33rd International Symposium on Automation and Robotics in Construction and Mining (ISARC 2016)

Auburn, Alabama, USA
18 - 21 July 2016

ISBN: 978-1-5108-2992-3

Printed from e-media with permission by:

Curran Associates, Inc.
57 Morehouse Lane
Red Hook, NY 12571

Some format issues inherent in the e-media version may also appear in this print version.

Printed by Curran Associates, Inc. (2017)

For permission requests, please contact
International Association for Automation and Robotics in Construction (IAARC)
at the address below.

International Association for Automation and Robotics in Construction (IAARC)
Slovak University of Technology in Bratislava
Faculty of Engineering
Radlinského 11
813 68 Bratislava
SK - Slovakia

Phone: +421 2 5927 4599
Fax: +421 2 5296 8811

secretariat@iaarc.org

Additional copies of this publication are available from:

Curran Associates, Inc.
57 Morehouse Lane
Red Hook, NY 12571 USA
Phone: 845-758-0400
Fax: 845-758-2633
Email: curran@proceedings.com
Web: www.proceedings.com

TABLE OF CONTENTS

Analytical Hierarchy Process Decision Support System (AHP-DSS) for Trenchless Technology

Mohamed Salah[a], Soliman Abu Samra[b], Ossama Hosny[a]

[a] Construction Engineering Department, The American University in Cairo, Egypt
[b] Department of Building, Civil and Environmental Engineering (BCEE), Concordia University, Canada
Email: msahmed@aucegypt.edu, solimanamr_8@aucegypt.edu, ohosny@aucegypt.edu

Abstract

Among the increasing complexities and surface development, underground utilities installation, renewal and repair remain one of the most challenging projects worldwide. In addition, the crucial need for a minimal surface disruption is what even makes it more thought provoking for contractors/specialists to maintain. That is why, trenchless technology has been an economical choice for many contractors/specialists, especially in urban areas, to guarantee less restoration costs, social, and environmental impact and higher accuracy with less time compared to the open cut and cover method. This paper aims to introduce a framework, utilizing a fully automated Analytical Hierarchy Process engine, which supports the contractors in their selection for the most appropriate trenchless method, taking the project characteristics and site conditions into consideration. The framework features through four different modules as follows: (1) Input Module where the user enters the project attributes through the AHP-DSS user interface. (2) Central Database Module that contains the considered trenchless methods, project attributes limits & their weights and trenchless methods & their scores. (3) Analytical Hierarchical-based Engine that runs simultaneously with the central database module to provide the user with the most suitable construction method. (4) Trenchless Technology Method Module that shows the most suitable method that suits the pre-defined user inputs. Spreadsheet modelling has been used for developing the Analytical Hierarchal Process Decision-Support System (AHP-DSS). A case study composed of 20 projects with various characteristics and conditions has been used for validating and verifying the model. The results showed a percentage of error less than 10% compared to the actual executed results

Keywords – AHP, Decision support system; Modeling, Trenchless technology

1 Introduction

Underground utilities e.g. (pipes, cables, conduits, force mains, etc.) can be executed using traditional methods, known as open cut convention methods where a trench is excavated using excavator or labour then the required utility is placed followed by backfilling. Open cut method requires cautious contractors to excavate and maneuverer around existing under-ground utilities in order to avoid damaging any existing underground utility; consequently, it may lead to slower production rates. In addition, they might be more expensive compared to trenchless methods in congested areas, especially when considering the restoration costs i.e. costs of restoring back the sidewalks, landscapes, road pavement, etc. [1].

Moreover, open cut method proved to result in dramatically higher social costs as it leads to traffic disruption, noise, and damage of existing underground utilities in many occasions [2]. Ariaratnam et al [3] has developed sustainability index "USIR", which considers three factors "costs, social impact and environmental impact" to measure how sustainable open cut and trenchless methods are. The results showed that open cut has higher USIR i.e. (higher adverse impact economically, socially and environmentally) compared to the other trenchless methods in the study.

On the other hand, trenchless methods according to North America Society of Trenchless Technology (NASTT) is defined as "a family of methods, materials, and equipment capable of being used for the installation of new or replacement or rehabilitation of existing underground infrastructure with minimal disruption to surface, traffic, business, and other activities" which makes it more suitable in congested areas. However, some contractors and decision makers are not familiar with the trenchless methods, when it shall be used, and which method suits the project being approached. Careful planning and selection of the appropriate trenchless method is crucial to aid the stakeholders in taking their decisions and reduce any risks stemming from the inappropriate trenchless method selections.

2 Literature Review

Hastak and Gokhale [4] developed AUTOCOP that aids the user in selecting between pipe ramming, pipe bursting and open cut method through the application of an AHP-based engine through evaluating groups of criteria and sub-criteria with respect to the knowledge based preference of the user and the project specific situations.

Ueki et al [5] developed a decision tool that utilizes a rule-based expert system for selecting between open cut method and microtunneling methods (Slurry and Earth Pressure Balance "EPB"). The decision was based on depth of invert below the ground, pipe diameter, driving length, ground water level, site conditions, soil information and boulders existence and size.

Baik et al [6] developed a DSS that selects the most appropriate horizontal directional drilling method for a certain project based on pipe diameter, depth of installation, driving length, soil type, pipe material, pullback load, machine selection, and productivity.

Chung et al [7] further modified Ueki's model by including cost, duration estimation and the selection of shaft's construction method.

Bottero and Peila [8] developed a DSS for open cut and microtunneling methods using AHP based on construction time, construction cost, and environmental problems.

As highlighted above, it was recognized that most of the previous research was focusing on certain types of trenchless methods and disregarded the others such as; auger boring, pipe ramming, open face TBM, etc. In addition, some researchers did not vitalize the impact of crucial attributes such as; social and/or environmental surroundings, required level of accuracy, and type of utility installed e.g. (sewer, cables, etc.).

3 Objective

This paper aims to develop a computerized trenchless technology decision support system that aids decision-makers in selecting the most appropriate trenchless method based on various project characteristics and conditions (soil condition, depth, diameter…etc.) and recognizing the social, environmental attributes, accuracy level required and type of utility installed.

4 Methodology

There are various methods and approaches for developing a decision support system. However, Analytical Hierarchy Process "AHP" was used in this research. Saaty developed AHP in 1980 where: it structures the decision problem in various levels, corresponding to the situation understanding, objectives, criterion and sub-criterion, and solution alternatives. Through this structure, the decision maker is able to focus on a smaller set of decisions. The proposed model was performed using both MS-Visual Basic and MS- Excel.

The model considers seven different trenchless methods in addition to the open cut method, which are as follows: (1) Horizontal Auger Boring; (2) Pipe Ramming; (3) Mini horizontal directional drilling (Mini HDD); (4) Maxi HDD; (5) Pipe Jacking Open face TBM; (6) Earth Pressure Balance (EPB) microtunneling and (7) Slurry microtunneling. On the other hand, the model considers various attributes as follows: (1) Project type; (2) Depth of installation; (3) Soil type; (4) Driving length; (5) Pipe diameter; (6) Pipe material; (7) Social consideration; (8) Environmental consideration; (9) Utility type installed; (10) Accuracy required; (11) Ground water table level (12) Boulders size (if any).

5 AHP-DSS Framework

The AHP-DSS framework features through four consecutive modules as follows: (1) Input Module (2) Central database; (3) AHP engine and (4) Trenchless technology interface. Figure 1 displays the framework of the model and its modules.

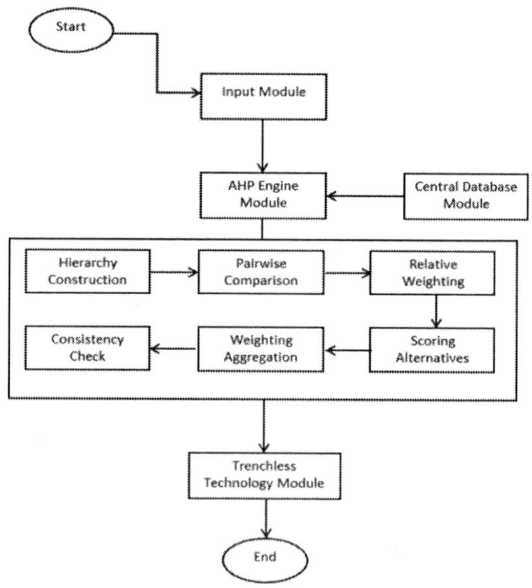

Figure 1: AHP-DSS Framework

5.1 Module 1: Inputs Module

The user enters the project attributes through the AHP-DSS user interface. The model placed list of options to some attributes where the user has to select one of them. This is applied to pipe material, soil type, boulders availability, social & environmental issues, accuracy required & type of utility installed.

5.2 Module 2: Central database

The central database module consists of three sub-modules as follows: (1) Trenchless methods in which the methods, considered in this research, are defined; (2) Attributes limits and weights where; the attributes weights and limits are defined; and (3) Methods vs attributes scoring where; the methods are scored against the attributes. The data was gathered from industry experts and previous researches and papers (Baik et al [5] and Allouche [9]).

5.3 Module 3: AHP Engine

AHP engine runs through consecutive modules as follows: (1) Hierarchy construction, (2) Pair wise comparison, (3) Relative Weight, (4) Scoring alternatives, (5) Aggregation of weights and, (6) Consistency check.

5.3.1 Hierarchy Construction

Involves decomposing the complexity where; decision factors are organized in a hierarchy-type structure. As shown in Figure 2, the primary goal of the problem "Selection of Trenchless technology" occupies the highest level of the structure, followed by "sets of attributes" that are the factors affecting the selection of trenchless methods Then, a third level is creating "Sub factors" that explain the attributes.

Finally, the feasible alternatives evaluation takes place at the lowest AHP hierarchy level.

5.3.2 Pair Wise Comparison

Once the hierarchy is constructed, comparing the attributes needs to be conducted. Hence, after, the following questions need to be answered, which attribute is more important or has greater influence on the attribute one level higher in the hierarchy. What is the intensity of that importance?

Saaty [10] translated the intensity assessment into numbers as shown in table 1. It should be noted that the pairwise comparison method is perhaps the cornerstone of the entire AHP philosophy, as it allows the user to systematically determine the intensities interrelationships of a practically great-unlimited number of decision factors.

Table 1: Random Index (Saaty [10])

Relative Intensity	Definition	Explanation
1	Of equal value	Two requirements are of equal value.
3	Slightly more value	Experience slightly favours one requirement over another.
5	Essential or strong value	Experience strongly favours one requirement over another.
7	Very strong value	A requirement is strongly favored and its dominance is demonstrated in practice.
9	Extreme value	The evidence favouring one over another is of the highest possible order of affirmation.
2,4,6,8	Intermediate values between two adjacent judgments	When compromise is needed.

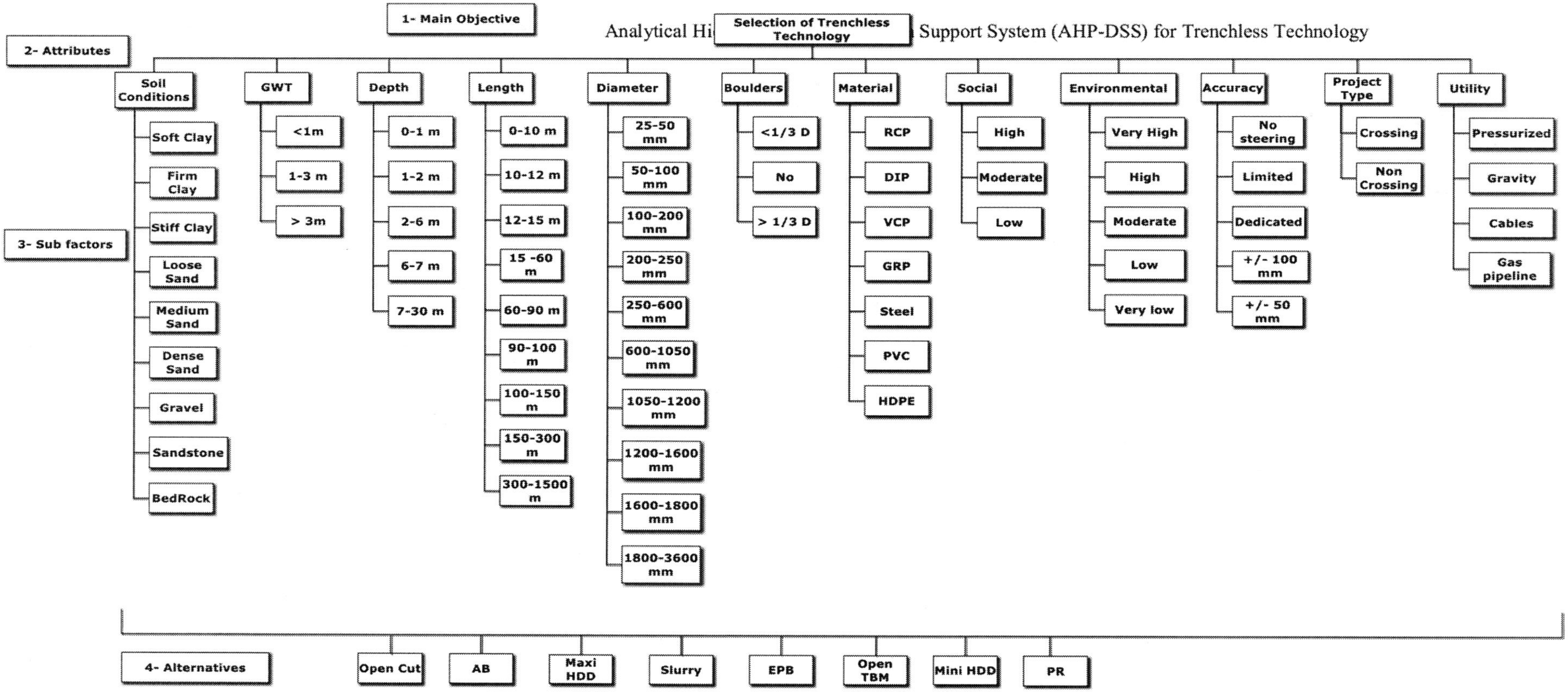

Figure 2: Hierarchy Construction

5.3.3 Relative Weight

Relative weights are calculated for the attributes called "Eigen Vector" or priority weight. This can be achieved by using "Column Normalization" or "Nth Root Method". In this paper, Nth root method was used where; it is calculated by multiplying the n elements in each row by each other and taking the nth root to form a new column vector. Then, the priority vector (W) is calculated by normalizing the column vector. The example in table 2 illustrates the calculations procedure for a pairwise comparison.

Table 2: Priority Weight Calculation

	A	B	C	Product	Nth root	Priority Vector (W)
A	1	3	1	3	1.44	0.44
B	0.333	1	0.5	0.166	0.55	0.18
C	1	2	1	2	1.26	0.38
					3.25	1.0

5.3.4 Scoring Alternatives

In this part, the same methodology, adopted for attributes relative weight calculation, is used for scoring the alternatives. It determines the extent to which each trenchless method satisfies each sub-factor, e.g. to what extent EPB work in soft clay or to what extent auger boring works with driving length more than 120 m, etc.

5.3.5 Aggregation of Weights

Results obtained from the two previous phases; the pairwise comparisons and the scoring of the alternatives, are aggregated to produce a quantitative measure of the benefits, and offered by each trenchless method, as per the equation below.

$$TT_{Score} = \sum(A_s * RW_c) \tag{1}$$

Where;

TT_{Score} : Is the trenchless method quantitative measure.

A_s : The score of each alternative with respect to each attribute.

RW_c : The relative weight of each attribute.

The total AHP score obtained for each alternative represents its relative value with respect to all attributes in the hierarchy.

5.3.6 Consistency Ratio Check

Consistency check has been developed to guarantee proper implementation of the pair-wise check comparison. For instance, let us consider that x is more important than y by factor of two and y is more important than z by factor of three then x should be six times z. Based on numerous empirical studies, it is acceptable to have inconsistency with a Consistency Ratio (CR) of 10% or less [10]. The consistency is checked using the following equations.

$$CR = \frac{CI}{RI} \tag{2}$$

$$CI = \frac{\lambda_{max-n}}{n-1} \tag{3}$$

Where;

CI is the consistency index.

λ_{max}: denotes the maximum principal eigenvalue of the comparison matrix.

RI: is random indices, as identified in Table 3.

N: is the matrix size.

To calculate λ_{max}, the weighted sum vector (W') is calculated by multiplying pair-wise comparison (A) by the priority vector (W). After that, the consistency vector (W'') is calculated by dividing each element in W' by its corresponding element in W, as per the equations below:

$$W' = W * A \tag{4}$$

$$\lambda_{max} = \frac{\sum_{i=1}^{n} W''}{n} \tag{5}$$

Table 3: Random Index (Saaty, 1980)

n	*1*	*2*	*3*	*4*	*5*	*6*	*7*	*8*	*9*	*10*	*11*	*12*	*13*	*14*	*15*
RI	0.00	0.00	0.58	0.9	1.12	1.24	1.32	1.41	1.45	1.49	1.51	1.48	1.56	1.57	1.59

6 Trenchless Technology Module

This module provides the suggested appropriate method or methods to the user that suits the project. In addition, data validation rules have been introduced to guarantee both logical data and reliable output. Figures 3 and 4 show screenshots from the interface showing the project attributes, defined in cases 1 and 2 in the respectively in the following sections and their selected trenchless method.

7 Model Validation

AHP-DSS model was tested by comparing the results obtained from the model and the actual decisions made, based on past projects. Twenty different projects were selected and used for validation, the results obtained indicates that the percentage error was not exceeding 5%. Two cases were selected to be discussed in details while the rest will be summarized in table 6.

7.1 Case Study 1

The first case was "Metro line crossing – New Cairo pipeline". This project involved the construction of two shafts (jacking and receiving) with 14 m depth and the installation of 2 x 3200 mm reinforced concrete pipes for a driving length of 80 m under a metro line, as a part of the New Cairo pipeline project. This project aimed to discharge 500,000 m3/day of potable water to the New Cairo water treatment plant. The crossing was in an urban congested area near the Nile corniche where; the intake is located. In addition, the ground water table level was high. The pipes are almost under the piers and therefore high accuracy during installation is required. Table 4 summarizes the input data for the model.

Table 4: Summary data for case 1- Metro line crossing – New Cairo pipeline

Parameters	*Condition*
Project type	Crossing
Soil condition	Medium Sand
Depth of Installation	14.0 m
Driving length	80 m
Pipe diameter	3200 mm
Pipe material	RCP
Utility installed	Pressurized flow
Boulders (if any)	No
Social consideration	High
Environmental consideration	Not mentioned (low)
GWT	7m
Accuracy required	+/- 50 mm

The AHP-DSS chose the "Slurry Microtunneling", which matched the actual construction method used by the contractor, accountable for this project, as shown in figure 3

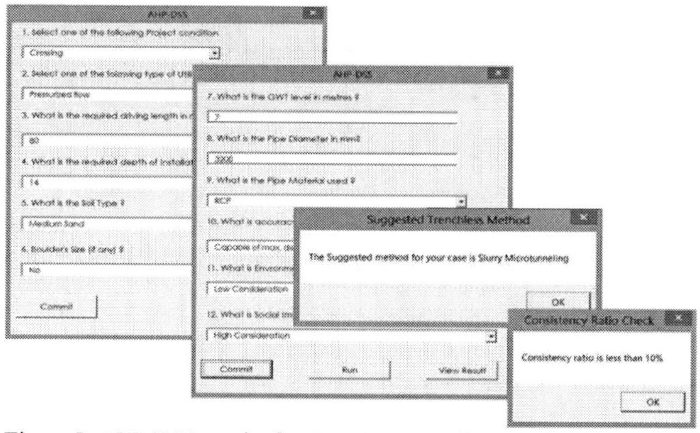

Figure 3: AHP-DSS results for Case 1 - Metro line crossing – New Cairo pipeline

7.2 Case Study 2

The second case was "Gas-line project in Zionsville, IN" which was retrieved from (Abraham et al [11]). This project included the installation of a gas pipeline of 100 mm diameter in Zionsville, Indiana. The total driving length for this project was 1,470 m and one of the drives included the installation of 87 m of pipeline. The average depth of installation was 1.1 m. The soil type was dry sand and the pipe material used was HDPE Nothing was mentioned about the social or environmental impact consideration or accuracy thus it would be assumed accordingly. Table 5 summarizes the input data for the model. The AHP-DSS chose the "Mini HDD", which matched the actual project construction method as shown in figure 4.

Table 5: Summary data for case 2 - Gas-line project in Zionsville,

Parameters	*Condition*
Project type	Crossing
Soil condition	Medium Sand
Depth of Installation	1.1 m
Driving length	87 m
Pipe diameter	100 mm
Pipe material	HDPE
Utility installed	Gas pipeline
Boulders (if any)	No
Social consideration	Not mentioned
Environmental consideration	Not mentioned
GWT	0
Accuracy required	Not mentioned

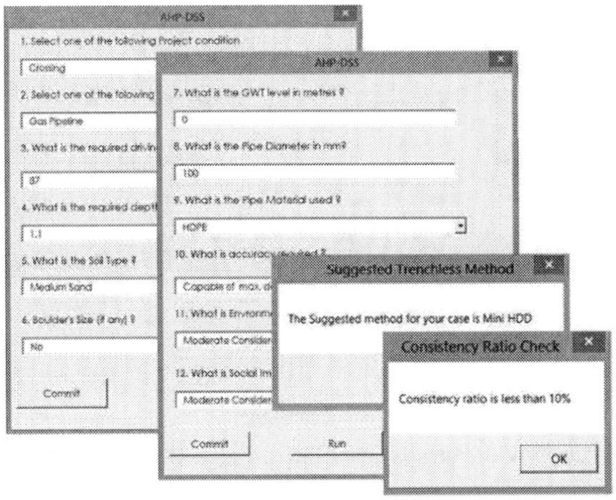

Figure 4: AHP-DSS results for case 2 - Gas-line project in Zionsville, IN

The following table summarizes the results obtained from the model for the remaining cases and the actual method used and selected to be implemented. The AHP-DSS in one case provided a different solution from the actual method used (Case 12). The model selected the Horizontal Auger Boring while the contractor used Pipe Ramming. The difference was due to the soil type "stiff clay", according to the experts Horizontal Auger Boring is able to work in stiff clay while Pipe Ramming is possible to work in it. The model gave the auger boring higher score than the Pipe Ramming and that is why Horizontal Auger Boring was selected. Both methods got the same scores for the remaining attributes.

Table 6: AHP-DSS Results

Cases	Project Type	Diameter /mm	Depth/m	Soil type	Pipe material	Utility Installed	Social	Environmental	Accuracy required	Boulders	Length	GWT/m	AHP-DSS Output	Actual method
Case 1	Crossing	900	3	Med.Sand	Steel	Pressure	Low	Low	Limited steering	No	23	0	HAB*	HAB
Case 2	Crossing	100	1.1	Dense Sand	HDPE	Gas	Low	Low	(+/- 100 mm)	No	87	0	Mini HDD	Mini HDD
Case 3	Crossing	1000	6	Firm Clay	GRP	Gravity Sewer	Low	Low	(+/- 100 mm)	No	120	0	Slurry MT*	Slurry MT
Case 4	Crossing	600	6	Med.Sand	VCP	Gravity Sewer	Low	Low	(+/- 100 mm)	No	122	4.5	Slurry MT	Slurry MT
Case 5	Crossing	750	3.2	Firm Clay	Steel	Pressure	Low	Low	Limited steering	No	18	0	Pipe Ramming	Pipe Ramming
Case 6	Crossing	1300	7	Loose Sand	GRP	Gravity Sewer	High	Low	(+/- 100 mm)	No	200	1.5	Slurry MT	Slurry MT
Case 7	Crossing	112	3.1	Firm Clay	HDPE	Gas	Mod*	Low	(+/- 100 mm)	<1/3 D	87	0	Mini HDD	Mini HDD
Case 8	Crossing	3200	14	Medium Sand	RCP	Pressure	High	Low	(+/- 50 mm)	<1/3 D	80	7	Slurry MCT	Slurry MCT
Case 9	Non-Crossing	1200	4	Gravel	GRP	Gravity	Low	Low	(+/- 100 mm)	No	6	0	Open Cut	Open Cut
Case 10	Crossing	3000	7	Dense Sand	RCP	Pressure	Low	Low	(+/- 100 mm)	No	24	0	Open face TBM	Open face TBM
Case 11	Crossing	1200	3	Gravel	Steel	Pressure	Low	Low	Limited Steering	No	60	0	Pipe Ramming	Pipe Ramming
Case 12	Crossing	800	5.4	Stiff Clay	Steel	Pressure	Low	Low	Limited Steering	No	50	0	HAB	Pipe Ramming
Case 13	Crossing	3000	15	Stiff Clay	RCP	Pressure	Low	Low	(+/- 100 mm)	No	65	2	EPB MT	EPB MT
Case 14	Crossing	1000	10	Medium Sand	HDPE	Gravity	Low	Low	(+/- 100 mm)	No	270	0	Maxi- HDD	Maxi HDD
Case 15	Crossing	800	8	Stiff Clay	VCP	Gravity	Med	Med	(+/- 100 mm)	No	150	2	EPB or Slurry MT	Slurry MT
Case 16	Crossing	1100	3.2	Soft Clay	Steel	Pressure	Mod	Mod	Limited Steering	<1/3 D	45	2	HAB	HAB
Case 17	Non-Crossing	900	2	Gravel	HDPE	Cables	Low	Low	(+/- 100 mm)	No	20	3	Open Cut	Open Cut
Case 18	Crossing	1200	10	Soft Clay	Steel	Gravity	Low	Low	(+/- 100 mm)	No	210	4	Maxi- HDD	Maxi HDD
Case 19	Crossing	150	4	Sand-stone	HDPE	Cables	Low	Low	(+/- 100 mm)	No	25	0	Mini- HDD	Mini HDD
Case 20	Crossing	3200	20	Medium Sand	RCP	Pressure	High	Low	(+/- 50 mm)	No	30	4	Slurry MT	Slurry MT

HAB: Horizontal Auger Boring, MT: Microtunneling and Mod*: Moderate

8 Analysis & Discussion

For the first case "Metro line – New Cairo pipeline", the AHP-DSS output was identical to the actual construction method implemented by the contractor (i.e. slurry microtunneling, which seems to be the most suitable method). Accordingly, the second best method after the slurry microtunneling was the pipe jacking open face TBM, which has satisfied all the project attributes and parameters except the GWT as it was high beyond its capabilities. As shown in figure 5, the highest weights were assigned to the project type, pipe diameter and GWT. Hence, as it was a crossing project in a congested area, open cut was excluded. Additionally, the trenchless methods that can satisfy a 3.2 m reinforced concrete pipe diameter are limited to the open cut (excluded due to the project type), pipe jacking open face TBM and slurry microtunneling. Finally, the 7.0 m GWT is limited to work with slurry microtunneling, open cut (excluded due to the project type) and HDD (excluded due to insufficiency for the pipe diameter). Therefore, project type, GWT and pipe diameter were assigned the highest weights since they were satisfied by few number of methods for this case.

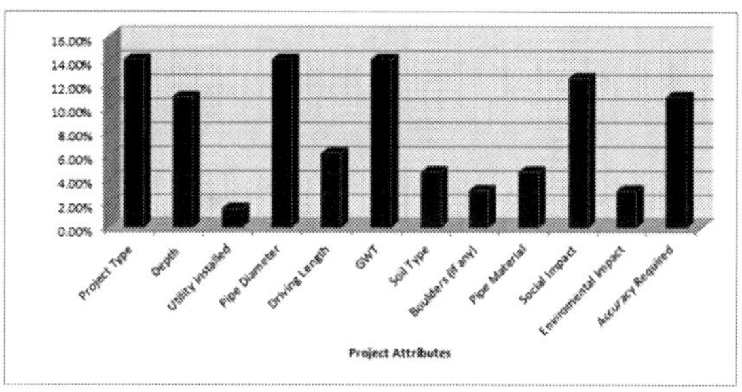

Figure 5: Attributes weights for case 1 - Metro line crossing – New Cairo pipeline

Regarding the second case "Gas-line project in Zionsville, IN", the AHP-DSS output was identical to the actual trenchless construction method, (i.e. Mini-HDD). As shown in figure 6, the governing attributes that gained the highest weights were project type, and pipe diameter. Hence, as it was a crossing project, open cut was excluded. Then, the 0.9 m pipe diameter can only work with pipe ramming, open cut (excluded due to the project type) and Mini HDD. Finally, the HDPE pipe material can only work with Mini HDD, open cut (excluded due to the project type). Since, few numbers of methods satisfied these attributes, they gained higher weights compared to the others. Therefore, Mini HDD was chosen, based on the below weights defined in figure 6, as the most suitable construction method for this project.

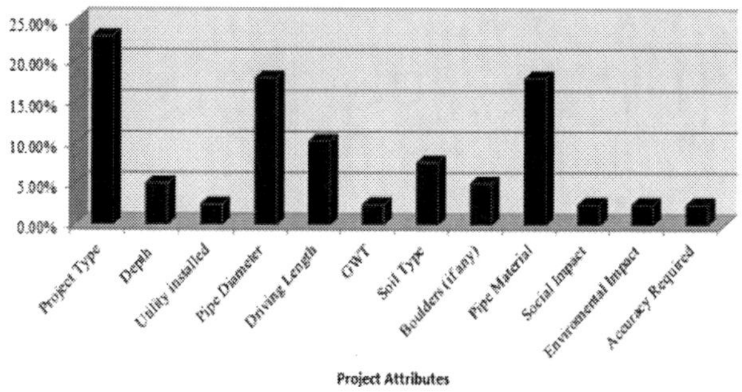

Figure 6: Attributes weights for case 2 - Gas-line project in Zionsville, IN

9 Concluding Remarks

The trenchless technology methods are more favourable compared to the open-cut method in urban areas. Due to the variety of trenchless methods, it was crucial to present a DSS that selects the most appropriate construction method that fits the project attributes and surrounding conditions. The proposed model was built using an AHP engine, incorporating wide spectrum of parameters to provide precise decisions for the construction method. Further work is required to include a cost estimation module, which incorporates the cost as a key attribute in deciding on the construction method, after being technically accepted by the existing AHP-DSS.

Some other parameters can be considered for future research that include safety and health issues that could affect the selection of different trenchless methods.

10 Acknowledgment

Special acknowledgement to Mr. Nabil Abdel Salam, the project manager of New-Cairo pipeline project, and Mr. Ahmed Abo Deshish, the head of tunnelling department in Arab Contractors; for their provision of data and insightful contributions in performing the AHP-DSS model.

References

[1] Ariaratnam, S. T., Member, A., Asce, A., Lueke, J. S., and Allouche, E. N. Utilization of Trenchless Construction Methods by Canadian Municipalities. Journal of Construction Engineering and Management, 125(2), 76, 1999.

[2 Pau, S. H., Chau, K. W., and Wong, W. G. Prospects for No-Dig Technology in Hong Kong Construction Industry. Journal of Construction

Engineering and Management, 119(3), 550, 1993.
[3] Ariaratnam, S. T., Piratla, K., Cohen, A., and Olson, M. Quantification of Sustainability Index for Underground Utility Infrastructure Projects. Journal of Construction Engineering and Management, 139(12), A4013002, 2013.
[4] Gokhale, S., and Hastak, M. Decision aids for the selection of installation technology for underground municipal infrastructure systems. Tunnelling and Underground Space Technology, 15, 1-11, 2000.
[5] Ueki, M., Haas, C. T., and Seo, J. Decision Tool for Microtunneling Method Selection. Journal of Construction Engineering and Management, 125(2), 123, 1999.
[6] Baik, H., Abraham, D., and Gokhale, S. A decision support system for horizontal directional drilling. Tunnelling and Underground Space Technology, 99-109, 2003.
[7] Chung, T. H., Abraham, D. M., and Gokhale, S. B. Decision Support System For Microtunneling Applications. Journal of Construction Engineering and Management, 130(6), 835, 2004.
[8] Bottero, M., and Peila, D. The use of the Analytic Hierarchy Process for the comparison between microtunneling and trench excavation. Tunnelling and Underground Space Technology, 20(6), 501-513, 2005.
[9] Allouche, e. (n.d.). A decision-support model for selection of a trenchless construction method, 2001.
[10] Saaty, T. L. *The analytic hierarchy process: planning, priority setting, resource allocation*. New York: McGraw-Hill International Book Co, 1980.
[11] Abraham, D., Baik, H., Gokhale, S. Development of a Decision Support System for Selection of Trenchless Technologies to Minimize Impact of Utility Construction on Roadways. Joint Research Study, School of Civil Engineering, Purdue University, West Lafayette, Indiana, 2002.

Time-Cost Tradeoff Model for Multiple Asphalt Paving Projects Using Genetic Algorithms

Mohamed Salah, Tarek Zaki, Ossama Hosny

Construction Engineering Department, The American University in Cairo, Egypt
Email: msahmed@aucegypt.edu, tarekzaki@aucegypt.edu, ohosny@aucegypt.edu

Abstract

Road construction companies work on delivering multiple projects at the same time. Careful planning and allocation of each project's resources need to be determined and evaluated in order to minimize their costs and maximize their profits. In this paper, a model was developed with the objective of trading off between time and cost for Multiple Paving Projects using Genetic algorithms (MPP-GA). The MPP-GA was designed with four different modules: (1) an input module that requires the number of multiple projects in hand and their related information; including each project's contractual milestones and construction specifications. (2) An equipment database module that contains a list of the available different types of equipment used on paving projects including their production rates and their cost. (3) an Optimization engine module that works to determine the optimum equipment fleet required to be allocated on each paving project based on the total equipment cost. The optimization module performs a trade-off between the cost of resources allocated for each project and the delay damages and/or incentives calculated as a result of late, early or on time delivery of each project. and finally, (4) an output module that reports the fleet configuration, delay damages and/or incentives, total cost and expected finishing dates for each project. A case study was presented to illustrate a number of practical features of the proposed model and to demonstrate its capabilities in selecting the near optimum fleet configuration.

Keywords –

Paving Projects; Multiple Project; Resource Allocation; Optimization; Genetic Algorithms; Construction Planning.

1 Introduction

Asphalt paving projects require the use of a set of heavy equipment including asphalt batching plants, mechanical pavers, vibratory rollers, pneumatic rollers and haulers. The typical construction sequence for paving asphalt roads starts by first producing hot-mix asphalt at an asphalt batch plant. For large jobsites, contractors set up the asphalt batch plants on/near the jobsite to minimize the hauling time and, hence, the cost from the production to the discharge location. Second, produced hot-mix asphalt is discharged into trucks which are hauled to the jobsite. Trucks loaded with asphalt feed the paver finisher which spreads uniformly thick layers of asphalt mix. Finally, vibratory rollers apply compaction to reach the required density of the road layers and then pneumatic rollers are used to smooth the finalized asphalt road.

The majority of contractors work on multiple paving projects, where every project has its own specifications, conditions and requirements. However, many contractors fail to plan and allocate their resources on their concurrent projects and hence they fail to deliver their project on time which leads the employer to impose delay damages on the contractor, or they allocate extra resources to avoid late delivery and hence extra cost of resources which in turn reduce their profit. Optimization models are one of the widely used techniques for approaching resources allocation problems. Optimization models can be developed using either traditional methods like linear programming, integer programming, dynamic programming…etc. or un-traditional methods like genetic algorithms, particle swarm, ant colony…etc. Traditional optimization models are inappropriate for resources allocation in paving projects due to their complexity and the presence of many variables and constraints, therefore un-traditional optimization methods are suitable for this problem.

Genetic algorithms (GAs) are widely used technique in optimizing the allocation and utilization of resources in construction industry [1].

2 Literature Review

Efficiently selecting the number of equipment to be used on each highway project or section is a resource allocation optimization problem. Different studies throughout the literature have tackled the use of different optimization techniques to provide near optimum solutions to resource allocation.

Hegazy [1] introduced a technique for solving the resource allocation and levelling problem using Genetic Algorithms (GAs) by searching for near-optimum solution. The developed technique defines priorities to selected tasks based on their impact on the schedule. The GA searches for an optimum set of tasks' priorities that produce shorter project duration and better-levelled resource profiles. Hegazy's work concluded that the use of GA in solving the resource allocation problem can be done by incorporating multi-objectives to the model; such as: selecting the appropriate methods of construction for each activity, considering the daily penalty of exceeding the deadline as well as the incentive for early completion. The work was generic and it does not consider the idea of allocating the resources on multiple projects.

Hassan and Gruber [2] developed a simulation model for simulating an asphalt paving project to optimize costs & time using STROBOSCOPE software. A sensitivity analysis was performed to estimate the effect of resources on costs & productivity. The model considered only one project, in addition sensitivity analysis is not an efficient method for determining the optimum fleet configuration.

A study by Heon Jun and El-Rayes [3] developed a multi-objective optimization model using GA that is capable of measuring and minimizing the undesirable resource fluctuations. The model's objective was set to maximize resource utilization efficiency and minimize project duration while complying with all precedence relationships and resource availability constraints. The authors concluded that the use of a multi-objective resource utilization function generates a trade-off. In this case, between minimizing the overall project duration while maximizing the resource utilization efficiency.

Kane and Tiesser [4] developed a mathematical optimization model that was used in the planning of a multi-project program. The working principle of their model was that the costs of completing a group of linked projects were optimized by speeding up the realization of the whole program to save time. The model was validated via a case study of three concurrent projects, each with a different duration, budget, number of operators and quantities of work. Resources were effectively allocated to minimize the costs and accelerate the project time. However, the model was limited to the allocation of human resources only.

Sarkar and Shah [5] developed a framework for the application of genetic algorithms to optimize the productivity of the site excavation activities for highway construction project using two types of equipment: hydraulic excavators and tandem vibratory rollers. The model was coded on MS-Excel and Evolver add-in was used as the GA optimization engine. They compared the actual resource allocation done on the project with the allocation output from the GA model and observed the results. It was concluded from their model that the resource allocation done by the GA produced an increase of 15.7% of the total productivity of the project compared to the manual allocation. However, the model did not consider the paving activities for road construction.

After reviewing the previous studies, it can be concluded that the previous studies were mainly discussing resource utilization for construction projects in general. In addition, the optimization models for asphalt paving projects were done on a single project scale and not applied on multiple one.

3 Objective & Scope

Since Highway projects is comprised of several and wide range of activities that include excavation, backfilling, grading, paving, compacting...etc. This paper will highlight and focus on the paving package as an activity of paramount importance for the project completion.

This paper presents a time-cost trade-off model to aid contractors in allocating their resources on multiple paving projects. The proposed model will report the fleet configuration for each project, finishing date, equipment costs, delay damages and incentives (if any).

4 MPP-GA Framework

A model was developed that is comprised of a four modules: input, equipment database, optimization engine and output as shown in figure 1. The developed model was coded on MS-Excel & Visual basic for Applications (VBA) to facilitate the user inputs and the output reports.

The input module compiles information such as project description, contract information...etc. such inputs are fed along with a pool of equipment with their information and specifications into the optimization engine module. The model's engine works on providing the optimal allocation of resources on each project by providing a time-cost trade-off.

Finally, the results are transferred to the output module which generates a number of outputs per project including: (1) equipment fleet configuration, (2) actual project finishing dates, (3) total projects cost and (4) amount of incentives/delay damages if any.

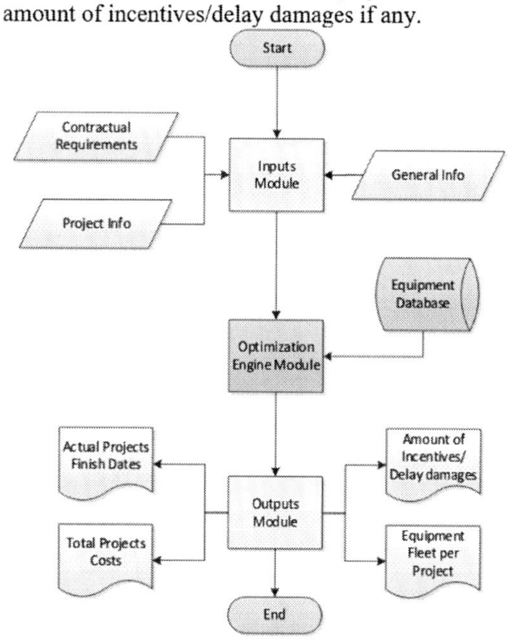

Figure 1: MPP-GA Framework

4.1 Module 1: Inputs Module

4.1.1 Project Information

In this part the user specifies some inputs including the number of the multiple projects in hand and their related information.

The volume of work required for each project (n) was calculated based on its designed specifications in terms of Length (Ln), Width (Wn), Thickness (Thn) and Density of Asphalt (Dn), which is assumed to be 2.4 ton/m3.Thus, the quantity of work was calculated (see Equation (1)).

$$Qn = Ln \; x \; Wn \; x \; Thn \; x \; Dn \qquad (1)$$

4.1.2 Contract Information

Each project has to be set with a target completion date expressed in number of weeks to complete the required work. In the event of surpassing the target completion duration for any project a penalty ($/week) is incurred for each day surpassed by the completion date. On the other hand, in the event of concluding the project prior to the completion date, an incentive ($/week) can be given to the appointed contractor. Based on the contract terms and conditions for each project, the end user defines the amount of delay damages and incentives for as well as the target competition duration for each project.

4.1.3 General Information

Other inputs include the hauling and return speed of trucks as well as the distance between the nearest asphalt batch plant and the project. However, several assumptions were considered in MPP-GA that includes the following: (1) grade resistance for the roadway between the asphalt plant and the project location are not considered in trucks cycle time and hence in the trucks production rates, (2) the efficiency of each construction equipment are determined by the user and considered constant throughout the project duration.

4.2 Module 2: Equipment Database

A database was designed to include five different equipment used in any paving project including: asphalt batch plants, trucks, mechanical asphalt paver, tandem vibratory rollers, and pneumatic rollers. An extract for the equipment database module in this model is demonstrated in table 1.

Table 1: Equipment Database Extract

ID	Equipment Type	Available No.	Production rate (ton/hr)	Cost/hour ($/hr)	Model no.
TR1	Truck	7	25.2	$65	Mack CH613
TR6	Truck	8	30.1	$70	Mack RD688
M1	Asphalt plant	2	168	$680	Tarmac HRODC_AP1
M3	Asphalt plant	3	67	$530	Tarmac HRODC_AP10
PAV-1	Asphalt paver	3	425	$280	DYNAPAC F800T
TV-1	Tandem Roller	5	224	$57	CAT-CB-534D XW
PNR-1	Pneumatic Roller	6	251	$55	CAT-PS150C

The production rates for each type of equipment were retrieved from either manufacturer's catalogs or field experts. The ownership cost & operating cost were calculated in order to end up with the rate per hour [6]. The production rates for rollers were converted from cubic yard per hour (see Equation (2)) to ton/hour (Equation (3)) [7].

$$Cubic\ Yard/hr = (16.3\ \times W \times S \times L \times E) \div P \qquad (2)$$

Where, (P) is the number of passes required, (W) is the width compacted in feet, (S) is the Roller speed in mph, (L) is the Lift thickness in (in) and E is the Efficiency.

$$Ton/hr = Cubic\ Yard/hr \times 0.7645 \times Dn \quad (3)$$

Where, (0.7645) is a constant for converting to cubic meters/hr.

4.3 Module 3: Optimization Engine

This module is the engine of the model where all the calculations are made. Since the resource allocation problems for paving projects is quite complex problem, genetic algorithm was used to solve our problem since it is a powerful optimization algorithm that deals with combinatorial-in-nature problems.

Genetic algorithms (GAs) are search algorithms developed by Holland in 1975, which are based on the mechanics of natural selection and genetics to search through decision space for optimal solutions [8]. Over the last two decades, these search algorithms have gained significant popularity in engineering fields [9].

Palisade (2015) Evolver™ v7 is a MS-Excel add-in that utilizes genetic algorithms in optimization. The GAs engine performs operations in sequential steps: (1) an initial population of solutions is created and is composed of a number of chromosomes as shown in table 2

Table 2: Chromosome Representation

Plants					Trucks						Pavers				Tandem Rollers			Pneumatic Rollers		
M1	M2	M3	M4	M5	TR1	TR2	TR3	TR4	TR5	TR6	PAV-1	PAV-2	PAV-3	PAV-4	TV-1	TV-2	TV-3	PNR-1	PNR-2	PNR-3
1	1	0	0	0	4	3	0	1	0	0	1	0	0	0	1	1	0	1	0	0

Each gene in the chromosome structure represents the number of equipment required per project duration. The domain of variables for the genes is a set of integer numbers that have a range equal to the number of equipment per type from Module 2: Equipment Database. (2) A fitness function is used to evaluate the pool of chromosomes (see Equation (4)).

$$Minimize.\ Total\ Cost = \sum_{j=1}^{w}\sum_{i=1}^{p}(EHC\ x\ NH\ x\ NE) + DLn - In \quad (4)$$

Where p= total number of projects, i=project number, w= total number of weeks, j=week number, (EHC) is the equipment hourly cost ($/hour), (NH) is the number of hours worked by each equipment, (NE) is the number of equipment for each type, (DLn) is the delay damage per project and (In) is the incentive per project.

The executed quantities at the end of each week are recorded, if the executed quantity was equal to the required quantity at the planned finishing date then the project was finished as scheduled and hence neither delay damage nor incentive will imposed or gained respectively. However, if the executed quantity was more than the required quantity at the planned finishing date, then the project was finished earlier and the model will provide the actual finishing date. On the other hand, if the executed quantity was less than the required quantity then the project is behind the schedule & extra duration will be calculated based on equation 5.

$$ED = rQ - eQ/\ Pa \quad (5)$$

Where (ED) is the extra duration in (weeks), (rQ) is the required quantity of asphalt to deliver per project in (tons), (eQ) is the actual executed asphalt quantity as scheduled by the model in (tons/week) and Pa is the average production rate during the project execution by the governing resource which is the asphalt batch plant in (tons/week).

The total delay damage incurred will be calculated using equation 6

$$DLn\ (\$) = ED\ \times\ D \quad (6)$$

Where (DLn) is the total damages incurred, (ED) is the extra duration in (weeks), (D) is the contractual delay damage per week per project.

The total incentive earned in case of early finish will be calculated using equation 7.

$$In(\$) = (PD - FD) * I \quad (7)$$

Where (In) is the total incentive earned, (PD) is the planned finishing date specified by the user in weeks, (FD) is the actual finishing from MPP-GA model in weeks, (I) is the contractual incentive specified in the contract entered by the user.

Then the actual finish date for delayed projects can also be calculated using Equation (8).

$$aF = ED + pF \quad (8)$$

Where (aF) is the actual finish date and (pF) is the planned finish date of the project.

(3) Genetic operators (selection, crossover, and mutation) are performed on the number of chromosomes and evaluating the fitness function to determine the fit solutions replace the weak solutions; thus a new population of solutions is formed and so on.

A group of constraints were considered in developing MPP-GA including: (1) the cumulative production rate for trucks, pavers and rollers should cover that of asphalt plants. (2) The number of equipment used should not exceed the available number. (3) A soft constraint was included that makes sure that cumulative executed quantity delivered should be equal or greater than the required quantity to finish the project.

4.4 Module 4: Output Module

This module produces the outputs generated by the GA engine on the developed reporting user interface which includes: (1) the set of equipment fleet to be used per project per week, the total projects costs, the actual finish dates per project and the amounts of incentive or delays per project that were expected to be incurred.

5 Case Study

A hypothetical example is considered to illustrate the use of the developed model in selecting near-optimum fleet configurations from a set of scenarios. It also demonstrates the ability of the model in conducting time–cost trade-off analysis. Figure 2 & 3 show the projects location and description respectively.

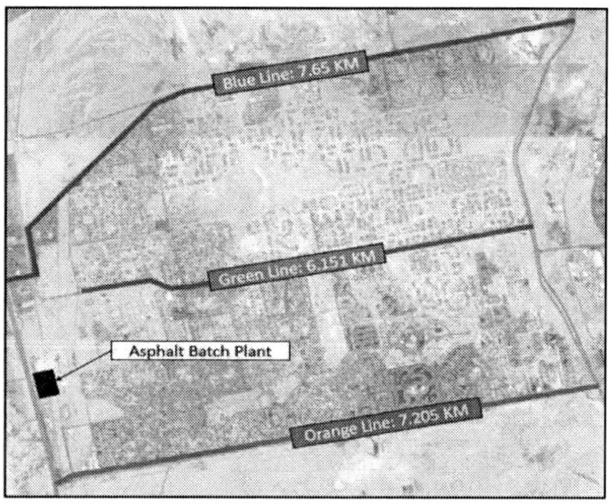

Figure 2: Projects Location and Description

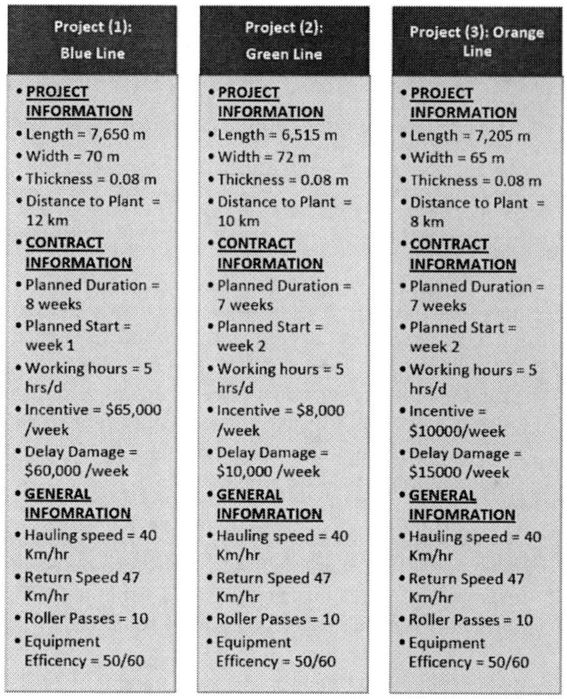

Figure 3: Case Study inputs for the three projects

The user will enter the projects details using the MPP-GA interface that include the roads dimensions, contract duration, commencement date, working hours, equipment efficiency…etc. Figure 4 shows a sample for the user's input for project 1 using MPP-GA interface.

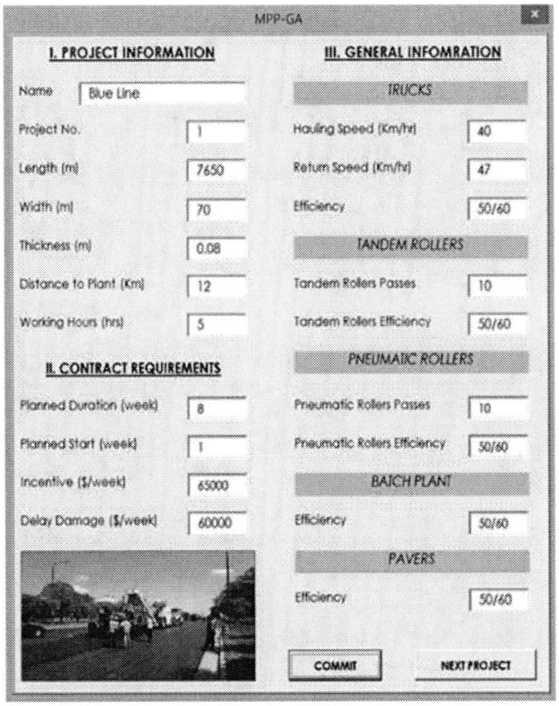

Figure 4: MPP-GA inputs module with Project (1) inputs defined by the user.

The fitness function was originally set to minimize the total cost of all projects while satisfying the set of constraints as defined in the model engine. The GA initial solutions' pool was set to be 1000 solutions, the termination criteria was set to be completing 200,000 trials, the crossover rate was set to 90% and the mutation rate was set to 10%.

6 Results and Discussion

The total cost after optimization was almost $ 2.07 million which includes the cost of resources allocated to the three projects, incentives earned and delay damages incurred. Figure 5 shows Evolver's progress after 50,000 trials.

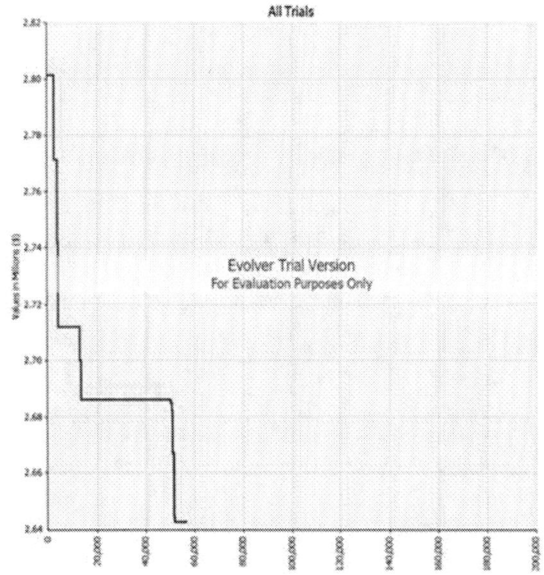

Figure 5: Solution convergence on Evolver®

Figure 6 depicted the results of MPP-GA output module for Project 1, which shows project cost, the actual finishing date, planned finishing date, delay damage, incentive & fleet configuration for Project 1 in week 4, in addition to the total projects cost.

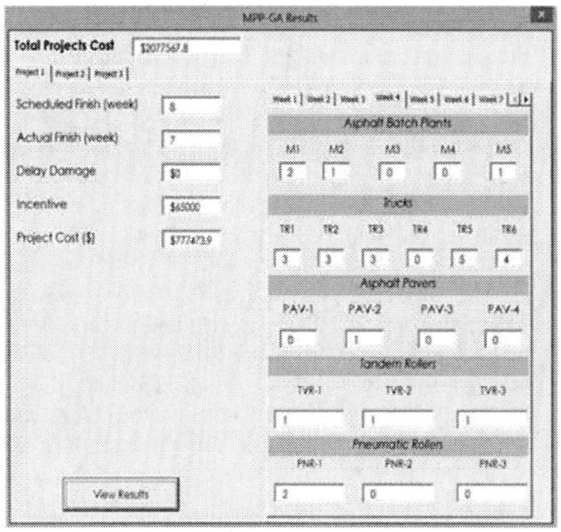

Figure 6: Model output sample for Project 1, week 4

Table 3 summarizes the output results of MPP-GA for the three projects showing their costs, planned finishing date, actual finishing date, delay damages & incentives.

Table 3: MPP-GA Results

	Project [1] Blue Line	**Project [2] Green Line**	**Project [3] Orange Line**
Planned Finish (weeks)	8	8	7
Actual Finish (weeks)	7	10	8
Delay Damages ($)	0	$20,000	$15,000
Incentives ($)	$65,000	0	0
Project Costs (including Incentives/delay damages)	$777,473.9	$651,952.9	$648,141

It was expected that due to limited resources available, the three projects cannot be delivered on schedule. Project 1 was delivered one week earlier, while Project 2 and Project 3 were delivered later than the contract duration by two weeks and one week respectively. Project [1] has the highest delay damage and incentive compared to project [2] and [3]; consequently, avoiding delay in project [1] was preferable where the amount incurred in case of delay will be $60,000/ week while the amount incurred as result of the late delivery of project [2] & [3] was $ 35000.

The total cost for project [1] before considering incentive was $ 842,473. By including the incentive, the total cost was reduced to be $ 777,473. If project [1] was finished on week 8 as planned and the other projects [2] & [3] were finished on week 10 & week 8 respectively, the total cost for project [1] will be $ 806,465 as shown in Figure 7. The cost of allocating extra resources (mixers, haulers…etc.) to project [1] for finishing earlier was $ 36008. Consequently, the incentive earned in project [1] covered the extra resources allocated to finish earlier & contributed to cover damages incurred in the other projects. In order to finish project [2] on time (two weeks earlier), this requires allocating extra resources which will increase the cost by almost $70,000. While for project [3], extra resources that cost almost $ 45,000 to the project cost in order to deliver the project on time. Consequently, MPP-GA effectively traded of between cost and time to obtain near optimum results as explained.

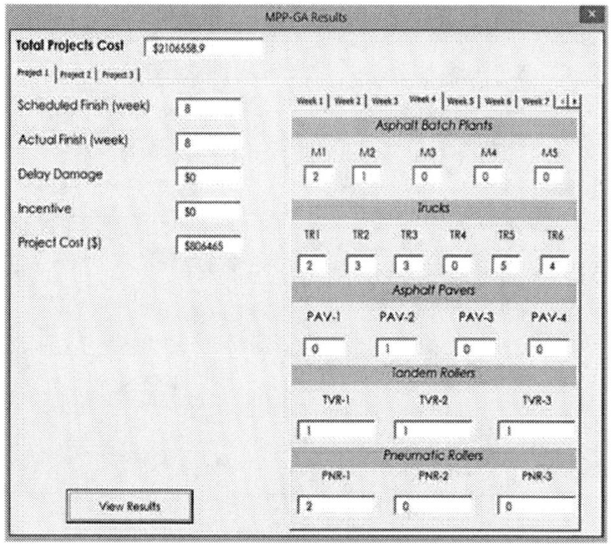

Figure 7: Model output for finishing Project 1 on time

7 Conclusion

Road construction agencies procure contractors to work on multiple projects or section simultaneously to save construction time and minimize costs. In this paper a model was developed to aid contractors in determining the optimum equipment fleet configuration to use in order to finish each project on time, within its set budget and avoid delay damages incurred as a result of surpassing the contractual finish dates. An equipment database module was incorporated into the model which included different types of paving equipment. The model uses an optimization engine coded with GA to provide the optimum equipment utilization configuration while preforming a time-cost trade-off to minimize the total costs and finishes each project on its contractual time. A case study was used to demonstrate the model's essential features.

For further studies, a second step optimization module can be added to optimize the allocation of resources for the delayed projects during extra time, since some of the resources used for project [1] will be back to the warehouses, hence there is chance to use cheaper resources (mixer, trucks, pavers…etc.) and get the work done as required.

Moreover, 3D simulation module can be considered to simulate the construction sequence and resources movement, the 3D simulation will be beneficial to demonstrate some complexities like change in elevation, bridges, turns...etc. consequently, the model will be able to figure out the change in productivity of resources together with changes in cost.

References

[1] Hegazy, T. Optimization of resource allocation and leveling using genetic algorithms. *Journal of construction engineering and management*, 125(3), 167-175, 1999.

[2] Hassan, M. and Gruber, S. Application of discrete-event simulation to study the paving operation of asphalt concrete. *Construction Innovation: Information, Process, Management*, 7-22, 2008.

[3] Heon Jun, D. and El-Rayes, K. Multiobjective optimization of resource leveling and allocation during construction scheduling. *Journal of Construction Engineering and Management*, 137(12), 1080-1088, 2011.

[4] Kane, H. and Tissier, A. A Resources Allocation Model for Multi-Project Management. In *proceedings of the 9th International Conference on Modeling, Optimization & SIMulation*, 2012.

[5] Sarkar, D. and Shah D. A Framework for Application of Genetic Algorithm in Productivity Optimization of Highway Equipment Using Evolver Software. *European International Journal of Science & Technology*, 2(5), pp. 151 –. 163, 2013.

[6] Peurifoy, R. L., and Oberlender, G. D. *Estimating construction costs*. McGraw-Hill, Boston, 2002.

[7] Nunnally, S. *Managing construction equipment*. Englewood Cliffs, N.J.: Prentice-Hall, 1977.

[8] David E. Goldberg. *Genetic Algorithms in Search, Optimization and Machine Learning*, Addison-Wesley, New York, NY, 1989.

[9] Georgy, M., & Basily, S. Y. Using genetic algorithms in optimizing construction material delivery schedules. *Construction Innovation: Information, Process, Management, 8(1), 23-45* 2008.

Fast and Robust 3D Terrain Surface Reconstruction of Construction Site Using Stereo Camera

Changhun Sung[a], Sang Hoon Lee[a], Young Min Kwon[a] and Pan Young Kim[a]

[a] Hyundai Heavy Industries, Republic of Korea
E-mail: sungch@hhi.co.kr, lshleee@hhi.co.kr, ymkwon@hhi.co.kr, projjang@hhi.co.kr

Abstract –

This paper presents a fast and robust three-dimensional (3D) terrain surface reconstruction system using a stereo camera. The local feature-based 3D terrain surface reconstruction algorithm consists of two major steps: matching and terrain surface reconstruction. In this paper, extracted corners are described by multi-scale descriptors (MSDs), and the matching precision is increased by a quadratic interpolation method. In the terrain surface reconstruction step, the construction terrain surface is modeled using a 3D regular grid plane for its computational efficiency and robustness. The precision of the 3D regular grid plane is improved by inferring the 3D grid vertices from a robustly matched nearest point cloud. The computational complexity of the proposed terrain surface is simplified using 2D and 3D triangle mesh structures.

In order to evaluate the proposed method, different types of 3D reconstruction methods were compared under the conditions of a practical construction site. The results of the experiment show that the MSD-based 3D terrain surface algorithm performed well in terms of robustness and computational efficiency.

Keywords –

3D terrain surface, Stereo Camera, Interpolation, Triangle mesh, Regular grid plane

1 Introduction

Recently, the development of automated construction systems on construction sites for safety and convenience has attracted interest [1,2,3]. In order to operate an automated construction system, it is essential to understand the environment in which construction equipment operates, and the first step of environment recognition is terrain reconstruction. To reconstruct construction site terrain, there are two general approaches: vision-based and Light Detection and Ranging (LiDAR)-based [5].

Figure 1. Example of a construction environment

LiDAR can provide highly precise distance information about the observed terrain. Therefore, when surveying construction sites, LiDAR based terrain reconstruction is normally used [4,5]. However, the use of LiDAR is limited in real-time applications. This is because the post-processing of the three-dimensional (3D) point cloud acquired from LiDAR is a time-consuming task. In addition, LiDAR consumes a large amount of power and it is expensive to mount these devices on real construction equipment.

A vision sensor can provide construction site environment images quickly, at over 30 Hz. Furthermore, it is possible to quickly calculate terrain surface information using current computer vision techniques [6,7,8]. However, there remain some problems when using vision to reconstruct construction site conditions because of its challenging environment. An image of a construction site generally comprises an irregular-appearing surface and textureless ground (see Figure 1). This makes it difficult to use vision-based methods to reconstruct 3D terrain surfaces in practical applications.

In order to deal with this problem, in this paper, we propose a fast and robust 3D terrain surface reconstruction algorithm suitable for construction sites. In the matching step, the quadratic interpolation method is applied to increase the precision of sparse 3D point location. In the 3D terrain surface reconstruction process, regular a grid vertex-based 3D terrain surface reconstruction algorithm is presented. This method efficiently estimated terrain surfaces from 2D and 3D triangle meshes while improving accuracy.

2 Related Work and Proposed Method

2.1 Related Work

In order to infer 3D terrain surface information using a stereo vision sensor, the first step is to find correspondence points between stereo images. There are many approaches to searching for correspondence points. One popular method is to find correspondence points by minimizing global constraints. Graph-cuts [9] and the belief propagation algorithm [10] are well-known global methods for stereo matching. These algorithms have shown good performance [10,11]. However, the selection of parameters such as disparity range and kernel size critically affects the result [9,15].

A local feature-based reconstruction algorithm is another general method. This method finds correspondence points based on interest points represented by a descriptor. In local feature-based reconstruction, how to represent extracted interest points is important, because descriptor distinctiveness significantly affects matching precision. The most widely used local features are SIFT [12] and SURF [14], which are based on gradient information. In various applications, these features have shown good performance. However, their high computational complexity restricts their use in real time applications [15,21].

Recently, a robust and fast feature descriptor, the multi-scale descriptor (MSD), was proposed by Sung et al. [13] for outdoor practical applications. This descriptor already has demonstrated its performance in outdoor applications such as motion estimation and dense 3D reconstruction [3,13]. Therefore, we also used the MSD to describe extracted corner points because of its distinctiveness and computational efficiency.

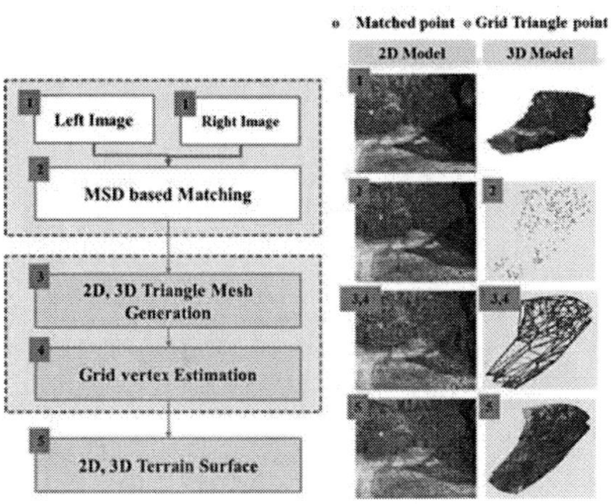

Figure 2. Overall procedure for the proposed 3D terrain surface reconstruction algorithm

2.2 Proposed work

The proposed terrain surface reconstruction method comprises three major steps: MSD-based matching, matched point-based 2D and 3D triangle mesh generation, and a regular grid-based terrain surface creation (see Figure 2). In the matching step, the distribution of the corners and the precision of the 3D point cloud are important, as they significantly affect the accuracy of reconstructed 3D terrain surface. In order to achieve these requirements, an equal number of corners are selected from each sub-region to ensure that the distribution of the interest points is uniform. The precision of the 3D point cloud is improved by quadratic interpolation techniques.

In this paper, the construction site's terrain surface is represented by a regular grid plane. Regular grid planes can dramatically reduce computational burden by inferring regular grid vertices from a robustly estimated triangle mesh. The accuracy of the terrain surface plane is improved by precisely calculated closest 3D points. Furthermore, when reconstructing a 3D terrain surface, we also generate a 2D terrain surface to increase the accessibility to the target surface distance and as information for the operator of the construction equipment.

3 Matching

Various methods have been proposed to extract interest points. One popular detector is the Harris corner detector [17]. This corner detector extracts corner points by eigenvalues analysis. For computational efficiency, the Features from Accelerated Segment Test (FAST) detector has been proposed [18]. This detector finds corner points by comparing the pixel intensity around a candidate point with machine learning techniques. The FAST detector has performed well in various practical tasks [18,21]. In this proposed method, we also find corner points using the FAST detector.

The proposed terrain surface reconstruction algorithm is based on a triangle mesh formed from the nearest extracted corner points. Therefore, in order to reconstruct the terrain surface more efficiency, it is important that the distribution of the corner points is uniform. In order to achieve this goal, in the proposed algorithm, corner points are independently extracted from regularly divided sub-regions. This enables a triangle mesh to be generated evenly over the input image.

After detecting the corner points, each extracted point must be described distinctively for finding correspondence points. Several ways to describe corner points have been proposed [15]. The well-known local feature descriptors are SIFT and SURF [12,14]. These descriptors have shown good performance in many tasks [15,21]. However, there is a limit to their practical applications because of their computational complexity. In order to overcome this problem, MSD has been proposed for real practical vision application in challenging outdoor conditions [3,13]. Therefore, in order to find correspondence points, each extracted corner point is described by MSD in the proposed method.

In order to keep this paper self-contained, we briefly discuss the concept of MSD. As shown in Figure 3, MSD describes each corner point with three pre-defined scale descriptors $(s1, s2, s3)$. These multiple scale descriptors improve the distinctiveness of the descriptor, as combining the different scales represents the corner point characteristics more efficiently. Furthermore, the computational complexity of MSD is significantly reduced using an integral image.

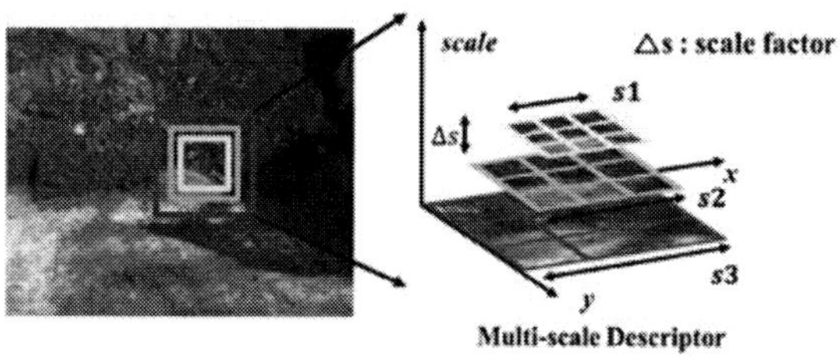

Figure 3. Concept of MSD

Once a pair of correspondence points $x = \{u, v\}$ is found using the MSD descriptor and matching process [3,13,16], the 3D point $X = (u, v, d)^T$ of x is estimated from the difference d, between the matched points, which is called the disparity value. In general, the location of a pixel is presented by an integer value. Therefore, the data type of d is also an integer because it is computed by subtracting corresponding integer-type pixel point locations. This leads to inaccuracy in the 3D point position.

In this paper, to enhance the precision of the 3D point position, a quadratic interpolation method is applied as follows:

$$S(x_{r,t}) = ax_{r,t}^2 + bx_{r,t} + c \quad (1)$$

where $S(x_{r,t})$ indicates the similarity between the target corner location in left image x_l and the rth candidate corner location in right image $x_{r,t}$. The interpolation $S(x_{r,t})$ is computed by comparing the MSD descriptor similarity between x_l and $x_{r,t}$ (see Figure 4).

The quadratic interpolation equation parameters $\{a, b, c\}$ are computed by solving the following linear equation.

$$\begin{bmatrix} x_{r,t-1}^2 + x_{r,t-1} + 1 \\ x_{r,t}^2 + x_{r,t} + 1 \\ x_{r,t+1}^2 + x_{r,t+1} + 1 \end{bmatrix} \begin{bmatrix} a \\ b \\ c \end{bmatrix} = \begin{bmatrix} s(x_{r,t-1}) \\ s(x_{r,t}) \\ s(x_{r,t+1}) \end{bmatrix} \quad (2)$$

Once the quadratic interpolation parameters have been estimated, we can find the optimal disparity value d_{op} easily by solving the quadratic equation.

Note that interpolation is a highly important step in the proposed terrain surface reconstruction algorithm. Regular grid plane vertices are inferred based on the triangle mesh estimated from the sparse 3D point cloud. Therefore, once the precise sparse 3D point cloud is calculated, accurate grid plane vertices can be obtained. This also positively influences the precision of the regular grid plane.

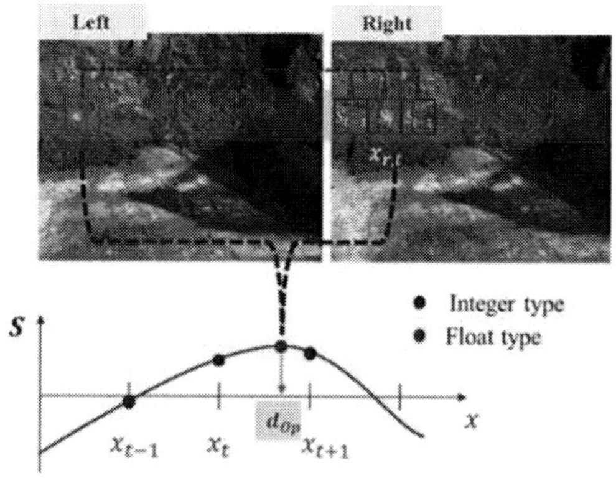

Figure 4. Quadratic interpolation for sparse 3D point cloud improvement

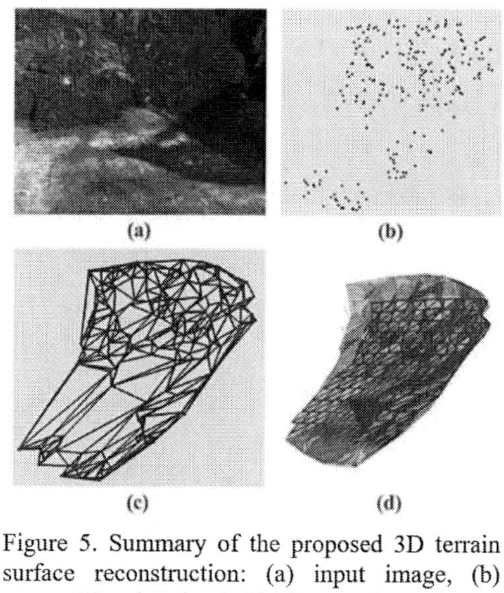

Figure 5. Summary of the proposed 3D terrain surface reconstruction: (a) input image, (b) sparse 3D point cloud, (c) 3D triangle mesh, and (d) reconstructed 3D terrain mesh

4 3D Terrain Surface Reconstruction

The proposed 3D terrain surface reconstruction step comprises three major steps:

(1) sparse 3D point cloud estimation
(2) 2D and 3D triangle mesh generation
(3) 2D and 3D terrain surface creation

Figure 5 shows each step in detail.

Once the disparity value has been estimated by subtracting the location of the corresponding point pair, the 3D point position can be computed based on the stereo camera geometry. The calibration and rectification of the stereo camera is an essential step for obtaining the camera's geometric information. Camera calibration is related to knowing the intrinsic and extrinsic parameters of camera, and the stereo camera rectification step involves horizontally aligning the epipolar line of the stereo images for matching efficiency. In this paper, the well-known open-source library OpenCV [21], was used to calibrate and rectify the stereo images.

In order to infer a regular grid vertex for the 3D terrain surface, we first establish a 2D triangle mesh with the matched point cloud using Delaunay triangulation. Once the 2D triangle mesh set is constructed with the three nearest matched 2D points in the left image, $x_{l,t} = \{u_{l,t}, v_{l,t}\}$, $t = 1, 2, 3$, we then construct the 3D triangle mesh $Ms(x_l^n)$ using the estimated 3D point $X_{l,t} = \{u_{l,t}, v_{l,t}, d_{l,t}\}$ corresponding to the 2D point $x_{l,t}$, where $d_{l,t}$ is the disparity value of 3D point, $X_{l,k}$ calculated from the camera parameters and quadratic interpolation. The mesh is calculated as follows:

$$Ms_i(X_{l,k}) = a_i u_{l,k} + b_i v_{l,k} + c_i \qquad (3)$$

where Ms_i indicates the ith 3D triangle mesh and (a_i, b_i, c_i) are the mesh parameters. The mesh parameters are easily acquired by solving a linear equation with the given matched 3D point set.

Note that construction site can be assumed to be a collection of continuous small surfaces. Construction terrain can be approximately described as a combination of small planes. Therefore, a small-sized triangle mesh reflects the real terrain surface well. Furthermore, we efficiently construct a 3D triangle mesh from the 2D triangle mesh. It is possible to further reduce the computational complexity of the 3D triangle mesh.

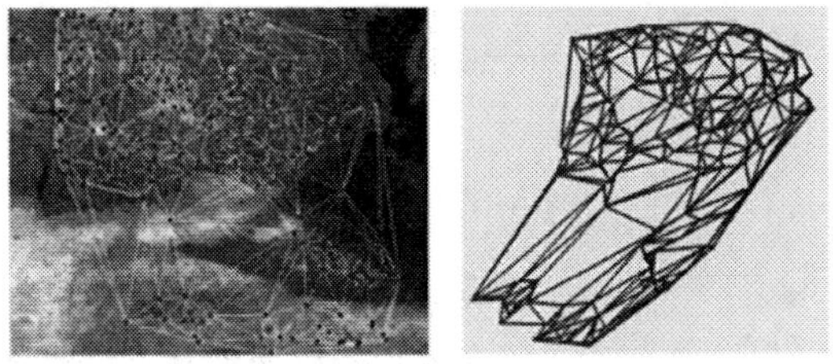

(a) (b)

Figure 6. (a) 2D triangle mesh and (b) 3D triangle mesh structure

After building the triangle mesh, we calculate a regular grid of vertices from the 3D triangle mesh. If the kth regular grid vertex $x_k = \{u_k, v_k\}$ in the image plane is given, we search for the corresponding 3D triangle mesh Ms_i that contains the target grid vertex using the relationship between the 2D and 3D triangle meshes. We then compute the disparity value d_k of the target grid vertex using the 3D triangle mesh model.

$$d_k = a_i u_k + b_i v_k + c_i \qquad (4)$$

Once the disparity value d_k of x_k is calculated, we can define the 3D position of x_k to be $X_k = \{u_k, v_k, d_k\}$, as previously explained, it is easy to convert the disparity-based 3D position into to Cartesian coordinates using the camera parameters.

Note that proposed regular grid-based 3D terrain surface reconstruction has two major advantages. The first one is that regular grid vertices are inferred from robustly estimated nearest 3D points using the triangle mesh model. This improves the result of the proposed 3D terrain surface. The other one is that the grid vertices are computed by simply solving a linear equation from the triangle mesh model. It is hence possible for the proposed algorithm to be applied in practical applications.

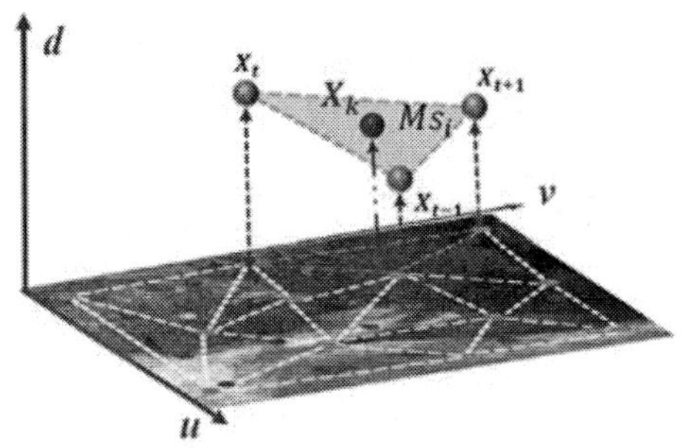

Figure 7. Regular grid vertex estimation

The 3D terrain surface is represented by regular grid plane Gr using the three adjacent grid vertices X_k, k=1, 2, 3 as follows:

$$Gr_i(X_k) = p_i u_k + q_i v_k + r_i \quad (5)$$

where Gr_i indicates the ith regular grid plane containing the three adjacent 3D points, X_k, k=1,2,3. The grid plane parameters (p_i, q_i, r_i) are easily obtained by solving a linear equation with three given grid vertices. Once the grid plane parameters are estimated, we can easily calculate the target area information such as surface slope, area, and distance from the equipment.

Note that the proposed 3D terrain surface is also connected to the corresponding 2D image, similarly to the 2D and 3D triangle meshes. It is a very convenient way to describe the 3D terrain surface. This is because the operator of a piece of construction equipment such as an excavator or wheel loader can intuitively access the 3D information of a certain task area simply by touching the 2D input image using the instrument displays in the cabin.

5 Experiments and Analysis

To evaluate the proposed method, we captured test images from real construction sites to take into account practical conditions. The test images were acquired from a stereo camera mounted on an excavator that consists of two Flea3 USB 3.0 cameras (see Figure 8) [3]. We evaluated each algorithm with four test images Image I (gravel slope), Image II (soil slope), Image III (mud hole), and Image IV (mud bumps). Each test image reflects typical construction environments such as complex surface appearances and various solid materials.

We compared the MSD-based 3D terrain surface reconstruction (MSD-TS) with three popular 3D terrain reconstruction methods: semi-global block matching (SGBM-TS), block matching (BM-TS), and SURF descriptor-based matching (SURF-TS) methods.

Figure 8. Stereo camera system mounted on an excavator

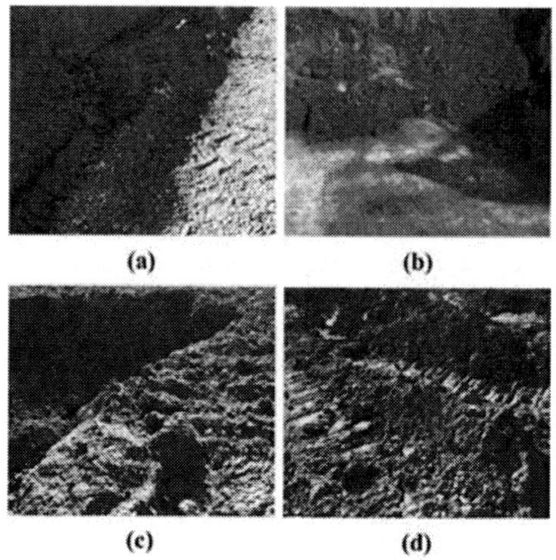

Figure 9. Evaluation image, (a) Image I (gravel slope), (b) Image II (soil slope), (c) Image III (mud hole), and (d) Image IV (mud bumps)

The three comparative algorithms were implemented with the latest OpenCV library [19]. In order to fairly test each algorithm, we used the same 3D terrain surface reconstruction process after finding the correspondence points with each original method. All experiments were performed on a 2.6 GHz single core processor with 8 GB memory.

5.1 Experimental results

The three major steps of the MSD-TS method are shown for each of the four evaluation images in Figure 10. For each test image, the shape of the reconstructed surface is close to the actual slope of the real ground surface.

Note that construction site terrain surfaces consist of textureless and irregular shaped surfaces with different type of solid material such as gravel, mud, and sand. This makes it difficult to reconstruct 3D terrain surfaces. However, given these challenging conditions, the proposed method performed well.

This is explained by the fact that proposed terrain surface model is built from a triangle mesh using a sparse 3D point cloud. Therefore, the accuracy of the sparse 3D points and robustness of triangle mesh plays an important role in the proposed method. The accuracy of the sparse 3D point cloud is improved by quadratic interpolation. The triangle mesh is robustly established with the previously matched nearest three points. This combination of a matched 3D point cloud and robust triangle mesh set has a positive effect on the result of the 3D terrain surface model.

To show the performance of the proposed method clearly, we also compared it the three algorithms, BM-

TR, SGBM-TR, and SURF-TR. As can be seen Figure 11, MSDTR outperformed the BM-TR and SGBM-TR methods for all test images. This is because the matching of patch-based descriptors such as BM and SGBM perform poorly in these kinds of textureless and periodic conditions. This also affected the result of the 3D terrain surface.

SURF-TR achieves a result that was similar to those of MSD-TR because of its good matching performance. However, as mentioned in Section II, SURF-TR has limited suitability for practical applications because of its computational complexity. In contrast, the MSD-TR was fast to calculate in the real 3D terrain surface reconstruction application. We discuss this fact in detail in Section 5.2.

5.2 Computation Time

In Tables I and II, we evaluated the computation time of each method. Although BM-TR and SGBM-TR reconstructed the noisy terrain surface poorly in terms of accuracy, with respect to computation time, BM-TR and SGBM-TR performed well because of the descriptor characteristics of patch-based descriptors.

The average computation time of MSD-TR was about 360 ms. This is far faster than SURF-TR. Furthermore, MSD-TR is 1.2 times faster than SGBM-TR. Note that the operators of construction equipment are generally recommended to operate their machine slowly (under 15 km/h) for safety reasons. In this environment, the computation time of MSD-TR is sufficiently fast to provide ground surface information for automated equipment.

Table 1. Computation Time (ms)

Img		SGBM-TR	SURF-TR	MSD-TR
I	Matching	429.9	17,199	358.8
	Terrain Surface	10.7	10.9	10.6
	Total	440.6	17,209.9	369.4
II	Matching	415.3	17,131	348.2
	Terrain Surface	9.0	9.4	9.2
	Total	424.3	17,140.4	357.4
III	Matching	423.3	17,080	327.2
	Terrain Surface	11.3	11	11.2
	Total	434.6	17,091	338.4
IV	Matching	426	17,371	356.8
	Terrain Surface	10.5	10.3	10.2
	Total	436.5	17,381.3	367

Table 2. Total Computation Times (ms)

Img	BM-TR	SGBM-TR	SURF-TR	MSD-TR
I	209.0	440.6	17,209.9	369.4
II	211.1	424.3	17,140.4	357.4
III	212.5	434.6	17,091.0	338.4
IV	209.6	436.5	17,381.3	367.0

6 Concluding Remarks

This paper proposed a robust and fast 3D terrain surface reconstruction algorithm combined with a regular a grid plane for computational efficiency and robustness. In order to enhance the precision of the 3D terrain surface, the vertices of a 3D grid are calculated from a robustly interpolated nearest 3D point set. Computational efficiency is achieved by constructing 2D and 3D triangle mesh structures. The 3D grid vertex is calculated by simply solving a linear equation using the triangle mesh structure.

The experimental results show that MSD-based 3D terrain surface reconstruction performed well in terms of robustness and computation time. The experimental result demonstrate that MSD-TR is suitable for a terrain surface reconstruction system where accuracy and operating time are vital.

References

[1] Yang, M.-D., Chao, C.-F., Huang, K.-S., Lu, L.-Y., & Chen, Y.-P. (2013). Image-based 3D scene reconstruction and exploration in augmented reality. Automation in Construction, 33, 48–60. doi:10.1016/j.autcon.2012.09.017

[2] Brilakis, I., Fathi, H., & Rashidi, A. (2011). Progressive 3D reconstruction of infrastructure with videogrammetry. Automation in Construction, 20(7), 884–895. doi:10.1016/j.autcon.2011.03.005

[3] Sung, C., & Kim, P. Y. (2016). 3D terrain reconstruction of construction sites using a stereo camera. Automation in Construction, 64, 65–77. doi:10.1016/j.autcon.2015.12.022

[4] Arayici, Y. (2007). An approach for real world data modelling with the 3D terrestrial laser scanner for built environment. Automation in Construction, 16(6), 816–829. doi:10.1016/j.autcon.2007.02.008

[5] Du, J.-C., & Teng, H.-C. (2007). 3D laser scanning and GPS technology for landslide earthwork volume estimation. Automation in Construction, 16(5), 657–663. doi:10.1016/j.autcon.2006.11.002

[6] Davison, A. J., Reid, I. D., Molton, N. D., & Stasse, O. (2007). MonoSLAM: Real-Time Single Camera SLAM. IEEE Trans. Pattern Anal. Machine Intell., 29(6), 1052–1067. doi:10.1109/tpami.2007.1049

[7] Brostow, G. J., Shotton, J., Fauqueur, J., & Cipolla, R. (2008). Segmentation and Recognition Using Structure

from Motion Point Clouds. Computer Vision – ECCV 2008, 44–57. doi:10.1007/978-3-540-88682-2_5
[8] Golparvar-Fard, M., Bohn, J., Teizer, J., Savarese, S., & Peña-Mora, F. (2011). Evaluation of image-based modeling and laser scanning accuracy for emerging automated performance monitoring techniques. Automation in Construction, 20(8), 1143–1155. doi:10.1016/j.autcon.2011.04.016
[9] Kolmogorov, V., & Zabih, R. (2001). Computing visual correspondence with occlusions using graph cuts. Proceedings Eighth IEEE International Conference on Computer Vision. doi:10.1109/iccv.2001.937668
[10] Felzenszwalb, P. F., & Huttenlocher, D. P. (2006). Efficient Belief Propagation for Early Vision. International Journal of Computer Vision, 70(1), 41–54. doi:10.1007/s11263-006-7899-4
[11] Hirschmuller, H. (2008). Stereo Processing by Semiglobal Matching and Mutual Information. IEEE Trans. Pattern Anal. Mach. Intell., 30(2), 328–341. doi:10.1109/tpami.2007.1166
[12] Lowe, D. G. (2004). Distinctive Image Features from Scale-Invariant Keypoints. International Journal of Computer Vision, 60(2), 91–110. doi:10.1023/b:visi.0000029664.99615.94
[13] Changhun Sung, & Myung Jin Chung. (2013). Multi-Scale Descriptor for Robust and Fast Camera Motion Estimation. IEEE Signal Process. Lett., 20(7), 725–728. doi:10.1109/lsp.2013.2264672
[14] Bay, H., Tuytelaars, T., & Van Gool, L. (2006). SURF: Speeded Up Robust Features. Lecture Notes in Computer Science, 404–417. doi:10.1007/11744023_32
[15] Mikolajczyk, K., & Schmid, C. (2003). A performance evaluation of local descriptors. 2003 IEEE Computer Society Conference on Computer Vision and Pattern Recognition, 2003. Proceedings. doi:10.1109/cvpr.2003.1211478
[16] Veksler, O. (2005). Stereo Correspondence by Dynamic Programming on a Tree. 2005 IEEE Computer Society Conference on Computer Vision and Pattern Recognition. doi:10.1109/cvpr.2005.334
[17] Harris, C., & Stephens, M. (1988). A Combined Corner and Edge Detector. Procedings of the Alvey Vision Conference 1988. doi:10.5244/c.2.23
[18] Rosten, E., & Drummond, T. (2006). Machine Learning for High-Speed Corner Detection. Lecture Notes in Computer Science, 430–443. doi:10.1007/11744023_34
[19] G. Bradski, A. Kaehler. Learning OpenCV. O'Reilly Media Inc., 2008.
[20] Beis, J. S., & Lowe, D. G. (1997). Shape indexing using approximate nearest-neighbour search in high-dimensional spaces. Proceedings of IEEE Computer Society Conference on Computer Vision and Pattern Recognition. doi:10.1109/cvpr.1997.609451
[21] Gauglitz, S., Höllerer, T., & Turk, M. (2011). Evaluation of Interest Point Detectors and Feature Descriptors for Visual Tracking. Int J Comput Vis, 94(3), 335–360. doi:10.1007/s11263-011-0431-5

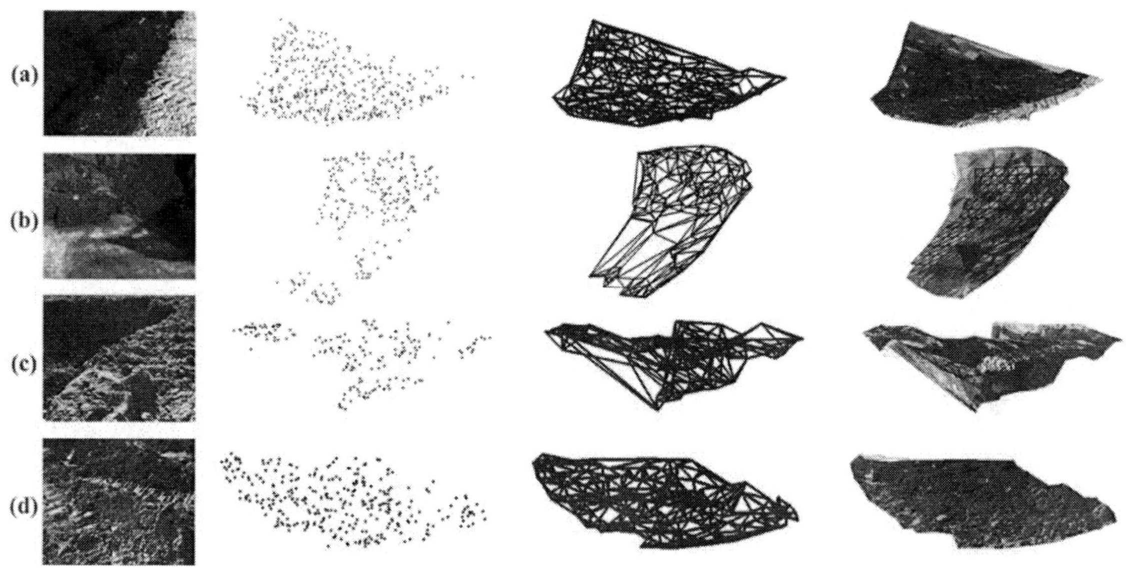

Figure 10. Results of the MSD-TR 3D terrain surface reconstruction method: test images (column 1), sparse 3D reconstruction (column 2), triangle mesh (column 3), and dense reconstruction (column 4)

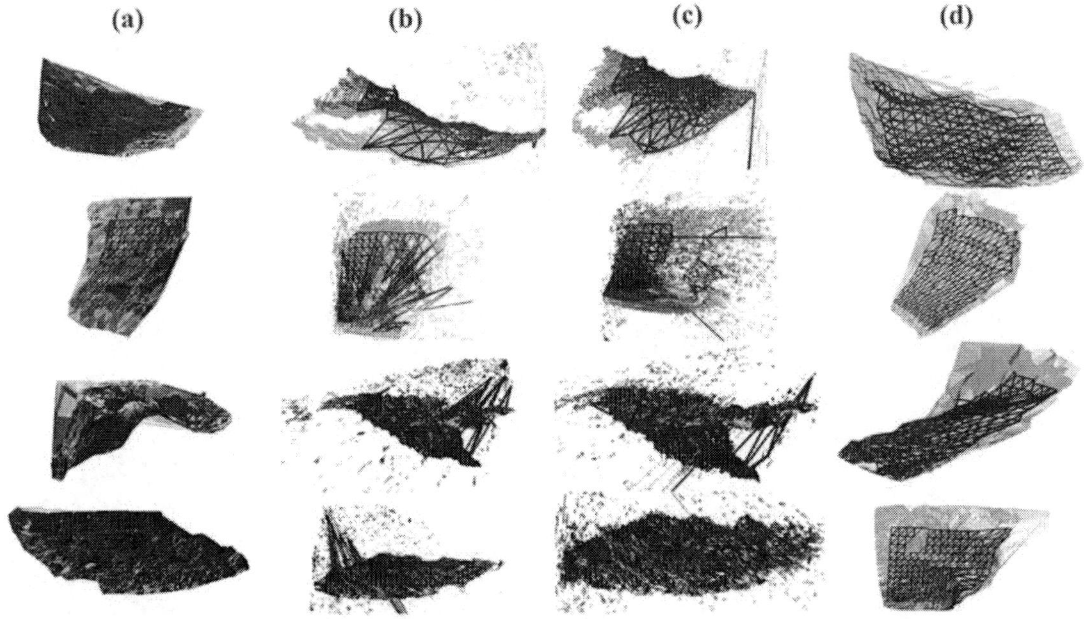

Figure 11. Comparison of the results of different 3D terrain surface reconstruction algorithms: (a) MSD-TR, (b) BM-TR, (c) SGBM-TR, and (d) SURF-TR

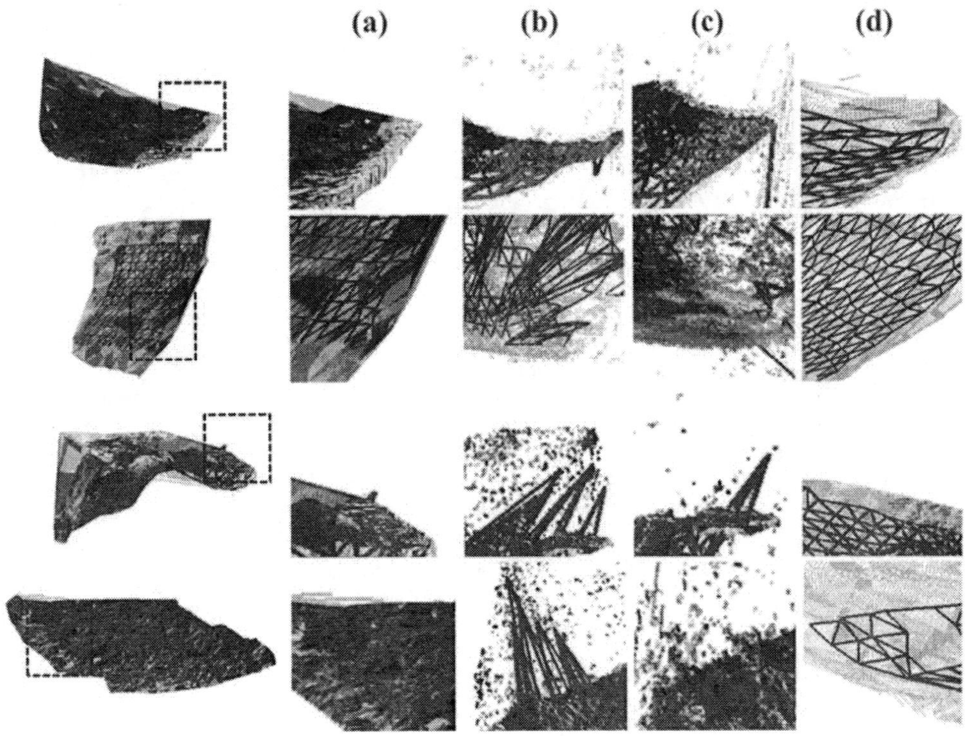

Figure 12. Magnified results of (a) MSD-TR, (b) BM-TR, (c) SGBM-TR, and (d) SURF-TR

Sampling-based floor plan analysis on BIMs

Christian Schlette and Jürgen Roßmann

Institute for Man-Machine Interaction (MMI), RWTH Aachen University, Germany
Email: schlette@mmi.rwth-aachen.de, rossmann@mmi.rwth-aachen.de

Abstract –

This contribution explores a novel combination of methods from robotics and BIM by working on Building Information Models (BIMs) with the Open Motion Planning Library (OMPL), an extensible programming interface and reference collection of various algorithms for sampling-based motion planning. We present a seamless workflow from industrial BIM generation to sampling-based planning of collision-free paths for mobile robots on 3D floor plans extracted from BIMs.

Beyond robot motion planning, we apply the workflow to generate huge amounts of paths for floor plan analysis. We assume the paths to resemble walking patterns where many paths crossing a given floor tile match with heavy usage of this space. The resulting paths are sampled in heatmaps, where pixels then correspond to floor tiles and the color value corresponds to the number of crossing paths, yielding a semi-automatic, easy-to-use method to analyze floor plan layouts in different scenarios.

Keywords –

BIM, Sampling-based Motion Planning, Floor plan analysis

1 Introduction

In robotics, we are used to see CAD data for mechanisms, machines and plants from industrial sectors such as automotive and aerospace, but based on the underlying approaches in computer aided engineering, the construction industry developed a similar concept called Building Information Modeling (BIM). At their core, Building Information Models (BIMs) are databases with well-defined layers for modeling and managing the relevant information for a wide range of share- and stakeholders of modern construction enterprises.

As the main application of BIMs is to hold the necessary data for construction, BIMs contain central parameters and semantic information about all structural elements, e.g. classes, dimensions and materials of walls and doors. Using Constructive Solid Geometry (CSG), the elements can be evaluated to generate full 3D geometric models (see Figure 1), yielding semantically labeled 3D models.

Figure 1. Semantically labeled 3D model of a multi-floor office building from BIM

Due to its significance in the construction industry, architects and construction engineers already explore new applications of BIMs, e.g. by using BIMs for building simulation and analysis [1][2]. As an example of how methods from robotics can contribute to such developments, we demonstrate the application of sampling-based algorithms for robot motion planning as well as floor plan analysis in this contribution.

Many well-established methods for motion planning are available for mobile platforms, with sampling-based motion planning being the most prominent approach for high-performance, model-based path generation in complex environments. Here, the Open Motion Planning Library (OMPL) provides a thoughtful Application Programming Interface (API) as well as an extensible set of reference implementations of historical and current algorithms (see Section 2). The models for collision testing in these algorithms are usually provided as CAD data for known environments, sensor data to enable planning in (partially) unknown environments [3][4] or plans sketched by human operators [5].

In section 3, we describe the workflow and prerequisites for making BIMs available in our software environment for robot simulation and control. Since motion planning for mobile platforms is a central task in indoor working environments, we demonstrate the usage of BIMs in section 4 by running OMPL on the model data. In section 5, we present first results of this combination of two strong concepts – BIM from the construction industry, OMPL from the robotics community – and discuss specific aspects of our solution for motion planning on BIMs and floor plan analysis.

2 State of the Art

Early approaches in motion planning for manipulators and mobile robots started planning in sandbox environments with fully known semantics and geometries, e.g. [6][7]. But since the a priori availability of detailed plans and models was not realistic and thus doubted at that time, research on mobile platforms turned towards "bottom-up", sensor-based localization and navigation in unknown environments. With BIM, this situation changes dramatically, as semantically labeled 3D models are becoming the standard data basis for modern construction enterprises. So far, only a few attempts have been made to link BIMs with navigation and motion planning. Isikda et al are extending BIM with a new data layer for (human) indoor navigation, e.g. to support emergency operations [8]. Borkowski et al are correlating semantic information from BIMs with sensor findings from mobile platforms in order to support "semantic navigation" [9], an approach which is later extended by Siemiatkowska et al towards "semantic localization" [10].

2.1 Open Motion Planning Library

Sampling-based methods are the most prominent approach for solving problems in fully defined, tight and cluttered workcells [11][12]. OMPL [13] is a C++ library developed by Kavraki et al. [14]. The development is visible since 2012 and offers many versions of "Rapidly-exploring Random Trees" (RRT) [15], "Probabilistic Roadmap Method" (PRM) [16], Sparse Roadmap Spanner algorithm" (SPARS) [17], "Expansive Space Trees" (EST) [18] as well as many other algorithms, in parts as contributions from OMPL users. The library is a Linux-driven development with the option to run on Windows and Mac. OMPL is the library behind "MoveIt!", the motion planning package for the "Robot Operating System" (ROS) [19].

2.2 Building Modeling Information

The BIM methodology encompasses processes and methods to create and manage virtual 3D building models (BIMs). As a central database for construction projects, BIMs contain semantic information, parameters and geometries of all structural elements, but also tracking information to share developments, changes and comments with share- and stakeholders. Although BIM started as a computer aided engineering approach for architects, construction engineers and civil engineers, BIMs are managed and applied today throughout the whole lifecycle of buildings – beyond design, planning and construction also for energy and facility management, automation and destruction [21][22].

BIM generation is currently limited to specialized 3D construction tools. Most of the tools are developed as parts of bigger software packages for 3D construction, which also offer solutions for managing and rendering BIM data. Today, the integration and usage of BIM data beyond the given capabilities of these dedicated software packages requires exporting the data to an intermediate format – e.g. exporting to VRML for rendering 3D views of the building [23].

A more general approach to exchange BIMs independently from specific software packages is provided by the "Industry Foundation Classes" (IFC) [24]. IFC is a neutral, ISO-certified (ISO/PAS 16739) data format standardized by buildingSMART, an international alliance of the developers of the leading BIM software packages. IFC data contains three major subsets, so-called "Views", *CoordinationView* for building design and construction management, *StructuralAnalysisView* for building simulation and analysis, and *FMHandOverView* to link buildings to computer aided facility management software. The basic structure of IFC data follows the "Standard for the Exchange of Product Model Data" (STEP) for the file-based exchange of geometric data (ISO 10303).

3 Concept

Our approach of running OMPL on BIMs is based on the software environment VEROSIM® for robot simulation and control, which we co-design and co-develop at the Institute for Man-Machine Interaction (MMI) for our research on eRobotics [25]. The eRobotics methodology provides a development platform for roboticist to exchange ideas and to collaborate with experts from other disciplines. The central method in eRobotics are Virtual Testbeds, where complex technical systems and their interaction with prospective working environments are first designed, programmed, controlled and optimized in 3D simulation, before commissioning the real system. Typical working environments are open landscapes, forests, cities, buildings, factories etc. In consequence, BIM on the one hand provides an important standard for modeling relevant environments and simulating their interdependencies with complex technical systems. On the other hand, modern buildings themselves are complex, highly automated, technical systems, which can profit from the simulation-driven development processes in Virtual Testbeds. We described this idea of "Virtual BIM Testbeds" in [23].

3.1 Prerequisites

The general idea of VEROSIM is to provide a micro-kernel architecture for simulation, where a slim core provides the central functionality to structure and manage a database which represents the simulated scenario. Individual simulation methods, datatypes and UIs are added via plugins. We briefly outline the system features which contribute most to our approach of running OMPL on BIMs.

Following the principles of Object-Oriented Modelling (OOM), the central database of VEROSIM is an active database, that contains all simulation entities and information of the environment as well as extensive meta data/ meta type information. The database is active, in the sense that it contains functionalities to connect and define the behavior of simulation entities directly in the database. This behavioral information differentiates the database from simple scene graph representations. The ability to interpret and manage meta data and meta type information enables the generic import, mapping and access of external database formats, such as IFC data.

For motion planning, the Kinematics plugin of VEROSIM provides the underlying functionalities to control robots in Cartesian or configuration space to sample the validity of given states. Collision detection in our system is carried out based on the simple approach to hierarchically compare the intersection of 1) axis-aligned bounding boxes, 2) user-defined simplified collision geometries, and 3) full polyhedrons. In case of BIMs, the geometries are simple enough to sustain performant collision detections without the need to introduce additional simplified geometries for level 2) tests.

3.2 Workflow

Figure 2 depicts the basic workflow of our approach. BIMs are typically modeled in 3D construction tools such as Autodesk Revit®, which have much in common with 3D CAD tools for other disciplines. The interaction with BIMs at this stage is mainly based on manufacturer-specific, proprietary database formats. As described in section 2, the major exchange format to transfer BIM data between the construction tools of different manufacturers is IFC. Using the IFC Engine DLL [26], our software is able to import and map IFC data directly into our own database (see Section 4 for details). As part of the import process, the IFC Engine DLL generates 3D geometries from CSG evaluations of the structural elements and their properties. In combination, the IFC data and the geometries yield semantically labeled 3D models, in which geometries are associated with their generator classes, the structural elements. The direct accessibility of the IFC data in our own database allows us to a) use the methods and tools in our software environment to work on and interact with the BIM, and b) to refer to and mix the IFC elements with our own Versatile Simulation Database (VSD) elements for robot simulation and control.

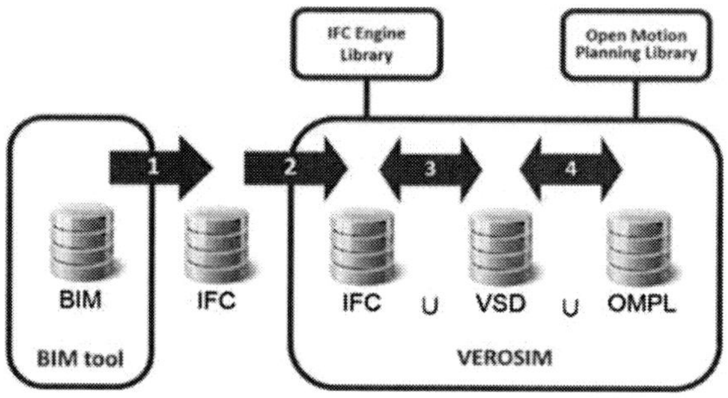

Figure 2. Workflow of running OMPL on BIM: 1) Export of IFC data from BIM tool, 2) Import of IFC data into VEROSIM database, 3) Preparation of IFC data with additional VSD elements, 4) OMPL queries and solution based on OMPL adapter classes

In particular, we use this access to prepare the building data for collision detection. Based on the semantics and a database search, we can identify doors directly from their class names and remove them in order to open rooms and corridors for passages. Also based on the semantics and a database search, we can collect all geometries which belong to a given entity (e.g. a selected floor or wing of a building) and connect them to the collision detection methods in our system.

In order to use OMPL on any 3D model in VEROSIM, we developed a set of adapter elements to make OMPL functionalities available in our database (see Section 4 for details). The adapter classes allow us to parameterize, trigger and store any number of OMPL queries and solutions directly by referring to existing elements in the database, e.g. by referring to robots, workcell geometries and 3D poses in the VSD or the IFC data. Also via the adapter classes, paths generated with planners in OMPL are directly available for robot simulation and control.

4 Running OMPL on BIMs

VEROSIM provides an importer based on the IFC Engine DLL by RDF Ltd. [26] for integrating and accessing IFC data [23]. The importer is set up by reading the meta data/ meta type schemas of the IFC language, thus being fully and generically in sync with a given version of the IFC data format. Then, the IFC schema is bidirectionally mapped to the VSD schema for associating incoming IFC data types with corresponding representations in the VSD (and vice versa). The import of concrete IFC data makes use of this so-called "schema mapping" by instantiating and initializing corresponding VSD types for the incoming IFC instances, thereby representing structure and details of the given IFC data in our own database. After this so-called "instance mapping", we are able to refer to and mix the IFC elements with our own VSD elements in a third step called "functional mapping". More details on this three-step mapping process are available in [27].

For interfacing OMPL, we developed a set of adapter classes to represent and map the main classes and parameters for formulating single query planning problems for the OMPL library (see Figure3 for details).

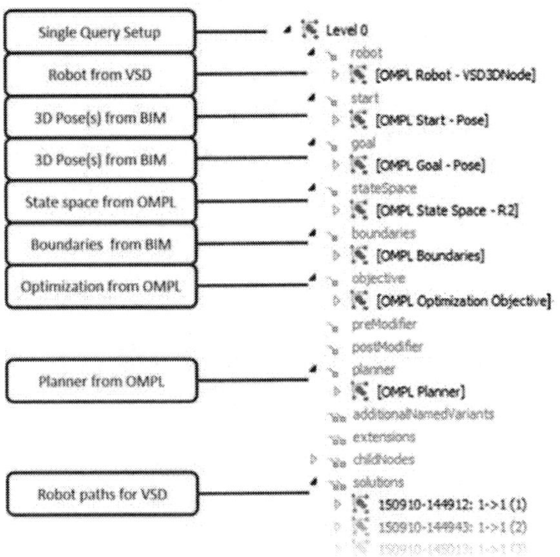

Figure 3. Setup with OMPL adapter classes

For a basic single query setup, we select a state-space model (e.g. *SE(3)*) and a planner (e.g. RRT*), and define start and goal states of the planning problem by selecting 3D poses from the database. In addition, we select a robot (including geometries, forward and inverse kinematics) and link it to an object derived from *ompl::base::StateValidityChecker*. During planning, this object signals OMPL requests to check the validity of proposed states in order to advance the sampling-based building of a collision-free path to reach the goal. For responding to the request, our derived implementation of the validity checker converts states proposed by the planner to our internal pose representation for Cartesian frames or joint configurations. The robot is set to this pose for a moment with the (invisible to the user) effect of rearranging the 3D workcell accordingly. In this rearranged situation, which represents the state proposed by the planner, we call the collision detection in VEROSIM and store the result. We then set the robot back to its original pose and respond to the request of the planner: Only if the rearranged situation did not exhibit any undesired collisions, the state is assessed to be valid (and invalid otherwise). The communication between OMPL and our system is depicted in Figure 4.

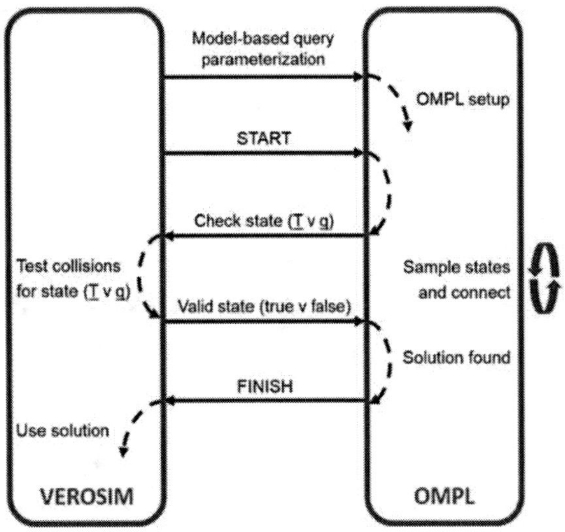

Figure 4. Interfacing OMPL with our system

5 Results

Following our workflow, running OMPL on BIMs is easily set up by importing a building model into our system, preparing it for collision detection and defining OMPL queries. As we are looking into motion planning for a mobile platform working on indoor tasks, we reduce the complexity of the full 3D problem in *SE(3)* by planning on 2D planes on individual floors in R^2 (see Section 5.1). In order to compensate for the missing transfer connections between floors, we then introduce "portals" on each floor, i.e. pre-defined sets of start resp. goal poses for entering and leaving a given floor (see Section 5.2). In section 5.3, we demonstrate how OMPL can be applied for BIM-based building simulation and analysis by evaluating floor plans by compiling numerous paths between arbitrary poses to learn about often used, thus critical passages.

In our experiments, we are using RRT*, an asymptotically optimal derivative of RRT by Karaman et al [20] which converges on the optimal path resp. shortest path in our experiments. This allows us to allocate a fixed runtime of 30 [s] for queries, in which RRT* will first find a collision-free path and then continue to minimize its path length to provide a high quality solution. Beyond a certain threshold of about 3 [s] for the construction of a first connection between start and goal pose, runtimes in our experiments can thus be assumed to scale linearly with the number of queries.

5.1 Single Floor Planning

For each floor $j \in \{-1, 0, 1, 2\}$, we select one representative start pose per room as well as one goal pose per exit resp. entry to the floor. In the given example, the floors are connected by central stairs, yielding one goal pose per floor, with only one additional exit on ground level. For each floor, we run RRT* three times to generate solutions from each start pose to the goal pose resp. both poses on Floor 0. The result is depicted in Figure 5.

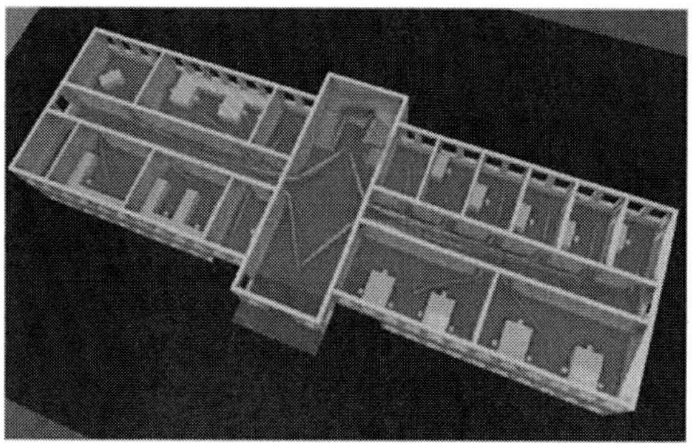

Figure 5. RTT* on Floor 0: 15 start poses (1 per room), 2 goal poses (portal "Stairs" and Exit), avg. solution length: 17.34 [m], runtime: 4.5/45.0 [min] (low/high quality)

5.2 Multi Floor Planning

We introduce "portals" for evaluating motions between individual floors and for describing the topology of multiple floors (see Figure 6). Portals are pre-defined poses at exits resp. entries to a floor and thus the essential start and goals poses for single floor motion planning. Using a portal for transfer between floors is assumed to add specific costs P to the solution, which represent average times for waiting and usage, e.g. for stairs and lifts. Equation (1) formulates a simple cost function for entering a portal i on floor j_{in} and leaving it on floor j_{out}:

$$P_i = c_{i_{j_{in}}} + \sum_{j_{in}}^{j_{out}} p_i w_{i_j} + c_{i_{j_{out}}}, \qquad (1)$$

where $c_{i_{j_{in}}}$ resp. $c_{i_{j_{out}}}$ represent the costs to enter resp. leave this specific portal on these specific floors. p_i is a portal-specific factor to ascend resp. descend one floor. The additional weight w_{i_j} allows for considering floor-specific variations from p_i as well as priorities and penalties.

The costs to use a portal can be expressed in terms of additional solution length, which allows for comparing multi floor planning solutions by assessing various combinations of average solutions lengths from single floor planning and portal costs.

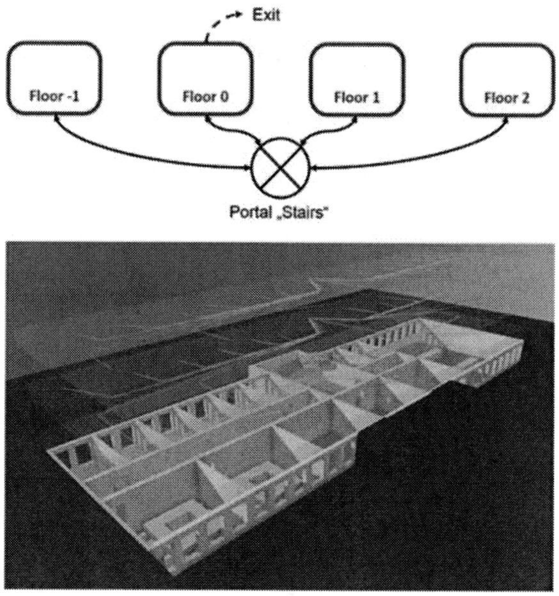

Figure 6. **(Top)** Floor topology with portals; **(Bottom)** RTT* on all floors: 64 start poses (1 per room), 1 goal pose (portal "Stairs") plus 1 goal pose on Floor 0 (Exit), avg. solution length: 16.29 [m], runtime: 9.6/96.0 [min] (low/high quality)

5.3 Floor Plan Analysis

So far, we demonstrated the combination of OMPL and BIM for robot motion planning in building maintenance tasks. On the other hand, the combination can also contribute to BIM-based building simulation and analysis. For floor plan analysis, we span a grid of 6 × 21 poses over a given floor (see Figure 7), which may well include invalid poses permanently colliding with walls and furniture. In a basic approach, we run RRT* to generate solutions from all grid poses to a given goal pose. The resulting solutions are then sampled in a heatmap, where a pixel position corresponds to a floor tile and the color value corresponds to the number of solutions crossing this floor tile. Figure 8 depicts the solutions "grid to exit" and the corresponding heatmap. We refine the basic approach by using the grid to define not only the start poses but also the goal poses, yielding a total of 15750 combinations in our example (however, many invalid due to collisions). Figure 9 depicts the solutions "grid to grid" and the corresponding heatmap. Clearly visible now are heavily used floor tiles in the right corridor in front of the larger offices, which may indicate critical passages, e.g. in case of emergency.

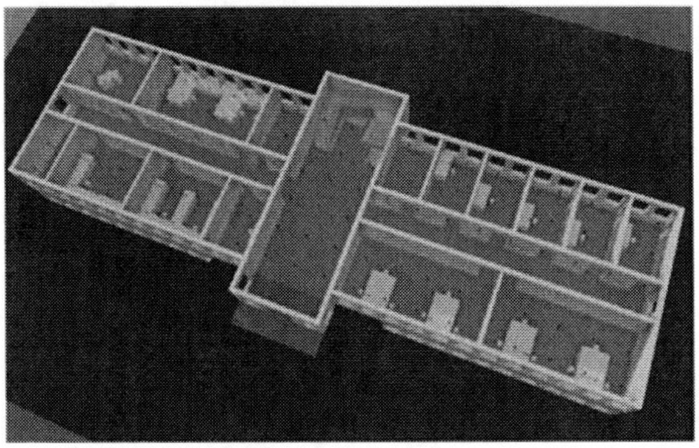

Figure 7. Grid of 126 poses on Floor 0

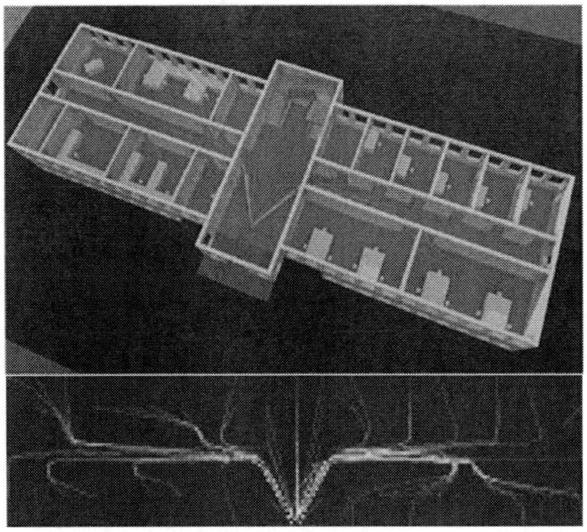

Figure 8. (**Top**) RTT* on Floor 0: 126 start poses from grid (69 invalid), 1 goal pose (Exit), avg. solution length: 14.50 [m], runtime: 8.5/85.0 [min] (low/high quality); (**Bottom**) Heatmap: 200 x 50 [px], 10 samples per [m]

6 Conclusions

In this contribution, we are combining two strong concepts – BIM from the construction industry, OMPL from the robotics community – to allow for sampling-based motion planning directly on semantically labeled 3D building models.

In robotics, the availability of semantics and geometries of working environments has been assessed to be unrealistic a priori knowledge for many years. Now BIM has the potential to change future developments in robotic indoor applications significantly, as professionally generated, semantically labeled 3D models will become more and more available. For robotic applications such as building maintenance, BIM thus provides profound data for navigation and motion planning, which of course has to be additionally enriched and correlated with sensor data to cope with time-variant obstacles such as furniture and human co-workers.

Beyond motion planning for mobile platforms in indoor environments, we demonstrate the application of algorithms from motion planning for floor plan analysis as an example of how methods from robotics can contribute to BIM-based building simulation and analysis. In particular, the proposed floor plan analysis allows for identification of "shortest path" movement patterns between rooms and consequential heavily used floor segments, which may indicate critical passages that need closer attention.

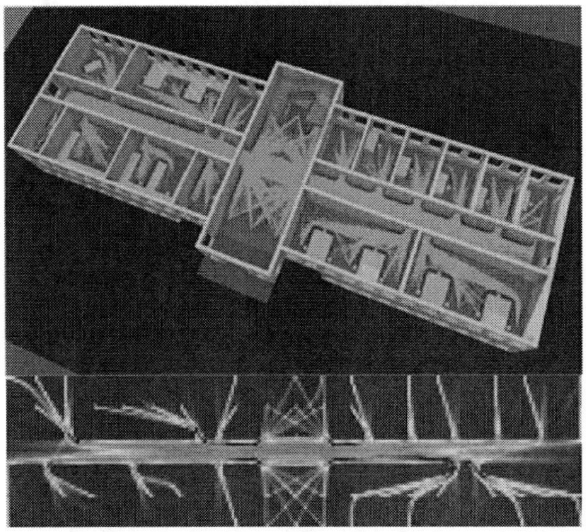

Figure 9. (**Top**) RTT* on Floor 0: 126 start poses from grid (69 invalid), 126 goal poses from grid (69 invalid), avg. solution length: 18.18 [m], runtime: 0.3/3.3 [d] (low/high quality); (**Bottom**) Heatmap: 200 x 50 [px], 10 samples per [m]

Acknowledgment

We would like to thank Peter Bonsma, main developer of the IFC Engine DLL at RDF Ltd., for sharing experiences and giving advice on using the engine on Windows 8.

References

[1] Bazjanac, V. IFC BIM-based methodology for semi-automated building energy performance simulation. In *Proc. CIB W78: Improving the management of construction projects through IT adoption*, pp. 292–299, 2008.

[2] Li, J., Hou, L., Wang, X., Wang, J., Guo, J., Zhang, S. and Jiao, Y. A project-based quantification of BIM benefits. *Int. Journal of Advanced Robotic Systems*, 11(123), 2014.

[3] Um, D., Gutierrez, M., Bustos, P. and Kang, S. Simultaneous planning and mapping (SPAM) for a manipulator by Best Next Move in unknown environments. In *Proc. IEEE/RSJ Int. Conf. Intelligent Robots and Systems (IROS)*, pp. 5273–5278. 2013.

[4] Masehian E. and Katebi, Y. Sensor-based motion planning of wheeled mobile robots in unknown dynamic environments. *Journal of Intelligent & Robotic Systems*, 74(3-4):893–914, 2014.

[5] Li, J., Xu, A. and Dudek, G. Graphical State Space Programming: A visual programming paradigm for robot task specification. In *Proc. IEEE Int. Conf. Robotics and Automation (ICRA)*, pp. 4846–4853, 2011.

[6] Nilsson, N. *Shakey the robot.* SRI International, Menlo Park, CA, US, Tech. Rep. Technical Note no. 323, 1984.

[7] Lozano-Perez, T.A simple motion-planning algorithm for general robot manipulators. *IEEE Journal of Robotics and Automation*, 3(3):224–238, 1987.

[8] Isikdag, U., Zlatanova, S. and Underwood, J.A. BIM-oriented model for supporting indoor navigation requirements. *Computers, Environment and Urban Systems*, 41:112–123, 2013.

[9] Borkowski, A., Siemiatkowska, B. and Szklarski, J. Towards semantic navigation in mobile robotics. In *Graph transformations and model-driven engineering*. Springer, Heidelberg, Germany, pp. 719–748, 2010.

[10] Siemiatkowska, B., Harasymowicz-Boggio, B., Przybylski, M., Rozanska-Walczuk, M., Wisniowski, M. and Kowalski, M. BIM based indoor navigation system of Hermes mobile robot. In *Romansy 19 - Robot Design, Dynamics and Control*. Springer, Vienna, Austria, pp. 375–382, 2013.

[11] LaValle, S.M. *Planning algorithms*. Cambridge University Press, New York, NY, USA, 2006.

[12] Choset, H., Lynch, K.M., Hutchinson, S., Kantor, G.A., Burgard, W., Kavraki, L.E. and Thrun, S. *Principles of robot motion*. MIT Press, Cambridge, MA, USA, 2005.

[13] Kavraki Lab, Rice University. Houston, TX, USA. Open Motion Planning Library. Online: http://ompl.kavrakilab.org, Accessed: 03/2016.

[14] Sucan, I.A., Moll, M. and Kavraki, L.E. The Open Motion Planning Library. *IEEE Robotics & Automation Magazine*, 19(4):72–82, 2012.

[15] Kuffner J. and LaValle, S.M. RRT-connect: An efficient approach to single-query path planning. In *Proc. IEEE Int. Conf. Robotics and Automation (ICRA)*, pp. 995–1001, 2000.

[16] Kavraki, L.E., Svestka, P., Latombe, J.-C. and Overmars, M.H. Probabilistic roadmaps for path planning in high-dimensional configuration spaces. *IEEE Transactions on Robotics and Automation,* 12(4):566–580, 1996.

[17] Dobson, A., Krontiris, A. and Bekris, K.E. Sparse roadmap spanners. In *Proc. Workshop Algorithmic Foundations of Robotics (WAFR)*, pp. 279–296, 2013.

[18] Hsu, D., Latombe, J.-C. and Motwani, R. Path planning in expansive configuration spaces. In *Proc. IEEE Int. Conf. Robotics and Automation (ICRA)*, pp. 2719–2726, 1997.

[19] Sucan, I.A. and Chitta, S. MoveIt! Online: http://moveit.ros.org, Accessed: 03/2016.

[20] Karaman S. and Frazzoli, E. Sampling-based algorithms for optimal motion planning. *Int. Journal of Robotics Research*, 30(7):846–894, 2011.

[21] Eastman, C., Teicholz, P. and Sacks, R. *BIM Handbook: A guide to Building Information Modeling for owners, managers, designers, engineers and contractors*. Wiley, Hoboken, NJ, USA, 2011.

[22] Hardin B. and McCool, D. *BIM and construction management: Proven tools, methods, and workflows*, Wiley, Hoboken, NJ, USA, 2015.

[23] Roßmann, J., Schluse, M., Hoppen, M., Losch, D., Hempe, N. and Schlette, C. Virtual BIM Testbeds. In *Proc. IEEE Int. Conf. Developments in eSystems Engineering (DeSE)*, 2015.

[24] buildingSMART International Ltd. Industry Foundation Classes (IFC). Online: http://www.buildingsmart-tech.org, Accessed: 03/2016.

[25] Schluse, M., Schlette, C., Waspe, R. and Roßmann, J. Advanced 3D simulation technology for eRobotics. In *Proc. IEEE Int. Conf. Developments in eSystems Engineering (DeSE)*, pp. 151–156, 2013,

[26] RDF Ltd. Bankya, Bulgaria. IFC Engine DLL. Online: http://rdf.bg, Access: 03/2016.

[27] Hoppen, M., Schluse, M., Roßmann, J. and Weitzig, B. Database-driven distributed 3D simulation. In *Proc. Winter Simulation Conference (WinTeSys)*, pp. 1–12, 2012.

Site Plan Development: Tower Crane Placement Based on Data Obtained from IFC File

T. Funtík[a] and J. Gašparík[b]

[a]Slovak University of Technology in Bratislava, Faculty of Civil Engineering
[b]Slovak University of Technology in Bratislava, Faculty of Civil Engineering
E-mail: tomas.funtik@stuba.sk, jozef.gasparik@stuba.sk

Abstract -

One of the fundamental issues of site plan development is the location of a tower crane which needs to be designed to meet specific criteria and specific needs of structures. With today´s traditional techniques it is quite difficult to consider optimal position and to select adequate technical parameters of a tower crane. This task is rather time consuming and complicated. With use of Building Information Modeling (BIM) this process can be simplified and the tower crane can be designed more effectively considering various criteria.

This paper focuses on obtaining the data of all relevant parametric objects from IFC file and using them to compute optimal position of a tower crane. A suggested position is evaluated from the perspective of travel time of a hook, travel distance and length of jib needed to deliver all elements and materials. Data obtained from IFC model are compared with workload curve of a tower crane to ensure constructability and safety.

As a result, map of criteria is done according to the coordinates of a tower crane arranged together with above-mentioned parameters to indicate differences between all possible locations.

Keywords –

BIM; Tower Crane; Building Technology

1 Introduction

During pre-construction stage we need to design the site plan according to a chosen building technology and selected facilities. The issue of a suitable location of tower cranes for the construction is up to date almost within all ranges of buildings. Currently, the cranes are mostly designed according to personal experience of the processor of construction site plan (POV), resp. the planner who has the task to develop a plan of construction site equipment within preparation phase of a contractor. The costs and also the smoothness and speed of construction directly depend on appropriate selection of the type, number and location of cranes. Today's computer technology allows for simplification of work in planning, and therefore it would be benefitial to establish a mathematical model and a computer program that would make this task easier.

2 BIM employment

The goal of BIM is not to create a model itself, but to gain complete, reliable, accessible and easily exchangeable information to anyone who might need it throughout the whole life cycle of the building. One of the essential prerequisites for BIM is to set cooperation between all stakeholders at different stages of a life cycle. Within the workflow, it is necessary to allow users to enter, change, update or correct any information in the model that is relevant to their roles in the project.

Virtual model is composed of 3D parametric objects that allow to bear information by using a specified data format. By assembling them together, we make so-called virtual construction resulting in improved quality of a project. While using traditional CAD (Computer Aided Design) it happens quite often that just a limited part of a building is designed and most of architects select the simplest sections to make their job easier. They do not care what is 200mm behind the cross section is and this may lead to delay in construction and makes the coordination of a project impossible. On the other hand, by using BIM, architect is obliged to design all details in a building so this certainly results in a coordinated project of high quality which is moreover much easier to check. Key to interoperability, which is another fundamental prerequisite of this approach is an IFC (Industry Foundation Classes) file type – a neutral data format used to describe, exchange and share information between stakeholders and various software.

Not many people are aware that BIM is not just about buildings and civil engineering, but of course highly suitable also for infrastructure - railways, highways, bridges, etc.

3 Proposed methodology

3.1 Actual Developments

In todays practice cranes are often designed without precise conceptuality, which reduces the efficiency of the entire construction site traffic and results in complications in works that cannot be done without a crane. Besides, with increasing pressure to reduce operating costs contractors must emphasize not only savings in management, but also shortening the time of deployment techniques, mechanisms and workers and thus saving resources.

BIM in this issue offers an opportunity to obtain input data directly from a virtual model without extensive manual analysis of the object. By reducing the number of manual tasks, construction planner can devote more time to optimization and alternatives. Depending on the size of the object, this time savings can also be a few days of work.

Although we are able to use sophisticated methods developed by previous authors, there was still an issue with data extraction from traditional plans. Authors (Al-Hussein, Niaz, Yu, & Kim, 2005; Lee, 2009, 2012; Irizarry & Karan, 2012) presented their studies how to take the advantage of parametric modelling in tower crane design but they did not address the question of obtaining information from the object. Although it is clear that all the information we need to design a crane are in a virtual model, currently there is no tool available which can extract this data from virtual parametric model. Model itself contains a lot of information and it is necessary to provide filtering of relevant data for the design of the crane.

3.2 Decision flowchart of methodology

As defined in a scheme shown on Figure 2 decisions are taken according to data obtained from BIM model represented by IFC file.

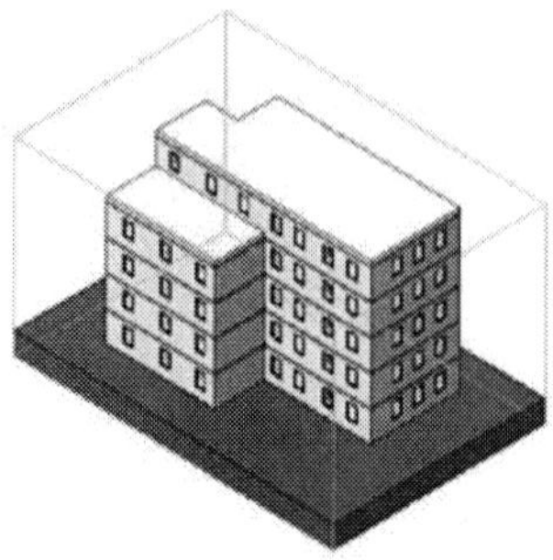

Figure 1 Investigated BIM model, Revit

Primary focus is on the individual data about the position, volume and material type of elements. These data are clear to read and therefore usable for further processing.

Individual steps of methodology are described in the following chapters.

The methodology was further verified on a building with more complicated floor plans. The more complicated building is, the use of automation helps more significantly. Finding the optimal position is not trivial and cannot be determined only intuitively as the mass of the building can be variable from bottom up.

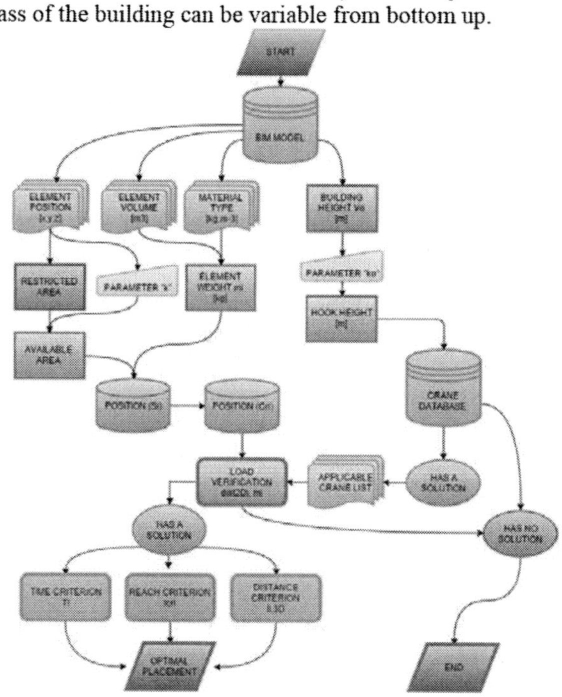

Figure 2 Decision flowchart of methodology

3.3 Model Analysis

To make the whole design process of cranes automatic, it is necessary to ensure that the program is able to obtain information from the virtual model of the building, which was created in an environment using parametric BIM objects.

Table 1 Data export from parametric model

i	x	y	z	Název prvku	Objem [mm3]	Materiál	Hmotnost [kg]	Počet dodávek	Hmotnost dodávky - m [kg]
1	1,92	7,058	-7,803	Wall Foundation:600/1200 na osu:966472	13608000000	Beton prostý	34020	14	2500
2	-7,23	4,758	-7,803	Wall Foundation:600/1200 na osu:966473	2880000000	Beton prostý	7200	3	2500
3	5,685	-1,742	-7,803	Wall Foundation:600/1200 na osu:966474	8172000000	Beton prostý	20430	9	2500
4	11,67	2,658	-7,803	Wall Foundation:600/1200 na osu:966475	5904000000	Beton prostý	14760	6	2500
5	-5,155	3,058	-7,803	Wall Foundation:600/1200 na osu:966476	2340000000	Beton prostý	5850	3	2500
6	-10,48	-3,142	-7,803	Wall Foundation:600/1200 na osu:966477	7056000000	Beton prostý	17640	8	2500
7	-4,78	-6,742	-7,803	Wall Foundation:600/1200 na osu:966478	7776000000	Beton prostý	19440	8	2500
8	0,32	-4,242	-7,803	Wall Foundation:600/1200 na osu:966479	3168000000	Beton prostý	7920	4	2500
9	-3,048	3,158	-7,603	Základová patka s pilotou:1200/1200/800:966779	1152000000	Beton prostý	2880	2	2500
10	1,645	3,158	-7,603	Základová patka s pilotou:1200/1200/800:966780	1152000000	Beton prostý	2880	2	2500
11	3,82	3,158	-7,603	Základová patka s pilotou:1200/1200/800:966781	1152000000	Beton prostý	2880	2	2500
12	6,677	3,158	-7,603	Základová patka s pilotou:1200/1200/800:966782	1152000000	Beton prostý	2880	2	2500
13	8,108	3,158	-7,603	Základová patka s pilotou:1200/1200/800:966783	1152000000	Beton prostý	2880	2	2500
14	9,539	3,158	-7,603	Základová patka s pilotou:1200/1200/800:966784	1152000000	Beton prostý	2880	2	2500
15	-14,266	-10,448	-7,003	Obdélníkový Sloup 400x200:961468	234000000	Železobeton - prefa	627,2	1	627,2

The first step is the analysis of the object in 3D environment where we can extract important information about the object itself, its shape, characteristics, and in particular the position and weight of the individual elements. This information is needed for the proposal and it is available and commonly used by structural engineers in the design of the individual sections and elements. The position information must be kept in a coordinate system (x, y, z) linked together with the information about object weight. The transfer of information is maintained through IFC file of a model.

As a result of the analysis, following information is extracted from a model:

- Weight of individual elements (calculated from volume and material type),
- Location of elements,
- Transport paths of elements.

Essential part of the collection of data from the project is programing of suitable algorithms and the use of C # programming language, to allow the automation of the process.

3.4 Analysis of Extracted Parameters of an Object

After analyzing the virtual data model, from which we obtain the necessary information - critical parameters of elements, it is necessary to evaluate them properly. Based on these parameters, it is possible to calculate the necessary radius to cover all construction processes – length of a jib, the required bearing capacity of the crane and the optimal position in relation to the distribution of load and location of individual elements in 3D space.

3.5 Defining the Length of the Jib

For each position, it is necessary to obtain the information about the distance from the centre of mass to the individual elements, and use this data for the design of the crane in working load diagrams of different types of cranes.

Value l_{cri} is the desired length of the jib, which is calculated based on the distance of the crane (C_{ri}) and the location of individual elements (D_i), respectively the location of supply point (S_i).

$$|l_{cri}| = \sqrt{(x_{cri} - x_{Di})^2 + (y_{cri} - y_{Di})^2}$$

Equation 1 Calculation of length between the crane and demand point

$$|l_{cri}| = \sqrt{(x_{cri} - x_{Si})^2 + (y_{cri} - y_{Si})^2}$$

Equation 2 Calculation of length between the crane and supply point

The maximum value of the $l_{cr,max}$ is the maximum distance of the element from the crane position. This value is considered as the minimal length of jib needed to cover all locations of demand and supply from selected crane location.

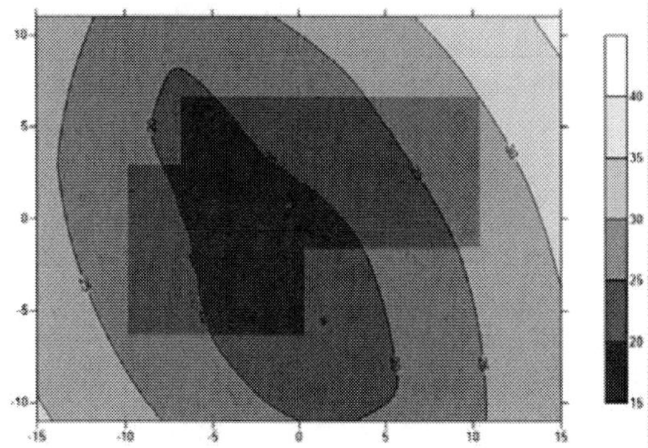

Figure 3 Map of the length of a jib according to its position

3.6 Defining the Bearing Capacity

In many cases, verification of a bearing capacity of a crane is done solely based on the information of the heaviest and furthest element. It is caused mainly by difficulty to obtain information about all elements manually, mainly in larger scale projects. In certain cases it might be misleading resulting in improper selection of a crane. Using traditional methods it is almost impossible to ensure the weight of all elements transported to its demand point will be within load curve of a selected crane. Therefore it is necessary to compare the set of all ordered pairs [di, mi], where d_i is the distance from the position of the crane i-th element and m_i is the weight, with working load diagram. This criterion ensures that all elements will be transported to the point of demand by specified type of crane.

3.7 Optimal Location of a Tower Crane

The proposed methodology approach consists of the steps that lead to the effective design of tower crane, including its location and the point of supply. As a result, a list of suitable cranes associated with critical parameters ordered from the most optimal location to the least optimal is composed. Following parameters are taken into consideration:

- Time of transfer of materials from (S_i) to (D_i),
- Required length of boom (L_{CR}),
- Sum of the distances between all (S_i) and (D_i).

An important criterion for placing the crane is to have a suitable assessment of traffic time of elements. Used formula was developed by the authors (Zhang, P., Harris, Olomolaiye, & Holt, 1999). Calculation of a time of transport consists of vertical, horizontal and radial movements from (S_i) to (D_i). Two parameters are taken into account - α and β. Alfa reflects the level of coordination within radial and tangential movement in horizontal direction and Beta parameter sets time overlap between horizontal and vertical movement. Both parameters are taken from the research of (Kogan, 1976). By observing transport paths estabilished α =0,25 and β = 1.

Figure 2 shows the range of transport times which differs by location of a crane. Contour map is created in order to help the planner to select optimal location according to site conditions. This map makes the decision faster and easier mainly in cases that some of the areas are not suitable for crane location.

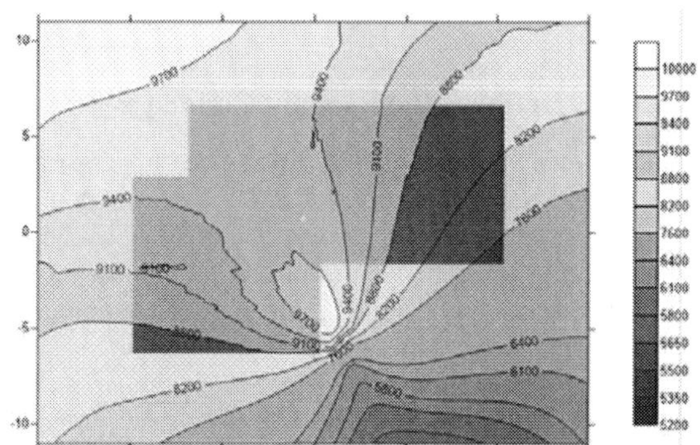

Figure 4 Time of transport according to the crane location

After precise calculation of all criteria, they are evaluated in order to define the ordered list of optimal positions. Dark indicated parts at Figure 3 are preferred in terms of location.

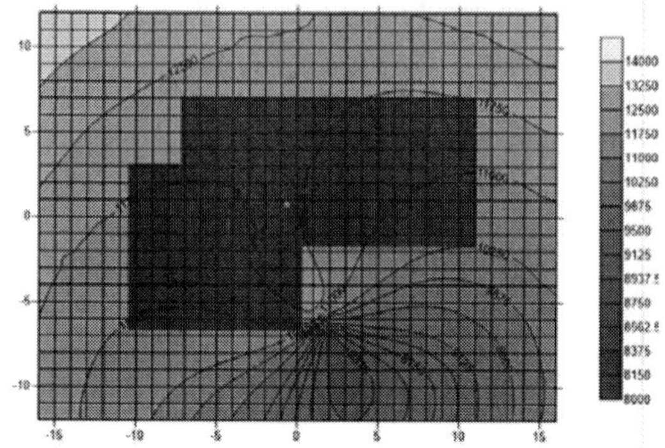

Figure 5 Contour map of the sum of all criteria

The basis of the methodology is an export of the data from a virtual model, which are crucial in designing the position and type of a crane. Location of a crane is further analyzed in each variant position in terms of capacity of selected types of cranes and their bearing capacity. The first step is verified by a particular type of crane capacity required for reaching the supply area and the location of all elements – demand point. Based on the distance of the point of supply, optimal jib is defined taking into consideration the module of a manufacturer.

Individual elements are defined by location and weight and by comparison with the database of the crane load it is verified whether it is possible to use selected type of crane for construction, i.e. whether a crane is capable of delivering all the elements from point of supply to all demand points and therefore it is guaranteed that selected crane is able to reach and deliver all elements.

Figure 6 Optimal position of tower crane after evaluation of all criteria

Based on a comparison of crane capacity data, the reserve between required and available capacity of a crane is calculated. Total reserve is the sum of all reserves and it allows for the optimal crane selection within the closest range of weights.

4 CONCLUSION

Due to complexity of projects and the resulting amount of manual work needed to obtain necessary information relevant for the proper selection of a tower crane type and its location by traditional methods it is common to examine only the heaviest and farthest element. This may lead not only to improper crane selection but also it makes it impossible to automate the whole process. The proposed methodology addresses the benefits of a functional link building information modeling and deployment issues of tower cranes. Methodology itself is more complex than other methods that are currently in use because it is possible to evaluate all elements distributed in 3D space of a construction and verify that they are within the capacity diagram of a crane. The methodology also defines the procedure for choosing the appropriate crane for fast, smooth and cost optimal solution.

References

[1] Abdel-Khalek, H., Shawki, K., & Adel, M. (2013). A Computer-based Model for Optimizing the Location of Single Tower Crane in Construction Sites. *International Journal of Engineering Science and Innovative Technology, 2*(2), 438 - 446.

[2] Abeid, J., & Arditi, D. (1999). Time series analysis for construction productivity experiments. *Journal of Construction Engineering and Management*, 87 - 95.

[3] Al-Hussein, M., Niaz, M., Yu, H., & Kim, H. (2005). Integrating 3D visualization and simulation for tower. *Automation in construction, 15*, 554-562.

[4] Alkriz, K., & Mangin, J. (2005). A New Model for Optimizing the Location of Cranes and Construction Facilities Using Genetic Algorithms. *ARCOM 21st Annual conference* (s. 981-992). Salford: Association of Researchers in Construction Management.

[5] Irizarry, J., & Karan, E. P. (September 2012). Optimizing location of tower cranes on construction sites through GIS and BIM integration. *ITcon*, 16.

[6] Funtík, T. (2013). Automatizácia umiestnenia vežového žeriava pomocou BIM. *Juniorstav* (s. 4). Brno: VUT Brno.

[7] Funtík, T. (2013). Automatizovaný proces návrh umiestnenia vežového žeriava pomocou BIM - export údajov z virtuálneho modelu. *Advances in Architectural,Civil and Environmental Engineering* (s. 261-266). Bratislava: STU v Bratislave.

[8] Lee , G., Kim, H., Lee, C., Ham, S., Yun, S., Cho, H., . . . Kim, K. (2009). A lasertechnology-based lifting-path tracking system for a robotic tower crane. *Automation in construction, 18*, 865-874.

[9] Lee, G., Cho, J., Ham, S., Lee, T., Lee, G., Yun, S., & Yang, H. (2012). A BIM- and sensor-based tower crane. *Automation in construction*(26), 1-10.

[10] Zhang, P., Harris, F., & Olomolaiye, P. (1996). A computer-based model for optimizing the location of a single tower crane. *Building Research & Information, 24*(2), 113 - 123.

[11] Zhang, P., Harris, F., Olomolaiye, P., & Holt, G. (1999). Location Optimization for a Group of Tower Cranes. *Journal of Construction Engineering and Management, 125*(2), 115 - 122.

[12] Building Information Modeling, A Strategic Implementation Guide, Dana K. Smith, Michael Tardif, Published by John Wiley & Sons, Inc. New Jersey 2009, ISBN 978-0-470-25003-7

Application of self-location system using a floor of random dot pattern to an automatic guided vehicle

Yutaro Fukase, Hiroshi Kanamori, Hiroshi Taga
Shimizu Corporation, 3-4-17, Etchujima Koto-ku, Tokyo, 135-8530, Japan
fukase@shimz.co.jp, kanamori@shimz.co.jp, htaga@shimz.co.jp
Kouichi Saito and Eiichi Nagata
UniCarriers Corporation, 1-2, Shin-Ogura,Saiwai-ku, Kawasaki, Kanagawa, 212-0031, Japan
saitokouichi@unicarriers.co.jp, nagataeiichi@unicarriers.co.jp

Abstract –

Application of the AGV (automatic guided vehicle) has broadened in factories and warehouses. Although most AGVs use magnetic tape for the guide path, the guided method has some problems. Once a path has been set, you require re-installation of the magnetic guide tape on a floor in order to reset the path. Also it is difficult to set intricately branched. For this reason, the application of AGV in the field of logistics, such as day-to-day transport route changes, has been difficult. In recent years, in order to solve these problems, several navigation methods using various self-location systems have been developed.

We have been developed a self-localization system which uses dots patterns on a floor. At first we construct the floor on which there are random dots patterns and make a database of the all dot positions on the floor. A camera captures an image of the floor and several dots are cropped in the image. The cropped dots are matched in the database of all the dot positions on the floor and then we can detect position and direction of the captured image on the floor.

We have set the floor which has dots pattern for detecting position at a factory where several forklifts move around and tire marks of those forklifts are attached. And an AGV using the developed self-localization system has run autonomously in the factory. In this paper we report the interim results of the experiment.

Keywords –

Automatic Guided Vehicle; Self-localization System

1 Introduction

A mobile robot programmed to move autonomously in indoor must be equipped with a self-localization system to operate effectively. Several types of self-localization systems are therefore being pursued.

A number of strategies using electric devices installed inside buildings have been developed. One system deploys a pseudo satellite (Sakamoto, Niwa, Ebinuma, Fujii, & Sugano, 2010 [1]); another detects differences in radio field intensity among several wireless LAN access points (Umetani, Yamashita, & Tamura, 2011 [2]); the third one uses an ultrasonic 3D position sensor (Nishida & Takeda, 2010 [3]). Yet in each case, non-uniformity of the electromagnetic or acoustic environment causes instability the detection position.

Other methods rely on ceilings or floors for position detection. One, for example, uses arrayed marks on ceilings (Nakazato, Kanbara, & Yokoya, 2008 [4]), while another tracks communications transmitted LEDs on ceilings (Uchiyama, Haruyama, & Nagamoto, 2009 [5]). Both of these methods fail when the ceilings in a room are tall. Kodaka, Niwa, and Sakamoto (2009) developed a floor-based system by placing a series of RFID tags on a floor and reading them to detect positions [6]. At about the same time, Nishizaka, Hiyama, Tanikawa, and Hirose (2009) printed position-coded patterns on a floor and read them with a camera [7]. These latter floor-based systems are costly, however, as the first requires many RFID tags and the second requires the printing of code patterns on a floor. Another floor-based system relies on cracks in a floor as landmarks (Kelly, 2000 [8]). This method is difficult to adopt for floors with finishing materials, though it is effective for factories or warehouses with solid concrete floors.

Our group is attempting to surmount these challenges by devising a method using random dot patterns on floors. Floors are often paved with vinyl or coated with epoxy resin in structures such as hospitals, factories, or office buildings. To improve the appearance and durability and to recycle materials, some floors are finished with dot patterns formed by mixing milled plastics into the flooring materials at the manufacturing or construction phase. These dots spread in a random manner, hence one pattern at one position is statistically different from all of the other patterns. Our group has taken advantage of this property to develop a practical self-localization system for mobile robots with a focus on low cost and high detection stability.

We have applied our developed self-localization system to an automatic guided vehicle. A problem, which was caused by changing the height of the sensor camera, had happened. We have solved this problem and presently we are having an experiment of the AGV control in a machinery factory. An interim report of the experiment is shown in this paper.

2 Self-localization System Using a floor with random dot pattern

The basic configuration of our developed location-system is shown in Figure 1. In a floor manufacturing process random dot patterns are generated. After that the floor is scanned to make a dot-position database. As the robot moves, a sensor-camera on the bottom captures images of the dot patterns. To determine the robot's position and direction of travel, the system matches the position and orientation of the dot pattern in the captured image with the data in the dot-position database of the entire floor.

The dot patterns in the image are matched with those in the dot-position database using an attitude-detection technology applied for satellites. In a satellite, an on-board camera captures an image of the space and the camera orientation is identified by matching the stellar constellation in the captured image with a known star map. This technology, otherwise known as the *Star sensor* technology, is illustrated schematically in Figure 2. Our detection method shares the following elements with the Star sensor method.

1. Naturally occurring random pattern

Both the star constellation and floor dot pattern are random dot patterns.

2. Foreign dots and disappearing dots

Countless stars may appear in a satellite image, hence the stars used for matching are limited to those above a preset threshold for brightness. Yet in the brightness range just above and below this threshold, non-target stars may inadvertently appear or target stars may disappear. The same principal applies in our system, as previously unregistered dots (hereafter, "foreign dots") may inadvertently appear or database-registered dots may disappear.

To cope with this problem we apply the Polestar algorithm (Silani & Lovera, 2006 [9]), an attitude-detection algorithm with a robust ability to handle the blending of non-target stars or the disappearance of target stars. We call this method "database matching method" in this paper.

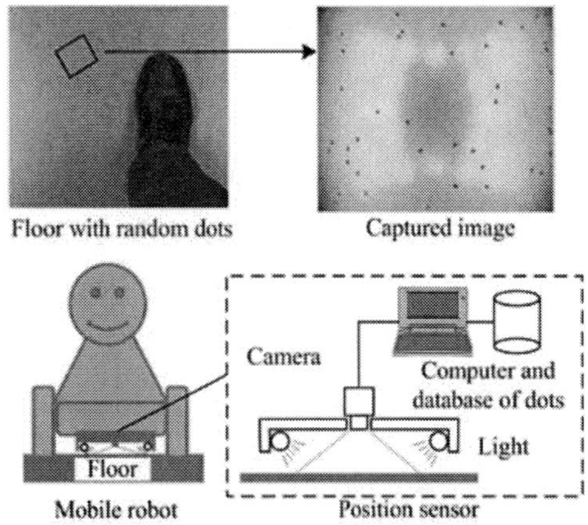

Figure1. Developed Self-localization system

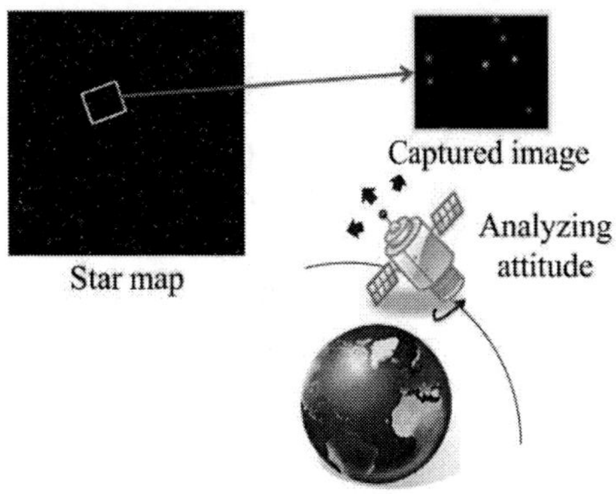

Figure 2. Pole Star algorithm

Sometimes, there are too many foreign dots to match the dot-position database on the floor. In that case, when we apply the database matching method to control a mobile robot, the robot will stop. In order to continue detecting position, we have devised another method. By detecting the movement of dot patterns in two images back and forth in time and accumulating it, we can calculate the position. In this paper, we call the method "detecting movement method."

Even if the database matching method fails, the position will be detected by the detecting movement method. And while it spends shorter time than the database matching method, it generates errors. Thus the two methods run in parallel. The detecting movement method calculates the position in a short cycle and when the matching database method calculates the position, the position calculated by the detecting movement method is modified. Thus, if the self-localization device goes into an area where there are many foreign dots, the position could be detected continuously and quickly. The device, which is showed in figure 4, can detect position in 14[ms] interval and with an accuracy of 1[mm] and 0.5[deg].

We have mounted the self-localization device on a mobile robot and properly controlled it using the detected position. Figure 5 shows the robot.

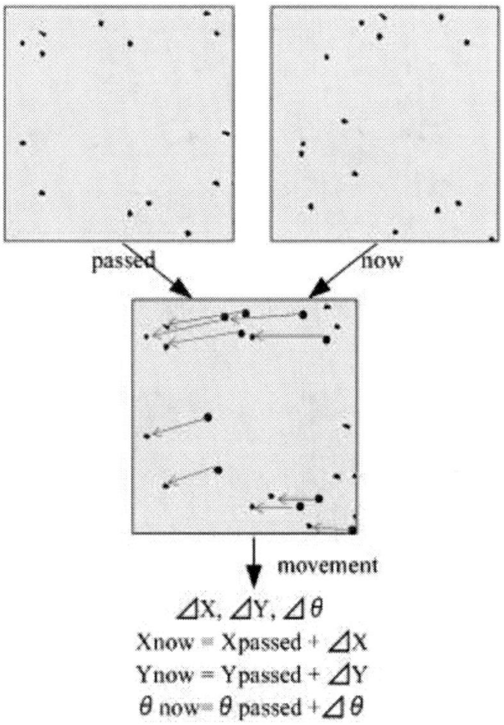

Figure 3. Detecting Movement Method

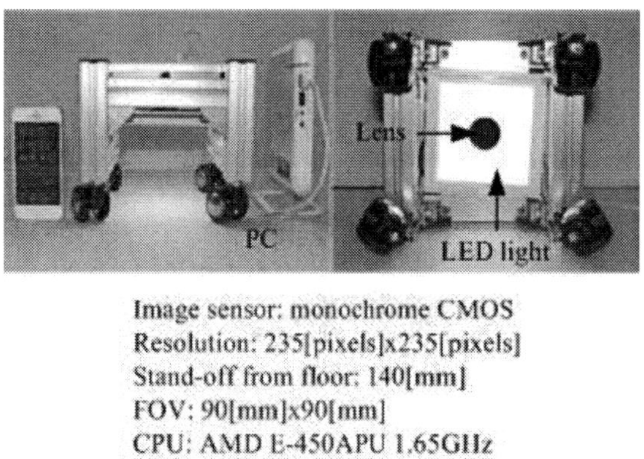

Figure 4. Self-localization Device

Figure 5. Mobile Robot

3 Automatic Guidance Vehicle

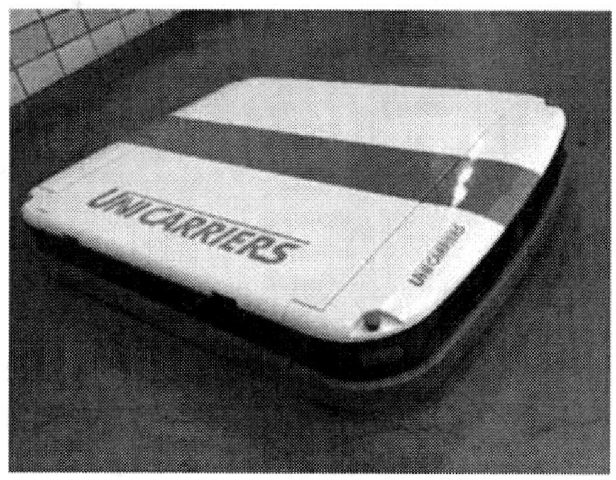

Figure 6. Automatic Guidance Vehicle

Figure 6 shows the automatic guided vehicle (=AGV) which is applied our self-localization system. Table1 shows a specification of the AGV. The AGV has been designed to be navigated with a magnetic guide tape on a floor. It can move omnidirectionally. So we have mounted our developed position sensor on the AGV to detect the position and control it not only on the tape but also on the all floor.

Table 1. Specification of the AGV

Navigation	Magnetic Guide Tape
Communication	Wireless communication
Driving mode	Forward-Back Traverse Spin turn
Tranport weight	1,000[kgf]
Speed	1,000[mm/s] max.
Stopping accuracy	$\pm$10[mm]
Size	Length:1,600[mm] Width: 1,250[mm] Height:238[mm]
Weight	450[kgf]
Power	Automatic quick charge Lithium ion battery

4 Undetected Case with AGV

The self-localization device is mounted on the AGV showed in Figure 6. We spread out a random dot pattern vinyl flooring (5[m] by 5[m]) on a concrete floor. When we moved the AGV on the whole floor, the device couldn't detect the position in several parts of the floor.

Therefore we applied a self-localization device shown in Figure 4. It could detect the positions in most parts of the floor where the device on the AGV could not detect previously.

The difference between these two devices is a length of the wheel bases. When the device is on a carrier with a long wheel base, the height of a sensor camera is variable on a corrugated surface (Figure 7). Matching algorithm was based on the premise that the height is constant.

Changing a distance from the sensor-camera to the floor makes the distances between each dot in a captured image different. In this case, because the matched database is based on the distances between each dot captured from determinate height, the matching process doesn't work. If the image captured at the location was adequately rescaled, we could detect the position. Therefore failure in detecting positions is thought to be caused by the changing height of the sensor-camera.

Figure 8 shows a captured image. Figure 9 shows a cropped dot pattern. Figure 10 shows difference between cropped dot pattern (white dots) and dot pattern in database (red dots).

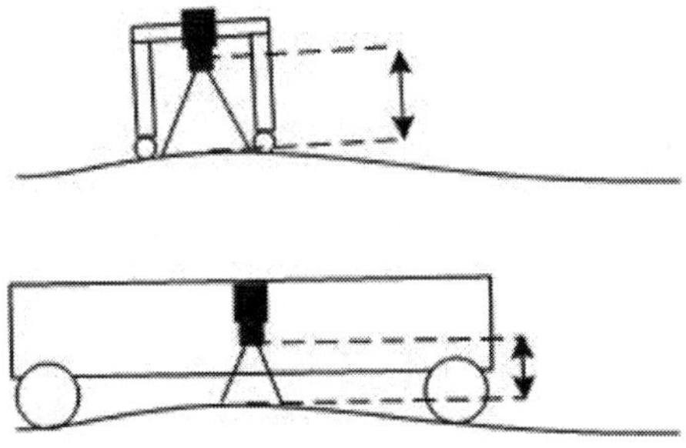

Figure 7. Height of sensor camera on AGV is variable on a corrugated surface

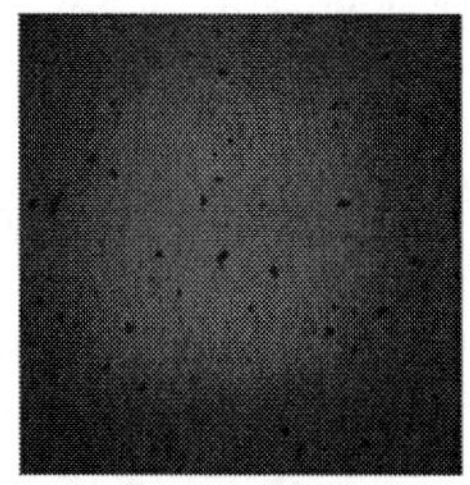

Figure 8. Captured image

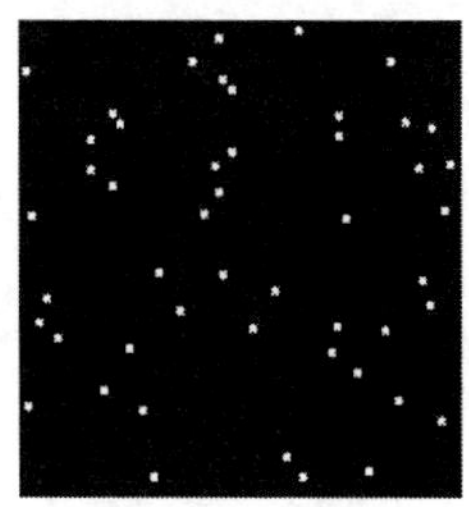

Figure 9. Cropped dots

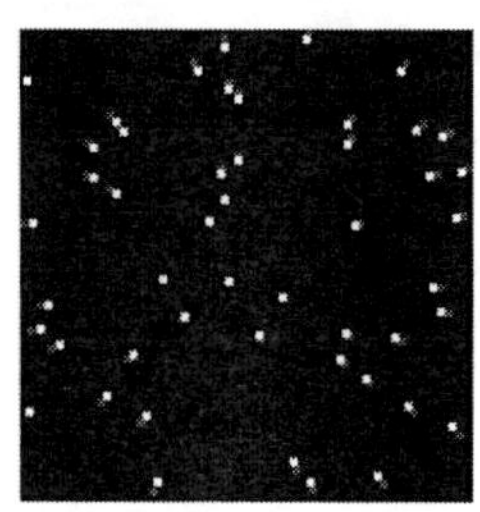

Figure 10. Red dots are pattern in database

5 Experiment for Changing Height

In order to improve the detecting performance, we measured a height of the sensor-camera and rescaled an image in proportion as the height.

We spread out a random dot pattern flooring whose size was 3640[mm] by 9000[mm] on a concrete floor, and set a magnetic guide tape on the floor.

We navigated the AGV by using the guide tape (Figure 11), set a Laser Range Finder (=LRF) near sensor-camera, and measured the distance between the sensor-camera and the floor (Figure 12). Using the measured distance, the image was rescaled in order to detect position. Figure 13 shows a trajectory of the position of the sensor camera. Figure 14 shows the height of the sensor-camera at each travel distance. We can see the changes in approximately 5[mm] range. Figure 15 shows the result of the detection when not using the height measured by the LRF. Figure 16 shows the result of the detection when using the measured height. Red points indicate undetected positions. We can see that the red points decrease from 5.1% to 0.8% on the trajectory.

Figure 11. Status of Experiment

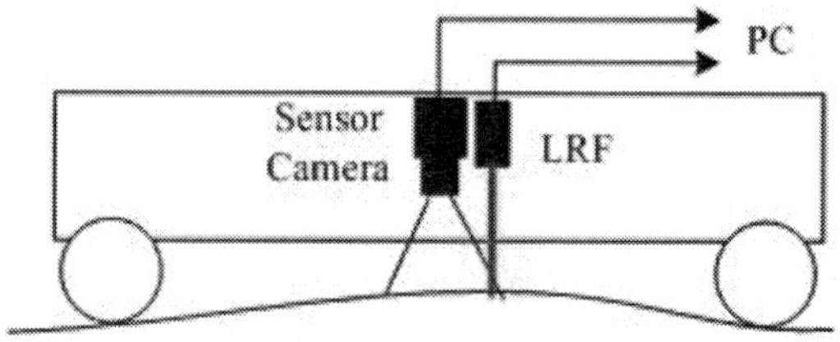

Figure 12. Device with LRF

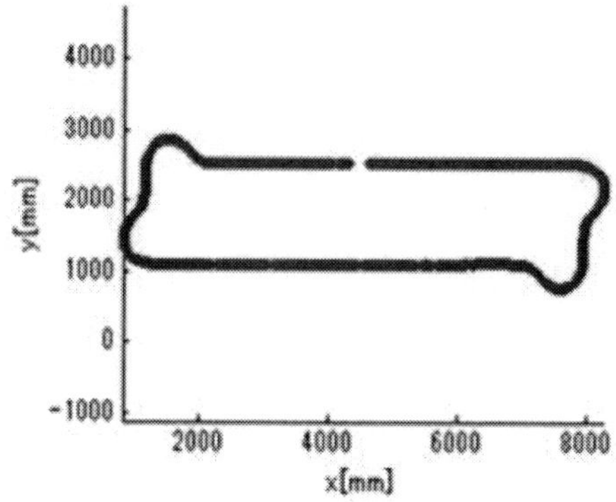

Figure 13. Trajectory of sensor-device

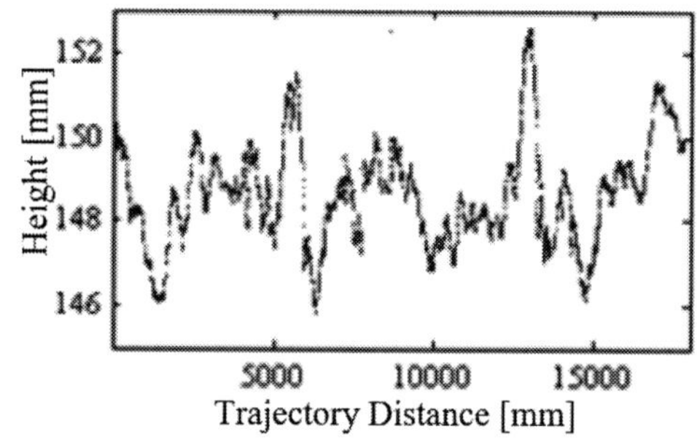

Figure 14. Height of sensor-device

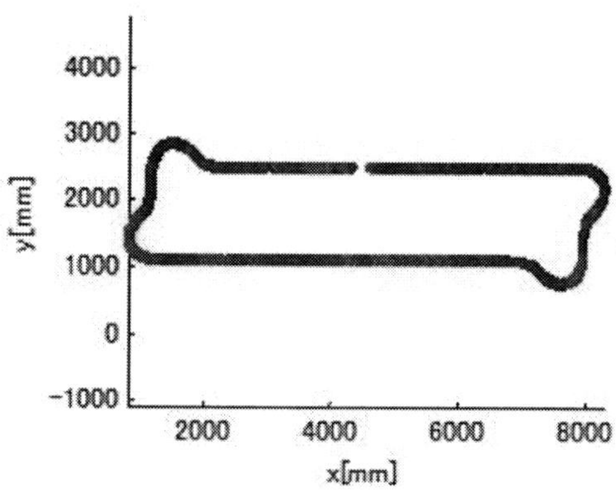

Figure 15. Case of fixed scale
The red points indicate undetected position and the blue points indicate detected position.

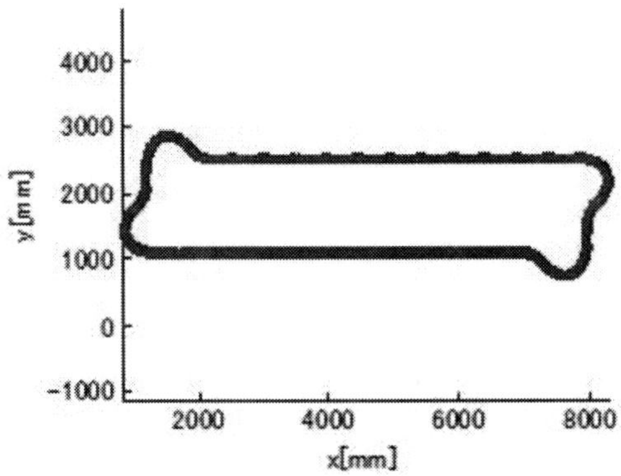

Figure 16. Case of scale changed by height.
The red points (=undetected points) decreased.

6 Experiment In a Factory

Presently we are having an experiment of an AGV controlling with our self-localization system in a mechanical factory. Several folk-lifts are moving around inside and outside the factory. The folk-lifts may put dirt on the floor or damage to the floor. We put a dot pattern floor and three months later we moved the AGV on the floor. In this section, an interim report is shown.

6.1 Experimental Condition

We spread out a flooring in a part of the factory (Figure 17). Figure 18 is a picture of the factory. In the factory there are many numerical control lathes and racks. The folk-lifts carry the racks near the lathes. Oil and cutting chips, which are attached to the tires of folk-lifts, cause the floor dirty (Figure 19). The AGV is controlled autonomously by the position which is detected by the self-localization system. It travels repeatedly approximately 40 [m] from one end to the other end of the floor at 400 to 500[mm/sec].

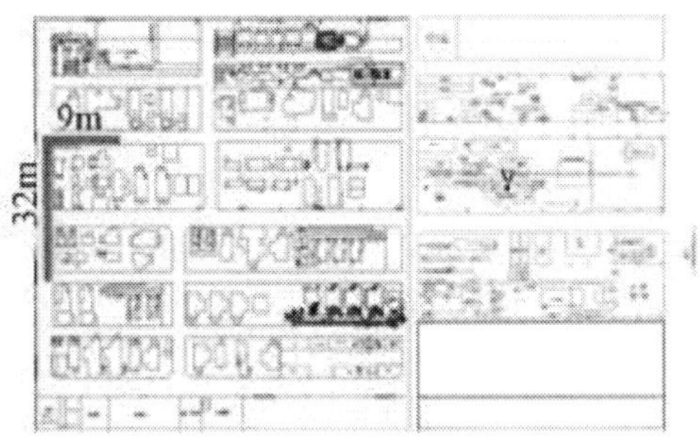

Figure 17. Floor in part of the factory

Figure 18. Mechanical factory

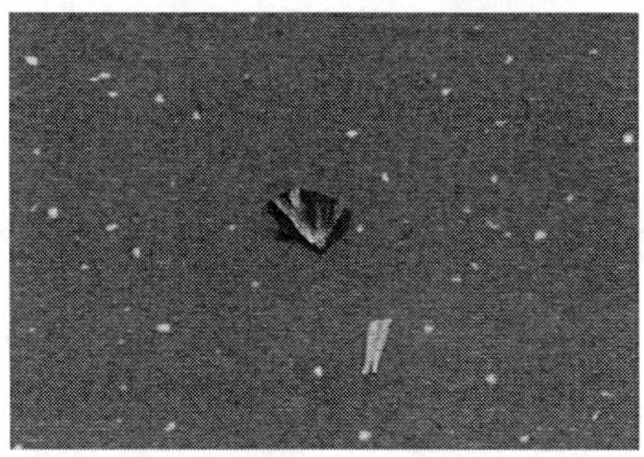

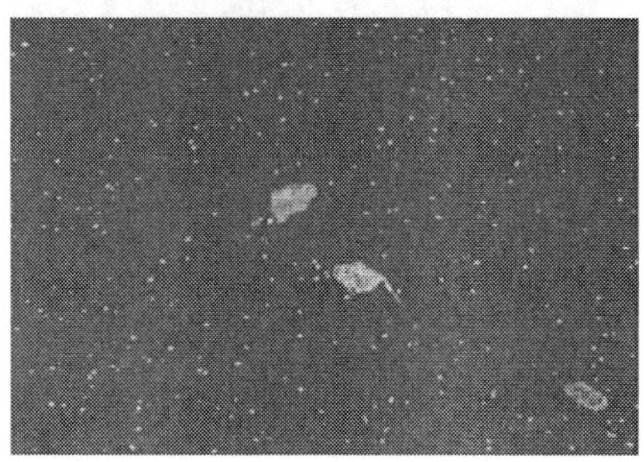

Figure 19. Dirt on the floor

6.2 Result

Followings are the result of detection of the AGV running 216 [m] in 8 [min.].

Output Interval (average) : 14 [msec]

Movement detection success rate: 99.1[%]

Matching database success rate: 92.4[%]

Figure 20 shows a trajectory of the AGV. The blue points are detected positions and the red circles are points where the matching database error occurred. The matching database failure rate is 7.6%.

Pictures in figure 21 are the images with which matching database failure occurred. In some cases dots were disappeared because of oil spots and in the other cases cutting chips became foreign dots. Even, in those cases, self-localization system could be continuously working using the detecting movement method in parallel with the matching database method and the AGV was controlled without problems.

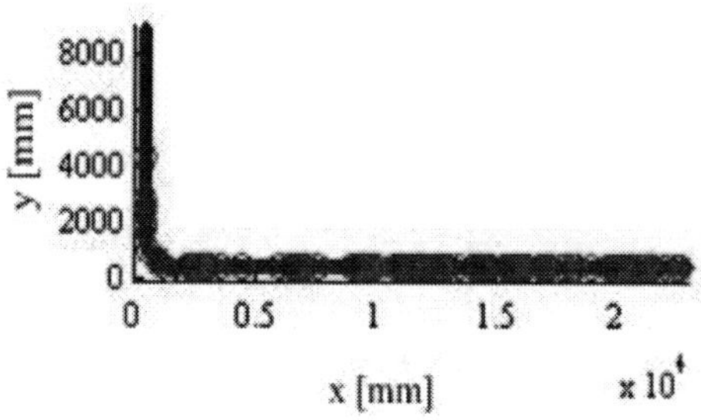

Figure 20. Trajectory of AGV

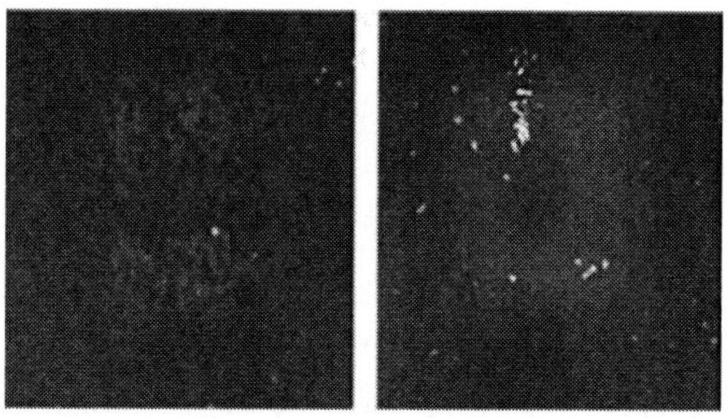

Figure 21. Mactching Failure Images

7 Summery

We applied the self-localization system which we developed using a floor of random dot pattern to an AGV. At first the system did not work properly because of the height variation of the sensor camera. We solved the problem by using the magnification adjustment of the image by measuring the height with a laser measuring instrument. Currently, laying the point cloud pattern floor in the machine factory, we have implemented a running experiment of AGV. After about three months since spreading out the floor, the self-localization system could detect the position properly and the AGV was controlled autonomously without problem. We plan to have the same experiment after six months and after one year in order to evaluate the effectiveness of the self-localization system.

References

[1] Yoshihiro Sakamoto, Haruhiko Niwa, Takuji Ebinuma, Kenjiro Fujii, Shigeki Sugano (2010). Indoor Positioning with Pseudolites. SICE Annual Conference 2010, CD-ROM

[2] Tomohiro Umetani, Tomoya Yamashita and Yuichi Tamura (2011). Probabilistic Localization of Mobile Wireless LAN Client in Multistory Building based on Sparse Bayesian Learning. Journal of Robotics and Mechatronics (vol. 23, no. 4, pp. 475-483)..

[3] Yoshifumi Nishida, Hideki Takeda (2010). Ultrasonic Three-Dimensional Location System. Journal of the Society of Instrument and Control Engineers(vol. 49, no. 1, pp. 56-59).

[4] Yusuke Nakazato, Masayuki Kanbara, Naokazu Yokoya (2008). A User Localization and Marker Initialization System Using Invisible Markers. Journal of the Virtual Reality Society of Japan (vol. 13, no. 2, pp. 257-266).

[5] Hideaki Uchiyama, Shinichiro Haruyama, Naoki Nagamoto (2009). *Construction Machinery and Equipment* (vol. 45, no. 2, pp. 50-53)

[6] Kenri Kodaka, Haruhiko Niwa , Yoshihiko Sakamoto (2009). *Robot Pose Estimation on Floor Equipped with a Lattice of RFID Tags*. Transactions of the Society of Instrument and Control Engineers (vol. 45, no. 7, pp. 379-387)

[7] Shinya Nishizaka, Atsushi Hiyama, Tomohiro Tanikawa, Michitaka Hirose (2009). Robustness Enhancement in Localization using Interior Decoration with Coded Pattern. Technical Report of the Institute of Image Information and Television Engineers (vol. 33, no. 21, pp. 43-48)

[8] Kelly. A. (2000). Mobile Robot Localization from Large Scale Appearance Mozaics. International Journal of Robotics Research (vol. 19, no. 11, pp. 1104-1125)

[9] Enrico Silani, Marco Lovera (2006). Star Identification Algorithms: Novel Approach & Comparison Study. IEEE Transactions on Aerospace and Electronic Systems (vol. 42, no. 4, pp. 1275-1288)

Automatic Surface Flatness Control using Terrestrial Laser Scanning Data and the 2D Continuous Wavelet Transform

E. Valero and F. Bosché

Centre of Excellence for Sustainable Building Design, Heriot-Watt University, UK
E-mail: e.valero@hw.ac.uk, f.n.bosche@hw.ac.uk

Abstract –

Evaluating flatness defects in surfaces during the construction of a building can prevent problems during subsequent construction tasks or the operation phase. An area of particular interest is that of the flatness of warehouse slabs on which pallet-transporting equipment is operated (nowadays including robots). Flatness control has typically been done with extensively manual methods, such as the Straightedge, F-Numbers or Waviness Index methods. However, these are time-consuming, based on very sparse measurements, and conduct assessment in linear ways only (i.e. in 1D along survey lines, as opposed in 2D on the entire surface), which can lead to inaccurate assessment. This paper presents a novel approach that takes advantage of (1) Terrestrial Laser Scanners (TLS) to speed up data acquisition and provide precise and dense 3D measurements of surfaces; and (2) the 2D Continuous Wavelet Transform (CWT) to deliver a 2D wavelength analysis of surface data with high resolution both spatially and in the frequency domain. The value of the proposed method over existing approaches is discussed and demonstrated with a first experiment conducted with a real concrete slab.

Keywords –

Surface; Flatness; Control; Wavelet Transform; 2D

1 Introduction

Construction is about the building and installation of components in the three-dimensional (3D) world. An important aspect of the control of construction quality is the verification that all components have indeed been built or installed in their specified location and shape [1] [2]. This may be collectively referred to as *dimensional quality control.*

Dimensional quality control is interested in many kinds of geometric measures, and the one focused on here is *surface flatness*. Surface flatness is an important part of dimensional quality control because the built environment remains largely made up of components with planar surfaces. This is due to the relative ease of constructing such surfaces, but in the case of floors this is even a matter of functionality requirement.

Ensuring that planar floors are indeed flat is important in houses for comfort and for preventing warping of floor finishing or furniture. Local planarity is also an important requirement of road surface to ensure comfortable driving [3]. And the requirements for flatness can be even greater in contexts such as television studies (where the travelling motion of cameras must be as smooth as possible) or warehouses where forklifts (now increasingly robotic ones) must travel fast and safely, without the risk of losing their loads due to vertical vibrations resulting from uneven floors [4].

1.1 Traditional Flatness Control Methods

Various flatness specifications and control procedures have been developed over time, with new methods typically being proposed as a result of the availability of improved measurement technologies.

The *Straightedge* method [5] is the oldest flatness measurement method. It is based on the measurement of deviations under a 3-meter straightedge that is manually and randomly laid on the floor. The floor is within tolerance if none of the deviations exceeds a value specified. This method is simple to understand and apply, and requires basic, inexpensive tools. However, its implementation is time consuming, prone to errors, and generally provides a partial assessment of flatness in both the spatial domain (due to the sparsity of measurements typically conducted) and frequency domain (only one surface wavelength, 3m, is considered).

The *F-Numbers* method [6] and later the *Waviness Index* method [7] emerged from the development of measurement tools like profilometers. Both methods require the delineation of survey lines on the floor and the measurement of the floor elevation at one-foot intervals along them. Formulas are then applied to the measured data to that calculate a few metrics summarizing the level of flatness. For the F-Numbers methods, these metrics include the F_F (floor flatness) and F_L (floor levelness) numbers, for each line and subsequently for the entire floor. For the Waviness Index method, five Waviness Indices (WIs) are calculated for each line and for the floor.

By their reliance on more modern measurement methods, the F-Numbers and Waviness Index methods are more efficient and precise than the Straightedge method. While the F-Numbers provides waviness information for two different wavelengths (approx. 60cm and approx. 500cm), the Waviness Index method extends those to five (60, 120, 180, 240 and 300cm). The Waviness Index method is thus the one that currently covers the frequency domain the best, albeit for only 5 spatial wavelengths. It should be noted though that these wavelengths were selected specifically to cover the range of wavelengths from 60cm to 300cm that are most likely to impact the operation of forklifts typically used in warehouses (the range is selected to correspond to 50% to 200% of their wheelbase length) [3]. The Waviness Index method was therefore designed to control a range of distances not covered by the F-Numbers method.

Nonetheless, despite its superiority, the Waviness Index method still presents four main limitations:

1. *Partial spatial analysis*: Its measurement process remains fairly tedious with the user having to mark survey lines on the floor, then carefully roll the profilometer along the lines and finally compute the WIs from all measurements. As a result, the number of measured survey lines is typically small, leading to spatially sparse results which may not be representative of the true level of flatness of the floor.
2. *Partial frequency analysis*: The method is limited to the study of five specific wavelengths within the [60cm;300cm] range – in fact experimental results previously reported by the authors even suggest a weakness of the method for the study of the wavelength 60cm [8]. It would be of interest to develop a method that considers more wavelengths within that range but also outside that range.
3. *No direct localisation of defects*: The WIs enable the detection of discrepancies but not directly their localization, although this information is important for remedying the defects. Further manual analysis of the results is required to localise the problematic areas.
4. *1D instead of 2D analysis*: The method is still based on measurements along lines. Floors are inherently 2D surfaces and so flatness should be assessed in 2D.

1.2 Terrestrial Laser Scanning and State-of-the-art Research on Flatness Control Methods

Terrestrial Laser Scanning (TLS) is revolutionizing geometric surveying in construction by its capacity to provide both accurate and dense point measurements very rapidly [9] [10] [11].

Early works on the application of TLS to dimensional control aimed at color-mapping the deviations of the measured points from a reference surface, which enabled a visual detection of areas of potential concern [12]. However, these first works did not aim to automatically detect these areas or quantify the deviations. Tang et al. [1] may be the first to have tried to address this for planar surfaces. They propose an algorithm (with two variants) to detect flatness deviation peaks in 2D TLS data (from a planar reference surface). However, this approach is limited by the fact that it focuses on detecting the maximum(s) of the signal amplitude (i.e. deviation from the reference surface), while existing standards on flatness control clearly highlight the need to characterize surface waviness, not just amplitude.

Bosché and Guenet [2] show how the Straightedge, F-Numbers and Waviness Index methods can be encoded for automated application to TLS point clouds of floors. The main advantage of that approach is that it is fully automated and thus very efficient. As a result, they show how the density of measurements (i.e. number of straightedge measurements, or number of survey lines assessed in the F-Numbers and Waviness Index methods) can be increased significantly at no cost (time-wise), thereby addressing the limitation (1) above (*'partial spatial analysis'*). Nonetheless, the other three limitations still apply.

In [8], Bosché and Biotteau propose a novel approach to flatness control based on the automated analysis of wavelength along survey lines virtually surveyed in the dense TLS point clouds of floors (similarly as in [2]) using the Continuous Wavelet Transform (CWT). This approach is shown to be powerful as it addresses both limitations (2) and (3) above. By measuring points at shorter increments along the survey lines (1cm instead of 30cm used by the F-Numbers and Waviness Index methods; this is made possible by the density of measurement provided by TLS), the authors show that the wavelengths present in a survey line elevation profile can be automatically detected with great precision in both the frequency and spatial (i.e. along the line) domains.

Figure 1 shows an example of result obtained for a survey line. The authors validate the approach by demonstrating a correlation of the results obtained by their approach with those obtained with the Waviness Index method. In fact, the results even suggest a potential weakness of the Waviness Index method for the 60cm wavelength. Nonetheless, the CWT method in [8] still presents two limitations:

1. The method analyses flatness along survey lines (1D) as opposed to over the entire surface (2D), which makes the overall analysis of a surface somewhat complex. Surface waviness is inherently a 2D matter and so should be studied in 2D; and
2. While Bosché and Biotteau show visually how their method is able to detect and localize any wavelength along survey lines, they did not actually present a method for the automated localization.

A method is still needed that achieves all this.

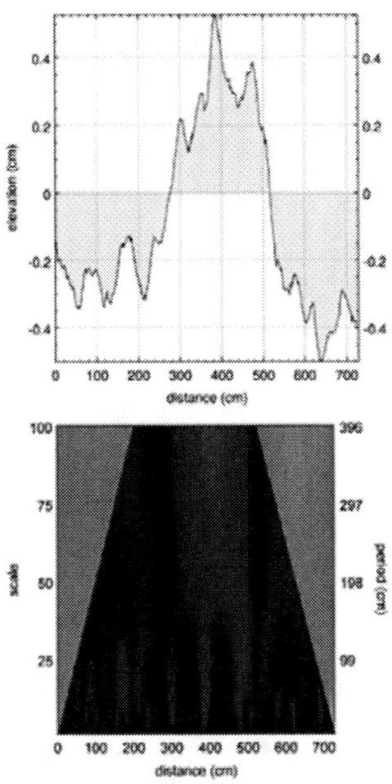

Figure 1. Example of result obtained for the analysis of the elevation profile along a survey line using the 1D CWT method presented in [8].

1.3 Contribution

This paper presents a method that extends that of [8] to 2D and that automatically feeds back to the user the areas of concern regarding the overall flatness. This method is thus the first that addresses all the limitations found in the current state of the art.

Section 2 presents the 2D CWT-based detection and localization method. Preliminary results are then reported in Section 3 with experiments that use the same dataset as that in [8]. The paper is concluded with an overall discussion in Section 4.

2 2D Continuous Wavelet Transform for Surface Flatness Analysis

The Wavelet Transform is a signal analysis method that is based on the convolution of the input signal with a wavelet function at different locations along the signal and at multiple scales. This enables the detection of the signal pattern of the wavelet function at potentially any scale and at any location. Wavelets take their name from the fact that their energy is contained within a short period, and they typically have one main center frequency f_c. Therefore, the convolution of the wavelet at multiple scales and locations along the input signal can be used not just to detect the type of pattern represented by the wavelet, but to detect specific frequencies [13]. The Continuous Wavelet Transform (CWT) is one of the several variants of the Wavelet Transform that is commonly considered for pattern/frequency detection in a signal (the pattern/frequency being that of the selected wavelet). As previously shown by the authors in [8] and also [3], the CWT is well-suited to the problem of surface waviness characterization. An interesting property of the CWT (WT in general) is that it is applicable not just to a 1D signal, but also to 2D signals (and signals of higher dimension) [13].

Applying the CWT, like any other WT, requires the selection of the mother wavelet. One common CWT wavelet is the Mexican Hat wavelet. As show in 2D in Figure 2, this wavelet is composed of one main undulation with center frequency f_c that is the same for both dimensions. The center frequency of the Mexican Hat wavelet is f_c=0.252. By convolving an input 2D signal with the Mexican Hat wavelet at a given scale a, undulations of characteristic frequency f can be detected; f is simply calculated as [13]:

$$f = \frac{f_c}{\delta_p a} \quad (1)$$

where δ_p is the point sampling period in the input signal along the given dimension.

Table 1 gives examples of wavelengths which the CWT will respond to when applying the Mexican Hat to a signal with point sampling period δ_p=1cm for 10 different scales a. The last column shows the five k levels evaluated in the Waviness Index method that correspond to each scale a.

Figure 2: 3D view of the 2D Mexican Hat wavelet.

Table 1: Characteristic frequency f and corresponding wavelength λ at which the CWT responds when applying the Mexican Hat to a signal with point sampling period δ_p=1cm for 10 different scales a.

a	f [cm^{-1}]	λ [cm]	k
7.5	0.034	29.8	
15	0.017	59.5	1
22.5	0.011	89.3	
30	0.0084	119	2
37.5	0.0067	148.5	
45	0.0056	178.6	3
52.5	0.0048	208.3	
60	0.0042	238.1	4
67.5	0.0037	267.9	
75	0.0034	297.6	5

The results obtained using the 2D CWT analysis are typically presented in the form of a 3D *scalogram*, showing the CWT responses at each sampled point *(x,y)* on the 2D surface and each characteristic frequency f (i.e. scale a), as shown in Figure 3.

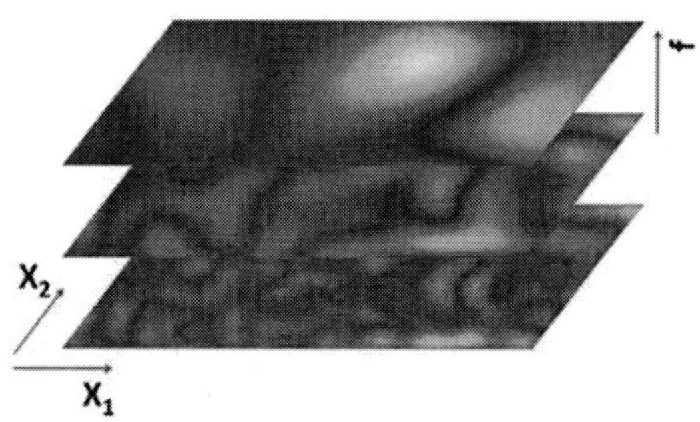

Figure 3. Scalograms for different values of characteristic frequency f.

By sampling TLS point clouds of a floor into a 2D grid with grid cell size δ_p, the 2D CWT can be applied to that transformed dataset to detect and precisely locate on the floor 'bumps' with a wide range of wavelengths. This is demonstrated with the experimental results shown in Section 4.

3 Proposed Method

A general overview of the method presented in this paper is shown in Figure 4. This method has four steps detailed in the following.

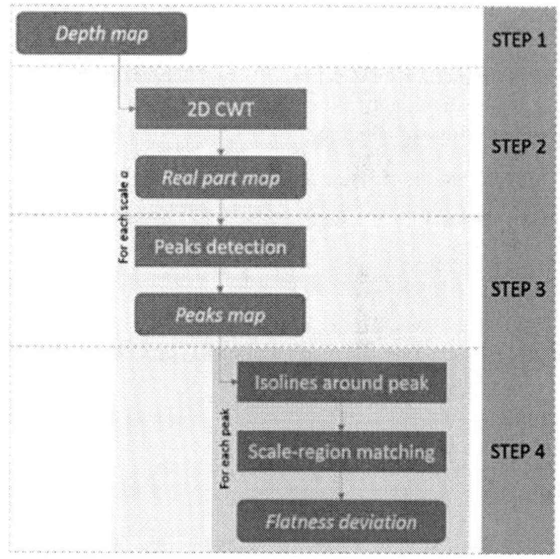

Figure 4. Overview of the method.

- Step 1: The point cloud obtained by means of the TLS device is pre-processed in order to segment the data corresponding to the floor. This data can then be processed as a *depth map*, by using a regular sampling δ_p of the slab surface (here we use δ_p=1cm) and associating to each sample the mean distance of the set of cloud points that can be locally associated to that sample location. Note that this process can be conducted automatically using scan-vs-BIM methods, as suggested in [14] [15] [16].
- Step 2 (Figure 3): The 2D CWT is applied to the depth map for each scale a of interest. This leads to the generation of a 3D scalogram, such as those shown in Figure 3, containing the CWT responses.
- Step 3 (Figure 5): For each 2D CWT scalogram obtained for each scale a of interest (Figure 5a), the peak responses (local maxima) are first detected, and peaks with values below 10% of the value of the maximum peak are filtered to remove 'noise' and non-critical responses (Figure 5b). Next, the closed *CWT response isolines* (i.e. connected sets of pixels with the same CWT response) that surround each detected peak are retrieved (Figure 5c) and the smallest ellipse that encloses the area defined by each isoline is calculated (Figure 5d).
- Step 4 (Figure 6): The lengths of the two main axes of each ellipse are calculated (Figure 6a). If any of

the two axes matches within ±1cm the wavelength λ corresponding to the scale a at which the CWT is applied, a flatness deviation is considered to be detected in that region for that wavelength (Figure 6b).

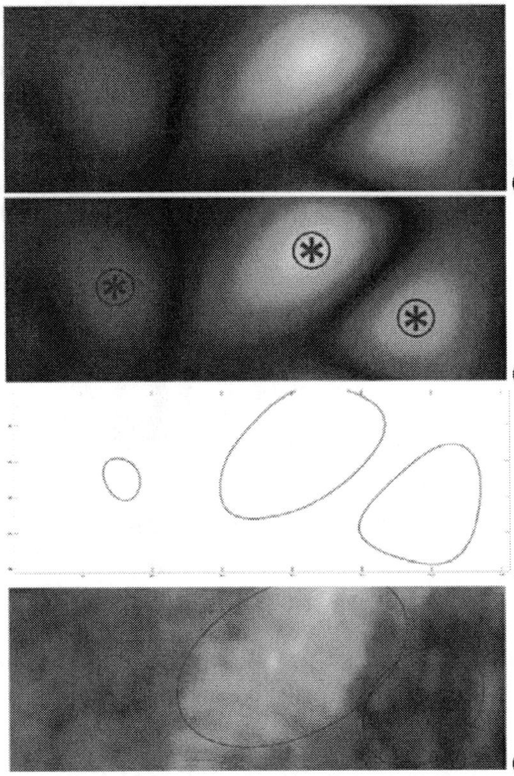

(a) (b) (c) (d)

Figure 5. Step 3 of the process carried out for the detection of potential defects in a slab with λ=297.6cm (i.e. a =75).

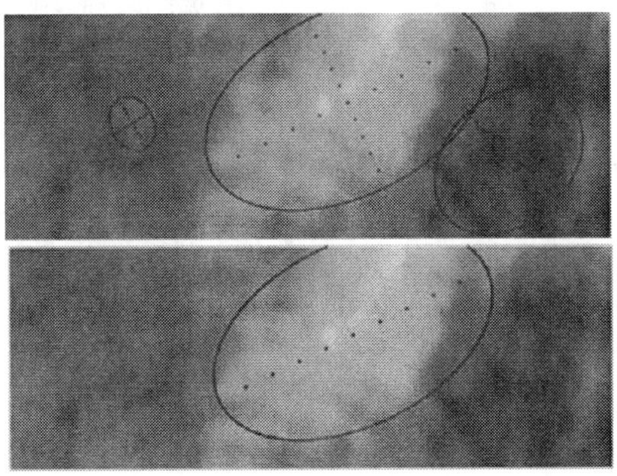

(a) (b)

Figure 6. Step 4 of the process carried out for the detection of potential defects in a slab with λ=297.6cm (i.e. a =75).

4 Experimental Results

4.1 Dataset

A first set of experiments have been carried out using the concrete floor slab of the Drainage Lab of the School of Energy, Geoscience, Infrastructure and Society at Heriot-Watt University whose dimensions are 2.75m x 7.25m. This slab is shown in Figure 7 and was already used in [8].

For the data acquisition, a Faro Focus 3D TLS was used. The final point cloud for the slab section of interest contained approx. 200,000 points (see Figure 8). As illustrated in Figure 8, the slab presents important deviations: the difference between the maximum and minimum height is 14 mm.

Figure 7. Drainage Lab slab studied in this work.

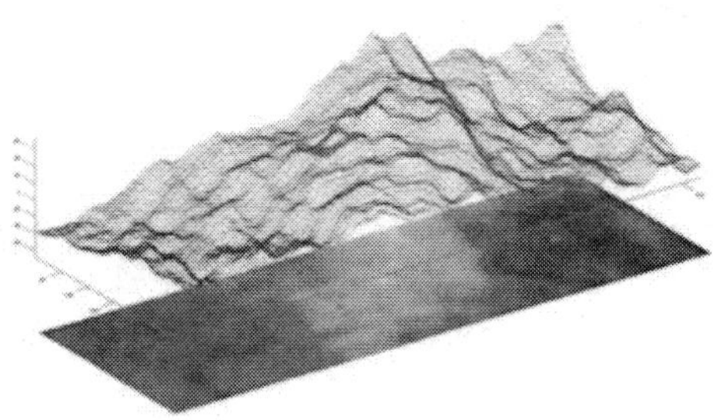

Figure 8. Point cloud and depth map of the Drainage Lab slab. Note that in this figure, the vertical axis presenting the point cloud elevation has been magnified to highlight the floor waviness.

4.2 Results

Figure 9 shows, enclosed by ellipses, the areas in which potential defects are found by the proposed method for the five wavelengths: 60, 120, 180, 240 and

300 cm (we show these wavelength because we later compare those results with those obtained with the Waviness Index method). For each ellipse, the axis with length similar to the studied wavelength, is plotted to illustrate the main direction of the detect planarity deviation. Note that no potential deviation is found for the λ=180 cm.

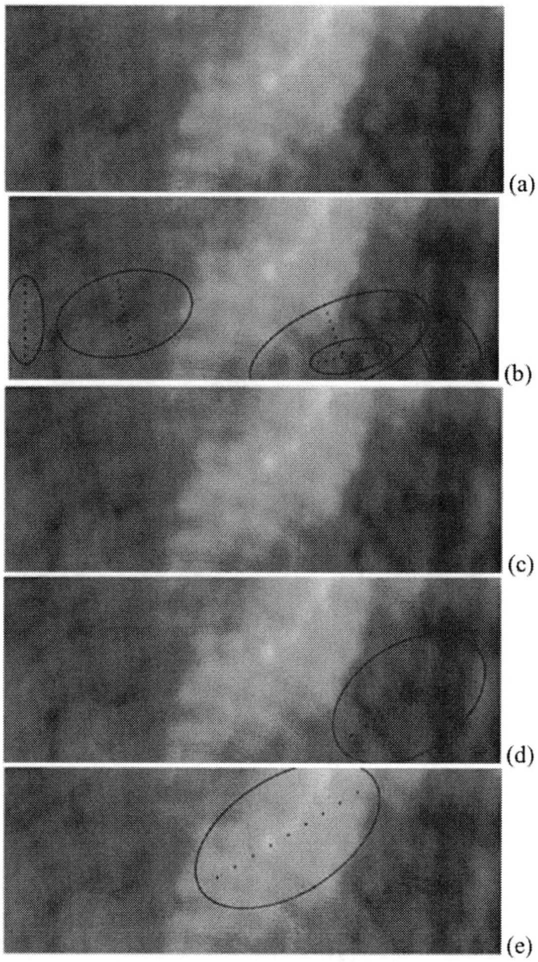

Figure 9. Potential deviations found for (a) 60, (b) 120, (c) 180, (d) 240 and (e) 300cm.

As shown in the previous images, several areas of deviation have been identified in the depth map of the slab for the considered wavelength. In previous works [7], this information had to be extracted from the scalogram by the reader in a manual manner. This task, which is time consuming and somehow complex, is now done completely automatically by the proposed new approach. This makes the whole process faster and more effective.

Figure 10 visually summarises the detected planar deviations for a study considering a continuous range of wavelengths between 60cm and 300cm. That information easily communicates to ground crews where planarity correction actions should be taken.

4.3 Comparison of with Waviness Index

Considering the Waviness Index (WI) method as the current state-of-the-art method, a comparison between WI and the proposed 2D CWT method is carried out.

Following the WI procedure specified in the standard ASTM-E1486 [7], survey lines are defined in directions parallel to the principal axes of the surface. Along these lines, survey points are defined every 30 cm and their elevation is measured. Then, the WI response [7] for each of the five levels k, corresponding to the wavelengths 60, 120, 180, 240 and 300 cm, is calculated using the equation:

$$LAD_k = \sqrt{\frac{\sum_{i=1}^{imax} LAD^2{}_{k,i}}{imax_k}} \tag{2}$$

where $LAD_{k,i}$ is the WI response at the level k at the ith evaluated location along the survey lines. The coefficient $imax_k$ is the number of locations where the response has been calculated at level k.

A comparable metric CWT_a is calculated to evaluate the 2D CWT response at each point evaluated at scale a. These values correspond to the previously mentioned LAD_k and are calculated as follows:

$$CWT_a = \sqrt{\frac{\sum_{j=1}^{jmax} CWT_{a,j}^2}{jmax_a}} \tag{3}$$

where $CWT_{a,j}$ is the 2D CWT response at the scale a at the jth sample location on the surface. The coefficient $jmax_a$ is the number of locations where the response has been calculated at scale a.

If the 2D CWT works well, then it would be expected that its response correlated with that of the WI method when for comparable scales a and levels k (i.e. the same wavelength), as defined in Table 1. Figure 11 shows the four points (CWT_a, LAD_k) obtained for the wavelengths 60, 120, 180 and 240cm (the slab is actually too small to assess the wavelength 300cm). Although it is hard to draw a full conclusion, the results show a remarkably strong correlation between CWT_a and LAD_k values, with R^2=0.9469. This result is supportive of the approach proposed here that can actually efficiently evaluated waviness with many more wavelength and can accurately report the location of where deviations are detected.

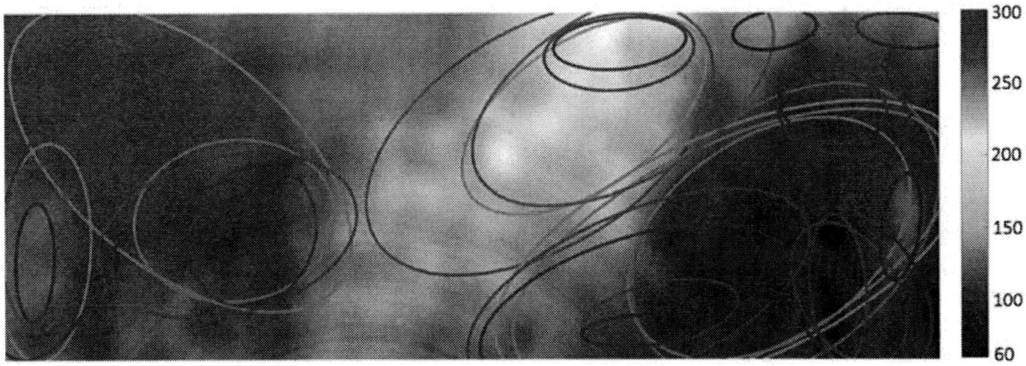

Figure 10. Map of defects for a continuous wavelength study from 60 to 300 cm.

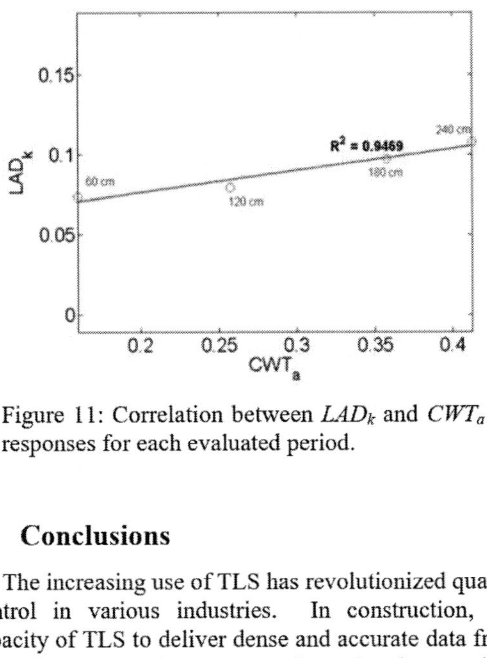

Figure 11: Correlation between LAD_k and CWT_a responses for each evaluated period.

5 Conclusions

The increasing use of TLS has revolutionized quality control in various industries. In construction, the capacity of TLS to deliver dense and accurate data from surfaces can greatly improve dimensional control in general, and flatness control particularly. Just like the introduction of profilometers has led to the emergence of the F-Numbers and Waviness Index methods (the current state of the art), the introduction of TLS offers an opportunity to review those methods once again and propose more robust and powerful ones.

The proposed 2D CWT approach makes full use of the density of measurements provided by TLS, and can assess waviness with levels of precision in the spatial and wavelength domains that more than surpass what was achievable with prior methods. The experimental results reported here positively demonstrate the potential value of the approach. Furthermore, a method is proposed to effectively detect deviations and communicates them effectively.

Nonetheless, the authors acknowledge that further work remains to be conducted to further validate the method. In particular, the authors will now acquire data from various slabs with various levels of specified flatness, ranging from housing foundation slabs to very flat slabs of industrial warehouses.

References

[1] Tang P., Akinci B. and Huber D. Characterization of laser scanners and algorithms for detecting flatness defects on concrete surfaces, *ASCE Journal of Computing in Civil Engineering*, 25:31-42, 2011.

[2] Bosché F. and Guenet E. Automating surface flatness control using terrestrial laser scanning and building information models, *Automation in Construction*, 44:212-226, 2014.

[3] Alhasan A., White D. J., De Brabanter K. Continuous wavelet analysis of pavement profiles, Automation in Construction, 63:134-143, 2015.

[4] Ytterberg C.N. *Flatness tolerances for random-traffic floors*, Technical Report, The Aberdeen Group, 1996.

[5] British Standards Institution (BSI), *BS 8204 - Screeds, bases and in situ flooring*, 2009.

[6] ASTM International, ASTM E 1155-96 - Standard test method for determining FF floor flatness and FL floor levelness numbers, 2008.

[7] ASTM International, ASTM E 1486 - Standard test method for determining floor tolerance using Waviness, wheel path, and levelness criteria, 2010.

[8] Bosché F. and Biotteau B. Terrestrial Laser Scanning and Continuous Wavelet Transform for Controlling Surface Flatness in Construction - A First Investigation, *Advanced Engineering*

Informatics, 29:591-601, 2015.

[9] Cheok G.S., Stone W.C., Lipman R.R. and Witzgall C. LADARs for construction assessment and update, *Automation in Construction,* 9:463-477, 2000.

[10] Akinci B., Boukamp F., Gordon C., Huber D., Lyons C. and Park K. A formalism for utilization of sensor systems and integrated project models for active construction quality control, *Automation in Construction*, 15:124-138, 2006.

[11] Bosché F., Haas C.T. and Akinci A. Automated recognition of 3D CAD objects in site laser scans for project 3D status visualization and performance control, *ASCE Journal of Computing in Civil Engineering*, 23:311-318, 2009.

[12] Schafer T. and Weber T. Deformation measurement using terrestrial laser scanning at the hydropower station of Gabcikovo, In: *INGEO 2004 and FIG Regional Central and Eastern European Conference on Engineering Surveying, FIG*, Copenhagen, Denmark, 2004.

[13] Addison P.S. *The Illustrated Wavelet Transform Handbook: Introductory Theory and Applications in Science, Engineering, Medicine and Finance*, 1st edition, CRC Press, 2002.

[14] Bosché, F., Guillemet, A., Turkan, Y., Haas, C.T., Haas, R. Tracking the Built Status of MEP Works: Assessing the Value of a Scan-vs-BIM System, *ASCE Journal of Computing in Civil Engineering*, 28:4, 2014.

[15] Golparvar-Fard, M., Peña-Mora, F., Savarese, S. Automated progress monitoring using unordered daily construction photographs and IFC-based Building Information Models, ASCE Journal of Computing in Civil Engineering, 29:1, 2014.

[16] Kim, C., Son, H., Kim, C. Automated construction progress measurement using a 4D building information model and 3D data, Automation in Construction, 31, 75-82, 2013.

Inspection Robot with Low Cost Perception Sensing

Harriet Peel*, George Morgan, Colin Peel, Anthony Cohn, **Raul Fuentes,
School of Civil Engineering, University of Leeds, Leeds, LS2 9JT
*Email addresses *cnhap@leeds.ac.uk, **r.fuentes@leeds.ac.uk*

Abstract - This paper presents the use of off-the-shelf products as a low cost solution to bridge bearing inspection. A commercial product, known as a DiddyBorg, is a robot designed for use with a Raspberry Pi as the on-board computer. The DiddyBorg is used as a robotic platform to make a photogrammetric survey of the bearing area of a bridge. The images collected from this survey are then used to make a 3D reconstruction using Structure-from-Motion (SfM) and software 3DFlow Zephyr Aerial (Zephyr). The quality of the 3D reconstruction had an accuracy of +/- 30 mm when compared to the known dimensions of the area. The resulting point cloud was then used as a map that the robot can use for navigation purposes. In particular we present a simple localization algorithm based on distance three readings measured from the robot.

Keywords-

Structure-from-Motion, robot, inspection, point cloud

1 Introduction

Bearings are critical for the performance of bridges and yet surveys are carried out infrequently, in many cases due to difficult access.

A solution to this problem is the use of autonomous robots which allows carrying out inspections with more repeatability than the traditional visual inspections that engineers undertake.

In particular, the geometry of the bridge bearing is of great importance. Although this paper does not focus on a methodology to achieve high accuracy, we present the use of a commercially available robot that uses low-cost cameras to undertake a full 3D reconstruction of the bearing area using Structure-from-Motion (SfM). There have been several applications of SfM to geosciences [1], archaeology [2] and for assessing the progression of work on construction sites [3]. One advantage of using SfM is that data collection using digital photography already occurs in many instances. Therefore, it is possible to use the images that have already been collected for other purposes. The images collected by the camera for the reconstruction can also be used to make a record of the status of the bridge and to observe damage using other techniques such as digital image correlation.

The focus of this paper is on the important issue of autonomous navigation. Specifically we present a solution to localization in near real-time using distance measurements and the initially recorded point cloud.

2 Structure-from-Motion (SfM)

SfM is now a relatively mature 3D reconstruction technique, much literature exists on the subject, and several commercial products exist. Zephyr Aerial is one such commercial software made by 3DFlow [4] and uses a proprietary algorithm known as SAMANTHA [5].

SfM uses multiple 2D views to find the 3D geometry (i.e. the structure) of a scene or an object by taking a series of images from different viewpoints (i.e. the camera has motion). The images collected do not need to be organised/ordered, nor do the camera locations need to be planned. Distinctive features, known as key points, are extracted from the images using feature detectors such as the scale-invariant feature transform (SIFT) [6], or in the case of SAMANTHA, a feature detector with automatic scale selection based on the work by Lindeberg in [7].

In SAMANTHA, keypoints are matched between images in two stages: broad stage matching and narrow stage matching [5]. Broad stage matching uses keypoints that are ranked according to the scale value obtained in keypoint detection: keypoints with the strongest response are used for matching with keypoints in other images. Once they are matched between images, the corresponding images are connected to the images with which it shares the greatest number of keypoint matches to form the epipolar graph. The epipolar graph describes how views from different cameras are geometrically related.

In narrow stage matching, some keypoints are discarded depending on nearest neighbour relationships. The fundamental matrices, which describe the location and rotation of the cameras in a global (but unscaled) reference frame, are then calculated using a subset of the key point matches. To find the fundamental matrix between two camera views, at least 7 keypoints are required. The fundamental matrix is commonly calculated using RANSAC [8], although SAMANTHA uses a variation of RANSAC called M-estimator sample consensus (MSAC) [9] which applies a set penalty to outliers.

Once the fundamental matrix has been calculated, the relative positions of all the cameras can be computed. Some SfM methods, such as [10], use an incremental approach, adding one new camera in at a time and using triangulation to find the 3D geometry, then bundle adjustment is carried out using Bundler [11]. SAMANTHA uses a 'hierarchical' approach [5]. The advantage of the hierarchical approach is that the calculations are parallelisable. Initially, images are clustered based on the amount of overlap between them. Partial stereo models are formed between the individual clusters of images. These partial stereo models can then be merged with other partial models, or can be expanded with other individual images. Bundle adjustments are carried out at each stage. Local bundle adjustment may also be used [5].

It is important to note that for all SfM methods the absolute scale of the scene cannot be recovered and that some method of scaling is required to relate the point cloud to a global scale.

A technique known as Multi-View Stereo [12], [13], can then be used to enhance the point cloud generated using SfM to make a dense point cloud.

3 Robot Description

A robot was selected to fill the following criteria;

- To be small and compact enough to navigate through limited spaces.
- Could be programmed.
- Could be adapted for autonomous navigation.
- Could be controlled remotely.
- Could capture photographs of a suitable standard for use in 3D reconstructions.

A variety of commercial products were considered that fulfilled these criteria, but the one that was selected was the DiddyBorg, a self-build kit made by PiBorg [14]. The DiddyBorg is a six wheeled vehicle approximately 182 x 220 x 95mm. Each wheel is driven by 5V motor, which is required to be used outside and required an adequate level of torque. The DiddyBorg is built around a Raspberry Pi Model 2 B motherboard with the operating system Raspbian Jessie [15].

The DiddyBorg with a Raspberry Pi was chosen for this application due to its low cost (approx. £180 / $260) and ease of programming, with many open source projects available online that can be integrated into a project. The Raspberry Pi adds the ability to mount multiple sensors to the DiddyBorg. For the experiments carried out in this paper, three ultrasound sensors were mounted to the DiddyBorg platform, one at the front, one on the left and one on the. A further description of these sensors is provided in Section 7.

The camera used is a Raspberry Pi Camera Board [16], it produces 5MP pictures and 1080p HD video at 30fps. The software required to control the camera is also open-source and readily available.

4 Site Description

The site considered in this paper is the Centenary Bridge, in south Leeds, UK, opened in 1993. It crosses The River Aire, connecting The Calls to Brewery Place. It is a cable suspension bridge spanning approximately 57m. The bridge was chosen because it was accessible and the bearings are located in an enclosed region (Figure 1) in which the DiddyBorg system could be tested easily. One disadvantage of this area being accessible, is that there was a build-up of litter inside the bearing enclosure, the quantity of this litter is likely to change from inspection to inspection and cause anomalies in the point clouds that are generated. However, not being able to control the environment highlights one advantage of using camera as a sensor over methods such as laser scanning – a visual assessment can easily be made of the data that is obtained.

The bearings in question are the north side bearings. The maximum height of the bearings, and subsequently of the enclosure, is 400mm. The top bearings are seated on the bridge by means of a machined steel plate bolted to the bearing. The bottom bearings are seated on the abutment by means of bedding mortar.

The top of the abutment is a space approximately 2.8m by 1.2m (Figure 3). There are few restrictions to the site. Firstly, there is a 155mm wide trough that runs along one side on the site. This is too wide for the robot to cross so the robot's movements were restricted to one side of this trough. Secondly, there were various pipes and electrical cables that occupied the centre of the site. The robot could travel over these obstacles, but with risk of getting stuck, so these regions were avoided. There was also a gate at street level that had a top support within the site, which was avoided.

5 Survey

For reconstructions of enclosed areas, such as a room or a town square, the best practice, as recommended by [17], is to take a panoramic sequence of images from at least 2 corners of the region being captured and to obtain at least 60% overlap between images. To meet these recommendations, the DiddyBorg platform was rotated by small increments up to 360 degrees in three locations – refer to Figure 4. Each time the DiddyBorg was rotated and stopped a photo was captured. Data capture took in the region of 20 minutes to complete.

5.1 Reconstruction

The photographs were then reconstructed using Zephyr Aerial. Due to computational requirements, and since reconstruction on-board the robot was not necessary, the reconstruction was performed on a desktop PC. Since Zephyr Aerial produces a reconstruction automatically from the data it is given, no changes in the default settings were made to the Structure from Motion or Multiview stereo stages of the reconstruction. The reconstruction took approximately 20 minutes running on a desktop computer with 32.0 GB RAM, a 3.60 GHz CPU and two 2.0GB NVIDA video cards with CUDA capabilities. Cloud computing services for reconstruction are also commercially available from 3DFlow [18].

There was some noise from the changing lighting conditions and the litter, and for this reason a confidence calculation was performed in Zephyr. The confidence score is calculated by considering how well a point is matched between photos. The matching score is from 0 to 1 and is summed over the photos. The confidence calculation function in Zephyr can be used to remove points where a certain matching score has not been reached. In this case, points with a confidence value less than 1 were removed. The resulting point cloud, after updating the confidence, is presented in Figure 4.

Since SfM creates a point cloud at an arbitrary scale, a method is required to scale the point cloud. Control points were picked from the photographs. The point cloud was then scaled using the scale drawings from the bridge design, this is a feasible approach for applications in inspection of civil infrastructure.

To scale the bridge, four control points were used, more control points would give a more accurate reconstruction, but since the environment surveyed here is broadly of one texture it was difficult to find points that clearly stood out that also corresponded to the scale drawings. For this reason, points such as corners were used. Once the control points are picked (control points appear as dots marked in and Figure 4), a further bundle adjustment is carried out and an RMS error is calculated by Zephyr showing the difference between the values used and their locations within the Zephyr model. The RMS error for this scaled reconstruction was 30mm.

Figure 1: Photograph showing the location of the north-side bearings of Centenary Bridge, Leeds, UK.

Figure 2: Photograph showing the enclosure in which the bridge bearing survey was carried out using the DiddyBorg robotic platform.

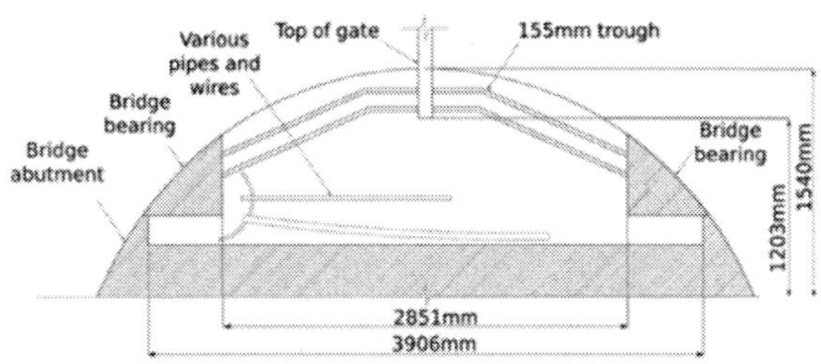

Figure 3: Schematic showing the dimensions and features of the enclosure from Figure 4.

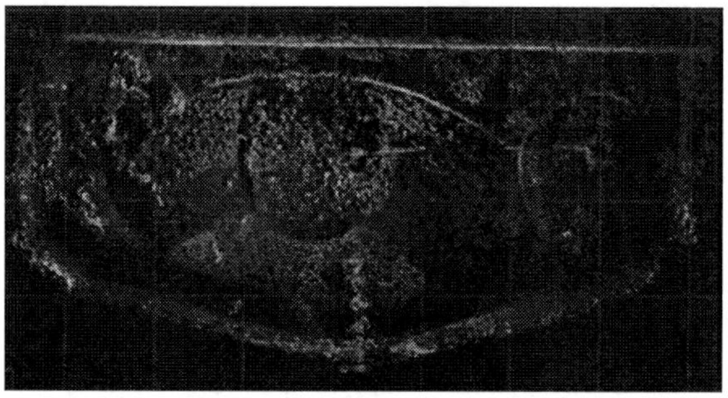

Figure 4 : Top view of the SfM reconstruction of the bridge bearing enclosure generated using Zephyr Aerial with confidence of points updated. Control points are marked by red dots. The blue shapes in the middle of the cloud show the locations where the robot took pictures.

5.2 Platform Integration

To interface with the Raspberry Pi, a network development environment known as Node.js [19] was used. The open-source nature of the community developing with Node has led to the development of many exciting projects, and many libraries exist that can easily be installed and integrated into a project. The work here to integrate the Raspberry Pi with the DiddyBorg robotic platform with several different sensors uses and brings together work from open-source libraries including: r-pi-usonic [20], picoborgrev [21] as well as taking some inspiration from the web Graphical User Interface developed by PiBorg [22] specifically for manipulating the motion of the DiddyBorg.

Requests were sent from a webpage at the front end to a web application program interface (API) which interfaces with the Raspberry Pi. Commands are sent to control the motors and the camera, but this interface can be expanded easily to incorporate any other sensors that are on-board. The algorithm for detecting the location of the DiddyBorg using distance measurements was also integrated into this system and the results can be viewed directly from the webpage. The design criteria when creating this system was to use familiar technology, like a webpage, for ease of use when working in the field. The user can manipulate the motion of the robotic platform, take video footage or camera stills, distance measurements and calculations from on-board sensors (e.g. ultrasound, infrared) and calculate the location of the platform through a simple and straightforward interface.

6 Algorithm

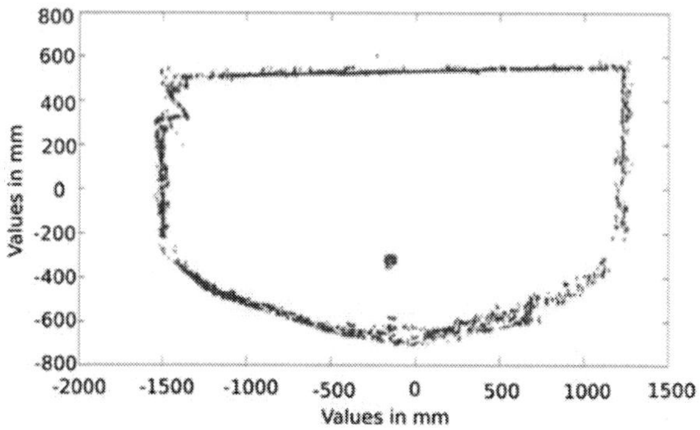

Figure 5: Convex hull point cloud generated using Cloud compare (plotted here as (x,y) coordinates with values in mm).

Once a point cloud had been generated, it had to be reduced in a way that would be useful for the navigation of the robot. The proposed solution was to produce a convex hull of the point cloud (see Figure 5), as an outline of the region the DiddyBorg is in. This outline was taken at the same height as the sensor mounts on the DiddyBorg. This reduction was completed using CloudCompare [23]. A localisation algorithm was then required to obtain the location of the DiddyBorg within the region using distance measurements. The steps of the localisation algorithm are as follows (also see Figure 6):

1. Take a reading from each of the sensors: front, left and right.
2. Find pairs of points where the distance between the points is equal to the sum of the left and right readings within a given error.
3. Using trigonometric relations and similarity find a guess location for the location of the DiddyBorg.
4. This method cannot calculate which direction the DiddyBorg is facing, so
5. Guess locations for two directions must be considered.
6. Calculate the distance from each guess point to all other points in the point cloud.
7. Find whether any of the distances match the front sensor reading within a given error.
8. Narrow down the number of possible locations by calculating the gradient of the lines connecting the points joining the side measurements and the gradients connecting the guessed point and the points that match the front readings. If the gradient of one is equal to -1/gradient of the other then the points are perpendicular, this is a requirement since the sensors are perpendicular on the DiddyBorg platform.

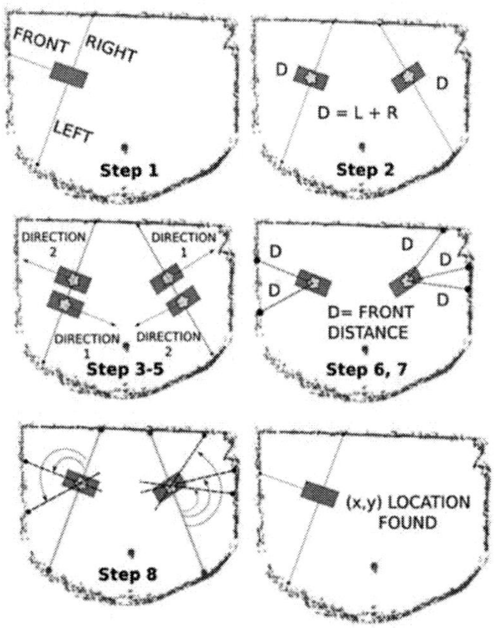

Figure 6: Schematic showing the steps of the algorithm described in Section 6.

It is highlighted that this algorithm will not give the unique location of the DiddyBorg platform when the point cloud used for navigating has axes of symmetry. However, the aim at this stage of the project was to reduce the possible number of locations where the robot could be located. Improvements to this algorithm and the methods used to detect potential locations will be the target of future work.

7 Validation

The algorithm was validated from tests within a simulated rectangular convex hull point cloud. Test locations for the DiddyBorg were chosen inside this location, the corresponding sensor readings were calculated and the algorithm was run.

A rectangular area was chosen to test the algorithm because it is a case where multiple locations will be returned due to symmetry. If the algorithm was successful, it was expected that four predicted locations would be returned.

Within the algorithm, (listed above in Section 6) an error value is used to account for the error in the sensor readings. This value is a constant and can be varied. When the error value is set to be very low (below 0.05 mm), i.e. the accuracy of the sensor is very high, the algorithm returns 4 possible locations for the DiddyBorg including the test value set at the beginning and showed that the logic of the algorithm is correct. However, this error value is unrealistic for what can be expected from a real sensor, and as the error value was increased more possible locations for the robot were detected, such that it would not be possible to say for certain where the robot was.

To determine the reliability that might be expected from using real sensors, an ultrasound sensor HC-SR04 Ultrasonic Module Distance Measuring Transducer Sensor [24]) that returns distance from an object was tested. The sensor range from the sensor datasheet [24] is 2cm to 4m, with an accuracy of 3mm. This sensor has been used with the Raspberry Pi [25] and the Arduino [26] open-source electronics platform.

The ultrasound sensors were wired in a similar way to [25], but by daisy chaining the ultrasound sensors from the battery pack that also powers the Raspberry Pi and the motors. The system and user interface developed in Section 5.2 were easily adapted to incorporate the functionality of these sensors. The algorithm used to take readings from the sensor was based on the work of [20], expanding it to incorporate multiple sensors. The sensors were tested by taking readings when the robot was stationary and then checking the readings against manual measurements using a tape measure. When the robot was facing a large planar surface, such as a wall at medium range, the readings were consistent and within the accuracy stated in [24]. However, the distance readings from the sensor became unreliable with variable surfaces and obtructions with readings fluctuating by up to 20cm in some cases. This variability is likely due to measurement angle of the sensor being 15 degrees, so a different object to the one the ultrasound sensor is facing may have been detected. Therefore, if this sensor was to be used with the algorithm in Section 6 it would, at best, be able to slightly narrow down the number of potential locations, but not adequately enough for navigation purposes. It was concluded that this sensor is inadequate for use on its own for the purpose of navigation, but perhaps if localisation is carried out in with other sensors, the ultrasound sensors could be used to detect unexpected objects such as the litter found in the bridge bearing enclosure.

8 Future Work

Given that the ultrasound sensors were not accurate enough to use for navigation purposes, in the real case of the bridge bearing enclosure, alternative sensors will be identified and tested for the purpose of navigation. Further work will also be carried out to investigate whether the ultrasound sensors can be used effectively to detect unwanted objects in the region being inspected such as litter.

In order to deal with uncertainties encountered in the

field, further development of the algorithm presented in Section 6 will be carried out.

A method for independently scaling point clouds without the use of scale drawings (although the drawings may be used as a measure of accuracy) is to be developed. This method should be easily integrated into the surveying process detailed in Section 5.

9 Conclusions

The DiddyBorg platform was used to successfully survey a bridge bearing through photographic data collection. These photographs were then used to reconstruct the enclosure where the bearings are situated using the SfM software Zephyr Aerial. The point cloud genereated in this reconstruction was used to create a convex hull point cloud suitable for use with the localisation algorithm developed for navigation (Section 6), which, in principle, can be used with the DiddyBorg robotic platform that was used to gather the original photographic data. The robot was successfully controlled using the web-based API detailed in Section 5.2.

10 Acknowledgements

We would like to acknowledge funding from ESPRC for the studentship of H.Peel.

References

[1] J. Westoby, M. Brasington, M. Glasser, N. Hambrey and M. Reynolds, "Structure-from-Motion' photogrammetry: A low-cost, effective tool for geoscience applications," *Geomorphology,* vol. 170, pp. 300-314, 2012.

[2] C. Balletti, F. Guerra, V. Scocca and C. Gottardi, "3D Integrated Methodologies For The Documentation and the Virtual Reconstruction of an Archaeological Site," in *The International Archives of the Photogrammetry, Remote Sensing and Spatial Information Sciences, Volume XL-5/W4,* Avila, Spain, 2015.

[3] M. Golparvar-Fard, F. Peña-Mora, S. Savarese and e. al, "Automated Progress Monitoring Using Unordered Daily Construction Photographs and IFC-Based Building Information Models," *Journal of Computing in Civil Engineering,* vol. 29, no. 1, p. 147, 2012.

[4] 3DFlow, "3DFLOW," 3Dflow, 2016. [Online]. Available: http://www.3dflow.net/. [Accessed 03 04 16].

[5] R. Toldo, R. Gherardi, M. Farenzena and A. Fusiello, "Hierarchical structure-and-motion recovery from uncalibrated images," *Computer Vision and Image Understanding,* vol. 140, pp. 127-143, 2015.

[6] D. Lowe, "Distinctive image features from scale-invariant keypoints," *International Journal of Computer Vision,* vol. 60, pp. 91-110, 2004.

[7] T. Lindeberg, "Feature detection with automatic scale selection," *International Journal of Computer Vision,* vol. 30, pp. 79-116, 1998.

[8] M. A. Fischler and R. C. Bolles, "Random Sample Consensus: A Paradigm for Model Fitting with Applications to Image Analysis and Automated Cartography," *Communications of the ACM,* vol. 24, pp. 381-395, 1981.

[9] P. H. S. Torr and A. Zisserman, "MLESAC: A new robust estima-tor with application to estimating image geometry," *Computer Vision and Image Understanding,* vol. 78, no. 1, p. 138–156, 2000.

[10] N. Snavely, S. M. Seitz and R. Szeliski, "Photo tourism: exploring photo collections in 3D, in: SIGGRAPH," in *International Confer-ence on Computer Graphics and Interactive Techniques, pp 835–846,* 2006.

[11] M. Lourakis and A. Argyros, "SBA: a software package for generic sparse bundle adjustment.," *ACM Transactions on Mathematical Software,* vol. 36, p. 1–30, 2009..

[12] Y. Furukawa and J. Ponce, "Accurate,dense, and robust multi-viewstereopsis," in *Proceedings, IEEE Conference on Computer Vision and Pattern Recognition (CVPR),* Minneapolis, USA, 2007.

[13] Y. Furukawa, B. Curless, M. Seitz and R. Szeliski, "Clustering view for multi-view stereo," in *Proceedings, IEEE Conference on Computer Vision and Pattern Recognition (CVPR),* San Francisco, USA, 2010.

[14] PiBorg, "DiddyBorg," 03 April 2016. [Online]. Available: https://www.piborg.org/diddyborg.

[15] R. P. FOUNDATION, "Download Raspian for the Raspberry PI," [Online]. Available: https://www.raspberrypi.org/downloads/raspbian/ . [Accessed 03 04 2016].

[16] R. P. FOUNDATION, "Raspberry Pi Camera Module," [Online]. Available: https://www.raspberrypi.org/products/camera-module/. [Accessed 03 April 2016].

[17] 3Dflow, "3DFLOW," 2016. [Online]. Available: http://www.3dflow.net/technology/documents/photogrammetry-how-to-acquire-pictures/. [Accessed 03 04 2016].

[18] 3DFlow, "SUNNY Beta - the Cloud service for your reconstruction," [Online]. Available: https://sunny.3dflow.net/login.php. [Accessed 03 April 2016].

[19] Node.js.Foundation, "node JS," 2016. [Online]. Available: https://nodejs.org/en/. [Accessed 03 April 2016].

[20] clebert, "A high performance, memory mapped, Node.js API for the HC-SR04 ultrasonic sensor connected to a Raspberry Pi.," 2015. [Online]. Available: https://github.com/clebert/r-pi-usonic. [Accessed 01 April 2016].

[21] CodyErekson, "picoborgrev: A controller library for PiBorg's PicoBorg Reverse motor driver board designed for use with the Raspberry Pi.," 2014. [Online]. Available: https://github.com/CodyErekson/picoborgrev. [Accessed 03 April 2016].

[22] PiBorg, "diddyborg-web: Web based interface for controlling the DiddyBorg from a phone or browser," 2015. [Online]. Available: https://github.com/piborg/diddyborg-web. [Accessed 15 March 2015].

[23] D. Girardeau-Montaut, *Cloud Compare—3D Point Cloud and Mesh Processing Software,* Open Source Project - http://www.danielgm.net/cc/, 2015.

[24] E. Freaks, "Ultrasonic Ranging Module HC - SR04 Product Features," [Online]. Available: http://www.micropik.com/PDF/HCSR04.pdf. [Accessed 03 04 16].

[25] P. Staff, "Building a Raspberry Pi Motion Sensor with Realtime Alerts," PubNub, 16 June 2015. [Online]. Available: https://www.pubnub.com/blog/2015-06-16-building-a-raspberry-pi-motion-sensor-with-realtime-alerts/. [Accessed 03 April 2016].

[26] Arduino, "Genuino," 2016. [Online]. Available: https://www.arduino.cc/. [Accessed 03 April 2016].

[27] R. P. FOUNDATION, "Raspberry Pi - Teach Learn and Make with Raspberry Pi," [Online]. Available: https://www.raspberrypi.org/products/raspberry-pi-2-model-b/. [Accessed 01 04 2016].

[28] Itseez, "OpenCV," 2016. [Online]. Available: http://opencv.org/. [Accessed 04 April 2016].

How can Lean, IPD and BIM Work Together?

A.H. Fakhimi[a], J. Majrouhi Sardroud[b] and S. Azhar[c]

[a]Department of Civil Engineering, Kashan Branch, Islamic Azad University, Kashan, Iran
[b]Department of Civil Engineering, Central Tehran Branch, Islamic Azad University, Tehran, Iran
[c]McWhorter School of Building Science, College of Architecture, Design and Construction, Auburn University
E-mail: Ahfakhimi@iaukashan.ac.ir,J.Majrouhi@iauctb.ac.ir,Salman@auburn.edu

Abstract

As buildings have become more complex, the construction industry has become more specialized, but it is under greater pressure to become more efficient, productive and integrated. Thus, all recent and novel attempts including Lean Construction (LC) philosophy, Integrated Project Delivery (IPD) method and recently Building Information Modeling (BIM) technology are greatly aiding to improve productivity and efficiency via enhanced collaboration and integration. They have been very useful and resulted many benefits, but their drawbacks and challenges prevent them to take fully satisfying results from their individual implementation.

The objective of this paper is to investigate if LC, IPD and BIM can play all together complementary and synergistically, then to draw benefits and challenges of their collaboration. The research is archival in nature and an extensive literature review has been conducted on the implementation of LC, IPD and BIM in the complex project settings. No trilateral collaboration is found and so this study is focused on bilateral collaboration experiences to perceive the demarche of their combination. This paper presents a combination scheme for LC, IPD and BIM in an effort to provide pragmatic solutions to the complex project problems. Challenges, which need to be addressed, are also highlighted.

Keywords

BIM, Integrated Project Delivery, Lean Construction, collaboration, benefit and challenges

1 Introduction

The architecture, engineering and construction (AEC) industry does not have a reliable reputation as a protagonist in quality, efficiency, productivity, cost and time management due to recent more complex and multi-discipline projects. Many researchers believe project success is dependent on the complexity of a project and having a direct effect on the overall project performance [1].

It is a commonly held opinion that the reason for the poor performance is the design and construction processes being particularly complex. Therefore, an understanding of project complexity and its management is of significant importance. Gidado K. [2] suggests that there seem to be two perspectives of project complexity in the industry: The managerial perspective, which involves the planning of integration to form workflow and the operative and technological perspective, which involves the technical intricacies or difficulties of executing individual pieces of work. This may originate from the resources used and the environment in which the project was carried out [2].

Construction industry due to their higher complexity than many other industries needs new theories of production or construction to support a renewed drive for greater performance improvement. Therefore, it has been revived growing debate on flame ideas regarding construction complexity including, managerial perspective (process, LC) and operative and technological perspective (people, IPD and tools, BIM).

Many organizations and researchers pay attention to concept of these novelty ideas/approaches (LC, IPD, BIM) and direct their specific research to develop and implement these ideas individually or even bilaterally for overtopping the mentioned complexities as well as low efficiency and productivity. This paper strives to present a quick review on individual and bilateral collaboration concepts, benefits, barriers and challenges and suggests a trilateral collaboration idea for complex construction projects. The minority of bilateral collaboration researches and lack of trilateral collaboration of LC, IPD and BIM show us the necessity of this kind of research to make construction industry more integrated.

2 Methodology

The research data are comprised of journals, conference papers, thesis, dissertations, and internet articles collected through databases such as Science

Direct, Scopus, Emerald, universities libraries. Instead of comparing or evaluating the strength of reported studies, this paper seeks to reflect the interest of research community on LC, IPD and BIM collaboration topic, so conference papers, thesis, dissertation, and internet article are included in the pool.

The purpose of this Study is to discover the benefits, barrier and challenges of LC, IPD and BIM collaborations for complex construction. As trilateral collaboration did not find, analysis and synthesis of their individual concepts and bilateral collaborations are used to draw trilateral collaboration scheme through literatures and published case studies in AEC industry. The literature review includes the concept, definition and the use of individual and bilateral collaboration of LC, IPD, BIM and their benefits, barrier and challenges in three main aspects: process, people and tools. In addition, the study looks for direction of recent research studies on LC, IPD and BIM bilateral collaboration and their probable trilateral collaboration as an important gap found in this study especially in AEC industry.

3 Lean Construction(LC) project

Concept: Since the 1950s, Toyota Motor Company has implemented Toyota Production System (TPS) principles successfully. TPS had two pillar concepts: (1) Just In Time flow (JIT) and (2) Autonomation (smart automation) [3]. Lean construction is a philosophy based on the concepts of TPS. It is about managing and improving the construction process to deliver profitably what the customer needs. Koskela L. introduced the basic theoretical innovation idea of understanding construction as production, International Group for Lean Construction (IGLC) stated. Further, Koskela and Howell [4] showed the need for a broader foundation for project management in order to getting more efficiency and productivity via elimination of waste. Koskela [5] states that lean construction shares the goals of lean production: elimination of waste, cycle time reduction, and variability reduction. In fact, workflow reliability and labor flow are regarded as key determinants of construction performance [6].

Benefits: Shorter order fulfilment lead times, fewer projects down time, more innovation and true cost reduction are reported as benefits of the successful application of LC. This paper suffices to illustrate the result of McGraw Hill Construction finding in this regards (Figure 1).

Barriers and Challenges: Many studies focused on investigating barriers that prevent the diffusion and implementation of LC concept and identifying barriers that emerge during the execution of LC practices. These barriers could affect the application process of LC and hinder the project performance, if not properly managed. As mentioned, LC is a philosophy and focuses on process and people and it does not uses specific tools for its implementation. LC implements via a few well-known techniques like last planer system (LSP), Five S's, Huddle meetings, First-run studies (PDCA) and so on. This study assessed a number of structural (process) and cultural (people) barriers hindering the progress of the LC approach. Process barriers such as lack of adequate lean awareness and understanding, incorrect interpretation, lack of the use of process based PMSs, traditional management organization, contractual structure, financial issues and lack of exposure on the need for lean construction and people challenges such as lack of top management commitment, governmental aspects, culture and attitudinal issues, the fear and resistance to change, educational issues, lack of training and lack of human skills and experience are reported as main LC implementation's barriers and challenges [7,8].

Figure 1. Potential Benefits with a high influence on Non Practitioners for the adoption of lean practices (*McGraw Hill Construction*, 2013)

4 Integrated Project Delivery(IPD)

Concept: In 1990, BP in-house team formed an integrated group that combined engineering, subsurface and commercial interests. They called it Alliancing and were spectacularly successful. It continued to improve the concept and emerged with new face called IPD. Based on American Institute of Architects, AIA's most recent definition of IPD is defined as "a project delivery method that integrates people, systems, business structures and practices into a process that collaboratively harnesses the talents and insights of all participants to reduce waste and optimize efficiency through all phases of design, fabrication and construction" [9]. IPD brings all participants together early with collaborative incentives to maximize value for the project. Many researchers believe IPD has parallels with Koskela's lean construction movement, which has aimed to translate product manufacturing and production methods to construction. Mossman et al. do - not recognize a distinction between IPD and LC in their particular research [10]. IPD approach built around six characteristics that differentiate it from traditional project delivery; (1) a multi-party contract, (2) early involvement of key participants, (3)collaborative decision making and control, (4) shared risks and rewards, (5) liability waivers among key participants, and (6) jointly developed project goals [11].

Benefits: AIA explains that, achieving the benefit that IPD offers requires the participants of the project to follow some key important principles such as mutual trust, mutual reward, early involvement of key participants, early goal definition and leadership. AIA claims some benefits in this regards as bellow:

- Reduce or eliminate conflict on the project team.
- Optimize the efforts of the workforce.
- Improve communication and understanding among project stakeholders.
- Result in a clearer definition of project goals.
- Create incentives for exceptional results.
- Reduce waste by better planning and shared costs.
- Improve project delivery timelines.
- Reduce operational and maintenance costs of the finished project.
- Reduce requests for interpretation (RFI) from the contractor.
- Facilitate sharing of rewards and risks among stakeholders.
- Encourage the team to take a broader, more creative approach to addressing the client's needs.

Barriers and Challenges: Despite the emergence of IPD as a comparably advantageous project delivery approach, literature on this topic presents numerous obstacles. AIA with a group pioneer in IPD [12] defined the different level of collaboration achieved through contractual (process) and organizational (people) from IPD philosophy to IPD delivery method. The group stated, "Once an owner decides that change needs to occur and IPD is the desired direction, the bigger the organization, the bigger the challenge".

IPD same as LC does not introduces specific tools for its implementation. IPD introduces BIM as its powerful tool. Main IPD implementation barriers and challenges regarding process are legal (Liability , insurance and risk) issues, skill sets and communication protocols, technological barrier and financial issues, and regarding people are fear and resistance to the changes, willingness and knowledge of owner organization, overcoming decades of mistrust and training issues [11,13].

5 Building Information Modeling(BIM)

Concept: Professor Charles M. Eastman [14,15] has introduced BIM concept since 1970. As a solution to inefficiencies in the industry, BIM technology has been gradually developed and practically used in the AEC industry in projects starting from the mid-2000s [16].

The USA is the first country to implement BIM [17]. Nowadays, BIM is implemented in many countries [14]. A commonly accepted definition of BIM is: "Building Information Modeling (BIM) is a digital representation of physical and functional characteristics of a facility. A BIM is a shared knowledge resource for information about a facility forming a reliable basis for decisions during its life-cycle; defined as existing from earliest conception to demolition" [17].

BIM technology creates a virtual model of a building's with quantitative and qualitative characteristics. Cooperation between the different parties involved in the project and supporting the project through its lifetime in the design, construction, fabrication, procurement and maintenance phases emphasized by LC and IPD, facilitate by this concept.

Benefits: BIM offers advantages to those businesses that embrace it. The most cited benefits expected BIM to grant AEC industry are: better production quality, faster and more effective processes and data transmission, reduction in project duration, better customer / client / stakeholder services and satisfaction, more productivity and efficiency, earlier and better coordination and collaboration between all partners, cost tracking possibilities, early design assessment to ensure project requirements are met, mitigating litigations, reduced number of RFI and change orders, elimination of rework via error free drawing production, achieve better logistics and procurement planning, and finally improved Commissioning and delivery of project [14-16,18-24].

Barriers and Challenges: By LC and IPD contrast, because BIM introduces powerful tools for its implementation individually or in collaboration with LC and IPD, BIM barriers and challenges divided into three categories: Process, People and Tools. Table 1 illustrated some recent references of main BIM implementation barriers and challenges regarding these three categories.

6 Collaboration and Integration

Collaboration is a data-centric activity while integration is a knowledge-centric activity. Integration occurs best when the participants view themselves as equal in the process and when the initial collaboration focuses on exploring and defining the problem, rather than commenting on another's proposed solution. The building process cannot be optimized without full collaboration and finally integration among all members.

Table 1 Main reported BIM barriers and challenges

BIM Barriers and Challenges		Azhar, 2011	Azhar et al, 2012	Both and Kindsvater, 2012	Erezi Utiome, 2010	Panaitescu, 2014	Stanley and Thurnell, 2014	Ning and Landon, 2010	Nurain, 2013	Elmualim and Gilder, 2014
Process	Legal Changes/Absence of Standard BIM Contract Documents	√	√	√		√				
	Implementation Issues/Need for a New Business Model	√		√		√		√		√
	Changes in Practice and Use of Information/Developing BIM modeling skills and redefining staff roles and responsibilities		√			√	√	√	√	
	Need to change information flow management / Integration of meaningful information/Data Organization/ data Validation	√	√			√	√	√	√	
	Required more time spent inputting and reviewing BIM data/cost of implementation	√	√			√	√		√	√
	Lack of collaborative work processes and modeling standards		√		√	√	√			
	complex nature of the AEC						√			
	lack of understanding of implementation processes									√
	Risk distribution (Liability and legal issues)					√			√	√
People	Collaboration and Teaming	√	√							
	Inertia/Resistance to change (Employee, Organization, Owner)			√	√	√	√			
	Absence of qualified BIM implementation team/users/operators		√						√	√
	Absence of owner's interest in using BIM technologies and further ownership of the model	√				√	√		√	
	Intellectual Property / Security /(IP and copyright issues)	√	√					√	√	
Tools	Need for standard data exchange languages/ Interoperatability	√	√	√	√	√	√	√	√	
	Training issue/ In adequate training/ Cost and time of training		√	√			√	√	√	√
	Taking responsibility for data exchange (Liability and legal issues)/Registry of communication and information exchange		√		√	√		√		
	growing propagation of non-standardised BIM applications				√		√			

However, with reviewing concept, benefits, barriers and challenges of the novel approach in the AEC industry (LC, IPD and BIM) and deep looking at their originations, their overlaps are clear. Nowadays, researcher and practitioner going to consider integration of them and introducing integrated framework via their reciprocal synergetic collaborations. Collaboration is the first step of integration and the AEC industry has to implement and experience synergetic collaboration of novel construction methods before integrating them.

7 Discussion, Finding and Areas for further research

LC, IPD and BIM are making fundamental changes in the AEC industry. While all of them conceptually independent, separate, all different from one another, address different aspects of professional practice and do not require together for running and implementation, but they are great complements, each can empowers the others and will have less effect on successful project collaboration if done in the absence of its complementary components. In the following, according to the previous description about benefits and challenges of LC, IPD and BIM, necessity of their collaboration for empowering them are stated.

LC focused its limit to attitudes, processes, techniques for continuous improvement, increasing value, eliminating waste inside a project [8], loose supply chains and interactions with third parties besides main barriers mentioned before. LC may be empowered by IPD, because IPD instead of introducing appropriate processes of how to reduce waste and optimize efficiency, concentrated on multi-party contract and collaboration between all parties [11]. IPD help the project team to apply the LC to maximize value and minimize waste in the production process. Some researchers (such as Mossman et al [10]), guidelines and standards (e.g. ConsensusDOCS 300, AIA) do not recognize a distinction between LC and IPD so that they introduce a combined name for their collaborations, Lean Integrated Delivery Project (LIPD).

One of the main drawbacks of LC and IPD is that, they focus on process and people via some techniques and they do not use specific tools for implementation. By contrast, BIM as a tool provides technological territory of information sharing between all parties and does not introduces any framework for cooperation between the different parties in the project. Looking at the overlaps of LC and BIM benefits show that the application of BIM produces almost the same benefits that LC is supposed to generate [15]. Sacks et al [25] uncovered 56 interactions of both approaches revealing significant synergy effects. They proposed a framework of BIM functionalities and associated Lean principles of such systems beyond construction. However, little has been published on its materialization in practical works. In this regards, The University of Salford has developed a maturity model for Lean and BIM. This model helps to assess the project's Lean and BIM implementation against the 10 identified main criteria [26].

AIA recommend that BIM should be used to achieve required collaboration for IPD. Through early collaboration and the use of BIM technology, a more integrated, interactive, virtual approach to building design, construction and operation is emerging [27,28]. The coupling of BIM with IPD enables a level of collaboration that not only improves efficiency and reduces errors, but also enables exploration of alternative approaches. Among other applications, IPD is materialized as a delivery method that could most effectively facilitate the use of BIM for construction projects [29].

Any collaboration is created to answer the concern, overcome the barrier, challenges and issues and pool the benefits. Based on this and considering the bilateral collaboration of LC, IPD and BIM, the main points should be taken into account for their collaboration and integration are [9]:

- Mutual trust and open communication among parties,
- Understanding each other's objectives and alignment of objectives
- Equitable and clear allocation of foreseeable and quantifiable risks
- Attitude of the Project Participants by develop a partnering culture and replacing individual responsibility by collective responsibility
- Readiness to compromise on unclear issues, awareness of risks and rewards
- Legal issues, What types of contracts should be used for optimum collaboration between all parties
- Develop a partnering culture

As a finding, this study clarified LC, IPD and BIM can work separately and appropriately, but achieving their full capacity is not possible without their collaborations. Until now, many researchers and practitioners define frameworks and methods for their reciprocal synergetic collaborations to issue their findings (Table 2, Table 3).

Table 2 Main reported bilateral collaboration barriers and challenges

	Bilateral Collaboration Barriers and Challenges	Bilateral: BIM+LC	Bilateral: LC+IPD	Bilateral: IPD+BIM	References
Process	•Need to qualified contractors			√	[38]
			√		[36]
	•Requires profound process changes of the involved parties			√	[29] [38] [39]
		√			[26]
	•Risk distribution between parties			√	[27]
	•Absence of standard collaborative contract documents			√	[29] [30] [38]
	•Responsibility, liability and model ownership(legal issues)			√	[29] [38] [39]
People	•Need to adequate training and education		√		[38]
	•Need to trusted and capable members		√		[36] [38]
	•team communication			√	[36] [38]
	•team's prior experience as a unit	√			[42]
	•cultural change			√	[38]
	• Fair of Changes		√		[29] [43]
	•commitment by management		√		[43]
		√			[44]
Tools	•Interoperability issues			√	[27]
		√			[42]
	•Lack of BIM product			√	[29]

Table 3- Main reported bilateral collaboration benefits

	Bilateral Collaboration Benefits	Bilateral: BIM+LC	Bilateral: LC+IPD	Bilateral: IPD+BIM	References
Process	• Better Planning and collaborative, integrated and transparent construction process			√	[9] [12] [30] [25] [7] [31] [32] [33]
	• Facilitate/catalyze the optimum project delivery process	√			[22]
	• Effectively facilitate the use of each other for construction projects		√		[34] [35] [35] [36]
	•Solve many of the issues (cost, constructability, schedules, quality, sustainability, waste, HSE, etc.) during project life cycle			√	[30] [31] [29] [32] [33]
		√			[37]
	•Increase productivity and efficiency and generates more added value to the client		√		[35]
	•Reduce amount of redundant data •Facilitate sharing of information			√	[30] [34] [38] [39]
	•Improve project relationships		√		[35]
	• Offer significant improvements in contractual and legal matters.			√	[35]
			√		
People	• Risk and reward could distributed amicably amongst project team			√	[39]
	•Facilitate closer collaboration from early stages of the Project •Suppliers will be integrated into the processes of the construction industry			√	[15] [27] [30] [36] [35] [40]
	• Help work teams to do the work with higher performance	√			[41]
Tools	• BIM acts as an excellent team building tools and accelerated the formation and strategies of IPD			√	[30]
		√			[42]
	• BIM provides data storage exchange service				[38]

However, their individual strength and weakness and bilateral collaborations for covering individual weakness to emerge their strength, direct us to suggest a trilateral collaborations for more empowerment by overruling some individual and bilateral barriers and challenges. It could be stated that IPD structures people's interactions and incentives, LC increases value and efficiency and BIM provides transparency and a single source of truth. Full benefits of this collaboration appear whenever collaboration tabernacles with integration. When BIM as a tool and LC as a process connect all people of project via IPD, the synergetic trilateral collaboration appears and all partners (owner, client, designer,..) get their benefit using synergies of them as a whole. Based on this study and inspiring existing experienced bilateral collaborations frameworks and models specially lean project delivery system (LPDS) [30], we propose a trilateral collaboration model (Figure 2) in which all parties could be collaborated and finally integrated in co-location, trilateral collaborative LC, IPD and BIM circumstance. It could be called the project delivery "hat trick", a state-of-the-art approach to design and construction.

8 Conclusion

To date, to realise the full potential benefit of novel methods, it needs to be used collaboratively in a project. The necessary trilateral collaborative circumstance does not exist but, rather, is best created when all parties enter into a relationship based contractual arrangement(IPD) for delivering a lean project via a strong technological tools (BIM). In this regards, a trilateral collaboration scheme is proposed to emerge their synergetic use in a unique environment (Figure 2).

This trilateral collaboration enables a level of collaboration that not only improves efficiency and reduces errors but also enables exploration of alternative approaches and expansions of market opportunities. It is expected this scheme could help AEC to surmount challenges of bilateral collaboration such as team communication and synergise all benefits of bilateral collaboration.

The findings of this study could be used to help researchers, practitioners, general contractors and companies in the AEC industry to focus their attention and resources on the significant issues necessary to support the trilateral collaborative implementation of LC, IPD and BIM. Future research is recommended to investigate for overcoming barriers to implementing trilateral collaborations.

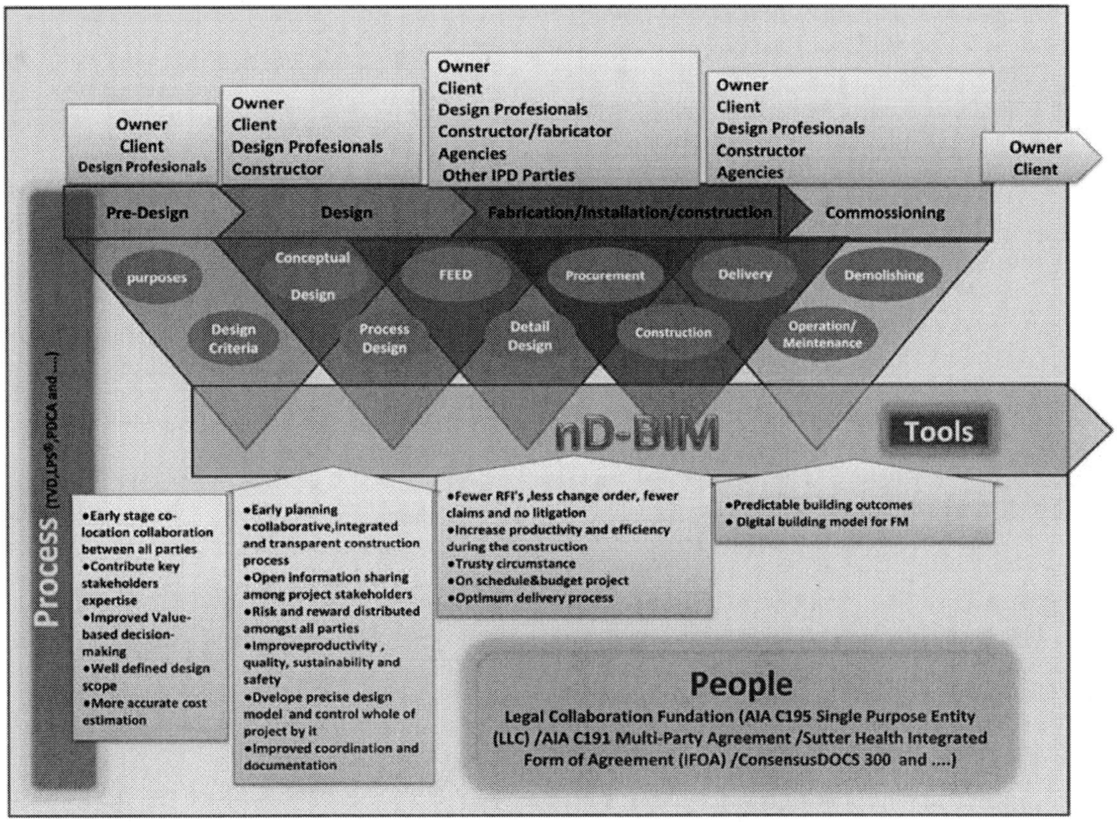

Figure 2- Proposed trilateral collaboration scheme

9 References

[1] Ochieng E. and Hughes L. Managing Project Complexity in Construction Projects: The way Forward. *Architectural Engineering Technology* 2(1): 1-2, 2013

[2] Gidado K. Project complexity: The focal point of construction production planning. *Construction Management and Economics*, 14:213-225, 1996.

[3] Aziz F.R. and Hafez SH.M. Applying lean thinking in construction and performance improvement. *Alexandria Engineering Journal*, 52:679-695, 2013.

[4] Koskela L. and Howell G. The theory of project management: Explanation to novel methods. *Proc. IGLC-10, 10th Conf. of Int. Group for Lean Construction.* pages 1-11,Porto Alegre, Brazil,2002.

[5] Koskela L. Management of production in construction: a theoretical view. *IGLC-7, 7th Conf. of Int. Group for Lean Construction.* pages 241-252, Berkeley, CA,1999.

[6] Thomas H., Horman M., Minchin R.JR. and Chen D. Improving labor flow reliability for better productivity as lean construction principle. *Journal of Construction Engineering and Management*,129(3): 251-261, 2003.

[7] Porwal V., Fernandez-Solis J., Lavy S. and Rybkowski Z.K. Last Planner System Implementation Challenges. *IGLC-18, 18th Conf. of Int. Group for Lean Construction.* pages 548-556. Haifa, Israiel,2010.

[8] Wandahl S. Lean Construction with or without Lean – Challenges of Implementing Lean Construction. *IGLC-22, 22th Conf. of Int. Group for Lean Construction.* pages97-108. Oslo, Norway, 2014.

[9] American Institute of Architects, California Concile. Integrated Project Delivery - An Updated Working Definition. AIA , 2014.

[10] Mossman A., Ballard G. and Pasquire C. The growing case for lean construction. *Construction Research and Innovation*, 2 (4):30-34,2011.

[11] Ghassemi R.and Becerik-Gerber B. Transitioning to Integrated Project Delivery: Potential barriers and lessons learned. *LCI Lean Construction Journal*, Lean and integrated project delivery special issue:32-52, 2011.

[12] NASFA, COAA, APPA, AGC AND AIA. Integrated Project Delivery For Public and Private Owners. pages 45, 2010.

[13] Rached F., Hraoui Y., Karam A. and Hamzeh F. Implementation of IPD in the Middle East and its Challenges. *IGLC-22, 22th Conf. of Int. Group for Lean Construction*, pages 293-304,Oslo, Norway, 2014.

[14] Aryani A.L. ,Suzila M. ,Narimah K. and Fathi M.S. Building InformationModeling (BIM) Application in Malaysian Construction Industry. *International Journal of Construction Engineering and Management*, 2(4A): 1-6,2013.

[15] Eastman C. et al. *BIM Handbook, A guide to building information modellin for owners, managers, designers, engineers and contractors.* Second edition. Wiley, 2011.

[16] Azhar S., Khalfan M. and Maqsood T. Building information modelling (BIM): now and beyond. *Australasian Journal of Construction Economics and Building*, 12(4):15-28, 2012.

[17] National Institute of Building Sciences (NIBS), United States National Building Information Model Standard, Version1,NIBS,2013.

[18] Arayici Y. and Tah J. Towards building information modelling for existing structures. *Structural Survey*,26(3):210-222, 2008.

[19] Succar B. The five components of BIM performance measurement. *18th CIB World Building Congress.* Salford,UK, 2010.

[20] Joannides Maya M. et al. Implementation of Building Information Modeling into Accredited Programs in Architecture and Construction Education. *International Journal of Construction Education and Research*, 8:83-100, 2012.

[21] Azhar S. Building Information Modeling (BIM): Trends, Benefits, Risks, and Challenges for the AEC Industry. *Leadership Manage. Engineering*, 11(3):241-252, 2011.

[22] Panaitescu R. *Building Information Modeling,Towards a structured implementation process in an engineering organization.* Delft University of Technology,Faculty of Civil Engineering. Delft, Netherland. 2014.

[23] Stanley R. and Thurnell D. The Benefits of, and Barriers to, Implementation of 5D BIM for Quantity Surveying in New Zealand. *Australasian Journal of Construction Economics and Building*, 14:105-117, 2014.

[24] Kalinichuk S. and Tomek A. Construction Industry Products Diversification by Implementation of BIM. *International Journal of Engineering and Technology Innovation*, 3(4):251-258, 2013.

[25] Sacks R., Koskela, L. Dave, B. and Owen R. Interaction of Lean and Building Information Modeling in Construction. *Journal of Construction Engineering and Management*, 136(9):968-980, 2010.

[26] Dave B., Koskela L., Kviniemi A., Owen R. and Tzortzopoulos P. *Implementing Lean in construction: Lean construction and BIM.* London: CIRIA, C725, 2013.

[27] Wright, J.A. The Integration of Building Information Modeling and Integrated Project Delivery into the Construction Management Curriculum. *ASEE Annual Conference*, 2012.

[28] Zallan Jay B et al. Restructuring The Achitechtural Practice: AEC Success in BIM, IPD and Beyond. *PRACTICAL BIM 2012:Building Information Modeling – Management,*

Implementation, Coordination and Evaluation. pages 257-262.California, USA: University of Southern California, 2012.

[29] Porwal A.A. and Hewage N.K. Building Information Modeling (BIM) partnering framework for public construction projects. *Automation in Construction,* 31:204-214, 2013.

[30] Ballard G. The Lean Project Delivery System: An Update. *Lean construction Journal,*pages 1-19, 2008.

[31] Elmualim A. and Gilder J. BIM: innovation in design management, influence and challenges of implementation. *Architectural Engineering and Design Management,*10(3-4):183-199, 2014.

[32] Dave B., Boddy S. and Koskela L. Challenges and Opportunities in Implementing Lean and BIM on an Infrastructure Project. *IGLC-21, 21th Conf. of Int. Group for Lean Construction.* pages 741-749, Fortaleza, Brazil,2013.

[33] Jones B. Integrated project delivery (IPD) for maximizing design and construction considerations regarding sustainability. *2nd International Conference on Sustainable Civil Engineering Structures and Construction Materials.* pages 528-538. Yogyakarta, Indonesia, 2014.

[34] Volk R., Stengel J. and Schultmann F. Building Information Modeling (BIM) for existing buildings — Literature review and future needs. *Automation in Construction,*38:109-127, 2014.

[35] Halttulla H., Aapaojab A. and Haapasaloc H. The contemporaneous use of building information modeling and relational project delivery arrangements. *8th Nordic Conference on Construction Economics and Organization.* pages 532 – 539, Tampere, Finland, 2015.

[36] Ilozor B.D. and Kelly D.J. Building Information Modeling and Integrated Project Delivery in the Commercial Construction Industry: A Conceptual Study. *Journal of Engineering, Project, and Production Management,*2(1): 23-36, 2012.

[37] Rokooei, S. Building Information Modeling in Project Management: pages 87-95, 2015. *Procedia - Social and Behavioral Sciences, Elsevier,*210(2):87–95, 2015.

[38] Sarkar D. A framework for development of Lean Integrated Project Delivery Model for infrastructure road projects. *International Journal of civil and structural Engineering,* 5(3):267-271, 2015.

[39] Hall D., Algiers A.,Lehtinen T. et al. The role of Integrated Project Delivery Elements in Adoption of Integral Innovation. *Engineering Project Organization Conference.* Devil's Thumb Ranch, Colorado, 2014.

[40] Suttie J.B. The Impacts and Effects of Integrated Project Delivery on Participating Organisations With a Focus on Organisational Culture. *IGLC-21, 21th Conf. of Int. Group for Lean Construction.* pages 267-276, Fortaleza, Brazil,2013.

[41] Rodewohl, C.F. *The presence of Lean Construction principles in Norway's transport infrastructure projects.* Norwegian University of Science and Technology,Department of Civil and Transport Engineering. Trondheim, Norway. 2014.

[42] Oskouie P., Gerber D.J. , Alves T. and Gerber B.B. Extending the interaction of building information modeling and lean construction. *IGLC-20, 20th Conf. of Int. Group for Lean Construction,* San Diego, California, 2012.

[43] Ma C., Li X. and Meng Y. Study on the Application of BIM Technology in Construction Projects under IPD Mode. *ICCREM 2014: Smart Construction and Management in the Context of New Technology,* pages 229-236.Kunming, China, 2014.

[44] Miettinen R. and Paavola S. Beyond the BIM utopia: Approaches to the development and implementation of building information modeling. *Automation in Construction,* 43: 84-91, 2014.

[45] Clemente J. and Cachadinha N. BIM-Lean Synergies in the Management on MEP Works in Public Facilities of Intensive Use - a Case Study. *IGLC-21, 21th Conf. of Int. Group for Lean Construction.* pages 751-760, Fortaleza, Brazil,2013.

[46] Hamd O. and Leite F. BIM and Lean Interactions from the BIM Capability Maturity Model Perspective: A Case Study. *IGLC-20, 20th Conf. of Int. Group for Lean Construction,* San Diego, California, 2012.

[47] Both P. and Kindsvater A. Potentials and barriers for implementing BIM in the German AEC market- Results of a current market analysis. *14th International Conference on Computing in Civil and Building Engineering.* Moscov, Rusia, 2012.

[48] Ning G. and London K. Understanding and facilitating BIM adoption in the AEC industry. *Automation in construction,* 19(8): 988-999, 2010.

[49] Nurian H.I. Reviewing the evidence: use of digital collaboration technologies in major building and infrastructure projects, *Journal of Information Technology in Construction,* 18: 40-63, 2013.

Direction Aware Bluetooth Low Energy Based Proximity Detection System for Construction Work Zone Safety

J. Park[a], Y.K. Cho[b] and S.K. Timalsina[c]

[a] School of Civil and Environmental Engineering, Georgia Institute of Technology, USA
[b] School of Civil and Environmental Engineering, Georgia Institute of Technology, USA
[c] School of Civil and Environmental Engineering, Georgia Institute of Technology, USA
E-mail: jpark463@gatech.edu, yong.cho@ce.gatech.edu, stimalsina3@gatech.edu

Abstract

Over the last decade, the rate of fatality of workers in construction work zones has been of serious concern. One of the most frequent accidents happens due to a ground worker being struck by construction equipment. A major reason behind this is the inability of equipment operators to be aware of the existence of the nearby pedestrian worker. In this paper, we propose a direction aware proximity detection and alert system. The proposed system can detect hazardous situations caused by proximity between equipment and the ground workers. Upon detection, it notifies both the ground worker and equipment operator with audible alerts and visualization of the relative location of the hazard. This research utilizes a recent Bluetooth low energy technology to build a proactive proximity and alert system. To evaluate the effectiveness of the system, outdoor tests were conducted. The obtained results show that the equipment operators can efficiently obtain the information on potential hazardous situations to avoid collision with the ground worker.

Keywords –

proximity sensing; direction awareness; work-zone safety; Bluetooth low energy; construction

1 Introduction

Workers in the construction industry are prone to various hazardous situations, and most of these situations lead to severe and fatal accidents. In spite of the serious attention paid to accidents in the construction industry, significant occupational injuries at construction sites prevail. During 2008 to 2013, a total of 683 worker deaths was reported on road construction sites [1]. The dynamic nature of work zones comprises of situations where ground workers operate in close proximity to construction equipment. This characteristic of work zones puts the personnel in a potentially hazardous situation. Most of the fatal work zone accidents were identified as the "Worker being struck by a vehicle or equipment in a work zone". It signifies that special attention needs to be given to the safety of ground workers working in close proximity to vehicles or equipment in construction sites.

In past research, efforts have been made to detect and alert the proximity of ground workers to construction equipment [2; 3]. The hazardous proximity situation can thus be avoided to prevent worker injuries and fatalities. However, their measures fail to provide the information on direction of the event occurrence. As noted in [4], many collision accidents are due to the blind area in the field of vision of an equipment operator. Depending on the construction equipment type and operator's position, the blind area can include front, back and sides of the equipment [5]. Therefore, information on relative direction of hazardous situations is crucial to take action to avoid them. This paper reviews the current proximity detection solutions and describes the proposed direction aware proximity detection system followed by the explanation of the testing scenario and its emulation. Finally, the observed results are analyzed to evaluate the performance of the system.

2 Background

In concern of the current work zone safety status, Occupational Safety and Health Administration(OSHA) and Manual on Uniform Traffic Control Devices (MUTCD) have made efforts on implementing regulations in work zone safety [6; 7]. The imposed regulations consist of requirements, such as wearing hard hats and shiny vests, signs and signals, flaggers, traffic controlling systems and other conventional measures. However, in spite of the application of these rules, it is affected by several pitfalls. It is found with insignificant improvement in the concerned safety hazards and prevailing fatalities. A major cause of this unsatisfactory result is due to the inability to recognize in advance and avoid the potential hazard; for example, due to blind areas in the equipment operator's vision or lack of knowledge of the ground worker in proximity to the equipment [4; 8].

2.1 Proximity Detection

Over the last decade, a considerable amount of research has been conducted in applying sensing and wireless technology into the work zone safety. In [9], the sensing technologies such as Radar, video camera and Radio Frequency Identification (RFID) have been discussed as measures to implement the collision warning system by monitoring. In [3], magnetic, RFID, and Bluetooth Low Energy (BLE) technologies are discussed as potential candidates for proximity detection. These technologies were compared in terms of their performance in average distance estimation and proximity detection. An coverage area of each of the technologies was also evaluated. As mentioned in the previous section, the current state-of-the-art technologies perform efficiently in sensing the proximity but lack on identifying and providing direction information of the potential hazard. This information is critical for the equipment operator who needs to overcome a potential collision by avoiding proceeding to the direction of potential hazard. In other words, it is required not only to find out "if" there is a potential hazard, but also "where".

In [2; 8], research have been performed on the application of laser and imaging technologies in construction to mitigate the effect of blind spots in equipment operator's vision. This research utilized the Flash Laser Detection and Ranging (Flash LADAR) and a low-resolution image ranging method. Using these techniques together with learning algorithms, the researchers have attempted to scan through subject area and predict the orientation of the equipment operator. However, using the image-ranging method has a main drawback as it needs to go through several preliminary phases before the system implementation. It should be noted that the image-ranging methods additionally have requirements of complex backend to run algorithms and store the point cloud. In addition, such required process can lead to the system unable to detect the hazardous situation when applied in a new environment. Besides, as mentioned in [8], the performance of the image ranging method degrades in an outdoor scenario, which is typical to construction. Further, the quality of an image can be significantly affected by many external sources, including lights, and angles; for example, images captured at night may be of too low quality to properly process. More importantly, although this method detects the orientation of operators head and dynamically updates the blind area of the operator, it still fails to eliminate the existence of blind areas.

In order to address the issues discussed above, our study focused on building a system capable of detecting proximity and its direction relative to the construction equipment with the ground worker in a dynamic construction environment. Our primary objective was to build a system, which can cover the entire region surrounding the construction equipment to eliminate the blind areas. This enables more interactive communication about hazard incidents between associated workers; it alerts an equipment operator upon the detection of potential collision with the ground worker. The system should be minimally affected by changes in its environment. It should have least or no preliminary installation requirement.

2.2 Bluetooth Low Energy (BLE)

For fulfilling our objective, BLE has been studied in terms of its application in the proximity detection with direction awareness. BLE is a relatively new technology that differs from the classical Bluetooth technology with the capability of functioning by consuming comparatively low energy. It operates in the 2.5 GHz industrial, scientific and medical (ISM) band and was introduced in 2010 by Bluetooth Special Interest Group (SIG) along with the Bluetooth 4.0 specification [10]. The low energy consumption attribute of BLE enables it to stay alive on a coin-sized battery for months if not years. Compared to the classical Bluetooth technology, it is able to send smaller data packets, thus having specific applications, which require long-lasting connectivity but small or no data communications. Furthermore, the devices capable of detecting the classical Bluetooth can easily detect BLE as well.

With relatively small sizes (see Figures 1 and 3), low power consumption, low cost and durability, BLE can be considered an ideal tool for proximity detection for construction applications. Construction sites are very dynamic and demand not only proactive but also easy-to-use technology, which can adapt to the changes with minimum overhead. As an accessory to the construction site, its cost is reasonable, and its size is negligible to pose hassle in undergoing the construction projects. Nevertheless, incurred device and instalment costs for these devices are also low.

The devices able to detect multiple BLE sensors within a region can obtain the Received Signal Strength Indicator (RSSI) for each of the BLE sensors. The RSSI can in turn be used to estimate the relative distance to the detector. In addition, the detectors can uniquely identify each BLE sensor. The information obtained by combining these capabilities can be utilized to detect the proximity hazardous situations and obtain the relative direction of the detectors.

Figure 1. Commercially available BLE transmitters.

3 Experiment

3.1 System Architecture

The key objective of this study was to create a reliable proximity detection and alert system capable of 1) identifying proximity hazardous situations between ground workers and equipment and 2) providing, in addition to warning alerts, the associated personnel with the relative direction of the hazard with respect to the construction equipment. The setup requires less overhead in terms of the installation and maintenance, and thus, it can work as plug and play. A typical scenario of the system in construction is shown in Figure 2. The system consists of three major components, such as BLE sensors, ground worker's personal protection units (PPUs) and equipment operators' PPUs. However, there can be other auxiliary components to enhance the feature and usability of the system.

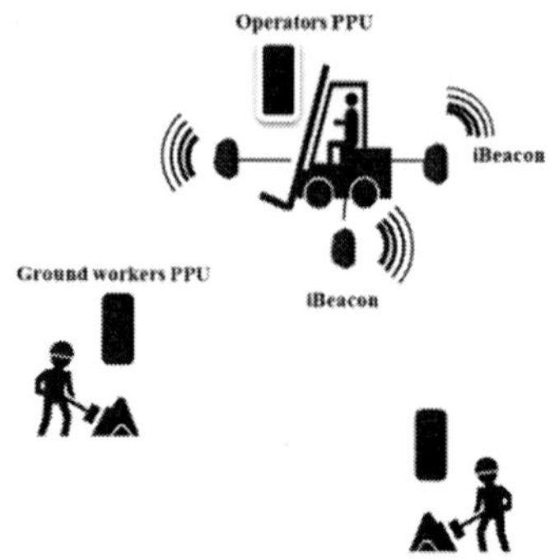

Figure 2. Typical scenario of the proposed system.

3.1.1 Location Broadcasters

Location broadcasters are the devices deployed throughout the region in order to get the knowledge on location information of object of concern. Our system uses BLE sensors to act as the location broadcasters. As described in the previous section, a BLE sensor is the radio signal transmitter, which advertises micro-location related data. The advertised data is broadcasted in Omni-direction, which can be utilized by any device capable of reading it. It is easy to install in the system by simply attaching to the equipment, whose location one wants to monitor. With its small size, the BLE sensor is easily attachable to the construction equipment while not requiring cumbersome infrastructure, including external power lines, nor interfering with the working environment (Figure 3). The sensor does not require any pre-setting to incorporate with the system as they continuously broadcast characteristic Bluetooth signals. The proposed system uses the Received Signal Strength (RSS) from these devices to estimate the relative distance to equipment.

Figure 3. A BLE sensor attached to construction equipment.

3.1.2 Personal Protection Unit/s (PPU/s)

PPU is a device, which monitors and processes the data transmitted by location broadcasters, and interprets the valuable information. The proposed system utilizes widely available smartphones to act as PPUs. Smartphones are capable of detecting the BLE signals and process it reliably. We developed software that converts the broadcasted RSSI to distance estimation. Further, the system monitors for proximity related incidents and provides alerts upon detection of such incident. This software can be configured to define the alert zones as well. The PPU can therefore alert workers through various forms such as sound and vibration when they enter the predetermined alert zones. Additionaly, PPUs can also communicate between themselves to share the local information.

3.1.3 Workers

Workers, including ground workers and equipment operators, who are prone to potential hazard situation are equipped with PPUs. The ground workers are alarmed through alerts such as sound and vibrations with intensity proportional to proximity to the equipment as shown in Figure 4. The equipment operators are also provided with the visual information regarding the hazard situation; direction of potential collision with reference to the equipment (Figure 5). This could help the operator to manoeuvre the equipment to a safer direction.

Figure 4. Ground worker with PPU.

The system comprises of the dynamic interaction between the components described previosly. Ground workers are equipped with a PPU with software that alerts upon reaching proximity to hazardous situation. Similarly, the equipment operators also have a PPU, which visualizes the proximity of the ground worker in case of potential danger. The location transmitters can be attached on different parts of the equipment such that PPUs could receive signals emitted from different regions. PPU is able to uniquely identify each of the location broadcasters. Therefore, this can provide visual information with the direction of the ground worker in proximity relative to the equipment. As shown in Figure 6, the system works by continuously receiving BLE signals and providing the information to workers through their PPUs.

With the complete knowledge on proximity of the ground workers relative to the equipment in all the directions, the equipment operator can efficiently undergo collision avoidance through the direction aware alert system. In this manner, with the proposed system, the blind area of equipment operator can be virtually eliminated.

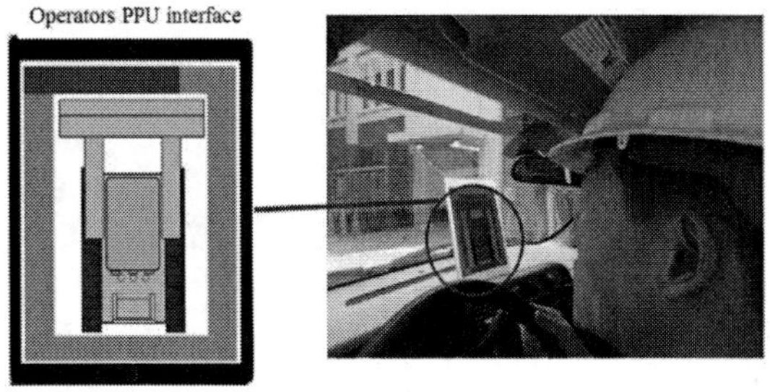

Figure 5. Equipment operator observing the PPU.

3.2 Test Setup

In order to evaluate the performance of the direction aware alert system, a simulation of scenario with a ground worker approaching the equipment was conducted. To measure the signal communication and warning responsiveness effectively, we conducted the one direction approach. The ground worker with a PPU approached stationary equipment in forward direction and raised hand to indicate the initiation of alert (Figure 4). The equipment operator then observed visualization of alert in the PPU (Figure 5), and confirmed the directional information. On moving out of the hazard zone, ground worker lowered the hand to signal the event. This was again observed by the equipment operator along with the disappearance of visual information on the PPU. Occurrences of each of these events were being logged against time. The time delay between initiation of alert and actual visualization on operators PPU (start delay) and that between termination of alert and disappearance of visual information (end delay) was also observed by the equipment operator. Observed delays were then evaluated for effectiveness of the system.

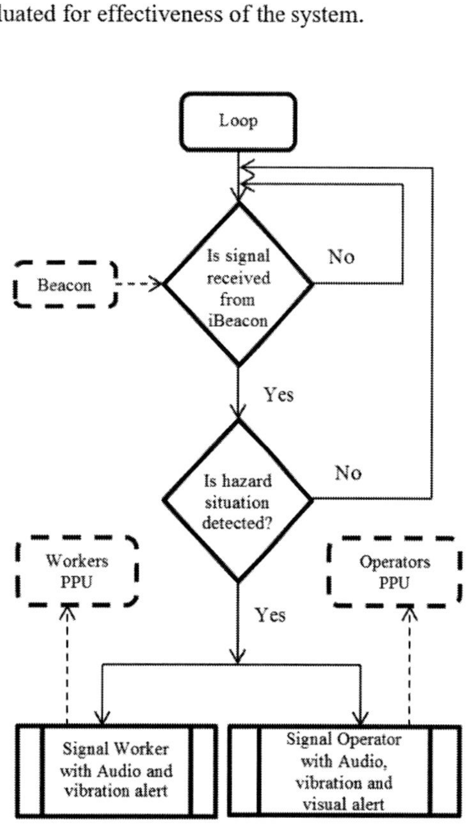

Figure 6. Working methodology of the system.

4 Results and Discussion

Figure 7 shows the result of the alert responsive rate test. The responsiveness of the system to an alert is an important metric to measure the performance of the system. The visual information in addition to the audible alerts can get the attention of the equipment operator to avoid potential collisions caused by blind spots or unawareness.

On a hazardous situation being originated at the ground worker's end, system signals operator's PPU about it. Due to the inherent nature of the system, a delay is incurred between these events. This delay in time is represented as start delay. Similarly, lag in time between a ground worker leaving the hazard zone and respective information conveyance by PPU to the operator was observed, which is represented as end delay.

Thirty trials in total were conducted on the test setup, and the obtained result is shown in Figure 6. The start delay was found to be approximately 0.1 second. However, the value of observed end delays varied in respective trials. The average end delay for the conducted trials was 8.03 seconds. Compared to start delay, end delay was found to be large. This can be attributed to the nature of radio signal transmission; the transmitted signals tend to reach the PPU in multipath. However, this outcome can be considered to have a practical importance as the operator is provided with additional time to take appropriate action to confirm the safety of workers. Importantly, the equipment operator was able to promptly receive proper information on

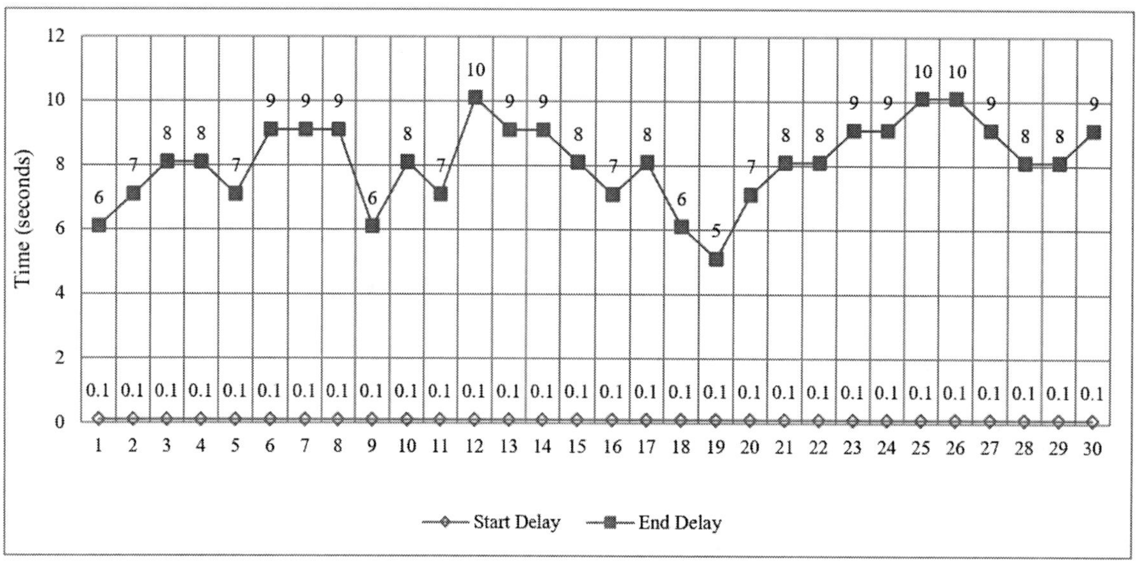

Figure 7. Alert responsiveness between ground worker and operator.

relative distance and direction of the ground worker and avoid potential collision.

5 Conclusion and Future Directions

Utilizing various proximity detection devices, many research efforts have been made to alert both the ground workers and approaching construction equipment operators regarding hazardous situations. This study focused on building a novel proximity detection system which is able to provide direction information in addition to alerts to workers and vehicle operators in vicinity. The BLE technology was used to build the direction aware proximity detection and alert system. The proposed system can synchronously provide information to the construction equipment operators regarding the potential hazard situation and its relative direction. In this way, an operator is able to avoid collisions with ground workers. Due to the simple hardware requirements, the proposed system is simple, reliable and easily adjustable to a dynamic construction environment.

In future, the system providing more granularities in the estimation of location and direction will be developed. Additionally, the improved system will be tested in more realistic and complex construction environments with a considerable number of ground workers and vehicles interacting with each other.

References

[1] National Institute for Occupational Safety and Health. CDC - Highway Work Zone Safety - NIOSH Workplace Safety and Health Topic. onlineavailable athttp://www.cdc.gov/niosh/topics/highwayworkzones/accessed2015

[2] Marks, E., and Teizer, J. Real-time proactive equipment operator and ground worker warning and alert system in steel manufacturing. *Iron and Steel Technology*, *9*(10), 56–69., 2012.

[3] Park, J., Marks, E., Cho, Y. K., and Suryanto, W. Performance Test of Wireless Technologies for Personnel and Equipment Proximity Sensing in Work Zones. *Journal of Construction Engineering and Management*, *142*(1), 04015049. doi:10.1061/(ASCE)CO.1943-7862.0001031, 2015.

[4] Teizer, J., Allread, B. S., and Mantripragada, U. Automating the blind spot measurement of construction equipment. *Automation in Construction*, *19*(4), 491–501. doi:10.1016/j.autcon.2009.12.012, 2010.

[5] Ruff, T. *Recommendations for evaluating and implementing proximity warning systems on surface mining equipment*. onlineavailable athttp://stacks.cdc.gov/view/cdc/8494/Printaccessed2007

[6] Federal Highway Administration. Manual on Uniform Traffic Control Devices (MUTCD) - FHWA. onlineavailable athttp://mutcd.fhwa.dot.gov/accessed2015

[7] OSHA. Regulations (Standards - 29 CFR).

[8] Ray, S. J., and Teizer, J. Coarse head pose estimation of construction equipment operators to formulate dynamic blind spots. *Advanced Engineering Informatics*, *26*(1), 117–130. doi:10.1016/j.aei.2011.09.005, 2012.

[9] Ruff, T. M. MONITORING BLIND SPOTS: A MAJOR CONCERN FOR HAUL TRUCKS. onlineavailable athttp://www.cdc.gov/niosh/mining/UserFiles/works/pdfs/mbsam.pdfaccessed2001

[10] Bluetooth Special Interest Group. Specification of the Bluetooth System Covered Core Package Version 4.0, *0*(June), 2302., 2010.

Field Construction Management Application through Mobile BIM and Location Tracking Technology

J. Park[a], Y.K. Cho[b], and K. Kim[c]

[a] School of Civil and Environmental Engineering, Georgia Institute of Technology, USA
[b] School of Civil and Environmental Engineering, Georgia Institute of Technology, USA
[c] School of Civil and Environmental Engineering, Georgia Institute of Technology, USA
E-mail: jpark463@gatech.edu, yong.cho@ce.gatech.edu, kkim369@gatech.edu

Abstract

While Building Information Modeling (BIM) has been increasingly accepted in the construction industry, the rich project information in BIM is still not well utilized to support the field construction management. Project managers need to spend significant time and effort to find out useful project information (such as plans and specifications) for field management, and communicate with construction participants in the site and offices. To avoid such inefficiency caused by separation between digital project information and field management, this study proposes an innovation that integrates BIM and location tracking technology for construction field management. We created a new management tool by integrating a Bluetooth Low Energy (BLE) location tracking system and a BIM environment. In addition to the visualization of a mobile BIM in hand, the developed system provides the users with real-time access to more contextual information based on locations of construction resources. The system has been demonstrated with a field trial at an indoor construction site. The result indicated that the integration of a tracking system in BIM 1) enabled timely visualization of a BIM model in hand, 2) provided immediate access to BIM data on site, and 3) showed the capability to interact with the model in real-time based on location information. Overall, the developed system showed the potential to improve the construction field management by bridging the BIM to the field through the BLE tracking system. Further, the real-time coordination of tracked resources and seamless communication of project information in BIM can significantly reduce the manual efforts made by the site managers and improve the project delivery efficacy.

Keywords –

Building Information Modeling; Bluetooth; Cloud communication; Construction management; Location tracking

1 Introduction

Building Information Modeling (BIM) is changing the way construction management is conducted [1]. Based on accurate digital modeling of BIM, construction plans can be created virtually and realistically evaluated. However, even with the increasing use of BIM, utilization of rich project information in BIM is still limited in terms of construction field management. Current construction field management practices are ineffective in several aspects. For example, many methods for construction progress monitoring, safety field safety monitoring, resource tracking, and management, and inspection are still manual and paper-based, which are considered unreliable, labor intensive, and costly [2]. Significant amount of effort is also needed to find proper information from BIM while conducting these field management tasks. Field managers need to find information, such as plans and specifications, take notes while in the field and later log in such information in the office. Furthermore, additional problems are associated with this manual process. They include time and space discrepancies between a construction site and an office, insufficient number of on-site managers, and inefficient communication among project stakeholders [3; 4]. The paper or tablet-based walk-through monitoring and data collection without locational information attached to data would require extra efforts to link the data into the monitored components. This labor-intensive, manual process and separation between the office and construction site can affect the quality of the work, and thus may decrease the productivity. In addition, monitoring and controlling personnel and keeping track of their log-in information is time-consuming and tedious and requires extra human resources [5].

Over the last decade, the development of 3-D digital modeling of construction projects has been of great interest. Modeling technologies including BIM have rapidly grown, and made a big shift from the conventional construction method. As one of the most critical tools, BIM helps to facilitate various construction related works, including planning, design, construction, and operation and maintenance stages. Advancement in the mobile technology triggered the research and development of mobile BIM technology [6; 7]. Although mobile BIM for collaborative construction planning and management, such as BIM 360, has started to gain in popularity, construction field managers are still required to be fully aware of the important information in BIM, manually localize their positions, and control the user interface to navigate through the model. In order words, the current mobile BIM technology lacks automated location awareness; thus it is incapable to find the user's location, visualize and share nearby contextual information in real-time. This is a major bottleneck to effectively utilize the mobile BIM and improve efficiency in the construction field management.

2 Related Work

With the development of BIM over the last decade, many researchers have studied its potential in various aspects: safety planning for temporary structures [8] and safety rule checking for fall related hazard were investigated, a prototype study for a large scale BIM model visualization was conducted [9], construction schedule optimization [10], and productivity study [11] were conducted. In addition, BIM has also been explored together with other resources to improve other aspects of construction. For instance, 3D laser scanning technology was used together with as-built BIM from a set of point clouds [12; 13; 14].

Availability of location awareness has also been recognized to offer a number of potential improvements to construction. They include decision making [15; 16], saving cost and time, thus increasing productivity [2; 15; 17; 18], enhancing safety and security level [19; 20]. Although such potential has been conceived by the construction industry, limited research efforts have been made to investigate their usages and the potential effects in construction.

Recent research [21] developed a RFID based tracking system and BIM visualization of the target. This research, however, performed course tracking based on proximity sensing, and BIM was used only for visualization purpose without making contextual interaction. A similar work was found in [22] but with a limited verification of the system; their testbed was a very short straight path, which is insufficient to present a real-world situation.

A BIM model contains an abundance of project information from individual components to aggregated details and summaries. If a tracking system utilized such information to communicate with stakeholders in real-time, it would improve work efficiency and add new capabilities to not only construction sites but also indoor offices. This paper presents our research work that integrated a Bluetooth Low Energy (BLE) based in door tracking system with the mobile BIM that we created. This system uses extracted BIM information to enable communication within a 3D BIM environment.

3 Objective and Scope

The objective of this study was to add location-awareness to BIM in order to assist in field construction management. We developed a system that integrates a BLE-based tracking system within a BIM environment. This integration should provide 1) visualization of the target within a BIM model, 2) nearby context aware information that are extracted from the BIM model, and 3) a communication tool for more efficient project management. The developed system extracts project specific information from a BIM model and integrates it into a BLE tracking system. The BLE tracking system used in this integration was developed in this research. Details of implementation of the BLE tracking system is discussed and can be found in [23]. A field trial was conducted in an indoor construction site to test the system.

4 Approach

4.1 System Overview

The developed BIM-integrated tracking system is composed of three components: 1) BLE tracking, 2) a BIM model, and 3) a cloud server. Figure 1 depicts the overall system architecture. The purpose of the BLE tracking component is to obtain location information of the target. A BIM model is used to extract project specific contextual information, including building geometry and object information for building components. By utilizing an as-built BIM model, the up-to-date state of the site condition can be collected and serve as important information, such as map knowledge for the tracking system and project detail data for quality inspection. This information is exported into a XML formatted file directly from a BIM model and then imported into the mobile system. With the conversion of a BIM model to a mobile device compatible BIM format, the extracted BIM information can be combined with the tracking system. This enables us to realize a mobile BIM platform to visualize, explore, and interact with a BIM model in the field.

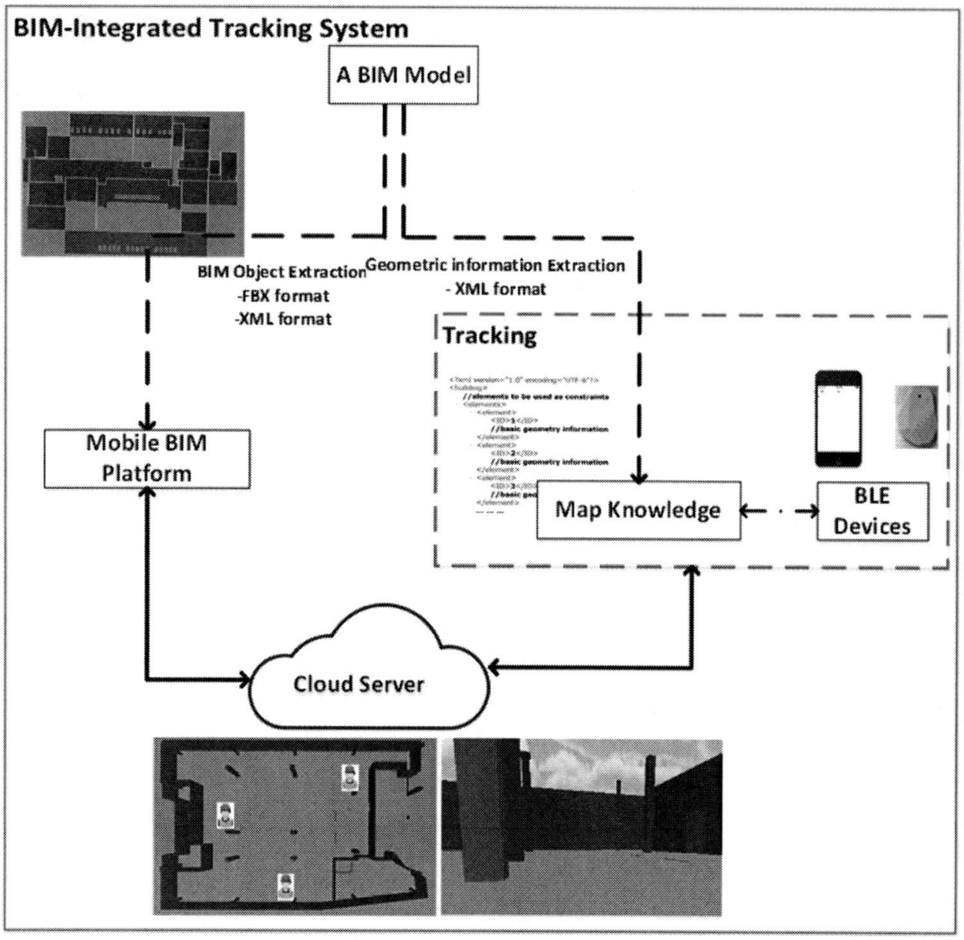

Figure 1. System architecture

This capability becomes more useful when it enables real-time communication of the valuable data collected at the site. This last component can bridge the onsite personnel with the remote stakeholders and thereby close the gap between them. For this purpose, a cloud server is configured to provide a safe storage for sharing and processing information collected by various stakeholders; for example, an inspection document that is digitally recorded and stored though the mobile system can be immediately transmitted to the server and become ready to be shared with the relevant stakeholders.

5 Validation

To test the developed BIM-integrated BLE tracking system, an indoor construction site was selected. The test area was located on the second floor of a concrete-frame building located in Atlanta, GA. Figure 2 (a) shows an extracted as-built floor view of the testbed. An as-built model was used to reflect the current stage of construction in the visualization and interaction with the model. Figures 2 (b) and (c) show the actual site condition of the site on the test day and the test subject, respectively. A test scenario was designed to evaluate the capability of the system to provide an accurate visualization of the BIM view and contextual information of the nearby objects on the site. In addition, the real-time information recording, processing, and sharing capability of the system is checked within the pre-configured cloud server. Figure 3 presents the designed path of the test; this path was particularly designed for the test subject to travel a long, complicated path with a number of turns. By testing a complex scenario, rather than a simple one, we validated the reliability of the system to provide proper visualization and nearby object information in a more realistic situation.

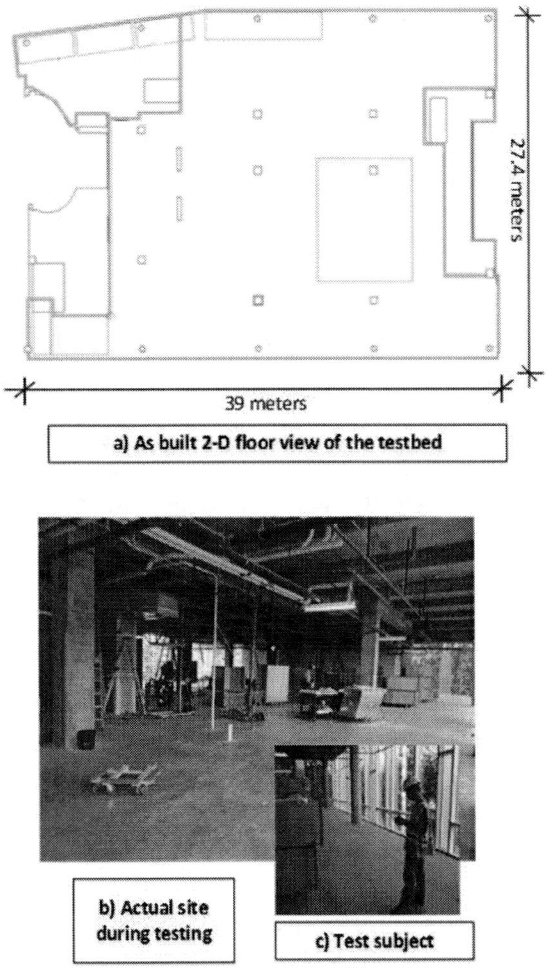

Figure 2. Floor view of the testbed, actual site condition, and test subject

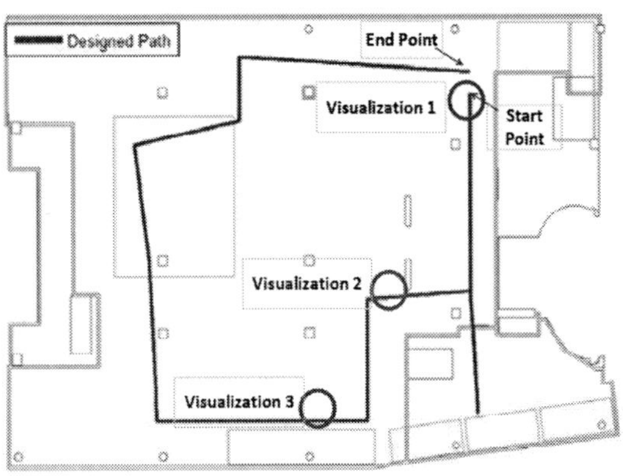

Figure 3. Designed path

The three locations shown in Figure 3 correspond to Figure 4 a), b) and c), respectively. As the test subject traveled, the location is obtained by the BLE tracking system, and the visualization of the surrounding is realized through the mobile BIM. Each visualization also indicates the current location in a two dimensional map at the top right corner. Availability of both the as-built model in hand and actual site condition can offer a new level of capabilities for project management in the field.

Figure 5 shows the developed user interface that displays the details of the selected building component. In this case, the user selected a column that is highlighted, and its property information, that was created in the BIM model, is listed in the interaction panel on the left. Utilization of this interaction panel enables more efficient communication of any issues found at the site. It allows the user to take notes, report safety issues, mark the level of urgency of the issues, take a photo, and inspect the detailed information of building components of concern. Transmission of this information is then logged with a time stamp, allowing a historical analysis.

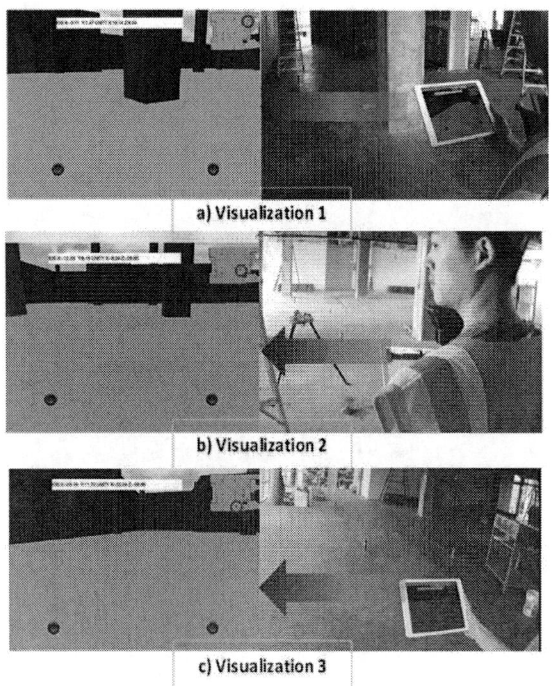

Figure 4. Visualization of as-built BIM model on site

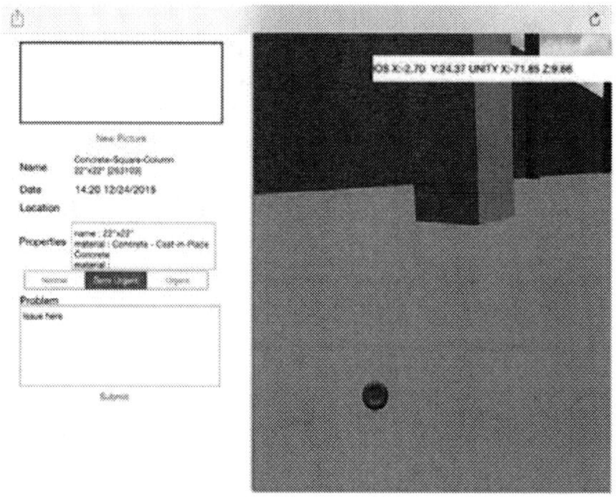

Figure 5. Incident reporting interface

As an example of such communication, Figure 6 shows a temporary storage area that was not found in the as-built model. Figure 6 displays the logged documentation as well as a picture of the scene, which all was transmitted to the cloud. This automation showed that the data collected at the site by quality inspection and / or safety inspection can be directly registered in the system and shared with the relevant stakeholder.

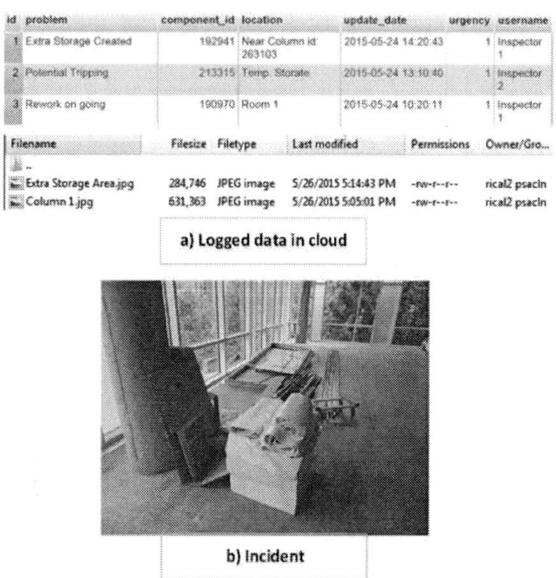

Figure 6. Cloud server data storage

6 Conclusion and Discussion

This study demonstrated the integration between BIM and location tracking to assist in field construction management. Beyond realization of the visualization of a mobile BIM in hand, the developed system provides the user access to location-aware and further context-aware information of BIM data as well as a communication tool that can be used in real-time. The system has been demonstrated with a field trial at an indoor construction site. The result indicated that the integration of a tracking system in BIM 1) enabled visualization of a BIM model in hand, 2) provided immediate access to BIM data on site, 3) showed the ability to interact with the model in real-time, and 4) report and share the results of the site investigation. This capability shows the potential to 1) improve effective coordination of resources, 2) provide a seamless communication tool among stakeholders, and 3) allow an instant share of information and documentation, thus overall reducing the amount of manual efforts made by the site managers and improving the quality of project delivery.

References

[1] Eastman, C., Teicholz, P., Sacks, R., and Liston, K. *BIM Handbook: A Guide to Building Information Modeling for Owners, Managers, Designers, Engineers and Contractors. Building* (Vol. 2). doi:10.1002/9780470261309, 2011.

[2] Wang, L. C. Enhancing construction quality inspection and management using RFID technology. *Automation in Construction*, *17*(4), 467–479. doi:10.1016/j.autcon.2007.08.005, 2008.

[3] Kimoto, K., Endo, K., Iwashita, S., and Fujiwara, M. The application of PDA as mobile computing system on construction management. *Automation in Construction*, *14*(4), 500–511. doi:10.1016/j.autcon.2004.09.003, 2005.

[4] Kim, Y. S., Oh, S. W., Cho, Y. K., and Seo, J. W. A PDA and wireless web-integrated system for quality inspection and defect management of apartment housing projects. *Automation in Construction*, *17*(2), 163–179. doi:10.1016/j.autcon.2007.03.006, 2008.

[5] Fiatech. *RFID for Access Control in Construction Sites*. doi:10.1016/S0022-3913(12)00047-9

[6] Shen, Z. Z., and Jiang, L. An Augmented 3D iPad Mobile Application for Communication, Collaboration, and Learning (CCL) of Building MEP Systems. In *Computing in Civil Engineering* (pp. 204–212)., 2012.

[7] Larsen, B. Accessing large 3D BIMs from mobile devices. *eWork and eBusiness in Architecture,*

Engineering and Construction, 505–508. doi:doi:10.1201/b12516-80, 2012.

[8] Kim, K., and Cho, Y. K. Automated Safety Planning of Scaffolding-Related Hazards in Building Information Modeling (BIM). In *Proceedings of The 6th International Conference on Construction Engineering and Project Management*. Pusan., 2015.

[9] Johansson, M., Roupé, M., and Bosch-Sijtsema, P. Real-time visualization of building information models (BIM). *Automation in Construction*, *54*, 69–82. doi:10.1016/j.autcon.2015.03.018, 2015.

[10] Kim, K., Walewski, J., and Cho, Y. K. Multiobjective Construction Schedule Optimization Using Modified Niched Pareto Genetic Algorithm. *Journal of Management in Engineering*, 04015038. doi:10.1061/(ASCE)ME.1943-5479.0000374, 2015.

[11] Kim, S. A., Chin, S., Yoon, S. W., Shin, T. H., Kim, Y. S., and Choi, C. Automated building information modeling system for building interior to improve productivity of bim-based quantity take-off. *2009 26th International Symposium on Automation and Robotics in Construction, ISARC 2009*, 492–496., 2009.

[12] Hajian, H., and Becerik-Gerber, B. Scan to BIM: factors affecting operational and computational errors and productivity loss. *27th International Symposium on Automation and Robotics in Construction*, (Isarc), 265–272., 2010.

[13] Wang, C., and Cho, Y. Automated As-is BIM Extraction for Sustainable Simulation of Built Environments. In *The Fifth International Conference on Construction Engineering and Project Management (ICCEPM-2013)*. Anaheim, CA., 2013.

[14] Wang, C., Cho, Y., and Kim, C. Automatic BIM Component Extraction from Point Clouds of Existing Buildings for Sustainability Applications. *Automation in Construction*, *56*, 1–13. doi:10.1016/j.autcon.2015.04.001, 2015.

[15] Khoury, H. M., and Kamat, V. R. Evaluation of position tracking technologies for user localization in indoor construction environments. *Automation in Construction*, *18*(4), 444–457. doi:10.1016/j.autcon.2008.10.011, 2009.

[16] Torrent, D. G., and Caldas, C. H. Methodology for Automating the Identification and Localization of Construction Components on Industrial Projects. *Journal of Computing in Civil Engineering*, *23*(1), 3–13. doi:10.1061/(ASCE)0887-3801(2009)23:1(3), 2009.

[17] Grau, D., Caldas, C. H., Haas, C. T., Goodrum, P. M., and Gong, J. Assessing the impact of materials tracking technologies on construction craft productivity. *Automation in Construction*, *18*(7), 903–911. doi:10.1016/j.autcon.2009.04.001, 2009.

[18] Cho, Y. K., Youn, J. H., and Martinez, D. Error modeling for an untethered ultra-wideband system for construction indoor asset tracking. *Automation in Construction*, *19*(1), 43–54. doi:10.1016/j.autcon.2009.08.001, 2010.

[19] Park, J., Marks, E., Cho, Y., and Suryanto, W. Mobile Proximity Sensing Technologies for Personnel and Equipment Safety in Work Zones. In *Computing in Civil Engineering* (pp. pp. 41–48), 2015.

[20] Cho, Y. K., and Youn, J.-H. Wireless sensor-driven intelligent navigation robots for indoor construction site security and safety. In *23rd International Symposium on Automation and Robotics in Construction (ISARC)*, 2006.

[21] Fang, Y., Cho, Y. K., Zhang, S., and Perez, E. Case Study of BIM and Cloud–Enabled Real-Time RFID Indoor Localization for Construction Management Applications. *Journal of Construction Engineering and Management.*, 2016.

[22] Costin, A., Pradhananga, N., and Teizer, J. Passive RFID and BIM for Real-Time Visualization and Location Tracking. In *Construction Research Congress* (pp. 169–178). Proceedings of the Construction Research Congress., 2014.

[23] Park, J., Cho, Y. K., and Ahn, C. Wireless Tracking System Integrated with BIM for Indoor Construction Applications. In *Proceedings of The 2016 Construction Research Congress (CRC)*. San Juan, Puerto Rico., 2016.

ADAPTIVE CONTROL OF BULLDOZER'S WORKFLOWS

Alexej Bulgakov [1]*, Sergei Emelianov[1], Thomas Bock [2], and Georgy Tokmakov [3]

[1] South West State University, Kursk, Russia

[2] Technical University of Munich, Germany

[3] South Russian State Technical University, Novocherkassk, Russia

* Corresponding author (a.bulgakow@gmx.de)

ABSTRACT

The most important task for bulldozer's traction mode control is to use its traction capacity in full by means of its end-effectors control. To keep traction mode at maximum or at a given resistance value applied to end-effectors automatically is difficult due to a great number of stochastic factors affecting the bulldozer. Bulldozer is taken as a mechatronic system [1, 2]. The study presents analytic dependences for the sub-processes where analytic modeling based on bulldozer's parameters correlation knowledge is applicable. Models of the sub-processes are included into the general structure of bulldozer's workflow simulation model.

Simulation technique is demonstrated through model development of the bulldozer as a universal machine operating in modes of soil movement and subgrade surfacing.

In developing the models mathematical apparatus of the theory of random processes, transfer functions, table interpolation, numerical solution of algebraic equations and ordinary differential equations in the Cauchy form was used.

A dynamic model of the drawing prism formation was developed describing the dependence of the volume of prism on the variable digging depth and variable bulldozer speed. A general structure of the model of bulldozer's workflows due to the working process control objectives was developed.

KEYWORDS

Robotics and mechatronics, Automation and control, Bulldozer, Neural network technologies.

1. INTRODUCTION

Bulldozers equipped with modern navigation and information systems are mobile mechatronic objects, and they can be integrated into general process of intellectual construction [3]. The integration will provide optimal efficiency of the construction cycle and will ensure lean production process [4,5].

On the basis of bulldozer's workflow dynamics modeling and analyses described in a variety of works, we have concluded that the models to describe kinematics and dynamics of its working equipment, hydraulic and transmission features tend to be analytical formulas derived from well-known laws of physics and from information on bulldozer's structure and mechanisms. If some parameters of the workflow are unknown or constantly changing, the models are either statistical tables or empiric dependences summarizing experimental data. The models depict interaction of end-effectors, engines and environment as well as statistic features of bulldozer's complex units.

Application of regulators based on classical control theory is difficult due to the frequent changes in workflow conditions. Thus, it is necessary to develop adapted control systems to eliminate the difficulties described. The system includes both the bulldozer's dynamics modeling and bulldozer's workflow control method to take into consideration the complex non-linear dependencies between workflow parameters and incomplete information on its working conditions changes.

Having reviewed adaptive and intellectual control methods [6, 7], we propose to create an adaptive control system for technological processes to increase efficiency of bulldozer's control in comparison with traditional control methods.

2. MOBILE MECHATRONIC OBJECT MATHEMATICAL DESCRIPTION TO PERFORM EXCAVATION WORKS ON THE BASIS OF A DOZER

When researching a dozer's working process usually a number of design schemes are considered – straight line, thread milling, wedge and exponential cutting. Meanwhile, a dozer moves along the surface that is formed by its blade. Therefore, when driving onto any surface roughness resulting from the dozer blade control or the change in its position due to any reason, causes position changes of the machine frame and along with the cutting edge that is any face deviation from a straight line in some extent is copied by the dozer.

Observations [1,2] show that quite often while designing a face its roughness is progressing, reaching a size at which the control over the workflow is lost. In this case, the operator has to align the face deliberately, trying to ensure its "tranquil" profile that allows doing excavation works smoothly, without frequent control system switching and reducing the dozer's operating speed that causes a slowdown and shows inferiorities of the blade control system. Obviously, if the control system operates in the antiphase towards deviations of the tractor frame with sufficient accuracy, the initial face roughness will not evolve and will be gradually cut. One of the most likely causes of the opposite phenomenon observed in practice, is the disparity between the velocity of the dozer V_p and actual conveying speed of the working body V_{ot} required in certain areas Si of the digging operating cycle, where i- is the number of the speed change V_{ot}. Speed ratio depends on the dozer's geometrical dimensions (Figure 1) and its control system.

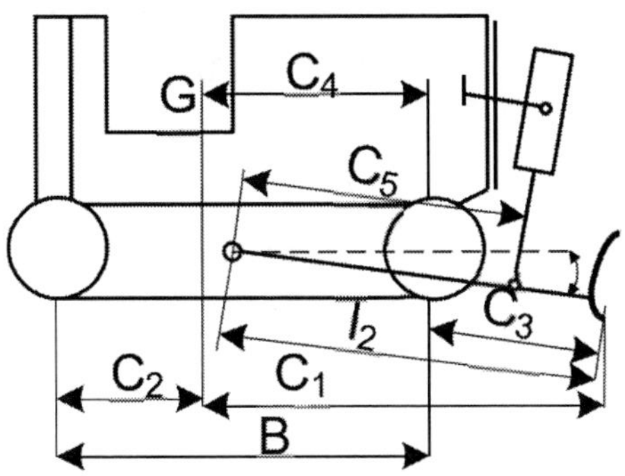

Figure 1 Dozer's geometrical dimensions

Mathematical model of the dozer's movement on a straight line tracking (frame alignment) is built using the Lagrange equations of the 2nd kind, under the assumption that the contribution to the dynamics of the drive gears and a track is small, compared with the contribution of the remaining parts of the dozer[8,9].

At the beginning of digging (Figure 2), the frame of the tractor makes a strictly forward movement over a distance of S_1+S_2 without hesitation relatively its mass center. The blade cutting edge in the area S_1 dives into the soil to a depth equal to a predetermined cutting thickness h. Thus, the control action a_1 may be determined by the formula:

$$a_1 = \frac{30 i_{tr} m\, l_2}{\pi r_k F_z i_{pr} C_5 n} \quad ; \qquad (1)$$

where i_{tr}, i_{pr} - tractor transmission and hydraulic pump ratios;

n - number of hydraulic cylinders;

m - fluid mass in the hydraulic cylinders.

In the area S_2 the movement is made with $a_2=0$ until the mass center of the tractor won't move to the buttomhole edge.

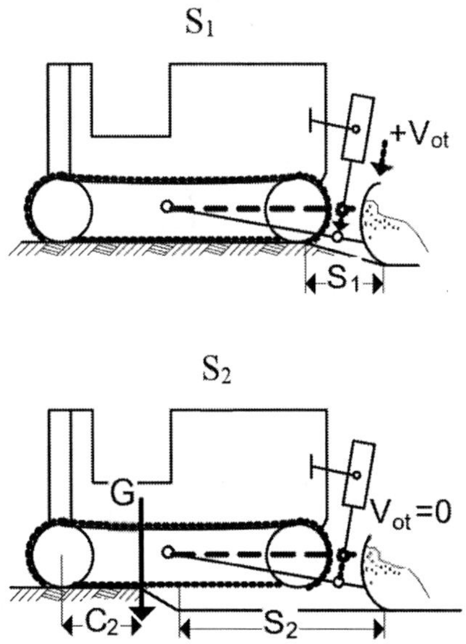

Figure 2. The movement of the tractor frame the beginning of digging.

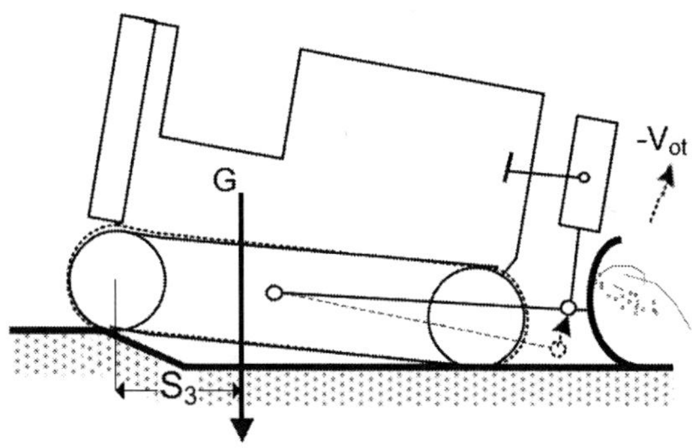

Figure 3 "Dives" in the drawn buttonhole

On further movement the dozer "dives" in the drawn buttonhole (Figure 3), so in the area S_3 it is necessary to lift the blade at a rate of V_{ot}, determined by the coefficient a_3:

$$a_3 = \operatorname{tg}\beta \left[e^{\frac{aV_n t}{C_1+V_n t}} \left(1 + \frac{aC_1}{C_1+V_n t}\right) - 1 \right]; \quad (2)$$

The area S_3 ends after the dozer's back gear hits the edge of the face and reverse alignment of tractor frame starts. Length of the alignment area is $S_4 \approx S_1$. Obviously, during this period it is necessary to start dropping the blade. The a_4 determines the rate of dropping the blade in the given area:

$$a_4 = \frac{C_3 S_1}{(C_4+S_3+V_n t)^2}; \qquad (3)$$

To implement control actions $a_i=f(S_i, t, h)$ the dozer must be equipped with a vertical blade control system.

3. GENERAL STRUCTURE OF BULLDOZER WORKING PROCESS MODEL

The main goals for analytic simulation modeling of bulldozer workflow are [1.2]:

- Bulldozer simulation as a controlled object to realize bulldozer's workflow parameters for using them at workflow neural network identification;

- Efficient traction modes parameters definition to be supported by the control system;

Simulation tasks:

- To single out the main sub-systems in bulldozer's structure and interrelations between the sub-systems;

- To develop analytic and simulation models for workflow elements and to include them into the general structure of the model.

The structure meets the goals of workflow control. When moving soil by the bulldozer, it is necessary to utilize bulldozer's traction capacity in full keeping the nominal traction value; when surfacing, the altitudes of the right and left side of the blade are to correspond the design marks. The key element at the scheme (Figure 4) shows the choice for the first or the second operational mode.

At developing the models, we use mathematical apparatus of the random processes theory, transfer functions, table interpolation, numerical solution of algebraic equations and ordinary differential equations in the Cauchy form. Random changes in the coordinates of untreated soil surface, as well as normalized fluctuations in the resistance forces on the working organ , caused by the heterogeneity of the soil are highlighted among the disturbing effects on the working organ of the bulldozer from soil conditions. Disturbance cause unwanted vertical movement of the working organ that affects both the coordinates and the change in the digging depth. Dependence of the blade position and dig depth from disturbances reflects the intricate relationship between the geometric parameters of the bulldozer in space. Loading conditions on the working organ are due to random variation in the dig depth and heterogeneity of soil properties. Soil digging process with bulldozer working organ is studied on the base of the finite element model of the soil mass, a mathematical model of random forces of resistance on the working organ being developed. The actual bulldozer velocity depends on the strength and the properties of the mover, transmission and the power unit. In its turn, disturbance parameters, movement of the working organ and the formation of stress depend on the velocity. Bulldozer drive model and mover interaction with the soil include engine model, mechanical and hydro mechanical transmission, as well as slipping. Control system regulator depending on the objectives, control algorithm and the incoming data from the bulldozer as a control object produces electrical signals to the electro- hydraulic distributors being part of the working organ hydro drive. Lifting or burying the blade is done to control either the pulling power, or the blade coordinates.

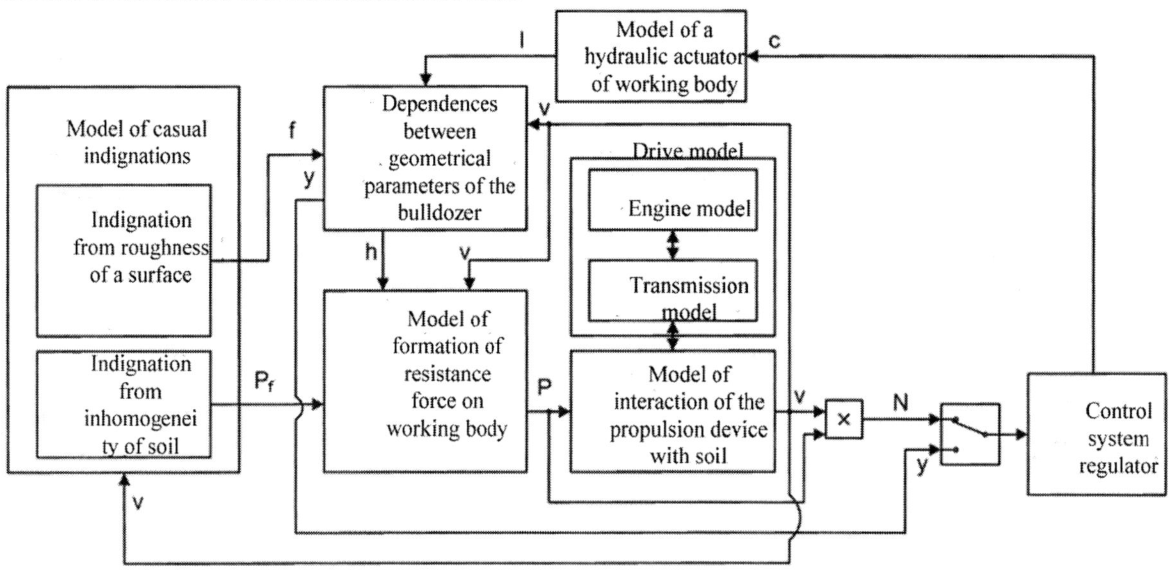

Figure 4. General structure of bulldozer working process model

4. NEURO-FUZZY CONTROL OF BULLDOZER'S WORKFLOWS

Applying a hybrid neural network consisting of a combination of traditional neural networks and neural networks of higher order (Figure 5). Thus, the neural network has the ability to switch between linear connections and connections of high order that can be described by the following dependencies [10-12].

Linear coupling:

$$y_i = \sum(w_{ij}x_i + b_{jo}); \quad (4)$$

High order coupling:

$$y_i = f\left(\prod_i x_i^{p_{ij}} * 1^{b_{jo}}\right); \quad (5)$$

Activation function:

$$f(x) = \frac{1}{1+e^{-ax}}; \quad (6)$$

where w_{ij} – coupling weight coefficients; y_i – output neuron signal; x_i – input neuron signal.

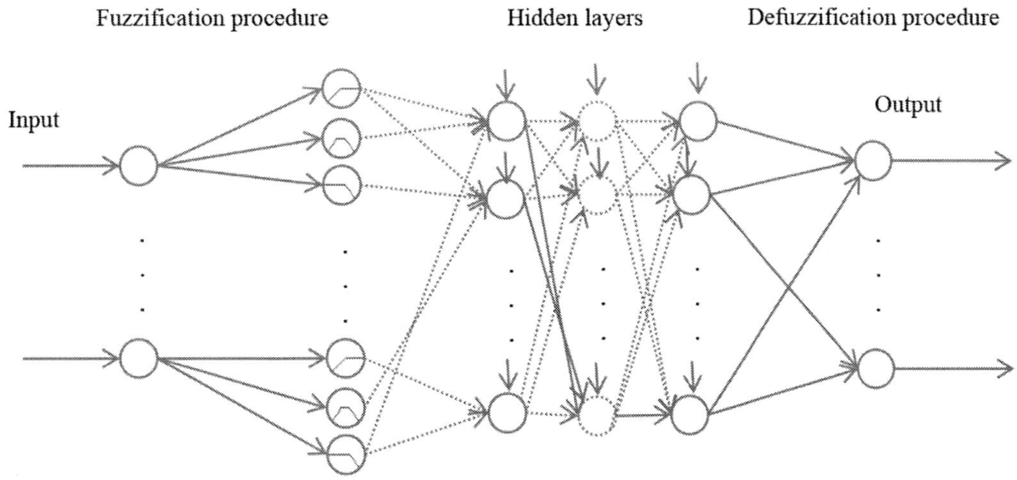

Figure 5. Hybrid Neural Network Structure

This implies that each layer depending on the operating mode may change the type of connection between neurons. For example, for a neural network consisting of 3 layers, the following options are possible (linear - L, higher order - HO): L-L; L-HO; HO-L and HO-HO.

To optimize the created neural network is possible with the help of the genetic algorithm adaptation [13-15] (Figure 6).

It is a method of random search with elements of adaptation, which is based on principles similar to the Darwin's evolution process of biological organisms. In this case, three types of operations are performed: crossing, mutation, selection. The fitness degree (how the population corresponds to the given task) is defined through the fitness function that can also include penalty functions for violation of additional restrictions on variable variables. There are various forms of crossing [16]. They make a selection of the fittest specimen, which constitute a parental pair and the crisscrossing of the chromosomal chains takes place, i.e. the descendant line code inherits fragments of codes of parental chromosomes. The mutation operator produces a local change in the line code of chromosomes with a given probability, which is one of the configurable parameters of the genetic algorithm [17, 18].

The selection operator allows creating a new population from a set of specimen, generated and modified descendants of specimen after mutation. The genetic algorithm is used to adjust the membership functions that are defined within the accuracy of a few changeable parameters, such as triangular, trapezoidal, radial functions. When simultaneously configuring several membership functions, the parameters of each of them are coded by their own segment of the chromosome, so that during the process of crossing the code sharing occurs only between chromosome segments of the same type. To configure a rule base to a specific chromosome fragment, some variant of the rule base is corresponded and in accordance with the accepted coding the choice of the genetic operators' type is performed. Thus, the architecture of the management and control system can be represented as follows (Figure 7):

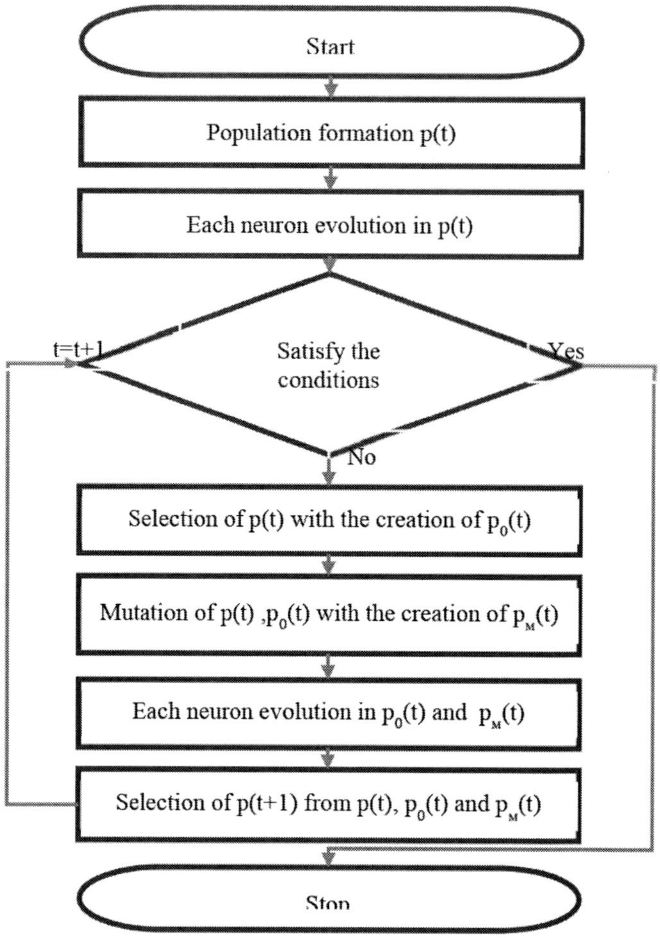

Figure 6. Hybrid neural network optimization algorithm

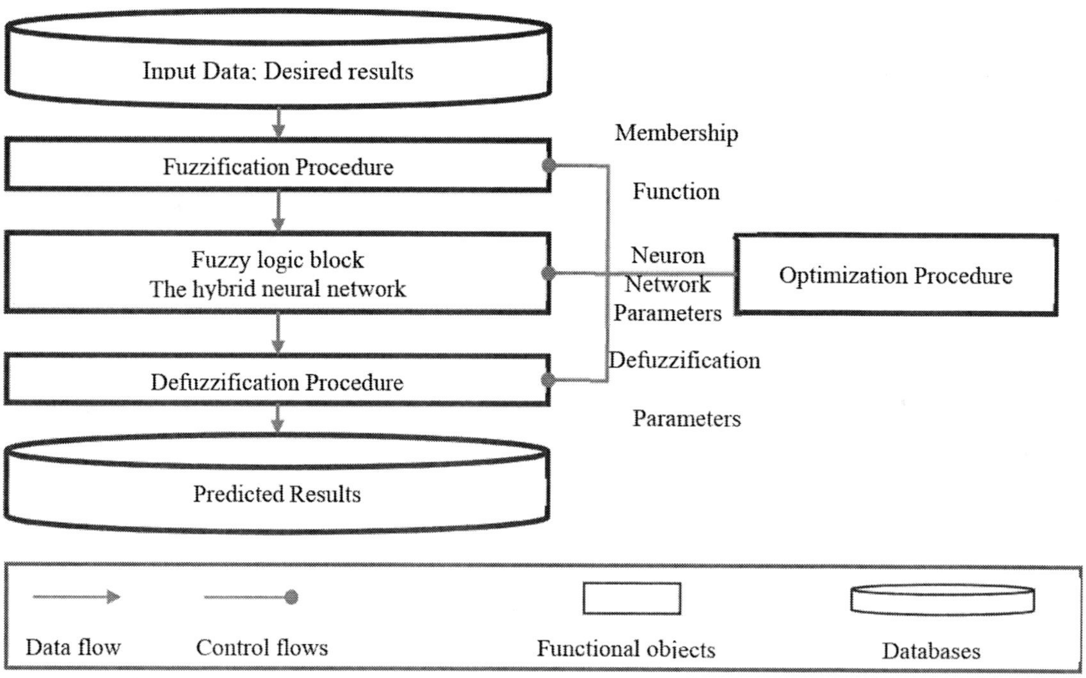

Figure 7. Architecture of the management and control systems

To conduct researches on the basis of fuzzy modeling, quite versatile software have been developed that greatly simplify the creation of new control systems using neural networks and fuzzy models. The use of different aspects of evolutionary modeling as a new direction of computing technology allows applying principles of learning for the management and control system and adaptation of the described hybrid neural network allows getting the following results.

CONCLUSIONS AND RESULTS

Automatic control function of blade positions precisely adjusts the cutting edge. Depending on the content of correction signals, the regulating dual hydraulic valve automatically lifts or drops the cutting edge of the blade, constantly keeps it in position that ensures the accuracy of work and ensures an optimum level of productivity.

Identification technique of the dozer's working processes and models obtained on its base, are intended to be used in the development of adaptive systems of automatic control of the dozer's working process.

Methods of development of adaptive systems of control of the dozer's working process, is based on neural network technology. For the formation of control actions on a dozer, and of the electrical switch signals of the hydraulic directional valves of the lifting and dropping hydraulic cylinders of the working body, in particular, the structure and functioning algorithms of the adaptive neural network controller have been designed.

Use of the presented approach can be used not only for earthmoving machinery, but also for the machinery does not require high speed of movement, but dependent of resistance on work body, such as agro-machinery.

REFERENCES

[1] Krapivin D.M., Nefedov V.V., Tokmakov G.E. Mathematical model for the movement of mechatronischen devices for the intelligent building site, Mechatronik, Lik, Nowotscherkassk, 2010.- S. 50-54.

[2] Bulgakov A.G., Tokmakov G.E. The analysis of the control systems for building site, problems, possible solutions. VII Internationally scientific conferences: Realisation of the European scientists; 17th-25th of July, 2011. Sofia. S. 43-44.

[3] Shestopalov K.K., Hoisting-and-transport, building and road machines and equipment. M .: Academy, 2009. P. 67-72.

[4] Bulgakow A.G., Jehle P.,Tokmakov G. SCM-logistic and mechatronics systems for ensuring the smooth construction process // Innovation in Mechanical Engineering - Shaping the Future: 56-th International Scientific Colloquium (12-16 September 2011: Conference Proceedings /Ilmenau University of Technology).Ilmenau, 2011.

[5] Hughes S.W., Tippett D.D., Thomas W.K. Measuring project success in the construction industry // EMJ – Engineering Management Journal. 2004. 16(3). P. 31–37.

[6] Min-Yuan Cheng, Hsing-Chih Tsai, Erick Sudjono. Evolutionary fuzzy hybrid neural network for construction industry. Automation in Construction 21 (2012) S. 46-51.

[7] Meshcheryakov V.A., Identification of earth-moving machines as a dynamic objects based on neural network technology // Problems of development and operation of vehicles, special machinery and technology in Siberia and the Far North. Proceedings 43 th International Scientific Conference of the Association of Automotive Engineers. - Omsk: Publishing House of the "LEO", 2004. - S. 176.

[8] Denisov V.P., Automatic speed control system for grader // Building and road machines. - 2003.- № 5.- pp 39-41.

[9] Weber V.V. Dynamic analysis grader equipment on base of CAD / CAE technology // Interuniversity collection of works of young scientists and students. - Omsk: SibADI, 2004. - Vol. 1, Part 1. - pp. 141-146.

[10] Kruglov V.V., Borisov V.V., Neural networks. Theory and practice. M: Hotline Telecom 2002.

[11] Gorban A.N., Dunin-Barkovskii V.L., Kirdina A.N., Neuroinformatics / Novosibirsk: Nauka, 1998.

[12] Gorban A.N., Russiev D.A., Neural networks on personal computer. Novosibirsk: Nauka, 1996.

[13] Kureychik V.M. Genetic algorithms. State problems, prospects., Izvestiya RAN, Theory and control system., 1999. № 1. C. 144-160.

[14] Rutkovskij L., Pilinskij M. Neural networks, genetic algorithms and fuzzy systems., Gorzachaza Liniza TELEKOM, 2004.

[15] M.E. Georgy, L.M. Chang, L. Zhang, Prediction of engineering performance: a neurofuzzy approach, Journal of Construction Engineering and Management 131 (5) (2005) 548–557

[16] Mecheryakov V.A. Recurrent algorithm neural identification working process earthmoving machinery. Siberian State Automobile and Highway Academy, Omsk, 2007.- S. 63-66

[17] Leonenkov A.V., Fuzzy modeling in MATLAB environment and fuzzyTECH. SPb .: BHV - St. Petersburg, 2003. 736 p.

[18] T.M. Cheng, C.W. Feng, M.Y. Hsu, An integrated modeling mechanism foroptimizing the simulation model of construction operation, Autom. Constr. 15 (2006) 327–340.

Mathematical Representation of Haptic Robotic Realization for Artefacts Maintenance

Alexey Bulgakov [a] Andrey Volkov [b] and Daher Sayfeddine [c]

[a] Professor, Dr.-tech-sc., South West State University, Kursk, Russia
[b] Professor, Dr.-tech.sc., Rector, Moscow State of Civil Engineering, Moscow, Russia
[c] Ph.D., South Russian State Polytechnic University, Novocherkassk, Russia
E-mail: [a] a.bulgakow@gmx.de, [b] volkov@mgsu.ru, [c] daher@live.ru

Abstract –

Originally generated from Greek world "*haptikos*", haptic technology is used in many grasping and robotic applications. Commercially, it is widely known solution in videogames and mobile technology. Essentially, haptics is used in medical robots, where the surgeon is located outside of the sterile area and remotely operates vide a direct video connection the robot manipulators, which are holding endoscopes and working elements. Haptics has proven its worth especially for minimal and non-invasive surgical procedures, where the patient care coefficient has increased. As it is directly related to human existence, artifacts may be as sensitive as human life and need extreme care to last for the next generations. Here we could see a possibility to combine haptics with the maintenance of the monuments. In this paper we represent mathematically a haptic robotic concept for artefact maintenance purposes.

Keywords –
Haptics; Telerobotics; Artefact Maintenance

1 Introduction

Haptics is human-machine interface technology aiming to guide a human operator-controller by recreating his tactile sense signals. The realization process is based on applied forces, vibrations or motions fed back to the operator in order to perceive the surrounding world [8]. This terminology is widely used in telerobotics, essentially for semi-autonomous distantly guided mobile robots or manipulators. The biggest success of the haptic implementation is observed in medical field (surgical robots) [6,9]. Therefore it is believed that haptics can be used where machine infinite precision, human flair and passion need to be combined in an integrated work.

Historically, polymaths and inventors were blessed with such gift: flair and precision. We appreciate their work by contemplating monuments, old machineries and inventions. Artefacts are very important to humanity. In fact, a small piece of stone can tell a story of a whole generation or race. Human being has appreciated the value of the artefacts and started maintaining them periodically. In a refurbishment workshop, several artists, architects and engineers team up together to conserve the masterpiece.

Looping back to haptics, we see possible integration for this technology in artefacts maintenance. The remote control allows the artist to have a wide-view of the masterpiece and improve its simulation in his mind. In parallel, the mechanical feedback and the robust robotic arms achieve better stabilization of the working tools than the human hand. To proceed with the simulation of this idea, we have firstly to describe the haptic approach in telerobotics.

There exist two major haptic mechanical categories: impedance-type and admittance-type devices. The first is drivable, open loop operated mechanism that has less output force. In contradiction, the latter is non-drivable, closed-loop operated device and has high output force. The effectiveness of each category is evaluated through a factor called transparency or the output impedance. The physical meaning of this factor is the ability to render zero forces with the disturbance of the existing user motion.

Developing criteria and simulators to avoid costly prototyping and experimental comparisons and analyzing each approach are yet to be perfected. However, we can bring to your notice that the impedance-type devices are easier to simulate and more straightforward in opposition to the admittance-type. This is because the latter is inherently non-drivable and particularly because the output impedance is very high in the frequency range above bandwidth of the admittance controller.

2 Related work

Many literatures are oriented to study the haptic mechanism and simulation models. Some researchers have developed methods to analyze the transparency factors. While some of them are difficult to implement due to experimental dependencies [1,2,3], others are more simulation-friendly [4,5].

Choosing a haptic device type depends on the control task to be performed. From this point of view we can characterize these two categories as follows:

a- An impedance control system senses motion developed by the operator (position, orientation) and controls the force on the haptic device. While an admittance control system senses forces commanded by the operator and accordingly controls the motion of the haptic device.

b- For impedance-type devices, any small change in position will cause a very high rise in actuator reaction force. To minimize this effect, we require very high control gain from measured device position to actuator force [8]. At the same time, the high gain value will have implication on the stability, as control gains cannot become infinity high.

On the other hand admittance control was used to manipulate rigid objects. Therefore, the admittance-type device is highly geared. This guides to non-backdrivable effect and generates forces at the end-effectors.

For the paper purposes, we choose to simulate impedance-type haptic device as it can render small masses necessary to work with artefacts. Using the abovementioned definitions developed earlier, we can draw the following comparison that approves our choice.

Table 1. Impedance-type (IT) vs. Admittance-type (AT)

Criterion	IT	AT
Input	Motion	Force
Output	Force	Motion
Render mass	Yes	No
Stiffness	Poor	Capable

Additionally, we would like to mention that the most frequent problem addressed in the literature on haptic devices is the stability problem. Listed below are some of the possible reasons for the stability issue of the haptic devices:

- Coulomb friction;
- Actuator saturation;
- Sensors fidelity;
- Discretization and sampling;
- Joints flexibility;
- Virtual environment dynamics;
- Human hand dynamics.

Addressing all these points is a challenging task. Some of these factors can be neglected for simulation and analysis purposes.

3 Design and control of the haptic device

A haptic device interface represents a challenging design task, as it is required to be light back drivable. In such scenario, the operator feels no resistance in case of free motion of the robotic manipulator. At the same time, it is important that the operator feels the force generated from the virtual environment and not that of the mechanism weight. Hence, the design should address the following criteria:

- Free space is a free motion;
- Solid virtual object must be felt stiff enough;
- Haptic dynamics must not be additional noise on the operator perception of the virtual world.

The minimum stiffness calculated experimentally is calculated around 20 N/cm in order to feel rigid bodies [6].

In table 2 are given solutions addressing the abovementioned design criteria

Table 2. Control solution for haptic design

Criterion	IT
Freedom	Active control
Stiffness	High-bandwidth controller
Dynamic noise	High computation speed

A mechanical requirement to be added to the control task is the saturation requirement. By that, we understand the higher the force applied on the haptic device the larger the force that must be supplied by the actuator.

Typical block diagram for haptic controller is illustrated in fig.1

3.1 Impedance Control Architecture

There are several approaches for impedance control: open loop impedance control, impedance control with feed-forward and feedback terms and impedance control with hybrid compensation. The idea in impedance control is to make the robot to behave as a spring-mass- damper system, which can be described mathematically using the following equation

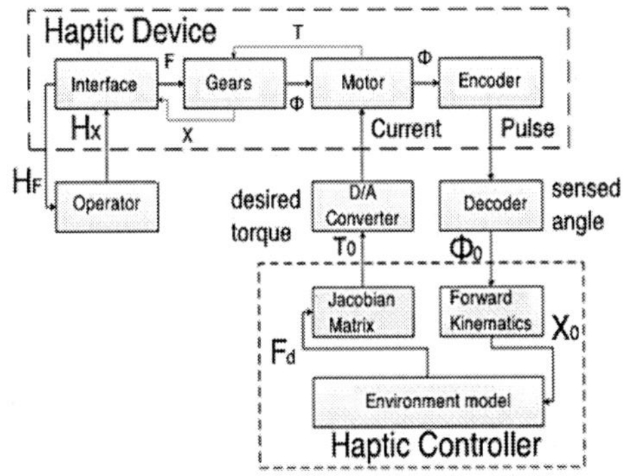

Figure 1. Typical block diagram for haptic controller

$$F = M\ddot{x_{ref}} + D(\dot{x} - \dot{x_d}) + K(x - x_d) \quad (1)$$

Where F is the force of the environment affecting the robot, *M* is the mass, *D* and *K* are the damping and spring stiffness coefficients respectively. x_d and $\dot{x_d}$ are the desired position and velocity of the robot, while x and $\dot{x}$ are the expected values.

An illustration of the Spring-Mass-Damper is given in figure.2.

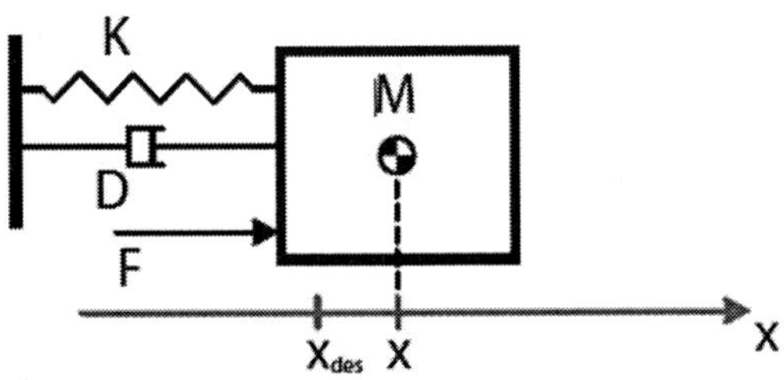

Figure 2. Spring-Mass-Damper concept

If we assume that the acceleration of the robot is equal to the referential acceleration, then using equation (1) we can say:

$$\ddot{x_{ref}} = \frac{F - D(\dot{x} - \dot{x_d}) - K(x - x_d)}{M} \quad (2)$$

By integrating the acceleration we can obtain the referential velocity and position. Since we have to implement control law, the values of the $\dot{x_{ref}}$ and x_d have to be discretized with reference to a sampling rate *h*. This guides us to write the following equations

$$\dot{x_{ref}}[k] = \sum_{i=0}^{k} \ddot{x_d}[i] * h; \quad (3)$$

$$x_{ref}[k] = x[0] + \sum_{i=0}^{k} \dot{x_{ref}}[i] * h; \quad (4)$$

Where the discrete integration of a signal *u[k]*is defined as $\sum_{i=0}^{k} u[k] * h$;

By achieving the discrete integration, it is now possible to design a control system, which is able stabilize a Spring-Mass-Damper system, representing here the haptic robot. The task doesn't seem to be smooth, as it is difficult to imagine how a system with chosen damping and stiffness coefficient will behave. Nevertheless, it is possible to compare the characteristic polynomial of the system transfer function $W_{DS}(p)$ with characteristic polynomial of an ideal aperiodic second order transfer function $W_{2s}(p)$; where:

$$W_{DS}(p) = Mp^2 + Dp + k; \quad (5)$$

$$W_{2s}(p) = p^2 + 2\zeta\omega p + \omega^2 \quad (6)$$

Where ω – is the eigenfrequency of the sring-mass-damper system; ζ – is the damping ∈ [0;1] and p – is the Laplace operator.

From equations (5) and (6), we can obtain the following equalities:

$$D = 2\zeta\omega; \quad (7)$$

$$k = M * \omega^2; \quad (8)$$

3.2 Simulation Results

The haptic device is usually a robotic arm manipulating tools. Therefore, the end effector mechanism or the grasping body are important to us. Nevertheless, the behaviour of the remaining joints is essential to have stable grasping or precise execution [7]. In this paragraph, we illustrate the robotic arm with 5 joints. This design is called an Elbow-Manipulator (figure.3.). It consists of the following joints: base fixation on wheels, body, shoulder, elbow and forearm.

Based on the Force/Torque relationship, the interaction of the manipulator with the surrounding environment will produce forces and moments on the grasping element. We designate F as the vector of forces and torques. Hence we can write:

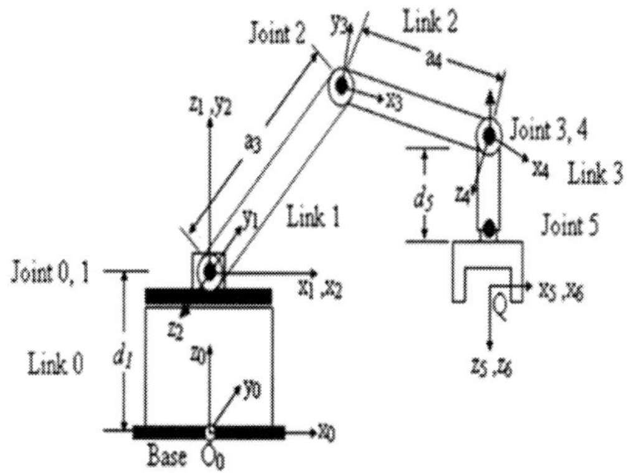

Figure 3. Representation of kinematic structure of elbow-manipulator

$$F = [F_x \quad F_y \quad F_z \quad n_x \quad n_y \quad n_z]^T; \qquad (9)$$

Where $F_x \quad F_y \quad F_z$ – are the components of the force and $n_x \quad n_y \quad n_z$ – components torques on the grasping tool accordingly.

We donate τ for vector of joints torques, δ is the displacement of the end effector caused by the vector F and γ is the corresponding virtual joints displacement. We can write:

$$\delta = J(q) * \gamma; \qquad (10)$$

Where $J(q)$ – is the Jacobian of the manipulator.

The virtual work of the system is given by:

$$\omega = F^T * \gamma - \tau^T * \gamma; \qquad (11)$$

Substituting (9) into (8) we obtain:

$$\omega = (F^T * J - \tau^T)\gamma; \qquad (12)$$

Since the generalized q is independent, we have the equality

$$\tau = J(q)^T * F; \qquad (13)$$

From (11), we can write:

$$\begin{bmatrix}\tau_1\\ \tau_2\\ \tau_3\\ \tau_4\\ \tau_5\end{bmatrix} = \begin{bmatrix}J_1\\ J_2\\ J_3\\ J_4\\ J_5\end{bmatrix} * \begin{bmatrix}F_x\\ F_y\\ F_z\\ n_x\\ n_y\\ n_z\end{bmatrix}; \qquad (14)$$

By computing the mathematical model in MATLAB, we obtained the following results shown in figure 4. The results represent the behaviour of the joints, i.e. angles, velocity and acceleration, with reference to the motion of the human operator. This task was resolved using inverse kinematics with reference to the geometry given in figure 3.

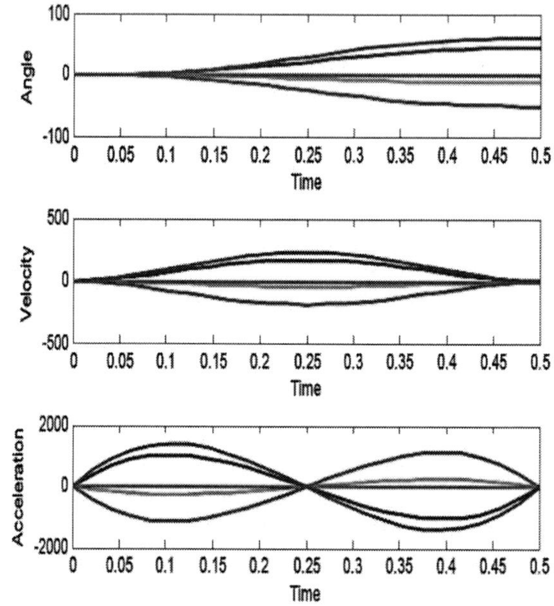

Figure 4. Robotic arm joints angles, velocities and accelerations with reference quantic polynomial trajectory

3.3 Discussion

By implementing the impedance control design using the obtained mathematical models in equations (3-6), we rearranged the system to focus on single input- single output. This permits evaluating whether the end effector is reaching the target area and applying the necessary force to execute the task. Therefore the simulation results as single input-single output presentation should highlight the position and the speed of movement of the end effector with reference to a desired position: tracking desired position will show the precision of the control system, while controlling the speed variation will generate enough power to gently execute the job.

Figure 5. represents the behaviour of the joints (thin blue curves) with reference to desired position (green line) according to the operator motion (red step signals).

From the results, it can be easily seen that the impedance controller has successfully driven the position of the end effector to the desired spot without overshooting. Taking into consideration that variable step signals are difficult to handle, the obtained results are satisfactory.

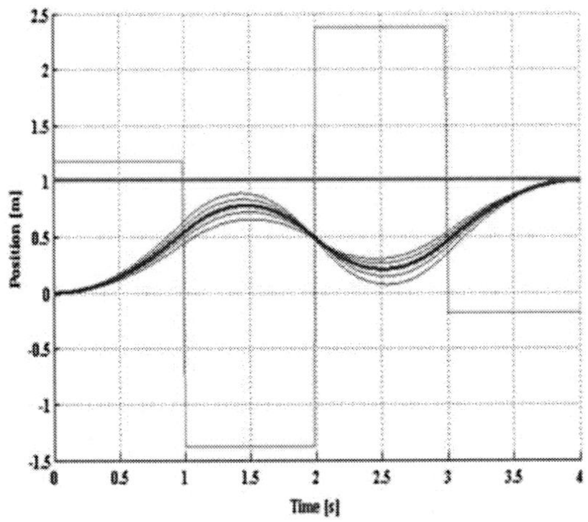

Figure 5. Position control of end effector using haptic impedance control method

Similarly, the control approach permitted to stabilize the velocity of the end effector without overshooting. This is a very important result. As was mentioned earlier impendace-type devices are very sensitive to motion. Hence any small variation in speed will generate more acceleration, which will produce more force. This will cause the end effector to exert more pressure on the structure being maintained. Figure 6. illustrates results of speed stabilization using hapric impedance control approach.

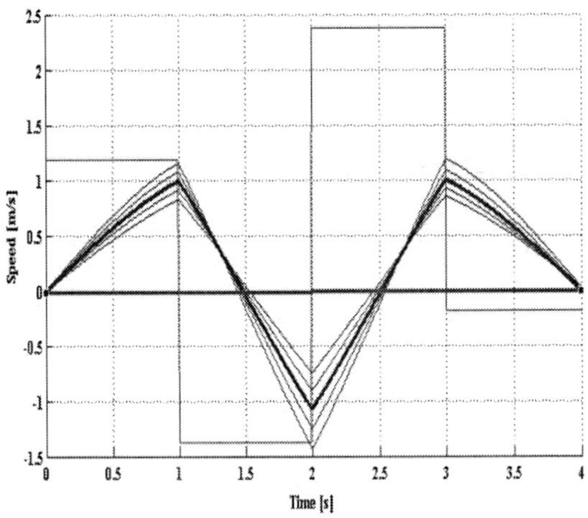

Figure 6. Speed control of end effector using haptic impedance control method

4 Future work

The global haptics financial studies indicate that the market actuator reached 2.7 billion dollars in 2015. It is also forecasted that the haptic market share can potentially reach 10.3 billion dollars with a growth rate of 37.3% annually [10].

With the new development in technologies, it is forcasted that haptics will reach new heights and be integrated in different intelligent products. Along with the wordwide tendency to build trusted fully autonomous robotic systems, telerobotics is still acquiring lots of researchers attention. This is due to the simplicity of the intelligent systems of the telerobotics in comparasion with the autonomous robots. One more attracting point that serves this field is the human ability to learn sometimes faster and percieve better than the machine. For instance, this is one of reasons why we didn't see until now fully autonomous surgical robot. Until having fully trusted autonomous systems, it is always good to have a human operator, as an observer or indirect executer for sensitive tasks. Telerobotics allows this option.

Referring to the presented paper here, we see room for improvement, especially when combining impedance and admittance haptic control into one algorithm. Although this was addressed by reseachers as hybrid haptics, the topic is not yet closed as the results are not close to perfection.Therefore, we dedicate our future work to study the hybrid approach, design intellectual control haptic system enabling to render small masses and generate better net force and accurate positioning of the end effector.

References

[1] Vipul G. Gohil, S.D. Bhagwat, Amey Raut and Pratik Nirmal. Robotic arm control using haptic technology. International Journal of latest research in science and technology, Vol.2, issue 2, pp. 98-102, 2013.

[2] A. Reshamwala, R. Singh. A review on robot arm using haptic technology. International Journal of advanced research IJAREEIE, Vol.4, issue 4, pp.2204-2211, 2015.

[3] Y.Jung, J. Choi, S. Lee and H. Hahn. Robot mobile control technology using robot arm as haptic interface. Recent advances in circuits, systems, electronics, control and signal processing, pp. 133-136.

[4] A. Das, S.K. Samantaray, A.K. Patra. Operative percept of robotic arm expending haptic virtual system. IOSR Journal of Engineering, Vol. 2, issue 11, pp.37-43, 2012.

[5] O. Khatib, O. Brock, K.S. Chang, D. Rupsini, L. Sentis and S. Viji. Human Centered Robotics and Interractive Haptic Simulation. 27 pages.

[6] N. Diolaiti, C, Melchiorri. Tele-Operation of a mobile robot through haptic feedback. IEEE international workshop on haptic environments and their application. 6 pages, 2002.

[7] M.A. Hussein, A.S. Ali, A.B. Sharkawy and A.M. Mohamed. Haptic control development of robotic arm. International Journal of control, automation and systems, Vol.3, No. 3, pp.16-22, 2014.

[8] O. Khatib, O. Brock, K.S. Chang, D. Rupsini, and S. Viji. Robots for the human and interractive simulations. Proceedings of the 11th world congress in mechanism and machine science. 5 pages, 2003.

[9] Y. Bar-Cohan, C. Mavroidis, M. Bouzit, B. Dolgin et al. Virtual reality robotic telesurgery simulations using MEMICA haptic system. Proceedings of the SPIE Smart Structure Conference, Vol. 4329, 7 pages, 2001.

[10] Haptic technology: Application and global market. BCC reseach report. 153 pages, 2015.

Innovative Assessment of Selected Properties of Industrial Floors

T. Funtík[a], J. Erdelyi[b], M. Dubek[c], J. Gašparík[d]

[a,c,d] Slovak University of Technology in Bratislava, Faculty of Civil Engineering, Dept. of Building Technology
[b] Slovak University of Technology in Bratislava, Faculty of Civil Engineering, Dept. of Surveying
E-mail: tomas.funtik@stuba.sk, jan.erdeyi@stuba.sk, marek.dubek@stuba.sk, jozef.gasparik@stuba.sk

Abstract -
The quality and precision of casted industrial floor is often a dispute between an investor and a contractor. Its geometrical and physical properties divert from values defined by the project and also often do not meet the criteria of National Standards requirements. The aim of this paper is to evaluate the practical use of the Terrestrial Laser Scanning (TLS) in non-destructive testing of selected industrial floors. Our focus is on the geometric flatness evaluation. With TLS technology we assess the surface quality and flatness more precisely, faster and moreover, investigate 100% of area. This may lead to better evaluation of the relative position of investigated areas, points and constructions. However, a further investigation, which would define the methodology for evaluation the geometric tolerance from point clouds is needed.
Keywords –
TLS; Non-destructive testing; Industrial floor

1 Introduction

Construction industry deals almost daily with various requirements for geometric accuracy of any structure or surface. As the ideal shape of any structure is technically not achievable in most of the cases, various standards define the limits for the deviations from projected shape. These geometric tolerance states how much an element, a structure or a surface can differ from its ideal, designed geometry and proposed location. All surfaces are usually defined as planes, cylinders, spheres, etc, so we are able to use common techniques to verify above mentioned parameters. Those techniques don´t reflect actual developments in digital age e.g. TLS and we assume that by setting a clear methodology, geometric tolerance verification may be easier, faster, more accurate and may also eventually leads to automation.

Further investigation may include BIM based evaluation of potential construction errors. Main benefit will include faster spatial coordination and better location determination and automatic evaluation especially if there is planned cross slope in order to drain water. Information from BIM model would speed up and ease the proces and provide precise information regarding the cause of possible construction or technological errors.

The spots with depression could be evaluated by destructive testing to determine cause of local inaccuracy. To evaluate the influence of fiber distribution CT scan of drilled cylinder may be used. This information could result in determination if the compression of floor or technological discipline failure has taken place.

For a purpose of this paper, only the geometric tolerance of finished surface of industrial floor is considered as this property is often a source of claims between Investor and Constructor in the construction industry.

2 Related work

2.1 Geometrical precision according actual standards

The challenge is to apply the standards and geometrical precision in construction practice. In doing so, it is still necessary to proceed in accordance with the standards and properly define the variables with which it comes into contact with only a small proportion of professionals in the construction industry. Within handover of the construction all structures and surfaces should be measured in accordance with prescribed procedures which describe data collection, evaluation and correct interpretation.

Standards that deal with geometric accuracy in the construction industry can be divided into following groups:

- *Standards for design*: STN 73 0205 Geometrical accuracy in construction. Designing geometric accuracy;

STN 01 3405 Construction drawings, labeling accuracy characteristics.

- *Standards for construction*: STN 73 0210-1 Geometrical accuracy in construction. Part 1: The accuracy of the placing; STN 73 0210-2 Geometrical accuracy in construction, Part 2: The accuracy of monolithic concrete structures (canceled after the release of STN EN 13 670 (73 2400) Execution of concrete structures).
- *Standards for verification and evaluation*: STN 73 0212-1 Geometrical accuracy in construction. Verification of accuracy. Part 1: Basic provisions; STN 73 0212-3 Geometrical accuracy in construction. Verification of accuracy. Part 3: Civil engineering objects; STN 73 0212-5 Geometrical accuracy in construction, Verification of accuracy, Part 5: Checking the accuracy of building components; STN 73 0212-6 Geometrical accuracy in building, Verification of accuracy, Part 6: Statistical analysis and handover; STN ISO 7737 Geometrical accuracy in construction. Tolerances in construction. Data logging for dimensional accuracy. STN ISO 7077 Geometrical accuracy in construction. Measurement methods in construction. General principles and procedures for verifying the accuracy of dimensions.

To assess local flatness of investigated industrial floor following standard has been used: STN 74 4505 Floors, Common provisions, Design and Execution. This standard specifies requirements for the design, execution and testing of floors both in internal and external environment of buildings. The standard defines maximum deviation values.

2.2 Terrestrial Laser Scanning

The technology of Terrestrial Laser Scanning (TLS) is a non-selective method of spatial data acquisition. TLS determines the 3D coordinates of the measured points on the surface of the measured object in a grid, which is defined by regular angular spacing in the horizontal and vertical directions [1]. The result of TLS is an irregular raster of measured points, the so-called point cloud, which documents the measured object (fig. 1). The difference between TLS and conventional surveying methods is that the coordinates of characteristic points are obtained by modelling, respectively by generalization of the main elements of 3D models or the resulting point cloud [2].
Most of the current TLS works on the principle of spatial polar method. The spatial position of measured points are calculated from the measured horizontal and vertical angles and from the measured slope distance (fig.2 left).

Figure 1 The point cloud of The bridge No. 137, Bojnicka street in Bratislava

The optimal source of radiation of electromagnetic waves for scanning systems are lasers. These are used for contactless measuring of distances. Laser beams are highly monochromatic and have a narrow spectral line width compared to other sources of radiation.
The deflection of the laser beams is provided by oscillating mirrors, rotating prism, by rotation of the laser source around horizontal and vertical axis of the instrument or by fiber optics, resp. by combination of the methods above mentioned [1]. The most common used is the combination of rotation of instrument around the vertical axis, and an oscillating mirror.

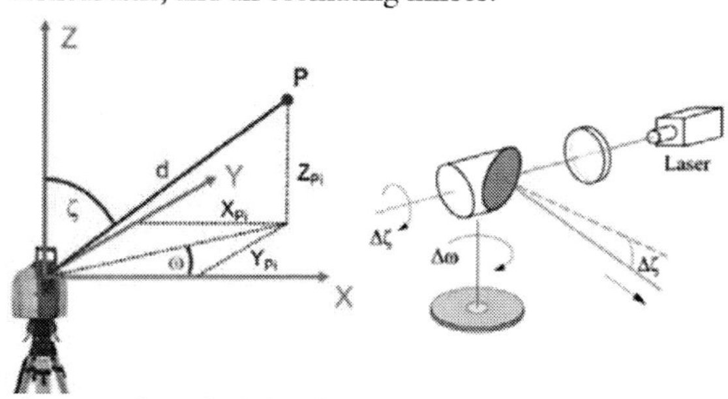

Figure 2 The principle of spatial polar method (left), and the combination of rotating source and oscillating mirror (right)

The process of data acquisition and the modelling using TLS can be divided into three main steps. The first one is the preparation for the measurement (for scanning), recognition of the measured object, the choice of positions for the instrument, signalization of the control points. The second step is the process of scanning and the third one is the processing of data obtained by TLS. The data processing contains:

- Preparation of the point cloud for data processing. This includes initial adjustments of point cloud: error elimination, filtering and data reduction, transformation (between different coordinate systems), elimination of unnecessary points, and coloring of points (assigning the colors according to the intensity of measuring signal or from photographs).
- Processing of data obtained by TLS. Spatial model creation of the measured object or its parts,

determination of geometric parameters (e.g. dimensions) and deformations of chosen parts of the measured object.

- Visualization. Rendering of the created model, and creation of animations.

2.3 Determination of the quality of as-build state of structures

Common mistakes and faults occurring on industrial floors are cracking, flatness deviation, failure of surface inclination, incorrect execution of expansion joints and delamination. As stated above, flatness and surface inclination can be easily evaluated by TLS.

The determination of the flatness of structures from laser scanning data is relatively simple when the flatness is represented by mesh surfaces (complex meshing or TIN), or by measuring the coordinate difference between discrete measured points. In both cases the accuracy of the results depending on the accuracy of the position of scanned points (several millimetres). To increase the accuracy of the results, the measured points (monitored parts of the scanned structure) have to be modelled using regression. Using regression models the noise of the point cloud can be significantly reduced.

The flatness of industrial floors can be determined as the difference between the height of measured points (lying on the surface of the floor) and a horizontal plane. It also allows to check the slope of the floor surface. The height of the points can be calculated by modelling small planes using orthogonal regression. The position of the measured points in the XY plane can be defined by their coordinates in the plane (in a regular grid) (fig. 3).

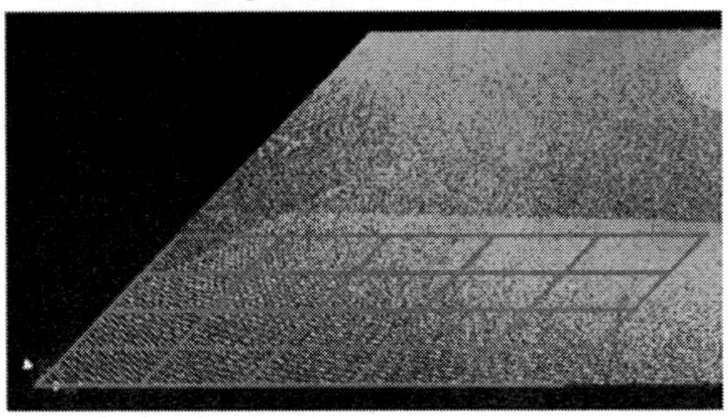

Figure 3 Modelling the position of the measured point using small planar surfaces

In this case the position of the measured points is calculated from tens or hundreds of scanned points, which increasing the accuracy of the results.

The deviation of the measured surface from a horizontal plane can be represented by isolines (contours) or by mesh, where the deviation is drawn on the Z axis of the figure.

Orthogonal regression is calculated from the general equation of a plane:

$$a \cdot X + b \cdot Y + c \cdot Z + d = 0 \quad (1)$$

where: a, b and c are the parameters of the normal vector of the plane,

X, Y and Z are the coordinates of the point lying in the plane,

d is the scalar product of the normal vector of the plane and the position vector of any point of the plane.

The orthogonal distance of a point from the plane is calculated by:

$$d_{P,\rho} = \frac{\left| a \cdot X_P + b \cdot Y_P + c \cdot Z_P + d \right|}{\sqrt{a^2 + b^2 + c^2}} \quad (2)$$

The requirement of orthogonal regression is that the sum of the squares of orthogonal distances have to be minimal, so:

$$\sum_{i=1}^{n} \frac{\left| a \cdot X_i + b \cdot Y_i + c \cdot Z_i + d \right|^2}{a^2 + b^2 + c^2} = \min \quad (3)$$

where: n is the number of points used for the calculation of the plane.

Partial derivation of (3) with respect to d leads to:

$$2 \cdot \sum_{i=1}^{n} \frac{\left| a \cdot X_i + b \cdot Y_i + c \cdot Z_i + d \right|}{a^2 + b^2 + c^2} = 0 \quad (4)$$

According to the previous formula, the parameter d can be formulated as:

$$d = -(a \cdot X_0 + b \cdot Y_0 + c \cdot Z_0) \quad (5)$$

And the formula for the general equation of a plane becomes:

$$a \cdot (X_i - X_0) + b \cdot (Y_i - Y_0) + c \cdot (Z_i - Z_0) = 0 \quad (6)$$

where: $(X_i - X_0)$, $(Y_i - Y_0)$ and $(Z_i - Z_0)$ are the coordinates of the point cloud reduced to a centroid.

For each point of point cloud is possible to write a formula according to (6),then the design matrix of the system of equations has the form:

$$\mathbf{A} = \begin{pmatrix} (X_1 - X_0) & (Y_1 - Y_0) & (Z_1 - Z_0) \\ (X_2 - X_0) & (Y_2 - Y_0) & (Z_2 - Z_0) \\ \vdots & \vdots & \vdots \\ (X_n - X_0) & (Y_n - Y_0) & (Z_n - Z_0) \end{pmatrix} \quad (7)$$

Orthogonal regression is calculated by applying Singular Value Decomposition:

$$\mathbf{A} = \mathbf{U \Sigma V}^T \quad (8)$$

where: **A** is the design matrix, with dimensions *nx3*, and *n* is the number of points used for the calculation. The column vectors of $\mathbf{U}^{nxn}$ are normalized eigenvectors of matrix $\mathbf{AA}^{T}$. The column vectors of $\mathbf{V}^{3x3}$ are normalized eigenvectors of $\mathbf{A}^{T}\mathbf{A}$. The matrix $\mathbf{\Sigma}^{nx3}$ contains eigenvalues on the diagonals. Then the normal vector of regression plane is the column vector of **V** corresponding to the smallest eigenvalue from **Σ** [3], [4].

The position of the observed points in XY plane is defined as fixed in a defined grid (e.g. 100 m x 100 m). The Z coordinates (heights) of the measured points are calculated by projecting the points onto regression planes (fig.4) using formula:

$$Z_P = -\frac{a \cdot X + b \cdot Y + d}{c} \tag{9}$$

In case of large floor surfaces, when it is not possible to scan the whole floor from a single position of the instrument, the transformation of the point clouds from each instrument position is needed to obtain data in a common coordinate system.

The standard deviation of the results is calculated using uncertainty propagation law, from the standard deviation of the vertical component of the transformation error and the standard deviation of the regression planes:

$$\sigma_{Z_P} = \sqrt{\sigma_{T_Z}^2 + \sigma_\rho^2} \tag{10}$$

where: σ_{T_Z} is the vertical component of the error of the data transformation and σ_ρ is the standard deviation of the calculated regression plane.

The accuracy of the transformation is given by the differences (ΔX, ΔY, ΔZ) between the identical reference points after the transformation of the scanned point cloud from each instrument position into a common coordinate system. The standard deviation of the regression planes is calculated from the orthogonal distance of the points of point cloud from these planes. Dispersion of the points around the plane reflects the random error (noise) of the distance measurement by TLS (coordinate determination) mainly.

An application based on software MATLAB® was developed for automated data processing. The above mentioned computational procedure is performed and controlled with help of this application. The application was created as a standalone app; however, the Matlab Runtime is necessary to be installed. The work with the app is as follows: In the first step the user can choose a work directory in which the resulting files will be saved. The second step is the point cloud file loading in *.txt or *.xyz file format which contains the coordinates of scanned points. The position of measured points is defined by the step of a grid in XY plane. The deviation of measured points from the horizontal plane (Z coordinates) is calculated in the points defined by the mentioned grid definition.

According to the dimensions of the grid the application defines fencing boxes (selects part of the point cloud around the grid points). Using the above mentioned calculation procedure, calculates the Z coordinates and their standard deviation for each measured point using the uncertainty propagation rule. The results are saved into an *.xlsx file in the work directory. Flatness is represented by isolines and by mesh surface. Both are created and saved into a *.png file.

The accuracy of the results is defined by the standard deviation of the flatness determination. Actually it is difficult to define one scalar value for the whole floor, but it can be calculated from the standard deviation of the Z coordinates of measured points (grid points) using quadratic mean.

3 Case study

Precision of works was evaluated on selected type of industrial floor located in Bratislava, Slovakia.
Within this case study, we have measured two dilatation units, A and B (see Figure 4). They were further evaluated according to above mentioned methodology. The dimensions of A and B are equal, length is 6,0m and width is 6,0m. To eliminate imperfection of works, we have selected internal dilatation units.

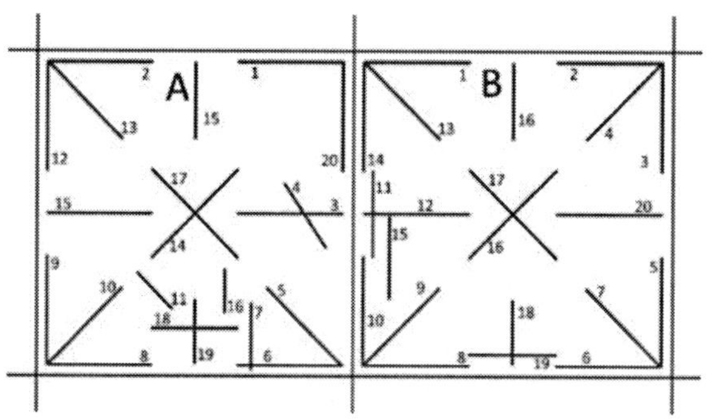

Figure 4 Measured area of industrial floor

Due the fact that investigated industrial floor was recently finished and has not been handed over for use, data collected by TLS are stored and might be used in future for comparison and further investigation of deformations caused in time.

3.1 Standard Measurement

The standard allows maximum deviation of 5 mm for industrial floors. Measurement was carried out using a standard pads at the ends of the measuring spirit level meter. Defined height was 10 mm.

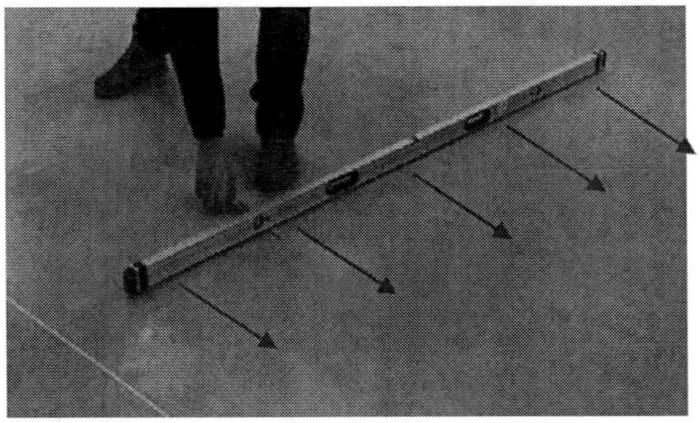

Figure 5 Local flatness measurement showing the position measurement

Measurements were performed on two expansion sections, the two-meter spirit level within the meaning of the abovementioned standard was used. The method of measurement is captured in the image below. Number of sets and locations on the floor was 20 times per section. Each spot was measured by spirit level measure and calibrated wedge in five positions. The measurement result is a set of 100 measurements which are in accordance with the standard STN 73 0212-6 Geometrical accuracy in building. Control accuracy. Part 6: Statistical analysis and thinning, - Table No. 10, when considering the same risk of the investor and the contractor with a set of permitted levels.

Evaluation for a set of measurements n = 100 is then the result for section No. 1 is shown in Table 1 below. In the evaluation, it is therefore considered PR = CR = 5%, PRQ = 1.5%, CRQ = 5%, n = 100, and k = 1.91.

Table 1 Evaluation of Section n.1

n.	Deviation +10mm	Measured deviation abs(x_n)	x_n - x	$(x_n - x)^2$
1	10,5	0,5	-0,265000	0,0702250
2	9,0	1,0	0,235000	0,0552250
3	8,5	1,5	0,735000	0,5402250
4	10,0	0,0	-0,765000	0,5852250
100	10,5	0,5	-0,265000	0,0702250

Σx_n	76,5
x	0,765
$\Sigma(x_n - x)^2$	50,05705
x_m	0,505626
s	0,711074
x + k . s	2,123152

n	100
k	1,91
PRQ	1,50%
RQ	5%

According to the standard, calculated deviation is at section no. 1 equal to 2.12 mm and the deviation of section no. 2 equals 1.7 mm.

In search of extremes according to STN 74 4505, as well as statistical analysis and handover according to 730212-6, evaluated industrial floor is within 5 mm tolerance for the category of manufacturing warehouses and garages.

Table 2 Evaluation of Section n.2

n.	Deviation +10mm	Measured deviation abs(x_n)	x_n - x	$(x_n - x)^2$
1	10,5	0,5	-0,180000	0,0324000
2	10,0	0,0	-0,680000	0,4624000
3	10,0	0,0	-0,680000	0,4624000
98	10,5	0,5	-0,180000	0,0324000
99	10,5	0,5	-0,180000	0,0324000
100	10,0	0,0	-0,680000	0,4624000

Σx_n	68
x	0,68
$\Sigma(x_n - x)^2$	28,2600
x_m	0,28545
s	0,53427
x + k . s	1,700473 7

n	100
k	1,91
PRQ	1,50%
RQ	5%

3.2 TLS Measurement

TLS was performed from two positions by Leica ScanStation2 device to cover the area underneath the device. The device was directly connected to the Notebook. Data from point cloud were then processed using specialized software tool developed by our research team.

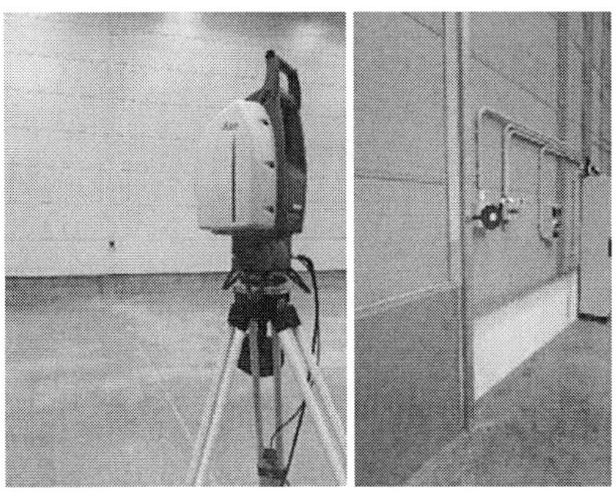

Figure 6 Laser scanner Leica ScanStation2

The point cloud file was loaded into the software in order to obtain the coordinates of scanned points. The extracted position of measured points is shown in the Table 3 below. The deviation of measured points from the horizontal plane (Z coordinates) is calculated using the uncertainty propagation law.

Table 3 Data output after processing

	A	B	C	D	E	F	G	H
1	X[m]	Y[m]	Z[mm]	a	b	c	d	std[mm]
2	0,00	0,00	0	-0,0723	-0,06652	0,99516	0,00861	0,2694
3	0,10	0,00	1	0,00456	0,00327	-0,99998	-0,00547	0,43418
4	0,20	0,00	1	0,0048	0,01072	0,99993	0,004	0,49058
5	0,30	0,00	1	0,01134	0,00754	-0,99991	-0,00869	0,37806
6	0,40	0,00	2	0,00808	0,02324	0,9997	0,00105	0,53103
7	0,50	0,00	1	-0,00659	-0,00624	0,99996	0,00794	0,71579
8	0,60	0,00	2	0,0041	0,00965	0,99994	0,00189	0,58929
9	0,70	0,00	1	-0,00078	0,00748	-0,99997	-0,00427	0,59528
10	0,80	0,00	1	0,00129	-0,00282	-1	-0,00603	0,56227

Column A – defined coordinate X [m], Column B – defined coordinate Y [m], Column C – computed height, coordinate Z [m]. Columns A, B define the grid.

The accuracy of the results in this case study defined by the standard deviation of the flatness determination calculated from the standard deviation of the Z coordinates of measured grid points using quadratic mean is represented by the value 0.7 mm.

Graphical representation based on output data is presented on the Figure 7 where overall flatness is represented by isolines and on the Figure 8 where by the flatness is represented by mesh surface.

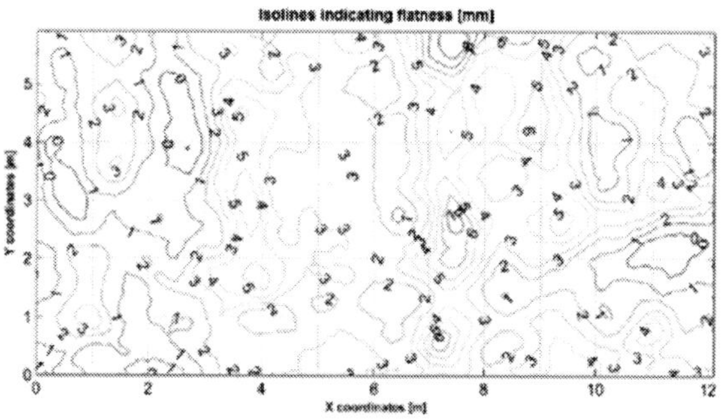

Figure 7 Isolines indicating flatness

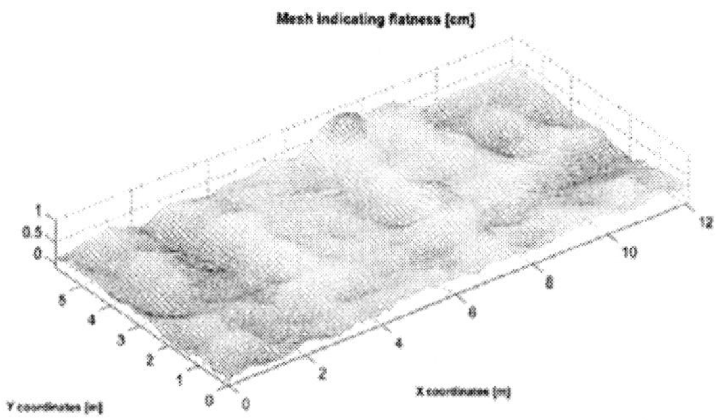

Figure 8 Mesh indicating flatness

3.3 Comparation of two methods

Comparation of mentioned methods for flatness evaluation (standard methodology and non-standard methodology using TLS) provided several facts and factors to be taken into account developing the process of new methodology.

Evaluation according to actual standards could provide information about local flatness without the possibility of including individual points from wider area than 2m. Total deviation of individual points of industrial floor from defined plane is not possible neither the verification of inclination nor declination defined by project. Moreover, this process is very laborious, time consuming and it requires a high degree of attention. This may lead to significant errors in reading of measures and result in inaccurate records.

The measurement is directly dependent on the choice of places where the spirit level was placed and also the direction in which the spirit level is placed on floor, therefore different interpretation of the results are possible. Under certain circumstances, the very same floor can be evaluated both to meet and not to meet the criteria of standards for flatness. On the Figure 9 are shown various positions of spirit level and influence to measurement. By placing spirit level in position A and B measured data provide an information that floor has almost no deviation from plane and local defects. By placing the spirit level into position C, data obtained from a measure proves significant deviations.

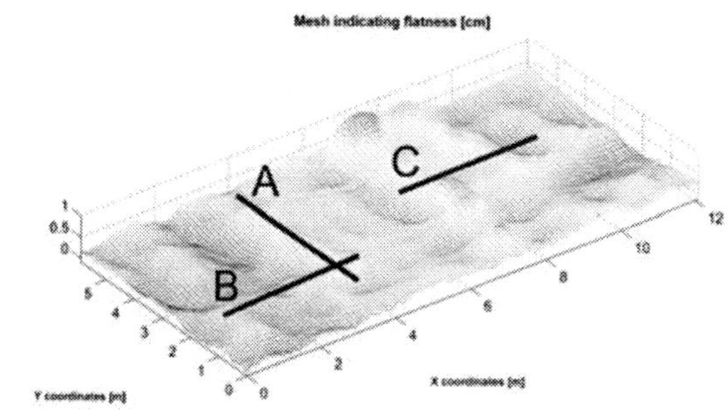

Figure 9 Various positions of spirit level meter

One of major disadvantage is the fact that the measurement cannot be automated.

Measurement using TLS can be evaluated from perspective of the local flatness and at a same time total deviation from the plane or verification of inclination nor declination defined by project can be assessed. The measurement by TLS is very accurate and fast. Measurements record exact data in defined grid, i.e. it could be used also for the evaluation of the data in accordance with standard methodology.

Proposed methodology is able to provide clear answer what degree of the flatness of the floor has been reached.

Significant advantage is that the measurement and evaluation can be automated and therefore will not depend on personal judgement of evaluator.

3.4 Prospective future research

Destructive tests were not allowed on this case study as the deviations were not significant and investor has not raised a claim. Local defects might be better assessed by drilling cylinders and performing tests e.g. determining the number of fibres in this spot by CT scan. The declined spot may be a result of less fibres or improper distribution of fibres.

By providing 3D BIM model of a project, other possible causes of defects may by easily evaluated and rejected or confirmed. By comparing the planned 3D model vs model obtained from actual measurements, differential models will be automatically generated (processed). Analysis of difference models will be the part of new tools and the final assessment of the quality of the work, its correct location in space, as well as the fulfilment of quality requirements (construction deviations), will be determined in the BIM environment.

These data could be easily used to discover local defect that can be caused by failure of underfloor horizontal piping system (leakage, inadequate compaction, etc.). Without having precise and complete data, technological uncertainty is higher, investigation is more complicated, time consuming and the result is ambiguous.

By continuing on this research we assume that more complex testing will be included into methodology.

CONCLUSION

Geometric accuracy measurement of flatness of industrial floors is rather time consuming activity and the quality of works cause many legal disputes between investor and contractor. Standards that are currently in use don't reflect to actual digital technology developments and advanced technologies as many of processes are manual which leads to significant errors e.g. error in measuring, error in evaluation, error caused by low complexity of operation. Moreover, such a manual process cannot be automated. Proposed methodology by our research team is able to provide clear answer what degree of the flatness of the floor has been reached.

Methodology could result in new standards but further study and research in this field needs to be performed.

REFERENCES

[1] KAŠPAR, M. - POSPÍŠIL, J. - ŠTRONER, M. - KŘEMEN, T. - TEJKAL, M.: Laser Scanning in Civil Engineering and Land Surveying. 1. ed. Hradec Králové: Vega, 2004. 103 s. ISBN 80-900860-7-1.

[2] VOSSELMAN, G. – MAAS, H-G. 2010. Airborn and Terrestrial Laser Scanning. Dunbeath: Whittles Publishing, 2010. 318 s. ISBN 978-1904445-87-6.

[3] Čepek, A. and Pytel, J. 2009. A note on numerical solutions of least squares adjustment in GNU project gama, In Pilz J., editor, Interfacing Geostatistics and GIS, Springer Berlin Heidelberg, pp. 173-187. doi:10.1007/978-3-540-33236-7_14.

[4] LACKO, V. 2008. Singular Value Decomposition and Difficulties of Software Implementation of Golub Algorithm and its Determination: Student sciense conference. Bratislava: Comenius University in Bratislava, 2008. 69 p.

[5] STN 73 0270 The accuracy of geometrical parameters in construction. Control of building works (effective date 01.06.1991)

[6] STN EN 13914-2 / 3710 73 / Design, preparation and application of external and internal plastering - Part 2: Preparation of design and basic procedures for plasters (01, 2006)

[7] CSN 73 3714 / IDT CEN / TR 15124: 2005 / Design, preparation and execution of internal gypsum plastering systems (07, 2006)

[8] ISO 7078 STN Civil Engineering. Measurement procedures and setting out. Dictionary and explanations

[9] STN ISO 1803 Building construction - Tolerances - Expression of dimensional accuracy - Principles and terminology

[10]STN 73 0205 Geometrical accuracy in construction. Designing of geometric accuracy

[11]STN 01 3405 Construction drawings, labeling accuracy characteristics.

[12]STN 73 0210-1 Geometrical accuracy in construction. Part 1: Placing Accuracy

[13]STN 73 0210-2 Geometrical accuracy in construction, Part 2: The accuracy of monolithic concrete structures.

[14]STN EN 13670 (73 2400) Execution of concrete structures.

[15]STN 73 0212-1 Geometrical accuracy in construction. Verification of accuracy. Part 1: Basic provisions

[16]STN 73 0212-3 Geometrical accuracy in construction. Verification of accuracy. Part 3: Building construction objects

[17]STN 73 0212-5 Geometrical accuracy in construction, Verification of accuracy, Part 5: Checking the accuracy of structural components

[18]STN 73 0212-6 Geometrical accuracy in construction, Verification of accuracy, Part 6: Statistical analysis and thinning

[19]STN ISO 7737 Geometrical accuracy in construction. Tolerance under construction. Data logging for dimensional accuracy.

[20]STN ISO 7077 Geometrical accuracy in construction. Measurement methods in construction. General principles and procedures for verifying the accuracy of dimensions.

[21]STN 73 2901 Execution of external thermal insulation composite systems (ETICS), Bratislava 2008

[22]Stefano PETRO, Geometric tolerances verification: strategy optimization for CMM measurement, Politecnico di Milano, Italy, 2005

[23]Makys P., Funtik T., Verification of BIM Model using TLS, In: Eurostav, 4/2015, Bratislava, ISSN 1335-1249

[24]FUNTÍK, Tomáš - ĎUBEK, Marek - ERDÉLYI, Ján. Geometric Tolerance Verification - Innovative Evaluation of Facade Surface Flatness Using TLS (Terrestrial Laser Scanning). In Applied Mechanics and Materials: Advanced Architectural Design and Construction. Vol. 820, (2016), s. 81-89. ISSN 1660-9336

[25]Methods for laser scanning data processing / Šoltés Tibor, Kozlovská Mária, 2014. In: SGEM 2014: International multidisciplinary scintific conferences on social sciences & arts: Arts, Performance arts Architecture and Design: 1. - 10.9.2014: Albena, P. 795-800, Sofia : STEF92 Technology Ltd., 2014 /978-619-7105-30-08

[26]The Methodology of Interactive Parametric Modelling of Construction Site Facilities in BIM Environment / Mária Kozlovská, Jozef Čabala, Zuzana Struková In: SSP - Journal of Civil Engineering: Scientific Selected Papers Roč. 9, č. 2(2014), s. 85 - 96 2013

Automatic Reconstruction of As-built BIM from Laser Scanned Data of Precast Concrete Elements for Dimensional Quality Assessment

Q. Wang[a,b], J. C. P. Cheng[a] and H. Sohn[b]

[a]Department of Civil and Environmental Engineering,
The Hong Kong University of Science and Technology, Hong Kong
[b]Department of Civil and Environmental Engineering,
Korea Advanced Institute of Science and Technology, South Korea
E-mail: qwangau@ust.hk, cejcheng@ust.hk, hoonsohn@kaist.ac.kr

Abstract –

Precast concrete elements are popularly used in the construction of buildings and infrastructures because they enable higher construction quality, less construction time, and less environmental impact. To ensure the performance of complete precast concrete systems, dimensional quality of individual precast concrete elements must be assessed before they are transported to the construction sites. However, the current quality assessment methods mainly rely on manual inspection with traditional measurement devices, which are inefficient and inaccurate. Besides, the quality assessment results are not well managed. To realize efficient and accurate quality assessment of precast concrete elements and to facilitate the management of quality assessment results, this study proposes a technique which can automatically reconstruct the as-built Building Information Models (BIM) from laser scan data of precast concrete elements for dimensional quality assessment. The proposed technique firstly performs a scan planning to determine the number of scans and the locations of scanners. Then, the pre-processing of scan data removes noisy data and registers multiple scans in a global coordinate system. Afterwards, the as-built geometries of the element are extracted from the registered scan data, and finally the as-built BIM is reconstructed. To validate the proposed technique, a scanning experiment was conducted on a small-scale test specimen. The experimental results demonstrate that the proposed technique can accurately and efficiently create as-built BIM of precast concrete elements.

Keywords –

Building Information Models; Laser scanning; As-built BIM reconstruction; Precast concrete elements; Quality assessment

1 Introduction

Precast concrete elements are popularly used in the construction of buildings and infrastructures because they enable higher construction quality, less construction time, and less environmental impact, compared to cast-in-place concrete [1, 2]. To ensure the performance of complete precast concrete systems, the dimensions of individual precast concrete elements must be assessed before they are transported to the construction sites. However, the current quality assessment of precast concrete elements is mainly performed manually using traditional devices such as measuring tapes. Such traditional quality assessment methods are proven to be time-consuming and error-prone [3]. Besides, as the quality assessment results are traditionally recorded in paper sheets or electronic spreadsheets, the as-built conditions of the elements are not well managed in a long term.

These days, 3D laser scanners have become popular in civil engineering because they can acquire 3D range measurement data, known as point cloud data, at a high speed of million scan points per second and with a high accuracy up to a few millimeters. Utilizing laser scan data, a variety of applications have been reported, including 3D model reconstruction [4, 5], construction progress monitoring [6, 7], structural quality assessment [8, 9], etc. On the other hand, Building Information Models (BIM) have become a trend in the Architecture, Engineering, and Construction/Facility Management (AEC/FM) industry because BIM facilitates the storage, visualization, and management of building information throughout the whole lifecycle of a facility.

To achieve efficient and accurate quality assessment of precast concrete elements and to facilitate the management of quality assessment results, this study proposes a technique that can automatically reconstruct the as-built BIM of precast concrete elements, particularly focused on precast concrete bridge deck panels. The as-built BIM representing the as-built conditions of an element can be compared to the corresponding as-design blueprints or models for dimensional quality assessment. Furthermore, the as-built BIM can facilitate the assemblage, maintenance, and replacement of the precast concrete element. This paper is organized as follows. Section 2 provides background information on (1) the existing research on BIM reconstruction from laser scan data, and (2) the precast concrete bridge deck panels. Then, the proposed technique for automatic as-built BIM reconstruction is presented in Section 3, and an experimental validation of the proposed technique is described in Section 4. Finally, Section 5 concludes this study and suggests future work.

2 Background

2.1 BIM reconstruction from laser scan data

While manual BIM reconstruction from laser scan data is time-consuming and labor-intensive, a few studies have developed algorithms for automatic BIM reconstruction. The main challenges and research topics for automatic BIM reconstruction include (1) the registration of multiple scans, (2) geometric modeling, and (3) object recognition.

To generate a complete point cloud covering all the surfaces of a target, multiple scans must be conducted at different locations. Then, the scan data obtained from different scans are aligned in a global coordinate system during the registration process of multiple scans. Although automatic registration methods have been developed in computer vision area, they are not applicable to the AEC/FM industry [10]. Instead, considering that most surfaces are planar in the AEC/FM context, some studies have developed automatic or semi-automatic plane-based registration methods, which register two scans by matching their common plane [11, 12].

Geometric modeling refers to the process of detecting and modeling certain geometry shapes from the scan data. The most common shape in the AEC/FM context is plane. A number of algorithms have been adopted to detect planes, including the RANSAC algorithm [11], the Hough transform [13], and the region-growing algorithm [5]. Besides planar surfaces, cylindrical shapes, which can represent pipes and some conduits, have also been studied [14].

Object recognition aims to recognize specific construction components, e.g., wall, roof, column, etc., from the detected geometry shapes. Several approaches for automatic object recognition have been investigated, including machine learning algorithms [5], context-based recognition [15], and the use of prior knowledge [16].

2.2 Precast concrete bridge deck panels

Precast concrete bridge decks have been used successfully for several decades. Along with the reduction of construction time, the quick replacement of precast concrete bridge decks also reduces the impact on traffic flow significantly, which is essential in urban areas. A precast concrete bridge deck consists of a series of bridge deck panels, which are connected one by one. To connect panels with the girders and to connect adjacent panels, three structural features, namely shear pockets, shear keys and flat ducts, are usually provided on the bridge deck panels, as shown in Figures 1(a)-(b).

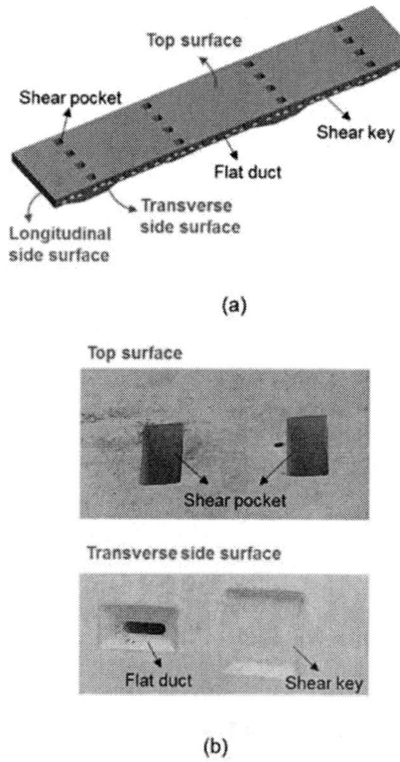

Figure 1. Precast concrete bridge deck panels. (a) A 3D model of a bridge deck panel. (b) Examples of shear pocket, flat duct, and shear key on an actual bridge deck panel.

Shear pockets are provided on the top surface of a panel and they usually have rectangular shapes. They serve as panel-to-girder joints after grout is filled inside. Shear keys are distributed along the transverse side surfaces of panels and serve as transverse panel-to-panel joints. They are designed to eliminate relative vertical movement between adjacent panels. Shear keys are mainly in two categories, non-grouted match-case shear keys using epoxy adhesive and grouted female-to-female shear keys, and the one shown in Fig. 1(b) is the latter. Female-to-female shear keys use grout to fill the joints between adjacent panels to achieve the required strength. Last, ducts are also distributed along the transverse side surfaces of panels. They are used to place post-tensioned longitudinal reinforcements, which put the transverse panel-to-panel joints under compression and eliminate the tensile stress resulting from the live load. Ducts are usually rounded or flat in shape, and the one shown in Fig. 1(b) is a flat duct.

3 Proposed technique for automatic reconstruction of as-built BIM

Figure 2 illustrates the procedures of the proposed technique based on laser scanning for automatic as-built BIM reconstruction. The proposed technique includes four steps: (1) scan planning, (2) scan data pre-processing, (3) extraction of geometry, and (4) reconstruction of as-built BIM. The details of the four steps are illustrated in Sections 3.1-3.4, respectively.

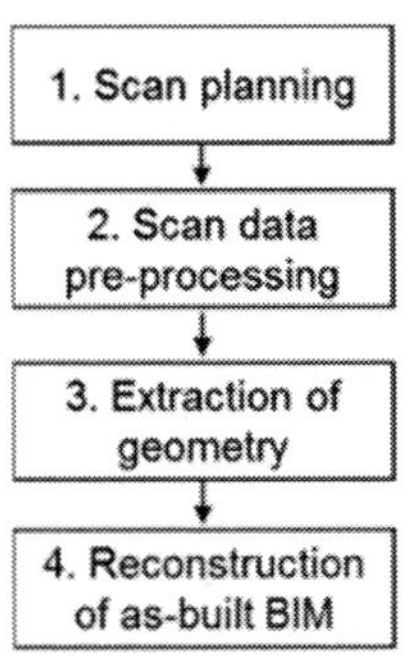

Figure 2. Procedures of the proposed technique based on laser scanning for automatic as-built BIM reconstruction

3.1 Scan planning

To acquire scan data covering all the surfaces of an element, multiple scans must be conducted at different locations. To obtain high resolution scan data and to save scanning time, scan planning is performed to minimize the number of scans and to optimize the locations of the scanner. The set of scans, namely scan set, must meet the following two requirements.

(1) For any arbitrary point on the target, there should exist at least one scan that provides high resolution scan data surrounding this point. Here, high resolution scan data require that the spacing s between two adjacent scan points is less than a threshold value S_T.

(2) Scan data from all the scans should be able to be registered based on common planes. This is motivated by that a plane-based registration will be adopted in the proposed technique to register multiple scans. Assuming that a total of four planes, P1, P2, P3, and P4, need to be scanned, and a total of three scans, S1, S2, and S3, are conducted, two different cases are illustrated in Figure 3. In case 1, S1 covers P1 and P2, S2 covers P2 and P4, and S3 covers P3 and P4. Therefore, firstly S1 and S2 can be registered based on their common plane P2; and then the registered scan data can be further registered with S3 based on their common plane P4. However, in case 2, although S1 and S2 can be registered, S3 cannot be registered with any other scan. Therefore, the scan set in case 2 does not meet requirement (2).

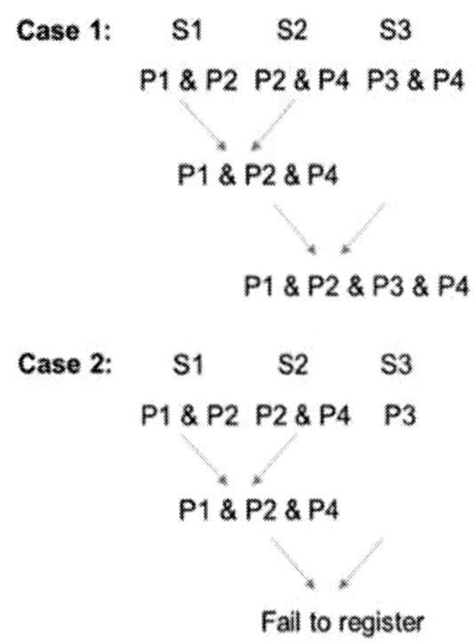

Figure 3. All the scans should be able to be registered by common planes: case 1 is a success case and case 2 is a failure case

Among all the scan sets that fulfill the above requirements, the ones that contain the minimum number of scans are firstly found. Here, the minimum number of scans is denoted as N_{min}. Then, among all the scan sets with N_{min} scans, the one that brings the smallest averaged s value is selected.

3.2 Scan data pre-processing

After the scan data are acquired at the planned

scanning locations, the scan data pre-processing is performed, which consists of filtering of noisy data and registration of scans, as illustrated in Sections 3.2.1 and 3.2.2, respectively.

3.2.1 Filtering of noisy data

Raw scan data contain three types of scan points – (1) valid points belonging to the target object, (2) background points belonging to the background objects, and (3) erroneous data points called mixed pixels. A mixed pixel occurs when a laser beam emitted at the edge of an object is split into two parts and falls on two different surfaces. The two reflected signals are both received by the scanner and the resulting scan point becomes a mixed pixel. Hence, the filtering of noisy data aims to remove both mixed pixels and background points, while retaining valid points only.

Although mixed pixels can be anywhere along the direction of the laser beam, they usually have a lower spatial density compared to valid points or background points [17]. Therefore, a density-based clustering algorithm, namely the density-based spatial clustering of applications with noise (DBSCAN) [18], is adopted to filter out mixed pixels and background points, which is proven to be effective in [19]. After applying the DBSCAN algorithm, mixed pixels are classified as noise due to the low density, valid points become one cluster due to the high density, and background points become one or multiple clusters, depending on the specific background objects. Finally, among all the clusters of points, the one that has the least average distance to the scanner is identified as valid points.

3.2.2 Registration of scans

The registration of scans includes two steps – (1) a coarse registration which provides a rough alignment of scans, and (2) a fine registration which further improves the initial solution.

The coarse registration aligns two scans based on their common plane. For a typical bridge deck panel, five planes (the top surface, two transverse side surfaces, and two longitudinal side surfaces, as shown in Figure 1(a)) are used for coarse registration. Firstly, for each scan, all the planes are detected from the scan data using the RANSAC [20] algorithm. Then, the common plane of the two scans is identified. If two scans have more than one common plane, the one with the largest size is used for registration. In case that one scan covers two or three planes, the correct common plane is identified based on the sizes of planes, because the top, the transverse side, and the longitudinal side surfaces of a panel have substantially different sizes. Thirdly, for the common plane, its mass center (O) and three eigenvectors (e_1, e_2 and e_3) are obtained from both scans, as shown in Figure 4. Lastly, the two scans are aligned by making the mass center and eigenvectors of their common plane coincident.

After coarse registration, the point-to-plane Iterative Closet Point (ICP) [21] algorithm is used for fine registration, which optimizes the registration results by minimizing the point-to-plane distances between the two scans.

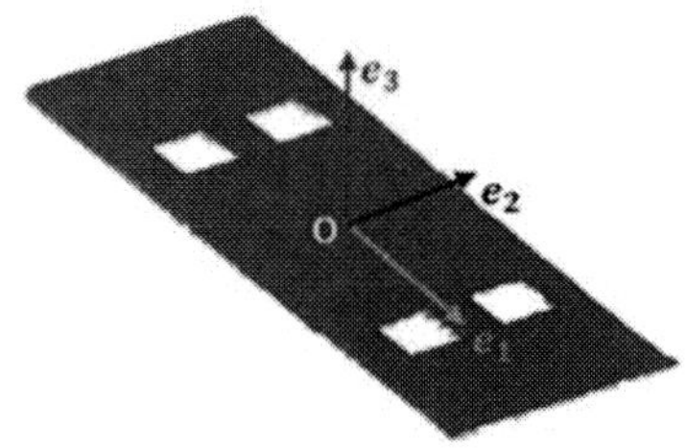

Figure 4. The mass center (O) and the eigenvectors (e_1, e_2 and e_3) of a plane

3.3 Extraction of geometry

To create the as-built BIM, the necessary geometric information of the element needs to be extracted from the registered scan data, including the dimensions and locations of shear pockets, flat ducts, and shear keys, and the dimensions of the outer boundaries of the element.

3.3.1 Extraction of shear pockets and flat ducts

Shear pockets and flat ducts are basically holes that go through the element. Hence, in the scan data, they are characterized by areas without any scan point. In the following explanations, the shear pockets on the top surface are taken as the example, and the flat ducts can be extracted in a similar way.

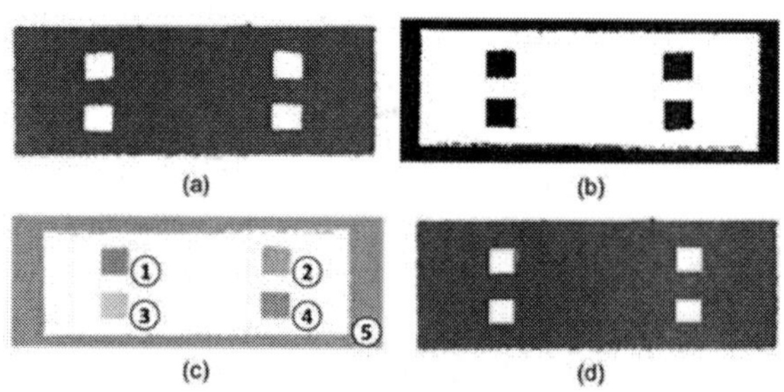

Figure 5. Extraction of shear pockets from scan data. (a) The scan data of the top surface. (b) Binary image of the top surface. (c) Detection of shear pockets. (d) Extraction of the edges of the

shear pockets.

First, the scan data of the top surface (Figure 5(a)) are transformed into a binary image (Figure 5(b)), where white pixels represent areas with scan points and black pixels represent areas without any scan point. Then, the black areas are clustered by the DBSCAN algorithm and the areas are classified into five clusters, which are denoted as clusters 1-5 and shown in different colors in Figure 5(c). Among all the five clusters, only the ones with proper sizes are identified as shear pockets. Such proper sizes are determined based on an estimation of the actual sizes of the shear pockets. Therefore, clusters 1-4 are identified as shear pockets, whereas cluster 5 is ignored due to its too large size. Last, based on the locations of the shear pockets, the four edges of each shear pocket are estimated using the edge line estimation algorithm proposed in [19], as shown in red lines in Figure 5(d).

3.3.2 Extraction of shear keys

A typical shear key has six surface in its local area, as shown in Figure 6(a). Among them, surfaces 5 and 6 have the same normal vector $\vec{n}$, whereas surfaces 1-4 have normal vectors which have 45° difference from $\vec{n}$. Based on this observation, shear keys on the transverse side surface are extracted as follows.

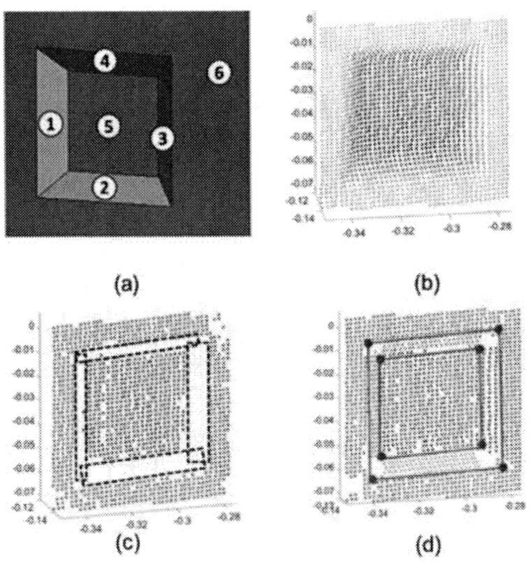

Figure 6. Extraction of shear keys from scan data. (a) The 6 surfaces in the local area of a shear key. (b) The scan data surrounding a shear key. (c) The scan points for surfaces 5 (blue dots) and 6 (orange dots), and the estimated areas of surfaces 1-4 (dashed boxes). (d) The scan points for all six surfaces (in different colors) and the extracted corner points (red dots) of the shear key.

First, the normal vector $\vec{n}$ of surface 6 is estimated as the normal of the least-squares fitting plane of all the scan points (Figure 6(b)) on the transverse side surface. Then, the scan points, the local normal vectors of which have large differences from $\vec{n}$, are filtered out. Here, scan points representing surfaces 1-4 are expected to be removed. The remaining scan points are further classified by the DBSCAN algorithm into two clusters, which represent surfaces 5 and 6, respectively, as shown in Figure 6(c). Once the scan points for surfaces 5 and 6 are obtained, the areas of surfaces 1-4 can be estimated between them, as shown in dashed boxes in Figure 6(c). Last, for each surface among surfaces 1-4, a plane is detected by the RANSAC algorithm using scan points within its estimated area. As shown in Figure 6(d), the scan points representing all the 6 surfaces are shown in different colors and each corner point (shown in red dots in Figure 6(d)) of the shear key is obtained by intersecting three surfaces surrounding this corner point.

3.3.3 Extraction of outer boundaries

Although the outer boundaries are designed to be straight lines, the as-built conditions can be different. To reflect the actual conditions, a set of points are extracted along each boundary to represent it. As shown in Figure 7, surfaces SF1 and SF2 intersect at the boundary B1. With an interval of q, a set of boundary points are obtained to represent B1, as shown in red dots. Each boundary point is obtained as the intersection point of 3 planes, i.e., the local fitting plane of scan points on SF1 (shown in green), the local fitting plane of scan points on SF2 (shown in blue), and the cross section plane perpendicular to both SF1 and SF2.

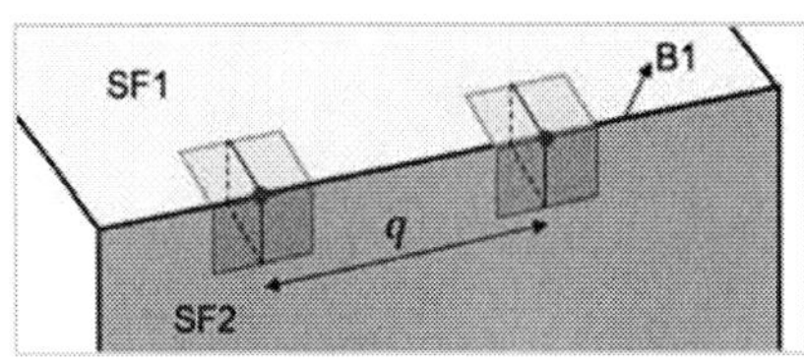

Figure 7. Extraction of a set of boundary points (shown as red dots) representing the actual conditions of boundary B1

3.4 Reconstruction of as-built BIM

To facilitate data exchange among different stakeholders throughout the lifecycle, a neutral BIM data model, namely Industry Foundation Classes (IFC), is adopted to create the as-built BIM. Because the geometry of bridge deck panels is complex, the panel is divided into several units and modelled separately. Each shear pocket, flat duct, or shear key becomes one unit, and the panel itself excluding all the above-mentioned structural features becomes another unit. First, for each single unit, its geometry is triangulated and represented by an instance of *IfcFacetedBrep*, which represents a geometry by a set of connected surfaces. Then, an instance of *IfcBooleanResult* is created, which can represent the difference (or other Boolean operators) of two *IfcFacetedBrep* instances. For example, Figure 8(a) shows two *IfcFacetedBrep* instances and Figure 8(b) shows their difference using an *IfcBooleanResult* instance. Therefore, when one instance is set as the unit of the panel itself and the other is set as the units of all structural features, the panel with all structural features can be represented by the *IfcBooleanResult* instance.

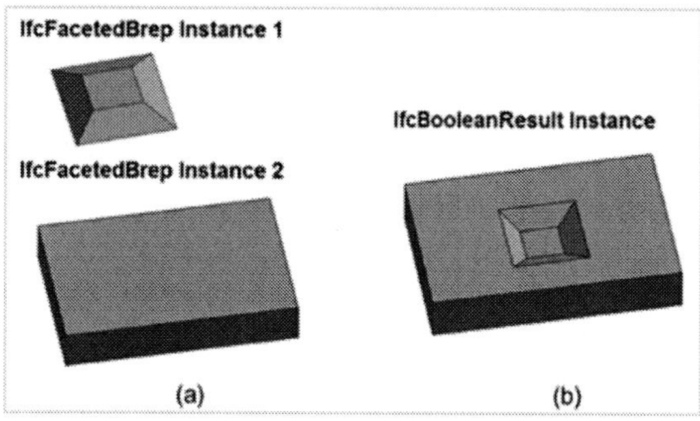

Figure 8. Geometry representation using IFC format. (a) Two *IfcFacetedBrep* instances. (b) An *IfcBooleanResult* instance representing the difference of the two *IfcFacetedBrep* instances.

4 Experimental validation

4.1 Test specimen

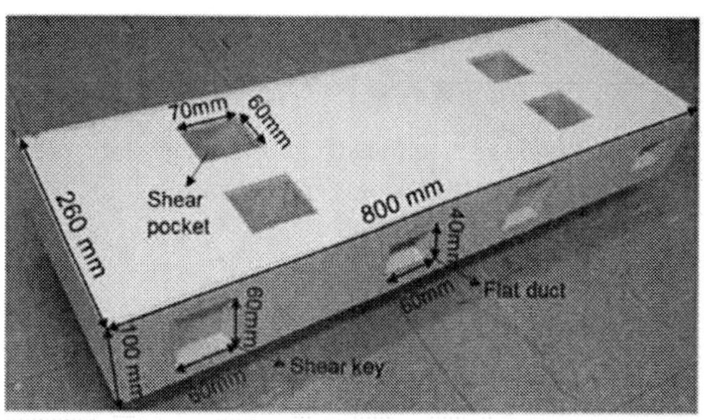

Figure 9. Dimensions of the test specimen

To validate the proposed technique of automatic as-built BIM reconstruction, a small-scale test specimen was manufactured in the laboratory using Styrofoam materials. As shown in Figure 9, the test specimen had dimensions of 800 mm (length) × 260 mm (width) × 100 mm (height). On the top surface, there were a total of 4 shear pockets, with identical dimensions of 70 mm × 60 mm. On each transverse side surface, there were a total of 2 flat ducts and 2 shear keys, with identical outer dimensions of 60 mm × 60 mm and 60 mm × 40 mm, respectively.

4.2 Applying the proposed technique

The proposed technique was applied to reconstruct the as-built BIM of the test specimen. In this experiment, all the surfaces, except the bottom surface, of the specimen need to be scanned.

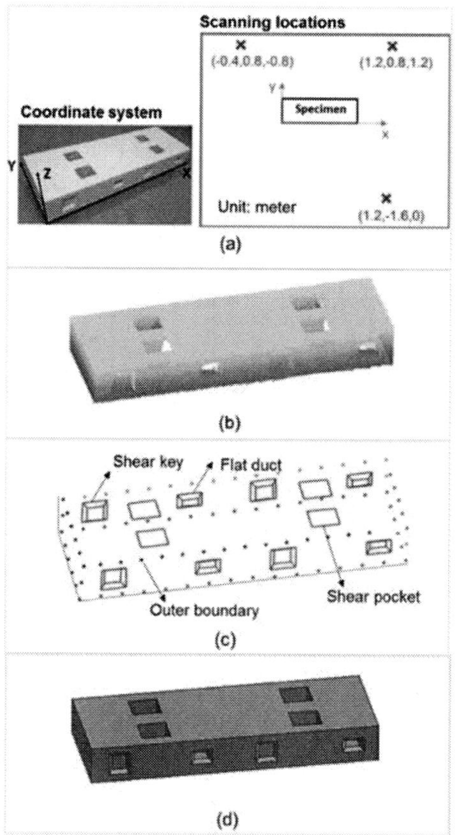

Figure 10. Applying the proposed technique on the test specimen. (a) The XYZ coordinates of the optimal set of scanning locations. (b) The scan data after data pre-processing. (c) The extracted geometries. (d) The reconstructed as-built BIM of the test specimen.

First, in the scan planning, the angular resolution was set as 0.036° and the spacing threshold S_T was set as 2 mm. Then the minimum number of scans was solved as 3 and the XYZ coordinates (unit: meter) of the optimized set of scanning locations are (-0.4, 0.8, -0.8), (1.2, 0.8, 1.2), and (1.2, -1.6, 0), as shown in Figure 10(a). Since the scanning locations are solved from an exhaustive search among a set of possible scanning locations, the solution is approximately optimized. In this study, a FARO Focus 3D terrestrial laser scanner was used to acquire the scan data. Then, by data pre-processing, the scan data from all the three scans were filtered and registered, and the obtained scan data are shown in Figure 10(b). Afterwards, the geometries of the test specimen, including shear pockets, flat ducts, shear keys, and the outer boundaries, were extracted from the registered scan data, as shown in Figure 10(c). Lastly, based on the extracted geometries, the as-built BIM of the specimen was created with the IFC data model, as shown in Figure 10(d).

4.3 Accuracy of the as-built BIM

To examine the accuracy of the created as-built BIM, the dimensions of the specimen in the as-built BIM were compared to the actual dimensions, which were measured using a measuring tape with the smallest division of 1 mm. The dimensions selected for comparison include (1) the dimensions of the structural features, i.e., shear pockets, flat ducts, and shear keys, (2) the dimensions of the outer boundaries of the specimen, and (3) the locations of the structural features, e.g., the location of a shear pocket is measured as the distance from the center of the shear pocket to the outer boundaries of the top surface. The comparison results show that, for the above-mentioned three categories of dimensions, the average discrepancies between the as-built BIM and the actual dimensions are 0.8 mm, 1.6 mm, and 1.3 mm, respectively. Considering that the tolerance values for constructions are usually more than 5 mm, the as-built BIM can provide accurate as-built dimensions for dimensional quality assessment.

5 Conclusions

To conduct quality assessment of precast concrete elements in an efficient and accurate manner and to facilitate the management of the quality assessment results, this study develops a technique for automatic reconstruction of the as-built BIM of precast concrete elements from laser scan data. Particularly, this study focuses on the precast concrete bridge deck panels, which have structural features with complex geometries, including shear pockets, shear keys, and flat ducts. The proposed technique first determines the minimum number of scans and the optimal locations of scanners. Then, data pre-processing removes noisy data and registers scan data using a plane-based approach. Third, the geometric information of the element is extracted from the scan data. Finally, the as-built BIM is created based on the extracted geometric information using the IFC data model. A small-scale test specimen was manufactured and the proposed technique was applied to the specimen to validate the proposed technique. After the as-built BIM of the specimen was reconstructed, the dimensions of the specimen stored in the as-built BIM were compared to the actual dimensions, showing an average discrepancy of around 1 mm. Therefore, it is demonstrated that the proposed technique can create the as-built BIM of precast concrete elements efficiently and accurately.

However, this study still has the following limitations. (1) Though the proposed technique is validated through small-scale specimen, its applicability needs to be further examined on real precast concrete elements. (2) This study currently focuses on a particular type of precast concrete elements, i.e., precast concrete bridge deck panels, thereby future research is needed to extend the proposed technique to other types of elements.

Acknowledgement

This research was supported by a grant (13SCIPA01) from Smart Civil Infrastructure Research Program funded by Ministry of Land, Infrastructure and Transport (MOLIT) of Korea government and Korea Agency for Infrastructure Technology Advancement (KAIA).

References

1. Glass, J., B.P.C. Federation, and B.C. Association, *The future for precast concrete in low-rise housing*. 2000: British Precast Concrete Federation Leicester.
2. Tadros, M.K. and M.C. Baishya, *Rapid replacement of bridge decks*. Vol. 407. 1998: Transportation Research Board.
3. Phares, B. M., Washer, G. A., Rolander, D. D., Graybeal, B. A., & Moore, M. (2004). Routine highway bridge inspection condition documentation accuracy and reliability. *Journal of Bridge Engineering*, *9*(4), 403-413.
4. Tang, P., Huber, D., Akinci, B., Lipman, R., & Lytle, A. (2010). Automatic reconstruction of as-built building information models from

laser-scanned point clouds: A review of related techniques. *Automation in construction,19*(7), 829-843.

5. Xiong, X., Adan, A., Akinci, B., & Huber, D. (2013). Automatic creation of semantically rich 3D building models from laser scanner data. *Automation in Construction, 31*, 325-337.
6. El-Omari, S., & Moselhi, O. (2008). Integrating 3D laser scanning and photogrammetry for progress measurement of construction work. *Automation in construction, 18*(1), 1-9.
7. Turkan, Y., Bosche, F., Haas, C. T., & Haas, R. (2012). Automated progress tracking using 4D schedule and 3D sensing technologies. *Automation in Construction, 22*, 414-421.
8. Kim, M. K., Sohn, H., & Chang, C. C. (2014). Automated dimensional quality assessment of precast concrete panels using terrestrial laser scanning.*Automation in Construction, 45*, 163-177.
9. Bosché, F. (2010). Automated recognition of 3D CAD model objects in laser scans and calculation of as-built dimensions for dimensional compliance control in construction. *Advanced engineering informatics, 24*(1), 107-118.
10. Makadia, A., Patterson, A., & Daniilidis, K. (2006, June). Fully automatic registration of 3D point clouds. In *Computer Vision and Pattern Recognition, 2006 IEEE Computer Society Conference on* (Vol. 1, pp. 1297-1304). IEEE.
11. Bosché, F. (2012). Plane-based registration of construction laser scans with 3D/4D building models. *Advanced Engineering Informatics, 26*(1), 90-102.
12. Dold, C., & Brenner, C. (2006). Registration of terrestrial laser scanning data using planar patches and image data. *International Archives of Photogrammetry, Remote Sensing and Spatial Information Sciences, 36*(5), 78-83.
13. Tarsha-Kurdi, F., Landes, T., & Grussenmeyer, P. (2007, September). Hough-transform and extended ransac algorithms for automatic detection of 3d building roof planes from lidar data. In *Proceedings of the ISPRS Workshop on Laser Scanning* (Vol. 36, pp. 407-412).
14. Bosché, F., Ahmed, M., Turkan, Y., Haas, C. T., & Haas, R. (2015). The value of integrating Scan-to-BIM and Scan-vs-BIM techniques for construction monitoring using laser scanning and BIM: The case of cylindrical MEP components. *Automation in Construction, 49*, 201-213.
15. Pu, S., & Vosselman, G. (2009). Knowledge based reconstruction of building models from terrestrial laser scanning data. *ISPRS Journal of Photogrammetry and Remote Sensing, 64*(6), 575-584.
16. Vosselman, G., & Dijkman, S. (2001). 3D building model reconstruction from point clouds and ground plans. *International archives of photogrammetry remote sensing and spatial information sciences, 34*(3/W4), 37-44.
17. Hebert, M., & Krotkov, E. (1992). 3D measurements from imaging laser radars: how good are they?. *Image and Vision Computing, 10*(3), 170-178.
18. Ester, M., Kriegel, H. P., Sander, J., & Xu, X. (1996, August). A density-based algorithm for discovering clusters in large spatial databases with noise. In *Kdd* (Vol. 96, No. 34, pp. 226-231).
19. Wang, Q., Kim, M. K., Cheng, J. C., & Sohn, H. (2016). Automated quality assessment of precast concrete elements with geometry irregularities using terrestrial laser scanning. *Automation in Construction*.
20. Fischler, M. A., & Bolles, R. C. (1981). Random sample consensus: a paradigm for model fitting with applications to image analysis and automated cartography. *Communications of the ACM, 24*(6), 381-395.
21. Chen, Y., & Medioni, G. (1992). Object modelling by registration of multiple range images. *Image and vision computing, 10*(3), 145-155.

Active Control of a Pan-Tilt-Zoom Camera for Vision-Based Monitoring of Equipment in Construction and Surface Mining Jobsites

E. Rezazadeh Azar[a]

[a] Department of Civil Engineering, Lakehead University, Canada
E-mail: eazar@lakeheadu.ca

Abstract –

Automated monitoring systems have proven to be effective in improving the productivity of equipment-intensive operations in the construction and mining sectors. Vision-based systems are the most recent methods employed to detect, track, and monitor construction equipment. The currently developed systems, however, analyze video frames captured by a stationary camera, which dramatically limits their coverage area and requires manual adjustment of the viewfinder. This research paper introduces methods to proactively steer a pan-tilt-zoom camera to localize, track, and identify objects of interest in construction jobsites. This automated camera control system uses a number of image and video processing algorithms to detect objects and estimate their trajectory and velocity, and then uses the extracted information to set the camera movement parameters, including direction and magnitude. The experimental results of this system showed promising performance for equipment monitoring in construction and mining jobsites.

Keywords –

Computer vision; equipment monitoring; pan-tilt-zoom camera; automated control

1 Introduction

There is a growing trend in the construction and mining industries to use cameras to monitor jobsites [1], [2], [3]. The resulting videos, however, are mainly processed manually, which is cumbersome, labor-intensive, and hinders the potential use of these rich resources. Therefore, there has been considerable interest in the construction research community to apply computer vision methods to automate video analysis processes. In particular, research efforts have focused on detection, localization, and tracking construction equipment for productivity estimation, safety, and fleet management purposes. For example, projects investigated equipment detection [2], [4], [5], [6], tracking [3], [7], [8], [9], activity interpretation [3], [7], [10], and safety assessment [17], [18]. These investigations mentioned some advantages to using vision-based monitoring systems, including low cost and the non-intrusive nature of this monitoring approach. Despite all the achievements, a number of shortcomings exist in the developed vision-based methods that prevents their practical applications. These limitations include those described below.

Low level of reliability: the rates of false positives and false negatives are still high compared to the radio-based sensing systems.

Line of sight and occlusion: vision-based systems require clear sightlines for the successful detection and tracking of operations, and obstructions can hinder the performance of the system. However, methods have been developed to improve the detection and tracking of semi-obscured targets.

Limited coverage: the developed systems use videos captured by stationary cameras. This limits the coverage area and will be problematic for large construction and open-pit mining fields. In those types of settings, site personnel should be able to set the camera's view as the work progresses.

This research paper presents some methods for automated steering of a pan-tilt-zoom (PTZ) camera to overcome the coverage limitations of stationary cameras. In particular, it introduces intelligent steering frameworks for three main purposes: First, it demonstrates an algorithm to automatically scan the site by steering the camera and then finding active operations. Second, an algorithm is introduced to continue tracking an object when it is about to exit from the static view. Third, it discusses methods for identification and localization of equipment. All of these methods use image and video processing techniques to extract spatiotemporal data from 2D video frames, which are then used to steer the PTZ unit. Finally, opportunities and future research directions are discussed.

2 Research Methods

This section introduces methods developed to steer a PTZ camera for various purposes, such as resource tracking and automated operation localization in jobsites. These frameworks employ combinations of computer vision algorithms, such as image stitching, object recognition and feature tracking; active control of a PTZ camera; and principles of camera geometry to achieve the expected performance. The following subsections explain the technical approaches used in these systems.

2.1 Automated Search for Active Operations

Stationary cameras cover a limited area, which is especially problematic for large construction jobsites. For full coverage, the viewfinder of the camera must be adjusted manually as the work progresses. However, this issue could be tackled using a self-steering camera that searches the jobsite for an active operation. The first step of this process is to determine the boundaries of the camera's coverage area. In this step, the position of the camera, including its 3D coordinates and the initial yaw and pitch angles of the camera mount, is mapped to the jobsite plan. Then the boundaries of the site are considered to avoid scanning areas outside of the site boundaries (see Figure 1). In this arrangement, the optical axis is not allowed pass the boundaries of the jobsite.

Figure 1. Mapping the camera in the site plan

The next step is the systematic search within the assigned boundary for active operations. An image stitching technique was used to create a panorama of the jobsite from several overlapping images. In this approach, the system changes the yaw angle of the mounting unit in a certain step and captures a frame. Then it steers the camera to the next viewpoint, in which it has an overlapping field of view (FOV) with the previous frame.

After the required frames are captured, invariant local features are used to find matches between frames and stitch those together [11]. This algorithm finds and matches SIFT features [12] in frames, and then employs a probabilistic method to verify the matches. Afterwards, it solves and refines camera parameters, compensates for errors, and blends the images (Figure 2. a).

The next step in this process is to search the panorama for construction equipment (Figure 2 b). Methods for recognizing rigid [4] and articulated equipment (such as excavators) [5] have been discussed. If detectors locate potential targets in the panorama, the system steers the PTZ camera to the located targets to verify detection. Sometimes the detectors provide false positives, so spatiotemporal information is used to help eliminate them. Some consecutive frames are searched for the target, then the spatiotemporal data of the detections are compared with the movement pattern of active equipment to reject false positives [5] [13].

When the detectors find different types of equipment, stationary plants have priority over highly mobile equipment in identifying jobsite tasks. For example, an excavator is an indicator for an excavation/loading operation, whereas a dump truck could be seen in different parts of a construction site.

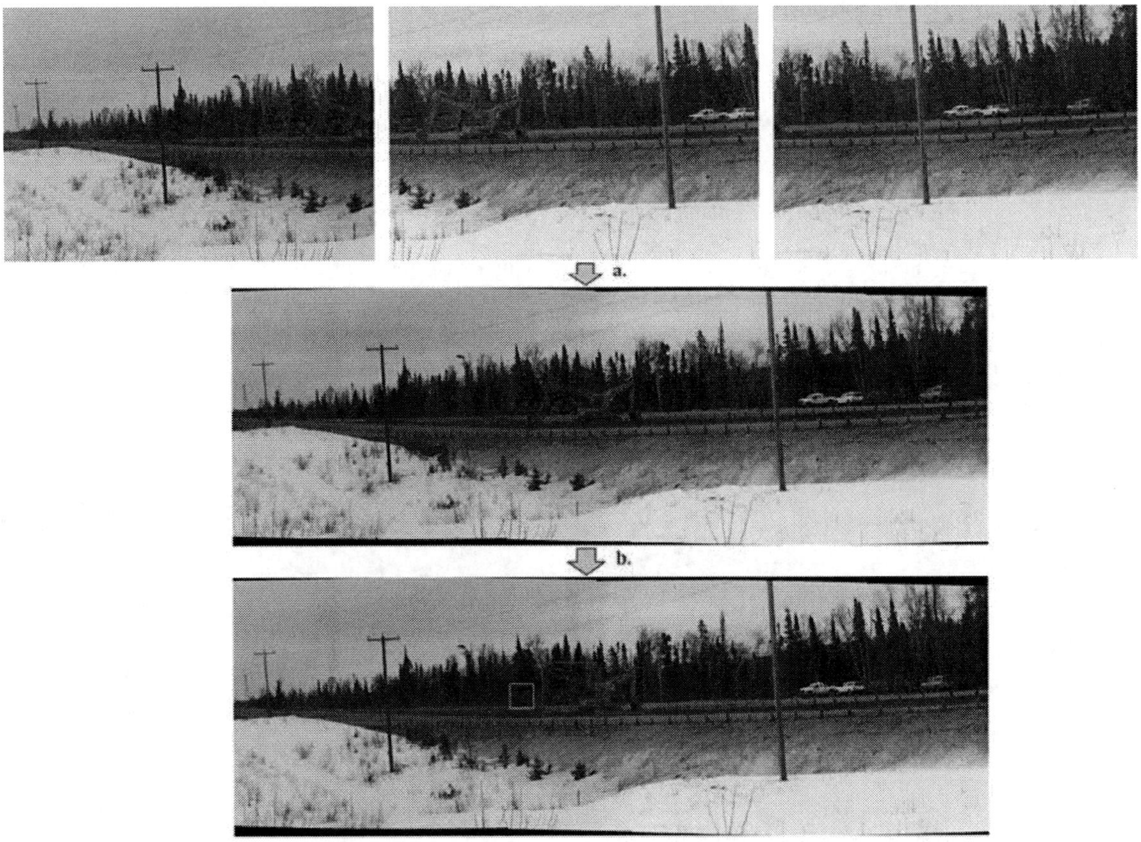

Figure 2. Steering of a PTZ camera to locate operations: a) stitching of images to create a panorama, b) object detection

2.2 Active Camera Control to Track Equipment

The tracking performance of the existing vision-based systems is limited to the camera's FOV. But some applications, such as security and safety monitoring systems, might need to track a target beyond the FOV of a stationary camera. Thus, it is useful to benefit from the physical panning and tilting capabilities of a PTZ camera to extend its tracking range.

A key issue for continuous tracking of construction resources is to estimate their motion vector, which includes velocity and trajectory. The velocity of the object is measured in the frame coordinate system, thus it would be expressed in terms of pixels per second. Then the motion vector of the target is used to predict the upcoming 2D coordinate of the target. The trajectory and velocity of the detected objects could be estimated using a tracking algorithm. The KLT feature tracker [14] was utilized to track certain features of the target object. The KLT is a differential method that uses spatial intensity data to search for the best match in the next frame. Successful motion estimation requires a number of specific interest points that are mathematically well-founded. Corners (the intersection of the edges) are suitable visual features for tracking, and Harris corner detection [15] was used to detect a number of corners for tracking (see Figure 3.b). Tracking each of these features yields a motion vector (see Figure 3.c). The overall motion vector of the target is calculated as the average of the valid motion vectors. The random sample consensus method (RANSAC) was used to remove the outlier vectors, vectors that indicate a direction inconsistent with that of most of the other optical vectors. Since construction equipment provides slow-moving targets in videos, the system estimates the upcoming coordinates of a piece of machinery using its current coordinates in the frame and its speed and trajectory. Then the system should steer the camera mount to maintain the target in the center of the frame. Given the pixel distance between current and predicted positions of the target, it is possible to determine also the pan and tilt angles to reach the projected position. This could be calculated using the horizontal and vertical angles of camera views. These angles are determined based on the size of the sensor and focal length of the camera's lens as presented in Equation (1):

$$\alpha = 2arctan\frac{d}{2f} \quad (1)$$

α is the angle of view (horizontal, vertical, or diagonal)

d is the length of the sensor in the direction measured (horizontal, vertical, or diagonal)

f is the focal length

Given the horizontal and vertical angles of camera views, the required displacement of the frame's center, and frame dimensions in pixels, the camera steering pan and tilt angles are calculated and executed (see Figure 3.d).

This algorithm, however, could consider the speed of the camera's panning and tilting system to achieve the best results. Most camera mounting systems are quite fast (more than 180°/Sec in pan and 60°/Sec in tilt movement) and don't cause any latency.

a. b. d. c.

Figure 3. Camera steering process: a) detection of the target, b) generation of suitable features to track, c) optical flow estimation, d) panning of the camera to a new position

2.3 Zooming for Equipment Identification

Sometimes the zoom of the camera needs to be changed to obtain a suitable resolution of the targets. For example, the visual markers attached to pieces of equipment for identification and localization purposes should have a specific minimum size in the frame (in pixels) to be robustly detected and provide useful localization results. On the other hand, sometimes it is necessary to have a wide-angle view of all of the processes for tracking and monitoring purposes.

One of the main shortcomings of the model-based object recognition methods is that they cannot differentiate individual pieces of equipment within a same class. For example, they are not able to identify individual dump trucks within a fleet of similar machines. Identification of individual equipment is an important parameter for productivity estimation. The solution for this problem is to attach unique visual markers to every piece of equipment.

For detection of visual markers, such as AprilTag [19], attached on the equipment, the size of the marker in the frame (in pixels) is the key factor. Therefore, an automated zooming feature was developed to control the focal length of the lens. In particular, this module was developed to reliably detect markers attached to the equipment. Experiments revealed that an AprilTag marker should be at least 18x18 pixels to be reliably detected (>98%) [16].

An automated zoom control algorithm was developed to change the focal length of the camera's lens to obtain the expected marker resolution. First, the linear relationship between the focal length and the magnification factor of the camera lens was determined using a set of experiments. For this purpose, a specific target was captured in frames from the same viewpoint but with altered focal lengths. Then the linear relationship was established by measuring the changes of the target size (in pixels) versus the focal length. Second, the algorithm calculates the magnification factor required to obtain expected marker size (in pixels) based on the size of the detected equipment (in pixels), and sets the focal length of the lens to achieve the expected resolution. This algorithm calculates the magnification factor using the ratio of the actual length of the visual tag to the actual length of the equipment (e.g. 0.6m label and 7m dump truck length) and the expected marker size (e.g. 20x20 pixels). For example, a dump truck is detected with a size of 190x119 pixels in a frame, thus the size of the marker is about 16x16 pixels in the frame (190x(0.6/7) = 16). Assuming that the expected marker resolution is 20x20 pixels, the required magnification factor would be 20/16=1.25.

In addition, to avoid useless zooming, the algorithm checks whether the target will remain in the view after zooming. This test is done using a scale affine transformation, in which the coordinates of the object after magnification are calculated. The new coordinates indicate whether the target will appear within the zoomed frame.

Since the camera lens zooms toward the center of the view, the center of the frame is set as the origin of the 2D coordinate system. Then the coordinates of the detected objects are multiplied by the calculated magnification factor to check whether the target will appear in the frame after zooming. An autofocus process (usually provided by the camera manufacturer) is employed to correct the usual blurriness resulting from change of the focal length.

The zooming module was used to identify dump trucks and excavators that were labelled with unique fiducial markers, and the reported results are presented in Table 1. Due to the robust performance of the marker recognition algorithm, no false positives were observed [16].

Table 1. Performance of the equipment identification system using fiducial markers and active zooming

Experiment	No. appeared machines	No. identified machines	Precision	Identification rate
Dump truck	32	27	100%	84.4%
Excavator	18	16	100%	88.9%

2.4 Active Camera Control to Localize Equipment

Automated camera steering could be used to estimate 3D location of the targets. In this approach, the PTZ camera is the origin of a spherical coordinate system. The three parameters—the radial distance of the target from the camera, its tilt (polar) angle measured from the zenith direction, and pan (azimuthal) angle measured by its projection on the horizon plane from a fixed reference direction—are required to estimate the 3D coordinates of an object with respect to the PTZ camera. The pan and tilt angles are provided by the motorized camera mounting systems, thus the only parameter requiring estimation methods is "the radial distance"—the distance of the target to the camera.

Depth estimation using a single camera could be resolved using two main approaches: one based on the geometric principles of a pinhole camera and one based on a pinhole camera model. These methods also rely on detection of a fiducial marker attached to the equipment. So the system needs to zoom on the target for localization of the target.

The first step in both approaches is to calibrate the

camera to find the focal length(s) and principal point of the camera. Then, the vision system searches the video frames for the objects of interest, i.e. construction equipment. The object recognition methods were discussed in previous works [4], [5]. As soon as a target is spotted in the frame, the system steers the camera to capture the target in the center of the frame to achieve the best depth estimation result. The process of the camera steering to capture the target in the center of the frame is similar to the method described in section. 2.2. Then the system zooms (by changing the focal length) on the target to detect the marker attached on the equipment. The following subsections describe the two depth estimation approaches.

2.4.1 Camera Geometry

The first technique is to use the geometric principle of a pinhole camera. In Equation (2):

$$d = \frac{f \times rho \times imh}{rhi \times seh} \tag{2}$$

d is distance to the camera in mm
f is the focal length in mm
rho is the real height of the object in mm
imh is the image height in pixels
rhi is the object height in image in pixels
seh is the camera's sensor height in mm

This method achieves the best result when the target is parallel to the image plane, and it is less successful when the planar target is rotated with respect to the image plane.

2.4.2 Pinhole Camera Model

This approach is based on the pinhole camera model, in which the intrinsic and extrinsic parameters of the camera are used to estimate the distance from a planar target to the camera (see Equation (3)). The intrinsic matrix includes the camera's focal length, principal point, and distortion parameters, which is extracted using a well-known camera calibration process.

$$sm = A[R|T]M \tag{3}$$

s is a scale factor. This coordinate system is a homogenous system and scale invariant.

m is a 3x1 matrix of the pixel coordinates of the projected point in the image. The third element (z) is equal to 1.

A is the camera's intrinsic matrix.

[R|T] is the 3x4 extrinsic matrix. R is a 3x3 rotation matrix demonstrating the orientation of the camera with respect to the planar target, and T is a 3x1 translation matrix which denotes the camera's position.

M is a 4x1 matrix which includes the 3D world coordinates of the point. The fourth element is equal to 1. To simplify the calculations, a point on the planar object is usually set as the origin of the world coordinate.

This approach handles rotation of the targets theoretically, by considering the rotation of the planar target. In this method, a corner of the target (i.e., the visual marker) will be the origin of the world coordinate and the target plane is set as the world's x-y plane; given the dimensions of the target, the world coordinates of the rest of three corners are calculated. Then four equations (based on Equation (3)) are established for the four corners, in which all parts except [R|T] matrix, including m, A, and M, are known. Using a set of linear equations, the [R|T] matrix is extracted [20].

Afterwards, the [R|T] matrix is used in Equation (3) to estimate the 3D coordinates of the camera origin in the 3D coordinate system (where its origin is on the target). Lastly, the norm of the camera's 3D coordinate yields the distance between the target and camera.

2.4.3 Experimental Results

This localization framework was evaluated using a number of real-time videos of different earthmoving equipment, including excavators and off-highway and urban dump trucks. A total station was set up adjacent to the PTZ camera for accurate measurement of the location of the targets. These videos included 67 test samples, in which the distance of the test subjects to the camera varied from 30.44 to 100.50 meters. A few samples are presented in Figure 4. Table 2 provides the performance of localization methods. As it is presented, the results are relatively close, but the pinhole camera model performed slightly better. Comparison of the results demonstrated that the performance of the pinhole camera model approach was better than the camera geometry method in localization of the targets that were oriented with respect to the image plane.

Table 2. Performance of the localization algorithms

Localization method	Average localization error (m)	Standard deviation of error (m)	Maximum error (m)	Upper limit at 95% confidence level (m)
Camera geometry	1.389	0.992	4.18	1.631
Pinhole camera model	1.346	1.031	4.11	1.597

Figure 4. Samples of marker localization

2.5 Conclusion

Vision-based systems are among the sensing technologies that are increasingly being tested to monitor the productivity and safety of equipment-intensive operations in the construction and surface mining industries. The developed systems, however, analyze videos captured by a motionless camera, which drastically limits their coverage area. This research paper introduces some algorithms to automatically control pan, tilt, and zoom features of a PTZ camera. These camera steering algorithms command the PTZ unit based on the spatiotemporal data extracted from the video frames. The methods discussed include automated scanning of the site for active operations, continuous tracking, and identification and localization of the labelled equipment.

The discussed methods, however, are able to control only a single camera, which is not sufficient for large construction and mining fields. Future research will investigate autonomous control of a network of connected PTZ cameras to monitor large jobsites.

References

[1] Yang J., Park M.W., Vela P.A. and Golparvar-Fard M. Construction performance monitoring via still images, time-lapse photos, and video streams: Now, tomorrow, and the future. *Advanced Engineering Informatics*, 29(2):211-224, 2015.

[2] Memarzadeh M., Golparvar-Fard M. and Niebles J. C. Automated 2D detection of construction equipment and workers from site video streams using histograms of oriented gradients and colors. *Automation in Construction*, 32:24–37, 2013.

[3] Gong J. and Caldas C.H. An object recognition, tracking, and contextual reasoning-based video interpretation method for rapid productivity analysis of construction operations. *Automation in Construction*, 20(8):1211–1226, 2011.

[4] Rezazadeh Azar E., and McCabe B. Automated visual recognition of dump trucks in construction videos. *Journal of Computing in Civil Engineering*, 26(6):769-781, 2012.

[5] Rezazadeh Azar E. and McCabe B. Part based model and spatial-temporal reasoning to recognize hydraulic excavators in construction images and videos. *Automation in Construction*, 24:194-202, 2012.

[6] Chi S. and Caldas C.H. Automated Object Identification Using Optical Video Cameras on Construction Sites. *Journal of Computer-Aided Civil and Infrastructure Engineering*, 26(5):368–380, 2011.

[7] Rezazadeh Azar E., Dickinson S. and McCabe B. Server-Customer Interaction Tracker: Computer Vision–Based System to Estimate Dirt-Loading Cycles. *Journal of Construction Engineering and Management*, 139(7):785–794, 2013.

[8] Park M., Koch C. and Brilakis I. Three-Dimensional Tracking of Construction Resources Using an On-Site Camera System. *Journal of Computing in Civil Engineering*, 26(4):541–549, 2012.
[9] Park M.W., Makhmalbaf A. and Brilakis I. Comparative study of vision tracking methods for tracking of construction site resources. *Automation in Construction*, 20(7):905-915, 2011.
[10] Golparvar-Fard M., Heydarian A. and, Niebles J.C. Vision-based action recognition of earthmoving equipment using spatio-temporal features and support vector machine classifiers. *Advanced Engineering Informatics*, 27(4):652–663, 2013.
[11] Brown M. and Lowe D. G. Automatic panoramic image stitching using invariant features. *International Journal of Computer Vision*, 74(1):59-73, 2007.
[12] Lowe D. Distinctive image features from scale-invariant keypoints. *International Journal of Computer Vision*, 60(2):91–110, 2004.
[13] Rezazadeh Azar E. and McCabe B. A visual sensing approach to estimate material hauling cycles in heavy construction and surface mining jobsites. In *Proceedings of the ISARC 2013 conference*, paper 229, Montreal, Canada, 2013.
[14] Tomasi C. and Kanade T. Detection and tracking of point features. *Technical Report CMU-CS-91-132*, Carnegie Mellon University, 1991.
[15] Harris, C. and Stephens, M. A combined corner and edge detector. In *Proceedings of the Alvey vision conference*, 15, page 50, 1988.
[16] Rezazadeh Azar E. Construction Equipment Identification Using Marker-Based Recognition and an Active Zoom Camera. *Journal of Computing in Civil Engineering*, 04015033, 2015.
[17] Chi S. and Caldas C. H. Image-based safety assessment: Automated spatial safety risk identification of earthmoving and surface mining activities. *Journal of Construction Engineering and Management*, 138(3):341–351, 2012.
[18] Kim H., Kim K. and Kim H. Vision-Based Object-Centric Safety Assessment Using Fuzzy Inference: Monitoring Struck-By Accidents with Moving Objects. *Journal of Computing in Civil Engineering*, 04015075, 2015.
[19] Olson E. AprilTag: a robust and flexible visual fiducial system. In *Proceedings of IEEE International Conference on Robotics and Automation (ICRA'11)*, pages 3400–3407, Shanghai, China, 2011.
[20] Zhang Z. A flexible new technique for camera calibration. *Pattern Analysis and Machine Intelligence, IEEE Transactions on*, 22(11):1330-1334, 2000.

A Face Recognition System for Automated Door Opening with parallel Health Status Validation Using the Kinect v2

A. Ogawa[a], A. Mita[a], C. Georgoulas[b] and T. Bock[b]

[a]Department of System Design Engineering, Keio University, Kanagawa, Japan
[b]Chair for Building Realization and Robotics, Technical University of Munich, Germany
E-mail: ami_ogawa@keio.jp, mita@keio.jp, christos.georgoulas@tum.de, thomas.bock@tum.de

Abstract

Nowadays it is said that extending the healthy life expectancy is important in the aging society target group. Of course it is important for every single-person household to be regularly checked upon their safety, but especially for elderly people the importance is higher at the point of healthy living, because this particular target group comprises a higher accidents rate. Therefore, the authors here propose the design and implementation of an unobtrusive vision system for single-person households, particularly for elderly people, based on the Microsoft Kinect v2. The entrance area of the home environment is here considered, where the monitoring system is activated upon while the user approaches, in order to first detect and calculate the user's fatigue levels, as well as to secondly identify the user by facial recognition, in order to actuate a door opening mechanism. The proposed system algorithm is divided into two distinct phases. The first phase concerns the acquisition of the user knee position while walking up a staircase in order to calculate fatigue. The second phase comprises the face recognition based door opening method. The proposed system has been successfully tested, and it could undoubtedly comprise an unobtrusive health status validation and automated door opening solution for elderly people in single-person households.

Keywords –

Computer Vision, Kinect v2, Face Recognition, Fatigue Detection, Ambient Assisted Living

1 Introduction

Nowadays the number of single-person households is increasing and it is expected to rapidly grow within the following years [1,2]. In Norway Finland and Denmark the single person households already reached more than 35%. In Japan this percentage is currently at 33%. One of the concerning things is that the single-person households have a higher risk of exacerbation of the accidents which happen in living spaces than the others. It is because in the case of the single-person households, normally there are no other people present in their living spaces, so they cannot retrieve any assistance or care in case of an accident.

This is a seriously concern, especially for elderly people. Elderly people are affected heavily because they have less ability of recovering. For example, if they fall down, many of them might not be able to walk and this could psychologically affect them. Therefore, an unobtrusive monitoring system could increase their safety.

To decrease the risk of accidents in living space for single-person households, the detection of the resident's physical and mental information in real-time is required, in order to be able to recognize risk situations. Considering the time flow of the monitoring system in the living space, the sensing is activated when the resident enters the house. Therefore, it is more effective to position the sensing system at the entrance part so that we can retrieve as early as possible the resident's initial condition.

The authors eventually aimed to predict the resident's physical and mental fatigue. As the first step of it, in this research, it was suggested the combined system which consists of the knee position acquisition of stair walking and face recognition based door opening. The knee position acquisition is expected to be extended to get the feet joint parameters for physical fatigue evaluation. And face recognition is expected to be a tool for aspect prediction for mental fatigue evaluation. In this study, we used the Kinect v2 which provides RGB-color, IR-depth and IR images with 1920x1080, 512x424 and 512x424 spatial pixel resolution respectively, and can detect the human body and track 25 joints without putting any markers [3]. The proposed system has been implemented and realized in the experimental ambient assisted living laboratory of the authors at the Technical University of Munich.

2 Background

There are various attempts addressing smart home environments embedded with many sensors and actuators inside walls, floors, ceilings, etc, which retrieve information about the environment accumulate into a sort of live log, to optimally control smart devices, and to make residents comfortable [4-7]. However, all these attempts also impose limitations due to the necessary and expensive reconstruction costs in order to install and embed all sensors and actuators. In some situations large scale reconstruction is required or even to move to a brand new smart home construction. Moreover, even if we can live in such smart house once, we have to change the sensors and devices often, because nowadays the technology advances with higher rates compared to a buildings lifecycle. Thus, instead of focusing on a smart wall, or smart floor, or smart ceiling approach, the authors chose the so-called "Terminal" approach[8].

The "Terminal" comprises a modular furniture containing specific functions and services depending on the room or life center it is installed. It can be said a kind of furniture with several sensors in simple words, so that it is easy to install into existing buildings. This concept saves us from renovating or reconstructing the whole house. It is designed and manufactured by combining standard products to save the cost of production. Moreover due to the fact that it is "modular", it can straightforwardly be adapted to any kind of house. Every house has its unique features and arrangement. Additionally every user might require a different level of intelligence (considering the terminal sensors and actuators), so modularity here allows for user customization.

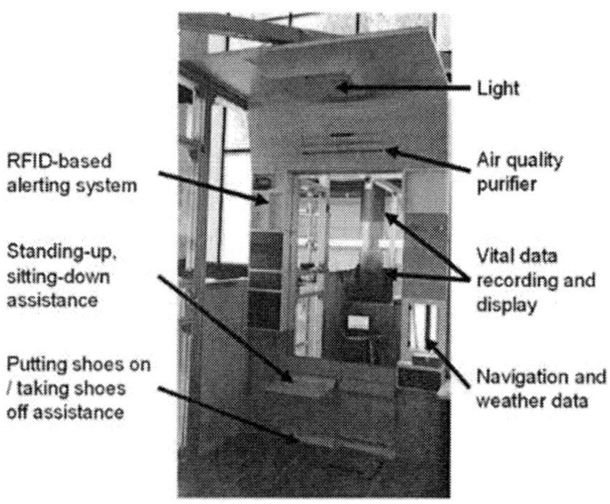

Figure 1. The entrance "Terminal", LISA smart Wall [8]

An example of such terminal can be seen in Figure 1, which was developed under the research project LISA, investigating the possibilities to embed mechatronic, assistive functions and services into compact wall "terminal" elements thereby enabling autonomous and independent living upon performing Activities of Daily Living (ADLs) by means of generated structured environments and robotic micro-rooms (RmRs) 8. The proposed system in this paper comprises an add-on for the terminal shown in Figure 1, since its output can be fed to the terminal onboard screens, to display the resulting user fatigue condition at the time, and provide with necessary advices. Also, it is possible to control the lighting and air conditioning devices on the L.I.S.A. wall based on the results.

3 Proposed Implementation

The proposed system has been implemented and realized in the experimental ambient assisted living laboratory of the authors at the Technical University of Munich. A full scale home environment has been built, comprising entrance space, living room, bed room and bathroom, kitchen. There are eight steps towards the entrance door, and the entrance door is simply composed of steel frames (Figure 2). The Kinect was used to acquire RGB, depth, and also IR consecutive images as an animated film. A 3D models was created by combining these data. Moreover, Kinect v2 (Figure 3) has a skeleton tracking function, face tracking function, and microphone arrays. In the proposed implementation, the skeleton tracking and depth data were used for knee position acquisition, and face tracking and IR images for face recognition respectively (Figure 4).

Figure 2. Experimental ambient assisted living laboratory

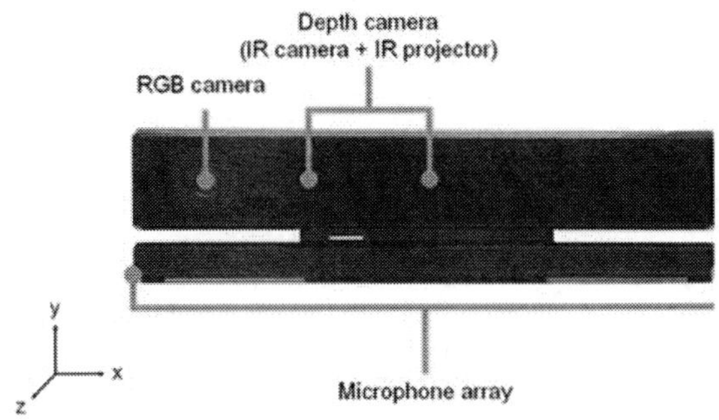

Figure 3. MS Kinect v2

3.1 Proposed Knee Position Acquisition Method

This knee position acquisition while walking up the stairs was proposed as a prediction of the user fatigue level upon entering the home environment. It is meaningful to know the user condition upon arriving home. If the user fatigue level is high, the system can warn to notice upon the condition, in order to prevent any potential accidents due to it. Kinect v2 can detect up to 6 people's bodies. It can not only detect human body, but also mark the 25 different joint points. The available range of this function is limited in 4.5m from Kinect, but the entrance approach of experimental house was within the range, so the efficiency of the system was not questioned.

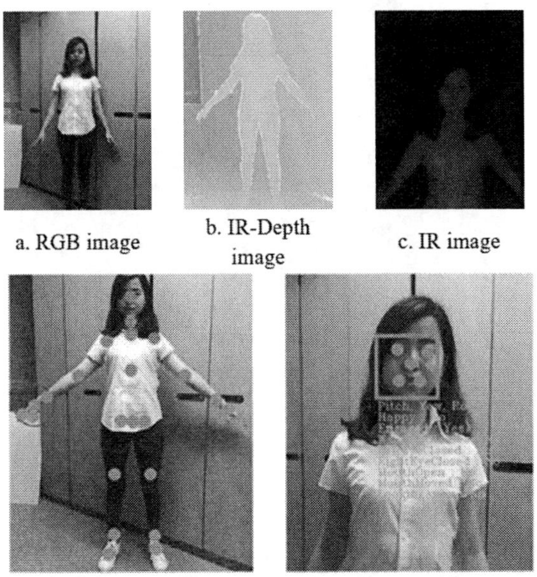

a. RGB image b. IR-Depth image c. IR image

d. Skeleton tracking e. Face tracking

Figure 4. The main function of Kinect v2

Firstly, the knee position while walking up the stairs was detected and retrieved using the skeleton tracking function directly. The visualized joint position marks are shown in Figure 5. According to the original knee position, it is evident that the mark does not reside with the exact knee position. Then knee position correction was considered based on the following two facts about the features of stair walking.

- Fact 1: The true knee position is always between the knee position and the ankle position which are given by Kinect v2 (Figure 6).
- Fact 2: The true knee position is always the forefront of forward movement in the part of the foot (Figure 7).

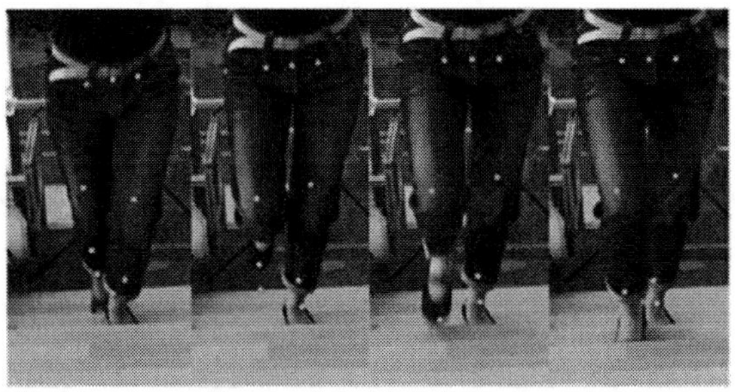

Figure 5. The original joint position of Kinect v2

Then the knee position correction algorithm was implemented which is based on the knee and anckle position provided by the Kinect. Firstly, the 3-D value of the knee and ankle position is obtained. Then the straight line from the knee to ankle position is calculated according to equations (1), (2) and (3). At this time, only y-z plane is considered:

$$y = mz + n \tag{1}$$

$$m = \frac{y_{Knee} - y_{Ankle}}{z_{Knee} - z_{Ankle}} \tag{2}$$

$$n = y_{Knee} - \frac{(y_{Knee} - y_{Ankle})z_{Knee}}{z_{Knee} - z_{Ankle}} \tag{3}$$

After the points along the line on the x-y plane are examined one by one. Then the distance between the depth, z value, of each point and the straight line which connects knee and ankle position in the y-z plane is calculated. The equation of the distance between line and each position is expressed in equation (4):

$$d = \frac{|y_i - mz_i - n|}{\sqrt{1 + m^2}} \tag{4}$$

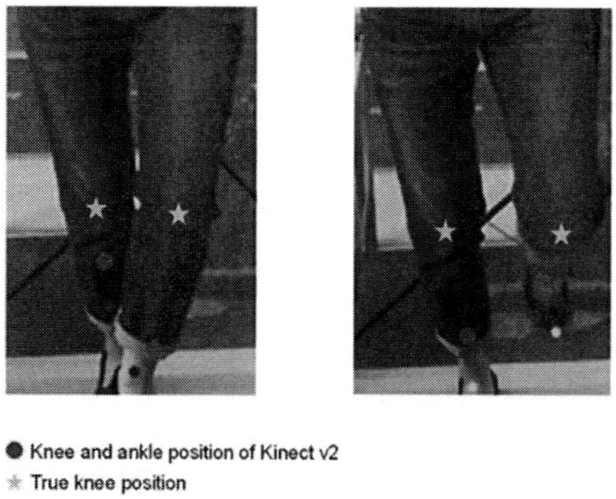

Figure 6. True knee position is always between the original position of knee and ankle

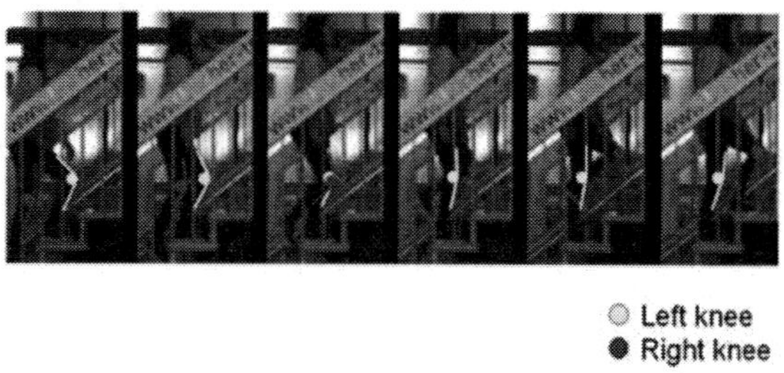

Figure 7. True knee position is always the fore-front of forward movement in the part of the foot

Finally, we can detect the point which has the maximum distance from the straight line as the true knee position (Figure 8).

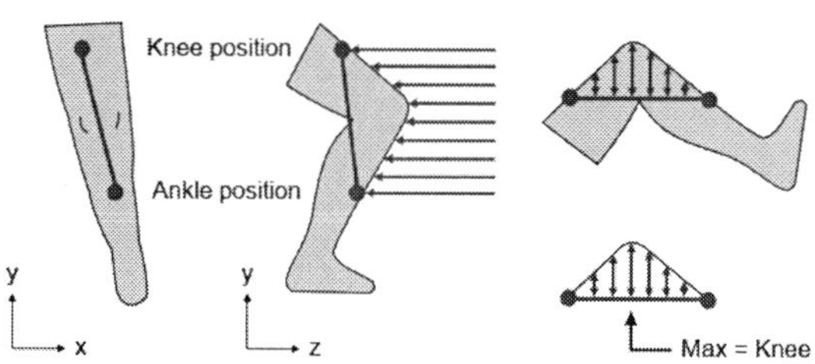

Figure 8. The proposed knee position detection algorithm

Applying this method, the accuracy of knee position was seriously improved. However, it is not always accurate because it is not sure that we can always get the knee and ankle position by Kinect v2. For this reason another 2 post-processing steps to accurately calculate the true knee position were developed. In case the calculated value was incorrect, i.e. extremely small, extremely large, or out of range, the value was rejected by filtering. In particular, the value will be rejected when the z value difference of calculating knee and spine base is more than 1000mm. Sometimes we cannot have the value of the specific position. Especially those kind of error is happening with hands and feet and the accuracy is unstable. On the other hand, the most stable joint position is spine base. Therefore, we used the difference with spine base. And the true knee will be the smallest value selected in the area of the rectangle which consist of knee and ankle x and y position of Kinect v2 instead (Figure 9).

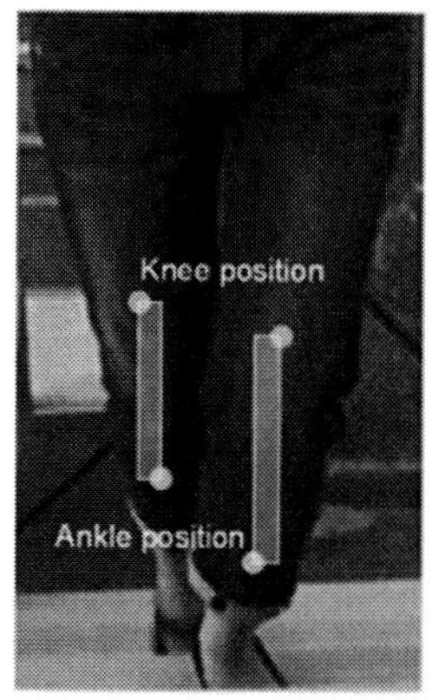

Figure 9. The range of knee detection for 1rst post processing step

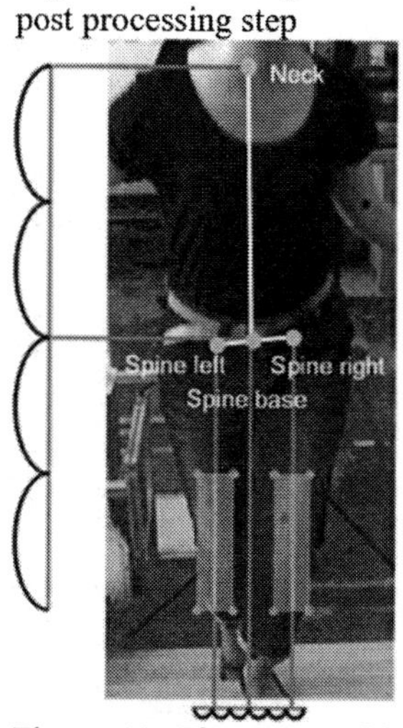

Figure 10. The range of knee detection of 2nd post processing correction

Finally the last post-processing step was applied in the case of no or incorrect value of knee and ankle position given by the Kinect v2. Same as the previous step, the value will be recognized as an incorrect value when the z value difference of the knee or ankle and spine base is more than 1000mm. The program is written to find the true knee position by using spine base, spine left, spine right and neck to define the range of the rectangle, and extract the minimum depth value inside of this range. In particular, the definition of the range of rectangle is shown in Figure 10. The y distance is defined as half of the y distance between the neck and the spine base. And the x distance is defined as the x distance between spine base and spine left or right centering on spine left or right. The total algorithm flow is shown in Figure 11.

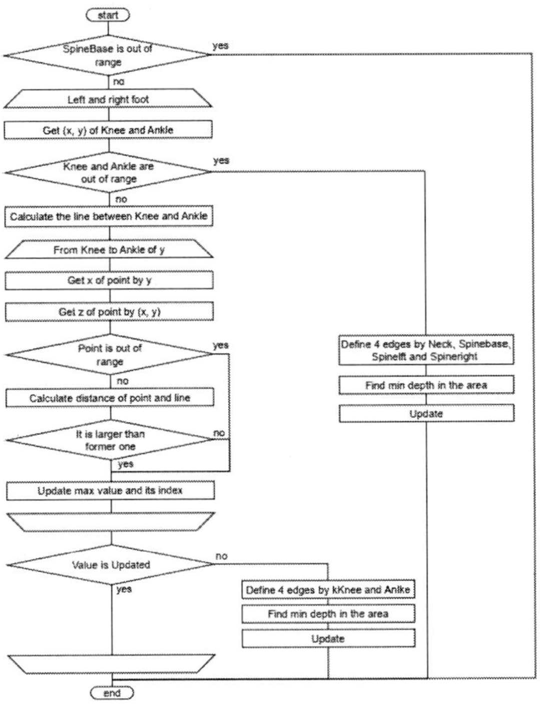

Figure 11. Flowchart of the proposed algorithm for knee position detection

3.2 Proposed Face Recognition Method

The face recognition based door opening system was placed on the entrance door of the experimental house. Thus, it will be held after the knee position acquisition. Normally depth data should be used for sensing because of the privacy protection in living space where is a very personal place and hard to accept putting cameras inside. However, this face recognition system is supposed to be used in front of the entrance, which means outside of the house just like an intercom, so that there is no need to consider about resident's privacy because outside of the house is no longer public space. Therefore, we can use any of the RBG, IR-depth or IR images. In the proposed approach only RGB and IR images were used for the face recognition function. Firstly, RGB images were used, but in the feature extraction phase, due to the fact that the features in the images depend on the lighting condition, the accuracy was unstable (Figure 12). Therefore, IR images were used instead. The proposed face recognition flow consists of four phases: 1) Face detection, 2) Feature extraction, 3) Read of database, and 4) Identification of the person.

a. Color image in the dark b. IR image in the dark

c. Color image in the light d. IR image in the light

Figure 12. RGB images are affected by the light condition but IR images are not

3.2.1 Face Detection

As we mentioned before, Kinect v2 has the face tracking function which actually works based on the skeleton tracking function, i.e. it can detect the face automatically. Therefore, we used this function directly.

3.2.2 Feature Extraction

Here the Local Binary Pattern Histogram [9, 10] was used as the feature extraction method of the face recognition. Local Binary Pattern Histogram is one of the local feature descriptor and is widely used for face recognition [11]. Also, it is evaluated as a balanced feature extraction method at the point of accuracy and processing speed [12]. Therefore, the authors used LBPHFaceRecognizer provided by OpenCV.

3.2.3 Image Retrieval Database

Several JPG images were obtained of various subjects for the generation of the database and an Excel file was created listing the image file names. The size of the database was also concerned. The larger the number of images in the database, the higher the performance of the face detection algorithm, but the processing speed is slower, and vice versa. Eventually we collected three subject's images in the database.

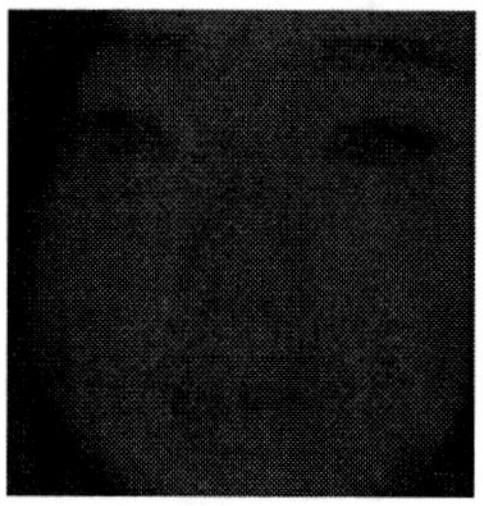

Figure 13. Example of JPG images in the database

3.2.4 Person Identification

If the input image identified specific person who were saved in database previously, it means face recognition is done. The threshold of the classification after a series of experiments was empirically set to 50.0. If the distance from the feature value of the input image to any prototypes is larger than the threshold, the input data is rejected and the person is recognized as a "Stranger".

3.3 Proposed Door Opening System

Using and Arduino Uno microcontroller board interfaced with the door opening mechanism (Figure 14) as well as with the Kinect, following a successful face recognition result, the result could be used to enable a signal on the microcontrollers output to actuate the door opening mechanism. When the user's face is recognized as the resident, a logical '1' is enabled on a specific pin on the Arduino GPIO, which is used as a signal to drive the door opening mechanism. Also a green LED is lit for visual response. On the other hand, when the user's face is not recognize as any of the residents stored in the database, meaning the user is recognized as a stranger, a logical '0' is enabled on the GPIO pin. Additionally a Red LED is lit accordingly for visual purposes. With a logic '0' the door opening mechanism remains idle, i.e. the door remains closed. The door operating system is programmed to close the door several seconds after the door opening.

Figure 14. ARDUINO Uno Board

4 Testing

The system performance was tested following the installation of the Kinect sensor in the entrance of the flat (Figure 15). The Kinect was set at an angle of 34.3 degrees. Both subsystems, knee position acquisition in stair walking and face recognition based door opening, were implemented and developed in C++. The testing followed the steps below:

- When the Kinect detects a subject, the knee position acquisition of stair walking is initiated. The calculated data is accumulated temporarily.
- When the user terminates the walking sequence i.e. arrives at doorstep, the knee position acquisition is interrupted and the face recognition peocess starts.
- If the subject is recognized as a resident within the database the door automatically opens.

The trigger to switch from stair walking acquisition to face recognition is the distance between subject and the Kinect. The distance is defined as the depth value of spine base position which is given by Kinect and it was 1150 mm.

5 Conclusions

In this paper, a combined system is proposed offering knee position acquisition while walking on stairs as well as face recognition based automatic door opening. These combined system aim to predict both the mental and the physical fatigue of the user upon returning to the home environment.

In the knee position acquisition phase, a novel knee position correction algorithm is proposed. According to the acquired results, the accuracy is undoubtedly improved. In the face recognition phase, images depicting the faces of three users are stored in a database. The number of images per user, as well as the number of users has been considered for the trade off between face recognition accuracy and processing speed. The system was installed and tested under realistic 1:1 scale in order to prove its possibility for realization. The position of the Kinect sensor in the experimental setup was extensively tested.

Currently the authors are conducting accuracy verification experiments for both subsystems. As future development, the face recognition phase can be expanded to detect the mental fatigue level according to the facial expressions of the user. It is expected that the Kinect v2 detailed face tracking function can be exploited towards this goal. Considering the knee position acquisition phase, it is expected to first predict the fatigue levels by comparing the acquired data and accumulated data within the database, and then enable warnings and advices depending on the calculated fatigue level.

The recognition can be used not only to actuate the door opening mechanism, but also to access to the user personal database including gait data for this study. This can enable a personalized actuated environment since lighting, air conditioning, aroma diffuser, music player, etc can be automatically controlled for the specific resident and his / her condition which is predicted from the fatigue level etc.

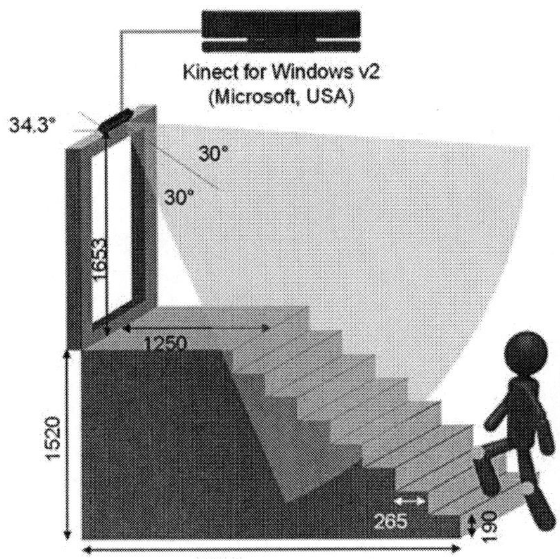

Figure 15. System Setup Arrangement

a. A subject is detected and knee position acquisition starts

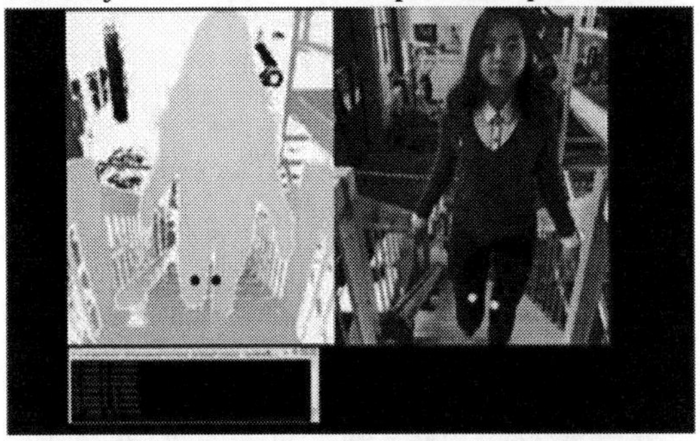

b. Knee position acquisition finished

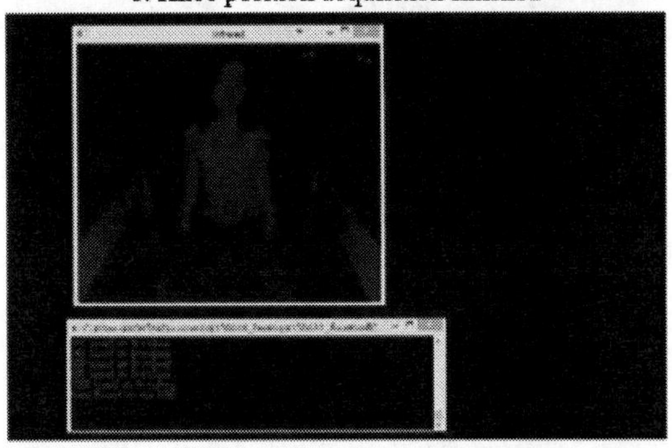

c. Face recognition starts

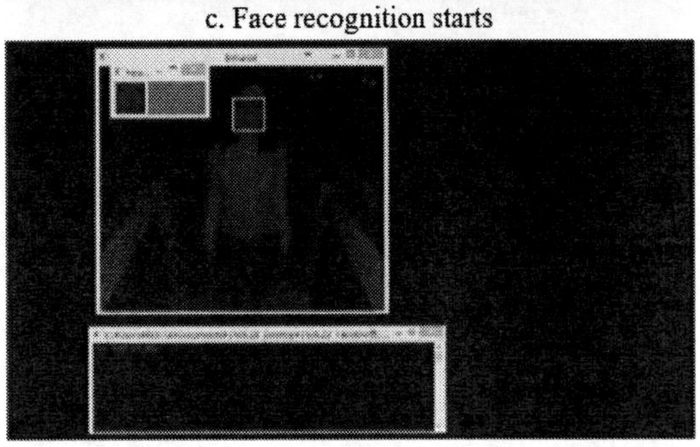

d. The subject is recognized as a resident, and the door will open

Figure 16. Execution screen of the testing sequence

References

[1] Yeung W.J.J. and Cheung A.K.L. Living alone: One-person households in Asia. Demographic Research, 32(40):1099-1112, 2015.

[2] Podhisita C. and Xenos P. Living alone in South and Southeast Asia: An analysis of census data. Demographic Research, 32(41):1113, 2015.

[3] Amon C and Fuhrmann F. Evaluation of the spatial resolution accuracy of the face tracking system for kinect for windows v1and v2. In Proceedings of the 6th Congress of the Alps Adria Acoustics Associa-tion, Graz, Austria, 2014.

[4] Waseda Univ, Wabot house, On-line: http://www.wabot-house.waseda.ac.jp/html/e-top.htm, Accessed: 15.03.2016.

[5] Murakami K., Hasegawa T., Kimuro Y., Karazume R. A structured environment with sensor networks for intelligent robots. In Proceedings of IEEE Sensors, pages 705-708, Lecce, Italy.

[6] Initlle S., Larson K., Tapia E.M., Beaudin J., Kaushik P., Nawyn J., Rockinson R. Using a live-in laboratory for ubiquitous computing research. In Proceeding of the 4th International Conference on Pervasive Computing, pages 349–365, Dublin, Ireland, 2006.

[7] Larson K and Stephen I. MIT Open Source Building Alliance — A house_n Initiative, Position Paper, MIT House_n, 2003.

[8] Linner T., Güttler J., Bock T. and Georgoulas, C. Assistive robotic micro-rooms for independent living. Automation in Construction, 51(2015):8-22, 2015.

[9] Ojala, T., Pietikäinen, M. and Mäenpää, T. Multiresolution gray-scale and rotation invariant texture with local binary patterns. IEEE Trans. Pattern Analysis and Machine Intelligence, 24(7):971-987, 2002.

[10] Ahonen T., Hadid A. and Pietikäinen, M. Face description with local binary patterns: Application to face recognition. IEEE Trans. Pattern Analysis and Machine Intelligence, 28(12):2037-2041, 2006.

[11] Aoyama S.K. and Aoki T. A study of biometric recognition Algorithm based on local phase features. In Proceedings of Biometrics Workshop, The Institute of Electronics, Information and Communication Engineers, pages 92-98, 2012.

[12] Terashima H. and Kida T., "勾配情報を用いた Local Binary Pattern の改良", DEIM Forum 2014, F5-4, 2014.AuthorSurname A. and AuthorSurname B. A journal article. *The Journal*, 1(2):123–345, 2014.

Visual and 3D Mapping for Steel Bridge Inspection Using a Climbing Robot

[1]Nhan H. Pham, [1]Hung M. La*, [2]Quang P. Ha, [3]Sy N. Dang, [3]Anh H. Vo, [3]Quang H. Dinh

[1]Department of Computer Science and Engineering, University of Nevada, Reno, USA
[2]Faculty of Engineering and IT, University of Technology, Sydney, Australia
[3]Robotics Center, Duy Tan University, Da Nang, Vietnam
*Corresponding Author: hla@unr.edu

Abstract –

There are over six hundred thousand bridges in the U.S. which require great amount of human effort along with expensive and specialized equipment for maintenance. Current bridge inspections are manually performed by inspectors which are inefficient and unsafe. This paper presents visual and three dimensional (3D) structure inspection for steel bridges and steel structures using the developed climbing robot. The robot can move freely on steel surface, carry several sensors, and collect data then send to the ground station for real-time monitoring as well as further processing. Steel surface image stitching and 3D map building are conducted to provide a current condition of the structure.

Keywords –

Field robotics; Climbing robots; Steel bridge inspection.

1 Introduction

There are currently over six hundred thousand bridges in the U.S. [1] and twenty percent of them are steel bridges. Among those, more than fifty thousand bridges are either deficient or functionally obsolete which are likely a growing threat to human's safety. Collapse of numerous bridges recorded over past 15 years has shown significant impact on the safety of all travellers. For instance, the Minneapolis I-35W Bridge in Minnesota, collapsed in 2007 due to undersized gusset plates, increased concrete surfacing load, and weight of construction supplies/equipment [2]. The accident along with others have grown a demand of frequent and adequate bridge inspection and maintenance. However, current activities require great amount of human effort along with expensive and specialized equipment. Besides, most bridges inspections are manually inspected by human inspectors with visual inspection or hammer tapping and chain dragging for delamination and corrosion detection which are very time consuming. Moreover, it is difficult and dangerous for the inspectors to climb up or hang on cables inspecting large bridges with high structures as shown in Figure 1. In addition, reports from visual inspection may vary among inspectors, hence the bridge's condition cannot be assessed precisely. Therefore, there should be a solution which provides consistent and accurate bridge condition report with high efficiency along with safety assurance. If a robot carrying several advanced evaluation sensors is able to climb on those bridges to collect data for condition assessment, it will significantly improve the inspection efficiency as well as enhance inspectors' safety since they no longer have to work under dangerous conditions.

Figure 1. Dangerous bridge inspection (source: icaeng.com, hireuavpro.com).

The number of studies related to utilizing advanced technology for bridge inspection and evaluation has significant increased recently. In one hand, R.S Lim et al. [3], N. Gucunski et al. [4], and H.M. La et al. [5] introduce a novel robotic systems for bridge deck inspection, while the NDE technology and NDE data processing for bridge deck inspection and evaluation are presented in [6] and [7]. In [8] and [9], an autonomous robotic system integrated with advanced non-destructive evaluation (NDE) sensors for high-efficiency bridge deck inspection and evaluation. B. Li et al. [10] also utilized NDE technique to perform automatic inspection on bridge deck and record bridge's health condition. In the other hand, there are several works focusing on cable inspection. F. Xu et al. [11] introduced the design and experiments of a wheel-based cable inspection robotic system consisting of CCD cameras for visual inspection. Similar robot from K.H. Cho et al. [12] enables effective visual inspection of the cable on suspension bridge. Besides, there were several initial implementations of climbing robots for bridge inspection including steel bridges. A. Mazumdar et al. [13] proposed a legged robot that moves across a steel structure for steel bridge inspection. Powerful permanent magnets embedded in each foot allow the robot to hang from a steel ceiling powerlessly while the attractive force is modulated by tilting the foot against the steel surface. A robot with magnetic wheels was developed by R. Wang, et al. [14] carrying Giant Magneto Resistive sensor array for crack and corrosion detection. Q. Liu et al. [15] and Y. Liu et al. [16] also proposed a bridge inspection method using a wall-climbing robot based on negative pressure adhesion mechanism which collects crack images with a high-resolution camera so that the crack can be extracted and analysed precisely. Based on attraction force created by permanent magnets, A. Leibbrandt et al. [17] and H. Leon-Rodriguez et al. [18] developed different wall-climbing robots carrying NDE devices to detect weld defects, cracks, corrosion testing that is capable of inspecting oil tanks or steel bridges. A. San-Millan [19] presented the development of a teleoperated wall climbing robot which can be equipped with various testing probes and cameras for different inspection tasks. D. Zhu et al. [20] used a magnetic wall-climbing robot capable of navigating on steel structures, measuring structural vibrations, processing measurement data and wirelessly communicating information to investigate field performance of flexure-based mobile sensing nodes and identify minor structural damage, and illustrate a high sensitivity in damage detection.

In this paper, the inspection technique using visual and three dimensional (3D) images integrated on our developed climbing robot [21] is introduced. The robot is able to strongly adhere and freely move on steel surface to collect images from visual camera as well as 3D camera. The captured images from visual camera are then stitched together to provide a whole image of steel surface that has been inspected. Moreover, a 3D map can be built from 3D data in order to assist robot navigation as well as inspecting bridge structures. With an advanced mechanical design, the robot is able to carry a heavy load (approximately 15 pounds) while climbing on both inclination and upside-down surfaces. The robot can also transit from one surface to another surface with up to 90 degrees change in orientation. Collected data are sent to a ground station in real time for further processing. Steel surface image stitching and 3D structure map are generated from collected data to provide a condition of the bridges.

The rest of this paper is organized as follows. Section 2 describes the overall robot design. Section 3 presents data collection and processing. Section 4 shows various experiments to verify robot design and data processing. Finally, Section 5 gives the conclusion and future work.

2 Overall Design

2.1 Mechanical Design

A four motorized wheels robot design as shown in Figure 2 is proposed which takes advantage of permanent magnets for adhesion force creation. Therefore, it allows the robot to adhere to steel surfaces without consuming any power. In addition, cylinder shaped magnets are used for convenience and flexibility in adjusting the attraction force. Moreover, eight high torque servo motors (525g/mm) are modified to use for driving four wheels and four shafts – which are used for robot lifting to avoid being stuck in rough terrain.

36 Neodymium magnet cylinders embedded in each wheel can create magnetic force up to 14 kg/wheel if there is no air gap between the magnets and the steel surface [21]. However, each wheel is covered by a thin layer of cloth to enhance the friction with steel surface, hence the magnetic force created is reduced which is approximately 6 kg/wheel. Additionally, a mechanism has been added in order to lift either front or rear wheels off if the robot is stuck on rough terrains.

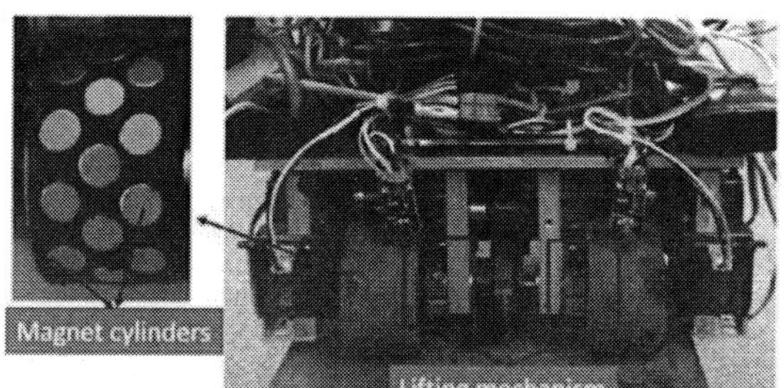

Figure 2. Embedded magnet cylinders and lifting mechanism.

When moving on steel surfaces, there are problems needed to be addressed in order to maintain stability of the robot which are sliding and turn-over failure. Our previous work [21] shows that in order to avoid two types of failure, the magnetic force created by all wheels should satisfy

$$F_{mag} > max\left\{2.237P; 2\frac{Pd}{L}\right\}, \quad (1)$$

where F_{mag} is the total magnetic force, P is the robot's weight ($P = mg$ where m is robot's mass and g is gravitational acceleration), d is the distance between the center of mass to steel surface, and L is the distance between front and rear wheels.

2.2 System Integration

Overall, the structure of the system is shown in Figure 3 and the robot with fully installed sensors and other components are depicted in Figure 4.

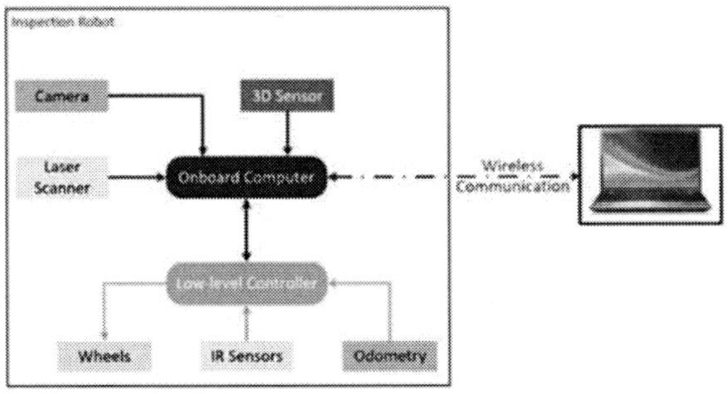

Figure 3. System architecture.

Regarding sensing devices, the robot is equipped with multiple imaging sensors: two video cameras for images capturing and video streaming, and a time-of-flight (ToF) camera for capturing 3D data. The USB camera in the back of the robot is installed facing downward to the steel surface in order to take images of steel surfaces to feed the image stitching technique. Besides, the ToF camera is placed so that 3D images received from the camera can assist navigation as well as 3D structure construction.

Apart from cameras, eight Hall Effect sensors which are able to detect the presence of magnetic field are also used. We use two sensors which are mounted next to each other and close to robot's wheel. By taking advantage of the fact that magnet cylinders inside each wheel will move when the robot moves, we can extract the velocity and traveling distance of each wheel after combining the data from these two sensors.

Besides, the robot has four IR range sensors mounted at four corners of the robot, which can detect whether there exists a surface underneath. Consequently, an edge avoidance algorithm can be implemented using this input to make sure that the robot can safely travel on steel surfaces.

The robot is controlled by two controllers: a microcontroller (MCU) based controller handling low-level tasks, and a powerful on-board computer for complex processing and communication with ground station. The low-level controller has the capability of receiving commands from on-board computer via serial connection including motion control, sensors data acquisition. The on-board computer is a NUC Core i3 computer from Intel which captures video camera frames and 3D camera images then sends them to ground station over wireless LAN connection for data post processing and logging. It also executes the edge avoidance algorithm with sensors data received from the low-level controller to ensure safe traveling on steel surfaces.

2.3 Robot Navigation

While moving on steel surfaces, there can be a circumstance that the robot moves far away toward the edge of the surface which may cause it to fall off. Therefore, an algorithm using input from IR range sensors is incorporated to prevent this failure. Denote $r_{cal_i}(i = 1, \ldots, 4$ is the index of four IR sensors of the robot) as the calibrated ranges before robot starts moving, $r_i(i = 1, \ldots, 4)$ as IR sensor reading corresponding to $sensor_i$ and $d_i(i = 1, \ldots, 4)$ as travel distances calculated from Hall Effect sensors. When sensor reading is out of the range between $[r\_cal_i - \varepsilon; r_{cal_i} + \varepsilon]$ where ε is a predefined threshold, the robot considers that there is no surface below $sensor_i$.

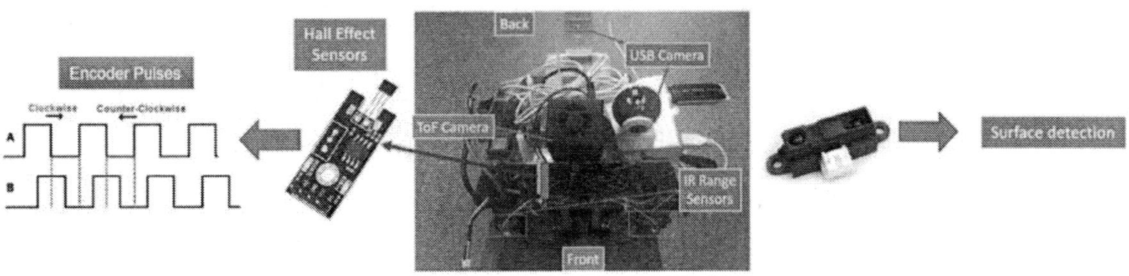

Figure 4. Robot prototype with integrated sensors.

Algorithm 1: Edge Avoidance.

Input: $(r_{cal_1}, r_{cal_2}, r_{cal_3}, r_{cal_4}), (r_1, r_2, r_3, r_4), \varepsilon, (d_1, d_2, d_3, d_4)$

```
for i=1:4 do
    if only one (r_i) ∉ [r_cal_i − ε; r_cal_i + ε] then
        if i:=front right IR sensor then
            Stop
            Go backward with a distance of 5 cm (Δd_i ≈ 3).
            Rotate left when travel distance of either right wheels reach 3 cm (Δr_i ≈ 2).
            Keep moving.
        Check other sensors and take similar actions.
    else stop and wait for commands.
```

Moreover, the velocity and distance calculated from Hall Effect sensor reading can help implement the automatic images capturing so that we can stitch all collected images to detect corrosion or defect on steel surface. In this mode, the robot need to stop after traveling a particular distance to ensure that the two consecutive images are overlapped.

3 Data Collection and Processing

3.1 Image Stitching

In order to enhance steel surface inspection, we combine image sequences. Captured images are saved to the memory so that they can be processed later when the robot finished its operation.

In this technique, it is required that two consecutively taken images are at least 30% overlapped as shown in Figure 5. The camera can cover an area of 18cm x 14cm which means the robot should stop and capture images every 12 cm.

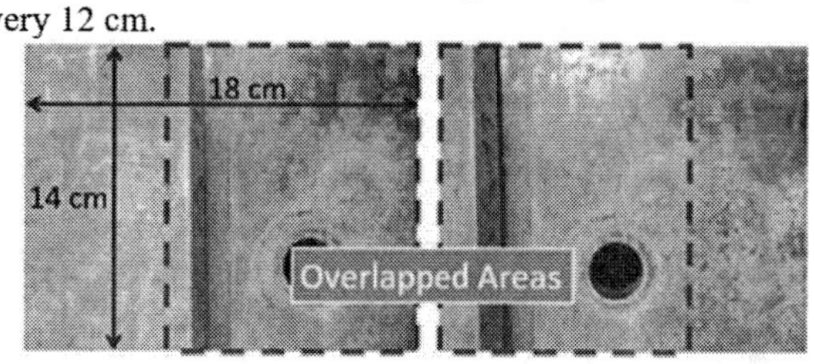

Figure 5. Overlapped images used for stitching.

While stitching all of the images, we can also transform the stitched image's coordinate to the world's coordinate. Denote (X_{im}, Y_{im}, Z_{im}) as the coordinates of image frame, $(X_{cam}, Y_{cam}, Z_{cam})$ are coordinates of camera frame, (X_r, Y_r, Z_r) are coordinates of robot frame and (X_w, Y_w, Z_w) are coordinates of the world fix frame as shown in Figure 6.

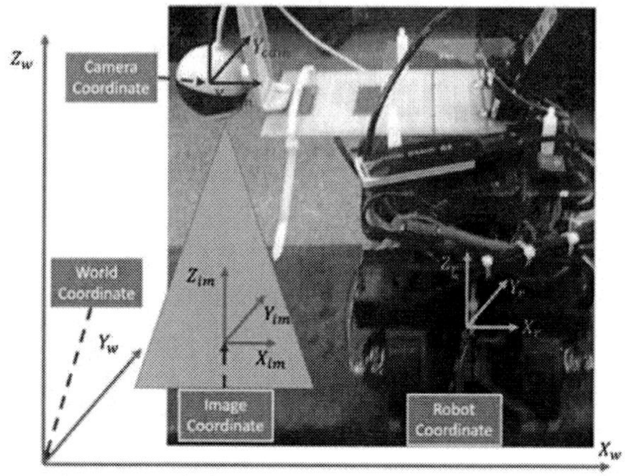

Figure 6. Relationship between multiple frames.

The following series of transformation should be done to convert coordinates in image frame to world frame:

$$\begin{array}{ccc} (X_{im}, Y_{im}, Z_{im}) & \xrightarrow{T_{ic}} & (X_{cam}, Y_{cam}, Z_{cam}) \\ & & \downarrow T_{cr} \\ (X_w, Y_w, Z_w) & \xleftarrow{T_{rw}} & (X_r, Y_r, Z_r) \end{array}$$

where T_{ij} are the transform matrix from frame i to frame j. Since the camera mounting location is fixed on the robot, T_{ic} and T_{cr} can be easily calculated by measuring the distance between the camera and the steel surface and the robot's centre of mass while T_{rw} can be extracted from odometry. For more details of this transformation, please refer to [23].

3.2 3D Construction

The concept of 3D construction or 3D registration is based on the Iterative Closest Point (ICP) algorithm introduced by Besl et al. [24] and Y. Chen [25] in the early 1990s. The algorithm was used to construct 3D model of objects and has later been employed for robotic applications including 2D, 3D mapping and localization [26]. The goal of the ICP is to find the transformation parameters (rotation and translation) that align an input point cloud to a reference point cloud. Those transformation parameters are presented in the coordinate frame of the reference point cloud. Figure 7 shows the process of 3D registration including ICP algorithm consisting of following steps [26], [27]:

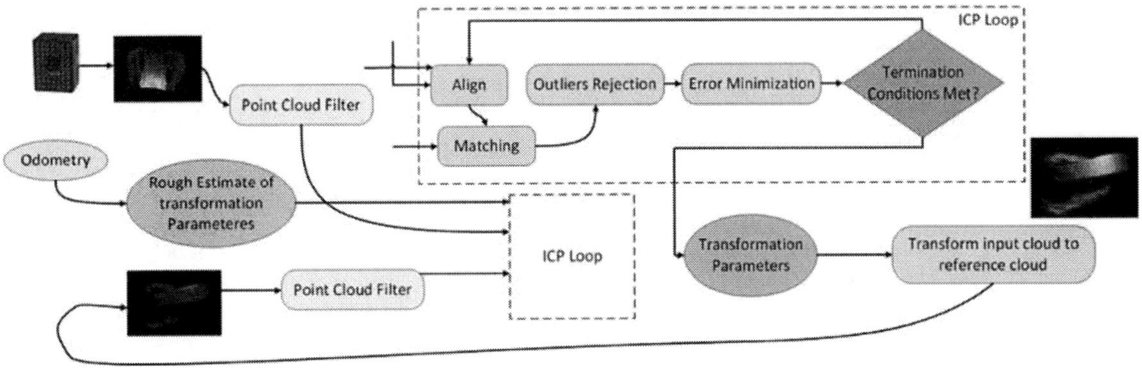

Figure 7. 3D map construction process employing ICP algorithm.

1. Selection: Input point cloud captured from ToF camera are pre-sampling for higher efficiency.
2. Matching: Odometry data can be used to estimate the correspondences between the points in the subsampled point clouds considered as the initial alignment.
3. Rejection: Filtering the correspondences to reduce the number of outliers, multiple points with same corresponding point are rejected.
4. Alignment: Calculating assessment criteria which are normally point-to-point or point-to-plane error metrics, then minimizing it to find the optimal transformation.

The algorithm stops when one of these case happens:

1. Convergence: the error metric falls under a threshold or remain constant; or the transformation parameters become unchanged.
2. The algorithm does not converge after a given number of iterations.
3. Transformation parameters are out of bound.

4 Experimental Results

In order to verify the edge avoidance algorithm as well as the data collection and processing technique, we have conducted multiple experiments under lab environment utilizing different steel bars. During the test, one 2S1P (2 cells) 7.4V 5000 milliampere-hour (mAh) and one 3S1P 11.1V 5000 mAh batteries are used to power the robot. One laptop which can connect to a wireless LAN is used as a ground station.

First of all, the full robot with all sensor integrated is tested to climb on a vertical steel surface in order to verify the load carrying capability. During the test, the robot strongly adheres to steel structure as shown in Figure 8. Another test is conducted when robot moves on a constructed steel structures from one end to the other end as depicted in Figure 9. The robot successfully reaches the destination as well as using lifting mechanism to overcome stuck condition as shown in Figure 9.b.

Figure 8. Robot moves on vertical surface while carrying all sensors.

Figure 9. Robot moves on a bridge-like steel structure with Edge Avoidance.

During the test, data collected from both video camera and depth camera are transmitted over wireless connection to the laptop acting as ground station. Received data are illustrated in Figure 10.

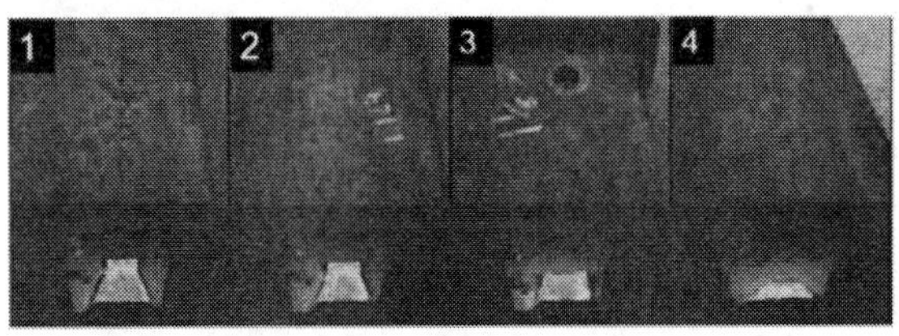

Figure 10. Visual and depth images from cameras showing the condition of the bridge surface. Top images show visual condition, and bottom images show the 3D structure maps.

Regarding 3D construction capability, the robot moves on a flat steel structure as shown in Figure 11. While the robot is traveling on the structure, 3D images are captured from the ToF camera. The data is both transferred to the ground station and saved to the on-board computer's hard disk. The 3D images are taken every 100 milliseconds in order to obtain enough overlapped area between two consecutive images. After the robot successfully reach the other end of the steel structure, we apply ICP algorithm to align all stored 3D images to form a 3D map of the environment including the steel structure. The 3D construction results are presented in Figure 12 and Figure 13.

Figure 11. Robot setup to test 3D construction capability.

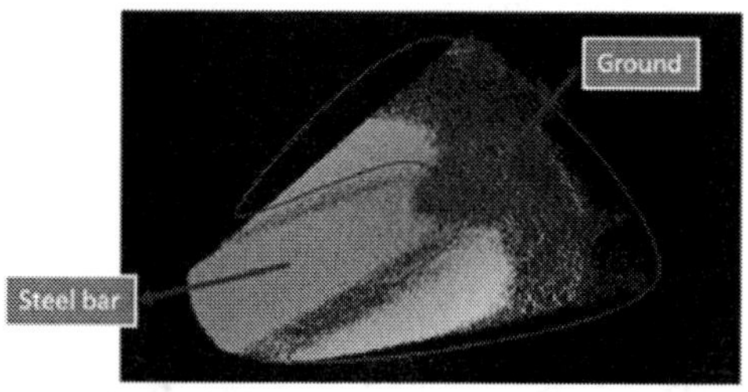

Figure 12. Resulting map from 3D images.

Furthermore, from the images collected through the camera facing downward to the surface, we are able to stitch the images together. From the stitched image, we can spot some areas of steel structure which have been greatly or slightly deteriorated as shown in Figure 14.

5 Conclusion and Future Work

This work describes the visual and 3D structure inspection for steel bridges and steel structures using the developed climbing robot. The robot is able to carry several sensors and move safely on steel surfaces to collect data. The acquired visual and 3D images are transferred to ground station for real-time monitoring and processing. An image stitching technique has been implemented to create an entire image of steel surface from captured images. Moreover, various sensors are integrated to ensure safe navigation on steel surfaces.

Further work needed to be done including localization using combined odometry and camera data, implementation of map construction method as well as visual crack detection algorithm from stitched images. Additionally, a cooperative process can be used to employ multiple robots for faster inspecting a steel large bridge.

Acknowledgments

This work is supported by the University of Nevada, Reno and the National Science Foundation under the grant: NSF-ICorps#1559942. The authors gratefully acknowledge contributions of: Luan Nguyen, Jesus Sanchez, and Tuan Le during the implementation and test of the robot at the Advanced Robotics and Automation (ARA) Lab.

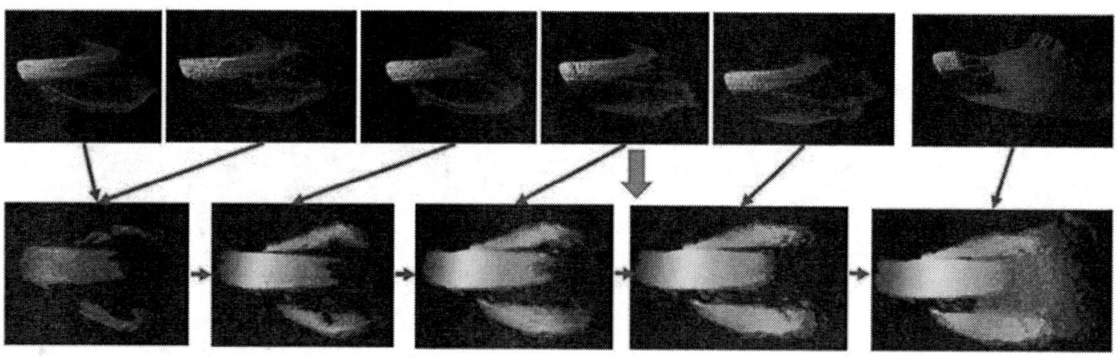

Figure 13. 3D registration from point cloud data.

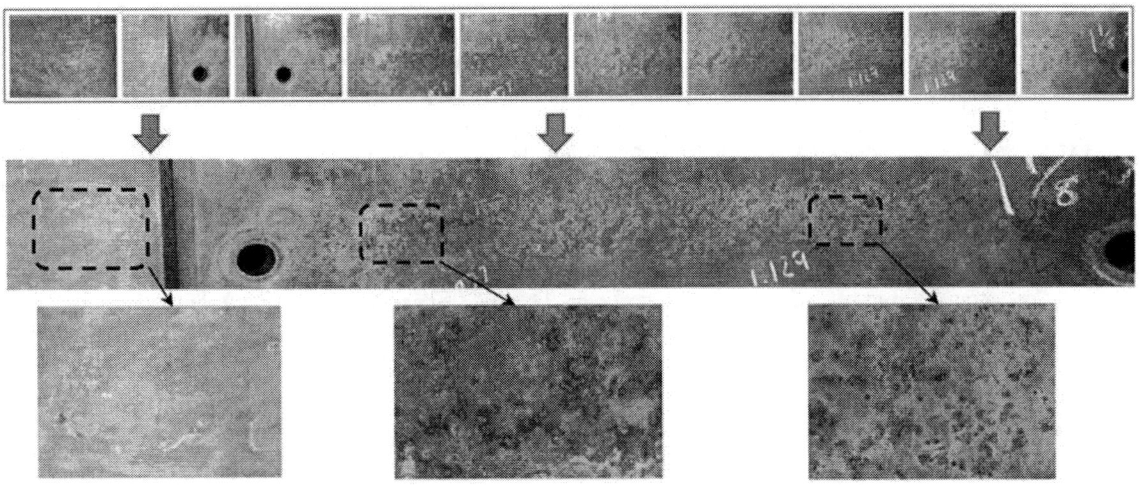

Figure 14. Images stitching result: (Top) 10 individual images taken by the robot; (Middle) Stitching image result from those 10 individual images; (Bottom) Closer look (zoom-in) at some areas, from left-to-right, showing "good condition", "serious deteriorated condition", and "light deteriorated condition" of the steel surface, respectively.

References

[1] U.S. Department of Transportation Federal Highway Administration. National Bridge Inventory Data. On-line: http://www.fhwa.dot.gov/bridge/nbi.cfm, Access: 01/30/2016.

[2] I-35W St. Anthony Falls Bridge collapse. Minnesota Department of Transportation. On-line: http://www.dot.state.mn.us/i35wbridge/collapse.html, Access: 01/30/2016.

[3] R.S. Lim, H.M. La and W. Sheng. A robotic crack inspection and mapping system for bridge deck maintenance. *IEEE Transactions on Automation Science and Engineering*, 11(2):367–378, April 2014.

[4] N. Gucunski, S.H. Kee, H.M. La, B. Basily and A. Maher. Delamination and concrete quality assessment of concrete bridge decks using a fully autonomous RABIT platform. *International Journal of Structural Monitoring and Maintenance*, 2(1):19–34, 2015.

[5] H.M. La, N. Gucunski, S.H. Kee, J. Yi, T. Senlet and L. Nguyen. Autonomous Robotic System for Bridge Deck Data Collection and Analysis. In *IEEE International Conference on Intelligent Robots and Systems (IROS)*, pages 1950–1955, Chicago, USA, 2014.

[6] H.M. La, N. Gucunski, S.H. Kee, and L. Nguyen. Visual and Acoustic Data Analysis for the Bridge Deck Inspection Robotic System. In *31st International Symposium on Automation and Robotics in Construction and Mining (ISARC)*, pages 50–57, Sydney, Australia, 2014.

[7] N. Gucunski, B. Basily, S.H. Kee, H.M. La, H. Pavardeh, A. Maher and H. Gashemi. Multi NDE Technology Condition Assessment of Concrete Bridge Decks by RABITTM Platform. In *NDE/NDT for Structural Materials Technology for Highway & Bridges*, August 25, 2014.

[8] H.M. La, R.S. Lim, B.B. Basily, N. Gucunski, J. Yi, A. Maher, F.A. Romero, and H. Parvardeh. Mechatronic systems design for an autonomous robotic system for high-efficiency bridge deck inspection and evaluation. *IEEE/ASME Transactions on Mechatronics*, 18(6):1655–1664, Dec 2013.

[9] H.M. La, N. Gucunski, S.H. Kee, and L.V. Nguyen. Data analysis and visualization for the bridge deck inspection and evaluation robotic system. *Visualization in Engineering*, 3(1):1–16, 2015.

[10] B. Li, J. Cao, J. Xiao, X. Zhang, and H. Wang. Robotic Impact-Echo Non-Destructive Evaluation Based On FFT and SVM. In *2014 11th World Congress on Intelligent Control and Automation (WCICA)*, pages 2854–2859, Shenyang, China, 2014.

[11] F. Xu, and X. Wang. Design and Experiments on a New Wheel-Based Cable Climbing Robot. In *IEEE/ASME International Conference on Advanced Intelligent Mechatronics (AIM)*, pages 418–423, Xian, China, 2008.

[12] K.H. Cho, H.M. Kim, Y.H. Jin, F. Liu, H. Moon, J.C. Koo, and H.R. Choi. Inspection robot for hanger cable of suspension bridge: Mechanism Design and Analysis. *IEEE/ASME Transactions on Mechatronics*, 18(6):1665–1674, Dec 2013.

[13] A. Mazumdar, and H.H. Asada. Mag-Foot: A Steel Bridge Inspection Robot. In *IEEE/RSJ International Conference on Intelligent Robots and Systems (IROS)*, pages 1691–1696, Missouri, USA, 2009.

[14] R. Wang, and Y. Kawamura. A Magnetic Climbing Robot for Steel Bridge Inspection. In *2014 11th World Congress on Intelligent Control and Automation (WCICA)*, pages 3303–3308, Shenyang, China, June 2014.

[15] Q. Liu, and Y. Liu. An Approach for Auto Bridge Inspection Based On Climbing Robot. In *2013 IEEE International Conference on Robotics and Biomimetics (ROBIO)*, pages 2581–2586, Shenzhen, China, Dec 2013.

[16] Y. Liu, Q. Dai, and Q. Liu. Adhesion-Adaptive Control of a Novel Bridge-Climbing Robot. In *2013 IEEE 3rd Annual International Conference on Cyber Technology in Automation, Control and Intelligent Systems (CYBER)*, pages 102–107, Nanjing, China, May 2013.

[17] A. Leibbrandt, G. Caprari, U. Angst, R.Y. Siegwart, R.J. Flatt, and B. Elsener. Climbing Robot for Corrosion Monitoring of Reinforced Concrete Structures. In *2012 2nd International Conference on Applied Robotics for the Power Industry (CARPI)*, pages 10–15, Zurich, Switzerland, Sept 2012.

[18] H. Leon-Rodriguez, S. Hussain, and T. Sattar. A Compact Wall-Climbing and Surface Adaptation Robot for Non-Destructive Testing. In *2012 12th International Conference on Control, Automation and Systems (ICCAS)*, pages 404–409, Korea, Oct 2012.

[19] A. San-Millan. Design of a Teleoperated Wall Climbing Robot for Oil Tank Inspection. In *2015 23th Mediterranean Conference on Control and Automation (MED)*, pages 255–261, Torremolinos, Spain, June 2015.

[20] D. Zhu, J. Guo, C. Cho, Y. Wang, and K.M. Lee. Wireless Mobile Sensor Network for the System Identification of a Space Frame Bridge. *IEEE/ASME Transactions on Mechatronics*, 17(3):499–507, June 2012.

[21] N.H. Pham and H.M. La. Design and Implementation of an Autonomous Robot for Steel Bridge Inspection. Manuscript submitted to *2016 IEEE/RSJ International Conference on Intelligent Robots and Systems (IROS)*, 2016.

[22] K&J Magnetics Inc. Orignial magnet calculator. On-line: http://www.kjmagnetics.com, Accessed: 01/30/2016.

[23] R.S. Lim, H.M. La, Z. Shan and W. Sheng. Developing a crack inspection robot for bridge maintenance. In *IEEE International Conference on Robotics and Automation (ICRA)*, Shanghai, China, May, 2011.

[24] P. Besl and N. McKay. A method for registration of 3-D shapes. *IEEE Transaction on Pattern Analysis and Machine Intelligence*, 14(2):239–256, 1992.

[25] Y. Chen and G. Medioni. Object modelling by registration of multiple range images. In 1991 Proceedings of the IEEE International Conference on Robotics and Automation, 3:2724–2729, 1991.

[26] T. Candia. ICP optimization for 3D robot mapping. Master Thesis, Swiss Federal Institute of Technology in Zurich, 2011.

[27] D. Holz, A. E. Ichim, F. Tombari, R. B. Rusu and S. Behnke. Registration with the point cloud library: a modular framework for aligning in 3-D. *IEEE Robotics & Automation Magazine,* 2(4):110–124, Sept 2015

Improving Construction Labor Productivity Using Automatic Rebar Tying Gun

M. Safa1[a], J. G. Cardenas2[a], D. G. Leblanc3[a], D. Rose4[a], and A. Shahi5[b]

[a]Reese Construction Management, Lamar University, US
[b]Department of Civil Engineering, University of Toronto, Canada
E-mail: msafa@lamar.edu, j_cardenas60@yahoo.com, bbmanjr889@yahoo.com, dannyrose326@yahoo.com
arash.shahi@utoronto.ca

Abstract – Rebar tying is labor-intensive and expensive, with a high learning curve that restricts proficiency at the craft to a great deal of construction field experience. Obtaining this experience requires tasks that are taxing on the human body, often resulting in muscular and skeletal injuries. Previously, research and engineering controls for rebar tying/rod busting have been reasonably limited. The main contribution of this study is investigating the use of a new advanced technology to reduce the time, cost, and hazard associated with rebar tying. Hence, several experiments were conducted on the effectiveness of an automatic rebar tying gun by comparing the speed and cost effectiveness of this tool with the common industry practice of manually tying rebar with a spool of wire and pliers. The study results show that the automatic rebar tying gun could lower the learning curve, thereby reducing training times and minimizing redundant routines. By minimizing the amount of time invested in training employees, the possibility of saving time and money is apparent. However, risk reduction is also implied by the lessening of time invested in training employees. This is because resignations frequently happen before a return on the investment of training is produced, resulting in sunk costs. Various repetitive, often manual, movements required of workers have been observed to be reduced from the use of the rebar tying gun. The results of this study also shows that the automatic rebar tying gun has a potential to save time, money, and ergonomic liabilities without risking productivity.

Keywords – Automatic Rebar Tying; Labor Productivity; Rod Busting Process; Musculoskeletal Disorders; Construction Project

1 Introduction

Rebar tying is one of the most labor-intensive jobs in construction projects; requiring repetitive, trunk-bending motions to complete a single tie. The most common type of construction worker injury, a musculoskeletal injury, results from conditions experienced by workers tying rebar. In fact, this type of injury is predominantly seen in the rod working field [1, 2]. This can be reduced, if not eliminated, through use of state-of the-art technology [2, 3]. Hence, the primary contribution this study offers is the investigation of using an innovative technological solution to reduce the time, cost, and hazard associated with rebar tying. This rebar tying technology will not only reduce the occurrences of injury to workers, it will also increase the productivity of the time spent working.

Increasingly risky exposure to heavy material handling, repetitive movements, awkward postures, contact stresses, vibrations, and forceful exertions insidiously incur musculoskeletal damage to workers over the course of a rebar tying career. The damage experienced by the seasoned workers of today have generated a sense of urgency to innovate ways of tying rebar that prevent or reduce injuries to the seasoned workers of tomorrow (those now entering or freshly entered into the workforce).

Workers are inevitably required to assume many awkward postures while tying rebar. A University of Wisconsin study conducted by the Department of Occupational & Environmental Safety & Health showed that "25% of all workers' compensation costs are associated with construction injuries," even though construction industries account for a mere 6% of construction work. Data shows that workers in this field can have "significant risk of musculoskeletal injury." Estimates of damages related to musculoskeletal disorders in the construction industry are "between $45 and $54 billion annually" in the United States alone [5].

With such high estimates, Choi initiated an investigation to uncover the causalities for this phenomenon. The investigation consisted of studying the physical effects rebar tying had on eleven rod workers by examining stress levels experienced by different parts of the body during traditional rebar tying. This study confirmed the "high potential risks for musculoskeletal disorders and injuries" in the rebar work of construction projects; the lower back and wrists of workers being among the most injurious areas for concern. "A guideline suggesting 15% - 30% of the maximum capability to avoid fatigue" exists, however, inevitably "65% - 92% of [rod worker's] maximum capability exerted results in long-term injuries. Naturally, monetization is a priority in any change implemented in business, however, the health and wellness of workers in this field should also serve as a bottom line [6]. This double-bottom line philosophy has been adopted in many other industries to improve customer and employee satisfaction, ultimately resulting in increased monetization. The effects of rapid, repetitive motions of the wrists and forearms while using pliers for the traditional tying method were observed. The long-term musculoskeletal effects that traditional rebar tying can have on the trunk of a person is tested as the basis for this study. [7].

Additional research on the Biomechanical Assessment of Three Rebar Tying Techniques states in an article that construction workers specializing in highway, street, or bridge construction have incident rates "one third higher than the national average." A study of a similar nature was conducted to see the amount of these workers that had experienced any type of musculoskeletal disorders (MSD). Of the 1,000 ironworkers, a mere 12% had no record of reporting MSD in their career. Observations revealed that while manually having to tie rebar using pliers, workers "sustained deep trunk bending and rapid and repetitive hand and wrist movements." 94% of the time rod workers using the conventional method experienced extreme trunk flexion. Whereas, once the automated tier, extension pole included, was introduced, 83% of the time neutral flexion was experienced and moderate forward flexion was experienced 16% of the time, leaving workers in extreme flexion only 1% of the time. These problems were "greatly reduced" while using an automated tier, yet their elimination is possible by attaching the extension handle to the automatic rebar tier, allowing the worker to tie standing erect [8].

A potential reduction in the risk of developing a MSD in the trunk from rebar tying is demonstrated in this study through the analysis of workers using of a rebar tying machine versus those using conventional methods. The study made evident the hypothesized benefits of the rebar tying machine. The following section is an overview of the experiment and results.

2 Methodology and Results

Our experiment consisted of a prefabricated rebar cage with evenly spaced rebar. A technician tested two different rebar tying guns alongside the performance of an experienced traditional rod buster. The time spent by each the gun users and the rod buster was compared to create a cost analysis to determine any benefits from the rebar tying gun. The performance resulting from use of the gun was examined along with the performance demonstrated by the rod buster to provide evidence for the effect on performance resulting from the use of the gun.

The rebar cage has been designed and made for conducting this experiment in the construction lab at Lamar University (Figure 1). We prepared the rebar cage by cutting four 2X4's to 5ft lengths, then connecting them at each end with screws. Holes were drilled on each of the cage's sides to provide a space for the rebar to go through. The 5ft lengths of rebar were inserted into the pre-drilled holes aforementioned.

Figure 1. Rebar Cage

The rebar cage we used consisted of 2X4's with #3 rebar spaced evenly in order to create 36 tie points. The two different types of rebar tying guns used were the RB655 with 16 gauge tie wire and the RB397 with 21 gauge wire. Our experienced rod buster used 36 gauge wires with pliers to create his ties while a timer was used to accurately record how long it took to tie the predetermined number of ties (Figure 2).

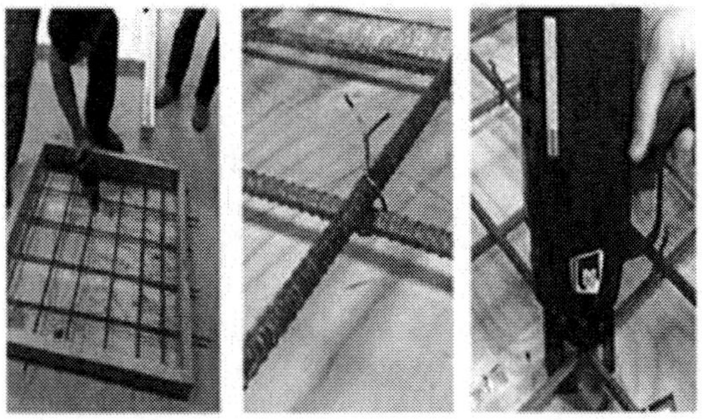

Figure 2. Lab experiment at Construction Lab (Lamar University)

The rebar tying guns were prepared to tie the allotted ties in the rebar cage, and all necessary tools and materials, such as wires and pliers, were provided for our experienced rod buster to do the same.

Our research design consisted of recording the amount of time that it took the rebar guns and the rod buster to tie 36 ties (not including actual prep time). The rebar cage was placed at ground level for all trials, and durations for each the gun users and the rod buster were recorded using a digital timer. Qualitative factors, such as each tie's tightness, were considered in performance observations as well. Thus, cost analysis calculations were performed by processing the raw quantitative and qualitative data with the wage rates listed by the Bureau of Labor Statistics to produce a numerical value that could represent the "cost" basis for our research [9].

The 16 gauge wire gun was able to tie the allotted ties in 1 minute 13 seconds, while the 21 gauge wire gun was 17 seconds quicker, recorded at 54 seconds. The experienced rod buster took 4 minutes 31 seconds to tie the same amount of ties each of the guns tied in an approximately one minute average. This shows well over a 4:1 ratio in the productivity of the guns versus that of the rod buster. Experiment conditions were the same for both gun users and the rod buster, ensuring further confidence that the deviations between the times recorded resulted directly from the tying speed of each.

However, although preparation time wasn't taken into account, it must be noted that the amount of time it takes to feed a new roll of wire into the guns can be reasonably large if operators do not have proper experience. The time variance for feeding the roll into the gun ranged from 1 minute to 6 minutes depending on the experience of the feeder. Another problematic observation of the guns' use was that the feeding process produces increasing waste with decreasing feeder experience, varying from 2 inches to almost 2 feet. These variance values were observed visibly during the experiment. Furthermore, it was noted that increased tightness and waste reduction achieved by rebar guns was visibly lower (excluding the waste produced during feeding).

Both the cost comparison (Figure 4) and time comparison (Figure 3) charts were developed using data for one rod buster manually tying two #3 rebar, and a rod buster operating each gun tying the same.

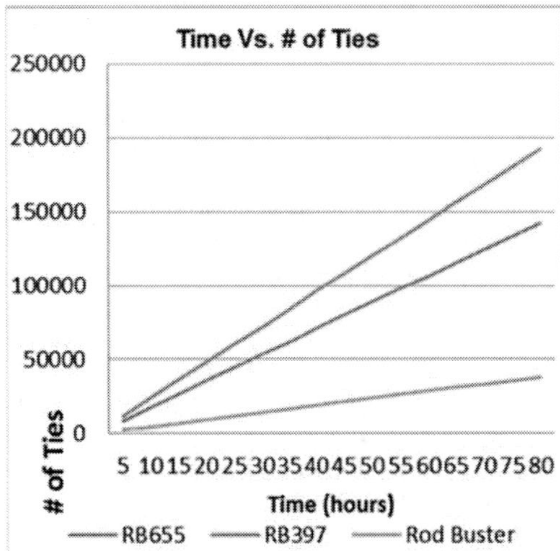

Figure 3. Time Comparison

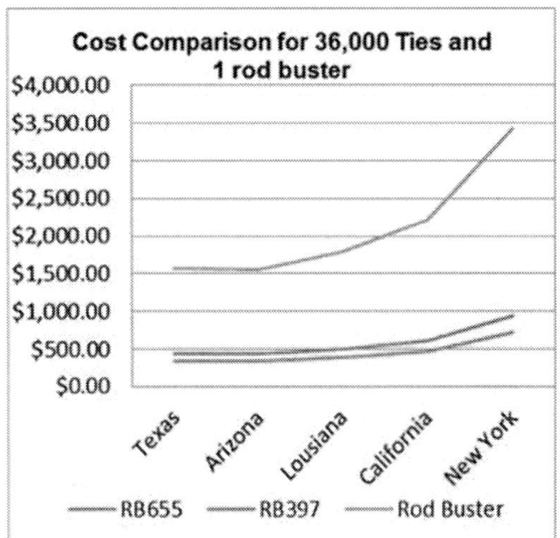

Figure 4. Cost Comparison

As can be observed from the graphs, the rebar tying guns offer a clear advantage when it comes to cost and time savings. The average cost of using the rebar tying guns for tying the set number of 36,000 ties is approximately 4 times less than the traditional method of tying rebar with pliers. Also, the amount of ties that can be made with the rebar tying guns is drastically higher per a set time. For example, at the 40 hour mark a rod buster operating an RB655 gun will tie 71,013 ties, and one with the RB397 will tie 96,000 ties. On the other hand, a rod buster following the traditional method of using pliers will only tie 19,129 ties.

As mentioned earlier, preparation times were not counted towards our calculations, and were found to increase with lacking experience. Thus, to achieve the results shown in this study there should be some training before rod busters use a rebar tying gun. This will reduce time consuming problems associated with using the gun in operations. In our experiment, one of the attempts at feeding a new coil into a gun posed problems that took the demonstrator approximately 6 minutes to finish.

The rebar tying guns also demonstrated limited capability in tight spaces. However, our observations have led us to conclude that tying rebar in spaces of such size is not a typical demand of most construction projects. In any case, the rod buster using the gun may need to stop and tie these few ties with unusually difficult accessibility.

The average price for an automatic rebar tying gun is approximately $2500. The rebar tying gun training takes approximately 30 minutes, which includes to train a person to replace the wire and disassemble any parts that does not void the 2-year warranty period. The initial costs of the guns were not calculated into our data, largely because this cost will be negligible over the useful life of the gun. The medical benefits for the rod buster and rod buster's employer alone justify the adoption of using these guns as a standard for the construction industry; if the optional extension is incorporated with the guns, the rod busters will need to bend over less which further reduces the risk of back injuries.

3 Discussion

Through our research it appears that the rebar tying gun is more efficient than traditional rebar tying. The costs of the rebar gun is easily offset by the productivity achieved by larger rebar tying jobs, such as those done in roadwork or concrete piling construction, where a large number of ties are required. The higher the volume of ties, the greater the benefits of the rebar tying gun will be. This implies that the gun may not be justifiable economically for jobs of a smaller scale. However, the safety benefits remain a motivating factor for these smaller project contractors to consider. The cost of the gun can seem unconventional, but this product has a high potential for value-added with a reasonable payback period.

The advantages of using an automatic rebar tying gun are identified in both road and beam construction. The average rodman can tie approximately 1,775 ties and use 13.7 pounds of tying material in an hour while using an automatic tying gun in road construction. The average rodman in road construction using the traditional method can only tie approximately 448 ties and use 21.2 pounds of tying material in an hour. This considerable difference is due both to the ease of creating a tie with the automatic rebar tying gun. In the traditional method the rodman has to actually lift the rebar at the point where the tie will be located and then commence the manual tying, taking up extra time.

Based on our results, the rebar tying guns are more efficient than the traditional tying method. However, our results came from experimentation in a relatively trivial and controlled environment, a small prefabricated box with evenly spaced and elevated rebar. Hence, the box could have failed to simulate the real world applications of the rebar tying gun. The test did not take into consideration larger scaled rebar tying jobs that would incorporate movement and fatigue.

Research was limited by this factor of size, as they were not conducted on a large scale. Another limiting factor was the level of realism in the conditions in which the experiment was conducted; a more realistic trial is necessary to truly evaluate the viability of these guns on a larger scale. This inhibited results from reflecting long-term usage and effectiveness of the rebar tying gun. Future tests should include a longer trial period in a real world setting with a full project or foundation to test the timeliness and effectiveness of each tie.

The research that we conducted did not determine the strength of each tie of the rebar. Due to the different types of ties that can be accomplished with hand tying, the traditional rebar tying method supposedly has a tie with greater strength than that of a rebar tying gun, which is limited to one type of tie. This leads to another disadvantage of using the gun encountered in certain types of formwork where regulations restrict the use of the gun due to the requiring of a specific tie that only hand tying it is capable of. Future tests should confirm the strength of each tie done by hand or the gun, as well as the failure rate of each tie.

It is recommended to perform a weekly cleaning of the debris such as dirt or dust from the prongs of the rebar tier with a simple brush to prevent build up and clogging of the wire feeding mechanism. Cleaning could be required more frequently depending on the amount used and the conditions that the gun is exposed too.

Limitations to the rebar guns were not tested. The experts noted that the gun is unable to tie rebar consistently and tightly if the tie has to be tied on an upright surface or overhead. This was not tested in our research, and also should be included in future testing.

4 Conclusion

An automatic rebar tying gun could eliminate some time spent training and executing repetitive motions. It also has the possibility of saving time and money by reducing man hours while increasing productivity. Also, since it eliminates much of the repetitive, often heavy, exertions that are normally employed by rod busters, it could reduce health risks in the rod busting community (especially if it is used with the optional extension bar). We will conduct our experiment by comparing the speed and cost effectiveness of this tool with the normal area practices, which include manually tying the rebar with pliers and a wire spool. In our study, we found that the rebar tying machine could tie two ties to an experienced rod buster's one. This advantage can be observed in day-to- day activities in the work place. The machine exhibits an optimal utility of its performance to be in the afternoon when rod busters are burnt out after intensive physical activity. This product could not only save time and money, it also has the potential to increase productivity, and therefore profitability, all the while reducing risks of injury to the long-term and short-term health of employees.

5 Acknowledgement

The Authors thank Mathew Johnson; Dillon Bergstrom; Eric T. Payton; Travis Clarke for assistance with the rebar tying experiment, and Matthew Chance for comments that improved the manuscript.

References

[1] Choi, Sang D., Lu Yuan, and James G. Borchardt. Critical Analyses of Work-related Musculoskeletal Disorders and Practical Solutions in Construction. *Proceedings of the Human Factors and Ergonomics Society Annual Meeting*. Vol. 58. No. 1. SAGE Publications, 2014.

[2] Vi, P. A field study investigating the effects of a rebar-tying machine on trunk flexion, tool usability and productivity. *Ergonomics 49*.14, 1437-1455, 2006.

[3] Albers, James T., and Stephen D. Hudock. Biomechanical assessment of three rebar tying techniques. *International Journal of Occupational Safety and Ergonomics,* 13.3, 279-289, 2007.

[4] Vi, P. Reducing risk of musculoskeletal disorders through the use of rebar-tying machines. *Applied occupational and environmental hygiene, 18.9,* 649-654, 2003.

[5] Kramer, D.M., Bigelow, P.L., Carlan, N., Wells, R.P., Garritano, E., Vi, P. and Plawinski, M. Searching for needles in a haystack: Identifying innovations to prevent MSDs in the construction sector. *Applied ergonomics*, 41(4), pp.577-584, 2010.

[6] Choi, Sang D. Ergonomic assessment of musculoskeletal discomfort of iron workers in highway construction. *Work 36.1*, 47-53, 2010.

[7] SAFETY AGENCY COMPARES INJURY TRENDS IN REBAR TYING METHODS, Concrete Products, 109, 4, p. PC26, Small Business *Reference Center, EBSCOhost*, 2016.

[8] Albers, J.T. and Hudock, S.D. Biomechanical assessment of three rebar tying techniques. *International Journal of Occupational Safety and Ergonomics*, 13(3), pp.279-289, 2007.

[9] Bureau of Labor Statistic. Occupational Employment and Wages. *47-2171 Reinforcing Iron and Rebar Workers*, 2014.

HYBRID SIMULATION MODELING OF HOIST DOWN-PEAK OPERATIONS IN CONSTRUCTION SITES

Shousheng Xiang[a], Mehrdad Arashpour[b] and Ron Wakefield[b]

[a]School of Management Engineering, Xi'an University of Finance and Economics, Xi'an, China
[b]School of Property, Construction and Project Management, RMIT University, Melbourne, Australia
E-mail: xiangshousheng@126.com, mehrdad.arashpour@rmit.edu.au, ron.wakefield@rmit.edu.au

Abstract – The **down-peak period is an important aspect of hoist operations in construction projects. To make the vertical transportation service more efficient, the operation parameters of down-peak should be modeled precisely in the process of planning. A simulation model is developed to analyze the operations of elevator in down-peak. The model incorporates hybrid use of discrete-event simulation and agent-based modeling to provide a robust methodology to analyze the two most important parameters for elevator operation planning: time spent and average waiting time. The developed model is validated under several conditions, and its usage is expanded to random situations.**

Keywords –

Agent-based modeling, Automation, Building projects, Construction management, Discrete event simulation, Elevator planning, Robustness, Process visualization, Validation

1 Introduction

Research on the physics and dynamics of construction hoists has attracted the attention of many researchers [1-2]. Now with the increase in high-rise construction, hoists have become a key important vertical transport vehicle for workers. So planning of the hoist capacity for a building construction has become very important. Each building has its own critical traffic period which is the busiest time of the hoist [3]. For each kind of building, there may have three critical traffic periods: up-peak (morning rush hours), down peak (evening rush hours) and down-up peak (lunchtime rush hours) [4]. If a hoist is sufficient to serve the critical traffic periods of the building, the rest of the time will not be a problem [3]. Two commonly used parameters to select hoists for a building are total time needed to send all the workers to their destinations and average waiting time for workers [5-6].

During the period of design and construction of a tall building, some key parameters for the hoist operation have to be considered such as average waiting time for workers, or total time needed to send all workers to their destinations [6-8]. Some authors use mathematical methods to determine these parameters for hoist planning [9-10]. These methods while useful, have some practical problems; it is not easy for those with limited mathematical knowledge to understand and use these methods to analyze the time spent and waiting time criteria when designing hoists arrangements for building design or construction site.

In this paper, a simulation model is proposed to analyze the operation of hoists in down-peak periods for a moderately tall building (for example 15 floors) with one hoist which have a maximum capacity of 14 workers. The model is a hybrid of agent based methodology with discrete event methodology to easily get parameters discussed above for hoist operation.

In the paper, we first describe the down-peak, next build a model for this situation; then the validation of the model is tested, and lastly the base model is expanded to random situation with any arrival distribution.

2 Description of the down-peak mode of construction hoists

In the lunch break or after work time of workers on floors above the ground want to go down to the ground floor (floor1). Workers from a given floor will first leave their work site and arrive at the waiting area of the floor, then call the hoist, if the hoist is just there taking the hoist and leaving or waiting for the hoist. If we assume the hoist first stays idle at the ground floor (floor 1). If there is user calling for it from a given floor above, the hoist will go upward to pick the user. During the going upward process, if someone whose floor number is bigger than the floor number of the user who just called the hoist, the hoist will go to the user with the highest floor number. During the downward process, when the hoist arrives at each floor, if it has spare capacity and there are workers waiting on this floor, it will load workers on this floor, if the hoist is full loaded or there is no user waiting on this floor, it will go down directly to the next floor. When the hoist arrives at the ground floor, it will unload all the workers in it and wait for another call. During the down-peak, it is assumed that there is no user who wants to go upward. And also there is no user who wants to go upward or downward between the floors. The situation can be schematically illustrated as Fig.1.

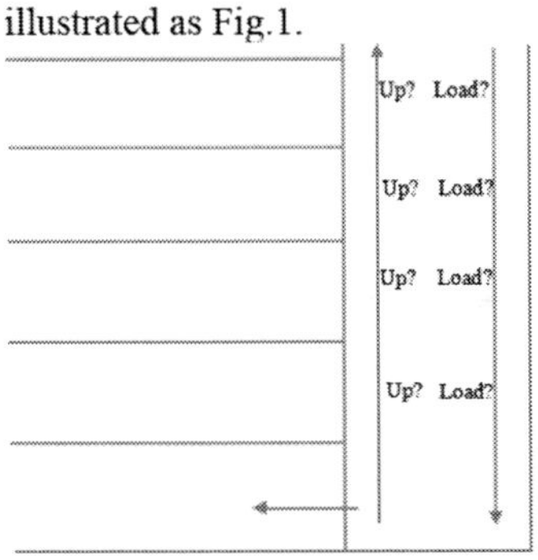

Fig. 1. Hoist's operation in down-peak

3 Base model of the down-peak situation

3.1 The behaviour of user

Each user leaves his work site in the building and arrives at the waiting area of his floor waiting for the hoist. The workers' behavior can be treated as a process. A process flow chart is used to express the behavior of workers on each floor (see fig.2).

The behavior of workers on each floor is expressed with two blocks. The first block (for example *f2*) is a source block to express workers' behavior of leaving his work area on a given floor (floor2) and arriving at the waiting area. So when a user arrives at the waiting area of the floor, the source block will generate an entity. The second block (for example q2) is a queue block to express the workers' behavior of waiting on a given floor (floor2). The function of the block is to store the entity generated by the corresponding source block. A *person* agent is predefined to represent the user entity.

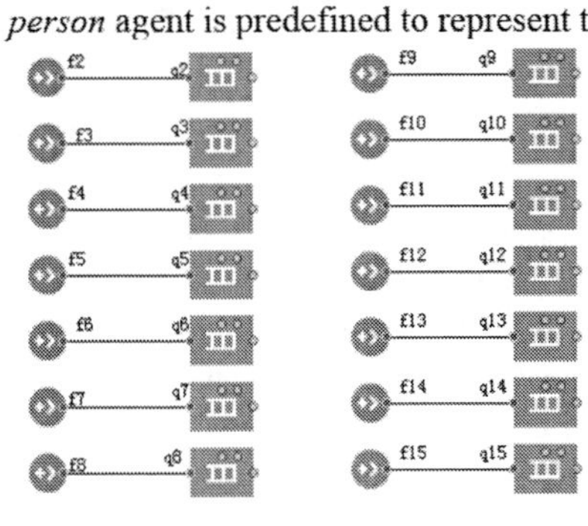

Fig. 2. User's behavior in down-peak

3.2 The behaviour of hoist system

The hoist's behavior can be represented as a series of events that can occur in one or more possible states. A state chart of agent-based modeling methodology is used to define the behavior of the hoist system (see fig.3).

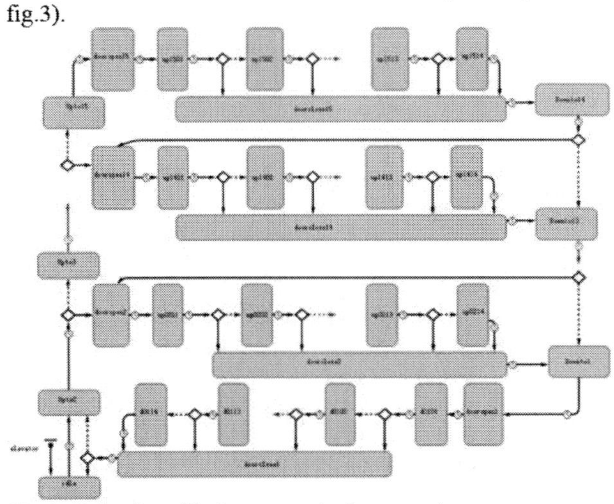

Fig.3. Behavior of hoist system in down-peak

The hoist first stays idle on the ground floor (state *idle*). If there are workers calling for it, it will go upward (transiting to state *Upto2*) to pick them up, or the hoist will stay idle, so the transition from the state *idle* to state *Upto2* is a condition transition. It will take the hoist some time to arrive at floor2, so the state *Upto2* is followed a time transition, and all states in fig.3 are followed by time transition except state *idle* because time are needed to transit from these states to other states. When the hoist arrives at floor2, it will either keep going upward or opening door and loading workers on this floor. So the time transition coming from state *Upto2* is followed by two branches, one goes to state *Upto3*, the other goes to state *dooropen2*.

If the hoist opens the door and loads on this floor, its state will transit to *dooropen2*, it will then load the first user on the floor (transiting to state *up0201*), after loading the first user in the waiting queue of the floor, it will then either load the second user (transiting to state *up0202*) or close the door (transiting to state *doorclose2*), because after the loading the hoist may not have spare capacity to hold user or there may not have workers waiting on the floor. The hoist can load 14 workers in maximum because the maximum capacity of the hoist is set to 14 workers. So after loading the 14th user, the hoist will close the door. If the hoist keeps going up (transiting to state *Upto3*) but not to load on the floor, when it arrives at floor3, it will have similar behavior as behavior on floor2. These similar behaviors on floor3 to floor13 are omitted in fig.3.

When the hoist loads workers on a given floor, after closing the door, it will go down to the next floor (transiting to state *Downto#*). When it arrives at the next floor, if the hoist has some spare capacity and there are workers waiting on the floor, the hoist will open the door and load workers, if the hoist has no spare capacity or there is no user waiting on the floor, the hoist will keep going down to another floor. That means when the hoist goes down to a given floor above the ground, it either loads workers on the floor or keeps going down to another next floor. So, each of the time transition of state *Downto#* is followed by two branches. For example, if the hoist closes the door on floor15 (state *doorclose15*), it will go down to floor 14 (transiting to state *Downto14*). When the hoist arrives at floor14, it will make a decision about opening the door (transiting to state *dooropen14*) and load workers or going down to floor13 (transiting to state *Downto13*).

When the hoist goes down to ground floor (floor1), it will open the door (state *dooropen1*), and unload the first user (state *d0101*) in the hoist, after unloading the first user, the hoist will check if there are workers in the hoist, if there are, the hoist will keep unloading the second user until the last user is unloaded, if there is not, the hoist will close the door. There are 14 possible workers to be unloaded (states *d0101* to *d0114*) because the hoist's maximum capacity is 14 workers. The time transition of each unloading state (states *d0101* to *d0113*) is followed by two branches except the state *d0114*. One is pointing to the next unloading state (one of the states *d0102* to *d0114*) and the other is pointing to state *doorclose1*.

After unloading workers and closing the door, if there are workers waiting on the floors, it will choose to go up again, if there is not, it will stay idle on the ground floor. So the time transition of state doorclose1 is followed by two branches, one is pointing to state *Upto2*, the other is pointing to state *idle*.

3.3 Parameters, collection, variables, Java codes or expressions for the base model

The behaviors of workers and hoist system are expressed with different methodologies. To construct the model , the *Anylogic* simulation tool (education version) is used to incorporate the discrete-event process flows chart with the agent based state chart. It is said that *AnyLogic* is the only simulation tool that supports all the most common situation methodologies in place today: system dynamics, process-centric, and agent based modeling [11-13]. The discrete-event process flow chart is put in the *main* agent of AnyLogic with a *hoist* agent collection named hoists. The state chart of the hoist system is embedded in the *hoist* agent of collection hoists.

Some parameters, collection, variables, events, Java codes and expressions are used to construct the model. The hoist first stays idle on the ground floor (floor1), when there is any user calling for it, it will leave state *idle*. So the trigger condition for the transition from the state *idle* is that at least one of the queues on the floors is not empty (see fig.4).

Condition:

```
get_Main().q2.size()>0||
get_Main().q3.size()>0||
get_Main().q4.size()>0||
get_Main().q5.size()>0||
get_Main().q6.size()>0||
get_Main().q7.size()>0||
get_Main().q8.size()>0||
get_Main().q9.size()>0||
get_Main().q10.size()>0||
get_Main().q11.size()>0||
get_Main().q12.size()>0||
get_Main().q13.size()>0||
get_Main().q14.size()>0||
get_Main().q15.size()>0
```

Fig.4. Trigger condition for the hoist leaving state idle

A predefined parameter in the *hoist* agent named *Onefloortime* is used to represent the time needed for the hoist to go through one floor (see fig.5). The timeout values of the transitions from states *Upto#* and *Downto#* are defined by *Onefloortime* respectively.

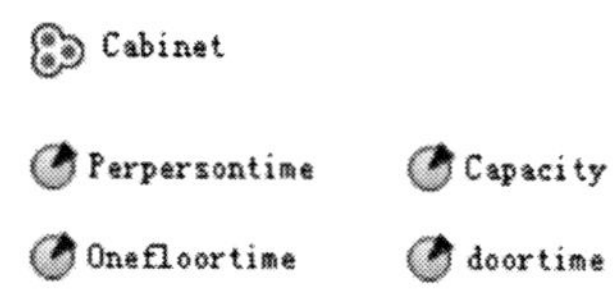

Fig.5. Parameters and collection in the hoist agent

When the hoist arrives at floor2, the condition for the hoist to open the door and loads workers on this floor is that there are workers waiting for the hoist on floor2 but if there is not any user waiting for the hoist above this floor at this moment. That means the queue on floor2 is not empty, but all the queues above floor2 are empty (see fig.6), the condition for the hoist to to go up to floor3 (state *Upto3*) can be set to default.. If the hoist goes up to a given floor, the condition for the hoist choosing to open the door can be similarly defined. If the queue of the floor is not empty, but the queues above the floor are empty, the condition for the other branch is set to default and all the condition for the branches with dashed lines are set to default.

Condition:

```
get_Main().q2.size()>0&&
get_Main().q3.size()==0&&
get_Main().q4.size()==0&&
get_Main().q5.size()==0&&
get_Main().q6.size()==0&&
get_Main().q7.size()==0&&
get_Main().q8.size()==0&&
get_Main().q9.size()==0&&
get_Main().q10.size()==0&&
get_Main().q11.size()==0&&
get_Main().q12.size()==0&&
get_Main().q13.size()==0&&
get_Main().q14.size()==0&&
get_Main().q15.size()==0
```

Fig.6. Condition for the hoist to transit from state Upto2 to dooropen2.

A parameter named *doortime* is predefined in the hoist agent to represent the time needed to open or close the door (see fig.5). All the timeout values of states *dooropen#* and *doorclose#* are defined by the *doortime* parameter. Another parameter named *Perpersontime* is also predefined in the hoist agent, this parameter is used to represent time needed to load or unload a user. The timeout value of each state to express the time need to load or unload a user is defined by this parameter.

When the hoist begins to load a user, the first user in the queue of the floor will go into the hoist. The user's positive action is treated as passive action executed by the hoist. The action of a user going into the hoist is treated as the hoist removing a user from the waiting queue of the floor and adding it into the hoist. So when the hoist enters into a loading state, it will execute these actions. For example, if the hoist is loading user on floor2, the hoist will execute the actions in fig.7. *Cabinet* is an agent collection predefined in the *hoist* agent to represent the car of the hoist (see fig.5). The Java codes in fig.7 show that the hoist removes the first user (person agent) from queue of *q2*, define this user as *man*, and add this *man* into the hoist (*Cabinet*).

Entry action:

```
Person man=get_Main().q2.removeFirst();
Cabinet.add(man);
```

Fig.7. Action executed by loading state on floor2

After loading each of the workers on a given floor, the condition for the hoist to choose to close the door is that the queue on the floor is empty or the hoist is full loaded now. For example, if the hoist is now on floor2, after loading a user, the condition for the hoist to close the door is `get_Main().q2.size()==0||(Capacity-Cabinet.size()==0)`. Where *Capacity* is a parameter predefined in the *hoist* agent to represent the maximum capacity of the hoist, its value is 14 in our model. (See fig.5).

If the hoist has loaded workers and closed the door, it will go down to the next door (transiting to state *Downto#*), when it arrives at the next floor (floor#). The condition for the hoist to open the door to load user on the floor (state *dooropen#*) is that the hoist still has some spare capacity and there are workers waiting on floor# (`get_Main().q#.size()==0&&(Capacity-Cabinet.size()==0)`). But when the hoist arrives at ground floor (floorl), the hoist opens the door and loads workers.

When the hoist begins to unload a user it will remove a user from its car, the action of unloading state on the ground floor is to remove a *person* agent from the *Cabinet* collection (`Cabinet.removeFirst();`). When a user is unloaded, the total number of workers moved down is recorded by a predefined variable named *totalpersonmoved* in the *main* agent. So when the hoist leaves an unloading state (states *d0101* to *d0114*), it will execute an action to make the parameter *totalpersonmoved* increased by one (`get_Main().totalpersonmoved++`). After unloading a user on the ground floor, the condition for the hoist to close the door is that there is not any user in the hoist (`Cabinet.size()==0`).

If all the workers on the floors are moved down to the ground floor, the hoist will go into state idle, the moment is set as the time spent to move all the workers. So when the hoist enters into state *idle* and the total workers moved are equal to the number of workers needed to be moved, the hoist will record the time. The entry action of state idle will be like action in fig.8 if the total number of workers needed to be moved is 336. *Timespent* is a predefined variable in the main agent to represent the time needed to remove all workers down to ground floor.

Entry action:

```
if(get_Main().totalpersonmoved==336)
get_Main().Timespent=time()
```

Fig.8. Entry action of state idle

To get the waiting time and average waiting time of workers on each floor, eight variables are used for each queue of a given floor. For example, eight variables in fig.9 are used for the queue on floor2 to record the waiting time and average waiting time of workers on the floor. Number 2 is used to represent the floor number; variables for the queues on other floors can be expressed in the similar way. The variables *Intime2*, *Outtime2* and *waitingtimearray2* are array variables, the others are plain variables. When a user enters into the queue, an index is tagged to him to represent the entering order of the user and when a user goes out of the queue, an out index is tagged him to the out order of the user. For example, the first user entering the queue is tagged by number 1 and the second by number 2. The first user out of the queue is tagged by number 1 and the second by number 2, and so on. When a user enters into the queue, the entering time is recorded by *Intime2* variable (see the on enter action in fig.9a). And when a user goes out of the queue, the *outtime* variable is used to record the out time of the user (see the on remove action in fig.9a). The waitingtimearray2 is used to record the waiting time of the user going out of the queue. The other variables are used to get the average waiting time of workers having gone out of the queue (see the on remove action in fig.9b).

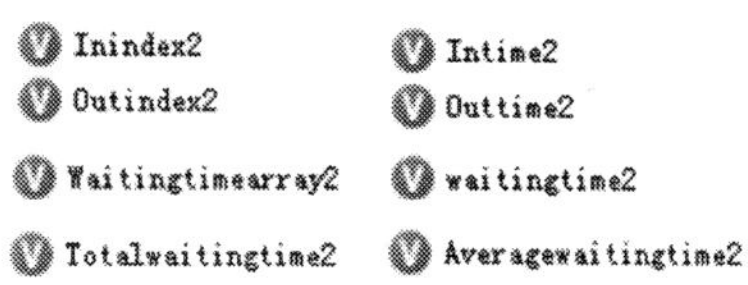

Fig.9a. Variables used for the queue on floor2

On enter:

```
Intime2[Inindex2]=time();
Inindex2++;
```

On remove:

```
Outtime2[Outindex2]=time();
Waitingtimearray2[Outindex2]=Outtime2[Outindex2]-Intime2
waitingtime2=Waitingtimearray2[Outindex2];
Totalwaitingtime2+=waitingtime2;
Outindex2++;
Averagewaitingtime2=Totalwaitingtime2/(Outindex2);
```

Fig.9b. Actions for the queue on floor2.

4 Validation test for the base model

To test the model's validation, the values of parameters for the hoist are assumed to be certain, the values of door time, per person time and one floor time are 1.5, 1, and 1.5 seconds respectively. The capacity of the hoist is 14 workers. The total number of workers that need to be moved is 168, and all the workers are waiting in the queue on a floor when the simulation starts.

4.1 Testing the time spent parameter

All the workers are first put on floor15 then floor14 and finally on the floor2. To put all the workers on a given floor at the beginning of the simulation, a system function `inject()` is used in the *on startup* field of *main* agent. For example, if all the 168 workers are on floor15, `f15.inject(168)` is used in the *on startup* field of agent Main to put all the 168 workers in the queue on floor15. All the outcomes for each situation are show in fig.10. If all the workers are on floor15, the total time spent is 912 seconds, if all the workers are floor9; the total time spent is 696 seconds and if all the workers are on floor2, the total time spent is 444 seconds.

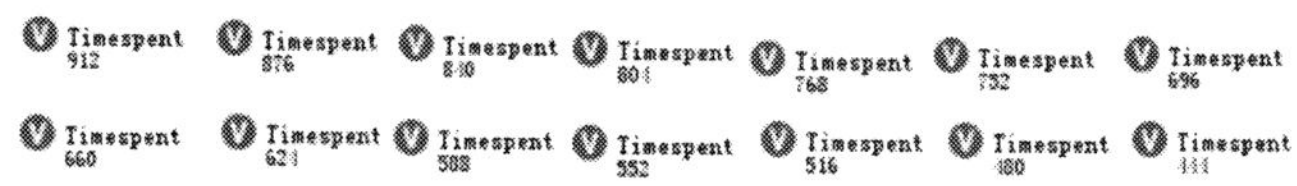

Fig.10. Time spent outcome for the time spent parameter testing.

To check if these outcomes are correct, the values of time spent parameter for each of these situations are calculated (see table1). The total workers are 168; it will take the hoist to go 12 runs (168/14) for each of the situation. The time spent for one run will be decided by the door time, one floor time, per person time and workers' floor number (*doortime**4+*perpersontime**14*2+*onefloortime**(floor number -1)*2). If all the workers are on floor15, the time spent for run is 76 seconds (1.5*4+1*14*2+1.5*14*2). The total time needed for moving down all the workers on floor15 is 912 (12*76). Total time needed for moving all the workers down to floor1 in other situation are calculated in table1, the outcomes of simulations (see fig.10) are the same with the calculations values (see table1).

Table 1. the values of time spent and average waiting time for 1-3lifts situation.

Workers' floor number	Time needed for one run	Total time needed to move all the workers down to floor1
15	76	912
14	73	876
13	70	840
12	67	804
11	64	768
10	61	732
9	58	696
8	55	660
7	52	624
6	49	588
5	46	552
4	43	516
3	40	480
2	37	444

4.2 Testing the average waiting time parameter

If all the workers are on floor15, the simulation outcome of the workers' average waiting time is 447 seconds (see fig.11). The calculation outcome of this situation is 447 seconds too (see table2).

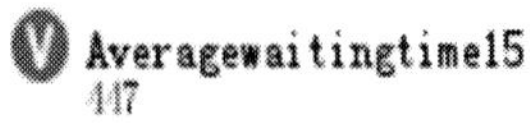

Fig. 11. Average waiting time outcome of simulation for workers all on floor15

In this situation, for the first run, the waiting time of the first user in the queue is the time spent by the hoist to move itself from floor1 to floor15 and open the door. It is equal to 22.5 seconds (1.5*14+1.5), the second workers waiting time will be increased by one second because he will wait for the first user to go into the hoist. The waiting times of the first workers in the second run, third run will be 98.5 (76+22.5), 174.5(76*2+22.5). And the waiting time of the first user in the last run will be 858.5 (76*11+22.5). The average waiting time for all the workers is 447 seconds.

Table 2. The calculation values of average waiting time for all workers on floor15.

Number of run	1	2	3	4	5	6	7	8	9	10	11	12
Workers moved	14	28	42	56	70	84	98	112	126	140	154	168
1st user's waiting	22.5	98.5	174.5	250.5	326.5	402.5	478.5	554.5	630.5	706.5	782.5	858.5
2nd user's waiting	23.5	99.5	175.5	251.5	327.5	403.5	479.5	555.5	631.5	707.5	783.5	859.5
3rd user's waiting	24.5	100.5	176.5	252.5	328.5	404.5	480.5	556.5	632.5	708.5	784.5	860.5
4th user's waiting	25.5	101.5	177.5	253.5	329.5	405.5	481.5	557.5	633.5	709.5	785.5	861.5
5th user's waiting	26.5	102.5	178.5	254.5	330.5	406.5	482.5	558.5	634.5	710.5	786.5	862.5
6th user's waiting	27.5	103.5	179.5	255.5	331.5	407.5	483.5	559.5	635.5	711.5	787.5	863.5
7th user's waiting	28.5	104.5	180.5	256.5	332.5	408.5	484.5	560.5	636.5	712.5	788.5	864.5
8th user's waiting	29.5	105.5	181.5	257.5	333.5	409.5	485.5	561.5	637.5	713.5	789.5	865.5
9th user's waiting	30.5	106.5	182.5	258.5	334.5	410.5	486.5	562.5	638.5	714.5	790.5	866.5
10th user's waiting	31.5	107.5	183.5	259.5	335.5	411.5	487.5	563.5	639.5	715.5	791.5	867.5
11th user's waiting	32.5	108.5	184.5	260.5	336.5	412.5	488.5	564.5	640.5	716.5	792.5	868.5
12th user's waiting	33.5	109.5	185.5	261.5	337.5	413.5	489.5	565.5	641.5	717.5	793.5	869.5
13th user's waiting	34.5	110.5	186.5	262.5	338.5	414.5	490.5	566.5	642.5	718.5	794.5	870.5
14th user's waiting	35.5	111.5	187.5	263.5	339.5	415.5	491.5	567.5	643.5	719.5	795.5	871.5
Total waiting	406.0	1470.0	2534.0	3598.0	4662.0	5726.0	6790.0	7854.0	8918.0	9982.0	11046.0	12110.0
Accumulated total waiting time	406	1876	4410	8008	12670	18396	25186	33040	41958	51940	62986	75096
Average waiting time	29.00	67.0	105.0	143.0	181.0	219.0	257.0	295.0	333.0	371.0	409.0	447.0

5 Expand the base model to random situations

In reality, the parameters of the hoist and arrival distribution of workers may be random. To expand the base model to the random situation, the arrival distributions of workers should be expressed as random. The other is that enough runs of simulation should be done to get robust simulation outcomes.

Some assumptions are made to explain the expanding. The one floor time is assumed to be between 1.5 and 2 seconds uniformly, the door time is between 1and 1.5 seconds uniformly, the per person time is between 0.5 and 1 uniformly. There are 24 workers on each floor. Their arrival distribution is one user every 20 to 30 seconds.

5.1 Random variables in the model

In the random situation, some of the parameters or variables are random. The system function of the *AnyLogic* software can be used to define these random parameters and variables. For the assumption numerical example, function `uniform()` is used to express the parameters of the hoist. The default values of parameters *doortime*, *onefloortime* and *perpersontime* are set to `uniform(1,1.5)`, `uniform(1.5,2)` and `uniform(0.5,1)` respectively.

In the example, the arrival distributions of workers are random too; events can be used to define the arrival distribution. There are three kinds of events: Timeout, Rate, and Condition. A Timeout event with cyclic mode is used to define the arrival distribution of workers on each floor in our example. The Event in fig.12 is used to define the arrival distribution of workers on floor2. In 20 to 30 seconds, a user will come, and the number of user on this floor should not more than 24. `count()` is a system function used to count the number of user that have been injected into the source block *f2*.

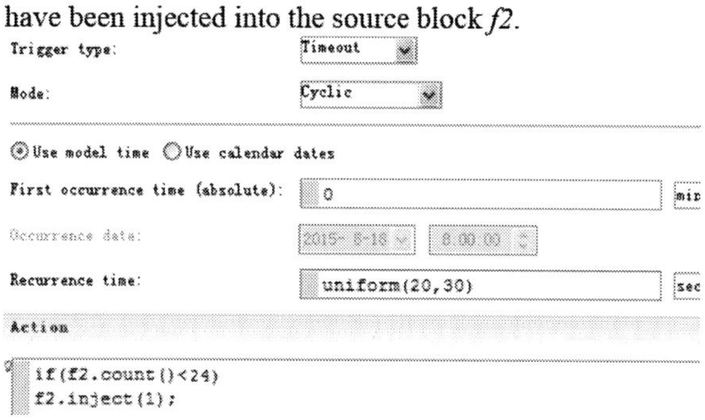

Fig.12. Arrival distribution defining with an event

5.2 Run the simulation for enough time

To get robust simulation outcomes, the model should be able to run enough times. The simulation is repeated. A variable named N is first defined in the agent of *main* to represent the number of runs that has been finished setting the initial value of N to 1. Then the finishing condition of each run is defined in the Entry action field of state *idle* of the hoist. For the example, the finishing condition is defined as when the simulation is on the Nth run, its finishing condition is that the workers that have been moved (*totalpersonmoved*) is N times of total workers needed to be moved (variable *Totalpersonneedtomove*) , variable is a predefined variable in the *main* agent, one run is finished, the value of N will be increased by one (see fig.13).

```
Entry action:
if(get_Main().totalpersonmoved==get_Main().N*get_Main().Totalpers
get_Main().N++;
}
```

Fig.13. Finishing condition for each run.

When the simulation runs are repeated, the arrival distribution is also repeated. The change in code is shown in fig.14.

```
Action
if(f2.count()<N*24)
f2.inject(1);
```

Fig.14. Repeating workers' arrival distribution for workers on floor2

When the simulation has finished the specified runs, average time spent of all runs and average waiting time for all the workers are calculated. An event named *Pausesimulation* is used to pause the simulation. Two variables, *Averagetimespent* and Averagewaitingtime are used to calculate the two time parameters; these two variables are predefined in the *main* agent. The condition of the event is defined by the variable N and the specified runs. If the specified runs are 100, then the condition and the action of the event can be expressed as condition and action in fig.15.

```
Condition: N>=101
Action
pauseSimulation();
Averagetimespent=time()/(N-1);
Averagewaitingtime=(
Averagewaitingtime2+
Averagewaitingtime3+
Averagewaitingtime4+
Averagewaitingtime5+
Averagewaitingtime6+
Averagewaitingtime7+
Averagewaitingtime8+
Averagewaitingtime9+
Averagewaitingtime10+
Averagewaitingtime11+
Averagewaitingtime12+
Averagewaitingtime13+
Averagewaitingtime14+
Averagewaitingtime15)/14
```

Fig.15. Condition and action of the event Pause simulation

For the assumed example, after 100 runs of simulation, values of variable *Averagetimespent* and *Averagewaitingtime* can be determined. The average time spent is 1303 seconds and average waiting time is 473 seconds (see fig.16).

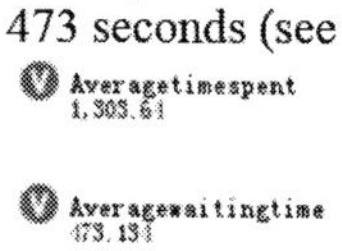

Fig.16.Outcome of the simulation after specified runs

6 CONCLUSION

In this paper we propose a down-peak model with single hoist. The model hybrids discrete-event methodology with agent based methodology. The model can be used in certain condition and random situations. Using this model we can analyze the time spent parameter and average waiting time parameters for hoist planning. If there are two more hoists but each hoist is with its independent waiting queue, it can also be treated as independent single hoist situation.

In this paper, a down-peak model with single hoist is proposed. The model is a hybrid of discrete-event methodology with agent based methodology to simulate the operation of hoist in the down peak. The discrete-event methodology is used to simulate the behavior of the workers on each floor, and the agent-based methodology is used to simulate the behavior of the hoist system. The model's validation is tested under deterministic conditions and then expanded to handle random situations. The two commonly used operation parameters (time spent and average waiting time) can be determined for hoist planning through model simulation. With two or more hoists, during the rush hours, if each hoist has a separate waiting line, each hoist can be treated as an independent hoist. If all the hoist s only have one waiting line, hoists' behavior will be affected by each other, in that case, the model will need to be modified. That will form the next stage of this work.

REFERENCES

[1] Bao Ding, Yong-Ming Zhang, Xi-Yuan Peng (2013). A hybrid approach for the analysis and prediction of hoist user flow in an office building. Automation in construction, 35 69-78 DOI 10.1016/j.autcon.2013.03.003

[2]M.L. Siikonen (1997). Customer service in an elevator system during up-peak, Transportation Research 31(2) 127-139

[3] G.F. Newell(1998). Strategies for serving peak hoist traffic, Transportation Research 32(8) 583-588

[4] T. Nagatani(2003). Complex behaviour of hoists in peak traffic, Physica A 326 556-566 DOI:10.1016/S0378-4371(03)00278-4

[5] T. Nagatani(2004). Dynamical transitions in peak hoist traffic, Physica A 333 441-452 DOI 10.1016/j.physa.2003.10.001

[6] Yutae Lee, Tai Suk Kim, Ho-Shin Cho, Dan Keun Sung, Bong Dae Choi (2009). Performance analysis of an elevator system during up-peak, Mathematical and Computer modelling, 49 423-431DOI 10.1016/j.mcm.2008.09.006

[7] Yoonseok Shin, Hunhee Cho, Kyung-In Kang (2011). Simulation model incorporating genetic algorithms for optimal temporary hoist planning in high-rise building construction, Automation in construction, 20 550-558 DOI 10.1016/j.autcon.2010.11.021

[8] Stefan Heinz, JÖrg Rambau, Andreas Tuchscherer (2014). Computational bounds for hoist control policies by large scale linear programming, Math Meth Oper Res 79 87-117 DOI 10.1007/s00186-013-0454-5

[9] U Beißert, M König, H-J Bargstädt (2010). Soft constraint-based simulation of execution strategies in building engineering, Journal of simulation, 4 222-231 DOI 10.1057/jos.2010.8

[10] Strakosch, G. R. (2010). *The vertical transportation handbook* (Vol. 4). Hoboken, NJ, USA: John Wiley & Sons.

[11] Arashpour, M. and M. Arashpour (2015). "Analysis of Workflow Variability and Its Impacts on Productivity and Performance in Construction of Multistory Buildings." Journal of Management in Engineering 31(6): 04015006.

[12] Arashpour, M., R. Wakefield, N. Blismas and T. Maqsood (2015). "Autonomous production tracking for augmenting output in off-site construction." Automation in Construction 53: 13-21.

[13] Andrei Borshchev (2013), The big book of simulation modeling: multi-method modeling with AnyLogic 6, Lisle,IL:Anylogic North America.

[14] Arashpour, M., R. Wakefield, N. Blismas and B. Abbasi (2016). "Quantitative analysis of rate-driven and due date-driven construction: Production efficiency, supervision, and controllability in residential projects." Journal of Construction Engineering and Management 142(1): 04015006.

Alternative Scheduling and Planning Processes for Hybrid Offsite Construction

Tarek Salama, Ahmad Salah, and Osama Moselhi
Department of Building, Civil and Environmental Engineering, Concordia University, Montréal, Canada

E-mail: salamatarek@gmail.com, aaasalah@hotmail.com, osama.moselhi@concordia.ca

Abstract

Offsite construction has gathered momentum in recent years due to its improved performance in terms of project cost, schedule, quality, environmental impact, and safety. Offsite prefabricated systems vary depending on the size of prefabricated components, which affect the need for onsite construction. These systems include categories such as modular, panelized, prefabricated, and processed materials construction. Each category has its own unique practices and can be blended with other categories in a "hybrid" offsite construction system. Several research studies have introduced scheduling and planning techniques for panelized and hybrid offsite construction using BIM and simulation tools. This paper thus presents an alternative BIM-based integrated framework for modelling and planning of hybrid offsite construction projects. BIM software (eg. Vertex BD) is used in the proposed framework for automating the data exchange between BIM model and the proposed method for scheduling of hybrid construction system. The objective of the proposed method is to develop a practical planning and scheduling alternative at the components level for offsite hybrid construction project. The proposed method integrates the linear scheduling method (LSM) and the buffering techniques of the critical chain project management (CCPM) into a comprehensive BIM-based framework while considering uncertainty associated with activities durations. This paper first introduces the literature pertaining to planning, and scheduling techniques for offsite construction; then, it presents the proposed computational along with a case study to verify the efficiency of the proposed integrated framework. Finally, conclusions related to implementation of the proposed framework are summarized.

Keywords – BIM, Panelized Construction, Planning, Hybrid Offsite Construction Introduction.

1 Introduction

Offsite construction and prefabrication facilitate the work simultaneously on offsite and on-site deliverables which may reduce considerably project schedule. Reduction in schedule leads to significant cost savings due to the reduced need for expensive and labor intensive onsite operations.

McGraw-Hill smart market report [1] asserts that 67% of firms reduced project schedules using prefabrication and modularization, and 35% of firms experiencing decreases of four weeks or more. Offsite construction provides valuable assistance for tight scheduling where project deadlines are usually inflexible as the higher education buildings sector [1].

Two features of offsite and modular construction are controlling the determination of its appropriate scheduling techniques. . The first is the repetitive nature of manufacturing for offsite construction, and the second is the uncertainty and variability of these operations. Thus, the traditional scheduling techniques that depend on the critical path method (CPM) cannot be applied due to lack in satisfying the resource continuity constraint in repetitive projects [2].

The linear Scheduling Method (LSM) is an advancement of the traditional line of balance (LOB) technique [3] that is commonly used for scheduling of repetitive projects. LSM uses similar mathematics to LOB though it represents the activity as a single line rather than dual lines as in LOB. However unlike LOB, LSM allows for representing non-repetitive and non-typical activities in complex construction projects.

The (LSM) diagram outlines time along the horizontal axis and the number of repetitive units along the vertical axis. Hence, it would be a great alternative to schedule offsite and modular construction though a few methods incorporate uncertainty with LSM in the literature. The other alternative for LSM is the use of simulation engines due to its capabilities for considering the cycles of repetitive activities and uncertainty.

However, the simulation models require dedicated simulation professionals [4] and it is usually tailored based on the special needs and requirements of the project being considered. Other studies [5], [6] and [7] integrate simulation with Building Information Modelling (BIM) to automate quantity take-off data and to visualize offsite and modular construction activities. BIM assists in automating the quantity-take off for modular construction components including the properties of modules, openings, and framing. Such automation improves the accuracy of schedule that can be integrated with other techniques to produce 4D and 5D schedules.

2 Literature Review

Limited number of scheduling techniques has been developed for off-site or modular construction projects. In this respect, the review focuses on three perspectives; linear scheduling, critical chain project management, and off-site construction scheduling.

2.1 Linear Scheduling Method (LSM)

The linear graphical techniques were used as early as 1929 on the empire state building, and then it was developed by the Goodyear Company in the 1940s and expanded by the US Navy in the 1950s [3]. Since the early 1970's, many variations of linear scheduling methods have been proposed with different names [2], [8], [9], [10], [11], [12], and [13].

LSM has a different approach to identify the critical path of activities due to resource continuity constraint.

Several methods have been introduced for identifying the controlling activity path (e.g. critical path) for linear schedules based on: controlling sequence of activities [14], controlling points concept [2], resource continuity constraint [15], and time and distance constraints [16]. However, these methods didn't consider the uncertainty associated with linear scheduling. In this respect, Schoderbek and Digman [17] integrated the LOB and the program evaluation review technique (PERT) into one scheduling method to account for repetitive activities and uncertainty. Dressler [18] considered uncertainty for a project by assigning probability distributions to each project segment however this approach requires complex linear programming methods to develop a schedule. Other researchers developed simulation-based methods [19], [20], and [21] and fuzzy set-based methods [22] and [23] to account for uncertainty associated with linear scheduling. However, these methods either depend on data availability, third party software, or tailored for a specific type of repetitive projects.

In a recent study, Slorup [24] suggested the application of critical chain buffer management theory in location-based management to account for uncertainty in linear scheduling without providing any formulation for this concept. The critical chain project management (CCPM) incorporates the critical chain scheduling technique which was introduced in 1997 as an application of the Theory of Constraints (TOC) by Goldratt [25]. The critical supply chain is first identified as the longest chain of activities that determines the project total duration based on availability of resources and durations of activities. Goldratt [25] introduced an aggressiveness procedure that reduces the activities durations confidence level from 95 to 50 percentile. The project buffer consists of the deducted durations of critical chain while, the feeding buffer consists of the deducted durations of non-critical chains as shown in Figure 1. A recent comprehensive review conducted by Ghaffaria et al. [26], highlights 21 aspects related to CCPM and outlines the research potentials in CCPM area.

2.2 Scheduling of offsite construction

Lack of tools that allows for modeling of uncertainty associated with linear scheduling led researches to develop simulation based methods for off-site construction [5], [27], and [28]. Other studies [6] and [7] integrate the Building Information Modelling (BIM) with simulation to facilitate the quantity take-off data for construction project scheduling.

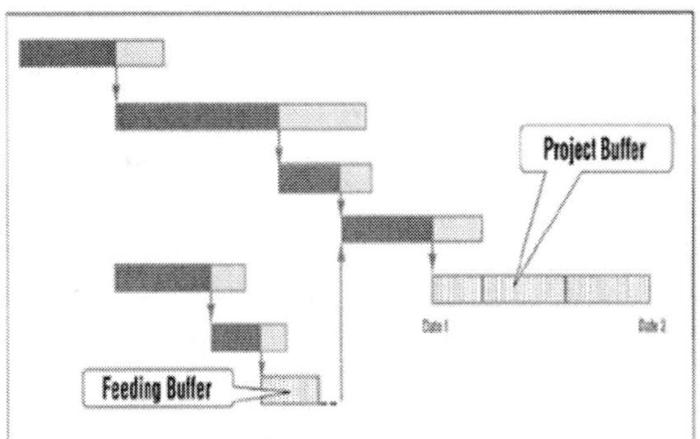

Figure 1: CCPM network with project and feeding buffers [24].

The main limitation identified based on the literature review is the lack of scheduling tools that utilize the linear scheduling under uncertainty for offsite construction projects. Thus, this paper presents a generic scheduling methodology for manufacturing of hybrid and modular offsite construction based on the LSM while considering the uncertainty associated with productivity of crews involved in the manufacturing process using the CCPM buffering technique.

3 Proposed Methodology

The proposed methodology utilizes the critical chain buffer management theory proposed by Slorup [24] to model the uncertainty associated with linear scheduling for manufacturing of offsite construction wall panels. The aggressiveness of CCPM eventually leads to shorter schedules while LSM visualizes the repetitive manufacturing processes.

The proposed framework depends on BIM for automating the data collection, then the collected data is used to generate the linear schedule. A set of assumptions are considered in the proposed methodology as follows:

1- Each of the manufacturing processes (or stations) has one fixed crew with a given productivity rate.
2- Each of the manufacturing processes is considered continuous due to the use of racks that reduce the variability in productivity rates of panels.
3- The manufacturing of panels follows the sequence of on-site erection.

The proposed methodology consists of four main phases: 1) Building Information Modelling (BIM) , 2) Automated data collection, 3) Calculation of activity (i.e. panel manufacturing) duration under uncertainty, and 4) Generation of linear schedules.

3.1 Building Information Modelling (BIM)

The project is modelled using BIM software (e.g. Vertex BD Pro 2016 22.0) that facilitates the modelling with automatic framing capabilities. The BIM model allows for generation of database that includes; property (e.g. Dry wall), dimensions, list of components (e.g. Openings), and the sequence of on-site erection of each panel. Figure 2 presents the interconnections between the BIM model and the project database.

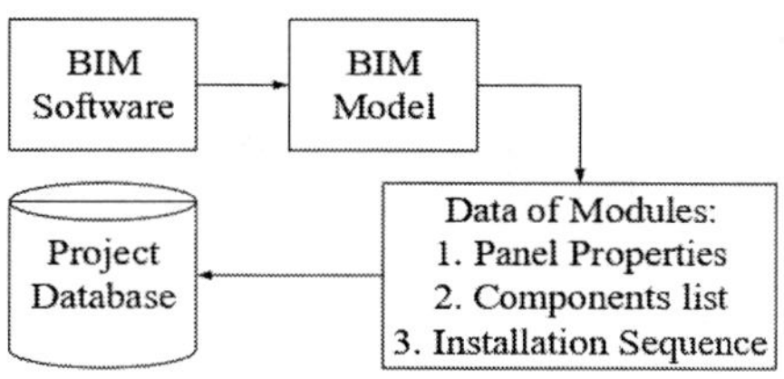

Figure 2: The interconnections between the BIM model and the project database.

3.2 Data Collection

The data collection has two main levels; 1) Project database generated from BIM model and 2) Data for productivity rate of crews involved in the manufacturing processes. The framework of the data collection is presented in Figure 3.

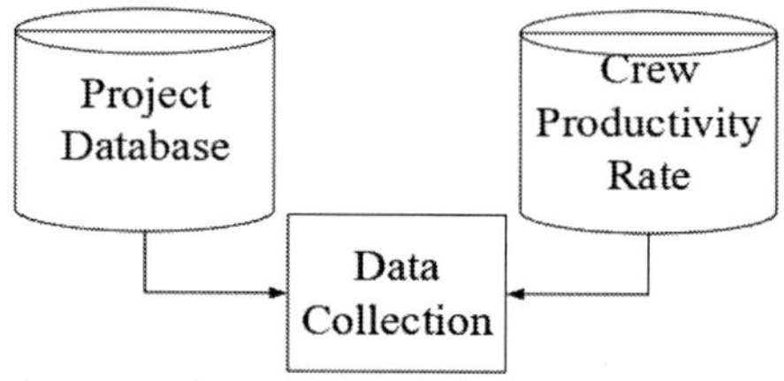

Figure 3: The framework of the data collection.

3.3 Calculation of Activity Duration

Each manufacturing process has "n" number of activities where n represents the number of panels. Each process has one fixed crew with a given productivity rates per the net area of panel which is referred to Average Productivity Rate (APR) in the remaining part of this paper. The activity duration for a given confidence level (CL) percentile (e.g. 50 percentile) is calculated as follows:

$$D_{i,j}(CL) = Area_i \,/\, APR_{i,j}(CL) \qquad (1)$$

Where,
Di,j(CL), represents the duration of panel "i" in process "j" with given confidence level "CL"
Area i, represents the area of the panel i
APRi,j(CL), represents the average panel productivity rate .

3.4 Generation of Linear Schedule under Uncertainty

The proposed method considers the linear schedule at 50 percentile as the basic schedule. However, the uncertainty associated with the linear schedule is modelled using time buffer for each activity at each process. The buffer formula, as shown in Eq. (2), follows the Root Square Error Method (RSEM) [26]. The buffer of each panel "i" at process "j" with a confidence level "CL" is calculated as follows:

$$\text{Buffer}_{i,j}(CL)=\sqrt{\sum_{1}^{i}(D_{i,j}(CL)-D_{i,j}(50))^{2}} \quad (2)$$

Where,
Bufferi,j(CL), represents the uncertainty associated with the duration of panel "i" in process "j" with a confidence level CL.

It should be noted that the CL=50 results of buffer =0 which means that the scheduling with 50 percentile represents the basic linear schedule. However, a higher confidence level generates a buffer to be added to the duration of 50 percentile as presented in Eq. (2).

The linear schedule of a given process "j" can be generated using start and finish dates of each panel "i" which can be calculated using Eq. (3) and (4) respectively. Figure 4 shows the schedule of one process with multiple level of confidence (e.g. 85 percentile).

$$SD_{i,j}(CL) = FD_{i-1,j}(CL) \quad (3)$$

$$FD_{i,j}(CL) = FD_{i,j}(50) + \text{Buffer}_{i,j}(CL) \quad (4)$$

Where,
FDi,j, SDi,j represent respectively start and finish dates of panel "i" at the process "j"

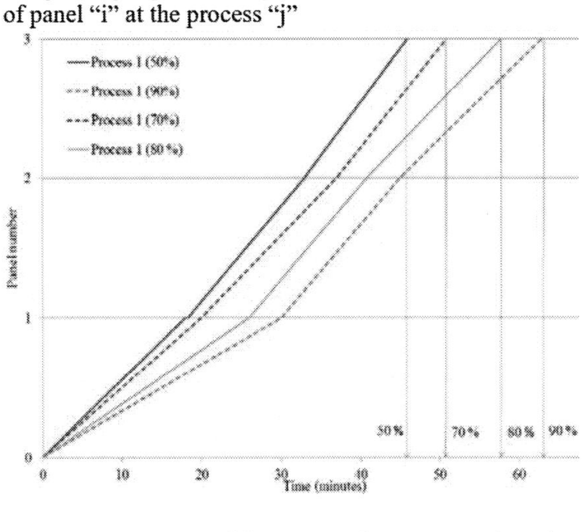

Figure 4: Various confidence rates for one manufacturing process.

In case of a manufacturing process with multiple stations, start and finish dates of activities with a confidence level of 50 percentile are calculated using Eqs. (5) and (6) respectively.

$$SD_{i,j}(50) = \text{Max}\,[FD_{i,j-1}(50), FD_{i-1,j}(50)] \quad (5)$$

$$FD_{i,j}(50) = SD_{i,j}(50) + D_{i,j}(50) \quad (6)$$

Start date of first activity within process "j" is calculated using Eq. (7) as follows:

$$SD_{1,j}(CL) = \text{Max}\,[SD_{1,j}(50), FD_{1,j-1}(CL)] \quad (7)$$

For the succeeded activities of process "j", start and finish dates with a given confidence level (e.g. 80 percentile) are calculated using Eqs. (8) and (9) respectively.

$$SD_{i,j}(CL) = \text{Max}\,[FD_{i-1,j}(CL), FD_{i,j-1}(CL)] \quad (8)$$

$$FD_{i,j}(CL) = FD_{i,j}(50) + \text{Buffer}_{i,j}(CL) \quad (9)$$

The proposed methodology generates two linear schedules for each process as shown in Figure 5. The first represents the linear schedule of each process at confidence level of 50 percentile, while the second represents the linear schedule of each process at a given confidence level. Thus, the proposed methodology illustrates graphically the manufacturing processes of all panels (or modules) at any confidence level (e.g. 85 percentile).

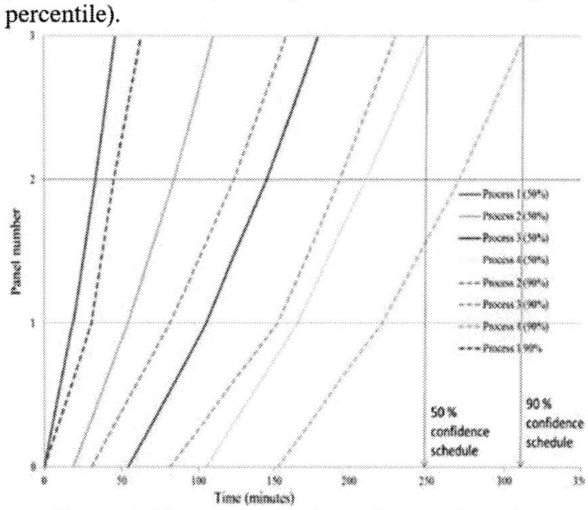

Figure 5: The two curves for each manufacturing process.

The proposed method utilizes the procedure developed by Harmelink [14] for identifying the critical segments within each process. This procedure is based on two main steps, the upward pass and the downward pass.

Step1: Upward pass

The upward pass begins with "the origin activity" which is the first activity/process. While the second activity/ process is named "the target activity" as shown in Figure 6. Harmelink [14] indicated that the controlling activity path occurs where activities/processes are closest to each other. The upward pass determines the potential controlling segments by identifying the least distance (LD) between the origin activity/ process and the target activity/process. The intersection between LD and origin activity is named critical vertex as shown in Figure 6. The LD interval is a potential controlling link between the origin and target activities/ processes as shown in figure 6. This step is repeated until all the potential controlling segments and links are identified.

Step2: Downward pass

The downward pass is equivalent to the backward pass in CPM scheduling, which identifies the activities that in case of delay extend the duration of the project. It starts at the end of last activity/process in the project until the potential controlling link is reached. The segments between these two points belong to the controlling activity path, and the potential controlling link between the last two activities/processes is a controlling link.

The project buffer is then added at the end of controlling activity path by considering the variability of the critical activities/panels only. The project buffer can be calculated using Eq. (8).

$$\text{Project Buffer} = \sqrt{\sum_{k=1}^{k=n} \left(D_k(CL) - D_k(50)\right)^2} \qquad (10)$$

Where,
n, represents the number of activities on the critical path
k, represents the couple (i, j) that indicates the critical activity "i" within process "j".

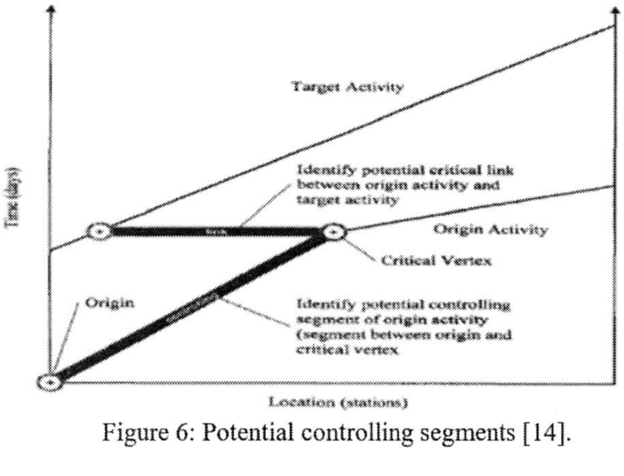

Figure 6: Potential controlling segments [14].

4 Case Study

A hypothetical case study was modelled on Vertex BD for three identical hybrid construction units including 45 panels with different dimensions and properties. A hybrid unit consists of bathroom modules combined with panelized construction forms the hybrid units shown in Figure 7. The bathrooms modules are fabricated as separate units of four panels, while the rest of units are panelized construction.

Vertex BD was used to generate the BIM model that includes panels of the three hybrid units as shown as an isometric view in Figure 8. The project database that includes list of components (panels and modules), properties of each components, and the sequence of installation on site. The net area is calculated based on the dimensions generated from Vertex BD. Figure 9 shows panel framing generated by Vertex BD which illustrate panel dimensions, components, and openings. The productivity rates for four different manufacturing processes (e.g. assembly, framing, etc.) for 3 different groups of panels based on their respective net areas as shown in Table 1. The panels are categorized into three categories; the first represents the panels which have net area less than 60 square feet, the second represents the panels which have net area between 60 and 90, and the third represents the panels which have net area higher than 90 as shown in Table 1. Accordingly, the productivity rates vary from panel to another as shown in Table 1.

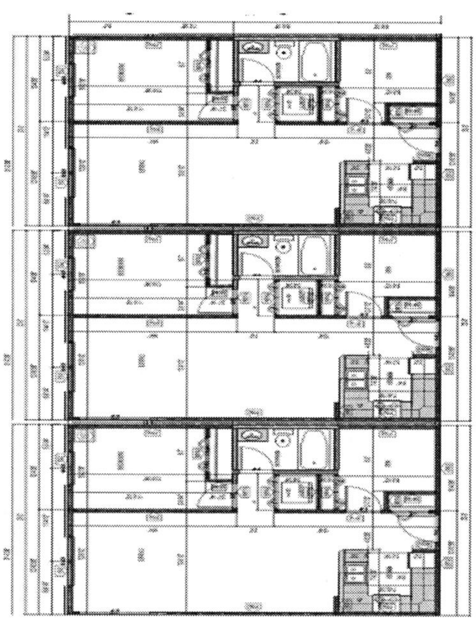

Figure 7: BIM model plan view using Vertex BD.

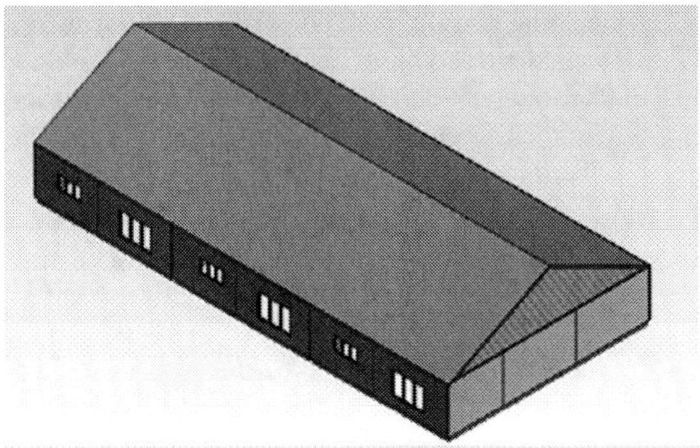

Figure 8: Isometric view of the BIM model

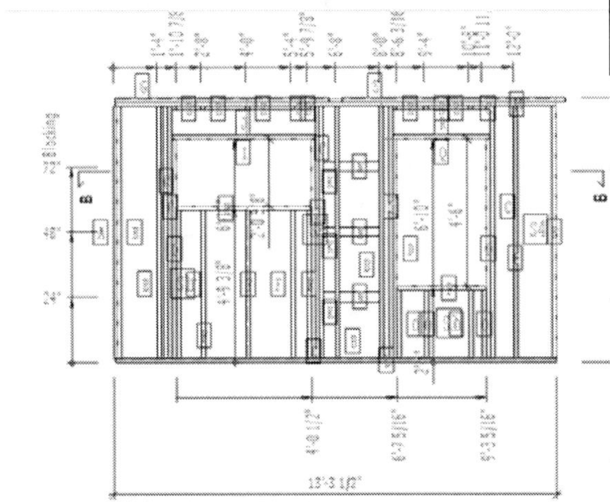

Figure 9: panel layouts generated by Vertex BD.

The net area is calculated based on the panels dimensions acquired from the project database generated by Vertex BD. An EXCEL based tool is generated to facilitate the calculations of duration, start date, finish date and time buffer of each module or panel (i.e. activity). The required calculations for panel durations and buffers are shown in Figure 10. Then, the basic linear schedule for all manufacturing processes is generated using Eqs. (1) - (6) as shown in Figure 11. Then the second linear schedule with 90% confidence rate is generated for all manufacturing processes using Eqs. (7) - (9) by considering all the necessary buffers to account for uncertainty associated with production of panels. The 50 percentile basic schedule and the 90 percentile schedule are shown in Figure 11as the solid and dashed lines respectively.

Table 1 Panel groups productivity rates

Panel Group	Net area (S.F)	Productivity rate (50 %) S.F/min		Productivity rate (90 %) S.F/min	
1	0:60	Process 1	4	Process 1	2.5
		Process 2	2	Process 2	1.1
		Process 3	1.5	Process 3	1
		Process 4	1.3	Process 4	0.9
2	60:90	Process 1	5	Process 1	3
		Process 2	2.5	Process 2	1.5
		Process 3	1.75	Process 3	1
		Process 4	1.5	Process 4	1
3	90:150	Process 1	6	Process 1	3.5
		Process 2	3	Process 2	2
		Process 3	2	Process 3	1.5
		Process 4	1.8	Process 4	1.6

Panel#	PanelName	Aggressive Productivity rate 50 %				Safe Productivity rate 90 %				Aggressive Duration				Safe Duration				NetSF	Buffer			
		Process1	Process 2	Process 3	Process 4	Process1	Process 2	Process 3	Process 4	Process1	Process 2	Process 3	Process 4	Process1	Process 2	Process 3	Process 4		Process 1	Process 2	Process 3	Process 4
1.0	bath	5.0	2.5	1.8	1.5	3.0	1.5	1.0	1.0	18.08	36.17	51.67	60.28	30.14	60.28	90.42	90.42	90.42	12.0	27.0	47.2	56.0
2.0		4.0	2.0	1.5	1.3	2.5	1.0	1.1	0.9	15.03	30.06	40.08	46.24	24.05	60.12	54.65	66.80	60.12	15.1	40.4	49.4	59.7
3.0		4.0	2.0	1.5	1.3	2.5	1.0	1.1	0.9	12.92	25.83	34.44	39.74	20.67	51.67	46.97	57.41	51.67	16.9	47.9	51.0	62.2
4.0		4.0	2.0	1.5	1.3	2.5	1.0	1.1	0.9	12.92	25.83	34.44	39.74	20.67	51.67	46.97	57.41	51.67	18.6	54.5	52.5	64.7
5.0	bath	5.0	2.5	1.8	1.5	3.0	1.5	1.0	1.0	18.08	36.17	51.67	60.28	30.14	60.28	90.42	90.42	90.42	22.2	59.6	65.2	71.4
6.0		5.0	2.5	1.8	1.5	3.0	1.5	1.0	1.0	12.02	24.05	34.35	40.08	20.04	40.08	60.12	60.12	60.12	23.6	61.7	70.1	74.1
7.0		4.0	2.0	1.5	1.3	2.5	1.0	1.1	0.9	12.92	25.83	34.44	39.74	20.67	51.67	46.97	57.41	51.67	24.8	66.9	71.3	76.2
8.0		4.0	2.0	1.5	1.3	2.5	1.0	1.1	0.9	12.92	25.83	34.44	39.74	20.67	51.67	46.97	57.41	51.67	26.0	71.7	72.3	78.2
9.0	bath	5.0	2.5	1.8	1.5	3.0	1.5	1.0	1.0	18.08	36.17	51.67	60.28	30.14	60.28	90.42	90.42	90.42	28.7	75.6	82.1	83.8
10.0		5.0	2.5	1.8	1.5	3.0	1.5	1.0	1.0	12.02	24.05	34.35	40.08	20.04	40.08	60.12	60.12	60.12	29.8	77.3	86.0	86.2
11.0		4.0	2.0	1.5	1.3	2.5	1.0	1.1	0.9	12.92	25.83	34.44	39.74	20.67	51.67	46.97	57.41	51.67	30.8	81.5	86.9	85.5
12.0		4.0	2.0	1.5	1.3	2.5	1.0	1.1	0.9	12.92	25.83	34.44	39.74	20.67	51.67	46.97	57.41	51.67	31.7	85.5	87.8	85.5
13.0	PT4 B	6.0	3.0	2.0	1.8	3.5	2.0	1.5	1.6	23.68	47.36	71.04	78.94	40.60	71.04	94.72	88.80	142.08	35.9	88.7	91.0	90.3
14.0	PT5 B	5.0	2.5	1.8	1.5	3.0	1.5	1.0	1.0	17.39	34.79	49.70	57.98	28.99	57.98	86.97	86.97	86.97	37.8	91.7	98.3	94.8
15.0	B10	6.0	3.0	2.0	1.8	3.5	2.0	1.5	1.6	24.25	48.51	72.76	80.85	41.58	72.76	97.02	90.95	145.53	41.6	94.9	101.2	95.3
16.0	B11	5.0	2.5	1.8	1.5	3.0	1.5	1.0	1.0	16.09	32.19	45.98	53.64	26.82	53.64	80.46	80.46	80.46	42.9	97.3	107.0	99.0
17.0	PT4 B	6.0	3.0	2.0	1.8	3.5	2.0	1.5	1.6	23.68	47.36	71.04	78.94	40.60	71.04	94.72	88.80	142.08	46.1	100.1	109.5	99.5
18.0	PT5 B	5.0	2.5	1.8	1.5	3.0	1.5	1.0	1.0	17.39	34.79	49.70	57.98	28.99	57.98	86.97	86.97	86.97	47.6	102.7	115.7	103.7
19.0	B10	6.0	3.0	2.0	1.8	3.5	2.0	1.5	1.6	24.25	48.51	72.76	80.85	41.58	72.76	97.02	90.95	145.53	50.6	105.6	118.2	104.2
20.0	B11	5.0	2.5	1.8	1.5	3.0	1.5	1.0	1.0	16.09	32.19	45.98	53.64	26.82	53.64	80.46	80.46	80.46	51.7	107.7	123.2	107.6
21.0	PT8	5.0	2.5	1.8	1.5	3.0	1.5	1.0	1.0	17.22	34.44	49.21	57.41	28.70	57.41	86.11	86.11	86.11	53.0	110.2	128.6	111.3
22.0	PT9	6.0	3.0	2.0	1.8	3.5	2.0	1.5	1.6	24.69	49.37	74.06	82.28	42.32	74.06	98.74	92.57	148.11	55.9	112.9	130.9	111.8

Figure 10: The calculations for panel durations and buffers

The generated schedules in Figure 11 presents the uncertainty associated with scheduling of panel manufacturing in a manner that enables project stakeholders to monitor and control each manufacturing process. Alternatively the project duration can be determined without the need for generating all the schedules at 90 percentile confidence level. The critical controlling activity path shown as a red arrow on Figure 10 was identified based on the procedure described in the methodology. Eq. (10) is used to calculate the project buffer which equals 147 minutes due to the variability of production rates of panels that belong to controlling activity path.

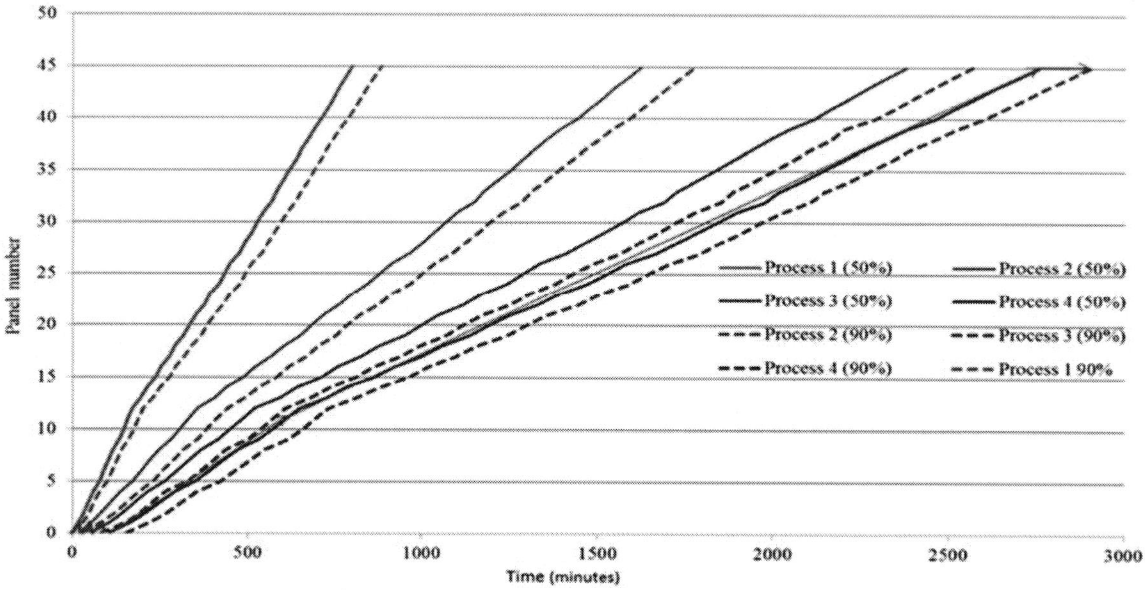

Figure 11: The 50% and 90% percentile schedules

All panels in the last process are critical because this process has the slowest productivity rate among all the four processes. By adding the calculated project buffer to the identified finish date for the last panel at the basic schedule (i.e. 50 percentile), the project total duration with 90 percentile confidence level is 2902 minutes. While, duration of the schedule at 50 percentile confidence level is only 2755 minutes.

5 Conclusion

The developed methods accounts for the uncertainty associated with the schedule of each activity. It also integrates between BIM and the linear scheduling method (LSM) to automate the data collection and the generation of linear schedule. Visualization capabilities of LSM allow the identification of controlling path. The developed method presents a generic alternative for scheduling of offsite construction projects that can be used by the manufacturing industry without the need for a specific simulation model for each different production line. The developed method allows for generation of schedules at various confidence levels in a manner that supports the decision making procedure in selecting the best schedule for each project. The results of a case study shows that the linear behavior of manufacturing processes for offsite construction panels can be scheduled using LSM while considering the uncertainty associated with productivity rates. It also concluded that the automating procedure of linear scheduling method provides the manufacturers with an easy to use approach of planning for their manufacturing processes.

This methodology assumes the continuation of production for all stations based on unlimited storage capacity of racks between stations. This assumption prevents the identification of bottlenecks and starvations on any manufacturing production line. In this respect, future research in identification of bottlenecks and starvations in linear scheduling techniques is recommended.

Acknowledgements

The authors wish to thank Dr. Mohamed Al-Hussein from the University of Alberta, and the management team at Fortis Company in Edmonton, Alberta for their guidance and support in this research project.

References

[1] McGraw-Hill Smart Market Report: Prefabrication & Modularization at World of Modular, 2011.

[2] Harris, R. and Ioannou, P. Scheduling Projects with Repeating Activities." *J. Constr. Eng. Manage.*, 10.1061/(ASCE)0733 9364 (1998) 124:4(269) , 269-278,1998.

[3] Russell Kenley and Olli Seppänen. Location-Based Management for Construction Planning, scheduling and control, Spon Press, 2010.

[4] Hajjar, D., and AbouRizk, S. Unified modelling methodology for construction simulation." Journal of Construction Engineering and Management, 128(2), 174-185, 2002.

[5] Ajweh, Z. A Framework for Design of Panelized Wood Framing Prefabrication Utilizing Multi-panels and Crew Balancing", Master Thesis, Department of Civil and Environmental Engineering, University of Alberta, 2014.

[6] Liu, H., Altaf, M. S., Lei, Z., Lu, M., & Al-Hussein, M. "Automated production planning in panelized construction enabled by integrating discrete-event simulation and BIM, ICSC15 – The Canadian Society for Civil Engineering's 5th International/11th Construction Specialty Conference, 2015.

[7] Shafai, L. Simulation Based Process Flow Improvement for Wood Framing Home Building Production Lines, Master Thesis, Department of Civil and Environmental Engineering, University of Alberta, 2012.

[8] Peer, S. Network Analysis and Construction Planning. Journal of the Construction Division, 100(CO1), 203-210, 1974.

[9] O'Brien, J. VPM Scheduling for High-Rise Buildings. Journal of the Construction Division, 101(CO4), 895-905, 1975.

[10] Birrell, G. S. Construction Planning - Beyond the Critical Path. Journal of the Construction Division, 106(CO3), 389-407, 1980.

[11] Johnston, D. Linear Scheduling Method for Highway Construction. Journal of the Construction Division, 107(CO2), 247-261, 1981.

[12] Stradal, O., and Cacha, J. Time Space Scheduling Method. Journal of the Construction Division, 108(CO3), 445-457, 1982.

[13] Kenley, R. Project micromanagement: practical site planning and management of work flow. The 12th Annual Conference on Lean Construction, 2004.

[14] Harmelink, D. J., and Rowings, J. E. Linear scheduling model: Development of controlling activity path. J. Constr. Engrg. And Mgmt., ASCE, 124(4), 263–268, 1998.

[15] Ammar, M. and Elbeltagi, E. Algorithm for Determining Controlling Path Considering Resource Continuity. J. Comput. Civ. Eng., 10.1061/(ASCE)0887-3801(2001)15:4(292), 292-298, 2001.

[16] Kallantzis and Lambropoulos, Critical path determination by incorporating minimum and maximum time and distance constraints into linear scheduling, Engineering, Construction and Architectural Management, Vol. 11 Iss 3 pp. 211-222, 2004.

[17] Schoderbek P.P and Digman L.A. Third Generation in PERT Systems. ACAD MANAGE PROC, 195-200, 1967.

[18] Dressler, J. Stochastic Scheduling of Linear Construction Sites." Journal of the Construction Division, 100(CO4), 571-587, 1974.

[19] Crandall, K. C. Probabilistic Time Scheduling Journal of the Construction Division, 102(CO3), 415-423, 1976.

[20] El-Sayegh, S. M. Linear Construction Planning Model (LCPM): A New Model for Planning and Scheduling Linear Construction Projects. Doctor of Philosophy. Texas A&M University, 1998.

[21] Trofin, I. Impact of Uncertainty on Construction Project Performance Using Linear Scheduling. Master of Science. University of Florida, Florida, 2004.

[22] Maravas, A. and Pantouvakis, J.P. Fuzzy repetitive scheduling method for projects with repeating activities. Journal of Construction Engineering and Management, 137(7), 561–564, 2011.

[23] Bakry, I., Optimized Scheduling of Repetitive Construction Projects under Uncertainty. PhD thesis, Department of Building, Civil and Environmental Engineering, Concordia University, 2014.

[24] Slorup, R.B. Applying critical chain buffer management theory in location-based management, Construction Management and Economics, 32:6, 506-519, 2014.

[25] Goldratt, E. M. Critical chain. North River Press Great Barrington, MA, 1997.

[26] Ghaffaria, M., Emsley M. W. Current status and future potential of the research on Critical Chain Project Management. Journal of Surveys in Operations Research and Management Science, 43–54, 2015.

[27] Moghadam, M., Lean-Mod: An Approach to Modular Construction Manufacturing Production Efficiency Improvement." PhD thesis, Department of Civil and Environmental Engineering, University of Alberta, 2013.

[28] Taghaddos, H., Hermann, U., AbouRizk, S., Y., Mohamed, A. Simulation-based multiagent approach for scheduling modular construction, J. Comput. Civ. Eng. 28 (2) 263–274, 2012.

A Sequential Pattern Mining Approach to Extract Information from BIM Design Log Files

Saman Yarmohammadi[a], Reza Pourabolghasem[b], Arezoo Shirazi[a], Baabak Ashuri[c]

[a] Ph.D. Student, Economics of the Sustainable Built Environment (ESBE) Lab, School of Building Construction, Georgia Institute of Technology, USA
[b] Ph.D., School of Electrical & Computer Engineering, Georgia Institute of Technology, USA
[c] Associate Professor, Economics of the Sustainable Built Environment (ESBE) Lab, School of Building Construction, Georgia Institute of Technology, USA
E-mail: Saman.yar@gatech.edu, pourabolghasem@gatech.edu, Arezoo.shirazi@gatech.edu, Baabak.ashuri@coa.gatech.edu

Abstract –

Building Information Modeling (BIM) and its implementation in Architecture, Engineering and Construction (AEC) industry is rapidly growing all over the US. The growing utilization of BIM application has resulted in accumulation of tremendous volumes of computer-generated design data. Such vast datasets provide practitioners with an opportunity to extract valuable information regarding the process of design. Design log files, in particular, are rich data sources that contain model development data automatically recorded throughout the design process. However, the existing studies are mostly restricted to mining structured data and hence, lack the capabilities to handle unstructured temporal data sources such as design log files. The main objective of this paper extract implicit process information from design log data by implementing a tailored sequential pattern mining approach. For the purpose of this research, we examined the feasibility of utilizing Revit journal log files as a non-intrusive data collection mechanism to capture modelers' interactions with the software and detect common command execution structures during model development sessions. To this end, user-model interaction data such as modeler characteristics, command type, and command time were extracted from the data sample's journal files using a text file parser. After careful data cleaning, the final set of temporal data were transformed into sets of multiple character strings. We used an efficient implementation of Generalized Suffix Trees (GST) data structure to identify common command execution sequences among several modelers. The results of the study identified several shared command sequences among five modeler. This study proposes a novel pattern mining methodology to extract useful information from time-stamped design log data and enables project managers to obtain valuable insight into design development processes.

Keywords –

Building Information Modeling; Data Mining; Sequential Pattern Mining; BIM Log Files

1 Introduction

Design log data in general and Autodesk Revit© journal files, in particular, are unstructured text files that BIM tools create when modelers use the application. These files capture all modeling activities that occur during a design session, as well as system information, such as memory performance and operating system [1]. Revit journal files are largely used to diagnose and troubleshoot technical problems with the software. In this research, however, we examine the feasibility of utilizing these log files as a non-intrusive data capturing mechanism for documenting modeller-software interactions and recording model development events. The information extracted from Revit journal files will be utilized to examine our research hypothesis that there are common command execution patterns among BIM modelers who work on similar projects. In the context of this research, a journal log file is regarded as a database of ordered modeling events (commands) recorded with a concrete notion of time. A frequent pattern is an ordered set of individual commands that occurs more than a threshold number of times (i.e., minimum support) in the original sequence database. The following questions are of particular interest in this research: What types of commands sequences do modelers execute frequently? What structures are formed from various commands at each stage of modeling and how? What command pattern sequences are common among different modelers? Is there a resemblance between the modeling behaviour of BIM users who work on similar types of projects?

In recent years, there has been an increasing amount

of studies on the topic of mining information from BIM-generated documents. In particular, a considerable literature has accumulated around the topic of extracting patterns from geometric and semantic information stored in virtual building models. For instance, Hu et al. developed a multivariate linear regression model to predict man-hour quantity for steel fabrication projects in the planning phase [2]. The utilized data was extracted from historical Tekla BIM models. In another study, Abdelmohsen et al. developed a cost analysis and reporting system that utilizes elements' dimensions extracted from BIM models [3]. The researchers used Industry Foundation Classes (IFC) formats and Solibri Model Checker (SMC) to obtain quantity take off data, and integrate BIM and cost models. The main emphasis of these research efforts was on utilizing data stored in BIM models to enhance non-design practices, such as production planning, cost estimation, and site layout analysis. However, these studies are limited to utilizing information stored as physical and functional characteristics of elements after the model is already created. As a result, little is known about the implicit information that can be obtained by studying BIM model development data throughout design processes.

Current applications of textual data mining techniques in AEC literature can effectively extract patterns from information stored in structured relational datasets or unstructured documents [4]. However, these approaches are restricted to data that lack time dimension [5]. Because of this shortcoming, these methods are inherently limited to analyze sequential design log data [6]. Soibelman and Kim outlined the steps necessary to apply data mining and Knowledge Discovery in Databases (KDD) as tools to extract novel patterns in design and construction fields [7]. In two consecutive studies, Caldas et al. [8] and Caldas and Soibelman [9] proposed text mining-based approaches to automatically classify unstructured construction documents. The presented approaches were implemented to classify several product description documents into categories defined based on CSI MasterFormat classification hierarchy. Another example of implementing data mining algorithms to analyze unstructured textual data is construction cost overrun prediction by William and Gong [10]. The authors used a stacking ensemble model of several classifiers to forecast the level of cost overruns. These approaches present valuable insights on how to overcome several challenges of handling unstructured textual information such as dealing with the high dimensionality of data, removing incorrect and irrelevant data points, and identifying features to represent the data in an analysis-friendly format. However, the existing methods fail to address how to incorporate the chronological dependencies of temporal data and hence, are not suitable to analyze design log files.

The primary objective of this study is to extract implicit process information from design log data by implementing a tailored sequential pattern mining approach. For the purpose of this research, we examined the feasibility of utilizing Revit journal log files as a non-intrusive data collection mechanism to capture modellers' interactions with the software and detect common command execution structures during model development sessions. Throughout this paper, several steps necessary to collect, prepare, and analyze data to extract frequent command patterns are elaborated. A major international design firm provided the data for this exploratory study. This study contributes to the state of practice by providing new insights into what additional design process information can be retrieved from Revit journal files. The novel method created in this research contributes to the body of knowledge by incorporating chronological dependencies of textual records into the existing pattern matching models. Section 2 describes the methodology used in this paper to analyze time-stamped design log data. We outline the implementation details of the proposed approach in section 3. In section 4, we provide results obtained from conducting an experiment on real design project data. Section 5 is conclusions and future research directions.

2 Research Approach

"Figure 1" presents the flowchart of the process to extract necessary information from Revit journal files and identify common frequent command execution sequences. This process consists of three major parts. First, a large number of Revit journal files that belong to a design project are collected. These text files are parsed to extract and store necessary data items. Second, we transform the obtained data to construct long strings of characters and generate input vectors. Finally, we construct Generalized Suffix Tree data structures for each user. A Depth-First Search (DFS) is later conducted on the GST structures to find common command sequences for each and among the five users.

2.1 Data Collection

We utilized modeling log data provided by an international design firm. The research team had access to data from 10 healthcare project designed in 2013 and 2014. The provided database consists of over 5,000 Revit journal files that were later parsed to extract useful entries, amounting to over 10 Gigabytes of structured data. In addition to the modeling events, journal files contain information about the user and the project to which the model belongs. The steps that were followed to extract required information from the journal files are outlined in section 3.

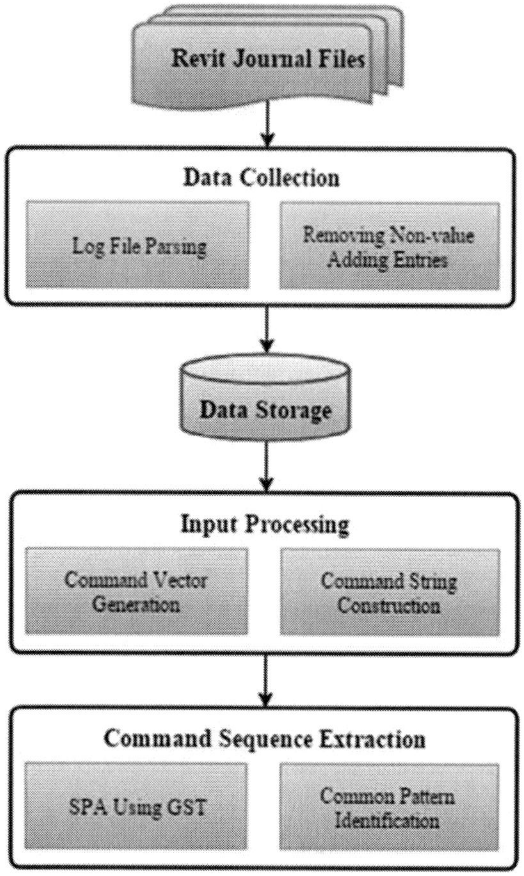

Figure 1. Flowchart of Common Command Sequence Extraction from Revit Journal Files

2.2 Input Processing

The records of five modelers that each had over 100,000 log entries are used to conduct further analysis. We treated the log entries for each user as a long string. The commands that each user issued were modelled as characters. The final string for each user was constructed by putting these characters in the original order they were recorded. Input vectors of modelers were then generated by joining the constructed command strings in chronological order.

2.3 Command Sequence Extraction

Log files (or transaction logs) have for long been studied in the data mining community. Log files can be generated in different applications, such as retail transactions data, web access logs, etc. Agrawal and Srikant introduced Generalized Sequential Pattern (GSP) algorithm to mine shopping patterns [11]. The authors studied a large database of customer transaction data, where each transaction consisted of customer-id, transaction time, and items bought in the transactions. Utilization of downward closure property (i.e., Apriori) provides GSP with the advantage of reducing search space compared to an exhaustive search approach. However, the time and memory performance of GSP is relatively low since a huge number of candidates must be generated and stored in each repetition for evaluation [12]. PrefixSpan algorithm overcomes this issue by taking advantage of pattern-growth methodology [13]. This algorithm greatly reduces the candidate sequence generation efforts by following a divide-and-conquer approach called prefix-based projection. The approaches proposed in these studies can extract valuable information from sequential datasets. However, general Sequential Pattern Mining (SPM) methodologies do not preserve the exact order of elements in a sequence. As a result, some of the elements in extracted sequences may not necessarily be consecutive in the original string of transactions.

For the purpose of this study, a special GST-based string processing algorithm was used. This method is particularly useful as it maintains the order in which the executed commands are recorded. Xiao and Dunham first proposed applying GST data structure to mine web access log data [14]. The authors analyzed clickstream data generated based on the access made by internet users to find frequent web page traversal patterns. The technique proposed in this study achieved a high level of adaptability to large databases through dynamic compressions and effective pruning. Guerbas et al. introduced an improved version of GST algorithm with an optimized data structure to extract page visit patterns of internet users [15]. The primary objective of the authors was to improve the search experience of users by predicting what pages they intend to view next. To this end, the research team utilized GST algorithm to find common web navigational patterns among users with similar interests. The methodology utilized in our study is a modified version of Guerbas et al.'s algorithm tailored for mining journal log files. The algorithm's implementation details are outlined in the following section.

3 Implementation of the Proposed Design Log Files Mining Approach

In this section, we present a detailed explanation of our proposed approach to process and analyze information embedded in Revit journal files using

Table 1 Examples of data items as recorded in Revit journal files

Column name	Example
Date and Time	'E 19-Apr-2014 22:22:53.175; 0:< Jrn.Command "Internal" , "Save the active project with a new name , ID_ ' 1:< System (MB) [Available / Total] [Revit Memory Usage (MB)]
Project Path	'H 19-Apr-2014 22:23:05.063; 2:< Jrn.Data "File Name" , "IDOK", "..\studentunion\Atlanta\gatech.rvt"
General Command	Jrn.Command "AccelKey" , "Cancel the current operation , ID_CANCEL_EDI 'C 19-Apr-2014 22:22:48.142; 0:< idle0_doc ' 0:< Setup Tool transaction group (wall)
Specific Command	Jrn.Command "AccelKey" , "Cancel the current operation , ID_CANCEL_EDITOR" 'C 19-Apr-2014 22:22:48.142; 0:< idle0_doc ' 0:< Setup Tool transaction group (wall)
Revit Version	' 5:< File was saved in Autodesk Revit 2014 (Build: 20130123 1540(x64)) ' 6:< ::8:: Delta VM: Avail -20 -> 8386966 MB, Used +16 -> 334 MB, Peak +8 ' 0.000370 7:<<<ET::repairModificationDatesOutOfSequence
View Type	Jrn.Directive "WindowSize" _ , "[Project1]", "Floor Plan: Level 1" _ , 1322, 625

GST string mining algorithm. To begin this process, we need to parse journal files and extract necessary data items. This step is particularly challenging as there is no documentation available on public domain that specifies how and where different data components are recorded. The data items extracted for the purpose of this study are modeller name, date of the modeling session, project name, command execution time, general and specific command description, view type, and the Revit version utilized by the modeller. To find this information in journal files, we use regular expression matching operations. "Table 1" shows examples of each data item as they appear in journal files. We manually search the files to identify the local format using which each data instance (e.g., project name, command, and view type) is recorded. Once patterns are identified, we use a text processor to extract and store information in a Comma-Separated Values (CSV) file as shown in "Table 2". Non-value adding commands, such as "cancel the current operation", are removed from stored entries to improve the quality of obtained command patterns.

Testing whether a sequence of commands occurs frequently in a database needs to be performed in an efficient manner. That is why we opted for GSTs. Suffix trees are very efficient data structures and often provide linear time solutions to challenging string problems [16]. "Figure 2" shows the algorithm utilized to search for frequent command execution sequences where the original order of data is conserved. First, we assign unique characters to each specific command name to generate the necessary input strings. We generate long strings by joining these characters in accordance to the original order in which corresponding commands are recorded in each modeling session. Once these strings are constructed, we use suffix trees to represent all suffixes of the string set. In our

Table 2 Sample of structured processed data

Modeller Name	Time	Project Name	General Command	Specific Command	Revit Version	View Type
modeller	10:30:07	StdUnion	"Internal"	"Show or hide recent files	2014	Floor Plan
modeller	10:33:50	StdUnion	"StartupPage"	"Open an existing project	2014	Floor Plan
modeller	10:36:51	StdUnion	"Internal"	"Print the active window	2014	sheet
modeller	10:45:25	StdUnion	"Internal"	"Activate this viewport	2014	sheet
modeller	10:48:04	StdUnion	"Internal"	"Modify View Templates	2014	sheet
modeller	10:48:07	StdUnion	"Internal"	"Manage Links	2014	sheet
modeller	10:48:09	StdUnion	"KeyboardShortcut"	"Steering Wheels	2014	3D view

```
Input: a csv file containing recorded entries, minimum length of the common command patterns
l, minimum number of strings sharing the pattern m
Output: common command patterns
Construct input vectors for each by transforming command sequences
Build generalized suffix tree for all strings
For each leaf node f
    current node ← f
    While root is not reached
        p ← parent (current node)
        add string id of the leaf node f to the node p if this id has not been added to it yet
        if the number of string ids added to p>=m and ||path(p, root)||>l then
            return path(p, root)
        end if
        current node ← p
    end While
end For
```

Figure 2 Using generalized suffix trees to detect frequent command execution sequences (Adapted from Guerbas et al. [15])

implementation, the DFS step generates an ordered list of leaf nodes of the suffix tree. The leaves corresponding to each internal node are, therefore, a consecutive sub-list of this ordered list of leaves. The DFS saves the start and end position of the leaves (based on their DFS order) for each pattern (i.e., internal node) in a helper hash table. The subtraction of the end position and the start position will give the number of repetitions of the specific pattern. Additionally, this way we can access all the instances of that pattern in the original string by having access to the leaf nodes and their corresponding suffix index. This technique enables us not only to identify patterns, but also calculate their frequency. These substrings, then, are filtered based on simple heuristics (e.g., minimum length of the substring, minimum frequency, etc.). At the end of this step, we have a limited number of most frequent substring of commands that is common between all the users. We call these substrings as primitives. We repeat this process for all users to extract their frequent command patterns. Finally, the retrieved patterns are compared against each other to identify the ones that are common among different users. The results obtained from implementing this approach on the provided Revit journal files are presented and discussed in the following section.

Table 3 Statistics of the dataset used

Data set	Period	Number of entries
Modeler 1	2013	126,815
Modeler 2	2014	122,813
Modeler 3	2014	114,290
Modeler 4	2013	112,082
Modeler 5	2014	106,887
All records	2013-2014	5,568,243

4 Experiments and Results

This section describes the results of experiments conducted and discusses the patterns obtained. We present some statistics about the dataset used. We comment on the patterns discovery results obtained using GST and how they support our initial hypothesis regarding the existence of common command execution patterns among BIM modelers.

A dataset containing 5,000 Revit journal files was used to implement our approach. We started the analysis by processing these files using a text parser. We also filtered out all noisy data (e.g., "cancel the current operation", "delete") and entries related to errors. The processed data was stored in .csv format amounting to over 5.5 million records. Data from five architects that model interior systems of healthcare projects were selected to conduct analysis. "Table 3" provides some statistics about the utilized data.

Table 4 Top three most executed individual commands

	Frequency		
Modeler	Move selected objects or their copies	Align references	Create a line
Modeler 1	19.30%	11.92%	7.89%
Modeler 2	15.27%	13.56%	14.23%
Modeler 3	18.19%	6.21%	11.07%
Modeler 4	15.57%	15.98%	13.07%
Modeler 5	14.31%	6.94%	10.00%

Prior to using GST, we conducted some preliminary analysis to identify the most frequently executed individual commands. "Table 4" shows the top three commands for each modeller. These commands were consistently found to be the most frequent events for all users.

In the next step, we constructed a GST data structure for each modeller's command string to identify shared sequences. Several arbitrary minimum frequency threshold values were tested, among which we selected 250 and 500. The minimum length of extracted common command patterns was also set to three. Interesting patterns obtained for the two threshold values are presented in "Table 5".

The primitives extracted for minimum threshold of 250 are longer and represent meaningful modeling activities. Pattern 1 seems to correspond to commands executed to create and extend multiple lines. In this case, modelers have hidden a number of objects to gain easier access to the elements they want to modify. The second pattern appears to show cases where modelers make copies of different elements and visualize their dimensions. The third pattern captures commands used to make copies of a specific object in the model and modify them.

The obtained patterns became shorter when we increased the minimum frequency threshold. This observation was expected since longer sequences tend to match less frequently. The first pattern corresponds to cases where modellers navigate through different viewpoints to select and copy certain objects. The third pattern also captures command sequences used to change visibility of different layers. Contrary to these two patterns, it is not clear what specific activity the second command sequence represents. We saw more examples of such repetitive sequences for both thresholds. Presence of such patterns may be because of noisy input data or consecutive execution of similar commands by modelers.

5 Conclusion and Future Work

In this paper, we proposed an efficient approach to extract modeling development information embedded in design log files produced by Autodesk Revit. We have reviewed all steps of this process and made an effort to make a contribution at each step. The first step in Revit journal mining consists of different phases. The first phase is to identify the format in which different information items are stored. Our code accepts the raw journal files and produces structured csv files as shown in "Table 2". We also clean the obtained data by removing non-value adding entries, such as cancel and error messages. The conclusion we derive from our investigation at this level is that using the suggested approach will help in processing unstructured journal log files and producing good quality input data for mining algorithm.

In the second step, we use GST data structures to find common command sequences among BIM users. First, we transform command sequences into character-based input vectors. Then, the transformed data is utilized to construct GST. Frequent command patterns are identified by conducting DFS on the trees ("Figure 2"). Extracted patterns for different users are compared against each other to identify shared sequences. The conclusion at this step is that GST-based string mining approach is an efficient method to extract common command patterns among several modelers.

Table 5 Common command sequences extracted using GST

	Pattern 1	Pattern 2	Pattern 3
Extracted Pattern 250	1. Select objects to modify 2. Hide selected elements 3. Create a straight detail line or a detail arc 4. Rotate selected object(s) 5. Trim/Extend two lines or walls to make a corner	1. Copy the selection and put it on the Clipboard 2. Move copies of selected objects 3. Create aligned Dimensions	1. Select objects to modify 2. Create an object similar to selected object 3. Move selected objects or their copies 4. Align references 5. Finish Sketch
Extracted Pattern 500	1. Activate this viewport 2. Copy the selection and put it on the Clipboard 3. Deactivate the currently active viewport	1. Move selected objects or their copies 2. Move selected objects or their copies 3. Move selected objects or their copies	1. Deactivate the currently active viewport 2. Activate this viewport 3. Control visibility and appearance of objects (applied only in the active view)

Finally, we conducted experiments on Revit journal files provided by a design firm to verify our proposed approach. Our code successfully processed the raw log files and extracted common command patterns for five architects who model interior systems of healthcare projects. The obtained results also confirmed our initial hypothesis that there are frequent command execution sequences shared among BIM modelers who work on similar projects.

This study contributes to the state of knowledge by proposing a tailored string mining algorithm capable of extracting meaningful information from time-stamped design development data. We contribute to the state of practice by enabling design project managers to gain an unprecedented insight into the evolution of design. It is suggested that the possibility of using command patterns to characterize modelers is investigated in future studies. For instance, it might be a good idea to modify the proposed approach to calculate the average time it takes BIM users to executed different command patterns. Such empirical assessment of modellers' behaviour can potentially be used to provide customized training to improve designers' performance. Moreover, calculated average times can help design manager to improve project outcomes by choosing an optimal team configuration.

Acknowledgement

The authors take this opportunity to thank Professor Charles M. Eastman, who leads Digital Building Laboratory (DBL) at the Georgia Institute of Technology, for his support of this research project. In addition, we are grateful to Dr. John Haymaker, Mr. Dan Chasteen, and Mr. Josh Emig who kindly helped us throughout the progress of this study.

References

[1] Autodesk Revit. On-line: http://www.autodesk.com, Accessed 15/03/2016.

[2] Hu X. and Lu M. and AbouRizk S. BIM-based data mining approach to estimating job man-hour requirements in structural steel fabrication. In *Proceedings of the 2014 Winter Simulation Conference*, pages 3399–3410, IEEE Press, 2014.

[3] Abdelmohsen, S. and Lee J. and Eastman C. Automated cost analysis of concept design BIM models. *Designing Together: CAAD Futures*, 2011.

[4] Manyika J. and Chui M. and Brown B. and Bughin J. and Dobbs R. Big data: The next frontier for innovation, competition, and productivity, 2011.

[5] Baars H. and Kemper H. Management support with structured and unstructured data—an integrated business intelligence framework. *Information Systems Management*, 25(2): 132-148, 2008.

[6] Doan A. and Naughton J. and Ramakrishnan R. and Baid A. Information extraction challenges in managing unstructured data. *ACM SIGMOD Record*, 37(4): 14–20, 2009.

[7] Soibelman L. and Kim H. Data preparation process for construction knowledge generation through knowledge discovery in databases. *Journal of Computing in Civil Engineering*, 16(1): 39–48, 2002.

[8] Caldas C. and Soibelman L. and Han J. Automated classification of construction project documents. *Journal of Computing in Civil Engineering*, 16(4): 234–243, 2002.

[9] Caldas C. and Soibelman L. Automating hierarchical document classification for construction management information systems. *Automation in Construction*, 12(4): 395–406, 2003.

[10] Williams T. and Gong J. Predicting construction cost overruns using text mining, numerical data and ensemble classifiers. *Automation in Construction*, 43: 23–29, 2014.

[11] Agrawal R. and Srikant R. Mining sequential patterns. In *Data engineering 1995. Proceedings of the Eleventh International Conference on*, pages 3–14, IEEE, 1995.

[12] Verma M. and Mehta D. A comparative study of techniques in data mining. *International Journal of Emerging Technology and Advanced Engineering*, 4(4), 2014.

[13] Pei J. and Han J. and Mortazavi-Asl B. and Pinto H. and Chen Q. and Dayal U. Prefixspan: Mining sequential patterns efficiently by prefix-projected pattern growth. In *iccen*, page 0215, IEEE, 2001.

[14] Xiao Y. and Dunham M. Efficient mining of traversal patterns. *Data & Knowledge Engineering*, 39(2): 191–214, 2001.

[15] Guerbas A. and Addam O. and Zaarour O. and Nagi M. Effective web log mining and online navigational pattern prediction. *Knowledge-Based Systems*, 49:50–62, 2013.

[16] Gog S. and Beller T. and Moffat A. and Petri M. From Theory to Practice: Plug and Play with Succinct Data Structures. In *sea*, pages 326–337, 2014.

HBIM AND FIRE PREVENTION IN HISTORICAL BUILDING HERITAGE MANAGEMENT

Frosini, Giulia[a] - Biagini, Carlo[b] – Capone, Pietro[c] – Donato, Vincenzo[b] – Giusti, Tommaso[c]

[a]University of Florence. Florence – Italy
[b]University of Florence. DIDA – Department of Architecture. Florence – Italy
[c]University of Florence. DICeA – Department of Civil and Environmental Engineering. Florence – Italy

E-mail: frog.giulia@gmail.com, cbiagini@dicea.unifi.it, pcapone@dicea.unifi.it, vincenzo.donato@unifi.it, tommasog@dicea.unifi.it

Abstract

Italy possesses a unique cultural heritage that is constantly under threat from natural and anthropic risks: fire vulnerability is one of the main causes of architectural heritage destruction.

The goal of this work is to propose a methodological approach to the management and preservation of historical building heritage with respect to fire safety. This approach aims to facilitate the process of analysis, design and fire management in historical buildings, using automated procedures based on HBIM (Historical BIM) processes.

BIM can be used effectively in the digitization of cultural heritage: HBIM enables the three-dimensional modeling of the building by means of parametric objects, to which infographic information is associated, related to an appropriate Level of Development (LoD).

A workflow process has been defined to set up a procedure for best practices which can guarantee the correct interoperability between BIM programs and fire safety design.

The procedure, based on the exchange of IFC files, shows result applying analysis through Model Checking software of Model Checking and shows limited interoperability with fire safety software. The procedure has been validated through real tests and measurements, by comparing data obtained from a traditional fire prevention analysis with the data gathered from the automated procedure.

The real case study is the "Command Building" of the School of Air Warfare in Florence. Results mainly indicate that fire safety in historical buildings should always be managed using a performance-based approach (interoperability issue between HBIM and fire safety software).

Keywords – HBIM, interoperability, fire safety, historical buildings.

1 Introduction

In the past a high number of fire events, with serious consequences in terms of damage, have occurred both in Italy and abroad. The knowledge of these events and their analysis is an important source for their prevention and for the identification of specific guidelines to improve standard safety conditions. In order to protect historical building heritage it is useful to define a methodology based on an integrated digital design procedure aimed at fire safety management.

One possibility could be to use BIM (Building Information Modeling) and interoperable procedures, not only to coordinate the project during the whole building process, but also to obtain numerical data by analysing the virtual model through specialist simulations [1,2,3]. In fact, Building Information Modeling has been used for years during the renovation and restoration phases, also through collaborative platforms for the "design" of existing buildings (HBIM) and with different aims [4,5,6]: the virtual model can be used as a tool to highlight criticisms in the existing fabric and drive the decision making process [7].

This article sets out an approach for improving the interoperability between HBIM models and open source software for fire prevention analysis, validating the procedure through a comparison between the results achieved by the automated procedures and those derived from a traditional analysis.

The reported case study, the "Command Building of the School of Air Warfare" in Florence [8], deals with a building that belongs to Italian artistic and cultural heritage. The main purpose of this study was to evaluate the use of BIM for historic buildings and, specifically, the possible correlation between the process of fire prevention and design.

Figure 1. The Command Pavilion, aerial photo [8]

2 State of the art

The eBIM (Existing Building Information Modeling) and HBIM (Historical Building Information Modeling) focus attention on a different design process to the one applicable to new buildings, but at the same time guarantee the database definition of various info-graphic information capable of transmitting specific meaningful content [9]. The model assumes significance not only as a geometric representation of the architectural object, but rather as a database for all the data collected during the survey phase and construction analysis [2,10].

In contrast to the modeling of ex-novo artifacts, where the parametric entities (BOMs) that define the model are collected within BIM platforms, in a cultural and historical building the definition of the "Historic Building Object Models" (H-BOMs) is subjected to a survey phase, which must possess specific features for the development and management of historic building models [5,6]. The difficulties of implementing projects for historical building heritage are not only related to the definition of new architectural elements that complement the existing database, but also to the need to define a Level of Development (LoD) as a function of the analysis that is going to be performed. In that sense, it is possible to calibrate the quantity and quality of necessary information and avoid time-consuming procedures during the reconstruction of the monument under analysis [5].

Also in the field of Fire Prevention an integrated approach based on BIM technology is useful, aiming to guarantee the interoperability between BIM software and specific software packages aimed at the planning phase and/or at fire testing [4,11,12].

From the fire safety point of view, in order to protect the buildings and their contents it goes towards a performance-based approach and it's therefore able to adapt the design solutions to the reference building context. This is a very important aspect for fire prevention and for protecting historical and artistic heritage, in which there is considerable difficulty in applying the rules, codes and prescriptive regulations: the identification of *ad hoc* solutions to this case makes it possible to reach a level of safety for buildings equivalent to that regulatory ensuring a lesser impact on the structure and convenient economic implications [13,14]. Nowadays there is a wide range of software, for example CFAST and FDS [13,15], that can both simulate a fire in order to determine the spread of fire and smoke, the response of a building at high temperatures and the movement of users leaving the building, and check the regulatory requirements concerning divisions and escape routes.

The development of this software provides high design flexibility and the opportunity to analyse aspects related to both the protection of human life and the protection of artistic goods [13,16].

3 Methodology

The main issues related to the historical heritage of artefacts relate to management as a whole; an important aspect is the fire vulnerability of structures which are sometimes subject to a functional change (such as office buildings, hotels or hospitals) with the inclusion of activities which do not provide specific rules for these types of buildings.

The use of an HBIM approach, still to be finalized, makes it possible to address the problem of the management of historic buildings by allowing interoperability between the software in order to facilitate the exchange of information in a digital format.

It is therefore essential to define from the outset the level of detail desired, so as to calibrate the information necessary for the realization of the various parametric objects that will compose the model. Once the HBIM model is completed, we propose an approach to fire prevention analysis by performing the relevant checks with a validation software program, such as Solibri Model Checker, to identify problems and/or design errors. Parallel analysis was carried out with traditional techniques to have a parameter for comparing the obtained results.

4 H-BIM and LoD

Applying the HBIM techniques means applying different approaches that can change in accordance with the building typology and its cultural relevance (for instance, if the building is a monument and if it has a certain aesthetic and architectural value) [5]. The BIM model has to be modelled with a certain LoD that must be chosen in accordance with the final purpose of the model. Possible purposes are discussed in the following list

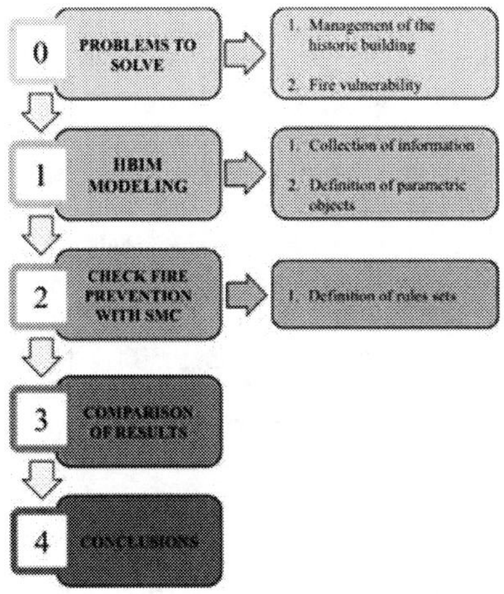

Figure 2. Scheme of the proposed method

1. architectural restoration for the preservation of historical heritage through the simulation of conservative, supplementary or reconstructive assumptions and the creation of a visual data-base where they are collected, and the various preliminary investigations, contextual and post intervention;
2. recovery and reuse interventions in order to restructure and retrain the existing buildings, possibly by proposing a change of intended use;
3. structural analysis aimed at the rehabilitation of existing buildings through local interventions, adjustment or seismic improvement;
4. upgrading of plants through energy design;
5. to design facility management interventions aimed at the ordinary and extraordinary maintenance of buildings, of safety in the workplace and of fire prevention.

In the case of renovation projects and/or restoration of existing buildings, the experts involved are almost always forced to create new elements reproducing the objects belonging to the building's historical repertoire that complement the existing database.

Our methodology refers to Common BIM Requirements [17,18] that describe the requirements for the implementation of a BIM model, both in relation to data collection procedures and info-graphic modeling. In the case of existing or historical buildings, series 2 [18] describes how to start with an "inventory model", understood as the modeling of the state of information quality related to the building at the time of the recovery and/or restoration project. The COBIM identifies three methodologies for data collection which involve the use of specific tools that are not always available; therefore, this highlights the need for a new classification that includes printed material about the original design. The means for the reconstruction of an existing building, through a BIM model, can then be listed in the following way:

- Level 0 - Original documents and archival sources;
- Level 1 - Laser distance meter and total station;
- Level 2 – Photogrammetry;
- Level 3 - Laser Scanner.

For each of these levels, it is possible to associate a "precision degree" and an "information accuracy", which can influence the precision of the overall BIM model.

If we refer to the lowest level of accuracy, Level 0, the precision of the BIM model is connected to subordinate to the graphic errors and tolerances expressed by the source documents. For instance, if the document is represented with an architectural scale of 1:200 and the elements are represented with a weight line of 0.2 mm, the error can be quantified as 4 cm and the tolerance is expressed around 4/10 cm. To drastically reduce this inaccuracy, it is necessary to integrate the historical sources with a direct survey and, even better, use photogrammetry and a laser scanner. Figure 3 shows the relationship between the degree of accuracy and the representation error derived from the different methods used to perform the survey.

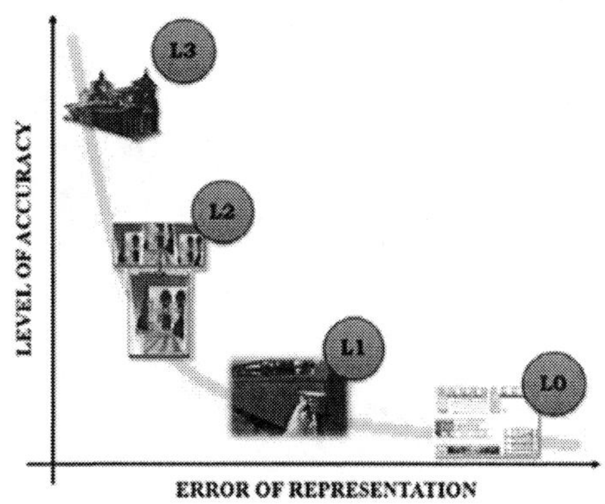

Figure 3. Relationship between error of representation and level of accuracy of different data collection techniques of a BIM model.

When there is the need to create an HBIM model it is therefore essential to obtain a sufficient amount of geometric and topological information, qualitatively suitable for the reproduction of each construction element that characterizes the object of study: the more data available to the designer, the more the result will correspond to reality.

Often the choice regarding the use of the best

available technology for survey (i.e. the laser scanner), which involves minimum representation errors, is not always the most appropriate solution: the degree of accuracy provided by each data collection methodology is closely related to the objectives which must be achieved. Depending on the purpose for conducting the survey, and the reference scale representation with which the measured material is to be processed, it is clear that direct detection methods can be used, bearing in mind the increase of possible errors that may occur. An accuracy level of the detail needed for the reproduction of the building is associated to each specific purpose of the survey realisation. Therefore, we can find three categories:

- High detail level of accuracy, which requires an almost realistic representation of the object;
- Average detail level of accuracy, in which the loss of precision does not create excessive problems;
- Lowest detail level of accuracy, for which a less detailed reproduction of the components of the object is sufficient.

In parallel, according to the type of representation that is needed (and is possible to obtain), each level corresponds to the most appropriate type of tool, considering advantages and disadvantages resulting from its use.

The higher the level of detail required, the greater the degree of achievable accuracy with which the survey must be performed.

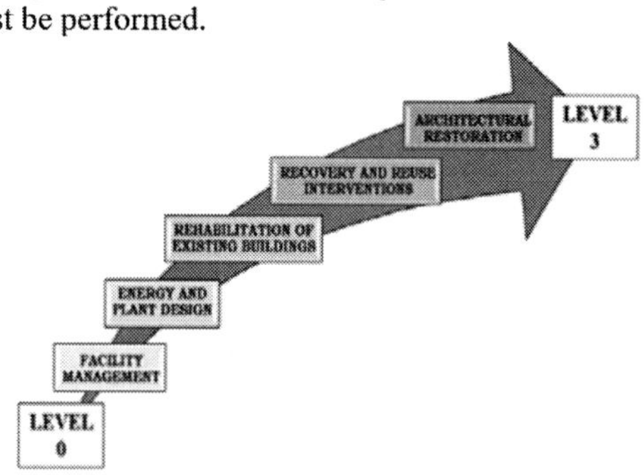

Figure 4. Relationship between a survey proposed for historical-monumental buildings and the level of accuracy obtained with the different methods.

5 Risk assessment through BIM processes

Although there is a lot of fire simulation software, in existing buildings a traditional approach to fire prevention is still one of the most commonly-used methods.

The implementation of risk assessment through BIM methodology allows a simplification of procedures between the design and the analysis, avoiding the need to refer to paper material or CAD files which allow the passage of a small amount of information that is mostly related to dimensional data. Starting from a BIM model and designed with an appropriate LoD for the preparation of a preliminary evaluation aimed at fire events, the user may be able to develop the necessary evaluations which can handle not only the geometrical aspects of the construction but also all the data related to the intrinsic properties of the materials. The main applications for fire safety verification adopt the IFC standards as a tool for interoperability, for the sharing of design information, but there are still issues related to compliance with the interoperability standards adopted by various implementation software programs and by intrinsic conformity to the IFC model that always needs integration.

6 Case study: the "Command Building" of the School of Air Warfare in Florence (Italy)

The proposed case study involves the application of BIM technology for the "Command Building" of the School of Air Warfare in Florence, one of the most avant-garde architectural and engineering masterpieces of 1930s.

The construction began in 1936 with the arch. Raffaello Fagnoni, who was commissioned to build the "Application for the Royal Air Force School", now the "Institute of Military Aeronautical Sciences" [8].

One of the buildings that is part of the School of Air Warfare is the Command Building, where office activities take place: the exception is the top floor where there the apartments of the Commander are located. The building looks like a compact parallelepiped with a square plan that essentially consists of two parallel buildings placed in an east-west direction, connected to each other by two transverse bodies, realized with characteristic Vierendeel beams, that on the top floor open into a loggia. The case study can be summarized in three steps:

1. Modelling of the building according to the principles of HBIM;
2. traditional fire prevention analysis by checking the rules on fire prevention and later defining the possible intervention solutions;
3. application of specific software (Solibri Model Checker), to evaluate the fire safety of the building.

7 The modelling phase with Autodesk Revit Architecture

In this case study Autodesk Revit Architecture 2015

was used, as one of the possible parametric software programs for the construction industry based on the BIM platform for the modeling of new and historic buildings.

The "construction" of the Command Building began with the reconstruction of maps of various levels using AutoCAD in order to have a frame of reference from which to begin the modeling. Subsequently, through the analysis of the topological information of the various technical components the appropriate families for walls and floors were defined, thereby obtaining a first and schematic project.

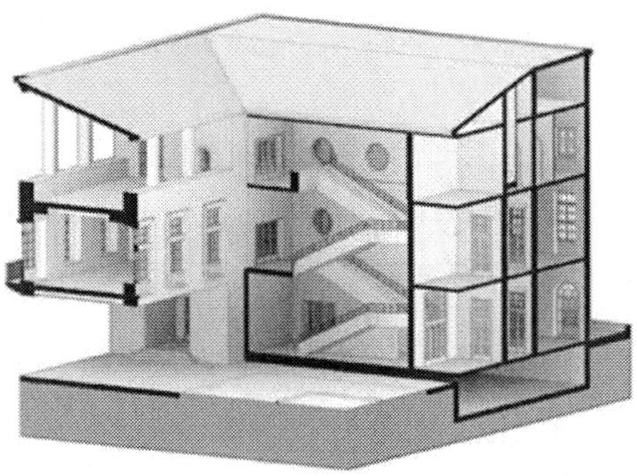

Figure 5 .Axonometric section of the Command Building.

Following this, the model was gradually refined with "in-place" family, representing the various internal and external fixtures, stairs and all the architectural elements that characterize it. An identification code was assigned to each architectural and/or structural component, generally composed of four parts:

1. generic item category and its location;
2. material and/or type of element;
3. characterizing size (for example, the thickness of walls and floors, or size for fixtures);
4. finishing material.

Proceeding in that way, by a simple request to the software, it was simple to go back immediately to family and type of each technical element that compose the building. It is important to consider that the preliminary realization of the 2D project occurred through a survey the paper material available without having the possibility to execute a direct validation of the building. This certainly implies a very high degree of approximation, directly proportional to the error of the representation expressed in centimetres. Compared to the actual size of detected objects, bigger mistakes will be tolerated on small scales, and smaller errors on big scales. Despite being in possession of limited and not very detailed information, it was possible to reconstruct the building in order to perform assessments on the fire safety of the structure.

Figure 6. 3D building.

8 Solibri Model Checker

Solibri Model Checker (SMC) is a software for "model checking" that allows the analysis of models through rule sets [19]. SMC is able to open both .dwg and IFC files: the hierarchical structure, the topological relation between elements and their dimensional and material characteristics are automatically recognized by the software. The SMC software applies a "Quality Assurance" process for each element (for instance, data of the project, products and components, services, management, production and control processes) by verifying the element's compliance with specific requirements by means of pre-set rules. The user can create customizable rules and run specific analyses in accordance with regulatory requirements, which can vary for each country.

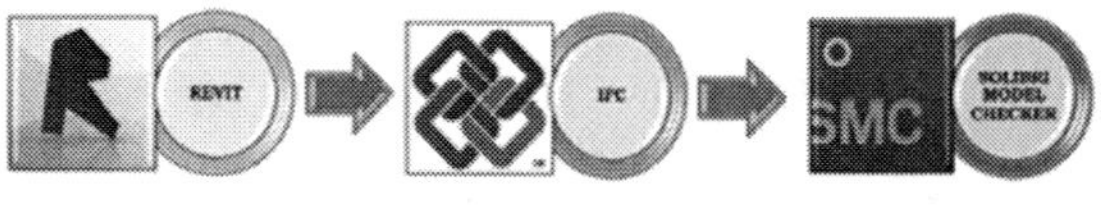

Figure 7. Export process of the Revit model on Solibri Model Checker.

These rules can be employed for the evaluation of the quality of the model and for the interference checking that makes it possible to highlight criticisms and shortcomings of projects, contributing to their validation. Among the rules proposed by SMC software is a set concerning the checking and control of escape routes. The system uses a sophisticated algorithm to identify all the possible escape routes, calculating their length: this is then compared with the user-entered limit and defined by a specific legislation. At the same time it is also possible to analyse the fire compartments, through a check on the various building components, and the minimum dimensions of the doors that define the various spaces. In this way, it provides a secure method to ensure the safety criteria for a proper fire design. Since the primary purpose of the case study is the evaluation of fire safety, we can identify three sets of control rules within the Architectural Checking "package":

- BIM VALIDATION - ARCHITECTURAL: set of rules that includes laws to generally validate the BIM model by means of audits of all structural components;
- GENERAL SPACE CHECK: rules for the setting controls on the minimum surfaces of the environments' space.
- EGRESS ANALYSIS: set of rules that control the fire compartments and the building escape routes.

In this way, the program identifies a number of actions to be taken in order to refine and detail the model. The software automatically suggests adding missing data, if not previously defined in the BIM software. The analysis begins with the classification of the doors between Internal Door, Exit Door or Balcony Door, and vertical connections between Lift, Ramp or Stair. A data consistency check of the various construction elements is automatically given by the software, verifying if the elements are classified in accordance with the IFC standards. It is necessary to associate the corresponding classification to the walls and to the exterior doors, as SMC automatically associates functions of internal elements with these.

The next step is to assign the corresponding intended use for each space detected (IFC space) distinguishing between: office, storage, residence, space group, elevator, technical, vertical and horizontal paths (circulation).

As a last step, the compartments must be defined according to various methods by choosing the most appropriate in this case, specifying the maximum area laid down by legislation. Last step for the rules-et "Egress Analysis" approach is the definition of the parameters for the escape routes: according to the classification of the various rooms, you set the maximum path length, the minimum acreage for each occupant and the point of from which to calculate the escape route. It also needs to specify the methodology for measuring horizontal and vertical paths and the minimum size of corridors and doors. After starting the analysis, the software checks the various "packages" of rules according to the limits set by the user, reporting problems and validating the parameters.

9 Test results and problems of Solibri Model Checker

From the data consistency analysis of the BIM model there is are no errors in its structure. With regard to the analysis of fire prevention, one of the main issues is the lack of elements with adequate fire resistance, such as doors, walls and windows: this discrepancy appears to be in agreement with what had been highlighted during the traditional fire analysis.

To verify the size of escape routes the rule "Escape Route Analysis" was analysed, through which it is possible to locate the escape routes, verifying that the length does not exceed the imposed limit. In addition to this, having set the size of the passages and doors, the number of planned occupants and the area assigned to each user, SMC is able to verify if the width of each frame is sufficient for evacuation, according to the crowding expected for any given space.

From this point of view, there are no problems related to the sizing of the passages and doors, so attention must be focused on the excessive length of the paths of some environments.

Performing an additional comparison with the results obtained from the calculation of the lengths through measurement made using the traditional method, (using AutoCAD 2012), we can highlight a deviation between the two values of about ten meters.

This is mainly due to the method of calculation chosen on SMC, especially in regard to the measurement along the stairs.

A further verification was carried out with the aim of reducing the escape routes, by transforming the stairwells into protected environments equipped with doors with adequate fire resistance and smoke filters so as to define a new dynamic safe place for evacuation. In this case it is not necessary to load the template again, but the classification of doors of the stairwells can be changed. In this way, the software directly modifies the extreme of the calculation of the paths.

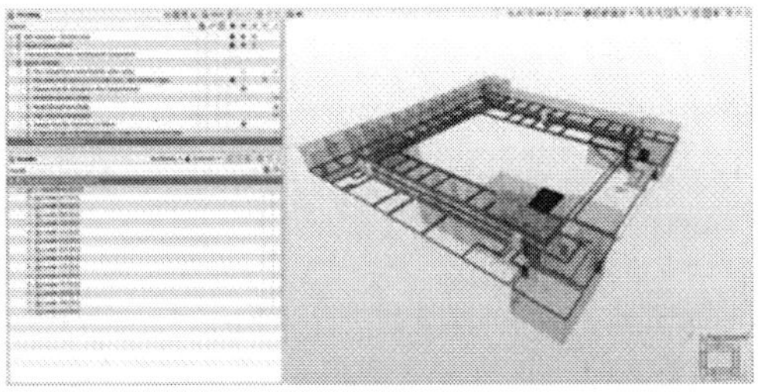

Figure 8. Screen of analysis results with Solibri Model Checker on the length of the escape routes and relevant details.

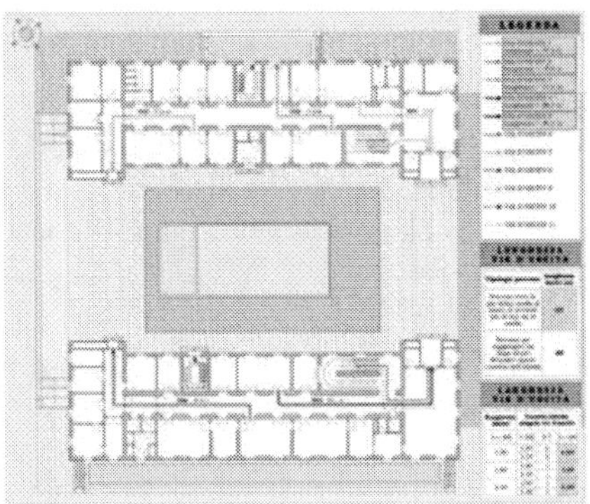

Figure 9. Graphic explanation of the traditional fire testing.

Then, the software calculates the new distance and highlights those environments which exceed the regulatory limits; graphically, the entire escape is highlighted, from the internal selected space to the outside. Comparing the two proposed solutions, we can notice a reduction of the gap between the two measurements, on AutoCAD 2012 and on SMC files: this is due to the absence of the stairs in the distance calculation.

Despite the considerable advantages of using SMC, we must take into account some gaps in the software. First of all, it is necessary to emphasize that any standards imposed, and the regulatory limits required, refer to parameters that are sometimes not the same as those set out in Italian laws: if it can, on occasion, be difficult to adapt the program requests to the user's needs, another problem is that some checks on relevant national rules could be missed.

Another important issue is the difficulty of importing the model between Autodesk Revit and SMC. This is not due to the creation of the file in IFC format and its loading, but rather to the loss of information and already-defined parameters that must be re-assigned every time. SMC assigns precise classifications to the various structural elements that must be carefully checked and modified if necessary, every time a new model is defined. If this aspect can be considered of secondary importance for an existing building, the same can hardly be said of building the new design, especially in the case of relatively complex buildings: the template cannot be directly edited. Instead, the user can only change the options or report any issues and their solutions. Each time the original project undergoes a variation, the same process of importing must be applied, carefully controlling the specific classification.

10 CONCLUSION

The case study allowed us to analyse a building belonging to the Italian historical-cultural heritage.

HBIM allowed us to perform a full survey of the building that, thanks to its three-dimensional restitution, makes it possible to consider lots of potential conservation measures, recovery, rehabilitation or functional re-use, and to perform structural, energetic and plant design analysis as well as assessments regarding the safety of the building.

However, the technology available does not allow a true representation of many features and of specific details that are beyond the seriality and prefabrication which affects new construction: for this type of modeling we can find groups of buildings that are difficult to model and we need to define *ad hoc*.

It is still essential to establish, from the earliest stages of modeling, the correct level of development to be obtained in order to streamline the resources available: according to the survey purposes it is possible to define a relative degree of accuracy of the available data in order to calibrate the quality and quantity of information for a faithful reconstruction of the buildings, and to reliably perform the analysis that we intend to carry out.

The analysis with SMC has also highlighted a number of errors that has affected the final results, if we compare to the results achieved by traditional method performed with spreadsheets and AutoCAD files. The use of automatic control, with respect to the prescriptive indications of fire prevention, would lead to considerable advantages from the economic point of view and also with regard to the management of the design for users' safety.

To solve the interoperability issue (importing data from Revit into SMC or any other model checking software), is necessary a redefinition of the input data. This interoperability problem arises from the missing semantic information in the IFC file exported from Revit. This issue is best dealt with using "Semantic

Enrichment" of the BIM model files. Further developments of this research could be done considering the semantic enrichment approach.

In conclusion, we can state that Fire Safety Engineering is the way to manage historical and artistic heritage in fire prevention. Thanks to its design flexibility, a proper use of FSE ensures a level of safety for buildings which is equivalent to that provided by means of performance standards. This approach is made easier by having a complete BIM model, so research must be directed towards the development of an appropriate level of interoperability between BIM and FSE, in order to optimize the results and to better manage the information during both the design and management phase.

References

[1] Bernstein P.G. and J.H. Pittman. Barriers to the Adoption of Building Information Modeling in the Building Industry, *Autodesk BUilding Solutions, White Paper.* :1–14, 2004. http://academics.triton.edu/faculty/fheitzman/Barriers to the Adoption of BIM in the Building Industry.pdf.

[2] Eastman C., P. Teicholz, R. Sacks, and K. Liston. *BIM Handbook: A Guide to Building Information Modeling for Owners, Managers, Designers, Engineers and Contractors*, 2° Edition, John Wiley & Sons Inc., New Jersey, 2011.

[3] Miettinen R. and S. Paavola. Beyond the BIM Utopia: Approaches to the Development and Implementation of Building Information Modeling, *Automation in Construction.* (43) :84–91, 2014.

[4] SFPE - society of fire protection engineers. *Building Information Modeling and Fire Protection Engineering*, 2011.

[5] Biagini C. and V. Donato. Building Object Models (BOMs) for the documentation of historical building heritage, in: H.C. Lomonaco, S. Barba (Eds.), EGraFIA 2014: Revisiones Del Futuro, Previsiones Del Pasado. V Congreso Internacional de Expresión Grá ca En Ingeniería, Arquitectura Y Carreras A Nes Y XI Congreso Nacional de Profesores de Expresión Grá ca En Ingeniería, Arquitectura Y Carreras A, EGraFIA 2014, Rosario (Argentina), 2014.

[6] Volk R., J. Stengel, and F. Schultmann. Building Information Modeling (BIM) for Existing Buildings — Literature Review and Future Needs, *Automation in Construction.* (38) :109–127, 2014. doi:10.1016/j.autcon.2013.10.023.Abstract.

[7] Donato V. *Transformability of the Existing Buildings: an approach based on BIM technologies*, Ph.D. Thesis, 2015.

[8] Raffaello F. *Architettura della Scuola di Applicazione Aeronautica di Firenze*, Electa Firenze, Firenze, 1988.

[9] Oreni D., R. Brumana, A. Georgopoulos, and B. Cuca. Hbim for Conservation and Management of Built Heritage: Towards a Library of Vaults and Wooden Bean Floors, *ISPRS Annals of Photogrammetry, Remote Sensing and Spatial Information Sciences.* (II-5/W1) :215–221, 2013.

[10] Osello A. *Building Information Modeling - Geographic Information System - Augmented Reality per il Facility Management*, Dario Flacovio Editore, Torino, 2015.

[11] Dimyadi J.A.W., M. Spearpoint, and R. Amor. Generating Fire Dynamics Simulator Geometrical Input Using an IFC-Based Building Information Model, *Electronic Journal of Information Technology in Construction.* (12) :443–457, 2007.

[12] Dimyadi J., M. Spearpoint, and R. Amor. Sharing building information using the IFC data model for FDS fire simulation, in: Fire Safety Science, 2008: pp. 1329–1340. doi:10.3801/IAFSS.FSS.9-1329.

[13] Marsella S. and L. Nassi. *L'ingegneria della sicurezza antincendio e il processo prestazionale. Guida alla fire safety engineering ed esempi applicativi*, Ed, EPC Libri, Roma, 2006.

[14] Marsella S. and L. Nassi. *Sicurezza antincendio per i beni culturali*, UTET, Torino, 2008.

[15] Balocco G., A. Carpignano, and L. Nassi. Approccio "prestazionale" Alla Sicurezza Antincendio in Edifici Di Interesse Storico Artistico: Il Caso Di Santa Maria Della Scala in Siena, *Quaderni Di Scienza & Tecnica.* :26–56, 2007.

[16] Nassi L. I Modelli Automatici Di Simulazione Dell'incendio - Come Vengono Utilizzati Nella Progettazione Antincendio E Come Si Valutano Le Ipotesi Di Base E I Loro Limiti, *Ordine Ingegneri Di Latina.* :10, 2010.

[17] BuildingSmart Finland. COBIM - Common BIM Requirements 2012 Series 1: General Requirements, *Cobim: Common BIM Requirements.* (1) 2012.

[18] BuildingSmart Finland. COBIM - Common BIM Requirements 2012 Series 2 - Modeling of the Starting Situation, *Cobim: Common BIM Requirements.* (2) 2012.

[19] Khemlani L. *Solibri Model Checker v7*, 2011. www.aecbytes.com/review/2011/SolibriModelCheckerv7.html.

The Role of Automation in Right-time Construction Safety

J. Teizer [a]

[a] RAPIDS Construction Safety and Technology Laboratory, Ettlingen, Germany
E-mail: jochen@teizer.com

Abstract –

Construction organizations continue to be challenged to adequately prevent accidents. Although the five C's (culture, competency, communication, controls, and contractors) have been focusing for many years on compliance, good practices, and best in class strategies, even industry leaders have only marginal improvements in recorded safety statistics. Right-time vs. real-time construction safety and health identifies three major focus areas to aid in the development of a strategic – as opposed to a tactical – response. (1) Occupational safety and health by design, (2) real-time safety and health monitoring and alerts, and (3) education, training, and feedback leveraging state-of-the-art technology provide meaningful predictive, quantitative, and qualitative measures to identify, correlate, and eliminate hazards before workers get injured or incidents cause collateral damage. Based on the current state-of-the-art of existing innovative initiatives in the occupational construction safety and health domain, a framework for right-time vs. real-time construction safety and health presents the specific focus on assisted safety and health data gathering, analysis and reporting to achieve a better safety performance. The practical as well as social implications in conducting a rigorous right-time safety culture in a construction business and its entire supply chain are tested in selected application scenarios and results are presented.

Keywords –

Right-time alert and warning technologies, automation and control, construction safety and health, historical safety statistics, lagging and leading indicators, performance measures, prevention through design, pro-active decision making, real-time monitoring and feedback, remote sensing, safety culture and climate, worker education and learning.

1 Introduction

Many safety studies that were led by industry or research experts focused greatly on traditional safety assessment and improvement techniques. In fact, these efforts have been successful and have changed the culture of safety. Today, many owners take safety very seriously and demand from contractors or subcontractors – and slowly from the entire supply chain in construction – to provide excellent track records and safety management on their jobsites. But even voluntarily reported safety performance statistics by industry leaders show that the real impact of safety best practices – which can be much stricter than existing federal or state safety rules regulations – hardly leads to further decreases in recordable incident rates. A valid question remains: How can the goal of zero accidents in construction and in any other hazardous workplace environments be achieved?

To review the magnitude of this problem, it is critical to consider construction with its significance to the economy of many countries. The U.S. construction industry, for example, accounts for $611 billion or 4% of the yearly gross domestic product (10% if equipment, furnishings, energy and other sectors are included), employs 7% of the total workforce, and invests $1.16 trillion every year in the economy [1]. A key message is that advances in emerging technologies offer significant opportunities to improve construction safety substantially and help meet other national challenges, such as becoming more efficient in work tasks. Thus, an investment in safety directly impacts the performance of other troubled industry sectors.

A key element that can help in objective data collection, processing, information, and knowledge generation of many of the mostly dynamic construction resources (personnel, equipment, and materials) [2] is automation [3]. Finding ways to achieve such automation and leveraging information that becomes available at the right-time vs. real-time is the topic of this paper. The author argues that such automation should be done at the right-time, enabling up-to-date safety data gathering, analysis and reporting when needed and with the ultimate goal of achieving better safety performance.

The paper outline has the following structure: the background reviews occupational safety and health literature; second, up-to-date safety performance statistics and performance measurement techniques are reviewed; third, the motivation for right-time vs. real-time pro-active construction safety and health is introduced by creating a framework and a strategic roadmap for implementation; fourth, examples of state-of-the-art research illustrate the feasibility of implementing the framework; fifth, the paper concludes with a summary of its contributions.

2 Background

2.1 Safety as a Key Problem in the World

More than a decade ago and varying by country, up to 50% of all the fatalities in work related accidents occurred in construction [4]. While today in Europe more than one in four fatal accidents take place within the construction sector, Australia and the US report about one fifth of worker fatalities [5]. It is reality that construction accidents occur significantly more often than in any other industry sector [6-8]. In addition to the loss of life, construction accidents result in serious and debilitating injuries. One of the key reasons for incidents are unsafe worksite conditions and human error [9]; in particular, carelessness or lack of awareness on the part of workers [10]. Workers often do not have good perception of the risks involved in executing work tasks and according to Rasmussen's model, their or their management's motivation to achieve high levels of productivity pushes them to work 'near the edge' (in terms of their exposure to risk beyond the zone of control or recovery) [11]. A major limitation of current practices though is that safety and health data is most often collected after the fact – when loss of life, injury, near miss, or at-risk behavior have already occurred – with no chance of return.

2.2 Safety Education and Training

Safety educating and training construction workers is required off and on jobsites, but it is largely undertaken through frontal teaching and instruction of hazards. Very little, if any, experiential training or practical education is carried out on job sites. It is a rare scene that a safe training mock-up structure on the jobsite is provided to workers [5]. Few receive individualized safety training or continued education on smaller or mid-sized operations. Other successful efforts that are underway [12] involve construction safety experts early in the design and planning process of construction projects. Their goal is to identify and eliminate hazards before they appear during operations. Owner buy-in and funding, however, is required.

2.3 Time Dimension in Safety Performance Measurement

Effective monitoring and control of construction safety in modern projects require sufficient planning and management to succeed. Various measures such as lagging indicators, leading indicators and safety climate have improved safety [13]. These traditional approaches of measuring safety performance rely on manual means and subjective measures (e.g. surveys or manual counts), which are often costly or slow to conduct, error prone in accurateness, and infrequently performed for effective project safety control [14]. In addition, the availability and usefulness of indicator data diminishes over a short period of time rather quickly due to the quick progress on construction sites. Edirisinghe et al. [15] introduced a good definition on leading indicator lifespans: "the time period the indicator remains useful relative to a potential incident". They added, time-delays between indicator data collection, result reporting, and responsive action can undermine any possible advantages of having such information available. According to the same study physical hazard indicators have the shortest lifespan and – due to the dynamic nature of the work environment – should be collected immediately, if necessary in real-time, to derive safety performance evaluations. Perception and management leading indicators – the terms are all introduced later in this article – are typically collected on a regular basis over time. While in-depth information about the root cause of accidents or incidents is important to prevent the same type from happening again, the time dimension in analyzing the causality of accidents and incidents is vital, but is very often overlooked in research studies and practical world applications [15]. Dyreborg [16] therefore argues that the 'time-window' to understand the 'cause-and-effects relationships' is rather important. Although, automation in the safety and health performance measurement processes is one solution [3], to date, there exists no formalized approach for right-time automated safety monitoring, data analysis, information reporting, knowledge generation, and/or visualization in the construction industry.

2.4 Technology for Safety Data, Information, and Knowledge Generation and Sharing

Although the uniqueness and short duration of construction projects might be reasons for the slow and limited adoption of technology in construction, multiple emerging technologies are suited to be applied to the safety problem [2]. To improve the time-lag between data recording and change in safety performance, several improvements are necessary:

1. It is important to identify classes of technology, since a proposed system must be robust in operation and readily incorporate technological breakthrough to make it worthwhile for companies to invest.

2. Data that has been gathered and processed to information might be applied and disseminated as new knowledge with all project stakeholders immediately rather than limiting it to few.
3. Feedback of any necessary kind must be shared pro-actively at the desired level-of-detail at the right-time to identify or predict safety issues, perhaps in real-time when life is at imminent risk.

An abundance of research literature exists on sensing systems for resource location tracking, rapid and robust visual progress and motion tracking, limited availability of worker warning, feedback and actuation technology, and virtual reality as data visualization environment. Several of these have been reviewed in [17] and are not further discussed in greater detail in this paper. Based on the previous sections of the literature review, construction safety and health research and applications for the past few decades can be grouped into clusters. Figure 1 illustrates a few of the topics that build the base for right-time construction safety and health research. A trend towards safety engineering – the use of advanced sensing and data mining technologies – becomes visible, while all other domains are necessary to build a strong foundation of safety within an organization.

Figure 1. Existing domains in construction safety and health research

3 Method

3.1 Formalization of a Right-Time Pro-active Construction Safety System Architecture

Strategically-conducted safety research seeks to study the critical technological and algorithmic needs associated to the safety decision-making processes that require large-scale and data-intensive sensor networks for monitoring and visualization of infrastructure, human resources, equipment, and materials. Limited elements of this critical technology exist in hardware and software form to some degree. The design of a complete right-time pro-active safety system architecture has yet to be realized due to the:

1. impracticality of contemporary technology (data acquisition),
2. inefficiency or unreliability of existing data processing algorithms (data analysis),
3. lack of realistic methods for realizing a safety feedback system architecture (reporting and alerting), and
4. absence of proven safety management actions (safety culture).

Thus efforts are warranted into the identification and resolution of critical needs in right-time data acquisition, data processing, and reporting in the complex and data noisy construction environments. Implementation of right-time pro-active safety and health research requires addressing several open research questions: Which traditional safety information from accident causation models and safety indicators is useful for a right-time or real-time construction safety and health process? What type of data gathering sensors, processing techniques, and data visualization environments can provide efficient, effective, reliable, fast, and accurate safety information in a highly dynamic, unstructured, and/or cluttered environment? Even if such fast and accurate sensor, information, and visualization technologies are available, how can the proposed framework that is ultimately targeting safety applications, handle large amounts of data, reduce measurement errors, in what data file format and which existing open software interfaces, and perform all of that at the right time, eventually in real-time or near real-time? How are the potential construction hazards recognized early, preferably at the planning stage, and how can previously unrecorded safety data be gathered to assist safer construction design? What safety information provides the most relevant feedback to the project stakeholders and how can it be communicated to workers so they utilize it in their transition from skill-based to knowledge-based decision makers and how can organizations transform safety management actions? To address these challenges, the author proposes to use right-time data collection and processing techniques and immersive visualization environments as a catalyst for the conversation of safety and health information.

3.2 Linking Accident Causation Models and Safety Indicators

The main concept behind the accident triangle of [18] is that severe accidents can be prevented if one takes care of the more frequent unsafe acts first. Bellamy [19] noted that the investigation of accidents can help preventing similar ones of the same type. As Edirisinghe et al. [15] argue, however, both studies did not consider the time dimension in analyzing causations. They note "it is essential to consider the time dimension in causality analysis in order to undertake an evaluation of the time-sensitivity of the different types of indicators".

The author issues the following definitions to explain the context of data types in the developed framework:

- *Right-time (real-time) construction safety*: "The (latest) point in time when knowledge is required to prevent an injury or collateral loss." Right-time in real-world applications might frequently be real-time, for example, when a worker-on-foot requires real-time situational awareness and instant reaction time to avoid being struck by equipment traversing in too close proximity to a work area.
- *Physical accident/incident precursor indicators:* "Evaluation of the characteristics of the physical work environment." These include (pre-) work site conditions (e.g., weather, illumination, road conditions, and availability and condition of tools and materials), the presence and state of resources (e.g., static, moving, or interacting workers, equipment, and materials), work crew and equipment interfaces, and risk exposures and management approach.
- *Management leading indicators:* "Counting safety management activities" (e.g., frequency and number of inspections, safety walks or checks by contractor's and client's safety representative, training sessions, hazard reports, and the time it takes to address issues).
- *Perceptions or situational awareness leading indicators:* "Periodic measurement of worker's and management's perception of safety climate" (e.g., surveys measuring effectiveness of safety and health program, level of quality to commitment to safety culture).
- *Safety levels related to events:* "Measurement of accidents (e.g., injury or fatality), minor incidents (e.g., first aid treatment or small collateral damage), near-misses (e.g., unsafe act or event almost leading to accident/incident), and situational or personal issues (e.g., state of communication, supervision, and worker health and fatigue, behavioral factors of humans)".

It is important to understand the time-sensitivity associated with leading safety indicators. Some might even be dismissed if the time lag between data capture and analysis is too great. According to Lingard et al (2013), a great emphasis is on selecting the appropriate frequency of useful data capture and reporting.

3.3 Organization for Right-time Safety

Figure 2 depicts the complex time-dependent nature of safety performance indicator data that become available to different levels within an organization through the use of manual and/or automated recording methods. As construction stakeholders require different pieces of safety information at different time intervals, technology may assist in this task; in particular early in a project through automated safety in design checking. Construction safety planning is important, but less time-critical, since it typically occurs months or weeks (in other words 'at the right-time') in advance of construction. Perception and management indicator measurement frequencies vary accordingly from weeks to months depending on the organization's resources it has committed. Although these indicators identify gaps in an existing safety program, they have a limited ability to detect any potential hazards near real-time. Data from these indicators require though execution of appropriate actions once holes are detected. Once the severity of the hazard intensifies, injury, illness, or death may become more likely.

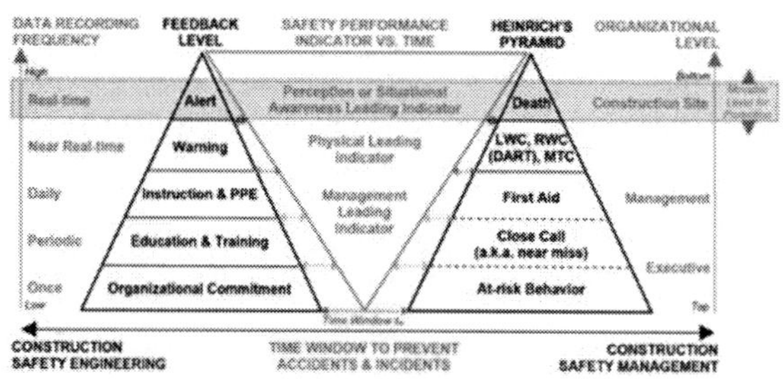

Figure 2. Right-time construction safety framework: 'movable layers for protection' address the critical time window for accident prevention and response

On-site real-time hazard monitoring is vital to detect physical hazard indicators. Real-time warning and alert feedback is required. As Edirisinghe et al. [15] noted, a good strategy is to use real-time captured leading indicators as a hazard precursor and execute appropriate actions immediately once holes are detected or intensify. A "movable layer for protection" – shown in an example as a layered barrier in Figure 2 – prevents a fatality using – at the latest – real-time data recorded on a construction site once the perception or situational awareness of site personnel fails. The time window to prevent accidents and incidents accordingly narrows the worse the outcome of one gets.

3.4 Techniques for Right-time Safety

To achieve the ambitious goal of building such a system, the roadmap builds on technologies for right-time safety data collection, processing, and reporting. Figure 3 depicts the proposed technological approach of a pro-active control system for construction safety.

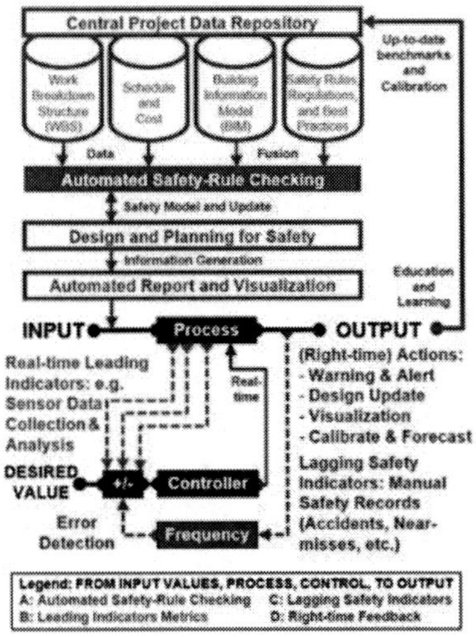

Figure 3. Technical right-time safety system architecture: Influence of safe design and indicators on control and output of a safe construction process

It consists of input (e.g., data collection), processing (e.g., data analysis), feedback (e.g., reporting and alerting), and output (e.g., management actions). The system includes several modules that are explained in more depth:

- *Pre-project and pre-task execution data collection, early rule checking, and reporting to prevent hazards:* This module captures project-related data from various sources in a central data repository, including important operations data such as work breakdown structure (WBS), schedule and cost, building information modeling (BIM), safety rules and regulations, and best practices data. Project data, if held in a central data repository and fused, can be checked using automated rule checkers [20]. These design and plan for construction safety and health during the pre-construction phase.
- *Lagging indicators vs. real-time leading indicators collection:* This module collects and processes lagging safety and health indicator data by applying conventional data collection approaches (e.g., manual use of smart phones). While many researchers noted that manually collected lagging indicators provide value detecting errors in the system, they cannot contribute from preventing hazards that appear in real-time. Therefore, Edirisinghe et al. (2014b) and Lingard et al. (2013) envision real-time benchmarks to be set across work groups, projects, organizations, and industry sectors and propose forecasting as a second type of analysis, which includes time series and trend analysis to identify deviations from normal conditions. The module on real-time leading indicators collection collects and processes leading indicator safety and health data in real-time. It takes place in the entire construction supply chain, before and throughout the construction operation duration, and eventually beyond.
- *Control through desired safety benchmark values:* They are responsible for accurate safety performance according to a-priori established safety and health benchmarks. Benchmarks are set up-to-date using various leading indicator values. Reporting and alerting mechanisms are based on real-time pro-active feedback technology or conventional management actions. Any control task creates new data that itself can calibrate existing data in the central data repository.
- *Right-time (real-time) feedback and management actions:* This module includes three steps: (a) real-time control, (b) near real-time or slower control, and (c) forecasting. As Teizer et al. [21] stated, "sensing and actuation technology provides an additional level of protection". They further noted, the specific application determines whether pro-active real-time hazard detection, identification, analysis and feedback techniques are required to prevent loss of life or severe collateral damage. Aside real-time control, near real-time or less frequent feedback enables reporting to various management or executive levels of an organization with the focus on changing mid- or long-term goals of established occupational safety and health policies. The last step is forecasting. All gathered information can be used to predict and rectify any potential future risk early in projects.

4 Results to Conceptual Proof of Concept

Several existing real-world application scenarios tested the right-time pro-active safety framework and its system architecture: (1) Safety rule checking to prevent hazards [20], (2) location data tracking and analysis to determine leading indicators [14, 22], (3) near real-time monitoring for rapid response [14], and (4) advanced education and training [23]. The first and third are explained while results to the second and fourth are discussed extensively in the mentioned literature.

4.1 Automated Safety Rule Checking to Detect and Eliminate Hazards Upfront

Zhang et al. [24-26] developed a safety rule-checker and safety ontology that integrate available project data (Building Information Model, work breakdown structure, schedule, etc.) and checks domain knowledge, e.g. official safety code, automatically using machine readable language. Necessary protective equipment will then appear for safety violations where and when needed, e.g. guardrails for building edges or covers for holes (see Figure 4). Similar to a structural clash detection tool, testing for safety violations in existing BIM designs (hazard model) uses parametric conditions of BIM objects. Visualization, simulation of alternative ways of construction over the project duration, optimization of work phases, and accurate quantity take-off of protective safety equipment becomes feasible. Visualization through BIM, however, will be limited to static protective safety equipment or hazard clash/location identification data. Currently, commercially-available BIM software is not able to engage users less on any of the data from dynamic resources that exist in the construction environment, such as workers and equipment movements. Real-time 3D immersive environments utilizing real field data, explained later, offer such capability, however, require further development towards safety applications.

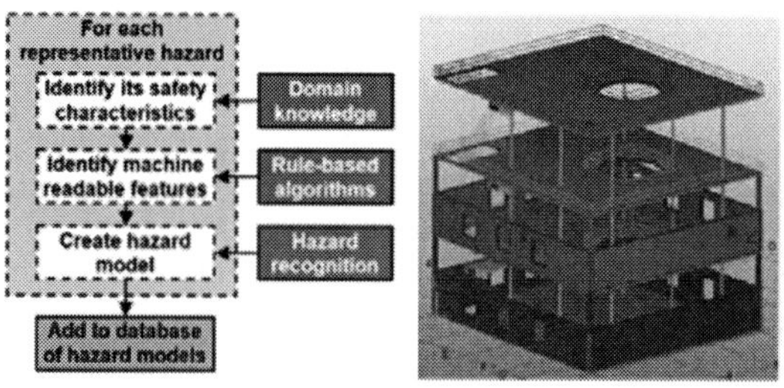

Figure 4. Rule-based safety checking (dashed lines indicate current manual practice) and result to automated rule checking that detects and mitigates fall hazards in BIM [20].

4.2 Near Real-time Leading Safety Performance Indicator Data Collection, Analysis and Visualization

As positioning sensors collect time-stamped location data, the spatio-temporal proximity of two (or more) resources can be found by cross-correlating data from both resources. User-defined proximity thresholds can be applied to define the terms of "near-miss", "warning", "alert", and "collision". Empirical data generated in this experiment, once analyzed automatically, creates valuable proximity graphs without any human interference. Mining such data further can lead to evaluation of truthfulness of safety best practices (e.g., rules established for work near dynamic hazards). Recorded site data can further be mined to result in personalized education and training [23]. The frequency of the proximity of workers can be quantified and mapped. See the location of fatal events of workers-on-foot with underground mining equipment in Figure 5.

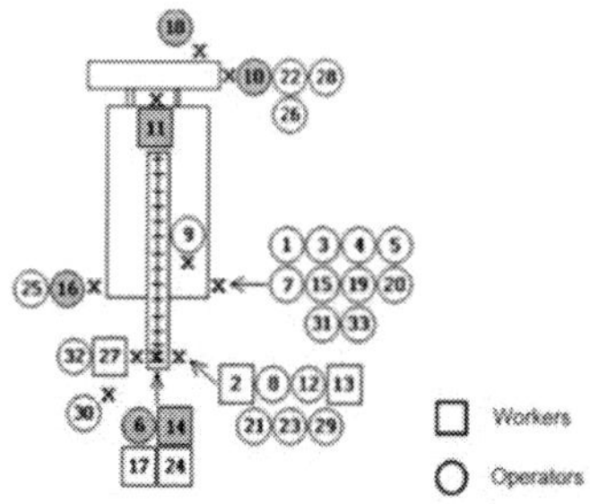

Figure 5. Number and location of fatalities around equipment recorded after the fact [27].

If needed, individual resources can be optimized, e.g. a worker can be taught to change his/her work behavior. Alternatively, a piece of equipment can be replaced with safer ones, if it is determined that it causes many near misses. Criteria on when to change will be set on leading indicators. The advantage of leading indicators (compared to lagging indicators) is that they can change the outcome of a process before it actually ends. The chance of improving the safety performance is subsequently high, as many incidents can be prevented if actions are taken rapidly enough. Thus, should a-priori safety thresholds or tolerable benchmarks be met, unfortunately, control or actions will be taken to resolve the issue immediately, if not automatically.

Such a scenario is novel as in the past, lagging indicators depended on incidents to occur to control a process after the fact. Lagging indicators offered no remedy since information was provided too late to have significant impact on changing the process (and safety performance). As important site safety metrics become measurable effortless and at low investment, immediate increase in safety performance while a process is running is possible.

Results to a real-world application are visualized in Figure 6. A construction site with many worker-on-foot and equipment near misses in and close to a ramp into an excavated pit is shown. Although the near miss data was collected in real-time and processed afterwards, timely information can be presented to the site safety management that then provides a second, safer entrance for workers-on-foot into the pit area or that orders placement of concrete barriers for a safe walking path for the workers-on-foot next to the traversing heavy equipment. After visualizing the proximity data, the contractor decided on installing a separate entry/exit into the pit. Even such alternatives (e.g., using potentially an unstable ladder) should be evaluated whether they are a good safety best practice.

Figure 6. Close proximity events visualization projected on satellite imagery [28].

5 Discussion and Conclusions

Safety in increasingly complex, large, and capital intensive construction projects is challenging, time consuming, and mostly a manual task for engineering and construction professionals. In such projects, communication of essential safety information among project stakeholders becomes key to save lives and cost. When technology is utilized to rapidly provide valuable information to decision makers, significant safety and efficiency improvements are envisioned to trigger tighter (safety and risk factor) integration of safety and construction process information. Several motivators have led this study on a right-time monitoring system for construction health and safety based on an intelligent safety indicator framework. Other approaches – based in safety culture and climate – exist as well [29].

First, enhancing workplace safety and health – in particular providing effortless and accurate data collection and reporting valuable information – has large potential to transform the industry's current best practices.

Second, what is considered a complex human capability to capture, process, interpret, and assimilate safety information, is yet a challenging task if human-designed systems are to be replicated automatically and in real-time as intelligent prototypes. Improving performance and robustness of tracking site resources for data recording, establishing leading indicator data fusion and inference algorithms, and information sharing through immersive visualization platforms are few of important examples to improve the current standards-of-practice.

Third, no research to date has established a formal link between the impacts of pro-active hazard prevention, right-time data collection, real-time processing, and reporting of safety-relevant information to decision makers. The effective use of advanced right-time techniques in monitoring and reporting construction workplace safety and health performance indicators allow rapid acting based upon of never before recorded events.

For these above mentioned reasons, a novel system for right-time risk identification and monitoring coupled with intelligent data analysis and reporting techniques has been introduced. The strategic importance of considering the time-dimension of a leading indicator and action framework to improve early pro-active feedback, provide advanced learning, and better safety and health decision-making has been successfully implemented and tested on various application scenarios. The framework and early testing show promising results that at some point in the future may lead to autonomous intelligent prediction and avoidance of hazards in the construction workplace. New concepts to solve these problems will likely need to use an interdisciplinary approach and utilize or adapt various tools from sensing, data fusion, machine learning, behavioral factors, and virtual reality research.

References

[1] Census Bureau, 2007 Economic Census, www.census.gov (Last accessed July 10, 2009).

[2] Teizer, J., Caldas C.H., and Haas, C.T., Real-Time Three-Dimensional Occupancy Grid Modeling for the Detection and Tracking of Construction Resources, *Construction Engineering and Management*, ASCE, 133(11):880-888, 2007.

[3] Navon, R. and Sacks R., Assessing research issues in automated project performance control (APPC). *Automation in Construction*, Elsevier, 16(4):474–484, 2006.

[4] Dupre, D., Accidents at work in the EU 1998-1999, KS-NK-01-016-EN-I, Eurostat, Luxembourg, http://epp.eurostat.ec.europa.eu/cache/ITY_OFFPUB/KS-NK-01-016/EN/KS-NK-01-016-EN.PDF, p. 9, 2001.

[5] Teizer, J. and Melzner, J., Sicherheitstechnische Planung von Hoch- und Ingenieurbauprojekten mithilfe von Bauwerksinformationsmodellen (BIM), VDI-Bautechnik, Jahresausgabe 2015/16, *Der Bauingenieur*, Springer, Heidelberg, 128-135, 2015.

[6] OSHA, Accidental Reports and Statistics, U.S. Department of Labor, Occupational Safety & Health Administration (OSHA), Washington D.C., http://www.osha.gov. (Accessed January 26, 2010).

[7] HSE. HSE press release E234:03A, Health and Safety Executive, HM Government, http://www.hse.gov.uk/press/2003/e03234a.htm, 2003.

[8] LOSH. Fatal Workplace Accident Statistics, Ministry of Trade, Industry and Employment, Jerusalem, http://www.moit.gov.il/NR/exeres/0FE9A855-90F2-4C15-8793-BF4D05159D08.htm, 2015.

[9] Saurin, T. A., Formoso, C. T., and Cambraia, F. B., A Human Error Perspective of Safety Planning and Control, *12th Annual Conference on Lean Construction*, Elsinore, Denmark, 2014.

[10] Abdelhamid, T.S., and Everett, J.G., Identifying root causes of construction accidents, *Construction Engineering and Management,* 126(1):52-60, 2000.

[11] Mitropoulos, P., Abdelhamid, T.S., and Howell, G.A., Systems Model of Construction Accident Causation, *Construction Engineering and Management*, ASCE, 131(7):816-825, 2005.

[12] CMD. The Construction (Design and Management) Regulations 2007, Office of Public Sector Information, National Archives, London, http://www.opsi.gov.uk/si/si2007/uksi_20070320_en_1.

[13] Lingard, H., Wakefield, R., and Blismas, N., If you cannot measure it, you cannot improve it: Measuring health and safety performance in the construction industry, *19th CIB World Building Congress*, Brisbane, Australia, 2013.

[14] Teizer, J. and Cheng, T., Proximity Hazard Indicator for Near Miss Location Recording and Mapping of Workers-on-Foot Interactions with Construction Equipment and Geo-Referenced Hazard Areas, *Automation in Construction*, Elsevier, 60: 58-73, 2015.

[15] Edirisinghe, R., Lingard, H., Blismas, N., and Wakefield, R., Would the time-delay of safety data matter? Real-time active safety system (RASS) for construction industry, *CIB W099 Achieving Sustainable Construction Health and Safety*, Lund, Sweden, 564-574, 2014a.

[16] Dyreborg, J., The causal relation between lead and lag indicators, *Safety Science*, Elsevier, 47:474-475, 2009.

[17] Teizer, J. Right-time vs. Real-time Pro-active Construction Safety and Health System Architecture, *Construction Innovation*, EmeraldInsight, 2016.

[18] Heinrich, H.W. *Industrial accident prevention, a scientific approach*, McGraw Hill, New York, 1931.

[19] Bellamy, L.J., Exploring the relationship between major hazard, fetal and non-fatal accidents through outcomes and causes, Safety Science, Elsevier, 71:93-103, 2015.

[20] Zhang, S., Teizer, J., Lee, J.-K., Eastman, C., and Venugopal, M., Building Information Modeling (BIM) and Safety: Automatic Safety Checking of Construction Models and Schedules, *Automation in Construction*, Elsevier, 29:183-195, 2013.

[21] Teizer, J., Allread, B.S., Fullerton, C.E., and Hinze, J., Autonomous Pro-Active Real-time Construction Worker and Equipment Operator Proximity Safety Alert System, *Automation in Construction*, Elsevier, 19(5): 630-640, 2010.

[22] Luo, X., Li, H., Huang, T., Rose, T. A field experiment of workers' responses to proximity warnings of static safety hazards on construction sites, *Safety Science*, Elsevier, 84: 216-224, 2016.

[23] Teizer, J., Cheng, T., and Fang, Y., Location Tracking and Data Visualization Technology to Advance Construction Ironworkers' Education and Training in Safety and Productivity, *Automation in Construction*, Elsevier, 35:53-68, 2013.

[24] Zhang, S., Sulankivi, K., Kiviniemi, M., Romo, I., Eastman, C.M., and Teizer, J., BIM-based Fall Hazard Identification and Prevention in Construction Safety Planning, *Safety Science*, 72:31-45, 2015a.

[25] Zhang, S., Boukamp, F., and Teizer, J., Ontology-Based Semantic Modeling of Construction Safety Knowledge: Towards Automated Safety Planning for Job Hazard Analysis (JHA), *Automation in Construction*, Elsevier, 52:29-41, 2015b.

[26] Zhang, S., Teizer, J., Pradhananga, N., and Eastman, C., Workforce location tracking to model, visualize and analyze workspace requirements in building information models for construction safety planning, *Automation in Construction*, Elsevier, 60:74-86, 2015c.

[27] MSHA - Mining Safety and Health Administration, *Presentation at a Workshop on Underground Mining Safety*, West Virginia, September 15, 2010.

[28] Teizer, J., Golovina, O., Wang, D., and Pradhananga, N., Automated Collection, Identification, Localization, and Analysis of Worker-Related Proximity Hazard Events in Heavy Construction Equipment Operation, *32nd ISARC*, Oulu, Finland, 2015

[29] Edirisinghe, R., Zhang, R., Lingard, H., and Boukamp, F., The development of an automated multi-level safety climate benchmarking tool for construction projects, *CIB W099 Conference on Achieving Sustainable Construction Health and Safety*, Lund, Sweden, 258-269, 2014b.

First Person Virtual Reality for Evaluation and Learning of Construction Site Safety

Thomas Hilfert[1], Jochen Teizer[2] and Markus König[1]

[1] Computing in Engineering, Ruhr-Universität Bochum, Bochum, Germany
[2] RAPIDS Construction Safety and Technology Laboratory, Ettlingen, Germany
E-mail: thomas.hilfert@rub.de, jochen@teizer.com, koenig@inf.bi.rub.de

Abstract –

Early hazard recognition in the construction environment plays a pivotal role in accident prevention. Modern technological advances in the field of visualization and display hardware enable unrivaled situational experiences when using head-mounted displays. Low-cost hardware, high refresh rates and increasing resolutions of devices enable immersion into a virtual world with useful safety applications for the architecture, engineering, and construction (AEC) industry. Realistic particle effects and audio simulation support the perception of human test subjects experiencing a construction site from a first person view. Such test infrastructure and environment generate more accurate reactions to dangers emerging from hazardous site conditions. The scope of the presented work is developing the framework for gathering scientific data that, once analyzed, defines new approaches towards situational awareness and learning. The particular application we investigate is the understanding of the safety behavior of pedestrian workers performing tasks near heavy equipment. To create a realistic virtual reality environment, we are importing building information modeling (BIM) data into the Unreal Engine 4 for visualization with modern, low-cost HMD hardware. The developed first person virtual reality infrastructure allows us to test human behavior in several unsafe work scenarios that are common in practice. While the test subjects experience the hazards in a safe test environment, their learning is made analyzable through the use of computational logging operating on a multitude set of data records collected in the virtual environment. Results are presented that show the potential impact of the developed method on existing construction safety applications, including but not limited to rapid hazard evaluation and learning.

Keywords –

first person view, construction safety, equipment, hazard recognition, learning, situational awareness, virtual reality.

1 Introduction

Construction site safety evaluation is an important part when planning projects. Since many workers are working in confined spaces for a manifold of different tasks, risks can be associated to the work tasks. Especially when operating in the vicinity of heavy machinery, such as mobile cranes, excavators and loaders, distracted workers are prone to serious and / or fatal injuries. Construction site safety should therefore be integrated from an early on point in the planning phase to minimize costs and overhead in the whole building project [1]. Ideally, safety engineers and workers should be able to review construction site plans at this stage. However, conventional 2D site layout plans are lacking intuitive information about safety and health considerations [2]. Virtual-reality (VR) with head-mounted displays (HMDs) enables users to view the construction sites from a First Person View (FPV) to intuitively perceive the impressions in a lifelike virtual environment.

While Virtual Reality (VR) hardware and specific software was very expensive and complicated to setup and use in the past, recent changes in commercialized products allow for a broader application of VR in engineering and construction. The rise of a new generation of head-mounted displays (HMDs) enables even private users to own and use them at a reasonable cost. Targeting home users and gamers is beneficial, as this means the products are fit for consumer hands-on experiences and are not as complicated to setup as developer versions.

In this paper, we propose a VR testbed that enables the users to assess the reactions of workers in simulated critical situations on construction sites while interacting with heavy machinery. Using realistic rendering in the Unreal Engine 4 (UE4, [3]), we are able to create construction site scenarios with the support of importing existing Building Information Modeling (BIM) geometry. Audio effects can be simulated to create an ambient and localized noise source for distraction and simulation of realistic environments. Data collection is available to further analyze the users' perception of heavy machinery and to test the availability of different warning systems for pedestrian workers and operators.

2 Related Research

Wang et al. [4] introduced the early concept and design of a virtual environment for excavator operator training using the Unity3D game engine. They mentioned the value of having a VR simulator for tasks that are "[...] expensive, logically difficult [or] dangerous [...]". They showed, among other researchers [5], the complexity and programming effort that needs to be taken in account when designing a construction VR simulator environment is highlighted. The benefit in having a realistic physical simulation engine is mentioned, when dealing with excavation tasks. However, they fall short on describing an actual implementation using VR hardware, such as HMDs or stereoscopic displays and their impact on the presence of the system to the user.

Li et al. [6] presented a virtual safety assessment system for training workers on construction sites. It highlights the use of a VR environment for simulating activities that are bound to high risks for the workers. The users are trained in certain scenarios and may answer questions about mistakes in their workflow and how to correct them. Especially tasks associated with unsafe site conditions, unsafe work behaviour or unsafe construction methods are candidates for training in construction. The visualization is done using Unity3D, but falls short on an actual implementation using VR hardware. Also, this system has its main focus on retraining work behaviour, instead of analysing workers reactions towards certain site layouts, similar to [7], where Nintendo Wii controllers are used as 3D input devices. Users have to fulfil certain tasks, while working on a virtual building platform. Physical errors, such as lifting objects with too much weight, will be relayed via optical cues. They also mention the potential of using open-source game engines to extend the potential beyond only the presented use case, as it allows developers for extension of the engine code.

Cheng et al. [8] described a system for tracking trainees' safety and productivity performance of ironworkers. They utilized Ultra Wideband (UWB) technology to track real workers in an instruction environment and map the results in a VR environment, enabling immediate analysis of the actions in the real world. On-site feedback, e.g. on the violation of defined hazard zones, has shown possible. Similar to [8], the authors in [9] show a system capable of real-time location tracking on a construction site. Laser scanning was used to create a topological realistic virtual scene and the Global Positioning System (GPS) and / or UWB were used to track pedestrian workers and machinery. The application can be used to enable stakeholders to make "informed decisions in shorter times and at lower cost". However, as the virtual environment is based on real actions in both cases ([8], [9]), simulation of potential dangerous activities is not possible.

A virtual crane simulation system is mentioned in Juang et al. [10], where stereoscopic displays are used. With the addition of kinaesthetic tracking, to alter the virtual view point of the user, it gives the impression of looking through a window. In the results of testing the performance on crane operation, only a marginally improvement was observed. This is due to users having to get used to the shutter glasses and / or refresh rates, according to the authors, and the complexity of the human vision system. While having a kinaesthetic system is an improvement over stereoscopic displays, modern HMDs are far more suitable for using such a system, as they translate head movements with high refresh rates to the virtual world. Also, displaying the world through a monitor is hindering, as static reference points around the screen exist and confuse the viewer, in contrast to a HMD.

While multiple concepts and applications of safety simulators, training tools and assessment exist ([11], [12]), few of them utilized modern, low-cost HMD technology. Therefore, we see benefit in creating a test bed for researching human behaviour with construction site safety and construction site layouts in mind, especially designed for HMD usage.

3 Concept

Our proposed concept for a testbed for construction site safety to analyse human behaviour consists of a hardware and software side.

As several major companies developing VR HMDs recently went out of development status with their products, consumer headsets are available from March / April 2016. The Oculus Rift Consumer Version 1 (CV1, [13]) and the HTC Vive ([14]) both promise to deliver visual fidelity and input solutions for our proposed approach. The Vive includes two hand controllers, which are tracked in a room-scale environment via passive laser diodes, while the Rift will sell their optical tracked Touch controllers ([15]) later in 2016. Both enable users to use grab gestures for picking up objects and interacting with their surroundings. Additionally, the Leap Motion controller ([16]) can be affixed to the front of the Oculus Rift to detect the hands via a stereoscopic camera setup. Leap's new Orion software release further improves hand detection and reliability on occluded parts of the hands to a usable degree for normal users. Support for both the HTC Vive and the Oculus Rift are included in most of the major game engines, including Unity3D ([17]) and UE4.

On the software side, our system relies on UE4 to create an interactable world for the user. As UE4 is free of charge (except commercial royalties, which apply upon exceeding a certain sale margin per year) and open source, we can modify the engine to our needs. Additionally, UE4 delivers an intuitive logic building block approach on programming, the visual programming language Blueprint. Therefore, it is easier to maintain for people not used to program in C++.

Figure 1 shows the proposed system in use. The user faces a traditional PC workplace, while wearing the Oculus Rift on his head. Scenario operators can watch a mirrored version of the VR rendering on the screen, so instruction is possible at any time. The Oculus positional tracker for absolute calibration of the headset position is mounted on top of the monitor. While the head tracking determines the view inside the VR world, a gamepad can be used to move around, interact with objects and override turning direction, if necessary.

Figure 1: VR test bed in use

3.1 World building

To create meaningful surroundings for construction site safety, the scene should resemble the later construction site in the real world. As the adoption of Building Information Modelling (BIM) in the AEC industry is rising internationally, we can leverage the existing models for our purpose. Existing map services, such as OpenStreetMaps can be used to further extract geometries of surrounding buildings and to gather the digital elevation model, relevant to the scene. Manual corrections are possible through the UE4 editor application, which is used to build environments in conventional games.

3.2 Data Logging

To analyse all interactions of a worker inside the VR scenarios, different log file information should be collected. As one of the main foci of this concept lies in the detection and perception of heavy machinery around pedestrian workers, data should be saved when a participant is able to visually process information about machinery in his / her vicinity.

As most game engines only render objects when they are inside the visible part of the world (the viewport), we can use this data to detect when certain objects are visible to the user in his peripheral vision (simplified by a grey arc in Figure 2). As the participant will wear a VR HMD, his head movements are translated to the viewport in the virtual world. The natural reaction is to turn one's head towards the location of a stimulus to further analyse the situation. Therefore, we are able to log when the user is able to react to avoid hazards. Non-reaction logging is also possible, enriching the learning process once data is processed to valuable safety information. While this only filters data on visual clues, users are expected to almost always react on auditive clues in their surroundings with looking towards them.

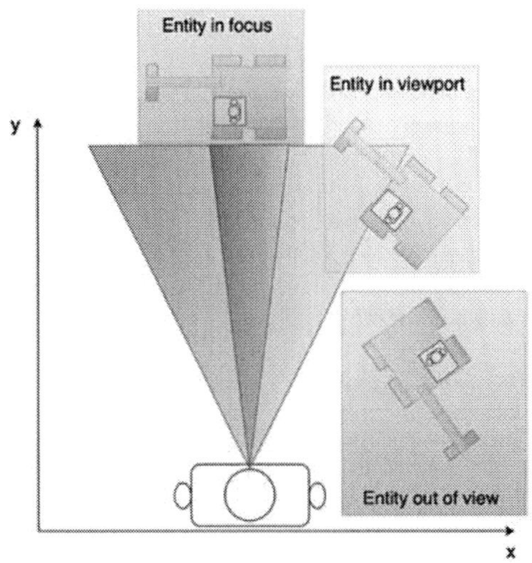

Figure 2: Schematic view of the user's perspective inside a VR environment

The second step of acquisition should be when the subject focuses on certain parts of the scenery that is associated with a-priori known hazards. As the VR HMDs contain Fresnel lenses, which will have a 'sweet spot' in the middle of the field of view (Figure 2, red arc), eye movements will be reduced in favour of head movements, such that the eyes maintain fixed in a forward looking position. We can use this downside of the HMDs to do an approximate 'eye tracking' with the middle of the camera frustum. If an object is in the middle of the view for multiple milliseconds, we can assume it has gotten the viewer's attention.

For further understanding the behaviour of the participant in the scene, it would be beneficial to have a replay functionality, which allows for the recording and playback of all events inside the scene. This will allow later in-depth analysis of all circumstances that led to certain decisions.

All of these data should be processable in an interchangeable format, for example character separated value (CSV) files, except the replay functionality. Logging all the scene data outside of the engine is not intuitive to analyse. Most game engines offer this feature in an own format, native to the engine.

3.3 Rule set

For users immersing themselves into a prepared scene, certain game logic rules have to be applied in order to deliver a realistic test bed environment.

Foremost, all of the interactions between dynamic participants (heavy machinery, physics bodies, the users) that lead to fatal or major injuries should result in the scene coming to a halt and giving feedback about the error. A supervisor should be in range to support the setup and resetting of the test run. If the users are given a specific task to fulfil, additional rules would apply.

One possible task for the participants would be to steer a wheelbarrow down a certain path. The game rule set would impose certain restrictions that will lead to failure, such as diverging outside of certain bounds away from the path or tilting the barrow above a predefined degree constraint. These restrictions would be specific to the feature set of the scene and therefore have to be programmed for each application.

3.4 Non-player character (NPC) setup

All interactive elements in the scenes, that will move and are relevant to the fulfilment of the task given, need to be prepared for the scenario. While 3D models of equipment are widely available on the internet (paid and free), most of them lack animations or the prerequisites to be animated. Usually, these preparations include binding the triangles of the meshes to certain handles (bones), which are then later animated in a meaningful fashion. This part of the setup cannot be skipped, as it will yield in a much more realistic environment for the users. Apart from using animations, textures are another requirement to be considered when designing realistic NPCs. For prototyping scenarios this may be neglected to a certain degree, but for the final presentation and test runs, animations and textures should be reworked and polished.

The movement of the NPCs inside the scene can usually be steered by setting up navigation nodes, which define the way a certain actor moves. Therefore, setting up a realistic testing environment additionally consists of planning vehicle movement throughout the construction site. Collisions between NPCs in this scenario should be avoided, as they may happen with superficial planning of navigation routes, to further increase immersion inside the scene.

As most of the models for NPC simulation should be prepared for physics interaction, it is easy to add player controls to them, enabling users to also use the test bed for vehicle driving. The same mechanisms used for processing the viewport to detect recognition of vehicles may be applied vice-versa to the detection of workers in a construction site environment. These versatile applications reduce the amount of preparation needed for multiple use cases, as the environment of the test bed would have to be prepared once and only additional scenarios have to be set up.

4 Implementation

As mentioned in the previous chapter, we are using UE4 for creating the software side of the VR test bed. To keep the system extensible to non-programmers, Blueprint visual scripting language is used for setting up the major interaction rules and scenarios. This also keeps the project compatible to additional C++ code, if needed, as projects support both languages at the same time.

4.1 General setup

UE4 comes with multiple project templates, enabling developers to swiftly create prototypes for game creation. These prototypes consist of control input schemes, a starting environment, and additional starter content, such as materials and elemental building blocks.

We chose the First Person template as a basis, as it includes most of the features needed for the creation of a VR test bed. Controls of the player character are done using the computer mouse and keyboard, or by using a gamepad, which is more intuitive to use while wearing a VR headset.

Setting up scenes in the UE4 editor is intuitive, as it has drag and drop functionality for adding models to the environment. Landscape sculpting tools are available to shape the ground of a construction site and painting different textures on the landscape is directly possible. As soon as basic placement of all necessary entities (scene starting point, lighting, landscape) is done, a preview version of the current scene can be tested inside the editor itself, or as a standalone running process.

4.2 Construction site layout

As the building to be constructed is the main focus on the site, attention should be set on modelling an authentic representation. Many buildings nowadays are already planned using BIM, therefore we are using the import plugin for UE4 described in [18]. This allows us to connect to an instance of the open source BIMserver to import geometry for the building. After importing a BIM model, textures need to be set in order for it to be realistically rendered. UE4 delivers some materials already included in the starter content, for example, concrete. Individual parts of the building can be later selected and hidden to reflect the current state of the construction process. Additionally, direct import of the Autodesk FBX file format is also possible.

Landscapes should be modelled according to the site layout plan. This can be done using UE4 landscape sculpting tools (see Figure 3), which allow on-the-fly modification of the ground. A landscape in UE4 is always assigned to a material, which can consist of different textures that blend into each other. During the design process of the scenario different textures can be painted directly on the landscape meshes in the editor. Raw height map files are also importable into an UE4 scene, which gives the possibility of automatically grabbing height information from geographical databases and use them as a basis for landscape building.

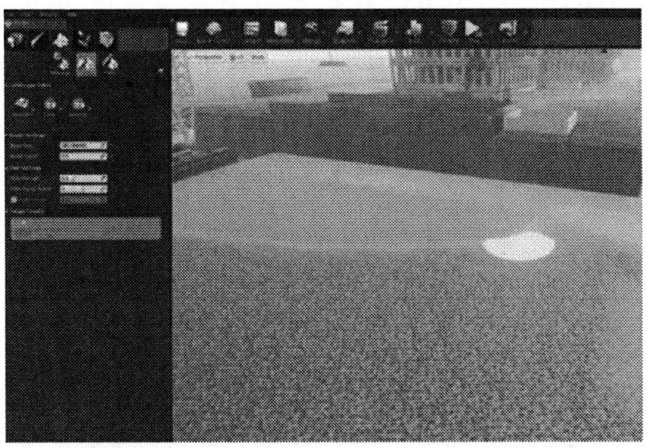

Figure 3: UE4 Landscape sculpting tools

Map services, such as OpenStreetMaps, deliver basic models of the buildings surrounding the construction site. Most of the times they are missing the height of the buildings in question, but the outline in relation to the streets and other buildings is correct. This gives some basic context to the construction site, in contrast to having only a flat landscape around. However, texturing these buildings should be considered regarding to the complexity of the scene setup, as this will be a time consuming process. For basic shapes it would be advisable to use distance fog to cover up the unrealistic textures and not break immersion, while keeping the setup time to a minimum.

4.3 Data logging components

For later assessment of the users' performance inside the VR test bed, data logging components, as mentioned in Concept, have to be implemented inside UE4. We are using Blueprints for modelling the logic, as this is easily extensible.

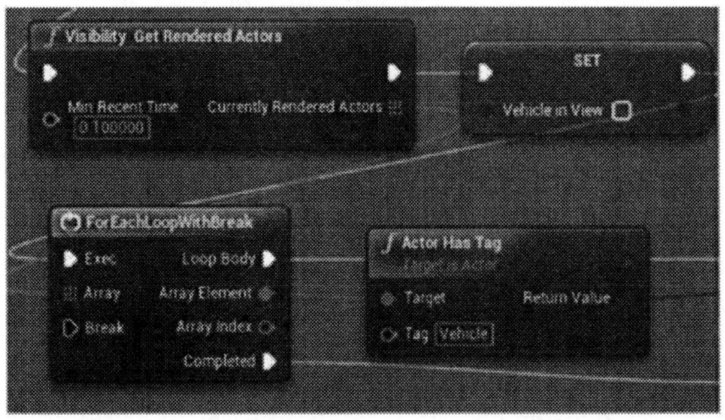

Figure 4: Checking if a vehicle is inside the viewport (Blueprint)

Figure 4 illustrates the Blueprint for detecting if a vehicle is inside the user's viewport. As this functionally is not included in the base UE4 classes, we are using the Blueprint extensions from [19]. We are then iterating over all actors, which are the base class for displayed entities, inside the viewport and check for an attached property tag. If it equals to "Vehicle", we can assume a vehicle actor has been rendered and can further process this input for analysis.

Figure 5 shows the process of detecting if a vehicle is inside the center of the viewport, which is more complicated. The first person camera is used as a reference and its location and rotation are extracted. Then we derive the forward vector from the rotation and extend it by a certain amount of distance, which varies from scene to scene. We can then use UE4's *LineTraceByChannel* function to check for a ray hit test result with any entity in the world. As the "visibility" channel is selected, only visible entities, not the ones occluded by other objects, will be returned. After this process, UE4 returns an array of actors to be checked for the aforementioned property tag "Vehicle".

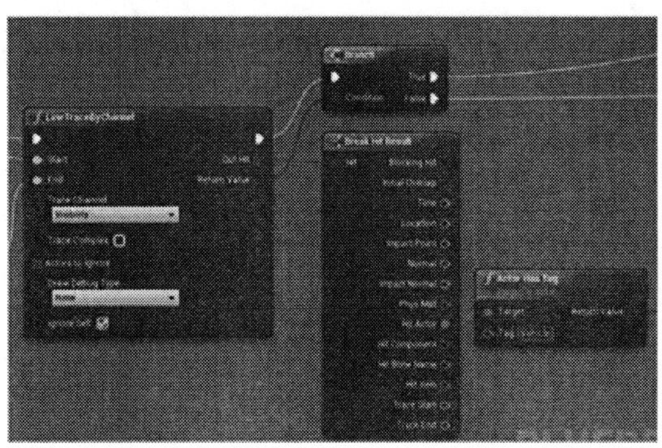

Figure 5: Line-tracing a vehicle in the center of the viewport

Both methods of detection deliver a reference to the actor in question, enabling to access all of its properties, e.g. the location, speed or other state variables. This information may be retained in the CSV log files, as it will enable complete analysis of the scene and events. As seen in Figure 6, we can use the FileIO Blueprint functions to write text to the file system. Normally the library allows only for overwriting the same file, so we did extend the functionality via C++ to being able to append text instead. As a reference value for the time elapsed, we can use the *GetRealTimeSeconds*, which references the time in milliseconds since the scene has been started. This is combinable with a custom console command, such that the time will be only measured from starting the scenario with the user already in place, instead of referencing only the time of the engine startup.

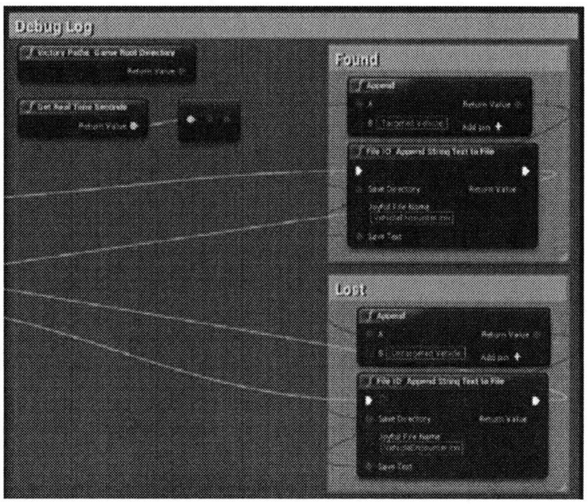

Figure 6: File logging to CSV

While logging the data about vehicles encountered in the viewport and center of view is essential for the benchmark of the performance inside the test bed, it misses some of the context information needed to fully assess the situational awareness of the environment. The test data will include the timing of the participant, but not all data about the view orientation, currently processed tasks at hand, and more Therefore, it is important to be able to record the scene execution in as much detail as possible. Creating a logging component for all of this would be time-consuming, but UE4 already has a replay and demonstration system available. When designing network-based player experiences and games, the information about all entities in the scene needs to be propagated to all connected clients, such that all of the distributed engine instances are in the same state at all times. This system, called replication, is available inside of UE4 and can be enabled for individual actors and entities, as, for example, static and unmovable entities do not need to be updated in their position. UE4's demonstration recorder will then use the information originally intended for network play, to store all needed information for later replay. When activating the replay feature, the demo can be paused and slowed down, as needed.

5 Case Study

To demonstrate the feasibility of the developed VR test bed system, we are using the layout, of a real world construction site. The users have to solve a simple task inside the test scenario to distract him/her enough from the surroundings of the construction site. For the selected application they have to carry spherical objects from one point to another. Objects can be picked up using the left mouse button and dropped in the same fashion. The two buckets (Figure 7, position A) are located at the intersection of a wheeled excavator driving path (Figure 7, position B). At a randomly triggered point in time, when the user is behind the building, the excavator will start to drive towards the user's path of operation. As most of the view is obstructed by the building, only audible clues are available to him/her at this time.

Figure 7: Case study scenario

Events to be logged in this case are: Start of the scenario, excavator in viewport, excavator in focus of viewport, excavator starting to drive, excavator striking user, excavator stopping. The scenario is stopped as soon as the excavator hits the user while on duty or the task is fulfilled. The scenario operator instructs the user a priori the test about the possibility of injuries and the construction site having safety hazards, but will not warn about the imminent danger of the driving excavator.

For the sound design we decided upon an ambient construction site recording which will be looped and is played omnidirectional. Additionally, excavators and cranes emit a motor sound at their respective positions.

Therefore, the user should be able to detect a vehicle with only audible clues. The excavator is set to drive with a top speed of 10 km/h and will start at a random point in time (between 10 and 180 seconds from start).

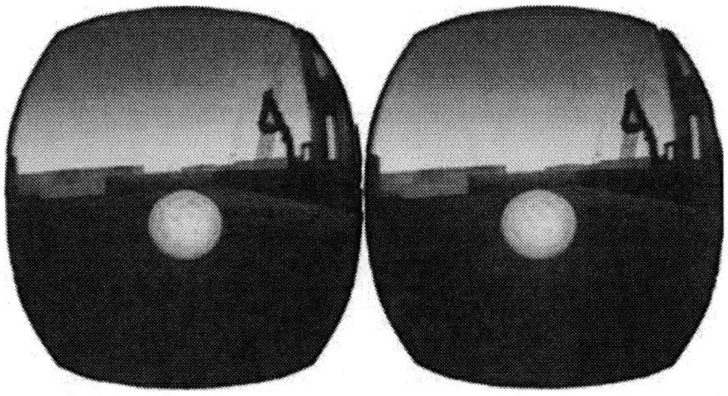

Figure 8: Stereoscopic view rendered to Oculus Rift

Figure 8 shows the view rendered to the user inside the VR headset. While s/he is carrying the spherical object for the task, s/he notices the approaching excavator around the building's corner. As the events on starting are random, it is not guaranteed, that the excavator will hit the participant, but identification on near misses are also important to consider when designing a safe construction site layout.

Research has shown that intelligent vehicle safety systems, for example the automatic deployment of an air bag, must work within 30 milliseconds of an event [20]. Table 1 shows the CSV results of a sample test run with a male doctoral student. Up to time index 10.413s the subject works on the task of carrying the spheres from basin to basin. The user is expected to see the excavator at its designated starting point, if the subject is looking towards its direction. As soon as the vehicle has started moving at around 11 seconds, the subject saw the excavator driving towards the subject and shifts his attention to it. This is seen by the "Vehicle in center" event that uses the line trace method to check for vehicle actors in the center of the view. He has, however, no time left to react and counter the struck-by event, which is seen at time index 13.297s. The subsequent "Vehicle out of center" was accompanied by an evasion reflex of the head, which was tilted away from the excavator.

The reaction time between seeing the vehicle in the viewport and focusing on it was 1612 ms, too late for a human to react in time. While this is above normal reaction time it has to be considered, that the subject was focused on the work task and it took time to realize the direction of movement of the vehicle. The narrow detection range in the center field is also difficult to interpret, as the movement towards the excavator may have started earlier. The time between "Vehicle in center" and "Vehicle hit" is merely around 87 ms and also not enough to counter the event. However, the time span between noticing the vehicle and moving the head away was well inside normal reaction margins of a visual stimulus (330 ms, [21]). For further analysis on this topic, the recoding functionality of UE4 can be used to replay the event in detail. The results should vary when modifying auditive and visual clues, such as emitting a beeper tone or signal light when driving, as auditive stimuli are faster reacted upon [21]. However, research has shown that many construction workers ignore them [22], making them less effective [22].

Table 1: CSV results of case study test run

Time (Seconds)	Event
1.036	Vehicle in viewport
1.985	Vehicle out of sight
9.557	Vehicle in viewport
10.413	Vehicle out of sight
11.064	Starting vehicle
11.598	Vehicle in viewport
13.210	Vehicle in center
13.297	Vehicle hit
13.336	Vehicle out of center

6 Conclusions & Outlook

Evaluation and analysis of human behavior at construction sites for safety considerations is an important task. With today's digital building modeling and the commercial availability of VR headsets it is possible to create the impression of on-site presence for the user to train and test inside a VR test bed. With our proposed concept, it is feasible to create such a system at a low cost and reasonable development effort. The resulting data sets are fit for later analysis and recording functionality can be used to examine and replay problems in greater detail than before.

We are looking forward to extend our test scenarios to vehicle driving and crane steering. First person interaction with the machinery may be done using hand detection and the Leap Motion, to enable realistic interactions with the steering controls. We will further elaborate on the presented and proposed scenarios and test the system with a reasonable amount of users, while trying to incorporate more complex tasks, such as building a brick wall or using machinery while trying to keep the user's environmental awareness high. Oculus CV1's Touch and HTC Vive's controllers may be used as an alternative to optical detection of the user's hands. Also, natural day/night light cycles and weather simulation can be incorporated, to create even more immersive scenarios, which distract the users with realistic environmental conditions. Also, it should be tested if an interface towards real-time location tracking using positioning systems can be implemented, to record and replay events in case of an accident.

While modern game engines like UE4 deliver a high visual fidelity, in terms of lighting, shadows, lens flares, etc., they may fall short in simulating complex physical effects, like windows fogging up. Also, the development of a scene can be complex, as animation and modelling take up a large amount of time. As soon as models are set up in a library, they can be reused for other test scenarios. However, this step should not be underestimated in project planning. Resource-wise the scenes can get also very costly to render fluidly, as the targeted frames-per-second (FPS) should be around 180 (90 FPS per eye), if wide open spaces are used without a level-of-detail mechanism in place.

Nevertheless, the current-gen HMDs for VR offer a feasible approach to run a test bed for a variety of applications, construction site safety being one of them. As more graphics card manufacturers will optimize their product to VR, entry costs will drop and the products will be affordable even for home and remote usage, priming and training users with experiences that overcome current limitations, such as motion sickness.

References

[1] Hinze J. Safety plus: Making zero accidents a reality, CII Research Rep. Univ. of Texas at Austin, 2002.

[2] Zhang S., Boukamp F., and Teizer J. Ontology-Based Semantic Modeling of Construction Safety Knowledge: Towards Automated Safety Planning for Job Hazard Analysis (JHA), *Automation in Construction*, 52:29-41, 2015.

[3] Epic Games, Unreal Engine 4. On-line: www.unrealengine.com, Acces 03/14/2016.

[4] Wang X., Dunston P.S., Proctor R., Hou L., and So J. Reflections on using a game engine to develop a virtual training system for construction excavator operators. In *Proc. 28th International Symposium on Automation and Robotics in Construction (ISARC)*, Seoul, Korea, 2011

[5] Kamat V. and Martinez J. Dynamic 3D visualization of articulated construction equipment, *Journal of Computing in Civil Engineering*, 19(4):356-368, 2005.

[6] Li H., Chan G. and Skitmore M. Visualizing safety assessment by integrating the use of game technology, *Automation in Construction*, 22:498-505, 2012.

[7] Guo H., Li H., Chan G. and Skitmore M. Using game technologies to improve the safety of construction plant operations, *Accident Analysis & Prevention*, 48:204-213, 2012.

[8] Teizer J., Cheng T. and Fang Y. Location tracking and data visualization technology to advance construction ironworkers' education and training in safety and productivity, *Automation in Construction*, 35:53-68, 2013.

[9] Cheng T. and Teizer J. Real-time resource location data collection and visualization technology for construction safety and activity monitoring applications, *Automation in Construction*, 34:3-15, 2013.

[10] Juang J.R., Hung W.H. and Kang S.C. SimCrane 3D+: A crane simulator with kinesthetic and stereoscopic vision, *Advanced Engineering Informatics*, 27(4):506-518, 2013.

[11] Chen A., Golparvar-Fard, M., Kleiner B. SAVES: A safety training augmented virtuality environment for construction hazard recognition and severity identification, CONVR 2013, 373-384.

[12] Sivanathan A. Abdel-Wahab M., Bosche F., Lim T. Towards a Cyber-Physical Gaming System for Training in the Construction and Engineering Industry, ASME 2014 International Design Engineering Technical Conferences and Computers and Information in Engineering Conference, 2014, doi:10.1115/DETC2014-34930.

[13] Oculus, Rift. On-line: https://www.oculus.com/en-us/rift, Accessed: 03/14/2016.

[14] HTC, Vive, On-line: http://www.htcvive.com, Accessed: 03/14/2016.

[15] Oculus, Touch. On-line: https://www.oculus.com/en-us/touch/, Accessed: 03/14/2016.

[16] Leap Motion, Leap Motion, On-line: https://www.leapmotion.com, Accessed: 03/14/2016.

[17] Unity, Unity3D, On-line: http://unity3d.com/, Accessed: 03/14/2016.

[18] Hilfert T. and König M., Low-cost virtual reality environment for engineering and construction, Visualization in Engineering, 4(2), 2016, http://dx.doi.org/10.1186/s40327-015-0031-5.

[19] Rama. Victory Blueprint Library. On-line: https://github.com/EverNewJoy/VictoryPlugin, Accessed: 03/14/2016.

[20] Air bags, https://en.wikipedia.org/wiki/Airbag, Accessed: 03/18/2016.

[21] Shelton J. and Kumar G., "Comparison between Auditory and Visual Simple Reaction Times," *Neuroscience & Medicine*, 1(1):30-32, 2010.

[22] Golovina, O., Teizer, J., Pradhananga, N. Heat map generation for predictive safety planning: Preventing struck-by and near miss interactions between workers-on-foot and construction equipment, *Automation in Construction* http://dx.doi.org/10.1016/j.autcon.2016.03.008.

[23] Hinze, J.W. and Teizer, J. Visibility-Related Fatalities Related to Construction Equipment", Journal of Safety Science, 49(5):709-718, 2011.

BIM Applications of Rule-based Checking in Construction Site Layout Planning Tasks

K. Schwabe[a], M. König[a] and J. Teizer[b]

[a] Department of Civil and Environmental Engineering, Ruhr-Universität Bochum, Germany
[b] RAPIDS Laboratory, Ettlingen, Germany
E-mail: kevin.schwabe@rub.de, koenig@inf.bi.rub.de, jochen@teizer.com

Abstract

Site layout and logistics planning generally plays an important role for the successful execution of construction activities. The allocation of the right amount and size and the timely use of resources play critical roles. Compared to other industrial sectors the construction industry shows a lack of technological progress in site logistics and fabrication planning. The automotive or ship building industries, for example, have stringent production planning methods in place to the point that almost every step in the planning and manufacturing processes is supported by digital simulation and optimization. On the other hand, construction planning appears to remain a manual process slowly taking advantage of building information modeling (BIM) processes and techniques. Our work investigates automated rule-based checking in construction site layout planning tasks to simplify the existing manual processes. A gap analysis of the traditional site layout planning process is performed to identify key areas that are of high concern to practitioners. A rule-based checking algorithm for site layout planning embedded in a commercially-available BIM-platform was created and tested on specific cases in the site layout planning process of a realistic building. Promising results and a discussion to the existing limitations of rule-based checking approaches for site layout planning are presented. A short outlook towards future research gives a potential path forward in advanced construction site layout planning.

Keywords –

building information modeling; construction schedule; logistics; rule-based checking; site layout planning

1 Introduction

Before construction starts, site layout planning must be performed in order to provide the necessary equipment and temporary facilities for the construction process. This includes the allocation and dimensioning of elements like tower cranes, containers or storage areas. Decisions during this planning phase have direct impact on cost development and occupational safety on site during construction. Meeting all requirements, such as safety standards and minimizing on-site overheads, is a complex process that follows specific rules or best-practices. Experience is needed for the time-consuming evaluation and checking of all relevant rules.

One of the most frequent problems during traditional construction and planning phases of a building is loss of data. Therefore, the most ambitious goal of implementing BIM (Building Information Modeling) - processes is to provide secure and consistent data management. This is achieved by adding relevant information into a parametric 3D model. Once stored in the 3D model, the information can be retrieved at any time in the future. This similarly applies to the phase of site layout planning. The 3D environment allows an exact representation of the actual equipment geometry. This can be used to easily derive site layout plans as 2D drawings from the 3D model, which are most commonly used in the field. The exact geometry and additional semantic data can also be used to retrieve required information for the dimensioning of site layout elements, where strict code compliance is mandatory. With the help of rule-based algorithms the process of checking the site layout model for code compliance can be automated. This could minimize human error, which increases economic benefit and occupational safety.

In this research we propose a concept for BIM-based site layout planning tasks. The concept includes predefined parametric site equipment models and rule-based evaluation of site layouts. Rule-checking algorithms are prototypically implemented to evaluate our findings in a case study.

2 Related Work

The steadily progressing digitalization of construction processes covers several construction phases, but cannot yet be described as exhaustive in every discipline. For example, for the phase of site layout planning no specialized BIM-based software is available. However, the idea of digitally assisted site layout planning is not new. Researchers around the world try to optimize the allocation of site equipment, e.g. tower cranes, in order to improve the workflow and logistics on construction sites. First approaches use optimization algorithms to find the optimal position of site equipment in a 2D site layout plan [1-3]. The next level of optimization and simulation approaches to site layout planning was to consider 3D and 4D environments [4-7]. The optimization of equipment positioning is highly attractive for BIM-based site layout planning, but basic requirements have to be met first. For example, the optimization of the crane's position requires characteristic parameters. But the basic process of site layout planning is the choice and dimensioning of equipment in order to define these parameters. Therefore, BIM-assisted dimensioning of site equipment will be the next step towards digital site layout planning. Choice and dimensioning of equipment is more than a geometric problem, as it is highly time-dependent [6]. For example, the size of storage areas is not calculated for all required material, but for a peak consumption. This peak consumption per day can be derived from the construction schedule linked with the 3D-model by dividing the volume of a construction element by the scheduled completion time. Another important parameter that can be derived at this point is the required workforce. Repeating this procedure for every construction phase takes the dynamic nature of workflows on construction sites into account. Furthermore, databases or object catalogs for parametric site equipment are proposed to simplify element selection and perform dynamic site layout planning [8].

Both optimization and dimensioning tasks are based on constraints. These constraints consist of specific rules and standards. In previous research the constraints are assumed, but their source is not defined any further. The process of defining and systematically evaluating rules is called *rule checking* or *code compliance checking* [9]. The automated process of rule-based model checking is still in early stages of research. The biggest obstacle is the transcription of existing mostly textual rules, standards and best-practices into an algorithmically readable form. The definition of such a form is called *rule language*. A consistent rule language that enables users to intuitively define and check rules in their 3D model does not yet exist. Simplified approaches to rule checking tasks are available in commercial software, e.g. Solibri Model Checker [10]. The software provides certain alterable rule sets, but the user is not able to define custom rules. This makes a comprehensive adoption of site layout planning rules impossible.

The application of rule checking in practice does not cover site layout planning tasks, but the similar topic of occupational safety can be found in recent research [11-12]. Safety-related conflicts are detected in a 3D model and proposals for possible solutions are provided. However, the calculation of possible solutions is very case-specific, but it gives an impression of what rule-checking can look like. Knowledge-based solution proposals belong to the field of artificial intelligence.

2.1 Rule checking

The process of rule checking in building models, as summarized by Eastman et al. [9], can be structured into four steps:

1. Rule interpretation
2. Building model preparation
3. Rule execution
4. Reporting of checking results

In the first step the naturally written rule texts need to be transcribed into an interpretable computer language. The form or grammar of an adequate rule language is still a matter of research, so that applications of rule languages remain to be customized. Eastman et al. [9] suggest parametric tables for rule interpretation.

The second step deals with objects and their parameters in a 3D model. The parameters in the parametric table must be equal to the parameters of the 3D objects, so that adequate filtering is guaranteed. The most common parameter type is *string*.

Step three checks the rules for compliance. Both positive (*true*) and negative (*false*) return values are added into the data structure. A false return value detects a rule conflict. The form of a conflict depends on the associated rule. Conflict information needs to be transferred to the next step.

The fourth and last step receives the conflict information from the rule execution und displays the results in two different ways: the textual report, e.g. IDs of intersecting elements, and the visual report, e.g. element selection or isolation in the model viewer.

This approach to rule checking is a subsequent procedure. This means, that the rule check takes place after design decisions are made. The afore mentioned commercial software Solibri Model Checker uses this principle to check 3D models for design quality. Predefined rules include element specific clash detection or element parameter comparison. The editing of these predefined rule sets is possible, but very limited. New rules cannot be defined.

3 Methods

Our general concept for BIM-supported site layout planning is depicted in Figure 1. Two of the four requirements are the building model and the construction schedule. The other two requirements have yet to be developed. They contain a database or so-called object catalog of parametric models of all existing site layout equipment and a comprehensive database of rules concerning site layout planning. These rules must be in form of an algorithmically readable rule language that enables the transcription of all existing rules, standards and best-practices. Company-specific expert knowledge or best-practices that have not yet been written down formally, can and have to be included as well. With this information the automated rule checking can support the dimensioning and placement of site layout elements. The 2D site layout plans that are needed in the field can be derived, costs including site equipment can be calculated and logistics can be planned.

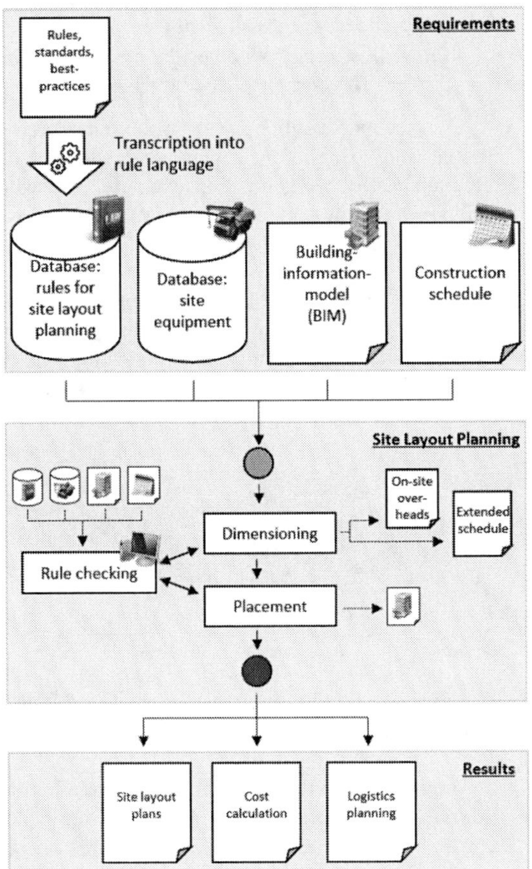

Figure 1. Concept of rule checking application

3.1 Interactive rule checking

An extended use of rule checking that is still in early stages of research describes an interactive process, where design decisions are supported by preprocessed rules. This means, that the supporting software visualizes rule-based constraints in advance. Figure 2 shows a conceptual example of interactive preprocessed rule checking. The light green area around the building indicates where the tower crane can be placed. Rules that would prohibit the placement are considered, for instance the installation of a site fence and its safe distances towards the surrounding terrain or inclined surfaces.

Although we consider the application of interactive preprocessed rule checking to be very attractive for design supporting software, the first step towards a rule-based evaluation of site layout plans is the implementation of subsequent rule checking. The problem with currently available rule checking software is the limited ability to edit the predefined rules. The solution is the development of a rule language, which can be operated intuitively by the user to create custom rules. This avoids the exclusion of the user from the rule algorithm and leads to a better understanding of the consequences of the self-made rule. Furthermore, the user is able to reedit the custom rule, if the result does not satisfy previous expectations.

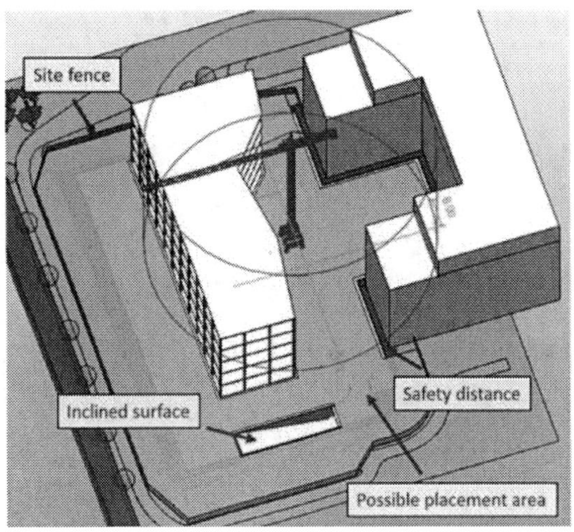

Figure 2. Interactive rule checking: possible placement area of a tower crane

3.2 Parametric site equipment

A required step of rule checking is building model preparation. This includes both the building model and the parametric 3D objects of site equipment. A database of predefined site equipment elements needs to be developed, serving as an object catalog during site layout planning to simplify model creation significantly. Parameters added to the 3D objects are required for adequate filtering and rule execution, e.g. crane radius or maximum load capacity. The digital site equipment catalog should include the following element categories:

- Large equipment (construction machinery, cranes, etc.);
- Social services and office equipment (containers, sanitary facilities, etc.);
- Traffic areas and transportation routes (construction road, storage areas, etc.);
- Supply and disposal (electricity, garbage containers, etc.);
- Construction site safety (site fence, scaffolding, etc.);
- Temporary pit system (excavation support, slope, etc.).

The commercial 3D modelling software Autodesk Revit [13] provides a small portion of these elements as prototype objects. So-called Revit-families of a tower crane, concrete pump vehicles, containers and a material storage area can be seen in Figure 3.

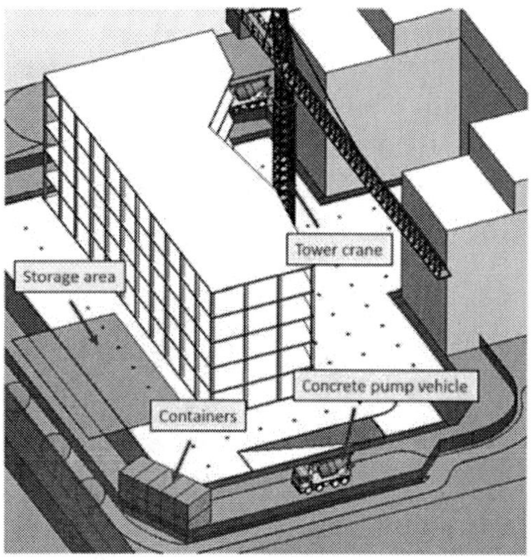

Figure 3. Examples of 3D site equipment in Autodesk Revit

4 Case Study

For the prototypical implementation of rule checking we chose the software Autodesk Revit. The 3D modelling software provides design of parametric objects and an API (application programming interface) that enables custom programming of so-called Add-Ins. Add-Ins can be designed for additional and custom interaction with the objects in the 3D model to extend the original software functionality. An evaluation or testing of similar modelling software for this particular purpose was not performed. The case study includes a realistic construction project of a five story office building with surrounding terrain and urban cityscape (see Figure 3). The implementation follows the four steps of rule checking mentioned earlier.

4.1 Rule interpretation

At first we defined a tabular structure for the transcription of the rules, as recommended by Eastman et al. [9]. The rule table consists of two different input types. The first type is the rule category. Every category invokes a specified algorithm. We defined three prototype categories for the implementation:

- *Radius rule:* Checks, whether an element cuts the pan radius of another element
- *Reach rule:* Checks, whether a set of elements can be reached by other elements
- *Lift rule:* Checks, whether the heaviest element of a selected type can be lifted by another element

The second input type are rule parameters. The amount and value of rule parameters depend on the associated rule category. For example, the category *Reach rule* requires information about which elements it affects. Both element types then serve as input parameters. Figure 4 shows the resulting rule table as a CSV-file. The first column always contains the category, so that the program can identify which algorithm to invoke. After identifying the category, the input parameters found in the subsequent columns are transferred to the invoked algorithm.

Radius rule	Tower_Crane	Container	False
Reach rule	Tower_Crane	Building	
Lift rule	Tower_Crane	Column	
Reach rule	Tower_Crane	Storage_area	
⇧	⇧	⇧ ...	⇧
Category	Parameter 1	Parameter 2	Parameter n

Figure 4. Prototypical rules transcribed into a rule table

The rule in line one transcribes the example rule text '*Containers* may *not* be placed within the load pan *radius* of *tower cranes*'. The keywords in italics signalize the input types for transcription, e.g. *radius* for Radius rule. The second rule represents another example rule.

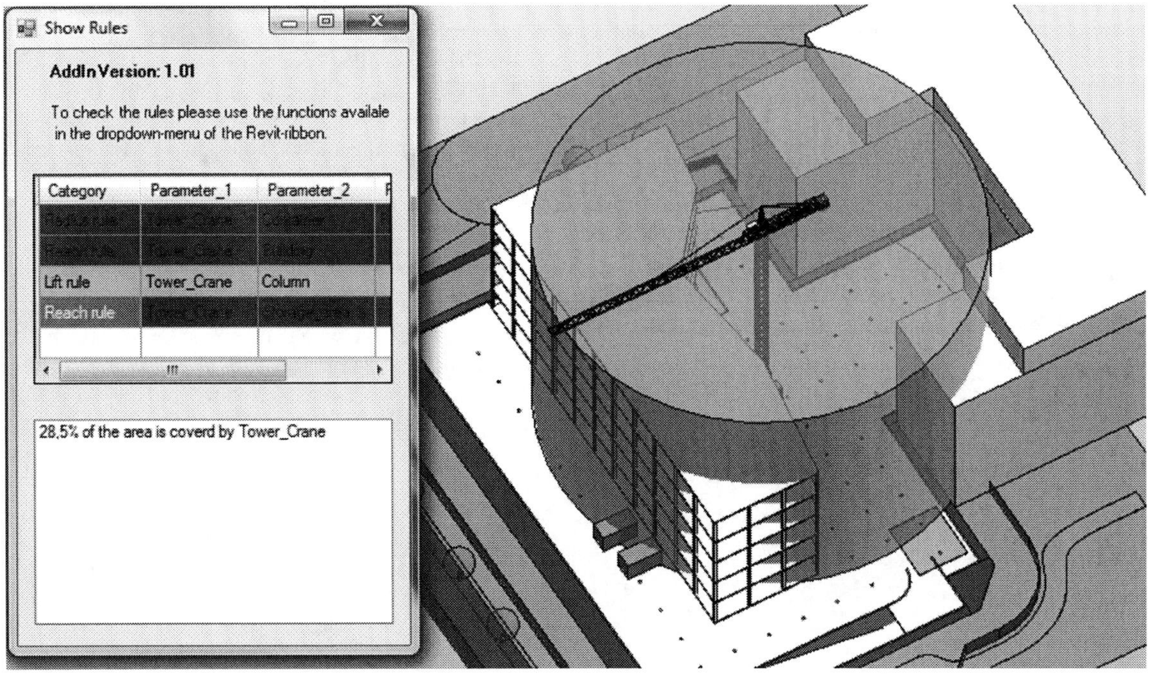

Figure 5. Rule execution: conflict appearance

'All building elements must be within reach of tower cranes'. Line three applies the example rule text 'Tower cranes must be able to lift the heaviest precast column'. In the fourth line a second reach rule is invoked with the example rule text 'All material storage areas must be within reach of tower cranes'. These example rules are derived from a German handbook for site layout planning [14]. Further rules will be implemented. This procedure can be repeated until every desired rule is categorized. If the user wants to include another rule and it can be assigned to one of the existing rule categories, another line must be added into the CSV-file. But if the user wants to define a new category new source code has to be written in order to include the new rule-based algorithm.

4.2 Model preparation

As seen in Figure 4 the rule parameters are in string format. For adequate filtering the exact string must be found among the parameters of the desired elements. For example, the keyword *Tower_crane* can be assigned to the name of the element. The keyword *Building* requires a different approach. Either the algorithm is taught which elements belong to a building, or the user assigns the string 'Building' to a custom parameter of the necessary elements manually. In order to check whether all necessary parameters have reasonable values a so-called pre-checking process must be carried out before the actual rule execution starts. However, pre-checking is not part of this research.

4.3 Rule execution & reporting of checking results

For the process of rule execution the three example rule categories mentioned above are implemented using the API of Autodesk Revit. At first the rule table must be read and saved into a data model. The data model mainly consists of two classes: *Rule* and *Conflict*, as seen in Figure 6. Other classes consider the user interface or event handling and play no role in the given data model.

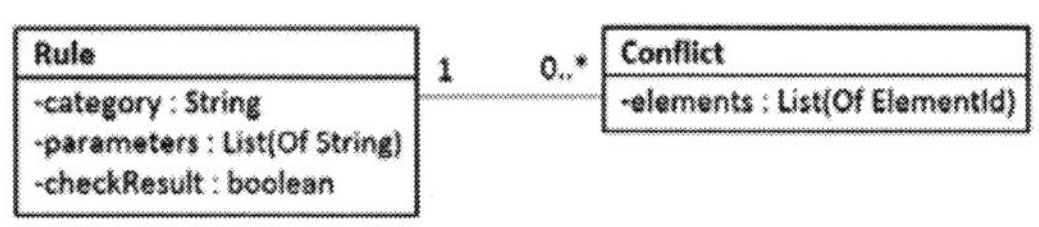

Figure 6. Classes in the data model

A rule can cause zero or an infinite amount of conflicts. But one conflict can only be associated with a specific rule. During the reading of the rule table the rule objects are created and the fields *category* and *parameter* are filled with values. During rule execution conflict objects will be created, if elements do not apply to the rule and added to the list of conflicts within the rule object. In addition the field *checkResult* will be set to *false*, if at least one conflict occurs. If no conflict is found the algorithm will return *true*.

The results of the first rule check can be seen in Figure 5. The 3D building and site layout model are depicted on the right, the Add-In window on the left. The site layout objects include two containers (grey cuboids at the bottom), a storage area (blue rectangle on the right) and a tower crane, whose radius is represented by a semitransparent green cylinder. The rule table introduced in Figure 4 can be found within the Add-In window.

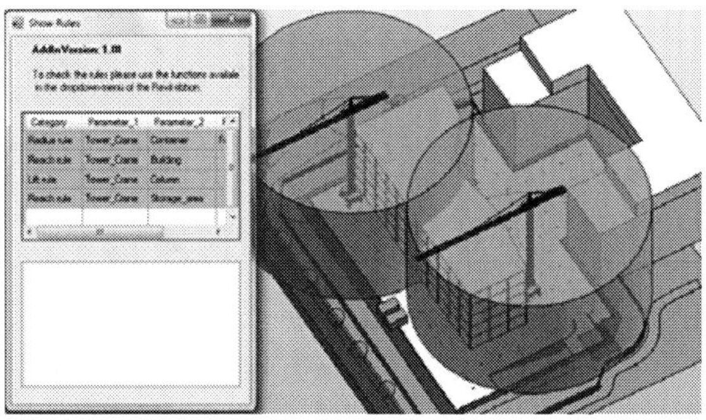

Figure 7. Rule execution: conflicts solved by layout changes

Below a text field reports potential conflicts. The first rule check marked three of the four rules as false (red rows) and one rule as true (green row). Clicking a red row adds the textual report of a conflict to the text field. The textual report of the selected Reach rule states, "28.5% of the area is covered by Tower_Crane". It returned false, because the rule dictates 100% storage area coverage. A click on the conflict text selects all objects involved in the conflict, in this case the tower crane. The Radius rule returned false, because both containers are within the load pan radius of the tower crane. The upper Reach rule returns false, because the tower crane does not cover 100% of the building. The next step is eliminating the conflicts. Therefore, a second tower crane is added to the layout model and the objects are rearranged. The conflict solution can be seen in Figure 7. Now the two tower cranes reach 100% of the building and storage area. The Lift rule still is marked as true and the containers are moved out of the load pan radius of the tower cranes.

The scenario described above illustrates our prototype implementation of subsequent rule checking. Although the displayed checking results do not report any conflict, not all imaginable rules have been considered in this case: for example, that both cranes pan at the same height or that the crane pan radius is not affected by the surrounding objects. This indicates how important a comprehensive rule data base is, so that even rules that can easily be missed will be considered. Another aspect that becomes obvious at this point is that a conflict cannot always be solved by relocating an object. In the case of the crane its load pan radius must be edited in the model and pan restrictions must be given to the crane operator.

4.4 Implementation

The Revit API provides the ability to automate features the user would otherwise carry out manually. Therefore the automation of rule checking in Revit is limited to the features provided by the API. One of those features for the analysis of object geometry is called 'SolidSolidCutUtils'. It allows intersections between two solid geometries and returns useful information, especially volume differences before and after the cut. This makes it easier to observe and evaluate object interactions. Otherwise the algorithm would have to use bounding boxes. A bounding box determines a cubic space that includes every geometrical point of an object. The problem with bounding boxes is that they remain cubic, even if the object does not. In Figure 6 this problem is visualized. Both spheres obviously do not intersect, but their bounding boxes do. Thus, the calculation of intersections via bounding boxes cannot guarantee complete accuracy. This problem can be avoided, by working with CSG (constructive solid geometry) - objects. CSG describes a method that uses boolean operators to interact with predefined mathematical geometries, e.g. cylinder, sphere etc. With this method the calculation of intersections is mathematically precise and time efficient.

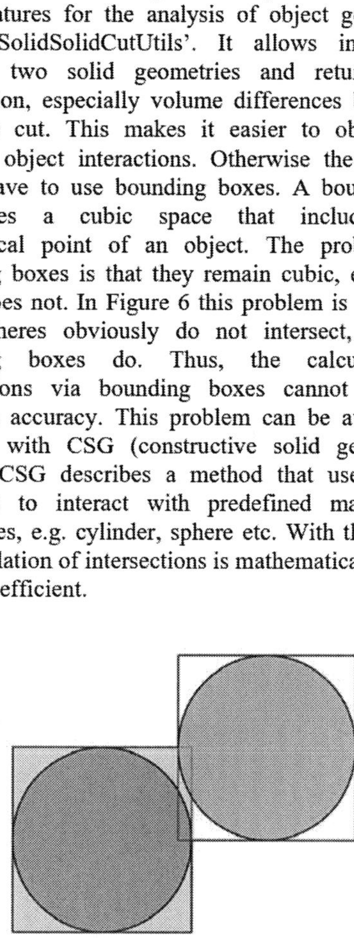

Figure 8. Intersections of objects and their bounding boxes

An obstacle of using 'SolidSolidCutUtils' is that it cannot be used for all family types. Standard family types, e.g. floors, are restricted by Revit. The solution to bypass these restrictions is to duplicate the element geometry. Each solid in the geometry needs to be read and rebuild. However, this procedure is not practicable in every case, so that each case must be handled individually.

The pseudo code of the algorithm implemented for the radius rule can be seen in Figure 7. First all elements in the BIM-model are filtered by the two element types received from the input parameters. Both element lists then get iterated over. If an intersection between the two elements is found they are added to a new element list that functions as the input parameter for the newly created conflict object.

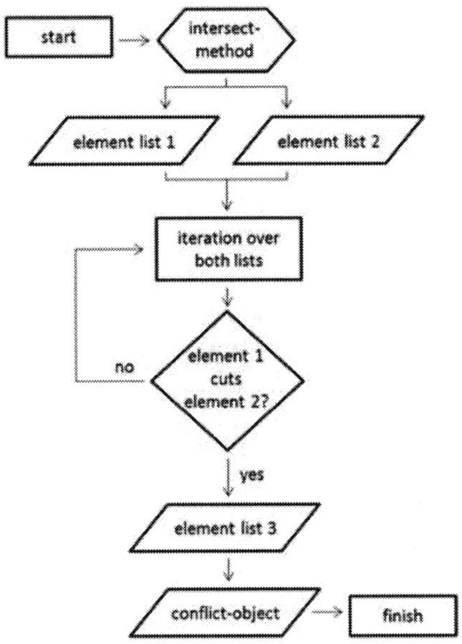

Figure 9. Schematic pseudo-code of the rule algorithm involving intersections

4.5 Dimensioning of site layout elements

The process of dimensioning site layout elements mostly includes time-dependent and logistic rules that can be derived from the construction schedule by calculating peak consumptions [6]. Figure 10 shows the schematic calculation of the size of a storage area for masonry in a simple example. The required information is derived from different sources. The first information is derived from the schedule. The completion time of the masonry works in this case is 2 weeks. The volume of masonry works is derived from the BIM-model: 80 m^3. This can be calculated to a peak consumption of 40 m^3/week. The best-practice for planning temporary storage of masonry material, found in the aforementioned German handbook for site layout planning [14], recommends a size of 2 m^3 per square meter of masonry work during a given week. Considering the masonry work tasks altogether and leveled by the available resources, the calculated storage area for masonry in the given project is 20 m^2. This information is used by the site layout planner to define a rectangular storage area with dimensions 5 x 4 m in the model. The use of peak consumptions for dimensioning of storage areas prevents the waste of useful space on the construction site. Time-dependent rules like these have not yet been implemented for this research.

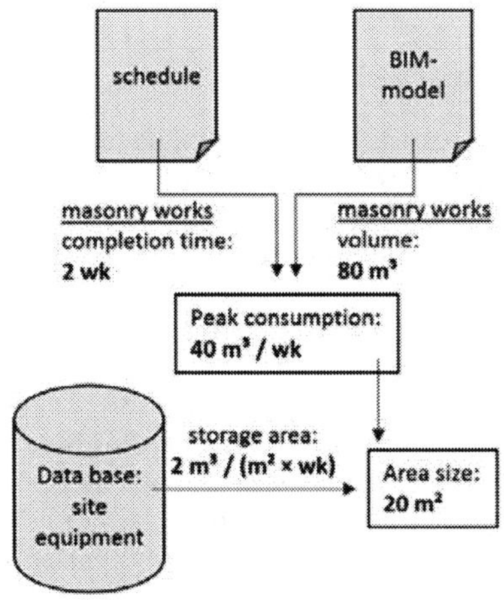

Figure 10. Schematic calculation of storage area size using peak consumption

5 Discussion

The implemented rules presented in this research demonstrate only a small number of prototype rules and raise no claim to be exhaustive. It is questionable, if more diverse and complex rules can be managed by a tabular approach to rule transcription. In addition, the exclusion of the user from the implemented rule algorithm further questions the use of rule tables. Nevertheless, it shows a realistic first application of subsequent rule checking in site layout planning tasks. The concept of interactive preprocessed rule checking introduced in this paper has not yet been tested, but should be addressed in future work. In order to calculate a proposed design solution the program requires a comprehensive database of rules. Another database including site layout equipment and machinery needs to be developed, so that choice and placement of elements is massively simplified, due to the connection of 3D and characteristic parameters.

The choice of the commercial software Autodesk Revit for implementation was arbitrary and served only for testing purposes. Although Revit provides some of the necessary requirements for the implementation, restricted or limited functions of the API leave implementing to be counterintuitive. Another important issue is the dependency on a single commercial software, which prevents the application from contributing to open-BIM. This could be solved by developing an external software dealing solely with rule checking and design optimization in site layout planning tasks using independent data exchange formats, e.g. IFC (Industry Foundation Classes) [15]. Early examples that require some human input and expert knowledge are shown in [16, 17].

6 Conclusion

The concepts presented in this research demonstrate that BIM- and rule based site layout planning is highly attractive for the engineers involved in this process. The most significant advantages are the intuitive 3D environment of the building model, where 2D drawings can easily be derived from. The permanent availability of additional relevant information enables integrated site layout scheduling (4D) and cost calculation (5D). Joining the relevant information leads to a better data management and prevents loss of data before and during construction. In addition, the development of automated rule checking could result in minimized human error during planning phase, which increases monetary benefit and occupational safety.

Future work should consider time-dependent rules, which are necessary for the dimensioning of site layout elements. Furthermore, the development of a designated rule language, that is able to comprehensively cover the diversity of existing rules and best-practices, should be addressed. If this is achieved, a rich rule data base and the corresponding object catalog of site equipment can be generated. In the process of implementing a rule checking application the idea of open-BIM using independent data exchange formats should be favored.

References

[1] Ning, X., Lam, K., and Lam, M., A decision-making system for construction site layout planning, *Automation in Construction,* 20, 459-473, 2011.

[2] Pradhananga, N. and Teizer, J., Congestion Analysis for Construction Site Layout Planning using Real-time Data and Cell-based Simulation Model, *Computing in Civil and Building Engineering,* 681-688, 2014.

[3] Yahya, M. and Saka, M., Construction site layout planning using multi-objective artificial bee colony algorithm with Levy flights, *Automation in Construction,* 38, 14-29, 2014.

[4] Andayesh, M. and Sadeghpour, F., The time dimension in site layout planning, *Automation in Construction,* 44, 129-139, 2014.

[5] Astour, H. and Franz, V., BIM-and Simulation-based Site Layout Planning, *Computing in Civil and Building Engineering,* 291-298, 2014.

[6] Cheng, J. and Kumar, S., A BIM Based Construction Site Layout Planning Framework Considering Actual Travel Paths, *The 31st International Symposium on Automation and Robotics in Construction and Mining,* 2014.

[7] Olearczyk, J., Lei, Z., Ofrim, B., Han, S., and Al-Hussein M., Intelligent Crane Management Algorithm for Construction Operation (iCrane), *The 32nd International Symposium on Automation and Robotics in Construction and Mining,* 2015.

[8] Hollermann, S. and Bargstädt, H., 4D Site Installation Planning in Virtual Reality for Multi-user, *Computing in Civil and Building Engineering,* 777-784, 2014.

[9] Eastman, C., Lee, J., Jeong, Y., and Lee, J., Automatic rule-based checking of building designs, *Automation in Construction,* 18, 1011-1033, 2009.

[10] Solibri Model Checker, On-line: http://www.solibri.com/products/solibri-model-checker/, Accessed: 11/08/2015.

[11] Kim, K. and Teizer, J., Automatic design and planning of scaffolding systems using building information modeling, *Advanced Engineering Informatics,* 28, 66-80, 2014.

[12] Zhang, S., Sulankivi, K., Kiviniemi, M., Romo, I., Eastman, C., and Teizer. J., BIM-based fall hazard identification and prevention in construction safety planning, *Safety Science,* 72, 31-45, 2015.

[13] Autodesk Revit, On-line: http://www.autodesk.com/products/revit-family/overview, Accessed: 12/06/2015.

[14] Schach, R., Otto, J., *Baustelleneinrichtung,* Vieweg+Teubner Verlag, Wiesbaden, Germany, 2011

[15] buildingSMART, On-line: http://www.buildingsmart-tech.org/ifc/IFC4/final/html/, Accessed: 15/07/2015.

[16] Schwabe, K., Liedtke, S., König, M., and Teizer, J., BIM-based Construction Site Layout Planning and Scheduling, *International Conference on Computing in Civil and Building Engineering (ICCCBE),* Osaka, Japan, 2016.

[17] Golovina, O., Teizer, J., Pradhananga, N. Heat map generation for predictive safety planning: Preventing struck-by and near miss interactions between workers-on-foot and construction equipment, *Automation in Construction,* http://dx.doi.org/10.1016/j.autcon.2016.03.008.

Automated Equipment Recognition and Classification from Scattered Point Clouds for Construction Management Applications

J. Chen[a], Y. Fang[b], and Y. K. Cho[b]

[a] School of Electrical and Computer Engineering, Robotics Institute, Georgia Institute of Technology
[b] School of Civil and Environmental Engineering, Georgia Institute of Technology
E-mail: jchen490@gatech.edu, yihaifang@gatech.edu, yong.cho@ce.gatech.edu

Abstract –

Recognizing construction assets from as-is point cloud data of construction environment provides essential information for engineering and management applications including progress monitoring, safety management, supply-chain management, and quality control. This study proposes a fast and automated processing pipeline for construction target assets recognition from scattered as-is point clouds. The recognition tasks can be subdivided into object detection, which involves computing the bounding box around each construction equipment, and object classification, which involves labelling point cloud clusters from discrete equipment categories, such as backhoe loader, bulldozer, dump truck, excavator and front loader. The object detection step consists of point cloud down-sampling, segmentation and clustering. For the object classification step, machine learning methods were employed to determine class membership probability using features derived from the ESF (Ensemble of Shape Functions) descriptor. The classifiers were trained on synthetic point clouds generated from CAD (Computer-aided Design) models. The method was validated using laser scanned point clouds from an equipment yard. The test results demonstrate promising advancements towards semantic labelling and scene understanding of point cloud data.

Keywords –

Object recognition; Object classification; Scattered point clouds; Laser scanning; Machine learning.

1 Introduction

In the past decades, construction practices and research have been actively embracing emerging technologies to improve project performance in productivity, quality and safety. The technical categories that have been extensively studied and tested in the construction industry include data sensing, simulation, information modeling, and visualization. From these categories, as-built data sensing and visualization are considered by many industry practitioners and academia experts as one of the most promising technologies that will greatly expand the utilization of site information [1][2]. For instance, 3D point clouds produced by laser scanners and other data acquisition technologies have been widely used for acquiring and generating as-built site information to support applications such as construction quality assessment and control [3][4], construction progress tracking [5][6], building energy analysis [7], construction hazard recognition [8], structural health monitoring [9] [10] and highway asset management [11] [12]. Each application has a different focus of target objects (e.g., building components, materials, equipment, workers, and traffic signs) on the as-built point clouds.

Regardless of the applications, recognizing objects from point clouds is often the very first step for post data processing and analyses. Although object recognition is a well-studied problem in the field of computer vision, recognizing objects from point clouds of an uncontrolled environment such as construction sites remains a challenging and unsolved task. Many previous algorithms for object recognition were validated based on complete scans of small objects (e.g. fruit, household objects) obtained in controlled lab settings. In uncontrolled settings with the presence of occlusion and other objects that share similar features with the target object, many state-of-the-art recognition algorithms fail to maintain their good recognition rate and speed [13]. These problems become even more serious when acquiring as-built 3D data from construction sites where the presence of huge building structure and a large amount of equipment and materials introduce potential occlusions to the point cloud acquired regardless of what technology is used. As construction projects become larger and more complex, obtaining complete 3D data becomes even more difficult [14]. In addition, similarities among different types of construction equipment (e.g., bulldozer vs. loader) present a formidable challenge to object recognition algorithms that aim to distinguish between them.

This study proposes an effective and efficient object recognition pipeline that is robust to incomplete point clouds and shape variances of construction assets including structures, materials, and equipment. This paper starts with a thorough literature review on as-built data acquisition and modeling, followed by a detailed description of methodology applied in the proposed approach. The validation of the proposed approach on two testbeds is introduced, and the limitations and future work on the proposed method are discussed.

2 Literature Review

2.1 As-built Data Acquisition in Construction

As-built data of building components and construction assets are important for various applications in construction engineering and management. Depending on different representations of as-built scene, approaches for acquiring as-built data can be categorized as 2D image or video based methods, and 3D range data or point cloud based methods. This section will review multiple technologies belonging to these two categories.

2.1.1 2D Image and Video-based Methods

Recognizing objects in 2D images or videos has been extensively studied in the field of computer vision. Various techniques are available for object recognition, pose identification, and location tracking from 2D images or videos captured from of-the-shelf cameras. Given its advantages of low cost and minimal infrastructure, 2D image or video based techniques have been applied in construction for various operation-level applications including construction progress monitoring [5], construction entity detection and tracking [15], and productivity and safety analyses [16][17]. Various techniques have been explored to recognize objects (e.g., equipment, workers) on construction sites from 2D image or video, such as hue, saturation, and value (HSV) color space [18], background subtraction [19], histogram of color (HOC) [20], and template matching [21].

2.1.2 3D Range Data and Point Cloud based Methods

Compared to 2D presentation of construction site, 3D as-built data provide additional information and flexibility for visualization, as-built modeling, and other applications that heavily depend on spatial data. The most popular technologies for obtaining as-built 3D data on construction sites are 3D range imaging camera, photogrammetry, and laser scanning. This section will briefly review the characteristics of these three technologies in their applications in construction environments.

Based on different techniques such as stereo triangulation, structured light, and time-of-flight, a 3D range imaging camera produces a range image with pixel values representing the intensity and range. Although the 3D range imaging camera has been widely applied in the fields of virtual reality and gaming, very few investigations of its benefits in construction have been undertaken. Such research focused on range data processing for real-time site spatial modeling [22]. Compared to laser scanners, 3D range imaging cameras are more portable and less expensive but they operate in a much closer range, and do not provide reliable range images when direct sunlight is present, which makes it difficult to apply on large construction sites [23][24]. While understanding the positive and negative aspects of each data acquisition technology in a construction environment is important for real-time applications, it is also vital to consider the difficulties of data processing and extracting useful semantic information out of as-built data, regardless of the data acquisition methods.

2.2 Object Recognition from a Point Cloud

The typical pipeline of point cloud processing for object recognition consists of the following steps. First, shape descriptors are computed for each object instance or class to be recognized and stored in an offline database. Second, point cloud data is examined by an object detection algorithm to identify the contained object instances or object classes and query descriptors are computed for each detected object. Finally, objects from the model database with high descriptor similarity are aligned with the objects recognized in the point cloud to produce a match [1]. A major disadvantage of instance-based object recognition is that it is not as applicable and effective when significant shape variance exists among objects [1]. The most common approach in class-based object recognition is to use global shape descriptors which are less discriminative than semi-local descriptors but better adapted to shape variance. However, a major challenge is that this approach cannot handle partial or incomplete point cloud caused by occlusion or clutter [25]. To resolve this problem, one solution is to use descriptors that are more robust to shape changes [25]. To recognize mobile construction assets such as workers and equipment, [7] proposed a projection-recognition-projection (PRP) method to automatically recognize construction equipment from the point cloud. The 3D point cloud is projected to a 2D space where the geometric features represented by a local SURF descriptor are compared to a prepared template database for recognition. This method is very effective and efficient for recognizing target objects that are known to be present on the construction site. For unknown objects with high shape variance, however, the performance of this method is limited.

2.3 Shape Distribution and Global Shape Descriptor

The concept of shape distribution was introduced by [26] to describe the geometric signature of an object. Shape distributions are sampled from multiple shape functions that measure global geometric properties of an object. Distance between two randomly chosen points (D2), for instance, is a robust shape function for distinguishing different object classes. It should be noted that the performance of this method can be compromised when the number of classes to be recognized increases or only a partial view of the object is available in the point cloud. [27] proposed a robust global shape descriptor named ESF (Ensemble of Shape Function) for recognizing a variety of object classes (e.g., mugs, fruits, cars) from point clouds. This descriptor specifically addresses the challenge of using partial views in incomplete point clouds for object recognition through a construction of three distinct shape functions describing distance, angle, and area distributions. Each shape distribution is represented by an ESF histogram that can be used to efficiently retrieve the k-closest matches from a pre-defined class database using the k-nearest neighbor algorithm (k-NN). Although the ESF descriptor shows fairly good recognition rates for small-scale objects such as apples (98.45%) and mugs (99.46%), it shows limited capability in recognizing large objects such as cars (43.64%).

2.4 Training in Object Recognition

In addition to a robust description of objects, another critical component in object recognition is training the object recognition algorithm. Typically, the training process consists of the following four steps: 1) gather a training set which consists of a number of point clouds of the object classes that need to be recognized and label the point clouds with desired classifications; 2) determine a descriptor to represent the objects in the training set; 3) select the structure of the learned function and corresponding learning algorithm (e.g., support vector machine, k-nearest neighbor, decision tree); and 4) determine training parameters and run the learning algorithm on the gathered training set. Thus, a new input point cloud can be recognized by mapping its descriptor to the function inferred by the previously labeled training data. Similar to other applications adopting supervised learning, the availability and collection of training set is a bottleneck due to the lack of a centralized repository for point clouds. This problem can be addressed by using synthetic point clouds generated from 3D CAD models [27].

3 Methodology

To address the limitations in existing object recognition techniques, this study proposes a fast and automated processing pipeline for construction target assets recognition from scattered as-is point clouds. Figure 1 shows the processing pipeline for our object recognition approach. The pipeline is divided into an object detection stage and an object classification stage. The two stages are further discussed in detail in the following sections.

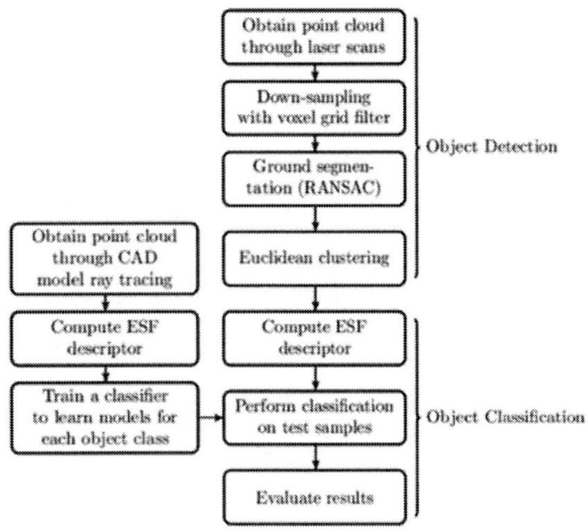

Figure 1: Flowchart for our object recognition approach

3.1 Object Detection in Laser Scanned Point Cloud

Laser scanned point clouds from construction sites often consist of millions of points, which renders the post-processing step computationally demanding. Thus, a voxel grid representation is used to store point cloud data in condensed form in order to achieve an efficient object detection pipeline. Next, the RANSAC (Random Sample Consensus) technique [28] is employed to filter out points belonging to the ground plane. In cases where the ground is a uniform flat plane or approximately flat, a single RANSAC pass is sufficient to obtain plane parameters while rejecting outliers. For cases where the ground consists of multiple planes with different gradients, multiple RANSAC passes are required to detect each plane. Finally, the Euclidean distance metric [29] is used to combine neighboring points into point cloud clusters for each object.

3.2 Object Classification

The object classification stage involves generating a training database of point cloud samples from objects with known categorization and matching query point cloud samples to the appropriate category based on computed shape descriptors. Further details concerning the classification process are discussed in the following sections.

3.2.1 Synthetic Point Cloud Generation

Sample 3D CAD models of our desired classification categories were downloaded online from 3D Warehouse. In this study, construction equipment samples were collected from *backhoe_loader*, *bulldozer*, *dump_truck*, *excavator (backhoe)*, and *front_loader* categories (Figure 2). A ray casting technique was used to sample points along the model surface with respect to virtual laser scanners placed at multiple view locations around the model. For each equipment model, point clouds were generated from multiple view angles and the points were perturbed with Gaussian white noise to simulate the data collection process in real world scenarios. This can also be viewed as a form of regularization which allows the classifier to be robust to data variability.

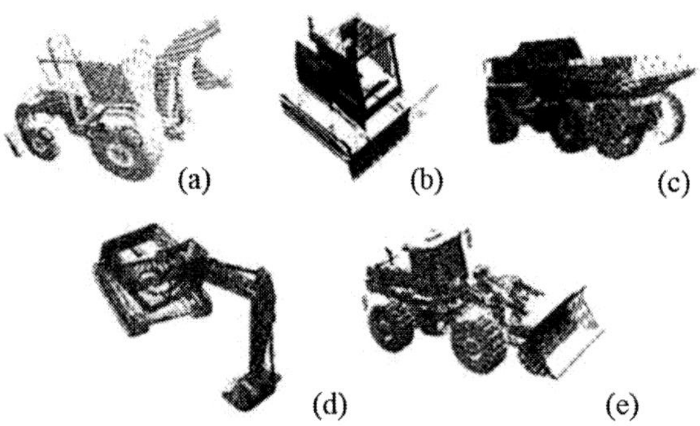

Figure 2: Synthetic point clouds of construction equipment: (a) backhoe loader, (b) bulldozer, (c) dump truck, (d) excavator, (e) front loader

3.2.2 ESF Descriptor

The ESF descriptor describes a 3D point cluster with a 640-dimensional set of features encompassing angle, area and distance shape functions. ESF is an example of a global shape descriptor since it encodes information about an object by considering points in the entire point cloud without making any prior assumptions about the object geometry. In our study, ESF was selected over competing descriptors such as VFH (Viewpoint Feature Histogram) or SHOT (Signature of Histograms of Orientations) due to its better performances in speed and robustness with respect to noise and partial views. Figure 3 shows the distribution of descriptor values of five construction equipment categories considered in this study. The x-axis describes each dimension of the descriptor whereas the y-axis represents the descriptor value at that dimension. ESF descriptor calculation is performed using the C++ implementation provided in the PCL (Point Cloud Library) software suite [30].

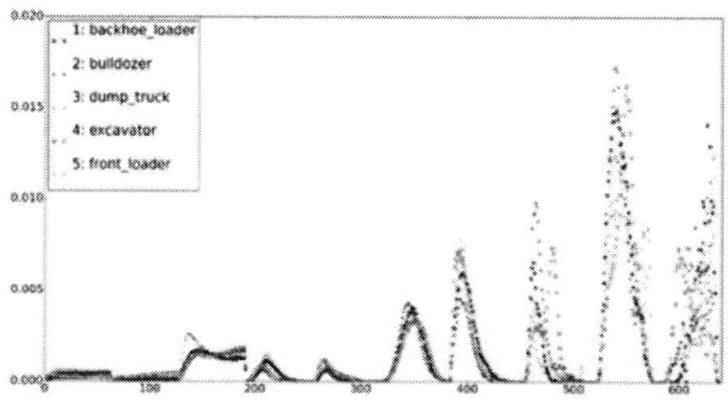

Figure 3: ESF descriptor distribution with separate plots representing the mean descriptor values for each category

3.2.3 Machine Learning

In the final stage of object classification, class labels are assigned to each input point cloud cluster by training and applying machine learning classifiers. In addition to the k-nearest neighbor (k-NN) algorithm used in the original ESF descriptor study [27], we also considered the approach of using discriminative classifiers such as logistic regression [31] and support vector machine [32]. The k-NN method involves picking k training samples closest to the test sample in terms of Euclidean distance in descriptor space. The output class label is then determined through uniform voting over class labels of the k nearest training samples. On the other hand, the logistic regression approach estimates the class label probability using logistic functions and learns a set of weights for descriptor elements from the training samples. Whereas, the support vector machine (SVM) method involves constructing a set of separating hyperplanes in high-dimensional space to perform classification. In our implementation of SVM, the linear kernel was used instead of the RBF (Radial Basis Function) kernel since the number of features in the descriptor is large. To achieve multi-class classification, the one-vs-all and one-vs-one scheme was used for logistic regression and SVM respectively.

After training and testing samples are obtained, the corresponding ESF descriptor is calculated for each point cloud cluster. For both training and testing, the descriptor values are scaled uniformly to a range of 0 to 1 and provided as input to each machine learning classifier. The output class labels and class probabilities are computed and stored for evaluation purposes. In this study, the training and testing processes were implemented using the Python toolkit *scikit-learn* [33].

4 Results and Validation

The validation dataset is composed of laser-scanned point clouds collected from a construction equipment yard. The data is collected from two sites with seven scans each and fused into a single point cloud of around 8 million points using registration techniques. The recognition results are shown in Figure 4. Point cloud clusters associated with recognized objects are identified with a bounding box labelled with the object class and computed class probability. Quantitative evaluation of the classification performance is discussed in the next section.

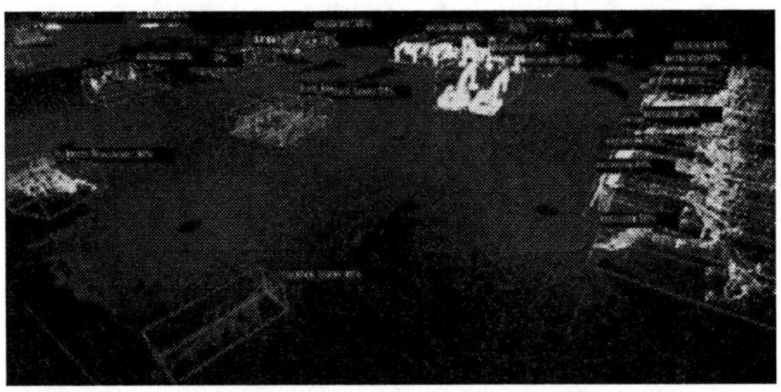

Figure 4: Testbench sample recognition results

In the object classification stage, recognition performance was quantified using the recall rate metric, which is the number of correctly identified samples divided by the total number of samples for each object class. Table 1 demonstrates that Logistic Regression and SVM classifiers [32] perform better compared to the original technique of k-nearest neighbors proposed by Wohlkinger and Vincze (2011) with respect to the number of correctly classified samples for each object category. The SVM classifier shows the highest overall recall rate for both and thus is used for all subsequent experiments.

Table 1: Number of correct classifications based on classification technique

Category	Sample number	K nearest neighbor	Logistic Regression	SVM
Backhoe Loader	3	1 (33%)	2 (67%)	2 (67%)
Bulldozer	3	0 (0%)	3 (100%)	1 (33%)
Dump truck	2	0 (0%)	1 (50%)	2 (100%)
Excavator	23	23 (100%)	19 (83%)	22 (96%)
Front Loader	9	3 (33%)	4 (44%)	4 (44%)
Total	40	27 (68%)	29 (73%)	31 (78%)

The effect of variables such as size of training data and point cloud resolution on the recall rate can also be investigated. Table 2 shows that the recall rate shows a general increase as the number of samples of training data increases. Table 3 shows that the recall rate increases with the resolution of both the training data and the testing data. This suggests that the object recognition outcome could theoretically be improved given a larger amount of training data and higher resolution laser scans.

Table 2: Recall rate for different sizes of training data

Training Size (Number of samples)	Recall Rate
48	7%
96	62%
144	60%
192	57%
240	68%
288	75%
336	78%

Table 3: Recall rate for different combinations of training data resolution and testing data resolution

Testing res. / Training res.	Downsampled by 20%	Downsampled by 5%	Original
40 scan lines	55%	53%	60%
200 scan lines	68%	70%	75%
400 scan lines	70%	73%	78%

5 Discussion

In terms of recognizing construction equipment, we demonstrated an overall classification recall rate of around 78% for our equipment yard testbench. Recall rates was improved using discriminative classifiers such as logistic regression and SVM compared to the original technique of k-nearest neighbors [27]. We further identified the number of training data and point cloud resolution as variables that could affect the recall rate. Classification performance for construction equipment was analysed using the confusion matrix shown in Table 4. Overall, the recall rate is high since there is less than one instance of wrong classification for all categories except for *bulldozer* and *front_loader*. The *excavator* category showed high classification performance due to its distinctive boom and bucket shape. On the other hand, the *front_loader* category showed less effective classification due to high shape variance and thus is often confused with other categories such as *bulldozer* and *excavator*. Overall, the classification method is shown to be theoretically sound and robust. The processed outcomes would be more viable given a larger amount of training data and more complete laser scans.

Table 4: Confusion matrix for construction equipment classes

Predicted / Actual	Loader	Bulldozer	Dump truck	Excavator	Front Loader
Backhoe loader	2	1			
Bulldozer	1	1		1	
Dump truck			2		
Excavator				22	1
Front loader		2	1	2	4

6 Conclusion

Our main contributions in this study are to integrate pre-processing, detection and classification steps of object recognition and classification from scattered point clouds into a single efficient recognition pipeline and to develop improvements over existing methods. We found that scan resolution, scan completeness and the availability of real world training data are key variables positively correlated with the classification recall rate. We were able to achieve promising recognition rates even without a large and comprehensive data set, as common machine learning applications often operate in the big data realm. This was achieved through a more rigorous processing of training data and better classification techniques. For future work, we envision expanding the training dataset, fine tuning the shape descriptor, and extending the method to attach multiple semantic labels such as coarse and fine categorization to each object in a construction site to aid the current construction management practices.

7 Acknowledgement

This material is based upon work supported by the National Science Foundation (Award #: CMMI-1358176). Any opinions, findings, and conclusions or recommendations expressed on this material are those of the authors and do not necessarily reflect the views of the NSF. This research was also supported by the Chung-Ang University Research Grants in 2015.

References

[1] P. Tang, D. Huber, B. Akinci, R. Lipman, and A. Lytle, "Automatic reconstruction of as-built building information models from laser-scanned point clouds: A review of related techniques," Autom. Constr., vol. 19, no. 7, pp. 829 – 843, Jul. 2010.

[2] F. Bosché, "Automated recognition of 3D CAD model objects in laser scans and calculation of as-built dimensions for dimensional compliance control in construction," Adv. Eng. Informatics, vol. 24, no. 1, pp. 107–118, 2010.

[3] L. Bu and Z. Zhang, "Application of Point Clouds From Terrestrial 3D Laser Scanner for Deformation Measurements," Int. Arch. Photogramm. Remote Sens. Spat. Inf. Sci., vol. XXXVII, pp. 545–548, 2008.

[4] M. Ahmed, C. T. Haas, and R. Haas, "Toward low-cost 3D automatic pavement distress surveying : the close range photogrammetry approach," NCR Res. Press, vol. 1313, no. November 2011, pp. 1301–1313, 2012.

[5] M. Golparvar-Fard, F. Peña-Mora, C. a. Arboleda, and S. Lee, "Visualization of Construction Progress Monitoring with 4D Simulation Model Overlaid on Time-Lapsed Photographs," J. Comput. Civ. Eng., vol. 23, no. December, pp. 391–404, 2009.

[6] Y. Turkan, F. Bosche, C. T. Haas, and R. Haas, "Automated progress tracking using 4D schedule and 3D sensing technologies," Autom. Constr., vol. 22, pp. 414–421, 2012.

[7] Y. K. Cho and M. Gai, "Projection-Recognition-Projection Method for Automatic Object Recognition and Registration for Dynamic Heavy Equipment Operations," J. Comput. Civ. Eng., vol. 28, no. 5, p. A4014002, Sep. 2014.

[8] S. Fekete, M. Diederichs, and M. Lato, "Geotechnical and operational applications for 3-dimensional laser scanning in drill and blast tunnels," Tunn. Undergr. Sp. Technol., vol. 25, no. 5, pp. 614–628, 2010.

[9] H. S. Park, H. M. Lee, H. Adeli, and I. Lee, "A new approach for health monitoring of structures: Terrestrial laser scanning," Comput. Civ. Infrastruct. Eng., vol. 22, pp. 19–30, 2007.

[10] D. Huber, "ARIA: the Aerial Robotic Infrastructure Analyst," 2014. [Online]. Available: http://spie.org/newsroom/technical-articles/5472-aria-the-aerial-robotic-infrastructure-analyst. [Accessed: 07-Oct-2015].

[11] J. Gong, H. Zhou, C. Gordon, and M. Jalayer, "Mobile Terrestrial Laser Scanning for Highway Inventory Data Collection," in Computing in Civil Engineering (2012), 2012, vol. Computing, pp. 545–553.

[12] S. Pu, M. Rutzinger, G. Vosselman, and S. Oude Elberink, "Recognizing basic structures from mobile laser scanning data for road inventory studies," ISPRS J. Photogramm. Remote Sens., vol. 66, no. 6 SUPPL., pp. S28–S39, 2011.

[13] B. Matei, Y. Shan, H. S. Sawhney, Y. Tan, R. Kumar, D. Huber, and M. Hebert, "Rapid object indexing using locality sensitive hashing and joint 3D-signature space estimation.," IEEE Trans. Pattern Anal. Mach. Intell., vol. 28, no. 7, pp. 1111–1126, 2006.

[14] C. Kim, C. Kim, and H. Son, "Automated construction progress measurement using a 4D building information model and 3D data," Autom. Constr., vol. 31, pp. 75–82, 2013.

[15] M. W. Park, A. Makhmalbaf, and I. Brilakis, "Comparative study of vision tracking methods for tracking of construction site resources," Autom. Constr., vol. 20, no. 7, pp. 905–915, 2011.

[16] J. Gong and C. H. Caldas, "An object recognition, tracking, and contextual reasoning-based video interpretation method for rapid productivity analysis of construction operations," Autom. Constr., vol. 20, no. 8, pp. 1211–1226, 2011.

[17] S. S. Han, S. S. Lee, and F. Peña-Mora, "Vision-Based Detection of Unsafe Actions of a Construction Worker: Case Study of Ladder Climbing," J. Comput. Civ. …, vol. 27, no. 6, pp. 635–644, 2013.

[18] J. Zou and H. Kim, "Using Hue , Saturation , and Value Color Space for Hydraulic Excavator Idle Time Analysis," vol. 21, no. August, pp. 238–246, 2007.

[19] S. Chi and C. H. Caldas, "Automated Object Identification Using Optical Video Cameras on Construction Sites," Comput. Civ. Infrastruct. Eng., vol. 26, no. 5, pp. 368–380, 2011.

[20] M. Memarzadeh, M. Golparvar-Fard, and J. C. Niebles, "Automated 2D detection of construction equipment and workers from site video streams using histograms of oriented gradients and colors," Autom. Constr., vol. 32, pp. 24–37, 2013.

[21] L. Cole, D. Austin, and L. Cole, "Visual Object Recognition using Template Matching," Australas. Conf. Robot. Autom., 2004.

[22] J. Gong and C. H. Caldas, "Data processing for real-time construction site spatial modeling," Autom. Constr., vol. 17, no. 5, pp. 526–535, 2008.

[23] M.-W. Park, C. Koch, and I. Brilakis, "Three-Dimensional Tracking of Construction Resources Using an On-Site Camera System," J. Comput. Civ. Eng., vol. 26, no. 4, pp. 541–549, 2012.

[24] F. Remondino and D. Stoppa, TOF Range-Imaging Cameras | Fabio Remondino | Springer. Springer, 2013.

[25] S. Ruiz-Correa, L. G. Shapiro, M. Meila, G. Berson, M. L. Cunningham, and R. W. Sze, "Symbolic signatures for deformable shapes," IEEE Trans. Pattern Anal. Mach. Intell., vol. 28, no. 1, pp. 75–90, 2006.

[26] R. Osada, T. Funkhouser, B. Chazelle, and D. Dobkin, "Matching 3D models with shape distributions," Proc. - Int. Conf. Shape Model. Appl. SMI 2001, pp. 154–166, 2001.

[27] W. Wohlkinger and M. Vincze, "Ensemble of shape functions for 3D object classification," 2011 IEEE Int. Conf. Robot. Biomimetics, pp. 2987–2992, 2011.

[28] R. Schnabel, R. Wahl, and R. Klein, "Efficient RANSAC for point-cloud shape detection," Comput. Graph. Forum, vol. 26, no. 2, pp. 214–226, 2007.

[29] R. B. Rusu, "Semantic 3D Object Maps for Everyday Manipulation in Human Living Environments," 2009.

[30] R. B. Rusu and S. Cousins, "3D is here : Point Cloud Library (PCL)," in IEEE International Conference on Robotics and Automation (ICRA), 2011.

[31] C. Lin, H. Yu, and F. Huang, "Dual Coordinate Descent Methods for Logistic Regression and Maximum Entropy Models," Mach. Learn., no. 85, pp. 41–75, 2011.

[32] C. Chang and C. Lin, "LIBSVM : A Library for Support Vector Machines," in ACM Transactions on Intelligent Systems and Technology, 2011, pp. 2:27:1–27:27.

[33] F. Pedregosa, G. Varoquaux, A. Gramfort, V. Michel, B. Thirion, O. Grisel, M. Blondel, P. Prettenhofer, R. Weiss, V. Dubourg, J. Vanderplas, A. Passos, D. Cournapeau, M. Brucher, M. Perrot, and E. Duchesnay, "Scikit-learn: Machine Learning in {P}ython," J. Mach. Learn. Res., vol. 12, pp. 2825–2830, 2011.

Real-time 3D Mobile Mapping for the Built Environment

Jingdao Chen[a], Yong K. Cho[b]

[a] School of Electrical and Computer Engineering, Robotics Institute, Georgia Institute of Technology
[b] School of Civil and Environmental Engineering, Georgia Institute of Technology
E-mail: jchen490@gatech.edu, yong.cho@ce.gatech.edu

Abstract –

Laser scanned point clouds are relevant for geometric analysis, building maintenance and energy modeling, yet obtaining fully registered point clouds is a labor-intensive process when targets or common registration features are absent. Thus, we propose a versatile mobile platform that forms an incremental 3D map of the environment in real time using an orthogonal pair of Lidar (Light Detection and Ranging) devices. The horizontal scanner aims to estimate the robot position and orientation with SLAM (Simultaneous Localization and Mapping) techniques whereas the vertical scanner recovers the building structure in the vertical plane. We also developed a real time point cloud visualization tool that allows an operator to track the mapping progress. The method was evaluated with walk-through laser scans of a complete building floor.

Keywords

3D, Lidar, point cloud, SLAM, real-time

1 Introduction

Three-dimensional models of building components in the as-built condition are important tools in building maintenance and restoration applications. The acquisition of data for creating building models has traditionally been done with tools such as measuring tapes and distometers [1]. However, advances in recent technology has led to use of a host of techniques such as robotic total station, stereo photogrammetry, depth cameras, Light Detection and Ranging (Lidar), as well as hybrid setups to survey and map various aspects of building infrastructure. The acquired data in the form of visual or range data can be incorporated into a single three-dimensional point cloud representation. The generated point cloud can be further post-processed to create a Building Information Model (BIM) [2], which describe relational characteristics and attributes of building elements. The generated model can also be used to monitor defects and estimate energy usage characteristics [3].

Static scanning solutions such as fixed camera setups and ground-based Lidar are widely used in the data collection step of infrastructure modelling. However, they suffer from the limitation that the registration step to combine data collected from multiple scan times is labor-intensive. For example, images taken in multiple locations for photogrammetry have to be combined in a feature point matching process. For the case of ground-based Lidar, multiple scans have to be registered together in a post-processing step using either Global Positioning System (GPS) information or iterative optimization techniques. These data collection methods offer limited feedback to the operator in terms of the final registered output during the scan process, so the generated data is often vulnerable to problems such as incomplete scans or occluding structures.

Thus, in this study we propose a mobile scanning solution that generates a registered three-dimensional point cloud output in real-time. The proposed method involves two Lidar devices mounted in an orthogonal position on a mobile platform. The two devices work concurrently in a Simultaneous Localization and Mapping (SLAM) setting to estimate the current position and map of the environment so that each scan can be automatically registered. The combined point cloud is presented to the operator through a visualization tool to determine scan progress. The proposed method aims to achieve a more efficient and accurate scanning process compared to conventional scan techniques.

2 Literature Review

Most instances of 3D mapping have been achieved by using photogrammetry, laser scanners or even a combination of both methods with some also using additional technologies such as sonar and infra-red (IR) [4]. Since most research has been focused on the photogrammetric and Lidar-based SLAM, we discuss some common practices of each approach and how close they are to solving the SLAM problem. We also discuss the pros and cons of each approach.

2.1 Application of SLAM

Simultaneous Localization and Mapping (SLAM) [5] is a significant problem in mobile robotics where map building and localization is carried out without a priori information. SLAM algorithms collect environment features and learn a mapping between the robot location and the environment. Thus, SLAM algorithms provide a useful tool to automatically obtain the scan origin and register sensor data onto a global map.

2.1. Photogrammetric Approach

In [6], Davison et al. were able to recover the 3D trajectory of a monocular camera moving through a previously unknown scene. Here the nature of the map depends on the initial and current positions of the camera, 3D position of the center of the feature and the orientation of its local surface. This algorithm is unable to handle sudden jerky movements and while it uses modest processing power, it has limited application in getting a detailed map of the environment. In [7], sixteen advantages are provided on the advantages of the photogrammetric 3D workflow over the directly measured laser point clouds with significant points being the inexpensive hardware available for vision based techniques and the density of surface points being higher but these require multiple cameras and hardware to run expensive algorithms to achieve real time SLAM. In [8], a Kinect sensor is used to generate a 3D model of a building and the result is compared to another model generated from a laser scanner. While the paper does make a case for detecting discontinuity points using a vision based system, the authors concluded that the measurements are highly influenced by the material and lighting conditions. As a consequence, the generated point cloud is noisy and not uniform. Khoshelham et al. [9] expounded on this in detail in their paper where they note that the random error of depth measurements increases quadratically with increasing distance from the sensor. Their technique also required accurate stereo calibration of the IR camera and visual camera.

Thus we can summarize the advantages of using visual based SLAM as being relatively less expensive, lighter and consuming less power per unit instrument than using laser scanners. Visual based systems are also able to detect colors and other features/objects which cannot be detected by Lidar in a cluttered environment. These approaches also have their disadvantages: 1) They can accumulate large errors if there are large distances, 2) They are very susceptible to even slight changes to lighting conditions, 3) They can get confused in a dynamic environment and 4) They are difficult to use at night [10].

2.2 Lidar Based Approach

There have been several attempts in the literature to map an indoor environment with laser scanners. Shohet and Rosenfeld [11] talk about two main factors affecting the precision of the map: 1) orientation of the carriage on which the scanner is mounted and 2) distance between the sensor and the walls being scanned. However advancements in Lidar technology have rendered the last point irrelevant with commercial Lidars now available with a range of 120 m [12]. Current research is focused on improving the accuracy of maps generated from laser scanners. Jung et al [13] proposed a kinematic 3D laser scanning system for indoor mapping which imposes some constraints on the environment where main structures are formed from straight lines and all structures are parallel or perpendicular. Another approach to solving the SLAM problem is having multiple robots on the field and then integrating the pose estimations of the robots and individual maps to create a more accurate model of the environment [14]. This approach, while successful, does not allow the operator to have a real-time view of the map as the off-line optimization step may require the robots to stop exploration and perform the optimization process before they resume the exploration task.

Kohlbrecher and von Stryk [15] proposed a 2D SLAM system known as Hector SLAM which performs laser scan matching to obtain a pose estimate and planar map of the environment. The SLAM system is flexible, scalable, and has been deployed on unmanned ground vehicles (UGV), unmanned surface vehicles (USV), and small indoor navigation systems. LOAM (Lidar Odometry and Mapping in Real time) [16] is another approach for real-time SLAM but requires a spinning Lidar which is more expensive to achieve than a planar Lidar approach like Hector SLAM. There have been several studies which built on Hector SLAM to produce more accurate maps of indoor environments using external sensors [17-20]. Khan et al [20] proposed a data driven method, built on Hector SLAM, which models laser intensities and measures the surface reflectivity to augment a geometric model of the surrounding environment.

In this study, the method of [15] was built upon to obtain full 3D Lidar mapping. The goal is to achieve a more mapping-focused SLAM compared to navigation [19] or exploration [17]. This study also utilizes less sensory requirements compared to [13] and [16].

3 Methodology

The hardware setup of our experiments consists of 2 Lidar devices as shown in Figure 1. The larger device is a SICK Lidar LMS511 which scans at a 0.5° angular resolution and 50 Hz sample rate. The smaller device is a SICK Lidar LMS151 which scans at a 0.25° angular resolution and 25 Hz sample rate. Each Lidar device only scans along a single plane so it is necessary to stack the Lidar devices in order to capture full 3D information. The Lidar devices are mounted orthogonally to each other on the mobile platform where the larger Lidar is used for scanning in the horizontal plane and the smaller Lidar is used for scanning in the vertical plane. The Lidar devices are connected to an NVIDIA Jetson TK1 processor running Ubuntu Linux. The Robot Operating System (ROS) [21] software suite is used to carry out sensor data input and output, coordinate timing, and perform frame transformation between the two Lidar devices. Point cloud processing modules from the Point Cloud Library (PCL) [22] are also used for point cloud filtering, storage and visualization.

Figure 1: Lidar devices used in the hardware setup

The following sections provide details on each step of the scan process. Each step is processed continuously to generate a real time scan output in the form of a point cloud.

3.1 Lidar Data Collection

The data collection process involves walk-through scans with the mobile platform equipped with two Lidar devices. The two sites utilized for this study are a laboratory room and corridors along a selected building floor since they contain features of interest to qualitatively evaluate the scan output. A laboratory room setting is challenging because it is relatively unstructured and contains occlusion in the form of furniture whereas mapping a series of corridors presents a challenge in terms of scanning long distances and mapping around corners.

3.2 Motion Estimation

The Lidar data from the horizontal device is processed with a SLAM algorithm to estimate the motion of the mobile platform on the horizontal plane. The SLAM algorithm used in this study is Hector SLAM [15], which performs scan matching between the current Lidar input and an incrementally built map. The SLAM algorithm estimates the optimal Lidar pose including translation and rotation to match the current scan with a map of the environment. The SLAM algorithm also outputs an occupancy grid map which indicates the grid locations that are likely to be obstacles and the grid locations that are unobstructed. An example of the generated planar map is shown in Figure 2 for the building corridor experiment. The map shows a series of grid lines representing the areas scanned by the horizontal Lidar and a single thick curve representing the trajectory of the mobile platform. Light areas of the map correspond to open space that may be traversed by the mobile platform whereas dark areas correspond to building structures. The trajectory is updated for each input scan from the horizontal Lidar with the calculated current position and orientation of the mobile platform.

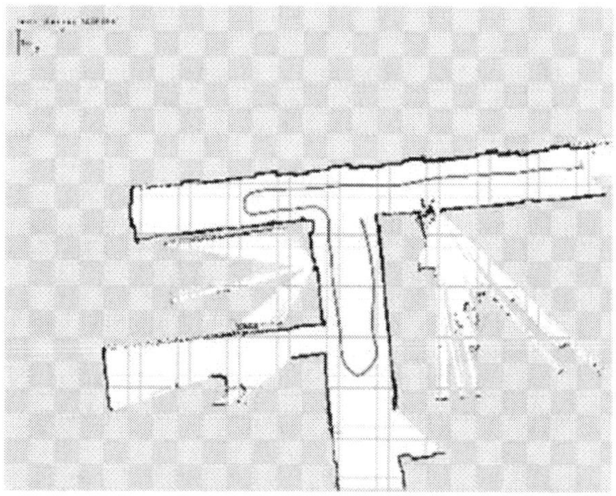

Figure 2: Two dimensional map on the horizontal plane with trajectory estimate of the mobile platform. Blue grid lines indicate scanned areas while the purple curve indicates the trajectory of the mobile platform.

3.3 Scan Registration

Based on information gathered about the mobile platform position from the previous step, the vertical scan input from the second Lidar device can then be registered onto a global three-dimensional map. For each vertical scan collected, the corresponding timestamp is obtained from the sensor and matched with the nearest timestamp from the horizontal sensor. Linear interpolation is then used to obtain the 2D position of the mobile platform at the time when the vertical scan is collected. In terms of position, the current estimate is obtained by fitting a linear motion trajectory through the preceding position estimates. In terms of rotation, the current estimate is obtained through Spherical Linear Interpolation (SLERP) [23] of the neighboring rotation quaternions. The vertical scan is added to a global map by horizontal translation with the calculated 2D vector. A three-dimensional map is thus incrementally built up through the stacking of multiple planar scans.

3.4 Scan Visualization

We also develop a visualization tool to aid the operator of the mobile platform in tracking the scan progress. Figure 3 shows an example of a scan in progress rendered using OpenGL. The visualization displays a set of larger points representing the current horizontal scan and a set of smaller points representing the registered vertical scans in 3D world coordinates. The visualization tool enables the operator to identify unscanned areas and detect potential occlusions during the scan process. This ensures a higher quality and more complete point cloud in the final output.

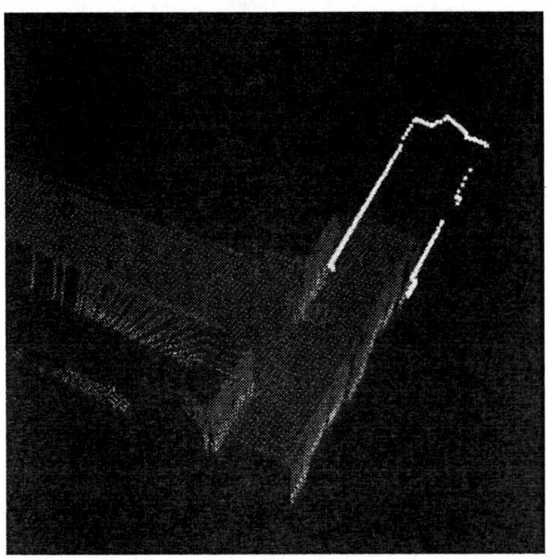

Figure 3: Mapping progress shown in user display. Bold white points indicate horizontal scans while small colored points indicate vertical scans.

4 Results

The proposed technique was applied to two separate scenarios and the scan results in the form of 3D point clouds are shown below. The first experiment was carried out in an indoor laboratory room as shown in Figure 4. The ceiling component of the point cloud was filtered out to allow a clearer view of the room contents. The scan trajectory can also be observed in the middle section of the point cloud. The scan result demonstrates the ability of our technique to capture shape profiles from furniture such as desks and cabinets even in a relatively cluttered environment.

Figure 4: Indoor lab mapping result

On the other hand, the second experiment was carried out over an entire building floor as shown in Figure 5. The result consists of scans along multiple corridors on the same floor. The scan result is largely extensive with a few missing areas corresponding to areas not accessible by the mobile platform. This demonstrates the ability of our technique to capture scans in a large area and register them in a single map.

Figure 5: Corridor mapping result

The accuracy of the generated point cloud map was validated against a commercially available laser scanning system [25]. The static Lidar was placed in 4 different scan points along the corridor shown in Figure 5 and the fully-registered point cloud was used as a reference. The reference point cloud is registered using manual coarse registration and Iterative Closest Point (ICP) fine tuning. Since the scans were taken in an indoor environment, the accuracy metric used is the Euclidean Root Mean Squared Error (RMSE) of point distances to three selected best-fit planes. The calculated Euclidean distance error of the point cloud data is shown in Table 1. The estimated scanning time involved is shown in Table 2. The results indicate that the proposed method achieves reasonable accuracy in the generated point cloud data but involves much faster scanning time.

Table 1: Euclidean distance error of point cloud

Plane	RMSE (single Lidar) (m)	RMSE (our method) (m)
1	0.03511	0.03534
2	0.03948	0.03883
3	0.04233	0.04215

Table 2: Scanning time comparison between different methods

Method	Scanning Time (minutes)
Single Lidar	30
2 Lidar mobile platform (our method)	5

5 Discussion

This section discusses the strengths and weaknesses of the proposed technique for generating a three-dimensional point cloud map of the built environment along with the associated challenges. From the scan results obtained from two test scenarios, we found that the hardware and software setup is capable of performing scan localization and mapping in real time with an update rate of 25Hz. The visualization tool is also effective in providing the operator with an informative view of the scan progress.

The proposed technique is able to generate detailed point clouds due to the high angular resolution of the Lidar devices. However, our methodology heavily relies on accurate localization of the mobile platform at each scan time. The sensor pose estimate may occasionally be erroneous due to jerky motion of the mobile platform which causes a mismatch in the scan registration process. One strategy is to employ a median filter to smoothen the estimated trajectory by eliminating outliers. We can also enforce geometric consistency conditions among vertical scans in a local region such as planarity constraints along the floor and walls. Another limitation of the proposed technique is non-uniform point cloud resolution in the generated 3D map. The point cloud tends to be denser for regions where the mobile platform is moving fast and smoothly and sparser for regions where the mobile platform is moving quickly or undergoing a rapid rotation. To overcome this problem, the operator can reduce the velocity when moving around corners and obstacles while obtaining feedback from the visualization tool to ensure a high resolution in the point cloud output.

Table 3 shows a comparison between our proposed mapping technique with alternative techniques. For example, generating a 3D map using pure visual input involves a less expensive hardware setup and is able to output color information. However, visual scene reconstruction is prone to scale inaccuracies because the process of translating between pixel values and metric units is complex and rely on accurate camera calibration. A depth camera, such as the Microsoft Kinect [24], is able to output both color and depth information but operates at lower ranges compared to a Lidar device. A single ground-based Lidar, such as the FARO Focus laser scanner [25], is capable of generating point clouds at a high resolution, but suffers from the limitation of requiring an additional step to register scans taken from multiple locations. Our proposed technique involving two Lidar devices has the ability to perform automatic registration of scan outputs in real time, but involves an expensive hardware setup.

Table 3: Comparison between scanning techniques

Technique	Advantages	Disadvantages
Visual Camera	Inexpensive hardware, has color information	Sensitive to lighting, vulnerable to scale inaccuracies
Depth camera	Has color and depth information	Limited range
Single Lidar	High resolution output	Require additional registration step
2 Lidar mobile platform (our method)	Real time, automatic registration	Expensive hardware

6 Conclusion

In conclusion, the results of this study demonstrate the viability of the proposed technique for creating three-dimensional maps of the built environment in real time using an orthogonal pair of Lidar devices mounted on a mobile platform. The proposed technique leverages state-of-the-art SLAM algorithms and scan registration to generate dense and accurate point cloud outputs. The main advantages to the proposed technique are the high update rate, high output resolution and effective progress visualization. The proposed technique has the potential to be utilized in infrastructure surveying applications such as geometric analysis, building maintenance and energy modeling. For future work, we hope to improve the accuracy of scan registration on the global three-dimensional map. This can conceivably be achieved through better handling of sensor error and identifying a more precise motion model for the mobile platform. We also consider including color information in the point cloud output by adding a visual sensor, which could be useful in providing a better visualization for the point cloud output as well as helping to identify color features in the post-processing stage.

7 Acknowledgement

This material is based upon work supported by the National Science Foundation (Award #: CMMI-1358176). Any opinions, findings, and conclusions or recommendations expressed on this material are those of the authors and do not necessarily reflect the views of the NSF.

References

[1] C. Eastman, P. Teicholz, R. Sacks, K. Liston, "BIM Handbook – A Guide to Building Information Modeling for Owners, Managers, Designers, Engineers, and Contractors", *John Wiley & Sons, Inc.* (2008)

[2] L. Barazzetti, F. Banfi, R. Brumana, G. Gusmeroli, M. Previtali, G. Schiantarelli, "Cloud-to-BIM-to-FEM: Structural simulation with accurate historic BIM from laser scans", *Simulation Modelling Practice and Theory*, Volume 57, September 2015, Pages 71-87,

[3] C. Wang, and Y. Cho. "Automated 3D Building Envelope Recognition in Point Clouds." *ASCE, Proceedings of Construction Research Congress* 2012, West Lafayette, IN, pp. 1155-1164

[4] Akshay Kumar Shastry, Sanjay Anand, V. Chaitra, K. Uma Rao, and D. R. Akshay, " SONAR Validation for SLAM," *International Journal of Innovation, Management and Technology* vol. 4, no. 5, pp. 498-501, 2013

[5] J. J. Leonard, J. J., and H. F. Durrant-Whyte. "Simultaneous map building and localization for an autonomous mobile robot." in *Proceedings of IEEE/RSJ International Conference on Intelligent Robots and Systems (IROS'91)*, pp. 1442–1447, New York, USA.

[6] A. J. Davison, I. D. Reid, N. D. Molton and O. Stasse, "MonoSLAM: Real-Time Single Camera SLAM," in *IEEE Transactions on Pattern Analysis and Machine Intelligence*, vol. 29, no. 6, pp. 1052-1067, June 2007.

[7] F. Leber, A. Irschara, T. Pock, P. Meixner, M. Gruber, S. Scholz, A. Wiechert. *Photogrammetric Engineering and Remote Sensing*. Vol 76, No.10, pp 1123-1134, 2010

[8] D. Roca, S. Lagüela, L. Díaz-Vilariño, J. Armesto, P. Arias, "Low-cost aerial unit for outdoor inspection of building façades", *Automation in Construction*, Volume 36, December 2013, Pages 128-135, ISSN 0926-5805

[9] Khoshelham K, Elberink SO. "Accuracy and Resolution of Kinect Depth Data for Indoor Mapping Applications." *Sensors* (Basel, Switzerland).2012;12(2):1437-1454. doi:10.3390/s120201437..

[10] Fuentes-Pacheco, J., et al. "Visual simultaneous localization and mapping: a survey." *Artificial Intelligence Review* 43(1): 55-81

[11] Igal M. Shohet, Yehiel Rosenfeld, "Robotic mapping of building interior—precision analysis", *Automation in Construction*, Volume 7, Issue 1, December 1997, Pages 1-12, ISSN 0926-5805,

[12] Velodyne Lidar. http://velodynelidar.com/hdl-64e.html

[13] Jung J, Yoon S, Ju S, Heo J. "Development of Kinematic 3D Laser Scanning System for Indoor Mapping and As-Built BIM Using Constrained SLAM". Passaro VMN, ed. *Sensors (Basel, Switzerland)*. 2015;15(10):26430-26456.

[14] H. Jacky Chang, C. S. George Lee, Y. Charlie Hu and Yung-Hsiang Lu, "Multi-robot SLAM with topological/metric maps," *Intelligent Robots and Systems, 2007. IROS 2007. IEEE/RSJ International Conference on*, San Diego, CA, 2007,pp.1467-1472.

[15] S. Kohlbrecher, J. Meyer, O. von Stryk, U. Klingauf. "A flexible and scalable SLAM system with full 3D motion estimation." in Proc. of the IEEE International Symposium on Safety, Security, and Rescue Robotics (SSRR'2011), pp. 50–55. Kyoto (2011)

[16] J. Zhang and S. Singh. "LOAM: Lidar Odometry and Mapping in Real-time". *Robotics: Science and Systems Conference* (RSS). Berkeley, CA,

July 2014

[17] F. Hoeller, A. Konigs and D. Schulz, "Autonomous reconnaissance and surveillance in urban structures - Eurathlon 2013," *Autonomous Robot Systems and Competitions* (ICARSC), 2014 IEEE International Conference on, Espinho, 2014, pp. 223-228.

[18] Hoeller, F.; Konigs, A.; Schulz, D. "Autonomous reconnaissance and surveillance in urban structures - Eurathlon 2013", Autonomous Robot Systems and Competitions (ICARSC), 2014 IEEE International Conference on, On page(s): 223 – 228

[19] Schueftan, D.S.; Colorado, M.J.; Mondragon Bernal, I.F. "Indoor mapping using SLAM for applications in Flexible Manufacturing Systems", *Automatic Control* (CCAC), 2015 IEEE 2nd Colombian Conference, page(s): 1 – 6

[20] S. Khan, D. Wollherr and M. Buss, "Modeling Laser Intensities For Simultaneous Localization and Mapping," in *IEEE Robotics and Automation Letters*, vol. 1, no. 2, pp. 692-699, July 2016.

[21] M. Quigley, B. Gerkey, K. Conley, J. Faust, T. Foote, J. Leibs, E. Berger, R. Wheeler, and A. Ng, "ROS: An open-source robot operating system," in *Proc. Open-Source Software Workshop Int. Conf. Robotics and Automation*, Kobe, Japan, 2009

[22] R. B. Rusu and S. Cousins, "3D is here : Point Cloud Library (PCL)," in *IEEE International Conference on Robotics and Automation (ICRA)*, 2011.

[23] K. Shoemake, "Animating rotation with quaternion curves." in *Computer Graphics: Proceedings of SIGGRAPH* '85, 245–254. San Francisco: ACM

[24] Microsoft Kinect. http://www.xbox.com/en-us/kinect/

[25] FARO Focus 3D X. http://www.faro.com/en-us/products/3d-surveying/faro-focus3d/overview

Parametric Library Components for BIM-based Curtain Wall Design Automation Module

Beomjun Kim[a] and Sangyoon Chin[b]

[a]Department of Convergence Engineering for Future City, Sungkyunkwan University, Korea
[b]School of Civil, Architectural Engineering and Landscape Architecture, Sungkyunkwan University, Korea
E-mail: skkueng09@skku.edu, schin@skku.edu

Abstract –

Recently, the use of curtain wall systems has been increasing. The architect should be able to intuitively select materials and design elements that are the most suitable for the building from among diverse curtain wall systems. However, in the current curtain wall design processes that are mostly implemented in 2D, lots of problems are occurring. Therefore, the BIM-based curtain wall design automation module should be developed. In this research, before developing BIM-based curtain wall design automation modules, general metal curtain wall systems are analyzed in order to classify components necessary for the fabrication of parametric design based libraries. Curtain wall library components are classified according to the locations, functions, and LODs. Finally, basic shape information of components according to LODs is defined.

Keywords –

Curtain Wall, Building Information Modeling (BIM), Parametric BIM Library, Level of Development (LOD), Basic Shape Information

1 Introduction

1.1 Background

Recently, as diverse forms of buildings including super high-rise buildings have been generalized, attention to the importance of curtain walls that are advantageous for construction period shortening, building lighting and quality homogenization has been increasing. In particular, aluminum curtain walls are currently the most widely used in super high-rise buildings because they are light weight and have excellent functions as filters to control and block the flow of elements that affect indoor environments [1].

However, in the current curtain wall design processes that are mostly implemented in 2D, problems such as excessive design drawings, errors in design drawings and documents, and redesign are occurring [2]. Key risk factors that delay the construction of super high-rise curtain walls include excessive unit type designs due to complicated designs, the occurrence of excessive design changes, and insufficient reflection of site conditions and constructability [3]. In addition, current curtain wall design highly tends to utilize data accumulated through existing projects or rely on consulting by enterprises specializing in curtain walls.

Curtain wall design should be performed with BIM by the architect from the beginning, and the architect should be able to intuitively select from among diverse curtain wall systems the materials and design elements that are the most suitable for the building. In addition, constructability should be reflected in design so that design and construction can be connected successively and to help with the preparation of shop drawings.

However, even if attempts are made to utilize BIM authoring tools that enable working in 2D and 3D in design, the utilization of BIM authoring tools is not easy because they have a limitation of not being able to reflect the materials and assembling methods for diverse curtain wall systems.

Therefore, BIM-based curtain wall design automation modules that will enable selecting those materials and assembling methods that are the most suitable for buildings from among diverse curtain wall systems and can help the preparation of shop drawings should be developed.

1.2 Research Purpose and Scope

In this research, before developing BIM-based curtain wall design automation modules, general metal curtain wall systems are analyzed in order to classify components necessary for the fabrication of parametric design-based libraries.

Since the kinds of curtain wall systems are diverse because of materials and assembling methods, the scope of the present study is limited to the analysis of the currently most widely used general metal curtain wall systems. In addition, curtain wall library fabrication through component classification and the development of BIM-based curtain wall design automation modules utilizing the curtain wall libraries will be performed in follow-up studies.

1.3 Research Procedure

The research was conducted using the following procedures.

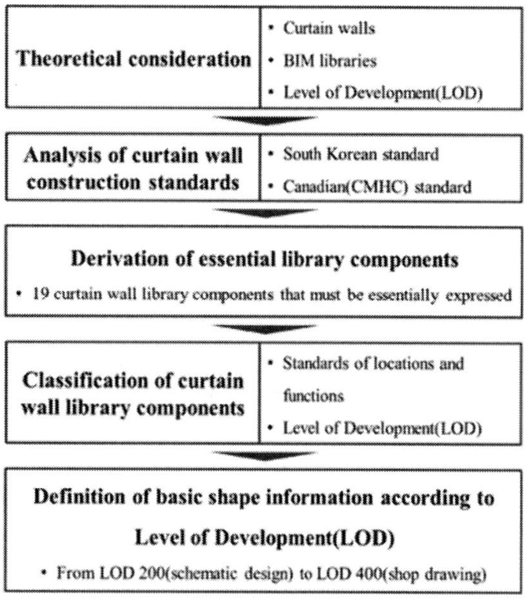

Figure 1. Research procedure

1. Theoretical consideration: Literature data and previous studies on curtain walls, parametric BIM libraries, and Level of Development (LOD) were examined.
2. Analysis of domestic and foreign curtain wall construction standards: The metal curtain wall construction standards in the South Korean building construction standard specification [4] and the Canadian metal curtain wall construction guide [5] were analyzed to derive components necessary for curtain wall library fabrication.
3. Classification of curtain wall library components according to locations and functions: The derived curtain wall library components were classified according to the standards, locations and functions.
4. Classification of curtain wall library components expressed according to LOD: Through LOD standards [6], curtain wall library components expressed in LOD 200, 300, 350, and 400 were classified.
5. Definition of basic shape information for curtain wall library components according to LOD: Basic shape information for curtain wall library components in LOD 200, 300, 350, and 400 were defined.

2 Theoretical Consideration

2.1 Curtain Wall

2.1.1 Overview of Curtain Walls

Curtain wall is a generic name for non-bearing outer walls and refers to those walls that compartmentalize and block spaces in the form of curtains within the structure of their construction. Curtain walls can be largely divided into metal ones and non-metal ones based on the materials used. Metal curtain walls include steel, aluminium, and stainless steel ones. In addition, the assembling methods can be divided into site assembling and shop assembling methods. The materials and assembling methods of curtain walls as such are selected according to building facade designs, construction periods, costs, and required performances.

Figure 2. Curtain walls used in building façade

2.1.2 Consideration of Curtain Wall Related Studies

Previous-curtain wall-related studies in South Korea include curtain wall design improvement-related studies and risk analysis-related studies. The curtain wall design improvement-related studies included studies that derived curtain wall structure design review processes and developed structural design understanding manuals [1]. In addition, there were studies that developed systems that enable sharing of curtain wall design information, cooperation, and support for decision making [2]. Furthermore, there were studies that presented improved curtain wall construction life cycle processes utilizing the VSM technique [7]. Curtain wall risk-analysis related studies include studies that derived core risk factors for delays in curtain wall construction using the FMEA technique and presented responding processes [3]. Studies related to the development of curtain wall design tools or curtain wall library fabrication using BIM were insufficient.

Among curtain wall-related previous studies conducted in foreign countries, many were related to curtain wall designs in terms of energy efficiency. In addition, there were studies that presented decision making support processes for selection of curtain wall systems in the design development stage [8]. In common with South Korea, studies related to curtain wall library fabrication using BIM were insufficient.

2.2 Parametric Design-based BIM Library

Parametric design is a methodology of architectural design that defines parameters as input values so that complicated shapes can be designed through many constraints based on the parameters. BIM libraries based on parametric design enable giving dimensions, creating diverse shapes from relational expressions defined using variables, and easily revising the shapes through the adjustment of variable values [9].

BIM libraries refer to the sets of individual members contained in BIM authoring tools or firsthand constructed by the user when necessary. Individual libraries include diverse pieces of information on certain members utilized in construction processes such as shapes, dimensions, materials, and prices. By utilizing the libraries in design processes, repetitive tasks can be reduced and data can be directly delivered so that design information can be successively connected to construction work. Constructed libraries can be utilized in drawing extraction, estimation, construction, and maintenance [10].

Currently, technologies to fabricate BIM libraries are largely divided into two types; those that are to fabricate libraries with dimensions and constraints while having library description information in their own formats such as the Family of Revit and those that are to fabricate libraries using languages that can describe parameters and even algorithms such as the Geometric Description Language (GDL) of ArchiCAD [11].

2.3 Level of Development(LOD)

Before fabricating BIM based curtain wall libraries, plans for detailed levels of models are necessary. To this end, standards for Level of Development (LOD) that indicates the level of BIM model preparation should be checked first.

Represent examples of standards that define LOD include The American Institute of Architects (AIA) Contract Documents and BIMForum LOD Specification [6]. AIA first defined LOD in AIA Document E202™-2008 and has developed the contents through E203TM-2013, G201™-2013, and G202™-2013. AIA divides LOD into five levels from 100 through 500 based on the contents of major tasks and model preparation levels.

The contents of the LOD Specification of BIMForum were developed based on AIA Contract Documents. BIMForum more concretely defines the standards for preparation of construction members classified by Uniformat™ into BIM models in individual stages of projects. In addition, BIMForum did not add any separate content to the LOD 500 defined by AIA but added 350 between 300 and 400 to divide the specification into five levels.

3 Analysis of Domestic and Foreign Curtain Wall Construction Standards

3.1 Standard Specifications for Building Construction Work in South Korea

The Ministry of Land, Infrastructure and Transport of South Korea provides standard specifications that can be applied to building construction work performed in South Korea. As sub-items of exterior wall construction work (14000), the specifications describe in detail, the require performance of metal curtain wall construction work (14015), materials, and contents related to construction work. The materials that constitute metal curtain walls specified in the standard specifications are listed in Figure 3.

3.2 CMHC Glass and Metal Curtain Walls Guide

Canada Mortgage and Housing Corporation (CMHC) is a crown corporation of the government of Canada. The materials that constitute metal curtain walls specified in chapters 3 (Components and Materials) and 8 (Specifications) of the glass and metal curtain walls guide provided by CMHC are listed in Figure 3. The glass and metal curtain walls guide describes the require performance of metal curtain wall construction work, materials, and contents related to construction work in more details than the standard specifications for construction work of South Korea along with detailed drawings.

3.3 Derivation of Essential Curtain Wall Library Components

The materials that constitute metal curtain walls specified in the standard specifications for construction work of South Korea and Canada were compared with each other and nineteen curtain wall library components that must be essentially expressed were listed in Figure 3. The insulations among the derived components include both those insulations that are used in aluminum bars and panels and cavity insulations. In addition, since sealants are used in many areas and are small in size, they were not included in component classification in order to express them as attribute information instead of shape information when curtain wall libraries are fabricated.

Korean Standard	Canadian Standard
Extruded aluminum and panel	Extruded aluminum
Steels	Sheet aluminum
Fasteners	Sheet steel
Insulation	Sheet sections
Insulation using in bar	Fasteners
Back panel	Spandrel Panel
Glass and glazing material	Bituminous paint
Sealant	Vertical glass units
Gasket	Sealant
Setting block	Waterproofing/air barrier membrane
Fire stop	Shims, spacers, tapes, glazing, gaskets
Anchors	Setting blocks, sealants
Flashing and gutter	Spandrel glass
Extractor	Spandrel and cavity insulation, stick pins, adhesives
Hardware of openings	Joint backers
Screen, louver, cavity insulation	Structural sealant

Components
Embedment
Anchor clip
Fastener
Mullion
Transom
Spandrel panel
Back panel
Insulation
Glass
Gasket
Setting block
Flashing
Gutter
Fire stop
Extractor
Sun shade
Window
Door
Louver

Figure 3. Derivation of essential curtain wall library components

4 Classification of Curtain Wall Library Components

4.1 Classification of Curtain Wall Library Components according to Locations and Functions

The nineteen essential library components are divided into eight groups as shown in Figure 4 based on their locations and functions. The components are classified into upper eight items which are *Anchor, Frame, Panel, Insulation, Glazing, Waterproofing, Firefighting, and Accessory* and sub-elements are included in each item. As mentioned above, sealants will be expressed as attribute information instead of shape information.

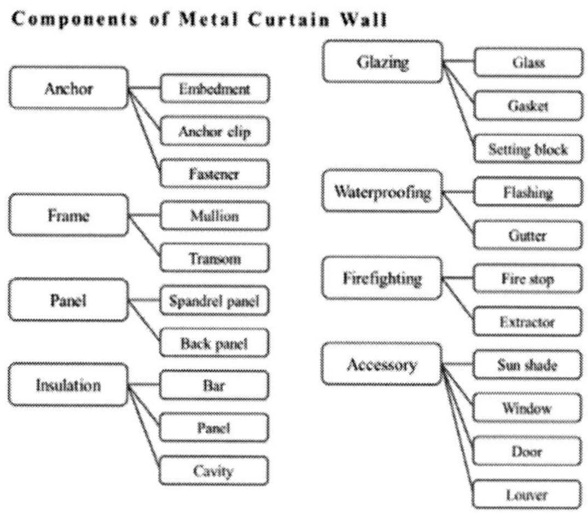

Figure 4. Classification of curtain wall library components according to locations and functions

4.2 Classification of Curtain Wall Library Components Expressed according to LOD

Referring to the newest LOD specification of BIMForum, curtain wall library components expressed according to LOD are classified. BIMForum 2015 LOD specification [6] expresses curtain walls as exterior window walls pursuant to Uniformat^{TM}. The detailed model levels from LOD 100 to 400 are not clearly expressed with materials to be used but are shown with descriptions and figures.

The curtain wall library components classified earlier can be prepared as a classification table as shown in Figure 5 based on the LOD standard of BIMForum. Since LOD 100 is the conceptual design level and expresses the shapes and volumes of masses instead of components, it was not expressed in the classification table. The classification table shows the curtain wall library components expressed differently according to individual LODs ranging from LOD 200 which is a schematic design level to LOD 400 which is a shop drawing level.

LOD 200(1)	LOD 300(9)	LOD 350(12)	LOD 400(19)
Metal curtain wall	Anchor	Embedment	Embedment
		Anchor clip	Anchor clip
		Fastener	Fastener
	Mullion	Mullion	Mullion
	Transom	Transom	Transom
	Panel	Spandrel panel	Spandrel panel
		Back panel	Back panel
			Insulation
	Glass	Glass	Glass
			Gasket
			Setting block
			Flashing
			Gutter
			Fire stop
			Extractor
	Sun shade	Sun shade	Sun shade
	Window	Window	Window
	Door	Door	Door
	Louver	Louver	Louver

Figure 5. Classification table of curtain wall library components expressed according to LOD

LOD 200 expresses metal curtain walls as a single model object and LOD 300 expresses a total of nine components; *Anchor, Mullion, Transom, Panel, Glass, Sun shade, Window, Door, Louver*. LOD 350 expresses anchors and panels in more details compared to LOD 300. Anchors are divided into three components; embedment, anchor clips, and fasteners and panels are divided into two components; spandrel panels, and back panels so that a total of twelve components are expressed; *Embedment, Anchor clip, Fastener, Mullion, Transom, Spandrel panel, Back panel, Glass, Sun shade, Window, Door, Louver*. LOD 400 expresses a total of nineteen components including insulations, glazing, waterproofing, and firefighting components; *Embedment, Anchor clip, Fastener, Mullion, Transom, Spandrel panel, Back panel, Insulation, Glass, Gasket, Setting block, Flashing, Gutter, Fire stop, Extractor, Sun shade, Window, Door, Louver*.

BIM-based curtain wall design automation modules should be developed so that LOD 300 can be selected to make up to design development levels, LOD 350 can be selected to make up to construction design levels, and LOD 400 can be selected to help even shop drawing preparation.

4.3 Definition of Basic Shape Information for Curtain Wall Library Components according to LOD

In LOD 200, information on approximate dimensions (width, depth, and height), shapes, locations, quantities, orientations, and major types of single model objects combining frames, panels, and glasses is necessary.

The nine curtain wall library components expressed in LOD 300 basically require information on design development level dimensions, shapes, locations, quantities, orientations, and materials. In the case of anchors and panels, since sub elements included in them are expressed in LOD 350, information on the shapes of single model objects before segmentation should be necessary in LOD 300. In particular, in LOD 300, mullions and transoms additionally require information on spacing.

The twelve curtain wall library components expressed in LOD 350 basically require information on construction design level dimensions, shapes, locations, quantities, orientations, and materials. Since anchors in LOD 300 are expressed as embedment, anchor clips, and fasteners in LOD 350, shape information on each of embedment, anchor clips, and fasteners should be necessary. This is also the case with spandrel panels and back panels. In particular, in LOD 350, mullions and transoms require accurate information on shapes and spandrel panels require accurate information on dimensions.

The nineteen curtain wall library components expressed in LOD 400 require information on shop drawing level dimensions, shapes, locations, quantities, orientations, and material for actual fabrication stages. As with mullion extrusion profiles that must be perfectly expressed, all components require shape information for fabrication.

5 Conclusion

In this research, as the first stage to develop BIM-based curtain wall design automation modules, curtain wall library components were classified and basic shape information according to LOD was defined.

First, metal curtain wall constituting materials specified in curtain wall construction standards of South Korea and Canada were compared to derive nineteen curtain wall library components that must be essentially expressed. The derived nineteen essential library components were classified into eight groups; Anchor, Frame, Panel, Insulation, Glazing, Waterproofing, Firefighting, and Accessory based on their locations and functions.

Thereafter, referring to the newest LOD Specification of BIMForum, curtain wall library components expressed according to LOD were classified. The curtain wall library components expressed differently according to individual LODs ranging from LOD 200 which is a schematic design level to LOD 400 which is a shop drawing level were classified and organized into a table. Hereafter, BIM-based curtain wall design automation modules should be developed so that LODs can be selected according to design stages.

Finally, basic shape information for curtain wall library components in LOD 200, 300, 350, and 400 were defined. Information of dimensions, shapes, locations, quantities, orientations, and material for each design stages is necessary.

In future studies, for curtain wall library fabrication, not only shape information for components by design stage but also attribute information should be defined in more details. Using IFC property set is could be one solution. In addition, combination relationships between components should be defined.

Acknowledgement

This work is financially supported by Korea Ministry of Land, Infrastructure and Transport (MOLIT) as <U-City Master and Doctor Course Grant Program>.

References

[1] Y. W. Cho, M. C. Lee and J. H. Ock. A Study on the Development of the Structural Performance Evaluation Manual of the Skyscrapers Curtain Wall Systems for Construction Managers. *Korean Journal of Construction Engineering and Management*, 4(2):92-101, 2009.

[2] W. Kwon and J. Y. Chun. Collaborative Design Management System for Curtain Wall. *Journal of the Architectural Institute of Korea Structure & Construction*, 23(11):143-151, 2007.

[3] Y. M. Lee and Y. S. Kim. A Study for Major Delay Risk Factors in Curtain Wall Work of High-rise Building using FMEA. *Journal of the Architectural Institute of Korea Structure & Construction*, 27(1):189-196, 2011.

[4] South Korea Ministry of Land, Infrastructure and Transport, *Building Construction Standard Specification*, 2013.

[5] Canada Mortgage and Housing Corporation (CMHC), *Glass and Metal Curtain Walls*, 2004.

[6] BIMForum, *Level of Development (LOD) Specification*, 2015.

[7] C. D. Kim and S. H. Lee. The Improvement of Curtain Wall Design Process using Value Stream Mapping Tools. *Korean Journal of Construction Engineering and Management*, 7(5):128-137, 2006.

[8] M. Kassem, N. Dawood and D. Mitchell. A Decision Support System for the Selection of Curtain Wall Systems at the Design Development Stage. *Journal of Construction Management and Economics*, 30(12):1039-1053, 2012.

[9] J. K. Im and J. D. Park. A Study on Form Generation Methodology Using Parametric Constraints - Focused on the Irregular Shaped Skyscraper. *Journal of the Architectural Institute of Korea Planning & Design*, 29(08):43-50, 2013.

[10] C. W. Kim and B. S. Koo. Development of Parametric BIM Libraries for Civil Structures using National 2D Standard Drawings. *Korean Journal of Construction Engineering and Management*, 15(4):128-138, 2014.

[11] T. W. Kang, J. W. Lee, W. S. Lee and H. S. Choi. A Study on Civil BIM Description Neutral Library Development Direction for modeling the Construction Spatial Information. *Journal of the Korea Society for Geospatial Information Science*, 20(4):145-151, 2012.

A Framework for Developing an Estimation Model of Damages on Bridge Elements Using Big Data Analytics

S. Lim[a*], S. Chung[a], S. Chi[a], and J. Song[a]

[a]Department of Civil and Environmental Engineering, Seoul National University, Republic of Korea

E-mail: sorami@snu.ac.kr*, hwani751@snu.ac.kr, shchi@snu.ac.kr, junhosong@snu.ac.kr

Abstract

According to the Ministry of Land, Infrastructure and Transport in Korea, the number of bridges over 30 years old will be three times in 2025 than three thousands in 2015 so that proper maintenance enforcement has emphasized. The Korean government legislated on the Special Act on Safety Control for Infrastructure in 1995 and rated the condition grade of "A" to "E" to indicate that under the "C" grade of infrastructure requires critical repairs. Since the lack of maintenance budget and decrease of professional inspectors, more effective and efficient monitoring solutions for bridge conditions are required. The primary purpose of this study is to develop a model to predict deficiencies of bridge elements using big data analytics and the framework is proposed in this paper. The model analyzed a dataset of Bridge Management System (BMS) developed by the Korea Institute of Civil Engineering and Building Technology (KICT) (e.g., general bridge information, structural information, and inspection and maintenance history) and found combinations of significant factors causing damages of bridge elements. Data mining algorithms (Apriori and Ripper) were applied for the pattern analysis and the partitional clustering algorithm grouped similar patterns for more efficient and fast damage prediction. Machine learning concepts using bagging and boosting were also employed for resampling and upgrading the estimation model to enhance estimation performance. The expected results showed potential to predict causes, locations, current status, and the types of bridge damages that can be used for preventive bridge maintenance planning.

Keywords

Bridge Management; Infrastructure Maintenance; Damage Estimation; Big Data

1 Introduction

According to the World Economic Forum, infrastructure plays a key role for economic development since it can enhance national competitiveness. In addition, the condition of infrastructure is crucial because its deterioration can threat public safety and hinder economic activity[1]. Recently it is reminded owing to a case that a serious damage of steel wire to support main structure of the elevated bridge of Jeongneung stream in Seoul, Korea was found by emergency inspection and then the bridge was blocked for a month from February 2016. Bridge structures need enough time and budget for maintenance inspection considering its structural complexity and also direct and indirect impacts of deficiency, and thus preventive management have become significant.

A number of bridges were built with rapid economic growth of Korea in the 1970s and then the number of bridges over 30 years old is expected to rise over three times in 2025 from 3,094 in 2015[2]. Nowadays many of them have deteriorated and required proper maintenance[3]. After the collapse of Seongsu Bridge in Seoul in 1994, the Korean government legislated on the Special Act on Safety Control for Infrastructure in 1995 to designate major facilities as type Ⅰ and type Ⅱ facilities and manage them. The aim of the Act is to provide methods and regulations of periodic inspection for proper repairs based on infrastructures' condition grade of "A" to "E". A bridge under the state "C" requires critical maintenances.

Current bridge management systems in Korea only consider about ten thousand major targeted bridges based on its scale elements. Type Ⅰ bridges cover the bridges which have 1) a specific superstructure types including suspension bridges, cable-stayed bridges, arch bridges, and truss bridges, 2) maximum span length over 50m except one span bridge, 3) length over 500m, and 4) cut and cover structure with width over 12m and length over

500m. Type Ⅱ bridges are the second priority structures, and they include the bridges which are 1) one span bridges with span length over 50m, 2) length over 100m among not the type Ⅰ bridges, 3) cut and cover structure with width over 6m and length over 100m among not the type Ⅰ bridges. However, approximately twenty thousand smaller bridges which are under 100m in length, not targeted to be managed by the Act, are exposed to be in danger. Nonprofessional inspectors have been hired and even such an inspection has not been conducted regularly[4].

An uniform inspection cycle based on the condition grade also hinder efficient investigator arrangement and budget allocation for weak facilities. Since there are various causes of bridge damage and the time of occurrence are less predicted[5], there is a need for a model to support inspections at ordinary times.

Developing a model to support effective and efficient bridge inspection had been a critical issue in bridge management so many studies have been conducted not only in Korea, but also in many other countries. The bridge condition deterioration model have been one of main streams providing condition ratings. The models can be categorized into three groups which are not mutually exclusive: deterministic, stochastic, and artificial intelligence[6].

Deterministic models including regression models represent the relationship between the factors (e.g., bridge age, length, and width) that affect deteriorations and the facility functions based on mathematical or a statistical formulations. Stochastic models such as Markovian models focus on facility condition change from one state to another during one inspection period using probabilities of one or more random variable. Artificial intelligence (AI) models with data mining techniques comprise classification, clustering, and mining frequent patterns. Classification is to find a model that explains and distinguishes data classes to predict the class label of objects for which the class label is unknown. Artificial Neural Network have been utilized in many studies to predict condition ratings of bridge[7-10]. Clustering forms some groups based on similar characteristics measured by distance between data points. The latter analysis is advantageous to find frequent patterns which are a set of items that often appear together[6], [10-13]. Huang and chen (2012) clustered bridge data in similar characteristics and found mining association rules using the National Bridge Inventory in the United States[14], but this research did not show the relationships with bridge deficiency and just described the associations between conditions.

Regression models are simple but they have several constraint conditions. At first, they need to select several factors influencing bridge condition grade before making models and thus some factors can be ignored even if they are significant[15]. Second, the models are updated as calculating new coefficient, and therefore, it is hard to update frequently. Moreover, in the context of application, most regression studies have focused on the condition grade of the whole bridge rather than its elements or different types of elements' deterioration because a parsimonious model, a model that satisfies a desired level of explanation or prediction with as few independent variables as possible, is usually preferred[16]. Even though deterioration factors that cause bridge elements' problems were not much different by each type of elements and thus individual properties of elements did not properly reflected. For instance, the current deterioration model in the BMS uses combinations of factors among bridge age, span length, Average Daily Truck Traffic, average humidity, and amount of surface chloride as independent variables for different type of elements. Since the number of variables is small and the combinations are not much different by type of elements, the current model did not very much good at considering different type of elements[17].

Stochastic models such as Marcovian model are not suitable to develop a model to assist inspections since they have been applied for prediction of future conditions of bridges, not correspond with the objective of this research focusing on present condition predictions. Moreover, the Marcovian model based only on the current condition without considering historical records which make this model unrealistic according to [6], [18].

In this context, adopting the AI approach using big data analytics or data mining techniques to discover patterns is suitable for developing a model. Big data analytics enables the model to use as many as variables for more reliable model development. Easy updating of algorithms enhance the estimation performance and also can encourage frequent uses. Inspections supported by such functions will be more efficient by offering the elements which have high probability of deficiency.

The primary purpose of this study is to develop an easily updated estimation model for bridge elements' damage by considering numerous factors from the bridge element level with big data analytics. The model is to find a set of patterns such as sheet waterproofing as a deck waterproofing type and the maximum span length between 29.3 and 32.3 that can impact to each element's condition grade from A to E (e.g., deck delamination = "D").

This paper developed a model by using Bridge Management System (BMS) data collected by the Korea Institute of Civil Engineering and Building Technology (KICT). There are two types of bridge condition ratings in BMS which are a state condition grade and safety condition grade. The latter is assessed by structural analysis and load carrying capacity test and therefore, it is hard to be replaced. The estimation object is state condition grade and it will be mentioned as just "condition grade". BMS data covers bridges categorized as general national road, and smaller bridges as well as type Ⅰ and type Ⅱ bridges.

The word "prediction" in this research implies that the model estimates probable damages in target bridge elements. The prediction does not mean time series analysis.

This paper introduces the framework to develop the estimation model and preliminary results.

2 Research Methodology

2.1 Research Framework

The major steps to build the estimation model were described in Figure 1.

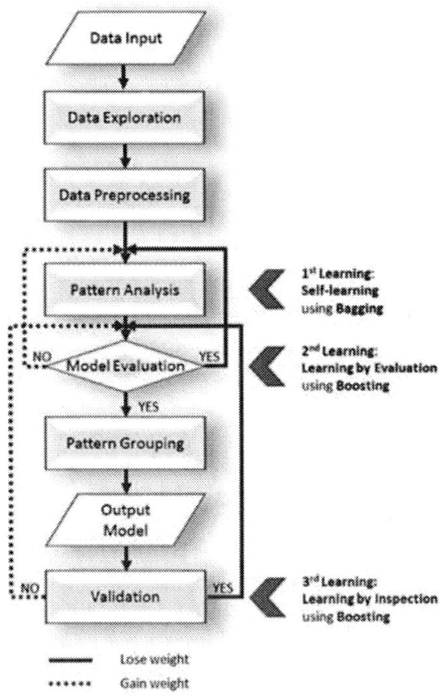

Figure 1. Research Framework

First of all, data exploration to understand data characteristics was performed. Data types (e.g., numerical variable and text variable) and data distributions were identified in this step. The next step was data preprocessing which covered data reduction, data transformation, data discretization, data integration, and data cleaning. Using the preprocessed data, pattern analysis including associations and classifications was carried out and this paper presents the preliminary results by this step. As one of machine learning algorithms, self-learning algorithm such as bagging, and an algorithm learning by evaluation such as boosting are applied to improve estimation performance. Significant patterns are then grouped by the similar input conditions or the same output damage. At the final stage, the model is validated using new inspection data by the boosting algorithm. All procedures were carried out in R software version 3.2.2 for statistical computing and graphics.

2.2 Data Characteristics

Accumulated inspection data in BMS were categorized into three parts: general factors (e.g., bridge class, locations, competent authorities, construction related factors, mileage, offset, detour, traffic volume, length, width, number of lanes, number of spans, main structure type, substructure type, design live load, and attached facilities), structural factors (e.g., span length, decks, girders, diaphragms, ribs for span and support types, abutments, piers, expansion joints, shoes, and stopper factors for support) and inspection factors (e.g., span or support number, element code, damage code, condition grade and maintenance record). Some text options were coded into numbers such as 11 indicated a RCS type of main structure. Factors were divided into discrete numerical variables (e.g. bridge class and main structure type), continuous numerical variables (e.g., mileage and traffic volume), and text variables (e.g., design firm, constructor, supervisor, and inspection descriptions). To utilize text string variables, they should be preprocessed into discrete values and they were not considered yet in this paper.

The size of BMS data were quite large enough to be called big data since it contained about 84 thousands tuples (i.e., records or rows). General data covered total 6,773 bridges and structural data included 19,625 rows for span and 25,729 rows for support. Inspection data was 10,655 which comprised 9,775 of detailed inspection and 900 of precise safety diagnosis. Detailed inspection data was used since it contained the condition grade in an element level. 834,815 rows inspection data in a level of different damage types of elements were composed of 3,701 bridges' span data and 3,000 bridges' support data.

The classes of 6,773 bridges in the data were type Ⅰ bridge (13.6%), type Ⅱ bridge (18.6%), the others (64.7%), and missing value (3.1%). A distribution of condition grade of the whole bridge showed grade "A" (27.9%) and "B" (64.7%) took large portion and "C" (3.6%), "D" (0.1%), "E" (0%), and the other values (3.7%). In the bridge element level, distribution is different from that of condition grade of the whole bridge:

the portion of grade "A" (34.4%) and "B" (37.0%) are smaller and "C" (11.2%), "D" (1.3%), "E" (0.1%), and the other values (16.0%).

2.3 Data Preprocessing

Variables used in the developed model consisted of predictors (i.e., input variables or independent variables) and targets (i.e., output variables or dependent variables), and preprocessing step was different according to the variable type and the role of each variables.

2.3.1 Predictors

Data reduction, data transformation, and data discretization were conducted for predictors. Predictors covered all general factors, all structural factors, and the condition grade and the maintenance record from inspection factors in order to consider condition grade. Data transformation was then performed to correct or erase typing errors and to have consistency of expression. For example, bridge elements' condition grade should comprised from "A" to "E" but number (e.g., 1, 2, and 3), some symbols (e.g.,] and '), and other characters (e.g., N, Q, V, and X) were sometimes included in a original dataset. Numbers and the other values were removed and small letters from "a" to "e" was changed into the capital letters.

Discretization was also carried out to make continuous numerical variable be categorized. It can be considered as if one dimensional partitioning and then a k-means clustering was utilized. By k-means clustering, the objects are distributed into k clusters based on a rule that minimizing within-cluster distances and maximizing inter-cluster distances[6]. The "discretize" function in the "arules" package was applied to make results on R software. To set the number of classes (k) before clustering, Sturge's rule was applied using size of sample (n) [19]. The calculated values were 20 for span and 19 for support when n was 560,272 for span and 240,418 for support.

$$k = 1 + 3.322(log_{10}\text{n}) \quad (1)$$

2.3.2 Targets

The major goal of this research is to find patterns of bridge elements' damages, so the related variables were restricted to represent target variables. An aggregated factor (e.g., 2B07100C) consisted of span or support number (2), element code (B07; expansion joint), damage code (100; crack), and condition grade (C) from inspection factors.

2.3.3 Generating Dataset

The preprocessed variables were combined by the bridge number which existed in the every data table. Span dataset (87 variables, 452,361 records) and support dataset (69 variables, 193,499 records) were generated to include not only predictors but also targets. Since there was not any common structural factor between span factors and support factors and thus two divided dataset were created.

2.3.4 Data cleaning

Handling missing values was critical because blank cells were randomly but frequently distributed in each dataset. There are two common approaches to deal with missing values: ignoring the tuple or filling in the missing values[11]. The former is very conservative and then the span dataset were shrunk as 66,542 tuples when it was applied. A global constant (e.g. 9999) or using a measure of central tendency (e.g., mean for numerical continuous variable and mode for numerical discrete variable) to fill in the missing value were used to find the best method that represented real population.

2.4 Data Mining

The core part of the estimation model was the pattern analysis and it had two axis in this research: descriptive mining and predictive mining. Descriptive mining is to comprehend essential characteristics of the data and answers to "What has happened?" and predictive mining deduces patterns from the data and replies to "What could happen?"[20], [21]. Two approaches used different techniques but the objective is same: finding patterns between predictors and targets.

The first descriptive axis was to find frequent patterns, which were association rules. The rule indicated only co-occurrence and did not mean causality. This process found significant factors that correlated with each bridge elements' damage. Association rules are sets of items (i.e. itemsets) which consist of "if" (i.e., antecedent) and "then" (i.e., consequent). Items indicate each condition such as bridge age under 20 years and a "frequent itemset" means the itemset can be found in the data more than a specific level called minimum support. "Support" is calculated as the number of occurrences with both antecedent and consequent itemsets divided by the number of the whole tuples. Another measure that shows more strong co-occurrences than support using conditional probability concept is "confidence". The "confidence" is calculated as support divided by the number of occurrences with only antecedent itemsets. It means how many occurrences which include consequents among the occurrences only with antecedents. The validity of the created rules is obtained by "lift" calculated as confidence divided by benchmark confidence. The benchmark confidence is computed by the number of occurrence with only consequent itemsets divided by the number of the whole tuples. Assume that a rule can be created as if six years old bridge and number of shoes under three then B16100C (i.e., pier, network of fine cracks, and grade "C"). In this case, benchmark confidence presents the frequency of B07100C. If lift ratio is smaller than 1, the obtained rule is not significant since there is a good possibility that B07100C can occur without antecedents [11], [21].

To mine association rules, Apriori algorithm proposed by Agrawal and Srikant in 1994[22] were used in this research since its efficiency for analyzing huge amounts of data. The basic concept of Apriori is that if an itemset is not frequent, each superset of the itemset does also not appear frequently. For instance, if {condition 1} → {condition grade C} is not frequent, then {condition 1, condition 2} → {condition grade C} is also not frequent. The contraposition for frequent cases is also true[11]. This concept prunes insignificant rules and then enhances the computational efficiency. Data discretization is needed to operate the Apriori algorithm. On R software, the "apriori" function in the "arules" package was utilized. Rules are created by adding conditions one by one until the number of conditions reaches the maximum length. The results contained a lot of combinations, and thus to obtain the desired outcome such as {condition 1, condition 2} → {condition grade C} an option was added that descendants should include the target variable column.

In the other axis, classification as performed to find patterns which combinations of certain conditions predict a particular condition grade. As one of the classification method, rule-based classification also finds rules (i.e., patterns) like mining association rules but its objective is to predict class label such as condition grade. For extracting predictive patterns, RIPPER (Repeated Incremental Pruning to Produce Error Reduction) as a rule producer[23] was applied. RIPPER directly extract rules from data without going through other classification models such as decision tree, so the computation time is faster.

In the RIPPER, one condition is created and the next condition is added with examination of how well the training data (i.e., data used for developing model) is explained. After making patterns, it sets a different training set from the whole data and compares rules from training sets. Through this step, the optimized combinations can be created[11], [24]. The "JRip" function in the "RWeka" package on R software was used.

After finding patterns, a partitional clustering algorithm groups similar patterns not only for efficiency but also for utility for users. The grouping includes two directions: one is to make clusters with common combinations of conditions, the other is to group using some component of the target such as condition grade, damage code, element code, and span or support number.

2.5 Updating

The developed estimation model is updated using ensemble algorithms of machine learning to obtain better prediction performance. Ensemble methods apply multiple classifiers which indicate an algorithms or a specific mathematical functions that implement classification[11].

In the pattern analysis phase, bagging, also called bootstrap aggregating proposed by Leo Breiman(1994) is utilized to reduce variance and to avoid overfitting of classification. The main concept of this method is creating multiple training sets by random sampling from the given training set, building estimation models in each sample space (i.e., bootstrap), and eventually combining estimation models using majority vote for classification[11], [24].

In addition, boosting algorithm is applied for model evaluation and validation. To decrease misclassification and maximize the classification performance, misclassified items gain weight and correctly classified items lose weight during boosting so that weak classifiers concentrate more on the items previously misclassified[11,] [24]. Model update occurs when a new data enters. Training population is changed and then bagging algorithm builds estimation model.

2.6 Validation

The developed model is validated by comparing new inspection data to the estimation results. After entering the data about newly inspected bridge into the model, the model matches the data to patterns found and evaluate whether inspected condition grade corresponds the predicted grade. The validation result will be an input data for boosting and the classifiers will be weighted.

3 Results

3.1 Preliminary Results

The 11 rules and 678 rules were found using the Apriori algorithm in the span and support dataset respectively with numerical settings that 0.001 for minimum support, 0.5 for minimum confidence, and four for maximum length of itemsets (Table 1). Four patterns from the support dataset were created using the RIPPER without any numerical settings. The number of patterns including only grade under C were selected since the bridge elements of condition grade under C were the targets to get proper maintenance.

Table 1. Obtained patterns of bridge span based on the Apriori algoritm

Combination of conditions			Estimated condition grade
Condition 1	Condition 2	Condition 3	
Inspection date =20071016	Maximum span length =[29.3,32.3)	City =Moon-Kyung	Expansion joint
		Road code =2003	Crack
		Millage =[274.64,316.48)	C
Inspection date =20080808	Type of cross =Normal (not elevated)	Effective width =[18.36,20.83)	
		Deck main rebar space =[27.51,37.21)	
		Girder space =[207.2,229.6)	Deck
	Allowable load=0	Substructure type =Round-shaped pier	Crack
		Effective width =[18.36,20.83)	C
	Total width =[18.91,20.11)		
	Barrier existence =Y		

The results can explain that some structural conditions determines specific grade and type of bridge elements. The results about span using the Apriori shows that specific location, time, and structural characteristics had an impact for C grade of crack on expansion joint. When a bridge which was located on the load code 2003 in Moon-Kyung city with maximum span length between 29.3 and 32.3 and millage between 274.64 and 316.48 was inspected in 2007 it frequently showed C grade of crack on expansion joint. For C grade of crack on deck was influenced by time, structural types, and structural sizes. A bridge which was not the elevation bridge and had barriers and rounded-shaped piers with widths between 18.36 and 20.83, rebar space between 27.51 and 37.21, or girder space between 207.2 and 229.6 frequently represented C grade of crack on deck. The support data presented similar format results.

Using the RIPPER algorithm patterns about not span but support were found as follows:

1) If bridge Age=21 and abutment high=[0, 1.41) then grade D of crack on bridge bearing.
2) If minimum distance=[21.21, 27.33) and bridge age = 16 then grade C of erosion on bridge bearing.
3) If bridge length = [43.2, 78.2) then grade C of erosion on bridge bearing.
4) If bridge age = 24 and number of up-lanes = 2 then grade C of deformation on bridge bearing

The patterns from the RIPPER algorithms were simpler than that from the Apriori but the explanation power was not good enough since the number of conditions was too short and it is less understandable such as only bridge age and the number of up-lanes leaded to bridge bearings' erosion.

3.2 Expected Outcomes

To provide interpretable information the output is designed as a portfolio type of each bridge span or support as illustrated in Table 2. Using confidence probabilities of each pattern will also be offered.

Table 2. Sample portfolio of a bridge span

1) Bridge Number: 03447
2) Bridge Name: Songgang
3) Span number: 3

Combination of conditions	Element	Damage	G	Prob.
{bridge length =(970.0, 1289.3), average traffic volume =(18291, 22064), deck pavement thickness =(7.75, 8.50)}	Deck	Crack	C	70%
{bridge width =(17.85, 20.08), deck depth =(23.7, 27.5)}	Girder	Crack	D	60%
{no. of lanes=4, girder type =PSCB}	Diaphra-gm	Exposed Steel bars	B	80%

4 Conclusions

This study proposed a framework to develop an estimation model to find patterns of bridge element's deficiencies by applying association, classification, and clustering in big data analytics. The preliminary results implied that specific conditions of factors can cause bridge elements' damages. The developed model uses the number of various factors and can be updated using machine learning algorithms which are strengths of this model compared to existing regression models.

The study used BMS data which covers not only type Ⅰ and Ⅱ bridges but also smaller bridges and therefore the model can be utilized widely. Identifying many factors related with specific condition grade of each bridge element provides useful information to be referred for inspection that enables effective and efficient monitoring. In short, this estimation model can enhance current inspection system by enabling a preventive bridge maintenance planning.

This paper suggested the framework to develop the estimation model and provide preliminary findings of rules. Other association rule algorithms and classification algorithms should be compared and conducted to find optimized level of performance as well as remaining steps from the research framework. The detailed results, portfolio and a set of critical factors influencing bridge elements' damage, will covered in our future research.

Acknowledgments

This work was supported by the National Research Foundation of Korea (NRF) Grant (No. 2015R1A5A7037372) funded by the Korean Government (MSIP) and the Promising-Pioneering Researcher Program through Seoul National University(SNU) in 2015.

References

[1] ASCE. 2013 Report Card for America's Infrastructure, 2013.

[2] Ministry of Land, Infrastructure, and Transport of Korea. *Yearbook of road bridges and tunnels*, 2015.

[3] Lee Y. H. Construction market prospects related to improvement of infrastructure safety, 2014.

[4] Seok I.H., Gong S. C. and Jung S. H. A consideration of safety inspection for small scale vulnerable facilities. *The journal of Infrastructure Safety*, 37:7-26, 2011.

[5] Lee Y. S. The improvement of precision safety diagnosis cycle for bridge. *The Magazine of the Korean Society of Civil Engineers*, 49(6):62-72, 2001.

[6] Morcous G., Rivard H. and Hanna A. Modeling bridge deterioration using case-based reasoning. *Journal of Infrastructure Systems*, 8(3):86-95, 2002.

[7] Li, Z. B., Shi, Z. Q., and Ososanya, E. T. Evaluation of bridge conditions using artificial neural networks. In *Proceedings of IEEE Southeast Conference 1996*, pages 366–369, Tampa, Florida, United States, 1996.

[8] Kawamura, K., and Miyamoto, A. Condition state evaluation of existing reinforced concrete bridges using neuro-fuzzy hybrid system. *Computers & structures*, 81(18):1931–1940, 2003.

[9] Elhag T. and Wang Y. Risk assessment for bridge maintenance projects: neural networks versus regression techniques. *Journal of Computing in Civil Engineering*, 21(6):402-409, 2007.

[10] Callow D., Lee J., Blumenstein M., Guan H. and Loo Y.-C. Development of hybrid optimisation method for Artificial Intelligence based bridge deterioration model — Feasibility study. *Automation in Construction*, 31:83-91, 2013.

[11] Han J., Kamber M. and Pei J. *Data mining: concepts and techniques*, Third Edition. Elsevier, Waltham, United States, 2012.

[12] Jiang Y., Saito M. and Sinha K. C. Bridge performance prediction model using the markov chain. *Transportation Research Record*, 1180: 25–32, 1988.

[13] Kobayashi K., Kaito K. and Lethanh N. Deterioration forecasting model with multistage weibull hazard functions. *Journal of Infrastructure Systems*, 16(4):282-291, 2010.

[14] Huang, R. Y. and Chen, P. F. Analysis of influential factors and association rules for bridge deck deterioration with utilization of national bridge inventory. *Journal of Marine Science and Technology*, 20(3), 336-344, 2012.

[15] Huang R.-Y., Mao I. S. and Lee H.-K. Exploring the deterioration factors of RC bridge decks: a rough set approach. *Computer-Aided Civil and Infrastructure Engineering*, 25(7):517-529, 2010.

[16] Forster M. Parsimony and simplicity. Available from: http://philosophy. wisc. edu/forster/220/simplicity. html, 1998.

[17] The Korea Institute of Civil Engineering and Building Technology. Development of bridge cost and performance database and module, 2012.

[18] Madanat S. M., Karlaftis M. G., and McCarthy, P. S. Probabilistic infrastructure deterioration models with panel data. *Journal of infrastructure systems*, 3(1): 4-9, 1997.

[19] Nie N. H., Hull C. H., Jenkins J. G., Steinbrenner K. and Bent D. H. *SPSS statistical package for the social sciences*, Second Edition. McGraw-Hill, New York, United States, 1975.

[20] IBM Descriptive, predictive, prescriptive: transforming asset and facilities management with analytics. On-line: http://www-01.ibm.com/common/ssi/cgi-bin/ssialias?infotype=SA&subtype=WH&htmlfid=TIW14162USEN, Accessed 08/03/2016.

[21] Slimani T. and Lazzez A. Efficient analysis of pattern and association rule mining approaches.

arXiv preprint arXiv:1402.2892, 2014.
[22] Agrawal R. and Srikant R. Fast algorithms for mining association rules. *Proc. 20th int. conf. very large data bases, VLDB,* 1215:487-499, 1994.
[23] Cohen, W. Fast effective rule induction. In *Proceedings of the twelfth international conference on machine learning*, pages 115-123, Tahoe City, California, Unites States, 1995.
[24] Shmueli, G., Patel, N. R., & Bruce, P. C. *Data mining for business intelligence: Concepts, techniques, and applications in Microsoft Office Excel with XLMiner*. John Wiley and Sons, 2011.

Use of Ultra-Wideband Sensor Networks to Detect Safety Violations in Real Time

J.D. Lucas[a], J.M. Burgett[a], A.W. Hoover[b] and M.X. Gungor[b]

[a]Department of Construction Science & Management, Clemson University, U.S.
[b]Department of Electrical & Computer Engineering, Clemson University, U.S.
E-mail: jlucas2@clemson.edu, jmburg@clemson.edu, ahoover@clemson.edu, mehmetg@g.clemson.edu

Abstract –

Safety is the highest priority for most construction professionals. However, enforcing the safety requirements and self-regulation of construction jobsites can be difficult. Often, one safety manager is responsible for multiple jobsites and must split their time appropriately depending on the type of work and size of job. This, coupled with large jobsites, makes consistent monitoring of safe working practices a challenge. This paper summarizes a preliminary study that examines the feasibility of utilizing Ultra-Wide Band (UWB) sensors for continuous monitoring of safety violations on a construction job site by placing UWB sensors on personal protective equipment (PPE) and common jobsite equipment. Laboratory experiments were conducted to test the feasibility of current system capabilities by developing a prototype system utilizing known algorithms for sensor localization with an application of relational parameters to test the systems capability of correctly identifying safe and unsafe working conditions. These preliminary experiments revealed the limitations of the current body of knowledge. This paper discusses the concept of continuous monitoring and the identified benefits that UWB sensor networks may be able to provide for construction safety monitoring. The preliminary prototype development and experiments are also discussed with an explanation of the identified limitations. Lastly, future research steps on how to overcome the identified limitations is discussed.

Keywords –

Ultra-Wide Band Sensors; Safety; Construction; Network Sensors; OSHA; fall protection

1 Introduction

Safety on a construction job site is a very high priority. A safe jobsite can lead to lower insurance rates, completing jobs more efficiently, and most of all allow everyone to go home safe at the end of the day. To help increase the safety monitoring capacity of a site, Ultra-Wide Band (UWB) Sensor Networks are being explored as a method for offering continuous safety monitoring in ensuring construction workers are following safe working procedures for common tasks.

UWB sensing is a promising new technology for tracking objects without requiring line-of-sight. Unlike cameras, which require the objects to remain within the field-of-view in order to be tracked, UWB signals propagate through walls and objects and allow non-line-of-sight tracking.

On a construction site, it may be feasible to track pieces of safety equipment and working equipment as they move around. The limitation of the technology is that UWB position measurements are noisy. Depending on the time resolution and spatial precision required, UWB position tracking may be suitable for detecting some safety violations. This research is a preliminary investigation of this area.

This paper summarizes the preliminary study that examines the feasibility of utilizing UWB sensors for continuous monitoring of safety violations on a construction job site. A prototype monitoring system, preliminary laboratory experiments, and identified limitations to the current body of knowledge are also discussed.

2 Background and Literature Review

Construction projects are very dynamic in nature. The dynamic work environment is inherently dangerous, making safety of those who work in construction a priority for those involved. Each company is responsible for safety on the job-site and most have a formalized safety training program. A site superintendent or safety officer, for larger jobs, is usually responsible for enforcing safety protocol. Because of job size and other job duties, these personnel can only observe a fraction of the activities each day so continuous monitoring by personnel is next to impossible and would be very costly. This research suggests that sensor networks may be utilized to help jobsite superintendents and safety officers identify possible unsafe working practices by personnel without having to witness it themselves. The UWB sensors can offer a continuous monitoring of site and worker conditions, especially for those who are working at height and have a risk of falling.

2.1 Safety in Construction

In 2014, OSHA reports that there were 8 million worksites with 130 million workers on them. During this same year there were 4,679 workers killed on the job. The construction industry makes up approximately 21% of these fatalities with 12 workers on average not making it home to their families every week [1]. It has been estimated that the cost of fatal and non-fatal injuries in the construction industry exceeds $11.5 billion annually [2]. Most of the construction fatalities involve what OSHA refers to as the "Focus Four" which include Falls, Electrocutions, Struck by Objects and Caught-in Between. Of these four, falls are by far the most significant accounting for 40% of all construction related fatalities. Despite Fall Protection (29 CFR 1926.501) being the most frequently cited standard by OSHA, it is clear that more can be done to protect the American work force [1]. It is for this reason that this research is examining the use of UWB sensors to help monitor conditions where fall is a risk in the preliminary study.

2.2 Ultra-Wide Band (UWB) Sensors

Ultra-Wideband (UWB) sensor networks use short pulse radio frequency waveforms over a large bandwidth for precise localization of tagged elements within an environment [3] and does not require line of site between the sensor and tag, allowing it to work both inside and outdoors [4]. UWB technology has its origins with time-domain electromagnetics or specifically base-band pulse technology of the early 1960s [3]. Baseband-pulses are short duration pulses concentrated between zero frequency and the microwave spectrum [5]. Early research was originally focused on understanding microwave networks and intrinsic material properties which then lead to the development of target-signature analytics in the time domain [5]. The term "UWB" originated with the Defense Advanced Research Project Agency (DARPA) in the early 1990s as a means of distinguishing between more traditional radar technologies and that of low probability of detection (LPD) (covert) radar systems [6]. UWB systems have primarily been used in LPD radar; however, is now more frequently being researched for civilian applications.

One of the areas that UWB is being studied, and is the topic of this paper, is in precise 3D object localization in real time. With these systems, short-pulse waveforms are transmitted by mobile tags and received by three or more stationary receiving boards. The receiving boards transmit the data to a central processing hub via standard CAT-5 cable (tethered) or wireless gateway signal (untethered). The time of arrival (TOA) to each of the receivers is used to triangulate the tags to specific spatio-temporal locations. Depending on the commercial system used, the angle of arrival at each receiver board may also be used to determine the spatial location of the tags. Algorithm smoothers and filters are applied at the processing hub to filter out incomplete and noisy data [7]. UWB technology has several key advantages to other spatial location alternatives such as RFID tags, 3D video range camera and GPS. These advantages include immunity to multipath cancellation, low interference to legacy systems, high communication security and extended battery life with low pulse rate and short duty cycles tags [5,6,8]. While these advantages are important, what makes UWB particularly relevant for applications on a changing construction jobsite is that it has high signal fidelity inside the building (where satellite signal can be disrupted for GPS systems), where there is a high degree of variability in the density of the obstructions in the environment and when line of sight is difficult to maintain [7,9]. Past studies suggest that these advantages coupled with the relatively low costs of the tags may offer a high return on investment for UWB technology in the construction industry specifically in the areas of work zone safety, job site monitoring and resource tracking [7,8,10].

While the literature suggests that UWB sensor networks show promise in the construction industry, there are several limitations inherent with current technology. Under the best conditions, where the tags and receiving boards are in line of site, the tags are static and all receiver boards are tethered (hard wired), the accuracy is approximately 10 cm [7]. In a field trial at the University of Nebraska-Lincoln (UNL), the UWB inaccuracy increased to 59 cm as interferences were added [7]. Although not addressed in the UNL study, it has been suggested that there could be mutual interferences between UWB sensor tags reducing the fidelity of the network [11]. The UNL field trial also confirmed an experiment by Welch et al. which concluded that the human body interacts with the UWB antenna creating a sharp and pronounced null in the signal [9]. The Welch et al. [9] experiment concluded that the system performance when using UWB sensors near human targets will depend significantly on the signal angle of arrival; however, the effect was reduced in a dense multipath environment common with active construction jobsites. Welch et al. notes that in dense multipath environments where more traditional communications system struggle to perform, UWB systems excel further suggesting that UWB systems have a niche in asset location in the construction industry.

3 UWB Safety Monitoring System

This research is looking at the feasibility of using UWB sensor networks for tracking the position for critical equipment and personnel to identify potentially unsafe conditions. The proposed system would be designed to assist safety officers and superintendents in monitoring safe and unsafe working habits of their employees. The idea is to place UWB sensor tags at strategic locations on equipment and on personnel protective equipment worn by workers to identify safe and unsafe conditions based on the proximity, location, and movement of the tags.

The research set out with three specific aims: (1) examining the feasibility of UWB sensor networks use in constructions safety, (2) based on those tests, generate a list of applicable safety violations that can be monitored using the sensor networks, and (3) identify limitations and barriers of the sensor networks that need further examination.

3.1 Mock-up Testing

The mock-up testing was used to test the feasibility of the UWB sensor networks use in construction safety. The laboratory based experiment utilized the simple equipment and work activities related to the "falls" risk. Specifically, working on a bakers-style scaffold (an easily moveable elevated platform) and a-frame ladder were used.

Basic activities and violations for the bakers scaffold and ladder were identified. For the bakers scaffold, a common violation would be the worker not locking the wheels to prevent it from moving every time they needed to move the scaffold. This also commonly leads to the worker being able to pull the scaffolding along the ceiling when placing ductwork or ceiling panels. This is a clear safety violation and fall risk, however many workers may be tempted to do this in order to speed up their productivity because getting on and off the scaffolding and properly locking the wheels can take significant time if shorter, simpler tasks are required at height over a large area. For the ladder, a fall risk is present when the worker steps on the last rung or top of the a-frame ladder. This causes the worker's center of mass to be higher than the ladder can safely support.

3.1.1 Bakers Scaffold Violation

To test the violation of the bakers scaffold, a UWB tag was placed on a boot of the actor who is simulating the movement of the worker and another tag is placed on the scaffold (Figure 1). Scripted testing included walking around on the floor pushing the scaffolding as well as standing on the stationary scaffolding and walking on top of it. Lastly, taking the boot off with the sensor and placing it on top of the scaffolding and pushing the scaffolding around was tested. This was done in order to test for the "unsafe condition" without actually putting the actor in danger of falling.

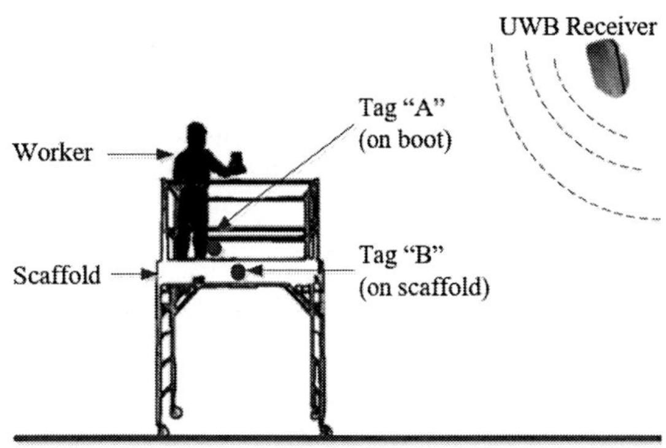

Figure 1. Safety Violation Mock-up

3.1.2 Ladder Violation

When working on a ladder it is important for the worker to not step on the top or next to top rung of the ladder. In order to script this scenario and test the UWB feasibility a UWB tag was placed on the rung of the ladder that should not be stepped on. The other tag was left on the actors boot. The test was then run to see if the UWB sensors could detect when the boot was on the rung with the tag, thus signally a violation.

3.1.3 Mock-up Testing Goals

The goal of these mock-up tests is to identify if conditions can be detected by the UWB sensor network that would signal to a superintendent or safety officer that a problem has occurred which could signal additional safety training is needed by an employee. The goal of the system is to offer a tool for those in charge of safety to better understand the habits of the workers and potentially offer additional training. It is not intended to be used as a means for enforcing violations through punishment. The recorded information can then potentially be used one of two ways depending upon the processing time needed for the system to adequately and accurately detect unsafe conditions. The first method would be to send out an immediate alert via text message or email that will notify the superintendent or safety officer that an unsafe condition is present. This would require a close to real time calculation and notification that the event is occurring. Alternately, the system could take readings over a longer period of time, thus offering a larger lag of notification but potentially higher accuracy because of the use of more data. This would allow for a log to be created over a time period, for instance the course of one day, which can then be reviewed by the safety officer or superintendent. Since the system is to be used as a tool to notify the superintendent and safety officer that additional training is needed or someone may need to be observed, and it is not intended to be a method for automatic reprimand to workers committing the potential violation, the system log could be a better indicator of an employee's actions over a period of time and highlight actual trends of behaviour instead of one incident that has the potential to be a false positive.

3.2 Laboratory Experiment Setup

The testing laboratory is equipped with a UWB tracking system manufactured by Ubisense, Inc. Figure 2 shows a UWB transmitter, commonly referred to as a tag, and a UWB receiver. Transmitters are encased in small rugged plastic containers. Eight receivers are distributed around the lab and are mounted above ceiling tiles. Positions of tags are measured by how long the UWB signals take to propagate from the transmitters to receivers. The Ubisense system uses the 5 receivers that measure the strongest signal strength for each position calculation.

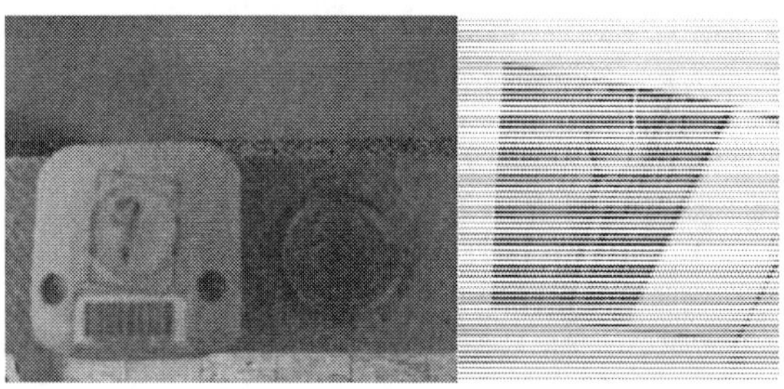

Figure 2: UWB transmitter (left) and receiver (right).

Figure 3 shows a ladder and scaffolding that were used for testing. A tag was mounted below the top surface of the scaffolding and another tag was mounted below the highest step on the ladder. A tag was placed on the boot of a person to track their motion to determine if they were using the ladder or scaffolding in an unsafe manner. Each tag has a unique identifier in the system allowing for the equipment and the person interacting with the equipment to be distinguished from each other. Each piece of equipment is linked to a specific set of parameters defining safe and unsafe conditions (based on sensor location, proximity, and movement).

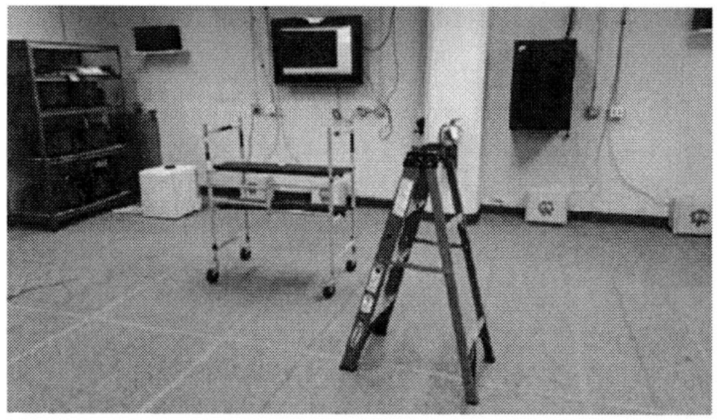

Figure 3 shows a floorplan and image of the laboratory space.

Figure 4 shows an overview of the algorithm. There are four stages. In the first stage raw UWB position measurements are taken for each tag. In the second stage the raw measurements are smoothed using a Kalman filter. In the third stage we developed a custom algorithm to determine if a tag is stationary or in motion. In the fourth stage, safety violations are detected depending on the relative positions and stationary/motion conditions of the tags related to their movement and proximity to the "actor" tag. The following provides the details for each stage.

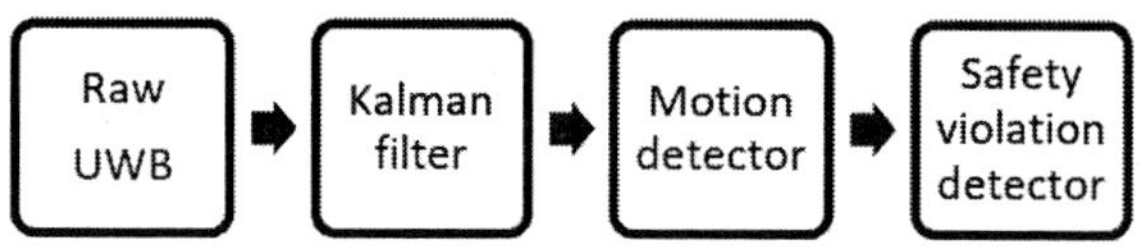

Figure 4: Overview of algorithm.

Given a set of N UWB tags we denote each tag using a subscript i where i=1...N. Let the raw position measurement for a UWB tag to be denoted $(\tilde{x}_i, \tilde{y}_i, \tilde{z}_i)$. These are sampled at 10 Hz. On occasion samples are missed due to the UWB receivers not receiving sufficient power to detect a tag and thus make a measurement. This can happen when a tag is obstructed by large amounts of dense materials that diminish the power received. In such a case, the tracked position of the tag does not change until a new measurement is received.

The purpose of the second stage is to smooth the raw measurements. Raw UWB measurements have an average error of 30 cm but the noise distribution is non-Gaussian [12]. Sources of noise include the transmission of signals through materials obstructing the direct line-of-sight path, reflections of the signals causing non-line-of-sight estimates, and changes in the sets of receivers obtaining the strongest signal strengths [13,14]. To smooth the data we apply a Kalman filter [15]. Our filter uses a constant velocity model. The state is defined as

$$X_t = \begin{bmatrix} x \\ y \\ z \\ \dot{x} \\ \dot{y} \\ \dot{z} \end{bmatrix}$$

where (x, y, z) gives the position and $(\dot{x}, \dot{y}, \dot{z})$ gives the velocity. The velocity is assumed to be constant and accelerations are modeled as dynamic noise using the state transition equation

$$X_{t+1} = \Phi X_t + A_t$$

where Φ is the state transition matrix

$$\Phi = \begin{bmatrix} 1 & 0 & 0 & \Delta T & 0 & 0 \\ 0 & 1 & 0 & 0 & \Delta T & 0 \\ 0 & 0 & 1 & 0 & 0 & \Delta T \\ 0 & 0 & 0 & 1 & 0 & 0 \\ 0 & 0 & 0 & 0 & 1 & 0 \\ 0 & 0 & 0 & 0 & 0 & 0 \end{bmatrix}$$

and A_t is the dynamic noise matrix

$$A_t = \begin{bmatrix} 0 \\ 0 \\ 0 \\ N(0, \sigma_d) \\ N(0, \sigma_d) \\ N(0, \sigma_d) \end{bmatrix}$$

Observations are denoted as

$$Y_t = \begin{bmatrix} \tilde{x} \\ \tilde{y} \\ \tilde{z} \end{bmatrix}$$

and are modeled by the observation equation

$$Y_t = MX_t + N_t$$

where M is the observation matrix

$$M = \begin{bmatrix} 1 & 0 & 0 & 0 & 0 & 0 \\ 0 & 1 & 0 & 0 & 0 & 0 \\ 0 & 0 & 1 & 0 & 0 & 0 \end{bmatrix}$$

and N_t is the measurement noise matrix

$$N_t = \begin{bmatrix} N(0, \sigma_n) \\ N(0, \sigma_n) \\ N(0, \sigma_n) \end{bmatrix}$$

Using these matrices, the standard Kalman filter algorithm is applied. The values σ_d and σ_n determine the amount of smoothing. We set $\sigma_d = 0.001$ and $\sigma_n = 1$. This tuning of the filter provides a large amount of smoothing so that the jitter-like Gaussian noise is completely eliminated and the occasional outlier causes only a small bump. This reduces the speed at which actual changes in velocity can be detected. However, the intent is not to provide instant real-time tracking, but to detect periods of time during which a piece of equipment is being used in a non-safe manner. Thus it is important to smooth out false detections that may be caused by the large measurement noise at the cost of not detecting violations caused by motion that occurs for a brief period of time (e.g. 1 second).

The purpose of the third stage is to determine if each tag is in motion or stationary. Let the motion state for tag i be defined as $m_i = 0$ if the tag is stationary and $m_i = 1$ if the tag is in motion. A buffer of the most recent 10 filtered measurements is kept for each tag:

$$\begin{bmatrix} x_t & y_t & z_t \\ x_{t-1} & y_{t-1} & z_{t-1} \\ \vdots & \vdots & \vdots \\ x_{t-9} & y_{t-9} & z_{t-9} \end{bmatrix}$$

The average $(\bar{x}, \bar{y}, \bar{z})$ of the values in the buffer is calculated. If the tag is stationary, its position is tracked as $(\bar{x}, \bar{y}, \bar{z})$. If the tag is in motion, its position is tracked as (x_t, y_t, z_t) which is the most recent filtered measurement in the buffer. The effect is to provide even more smoothing while a tag is considered stationary but a quicker response to motion if the tag is considered in motion. To change state from stationary to in motion, the count of positions within the buffer that are greater than 20 cm from $(\bar{x}, \bar{y}, \bar{z})$ must be greater than 6. This allows for a few outliers without triggering a state change. Similarly, to change state from in motion to stationary, the standard deviation σ_{buffer} of the values in the buffer must be less than 15 cm. This allows for a small amount of motion in an area without triggering a state change.

The purpose of the fourth stage is to check the

positions and motion states of the tags for safety violations. The algorithm checks each tag mounted on a boot against each tag mounted on a piece of equipment. For each type of equipment, a separate set of conditions is tested. Two examples are detailed below. If the conditions fail, then a counter is increased; if the conditions pass, then a counter is decreased. A warning for a safety violation is triggered if the counter rises above a threshold. The up/down operation of the counter helps prevent false alarms by requiring unsafe conditions to persist for a period of time.

For experimental purposes, two types of violations were tested. The first occurs when a person stands too high on a ladder. The second occurs when a person stands on scaffolding while it is in motion. The ladder violation is detected by comparing the position of the tag mounted on a boot to the position of the tag mounted on a ladder. If the z coordinate of the boot is higher than the z coordinate of the ladder, and if the x,y coordinates of the boot are within 40 cm of the x,y coordinates of the ladder, then a violation is detected. The scaffolding violation is detected by checking the motion state of the scaffolding and comparing the position of the tag mounted on a boot to the position of the tag mounted on the scaffolding. If the scaffolding tag is in motion, the z coordinate of the boot is above the z coordinate of the scaffolding, and the x,y coordinates of the boot are within 50 cm of the x,y coordinates of the scaffolding, then a violation is detected.

Upon initialization, all the relevant position variables are set to the center of the tracking area. It takes a few seconds for the filtered estimates and motion variables to settle to their actual values. All the motion states are initialized to in motion and safety violation counters are initialized to zero.

4 Findings and Limitations

The experiment findings identified two areas related to the use of UWB sensors to construction safety monitoring. This includes that the protocols for identifying safety violations, though relevant for self-regulations and personal supervision, are not definitive enough for defining appropriate computer based algorithms for monitoring. Another issue is that the accuracy of sensor localization is directly linked to the period of time that the data is collected, so certain movements may be hard to detect if the collection period is increased to gain a more accurate location. The short movements may be seen as outliers in the data instead of short, legitimate movements.

4.1 Future Research

In order to more clearly define the parameters, the researchers are proposing a four step grounded theory methodology to identify better tracking with the UWB system. This includes (1) video documentation of workers performing typical tasks. This will be a combination of scripted activities in the lab that will include scripted violations or errors in methods. (2) Data analysis of the videos that will include coding of the workers movements, times, length of movements, etc. (3) System Coding based on parameters identified with the video coding. (4) Prototype validation through the monitoring of scripted activities in the laboratory to identify the system's ability to positively identify safety violations. (5) Finalize protocol for documenting and coding additional violations for easy inclusion in an expanded safety monitoring system. Activities related to fall protection will be the focus of the next research steps.

5 Discussion

In studies conducted in other fields, employee performance, when they know they are monitored has increased [16]. This leads to others also identifying benefits of electronic monitoring systems [17]. However, electronic monitoring can also have negative effects on job attitude and satisfaction especially when using close monitoring techniques such as camera recording, data entry, and chat/phone recording that lead to the employees feeling as if they lose control [17]. Attitudes also go down when monitoring is focused on individuals and unpredictable [17]. Monitoring can also be seen as an invasion of privacy, but McNall and Stanton [18] identified within a study that employee control over monitoring, such as providing areas where monitoring does not take place, can help those being monitored feel more comfortable. The study also stated there might be a link to reasons for monitoring and transparency of monitoring being a factor but could not definitively say.

The fact of the matter is that employees' concerns will need to be addressed and they will need to be comfortable with the technology. There are clear applicable benefits to the contractor in terms of ensuring a safer work environment and monitoring safe work practices that can lead to improved safety training, however it can also lead to employers taking advantage of the system to monitor productivity and utilize it as a disciplinary tool. A positive, transparent environment, focusing on improvement and not discipline may help overcome potential challenges of employee acceptance of the technology.

The benefits of better trained employees in terms of safe working practices can have a large impact on the company's bottom dollar. If safety incidents onsite decrease, insurance rates and worker's compensation payments will also decrease, this will provide a monetary savings to the company. Safety also helps with a company's reputation. The use of an electronic monitoring system such as a UWB sensor network for continuous safety monitoring can be a positive tool within the construction industry.

References

[1] Occupational Safety and Health Administration (OSHA). Commonly Used Statistics United States Department of Labor. Online: www.osha.gov/oshstats/commonstats.html, Accessed: 0/15/2016.

[2] Waehrer G. M., Dong X. S., Miller T., Haile E. and Men Y. Costs of Occupational Injuries in construction in the United States. *Accident Analysis & Prevention.* 39(6): 1258-1266, 2007

[3] Saidi K. S., Teizer J., Franaszek M., and Lytle A. M. Static and Dynamic Performance Evaluation of a Commercially-Available Ultra Wideband Tracking System. Automation in Construction 20: 519-530, 2011.

[4] Shahi, A., Aryan, A., West, J., Haas, C., & Haas, R. Deterioration of UWB positioning during construction. *Automation in Construction,* 24: 72-80, 2012.

[5] Bennett L. Time-Domain Electromagnetics and Its Applications. *Proceedings of the IEEE* 66(3), 1978.

[6] Fontana R. Recent System Applications of Short-Pulse Ultra-Wideband (UWB) Technology. *IEEE Transactions on Microwave Theory and Techniques*, 52(9), 2004.

[7] Cho Y. K., Youn J. H., and Martinez D. Error Modeling for an Untethered Ultra-Wideband System for Construction Indoor Asset Tracking. *Automation in Construction*, 19(1):43-54, 2010

[8] Teizer J., Lao D., and Sofer, M. Rapid Automated Monitoring of Construction Site Activities Using Ultra-Wideband. In *24th International Symposium on Automation & Robotics in Construction,* Madras, India, 2007.

[9] Welch T., Musselman R., Emessience B., Gift P., Choudhurry D., Cassadine D., and Yano S. The Effects of the Human Body on UWB Signal Propagation in an Indoor Environment, *IEEE Journal on Selected Areas Communications.* 20 (9):1778-1782, 2002.

[10] Cheng T., Venugopal M., Teizer J., and Vela P. (2011). Performance Evaluation of Ultra Wideband Technology for Construction Resource Location Tracking in Harsh Environments. *Automation in Construction,* 20(8): 1173-1184 2011.

[11] Porcino D. and Hirt W. Ultra-Wideband Radio Technology: Potential and Challenges Ahead. *IEEE Communication Magazine.* (July) 2003.

[12] Suski, W., Banerjee, S., and Hoover, A., System-Level Noise of an Ultra-Wideband Tracking System, in the *proc. of IEEE Int'l Conf on Information Science*, Signal Processing and their Applications, July 2012.

[13] Suski, W., Banerjee S., and Hoover, A. Using a Map of Measurement Noise to Improve UWB Indoor Position Tracking, in *IEEE Trans. on Instrumentation and Measurement, 2013*, 62(8): 228-2236, 2013.

[14] Banerjee, S., Suski, W. and Hoover, A. Identifying and Filtering Noise Caused by Sensor Set Switching in UWB Indoor Position Tracking, *International Journal of Ultrawideband Communications and Systems*, 3(1): 8-18, 2014.

[15] Kalman, R. E. A new approach to linear filtering and prediction problem, *Journal of basic Engineering* 82(1):35-45, 1960.

[16] Bhave, D.P. The Invisible Eye. Electronic Performance Monitoring and Employee Job Performance, *Personnel Psychology*, 67(3): 605-635, 2014.

[17] Jeske, D. and Santuzzi, A.M. Monitoring what and how: psychological implications of electronic performance monitoring, *New Technology, Work and Employment*, 30(1): 62-78, 2015.

[18] McNall, L. A. and Stanton, J. M. Private Eyes are Watching You: Reactions to Location Sensing Technology. *Journal of Business and Psychology*, 23 (6): 299-309, 2011.

Ergonomic Assessment of Residential Construction Tasks Using System Dynamics

Hossein Abaeian[a], Ndukeabasi Inyang[b], Osama Moselhi[a], Mohamed Al-Hussein[b], Marwan El-Rich[b]

[a] Department of Building, Civil and Environmental Engineering, Concordia University, Canada

[b]Department of Civil and Environmental Engineering, University of Alberta, Canada

E-mail: hosseinabaeian@ymail.com , ndukeaba@ualberta.ca , elrich@ualberta.ca , malhussein@ualberta.ca moselhi@encs.concordia.ca,

Abstract-Modular residential construction activities often require prolonged standing, bending, stooping, and material handling, while working in crowded spaces; these activities increase the potential risk of work-related musculoskeletal disorders (WRMSDs), which may worsen over time, resulting in permanent disability and, consequently, the loss of ability to work. The use of system dynamics (SD) modeling to assess ergonomic risks provides a decision support tool for job managers and job designers and delivers a powerful graphical illustration, showing the logical links between cause and effects and helps illustrate how ergonomic risks may lead to WRMSDs. This paper presents a SD model for ergonomic analysis of residential construction tasks. A case study is presented and used to evaluate variations in risk exposure to identify most contributing factors to potential ergonomic injury. Also a literature review is performed to identify the main ergonomic risk variables. The results are expected to assist project participants in controlling and assessing ergonomic risks leading to improved work efficiency, safety and reduced lost time injuries and related cost, insurance premium (WCB) and claims caused by WRMSDs.

Keywords: residential construction task; system dynamics; ergonomic risk

1 Introduction

Construction productivity correlates to how well, how quickly and at what cost buildings and infrastructure can be constructed and can influence prices of homes, consumer goods and the strength of the national economy [13]. While only 63% of traditional building projects are completed on time [15], the off-site fabrication method has made significant advances in the last two decades, offering benefits in several major areas including productivity improvement [8] by shifting house construction to a manufacturing process [7]. Although the manufacturing process brings many benefits to construction projects; it is also challenged with different internal and external risk factors [9]. By moving the construction process from on-site to factory, most of the tasks are performed in manufacturing facilities and include physically demanding tasks, that expose construction workers to a number of ergonomic risks leading to (WRMSDs).

WRMSDs have been a major cause of non-fatal injuries in construction in the United States accounting for almost 33% of all occupational injuries [4]. In Canada the total cost of occupational injuries to the economy, considering direct and indirect costs, is more than $19 billion annually [10]. WRMSDs develop over time as a result of exposure to factors such as, awkward posture, contact stress, hand/arm vibration, force/static load, and repetition. Also the work pace (organizational factors) and environmental factors contribute to the risk of WRMSDs. Symptoms of WRMSDs may include pain, aching, discomfort, numbness, tingling, and swelling, and normally occur in the back, shoulders, neck, legs, wrists, fingers, elbows, and arms [4]. WMSDs are attributable to a combination of interacting risk factors [2]. The main concept underlying the application of ergonomics to reduce musculoskeletal injury is to match the job to the worker capacity based on the characteristics of the worker, rather than requiring the worker to adapt to the job [5]. Thus proactive ergonomic practices and periodical ergonomic risk assessments can help to identify and eliminate exposure to risk factors and ensure safer construction work conditions, reduces occurrence of WRMSDs and consequently improves quality and productivity of the process. A number of different methods have been developed to evaluate the existence of ergonomic risks. Rapid Upper Limb Assessment (RULA) is a survey method consisting of a single page worksheet which generates only a single risk score associated with upper limbs [3]. Rapid Entire Body Assessment (REBA), just like RULA, is a single worksheet but it evaluates body posture, forceful exertions, type of movement, repetition, and rapid changing associated with the task [12]. Quick

Exposure Check (QEC) assesses work risks related to the back, wrist, neck, and shoulder/arm and is based on epidemiological evidence of the observer's ability to differentiate between different levels of exposure [6]. The Ergonomic Workload Stress Index (EWSI) model is based on the concept of fuzzy set theory to predict the existence and level of ergonomic workload stress; however this model is challenged with the problem of fitting fuzzy logic to human judgment data and accommodating the issue of vagueness of human language [11]. Other methods include the Ovako Working Posture Analysis System (OWAS) which is mainly developed for use in medium to heavy assembly tasks in steel industry to identify and evaluate poor working postures [14]. The Three-dimensional Static Strength Prediction Program (3DSSPP), which is based on biomechanical analysis concepts, considers the worker's posture, gender, anthropometry factors (weight and height), and forces on the hand. It calculates spinal compression and shear forces at L5/S1 disc [16]. However none of the aforementioned methods integrates hand arm vibration exposure, effect of organizational and environmental factors. ErgoCheck is an ergonomic assessment framework which quantifies risks associated with modular and panelized residential construction tasks considering awkward posture, contact stress, force and static loading, repetitive tasks, hand arm vibration and environmental factors as well as the organizational factors [22].This method is selected as the basis of this study for the purpose of developing an integrated ergonomic tool using system dynamics (SD) as it offers a body part based analysis and rating of ergonomic risk.

System thinking, which is a compelling solution for many real world problems, refers to the paradigm in which the world is seen as a complex system, in which everything is connected to everything else [17]. SD, as a powerful graphical illustration tool, is capable of showing the logical links between cause and effects, and has been implemented where a holistic view is essential and feedback loops are critical in understanding the interdependencies among the variables included in the system. SD models are capable of addressing a variety of the problems identified in the literature, ranging from environmental or public policy, to corporate strategy, security, healthcare, and operations management.[19].However there has been no extensive application of SD reported in the ergonomic analysis literature. SD modeling is suitable for any dynamic system characterized by mutual interactions, information feedback, and interdependencies among variables. The behavior of complex systems and consequences of different policies involved in the system can be examined throughout the process by means of a philosophy that reviews problems from a globally perspective [20]. The application of SD allows a four stage approach: (1) recognizing the problem. To have a successful model the modeler must have a clear understanding of the problem. A full understating of the models is often beyond the capacity of human cognition therefore, it should be broken down into smaller systems without violating the holistic concept of SD; (2) Describing the system by means of causal loop diagrams also known as influence diagram; (3) qualitative analysis that involves analyzing the causal loops closely; (4) construction of simulation model. This step covers specification of the structure and decision rules. The estimation of variables, behavioral relationships, and initial conditions should be quantified at this stage [21]. While all the existing SD computer packages use the same modeling concept, the three most widely used packages are STELLA/iThink, Powersim, and Vensim (which is the package used in this research). Different types of components as well as the modeling process are explained more fully in the methodology section below.

The objective of this research is to develop an SD model application for ergonomic analysis of residential construction tasks. The proposed methodology focuses on; (i) identifying principal components of the system, (ii) providing comprehensive ergonomic risk factors based on the existing literature, (iii) developing causal loop diagrams to illustrate the relationships among the variables in the system, and (iv) exploring different what-if scenarios and policy tests as a result of changes in variables which eventually will lead to increase confidence in particular job design strategies, task cycle strategies and policies. The proposed methodology is represented in figure 1.

2 Methodology

The application of SD is a four stage approach in which the first step is recognizing the problem. The model designer must develop a reference model composed of descriptive data illustrating the behavior of the problem. The reference model assists the modeler in obtaining holistic view of the problem. As the primary goal of this study is to build an SD model application for ergonomic analysis of residential construction tasks, the hazard quantification rating system developed by Inyang et al [22] is selected as the basis reference model in the process of ergonomic risk assessment in this research. This is because other existing ergonomic assessment methods mainly focus on awkward posture, repetition, force and static loading, while the selected reference model considers all of the aforementioned factors as well as hand/arm vibration, contact stress, and organizational and environmental factors. After selecting the reference model, it is important to choose a proper time horizon for the SD model, and to define variables and impressions that the modeler deems significant for understanding and discussing the problem[17].

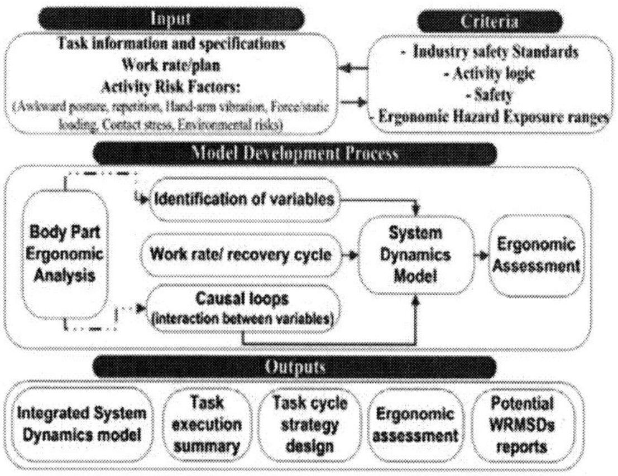

Figure 1. Methodology of Integrated System Dynamics Model Development

2.1 List of variables

There are four types of variables in SD: (1) level (those representing accumulation in the system); (2) flow (those the value of which changes over time); (3) constant (variables the value of which doesn't change over time); and (4) auxiliary (any variables computed from other variables at a given time). The causal loop diagram is a good starting point for developing an SD model; however, the list of variables and their type should first be identified. A hazard is anything that can cause harm or adverse effects and may be biological, physical, ergonomic, chemical, or psychosocial. Having identified the variables, they are categorized into three main groups: (i) hazard score which includes postural hazard, force/static load risk, hand/arm vibration, contact stress and environmental risks. (ii) Organizational risk factor and, (iii) impact of time (repetition and cycle time).

2.2 Relationships among the variables

Developing a solid causal loop diagram can be viewed as a foundation block for a continuous simulation model. A causal loop diagram illustrates the relationships among the variables in a system and can be useful during the early stages of model conceptualization and identifying principal components of the system [21]. In this study, factors such as postural hazard, force/static load risk, contact stress, hand/arm vibration and environmental risks are assumed to be constant during the simulation process. The developed model evaluates the existence of ergonomic risks while a specific posture is held during the performance of tasks under definite working conditions. As Figure 2 illustrates, a hazard score results from the evaluation of six ergonomic risk factors which are evaluated separately for each body part. (1) Postural hazard, considers the relative position of the body segments while performing work activities and is measured and valued in terms of the angle by which a specific joint deviates from the neutral position. (2) Forceful/ static exertion, represents the amount of muscular effort required to perform a task based on the weight of object, the location of object compared to shoulder and knee, and the distance of carrying the load, as well as the load characteristic, which shows how manageable the load is to be carried; (3) Contact stress risk, implies the repeated contact of the body with a hard surface or edge and level of pressure on the skin; (4) Hand arm vibration, pertains to vibration applied to the hand/arms through a tool or piece of equipment [22].

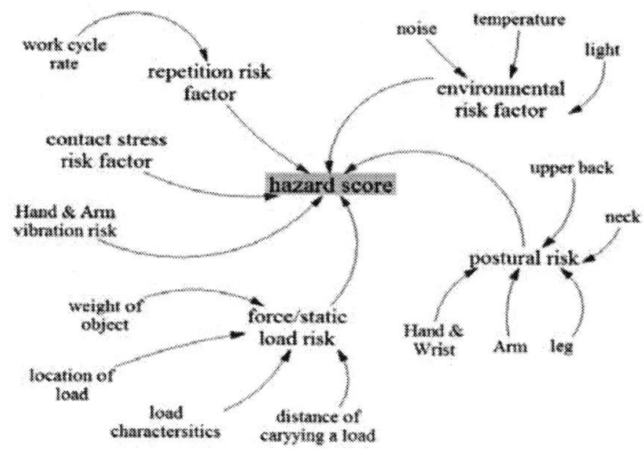

Figure2. Overview of hazard score variables

(5) Environmental factors, refers to the prevailing conditions of the work environment and their adverse effects on the worker's health including sources and levels of light that provide too much or too little illumination, cold and excessively warm temperatures and level of noise distraction; (6) Repetitive factor refers to the frequency or number of similar exertions performed during a task. Repetitive tasks are those with cycle times lasting less than 30 seconds or in which 50% of the cycle involves performing the same fundamental activities [22]. Generally, the greater the number of repetitions, the greater the degree of risk of cumulative trauma injuries; however, there is no specific repetition limit or threshold value (cycles/unit of time, movements/unit of time) associated with injury, which is why this study aims to highlight the impact of cycle time and its frequency on the development of WRMSDs.

Organizational risk factors refer to aspects of how a job is organized; these factors are based on work recovery cycle, work rate and degree of difficulty in keeping pace, workers control over work and level of mental stress [22]. Although existing statistical analysis shows a moderate correlation between mental stress and physical exhaustion, and also the development of mental stress as task duration and repetition increase, in this study these levels have been assumed to be constant through different job cycles. Further research is required to quantify the interaction between psychosocial risks and physically hazardous risks as well as development of mental stress over time [23]. Figure 3 illustrates the organizational factor as well as its associated sub-factors, which have been developed according to the organizational risk scoring system by Inyang and Al-Hussein [22]. The symbols "+" and "–"included in the diagram, represent the causal direction. While the sign '-'in the diagram represents the change in the opposite direction, the sign '+' implies a change in the same direction.

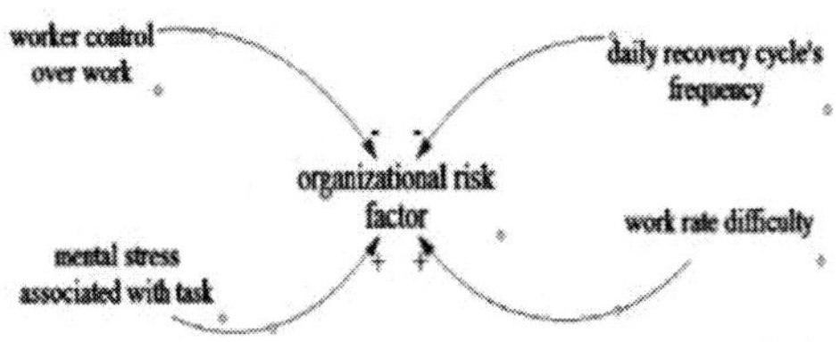

Figure 3. Organizational risk factors and related variables

It is imperative to identify a proper time horizon for the simulation process and to define those variables and impressions that seem to be significant for understanding the problem, and for planning policies to rectify it. The time horizon should meet two criteria: (1) it should begin far enough in the past to indicate how the problem occurred; and (2) it should span far enough into the future to cover the delayed and indirect effects of the potential policies [17].Since this study evaluates the ergonomic risks associated with tasks during the work day, the selected time horizon is eight hours (total daily working hours) and the time interval is one minute. Based on the ergonomic model presented by Inyang and Al-Hussein [22], the resultant hazard score of each body part during assessment (R_S) is calculated by Equation 1:

$$R_S = H_S * M_d * M_o \qquad (1)$$

where, H_S is the hazard score, M_o is the organizational risk factor, and M_d is the total daily exposure duration calculated based on the number of accomplished cycles in the day, activity cycle time and percentage of activity time during which a specific body part is exposed to risk factors. Figure 4 shows the causal links for the duration exposure factors including major variables in the developed system.

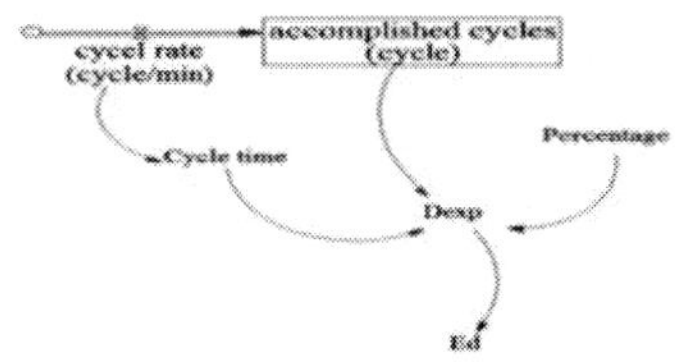

Figure 4. Graphical illustration of duration factors

2.3 Mathematical modeling process

SD simulations are developed based on a set of mathematical equations that represent interactions among variables. It should be mapped based on defined cause and effect relationships in the system using simple mathematical forms and can be either inserted into the system to represent the interactions among variables or added to the model using other methods such as tables of function. Tables of function also known as lookup tables, are typically used in SD to show nonlinear relationships between two variables. A table of function is easier to interpret and visualize and can be modeled as a list of numbers whereby input values to a function are positioned relative to the x axis and output values are read from the y axis. By employing these tables, a user can control the shapes, slopes and saturation points to represent the relationship more accurately [18].These tables are utilized in this study in order to model interactions among variables based on the ErgoCheck ergonomic risk scoring system [22]. The developed model includes separate normalized inputs for major variables rather than normalizing the input variable within itself as normalization helps examine variables without a need to redesign functions for each change in a given variable. For instance, the lookup table below (Figure5) shows the relationship between daily hazard exposure (M_d) and total daily exposure duration (D_{exp}). D_{exp} is calculated based on the number of cycles of activity during the day (n), activity cycle time (CT), and percentage of activity time during which a body part is exposed to a risk factor [22]:

$$D_{exp} = n * CT * P_{ex} \qquad (2)$$

After defining the structure of the SD model, continuous simulation is conducted in order to model all the variables included in the model and to evaluate the ergonomic risks associated with performance advancement over a time horizon of eight hours with time intervals of 1 minute.

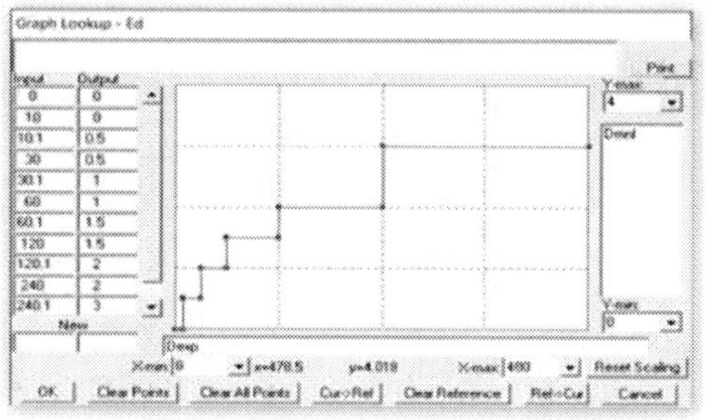

Figure 5. Table of function for daily duration risk factors

3 Case Study

Data obtained from 25 hourly observations of the sanding workstation at the Nava Cabinet Solutions production facility, a kitchen cabinet manufacturing company based in Edmonton, Canada, is used to assess the applicability of the developed simulation model. Manufacture of kitchen cabinets is a complex process that consists of many operations and requires many skills, which include shaping, tasks involving the operation of machinery such as cutting and edge banding; sanding, repetitive tasks involving excessive bodily motions; and cabinet assembly. The sanding operation consists of smoothing the surface of wooden sheets using a hand-operated power sander while both hands are used to carry out this task. The dominant hand is used to operate the sander while the other hand is used to hold the sheet of wood (Figure 6).

Figure 6. Body posture observation during sanding process

The developed SD model is implemented to assess the existence of ergonomic risk at the sanding station based on observed body postures, activity rate, cycle time, and related organizational risk factors.

4 Results and Validation

Having identified the body parts and range of major ergonomic risk factors involved in performing tasks through observations, the proposed SD model is applied to assess the hazard score over time cycles for each body part. Sanding operators are exposed to the risk of repetitive motion injuries and hand/arm vibration due to induced stress, and forceful movements. Significant body parts identified as "exposed" include hands, neck, and arm/shoulder. To complete a task, the sanding operator must pick up the sheet, place the sheet on the work bench, sand the sheet (which may involve either manual sanding or using a vibrating sander), and then pile the sheet for any further required operation (Figure 6). Since the sanding operation itself is a major task in the work station, it is selected as the main task for the purpose of ergonomic evaluation. The cycle time as well as the cycle rate of the sanding operation can vary depending on the size of the sheet. An average of 0.54cycles/minute and a 55% exposure rate to repetitive risk during activity are estimated based on video observations. The evaluation conditions for organizational risk factor, including mental stress, work rate difficulty, and worker's control over work are observed. Table 1, developed based on the study by Inyang and Al-Hussein [22], shows the risk classification.

Table 1. Risk range classification

Risk Score Range	Risk Classification
$R_S < 6$	Low risk
$6 <= R_S < 13$	Medium risk
$13 <= R_S < 15$	High risk

Figure 7, illustrates the development of R_S resulting from repetition risk over 480 minutes (eight working hours). This can assist job designers and managers to design the recovery cycle in a manner which mitigates the risk related to task repetition. Figure 10 provides an overview of the developed SD model for ergonomic evaluation resulting from repetition risks.

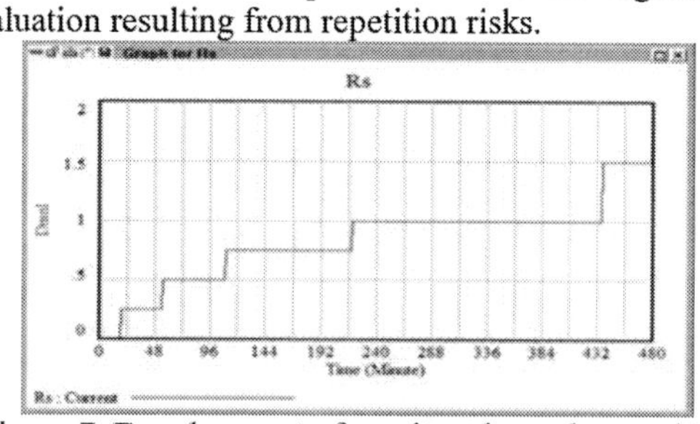

Figure 7. Development of resultant hazard score (Rs) over time

Exposure to vibration puts the hand and wrist of the operator at risk. To evaluate the impact of vibration risk, sanding operations using the vibrating sander are selected. Based on the tool vibration data, the vibration magnitude is 1.9 m/s^2, and video observations show an average of 0.56 cycle rate and 57% of exposure during activity. Figure 8 shows the resultant risk score for vibration risk factor in comparison to vibration exposure limit recommended by the American Conference of Governmental Industrial Hygienists [2]:

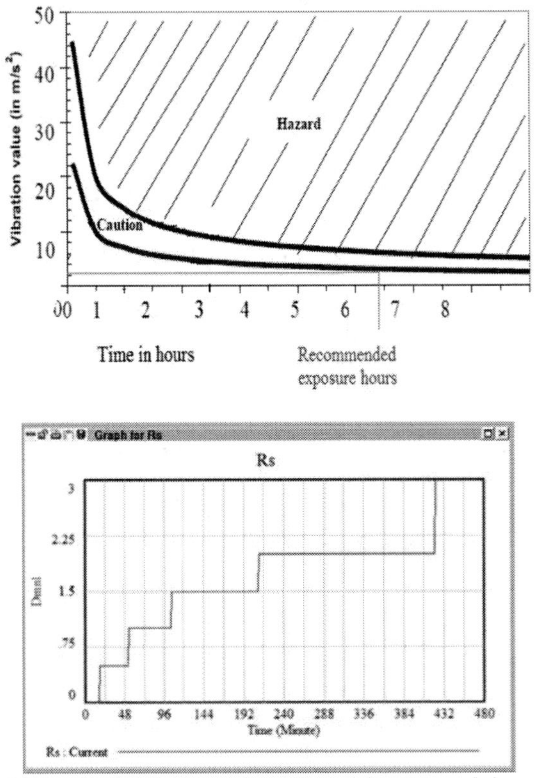

Figure 8. Comparison of resultant risk score and time hours recommendation by American conference of governmental industrial hygienists

A postural analysis is performed for major body parts involved in the task including neck, hand/wrist, and arm/shoulders. Considering the different ranges of movements observed through the operation as well as exposure duration to the posture, cycle rate and percentage of exposure, the resultant risk scores related to postural hazard for each body part are identified from the developed SD model. Table 2 summarizes the resultant risk score associated with each body part; showing the summary of daily exposure durations and risk classification for the neck, hand/wrist, and arm/shoulder. Workers are exposed for about 2.7 daily hours to 20-90° flexed neck posture, 3.16 hours hands and wrist flexion, and 4.56 hours of upper arm flexion > 45°. While methods such as REBA and RULA show risk scores of 3 and 4 respectively, representing low risk associated with the task, the proposed model clearly shows a breakdown of risk classification by body part. As illustrated in Figure 9, R_S resulting from neck posture shows a medium risk after 350 minutes while prior to that, the risk rate falls within the low risk category. Hand/wrist postural risk is classified as "medium" after 300 minutes (5 hr). Likewise, arm/shoulder shows a medium risk level after 211 minutes (3.5 hr). The postural risk for the hand/wrist is classified as "medium" after 300 minutes of work, while that of the arm/shoulder becomes medium after 211 minutes; also postural risk for the neck is classified as "medium" after 350 minutes of a task (5.8 hr of work). These findings can be useful for ergonomists and work designers, as they can modify the work to ensure low risk by controlling the exposure duration in order to secure the desired risk classification.

Table 2. Postural hazard assessment

Body part	Observed posture range	Daily exposure (8 working hr)	body part exposure %	Rhs	Risk Class
Neck	< 20	1.848	23.11%	1.5 after 264 min	Low
	20 - 90	2.706	33.83%	6 after 350 min	Medium
Hand & Wrist	0 - 15	3.163	39.55%	6 after 300 min	Medium
	> 15	1.395	17.45%	9 after 348 min	Medium
Arm & Shoulder	45 - 90	4.56	57%	6 after 211 min	Medium
				9 after 422 min	Medium

Table 3 summarizes the ergonomic risk scores and risk classification for the postural hazard, force/static load, hand/arm vibration, contact stress, repetition risk, and environmental risk factor. As it can be inferred from Table 3, workers are exposed to medium postural risk for the arm/shoulder, and hand/wrist. Ultimately it demonstrates an overall medium risk classification for the observed task.

Table 3. Resultant ergonomic risk scores

	Hazard Score							
Body part	Postural hazard	Force /static load risk	Hand/ arm vibration	Contact stress	Repetition risk	Environmental risk	Resultant body part risk score	Risk class
Neck	6						6	Low
Arm/ shoulder	9		3		1.5		9	Medium
Hand/ wrist	9		3				9	Medium

Based on the above results work can be rotated or changed within safe limits to ensure low exposure to ergonomic hazards.

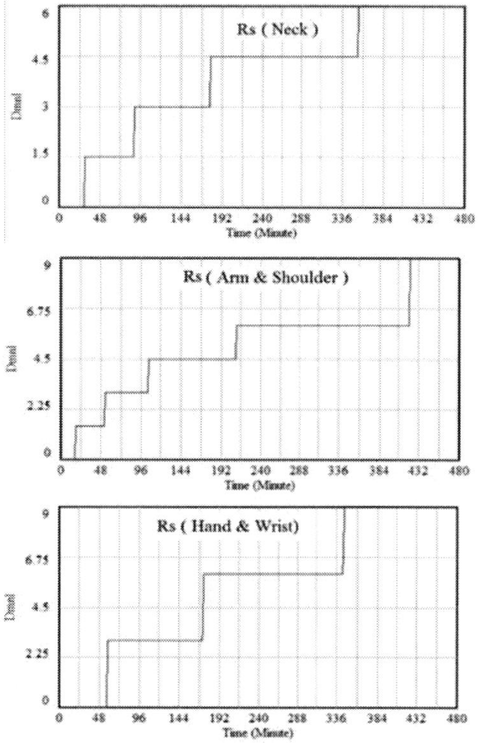

Figure 9. Resultant risk score of awkward posture including neck, arm/shoulders, and hand/wrist

5 Conclusion

The presented work shows an application of system dynamics (SD) to assess the ergonomic risk of residential construction tasks. The results obtained from the case study show that SD modeling not only can provide a graphical illustration, showing the logical links between cause and effects, but also increases the knowledge of how ergonomic risk can develop while performing task cycles. This can assist job designers and managers to have a clear understanding of possible ergonomic risks associated with the tasks as well as the major body parts affected. In order to address the ergonomic risks, recovery cycles should be designed in a way to avoid potential ergonomic risks. SD, as a powerful tool, can be used to explore several what-if scenarios caused by a range of variations in risk exposure, and can identify most contributing factors to potential ergonomic injury. While various factors, such as environmental and organizational, are typically assumed to be constant over time, they may be found to change as a result of their inter-relationships with other factors. Further investigation is thus recommended to address interactions among governing variables and their variations over time. The results identified in this study are expected to assist project work designers and ergonomists in controlling and assessing ergonomic risks, thereby leading to improved work efficiency, safety and reduced lost time injuries caused by WRMSDs.

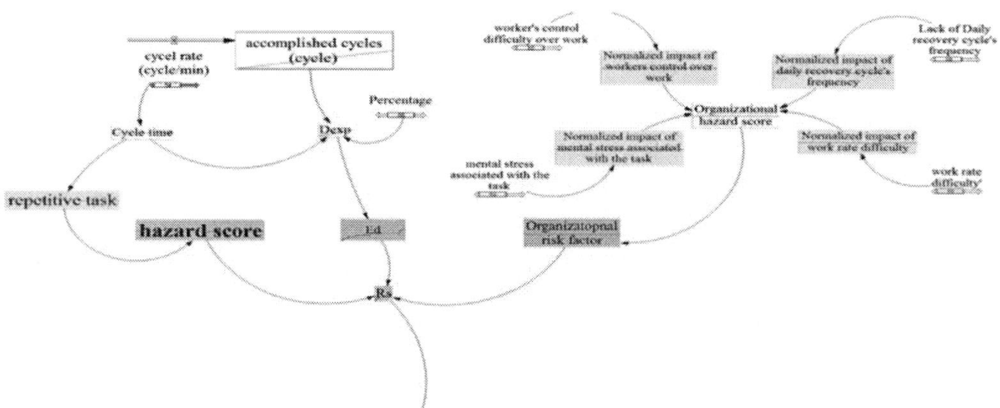

Figure 10, SD model of repetition risk factor

Acknowledgements

The authors are grateful to the management of Nava Cabinet Solutions for allowing their facility to be used as the model for this study.

References

[1] Bidiawati A. J and Suryani E "Improving the Work Position of Worker's Based on Quick Exposure Check Method to Reduce the Risk of Work Related Musculoskeletal Disorders". *Procedia Manufacturing*, 496-503, 2015.

[2] CCOHS "Canadian Centre for Occupational Health and safety". <Retrieved on Feb 2015 from http://www.ccohs.ca/oshanswers/ergonomics/>.

[3] McAtamney L. and Corlett E.N "RULA: a survey method for the investigation of work-related upper limb disorders. *Applied Ergonomics* , 91-99, 1993.

[4] Wang D. and Dai F. and Ning X "Risk Assessment of Work-Related Musculoskeletal disorders in construction: State-of-the-art review". *Journal of Construction Engineering and management*, ASCE, ISSN 0733-9364/04015008(15), 2015.

[5] Tompa E. and Dolinschi R. "A Systematic Review of Workplace Ergonomic Interventions with economic analysis". *Journal of occupation, rehabilitation* , 220-234, 2010.

[6] David G. and Woods V. and Li G. and Buckle P. "The development of the Quick Exposure Check (QEC) for assessing exposure to risk factors for work-related musculoskeletal disorders". *Applied Ergonomics* , 57-69, 2008.

[7] Yu H. and Al-Hussein M. and Nasiri R. "Process Flowcharting and Simulation of House Structure". In Proceedings of the 2007 Winter Simulation Conference, *2007.*

[8] SAID H. and ALI A.R. and ALSHEHRI M. "Analysis of the Growth Dynamics and Structure of Modular Building Construction Industry". *Construction Research Congress ASCE, 2014.*

[9] Li H.X. and Al-Hussein M. and Lei Z. and Ajweh Z. "Risk identification and assessment of modular construction utilizing Fuzzy Analytic Hierarchy Process (AHP) and Simulation". *Canadian Journal of Civil Engineering*, 2013.

[10] HRSD Canada, Human Resources and Skills development Canada. "*Injury Rates and Cost to the economy". 2010.*

[11] Chen J.G. and Jung H.S. and Peacock B.J. "A fuzzy sets modelling approach for ergonomic workload stress analysis". *International Journal of Industrial Ergonomics*, 189-216, 1994.

[12] McAtamney L. and Hignett S. "Rapid Entire Body Assessment (REBA)". *Applied Ergonomics* , 201-205, 2000.

[13] NRC (National Research Council). "*Advancing the Competitiveness and Efficiency of the U.S. Construction Industry".* The National Academies Press, Washington,D.C. , 2009.

[14] Karhu O. and Kansi P. and Kuorinka I. "Correcting working postures in industry: A practical method for analysis". *Applied ergonomics*, 199-201, 1977.

[15] Rathnapale D.G. "*Incorporating Prefabrication Process into Building Information Modelling".* thesis submitted to the University of Wolverhampton, 2009.

[16] Feyen R. and Liu Y. and Chan D. and Jimmerson G. and Joseph B. "Computer-aided ergonomics: a case study of incorporating ergonomic analyses into workplace desgin". *Applied Ergonomics* , 291-300, 2000.

[17] Sterman, J. D. "Systems dynamics modeling: Tools for learning in a complex world". *Engineering Management Review* , 30(1), 42-42, 2002.

[18] Vensim. "Reference Manual, Version 6.3E, Chapter 8 Building a Function with Lookups". Retrieved November 10, 2015, from online:https://www.vensim.com/documentation/index.html?vensim_help.htm.

[19] Richardson G.P. and Otto P. "Applications of system dynamics in marketing: Editorial". *Journal of Business Research*, 61, 1099–1101, 2008.

[20] Chuang K.W. "*Building system dynamics simulation model in support of ERP project implementation".* Purdue University ,West Lafayette, Indiana: PhD thesis, , 2011.

[21] Sonawane R. "*Applying system dynamics and critical chain methods to develop a modern construction project management system".* Texas A&M University-Kingsville. Master of science thesis,. 2004.

[22] Inyang N. and Al-Hussein M. "Ergonomic hazard quantification and rating of residential construction tasks". *CSCE 2011 General Conference*, (pp. 193-1-11), 2011.

[23] Bao S. and Kapellusch J.M. and Merryweather A.S. and Matthew S. "Relationships between job organizational factors, biomechanical and psychosocial exposures". Ergonomics, 1-17, 2015.

Coupling Wireless Sensor Networks and Unmanned Aerial Vehicles in Bridge Health Monitoring Systems

H. Izadi Moud[a] and M. Gheisari[a]

[a] M.E. Rinker, Sr. School of Construction Management, University of Florida, USA
E-mail: izadimoud@ufl.edu, masoud@ufl.edu

Abstract

Bridges are exposed to different types of damages within their lifetime. Bridge maintenance is an essential procedure that can significantly affect lifetime and serviceability of bridges. First step in bridge maintenance is inspection. Visual inspection has been traditionally the dominant method of bridge inspection. Although visual inspection is still used by several authorities to inspect bridges, it is prone to errors, due to human biased judgment. Using wireless sensor networks is a non-destructive method that can help bridge maintenance teams to monitor bridge health. Although there are many advantages associated with using wireless sensor networks in bridges, there are some downsides such as battery failure and large volume of collected and transmitted data. Unmanned aerial vehicle is a mobile robotic agent that can help wireless sensor networks to overcome some of its shortcomings. This paper initially explores the advantages and disadvantages of using wireless sensor networks in bridge health monitoring systems. Then, the advantages of using unmanned aerial vehicles in bridge health monitoring systems are discussed. And finally, this paper investigates possible synergies of coupling unmanned aerial vehicles and wireless sensor networks in bridge health monitoring systems that might help bridge health monitoring systems to overcome some of its traditional limitations. The contribution of this paper is to identify and discuss the possible synergies between unmanned aerial vehicles and wireless sensor networks in bridge health monitoring systems.

Keywords

Bridge Health Monitoring Systems; Unmanned Aerial Vehicles; UAV; Wireless Sensor Networks

1 Introduction

Human-made structures are exposed to different kinds of damages. Damages may happen due to the severe structure ambient factors, e.g. weather condition, which may frequently happen to the structure or due to the less frequent factors, e.g. natural disasters, which may or may not happen to the structure during its lifetime. In order to maintain an acceptable level of service for a structure, it is necessary to monitor the health condition of the structure and take the necessary actions to maintain its health – e.g. repairing damaged parts or redesigning. Ponte Fabricio Bridge, which was made in 62 BC in Rome, and still in service, is a good illustration of the importance of the structural maintenance. It shows that a good design accompanied with good maintenance service can expand the lifetime of a structure to more than two thousand years [1].

First step in maintenance is inspection. Amongst the numerous ways to inspect structure health, visual inspection, traditionally, has been the common method. However, by having massive structures – in terms of dimensions – in the twentieth century, mostly after the Second World War, visual inspection was not able to satisfy the requirements for structure health inspection. Cabled (wired) sensors replaced the visual inspection method gradually. Nevertheless, the visual inspection is remained as a common and, most importantly, low-priced method for structure health monitoring usually in small and medium scale structures [2].

Bridges, as highly loaded structures, also need proper inspection. Researchers have investigated the use of new technologies in bridge health monitoring systems (BHMS) for more than a decade. Pines and Aktan [3] highlighted the lack of accuracy in visual inspection. A study by Federal Highway Administration revealed that around 56% of the average condition ratings by visual inspection were incorrect [3]. Oh et al. [4] noted that the majority of systems that have been used for bridge health monitoring rely on manual processes of counting the cracks, measuring cracks' lengths and widths and using static images that have been taken from bridges. The reliability of these methods depends on inspectors experience in bridge inspection. Consequently, sensors have been used in bridge health monitoring to overcome drawbacks of visual inspections.

Using sensors in structural health monitoring instead of visual inspection has brought many advantages to structural maintenance. Although sensors are more accurate than visual inspections there are a few disadvantages associated with them. Sensor networks in structural health monitoring were cabled sensor networks and there was traditionally an important downside to the cabled sensors: the cost of the system. Since cabling and instrumentation in cabled sensors are expensive, gradually there was a trend to find a more economically reasonable method for structural health monitoring. In last decade, the wireless sensor technology became available and was used in structural health monitoring but the price was still high at that moment. Having lower-priced sensors with more powerful wireless technologies increased the interest in using wireless sensors. Not only lower prices make wireless sensors attractive in structural health monitoring, but also having less cabling and instrumentation makes them more favorable to construction industry. These factors combined with faster communication technology, fast and low-price computation technology, and a government regulated playground play a vital role in making wireless sensor technology promising to construction and facility managers [5-6].

Safety concerns, inaccessibility to some of bridge sections and the hardness of bridge inspection on heavy traffic situations arise the question of the possibility of bridge inspection without use of workforce. Metni and Hamel [7] discussed the use of unmanned aerial vehicles (UAVs) in bridge inspection. Metni and Hamel [7] mentioned that aging infrastructure is a constant concern to the societies in western European countries as it is estimated that half of the bridge life cost is due to its long lifetime maintenance cost. As Metni and Hamel [7] argued, use of drone in bridge inspection would have advantages over traditional inspection methods like visual inspection. UAVs bring many advantages in to bridge inspection process including reducing work accidents, budget reduction, less bridge lane closure and possibility of using other non-destructive crack detection techniques [7].

Although it has been more than a decade since Lynch et al. [8] introduced the concept of using wireless sensor systems in infrastructures, the construction industry still struggles to have a widely acceptable and implementable smart wireless sensor networks (WSN) for BHMS. This study presents a study of critical elements and downsides of WSN in BHMS. The use of UAVs in BHMS is also explored. A summary of previous applications and potential areas that UAVs can be coupled with WSN to provide a faster, safer, and more efficient BHMS is presented [Figure 1].

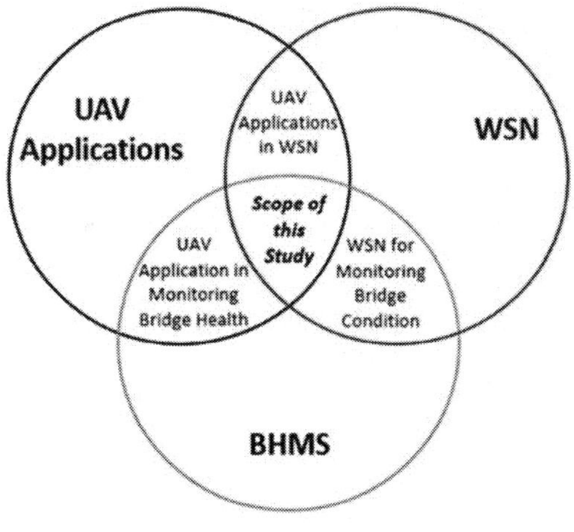

Figure 1. Scope of this study.

2 Wireless Sensor Networks in Bridge Health Monitoring Systems

There are three important elements in designing any BHMS: (1) identifying the inspection elements in bridge, (2) choosing a suitable sensor (hardware) and sensor networks platform, and (3) planning to overcome the common, and known, challenges that may be associated with selected sensor network platforms. This section focuses on these three aforementioned factors and discusses challenges associated with them. Moreover, a brief overview of general challenges associated with any WSN is also presented.

2.1 Identifying the Right Inspection Elements in Bridge Health Monitoring Systems

Finding the elements that need to be inspected is the first step in designing an inspection system for any kinds of infrastructure. In bridge health monitoring, recognizing the must-investigable elements plays a crucial role in designing a proper BHMS. In order to recognize bridge must-investigable elements, it is important to have a thorough look to the bridge inspection manuals. It is worth noting that although recognizing the must-investigable bridge elements is a crucial step in designing a BHMS, the ability of wireless sensor in monitoring any bridge element is still debatable.

NYDOT [9] breaks down bridges to different bridge components and discusses the process of inspection regarding specifications of each component. NYDOT [9] considers stream channel, scour and erosion, abutments and wing-walls, piers, deck and superstructure as the elements that need to be inspected. NYDPT [9] is not only consider bridge elements to be inspected but also focus on some features like stream channel, scour and erosion as the features which need to be inspected. The reason that there is a focus by NYDOT [9] on these environmental features is that environmental factors can seriously affect bridge health conditions. Environmental factors are explicitly highlighted in bridge inspection manuals. For instance, NYDOT [9] mentioned the importance of stream channel as *"the most dynamic system affecting the condition of bridges."*

Bridge inspection manuals help bridge inspectors to design the most appropriate BHMS for any bridge regarding its specifications, though it is not the only tool that guides to design a useful BHMS. Having a methodical view to the failure causes in failed bridges, assist construction managers to design effective BHMS [10].

Wardhana and Hadipriono [10] studied bridge failures in the United States from 1989 to 2000. Their study revealed that there were more than 500 reports of bridge failures in the United States from 1989 to 2000 [10]. The average age of failed bridges was 52.5 years with bridges that were in the construction phase to a 157 year-old one. Another finding by Wardhana and Hadipriono [10] study points out that steel beam/girder bridges are the dominant type of failed bridges with 29% of total failed bridges. The other common types of failures happens in steel truss, concrete beam/girder, and concrete slab bridges with 21%, 6%, and 5% of total failed bridges, respectively [10].

Most of failed bridges have steel or concrete structures. It is important to focus on structural materials of bridges in order to see what the critical factors are in inspection phase. It is believed that accelerations, temperature, wind, inclination and cable tension are among the most important structural-related factors in BHMS [11].

2.2 Sensor Platforms and Sensor Networks in BHMS

After identifying the inspection elements in BHMS, the appropriate sensor platform should be chosen. Numerous motes have been used in structural and bridge health monitoring. Lynch et al. [8] used WiMMS to prove the wireless system proof of concept in in structural health monitoring. MicaZ, Imote2 and some high sensitivity accelerometer board, such as SHM-DAQ board, have been also used in the area of structural and bridge health monitoring systems. Choosing the most suitable sensor network architecture is an important aspect of designing a BHMS, however discussing sensor platform and network architecture are out of the scope of this paper.

2.3 Challenges Associated with Using Wireless Sensor Networks in Bridge Health Monitoring Systems

Apart from the fact that choosing the appropriate sensor platform is important for BHMS, sensors energy dependency should also be appropriated chosen in any WSN [12]. Yoo et al. [12] argued that one of the most important issues in deploying wireless sensors in structures is energy limitation due to the high-energy consumption nature of the wireless sensors. Mainly because of huge portion of energy needed to transmit the data to the base station - hub. In order to overcome high-energy consumption nature of wireless sensors, Yoo et al. [12] proposed that wireless sensors should be more energy dependable. In other words, wireless sensors should be able to work with their energy supplies – batteries – for a reasonable period of time or they should use undependable, and potentially more sustainable, rechargeable batteries.

Not only high energy consumption nature of wireless sensors may act as an obstacle in the process of sending and receiving the data between base station and sensors, Yoo et al [12] indicated that traffic overload caused by huge volume of transferred data between base station and sensors can also lead to a problem in wireless sensor networks of structural health monitoring systems. Power management and data transfer problems in WSN are not the only downsides of implementing WSN in BHMS. Hoult et al. [13] identifies other major downsides of WSN such as data transmission bandwidth, reliability, and time synchronization. Another challenge indicated in the literature was wireless networks' incapability to achieve real time transmission of data; however, this can be achieved by lowering sample rates [13-14].

3 Unmanned Aerial Vehicles and Wireless Sensor Networks

Investigating the possibility of using UAVs in sensor networks appeared for the first time, in 2001, in a white paper by United States Department of Energy (USDOE). In 2001, USDOE investigated the feasibility of using cooperative UAV-based communication as a backbone for sensor networks. The feasibility of using UAVs to provide communication connectivity to the sensors that cannot communicate with each other due to distance and geographical constraints was the objective of this white paper. Since the focus was on the cooperative UAV-base communication, authors of USDOE [15] tried to explore whether or not a group of UAVs can possibly divide a ground-based sensor network into a few sub-networks (subnets) and help the sensor networks by uplinking sensors-collected data and transmit to the base ground station.

But the application of UAVs in WSN is not limited to relaying features. UAVs can act as mobile base station, which can improve the WSN status in terms of data traffic load balancing and energy consumptions. Also UAVs can act as mobile sensing nodes, which can potentially act as extra mobile nodes augmenting the network by bringing additional information [16]. Teh et al., [16] argues that a combination of UAVs and WSN may result in unique applications such as:

1. Remote data mulling: UAVs act as data agents, which can upload data to the network and/or send the data back to base stations.
2. 3D sampling of sensor data for urban pollution monitoring: Consider having ground-deployed WSN in an area, UAVs can potentially provide three-dimensional sampling, which can be used for further analysis.
3. Traffic monitoring

The applications of UAVs in WSN are not limited to the above-mentioned applications. Corke et al. [17] used UAVs to deploy sensors on specified ground locations. This application improves the robustness in designing any WSNs. It brings the flexibility of flying UAVs to any location to design more flexible and location-independent WSNs. However, this application is only a concept and fully autonomous use of UAVs for sensor deployment has not been fully applicable due to the current UAVs limitation on carrying loads to far distances.

4 Unmanned Aerial Vehicles Applications in Bridge Health Monitoring Systems

UAVs have been conventionally used for enhancing visual inspection of bridges by providing detailed images of hard-to-reach areas of bridges. Although taking photos and recording videos for post-inspection uses are some of the traditional uses of UAVs in BHMS, other visual uses of UAVs in BHMS like developing 3D models of existent bridges has also been investigated [18].

It is argued that using UAVs has some advantages compare to the traditional bridge inspection methods; including reducing work accident risk, and ultimately increasing safety, budget reduction, robustness and possibility of using other non-destructive methods [7]. UAVs have also been used for transmitting wireless power to wireless sensor nodes, installed in a bridge [19].

Although numerous advantages of using UAVs for bridge inspection purposes have been counted, there are some disadvantages associated with using UAVs in bridge inspection systems. Wind gusts and turbulences, no-fly zones, hazardous zones and difficulty in path planning are some of the issues that need to be considered in using UAVs for bridge inspection purposes [20].

5 Discussion and Conclusion

Traditionally, infrastructures health conditions have been monitored with visual inspection and also other forms of non-destructive inspection method including cabled (wired) sensors. Due to many disadvantages of wired sensors, most importantly high cost, using wireless sensors in structural health monitoring is becoming a trend nowadays. Numerous elements have been measured regarding the infrastructure health monitoring; including accelerations, and temperature as the most common sensed elements. Choosing the right elements to measure in a structure depends on the necessity of measuring these elements in terms of their effects on structural lifetime health. Despite the effort that have been placed in implementing vibration-based WSN in bridge health monitoring, some important elements, which are already highly emphasized in bridge inspection manuals, have been forgotten to take into account in the process of implementing WSN in bridge health monitoring. Wardhana and Hadipriono [10] study of failed bridges in the United States revealed that 53% of total causes of bridge failures are associated with flood/scour. It raises the importance of bridges environmental condition monitoring. Deploying WSN in order to measure some of these ambient factors, such as level of water, which might contribute to predicting flood and scour, could be a novel approach in measuring bridges ambient factors. UAVs can be used in order to deploy sensors in bridge surrounding environment to measure bridge ambient factors. UAVs might be good agents for deploying wireless sensors in bridge surrounding environment due to flexibility that they provide for deployments.

Although WSN have been implemented and used for BHMS purposes, there have been some technical issues concerning widespread use of them including: time synchronization and sensors energy dependency. Energy dependency and sensors' battery lifetime is one of the most critical issues in WSN. It is been concluded that UAVs can help to solve sensor battery problems. Lifetime of wireless sensors are dependent on sensors' battery lifetime. Wireless sensors batteries need to be charged after a while if network administrator wants to keep sensors working. Charging wireless nodes may not be a problem in simple scenario but recharging sensors batteries that are deployed in hard-to-reach environment, such as some parts of bridges that are hardly accessible by maintenance team, makes it hard and expensive for network administrator to recharge wireless sensors. UAVs can act as power transfer agents to (re) charge the wireless sensors. The concept of using wireless power to supply energy to another agent goes back to 1964, when it has been tried to supply a flying helicopter with wireless power. Griffin and Detweiler [21] and Johnson et al. [19] proved the concept of wireless magnetic resonant power transfer to wireless nodes in a WSN. BHMS can use the mobility of UAVs and use UAVs as mobile battery chargers in order to charge sensor batteries in WSN in bridges.

Wireless sensor failure is another common phenomena that might happen in any WSN. Sensors (nodes) can fail due to any reason including battery or transmitter failure. Wireless network planners usually deploy more number of sensors than the estimated number of needed sensors in order to avoid the effects of any future sensor failure. Node failure could potentially slow down the data transfer in the network [6]. UAVs can be used to deploy wireless sensors in any area in bridges that are hardly accessible by humans.

Data transfer is another critical issue in WSN. Volume of data might be large, which makes receiving live data hard. UAVs can be potentially used as temporary hubs to ease the data transfer in WSN.

References

[1] O'Connor, Colin. *Roman bridges*. Cambridge University Press, 1993.

[2] Lynch, Jerome Peter. "An overview of wireless structural health monitoring for civil structures." *Philosophical Transactions of the Royal Society of London A: Mathematical, Physical and Engineering Sciences* 365.1851: 345-372, 2007.

[3] Pines, Darryll, and A. Emin Aktan. "Status of structural health monitoring of long-span bridges in the United States." *Progress in Structural Engineering and materials* 4.4: 372-380, 2002.

[4] Oh, Je-Keun, et al. "Bridge inspection robot system with machine vision. "*Automation in Construction* 18.7: 929-941, 2009.

[5] Townsend, Chris, and Steven Arms. "Wireless Sensor Networks."*MicroStrain, Inc,* 2005.

[6] Akyildiz, Ian F., and Mehmet Can Vuran. *Wireless sensor networks*. Vol. 4. John Wiley & Sons, 2010.

[7] Metni, Najib, and Tarek Hamel. "A UAV for bridge inspection: Visual servoing control law with orientation limits." *Automation in construction* 17.1: 3-10, 2007.

[8] Lynch, Jerome P., et al. "Field validation of a wireless structural monitoring system on the Alamosa Canyon Bridge." *Smart structures and materials*. International Society for Optics and Photonics, 2003.

[9] NYDOT (New York State Department of Transportation). Bridge inspection manual. 2014.

[10] Wardhana, Kumalasari, and Fabian C. Hadipriono. "Analysis of recent bridge failures in the United States." *Journal of Performance of Constructed Facilities* 17.3: 144-150, 2003.

[11] Pakzad, Shamim N., et al. "Design and implementation of scalable wireless sensor network for structural monitoring." *Journal of infrastructure systems*14.1: 89-101, 2008.

[12] Yoo, H. S., J. R. Kim, and M. Y. Cho. "BRIDGE CONDITION MONITORING SYSTEM USING WIRELESS NETWORK (CDMA AND ZIGBEE)."

[13] Hoult, Neil A., et al. "Long-term wireless structural health monitoring of the Ferriby Road Bridge." Journal of Bridge Engineering 15.2: 153-159, 2010.

[14] Whelan, Matthew J., et al. "Development of a wireless bridge monitoring system for condition assessment using hybrid techniques." The 14th International Symposium on: Smart Structures and Materials & Non-destructive Evaluation and Health Monitoring. International Society for Optics and Photonics, 2007.

[15] USDOE (United States Department of Energy), white paper on wireless power transfer. 2001.

[16] Teh, Steven K., et al. "Experiments in integrating autonomous uninhabited aerial vehicles (UAVs) and wireless sensor networks." (2008).

[17] Corke, Peter, et al. "Autonomous deployment and repair of a sensor network using an unmanned aerial vehicle." *Robotics and Automation, 2004. Proceedings. ICRA'04. 2004 IEEE International Conference on*. Vol. 4. IEEE, 2004.

[18] Hallermann, Norman, and Guido Morgenthal. "Visual inspection strategies for large bridges using Unmanned Aerial Vehicles (UAV)." *7th International Conference on Bridge Maintenance, Safety and Management, IABMAS 2014, July 7, 2014-July 11*. 2014.

[19] Johnson, Jamie, Elizabeth Basha, and Carrick Detweiler. "Charge selection algorithms for maximizing sensor network life with UAV-based limited wireless recharging." *Intelligent Sensors, Sensor Networks and Information Processing, 2013 IEEE Eighth International Conference on*. IEEE, 2013.

[20] Guerrero, Jose Alfredo, and Yasmina Bestaoui. "UAV path planning for structure inspection in windy environments." *Journal of Intelligent & Robotic Systems* 69.1-4: 297-311, 2013.

[21] Griffin, Brent, and Carrick Detweiler. "Resonant

wireless power transfer to ground sensors from a UAV." *Robotics and Automation (ICRA), 2012 IEEE International Conference on*. IEEE, 2012.

The Impact of Eco-Feedback on Energy Consumption Behavior: A Cross-Cultural Study

Jing Lin [a], Nan Li [b], Guofeng Ma [a] and Junjie Zhou [a]

[a] Department of Construction Management and Real Estate, Tongji University, Shanghai 200092, China
[b] Department of Construction Management, Tsinghua University, Beijing 100084, China
E-mail: 5_linjing_jocelyn@tongji.edu.cn, nanli@tsinghua.edu.cn, guofengma@tongji.edu.cn, 1432701@tongji.edu.cn

Abstract –

To reduce energy consumption in buildings, researchers have in the recent decade explored the potential of changing occupants' energy consumption behaviors using eco-feedback technologies. Energy consumption behavior is a type of consumer behavior, which has been proven cultural-specific in prior research. This paper aims to examine the impact of cultural differences on the effectiveness of eco-feedback technologies in reshaping building occupants' energy consumption behaviors. Energy consumption monitoring instruments were installed in four student dormitories, three resided by Chinese students and one by international students. The daily energy consumption of every unit in the dormitories was monitored. A web-based eco-feedback system was developed, which was responsible for sending weekly email reminders to students participating in the study and, upon their logins, providing them with their detailed energy consumption data as well as those of their peers. Changes in the students' energy consumption behaviors were analyzed, and correlated with their cultural background, which was assessed using a survey instrument designed based on Hofstede's dimensions of culture. Results proved that cultural background assessed in certain dimensions were significantly correlated with the effectiveness of eco-feedback technologies. The results suggested that eco-feedback technologies should be tailored to specific cultural context to maximize their effectiveness in building energy conservation.

Keywords –

Cross-cultural; Behavior; Eco-feedback; Energy consumption

1 Introduction

Global carbon dioxide (CO_2) emissions have been continuously increasing in the past decades [1], 11% of which was accounted by electricity and heat production consumed in the residential sector [2]. Thus, reducing residential energy consumption can significantly reducing global CO_2 emissions. Many energy saving technologies, such as energy-efficient lighting and thermal insulation materials, have been developed in recent years to improve residential energy efficiency. However, as these technologies are adopted, a "take-back" effect is observed. This effect refers to the phenomenon that effectiveness of the energy-saving technologies decreases due to inefficient energy consumption behaviors by users [3]. Therefore, more recent research has indicated that changing users' energy consumption behaviors is of significant importance in achieving long-term energy conservation [4]. Based on the premise that awareness of actual consumption will motivate energy conservation behavior of occupants [5], many researchers have studied eco-feedback systems which, designed to monitor and present occupants' energy consumption data [6], are believed to have the potential of changing occupant's energy consumption behaviors. The effectiveness of eco-feedback systems has been validated in a number of studies. However, the effectiveness is observed to vary across different settings [7,8], and is reportedly affected by factors such as interface design [9], information representation [7], social influence [10], and so on. One factor that is potentially important but has yet to be examined in the literature is the culture context. Energy consumption behavior by its nature is a type of consumer behavior, which has been proven culture specific in prior research [11]. This paper aims to examine whether the effectiveness of an eco-feedback system is impacted by the cultural background of building occupants whose behaviors the system aims to change. The remainder of this paper is organized as follows: section 2 presents related research, followed by section 3 that introduces the settings of an eco-feedback system developed in this study. Section 4 describes the design and implementation of an eco-feedback experiment. Section 5 then presents findings of this study. Section 6 concludes the paper.

2 Related Research

2.1 Factors Influencing the Effectiveness of Eco-Feedback Systems

The use of eco-feedback system has been reported in prior research to result in reductions in energy consumption that varied between -18% and 55% [7,8]. Such wide range has motivated many researches that focus on identifying and assessing factors that may influence the effectiveness of eco-feedback system, and developing a more effective way of designing and implementing eco-feedback systems. Peschiera et al. [12] reported in their study that weekly email reminders would cause response–relapse behavioral pattern that were impactful on the energy-saving effectiveness. They reported significant energy conservation when test subjects received weekly email reminders. However, their energy consumption quickly relapsed to pre-study levels after they stopped receiving the reminders. The format of information representation could also impact the effectiveness of eco-feedback systems [7]. Another way to improve the effectiveness of eco-feedback is to include users' energy consumption history and their peers' energy consumption data, which could serve to motivate energy-saving behaviors [9,10]. Computerized methods are considered the most effective for delivering eco-feedback information [13]. For instance, interface websites could be used to enable better interactivity of eco-feedback than traditional energy bills, and emails could be used to enable higher frequency of the eco-feedback information being read by energy users. These two methods of delivering eco-feedback information could also be combined to further improve the effectiveness of eco-feedback systems [9].

One factor that has not been thoroughly examined is the characteristics of energy users. When studying the effectiveness of in-home-displays, a typical type of eco-feedback system, Van Dam et al. [14] argued that "a one-size-fits-all approach for in-home-displays cannot be justified", which highlights that the performance of eco-feedback systems is dependent on the characteristics of the users. In fact, it has been reported that not all consumers have the potential of saving energy consumption by using eco-feedback system [15]. Therefore, to make eco-feedback systems work, it is necessary to consider the characteristics of their users, for instance, their cultural background, and to provide a certain level of customization.

2.2 Cross-Cultural Study of Consumer Behaviors

It is a proven theory that culture "is a determinant of certain aspects of consumer behavior " [16]. Culture can impact consumer behaviors through values, heroes, rituals, and symbols [17]. In addition, there are factors that can affect both culture and consumer behavior. Examples include personality, social processes and mental processes [16].

There are two types of approaches to analyzing culture, including etic approaches and emic approaches [18]. Etic approaches define culture as "the collective programming of the mind which distinguishes the members of one group or category of people from another" [19]. This definition emphasizes the comparison among different cultures. Emic approaches, to the contrary, focus on understanding within the culture, where culture is defined as "the 'lens' through which all phenomena are seen". It determines how these phenomena are apprehended and assimilated, and considers values the central views of culture. Hofstede's cultural dimensions model, which adopts the etic approach, is usually used to measure the cultural values for cross-culture comparison purpose [19].

3 Eco-Feedback System

An eco-feedback system was developed for this study. The system consisted of two components, including a data capture component and a data processing and delivery component, as demonstrated in Figure 1. The data capture component included electric meters, concentrators, cables and two servers. Electric meters, each responsible for monitoring one unit in a building, were connected to a concentrator through cables in every building. The concentrators reported to a server to upload the latest readings of energy consumption in a MySQL database, where the data were analyzed and prepared for delivery to the occupants. The delivery component was composed of an interface website that allowed for online checking of current eco-feedback, and an email portal for sending automatic weekly emails to the occupants reminding them of checking eco-feedback information through the website. The following information was presented to the occupants once they logged into the website: (1) an overview of their energy consumption of the previous day, the last seven days and the last thirty days, all measured in kWh; (2) a list of navigation options, including reviewing the charts of daily consumption of the last seven days and thirty days, changing display language, and changing account settings; (3) charts of the users' daily energy consumption of the last seven days and thirty days, as well as of the average of their peers whose rooms were on the same floor and with the same orientation (south or north); (4) contact information of the research team for assistance and feedback.

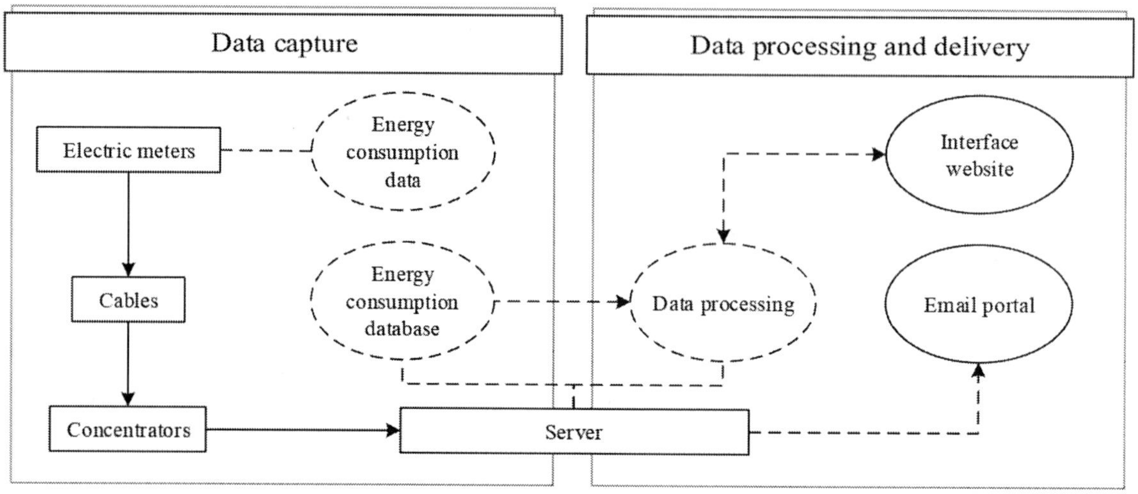

Figure 1. Architecture of the eco-feedback system developed in this study

4 Cross-Cultural Eco-Feedback Experiment

4.1 Test-bed Buildings and Participants

For empirically comparing the effectiveness of eco-feedback system in changing the energy consumption behaviors of occupants having different cultural backgrounds, an experiment was carried out in this study. The test-bed buildings used in the experiment were four six-story dormitories on the campus of Tongji University in Shanghai, as shown in Figure 2. Each unit in the dormitories had an area that varied between $8m^2$ and $13.77m^2$, included a bathroom, a bedroom and a balcony, and could accommodate one or two students. Chinese students lived in buildings 1, 2 and 3, and international students lived in building 4. All units involved in the study had stand-alone air conditioning units installed during the experiment period.

Figure 2. Test-bed Building

A total of 39 occupants in the test-bed buildings participated in the experiment, including 10 Chinese students and 29 international students who were from Germany (16), Italy (5), France (2), USA (1), Ethiopia

(1), Montenegro (1), Portugal (1), Sweden (1) and Switzerland (1). These participants all lived in single units, so that each electricity meter was associated with one participant. All other single units in the test-bed buildings were also monitored by the eco-feedback system, and their occupants, though not directly invited to participant in the experiment, were considered in control groups. Participants were recruited with face-to-face interviews, during which they were explained the purpose of the study, informed that they would receive weekly emails to remind them of checking the energy consumption information through the website, and provided with an individual login account to the website. Participants filled in a questionnaire about their personal information, such as email address, gender, education, age and so on. Gift cards were given to thank them for participating in the study. All single unit occupants in the test-bed buildings were divided into the following four groups:

- Study Group A (SG A) – Chinese occupants invited to participate in the experiment and provided with online eco-feedback and weekly email reminder.
- Study Group B (SG B) – International occupants invited to participate in the experiment and provided with online eco-feedback and weekly email reminder.
- Control Group A (CG A) – Chinese occupants having no access to eco-feedback.
- Control Group B (CG B) – International occupants having no access to eco-feedback.

The experiment lasted for 13 weeks from October 17 , 2015 to January 15, 2016. It included two periods, a pre-study period and a study period. The pre-study period lasted for 6 weeks from October 17 to November 27, 2015, before the participants were recruited. Daily energy consumption data were collected during this period and used for offsetting the effect of inherent difference between groups. The study period lasted for 7 weeks from November 28, 2015 to January 15, 2016, and daily energy consumption data were collected during this period for analyzing the effectiveness of eco-feedback system upon study groups. During the study period, participants received emails every Monday reminding them of logging into the website to check their the eco-feedback information.

4.2 Six Dimensions of Culture

The Hofstede's cultural dimensions model was applied in this study to compare participants' cultural background. Hofstede defined six dimensions of national culture, including Power Distance Index (PDI), Individualism vs. Collectivism (IDV), Masculinity vs. Femininity (MAS), Uncertainty Avoidance Index (UAI), Long Term Orientation versus Short Term Normative Orientation (LTO), and Indulgence vs. Restraint (IND) [17]. PDI represents the degree to which the less powerful members of a society accept and expect that power is distributed unequally. The fundamental issue here is how a society handles inequalities among people. IDV represents the contrast between individualism, a preference for a loosely-knit social framework in which individuals are expected to take care of only themselves and their immediate families, and collectivism, a preference for a tightly-knit framework in society in which individuals can expect their relatives or members of a particular in-group to look after them in exchange for unquestioning loyalty. MAS represents the contrast between masculinity, a preference in society for achievement, heroism, assertiveness and material rewards for success, and femininity, a preference for cooperation, modesty, caring for the weak and quality of life. UAI represents the degree to which the members of a society feel uncomfortable with uncertainty and ambiguity. Societies exhibiting strong UAI tend to maintain rigid codes of belief and behaviors and are intolerant of unorthodox behavior and ideas. LTO represents the degree of consideration of the future when a society deals with challenges. IND represents the attitude of a society in meeting needs of life, indulgence or restraint.

The VSM 2013 survey instrument was employed in this study to assess the cultural dimensions of the experiment participants [20]. This survey instrument is a 30-item paper-and-pencil questionnaire developed for comparing culturally influenced values and sentiments of similar respondents from two or more countries, or sometimes regions within countries [21]. For each of the six dimensions of culture as defined by Hofstede [17], the survey presents four relevant questions, and provides a technique for computing a score for the dimension based on answers of survey respondents. Six additional questions are about demographic information of the respondents, including their gender, age, education level, kind of job, present nationality, and nationality at birth. All 39 participants were surveyed in this study. Scores of each participant in each cultural dimension were calculated, then averaged per nationality, and related to the participants' behavioral responses to the eco-feedback system, in order to examine the correlation between each cultural dimension and the effectiveness of the eco-feedback system. Details of the analysis and findings are presented in the next section.

5 Findings

A total of 20,304 daily energy usage data points were collected, including 897 data points for SG A, 2,745 data points for SG B, 10,029 data points for CG A, and 6,633 data points for CG B. The statistical method used for measuring the effectiveness of eco-feedback was similar to the one used in prior research [7]. The statistical difference in energy consumption resulting from the eco-feedback information, which was an indicator of the effectiveness of the eco-feedback system, was calculated as follows:

$$\delta_{pre-study} = \frac{\sum_{i=1}^{n}((P-C)/C)}{n} \quad (1)$$

$$\delta_{study} = \frac{P-C}{C} \quad (2)$$

$$\Delta_{consumption}(\%) = \delta_{study} - \delta_{pre-study} \quad (3)$$

where P denotes the energy consumption of a given room on a given day, C denotes the average daily consumption per unit in CG, $\Delta_{consumption}$ denotes the statistical changes of a SG in energy consumption, relative to the corresponding CG, between the pre-study period and the study period, $\delta_{pre-study}$ denotes the average percentage difference between P and C for the pre-study period, δ_{study} denotes the percentage difference between P and C for the study period, and n denotes the number of days of the pre-study period..

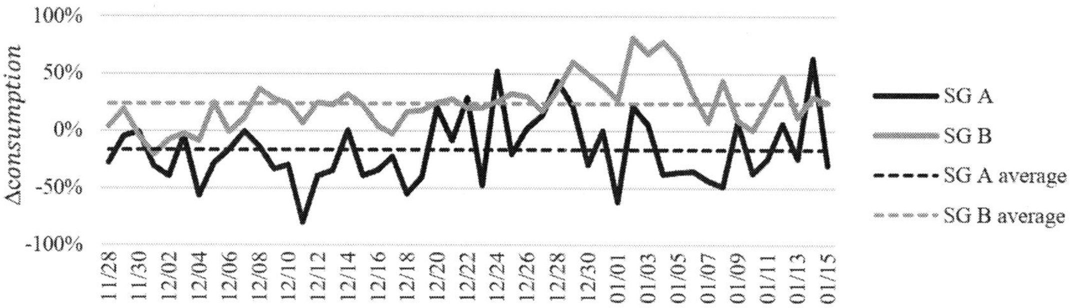

Figure 3. Plot of changes in daily energy consumption between SG A and SG B

The effectiveness of the eco-feedback system was thus calculated for both SG A and SG B, and these two groups were compared. Figure 3 shows the changes in consumption relative to the corresponding control group ($\Delta_{consumption}$) for each day during the study period for both SG A and SG B. As can be seen in Table 1, SG A reduced their energy consumption by 16.7% after using the eco-feedback system, whereas SG B increased their energy consumption by 23.7% during the same period. The results suggested that cultural background significantly influenced the effectiveness of eco-feedback system, which was also supported by a t-test carried out between these two groups, with a p-value of 0.000, rejecting the null hypothesis that SG A and SG B had the same behavioral response to the eco-feedback system. It need to be noted that it was interesting to observe an increase in the average energy consumption of international students after the eco-feedback system was implemented. A possible explanation was that, unlike Chinese students, energy bills of international students were paid by the university, so it was difficult to motivate them to save energy by simply presenting the eco-feedback information.

Table 1 Results of comparison between SG A and SG B

$\Delta_{consumption}$	SG A	SG B
Mean	-16.7%	23.7%
Standard deviation	1.403	0.772
p-value	0.000	

Additional analysis was performed to examine which dimensions of culture had impact on the effectiveness of eco-feedback. All 39 participants were divided into ten groups based on their nationality, and Kendall's tau-b correlation analysis was conducted between the groups' average scores in each cultural dimension and their average $\Delta_{consumption}$ values over the study period. At a 95% confidence level, a p-value less than 0.05 indicates statistically significant correlation. For influential cultural dimensions, correlation coefficient represented the extent of the influence. Table 2 shows the results of correlation analysis between the changes in energy consumption behaviors and six cultural dimensions.

Table 2 Correlation between the changes in energy consumption behaviors and cultural dimensions

Cultural dimension	PDI	IDV	MAS	UAI	LTO	IVR
Correlation coefficient with $\Delta_{consumption}$	-.231	.211	.254	.186	-.063	-.138
Sig.	.000	.000	.000	.000	.054	.000

The results indicated significant correlation between five cultural dimensions (PDI, IDV, MAS, UAI and IVR) and the effectiveness of eco-feedback. Specifically, PDI and IVR had negative correlation, which suggested that occupants in societies exhibiting a large degree of power distance who accept a hierarchical order in which everybody has a place and which needs no further justification would more positively respond to eco-feedback systems. The results also suggested that occupants in societies where people tend to accept a hierarchical order in which everybody has a place and which needs no further justification would more likely reduce energy consumption when provided with eco-feedback information. Furthermore, the results suggested that occupants in societies with low IVR scores, where people are less inclined to satisfy their needs and enjoy life, would behave with more restraints and conserve more energy when interacting with the eco-feedback system. In addition, IDV, MAS and UAI had positive correlation with $\Delta_{consumption}$. Occupants in societies with collectivist culture would respond to eco-feedback information, which contains a comparison with their peers, more positively, as by definition collectivist occupants care about not only themselves but also groups, so they would like to adjust their energy consumption behaviors as a gesture to contribute to the environment shared by others. High scores in MAS represent the society at large is more competitive and preference for success, which may decrease concern of occupants about energy. Conversely, feminine culture making society at large more consensus oriented cares the quality of life and concerns more about energy. Occupants with high scores in UAI would try to control the future and any uncertainty. They feel strongly uncertain about energy consumption without eco-feedback and will take actions to control and save energy consumption. Therefore, when they receive eco-feedback, they lose the incentive to save energy because they are sure about their energy consumption. As there were 6 countries only including one participant in the experiment, the behavior and cultural background may be influenced by individual characteristics. We acknowledge that our study could have benefited from a larger sample size. However, current sample size was adequate to obtain statistically significant results.

6 Conclusion and Implications

This study examines the impact of culture on the effectiveness of eco-feedback systems. An eco-feedback system was developed and implemented in test-bed buildings, and an experiment was conducted among students from ten different countries. By analyzing the energy consumption behavioral changes of the participants and assessing the cultural characteristics, examined in six dimensions, of participants from each country, the correlation between cultural dimensions and the effectiveness of eco-feedback systems was analyzed. The results showed that Chinese students and international students exhibited noticeably different behavioral responses to the eco-feedback system, and that five cultural dimensions, including PDI, IDV, MAS, UAI and IVR, had statistically significant impact on the effectiveness of the eco-feedback system. The results implied that the performance of eco-feedback systems is dependent on the cultural background of the users, and that the systems require a certain degree of adaptation to the cultural context in which they are implemented, in order to improve their effectiveness in building energy conservation.

References

[1] BP. BP Statistical Review of World Energy 2015. On-line: http://bp.com, Accessed: 20/01/2016.

[2] Statistics, I. E. A. CO2 emissions from fuel combustion-highlights. On-line: http://www.iea.org, Accessed: 20/01/2016.

[3] Haas R., Auer H., and Biermayr P. The impact of consumer behavior on residential energy demand for space heating. *Energy and Buildings*, 27(2): 195–205, 1998.

[4] Gardner G. T. and Stern P. C. The short list: The most effective actions US households can take to curb climate change. *Environment: Science and policy for sustainable development*, 50(5): 12–25, 2008.

[5] Attari S. Z., DeKay M. L., Davidson C. I., and De Bruin W. B. Public perceptions of energy consumption and savings. *Proceedings of the National Academy of sciences*, 107(37): 16054–16059, 2010.

[6] Jain R. K. *Building Eco-Informatics: Examining the Dynamics of Eco-Feedback Design and Peer Networks to Achieve Sustainable Reductions in Energy Consumption*, , Columbia University, New York, 2013.

[7] Jain R. K., Taylor J. E., and Culligan P. J. Investigating the impact eco-feedback information representation has on building occupant energy consumption behavior and savings. *Energy and Buildings*, 64: 408–414, 2013.

[8] Petersen J. E., Shunturov V., Janda K., Platt G., and Weinberger K. Dormitory residents reduce electricity consumption when exposed to real-time visual feedback and incentives. *International Journal of Sustainability in Higher Education*, 8(1): 16–33, 2007.

[9] Jain R. K., Taylor J. E., and Peschiera G. Assessing eco-feedback interface usage and design to drive

energy efficiency in buildings. *Energy and Buildings*, 48: 8–17, 2012.
[10] Jain R. K., Gulbinas R., Taylor J. E., and Culligan P. J. Can social influence drive energy savings? Detecting the impact ofsocial influence on the energy consumption behavior of networkedusers exposed to normative eco-feedback. *Energy and Buildings*, 66: 119–127, 2013.
[11] Bedwell B. and Shao J. C-Aware: Introduction to a Case-Study of Cross-Cultural Participatory Design of Persuasive Visualizations. In *Workshop on Transnational HCI at CHI 2011*, Vancouver, Canada, 2011.
[12] J Peschiera G., Taylor J. E., and Siegel J. A. Response–relapse patterns of building occupant electricity consumption following exposure to personal, contextualized and occupant peer network utilization data. *Energy and Buildings*, 42(8): 1329–1336, 2010.
[13] Heras D. B., Otero D., and Argüello F. An eco-feedback system for improving the sustainability performance of universities. In *Virtual Environments Human-Computer Interfaces and Measurement Systems (VECIMS)*, pages 1–6, Ottawa, Canada, 2011.
[14] Van Dam S. S., Bakker C. A., and Van Hal J. D. M. Home energy monitors: impact over the medium-term. *Building Research & Information*, 38(5): 458–469, 2010.
[15] Buchanan K., Russo R., and Anderson B. The question of energy reduction: The problem (s) with feedback. *Energy Policy*, 77: 89–96, 2015.
[16] Henry W. A. Cultural Values Do Correlate with Consumer Behavior. *Journal of marketing research*, 121-127, 1976.
[17] Hofstede G., Hofstede G. J., and Minkov M. *Cultures and organizations: software of the mind: intercultural cooperation and its importance for survival*. McGraw-Hill, New York, 2010.
[18] Luna D. and Forquer Gupta S. An integrative framework for cross - cultural consumer behavior. *International Marketing Review*, 18(1): 45–69, 2001.
[19] Hofstede G. *Cultures and organizations: Software of the Mind*. McGraw-Hill, New York, 1997.
[20] Hofstede G., Hofstede G. J., Minkov M., and Vinken H. Values survey module 2013. On-line: http//www.geerthofstede.nl/vsm2013, Accessed: 30/11/2015.
[21] Hofstede G. and Minkov M. Values Survey Module 2013 Manual. On-line: http//www.geerthofstede.nl/vsm2013, Accessed: 30/11/2015.

Eye-Tracking Technology for Construction Safety: A Feasibility Study

R. B. O. Pinheiro[a], N. Pradhananga[b], R. Jianu[c] and W. Orabi[b]

[a]Centro Universitário Jorge Amado, Paralela Salvador, BA, Brazil
[b]OHL School of Construction, Florida International University, Miami, FL, USA
[c]School of Computing and Information Sciences, Florida International University, Miami, FL, USA

E-mail: renatabritoo@gmail.com, npradhan@fiu.edu, rdjianu@cis.fiu.edu, worabi@fiu.edu

Abstract

A construction site is a harsh environment demanding entire human senses and attention, even for regular scheduled tasks. Many accidents occur on construction sites because of inability of workers to identify hazards and make timely decisions. To understand why some hazards go unseen, it is crucial to study how workers perceive the site. The objective of this research is to leverage eye-tracking technology to study workers' gazing pattern in a construction environment. A real picture from an active construction site is modified to introduce hazards and a desktop experiment is conducted, in which, subjects are asked to identify the hazards. A different group of subjects are made to make similar observations on a 2D sketch-representation of the same construction scenario. Eye-tracking data gathered from their observations is analyzed to understand when, how, and which hazards do they recognize and the pattern of recognition is studied. The results of this study will enhance our understanding on the visual factors that govern attention and help workers recognize potential hazards in a construction site. The comparison between the observation pattern in real and sketch-representation is done to assess how subjects respond to artificial images compared to real images. This comparison will test the feasibility of using virtual reality for safety training and simulations.

Keywords –

Construction Safety; Eye-tracking; Virtual Reality; Site Perception; Hazard Recognition

1 Introduction

Despite safety regulation efforts to reduce numbers of injuries and deaths in construction industry, it remains responsible for most of the work-related injuries and deaths consistently [1, 2]. There is still no adequate investment and research in developing new methods and technologies to effectively improve safety within the construction industry.

It is crucial that construction workers are able to identify possible hazards in work sites in order to take preventive actions to reduce or eliminate the risk of injuries and illnesses. Therefore, it is important to understand their perception of the construction environment and what kind of hazards they are able to recognize when exposed to different circumstances [3]. By evaluating which aspects need improvement it is possible to develop more efficient approaches, increasing worker and student assimilation of site hazards and construction safety practices. This study aims to explore how eye tracking can best contribute in analyzing the perception of hazards in construction sites and utilize it as a potential educational/training tool.

2 Literature review

2.1 Construction Safety Scenario

Even with the rapid advance of new technologies, methods and materials in construction industry, safety tends to receive insufficient attention. According to the Occupational Safety and Health Administration (OSHA) [1], in the United States, five work-related deaths in private industry, one was in construction, which represents 20.5% of work-related accidents occurred in 2014. The Census of Fatal Occupational Injuries, conducted by the U.S. Bureau of Labor Statistics [2], shows that this scenario has hardly improved from the past years. In 2014, construction industry presented the highest rate of fatal work injuries since 2008, rising from 828 in 2013 to 874 in 2014. Heavy civil engineering was the only segment that showed some improvement, decreasing from 165 fatal injuries in 2013 to 138 in 2014.

Different factors are responsible for these high accident and fatalities rates, such as non-use of Personal Protective equipment (PPE), lack of supervised working environment, lack of a site safety plan and insufficient training. A study conducted by Hinze and Teizer [3], analyzed 659 fatality accidents investigated by OSHA. It was observed that blind spots, obstructions and lighting conditions were the most common factors contributing to vision-related fatalities. Equipment or vehicles, especially heavy equipment were involved in most of the identified accidents. Struck-by, collisions, rollovers and other equipment incidents represented one-fourth of construction workers deaths. The lack of visibility or a worker being present in the blind spot of equipment was the main cause of struck-by accidents, which represents 22% of all construction fatalities. It was also noted that 72.6% of the incidents occurred when the equipment were traveling in reverse.

The increasing cost of injuries has a direct financial impact on medical, legal and insurance expenses [3]. Therefore, the implementation of an efficient construction safety plan can reduce costs in construction industry besides improving employee health.

2.2 Safety Training

Planning and design of construction projects has seen revolutionary changes in recent years while safety training still follows traditional methods. New methods and technology applications are being considered and studies are being performed to encourage the use of more dynamic and precise methods, such as 3D training environments or personnel tracking and monitoring. Global Positioning System (GPS) and radio frequency technology, for instance, can be used to monitor workers distance from risky areas, fall prevention from height and collision avoidance [4]. Studies have shown sensing devices as a useful tool for safety managers identify, monitor and assist risky situations involving workers and equipment in construction sites [5].

As stated by Rosch and Vogel-Walcutt [6], the applications of eye tracking devices in safety training are wide. Physical reactions, such as gaze and pupil response measurements provided by the eye tracker can help to evaluate the effectiveness of training, measuring attentional focus and reading comprehension users, towards given image stimuli or tasks. By customizing e-learning environment based on eye tracking data, the content is adapted to the needs of each user, according with individual learning levels. By analyzing cognitive load, distraction and attention levels and user's perceptions and comprehension towards instructional materials it is possible to improve the user interface of courseware, resulting in a more interactive and efficient learning method.

2.3 Applications of Eye Tracking Technology

Murray et al. [7] defines eye-tracking as the process of measuring the point of gaze, and an eye tracker as a portable device that measures eye movement and eye positions. This technology can assess the user's mental state, exploring visual pattern and allowing the access to information that, otherwise, could not be accessible through any different approach. Through eye movement recording, it is possible to understand the behavior and cognitive processing of users and interpret their response to different image stimuli.

Eye tracking has become a multifunctional technology that can be implemented in many different fields of study. Human-computer interaction, virtual and augmented reality, usability tests, e-learning, assistive technology, pilot training assistance are some of the many eye-tracking applications [8]. Medicine, psychology and neuroscience can also benefit from this technology in studies, accessing cognitive abilities, attention and recognition memory in patients [9]. Assistive technology has been helping individuals with disabilities to improve their functional capabilities, enabling them to perform activities conducted by eye movements. A computer controlled by an eye tracker makes communication possible and also entertainment activities, which significantly improving patient quality of life [10].

Published studies relating eye tracking and construction are very limited so far. Driver Safety System (DSS) is one of the very few examples found regarding tracking applications used in construction industry. Developed by Caterpillar, this system can be used in construction and mining equipment in order to reduce the risk of drowsy driving-related accidents involving heavy machinery. The equipment consists a console-mounted camera to track the driver's head and eye movement, monitoring the driver's behavior in real time. When signs of sleepy or distraction occurs, the system sounds an alarm and produces a loud vibration in the driver's seat, alerting the driver to a possible risk situation. It also helps the company to track and monitor driving practices.

Eye tracking can also be used as a tool for education. In the 3D game "Safety Inspector", students act as a controlling contractor's safety inspector on the job site, identifying on-site hazards during a virtual walk-thru [11]. There are different levels of challenge, according with the player safety background. For every point obtained, the game provides the player with valuable information such as corrective actions, rules and regulations. This game can be used as a supplemental educational resource in the classroom and as an alternative to walk-throughs. Even though this project does not support eye tracking, it could work as an excellent future application for this technology, since the use of eye tracking in a 3D construction environment would provide a more accurate study of the student's perception of a jobsite. Besides, for students, it can be used with construction workers such as heavy equipment operators, laborers, electricians, foremen, drivers or inspectors. In this case, the challenges would be related with the occupation and activities performed by each worker. This approach can also be used for training, helping to facilitate the comprehension of the workers perception of construction sites, potential risks and its relationship with their visual attention [12]. Games in a virtual environment also increase the learning interest of the participant and potentially lead to a more critical awareness of safety hazards [4].

2.4 Eye Tracking Data

Eye tracking experiments can provide a wide variety of data. The results can be displayed according to the study requirements. The principal measurements obtained through eye-tracking applications are fixations, saccade and gaze. Yousefi et al, [12] provide the following definitions for these metrics:

- Fixations: Eye movements with a series of short stops pauses in specific positions
- Saccade: Eye movement from one position to the other
- Gaze: The eye gaze pauses in a certain position

In function of these data, heat maps or gaze plots are generated, allowing the measurement of attention allocation, the number and duration of eye fixations, as well the amount of re-fixations, establishing patterns of the user's attention.

3 Objectives and Methodology

The main objectives of this research are as follows

- to test the feasibility of using eye-tracking technology for construction safety training
- to assess the difference in gazing pattern on a real image and a virtual scene in regards to construction hazard identification
- to understand the difference in gazing pattern of individuals based on their knowledge about construction processes and operations

To achieve these objectives, the first step was to get an overview about safety scenarios in the construction industry and what has been done in terms of training to prevent injuries and fatalities on constructions sites. The literature review also consisted of the study of what has been published about eye tracking and what has been marketed about this technology in the last few years.

Then, an experiment was designed to evaluate the perception of subjects towards a construction scene. A picture taken from an active construction site was selected based on the level of activity, areas that require attention and potential of making adjustments. Two groups were defined and the selected image of a construction site was adjusted to contain a variety of hazards, what will be explained in detail in the next section. Beside the results generated by the eye tracker, a checklist was designed to be completed by the subjects to reinforce the data obtained from this study.

This research isolates visual gazing pattern and ignores the effects of prior training, cognitive and learning capabilities and other potential factors.

4 Experiment

The initial purpose of this experiment is to analyze the visual perception of an image of a construction site based on recorded eye tracking data. For this test, a picture of a construction site is displayed at the monitor to volunteers. The goal is to identify patterns of visual fixation, visual hierarchy, and perception of the displayed picture.

4.1 Eye-Tracking System Selection Process

The required level of data precision and relevant metrics must be considered in order to select the more appropriate eye tracking system for each type of experiment. There are two types of eye tracker: static or head-mounted (mobile). The static eye tracker consists of a portable device that can be mounted on computer monitors, laptops, iPads or other equipment. A head-mounted eye tracker is a more flexible alternative that consists of a helmet or glasses with a built-in eye tracker and a video camera. This option is recommended when the participant needs freedom of movement or in outdoor environments. For the purpose of this study, an indoor experiment using laptop monitors, the more suitable option is the use of a static eye-tracker. The equipment used in the present study was a Tobii EyeX eye tracker for data collection and a 15" computer monitor.

4.2 Participants

Sixteen student participants were recruited for this experiment. The subjects, undergraduate and graduate students, were divided in two groups: construction students and students of a different major (no construction background). No prior training was given to the participants. They went through a pre-written briefing.

4.3 Image Stimuli

Two different images were used on the experiment. A picture taken of a construction site and a 2D representation of the same picture. The photograph used was manipulated in Photoshop to simulate different situations of hazards in the same image. The hazard selection for the scene was based on the major causes of injuries and fatalities that occurred in construction sites for the past few years.

Figure 1. Original image from a construction site

Figure 2. Photoshopped image based on Figure 1

Some elements were modified or added to the scene, such as a bulldozer and part of a crane and a loader, representing lifting and hoisting operations and an operator, moved to right behind an equipment, representing a struck-by hazard. Two workers were positioned in the area within the crane's swing radius, representing the risk of being struck by an overhead load. Two other pieces of equipment were removed from the scene. The perimeter protection was edited and part of the safety net was removed, representing a possible risk of fall. In the same area, a construction safety sign was added. Some operators had their positions changed. The worker with the orange vest was positioned behind the equipment, which might be a blind spot for the driver that could lead to a struck-by accident. In addition, the helmet of one of the workers was removed.

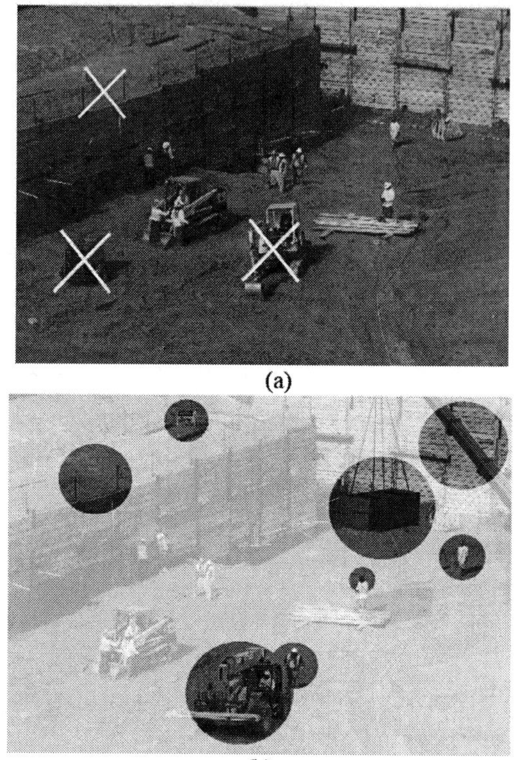

(a)

(b)

Figures 3: (a) Items removed from Figure 1, (b) Items added to Figure 1

Using the manipulated picture as a base, a 2D drawing was created. The purpose of using a photograph and a 2D image of the same scenario was to compare the perception of the subjects when looking to different types of image stimuli, real and artificial.

Figure 4. Artificial 2D representation of Figure 1

Through this analysis, it is possible to evaluate if the use of 2D images or even 3D non-realistic models on a test, experiment or training have the same effectiveness in regards to gazing pattern.

Before performing the experiment, Areas Of Interest (AOI) were defined by the research team. It is important to define the AOI prior to the tests in order to evaluate gaze data and compute the number of times the subjects visit the AOI and the amount of processing time spent on each area. There were seven AOIs identified in the scene. Figure 5 shows the AOIs in different color. Figure 8 and Figure 9 will use the same color code that is assigned to the AOIs in Figure 5.

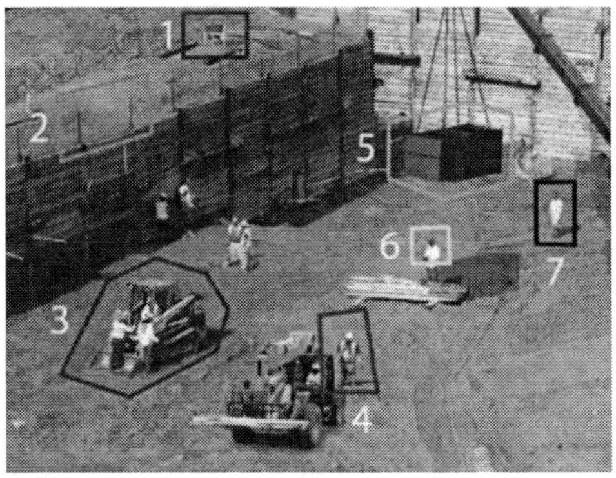

Figure 5. Color-coded Areas of Interests (AOIs)

4.4 Procedure

Each student looked at one image individually. The participants were randomly divided in two groups. Half of them were assigned to observe the photoshopped image (Figure 2), and the other half, the 2D representation (Figure 4). A list with construction hazards, and brief explanation about each item, was provided to the subjects prior to the test. The participants were seated approximately 30 cm from the computer screen.

Calibration of the eye tracker is a crucial procedure toward accurate data collection and analysis. The calibration was performed based on the Tobii's default calibration settings, where the subject is asked to look at specific points on the screen (calibration dots). This process optimizes the eye tracker settings for the current light conditions and the user's eye characteristics. After calibration, the subjects were asked to look to the image and fill out the checklist to identify potential hazards. The identified items should be numbered so it would be possible to observe the sequence of recognition of the items. The list also contained extra items (hazards), not included on the image. The checklist can be found in Figure 9. The subjects were free to take as much time as they want to view and analyze the scene. Figure 6 shows the experimental setup.

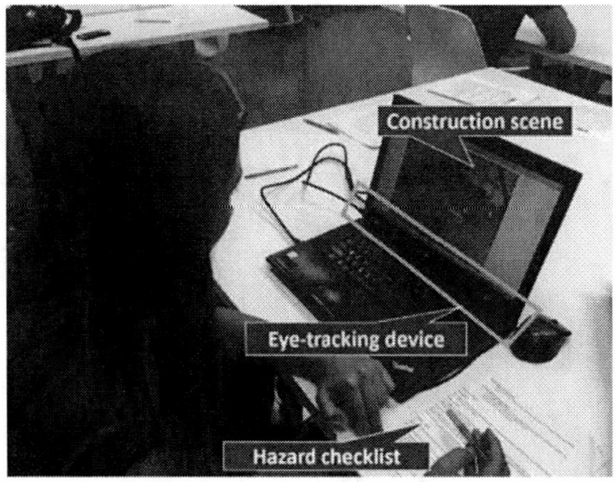

Figure 6. Experimental setup

5 Preliminary Results

Sixteen subjects participated in this experiment, eight construction students and eight non-construction major students. Four subjects from each groups were assigned to view photoshopped image and the other four were supposed to view the 2D representation. This assignment was also made randomly.

The results are, hence, presented and categorized by

- Non-construction students / Construction students
- Photoshopped image (PS) / 2D image (2D)

Based on horizontal and vertical coordinates for each sample, it was possible to compare the time spent on each AOI or something other than the marked AOIs. The device measured 137,256 gaze samples during the experiment, with a frequency of 60Hz, which means that 60 gazes were recorded each second. The total gaze samples was equivalent to 2,287.60 seconds, or 38.13 minutes of eye-tracking data.

Gaze plots indicate the eye movement sequence through lines and the fixations illustrated with dots. The size of the dots indicates the fixation duration and shows which elements of the picture holds user attention. A gaze plot was generated in order to observe the sequence of visualization of each subject. However, since we had many AOIs and the time the subjects were exposed to the pictures was considerably long for this type of experiment, the gaze plots will need further in-depth analysis to provide any meaningful information. Figure 7 shows a sample gaze plot for a subject.

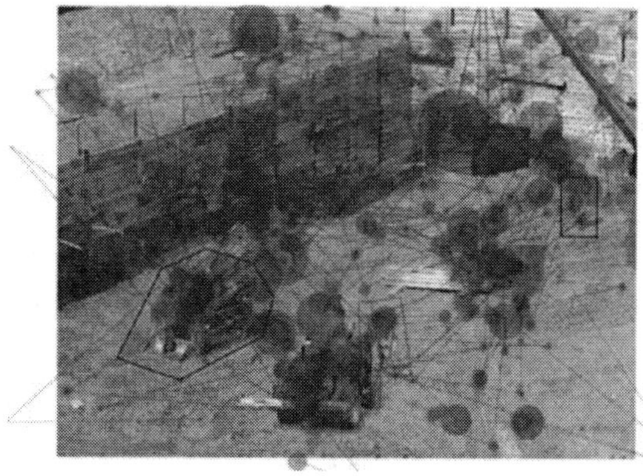

Figure 7. Gaze Plot of a subject

5.1 AOI Analysis

By measuring the two-dimensional eye movement pattern (x,y) the eye tracker software provided the percentage of gaze for each AOI, numbered as one to seven and identified by colors. The scarf plot in Figure 8 shows the gazing pattern of all the subjects as they move from one AOI to another. The colors in vertical bars represent the time spent in the AOIs coded by color shown in Figure 5. Grey part on the bars represent the times spent outside the AOIs. Figure 8 only shows the data pertaining to initial part of the experiment and the vertical bars (time) continues in a similar manner throughout the experiment. Entire scarf plot is not presented in the paper because the colors becomes indistinct as the vertical bars get longer.

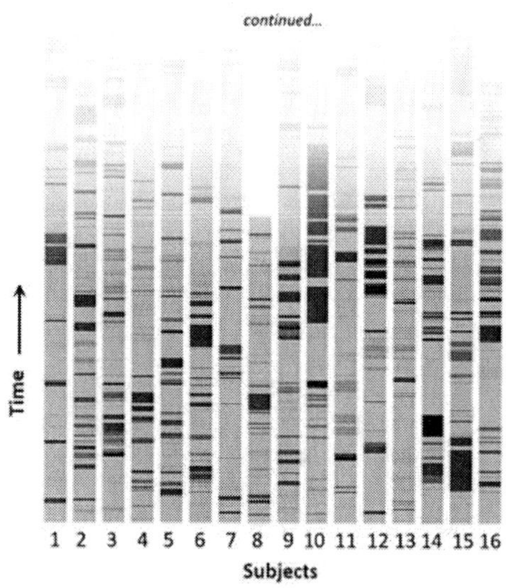

Figure 8: Scarf plot of subjects' gazing pattern color-coded by AOI

During the experiment, the subjects looked outside the AOIs 74% of the time and 26% inside AOI's. All the grey region in Figure 8 contribute to time outside AOI while time inside AOI is the summation of all the colored part of the scarf plot. AOIs only occupied a small portion of the entire picture so it is understandable that the subjects spend more time outside AOI.

Among all seven pre-defined AOIs, AOI 3 was the most viewed, with 32% of total visualizations inside AOIs. The least viewed was AOI 2, with 7% of gaze samples. The position of the AOI in the image and its size must influence the order of visualization, since elements exposed in foreground are usually bigger so they can be noticed first, rather than a background element. The two most visualized elements, the crane and the loader, for example, are considerably bigger compared to the other elements in the image.

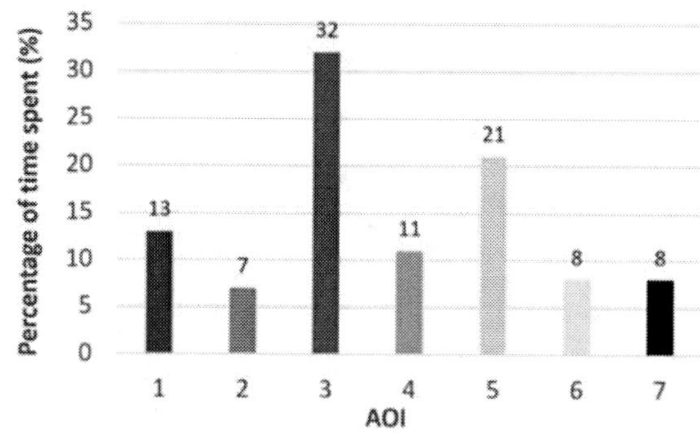

Figure 9. Proportion of gaze duration on each AOI

From the checklist, it was possible to identify the order of recognition of each hazard. Among all the hazards on the checklist, "Personal protective equipment" was the most recognized and fastest identified item in the pictures, by all participants (Figure 10). The fact that the helmet is one of the most well-known personal protective equipment, used in any construction site, might have facilitated its identification before other items. "Dust exposures", on the other hand, was the least recognized item, which was expected, since this item was listed on the checklist, but it was not represented on the images. A study lead by Bhoir et al [13] indicated low levels of attention on danger signs and most of the participants did not even look at it.

5.2 Non-construction Students vs Construction Students

Heat map is another method that can be used to analyze the results. It shows the areas of most interest and the cumulative eye fixations of those areas. The scale

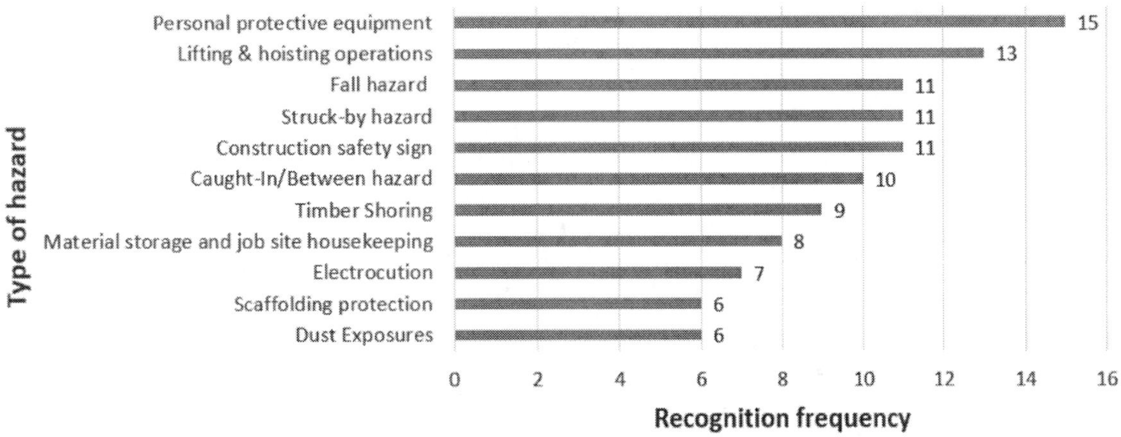

Figure 10. Frequency of hazard recognition

color changing from green to red represents the amount of fixations while the red spots indicate increased duration of fixations. The heat maps generated by the eye tracking software combined data from all participants, divided into four groups: construction and non-construction students, 2D representation and photoshopped image. Visual analysis of the maps show that non-construction students spent more time looking at areas out of the AOI, showing a more spread visualization distribution in the map and focused in the workers, whereas construction students have focused more on the AOI's.

Figure 11. Non-construction students' heat map

Figure 12. Construction students' heat map

The heat maps generated for all subjects who observed the photoshopped image and 2D representation, illustrate that user's attention is considerably spread out in the 2D representation, while in the photoshopped image the attention is more directed to the AOI. Also, during the experiment, the subjects spent 18% less time observing the 2D representation than the photoshopped image. The subjects were not restricted by time to identify the hazards in the image. In this context, data collected from photoshopped image constituted to 55% of the total gazing data while only 45% of the data came from 2D representation as shown in Figure 13. Despite the real picture being more realistic and detailed, identification of elements were faster than in the 2D representation. One possible reason might be the minimal level of shadow/highlight and distinctly defined element boundaries, which simplifies shape, contributing to a faster recognition.

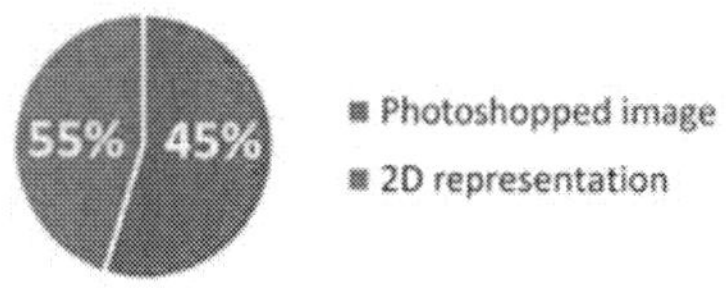

Figure 13. Proportion of time spent on photoshopped image and 2D representation

6 Conclusion

Even though varieties of new technologies have been increasingly being utilized in construction industry, safety scenario still yields unsatisfactory results. When it comes to training, it is very essential to apply something beyond the conventional approach, investing on more dynamic learning methods, such as virtual environments. Eye-tracking technology, still under-utilized in construction industry, is a promising alternative to analyze the efficiency of safety training. The analysis of workers' perception can determine the understanding level of safety requirements and guidelines adopted by a company or even indicate layout problems in the construction site planning. The use of real pictures or ultra-realistic 3D models can produce valuable results. As observed in these results, there was a slight difference between the perception of the photograph and the 2D image, implying more research needs to be done to use of 2D drawings or basic 3D models for scene-perception studies. The camera position and viewing angle of the picture might also influence the perception of an element instead of another.

An eye-tracking experiment performed in a 3D environment simulating a construction jobsite, would allow students and workers to be exposed to different types of hazards prior to going to a real jobsite. Furthermore, head-mounted eye trackers, such as glasses, can potentially lead to walkthrough experiments in real construction sites. From eye-tracking data, it would be possible, during and after training sessions, to assess the safety knowledge level of each individual, analyze group patterns and develop new prevention measures.

Interactive methods as video games also can be integrated in these experiments. Further research in this area may include eye movement combined with other types of data such as think aloud method. Equipped with a microphone, the participants would provide a verbal feedback, describing their experiences and thoughts while doing the tasks or after tasks are completed.

It is necessary to provide effective learning methods and training programs, encouraging protective measures to ensure a safe working habit. Due to its dynamism, eye-tracking experiments and tests can be a helpful tool to complement traditional training approaches and evaluation methods.

7 References

[1] U.S Department of Labor. *Commonly Used Statistics*. On-line: https://www.osha.gov/oshstats/commonstats.html, Accessed: 25/06/2015.

[2] Bureau of Labor Statistics U.S Department of Labor. *National Census of Fatal Occupational Injuries in 2014*, 2014.

[3] Hinze, J. W., and Teizer, J. *Visibility-related fatalities related to construction equipment.* Safety science, 49 (5):709-718, 2011.

[4] Teizer J., Cheng T. and Fang Y. *Location tracking and data visualization technology to advance construction ironworkers' education and training in safety and productivity.* Automation in Construction, 35: 53–68, 2013.

[5] Pradhananga, N., & Teizer, J. (2013). *Automatic spatio-temporal analysis of construction site equipment operations using GPS data.* Automation in Construction, 29, 107-122.

[6] Rosch, J. L. and Vogel-Walcutt, J. J. *A review of eye-tracking applications as tools for training.* Cognitive Technology, 15:313–327, 2012.

[7] Murray, I. C., Fleck, B. W., Brash, H. M., MacRae, M. E., Tan, L. L., and Minns, R. A. *Feasibility of saccadic vector optokinetic perimetry: a method of automated static perimetry for children using eye tracking.* Ophthalmology, 116 (10), 2017-2026, 2009.

[8] Lupu, R. G and Ungureanu, F. *A survey of eye tracking methods and Applications.* Gheorghe Asachi, Technical University of Iaşi, Faculty of Automatic Control and Computer Engineering, 2013.

[9] Rose SA, Djukic A, Jankowski JJ, Feldman JF, Fishman I, et al. *Rett syndrome: an eye-tracking study of attention and recognition memory.* Dev Med Child Neurol 55: 364-371, 2013.

[10] Majaranta P et al. *Gaze Interaction and Applications of Eye Tracking: Advances in Assistive Technologies.* IGI Global, Hershey, 2012

[11] Lin. K, *A pilot study of a 3D game environment for construction safety education.* Journal of Information Technology in Construction (ITcon), Vol. 16, pg. 69-84, 2011

[12] S Yousefi, M. V., Mohammadpour, A., Karan, E. P., and Asadi. *Implementing Eye Tracking Technology in the Construction Process.* 51st ASC Annual International Conference Proceedings, 2015.

[13] Bhoir, Sneha A. et al. *Measuring construction workers' attention using eyetracking technology.* 5th International/11th Construction Specialty Conference, 2015.

A Framework of Lift Virtual Prototyping (LVP) Approach for Crane Safety Planning

Y. Fang[a] and Y. K. Cho[a]

[a] School of Civil and Environmental Engineering, Georgia Institute of Technology
E-mail: yihaifang@gatech.edu, yong.cho@ce.gatech.edu

Abstract –

Many crane accidents occur due to a lack of understanding of potential constraints present on lift sites, especially the spatial conflicts with inadequate clearance to surrounding obstructions. Even the most thorough lift plans can be compromised by changes in lift sites such as a newly-erected frame in the crane workspace or a sudden presence of another equipment. Therefore, it is critical for the lift crew (i.e., planner, rigger, signalman, operator) to be aware of the condition and potential constraints in the as-is lift settings prior to conducting the actual lift job. This study proposes a framework of Lift Virtual Prototyping (LVP) approach to assist quick lift planning based on actual site settings. This is achieved by rapid modeling, simulating, and analyzing crane lifting operation by reconstructing as-is site conditions and incorporating the characteristics of individual operator. The proposed LVP approach was implemented in a case study and the results showed that the LVP approach is effective in familiarizing the lift crew with potential constraints existing in the lift site and improving the operator's confidence in conducting the lift job safely and efficiently. Implementing this approach in daily lifting operations will facilitate the recognition of operation constraints that are difficult to be identified in the planning phase, and thus advance the crane operational safety and efficiency.

Keywords –

Lifting Virtual Prototyping (LVP); obstruction recognition; spatial conflicts; Crane safety planning; Lift safety and efficiency

1 Introduction

A crane is one of the most essential and commonly used construction machinery present on the various construction sites of buildings, plants, and bridges. Given the magnitude of a crane's weight and the workspace it occupies, the potential loss of property and life due to incidents related to cranes is tremendous. From 1997 to 2013, 1171 fatalities were observed in crane-related accidents in all industry sectors in the US, among which construction industry alone was responsible for 544 fatalities accounting for nearly half of the deaths in all sectors combined (Bureau of Labor Statistics (BLS) 2015). Many of these accidents are were due to inadequate lift planning or failures to identify lift constraints existing on complex and changing construction sites.

Virtual prototyping (VP) originated from the manufacturing industry especially in the automobile and aerospace fields. During product development, VP helps to validate a design before committing to making a physical prototype. Through computer-aided design (CAD) and computer-aided engineering (CAE) software, the designers and engineers can test the mechanical motions and functions, and simulate the behavior of the product in the real world. The manufacturing industry and the construction industry share many similarities and thus the construction industry has benefited from many concepts adopted from manufacturing industry such as lean construction and prefabrication. Recently, the VP method has drawn much attention from the academics and practitioners in the construction industry.

As an early attempt in this direction, Huang et al. introduced the concept of VP in the construction industry, termed construction virtual prototyping (CVP) [1]. They envisioned that CVP will enable project teams to model, simulate, and analyze the construction processes before the commencement of the project. The CVP, therefore, allows them to check constructability and safety issues in advance so that rework and construction defects can be minimized. Li et al. introduced a framework for efficient implementation of CVP in the construction planning phase [2]. They reported the application of the CVP in an office building project and the findings shows that the CVP can reduce construction risks and enhance the communication and collaboration between construction stakeholders. Dong et al. adopted a virtual prototyping environment as it provides a configurable setting to evaluate the augmented reality-assisted method for building damage reconnaissance [3].

In light of the CVP concept, this study proposes a lift virtual prototyping (LVP) approach to enhance lift safety and efficiency on construction sites. This paper starts with a review of the practices and techniques used in traditional lift planning. Then, a framework of the LVP approach is introduced with details in the steps of modeling, simulation, and analysis. It is followed by a case study and validation of the LVP approach in a real world implementation. The last sections discuss the limitations and envision the future work in the LVP approach.

2 Related Work

Within the huge workspace of cranes, the presence of built structure components, overhead power lines, and the workspace of other equipment induce massive spatial conflicts during crane lift operations. Suruda et al. reported that 40% of the deaths in crane-related accidents were related to spatial conflicts [4]. Among 632 crane-related deaths from 1992 to 2006, 61% of all fatalities involved crane parts or loads colliding with obstructions (e.g., power lines, personnel, structure) [5]. Therefore, designing crane locations [6], calculating the lift range and crane boom reach based on load chart [7], and planning a collision-free lift path [8] are important planning tasks in order to mitigate the collision hazards and improve lift efficiency in crane operation.

In crane lift planning, lift path planning is an important task to ensure lifting safety and efficiency. The goal of path planning is to plan a collision-free and efficient lift path based on load data and site geometry constraints. Traditionally, the lift path was planned largely depending on the planners' intuition and experience and thus this process is often labor intensive and error prone [9]. Recently, researchers have adopted the knowledge and techniques in robotic motion planning in lift path planning trying to automate the lift planning process. In this approach, a crane was treated as a multi degree-of-freedom robotic manipulator and various searching algorithms were tested to optimize the lift path to ensure a collision-free travel for the lifted load while guaranteeing a respectively short lift path [9]. To validate the advantage of computer-aided path planning , researchers have extensively investigated different algorithms including Heuristic Search [10], Ant Colony Algorithm [11], Probabilistic Road Map (PRM) [12], and Rapidly Exploring Random Tree (RRT) [13]. Findings from these research efforts have demonstrated that using computer-aided path planning techniques can greatly improve the accuracy and efficiency in path planning.

It should be noted that computer-aided path planning only works when provided with accurate and comprehensive site geometric information. Site geometric information, however, is not always available or accurate in the early planning phase. In addition to static obstructions, when analyzing crane-related spatial conflicts, one must consider the dynamic crane motions and condition changes in the surrounding environment as the project proceeds [14]. Design changes occur throughout the planning phase and even during the construction phase. It is not practical, however, to re-plan the lift path for every design change. In addition, site drawings or models used for lift path planning can hardly represent all the geometric constraints existing in the actual lift site. For instance, temporary structures such as scaffoldings and formworks are often not included in the design model and thus won't be considered in the path planning. These variances between the planned and actual site conditions will inevitably compromise the validity of the planned lift paths. Furthermore, the lift job will be eventually performed by a crane operator who might not prefer or not be able to follow the exact planned path. Although the operator is required to be involved in the lift planning and be familiar with the lift plan, he or she will most likely execute the lift job based on their judgements in the actual situations. Therefore, it is critical for the lift planning to consider the actual as-is condition of the lift site and to involve the operator in the planning by incorporating their individual operation preference and tendency.

3 Lift Virtual Prototyping (LVP) Framework

The objective of the Lift Virtual Prototyping (LVP) approach was to plan the lift based on actual site settings and the actual capabilities of the lift crew. This was achieved by rapid modeling, simulating, and analyzing crane lifting operations with as-is site conditions and the characteristics of an individual operator. Figure 1 shows the framework of the proposed LVP approach and the following sections will introduce the details in each part of the framework.

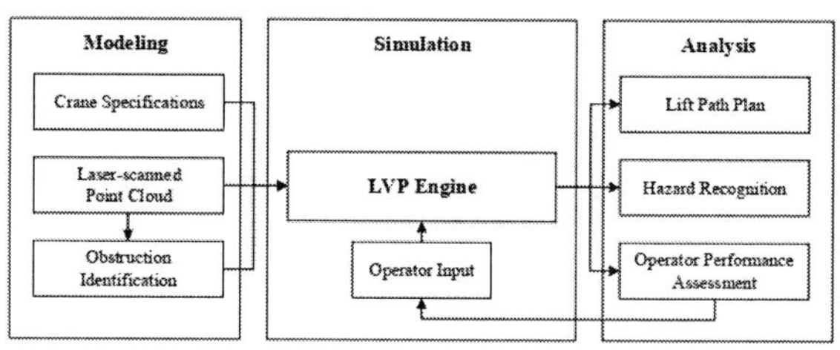

Figure 1: Framework of the Lift Virtual Prototyping (LVP) approach

3.1 Lift Scenario Modeling

Cranes are different in types, size, and capacity, and

thus it is important in the LVP to model the critical crane configurations according to the actual crane. This can be accomplished by modifying an existing crane model according to specification of the actual crane that will be used for the job. A major limitation in traditional lift planning approach is the lack of information to represent the as-is lift site condition. Recent advances in rapid geometry data acquisition makes it possible to obtain massive 3-dimensional (3D) geometric data in a short amount of time with minimal effort. This study adopted laser scanning technology that can rapidly and accurately capture the shape and geometry of a physical scene. The scanned scene is represented by a dense point cloud comprised of millions of points, each of which contains position (XYZ) and color (RGB) data. This technology, therefore, is suitable for wide range measurement given only a limited amount of time.

To reconstruct as-is lift scenario, the laser scanned point cloud need to be converted to bounding boxes to represent these site obstructions. The pipeline of obtaining oriented bounding boxes for various objects in a point cloud involves the steps of segmentation, clustering, and orientation estimation. The input point cloud is first down-sampled to 10% of its original size by performing voxel grid filtering. Next, ground plane segmentation is performed using Random Sample Consensus (RANSAC) [15]. Points from the ground plane are filtered out so that object points can be separated nicely in the clustering step. The ensuing clustering step involves grouping points that are contained within the same neighborhood together and labeling each group as an individual cluster. The threshold parameter can also be manually adjusted so that points belonging to the same object would not be split into separate clusters and points belonging to different objects would not be grouped into the same cluster. Finally, an oriented bounding box is computed for each point cluster by considering the physical spread of points in the z-axis (vertical axis) and the x-y-plane. Principal Component Analysis (PCA) [16] is used to determine the two principal directions in which points in the cluster vary the most in the x-y-plane. The principal directions are used to determine the orientation of the object's bounding box in the x-y-plane. The final bounding box is then computed by determining the maximum and minimum length along the vertical axes and the horizontal principal axes. The total computation time for this process largely depends on the down-sampling ratio but it only takes a couple of minutes for a down-sampled point cloud with around 10,000 points.

3.2 Lift Simulation

The crane model, site point cloud, and the bounding boxes of the identified obstructions were imported into a LVP engine powered by Unity 3D game engine. The LVP engine can simulate actual lighting and wind conditions, customize data input/output, and support close-to-reality physics simulation, such as load pendulum motion. A major advantage of using LVP in lift planning is that the variances among the skill and experience of different operators are considered in the prototyping process so that the results can reflect actual proficiency of individual operator in executing a particular lift task. The operator can practice and rehearse the lift task in the LVP engine with the presence of actual constraints. As such, their ability to recognize and react to hazards is incorporated and will influence the prototyping results. From a training perspective, the operator can continuously rehearse the lifting until a satisfying safety and efficiency results is achieved.

3.3 Safety and Efficiency Analysis

One major objective of the LVP approach was to introduce as-is site conditions, especially geometry constraints, in order to advance the accuracy and practicality of traditional lift planning. The results of the LVP approach was analyzed from different safety and efficiency perspectives. The safety analysis is enabled by continuously monitoring and measuring the proximity of the lifted load to surrounding obstructions and lift capacity by comparing the actual lifted load to the rated allowable capacity. Two measurements (i.e., $I_{clearance}$, $I_{overload}$) are automatically recorded during the LVP process to reflect the ability to maintain a safety clearance to obstructions and the ability to maintain a safe load capacity:

$I_{clearance}$ – Incidence that the load is within dangerous clearance (2m) to obstructions.

$I_{overload}$ - Incidence that the capacity is beyond 75% of the rated capacity.

The efficiency of the lift job is evaluated by the overall ability to efficiently complete a task and the ability to select an efficient lifting path. They are indicated by the following two measurements:

T_{total} - Total time spent for the lifting task.

D_{total} - Total distance the hook travels during the lifting task.

4 Case Study

The proposed LVP approach was implemented in a real world lift task to evaluate its effectiveness in lift planning. This lift task adopted a 70-ton telescopic boom mobile crane to lift a 25 ton load from its original location to a trailer. A telescopic boom crane model downloaded from 3D Warehouse was modified according to the specifications of the actual crane in order to model critical geometries such as boom length, boom slewing pivot point, and load shape and size.

As-is geometric data of the lift site was acquired by using a commercially available laser scanner. Six scans in low resolution setting was conducted around the site which took 40 minutes in total. These scans were then registered as a single point cloud and converted to PTS format. This point cloud was processed by the obstruction identification algorithm and as a result 92 bounding boxes were automatically generated. Finally, the crane model, point cloud, and the bounding boxes of the identified obstructions were visualized in the LVP engine (see Figure 2).

Figure 2: Reconstructed virtual lift scenario with laser scanned point cloud and bounding boxes generated for identified obstructions.

Operator involvement in the prototype was realized by providing the operators an immersive and interactive game type virtual environment where they can rehearse the lift in the simulated as-is lift site condition. To maximize the immersive feel of the operator in the prototyping, a head-mounted display (HUD) was adopted as the primary user interface (see figure 3). A gamepad was used to simulate the actual crane control where two joysitcks replace the levers and trigger buttons replace the pedal (see Figure 4).

Figure 3: Simulated lift scenario in an HUD interface

Figure 4: Operator input in the VLP process

To validate the effectiveness of LVP approach in identifying constraints existing in the lift job, the analysis results in the LVP approach were compared to the results of analysis based on traditional 2D drawings and experience. Two crane operators were asked to perform the lift. They had similar levels of training and

Table 1: Planning result comparison between traditional planning and LVP approach

	Traditional planning	LVP approach
Safety Analysis	• Potential proximity hazards: crawler crane, spare jibs, trailers • Boom reach < 32m	• Time and location of proximity and overload incidents • Recorded incidents for review
Efficiency Analysis	• Rough lift time estimation: 4 min	• Simulated lift time: 3 min 35 sec • Simulated lift path: 95.3m • Recorded lift path for review

experience in operating this particular type of crane. As such, we assumed that they hadsimilar proficiency in crane operation, hazard recognition, and decision making. The hazards to be identified were limited to proximity between the load and surrounding obstructions and the cases when load was beyond 75% percent of the rated capacity. Table 1 shows the comparison between the LVP and traditional planning results.

Based on limited site information, traditional planning provides subjective judgements and rough estimations of potential hazards that might occur during the lift operation. LVP approach, taking advantage of as-is site information, provides a more accurate simulation of the actual site conditions. As such, the lift simulation is closer to real lift performance and thus provides more detailed planning results such as the exact time and location that potential hazards might occur, which offers the lift crew an opportunity to take action in advance or pay more attention during the lift. In addition, the hazards and efficiency constraints identified in the simulation can be reviewed in a virtual environment that allows further investigation (see Figure 5).

Figure 4: Hazard identification in LVP

5 Conclusions

Traditional lift planning is a manual process based on outdated information and thus can hardly address the actual lift site constraints. This study proposed a Lift Virtual Prototyping (LVP) approach to plan the lift based on actual site settings. This is achieved by rapid modeling, simulating, and analyzing crane lifting operation by incorporating as-is site condition and the characteristics of individual operator. The benefits of the LVP approach over conventional crane lift simulation are in twofold. First, it takes the as-is environment into the prototyping process so that constraints, which are not identified by conventional simulation, can be recognized. Second, it emphasizes more on human-operation interaction by involving the same operator that will perform the lift in the prototyping. The simulation, therefore will not only identify hazards that are more prone to this operator, but at the same time familiarize the operator with the lifting process in advance. The LVP approach was implemented in a real world lift operation and its effectiveness was validated by comparing the prototyping results to conditional planning results. Although the test results show a great potential of the LVP approach in crane lift planning, the current state of the LVP approach is not fully-automated and the preparation of the LVP engine requires an expensive laser scanner that might not be available for small projects and personnel with limited knowledge and experience in data acquisition and processing. Future work will be directed toward exploring technologies that are less labor-intensive in data collection and automating the process of LVP engine preparation.

6 Acknowledgement

This material is based upon work supported by the National Science Foundation (Award #: CMMI-1358176). Any opinions, findings, and conclusions or recommendations expressed on this material are those of the authors and do not necessarily reflect the views of the NSF.

References

[1] T. Huang, C. W. Kong, H. L. Guo, A. Baldwin, and H. Li, "A virtual prototyping system for simulating construction processes," Autom. Constr., vol. 16, no. 5, pp. 576–585, 2007.

[2] H. Li, T. Huang, C. W. Kong, H. L. Guo, A. Baldwin, N. Chan, and J. Wong, "Integrating design and construction through virtual prototyping," Autom. Constr., vol. 17, no. 8, pp. 915–922, 2008.

[3] S. Dong, C. Feng, and V. R. Kamat, "Sensitivity analysis of augmented reality-assisted building damage reconnaissance using virtual prototyping," Autom. Constr., vol. 33, pp. 24–36, 2013.

[4] A. Suruda, M. Egger, and D. Liu, "Crane-Related Deaths in the U.S. Construction Industry, 1984-94," J. Constr. Eng. Manag., vol. 132(9), pp. 901–910, 1997.

[5] CPWR, "Crane-Related Deaths in Construction and Recommendations for Their Prevention," no. November, pp. 1–13, 2009.

[6] K. Tantisevi and B. Akinci, "Simulation-Based Identification of Possible Locations for Mobile Cranes on Construction Sites," J. Comput. Civ. Eng., pp. 21–30, 2008.

[7] L. K. Shapiro and J. P. Shapiro, Cranes and Derricks. New York: McGraw-Hill, 2011.

[8] H. R. Reddy and K. Varghese, "Automated Path Planning for Mobile Crane Lifts," Comput. Civ. Infrastruct. Eng., vol. 17, no. 6, pp. 439–448, Nov. 2002.
[9] Z. Lei, H. Taghaddos, U. Hermann, and M. Al-hussein, "A methodology for mobile crane lift path checking in heavy industrial projects," Autom. Constr., vol. 31, pp. 41–53, 2013.
[10] P. L. Sivakumar, K. Varghese, and N. R. Babu, "Automated Path Planning of Cooperative Crane Lifts Using Heuristic Search," J. Comput. Civ. Eng., vol. 17(3), pp. 197–207, 2003.
[11] X. Wang, Y. Y. Zhang, D. Wu, and S. De Gao, "Collision-Free Path Planning for Mobile Cranes Based on Ant Colony Algorithm," Key Eng. Mater., vol. 467–469, pp. 1108–1115, Feb. 2011.
[12] Y.-C. Chang, W.-H. Hung, and S.-C. Kang, "A fast path planning method for single and dual crane erections," Autom. Constr., vol. 22, pp. 468–480, Mar. 2012.
[13] C. Zhang and A. Hammad, "Improving lifting motion planning and re-planning of cranes with consideration for safety and efficiency," Adv. Eng. Informatics, vol. 26, no. 2, pp. 396–410, Apr. 2012.
[14] K. Tantisevi and B. Akinci, "Automated generation of workspace requirements of mobile crane operations to support conflict detection," Autom. Constr., vol. 16, no. 3, pp. 262–276, May 2007.
[15] R. Schnabel, R. Wahl, and R. Klein, "Efficient RANSAC for point-cloud shape detection," Comput. Graph. Forum, vol. 26, no. 2, pp. 214–226, 2007.
[16] N. Locantore, J. S. Marron, D. G. Simpson, N. Tripoli, J. T. Zhang, K. L. Cohen, G. Boente, R. Fraiman, B. Brumback, C. Croux, J. Fan, A. Kneip, J. I. Marden, D. Peña, J. Prieto, J. O. Ramsay, M. J. Valderrama, A. M. Aguilera, N. Locantore, J. S. Marron, D. G. Simpson, N. Tripoli, J. T. Zhang, and K. L. Cohen, "Robust principal component analysis for functional data," *Test*, vol. 8, no. 1, pp. 1–73, 1999.

Analytical Model and Data-driven Approach for Concrete Moisture Prediction

Karthick Thiyagarajan and Sarath Kodagoda

Centre for Autonomous Systems, University of Technology Sydney, Australia.
E-mail: Karthick.Thiyagarajan@student.uts.edu.au, Sarath.Kodagoda@uts.edu.au

Abstract –

The advent of smart sensing technologies has opened up new avenues for addressing the billion dollar problem in the wastewater industry of H_2S corrosion in concrete sewer pipes, where there is a growing interest in monitoring the environmental properties that govern the rate of corrosion. In this context, this paper proposes a methodology to predict the moisture content of concretes through data-driven approach by using Gaussian Process Regression modeling. The experimental program in this study practices measurements during wetting and drying phases of concrete. The obtained moisture data is used to train the prediction model against interpreted electrical resistivity data. The data of analytical model formulated from Archie's Law is then analyzed with experimental and Gaussian Process prediction data.

Keywords –

Analytical Model; Data-driven approach; Drying-wetting process; Electrical resistivity data; Gaussian Process prediction model; Moisture data.

1 Introduction

Sewer concrete corrosion is now widely recognized as a significant global problem by water utilities. This problem is vastly influenced by the gas concentration of Hydrogen Sulphide (H_2S) and bacterial activity on the surface of concrete pipes, resulting in the failure of sewer infrastructure. Since the atmospheric conditions inside confined sewers are not sensor friendly, there is a need for an adaptation of current sensor technologies or development of new innovative sensor technologies. For the reason that highly toxic components in sewer gas induces Occupational Health and Safety (OHS) issues, a custom-built smart sensing framework with capabilities of deploying an intelligent autonomous robot can be significantly beneficial in tracking the rate of corrosion.

Several models have been developed previously to predict the possible failure of sewer pipes due to H_2S influenced corrosion. However, those models lack the reliable data of key parameters [1]. It has been identified that one of the main factors that govern the high rate of H_2S corrosion is moisture content on the walls of sewer pipes [2][3]. In context to aforementioned factor, the objective is to devise a sensor system with reliable sensing performance for measuring the moisture of concrete on the surface of interest. Those sensor data will provide salient information to the existing models of predictive analytics for sewer corrosion.

Current research has reported the relative humidity measurements of the sewer atmosphere by deploying the sensors that are commercially available in the market [4]. However to date, no sensor system has been reported which is capable of measuring moisture on the sewer walls inside confined sewers. As the sewer concrete corrosion incur losses that are estimated to be in the order of billions of dollars per year [5], the wastewater governing bodies are seeking an application-specific sensor framework that performs the measurements of interest in the given ambiance.

The work reported in this paper focuses on presenting an analytical model and data-driven approach using Gaussian Process (GP) modeling for predicting the concrete moisture. In this study, the experimental measurements of the electrical resistivity were used to determine the moisture conditions of a concrete. The experimental measurements data were used to train the GP model for high-quality prediction. The data from the analytical model and prediction model was also studied. The work described in this paper is a furtherance of work carried out by [6][7], where the behavior of pH, resistance and moisture at various depths was studied.

The remainder of the paper is organized as follows. Section 2 provides the theoretical background. Section 3 elucidates the analytical model. Section 4 illustrates the experimental data collection methods. Section 5 explains the interpretation of data. Section 6 formulates the GP model. Section 7 presents the results with analyses, followed by conclusions in Section 8.

2 Theoretical Background

A recent research study has established the reliance on the rate of corrosion (mm $year^{-1}$) as a functional relationship with the concentration of H_2S (ppm), relative humidity (%) and temperature (Kelvin) in the sewer atmosphere [8]. The bacterium that dwells on the sewer concrete surface oxidizes the gas phase H_2S into aqueous sulphuric acid (H_2SO_4), which in turn reacts with the cement component of concrete [9]. The aqueous by-product that was produced as a result of the bacterial activity is responsible for filling the capillary pores of the sewer concrete. As the bacterial activity increases, the contained liquid within the pores of concrete penetrates deeper and eventually deteriorates embedded rebar. Since the production of aqueous by-product occurs on the surface of the sewer concrete, it is vital to measure the surface properties such as temperature and moisture of the sewer concrete. The measured data can be used to fit into the corrosion prediction model proposed by [8] as an alternative to relative humidity and gas-phase temperature data.

The changes in the moisture content of a concrete can be monitored by measuring changes in the electrical resistance of the concrete [10]. The electrical resistivity, which is a reciprocal of electrical conductivity, is an intrinsic property of a concrete. In [11], it was established that the electrical conduction of a concrete is dependent on both porosity of the concrete and the pore solution conductivity. Also, the electrical properties of the concrete are concurrently affected by the environmental parameters such as temperature variation [12] and moisture variation [13][14]. A new concrete health monitoring system was developed by [15] was used to perform in-situ measurements for evaluating the durability of concretes by measuring the electrical conductivity of concrete, an electrical conductivity of pore solution and temperature variations to calibrate electrical sensors. The study in [16] examined the electrical surface resistivity and bulk electrical resistivity for various concrete mixtures and established a correlation between bulk and surface resistivity of concrete.

Archie's Law described in [17] was developed for a sedimentary rock, which correlates the electrical conductivity to its porosity and brines solution. The Archie's law model that includes properties such as electrical resistivity, porosity and tortuosity of the pore network is used in [18] to estimate the appropriate composition of a concrete mixture for attaining recommended electrical resistivity for guaranteeing the durability of the structure, which is exposed to aggressive environments. By using Archie's Law, a correlation between electrical resistivity and volumetric moisture content of municipal solid waste was expressed in [19].

3 Formulation of an Analytical Model

In [17], the classical Archie's Law correlates the in-situ electrical conductivity, which is the inverse of electrical resistivity of a rock to its porosity and brines solution resistance. Archie's Law is defined in equation (1) shown below.

$$R_t = a\, \varnothing^{-m}\, S_w^{-n}\, R_w \tag{1}$$

where R_t is the liquid saturated rock resistivity, R_w is the brine resistivity, a is the tortuosity factor, m is the cementation exponent of a rock, n is the saturation exponent and S_w is the brine saturation.

The research in [18] has proposed a relationship between the total porosity of the concrete material with the electrical resistivity of the concrete by again using the classical Archie's Law, which is shown in equation (2).

$$R_t = a\ R_w\ \varepsilon^{-m} \tag{2}$$

where ε is the total porosity of the concrete material and the exponential in the equations was taken as the other aspects of the microstructure. In [18], R_w was taken as the pore solution resistivity. From the literature, the volumetric water content θ_V relationship with the porosity, which was expressed as in equation (3)

$$\theta_V = \varepsilon\ R_S \tag{3}$$

where R_S is the water saturation ratio defining the ratio of the volume of water to the volume of pore space. The literature [19] shows the relationship of volumetric moisture content and electrical resistivity of the solid waste by rewriting the Archie's Law as shown in equation (4).

$$R_t = a\ R_w\ \theta_V^{-m} \tag{4}$$

The relationship of R_t and θ_V was adopted from [19] in this study for the concrete material to formulate an analytical model. Rewriting the above equation (4) for mathematical simplification in the following equations:

$$\theta_V^{-m} = [\,(R_t)\,/\,(a.\,R_w)\,] \tag{5}$$

$$\theta_V = [\,(\mathrm{R_t})\,/\,(\mathrm{a.\,R_w})\,]^{[-(1/m)]} \tag{6}$$

$$\theta_V = [\,(a.\,R_w)\,/\,(R_t)\,]^{[(1/m)]} \tag{7}$$

From [18], the value of tortuosity factor a was approximately equal to 1.

4 Experimental Data Collection Methods

The experimental preparation of concrete samples and methods for measuring environmental variables that were used in this work for interpreting moisture conditions of the concrete are described in the following subsections. All the experiments were carried out in laboratory conditions at CAS Robotics Research Lab, UTS.

4.1 Preparation of Concrete Samples

Five concrete samples were used for this experimental program. The dimensions of each concrete sample were 20 cm (length) x 10 cm (height) x 5 cm (width). All the concrete blocks were dried using a microwave oven and left to cool for a few hours to ambient temperature. Two holes were drilled on each concrete sample for placing the electrodes to facilitate electrical resistivity measurements. Before placing the electrodes in the drilled holes of the concrete samples, the dry weight of each concrete sample was recorded. As shown in Figure 1, each concrete sample was placed in a plastic container that was filled with a liquid to a level such that the base of the concrete sample is in contact with the liquid. The liquid level of each container was monitored frequently and kept adding the necessary amount of liquid throughout the experimental program.

The electrodes were placed at a depth of 4.5 cm from the top surface of the concrete, i.e., 0.5 cm height from the surface that was exposed to the liquid. The electrodes that were used in the experiment were tampered resistance with dimensions of 75 mm (Height) x 4 mm (Width). These electrodes were made up of carbon steel and plated with zinc for corrosion protection. Two electrodes were drilled onto each concrete block with the spacing of 1 cm apart and each electrode weighed approximately 7.55 grams.

Figure 1. Experimental setup during wetting process

Figure 2. Experimental setup during drying process

4.2 Approach to Experimental Measurements

In this subsection, the experimental approach for measuring the variables such as moisture content and electrical resistivity of the concrete samples are described in detail. The measurements of all the variables were taken simultaneously.

The moisture content of the concrete sample was measured by means of gravimetric analysis. Two different methods were practiced to determine the moisture content of concrete samples. The first method to determine the concrete moisture content was through the wetting process. In this method, each concrete sample surface was exposed to a container containing water. As the water ingresses into the concrete from the exposed surface, there is an increase in the mass of the concrete sample. Then, the concrete sample was taken out of the container with care and its surface was wiped gently with paper towel to remove excess water. Subsequently, the mass of the concrete sample was recorded by using a weighing scale (OHAUS SP-6001) that has a precision of 0.1 grams. The second methodology used in the experimental program for determining moisture content was through the drying process. In this method, the concrete sample was removed from the container and allowed to dry in laboratory conditions. While drying, the mass of wet concrete sample decreases and its decreasing mass were recorded. The value of m_w increases during the wetting process and it decreases while drying. By using the gravimetric method, the determination of moisture content θ_G of the concrete sample for wetting and drying process can be defined as in equation (8):

$$\theta_G = \frac{m_w - m_d}{m_d} \text{ x } 100 \quad (\%) \qquad (8)$$

where θ_G is the gravimetric moisture content, m_w is the mass of the concrete sample in wet condition and m_d is the mass of the concrete in dry condition.

An electrical resistivity test was conducted on all the concrete samples. As covered in section 4.1, a pair of electrodes was used to facilitate electrical resistivity measurements while recording moisture content measurements during wetting and drying process. The electrodes were inserted in a way such that the tips of the electrodes were in contact with the concrete depth of 0.5 cm height from the surface that was exposed to water.

5 Interpretation of Experimental Data

This section explains the interpretation and analysis of data that were obtained during the experimental program. The moisture content measurements of the concrete samples that were obtained through the wetting and drying process by using gravimetric method were used to determine the volumetric moisture content. As in [20], the volumetric moisture content θ_V of the material can be determined by using the equation (9):

$$\theta_V = \frac{\rho_d}{\rho_w} \text{ x } \theta_G \quad (\%) \qquad (9)$$

where θ_G is the gravimetric based moisture content, ρ_w is the density of the liquid and ρ_d is the density of concrete sample in a dry condition, which is determined using the formula in equation (10):

$$\rho_d = \frac{m_d}{volume} \quad (\text{g/cm}^3) \qquad (10)$$

The volume of each concrete sample was 1000 cm^3 (20 cm x 10 cm x 5 cm). Since the electrodes were drilled and placed at a height of 0.5 cm from the concrete surface that was exposed to aqueous solution, the height of the concrete 5 cm is reduced to 0.5 cm and then the new volume of the concrete was calculated to be 200 cm^3 and it was used to calculate the ρ_d values that are presented in Table 1. The liquid used for the experiment was the water and the value of ρ_w was taken to be 1.0 g/cm^3.

Table 1 Dry Density of Concrete Samples

Concrete Number	Dry Density (ρ_d)
1	21.961 g/cm^3
2	21.567 g/cm^3
3	22.121 g/cm^3
4	21.808 g/cm^3
5	21.648 g/cm^3

The determined data of volumetric moisture content θ_V of concrete samples from θ_G were interpreted with the data of electrical resistivity measurements of concrete samples during the wetting and drying process of moisture measurements. The concrete samples that were numbered from 1 to 4 were used for training, while the data-driven model and concrete sample numbered 5 was used to test the prediction model. Figure 3 and Figure 4 presents the volumetric moisture profile during wetting and drying process and the interpreted electrical resistivity data.

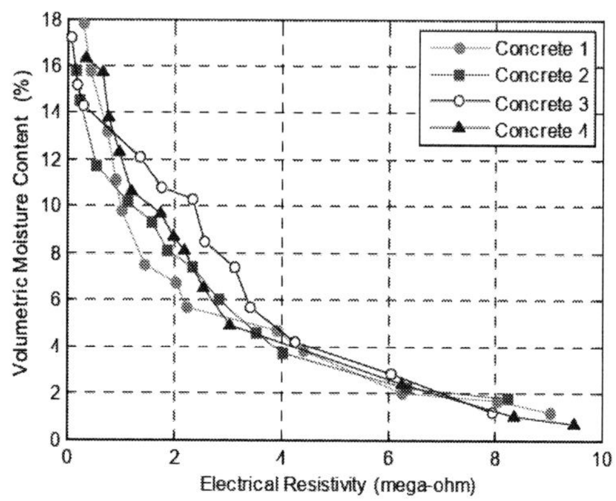

Figure 3. The plot of electrical resistivity data versus volumetric moisture content profile during the wetting process.

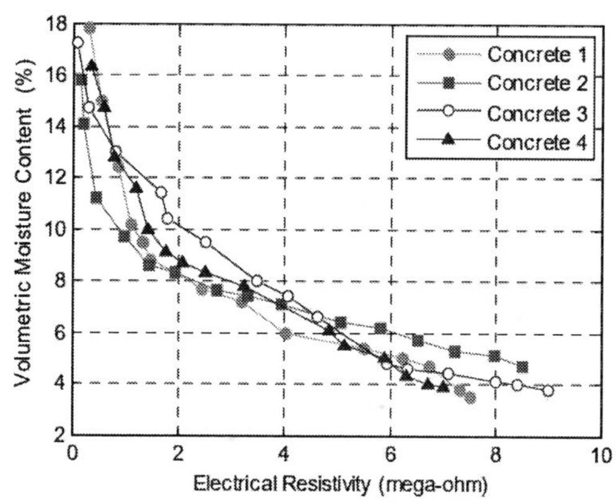

Figure 4. The plot of electrical resistivity data versus volumetric moisture content profile during the drying process.

It can be observed from the interpreted experimental data of electrical resistivity and volumetric moisture content shown in Figure 3 and Figure 4 that there is a decrease in electrical resistivity measurements and increases in volumetric moisture content, during both wetting and drying process and it is evident that the trend was similar for all the concrete samples.

6 Data-driven Approach Using Gaussian Process Regression (GPR) Modeling

This section describes the data-driven machine learning approach using Gaussian Process Regression (GPR) for predicting the moisture conditions on the concrete. The Gaussian Process (GP) modeling approach in [21] was used in this study to train the non-parametric model that obtains electrical resistivity (R) values as inputs and predicts the percentage volumetric moisture content (VMC). The motive is to learn the function f in the form as in equation (11)

$$VMC = f(R) + \xi \tag{11}$$

where ξ is the uncertainty. Let X_1, X_2, X_3 and X_4 be the electrical resistivity data and Y_1, Y_2, Y_3 and Y_4 be the volumetric moisture content of four different concrete samples from the experimental program during wetting and drying process. The data from each concrete was combined together to increase the training data points for GP modeling. Now, $X = [X_1 + X_2, + X_3 + X_4]$ and $Y = [Y_1 + Y_2, + Y_3 + Y_4]$.

Let $[X, Y]$ be the training data set input to GP model, where $X = [x_1, x_2, x_3 .... x_m]^T$ and $Y = [y_1, y_2, y_3 .... y_m]^T$. The X is a matrix of training input extracted from electrical resistivity measurements having $x_i = [R_i]^T$ and i ($1 \leq i \leq$ m), where is i is an integer and m is number of captured data points. The Y is the training vector label data of volumetric moisture content corresponding to the training targets having $y_i = VMC_i$. Likewise, $[X^*, Y^*]$ be the testing data set to the GP model, where $X^* = [x^*_1, x^*_2, x^*_3 .... x^*_n]^T$ and $Y^* = [y^*_1, y^*_2, y^*_3 .... y^*_n]^T$. The X^* is a matrix containing testing data and Y^* is a vector containing predicted outputs corresponding to X^*. Once the function has been learned using $[X, Y]$, the f can be used to predict Y^* for testing data X^* as in equation (12).

$$VMC^* = f(R^*) + \xi^* \tag{12}$$

From [21], a kernel $K(X, X)$ having elements $k_{i,j} = k(x_i, x_j)$ was selected to apply GP modeling technique for the non-linear regression problem in this study. After assessing number of generally used kernels, a squared-exponential kernel having a one-dimensional form with free parameters as in (13) was chosen for this work.

$$k(x_i, x_j) = \alpha^2 exp\left\{-\frac{1}{2\beta^2}\left(x_i - x_j\right)^2\right\} \tag{13}$$

where the α and β in (13) are free parameters that were learned from the training input data set $[X, Y]$. Normally, the free parameters are the hyper-parameters. The technique of training GPR model in [21] by minimizing the negative log marginal likelihood in (14) with respect to the hyperparameters $\theta = \{\alpha, \beta, \sigma_n\}$ was performed while formulating the prediction model.

$$-log\, p(Y|X, \theta) = \frac{1}{2} Y^T \Sigma^{-1} Y + \frac{1}{2} log\, |\Sigma| + \frac{m}{2} log\,(2\pi), \tag{14}$$

The covariance function denoted by Σ in (14) is expressed in (15)

$$\Sigma = K(X, X) + \sigma^2_n I \tag{15}$$

The basic GPR model equations are given by (16) and (17).

$$\mu^* = K(X^*, X)\{K(X, X) + \sigma^2_n I\}^{-1} y \tag{16}$$

$$\Sigma^* = K(X^*, X^*) + \sigma^2_n I - K(X^*, X)\{K(X, X) + \sigma^2_n I\}^{-1} K(X, X^*) \tag{17}$$

The predicted volumetric moisture content (Y^*) for the testing input vector (X^*) will be therefore given by the mean of the posterior distribution μ^* and the associated uncertainty will be given by the covariance Σ^*. The functional block diagram in Figure 5 describes the formulation of GPR model for the data-driven approach.

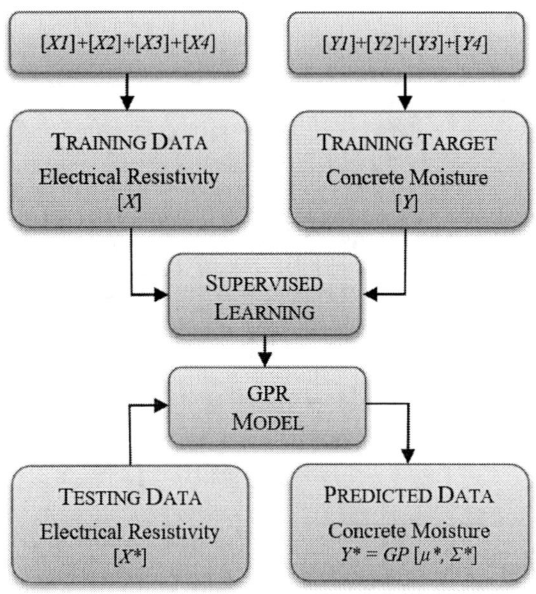

Figure 5. Functional Block Diagram: Training GPR Model and Predicting Output.

7 Results and Analysis

This section presents the results of data-driven prediction modeling using GPR and analysis of the formulated analytical model. The experimental data from the wetting and drying phase of concrete were combined together to increase the number of training data input set feature for GP modeling. Figure 6 shows the obtained input training data in 2D for GP modeling.

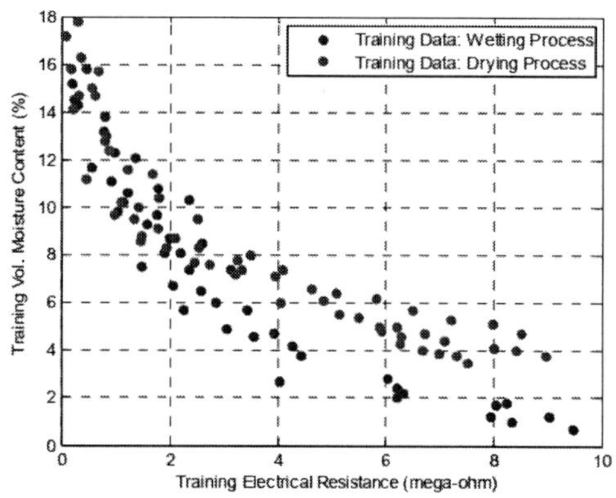

Figure 6. Learned GP model with training data.

For the given electrical resistivity testing data from wetting and drying process, the volumetric moisture content was predicted. Figure 7 shows the Gaussian Process model predictions in 2D.

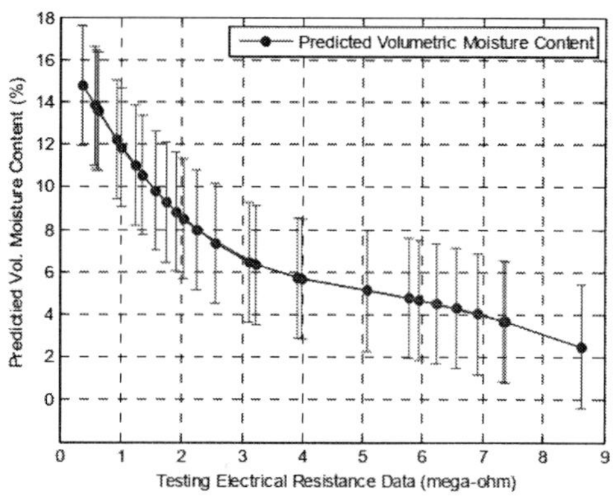

Figure 7. Gaussian Process predictions in 2D

The behavior of predicted volumetric moisture content with respect to training volumetric moisture content data is shown in Figure 8. It can be observed that the Gaussian Process predicted data and the training data fits within the 2σ uncertainty bounds as it learns the function f from equation (12).

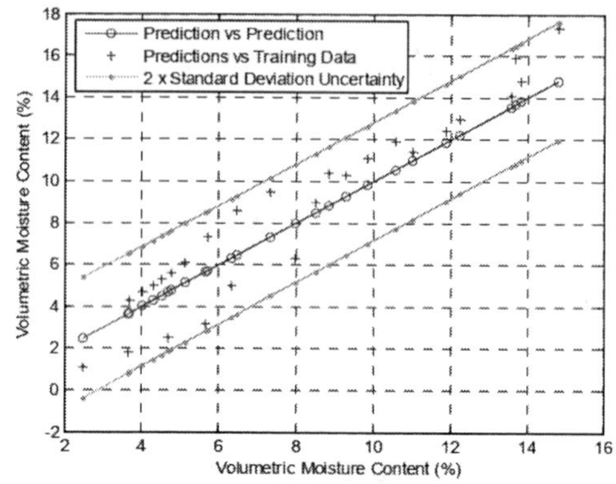

Figure 8. The behavior of predicted values corresponding to training data.

The statistical metric, Root-Mean-Square (RMS) error metric was used as a criterion to compute prediction performance. The RMS error between the GP predicted data and measured data (testing data) is 1.5020 % while the mean absolute error is 1.3412 %. Figure 9 compares predictions to measured values (testing data) from the concrete that was used for testing purpose.

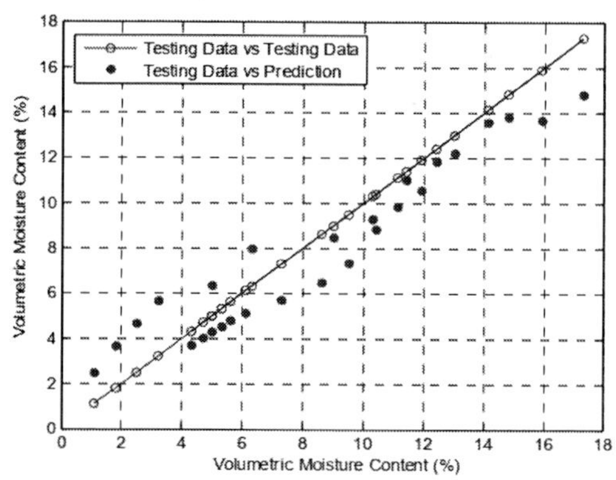

Figure 9. The behavior of Predictions vs. Measured values (testing data).

The analytical model curve in Figure 10 is based on the measurements of values that were substituted in the analytical model equation (7). The aqueous solution that was used to soak the concretes has a resistance value that was measured to be R_w=35KOhm. As from the literature, the tortuosity factor (a) was chosen as 1 and the cementation exponent was adjusted to be 1.65 in this study. The cementation exponent m for rock varies from 1.3 to 2.2 [22]. By choosing arbitrary values of resistivity from 0.5 MOhm to 10 MOhm, the volumetric moisture content of concrete material was determined using analytical model equation (7). Figure 12 presents the plot of analytical model data with experimental data that was obtained during the wetting-drying process, and Figure 11 presents the curve of predicted moisture profiles and determined curve from the analytical model equation.

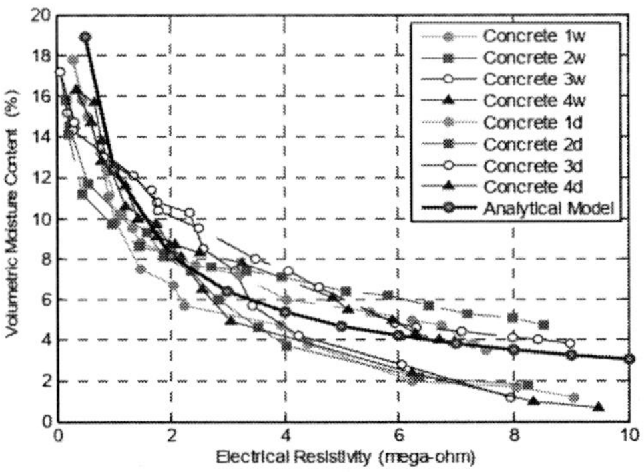

Figure 10. The plot of analytical model data with the experimental measurement data.

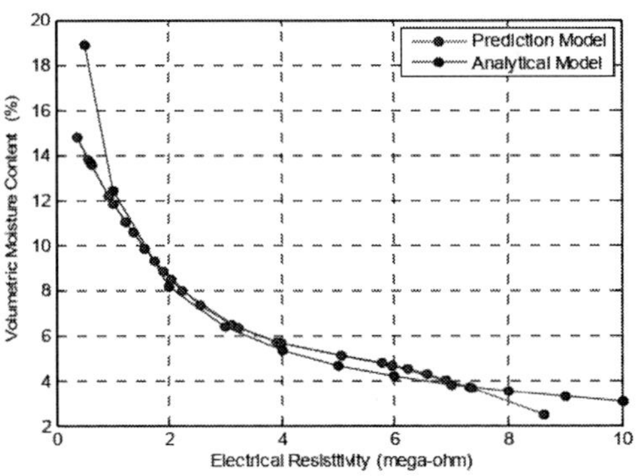

Figure 11. The plot of predicted model data and analytical model data.

From Figure 7, it can be observed from the predicted data has a high value of error bounds. This is primarily because of the experiment that was carried out on various concrete samples, which may differ in porosity, concrete material composition and water-cement ratio. The key element of the research work was to estimate the levels of moisture conditions (0%, 25%, 50, 75 and 100%). Having the high value of error bound in predicted data may not significantly affect accuracy in predicting moisture levels. As mentioned in the previous section, the majority of concrete corrosion happens on the crown of sewer concrete pipe. As we intend to mount the sensory unit in the crown of the sewer pipe, we may need corrosion resistant bolts to mount sensing enclosure unit. These bolts may be also used as electrodes to obtain the moisture conditions of the concrete. The resistivity based measurements need calibration, as it is susceptible to errors due to other chemicals that are present in sewer walls.

8 Conclusion and Future Work

The study in this paper presented a data-driven machine learning approach for predicting the volumetric moisture conditions of concrete samples based on the interpreted electrical resistivity measurements data that were obtained by embedding the electrodes near to the concrete surface, which was exposed to the aqueous solution. Also, the study investigated the use of electrical resistivity measurements for determining the moisture conditions of concrete samples. The results of experimental data show that the electrical resistivity of concretes increases with the decrease in volumetric moisture content data during both wetting and drying process. The GPR modeling by training data from both wetting-drying processes measurements was presented, analyzed and their prediction performance was assessed via RMS error. The data obtained from the analytical model was compared and analyzed with the data-driven predicted data and experimental data. The experimental methodology carried out in this paper provides the preliminary results on moisture content profile on concretes during wetting and drying process. Further study needs to be conducted for evaluating the reliability of resistance based measurements on varying surface pH and ambient H_2S levels.

9 Acknowledgement

This publication is an outcome from the Data Analytics on Sewers Project funded by Sydney Water Corporation, Melbourne Water Corporation, Water Corporation (WA) and South Australian Water Corporation. The research participants are NICTA (lead), University of Technology Sydney and The University of Newcastle. Karthick Thiyagarajan receives UTS President's Scholarship (UTSP) and International Research Scholarship (IRS) from The University of Technology Sydney.

References

[1] Romanova A., Mahmoodian M. and Alani M.A. Influence and interaction of temperature, H2S and pH on concrete sewer pipe corrosion. *International Journal of Civil, Architectural, Structural, Urban Science and Engineering*, 8(6): p. 592-595, 2014.

[2] Mori T., Nonaka T., Tazaki K., Koga M., Hikosaka Y. and Noda S. Interactions of nutrients, moisture and pH on microbial corrosion of concrete sewer pipes. *Water Research*, 26(1): p. 29-37, 1992.

[3] Joseph A.P., Keller J. Bustamante H. and Bond P.L. Surface neutralization and H2S oxidation at early stages of sewer corrosion: Influence of temperature, relative humidity and H2S concentration. *Water Research*, 46(13): p. 4235-4245, 2012.

[4] Wells P. and Melchers R.E. Microbial Corrosion of Sewer Pipe in Australia–Initial Field Results. In *Proceedings of the 18th International Corrosion Congress*, November 2011.

[5] Jiang G., Wightman E., Donose B.C., Yuan Z., Bond P.L. and Keller J. The role of iron in sulfide induced corrosion of sewer concrete. *Water research*, 49: p. 166-174, 2014.

[6] Thiyagarajan K. and Kodagoda S. SMART monitoring of surface temperature and moisture content using multisensory data fusion. In *Proceedings of the Cybernetics and Intelligent Systems (CIS) and IEEE Conference on Robotics, Automation and Mechatronics (RAM), 2015 IEEE 7th International Conference on*, pages 222-227, 2015.

[7] Thiyagarajan K., Kodagoda S. and Ulapane N. Data-driven Machine Learning Approach for Predicting Volumetric Moisture Content of Concrete Using Resistance Sensor Measurements. In *Proceedings of the Industrial Electronics and Applications (ICIEA), 2016 IEEE 11th Conference on*, pages 1288-1293, 2016.

[8] Wells T. and Melchers R. Modelling concrete deterioration in sewers using theory and field observations. *Cement and Concrete Research*, 77: p. 82-96, 2015.

[9] Zhang L., Schryver P.D., Gusseme B.D., Muynck W.D., Boon N. and Verstraete W. Chemical and biological technologies for hydrogen sulfide emission control in sewer systems: a review. *Water research*, 42(1): p. 1-12, 2008.

[10] Parrott L. A review of methods to determine the moisture conditions in concrete. 1990.

[11] Xiao L. and Li Z. Early-age hydration of fresh concrete monitored by non-contact electrical resistivity measurement. *Cement and Concrete Research*, 38(3): p. 312-319, 2008.

[12] Whittington H.W., McCarter J. and Forde M.C. The conduction of electricity through concrete. *Magazine of concrete research*, 33(114): p. 48-60, 1981.

[13] McCarter W., Ezirim H. and Emerson M. Properties of concrete in the cover zone: water penetration, sorptivity and ionic ingress. *Magazine of Concrete Research*, 48(176): p. 149-156, 1996.

[14] Rajabipour F., Weiss J., Shane J.D., Mason T. and Shah S.P. Procedure to interpret electrical conductivity measurements in cover concrete during rewetting. *Journal of Materials in Civil Engineering*, 17(5): p. 586-594, 2005.

[15] Rajabipour F., Sant G. and Weiss J. Development of electrical conductivity-based sensors for health monitoring of concrete materials. In *Proceedings of the TRB Annual Conference*, 2007.

[16] Ghosh P. and Tran Q. Correlation Between Bulk and Surface Resistivity of Concrete. *International Journal of Concrete Structures and Materials*, 9(1): p. 119-132, 2015.

[17] Archie G.E.The electrical resistivity log as an aid in determining some reservoir characteristics. *Transactions of the AIME*, 146(01): p. 54-62, 1942.

[18] Andrade C. and D'Andréa R. Concrete mixture design based on electrical resistivity. In *Proceedings of the Second International Conference on Sustainable Construction Materials and Technologies*. 2010.

[19] Grellier S., Reddy K.R., Gangathulasi J., Adib R. and Peters C. Correlation between electrical resistivity and moisture content of municipal solid waste in bioreactor landfill. *Geotechnical Special Publication*, 163: p. 1-14, 2007.

[20] Alam M.N., Bhuiyan R.H., Dougal R. and Ali M. Concrete moisture content measurement using interdigitated near-field sensors. *Sensors Journal, IEEE*, vol. 10, pp. 1243-1248, 2010.

[21] Rasmussen C.E. Gaussian processes for machine learning. 2006.

[22] Breysse D. Non-Destructive Assessment of Concrete Structures: Reliability and limits of Single and Combines Techniques. *Springer*, 2012.

Automated Classification of Detected Surface Damage from Point Clouds with Supervised Learning

Burcu Guldur[1] and Jerome F. Hajjar[2]
[1]Assistant Professor, Department of Civil Engineering, Hacettepe University, Ankara, Turkey
burcuguldur@hacettepe.edu.tr
[2]CDM Smith Professor and Chair, Department of Civil and Environmental Engineering, Northeastern University, Boston, MA
JF.Hajjar@neu.edu

Abstract –

Recent advances in sensing technologies provide opportunity to utilize advanced data collection equipment such as laser scanners, high-resolution cameras, etc. for both short- and long-time monitoring of structures. Laser scanners, which are capable of collecting up-to two million points per second along with high-resolution images, are especially evolving rapidly and their usage for capturing and documenting the current condition of varying structural types is becoming increasingly feasible. The authors' previous research has focused on developing generalized algorithms for detecting surface damage on structures from captured laser scans and images. These algorithms are capable of automatically detecting surface damage on varying structural types as well as several construction materials since they use underlying surface geometry for performing damage detection. These damage types include concrete cracking, concrete spalling, steel delamination, steel section loss, bent members, and other misalignments. This paper investigates the use of supervised learning methods for determining whether a detected region represents an actual damage location. First, the input feature representations of the learning algorithms for each damage type are determined and the associated training sets, which are representative of real-world use of the algorithms, are prepared. Once the training and validation are completed, the learning algorithms are tested on the detected damage data, which is obtained by running the damage detection algorithms on both laser scans and texture-mapped images from a concrete test setup. Finally, the accuracy of these learning algorithms is investigated.

Keywords –

Terrestrial Laser Scanning; Surface Damage Detection; Damage Classification; Supervised Learning

1 Introduction

In the past two decades, researchers have developed several methodologies for both monitoring structures and detecting damage by using laser-scanning technology. The common methodologies used for structural monitoring include measuring high accuracy displacements, strains, pressures, or using a small number of points for computing related quantities or collecting visual inspection information [1]. Since the texture-mapped point cloud-capturing laser scanning is a recent technology, its usage for damage detection has not been investigated broadly. A couple of novel advances include the usage of point clouds to generate Building Information Model (BIM) of reinforced concrete walls and performed BIM–based earthquake damage assessment [2] and the utilization of LiDAR data and imagery for performing post-earthquake assessment, in order to determine the building damage degree [3].

Laser scanning technology provides a 3D representation of the entire structure under investigation. As has been discussed, with the current developments in this technology, it is possible to collect high-density point clouds that capture the surface properties accurately for structures. Authors' previous research focuses on the surface damage detection capabilities of the laser scanning technology that captures 3D point clouds with color information [4; 5; 6; 7]. The developed damage detection algorithms provide an adaptive system for damage detection, since the parameters used for the detection algorithms are computed automatically using the associated surface properties. The strategies developed for laser-based structural sensing and surface damage detection have been discussed extensively in [5].

For this paper, these developed damage detection algorithms are used to determine the damage locations on a concrete testing frame, and then, the detected surface damage is quantified. Since the main focus of this paper is to use supervised learning algorithms for damage classification, the detailed damage detection results obtained from the concrete testing frame are not included in the text, however, these results could be found in [5].

Once the damage detection and quantification are completed, the damage detection results obtained from the concrete testing frame are used for both training and validation of the learning algorithm, and finally, its accuracy is investigated.

The brief information on the methodologies used for surface damage detection and the detailed explanation on the use of supervised learning algorithms for damage classification are discussed in the following sections.

2 Laser-based Surface Damage Detection

Two main surface damage detection methodologies were developed in authors' previous research. The first methodology is a surface-normal based damage detection method that only uses the 3D coordinate information for locating rupture, spalling, delaminations. This method is later improved by using intensity values along with the 3D point information for locating small deformations such as cracks, corrosion.

The second methodology, on the other hand, is a graph-based damage detection method that is used for detecting alignment issues and points of discontinuity. This method is an extension of the graph-based object detection method that generates skeletons from cross-section cuts of a voxelized cluster, where a voxel represents a single sample or data point on a regularly spaced, three-dimensional grid, through extracting skeleton of an object in order to detect common structural members. The deviations from the predicted object alignments are used for extracting problematic locations on structures. Then, another method is introduced that converts cross-section voxel representation automatically into a polygon for computing the changes in the cross-section through area calculation and determining the total volume change on the investigated member.

Detected defects are automatically clustered and a mesh grid-based defect area and volume extraction method is developed in order to obtain quantifiable defect outputs for further investigation. For smaller defects such as cracks, an additional methodology is proposed for automated crack length and width extraction.

2.1.1 Surface Normal-based Damage Detection

The surface normal-based damage detection method relies on the modal properties of the detected surfaces and/or objects. The relative orientation of the estimated surface normal with respect to a reference normal is used to locate the defected areas on the surface of structures. The reference normal can be a surface normal computed via surface patches; the normal representing the skeleton of the detected object; or the normal vector between a reference point and the current query point.

The usage of only the normal variations on the surface would be sufficient for detecting large surface defects. However, another parameter should be introduced if certain defects with sizes close to the resolution of the scanner, such as cracks, are to be detected. Thus, the pixel information (intensity) obtained from texture-mapped point clouds is used for enhancing the detection capabilities for smaller defects.

In order to calculate both area and volume of a detected defect, a mesh-grid is fitted to the defect surface. Later, each part of the grid (for the area) and the corresponding rectangular prism (for the volume) are used for quantifying the detected damage. This damage quantification method is used to compute the area and volume associated with each detected defect.

For the concrete testing frame, the developed damage detection methods are used for both locating and quantifying concrete spalling. Several point cloud patches, which are extracted from the surface of the concrete testing frame, are processed. A set of representative dimensions are extracted from the detected damage areas, and these dimensions are compared with the corresponding hand-measurements in order to validate the efficiency of the detection algorithms.

Even though the mentioned quantification method is capable of both locating and quantifying relatively large defects, it is not suitable for recording required crack dimensions, which can be listed as length and thickness. Thus, a new methodology is developed for computing the necessary crack dimensions automatically.

2.1.2 Improvements for Crack and Spalling Detection

In most of the previous crack detection studies, which are predominantly image-based, some of the important parameters, such as camera-object distance, are not considered or are assumed to be constant. This prevents most of the current approaches from being used for crack quantification, since these methods are specifically developed for crack detection rather than quantification. For current approaches, it is required to maintain a constant focal length, resolution, or distance to the object in order to be able to extract crack dimensions ([8]; [9];

[10]; [11]; [12]; [13]). The damage detection method discussed in authors' previous research eliminates the requirement for prior knowledge on focal length, resolution, or distance to the investigated object, since all the required parameters for the defect detection are extracted from the point cloud automatically. However, it should be noted that the proposed method, similar to many other damage detection methods, especially crack detection methods, results in false positives along with actual defects.

The first step towards automated crack dimension extraction is to adjust the existing clustering algorithm, cluster tuning, which separates detected defect regions into individual defects, for cracks, and then quantify the dimensions of each separated point cluster with the proposed crack dimension extraction method. The crack dimensions are computed by defining a bounding box around each crack and performing a dimension extraction procedure as explained in [5].

The results obtained through defect detection and dimension extraction algorithms prove that these developed methods provide an opportunity to use laser scanning technology for detecting small defects, such as cracks and spalling regions, effectively. However, it should be noted that since the defect detection is achieved through investigating the local variations at a point of interest on the surface, the developed detection algorithms are sensitive to surface impurities. These impurities and/or any reoccurring patterns result in false positives when the damage detection algorithms are executed. Thus, the obtained results include several falsely detected clusters, which do not represent an actual defect. In order to avoid detecting these false positives, to improve the accuracy of the developed algorithms and to perform automated classification, a neural network classifier is introduced. The details of this classification strategy and the classifier's overall performance on the concrete testing frame are discussed in the following section.

3 Damage Classification with Supervised Learning

The detected false positives for damage results could be an issue due to a repeated pattern, any surface impurity, and etc.; essentially, any surface variation that causes the surface normal to deviate from the computed reference normal and/or results in significant variations in local intensity values. Two significant example cases that are encountered for this dataset are shown in Figure 1(a) and (b). In Figure 1(a), the locations of small surface holes are shown with black circles, and a repetitive pattern for intensity variation, resulted from inefficient texture-mapping, is represented in Figure 1(b). It is required to eliminate the false positives from the detected damage clusters in order to improve the overall efficiency of the proposed defect detection algorithm. To perform this, a trained artificial neural network classifier is used to differentiate the real defects (cracks and/or small spalling regions) from false positives.

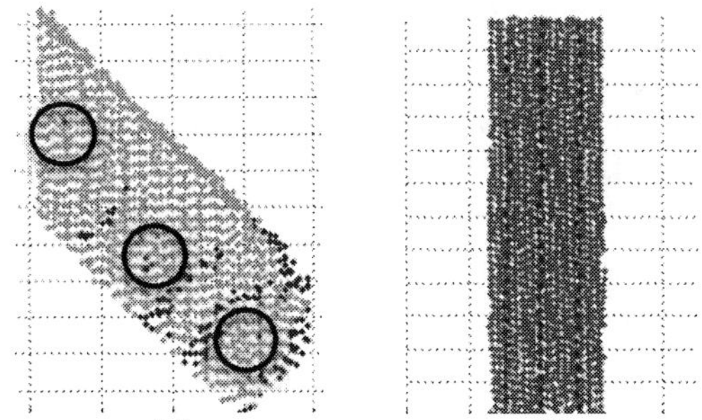

(a) (b)

Figure 1. Sample false positive patterns: (a) small holes existing on the surface and (b) lines with significant intensity variation

In [14], an image-based study, which uses 2D images to create 3D surface representations, was performed, in order to eliminate the dependency on the previously listed parameters: constant focal length, resolution, or distance to the object. The first step of this study is segmentation, which is used for isolating the patterns that can be potentially classified as defects. The useful information about the scene objects is extracted by using morphological image processing and then, structuring elements are used to complete the segmentation. Segmentation is followed by feature extraction and finally, the study is completed by classification.

In this paper, a similar methodology for classification is followed. First, the defects are segmented as objects by using the developed clustering methods. Second, a feature set that stores quantitative information on each defect cluster is formed. Finally, this feature set is used to train, validate and test the neural network classifier.

3.1 Feature Extraction

A feature is defined as a set of finite values that represents the quantitative attributes or properties of any segmented object; in our case, clusters. It is crucial to generate an appropriate feature set that includes all the important characteristics that helps identifying similar patterns. In this paper, a feature set, which is similar to the one proposed in [14], is used, with two additions representing the properties of intensity distribution, since both application have a similar nature. The feature set defined in [14] is extracted by using only the 2D properties of the segmented objects; however, since the detected defect clusters represent a 3D dataset, for this research, some of the described features are extracted by using 3D information.

The features that are included in the feature set can be listed as eccentricity, area of the projected cluster divided by the area of the bounding ellipse, solidity, absolute value of the correlation coefficient, compactness, and mean and standard deviation of intensity values. The definitions of each item in the listed feature set could be found in both [14] and [5].

These features listed above are computed for each defect cluster, and the values are stored in a feature matrix. This feature matrix is then used for training, validating and testing the neural network classifier.

3.2 Classification

There are several classifier options that can be used for this application. Some of the possible classifiers can be listed as Bayes classifier [15], k-nearest neighborhood classifier [16], multilayer feed-forward artificial neural networks [17], support vector machines [18], and etc. However, in most of the recent studies performed for crack detection, where several classifiers are compared, the accuracy of the results obtained by using neural network classifiers is shown to be higher than the other listed methods ([8]; [9]; [14]; [19]). Thus, for this research, the neural network classifier is selected to be used for classification.

An artificial neural network, which is composed of processing elements that are interconnected via synaptic or weighted connections, is a parallel processing optimization system. In a neural network, weighted interconnections are used to process inputs received at a processing element in a layer, and then the resulting outputs are transmitted to the following set of processing elements. For nonparametric pattern classification, feed-forward neural networks, which are trained by using a back-propagation algorithm, are the most commonly used neural approaches [20]. This type of neural network can be composed of three or more layers of processing elements: the input layer, hidden layer/layers, and output layer. Through weighted interconnections, each processing element in a layer is connected with all processing elements in the preceding and following layers. For pattern classification applications, the input layer consists of a set of feature vectors; thus, the size of the input layer is always equal to the number of features. However, it should be noted that the number of processing units in hidden layers is completely dependent on the complexity of the pattern recognition problem. The number of neurons in the hidden layer is generally empirically determined by trial and error [21]. The number of layers and the number of neurons in the hidden layers of neural networks have to be chosen carefully in order to obtain good classification results. Since the computed feature matrix is composed of similar entities, in this research, the neural network configuration given in [14] is used. The classifier used for this application is a three-layer feed forward neural network, which has 2 output neurons and 10 neurons in the hidden layer.

4 Results

The developed damage detection algorithms are used to process 106 randomly selected surface patches from the concrete testing frame. As a result, 201 candidate defect clusters are detected and separated for further analysis. These defect clusters are plotted, and the actual defects and the false positives are manually separated for every patch.

The generated feature set is composed of 201 damage and non-damage feature vectors. Out of 201 feature vectors, 74 vectors represent the properties of actual defects, whereas 127 vectors are composed of features extracted from false positives. For training, % 70 of the complete feature set is used; 15% is used for validation and finally, 15% is used for testing.

The performance of the selected classifier is shown by using four items: accuracy, precision, sensitivity, and specificity. Accuracy shows the proportion of true classifications in the test set (15% of the entire feature set). Precision is defined as the proportion of the true positive classifications against all positive classifications. Sensitivity is the proportion of actual positives that were correctly classified, and specificity is the proportion of negatives that were correctly classified. The results are shown in Table 1.

Figure 2(a) shows a portion of the concrete testing frame, and Figure 2(b) presents the post-classification damage detection results on the corresponding region of the point cloud.

Table 1 Performance results for the neural network classifier

	Percentage (%)
Accuracy	93.51
Precision	94.45
Sensitivity	95.89
Specificity	90.38

5 Conclusions

The results show that the accuracy of the neural network classification is high for this specific application.

However, it should be noted that the detected defect clusters sometimes fail to represent all the surface damage that is present on an investigated patch. Some of the defects may not be detected depending on the properties of the point cloud. For these cases, even though the accuracy of the classifier is high for the generated feature set, the obtained results may not represent the overall efficiency of the developed damage detection algorithm. At the same time, it should be mentioned that the performance of the trained neural network on another structure cannot be estimated from the obtained results; however, in literature, there are several examples that show the accuracy of a trained neural network reduces when the classifier is tested on different structures.

Further study on the subject, which includes the application of the learning algorithms on larger datasets with varying surface properties, needs to be performed, in order to build a comprehensive understanding on the use of supervised learning algorithms for laser scanning-based damage detection and classification applications.

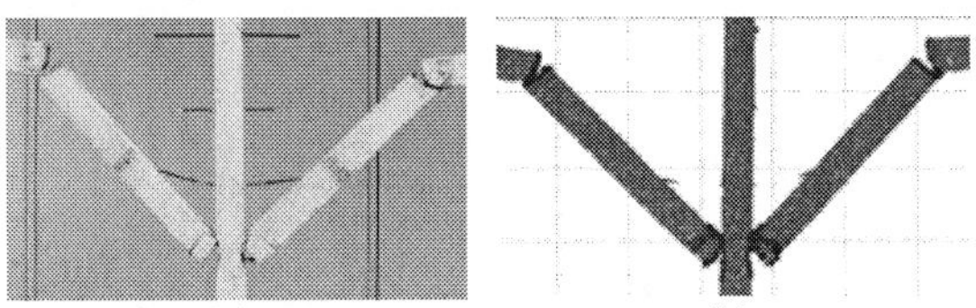

(a) (b)

Figure 2. (a) Image of a portion of the concrete testing frame, and (b) defect detection results shown on corresponding portion of the point cloud

Acknowledgements

This material is based upon work supported by the National Science Foundation under Grant No. IIS-1328816 and Northeastern University. Any opinions, findings and conclusions or recommendations expressed in this material are those of the authors and do not necessarily reflect the views of the National Science Foundation.

References

[1] Chang, P. C., Flatau, A. and Liu, S. (2003). "Review Paper: Health Monitoring of Civil Infrastructure." *Structural Health Monitoring*, 2(3), 257-267.

[2] Anil, E. B., Akinci, B., Kurc, O. and Garrett, J. H. (2015). "Building-Information-Modeling–Based Earthquake Damage Assessment for Reinforced Concrete Walls." *Journal of Computing in Civil Engineering*, 04015076.

[3] Li, M., Cheng, L., Gong, J., Liu, Y., Chen, Z., Li, F., Chen, G., Chen, D. and Song, X. (2008). "Post-earthquake Assessment of Building Damage Degree using LiDAR Data and Imagery." *Science in China Series E: Technological Sciences*, 51(2), 133-143.

[4] Guldur, B. and Hajjar, J. F. (2013). "Laser-based Automatic Cross-Sectional Change Detection for Steel Frames." *Proceedings of the 9th International Workshop on Structural Health Monitoring 2013*, Stanford University, Stanford, CA.

[5] Guldur, B. (2014). "Laser-based Structural Sensing and Surface Damage Detection." Ph.D. Dissertation, Northeastern University, Boston, Massachusetts.

[6] Guldur, B. and Hajjar, J. F. (2014). "Damage Detection on Structures Using Texture Mapped Laser Point Clouds." *Proceedings of the ASCE/SEI Structures Congress 2014*, Boston, MA.

[7] Guldur, B. and Hajjar, J. F. (2014). "Automated Damage Assessment from 3-D Laser Scans." *Proceedings of the Istanbul Bridge Conference*, Istanbul, Turkey.

[8] Kaseko, M. S. and Ritchie, S. G. (1993). "A Neural Network-based Methodology for Pavement Crack Detection and Classification." *Transportation Research Part C: Emerging Technologies*, 1(4), 275-291.

[9] Cheng, H., Chen, J.-R., Glazier, C. and Hu, Y. (1999). "Novel Approach to Pavement Cracking Detection based on Fuzzy Set Theory." *Journal of Computing in Civil Engineering*, 13(4), 270-280.

[10] Abdel-Qader, I., Abudayyeh, O. and Kelly, M. E. (2003). "Analysis of Edge-detection Techniques for Crack Identification in Bridges." *Journal of Computing in Civil Engineering*, 17(4), 255-263.

[11] Abdel-Qader, I., Pashaie-Rad, S., Abudayyeh, O. and Yehia, S. (2006). "PCA-based Algorithm for Unsupervised Bridge Crack Detection." *Advances in Engineering Software*, 37(12), 771-778.

[12] Choudhary, G. K. and Dey, S. (2012). "Crack Detection in Concrete Surfaces using Image Processing, Fuzzy Logic, and Neural Networks." *IEEE Fifth International Conference on Advanced Computational Intelligence*, Nanjing, Jiangsu, China.

[13] Rabah, M., Elhattab, A. and Fayad, A. (2013). "Automatic Concrete Cracks Detection and Mapping of Terrestrial Laser Scan Data." *NRIAG Journal of Astronomy and Geophysics*, 2(2), 250-255.

[14] Jahanshahi, M. R. and Masri, S. F. (2012). "Adaptive Vision-based Crack Detection using 3D Scene Reconstruction for Condition Assessment of Structures." *Automation in Construction*, 22, 567-576.

[15] Rish, I. (2001). "An Empirical Study of the Naive Bayes Classifier." *IJCAI 2001 Workshop on Empirical Methods in Artificial Intelligence*, Seattle, WA.

[16] Keller, J. M., Gray, M. R. and Givens, J. A. (1985). "A Fuzzy k-nearest Neighbor Algorithm." *IEEE Transactions on Systems, Man and Cybernetics*, 4, 580-585.

[17] Hornik, K., Stinchcombe, M. and White, H. (1989). "Multilayer Feed Forward Networks are Universal Approximators." *Neural Networks*, 2(5), 359-366.

[18] Hearst, M. A., Dumais, S., Osman, E., Platt, J. and Scholkopf, B. (1998). "Support Vector Machines." *Intelligent Systems and their Applications, IEEE*, 13(4), 18-28.

[19] Adhikari, R., Moselhi, O. and Bagchi, A. (2014). "Image-based Retrieval of Concrete Crack Properties for Bridge Inspection." *Automation in construction*, 39, 180-194.

[20] Rumelhart, D. E., McClelland, J. L. and Group, P. R. (1995). *Parallel Distributed Processing*, Cambridge, MA, MIT Press.

[21] Kaseko, M. S., Lo, Z.-P. and Ritchie, S. G. (1994). "Comparison of Traditional and Neural Classifiers for Pavement-crack Detection." *Journal of Transportation Engineering*, 120(4), 552-569.

Construction Progress Management and Interior Work Analysis Using Kinect 3D Image Sensors

Kosei Ishida[1]

[1] School of Architecture, Kogakuin University, Tokyo, Japan
E-mail : k_ishida@cc.kogakuin.ac.jp

Abstract –

Construction progress management and work analysis are often difficult in building interiors. In order to make progress management easier, the authors developed methods for recording the interior work environment. These methods record data using Microsoft Kinect sensors.

In this paper, the authors describe methods for recording work data utilizing several Kinect sensors. The authors recorded the following types of data:

1. **The work environment at the construction site**
2. **The shape of the building at the construction site**
3. **The work efficiency at the site derived from the motion capture data**

To record the three types of data, the authors developed a method for performing motion capture using several Kinects.

In addition, a shape recognition method for identifying building materials was developed. The shape recognition method utilizes point cloud data and camera image data. The authors also studied methods for analyzing the work efficiency data on the basis of skeletal tracking information.

Keywords –

Work efficiency; Point cloud; Kinect; Skeletal tracking; Construction management; Interior work

1 Introduction

Construction work is broadly divided into outdoor work such as building frame construction and indoor work such as interior finish work. The progress management and observation of outdoor work at a construction site are relatively easy to implement using methods such as employing the Global Positioning System (GPS) to identify worker positions and trajectories.

However, it is difficult to observe indoor work such as interior finish work because interior workers are located in buildings with room and floor divisions. It is generally not possible to use the GPS inside buildings to record worker positions and trajectories. Construction progress management and work analysis in indoor environments require a great deal of effort.

According to Japan's Ministry of Land, Infrastructure, Transport and Tourism, the cost of renovation and repair work in Japan was 11.02 trillion yen in 2013, and the cost is expected to increase in the future.

Moreover, if humans and robots work together on the construction site, we need the methods of collaborating safely with human workers in the same floor. For safety of construction workers, it is important to record pedestrian trajectories inside buildings.

In order to reduce construction costs with improved progress management, this paper focuses on methods for determining construction-worker positions and trajectories inside buildings. This requires shape recognition and motion capture devices to identify the 3D shape of the building and worker movements inside the building.

Several researchers have developed methods for recording pedestrian trajectories inside buildings. Some researchers have recently shown interest in 3D image sensors such as Kinect. A method for collecting trajectories with multiple Kinect sensors has been developed by Stefan Seera [1].

We also developed methods using Microsoft Kinect sensors [2]. The Kinect sensors collect camera data and calculate a 3D depth map. In addition, a Kinect sensor is able to track the skeletal images of one or two moving people.

In this paper, we describe methods for recording the work environment and worker trajectories using several Kinect sensors. The authors recorded the following types of data:

1. The work environment at the construction site
2. The shape of the building at the construction site
3. The work efficiency at the site derived from the motion capture data

2 Measurement Technique Using Multiple Kinect Sensors

Kinect sensors can maintain tracking through a range of a few meters; however, this is an insufficient range for measurements inside buildings. One Kinect sensor cannot measure all of the intricacies of building spaces, such as stairwells. In order to measure a wide area inside buildings, we have developed a measurement technique using multiple Kinect sensors.

2.1 Comparison of the Precision of 3D Point Clouds Acquired from a Kinect Sensor and 3D Laser Scanner

A Kinect collects camera data and calculates a 3D depth map. We developed a system to create 3D color point clouds from camera data and depth maps.

We then compared the precision of 3D point clouds acquired by a Kinect sensor and a Leica HDS3000 3D laser scanner. Figure 1 shows a Kinect sensor and 3D laser scanner setup in a measurement room with spherical targets. The distance between the Kinect sensor and the wall is 4 m. Figure 2 shows the 3D point clouds acquired by the Kinect (in red) and 3D laser scanner (in green and blue). As can be seen in Figure 2, the precisions of the 3D laser scanner and Kinect sensor are approximately ±3 and ±10 mm, respectively.

Figure 1. Kinect sensor and 3D laser scanner setup in a measurement room with spherical targets

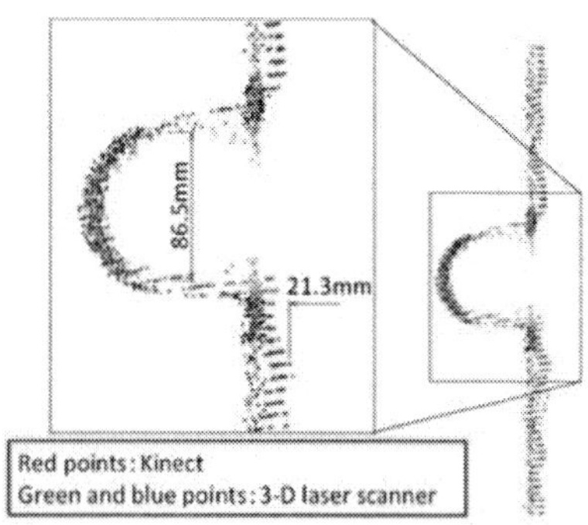

Figure 2. Comparison of the precision of 3D point clouds acquired from a Kinect sensor and 3D laser scanner

2.2 Shape Recognition of Circular and Spherical Targets

Two 3D depth maps with separate individual coordinate systems were used as inputs to perform the registration of point-cloud data based on multiple targets. We used spherical and circular targets to register multiple individual scans on a single coordinate system. Figure 3 shows the results for the shape recognition of circular targets based on 3D images and the measurement of the 3D coordinates of the circle center. Figure 4 shows the point clouds on spherical targets acquired by a Kinect sensor.

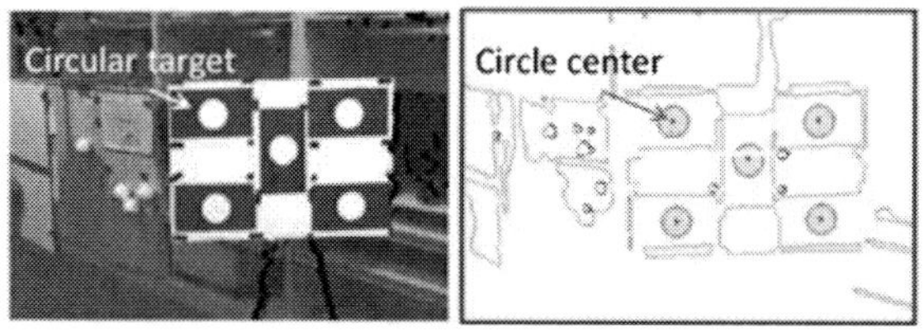

Figure 3. Shape recognition of circular targets based on 3D images and the measurement of the 3D coordinates of the circle center

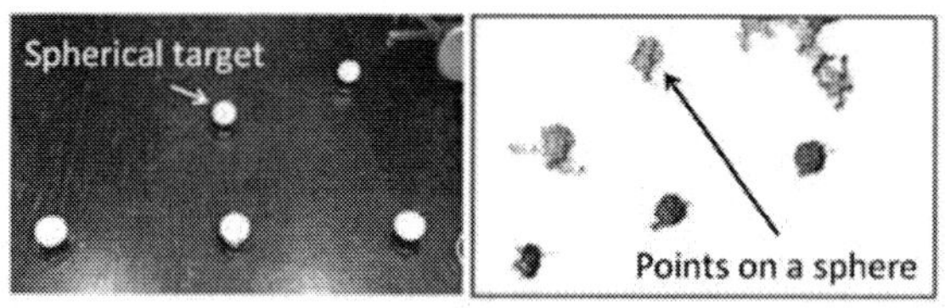

Figure 4. Point clouds of the spherical targets

2.3 Procedure for Registering Multiple Individual Point Clouds on a Single Coordinate System

In order to register multiple individual point clouds acquired by Kinect sensors on a single coordinate system, we developed a procedure for Kinect registration. This procedure consists of the following five steps:

1. Install more than four spherical targets for each Kinect.
2. Map the depth information from each Kinect.
3. Import point-cloud data and extract the points on the spherical targets.
4. Calculate the coordinates of the centers of the spherical targets using the least-squares method.
5. Register the target center coordinates into a CAD coordinate system.

Figure 5 shows spherical targets and the importation of point-cloud data from two Kinect sensors. Figure 6 shows multiple 3D depth images in a complete 3D model.

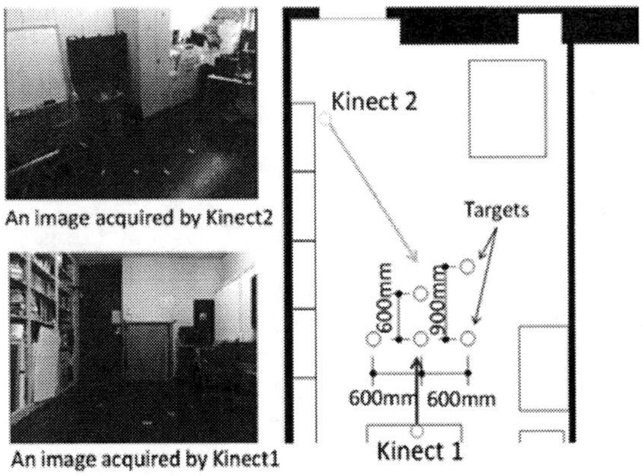

Figure 5. Measuring the position of the Kinect based on the spherical targets

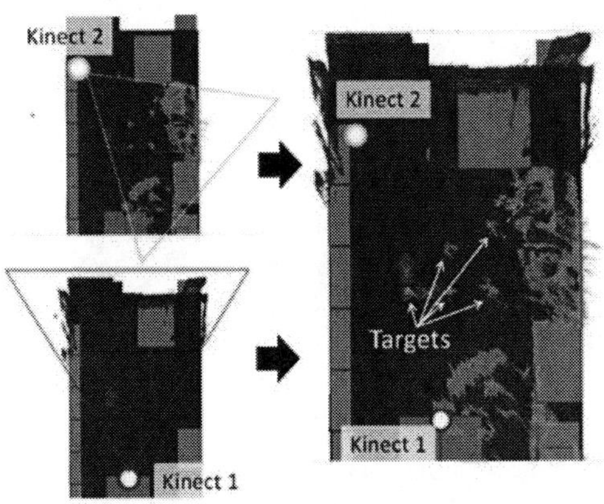

Figure 6. Multiple images in a complete 3D model

2.4 Evaluation of Pedestrian Trajectories Acquired by Kinect Sensors

We evaluate the pedestrian trajectories acquired by two Kinects. We collected skeletal tracking data in the room using two Kinect sensors.

Our evaluation data consist of two trajectory sets: the first data set was acquired by Kinect1 shown in Figures 6 and 8. The second data set was acquired by Kinect2. Two Kinect sensors measure the same pedestrian for 7668 frames ($\approx$ 259 s). The pedestrian walked around the perimeter of a triangle with a side length of 1300 mm. We analyzed the distance of the two trajectories.

Figure 7 shows the triangle on the floor and the positions of Kinect sensors. Figure 8 shows the point clouds and student trajectories acquired by the two Kinect sensors. Figure 9 shows the evaluation results of the pedestrian trajectories acquired by the Kinect sensors, and Table 1 summarizes the statistics of the measurement precision of the pedestrian trajectories.

Figure 9 shows the distribution of the Euclidean distance between the Kinect1 and Kinect2 trajectories acquired simultaneously. The mean and median distances are 82.9 and 70.5 mm, respectively. The deviation in the two trajectories is a few centimeters. The reason for the deviation is considered to be the deviation due to registering multiple individual point clouds on a single coordinate system. In addition, the reason for the deviation is due to the student's orientation in the camera image. Owing to the differences in the acquisition conditions such as the image in the pedestrian facing forward or backward, it is assumed that the two skeleton positions have different positioning of approximately 70 mm.

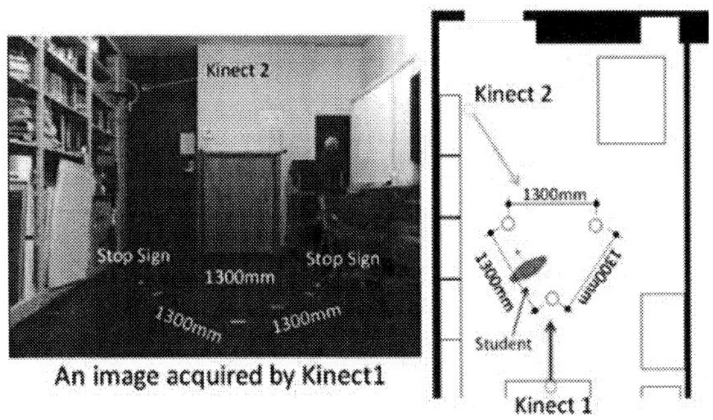

Figure 7. Collecting the skeletal tracking data

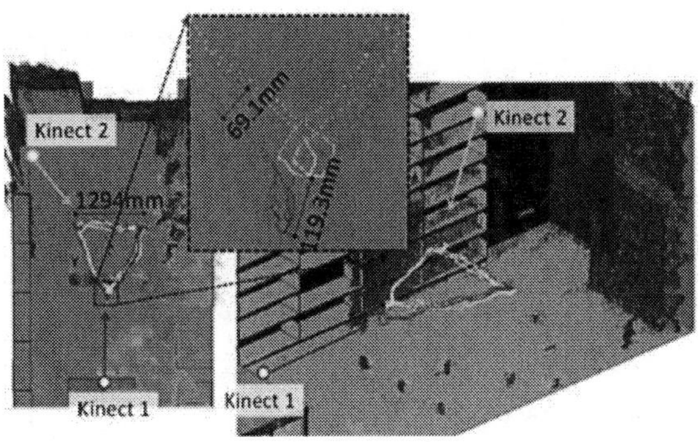

Figure 8. Point clouds acquired by Kinect1 (in blue) and Kinect2 (in green) and the student trajectories obtained by Kinect1 (in red) and Kinect2 (in yellow)

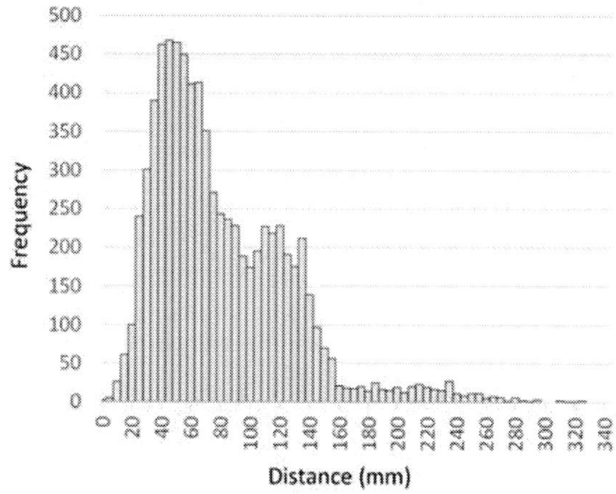

Figure 9. Evaluation results of the pedestrian trajectories acquired by the Kinect sensors.

Table 1 Statistics for the measurement precision of the pedestrian trajectories

Variable	Value
n	7668
Mean (mm)	82.9
SD (mm)	46.3
Maximum (mm)	325.7
Median (mm)	70.5

3 Method for Registering Skeletal Tracking Data from Multiple Kinect Sensors on a Single Coordinate System

The registration procedure discussed in the previous section is used with point-cloud data and 3D depth maps. The registration procedure for skeletal tracking data from multiple Kinect sensors is different because the skeletal tracking data include a time axis. We developed a method of registering skeletal tracking data from multiple Kinect sensors on a single coordinate system.

3.1 Procedure for Registering Skeletal Tracking Data from Multiple Kinect Sensors on a Single Coordinate System

In order to register skeletal tracking data from multiple Kinect sensors on a single coordinate system, we developed the following five-step procedure:

1. Register skeletal tracking data from each Kinect on individual coordinate systems (using the procedure discussed in the previous section).
2. Set up the measurement space for each Kinect.
3. Integrate the skeletal tracking data from all Kinect sensors into one dataset according to the data acquisition time.
4. Perform noise reduction on the skeletal tracking data.
5. Perform smoothing of the skeletal tracking data using a simple moving-average technique.

3.2 Measurement Space Setup for Kinect Sensors

Skeletal tracking data are generated from camera images and 3D depth maps obtained from the Kinect sensor. The shape of the skeleton is different in images obtained from the front and back camera positions. The data precision deteriorates at the point of camera switching. In order to prevent precision deterioration, the measurement space should be set up for each Kinect. Figure 10 shows the setup for four Kinect measurement spaces. The skeletal tracking data that are outside the measurement spaces need to be deleted.

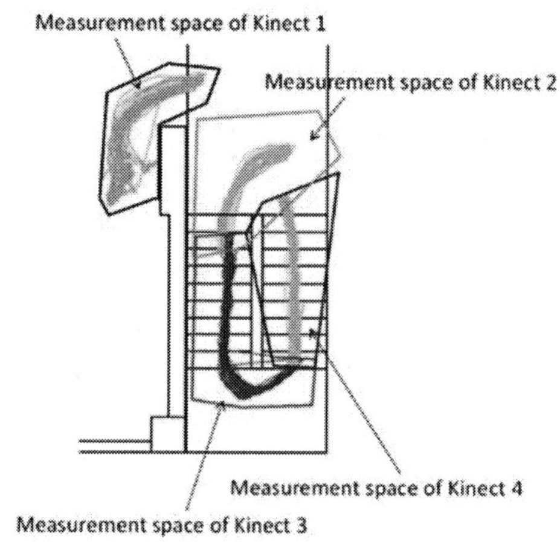

Figure 10. Setup of the Kinect measurement spaces

3.3 Integration of Skeletal Tracking Data into One Dataset According to the Acquisition Time

Skeletal tracking data have XYZ coordinates and Kinect acquisition times. It is necessary to integrate skeletal tracking data into one dataset according to the Kinect acquisition time. The method used to integrate the skeletal tracking data from multiple Kinect sensors into one dataset is shown in Figure 11.

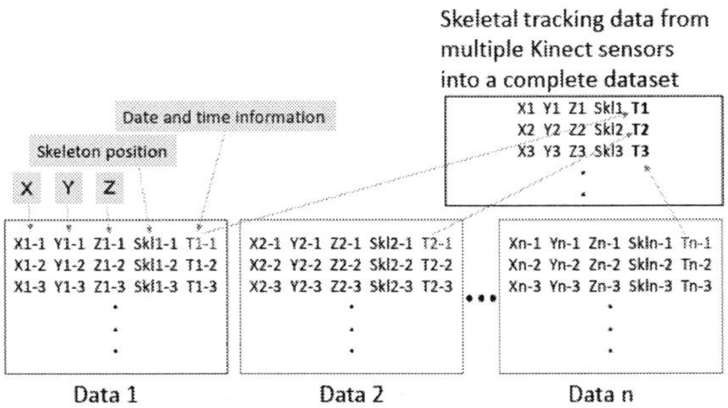

Figure 11. Method for integrating skeletal tracking data from multiple Kinect sensors into one dataset

3.4 Noise Reduction of Skeletal Tracking Data

The advantage of motion capture based on a variant of image-based 3D reconstruction is that it does not require a special marker for motion capture. Owing to the occasional erroneous recognition of skeletal positions, we have developed a noise reduction method for skeletal tracking data.

If the skeletal points exceed a speed of 6 m/s, noise is recognized, and points are deleted as errors. Figure 12 shows the noise reduction for skeletal tracking data.

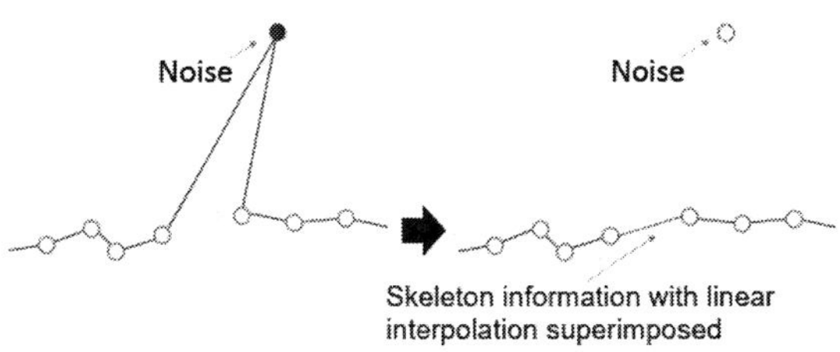

Figure 12. Noise reduction for skeletal tracking data

3.5 Data Smoothing of Skeletal Tracking Data Using a Simple Moving Average Technique

The skeletal tracking data obtained by a Kinect sensor are acquired at an approximate frame rate of 30 Hz. There is a large variability in the accuracy of the individual points. To reduce this variability, we have developed a method to smooth the skeletal tracking data using a simple moving average technique. Figure 13 shows the method for smoothing the skeletal tracking data. We normalize using two data points on both sides of a central point.

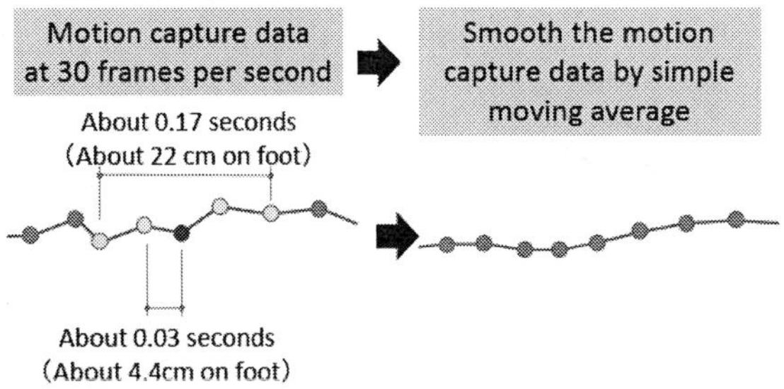

Figure 13. Method for smoothing the motion capture data using a simple moving average technique

4 Collection of Skeletal Tracking Data obtained by a Kinect Sensor in a Building Stairwell

We collected skeletal tracking data in a building stairwell using four Kinect sensors. In this experiment, a student (similar to a worker) hand-carried luggage on building stairs located between the eighth floor and the ninth floor. The student went up and down the stairs 20 times, acquiring data that was subsequently analyzed.

4.1 Kinect Setup in a Stairwell

To collect skeletal tracking data in a stairwell, we installed four Kinect sensors. Figure 14 shows the stairwell used in the experiment with the Kinect sensor set up on the wall. Using the spherical targets shown in Figure 14, we registered multiple Kinect sensors on a single coordinate system. The locations of the four Kinect sensors are shown in Figure 15.

Figure 14. Stairwell with Kinect sensors set up on the wall

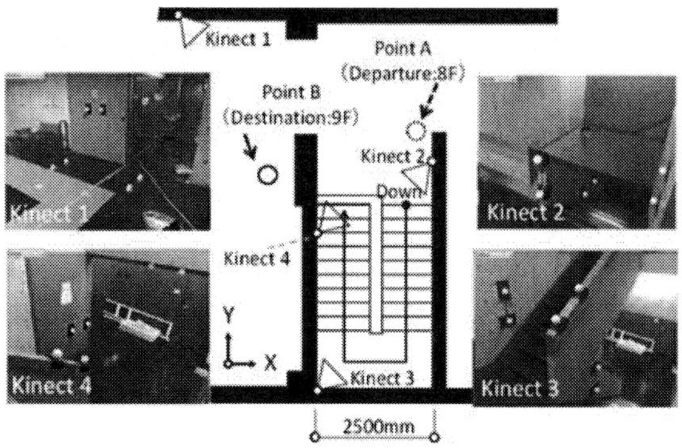

Figure 15. Layout of the stairs and the positions of the Kinect sensors

4.2 Collection of Point Cloud Data and Skeletal Tracking Data Obtained by Kinect Sensors in the Stairwell

We collected point cloud data and skeletal tracking data using Kinect sensors in a stairwell. Figures 16 and 17 show the results of this experiment. The raw 3D depth data (the point cloud data for the stairwell walls and stairs) are shown in gray, and the pedestrian trajectories of the student are shown in black. The pedestrian trajectories represent 20 round trips. Points A and B are on the eighth and ninth floors, respectively. The student traversed the stairs 20 times from Point A to Point B. We succeeded in measuring the shape of the stairs and collecting pedestrian trajectories.

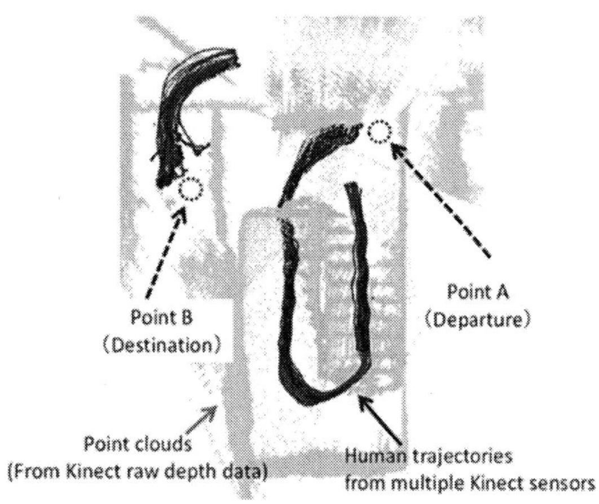

Figure 16. Point clouds on walls and stairs (in gray) and pedestrian trajectories (in black)

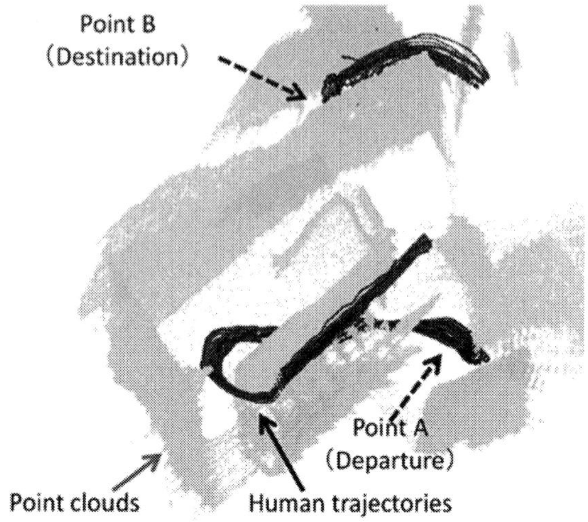

Figure 17. Isometric projection of the point clouds on the walls and stairs (in gray) and pedestrian trajectories (in black)

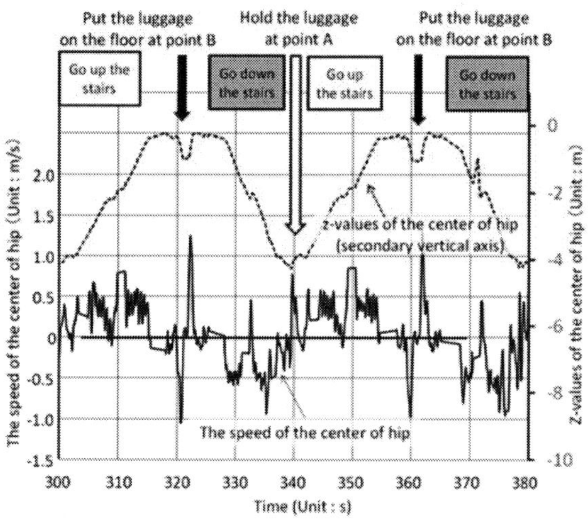

Figure 18. Results of the pedestrian position and velocity tracking

4.3 Analysis of the Skeletal Tracking Data in the Stairwell

The skeleton data consist of a set of 20 joints. We use the joint that is located at the center of the hip. The skeleton data encode a set of three-dimensional points $p_i = [x_i, y_i, z_i, t_i]$ including the acquisition time t_i. In this study, the Z-values of the center of the hip are z_i of the skeleton data.

The results of skeletal position and velocity tracking are shown in Figure 18. The horizontal axis represents the acquisition time of the skeletal tracking data. The left axis represents the speed (of the center) of the skeletal hip. The right axis of the graph represents the Z-values of the center of the hip.

In this experiment, the student pedestrian holds luggage at Point A and walks up the stairs. The pedestrian then places the luggage on the floor at Point B and walks down the stairs back to Point A.

From Point A to Point B, the Z-value of the center of the pedestrian's hip increased from −4.0 m in 300 s to −0.5 m in 315 s. From Point B, where the pedestrian places the luggage on the floor, back to point A, the Z-value of the hip center decreased from −0.5 m in 320 s to −1.0 m in 322 s. There is also a sudden decrease in the speed between 320 and 322 s.

4.4 Human Pose Recognition Using Skeletal Tracking Data

We developed a method for human pose recognition using skeletal tracking data. Human pose recognition is based on the pedestrian's speed and movement. This section describes the evaluation criteria used for human pose recognition.

The variables used in evaluation criteria for movement are the speed of the pedestrian in the X–Y plane, V_{xy} [m/s]; the total distance traveled after the state has changed, D_t [m]; and the duration of a human movement behavior, t [s].

The evaluation criteria for movement are defined in Table 2.

Table 2. Evaluation criteria for movement

State	Evaluation criteria
Stopped	$V_{xy} \leq 0.3$, $t \geq 1$, and $D_t \leq 0.4$
Walking slowly	$V_{xy} \leq 0.3$, $t \geq 1$, and $D_t \geq 0.4$
Walking	$V_{xy} \geq 0.5$ and $D_t \geq 0.4$
Transition state	None of the above conditions

The variables used in the evaluation criteria for standing and sitting are the Z component of the pedestrian's speed, S_z [m/s], and the height between the floor and the pedestrian's hip, H_{hip} [m].

The evaluation criteria used for standing and sitting are defined in Table 3. H_{hip} is calculated from the distance between the center of the pedestrian's hip and the floor, as shown in Figure 19.

Figure 20 shows the results of the height between the floor and the pedestrian's hip.

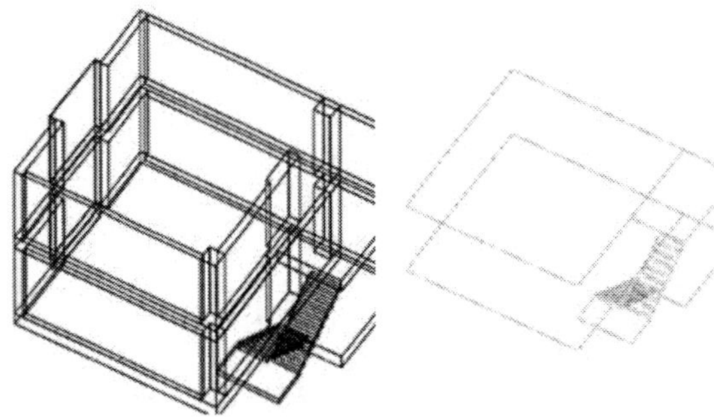

Figure 19. 3D CAD models of the stairwell and floor

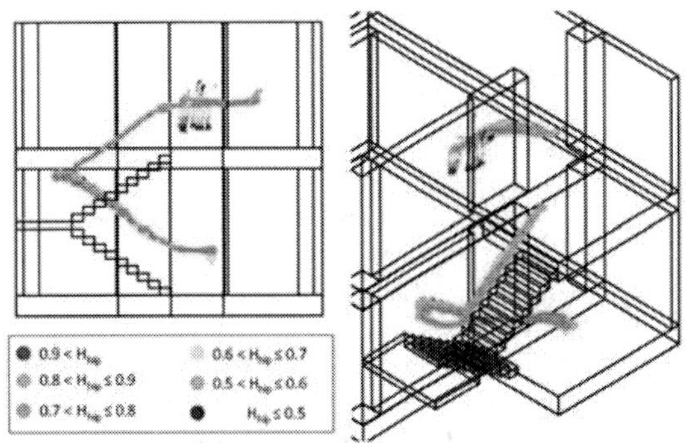

Figure 20. Height between the floor and the pedestrian's hip

Table 3. Evaluation criteria for standing and sitting

State	Evaluation criteria
Trying to stand up	$S_z \geq 0.8$
Trying to sit down	$S_z \leq -0.8$
Standing	$-0.8 < S_z < 0.8$ and $H_{hip} \geq 0.8$
Sitting	$-0.8 < S_z < 0.8$ and $H_{hip} \leq 0.6$
Transition state	None of the above conditions

Using the conditions in Table 2, we analyzed the skeletal tracking data with automatic human pose recognition. Figure 21 shows the results of the automatic evaluation of walking and stopping.

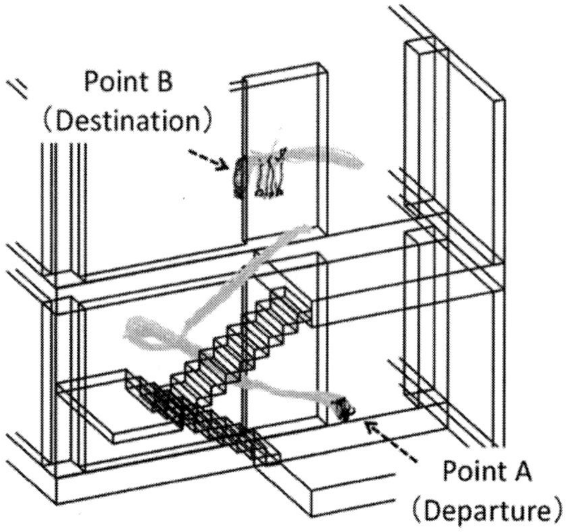

Figure 21. Results of the automatic evaluation of walking (in gray) or stopping (in black)

4.5 Automatic Acquisition of Repetitive Tasks

Construction work often includes repetitive tasks. In this experiment, a student pedestrian performed the repetitive task of climbing stairs 20 times from Point A to Point B. We developed a method for repetitive task recognition. This method is based on pedestrian positions and the evaluation criteria for walking and stopping. If the pedestrian worker has stopped, a task is likely finished.

Therefore, repetitive tasks are identified when a pedestrian has stopped by the time and location. Figure 22 shows the automatic acquisition of pedestrian trajectories for the first round-trip on the stairs.

Figure 23 shows the amount of time taken for each task repetition.

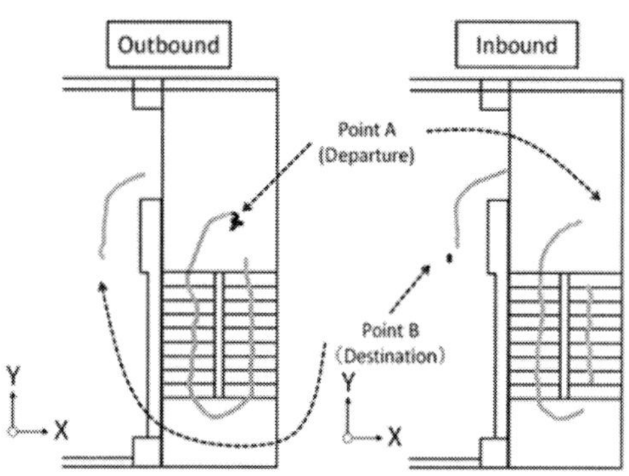

Figure 22. Automatic acquisition of pedestrian trajectories for the first round-trip

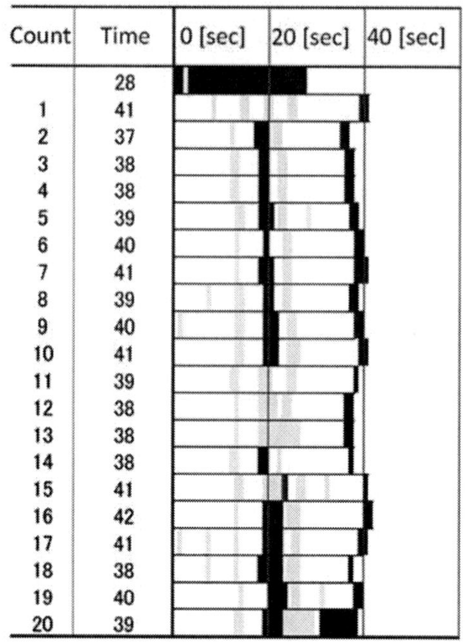

Figure 23. Time used for each task repetition

Black and white horizontal bars represent nonmoving and walking workers, respectively. Gray bars represent unmeasured time or transition states.

5 Conclusion

In this study, we have developed methods of registering skeletal tracking data from multiple Kinect sensors on a single coordinate system. We have analyzed the skeletal tracking data obtained in stairwells and have developed methods for human pose recognition using skeletal tracking data. We have also developed methods for the automatic acquisition of repetitive task data.

References

[1] Seera S., Brändlea N., and Rattib C. Kinects and human kinetics: A new approach for studying pedestrian behavior. *Transportation Research Part C: Emerging Technologies*, 48:212–228, 2014.

[2] Kosei I., and Yusuke O. Study on Analytical Method of the Workers Movements Monitor on Construction Site - Measurement Methods of Transport Route of the Worker -. *Proceedings of Thirty Symposium on Building Construction and Management of Project (Architectural Institute of Japan)*, 135-140, 2014.7.31

Immersive Virtual Environments for Investigating Building Emergency Evacuation Behaviors: A Feasibility Study

Hao Zou [a], Nan Li [a] and Lijun Cao [a]

[a]Department of Construction Management, Tsinghua University, China
E-mail: zouh14@mails.tsinghua.edu.cn, nanli@tsinghua.edu.cn, caolj12@mails.tsinghua.edu.cn

Abstract –

Evacuee behaviors have significant influence on the efficiency of emergency evacuation in buildings. While crowd evacuee behaviors have been extensively examined with various simulation models, the understanding of individual evacuee behaviors has to date mostly relied on sociological models, which are usually subjective and descriptive. Using latest advances in virtual reality (VR) technologies, it is possible to create immersive virtual environments (IVEs) that can be used for behavioral experiments in controlled setups to examine individual evacuee behaviors. One challenge in conducting such experiments is to ensure their ecological validity, so that the decisions and actions made by the experiment subjects are what they would actually make in reality. As a first step to achieve sufficient ecological validity, this study examines the feasibility of using a combination of subjective and objective measures, including an emotion scale and a physiological indicator, to assess the emotional responses of subjects in IVE-based evacuation experiments. Two IVEs were developed in this study, both representing a fire emergency scenario in an apartment but having different levels of realism and hence different levels of ecological validity. Subjects were asked to perform an evacuation task in both IVEs, and their emotional responses were monitored and analyzed throughout the experiments. Statistical difference was observed between the subjects' emotional responses when exposed to the two IVEs. The results suggested it was feasible to assess the ecological validity by assessing the emotional response of the subjects in IVE-based evacuation experiments.

Keywords –

Emergency evacuation; Virtual reality; Evacuation experiment; Sense of presence; Feasibility study

1 Introduction

When people evacuate in buildings during emergencies such as fire and explosions, their individual and collective behaviors have significant impact on the overall efficiency of the evacuation process, which in turn impacts the chances of occurrence of fatalities and injuries. The crowd behaviors in evacuation has been extensively examined in prior research, resulting in a variety of simulation models such as those based on social forces, multi-agent modeling and cellular automata. These models, however, have largely relied on various assumptions and over simplification about behavioral patterns of individual particles and agents, which represent individual evacuees in reality, limiting to a large extent fine-grained simulation and prediction of evacuation processes and assessment of evacuation management measures. Despite their importance in evacuation studies, the behaviors of individual evacuees have to date not received the attention they deserve, as evidenced by the absence of computable and verifiable behavioral models. Prior research on individual evacuee's behaviors mostly employs one of the following approaches, including evacuation drills, incident investigation and questionnaire-based psychological experiments. However, these approaches all bear certain limitations, such as high costs, lack of details of emergency scenarios, and difficulty in isolating controlled variables, which limit the collection of adequate behavioral data that are needed for in-depth behavioral analysis of individual evacuees.

Recent development of virtual reality (VR) technologies has provided a promising solution to these limitations, allowing for the creation of low-cost, low-risk, and controllable immersive virtual environments (IVEs) for virtual evacuation experiments. The advantages of IVEs have been well demonstrated in various behavioral psychology research [1], including studies on evacuation behaviors [2]. A number of IVE-based evacuation systems have been proposed in the literature, such as Vegas [3] and interFIRE VR [4], and used for firefighting training or evacuation simulation. IVE-based behavioral experiments, however, are usually challenged by its ecological validity, namely the extent to which experiment subjects' perception and responses can be generalized to real-life settings [5]. The ecological validity of IVE-based studies can be regarded satisfactory if experiment subjects' behavioral reactions in the IVEs are similar to those in the real world [6]. The level of similarity is usually measured by comparing subjects' behaviors in IVEs to their behaviors in corresponding real-world settings in studies about e.g. building design optimization [7], driving behavior analysis [8], and interpersonal behavior analysis [9]. However, this direct measurement of ecological validity is much less applicable in evacuation behavioral studies, because it is highly difficult, if possible at all, to create real building emergency scenarios, without causing significant damages and public panic, to match the scenarios in the IVEs for benchmarking purpose. Therefore, alternative approaches are needed for ecological validity assessment of IVE-based evacuation behavioral studies.

Prior research has pointed out that the sense of presence of the subjects is a fundamental factor that determines the ecological validity of IVE-based studies [10]. There is also evidence that shows noticeable correlation between the sense of presence of experiment subjects and their emotions when immersed in virtual environments [11]. Therefore, assessing the emotional response of the subjects in IVE-based experiments provides a possible way of assessing the subjects' sense of presence and hence the overall ecological validity of the experiments and ultimately the level of ecological validity of IVE-based evacuation experiments. This paper examines the feasibility of using a combination of subjective and objective measures, including an emotion scale and a physiological indicator, to assess emotional responses, which are comprised of emotional valence and emotional arousal, of the subjects in IVE-based evacuation experiments. Two IVEs were developed in this study. They both represented a fire emergency scenario in an apartment, but they had different levels of realism, and were expected to cause different emotional responses. The combined emotional assessment was applied to distinguish the emotional responses of experiment subjects elicited by the two IVEs when performing virtual evacuation tasks. The remainder of the paper is organized as follows: section 2 reviews relevant literature, followed by section 3 that introduces a combined method for emotional assessment. Section 4 presents the development of two building fire emergency IVEs with two different levels of realism. The design and implementation of a simplified IVE-based evacuation experiment is explained in section 5 and the results are presented in section 6. Section 7 concludes the paper.

2 Literature Review

Emotional response assessment in general includes the assessment of emotional valence, which is the hedonic value of a specific emotion, and the assessment of emotional arousal, which is the intensity of a specific emotion. Self-report emotion scale is a typical subjective measure for emotional assessment in psychological studies. It allows subjects to assess the type and magnitude of their own emotions by filling a deliberately designed survey. Several self-report emotion scales have been proposed and tested in prior research [12]. While they are convenient to use, emotion scales are challenged due to its subjectivity, as different subjects may have different comprehension about the evaluation standard [13], and their lack of sensitivity to subtle emotional fluctuations [14]. On the other hand, certain types of physiological reactions, such as brain activations and facial muscle activity, have found to be correlated with and sensitive to human emotions, and have hence been adopted in prior research for assessing the sense of presence of subjects in IVE-based studies [15]. For instance, autonomic nervous system (ANS) activity is an autonomic activity that is not controlled by subjective factors, and it has been proven a direct reflection of subjects' emotional response [16]. There is a number of types of physiological indicators, such as cardiovascular, respiratory and electrodermal measures, that can be used to measure ANS activities. Kreibig [17] summarized the use of these physiological indicators and their varying patterns under different emotions. It was reported that distinguishing different types of positive and negative valence based on ANS activities remains a challenging task. However, the magnitude of emotional arousal has been proven to have clear relevance with certain types of physiological data. For instance, skin conductance (SC) [18] and minute ventilation (VE) [19] were reported to have positive correlation with the magnitude of emotional arousal. The low frequency (LF) and high frequency (HF) power of heart rate variability (HRV) were reported to respectively reflect the activity of two parts of ANS including sympathetic and parasympathetic nervous system [20]. Despite these findings, there still lacks a widely adopted approach to analyzing physiological data for ANS activity assessment and ultimately emotion responses assessment of subjects in IVE-based experiments.

In short, various approaches have been proposed in literature for assessing emotional valence or emotional arousal. However, there lacks an integrated approach to assess both elements of emotions applicable to typical emotional responses observed in virtual evacuation experiment settings, where further research is needed.

3 A Combined Method for Assessing Emotional Responses

As discussed above, self-report emotion scale is the most widely used measure of emotional responses, especially for assessing emotional valance. However, it bears certain limitations including high subjectivity, and low sensitivity to emotional changes. Physiological indicators, on the other hand, have proven relevance with emotional arousal, and are used in prior research as an alternative measure of emotional responses. This study proposes to use a combination of subjective measure (emotion scale) and objective measure (physiological indicator) to assess the emotional responses of subjects in IVE-based evacuation experiments. It is hypothesized that these two measures, when used together, can effectively distinguish specific types of positively and negatively valenced emotions, and accurately capture subtle changes in the magnitude of emotional responses.

Specifically, the emotion scale used in this study is the Positive Affect and Negative Affect Scale (PANAS) scale, which is widely adopted in the field of psychology. Designed by Watson and Clark in 1980s [21], PANAS scale is composed of twenty adjectives, including ten adjectives describing positive emotions and the other ten adjectives describing negative emotions. Subjects are asked to evaluate the extent of each of these emotions using a scale from 1 to 5, with 1 meaning the least and 5 the strongest, according to their present emotional state. The total scores of all positive and all negative emotions respectively represent the extent of the subjects' overall positive and negative emotional state. A Chinese translation proposed in [22] was presented to experiment subjects in this study along with the original PANAS survey to aid precise comprehension of the adjectives. The physiological indicator used in this study is the SC, which is used to describe the conductivity of skin. The SC can be characterized by skin conductance level (SCL), which is a slowly varying tonic activity, and skin conductance response (SCR), which is a fast varying phasic activity, and the SC score is equal to SC tonic score plus SC phasic score. The SCL and SCR were measured using a multichannel physiological recorder in the experiments.

4 Development of Two Building Fire Emergency IVEs

Two IVEs were developed in this study to validate the feasibility of using the proposed method for ecological validity assessment. Both IVEs simulated a fire emergency scenario in a one-bedroom apartment. Basic geometries of the apartment were modeled in Revit, and exported to 3ds Max for rendering, by adding photorealistic materials, texture, lighting and furniture. The rendered model was imported into Unity 3D, in which the components of fire hazards, including flame, smoke and glow effect, were created using the embedded particle system. A first-person controller was also created in Unity 3D that enabled user-IVE interactions such as navigation and door opening. Stereo combustion sound effect and fire alarms were also added using audio source and listener system in Unity 3D. An Oculus Rift DK2 head-mounted display (HMD) and a Bose QC20 noise cancelling headphone were used to immerse subjects into the virtual environment, and a Microsoft Xbox joystick was used by the subjects to navigate in the virtual environment. The only difference of the two IVEs was their level of realism: for all components of the fire hazards, one IVE (hereafter referred to as IVE 1) contained flame effect only and the other one (hereafter referred to as IVE 2) contained flame, smoke, glow effect as well as the sound of combustion and fire alarm. Screenshots of both IVEs are shown in Figure 1. The computer used for IVE creation and evacuation experiments was Dell Precision T7800 workstation with Microsoft Windows 7 operating system, Intel Xeon E5-2603 processors, NVIDIA Quadro K620 graphics card and 16GB of installed memory.

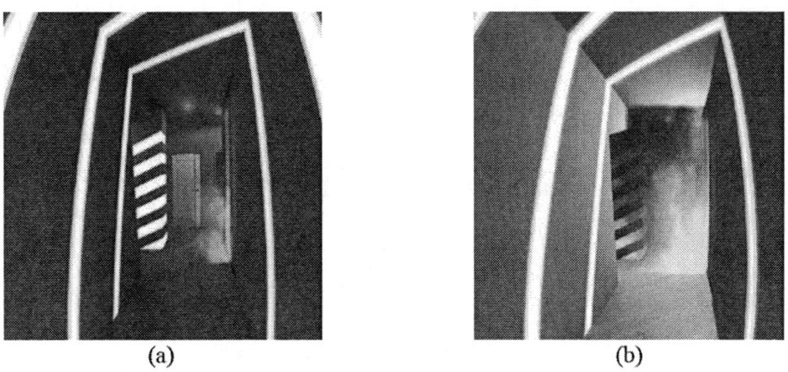

(a) (b)

Figure 1. IVEs with low (a) and high (b) level of realism

5 Virtual Evacuation Experiment

A total of 35 subjects were recruited to participate in the experiments. The subjects were all engineering students from Tsinghua University, including 25 male and 10 female between 18 and 28 years old. The experiments were conducted in a sound-attenuated room. Every subject went through the following procedure: at the beginning of the experiment, the subject wore SC data acquisition sensors by binding two electrodes around his/her index finger and middle finger, and filled in a PANAS survey. The subject then took a three-minute rest during which his/her baseline physiological data were recorded. Next, the subject put on the HMD, and took a training in the IVE, which contained no fire hazards, to learn how to navigate around. After another three-minute rest, the subject was asked to complete an evacuation task in IVE 1 and then IVE 2, with a three-minute rest between the two evacuation tasks. The subject's physiological conditions, interactions with IVE, and first-person view from the HMD were monitored and recorded throughout the experiment using ErgoLAB platform (Figure 2). After each task, the subject was asked to fill up a PANAS survey. Since the investigation of the subject's behaviors was not the objective of this feasibility study, the evacuation task was purposely designed to be simple: the subject would find him/herself in a bedroom at the beginning of the experiment, with a health warning message about use of IVE displayed for a few seconds. After the message disappeared, he/she would be instructed to evacuate from the apartment, by leaving the bedroom, going through a short burning hallway, and exiting from a door across a living room. Figure 3 shows some of the subjects performing the assigned tasks (training, evacuation, filling up the survey) during the experiments. If a subject felt any sickness in the IVE, he/she could ask to terminate the experiment at any time.

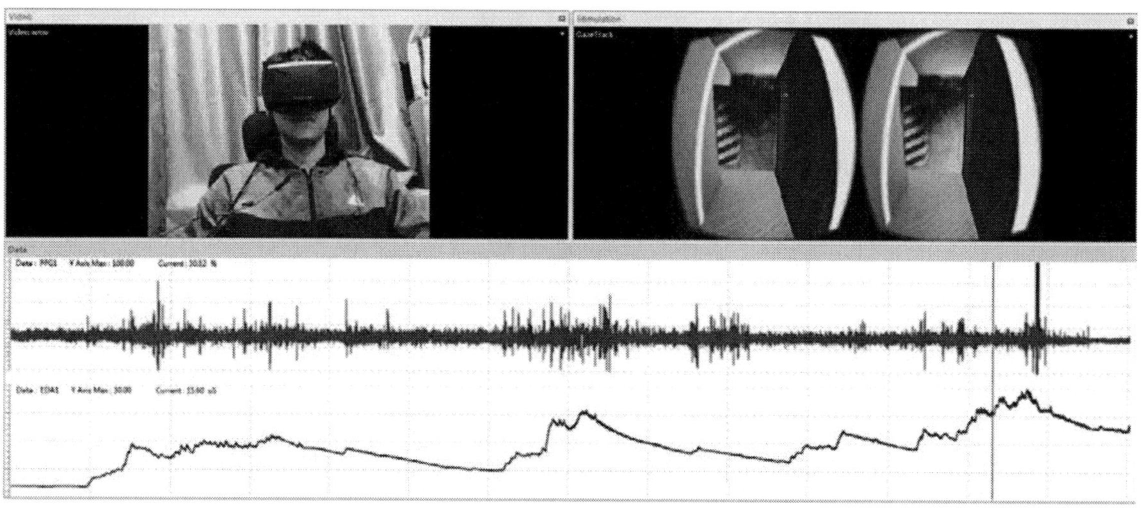

Figure 2. Interface of the experiment data collection platform

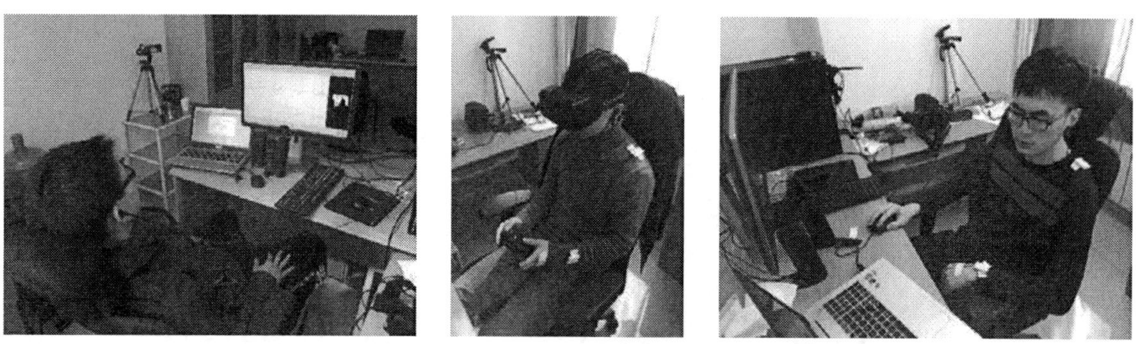

Figure 3. Subjects performing the assigned tasks

6 Findings and Discussions

Out of the 35 subjects, 33 subjects completed the experiments. Two subjects chose to quit due to motion sickness in the IVEs. One data set was incomplete due to technological problem of the physiological sensors. A total of 32 complete data sets were collected and used for analysis. The difference between subjects' respective emotional responses to IVE 1 and IVE 2 were compared, in order to examine the feasibility of using the proposed approach to assess the level of sense of presence. T-tests were conducted for negative affective scores and positive affective scores of PANAS, and SC mean scores. The findings are presented in this section.

6.1 Analysis of PANAS Scores

A pairwise comparison was conducted between three sets of PANAS scores, which were collected at the baseline (training) stage and after the two evacuation tasks. The results are summarized in Table 1. At a 95% confidence level, t-test results showed that the negative scores of IVE 2 was statistically larger compared with the negative scores of the baseline ($t = 1.796$, $p = 0.041 < 0.05$) and IVE 1 ($t = 5.065$, $p = 0.000 < 0.05$). It suggested that the higher level of realism in IVE 2 did cause subjects to experience sensibly higher negative emotions, and that such difference in emotional responses could be captured using the PANAS emotion scale. It needs to be pointed out, however, that there was noticeable decrease in the negative scores from the baseline stage to IVE 1 ($t = -2.204$, $p = 0.982 > 0.95$). It was possibly because that the subjects were not familiar with the feeling of being immersed in a virtual environment during the baseline period and therefore felt uncertain and dubious, which faded away to a larger extent than that of negative emotions elicited by IVE 1 when the subjects performed the evacuation task. For positive scores, t-test results suggested that there were no significant changes ($t = 0.909$ and $p = 0.185 > 0.05$ between baseline and IVE 1; $t = 1.155$ and $p = 0.128 > 0.05$ between baseline and IVE 2; $t = 0.740$ and $p = 0.232 > 0.05$ between IVE 1 and IVE 2), suggesting that the emotional responses to virtual fire emergency scenarios were mainly negative while positive emotions mostly remained the same. Furthermore, a detailed comparison between the average scores of each type of negative emotions collected during the baseline stage and IVE 2, as summarized in Table 2, revealed that scared (1.394 to 2.152), nervous (2.061 to 2.545) and afraid (1.485 to 2.000) emotions were the main negative emotions elicited by the fire emergency IVEs. These types of emotions were in accordance with the nature of fire emergency scenarios, which demonstrated the feasibility of using PANAS emotion scales to assess emotional valence and changes in their magnitude.

Table 1. Comparison between three sets of PANAS scores

	Baseline and IVE 1		Baseline and IVE 2		IVE 1 and IVE 2	
	Positive	Negative	Positive	Negative	Positive	Negative
t-score	0.909	-2.204	1.155	1.796	0.740	5.065
p-value	0.185	0.982	0.128	0.041	0.232	0.000

Table 2. Comparison between average scores of different types of negative emotions

	Upset	Distressed	Guilty	Scared	Hostile	Irritable	Ashamed	Nervous	Jittery	Afraid
Baseline	1.727	1.879	1.242	1.394	1.152	1.394	1.758	2.061	1.515	1.485
IVE 2	1.727	1.939	1.333	2.152	1.333	1.303	1.152	2.545	1.485	2.000
Δ	0	0.061	0.091	0.758	0.182	-0.091	-0.606	0.485	-0.030	0.515

6.2 Analysis of SC Scores

The SC scores were calculated by adding average SC tonic score and average SC phasic score. A typical SC curve of a subject throughout the entire experiment is shown in Figure 4. A pairwise comparison was conducted between the SC scores collected at the baseline stage and during the two evacuation tasks. The results are summarized in Table 3. At a 95% confidence level, t-test results showed that there was significant pairwise difference between the SC scores of the baseline, IVE 1 and IVE 2. Specifically, the SC scores of the subjects first increased by 120.84% on average from the baseline to IVE 1, and further increased by another 14.93% from IVE 1 to IVE 2. The comparison of the SC scores is shown in Table 4.

Moreover, it was observed that the SC score at the beginning of each evacuation task was slightly different than each other and both were higher than the SC score at the baseline stage. It showed that the subjects' emotion did not return to the original state after the three-minute rest. To take into consideration such residual impact of the preceding training or evaluation task, the relative score of SC, namely the increase of SC score over the period of each evaluation task, was calculated to compare between the two IVEs. The relative SC scores of the two IVEs were denoted as ΔIVE 1 and ΔIVE 2, respectively, and summarized in Table 5. It can be seen the table that ΔIVE 2 (5.351) was 23.52% higher than ΔIVE 1 (4.332), which indicated that the experiment in IVE 2 resulted in more significant emotional response than the experiment in IVE 1.

In short, the analysis of SC scores showed, in accordance with the assessment results by the emotion scale, that the subjects experienced higher emotional responses and hence higher sense of presence in a more realistic IVE. Moreover, unlike the PANAS scores, the SC scores captured statistically noticeable difference in the emotional arousal between the baseline and IVE 1. The results indicated that the SC score was an effective indicator, with higher sensitivity than the emotion scale, for assessing the emotional arousal in IVEs.

Table 3. Comparison between three sets of SC scores

	Baseline and IVE 1	Baseline and IVE 2	IVE 1 and IVE 2
t-score	7.307	7.503	6.929
p-value	0.000	0.000	0.000

Table 4. Increase of SC scores in IVE 1 and IVE 2 compared to their preceding stages

	Baseline	IVE 1	IVE 2
SC mean (μS)	7.366	16.267	18.696
Percentage increase	/	120.84%	14.93%

Table 5. Comparison between two sets of relative SC scores

	Rest 1	IVE 1	Rest 2	IVE 2
SC mean (μS)	11.936	16.267	13.345	18.696
Relative SC scores (μS)	/	4.331	/	5.351

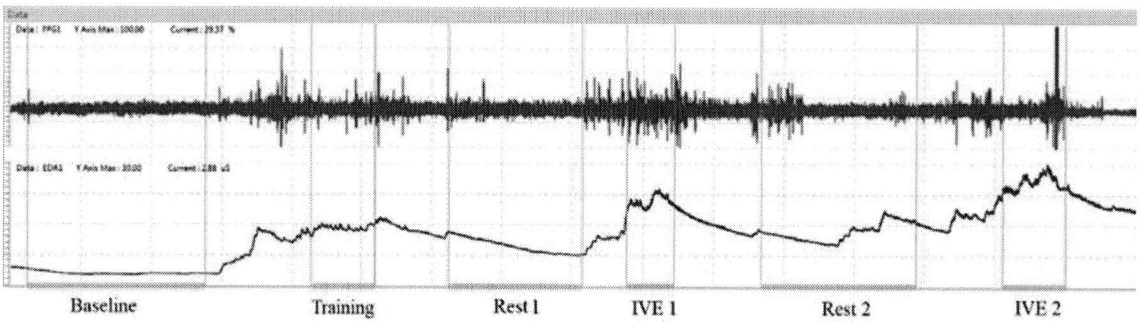

Figure 4. A typical SC curve in the experiments

7 Conclusions and Future Work

This study examines the feasibility of using a combination of subjective measure (an emotion scale) and objective measure (physiological indicators) for both emotional valance and emotional arousal to assess subjects' sense of presence, a critical factor that determines the overall level of ecological validity of evacuation experiments conducted in IVEs. Two IVEs were developed in this study, both representing a fire emergency scenario in an apartment but having different levels of realism. Experiment subjects were asked to perform an evacuation task in both IVEs, and their emotional responses were monitored throughout the experiments using the subjective and objective measures. The results showed that the PANAS emotion scale and the SC score were effective measures of emotional responses of the subjects to fire emergency IVEs. Assessment reported by both measures were generally in accordance with each other. The emotion scale was able to distinguish specific types of emotion responses elicited by the IVEs, especially scared, nervous and afraid, whereas the physiological indicators were more sensitive to subtle changes in the magnitude of emotional responses. When used together, the subjective and objective measures provided reasonable assessment of the sense of presence that subjects experience in emergency evacuation IVEs.

To advance this line of research, the authors plan to examine the effectiveness of other types of physiological indicators in assessing the sense of presence in the IVEs, and establish the relationship between these indicator and the emotional responses based on the recognition of physiological signal patterns. Moreover, the long-term objective of this study is to build IVEs with high sense of presence so that they can be adopted in evacuation behavioral experiments with high ecological validity for investigating the sensing, decision making and behaviors of individual evacuees during emergencies. Therefore, further work is needed to determine the threshold of the sense of presence needed to ensure sufficient ecological validity of virtual evacuation experiments, and to explore new approaches and technologies to develop IVEs that can satisfy the threshold of the sense of presence.

Acknowledgements

This material is based upon work supported by Tsinghua University Initiative Scientific Research Program under Grant No. 2014z21050. The authors are thankful for the support of Tsinghua University. Any opinions, findings, and conclusions or recommendations expressed in this paper are those of the authors and do not necessarily reflect the views of Tsinghua University. The authors would also like to thank Kingfar International Inc. for providing the ErgoLAB platform that was used in this study, and Shengxing Ban, Lizhi Bao and Po Tian for their technical support in the acquisition of physiological data and development of IVEs.

References

[1] Ioannou I., Avery A., Zhou Y., Szudek J., Kennedy G. and O'Leary S. The effect of fidelity: How expert behavior changes in a virtual reality environment. *Laryngoscope*, 124(9):2144-2150, 2014.

[2] Mc Daid A. Study into the use of signage on exit choice behaviour in a virtual environment. *Applied Mechanics & Materials*, 796:1-11, 2015.

[3] Xi M. and Smith S.P. Simulating cooperative fire evacuation training in a virtual environment using gaming technology. In *Virtual Reality (VR)*, IEEE, 2014.

[4] Putorti Jr A. and McElroy J. Full-scale house fire experiment for InterFIRE VR. *National Institute of Standards and Technology Report of Test FR4009*, US Dept.of Commerce, 2000.

[5] Brewer M.B., Reis H. and Judd C. Research design and issues of validity. *Handbook of research methods in social and personality psychology*, 3-16, 2000.

[6] Anderson C.A. and Bushman B.J. External validity of "trivial" experiments: The case of laboratory aggression. *Review of General Psychology*, 1(1):19, 1997.

[7] Heydarian A., Carneiro J.P., Gerber D., Becerik-Gerber B., Hayes T. and Wood W. Immersive virtual environments versus physical built environments: A benchmarking study for building design and user-built environment explorations. *Automation in Construction*, 54:116-26, 2015.

[8] Shechtman O., Classen S., Awadzi K. and Mann W. Comparison of driving errors between on-the-road and simulated driving assessment: a validation study. *Traffic injury prevention*, 10(4):379-85, 2009.

[9] Iachini T., Coello Y., Frassinetti F., Senese V.P., Galante F. and Ruggiero G. Peripersonal and Interpersonal Space in Virtual and Real Environments: Effects of Gender and Age. *Journal of Environmental Psychology*, 45:154–164, 2016.

[10] Slater M., Lotto B., Arnold M.M. and Sánchez-Vives M.V. How we experience immersive virtual environments: the concept of presence and its measurement. *Anuario de Psicología*, 40:193-210, 2009.

[11] Riva G., Mantovani F., Capideville C.S., Preziosa A., Morganti F. and Villani D. Affective interactions using virtual reality: the link between presence and emotions. *CyberPsychology & Behavior*, 10(1):45-56, 2007.

[12] Luigi M., Massimiliano M., Aniello P., Gennaro R. and Virginia P.R. On the Validity of Immersive Virtual Reality as Tool for Multisensory Evaluation of Urban Spaces. *Energy Procedia*, 78:471-6, 2015.

[13] Mardaga S., Laloyaux O. and Hansenne M. Personality traits modulate skin conductance response to emotional pictures: An investigation with Cloninger's model of personality. *Personality and individual differences*, 40(8):1603-14, 2006.

[14] Wilson G.M. and Sasse M.A. Do users always know what's good for them? Utilising physiological responses to assess media quality. In *People and Computers XIV—Usability or Else! Springer*, pages 327-339, 2000.

[15] Rodríguez A., Rey B., Clemente M., Wrzesien M. and Alcañiz M. Assessing brain activations associated with emotional regulation during virtual reality mood induction procedures. *Expert Systems with Applications*, 42(3):1699-709, 2015.

[16] Stephens C.L., Christie I.C. and Friedman B.H. Autonomic specificity of basic emotions: Evidence from pattern classification and cluster analysis.

Biological Psychology, 84(3):463-473, 2010.

[17] Kreibig S.D. Autonomic nervous system activity in emotion: A review. *Biological Psychology*, 84(3):494-421, 2010.

[18] Kallinen K. Emotion related psychophysiological responses to listening music with eyesopen versus eyes-closed: electrodermal (EDA), electrocardiac (ECG), and electromyographic (EMG) measures. In *Proceedings of Music Perception & Cognition*, pages 299-301, 2004.

[19] Gomez P., Shafy S. and Danuser B. Respiration, metabolic balance, and attention in affective picture processing. *Biological Psychology*, 78(2):138-149, 2008.

[20] Appelhans B.M. and Luecken L.J. Heart rate variability as an index of regulated emotional responding. *Review of general psychology*. 10(3):229-240, 2006.

[21] Watson D., Clark L.A. and Tellegen A. Development and validation of brief measures of positive and negative affect: the PANAS scales. *Journal of personality and social psychology*, 54(6): 1063-1070, 1988.

[22] Zuoji Z. Behavioral medicine inventory manual. *Chinese Journal of Behavioral Medicine Science*, 10(10):19-24, 2001.

Automated Construction of Masonry Buildings using Cable-Driven Parallel Robots

T. Bruckmann[a], H. Mattern[b], A. Spengler[c], C. Reichert[a], A. Malkwitz[c] and M. König[b]

[a] Chair of Mechatronics, University of Duisburg-Essen, Germany
[b] Chair of Computing in Engineering, Ruhr University Bochum, Germany
[c] Institute of Construction Management, University of Duisburg-Essen, Germany
E-mail: tobias.bruckmann@uni-due.de, hannah.mattern@rub.de, arnim.spengler@uni-due.de, christopher.reichert@uni-due.de, alexander.malkwitz@uni-due.de, koenig@inf.bi.rub.de

Abstract

This paper presents an approach of using cable-driven parallel robots for automated construction. Masonry buildings are considered as appropriate application case due to repetitive construction procedures, high demands on quality as well as accuracy of construction, and handling of small and lightweight material. In this context, the application of Building Information Modeling (BIM) as data source is proposed. Furthermore, a simulation model representing the operation method of a wire robot is created to analyze the process of assembling a masonry wall. Special focus lies on creating collision-free motion profiles that can be exported to the robot control system.

Keywords – Automation and Robotics, Building Information Modeling (BIM), Mechatronics, Simulation

1 Introduction

Due to the potential to improve productivity, quality, safety and to reduce cost, automated production has been established in many industrial sectors. Today, Computer Aided Design (CAD) system modeling as well as automatic assembly methods are frequently applied in industries such as aerospace, ship building or automobile [1]. Concerning the construction sector, however, automated technologies have emerged in the 1990s. One of the most attractive aspects of automated production is the opportunity to reduce personnel costs, while at the same time achieving a constantly high production quality. In many cases, robots or electronically controlled machines are applied to increase production efficiency. Focusing on the construction sector, however, automated production techniques are still limited to the sector of prefabrication (e.g., the production of precast concrete elements). With the development of Building Information Modeling (BIM) in the last decades, the application of automated construction techniques needs to be reconsidered. In this context, BIM represents a comprehensive data source by enabling a high amount of precise information concerning a building's construction phase.

The intention of this work is to present a concept for applying cable-driven robots for automated construction. The results of a literature research show possible application cases of BIM in the context of automated production, site layout planning and work safety (Section 2). In addition, several projects that combine BIM and robotics are presented. Recent approaches to use wire robots in large scale manipulation projects are examined and provide first ideas for possible application cases in the construction industry (Section 3). A simulation model is created to validate the concept concerning feasibility, efficiency, and work safety (Section 4). In the following, the background of developed concept is presented in more detail.

2 Related Work in the Context of BIM

On the basis of digital building models, the design, construction, and operation processes can be significantly improved. The focus is on three-dimensional digital models that consist of individual building components characterized by rich semantics. BIM models can be applied and extended for various planning and analyzing tasks. During project execution, typical application cases comprise clash detection, cost estimation and construction scheduling. Further applications also deal with the simulation and control of detailed construction and logistics processes. In this context, BIM can also be an enabler for developing new approaches to use robots in construction projects frequently.

In the following, current applications of BIM for automated construction processes are presented. With regard to the purpose of this paper, special focus lies on prefabrication and automated assembly of building elements, site layout planning and work safety. Additionally, several projects that combine BIM with robotics are described.

2.1 BIM for Prefabrication and Automated Production

The growing application of BIM contributes to the potential for increased use of prefabrication and modularization [2]. Costa et al. connected building components catalogues with BIM models by using semantic technologies and tested the concept on concrete components [3]. Thus, they facilitated the participation of manufacturers in the design and building process. Buswell et al. focus on large scale processes for construction applications based on Rapid Manufacturing principles [4]. They state that BIM is likely to become a key element for information delivery in this context. An exploration of a new BIM-based automation construction system (BIMAC) can be found in [5]. Based on the technique of Additive Manufacturing (AM) techniques, BIM model layer data are converted to specific Computerized Numerical Control (CNC) codes enabling an automated manufacturing of building components. Babič et al. aimed at integrating mass production prefabrication processes with construction site activities [6]. BIM was used as a link between an enterprise resource planning (ERP) information system that supports manufacturing processes and construction object related information. By introducing their approach to a construction company, they observed an increase in transparency between production and construction for the purpose of project monitoring and material flow tracking. Moghadam et al. developed an automated model for the design and manufacture of modular construction that applies both Building Information Modeling (BIM) and lean concepts on a modular construction manufacturing (MCM) process [7]. The described projects prove the possibility to use data from BIM models for prefabrication and automated production of single building elements. In contrast to the proposed robot application, assembly work needs to be performed manually resulting in decreased working precision.

2.2 BIM for Site Layout Planning and Work Safety

Besides the application of BIM for automating construction processes, many researches concentrate on BIM-supported site layout planning. When using wire robots for construction tasks, the site layout needs to be modified accordingly. In the best scenario, basic information on site layout are directly derived from the BIM model. Several projects of the past focused on this aspect. For example, Kumar et al. generated an automated framework for creating dynamic site layout models by utilizing information from BIM models [8]. Taking into account actual travel paths of on-site personnel and equipment allowed optimal utilization of the available space. Wang and Zhang developed an integrated approach which combines BIM and Firefly Algorithm (FA) to automatically generate an optimal tower crane layout plan [9]. A BIM-based simulation can be created to visualize and evaluate the chosen layout.

Using a wire robot results in high demands concerning work safety. Collisions between robot parts and workers may result in an increased risk of injuries. Consequently, checking work safety represents an important step when validating the proposed concept. In addition to the above described application areas, BIM represents a viable tool to check work safety and identify possible hazards. Schwabe presented concepts for possible software solutions using rule checking algorithms for construction site layout planning [10]. Prototypical implementations demonstrate to what extent currently existing BIM software is able to realize the existing concepts. Zhang, Teizer et al. developed an automated safety rule checking platform based on BIM models including automated hazard identification and correction [11]. In [12], a BIM-based approach for activity-level construction site planning is presented enabling an improvement of construction safety. An automated hazardous area model was proposed by Kim, Lee et al. [13]. It is based on the deviation between the optimal route (shortest path) which is determined by extracting nodes from objects in a BIM model and the actual route of on-site personnel collected from a real-time location system. Based on this information, the developed identification module identifies potentially hazardous areas in laborers' paths. Combining information form the BIM model (e.g., 4D construction schedules) with the path planning of the robot may help to increase work safety at all project stages.

2.3 BIM and Robotics

Vähä, Heikkilä at el. conducted a survey on potential sensor technologies and robotic applications for construction projects [14]. They distinguished between manufacturing of prefabricated components and assembly work. In this context, BIM can be used for guiding the assembly work by providing component reference values and other assembly related information. Lee, Kim et al. developed a robotic crane system deploying a laser-technology-based lifting-path tracking system which requires the application of BIM [15]. The tracking system receives an identifier for material to lift from a central database that stores a construction schedule and a 3D BIM model. Lee, Cho et al. present a tower crane navigation system that provides three-dimensional information about the building and surroundings and the position of the lifted object in real time [16]. Besides various sensors, a BIM model is applied and linked to a real-time construction progress management system. The results of a case study indicated that crane operators preferred the crane navigation system to a text-based anti-collision system.

The results of the literature research prove the wide range of possible applications of BIM. This ongoing development can be regarded as key motivation for examining the application of robots for automated construction. Automatization techniques may now benefit from detailed and precise information provided in BIM models which can be used as digital construction plans. Thus, the application range of the presented concept is extremely increased. The following section contains a brief overview concerning existing concepts to use robots in the field of construction.

3 Application of Wire Robots for Construction

When examining the typical processes of construction projects, the size of a building lot makes the application of robots extremely demanding. Thus, the integration of robots in the construction of buildings has always been challenging. At the same time, in mechanical engineering, the automation of processes has reached a high level thanks to widely used series production. Consequently, both development and production processes have been established towards standards, tools and interfaces that allow for the application of robots and automated production lines. Nowadays, the introduction of Industry 4.0 even aims at the fully automated production of individual items and products.

Today's standard industrial robots hardly cover workspaces of more than five meters in radius. Accordingly, even fundamental processes like the stacking of bricks to rise a wall are impossible to realize without the need to move the whole robot. This process requires recalibration, takes time and drives the costs. These are only some of the reasons why many approaches failed, most of them during the first wave of automation attempts in the 1990s.

Today, some novel approaches for bricking based on serial robot manipulators are under development. The Australian company Fastbricks Robotics currently develops a large-scale manipulator called Hadrian that could be large enough to move bricks along the manipulator arm to any desired position on the construction site. In parallel, the American company Construction Robotics is developing the SAM robot that is basically a conventional serial robot on a moving platform, equipped with a wide range of pose sensors. Both approaches still need their broad applicability. On the level of innovative fundamental research, Kohler presented approaches based on flying drones and moving industrial robots [17, 18].

But besides the development of BIM, also robotics research explored revolutionary concepts and new types of robots. One of them is the so-called cable-driven parallel robot or simply wire robot, similar to a Steward-Gough platform. This type of robot completely eliminates the aforementioned drawbacks. A wire robot uses a set of wires (or cables) in a tensed configuration to guide a payload along a predefined path.

3.1 Overview on Wire Robots for Large Scale Manipulation

The concept of the Steward-Gough platform – a parallel kinematic machine – has been used since the 1950s. In the last decades, several approaches were taken to transfer some of its advantageous properties like high stiffness and good precision into industrial applications. However, its drawbacks like high manufacturing costs and limited workspace appeared as major issues and prohibited its success on a broader level. Only within niches like simulator motion systems, the Steward-Gough platform became a popular solution. To overcome the described drawbacks, in 1985 Landsberger [19] proposed to replace the stiff linear actuators of the conventional Stewart-Gough platform by cables. These cables are attached to a moving platform (the payload), guided by pulleys on a supporting frame and wound up by motor-driven winches on the ground. This results in a so-called cable-driven parallel robot, often simply called wire robot. In a certain sense, a wire robot is a combination of computerized cranes that act in a synchronized manner, keeping the appropriate cable lengths and cable tensions to precisely move materials to the desired position. From a certain perspective, the concept resembles a 3D crane system. A wire robot shows some remarkable advantages:

- Extremely large workspace: Since cables can be easily wound up for dozens of meters, a wire robot covers the complete building lot.

- Simple and cheap mechanical elements: Only winches, pulleys and cables are needed. These elements are traditionally widely used on building lots.
- Easy to install: The wire robot just needs a supporting cuboid frame which can be easily erected in place.
- Extremely lightweight and fast: Since a wire robot has nearly no mass except from the payload, it can move very fast as long as human safety is not affected. At large dimensions, the frame may be constructed by lattice girders.

Starting with the 1990s [19], the wire robot has been tested for several applications. Recently, the application of wire robots for large-scale applications has been extensively addressed.

Figure 1: Storage Retrieval Machine based on Wire Robot Technology

In 2010, the project "Storage and Retrieval machine based on the Stewart-Gough-Platform" – funded by the German Federal Ministry of Education and Research – started to develop a storage retrieval machine that effectively uses the advantages of a parallel wire robot [20]: While conventional storage retrieval machines have a mass of one to two tons to move a payload of 20-50kg, the wire robot only moves 100kg including payload. Besides the University of Duisburg-Essen (Germany), the project included eight globally leading industrial partners from the logistics industry to realize a full-scale demonstrator (see Figure 1).

In 2009, the Chair of Mechatronics at the University of Duisburg-Essen and the Technical University Hamburg-Harburg realized an active suspension system for wind tunnels based on wire robots [21]. The system is currently in normal operation and performs well. In 2011, a large initiative on European level started: Within the European Union Seventh Framework Program FP7/2007-2013, the project CableBOT was realized. Its objective was the investigation of wire robots for large goods handling and manipulation. In the frame of the project, wire robots were explored for two key application examples: First, the maintenance of airplanes together with EADS Innovation Works (now Airbus group). Second, the production of large prefabricated steel parts for facades. Within the project, the Chair for Mechatronics was involved in the development of an advanced control system.

Nowadays, wire robots are used commercially to move cameras e.g. above a sports stadium. This is a simple application example from a robotics point of view, but very demanding from the requirements for safety and reliability.

3.2 Robot Design

Wire robots are intrinsically modular machines: They are composed by identical winches that share the weight of the payload. This allows to adapt wire robots to a given task in a very wide range.

As cables are flexible, they can only pull but never push. Based on this, researchers fundamentally distinguish between two types of wire robots that are composed by the same principle, but show very different properties. These two types are explained in the following.

3.2.1 Suspended Wire Robot

This type of wire robot needs external forces like gravity to tense the wires and is e.g. used to carry cameras. The options to vary the tension in the cables for a given robot pose are very limited. As an advantage, usually the cables come "from above", i.e. the area below the robot is not occupied (see Figure 2). This dramatically reduces the problem of collisions.

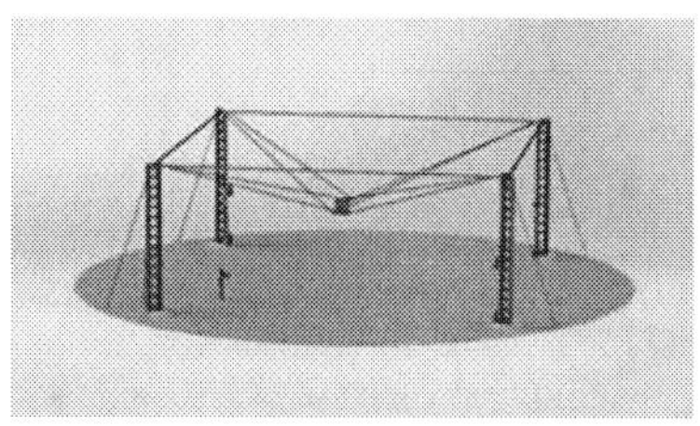

Figure 2: Suspended Wire Robot

3.2.2 Fully-tensed Wire Robot

Here, the cables needed to suspend the payload are completed by a set of cables that counteract the others. This allows to actively vary the tension in the system: If one cable increases the tension, the others can react with an appropriate force and establish the force equilibrium again. Additionally, this allows to fully constrain the payload and to suppress vibrations effectively. Assuming the payload needs to be moved in n degrees-of-freedom, at least m cables are needed, i.e. $m \geq n + 1$. Since usually also cables below the moving platform are employed, collisions might be an issue. As an option, sliding pulleys in rails can help to avoid this.

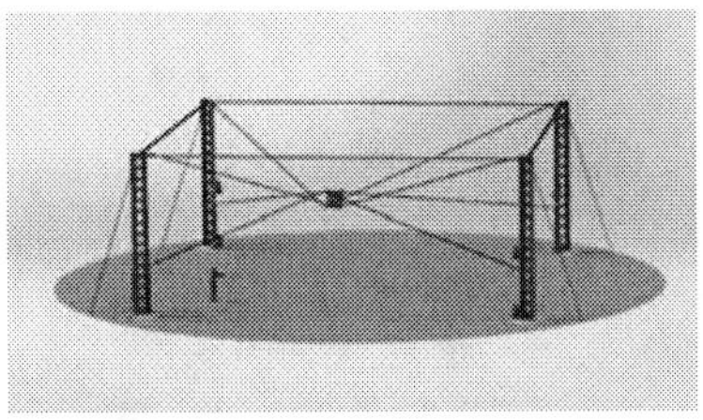

Figure 3: Fully-tensed Wire Robot

For first experiments, a fully-tensed robot is proposed due to its superior mechanical properties and its higher stiffness. For the production of buildings, a certain level of accuracy is required. Since the cables, the drivetrains of the winches and even the controller show an elastic behavior (with possibly changing elasticities over time), the stiffness of the platform might be an issue for accuracy and therefore needs to be carefully investigated and optimized. Otherwise, all disturbances like e.g. wind forces lead to bad accuracy results. Based on numerical optimization, a design with a cuboid supporting frame is proposed. It shows a remarkably large workspace of the wire robot, see Figure 4. Currently, the design is optimized with respect to workspace shape, volume and especially stiffness to increase the accuracy of the system. To avoid the danger of collisions, movable pulleys should be used in the future.

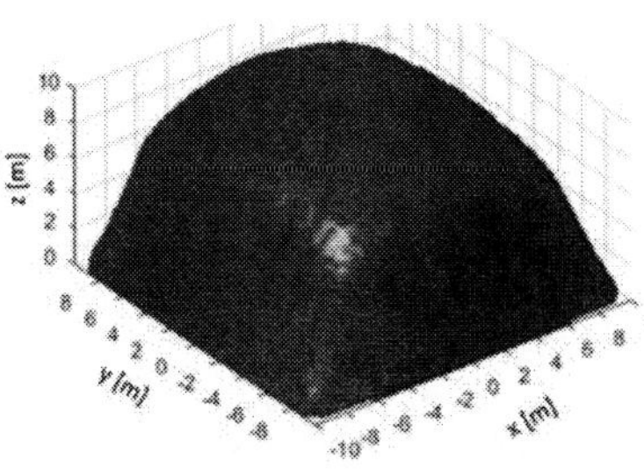

Figure 4: Workspace of the Wire Robot

3.2.3 Planning and Trajectories

The planning of paths and the generation of motion profiles over time – called trajectories – is a key issue of the planned wired robot. The robot control system needs to generate all necessary motions to create buildings using material elements like bricks, steel parts and prefabricated modules. Path planning needs to include a number of features:

- Completeness: For each single material element, a trajectory must be generated that moves it from its storage or delivery location to its mounting position.
- Efficiency: To successfully enable the automated wire robot approach, the production of a building must become cheaper. Usually, this is related to the reduction of the construction time, i.e. the robot must be fast or should move large material elements. Accordingly, the sum of the generated trajectory times must be minimal and tweaked. This can be solved e.g. by numerical optimization.
- Sequences and Collisions: The path planning must include the position of all material elements at all times. This is required due to two facts:
  - Building material is often delivered in a stacked manner, e.g. palletized. Only the top layer is usually accessible for the robot.
  - Many building structures can only be created bottom-up in layers. Accordingly, the production sequence needs to be considered.
- During the production process, the building intrinsically changes its shape. The wire robot and its payload may not collide with storage, delivered parts, the building, other machinery or itself (self-collisions). As most of these obstacles also change dynamically their size and/or position, this becomes a complex task that usually cannot be solved in real-time and therefore needs to be solved in advance of the production process.

4 Concept of BIM-based Construction Planning using Wire Robots

The planning of the construction process is based on the virtual 3D-model of the building. By providing detailed information concerning building structure, geometry and materials, BIM models represent a valuable source of information. Furthermore, the BIM-based approach allows the consideration of complex restrictions making each construction project unique. Connecting the digital model with information from the construction schedule enables the detection of processes that can be performed by wire robots. Those are described in Section 4.1.

One of the most important steps of this project is proving the feasibility of the developed concept. A simulation model is created for analyzing the robot operation method instead of conducting expensive and time-consuming practical tests. To achieve reliable results, process-relevant objects and interactions with the robot need to be implemented. Following a modular structure, the model can be used for different application cases (e.g., different building geometries or robot designs). The implementation is mainly based on the developed catalogue of automated processes (cf. Section 4.1) as well as the BIM model of the building. Restrictions considered in the simulation model include site and robot layout. Conventionally performed work steps are not considered explicitly. In the following, processes and elements implemented in the model are described in more detail. Furthermore, the data flow between BIM model, simulation and robot control is presented.

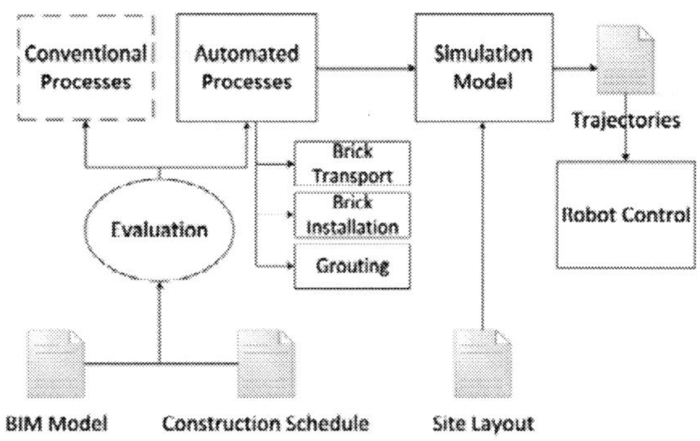

Figure 5: Basic Concept and Data Flow

4.1 Automated Processes

The first step comprises creating a catalogue of building processes and associated materials that can be automatized and performed by wire robots. Processes are examined concerning materials to be moved, required equipment and desired working precision. Due to its high potential for automatization, the creation of masonry buildings represents the focus of this project. In this context, a fully-tensioned wire robot transports and assembles bricks to create a masonry building. After having completed one row of bricks, a special grouting system is applied to create a firm bonding between the single bricks. This work step is also performed by the robot. Processes that are unsuitable for automatization need to be performed in a conventional manner and are not regarded explicitly. Figure 6 shows the operation method of the wire robot which comprises the automated processes of brick transportation, brick installation and grouting.

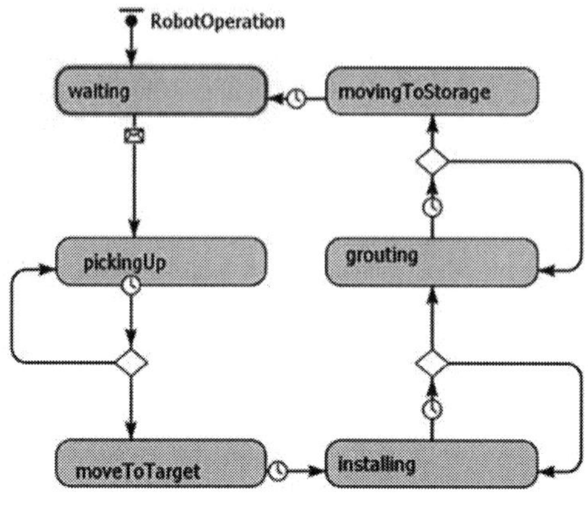

Figure 6: State chart: Robot Operation Method

4.2 Data Exchange

The BIM model represents the major data source concerning building geometry, material and installation position of the single components. Focusing on the automated construction of masonry buildings, wall elements are extracted from the BIM model which is provided in the neutral IFC format. Relevant information on walls (material, position, length, height and thickness) are imported to the simulation model. However, within BIM models, walls are defined as coherent elements. To simulate the assembly procedure, they need to be divided into single bricks. Furthermore, a formal language is developed and implemented to define transport units moved by the robot. The installation sequence of the single walls corresponds to the construction schedule of the project. As the transportation of building material represents the major part of the robot operation, moving paths are recorded paying attention to the requirements mentioned in Section 3.2.3. The resulting trajectories can be used as input source of the crane control system and to check the robot operation for possible collisions.

4.3 Simulation Model

The described approach was implemented using the commercial simulation software AnyLogic. This Java-based software offers state charts to model processes for discrete event simulation. To create a realistic model, system-relevant elements are implemented as AnyLogic "Active Objects" which can be easily modified by adding Java code. Figure 7 shows the most important elements of the simulation after the completion of the first masonry wall: Wire robot, workspace, cables, and storage area.

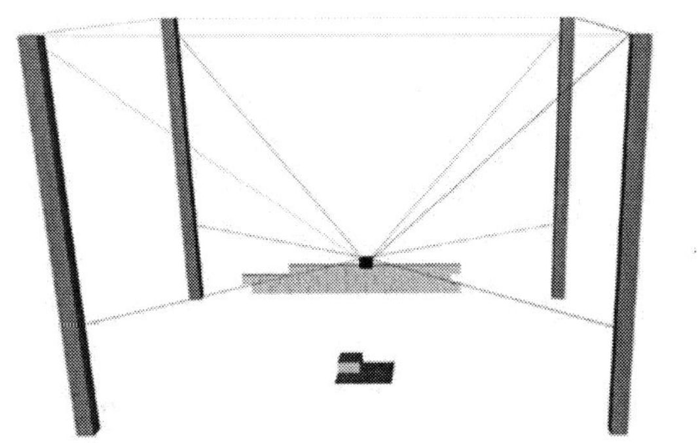

Figure 7: Screenshot of the Simulation: Creation of a Masonry Wall

The design of the wire robot and the positioning of the frame corresponds to the chosen robot type and setup as well as the given construction site layout. Position and geometry of the single bricks derive from importing wall elements provided in the BIM model. In a second step, the exact installation position of each brick is calculated in accordance with brick format (height, width and thickness) as well as grid design. In order to avoid cross joints, every second brick row is displaced by a half stone. Storage position and capacity correspond with the prevailing site layout. As the main focus lies on analyzing the robot operation method, it is assumed that the required amount of bricks and grout is available at any time of the construction phase.

4.4 Results

Analyzing the robot operation method represents the main intention of conducting simulation experiments. By conducting case studies on simple building structures, the robot motion profile can be examined concerning feasibility, efficiency and work safety. Figure 8 shows the moving paths in x- and y-direction while creating the first row of a masonry wall. The single bricks measure 1000mm x 500mm x 240mm and the center of the storage is set to be in the point (0;0). Concerning the aspect of feasibility, for example, the simulation can help to check whether all building materials are positioned within the workspace of the robot. In addition, the model can be applied to record movement paths required to create the exemplary building structure. The feasibility of the developed concept is validated if the resulting trajectories are collision-free. Collision-free motion paths can be used as input for the robot control system and thus, support an efficient and safe work progress.

The current underlying model is purely based on geometric assumptions and thus limited to kinematic properties. Accordingly, errors arising from dynamic effects (including forces, elasticities, inertia) cannot be covered, but the resulting paths can serve as input for first experiments on a prototype that will indicate the accuracy for practical applications.

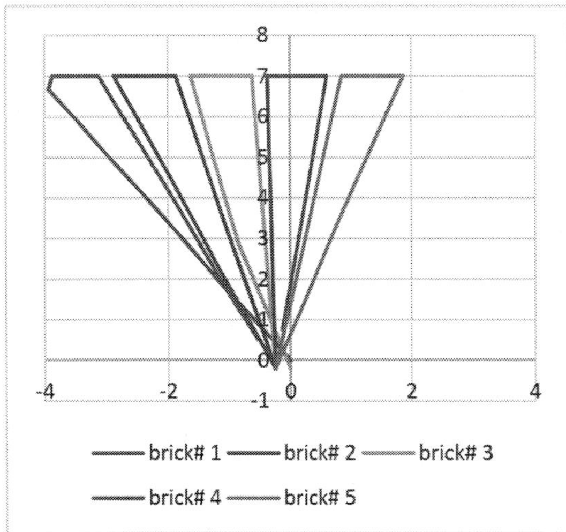

Figure 8: Resulting Moving Paths for the First Row of a Masonry Wall

5 Summary and Outlook

This paper presents an approach of using wire robots for the automated creation of masonry walls. The developed concept represents a first step towards automatizing construction work by the use of robotics. By combining the advantages of BIM and automated production techniques, quality as well as production rates can be increased. In a first step, construction processes which are suitable to be performed by wire robots were detected. This evaluation mainly depends on the BIM model and construction schedule of a construction project. Furthermore, recent applications of robots, especially in the construction industry, were analyzed. Due to its superior mechanical properties, a fully-tensioned wire robot was chosen for the transportation of bricks. A simulation model was created to represent the robot operation method in detail. In this context, BIM was included as main data source for building geometry and material while the construction schedule provided further information on the installation sequence. Next steps comprise the development of a rule checking language to develop a collision-free motion profile. Subsequently, a common data format for exporting moving paths to the control system of the wire robot needs to be developed. The accuracy of the developed approach will be examined by conducting experiments with a prototype of the robot. Additional application cases of the simulation might include optimizing storage position and size and thus, leading to a reduction in project makespan and costs. Focusing on the application of wire robots, the achievable level of working precision needs to be investigated which might result in higher demands concerning platform stiffness.

References

[1] Buswell R.A., Soar R.C., Gibb A., Thorpe A., The Potential of Freeform Construction Processes. In: *Proceedings of the 16th International Symposium on Solid Freeform Fabrication*, pages 503–512, 2005.

[2] McGraw-Hill Construction, Prefabrication and Modularization: Increasing Productivity in the Construction Industry. In: *SmartMarket Report*, .

[3] Costa G., Madrazo L., Connecting Building Component Catalogues with BIM Models Using Semantic Technologies. *Automation in Construction*, 57 :239–248, 2015.

[4] Buswell R.A., Thorpe A., Soar R.C., Gibb A., Design, data and process issues for mega-scale rapid manufacturing machines used for construction. *Automation in Construction*, 17 (8):923–929, 2008.

[5] Ding L., Wei R., Che H., Development of a BIM-based Automated Construction System. *Procedia Engineering*, 85 :123–131, 2014.

[6] Babič N.Č., Podbreznik P., Rebolj D., Integrating resource production and construction using BIM. *Automation in Construction*, 19 (5):539–543, 2010.

[7] Moghadam M., Al-Hussein M., Umut K., Automation of Modular Design and Construction Through an Integrated BIM/Lean Model. *Gerontechnology*, 11 (2):, 2012.

[8] Kumar S.S., Cheng J.C., A BIM-based automated site layout planning framework for congested construction sites. *Automation in Construction*, 59 :24–37, 2015.

[9] Wang J., Zhang X., Shou W., Wang X., Xu B., Kim M.J., Wu P., A BIM-based approach for automated tower crane layout planning. *Automation in Construction*, 59 :168–178, 2015.

[10] Schwabe K., BIM-basierte Baustelleneinrichtungsplanung., Bochum., 2015.

[11] Zhang S., Teizer J., Lee J.-K., Eastman C.M., Venugopal M., Building Information Modeling (BIM) and Safety. *Automation in Construction*, 29 :183–195, 2013.

[12] Zhang S., Teizer J., Pradhananga N., Eastman C.M., Workforce location tracking to model, visualize and analyze workspace requirements in building information models for construction safety planning. *Automation in Construction*, 60 :74–86, 2015.

[13] Kim H., Lee H.-S., Park M., Chung B., Hwang S., Automated hazardous area identification using laborers' actual and optimal routes. *Automation in Construction*, 65 :21–32, 2016.

[14] Vähä P., Heikkilä T., Kilpeläinen P., Järviluoma M., Gambao E., Extending automation of building construction — Survey on potential sensor technologies and robotic applications. In: *Automation in Construction*, pages 168–178, 2013.

[15] Lee G., Kim H.-H., Lee C.-J., Ham S.-I., Yun S.-H., Cho H., Kim B.K., Kim G.T., Kim K., A laser-technology-based lifting-path tracking system for a robotic tower crane. *Automation in Construction*, 18 (7):865–874, 2009.

[16] Lee G., Cho J., Ham S., Lee T., Lee G., Yun S.-H., Yang H.-J., A BIM- and sensor-based tower crane navigation system for blind lifts. *Automation in Construction*, 26 :1–10, 2012.

[17] Willmann J., Gramazio F., Kohler M., Roboter und Megastrukturen: Neue Masstäbe in der digitalen Fabrikation. In: *Technik in Bayern 04*, 2014.

[18] Willmann J., Gramazio F., Kohler M., If Robots Conquer Airspace: The Architecture of the Vertical City. In: *Springer Optimization and Its Applications*, pages 1–11, Switzerland, 2015.

[19] Landsberger S. E., Sheridan T.B., A new design for parallel link manipulator. In: *Proceedings Systems Man and Cybernetics Conference*, pages 812–814, 1985.

[20] Bruckmann T., Lalom W., Nguyen K., Salah B., Development of a Storage Retrieval Machine for High Racks Using a Wire Robot. In: *Proceedings of the ASME International Design Engineering Technical Conferences and Computers and Information in Engineering Conference*, pages 771–780, 2012.

[21] Sturm C., Lalo W., Bruckmann T., Wire Robot Suspension Systems for Wind Tunnels. In: *Wind tunnels and experimental fluid dynamics research*, Rijeka, 2011.

Integration of NDE Measurements and Current Practice In Bridge Deterioration Modeling

Marwa Ahmed[1], Osama Moselhi[2] and Anjan Bhowmick[3]
1Phd student, Building, Civil, and Environmental Engineering Dept., Concordia University, Canada.
marwa.bella@hotmail.com
2Professor, Building, Civil, and Environmental Engineering Dept., Concordia University, Canada,
moselhi@encs.concordia.ca
3Assistant Prof., Building, Civil, and Environmental Engineering Dept., Concordia University, Canada,
anjan.bhowmick@concordia.ca

Abstract –

Deterioration models are required and used in Bridge Management System (BMS) to predict the condition and performance of bridges. Effective maintenance of bridge structure relies on the accuracy of deterioration models used to predict bridge performance. Markov Model is a deterioration forecasting model that is widely used in BMS. However, research showed that Markov Model has many shortcomings. This research provides a review of bridge deterioration modeling with emphasis on accuracy improvement of the generated transition probability matrix used in the model. Dynamic Bayesian Network (DBN) technique is utilized to predict the future conditions of bridge decks. Variables such as, factors affecting deterioration process and inspection measurements from Non-Destructive Evaluation (NDE) methods are incorporated to increase the accuracy of the developed deterioration model. The impact of these factors is extracted from the literature and the DBN model is built using a numerical example. Measurements of NDE for years 2008 and 2013 for a case of a bridge deck are used to apply the model. The developed method is expected to improve current practice in forecasting bridge deck deterioration and in estimating the frequency of inspection.

Keywords –

NDE; Deterioration Modeling; Bridges; DBNs

1 Introduction

Deterioration models are used in Bridge Management System (BMS) to predict the future conditions and performances of bridges. Large number of historical data is required for deterioration modeling. The deterioration models are influenced by: 1-Bridge age, 2-Bridge type, 3-Bridge environment, 4-Material properties, 5-Bridge design, 6-Bridge loading and 7-Bridge Capacity. Bridge deterioration rate is a decrease in condition rating per year. Bridge age and daily traffic load are the most critical factors that cause bridge deterioration. Bridge service life can be determined by defining the correlation between bridge age and condition rating. Therefore, the effective maintenance of bridge structure relies on the quality, accuracy of deterioration models that are used to predict bridge performance and service life [1; 2; 3].

Currently, there are two major types of deterioration models:

(i) Deterministic Models:

Deterministic models describe relationships between factors affecting bridge deterioration. However, it ignores random errors in prediction. Some of the limitations of deterministic deterioration models are as follows:

1- Deterministic models neglect uncertainty

2- They predict the average condition of group of (bridges) without focusing on individual bridge. These models provide less focus on current condition and the history of the bridge.

3- It is always difficult to estimate the impact of maintenance actions on deterioration when deterministic deterioration models are used.

4- These models neglect the interaction between bridge components.

(ii) Stochastic Models:

Stochastic models deal with deterioration process as random variables that incorporate uncertainty. Markov models are the most widely used deterioration models used to predict the condition of infrastructure facilities. It covers two limitations of deterministic models as it incorporates uncertainty and account for the current facility condition. Markov Chain Model forecasts bridge condition rating based on the concept of defining states of bridge condition from one to another during transition period. Markov approach is a discrete time stochastic process that takes number of possible discrete states. Markov model has the following limitations:

1- Markov models assume discrete transition time intervals.

2- Future condition of a facility depends only on current facility condition and not on a history of the bridge, which is unrealistic.

3- Markov models assume that the condition of a bridge can stay the same or reduced to avoid the complexity to consider the treatment process and its impact.

4- Markov models cannot determine the interaction between different components of bridges.

5- In these models, transition probabilities require update when new information is available.

It has been suggested that integrating NDE methods into Markov model will reduce its limitation [4]. Also, the accuracy of Transition Matrix increases the accuracy of Markov-deterioration model [5, 6].

Another type of stochastic models available are Bayesian Networks (BNs). These models consist of a graph that includes nodes and arcs. The arcs connecting two nodes represent the dependences relationships between random variables nodes. BNs has many application in medicine diagnostics and in engineering predictions [7, 8, 9]. Few researchers have applied BNs in deterioration modeling. By using BN, dependencies among variables is easy to interpret. Variables are considered independent if there is no edge connecting those variables.

According to Weber et al. [10], BN has the capability of modeling complex system. It makes prediction and diagnostics. It computes the probability of event occurrence. It updates beliefs based on new evidence. It integrates qualitative information and the quantitative ones. BN merges experience, past knowledge, impacting factors and measurements. So far, according to the literature review, BN has limited applications in maintenance and in bridge deterioration modeling. Dynamics Bayesian Network (DBN) is a class of BNs which represent stochastic process. DBN consists of sequence of slices; each consists of BN nodes. These slices are connected by direct arc from slice at time T1 to slice at time T2. DBN provides computational framework that allows accurate and efficient prediction of deterioration based on observations and deterioration parameters [11, 9]. These DBNs are expected to alleviate the main limitations of current Markov model.

In this research, the main objective is to develop a model for bridge deterioration using multiple NDE methods. The integration of NDE methods and current practice is used to improve the accuracy of forecasting bridge condition.

2 DBN For Bridge Deck Deterioration

Dynamic Bayesian networks are a special class of BNs to analyze problems of bridge deterioration with time variation. It consists of a sequence of time slices (T1, T+1, ..., T+n). In each slice, there are one or more BN nodes. Time slices are connected with direct link, these links present probabilistic dependences.

In DBNs, bridge deterioration can be predicted from past experience. The knowledge from experts are used to build conditional probability table (CPT) directly. According to Wang et al. [13], this task is performed through 5 steps; expert selection, expert training, questions preparation, expert judgment, and results verification.

CPT can be determined directly from visual inspection or from NDE methods. In DBN discrete units of time is modeled, where each unit defines a time slice. These time slices are connected through links. The probabilities associated with links connecting these slices are defined as transition probabilities. In DBN, the basic network is repeated over time [7, 9, 14, 13]. DBNs utilize the Markov model process and allow taking into consideration the prior probability distribution of random variables considered in the deterioration process.

According to Rafiq et al. [14], deterioration of bridge element leads to reducing level of service and bridge safety level. In current practice, deterioration models are presented by Markov stochastic process. For simplicity in existing BMS, discrete time stochastic process is employed to model bridge deterioration at T+1 by prior knowledge about deterioration at T1. Rafiq et al. [14], applied DBN model for modeling the deterioration of masonry arch bridge. The authors utilized DBN to address the interdependency between main element and sub element.

Wang et al. [13] focused and studied BN. Wang et al. [13], used dynamic Bayesian Networks for prediction of structural reliability of steel bridge elements. The authors developed an approach that is able to update information from the observed measurements, and then corrosion process is modeled. Straub [9] Proposed DBN to model that updates variables based on information from inspection. Faddoul et al. [11] Presented DBN for maintenance action of roads taking into account updated information to improve the existing inspection, maintenance and rehabilitation actions.

3. Model Development of Bridge Deck Deterioration with DBNs and NDE Methods

The bridge deterioration model utilizes the inspection measurements acquired during bridge inspection. It includes the measurements from multiple NDE methods that usually used in the advanced inspection. These measurements are combined and their outputs are used to determine bridge deck condition rating. NDE measurements are used to detect bridge deck defected area. Bridge deck condition rating is assigned based on the percentage of the defected area. According to Minnesota department of transportation, 2013, there are five condition states used to assign bridge deck condition rating. These five condition states are defined as follows:

Condition state 1: deck shows little or no deterioration
Condition State 2: combined deterioration of deck areas are less than 2%.
Condition State 3: combined deterioration of deck areas are between 2% and 10%.
Condition State 4: combined deterioration of deck areas are between 10% and 25%.
Condition state 5: combined deterioration of deck area are more than 25%.

Many attributes are impacting bridge deck deterioration. These factors are; bridge age, bridge design, environmental factors and excessive loading. The impact of these factors are stochastic. This impact is incorporated in the developed model.

The basic Bayesian network is modeled as illustrated in Figure 1. F1, F2,...... Fn are the deterioration factors impacting the bridge deck. Factors nodes are parents' variables contributing their impact to the condition states of the bridge deck. The bridge deck condition contribute the information to impact and cause the inspection measurements using NDE methods. Accordingly, the multiple measurements from NDE methods that are collected after the bridge deck inspection are child nodes of the bridge deck condition. These NDE measurements are considered parents nodes contributing and causing the information to their child node which is the bridge deck condition rating. Bridge deck condition is assigned based on the combined defected area. The qualitative part of the basic Bayesian network is illustrated as a graph as shown in Figure 1. The quantitative part is defined using conditional probability table (CPT) between each parent node and its child node. CPT measures the strength of the relationship between them. In this research, the CPT is defined by varying the impact and occurrence of factors on bridge deck condition. CPT between bridge deck condition rating and NDE measurements is defined by Varying the measured areas through five states, each state has specific range of the % of the measured defected area.

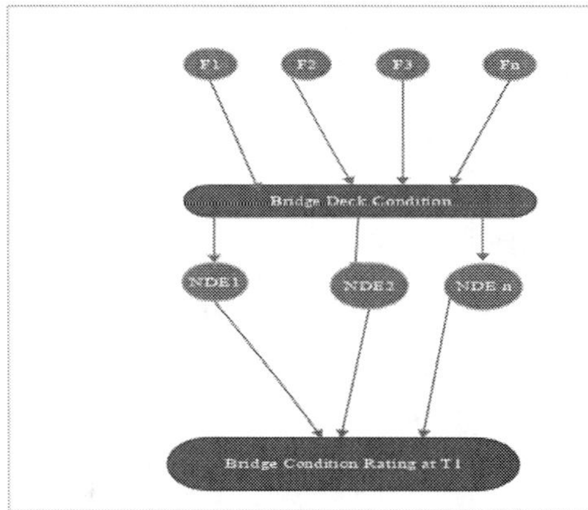

Figure 1. Bayesian Network for bridge deck deterioration modeling

The bridge deck condition state and the bridge deck condition rating are variables changing over the time. So, the basic Bayesian network is modeled as dynamic Bayesian network (DBN) through time slices. Each time slice includes the basic Bayesian network at specific time. Direct arrow is linking nodes of bridge deck condition at different times. The probabilities associated with links connecting the nodes of bridge deck condition at different time slices are defined as transition probabilities. As illustrated in Figure 2. The basic Bayesian network is repeated over the time. The bridge deck condition and condition rating are variables changing over the time T1, T2,....Tn. Temporary arc is used to link the change of bridge condition rating over the time to build the transition probabilities of the bridge condition.

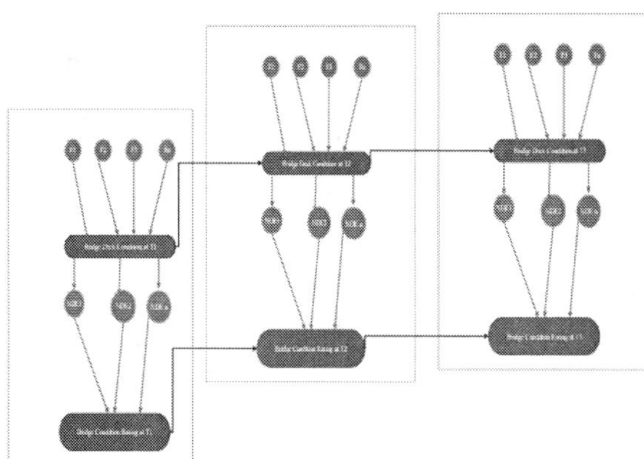

Figure 2. DBNs Model for bridge deterioration

The arc linking bridge deck condition at different times ensures that current bridge condition T2 depends on previous history of bride deck condition at T1. Modeling bridge deck deterioration this way, incorporates the maintenance action in previous time units. Incorporating the stochastic impact of deterioration factors at each time unit, helps to accurately forecast bridge deck condition. In the developed model, the experience, past knowledge, measurements from different sources of NDE and deterioration factors are combined. The model can be updated with new information from NDE measurements. It will be updated if more NDE methods are incorporated and their results are fused. More extra factors can be incorporated as well.

4. Numerical Example As Application of Dynamic Bayesian Networks Model

Factors impacting bridge deterioration are incorporated in Dynamic Bayesian network model. The impact of these factors are extracted from the literature review. Huang [16] identified 8 attributes that extracted from the inventory data of decks record from Pontis BMS. The author analyzed the five factors that has great impact on transferring bridge deck condition from state 1 to state 2, A12. These factors are: District, Design Load, ADT (Vehicle/Day), Environment, and Degree of Skew. The author analyzed the five factors that have great impact on transferring bridge deck condition from state 2 to state 3, A23. These five factors are: Design Load, Deck Length (m), Deck Area (m2), Environment, and Number of Spans. Huang [16] listed the 11 factors that did impact the bridge deck deterioration as illustrated in Table 1. Table 1 illustrates the impact of factors on the transition of bridge deck condition from state 1 to state 2 (A12), from state 2 to state 3 (A23) and from state 3 to sate 4 (A34).

Table 1. The impact of factors on the transition of bridge deck condition

States	*P-value*	ATTRIBUTES
A12	0.05	Degree of Skew
A34		Maintenance History
A34		Age
A34		Previous Condition
A12	0.0014	District
A12, A23	0.0001, 0.0211	Design Load
A23	0.0158	Length
A23	0.0019	Deck area
A12	0.0158	ADT
A12, A23	0.0005, 0.0053	Environment
A23	0.0149	Number of Spans

Measurements for years 2008 and 2013 of GPR for a bridge deck are used in this study. This data were extracted from condition mapping of bridge deck in years 2008 and 2013 [17]. The % of delamination for years 2008 and 2013 are calculated using generic model of Martino [18]. The amplitude values for years 2008 and 2013 of bridge deck are illustrated on Table 2. The amplitude values are extracted approximately from the deterioration mapping of previous research efforts [17] for years 2008 and 2013. Table 2 illustrates % of delaminated areas that are calculated using the generic model. % of delamination is calculated in the following steps because of lack of the raw data:

1- Dinh et al.[17] plotted two deterioration mapping for years 2008 and 2013. These deterioration mapping were built based on GPR signal attenuation. Amplitude values for the bridge deck are obtained from the deterioration maps.

2- Dinh et al.[17] plotted the results based on the rebar reflection amplitude and related the deterioration of bridge deck with this reflection.

3- The GPR amplitude ranges that appears as values of GPR in [17] are utilized to apply [18] generic linear model to get % of delamination in the bridge deck for each year.

4- Martino [18] indicated that the model can be used for bridge deck with moderate corrosion and with threshold -1.6dB) by using equations (1), (2) and (3)

$$Y = 7.051725 * X + 1.78044 \quad (1)$$

$$Y = \%\ Delamination \quad (2)$$

$$X = Skew * Mean\ GPR\ Amplitude \quad (3)$$

5- By Following this way, % delamination for each year (2008 and 20013 of the case study) are obtained.

Table 2. Amplitude values of Bridge Deck Years

2008	2013	2008	2013	
-9	-7	-18	-13	
-7	-7	-16	-14	
-10	-8	-7	-16	
-11	-9	-17	-8	
-16	-11			
-16	-11			
-17	-13			
-18	-18			
-10	-10			
-11	-14			
-7	-4			
-6	-5	-25.667	-22.222	**Mean**
-9	-9	-0.0608	-0.1694	**SKEW**
-12	-13	1.56151	3.7651	**Mean *SKEW**
-14	-10	12.7105	28.1357	**%Delamination**

Figure 3 illustrates the basic network that consists of factors impacting on the transition of bridge deck condition. NDE measurements are child node of A12, A23, A34 and A45. From NDE measurements at different times, condition assessment nodes are determined with different time. Figure 3 is considered the qualitative part of the network as it shows the relationship between different nodes.

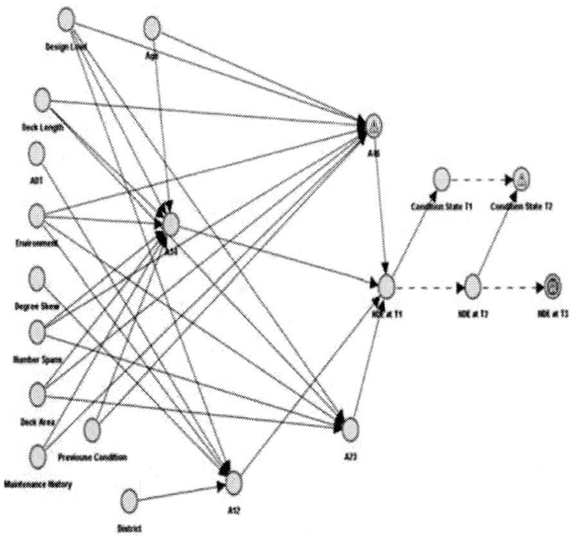

Figure 3. Dynamic Bayesian Network of Bridge Deck Deterioration Model

The relationship between nodes is quantified by defining the conditional probability table. Table 3 shows 32 conditional probabilities on node A12. Table 3 utilized to define the relationship between the factors and bridge state transition from state 1 to state 2 (A12). It measures the true and false percentage of A12 occurrence with high and the low impact of the factors. As illustrated in Table 3, the factors impacting A12 are District, Design Load, ADT, Environment and Degree of Skew. The 32 probabilities are generated based on varying the impact of the factors. For example, the first conditional probability is generated based on the low or false impact of factors District, Design Load, ADT, Environment and Degree of Skew. So, the first conditional probability is assigned true and false percentage values of 0% and 100% respectively. The second conditional probability is generated based on the false value of factors District, Design Load, ADT, Environment and the true value of the factor Degree of skew. So, the second conditional probability is assigned true and false percentage values of 50% and 50% respectively because Degree of skew factor is one of the significant factors with high impact comparing to the other factors. The other 30 conditional probabilities are assigned by the same way. Table 4 defines the conditional probability table of 32 conditional probabilities that measure the strength of the relationship between factors Deck Length, Deck area, Number of Spans, Environment, Design Load and node A23.

Table 3. Conditional Probability table of transition from state 1 to state 2 (A12)

District	Design Load	ADT	Environment	Degree Skew	FALSE	TRUE
FALSE	FALSE	FALSE	FALSE	FALSE	100	0
FALSE	FALSE	FALSE	FALSE	TRUE	50	50
FALSE	FALSE	FALSE	TRUE	FALSE	95	5
FALSE	FALSE	FALSE	TRUE	TRUE	75	25
FALSE	FALSE	TRUE	FALSE	FALSE	70	30
FALSE	FALSE	TRUE	FALSE	TRUE	20	80
FALSE	FALSE	TRUE	TRUE	FALSE	65	35
FALSE	FALSE	TRUE	TRUE	TRUE	15	85
FALSE	TRUE	FALSE	FALSE	FALSE	99	1
FALSE	TRUE	FALSE	FALSE	TRUE	51	49
FALSE	TRUE	FALSE	TRUE	FALSE	94	6
FALSE	TRUE	FALSE	TRUE	TRUE	44	56
FALSE	TRUE	TRUE	FALSE	FALSE	49.286	50.714
FALSE	TRUE	TRUE	FALSE	TRUE	19	81
FALSE	TRUE	TRUE	TRUE	FALSE	64	36
FALSE	TRUE	TRUE	TRUE	TRUE	14.286	85.714
TRUE	FALSE	FALSE	FALSE	FALSE	86	14
TRUE	FALSE	FALSE	FALSE	TRUE	64	36
TRUE	FALSE	FALSE	TRUE	FALSE	19	81
TRUE	FALSE	FALSE	TRUE	TRUE	31	69
TRUE	FALSE	TRUE	FALSE	FALSE	44	56
TRUE	FALSE	TRUE	FALSE	TRUE	80	20
TRUE	FALSE	TRUE	TRUE	FALSE	51	49
TRUE	FALSE	TRUE	TRUE	TRUE	1	99
TRUE	TRUE	FALSE	FALSE	FALSE	85	15
TRUE	TRUE	FALSE	FALSE	TRUE	35	65
TRUE	TRUE	FALSE	TRUE	FALSE	80	20
TRUE	TRUE	FALSE	TRUE	TRUE	30	70
TRUE	TRUE	TRUE	FALSE	FALSE	55	45
TRUE	TRUE	TRUE	FALSE	TRUE	5	95
TRUE	TRUE	TRUE	TRUE	FALSE	50	50
TRUE	TRUE	TRUE	TRUE	TRUE	0	100

Table 4 shows the relationship between factors and A34 through different probabilities and varying of incorporating the impact of the factors. It measures the true and false percentage of A34 occurrence with the high and the low impact of the factors. For example, the first conditional probability is generated based on the low or false impact of factors Deck Length, Deck area, Number of Spans, Environment and Design Load. So, the first conditional probability is assigned true and false percentage values of 0% and 100% respectively. The second conditional probability is generated based on the false value of factors District, Design Load, ADT, Environment and the true value of the factor Design load. So, the second conditional probability is assigned true and false percentage values of 28% and 72% respectively.

Table 4. Conditional Probability table of transition from state 2 to state 3 (A23):

Deck Length	Deck Area	Number Spans	Environment	Design Load	FALSE	TRUE
FALSE	FALSE	FALSE	FALSE	FALSE	100	0
				TRUE	72	28
			TRUE	FALSE	86	14
				TRUE	58	42
		TRUE	FALSE	FALSE	77	23
				TRUE	49	51
			TRUE	FALSE	63	37
				TRUE	35	65
	TRUE	FALSE	FALSE	FALSE	90	10
				TRUE	37	63
			TRUE	FALSE	76	24
				TRUE	48	52
		TRUE	FALSE	FALSE	67	33
				TRUE	39	61
			TRUE	FALSE	53	47
				TRUE	25	75
TRUE	FALSE	FALSE	FALSE	FALSE	75	25
				TRUE	47	53
			TRUE	FALSE	61	39
				TRUE	33	67
		TRUE	FALSE	FALSE	52	48
				TRUE	24	76
			TRUE	FALSE	38	62
				TRUE	10	90
	TRUE	FALSE	FALSE	FALSE	65	35
				TRUE	37	63
			TRUE	FALSE	51	49
				TRUE	23	77
		TRUE	FALSE	FALSE	42	58
				TRUE	14	86
			TRUE	FALSE	28	72
				TRUE	0	100

Table 5 shows the conditional probabilities of GPR measurements at year 2008. Table 5 measures the strength of transition for condition states between A12, A23, A34 and A45 and the probability of existence of defected areas. For example, the first conditional probability is generated based on the true occurrence of A12, A23, A34 and A45. So, the first conditional probability is assigned 0 values if there is no area defected. The first conditional probability is assigned 0.25 if the defected area is less than 2%, less than 10%, more than 10% and more than 25 %. The second conditional probability is generated based on the true occurrence of A12, A23, A34 and the false occurrence of A45. So, the second conditional probability is assigned 0 value if there is no area defected or the area defected is more than 25%. It is assigned a value of 0.1 if the defected area is less than 2%. It is assigned a value of 0.45 when the defected area is less or more than 10%.

Table 5.Conditional Probabilities of node NDE measurements

A12	TRUE				
A23	True				
A34	True		False		True
A45	True	False	True	False	True
No Area Defected	0	0	0	0	0
Area Less 2%	0.25	0.1	0.1	0.5	0
Area Less 10%	0.25	0.45	0.45	0.5	0.333
Area More 10%	0.25	0.45	0	0	0.333
Area More 25%	0.25	0	0.45	0	0.333

The dynamic Bayesian network of the basic network at different times T0, T1 in 2008 and T2 in 2013 is built. The basic Bayesian network is repeated within the time and at each time slice the networks are connected thorough temporary arcs. Modeling deterioration this way, ensures that future condition depends mainly on current condition, previous condition and related factors.

The final results of the developed NBNs show the probability of condition states at different times. As illustrated in Figure 4, the vertical axis represents the probability of different condition states. The horizontal axis represents the time steps. The spacing between time steps is 5 years. From time t0 to time t1, the bridge deck falls under condition state 1. From time t1 to time t2 (year 2008) and from time t2 (year 2008) to t3 (year 2013), the bridge deck fall under condition state 3. After 5 years (year 2018), the bridge deck will fall under condition state 5.

The results also show the probability of the existence of the defected areas measured by NDE at different time steps, from t0 to t1, the defected area is falling under the category of no area defected. From t0 to t2, the defected area is falling under the category of area defected more than zero and less than 2%. From t1 to t4, the defected area is falling under the category of area defected is more than 10% and less than 25%. From t3 to t4, the defected area is falling under the category of area defected is more than 25%.

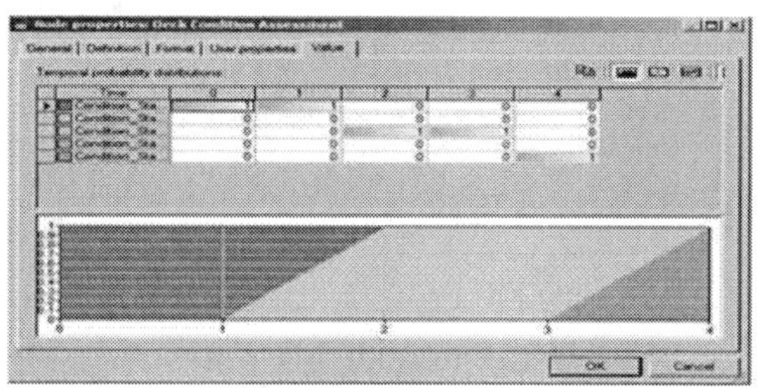

Figure 4. The Result of bridge deck condition

5. Prediction of Bridge Condition Using Markov Model

Markov Model is used to predict the future condition of the bridge deck for the same case study. So, condition state C0 at time 0 is taken at 1978 year, In order to predict the future condition of the deck at different times every 5 years Equation (4) is applied, where t is the number of transactions

C (t) = C (0) * TPM power t ....(4)

= [1 0 0 0 0] *

[0.513	0.487	0	0	0] Power t
0	0.492	0.507	0	0
0	0	0.5	0.5	0
0	0	0	0.513	0.487
0	0	0	0	1

At year 1993 which means 15 years after the initial condition and the time interval between inspection measurements are 5 years. So, the number of transactions are 3

Condition States = **[0.135005697 0.368965323 0.274562808 0.220489737 0]**

At year 2008 which means 30 years after the initial condition and the time interval between inspection measurements are 5 years. So, the number of transactions are 6

Condition States = [0.018227 0.093755 0.23468386 0.3171046 0.3345]

At year 2023 which means 45 years after the initial condition and the time interval between inspection measurements are 5 years. So, the number of transactions are 9

Condition States= [0.0024607 0.01756 0.0698756 0.16798167 0.708455]

The results of the probabilities of the five condition states at different years are summarized in Table 6.

Table 6: Probability of bridge condition states at different transactions

Years	1978	1993	2008	2023
Condition 1	**1**	0.135	0.01823	0.0024
Condition 2	0	**0.3689**	0.09375	0.0175
Condition 3	0	0.27456	0.2347	0.06988
Condition 4	0	0.22049	**0.3171**	0.1679
Condition 5	0	0	0.3345	**0.70846**

6. Comparing The results of DBNs and Markov Model

Table 7 illustrates the final results of modeling bridge deterioration using Dynamic Bayesian Networks and Markov model techniques. Although, transition probabilities matrices for Markov model were built using some of the information from Bayesian networks, it doesn't consider the impact of the deterioration factors. Also, it doesn't take into consideration the previous condition at previous time. It is very clear from the result of Markov model, that it doesn't consider the impact of maintenance action. So, at years 1993 and 1998, the bridge was deteriorated faster to reach condition 2. Starting from year 2008, the bridge deteriorated faster to reach condition 5. In the DBNs model, factors impacting bridge deterioration are incorporated. It is very clear from the results that bridge deck will start to deteriorate and reach condition state 5 at year 2018.

Table 7. The Results Comparison between Markov and DBNs models

Years	Markov Model	DBNs Model
1993	2	1
1998	2	1
2003	3	3
2008	5	3
2013	5	3
2018	5	5
2023	5	5

7. Conclusion

This research provides a method to predict bridge deck condition states using Dynamic Bayesian Networks. The model is built using limited inspection records for two years at 2008 and 2013. The model incorporates the impact of deterioration factors extracted from the literature. Modeling bridge deck deterioration this way, ensures that future condition depends mainly on current condition, previous condition and factors impacting bridge deck deterioration. The model circumvents the limitations of current practice which is based on traditional Markov model. The final results of Dynamic Bayesian Networks are compared with the results of Markov model. These results show that incorporating deterioration factors improve the forecasting accuracy and its impact on forecasting inspection frequency and maintenance action required. The main contribution of the developed model lies in building an advanced deterioration modeling for bridge deck by using measurements of NDE methods and incorporating related factors. The model is generic, it can be updated when new observations are incorporated.

References

[1] Agrawal, A., Kawaguchi, A., and Chen, Z. "Deterioration Rates of Typical Bridge Elements in New York." J. Bridge Eng., 15(4): 419-429, 2010.

[2] Cesare, M., Santamarina, C., Turkstra, C., and Vanmarcke, E. "Modeling Bridge Deterioration with Markov Chains." Journal of Transportation. Engineering, 118(6): 820-833, 1992.

[3] Robelin, C. and Madanat, S. "History-Dependent Bridge Deck Maintenance and Replacement Optimization with Markov Decision Processes." J. Infrastructure System, 13(3):195-201, 2007.

[4] Frangopol, D. M., Kallen, M. J., and Van Noortwijk, J. M. "Probabilistic models for life cycle performance of deteriorating structures: review and future directions". J. Progress in Structural Engineering and Materials, 6(4): 197-212, 2004.

[5] Madanat, S., Mishalani, R., and Ibrahim, W. H. W. "Estimation of infrastructure transition probabilities from condition rating data". Journal of infrastructure systems, 1(2), 120-125, 1995 .

[6] Roelfstra, G., Hajdin, R., Adey, B. and Brühwiler, E., "Condition evolution in bridge management systems and corrosion-induced deterioration." Journal of Bridge Engineering, 9(3): 268-277, 2004.

[7] Murphy, K. P. "Dynamic Bayesian Networks: representation, inference and learning" (Doctoral dissertation, University of California, Berkeley), (2002).

[8] Jha, M. (2009). "Dynamic Bayesian Network for Predicting the Likelihood of a Terrorist Attack at Critical Transportation Infrastructure Facilities". Journal of Infrastructure Systems, 15(1): 31-39, 2009.

[9] Straub, D. "Stochastic modeling of deterioration processes through dynamic Bayesian networks". Journal of Engineering Mechanics, 135(10): 1089-1099, 2009.

[10] Weber, P., Medina-Oliva, G., Simon, C., and Iung, B. "Overview on Bayesian networks applications for dependability, risk analysis and maintenance areas". Journal Engineering Applications of Artificial Intelligence, 25(4): 671-682, 2012.

[11] Faddoul, R., Raphael, W., Soubra, A. H., and Chateauneuf, A. "Incorporating Bayesian Networks in Markov Decision Processes". Journal of Infrastructure Systems, 19(4): 415-424, 2013.

[12] Jensen, F. V., and Nielsen, T. D. "Bayesian network and decision graphs" Springer-Verlag. NY, USA, 2001.

[13] Wang, R., Ma, L., Yan, C., and Mathew, J. "Condition deterioration prediction of bridge elements using Dynamic Bayesian Networks (DBNs)". In Proceedings of the conference Quality, Reliability, Risk, Maintenance, and Safety Engineering (ICQR2MSE), International Conference pages. 566-571. IEEE, 2012

[14] Rafiq, M. I., Chryssanthopoulos, M. K., and Sathananthan, S. "Bridge condition modelling and prediction using dynamic Bayesian belief networks". Journal of Structure and Infrastructure Engineering, 11(1): 38-50, 2015.

[15] Wang, R., Ma, L., Yan, C., and Mathew, J. "Structural reliability prediction of a steel bridge element using Dynamic Object Oriented Bayesian Network (DOOBN)". In *Proceedings of the conference*, Quality, Reliability, Risk, Maintenance, and Safety Engineering (ICQR2MSE) pages. 7-12, International Conference on). IEEE, 2011.

[16] Huang, Y. H. (2010). "Artificial neural network model of bridge deterioration". Journal of Performance of Constructed Facilities, 24(6): 597-602, 2010

[17] Dinh, K., Zayed, T., Romero, F., and Tarussov, A. "Method for Analyzing Time-Series GPR Data of Concrete Bridge Decks". Journal of Bridge Engineering, 20(6): 04014086-1- 04014086-8, 2014

[18] Martino, N. Quantifying reinforced concrete bridge deck deterioration using ground penetrating radar (Doctoral dissertation, Dissertation, Northeastern University), 2013

Discrete and Continuous Simulation Approach to Optimize the Productivity of Modular Construction Element

Mona Afifi [a], Mohamed Al-Hussein [b], Simaan Abourizk [c], Ahmed Fotouh [d], Ahmed Bouferguene [f]

[a, b, c] Department of Civil & Environmental Engineering, Alberta University, Canada
[d] Department of Quality, KBR Canada, Canada
[f] Campus St. John, Alberta University, Canada
E-mail: mkafifi@ualberta.ca, malhussein@ualberta.ca, abourizk@ualberta.ca, fotouh@ualberta.ca, ahmed.bouferguene@ualberta.ca

Abstract

Bolstered by significant productivity improvement in manufacturing technology, the construction paradigm has shifted toward industrialized construction methods, such as modular construction. Modular construction has benefitted from the long term development of the manufacturing technologies and core techniques used in the assembly production lines of construction prefabrication plants. Lean production is a methodological approach for assembly line production that aims to eliminate waste, minimize inventory, and maximize production flow.

This research proposes implementation of lean concepts through combined discrete-event and continuous simulation approaches with the aim of increasing the productivity of a modular construction element. The proposed model incorporates a simulation of the system reliability by modeling various problems detected within the production line, such as double-handling. Applying continuous simulation facilitated attaining optimal threshold by detecting the inventory level. In addition, improvements are implemented within the model, such as introducing advanced alternatives for the automated stations and adding parallel stations to increase productivity. The effect of proposed production line improvement, along with each station improvement are examined and investigated.

Keywords

Lean; construction; productivity; simulation

1 Introduction

In the early 1990s, lean production emerged as a new manufacturing paradigm. Studies by Koskela (1992; 2000) were instrumental in integrating lean concepts to the construction process as TFV (transformation, flow, and value) theory of production in construction. This was followed by multiple research efforts to characterize this paradigm shift toward lean construction: (1) introduction of lean contraction workflow planning and Last planer system (LPS) (Ballard 1993); (2) establishment of Poka-yoka and Kanban systems in construction (Milberg and Tommelein, 2003; Arbulu, Ballard and Harper, 2003); and (3) formulation of a cohesive framework of lean construction paradigm and its relationship with lean production (Salem, et al., 2006). In this context, modularization, where units are designed, produced, and assembled off site that consist of various construction elements such as walls, frames, doors, ceilings, and windows, has garnered increasing attention.

Production optimization and plant layout optimization based on lean principles using simulation has been widely implemented in the construction industry (Senghore, et al., 2004; Mehrotra, Syal and Hastak, 2005; Jeong, Hastak and Syal, 2006; Yu, et al., 2011). As the construction industry seeks to increase productivity, reduce costs, and increase quality based on lean principles, Simphony has proven to be an effective simulation tool by which to model construction processes or production lines (Yu, Al-Hussein and Nasseri, 2007). Simphony is software developed at the University of Alberta which supports both general-purpose and special-purpose real-time simulation (Ekyalimpa, AbouRizk and Farrar, 2012; AbouRizk and Hajjar, 1998). This software allows users to model different work stations with their associated resources in a manner that resembles the actual production process. In addition, Simphony provides the opportunity to modify the production process, adjust the parameters of the model, and examine the results by running the simulation as many times as needed. Simphony has been used within the construction domain for many purposes, such as project duration forecasting, resource utilization, and general decision making support (Moghani, et al., 2011; Farrar, AbouRizk and Mao, 2004; Fernando, et al., 2003;

AbouRizk, 2010).

This paper represents the development of a Simphony model (Simphony.NET 3.5) to examine a modular building component prefabrication that is being built to satisfy lean production principles. The aim is productivity increase of the single-door line for All Weather Windows Company. A combined simulation model is implemented using discrete-event simulation for different work stations, and continuous simulation in the distribution area s between the stations. The current production line is modeled to satisfy lean principles based on pull system techniques using the combined simulation approach. The issues affecting the reliability of the current system are modeled. Recommended improvements are proposed to minimize production waste throughout the current lean production line. Thus, waste events are detected and eliminated by fixing the double-handling problems, and investigating the optimal threshold level in order to increase the productivity for the continuous simulated part. All the proposed improvements are implemented and tested using the developed combined simulation model generated within a Simphony environment.

2 Lean Approach of the Single-Door Line

2.1 Project Description

The aim of this research is to increase the productivity of the single-door line and increase the efficiency of various stations within the assembly production line. Therefore, a work study is developed to monitor each stations and its cycle time, in addition to investigate the causes of waste in each station. As illustrated in Figure 1, the current assembly door line consists of various stations that includes an automated door slab picker; door measuring table; door insert/cut-out station required for glass opening; the doors then moves through a conveyer that can install up to 3 doors; install and assembly door sweep; cell and jamb assembly preparation station; doors are then moves through a conveyer to the final assembly station, where doors is enveloped for delivery.

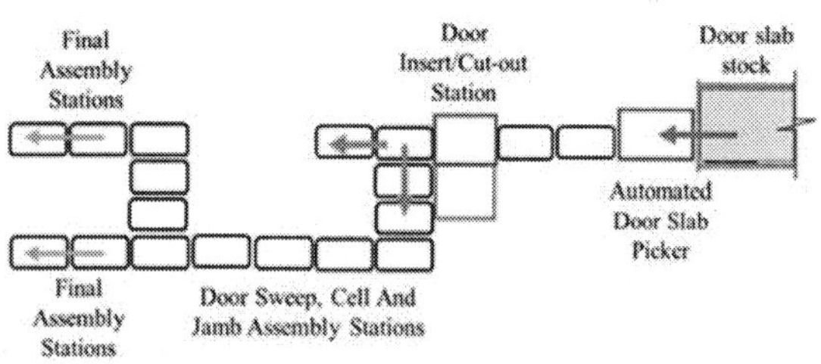

Figure 1. Schematic floor plan of the proposed lean production line

2.2 Lean Implementation

Lean principles were implemented throughout the current construction element of the door production line when it was first built, targeting a total elimination of waste. The floor layout has been improved to eliminate transportation waste by moving all the necessary materials as close as possible to the relevant station. Different Standard Operating Procedures (SOPs) have been developed to eliminate processing waste. However, based on the conducted work study, some areas of potential improvement in the current production line have been identified.

Double handling problems are observed after the door insert/cut-out station, while the work pieces (door slabs) are moving toward the door sweep station. The worker assigned to the door insert/cut-out station and door sweep station leave their stations for an average of 1.00 min for each work piece going to the double-door line in order to adjust the position of the work pieces moving through the first distribution area (from the door insert/cut-out station to door sweep station). Another double handling problem is detected in the automated door slab picker, by which the worker has to leave the work station for 0.25 min in order to adjust the door so the automated slab machine can successfully pick up the slab. In addition, a waste of motion is detected in the door production line.

A waste of motion is identified as the waste caused by a worker's movements, such as bending, twisting, or stretching, that are not necessary for the process. In the door production line, waste of motion is detected when a worker bends over to pick up door shipping blocks to be used for the final assembly station (see Figure 2). This qualifies as a waste of motion because the worker could have bent less and completed the task in less time. In addition, ergonomically it is not recommended to bend over in this manner, as illustrated in Figure 2(a), since this case is more likely to exert high pressure on the spine (90 degree bending). A solution is provided by lifting the entire load of the door shipping blocks to a higher level; this will enable safer lifting of the shipping blocks by workers (a max of 30 degree bending), as illustrated in Figure 2(b).

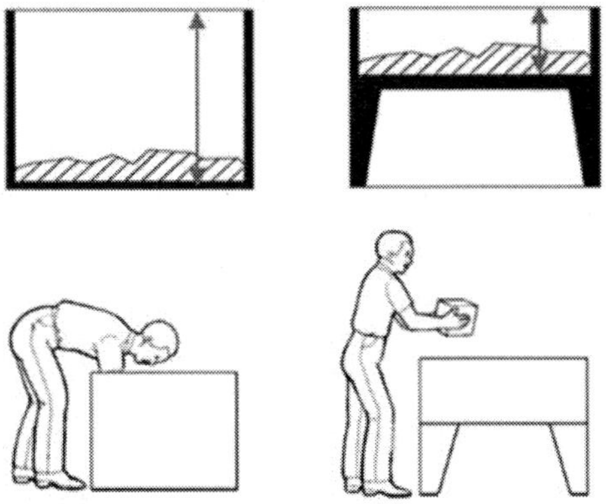

(a) (b)

Figure 2. Picking up the door shipping blocks required for the final assembly station

3 Current Lean Production Line modeling and analysis

3.1 Model Description

Based on the work study, the double handling problems in addition to the waste of motion problems are modeled and activated through valves. Door slabs are classified as either insert (doors that require glass insert) or non-insert (doors that do not require glass insert) before the door insert/cut-out station, as the processing time for non-insert doors is 2.06 min/door, whereas the processing time for insert doors is 5.03 min/door. Door slab ratio for average production of insert to non-insert doors is 46 to 54. A snapshot of parts of the Simphony simulation model is illustrated in Figure 3.

After the cut-out station, the door slabs are moved to the first distribution area, after which the slabs are distributed as either single-door slabs or double-door slabs. The area is simulated as a continuous simulation process. Door slabs state variable is modeled to continuously vary based on the real time simulation of the door slabs movement. The insert assembly station, the cell preparation station, and the jamb assembly station are modeled based on the pull system to represent the current lean production line. A chart collector has been used after the final two modeled branches to monitor the number of insert and non-insert doors.

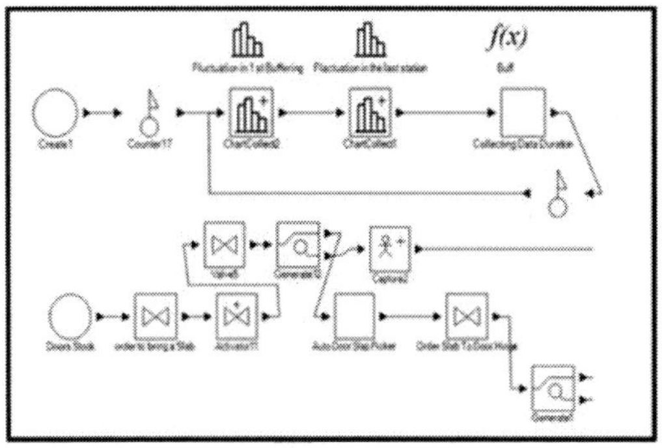

Figure 3. Snapshot of parts of the devolved simulation model by Simphony program

3.2 Production Line Model Analysis

As the actual total current productivity has been found to vary from 95 to 105 door slab/shift, the total productivity based on the developed model in Simphony has been found to vary from 95 to 106 (see Table 1), which indicate the ability of the developed model to accurately model the current production line. Table 1 illustrate the current production line data collected from different runs, along with the total production of the single-door and double-door line. Based on run 1 and 3 in Table 1 that represent the highest and lowest production, the number of single-door and double-doors produced within the lean production line have an inverse relationship. Single-door production is thus affected by the double-door production.

Table 1. Single-door production in relation to double-door production of the developed mode

Number of random runs	Total doors to double-door line	Total doors to single-door line
1	17	106
2	22	102
3	36	95
4	28	99

Based on the continues simulation data in run 1 that extracted from the developed model of the first distribution area, the lowest single-door fluctuation happens when low number of door slabs is going to double-door production line, as shown in Figure 4(a). When number of slabs going to double-door are high, and the number of insert slabs going to single-door production are also high, the fluctuation increase to 3 doors, as shown in Figure 4(b) for run 2. This fluctuation resulted from longer cycle time of insert doors. In the second distribution area, door fluctuation did not seem to be affected by insert door productivity because slabs number did not exceed one slab, as shown in Figures 5(a) and 5(b). This means that no door is waiting in the second distribution zone except the one door moving to the first or second branch of the last station.

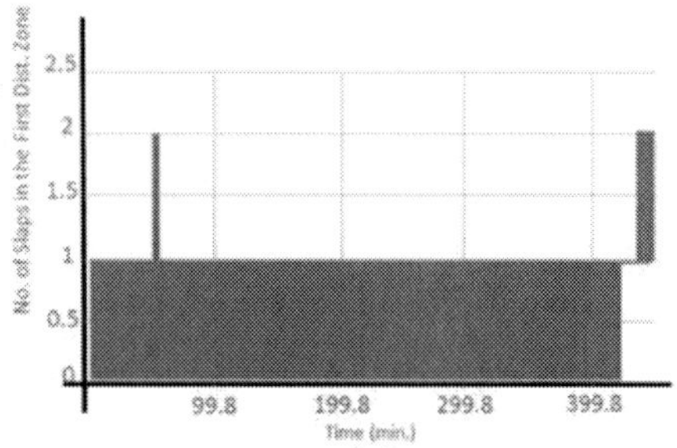

(a)

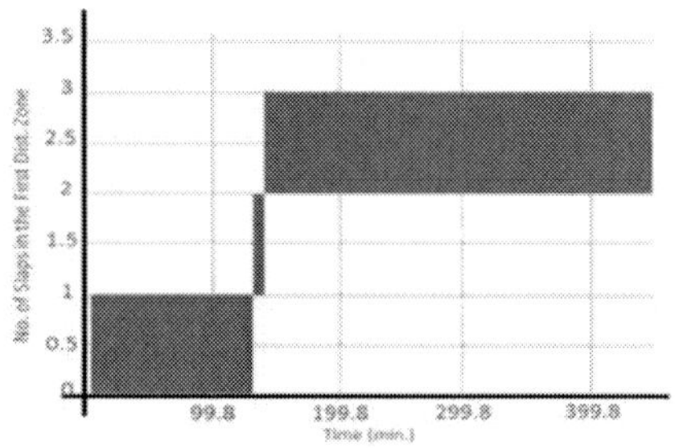

(b)

Figure 4. Bandwidth fluctuation of doors in 1st distribution area for: (a) run 1 and (b) run 2 in Table 1

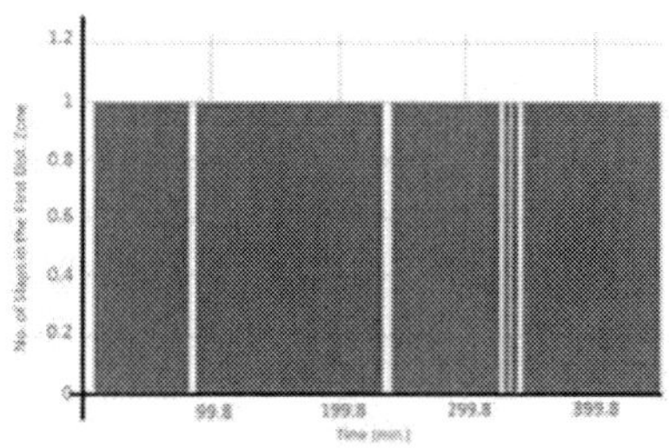

(a)

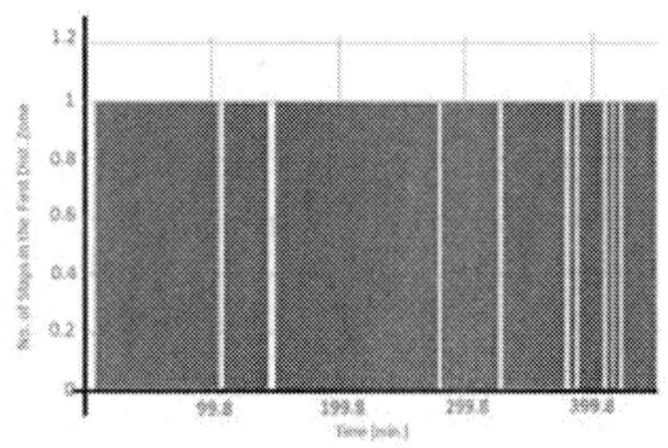

(b)

Figure 5. Bandwidth fluctuation of doors in 2nd distribution area for: (a) run 1 and (b) run 2 based on Table 1

3.3 Modifications of Single-Doors Production Line

In the modified model, the waste detected through the work study was eliminated. Both double handling problem and the waste of motion problem was eliminated. Also, a source of imbalance was detected from the doors accumulation problem in the first distribution area, which is due to the longer processing time of the insert doors in relation to non-insert doors. Therefore, a parallel station is created beside the inserting station in order to process insert doors while the production line of non-insert doors continues through the original station.

Another source of unbalance is identified in the predecessor station to the first distribution area door insert/cut-out station, which is based on the time difference between the processing of insert and non-insert doors: insert doors are processed in 5.03 min while non-insert doors are processed in just 2.06 min. To overcome this unbalance in the processing time of the door insert/cut-out station, a new Computer Numerical Control (CNC) machine is proposed to replace the current machine. This new machine can reduce the cut-out processing time of the insert doors to 3.60 min. Although the increase in production due to this improvement may cause some accumulation in the first distribution area, the previous improvement of adding a parallel station is expected to reduce this potential accumulation.

3.4 Analysis of the Improved Production Line

Based on the proposed improvements, the production of single-door line has been increased, as shown in the random runs in Table 2. Although the objective of adding a parallel station was to minimize the inventory to be one work-piece in the first distribution zone waiting for the work in progress, the number of work-pieces was increased to 3 doors in the last quarter of the shift, as illustrated in Figure (6)a. This can be due to the fact of installing the new "door cut-out station." However, the number of work-pieces in the first distribution zone was reduced back to one when slabs going to the double-door is reduced from 39 to 31 doors, as illustrated in Figure 6(b). For the second distribution zone, the number of work-pieces waiting in the second distribution zone did not get affected as it was a max of one door, as shown in Figure 7 and Figure 8.

Table 2. Number of doors moving to the double-door production line, and the production of modified single-door

Number of random runs	Total doors to double-door line	Total doors to single-door line
1	39	115
2	31	130
3	38	122
4	31	124

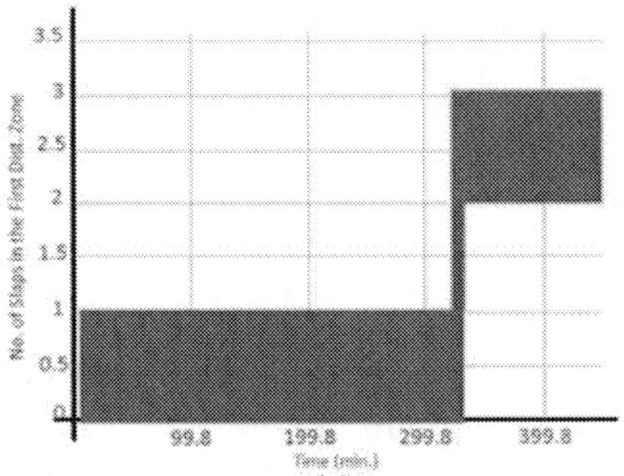

(a)

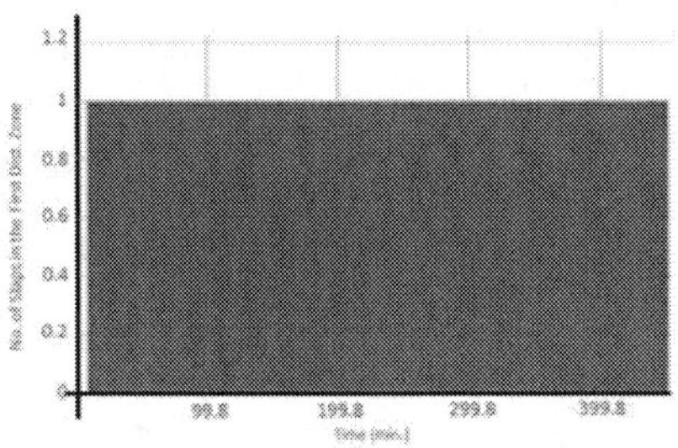

(b)

Figure 6. Bandwidth fluctuation of doors in the first distribution area for (a) run 3 and (b) run 4 in Table 2

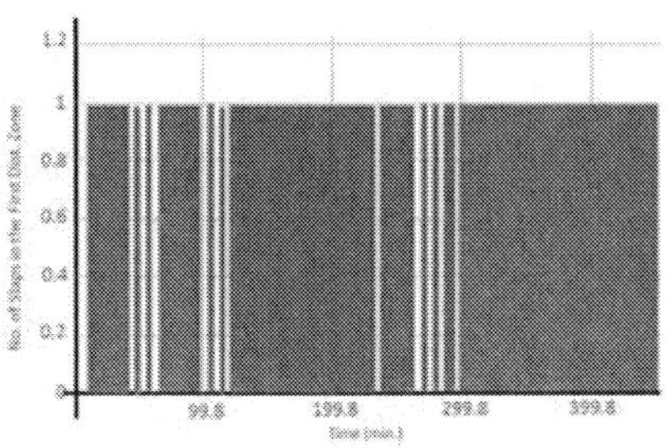

Figure 7. Bandwidth fluctuation of doors in the second distribution area for run 3 in Table 2

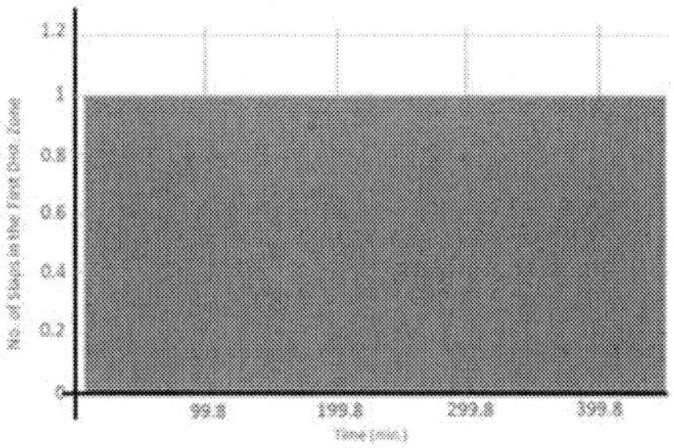

Figure 8. Bandwidth fluctuation of doors in the second distribution area for run 4 in Table 2

4 Conclusion

Using the proposed methodology of a production line improvement based on a real-time simulation approach, the strategy of implementing any improvements of modular construction element can be investigated and refined. The existing single-door production line has been modeled using real-time continuous/discrete-event simulation, and the reliability issues, such as double handling and waste of motion problems, in the current production line have been modeled as well. Using the developed model, the performance of the current production line has been assessed.

Two stations have been found to be the sources of unbalance in the production line, as the unit processing times of these two stations were found to have two separate values depending on the type of the processed doors, whether non-insert or insert. Improving the automatization of one station along with adding another parallel station have been proposed as alternative strategies by which to reduce the sources of the imbalance in the production line. All of the proposed alternatives have been implemented in the modified model. Based on the developed simulation model, single-door production has been increased, including production increasement in "insert", "non-insert", and double-door production.

Acknowledgements

The authors are grateful for the support of the Natural Sciences and Engineering Research Council of Canada (NSERC), as well as the contributions of All Weather Windows Company in funding and supporting this study.

References

[1] AbouRizk, S. M. Role of Simulation in

Construction Engineering and Management. *Journal of Construction Engineering and Management*, 136 (10): 1140-1153, 2010.

[2] AbouRizk, S. M. and Hajjar, D. A Framework for Applying Simulation in Construction. *Canadian Journal of Civil Engineering*, 25(3):604-617, 1998.

[3] Arbulu, R. J., Ballard, G. and Harper, N. Kanban in Construction. In *Proceedings of the 11th Annual Conf. of the Int. Group of Lean Construction, IGLC-11.*, pages 16-17, Blacksburg, Virginia, USA, 2003.

[4] Ballard, G. Lean Construction and EPC Performance Improvement. In *Proceedings of the Lean Construction.* August, pages 79–91, Espoo, Finland, 1993.

[5] Dennis, P. *Lean Production Simplified: A Plain Language Guide to the World's Most Powerful Production System,* New York: Productivity Press, 2002.

[6] Ekyalimpa, R., AbouRizk, S. M. and Farrar, J. Effective Strategies for Simulating One-Of-A-Kind Construction Projects. In *Proceedings of the Winter Simulation Conference*, pages 260-267, Berlin, Germany, 2012.

[7] Farrar, J. M. , AbouRizk, S. M. and Mao, X. Generic Implementation of Lean Concepts in Simulation Models. *Lean Construction Journal*, 1 (1):1-23, 2004.

[8] Fernando, S., Er, K. C., Mohamed, Y., AbouRizk, S. M. and Ruwanpura, J. A Review of Simulation Applications for Varying Demands in Tunneling. *Bridges*, 10(40671): 35, 2003.

[9] Hassan, K. and Kajiwara, H. Application of Pull Concept-Based Lean Production System in the Ship Building Industry. *Journal of Ship Production and Design,* 29 (3):105-116, 2013.

[10] Jeong, J. G., Hastak, M. and Syal, M. Supply Chain Simulation Modeling for the Manufactured Housing Industry. *Journal of urban planning and development*, 132(4):217-225, 2006.

[11] Koskela, L. *Application of the New Production Philosophy to Construction.* Technical Rep. Center for Integrated Facility Engineering, Dept. of Civil Engineering, Stanford University, CA, 1992.

[12] Koskela, L. *An Exploration towards a Production Theory and Its Application to Construction*, VVT, Technical Research Centre of Finland, Finland, 2000.

[13] Mehrotra, N., Syal, M. and Hastak, M. Manufactured Housing Production Layout Design. *Journal of Architectural Engineering,* 11(1):25-34, 2005.

[14] Milberg, C. and Tommelein, I. Role of Tolerances and Process Capability Data in Product and Process Design Integration. In *Proceedings of the Construction Research Congress,* Honolulu, Hawaii, ASCE, 2003.

[15] Moghani, E., Sander, H., AbouRizk, H. and AbouRizk, S. M. Analyzing Transit Tunnel Construction Strategies Using Discrete Event Simulation. *In Proceedings of the Winter Simulation Conference*, pages 3505-3515, Phoenix, Arizona, USA, 2011.

[16] Nee, L. S., Juin, G. W., Yan, F. J., Theng, C. Y. and Kamaruddin, S. Lean Production in Manual Assembly Line - A Case Study. *Humanities, Science and Engineering Research (SHUSER), 2012 IEEE Symposium*, pages 815-820, Kuala Lumpur, Malaysia, 2012.

[17] Salem, O., Solomon, J., Genaidy, A. and Minkarah, I. Lean Construction: from Theory to Implementation. *Journal of Management in Engineering*, 22(4):168–175, 2006.

[18] Senghore, O., Hastak, M., Abdelhamid, T. S., AbuHammad, A. and Syal, M. G. Production Process for Manufactured Housing. *Journal of construction engineering and management,* 130(5):708-718, 2004.

[19] Tapping, D., Luyster, T. and Shuker, T. *Value Stream Management: Eight Steps to Planning, Mapping, and Sustaining Lean Improvements.* New York: Productivity press, 2002.

[20] Yu, H., Al-Hussein, M., Al-Jibouri, S. and Telyas, A. Lean Transformation in a Modular Building Company: A Case for Implementation. *Journal of Management in Engineering*, 29(1):103-111, 2011.

[21] Yu, H., Al-Hussein, M. and Nasseri, R. Process Flowcharting and Simulation of House Structure Components Production Process. *In Proceedings of the 39th Conference on Winter Simulation conference,* pages 2066-2072, Washington, D.C., USA, 2007.

Automated Removal of Planar Clutter from 3D Point Clouds for Improving Industrial Object Recognition

T. Czerniawski [a], M. Nahangi [a], S. Walbridge [a], C. Haas [a]

[a] Ralph Haas Civil Infrastructure Sensing Laboratory, University of Waterloo, Waterloo, Canada.
E-mail: tczernia@uwaterloo.ca, mnahangi@uwaterloo.ca, swalbridge@uwaterloo.ca, chaas@uwaterloo.ca

Abstract

The industrial construction industry makes use of prefabrication, preassembly, modularization and off-site fabrication (PPMOF) for project execution because they offer a superior level of control as compared to on-site operations. This control is enabled by systematic and thorough performance feedback loops. Improvement of the feedback systems within these facilities will require a transition away from suboptimal manual data collection to more reliable automated data collection and processing. Laser scanners are an effective tool for automatically gathering dimensional data but extraction of useful information from point clouds remains a challenge. The speed of 3D object recognition methods depends on the size of the search space. Methods for reducing this search space are needed in order to improve the performance of 3D object recognition and subsequent information extraction. Large planar objects (e.g. floors and walls) constitute a large portion of the search space in fabrication facilities, yet are rarely the objects of interest for analysis. In this paper, an automated framework for detecting and removing large planes in point clouds is presented to speed up object recognition. The raw point cloud is first Guassian mapped to normal vector space by calculating normal vectors at each point. The Gaussian sphere is clustered using a density-based clustering algorithm and major parallel planes are segmented from the rest of the point cloud. The major planes are removed and the remaining objects in the scene continue on to 3D object recognition. Results show the algorithm for automatic plane removal can reduce the search space for object recognition by as much as 60% or 70%.

Keywords:

Fabrication and process control, industrial fabrication, pipe spool, 3D laser scanning, 3D point cloud, 3D object recognition, clutter removal.

1 Introduction

Laser scanning has been profoundly affecting project surveying in the construction industry. These 3D imaging sensors capture existing structural and terrestrial conditions accurately, objectively, and with greater continuity than any manual metrology methods. Despite these benefits, the potential of 3D imaging for applications like automated progress tracking [1] and automated dimensional compliance control [2-9] remains limited because isolating the object of interest from the collected data remains a manual process. Manually extracting information from the raw 3D images and running analysis is painstaking, requires many man-hours and specialized personnel training, and therefore discourages adoption by industry. Industrial object recognition is the fundamental enabler for further developments in automated spatial analysis and information flow in construction applications.

The search space in many industrial scans is largely comprised of massive planer objects (i.e. walls, floor, and ceiling). For cases where these planes are not the focus of the analysis, they clutter the search space and substantially slow the recognition of the object of interest. Therefore, quick removal of these planes before the object recognition process begins is desirable.

2 Background

2.1 Industrial Object Recognition

3D object recognition is the process of detecting the presence of an object in a 3D image with similar characteristics to a reference image or model and mapping the 3D coordinates of the reference to the 3D coordinates (or world coordinates) of the detected object in 3D space [10,11]. For industrial specific object recognition, some of the popular methods found in the literature are based on voxel connectivity [12-14], 2D primitive fitting [15-18], classifier training [19], and feature and shape descriptor matching [20-23]. Although a variety of approaches exist, all are negatively impacted by unnecessarily cluttered search spaces.

2.2 Plane Recognition

A few plane removal algorithms exist in the literature, each with varying degrees of performance.

2.2.1 RANSAC

Basic RANSAC [24] is comprised of two repeating steps: (1) minimal set selection and (2) minimal set evaluation. The minimal set for plane removal is a single point along with its normal vector, as this provides a complete description for a plane. RANSAC randomly samples minimal sets from the scan data, fits a plane using their description, and counts the number of points in the scan that are consistent with the fitted plane. After a given number of trials, a plane is considered to be recognized at the locations defined by the minimal set which achieved a score higher than a predefined threshold. Although basic RANSAC is conceptually simple, a direct application to plane recognition is computationally intensive. Methods for speeding up RANSAC have been explored [25].

2.2.2 Hough Transform

The general Hough transform [26] can be used to recognize planes within noisy data. It is comprised of three steps, (1) repeated transform mapping, (2) application of a voting rule and (3) finding the shape parameters within the accumulated array of votes. Use of the 3D Hough transform for extraction of planar faces from point clouds was investigated by Vosselman and Dijkman. Randomized Hough Transform is a variant of the 3D Hough transform that has proven to be especially effective for plane detection in point clouds [27,28].

2.2.3 Gaussian Mapping

An elegant solution for identifying major planes within point cloud data includes mapping normal vectors to a Gaussian sphere [29,30]. Each cluster on the Gaussian sphere represents a direction that is perpendicular to major sets of parallel planes. Gaussian mapping will be used by the framework presented in this paper because of its superior speed of execution on large point sets.

3 Research Methodology

The framework for automatic plane removal is a pre-processing step for 3D object recognition. The method for planar clutter removal has five steps as outlined in Figure 1. Following plane removal, 3D object recognition is then performed using a novel method based on local data level curvature estimation, clustering, and bag-of-features matching [31].

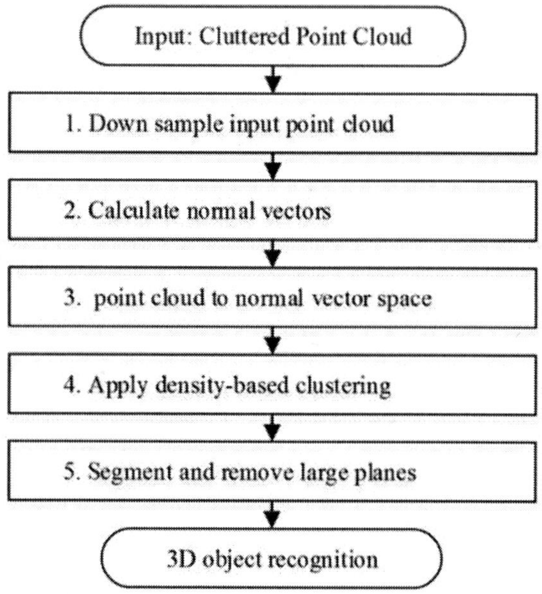

Figure 1: Framework for automatic removal of planar clutter objects.

3.1 Down Sampling

The point cloud is first down sampled because the original point set is usually massive in size with a density of points exceeding the amount necessary for accurate spatial analysis of any objects of interest. For down sampling, a random filter is applied on the original point cloud to reasonably reduce its size. The down sampled point cloud still adequately represents the objects and components without containing an excessive amount of points. Down sampling resolves the problem of redundancy that causes expensive computation with no additional benefit.

3.2 Calculating Normal Vectors

In order to detect planar clutter and filter it out from the point cloud, normal vectors are calculated and provide the primary feature for clustering and segmentation. A three step procedure is applied to each point to calculate its normal vector:

1. Identify the nearest neighbours of a point using K-Nearest Neighbours (KNN) with K=12. The 12 closest points to each point constitute the nearest neighbourhood for that point and are the set of points used for plane fitting.
2. Plane fitting is performed on the point and corresponding neighbourhood identified in the previous step. For fitting a plane to a dataset, least-square adjustment is employed.
3. The normal vector of the plane, which also represents the normal vector of the point being investigated, is extracted from the plane fitted to the investigated point.

3.3 Transforming the Point Cloud into the Normal Vector Space

Once a normal vector has been calculated for each point in the down sampled point cloud, the point cloud is again down sampled and mapped to vector space in a process called Gaussian mapping. So rather than each point being represented using (x,y,z) in Cartesian space, each point is represented by their normal vector (n_x, n_y, n_z) on a Gaussian sphere (Figure 2). This coordinate space transformation will result in identification of major planar regions existing in the point cloud. A similar transformation is used for clustering parallel branches of a pipe spool in a 3D point cloud [32]. The transformation will results in a sphere with scattered points.

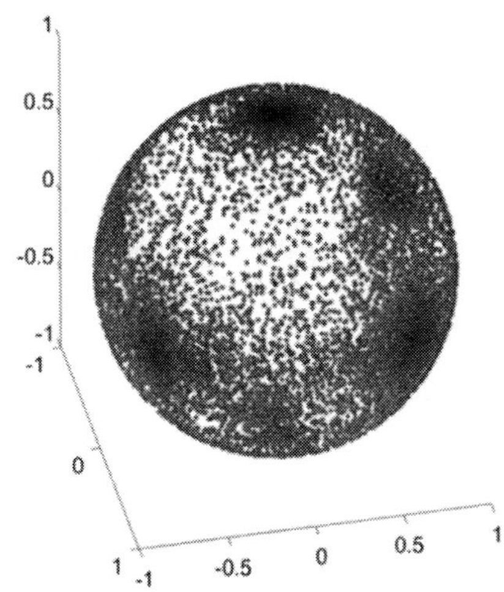

Figure 2: A typical Gaussian sphere of a point cloud. Dense point clusters on the Gaussian sphere correspond with directions perpendicular to major planar objects.

3.4 Clustering

For extracting the planar regions in the point cloud, a density-based clustering algorithm is used. In the normal vector space, planar regions are all concentrated around a point, which is the normal vector of the plane. In this paper, density-based clustering of applications with noise (DBSCAN) is employed [33]. The advantage of using density-based methods rather than centroid-based methods is that the noise and outliers are removed simultaneously using DBSCAN. While in centroid-based methods (e.g. k-means clustering), outliers and noise is also clustered with the inliers, which causes errors in the plane detection phase [34]. In addition, DBSCAN does not require the number of clusters to be known as a-priori, while in k-means clustering, the value of k is to be known to initiate the clustering process. : (a) Clustered points in the point cloud based on DBSCAN clustering. (b) Corresponding points are shown with similar colours in the down sampled point cloud.

Figure 3 shows typical results of major parallel plane set detection using DBSCAN on a point cloud.

(a)

(b)

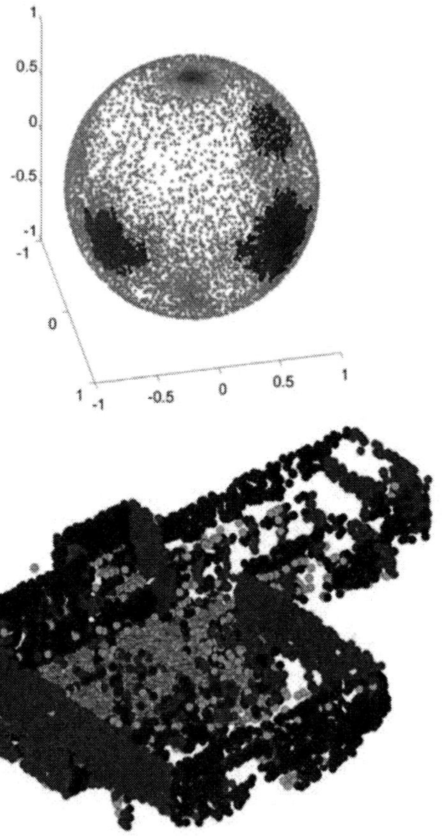

Figure 3: (a) Clustered points in the point cloud based on DBSCAN clustering. (b) Corresponding points are shown with similar colours in the down sampled point cloud.
As seen in

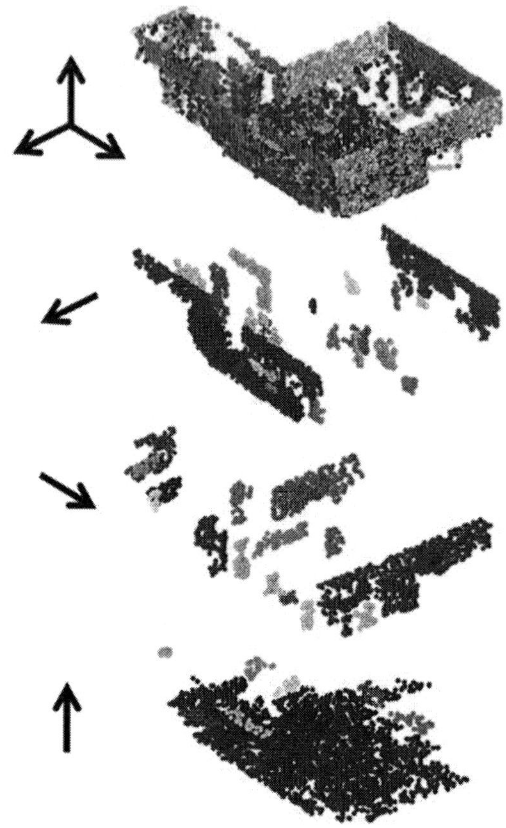

(a) (b)

Figure 4: Segmenting and removing planar objects. (a) Normal vectors to major parallel plane sets (b) segmented major parallel plane sets

Figure 3, major parallel plane sets are clustered, because parallel planes have the same normal vectors. The major parallel plane sets can then be segmented based on Euclidean space separation using an appropriate approach such as hierarchical clustering (Figure 4). The same approach is used to separate parallel branches of a pipe spool that are clustered similarly [32].

3.5 Filtering out the Planar Clutter

The segmented planes or clusters, which represent the major planes in the point cloud are then removed from the original point cloud along with all other points within their bounding box. In other words, once the planes are detected as the major clusters, the outliers which represent the points that are not belonging to the planes are kept in the point cloud.

4 Experimental Validation

The automated plane removal framework proposed is tested on two case studies: a laser scan of the Civil Infrastructure Sensing (CIS) laboratory at the University of Waterloo (Figure 5) and a scan of an industrial pipe spool fabrication facility (Figure 6).

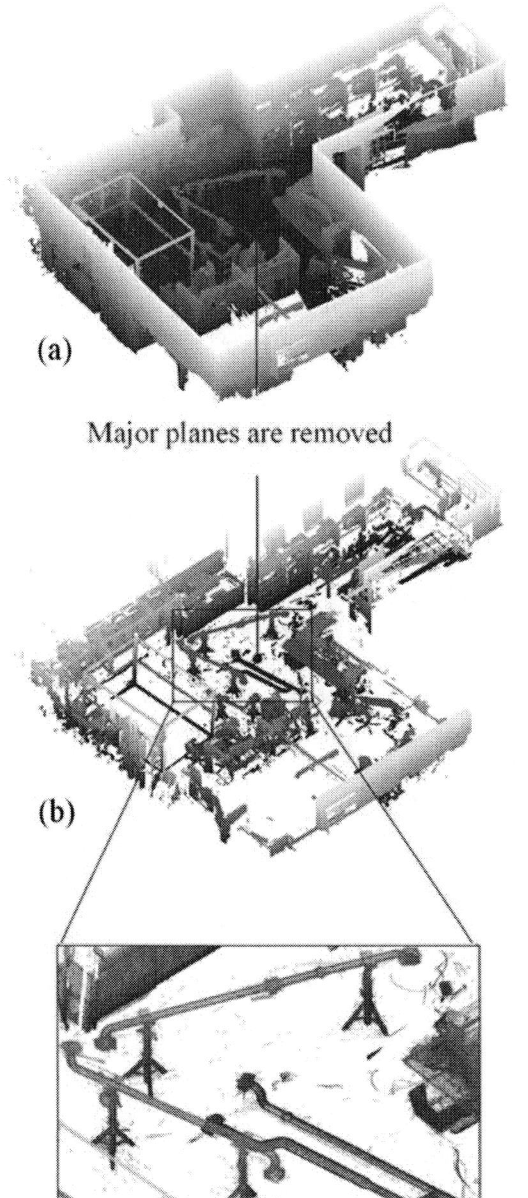

Figure 5: : Results of automated plane removal case 1, (a) original scan with 8.2 million points, (b) scan with major planar clutter removed with3.3 million points maintains rich information on the pipe spools of interest.

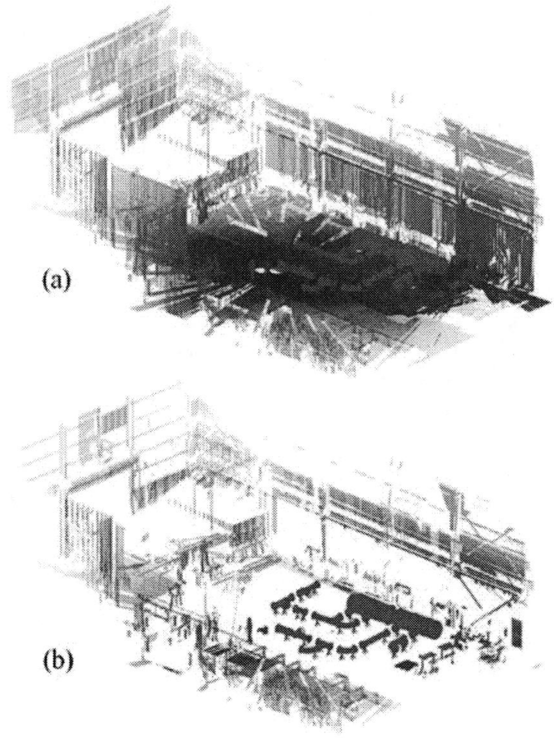

Figure 6: Results of automated plane removal case 2, (a) original scan with 7 million points, (b) scan with major planar clutter removed with 2.1 million points.

In both case studies, a set of pipe spool objects must remain in the scans for 3D object recognition and must not be affected by the plane removal. During the data acquisition, walls and floors are captured by the indiscriminant laser scanner and clutter the point cloud search space. Such clutter and unwanted objects are to be automatically removed, in order to improve the 3D object recognition step.

The object recognition framework used by the authors [31] relies on curvature calculation. Knowing mechanical/electrical/piping (MEP) components are usually non-planar, removing planar regions will reduce the processing time for curvature calculation.

As seen in Figures 5 and 6, major planes captured during laser scanning were successfully removed automatically using the framework for planar regions removal. This removal will result in a significant reduction of the point cloud search space. In case study 1, the original size of the point cloud was approximately 8.2 million. The plane removal framework reduced the size of the point cloud to approximately 3.3 million points, a 60% reduction. In case study 2, the original size of the point cloud was approximately 7 million points. The plane removal framework reduced the size of the point cloud to approximately 2.1 million points, a 70% reduction Applying the automated plane removal framework will down sample the point cloud in a supervised way. In other words, down sampling is performed without losing the required details in the regions of interest (**Error! Reference source not found.**). The clutter and unwanted regions are removed automatically. Applying indiscriminant random down sampling methods on the original point cloud will also reduce the size; however, such algorithms have no control on keeping the regions of interest. This will impact the accuracy of object recognition and subsequent object analysis. Using this plane removal framework will resolve this issue.

The plane removal execution time for both the CIS laboratory and industrial fabrication facility is approximately 3 minutes. The analyses are benchmarked on a computer with a 64 GB RAM and a 2×1.90 GHz×12 cores. This 3 minute execution time consequently reduces the execution time of object recognition or object search. The execution time for object recognition is proportional to the point cloud size. Therefore, the execution time for object recognition in case 1 and case 2 were reduced by 60% and 70% respectively.

5 Conclusions and Recommendations

Unsupervised down sampling of massive point clouds resolves the problem of processing time to some extent; However, it removes critical information about the object(s)-of-interest, and will therefore impact the accuracy of extraction of semantic information. A supervised down sampling algorithm was proposed. The proposed algorithm removes major planar regions from the massive laser scans acquired on fabrication shops and construction sites. Knowing that the majority of the industrial objects (focused in this paper), are non-planar, removing planar regions will reduce the point cloud size without losing detail and point density in the regions that contain the objects to be recognized and analyzed. The algorithm requires calculating normal vectors on a sub-sampled point cloud. Mapping the point cloud into the normal vector space will result in concentration of the planar regions around a point, which is normal vector to the estimated planar region. Applying a density-based clustering step will automatically segment the points that are belonging to a major plane in the point cloud. The inliers of the fitted plane are then removed to reduce the point cloud search space. The key observations and results are summarized as follows:

- Density-based clustering method is suitable for planar regions segmentation without knowing the number of planes a-priori. Specifically, DBSCAN approach is employed to automatically segment the planar regions on the normal vector space.
- The object recognition accuracy is not reduced because the down sampling framework only removes the planar regions.
- Experimental tests showed that the proposed framework can typically reduce the processing time for cylindrical object recognition down by as much as 60% or 70%.

Evaluating effective variants on the validation metrics can be a potential extension for future research. In this paper, effective variants such as the search region for plane fitting, and the thresholds for DBSCAN clustering are fixed. Obviously, these parameters must be calibrated to generalize the framework. Integrating the developed framework with some other processing steps, such as automated discrepancy analysis and automatic realignment of defective assemblies are other potential applications that the authors are currently investigating.

Acknowledgement

The authors would like to thank Natural Science and Engineering Research Council (NSERC) and Aecon for the financial support.

References

[1] Y. Turkan, F. Bosche, C.T. Haas, R. Haas, Automated progress tracking using 4D schedule and 3D sensing technologies, Autom. Constr. 22 (2012) 414-421.

[2] M. Nahangi, C.T. Haas, Automated 3D compliance checking in pipe spool fabrication, Advanced Engineering Informatics. (2014).

[3] F. Bosché, A. Guillemet, Y. Turkan, C. Haas, R. Haas, Tracking the Built Status of MEP Works: Assessing the Value of a Scan-vs-BIM System, J. Comput. Civ. Eng. 28 (2014) 05014004.

[4] F. Bosché, M. Ahmed, Y. Turkan, C.T. Haas, R. Haas, The value of integrating Scan-to-BIM and Scan-vs-BIM techniques for construction monitoring using laser scanning and BIM: The case of cylindrical MEP components, Autom. Constr. 49 (2015) 201-213.

[5] P. Tang, E. Anil, B. Akinci, D. Huber, Efficient and effective quality assessment of as-is building information models and 3D laser-scanned data, (2011) 19-22.

[6] F. Bosche, C. Haas, Automated retrieval of 3D CAD model objects in construction range images,

Autom. Constr. 17 (2008) 499-512.
[7] F. Bosche, C. Haas, B. Akinci, Automated Recognition of 3D CAD Objects in Site Laser Scans for Project 3D Status Visualization and Performance Control, J. Comput. Civ. Eng. 23 (2009) 311-318.
[8] F. Bosché, Automated recognition of 3D CAD model objects in laser scans and calculation of as-built dimensions for dimensional compliance control in construction, Advanced Engineering Informatics. 24 (2010) 107-118.
[9] M. Nahangi, T. Czerniawski, T.C. Haas, S. Walbridge, J. West, Parallel systems and structural frame realignment planning and actuation strategy, Journal of Computing in Civil Engineering. (2015).
[10] D. Hoiem, S. Savarese, Representations and Techniques for 3D Object Recognition and Scene Interpretation, San Rafael, Calif. (1537 Fourth Street, San Rafael, CA 94901 USA) : Morgan & Claypool, c2011, 2011.
[11] L. Brown, A survey of image registration techniques, ACM Comput. Surv. 24 (1992) 325.
[12] D. Gilsinn, G.S. Cheok, C. Witzgall, A. Lytle, Construction Object Identification from LADAR Scans: An Experimental Study using I-Beams, US Department of Commerce, Technology Administration, National Institute of Standards and Technology, 2005.
[13] J. Teizer, C.H. Caldas, C.T. Haas, Real-time three-dimensional occupancy grid modeling for the detection and tracking of construction resources, J. Constr. Eng. Manage. 133 (2007) 880-888.
[14] G. Erdos, T. Nakano, G. Horváth, Y. Nonaka, J. Váncza, Recognition of complex engineering objects from large-scale point clouds, CIRP Ann. Manuf. Technol. 64 (2015) 165-168.
[15] M. Ahmed, C.T. Haas, R. Haas, Autonomous modeling of pipes within point clouds, Proceedings of the 30th ISARC, Montréal, Canada. (2013) 1093-1100.
[16] M. Ahmed, C. Haas, R. Haas, Automatic Detection of Cylindrical Objects in Built Facilities, J. Comput. Civ. Eng. 28 (2014) 04014009.
[17] R. Qiu, Q. Zhou, U. Neumann, Pipe-run extraction and reconstruction from point clouds, in: Anonymous Computer Vision–ECCV 2014, Springer, 2014, pp. 17-30.
[18] Y. Liu, J. Zhang, J. Hou, J. Ren, W. Tang, Cylinder detection in large-scale point cloud of pipeline plant, Visualization and Computer Graphics, IEEE Transactions on. 19 (2013) 1700-1707.
[19] G. Pang, U. Neumann, Training-based object recognition in cluttered 3d point clouds, (2013) 87-94.
[20] B. Akinci, F. Boukamp, C. Gordon, D. Huber, C. Lyons, K. Park, A formalism for utilization of sensor systems and integrated project models for active construction quality control, Autom. Constr. 15 (2006) 124-138.
[21] A.E. Johnson, M. Hebert, Using spin images for efficient object recognition in cluttered 3D scenes, Pattern Analysis and Machine Intelligence, IEEE Transactions on. 21 (1999) 433-449.
[22] H. Son, C. Kim, C. Kim, Fully automated as-built 3D pipeline extraction method from laser-scanned data based on curvature computation, J. Comput. Civ. Eng. (2014) B4014003.
[23] H. Date, Y. Kaneta, A. Hatsukaiwa, M. Onosato, S. Kanai, Object Recognition in Terrestrial Laser Scan Data using Spin Images, Computer-Aided Design and Applications. 9 (2012) 187-197.
[24] M.A. Fischler, R.C. Bolles, Random sample consensus: a paradigm for model fitting with applications to image analysis and automated cartography, Commun ACM. 24 (1981) 381-395.
[25] R. Schnabel, R. Wahl, R. Klein, Efficient RANSAC for Point-Cloud Shape Detection, Comput. Graphics Forum. 26 (2007) 214-226.
[26] D.H. Ballard, Generalizing the Hough transform to detect arbitrary shapes, Pattern Recognit. 13 (1981) 111-122.
[27] D. Borrmann, J. Elseberg, K. Lingemann, A. Nüchter, The 3D Hough Transform for plane detection in point clouds: A review and a new accumulator design, 3D Research. 2 (2011) 1-13.
[28] L. Xu, E. Oja, P. Kultanen, A new curve detection method: randomized Hough transform (RHT), Pattern Recog. Lett. 11 (1990) 331-338.
[29] Y. Liu, Y. Xiong, Automatic segmentation of unorganized noisy point clouds based on the Gaussian map, Comput. -Aided Des. 40 (2008) 576-594.
[30] Y. Wang, W. Hao, X. Ning, M. Zhao, J. Zhang, Z. Shi, X. Zhang, Automatic Segmentation of Urban Point Clouds Based on the Gaussian Map, The Photogrammetric Record. 28 (2013) 342-361.
[31] T. Czerniawski, M. Nahangi, C. Haas, S. Walbridge, Pipe spool recognition in cluttered point clouds using a curvature-based shape descriptor, (Submitted (2016)).
[32] M. Nahangi, C.T. Haas, Skeleton-based discrepancy feedback for automated realignment of industrial assemblies, Autom. Constr. 61 (2016) 147-161.
[33] M. Ester, H. Kriegel, J. Sander, X. Xu, A density-based algorithm for discovering clusters in large spatial databases with noise. 96 (1996) 226-231.
[34] A.K. Jain, M.N. Murty, P.J. Flynn, Data clustering: a review, ACM computing surveys (CSUR). 31

(1999).

Arbitrary 3D Object Extraction from Cluttered Laser Scans Using Local Features

M. Nahangi [a], T. Czerniawski [a], C. Rausch [a], C. T. Haas [a]

[a] Department of Civil and Environmental Engineering, University of Waterloo, Waterloo, ON, Canada.
E-mail: mnahangi@uwaterloo.ca, tczernia@uwaterloo.ca, ctrausch@uwaterloo.ca, chaas@uwaterloo.ca

Abstract

Determination of construction performance metrics requires intensive processing of large amounts of data collected on construction sites including cluttered laser scans. For example, for quality control of construction components using 3D laser scans, the acquired point cloud should be cleaned and the object-of-interest should be extracted for measuring the incurred deviations. Such a procedure is tedious, time consuming and inaccurate due to intensive manual user operations. Although automatic extraction of rough and simple 3D shapes and features is performed by applying techniques such as Hough transform, automatic extraction of construction components with complex geometry is a challenging research need that must be addressed for fully automated modelling and processing. This paper presents a framework for automated extraction of 3D objects with arbitrary shapes and geometry. A new local feature set, which is globally invariant, is created in order to represent 3D models. The feature space created is then searched for in the cluttered laser scan by hashing from a hash table created for the 3D model. The best match is then extracted automatically by applying a post-processing RANSAC loop. The framework is then followed by an ICP-based registration in order to refine the best match identified. The results show that the method is sufficiently robust and quick to be applied for effective and efficient post processing of the laser scans acquired on construction sites.

Keywords:

Local feature descriptors, hashing, hash table, RANSAC, 3D object recognition, clutter, laser scanning

1 Introduction

3D imaging systems have become effective and accurate tools for acquiring the as-built status on construction sites. As-built status acquisition is critical due to continuous change and therefore the continuous monitoring required for construction components. 3D imaging in the construction industry is often referred to a set of processes that the built environment containing the construction elements are represented by a dense 3D point cloud acquired using the appropriate sensor. 3D imaging sensors enable the capture of existing structural and terrestrial conditions objectively, accurately, quickly, and with greater detail and continuity than any manual methods. Current applications of laser scanners by construction firms include schedule and progress tracking [1], creating complex as-built construction documents and 3D models [2], quality control and assurance, deviation and discrepancy quantification, steel column plumbness, etc.

Despite the capabilities that 3D imaging provides, its potential is still limited and challenging, because extracting usable information from the collected data remains primarily a manual process. Manual extraction of semantic and meaningful information from the raw 3D images and running analyses is painstaking, requires many man-hours and specialized personnel training, and is therefore not well suited for real-time or rapid decision making on a large scale. Automating this process is key for further developments in automated spatial analysis as well as automated, integrated and electronic information flow. Search and extraction involves many technical challenges that stem from the variability of spatial data and other operational realities such as non-uniform point density, clutter which may be in contact with the object-of-interest or of similar shape and size to the object-of-interest, occlusion, missing data, and sensor noise and inaccuracy [3].

This paper presents a method for automated extraction of construction components from cluttered laser scans acquired on construction sites. A 3D CAD model integrated in the building information model (BIM) represents the designed status. A local feature is created to describe the geometry of the model. This feature-based descriptor is generated and stored in an array for further analysis and search in the laser scan. The 3D point cloud is then searched for the feature created in the first step. Potential matches are tested and the best hypothesis is selected for the match between the built and designed state. Although there is no limitation for the type of the assembly, this paper focuses on industrial components.

2 Background

The related background is investigated from two different perspectives: (1) Point cloud analysis for as-built status assessment, and (2) 3D object recognition and segmentation.

2.1 Point Cloud Analysis for As-Built Status Assessment

There has been relatively little work on construction specific object recognition from laser-based 3D imaging systems. The primary application areas for construction are automated dimensional compliance control, progress tracking, and equipment guidance. In particular, many methods for visual inspection outlined in the literature fall short of full automation because of the absence of a reliable object recognition method. Fully automated frameworks have the capability of being integrated with construction product management systems. This will improve the construction process performance by minimizing the rework associated. In the construction literature, rework is defined as the wasteful effort involved in redoing work that has not yet yielded a product adequately conforming to contractual requirements [4,5]. Rework directly and significantly contributes to cost and schedule overruns on construction projects [6]. The function of QA/QC personnel is to perform lifecycle inspections to mitigate these rework situations. Inspection is the process of determining if a product deviates from a given set of tolerance specifications. The predominant processes for monitoring the critical dimensions of an assembly involve a temporary production stoppage and manual direct contact measurement devices such as measuring tapes, calipers, custom gauges, squares, and straightedges. These processes can help fabricators evaluate whether basic assemblies are compliant with design specifications, but their effectiveness deteriorates as the assembly's geometric complexity increases. Automated inspection is desirable because manual inspection by humans is time-consuming, and can be excessively subjective, unreliable, and boring for humans to perform.

Using 3D imaging for dimensional compliance assessment of construction components has proven to have potential for mitigating costly repair and rework while tracking progress. 3D image-to-BIM comparison requires the superposition of the BIM onto the object of interest within the 3D image, i.e. registration of the object centred coordinate systems. However, unwanted clutter in the 3D image makes automated registration a challenge. Within the construction literature, this initial registration step has, predominantly, remained a manual process and must be resolved before the enormous amount of geometric data that 3D imaging makes available can be fully utilized.

2.2 3D Object Recognition and Segmentation

3D object recognition is the process of detecting the presence of an object in a 3D image with similar characteristics to a reference image or model and mapping the 3D coordinates of the reference to the 3D coordinates (or world coordinates) of the detected object in 3D space [7,8]. This process requires recognition of object and non-object components, which is difficult if the position of the object is unknown.

In order to minimize required computation, the 3D image and reference model are abstracted using features or descriptors. Then point descriptions from the 3D image and the reference model are matched to determine the pose of the object in the scene. Based on the literature [9,10], the type of abstraction method may vary but for object recognition in cluttered scenes, it must demonstrate (1) discrimination between objects of dissimilar geometry, (2) insusceptibility to noisy data, (3) invariance under transformation and rotation, (4) conciseness and ease of indexing, and (5) the ability to perform partial matching i.e. describe parts of a point set (object of interest) independently of the rest of the point set in order to enable recognition of those specific parts.

Global descriptors consider the point set in its entirety and produce contaminated descriptions if clutter is present. Conversely, local descriptors abstract a 3D point set by considering the point set in subsets or regions. A popular subset type is the nearest neighbourhood, which is a collection of points in a spherical volume in 3D Euclidean space surrounding a query point. Determining the optimal nearest neighbourhood size to use for applying the shape descriptor is a critical problem in obtaining useful results from the abstraction process. Local descriptors allow for partial matching, and therefore, are ideal for object recognition in cluttered scenes. For a detailed review of the state of the art in shape descriptors and feature based object recognition see [9-11].

3 Research Methodology

As shown in Figure 1, the proposed method for automated object isolation has two primary steps: (1) feature space creation, and (2) feature matching. 3D CAD models are first represented by a local feature descriptor. The descriptor is stored in a hash table for further search and match in the following step. The first step is performed in the offline phase. Two arbitrary points are then selected by users and the feature is calculated for the selected pair. The potential matches are then tested to find the best match for the pair. This is performed using a RANSAC (Random Sample Consensus) loop followed by an ICP-based registration for refining the match.

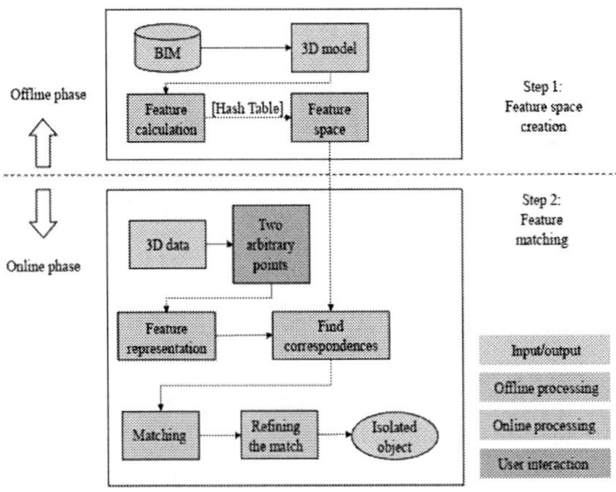

Figure 1: General framework for arbitrary shape isolation from cluttered and incomplete laser scans

3.1 Calculation of model descriptors

Our approach requires a point pair to create the feature space. The feature created here, identifies the pose and orientation of the point pair. illustrates the parameters for creating the feature space for a point pair (p_1, p_2).

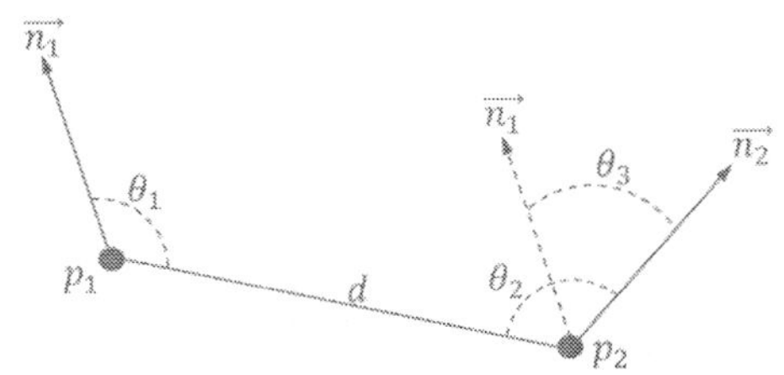

Figure 2: The four-dimensional feature space for a point pair (p_1, p_2)

As shown in , for a point pair (p_1, p_2), normal vectors (n_1, n_2) are first calculated (detailed in Section 3.1.1). The feature of the pair is therefore calculated as follows:

$$F = [f_1 = \|d\|, f_2 = \angle(n_1, d), \\ f_3 = \angle(n_2, d), \\ f_4 = \angle(n_1, n_2)] \qquad \text{Eq. 1}$$

in which, d is the Euclidean distance between p_1 and p_2, and $\angle(v, w)$ returns the angle between two vectors v and w. In summary, the feature is represented by four elements as: $F = (f_1, f_2, f_3, f_4)$. Our feature is similar to the feature used by [12], however, we use a different representation for the matching. In addition to storing the four parameters explained, (p_1, p_2) and (n_1, n_2) are stored for the matching step. The representation of the feature is explained is Section 3.1.2.

3.1.1 Feature elements calculation

The calculation of feature elements (f_1, f_2, f_3, f_4) is described in this section. The key parameter for creating the feature space is to calculate normal vector at each point in the point pair. An illustrative summary of the normal vector extraction is shown in .

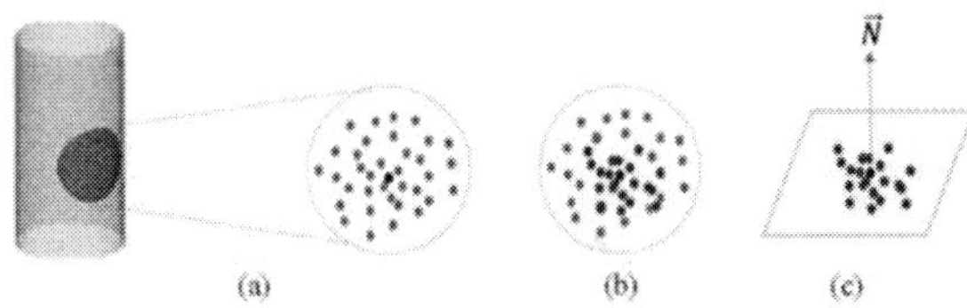

Figure 3: Normal calculation procedure. (a) A point p is selected in a point cloud $p \in P$. (b) K-nearest neighbors (KNN) of point p, denoted as $NN_p \subseteq P$, are identified using kd-tree. (c) A plane is fitted to NN_p and the normal vector N_p of the plain is calculated.

The proposed algorithm for normal vector calculation has three primary steps, described as follows:

1. Finding nearest neighbours

As discussed earlier and shown in , each point should be grouped with its nearest neighbours in order to calculate a plane. For this purpose, a k-nearest neighbour (KNN) algorithm [13,14] is employed with k=10. KNN algorithm requires a distance calculation step for finding the nearest points. A kd-tree searcher is employed for the range searching step involved here. The point and its nearest neighbours are then stored in an array for the further manipulations required in normal vector extraction.

2. Fitting a plane to a neighbourhood

Given a dataset including a point, which is being investigated, and its k-nearest neighbours, a plane that best fits the dataset is calculated. In general, a plane p is defined as $p: ax + by + cz + d = 0$. Without loss of generality, let's assume that $c = 1$; thus, the plane equation can be rewritten as $p: z = -ax - by - d$. Rewriting the plane equation in matrix form results in $p: z = -[x \;\; y \;\; 1]\langle a \;\; b \;\; d\rangle^T$. Given a dataset, the plane fitting step results in a least-square adjustment problem in the form of a linear matrix equation as:

$$\begin{bmatrix} x_1 & y_1 & 1 \\ x_2 & y_2 & 1 \\ \vdots & \vdots & \vdots \\ x_n & y_n & 1 \end{bmatrix} \begin{bmatrix} a \\ b \\ d \end{bmatrix} = \begin{bmatrix} -z_1 \\ -z_2 \\ \vdots \\ -z_n \end{bmatrix} \qquad \text{Eq. 2}$$

Where, n is the total number of point for plane fitting that equals to $k + 1$ (k= number of nearest neighbours). The plane equation can be rewritten in the form of a standard linear matrix equation $AX = B$, where X is the matrix of unknowns. The solution of the linear matrix equation is as follows:

$$X = (A^T A)^{-1} A^T B \qquad \text{Eq. 3}$$

in which, $(A^T A)^{-1} A^T$ is also known as the Moore-Penrose pseudoinverse of matrix A. More details on pseudoinverse can be found in [15-17].

3. Calculating normal vectors

Given an equation of a plane in the form of a vector $X^T = \langle a, b, c = 1, d\rangle$, the normal equation can be calculated as follows:

$$\vec{n} = \left(\frac{a}{\sqrt{a^2 + b^2 + 1}}, \frac{b}{\sqrt{a^2 + b^2 + 1}}, \frac{1}{\sqrt{a^2 + b^2 + 1}}\right) \qquad \text{Eq. 4}$$

The explained feature detection step is applied on the CAD model $\{M\}$, which is being detected in the acquired laser scan $\{S\}$. The output of the feature detection step are the descriptors stored in $F_M(m_1, m_2) \subseteq M^2$ and $F_S(s_1, s_2) \subseteq S^2$ that describe the point clouds $\{M\}$ and $\{S\}$, respectively. The calculated descriptors are then matched in the matching step (Section 3.2.3). Summary of the normal vector calculation step explained here, is shown in Algorithm 1.

Algorithm 1: Normal vector calculation

Input(s): Point cloud $\{P\}$
Output(s): Normal vector features $\{F\}$
for each point $p \in \{P\}$
(1) Finding nearest neighbours
  Apply KNN and find the nearest neighbours
  Store each point with its nearest neighbour in an array $\{N\}$
(2) Fitting a plane
  Build up matrix A and B for N using Eq. 2.
  Calculate plane parameters (X) using Eq. 3.
(3) Calculating normal vectors
  Find normal vector parameters using Eq. 4.
  $\{F\} \leftarrow$ (point coordinates) **and** (normal vector)
End for
Return $\{F\}$

3.1.2 Model representation

Once the feature $F_M = (f_1, f_2, f_3, f_4)$ is created for all possible pairs in the 3D CAD model $\{M\}$, it is used as a global descriptor for representing the model. In other words, the feature created is mapping transformation from the model space to the feature space as: $F_M \rightarrow M^2$. For the 3D CAD model $\{M\}$, all features are calculated and then stored in an array. In order to be efficiently searchable, the point pairs with identical features are stored in the same cell in a four-dimensional array, of which each dimension represents the feature elements (f_1, f_2, f_3, f_4). For this purpose, a four dimensional hash table is created and the point coordinates along with the normal vectors are stored. Figure 4 illustrates how point pairs are stored in the hash table, and how they can be searched efficiently.

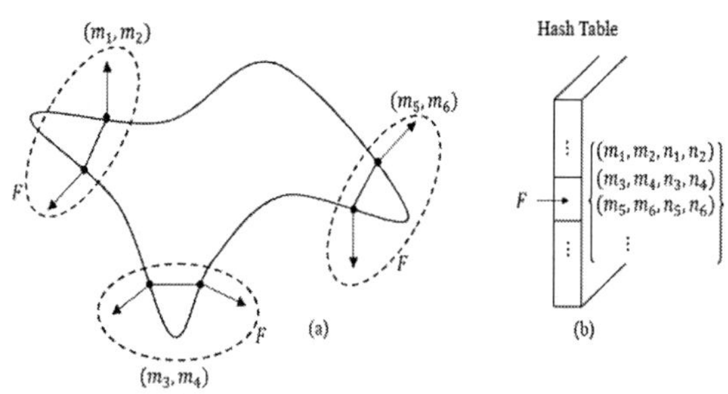

Figure 4: Model representation. (a) Feature elements are calculated for point pairs. (b) A 4D hash table is created. Each dimension represent an element in the feature space (f_1, f_2, f_3, f_4). The points having similar feature elements, along with the corresponding normal vectors are stored in the same cell of the hash table created. The feature is then the key to the table and similar pairs can be

efficiently searched for the matching step.

This algorithm for feature calculation and model representation is applied on all possible point pairs in the model. Such a procedure is performed in the offline mode for the arbitrary objects being searched in a cluttered laser scan. In the following section, the matching algorithm is explained.

3.2 Feature Matching

Once the model features are calculated, they are being searched by the features calculated on the laser scan F_S. Similar pairs are then matched geometrically and the best pair is detected by employing a RANSAC loop on the matching pairs. The matching algorithm has three primary steps: (1) matching initiation for speeding up the search process, (2) finding corresponding features by mapping the initial pair being searched in the model and the hash table created in the previous step, and (3) matching and refining that finds the best pair globally by employing a RANSAC loop. These steps are detailed in the following sections.

3.2.1 Matching initiation

Once feature space on 3D CAD model is created and stored in the hash table, the search phase, in the online mode, commences. Two arbitrary points $s_1, s_2 \in S$ are arbitrarily selected from the scanned point cloud. Once these two points are selected, the previously defined feature is calculated for the pair (s_1, s_2). This feature F_S is the key to the hash table for finding the matching pairs in the model (Figure 4).

3.2.2 Finding the potential matching pairs

All matching pairs stored in the cell with the same scan feature values F_S are hashed to test the hypothesis. The hypothesis is finding the pair that best matches F_S. For this purpose, the distances and the angles representing the hash table are sampled in steps of d_{dist} and d_{ang}, respectively. For finding the best match, the algorithm searches a set of possible matching pairs $m_{match} \subseteq M^2$. The matching pairs are represented as: $m_{match} = \{(m_1, m_2)\}$. This will improve the accuracy of finding the correct match, and it therefore increases the robustness of the algorithm.

3.2.3 Matching and Refining

For matching the scene pair with the model, $\{s_1, s_2\}$ are projected along the normal vectors. A set of four points is therefore created. This set is aligned with the set of four derived from the model, knowing the potential pairs and the corresponding normal vectors (see Figure 5). In order to avoid incorrect matching for symmetric sets k_1 and k_2 must be unequal ($k_1 \neq k_2$).

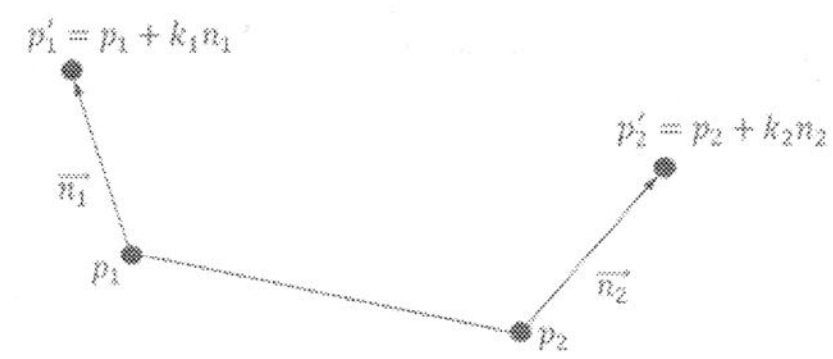

Figure 5: Matching the scene and model. A set of four points are calculated for each pair in m_{match} and it is matched with the set resulted from the scene.

In order to match each set of four from the CAD model and the scene, Principal Component Analysis (PCA) is employed. Principal axes of the two datasets are first calculated using singular value decomposition (SVD). The principal axes are then aligned by a rotational transformation. The centroids of the dataset are then coincided with each other. For detailed explanation about the rough alignment of point sets see [18].

The algorithm is followed by a RANSAC loop to find the best set from the model matching the scanned point cloud. Once the transformation $T = (\vec{r}, \vec{t})$ is calculated, it is applied on the entire model point cloud. The matching closest points with a threshold distance value are recorded as the number of inliers for the transformation calculated. A certain number of iterations will find the best match that has the maximum number of inliers. An iterative closest point (ICP)-based registration will finally refine the match.

4 Verification and Validation

The proposed method for 3D object isolation from cluttered laser scans is tested by setting a set of experiments. A typical scene containing industrial components is scanned (Figure 6). A branch of a pipe spool is desired to be isolated from laser scans for further analyses and processing such as discrepancy analysis, as shown in Figure 6. The explained algorithm is implemented in a MATLAB environment. Table 1 shows the parameters considered for implementing the algorithm.

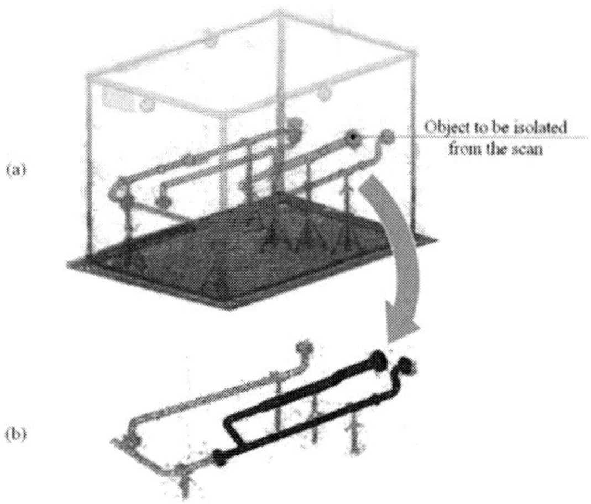

Figure 6: Isolation and extraction of the object-of-interest from a typical laser scan. (a) A typical laser scan containing industrial objects is captured. (b) The desired branch is isolated using the proposed method.

Table 1: Values of the effective parameters on the algorithm

Parameter	Value
d_{dist}	5 cm
d_{ang}	12 °
k_1	1
k_2	0.5
Inlier threshold for RANSAC	1 cm
ICP iteration	5

For a typical scene shown in Figure 6, the feature space creation step took approximately 26 min. The 3D CAD model converted to a point cloud had 5000 points. The point cloud was resampled to expedite the algorithm and avoid over-interpretation. The resampling factor was selected as the maximum value that the distance between furthest points in the resampled point cloud is not greater than d_{dist}. Similar criterion is used for down sampling the laser scan representing the as-built status. The search and match step for the example provided here, took approximately 16 sec.

5 Conclusions and Recommendations for Future Research

A framework was developed for automated isolation of an object-of-interest from a cluttered laser scan. Isolation of specific objects is critical because the as-built status needs to be automatically assessed and compared with the as-designed state integrated in the BIM. The method uses a local feature-based descriptor for representing the 3D CAD models. The feature space is stored in a hash table in order to better search and match with the laser scan. This step was implemented in the offline mode. The created feature is then searched for in a given laser scan. A pair in a given scene is characterized using the described feature space, and it is then searched from the potential matches in the created hash table. A PCA-based rough alignment is first employed to identify the matching inliers based on a distance threshold between the two point clouds representing the two states. The pose with maximum number of inliers is therefore selected as the initial match. The results are finally refined using an ICP-based fine alignment. Key findings are summarized as follows:

- The feature space creation can be performed in the offline phase; meaning that, all 3D objects can first be characterized and represented by the local feature space explained in the method. The results will be stored in a database for further use in the framework. This improves the efficiency of the framework.
- The algorithm is capable of finding a suitable match in a time effective and cost efficient manner. Therefore, it can be integrated with real-time frameworks for fabrication and process control.

Effective parameters such as the threshold values for inliers identification and down sampling factors should be investigated, and optimum values must be calibrated. These values depend on the size of the objects and resolution of the scan for capturing the as-built status. Improving processing time for real-time applications is also a potential for future work, on which the authors may expand the work.

Acknowledgement

The authors would like to thank Natural Science and Engineering Research Council (NSERC) Canada and Aecon for funding this research.

References

[1] Y. Turkan, F. Bosche, C.T. Haas, R. Haas, Automated progress tracking using 4D schedule and 3D sensing technologies, Automation in Construction. 22 (2012) 414-421.

[2] V. Pătrăucean, I. Armeni, M. Nahangi, J. Yeung, I. Brilakis, C. Haas, State of research in automatic as-built modelling, Advanced Engineering Informatics. (2015).

[3] T. Czerniawski, M. Nahangi, C. Haas, S. Walbridge, Pipe spool recognition in cluttered point clouds

using a curvature-based shape descriptor, Submitted (2016).
[4] S. Du, N. Zheng, S. Ying, J. Liu, Affine iterative closest point algorithm for point set registration, Pattern Recognition Letters. 31 (2010) 791-799.
[5] J. Lim, J. Park, G.G. Medioni, Text segmentation in color images using tensor voting, Image and Vision Computing. 25 (2007) 671-685.
[6] P. Love, Influence of Project Type and Procurement Method on Rework Costs in Building Construction Projects, Journal of Construction Engineering and Management. 128 (2002) 18-29.
[7] D. Hoiem, S. Savarese, Representations and techniques for 3D object recognition and scene interpretation, Synthesis Lectures on Artificial Intelligence and Machine Learning. 5 (2011) 1-169.
[8] L.G. Brown, A survey of image registration techniques, ACM computing surveys (CSUR). 24 (1992) 325-376.
[9] J.W. Tangelder, R.C. Veltkamp, A survey of content based 3D shape retrieval methods, Multimedia Tools and Applications. 39 (2008) 441-471.
[10] M. Yang, K. Kpalma, J. Ronsin, A survey of shape feature extraction techniques, Pattern Recognition. (2008) 43-90.
[11] Y. Guo, M. Bennamoun, F. Sohel, M. Lu, J. Wan, 3D object recognition in cluttered scenes with local surface features: a survey, Pattern Analysis and Machine Intelligence, IEEE Transactions on. 36 (2014) 2270-2287.
[12] B. Drost, M. Ulrich, N. Navab, S. Ilic, Model globally, match locally: Efficient and robust 3D object recognition, Computer Vision and Pattern Recognition (CVPR), 2010 IEEE Conference on, 2010, 998-1005.
[13] T. Cover, P. Hart, Nearest neighbor pattern classification, Information Theory, IEEE Transactions on. 13 (1967) 21-27.
[14] A.K. Jain, R.P.W. Duin, J. Mao, Statistical pattern recognition: A review, Pattern Analysis and Machine Intelligence, IEEE Transactions on. 22 (2000) 4-37.
[15] Eric W. Weisstein. (2014). "Moore-Penrose Matrix Inverse". <http://mathworld.wolfram.com/Moore-PenroseMatrixInverse.html>. July 10, 2014.
[16] R. Penrose, A generalized inverse for matrices, Mathematical proceedings of the Cambridge philosophical society, 1955, 406-413.
[17] M. Nahangi, C.T. Haas, J. West, S. Walbridge, Automatic Realignment of Defective Assemblies Using an Inverse Kinematics Analogy, ASCE Journal of Computing in Civil Engineering. (2015).
[18] M. Nahangi, C.T. Haas, Automated 3D compliance checking in pipe spool fabrication, Advanced Engineering Informatics. 28 (2014) 360-369.

Applying the Concept of Selective Assembly to Modular Construction to Mitigate Impacts of Component Variability

Christopher Rausch [a], Mohammad Nahangi [a], Melanie Perreault [a], Carl Haas [a], and Jeffrey West [a]

[a]Ralph Haas Infrastructure and Sensing Analysis Laboratory, University of Waterloo, Canada
E-mail: ctrausch@uwaterloo.ca, mnahangi@uwaterloo.ca, mel29per@gmail.com, chaas@uwaterloo.ca, jswest@uwaterloo.ca

Abstract –

As adoption of offsite methods of production continues to grow within the construction industry, optimization techniques from manufacturing are increasingly being utilized analogously for increasing productivity, reducing rework, and improving assembly processes. This paper demonstrates how the concept of selective assembly can be applied in modular construction as a potential assembly optimization technique. Rather than specifying and controlling tight fabrication tolerances, the selective assembly process groups components into bins or categories of compatible dimensional and geometric properties in order to find an optimal arrangement of interchangeable components in an assembly. This concept has traditionally proven to be more cost effective in certain manufacturing applications than using rigorous specification and control of tight fabrication tolerances. Using a laser scanner for as-built data acquisition, a modular steel bridge is analyzed as a case study to demonstrate how the concept of selective assembly can be applied in modular construction. The results of this case study show that selective assembly has potential to reduce rework in certain modular construction applications.

Keywords –

Selective Assembly; Tolerance Control; Modular Construction; Geometric Variability; Assembly Planning; Interchangeable Components

1 Introduction

The paradigm shift in the construction industry towards offsite methods of production has raised the question about how to specify adequate component tolerances to ensure that fit-up and onsite installation are not met with serious challenges such as cost overruns, delays and extensive rework. While the traditional approach of benchmark tolerances has been used in stick-built construction [1], these same component tolerances are often not strict enough to ensure problem-free fit-up of components and modules in modular construction projects [2-4]. Dimensional coordination is thus a vital aspect of production in modular construction.

Two comprehensive approaches can be taken to develop adequate component tolerances for modular construction: (1) development of a new set of benchmark tolerances specifically for use in modular construction projects, or (2) creation of a systematic tolerance design through utilization of tolerancing theory employed in manufacturing (i.e., tolerance analysis, tolerance synthesis, vector loop models, etc.). These two solutions are non-trivial and can be very intensive to implement. Both of these approaches tackle the challenges faced in component fit-up and installation during the design phase by either proper selection or adequate design of critical tolerances. However a third approach is also viable, by solving component fit-up and installation challenges during the assembly phase through collection and analysis of 3D as-built data. This third approach uses a technique found in the manufacturing industry called selective assembly. The basic premise behind this technique is to measure and organize parts which are to be mated together into classes or bins of dimensional compatibility. Selective assembly is traditionally used to achieve tight clearance tolerances between mating parts that are manufactured using imprecise techniques; leading to high quality assembly using inexpensive techniques [5-7].

Modular construction has great potential for implementing selective assembly, since numerous components are often manufactured and assembled into repetitive modules. Even with the use of precise methods of dimensional control during production (i.e., framing tables, fixtures and jigs), offsite construction modules can experience geometric conflicts during fit-up and erection on site. Furthermore, not only are modular construction components often manufactured to be interchangeable (i.e., utilization of economies of scale), but fabrication and assembly processes themselves are also intrinsically highly repetitive. Process and product repetition are essential requirements for implementing selective assembly. Due to its intrinsic product and process repetition, modular construction has the potential to successfully use selective assembly to optimize assembly processes.

This paper investigates the use of selective assembly in an ad-hoc manner by optimizing the aggregation of mating parts in a small-scale modular steel bridge. The scope of this paper is limited to selective assembly on a simple 'bare-bones' structural assembly. However, the methodology can be applied to more general structural systems such as modular steel frames (Figure 1), where there is sufficient repetition of processes (e.g., cutting, fit-up, alignment, welding), and quantity of modules.

(a)

(b)

Figure 1: (a) case study in this paper examines the mating parts in modular steel bridge, (b) methodology can be generalized for mating parts in prefabricated steel systems such as modular steel frame assemblies.

2 Background

2.1 Overview of Selective Assembly

The need for tolerances within the production of parts arises because variability is inevitable. Regardless of the amount of effort placed into controlling the dimensional and geometric properties of production processes, some degree of variability cannot be avoided. However, there are certain types of variability which have larger impacts than others on ease of assembly. These specified limits of variability, known as tolerances, are used to target critical sources of variability in order to control certain dimensional and geometric attributes of parts so that production goals can be met in way which balances cost, quality and customer satisfaction [8].

Production variability is an issue that emerges in many industries, including manufacturing. While the use of expensive equipment and precise production methods is a common approach for controlling critical variability in manufacturing, selective assembly is a valuable quality improvement tool which sidesteps the use of expensive equipment and precise production methods. Select assembly is a dimensional quality improvement tool used to determine optimal pairs of mating parts from stockpiles which are nominally identical [6,7,9].

One of the simplest examples used to demonstrate selective assembly is the aggregation of shaft and sleeve parts. In this example, the critical dimensions for aggregation are isolated (e.g., the radius of each part). Assuming the critical aggregation dimension of a given part can be modelled by a normal distribution with a certain tolerance threshhold (Figure 2-a), then sample distributions will vary within an allowable variability region (Figure 2-b). Then, as the stockpile for each part is populated, critical dimensions can be quantified and parts organized into bins which enable optimal aggregation. As seen in Figure 2-c, if four randomly selected sleeve and shaft parts are selected from their respective stockpiles, it is clear that based on their respective dimensional distributions, that an optimal set of best-fit pairs does exist.

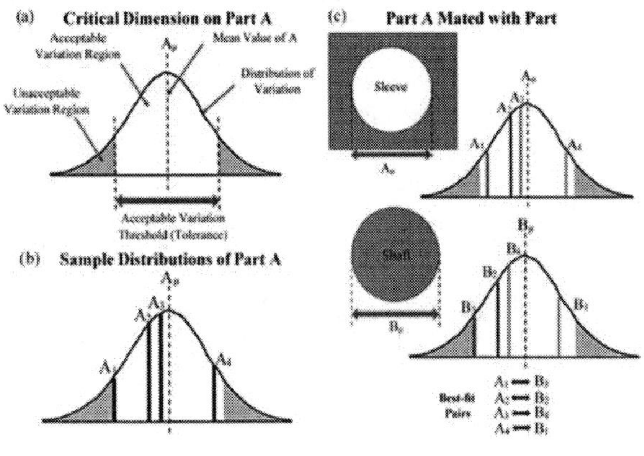

Figure 2: Demonstrating the concept of selective assembly: (a) dimensional distribution of critical feature for aggregation, (b) sample distributions of feature and (c) optimal matching of sleeve and shaft parts based on the distribution of critical aggregation features.

2.2 Current State of Selective Assembly

Selective assembly has been used in the manufacturing industry for years. Typical applications include the assembly of ball bearings and joints [10], production of pistons and cylinders [7], production of scroll compressor shells [11], and is still used in the assembly of engines, transmissions, and compressors [12].

The use of statistical selective assembly was introduced as a way of optimizing assembly based on the dimensional distribution of mating components, which for a number of years focused on parts with similar dimensional attributes that follow the Normal distribution [7]. In order to generalize statistical selective assembly, researchers began introducing novel grouping methods to reduce the dependency on the constraint that parts must have similar distributions [13,14]. The efforts to generalize statistical selective assembly opened a new area of research surrounding optimal binning strategies for a range of applications.

Binning strategies can either be designed before production, during production, or post-production (note that in the case of post-production, parts have been manufactured, but mating parts still await aggregation). Since selective assembly encompasses measuring part dimensions after they have been produced, it is often preferable to design the binning strategy after the design stage, in a prototypical manner. Performing selective assembly during production (before all parts have been manufactured), often utilizes a buffer system where bins are populated and depleted in real time [11]. In terms of the manufacturing environment, selective assembly is typically better suited for batch production rather than mass production, since the extra steps to utilize selective assembly in mass production can create congestion and bottlenecks, which is less likely to occur in batch production due to the lower production rate [9].

While research on selective assembly has focused exclusively on manufacturing applications, the proposed methodology demonstrates how selective assembly can be used in modular construction. Reasons why selective assembly can be used as a quality improvement tool in modular construction include (1) the ability to achieve tight tolerance requirements without the use of a rigorous tolerance design, and (2) the use of production techniques currently employed in stick-built construction rather than adoption of highly precise equipment which can be very expensive. Modular construction also often resembles batch production manufacturing more so than mass production, making it very favourable for application of selective assembly.

2.3 Production Optimization Methods in Modular Construction

In the area of production optimization for modular construction, numerous methods can be employed. This section briefly introduces two production optimization techniques which achieve similar goals as the proposed methodology: application of lean principles, and design for manufacturing and assembly (DfMA).

Lean production principles, which first emerged in the Japanese manufacturing industry, are a collection of practices that aim to increase productivity, maximize value, and minimize waste [15]. While the application of lean principles in modular construction can significantly improve project performance [16-18], there still exists some key challenges with respect to the dichotomy between design and construction. For instance, Sarhan and Fox [19] found that in lean construction, there is often a conflict border between design and construction, resulting in inaccurate designs, rework, lack of constructability, and final products with significant variation from specified design values.

Design for manufacturing and assembly (DfMA) is a common design approach used in the manufacturing industry for optimizing production costs by addressing the trade-off between manufacturing and assembly that exists for various processes [20]. With respect to component tolerances, strict tolerances allow for easy assembly, however often come at an increased manufacturing cost. As such, DfMA is an effective design approach to evaluate tolerance decisions with respect to both manufacturing and assembly processes in order to determine an optimal component tolerance value. While DfMA is primarily only used in the manufacturing industry, recent efforts have been made to adopt its use for construction. Design for construction (DFC) is one of these efforts, and utilizes three key ideas: (1) design for manufacturing and assembly (DfMA), (2) lean production and (3) constructability [21]. DFC aims to capture tacit knowledge learned from previous production problems, so that unnecessary rework is minimized in future projects.

Some of the key goals of lean production and DfMA in modular construction are increased productivity, greater constructability and rework reduction. Current production optimization techniques in modular construction do not offer a systematic approach for managing problems that are often encountered in the assembly of modular components. To address this, selective assembly tackles the problem from a very unique perspective – by measuring and grouping components in order to optimize production processes.

3 Proposed Methodology

This paper demonstrates how the principles of selective assembly can be applied to modular construction to optimize the aggregation of repetitive pairs of components. Since selective assembly relies on sorting pairs of components based on the dimensional distribution of features that make up the physical interface between components, it is easier to develop an optimal strategy using as-built data rather than theoretical predictions. As such, the proposed use of selective assembly is applied after production has finished (i.e., parts are manufactured, but await aggregation). The proposed framework (Figure 3), is comprised of three key steps: (1) identify critical interfaces between components and tolerances, (2) calculate the minimum number of bins required, and (3) organize parts into bins based on a binning strategy.

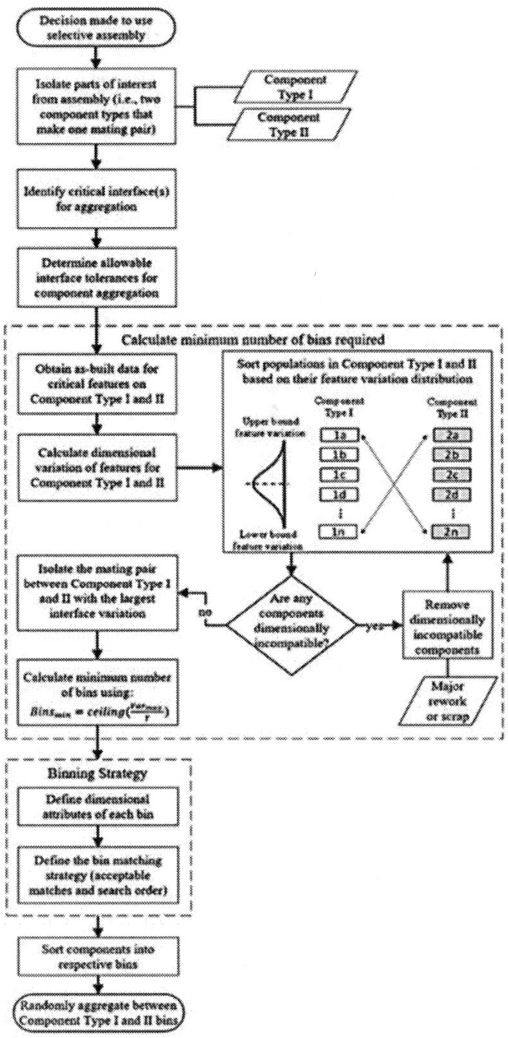

Figure 3: Framework of proposed methodology

3.1 Isolating the Critical Interfaces and Determining Aggregation Tolerances

The critical interfaces between components are the physical regions or points on components in direct contact with each other. The dimensional variation of features that make up the critical interfaces on the as-built components must be determined. Laser scanning is proposed for this purpose based on its ability to yield rich and accurate data [22]. One of the most challenging aspects of applying selective assembly lies with determining tolerances that govern adequate component aggregation. For this purpose, three approaches can be taken: (1) measure variation on prototyped pairs (2) use previous experience or benchmark tolerances, or (3) conduct a systematic tolerance analysis.

3.2 Calculating the Minimum Number of Bins Required for the Binning Strategy

Before calculating the minimum number of bins required in a binning strategy, it is important to ensure that all components are dimensionally compatible based on the allowable aggregation tolerances. For checking dimensional compatibility, each isolated component must have at least one possible mating component, otherwise it will not be able to aggregate properly. In the event that dimensional incompatibility exists (i.e., the geometry variability of a particular component is such that it cannot connect with any other component), major rework or component scrapping is required (depending on which option is least expensive or time consuming). Finally, the minimum number of required bins ($Bins_{min}$) is calculated using:

$$Bins_{min} = ceiling(\frac{Var_{max}}{T}) \quad (1)$$

where Var_{max} is the maximum measured interface variation between all possible pairs, T is the allowable aggregation tolerance and *ceiling* is a function that rounds up to the nearest whole integer. Although the minimum required bins is calculated here, a larger number may be used, depending on desired accuracy. In general, increasing the number of bins decreases assembly variations, but also increases the likelihood of having surplus parts, disproportional bin populations, and decreased overall effectiveness [11]. Selective assembly in manufacturing usually aims to achieve low assembly deviations since most mating parts are moving (i.e., pistons in automotive engines). However for modular construction, the level of assembly deviations only needs to ensure adequate component aggregation, and parts are typically not designed to experience movement after aggregation. As such, it is preferable to minimize the number of bins for selective assembly in modular construction.

3.3 Developing the Binning Strategy

The next step for applying selective assembly is determining the binning strategy, which outlines how components are organized into bins and how bins are matched together. Two common ways to partition the dimensional attributes of bins are (1) equal dimensional width or (2) equal probability. Equal width partitioning divides the total interface variability equally between bins, while equal probability partitioning ensures that each bin has equal populations of components [5]. Matching criteria defines how components are matched between bins. Traditional methods include one-to-one matching (each bin has exactly one other matching bin), or one-to-three matching (each bin has one matching bin but can pull from adjacent bins to the matching bin if need be) [11]. The methodology for developing the a binning strategy is outside of the scope of this paper, however based on the results of the case study, the authors recommend using equal probability bin partitioning and one-to-one matching due to its simplicity. Since modular construction typically has a lower number of mating pairs of components than in manufacturing, and since part inventories are not common, everything should be matched on each project. These factors lend themselves to have equal probability bins (avoiding surplus parts), and one-to-one matching (to ensure that every part is matched). After the binning strategy has been determined, components between matched bins can be randomly aggregated.

4 Modular Bridge Case Study

A 6 m long modular steel bridge designed and built by the University of Waterloo Steel Bridge Team is examined. The bridge is comprised of hollow steel section members, and has five types of assemblies or modules that are bolted together (Figure 4). As part of a modularization strategy, 24 assemblies were designed as interchangeable top and bottom pairs (A2 and A3 in Figure 4-c). During construction, selective assembly was utilized as an approach for mitigating the impact of fabrication error (accumulating effects of cutting, milling, fit-up, measurement, welding distortion, and inspection). Selective assembly was applied for the aggregation of the top and bottom pairs.

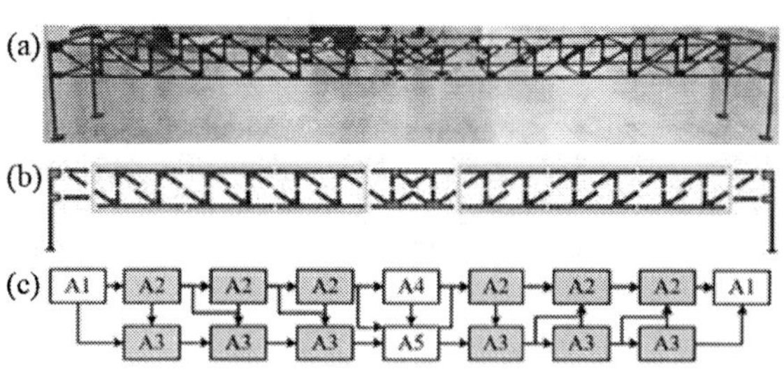

Figure 4: (a) Fully assembled steel bridge, (b) deconstructed single truss, (c) assembly diagram of single truss showing the five main assembly types.

4.1 Critical Aggregation Interfaces

For each top and bottom assembly pair (A2 and A3), there are three direct contact points that make up the aggregation interfaces. Of these interfaces, two critical dimensions are extracted: (1) an angular dimension, ϴ, and (2) a linear dimension, X, as illustrated in Figure 5. Aggregation between top and bottom pairs is assumed to rely on the compatibility of these critical dimensions.

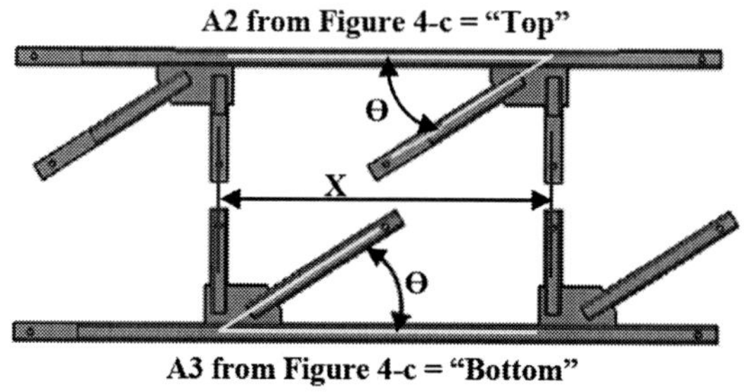

Figure 5: Repetitive assembly pair in case study, with critical interface dimensions as ϴ (yellow) and X (red).

The as-built data of the critical interface dimensions were obtained by conducting coordinate probing using a laser scanner (Faro Edge Arm). This device has an accuracy of 0.024 mm for the working length employed [23]. Coordinate probing was used since each part can be reduced from its as-designed model into a centreline model, and then to a series of critical coordinates at the interface points (Figure 6).

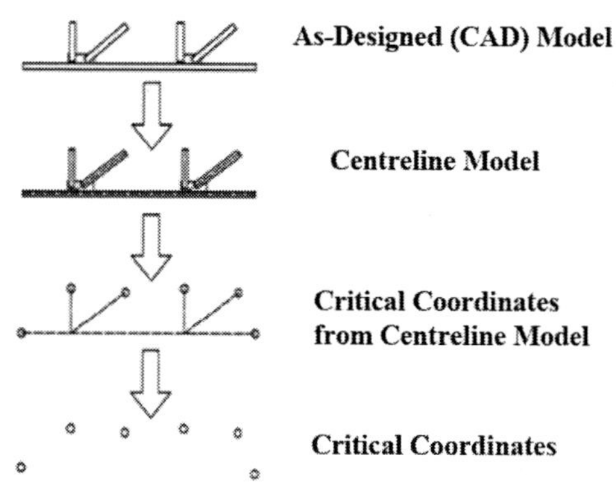

Figure 6: Process of Extracting Critical Coordinates from Original As-Designed Model

Selective assembly in this case study depended solely on the centreline alignment of the physical interfaces for top and bottom pairs. Rework during aggregation is assumed to be constrained to shimming (i.e., extending the length of a member at an interface) and grinding (i.e., reducing the length of a member at an interface). This type of rework is much less intensive than having to cut, realign and weld a member into proper position (which does not necessarily eliminate the need for grinding and shimming at interfaces). After critical interface coordinates were obtained, populations of top and bottom components were sorted based on the distribution of each critical dimension (Table 1).

Table 1: Sorted Populations of Top and Bottom Components based on Distribution of Critical Aggregation Dimensions

Dim Θ (degrees)				Dim X (mm)			
ID	A2 (top)	ID	A3 (bot)	ID	A2 (top)	ID	A3 (bot)
10	30.67	1	31.44	8	393.17	6	392.48
6	31.62	8	31.88	9	393.35	9	393.79
9	31.67	3	31.95	11	393.53	8	394.58
2	31.90	11	32.02	2	395.06	12	394.58
5	31.94	7	32.21	12	395.27	11	394.82
7	32.29	4	32.33	4	395.59	1	395.32
4	32.32	12	32.37	6	395.62	4	395.44
12	32.51	5	32.55	10	395.66	5	395.97
11	32.97	2	32.58	5	395.71	10	396.04
8	33.07	9	32.80	1	396.07	2	396.97
3	33.51	6	33.09	3	396.26	3	397.10
1	33.60	10	33.40	7	398.89	7	399.37
Nominal Dim – 32.9				Nominal Dim – 396.88			

4.2 Binning for Critical Angular Dimension

Using the sorted populations of top and bottom components, the first iteration of binning was carried out for the critical angular dimension, Θ (Figure 5). Since the allowable tolerance for this dimension was not specified in the design, it was determined through prototyping. The maximum angular discrepancy between Θ values of top and bottom components for a successfully aggregated pair was measured as 2.72° from testing all possible component pairs. Since all possible top and bottom pair combinations result in angular discrepancies equal to or less than 2.72°, a single bin can be used for each top and bottom population for aggregation based on Θ. As such, a binning strategy was not required for the critical angular dimension, since random aggregation can proceed between all possible top and bottom pairs.

4.3 Binning for Critical Linear Dimension

Using the sorted populations of top and bottom components, the second iteration of binning was carried out for the critical linear dimension, X (Figure 5). The allowable tolerance for this dimension was specified in the design as +/- 1/16" (1.588 mm), since the bolt hole diameters are 1/16" (1.588 mm) larger than the bolt diameter used. As such, the tolerance range is equal to 1/8" (3.175 mm) to account for the case where an upper bound deviation is matched with a lower bound deviation. One final check was performed before determining the minimum number of bins to ensure that no components would be rejected based on dimensional incompatibility: for every bottom component there is at least one top component that does not exceed the allowable tolerance, and for every top component there is at least one bottom component that does not exceed the allowable tolerance. This check yielded no dimensionally incompatible components. Using the largest deviation between critical linear dimensions for all possible pairs (6.41 mm) and the tolerance (3.175 mm), the minimum number of bins was calculated using (Equation 1): $Bins_{min} = ceiling\left(\frac{6.41\ mm}{3.175\ mm}\right) = 3$. Since the minimum number of bins is greater than 1, random aggregation cannot proceed between all top and bottom pairs, and a binning strategy is required.

The binning strategy uses a one-to-one strategy, where every bin for top components has exactly one matching bin for bottom components. Furthermore, equal probability partitioning is employed to avoid having surplus components. Since the combined width of all bins (3 x tolerance = 9.525 mm) is larger than the largest deviation between all possible pairs (6.41 mm), several bin arrangements are possible. A script was compiled in MATLAB to find that there 8 possible bin arrangements (Figure 7).

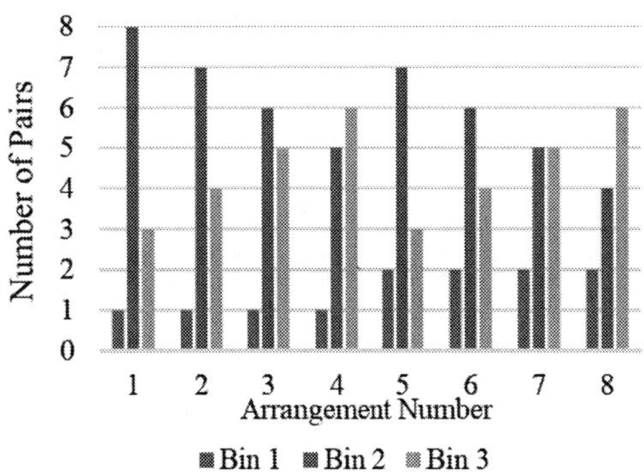

Figure 7: Bin Populations for all Possible Bin Arrangements Based on the Critical Linear Dimension

4.4 Selection of Optimal Bin Arrangement

While each of the 8 possible bin arrangements yield dimensionally compatible component pairs, there exists a specific arrangement with a statistically optimal amount of rework avoidance. There are two approaches for selecting the optimal bin arrangement in this case study: (1) absolute rework avoidance or (2) least expected rework. Absolute rework avoidance finds the pairs of components in each bin with the largest possible deviation. Then, these values are summed together and used for comparison to find the bin arrangement with the least overall deviation (Table 2).

Table 2: Maximum Deviations for Critical Linear Aggregation Interface for all Possible Pairings in Each Bin Arrangement (all deviations in mm)

Arrangement	1	2	3	4	5	6	7	8
Bin 1	0.48	0.48	0.48	0.48	3.11	3.11	3.11	3.11
Bin 2	2.04	1.83	1.51	1.48	1.91	1.70	1.38	1.35
Bin 3	1.41	2.58	2.79	3.11	1.41	2.58	2.79	3.11
Sum	3.93	4.89	4.79	5.07	6.43	7.39	7.29	7.57
Average	1.31	1.63	1.60	1.69	2.14	2.46	2.43	2.52
Std. Dev.	0.64	0.87	0.95	1.08	0.71	0.58	0.75	0.83

Least expected rework associates a probability of selecting a random pair with its deviation value. As such, this approach finds the most probable deviation to be expected between pairs in each bin. The expected deviations for each bin are summed together and compared to find the bin arrangement with the least expected overall deviation (Table 3).

Table 3: Expected Deviations for Critical Linear Aggregation Interface for all Bin Arrangements (all deviations in mm)

Arrangement	1	2	3	4	5	6	7	8
Bin 1	0.48	0.48	0.48	0.48	1.56	1.56	1.56	1.56
Bin 2	0.79	0.72	0.65	0.66	0.67	0.59	0.51	0.52
Bin 3	0.85	1.01	1.08	1.14	0.85	1.01	1.08	1.14
Sum	2.11	2.20	2.21	2.28	3.07	3.15	3.14	3.21
Average	0.70	0.73	0.74	0.76	1.02	1.05	1.05	1.07
Std. Dev.	0.16	0.22	0.25	0.28	0.38	0.40	0.43	0.43

As seen from both approaches, bin arrangement 1 is optimal since it has the least absolute rework and least expected rework. As such, it was selected for as the bin arrangement for this case study.

5 Discussion

The binning strategy in this case study was conducted by isolating repetitive pairs of component in a modular assembly, classifying them by distinct dimensional attributes and sorting them into bins that yielded component pairs that could be correctly assembled. Laser scanning was employed for as-built data collection primarily for its ease of capturing rapid and accurate data. Currently, the method of determining critical aggregation interfaces, and tolerances is manual and requires proper user judgement. Although the critical angular dimension was not directly used in this case study, it should be noted that a binning strategy needs to incorporate all critical dimensions. For instance, the ID numbers shown in Table 1 do not match up for a given component between each dimension. The distribution for the angular dimension does not match with the distribution for the linear dimension. This means that as the number of critical dimensions increases, it becomes increasingly more difficult to decrease the minimum number of possible bins.

As a result of the binning strategy shown in this case study, the component aggregation of mating pairs in the modular steel bridge proceeded with no major rework or wasted components. Before applying selective assembly, the fabrication team of this bridge attempted to apply random aggregation of top and bottom components without a binning strategy. Due to compounding effects of fabrication error, there were two instances of extensive rework, where members had to be cut, realigned and welded into proper alignment. While the exact quantitative impact of this rework is unknown, the team found the results of binning components to yield component pairs that could be successfully aggregated.

The primary limitation of this case study lies in the assumption that component aggregation is solely based on the three direct contact points between top and bottom components (Figure 5). In this regard, the binning strategy finds optimal pairs locally, but does not consider the impact that a given assembly has on its adjacent neighbours (i.e., assemblies on either side).

6 Conclusions

This paper demonstrates how selective assembly can be used in modular construction to reduce the adverse effects of component variability. Variability is an unavoidable aspect of component manufacture and if not controlled properly, it can lead to severe conflicts during component aggregation. This was observed by the fabrication team in the case study, but also occurs in modular construction projects. Instead of conducting comprehensive tolerance designs, which is one approach used in manufacturing, selective assembly is a technique that can be applied during fabrication to avoid aggregation conflicts. The proposed binning strategy is shown to yield an optimal set of component pairs based on least overall deviation. It is reasonable to assume that rework is directly related to the amount of deviation between components, which means that the proposed framework for selective assembly results in an optimal rework mitigation strategy.

Future work of this research aims to extend selective assembly to include a more systematic approach for optimal assembly planning.

Acknowledgements

The authors acknowledge the financial support of the Natural Science and Engineering Research Council (NSERC) and PCL-Permanent Modular Construction.

Furthermore, the authors acknowledge the University of Waterloo Steel Bridge Team and their faculty advisor Dr. Scott Walbridge for providing access to the modular steel bridge shown in this paper.

References

[1] D.K. Ballast, Handbook of Construction Tolerances, John Wiley & Sons, 2007.

[2] A.G. Gibb, Off-Site Fabrication: Prefabrication, Pre-Assembly and Modularisation, John Wiley & Sons, 1999.

[3] CII, Transforming modular construction for the competitive advantage through the adaption of shipbuilding production processes to construction, Research summary 255-1 (2011).

[4] N. Boyd, M.M. Khalfan, T. Maqsood, Off-site construction of apartment buildings, J. Archit. Eng. 19 (2012) 51-57.

[5] D. Mease, V.N. Nair, A. Sudjianto, Selective assembly in manufacturing: statistical issues and optimal binning strategies, Technometrics. 46 (2004) 165-175.

[6] A. Asha, S. Kannan, V. Jayabalan, Optimization of clearance variation in selective assembly for components with multiple characteristics, The International Journal of Advanced Manufacturing Technology. 38 (2008) 1026-1044.

[7] G.A. Pugh, Partitioning for selective assembly, Comput. Ind. Eng. 11 (1986) 175-179.

[8] C.M. Creveling, Tolerance Design: A Handbook for Developing Optimal Specifications, Prentice Hall, 1997.

[9] M.H. Tan, C.J. Wu, Generalized selective assembly, IIE Transactions. 44 (2012) 27-42.

[10] E. Mansoor, Selective assembly—its analysis and applications, Int J Prod Res. 1 (1961) 13-24.

[11] A. Thesen, A. Jantayavichit, Design and evaluation of a selective assembly station for high precision scroll compressor shells, (1999) 694-700.

[12] MARPOSS, Gauging stations for selective assembly, <http://www.marposs.com/family.php/eng/selective_assembly >. 2016 (2016).

[13] G.A. Pugh, Selective assembly with components of dissimilar variance, Comput. Ind. Eng. 23 (1992) 487-491.

[14] K.C. Chan, R.J. Linn, A grouping method for selective assembly of parts of dissimilar distributions, Quality Engineering. 11 (1998) 221-234.

[15] R. Shah, P.T. Ward, Lean manufacturing: context, practice bundles, and performance, J. Oper. Manage. 21 (2003) 129-149.

[16] J. Meiling, F. Backlund, H. Johnsson, Managing for continuous improvement in off-site construction: Evaluation of lean management principles, Engineering, Construction and Architectural Management. 19 (2012) 141-158.

[17] M. Höök, L. Stehn, Applicability of lean principles and practices in industrialized housing production, Constr. Manage. Econ. 26 (2008) 1091-1100.

[18] H. Yu, M. Al-Hussein, S. Al-Jibouri, A. Telyas, Lean Transformation in a Modular Building Company: A Case for Implementation, J. Manage. Eng. 29 (2013) 103-111.

[19] S. Sarhan, A. Fox, Barriers to implementing lean construction in the UK construction industry, The Built & Human Environment Review. 6 (2013).

[20] A.K. Kamrani, M.S. Sa'ed, Product Design for Modularity, Springer Science & Business Media, 2002.

[21] R. Gerth, A. Boqvist, M. Bjelkemyr, B. Lindberg, Design for construction: utilizing production experiences in development, Constr. Manage. Econ. 31 (2013) 135-150.

[22] F. Bosché, Automated recognition of 3D CAD model objects in laser scans and calculation of as-built dimensions for dimensional compliance control in construction, Advanced engineering informatics. 24 (2010) 107-118.

[23] FaroARM, Measuring arm FaroArm - overview - portable measuring solutions from FARO, <http://www.faro.com/en-us/products/metrology/measuring-arm-faroarm/overview#main>. 2016 (2014) 2.

A Preliminary Study on Text Mining–Based Human Resource Allocation in a Construction Project

Sangil Han [a] and Ghang Lee [a]

[a] Yonsei University, Republic of Korea
E-mail: hancowr@gmail.com, glee@yonsei.ac.kr

Abstract –

Even engineers who have the same job title and are working on the same project have different skills and backgrounds. However, despite the importance of assigning the right person to the right job to ensure successful project delivery, current human resource (HR) allocation practices are only concerned with the list of projects in which a candidate has been involved, years of experience in a discipline, and their previous job titles. As a result, some engineers lack the knowledge and experience to perform the tasks they are assigned on a project. This study demonstrates the need for a long, descriptive résumé rather than the commonly used short, brief résumé. When the job candidates who participated in this study were asked to describe their work experience in several sentences on their curriculum vitae (CV) instead of describing them using a few words composed of just the job title and roles, the contents of the CVs were clearer and showed less bias. This paper therefore presents a text-mining algorithm and a semantic résumé analysis system that we developed to automatically extract and analyze the work experience candidates write about on their résumés. This algorithm and analysis system can help HR managers control the significant amount of information found on prospective employees' long résumés. To validate the algorithm and the system, six sample résumés were collected anonymously. These résumés were then reviewed both manually and by the developed system. In a subsequent study, we expect to apply the text mining–based HR resource allocation algorithm to select construction engineers for a real project and measure the job-matching rate.

Keywords –

Construction engineer; text mining; human resource allocation; résumé; KNIME

1 Introduction

The purpose of this study is to develop an approach to resolve the problems occurred during the allocation of human resources (HR) in construction organizations by applying the text-mining method to long descriptive résumés. Searching for the best candidate for a specific job at a construction project is one of the main challenges of HR managers in the architecture, engineering, and construction industries. Although large companies have their own HR management systems that include information technology (IT) solutions (such as web-based management systems) to facilitate job allocation, their current usefulness and accuracy are questionable. The reason for this is that most HR management (HRM) systems only allow HR managers to search for appropriate engineers or managers for a new project based on previous team names or the names of project that the engineers/managers worked on, job titles, durations of each project, etc. Nevertheless, how candidates perform on a new project greatly differs depending on the specific tasks that each candidate performed for previous projects and the experience and skills that they gained through these tasks. For example, if a new project needs a site manager for a building construction project (referred to as an architectural site engineer, hereafter), an HR manager generally looks for potential candidates from an HR system by searching for employees who previously worked as an architectural site engineer. However, architectural site engineers perform highly diversified tasks throughout the entire period of a project and also work with different types of subcontractors and vendors and on different contractual systems, but the employee information in an HR system does not include these data, which therefore hinders the HR manager from finding the best person for the new project.

Similar problems exist when hiring new employees. Résumés or curriculum vitae (CVs) are often composed of discrete words or short sentences and thus are often vague and unclear. Some companies ask for a statement of purpose (SOP) in addition to a CV, but long SOPs are very time-consuming to analyze, especially when there are many applicants (for the more popular companies in South Korea, it is common to review hundreds of applicants for a single position). In addition, with new skill sets required due to the increasing globalization of the construction market and the rapid development of computer integrated construction and building information modeling, the need for an improved and detailed job classification system is increased [1]. To overcome these limitations, construction companies have tried to develop an HR system that can support detailed job classifications. However, a drawback of the new systems is that they ask applicants and employees to provide too much information, sometimes by selecting an item from a multi-layered dropdown menu and sometimes by filling in blank text boxes. A more critical drawback of such systems is that they will always be limited in their ability to capture all the details of various types of construction tasks that the applicants and employees feel are important.

Therefore, this study has developed an improved HR solution for finding the best person for a job and for maintaining the records of project staff by collecting and analyzing long sentence-based job descriptions related to the construction industry using a text-mining method instead of keyword-based job descriptions. The text-mining system compensates for the natural shortcoming of long descriptions by refining and stemming key factors from the descriptions.

2 Previous studies

2.1 Literature Review

The importance of HRM has been emphasized by many researchers. As intellectual capital has become one of the most strategic assets of successful organizations, the ability to manage the expertise, skills, and experience of employees has become a key factor in overcoming the increasing competitiveness of the global market [2]. In today's competitive business environment, companies need to accurately grasp the competency of their HR in order to be successful [3]. However, technology companies often suffer from high turnover rates and find it hard to recruit the right talents [4]. Due to the current trend of pursuing better HRM options, an increasing number of publications concerning data mining in the context of HRM are providing new research data [5]. For example, one study investigated the relationship between web data mining and strategic HR planning by developing a practical model [6]. Another study focused on the HRM performance evaluation process by introducing a data-mining method termed knowledge discovery in databases [7].

Thus far, filtered résumés have been the most important resource for HR departments when hiring new employees [8]. One study introduced an approach to improve the matching of profiles by searching job descriptions and applicant profiles using filters that represent the relevant skills and competencies [9]. Nevertheless, several studies have found that résumés and work experience lists, which are composed of brief words or short sentences, are limited and have tried to improve HR solutions by adopting a semantic system approach, such as ontologies and text-mining methods. An example is the On-To-Knowledge project. The On-To-Knowledge project focuses on the application-driven development of ontologies during the introduction of ontology-based knowledge management systems [10]. One research study has suggested that a possible approach could be addressing an intelligent decision support system composed of case-based reasoning and ontology [11]. A case study focused on the electronic market introduced a unique configurable framework that was designed to be adapted for electronic marketplace applications [12]. Another research stressed the importance of HR recruiting, selecting individuals for teams based on different skills and qualifications, determining who to train and what training programs to offer, and recommending the right expert to individuals for acquiring information or learning from within the organization [3]. In addition, previous research has focused on the overall HR domain, whereas researchers studying the construction industry specifically have not attempted to actively apply text mining to HRM.

In addition, to fully understand the current status of HRM in the construction industry, we reviewed the relevant articles and papers relating to the inherent problems of construction HRM. One study pointed out the importance of an automated HR information system using a case study of Australian construction companies to prevent chronic problems like high turnover, lower job satisfaction, and motivation [13]. Another study stressed that a typical Korean construction organization has been simply classified by terms such as architect, safety, quality, administration, etc., even though it should be categorized with more detailed activities, such as claim/contract management, design management, HRM, document management, information management, risk management, etc. [14]. A report issued by a Korean research institute pointed out that construction companies' job classifications have not been set up systematically to trigger inefficient HRM and material resource management [15].

2.2 Existing Web-Based HR Systems

Probably the most widely well-known web-based HR systems are Linked-In and OilandGas. Their search systems are designed so that a recruiter can input a few words, and within milliseconds, the web page shows the search results of résumés through keyword matching, similar to portal web searching like Google. With the special services of Linked-In, recruiters can also search for specialists and candidates from various industries, not only for the construction industry. On the other hand, the OilandGas service is specialized for searching for experts in the oil and gas industry. These services use basically the same concept of discrete keywords matching. Hence, users can put in keywords, like a position or job title, add some more information like nationality, and then search. The results can be anywhere from just a few to thousands of CVs. However, once this search process is completed, the recruiter has to download all the search results and résumés and then read through all the documents to pick the most suitable candidates.

3 Analysis of the Current Practices

To fully understand current practices, we conducted an interview with a HR manager at one of the largest construction companies in Korea. He has been working on the HR team at the company's headquarters for about 5 years and is in charge of recruiting experts for international construction projects. Due to the growing needs for the international construction market—especially for the oil and gas plants and the power plants, the HR overseas department target-recruits experts suited for those projects from outside the company. We asked the HR manager several questions related to our research subject:

1. How do you find a person who is suitable for a particular job?
2. How do you find new employees from outside the company?
3. Do you keep detailed records of each employee's job experience?
4. If so, how do you collect and manage detailed job experience information? Do you use long descriptions or the more traditional keyword-based HR database?
5. How important do you think are long descriptions about job experience for job allocation?
6. Do you use an automated HR resource allocation or recommendation system?

For the job allocation of employees, this particular company uses its own web-based HRM system that had fields to input detailed and descriptive records of employees' past job experiences, but these fields are optional. Thus, according to the HR manager, most staffs do not input these descriptions. In addition, the HR department does not consider these descriptions to be important, especially for low-profile staff. Internal job allocation of employees relies heavily on recommendations from previous project managers. For the recruitment of experienced employees from outside the company, the interviewee responded that résumés and SOPs were essential to filter out unqualified personnel and to allocate an individual most suitable for a particular job. Nevertheless, the analysis process is very time-consuming and challenging, as the number of applicants for each position has grown rapidly. According to the HR manager, the company uses its own HR system for controlling and managing all the employees. This system handles various aspects of individual staff, from private information like family members to involved projects of the company or of previous companies. But, like most Korean companies, work experience and job descriptions are expressed by a single word such as 'Architectural (construction)' or 'safety,' along with the involved project or team name.

For comparison, we collected other cases of Korean top-ranking construction companies and found that their HR systems were very similar. For companies A and B, their web-based HR systems showed only job position and its domain, such as architecture or infrastructure, and even such categories were generally left blank. This shows that most Korean construction companies do not seem to consider individuals' detailed experiences as important factors for HRM and that they prefer simple job categories due to ease of management. Also, we can assume that the Korean construction companies' HRM relies on simple facets of each employee, like the level of their position (staff, assistant manager, senior manager, etc.), their job title, and their reputation and references.

However, the interviewee's company has a unique system for collecting long descriptions of each staff. The staff members write their own job descriptions freely, using their own judgement to rank their skills and experience. It may seem like the most useful type of information for depicting each staff's career and abilities for certain tasks and positions; however, the company does not view this as a beneficial asset for HRM. The input guidelines consist of two sentences only, so the staffs also do not want to put their energy and time into filling out long descriptions to express their experience and major job skills.

From exploring the current status of HRM in common Korean construction companies, we can conclude that the HRM mainly relies on fragmentary factors of individuals or on their reputation that is transferred verbally. Hence, the purpose of this study to improve the current shortcomings of the HR systems of construction companies seems to be reasonable.

4 Introduction to the Text Mining–Based HR Allocation System

This section explains the text mining–based HR allocation system that we propose. It briefly explains the basic concept of the proposed text-mining method and then demonstrates the applicability of the proposed method using one of commonly used text-mining tools and a sample case.

4.1 Overview of the Proposed Method

The first step to accomplish the method is setting up the basic concept algorithm as follows:

1. The main objective is to use text mining to efficiently and effectively analyze the résumés of candidates for allocating positions for a construction project.
2. Setting the basic requirements of each position by searching for typical job requirements or applying an organization's own requirements.
3. The main usage of the method is (1) for in-house employees and (2) for new job seekers from outside the company.
4. The natural languages consist of long descriptive sentences in résumés that are modified and refined by the flow, such as 'input,' 'parsing,' 'filtering,' 'collecting,' 'matching,' and 'result showing'.

By using the concept of text mining to build the proposed system, we constructed a flow chart of the entire process for the final judgement of whether a candidate is the best fit for a designated position. The flow and its step-by-step explanation are shown in Figure 1.

- The construction domain must be selected before starting the text-mining process among industry divisions such as 'building,' 'infrastructure,' and 'plant' since each sector's expertise is defined specifically for its own purposes and characteristics.
- The input data is divided into two major categories: (1) for in-house use and (2) for hiring external candidates. The input data to be validated has to fully contain sufficient information of the personnel semantically before entering through the algorithm.
- The mining process consists of 'input,' 'filtering & parsing,' 'classification,' and 'matching.' This process enables the natural language used in the long descriptions in a CV to be converted into strings that are readable and comparable by a computing system. The last step 'matching,' is implemented after the requirements of the position are received, which was derived from a separate source. The requirements can be defined by any source, but they should cover the position as exhaustively as possible using essential keywords.
- The end of process is position matching; e.g., how well a candidate is suited for a position or a recommendable position for a particular candidate.

We would like to emphasize that this text-mining process requires that candidates' CVs are composed of long descriptive sentences that explain the candidates' work experience in detail. Otherwise, there is no reason to develop this much more complicated system as compared to the traditional HRM system that only uses short keywords.

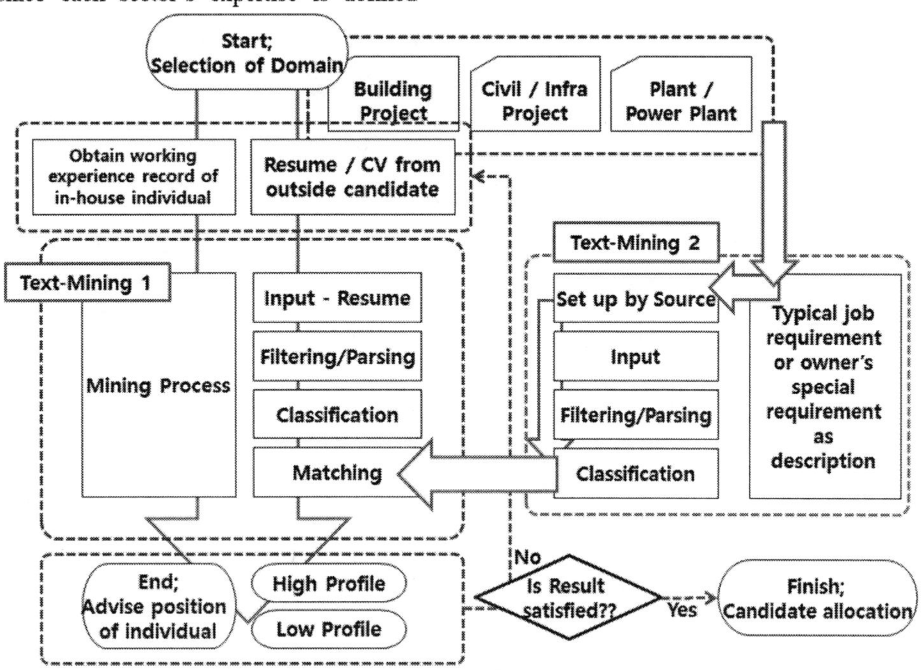

Figure 1. Text-mining flow

4.2 Applying the text-mining tool

The concept of text mining has been used since it was introduced, and many research teams and organizations have created text-mining platforms for various purposes. The text-mining tool employed for this study was KNIME, which is one of the more widely used open platforms (Figure 2). Since 2006, KNIME has been used in pharmaceutical research, but it is also used in other areas like CRM customer data analysis, business intelligence, and financial data analysis. This platform has intuitive visual, so users can build the process with a visualized work flow.

4.2.1 Main Functional Nodes Set-Up

In the introduced text-mining platform, 9 functional nodes were dragged and linked to conduct the mining process for résumé and requirement in accordance with the framework. In KNIME, we numbered the nodes from 1 to 9 so that the input resources—résumés and job requirements—were transformed when they went through the nodes in the system.

- #1. Document parser: This node reads the documents and creates a document for each file. The full text is extracted.
- #2. POS tagger: This node assigns a part of speech to each term in the document (tag).
- #3. Bag of words (BoW) creator: This node creates a BoW for a set of documents.
- #4. Number filter: This node filters all terms consisting of numbers only.
- #5. Punctuation erasure: This node erases the punctuation characters from terms.
- #6. Stop word filter: This node filters all terms contained in the specified stop word file (such as 'a,' 'about,' 'but,' 'could,' 'the,' 'from,' etc.).
- #7. Stemmer: This node stems terms by stemmer algorithms.
- #.8. Keyword extractor: This node extracts relevant keywords based on frequency and its own scoring method.
- #9. String matcher: This node compares two lists of strings, computes the distance between the strings, and lists the most similar strings. (These functional node descriptions are cited from the KNIME platform.)

The functional nodes were extracted from the Node Repository sector of the KNIME platform categorized as 'Text Processing.' Each node has 3 color signals similar to traffic lights so a user could easily recognize when an error occurs and amend the process effectively. In addition, a user can adjust the options of each node, which affects the level of function, to control the outcome level of the process.

5 Test Implementation of Text Mining

5.1 Collection of Test Résumés and Job Requirements

To assess the usability and reliability of the structured text-mining process, we collected six résumés of different candidates who were willing to be assigned a construction manager construction manager (CMr) position. We also obtained the job requirements of the CMr position from the Civil Service Commission of State of New Jersey, USA, website

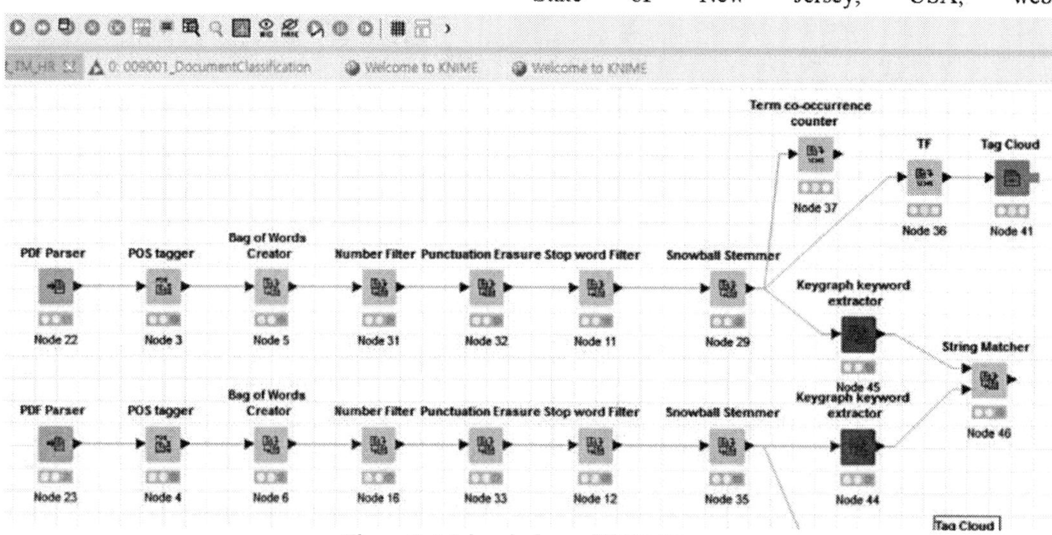

Figure 2. Main window of KNIME.

(http://www.state.nj.us), which provides official descriptions of construction job positions. Job requirements can be obtained from various channels, so we have searched many sources through web searching and our literature review. We selected the aforementioned website from this search process because it offered sufficient descriptions of construction job positions for our text-mining test. The candidates' CVs for our test were collected anonymously from a commercial service provider. The résumés and requirements were written in English and in long sentences. All of these input resources were in a PDF format, and no artificial amendments were added. We gathered samples from only six individuals, although we could have collected more CVs from various sources because we needed to verify each résumé carefully and compare it with the text-mining results within the limited research time, which is a potential limitation of this study.

5.2 Conceptual Design of Weight by Distance

To measure the degree of accuracy that the candidates were matched to the job position of construction project manager, we used the matched results of the 'String Matcher' node, which showed how many keywords were coincidental or related between two input resources—the job requirements and candidates' résumés—as a result of the text-mining process. Two important indices were extracted for assessment, one was the number of matched strings and the other was the distance between each string. The distance indicated the accuracy of two matched keywords; for instance, '0' was perfectly coincidental and '1' showed a small, but ignorable, difference between two keywords. Two strings with a distance of less than '1' (e.g., 0 and 1) were considered matched strings, while strings with a distance of more than '2' were assumed to not be paired words. Thus, in the formula for calculating the matching score, a distance of more than 2 was not considered.

Hence the six different results were compared by the number of matched strings under a distance of '1.' To make this comparison, we decided to input the weight for '0' distance and '1' distance only. '0' had a weight of 1, and '1' had a weight of 0.9.

$$\text{Score} = W0 \times n(D0) + W1 \times n(D1),$$

where
D0(Distance 0) has weight W0 = 1,
D1(Distance 1) has weight W1 = 0.9,
Distance (≥2): Weight = 0 (excluded in formula),
n(D0): number of DO, and
n(D1): number of D1.

5.3 Running and Validating the Results of the Text Mining

Candidate A's score was 3 matched keywords, 'respons', 'project,' and 'manag,' with a distance of 0. Candidate B's score was 4 matched keywords with distances of 0 and 1. Candidate C and candidate D had a score of 5 keywords within less than a distance of 1. Candidate E had 4.9 points. Finally, candidate F scored 8 matching keywords with a distance of 0 and 1 (Figure 3). In terms of the validation rule we set up, candidate F received the best score and was therefore the most qualified for the CMr position (Table 1).

To validate the results of the text-mining process, we asked an HR manager who was not informed of the results of our text mining to independently assess the six résumés. Similar to our results, the HR manager judged that candidate F was the best option for the CMr position among candidates, and he mentioned he would disqualify candidates A and B. He judged candidates C and D to be acceptable but not fully qualified and candidate E to be relatively qualified (he didn't have experience as a project manager). We had a short discussion after the manger's appraisal and concluded that candidates A and B did not have enough experience to qualify for the CMr position. Candidate A had experience as a quality control manager, and candidate B had worked as a labor control manager at the

Table 1. Text mining result

Candidate sample	Main Experience	Match Result Score
A	Quality Control Manager	3 D0 x W0 + 0 D1 x W1 = 3 x 1.0 + 0 x 0.9 = 3.0
B	Resource Management	2 D0 x W0 + 2 D1 x W1 = 2 x 1.0 + 2 x 0.9 = 3.8
C	QM & CM	2 D0 x W0 + 3 D1 x W1 = 2 x 1.0 + 3 x 0.9 = 4.7
D	Construction Manager	3 D0 x W0 + 2 D1 x W1 = 3 x 1.0 + 2 x 0.9 = 4.8
E	Construction Manager	4 D0 x W0 + 1 D1 x W1 = 4 x 1.0 + 1 x 0.9 = 4.9
F	Construction Manager	6 D0 x W0 + 2 D1 x W1 = 6 x 1.0 + 2 x 0.9 = 7.8

subcontract level. Candidates C and D were relatively better than A and B, but both résumés did not contain enough description of roles and duties relevant to the CMr position. Candidate E was also better than candidates A and B and a little bit above candidates C and D, but his experiences were mainly related to contract and QS part. The last candidate F, who achieved the best score through text mining, had more experience as a CMr listed on his résumé than the other candidates.

Based on these results, we tentatively concluded that the deployed text-mining method has potential to be used for HRM of the construction industry, especially for allocating candidates to proper positions when the HR manager needs to screen a large amount of résumés in a short amount of time. If the algorism of text mining is amended and elaborated further, the results can be even more accurate and satisfactory.

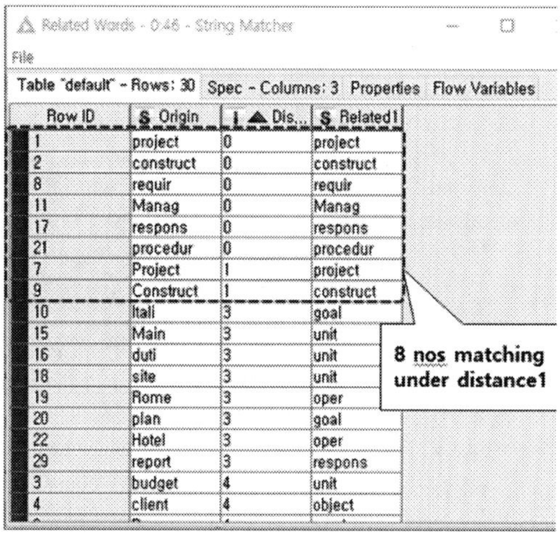

Row ID	S Origin	I Dis...	S Related1
1	project	0	project
2	construct	0	construct
8	requir	0	requir
11	Manag	0	Manag
17	respons	0	respons
21	procedur	0	procedur
7	Project	1	project
9	Construct	1	construct
10	Itali	3	goal
15	Main	3	unit
16	duti	3	unit
18	site	3	unit
19	Rome	3	oper
20	plan	3	goal
22	Hotel	3	oper
29	report	3	respons
3	budget	4	unit
4	client	4	object

Figure 3. Matching results of candidate D in KNIME.

6 Conclusion

This study started with a question regarding the current methods that HR managers in construction companies use. It was also motivated by the struggle to find a better way to manage HR, especially position designation for construction projects for those with different, unique experiences and training. We based our study on the premise that long descriptive résumés or CVs are essential for exactly figuring out a candidates' capabilities and qualifications in view of the roles and responsibilities from his or her career history, but this information has not been used well by construction companies due to the difficulties of management. This study does not claim that a long descriptive résumé is superior to a set of short keywords for HRM, but this was the premise of this study. From this premise, in order to overcome the drawback that the long descriptive résumé cannot be easily controlled and classified in practical level, we introduced the text-mining method that is commonly used to treat large amounts of semantic data and also is used by the HRMs of other industry domains. The purpose of the text-mining process was to compare and match the keywords taken from a candidate's résumé and the job requirements, and the process was implemented by the widely used platform, KNIME. Six candidates' résumés and one document listing the CMr position requirements were used as input data for the text mining. Based on the results, we tentatively conclude that the text-mining method has potential for screening and selecting proper candidates for a certain position by comparing the results with an HR manager's manual judgement, which was similar. The text-mining method, however, only took less than a minute for inputting the data and receiving results, while the HR manager took around 30 minutes—5 to 7 minutes for each resume—for screening and 1 hour to fully review the six résumés. This means that this kind of system could improve current HRM of construction companies when searching for an appropriate person to assign to a position by automatically screening out many inappropriate candidates, at least in regard to time. If an HR manager needs to review a bunch of CV documents, then the time gap between the manual and automatic methods will be far more than even we have expected during the screening stage.

We did encounter several limitations as the research proceeded. The text-mining algorithm was not built perfectly with full consideration of various alternatives and conditions because it was challenging to gather enough résumés that contained long-enough descriptions. In addition, the algorithm was not designed to consider unique characteristics of projects and the importance of a candidate's recent project experience, but rather was focused on key word parsing, extracting, and matching from typical requirements using only the basic algorithm of text mining. However, it has led to the question of whether the text-mining method can be practical in real situations of job positioning and target recruiting of experts. Methods to effectively gauge the level of accuracy of text mining and to determine the proper keyword extraction and matching ratio must be addressed in future studies. In addition, the introduced text-mining platform needs to be validated by comparing it to other platforms. Moreover, this study has only concentrated on manager positions that are directly involved in construction management in a construction company. Thus, it has not been shown to be applicable to the entire domain of HRM in the construction industry. And lastly, this text-mining method might not seem different from existing web-based recruiting systems based on keyword input and matching mechanisms in terms of aim and function, even though the basic concept and the process itself are different.

Although these shortcomings and limitations should be taken into account, this study has made a meaningful step forward as a preliminary study to reach practical use and to extend various domains for improved HRM by applying a text-mining methodology to the construction industry, which encompasses relatively diversified positions that require various training and experiences compared to other industries. Specially, the text-mining method provides an opportunity to dramatically save time compared with the traditional method of reviewing CVs. We hope that this text-mining methodology can be used to more effectively match candidates with the best positions as a very helpful tool for HR managers in construction projects in a shorter amount of time.

References

[1] A. Ekholm, S. Fridqvist, A conceptual framework for classification of construction works, Electronic Journal of Information Technology in Construction, 1 (1996) 1-25.

[2] S. Colucci, T. Di Noia, E. Di Sciascio, F.M. Donini, M. Mongiello, M. Mottola, A formal approach to ontology-based semantic match of skills descriptions, Journal of Universal Computer Science, 9 (2003) 1437-1454.

[3] M. Fazel-Zarandi, M.S. Fox, Semantic matchmaking for job recruitment: an ontology-based hybrid approach, in: Proceedings of the 8th International Semantic Web Conference, 2009.

[4] C.-F. Chien, L.-F. Chen, Data mining to improve personnel selection and enhance human capital: A case study in high-technology industry, Expert Systems with applications, 34 (2008) 280-290.

[5] S. Strohmeier, F. Piazza, Domain driven data mining in human resource management: A review of current research, Expert Systems with Applications, 40 (2013) 2410-2420.

[6] H. Al-Zegaier, H.A. Al-Zu'bi, S. Barakat, Investigating the link between web data mining and strategic human resources planning, Computer and Information Science, 4 (2011) 67.

[7] X. Zhao, An empirical study of data mining in performance evaluation of HRM, in: International Symposium on Intelligent Information Technology Application Workshops, IEEE, 2008, pp. 82-85.

[8] D. Çelik, A. Elçi, An ontology-based information extraction approach for résumés, Pervasive Computing and the Networked World, 7719 (2013) 165-179.

[9] A.L. Paoletti, J. Martinez-Gil, K.-D. Schewe, Extending Knowledge-Based Profile Matching in the Human Resources Domain, in: Database and Expert Systems Applications, Springer, 2015, pp. 21-35.

[10] T. Lau, Y. Sure, Introducing ontology-based skills management at a large insurance company, in: Proceedings of the Modellierung, Citeseer, 2002, pp. 123-134.

[11] I. Zhukova, M. Kultsova, M. Navrotsky, A. Dvoryankin, Intelligent support of decision making in human resource management using case-based reasoning and ontology, in: Knowledge-Based Software Engineering, Springer, 2014, pp. 172-184.

[12] D. Veit, J.P. Müller, M. Schneider, B. Fiehn, Matchmaking for autonomous agents in electronic marketplaces, in: Proceedings of the fifth international conference on Autonomous agents, ACM, 2001, pp. 65-66.

[13] S.T. Ng, R.M. Skitmore, T. Sharma, Towards a human resource information system for Australian construction companies, Engineering Construction and Architectural Management, 8 (2001) 238-249.

[14] JY. Cho, YS. Kim, Improving Construction Management Functions of the Korean Construction Companies through the Reformation of the On-site Organization , Educational Program and Career Management Systems JOURNAL OF THE ARCHITECTURAL INSTITUTE OF KOREA Structure & Construction, 17 (2001) 87-94.

[15] J.S.K. T.H. Kim, Improvement way for job classification of construction company, in, Construction Economy Research Institute of Korea, Korea, Republic of, 2001, pp. 86.

Incorporating BIM into Architectural Precast Concrete Fabrication

Jeffrey Collins
Georgia Institute of Technology
jcollins36@gatech.edu

Abstract

Digital models have many benefits for construction; for example, clash detection prior to approval, ability to confirm assembly sequencing, verification of as-built dimensions, and more. Customizing and preparing such models for project delivery takes a large amount of time and, unfortunately, cannot guarantee that the physical building or part can be built as modeled; a disconnect between design intent and constructability can often occur. One reason for this is that design team is unaware of certain details that are involved in fabricating and assembling building components; an understanding that craftsmen take for granted.

This paper presents the process of incorporating Autodesk Revit into the workflow of an architectural precast concrete manufacturer, tracking a real-world project from design intent model through the incorporation of industry-specific fabrication details. Methods used to generate a novel digital model, as well as to produce shop drawings and shop tickets from that model are documented. Three approaches to creating custom Revit models are described: using parametric Revit families; through Dynamo scripting; and via Excel spreadsheet input. Future potential workflows which could extend the findings even further are discussed, setting the groundwork for "Design-Assist", an emerging project delivery approach wherein subcontractors are engaged early in project development to provide advice to the design team, allowing for more informed decisions, conversations, and digital models for construction.

Keywords

Building Information Modeling, BIM-based workflow; Precast concrete; Parametric modeling; Fabrication; Computer Aided Design

1 Introduction

Building Information Modeling (BIM), the process, and Building Information Models (BIM), the document, have become such a fundamental part of the building of buildings that owners are now requiring BIM processes and BIM documents as part of their contractors' contract. BIM "support[s] design through its phases, allowing better analysis and control than manual process." [1] This modeling approach "contain[s] precise geometry and data needed to support the construction, fabrication, and procurement activities through which the building is realized." [2] The benefits for construction schedule and budget are varied and vast – clash detection prior to approval, ability to confirm assembly sequencing, verification of as-built dimensions, and more. Not too long ago, contractors who took on the additional software and technical personnel expenses that BIM incorporation requires had a competitive advantage to those who could or would not. "Concerns that impede [a more] widespread implementations of BIM are…primarily [due to] a perceived inability of subcontractors to adopt or work with…technology." [3] However, "resistance disappears once the efficiencies accruing to the bottom line are evident." [4] Indeed, over time, the Architectural Engineering and Construction (AEC) industry as well as the perspective of the owners funding their buildings have converted to accept BIM so widely that those contractors who can't or won't evolve their "traditional" processes are in danger of being left behind.

1.1 Precast/Prestressed Concrete Industry

The Precast/Prestressed Concrete Industry (PCI) is one group that has consistently tried to stay ahead of advancing technology. The PCI was "founded in 1954 to advance the design, manufacture and use of precast and prestressed concrete." [5] Initially producing a series of graphic standards for use in hand drafting, and then computer aided drafting, "CADD drawings [were] only readable as graphics, so that information transfers for

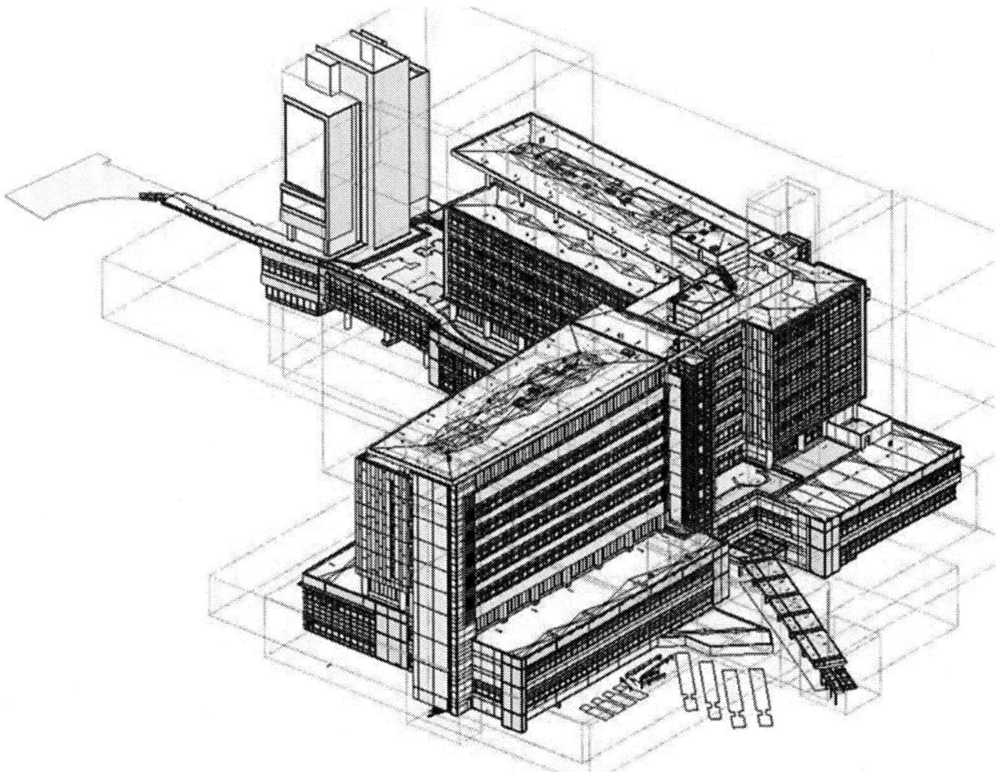

Figure 1. Shands digital model from the design team

process activities such as structural analysis, bills of material, coordination between building systems, quality control, rebar fabrication and piece production, [had to] be done by people." [6] A desire for a common language, procedures, and automation appears to have been the PCI's aim for some time.

Advances in digital technology led to research in parametric definitions of precast pieces [7], developing solution to concerns of software interoperability and data exchanges [8, 9], and improvements in process modeling. [10] Today, BIM software has now developed wherein achieving many goals, from the point of view of many stakeholders, including the precast concrete industry, is within reach.

1.2 Architectural Precast Concrete

Precast concrete, though rather specific in material and method, still includes an enormous catalog of building products. Just look around our built environment every day. Precast concrete is everywhere, used for many purposes, and specified by many professions – structural, civil engineers, architects, and more. One specific category of precast concrete called "architectural precast concrete" is defined by the PCI as "any precast concrete unit employed as an element of architectural design." [11] More particularly, it is assumed that such pieces are critical components of a building skin – they have a high-quality finish and are integral to the overall aesthetic of the design.

Advantages of architectural precast concrete listed in the Architectural Precast Concrete Design Manual include: design freedom, quality control, plasticity (unlimited shape and configuration), economical erection and rapid enclosure, and trade scheduling.

1.3 Castone Corporation

Castone Corporation (Castone), an architectural precast concrete manufacturer, was founded in Opelika, Alabama in 1962 by David M. James, Sr. Mr. James' three sons, David Jr, Michael, and Cooper now lead the company. They carry on their fathers' "deep commitment for serving [their] customers, employees, and communities." [12] Castone manufactures two products: cast stone, or "dry tamp" concrete and architectural precast concrete, or "wet pour" concrete. The focus of this paper is wet pour. Castone has an in-house engineer that provides specifications regarding the structural capacity of their architectural precast panels, including concrete mix, panel thickness, rebar sizing and spacing, and embed and other associated steel hardware design.

The facility in Opelika is housed in two main buildings; one for administration and the other for manufacturing. Administration manages a project from initial RFQ, estimating, contract negotiation, shop drawings, engineering, and through shop tickets production. It is only when shop tickets are completed that the manufacture of pieces required for a project will begin. The manufacturing facility handles each piece from the construction of formwork, assembly of rebar, concrete pouring, curing, and finishing. In addition to the building proper, the facility also has an extensive "yard" for the storage, staging, and loading of pieces onto trucks for shipping to job sites.

A more recent development in the history of the firm is that Castone is a General Contractor for building envelopes, certified by the Architectural Precast Association. This allows the company to offer "pre-construction design consulting, drafting and engineering, custom precast manufacture, in-house testing and Quality Control, masonry, installation, cleaning, and waterproofing of...building skin[s]." [13] One potential implementation of these services, "Design Assist," will be discussed.

In 2014, Castone contacted Georgia Tech's Digital Building Lab to start a conversation about the possibilities of incorporating digital modeling software into their existing workflow. That conversation developed into research of methods of incorporating BIM into architectural precast concrete fabrication. A report of the processes, current state, and results of this ongoing research follows.

2 Project Description and Goals

The building that serves as the case study for this project is a hospital designed by Flad Architects which expands the University of Florida Health Shands complex, located in Gainesville, Florida. The General Contractor for the 500,000 square foot building is Skanska. Construction began in 2014 and is expected to be completed in the fall of 2017. [14, 15]

The primary goal for this project is to aid Castone in building a digital model of the architectural precast concrete pieces to be fabricated for the new Shands building, specifically through the use of the software program Autodesk Revit. Submitting this model is a contractual obligation. On previous projects, Castone had been fulfilling such obligations by paying for the model to be built out of house. While this method fulfills the letter of the contract, other benefits of creating a digital model are squandered. Creating the model in-house with

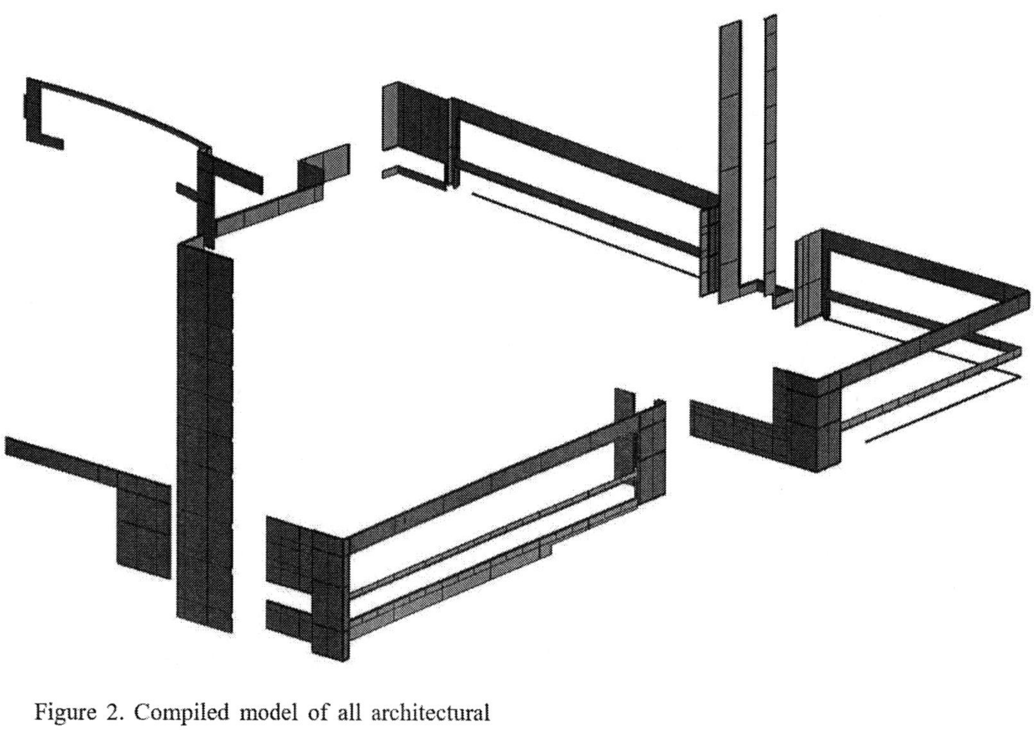

Figure 2. Compiled model of all architectural precast concrete pieces

this project, Castone hopes to be able to use the model for these additional goals:

- Clash detection with building superstructure and other exterior wall assembly trades – during the shop drawing phase, Skanska held weekly virtual coordination meetings
- Use the model to create shop drawings and shotickets – track time savings of production compared to current processes, access the process for potential future application and expansion
- Incorporate BIM into existing process – compare existing and future processes
- Generate families of panels for future use – create a limited set of model families that can capture as many variations of architectural precast concrete pieces as possible
- Use model for material take-offs currently done by hand – all accountable on spreadsheets, time-consuming and error-prone

3 Methods

The first step before starting to build any models or model families is an assessment of the whole building. The design team shared a Revit model with the project manager, Figure 1. Along with printed Construction Documents, the drawings and model are examined by Castone to determine quantity and variety of architectural precast concrete pieces and what parameters or alternatives for each piece will need to be obtainable. Based on the size of their facility, transportations costs, and scheduling, Castone determines that the completed building will require 313 individual pieces of architectural precast concrete.

Figure 3. Piece generated using Revit parameters

3.1 Revit

Each of the piece types is modeled in Revit as a *Family* with parameters. For example, the piece shown in Figure 3 belongs to the "W Family" (W for wall) of models and has a long list of parameters controlled its form. Modeling in this manner allows many different variations and saves modeling time; each time a similar piece with a slight variation is needed one does not need to model completely from scratch, but use the Family model and make adjustments. Each adjusted model can be saved as a "Type" and a catalogue of piece varieties is soon collected. For the Shands project, Castone developed 56 Families.

As each piece is completed, they are assembled together in the *Project File*, Figure 2. When all pieces are present, this compiled model is federated into the General Contractor file for coordination with the building superstructure and other exterior components and materials. Models from Castone and the intended building superstructure are brought together in Navisworks for clash detection and other details of coordination. In one particular example, viewing the federated model revealed a clash of panel embeds with a column. During the coordination phase, the embed could easily be moved. Without the use of the digital model in this way, it would have been far less likely to catch this conflict, leading to costly adjustments in the field long after the pieces had already been manufactured.

The compiled model is also used to generate shop drawings sheets. Shop drawings are "detailed plans prepared by a fabricator." [16] They are based on the Construction Documents – drawings, specifications, and digital model supplied by the design team and construction manager – but are produced by the fabricator to validate the design intent through the additional details required by each specific trade. The shop drawing process anticipates the potential of revisions to the drawings as pieces are coordinated to model the design intent and coordinate with related components and other trades.

On the shop drawing sheets, Castone notes the following timeline as the history of submissions, changes, and approvals:

1.	For approval	4/14/15
2.	Revised for approval	4/27/15
3.	Revised for approval	6/10/15
4.	Embeds for field use	6/24/15
5.	Revised for approval	7/10/15
5.	For field use	10/7/15

Because the shop drawings where generated from a digital model of assembled parametric *Family Types*, adjustments can easily be made to individual pieces through each of these drawing iterations and the shop drawing sheets update automatically.

Following the approval of the July 10, 2015 set of shop drawings, shop ticket production can commence for each of the panels. Shop tickets are different from shop drawings in specificity and detail. Each sheet will show one piece of architectural precast concrete only, as opposed to the shop drawings sheets which show an entire façade composed of many pieces. Shop drawings show the big picture. Shop tickets detail and dimension every single aspect of each architectural precast concrete piece that the fabrication team will need to produce it – the shape, details, rebar, locations and types of various embeds and lift points, and any other requirements.

Castone uses the same Revit model, updated through the coordination and shop drawing approval phase, to generate the shop tickets. This assures consistency of the fabricated panels and the approved design.

5.2 Dynamo

Parallel to the effort Castone made to model the architectural precast concrete pieces in Revit, the author extended the logic which developed the model parameters into Dynamo, an open-source graphical algorithm editor. [17] Dynamo models can be instantiated directly into Revit, thereby also capturing the beneficial abilities of Revit models including ease of coordination and drawing sheet production. Described below, similar parametric qualities of the above Revit models are coded into Dynamo. Five types of parametric features are modeled: Top cap, Turn back, Reveal (front), Notch (back), and Embed locations.

As with the Revit process, the model begins as a rectangular prism (*Cubiod.ByLengths*) to which additional features are successively added.

3.2.1 Top cap

Panels at the tops of walls at Shands have a cap detail. Another *Cuboid.ByLength* is added to the front face of the original rectangular prism. *Integer Sliders* allow flexibility of the cap's height and depth. *Solid.Union* joins the pieces together.

3.2.2 Turn back

A "turn back" is an extension of a wall panel perpendicular and away from the face of the building. This detail allows adjacent materials to be attached to or sealed to the panel. Assuming a rectangular panel, the turn back could occur on any of the four sides (top, bottom, right, or left) or any combination of the four sides. Again, *Cuboid.ByLength* is used to add a rectangular prism, though here additional scripting is required to allow the turnback to be able to occur on any side. *Integer Sliders* allow flexibility of the turn backs thickness and depth. *Solid.Union* joins the pieces together.

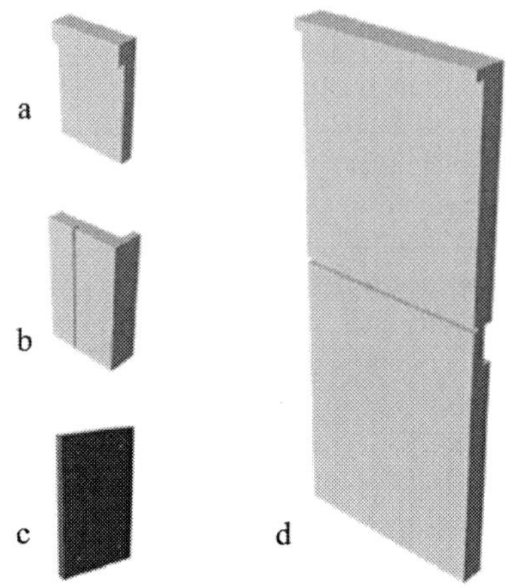

Figure 4. Pieces generated using Dynamo script

3.2.3 Reveal (front)

Perhaps the most "architectural" for its effect on the aesthetic of the overall façade, reveals are linear voids in the surface of a piece. Cuboid.ByLength and Integer Sliders control the thickness and depth of the reveals and Solid.Difference removes that part of the model. An added Dynamo node, Geometry.Translate, is used to allow adjustability of the location of the reveal.

3.2.4 Notch (back)

A back notch would be used to coordinate with the building superstructure, for example, the edge of a structural floor slab or beam. Node scripting is similar to the "Reveal" logic, just selecting the back face instead of the front. Again, thickness, depth, and location of the notch are parametric.

3.2.5 Embed locations

Points can be defined on the panel which serve later as insertion points for other related models, such as embed plates. *Curve.Offset* to offset each of the edges of the panels by a certain distance, thus locating each *Curve.EndPoint*. If more or less than four embeds where required or the offset distance on the sides differed from that at the top or bottom, alternate script logic could be used.

Each of the above described parameters can be combined with one another, used multiple times in each model instance, or not at all. The example shown in Figure 4d shows one possible combination. This model instance is intended to replicate the piece generated in Revit by Castone in Figure 1. This example is shown to demonstrate that the Dynamo model can create any and all of the panels that the previous Revit model could.

Figure 5. Navisworks views of coordination model

3.3 Excel

Many of the above Dynamo nodes work on the premise of (1) defining a parameter and (2) applying a value to that parameter. Once the scripting of nodes has been defined, the model can be flexed by simply adjusting the values. Taken a step further, just as Revit writes spreadsheets of numerical information, Dynamo can read the same from Microsoft Excel. Each Excel "cell" can represent one parametric value of the Dynamo model. When the files are linked, numbers in the spreadsheet can be modified and the Dynamo model updates accordingly. The node *ImportInstance.ByGeometries* can then instantiate the geometry into Revit.

Dynamo reads the Excel file as a list of lists – each column of information is a list of the items in each row of that column. Through successive *List.GetItemAtIndex*, one is able to connect the value in each specific cell to the appropriate Dynamo node that it is meant to control. This process of controlling Revit geometry through Excel and Dynamo is comparable to the process of manipulating the parameters set in Castone's Revit model.

This strategy of "modeling" through Excel, once scripting and actuated, is much quicker and potentially more user-friendly for novice software users. However, creating models this way does not yet offer the continual updating that is described in the Shop Drawing process above. Using Excel and Dynamo, once a model is placed into Revit, it must always be linked to that file if future changes are desired and a copied instance will no longer be parametric.

On the other hand, once a catalog of architectural precast model families is established and organized (discussed below in Section 4.2), these models could easily be linked into and through an Excel file.

4 Results

4.1 Previous work flow

The traditional work flow of architectural precast concrete manufacture, from receipt of Contract Documents through completion of pieces ready for shipping to the construction site, is described here in four stages: Estimating, Shop Drawings, Shop Tickets, and Production.

The estimating stage sets the road map for future stages of the project. The first task is interpreting the design intent and identifying architectural precast parts in the received Contract Documents that will be a part of the completed building. The more involved and critical aspect is translating this list of pieces into a bid. This time consuming process requires experience and knowledge regarding fabrication details, installation sequencing and strategy, production man-hours, and material unit costs. Furthermore, the assumptions made during this stage have lasting implications to the outcome of the completed project, not to mention profitability for the business.

Moreover, all of the work put into the estimating stage does not guarantee that the fabricator will be awarded the contract. If they do move forward with the project, the next stage is shop drawings, discussed in Section 3.1. The shop drawing stages involves implementing all of the assumptions made during the estimate (and related negotiation) phase, based on the described design intent, into drawings for approval by the design team. This iterative process can take several attempts; recall the Shands shop drawing process described previously.

Upon approval of shop drawing, the next stage is generation of shop tickets. This stage translates design intent drawings into drawing for production. Each individual architectural precast piece is carefully planned, considering, to name a few constraints: control joint locations, the maximum bed size in the production facility, the maximum truck size allowed on the roads from the production facility to the construction site, site limitations in terms of installation each panels' internal structural capabilities and weight, which side will be "up" during the concrete pour, whether formwork will be built traditionally or digitally, and more. Completed shop tickets are taken to the production facility for the production of each architectural precast piece.

4.2 Results relative to Castone goals

As described above, Castone was able to use Autodesk Revit to develop a novel digital model for a real-world project and to use that digital model to:

- Coordinate with other exterior wall assembly trades and components
- Explore construction sequencing
- Generate updatable shop drawings
- Generate updatable shop tickets
- Calculate material quantities, for example; concrete volume, number and type of embeds

These factors alone characterize this project as a success for Castone. As this is the first time that Castone is fully incorporating BIM into their work flow, time savings benefits were not yet achieved; there is a large learning curve when integrating any new software. Nath et al [18] describe a similar process of incorporating BIM into a "technologically-enhanced workflow." Their results show "overall productivity improvement ...of approximately 36% for processing time and 38% for total time." *The Rosewood Experiment* [19], submits even more positive results – "production of the same set of general arrangement drawings and shop tickets that required 830 [man hours] in the 2D CAD process was accomplished within 358 [man hours], representing a productivity gain of 57%." Future Castone projects will no doubt capitalize on the teams' experience, improving productivity and efficiency and realizing similar productivity results. Furthermore, as more projects successfully incorporate and profit from BIM, Castone hopes to be able to expand incorporation of digital modeling to other phases such as estimating. If a digital model could be made very quickly, it could be used to improve accuracy in estimates before projects are even contracted. One method for this is to create a limited set of model families for future use that capture as many variations of architectural precast pieces as possible.

4.3 Advanced operability

It is worth noting that the building and, consequently, the architectural precast concrete pieces created for the case study described in this paper are somewhat standard in form. The digital modeling practices and processes described here are intended to be extensible to more complex forms. The overall shape, flatness, or verticality of the panels, texture of the formwork, and incorporation of voids through panels are all aspects of architectural precast concrete currently being explored by various design teams. Future research will explore how these features can be incorporated into the modeling process described above. Discussed below, it is particular interest to study how fabrication details from industry experts like Castone can be incorporated into the design process in order to make such features constructible.

5 Discussion

Project delivery in AEC refers to the handing over of a set of building descriptions from the designers to those who will construct the building. Procedures for how exactly to do this are fast evolving. Furthermore, as modes of designing and modes of representing designs increasingly utilize and rely upon digital technology, the boundary line between who designs and who builds is becoming blurred.

Regardless of the project type or size, "a building...is a complex venture encompassing many tasks and involving many different persons, firms, and organizations in these tasks. These tasks, roles, and responsibilities can be put together in numerous ways to deliver a building project." [20] For example:

> One... way is for the owner to contract with an architect for design and construction documents; when these documents are fully completed they become the basis for soliciting contractor proposals and awarding construction contracts. The architect administers the contract between owner and contractor and, when the work is completed and certified to, the project is occupied and closed out Although there are many options and variations... it is basic and widespread, and over the years, standard contract forms and procedures have evolved to guide its use. [21]

While this Design-Bid-Build process removes much liability from the design team and emphasizes competitive pricing, it does not allow any feedback to the design team on the proposals' feasibility, budget, or schedule. This process does not guarantee that the physical building or part can be built as designed; a disconnect between design intent and constructability can easily occur. One reason for this is that the designer is simply unaware of certain design details that are involved to fabricate building components; an understanding that craftsmen take for granted. This is in part due to above-referenced "standard contract forms and procedures" of the profession; the designer is not required to provide the means and methods of the fabrication, but a general direction for the design intent at which point various contractors fill in the gaps.

"Design intent has to do with the desired outcome, not the means by which it is achieved, however, as any designer knows, the design process can heavily influence that outcome." [22] Coordinating the various contractors, and therefore the built outcome, is the role of the Construction Manager (CM). The CM "use[s] the interaction of their design and construction staff to generate cost effective construction details (constructability)." [23] Though, as George Heery wrote in *Time, Cost, and Architecture*, "many would point out, and with some validity, that the evolution of construction management as a profession, or definable professional service, has taken place within and because of a management void left by the architectural and engineering professions." [24]

The filling of the "management void" by "these [new CM] specialists – often trained as architects or professional engineers themselves – was meant to streamline the building process by making others accountable for aspects of construction, changing costs of materials, managing leads times for materials or equipment and making substitutions, or even challenging design decisions made in the construction documents, but it did little to increase trust between owner, architect and contractor – in fact often it made the building process more contentious." [25]

Heery was aware of this downside as well as the potential effect, or perceived effect, on the design outcome, pondering:

> "[C]an something so intimately related to a building's design as the control of its cost and time of delivery be successfully separated from the design approach? It would seem clear that the two cannot be separated if that control is to be as effective as it can be...Still another view it that when there is strong control of cost and time, the design must suffer...What, then, is the relationship between architectural design and construction management, and what should it be? That relationship would seem to be, basically, that both are indigenous parts of the process that creates architecture, and that they both have to be carried out well and thoroughly intertwined to produce truly successful architecture. [26]

Furthermore, he knew that "the greatest savings in time and cost... can be achieved during the design phase. It is during the design phase that the quantity and quality of the building are established, the systems which will affect construction procedures are selected, and the start time for construction is determined. These activities almost invariably have more control over time and cost than management activities not initiated until bid/award or construction phases." [27] Shown in graphic form, this phenomenon has become known as the "MacLeamy Curve", named after Patrick MacLeamy, HOK's chairman and CEO, who invented it. [28] The graph shows that as design phases progress, teams are decreasingly able to impact cost and functional capabilities through changes to the design because the potential cost of such changes increases. The MacLeary curve depicts a scenario of the "traditional design process" wherein "by the end of Construction Documents, the only option to control cost is to degrade finishes – a terrible option." [29, 30] MacLeamy proposes shifting to an earlier "preferred design process," focusing "more effort to develop and test design alternatives" sooner.

In order to bolster the design teams' confidence in their proposals' proposals' feasibility, budget, and schedule, consultation with construction experts during this "preferred design process" time period is necessary. Several project delivery methods employ this approach, namely; Fast Track (also called Accelerated Scheduling), Design-Build, and Integrated Project Delivery (IPD).

Yet another emerging technique in which design teams and owners are supplementing the design phase with industry knowledge is called "Design Assist," defined by the AIA as "the procurement method by which, prior to completion of design... a contractor provides design assistance to the architect or engineer." [31] Design Assist offers a method for designers to have a similar amount of building control – of the design as well as the construction feasibility, budget, and schedule. Further, while Design Assist is not dependent on BIM, there does appear to be something in the way that BIM has been developed and is used that facilitates Design Assist – "innovation in construction delivery methodology is clearly trending toward collaborative, teamwork approaches." [32] Cavieres et al [33] provide an overview of these and other early conceptual design strategies and develop a case study of parametric templates based on rules and requirements from the concrete masonry industry. Future research will aim towards similar goals for based on industry knowledge of architectural precast concrete.

6 Conclusion

Advances in digital technology have led to a revolution in the way that buildings are documented and constructed. For some time, many AEC teams have been organizing their processes so that the transfer of design documents to contract documents occurs online and via

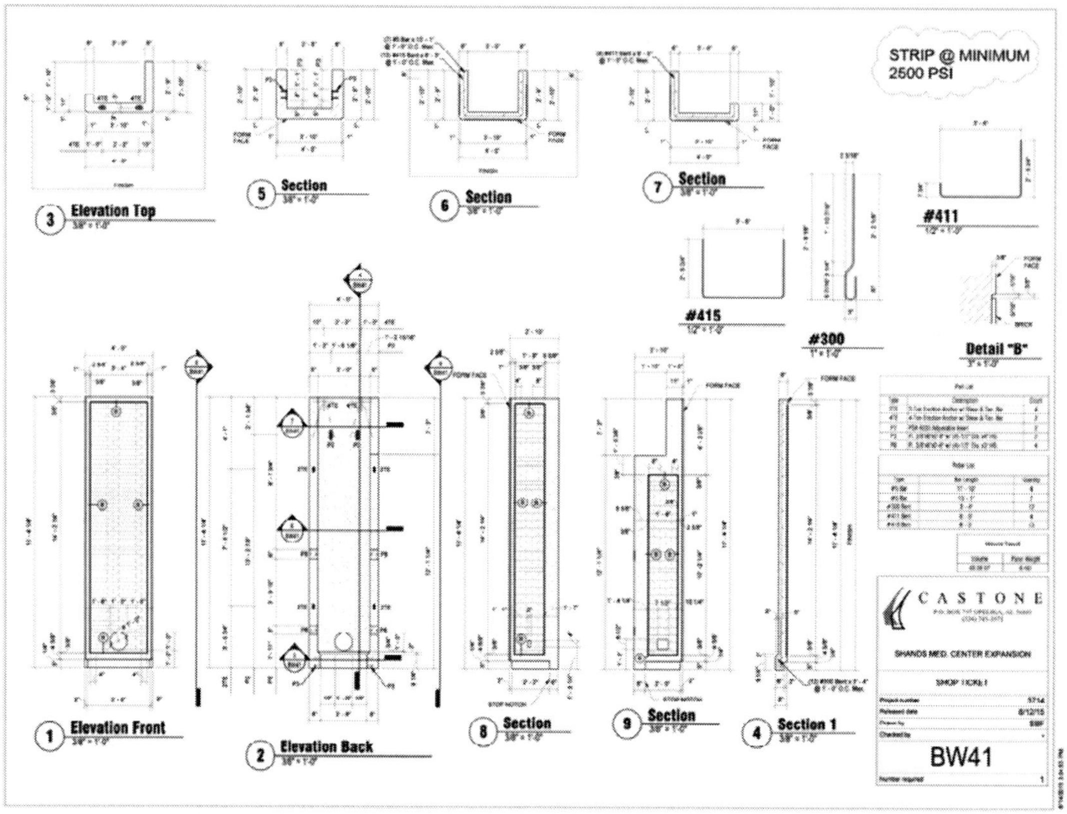

Figure 6. Example Shands shop ticket

digital model. This revolution has huge benefits for construction; for example, clash detection prior to approval, ability to confirm assembly sequencing, verification of as-built dimensions, and more. Castone was able to achieve all of these benefits through their Revit model described in Section 3. Design teams also find many advantages to digital modeling, particularly in the sharing and documenting of architectural representations. Various software programs propagate changes to designs to relevant model views and drawing sheets automatically, reducing the likelihood of errors. Further, digital models can be shared among team members for quicker, easier, and more accurate coordination. However, manipulating and customizing a digital model takes a large amount of time; perhaps equivalent to the time that goes into conceiving the design itself. Furthermore, such manipulation of the digital model does not guarantee that the physical building or part can be built as modeled. Many other factors – constructability, budget, and schedule commitments – may force changes to the design, resulting in time consuming re-modeling.

The potential need to remodel highlights the disconnect between design intent and industry knowledge. BIM offers the opportunity for AEC teams to work across disciplines and through various phases, allowing for more informed design decisions. This is true for "standard" designs such as the Shands buiding, and for more complex designs eluded to in Section 4.3. Engaging trades in conversation early in the design phases gives the design team access to industry knowledge regarding – to use Castone's work as an example – parameters effecting shop tickets, and therefore the outcome of the architectural precast pieces, during the design phases. Working together through emerging project delivery approaches such as Design Assist could imbed design intent models with fabrication knowledge that allows them to mature directly into models for construction, expediting much of the back and forth of traditional work flows discussed in Section 4.1. These steps work in concert to unleash further benefits of BIM, the process and the document.

Acknowledgements

All modeling of architectural precast concrete pieces

in Revit (Figures 2, 3 and 6) was completed by Castone, with some technical advice from the author. Modeling in Dynamo and Dynamo via Excel was completed by the author.

This project would not have been possible without the interest and dedication of Castone Corporation, especially the outlook and knowledge shared by Mo Wright, Stephen Farr, Chris Cox, and Robert Adkins; nor without the support of Georgia Tech's Digital Building Lab, specifically Russell Gentry and Chuck Eastman. Special thanks also to Ratnakar Sonthi at Autodesk.

References

[1] Eastman C, Teicholz R, Sacks R, Liston K. *BIM Handbook: A guide to building information modeling for owners, managers, designers, engineers and contractors*. Wiley. 2011.

[2] Ibid.

[3] SHoP Architects, *Shop: Out of Practice*. The Monacaelli Press. 2012.

[4] Ibid.

[5] Precast/Prestressed Concrete Institute. *Architectural Precast Concrete*. Precast/Prestressed Concrete Institute. 1989.

[6] Ibid.

[7] Sacks R, Eastman C, Lee G. "Process Improvements in Precast Concrete Construction Using Top-Down Parametric 3-D Computer Modeling." *PCI Journal, Vol. 7, Issue 3*. p46-55

[8] Belsky, M., Eastman, C., Sacks, R., Venugopal, M., Aram, S. & Yang, D. "Interoperability for precast concrete", *Journal of Precast Concrete Institute*, 59(2), p144–55.

[9] Jeong, Y.-S., Eastman, C.M., Sack, R., Kaner, I. "Benchmark test for BIM data exchanges of precast concrete." *Automation in Construction 18*. 2009.

[10] Sacks R, Eastman C, Lee G. "Parametric 3D Modeling in Building Construction with Examples from Precast Concrete." *Automation in Construction, Vol. 13, No. 3*. 2004. p291–312

[11] Precast/Prestressed Concrete Institute. Op Cit.

[12] Castone Precast Concrete Panels and Cast Stone Products. Retrieved from website October 8, 2015. http://castonecorp.com/

[13] Ibid.

[14] Skanska contracted for Shands expansion. *The Business Report of North Central Florida*. April 17, 2014. Retrieved from website October 23, 2015. http://gainesvillebizreport.com/skanska-contracted-for-shands-expansion/

[15] Blouin, M. *UF Health Shands board approves funding for new hospital*. Retrieved from website October 23, 2015. https://ufhealth.org/news/2014/uf-health-shands-oard-approves-funding-new-hospital

[16] Allen, E. *Fundamentals of Building Construction: Materials and Methods*. John Wiley & Sons Inc. 1990.

[17] Open source graphical programming for design. Retrieved from website October 23, 2015. http://dynamobim.com/

[18] Nath, Attarzadeh, Tiong, Yu. "Productivity improvement of precast shop drawings generation through BIM-based process re-engineering" *Automation in Construction*. 2015.

[19] Sacks, R., Kaner, I., Eastman, C. & Jeong, Y. "The Rosewood experiment—building information modeling and interoperability for architectural precast facades" *Automation in Construction 19(4)*. p419–32.

[20] Haviland, D. *Project Delivery Approaches: An AIA Guide*. The American Institute of Architects. 1976.

[21] Ibid.

[22] Garber, R. "The C(reative)onstruction Process, Then and Now." *BIM Design*. Wiley. 2014.

[23] Haltenhoff, C. *The Construction Management Project Delivery System: Professional CM*. Michigan Technological University. 1978.

[24] Heery, G. *Time, Cost, and Architecture*. McGraw-Hill Book Company. 1976.

[25] Garber, R. Op Cit.

[26] Heery, G. Op Cit.

[27] Ibid.

[28] Video. Retrieved from website December 1, 2015. http://www.hok.com/thought-leadership/patrick-macleamy-on-the-future-of-the-building-industry/

[29] Ibid.

[30] Light, D. *BIM Implementation - HOK Buildingsmart: BIM As A Business Platform*. Presentation. https://www.thenbs.com/are-you-ready/delegate-presentations/BIM-Implementation_HOK-buildingSMART.pdf

[31] Hart, D. "The Basics of Design-Assist Contracting." *AIA Best Practices*. http://www.aia.org/aiaucmp/groups/secure/documents/document/aiap029127.pdf

[32] Andres, G. "Design-Assist: Getting Contractors Involved Early." *Legal Insight*. September 19, 2012.

[33] Cavieres, A, Gentry, R, Al-Haddad, T. "Knowledge-based parametric tools for concrete masonry walls: Conceptual design and preliminary structural analysis." *Automation in Construction 20* (2011) 716-728.

Automated 3D Image-Based Section Loss Detection for Finite Element Model Updating

Abstract –

Accurate and rapid condition assessment of in-service infrastructure systems is critical for system-wide prioritization decisions. One major assessment consideration is structural section loss due to deterioration, for instance from corrosion. Modern 3D imaging, which generates high-resolution 3D point clouds, is capable of measuring this degradation. Despite growth in the fields of point cloud analysis, few studies have addressed the potential of using such data for creating and updating numerical finite element models of structures. This paper presents a framework for automatic and systematic 3D section loss detection in structural components, followed by a corresponding update to a finite element model. Point cloud data of a targeted structure is obtained by using recently developed Dense Structure from Motion (DSfM) algorithms. Section loss damage is then located and identified by using computer vision techniques. In order to preserve data integrity and resolve localized high fidelity details, direct 3D point cloud comparisons are applied instead of 3D surface reconstruction or curve fitting techniques that limit the accuracy of the structural analysis. An experimental case study validating the developed approach is presented, along with a discussion of potential uses for the analysis framework.

This study aims to prototype a method for fast and reliable detection, structural model updating, and tracking of deterioration in structures, for use by infrastructure managers and engineers. The proposed methodology will enable engineers to use the updated structural model to determine the reserved capacity and remaining service life of structural elements in both in-service structural systems and under severe loading conditions.

Keywords –

3D imaging; Condition assessment; Computer vision; Damage detection; Finite element model; Computational mechanics; Structural damage

1 Introduction

Existing infrastructure systems are faced with serious durability problems due to increasing service demand and natural aging. Current field inspection protocols are subjective, primarily based on visual inspection, and cannot be directly incorporated into existing numerical models of structures. Thus, there is a growing need for accurate, low cost, and consistent inspection and evaluation processes that enable more sustainable decision-making about how to repair or replace infrastructure.

3D imaging techniques are increasingly used in civil engineering applications for making 3D geometric measurements [1-3]. The fast acquisition speed, high accuracy, and portability of these modern remote-sensing tools make them an efficient and economically justifiable alternative to conventional methods. While laser scanning and photogrammetry have been used as efficient 3D measurement techniques for decades, the idea of using the acquired 3D point clouds to generate numerical structural models is relatively new. Despite the fast growth in the fields of point cloud analysis and surface reconstruction algorithms, few studies have addressed the potential of using such data for creating and updating computational models, e.g. finite element (FE) models, for structural analysis. A review of the literature [4,5] in this field not only presents a wide range of the possible applications for 3D imaging in structural engineering, but also highlights the need to develop robust methodologies for creating and updating 3D structural models. Further research is needed into the use of reconstructed 3D models of a damaged structure for updating a finite element model for the strength analysis of structural elements under in-service and severe loading conditions. The current practice of converting 3D point clouds to watertight surfaces using commercially available software in order to generate structural models may fail to preserve highly localized details of the damaged component and also limits the scope of the anticipated structural analysis by imposing a specific type of finite element mesh.

In the current study, image-based 3D reconstruction is used to generate point cloud models of structural elements. Computer vision algorithms are then utilized to automatically locate and identify the damage in the inspected element, and finally, the structure's finite element (FE) model is updated accordingly by using an integrated methodology. The objectives of this study are 1) to evaluate the efficiency of 3D computer vision algorithms in detecting section loss in various structural element shapes, 2) to compare and evaluate the efficiency of surface reconstruction techniques in restoring the geometry and fine details of damaged regions, 3) to use the findings from Objective 1 and 2 to develop a methodology to update the structural 3D model that preserves highly localized damage on structural elements.

2 Methodology Description

The proposed methodology (Figure 1) is composed of several steps that are described in the following subsections. This methodology requires an initial solid model and point cloud data of both initial (as-built) and current conditions of the structure (i.e. undamaged and damaged) as input data. The point cloud data can be obtained by using Dense Structure from Motion (DSfM) algorithms or LIDAR. In absence of as-built condition point cloud data, a 3D solid model of the initial structure (ISM in Figure 1) can be used as a basis for sampling surface points so as to generate a synthetic 3D point cloud model. In such cases, the ISM is created based on structural drawings or the building information model (BIM). The damaged and undamaged state point clouds are then compared on a point-to-point scale to detect structural deformations in the damaged component.

2.1 Data Collection

3D point cloud data of the current condition of the structure (CPC in Figure 1) is collected via a 3D laser scanner or photogrammetry (or a combination of both). Another set of point cloud data for the initial or as-built condition of the structure (IPC in Figure 1) is also generated based on the initial condition or the original structural detail drawings of the structure, respectively. The latter can be done based on the computer-aided design (CAD) model of the structure in order to generate a synthetic point cloud. The goal here is to directly compare the point cloud data of the damaged and undamaged states of the structure on a point-to-point scale and plot a deviation map. This will help to visually represent the visible deformations in the damaged structure, detected by using 3D reconstruction and computer vision methods.

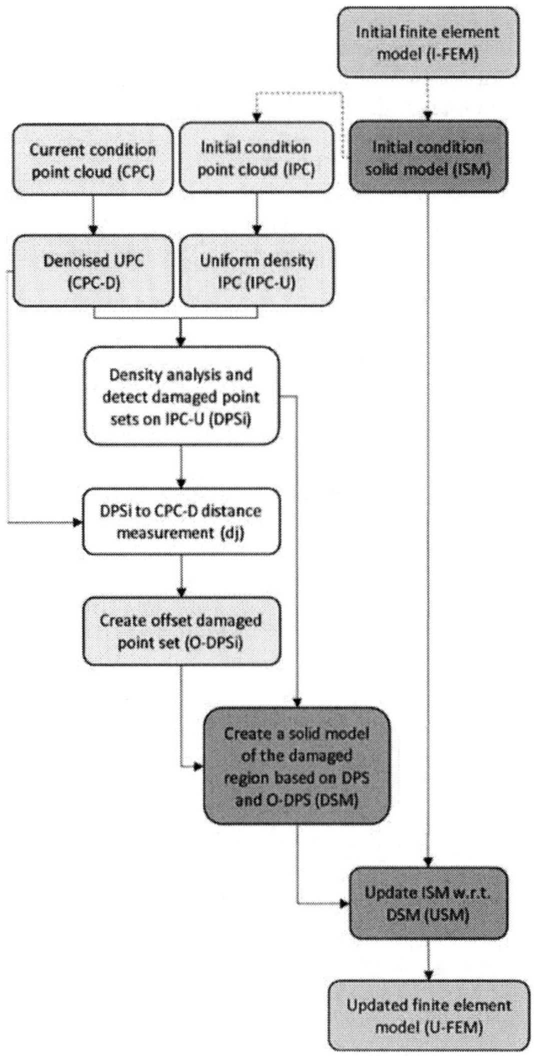

Figure 1. Flowchart of the proposed methodology

2.2 Data Preparation

The generated initial (IPC) and current (CPC) point clouds then undergo pre-processing. The goal is to create a subsampled point cloud from the initial model (IPC-U in Figure 1) in which the points are approximately spaced at an equal distance of r. In other words, the nearest neighbor point for each point in IPC-U is located at a distance equal to or slightly greater than r. One of the most efficient ways to achieve this uniform sampling of the point cloud is by building a 3D voxel grid with the input data and taking the centroid (average point inside a voxel) of the voxel grid as a key point. The number of points in the resulting point cloud is then reduces and down-sampled uniformly. In this paper, the open source software package *CloudCompare* [6] is used for this purpose. The value for r is empirically determined and should be set by considering the desired flaw detection precision, the UPC density, and the IPC surface density. For an initial value, a radius corresponding to 24 points per square unit for the IPC model is suggested. To obtain the local point density ρ at a point **p**, $\rho = k / r^2$, where r is the radius of the enclosing sphere of the k-nearest neighbors of **p**, denoted by the index set N_p, given by:

$$r = \max_{i \in N_p} \|p - p_i\| \quad (1)$$

After determining the r value and generating the IPC-U model, a de-noised version of the current condition point cloud (CPC-D) is created by deleting all the points that have only one or two neighboring points within their specified r distance.

2.3 Density Analysis for Damage Detection

Intuitively, surfaces that are present in IPC-U but are not present in CPC-D correspond to deformed regions or section loss. These regions can be detected by the merging-and-isolating procedure described as follows. First, IPC-U and CPC-D are merged and, consequently, all of the points in IPC-U will find at least one neighboring point in CPC-D with a distance less than r, except for those points corresponding to the damaged regions. By isolating the points with no neighbor within their r distance, point sets corresponding to the damaged surfaces (DPS) can be found. In case of multiple scattered damaged regions, a simple segmentation procedure can be used to break DPS into i distinct separated regions (DPS_i) where each DPS_i represents a deformed surface in IPC, each composed of j points.

2.4 Cloud-to-cloud Distance Analysis for Damage Quantification

Typically, another surface is present in CPC-D for each DPS_i within a damaged region. An exception is a through hole for which there are two DPS_i on two opposite sides of IPC-U. Thus, the distance from each DPS_i to its corresponding surface in CPC-D can be used as a quantification measure for the extent of the damage. There are different definitions and methods to calculate the distance between two point clouds (or point sets). In the current study, a definition based on [7] is used. For each DPS_i, the distance (d_j) from each point to CPC-D is then calculated. Using the method in [7], the *Hausdorff* distance for each point p in a reference cloud S is the distance to the nearest point in the other cloud S':

$$d(p, S') = \min_{p' \in S'} \|p - p'\|_2 \quad (2)$$

The octree-based *C2C* plugin implemented in *CloudCompare* is used in this study to calculate vector d_j for each point in each DPS_i. In addition, the normal is estimated for each point within the point clouds using Principal Component Analysis in a local neighborhood of p as described in [8].

2.5 Reconstruct Damaged Surface

Each damaged surface in CPC-D can be estimated by offsetting the points in each DPS_i by each point's displacement vector d_j. This can be used to create an offset point set (O-DPS_i) for each DPS_i. The implementation of the code for this purpose was done in *Matlab*. At the end of this stage, there are two point sets for each damaged regions: a DPS_i corresponding to the original surface and an O-DPS_i corresponding to the estimated new surface in the current condition (see Figure 4h). If DPS and O-DPS are converted into their corresponding surfaces, the volume encompassed between these two surfaces will correspond to a section loss damage or surface expansion (depending on the direction of vector d_j) in the initial solid model (ISM).

2.6 Create Solid Model of the Damaged Region

The goal of this step is to use O-DPS_i and DPS_i to create a watertight volume for each of the i damaged regions. This can be done by generating a watertight surface (mesh) encompassing the volume created between each O-DPS and DPS. The created watertight surface will be then converted to a solid model. Fundamentally, surface reconstruction algorithms generate a polygonal mesh from a dense point cloud model in order to recover the original surface on which those points lie. The method developed in [9] is used in this study. The basic idea behind this method is to reconstruct triangulated surfaces that are formed by a subcomplex of Delaunay triangulation. The created mesh is then converted into a solid model (DSM) using *Autodesk Inventor* software.

2.7 Update the Initial Solid Model

Prior to updating the FE model, the last step of the

procedure involves subtraction (or addition in the case of expansion) of the solid model of the damage (DSM) from the initial solid model (ISM) via *Boolean* operators. This operation can be performed in several ways. In the current study, *Autodesk Inventor* software is used for this purpose. The final updated solid model (USM in Figure 1) can then be exported into a finite element program for further analysis.

3 Application and Results

For experimental validation, the presented methodology was tested using two specimens. The specimen geometry and material were chosen in order to easily fabricate proof-of-concept specimens and controlled damage patterns. The specimens (shown in Figure 2) included a built-up section made from medium-density fiberboard (MDF) material, and a cardboard tube. The 19 mm thick I-profile specimen represented a structural component with planar surfaces and several controlled flaws were machined using a CNC machine. The 12.7 mm thick tube specimen had an outer diameter of 127 mm and represented a structural component with a curved surface. The damage in the tube specimen included a randomly shaped hole.

In order to create accurate and highly dense 3D point cloud data for the targeted specimens before and after being damaged, an image-based 3D reconstruction technique called Hierarchical Point Cloud Generation (HPCG) [10], an adaptation of the well-known Semi-Global-Matching (SGM) algorithm [11], was employed. One of the key features of this multi-scale 3D reconstruction process is its ability to control the resolution of the point clouds at different locations, which was essential in capturing localized details of the damage in each of these specimens. Throughout this study, a Nikon D-800E (36.3 mega pixels resolution) D-SLR camera with a Nikon AF-S 50mm lens was used to capture images. All images were taken with a sensitivity (ISO) of 200 and an aperture of *f*/8.

The undamaged and damaged point clouds for the I-profile were generated using 165 and 293 images that resulted in dense 3D point clouds with 27,120,412 and 35,610,062 points, respectively. For the tube specimen test, there were 92 images and 2,405,713 points for the undamaged condition and 83 images and 11,655,611 points for the damaged condition. The average local density of points for the damaged specimen point clouds was approximately 600 and 550 points per bounding box volume (cm^3) for the I-profile and tube specimens, respectively.

3.1 I-Profile Specimen

In total, 9 defects of different shapes and sizes (DPS_i) were machined in the I-profile specimen's web. These defects are listed in Table 1 and are shown in Figure 4n. The ISM was created in *AutoCad Inventor* based on the actual dimensions of the specimen. Each defect was modeled separately following the methodology described in the previous section and the ISM was updated with one defect at a time.

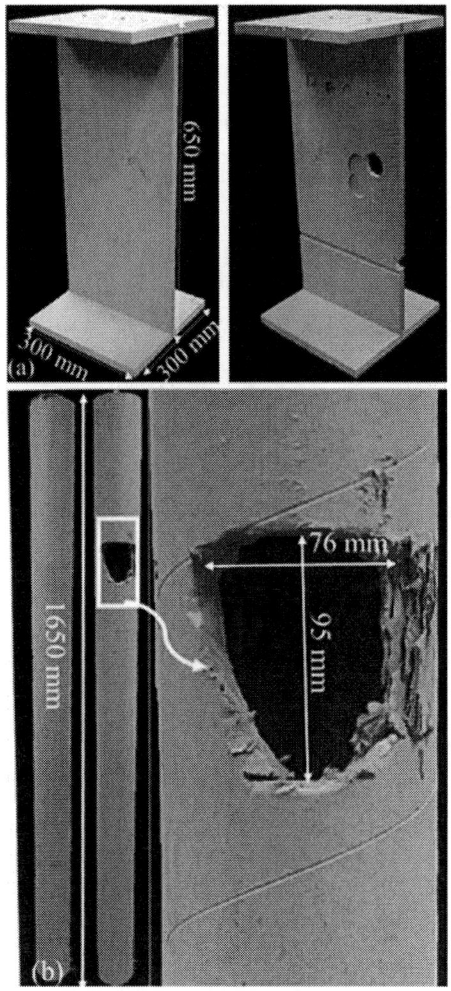

Figure 2: Undamaged and damaged specimens: (a) MDF I-profile; (b) cardboard tube

As can be seen in Table 1 and Figure 4, the first part of the methodology (i.e. the point cloud density analysis) successfully detected all of the 9 defects (DPS in Section 2.3) and the second part of the methodology (i.e. the cloud-to-cloud distance analysis) was able to provide a quantification estimate for 6 of them (O-DPS in Section 2.4). Generally, better damage extent estimations were achieved with an increase in the defect dimensions and volume. For the defects with a minimum dimension of less than 10 mm, it was observed that the captured volumes were systematically lower than the measured values. For instance, the captured defect volume for DPS_9 (5 mm deep transverse groove) was significantly lower than the actual volume.

To further study the correlation between the captured and actual volume of the defects, the point cloud-based measurements are plotted versus the actual measurements in Figure 3. According to this figure, the correlation between the measurements improves for 10,000 mm^3 and larger defects; however, the point cloud-based measurements were significantly lower than the direct measurements for smaller volumes.

Table 1: DPS_i on I-profile web

i	Defect	Depth (mm)/ Area (mm^2)	Actual Volume (mm^3)	Captured Volume (mm^3)
1	Circular hole	19/ 30.66	583	0 (detected)
2	Circular groove	4.1/ 30.66	126	0 (detected)
3	Circular groove	3.1/ 30.66	95	0 (detected)
4	Circular groove	5/ 449	2,242	607
5	Circular groove	10/ 433	4,335	1,163
6	Square groove	3.2/ 351	1,124	133
7	Connected circles	3, 5, 7, 10,19/	98,318	88,490
8	Edge defect	19/ 288	5,472	3,214
9	Transverse groove	5.1/ 7437	37,929	6,920

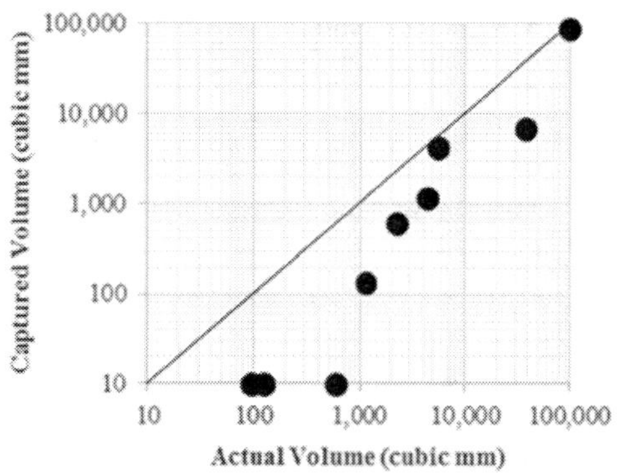

Figure 3: Captured vs. actual defect volumes

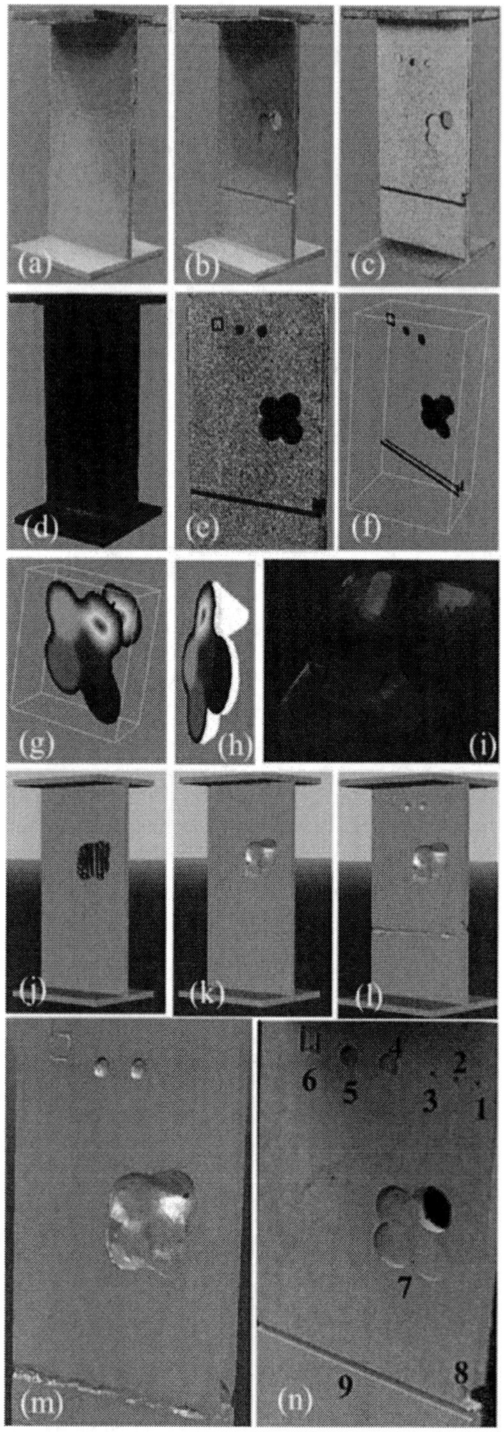

Figure 4: Updated I-section solid model: (a) IPC; (b) CPC; (c) CPC-D; (d) IPC-U; (e) merged IPC-U and CPC-D (web only); (f) detected DPS_i's; (g) calculated d_j for DPS_7; (h) O-DSP_7; (i) DSM_7; (j) ISM and DSM_7 alignment; (k) first ISM update; (l) and (m) USM; (n) DPS_i's

The employed cloud-to-cloud distance measurement technique was found to be the main reason for the systematic underestimation of the defect volumes. As can be seen in Figure 4g, the current distance definition intrinsically results in varying depth estimation (rendered as a heat map) for all of the defects. This leads to an inherently built-in defect damage threshold (here about 10 mm) that can be quantified. Consequently, the volume encompassed between the initial and offset point sets for each defect (DSP and O-DSP, respectively), will be automatically smaller than the real volume (see Figure 4h). This was particularly the case for DSP_7 where the captured volume for the through hole portion of the defect was in the form of a cone rather than a cylinder (Figure 4i and m).

3.2 Tube Specimen

To simulate a field condition assessment of a structural component, the point cloud data for the undamaged and damaged conditions of the tube specimen were collected at different times and light conditions. The undamaged data (IPC) was collected on a sunny day in November 2015 and the damaged data (CPC) was collected on a cloudy day in February 2016. This change in the condition of data collection can be seen in the RGB color information of the point clouds shown in Figure 5a and b. The defect for this specimen was a randomly shaped through hole in the pipe to simulate a burst pipe. The methodology was applied to the collected data and a summary of the results is presented in Figure 5.

Similar to the I-profile specimen, the damaged region on the specimen was detected using the density analysis described in the previous sections. As can be seen in Figure 5d, the cloud-to-cloud (C2C) distance pattern for the detected hole in the tube was similar to the pattern to the through hole in the I-profile (Figure 4g). In other words, the employed C2C distance definition provided a gradual DSM.

4 Finite Element Model Updating

Once updated, the updated solid model (USM) can be exported to a finite element program to study the structural behavior of the damaged component. In this study, the finite element (FE) program *Abaqus* was used to create the FE models of the ISM and USM.

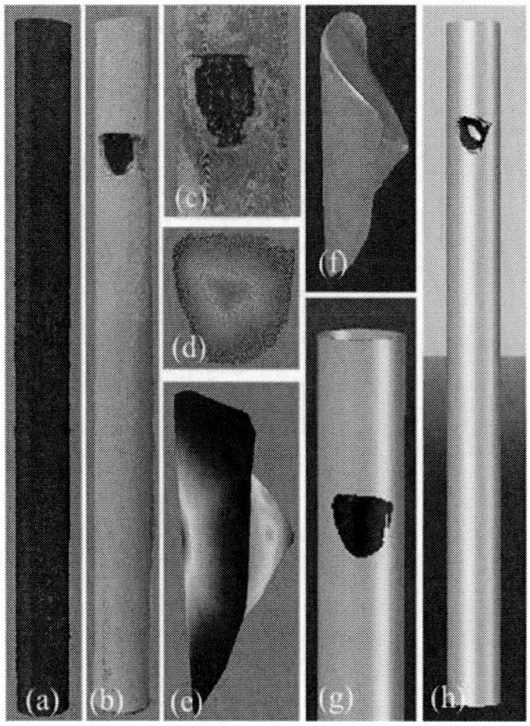

Figure 5: Updated solid model for the tube specimen: (a) IPC; (b) CPC; (c) merged IPC-U and CPC-D; (d) calculated d for DPS; (e) surface reconstruction based on DPS and O-DSP; (f) DSM; (g) ISM and DSM alignment (close up); (h) USM

The initial and updated solid models were exported in the form of an *.igs* file into *Abaqus*. After importing the geometry of the model, other model parameters including material properties, boundary conditions, and loading conditions can be defined for the FE model. Prior to the analysis, however, the model must be discretized into a finite element mesh. Typically, a finer mesh (i.e. smaller element size) is required at locations with complex geometry to capture the induced stress concentration effects. Figure 6 presents the FE mesh for the undamaged and damaged conditions of the tube specimen. As can be seen in this figure, a uniform, structured FE mesh was created for the ISM whereas a rather complex FE mesh was required for the USM to facilitate the FE analysis. In particular, a biased, fine FE meshing scheme was employed around the damaged region (through hole) as can be seen in Figure 6c.

Defining the desired type of analysis (e.g. linear or nonlinear analysis), assigning material type and properties, and defining the boundary and loading conditions will be the final steps in completing the developed point cloud-FE model updating pipeline that was presented in this study.

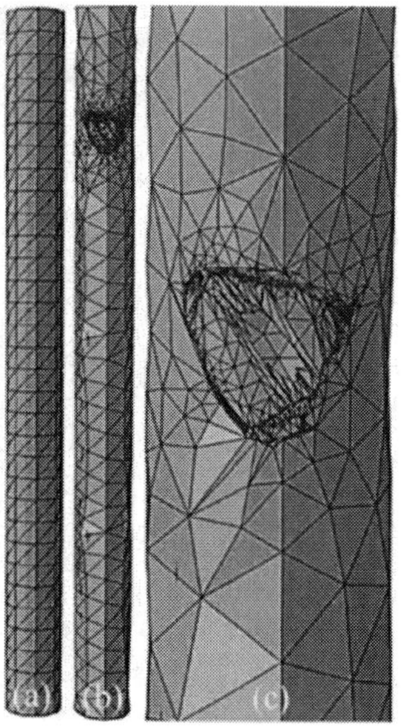

Figure 6: Finite element mesh in *Abaqus*: (a) undamaged model (I-FEM); (b) damaged model (U-FEM); (c) refined mesh in the damaged region

5 Comparison with the Current Practice

The proposed methodology has several advantages over the current practice of using commercial software for surface reconstruction (also called meshing). A number of these advantages are:

- The proposed methodology avoids applying a surface reconstruction scheme on a global scale. Instead, meshing techniques are applied at a local scale that reduces the computation time and cost significantly. For instance, the damaged I-profile specimen was reconstructed using the Screened Poisson surface reconstruction algorithm described in [12] with spatial octree depth equal to 12 and is shown in Figure 7. This model consists of more than 7 million triangles whereas the USM shown in Figure 4l had less than 15,000 triangles. The key reason for this substantial difference is that most of the triangles in the reconstructed ISM were used to model essentially flat and flawless parts of the specimen (Figure 7). The inevitably large number of triangles used in the reconstructed model makes it very difficult to perform the subsequent structural analysis. Still, the final reconstructed model will fail in preserving some basic features of ISM such as the flat surfaces and sharp edge boundaries, which due to the intrinsic definitions of these methods, will be transformed into wavy surfaces.
- Reducing the number of triangles (decimation) and/ or smoothing the surfaces in the reconstructed model will result in the loss of important features that were already captured in the point cloud data. Using a point cloud-based approach, such as the one introduced in this paper, helps to accurately represent the localized details of the model through the model updating procedure.
- It is not easy to further update a reconstructed model based on the subsequent point cloud data obtained from future field inspections. Considering that civil structures are inspected on a regular basis, a new reconstructed model is required for each inspection. In the proposed methodology, however, the model (USM) is updated one defect at a time and, thus, damage propagation as well as new damage can be easily accommodated and tracked based on new inspection data.

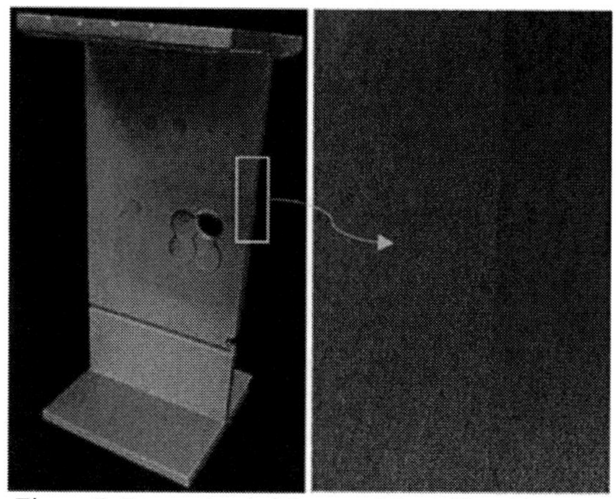

Figure 7: Reconstructed damaged I-profile with more than 35 million vertices and 7 million triangles

6 Conclusions and Future Work

In this paper, a methodology to update the finite element model of a damaged structural component based on comparative point cloud analysis was introduced. The methodology was successfully used to detect and quantify the extent of different types of damage and to then update the solid model of the member, for further finite element analysis. The proposed point cloud-based methodology has a number of advantages over the current practices, which are mainly based on localized surface reconstruction approach taken. These advantages include a significantly lower computation time and cost, the ability to preserve highly localized details of the original point cloud data, providing quantitative metrics for the extent of damage, and the capability to further update the model based on future inspection data and for future referencing.

Based on the measurements conducted in this study, the following observations and consequent recommendations are made, with regard to FE model updating based on point cloud data:

- Good agreement was observed between the point cloud-based and direct measurements when all of the dimensions for the defect were about 10 mm or larger. The results generally improved as the magnitude of the measured dimensions increased.
- The methodology in its current form is capable of modeling section loss and corrosion expansion damage.
- The employed cloud-to-cloud distance definition caused a systematic underestimation of defect size and volume.
- A distinct distance deviation pattern was observed for the through hole defects that resulted in a cone-shaped defect rather than a cylindrical shape one.

This study was part of an on-going research program and several aspects of the presented methodology are being considered for further improvement. These aspects include:

- More advanced and robust direct cloud-to-cloud comparison techniques such as the Multiscale Model to Model Cloud Comparison (M3C2) proposed in [13].
- Using Non-Uniform Rational Basis Spline (NURBS) to create a DSM instead of using surface reconstruction methods.
- Evaluating and incorporating different registration, alignment, and de-noising algorithms to facilitate automatic updating of the FE model.
- Designing and fabricating large-scale specimens to assess the performance of the proposed methodology on large-scale specimens and possibly full-sized structural components.

This study also shows that 3D vision-based techniques offer an appropriate and practical means for structural inspection because of their portability, the relative ease of use, and the ability to archive the data for future referencing.

References

[1] Fathi H., Dai F., Lourakis M. Automated as-built 3D reconstruction of civil infrastructure using computer vision: achievements, opportunities, and challenges, *Advanced Engineering Informatics*, 29(2):149-161, 2015.

[2] Koch C., Paal S., Rashidi A., Zhu, Z., König M., Brilakis I. Achievements and challenges in machine vision-based inspection of large concrete structures. *Advances in Structural Engineering*, 17(3):303-318, 2015.

[3] Koch C., Georgieva K., Kasireddy V., Akinci B., Fieguth P. A review on computer vision based defect detection and condition assessment of concrete and asphalt civil infrastructure. *Advanced Engineering Informatics*, 29(2):196-210, 2015.

[4] Barbieri G., Biolzi L., Bocciarelli M., Fregonese L., Frigeri A. Assessing the seismic vulnerability of a historical building. *Engineering Structures* 57:523-535, 2013.

[5] Conde-Carnero B., Riveiro B., Arias P., Caamaño J.C. Exploitation of Geometric Data provided by Laser Scanning to Create FEM Structural Models of Bridges. *Journal of Performance of Constructed Facilities*, 04015053, 2015.

[6] Girardeau-Montaut, D. Cloud Compare: 3D point cloud and mesh processing software, open-source project. On-line: http://www.danielgm.net/cc, Accessed: 04/03/2016.

[7] Girardeau-Montaut D., Roux M., Marc R., Thibault G. Change detection on points cloud data acquired with a ground laser scanner. *International Archives of Photogrammetry, Remote Sensing and Spatial Information Sciences*, 36(part 3) W19, 2005.

[8] Hoppe H., DeRose T., Duchamp T., McDonald J., Stuetzle W. *Surface reconstruction from unorganized points*, 26(2):71-78, ACM, 1992.

[9] Cazals F. and Giesen J. Delaunay triangulation based surface reconstruction. In *Effective computational geometry for curves and surfaces*, pages 231-276. Springer Berlin Heidelberg, 2006.

[10] Khaloo A. and Lattanzi D. Extracting structural models through computer vision, In *Proceedings of the Structures Congress 2015*, pages 538–548, Portland, USA, 2015.

[11] Hirschmuller H. Stereo processing by semiglobal matching and mutual information, *IEEE Transactions on Pattern Analysis and Machine Intelligence*, 30(2):328-341, 2008.

[12] Kazhdan, M. and Hoppe H. Screened poisson surface reconstruction. *ACM Transactions on Graphics (TOG)*, 32(3), 29, 2013.

[13] Lague D., Brodu N., Leroux J. Accurate 3D comparison of complex topography with terrestrial laser scanner: Application to the Rangitikei canyon (N-Z), *ISPRS Journal of Photogrammetry and Remote Sensing*, 82:10-26, 2013.

Automation of modular assembly of structural frames for buildings

Benny Raphael[1], K Sai Chowdeswara Rao[2], Koshy Varghese[3]

[1]Associate Professor, [2]Under graduate student, [3]Professor,
Department of Civil Engineering, IIT Madras, Chennai, India.
E-mail: benny@iitm.ac.in, saichowdeswararao@gmail.com, koshy@iitm.ac.in.

Abstract:
On congested sites in many cities around the world, there is no space available for installing heavy tower cranes and the approach roads are narrow. On such sites, it is desirable to have an automated construction system that can assemble the required components without the use of heavy lifting equipments. This paper introduces a new automation technique which aims to reduce the usage of heavy equipment's in construction projects. The construction is done using automated lifting and positioning systems at the ground level which help in completing the project cheaper and faster. The concept proposed here is a top to bottom construction method using modules of standard sizes. A semi-automatic demonstration prototype of the concept has been implemented in a laboratory. In order to evaluate the potential for saving construction time, experiments were conducted to estimate the time for various operations and these were projected to estimate the time for a typical residential building. It is shown that the structural frame of small buildings can be constructed in a short time, with significantly reduced labor and at low cost. The experimental results show that the total time taken by different people to complete the activities are very consistent and the use of skilled manpower is not essential. With further automation, the time is expected to be reduced considerably.

Keywords –
Automation, Modules, Assembly, Optimization, Automated Construction.

1 Introduction

Automation techniques have been used for both large and small scale construction. An example of a large scale system using heavy machinery is the construction factory system described by Ikeda and Harada [15]. In this system, a factory like structure that surrounds a building under construction includes various robotic systems to automate construction. Very few construction factory systems have been implemented in real construction projects [15]. Another example is the Shimizu Manufacturing system by advanced Robotic Technology (SMART) which is another fully automated construction system for use in high rise buildings [10]. The SMART system was developed until its third generation but its deployment was limited because the system could be used only for rectangular shaped buildings and entire system weighs 1200 tons. It takes more than 2 months for installation and is reported to have low productivity [3]. Most of the above automation techniques rely on heavy machinery like tower cranes. Many authors have reported automation using Tower Cranes [8] [9] [12] to increase the productivity [11] [13]. Even though the construction time can be reduced with these heavy equipment's, many accidents during their use have been reported [1] [2] [14]. There are also several examples of the use of small robots in buildings and construction. A robot used to construct a masonry wall is reported in [7]. A wall climbing robot is described in [16]. Robots have been used to fix glass in buildings. These robots enhance safety by eliminating the need for human beings to do these works [18], but the productivity of these robots are less and cannot be used for large scale construction. Many of these robots are heavy and it is difficult to move them to greater elevations.
Even though, many robotic applications are found in the literature, robotic systems that are appropriate for fast construction of small buildings (up to three floors) could not be found. On congested sites in Indian cities there is no space available for installing heavy tower cranes and the approach roads are narrow. In such sites, it is desirable to have an automated construction system that can assemble the required components without the use of heavy lifting equipment's.

This paper introduces a new automation technique which aims to reduce the usage of heavy equipment's in construction projects. The construction is done using small robots on ground which helps in completing the project cheaper and faster. A demonstration prototype has been implemented using timber modules and the assembly system is controlled by Arduino microcontrollers and sensors. In order to bring out the potential for saving construction time, experiments are conducted to estimate the time for various operations and these are projected to estimate the time for a typical residential building

2 Construction Technique

The usual construction procedure for small residential buildings is bottom up - the ground floor is constructed first, then upper floors are constructed one after the other. As the height increases, the construction cost escalates because of the need to lift materials. In the case of small scale construction, large amount of manual effort is wasted in carrying material to the top floors. The risk of construction at large heights is higher and there are frequent accidents [1].

The idea proposed here is to construct the top floor first and then lift the top part in small steps in order to assemble the modules for lower floors. This will help in performing all the construction activities at the ground floor. This will also support automation since the assembly system can be permanently installed on the ground. These are controlled by using sensors and microcontrollers.

The structure is constructed using small modules that can be easily assembled. For the final automation system, modules will be made of steel bars. Interlocking mechanisms are used to make automation simpler since it requires simple robotic movements for pushing parts and connecting them. However, this requires the parts to be fabricated with much stringent tolerance requirements. In the initial demonstration prototype timber modules have been used because of the ease of fabrication to the required tolerance.

The construction scheme is illustrated in Figure 1. In Step 1, the structural frame for the roof is assembled at the ground level, supported on a series of lifting platforms (Figure 1a). For simplicity, only two sides of the frame are shown in the figure (figure 1a) and only three supports S1, S2 and S3 are shown. These supports are located below the modules M1, M2 and M3, where structural columns will be inserted. In Step 2, all the supports are lifted up by 400 mm which is the height of one module (Figure 1b). In Step 3, the first support is lowered, leaving the frame to be supported by all the remaining lifting platforms. In Step 4, the next module is inserted under M1 (Figure 1c). In Step 5, support S1 is lifted up until the load is transferred to the lifting platform (Figure 1d). Then the next support S2 is lowered (Figure 1e). The next module is inserted under M2 and the process is repeated for all the supports. After the second layer of modules is inserted at all the support locations, the whole frame is lifted up by another 400 mm and the process is continued (Figure 1f).

Figure 1: Construction Scheme

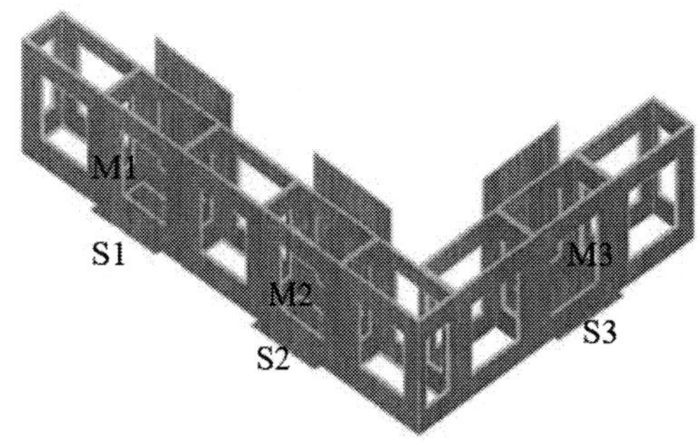

Figure 1a: The top layer of modules are assembled on the ground

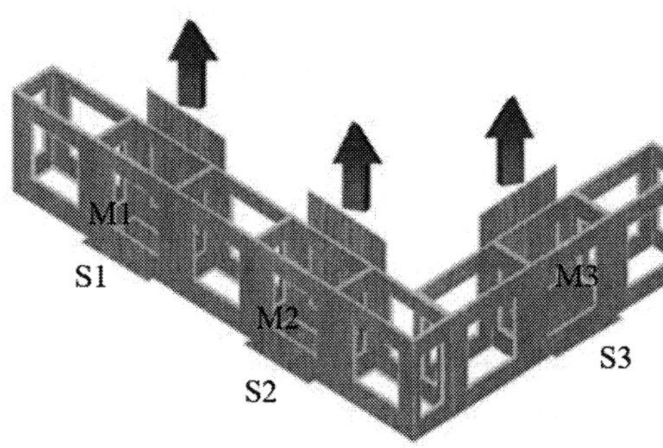

Figure 1b: The assembled layer is lifted up by 400 mm

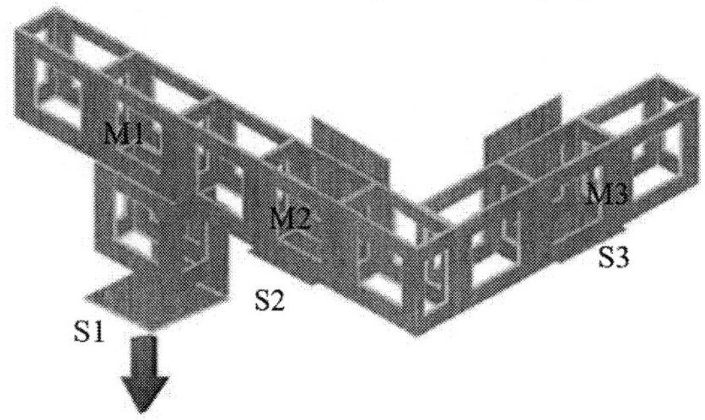

Figure 1c: The lifting platform of the first support is lowered and a new module is inserted underneath

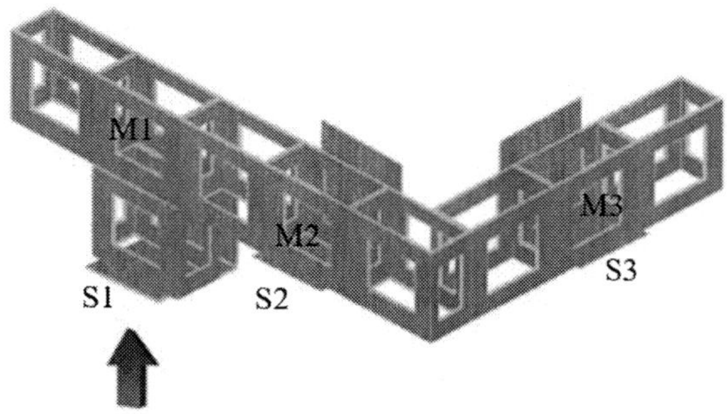

Figure 1d: The lifting platform of the first platform is raised till it takes the load

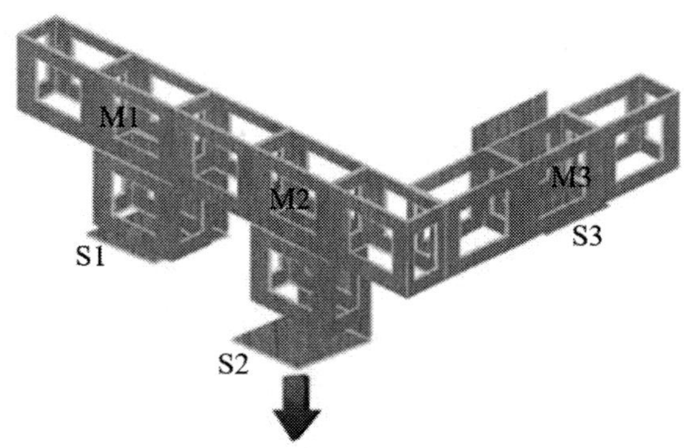

Figure 1e: The lifting base for the second support is lowered and a new module is inserted underneath

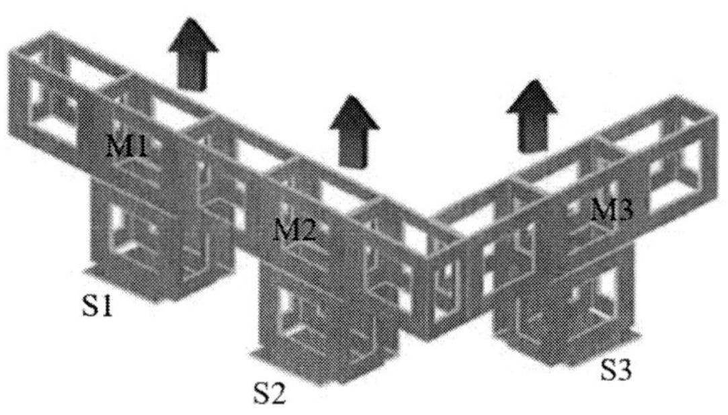

Figure 1f: The process is repeated for all the supports and then the whole frame is lifted by another 400 mm height

3 Initial prototype

The initial prototype is constructed using timber modules and once the feasibility is established, the system will be adapted to use steel modules. The automation system consists of five important parts. They are:

- Frame modules
- Interlocking mechanism
- Lifting system
- Alignment robots
- Assembly robots

3.1 Frame Modules

These are the building blocks of the structural frame and are of standard size. These are prefabricated in factories or workshops under controlled conditions where the required tolerance can be achieved. The strength of the structural frame depends on the strength of the frame modules and the connections. Detailed design of these modules will be performed for buildings of standard dimensions before the next prototype is built. For the initial prototype, predominantly constructability issues are investigated. The wooden modules that are currently used have not been designed to take the full load expected for the building. The geometry of the frame module fabricated for the initial prototype using timber is shown in Figure 2. The dimensions of the module are 400 mm length, 200 mm width and 400 mm height and are made with 20 mm thick timber pieces. The structural frame constructed with these modules are expected to be concealed within the walls of standard thickness used in residential buildings.

Figure 2: Module made of Timber

3.2 Interlocking Mechanism

Interlocking mechanism for connecting modules was designed to make the assembly easier through the use of robots. The "lock" used for interconnecting the timber frame modules is shown in Figure 3. The lock consists of a flat base (flange) with two projections (web) meant to tightly hold together the bars of adjacent frame modules. Pins are designed to hold the locks in position, however, in the initial prototype, the web of the lock is directly screwed to the frame modules. Four frame modules connected using locks are shown in Figure 4.

Figure 3: Lock for interlocking. The space between the flanges of the lock hold together the web of adjacent modules.

Figure 4: Assembled Modules with interlocks; 1. Module; 2. Lock

3.3 Lifting System

The purpose of the lifting system is to raise the partially assembled frame to the next level so that the next layer of modules could be connected under it. Lifting could be done using hydraulic jacks, linear actuators or hoists. Since the height of the modules is 400 mm, jacks should have a stroke length of at least 400 mm. Jacks with 400 mm stroke length turned out to be more expensive and heavier than motorised hoists. Hence light weight hoists were used in the prototype. Lifting is done using multiple hoists which operate in a coordinated manner. These are placed at regular intervals along the walls to be constructed, within the building foot print as shown in Figure 5. The hoists are supported on a metal platform of height 1.6 m and the platform is designed to ensure stability during the lifting process. Currently six hoists of 2 Tonne capacity are used to construct a wall enclosing an area of 2 m X 1.6 m. Each hoist weighs about 15 kg. Two hoists are installed in the front, two at the back and two on the sides. Their positions have been decided such that the assembled structure is stable with support from five hoists at any time, while the sixth one is being lowered.
During lifting, all the hoists support the currently assembled frame through a lifting base pulled by the steel rope of the hoist. After the frame is lifted to a height equal to the height of the module, one hoist is lowered in order to insert the next module below. After the next module is inserted, the hoist is raised until the load is fully supported at this point. Then the next hoist is lowered and the process is repeated.
The steel rope of the hoist is connected to a lifting base by a metal hook. The lifting base is guided through C-channels on rollers which help in moving the module vertically. The guides in the C-channel help in resisting the horizontal movement of the module and providing moment reaction to the eccentric load. A load cell is placed on the lifting base which helps in detecting the contact between modules and the lifting system and also measures the load taken by each lifting base. The lifting base also has a height sensor. Height sensor is used to control the height to which the module is lifted. These sensors are connected to a microcontroller which controls the actuator which consists of contactors and relays that provide power to the motor of the hoist.
All the actuators are simultaneously operated to lift the structure to a height which is sufficient enough to assemble the next layer of modules below. Then one of the hoists is lowered leaving the structure supported on five remaining hoists. One module is added below the released hoist. This hoist is then operated such that the support touches the newly added module. This contact can be detected with the help of load cell. Then the next hoist is lowered and the process is repeated until all the

modules are added below. Then the scheme is repeated till the structure is constructed.

According to the construction scheme, the beams along the roof of the building are assembled first. This forms the first layer of modules as shown in Figure 6. Then the columns supporting the beams are assembled below by inserting the modules one at a time. After the frame is raised to a height above the platform supporting the hoists (1.6 m), the roof slab can be constructed supported on purlins and rafters. Similarly other floor slabs can be constructed at the appropriate stages of lifting. The walls are constructed either immediately after the floor slab is completed or after the entire structural frame is constructed.

Figure 5: Lifting module and Actuator; 1. Metal Frame; 2. Actuator; 3. Steel Rope; 4. Metal Hook; 5. Lifting Module; 6. Load cell sensor

Figure 6: Two levels of Assembled modules; 1. Level 1; 2. Level 2 Modules connected to Level 1

3.4 Alignment robots

The Robot transports the modules and places them precisely under the existing assembled frame. This robot has three degrees of motion. It moves horizontally and vertically for aligning the new module with the existing frame. The robot designed for the initial prototype is shown in Figure 5 and is currently remotely controlled. Current challenges include, installing sensors for aligning the module and controlling its motion automatically. These challenges might be overcome by using image processing sensors. Future work also involves automatically inserting the locks and pins. Currently the modules are manually screwed to the existing structure.

Figure 7: Robot for aligning modules; 1. Robot; 2. Module; 3. Lifting Module

3.5 Assembly robots

Connecting the modules require aligning them precisely side by side and pushing the locks such that adjacent modules are tightly connected. Then pins have to be pushed into the locks to keep them in position. These operations are expected to be challenging because of the uncertainties involved in the fabricated modules. This is topic of future work.

Currently the connections are done manually through screws; work is ongoing to automate this task. Once a layer of modules is assembled, it is lifted to the next level using hoists. Height sensors and load cells are used to control the lifting process. The next layer of modules are assembled under the lifted structure. Custom made robots are used to transport the modules to the required location and align them under the existing assembly. The parts and operations of the initial prototype are described in detail in the next Section.

3.6 Control system

The lifting mechanism is controlled by Arduino controllers. Sensors measure the height through which the lifting device is lifted. Load cells are used to detect contact and the magnitude of the load carried by each support. The output channels of the controller is connected to relays which start and stop the motor in the desired directions.
While the assembly is being lifted, the readings from the height sensor are recorded continuously. The controller stops the motors by sending commands to the relays when the lifting base reaches the required height. Similarly after a new module has been added and the lifting base is raised, the load cell reading is used to detect the contact between the lifting base and the module. When the module is placed on the lifting base and the load cell reading confirms that the load has been correctly transferred to the lifting base, the controller stops the motor of the hoist.

4 System Evaluation

The short term objective of present work is to evaluate the feasibility of the scheme and to establish the potential for saving construction time. The feasibility was demonstrated by constructing the structural frame for a test room of area 2 m x 1.6 m and 2 m height using timber modules.
An experiment was conducted to obtain a realistic estimate of the construction time. Different operations involved in the construction of the test room with the help of 2 workers were repeated several times and the time was recorded.
The experimental procedure involved the following steps:
1. The metal platform supporting the six hoists is installed. The frames for the platform were fabricated in a welding shop and assembled on site with screws. This step was done only once and the time taken for this was noted to be 15 minutes.
2. The initial layer of modules were connected to form the roof beams as shown in Figure 8. This step was also not repeated and it took 45 minutes.
3. The assembled structure consisting of the initial layer of modules is lifted to a height of 0.45 m by the hoists.
4. The lifting base where the next module is to be added is brought down (starting with the first hoist).
5. The spot for loading the timber frame modules is fixed nearly 1 m from the lifting base. The frame module is brought to the lifting base with the robot and the module is placed on the lifting base. The time taken by the robot to reach each lifting base varies since the distances were different. So the average time is taken.
6. The frame module brought by the robot is connected to the module vertically above in the assembled frame. The connection is made manually with screws using a mechanical drill.
7. The lifting base is raised until it touches the newly added frame module.
8. Steps 4-7 are repeated for each column.
9. Steps 4-8 are repeated to fill the next 4 layers resulting in a height of 2 m.
The total time for the completed procedure is calculated by summing up the time taken for all the activities. The short duration activities that are repeated many times during the construction are estimated more accurately by repeating the experiment with different people. The activities that are repeated to estimate the time are shown in Table 1.

Table 1: Activities considered for calculating total time

S.No	Activity
1	Time taken to lift the structure (Step 3)
2	Time taken to lower the first lifting base (Step 4)
3	Time taken to transport a new module using the robot (Step 5)
4	Time taken to connect the module to the frame (Step 6)
5	Time taken to raise the lifting base and make contact with the new module (Step 7)

The experiment was repeated with seven different people who are male college going students who have limited experience with mechanical tools. The total time taken by each person was recorded and is shown in Table 2.

Table 2: Experimental values

S.No	Experiment	Total time calculated
1	Experiment 1	4635 sec
2	Experiment 2	4575 sec
3	Experiment 3	4935 sec
4	Experiment 4	4305 sec
5	Experiment 5	4665 sec
6	Experiment 6	4981 sec
7	Experiment 7	4245 sec

The average time taken to complete the activities in the experiment was 4620 sec (77 minutes). When the time taken by each activity was analysed, it was found that connection and alignment takes most of the total time. Therefore, further automation of these activities are necessary.

The estimated time for activities can be projected to estimate the time for constructing a complete building frame and floor slabs. For illustration, a 3 storey building of plan area (8 m x 8 m) height 12 m is considered. The structural elements consisting of beams, columns, slabs and roof are included in the estimate. The typical time taken using traditional RCC construction was obtained from a local contractor. The time taken and number of labourers required by traditional construction and proposed mechanism are given in Table 3. These estimates show significant reduction in construction time using the automated procedure.

Table 3: Comparison between Traditional construction and proposed mechanism

	Proposed Automation Model	Typical Construction
Time of Construction	30 Hours	45 Days
Number of Labours	2	10

In the real time construction, the building frame is made of steel modules. The steel components are welded into the module shape in a pre-fabrication site. These modules can be directly used in the construction of the building frame, same as the timber modules. The modules can be connected by welding or by connecting them with steel locks using bolts. The likely weight of each module is around 3.9 Kg and the weight of each steel lock is around 2 Kg. The same proposed scheme of mechanism can be used to build the building frame. The above illustrated 3 storey building of plan area (8 m x 8 m) height 12 m require around 1248 modules and 2496 steel locks. The total weight of structure will be around 9859.2 Kg.

Because of the scaling of the modules the scheme of the mechanism will not be changed, only the number of lifting bases required will only change depending on the scale of construction.

Figure 8: Assembled layer of modules

5 Concluding Remarks

The proposed automation system can easily be dismantled into small components and can be transported by one or two people in small vehicles. The cost of the system is very less when compared to existing automation equipment's. It has been demonstrated that the structural frame of small buildings can be constructed in a short time, with less manual labour and at low cost. The experimental results show that the total time taken by different people to complete the activities are very consistent and the use of skilled manpower is not essential.

6 References

[1] H.H.Kim, G.Lee, A quantitative analysis of fatal accidents related to cranes using the FMEA method, Journal of Korean Institute of Building Construction 7(3):115-122, 2007.

[2] G.Mistikoglu, I.H.Gerek, Ercan Erdis, P.E.M.Usman, Hulya Cakan, E.E.Kazan, Decision tree analysis of construction fall accidents involving roofers, Expert Systems with Applications 42: 2256-2263, 2015.

[3] Baeksuk Chu, Kyoungmo Jung, Myo-Taeg Lim,

Daehie Hong, Robot based construction automation: An application to steel beam assembly (Part I), Automation in Construction 32: 46-61, 2013.

[4] E.Gambao, C.Balaguer, F.Gebhart, Robot assembly system for computer-integrated construction, Automation in Construction 9: 479-487, 2000.

[5] Ger Maas, Frans van Gassel, The influence of automation and robotics on the performance construction, Automation in construction 14: 435-441, 2005.

[6] Pentti Vaha, Tapio Heikkila, Pekka Kilpelainen, Markku Jarviluoma, Ernesto gambao, Extending automation of building construction - Survey on potential sensor technologies and robotic applications, Automation in construction 36:168-178, 2013.

[7] T.Bock, D.Stricker, J.Fliedner, T.Huynh, Automatic generation of the controlling system for a wall construction robot, Automation in construction 5: 15-21, 1996.

[8] Kyoungmo Jung, Baeksuk Chu, Daehie Hong, Robot-based construction automation: An application to steel beam assembly (Part II), Automation in construction 32:62-79, 2013.

[9] Lee, H.-H. Kim, C.-J. Lee, S.-I. Ham, S.-H. Yun, H. Cho, B.K. Kim, G.T. Kim, K. Kim,A laser-technology-based lifting-path tracking system for a robotic tower crane, Automation in Construction 18 (7) : 865-874, 2009.

[10] Y. Yamazaki, J. Maeda, the SMART system: an integrated application of automation and information technology in production process, Computers in Industry 35 (1): 87-99, 1998.

[11] J.G. Everett, A.H. Slocum, CRANIUM: device for improving crane productivity and safety, Journal of Construction Engineering and Management 119 (1): 23-39, 1993.

[12] U.-K. Lee, K.-I. Kang, G.-H. Kim, H. Cho, Improving tower crane productivity using wireless technology, Computer-Aided Civil and Infrastructure Engineering 21 (8): 594-604, 2006.

[13] Lee,J.Cho,S.Ham,T. Lee,G.Lee,S.-H.Yun, H.-J.Yang,A BIM- and sensor-based tower crane navigation system for blind lifts, Automation in Construction 26: 1-10, 2012.

[14] In Jae Shin, Factors that affect safety of tower crane installation/dismantling in construction industry, Safety Science 72: 379-390, 2015.

[15] Y.Ikeda, T.Harada, Application of the automated building construction system using the conventional construction method together, Proceedings of the 23rd international Symposium on Automation and Robotics in Construction, pp.722-727, 2006.

[16] Daniel Scmidt, Karsten Berns, Climbing robots for maintenance and inspections of vertical structures-A survey of design aspects and technologies, Robotics and Autonomous systems 61:1288-1305, 2013.

[17] Thomas Linner, Jorg Gutter, Thomas Bock, Christos Georgoulas, Assistive robotic micro-rooms for independent living, Automation in construction 51: 8-22, 2015.

[18] Seungyeol Lee, Seungnam Yu, Junho Choi, Changsoo Han, A methodology to quantitatively evaluate the safety of a glazing robot, Applied Ergonomics 42: 445-454, 2011.

Suggestion of the ground stiffness estimative method with the running speed of a plate compactor

Takeshi HASHIMOTO[a], Kenichi FUJINO[b] and Kazuyoshi TATEYAMA[c]

[a] Public Works Research Institute, 1-6, Minamihara, Tsykuba-shi, Ibaraki-ken, 305-8516, Japan, t-hashimoto@pwri.go.jp
[b] Public Works Research Institute, 1-6, Minamihara, Tsykuba-shi, Ibaraki-ken, 305-8516, Japan, fujino@pwri.go.jp
[c] Ritsumeikan University, 1-1-1, Nojihigashi, Kusatsu-shi, Shiga-ken, 525-8577, Japan, tateyama@se.ritsumei.ac.jp

Abstract

A reversible plate compactor, that is one type of small compaction machine, travels by performing small jumps forward or backward according to its own vibration and resistance from the ground. So it is assumed that its running speed is affected by the stiffness of the ground. Therefore it is thought that it would be possible to estimate ground stiffness by measuring the running speed of a reversible plate compactor using an inexpensive method (stopwatch etc.).

Keywords –

soil compaction, ground stiffness, reversible vibratory plate compactor, running speed

1 Introduction

The vibration behaviour of the vibratory compaction machine is impacted by the stiffness of the ground it contacts, so it is possible to estimate the stiffness of the ground by analysing the signal of the acceleration sensor which equipped on the machine [1][2][3][4]. This technology has attracted interest as far more rational ways of performing quality control of compaction. But this requires instruments called acceleration sensors, calculation modules, and display instruments in addition to the normal compaction machine, increasing the cost. Another problem obstructing their popularization is the difficulty in incorporating them in existing machines already in use at construction sites.

On the other hand, a reversible plate compactor, that is one type of small compaction machine, travels by performing small jumps forward or backward according to its own vibration. So it is assumed that its running speed is affected by the stiffness of the ground too. Therefore it is thought that it would be possible to estimate ground stiffness by measuring the running speed of a reversible plate compactor using a simple method (stopwatch etc.).

So this research was done in an attempt to establish a method of estimating ground stiffness value based on the running speed of a reversible plate compactor, and evaluating whether or not the value obtained satisfies compaction quality control standards. (Japanese compaction quality control standard value for railway embankments.)

First, a numerical calculation was done to study the relationship of the running speed of the reversible plate compactor with the ground stiffness. Then an actual reversible plate compactor was used to perform an experiment in a laboratory test pit to verify the result of numerical calculation.

2 Numerical calculation model

The study of the running speed of the reversible plate compactor based on the numerical calculation was performed separately for vertical direction movement and horizontal direction movement.

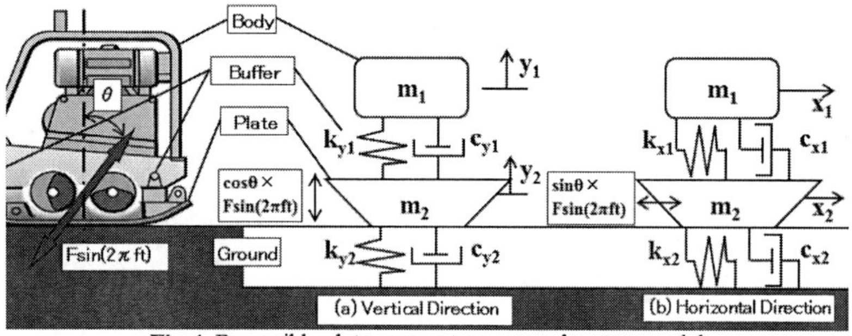

Fig. 1. Reversible plate compactor – ground system model

For the vertical direction, the reversible plate compactor and the ground were replaced with the 2-degrees of freedom vibration model like that shown in Figure 1(a). The equation of motion of this model is shown as formula (1), and solving this using the Runge-Kutta method calculated the vertical displacement accompanying change over time.

$$\begin{aligned} m_1\ddot{y}_1 + c_{y1}(\dot{y}_1 - \dot{y}_2) + k_{y1}(y_1 - y_2) &= -m_1 g \\ m_2\ddot{y}_2 + c_{y2}\dot{y}_2 + k_{y2}y_2 - c_{y1}(\dot{y}_1 - \dot{y}_2) - k_{y1}(y_1 - y_2) & \\ = -m_2 g + \cos\theta \times F\sin(2\pi f t) & \end{aligned} \quad (1)$$

m_1	Mass of upper body	(kg)
m_2	Mass of plate	(kg)
k_{y1}	Vertical spring constant of rubber buffer	(MN/m)
k_{y2}	Vertical spring constant of soil	(MN/m)
c_{y1}	Vertical viscous damping coefficient of rubber buffer	(Nsec/m)
c_{y2}	Vertical viscous damping coefficient of soil	(Nsec/m)
F	Max centrifugal force	(kN)
f	Frequency	(Hz)
θ	Angle of centrifugal force (Angle between resultant force of two eccentric shafts and vertical line)	(deg)

Because the ground and exciter plate are connected in the model in Figure 1(a), the subgrade reaction N (N) represented by formula (2) is also calculated in cases where it is negative. But, because these are not actually connected, the subgrade reaction N cannot become a negative and represents the exciter plate separating from the ground, or in other words, jumping. So in this calculation, the change of the subgrade reaction N from positive to negative is considered to represent the start of a jump, and the displacement in the air was calculated, by later applying $k_{y2} = c_{y2} = 0$ to formula (1) by the time it landed.

$$N = -c_{y2}\dot{y}_2 - k_{y2}y_2 \quad (2)$$

To judge that it has landed, because if, at the same time, $N = 0$ in formula (2), this formula represents displacement of the ground, solving this calculates the ground displacement while it is in the air and the time when the exciter plate displacement and the ground displacement are reversed is judged to be the landing time.

For the horizontal direction, the reversible plate compactor and ground are replaced by the 2-degrees of freedom vibrating model such as that shown in Figure 1(b). The equation of motion of this vibrating system is represented by formula (3), and like formula (1), solving it using the Runge-Kutta method calculated the horizontal displacement accompanying change over time.

$$\begin{aligned} m_1\ddot{x}_1 + c_{x1}(\dot{x}_1 - \dot{x}_2) + k_{x1}(x_1 - x_2) &= 0 \\ m_2\ddot{x}_2 + c_{x2}\dot{x}_2 + k_{x2}(x_2 - x_0) & \\ - c_{x1}(\dot{x}_1 - \dot{x}_2) - k_{x1}(x_1 - x_2) &= \sin\theta \times F\sin(2\pi f t) \end{aligned} \quad (3)$$

k_{x1}	Horizontal spring constant of rubber buffer	(MN/m)
k_{x2}	Horizontal spring constant of soil	(MN/m)
c_{x1}	Horizontal viscous damping coefficient of rubber buffer	(Nsec/m)
c_{x2}	Horizontal viscous damping coefficient of soil	(Nsec/m)

To judge ground contact or jump, the results of the above vertical direction calculation were used to calculate displacement in air by applying $k_{x2} = c_{x2} = 0$ to formula (3) while it was in the air. And it was presumed that while it was on the ground, no slippage between the exciter plate and ground occurred.

In the calculation, the vertical ground viscous damping coefficient c_{y2} was calculated from formula (4). Here, the damping ratio Dy2 was set at 0.4. [5]

$$c_{y2} = 2D_{y2}\sqrt{m_2 k_{y2}} \quad (4)$$

The vertical rubber buffer viscous damping coefficient c_{y1} can be calculated in the same way as formula (4). The damping ratio D_{y1} was set as 0.1. [3]

As the horizontal soil spring constant k_{x2} was calculated using the relational formula (5) with the vertical soil spring constant k_{y2}. [5] Here the Poisson's ratio ν of the ground was 0.4 as the normal value.

$$k_{x2} = \frac{8(1-\nu)^2}{7-8\nu} k_{y2} \quad (5)$$

The horizontal ground viscous damping coefficient c_{x2} can be calculated from formula (6). Here the damping ratio D_{x2} was set at 0.3. [5]

$$c_{x2} = 2D_{x2}\sqrt{m_2 k_{x2}} \quad (6)$$

The horizontal rubber buffer spring constant k_{x1} and viscous damping coefficient c_{x1}, were set equal to the vertical spring constant k_{y1} and viscous damping coefficient c_{y1} considering the shape of the rubber buffer and the way it is installed.

In the above numerical calculation, the ground stiffness is represented as the vertical soil spring constant k_{y2} based on rectangular loading (plate of compactor). On the other hand, the compaction quality control standard values are provided as the modulus of subgrade reaction K_{30} caused by circular loading (Plate loading test). Therefore, it is necessary to replace the value K_{30} with k_{y2} in order to consider whether or not the ground stiffness value satisfies the compaction quality control standard based on the running speed obtained by the numerical calculation. Formula (7) is used to transform the modulus of subgrade reaction K_{30} value to the vertical ground spring constant k_{y2}. [6]

$$k_{y2} = \frac{4\pi br}{L} \cdot K_{30} \qquad (7)$$

Here,

$$L = \ln\left(\frac{\sqrt{1+m^2}+m}{\sqrt{1+m^2}-m}\right) + m \cdot \ln\left(\frac{\sqrt{1+m^2}+1}{\sqrt{1+m^2}-1}\right)$$

$$m = \frac{b}{a}$$

$2a$	Width of plate	(m)
$2b$	Length of plate	(m)

3 Study of running speed by numerical calculation

Table 1 shows the machine specifications of the reversible plate compactor and Table 2 shows the ground conditions used for the calculation. Here, the machine specifications were the specifications for the reversible plate compactor that are now used most widely in Japan. The vertical direction soil spring constant k_{y2}, confirmed behavior of running speed for a broad range of spring constants from extremely soft ground to hard ground that has been compacted, so it was set at k_{y2} = 0.3 ~ 20.0 MN/m for this study.

Figure 2 shows the results of calculating under the conditions shown in Table 1 and Table 2. The horizontal axis represents the vertical soil spring constant k_{y2}, and the vertical axis represents the running speed calculated based on the horizontal displacement of the plate of the reversible plate compactor. The value obtained by converting the compaction quality control standard value in a railway embankment shown on Table 3 to the soil spring constant by formula (7) is represented by the broken line.

This shows that outside the diagonal line area in Figure 2, the running speed is almost constant even when the soil spring constant changes, and it is difficult

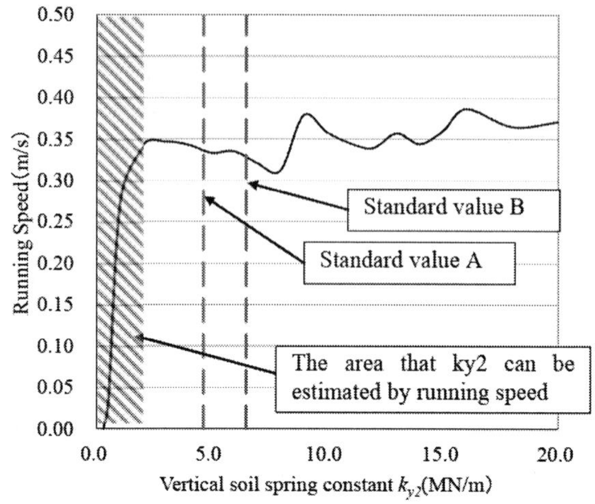

Fig. 2. Soil spring constant – running speed relationship according to numerical calculation

Table 1. Machine specification

θ(deg)	50
m_1 (kg)	200
m_2 (kg)	130
k_{y1} (MN/m)	1.0
F (kN)	45
f(Hz)	73
$2a$(m)	0.4
$2b$(m)	0.3
c_{y1} (Nsec/m)	$c_{y1} = 2D_{y1}\sqrt{m_1 k_{y1}}$ D_{y1} = 0.1
k_{x1} (MN/m)	$k_{x1} = k_{y1}$
c_{x1} (Nsec/m)	$c_{x1} = c_{y1}$

Table 2. Ground condition

k_{y2} (MN/m)	0.3 ~ 20.0
c_{y2} (Nsec/m)	$c_{y2} = 2D_{y2}\sqrt{m_2 k_{y2}}$ D_{y2} = 0.4
k_{x2} (MN/m)	$k_{x2} = \frac{8(1-\nu)^2}{7-8\nu} k_{y2}$ ν =0.4
c_{x2} (Nsec/m)	$c_{x2} = 2D_{x2}\sqrt{m_2 k_{x2}}$ D_{y2} = 0.3
ν	0.4

Table 3. Japanese compaction quality control standard value for railway embankments

Standard value A	Minimum K_{30} value : 50 MN/m
Standard value B	Minimum K_{30} value : 70 MN/m

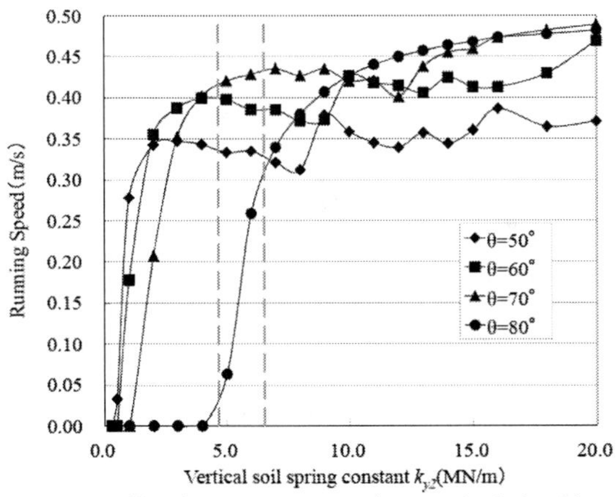

Fig. 3. Soil spring constant – running speed relationship according to numerical calculation (Difference in angle of centrifugal force)

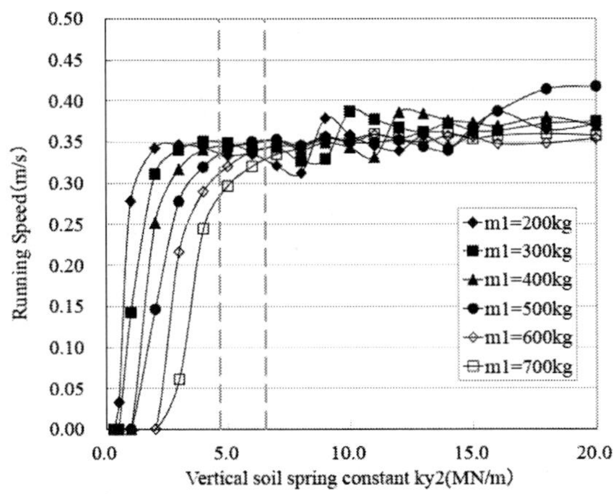

Fig. 4. Soil spring constant – running speed relationship according to numerical calculation (Difference in upper body weight)

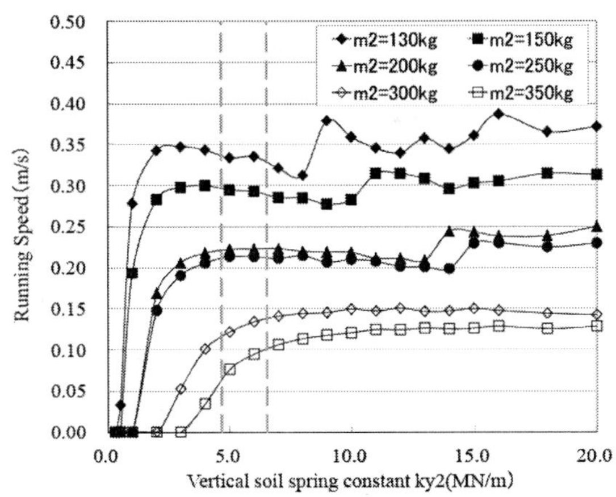

Fig. 5. Soil spring constant – running speed relationship according to numerical calculation (Difference in plate weight)

to estimate the soil spring constant from the running speed. In Figure 2, it is possible to estimate the soil spring constant based on running speed in only the diagonal line area, which is a range much lower than the quality control standard value. Therefore, it is impossible to judge whether or not the soil spring constant satisfies the quality control standard value by running speed.

So a study was done to learn if it is possible to judge whether or not the soil spring constant satisfies the quality control standard value by performing minor modifications to the machine.

Figures 3 to 5 shows the results of performing numerical calculations while changing the angle of centrifugal force θ from 50° to 80°, upper body weight m_1 from 200kg to 700kg, and the plate weight m_2 from 130kg to 350kg.

According to Figure 3 to Figure 5, in order to judge whether or not the soil spring constant satisfies the quality control standard value based on running speed, the angle of centrifugal force should be increased to 80° or the upper body weight should be increased to 700kg or the plate weight should be increased to 350kg. But, it is not realistic, in terms of both weight and space, to increase the upper body weight by 500kg or the plate weight by 220kg. Therefore, the most realistic and effective modification is increasing the angle of centrifugal force to 80°.

4 Verifying applicability by a test pit experiment

In order to investigate the applicability of the results of performing numerical calculations, an experiment was done in a laboratory test pit.

The experiment was done in a test field simulating an actual execution site prepared in a test pit with dimensions width 5m×length 44.8m×height 4m shown in Figure 6.

Table 4. Material properties of the test soil

	Soil(1)	Soil(2)	Soil(3)
Density of soil particle ρ_s(g/cm^3)	2.675	2.665	2.647
Maximum grain size D_{max}(mm)	9.5	4.75	9.5
Fine fraction content F_c(%)	15.3	57.1	4.3
Maximum dry density ρ_{dmax}(g/cm^3)	1.674	1.531	1.571
Optimum water content w_{opt}(%)	16.0	24.9	18.2

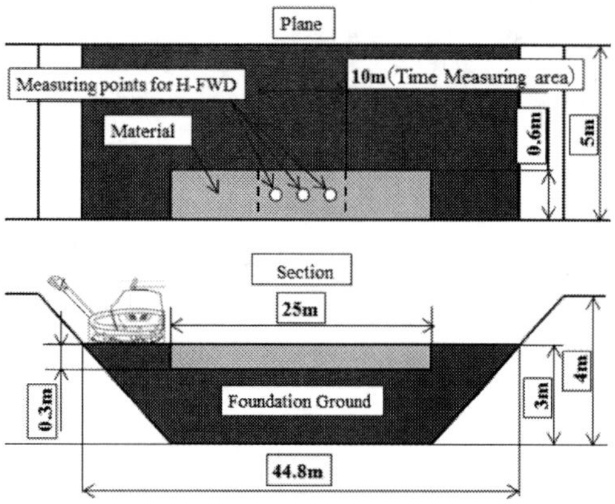

Fig. 6. Test field

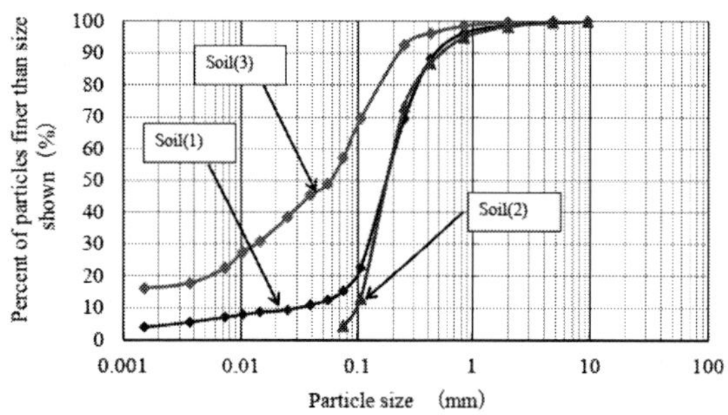

Fig. 7. Particle size accumulation curve

The test field was made according to following procedure. First, soil (1) shown in Table 4 was used to prepare enough compacted foundation ground with height of 3m in the test pit. Next, beside the wall of the foundation ground, a ditch with dimensions of width 0.6m×length 25m×depth 0.3m was excavated. Finally, test soil was embanked so that its finished thickness was 0.3m. Three kinds of test soil were used. The properties of the test soil are shown in Table 4 and Figures 7.

This test field was compacted with a reversible plate compactor with the specifications shown in Table 1 (Compactor1) and with a reversible plate compactor with the machine's angle of centrifugal force changed to 80° (Conpactor2). Therefore other specifications of both compactors are exactly same. And the time required for it to travel 10m in the field was measured with a stopwatch to calculate the running speed. At the same time, the modulus of subgrade reaction after passage by the reversible plate compactor was measured with a compact FWD. And the measurement by the compact FWD was done at three points in the 10m where the above speed was measured, and the average value of the measurements was used.

Figure 8 shows the results of the test. In Figure 8, the modulus of subgrade reaction obtained by the compact FWD (K_{30} value) was converted to the spring constant of the soil by formula (7).

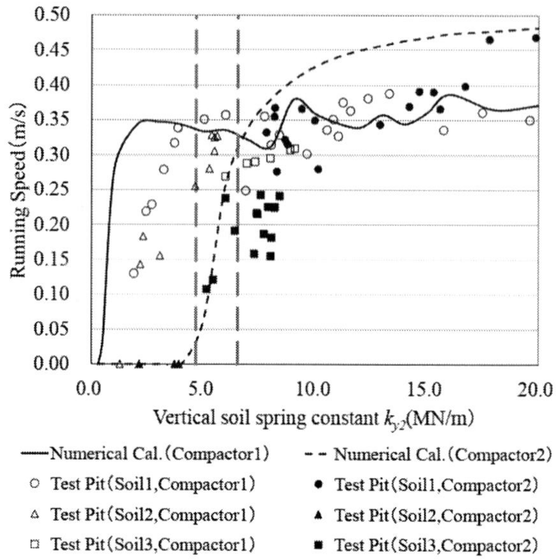

Fig. 8. Soil spring constant – running speed relationship according to numerical calculation

According to Figure 8, the test results and the calculation results show same tendency, when increasing the angle of centrifugal force from 50° to 80°. At angle of centrifugal force of 50°, it is difficult to judge whether or not the soil spring constant satisfies quality control standard values A and B respectively based on the running speed, but changing it to angle of centrifugal force of 80° permits judgments of the correctness of quality control standard values A and B.

However, it is expected that the compaction effect will be decreased because the vertical direction of the centrifugal force decreases when increasing the angle of centrifugal force from 50° to 80°. Therefore, a verification study was done.

Table 5 shows the compaction result (density and compaction degree) after 16 passes with compactor1 (angle= 50°) and compactor2 (angle= 80°) on soil(3).

Table 5. Compaction result after 16 passes

	Compactor1 (angle= 50°)	Compactor2 (angle= 80°)
Density (g/cm^3)	1.559	1.511
Compaction degree (%)	99.2	96.2

According to Table 5, the compaction degree of compactor2 is lower approximately 3% than the compaction degree of compactor1. Therefore, it is better to use compactor1 for compaction work, and to use compactor2 for measurement work only.

5 Conclusions

The relationship of the running speed with soil spring constant according to numerical calculation was considered. The results clarified the following points.

(1) In the case where the reversible plate compactor that was not modified (angle= 50°) was used, the range in which it is possible to estimate the soil spring constant using the running speed is in a range far lower than quality control standard values and it is difficult to use the running speed to judge whether or not the quality control standard value is satisfied.
(2) By changing the angle of centrifugal force to 80°, it is possible to judge whether or not the soil spring constant value satisfies the compaction control standard value in a railway embankment based on the running speed.

And the results of the numerical calculation were confirmed by an experiment using two reversible plate compacters. Compactor1 has the angle of centrifugal force as 50°. And compactor2 has the angle of centrifugal force as 80°. The results clarified the following points.

(3) Increasing the angle of centrifugal force to 80°, permits the estimation of whether or not the soil spring constant satisfies the compaction quality control standard value in a railway embankment based on the running speed.
(4) Because the compaction effect of compactor2 (angle= 80°) is lower than the compaction effect of compactor1 (angle= 50°), it is better to use compactor1 for compaction work, and to use compactor2 for measurement work only.

The method applied for this research is extremely simple if a stopwatch etc. is used, and it will help to easily rationalize quality control methods.

One major problem is the gap between the numerical calculation results and the test results. We wish to continue studies to bridge this gap in the future.

References

[1] Akiomi SHIMAZU, Kiyoshi MINAMI, Hiroki NAKATA, Isao SHIMADA, Kenichi ADACHI, 1985. The study of soil compaction using vibratory roller. Technical note of PWRI. No.2184, pp.37-76.

[2] Kazuyoshi TATEYAMA, Satoshi NAKAJIMA, Tetsuo FUJIYAMA, 1994. The evaluation of the degree of soil compaction by the behavior of the vibrating exciters. Journal of JCSA. No.487 / III-26. pp.237-245.

[3] Tetsuo FUJIYAMA, Kazuyoshi TATEYAMA, 2000. The evaluation of compacting ground stiffness by the acceleration of vibratory roller. Journal of JCSA. No.652 / III-51. pp.115-123.

[4] Kröber, W., Floss, E. and Wallrath, W., 2001. Dynamic soil stiffness as quality criterion for soil compaction, *Geotechnics for Roads, Rail Tracks and Earth Structures*, A. A. Balkema Publishers, pp.189-199.

[5] Toshio IWASAKI, Akiomi SHIMAZU, 1975. Vibrations of soils and foundations. Kajima Institute Publishing. pp.207-263. p.367.

[6] Takeshi KIMURA, 1978. Stress propagation of soil. Kajima Institute Publishing. pp.18-31.

Ambient Data Collection in Indoor Building Environments Using Mobile Robots

***Bharadwaj R. K. Mantha[1], Carol C. Menassa[2], and Vineet R. Kamat[3]**

[1]PhD Student, Dept. of Civil and Environmental Engineering, Univ. of Michigan, 2350 Hayward Street, Room 1350, G.G. Brown building, Ann Arbor, MI 48109, USA. *Corresponding Author: Tel: +1 (979)204-1981.
[2]Assistant Professor, Dept. of Civil and Environmental Engineering, Univ. of Michigan, 2350 Hayward St., Room 2322 G.G. Brown Building, Ann Arbor, MI 48109.
[3]Professor, Dept. of Civil and Environmental Engineering, Univ. of Michigan, 2350 Hayward St., Room 2340 G.G. Brown Building, Ann Arbor, MI 48109.
E-mail: baddu@umich.edu, menassa@umich.edu, vkamat@umich.edu

Abstract –

Building designers have increasingly looked at simulation to improve the performance of building systems with the primary objective of simultaneously maximizing comfort, and minimizing energy use. The quality of the simulations depends on the quality of the data being input into the models. The input data for simulation comprises the current real state of a building which encompasses several components that include the state of occupants, comfort parameters, and the building systems. Typically, such data is aggregated using preinstalled Building Automation Systems (BAS) with the help of stationary wired or wireless sensor networks. Such a process is cost-prohibitive, time consuming, and often impractical in existing buildings without BAS. This paper proposes mobile platform based robotic data collection for gathering energy and comfort related data in real-time which can be utilized for further simulation analysis and decision-making. The fiducial marker based navigation and drift correction algorithms developed to facilitate the robotic platform navigation in a building are discussed in detail. This method successfully achieves the navigation task by providing directional navigation information along with drift correction at critical discrete locations instead of the traditional continuous updating process, which is computationally intensive. An experimental study validating the statistical equivalence of the two data sets gathered by the traditional preinstalled fixed sensor networks and the multi-sensor fused robot was performed. The results demonstrate the feasibility of the proposed methodology in efficiently collecting large datasets in buildings using only a single set of sensors in contrast to the scads of similar sensors required by traditional data collection methods.

Keywords –

On-line Simulation, Data Collection, Indoor robots, Navigation and Drift correction algorithms.

1. Introduction

Buildings are responsible for 40% of the total energy consumption in the Unites States (US) [1]. According to US EIA 2014, future predictions show that these energy demands will continue to increase indicating a significant need for innovative efforts to optimize the energy consumption of buildings. The United States Green Building Council (USGBC) suggests that commissioning and/or retrofitting is the process through which buildings can be efficiently operated to achieve high performance [2].

Effective energy retrofits or commissioning process requires a comprehensive understanding of all the building systems, energy use behavior of the occupants, energy performance of the building, and changing weather patterns [3]. It is thus apparent that a significant amount of data needs to be continuously collected, managed, and analyzed in a building at the room and floor level in order to effectively optimize energy use in buildings while maintaining comfort of the occupants. Building energy simulation plays a key role in analyzing the aforementioned building energy data [4]. Several existing studies demonstrate how the data in real-time or near real-time or in offline/batch processing mode can be used for critical decision making in buildings [5,6].

Building energy and comfort parameter data collection has progressed significantly in the past few years because of the advancements in technology. In the early stages, inspectors used to collect data manually by inspecting different locations in the building, a process considered to be extremely tedious, time consuming and generates limited amounts of data which limits its practical applicability in large buildings [7-9].

With the advent of wired and wireless sensors, significant effort was saved by avoiding the need of manual inspection during the operation and maintenance phase of the buildings. Instead, in the newer buildings, sensors are installed, calibrated, and integrated with building systems before the operation and maintenance phase as part of a Building Automation System (BAS). However, this process is not very feasible for the existing buildings because of the challenges involved in installing the wiring, integrating with existing building systems, and calibrating the sensors.

Wired systems are most commonly used for video surveillance applications in residential or commercial buildings because of higher band width and power needs [10]. These are economically feasible for smaller buildings compared to larger buildings (e.g., dormitory buildings, academic institutions, and office spaces) where there are a large number of rooms that need to be monitored. This is because installing a wired system in an existing building requires additional costs associated with material (e.g., wiring) and labor intensive task of installation [11,12]. In addition, wired installation process is time consuming and sometimes limits the extent of space that can be monitored [12-14].

On the other hand, wireless systems have a capability to monitor, process, analyze the data locally, and eliminate the need of wires [13,15]. Wireless sensor networks are widely used for monitoring and controlling various indoor parameters such as temperature, humidity, CO2 levels, occupancy level, occupancy comfort, and water use [14]. However, they are expensive (due to a lot of initial upfront investment for large buildings), complex (due to installation of lot of sensors), time consuming (due to periodic calibration and maintenance requirements), and not feasible in existing buildings due to the need to install a BAS. In addition, existing buildings undergo retrofitting or retro-commissioning to improve the energy performance, which makes it even more time consuming, tedious, and expensive to uninstall, reinstall, recalibrate, and integrate with the BAS of the building. Wireless systems also suffer from power consumption, scalability, and limited information storage capacity issues limiting the extent and quality of the data that can be collected [7,16]. Further details about these data collection methods along with a thorough analysis of the respective characteristics can be found in [17].

In an aim to mitigate the aforementioned issues, this paper introduces a mobile platform based data collection process that uses mobile indoor robot equipped with single set of sensors. The mobile robot is capable of navigating in a known indoor environment with the help of onboard sensor suite (for data collection), onboard computing capabilities, and a RGB camera (for localization, navigation, and drift correction). The entire process of robotic navigation along with the localization, data collection, and geotagging are discussed in detail in the methodology section of the paper. One of the main advantages of the proposed data collection method is that it eliminates the need for instrumenting different locations in existing buildings with the same set of sensors [18]. The case study described at the end of this paper illustrates how the robot is capable of providing a rich data set for energy retrofit and/or retro-commissioning decision making purposes.

2. Research Objectives

The main goal of this study is to introduce a new data collection technique for gathering information regarding energy and comfort related parameters in buildings. The proposed approach is more economical and feasible compared to the existing techniques. The data collected using this approach has the potential to assist building managers in making timely and informed decisions about retrofitting and/or commissioning of the buildings which does not have the latest BAS installed. Thus, the primary objectives of this paper are to 1) Introduce a novel concept of using mobile indoor robots for collecting energy and comfort related data in buildings, 2) Discuss in detail the localization, navigation, and drift correction algorithms for two of the three robotic platforms developed. 3) Statistically compare the quality of the data sets collected using the traditional BAS and the mobile robot to illustrate the applicability of this framework for energy analysis in buildings.

3. Design of the Robot

One of the main contributions of this paper is the design of the robot along with its respective algorithms which are the basis for robots' navigation. Some of the other crucial aspects in the design of the robot are determining the type of data that needs to be collected (accordingly the type of sensors to be placed on the robot), frequency of data collection (how frequently the data needs to be collected at every location), waiting time at each location of data collection, algorithms that will help decide the number of mobile robots required to monitor (depending on the size of the buildings) the entire building, optimizing the travel time and path.

The TurtleBot robot platform, equipped with the iCreate base is chosen as the mobile data collection platform and sensors such as Cozir® CM 0199 (for temperature, humidity, and CO2 levels), HOBO U12 (for light and occupancy levels), Lutron (for natural light levels), NinjaBlocks (for air speed), Smart meters (for electricity consumption) is used for the data collection. Figure 1 shows the robot with the following components 1) TurtleBot – for navigating the indoor environment; 2) On-board netbook – to communicate with the TurtleBot; 3) RGB Camera - for the TurtleBot to detect fiducial markers, localize, and estimate its relative pose in an indoor environment; 4) Remote laptop for executing the corresponding navigation programs on the on-board netbook; and 5) Sensors – for monitoring and data collection of various occupant comfort and building energy parameters.

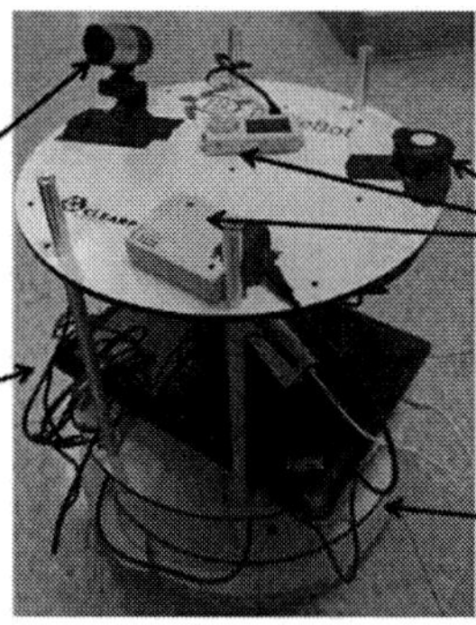

Figure 1. Components of the robot used for occupant comfort and indoor environmental data collection

4. Methodology

In order for the robot to autonomously navigate indoors and collect energy and comfort related data, it needs to 1) localize in the indoor environment, 2) navigate to the intended data collection locations, 3) collect the respective data (such as temperature, humidity, and light intensity), and 4) geo-tag (record the collected data with the physical location) for further analysis.

In the current context, localization of a robot means that the robot needs to be able to identify its current location in a known indoor environment setting. For example, a robot being able to recognize its current location to be in room 201, or knowing its xy location and orientation in the global coordinate reference system. The robot's navigation can be briefly defined as the robot's ability to plan a course of action to reach the destination location while accurately localizing itself in its frame of reference at strategic locations [19].

Many non-visual sensor based (non-vision based) techniques such as Global Positioning System (GPS) [20], Radio Frequency Identification (RFID) [21,22], Wireless Local Area Network (WLAN) [21,23], Ultra-Wide Band (UWB) [24], Bluetooth [25], and Inertial Measurement Unit (IMU) [26] used to be the primary focus of many studies for localization and navigation indoors. Though GPS and UWB have good accuracy, they require Line Of Sight (LOS) and tend to be costly. RFID is comparatively cost effective but for good accuracy a large number of tags are required. WLAN and IMU suffer from dynamic environment changes and drift accumulation respectively. Bluetooth based technologies require dense instrumentation of the beacons in the environment.

With the advent of efficient computing capabilities and developments in the field of computer vision, techniques such as Simultaneous Localization and Mapping (SLAM) [27] and visual registration which utilize visual sensors such as cameras, laser scanners [28,29] or LIght Detection And Ranging (LIDAR) were developed. Though 3D pose estimation is possible in most of the vision based techniques, they suffer some of the disadvantages such as requirement of high computational power, infrastructure dependency, and higher costs.

One of the vision-based methods reviewed (Fiducial Markers), however, is particularly immune to the aforementioned disadvantages afflicting other methods. Fiducial markers [30] offer high accuracy in determining and estimating their relative 3D pose in an environment, require relatively less computing capabilities, are cost-effective, and are easy to install [31].

4.1 Localization

In this section, the general computing framework that is developed as part of this research is described. This framework uses fiducial markers to link between actual physical locations and virtual information stored regarding those locations) for indoor robot localization.

Fiducial markers have the capability to store virtual information regarding multitude of things such as information regarding physical location (floor and room level information), emergency evacuation directions, indoor navigational information, and inspection related data regarding building systems helpful for facility managers [32]. For this study, unique fiducial markers are required to be placed along the navigational path of the robot (e.g., corridors, entrances to rooms etc.) as shown in Figure 4. More information on characteristics and properties of the markers is available in [30]. These markers, which are printed on regular paper, are used to store the physical location information (e.g., Room 101) and navigational information (e.g., take right) which is necessary to help the robot determine its current location and its targeted headed direction.

4.2 Navigation

Three different kinds of navigation techniques namely tele-operated, predefined path mode, and dynamically configurable path mode were developed by the authors. The former is discussed in [18] and the latter two are discussed in detail in this study. The complexity of the algorithm increases and human involvement decreases respectively. Tele-operated navigation (where human operator drives the robot remotely using a laptop or a wireless controller) is most suited for places which are highly reconfigurable and need sporadic data collection such as banquet halls, training rooms, and convention facilities. Predefined path mode navigation is mostly suited for places with regular occupancy and almost fixed spaces which require periodic data collection (Every 15 minutes or 30 minutes) such as offices, data centers, and ware houses. Dynamically configurable path mode is highly applicable for flexible/ dynamically changing occupancy buildings where data collection locations change dynamically depending on occupancy all through the day such as retail stores, schools, colleges, shopping malls, and airports. The navigation logic for each of these two types is described in detail in the following paragraphs. In addition, they are semi-autonomous because obstacle avoidance is not considered in the navigation.

4.2.1 Predefined path mode

In this mode, the robot is given a predefined path with a start location, end location, the path it needs to autonomously traverse to reach the end location, and also the data collection locations along the way. Several markers whose global positions and orientations are known in advance are placed at regular intervals along the navigational path as shown in Figure 4.

The overview of the navigational algorithm logic is represented as a flowchart in Figure 2. First, the aboard traditional RGB camera continuously captures images that might potentially contain a known fiducial marker. The images are processed by the marker recognition module (an algorithm which detects the presence of the marker). If a known marker is detected by the marker recognition module, the ID and relative pose of the robot with respect to the fiducial marker (in the camera's reference frame) is outputted by the module. Each ID is associated with a physical location in the indoor environment as shown in Figure 4. Current location of the robot based on the aforementioned information is estimated and pose correction is calculated based on the drift correction algorithms discussed later in this section of the paper. The current approach of navigation can be termed as treasure hunt based navigation because the robot traverses the path from one marker to another marker with the help of clues provided at each marker. This means that robot will follow the previously known navigational direction (or clue provided by the last seen marker) until it finds a new marker. To avoid the cases of obstruction (robot not being able to see one or two markers in the navigation path), it can be easily programmed to store information regarding next couple of markers and still traverse the navigational path accordingly.

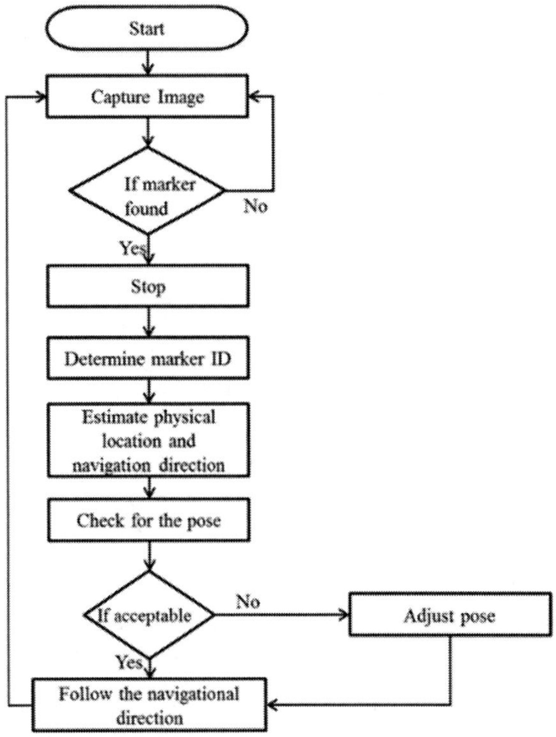

Figure 2. Indoor treasure hunt based navigation algorithm with the help of network of fiducial markers

Drift Correction (Predefined path): Drift is one of the major areas of concerns in the field of autonomous robots [33]. One of the most important factors that have a direct impact on whether the robot will traverse the entire path is drift accumulation of the robot. In this study, negative drift (-ve drift) means that the robot has drifted inwards (towards the marker) and conversely positive drift (+ve drift) means that the robot has drifted outwards (away from the marker).

The current drift correction algorithm developed is based on the relative pose information estimated with the help of the onboard camera, computer, the known marker's pose, and the known location in the fiducial markers network. The algorithm returns 3D relative pose information (as shown in Eq. 1) in camera reference frame with respect to the marker. From that, the lateral distance information (distance between the camera and marker), *d* as shown in Figure 3, is extracted and the drift at each location (δdi) is estimated. After that, the pose correction (turning angle (αi) in this case) is calculated based on the equations developed as shown in Eq. 2 and 3 along with a detailed description of various parameters are shown in Figure 3. Since the drift at each location might be different, the turning angle at each location also differs.

$$H = \begin{matrix} R11 & R12 & R13 & Tx \\ R21 & R22 & R23 & Ty \\ R31 & R32 & R33 & Tz \end{matrix} \text{ .... } Eq(1)$$

Where:

H is the part of Homogeneous transform matrix returned by the localization algorithm

R: Rotation matrix

T: Translation matrix

$$\alpha 1 = 2 * \tan^{-1}\left(\frac{\delta d}{M}\right) \text{ .... } Eq(2)$$

$$\alpha i = 180 - \tan^{-1}\left(\frac{\frac{M}{\delta di}}{1 + \left(\frac{\delta d(i-1)}{\delta di}\right)}\right) - \tan^{-1}\left(\frac{M}{\delta di}\right) \forall\, i \geq 2 \text{ .... } Eq(3)$$

4.2.2 Dynamically configurable path mode

In this mode, navigation is based on user inputted coordinates which also allows the user to configure the robot's path dynamically. This mostly resembles and matches the criteria for real-world applications where the robot might have to traverse different paths during different times of day.

The overarching logic of the entire algorithm is the same as previously discussed and shown in Figure 2. However, the drift correction technique differs in this algorithm when compared to the previously discussed (Predefined path mode) algorithm. The navigational direction is estimated with the help of user inputted coordinates by generating vectors based on adjacent pair of coordinates and calculating the angle between the pair of vectors

Drift Correction (dynamically configurable path): Similar to the drift correction algorithm discussed in the previous section, the current drift correction algorithm also works based on the relative pose information estimated with the help of the onboard camera, computer, the known marker's pose, and the known location in the fiducial markers network. However, instead of calculating the lateral distance between the camera and the fiducial marker detected, this system estimates the relative pose angle between the plane of the marker and the plane of camera in cameras reference frame. The angle determined is termed as adjusted angle and it is added to the calculated angle in the navigation logic discussed above. Thus, a net corrected angle (based on the calculated angle and adjusted angle) is calculated and the robot instead of turning the entire calculated angle, rotates only the amount of corrected angle thereby, correcting its drift. The same process is continued at every marker location until the destination/ targeted location is reached.

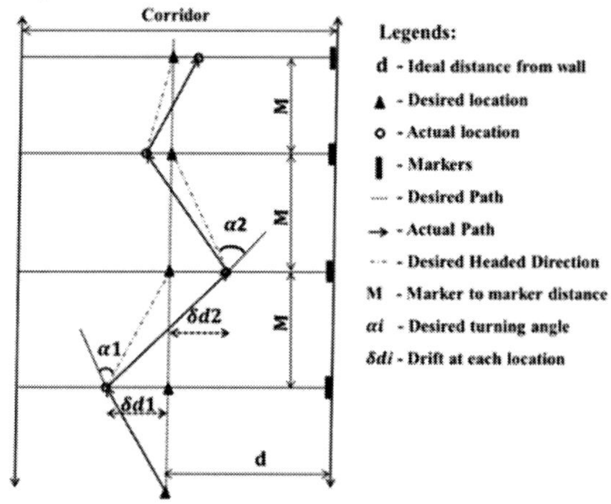

Figure 3. Visual representation of the drift error and respective terminology used in the drift correction algorithms.

4.3 Data Collection and Geo Tagging

The immediate next step once the robot is cognizant of its surroundings and capable of navigating indoors is collecting data. In the context of this paper, the parameters of interest are occupants comfort data in buildings such as temperature, humidity, indoor air quality, and light intensity.

In this study, geotagging is associating the building energy/ thermal comfort data collected along with its location information such as room number and floor number. To achieve this, a programmable interface is developed which bridges the communicated physical location information (as given by the fiducial markers) with the sensor data (obtained from the data collection). A python program is written which subscribes the published ROS data regarding the location of the robot (given by the fiducial marker), concatenates it with the retrieved sensor data along with the time stamp, and exports the data to an excel file locally stored in the on-board netbook. This data then can be stored in a local server or can be updated to an online big data set interface for further real-time processing/ analysis.

5. Case Study

A case study was performed to show the validity of all the four steps of methodology. Experiments were conducted in the Ross School of Business at the University of Michigan - Ann Arbor campus with the help of turtlebot as a mobile platform (as discussed in section 3 of this paper). The selected building is equipped with a BAS that collects different types of data at the room, system, and the building level. For example, different types of data gathered by the BAS of Ross are control temperature, supply air damper point, room temperature, hot water valve pint, and damper status operation. The basement floor comprising of an open study lounge (monitored by four thermostats) and the open corridor (monitored by three thermostats) were chosen as the test bed for the case study experiments. Figure 4 shows the locations of the thermostats and/or the locations in the basement where the temperature readings are recorded by the BAS. Since the case study location chosen consists of public spaces (corridors and open student lounge) with dynamically changing occupancy levels, dynamically configurable path mode technique as discussed in the navigation section of the methodology is chosen for the robot data collection path. However, the type of navigation technique used will not have any effect on the quality of the data collected.

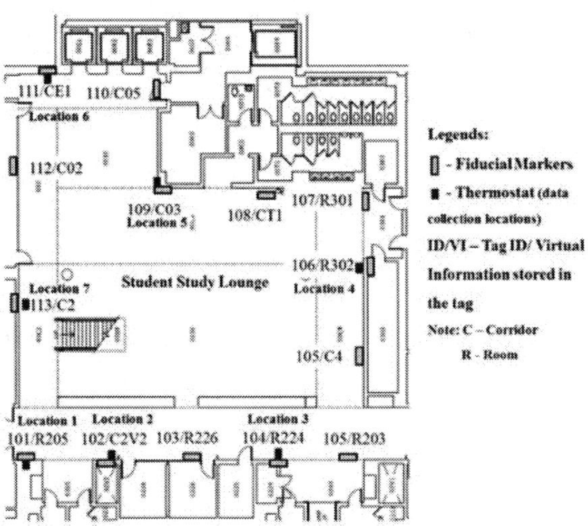

Figure 4. Fiducial marker network with the virtual information regarding the location stored in each of the markers along with the data collection locations in the basement floor of Ross School of Business, University of Michigan – Ann Arbor.

5.1 Validation

The data collected with the help of the proposed methodology is verified with data collected by existing technique using BAS. The BAS is programmed to collect data every 30 minutes in all the locations around the clock. For example, the BAS collects and time stamps data samples at 10:30:00 AM, 11:00:00 AM, 11:30:00 AM, and so on. Since the data collected with the proposed methodology is done with a mobile robot, it is not possible to sample data in all the locations in a time synchronized way. However, the data is collected at all the locations within a stipulated time range so that it can be compared to the BAS data. Also, it is assumed that there were no significant differences in temperature values within that time frame.

There are many statistical methods to assess two types of data sets. However, given the context of comparing two data sets, t-statistic hypothesis testing was done to compare the data collected by the BAS and the robot. The absolute difference between every pair of the readings was calculated and hypothesis testing was performed for the resulting data set. Prior to the data collection, experiments were also conducted in the same setting to find the difference in recorded temperature values from both the sensors (the sensor used on the robot and the thermostat in the BAS). The maximun absolute difference in the value observed was 0.8. Hence, for the statistical analysis, the null and alternate hypothesis were considered to be $|\mu_{BAS} - \mu_{Robot}| \leq 0.8$ and $|\mu_{BAS} - \mu_{Robot}| > 0.8$ respectively. The sample size of data at each location is 202 and hence the degrees of freedom are considered to be 201. The analysis and results are listed in Table 1.

Table 1 Results of paired test analysis done for comparing the data sets collected using BAS (1) and Mobile robot (2).

Location	$\lvert\mu1 - \mu2\rvert$	$\lvert\sigma1 - \sigma2\rvert$	t statistic	p value
1	0.283	0.184	-39.90	1
2	0.491	0.347	-12.67	1
3	0.466	0.262	-18.09	1
4	0.446	0.366	-13.76	1
5	0.647	0.491	-4.44	0.99
6	0.603	0.389	-7.20	1
7	0.493	0.453	-9.63	1

Considering α=0.05 (confidence level of 0.95), it can be noted from the p-values in Table 1that there is no evidence to reject the null hypothesis. Hence, it is evident that the data collected by the mobile data collecting platform is equivalent to the data collected by densely instrumented sensor network of BAS.

6. Conclusion

The proposed methodology of mobile indoor robotic monitoring and data collection of indoor environmental and occupant comfort parameters discussed in this paper offers an effective and economical method as compared to the traditional state of the art (Fixed stationary sensor network) data collection methods. This method involves using mobile indoor robots equipped with sensors to monitor and collect energy and comfort related data in buildings. It is particularly significant for old buildings that do not have an installed sensor network. With meager instrumentation of markers in buildings, the required data can be collected with the help of mobile robots.

Some of the main characteristics of the traditional (stationary/fixed) sensor networks and mobile robotic data collection are descried as follows. A) Upfront Costs: Fixed sensor networks need astronomical amount of sensors, while mobile based systems need only one set of sensors. B) Initial setup: Need to install and calibrate thousands of sensors in each room in a building. Though mobile based system requires installing markers in the environment, they are easily configurable and the calibration needs to be done only on one set of sensors. C) Operational and Maintenance Costs: For fixed systems, high manual and administrative costs are incurred for periodic battery replacement, maintenance, and calibration for each of these sensors. Locating these sensors may prove to be a challenge in complex buildings. On the other hand, it is comparatively a lot easier to perform the aforementioned tasks on one set of sensors. With regard to the runtime power consumption of the robot, it is envisioned that the robot will autonomously charge itself (similar to existing robotic platforms such as Roomba) when the battery is running low and resume the data collection. D) Frequency of data collection: Fixed systems are capable collecting data with any time interval, but mobile platforms are limited based on the area that needs to be monitored.

In addition, the limitations of the proposed system include requirement of marker instrumentation in the environment and occlusions. Careful consideration is required for determining the placement of the markers, as they might suffer from occlusions. However, with minor alterations in the algorithm, it can be programmed to account for the occluded markers. These markers are very easy to deploy and configure [32].

Future planned work includes addition of obstacle avoidance methods to the existing algorithms and subsequently testing the robot in more complex indoor environments.

7. Acknowledgments

The authors would like to acknowledge the financial support received for this research from the US National Science Foundation (NSF) CBET grants #1349921 and 1407908. Any opinions in this paper are those of the authors and do not necessarily represent those of NSF.

References

[1] US DOE/EIA (2014), "Annual Energy Outlook 2014 with projections to 2040", <http://www.eia.gov/forecasts/aeo/pdf/0383(2014).pdf>, Accessed : 23/3/15.

[2] Coyle (2014), "Green Building 101: Why is energy efficiency important?", < http://www.usgbc.org/articles/>,Accessed: 12/4/15.

[3] Mills, E. (2011). "Building commissioning: a golden opportunity for reducing energy costs and greenhouse gas emissions in the United States." *Energy Efficiency*, *4*(2), 145-173.

[4] Ma, Z, Cooper, P, Daly, D, and Ledo, L, (2012). "Existing building retrofits: Methodology and state-of-the-art." *Energy and buildings*,55(889-902)

[5] Menassa, C. C., Taylor, N., and Nelson, J. (2013). "A framework for automated control and commissioning of hybrid ventilation systems in complex buildings." *Automation in Construction*, 30, 94-103.

[6] Zhao, L., Zhang, J. L., and Liang, R. B. (2013). "Development of an energy monitoring system for large public buildings." *Energy and Buildings*, 66, 41-48.

[7] Swartz, R. A., Zimmerman, A. T., Lynch, J. P., Rosario, J., Brady, T., Salvino, L., and Law, K. H. (2012). "Hybrid wireless hull monitoring system for naval combat vessels." *Structure and Infrastructure Engineering*, 8(7), 621-638.

[8] Raftery, P., Keane, M., and Costa, A. (2011). "Calibrating whole building energy models: Detailed case study using hourly measured data." *Energy and Buildings*, 43(12), 3666-3679.

[9] Wang, W., Wang, N., Jafer, E., Hayes, M., O'Flynn, B., and O'Mathuna, C. (2010). "Autonomous wireless sensor network based building energy and environment monitoring system design." *Environmental Science and Information Application Technology (ESIAT), IEEE*, Vol. 3, pp. 367-372.

[10] Chandramohan, V., and Christensen, K. (2002). "A first look at wired sensor networks for video surveillance systems." *Local Computer*

Network ,IEEE,v27, pp.728-729.

[11] Nourollah, S., V. (2009). "Automated building monitoring using a wireless sensor network." *Doctoral dissertation*, Concordia University.

[12] Dermibas, M. (2005). "Wireless sensor networks for monitoring of large public buildings." *University at Buffalo, Tech. Rep.*

[13] Kim, J., and Lynch, J. P. (2011). "Autonomous decentralized system identification by Markov parameter estimation using distributed smart wireless sensor networks." *Journal of Engineering Mechanics*, 138(5), 478-490.

[14] Vlissidis, A., Charakopoulos, S., Kolokotsa, D., and Boian, I. (2008). "Wireless Sensor Networks Building Monitoring Application." *Bulletin of the Transilvania University of Brasov*, Series I: Engineering Sciences, 1(50).

[15] Ceriotti, M., Mottola, L., Picco, G. P., Murphy, A. L., Guna, S., Corra, M., and Zanon, P. (2009). "Monitoring heritage buildings with wireless sensor networks: The Torre Aquila deployment." *Information Processing in Sensor Networks (IPSN) 2009, IEEE Computer Society*, pp. 277-288.

[16] Bhadauria, D., Tekdas, O., and Isler, V. (2011). "Robotic data mules for collecting data over sparse sensor fields." *Journal of Field Robotics*, 28(3), 388-404.

[17] Mantha, B. R. K., Menassa, C., and Kamat, V, (2015) "A Taxonomy of Data Types and Data Collection Methods for Building Energy Monitoring and Performance Simulation", *Advances in Building energy research, Taylor and Francis,* 1-31, permanent link < http://www.tandfonline.com/doi/full/10.1080/17512549.2015.1103665>.

[18] Mantha, B. R. K., Feng, C., Menassa, C., and Kamat, V, (2015) "Real-time Building Energy And Comfort Parameter Data Collection Using Mobile Indoor Robots", *International Symposium on Automation and Robotics in Construction (ISARC) 2015*, 32(1), Oulu, Finland.

[19] Levitt, T. S., and Lawton, D. T. (1990). "Qualitative navigation for mobile robots." *Artificial intelligence*, *44*(3), 305-360.

[20] Khoury, H. M., and Kamat, V. R. (2009)." Evaluation of position tracking technologies for user localization in indoor construction environments."*Automation in Construction*, *18*(4), 444-457.

[21] Montañés, J. A. P., Rodríguez, A. M., and Prieto, I. S. (2013). "Smart indoor positioning/location and navigation: A lightweight approach." *IJIMAI*, 2(2), 43-50.

[22] López Escobés, P.(2009) "Summary of the State of the Art in indoor location systems.", *MSc ICT Research, University of Valladolid.*

[23] Torres-Solis, J., Falk, T. H., and Chau, T. (2010). "A review of indoor localization technologies: towards navigational assistance for topographical disorientation." *Ambient Intelligence*, pp. 51–84, 2010.

[24] Segura, M. J., Auat Cheein, F. A., Toibero, J. M., Mut, V., and Carelli, R. (2011). "Ultra Wide-Band Localization and SLAM: A Comparative Study for Mobile Robot Navigation." *Sensors*, 11(2), 2035-2055.

[25] Raghavan, A. N., Ananthapadmanaban, H., Sivamurugan, M. S., and Ravindran, B. (2010). "Accurate mobile robot localization in indoor environments using Bluetooth." *Int. Conf. on Robotics and Automation, IEEE* (pp. 4391-4396).

[26] Akula, M., Dong, S., Kamat, V. R., Ojeda, L., Borrell, A., and Borenstein, J. (2011). "Integration of infrastructure based positioning systems and inertial navigation for ubiquitous context-aware engineering applications." *Advanced Engineering Informatics*, 25(4), 640-655.

[27] Dissanayake, M. W. M. G., Newman, P., Clark, S., Durrant-Whyte, H. F., and Csorba, M. (2001). "A solution to the simultaneous localization and map building (SLAM) problem." *Robotics and Automation, IEEE* , *17*(3), 229-241.

[28] Thrun, S., Bennewitz, M., Burgard, W., Cremers, A., Dellaert, F., Fox, D., Hähnel, D., Rosenberg, C., Roy, N., Schulte, J., and Schulz, D. (1999). "MINERVA: A second generation mobile tour-guide robot." *ICRA*,pp. 3136-3141.

[29] Burgard, W., Cremers, A. B., Fox, D., Hähnel, D., Lakemeyer, G., Schulz, D, Steiner, S, and Thrun, S. (1998). "The interactive museum tour-guide robot." In *Aaai/iaai* (pp. 11-18).

[30] Olson, E., (2011). "AprilTag: A robust and flexible visual fiducial system." *ICRA*, pp. 3400-3407.

[31] Iwasaki, M., and Fujinami, K. (2012). "Recognition of Pointing and Calling for Industrial Safety Management." *ICT International Senior Project Conference IEEE Thailand* (pp. 50-53).

[32] Feng, C., and Kamat, V. R. (2012). "Augmented reality markers as spatial indices for indoor mobile AECFM applications." *CONVR*, pp. 235-240.

[33] Arleo, A., Millán, J. D. R., and Floreano, D. (1999). "Efficient learning of variable-resolution cognitive maps for autonomous indoor navigation." *Robotics and Automation, IEEE Transactions on*, *15*(6), 990-1000.

Fast terminal sliding mode control for gantry cranes

Ansu Man Singh, Van Truong Hoang and Q. P. Ha

School of Electrical, Mechanical and Mechatronic Systems,
University of Technology Sydney, Australia

E-mail: AnsuMan.Singh@uts.edu.au, VanTruong.Hoang@student.uts.edu.au, Quang.Ha@uts.edu.au

Abstract

Cranes remain a vital tool for the construction of infrastructure such as buildings, bridges, etc. Recently, there has been renewed interest in crane automation in dealing with concerns on safety and possible performance degradation due to system uncertainties and disturbances. One potential solution to the problem is the use of robust techniques based on the Sliding Mode Control (SMC) methodology. Much research has been conducted to design controllers based on linear sliding surfaces, aiming at achieving the desired control performance in finite time. In this context, this paper proposes a control method, based on the Fast Terminal Sliding Mode (FTSM), to guarantee finite-time stability of the crane. To do that, we have derived a mathematical model of the crane using Lagrangian formulation with uncertainties as bounding functions. Then, sliding surfaces based on the hierarchical sliding mode are defined, and a control law is derived using the Lyapunov stability theory. The hierarchical sliding surfaces consist of two layers. The first layer include sliding functions based on FTSM to enable faster convergence of the system to equilibrium. This can have advantages in high precision tracking applications. In the second-layer, the sliding surface is designed from the linear combination of the first layer sliding functions. Also, we have given a proof of the stability of the system in finite time. Extensive simulation results show the proposed controller based on FTSM can achieve higher performance in stabilizing the swinging load of a gantry crane. Laboratorial experiments have been conducted to verify the obtained results in terms of the superior convergence time and improved performance over conventional SMC.

Keywords –

Machine Control Systems, Mechatronics, hoist control, Automation and Robotics

1 Introduction

Cranes have a pivotal role in the construction of high-rise buildings, bridges, etc. Automation and control of cranes has been the subject of many studies to address safety and performance in the presence of parametric uncertainties and external disturbances.

Problems of uncertainties and external disturbances in the cranes can be solved by employing robust controllers, for which the Sliding Mode Control (SMC) method is very promising. Many published studies have shown the synthesis of such controllers. For instance, optimal sliding mode is proposed for trajectory tracking of offshore cranes [1] or second-order sliding mode is applied for vibration suppression in smart structures [2] to address the problem of uncertainties and external disturbances in these systems. Similarly, sliding surfaces incorporating the load swing dynamics along with an adaptive method to tune the controller's gain are presented in [3] for container cranes. Studies on the subject have mostly focused on the controllers that utilize linear sliding surfaces. By using these control approaches, convergence to the system equilibrium is asymptotic, and hence, the settling time of the crane operations may be not fast enough. Therefore, it is worth looking for a SMC scheme with a finite time reaching phase.

To enable finite time reachability of the system equilibrium while also guaranteeing robustness in the system, terminal sliding mode (TSM) has been proposed in [4] and applied to many mechanical systems [5]. However, there are issues with performance, particularly in the high-precision applications, due to the slower response of TSM when the system's states are far from equilibrium under a constrained control force. This paper presents a faster, finite-time stable, and robust controller for gantry cranes. The controller is based on Fast Terminal Sliding Mode (FTSM), a variant of TSM. The idea of FTSM is to provide faster convergence both at a distance from and at a close range of the equilibrium [6]. Many published studies have reported the application of FTSM on the fully-actuated systems in robotics and automation applications. However, to date, no studies have shown its design and implementation on underactuated systems such as cranes. The methodology used in the synthesis of the proposed controller is as follows: first, the two layer sliding functions are defined based on hierarchical SMC. The first layer consists of FTSM manifolds. The second layer is the linear combination of the first layer sliding functions. Then, the Lyapunov stability theory is used to derive the control law for the system.

The organization of this paper is as follows. A mathematical model for the gantry cranes is obtained in the next section, followed by Section 3 for the derivation of the proposed controller and its stability analysis. Simulation and experimental results are presented in Sections 4 and 5, respectively. Finally, the paper conclusion arrives with some recommendations.

2 Mathematical model of gantry cranes

The structure of a gantry type crane is depicted in Figure 1. It consists of a cart of mass m_c, a hoisting mechanism through a rope of length l, and a payload of mass m_p. The position of the cart and the swing angle of the payload are represented by x and θ, respectively. In this paper we have made following assumptions:

- the payload consists of a point mass
- the stiffness of the hoisting rope is negligible.

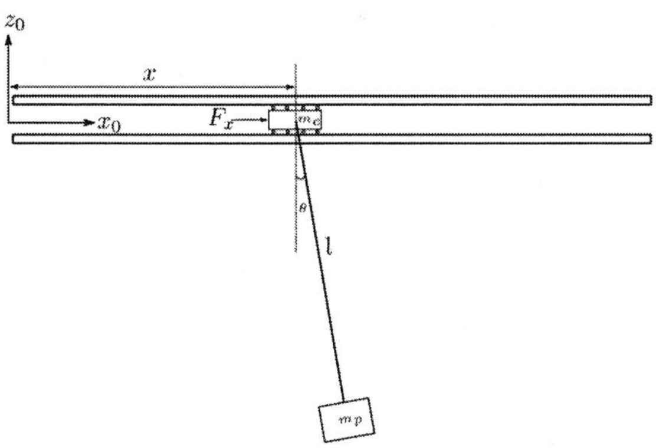

Figure 1. Structure of a gantry crane.

Using the Lagrangian formulation [7], the dynamic model of the gantry crane can be represented as,

$$M(q)\ddot{q} + C(q,\dot{q}) + g(q) = Bu, \tag{1}$$

where coordinates $q = [q_x \quad q_\theta]^\top = [x \quad \theta]^\top \in \mathbb{R}^2$,

$$M(q) = \begin{bmatrix} m_{11} & m_{12} \\ m_{21} & m_{22} \end{bmatrix} = \begin{bmatrix} m_c + m_p & m_p l \cos(\theta) \\ m_p l \cos(\theta) & m_p l^2 \end{bmatrix}$$

is the inertial matrix,

$$C(q,\dot{q}) = \begin{bmatrix} -m_p l \sin(\theta)\dot{\theta} \\ 0 \end{bmatrix}$$

is the Coriolis matrix, and

$$g(q) = \begin{bmatrix} 0 \\ -m_p g l \sin(\theta) \end{bmatrix}$$

is the gravitational term. The signal u is the control input with $B = [1 \quad 0]^\top$ being the input matrix. Equation (1) can be formulated in the following form

$$\begin{aligned} \ddot{q}_x &= f_x(q,\dot{q}) + b_x(q)u + \sigma_x(q) \\ \ddot{q}_\theta &= f_\theta(q,\dot{q}) + b_\theta(q)u + \sigma_\theta(q), \end{aligned} \tag{2}$$

where,

$$f_x(q,\dot{q}) = \frac{m_p^2 l^2 \sin\theta(l\dot{\theta} - g\cos\theta)}{(m_p + m_c)m_p l^2 - m_p^2 l^2 \cos^2\theta},$$

$$f_\theta(q,\dot{q}) = \frac{m_p l \sin\theta\left\{(m_p + m_c)g - m_p l \cos\theta\dot{\theta}\right\}}{(m_p + m_c)m_p l^2 - m_p^2 l^2 \cos^2\theta},$$

$$b_x(q) = \frac{m_p l^2}{(m_p + m_c)m_p l^2 - m_p^2 l^2 \cos^2\theta},$$

$$b_\theta(q) = \frac{-m_p l \cos\theta}{(m_p + m_c)m_p l^2 - m_p^2 l^2 \cos^2\theta}.$$

In (2), the lumped terms $\sigma_x(q)$ and $\sigma_\theta(q)$ represents the system uncertainties of the system as well as external disturbances. They are assumed to be bounded, i.e. $|\sigma_x(q)| \le \mu_x$ and $|\sigma_\theta(q)| \le \mu_\theta$.

3 Control system design

3.1 Fast Terminal Sliding Mode

Sliding function for FTSM is given by

$$s = \dot{x} + \beta x^{q/p} + \alpha x, \tag{3}$$

where $\alpha > 0$, $\beta > 0$, $q > 0$, and $p > 0$. Moreover, q and p are odd integers such that $q < p$. In the sliding surface $s = 0$ the state x in (3) reaches the equilibrium point $x = 0$ in finite time, also known as the reaching time, is given by

$$t_s = \frac{p}{\alpha(p-q)} \ln\left(1 + \frac{\alpha x_0^{(p-q)/p}}{\beta}\right), \quad (4)$$

where $x_0 = x(0)$ is the initial condition. The trajectory of the system in the sliding surface $s = 0$ can be obtained as

$$x(t) = \begin{cases} \mathrm{sgn}(x_0)\left\{\frac{\left(\alpha x_0^{(p-q)/p}+\beta\right)e^{-\frac{\alpha(p-q)}{p}t}-\beta}{\alpha}\right\}^{\frac{p}{p-q}} & t \le t_s \\ 0 & t > t_s \end{cases} \quad (5)$$

The difference between sliding surfaces of FTSM (3) and TSM is the additional linear attractor αx in addition to terminal attractor $\beta x^{q/p}$ [8]. The presence of the linear attractor is to make the state of the FTSM sliding surface converges to equilibrium at a faster rate compared to TSM [6] when the initial state is far from the equilibrium point.

3.2 Sliding surface design for the crane

The position error and the swing angle error of the crane are defined respectively as $e_x = x - x_d$ and $e_\theta = \theta$. Since the crane is an underactuated system, it is intuitive to design the first layer with two sliding functions:

$$\begin{aligned} s_1 &= \dot{e}_x + \beta_x e_x^{q_1/p_1} + \alpha_x e_x \\ s_2 &= \dot{e}_\theta + \beta_\theta e_\theta^{q_2/p_2} + \alpha_\theta e_\theta, \end{aligned} \quad (6)$$

where p_1, q_1, p_2, and q_2 are odd integers such that $\frac{q_1}{p_1} < 1$ and $\frac{q_2}{p_2} < 1$. Also, for coefficients of the terminal and linear attractors, scalars $\beta > 0$ and $\alpha > 0$. Now, the second layer sliding function is defined as,

$$S = k_1 s_1 + k_2 s_2, \quad (7)$$

where $k_1 > 0$ and $k_2 > 0$, which is the linear combination of first layer sliding functions. This strategy is known as the hierarchical sliding mode [9].

3.3 Stability analysis of the sliding surfaces

In the following, we design a control input for the system that guarantees the stability of the system and the convergence of the sliding surface S=0. Let us first define a Lyapunov function for S as,

$$V = \frac{1}{2}S^2 \quad (8)$$

Taking time derivative, one can get

$$\dot{V} = S\dot{S} \quad (9)$$

The time derivative of the of Equation (7) is given by

$$\begin{aligned} \dot{S} =& k_1\left(\ddot{e}_x + \beta_x \frac{q_1}{p_1} e_x^{q_1/p_1 - 1}\dot{e}_x + \alpha_x \dot{e}_x\right) \\ &+ k_2\left(\ddot{e}_\theta + \beta_\theta \frac{q_2}{p_2} e_\theta^{q_2/p_2 - 1}\dot{e}_\theta + \alpha_\theta \dot{e}_\theta\right) \end{aligned}$$

or,

$$\begin{aligned} \dot{S} =& k_1\{f_x(\boldsymbol{q},\dot{\boldsymbol{q}}) + b_x(\boldsymbol{q})u \\ &+ \beta_x \frac{q}{p} e_x^{q_1/p_1-1}\dot{e}_x + \alpha_x\dot{e}_x - \ddot{x}_d + \sigma_x(\boldsymbol{q})\} \\ &+ k_2\{f_\theta(\boldsymbol{q},\dot{\boldsymbol{q}}) + b_\theta(\boldsymbol{q})u \\ &+ \beta_\theta \frac{q}{p} e_\theta^{q_2/p_2-1}\dot{e}_\theta + \alpha_\theta\dot{e}_\theta + \sigma_\theta(\boldsymbol{q})\} \end{aligned}$$

or,

$$\begin{aligned} \dot{S} =& k_1\left\{f_x(\boldsymbol{q},\dot{\boldsymbol{q}}) + \beta_x \frac{q}{p} e_x^{q_1/p_1-1}\dot{e}_x + \alpha_x\dot{e}_x - \ddot{x}_d + \sigma_x(\boldsymbol{q})\right\} \\ &+ k_2\left\{f_\theta(\boldsymbol{q},\dot{\boldsymbol{q}}) + \beta_\theta \frac{q}{p} e_\theta^{q_2/p_2-1}\dot{e}_\theta + \alpha_\theta\dot{e}_\theta + \sigma_\theta(\boldsymbol{q})\right\} \\ &+ \{k_1 b_x(\boldsymbol{q}) + k_2 b_\theta(\boldsymbol{q})\}u \end{aligned} \quad (10)$$

Now, let us propose the control input

$$u = -\frac{1}{k_1 b_x(\boldsymbol{q}) + k_2 b_\theta(\boldsymbol{q})}\{u_1 + u_2 + u_3\}, \quad (11)$$

where,

$$\begin{aligned} u_1 =& k_1\left(f_x(\boldsymbol{q},\dot{\boldsymbol{q}}) + \beta_1 e_x^{q_1/p_1-1}\dot{e}_x + \alpha_1\dot{e}_x - \ddot{x}_d\right) \\ u_2 =& k_2\left(f_\theta(\boldsymbol{q},\dot{\boldsymbol{q}}) + \beta_2 e_\theta^{q_2/p_2-1}\dot{e}_\theta + \alpha_2\dot{e}_\theta\right) \\ u_3 =& \eta\,\mathrm{sgn}(S) + KS, \end{aligned}$$

in which $\eta > 0$ and $K > 0$. By applying the input (11) to Equation (10) and after some arrangement, we obtain

$$\begin{aligned} \dot{V} = S\dot{S} =& S\{k_1\sigma_x(\boldsymbol{q}) + k_2\sigma_\theta(\boldsymbol{q}) - \eta\,\mathrm{sgn}(S) - KS\} \\ =& -KS^2 - \eta|S| + \{k_1\sigma_x(\boldsymbol{q}) + k_2\sigma_\theta(\boldsymbol{q})\}S \\ =& -KS^2 - \eta|S| + \{k_1\sigma_x(\boldsymbol{q}) + k_2\sigma_\theta(\boldsymbol{q})\}|S| \end{aligned}$$

where $D_m = k_1\sigma_x(\boldsymbol{q}) + k_2\sigma_\theta(\boldsymbol{q})$.

If η is chosen such that

$$\eta > D_m = sup(k_1\mu_x + k_2\mu_\theta),$$

then

$$\dot{V} = -KS^2 - (\eta - D_m)|S| \le 0. \quad (12)$$

By applying the Lyapunov stability theory, it can be verified that the second layer sliding surface S is convergent to zero. In [10],Wang *et al.* have proved that if the second layer sliding surface is driven to the origin; then the first layer sliding functions, i.e., s_1 and s_2 , are also convergent to the origin, i.e. $s_1 = 0$and $s_2 = 0$. Thus, following the same argument we can conclude that e_x and e_θ reach the equilibrium in finite time as determined by (4).

4 Simulation Results

To validate the proposed method we used the gantry crane available in our control laboratory, as shown in Figure 2. The parameters of the crane and the design parameters of the proposed controller are provided in Table 1. To select values for p_1, q_1, p_2, q_2, α_x, α_θ, β_x, and β_θ we followed the conditions given in Section 3.3 and considered the reaching time expression for FTSM (4) for the required reaching time of the cart position and the swing angle. However, to optimize the values we did a number of trials in the simulation. Similarly, the values of k_1 and k_2 were chosen with some trial-and-error steps during the simulation.

Table 1

System parameters	Value
m_p	2.24 kg
m_c	2.27 kg
l	0.5 m
g	9.8 m/s^2
$\lvert\sigma_x(q)\rvert$	$\leq 0.5m/s^2$
$\lvert\sigma_\theta(q)\rvert$	$\leq 0.5rad/s^2$
Control parameters	$\alpha_x = \alpha_\theta = 0.1, \beta_x = \beta_\theta = 0.4$ $\eta = 0.1,\ p_1 = 11,$ $q_1 = 9,\ p_2 = 9,\ q_2 = 7$ $k_1 = 2,\ k_1 = 1,\ K = 1$

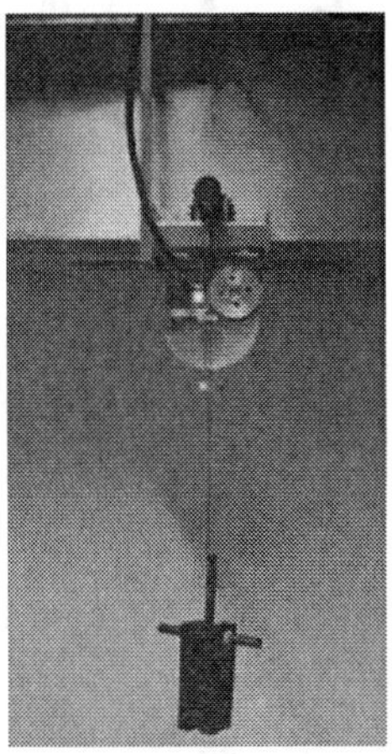

Figure 2: Gantry crane photo.

Figure 3 and Figure 4 show the simulation results of the cart position and the payload swing angle, respectively. The reference position for the cart is 1.5 m. Both results show achievement of the equilibrium point within 6 sec. To judge the control performance, Figure 5 represents the sliding surfaces. It is clear from the plot that the second layer sliding function S can reach $S = 0$ within 3 sec and the first layer sliding functions s_1 and s_2 approach the equilibrium asymptotically.

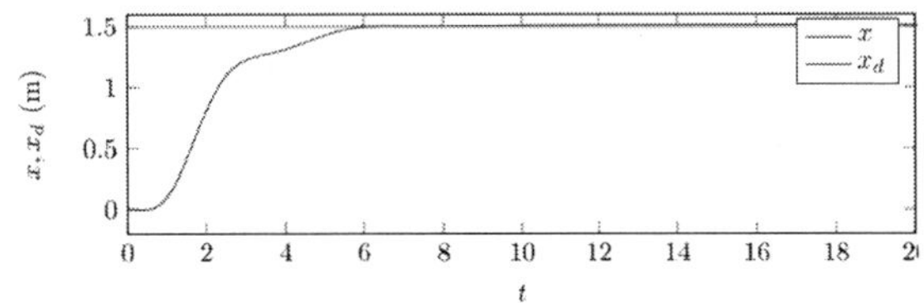

Figure 3: Cart position of the crane

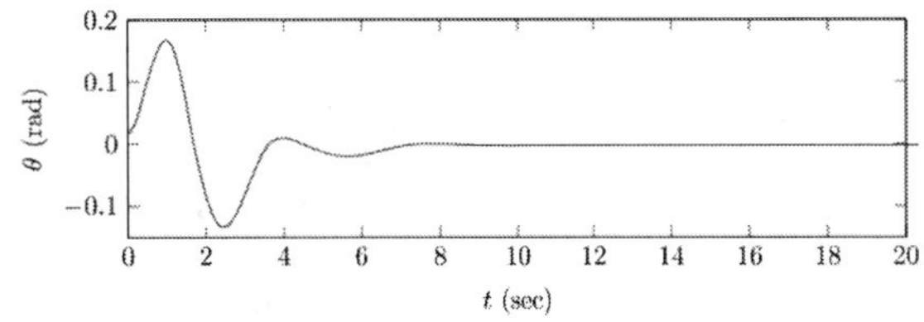

Figure 4: Swing angle of the payload

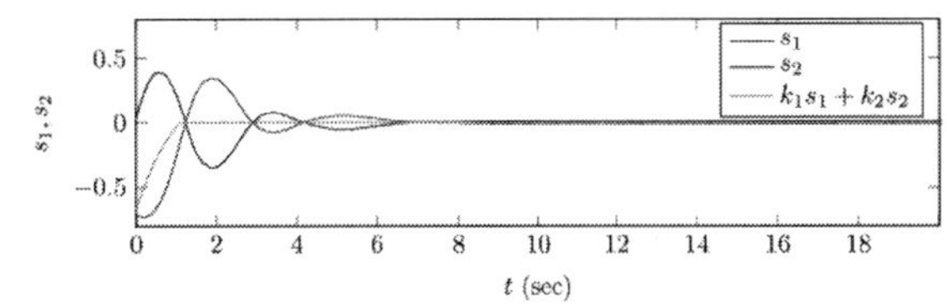

Figure 5: First and second layer sliding functions

Figure 6 shows the control input generated by the control system. The peak control force is around 12 N. Chattering can be observed after the system reaches its final position. This undesired effect can be reduced by using the well-known saturated function.

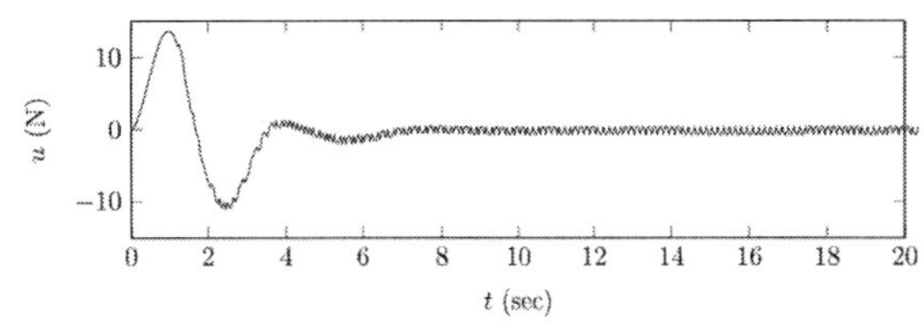

Figure 6: Control input to the plant

Figure 7 compares the result obtained from our proposed method and TSM. The cart reference for both the controller remains 1.5 m and the starting location is the origin. It is observed that the FTSM controller drives the cart to the set point in 6 sec whereas it takes 7 sec for the TSM. Therefore, a faster convergence is the clear advantage of the proposed method over TSM. Indeed, a faster settling of the control system would imply a higher efficiency of the construction or manufacturing automation processes that require pick-and-place operations of a crane.

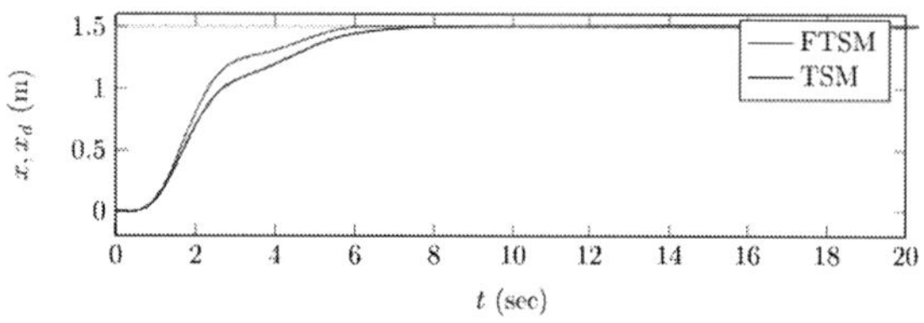

Figure 7: Comparison study of FTSM and TSM.

5 Experimental setup and results

To experimentally verify the proposed method, we implemented a control system for the crane of Figure 2 by using a STM32f411 Nucleo microprocessor board, as depicted in the Figure 8. The board consists of an ARM Cortex M4 processor with 512KB of flash memory and 128 KB of RAM. It receives the feedback signals, i.e. position of cart and payload's swing angle, from the crane through a signal conditioning circuit. The purpose of the circuit is to provide interface between the crane and the microprocessor board. Similarly, the control signal generated by the board is fed to crane through a Digital to Analog Converter (DAC) AD5726 IC. The DAC IC is connected with the board using SPI interface. The board is also connected to a PC via an UART interface to collect data from the crane's sensors for post processing.

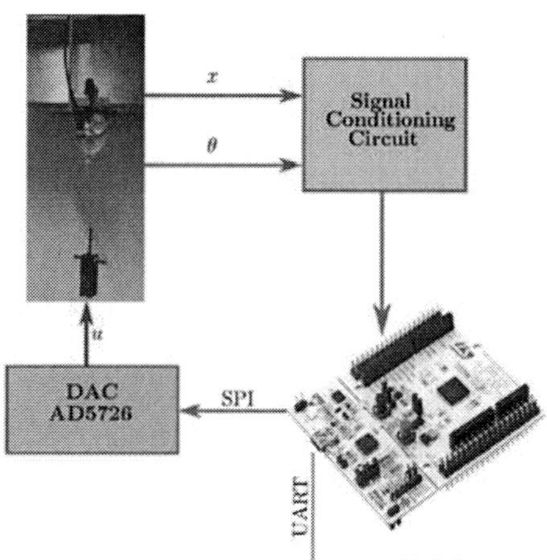

Figure 8: Experimental setup

The Nucleo board runs the FreeRTOS operating system (OS) and the proposed controller is implemented as an OS's stack. The feedback signals from the crane are sampled at the rate of 10 kHz. Similarly the controller generates the control signal every 0.01sec.

Figure 9 and Figure 10 show responses of the cart position and payload swing angle, respectively. The sampling frequency for the graphs is 10 Hz. The reference position for the cart, during the experiments, was set to 1.5m. The initial position of the cart is 0.1m, and reaches the desired location with the steady state error less than 0.05 m.

Initially the swing angle of the payload reaches up to 11 degrees. The oscillation, however, decreases to less than 3 degrees. The obtained results are quite coincident with those from the simulation. There are many reasons for the oscillation. One of the reasons is the uncompensated dynamics of the motor in the model of the crane, i.e. equation (1). Another is with the approximation. Indeed, the controller requires position and velocity feedback from the system, i.e. $\boldsymbol{q}$ and $\dot{\boldsymbol{q}}$. However, only position feedback is available. Therefore, for the experiment, we developed an estimator of the velocity using an approximate differentiator as

$$\dot{\boldsymbol{q}} = \frac{\boldsymbol{q}(t) - \boldsymbol{q}(t-T)}{T}, \tag{13}$$

where T is the sampling period of the controller. It should be noted that (13) does not provide an accurate estimation of the velocity, particularly of the swing angle. Therefore, by using another sensor or a more accurate estimation method, performance of our proposed control method can be enhanced.

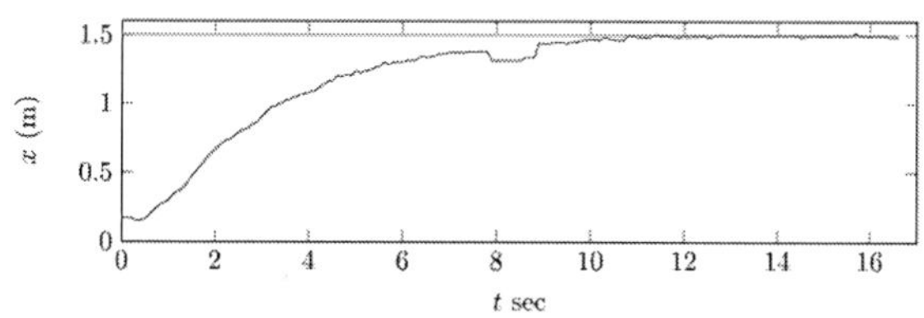

Figure 9: Position response of the cart

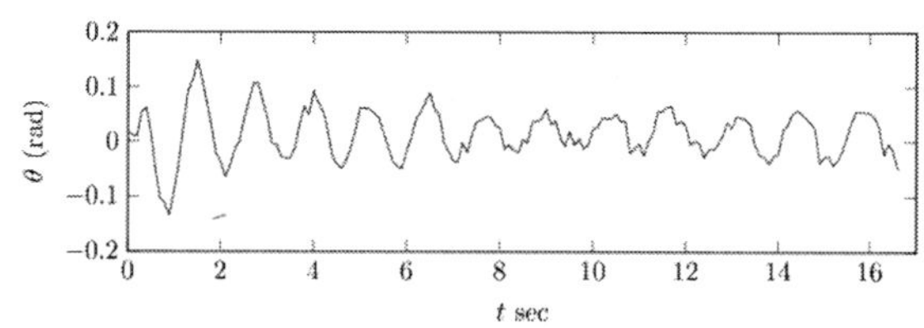

Figure 10: Swing angle response of the payload

Figure 11 and Figure 12 show respectively the control input generated by the controller and the sliding functions. Initially, the control force is around 2 N. The chattering effect can be observed after the system has reached the equilibrium. Figure 11 shows the sliding surfaces of our proposed method. It can be seen that the sliding surfaces $S = 0$ and consequently, s_1=0 and s_2=0 are controlled to reach the equilibrium in about 7 sec.

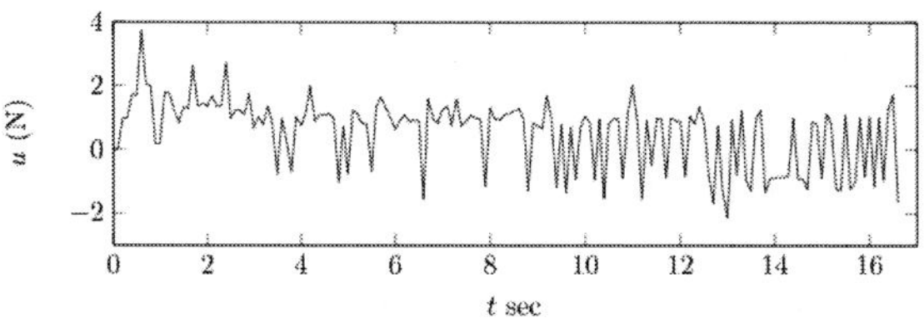

Figure11: Control input.

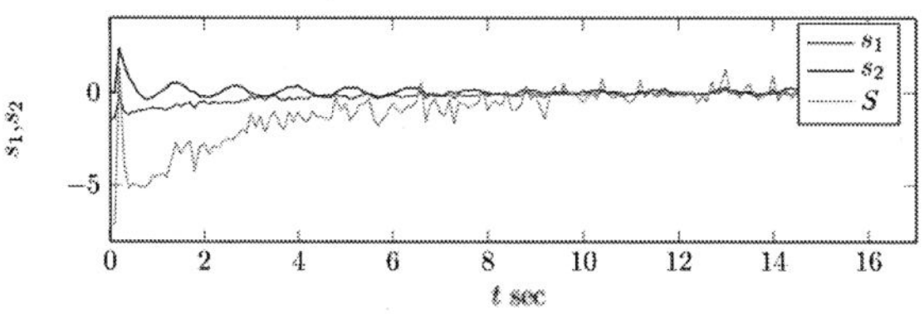

Figure 12: First and the second layer sliding surfaces

6 Conclusion

This paper has presented the design and implementation of a robust controller for gantry cranes. The control method is based on the Fast Terminal Sliding Mode. The proposed method provides enhanced convergence rate to the equilibrium in face of bounded uncertainties. Comparative study of the proposed method with the Terminal Sliding Mode verifies its advantage in simulation. Experiments were successfully conducted to validate the simulation results. The control performance obtained could be improved by using accurate sensors or estimation techniques. Furthermore, this method can be extended to other types of cranes such as those of boom type or offshore cranes.

References

[1] Raja Ismail, R. M. T., That, N. D. and Ha, Q. P. Modelling and robust trajectory following for offshore container crane systems. *Automation in Construction*. Vol. 59, pp. 179-187, 2015

[2] Ha, Q.P., Nguyen, M.T, Li, J. and Kwok, N.M. Smart structures with current-driven MR dampers: Modelling and second-order sliding mode control. *IEEE/ASME Transactions on Mechatronics*. 18-8, 1702-1712, 2013.

[3] Ngo Q. and Hong K. Adaptive sliding mode control of container cranes. *IET control theory & applications*. Vol. 6, 662-668, 2012.

[4] Yu X. and Zhihong M. Multi-input uncertain linear systems with terminal sliding-mode control. *Automatica*. Vol. 34, 389-392, 1998.

[5] Liu J. and Wang X. *Advanced Sliding Mode Control for Mechanical Systems*, Springer, 2011.

[6] Yu X. and Zhihong M. Fast Terminal Sliding-Mode Control Design for Nonlinear Dynamical Systems. *IEEE Trans. Circuits & Systems I.* Vol. 49, 261-264, 2002.

[7] Liu Y. and Yu H. A. Survey of underactuated mechanical systems. *Control Theory & Applications, IET*, Vol. 7, 921-935, 2013.

[8] Yu, X and Zhihong M. On finite time mechanism: Terminal sliding modes. *Proc. Int. Workshop on Variable Structure Systems*. pp. 164-167, 1996.

[9] Zak, M. Terminal attractors for addressable memory in neural networks. *Physics Letters A*. Vol. 133, 18-22, 1988.

[10] Wang W., Yi J., Zhao D. and Liu D. Design of a stable sliding-mode controller for a class of second-order underactuated systems. *IEE Proceedings: Control Theory and Applications*. Vol. 151(6), 683-690, 2004.

A Framework for Optimizing Lap Splice Positions within Concrete Elements to Minimize Cutting Waste of Steel Bars

Zahra S. Moussavi Nadoushani[1], Ahmed W. Hammad[2] and Ali Akbarnezhad[3]

[1] PhD candidate, School of Civil and Environmental Engineering, UNSW Australia, Sydney, NSW 2052, Australia;

[2] PhD candidate, School of Civil and Environmental Engineering, UNSW Australia, Sydney, NSW 2052, Australia;

[3] Lecturer, School of Civil and Environmental Engineering, UNSW Australia, Sydney, NSW 2052, Australia;

Email: z.moussavinadoushani@unsw.edu.au, a.hammad@unsw.edu.au, a.akbarnezhad@unsw.edu.au

Abstract

One of the material waste streams in the construction of concrete structures is the waste generated during cutting of steel bars to their final required length. The amount of waste generated to produce the required length of steel bars can be affected considerably by the cutting patterns adopted. Waste can, therefore, be minimized by optimizing the cutting pattern to be selected. Cutting patterns are a combination of bar lengths and, thus any change in the required lengths directly affects cutting patterns and the final amount of waste. This paper presents a framework for selection of the location of lap splices, through generating all possible lapping patterns, which define the bar lengths in cutting patterns. The focus is placed on steel bars used in concrete columns as their arrangement is affected by multiple parameters that make the manual preparation of shop drawings a tedious job. The developed framework is applied to a case study involving the construction of columns for an actual 6 storey building. The estimated waste produced after the adoption of optimal lapping patterns is compared with the actual waste generated based on the project shop drawings. The results indicate that considerable reduction in the steel bars wasted is achievable by optimizing the lapping patterns.

Keywords –

Minimizing rebar cutting waste, Lap splice, Cutting pattern generation, Linear Integer Programing

1 Introduction

Minimizing the material waste generated during construction is one of the major objectives in sustainable construction due to its economic and environmental benefits [1]. The economic benefits include reductions in material procurement and waste management costs. On the other hand, environmental benefits include decreasing the need for extraction of natural resources and reducing the carbon emissions and energy use associated with waste processing [2, 3]. One of material waste streams in the construction of concrete structures is related to reinforcing steel bars waste, accounting for up to 5 and 8% of the total waste in public and private residential construction respectively [4]. This waste is generated during the cutting of steel bars from standard lengths to required lengths [5-7].

The amount of the waste generated during cutting of steel bars can be affected considerably by the selected cutting patterns and thus may be minimized by selection of optimal cutting patterns. Cutting patterns are a combination of various steel bar lengths; therefore, any change in the required lengths directly affects cutting patterns and the final amount of waste. This is where the importance of engineering judgment, in terms of choosing alternative bar arrangements leading to different required lengths, becomes apparent [6-8]. However, due to the limitations of design software and Building Information Models (BIM) in producing comprehensive bar schedules, this process is mostly done by hand without using any special calculation tools, making the process a tedious and labour-consuming job [6, 7]. As Chen and Yang pointed out, even more recent developed software such as Tekla [9] which integrate design data and automatically generate design drawings still needs the engineer's experience and judgment. In their study, Chen and Yang picked concrete beams as one of the elements with the most complicated steel bar arrangement and developed an automated framework to generate shop drawings [6]. Porwal and Hewage also presented a framework integrating BIM with optimization techniques in order to minimize trim losses in the design stage [7]. Different optimization techniques such as Linear Programming (LP) [10], Integer Programming (IP) [11], Sequential Heuristic Procedure (SHP) [12] and Genetic Algorithm (GA) [5] have been proposed by previous studies to minimize the rebar trim loss referred to as one-dimensional Cutting Stock Problem (1D-CSP). Although the framework presented by Porwal and Hewage [7] can be the base of future development of BIM software, in the meantime a more practical methodology is needed to assist structural designers, in the task of generating steel bar arrangements that will lead to a reduction of trim loss.

The focus of this study is placed on steel bars used in concrete columns as their arrangement is affected by multiple parameters such as storey height, lap length and bar size that make manual preparation of shop drawings a tedious job. Including all these parameters in a model that yields the best steel bar arrangement in terms of the minimal final waste produced is a tedious job and time consuming to perform manually. In this study, a framework aiming at finding different cutting patterns and optimizing them based on variable lengths due to various possibilities for longitudinal lap splices positioning within structural columns is developed. The model is applied to a case project involving the construction of columns for an actual 6 storey building. The estimated waste produced after the adoption of optimal cutting patterns is compared with the actual waste generated based on the project's shop drawings.

2 Methodology

The proposed framework for trim loss analysis consists of four main modules; 1) Data collection module, a data entry interface for gathering all the required information needed to calculate the bar lengths, 2) Lapping pattern generation module that yields all the lap splice positioning within the permitted intervals specified by design codes, 3) Cutting pattern generation module that generates all the possible cutting patterns and 4) Trim loss optimization module that identifies the optimal cutting patterns to minimize the steel rebar waste generated during construction of concrete structures (Figure 1). Each module is described in the following sections. As shown in Figure 1, the trim loss analysis starts once the structural design phase is completed and all data regarding column sections, including the section dimensions, steel bar size and the number of steel bars, are finalized.

2.1 Data Collection Module

The data collection module is devised to provide the analyzer module with the data required to perform the trim loss analysis. As depicted in Figure 1, the information required in the data collection module comprises the number of stories of the building, storey heights, steel bar size and number of each column section, the overlap and hook length calculated based on design codes, the slab thickness of each storey and the foundation depth. These data are imported from the structural BIM model into a database which is used by the optimization code developed in MATLAB [13].

2.2 Lapping Pattern Module

Based on the ACI code, the lap splices of lateral load resisting columns are permitted only within the center half of the column length [14]. The overlap lengths are usually shorter than the columns half-length, therefore, the location of the center of lap can vary within the interval obtained from that difference (Figure 1). The aim of the developed framework is to identify the optimal rebar arrangements that will lead to a set of cutting patterns that will minimize the trim loss of the column steel bar and thus, minimize the steel trim loss of the whole project.

The lapping pattern generation module developed using MATLAB [13] yields all possible lapping patterns, considering the allowable intervals specified in design codes. The pattern generation code starts by importing the data made available by the data collection module. Then, the feasible region for the center of lap is identified based on the length of each column (storey height) and the overlap length required for each storey. It is assumed that bars are cut into lengths that are whole numbers. Thus, the length of the feasible region, which is basically a length in centimeter, represents the number of the center of lap positions. For instance, if the difference of the allowable upper and lower bound of the center of lap of a storey is 20 cm, there will be 20 different positions for the center of lap of that storey. Due to the difference in height and required overlap length of each storey of a column, the possible numbers of positions for the center of lap tend to be different. The role of the generation module is therefore to generate all the possible combinations of the center of lap positions for different stories of a column in a building. Based on the concept of permutations with repetitions [15], the total number of combinations for each column will be the product of the number of possible overlap positions for each storey height of that column. In other words, if there are 15, 20 and 30 allowable positions for storey 1, 2 and 3 of a three storey building respectively, there will be

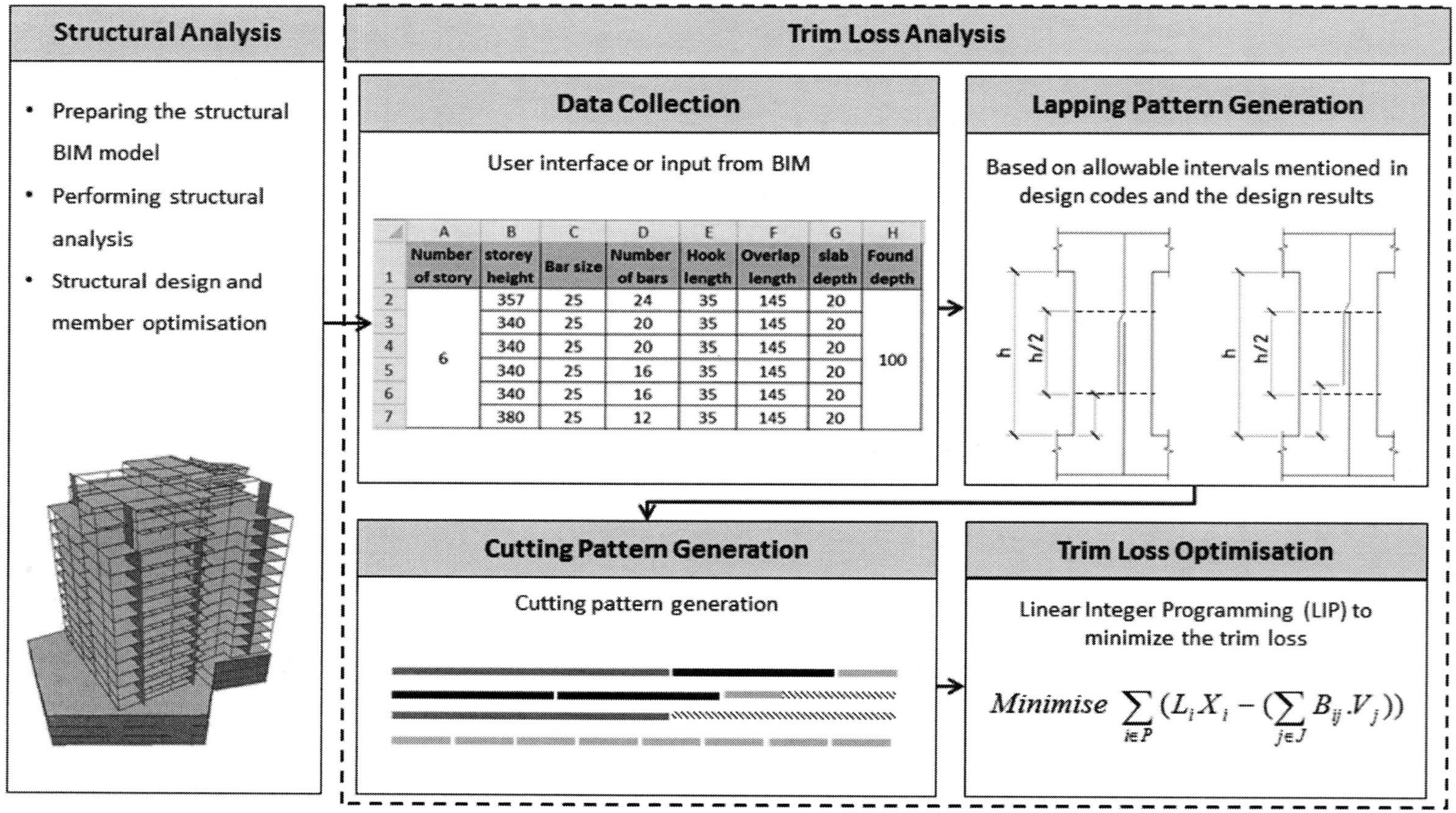

Figure 1. Framework for trim loss optimization of columns

15 × 20 × 30 equal to 9000 different arrangements for the lap splices of the columns of that building. The number of different arrangements for this simple example is indicative of the time and effort required for manual generation of all the possible arrangements for the columns in a typical building. To automate the process, once all the possible patterns are generated and stored in a matrix, the lengths of reinforcement bars of different stories are calculated based on the positions stored in that matrix. The length of dowels is calculated considering the depth of the foundation and hook length. Similarly, the length of steel bars of the last storey is calculated considering the slab thickness and hook length.

2.3 Cutting Pattern Generation Module

Cutting pattern generation module is responsible for the generation of all feasible cutting patterns for each set of lengths obtained in the previous step. In this study, this is achieved using the algorithm developed by Pierce [16]. First, the required lengths, l_1, l_2, l_3, ... , l_N, are arranged in descending order. The required number of each length is shown by d_1, d_2, d_3, ..., d_N. In the first pattern (i = 1), the number of bars corresponding to the longer required length (l_1) that can be cut from the standard length (L), A_{11}, is the minimum of the number of bars of that required length (d_1), and the quotient of the remaining length of the standard bar when divided by the required length. This is repeated for all other required bar lengths.

$$A_{i1} = \min\left(\left\lfloor \frac{L}{l_1} \right\rfloor, d_1\right) \tag{1}$$

$$A_{i2} = \min\left(\left\lfloor \frac{L - A_{i1}l_1}{l_2} \right\rfloor, d_2\right) \tag{2}$$

$$A_{iN} = \min\left(\left\lfloor \frac{L - \sum_{j=1}^{N-1} A_{ij} l_j}{l_N} \right\rfloor, d_N\right) \tag{3}$$

Efficient pattern number i is [A_{i1}, A_{i2}, . . . , A_{iN}], where A_{ij} is the number of demanded units of length l_j that are present in pattern number i. The algorithm proceeds as follows: If in pattern number i, there exists a j, $1 \le j \le N$, such that $A_{ij} > 0$ then let k be equal to the largest j and proceed as follow:

$$Set\ A_{i1} = A_{(i-1)1}, \ldots, A_{i(k-1)} = A_{(i-1)(k-1)} \tag{4}$$

$$A_{ik} = A_{(i-1)k} - 1 \tag{5}$$

$$A_{i(k+1)} = \min\left(\left\lfloor \frac{L - \sum_{j=1}^{k} A_{ij} l_j}{l_{k+1}} \right\rfloor, d_{k+1}\right) \tag{6}$$

This process continues until all the feasible patterns are generated. The Pseudocode of this pattern generation procedure can be found in Salem et al. [5].

2.4 Trim Loss Optimization Module

The next step is to select a combination of patterns that minimize the trim loss, thus generating the minimum waste possible. The optimization procedure selected is Linear Integer Programming (LIP). The model is formulated and solved using the linear solver CPLEX [17]. The objective function and constraints implemented are described next.

2.4.1 Objective Function

Equation below represents the objective function:

$$Minimize \sum_{i \in P} (L_i X_i - (\sum_{j \in J} B_{ij} . V_j)) \tag{7}$$

The set notation adopted is as follows: The set of patterns is indexed by i, such that the total number of patterns obtained from the cutting pattern generation module is P. The set of bar lengths demanded, which are cut from the standard steel bar lengths L_i, of each cutting pattern i, is indexed by the letter j. The total number of demanded bar lengths is thus represented by the letter J. Each pattern i, and each demanded length j is associated with a single value that represents the total number of bars of length j to be cut from L_i, and this value is represented by the notation B_{ij}. V_j denotes the length of demanded steel bar j. A single type of decision variable, an integer variable, X_i, specifies the number of times that pattern i is used. For instance, if in the final optimum solution, the value of X_3 equals to 8 and the value of X_6 equals to 5, the cuts that would minimize the generated waste are to use pattern number 3 a total of 8 times and pattern number 6 a total of 5 times. A total summation over the entire set of patterns is minimized to ensure that the solution produced is the one that corresponds to the minimum trim loss throughout the entire steel bar cutting phase in a project.

2.4.2 Constraints

Two constraints are defined to outline the shape of the feasible searching space for the algorithm. The first constraint requires that the total demand of bar length j is met by the solution. In particular, each demand d_j is set equal to the total number of instances a particular length is used within the solution pattern generated. This is to ensure that the set of lengths specified for each structural column as dictated by the design plans are met. The second constraint defines the domain of the decision variable. This constraint ensures the integrity of the decision variables, such that the variable x_i can take only integer values.

$$\sum_{i \in P} B_{ij} X_i = d_j \qquad \forall j \in J \tag{8}$$

$$X_i \in Z^+ \quad \forall i \in P \tag{9}$$

3 Case Study

The proposed framework was adopted to obtain the lapping patterns of the column's steel bars of an actual 6 storey building. The data collection Excel spread sheet was filled up based on the information displayed in Table 1, acquired from the structural drawings of the project (Figure 2). The standard length for the bars of this project is 12 meters. The analyzer code was then used to read the input data and perform the lapping pattern generation, cutting pattern generation and optimization operations.

Table 1. Structural data regarding stories and column sections obtained from the ETABS structural design file

Story height	Bar size	# of bars	Hook length	Overlap length	Slab depth	Found. depth
357	25	24	35	145	20	
340	25	20	35	145	20	
340	25	20	35	145	20	100
340	25	16	35	145	20	
340	25	16	35	145	20	
380	25	12	35	145	20	

Table 2 shows the cutting patterns and the number of times each pattern should be used (X_i) to fulfill the demand list while resulting in the minimum waste. A total of 1620 lapping patterns, and accordingly 49031 cutting patterns, were generated. Lapping pattern number 305 was found to produce the minimum amount of waste compared to the other 1619 lapping patterns. A total of 31 cutting patterns were generated for this lapping pattern. Among these 31 cutting patterns, only 6 were used in the optimum answer. As can be seen in Table 2, 48 bars with a standard length of 12 m, were cut to fulfill the required bar demand, shown in the Demand row of Table 2. The selected set of patterns and X_i resulted in 41.43 m of trim loss, generating 7.2% waste, considering the total length used (576 m).

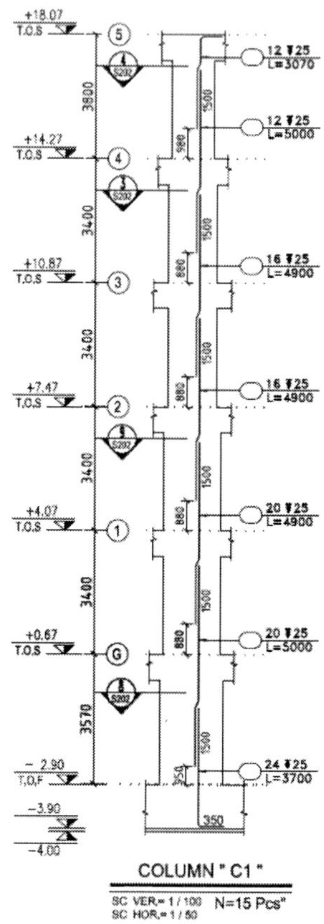

Figure 2. The original shop drawing prepared by the consultant

To show the effectiveness of the proposed framework, the bar demand list was prepared according to the structural drawings (Figure 2) and entered to the developed coded analyzer as the input for the third and fourth module, i.e. pattern generation and optimization modules. Table 3 shows the obtained cutting pattern along with the number of times that each of them is to be used (X_i) to fulfill the bar demand list. A total of 29 cutting patterns were generated, while only 7 patterns were included in the optimum solution. As can be seen in Table 2, 52 bars with a standard length of 12 m were cut to fulfill the required bar demand list. The selected set of patterns and X_i resulted in 83.56 m of trim loss, generating 13.4% waste, considering the total length used (624 m).

Comparing the total amount of material used and the

Table 2. Optimized cutting patterns based on lengths from actual shop drawing

	Patterns – number of bars (B_{ij}) of length V_j to be cut from the standard length (L)							Number of time each pattern is used in the final solution (X_i)	$X_i \times V_j \times \beta_{ij}$	$X_i \times L$
Patterns used in the final solution (i)	$V_1 = 5.35$m	$V_2 = 5.18$m	$V_3 = 5.05$m	$V_4 = 4.65$m	$V_5 = 4.65$m	$V_6 = 3.6$m	$V_7 = 2.7$m			
1	2	-	-	-	-	-	-	6	64.2	72
8	-	2	-	-	-	-	-	10	103.6	120
14	-	-	2	-	-	-	-	8	80.8	96
19	-	-	-	2	-	-	1	10	120	120
24	-	-	-	-	2	-	1	2	24	24
25	-	-	-	-	1	2	0	12	142.2	144
							Sum	48	534.8	576
							Trim loss	576-534.8=41.2		
							Total waste	41.2/576=7.2%		

Table 3. Optimized cutting patterns based on length generated from this framework

	Patterns – number of bars (B_{ij}) of length V_j to be cut from the standard length (L)							Number of time each pattern is used in the final solution (X_i)	$X_i \times V_j \times \beta_{ij}$	$X_i \times L$
Patterns used in the final solution (i)	$V_1 = 5.0$m	$V_2 = 5.0$m	$V_3 = 4.9$m	$V_4 = 4.9$m	$V_5 = 4.9$m	$V_6 = 3.7$m	$V_7 = 3.07$m			
1	2	-	-	-	-	-	-	6	60	72
8	-	2	-	-	-	-	-	10	100	120
14	-	-	2	-	-	-	-	8	78.4	96
19	-	-	-	2	-	-	-	8	78.4	96
23	-	-	-	-	2	-	-	4	39.2	48
24	-	-	-	-	1	1	1	12	140.04	144
26	-	-	-	-	-	3	0	4	155.4	48
							Sum	52	540.44	624
							Trim loss	624-540.44=83.56		
							Total waste	83.56/624=13.4%		

waste produced (Tables 2 and 3), reveals the effectiveness of the developed framework and the potential for reducing the rebar waste in construction projects through optimizing the cutting patterns. Optimizing the cutting patterns in the case project resulted in 49.5% waste reduction and a saving of 7.7% in the total amount of material used. Considering a unit weight of 3.85 kg/m for the bar with a diameter of 25 mm, used for these columns, and the fact that 15 of these columns were constructed in the case project, 2.77 ton of used steel bars and 2.43 ton of waste steel bars could be saved. This could in turn, result in considerable reduction in the cost of materials as well as the cost and energy consumed to recycle steel bars at the end of life of the structure. The proposed framework can also be extended to optimize the cutting pattern for other elements including shear walls.

4 Conclusion

Materials in the form of one-dimensional stocks such as steel bars generate a major fraction of the construction waste. This is due to the fact that these stocks are usually purchased in standard lengths and the required project lengths are shorter than the standard lengths. In this paper, a framework aiming at minimizing the trim loss, through automating the processes of lapping pattern generation, cutting pattern generation and trim loss optimization, was developed. A case study was presented to illustrate the effectiveness of the framework. The total amount of standard steel bars cut and the amount of generated waste produced after optimization were compared with those obtained from the actual shop drawings. A decrease of 7.7% and 49.6% in the total amount of material used and generated waste was observed respectively. The results highlight the potential for considerable reduction in rebar waste. Although, it should be noted that customizing the cutting pattern to minimize the rebar waste may in some cases add to the complexity of the rebar installation operation and thus reduce productivity. Therefore, careful consideration of costs and benefits by considering the project specific conditions is required.

References

1. Danatzko, J. and H. Sezen, *Sustainable Structural Design Methodologies*. Practice Periodical on Structural Design and Construction, 2011. **16**(4): p. 186-190.
2. Moussavi Nadoushani, Z.S. and A. Akbarnezhad, *Effects of structural system on the life cycle carbon footprint of buildings*. Energy and Buildings, 2015. **102**: p. 337-346.
3. Akbarnezhad, A. and Z.S. Moussavi Nadoushani, *A computational method for selection of optimal concrete recycling strategy*. Magazine of Concrete Research, 2014. **67**(11): p. 543-558.
4. Poon, C.S., A.T.W. Yu, and L. Jaillon, *Reducing building waste at construction sites in Hong Kong*. Construction Management and Economics, 2004. **22**(5): p. 461-470.
5. Salem, O., A. Shahin, and Y. Khalifa, *Minimizing Cutting Wastes of Reinforcement Steel Bars Using Genetic Algorithms and Integer Programming Models*. Journal of Construction Engineering and Management, 2007. **133**(12): p. 982-992.
6. Chen, Y. and T. Yang, *Lapping Pattern, Stock Length, and Shop Drawing of Beam Reinforcements of an RC Building*. Journal of Computing in Civil Engineering, 2013. **29**(1): p. 04014028.
7. Porwal, A. and K. Hewage, *Building Information Modeling–Based Analysis to Minimize Waste Rate of Structural Reinforcement*. Journal of Construction Engineering and Management, 2011. **138**(8): p. 943-954.
8. Benjaoran, V. and S. Bhokha, *Three-step solutions for cutting stock problem of construction steel bars*. KSCE Journal of Civil Engineering, 2014. **18**(5): p. 1239-1247.
9. Tekla, *Tekla Structures*. Kennesaw, GA.
10. Gilmore, P.C. and R.E. Gomory, *A Linear Programming Approach to the Cutting_Stock Problem*. Operation Research, 1961. **9**(6): p. 849-859.
11. Goulimis, C., *Optimal solutions for the cutting stock problem*. European Journal of Operational Research, 1990. **44**(2): p. 197-208.
12. Haessler, R.W., *Controlling Cutting Pattern in One-Dimensional Trim Problems*. Operations Research, 1975. **23**(3): p. 483-493.
13. MathWorks, I., *MATLAB*. 2011.
14. ACI, *Building Code Requirements for Structural Concrete (ACI 318-08) and Commentary*. 2008, American Concrete Institute: Farmington Hills, MI.
15. Gossett, E., *Discrete Mathematics with proof*. 2009, Hoboken, New Jersey: John Wiley & Sons.
16. Pierce, J.F., *Some large-scale production scheduling problems in the paper industry* in *Prentice-Hall*. 1964: Englewood, N.J.
17. IBM, *CPLEX optimizer*. 2009, IBM: NYC, USA.

Factors influencing BIM Adoption in Small and Medium Sized Construction Organizations

Ying Hong[1], Samad M. E. Sepasgozar[2], A. Ahmadian F.F.[1], and Ali Akbarnezhad[1]

[1]School of Civil and Environmental Engineering, University of New South Wales, Australia
[2]Faculty of the Built Environment, University of New South Wales Australia
E-mail: z5023853@student.unsw.edu.au, Sepas@unsw.edu.au, ahmadian@unsw.edu.au
a.akbarnezhad@unsw.edu.au

Abstract –

The advantages and disadvantages of Building Information Modelling adoption have been extensively investigated in available literature. However, the focus of previous studies has been placed mainly on BIM adoption in relatively large organizations and little effort has been made to investigate the effects that the size of the organizations may have on the BIM adoption process and its associated costs and benefits. This study aims to identify factors that affect the BIM adoption process in small and medium size construction organizations (SMOs) in terms of adoption motivations, ease of implementation and organizational competency. Quantitative and qualitative information are collected through survey and face-to-face interview of construction SMOs in Australia. Structural Equation Modelling is used to quantify the relationships between influential factors and organization's intention towards BIM utilization. The results presented in this paper aim to provide an insight into the needs and concerns of SMOs with regards to BIM adoption.

Keywords –

BIM adoption; small and medium sized construction organizations; influential factors

1 Introduction

Building Information Modelling has been advocated widely as a platform to enhance information communication, and reduce the project's costs and duration in the construction industry. Construction managers and engineers typically communicate with several parties involved in the project. Improved communication and exchange of technical information between the parties through BIM has been reported to affect project's productivity, quality and time [37]. National Institute of Standards and Technology reported an annual loss of $15.8 billion in the US, due to the lack of interoperability in the capital facilities. In addition, a variety of other benefits including project cost reduction and improved project team collaboration have been attributed to BIM adoption in construction [9,13,17,38]. However, apart from these associated benefits, BIM adoption in construction also faces a number of major challenges including difficulties in operating and maintaining BIM files, interoperability issues, and high implementation fee [3,25,28]. There is a great deal of literature on costs and benefits associated with BIM adoption process, the stages of the adoption process, and measures for effectiveness and maturity of BIM adoption and implementation [6,12,50]. However, previous studies have focused mainly on BIM adoption in relatively large organizations and little has been done to investigate the benefits and costs associated with BIM adoption as well as the BIM adoption motivations and challenges in small and medium sized organizations (SMOs). This is despite the fact that SMOs comprise the largest proportion of the construction industry. According to Australian government statistics [7], 97.8% construction firms are categorised as small business in Australia, and contributing 44.6% of the total market value. Despite the growth in clients' demand for BIM utilization, the adoption of BIM remains a big concern to SMOs [35,40].

This study contributes to the body of knowledge in construction technology adoption by identifying factors affecting BIM adoption decision in SMOs. This is achieved by analysing the results of a systematic survey and interview of 40 BIM professionals employed by SMOs across Australia. The results of this study can provide a basis for development of a decision making tool to assist SMOs managers to predict the outcome (i.e. adopt or reject) of BIM adoption in their organisation. This paper consists of four main sections: First, the available literature of advantages and disadvantages of BIM adoption in SMOs is reviewed. Second, the potential factors affecting the BIM adoption decision are categorised into three groups, viz. adoption motivation, ease of implementation, and organizational competency. Third, a generic BIM adoption model is presented and a number of hypotheses are developed. Finally, the results of a systematic survey conducted in Australia are analysed. SEM analysis and path analysis are presented and verifications are discussed.

2 Literature Review

2.1 Terms

This paper identifies factors which may affect BIM adoption in SMOs. The categorization criteria for SMOs vary from a country to another. European Commission (EU recommendation 2003/361) defines SMOs as companies with less than 250 full-time employees and an annual turnover of less than 50 million euros [18]. In China, the size of a construction company is determined by the number of its registered engineers (<60) and its total capital (<) [30]. Australian Bureau of Statistic [7] defines a small and medium-sized business as an activity trading business with 0-200 employees. In the present study the criterion used by Australian Bureau of Statistics is adopted to identify small and medium-sized contractors.

2.2 Advantages and Disadvantages

The costs and benefits associated with BIM implementation have been discussed in a number of previewed studies. However, the focus of such studies has been mostly placed on larger organizations. SMOs have considerable characteristic differences with larger organizations, which may affect the adoption process as well as the benefits and costs associated with the BIM adoption.

Based on the available literature, the advantages of BIM adoption by SMOs can be summarised in two main generic groups. 1) More organic organizational structure, which enables SMOs to have higher flexibility to change the existing working procedure. As Parida [19] pointed out, changing people's habits and getting them up to speed on a different technology is challenging especially for complex structured organizations. In larger organizations, all departments and levels have to face the challenge brought by working process change. 2) Smaller project size may help BIM implementation rate in SMOs to grow faster than in larger organizations [35]. Although SMOs are showing concerns about BIM's cost and applicability to their projects due to comparatively smaller size of projects, the size actually is a plus in driving higher levels of implementation. In larger organizations, a longer time is recognized to achieve complete and coordinate transition projects to BIM, because of the long project duration.

Furthermore, SMOs may not have enough budgets to afford BIM implementation and training expenses, as well as resources such as technical support [21,52]. Apart from above two disadvantages, the lack of understanding and knowledge of BIM in SMOs are obstacles to improve BIM implementation level within organizations [21,24].

3 Influential Factors

Based on the benefits and risks of BIM adoption, and theories of implementing innovations in construction industry, this study categorises influential factors in three groups – adoption motivation, organizational competency, and ease of implementation.

3.1 Adoption Motivation

During the innovation implementation process, the initial stage involves developing an awareness of the innovation and perceiving the needs to adopt the innovation [43,48]. Adriaanse, et al. [2] indicate that substantial benefits of implementing BIM in projects and external adoption motivations are key drivers for a company to approach BIM. In this study, in line with the existing literature on technology adoption the field of information technology, as discussed in the following, perceived usefulness of BIM implementation, organizational innovativeness, subjective norms, and awareness are categorised as major motivations of BIM adoption [33,52].

Perceived Usefulness (PU): This term was introduced by Davis [16], refers to the degree to which an individual believe that using a system would enhance his/her job performance. BIM utilization benefits not only individuals but also organizations [17,42]. Measures for PU include direct and indirect BIM implementation benefits in the organizational level. Direct benefits include reduced project costs and time, and improved team work; while indirect benefits mainly refer to intangible benefits, i.e. reputation, and benefits, which cannot be realized in short terms.

Innovativeness (IN): In a BIM acceptance study reported by Lee, et al. [33], this term was measured in two aspects: organizational innovativeness and individual innovativeness. A general survey of McGraw Hills Constructions [36] found that the innovativeness of an organization's leaders is a critical initiator of BIM adoption. Two proposed measures for innovativeness include the desire to build up organizational competiveness (IN1) and the need to streamline organization's workflow (IN2).

Subjective Norms (SN): Fishbein and Ajzen [19] defined this term as the person's perception that most people who are important to him think he should or should not perform the behaviour. Since in construction industry, activities are linked contractually, and contractual obligation is critical for an organization to accept new technologies [2]. Sexton and Barrett [47] found that one of motivations for small construction firms to implement innovations is survival in the market, which refers to the need for acceptance of technology as an operating necessity in the market. In this study, clients' requirement (SN1) and the need to share project information with other parties (SN2) are considered as key measures for SN.

Awareness (AW): Technology awareness is an essential component at early stages of implementation process [47,48]. A comprehensive understanding of innovations before decision making improves the efficiency of innovation implementation [43,48]. A study conducted by Parida, et al. [40] indicated that managers' knowledge of BIM has positive effects to BIM implementation. BIM awareness, in this study, is measured by the understanding of BIM definition (AW1), and knowledge of BIM applications (AW2).

3.2 Ease of Implementation

Technical issues have been highlighted by a number of studies as the main barrier affecting BIM adoption [1,25], especially in SMOs due to their limited resources [31,34]. The three primary factors contributing to ease of implementation include Ease of Operation, Ease of Maintenance, and Down Time.

Ease of Operation (EO): Also commonly termed as "ease of use" [27], this term refers to the degree to which an innovation is perceived as being difficult to use. The results of a survey reported by [52] and [38] indicate that difficulties in operating BIM are key barriers to BIM adoption. In this study EO is measured in terms of difficulties in learning process (EO1), model creation (EO2), and refining the project documentations (EO3).

Ease of Maintenance (EM): Maintenance for information systems generally refers to maintenance of applications, ongoing technical supports and upgrade (i.e. software and hardware). Maintenance is mentioned as an important factor for determining the cost of change management using BIM [29].

Down Time [27]: Down Time refers to a period of time that BIM fails to provide or perform its primary function either planned or unplanned [46]. The transition from traditional CAD drafting to BIM modelling causes higher requirements of hardware's capability. However, the tight budget may not allow SMOs to invest sufficiently in hardware upgrade [52]. Rogers, et al. [44] found that technical downtime contributed high probabilities to unsatisfied BIM implementation results. In this study, DT is measured in three aspects: the identification of DT as a risk (DT1), likelihood of DT occurrence (DT2), and consequence of DT occurrence (DT3)

3.3 Organizational Competency

Organizational competency can be measured in following areas: organizational support, expertise, and organization intention.

Organizational Support (OS): This term is defined as the degree to which an organization's policy supports BIM utilization [33]. Previous studies [40,49] show that top management support, including providing training (OS1) and encouraging staff to use BIM in daily work (OS2), is critical to BIM adoption.

Expertise (EP): BIM implementation is a complicated process, which requires professional technical skills [6,8,45]. The IT team plays a critical role in hardware selection, software installation, and ongoing support of the BIM implementation process. In project context, the variety of applied BIM software raise up issues in data interoperability [25,28]. Therefore, a technical support to shoot down relevant problems may improve the BIM implementation performance. In this study, expertise is measured by received professional support in hardware/software selection (EP1), and in BIM implementation (EP2).

Organizational Intention (OI): Fishbein and Ajzen [19], Davis [16] labelled this term as user acceptance, which is a prerequisite of actual use. Lee, et al. [33] estimated the organizational acceptance by the strength of willingness of an organization's intention to adopt, implement or recommend BIM to others. In the present study the organizational intention is considered to include intention to implement BIM into more projects (OI1) and to recommend BIM implementation to others (OI3).

4 Research Method

This study employed face-to-face interview and web based questionnaire to obtain quantitative and qualitative data. Interview helps to deeply understand industrial perspectives towards BIM adoption in SMOs. Questionnaire provides statistic basis to qualify relationships between influential factors and BIM adoption. The questionnaire uses 7-point Likert scale to measure each manifest variable. Research participants were chosen from senior personnel, managers, directors in SMOs with BIM knowledge. The research analysis was designed for a sample of 120 participants. At the first phase of this study, presented herein, a total of 40 responses were received.

The hypotheses of this study are tabulated in Table 1 and the hypothesises-based diagram is shown in Figure 1. This study employs Structural Equation Modelling to test the hypotheses. A two-step approach suggested by Anderson and Gerbing [5] is used to analyse SEM models, to reduce the potential model misspecifications. In the first step, the model's consistency, and the validate model's construct are assessed. The internal consistency of data is assessed using Cronbach's α value [15]. While the assessment of construct validate is performed by breaking the problem down to assessment of convergent validity and discriminant validity [14]. Convergent validity ensures that manifest variables measuring the same latent variable are demonstrating moderate inter-correlation; while an acceptable discriminant validity testing result requires independency among latent variables [20]. Convergent validity may be assessed using the following indicators: factor loading, composite reliability, and average variance extracted [10]. The most frequent way to examine discriminant validity is comparing the AVE value of a certain latent variable with the squared values of correlations between that latent variable and others latent variables [10]. The second step is hypotheses testing, based on proposed model (**Error! Reference source not found.**). Path analysis is employed in hypothesises testing. Path analysis decomposes effects into direct, indirect and total effects, [4].

Table 1 Hypothesises

Hypo thesis	Definition
H1	
a	Perceived Usefulness has a positive effect on Organizational Support
b	Innovativeness has a positive effect on Organizational Support
c	Awareness has a positive effect on Organizational Support
H2	
a	Perceived Usefulness has a positive effect on Organizational Intention
b	Innovativeness has a positive effect on Organizational Intention
c	Awareness has a positive effect on Organizational Intention
H3	
a	Ease of Operation has a positive effect on Organizational Intention
b	Ease of Maintenance has a positive effect on Organizational Intention
c	Down Time has a negative effect on Organizational Intention
H4	
a	Organizational Support has a positive effect on Ease of Operation
b	Organizational Support has a positive effect on Ease of Maintenance
c	Organizational Support has a negative effect on Down Time
H5	
a	Organizational Support has a positive effect on Organizational Intention

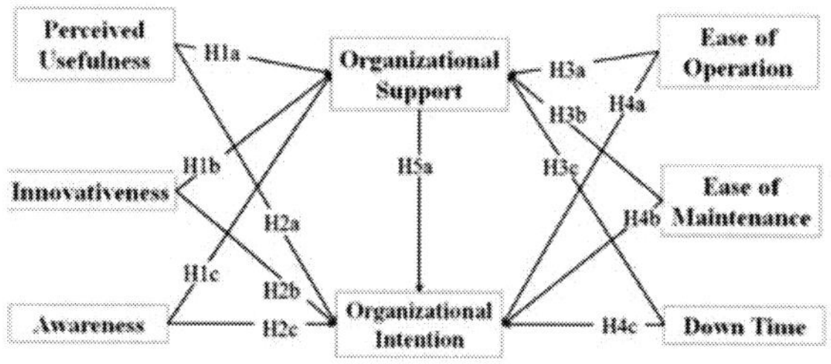

Figure 1. Organisation BIM adption model inclduing heypotheses

5 Results of Validity Tests

Error! Reference source not found. presents the profile of respondents. 60% of participants involve with commercial and residential building projects; 45% of them are junior BIM operators with less than a year experience. The respondents were roughly reported an average rate of 38% BIM usage in their technical documentation process. 3D visualization, clash detection and quantity take-off are three applications which most frequently were used.

Table 2 Respondents' Profile

Characteristic		Statistics (%)
Business Type	Contractors	90.0%
	Consultants	7.5%
	Architects	2.5%
Number of Full-time Employees	0-4	15.0%
	5-19	17.5%
	20-199	60.0%
	200-499	7.5%
Project Type	Commercial	60.0%
	Residential	60.0%
	Public	27.5%
Years of Using BIM	Non users	37.5%
	6-12 months	7.5%
	1-2 years	32.5%
	2-5 years	22.5%

BIM Applications	3D Visualization	88.0%
	Clash Detection	68.0%
	Quantity Take-off	48.0%
Projects using BIM	Average Percentage	38%

Table 3 shows the Cronbach's α value for variables investigated in this study. As shown all the values, except for OI, are higher than the minimum acceptable threshold of 0.60 [39].

Table 3 Reliability Test

Latent Variable	Cronbach's α	No. of items
Awareness	0.777	2
Innovativeness	0.694	3
Perceived Usefulness	0.754	4
Organizational Support	0.847	4
Ease of Operation	0.731	3
Ease of Maintenance	0.864	3
Down Time	0.744	3
Organizational Intention	0.572	2

Table 4 presents the results of the convergent validity test. As shown, factor loadings of manifest variables to latent variables range from 0.456 to 1.043, composite reliability ranges from 0.573 to 0.862, and AVEs are between 0.402 and 0.701. The minimum acceptable threshold of these indicators are 0.5, 0.6, and 0.5, respectively [26].

Table 4 Convergent Validity

LV	MV	Factor loading	CR	AVE
Awareness	AW1	0.974	0.819	0.701
	AW2	0.673		
Innovativeness	IN1	0.456	0.744	0.509
	IN2	0.723		
	IN3	0.892		
Organizational Support	OS1	0.777	0.857	0.604
	OS2	0.601		
	EP1	0.849		
	EP2	0.855		
Ease of Operation	EO1	0.597	0.767	0.553
	EO2	0.462		
	EO3	1.043		
Ease of Maintenance	EM1	0.942	0.862	0.684
	EM2	0.908		
	EM3	0.582		
Down Time	DT1	0.922	0.720	0.479
	DT2	0.506		
	DT3	0.576		
Organizational Intention	OI1	0.605	0.573	0.402
	OI2	0.662		
Perceived Usefulness	PU3	0.704	0.769	0.460
	PU4	0.565		
	PU5	0.808		
	PU6	0.609		

The results of the discriminant validity test are shown in [10]. As shown, the results indicate high correlations for PU and EM, SN and OS, EM and EO.

Table 5. The values on the diagonal of the matrix are AVEs of each latent variable, while off-diagonal values are square values of correlations between the latent variable and other latent variables in its corresponding column. An acceptable level of discriminant validity requires AVEs greater than off-diagonal elements [10]. As shown, the results indicate high correlations for PU and EM, SN and OS, EM and EO.

Table 5 Discriminant Validity

	PU	OI	IN	EM	DT	OS	EO
PU	**.68**						
OI	.47	**.72**					
IN	.43	.38	**.71**				
EM	.69	.59	.57	**.83**			
DT	.49	.23	.23	.45	**.69**		
OS	.46	.37	.71	.68	.24	**.78**	
EO	.49	.61	.46	.90	.45	.48	**.74**

Model fitness indices of independent model and developed model were tested using six indices. The values for these indices as well as the recommended values are shown in Table 6[11].

Table 6 Goodness of Fit

Measure	Measurement model	Structural Model	Recommend value
$\chi 2/df$	1.861	1.976	≤ 3.0
RMR	0.259	0.324	≤ 0.10
RMSEA	0.149	0.158	≤ 0.10
TLI	0.599	0.545	≥ 0.80
CFI	0.674	0.611	≥ 0.80
PNFI	0.422	0.396	≥ 0.50

Table 7 decomposes the internal effects of variables into total, direct, and indirect effects. According to analysis results, AW, IN and EO have significant effects on OI; OS is significant to EO and EM; IN has significant indirect effects on EO, DT and EM. Meanwhile, Ease of Operation has been identified as a mediator which fully mediates the effects between OS and OI.

In general, while indicating a nearly acceptable performance for a majority of indicators, the results indicate the need for a higher number of data to further validate the results of the present study. The data collection is currently ongoing and will continue to meet all the validity and convergence criteria tested above.

Table 7 Total, Direct and Indirect Effects

		AW	PU	IN	OS	EO	DT	EM
Total Effect	OS	.03	.34*	.58***	0	0	0	0
	EO	.02	.20*	.34***	.60***	0	0	0
	DT	.01	.10	.18**	.31	0	0	0
	EM	.02	.25	.42***	.74***	0	0	0
	OI	.48**	.08	.38*	.21	.52**	-.09	-.001
Direct Effect	OS	.03	.34*	.58***	0	0	0	0
	EO	0	0	0	.60***	0	0	0
	DT	0	0	0	.31	0	0	0
	EM	0	0	0	.74***	0	0	0
	OI	.47**	.01	.26	-.07	.52**	-.09	-.001
Indirect Effect	OS	0	0	0	0	0	0	0
	EO	.02	.20*	.34***	0	0	0	0
	DT	.01	.10	.18**	0	0	0	0
	EM	.02	.25	.42***	0	0	0	0
	OI	.01	.07	.12	.28	0	0	0

*p<0.1, **p<0.05, ***p<0.01

6 Findings and Discussions

Findings of this paper are different to other studies in two ways. First, it examines different factors influencing BIM adoption in both individual and organizational levels; whereas, previous studies are mainly focused on BIM adoption at individual levels Peansupap and Walker [41] and Lee, et al. [33]. Second, the paper specifically investigates SMOs in the Australian construction industry using both quantitative and qualitative methods.

The results show that three main factors of 'Awareness', 'Innovativeness', and 'Ease of Operation' have been identified as critical factors that have been influenced the respondents' decision when considering to adopt or reject BIM . The result presented in Table 7 particularly show that Ease of Operation (β=0.52, $p<0.05$) is critical for individuals who contribute in the BIM adoption decisions, while Innovativeness (β=0.38, $p<0.10$), and Awareness (β=0.48, $p<0.05$) are vital for organizations. Since individuals contribute in the decision process, and the organisations should follow their procedures, both individual and organisational factors should be considered for BIM adoption prediction. Previous studies [33] mainly focus on the individual attributes and intentions in the BIM adoption decision, while the result shows that organizational supports have significant positive effects on two factors Ease of Operation (β=0.60, $p<0.01$) and Ease of Maintenance (β=0.74, $p<0.01$) by individuals.

Table 7 shows that Ease of Operation positively and fully mediates the positive relationship between Organizational Support (β=0.60, $p<0.01$) and Organizational Intention (β=0.52, $p<0.01$) [22]. The relationship is critical for BIM technology vendors and facilitators, because it shows the organisation intention toward BIM adoption is mainly affected by Ease of Operation and Maintenance. Surprisingly, the Organisation Support does not directly effects on Organisation Intention to use BIM. This is an important finding and refers to the fact that organisation support will be an influential factor if it effectively increases perceive of Ease of Operation. This is new to the literature, as it has been ignored in previous relevant literature. The insignificant correlation between EM and OI as shown in Table 7 indicates that most respondents don't think difficulties in maintaining BIM models and files would affect their decision about BIM adoption. Although this result is different with literature [3,25,28], SMOs may not experience BIM maintenance issues yet, due to smaller projects' sizes [15], and respondents' limited experiences in using BIM.

The result of the qualitative analysis shows that SMOs in Australia concern about organizations' competiveness and satisfying clients' requirements. According to respondents' feedbacks, influences from clients are initiations for SMOs to approach BIM. These influences not only include clients' requirement of BIM involvement in projects, but also include showcasing company's capability.

Except satisfying clients' requirements, displaying visualized video clips on our website is a kind of testimonial of our capability and profession.

My previous firm lost ten projects due to not using BIM.

However, because of the financial limitation, BIM implementation stays at the very basic stage e.g. drafting. Some SMOs are focusing on remodelling their previous project by using BIM, by adding Sketch-up. In ongoing projects, SMOs were found to suffer from unavailability of sufficient financial support in maintaining BIM models and files throughout the process.

We have amount of budget to do models on projects, mainly use an add-ins on program to Sketch up. But if we use BIM through the building cycle, it's gonna cost more, especially we started using BIM, operation skills are not mature.

This basic level of BIM implementation leads to lack of interoperation within a project. Suppliers are unable to extract information from contractors' deliveries, project information from architects are not consumed by contractors. However, the results also indicate that SMOs tend to be fresh in BIM implementation (45% respondents are with less than one year experience in BIM utilization), and getting SMOs involved into a collaborative working circumstance based on BIM needs a long time.

7 Conclusion

This study was aimed at identifying the factors that influence the BIM adoption decisions in SMOs. The results show that three main factors, viz. 'Awareness', 'Ease of operation', and 'Innovativeness' are critical factors influencing organisations decision to adopt or reject BIM. Furthermore, the qualitative data cross validate the results of the paper, and additionally indicates that complexity and compatibility are the main specific BIM implementation barriers.,

The results of this study also contribute to the body of knowledge by examining key factors (e.g. ease of operation and maintenance) influencing the SMOs' decision in the BIM adoption process, rather than focusing on factors (e.g. playfulness and anxiety) related to the individuals perception of information systems usage. The operation costs associated with BIM were identified by respondents as one of the main barriers to BIM adoption. A higher emphasis was placed on this cost factor by non-users than users, which indicates the possibility of overestimation of real operation costs by non-users. In addition, the results can provide an enriched understanding of the BIM adoption process in SMOs in Australia, which account for the largest proportion of the construction industry. The main limitation of this study is the relatively small sample size used at this stage. The future work involves recruiting more participants to improve the reliability of the findings.

References

[1] A. Abuelmaatti, V. Ahmed, Collaborative technologies for small and medium-sized architecture, engineering and construction enterprises: implementation survey, ITCON special issue: emerging digital technologies and innovations. pg (2014).

[2] A. Adriaanse, H. Voordijk, G. Dewulf, Adoption and use of interorganizational ICT in a construction project, Journal of Construction Engineering and Management 136 (9) (2010) 1003-1014.

[3] A. Alabdulqader, K. Panuwatwanich, J.-H. Doh, Current use of building information modelling within Australian AEC industry, Proceedings of the Thirteenth East Asia-Pacific Conference on Structural Engineering and Construction (EASEC-13), The Thirteenth East Asia-Pacific Conference on Structural Engineering and Construction (EASEC-13), 2013, pp. C-3-1.

[4] D.F. Alwin, R.M. Hauser, The decomposition of effects in path analysis, American sociological review (1975) 37-47.

[5] J.C. Anderson, D.W. Gerbing, Structural equation modeling in practice: A review and recommended two-step approach, Psychological bulletin 103 (3) (1988) 411.

[6] Y. Arayici, P. Coates, L. Koskela, M. Kagioglou, C. Usher, K. O'Reilly, BIM adoption and implementation for architectural practices, Structural Survey 29 (1) (2011) 7-25.

[7] S. Azhar, Building information modeling (BIM): Trends, benefits, risks, and challenges for the AEC industry, Leadership and Management in Engineering 11 (3) (2011) 241-252.

[8] S. Azhar, J. Brown, R. Farooqui, BIM-based sustainability analysis: An evaluation of building performance analysis software, Proceedings of the 45th ASC annual conference, Vol. 1, 2009.

[9] R.P. Bagozzi, Y. Yi, On the evaluation of structural equation models, Journal of the academy of marketing science 16 (1) (1988) 74-94.

[10] R.P. Bagozzi, Y. Yi, Specification, evaluation, and interpretation of structural equation models, Journal of the academy of marketing science 40 (1) (2012) 8-34.

[11] K. Barlish, K. Sullivan, How to measure the benefits of BIM—A case study approach, Automation in construction 24 (2012) 149-159.

[12] B. Becerik-Gerber, S. Rice, The perceived value of building information modeling in the US building industry, Journal of information technology in Construction 15 (2) (2010) 185-201.

[13] D.T. Campbell, D.W. Fiske, Convergent and discriminant validation by the multitrait-multimethod matrix, Psychological bulletin 56 (2) (1959) 81.

[14] E. Commission, EU recommendation 2003/361, 2003.

[15] M.H. Construction, The business value of bim for infrastructure, The Business Value of BIM for Infrastructure Addressing Americas

Infrastructure Challenges with Collaboration and Technology. Bedford, MA: McGraw-Hill Construction Research & Analytics (2012).
[16] M.H. Construction, The business value of BIM in Australia and New Zealand: How building information modelling is transforming the design and construction industry, Bedford, MA: McGraw Hill Construction (2014).
[17] L.J. Cronbach, Coefficient alpha and the internal structure of tests, Psychometrika 16 (3) (1951) 297-334.
[18] F.D. Davis, Perceived usefulness, perceived ease of use, and user acceptance of information technology, MIS quarterly (1989) 319-340.
[19] C. Eastman, C.M. Eastman, P. Teicholz, R. Sacks, BIM handbook: A guide to building information modeling for owners, managers, designers, engineers and contractors, John Wiley & Sons, 2011.
[20] M. Fishbein, I. Ajzen, Belief, attitude, intention, and behavior: An introduction to theory and research, (1977).
[21] C. Fornell, D.F. Larcker, Evaluating structural equation models with unobservable variables and measurement error, Journal of marketing research (1981) 39-50.
[22] P. Forsythe, The Case for BIM Uptake among Small Construction Contracting Businesses, Proceedings of the 31st International Symposium on Automation and Robotics in Construction and Mining, University of Technology Sydney, 2014.
[23] J. Gaskin, "SEM", Gaskination's StatWiki, 2012.
[24] B.K. Giel, Return on investment analysis of building information modeling in construction, University of Florida, 2009.
[25] B. Gledson, D. Henry, P. Bleanch, Does size matter? Experiences and perspectives of BIM implementation from large and SME construction contractors, (2012).
[26] N. Gu, K. London, Understanding and facilitating BIM adoption in the AEC industry, Automation in construction 19 (8) (2010) 988-999.
[27] J.F. Hair, R.E. Anderson, R.L. Tatham, C. William, Black (1998), Multivariate data analysis, Upper Saddle River, NJ: Prentice Hall, 1998.
[28] J.F. Hair Jr, G.T.M. Hult, C. Ringle, M. Sarstedt, A primer on partial least squares structural equation modeling (PLS-SEM), Sage Publications, 2013.
[29] D. Haynes, Reflections on some legal and contractual implications of building information modeling (BIM), Construction Watch 2 (9) (2009) 1-9.
[30] D. Holzer, The BIM Manager's Handbook, Part 2: Change Management, John Wiley & Sons, 2015.
[31] JonesDay, China's Minstry of Construction Issues New Qualification Regulations for Construction, Design, and Supervision Enterprises, 2007.
[32] T. KOUIDER, G. PATERSON, Architectural Technology and the BIM Acronym, Architectural Technology: The Defining Features. Proceedings of the 4th International Congress of Architectural Technology, 2013, pp. 122-141.
[33] S. Lee, J. Yu, D. Jeong, BIM acceptance model in construction organizations, Journal of Management in Engineering (2013).
[34] K. Manley, Against the odds: Small firms in Australia successfully introducing new technology on construction projects, Research Policy 37 (10) (2008) 1751-1764.
[35] A. Mutai, Factors influencing the use of Building Information Modeling (BIM) within leading construction firms in the United States of America, Indiana State University, 2009.
[36] K. Newton, N. Chileshe, Awareness, usage and benefits of building information modelling (BIM) adoption-The case of the South Australian construction organisations, Association of Researchers in Construction Management, 2012.
[37] J. Nunnally, Psychometric Theory (2nd eel.), New York, Iv\ cGrow-Hill (1978).
[38] V. Parida, J. Johansson, H. Ylinenpää, P. Baunerhjelm, Barriers to information and communication technology adoption in small firms, Working paper, Swedish Entrepreneurship Forum, 2010.
[39] V. Peansupap, D. Walker, Factors affecting ICT diffusion: a case study of three large Australian construction contractors, Engineering, Construction and Architectural Management 12 (1) (2005) 21-37.
[40] E. Poirier, S. Staub-French, D. Forgues, Embedded contexts of innovation: BIM adoption and implementation for a specialty contracting SME, Construction Innovation 15 (1) (2015) 42-65.
[41] E.M. Rogers, Elements of diffusion, Diffusion of innovations 5 (2003) 1-38.
[42] J. Rogers, H.-Y. Chong, C. Preece, Adoption of Building Information Modelling technology (BIM) Perspectives from Malaysian engineering consulting services firms,

Engineering, Construction and Architectural Management 22 (4) (2015) 424-445.
[43] R. Sebastian, W. Haak, E. Vos, BIM application for integrated design and engineering in small-scale housing development: a pilot project in The Netherlands, International symposium CIB-W096 future trends in architectural management, 2009, pp. 2-3.
[44] S. Sepasgozara, S. Davisb, Modelling the Construction Technology Implementation Framework: An Empirical Study, America 30 (33) (2015) 63.
[45] M. Sexton, P. Barrett, Appropriate innovation in small construction firms, Construction Management and Economics 21 (6) (2003) 623-633.
[46] E.S. Slaughter, Implementation of construction innovations, Building Research & Information 28 (1) (2000) 2-17.
[47] H. Son, Y. Park, C. Kim, J.-S. Chou, Toward an understanding of construction professionals' acceptance of mobile computing devices in South Korea: An extension of the technology acceptance model, Automation in construction 28 (2012) 82-90.
[48] A.B.o. Statistics, Counts of Australian Business, June 2014, 2014.
[49] B. Succar, M. Kassem, Macro-BIM adoption: Conceptual structures, Automation in construction 57 (2015) 64-79.
[50] H. Yan, P. Damian, Benefits and barriers of building information modelling, 12th International conference on computing in civil and building engineering, Vol. 161, 2008.
[51] L. Zhou, S. Perera, C. Udeaja, C. Paul, Readiness of BIM: a case study of a quantity surveying organisation, (2012).

Simulation of HVAC Local Control Based on Occupants Locations and Preferences

Zheng Liu[a], Shide Salimi[b], and Amin Hammad[c]

[a] Department of Building, Civil, and Environmental Engineering, Concordia University, Canada
[b] Department of Building, Civil, and Environmental Engineering, Concordia University, Canada
[c] Concordia Institute for Information Systems Engineering, Concordia University, Canada (corresponding author)
Email: l_zhe28@encs.concordia.ca, Email: sh_sa@encs.concordia.ca, Email: hammad@ciise.concordia.ca

Abstract –

The Heating, Ventilation and Air Conditioning (HVAC) system consumes the major part of energy in buildings. By analyzing the gap between the actual energy consumed and the required energy to satisfy heating/cooling loads, it is estimated that an average of 38% of the energy can be saved with the adoption of more efficient control strategies. Occupancy-driven management has attracted great attention due to the potential energy savings. Therefore, the development and implementation of these strategies are of primary importance. On the other hand, future smart buildings will have the ability to detect and locate the occupants, and adjust the HVAC system to better satisfy their preferences, which is expected to result in considerable energy savings. This paper aims to simulate the energy performance that can be achieved by applying local control of the HVAC system. It proposes an HVAC localized control strategy based on the locations and preferences of occupants in an office open space with multiple zones. In order to evaluate the energy saving after applying the control strategy, DesignBuilder software is used to model one office at Concordia University and simulate its energy consumption.

Keywords –

Energy consumption; energy simulation; occupancy; locations; preferences; HVAC setpoint.

1 Introduction

It is estimated that the world energy consumption will increase by 56% from 2010 through 2040 [1]. Due to the catastrophic climate change, growing population and shortage in fossil fuels, the imbalance of increasing energy demand and limited energy resources has become a global environmental and economic problem and a driving force for researches on energy conservation techniques [2].

Buildings account for one-third of the total global energy consumption [3]. In Canada, approximately 85% of the total energy in institutional and commercial buildings is consumed by heating, cooling, lighting, and IT equipment [4]. In order to save energy in institutional and commercial buildings, a more efficient approach is to manage facilities based on actual occupancy locations and preferences.

Previous literature mainly contributed to developing occupancy models to predict the presence of the occupants. These models have been applied to controlling the HVAC and lighting systems. However, it did not consider local control combining the spatiotemporal variations of the space usage and the occupants' preferences.

This paper aims to simulate the energy performance that can be achieved by applying a localized control of the HVAC system. It proposes an HVAC localized control strategy based on the locations and preferences of occupants in an office open space with multiple zones. In order to evaluate the energy saving after applying the control strategy, EnergyPlus software and its graphical interface DesignBuilder are used to model one office at Concordia University and simulate its energy consumption.

2 Literature Review

Building energy management systems (BEMSs) are computer-based systems that can assist in the management, control and monitoring of HVAC, lighting, energy monitoring and targeting, etc. [5]. With programmed routines and repetitive functions, relatively simple operation, quick response to complaints, improved management information and graphical representation of operating conditions, BEMSs enable up to 10% reduction in energy consumption and costs [6]. On the other hand, occupants' behavior affects HVAC energy consumption with respect of the thermal loads, ventilation loads, and conditioning requirements of HVAC control settings to meet their thermal comfort and standard air quality [3].

There are four levels of occupancy models: building level, occupancy status of space, number of occupants in a space, and the location of occupants [7]. The higher level of occupancy information that can be extracted, the more accurate the simulation will be and the more energy the control strategy can save. To achieve occupancy-based control, common occupancy sensors are utilized to identify presence and advanced occupancy sensors (e.g. motion detectors) are utilized to identify the number of occupants in a zone and track their locations. Several occupancy detection and tracking technologies are utilized to sense the occupancy information. For example, optical cameras and thermal cameras are used to collect and analyze the images of the monitored areas including occupants; passive infrared (PIR) based motion detectors are used to sense movement of occupants; carbon dioxide based detection technology detects the concentration of CO_2 to measure occupancy; radio frequency technology such as Radio Frequency Identification Device (RFID) and Device Free Localization (DFL) detects location and activities of occupancy; Ultra Wideband (UWB) Real-Time Location Systems (RTLSs) use radio signals to detect the movement of occupants [8].

A report prepared for U.S. Department of Energy [9] proposes three cases to identify the energy savings using occupancy-based control in variable-air-volume (VAV) systems of large offices. The result shows that 17-23% of energy in HVAC and lighting can be saved by using advanced occupancy sensors. For individual offices, a group of researchers used synergy occupancy sensor nodes to detect occupancy and implemented the presence system in simulation [10]. They found that the daily HVAC savings can reach 10-15% after applying the occupancy system. Gunay et al. [11] developed an occupancy learning algorithm for terminal heating and cooling units to predict the arrival and departure time of occupants in private offices, leading to 10 -15% reduction in the annual heating and cooling loads. A versatile smart phone application is developed for the employees at the office building called Edge in Amsterdam to dial in their individual climate and lighting preferences. Packed with 28,000 sensors, The Edge is considered to be the greenest and most intelligent building in the world [12].

Due to the flexible work hours and large spaces that HVAC systems need to deal with, especially in office large open spaces, it is more energy-saving to apply localized heating and cooling control to specify the energy needed for occupied and unoccupied areas. Lo and Novoselac [13] developed a localized air flow with occupancy control strategy for cooling and simulated the Computational Fluid Dynamics (CFD) to examine the air quality and to calculate savings in energy consumption. It is concluded that 12% of energy can be saved. Budaiwai and Abdou [14] presented the energy saved by HVAC system operational strategies in the case of mosques, a unique type of buildings with intermittent occupancy in Saudi Arabia. To operate the HVAC system only when the space is occupied, they proposed the strategies of oversizing the HVAC system to balance the accumulated thermal load and zoning the conditioned area according to the schedule and location of occupants (e.g. one zone is occupied on all weekdays and another zone is only occupied on Friday). It is concluded that savings of 13% and 30% of energy consumption for a mosque with two zones (504 m^2) and another mosque with three zones (1050 m^2) can be achieved, respectively.

Selecting the suitable software for energy simulation is an important decision; Yang and Becerik-Gerber [3] compared available energy simulation programs in the market based on the accuracy and reliability to react to the effects of occupancy on HVAC energy consumption and response of HVAC system to occupancy-based control strategies. They concluded that the most qualified program is EnergyPlus, developed by U.S. Department of Energy, because of its capabilities for HVAC system simulation. DesignBuilder makes the simulation more user-friendly and flexible because of its graphical user interface [15]. In this paper, EnergyPlus and its graphical interface DesignBuilder are selected to simulate the proposed HVAC control strategy and to evaluate the energy savings.

3 Proposed Method

Figure 1 shows an example to demonstrate the proposed method. The figure shows a relatively large open office in a smart building equipped with Sensor Tags (STs), capturing the location of occupants at a certain frequency. To control the HVAC and lighting systems at a more detailed level, the office can be divided into two zones according to the number of HVAC terminal units, and six zones according to the number of lights. Knowing the location of a specific occupant, the corresponding HVAC terminal unit and corresponding light are activated.

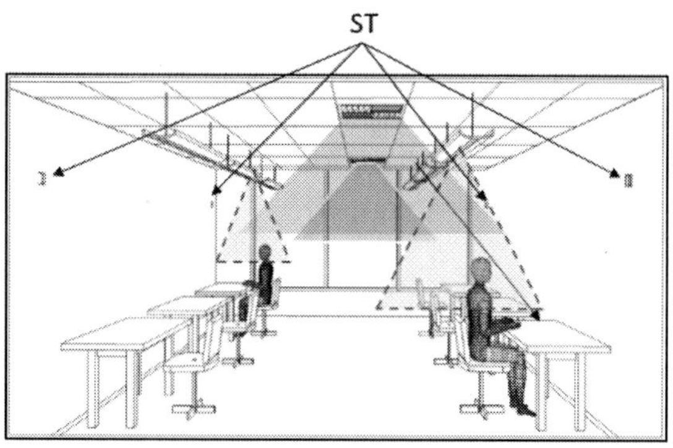

Figure 1. Proposed HVAC control strategy

In this paper, firstly, the identity and location of occupants over time are captured by UWB RTLSs [16]. The overall occupancy profiles can be created based on the collected data of the occupants on an hourly basis in one week. The probabilistic individual occupancy profiles can be used to identify the specific needs of each occupant and their preferred comfort settings. Assuming that the occupancy pattern is fixed, the resulting profiles can be used for long-term control of light and HVAC settings to avoid the problems of delays of system response that may be caused by depending only on real-time control. These profiles can be also used in the simulation of the energy consumption of the building, which can help forecast long term energy needs. It should be mentioned that providing control of the lighting and HVAC systems based on individual profiles depends on the availability of individual light units and HVAC outlets near each occupant location. As this condition may not be available in many cases, occupants' profiles should be integrated to match the actual physical layout of the lighting and HVAC systems in the specific space based on combined zones. To satisfy the occupants in one zone, when the occupants' profiles overlap, the preferences of the occupants are averaged.

3.1 Modeling of the Office Space

The open office with multiple zones is modeled in an energy simulation program. Before applying localized control, the building block model has to be divided into different zones. The criteria of zoning are: (1) different facade orientation for perimeter zones, because the wind and solar gain changes throughout the day will affect both thermal and daylight control performance; (2) different activity types in different areas, because the change in occupancy and equipment can affect the internal gains; (3) applying bespoke activity data including occupancy schedules, temperature setpoints etc.; (4) different HVAC system in different parts of the building; and (5) different lighting systems in different parts of the building [17].

The zones in the open office space are separated by virtual partitions. There are two types of partitions, standard and virtual partitions. Standard partitions are used to create enclosed spaces within the block using walls as a physical boundaries. Virtual partitions create a new zone without creating a physical boundary. They have no thermal mass and present no barrier to heat and air flow caused by the natural ventilation. Virtual partitions are used to separate open areas which have different HVAC or lighting systems [17].

3.2 Energy Simulation

Based on the occupants' schedules and preferences, the schedule of terminal heating and cooling units and lights can be adjusted. Then, the energy consumption of the proposed control strategy can be obtained by running the simulation. Three scenarios are generated to quantify the energy saving. Scenario 1 assumes the setpoint of HVAC system is always fixed to the condition that the area is occupied all the time. Scenario 2 assumes that when there is no occupant in the area, the setpoint of the HVAC system will be adjusted to save energy. When there is any occupancy in the office open space, the setpoint of the HVAC system will be changed to fit the occupants' preferences. Scenario 3 represents the case of localized control. It is assumed that the localized control can be implemented in the large open office with fixed number of users assigned to each zone. The HVAC system can be adjusted according to the occupancy in each zone. By comparing with the baseline (Scenario 1) and Scenario 2, the energy saving of Scenario 3 can be quantified.

4 Implementation and Case Study

A model is created in DesignBuilder, the graphical interface of EnergyPlus. After modeling the building, different schedules are used as input to run the simulation. Setpoint Managers are used to control the air loop of the HVAC system. By using Setpoint Managers, the setpoint of air handling units' (AHU) supply air temperature can be scheduled. The schedules of occupancy are imported to the model in DesignBuilder to quantify the energy saving after applying the local control.

To apply the proposed control strategy, a 3D model of an office space is created in DesignBuilder. The model represents an office at Concordia University in Montreal, Canada. The model is shown in Figure 2. To define the three scenarios, three schedules are generated as input for the model.

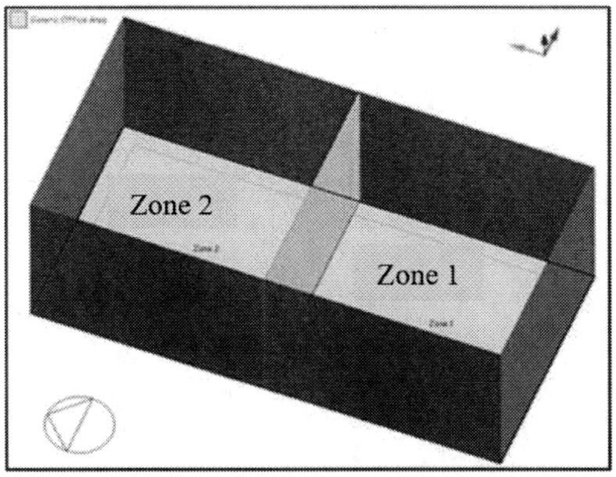

Figure 2. DesignBuilder model

4.1 Occupancy Schedules

Occupancy information can be obtained using occupancy tracking technique. The identity and location of four occupants in the room (M, N, S, L) over one week in June, 2014 are captured by an UWB RTLSs [16]. The readings of the tags are updated twice per minute. As Figure 3 shows, M and N are under the same terminal unit of HVAC system; therefore, M and N are assigned to Zone 1 of the space. Likewise, S and L are assigned to Zone 2. As an example of the occupancy data, the percentage of occupancy over time for one week is shown in Figure 4. The occupancy schedule is used for the calculation of internal loads for the HVAC system in the three scenarios.

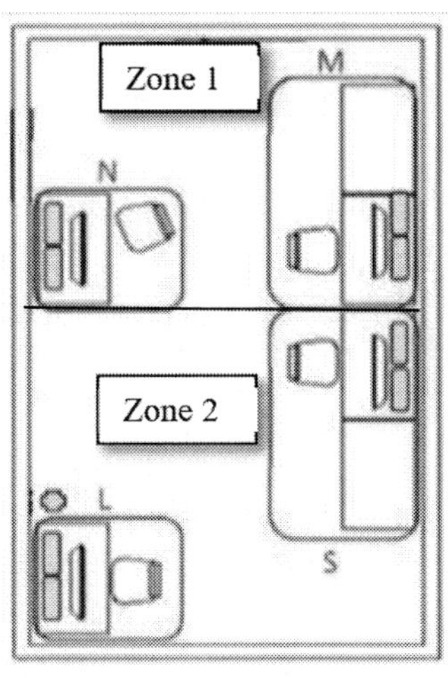

Figure 3. Distribution of occupants [16]

4.2 Defining HVAC Schedules

To achieve the proposed control, first, instead of using ideal loads in the modelling by applying the Simple HVAC model in DesignBuilder, it is necessary to apply the Detailed HVAC made by defining EnergyPlus HVAC systems graphically using components and customizing the setpoints in heating and cooling. There are two types of data in Detailed HVAC: (1) Simple HVAC data and (2) Detailed HVAC data. In Simple HVAC and Detailed HVAC/Simple HVAC data, the setpoint can be adjusted in each zone, but the value is constant. In this research, the setpoint should be specified by adding thermostat cooling and heating setpoint schedules according to the occupants' schedules and preferences for the corresponding zones. Therefore, Detailed HVAC/Detailed HVAC data is selected.

There are three schedule types for the schedule data in DesignBuilder, 7/12 Schedule, Compact Schedule and Day Schedule. Based on the collected data, the occupancy schedule for each day in the week can be customized. The text-based Compact Schedule is chosen to input the schedule. Compact Schedules are defined following a standard format which contains the elements Through (date), For (days), Interpolate (optional), Until (time of day) and Value (percentage of occupancy). Figure 5 shows an example of a compact schedule.

4.3 Defining Cooling Setpoint Schedule

As shown in Table 1, in Scenario 1, the setpoint for cooling is fixed (24 °C) on both weekdays and weekends. In Scenario 2, on weekdays, from 8:00 to 18:00, the setpoint is 24°C when there is occupancy in the space and 25°C when it is not occupied; before 8:00 and after 18:00 in the day, the setpoint is 26°C when the space is not occupied and 24°C when there is occupancy. On weekends, the setpoint is 26°C when the space is not occupied and 24°C when there is occupancy in the space. In Scenario 3, on weekdays, when there is no occupancy in the zone, the setpoint is 25°C from 8:00 to 18:00 and 26°C for other period of the day; when there is occupancy in the zone, the setpoint is adjusted to 24°C. On weekends, for the unoccupied zone, the setpoint is set to 26°C; when the zone is occupied, the setpoint is 24°C.

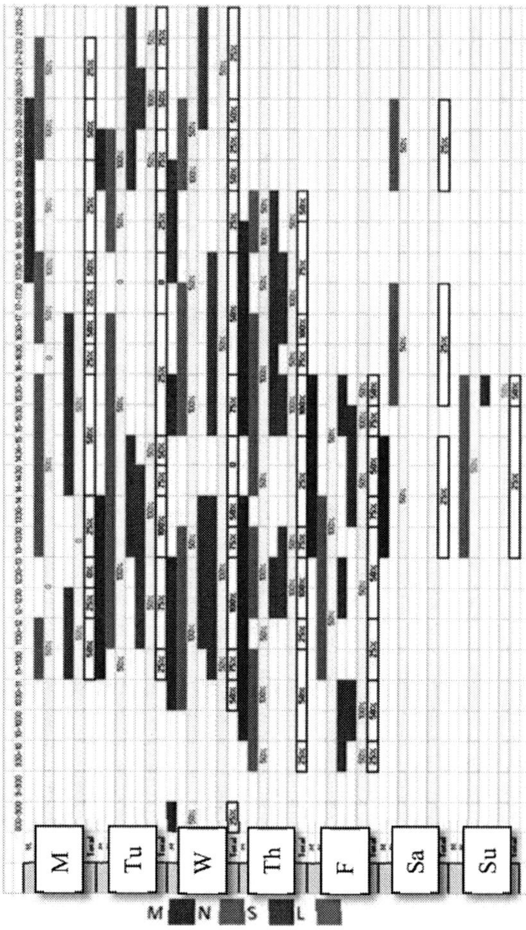

Figure 4. Occupants' schedule

```
Schedule:Compact,
Office_Occ_Z1,
Fraction,
Through: 13 Jul,
For: AllDays,
Until: 13:00, 0,
Until: 16:00, 0.5,
Until: 24:00, 0,
Through: 14 Jul,
For: AllDays,
Until: 11:00, 0,
Until: 12:00, 0.5,
Until: 13:00, 0,
Until: 16:00, 0.5,
Until: 16:30, 0,
Until: 17:30, 0.5,
Until: 18:00, 1,
Until: 19:30, 0.5,
Until: 20:30, 1,
Until: 21:30, 0.5,
Until: 24:00, 0,
```

Figure 5. Example of a compact schedule

Table 1 HVAC cooling setpoint schedule for three scenarios

		Not occupied			Occupied
		0:00-8:00	8:00-18:00	18:00-24:00	
Scenario 1	Weekdays	24 ^{0}C			
	Weekends	24 ^{0}C			
Scenario 2 (Space-level)	Weekdays	26 ^{0}C	25 ^{0}C	26 ^{0}C	24 ^{0}C
	Weekends	26 ^{0}C			24^{0}C
Scenario 3 (Zone-level)	Weekdays	26 ^{0}C	25 ^{0}C	26 ^{0}C	24 ^{0}C
	Weekends	26^{0}C			24 ^{0}C

4.4 Results and Discussion

After inputting the mentioned schedules in the model, the simulation is run for the three scenarios. The results are shown in Figure 6. By comparing Scenario 1 and Scenario 2, 1.5% of the energy can be saved at the space level without sacrificing the comfort of occupants. By comparing Scenario 1 and Scenario 3, 1.59% of the energy can be saved.

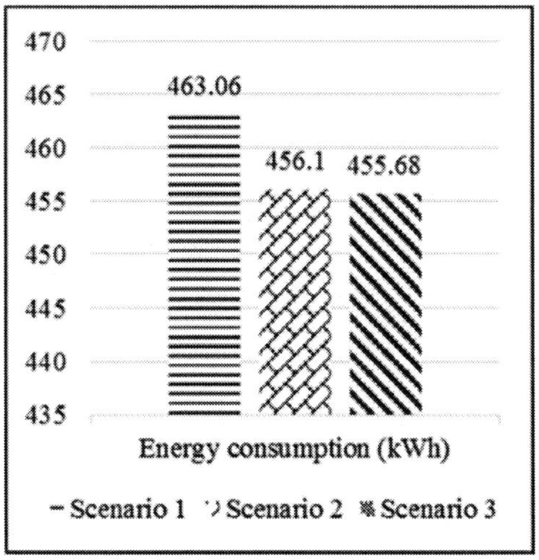

Figure 6. Energy consumption of the three scenarios

The results show only slight differences among the three scenarios. The probable reasons can be:

- The size of the model: It is expected that more significant energy saving can be achieved if the office space is large. As mentioned in [14], with strategies of oversizing the HVAC system and zoning the conditioned area, more energy can be saved in larger spaces.
- The lack of elements in the model: There is no windows in the proposed model, which play an important role in solar gain and energy consumption. Besides, the operation of equipment in the office is not considered, which can affect the heat gain in the space and the consumption of electricity.
- Number of occupants: In some office buildings, the distribution of occupants is generally based on the nature of their jobs. However, in some cases, it is possible to obtain more energy savings by the dynamic usage of the space. Reassigning the occupants with similar presence patterns or with similar preferences of heating and cooling setpoints to the same office can be one of the solutions to save energy without sacrificing the occupants' thermal comfort [18]. The zoning scheme can be applied to office buildings, which is called "hot desking" in the Edge building [12]. Without assigned seats for the occupants, a specific area in the space can be given precedence for dynamic assignment over others. With more occupants in that area, the preferred temperature for the heating and cooling setpoints can be averaged to satisfy the several occupants under same terminal unit.
- Setpoint: The setpoint can be adjusted within the range of the recommended values in ASHRAE standards. With greater difference in setpoints between the base case (Scenario 1) and other scenarios, the energy consumption difference can be more significant.
- Variety of occupants' schedules: In Scenario 2, the HVAC cooling schedule is based on the occupancy in the space; but the profiles of the four occupants are of little difference. Therefore, the schedules in Scenario 2 and Scenario 3 are very similar.

5 Conclusions and Future Work

In this paper, a localized control strategy of HVAC system is developed to adjust the setpoints and HVAC operation in an office open space with multi users according to the occupancy to save energy. According to the results of simulation, by setting different setpoints based on the occupancy presence, 1.59% of the energy can be saved.

Our future work will aim to validate the proposed control strategy by comparing the results of the simulation and the actual energy consumption. Furthermore, in this study, the occupancy profile is developed based on occupant detection using UWB RTLSs. Going forward, we will conduct tests using Bluetooth Low Energy (BLE) SensorTags with iBeacon [19]. Bluetooth is a set of wireless technologies for communication between mobile devices in short range. BLE is the enabler of the Internet of Things [20]. Compared with UWB RTLSs and other detection technologies, BLE is less expensive and more practical. Besides, it is much less energy consuming than the classic Bluetooth. The BLE hubs can be any device that implements the BLE standard, such as smart phones. The occupants do not have to wear tags as their smart phones can be tracked.

References

[1] SUSRIS. (2013, July 25). International Energy Outlook 2013. On-line: http://susris.com/2013/07/25/international-energy-outlook-2013/, Accessed: 06/06/2015.

[2] Bratt, N., Johansen, C. and Singh, M. P. Driving forces for energy demand. On-line: http://etreeprojects.com, Accessed: 02/05/2015.

[3] Yang, Z., and Becerik-Gerber, B. Coupling occupancy information with HVAC energy simulation: a systematic review of simulation programs. *Proceedings of the 2014 Winter Simulation Conference*, (pp. 3212-3223). Los Angeles, US, 2014.

[4] NRCan. 15th edition of energy efficiency trends in Canada. On-line: http;//www.nrcan.gc.ca, Accessed:

12/06/2015, 2011.

[5] Levermore, G. J. *Building energy management systems: applications to low-energy HVAC and natural ventilation control*. London: E & FN Spon, 2000.

[6] Defence Estates. *Building Energy Management Systems*. Stationery Office Books, 2001.

[7] Feng, X., Yan, D. and Hong, T. Simulation of occupancy in buildings. *Energy and Buildings*, 87:348-359, 2015.

[8] Priyadarshini, R. and Mehra, R. M. Quantitative Review of Occupancy Detection Technologies. International Journal of Radio Frequency Design, Vol.1 Issue 1, 2015.

[9] Zhang, J., Lutes, R., Liu, G., and Brambley, M. Energy savings for occupancy-based control (OBC) of Variable-Air-Volume (VAV) systems. Washington: Pacific Northwest National Laboratory, 2013.

[10] Agarwal, Y., Balaji, B., Gupta, R., Lyles, J., Wei, M. and Weng, T. Occupancy-driven Energy Management for Smart Building Automation. *Proceedings of the BuildSys, in conjunction with ACM SenSys Conference*, Zurich, Switzerland, 2010.

[11] Gunay, H. B., O'Brien, W. and Beausoleil-Morrison, I. Development of an occupancy learning algorithm for terminal heating and cooling units. *Building and Environment*, 93(2015) 71-85, 2015.

[12] BREEAM. The Edge, Amsterdam. On-line: http://www.breeam.com/index.jsp?id=804, Accessed: 14/09/2015.

[13] Lo, L. J. and Novoselac, A. Localized air-conditioning with occupancy control in an open office. *Energy and Buildings*, 1120-1128, 2010.

[14] Budaiwi, I., and Abdou. A., HVAC system operational strategies for reduced energy consumption in buildings with intermittent occupancy: the case of mosques. *Energy conversion and Management*, 73(2013) 37-50, 2013.

[15] DesignBuilder. DesignBuilder. On-line: http://www.designbuilder.co.uk/, Accessed: 20/06/2015.

[16] Masoudifar, N., Hammad, A., and Rezaee, M. Monitoring occupancy and office equipment energy consumption using real-time location system and wireless energy meters. *Proceedings of the 2014 Winter Simulation Conference*, pp. 1108-1119, 2014.

[17] DesignBuilder. DesignBuilder video tutorials. On-line: http://www.designbuilder.co.uk/content/view/105/184/, Accessed: 14/07/2015.

[18] Yang, Z., and Becerik-Gerber, B. The coupled effects of personalized occupancy profile based HVAC schedules and room reassignment on building energy use. *Energy and Buildings*, 78(2014) 113-122, 2014.

[19] Texas instruments. Texas instruments. On-line: http://www.ti.com/, Accessed: 05/09/2015.

[20] Bluetooth. Bluetooth. On-line; https://www.bluetooth.com/what-is-bluetooth-technology/bluetooth, Accessed: 12/09/2015.

Industrial timber house building – levels of automation

D. Popovic 1[a] and M. Winroth2[a,b]

[a] School of engineering, University of Jönköping, Sweden
[b] Chalmers University of Technology, Sweden
E-mail: djordje.popovic@ju.se, mats.winroth@chalmers.se

Abstract

Swedish industrial timber house building faces a constantly increasing housing demand. In order to respond to the demand, companies in this sector, have to improve their businesses in terms of productivity. At the same time, they need to meet customer requirements, i.e., offering flexible building solutions, thus creating a balance between productivity and flexibility. Off-site assembly of exterior walls is an essential part of the house production, and cutting down lead times at this phase hence, increases the competitiveness. If introduced in a right way, automation can contribute to a higher productivity. Thus, right levels of both physical and cognitive automation are necessary. The objective of this study is to measure the current levels of automation (LoA) within the off-site exterior wall assembly. Research design consisted of a literature study and a case study that was conducted at a Swedish company that is an engineer-to-order producer of single-family timber houses. A case study design was made according to the DYNAMO ++ framework. The framework was used in the assessment of LoA and designing flexible task allocation in many manufacturing industries, but there is a lack of knowledge on how to use this method in the industrial timber house building. The average physical and cognitive LoA of 124 identified tasks are 3 and 1 respectively. Increased physical and cognitive LoA for critical tasks would enable flexible task allocation between human operators and technology. It is believed that this type of flexibility can result in less production disturbances and higher productivity when a high variety of exterior walls is assembled.

Keywords –

Physical and cognitive levels of automation; hierarchical task analysis; exterior wall; single-family timber house; off-site assembly.

1 Introduction

Swedish industrial timber house building faces a constantly increasing housing demand. Compared to other materials, the use of wood for load-bearing structures is dominant in the Swedish single-family house production [1]. Moreover there is an increasing interest for timber load-bearing structures in multi-family house production [2]. Lately, 10 to 15 % of the total amount of houses that are built annually are timber frame multi-family houses [3]. In order to respond to the demand, companies in these industry sectors have to improve their businesses in terms of productivity. Nevertheless the balance between productivity and flexibility has to exist since many of the companies have customer oriented business strategies [4]. In such cases the type of supply chain is engineer-to-order (ETO) and the customer order enters the design phase of a supply chain [5]. Therefore, the lead time in ETO timber house building companies can be divided into three phases: design/engineering/process planning, off-site assembly and on-site assembly.

Timber house building, as a sub-division of the construction industry [6] has moved from the pure traditional on-site production towards the industrialized off-site prefabrication. [7] [8]. The level of off-site prefabrication varies depending on whether the house components, elements and/or volumes are delivered to the building site. [9] and [10] have made classification of the off-site assembly systems in this respect. Nevertheless customer oriented companies have understandably a lower level of the off-site prefabrication then the companies that offer standardized houses due to the transportation issues. Mostly non-volumetric house elements are prefabricated and delivered as such to the building site.

Off-site assembly of exterior walls is an essential part of the house production, in terms of the throughput time required. Assembly of these non-volumetric house elements is a bottleneck of the off-site prefabrication and cutting down lead times at this phase hence, increases the competitiveness of the company. One of the reasons to introduce automation as described by [11] is to increase productivity. Ironies of automation [12] have to be avoided therefore in order to reach higher productivity and maintain flexibility, it is necessary to establish right levels of both physical and cognitive automation [13]. The flexibility of the physical and cognitive task allocation between the operators and technology is needed due to the high number of variants of exterior walls that are pre-assembled. [14] developed a framework called DYNAMO ++ to measure levels of automation (LoA) and identify possible improvements in task allocation between humans and technology. There is a lack of knowledge about physical and cognitive LoA in the industrialized timber house building and currently there is no published material on these measurements. Therefore, it is of interest to put the developed framework for achieving flexible LoA into the context of industrialized timber house building.

The research question: What is the current state of levels of automation within the off-site exterior wall assembly? An additional objective is to compare the findings related to the research question to those of the previous case studies that used the same framework.

There are several delimitations of the study. Since the design variation in the case company is quite large, the research design for the data collection did not include all the types of exterior walls that are produced on the assembly line. The design variation refers to different types of siding on the outer side as well as different types of components such as windows, doors, electrical boxes and installations. In terms of the ETO supply chain lead time, the scope of the study was limited to the off-site assembly of exterior walls. The case study was done according to the part of the DYNAMO ++ framework. Therefore, the scope of the study is delimited only to the measurement phase of the framework.

Table 1. Reference scale for seven levels of physical and cognitive automation with exemplified explanations (adapted from [16])

LoA	Mechanical and Equipment (Physical)	Information and Control (Cognitive)
1	Totally manual - Totally manual work, no tools are used, only the user's own muscle power, e.g. the user's own muscle power	Totally manual - The user creates his/her own understanding for the situation, and develops his/her course of action based on his/her earlier experience and knowledge
2	Static hand tool - Manual work with support of static tool, e.g. screwdriver	Decision giving - The user gets information on what to do, or proposal on how the task can be achieved, e.g. work order
3	Flexible hand tool - Manual work with support of flexible tool, e.g. adjustable spanner	Teaching - The user gets instruction on how the task can be achieved, e.g. checklists, manuals
4	Automated hand tool - Manual work with support of automated tool, e.g. hydraulic bolt driver	Questioning - The technology question the execution, if the execution deviate from what the technology consider being suitable, e.g. verification before action
5	Static machine/workstation - Automatic work by machine that is designed for a specific task, e.g. lathe	Supervision - The technology calls for the user's attention, and direct it to the present task, e.g. alarms
6	Flexible machine/workstation - Automatic work by machine that can be reconfigured for different tasks, e.g. CNC-machine	Intervene - The technology takes over and corrects the action, if the executions deviate from what the technology considers being suitable, e.g. thermostat
7	Totally automatic - Totally automatic work, the machine solves all deviations or problems that occur. by itself, e.g. autonomous systems	Totally automatic - All information and control is handled by the technology. The user is never involved, e.g. autonomous systems

2 Theoretical background

Research on Levels of Automation (LoA), i.e. the allocation of functions or tasks between humans and technology, has been going on for more than half a century [15]. This early research focused very much on complex systems, mainly in aircrafts and traffic control. LoA in industrial applications was further developed by many authors, e.g. [16], who based their work on the automation strategy development by [17] and [18]. [19] defined 10 levels of automation, ranging from 1, i.e. human makes all the decisions and physical tasks, to 10, i.e. computer makes the decisions and the equipment carries out the task without humans being involved at all. In the DYNAMO project [16], reduced the number of LoAs to 7 and defined the reference scales according to Table 1. This reference scale is on a task level and so-far there is no methodology dealing with LoA on production systems level.

This method has been further developed into the DYNAMO++ framework [14]. The aim of the framework was to establish the accessible LoA present in the factory in order to create a range of possible LoA.

That would further enable a flexible task allocation by which production disturbances could be avoided and productivity increased when a high product variety is assembled at the factory [20]. The framework consists of twelve steps divided into four phases: pre-study, measurement, analysis and implementation [22]. Data is mainly collected in the measurement phase using three methods: value stream mapping, hierarchical task analysis, and levels of automation taxonomy.

The underlying question is why to automate, what to automate, and how to do it [17]. There is always a risk of creating a system that is too complex to manage and maintain, and too connected making everything stop in case of disturbances, and investing in sophisticated equipment that is not really needed.

3 Method/Case study

In order to answer the research question, case study was chosen for a research method. It involved several research techniques such as literature review, informal interviews with the operators working at the assembly line and video recording of the assembly process. The research design for data collection was done according to the measurement phase of the DYNAMO ++ framework.

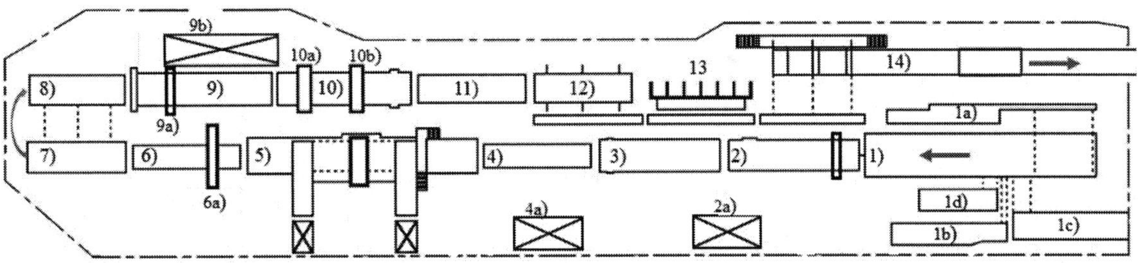

Figure 1. Assembly line layout: 1) basic wall frame assembly; 1a) horizontal studs; 1b) vertical studs; 1c) part units; 1d) rock-wool insulation; 2) wind protection sheet and air studs assembly; 2a) nailing and air studs inventory; 3) nailing studs assembly; 4) manual positioning of the horizontal siding wood panels; 4a) horizontal siding panels inventory; 5) nailing operation for vertical siding panels; 6) nailing operation for horizontal siding panels; 6a) nailing machine; 7) finalizing the outer side of the wall; 8) lifting belts assembly; 9) positioning of humidity protection plastic and gypsum/plywood boards; 9a) gypsum/plywood board positioner; 9b) gypsum/plywood board inventory; 10) nailing and routing operations for gypsum/plywood boards; 10a) nailing machine; 10b) CNC router; 11),12),13) final operations on the interior side of the wall; 14) quality control

The decision to use the DYNAMO ++ framework in this case study was based on the need to take both information and mechanical automation into consideration. A review of ten methods and models [20] that were used in redesigning, measuring or analysing production systems was used to justify the choice of the method. As the author indicated in the review, DYNAMO ++ was found to be an optimal choice where socio-cognitive and technical-physical aspects of a production system were equally taken into consideration. The other methods or models were developed for either analysis of socio-cognitive aspects:

- A model for types and levels of human interaction with automation.
- Complementary Analysis and Design of Production Tasks in Socio-technical Systems (COMPASS).
- Cognitive Reliability and Error Analysis Method (CREAM).
- Task Evaluation and analysis Methodology (TEAM)
- Taxonomy for Cognitive Work Analysis.

or technical-physical aspects of a production system:

- TUTKA production assessment tool.
- Systematic Production Analysis (SPA).
- Productivity Potential Assessment (PPA).
- Lean Customization Rapid Assessment (LCRA) [23].

The case study was conducted in the Swedish company that is an engineer-to-order producer of single-family timber houses. In terms of the level of off-site prefabrication the company pre-assembles non-volumetric house elements and does the final assembly on-site. The exterior walls are produced at the assembly line depicted in Figure 1. The assembly line consists of thirteen work stations.

The assembly of a single exterior wall was filmed with cameras. The unit of analysis was an exterior wall with windows and horizontal siding panels. This type of wall, shown in Figure 2, was chosen to be followed through the assembly process since it is the most commonly produced exterior wall at the moment and its complexity is rather intermediate.

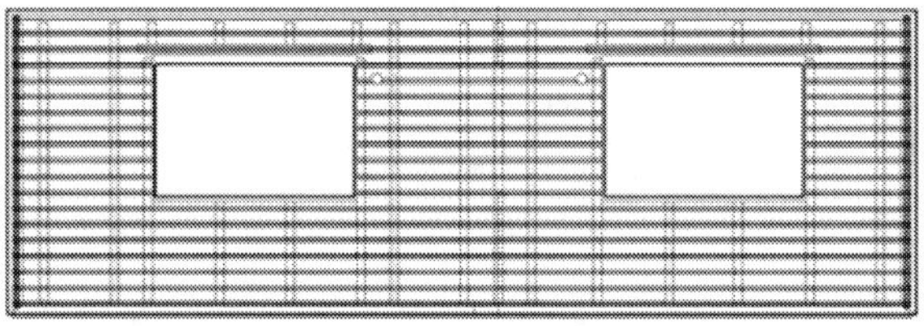

Figure 2. Part of the shop drawing of the type of wall that was followed and filmed through the assembly process

The operators working at the assembly line were given the information about the type of study and its purpose. This step was necessary for several reasons. It was important to make sure that there were no individuals who would not want to be recorded. By explaining the goals of the study operators were ensured that it is not their performance that would be assessed but rather the working tasks were under focus. This was particularly important in order to ensure that realistic current state of the work performed will be recorded. Furthermore the operators were assured that the filmed material would not be used for any other purpose other the one stated in the study objective.

The filmed material was sorted and analyzed with the purpose of developing a hierarchical task analysis map. The work stations were divided into operations/goals which were further divided into particular tasks. Every identified task was again analyzed from the video recordings and both physical and cognitive levels of automation were measured using reference scales (Table 1). For the tasks that appeared unclear in the recordings with respect to their cognitive part, i.e., which information the operators used when performing the tasks, informal interviews were conducted with several operators in order to get the full understanding of the tasks performed.

The levels of automation of all of the tasks within the assembly line were summarized according to their levels of physical and cognitive automation as shown in Figure 3. In order to put the assembly process back into the general perspective, it could be classified as either human assembling/monitoring, machine/technique monitoring or machine assembling depending on the spread of the tasks within the matrix.

A literature study was done in order to make the comparison between the exterior wall assembly and other manufacturing systems in terms of LoA. Published material from the previous case studies was analyzed to gather the data regarding measurements of LoA in other industries.

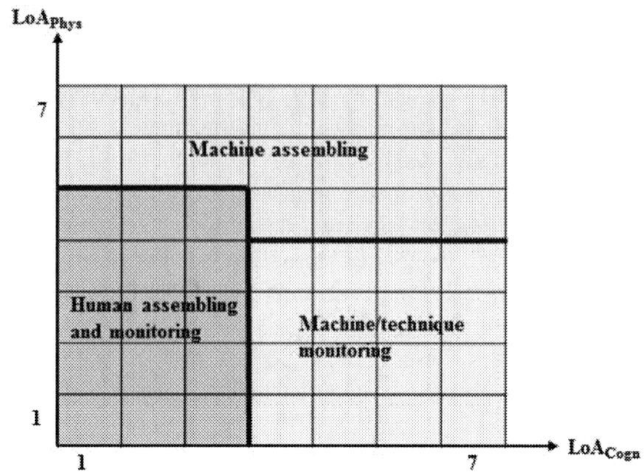

Figure 3. LoA matrix with three different regions tasks can be classified in depending on the allocation between humans and a machines (adapted from [13])

4 Findings

By implementing the hierarchical task analysis of the recorded material, 124 working tasks were identified. The working tasks are distributed over 24 operations which are further distributed over 13 working stations in the assembly line.

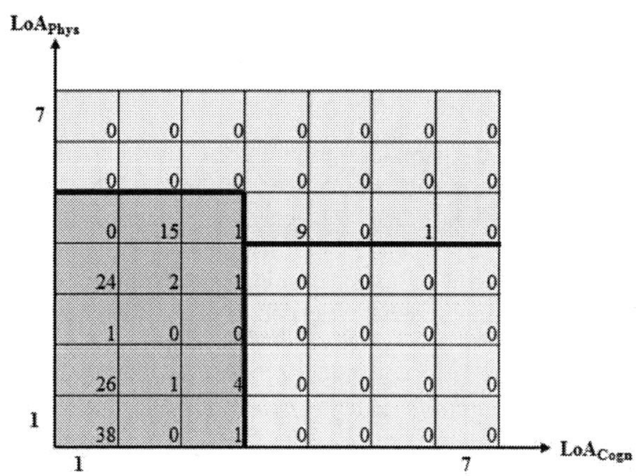

Figure 4. LoA of 124 tasks of the exterior wall assembly process

Most of the tasks, i.e., 114 out of 124, are regarded as human assembling and monitoring in terms of both physical and cognitive LoA. Aforementioned majority of tasks, the operators perform from their own knowledge and experience or using shop drawings which were the main information carriers at the assembly lines. Out of

124, 10 tasks are regarded as machine assembling. There were 9 tasks performed by the machines where information carriers were machine codes (LoA – 5:4). Only one task was performed by the machine that did not require a machine code but used sensors to take the course of the action (LoA – 5:6). Figure 4 shows a distribution of tasks according to physical and cognitive LoA.

The average physical and cognitive LoA is 3:1. The comparison between this case study and six previous studies [13] on LoA is shown in Table 2.

5 Discussion and conclusions

Some comments on research quality in terms of validity and reliability are made in the next three paragraphs. The validity of the study can be discussed from internal and external validity aspects. The internal validity, meaning as how well the researcher managed to measure and analyse what was meant to be measured and analysed, can be discussed from data collection and analysis point of view. Positioning of the cameras during the data collection had to be designed carefully in order to get all the tasks comprehensively recorded. Since the additional data collection for the missing information was done and completed through informal interviews, it can be concluded that the internal validity is quite strong in this respect. Nevertheless, the analysis of the recorded material with hierarchical task analysis and LoA measurements using LoA taxonomy, was based on researchers' judgements and subjective feeling. It can be argued that an internal validity can be threatened in this sense.

The fact that the case study was the research method in use where a single assembly process was studied in detail, the question of the external validity, i.e., how the findings can be generalized, has to be discussed. Obviously it is not possible to apply a statistical generalization and that is sometimes regarded as a drawback of the case study research. Nonetheless, an analytical generalization can be applied, as suggested by [24]. Existing theory is expanded by adding another context to it.

Reliability of the study was insured with the fact that the data collection was done through video recordings. By having recordings as a raw data, higher reliability is achieved than if the data was collected through regular observations.

Table 2. Comparison between the case study and previous LoA measurements in other industries. ATO - assemble to order; ATS - assemble to stock

	The case company	Previous case studies					
Companies	A	B	C	D	E	F	G
Production area	House building	Engine parts	Chemistry	Electronics	Cooling modules	Trucks	Vessels
Type of assembly	Line	U-cell	Line	U-cell	Job shop	Line	U-cell
Type of assembling	ETO	ATO	ATS	ATO	ATS	ATO	ATO
Number of stations	13	4	9	5	8	5	9
Average LoA_{Phys}	3	1	5	3	1	3	1
Average LoA_{Cogn}	1	1	5	5	1	-	1

The analysis of the recordings did not only give the figures about the LoA and the HTA map, but also revealed possible issues related to some tasks. The issues are related to both mechanical and information aspects of the tasks. Considering the mechanical part of the majority of tasks, it was noted that these are mostly pick and place, i.e., totally manual tasks (LoA – 1:1). The next level is the tasks performed using simple static tools such as measuring tape or hammer (LoA – 2:1). A number of tasks included automated hand tools such as hand nailing machine (LoA – 4:1). The pick and place tasks are mainly non-value adding activities related to material handling.

Solutions with more mechanized material handling can be applied and therefore help operators to focus more on value adding activities.

Based on the observations it was noted that for some tasks there is a lack of necessary information provided to the operators. The majority of tasks the operators perform from their own knowledge and experience without using shop drawings as main information carriers. In some cases the information that is provided to the operators through shop drawings is not good enough. The real measures, the dimensions and positions, do not exactly match those specified in the instructions. The operators often need to measure themselves, calculate and get the dimensions and positions for certain components. This problem reaches far beyond the scope of the study, towards the properties and heterogeneity of wood and dimensional tolerances of the off-site assembly. Nonetheless, one can think of control/scanning systems that can provide operators correct information and that implies the higher cognitive level of automation needed.

Conclusion is that it was possible to use the DYNAMO ++ method and quantify physical and cognitive LoA of the exterior wall assembly line. Compared to the other case studies, the current LoA state in the case company is rather intermediate. The average physical LoA is intermediate while the average cognitive LoA is the lowest possible. Both physical and cognitive LoA can be increased for certain critical tasks. The increased physical and cognitive LoA would further enable the flexible task allocation between human operators and technology which is believed to be a way to enable a better production flow with shorter assembly lead time and with less production disturbances and increased productivity.

Findings and conclusions from this case study together with a value stream map (VSM), will serve as an input information for the research continuation where the rest of DYNAMO ++ framework will be carried out. The information will be used and validated in the next phase of the framework which is the workshop organized in the company. At the workshop researchers involved in the project will meet the participants from the company. The participants chosen, cover all the enterprise levels from directors and top managers to the operators working at the assembly line. The goal of the workshop is to use the results of HTA, LoA and VSM and discuss the critical tasks. The discussion about the possible improvements should lead to the LoA range specification for the critical tasks. That, in turn, would hopefully result in a future state of flexible task allocation between human operators and technology.

6 Acknowledgments

This research project is part of the graduate school ProWOOD which is a collaboration between School of Engineering at Jönköping University, Linnaeus University, Nässjö Trächentrum, several companies, and two research institutes. The graduate school is financially supported by The Knowledge Foundation (KK-stiftelsen). The participating companies and institutes are: Södra Wood, Inwido, Flexator, Myresjöhus (OBOS), Yaskawa, Swerea IVF, and SP-trä.

References

[1] Von Platen, F. Mer trä i byggandet–Underlag för en nationell strategi att främja användning av trä i byggandet. Ds 1 (2004): 44.

[2] Gustafsson Å. Vessby J. and Rask. L.O. Identification of potential improvement areas in industrial housing: A case study of waste. *Lean Construction Journal*, 61-77, 2012.

[3] Nord T. and Brege S. Värden för världen – konsekvenser av ett ökat industriellt träbyggande. Linköpings universitet, 2013.

[4] Lu W. Olofsson T. Jensen P. and Simonsson P. BIM-based lean-agile supply chain for industrialized housing. In *Proceedings of the 11th International Conference on Construction Applications of Virtual Reality (CONVR 2011)*, Bauhaus-Universität Weimar, Germany, 2011.

[5] Gosling J. and Naim M. M. Engineer-to-order supply chain management: A literature review and research agenda. *International Journal of Production Economics,* 122(2): 741-754, 2009.

[6] Höök M. and Stehn L. Lean principles in industrialized housing production: the need for a cultural change. *Lean construction journal,* 2: 20-33, 2008.

[7] Lennartsson M. *The transition of industrialised house-building towards improved production control.* Luleå University of Technology, 2012.

[8] Höök M. Customer value in lean prefabrication of housing considering both construction and manufacturing. In *Proceedings of the 14th Annual Conference of the International Group for Lean Construction, Santiago de Chile,* 2006.

[9] Höök M. and Stehn L. Connecting Lean Construction to Pre-fabrication Complexity in Swedish Volume Element Housing. In *13th International Group for Lean Construction Conference: Proceedings. International Group on Lean Construction,* 2005.

[10] Gibb A. and Isack F. Re-engineering through pre-assembly: client expectations and drivers. *Building Research & Information,* 31(2): 146-160, 2003.

[11] Groover M. P. *Automation, production systems, and computer-integrated manufacturing*, Prentice Hall Press, 2007.
[12] Bainbridge L. *Ironies of automation,* Automatica, 19.6: 775-779, 1983.
[13] Fasth Å. and Stahre J. Does Levels of Automation need to be changed in an assembly system?-A case study. In *The 2nd Swedish Production Symposium (SPS), Stockholm, Sweden,* 2008.
[14] Fasth-Berglund Å. and Stahre J. Cognitive automation strategy for reconfigurable and sustainable assembly systems, *Assembly Automation,* 33(3): 294-303, 2013.
[15] Fitts R. M. *Human Engineering for an Effective Air-navigation and Traffic-control System,* National Research Council, London, 1951.
[16] Frohm J. Lindström V. Stahre J. and Winroth M. Levels of Automation in Manufacturing, *Ergonomia - International Journal of Ergonomics and Human Factors*, Vol. 30, Iss 3, 2008,
[17] Säfsten K. Winroth M. and Stahre J. The content and process of automation strategies, *International Journal of Production Economics*, No. 110, pp. 25-38, 2007.
[18] Winroth M. Säfsten K. and Stahre J. Automation strategies - Existing theory or ad hoc solutions? *International Journal of Manufacturing and Technology Management*, Vol. 11, No. 1, pp. 98-114, 2007.
[19] Sheridan T. B. Computer Control and Human Alienation, *Technology Review*, Vol. 83, No. 1, pp. 60-73, 1980.
[20] Frohm J. *Levels of Automation in production systems*, in Department of production system. vol. Doctorial Gothenburg: Chalmers University of technology, 2008.
[21] Fasth Å. *Quantifying Levels of Automation-to enable competitive assembly systems*, Chalmers University of Technology, 2012.
[22] Fasth Å. Stahre J. and Dencker K. Measuring and analysing Levels of Automation in an assembly system, *Manufacturing Systems and Technologies for the New Frontier*, 169-172, Springer London, 2008.
[23] Fasth Å. Reviewing Methods for Analysing Task Allocation in A Production System, *Journal of Logistics Management,* 1(1): 1-8, 2012.
[24] Yin R K. *Case study research: design and methods.* Sage publications, 2013

Research on Improving Work Efficiency of Unmanned Construction

Improving Work Efficiency of Unmanned Construction by Using Appropriate Visual Information and Necessity of Haptic Function for Construction Machine Operation Interface

Masaharu Moteki[*1] and Nishiyama Akihiko[*1], Shinichi Yuta[*2], Hiroyuki Mishima[*3], Kenichi Fujino[*1]

[*1] Advanced Technology Research Team, Public Work Research Institute,
[*2] Shibaura Institute of Technology
[*3] Faculty of Human Sciences, Waseda University

Abstract –

Unmanned construction is actively employed at dangerous locations where people cannot enter because of disaster damage.

Unmanned construction is a technology of remotely controlling construction machines as an effective system of construction to ensure the safety of operators. However, the work efficiency of unmanned construction is poorer than that of construction with manned operation.

Techniques for efficiently and remotely operating construction machines are necessary to realize swift rehabilitation activities after the occurrence of a disaster.

In the research, I decided to push forward a study by 2 approach.

As one, our research conducted a comparative experiment under different operation conditions (such as interfaces or operation environments) that represent the actual condition of work time related to the remote operation techniques of construction machines. This paper discusses the concept of operation conditions to improve work efficiency based on the results of the said comparative experiment.

As second, In general, when people do some act, they are considered to do so based on more accurate information they obtain from their visual sense as well as through their actions including touching things. When working with a construction machine, people generally cannot obtain haptic information, such as whether something is hard or soft, from their visual sense. Particularly, construction machine operators obtain haptic information such as whether something is hard or soft from some indirect information from the machines.

Our research reviews the validity of using haptic information in operating construction machines to meet the needs of the operators. The paper presents the results [1] of part of the said review, which is clarification based on fundamental experiments that machine operators will be able to indirectly perceive material differences, as expressed by "hard or soft," through media. It then discusses the potential of providing a haptic capability to the construction machine operation interface for the purpose of improving the work efficiency of unmanned construction.

Keywords –

Unmanned construction; construction machine; work efficiency; interface; visual information; haptic perception

1 Introduction

Unmanned construction with remotely operated construction machines is employed as the initial response in the event of a natural disaster (due to earthquakes or volcanic eruptions) and plays an active role in restoration work in situations of sediment disasters or volcano disasters.

The comparative experiments [1] so far conducted by the Public Works Research Institute (PWRI) revealed that unmanned construction has a work efficiency (cycle time) that is about 2.3 times greater than that of manned construction.

PWRI then conducted measurement experiments to examine the differences in interfaces, including differences in conditions, the operation environment, or operation methods between manned operation and remote operation and confirmed that the cycle time varied as the operation environment or operational interface changed, and did so in a step-by-step manner. (Figure 1)

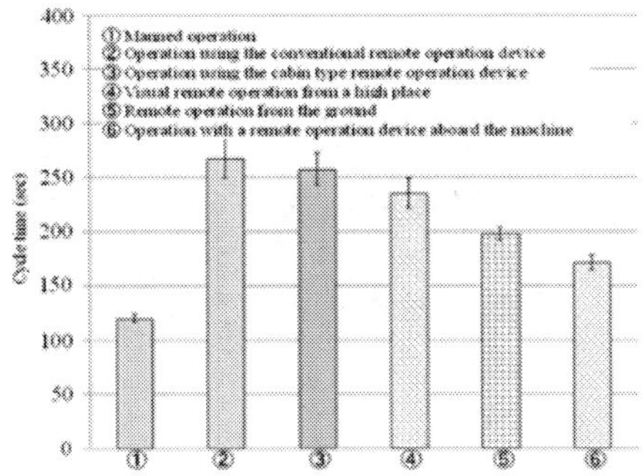

Figure 1 Comparative experiment in 2014

The authors hypothesized based on the findings obtained in the past experiments [1] that we can expect an improvement in work efficiency if we provide a lot of environmental information that is generally available during manned operation to remote control operators.

As one part of the discussion about the direction of future unmanned construction systems, the experiment clarified that remote operation of a construction machine can be improved in terms of work efficiency compared with the conventional unmanned operation by increasing the visual information obtained from monitors or other means, making it as close as possible to the volume of the visual information obtained by an operator who actually operates a machine. The details are described in Section 2.

In next research, when operating a construction machine, operators efficiently do their work using their perception including visual, auditory and haptic senses. Whereas, when construction machines are remote controlled at disaster restoration sites, the operation system is configured so that operators mainly depend on visual information. But, the operation-system does not deal with traveling on soft ground or breaking stones, which are the types of acts that cannot be judgment solely based on visual information.

In our study, we interviewed construction machine operators and confirmed the need for haptic information for remotely operating machines. This report describes the feasibility of indirectly perceiving differences in material texture such as "hard or soft" through media based on the fundamental experiment and discusses the need for a haptic sensing capability for a construction machinery operation interface for the purpose of enhancing the work efficiency of unmanned construction. The details are described in Section 3.

2 Comparative Experiment of Operation Interface

2.1 Purpose of study

A comparative experiment was conducted about visual information to be provided to operators during remote operation and about the operation device. The main purpose was to make remote operation much closer to manual operation, as part of our attempt to establish a basis of remote operation or an easy operating system to improve the work efficiency (cycle time) of remote operation and help reduce the fatigue and burden on operators.

2.2 Experiment task

A comparative demonstration experiment was conducted based on the experiment task [2]. The working environment of the experiment task is shown in Figure 2.

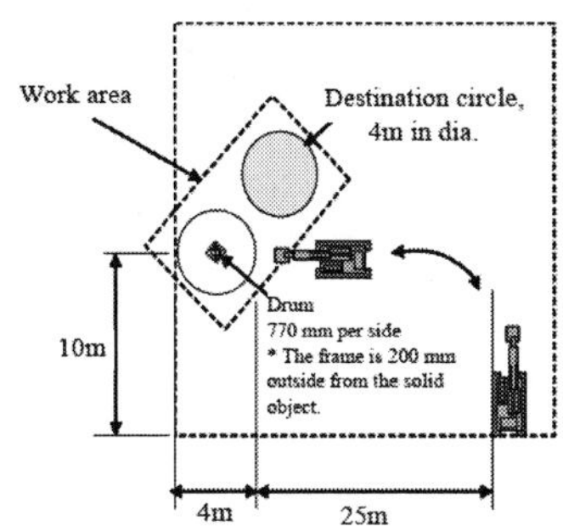

Figure 2 Plan view of the experiment task area

Figure 3 Operation procedure

This experiment task involves work elements of unmanned construction using a hydraulic excavator employed at disaster sites.

In the task, a hydraulic excavator travels along a route having a length of about 30 m, including a curve, to the work area, moves to the appropriate location, removes a solid object, as shown in Figure 3, takes it to a destination circle, and returns the object to a position in the original frame.

While this movement simulates excavation by a hydraulic excavator, when sediments are used, the difference in the volume of stacking with a bucket generally affects the work. It is difficult for operators to always scoop the same amount of sediments with a bucket, which makes it difficult to ensure reproducibility. In response, a similar work model, which is picking up a solid object as shown in Figure 4, was developed for this task.

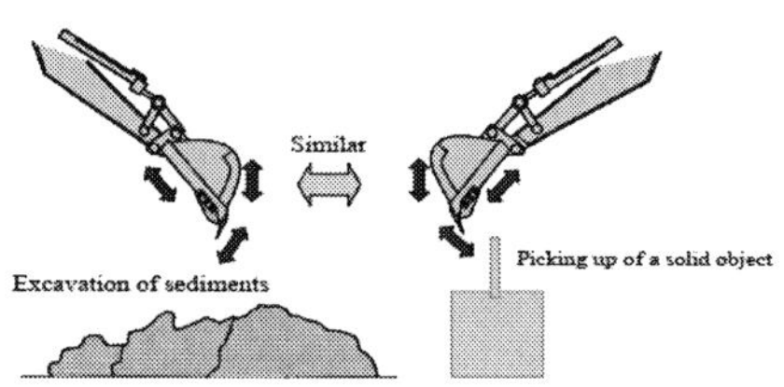

Figure 4 Work model

In this evaluation experiment, the hydraulic excavator used was a 0.5 m^3 class excavator, and the solid object selected was a drum-shaped object measuring 770 mm in width, 760 mm in height, about 100 kg in weight, and 2,000 mm in height at the holding handle.

2.3 Experiment cooperators and the flow of the experiment

Ten operators were selected as the experiment cooperators who conducted the experiment in 2014 for the purpose of comparative verification. The experiment measured the working time of operation by the conventional remote operation device (Figure 5) and by the operation cabin-type remote operation device (Figure 6).

Figure 5 Operation with the conventional remote operation device

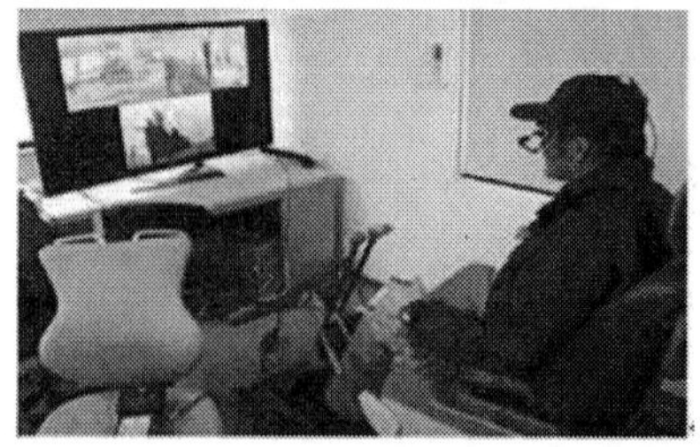

Figure 6 Operation with the cabin-type remote operation device

2.4 Determination of visual information

Image information is provided to the operator during the experiment through the monitors shown in Figure 5. The monitors show the image seen in front of the construction machine and the images of external points of view from the stationary cameras.

The study does not focus on the high definition of images or 3D feature of images but uses the monitor configuration generally used in conventional remote operation as the comparative criterion to review image information that is as close to that available for manned operation.

The reason why the study does not deal with high-definition telecommunication or imaging or 3D technology is that the authors realize such telecommunication or image technology is being studied by researchers in other fields and intend to use the outcome of their research in the future.

2.4.1 Necessary visual information

The viewing angle of a human is generally 60° upward, 70° downward, and about 200° laterally.

As shown in Figure 8 and 9, the effective viewing field is the field in which humans can receive information instantaneously only with their eyeball movement, which is generally within a horizontal angle of about 30° and a vertical angle of about 20°.

The viewing field of fixation is the field in which humans can naturally fix their eyes on something with their eyeball and head movement and effectively receive information, which is generally within a horizontal angle of 60 to 90° and a vertical angle of 45 to 70°. The surrounding visual field is a field in which a human can recognize the presence of an object, which is generally within a horizontal angle of 100 to 200° and a vertical angle from 85 to 130°.

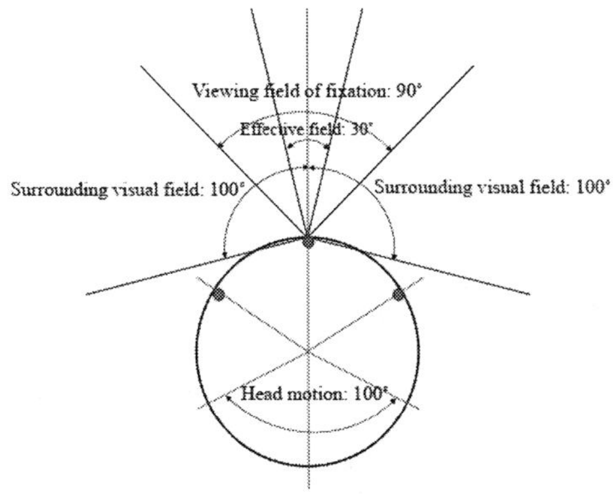

Figure 7 General viewing angle (planar)

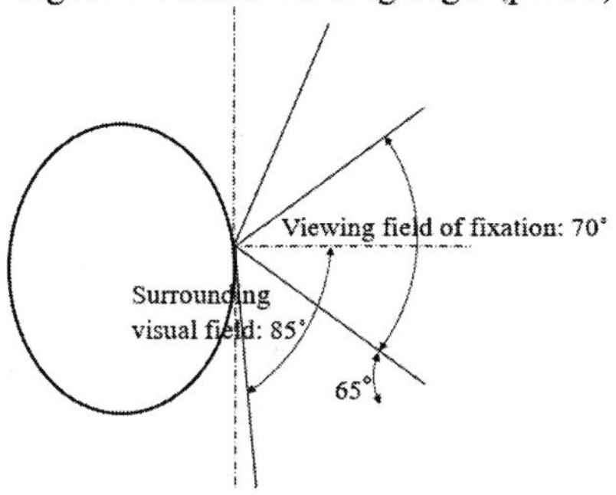

Figure 8 General viewing angle (lateral)

2.4.2 Setting of cameras and monitors

The camera arrangement was adjusted based on the general viewing angles and the measurement results.

1. Understanding of the working viewing field

The working viewing field of operators when they operate the machine in the operation cabin is determined as in Figure 9. Based on this working viewing field, the cameras were laid out.

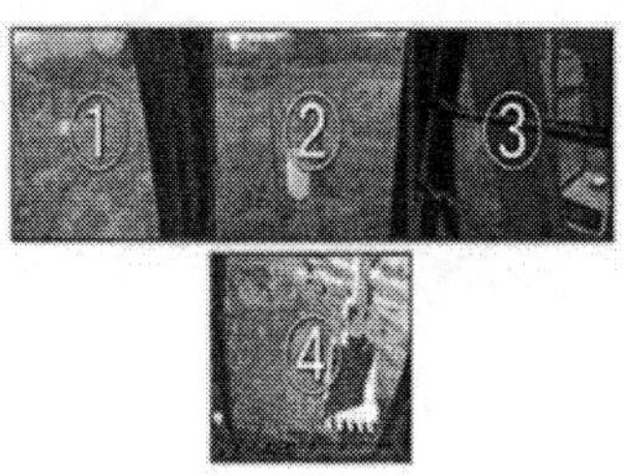

Figure 9 Working viewing field of the operator operating at the cabin

2. Camera positions

Cameras were installed as in Figure 10 based on the working viewing field of the operator operating the machine in the cabin.

Cameras used were those with a focusing capability close to the vision of a human (1/2.8 inch 2.8 – 8mm SONY Exmor CMOS 214 million pixels).

Figure 10 Camera arrangement (planar)

3. Monitor configuration

As shown in Figure 11, the monitor configuration that can provide a similar viewing angle as an operator based on the camera field angle was studied. The visual distance from the operator's viewpoint to the monitor was determined using Equation (1).

$$\tan(\theta/2) = d/2D \tag{1}$$

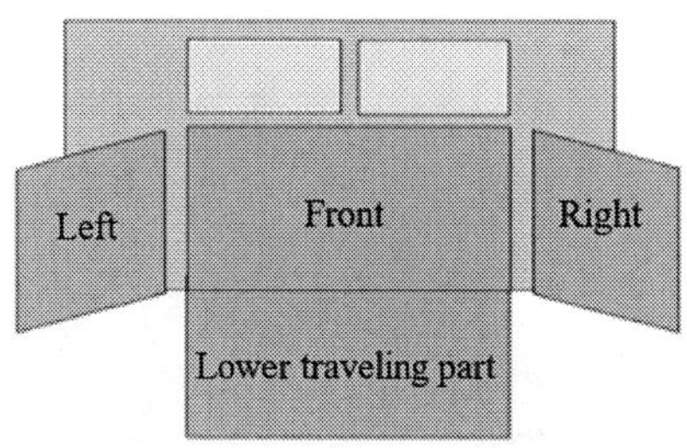

Figure 11 Monitor configuration

Monitors were laid out with the viewing field of fixation and head movement taken into consideration.

Monitors were laid out showing the front view and lateral views. The arrangement of the side monitors was determined by considering the surrounding visual field of 100° at the front as shown in Figure 12.

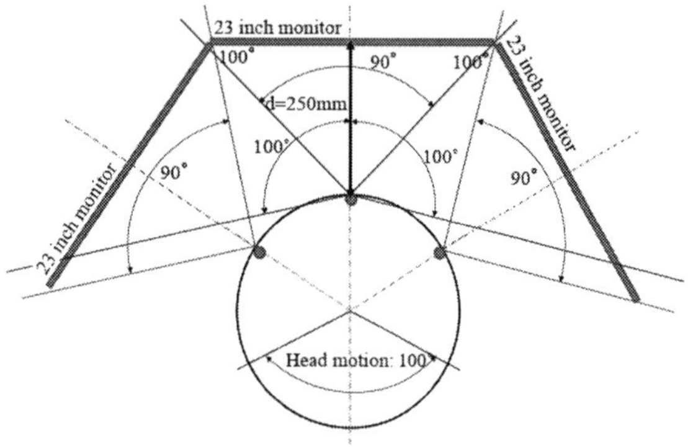

Figure 12 Monitor configuration (planar)

Monitors showing the frontal views were laid out to show the front viewing field of fixation of 70° and the lower surrounding field of 65° as shown in Fig. 13.

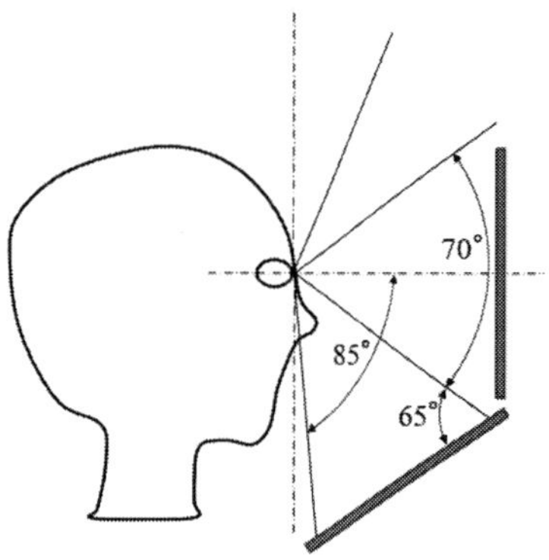

Figure 13 Monitor configuration (lateral)

Figure 14 Monitor layout

Six monitors were used, showing the front, the lower traveling range, left and right, and the two views of the conventional fixed cameras, based on the distance to the viewpoint of 250 to 300 mm. Each monitor has specifications of 23 inches, 16:9,720 × 480 and made by Iiyama.(Figure 14)

2.4.3 Image transmission

For the four cameras shown in Fig. 10 showing the front, left and right, and the lower traveling view of the operation cabin of the hydraulic excavator, the image data was transmitted using parallel transmission of wireless LAN, 2.4 GHz and 5 GHz.

In this experimental system, a serious image delay that may affect the work efficiency did not occur as in the case of the past experiments [1].

2.5 Operation device

A comparative review was made between an interface with a joystick operation device installed on the desk for remote operation and an interface that simulated the actual hydraulic excavator operation cabin.

2.6 Measurement

The cycle time was measured in the flow shown in Fig. 3 based on the experiment task [2]. Figure 15 and 16 show the condition of the measurement.

Figure 15 Operation using the conventional remote operation device

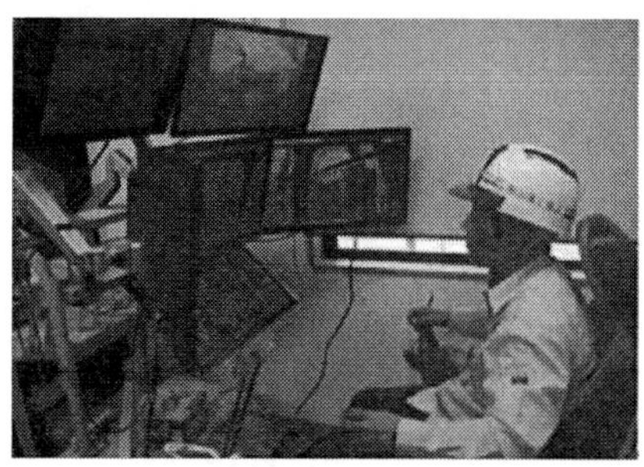

Figure 16 Operation with the cabin-type remote operation device

2.7 Experiment results

The results of the experiment with ten operators using the conventional remote operation device give an average cycle time of 232.7 sec ± 7.6. The operation with the cabin-type remote operation device resulted in an average cycle time of 194.8 sec ± 9.6.(Figure 17)

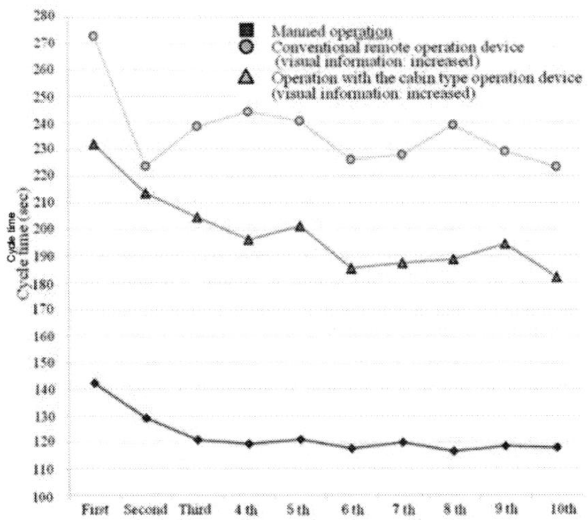

Figure 17 Cycle time tendency

2.8 Summary

The results indicate that the conventional remote operation system had a volume of visual information about 2.3 times that of the actual manual operation.

The authors considered the hypothesis that "giving a lot of environmental information, which is generally available for operators in the machine, to the operator may improve work efficiency," when comparing with the results of the experiment in which the volume of visual information was increased to that of the information for operators on the machine. The difference in cycle time turned out to be from 1.62 to 1.94 times greater than operation in the machine. This confirmed that increasing the volume of visual information is effective for improving work efficiency. (Figure 18 and 19)

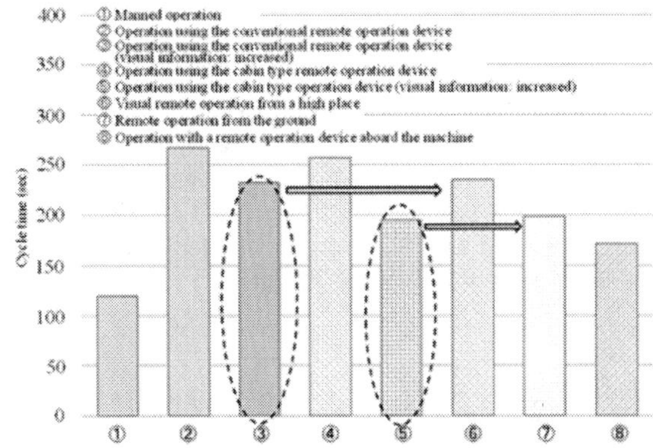

Figure 18 Comparison with the 2014 experiment results

As shown in Figure 18, when the volume of visual information is increased, the operation with the conventional remote operation device showed a tendency similar to the cycle time of visual remote operation from a high place as shown in Figure 19. The operation with the cabin-type remote operation device showed a tendency similar to the cycle time of remote operation on the ground as shown in Figure 20.

These results suggest that increasing the volume of visual information is greatly related to improvement in work efficiency.

Figure 19 Visual remote operation from a high place

Figure 20 Remote operation on the ground

As shown in Figure 17 and 18, comparison of the differences in cycle time revealed that the time difference between the conventional remote operation system with the increased visual information and the cabin-type remote operation device is 1.2 times. This is presumably because of the difference in operation interfaces. It is therefore suggested that the operation interface is one of the important elements to review in order to improve work efficiency.

3 Necessity of Haptic Perception in Construction Machine Operation Interface

3.1 Purpose and outline

Information on the hardness of materials, such as whether they are hard or soft, cannot be recognized with visual information and this is one of the essential elements for operators to efficiently remotely operate construction machines, for tasks such as traveling or excavation.

To improve the work efficiency of unmanned construction, a basic experiment on an interface with haptic capability was conducted to review the feasibility of having haptic capability in a construction machine's operation interface. The result of the experiment is explained in the following.

3.2 Needs for haptic information

Ten experienced operators who had previously remotely operated construction machines were interviewed.

The interview results clarified that the operators feel the need for haptic information on materials, such as their hardness or softness, when they conduct excavation at remote places when they cannot handle tasks with the available visual information, such as crushing of concrete blocks, or dismantling or removal of buildings.

3.3 Presence of indirect haptic sense and experiment on perceptive capability

It is assumed that a function is needed that lets an operator acquire haptic information to ensure safe and efficient remote operation of construction machinery.

It is however necessary to understand the current status about whether or not the operator can indirectly judge the necessary haptic feelings.

Preceding research by Katz verified by experiments that we can distinguish the differences in paper quality (14 kinds of different paper quality) through a bar having a length of 4 to 5 cm (a wooden bar attached to the tip of a pen shaft with the rounded front end) (D, Katz (1925))[3]. Based on this research, it is understood that we can indirectly perceive haptic feelings.

An experiment about the presence and accuracy of hardness perception via a human haptic sense through media was conducted based on the patterns (Figure 21) of actions actively conducted in haptic search in order to clarify the status of indirect haptic faculty.

In the experiment, 44 experiment participants including construction machine operators were randomly presented with six kinds of specimens, varying in degree of hardness, (namely rubber, urethane, plastic, wood, concrete, and steel), each measuring 200 mm × 200 mm × 100 mm (width, depth, height). The participants touched the specimens through a rigid material (bar made of ABS resin) as a medium, and were requested to judge which was which, in terms of the quality of the specimens. Then the right answer ratio was analyzed.

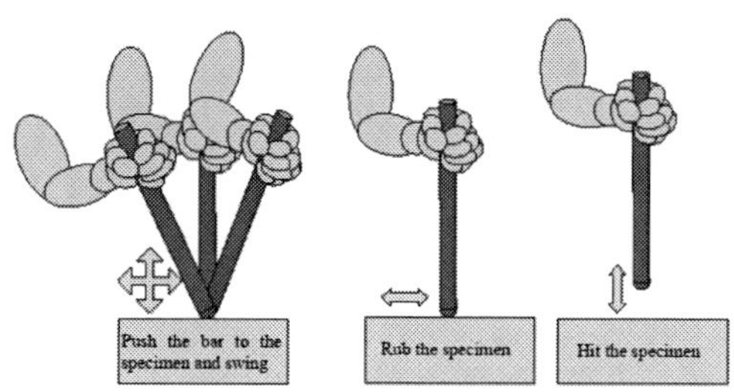

Figure 21 Action patterns

3.4 Summary

As shown in the experiment results of Figure 22, when the operator swung the search bar by pushing the bar to the specimens, the right answer ratio was 0.430 ± 0.167 on average, with a variable factor of 0.389. When the operator rubbed the specimens with the search bar, the right answer ratio was 0.648 ± 0.167 on average, with a variable factor of 0.257. For hitting the specimens with the search bar, the ratio was 0.494 ± 0.141 on average, with a variable factor of 0.286.

As indicated in Figure 22, the highest right answer ratio tends to be about 0.5 to 0.6 in the "search for back and forth and left and right by pushing the search bar to the specimens" and the "search by hitting the specimens with the bar." For "rubbing specimens with the search bar," the right answer ratio forms a mountain-shaped distribution pattern with 0.5 to 0.6 as the median.

In either case, the result exceeded the chance level of 0.167.

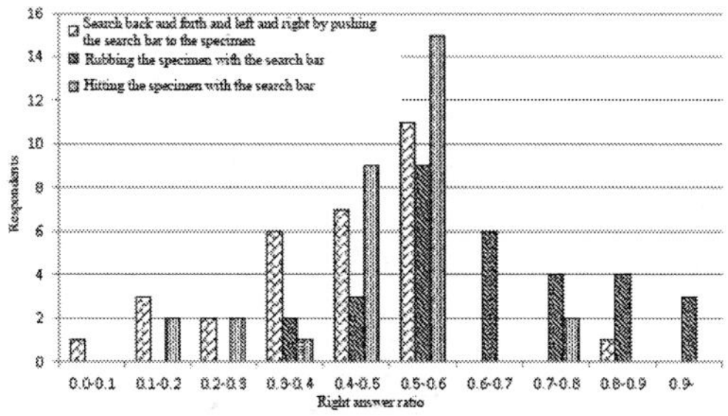

Figure 22 Right answer ratio histogram

Our study focused on an experiment with the feasibility of operators perceiving haptic information through media and distinguishing the differences in hardness of materials. The experiment conducted in our study confirmed that humans can distinguish differences of hardness through indirect haptic feelings. Although the detailed mechanism of haptic sensation through media is yet to be clarified, it is presumed that humans touch things to obtain information on the size, weight or hardness of materials in addition to visual information and that even when they indirectly touch things, they adjust the target actions based on the obtained information. In other words, it is presumed that operators actively make judgments of materials in terms of texture, such as hard or soft, through a haptic search (which may be referred to as "active tough" by Gibson, 1966) [4] and take appropriate actions.

4 Conclusion

4.1 Comparative Experiment of Operation Interface

The difference in cycle time turned out to be from 1.62 to 1.94 times greater than operation in the machine. This confirmed that increasing the volume of visual information is effective for improving work efficiency.

And, comparison of the differences in cycle time revealed that the time difference between the conventional remote operation system with the increased visual information and the cabin-type remote operation device is 1.2 times. This is presumably because of the difference in operation interfaces. It is therefore suggested that the operation interface is one of the important elements to review in order to improve work efficiency

4.2 Necessity of Haptic Perception in Construction Machine Operation Interface

One of the needs of remote control operators of construction machines is provision of haptic information for their operation of machines.

It is also understood to be one of the important factors to ensure safe and quick execution of unmanned construction.

Safe and quick execution of restoration work particularly at a disaster site will be realized by providing haptic information in such operations as excavation or removal of debris around buried water or gas pipes not visually accessible to workers.

Since it is important to provide haptic information to operational interfaces in the future, the authors intend to clarify the mechanism from the viewpoints of kinematics and dynamics and promote commercialization of an operation control system, such as drive-by-wire.

5 References

[1] Masaharu Moteki, Nishiyama Akihiko, Shinichi Yuta, Hiroshi Ando, Sadanori Ito, and Kenichi Fujino: Work efficiency evaluation on the various remote control of the unmanned construction, The 15th Symposium on Construction Robotics in Japan O-21 2015

[2] Masaharu Moteki, Shinichi Yuda, Kenichi Fujino: The Proposal of the Model Task for Efficiency Evaluation of the Construction Work by Remote Control of a Hydraulic Excavator, paper in the Journal of JCMA, Vol. 66, No. 8, pp.71 - 79, 2014

[3] David Katz (author), Atsuki Higashiyama and Kinuyo Iwakiri (translators): The World of Touch, Shinyosha, 2003

[4] Gibson, J. J. : The senses considered as perceptual systems, Oxford University Press, 1966. (authored by Gibson, J.J. and translated by Masato Sasaki, Norihiro Furuyama and Hiroyuki Mishima, University of Tokyo Press, 2011)

[5] Masaharu Moteki: Study on Human's Perception of Hardness by Haptic Sense through Media, master thesis for the Graduate School of Human Sciences, Waseda University, Waseda University, 2014

[6] Masaharu Moteki, Kenich Fujino, Akihiko Nishiyama: RESEARCH ON OPERATORS' MASTERY OF UNMANNED CONSTRUCION, The30th International Symposium on Automation and Robotics in Construction, USB No.186, 2013

Operation Analysis of Hydraulic Excavator with Different Operation System

Akihiko NISHIYAMA[a] and Masaharu MOTEKI[a], Kenichi FUJINO[a], Takeshi HASHIMOTO[a]

[a]Public Works Research Institute
E-mail: nishiyama@pwri.go.jp, moteki@pwri.go.jp, fujino@pwri.go.jp, t-hashimoto@pwri.go.jp

Abstract -

Unmanned construction with remotely operated construction machines is actively used a means of construction work in dangerous areas with no access to people such as rehabilitation work sites damaged by natural disasters including earthquakes or volcano eruptions. Unmanned construction is expected to progress in the field as a quick and safe technique for restoration from disaster damage.

However, unmanned construction with remotely operated hydraulic excavators is known to be much poorer in work efficiency than construction work with directly operated machines. The Public Work Research Institute (PWRI) has been promoting research on the actual status of, causes of, and solutions to reduced work efficiency of unmanned construction with remotely operated hydraulic excavators.

Our research focused on the impact that different operation systems of a remotely operated hydraulic excavator had on the work efficiency with the aim of obtaining knowledge that will help improve the work efficiency of unmanned construction.

To fulfil this purpose, a comparative experiment was conducted using a few operation systems with changed parameters of operation interface for a hydraulic excavator.

The data obtained from the experiment were analysed for each operation system and work element, with important points that express differences in work time and work efficiency analysed for clarification. Causes that resulted in differences were also discussed using video records.

Based on the experiment, the authors successfully clarified the status of the impact of interface operability for a remotely operated hydraulic excavator on the work efficiency and obtained knowledge that will lead to improved work efficiency of unmanned construction.

Keywords -

Remote control technology, Disaster, Construction machine, Hydraulic excavator, Operation interface Introduction

1 Introduction

Unmanned construction, construction with remotely operated construction machines, is actively used as means of construction work in dangerous areas that people have no access to, such as rehabilitation work sites damaged by natural disasters including earthquakes or volcano eruptions. Unmanned construction is expected to progress in the field as a quick and safe technique of restoration from disaster damage.

However, remote operation of hydraulic excavators conducted as unmanned construction sites is known to be much poorer in work efficiency than construction work with manned machines. The Public Work Research Institute (PWRI) has been promoting research on the actual status of, causes of, and solutions to reduced work efficiency of unmanned construction.

Preceding research on the efficiency of the remote operation of the hydraulic excavator were conducted in a flow of the following (1) to (3).

(1) The work of the provisions was carried out in two ways of remote operations and boarding operation of the hydraulic excavator. I grasped the reality of the work efficiency decreases by remote control[1].

(2) The work of the provisions were carried out multiple times by multiple operators. It was analyzed for the impact on the working efficiency decrease due to getting used to the work[2].

(3) Comparison of remote operations and boarding operation, a comparison of the novice and skilled person, were carried out by focusing on the gaze point in the operation and working time[3].

Based on these research results, we will further pursue the cause of the working efficiency decrease of the remote control. Therefore, in this study, we aimed to understand the impact on efficiency due to the difference in operation interface in the remote operation.

In our research, a comparative experiment was conducted based on a model task related to four different operation methods with different operation conditions (to be explained later). The time taken to complete a single run of the model task (hereinafter the cycle time) was measured, and the task work was filmed.

The time taken by a directly operated machine was used as the benchmark for comparison purposes. The cycle time recorded was divided into that of the excavation work and that of traveling, and each work efficiency was compared by the method of operation. We also discussed whether or not the skill of the operator (speediness of work execution) had any effect on the work efficiency depending on the operation interface.

2 Purpose of Research

The purpose of our research is to determine the actual status of interface operability as a factor that affects the work efficiency of unmanned construction and obtain knowledge that may help improve work efficiency.

3 Experiment Method

3.1 Experiment field

The experiment field was set up as shown in Fig. 1.

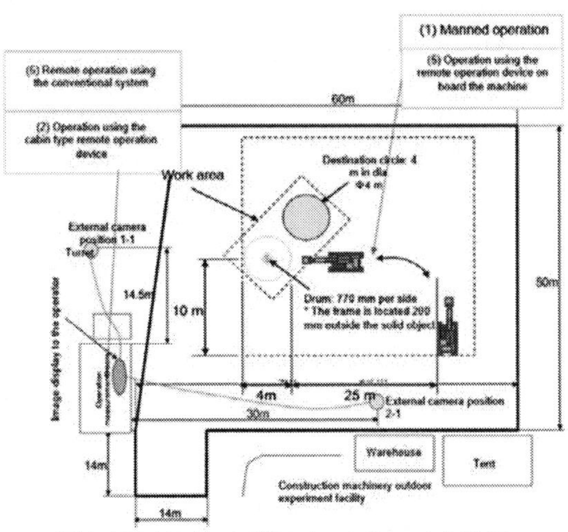

Fig. 1 Overview of the experimental field

3.2 Outline of the model task

The model task developed by preceding research was used for our study[1]:

(1) The operator drives a construction machine to the designated work position (near the work area) shown in the experiment field (Fig. 1);

(2) The machine grabs a solid object with its bucket and moves from the initial position to the destination circle, instead of conducting excavation work; and

(3) The machine grabs the solid object with its bucket and moves from the destination circle to the initial position. (The initial position is set as a square area with each side measuring 770 mm, and a solid object is placed in that square. When the operators fail to place the object inside the square, they lift it up again and place it in the square.) The solid object to be handled by the machine is a drum (570 mm in diameter, 760 mm in height, about 100 kg in weight, and 2,000 mm in handle height).

3.3 Equipment and materials for the experiment

3.3.1 Hydraulic excavator

A remotely operated hydraulic excavator (12-ton class with a bucket capacity of 0.5 m^3) was used.

3.3.2 Video equipment

Two cameras were placed in the field (Fig. 1) to film images for the operator's remote operation in an operation measurement room. Camera images (analog video, 30 fps, equivalent to 720 × 480) from a total of three cameras, including the said two and one installed on the hydraulic excavator (Fig. 1) were shown on monitors in the operation measurement room. Operators remotely operated the machine while viewing those monitors. Vertical and lateral rotation and zooming of the external cameras were performed by a dedicated camera operator.

3.4 Comparative experiment conditions

Four operation conditions were used in this experiment.

"Operation of a machine by an operator in the machine" (hereinafter "manned operation") and "remote operation by an operator in the machine" were compared to analyze the effect of the differences in the manned operation interface.

"Remote operation using the conventional system" and "remote operation using the cabin-type remote control system" were compared to analyze the effect of the differences in the remote operation interface.

Operation interfaces of manned operation were compared to aim at clarifying how the difference in operation interface operability will affect the work efficiency in case it is feasible in the future to provide the operator with an operation environment information equal to that available for manned operation (in terms of visual input, sound, vibration, etc.).

Operation interfaces of remote operation were compared with the aim of clarifying how use of the operating lever of the actual machine as the operation interface in current unmanned construction will affect the work efficiency.

3.4.1 Manned operation

An operator boarded a hydraulic excavator, used the operating lever of the actual excavator installed in the operation cabin, and conducted ordinary manned operation. (Fig. 2)

Fig. 2 Boarding operation

3.4.2 Manned operation with the remote operation device

An operator boarded a hydraulic excavator and operated the excavator with the remote operation device (to be explained later). (Fig. 3)

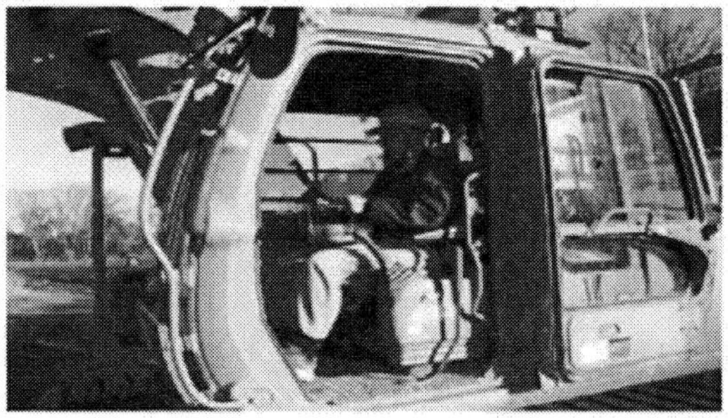

Fig. 3 Boarding operation using the remote control device

3.4.3 Remote operation using the conventional system

The conventional system as used here is the typical remote operation method currently used in unmanned construction. The operator conducted remote operation using the remote operation device (to be explained later) while checking the image shown on the video devices as explained in "3.3.2 Video equipment" taken by three cameras in the "operation measurement room" as shown in Fig. 1. (Fig. 4)

Fig.4 Remote control according to the conventional system

3.4.4 Operation using the cabin-type remote operation device

An operator conducted remote operation using the cabin-type remote operation device (to be explained later) while checking the images given by the cameras in the operation measurement room as in Fig. 1. (Fig. 5)

Fig.5 Cockpit-type remote device

3.5 Remote operation interface

In this experiment, two remote operation interfaces, were used in addition to the operating lever and traveling lever of the hydraulic excavator, and their effects on the work efficiency were measured.

3.5.1 Cabin-type remote operation device

The operator in the operation cabin, which is equivalent to that of an actual excavator, conducts remote operation using the operating lever and travel lever, the same as those of an actual machine. (Fig. 6)

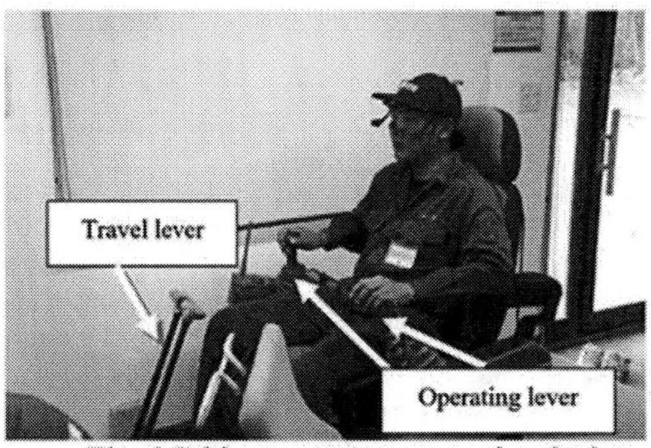

Fig. 6 Cabin-type remote operation device

3.5.2 Remote operation device

A typical remote operation device used in conventional remote operation (Fig. 7)

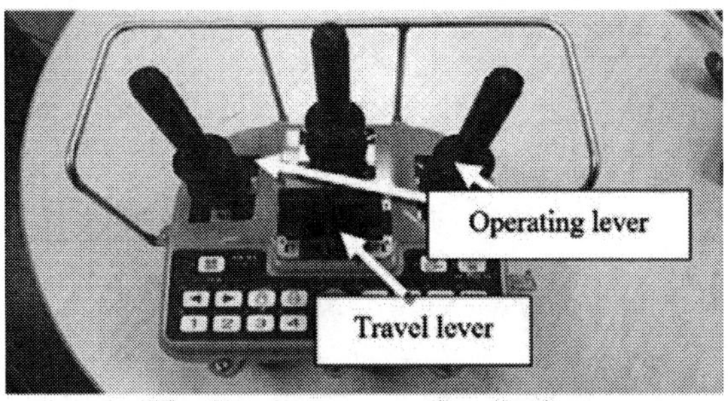

Fig. 7 remote operation device

3.5.3 Operating angle and neutral range of the lever

For the cabin-type remote operation device (and the actual lever) and remote operation device, the operating angle and neutral range of the lever were set as shown in Table 1 (default values by the manufacturer).

Table1 Operating angle of the operating lever

Operation lever specifications

Item		Remote control	Operating lever and remote operation cabin
Operating angle	[°]	20.00	21.70
Neutral range	[°]	2.50	0.81

Travel lever specifications

Item		Remote control	Operating lever and remote operation cabin
Operating angle	[°]	15.00	12.50
Neutral range	[°]	2.50	0.66

3.6 Operator

As shown in Table 2, 10 operators with 5 or more years of experience in manned operation of hydraulic excavators participated in the comparison experiment.

Table 2 Age and construction machinery years of experience of operator

Operator	Age	Experience	Operator	Age	Experience
No. 1	38 years old	20 years	No. 6	36 years old	15 years
No. 2	35 years old	10 years	No. 7	32 years old	10 years
No. 3	34 years old	10 years	No. 8	31 years old	15 years
No. 4	34 years old	10 years	No. 9	39 years old	15 years
No. 5	57 years old	30 years	No. 10	46 years old	20 years

3.7 Measurement and experiment times

Ten operators repeated the work run set in "3.2 Outline of the model task" ten times for each of the four operation conditions.

The measurement start timing was when the caterpillars of the hydraulic excavator started to rotate as the excavator began to move at the start point in Fig. 1. The end of measurement was when the operator completed the work, returned to the start point, revolved the upper part of the machine, and set the stationary posture.

The cycle time measured in a single operation from the measuring start to end was set as the indicator of work efficiency evaluation. (Work efficiency (%) = cycle time for manned operation / cycle time of the subject operation condition × 100)

Another fixed camera was installed separately from the cameras to film videos for remote operation to record the entire view of the work area as a record of the work (Fig. 8).

Fig. 8 Installation of work for the camera

4 Experiment Results

The experiment results were summarized based on data containing the results of the eight runs, or from the third to the tenth, in which the dispersion of cycle time was sufficiently settled[2]. There were 75 sets of measurement data for each of the four operation conditions, and the total number of data sets was 300 (75 × 4 cases).

4.1 Comparison of operation interfaces for manned operation

Based on the obtained results, the two operation methods, or "manned operation" and "remote operation in the machine" were compared.

When comparing the two work methods, one cycle time (measured in seconds) was divided into "work" and "travel," and the "work" was further divided into four elements, or "grabbing (of a solid object)," "lifting and rotating (of the solid object)," "lowering (of the solid object)," and "raising (of the boom and arm)" (see Fig. 9). The thus subdivided cycle time was defined as the "work time" of each division. Cycle time, work time and work efficiency using the manned operation as the benchmark were analyzed as shown in Table 3.

4.1.1 Comparison results

When the operation interface was switched from manned operation to the remote operation device, the work efficiency decreased by 30.1%. For "work" and "traveling," the efficiency of the "work" decreased by 36.2%, while that of "traveling" decreased by 25.3%.

Work time greatly differed particularly for some types of jobs, namely "grabbing" and "lowering."

4.2 Comparison of operation interfaces for remote operation

Two operation methods, or "remote operation using the conventional system" and "operation with the cabin-type remote operation device" were compared. The subdivided cycle times and the work efficiency with manned operation as the benchmark are analyzed in Table 4 as in the case of "4.1 Comparison of operation interfaces for manned operation" in comparison of the two work methods.

4.2.1 Comparison results

When the remote operation interface was switched from the remote operation device to the cabin-type remote operation device, the work efficiency remained almost the same (up 0.7%). Work efficiency differences differ between "work" and "traveling." There is a 2.9 % increase in efficiency for "work," while there is a 4.2 % reduction for "traveling."

In terms of cycle time difference, the largest cycle

Table3 Comparison of the interface in the boarding operation

Operation method	No. of data sets	Work time (unit: second)							Manned operation		
		Job					Travel	Total	Total	Work efficiency	Total
		Grab	Lift up and revolve	Lower	Raise	Total					
Manned operation	75	16.8	16.7	7.9	6.6	48.1	73.5	121.6	100.0%	100.0%	100.0%
Maximum		33.1	22.8	35.4	11.7	82.6	102.5	180.1	100.0%	100.0%	100.0%
Minimum		8.4	11.4	2.0	3.8	29.2	55.0	102.1	100.0%	100.0%	100.0%
Standard deviation		5.4	2.5	4.7	1.7	11.1	10.0	15.6	0	0	0
Operation using remote operation device on board the machine	75	26.8	22.0	16.9	9.7	75.4	98.5	173.8	63.8%	74.7%	69.9%
Maximum		65.8	35.9	49.1	21.2	131.1	134.0	245.9	130.7%	116.3%	116.2%
Minimum		12.7	14.8	5.5	2.4	40.0	63.5	115.4	52.7%	36.8%	48.5%
Standard deviation		9.1	4.5	8.3	3.8	18.6	19.1	33.4	14.3%	19.6%	13.9%
Work time (work efficiency) difference	75	10.0	5.4	8.9	3.0	27.3	25.0	52.3	-36.2%	-25.3%	-30.1%
Maximum		-10.9	-5.4	-4.5	-2.8	-9.1	-24.1	-21.9	-16.3%	-30.7%	-16.2%
Minimum		42.5	17.3	34.2	15.0	68.9	61.9	121.7	63.2%	47.3%	51.5%
Standard deviation		10.0	4.5	7.7	3.5	18.9	17.5	31.6	19.6%	14.3%	13.9%

Table4 Comparison of the interface in the remote operation

Operation method	No. of data sets	Work time (unit: second)							Manned operation		
		Job					Travel	Total	Total	Work efficiency	Total
		Grab	Lift up and revolve	Lower	Raise	Total					
Manned operation	75	50.8	35.6	31.4	20.4	138.3	124.9	263.2	37.0%	62.0%	48.1%
Maximum		162.7	132.9	125.0	71.9	279.5	195.2	429.1	79.4%	101.1%	86.6%
Minimum		21.3	19.9	13.6	9.2	81.9	73.8	155.7	16.1%	42.5%	27.9%
Standard deviation		21.8	15.7	16.6	9.0	42.9	32.7	61.8	11.6%	14.9%	10.1%
Operation using remote operation device on board the machine	75	50.7	33.7	22.5	19.6	126.6	132.4	258.9	39.9%	57.8%	48.7%
Maximum		108.0	61.7	59.4	46.6	212.0	206.3	399.4	77.0%	91.7%	69.5%
Minimum		20.8	18.9	3.1	7.2	68.1	90.9	165.8	18.0%	32.2%	26.4%
Standard deviation		15.9	8.2	12.3	7.3	31.5	30.7	54.5	11.9%	11.8%	9.9%
Work time (work efficiency) difference	75	0.1	1.9	8.9	0.8	11.7	-7.5	4.3	2.9%	-4.2%	0.7%
Maximum		113.8	95.5	107.4	46.4	157.4	68.8	134.2	27.4%	23.9%	23.3%
Minimum		-38.5	-24.9	-28.4	-25.6	-68.8	-90.4	-142.7	-31.5%	-48.6%	-29.4%
Standard deviation		19.7	16.2	19.8	8.7	38.3	36.4	54.0	10.0%	15.8%	9.5%

time difference occurred in the "lowering" job of the "work," while no major difference occurred for other jobs.

5 Discussion and Summary

5.1 Comparison of operation interfaces for manned operation

5.1.1 Relationship with operator skills

The swiftness of manned operation cycle time is defined as the skill of the operator. In order to clarify whether there is any correlation between the operator's skill and the work efficiency differences that occurred as a result of comparison, verification needs to be conducted.

To be specific, the cycle time of manned operation (operator's skill) and the work efficiency differences were subjected to regression analysis based on the scatter diagram prepared as in Fig. 9. The result of analysis gives a correlation factor of 0.23. Since this is low enough, it is judged there is no correlation.

5.1.2 Causes of work efficiency differences

Of 75 sets of data, three of those with the work efficiency differences closer to the median were picked up at random.

For "grabbing," "lowering," and "traveling," for which cycle time differences frequently occur among the three sets of data, the difference in behavior of the hydraulic excavator between "manned operation" with the actual lever and "remote operation by the operator on board the machine" using a remote operation device was checked based on the recorded video to examine what causes work efficiency differences.

It was revealed that manned operation with a remote operation device was a complex motion (that is, two or more motions, such as opening of the bucket and raising of the boom, simultaneously conducted) like the operation using the actual lever. It is presumed that differences in individual operators' adaptability to the difference in the layout or sensitivity of the lever affect the work efficiency differences.

For "traveling," the excavator followed almost the same track despite some lateral deviation regardless of the difference in operation, or manned operation with the actual lever and manned operation with a remote operation device. However, the time gap in "traveling" was attributable to the differences in time taken to adjust the traveling direction and in the traveling speed. What causes this difference is considered to be the adaptability of individual operators to the interfaces as in the case of "grabbing" and "lowering."

5.1.3 Summary

Information from the environment necessary for executing the model task shares the same conditions for any cases of manned operation. The difference in the operation interfaces, however, caused differences in cycle time.

According to the scatter diagram of Fig. 9, no correlation was found for the work efficiency differences between the differences in the operator's skill and the operation interfaces. At least according to what the dispersion of work efficiency differences indicates, there is a small difference in the skill about operation of the lever in manned operation. When the remote operation device is used, however, differences in work efficiency occurred. This is presumably because the operators failed to adapt to the operation interface depending on the difference in the layout of the lever or the motion reaction speed.

Therefore, experienced operators who were familiar with manned operation, the subject of the study, were affected in terms of work efficiency by changes in the operation interfaces even though they do the same work. The degree of such effect is assumed to vary depending on how each operator can adapt to the difference in operation interfaces.

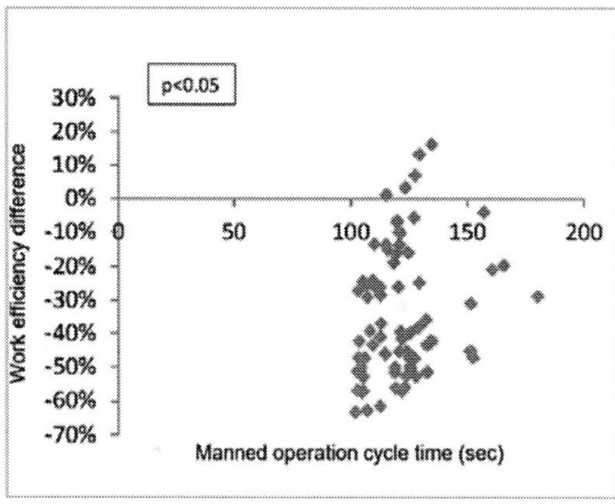

Fig. 9 Boarding operation time and work efficiency difference due to the difference in the operation interface

5.2 Comparison of operation interfaces for remote operation

5.2.1 Relationship with operator skill

The swiftness of remote operation cycle time is

defined as the skill of the operator. A scatter diagram was prepared as in Fig. 10 to conduct regression analysis in order to clarify whether or not there is any correlation between the operators' skill and the work efficiency differences determined by comparison. As the regression analysis gives 0.50 for the correlation factor, it is judged that there is a positive correlation of medium strength.

Since a decision function (R2) is low, at 0.24, the regression equation is not high enough in precision but is understood to be able to show the overall trend. Work efficiency differences caused by the difference in the operation interfaces for manned operation are presumed to be affected by the skill of the operator.

In order to clarify which job of the model task makes the operator's skill affect the work efficiency difference, 75 sets of cycle time data about "remote operation using the conventional system" were ranked based on the criterion that those with shorter cycle time are higher in rank. The 75 sets of data were categorized into three — higher, middle, and lower — with each having 25 sets, and the comparison results were rearranged as in Table 5.

For the "work" category, the longer the cycle time of "remote operation using the conventional system," a greater improvement is seen in work efficiency when the operation interface is switched to the cabin-type remote operation device. On the other hand, the shorter the cycle time of "remote operation using the conventional system," the less effective it is.

For "traveling," the work efficiency of the higher ranking data shows an 11.7% reduction, while the efficiency is almost the same for middle and lower ranking data.

5.2.2 Causes of work efficiency differences

To understand what caused the work efficiency differences as in Table 5, three sets of data whose "work" and "traveling" data were close to the median were picked up each from the higher ranking, middle ranking and lower ranking groups. A total of six sets of data were thus selected. The behavior of the hydraulic excavator in those six sets of data was checked from the video records about "remote operation using the conventional system" and "operation with the cabin-type remote operation device," which are subject to comparison in our study, in order to look for what caused the work efficiency differences.

The result revealed that "shut-down time during one- to two-second-long work suspension time" and "sudden start and stop" are what caused the cycle time differences in the lower ranking group. Not many delay cases by these causes are seen in the middle ranking group. Almost no such cases are seen in the higher ranking group. No operation by complex motion is almost seen in any group.

For "traveling," over 10% differences occurred only in the upper ranking group. There is, however, no major difference in operational behavior between operation with the cabin-type remote operation device and operation with the remote operation device in either the higher, middle or lower ranking group.

For the higher ranking group, the excavator moves slower when operated with the cabin-type remote operation device. The operator uses the actual traveling lever when operating the machine with the cabin-type remote operation device. In this case, the operators need to operate the lever located at their foot with a head-forward posture. As the work monitor is located at the front, the operators need to watch this front monitor. This means the operators have to operate without checking how they are holding the lever. This makes them feel careful about driving the machine, which presumably causes slower traveling of the machine.

It is estimated that no major influence occurred to the middle and lower ranking groups because the speed is lower in the first place for those groups.

5.2.3 Summary

According to the scatter diagram in Fig. 10, there is a tendency for a positive correlation in the difference in cycle time between operation with the remote operation device conventionally used in unmanned construction and that with the cabin-type remote operation device although the corroborating data are small in number.

It is found that operators who show a longer cycle time in operation with the remote operation device typically used in unmanned construction and find it difficult to adapt to remote operation shortened the cycle time and improved work efficiency when they switched to operation with the cabin-type remote

Table5 Comparison of the interface in the remote operation (divided by skill)

Remote operation cycle time	No. of data sets	Cycle time					Work efficiency difference		
		Grab	Lift and revolve	Lower	Raise	Grab	Work	Travel	Grab
Higher ranking	25	-5.8	-1.5	2.9	-0.5	-22.8	-1.4%	-11.7%	-5.2%
Middle ranking	25	-0.9	-3.1	9.1	-2.2	1.6	3.1%	1.3%	2.6%
Lower ranking	25	7.1	10.2	14.8	5.0	33.9	7.0%	-2.2%	4.6%
Grand total		0.1	1.9	8.9	0.8	4.3	2.9%	-4.2%	0.7%

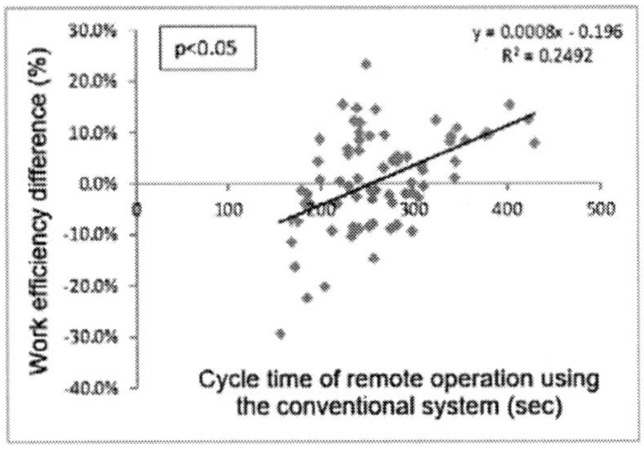

Fig. 10 Remote operation time and work efficiency difference due to the difference in the operation interface

operation device.

On the other hand, operators who show a shorter cycle time with operation with the remote operation device used in unmanned construction and can sufficiently adapt to remote operation turned out to show no reduction in cycle time when switched to operation with the cabin-type remote operation device.

The experiment results suggest that there are operators who do not depend on operation interfaces and others who depend on the operation interface similar to that of manned operation.

In conclusion, a certain work efficiency improvement or stability improvement may result from applying the shape or layout of the operating lever of an actual machine to the operation interface of remote operation.

It is also suggested that there are operators who have a significant capability to adapt to remote operation.

6 Conclusion

Switching the operation interface of current unmanned construction to the actual operating lever is effective for improving work efficiency only in a limited manner, but is considered to be effective as a remediation for operators slower in remote operation. There is still room for review about the layout of the traveling lever.

According to the results of our research that focused on the effect of operation interface operability for manned operation on work efficiency (a work efficiency reduction of 30% on average for the entire operation), improvement of the operation interface is considered to be an essential task to tackle in the future for unmanned construction that is predicted to become closer to manned construction in the years to come.

References

[1] Masaharu Moteki, Shinichi Yuta, Kenichi Fujino: The Proposal of the Model Task for Efficiency Evaluation of the Construction Work by Remote Control of a Hydraulic Excavator, paper in the Journal of JCMA, Vol. 66, No. 8, pp.71 - 79, 2014
Akihiko Nishiyama, Masaharu Moteki, and Kenichi Fujino: Research on the Comparison of Operator Viewpoints between Remote Control and Manned Operation in Unmanned Construction with Hydraulic Excavator, the 12th Symposium on Construction Robotics in Japan

[2] Masaharu Moteki, Kenichi Fujino, and Akihiko Nishiyama: Research on Proficiency of remote operation of Construction Machinery - Discussion on Work Efficiency and Accuracy for Manned Operation and remote operation of Construction Machinery, Proceedings and Summaries of the 2012 Construction Execution and Construction Machinery Symposium, pp. 19-24, 2012

[3] Akihiko Nishiyama, Masaharu Moteki, Kenichi Fujino, and Takeshi Hashimoto: Research On The Comparison Of Operator Viewpoints Between Manned And Remote Control Operation In Unmanned Construction Systems, the 30th International Symposium on Automation and Robotics in Construction, 2013

[4] Masaharu Moteki, Kenichi Fujino, and Takashi Otsuki: Diffusion of Unmanned Construction Technology to Construction Sites and Future Technical Needs - Validation Experiment Report for remote operation System, Proceedings of the 12th Symposium on Construction Robotics in Japan, pp. 89-96, 2010

Derivation of the Factors Influencing the Network Performance of Serverless Smart Exit Sign Systems based on Wireless Sensor Networks

Hyunoh Kim[a], Ghang Lee [a*], and Jehyun Cho[a]

[a]Dept. of Architectural Engineering, Yonsei University, South Korea
E-mail: hyunohkim7@gmail.com, glee@yonsei.ac.kr, jkun86@naver.com

Abstract

This paper presents the factors that could potentially affect the reliability of a serverless smart exit sign system, namely an evacuation guidance system based on a wireless sensor network (WSN) without a central server. Serverless smart exit sign system dynamically changes the directions of signs to indicate the shortest safe evacuation paths. Thus, the reliability of these systems is critical. Nevertheless, no research has been conducted to test the reliability of serverless smart exit sign systems. As a first step, we conducted a literature review to analyze the factors that could degrade the network performance of serverless smart exit sign systems. The identified factors were grouped into three categories: physical obstacles, environmental factors, and WSN properties. We plan to develop a prototype of a serverless smart exit sign system and then to conduct experiments to validate the influence of these factors on its network performance.

Keywords –
Serverless Evacuation System; Wireless Sensor Network; Smart Exit Sign System

1 Introduction

A smart exit sign system is an evacuation guidance system that dynamically changes the directions of the signs to indicate the shortest safe evacuation paths. A serverless smart exit sign is a smart exit sign that communicates data between the exit sign nodes using a wireless sensor network (WSN) without a central server (Figure 1) [6]. When a fire breaks out, evacuees are more likely to rely on exit signs when they are in a venue they do not know well than when they are in a familiar place [23]. Traditional exit signs show fixed direction signs to the nearest exit; however, this can lead evacuees to areas of greater danger if the fire has spread.

To solve this problem, several researchers have proposed smart exit sign systems; however, most of these studies have focused on the development of algorithms to find the shortest safe path to an exit based on the assumption that a central server exists [28,32,34]. Other studies have only proposed the concept of a smart exit sign system without elaborating on details [22,26,27].

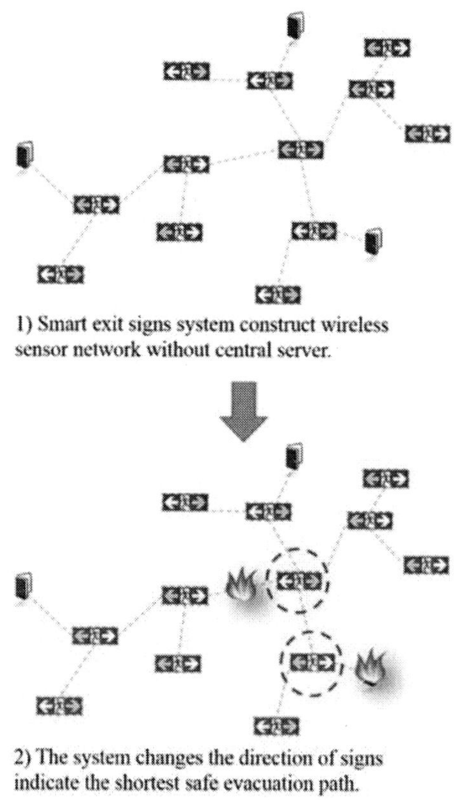

Figure 1 The concept of the serverless smart exit sign system

Serverless smart exit sign systems have several

advantages over the previous server-dependent smart exit sign systems. First, serverless smart exit sign systems are more reliable than server-dependent ones because the entire server-dependent system fails if the central server fails or any line between the server and the individual exit sign nodes is disconnected due to a fire, physical damage, or other reasons. On the other hand, exit sign nodes in serverless systems are not reliant on the status of the server because they do not use a server at all and can still communicate with neighboring nodes even if some nodes are damaged. The second advantage is that a serverless smart exit sign can be installed in an existing building without complex wiring work whereas a server-dependent smart exit sign system requires complex and expensive wiring.

Thus far, two groups of researchers have independently developed serverless smart exit sign systems, one in South Korea and the other in the UK [11,12,14]. Both groups have however focused only on the development of algorithms for serverless smart exit sign systems and have not considered the hardware aspects of serverless smart exit sign systems. This paper focuses primarily on the reliability of the wireless communication of serverless smart exit sign systems.

By reliability, we mean the extent to which a system stably and consistently operates under various internal and external conditions. For example, doors and walls that can potentially deteriorate network communication are external factors, and the specifications of network modules such as network packet size are internal factors. Reliability is an important requirement of any emergency evacuation system, including serverless exit sign systems, to prevent system failure [21]. The reliability of wireless communication (networking) in particular is key to the successful operation of a serverless exit sign system because, if individual exit sign units cannot send and receive data reliably, the entire system may malfunction.

As a preliminary study, we conducted a literature review to analyze and determine the factors that should be considered during the testing of the network reliability of serverless smart exit sign systems.

2 The reliability of serverless smart exit sign systems

The reliability we describe in this paper can be explained by two aspects: network reliability and data transmission reliability.

1. Network reliability refers to the stability of a system against external physical environments and indoor environmental variations. For example, potential physical obstacles are doors and walls while indoor environmental factors include temperature variations, humidity, and smoke variations. These factors may destruct the relay connection between motes (nodes in a sensor network) in a WSN.
2. Data transmission reliability refers to the extent to which a system stably and correctly communicates data between nodes. For example, packet size is one factor that may affect data transmission reliability. Network reliability can be improved by ensuring data transmission reliability.

Based on this classification, we conducted an analysis and derived three categories of factors: physical obstacles, environmental factors, and WSN properties.

3 Research method

The literature review has conducted keywords that Wireless sensor network, WSN performance, WSN experiment. The retrieved research are categorized in three fields that network reliability related physical obstacles, environmental factors and data transmission reliability related WSN properties. Among the cases, the only empirical research based on experiments using hardware are collated in this study. Table 1 summarizes collated cases.

Table 1 Collated cases

Index	Value
Collated cases	35
Published year	2001-2014
Database	IEEE, ScienceDirect, DBpia
Keyword	Wireless Sensor Network, WSN performance, WSN experiment
Derived factors	14

The collated result shows that physical obstacles include six factors that door, wall, floor, staircase, height, and distance between motes. WSN properties have five factors that nodal processing capacity, antenna orientation, RF power, packet size, other network interference. In addition, environmental factors include temperature, humidity and smoke. To sum up, serverless smart exit sign system has to consider both aspects of hardware and software hindrance, which may occurs indoor space. Figure 2 summarizes the derived factors of physical obstacles, environmental factors and WSN properties.

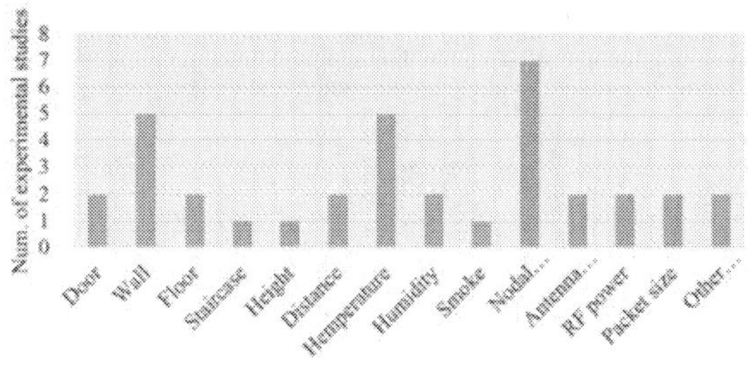

Figure 2. Derived factors influencing network performance

3.1 Physical obstacles

Many physical obstacles in a building may degrade the network performance of a WSN in an indoor environment. Jeon et al. [18] argued that signal transmission distance varies depending on building use and occupancy. For example, in office buildings, desk partition walls obstruct network communication, and in department stores, display stands and tables block signal transmission. It is however difficult to generalize the characteristics of physical obstacles by building use and occupancy. The reason is that many installations are temporary and are continuously being modified according to the occupants' changing needs. Moreover, the size of objects and their material properties may vary greatly.

Holland et al. [16] measured the degradation of WSN performance in an indoor environment. According to the study, the height difference between motes significantly affects network reliability. For instance, the packet transmission success rate between a ceiling-attached mote and a floor-embedded mote is lower than that between two ceiling-attached motes. The study showed that the vertical installation position of smart exit signs is another factor that should be taken into consideration.

Several other researchers [1,9,13,18,19,24,29] have conducted WSN performance tests while considering specific physical obstacles in indoor environments. The most commonly considered obstacle was a wall. Many studies [1,9,18,19,29] showed that the two layers of a 200 mm-thick concrete wall obstruct packet transmission. Doors, staircases, and the number of floors have also been identified as factors that degrade packet transmission [9,13].

In addition, the material type of a physical obstacle in a building is another factor that may affect network communication. For example, Di [9] and Fuhr et al. [9,13] demonstrated that motes cannot communicate with each other when they are installed over 15 m away with a steel door between them.

Although the findings of previous studies are useful in identifying the factors that generally affect network communication, the specific numbers (e.g., 200 mm concrete wall, 15 m) may change depending on the specifications of the network components and the physical obstacles. For example, Fuhr et al. [13] succeeded in reading data inside a metal food shipping container from outside the container using a radiofrequency (RF) device.

Table 2 summarizes the factors (i.e., doors, walls, floors, staircases, height, and distance between motes) that are potential physical obstacles in indoor environments.

Table 2 Derived physical obstacles

Factor	Previous research results	References	Considerations
Door	Steel doors degrade the packet success rate at a distance of 20 m between motes.	[9,13]	Door closure (half, fully) test with steel doors, glass doors, and fire shutters
Wall	Two layers of concrete walls with a total thickness of 200 mm block packet transmission.	[1,9,18,19,29]	A concrete wall test with a non-line of sight (NLOS) condition
Floor	Two stories of concrete floors block packet transmission.	[9]	Multi-story floors test with an NLOS condition
Staircase	20 m of stairs block packet transmission.	[9]	Additional mote installation test at the landing space
Height	Height differences between motes degrade the packet success rate.	[16]	Placement position test with radiofrequency (RF) power control
Distance	16 dBm over RF power is needed at a distance of 20 m between motes.	[24]	Effective distance test with RF power control

3.2 Environmental factors

In indoor spaces, a few environmental variations may influence WSN performance, for example, temperature and humidity variations. Indeed, fire causes sprayed water from sprinklers to raise the humidity level abruptly. Flames also increase the temperature of an indoor space immediately, which may affect WSN performance.

Several researchers conducted WSN performance tests in actual fired spaces to consider the temperature and humidity variations of the environment [3,5,10]. The motes read the temperature and humidity of the fired areas and continuously reported the status data using 400 MHz and 868 MHz multi-channel radio transceivers. The studies showed that WSN motes are able to communicate through penetrating flames and keep tracking the fire diffusion until they burn out. WSNs perform regardless of degradation caused by high-temperature environments.

Ruiz-Garcia et al. [31] conducted a WSN performance test in cooling environments (0–20°C). The researchers manipulated the temperatures in the cooling environments, which comprised a warehouse, a shipping container, and a vehicle for food logistics. Although the general indoor environment has a higher temperature than this range of venues, some corridors or spaces may be as cold in winter. A Zigbee-based WSN performance test showed that dynamic temperature variations have no effect on WSN performance.

Nevertheless, humid environments have a far greater effect on WSN performance than significant temperature variations. Anastasi et al. [2] and Holland et al. [16] conducted empirical research to test WSN performance in various humid environments. The studies both confirmed that rain and fog cause RF signal attenuation, which degrades WSN performance.

Hofmann et al. [15] investigated the impact of fire, smoke, and vapor on wireless communication in a tunnel. The study showed that fire and smoke do not severely affect communication performance; however, vapor significantly attenuates signal transmission.

Table 3 summarizes the environmental factors that could degrade WSN performance: temperature variations, a high humidity level, and smoke.

3.3 Wireless sensor network properties

The performance of a WSN is influenced by both its hardware and software properties. Hardware properties that could potentially affect WSN performance include the nodal processing capacity and antenna orientation. Software properties include RF power, packet size, and network interference.

Several researchers have investigated the influence of 1 to *n* connections [7,17,20,35]. When many motes transmit packets to a single mote, packet collision and channel interference occur. Kim et al. [20] found that increased data traffic and radio signal communication cause packet collision, which degrades network performance. Wilson et al. [35] argued that network expansion by multi-hop communication increases the packet error rate due to the increased number of packet collisions. Similar to maintenance work, security systems using sensor networks connect multiple nodes to a gateway node.

Choi et al. [7] determined that security system networks have an increased transmission time for inquiry and exchange packets due to the acknowledgement (ACK) responses from multiple motes. In their study, these ACK responses were used to check packet collision redundancy.

Smart exit signs also connect numerous sensor nodes in an entire building, which may cause packet collision. For instance, a hall or a lounge space may connect multiple corridors through which 1 to N connections are constructed.

Table 3 Derived environmental factor

Factor	Previous research results	References	Considerations
Temperature	1) Motes surrounded by fire can communicate with outside motes. 2) Cooling conditions (0–20°C) do not degrade the packet success rate.	[3,5,10,15,31]	This negligible factor depends on the WSN component specifications.
Humidity	1) Rain and fog attenuate radio signal. 2) Water vapor causes signal attenuation at a distance of 40 m from installed motes.	[2,15]	Humid environment test
Smoke	Smoke does not significantly attenuate the 2.4 GHz band of wireless communications.	[15]	This negligible factor depends on the WSN component specifications.

Therefore, the packet error rate at multiple connections should be considered in smart exit sign systems.

Furthermore, many other WSN properties are linked to the packet error rate. For example, the size of the packet, RF power, and other network interferences are included as well.

Bigger-sized packets can create sizable background traffic on a network, which then decreases the packet delivery ratio (%). Lee et al. [25] stated that larger-sized data payloads considerably degrade the networking performance of WSNs. When network traffic is increased, it causes packet collision due to the larger data payload. Moreover, a larger packet size also causes the battery life of the motes to shorten [30]. Therefore, packet size should be taken into account when considering network expansion and battery life.

Other networks could affect WSN performance, for instance, WiBro and satellite DMB WLAN. To avoid interference, many researchers have investigated the protocols of designated RF bands. Fuhr et al. [13] stated that an offset of 7 Mhz is required to avoid interference with other network communication. An et al. [1] conducted WiBro and satellite DMB interference tests on WSNs and suggested that an offset of 10 MHz is ideal.

Smart exit sign systems have the same network interference issues in buildings. A condensed network environment and other wireless networks may cause network interference with smart exit sign systems.

In terms of the hardware properties of WSNs, antenna orientation is important for connectivity performance [8]. In addition, the type of antenna and installation position have an influence on networking performance. Several researchers have conducted tests on antenna performance using different orientations and types of antennas [4,16].

Buckley et al. [4] pointed out that the type of antenna used influences networking performance significantly. The researchers conducted various experiments with chip internal antennas, external antennas, and planar-type, mono-type, and whip-type antennas. The results demonstrated that each antenna type performed significantly differently. The planar multilayer external antenna performed best in terms of distance and packet transmission success rate.

Holland et al. [16] conducted performance tests using different height conditions with antenna orientation. The study showed that different mote orientations with directional antennas degrade network performance.

Given the aforementioned findings, developers of smart exit sign systems should consider antenna orientation and type, particularly for floor-embedded types of smart exit signs.

RF power is another important factor influencing WSN performance. Higher RF power allows motes to communicate over longer distances; however, this does shorten battery life. Lee et al. [24] conducted a test to determine the effect of RF power over a fixed distance on packet transmission success rate. The study showed that, over a distance of 20 m, 16 dBm of RF power is needed to communicate correctly.

Table 4 Derived WSN properties

Factor	Previous research results	References	Considerations
Nodal processing capacity	The packet error rate is decreased when expanding connected motes.	[7,17,19,20, 25,29,35]	1 to *n* connection test
Antenna orientation	1) The direction of the antenna influences the packet transmission success rate by changing orientation. 2) Antenna type influences networking performance.	[4,16]	Omnidirectional antenna usage for floor-embedded smart exit signs
RF power	Higher RF power increases the packet transmission success rate.	[24,33]	RF power control test
Packet size	A larger-size packet decreases the packet delivery rate.	[24,35]	System algorithm and packet design
Other network interference	Analogue mobile communication, WiBro, Satellite DMB, and WLAN interference can be avoided if an offset of 10 MHz is used; a frequency offset of 7 MHz is demanded to avoid interference with other networks.	[1]	Other network interference tests

Sexton et al. [31] studied the physical layer protocol for use in industrial environments. The findings showed that 15 dBm may be required to achieve the desired range of 100 m NLOS to overcome physical obstacles and path loss uncertainty.

The available packet transmission distance may vary depending on the component specifications; however, these findings showed that the appropriate RF power needs to be configured to ensure optimal WSN performance in the installation environment.

Table 4 summarizes the network performance-related properties for WSNs: nodal processing capacity, antenna orientation, RF power, packet size, and other network interference.

4 Conclusion

An objective of the research is derivation of the factors considering the reliability of serverless smart exit sign systems. By conducting a literature review, we determined the factors that may influence the networking performance of serverless smart exit sign systems. The derived factors will be validated on further research using serverless smart exit sign system prototype.

So far, two groups of researchers have proposed algorithms for serverless smart exit sign systems [6,11,12,14]. No research has been done however to build a physical system while taking into consideration potential system failure. Reliability is key to the success of a serverless smart exit sign system and to prevent system failure. If individual exit signs cannot send and receive data reliably, the entire system will malfunction. We therefore defined reliability by considering two aspects of smart exit sign systems: network reliability and data transmission reliability.

Network reliability refers to the stability of a system in the presence of physical obstacles and indoor environmental variations as well as external factors. Data transmission reliability refers to the extent to which a system stably and correctly communicates data between nodes taking into account internal factors. Three categories of internal factors were derived:

1. Physical obstacles: doors, walls, floors, stairs, height differences, and distance
2. Environmental factors: temperature, humidity, and smoke
3. WSN properties: nodal processing capacity, antenna orientation, RF power, packet size, and interference from other networks

Although this paper analyzed and thus determined the factors that should be considered when testing network reliability, it has limitations. The identified factors are closely related to the installation environment and the component specifications of the WSN. Depending on the circumstances, specific numbers may change, for instance, if there is a 200 mm-thick concrete wall obstruction or a two-story floor obstruction. In addition, the importance level of the factors may vary depend on installation condition. To clarify this, we are planning to conduct experiment to validate the derived factors and its specific numbers.

Although the derived factors have the aforesaid limitations, this paper may make a meaningful contribution to future work on serverless smart exit sign systems. The identified factors should be considered when designing the network of disaster management system and testing the network reliability of a serverless smart exit sign system to fulfill the key requirements of the system. The factors can furthermore be implemented with ease in the design of a network system.

References

[1] T.-K. An, J.-R. Shin, G.-Y. Kim, S.-H. Yang, G.-B. Choi, B.-S. Sim, Performance Evaluation of Wireless Sensor Networks in the Subway Station of Workroom, Proceedings of Korean Society for Railway, Korean Society for Railway, 2011, 1701-1708.

[2] G. Anastasi, A. Falchi, A. Passarella, M. Conti, E. Gregori, Performance measurements of motes sensor networks, Proceedings of the 7th ACM international symposium on Modeling, analysis and simulation of wireless and mobile systems, ACM, Venice, Italy, 2004, 174-181.

[3] T. Antoine-Santoni, J.-F. Santucci, E. De Gentili, X. Silvani, F. Morandini, Performance of a Protected Wireless Sensor Network in a Fire. Analysis of Fire Spread and Data Transmission, Sensors 9 (8) (2009) 5878.

[4] J. Buckley, K. Aherne, B.O. Flynn, J. Barton, A. Murphy, C.O. Mathuna, Antenna performance measurements using wireless sensor networks, Electronic Components and Technology Conference, 2006. Proceedings. 56th, 2006, 6 pp.

[5] M. Chen, C. Majidi, D. Doolin, S. Glaser, N. Sitar, Design and construction of a wildfire instrumentation system using networked sensors, Network Embedded Systems Technology (NEST) Retreat, Oakland California. Retrieved April 5 (2004).

[6] J. Cho, G. Lee, S. Lee, H. Kim, Development of a Server-Independent Algorithm for Safe Evacuation Systems Utilizing Exit Signs,

32nd International Symposium on Automation and Robotics in Construction and Mining, Vol. 32, Vilnius Gediminas Technical University, Department of Construction Economics & Property, Oulu, Finland, 2015, 1-6.

[7] S.-H. Choi, B.-K. Kim, J. Park, C.-H. Kang, D.-S. Eom, An implementation of wireless sensor network, IEEE Transactions on Consumer Electronics 50 (1) (2004) 236-244.

[8] D. Culler, D. Estrin, M. Srivastava, Overview of Sensor Networks, Computer 37 (8) (2004) 41-49.

[9] X.M. Di, Byung-Hyun, Performance Analysis of Wireless Sensor Nodes over Indoor and Outdoor Environments, Journal of the Korea Industrial Information Systems Research 17 (2) (2012) 1-9.

[10] D.M. Doolin, N. Sitar, Wireless sensors for wildfire monitoring, Smart Structures and Materials 2005: Sensors and Smart Structures Technologies for Civil, Mechanical, and Aerospace Systems, Vol. 5765, San Diego, CA, 2005, 477-484.

[11] A. Filippoupolitis, E. Gelenbe, An Emergency Response System for Intelligent Buildings, in: N. M'Sirdi, A. Namaane, R.J. Howlett, L.C. Jain (Eds.), Sustainability in Energy and Buildings: Proceedings of the 3rd International Conference in Sustainability in Energy and Buildings (SEB'11), Vol. 12, Springer Berlin Heidelberg, Berlin, Heidelberg, 2012, 265-274.

[12] A. Filippoupolitis, G. Gorbil, E. Gelenbe, Autonomous navigation systems for emergency management in buildings, GLOBECOM Workshops (GC Wkshps), 2011 IEEE, Houston, Texax, USA, 2011, 1056-1061.

[13] P. Fuhr, R. Lau, Mesh radio network performance in cargo containers, Sensors 22 (3) (2005) 13-18.

[14] G. Gorbil, A. Filippoupolitis, E. Gelenbe, Intelligent navigation systems for building evacuation, Computer and Information Sciences II - 26th International Symposium on Computer and Information Sciences, ISCIS 2011, London, UK, 2012, 339-345.

[15] P. Hofmann, K. Kuladinithi, A. Timm-Giel, C. Goerg, C. Bettstetter, F. Capman, C. Toulsaly, Are IEEE 802 Wireless Technologies Suited for Fire Fighters?, Wireless Conference 2006 - Enabling Technologies for Wireless Multimedia Communications (European Wireless), 12th European, 2006, 1-5.

[16] M.M. Holland, R.G. Aures, W.B. Heinzelman, Experimental investigation of radio performance in wireless sensor networks, Wireless Mesh Networks, 2006. WiMesh 2006. 2nd IEEE Workshop on, 2006, 140-150.

[17] C. Jaikaeo, C. Srisathapornphat, C.C. Shen, Diagnosis of sensor networks, 2001 IEEE International Conference on Communications, Vol. 5, 2001, 1627-1632 vol.1625.

[18] J. Jeon, The study on safe management system of fire-fighter using sensor network, Proceedings of Korean Institute of Fire science & Engineering, KOREAN INSTITUTE OF FIRE SCIENCE & ENGINEERING, 2007, 118-125.

[19] M. Kang, Design of Multi-node Real-time Diagnostic and Management System Using Zigbee Sensor Network, Journal of the Institute of Electronics and Information Engineers 51 (6) (2014) 152-161.

[20] J. Kim, K. Chung, An Efficient Beacon Management Technique for Sensor Network-Based Indoor Location Systems, Information Networking 36(4) 36 (4) (2009) 330-338.

[21] S.-K. Kim, J.-Y. Won, A search feasilibilty of ubiquitous information technology in the Seoul Metropolitan Disaster Management, Journal of The Korean Regional Development Association 15 (4) (2003) 97-118.

[22] Y.W. Kim, Y.S. Lee, D.H. Kim, Design and implementation of autonomous refuge guide system based on wireless sensor network, Proceedings of Korea Institute Of Communication Sciences, Korea Institute Of Communication Sciences, 2008, 389-392.

[23] M. Kobes, I. Helsloot, B. de Vries, J.G. Post, N. Oberijé, K. Groenewegen, Way finding during fire evacuation; an analysis of unannounced fire drills in a hotel at night, Building and Environment 45 (3) (2010) 537-548.

[24] J.-h. Lee, I.-b. Jung, Analysis of Packet Transmission Performance for Construction of Wireless Sensor Networks in Indoor Environment, Journal of the Korea Institute of Information and Communication Engineering 13 (9) (2009) 1941-1946.

[25] J.-S. Lee, An experiment on performance study of IEEE 802.15.4 wireless networks, Emerging Technologies and Factory Automation, 2005. ETFA 2005. 10th IEEE Conference on, Vol. 2, 2005, 8 pp.-458.

[26] J.H. Lee, S.-j. Moon, Fire evacuation information system proposal, Proceedings of Korean Society For Internet Information,

Korean Society For Internet Information, Seoul, Korea, 2010, 323-326.

[27] M.G. Lee, K.K. Jung, W.-S. Lee, Design of fire emergency evacuation system using potential field, The Institute of Electronics Engineers of Korea -IE 48 (3) (2011) 26-32.

[28] N. Li, B. Becerik-Gerber, B. Krishnamachari, L. Soibelman, A BIM centered indoor localization algorithm to support building fire emergency response operations, Automation in Construction 42 (2014) 78-89.

[29] M. Paavola, J. Kemppainen, Wireless monitoring of a steam boiler - performance measurements in industrial environment, Industrial Electronics, 2008. ISIE 2008. IEEE International Symposium on, 2008, 1166-1171.

[30] R. Piyare, S.-r. Lee, Performance analysis of XBee ZB module based wireless sensor networks, International Journal of Scientific & Engineering Research 4 (4) (2013) 1615-1621.

[31] L. Ruiz-Garcia, P. Barreiro, J.I. Robla, Performance of ZigBee-Based wireless sensor nodes for real-time monitoring of fruit logistics, Journal of Food Engineering 87 (3) (2008) 405-415.

[32] Y. Seo, C. Lee, J. Jung, S. Shin, A guidance methodology using ubiquitous sensor network information in large-sized underground facilities in fire, JOURAN OF THE KOREAN SOCIETY OF CIVIL ENGINEERS D 28 (4D) (2008) 459-467.

[33] D. Sexton, M. Mahony, M. Lapinski, J. Werb, Radio Channel Quality in Industrial Wireless Sensor Networks, Sensors for Industry Conference, 2005, 2005, 88-94.

[34] L. Shen, Z. Andong, W. Xiaobing, C. Guihai, ERN: Emergence Rescue Navigation with Wireless Sensor Networks, 15th International Conference on Parallel and Distributed Systems (ICPADS), Melbourne, Australia, 2009, 361-368.

[35] W.T.H. Woon, T.C. Wan, Performance Evaluation of IEEE 802.15.4 Ad Hoc Wireless Sensor Networks: Simulation Approach, 2006 IEEE International Conference on Systems, Man and Cybernetics, Vol. 2, Taipei, Taiwan, 2006, 1443-1448.

A Multi-Objective Mixed Integer Programming Model for Minimising Obtrusive Effects and Installation Costs of Night-time Lighting on Construction Sites

A.W.A. Hammad[a], A.Akbarnezhad[a*] and D. Rey[a]

[a]School of Civil and Environmental Engineering, University of New South Wales, Australia
E-mail: a.hammad@unsw.edu.au, a.akbarnezhad@unsw.edu.au*, d.rey@unsw.edu.au

Abstract

The increase in the rate of urbanisation worldwide has led to a boom in the construction industry sector in most major cities. To cope with the associated rising demand for further services and facilities, contractors find themselves frequently obliged to extend working hours on construction sites. Construction of major infrastructure projects, such as road works, is conducted at times of the day when less disruption is likely to result to the affected population. As a result, night work is now a common sighting on many construction sites. To allow for better vision for construction personnel on sites, floodlights are deployed during night work to light up work zones. There is an expense however to the adoption of construction lighting, particularly when considering the social and environmental impacts arising due to light pollution. In a residential district, and especially when infrastructure projects are taking place, the lighting system utilised on a construction site can be a cause of major disruption to the people and wildlife in the vicinity of the site. This paper attempts to model the problem of construction lighting through an optimisation model that minimises both construction light set-up costs and the maximum light pollution perceived at light-sensitive receivers. At the same time, the model ensures an appropriate coverage level to avoid impairing workers' vision on site. The developed formulations take the format of a Mixed Integer Programming model. An illustrative case study is applied to a construction project to demonstrate the applicability of the model.

Keywords –
Construction Lighting; Light Pollution; Multi-Objective Optimisation; Mixed Integer Programming; ε - constraint Method; Sustainable Construction Operations.

1 Introduction

With the rising tendency of extending working times in the construction sector, in order to meet strict deadlines imposed on project completion, the sight of construction work taking place during the night time has become a norm in many cities around the world [1]. Along with the increasing rate of urbanisation, the steady increase in population of urban cities prompts contractors to race in the delivery of projects, in order to meet the generated demand. To allow for a smooth execution of tasks during the night hours on a construction site, an adequate lighting system needs to be installed. One of the main issues addressed by governing bodies, such as the International Commission on Illumination (CIE), which set out lighting regulations, is the provision of a reasonable level of lighting such that workers' vision is maintained properly. At the same time it has to be ensured that the negative impacts from the lighting system adopted, on the neighbouring vicinity is kept to a minimum. Thus, in Australia, obtrusive effects resulting from the lighting plan implemented on a construction site must adhere to the guidelines provided in outdoor lighting standards such as AS 4282 [2]. It is also necessary that the enforced plan produces an adequate light intensity, as stipulated in AS 1680.5 [3], to allow for optimum working conditions on site.

The operational impacts of the lighting plan deployed during night-time construction are wide and varied. Some of the factors reported to be influenced on construction projects include cost, productivity and safety of construction tasks [4], [5]. Dealing with excess lighting and its ramifications on human health, as well as its effect on wildlife and on stellar vision, is necessary, given that light pollution is one of the fastest growing forms of pollution [6]. When discussing the health impacts of lighting the most prevalent effect of light exposure is the reduction in the production and secretion of Melatonin [7], [8]. Among the plethora of ailments found to be associated with Melatonin suppression in humans, we list the following: sleep disruptions [9], psychiatric disorders [10], abnormal adrenergic functioning [11] and an increase in the risk of cancer progression [12]. Documented nocturnal lighting effects on wildlife include alteration of migration, communication, mating and predation behaviour [13].

Various ways of optimising outdoor lighting parameters have been studied in the literature. Noor- E - Alam et al., [14], proposed a grid-based lighting location problem where mixed integer programming (MIP) models were developed to minimise the amount of darkness and excess supply of light in an open park. Optimising the design of specific parts relating to the luminaire itself has been attempted in order to minimise energy consumption and optimise light distribution [15]. Other topics related to the optimisation of light photometric factors include the problem of optimising road lighting distribution [16], and the optimisation of the height of luminaires to improve energy efficiency and satisfy target illuminance of demand areas [17]. Gómez-Lorente et al. [18] presented a multi-objective evolutionary algorithm which maximises the illuminance uniformity and the installation efficiency of roadway luminaires. Lai et al. [19] proposed a non-linear optimisation model to minimise the luminous flux of light settings in a long tunnel.

When it comes to construction lighting, studies have mostly concentrated on the optimisation of lighting for highway construction tasks. The work of El-Rayes and Hyari [20] focused on developing a framework, labelled as CONLIGHT, which incorporates different parametric elements that concern the design of lighting plans for night-time highway construction and rehabilitation projects. The same authors then implemented the framework in a multi-objective optimisation problem that incorporated the use of genetic algorithms, where the authors maximised average illuminance and lighting uniformity, while cost and glare were minimised [21]. Said and El-Rayes, 2010 [22] incorporated a light system optimisation module in a proposed multi-objective framework to optimise the security of critical infrastructure projects through site layout planning. Corcione and Fontana [23] focused on outdoor lightings; in particular the use of GA to optimise the lighting requirements in sport fields. Excessive luminance, causing discomfort to an observer, is addressed in many of the standards, through specified restrictions on luminaire's aiming direction with respect to the observer's line of sight [24].

On a construction site, mathematical optimisation has been employed in previous studies to minimise noise pollution and transport costs [25]–[28], with no mentioning of light pollution as an objective. Optimisation studies accounting for light obtrusiveness are mostly related to road luminaires where observers are taken as vehicle drivers [29], with little or no attention directed at limiting light pollution at other types of light sensitive receivers. As such, we propose a location model to optimise the location of floodlights around a construction site while ensuring that light intensities at light sensitive receivers, in the vicinity of the construction site, are kept to a minimum. Light related formulations have been obtained from light design standards, including AS 4282 [2] and CIE 150 [30].

Location theory has been extensively studied and applied to various problems in a variety of fields. A class of models that we adapt, to suit our Construction Site Light Location Problem (CSLLP), are set covering location models, where the objective is to minimise the cost of meeting all demand through the servicing facilities [31]. We therefore consider the light poles as facilities. We present a multi-objective Mixed Integer Programming (MIP) model to positon the light poles that service work zones on the construction site during night-time, while minimising both light installation costs and light obtrusiveness.

The remainder of this paper is organised as follows. In section 2 we present the framework applied to solve the CSLLP. We describe the steps undertaken to compute the parameters of the model in a pre-processing phase, before introducing the objective functions and the constraints associated with our proposed model in Section 3. In section 4 we test our model on a case example and report the generated results. Finally, a conclusion is presented to summarise the major work contributions.

2 Framework

In this section of the paper, we explain the overall framework, depicted in **Figure 1**, for optimising the location of the light poles around the construction site.

As an initial step the site undergoes a pre-processing phase where work zones that require lighting are delineated as rectangular boxes. Next, we identify potential locations for the light poles on and around the construction site, based on the number and positioning of work zone areas. The demand for each work zone is next specified. Potential light pole locations are identified through their vertical and azimuth angles, B and β, respectively, measured with respect to the work zones and light sensitive receivers. The average horizontal maintained illuminance values at each of the work zones are computed to assess demand requirements, while vertical illuminance at each receiver, due to assumed light poles at each location, are used to assess the light obtrusiveness. The derived parameters are embedded in the optimisation model and a solution outlining the locations of the light poles is then obtained.

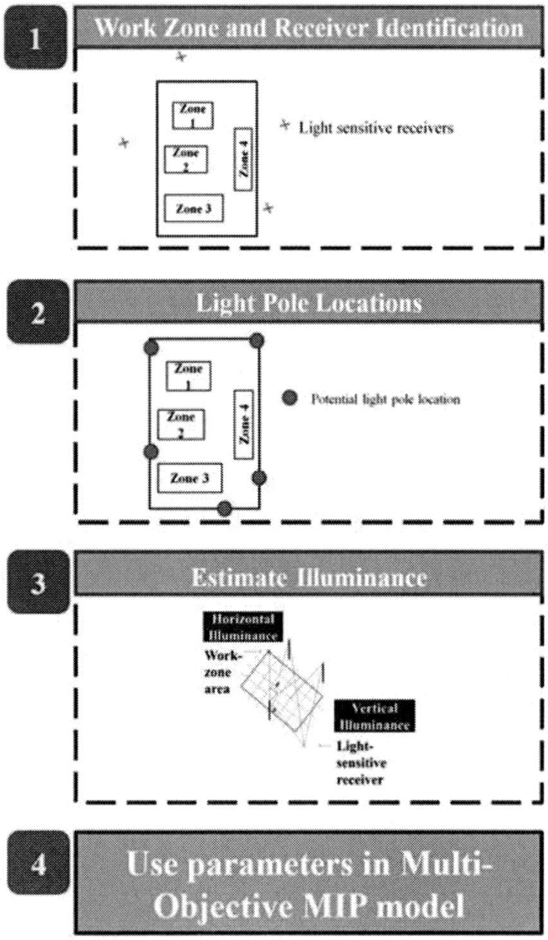

Figure 1. Framework of construction light location problem

2.1 Work zone and receiver identification

A work zone is defined as a rectangular area occupying places on the construction site where tasks are expected to be executed during night time. The reason for having to delineate the zones is to be able to discretise the overall space on a construction site, ensuring that the illuminance requirements of each relevant section of site are met. The area around the construction site needs to be surveyed to identify any light sensitive facilities, including residential dwellings, schools, etc. This then allows the planner to assess the impacts of obtrusive lights at these points, thus enabling prompt changes to the light plan to be adopted whenever needed. Lighting requirements for work zones are enacted in the form of constraints in the model that we propose.

2.2 Locating light poles

A location can only be specified if it is viable to mount a light pole on it. The location of light poles is modelled as a discrete problem and thus the potential locations appropriate for positioning the light poles need to be determined a priori. Locations should be chosen based on installation suitability, proximity to lighting area and distance away from sensitive receivers or nearby streets/roads [32], since all these factors affect the cost of light pole installations and its resulting light pollution levels.

2.3 Light parameter calculations

At the pre-processing phase, two main factors are computed for each of the potential light pole locations considered. The first of the factors, namely the horizontal maintained illuminance, E_{wl}^{H}, is used to assess the proportion of demand at work zone w that is met by the respective light pole at location l. The second factor is the vertical light illuminance, E_{rl}^{v}, used to assess impacts of light obtrusiveness on light sensitive receivers, r, in the vicinity of light installations at location l. We next detail how each of these factors is calculated

2.3.1 Illumination at work zones

To determine E_{wl}^{H} the lumen method is adopted [30]. Let ρ be the total lumen output from all luminaires within a single light pole. Let δ be the maintenance factor of the luminaires, which is dependent on the total months in use of the lamp, its burning hours and cleanliness of outdoor environment in which luminaires will be operating. Let α be the atmospheric loss factor, which varies depending on geographic location and on weather conditions. Let υ_l be the proportion of lamp flux reaching a specific area, also known as the utilisation factor, due to a pole at location l. Finally, let A_w be the total site area of work zone w that is to be lit. **Equation 1** is used to estimate the value of the average horizontal maintained illuminance at a given work zone w due to a light pole at location l.

$$E_{wl}^{H} = \frac{\rho\delta\alpha\upsilon_l}{A_w} \qquad (1)$$

To compute δ various lamp depreciation and surviving curves supplied by the luminaire manufacturers can be used [33]. For the purpose of our study we will assume that $\alpha = 1$ (i.e. no allowance for atmospheric loss is made). Calculations of υ_l are made based on the use of flux diagrams, reflecting the photometric characteristics of the considered luminaire, as will be described next.

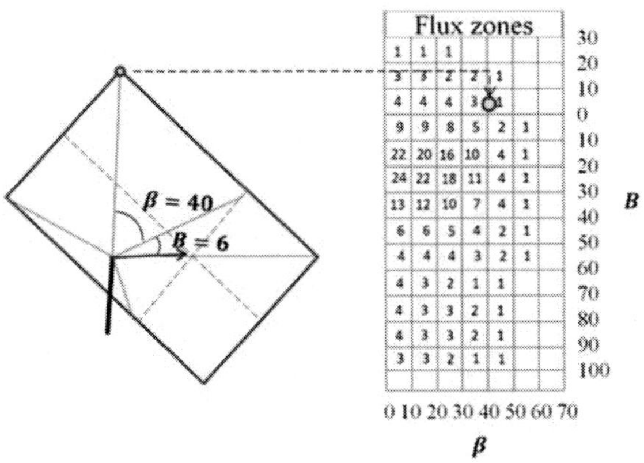

Figure 2. Plotting points of interest on the zonal flux diagram

2.3.1.1 Calculating Utilisation factor

To estimate the proportion of bare lamp flux from a light pole reaching a given area, we need to calculate the utilisation factor, using the zonal flux diagram, depicted in **Figure 2**. The zonal flux diagram, which can be obtained from the manufacturer of the luminaire under concern, corresponds to a coordinate system composed of two angles; the vertical and azimuth angles for the luminaires, expressed as B and β, respectively. We assume that the total lumen output of each light pole is given (hence we do not worry about the number of luminaires in each pole). We also assume for simplification that all light poles are identical, having fixed heights and with the peak intensity aimed at $\frac{2}{3}$ the width of the zone. The area being considered is divided, based on the aiming angle direction, into several panels; extreme points within each panel are plotted on the zonal flux diagrams. At each point we interpolate the flux emitted per 1000 lamp lumens, from a single potential pole location, making use the zonal flux diagram. Once all grid points resulting from the aiming direction of the pole are plotted, a summation of the flux values covered by the boundary drawn through the points on the flux diagram ensues. We are then able to calculate the utilisation factor corresponding to each light pole location. A single illustrative example of plotting the relevant angles on the zonal flux diagrams is highlighted in **Figure 2**.

2.3.2 Light intensity at light-sensitive receivers

For assessments carried out at light sensitive receivers, the vertical illuminance is computed. The point at which the intensity is calculated is usually chosen as the centroid of the body representing the receiving end. Light sensitive receivers mainly represent the window of a dwelling in the close vicinity of the site. **Equation (2)** is defined to compute the vertical illuminance, E_{rl}^{V}, at receiver r, due to a pole located at location l. Light intensity emitted from a source, I_{rl}, is assessed based on the angle coordinate system defined earlier, and these are plotted on an isocandela diagram. The angle between the normal to the surface of receiver r and the direction of the light emitted from the source is represented by the symbol θ_{rl}, while the distance between the light pole location and the receiver is given by d_{rl}. The rest of the symbols of **Equation (2)** have been defined earlier on.

$$E_{rl}^{V} = \frac{I_{rl}\cos\theta_{rl}}{1000 d_{rl}^{2}}\rho\delta\alpha \qquad (2)$$

3 Light pole location optimisation models

In this section, we present the multi-objective light pole location model, comprising of the objective functions and the set of constraints. One of the objective functions serves to minimise the maximum obtrusive effect of light, as measured at all light sensitive receivers in the vicinity of the constructions site. The second objective function is defined to minimise the cost of installing the light poles around the construction site. Constraints define the feasible search region; they are mainly specified to ensure full light coverage at all work zone areas. It is assumed that illuminance on a point is additive. Each objective function and the associated constraints of the model are described next. The notation adopted for the optimisation model is given in **Table 1**.

Table 1 Notation of sets, parameters and variables

Notation	Definition

Sets	W	Set of work zones, indexed by w
	L	Set of light pole locations, indexed by l
	R	Set of light-sensitive receivers, indexed by r
Parameters	D_w	Illuminance required at each work zone w
	E_{wl}^H	Average horizontal illuminance delivered to work zone w by a light pole positioned at l
	E_{rl}^V	Vertical illuminance delivered by a light pole positioned at l to receiver r
	c_l	Cost of installing a light pole at location l
	P	Maximum number of light poles that can be installed
Variables	x_l	Binary variable indicating whether a light pole is positioned at location l or not

3.1 Objective functions

The construction light location problem is modelled as a multi-objective MIP, where two objective functions are defined. Both objective functions warrant a sustainable implementation of a light plan, through targeting social, environmental and economic aspects.

3.1.1 Minimising Obtrusive effect of Light

The first objective function, defined in **Equation (3),** aims to minimise the maximum vertical illuminance experienced at any of the receivers near the construction area. **Equation (3)** thus acts as a social/environmental cost function that aims to reduce the obtrusiveness of the light plan adopted.

$$\text{Minimise} \quad \max_{r \in R} \left\{ \sum_{l \in L} E_{rl}^V x_l \right\} \tag{3}$$

3.1.2 Minimising Light pole installation costs

A second objective function, Equation (4) is incorporated into CSLLP, which targets the minimisation of the light pole installation costs. The location variable x_l controls the number of poles to be positioned within the construction site. Minimising this number will lead to cost savings. In addition, embedded in the equation is the cost parameter, which varies depending on the location of installation and any associated requirements needed for preparing that specific location for light pole fixing.

$$\text{Minimise} \quad \sum_{l \in L} c_l x_l \tag{4}$$

3.2 Constraints

Several hard constraints are defined to delineate the feasible search region based on criteria such as demand coverage, along with specifying the domain of the involved variables. The section below describes each formulated constraint.

3.2.1 Satisfying demand at work zones

An assessment of the average horizontal illuminance is carried out at each work zone, whenever a pole is chosen to be located at one of the given discrete locations. **Equation (5)** is specified to warrant full light coverage of each work zone, as is determined by illumination levels from all locations $l \in L$ that cover D_w.

$$\sum_{l \in L} E_{wl}^H x_l \geq D_w \qquad \forall w \in W \tag{5}$$

3.2.2 Number of light poles

When full coverage of an area is required, a natural tendency of the optimisation algorithm would be to associate as much light poles as is permitted, subject to the objective functions. In practice a limit is naturally enforced on the number of facilities that provide a service due to limited resource availability. As such, we define **Equation (6)** to limit the total number of permitted light poles to be positioned on the construction site.

$$\sum_{l} x_l \leq P \tag{6}$$

3.2.3 Variable Domain

A single group of variables is present in the proposed model, namely the location variable x_l . We require the variable to be binary; hence, Equation (7) is defined.

$$x_l \in \{0,1\} \qquad \forall l \in L \tag{7}$$

3.3 ε - constraint Method

For solving the multi-objective CSLLP, we adopt the ε - constraint method. This entails the re-formulation of the model to adhere to the format shown in **Equation (8)**, where $\mathbf{y}$ refers to the vectors of decision variables while Y denotes the feasible set. As a first step, both objective functions are optimised individually to obtain the extreme points of the Pareto frontier. Once the optimal values for both functions have been derived the model is formulated in such a manner that a single objective function is considered whilst the other function is incorporated as a constraint [34]. Initially, the objective function chosen as the constraint, $f_2(\mathbf{y})$, is upper bounded by its respective optimal solution, z_2^*. This requirement is slowly relaxed as the optimisation proceeds, through incrementing ε, until the constraint becomes no longer binding.

$$\begin{aligned} &\text{Minimise } f_1(\mathbf{y}) \\ &\text{such that} \\ &f_2(\mathbf{y}) < z_2^*(1+\varepsilon) \\ &g_u(\mathbf{y}) \le 0 \quad \forall u = 1,..,k \\ &\mathbf{y} \in Y \end{aligned} \tag{8}$$

4 Case Example

To highlight the applications of our proposed model, we implement the developed multi-objective MIP formulations on an illustrative case project to solve the CSLLP. The project, depicted in the plan view of **Figure 3**, relates to the construction of a railway line along the outskirts of an urban city. The project is split into various stages and our analysis is conducted over a single stage, namely the installation of concrete sleepers along the rail alignment.

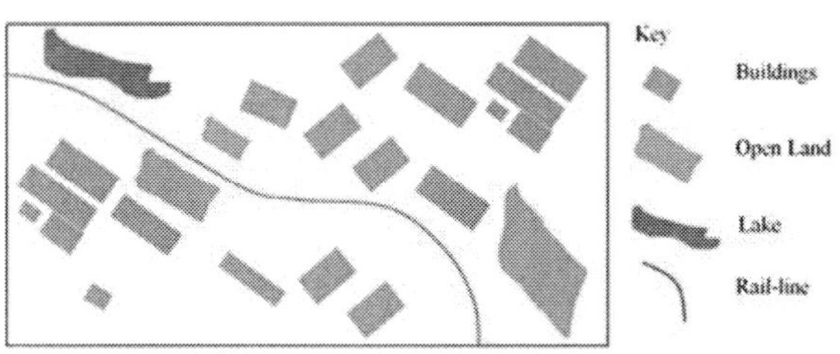

Figure 3. Plan view of rail line construction project

For installing the sleepers, it is required that work be conducted over a duration of two months, with several periods extending into the night time hours. Illuminance is measured at two main light sensitive receivers, namely residential dwellings and a nearby age care facility. The aim is then to provide a suitable light plan to account for the night work light requirements, while at the same time ensuring that light obtrusiveness at the receivers is kept to a minimal. We assume that 11 potential locations for installing the light poles, all satisfying electrical requirements etc., have been identified a priori, and these are shown by the red circles on **Figure 4**.

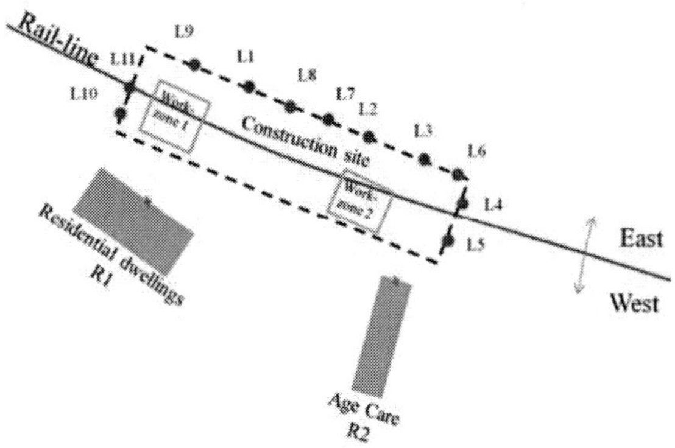

Figure 4. Construction site layout and proposed locations of floodlights

Two work zones are also delineated; area of work zone 1 is 12,000 m^2 while that of work zone 2 is 13,200 m^2. It is specified that for both work zones an average illuminance of 170 lx is required. The light luminaires to be installed are mounted 2000 W metal halide floodlights, beam type B, mounted at a height of 14 m, with 310,000 lumens of output. The associated zonal flux diagram is obtained from the manufacturer. All floodlights, once mounted, are assumed to have a fixed aiming angle of 62°. The calculation of the average horizontal illuminance, based on a grid point system as described earlier in the paper, was performed, and the results obtained for each potential location are reported in **Table 2**. Also, the vertical illuminance was calculated at the two receivers, the residential dwellings and the age care facility, shown in **Figure 4**. These parameters were then input into the model of **Equations (3) – Equation (7)**. Maximum number of light poles that can be installed was set to 8. The ε - constraint method was adopted for solving the proposed model, where the light obtrusiveness objective function was formulated as a constraint, while the light cost installation function was minimised. CPLEX was used as the linear solver to obtain the optimum solutions [35].

Table 2 also gives the cost of placing a floodlight at the discrete positions identified. The costs take into account the necessary electrical setup and site preparation required for installing the mounting poles. It is noticed from the cost figures of **Table 2** that the light pole locations on the eastern side of the line happen to have a higher cost than ones located on the western side. This has been attributed to differences in the topography of the land. The Pareto front of the optimisation problem is displayed in **Figure 5**, where the conflicting nature of the objective functions, for the particular project we consider, is apparent. Results highlight that having a low cost light plan configuration, at $ 27,289 would imply a high obtrusive light lux, measured at 118.76 lx, at one of the receivers. The opposite is true when a solution that minimises the light obtrusiveness is sought; this time the lowest of the maximum resulting light illuminance at a receiver is measured at 69.73 lx, with an associated light plan cost of $34,470. High costs are related to the locations chosen for positioning of the floodlights, as noted on the Pareto front of **Figure 5**.

Table 2 Model Parameters

Location	Work zone illuminance		E_{wl}^{V}		Cost (AUD)
	1	2	R1	R2	
L1	28.52	9.07	20.10	15.34	4903
L2	12.83	50.56	10.68	27.16	4785
L3	7.84	38.89	5.80	31.61	4941
L4	6.42	13.61	0.22	84.12	2478
L5	6.42	14.91	0.12	155.99	452
L6	7.13	25.28	3.31	32.01	5214
L7	14.26	25.93	12.81	22.11	4850
L8	17.11	12.32	14.81	18.28	5208
L9	47.77	7.78	25.04	12.89	5070
L10	57.75	5.83	49.85	0.29	369
L11	62.03	6.48	46.03	0.33	4200

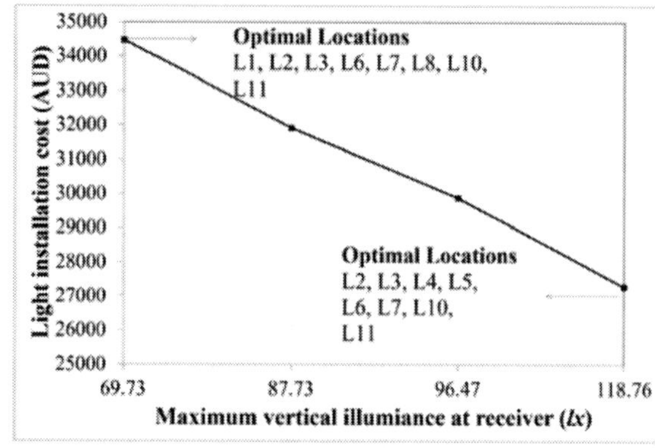

Figure 5. Pareto front highlighting efficient solutions to the multi-objective model

5 Conclusion

A multi-objective location set covering MIP was presented to solve the construction site light layout problem. The model involved two main objective functions; the first objective function minimised light obtrusiveness while the second function minimised light pole installation costs. Constraints were formulated to ensure full coverage of work zones on the construction site. In order to derive the Pareto front, the ε-constraint method was adopted, where globally optimal solutions were incrementally obtained. For the case study on which the model was applied, both objective functions happened to conflict, resulting in an efficient Pareto frontier. The results presented in this work highlight the potential for reducing light pollution through light location optimisation. Further work will focus on modelling the physics of cumulative light coverage at a point more precisely.

References

[1] O. A. Elrahman and R. J. Perry, “NIGHT-TIME CONSTRUCTION OPERATIONS,” Dec. 1994.

[2] Standards Australia, “Standards Australia 1997, Control of the obtrusive effects of outdoor lighting, AS4282,” 1997.

[3] Standards Australia, “Standards Australia 2012, Interior and workplace lighting - Outdoor workplace lighting, AS/NZS 1680.5,” 2012.

[4] R. D. Ellis, S. J. Amos, and A. Kumar, *Illumination Guidelines for Nighttime Highway Work.* Transportation Research Board, 2003.

[5] J. S. Shane, A. Kandil, and C. J. Schexnayder, “NIGHTTIME CONSTRUCTION IMPACTS ON SAFETY, QUALITY, AND PRODUCTIVITY,” 2011.

[6] F. Falchi, P. Cinzano, C. D. Elvidge, D. M. Keith, and A. Haim, "Limiting the impact of light pollution on human health, environment and stellar visibility," *J. Environ. Manage.*, vol. 92, no. 10, pp. 2714–2722, Oct. 2011.
[7] R. S. KNAUER, "Light Suppresses Melatonm Secretion in Humans," *Science*, vol. 210, p. 12, 1980.
[8] I. M. Mclntyre, T. R. Norman, G. D. Burrows, and S. M. Armstrong, "Human Melatonin Suppression by Light is Intensity Dependent," *J. Pineal Res.*, vol. 6, no. 2, pp. 149–156, Apr. 1989.
[9] D. E. Blask, "Melatonin, sleep disturbance and cancer risk," *Sleep Med. Rev.*, vol. 13, no. 4, pp. 257–264, Aug. 2009.
[10] C. Pacchierotti, S. Iapichino, L. Bossini, F. Pieraccini, and P. Castrogiovanni, "Melatonin in Psychiatric Disorders: A Review on the Melatonin Involvement in Psychiatry," *Front. Neuroendocrinol.*, vol. 22, no. 1, pp. 18–32, Jan. 2001.
[11] CIBA Foundation Symposium, *Photoperiodism, Metatonin and the Pineal*. John Wiley & Sons, 2009.
[12] B. De, D. Rt, S. La, K. Ja, and B. Gc, "Light during darkness, melatonin suppression and cancer progression.," *Neuro Endocrinol. Lett.*, vol. 23 Suppl 2, pp. 52–56, Jul. 2002.
[13] K. J. Navara and R. J. Nelson, "The dark side of light at night: physiological, epidemiological, and ecological consequences," *J. Pineal Res.*, vol. 43, no. 3, pp. 215–224, Oct. 2007.
[14] M. Noor-E-Alam, A. Mah, and J. Doucette, "Integer linear programming models for grid-based light post location problem," *Eur. J. Oper. Res.*, vol. 222, no. 1, pp. 17–30, Oct. 2012.
[15] X. Hu and K. Qian, "Optimal design of optical system for LED road lighting with high illuminance and luminance uniformity," *Appl. Opt.*, vol. 52, no. 24, pp. 5888–5893, Aug. 2013.
[16] W. Lai, W. Chen, X. Liu, and X. Lei, "Optimal design of light distribution of LED luminaries for road lighting," 2011, vol. 8123, p. 81231P–81231P-9.
[17] M. Akita, M. Miki, T. Hiroyasu, and M. Yoshimi, "Optimization of the Height of Height-Adjustable Luminaire for Intelligent Lighting System," in *Artifical Intelligence and Soft Computing*, L. Rutkowski, R. Scherer, R. Tadeusiewicz, L. A. Zadeh, and J. M. Zurada, Eds. Springer Berlin Heidelberg, 2010, pp. 355–362.
[18] D. Gómez-Lorente, O. Rabaza, A. Espín Estrella, and A. Peña-García, "A new methodology for calculating roadway lighting design based on a multi-objective evolutionary algorithm," *Expert Syst. Appl.*, vol. 40, no. 6, pp. 2156–2164, May 2013.
[19] W. Lai, X. Liu, W. Chen, X. Lei, and X. Cheng, "Light intensity distribution optimization for tunnel lamps in different zones of a long tunnel," *Opt. Express*, vol. 22, no. 19, p. 22952, Sep. 2014.
[20] K. El-Rayes and K. Hyari, "CONLIGHT: Lighting Design Model for Nighttime Highway Construction," *J. Constr. Eng. Manag.*, vol. 131, no. 4, pp. 467–476, 2005.
[21] K. El-Rayes and K. Hyari, "Optimal Lighting Arrangements for Nighttime Highway Construction Projects," *J. Constr. Eng. Manag.*, vol. 131, no. 12, pp. 1292–1300, 2005.
[22] H. Said and K. El-Rayes, "Optimizing the planning of construction site security for critical infrastructure projects," *Autom. Constr.*, vol. 19, no. 2, pp. 221–234, Mar. 2010.
[23] M. Corcione and L. Fontana, "Optimal design of outdoor lighting systems by genetic algorithms," *Light. Res. Technol.*, vol. 35, no. 3, pp. 261–277, Sep. 2003.
[24] R. H. Simons and A. R. Bean, *Lighting Engineering : Applied Calculations*. Jordan Hill: Taylor and Francis, 2008.
[25] A. W. A. Hammad, D. Rey, and A. Akbarnezhad, "A mixed-integer nonlinear programming model for minimising construction site noise levels through site layout optimisation," presented at the 31st International Symposium on Automation and Robotics in Construction and Mining, ISARC 2014 - Proceedings, 2014, pp. 722–729.
[26] A. Hammad, A. Akbarnezhad, D. Rey, and S. Waller, "A Computational Method for Estimating Travel Frequencies in Site Layout Planning," *J. Constr. Eng. Manag.*, p. 04015102, Dec. 2015.
[27] A. W. A. Hammad, A. Akbarnezhad, and D. Rey, "A multi-objective mixed integer nonlinear programming model for construction site layout planning to minimise noise pollution and transport costs," *Autom. Constr.*, vol. 61, pp. 73–85, Jan. 2016.
[28] A. W. A. Hammad, A. Akbarnezhad, D. Rey "A Comparison of Mixed Integer Programming Models for the Construction Site Layout Problem," *ISARC Proc.*, vol. 2015 Proceedings of the 32st ISARC, Oulu, Finland, pp. 1–9, 2015.
[29] A. Pachamanov and D. Pachamanova, "Optimization of the light distribution of luminaries for tunnel and street lighting," *Eng. Optim.*, vol. 40, no. 1, pp. 47–65, Jan. 2008.
[30] International Commission on Illumination, "Guide on the Limitation of the Effects of Obtrusive Light from Outdoor Lighting Installations, CIE 150," CIE Central Bureau, Vienna, Austria, 2003.

[31] C. Toregas, R. Swain, C. ReVelle, and L. Bergman, "The Location of Emergency Service Facilities," *Oper. Res.*, vol. 19, no. 6, pp. 1363–1373, Oct. 1971.

[32] D. Dilaura, K. Houser, R. Mistrick, and G. Steffy, *The Lighting Handbook: Reference and Application*, 10th edition. New York, NY: Illuminating Engineering, 2011.

[33] International Commission on Illumination, "Maintenance of indoor electric lighting systems," Central Bureau of the CIE, Vienna, Austria, Feb. 1993.

[34] V. Chankong and Y. Y. Haimes, *Multiobjective decision making: theory and methodology*. North Holland, 1983.

[35] "IBM CPLEX Optimizer - United States," 30-Dec-2014. [Online]. Available: http://www-01.ibm.com/software/commerce/optimization/cplex-optimizer/. [Accessed: 05-Feb-2015].

A Realisation of a Construction Scale Robotic System for 3D Printing of Complex Formwork

J.B. Gardiner, S. Janssen, and N. Kirchner

Laing O'Rourke Engineering Excellence Group, Chippendale, Australia
E-mail: JGardiner@laingorourke.com.au, SJanssen@laingorourke.com.au, NKirchner@laingorourke.com.au

Abstract – The cost of producing complex formwork or moulds for precast concrete and GRC (glass reinforced concrete) is perhaps the most significant limitation when constructing architectural intentions from prefabricated concrete elements. Our contribution, FreeFAB™ Wax[1], employs a 6-axis gantry robot with interchangeable end effector tooling to coarsely 3D print and subsequently mill finish complex moulds for concrete at a construction scale. Significantly, FreeFAB moulds can be up to 30m x 4m x 1.5m with significant time and cost savings relative to traditional formwork. Furthermore, FreeFAB's print material approaches 100% reusability, virtually eliminating waste and material costs within the process. This paper presents an overview of our method for construction scale 3D printing and its realisation. Our findings suggest a construction scale 3D printing is business-viable and significantly expands the feasibility of the wide spread use of bespoke moulds.

Keywords –
Robotic 3D Printing for Construction, Precast Concrete, Additive Manufacturing, Prefabrication

1 Introduction

The Crossrail commuter rail project, a new underground rail line for the London Underground network, is currently the largest construction project in Europe [1]. Laing O'Rourke are responsible for the construction of three underground stations for the project, these stations feature three dimensionally curved transitions for many of the tunnel intersections within these stations (example shown in Fig. 1), which results in a requirement for approximately 1400 unique moulds to create the Glass Reinforced Concrete (GRC) cladding. Laing O'Rouke's Operations Director of the Project Delivery Department identified the procurement of these moulds as a significant project cost and risk, due largely to the small number of UK mould suppliers.

Figure 1 - Photo of Crossrail tunnel intersection mockup (courtesy of Laing O'Rourke)

Further to this, there are a number of methods to fabricate moulds for precast concrete and GRC, significant examples include: moulds fabricated from timber & steel, milled from foam and formed with flexible materials such as polyurethane rubber. These methods require significant materials, which are generally not reusable and often not recyclable, and require significant machining times and/or manual labour. This yields relatively expensive moulds that must be used a number of times to ameliorate the costs. These limitations of the current approaches for mould production inherently limits design intentions as the need to ameliorate mould costs through mould reuse drives design standardisation and/or design repetition. FreeFAB was devised and realized in order to relieve these limitations. FreeFAB™ Wax (referred henceforth as FreeFAB) is a construction scale robotic system for 3D printing of complex wax formwork.

3D printing, also known as Additive Manufacturing and Rapid Prototyping, was first developed and commercialized in the 1980's [2]. The process generally works by incrementally adding or solidifying material in layers to build an object. There are three main categories of fabrication of any object; subtractive

[1] FreeFAB was developed by the Engineering Excellence Group Sydney – an innovation lab within Laing O'Rourke. Technologies incubated within this lab, are tailored to the needs of the company and the construction industry in general.

(carving an object from a block), formative (making a mould and casting a settable material) and additive (building an object with blocks) [3]. Construction 3D Printing, also known as Additive Manufacturing for Construction, was first developed in the mid 1990's [4], although more than 20 different technologies have been developed since (most after 2010), none to date have demonstrated a compelling advantage over existing construction techniques. Existing construction 3D printing techniques are either unable to deliver the unique capabilities of 3D printing (fabrication of complex three dimensional forms) or 3D printed materials do not match existing construction materials performance [5]. Only one technology Winsun concrete 3D printing technique has been demonstrated to be production ready to date [6], however this technique is limited to fabricating geometrically simple structures that do not fully take advantage of 3D printing.

The primary manner in which FreeFAB differs from these technologies is its intelligent use of an additive (3D printing) and subsequent subtractive (Milling) manufacturing method which highlights the advantages of each method whilst they inherently negate the limitations of the other. FreeFAB is a novel method for producing moulds suitable for use in standard precast and GRC concrete prefabrication that yields time, waste, and cost savings. FreeFAB aims to disrupt traditional mould production and drive a paradigm shift in precast concrete and GRC mould production.

This paper presents an overview of the realised FreeFAB system; both technical (Section 2.1) and business grounding (Section 2.2). Findings from a numerical and empirical exploration that probe the aforementioned potential time, waste, and cost savings are presented in Section 3. These findings are then leveraged to facilitate a discussion. Following this, conclusions are drawn and the potential of FreeFAB is extrapolated with mention of relevant future work.

2 An Overview of FreeFAB

FreeFAB (Fig. 2) aims to deliver a paradigm shift in the fabrication of moulds for precast concrete and GRC. This shift is enabled by the pairing of two technologies (3D printing & Milling) that each have particular have advantages and that's inherent limitations can be negated by the other's strengths. Specifically, it aims at significantly reducing the cost of producing a single mould for a single panel so that it is no longer prohibitively expensive and only possible for projects with excessive budgets. This would present a unique opportunity within the construction sector, as bespoke and one-off elements can become a common feature on buildings, within interiors, on bridges, noise walls and for structural elements such as columns and beams. The following sections present an overview of FreeFAB; Section 2.1 focuses on a technical overview and Section 2.2 focuses on the accompanying business grounding and context in which FreeFAB is deployed.

Figure 2 - Photo of FreeFAB Wax gantry installed at GRC UK plant, UK (courtesy of Laing O'Rourke)

2.1 The FreeFAB Wax technique

Figure 3 presents a conceptual overview of FreeFAB; FreeFAB first 3D prints a relatively coarse mould from wax (1) which is then milled to yield the required tolerances and surface finish (2). This mould is then used directly in place of a traditional mould (3), and without significant process change. Once set, the mould with panel in-situ is placed in a recycling station that melts the wax off the finished panel and collects it for use again in printing (4). The finished panel emerges from the recycling station ready for use (5).

The primary manner in which FreeFAB differs from the previously mentioned technologies is in its utilisation of a reliably reusable material; our specially devised and designed wax. By employing this material which is highly reusable the cost of materials and waste for each mould becomes negligible, significantly eliminating a major cost of mould manufacture. Furthermore, the wax we devised, designed, and realised for this technique, has excellent 3D printing properties allowing for very large beads of material (30mm x 10mm) to be deposited at high flow rates (up to 400L per hour). By 3D printing at this rate, while fabricating hollow tessellated internal structures – relatively large moulds can be fabricated at a fraction of the time relative to milling from a solid block of material, or traditionally constructed moulds.

However, whilst this high material throughput printing allows for significant amounts of material to be rapidly deposited in this additive process the surface finishes is not of a comparable quality to traditional techniques. A post printing complementary process is employed to counter this. Once the mould is fabricated using 3D printing the surface is then milled (removing approximately 5mm of wax at the surface) to achieve micron accuracy smoothness. By combining these two techniques the strengths of both are extracted and leveraged, while negating the limitations. For instance, 3D printing large moulds to micron accuracy would require very small layer heights, which would result in unpractically lengthy build times, by printing 'fast and dirty' we eliminate this limitation. Conversely, milling from a solid block often requires long roughing times to remove large volumes of excess materials. By removing the need for roughing and using 3D printing to fabricate the net shape, the accuracy of milling can be leveraged without the roughing time and waste costs.

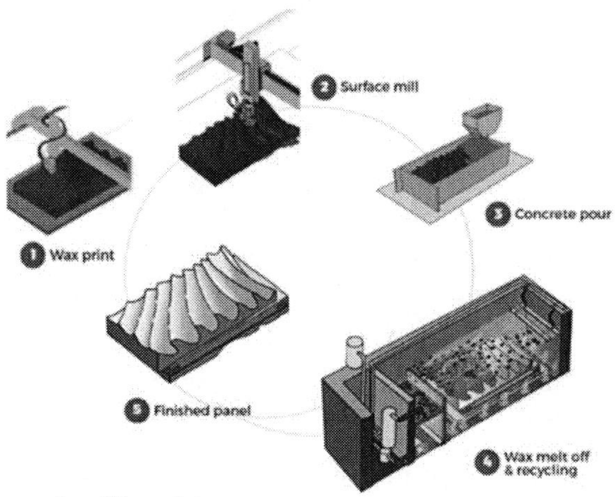

1. Wax print
2. Surface mill
3. Concrete pour
4. Wax melt off & recycling
5. Finished Panel

Figure 3 – A conceptual illustration of FreeFAB process (courtesy of Laing O'Rourke)

Contrasting FreeFAB with typical mould production methods provides further context and grounding. The previously identified, and typical, alternative mould manufacturing techniques predominately require a relatively large component of labour. For metal and steel moulds, sheets must be cut, handled and joined together and then the joints must be filed, sanded or ground down and then polished. With the milling and wire cutting of foam moulds the surface quality achieved in this relatively cheap material is generally not high enough to meet surface finish standards. Hence a further tooling paste must be applied to the foam to achieve a durable and high quality surface. This procedure requires a lot of manual labour for the application of the paste and subsequent polishing and sealing. The fabrication of polyurethane and silicone moulds requires the production of a positive 'buck' to pour the liquid mould material into, this process can be both machine and labour intensive.

2.2 Business Context and Grounding

This section builds the business context and grounding for FreeFAB – a realised effort in robotics and automation in construction. Figure 4 presents a simplified representation of the factors that drive mould cost. The three predominate factors are Process – the steps involved in the operation. Time – the time each process step consumes. These can include process steps and time consumers such as design time, order lead time, delivery time or may be focused on the fabrication time. Process and Time are often mistakenly considered as being entirely overlapping. This is not necessarily the case though, consider the case where a process step is removed and as a result the prior and post process steps proportionally increase in their Time cost. In this case the Process cost may have changed while the Time cost

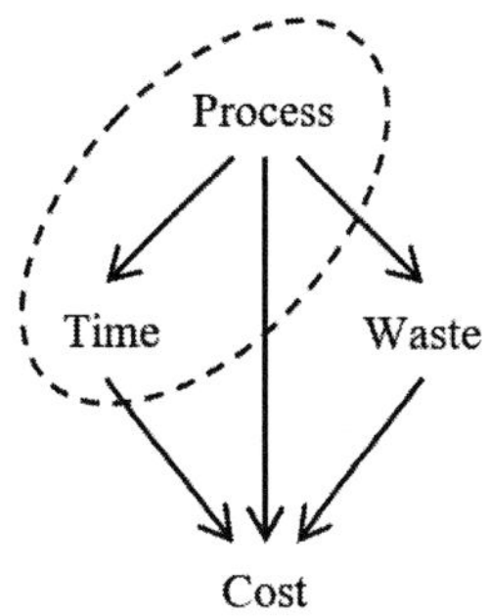

has not. The third cost factor is Waste – again this can be broad and consider a number of factors or focused on the cost of lost material for instance.

Figure 4 Factors affecting mould fabrication cost

Underpinning this model representation is a set of formations for both the traditional approach and the FreeFAB approach. The formations for the traditional approach as specified by the aforementioned UK suppliers are intended to provide benchmarks against which to evaluate FreeFAB. The formations are devised and described as follows:

$$Time_{trad} = E_{rough} + P_{app} + P_{set} + P_{finish} \quad (1)$$

where $Time_{trad}$ is the time taken to fabricate a mould using the traditional method, E_{rough} is the rough-machining time, P_{app} is the time to apply the tooling paste, P_{set} is the time required for the tooling paste to set, and P_{finish} is the finish-machining time.

$$Time_{FreeFAB} = W_{print} + W_{cool} + W_{finish} \quad (2)$$

where $Time_{FreeFAB}$ is the time taken to fabricate a mould using the FreeFAB method, W_{print} is the 3D printing time, W_{cool} is the cooling time of the printed wax, and W_{finish} is the finish-machining time.

$$Waste_{trad} = E + P \tag{3}$$

where $Waste_{trad}$ is the amount of waste produced while fabricating a mould using the traditional method, E is the amount of expanded polystyrene used in the mould, and P is the amount of tooling paste used in the mould.

$$Waste_{FreeFAB} = W \times \frac{R}{100} \tag{4}$$

where $Waste_{FreeFAB}$ is the amount of waste produced while fabricating a mould using the FreeFAB method, W is the total amount of wax used, and R is the percentage of reusable wax.

$$Cost_{trad} = \frac{Q}{A} \times A_{avg} \tag{5}$$

where $Cost_{trad}$ is the cost of a mould fabricated using the traditional method, Q is the quoted price for a mould by an external supplier, A is the surface area of the quoted mould, and A_{avg} is the average surface area of a range of typical moulds.

$$Cost_{FreeFAB} = L + M + I + O \tag{6}$$

where $Cost_{FreeFAB}$ is the cost of a mould fabricated using the FreeFAB method, L is the cost of labour, M is the machine maintenance cost, I is the electricity cost, and O is the overhead costs. Initial capital expenditure for the FreeFAB installation is not included.

3 Findings

This contextually specified model employs fiduciary insights from business, suppliers and logistics channels along with insights and numerical findings from piecemeal empirical evaluations and simulations to build a consolidated mechanism for comparison. The following sections exploit this model to build quantitive data to facilitate the exploration of the aforementioned primary comparisons of fabrication time, waste and cost.

3.1 Fabrication Process & Time

The model was first employed in order to investigate the relative mould fabrication times of traditional and FreeFAB methods. Specifically, each process was broken down into a sub-processes and computer simulations used to calculate the duration of each sub-process. The process does not include lead time, which would be dictated by the workload of the factory.

The typical method for fabricating a 3D curved mould involves machining a base of expanded polystyrene to a rough finish, applying a 40mm layer of tooling paste, allowing this paste to set and finally machining the top surface to the desired form. This process is costly, time consuming and wasteful. The FreeFAB process differs by 3D printing the rough form of the mould in wax, allowing this print to cool and then machining the top to create a high quality mould. Table 1 presents this breakdown of times for both traditional and FreeFAB mould fabrication methods.

Table 1 Fabrication Time Comparison

Traditional	Time (mins)
Rough-machining	58
Applying Paste	15
Setting Paste	720
Finish-machining	140
Total	**933 (15.6hrs)**

FreeFAB	Time (mins)
3D Printing Wax	24
Cooling Wax	96
Finish-machining	27
Total	**147 (2.5hrs)**

The data presented in Table 1 is derived from (1) for the traditional method, and (2) for FreeFAB. For clarity and to facilitate a depth of engagement, the time data is deconstructed into the key process steps. As can be seen from the table, the traditional mould fabrication method consists of four sub-processes totaling 15.6 hours. The first sub-process is rough-machining of the expanded polystyrene substrate to remove large quantities of material and reach the desired form. The second is the application of the viscous tooling paste. The third is the hardening of this tooling paste, and the final sub-process is the finish-machining of the tooling paste to obtain the casting surface of the mould.

Conversely, the FreeFAB mould fabrication method consists of three sub-processes totaling 2.5 hours. The first sub-process is the 3D printing of the wax, where a bead of semi-molten wax is extruded from a nozzle to create the rough form of the mould. The second is the cooling of this printed wax. The final sub-process is the finish-machining of the top surface of the wax mould to leave a high-quality finish for casting.

As can be seen, the FreeFAB process requires fewer sub-processes and yields a significant reduction in the overall mould fabrication time – a 6.5x reduction. The most significant time saving is achieved by using wax instead of tooling paste, thus reducing the setting time from 12 hours to only 1.5 hours.

Another factor affecting the mould fabrication time is the rough-machining time. The traditional method starts with a block of expanded polystyrene and must remove large quantities of material to reach the desired form. The FreeFAB method 3D prints material only where required, thus reducing the necessary initial rough machining time. The rough-machining is eliminated and replaced by the faster 3D printing process. As 3D printing only places material where it is actually needed, this cuts 49 minutes from the process. The finish-machining is also expedited by utilizing a 5-axis milling head, thus allowing it to more closely approximate the final surface with fewer passes compared to the 3-axis milling heads typically used with the traditional method.

3.2 Fabrication Waste

The model was subsequently used to investigate the amount of waste generated by traditional (on average over a range of comparable designs on file) and FreeFAB mould fabrication methods. Each was evaluated based on the weight and volume of material waste. Table 2 shows the amount of waste generated by both traditional and FreeFAB mould fabrication methods. The data presented is derived from (3) for the traditional method, and (4) for FreeFAB.

Table 2 Waste Comparison

Traditional	**Waste**
Expanded Polystyrene	5.8kg (291L)
Tooling Paste	12.8kg (20L)
Total	**18.6kg (312L)**

FreeFAB	**Waste**
Printed Wax	97.5kg (99.5L)
Loss (1%)	1.0kg (1.0L)
Total	**1.0kg (1.0L)**

As can be seen from the table, the traditional mould fabrication method uses expanded polystyrene for the substrate and tooling paste for the top surface of the mould, with a total of 18.6kg of material used. Once the mould has been used, it is discarded, thus all of this material is considered waste. Conversely, the FreeFAB mould fabrication method uses 97.5kg of wax, but is able to reuse ~99% of this material, a waste of ~1.0kg.

It can be seen that the FreeFAB method has 18 times less material waste than the traditional method by weight. If comparing waste volume the savings are even more significant. The FreeFAB mould uses three times less volume as it is able to create the mould with hollow internal support geometry. By reusing most of this material, the process reduces wasted volume to only ~1.0L; ~300 times less than the traditional method.

3.3 Fabrication Cost

The model was finally employed to investigate the relative cost between traditional and FreeFAB mould fabrication. The cost of both a single mould and five repeats of the same mould were compared. Figure 5 presents an illustrative comparison of traditional and FreeFAB mould fabrication methods for one-off and five-off conditions. The data presented is derived from (5) for the traditional method, and (6) for FreeFAB.

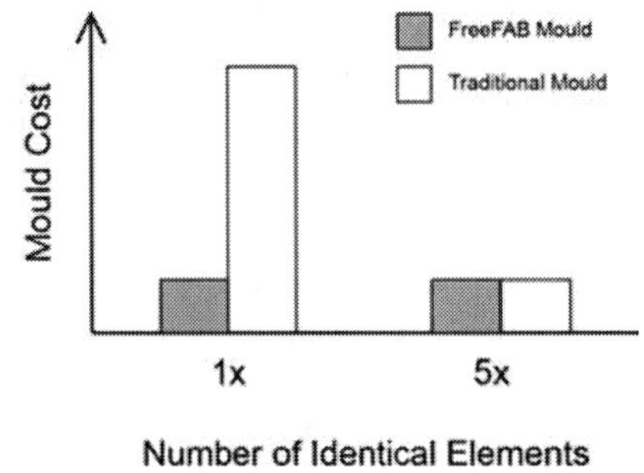

Figure 5 Illustrative comparison of mould fabrication costs using traditional and FreeFAB technologies for one repeat and five repeats.

As can be seen, the left most pair of bars compare the cost of a single-use mould using the FreeFAB and traditional fabrication methods. In this case, the FreeFAB method is approximately five times cheaper than the traditional method. The second two bars compare the cost of a mould that is reused five times. In this case the cost is comparable between the two methods as the FreeFAB mould must be fabricated and recycled for each casting. The significance of this is that there are no additional cost associated with one-off moulds when using the FreeFAB. This enables the designer with increased freedom as it relieves the requirement to standardise or repeat elements of the design to secure the cost benefit of ameliorating the mould cost over a number of castings.

Note: A significant portion of the high cost of traditional moulds is the cost of the raw materials. The tooling paste represents approximately 25% of the total cost of the mould using traditional methods. By comparison, the wax material used in the FreeFAB process is 2-3 times cheaper and can be reused hundreds of times with only 1% loss of material per cycle.

4 Discussion

In Section 2 we have devised and constructed a business-context placed and grounded model to provide a mechanism to benchmark FreeFAB with traditional approaches. This model was then utilised in Section 3 to produce a set of quantitive number to facilitate comparison. Our findings, presented in Section 3, support our claims that FreeFAB results in Time, Cost, and Waste savings relative to traditional mould approaches. Specifically, in Section 3.1 we showed that the FreeFAB fabrication method is significantly faster than traditional methods – 15.6hrs compared to 2.5hrs. With the FreeFAB system, a new mould can be created in hours, compared to the traditional method taking days (8-12hrs per work day). This means the production throughput of the FreeFAB system is much higher, increasing profit for the fabricator. Furthermore, this unlocks designers who are now able to implement a design that requires a relatively larger number of unique moulds. In Section 3.2 we showed that traditional mould fabrication consumes relatively large quantities of material to FreeFAB – 312L compared to 1L.

However, perhaps the most compelling finding was presented in Section 3.3 where we demonstrated the pressure that the expense of traditional mould fabrication places on designers – who are motivated to reuse moulds multiple times to reduce per casting cost – and how FreeFAB reduces the cost of moulds significantly, making it cost effective for designers to use moulds only once and significantly expanding the scope for feasible design. For instance, consider the enabling impact of the 5x lower cost FreeFAB on the design intentions described in the Crossrail project (Figure 1) where a large number of unique moulds are required to achieve the overall design intentions.

5 Conclusions

In this paper we have demonstrated that FreeFAB enables previously cost-prohibited design intentions to be achieved through its Time, Waste, and Cost savings compared to traditional approaches.

Further significant benefits lie in the use of the technology for projects that take advantage of the strengths of the technique in the production of highly bespoke buildings and infrastructure projects – two illustrative examples are given in Figure 6 and 7. This presents a unique opportunity within the construction sector, as bespoke and one-off elements can become a common feature on buildings, within interiors, on bridges, noise walls and airports. The cost of producing a single mould for a single panel is no longer prohibitively expensive, which frees the designers and engineers to add value (functional or aesthetic) where this was formerly out of reach.

Figure 6 - Rendering of non-repetitive design Freeway Noise Walls (courtesy of Laing O'Rourke)

In order to further explore the potential of FreeFAB and to deliver Crossrail, Laing O'Rouke has commissioned the GRC/FreeFAB factory in the UK's East Midlands (Figure 2); to be on line June 2016.

Figure 7 - Rendering of non-repetitive bespoke high rise building cladding (courtesy of Laing O'Rourke)

References

[1] Crossrail. Crossrail in Numbers. Online: http://www.crossrail.co.uk/news/crossrail-in-numbers, Accessed 12/03/16

[2] Chua, C K, and Kah F L. Rapid prototyping: principles and applications. Vol. 1. World Scientific, pages 7-11 2003.

[3] Onuh, Spencer O., and Yahaya Y. Yusuf. "Rapid prototyping technology: applications and benefits for rapid product development." Journal of intelligent manufacturing 10.3-4, pages 301-311 (1999):.

[4] Gardiner JB (2011) Exploring the Emerging Design Territory of Construction 3D Printing. PhD thesis, RMIT University, Pages 80-94. Melbourne, Australia

[5] Gardiner JB (2011) Exploring the Emerging Design Territory of Construction 3D Printing. PhD thesis, RMIT University, Pages 80-94, 362-364. Melbourne, Australia

[6] Kira (2015), Exclusive: WinSun China builds world's first 3D printed villa and tallest 3D printed apartment building. Online: http://www.3ders.org/articles/20150118-winsun-builds-world-first-3d-printed-villa-and-tallest-3d-printed-building-in-china.html, Accessed: 12/02/2016

Joint Segmentation and Recognition of Worker Actions using Semi-Markov Models

Jun Yang[a], Zhongke Shi[b] and Ziyan Wu[a]

[a]School of Mechanics, Civil Engineering and Architecture, Northwestern Polytechnical University, China
[b]School of Automation, Northwestern Polytechnical University, China
E-mail: junyang@nwpu.edu.cn, zkeshi@nwpu.edu.cn, zywu@nwpu.edu.cn

Abstract –

Vision-based automated recognition of worker actions has gain lots of interest during the past few years. However, existing research all requires pre-segmented video clips, which is not applicable in the real situation. Furthermore, pre-segmented videos abandon the temporal information of action transition. A joint action segmentation and recognition method, which can segment continuous video stream while recognizing the action type for each segment, is an urgent need. In this paper, we model the worker actions with a discriminative semi-Markov model. In the model, a set of features is defined to capture both the local and global characteristics of each action cycle. Then the semi-Markov model is formulated as an optimization problem and solved by the cutting plane method for simultaneous action segmentation and recognition. Scale-Invariant Feature Transform (SIFT) is applied to detect feature points in the region of interest in every frame. Two descriptors (Histograms of Oriented Gradients – HOG, Histograms of Optical Flow – HOF), are computed in the feature points to encode the scenario and motion flow simultaneously. Finally, the Bag-of-Feature strategy is adopted for feature representation. Experimental results from real world construction videos show that the proposed method is able to segment and recognize continuous worker actions correctly, resulting in a prospecting application in automated productivity analysis.

Keywords –

Action Segmentation; Action Recognition; Worker; Semi-Markov Models

1 Introduction

Worker activities have strong impact on productivity, progress and quality of construction projects. Traditional activity analysis usually involves foremen collecting data through onsite observations, survey or interview, and analyzing data offline [1, 2]. This procedure is labor intensive, cost sensitive and can be prone to error. Lacking of real-time information is another major concern.

With the prevalence of onsite cameras and the emergence of advanced computer vision technologies, video based construction operation analysis has become a new trend in recent years [3, 4]. Either workers or equipment are detected and tracked through image sequences for further activity analysis [5-9]. Or spatial-temporal features are extracted directly from videos for action recognition [10-12]. The latter, for being robust to changing view angle and moving cameras, has attracted lots of attentions.

Gong et al. [10] adopted the Bag-of-Feaures pipeline to recognize worker and equipment activities. Spatio-temporal features were extracted, clustered and quantized for motion pattern learning and prediction. Golparvar-Fard et al. [11] also adopted similar strategy for action recognition of earthmoving equipment. Yang et al [12] established a bigger worker action dataset and achieved a state-of-art performance on the new dataset by using dense trajectories for feature description.

Though the above mentioned research has made remarkable achievement, their limitation is obvious. They are all based on pre-segmented video clips by assuming that each clip only contains a single action cycle. In real world application, analyzing a continuous video stream with repetitive actions or sequential actions is often expected. What is more, pre-segmented video clips abandon the temporal information of action transition, which is valuable for operation flow analysis.

Hence, an automated action segmentation and recognition method is needed for continuous operation analysis.

In computer vision field, traditional solution is to treat action segmentation and recognition separately. Temporal segmentation is applied to partition videos into coherent constituent parts and recognition could then be simply carried out as categorization of the action classes corresponding to the segments. Typically three types of methods can be applied for segmentation: boundary detection, sliding windows and compositional approaches [13].

However, it is often difficult to segment a video into actions purely based on low-level cues. What is more, separating segmentation from classification may result in important loss of information related to the actions. A recent trend is to solve segmentation and classification jointly [14, 15].

Hoai et al [14] proposed a discriminative temporal extension of the spatial bag-of-words model for joint segmentation and recognition of human actions. A multi-class SVM was applied for classification and segmentation inference was done with dynamic programming. Shi et al [15] presented a discriminative semi-Markov model to define features over boundary frames, segments and also neighboring segments. The inference problem of simultaneous segmentation and recognition was solved efficiently using optimization algorithms. Their method exhibited good performance on repetitive action segmentation and recognition.

Inspired by Shi et al [15], in this paper, we adopt the discriminative semi-Markov model to represent continuous worker actions. Scale-Invariant Feature Transform (SIFT) is applied for feature point detection. Two descriptors (Histograms of Oriented Gradients – HOG, Histograms of Optical Flow – HOF), are computed in the feature points to encode the scenario and motion flow simultaneously. Then the semi-Markov model is solved by the cutting plane method for simultaneous action segmentation and recognition.

The rest of the paper is organized as follows. Section 2 illustrates the algorithm in detail. Section 3 gives out the experimental results. Section 4 concludes the paper.

2 Methodology

The paper aims at designing a joint action segmentation and recognition system for worker activity analysis. Some basic assumptions are as follows.

First, the system input is continuous video stream captured from construction site either by mounted surveillance cameras or hand-held cameras. Hence no static background is required. What is more, the algorithm is expected to handle various points of view. Second, cyclic worker actions, such as bricking or nailing, are the major concern. That is to say, the action should have clear starting point and ending point.

The overall workflow of the proposed system is shown in Figure 1. As can been seen, the system flow is divided into two pipelines: training and testing. First, feature points in labeled video clips are extracted by SIFT detector. Then, HoG and HoF descriptors are computed on the feature points. Feature vectors are quantized using the Bag-of-Features strategy. After that, feature functions are formed according to their definitions. Semi-Markov Model is learned through optimization. During testing, feature functions are formed using the codebook generated in the training stage. Then the trained SMM model is applied to infer the continuous action boundaries and types. More details are illustrated as follows.

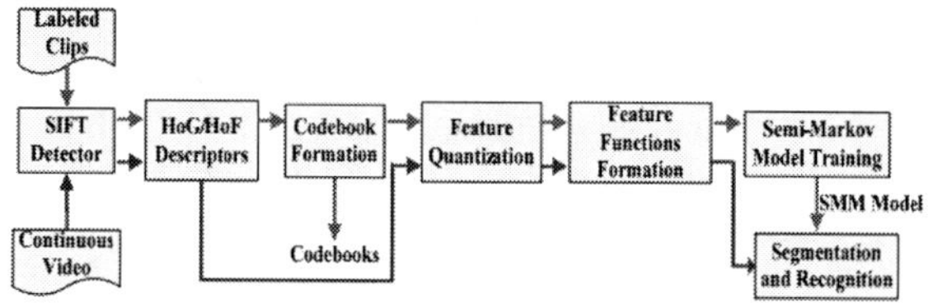

Fig 1. System Overview

2.1 Semi-Markov Model

The joint action segmentation and recognition problem is formulated as a convex optimization over a probabilistic semi-Markov model [15]. To make the paper self contained, a brief description of the model is provided as follows.

Considering a graph model, each node corresponds to a segment of video frames having the same action label, and each edge depicts the statistical dependency between adjacent segments. Let $C = \{1,\dots c\}$ represent the action labels. Given a video sequence X of length m, assume there are l segments with boundaries $\{n_k\}_{k=0}^{m-1}$. Then action sequence label Y can be represented as $Y = \{(n_k, c_k)\}_{k=0}^{l-1}$, where each pair (n_k, c_k) denotes the starting position and the corresponding action label for the kth segment.

Model parameter is denoted as W. And $\Phi(X,Y)$ represents a feature map over the joint input-output space. Assume that the conditional probability distribution over action sequence label Y given current observation sequence is a log-linear model,

$$\log p(Y \mid X, W) = \langle W, \Phi(X,Y) \rangle - A_W(X)$$

Where $A_W(X)$ is a normalization constant to ensure

$p(Y \mid X,W)$ is a valid probability distribution. And $\Phi(X,Y)$ is defined as,

$$\Phi(X,Y) = (\sum_{i=0}^{l-1}\phi_1(X,n_i,c_i), \sum_{i=0}^{l-1}\phi_2(X,n_i,n_{i+1,}c_i), \sum_{i=0}^{l-1}\phi_3(X,n_i,n_{i+1},c_i,c_{i+1}))$$

In which, ϕ_1 and ϕ_2 depicts the observation-label dependencies within the current action segment with ϕ_1 focusing on the segment's boundary frame and ϕ_2 describing the global characteristics of the current segment. While ϕ_3 encodes the interaction between two neighboring segments.

Given an unseen video sequence X, its action sequence can be labeled optimally by solving the following maximum likelihood decoding problem

$$Y^* = \arg\max_Y \log p(Y \mid X,W).$$

Learning is accomplished by solving a regularized optimization problem with respect to the parameter W. The goal is to make W be bounded to avoid over-fitting, while maximizing the minimum log ratio of the conditional probabilities,

$$\min_{W,\xi} \frac{\|W\|^2}{2} + \frac{\eta}{T}\sum_t \xi_t$$

$$s.t.\ \langle W, \Delta\Phi(X_t,Y)\rangle \geq \Delta(Y_t,Y) - \xi_t \quad \forall t,Y$$

for the set of video sequences $\{t : t \in 1,\ldots,T\}$. In the equation, $\Delta\Phi(X_t,Y) := \Phi(X_t,Y_t) - \Phi(X_t - Y)$ and ξ_t is a non-negative slack variable to account for the non-separable case.

For the sake of completeness, the dual program is given as

$$\max_\alpha \sum_{t,Y}\alpha_{t,Y}\Delta(Y_t,Y) - \frac{\eta}{2}\left\|\sum_{t,Y}\alpha_{t,Y}\Delta\Phi(X_t,Y)\right\|^2$$

$$\text{s.t.} \quad \alpha_{t,Y} \in M \quad \forall t$$

Where M represents the probability simplex constraints.

The above problem can be solved approximately using optimization techniques such as cutting plane [16] or the bundle method [17].

2.2 Feature extraction and description

A frame based feature representation is expected to encode the feature map $\Phi(X,Y)$. Considering the need for robustness to illumination, scale and view angle change, a local feature detector SIFT [18] is adopted. After feature points being detected, technically the SIFT descriptor can be applied for feature description. However, considering that action segmentation and recognition is not a task based on a single frame but usually requiring dynamic information from neighboring frames, we adopt HoG (Histograms of Oriented Gradients) [19] and HoF (Histogram of Optical Flow) [20] descriptors instead. HoG is used to encode static appearance information, while HoF is to capture the local motion information.

Descriptor are computed on a 3D video patch in the neighborhood of each detected SIFT point. The patch is partition into a grid with 3x3x2 spatio-temporal blocks. 4 bins HoG descriptors and 5-bin HoF descriptors are then computed for all blocks and concatenated into 72-element and 96-element descriptors respectively, finally forming a 168-dimensinal feature vector. For dimension reduction, the Bag-of-Features strategy is applied. A 50-dimensinal codebook is formed by clustering on all features. Then each feature vector is mapped to codebook centers, resulting in a quantized histogram.

Finally, feature functions ϕ_1, ϕ_2, ϕ_3 are computed. ϕ_1 describes the boundary frame features.

$$\phi_1(X,n_i,c_i) = \psi_1(X,n_i) \otimes c_i$$

Where $\otimes$ denotes the tensor products. ψ_1 is the concatenation of a constant 1 and the histogram vector on the boundary frame.

ϕ_2 captures features on the current segment.

$$\phi_2(X,n_i,n_{i+1},c_i) = \psi_2(X,n_i,n_{i+1}) \otimes c_i$$

Where $\psi_2(X,n_i,n_{i+1})$ contains three components: the length of the segment, the mean and the variance of the histogram vector of the segment (from frame n_i to $n_{i+1}-1$).

ϕ_3 is used to depict features on neighboring segments.

$$\phi_3(X,n_i,n_{i+1},c_i,c_{i+1}) = \psi_3(X,n_i,n_{i+1}) \otimes c_i \otimes c_{i+1}$$

It is the concatenation of four components: the mean of the histogram vector from frames n_i to $n_{i+1}-1$, and from frames n_{i+1} to $n_{i+1}+d$; the variance of the histogram vector from frames n_i to $n_{i+1}-1$, and from frames n_{i+1} to $n_{i+1}+d$. d is the minimum duration of a segment defined by the user.

3 Experimental results

This section describes experimental results on two groups of data: the repetitive actions and the sequential actions.

The original data is selected from the worker action dataset proposed by the authors previously [12]. Four types of worker actions are selected: 'Nailing', 'TieRebar', 'Shoveling' and 'LayBrick'. These actions are all cyclic with starting and finishing points for each cycle, offering a clear sign for segmentation. For each action type, we pick 3 workers. Each one performs the action 6 times. So there are totally 72 video clips (in resolution 320*240, frame rate 25 fps).

Figure 2 displays sample frames extracted from the dataset. As can be seen, various points of view are covered in the dataset. Furthermore, different workers may execute the same action differently. These all propose difficulties for action segmentation and recognition.

For algorithm evaluation, the segmentation performance and the recognition performance are measured separately. The algorithm will assign each frame a segment label and an action label. By comparing the segmentation and recognition results to the ground truth, the overall frame-level accuracy can be calculated as the ratio between the number of agreements over the total number of frames. Confusion matrix is computed to further evaluate the recognition performance.

Fig 2. Sample frames of the dataset (From top row to bottom: nailing, kntRod, shoveling and bricking, respectively)

3.1 Repetitive actions

Video clips from the same action type and the same worker are concatenated into longer video sequences. For each action type, two workers are used for training. One worker is left out for testing. Eventually, 8 video sequences are for training, 4 for testing. Each sequence contains 6 action clips.

The overall frame-level accuracy for action segmentation is 88.3%. The results for each sequence are shown in Fig. 3, in which different color represents different action type and black boundary of each bar gives the segmentation result. In each subfigure, 'Truth' represents the ground truth and 'Ours' represents the algorithm output. It can be seen that the number of segments is correct for all sequences. The segments boundaries are not very satisfying.

The recognition results are all correct for 'nailing' and 'bricking'. One segment in 'knitRod' is misclassified as 'nailing'. And all segments in 'shoveling' is misjudges as 'bricking'. The confusion matrix is shown in Fig. 4. The average recognition accuracy is 72.0%.

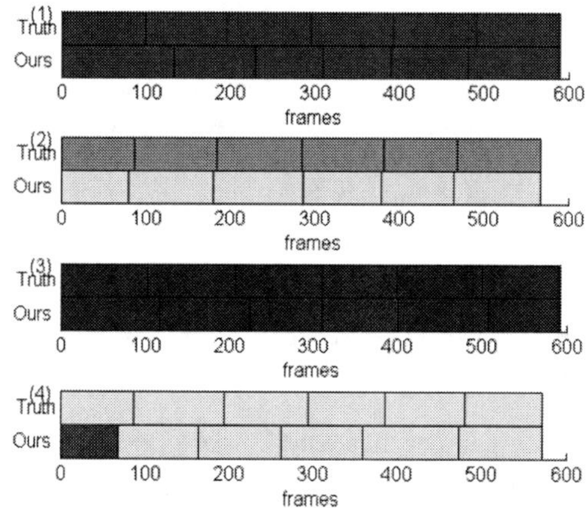

Fig.3 Automatic segmentation-recognition versus human labeled ground truth for repetitive actions. The segments are color coded; red, green, blue and yellow correspond to nailing, shoveling, bricking and knitRod classes, respectively. This figure is best seen in color.

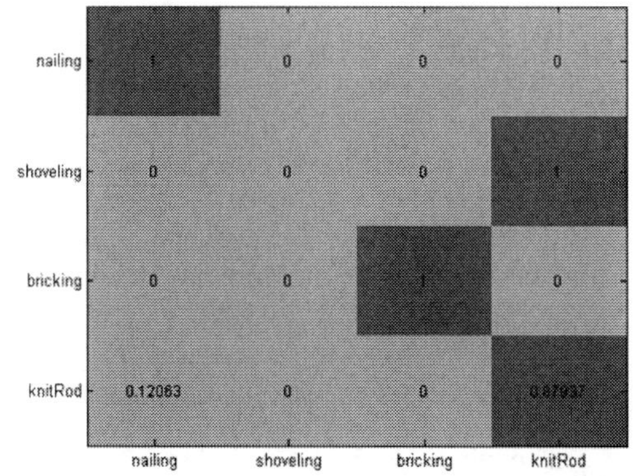

Fig. 4 Repetitive action - confusion matrix for action recognition in frame-level.

3.2 Sequential actions (Synthetic Dataset)

To simulate workers doing different task sequentially, video clips from different action types are concatenated into longer video sequences in the order of 'Nailing', 'Shoveling', 'LayBrick' and 'TieRebar', ending up 18 video sequences. 12 of them are training and the rest 6 are for testing.

The overall frame-level accuracy for action segmentation is 93.7%. The results for each sequence are shown in Fig. 5. It can be seen that the number of segments is correct for all sequences. The segments boundaries are close but not very accurate. Compared to the repetitive action result, the segmentation is slightly better, which is foreseeable because boundaries between two different actions are apparently easier to distinguish.

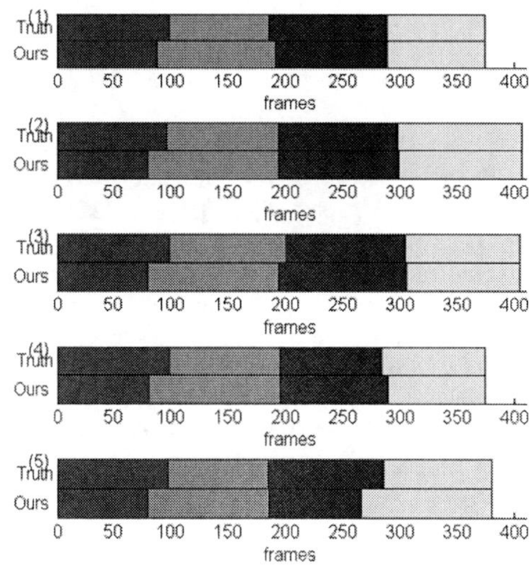

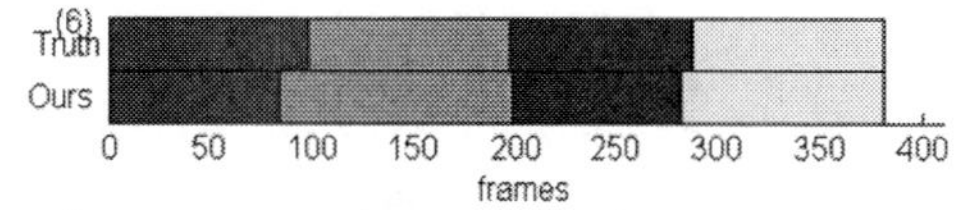

Fig. 5 Automatic segmentation-recognition versus human labeled ground truth for sequential actions. The segments are color coded; red, green, blue and yellow correspond to nailing, shoveling, bricking and knitRod classes, respectively. This figure is best seen in color.

All action type is recognized correctly per segments. The frame level confusion matrix for recognition is depicted in the confusion matrix in Fig. 6. The average accuracy is 93.7%.

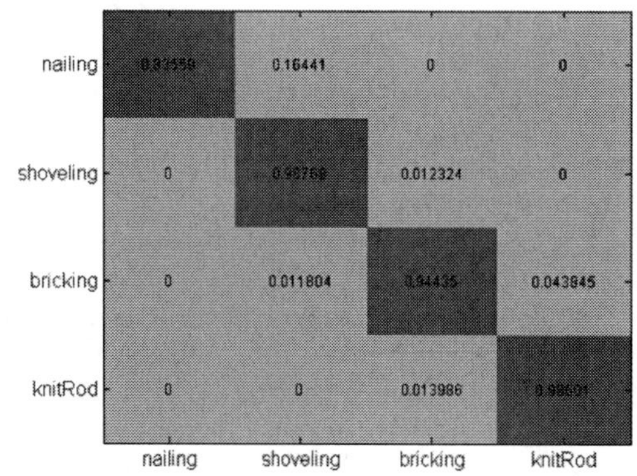

Fig. 6 Sequential action - confusion matrix for action recognition in frame-level.

3.3 Comments

From the two groups of experiments, we can see that the Semi-Markov Models method can segment continuous long construction videos and recognize worker action type for each segment. The action recognition rate is comparable to the state-of-art on pre-segmented videos [12]. The number of segmented are all correct in all testing sequences, which is important when applying the method to productivity analysis. However, the segmentation boundaries are not very accurate. One possible explanation is: worker actions are recorded from multiple points of view and different worker usually have different poses and various action length. These all introduce difficulties in segmentation.

Comparing the two groups of experiments, it can be concluded that the Semi-Markov Model can model the sequential action better than repetitive actions.

4 Conclusions

In this paper, we applied a Semi-Markov Model

method for joint action segmentation and recognition in construction videos. Experimental results showed that the method can recognize worker action type at a state-of-art accuracy while segmenting long videos into individual segments correctly in numbers. Joint action segmentation and recognition can be used for further productivity analysis and may supply useful information for workflow analysis. Future study may seek to test the methods on bigger datasets, improve the feature description method for better segmentation result.

References

[1] C. I. I. (CII) (Ed.), IR252.2a – Guide to Activity Analysis, Construction Industry Institute, Austin, TX, USA, 2010.

[2] M. C. Gouett, C. T. Haas, P. M. Goodrum, C. H. Caldas, Activity analysis for direct-work rate improvement in construction, Journal of Construction Engineering and Management 137 (2011) 1117–1124.

[3] J. Yang, M.-W. Park, P. A. Vela, M. Golparvar-Fard, Construction performance monitoring via still images, time-lapse photos, and video streams: Now, tomorrow, and the future, Advanced Engineering Informatics 29 (2015) 211–224.

[4] J. Teizer, Status quo and open challenges in vision-based sensing and tracking of temporary resources on infrastructure construction sites, Advanced Engineering Informatics 29 (2015) 225–238.

[5] M.-W. Park, C. Koch, I. Brilakis, Three-dimensional tracking of construction resources using an on-site camera system, Journal of Computing in Civil Engineering 26 (2012) 541–549.

[6] E. Rezazadeh Azar, S. Dickinson, B. McCabe, Server-customer interaction tracker: computer vision–based system to estimate dirt-loading cycles, Journal of Construction Engineering and Management 139 (2012) 785–794.

[7] J. Yang, P. Vela, J. Teizer, Z. Shi, Vision-based tower crane tracking for understanding construction activity, Journal of Computing in Civil Engineering 28 (2012) 103–112.

[8] J. Gong, C. H. Caldas, An object recognition, tracking, and contextual reasoning-based video interpretation method for rapid productivity analysis of construction operations, Automation in Construction 20 (2011) 1211–1226.

[9] M. Bugler, G. Ogunmakin, J. Teizer, P. A. Vela, A. Borrmann, A comprehensive methodology for vision-based progress and activity estimation of excavation processes for productivity assessment, in: Proceedings of the 21st International Workshop: Intelligent Computing in Engineering (EG-ICE), Cardiff, Wales.

[10] J. Gong, C. H. Caldas, C. Gordon, Learning and classifying actions of construction workers and equipment using bag-ofvideo-feature-words and bayesian network models, Advanced Engineering Informatics 25 (2011) 771–782.

[11] M. Golparvar-Fard, A. Heydarian, J. C. Niebles, Vision-based action recognition of earthmoving equipment using spatiotemporal features and support vector machine classifiers, Advanced Engineering Informatics 27 (2013) 652–663.

[12] J. Yang, Z. Shi, Z. Wu, Automatic recognition of construction worker activities using dense trajectories, in: Automation and Robotics in Construction and Mining, 32nd International Symposium on, pp. 75–81.

[13] Weinland D., Ronfard R. and Boyer E. A Survey of Vision-Based Methods for Action Representation, Segmentation and Recognition. Computer Vision and Image Understanding, 2011, 115(2): 224-241.

[14] Hoai M., Lan Z. and De la Torre F. Joint Segmentation and Classficiation of Human Actions in Video. IEEE Conference on Computer Vision and Pattern Recognition (CVPR), 2011, pp 3265-3272.

[15] Shi Q., Cheng L., Wang L. And Smola A. Human Action Segmentation and Recognition using Discriminative Semi-Markov Model. International Journal of Computer Vision, 2011, 93: 22-32.

[16] Tsochantaridis I., Joachims T., Hofmann T., and Altun Y. Large margin methods for structured and interdependent output variables. Journal of Machine Learning Research, 6, 1453–1484.

[17] Smola A., Vishwanathan S., and Le Q. Bundle methods for machine learning. Advances in neural information processing systems, 2007, pp. 1377-1384.

[18] Lowe, D. Distinctive image features from scale-invariant keypoints. International Journal of Computer Vision, 2004, 60(2), 91–110.

[19] Dalal N. and Triggs B. Histograms of oriented gradients for human detection. In Proceedings of IEEE Conference on Computer Vision and Pattern Recognition, Vol 1, pages 886-893, San Diego, USA, 2005.

[20] Laptev I., Marszelek M., Schmid C. and Rozenfeld B. Learning realistic human actions from movies. In Proceedings of IEEE Conference on Computer Vision and Pattern Recognition, pages 1-8, Anchorage, USA, 2008.

A Suggestion for Improvement of IFC Data in Energy Performance Index (EPI) based on Open BIM

Jungsik Choi[a], Inhan Kim[b], Minchan Kim[c] and Jeongwon Ryu[d]

[a]College of Engineering, Kyung Hee University, Republic of Korea
[b]Department of Architecture, Kyung Hee University, Republic of Korea
[c]BIM Business Div., U·MID SYSTEM, Republic of Korea
[d]Department of Architecture, Sunmoon University, Republic of Korea
E-mail: jungsikchoi@gmail.com, ihkim@khu.ac.kr, mchan_07k@naver.com, jwryu@sunmoon.ac.kr

Abstract

As many countries are becoming more aware of environmental issues, both the governmental and private sectors are making more of an effort to achieve eco-friendly construction. However, the assessment of circumstances for building energy performance and green building standards of legislation are still based on two dimensions and manual work. This takes time, costs, and labor and makes assessment inaccurate and inefficient. Therefore, this study suggests methods to improve the eco-friendly building permission process and to automatize the open BIM-based assessment of the EPI among several eco-friendly building standards.

Keywords –

Building Information Modeling (BIM); Building Permission; Energy Performance Index (EPI); Industry Foundation Classes (IFC); Open BIM

1 Introduction

Global warming and environmental problems are some of the most critical issues in modern life, which raises the importance of energy conservation and greenhouse gas emissions reduction, and most countries are endeavoring to achieve "low-carbon, green growth" [1]. As construction is a major cause of greenhouse gas emissions, which lead to global warming, the field gives considerable attention and effort to developing energy-saving, eco-friendly, and sustainable construction technology [2]. Many countries evaluate the sustainability of buildings and implement systematic and institutional practices to reduce the negative influence of construction on the environment as they develop eco-friendly building certification systems, such as GBCC, LEED, CASBEE, BREEAM, and EarthCheck. The Korean government has established the Green Building Development Support Act [3], which promotes eco-friendly construction and encourages energy efficiency and performance improvement. This act obligates buildings over a certain size to follow the Energy-Saving Design Standard and submit an Energy-Saving Plan and Energy-Saving Plan Design Review. In spite of these efforts on both governmental and private levels, the current assessment circumstances for performance, laws, and standards of eco-friendly building still depend on two dimensions and manual work, which is time consuming and inefficient because of the repetitive work involved and impeded feedback. The demands to solve this problem and improve the efficiency of eco-friendly building permission examination emphasize the importance of BIM technology. BIM is a technology that creates digital information on a building. In addition, the information needed for the life-cycle can be obtained and managed by using digital information.

Therefore, the BIM application on eco-friendly building permission examination analyzes necessary details effectively and accurately with the BIM-based integrated information model. It also establishes integrated and objective assessment circumstances for eco-friendly construction in various and complicated relations between buildings and the environment. Thus, the objectives of this study are as follows:

- To improve the eco-friendly building permission process based on the Energy-Saving Design Standard
- To provide a base study for the development of the Automatic Assessment System for the EPI

This study proposes an assessment method by connecting assessment criteria and IFC data and suggests a solution and a process of assessment.

To automatize the assessment of the EPI,

- It is necessary to check if a building satisfies assessment criteria through IFC data exported from the BIM authoring tool

- Required information and IFC data should be connected
- Assessment criteria should be analyzed in advance

2 Cases Studies

2.1 Analysis of Related Studies

Roh et al. used GBT, a BIM-based template, to suggest a use for the BIM application in eco-friendly building certification. GBT organizes necessary information and a series of procedures for eco-friendly building certification examination, and this study provides background work for BIM-based certification examination using GBT [5].

Yi and Park proposed a method to automatize open BIM-based Examination for requirements in the Energy-Saving Design Standard and some parts of the EPI. The method extracts necessary information from IFC data about the "thermal transmittance of exterior wall" in the requirements of the architecture part and "shading device" and "boiler efficiency according to type" items in the EPI [6].

Lee et al. saved the BIM model in the gbXML format and sent it to Ecotect Analysis, an energy performance analysis program. The study analyzed the annual energy acquisition and consumption according to geological information, direction, and elevation plan and pointed out problems of the EPI. The study applied BIM for energy performance assessment and presented possibilities for quantitative assessment and a clear certification standard [7].

Farzad and Ahmad provided an integrated BIM-based methodology for eco-friendly building certification, energy performance assessment, and the automation of estimation. They built a database by analyzing related information and established a BIM Library and predefined codes, which enable related information to be entered into BIM data. They also suggested an effective use of BIM data with the application, which connects BIM authoring tools and energy performance analysis tools [8].

2.2 Energy-Saving Design Standard of Building

The Green Building Development Support Act consists of Green Building Certification, the Energy Efficiency Rating System, and the Energy-Saving Design Standard of Building and defines obligations to submit an Energy-Saving Plan and Energy-Saving Plan Design Review when a building owner requests a building permit or a usage change.

The Energy-Saving Plan Design Review presents a checklist that confirms requirements and their credits. The Energy-Saving Plan Design Review is divided into Mandatory, EPI, and building energy requirement assessment, while the Energy-Saving Plan is divided into Architecture, Machine Equipment, Electrical Adiabatic, and New Renewable. Since the Mandatory in the Energy-Saving Design Standard of Buildings contains the obligation to follow the Chapter 2 Energy-Saving Design Standard of Buildings and EPI, these should be examined before deciding on the Mandatory.

3 Analyzing Energy Performance Index and Extracting Necessary Information

Analyzing assessment criteria is essential for the Automatic Assessment System for the EPI. It extracts the required information for the assessment criteria, and this information defines the necessary information for the BIM data.

The Parts of Architecture in the EPI consists of 14 categories, and the required information and necessary information from the BIM data are extracted to satisfy the requirements from each item. Necessary information is the information that satisfies the relevant required information and should be considered in the BIM model view. Additionally, related standards and necessary calculation methods regulated by the law should be examined. For example, the "Mean Thermal Transmittance of Exterior Wall," which is the first category in the EPI, needs not only the "Confirming Wall," "Relation between Wall and Exterior," and the "Thermal Transmittance of Wall," but also the "Azimuth and Area of Wall," 'Confirming Window," "Relation between Window and Exterior," "Azimuth and Area of Window," and "Thermal Transmittance of Window Information" because the Energy-Saving Design Standard of Buildings defines certain calculation methods, as shown in Table 1.

Table 1. Calculation Method of Mean Thermal Transmittance of Exterior Wall

Classification	Calculation Method
Mean Thermal Transmittance of Exterior Wall	Ue = [Σ(Thermal Transmittance of External Wall each Cardinal Point × Area of External Wall each Cardinal Point) + Σ(Thermal Transmittance of Window and Door each Cardinal Point × Area of Window and Door each Cardinal Point)] / (Σ Area of External Wall each Cardinal Point + ΣArea of Window and Door each Cardinal Point)

Table 2. Portion of Extracted Necessary Information

Evaluation Items	Requirements Information	Necessary Information	Necessary Object	Necessary Attribute	Necessary Relation
Mean Heat Transmission Coefficient of External Wall Ue (W/㎡.K) (Including Window and Door)	Mean Heat Transmission Coefficient of External Wall	Confirming Wall	Wall		Space
		Relation between Wall and Exterior	Wall	Exterior	
		Azimuth of Wall	Wall	Azimuth	
		Area of Wall	Wall	Area	
		Heat Transmission Coefficient of Wall	Wall	Heat Transmission Coefficient	
		Confirming Window	Window		Wall
		Relation between Window and Exterior	Window	Exterior	
		Azimuth of Window	Window	Azimuth	
		Area of Window	Window	Area	

Necessary information for each piece of required information can be extracted and categorized for BIM-based automatic assessment. The necessary information for the assessment of the Parts of Architecture in the EPI is shown in Table 2.

4 Plan for Open BIM-based Assessment Criteria Examination

4.1 Analyzing Relations between Necessary Information and IFC Data Structure

To improve the interoperability between BIM software in various parts of the construction industry, buildingSMART International has developed the IFC format as an international standard, and IFC-based BIM is the internationally used for standard open BIM [9]. IFC is an EXPRESS language-based data structure that expresses architectural information as Entity, Attribute, and Relation. This structure represents an object of a building, its attribute, and relations [1, 10].

Analyzing relations with the IFC structure determines the possibility of extracting necessary information from related IFC-format BIM data and provides an understanding of the relations [11].

This chapter introduces a method to analyze relations between necessary information and the IFC structure to extract the necessary information mentioned in the previous chapter.

4.1.1 General Decision Rules

To extract necessary information from IFC data, it is necessary to analyze the object, its attribute, and its relations and to judge the possibility, which leads to constructing decision rules. Based on the necessary information and categorized information defined in the previous chapter, the Entity, Attribute, and Relation of related IFC structure are mapped along with the process. For example, the "Thermal Transmittance of Wall," as shown in Figure 1, should be taken to confirm "IfcWallStandardCase," which means the wall in the IFC structure, and the thermal transmittance of the wall is extracted from the BIM data by finding the value of "Pset_WallCommon-ThermalTransmittance" through "IfcPropertySingleValue" in "PropertySet," which is defined for an additional property in the IFC structure.

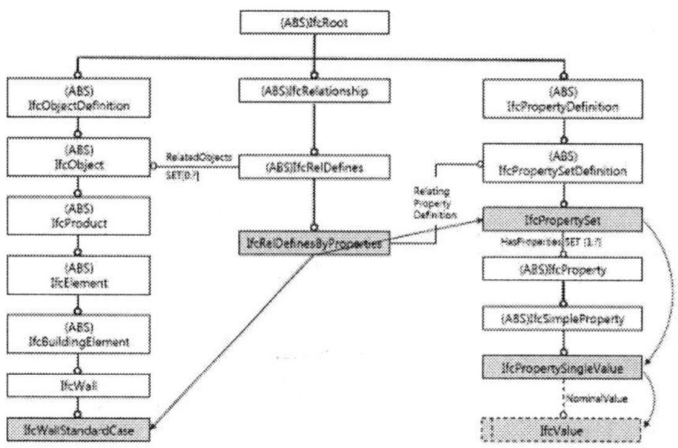

Figure 1. EXPRESS-G Diagram for Thermal Transmittance of Wall

4.1.2 Case for Extra Decision Rules

In spite of the presence of the IFC data structure related to necessary information, extra decision rules are required for precise information extraction. For example, the "Area of Wall and Slab" information can be extracted from the IFC structure because "IfcElementQuantity" defines the area of the object and "IfcRelDefinesByProperties" defines the relation between the object and the attribute in the IFC structure. However, this method is inappropriate for the calculation of spatial units because this information means the total area of an object. Likewise, slab is the same case. Part of the area related to a specific space is needed rather than the total area of the wall, as shown in Figure 2.

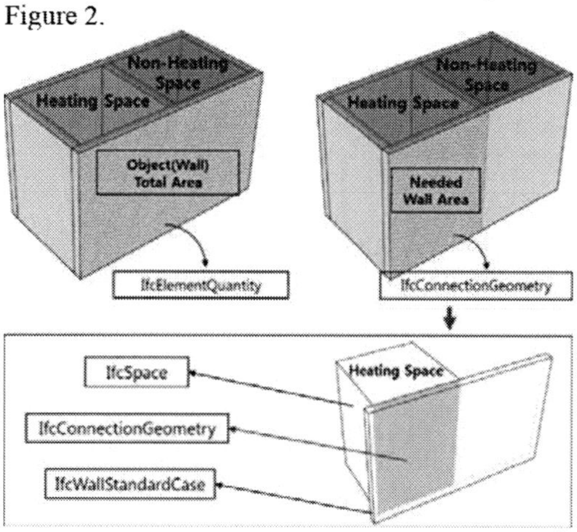

Figure 2. Extra Decision Rule for Area of Wall

Therefore, an extra decision rule is defined to extract that information from the IFC data. To extract the area of the wall in the IFC data, as shown in Figure 3, the wall related to a specific space can be found through "IfcRelSpaceBoundary," which means objects related to space in the IFC structure, by finding "IfcSpace," which means space in the IFC structure. After that, whether the wall is exterior is determined by confirming the value of "Pset_WallCommon-IsExternal" through "IfcPropertySingleValue" in "PropertySet." Finally, the relevant area can be extracted through "IfcConnectionGeometry," which means the connection surface between the space and wall and below the structure. This extra decision rule should be arranged separately to apply the development of the examination module.

4.2 Limitation of IFC Structure and Solutions

IFC is an international standard format used in various areas of the construction industry and is a neutral file format for the exchange of BIM data between different software and for interoperability. Thus, for effective information exchange, all the information in the lifecycle in various construction fields should be able to be stored in the IFC format. However, because the information related to the construction process is massive and complicated, there is a limitation of defining the data system, which can store all the information from each category. In addition, even though the IFC is constantly upgraded, it does not enable the data structure to store all the information. This causes data loss when data from different parts are exchanged. It also impedes data interoperability and interrupts effective uses of BIM data. Therefore, this chapter identifies necessary information that the IFC structure does not support and suggests a way to extract and store this information [12].

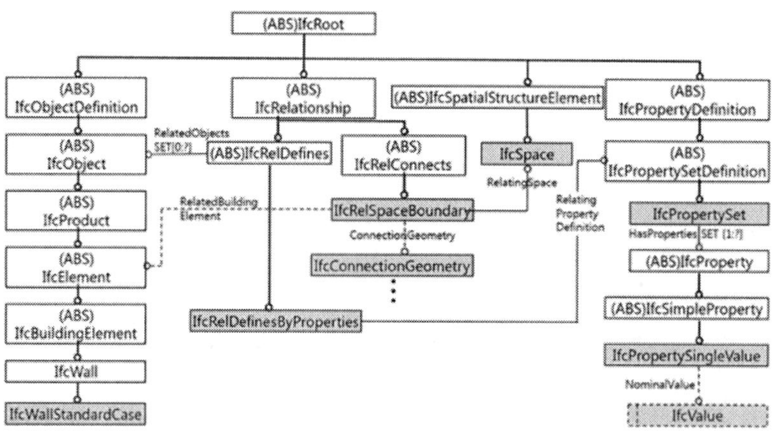

Figure 3. EXPRESS-G Diagram for Area of Exterior Wall

In the previous chapter, it was pointed out that some necessary information cannot be defined by the IFC structure after analyzing and mapping relations between necessary information and the related IFC data structure. This information can be omitted in the exchange of BIM data and cannot be evaluated because the examination system cannot identify it. Therefore, it is necessary to find an alternative to save this information as IFC-format BIM data.

4.2.1 Entering Additional Attribute

The additional attribute should be defined and entered in the object because the value of the thermal transmittance of the object is not included in the IFC structure. Although this attribute can be defined as "Thermal Transmittance" in "Pset_Common," it only regards the curtain wall, window, door, slab, and wall, but not the roof, which needs to be defined and entered additionally. When the model is made in the BIM authoring tool, the attribute of the thermal transmittance of the roof can be entered additionally by "IfcPropertySingleValue" in "PropertySet" and can be checked in "IfcPropertySingleValue" to extract necessary information.

4.2.2 Predefining Decision Rules

It does not need to be defined additionally, but should be predefined and can be inferred and judged according to the attributes saved in the IFC file. The living room can be judged by the name of the space, which is defined in the IFC structure. This space name should be categorized and organized previously, and the correct name for the space object should be entered in the BIM modeling process.

4.2.3 Using Proxy Object as Alternatives

Since the adiabatic shutter, adiabatic outer door, and shading are not defined in IFC, the related necessary information can be saved or extracted by "IfcBuildingElementProxy" from the IFC Entity. "IfcBuildingElementProxy" is an entity for saving the objects that are not defined in IFC. The adiabatic shutter object should be saved to "IfcBuildingElementProxy" after inputting the necessary information and name by using a general object when creating the adiabatic shutter in the BIM authoring tool. Afterward, information on the adiabatic shutter can be extracted from "IfcBuildingElementProxy" in the IFC data.

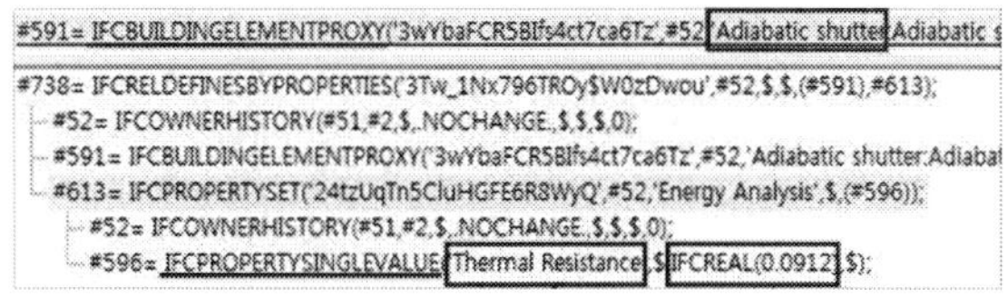

Figure 4. IFC Data for Adiabatic Shutter

Using these three solutions enables the conversion of the necessary information, which is not available in the IFC structure, into IFC data. It prevents data loss in the exchange of data and ensures accurate evaluation. These methods and standards should be managed in separate documents.

4.3 Problems and Solutions of IFC Conversion Function in BIM Authoring Tool

The BIM authoring tool helps a designer to build an architecture model and input architectural information with a 3D parametric object. In other words, it is a tool to connect the designer and the architecture model built in digital space. The model made by the BIM authoring tool is the digital information with the designer's plan, and the information of the architecture model should be converted to IFC without any loss. When the information is suitably converted to the IFC structure, the IFC data is reliable as an architectural information model and interoperability for a variety of software is ensured. In particular, the reliability of the IFC data is important because IFC data contains information for evaluations, and the reliability of IFC data affects the reliability of the evaluation. However, the IFC conversion function supported by the BIM authoring tool causes data loss and an information gap, which hinders the reliability of IFC data, effective use of BIM, and accuracy of evaluations. Therefore, this chapter analyzes the problems that occur in the conversion of necessary information to IFC data in the BIM authoring tool (Revit) and suggests solutions.

In the case of an operable window, as shown in Figure 5, although it can be defined by "IfcWindowPanelOperationalEnum" of "IfcWindowPanelProperties" in the IFC data structure, the information of the operable window in the BIM authoring tool is not exported as a correct IFC structure, and it is omitted. Thus, an alternative is needed to convert and save the information of the operable window to IFC data.

By entering the information of the operational window into Type Comments, one of the attributes for the window object, the operation of the window can be represented in IFC data. After exporting to IFC, the operation of the window can be identified by confirming the information previously entered in "IfcPropertySingleValue," named Type Comments, which is a sub-structure of "IfcPropertySet," which is related to "IfcWindowStyle" (Figure 5).

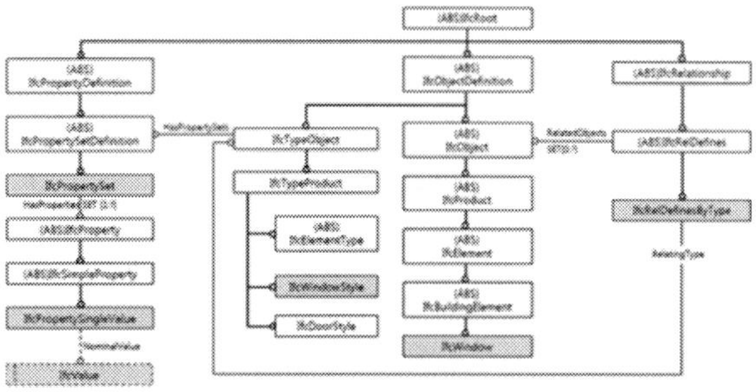

Figure 5. EXPRESS-G Diagram for Alternative

5 Suggestion and Application Scenario of Open BIM-based Automation Process for EPI

This chapter suggests a process to improve the permission process by using an open BIM-based automatic assessment system for the EPI and explains the procedure. In accordance with this method, the two-dimensional, manual work-based permission process can be improved for the BIM-based automatic assessment process, which extracts accurate and efficient results.

5.1 Creation of BIM Model and IFC Conversion

The designer builds a BIM model by considering related standards and requirements in the Energy-Saving Design Standard of Buildings with the BIM authoring tool. The designer should follow the BIM guidelines, which have standards and requirements, and these guidelines should contain the solutions mentioned in the previous chapter. The guidelines facilitate an accurate and smooth assessment process without any loss of required information in the conversion process.

Among the requirements for open BIM-based automatic assessment for EPI, this chapter focuses on the solutions mentioned in the previous chapter.

According to the solution, an additional attribute is entered for the thermal transmittance of the roof and entrance door, predefined attribute information for the living room, windbreak room, parking lot. For the adiabatic shutter, shading and proxy objects are used, and the attribute information is added. For the revolving door and operation of the window, the alternatives to compensate for the limitations of the BIM authoring tool are applied (Figure 6). Afterward, the IFC file is exported from the BIM authoring tool.

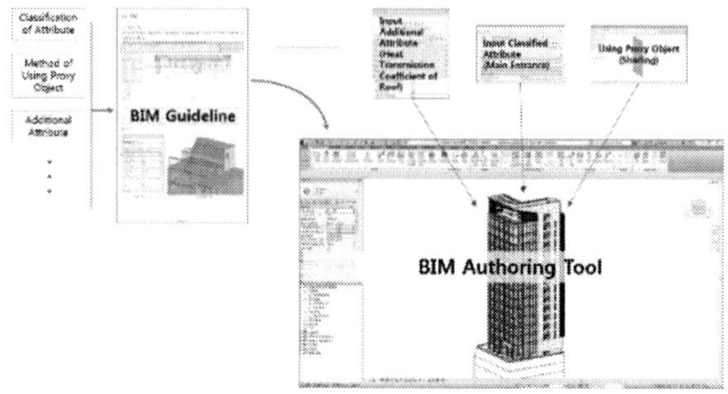

Figure 6. Creating BIM Model through Proposed Alternative

5.2 Assessment Module for IFC data

The assessment program examines if a building satisfies the EPI with IFC data converted by the BIM authoring tool. We developed an assessment module from the program function, which checks assessment criteria based on the rules applied when necessary information is extracted from IFC data (Figure 7). Moreover, the building assessment interface makes a user and an examiner to proceed with the EPI assessment with entering IFC data. Because some necessary information is not available in the IFC structure, the judgment rule should be predefined. External information can be used by entering information into the interface.

The visualization module is developed to make a user and an examiner to check if the building satisfies the assessment criteria and to give feedback for revising the design (Figure 7). The examination results and the total score are automatically input and calculated according to the results format in the Energy-Saving Design Standard, and this automatized process supports the permission process without changing the existing permission system. Because the results are automatically input into the document, the reliability and accuracy of the document and the evaluation can be ensured.

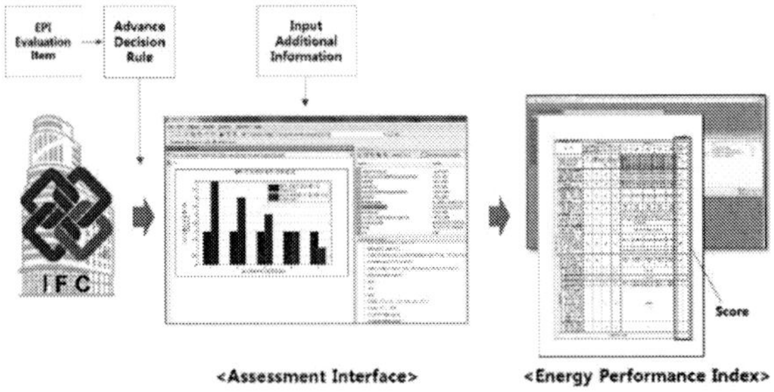

Figure 7. Development of Assessment Interface and Reporting Module

6 Conclusion

Changing two-dimensional manual work into a BIM-based three-dimensional information system in the green building permission process improves productivity, efficiency, and accuracy. In addition, the concept of open BIM in the BIM process ensures interoperability among a variety of software and the effective use of BIM data. Therefore, the objective of this study is to improve the eco-friendly building permission process based on open BIM. It is a base study for the development of the Automatic Assessment System for the EPI, proposes an assessment plan by connecting assessment criteria and IFC data, and suggests a solution and a process of assessment. This study

- extracts necessary information and required information for assessment criteria
- categorizes them into BIM information units by analyzing necessary information and connecting it with IFC data
- analyzes IFC structure and its relation to necessary data
- suggests solutions to the conversion problem and the limitations of the BIM authoring tool
- develops an automatic assessment system for the EPI that proceeds automatically from building the BIM model to visualizing the results by applying the solutions and methods suggested in this study

It is expected that time, costs, and labor will be reduced by the simplification and automation of the process, and not only the productivity and efficiency, but also the accuracy and reliability of the permission and design revision process can be improved.

This study focuses on the Parts of Architecture in EPI in the Energy-Saving Design Standard. A future study is expected to widen its range to the Energy-Saving Plan and Energy-Saving Plan Design Review and to find a method to automatize the open BIM-based evaluation in them.

Acknowledgements

This research was supported by a grant (15AUDP-C067809-03) from Architecture & Urban Development Research Program funded by Ministry of Land, Infrastructure and Transport of Korean government.

This research was supported by Basic Science Research Program through the National Research Foundation of Korea (NRF) funded by the Ministry of Education, Science and Technology (2014-01370001).

References

[1] Kim M., Choi J., and Kim I. Analysis of problems regarding BIM model interoperability and suggestion of improvement direction for openBIM based energy performance analysis. In *Proceedings of the Society of CAD/CAM Conference*, pages 313-315, Korea, 2015.

[2] Kim I., Choi J., and Kim H. Proposition and Application of Mapping System for Input Data between Open BIM Data and Building Energy Simulation Software-Focused on Building Material Library and Energy Plus-. *Journal of the architectural institute of Korea: Planning & design*, 29(4):3-12, 2013.

[3] Ministry of Land, Infrastructure and Transport. Green Building Development Support Act. On-line: http://www.law.go.kr, Accessed: 15/04/2015.

[4] Kim M., Kim I., and Choi J. A Study on the Composition of BIM Model Guideline for Architectural Design Quality Assessment based on OpenBIM. In *Proceeding of architectural institute of Korea*, pages 57-58, Korea, 2015.

[5] Roh M., Kim I., Kim M., and Jun H. A Study on Developing Evaluation Environment of Green Building Certification Criteria Using GBT (Green BIM Template). *Journal of the architectural institute of Korea: Planning & design*, 29(6): 117-126, 2013.

[6] Yi D. and Park C. Code Check for EPI Using Open-BIM. In *Proceeding of the Korea Institute of Ecological Architecture and Environment*, Pages 76-77, Korea, 2014.

[7] Lee K., Kim I., and Choo S. A Study on Improvement of Energy Performance Index in Green Building Certification System using BIM. *Journal of the architectural institute of Korea: Planning & design*, 27(9):13-21, 2011.

[8] Farzad J. and Ahmad J. Integrating BIM with Green Building Certification System, Energy Analysis and Cost Estimating Tools to

Conceptually Design Sustainable Buildings. In *Construction Research Congress 2014*, pages 140-149, 2014.
[9] Kim I., Yoo H., and Choi J. A Study on the Interoperability Improvement IFC Property Information for Energy Performance Assessment in the Early Design Phase. *Transactions of the Society of CAD/CAM Engineers*, 17(6):456-465, 2012.
[10] buildingSMART international. *IFC2XEdition 3 Technical Corrigendum1*. 2006.
[11] Kim I., Kim Y., and Choi J. Building Code Typology and Application for Open BIM based Code Checking. *Transactions of the Society of CAD/CAM Engineers*, 19(3):224-235, 2014.
[12] Lee S., Park C., and Kim I. Issues and Prospects of BIM-to-BEM (Building Energy Model). *Journal of the architectural institute of Korea: Planning & design*, 30(4):199-206, 2014.

Emissions Modelling of Earthmoving Equipment

K. Barati [a], **X. Shen** [a]

[a] School of Civil and Environmental Engineering, The University of New South Wales, Australia

E-mail: Khalegh.barati@unsw.edu.au, x.shen@unsw.edu.au

Abstract –

Earthmoving operations produce high amount of emissions and are one of the main sources of air pollutants in construction and mining industries. Modelling and quantifying the emissions produced by earthmoving equipment is the first step for developing emissions reduction schemes. Currently, emissions of construction and mining equipment are mainly estimated through simulation or laboratory tests which may not represent the real-world situations. This paper presents a comprehensive methodology to predict emission rates of carbon dioxide (CO_2), carbon monoxide (CO), hydrocarbon (HC) and nitrogen oxides (NO_x) of earthmoving equipment by considering operation modes and engine attributes. The developed framework includes three main processes of instrumentation, data collection and data analysis. Two instruments of portable emission measurement system (PEMS) and GPS aided inertial navigation system (GPS-INS) are proposed for conducting field experiments and collecting emission rates and operational parameters. Further, site observation is conducted to estimate the cycle time and time ratio of operation modes. An exploratory analysis method is then developed to process the gathered data and model emissions at operation and equipment level. The applicability of the developed methodology is finally verified through experimenting and modelling emissions of one loader and one excavator.

Keywords –

Emissions; Earthmoving Equipment; Field Experiments; Site Observation; Data Analysis

1. Introduction

In the previous two decades, climate change resulting from greenhouse gas (GHG) emissions has become a global concern [1]. The main compositions of GHGs are CO_2, CO, HC, NOx and general particulate matters (PM) pollutants. Construction and mining industries are two of the main sectors including more than 20% of all non-road equipment [2]. The machinery is mainly involved in large-scale earthmoving operations in which the emitted pollution is by far more than other vehicles. For example, the pollution of a 130 kW-power loader is nearly 500 times more than that of a private car [3]. Construction industry by itself is ranked the third highest emitted pollution industry behind oil and gas, and chemical manufacturing sectors, and also accounts for the third CO_2 emitter per unit of energy used just after cement and steel production industries [4]. Total GHG emission from construction and mining operations is estimated around 6.8% of all industrial emissions. According to the US Clean Air Act Advisory Committee (CAAAC), 32% of NO_x and 37% of PM emitted by all non-road engines result from construction and mining operations [5].

In spite of this significance, emissions of equipment involved in construction and mining sectors have not been precisely modelled yet. The main challenge in this field is that different researchers have been mainly focused on a certain aspect of issue. The lack of correlation and consistency amongst the conducted studies is the main barrier for modelling emission of equipment. Furthermore, the processes of instrumentation and data collection are challenging and error-prone to measure the exact amount of emissions produced by each type of equipment or operation mode. In addition, there are many parameters that their effects on emission have not been fully investigated yet.

This paper aims to develop a comprehensive methodology for emissions modelling of earthmoving equipment in construction and mining at operation and equipment levels through field data processing. This framework consists of three main steps of instrumentation, data collection and result analysis. After developing required instruments, field data need to be collected from in-use earthmoving equipment through experimentation. The methods of analysis are then presented for processing the collected data and developing emissions models at operation and equipment level. The case studies for emission modelling of one loader and one excavator are finally presented to verify the framework developed in the paper.

2. Literature Review

Many studies have been conducted by government agencies and scholars to model and estimate GHG emissions resulting from construction and mining industries at equipment, project and national levels. The NONROAD model developed by the United States Environmental Protection Agency (USEPA) can estimate emitted pollutions of equipment involved in construction industry at national level [6]. This model roughly estimates emission rates of six main pollutants of CO_2, CO, NO_x, PM and Sulphur oxides (SO_2) by considering some major affecting parameters, including engine size, equipment category, emission factors and activity hours [7]. Some of the inputs of this model such as emission factors have been determined based on laboratory tests which cannot provide the episodic nature of real-world operations [8]. Based on NONROAD model, the contributions to CO, NO_x and PM emissions by different types of equipment are compared in Table 1 [6]. The earthmoving machineries are found to be the main contributors to emissions in construction and mining industries. California Air Resources Board (CARB) developed OFFROAD model to estimate the emissions of non-road construction equipment in the California State. This model predicts five main GHG pollutants emitted from 94 machineries (17 categories) annually by considering engine power, annual activity and fuel-emission parameters [9]. URBEMIS is a project-level emission model developed by Sacramento Metropolitan Air Quality Management District (SMAQMD). This model takes into account of the project size, equipment specifications and fuel-emission factors developed by OFFROAD model to estimate the total air pollutants resulted from seven common construction projects [7]. Similar to URBEBIS, Pavement Life-cycle Assessment Tool for Environmental and Economic Effects (PaLATE) model was developed by the University of California-Berkeley for estimating the emissions of road projects at equipment and material level [10].

Numerous studies were conducted to collect and analyze real-world emission data of construction equipment. Through doing field experiments on construction equipment, the manifold absolute pressure (MAP) was identified as the main parameter affecting emissions [7]. Ahn and Lee developed operation efficiency parameter to consider idling and non-idling emission coefficients. The parameter was then defined as the criteria to determine the optimum fleet size for producing least amount of emissions [5]. Abolhasani and Frey conducted field experiments to determine the effects of different fuels on four pollutants emission rates of NO_x, CO, HC and PM [11].

3. Methodology

The framework for emissions modelling of earthmoving equipment is developed in the research, as shown in Figure 1. The parameters affecting emissions are first identified and categorized. Then, by considering the latest off-the-shelf technologies, an integrated instrumentation system is developed to collect real-world data. The data on emissions, operation modes and equipment cycle time were collected through conducting field experiments and site observations. Analysis methods are then developed for emission analyzing of earthmoving equipment. The equipment's cycle time needs to be estimated through analyzing the data collected by employed instruments, site observations and using manufacturer's performance handbooks. The equipment-level emission model is finally developed by considering the emission rates in each operation mode and estimated cycle times.

Table 1. Construction equipment contribution in NO_x, CO, PM emissions

Equipment	NOx		CO		PM	
	Contribution	Ranking	Contribution	Ranking	Contribution	Ranking
Front-end loaders	14.5%	1	11.5%	3	11.2%	3
Backhoes	9.2%	5	16%	1	15.1%	1
Bulldozers	12.5%	2	9.3%	4	9.1%	4
Skid-steer loaders	6.2%	6	14.5%	2	13.6%	2
Excavators	11.4%	3	7.4%	5	8.6%	5
Off-highway trucks	11.0%	4	7.3%	6	6.6%	6
Generators	4.7%	7	5.1%	7	6.0%	7
Forklifts	3.9%	8	4.9%	8	4.6%	8
Scrapers	3.4%	9	2.7%	11	2.3%	12
Cranes	3.2%	10	1.5%	15	1.9%	14

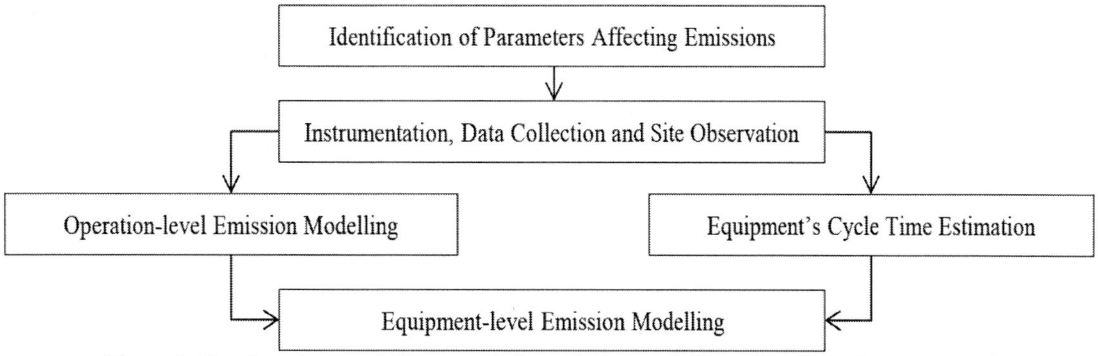

Figure 1. Developed research framework for emission modelling of earthmoving equipment

3.1. Affecting Parameters on Emissions

This section aims to identify and classify parameters affecting emissions of earthmoving equipment. Lewis investigated the effect of engine attributes including engine mode and engine size on emissions [7]. In NONROAD and OFFROAD emission models, three factors of engine size, average load factor and fuel-emission ratio were introduced as main affecting factors on emissions [6, 9]. Barati and Shen investigated the effect of acceleration, payload, speed and road slope as operational parameters for emission modelling of on-road construction equipment [12].

In this study, affecting parameters on emissions of earthmoving equipment are classified into two categories of engine attributes and operational parameters. As shown in Figure 2, engine size, engine load and fuel type are three main engine attributes affecting emissions. There is a direct relationship between engine size and emission rates. The engine load is defined as the amount of used power of engine over the maximum theoretical power as a percentage. The engine load of most construction equipment ranges approximately from 20% to 80% of the maximum engine power [13]. Barati et al. [14] showed a highly-correlated direct linear relationship between engine load and emission rates. Normally, construction equipment uses one certain type of fuel in their lifetime which has negligible changes in ingredients. So, engine fuel can be ignored as a parameter affecting emission rates for a certain piece of equipment. Since most of construction equipment consumes diesel, this research investigates the effect of diesel fuel on emissions modelling only.

Operation efficiency, cycle time and operator skill are the three main operational parameters affecting emissions of earthmoving equipment. Operation efficiency is defined by efficient operating time versus non-efficient idling time. It is obvious that as equipment operates more efficiently, the emission rates at equipment level will be higher. Total cycle time and the ratio of time in each operation mode are the other operational parameters affecting equipment level emission modelling of earthmoving operations. The effect of minor factors on emission rates, such as material type, swing angle and fleet position is considered in the parameter of cycle time. Operator skill is a main parameter affecting the time ratio and cycle time of equipment. Also, this parameter has significant effect on emission rates in operation modes.

3.2. Instrumentation and Data Collection

As shown in Figure 3, three instruments are required for emission modelling of earthmoving operations. The main instrument is a PEMS which measures real-world emissions of the equipment's exhaust. The instrument can measure CO_2, CO, HC and NO_x emission rates in each second using a sample probe inserted in the tailpipe. The model of PEMS is MEXA 584L automotive emission analyzer manufactured by the HORIBA Ltd. GPS-INS is the other multipurpose instrument used in the research. GPS-INS system combines accelerometers, gyroscopes and magnetometers with a commercial grade GPS receiver. GPS-INS is embedded to the boom or stick of earthmoving equipment and provides three-dimensional position and movement in each second. The GPS-INS system proposed for this study is SPATIAL-EK manufactured by Advanced Navigation Pty Ltd. The measured data by PEMS and GPS-INS instruments need to be transmitted through RS-232 serial data communication port to an industrial Tough Pad. Panasonic FZ-G1 is the industrial Tough Pad utilized to analyse the collected field data. Analysis of data collected by GPS-INS system determines the cycle time and operation efficiency of earthmoving equipment.

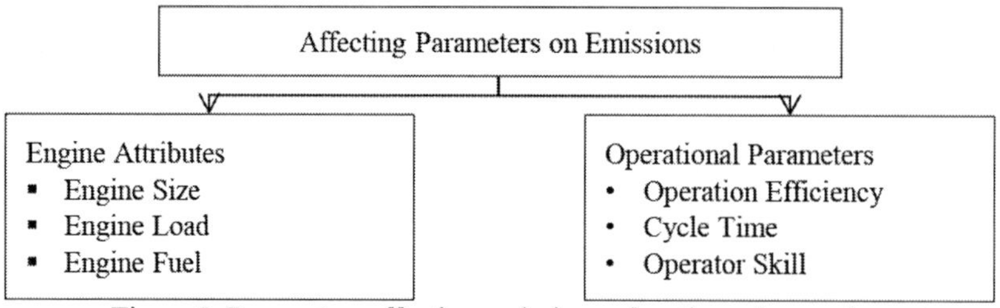

Figure 2. Parameters affecting emissions of earthmoving equipment

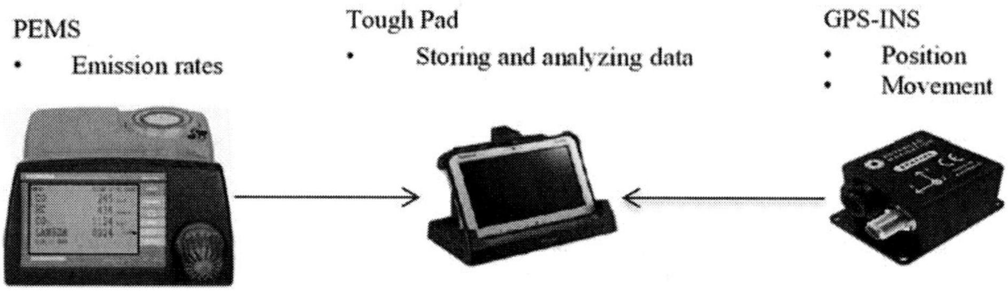

Figure 3. Instrumentation for field data collection

Due to main contribution of earthmoving equipment in producing NO_x, CO and PM_{10} pollutants [6], the aim of this study is to develop comprehensive methodology to model the emissions of two main types of construction equipment of loader and excavator. Data collection in this research is conducted in two steps of laboratory and field experiments. The laboratory experiments were conducted to verify the performance of instruments employed and synchronize the raw data. The data synchronization process shows that data measured by PEMS are around 8 seconds behind engine data retrieved from the engine data logger. This can be due to having 5 m long sampling tube and the time needed for gas analysis in PEMS. The data of the laboratory experiments can be used for developing the initial framework of emission model.

In the next step, field experiments were conducted to collect real-world data from in-use earthmoving equipment on construction and mining sites. Equipment with different brand, model year and size was experimented for representing the effect of engine technology, age and size on emissions. In this study, the field experimentation process was conducted on one loader and one excavator with approximately 3,000 data points collected from the PEMS and GPS-INS instruments. The data collected were then analyzed to model emissions of these two machineries at operation and equipment level.

4. Result Analysis

In this section, the method of data processing is developed for analyzing the emission data at operation and equipment levels. As discussed before, two main instruments of PEMS and GPS-INS are used to collect emission and operational data from earthmoving equipment. The data measured are transmitted and stored in the Tough Pad. The raw data were reviewed to identify any potential errors or problems occurred in the process of field experiments. Data filtering was also carried out to correct errors in certain points or remove invalid records [16]. In the next step, an exploratory analysis is conducted to model CO_2, CO, HC and NO_x emissions at operation level. Finally, the emissions of earthmoving equipment are estimated at equipment level by knowing cycle time and time ratio of operation modes. IBM SPSS Statistics V22 and Microsoft Excel software were used for data analysis.

4.1. Operation-level Emissions Modelling

The emissions analysis of two pieces of earthmoving equipment is presented in this section. The emission rates of CO_2, CO, HC and NO_x of the equipment are estimated in four operation modes: idling, loading, swing (moving), and dumping. It is anticipated that there is much variation in the amount of emissions in each operation mode due to many parameters such as operator skill, material type and engine conditions. By conducting statistical analysis on the raw data, it is shown the relationship between the value of emission rates in each operation mode and their corresponding occurrence frequency is similar to a Beta (β) distribution. Based on the variability relations developed, mean (M) and standard deviation (σ^2) for emission rates of pollutants in each operation mode is calculated by using Equations (1) and (2).

$$M_{ij} = (A_{ij} + 4B_{ij} + C_{ij})/6 \quad (1)$$
$$\sigma_{ij}^2 = ((C_{ij} - A_{ij})/6)^2 \quad (2)$$

Where:
M_{ij}: Mean of pollutant i emission rate in operation mode j (g/kWh)
σ_{ij}^2: Standard deviation of the value of pollutant i in operation mode j
A_{ij}: The minimum value of pollutant i in operation mode j
B_{ij}: The most likely value of pollutant i in operation mode j
C_{ij}: The maximum value of pollutant i in operation mode j.

4.2. Equipment-level Emissions Modelling

In this section, the emissions of earthmoving operations at equipment level are modelled. The cycle time of earthmoving equipment and time ratio of each operation mode were estimated first. GPS-INS instrument was attached to the stick of equipment and measured its three-dimensional position and movement second by second. By investigating the position and movement of stick, total cycle time and the time spent in each operation mode were estimated. Also, the cycle time of earthmoving equipment can be calculated through analyzing the films recorded from machinery operations on site. There are many parameters affecting the cycle time including bucket size, swing angle, operator skill and hauler position. Also, the main construction machinery manufacturers such as Caterpillar and Komatsu companies publish performance handbooks for presenting equipment specifications and estimating production rate and cycle times based on working conditions [15]. Tables 2 and 3 show cycle time of the loader and excavator used in the study based on manufacturers' performance handbooks. It was found the bucket size is the main parameter affecting the total cycle time and time ratio of operation modes for the earthmoving equipment. These two tables were prepared using specific swing angle and depth of cut, and minimum distance to hauler.

Operation efficiency of equipment (OEE) is another affecting parameter needs to be considered in emission modelling earthmoving machinery at equipment level. OEE is defined as the ratio of efficient operation time over total operation time of equipment. Actually, this parameter divides the efficient operating time from idling time of the equipment. In practice, this parameter varies from around 5% to 30% for earthmoving equipment involved in construction and mining sites. Equation (3) estimates emission rates of earthmoving equipment in efficient operation time by considering the time ratio and emission rates of different operation modes. Also, Equation (4) presents the equipment-level emission model based on OEE and P_{ni}.

$$P_{na} = P_{nl}*TR_l + P_{nsm}*TR_{sm} + P_{nd}*TR_d \quad (3)$$
$$P_n = PW*(OEE*P_{na} + (1\text{-}OEE)*P_{ni}) \quad (4)$$

Where:
P_n: Equipment-level emission rate of pollutant n (g/h)
P_{na}: Average emission rate of pollutant n in working time (g/kWh)
P_{nl}, P_{nsm}, P_{nd}, P_{ni}: Emission rate of pollutant n in loading, swing (moving), dumping and idling operation modes (g/kWh)
TR_l, TR_{sm}, TR_d: Time ratio of loading, swing (moving) and dumping operation modes in cycle time
PW: Power of equipment (kW).

Table 2. Average cycle time for loaders with different bucket size

Bucket Size (m^3)	Loading (sec)	Moving* (sec)	Dumping (sec)	Total Cycle (sec)
0.75-2.75	9	18	3	30
3-4	10	21	4	35

*: minimum distance between loader and hauler has been assumed.

Table 3. Average cycle time for excavators with different bucket size

Bucket Size (m^3)	Loading* (sec)	Swing** (sec)	Dumping (sec)	Total Cycle (sec)
<0.75	5	7	2	14
0.75-1	6	7	2	15
1.5-2	6	8	3	17
2.25	7	9	3	19
2.5	7	10	4	21
3	7	11	4	22
3.75	8	13	4	25

*: depth of cut is 50% of maximum cutting depth **: swing angle is 45°

Based on Equations (3) and (4), the influence of many affecting parameters including time ratio of operation modes and OEE on emissions can be measured. Also, different ways are suggested to reduce emissions of earthmoving equipment by considering the production rate. OEE is the main parameter used for reducing total emissions. Based on Equation (4), as OEE increases, the non-efficient operation time of equipment decreases which causes the reduction of total emissions. Minimizing the distance from hauler or swing angle decreases swing (moving) time and consequently total cycle time which decreases total emissions as well. Bucket size is another parameter affecting total cycle time and emissions at equipment level. By considering Tables 2 and 3, it is recommended to use equipment with larger bucket size for reducing total emissions.

5. Case Study

In this section, the methodology developed is verified through experimentation and emission analysis of two in-use pieces of equipment of a loader and an excavator. The total experiment time lasted around one hour and, after data mining and filtering, approximately 3000 data points were collected. Table 4 presents the specifications and engine attributes of the equipment used as in the case study. By developing Beta statistical distribution and using equations (1) and (2), the emission rates were estimated for different operation modes. Figure 4 presents the mean emission rates of CO_2, CO, HC and NO_x for the excavator and the loader in idling, loading, swing (moving), and dumping modes. Figure 5 shows the photos of the equipment took during the experiment.

Table 4. Equipment's specifications used for experimentation

Vehicle	Engine Size (kW)	Model	Bucket Size (m^3)	Experiment Time (min)
LCM LG966 loader	193	2010	3.4	24
Poclain 115 excavator	115	1998	0.8	27

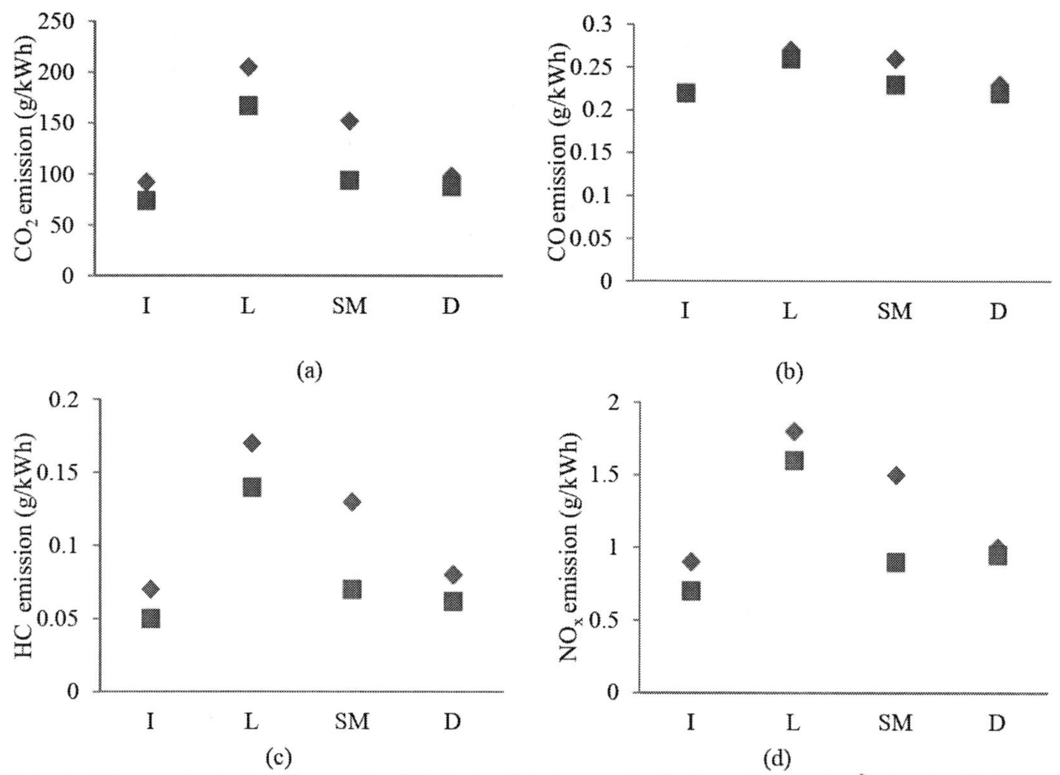

Note: I: Idling mode, L: Loading mode, SM: Swing/Moving mode, D: Dumping mode, ◆: Loader, ■: Excavator

Figure 4. Mean emission rates of two loader and excavator equipment types in different operation modes for (a) CO_2, (b) CO, (c) HC, and (d) NO_x pollutants

Table 5. Mean and standard deviation of emissions value in different operation modes

	Operation Modes							
	Idling		Loading		Swing/Moving		Dumping	
pollutions	LO	EX	LO	EX	LO	EX	LO	EX
CO_2	92,14	74,11	205,78	167,47	152.42	94,23	98,19	88,12
CO	0.23,2a	0.22,a	0.27,7a	0.26,5a	0.26,4a	0.23,2a	0.23,3a	0.22,a
HC	0.07,3a	0.05,2a	0.17,9a	0.14,6a	0.13,5a	0.07,3a	0.08,3a	0.06,2a
NO_x	0.9,b	0.7,b	1.8,7b	1.6,4b	1.5,4b	0.9,2b	0.9,2b	0.8,b

Note: LO= loader, EX= excavator, a = 10^{-5}, b = 10^{-3}

Table 6. Emission rates of experimented machineries at equipment level

Vehicle	CO_2 (g/h)	CO (g/h)	HC (g/h)	NO_x (g/h)
LCM LG966 loader	28,409	48.52	23.66	261.90
Poclain 115 excavator	12,972	27.14	10.10	125.28

(a)

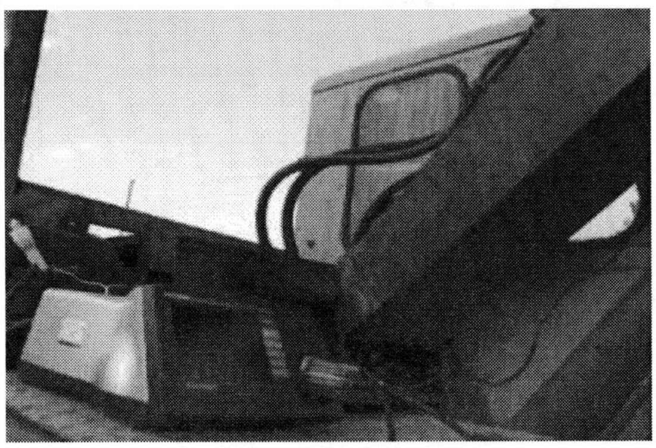

(b)

Figure 5. Equipment utilized for case study: (a) LCM LG966 loader, (b) Poclain 115 excavator

As shown in Figure 4, there is much variation in emission rates of earthmoving equipment in different operation modes. For loader and excavator, emission rates in loading mode are much higher due to more power of engine consumed. In the operation-level emission model developed by Barati et al., it was verified that there is a highly correlated relationship between the amount of used power of engine and emission rates [14]. It is also noteworthy that the emission rates in idling and dumping operation modes are relatively the same and minimum for both pieces of equipment considered in the study. Generally, excavators produce less emission per engine size in comparison with loaders, especially in swing/moving mode. It shows that the engine power of excavators is not used as much as loaders in earthmoving operations. The standard deviation of emission rates in different operation modes were calculated and given in Table 5. It is found the maximum standard deviation of all emissions for both excavator and loader is in loading operation mode. There are other parameters that may cause some variations in the amount of emissions in this operation mode, e.g. operator skill and material type.

In the next step, the emission rates of the loader and excavator are estimated at equipment level. Tables 2, 3 and 4 are used for estimating the cycle time and time ratio of different operation modes. The total cycle time for loader with 3.4 m^3 bucket size is 35 seconds, and loading, moving and dumping time take 10, 21 and 4 seconds respectively. Also, total cycle time of the experiment is 15 seconds (bucket size = 0.8m^3), and loading, swing and dumping modes take around 6, 7 and 2 seconds respectively. Site observations also found the experimented pieces of equipment were in idling mode for around 12 minutes per hour which means the OEE is 20%. By using Equations (3) and (4), the emission rates of two machineries are estimated. Table 6 presents the calculated emission rates of four pollutants investigated.

6. Conclusions

Earthmoving operations are one of the main contributors to GHGs emissions. The current models used for estimating emissions of such operations roughly estimate the pollutant emission rates at state or project level. This paper aims to develop a comprehensive methodology for emission modelling of earthmoving equipment at operation and equipment level. This framework involves three main processes of instrumentation, data collection and data analysis. The developed methodology is generic and can be used for other applications such as machinery maintenance and fuel use estimation.

Field data are necessary for modelling emission rates of earthmoving equipment and investigating the parameters affecting emissions. Three main instruments, a PEMS, a GPS-INS and a Tough Pad tablet PC were used in this study for collecting operational and emission data of in-use earthmoving equipment. Site observations were conducted to record videos of equipment operation for estimating cycle time and ratio of time spent in each operation mode. The developed methodology was successfully applied on one loader and one excavator with around 3,000 data point collected for data filtering and processing. IBM SPSS Statistics V22 and Microsoft Excel software were used to analyze the data gathered from the site.

Future studies will investigate the effect of operator skill on emissions of earthmoving equipment. This would provide better understanding of the significance of operation skill and also help developing guidelines and handbooks on how to operate equipment more efficiently to reduce emissions. More field experiments will be carried out to study the effect of engine tier on emissions as well.

References

[1] Intergovernmental Panel on Climate Change, Climate Change 2007. Synthesis Report, Contribution of Working Groups i, ii and iii to the Fourth Assessment Report of the Intergovernmental Panel on Climate Change, The IPCC Fourth Assessment Report, C.W. Team, R. K. Pauchauri, and Reisinger, IPCC, Geneva, 2007.

[2] Environmental Protection Agency (EPA). Clean Construction USA, *EPA-420-F-05-032*. Washington, D.C. 2005b.

[3] Kaboli A.S. and Carmichael D.G. Emission and Cost Configurations in Earthmoving Operations. *International Journal of Organization, Technology and Management in Construction*, 4 (1) (2012) 393-402.

[4] Avetisyan H.G., Miller-Hooks E. and Melanta S. Decision Models to Support Greenhouse Gas Emissions Reduction from Transportation Construction projects. *Journal of Construction Engineering and Management*, 138 (5) (2012) 631-641.

[5] Ahn C.R. and Lee S.H. Importance of Operational Efficiency to Achieve Energy Efficiency and Exhaust Emission reduction of Construction Operations. *Journal of Construction Engineering and Management*, 139(4) (2013) 404-413.

[6] EPA. User's Guide for the Final NONROAD 2005 model. *EPA-420-R-05-013*, Office of Transportation and Air Quality. U.S. 2005.

[7] Lewis M.P. *Estimating Fuel Use and Emission Rates of Nonroad Diesel Construction Equipment Performing Representative Duty Cycles.* (thesis) North Carolina State University, NC, USA, 2009.

[8] Kim B., Lee H., Park H. and Kim H. Greenhouse Gas Emissions from Onsite Equipment Usage in Road Construction. *Journal of Construction Engineering and Management*, 138 (8) (2012) 982-990.

[9] Lewis P., Leming M. and Rasdorf W. Impact of Engine Idling on Fuel Use and CO_2 Emissions of Nonroad Construction Equipment. *Journal of Management in Engineering*, 28 (1) (2012) 31-38.

[10] Melanta S., Miller-Hooks E. and Avetisyan H.G. Carbon Footprint Estimation Tool for Transportation Construction Projects. *Journal of Construction Engineering and Management*, 139 (5) (2013) 547-555.

[11] Abolhasani S. and Frey H.C. Engine and Duty Cycle Variability in Diesel Construction Equipment Emissions. *Journal of Environmental Engineering*, 139(2): 261-268, 2013.

[12] Barati K. and Shen X. Modelling Emissions of Construction and Mining Equipment by Tracking Field Operations. In *Proceeding of 32nd International Symposium on Automation and Robotics in Construction and Mining,* pages 478-486, Oulu, Finland, 2015.

[13] Nichols H. and Day D. *Moving the Earth: Excavation workbook.* McGraw-Hill, USA, 2005.

[14] Barati K., Shen X. and Carmichael D. Operational Level Emissions Modelling of On Road Construction Equipment through Field Data Measurement. *Journal of Automation in Construction*, 2016.

[15] Peurifoy R.L., Schexnayder C.J. and Shapira A. *Construction Planning, Equipment and methods.* (7). McGraw-Hill, USA, 2006.

[16] Frey H.C., Unal A., Chen J., Li S. and Xuan C. Methodology for Developing Modal Emission Rates for EPA's Multi-Scale Motor Vehicle and Equipment Emission Estimation System, *EPA420-R-02-027,* U.S. EP, Ann Arbor, MI, 2002.

Monitoring Apparent pH Value in Geopolymer Concrete Using Glass Electrode

J. Dang [a], X. Shen [a], A. Castel [a], and J. Aldred [a,b]

[a] School of Civil and Environmental Engineering, The University of New South Wales, Australia
[b] AECOM, Australia
E-mail: jin.dang@student.unsw.edu.au, x.shen@unsw.edu.au, a.castel@unsw.edu.au, James.Aldred@aecom.com

Abstract –

Geopolymer concrete (GPC) is estimated to reduce 80% carbon footprint in construction projects compared with ordinary Portland cement (OPC). Previous studies on GPC have been focusing on assessment of paste or mortar while the long-term durability of GPC has not been extensively investigated yet. This research aims to monitor carbonation process of GPC by measuring in-situ apparent pH value under accelerated carbonation test in laboratory environment. Glass pH electrode is typically applied for liquid medium and considered to be unsuitable for OPC due to high alkaline condition. We have endeavoured to test the glass electrode in GPC assuming the alkalinity in GPC would be much lower than OPC. Two pH probe sensors were embedded into GPC specimens with different depth and seal condition. After 18 hours heat cures and 7 days curing in ambient environment, one of GPC specimens was then relocated to a carbonation chamber with 1% carbon dioxide concentration. The glass electrodes were found to survive the GPC environment with real-time apparent pH value registered since the GPC specimens were poured. It is also observed that the apparent pH values were affected significantly by the level of humidity inside the concrete. Saturation experiments were then conducted to provide valid pH readings. As a result, we successfully monitored pH decreasing from 11.1 to 10.76 after three months carbonation experiment. The results are also validated by comparing against the data acquired by traditional concrete grinding test. The research features the first trial of its kind and has contributed to comprehensive understanding about corrosion mechanism of the new GPC material.

Keywords –

pH glass electrode, Geopolymer concrete, Carbonation, Long-term monitoring

1. Introduction

The contribution to carbon emissions from the ordinary Portland cement concrete (OPC) is second only to fossil fuels [1]. With growing environmental demands in terms of low carbon, geopolymer concrete (GPC) has emerged as environmentally construction material with great potential to substitute OPC [2]. The GPC utilises industrial waste as binder to cast concrete, e.g. fly-ash, slag and mud, etc. It potentially contributes to a maximum reduction of 80% carbon dioxide (CO_2) emission in construction projects compared with OPC. With similar mechanical properties to OPC, GPC also exhibits some superior characteristics such as stable state exposed to the high temperature and acid. Despite many advantages compared to the conventional concrete, there are also several challenges faced by this new material [4] [11]. There are still many fundamental issues to be addressed for GPC, e.g. lack of a complete standard or handbook commonly used globally; the corrosion mechanism of reinforcing steel in GPC is not well understood, which could be very different with traditional concrete [9]; and limited previous studies assessed long-term durability of GPC in terms of carbonation process.

Concrete carbonation is one of the primary reasons for steel corrosion in reinforced concrete structures [13]. Steel reinforcement in OPC is naturally protected by the high alkalinity pore solution (pH value being 12.5-13.5) in which induces an oxide passive film is formed on the reinforcement surface to protect it against corrosion. Carbonation process when occurs will cause a rapid decay and shorten the service life of concrete structure [1]. This can be evoked under the exposure to normal situation while carbon dioxide from atmosphere diffused into concrete and reacted with the hydrated cement in concrete [16]. The process is usually described as a simple chemical function, as given in Equation 1.

$$\mathrm{Ca(OH)_2} + CO_2 \rightarrow CaCO_3 + H_2O \qquad (1)$$

Carbonation itself doesn't cause the damage to concrete, which only lowers the alkalinity concentration of the pore solution and the passive film protection tends to be disappeared along with the decreased pH value in concrete. Therefore the pH value in concrete is well regarded as a key parameter indicating the level of corrosion [12]. As reported by Schießl, if pH value in concrete drops bellows 9, the passive film protection will lose in the greater surface area of reinforcement. Accordingly, long-term monitoring of pH value in the range from pH 9 to 13 with a resolution of about 0.5 pH units is relevant for early detection of a potential corrosion condition [10].

Carbonation is often assumed to be related to the diffusion gas though the porous concrete. The diffusion rate normally depends on the material properties, moisture and concertation of the carbon dioxide. As the properties of the concrete vary, the carbonation rate will also be different [13]. An easy model was developed for calculating the time-dependent depth according to Fick's first law of diffusion:

$$X=A\sqrt{t} \quad (2)$$

Where the X is the depth of the carbon dioxide front, A is the rate of carbonation, t is the time.

This research aims to assess performance and accuracy of the pH electrode in measuring apparent pH value GPC. The traditional electrode was embedded in GPC to measure the pH value throughout the entire life since concrete casting. It features the very first trial of a long-term pH monitoring system for GPC. It contributes to a comprehensive understanding about corrosion mechanism of the new GPC material and is potentially applicable for field measurement if the results of laboratory experiment prove to be successful and reliable.

2. Review on pH measurement

It is essential to measure the pH value to estimate the likelihood and start time of corrosion as the pH value is the only parameter that can directly indicate the corrosion state. Currently, most commonly used methods to measure the pH value in concrete are destructive, such as percussive drilling, X-ray and thermo-gravimetric analysis [18]. These methods require removals of concrete dust or sample from different depth of the testing structure and bring back to analysis in a laboratory. The construction industry uses a simple test, which involves spraying the phenolphthalein indicator solution onto the freshly cored concrete sample and measuring the carbonated depth of the concrete [3]. However, none of those methods is able to provide continuous measurement; the use of destructive methods to obtain pH information of the material is no longer accepted for long-term monitoring [5].

In-situ monitoring systems allow continuous monitoring over an extended time periods by embedding the sensing probe within permanent bodies. [20]. Due to the harsh condition of concrete in terms of the high pressure and high alkalinity condition, continuously measuring the pH value within concrete is always a very difficult task. The sensor has to be durable enough and requires many unique characteristics, such as high chemical and pressure resistance and a suitable measureable pH range. Since Werner reported the fibre optical pH sensor has the potential to measure the high pH value from 9~13 in 1990s, a few studies has been continued in developing the fibre optical pH sensor in concrete. This sensor is designed by use of an appropriate pH indicator attached to porous substrates allowing the ions exchange with the concrete structure. This matrix has the ability to respond the pH change as the colour difference and, consequently, the optical sensor will measure the light intensity. Although a few fibre optical pH sensors have been embedded in concrete to measure the pH value for several months, the development has not exceeded the status of prototyping and there is no commercial product available so far [6][21]. More importantly, it suffers from certain drawbacks, such as inconsistent response in different pH range [17]. As dye will easily leach leading to a signal drifting, a complex operation procedure in designing process is required. Meanwhile the fibre optical pH sensor has not been able measure pH value above 12 effectively [3] [15].

Usually, the pH value in solution is determined using an electrode which includes working electrode, reference electrode and temperature sensor. The pH electrode contacts with the solution by a special pH sensitive glass membrane which develops a voltage proportional to the pH value. Previous results in concrete showed that this type of sensor is not suitable for the OPC. The main problem is the sensor designed for the solutions might be not respond to the dry condition and easily leads to false readings. The high alkalinity concentration in concrete could break the sensor body in hours [15][21]. The conclusion is made based on the test results in conventional concrete while GPC tends to have a much lower alkalinity condition due to the different compositions. However, the pH electrode has never been tested in GPC even though the sensors have been commercially available for decades.

With technical advances in the pH electrode sensor, manufacturers use special glass membranes to measure high alkaline values which makes almost zero alkalinity error with pH value below 12. More importantly, periodically re-wetting the sensor by water can possibly eliminate the shift of readings [19].

An Emerson pH glass probe (Model: General

Purpose 3900) was evaluated in geopolymer concrete in this study. A rugged surface encloses the temperature, reference and working electrode into one sensor body, which provides a stable condition for electrode with high chemical and pressure resistance. The measurable pH range from 0-14 is suitable for entire carbonation process from the fresh concrete to cure and carbonation condition. The measurement unit is quite sensitive with a 0.01 unit of resolution, and therefore it is possible to monitor the pH change within the concrete throughout the service life.

3. Experimental Program

3.1. Design of Instrumentation

This pH sensor system consists of two pH probes, one data analyser and two data loggers, as shown in Figure 1. The probe is composed of a working electrode, reference electrode and temperature sensor included in one body. The data logger automatically records the pH readings in different environments from the two sensors. The overall potential of the measuring electrode equals the potential of the internal reference electrode plus the potential at the glass membrane surface. The working electrode is able to proportionally indicate the pH value while the potential of reference electrode is dependent. The overall voltage of the cell depends on the measured sample. Working and reference electrode will exhibit an iso-potential and this usually designed to be 0 mv at pH 7.0. Prior to the use of system, it is necessary to calibrate the sensor first before embedding it into concrete. As the cell voltage is a linear function of pH value, two buffer solutions were used with pH 4 and 10. The calibration slopes for sensor 1 and 2 were 57.8 and 57.6 respectively.

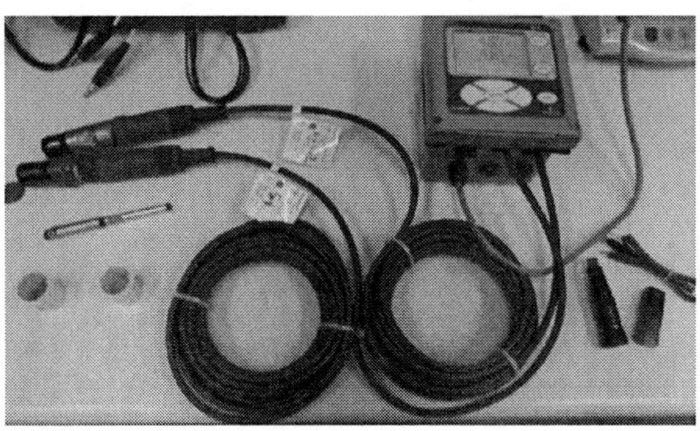

Figure 1. Experiment setup of the pH Sensor System (Emerson Ptd.)

3.2. GPC Material and Batching

The specimens are 50Mpa in design grade and produced by homogeneous paste from the mixture with fly ash and activator (Table 1). The fly-ash classification follows the ASTM standard C618. For concrete mixing, the priority procedure is to prepare the alkaline activator 24 hours before the casting day. The activator used was the mixture of 12M sodium hydroxide and sodium silicate solution while the aggregate used is 10 mm basalt and Sydney sand. The material mixing followed restrict sequence from coarse to fine, then the activator and free water.

Table 1 Concrete Mixing

Materials	Density ($Kg/^3m$)	Percentage
Coarse aggregate (10mm basalt)	1221.2	50.105
Fine aggregate	620.8	25.471
Fly ash	271.6	11.143
Kaolite	77.6	3.184
Slag	38.8	1.592
NaOH	55.3	2.269
Na2SiO3	138.7	5.691
Water	13.3	0.546
Total	N/A	100

3.3. Experiment Setup

After calibrating the pH sensors, the next step is the sensor embedment. Two concrete cylinders were built with same 100 mm diameter and 200 mm height. A magnet-stand was used to attach to the concrete steel mould and fix the sensor location, as shown in Figure 2. The sensors were embedded into different depths of 100 mm and 20 mm from the bottom surface respectively. Upon the sensor location is confirmed, the concrete was directly poured from the top and vibrated. To reach the expected strength shortly, both geopolymer concrete cylinders were heat cured at 75°C in a heat chamber for 18 hours.

Once the heat curing process finished, different treatments were applied into the individual concrete sample. Table 2 is shown the difference dispose methods between the 1st and 2nd specimen. The first specimen cylinder stayed at a plastic mould to isolate any contact with environment which intends to keep the inner moisture from evaporating quickly. The 2nd specimen was de-moulded and wrapped with the seal tap except bottom surface which was in contact with the environment. Both concrete cylinders were placed on the 25°C and 63% relative humidity for 7 days curing. But only the 2nd specimen was transferred to the carbonation chamber for investigating the carbon dioxide front at 20 mm depth. Environment in the chamber was set to 23°C temperature, 65% relative humidity with 1% carbon dioxide concentration. Meanwhile, another sample stayed in the controlled ambient condition with 23°C as a reference. The different measurement depth and seal condition in this experiment were designed to investigate the moisture and carbonation influence on the pH readings. It is obvious that the 1st specimen will provide a much more wet measurement environment for the pH sensor, while the moisture of 2nd specimen will decrease quickly due to the evaporation from bottom.

Figure 2. set-up of pH sensor embedment

Table 2 Experiment Plan

Items	Sealed condition	Embedment Depth	Carbonated
1st specimen	Sealed	100mm	No
2nd specimen	Leave bottom surface	20mm	YES

It should be noticed that the conventional pH glass sensors are designed to measure the pH in water or liquid solutions. When a non-water solvent is present in appreciable quantities, the pH reading will be shifted from the expected value by effects of the non-water solvent on the pH. Therefore, the concrete samples need to be wetted to reduce the errors of reading.

The compression strengths of 7 days and 28 days are 48.5 Mpa and 52.94 Mpa respectively. The specific test procedure followed the AS1012.9-1999 standard.

4. Results and Discussions

4.1. pH value in Curing process

Figure 3 shows the pH values of the two specimens during curing process. After concrete was poured, pH readings were 12.4 and 12.2 for the 1st and 2nd specimens respectively. Both specimens recorded a sharp decrease during the heating process as the temperature increased from 25°C to 75°C. Temperature compensation was already add into the data analyser to eliminate the error induced by the sensor. However, chemical equilibrium of the components affected the pH value in addition to the temperature effect.

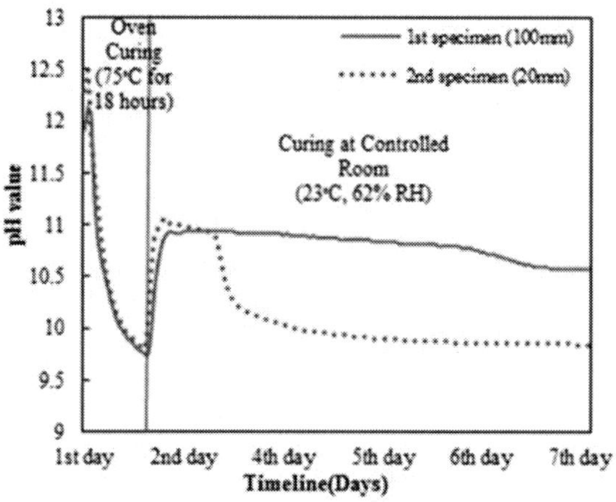

Figure 3. pH value during the heat and normal curing process

Both pH readings gradually recovered as the specimens cooled down to ambient temperature. The initial pH value of harden concrete was 11.1. In the next few days of curing process, a significant decrease was observed for the 2nd specimen in which the bottom can directly contact with the environment while the 1st specimen only experienced a slight drop at the same time due to different sealed condition. pH readings experienced a major change are not considered to be correct for the fresh concrete. The shift of pH readings is more likely caused by the moisture loss in terms of the evaporation and geopolymersation reaction within the concrete. Due to gradual loss of the water content within concrete, the resistance of the sensor increased substantially which induced a small current output and reflected a lower pH value. The specimens were then saturated to investigate and calibrate the moisture effect.

4.2. pH Value during Carbonation Process

After the pH reading of the 2nd specimen stayed constant in the dry condition, it was transferred to the accelerated carbonation chamber with 1% carbon dioxide and 65% relative humidity. The specimen was placed in the chamber for 6 weeks. To ensure the decreasing result is more reliable, another pH measurement method with percussive drilling was also introduced. The drilling test of concrete sample conducted every two weeks since the carbonation and all the samples were prepared from the same concrete batch. Figure 4 shows the pH value comparison between the sensor measurement and percussive drilling at the exact 20 mm depth of the concrete cylinder. The initial pH value tested by drilling test was 11.45 and the decreasing curve was presented mostly like a linear function and, eventually come to 11.04 after six weeks carbonated. By comparison, a very similar pattern was found in the raw data of sensor measurement result although the pH readings were not correct due to the loss of moisture. It decreased from 9.84 to 9.42, the decreasing values in two methods were almost the same being 0.41 and 0.42 pH unit respectively. The humidity within the concrete already kept constant and only had limited effect to the pH readings. Therefore the change of pH reading is mostly because of the carbonation.

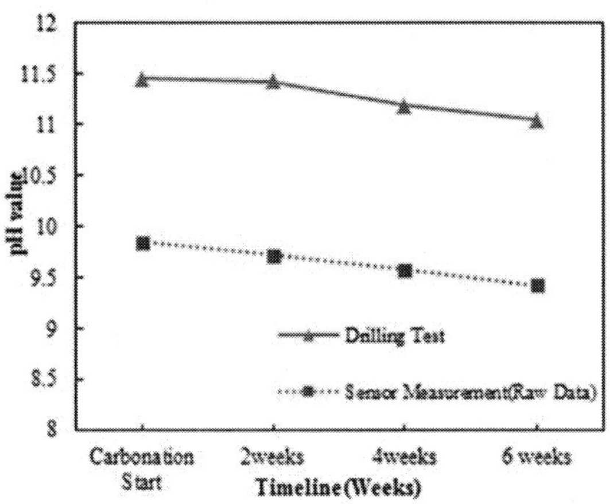

Figure 4. Comparison of 6 weeks carbonation value from raw pH sensor readings and drilling test

4.3. Saturation Test for pH Calibration

To verify the shift of pH readings due to moisture loss, a series of saturation tests was carried out on both specimens. The specimen was submerged into the tap water for 8 hours and then enclosed with the sealed tap. After that, the specimens were placed at controlled environment and let the humidity diffusion into the inner concrete gradually. Both specimens wre saturated into tap water twice; only hydrated for few hours of each time as a concern of the alkalinity leakage. Figure 5 shows the pH readings of specimens saturated after the carbonation test finished. Since the saturation begins, the pH readings of the both specimens increased apparently in the next few weeks. However, the increasing rate was gradually slowdown in the 2nd week and, therefore a second saturation was conducted to wet concrete more properly. The pH readings of the first specimen stabilised at 11.08 in the fourth week, which was exactly the initial pH value when concrete was harden. Meanwhile, the pH reading of second specimen was 10.74 remaining a 0.34 difference to 1st specimen. Considering the same initial values obtained from both specimens at the beginning of the test, the carbonation effect after 6 weeks was considered to be the main reason causing the gap of pH readings between the two specimens.

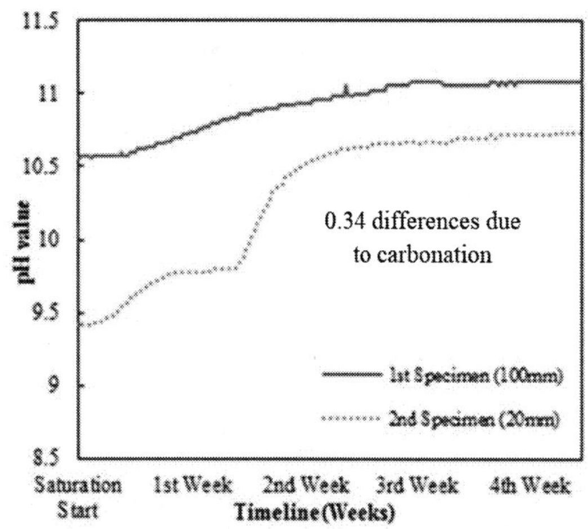

Figure 5. pH value of 2nd specimen during the saturation Test

It was also found from the saturation tests that the pH value shifting caused by the humidity loss could be reversible if the sensor probe was wetted properly. In this experiment, the pH value of 11.1 is regarded as the correct initial value. Meanwhile, the sensor measurement under the carbonation environment was considered as accurate.

Figure 6 shows the real pH value during 6 weeks carbonation. The pH readings of the sensor measurement decrease from 11.1 to 10.68 as an average weekly rate of 0.07 unit/week during the same duration.

The eventual value also contained small error compared to the value after 2nd specimen being well saturated (10.74). The difference of 0.3 pH unit was found between the two methods. This was considered to be in the acceptable range as the drop pH value due to carbonation is usually measured as a resolution of 0.5 pH unit.

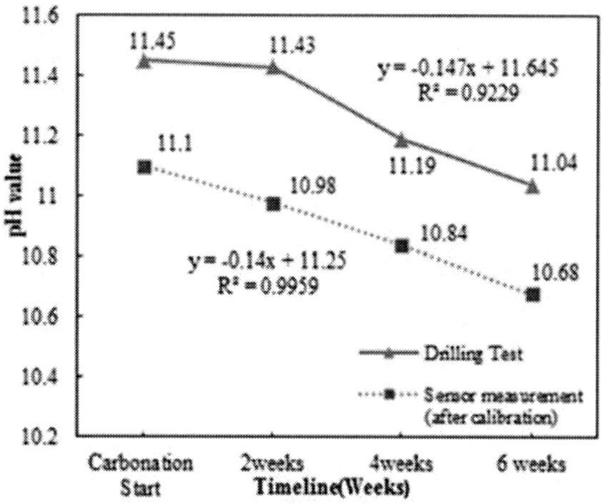

Figure 6. pH value comparison between sensor measurement and drilling test after calibration

5. Conclusion

This paper investigated continuous pH measurement by electrochemical method in geopolymer concrete for the very first time. The pH electrodes have survived the geopolymer concrete environment for 9 months since concrete casting. The pH value measured after 6 weeks carbonation test was validated by comparing to the result from drilling test. A decrease of 0.4 pH unit was recorded in both testing methods at cover depth of 20 mm. The pH readings in GPC were found to be affected by three factors: moisture, temperature and carbon dioxide diffusion rate. The dry condition induced a substantial resistance to the sensor circuit and resulted a lower pH reading. The pH readings were calibrated in this study by conducting saturation tests. Future research will investigate the relationship between the moisture loss and pH reading. A humidity monitoring sensor will be embedded in concrete to model pH drift due to humidity changes.

Acknowledgment

This research work is funded by the Cooperative Research Centre (CRC) for Low Carbon Living Ltd supported by the Cooperative Research Centres, an Australian Government initiative (project number RP1020). The experimental work was carried out in the materials and structures laboratory of the School of Civil and Environmental Engineering at the University of New South Wales. The assistance of the laboratory staff is acknowledged.

References

[1] Aguiar, J. B. and Junior, C. Carbonation of surface protected concrete. *Journal of construction and Building Materials,* 49, 478-483. 2014

[2] Badar. Md. Sufian. Corrosion of steel bars induced by accelerated carbonation in low and high calcium fly ash geopolymer concretes. *Journal of the Construction and Building Material*, 61,79-89.2014.

[3] Basheer M P, Grattan K T V, Sun T, et al. Fiber optic chemical sensor systems for monitoring pH changes in concrete. *Optics East. International Society for Optics and Photonics*, 144-153, 2004.

[4] Castel, Arnaud, Steve Foster, and James Aldred. Time-dependent behaviour of a class F fly ash-based geopolymer concrete. *The international journal of Research and Technology*. 13(3), 239-1163, 2014.

[5] Christiane, M. *Non-destructive evaluation of reinforce concrete structures*, Volumn 2, Non-destructive testing method, Woodhead Publishing. 2010

[6] Dantan, N., Habel, W. R., and Wolfbeis, O. S. Fibre optic pH sensor for early detection of danger of corrosion in steel-reinforced concrete structures." *Smart Structure and Materials*, 5758, 274-284, 2005.

[7] Dybko, A. Errors in chemical sensor measurements. *Sensors*, 1, 29-37. 2001

[8] El-Bagary, R. I., ElKady, E. F., and Kadry, A. M. Spectrofluorometric determination of certain antihyperlipidemic agents in bulk and pharmaceutical preparations. *International Journal of Spectroscopy*, 27(2), 83-92, 2012.

[9] Gourley J T, Johnson G B. Developments in geopolymer precast concret. *World Congress Ceopolymer.*: 139-143. 2005

[10] Habel,W.R and Kaebber,K. Fibre-optic sensor application in civil and geotechnical engineering.Photonic sensors,Vol ,262-280, 2010

[11] IPCC (Intergovernmental Panel on Climate Change). The fourth assessment report of the Intergovernmental Panel on Climate Change (IPCC). Climate change 2007: The physical science basis, S. Solomon, et al., eds., Cambridge University Press, Cambridge, U.K., 129–234, 2007.

[12] Juenger, M.C.G, Winnefeld, F., et al., Advance in alternative cementitious binders. Cement and Concrete Research, 41, 1232-1243, 2010.

[13] McPolin, D. O. New test method to obtain pH profiles due to carbonation of concretes containing supplementary cementitious materials. Journal of materials in Civil Engineering, 19(11),936-946,2007.

[14] McPolin, D. O. Preliminary development and evaluation of fibre-optic chemical sensors. Journal of Materials in Civil Engineering, 23(8), 1200–1210, 2011.

[15] Nguyen, T. H., Venugopala, T., Chen, S., et al. Fluorescence based fibre optic pH sensor for the pH 10-13 range suitable for corrosion monitoring in concrete structures. *Sensors and Actuators B*, 191, 498-507, 2014.

[16] Neville, A. *Properties of concrete*, 4th Ed., Longman, Harlow,1995

[17] Srinivasan, R, Phillips T. E., et, al, Embedded micro-sensor for monitoring pH in concrete structures, *Smart Structures and Materials: Smart Systems for Bridges, Structures, and Highways*, pp. 40-44, 2000.

[18] Schieb, P. and Yang, L. *Technique for corrosion monitoring*, Woodhead Publishing, 2007

[19] Rosemount Analytical, The theory of pH measurement, Online: http://www2.emersonprocess.com/siteadmincenter/PM%20Rosemount%20Analytical%20Documents/Liq_ADS_43-002.pdf, Accessed: 02/03/2015.

[20] Vimer, C and Yu,S et al. Probing pH levels in cvil engineering materials." ASCE, 21, 2(51), 2009

[21] Werner, T. and Wolfbeis, O. S. Optical sensor for the pH 10-13 range using a new support material. *Analytical Chemical*, 346, 564-568, 1992

Optimum Driving Pattern for Minimizing Fuel Consumption of On-road Vehicles

J. Klein[a], X. Shen[a], and K. Barati[a]

[a] School of Civil and Environmental Engineering, The University of New South Wales, Australia

E-mail: Jacob.m.y.klein@gmail.com, x.shen@unsw.edu.au, Khalegh.barati@unsw.edu.au

Abstract –

The consumption of fossil fuels by on-road vehicles is the greatest source of air pollution and the second largest source of greenhouse gases (GHGs) globally. The emitted pollutions have caused major problems such as respiratory diseases. Also, as the sources of these types of fuels are diminishing, their price is projected to steadily increase. This necessitates developing strategies and policies to ameliorate the environmental, economic and health impacts of such fuels. This paper aims to present a process to minimize fuel consumption through optimizing driving pattern. Four acceleration, speed, slope and payload parameters were investigated as operational parameters affecting fuel consumption. This paper developed engine load parameter as the surrogate of used power of engine linking operational parameters to fuel consumption. Through conducting ordinary least square and multivariate regression analysis on the field data collected from six on-road vehicles, the relations among operational parameters, engine load and fuel consumption were developed. In the next step, the optimum driving pattern was developed by investigating the effect of operational parameters in the fuel consumption developed model. The result analysis shows for driving a vehicle without payload and zero acceleration in a levelled route, the optimum speed is around 50km/hr for having minimum fuel consumption per kilometre. Also, the investigations on the road slope parameter shows by increasing this factor optimum speed decreases while the minimum used fuel increases.

Keywords –
Fuel Consumption; Statistical Modelling; Fuel Efficiency; On-road Vehicle; Instrumentation

1. Background

Today, there are over one billion vehicles in operation throughout the world consuming approximately five trillion litres of petroleum annually [1]. The rate at which petroleum is being consumed and the external costs that result from its use are extremely unsustainable [2]. Despite this, the demand for vehicles and petroleum is steadily increasing, the global on-road vehicle population is expected to reach two billion by 2050 [3]. With increasing demand and diminishing sources of fossil fuels, supplying the required fuels is one the main challenges in transportation field.

The fossil fuels are the main source of pollutants including carbon dioxides (CO_2), carbon monoxides (CO), hydrocarbons (HC), nitrogen oxides (NO_x) and particulate matters (PM). The pollution produced by on-road vehicles is one of the greatest killers today. The annual deaths associated with air pollution far exceed that of any currently ongoing conflict. Each year, 53,000 people die in the U.S alone due to vehicular air pollution [4]. Arguably of more concern is the fact that road transport accounts for more than 12% of the world's CO_2 emissions [5]. Based on the report prepared by Clean Energy Regulator Agency, by decreasing 10% of fuel consumed by on-road vehicles, only in Australia of over 3 billion litres of fuel is saved and over 8 million tonnes of CO_2 is produced less. By considering the increasing demand, reducing supplies and the impact of emitted pollution, developing strategies and schemes for decreasing fuel consumption is necessary.

Many studies have been conducted to develop strategies for optimising features of the road environment that are generally outside of a drivers control such as vehicle routing, platooning, road surface choice, intersection configuration, ambient weather conditions, vehicle load, vehicle condition and speed limit recommendations. However the greatest savings come from the studies that focus on driving patterns which are collectively known as eco driving. These studies have investigated the effect of acceleration rates, cruising speed, idling and speed variability on fuel consumption. It is estimated that as much as 70% of fuel economy is affected by driver decisions [6]. Eco driving strategies focus on driving behaviour and involve eliminating unnecessary sources of energy loss, they advise minimising time spent in acceleration as well as accelerating in a smooth and rapid manner, looking ahead to avoid unnecessary stops and reducing time spent in the wrong gear and idling [7]. A large number of studies on eco driving can be found however the reported savings differ substantially between experiments. For on road vehicles the savings from eco-driving vary from 5-30% [6, 7].

By considering the aforementioned studies, this paper aims to optimize driving pattern to minimize the fuel consumption of on-road vehicles. A comprehensive methodology is first developed to instrument, collect field data and analyse the results. Then, the effect of four operational parameters on fuel consumption is investigated through conducting regression analysis on the filtered data. The effect of operational parameters is finally considered on the presented fuel consumption model to develop optimum driving pattern.

2. Methodology

As was discussed, the main objective of this study is minimizing fuel consumption of on-road vehicle through optimizing driving pattern. In this section, a comprehensive methodology is developed for collecting and analysing field data. As Figure 1 shows, the required instruments are first developed for collecting required data. In the next step, the gathered field data are processed through data mining, and are analysed to develop fuel consumption model. The developed fuel consumption model is finally investigated to optimize driving pattern.

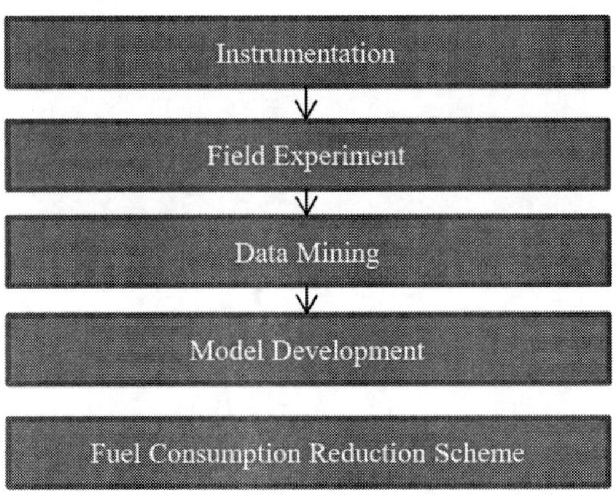

Figure 1. Methodology for developing fuel consumption reduction scheme

There are numerous parameters affecting fuel consumption. On the whole, these parameters can be classified to engine attributes and operational factors. Engine size, engine load are two main engine parameters having significant effect on fuel consumption. Also, engine age and engine tier are the other two engine attributes influencing fuel consumption, but due to their minor effect, they are not been investigated in this study. Operational parameters influencing fuel consumption can be categorized to four acceleration, speed, road slope and payload factors. Based on the research conducted by Barati and Shen [8], there is direct multi-linear relation between operational parameter and engine load. Figure 2 shows the schematic relations between operational parameters, engine attributes and fuel consumption.

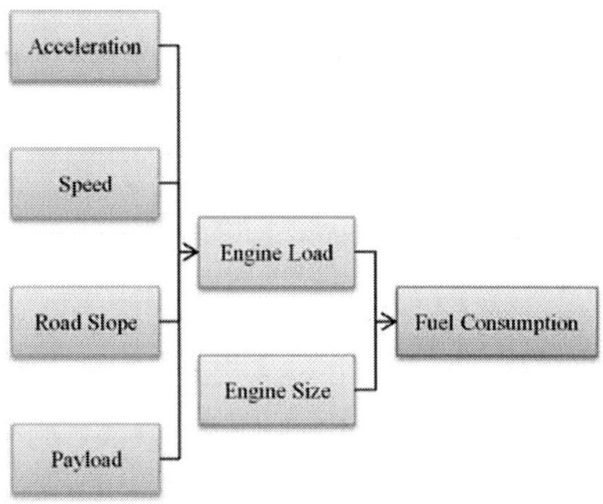

Figure 2. Schematic relations between operational parameters, engine load and fuel consumption

By identifying parameters affecting fuel consumption of on-road vehicles, required instrument needs to be developed to collect field data. Three engine data logger, GPS aided inertial navigation system (GPS-INS) and industrial Tough Pad instruments developed to collect and store required real-world engine and operational data. The main used instrument in this study was engine data logger. This instrument was connected to OBDII port of vehicle under the dashboard and collected real-world fuel consumption rate and engine load from engine control unit (ECU) in each second. The utilized engine data logger in this study is Bluetooth Vgate OBDII adapter. GPS-INS was another instrument used for collecting operational parameters investigated in this research. This tool combines accelerometers, gyroscopes and magnetometers with a commercial grade GPS receiver. By embedding this instrument on the dashboard, three-dimensional position, speed and acceleration of vehicle and slope of road were measured second by second. The GPS-INS system used for this study is SPATIAL-EK manufactured by Advanced Navigation Pty Ltd.

The measured data by engine data logger and GPS-INS instruments were transmitted through RS-232 serial data communication port to the industrial Tough Pad. Panasonic FZ-G1 is the industrial Tough Pad utilized to store and analyse the collected field data.

Using developed instrument, field experimentation were conducted on different on-road vehicles to collect real-world field data. Data collection in this research was conducted in two steps of laboratory and field experiments. The laboratory experiments need to be conducted to verify the performance of instruments, adopt the employed instruments and synchronizing the collected data. Around ten passenger and utility cars were experimented as laboratory test. The data of the laboratory experiments used for developing the initial framework of fuel consumption model.

In the next step, six vehicles were used as field data collection. The specifications of the vehicles used for field experiments are presented in Table 1. Also, Figure 3 shows some sample photos of the conducted field experiments on vehicles. Before starting experimentation, engine would be hot stabilised. To achieve this, the vehicle would be left in idling mode for the time that the cool engine indicator on the dashboard is deactivated. Each vehicle was driven for around 60 minutes, and the driving routes were designed so that each vehicle would spend a substantial amount of time in cruising, accelerating, decelerating and idling mode. Both engine data logger and GPS-INS instruments have a measurement rate of 1 Hz

Table 1. The specification of vehicle used for experimentation

Vehicle	Engine Size (kW)	Model	Tier	Fuel	Experiment Time (min)
Hyundai i30	107	2012	EU V	Petrol	68
Honda Civic VTi	103	2004	EU III	Petrol	57
Toyota Camry Atara	135	2013	EU VI	Petrol	55
Honda CR-V 4WD	120	2001	EU III	Petrol	66
BMW 120i	125	2012	EU V	Petrol	50
Toyota Echo	63	2004	EU III	Petrol	62

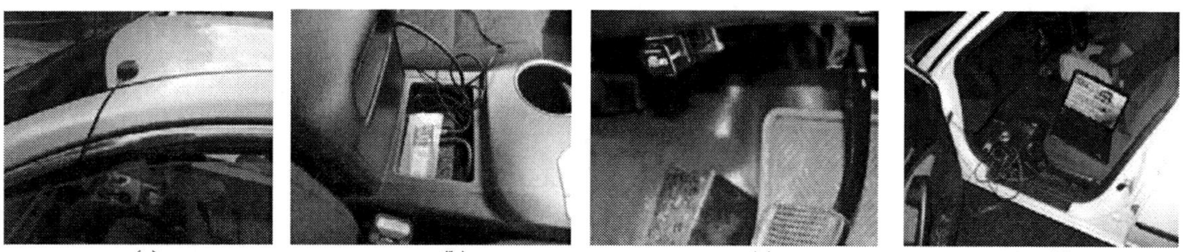

(a) (b) (c) (d)

Figure 3. Sample photos of vehicles used for data collection: (a) Toyota Camry Atara, (b) Honda CR-V 4WD, (c) LX Holden and (d) Honda Civic VTi

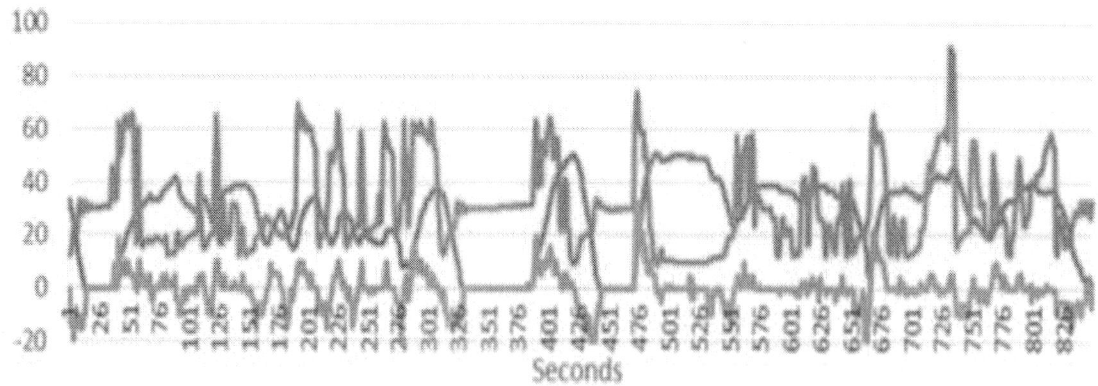

Figure 4. Samples of raw data collected by engine data logger and GPS-INS Instruments

generating 3600 data points per hour. The vehicles were driven for a portion of the time over planned drive cycles and for the remainder on unplanned routes extending over longer distances and traffic conditions. Over the course of the field experiment, a maximum speed of 116 km/hr was recorded. Acceleration rates ranged between -3 to +2.5 m/s^2 and the slopes encountered spanned -10 degrees to +10 degrees. It was ensured that a significant amount of time was spent in all driving modes and that a continuous array of data would be available for all conditions so that no gaps in the data would appear requiring interpolation. Figure 4 presents some sample data collected by GPS-INS and engine data logger instruments.

The collected data amounted to over 20,000 data points. The data was subsequently analysed and all data that did not meet a strict standard of quality was rejected. 25% of the data was rejected leaving 15,000 points deemed to be representative of the true conditions. The process of data filtering was done carefully and in a manner that would not bias the data. Sources of potential error were identified and the associated data points were filtered out.

3. Result Analysis and Discussion

In this section, an exploratory analysis is conducted on the field collected data to develop fuel consumption model and optimum driving pattern. As was discussed, two main engine data logger and GPS-INS instruments were used to collect fuel consumption, engine load and operational parameters. The collected data need to be synchronized and filtered in data mining process [9]. Conducting this procedure, the quality of raw gathered data was assured and outliers were removed. IBM SPSS Statistics V22 and Microsoft Excel Software were used to conduct an exploratory analysis on the gathered data.

The relation between engine load and instantaneous fuel consumption (IFC) was developed through conducting ordinary least square regression analysis on the collected data. Among different functions were correlated, it was found that exponential curve has the highest correlation coefficient (R^2) value for all experimented vehicles. Then, by combining the collected data from all vehicles, the Equation (1) was developed as the relation between IFC and engine load. The R^2 of developed model is 0.94 which shows there is high correlation between engine load and IFC, and developed model has high level of accuracy. Figure 5 shows the result of ordinary least square regression analysis conducted in Microsoft Excel Software.

$$IFC = 0.0126e^{0.036EL} \quad (1)$$

Where:
IFC: Instantaneous fuel consumption (l/kW.sec)
EL: Engine load of vehicle in (%)

As Figure 6 shows, there is exponential relation between IFC and engine load. This means that as engine load increases, the rate of IFC change per 1% variation of engine load increases. Based on developed curve, for decreasing IFC, we need to drive at the lowest engine load rate. It also shows the variation in engine load value must be minimized.

The effect of operational parameters needs to be considered in the next step. Barati and Shen developed multivariate linear relation between four acceleration, speed, road slope and payload parameters and engine

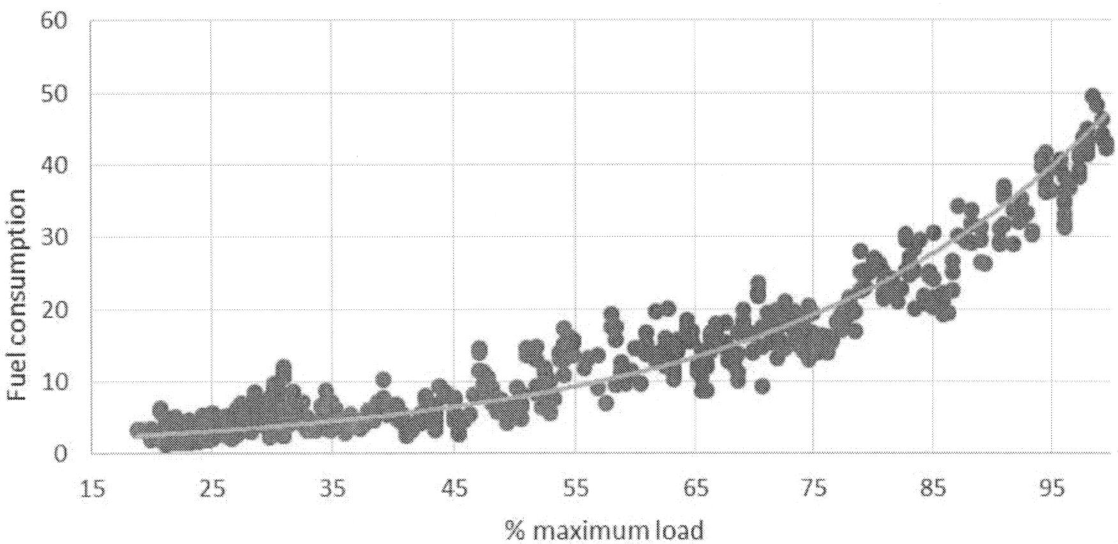

Figure 5. Ordinary least square regression analysis on IFC and engine load relation

load [8]. Actually, payload is a separate factor influencing the effect of the other three investigated parameters on engine load. So, this parameter does not have coefficient in the presented model, but its effect on engine load is considered in the coefficients of the other parameters. Also, the engine load of vehicles in idling mode needs to be considered in model which is around 18% for on-road vehicles. The developed engine load model is presented in Equation (2). Also, Table 2 presents the coefficients of acceleration, slope of road and speed parameters in the developed model. Load factor (LF) represents the payload parameter. The coefficients of this parameter show the current payload as a percentage of maximum allowable payload of vehicle.

$$EL = (C_{AC}*AC) + (C_{SL}*SL) + (C_{SP}*SP) + C \quad (2)$$

Where:
EL: Engine load of vehicle (%)
AC: Acceleration of vehicle (m/s^2)
SL: Slope of road (degree)
SP: Speed of vehicle (m/s)
C: Engine load of vehicle in idling mode which is around 18%.
C_{AC}, C_{SL}, and C_{SP}: The coefficients of acceleration, slope and speed respectively (unit less)

In the next step, we aim to develop optimum driving pattern to minimize the fuel consumption of on-road vehicles. As was noted, payload does not have any coefficient in the developed model, but its effect is considered on engine load through the coefficient of other parameters. The optimization process needs to be done for different payloads. For simplifying the analysis, as an example, we develop the optimization process for vehicles without any payload (LF=0). So, the fuel consumption model for vehicles without payload is presented as Equation (3). As can be seen, this is the combination of Equations (1) and (2).

$$IFC = 0.0126*e^{0.036(13.9AC + 1.2SP + 2.1SL + 18)} \quad (3)$$

Optimum driving pattern is developed by minimizing fuel consumption per moved distance for on-road vehicles. Based on Equation (3), three acceleration, speed and slope parameters affect fuel consumption at the same time. Equation (3) covers negative values of acceleration and slope of road as well which means the developed fuel consumption model is valid for deceleration and driving in downhills. In this step, we first focus to determine the effect of speed on fuel consumption to find optimum speed for driving with zero acceleration and payload in levelled route. For conducting optimization process, the value of three main parameters of IFC, instantaneous speed and moves distance is required for

Table 2.The coefficients of parameters in engine load model

Model	Coefficients			
LF	0	0.33	0.67	1
C_{AC}	13.9	16.10	19.15	24.30
C_{SL}	2.10	2.65	3.35	4.25
C_{SP}	1.20	1.40	1.60	1.90

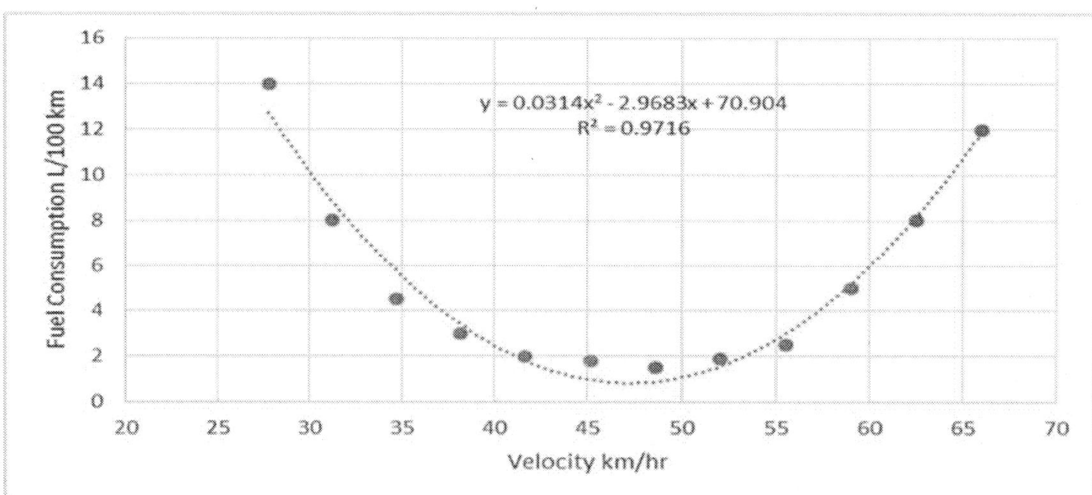

Figure 6. Relation between speed and fuel consumption

each second. IFC is measured by Equation (3), and speed and distance parameters are measured by GPS-INS instrument. As Figure 6 shows, the fuel consumption was calculated per 100 moved kilometers based on different fixed speeds. Ordinary least square analysis was conducted to develop this relation. As Equation (4) presents, there is highly correlated parabolic relation ($R^2 = 0.972$) between consumed fuel and speed of vehicle.

$$FC = 0.0314SP^2 - 2.968SP + 70.904 \quad (4)$$

Where:
FC: Fuel consumption (l/100km)
SP: Speed of vehicle (km/hr)

As Figure 6 indicates, for a vehicle with zero acceleration and payload driven in a levelled route, the speed of around 50 km/hr is optimum for having minimum fuel consumption. This figure indicates that driving with the speed of less than 50 km/hr, in spite of decreasing engine load, fuel consumption increases decreases due to having longer trip time. Also, for driving with the speed of higher than 50 km/hr increases the fuel consumption because of high change in the engine load value.

The result analysis on investigating the effect road slope shows this parameter has significant effect on the relation between optimum driving speed and fuel consumption. As Figure 7 indicates, by increasing the slope of road, the optimum speed decreases, but there is sharp increase in the amount of minimum fuel consumption. For example, in a road with 6 degree slope, optimum speed reduces to around 30 km/hr, while the minimum fuel consumption increases to about 6 l/100km.

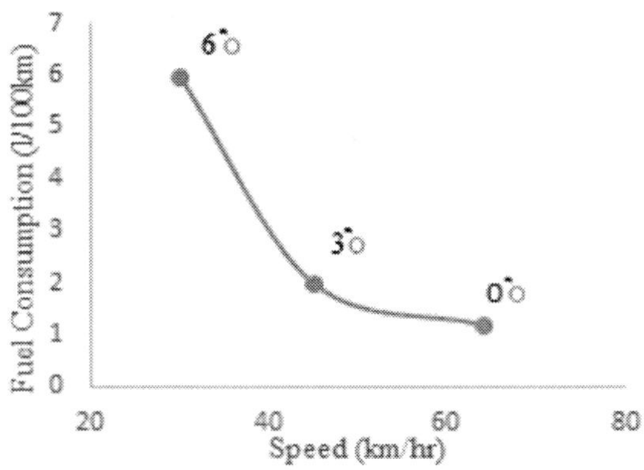

Figure 7. Effect of road slope on the relation between optimum speed and fuel consumption

4. Conclusion

The fuel used by on-road vehicles is the main source of pollution in urban area causing numerous respiratory diseases for inhabitants. Also, the sources of fossil fuels are diminishing because of huge use by vehicles globally. This paper optimized driving pattern to minimize fuel consumption used by on-road vehicles. A comprehensive methodology was first developed to collect required field data, model fuel consumption and optimize driving behaviour. Two main engine data logger and GPS-INS instruments were developed to collect real-world field data of six on-road vehicles through experimentation. After data mining and filtering, and removing outliers, ordinary least square and multivariate linear regression analysis were conducted on the data to investigate the relationship between operational parameters, engine load and fuel consumption. The result analysis indicates the developed relations have high accuracy ($R^2 > 0.90$). In the next step, by using developed fuel consumption model and field data measured by GPS-INS instrument, the driving pattern was optimized by investigating the role of speed and road slope parameters.

To simplify the optimization process, in the first step, acceleration, road slope and payload parameters were assumed to be zero, and the effect of speed was investigated. Based on the archived results, the speed of 50 km/hr is optimum for having minimum fuel consumption per distance. Also, the effect of road slope was investigated on the optimum speed and its correspondent fuel consumption. It was shown that as slope of road increases, the optimum driving speed decreases, while there is sharp increase in the amount of minimum fuel consumption.

For future studies, based on developed methodology, this research focuses on fuel modelling of off-road vehicles including construction and mining equipment. This is the first step for developing schemes and guidelines for minimizing the used fuel of these huge machineries. Also, the relation between fuel consumption and emission rates needs to be investigated for decreasing the emission produced by such equipment.

References

[1] Dargay J. Gately D. and Sommer M. Vehicle Ownership and Income Growth. *The Energy Journal*, 143-170, 2007.

[2] Khan M.A. Khan M.Z. Zaman K. and Naz L. Global estimates of energy consumption and greenhouse gas emissions. *Renewable and Sustainable Energy Reviews*, *29*: 336-344, 2014.

[3] Sperling D. and Gordon D. *Two billion cars: driving toward sustainability*. Oxford University Press, 2014.

[4] Caiazzo F. Ashok A. Waitz I.A. Yim S.H. and Barrett S.R. Air pollution and early deaths in the United States. Part I: Quantifying the impact of major sectors in 2005. *Atmospheric Environment*, *79*: 198-208, 2013.

[5] McCarthy J. J. *Climate change 2001: impacts, adaptation, and vulnerability: contribution of Working Group II to the third assessment report of the Intergovernmental Panel on Climate Change*. Cambridge University Press, 2001.

[6] Ho S.H. Wong Y.D. and Chang V.W.C. What can eco-driving do for sustainable road transport? Perspectives from Singapore City eco-driving programme. *Sustainable Cities and Society*, *14*: 82-88, 2015.

[7] Barkenbus J.N. Eco-driving: An overlooked climate change initiative. *Energy Policy*, *38*(2): 762-769, 2010.

[8] Barati K. and Shen X. Modelling Emissions of Construction and Mining Equipment by Tracking Field Operations. In *Proceeding of 32nd International Symposium on Automation and Robotics in Construction and Mining,* pages 478-486, Oulu, Finland, 2015.

[9] Nichols H. and Day D. *Moving the Earth: Excavation workbook*. McGraw-Hill, USA, 2005.

Reduced-order Models for Supporting Energy Audits of Buildings

M. Benedettelli[a], A. Carbonari[a], B. Naticchia[a] and M. Vaccarini[a]

[a]Università Politecnica delle Marche, DICEA Department, via Brecce Bianche, 60131 Ancona, Italy
E-mail: m.benedettelli@pm.univpm.it, alessandro.carbonari@univpm.it, b.naticchia@univpm.it, m.vaccarini@univpm.it

Abstract –

An energy audit is a standard process to support the decision-making in the area of energy management. Nevertheless, an energy auditor does not succeed as easily and quickly in identifying all the energy streams in a facility in order to decide whether retrofitting the audited object is beneficial. In fact, energy auditing of buildings is usually a time consuming and expensive process, due to efforts required for data collection and modelling of audited objects. Therefore, a decision-making tool should support the modelling phase and this purpose might be achieved by means of reduced-order modelling, after a quick data acquisition, through on-site measurements. As highlighted in literature, reduced-order models, also called grey-box models, showed their reliability to achieve a suitable description of the thermal response of buildings in a short time. Besides, they are more cost-effective and their thermal parameters can be usually extracted in short time. Consequently, deriving a lumped parameter model, by means of measurements collected in a rather short period, could allow the owners and managers of real estates to perform fast and cheap preliminary assessment on the opportunity to implement energy renovation actions. Starting from a test performed in the machine laboratory of our department at the Università Politecnica delle Marche, the investigated empirical procedure for deriving grey-box models will be provided and energy saving opportunities, within an energy retrofit, will be analysed as an example about the application of this procedure.

Keywords –

Energy audit; Reduced-order models; Energy retrofit

1 Overview of the energy management system

Buildings are responsible for almost half of the total primary energy use and the consequent greenhouse emissions worldwide. Even if current energy systems are improving, they do not satisfy yet the acceptable limits for the efficiency [1]. In this regard, the European Commission puts effort in promoting measures for the reduction of energy needs and greenhouse gas emissions, by recommending energy retrofitting projects. One of the targets of the European Directive 2012/27/UE is to retrofit 3% of the existing public stock, while upgrading it to current legislation [2]. Several actions in favour of the Energy Performance Contracting (EPC) market in Europe have been put into practice, e.g. within the Intelligent Energy - Europe (IEE) programme. The Energy service companies (ESCos) have been identified as key players in the implementation of EPC investments [3]. The EPC is a contractual arrangement between the beneficiary and the provider of an energy efficiency improvement, where investments in an energy efficiency project are paid in relation to the achievement of energy savings produced [2]. About this, contracts offered by ESCOs usually concern all the energy services, included energy audits. In such a way, facility owners and managers are supported in improving ageing and inefficient assets, transferring risks and reducing meanwhile energy costs [4]. The methodology of EPC is oriented to ensure the quality of performance. In fact, a strong commitment from the government to the use of EPC has been observed in several countries [5]. However, the success of EPC projects depends the most on the correct estimation of the expected energy savings. For the ESCOs, it is crucial to evaluate the feasibility of the proposed energy conservation measures (ECMs) to achieve an acceptable energy use baseline, according to a contractually agreed level of energy performance. A detailed energy audit is typically carried out at a first stage, in order to identify the energy saving opportunities.

Therefore, a decision support system for energy auditors might be beneficial in several scenarios, as for example:

- in case owners of multiple buildings must give a prority for refurbishment or every requalification;
- in case the energy enhancement of any existing buildings is conditioned upon the possibility of repaying back the initial investment with annual savings from energy reduction;
- in any case it is essential to select the most appropriate actions or strategies to reduce buildings' environmental impact.

Despite the fact that building simulation models are more effective to understand the implications in the energy use policies, they usually entail an expensive and time-consuming implementation process [6].

What makes the things worse, referring to an Ashrae level 3 of energy auditing [7], is that a detailed audit needs a substantial time to set the data and provide the detailed dynamic model of the audited object; besides, a considerable computational effort is necessary to run the large number of simulations, usually required. Furthermore, uncertainty in model inputs often does not guarantee an accurate prediction of the energy performance, particularly for existing buildings, in which a lot of information about intrinsic characteristics remain unknown. In order to develop quick and efficient diagnoses, reduced-order models represented by grey-box models are preliminary assessed in this paper. The reduced modelling can be applied in many fields in construction (i.e. safety, occupancy dynamics and structural damages) because of its reliability and usefulness for simulations, advanced controls, real-time predictions. Recently, it is widely used to improve energy management strategies. Relevant literature provides many examples and case studies about modelling the energy dynamics of buildings. In fact, parameters of the reduced models are extracted in real time; this makes the estimation of the thermal response of a building in its current state feasible, and this information reusable to make predictions about its expected behaviour in further analyses.

Our study started with a fact-finding survey on the reduced-order models and the use of identification techniques from literature. Then, we investigated the robustness of such techniques in a real case, with the purpose of defining experimentally an efficient decision support tool for the energy retrofitting. Therefore, we studied the possibility of identifying the unknown properties of a mathematical model based on partial observations of the heat dynamics of the building, that is, in our case, a small data collection of temperature measurements. To sum up, the modelling phase is discussed in Section 2. The preliminary test is described in Section 3, including the experimental set up and data generation. The results of the study are provided in Section 4. Conclusions are given in Section 5.

2 The modelling phase

Buildings modelling for advanced control, in particular to derive a total model for the heat dynamics of buildings, has been the purpose of several research groups in the last years, also related to the increasing attention for the energy savings. Therefore, several different approaches have been described and many different methods for the dynamic analysis of energy use in buildings have been developed. In literature, the problem of deriving a suitable reduced model is solved referring to two different approaches. The first one is a purely physical approach, in which it is possible to investigate the physical behaviour of a building. However, the inaccuracy of control strategies relying on these detailed models represents a clear weakness, since the real building parameters are often unknown in existing buildings [8]. Contrarily, the second approach concerns the statistical black-box models, in which the parameters result from a statistical relation between the input and the output of the system, because they are generally disconnected from physical systems. The later introduction of the grey-box models represents a "half-way" modelling to overcome the lacks of both modelling methods. In this work, the word "grey" is meant as defined by Reynders and Madsen [8-9], namely the combination of a prior knowledge of system dynamics, to define the model structure, and statistic methods, to estimate the unknown parameters.

In order to turn energy audits into fast and cheap processes, no effort must be employed for extensive simulations, thus our proposal is to train the model from a limited set of experimental data. Consequently, grey-box modelling represents an interesting solution because of the direct link between the physical laws of building dynamics and a strong framework of continuous stochastic differential equations formulated in a state space form [10]. The state-space representation provides a mathematical description to analyse dynamical systems with multiple inputs and outputs. A dynamical system is characterized by its state variables. The state variables are stacked in a time varying state vector, referred to as the system state.

Regarding grey-box models, the state-space form derives from the physical laws, which are formulated by first-order stochastic differential equations. Therefore, the model structure is a linear time-invariant model, represented by the following equations:

$$dX(t) = A(\theta)X(t) + B(\theta)U(t) + \sigma(\theta)d\omega \quad (1)$$
$$Y(t) = C(\theta)X(t) + D(\theta)U(t) + \varepsilon \quad (2)$$

The states of the system are represented by the vector $X(t)$ which correspond to the temperatures of the relevant building components, in the case of modelling the thermal response of buildings. $U(t)$ is a vector containing the measured inputs of the system, which can be either inputs or controlled outputs (heating and ventilation system, solar gains, outdoor temperature, etc.). The measured output $Y(t)$ is function of the states and the inputs. A, B, C, D are nonlinear functions, depending on the parameters' vector θ. The matrix A characterizes the dynamic behaviour of the system and the matrix B concerns the influence of input signals entering the system. ω is a random function of time, assumed as a standard Wiener process with independent increments; the mutual independent ε is a white noise process representing the measurement error.

To identify the unknown parameters of the vector θ, the ML estimation method is adopted, because it was validated by Madsen [9-11] and successfully applied in several applications. According to statistics, the MLE is the hypothetical population value that maximizes the likelihood of the observed sample, that is, the value most likely to have generated the sample actually observed. The maximum likelihood method thus provides a means of estimating a set of parameters characterizing a distribution of an observed phenomenon, if we know the form of this distribution.

The structures for modelling the heat dynamics are derived from the analogy with electric circuits. According to the model order, the thermal mass of the building is lumped to a discrete number of capacitances. In the thermal network, the capacitor C represents the active thermal capacity of the building zone for storing heat, while the thermal resistance R is linked to material properties and affects the heat flow across building layers at different temperatures. However, a given parameter is not identified with the same physical correspondence in each model. The models used in this work (Figure 1) are referred to the 1st, 2nd and 3rd-order models. To simplify the problem, in this early stage, the case study does not entail the internal and solar gains.

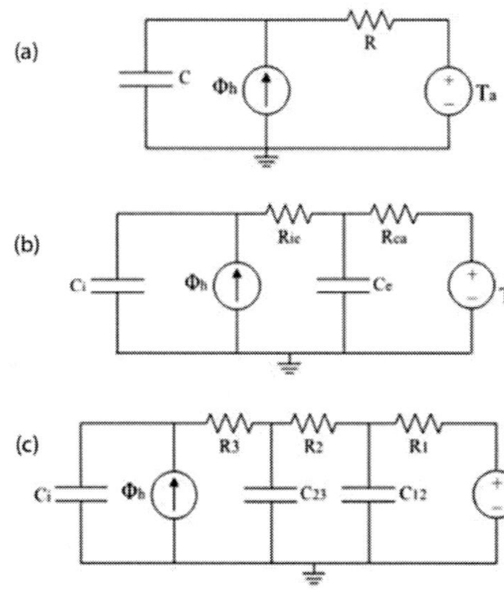

Figure 1. RC-networks of the grey-box models: 1st, 2nd, 3rd order models, from top to bottom

The stochastic differential equations, related to the state space form, are deduced from principles of energy balance at nodes and of heat transfer. For the first order model (Figure 1a), the only state is described by:

$$dT = \frac{1}{RC}(T_a - T)dt + \frac{1}{C}\Phi_h dt + \sigma d\omega \quad (3)$$

The two states grey-box model TiTe (Figure 1b), is defined by the following equations:

$$dT_i = \frac{1}{R_{ie}C_i}(T_e - T_i)dt + \frac{1}{C_i}\Phi_h dt + \sigma_i d\omega_i \quad (4)$$

$$dT_e = \frac{1}{R_{ie}C_e}(T_i - T_e)dt + \frac{1}{R_{ea}C_e}(T_a - T_e)dt + {} + \sigma_e d\omega_e \quad (5)$$

Consequently, the third order model (Figure 1c) is represented by these ordinary differential equations:

$$dT_i = \frac{1}{R_3C_i}(T_{23} - T_i)dt + \frac{1}{C_i}\Phi_h dt + \sigma_1 d\omega_1 \quad (6)$$

$$dT_{23} = \frac{1}{R_3C_{23}}(T_i - T_{23})dt + {} + \frac{1}{R_2C_{23}}(T_{12} - T_{23})dt + \sigma_2 d\omega_2 \quad (7)$$

$$dT_{12} = \frac{1}{R_2C_{12}}(T_{23} - T_{12})dt + {} + \frac{1}{R_1C_{12}}(T_a - T_{12})dt + \sigma_3 d\omega_3 \quad (8)$$

Whereas Φ_h [kW] and T_a [°C] constitute the input measured values of the power provided by the heater and the ambient temperature, the resistance/capacitance parameters, R [°C/kW] and C [kWh/°C], are estimated as well as the stochastic terms [°C] and the initial values of the other temperatures [°C]. The time is expressed in hours. For every system of differential equations, the output equation is represented by the discrete time measurement equation, which takes into consideration the data-driven:

$$Y_k = T_{ik} + e_k \quad (9)$$

where k is the point in time t_k of a measurement,
where Y_k [°C] is the measured interior temperature,
where T_{ik} [°C] is the state at the time t_k,
where e_k [°C] represents the measurement error, which is assumed to be a Gaussian white noise process with variance σ^2 [9]. In fact, a real context provides a number of disturbances to be taken into account in the modelling of the system, to not obtain largely distorted results.

3 The preliminary test

The main purpose of our experimental approach is the investigation of grey-box modelling in simulating real and complex cases, in order to implement it in a decision support tool for energy audits of existing buildings. To this aim, a first controlled situation was simulated in a machine laboratory at the Università Politecnica delle Marche and it will be described in this section.

3.1 Experimental set up

The entire laboratory of our department was used to create a first controlled situation. The changing room of our laboratory was chosen as the test-room, while the entire laboratory was one of the boundary conditions of temperature. The floor of the changing room, whose area is about 13 m^2, was covered with polystyrene sheets and bricks to insulate it from the ground. Besides, in this testing area an electric heater of nominal power 400 W and real measured power 377,6 W, generated the heat flux. This heat input was controlled by means of a timer, programmed with the sequence of PBRS signal, according to Madsen's procedure [11].

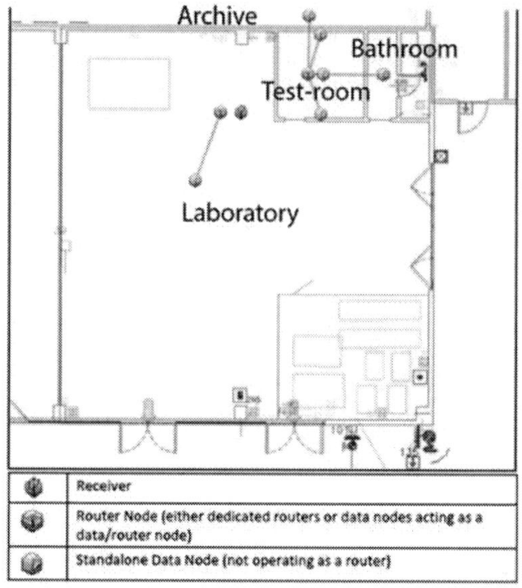

Figure 2. Map extracted from HoboNode Manager, representing the laboratory with sensors deployment

3.2 Data generation

A Hobo Zw Series Wireless Network was installed in the test room to measure temperatures of the various environments. The map in Figure 2, extracted from the hobo software interface (HoboNode Manager), shows the deployments of the data node, the routers and the receiver. The receiver was connected to a computer, which was used to set up the network and to offload the data from the receiver at the site. The six available sensors were installed in the area: three in the test-room, two on the internal walls and one hanging from the ceiling; one close to in the bathroom; another one in the laboratory and the last sensor was hung in the archive. This deployment of the sensors was dictated by the requirements posed by the adoption of numerical techniques (see section 2). In addition, a thermo-anemometer, VT200 Kimo, was placed in the changing room with two probes to measure the temperature of the internal air and of the floor. Finally, a Kestrel climate station was positioned on the roof of the changing room. These measurements were useful to test the network acquisition. Only the Hobo monitoring network provided the data for the identification process. The applied data covered a period of about 18 days, from 2nd to 20th July 2015. The number of total observations was equal to 27131, corresponding to an acquisition per minute. Prior to estimating the model parameters, the data were resampled at a sampling time of 15 minutes. This sampling time was chosen, according to Shannon's theorem, based on the smallest period of the electric heater that is 30 minutes. Temperature data from the sensors on the walls of the changing room were not considered in computing because too much affected by the convective motion of the air inside. At the same time, also temperature measurements of the bathroom were later excluded because they were too similar to the ones of the laboratory. Data of the first day of the test, July 2nd, were cancelled because corrupted. Some photos of the test set up are shown in Figures 3 to 6.

Figure 3. The testing area of the changing room

Figure 4. The sensor positioning in the laboratory

Figure 5. Overview of the laboratory

Figure 6. The archive bordering the laboratory

4 Results

In order to evaluate the robustness and reliability of the reduced-order modelling, in the data elaboration both the suitable structure of the lumped model and the influence of data period were investigated.

As found in the referenced literature, the simplest model, that is 1R1C network (Figure 1a), was analysed in the data period 3-10 July. As already known, the first reduced-order model was quite unrealistic, in fact it revealed insufficient to describe the heat dynamics of our test room. In fact, evaluating results of parameters estimation in CTSM-R, errors were meaningful and model validation [12] was not satisfied. After that, the 2R2C network (Figure 1b) was considered in the same data period of the first model, because measured data were more stable. The estimated results of this second-order model better fitted with measured plots. Three second-order models were tested, differentiated as for the boundary condition T_a in Equation 5. In the first case, the ambient temperature T_a was represented only by the laboratory temperature T_l. In the second case, the weighted average of the laboratory temperature (T_l) and the archive one (T_a), weighted on the respective surface, calculated it. In the last case, a coefficient (K_a) was introduced to weight the archive temperature as compared to the one of laboratory. This coefficient was estimated by CTSM-R improving estimation results and the Equation 5 was rewritten in this way:

$$dT_e = \frac{1}{R_{ie}C_e}(T_i - T_e)dt + \\ + \frac{1}{R_{ea}C_e}(K_a T_a + (1 - K_a)T_l - T_e)dt + \sigma_e d\omega_e \quad (10)$$

Then, to fix the best data period for the parameters estimation, a 3C3R model (Figure 1c) was considered; it was a natural extension of the second-order model (including the K_a parameter), for taking into account the more variable dynamics of the last test.

Four intervals were implemented for the ML estimation: 3-10 July, 8-15 July, 3-15 July, 3-20 July. The root of the mean squared error for the 1-step prediction and simulation of the identified model were investigated for each data set. The best data set was chosen depending both on the long period of acquisition and on the smallest RMSE value in simulation, that was equal to 0.1574. It corresponded to the period 3-15 July.

4.1 The reduced model validation

Finally, the last portion of data was used to validate the model of Equations 6-7-11. The validation was carried out by performing a simulation of the test-room indoor temperature from 16th until 20th July, using the 3C3R model with the estimated values in CTSM-R.

$$dT_{12} = \frac{1}{R_2 C_{12}}(T_{23} - T_{12})dt + \qquad (11)$$

$$+\frac{1}{R_1 C_{12}}(K_a T_a + (1 - K_a)T_i - T_{12})dt + \sigma_3 d\omega_3$$

The validation phase was alike performed for the second and the first reduced-order models seen before, using the best data set for both the parameter estimates (always including the K_a parameter). The first order model resulted as non-representative of the heat dynamics of the test room, while the third order model, which fitted better, presented some parameters with a Pr-value over 0.05 that indicated an over-parametrization of the model. Thus, the second order model was considered the best choice in terms of parameters validation in R and in terms of RMSE value in simulation, which was equal to 0.0745, as shown in Figures 7-8.

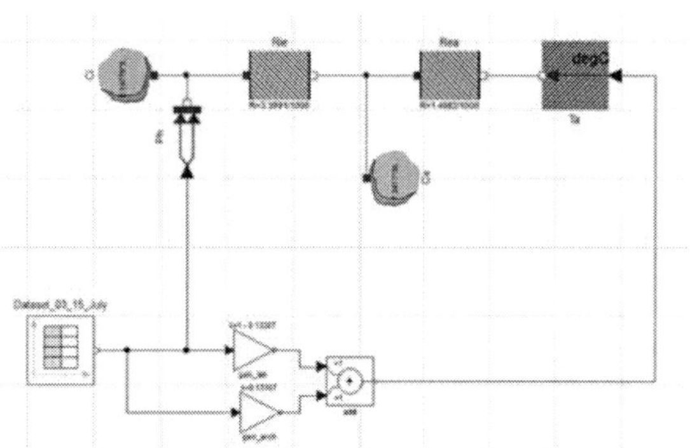

Figure 7. The 2nd-order model in simulation environment

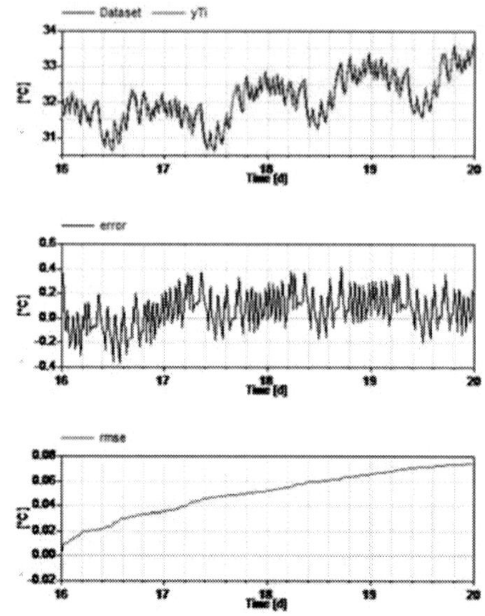

Figure 8. Results of the 2nd-order model for the validation phase

It was also possible to verify that replacing the K_a in the Equation 10 with its estimated value, that was 0.13307, the results of the estimation (Table 1) were the best for model validation, without any Pr-value over 0.05 (Table 2), that is, the model is not over-parametrized. The Pr-value represents the probability that the particular initial state or parameter is insignificant, i.e. equal to zero [12]. The variance of the system noise is represented in CTSM-R by the exponential function of the estimated parameters p11 and p22; while the variance of the measurement error in the output equation (see Equation 9) is the exponential function of the estimated parameter e11 (Tables 1-2).

Table 1. Parameter estimation in CTSM-R for the 2nd reduced-order model with fixed K_a

	Estimate	Std. error	t value
Ti0	2.9692e+01	4.9569e-02	5.9901e+02
Te0	2.9224e+01	5.7916e-02	5.0458e+02
Ce	7.4006e+00	9.8259e-01	7.5317e+00
Ci	1.7640e-01	1.6073e-03	1.0974e+02
e11	5.6405e+00	4.8150e-02	1.1714e+02
p11	1.0883e+01	1.6143e-01	6.7417e+01
p22	2.9891e+00	8.6550e-02	3.4536e+01
Rea	1.4677e+00	1.8143e-01	8.0896e+00
Rie	3.4447e+00	4.8751e-02	7.0658e+01

Table 2. Validation of the estimation results in CTSM-R for the second reduced-order model with fixed K_a

	Pr(>|t|)	dF/dPar	dPen/dPar
Ti0	0.0000e+00	-5.2349e-05	0

Te0	0.0000e+00	-4.8034e-06	0
Ce	9.7255e-14	-1.2173e-05	0
Ci	0.0000e+00	5.2610e-05	0
e11	0.0000e+00	-4.9432e-06	0
p11	0.0000e+00	8.7844e-05	0
p22	0.0000e+00	2.1533e-05	0
Rea	1.3323e-15	-1.3020e-05	0
Rie	0.0000e+00	-4.9613e-05	0

4.2 A validation of the reduced model for the energy audit

Grey-box modelling was applied to set reduced models and estimate thermodynamic properties of a building, starting from a data acquisition campaign. In the end, in order to provide the auditor a further support in decision-making, the influence of data acquisition period was investigated. Thus, identified the reduced model structure, a parametric study was carried out to evaluate the optimal season for a data acquisition campaign. In this validation for the energy audit, the annual energy consumptions of the heating system were analysed. Results were provided in terms of NMBE and CV-RMSE values, calculated on an hourly basis relatively to the reduced model and to the baseline. Therefore, they were compared and limits imposed by the ASHRAE guidelines were verified [13]. The followed procedure is summarized below.

Using our case study (see Section 3), the input data implemented in the identified 2nd-order model (see Paragraph 4.1) were derived from an annual simulation of the detailed building model (Figure 9), which was carried out by means of the Dymola software.

Then, the parameters estimation was run in CTSM-R, varying the data input period, that is a month of the year. In this case, the heat input was a PRBS signal. At this level, the reduced model was defined in its physical properties, as seen before. Subsequently, the dynamics of the model was simulated in Dymola environment, according to the different simulation periods for the comparison. In the simulations, the heat input was represented by a thermostatic control, in which the set-point temperature was fixed at 18°C (Figure 10).

Before comparing the energy consumptions, a comparison of temperatures between the reduced model and the baseline was carried out. ASHRAE limits of NMBE and CV-RMSE [13] were verified; moreover, it was observed that temperatures simulated in the model had the same trend of the baseline ones, even with a thermostatic control operating. Ultimately, the comparison of the energy consumptions in terms of NMBE and CV-RMSE showed that the optimal season for data acquisition campaigns must match the one of the heating system operating (Figure 11).

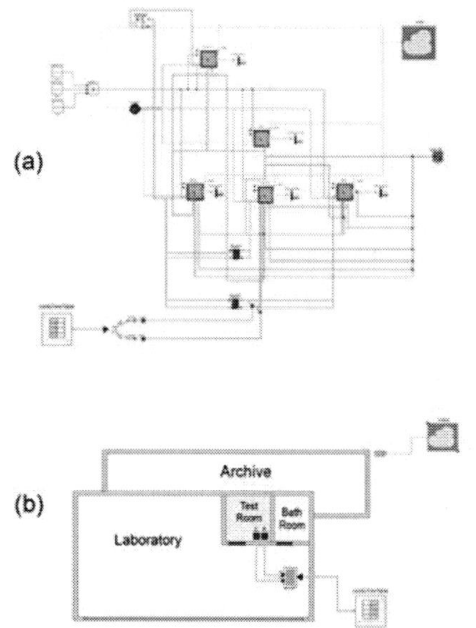

Figure 9. Detailed model of the building in Dymola environment (a) and the top layer of Dymola model (b)

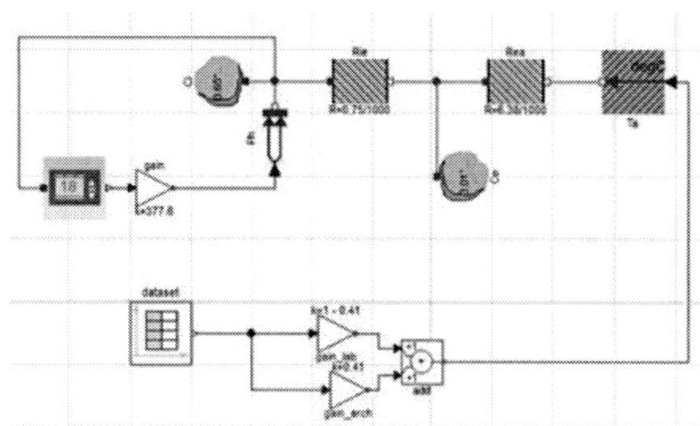

Figure 10. Identified reduced model of the building with thermostatic control

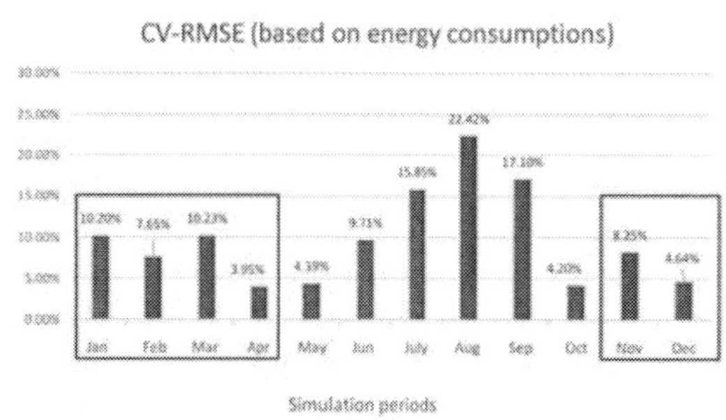

Figure 11. Comparison between CV-RMSE values in terms of energy consumptions for the different simulation periods

5 Conclusions

An empirical approach based on reduced-model strategy is provided to support energy audits of buildings. For this purpose, a preliminary test of a controlled real situation is performed, starting from a very little set of data. The influence of the data period is investigated and the results show the importance of measurements containing adequate dynamics, for the robustness of the identified reduced models.

The best predictions are obtained for the 2nd-order model; nevertheless, the reliability of the procedure is balanced on the comparison with real conditions. Subsequently, results of the parametric study prove that annual predictions on energy consumptions require a model structure of limited complexity; besides, the procedure shows its reliability in real operating conditions.

Further research should be able to generalize, as much as possible, this model identification approach. Future work should aim at testing other real and more complex cases, finding out saving opportunities of energy retrofits, in order to provide an efficient decision support tool for the energy audits of buildings.

References

[1] Krarti M. *Energy audit of building systems: an engineering approach*, Second edition. CRC press, Taylor & Francis Group, 6000 Broken Sound Parkway NW Suite 300, 2011.

[2] Directive 2012/27/EU of the European Parliament and of the Council of 25 October 2012 on energy efficiency, amending Directives 2009/125/EC and 2010/30/EU and repealing Directives 2004/8/EC and 2006/32/EC.

[3] Carbonari A. et al. Managing energy retrofit of acute hospitals and community clinics through EPC contracting: the MARTE project. *Energy Procedia*, 78:1033–1038, 2015.

[4] Sorrell S. The economics of energy service contracts. *Energy Policy*, 35:507–521, 2007.

[5] Marino A. et al. A snapshot of the European energy service market in 2010 and policy recommendations to foster a further market development. *Energy Policy* 39 (10):6190–6198, 2011.

[6] Lee P. et al. Risks in Energy Performance Contracting (EPC) projects. *Energy and Buildings* 92:116–127, 2015.

[7] Procedures for Commercial Building Energy Audits, 2nd Edition, ASHRAE 2011.

[8] Reynders G. et al. Quality of grey-box models and identified parameters as function of the accuracy of input and observation signals. *Energy and buildings*, 82:263-274, 2014.

[9] Bacher P. and Madsen H. Identifying suitable models for the heat dynamics of buildings, *Energy and Buildings*, 43:1511-1522, 2011.

[10] Jiménez M.J. and Madsen H. Models for describing the thermal characteristics of building components. *Building and Environment*, 43:152-162, 2008.

[11] Madsen H. *Short time determination of the heat dynamics of buildings*, Report 243, Technical University of Denmark, 1993.

[12] Juhl R. et al. CTSM-R User Guide. On-line: http://ctsm.info/pdfs/userguide.pdf, Accessed: 14/03/2016.

[13] ASHRAE guideline 14-2002, Measurement of energy and demand savings. Approved by the ASHRAE Standards Committee 22/06/2002, and by the ASHRAE Board of Directors 27/06/2002.

Human state estimation system implemented in an Office Deployable Getaway based on multiple bio information

Sayaka Nakaso [1], Jörg Güttler [2], Akira Mita [1], Thomas Bock [2]

1 Department of System Design Engineering, Keio University, Yokohama, Japan
2 Chair for Building Realization and Robotics, Technical University Munich, Germany
E-mail: sayakanakaso@keio.jp, joerg.guettler@br2.ar.tum.de, mita@sd.keio.ac.jp, thomas.bock@br2.ar.tum.de

Abstract –

This article is proposing a human state estimation system on the basis of multiple bio information implemented into a movable cover chair called "Office Deployable Getaway". The proposed system consists of 2 sub-systems; the first sub-system analyzes fatigue based on eye movement including blinking rate and eye closure time while the second sub-system analyzes breath frequency. These indicators are detected by applying image processing to IR images, using the Microsoft Kinect sensor. Hardware tests were conducted to determine the proper Microsoft Kinect (Kinect V1 or Kinect V2) and its position inside the chair to efficiently detect eye movement and respiration. The experimental results showed that these bio information could be measured appropriately under the condition that the Kinect V2 was set at the eye level opposite from the user. Based on the result, the implementation of the proposed system into the chair was carried out, installing the Kinect and a personal computer at appropriate positions. Finally, this system was tested in a real environment with test person sitting inside the chair; after the fatigue is analyzed over eye movement, breath frequency is analyzed. The results of these analyzations are indicated in the computer screen. This performance test showed that this system could be feasible for the realistic use. For future development, the authors will focus on audio system that triggers proper music based on estimated human state, which will be implemented into this chair.

Keywords –

Human-Machine-Interaction, Breath Detection, Fatigue Analyzation, Human Mood, Eye Recognition, Ambient Assisted Living

1 Introduction

It is considered that fatigue contributes to 20% of reported accidents across all sectors [1]. As the construction industry involves hazardous operations of machine tools, accumulation of fatigue might increase the risk of fatal accidents [2, 3]. Thus, getting enough rest is vital to staff safety especially in construction site. For this purpose, LIQUIFER Systems Group (in Vienna) has developed "Deployable Getaway for the Office" [4], which is a chair with a mobile, ergonomic, and transformable 'cocoon-like' structure that office staffs may utilize during the workday. It is expected that this chair could also provide construction staffs with private rejuvenation space. However, they sometimes fail to take enough rest due to increase in construction orders. Accordingly, this article is proposing a human state estimation system on the basis of multiple bio information implemented into "Deployable Getaway for the Office". The captured bio information will be used to trigger music, which will influence the mood of the person to improve his safety.

There has been a growing interest in human state estimation systems. Earlier studies proposed fatigue detection system for drivers using eye movement (eye closure time and blinking rate) [5 – 7]. Additionally, previous studies suggested that breath frequency or breathing capacity extracted from breath wave could be used as index to detect stress level [8, 9]. Therefore, the authors proposed the human state estimation system that consists of two sub-systems; the first sub-system analyzes fatigue based on eye movement including eye closure time and blinking rate, while the second sub-system analyzes breath frequency.

Regarding the method to capture bio information, methods using wearable sensors have been developed [10 – 12]. These methods provide high accuracy although their usage might be troublesome as the users must always remember to wear them. In order to solve this problem, image processing methods have been drawing intense research interest because of their contactless sensing [13]. However, the image processing is problematic in that it is easily affected by lighting condition. Given the fact that it is dark inside the chair, the image processing is applied to IR images instead of RGB images, obtained by the Microsoft Kinect sensor.

For the implementation, hardware tests were conducted where the proper Microsoft Kinect sensor (Kinect V1 or Kinect V2) and its proper position inside the chair were examined. In accordance with the results, the proposed system was implemented into Deployable Getaway Chair installing Kinect V2 and mini PC. Finally, the performance of the proposed system was examined with a test person sitting inside the chair to show its feasibility.

2 Proposed System

In order to run human state estimation system, the authors conducted hardware tests, where the proper device and its position inside chair were examined.

2.1 Hardware Tests

A device which could obtain images in low light condition is required for the human state estimation system. Accordingly, the author considered introduction of Microsoft sensor Kinect (Kinect V1 or Kinect V2) with IR image sensor that could function under the dark condition. In addition, they are easy in installation. The Kinect V2 could provide adequate performance for the implementation, though its cost is slightly higher compared to Kinect V1. Thus, the authors examined whether Kinect V1 could function through some tests. Kinect V1 mounts image sensor, distance sensor and microphone. Table 1 shows specifications of Kinect V1.

Experiments were conducted where a picture of subject was taken with Kinect V1. The obtained image included heavy noise which could cause difficulty in capturing bio-information. Accordingly, 2 kinds of image processing methods were applied to reduce the noise; "averaging method" and "erosion and dilation method". However, it was shown that these image processing methods were not sufficient to reduce the noise of image obtained by Kinect V1. The result showed that Kinect V2 should be introduced as a sensor for this proposed system.

For the installation of the device, experiments were carried out to find the best place to mount Kinect inside the chair. The authors attached importance to measure eye and chest parts with the same camera settings. In order to investigate the best angle and distance conditions for capturing bio information, experiments were conducted with several positions of the Kinect V2. Experimental results suggested that both eye movement and breath could be detected when Kinect V2 was set eye level within 0.8 m from the subject. According to the results, PC and Kinect V2 were installed into Deployable Getaway Chair as is shown in Figure 1.

Table 1. Specification of Kinect V1

RGB image resolution	640 × 480 pixel
Depth image resolution	640 × 480 pixel
Measurable distance	0.8 ~ 10 m
Infrared viewing angle	Horizontal: 57° Vertical: 43°
Frame rate	30 fps

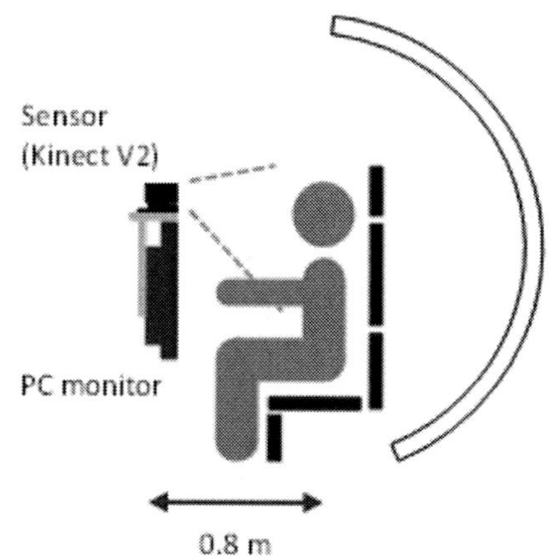

Figure 1. Position of Kinect for implementation (Created based upon data from [14])

2.2 Installation of Office Deployable Getaway

Depending on the obtained results, the devices such as PC and Kinect V2 were installed into the chair.

2.2.1 Deployable Getaway Chair

LIQUIFER SYSTEMS GROUP (in Vienna) has been taking on the "Deployable getaway" project [15], which promotes design of the space with space-efficient storage and a flexible set-up for more privacy; "Deployable Getaway for the international space station" and "Deployable getaway for the office".

"Deployable Getaway for the international space station" refers to the relaxation space for astronauts on long space trips. The design for relaxation space for astronauts who travel under space constraints might be applied to these for handicapped people on wheel chairs or elderly people who move in relatively small area inside their home.

For the implementation, the authors introduced

"Deployable getaway for the office". As illustrated in Figure 2 and Figure 3, this chair has transformable 'cocoon-like' structure. As can be seen from Figure 4, closing the cover provides dark space where employees may utilize in busy office environments for the purpose of rejuvenation.

Figure 2. Deployable Getaway for the Office; cover is open (Reprinted from [14])

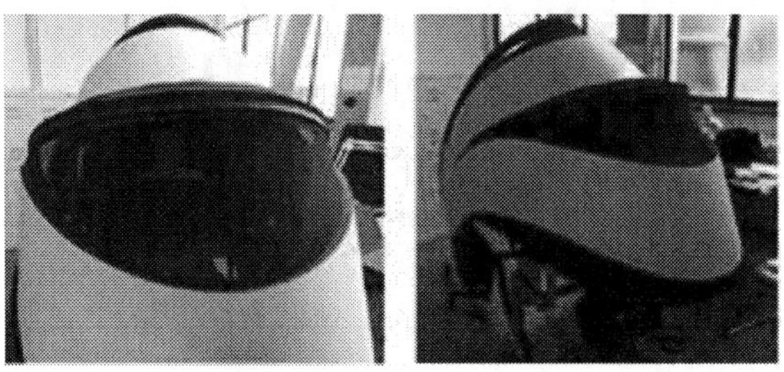

Figure 3. Deployable Getaway for the Office; cover is close (Reprinted from [14])

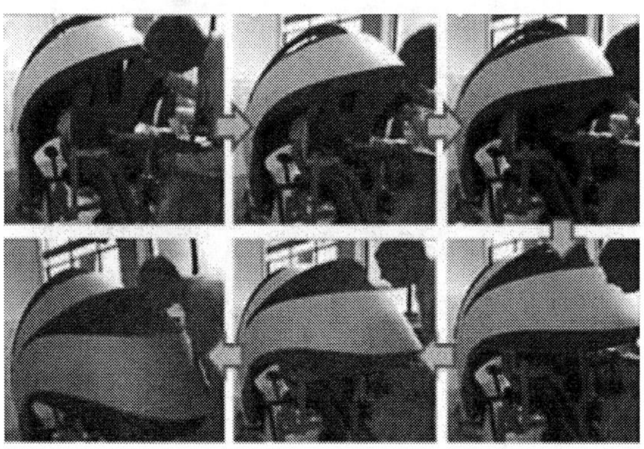

Figure 4. " Cocoon-like" structure (Reprinted from [14])

2.2.2 Installation

Based on the results obtained by hardware tests, the devices such as Kinect V2 and mini PC needed for the implementation were installed into the appropriate position inside Deployable Getaway Chair. The specifications of the devices are mentioned in Table 2 and Table 3. Table 2 shows specifications of mini PC. Kinect V2 mounts image sensor, distance sensor and microphone. The specifications of Kinect V2 are listed in Table 3. Figure 5 shows Deployable Getaway for the Office equipped with Kinect V2 and mini PC.

Table 2. Specification of mini PC

Windows edition	Windows 8.1Enterprise
Processor	Intel® Core(TM) i7-4770T CPU@2.50GHz
Installed memory RAM	8.00 GB
System type	64-bit Operating System, x64-based processor

Table 3. Specification of Kinect V2

RGB image resolution	1920 × 1080 pixel
Depth image resolution	512 × 424 pixel
Measurable distance	0.5 ~ 4.5 m
Infrared viewing angle	Horizontal: 70° Vertical: 60°
Frame rate	30 fps

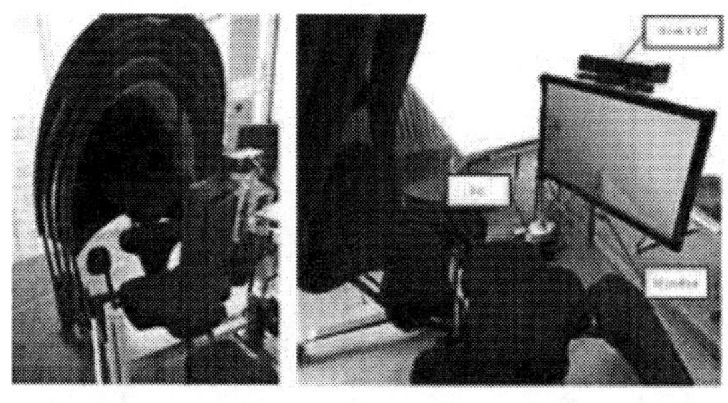

Figure 5. Chair with Kinect and mini PC (Reprinted from [14])

3 Software Architecture

Human state estimation system consists of 2 sub-systems; fatigue detection sub-system and breath detection sub-system. These sub-systems are as follows.

3.1 Fatigue detection sub-system

An overview of fatigue detection sub-system is shown in Figure 6. First, Kinect takes images of subject and stores the eye area. If the eye area is stored, image preservation starts. After enough images are stored, the eye is extracted from the images according to the stored eye area. Then, the extracted eye is classified as open eye or close eye to measure eye closure time and blinking rate. Based on the obtained data, fatigue is detected and the results is shown on the PC display.

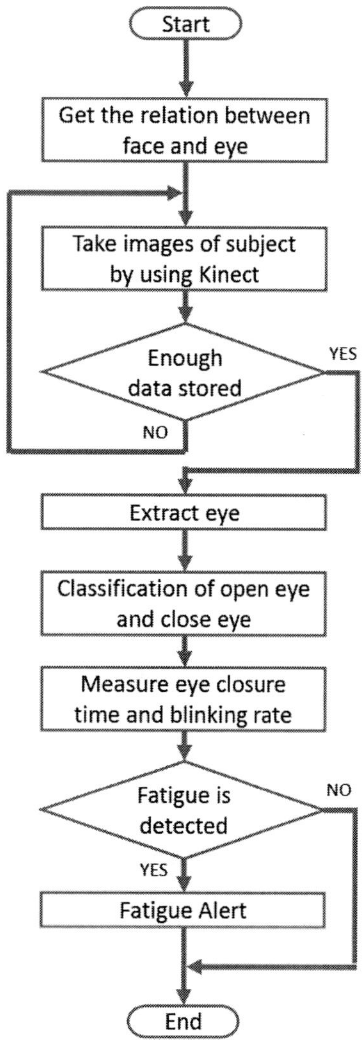

Figure 6. Flowchart of fatigue detection system

3.1.1 Eye segmentation

Eye segmentation was done by object detection based on Haar-like features using Open CV library [16 – 18]. Loaded xml file as object detector (face detector or eye detector) detects the desired object included in image. Accordingly, frontal face detector detects the face. Then, eye detector detects the eye from extracted face area. However, this eye detection algorithm tends to fail to detect eye, so that the relationship between eye and face was introduced for robust eye tracking algorithm; when the eye is detected for the first time, the relationship of face area and eye area are stored as r_x and r_y given in the equations (1) and (2) [14]. The variables used in these equations are illustrated in Figure 7. Numbers included in variables refer to frame number.

$$r_x = \frac{eye.x(1)}{face.width(1)} \tag{1}$$

$$r_y = \frac{eye.y(1)}{face.height(1)} \tag{2}$$

Assuming that subjects don't move forward or backward, *eye.height* and *eye.width* are defined based on preparatory experiments.

Drawing on the relationship, eye area is estimated when face area is detected as shown in equations (3) and (4) [14]. Here *n* refers to frame number.

$$eye.x(n) = r_x \cdot face.width(n) \tag{3}$$

$$eye.y(n) = r_y \cdot face.height(n) \tag{4}$$

This method could achieve accurate and stable eye area estimation because face detection could achieve higher accuracy and better robustness compared to eye detection based on Haar algorithm.

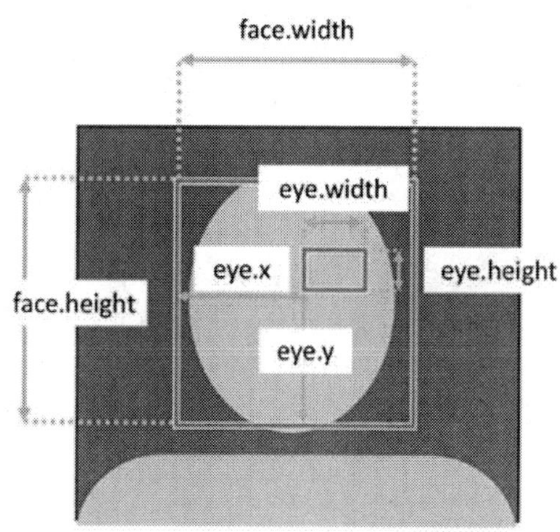

Figure 7. Eye segmentation (Created based upon data from [14])

3.1.2 Eye classification

In order to classify open eye and close eye, vertical projection curve was focused. Figure 8 shows the vertical projection curve of open eye image and close eye image. From this figure, it can be observed that the vertical projection curve of close eye image is flat compared to that of open eye image. Therefore, this difference in vertical projection could serve to classify open eye and close eye.

The valuable for the classification was calculated as *Change Factor* (represented as *CF*). *CF(n)* that refers to *CF* of nth frame was calculated as follows. First, vertical projection curve of 1st frame $V(1)$ and that of nth frame $V(n)$ were normalized as shown in equations (5) and (6) [14]. Here $f_{1dim}(1)$ represents normalized $V(1)$ while $f_{1dim}(n)$ represents normalized $V(n)$.

$$f_{1dim}(1) = \frac{V(1)}{max(V(1))} \cdot 100 \quad (5)$$

$$f_{1dim}(n) = \frac{V(n)}{max(V(n))} \cdot 100 \quad (6)$$

$CF(n)$ is calculated as the max value of difference between $f_{1dim}(1)$ and $f_{1dim}(n)$ given in equation (7).

$$CF(n) = \max(f_{1\text{dim}}(1) - f_{1\text{dim}}(n)) \quad (7)$$

By applying appropriate threshold to classify *CF* of open eye and that of close eye, eye classification is conducted. For more accurate classification, adjustment of threshold is required depending on the experimental situations.

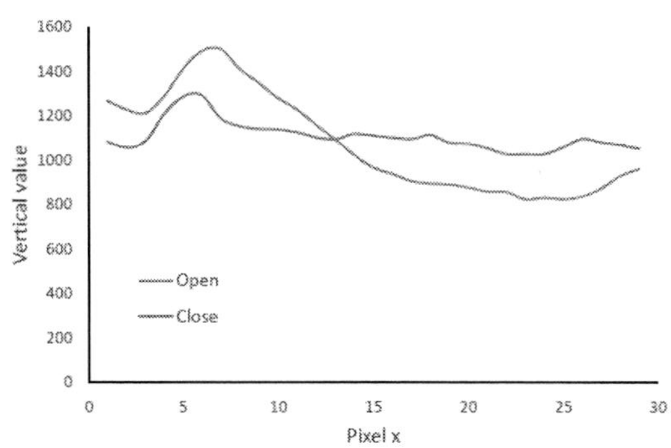

Figure 8. Vertical projection curve of open eye and close eye (Created based upon data from [14])

3.1.3 Fatigue detection

As mentioned above, eye was classified as open eye or close eye to calculate blinking rate and eye closure time. Previous research has suggested that blinking rate and eye closure time could serve as a baseline to detect fatigue [19]. Accordingly, the authors referred to this research to construct fatigue detection system using obtained eye information.

3.2 Breath detection sub-system

An outline of breath detection sub-system is illustrated in Figure 9. First, images of a subject is stored. After enough data are stored, chest part is extracted from the obtained image so that get rid of the effects caused by hand movements or head movements. Image processing was applied to extracted image to detect slight chest movement and obtain respiration wave. Then, Fourier transformation was applied to the detected respiration wave to calculate respiration rate.

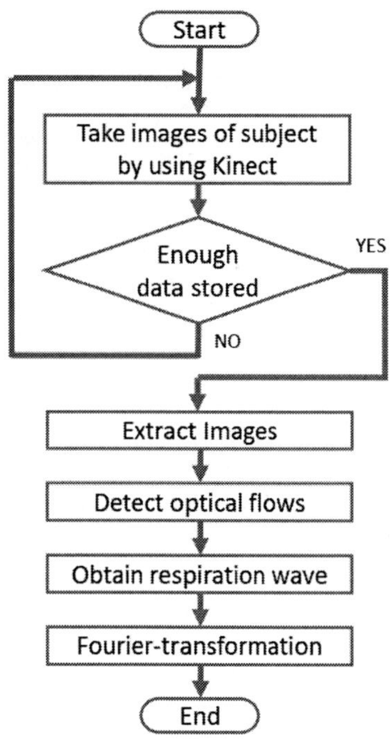

Figure 9. Flowchart of breath detection system

3.2.1 Breath detection using optical flow

As an image processing method to measure the slight chest movement, detection of optical flow method was introduced. Optical flow is vector that represents velocity field calculated from successive 2 frames. The detail of this method is shown below. First, optical flows are detected from successive 2 frames. One of the detected optical flows $v(i)$ is shown in Figure 10. In order to make a distinction between movement of inhalation and that of exhalation, the y direction of optical flow $v(i)$ is focused, represented as $v_y(i)$.

By averaging optical flows detected within extracted

area every frame, averaged optical flow $v_{ave}(i)$ is calculated shown in equation (8). Here n represents total optical flows detected per 1 frame.

$$v_{ave} = \frac{1}{n}\sum v_y(i) \tag{8}$$

The variable v'_{ave} is calculated given in equation (9) to obtain respiration wave.

$$v'_{ave} = \frac{v_{ave}}{dt} \tag{9}$$

For reduction of heavy noise included the obtained respiration wave, approximation based on least squares method is necessary. Then, Hanning Window and Fourier transformation have been applied to the obtained respiration wave to calculate respiration rate.

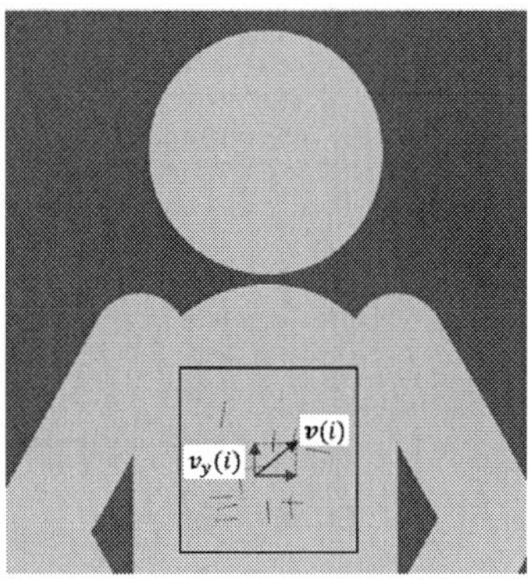

Figure 10. Detection of optical flow (Created based upon data from [20])

4 Measurement results and mood influence

In this chapter, verification experiments on human state estimation system have been carried out. Additionally, an audio system for mood control is scheduled for future development.

4.1 Verification experiments

After the proposed system was improved, further experiments have been conducted to test the performance of fatigue detection sub-system and breath detection sub-system. Experiments on fatigue detection system tested the detection of eye movement where the method to measure eye movement (see Section 3.1) was applied to data-set consists of 2000 images that equals to about 1 minute data. By comparing the obtained result and image data, accuracy of eye detection and eye classification were evaluated. As a result, the correct rate for eye detection could achieve 97.5%. Figure 11 shows the measurement result. From this figure, it can be seen that the proposed system has high accuracy of classification of open eye and close eye, which could lead to calculate eye closure time and blinking rate accurately.

Moreover, experiment on breath detection sub-system has been conducted; first, images of subject breathing for 1 minute were stored while the breath sound was recorded. Then, the breath detection sub-system was applied to the stored images. By comparing the obtained results (sound and image data), the accuracy of the breath detection sub-system was evaluated. Figure 12 illustrates an example of measurement result; the upper graph represents the recorded breath sound wave while the lower graph represents respiration wave obtained by the proposed system. From this figure, it was found that the proposed system could detect respiration wave properly.

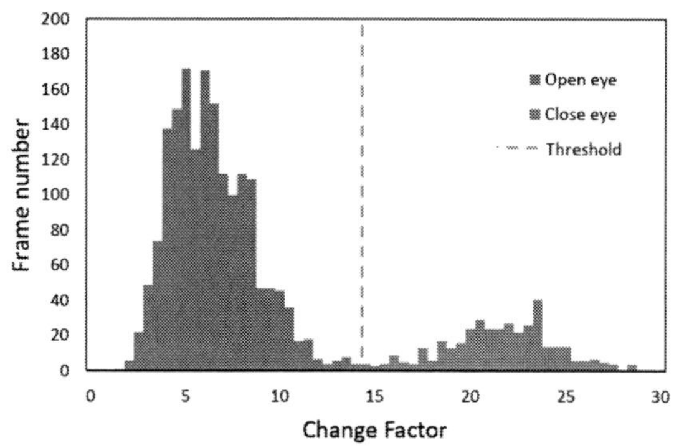

Figure 11. Measurement result of eye movement

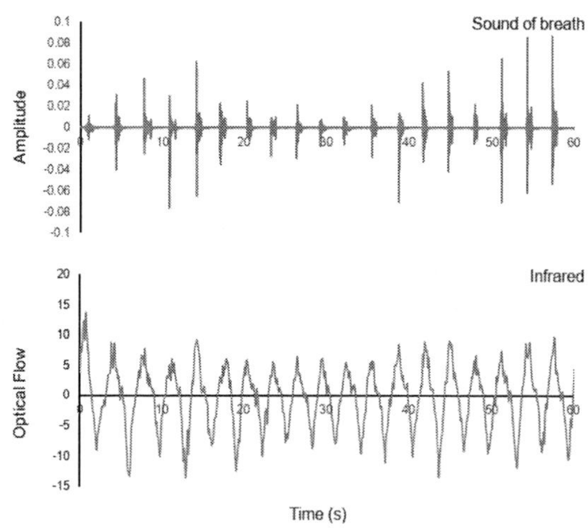

Figure 12. Measurement result of breath

4.2 Performance test on Deployable Getaway

Human state estimation system was implemented into Deployable Getaway Chair. The performance of the proposed system has been demonstrated with a test person sitting inside the chair; after the fatigue is analyzed over eye movement, breath frequency is analyzed. The results of these analyzations are indicated in the computer screen. Thus, this test suggested the feasibility of the proposed system.

4.3 Audio system for mood control

As a next step, it is required to construct an audio sub-system to trigger music for mood control based on the multiple bio-information captured by the human state estimation system. As this system is intended to let tired people get relaxed, fatigue detection is used as an input while breath detection is used for the audio system for mood control.

The authors focus on meditation as the method for mood control during rejuvenation. A previous research has suggested that *Tanden* respiration during Zen meditation could reduce negative mood and let people get relaxed, which might improve staff safety [21]. Accordingly, the authors take on the audio system that induces the proper respiration for meditation. An earlier study suggested that repeated phrase could be effective to control respiration [22]. Thus, sea wave sound is introduced for this audio system.

The detail of the audio system is described as follows. As soon as the fatigue is detected, the audio system is triggered, which influences the mood of the user at appropriate timing. It induces the desired respiration for the meditation starting from the breath rate obtained by the proposed system. This audio system will be implemented into the Deployable Getaway Chair with private space for meditation to increase staff safety.

5 Discussion

In this paper, the authors proposed a human state estimation system on the basis of multiple bio information implemented into "Office Deployable Getaway" which refers to a chair used as a work-station in busy office environments. The proposed system consists of 2 sub-systems; the first sub-system analyzes eye movement including blinking rate and eye closure time to detect fatigue while the second sub-system analyzes breath frequency. These indicators were detected by applying image processing to IR images obtained by Microsoft Kinect sensor. By so doing, bio information of the user could be captured non-contactly even under the dark environment.

Hardware tests were conducted to determine the proper Microsoft Kinect sensor (Kinect V1 or Kinect V2) and its position inside the chair to efficiently detect multiple bio-information including eye movement and respiration. The test results suggested that these bio information could be captured efficiently under the condition that the Kinect sensor V2 was installed at the eye level opposite from the user. Based on the result, the implementation of the proposed system into the chair was conducted, installing the Kinect and a personal computer at appropriate positions.

Performance tests on fatigue detection sub-system and breath detection sub-system were conducted to ascertain that both eye movement and breath could be captured properly. Accordingly, the proposed system was tested with test persons sitting inside the Deployable Getaway Chair; after the fatigue is analyzed over eye movement, breath frequency is analyzed. The results of these analyzations are indicated in the computer screen. Therefore, it was shown that this system could be feasible for the realistic use.

It remains a challenge for future research to improve the human state estimation system. Through this research, it was shown that there were 3 points of improvement. The first improvement point is long processing time; as the process to store images and the process to analyze images were conducted individually, the overall processing time tends to take long. In order to reduce the processing time, it is necessary to conduct these process at the same time by introducing multi thread program.

The second improvement point is influence caused by movements; it was observed that bio information could not be captured when the subject moved during measurement session. Hence, further work is on the way to reduce the influence caused by body movements.

The third improvement point is noise included in the obtained respiration wave; it was observed that some noises were included in the obtained respiration wave due to the distance between a subject and Kinect V2 or influence of experimental situations such as sun light. As a countermeasure, the authors are taking on the noise reduction system based on function approximation, the performance test of which provides promising results. It is expected that the noise reduction system will realize more robust respiration detection.

For future development, the authors will focus on audio system that triggers proper music based on estimated human state, which will be implemented into this Deployable Getaway Chair. It is expected that combination of Deployable Getaway Chair as an architectural environment and the system to control the environment based on the user's bio-information might contribute to safer construction work environment.

6 Acknowledgments

The authors are grateful to Mr. Andreas Bittner, Building Realization and Robotics, for his help in adjustment and installation of the devices into the Deployable Getaway Chair. This work is partially supported by MEXT Grant-in-Aid for the Program for Leading Graduate Schools.

References

[1] Civil Aviation Authority of New Zealand. Fatigue. Online: http://www.caa.govt.nz/Publications/Vector/Vector_Articles/Fatigue_NovDec00News.pdf, Accessed 26/4/2016.

[2] Krueger G. Sustained work, fatigue, sleep loss and performance: A review of the issues, *Work & Stress: An International Journal of Work, Health & Organizations*, 3(2):129-141, 1989.

[3] Chan M. Fatigue: the most critical accident risk in oil and gas construction, *Construction Management and Economics*, 29(4):341-353, 2011.

[4] LIQUIFER SYSTEMS GROUP. DEPLOYABLE GETAWAY for the Office. Online: http://www.liquifer.com/?p=727, Accessed 18/1/2016.

[5] Ogilvie R., McDonagh, D., Stone S., and Wilkinson R. Eye movements and the detection of sleep onset. *Psychophysiology*, 25(1): 81-91, 1985.

[6] Ogilvie, R., Wilkinson, R., and Allison S. The detection of sleep onset: Behavioural, physiological, and subjective convergence. *Sleep*, 12(5): 458-474, 1989.

[7] Zhang Z. and Zhang J. Driver fatigue detection based intelligent vehicle control. In *Proceedings of 18th International Conference on Pattern Recognition (ICPR 2006)*, pages 1262-1265, Hong Kong, 2006.

[8] Milic-Emili J. Recent Advances in Clinical Assessment of Control of breathing. *Lung*, 160(1):1-17, 1982.

[9] Umezawa A. Changes of respiratory activity during labolatory stress. *Japanese Journal of Physiological Psychology and Psychophysiology*, 9(1):43-55, 1991.

[10] Hoang L., Thanh D. and Feng L. Eye Blink Detection for Smart Glasses. In *Proceedings of 2013 IEEE International Symposium on Multimedia (ISM 2013)*, pages 306-308, Anaheim, USA, 2013.

[11] Knopp S. Bones P. Weddell S., Innes C. and Jones R. A wearable device for measuring eye dynamics in real-world conditions. In *Proceedings of 35th Annual International Conference on Engineering in Medicine and Biology Society (EMBC 2013)*, pages 6615-6618, Osaka, Japan, 2013.

[12] Grlica J., Martinović, T. and Džapo H. Capacitive sensor for respiration monitoring. In *Proceedings of Sensors Application Symposium (SAS)*, pages 1-6, Zadar, Croatia, 2015.

[13] Miyakawa T., Tsuruoka K. and Toda T. Involuntary-blink detection method robust against dynamically change of frame rate. In *Proceedings of 6th Biomedical Engi-neering International Conference (BMEiCON 2013)*, pages 1-5, Amphur Muang, Thailand, 2013.

[14] Nakaso S., Güttler J., Mita A. and Bock T. A Fatigue Detection System implemented in an Office Deployable Gateway based on Eye Movement. In *Proceedings of the CIB*IAARC W119 CIC 2015 Workshop,* pages 39-46, Munich, Germany, 2015.

[15] Imhof B. Hoheneder W. and Vogel K. Deployable getaway for international space station. In *Proceedings of 40th international conference on environmental systems*, pages 1-29, Barcelona, Spain, 2010.

[16] Viola P. and Jones M. Rapid object detection using a boosted cascade of simple features. In *Proceedings of 2001 IEEE Computer Society Conference on Computer Vision and Pattern Recognition (CVPR 2001)*, pages 511-518, 2001.

[17] Lienhart R. and Maydt J. An extended set of Haar-like features for rapid object detection. In *Proceedings of International Conference on Image Processing 2002,* pages 900-903, Rochester, New York, USA, 2002.

[18] OpenCV. Open Source Computer Vision Library Reference Manual, 2014.

[19] Tietze H. and Hargutt V. Zweidimensionale analyse zur beurteilung des verlaufs von ermüdung. Psychologisches institut, 2001.

[20] Nakaso S. and Mita A. Nap Environment Control considering respiration rate and music tempo using sensor agent robot. In *Proceedings of SPIE - The International Society for Optical Engineering,* 9435-100, SanDiego, USA, 2015.

[21] Arita H. Anterior prefrontal cortex and serotonergic system activation during Zen meditation practice induces negative mood improvement and increased alpha band in EEG. *Societas Neurologica Japonica*, 52(11): 1279-1280, 2012.

[22] Sato T., Ohsuga M., Moriya T. Increase in synchronization of respiration by repeatedly listening to the same piece of music. In *Proceedings of Communications and Information Technologies (ISCIT 2010)*, pages 1035-1038, Tokyo, Japan, 2010.

A Cloud-based BIM Platform for Information Collaboration

L. Ma[a] and R. Sacks[b]

[a]National Building Research Institute, Technion – Israel Institute of Technology, Israel
[b]Faculty of Civil and Environmental Engineering., Technion – Israel Institute of Technology, Israel
E-mail: lingma@technion.ac.il, cvsacks@technion.ac.il

Abstract

The Model View Definition (MVD) approach to interoperability of BIM models requires software vendors to write export functions that can map their internal representation to the whole scope of each of many domain-specific target MVDs. However, in everyday practical use of models, a user receiving an exchanged model often requires information that is outside the scope of the exporting application. Thus the need arises to enrich the model for different uses. We propose a cloud-based mode of working in which users can freely query and enrich model objects. The data structure uses the IFC schema but enables addition of dynamically user-defined properties. The approach has been implemented using a NoSQL cloud based database and its use is illustrated with an example of semantic enrichment of a BIM model representing a damaged reinforced concrete beam reconstructed from laser scanning data.

Keywords –

Building Information Modelling; Model View Definition; Cloud based database; Information collaboration; Semantic enrichment

1 Introduction

The Industry Foundation Classes (IFC) data schema has been developed to solve the information interoperability problem in the implementation of Building Information Modelling (BIM). To support exchange of BIM models across the full spectrum of architecture, engineering and construction (AEC) disciplines, IFC is designed to be generic and comprehensive in order to accommodate many different configurations, industry domains and levels of detail [1]. However, the IFC schema is not tailored to any individual processes within building construction, because in different business processes different users of building data have different information requirements, and authors of the building data will provide detail in different domains. Such discrete processes should be well defined in an Information Delivery Manual (IDM) [2], which maps the actors and information exchange requirements of any activities in the processes. Guided by the IDM, a set of Model View Definition (MVD) concepts is compiled to select a subset of the IFC schema and define allowable values at particular attributes of particular data types [3]. The MVD specifies which entities and attributes are mandatory or optional according to the needs of the specific business process.

The IDM and MVD approach has been used in subdomains like precast concrete construction [3-5], building energy analysis [6], and facility management handover [7]. A number of specialized applications have been developed for checking the compliance of IFC instance files to model view definitions [8, 9].

However, all BIM authoring applications have their own internal schema for building data modelling to suit their specific functional needs and information needs. Since no professional BIM software can support the full range of MVDs in IFC export subroutines, the need arises for tools that can supplement BIM exchanges with necessary information in a dynamic fashion according to the needs of importing applications. In addition, the file-based mode of most IFC exchanges is inadequate for dynamic data supplementation – a server solution is needed, preferably hosted in the cloud.

1.1 Semantic Enrichment

To address the need for supplementing information, Belsky et al. [10, 11] developed a rule-based semantic enrichment engine, which aims at upgrading a 'poorly-exported' BIM model to a 'high-standard' model that conforms to the MVD at the import end. The engine identifies undefined objects and missing object relationships in an IFC file based on encapsulated domain experts' (e.g., civil engineers) knowledge of geometrical and topological features of those objects. It automatically supplements the IFC file with information regarding objects' identification and relationships.

However, some information required by certain MVDs cannot be inferred from the geometrical and topological features the semantic enrichment engine depends on [10, 11]. Adding this missing information requires additional data entry, often in a collaborative fashion. In addition, users may also need to add information items that are beyond the scope of the current IFC schema.

1.2 BIM Servers and Cloud Systems

The semantic enrichment engine does not have an interface for users to supplement information, neither within the framework of the target MVD schema nor as part of an extended schema. A similar limitation also exists in current cloud-based BIM applications (such as Autodesk A360 [12] and Trimble Connect [13]) or BIM applications that can be deployed in the cloud (such as BIMServer [14] and GRAPHISOFT BIMcloud [15]). Most of these applications only allow users to query data rather than supplement the data, because they don't expose their local schema. None of them enables extension of the current IFC schema dynamically for enriching information items that are not defined in the schema.

To address these problems, we developed a cloud based BIM platform in which users could freely query, enrich and deliver model information in a working mode coincident with the IDM & MVD concepts. The platform enables both automated rule-based enrichment and user-friendly data entry. The platform's data structure uses the IFC schema but also enables dynamic addition of user-defined properties. The platform was built upon a non-SQL cloud based database. The following sections present the main components of the BIM platform and demonstrate its application.

2 The BIM Platform Components

The objective of the platform is to design a *data as a service* system in the cloud for collaboration on a BIM model. A system diagram is shown in Figure 1.

There are two major components in the BIM platform: a database system (MongoDB) for storing the BIM models and an interface for communication between the user (or other BIM applications) and the BIM models for different purposes, including uploading BIM models using IFC instance files, creating models from scratch, or updating a model. The interfaces follow a standard protocol to perform those actions upon the database system using CRUD operations.

The following sections explain why such a database system provides a good alternative for storing a BIM model, and how to work on a BIM model using database operations.

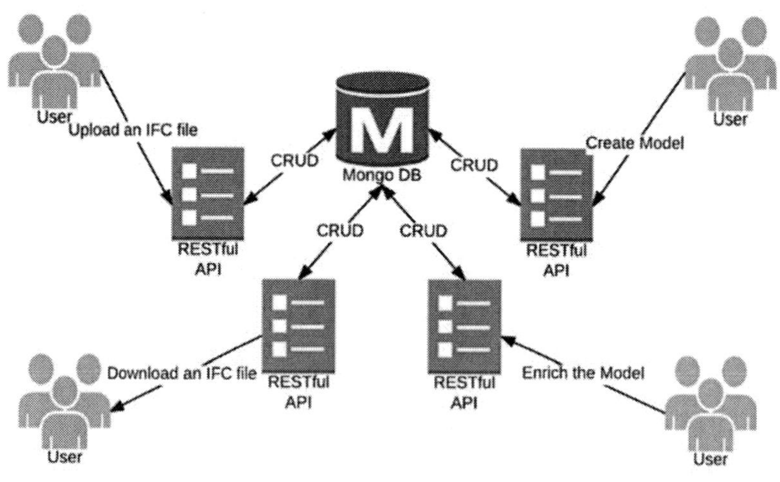

Figure 1 System Diagram of the Cloud Based BIM Platform

2.1 NoSQL Cloud Based Database

Database systems generally encapsulate the schema together with the data. IFC instance files, on the other hand, IFC separate the data from the underlying schema, which must be made available to any application that needs to import a BIM model stored in an IFC file. Database systems allow operations such as Create, Read, Update, and Delete (CRUD), which cannot be used directly to model data stored in IFC files.

However, IFC schema is mature in terms of representation of a building design. It can be used to model most of the information items required in subdomains, because it allows creating customized property sets, adding user defined element typing, and many other features. Hence, the IFC schema is a good starting point for the translation from an IFC instance file to a database storage of the BIM model.

Commonly-used relational databases use predefined tables and tabular relations, and as such are inappropriate for use with the IFC schema, which is an object-oriented schema coded using EXPRESS language. NoSQL is a new data management technology specifically designed to meet the requirements of data variety [16]. It does not require a predefined data schema, which means it not only allows mapping all the data definitions in IFC schema to the database structure, but also enables addition of new entities to the data structure. This conforms to the mode of development of the IFC schema, which is continuously adopting new modelling concepts into the schema in new releases.

One of the most popular NoSQL database is MongoDB, which is used in the proposed BIM platform. It is a document oriented database, and combines the best of relational database with the innovations of NoSQL technology. It stores the data in its featured Binary JavaScript Object Notation (BSON) format. The basic atom in MongoDB is called document. All documents are stored in collections, which are analogous to the data tables in relational databases. A document is an unordered set of field and value pairs, and it always contains a field of "_id" which is a unique index. For example, a document in the database could be like this:

```
{
    name : "concrete beam",
    dimension : [240, 300, 3900],
    _id: "12312366180a4ac15c7fdsdf"
}
```

A beam modelled in an IFC file that conforms to IFC Coordination View 2.0 is shown as follows:

```
#95=IfcBeam('2LyQvT49XF5xaTest6Jq_$',#5,'M_Concrete-Rectangular Beam:300 x 600mm:300 x 600mm:104323','',$,#92,#94,$)
```

A program for serialization of IFC instance file to MongoDB documents was developed. The above data entity of the beam can be mapped to MongoDB using this program and get the document as follow:

```
{
    _id: "56672266180a4ac15c7fb483",
    line_id: 95,
    type: "IfcBeam",
    GlobalId: "2LyQvT49XF5xaTest6Jq_$",
    OwnerHistory: "56672266180a4ac15c7fb429",
    Name: "M_Concrete-Rectangular Beam:300 x 600mm:300 x 600mm:104323",
    Description: "",
    ObjectType: null,
    ObjectPlacement:"56672266180a4ac15c7fb480",
    Representation:"56672266180a4ac15c7fb482",
    Tag: null,
    HasAssociations: [
        "56672266180a4ac15c7fba8e"
        ],
    IsDefinedBy: [
        "56672266180a4ac15c7fb4a4"
        ],
    ContainedInStructure: [
        "56672266180a4ac15c7fba81"
        ]
}
```

Note that the fields like ObjectPlacement and Representation link to other documents by using the latter's _id as their value. In addition, the program can also derive the inverse attributes after completely parsing the IFC file using the IFC schema. So fields like "HasAssociations", "ContainedInStructure" and "IsDefinedBy" are also derived in this example, although they did not exist in the IFC instance file. As a result, the data in the database is richer in terms of semantics and is more human readable.

Another benefit of using MongoDB is that it uses timestamp-based concurrency control, which is a non-locking control mechanism, so that it supports real-time read that will not conflict writes and high-speed logging. It has caching and high scalability, which allows running thousands of machines with distributed data, and is well-suited for deployment in the cloud. Sets of redundant servers called "replica sets" help maintain the system availability and data integrity even if individual cloud instances are taken offline.

As a result, the NoSQL-based MongoDB system serves a very good alternative for BIM application in terms of easy-to-use and concurrent collaboration among different users.

2.2 A RESTful API for CRUD Operations on Model Data

An Application Programming Interface (API) is used to allow one software/service talk with the other. Another term, i.e., Hypertext Transfer Protocol (HTTP) is the standard protocol that defines how a web browser and a web server communicate with each other. For example the client sends a request to the server and the server sends back a response. As technology moves forward, network-based software architectures have been widely adopted. Hence an architecture called REpresentational State Transfer (REST) [17] was proposed to define an API that uses HTTP for a series of constrained methods, such as get, put, post, or delete. Such kinds of API are usually called RESTful APIs.

Many RESTful APIs send responses in JSON, which is both human and machine readable. Another advantage of using RESTful APIs is that those HTTP methods can be easily mapped to the CRUD operations in a database like MongoDB.

In MongoDB, a CRUD operation targets a specific collection of documents. Generally CRUD operations can be grouped as queries and data modifications. A query specifies a projection (return which fields from the documents in response), query criteria, e.g., $eq (equal), $gt (greater than), $in (matches any of the values specified in an array), etc., and a result modifier, e.g, sort the data in certain order, limit the number of documents returned. For example the query for getting all the beams in a BIM model is simply:

db.records.find({type: "IfcBeam"})

Another group of CRUD operations is data modification, including create, update or delete data. For updating the names of all the beams to "Type A Beam", one only needs to run:

db.records. updateMany ({type: "IfcBeam"}, {$set: {name: "Type A Beam"}})

After associating the MongoDB operations to a customized RESTful API, one can perform the above two operations in a terminal as follows (it is supposed that the server is running on http://127.0.0.1):

$ curl –i http://127.0.0.1?where={"type": "IfcBeam"}

and

$ curl –X PATCH -i http://127.0.0.1?where={"type": "IfcBeam"} –d '{"name": "Type A Beam"}'

Note that PATCH is the HTTP method that links to the Update operation in MongoDB. As shown above, by using the RESTful API associated with MongoDB, performing operations on a BIM model is straightforward.

3 The BIM Platform Applications

Automated reconstruction of 'as-built' building models from remote sensing data, including point cloud data [18] and photogrammetry/videogrammetry data [19, 20] is the focus of much ongoing research around the world. To achieve the goal of creating a semantically rich BIM model from the sensing data, collaboration between experts in computer vision and in civil engineering is essential. The former contribute to reconstruction of 3D primitives from the redundant point cloud data, while the latter help to interpret and classify the 3D primitives using their domain knowledge.

The following sections utilize an example from this area of research as a vehicle to show how the customized cloud BIM platform can facilitate such collaboration. Specifically, we use the example of reconstruction of a reinforced concrete beam model taken from a research effort at the Technion–Israel Institute of Technology to develop a system for reconstruction of BIM models for earthquake damaged buildings [21-24] using terrestrial laser scanning.

3.1 3D Reconstruction of Remote Sensing Data

The first application of the platform shows how to create a 3D model from scratch using a computer vision algorithm.

A specimen of a damaged beam, shown in Figure 2 (a), was scanned using a terrestrial laser scanner - Leica ScanStation C10 at the Technion. The point cloud data, shown in Figure 2 (b), was processed using the Random Sample Consensus (RANSAC) algorithm [25]. RANSAC has become one of the most popular algorithms for model fitting from a set of data containing outliers. It is used to divide the point cloud into clusters, each of which contains points attached to a plane. The clustering result of the scan is shown in Figure 2 (c), in which each cluster of points is coloured differently.

In addition, we can define the edges as places where the geometry of the scene changes sharply/abruptly. Such features can be detected as points having high curvature values across a certain cluster of points. Those boundary points form polygons that represent 3D primitives in the scene.

For each polygon, the program sends a PUT request to insert documents for storing the vertices. For example, for a point with coordinates of (1400,120,150), a program will send a request as follows,

```
$ curl -d '[{
    "type": " IfcCartesianPoint",
    "Coordinates": [1400,120,150],
    "Dim": 3,
    "LayerAssignment":
    "56672266180a4ac15c7fb481",
    "StyledByItem": "56672266180a4ac15c7fbdfg"
    }]'
```

Note that, the field of "Dim" is a derived attribute calculated by the length of the array type field of "Coordinates"; "LayerAssignment" and "StyledByItem" are inverse attribute links to other documents for the purposes of grouping shapes and assigning presentation information (e.g., color) respectively. None of them is stored in an IFC instance file, but, for a rich sematic representation, they will be explicitly specified in the database documents by linking to other documents' object ids, which are accessible after insertion.

After defining the coordinates of vertices, a document for representing a polygon will be inserted with the object ids (like "56672266180a4ac15c7fb481") of these vertices using a command such as:

```
$ curl -d '[{
    "type": " IfcPolyLoop",
    "Polygon":    ["56672266180a4ac15c7fb481",
    "56672266180a4ac15c7fb482",
    "56672266180a4ac15c7fb483",
    "56672266180a4ac15c7fb484"],
    "LayerAssignment":
    "56672266180a4ac15c7fb481",
    "StyledByItem": "56672266180a4ac15c7fbdfg"
    }]'
```

(a) The damaged beam

(b) The scan of the beam

(c) The segmentation result of the scan

Figure 2 Scan of a damaged beam and the point cloud segmentation result

Usually, a polygon organizes the vertices in a certain sequence that makes the polygon's normal direction aligned with the face normal direction, which shows where the material is. In this case the value of the orientation of the face is defined as TRUE, otherwise is FALSE meaning that the polygon is reversed when used in the face. Hence the definition of the geometric representation is unambiguous. Suppose that the object id of the polygon is "56672266180a4ac15c7fb4db", then the insertion of the face geometry can be done as follows,

```
$ curl -d '[{
    "type": " IfcFaceOuterBound",
    "Bound": "56672266180a4ac15c7fb4db",
    "Orientation", TRUE,
    "LayerAssignment":
    "56672266180a4ac15c7fb481",
    "StyledByItem": "56672266180a4ac15c7fbdfg"
    }]'
```

In the IFC schema, besides the basic geometric information, a face can also be associated with texture maps through an inverse attribute "HasTextureMaps". As a result, adding a face can be done using the following command,

```
$ curl -d '[{
    "type": " IfcFace",
    "Bound": "56672266180a4ac15c7dsedb",
    "HasTextureMaps", TRUE,
    "LayerAssignment":
    "56672266180a4ac15c7fb481",
    "StyledByItem": "56672266180a4ac15c7fbdfg"
    }]'
```

The 3D reconstruction work can only derive the segmented faces, so only the geometric information of the faces is added to the model in the database at this stage. Generating a semantically rich BIM model requires the input of the civil engineers, not only computer vision.

3.2 Semantic Enrichment of a BIM model

The cloud-based BIM platform enables collaborative semantic enrichment of BIM models. The second application shows how to enrich the BIM model generated in the first application to meet the requirements of modelling damaged building components.

The normal direction of each face can be derived by computing the cross product the vectors of any two of its non-parallel edges, and can be added to the model using the following command:

```
$ curl -d '[{
        "type": " IfcDirection",
        "Axis": [1,0,0],
        "Dim", 3
    }]'
```

Thus pairs of perpendicular faces can be identified by the feature that the dot product of their normal directions equals zero. Each pair of perpendicular faces can be used to define a local coordinate system. The axes can be defined using:

```
$ curl -d '[{
        "type": " IfcAxis2Placement3D",
        "Location": "56672266180a4ac15c7fbtbc",
        "Dim": 3,
        "Axis": "56672266180a4ac15c7dsedb",
        "RefDirection": "56672266180a4ac15c7dsedb",
        "P": "56672266180a4ac15c7dsyuj",
        "LayerAssignment":
        "56672266180a4ac15c7fb481",
        "StyledByItem": "56672266180a4ac15c7fbdfg"
    }]'
```

where "Location" links to an origin (the end point of the face-face intersection line segment), "Axis" and "RefDirection" are the normal axes of the two faces, and "P" is a derived attribute calculated by the cross product of the two axes.

In this domain of concrete beams and columns, most of the building elements can be represented using extruded area solids with a rectangular profile. Hence, after aligning the point cloud to the local coordinate system of a pair of the faces, one can extrude a rectangle along the point cloud segments until all relevant points cover the visible faces of the extruded shape. The result is shown in Figure 3 (a). Insertion of this shape representation of object can be done using,

```
$ curl -d '[{
        "type": " IfcExtrudedAreaSolid",
        "Dim": 3,
        "SweptArea": "66672266180a4ac15c7fbtbc",
        "Position": "76672266180a4ac15c7dsedb",
        "ExtrudedDirection":
        "96672266180a4ac15c7dsedb",
        "Depth": "600",
        "LayerAssignment":
        "56672266180a4ac15c7fb481",
        "StyledByItem": "56672266180a4ac15c7fbdfg"
    }]'
```

where the "SweptArea" links to a profile definition (a rectangle in this case), and the "Position" links to the local coordinates system defined in previous step.

Another two solids can be derived and added to the model in a similar way, and the result is shown in Figure 3 (b). When a site engineer/damage inspector sees the three shapes, they will recognize that these shapes should be the solid segments that belong to a single beam from the original building, whose shape before the damage is as shown in Figure 3 (c). As a result, the correct representation of the beam should aggregate the three extruded area solids into one shape (using IfcProductDefinitionShape) and present this shape as a single beam using the following command:

```
$ curl -d '[{
        "type": " IfcBeam",
        "GlobalId": "3lB8HeqWX8Tfa2Zdze9mzo",
        "OwnerHistory":
        "3lB8HeqWX8Tfa2Zdze9m12",
        "Name": "Beam",
        "Description": "An earthquake damaged beam",
        "ObjectType": "Damaged beam",
        "ObjectPlacement":
        "76672266180a4ac15c7dsedb",
        "Representation":
        "546ds266180a4ac15c7fbdtt",
        "PredefinedType": BEAM
    }]'
```

In this way the model was progressively enriched for the purpose of representing a damaged beam in a suitable shape representation. This illustrates the ways in which the cloud-based platform can be used to dynamically compile a building model.

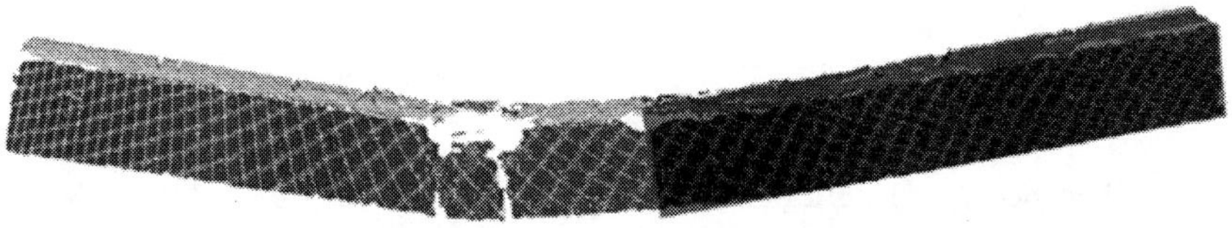

(a) Adjust the 3D model by fitting an extruded area solid

(b) Progressively create three solids in the scene

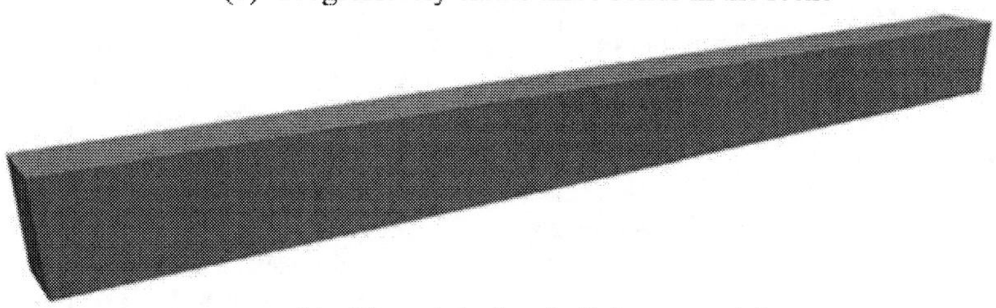

(c) The original as-built beam model

Figure 3 Enrichment of the Beam Model

4 Conclusions

The standard IDM and MVD approach was devised to specify the information exchange model schema for specific business processes in subdomains. However, BIM authoring applications serve a variety of purposes (architectural design, structural analysis, production management, etc.) and have specialized internal data schema. They therefore require extensive mapping functions in order to export customized instance models for export. To address this problem, an alternative cloud based approach to enriching and sharing BIM models in a collaborative way was proposed and tested.

The schema free feature of NoSQL allows users across different domains to share and collaborate on a single but comprehensive BIM model, where they can all contribute to the model enrichment. Instance models prepared according to different MVDs can be merged into this single model.

The major advantages of this platform are:

- Cloud based storage enables information sharing in a comprehensive model.
- The NoSQL databased is not limited to any predefined data schema, as would be required if a relational database was used.
- There is no redundancy of information storage as there is in file systems where new versions of files are saved periodically.
- The RESTful API is a lightweight, standard-based, and platform/language-independent API, which makes execution of operations on BIM model easy.

The cloud-based system developed can be used as the repository for semantic enrichment such as that implemented by the SeeBIM system, but it can also be accessed conveniently by other applications and BIM authoring tools.

References

[1] BuildingSmart. *Industry Foundation Classes Release 4 (IFC4)*. 2013 2013.

[2] Wix, J. and J. Karlshoej, Information delivery manual: Guide to components and development methods. *BuildingSMART International*, 2010.

[3] Venugopal, M., et al., Semantics of model views for information exchanges using the industry foundation class schema. *Advanced Engineering Informatics*, 2012. 26(2): p. 411-428.

[4] Sacks, R., et al., The Rosewood experiment — Building information modeling and interoperability for architectural precast facades. *Automation in Construction*, 2010. 19(4): p. 419-432.

[5] Eastman, C.M., et al., Exchange Model and Exchange Object Concepts for Implementation of National BIM Standards. *Journal of Computing in Civil Engineering*, 2010. 24(1): p. 25-34.
[6] Jeong, W., et al., Translating building information modeling to building energy modeling using model view definition. *TheScientificWorldJournal*, 2014. 2014: p. 638276-638276.
[7] East, E.W., N. Nisbet, and T. Liebich, Facility Management Handover Model View. *Journal of Computing in Civil Engineering*, 2013. 27(1): p. 61-67.
[8] Oh, M., et al., Integrated system for BIM-based collaborative design. *Automation in Construction*, 2015. 58: p. 196-206.
[9] Zhang, C., J. Beetz, and M. Weise, Interoperable validation for IFC building models using open standards. *Journal of Information Technology in Construction*, 2015. 20: p. 24-39.
[10] Belsky, M., R. Sacks, and I. Brilakis, Semantic Enrichment for Building Information Modeling. *Computer-Aided Civil and Infrastructure Engineering*, 2015. 31(4): p. 261-274.
[11] Belsky, M., R. Sacks, and I. Brilakis. A framework for semantic enrichment of IFC building models. in *Proceedings of the 30th CIB W78 International Conference*. 2013. Beijing, China.
[12] *Autodesk A360*. 2016; Available from: https://a360.autodesk.com/.
[13] *Trimble Connect*. 2016; Available from: http://connect.trimble.com/.
[14] Beetz, J., et al. BIMserver. org–An open source IFC model server. in *Proceedings of the CIP W78 conference*. 2010.
[15] *GRAPHISOFT BIMcloud*. 2016; Available from: http://www.graphisoft.com/bimcloud/overview/.
[16] Han, J., et al. Survey on NoSQL database. in *6th International Conference on Pervasive Computing and Applications (ICPCA)*. 2011. IEEE.
[17] Fielding, R.T., *Architectural styles and the design of network-based software architectures*. 2000, University of California, Irvine.
[18] Tang, P., et al., Automatic reconstruction of as-built building information models from laser-scanned point clouds: A review of related techniques. *Automation in Construction*, 2010. 19(7): p. 829-843.
[19] Brilakis, I., et al., Toward automated generation of parametric BIMs based on hybrid video and laser scanning data. *Advanced Engineering Informatics*, 2010. 24(4): p. 456-465.
[20] Fathi, H. and I. Brilakis, A videogrammetric as-built data collection method for digital fabrication of sheet metal roof panels. *Advanced Engineering Informatics*, 2013. 27(4): p. 466-476.
[21] Ma, L., et al., A Computational Procedure for Generating Specimens of BIM and Point Cloud Data for Building Change Detection, in *Computing in Civil Engineering 2015*. 2015, American Society of Civil Engineers. p. 684-691.
[22] Ma, L., et al., Preparation of Synthetic As-Damaged Models for Post-Earthquake BIM Reconstruction Research. *Journal of Computing in Civil Engineering*, 2015: p. 04015032.
[23] Ma, L., R. Sacks, and R. Zeibak-Shini, Information modeling of earthquake-damaged reinforced concrete structures. *Advanced Engineering Informatics*, 2015. 21(3): p. 396–407.
[24] Zeibak-Shini, R., L. Ma, and R. Sacks, *Mapping the Structural Frame of a Damaged Reinforced Concrete Building using As-Damaged Scans and As-Built BIM*, in *Proceedings of the 32nd CIB W78 International Conference*. 2015.
[25] Fischler, M.A. and R.C. Bolles, Random sample consensus: a paradigm for model fitting with applications to image analysis and automated cartography. *Communications of the ACM*, 1981. 24(6): p. 381-395.

Real-time Schedule Control

Shabtai Isaac[a]

[a]Department of Structural Engineering, Ben Gurion University of the Negev, Israel
E-mail: isaacsh@bgu.ac.il

Abstract –

Numerous previous studies have dealt with the use of automated technologies for on-site, real-time data collection to support construction project control. Most of these studies are based on the assumption, that the collected data can be compared with planning data to identify deviations and implement control actions. However, construction scheduling often focuses on pre-construction forecasting and logistics processes that are carried out by middle management. In contrast, the execution proper is governed by informal short-term scheduling, performed ad-hoc by site management. This reality creates discrepancies between the data provided by automated monitoring technologies and the information that can be obtained from project schedules, in terms of their granularity, scope and underlying assumptions. The aim of this paper is to review existing scheduling methods, and compare their outputs with the data provided by automated monitoring technologies. Ways are proposed in which scheduling methods can be enriched in order to better support monitoring and control processes.

Keywords –

Scheduling; Monitoring; Control; Real-time Data Collection

1 Introduction

These instructions will enable you to prepare your manuscript in an electronic format, ready for submission and peer review. It is therefore essential that these instructions be carefully followed. It is not widely known that Gantt charts, which are currently primarily used in construction projects for displaying schedules, were originally developed as monitoring tools – designed for supervisors to quickly know where production stood relative to the plan, and identify causes for reduced productivity. Regarding schedules, Gantt in fact warned: "Many shops have a very nice schedule system; they plan their work beautifully—at least, it looks very pretty on paper; but they have no means of finding out whether those schedules are lived up to or not. Usually they are not" [1]. This will not come as a surprise to practitioners involved in the management of construction projects, which typically suffer from significant deviations from their planned schedules [2, 3, 4]. However, Gantt's warning does not seem to have had much of an impact on those who use the chart that he developed for pre-project scheduling, given the prevalent gap between the scheduling that has been carried out offsite and the actual on-site execution management.

In many construction projects schedule control processes are implemented to identify the delays that occur, and reduce their impact on the project. Schedule control processes usually involve the following four steps:

1. A comparison of the actual performance on the construction site with the planned schedule
2. An identification of any differences between the actual and planned performance
3. An identification of those deviations that are likely to affect the date at which the project will be completed
4. The execution of controlling actions for the deviations identified in step 3.

However, such control processes, and the consequent updating of the project schedule, are often carried out manually, and thus infrequently and at a significant delay from when deviations actually occur. Consequently, many studies have focused on the use of automated tracking technologies to provide project management with detailed data on the actual progress in a project in real time, including the locations and production rates of different activities. Various technologies can be used for such a purpose, including global positioning system (GPS), radio frequency identification (RFID), 3D cameras and laser scanners [e.g. 5, 6, 7, 8]. Models have also been developed to convert automatically collected data into information regarding the project performance, which could then be compared with the project plan for the identification of deviations [9].

However, questions remain on how exactly automatically collected data should be compared with data from the schedule, in order to allow project management to implement control actions such as correcting and re-planning the schedule. In particular, it is unclear how the real-time monitoring data can be synchronized with planning data in terms of the granularity of the data, and the frequency at which it is provided. While the monitoring data can be collected in real-time, the scheduling data is, at best, accurate at a daily level. In practice, there is often a significant gap between the formal schedules prepared for construction projects prior to their execution, and the informal short-term planning performed by site management, which is often not well documented [10, 11].

The differences between formal schedules in construction projects, the informal methods used for onsite execution planning, and the detailed data that can be provided by automated data collection technologies, raise the question how scheduling practices should be adjusted to allow project management to fully utilize the real-time monitoring data for schedule control. This paper discusses how the current scheduling methods can be changed to support effective, real-time schedule control.

2 Existing Methods for Schedule Monitoring and Control

Many practitioners currently rely for the monitoring and updating of the schedule on the same methods that they use to define the schedule. Most use the Critical Path Method (CPM) for this purpose. However, traditional CPM network analysis has a number of limitations, which have been dealt with in numerous previous studies. These limitations include the inability to model continuous and simultaneous work flows of teams on the construction site, which need to be taken into account in case of overlapping, interdependent activities [12, 13]. Location-based scheduling methods have been developed to take into account the location of work on the site to avoid conflicts between overlapping activities [14]. However, these methods are limited to a novel visual representation of the activities, and are all defined using the same precedence diagram that is used in CPM,

Another important limitation of CPM and other traditional project scheduling methods is the assumption of unlimited resource availability, and the unexpected delays that often occur in projects due to insufficient resources [15]. A large number of methods have consequently been proposed to solve the resource-constrained construction scheduling problem. But despite these shortcomings of CPM and other traditional scheduling methods, few studies have addressed the basic precedence diagram that is used in all these methods. By using the same constructs and relationships as the existing activity network model, the effectiveness of alternative solutions that have been proposed so far has therefore been limited.

While most practitioners use for schedule control the same methods that they use for planning the schedule, some methods have been developed especially for this purpose. Earned Value Management (EVM) has been put forward as an alternative to the frequent updating of project schedules. EVM supports project control by forecasting the final project cost. The use of EVM to predict the final project duration has been found to be less reliable, since the EVM schedule indicators fail when projects continue execution past the planned end date [16]. Consequently, an extension to EVM, called earned schedule (ES), has been developed which uses time-based indicators for schedule performance. However, the reliability of ES has been criticized as well, since it fails to differentiate between critical and non-critical activities [17].

In the context of this study, both EVM and ES can be regarded as high-level approaches, based on a set of macro level indicators, which facilitate a simpler control process than detailed network analysis [18]. However, such an approach contradicts the need for a detailed schedule analysis when real-time data is used for schedule control, which addresses individual activities and continuously updates the status of the current critical path,. Similar to CPM and other planning methods, EVM and ES are thus not sufficient for real-time tracking and control of construction activities.

3 Proposed Approach

This paper discusses possible extensions to the existing precedence diagram used for project scheduling, to provide improved support for real-time schedule control. The precedence diagram used to produce construction schedules represents the relationships between the activities in the project network model. It is proposed that schedule control could be supported by adding new control points and milestones to the precedence diagram.

3.1 Control Points

Currently, only the start and finish points of an activity are represented in precedence diagrams. Accordingly, only four types of relationships between activities can be defined: finish-to-start (FS), start-to-start (SS), finish-to-finish (FF), and start-to-finish (SF) relationships [19]. These types are often insufficient for monitoring activities in real-time. Additional types of constructs and relationships are required for real-time control, given the granularity of the monitoring data and the frequency at which it is provided.

A solution for this can be to add internal control points to the definition of activities in schedules, in addition to their endpoints. For many activities in construction projects specific points in time can be defined as control points during the execution of an activity, so that an alert is provided when the activity reaches such a point. Those activities can be divided into discrete sections, such as apartments or floors. Naturally, the sections into which an activity is divided can differ in size when these sections represent different work-zones, and the size of these zones accurately represented in the schedule.

The mid-activity control points are differentiated from the start and finish of activities. Such control points can enable carefully tracking the status of activities, and the identification of deviations that occur as the activity is being executed. Currently, solutions for identifying deviations from such midpoints when they occur, and adjusting the execution of other planned activities accordingly, are mostly informal. Once the internal control points of two activities that are interdependent and overlapping have been defined, temporal relationships can be defined between those points in order to identify the implications of a deviation at a specific control point (Figure 1) [20]. For example, a time buffer that needs to be maintained between repetitive activities could be defined, and any change in the status of this buffer due to variations in the production rates of the activities identified in real-time.

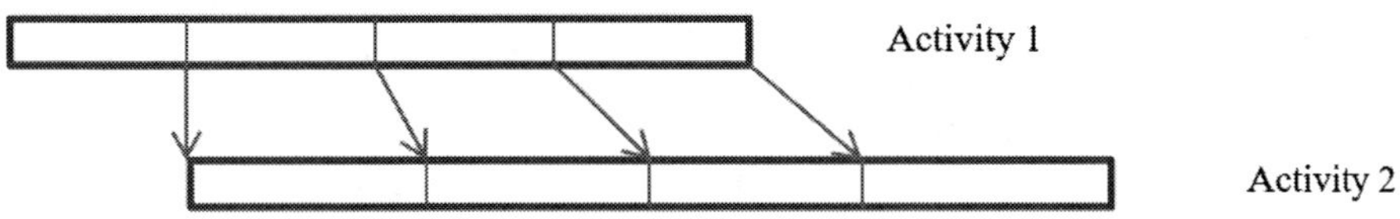

Figure 1: Temporal relationships between internal control points of activities

Internal control points can also be used for the management of limited resources, and when the allocation of those resources needs to be changed during the execution of an activity. Often, simultaneous activities in projects need to share limited resources such as equipment and manpower, and the project management reallocates those resources among the activities at certain points in their execution (Figure 2). Currently, there is no formal way to incorporate such requirements in schedules, to ensure that delays are taken into account when they occur. Internal control points can be used for this purpose as well.

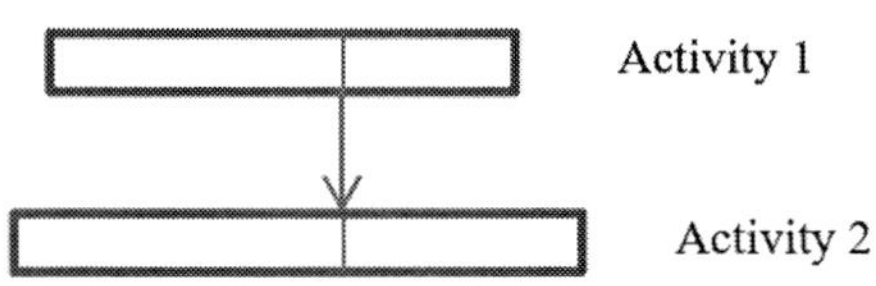

Figure 2: Reallocation of resources at control points

3.2 Milestones for Non-critical Activities

To control project execution, milestones can be defined as additional control points to provide an alert when a delay in a non-critical activity exceeds the free float (Figure 3). Float is currently considered to be a by-product of the CPM computation, and is often ignored in the daily planning and control of projects. In fact, it is not directly represented in most Gantt charts, and consumption of the float usually becomes evident only after the occurrence of delays in non-critical activities that exceeded the float. As a result, the critical path frequently changes in a ways that are not anticipated.

The status of milestones can automatically monitored, and used to help the project manager prevent delays from exceeding the free float and activities becoming critical. The milestones allow the project manager to identify potentially critical paths, and track the available float on them at any moment in the project. While the use of such control points does not directly resolve the question of who owns the float, it does increase the control the project manager has over its allocation and use. In those cases in which the project manager deems the slack between two activities to be insufficient in light of the uncertainties and risks involved in those activities, the float can be increased and the milestones adjusted accordingly, while ensuring that these don't violate technical or safety requirements that are reflected in maximum relationships.

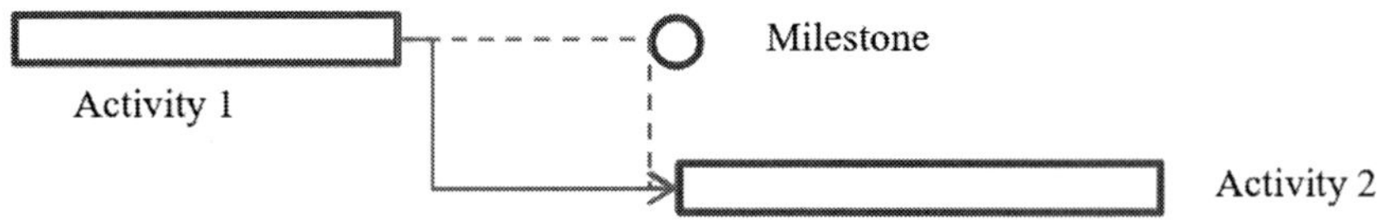

Figure 3: Milestone for non-critical activity (Activity 1)

3.3 Control Actions

Control actions are carried out in order to track and compare the actual progress with the planned progress using the predefined control points. When possible, the control actions make use of project buffers to absorb the impact of deviations from the schedule. Buffers are a gap between the (required) minimal and (actual) defined attributes of an activity. To this end, resources such as time, equipment and manpower are often allocated in the project plan beyond the minimum required for the planned activities. This is done to protect the activity from variability in other activities which produce or use the same resource, and upon which the first activity is therefore dependent. Several types of buffers are used in project plans [21]:

a. Inventory buffers consist of stocks of materials and finished components, and of plan buffers – backlogs of work for crews that are used to ensure reliable workflows [12].
b. Contingencies consist of time buffers (extra time allotted for activities) and reserve funds in the budget.
c. Capacity buffers consist of excess space, manpower and equipment capacities allocated for activities.

In order to carry out a quantitative assessment of the possible magnitude of the impact of a deviation that has been detected, and the consequent controlling action required to cope with this deviation, the status of buffers in the project has to be monitored, and the fraction of the deviation that can or should be absorbed by these buffers has to be assessed. Since the size of buffers that can be used to absorb the impact of changes may vary during a project, depending on external factors, it has to be continuously monitored in real time. In some cases, additional information has to be manually collected for this purpose, since project team members may have tacit knowledge regarding the actual buffers, when these are not explicitly known. For example, a sub-contractor can probably assess how much additional work he can carry out in the project.

4 Discussion and Further Research

To summarize, it is proposed that current scheduling methods can be changed to support effective, real-time schedule control, using automatically collected monitoring data, in the following way:

1. Additional control points are added to the project activity network. These include internal mid-activity points, and milestones for non-critical activities. New relationships are defined between those points to detect the implications of a deviation in one activity for other related activities in the project.
2. Deviations are automatically detected, in real time, at the control points that have been defined. This ensures that monitoring is carried out at any point that is important for project management, and not just at the end-points of activities.
3. The current status of buffers in the project is tracked, in order to detect the remaining slack, and consequently the control actions that can be taken.
4. Appropriate control actions are initiated to make use of available buffers in order to absorb the impact of deviations that have occurred and adjust the schedule. This will prevent a deviation from the project goals, in terms of the date of delivery and overall cost.

Further research is currently being conducted in order to develop new mathematical definitions that will allow the addition of the proposed control points to the precedence diagram, and to develop algorithms for updating the schedule in accordance with the data that is automatically collected.

References

[1] Gantt, H L (1913) *Work, wages, and profits*. Engineering Magazine Co.

[2] Ahsan, K. and Gunawan, I. (2010). Analysis of cost and schedule performance of international development projects. International Journal of Project Management, 28(1), 68–78.

[3] Hwang, B, Thomas, S, Haas, C and Caldas, C (2009) Measuring the impact of rework on construction cost performance. *Journal of Construction Engineering and Management*, 135(3), 187-98.

[4] Love, P E D, Watson, H and Davis, P (2010) Rework in civil infrastructure projects: Determination of cost predictors. *Journal of Construction Engineering and Management*, 136(3), 275-82.

[5] Cheng, T., & Teizer, J. (2013). Real-time resource location data collection and visualization technology for construction safety and activity monitoring applications. *Automation in Construction*, *34*, 3-15.

[6] Turkan, Y., Bosche, F., Haas, C. T., & Haas, R. (2012). Automated progress tracking using 4D schedule and 3D sensing technologies. *Automation in Construction*, *22*, 414-421.

[7] Song, L., & Eldin, N. N. (2012). Adaptive real-time tracking and simulation of heavy construction operations for look-ahead scheduling. *Automation in Construction*, *27*, 32-39.

[8] Vahdatikhaki, F., & Hammad, A. (2014). Framework for near real-time simulation of earthmoving projects using location tracking technologies. *Automation in Construction*, *42*, 50-67.

[9] Isaac S. and Navon R. (2014). Can project monitoring and control be fully automated?, Construction Management and Economics, 32:6, 495-505

[10] Docherty, P. (1972). *The management of Contingencies*. Stockholm, Sweden: The Economic Research Institute.

[11] Laufer, A., & Tucker, R. L. (1987). Is construction project planning really doing its job? A critical examination of focus, role and process. *Construction Management and Economics*, *5*(3), 243-266.

[12] Ballard, G and Howell, G (1995) Toward construction jit. In: *11th Annual ARCOM Conference*, Reading, UK, 304-12.

[13] Koskela, L. (1992). *Application of the new production philosophy to construction* (No. 72). Stanford, CA: Stanford University.

[14] Mahdi, I. M. (2004). A new LSM approach for planning repetitive housing projects. *International Journal of Project Management*, *22*(4), 339-346.

[15] Zhang, H., Li, H., & Tam, C. M. (2006). Particle swarm optimization for resource-constrained project scheduling. *International Journal of Project Management*, *24*(1), 83-92.

[16] Lipke, W., Zwikael, O., Henderson, K., & Anbari, F. (2009). Prediction of project outcome: The application of statistical methods to earned value management and earned schedule performance indexes. *International journal of project management*, *27*(4), 400-407.

[17] Book, S. (2006). Earned schedule" and its possible unreliability as an indicator. *The Measurable News*, 13(2), 24-30.

[18] Vanhoucke, M. (2013). An overview of recent research results and future research avenues using simulation studies in project management. *ISRN Computational Mathematics*, 2013.

[19] Crandall, K. (1973), Project planning with precedence lead/lag factors, *Project Management Quarterly*, 4, 18– 27.

[20] Francis, A., & Miresco, E. T. (2002). Decision support for project management using a chronographic approach. *Journal of decision systems*, *11*(3-4), 383-404.

[21] Alves, T and Tommelein, I D (2004) Simulation of buffering and batching practices in the interface detailing-fabrication-installation of hvac ductwork. In: *IGLC-12*, Elsinore, Denmark.

Human-Robot Collaboration for Creative and Integrated Design and Fabrication Processes

Shani Sharif [a], T. Russell Gentry [b] and Larry M. Sweet [c]

[a,b] School of Architecture, Georgia Institute of Technology, USA
[c] School of Interactive Computing, Georgia Institute of Technology, USA
Email: shani@gatech.edu, russell.gentry@coa.gatech.edu, lsweet7@cc.gatech.edu

Abstract

This paper proposes a new framework for digital design to fabrication workflow to create an effective collaboration process between human designers and industrial robots, in order to redefine the relationship between design and production. This framework aims at redefining the current one-directional workflow of design to robotic fabrication that is imposed by the current robot programming and control systems, originally developed for engineering product industries. Our proposed system employs machine perception, adaptive control, and design libraries to adapt fabrication processes in real time, in parallel with human perception, learning, and guidance, to act as a catalyst of innovation for design practices, accelerating the design and production of novel building and industrial components.

Keywords –

Robotic fabrication; Human-robot collaboration; Perception; Feedback; Adaptive control

1 Introduction

Over the last decade, industrial robots have been adopted by architects for fabrication and assembly, as they offer flexible and multi-purpose manipulators for a variety of additive, subtractive and deformative fabrication processes [1]. Combined with the potential of generative and parametric design and modeling, robots can fabricate complex three-dimensional geometries for building components and assemblies beyond the capabilities of conventional fabrication techniques. By using industrial robots in design to fabrication processes, architects can eliminate the need for costly fixtures and setup times that hinder the cost-effectiveness of automation, especially for physical prototypes in small lot sizes. Consequently, robotic fabrication, which can be enhanced by real-time process feedback, has created an opportunity to reconsider the entire design-to-production chain. In the current design to fabrication processes, many steps in the design process are speculative, dependent on estimation and prediction of the material, tool and design behavior in the production process. By parameterization of the design aspects, and measurability of the material/tool conditions with the aid of real-time sensors feedback has created an option to push the knowledge beyond speculation and into the realm of accurate and predictable architectural production.

However, the robot control and motion programming tools that are currently adopted by designers were all originally developed for engineering product industries [2]. These technologies have been developed for industrial mass production of components with known problems and processes, and predictable outcomes. The current industrial robot control system for fabrication requires that the designer have a comprehensive view of the design object and embed detailed design and machining data in the digital model before the start of the fabrication phase (i.e. a complete specification). As a result, the process of design to fabrication is mostly a one-directional workflow, in which the designer/maker has to predict material state, tool selection, fixture positioning and robot motion planning based on prior experience. Therefore, there is a high cost and time penalty for re-work (Figure 1). In the conventional workflow, while the design conception affects the process of part production, no feedback during the action of making affects the organization and content of design model and designer's intent. There is no formal mechanism for knowledge capture and reuse.

Recent research in the field of computational design calls for the new possibilities and potentials for human and machine interactions as complementing collaborators in design-making processes. These studies intend to unify design and fabrication processes and close the divide between the act of designing and making. As Gramazio and Kohler [3] argue: "Achieving a sophisticated building component ... can be compared to methods used by manufacturers from pre-industrialized ages. Despite the similarities, today the action of material handling is indirect through the use of CNC machines as opposed to the instant feedback about the work in progress the skilled manufacturer received through the tool in his hand. With CAM, the tool is controlled through explicit routing data, which leaves no room for interpretation and adaptation." Highlighting this gap, some studies have proposed the development of new interfaces for integration of physical and digital environments. These project-specific solutions augment actuators with sensors to incorporate the real time feedback on material state over the production process [4-7].

While these projects propose creative solutions for integration of sensor input and real-time feedback into the design and fabrication processes, the solutions are project-based. Focusing on this current technological gap, the goal of this research is to develop a "framework" for a comprehensive (computer and human) feedback system to be integrated into the design to fabrication workflow. This feedback system is aimed to change the current one-directional workflow to a complete closed-loop and interactive system between human designer, digital design environment, robot fabricator and the final product. In this new workflow for human-industrial robot collaboration, the design can evolve to meet the affordances of the production process, and at the same time, the production method can evolve to take advantage of feedback from computer sensing and simulation, coupled with human feedback into the process. Our proposed system will employ advanced machine perception, adaptive control, and design libraries to adapt fabrication processes in real time, in parallel with human perception, learning, and guidance, to act as a catalyst of innovation for design practices, accelerating the design and production of novel building and industrial components.

2 Role of Design Cognition in Human-Robot Interaction in Different Design Stages

The proposed framework for human-robot collaboration should enable cognitive technologies to essentially perform as complementary operations to those of human brain cognitive capabilities. Human brain has strong cognitive capabilities in tasks such as pattern recognition, extracting and encoding relational properties such as relative distance and parallelism, shape-based object identification and segmentation, or modeling simple dynamics of the world. However human cognition cannot perform strongly in other tasks such as recalling and executing of long or arbitrary sequences of operations [8, 9]. The main feature of human intelligence is that it can create tools and extended cognitive systems to complement its basic modes of processing and compensate for its less strong cognitive capabilities. A human-robot collaborative system is thus successful as if it can take advantage of human cognitive system strengths and compensate for its weaknesses with robotic and computer system strengths. To maximize the benefit for a successful design and fabrication system, human and robot should utilize their strengths to the fullest (Table 1).

In general, any creative design process consists of two main stages, conceptual design stage, followed by detailed design stage, leading to final design production. In the conceptual design stage, there are many factors that have to be decided by the designer, including aesthetic of design, functional adequacy, material selection and conditions, production techniques and procedures and tools, and financial and time constraints. At this stage of the design, the impact of design decisions is very high (Wang 2002). In this stage, design decisions are made based on the design intent that are matched selectively with designer's prior production experiences based on the closest correspondence with the current design situation. In addition, the act of making, the produced physical prototype, and the ongoing designer's perception of this action transform and enrich the designer's prior organization of knowledge. However, the designer's knowledge of a specific process is never adequately detailed and precise to predict all the possible situations and outcomes of the production process. In this

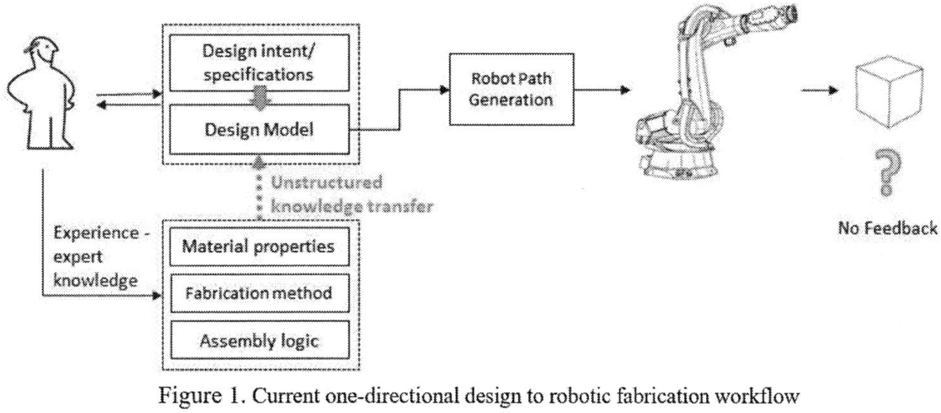

Figure 1. Current one-directional design to robotic fabrication workflow

process, design conception and artefact production co-evolve. The final design concept generated at this stage affects the product shape geometry, production material selection, as well as construction and manufacturing productivity and final product quality. In the following detailed design phase, it would be very difficult to compensate or correct the unanticipated aspects or shortcomings of the design concept developed at the conceptual design stage.

Table 1. Human Perception vs. Computer Perception, Weaknesses and Strengths [8-10]

Human perception	Robotic and Computer system
Strength	**Strength**
Creative design based on design intent	Accurate Form generation with numerical inputs (parametric form development)
Alternative Form development	Speed in numerical computation
Alternative fabrication technique development	Speed in physical operation
Synthesizing and interpolation	Accuracy in operation (computational and physical), complex sequence of operations
Detection of trends, patterns, or anomalies in visual data (digital and physical design model)	Endurance and reliability
Learning from limited test cases	Consistency
Collaboration and integration (people, tools, senses)	
Weakness	**Weakness**
Unable to detect minor deviations in the digital and physical prototypes	No creativity: difficult to synthesize new rules
Easily tired of repetitive operations	Unable to produce alternate and "out of the box" solutions to compensate for inadequate design solutions
Biased and inconsistent	No common sense
Erroneous	

Consequently, the proposed framework should be able to address different needs in design concept development and detailed design development. In the concept design development, the capability of the robotic system in facilitating design process for designer, and facilitating creativity has higher importance. In the detailed design development stage, capability of the system in building more accurate design prototypes, and avoiding potential fabrication errors is more essential.

3 Current Robotic Fabrication Workflow

As previously discussed, the current design to robotic fabrication workflow is one-dimensional. There are however, a number of relevant efforts that have made the design-to-robot programming section of the workflow more interactive. These efforts are discussed below.

3.1 Flexible Representation of Geometry

In the current workflow of design to robotic fabrication, designers start with computational representation of their design (geometric, parametric attributes and properties) using software tools such as Rhinoceros, Autodesk AutoCAD or Autodesk Revit. These environments are enriched by add-on tools that provide direct programming tools for design generation such as RhinoScript, or more commonly graphical programming and algorithm editors such as Grasshopper, running on Rhinoceros 3D CAD application, or Autodesk Dynamo for Revit. These design representation environments, in combination with computational programming logics provide flexible environments for designers to represent their complex geometrical forms with various parametric attributes and properties. The parametric representation of geometry – either through a script, or in a feature-based modeling environment – as apposed to a more conventional fixed representation of geometry, would allow for appending additional layers of information to different parameters of geometry. The appended data could range from robot control parameters to parameterized material or fabrication logic information.

3.2 Parametric Robot Control

As the conventional methods of robot control and motion programming were not developed based on the needs and skills of designers, researchers in these field have focused on the development of more flexible and intuitive robot control and programming tools. These new software tools have acquired graphical programming editors such as Grasshopper. Different plug-ins such as Kuka|prc [11], HAL [12] for programming and kinematic simulation of industrial robots such as KUKA, ABB, and Universal Robots have been developed as plug-ins for Grasshopper and Dynamo. These plug-ins for graphical robot programming provide the option for architectural designers to program and simulate industrial robots directly out of the parametric modelling environment based on the geometric parameters of their designs. However, the final physical prototype outcomes are still highly dependent on the architect's predictive capability of design, fabrication, analytical and process models. While robot motion can be derived from geometric parameters with these tools, they provide no option for integration of material properties or fabrication intelligence into the design model to be translated into robot motions.

3.3 Integration of design and fabrication

While theoretically, parametric and programming tools for digital modeling of the design, would create a platform for definition of material and fabrication knowledge as generative design factors in design to fabrication models [3, 13], in practice, the material and fabrication factors have to be selected and modified by designers. At the current state, the human expert knowledge (based on experience and past experiments) is required to define and embed material selection, material properties, manufacturing method, assembly, and installation logic as part of the parametric CAD model, a non-automatic and error-prone process. Consequently, any mis-estimation in the anticipated design, material, and fabrication behavior and their translation into parametric rules would result in rework with high cost and time penalty. In addition, while robots have sufficient repeatability and accuracy to produce desired tool paths in theory, in practice for many applications uncertainty is introduced through variability in the process depositing or deforming the work material. Even with highest level of consideration for integration of material and fabrication logic in the parametric CAD model, unpredictable material, tool, or environmental conditions might result in deviations from anticipated results for fabricated part, as there is no instant feedback about the work in progress [14].

4 Project Testbed

At the current state, the proposed framework is envisioned for adoption in design research labs or commonly known as fab-labs that are environments for development and assessment of new design and production knowledge [15]. Utilizing robotic technologies at fab-labs, designers can assess and improve their design intent through fast and accurate fabrication of full-scale physical prototypes, where there is a one to one mapping between a complete CAD-CAM model, fabricated model and anticipated actual building or industrial part. Industrial robots provide new techniques for prototyping of building and industrial components out of commonly-used materials of production such as sheet metal, as opposed to model materials and techniques such as 3D printing that is still constrained by high cost, long cycle time, materials with limited strength and environmental tolerance. Prototyping with materials of production would provide the designer with process and material feedback on the actual design's weaknesses and strengths in the final production process.

As the test bed, this research focuses on sheet metal robotic manufacturing methods due to their wide range of applications in architecture and manufacturing industries. With different developing and fabrication techniques, three-dimensional forms can be made out of two-dimensional planar geometries. Robotic sheet metal fabrication is most suitable for making test models, and prototypes in small batches.

Robotic fabrication eliminates the need for time consuming and expensive methods such as drawing, punching, or hydro forming into dies. The methods for robotic fabrication with sheet metals include:

1. **Bending/folding:**

- Bending or straight-line folding [16, 17]
- Curved folding resulting in more complex three-dimensional geometries [18]

Parts with complex geometries can be folded from a single planar sheet of metal without stretching, tearing or cutting, either along straight or curved lines [18-20]. Bending flat sheets along straight lines in press brakes is most common. Boxes, brackets, and similar shapes are possible by repositioning the workpiece in the brake, with subsequent bends perpendicular or oriented at other angles from prior bends. Only certain shapes are feasible, due to interference between bent shapes on the workpiece, or interference between workpiece and press brake, so human skill or advanced planning algorithms are required in design and bend sequence selection [21]. Collision detection algorithms based on segment intersection have been developed to create piecewise linear curves with a series of straight line bends, with bends all parallel to each other, with advanced planning possible for interference checking [16].

2. **Incremental sheet forming (ISF)** [22, 23]

A recent approach called "incremental forming" creates three-dimensional surfaces with continuous curvature, without constraints from straight line bending [24]. Single point and two-point incremental forming (SPIF and TPIF) tools are pressed into the metal, creating local plastic deformation. Moving the tools in discrete increments along paths digitally controlled through CNC or robots eventually creates a three-dimensional shape. Producing smooth surfaces may require very small, time consuming increments, limit material choice, and require careful characterization of material forming, lubrication, and tool shape properties, as shown by [25, 26], and [27]. Nimbalkar and Nandedkar explore the use of two six-DOF industrial robots instead of CNC machines, "Roboforming", with the potential advantage of forming "undercuts" compared to CNC produced shapes, which are concave in general. However, the resultant geometries produced deviate significantly from CAD models, due to inaccuracies in models for forming process, workpiece springback, and deflections in the robots themselves due to high forces applied by the incremental forming tools.

3. **Metal driving (shrinking and stretching)** [24, 28]

Driving is one of the older manufacturing techniques for sheet metal forming by which variety of geometries can be achieved. In driving process each tool set consists of a top and a bottom tool and has sub-methods, shrinking (compressive stress), and stretching (tensile stress) which results in deformation of sheet metal. The process has been mostly manual so far, in which parts are incrementally formed by a multitude of strokes into the sheet. The robotic and numerically controlled driving which requires automated positioning and handling of sheet metal which requires complex robot tool path generation techniques) [24, 28].

The main advantage of robotic sheet metal fabrication to other numeric controlled machines such as sheet metal CNC press brakes, shears and punching machines is the flexibility and adaptability of robotic arm for performing different fabrication and assembly methods with the appropriate end-effector. Utilizing robotic arms eliminates the need for production of dies, molds and presses. In addition, robotic arms offer fabrication methods for extremely complex geometries that otherwise could to some extend only be achieved through traditional metal craftwork in a slow and time-consuming process. The main factors in a systematic approach for robotic sheet metal fabrication includes geometric form development, analysis of material behavior during fabrication with finite element method (FEM) analysis, optimization of the fabrication process, operation sequencing, and finally robot toolpath generation.

Computer simulation of forming processes using finite element methods (FEM) are well established, so that for given shapes of forming tools (dies, molds) and material properties, the end shape can be predicted, including the effect of spring back. The more complex challenge is the inverse problem, i.e. given the desired shape, what tool geometry and material properties should be used? The first step is estimation of material properties, as shown in [29, 30]. If tool geometry is expressed parametrically, optimization methods can be developed for automated parameter identification [31], [32]. It is anticipated that based on the results of this research, the same method can be applied for other material systems and robotic fabrication methods based on the same principles.

5 Proposed Framework for Human-Robot Interaction

The proposed framework for human-robot interaction has two main elements (Figure 2):

1. **Adaptive robot control** (Based on Sensor feedback):
   - Matching sensor data (3D point cloud) with CAD model (in conceptual design stage)
   - Real time sensor data + robot motion program (in detailed design-fabrication stage)
2. **Design fabrication library:**
   - Design modeling (folding sequence, perforation shape, material type-thickness, fabrication method)
   - Tool path library (parameterized robot tool paths, parameterized fixture points)

Addition of of an extra layer of information gathering, processing and decision making would provide a comprehensive system for human designers to work with robotic arms in the process of digital design and fabrication. Most importantly, this layer of feedback-control ensures a comprehensive system for incorporation of both human and computer feedback into the design process for an effective human-robot collaboration in a bidirectional workflow from design to fabrication, and fabrication to design. The elements of this system are discussed in more detail in the text below. Note that for the successful development of this system, it is necessary to parameterize the representation of design model and the robot motion. This parametric model definition would facilitate the transformation of geometric data into fabrication data.

5.1 Parametric Model Definition

Developing algorithms to facilitate transformation of geometric data into fabrication data requires parameterized representation of design data. This parameterization of form would allow association of geometric information to material, robotic fabrication, and assembly attributes and properties through computational programming logics. A complete design to fabrication workflow requires association of geometric attribute to:

1. design intent (structural element, covering element, or connection part)
2. fabrication and assembly logic (folding sequence, assembly sequence, ...)
3. material properties (metal type, thickness, ductility, spring back, ...)
4. robot motion control

5. fixture positioning and material handling (relative to robot end-effector)
6. real-time data from scanner in point cloud format

In the last few years, there have been efforts for automation of some of these aspects, such as development of tools such as KukaPRC or HAL as discussed in Section 3.2 for motion programming and control of industrial robots directly from the geometric parameters of the design, in a single modeling environment. However, the other necessary factors that are related to materiality and fabrication logic are mostly not addressed. It has to be taken into account that while some of these factors such as design intent are known to the designer in the early design stage, many of the aspects such as material type, or assembly logic are not fully defined by the designer in the conceptual design development stage. Currently, decision-making process for these factors is a non-automatic and error-prone process, based on experience and trial and error. Although these factors are not decided at conceptual design stage, creating a framework for parameterizing the attributes related to materiality and fabrication would facilitate the development of an informed design model.

5.2 Adaptive control

The first element in the proposed framework is the *adaptive control* system of robot based on real time fabrication data from a sensor system (camera, 3D scanner, or force/torque sensors. The objective of this aspect of the research is development of a system that the data from a sensor system (one or a combination of 3D scanner, camera, depth sensor, force/torque sensor…) would feedback into the design modeling or robot path generation steps of the process; create a platform for informed decision-making process. There are two main parts of the system that will use the data received from the sensor system on the state of physical prototype:

- **Layer 1:** Matching sensor data (3D point cloud) with CAD model
- **Layer 2:** Real time sensor data + robot motion program correction

Layer 1 of the adaptive control is mainly intended for conceptual design development of the process. At this layer, the 3D point cloud (or other numeric data) from the sensor will be sent back to the CAD modeling environment. By aligning the point cloud model (with parametric feature matching) with the CAD model [33], the designer can compare the deviation of the physical prototype from the intended CAD model. Based on the comparison result, designer will have the chance to editing or changing the design model based on the limitation of the fabrication method.

Layer 2 of the adaptive control is mainly intended for detailed design and fabrication stage of the process. At this layer, system uses the defined adaptive control algorithms for adjusting tool path and fixture point locations, so that the actual shape converges to the designer's intent. Based on the sensor data, the gap between the scan points and the desired shape is calculated, and the adaptive control loop adjusts the robot planning vector parameters to achieve the fabrication result closest to the design intent, defined in the computer model. These algorithms that can be defined based on the acceptable tolerance, can be really helpful for fine-tuned fabrication of each single part, so that the geometric model and robot movements are refined to minimized the tolerance and error in the final assembly of all part.

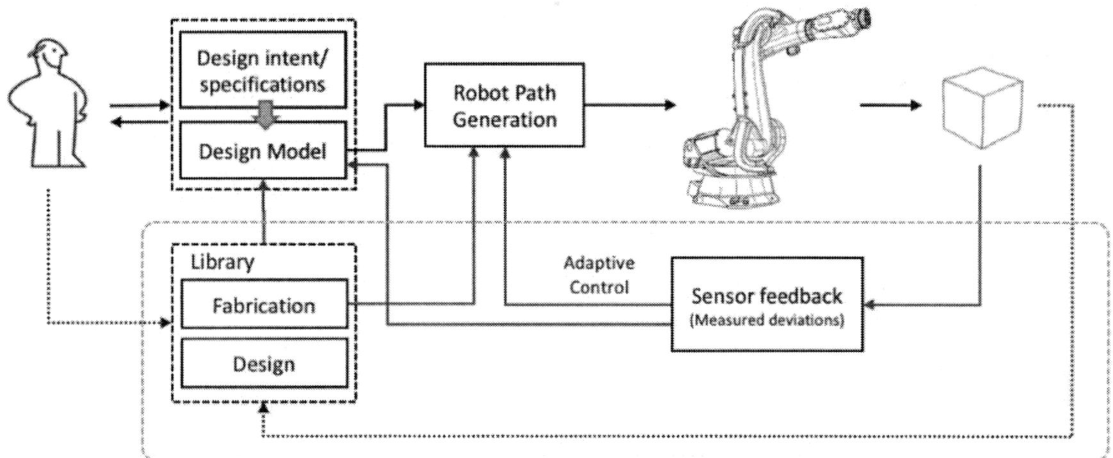

Figure 2. Proposed bi-directional design-fabrication workflow

5.3 Design Libraries

Design Fabrication Library is the next element of the system that utilized the infrastructure that is layout for parametric model definition. This library has two main parts:

1. Design-fab modeling (folding sequence, perforation shape, material type-thickness, fabrication method)
2. Tool path modeling (parameterized robot tool paths, parameterized fixture points)

In the conceptual design stage, this library will assist the designers with the decision making process on the adequacy of the design choices, detailing, material selection for the selected robotic production techniques and end-effectors. The correct choices in the computer-modeling environment at this step would extremely affect the success of the fabrication tests, and prevent rework and time and material waste. To achieve this goal, design library relates the mentioned parameters to sets of geometric shapes. Examples include compliant hinges for Origami-inspired sheet metal fabrications [34], and a database for optimal material and thickness identification for sheet metal folding [35].

The following example highlights one of the challenges that can solved with use of a design library. In most robotic and CNC controlled manufacturing processes, the tool path generally follows the surface of the part. This enables automatic tool path generation from CAD solid models, with adjustments made to the tool path allowing for cutting or forming tool geometry, approach vector, and process specific characteristics. Fixture points can be determined separately, their role primarily to assure stability of the workpiece. To fabricate three-dimensional sheet metal surfaces robotically, the tool path in general does not follow the surface, and location of fixture points has a strong influence on the shape. In the current metal folding process, the tool path is defined based on lots of experimentation with hand prototyping [18].

While for a given tool path, fixture locations, and material properties, the final shape can be determined through FEM simulation or experimentation; the main challenge is inverse planning problem. i.e. to go from desired end shape to tool path and fixture location [32]. Consequently, the design library will assist the designers with solutions stored in the library database. The designer selects 3D shape classes in the library that most closely matches the desired shape and material properties. If there is a close match with one of the stored classes, the stored parameterized tool path and fixture points may be used directly; if not the initial motion plan will be interpolated between adjacent classes. The designer has the option of modifying the initial tooling vector based on experience in working with the material. Using the initial tool path and fixture points, the first sample part is produced.

The design library will be developed incrementally based on both experimentation and human expert knowledge. Both real and virtual (FEM) experimentation along with statistical regression to interpolate between the results of these experiments would yield into a complete series of options for different design choices. The database will grow over time with data from continued experimentation by multiple users. For each initial and subsequent fabrication trial, scanned images for the surface produced will be added to the design library, so over time the library content will grow, filling in the gaps between adjacent designs.

6 Conclusion

This research addressed the main elements that are required for development of a new digital design to fabrication workflow, in order to create a structure for effective collaboration between human designers and industrial robots. We will implement these algorithms, controls and design libraries in an integrated test-bed and will test the test-bed with student designers in graduate and undergraduate level Robotic Fabrication courses at Georgia Tech.

We anticipate that by updating of design information based on fabrication sensor feedback would result in minimizing the errors through sensing and re-tooling/re-work. This would also result in increasing the body of work that can be accomplished with industrial robots. In addition, by developing techniques for sheet metal fabrication, this research would create the potential for extension of this system to other materials and system. Most importantly, this workflow would construct the required infrastructure for involvement of designers in robotics design and fabrication process.

References

[1] McGee, W., M.P. de Leon, and A. Willette, *Robotic Fabrication in Architecture, Art and Design 2014*. Springer, 2014.

[2] Kolarevic, B., *Architecture in the digital age: design and manufacturing*. 2004: Taylor & Francis.

[3] Gramazio, F., M. Kohler, and S. Oesterle, *Encoding material*. Architectural Design, 80(4), 2010.

[4] Mueller, S., P. Lopes, and P. Baudisch. *Interactive construction: interactive fabrication of functional mechanical devices*. in *Proceedings of the 25th ACM*. 2012.

[5] Willis, K.D., et al. *Interactive fabrication: new interfaces for digital fabrication*. in *Proceedings of the fifth international conference on Tangible,*

embedded, and embodied interaction. ACM, 2011.

[6] Raspall, F., F. Amtsberg, and S. Peters, *Material Feedback in Robotic Production*, in *Robotic Fabrication in Architecture, Art and Design 2014*. Springer, 2014.

[7] Johns, R.L., A. Kilian, and N. Foley, *Design Approaches Through Augmented Materiality and Embodied Computation*, in *Robotic Fabrication in Architecture, Art and Design 2014*. Springer, 2014.

[8] Clark, A., *Natural-born cyborgs: Minds, technologies, and the future of human intelligence*. Oxford University Press, 2004.

[9] Norman, D.A., *Being Analog*, in *The invisible computer: why good products can fail, the personal computer is so complex, and information appliances are the solution*. MIT press, 1998.

[10] Halim, S., *An integrated White+ Black box approach for designing and tuning stochastic local search algorithms (*in *Computer Science*. National University of Singapore, 2009.

[11] Braumann, J. and S. Brell-Cokcan, *Parametric robot control*. Integrated CAD/CAM for Architectural Design, ACADIA, 2011.

[12] Schwartz, T., *HAL*, in *Rob| Arch 2012*. p. 92-101. Springer, 2012

[13] Menges, A., *Material computation: Higher integration in morphogenetic design.*. 82(2): p. 14-21. Architectural Design, 2012.

[14] Kohler, M., F. Gramazio, and J. Willmann, *The Robotic touch: How robots change architecture*. Park Books, 2014.

[15] Valdes, F., A. Cavieres, and R. Gentry, *A Process-Centric Approach for Teaching Digital Fabrication*, in *SIGRADI*. p. 400-404, Valparaiso, Chile, 2013.

[16] Liao, X. and G.G. Wang, *Evolutionary path planning for robot assisted part handling in sheet metal bending*. Robotics and Computer-Integrated Manufacturing. 19(5): p. 425-430. 2003

[17] Aomura, S. and A. Koguchi, *Optimized bending sequences of sheet metal bending by robot*. Robotics and Computer-Integrated Manufacturing, 18(1): p. 29-39. 2002.

[18] Epps, G. and S. Verma. *Curved Folding: Design to fabrication process of RoboFold*. in *Shape Modeling International*. 2013.

[19] Tachi, T. and G. Epps. *Designing One-DOF mechanisms for architecture by rationalizing curved folding*. in *International Symposium on Algorithmic Design for Architecture and Urban Design (ALGODE-AIJ). Tokyo*. 2011.

[20] Kilian, M., et al. *Curved folding*. in *ACM Transactions on Graphics (TOG)*. ACM, 2008.

[21] Duflou, J.R., J. Váncza, and R. Aerens, *Computer aided process planning for sheet metal bending: A state of the art*. Computers in industry, 56(7). 2005.

[22] Kalo, A. and M.J. Newsum, *An investigation of robotic incremental sheet metal forming as a method for prototyping parametric architectural skins*, in *Robotic Fabrication in Architecture, Art and Design 2014*. p. 33-49. Springer, 2014.

[23] Ponticel, P. *Ford sheet-metal-forming technology on journey from lab to application*. 2013 [cited 2016; Available from: http://articles.sae.org/12385/.

[24] Opritescu, D. and W. Volk, *Automated driving for individualized sheet metal part production—A neural network approach*. Robotics and Computer-Integrated Manufacturing, 35: p. 144-150. 2015.

[25] Arshad, S., *Single Point Incremental Forming*. KTH Royal Institute of technology Stockholm, Sweden, 2012.

[26] Bouffioux, C., et al., *Identification of material parameters to predict Single Point Incremental Forming forces*. International Journal of Material Forming. 1(1): p. 1147-1150. 2008.

[27] Nimbalkar, D. and V. Nandedkar, *Review of incremental forming of sheet metal components*. Int J Eng Res Appl, 3(5): p. 39-51. 2013.

[28] Hoffmann, H., R. Hautmann, and R. Petry. *Studies for the development of a simulation basis for numerically controlled driving of sheet metal*. in *Advanced Materials Research*. Trans Tech, 2005.

[29] De-Carvalho, R., R. Valente, and A. Andrade-Campos, *Optimization strategies for non-linear material parameters identification in metal forming problems*. Computers & structures, 89(1). 2011.

[30] Chaparro, B., et al., *Material parameters identification: Gradient-based, genetic and hybrid optimization algorithms*. Computational Materials Science, 44(2): p. 339-346. 2008.

[31] Valente, R.A., et al., *Parameter identification and shape optimization: An integrated methodology in metal forming and structural applications*. Optimization and Engineering, 12(1-2). 2011.

[32] Ponthot, J.-P. and J.-P. Kleinermann, *A cascade optimization methodology for automatic parameter identification and shape/process optimization in metal forming simulation*. Computer Methods in Applied Mechanics and Engineering, 195(41). 2006.

[33] Ip, C.Y. and S.K. Gupta, *Retrieving matching CAD models by using partial 3D point clouds*. Computer-Aided Design and Applications, 4(5). 2007.

[34] Delimont, I.L., S.P. Magleby, and L.L. Howell, *Evaluating compliant hinge geometries for origami-inspired mechanisms*. Journal of Mechanisms and Robotics, 7(1). 2015.

[35] IndustrialOrigami. *Industrial Origami Database*. 2016 [cited 2016; Available from: http://www.industrialorigami.com/technology/proprietary_database.cfm.

Optimizing the trajectory of the painting robot

Ing. Michal Bruzl

Department of Construction Technology (K122), Faculty of Civil Engineering in Prague Thákurova 7/2077, 166 29 Prague 6 - Dejvice, Czech Republic

michal.bruzl@fsv.cvut.cz

Ing. Vyacheslav Usmanov

Department of Construction Technology (K122), Faculty of Civil Engineering in Prague Thákurova 7/2077, 166 29 Prague 6 - Dejvice, Czech Republic

vyacheslav.usmanov@fsv.cvut.cz

doc. Ing. Pavel Svoboda, CSc.

Department of Construction Technology (K122), Faculty of Civil Engineering in Prague Thákurova 7/2077, 166 29 Prague 6 - Dejvice, Czech Republic

pavel.svoboda@fsv.cvut.cz

Ing. Rostislav Šulc, Ph.D.

Department of Construction Technology (K122), Faculty of Civil Engineering in Prague Thákurova 7/2077, 166 29 Prague 6 - Dejvice, Czech Republic

rostislav.sulc@fsv.cvut.cz

Abstract –

The main objective of this research work was to create a technological and mathematical model of the robotic arm movement for conducting painting work. For experiment was chosen model of the robotic arm, which corresponds in its construction robots commonly used in industrial production. This model of the robot was adapted to the conditions of carrying out painting work. One of the objectives was to create a mathematical model, which is designed to find the optimal trajectory for moving of the robot arm. Thanks to the mathematical model, which allows to optimize the trajectory of the painting robot, it was possible to deal with technological aspects. One of the main aspects was the consumption of material, in this case the method of applying colour to the surface. The path of the endpoint of the robotic arm must apply uniform colour layers with minimal overlap of previously painted surfaces. This is related to the fluidity of movement of the robot, which is determined by the functionality of the articulated joints. In the model was monitored load of servomotors that are placed always in contact with the arm pads, arm with next arm and arm with the endpoint. The optimal load for the servomotor, in particular the total energy consumption during the movement, was designed to avoid overload and subsequent collision. The aim of optimizing the technological model of painting robot is shortening the period for which the robot, with defined consumption of materials and energy, is capable to make the optimal trajectory.

Keywords –

Mathematical model; Robotic arm; Painting work; Trajectory for moving; Energy of servomotors

1 Introduction

The topic of the given project goes hand in hand with the effort of implementing robotics and automation into the construction process. The process is completely different from automation of manufacturing due to the amount of performed operations, mobility and fluctuating work environment. Thanks to robotics in civil engineering we can achieve higher rate of safety, increase in productivity and maintaining the quality of the final product.

The project is divided into several phases. The first step is choosing the robotic machine (industrial robot with six axes) alternatively the construction of corresponding industrial robot (see Figure 1). The next step is a simulation of the movement itself, which has to be as economic as possible. The emphasis is put not only on ideal trajectory but also the final consumption of energy of every servomotor in the model. The last part of the research is suitable choice of color applicator with emphasis on fittingly chosen technology of painting in such a way that used material is saved.

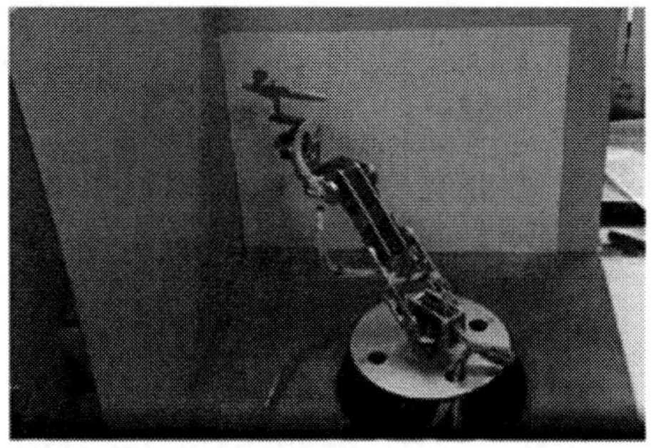

Figure 1 – First model of the painting robot

2 Optimal trajectory of spraying movement

The main objective is finding the optimal movement course of the painting robot's arm. To fulfill the main objectives it is necessary to accomplish following partial goals.

- Creation of mathematical model of the robot's movement;
- Choice of the applicator and nozzle;
- Specification of the optimal movement for color application,
- Identification of the optimal layer overlapping

2.1 Mathematical model of the robot's movement

For the correct choice of the model it is needed to perform a movement simulation which will abide by the following procedure:

- choice of mathematical method of optimization;
- suggestion of mathematical model (determination of the inputs, parameters and outputs);
- verification of mathematical model by physical model of the painting robot;

Suggested course of action has to be verified with the whole physical model of the robot.

To solve the task of robotic arm's optimal movement a model using a TSP (travelling salesman problem) algorithm was designed [1]. TSP algorithm is one of the NP-difficult (nondeterministic polynomial time) discrete optimization problems (heuristic algorithms giving approximate solution were used to solve this). Solution of algorithm mathematically displays and generalizes the task of finding the shortest possible path going through all the given spots on the map or knots in network graph. The goal of the task was to find the shortest path for the „business traveler" or for closed Hamilton's circle with the lowest edge rating. The task is one of the very important and widespread in graph theory. The difficulty of the task drastically rises with the increasing number of knots.

Solving methods:

- **Solving with brute force** – the whole area of solving is being investigated. The time needed for finding solution increases with factorial, so it is not possible to use this method in practice even for the small areas of possible solutions. For 20 knots of our task for painting robot it was necessary to evaluate 1,21 x 10^{17} variants of solutions [2].
- **Random solution** – does not guarantee any results, because the task is not convergent to any value [3].
- **Hunger algorithm** – choice of minimal closer value, often goes to a dead end [4].
- **Simulated annealing** - is a probability optimization method for searching of state area. Algorithm does not guarantee finding of global minimum [5].
- **Finding of minimal frame** (Jarnik's algorithm) – entry graph is not closed [6].
- **Method of branches and bounds** - the principle of this method is in systematic scanning of all potential solutions. The estimations get more precise during the calculation [7].
- **Genetic algorithm** - the principle of work is subsequent forming of generation of various solutions of given the problem [8].
- **Ant Colony Optimization** – algorithm uses set of agents, ants during calculation. ACO is a meta-heuristic technique used in artificial intelligence field for search of approximate solutions of combinatorial problems [9].

Choice of method: To find the optimal (minimal course) movement of the robotic arm the method of branches and bounds was used. This method was developed in 1960 by a group of mathematicians (A. H. Land, A. G. Doig) for searching optimal solutions of different optimization problems, especially in discrete optimization and combinatorial optimization.

Method of branches and bounds is used for solving optimization tasks and it searches the tree of solutions. Set of possible solutions of task T is marked by symbol S(T). By that the optimization task T with objective function *f* is expressed. The task T can be solved by creation of tasks $T_1,\dots,T_n$ with the same objective function so that (see Equation (1)):

$$S(T) = S(T_1) \cup \dots \cup S(T_n) \tag{1}$$

Optimal solution of task T can be obtained by choosing the best of all optimal solutions of subtasks, $T_1,\dots,T_n$. Individual subtasks could be considered as peaks of the root tree, that were called tree of solutions. During the calculation the tree grew larger and only the leaves of the root tree were considered interesting for the calculation. The leaves were divided into so called alive and dead. Solutions that were either optimal, did not have acceptable solution or a better solution existed were considered dead. The calculation ended when all the leaves were dead. To simplify the calculation for each minimization task the bottom bound $Pre(T_i)$ was set. During branching graph was searched in width or depth.

The shortest Hamilton cycle can also be searched for by method of branches and bounds. The task would be reformulated as follows:

Searching for the cheapest subgraph H of given graph G so that:

- Edges of the subgraph H would form a system of apex disjunctive cycles that would pass through all the apexes of the graph G;
- The subgraph H was coherent;

Subsequently a cycle where all the apexes were marked as a set M was determined. This led to creation of subtasks for each apex $v_i \in M$ (all the edges that led from apex v_i to set M were banned) [10].

After choosing suitable mathematical method, the starting entry conditions and parameters were set:

- placement of robot P_0 (point „zero") and $_0$ (vector „zero"): coordinates [x_0, y_0, z_0] in global coordinate system;
- definition of working plane: a 3-point method for dextrorotary orthogonal coordinate system (determination of the basis for robot „KUKA");
- decisive determination of the working plane (surface for painting without holes or permeation): matrix of critical points in working plane [x, y];
- setting of technical parameters of the robot's nozzle (effective stripe/width of slit of the nozzle, overlapping of color);

In the next phase the model was realized by the following steps [11]:

- Layout (division) of the working plane to active and passive rectangles (vertical, from the ceiling towards the floor, see Figure 2): rectangles (R) were decisively determined by points P_{i0} [x, y] – upper left corner, P_{i1} [x, y] – bottom right corner. Two matrixes were constructed (active R_a size 2 x n_a; passive R_p size 2 x n_p, where n_a and n_p is the number of active and passive rectangles);

- Setting of the robot's course during painting of the active rectangles (number of vertical stripes).
- Drawing of oriented network graph (see Equation (2) and (3)):
  - Number of knots:

$$N(k) = 2 \times n(a) + 1 \tag{2}$$

  - Number of edges:

$$N(e) = 3 \times n(a)^2 + 2 \times n(a) \tag{3}$$

- Construction of the matrix of oriented network graph containing the coordinates of all knots, edge orientation, edge length, determination of active and passive areas (respectively input/output points of active rectangles);
- Inspection of robot's arm reach (comparison of distance between point zero and points of working area with the robot's arm reach).
- Finding of the shortest path (closed Hamilton path) by several means, subsequent comparison and choice of optimal course.

In the final phase of the mathematical model of robotic arm, the optimal path (the shortest trajectory of movement of the painting robot's arm respectively Hamilton path - see Figure 3) was chosen and verified.

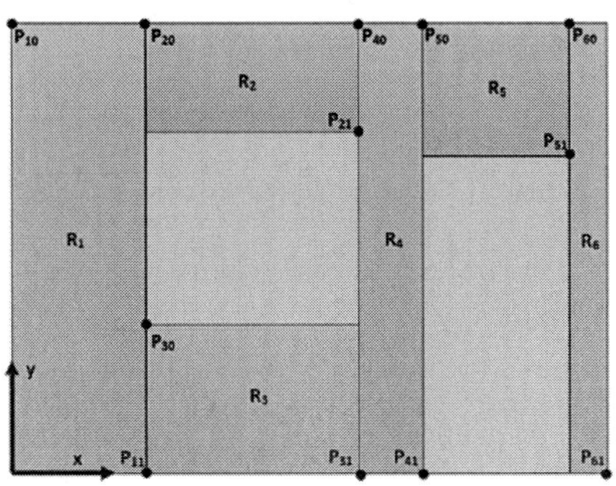

Figure 2 – Separation of the working surface to active and passive rectangles

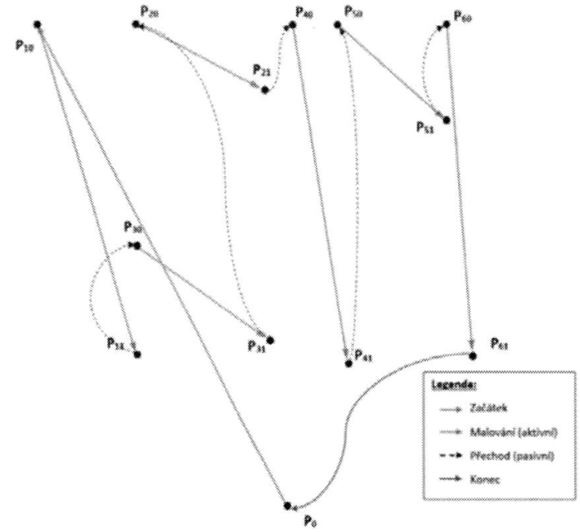

Figure 3 – Hamilton path

2.2 Choice of applicator and nozzle (pattern)

The spray gun was chosen from two technologies of spray (by air and by pressure) [12]:

- **By air:** air spraying is classic and so far most widespread. Compressed air from a compressor gets through a hose to the spray gun, where it mixes with the color from a container, which is located on the gun. The air pressure is depending on the type of the compressor in range from 2 to 7 bars.
- **By pressure:** less widespread (but gaining popularity) is so called airless spraying. It works without air, it means without using a compressor. It uses a pump. Strainer of the machine and a powerful pump, which depending on the type reaches pressure up to 230 bars is plunged to a container with the paint. It forces the paint or the glaze through the hose to the pistol. It eliminates the need for setting the air/paint ratio and complications caused by wrong adjustments. The only thing that is being adjusted is the pressure which is highly important. This way spraying is not only economical but also guarantees high quality of the spraying.

Subsequent choice of the nozzle and its pattern geometry is really important because it determines its shape and intensity of the spraying. For our applicator it was possible to choose from following patterns (see Figure 4) [13].

- **Solid stream** – One or more solid points (streams). Application with low pressures. The nozzles used for this type are exceptionally simple;
- **Flat spray** – Thin, long spraying. Almost unified pattern in its whole application. Useful for large areas;
- **Hollow cone** – Spraying is distributed in circle from the nozzle;
- **Full cone** – Spraying is applied in the whole circular area. The most widespread pattern;

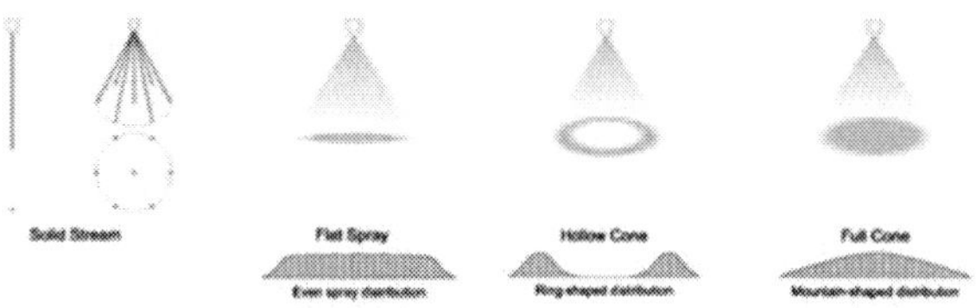

Figure 4 – Nozzles geometry [13]

2.3 Overlapping of layers

During the application of spraying the surface the most important criterion is keeping the suggested distance from the surface. The applicator should be in optimal distance of 30 centimeters (cca 12 inches) and it should always impact the surface perpendicularly. In order to secure equal amount of paint on the surface, the overlap ratio is suggested to be 50% and more in manual spraying. However, this solution is not economical and it causes unnecessary loses, which would have negative impact on the price when painting larger areas.

When using flat pattern (see Figure 5) we can achieve desired decrease in losses mainly thanks to reducing the overlapping which we modify from 50% to approximately 10%. The *a* area has in 90% of the spraying the same properties of overlapping and the two remaining areas *b* each occupying approximately 5% minimization their overlapping properties. It is due to the mutual overlapping of areas *b* while applying the paint we achieve economies.

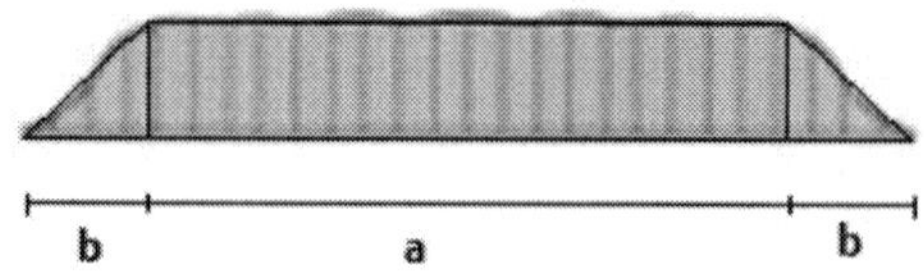

Figure 5 – Areas of coverage (*a*-constant, *b*-descending)

2.4 Delimitation of optimal movement for applying the paint

The last studied factor was the choice of appropriate movement of the trajectory of the end point where an emphasis on minimization of the energy consumption was made. For this part of the project a program presenting percentual workload of the servomotors was created. Basic trajectories (vertical and horizontal) were compared and each one of them was composed from three movements. The movement 1 of vertical course shifted from upper left downwards by distance x. Movement 2 followed from left to right by distance y. Furthermore, via movement 3 it returned back to the top by distance x. Horizontal movement started by shifting from right to left by the distance x, continued downwards by distance y and backwards from left to right by distance x (see Figure 6)

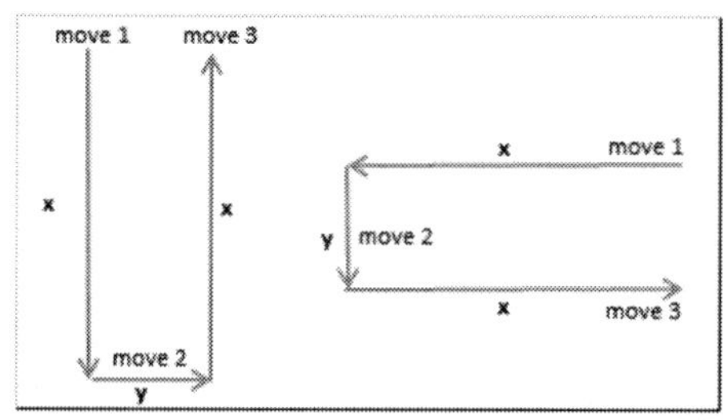

Figure 6 – Scheme of vertical and horizontal movement

Subsequently the times of each servomotors for both trajectories of robotic arm's movement were measured. Because of this the total energy consumption of each movement and their economic efficiency were expressed.

In the following Tables the servomotors (servo6, servo5, ...) are presented with their percentual workload while performing given movement. Due to percentual workload and measured time on servomotors, the individual work of both movements could be determined and compared.

Table 1 - Horizontal trajectory

servo	Position	Post 1	Post 2	Post 3	Post 4	Move 1	Move 2	Move 3
servo6	Rotate	225	132	124	225	40%	3%	44%
servo5	Bottom	149	149	175	175	2%	11%	2%
servo4	Middle	115	115	127	127	4%	5%	3%
servo3	Top	127	127	113	113	3%	6%	3%
servo2	Rotate	200	200	200	200	2%	2%	2%
servo1	Gripper	39	124	132	39	33%	3%	36%

Table 2 - Energy consumption (partial and total) for horizontal movement

W1[Ws]	*W2[Ws]*	*W3[Ws]*	*U[V]*	*I[A]*	*t1[s]*	*t2[s]*	*t3[s]*
43,67	3,76	47,43	12,00	2,00	1,82	0,16	1,98
2,16	12,21	2,16	12,00	2,00	0,09	0,51	0,09
4,32	5,63	3,24	12,00	2,00	0,18	0,23	0,14
3,24	6,57	3,24	12,00	2,00	0,14	0,27	0,14
2,16	2,16	2,16	12,00	2,00	0,09	0,09	0,09
36,00	3,39	39,39	12,00	2,00	1,50	0,14	1,64
TOTAL	222,886	Ws					
	6,19E-05	kWh					

Table 3 – Vertical trajectory

servo	Position	Post 1	Post 2	Post 3	Post 4	Move 1	Move 2	Move 3
servo6	Rotate	217	217	200	200	1%	7%	1%
servo5	Bottom	147	202	202	147	24%	1%	24%
servo4	Middle	145	75	75	145	30%	2%	30%
servo3	Top	154	56	56	158	43%	1%	44%
servo2	Rotate	200	200	200	200	2%	2%	2%
servo1	Gripper	50	45	61	61	2%	6%	1%

Table 4 - Energy consumption (partial and total) for vertical movement

W[J=Ws]	*W[J=Ws]*	*W[J=Ws]*	*U[V]*	*I[A]*	*t1[s]*	*t2[s]*	*t3[s]*
1,08	7,98	1,08	12,00	2,00	0,05	0,33	0,05
25,83	1,08	25,83	12,00	2,00	1,08	0,05	1,08
32,87	2,16	32,87	12,00	2,00	1,37	0,09	1,37
46,02	1,08	47,90	12,00	2,00	1,92	0,05	2,00
2,16	2,16	2,16	12,00	2,00	0,09	0,09	0,09
2,12	6,78	1,08	12,00	2,00	0,09	0,28	0,05
CELKEM	242,22	Ws					
	6,72E-05	kWh					

When abiding by all conditions of the spraying (optimal distance, ideal pattern of the nozzle and applicator) with the combination of optimization of movements (for spraying horizontal movement and for shifting the arm between the active areas by Hamilton path) the desired savings were achieved. Active width of spraying will be approximately 35 centimeters and the passive width on both sides around 2,5 centimeters. Sprayed stripe with total width around 40 centimeters can be put in area R1 y/37,5 times (2,5 centimeters is mutual overlap of the stripes). After finishing area R1 the robotic arm will move to another area based on the optimizing algorithm of ideal trajectory and here it will continue in horizontal movement for application of the paint again.

This option seems to be the most economical regarding the energy and paint consumption. When facing the request for multi-layer spraying, the very same method would have been used, the difference being that the new spraying would overlap with the old one by a half of the total width of applied paint.

3 Conclusion

The optimizing algorithm managed to quickly find at least partially optimal trajectory of the robotic arm's movement which saved engine time and wearing of the robot's servomotor. Algorithm can be extended by other optimizing parameters. The working area could be divided into rectangles also by using mathematical algorithm. The calculation could be extended also by optimizing moves of minimization of the paint and energy consumption. Evaluating edge would be evaluated not only by the distance but also by the time of robotic arm's travel and energy consumption respectively number of revs of servomotors.

The most suitable paint applicator was pressure spray gun. This method of spraying was not only economical but it guaranteed high quality of the spraying. The spraying was more comfortable also due to the pistol not having the container with paint. The arm could have moved without restrictions and spray from various positions. On top of that when used correctly the losses are minimal.

As the ideal pattern of the nozzle was chosen the flat pattern which was suitable for large areas too because of its geometry and coverage. Thanks to robotics the exact position of applied paint could be set and bound of coverage, which was reduced from 50% to approximately 10% determined.

The ideal trajectory of painting robot's end point was horizontal movement where the total energy consumptions 222 Ws was lower than the vertical one with 242 kWh. The differences were minimal but during the spraying of larger areas and applying the ideal movement, the energy savings will be increased too.

In the Figure (see Figure 7) the painting robot's model is depicted. Its paint transport to the final point is not yet completely solved. Next, the problems of tidiness and fluency of movement will be investigated.

In this case it was not taken into consideration because it would be unnecessarily complex due to limited software of the model.

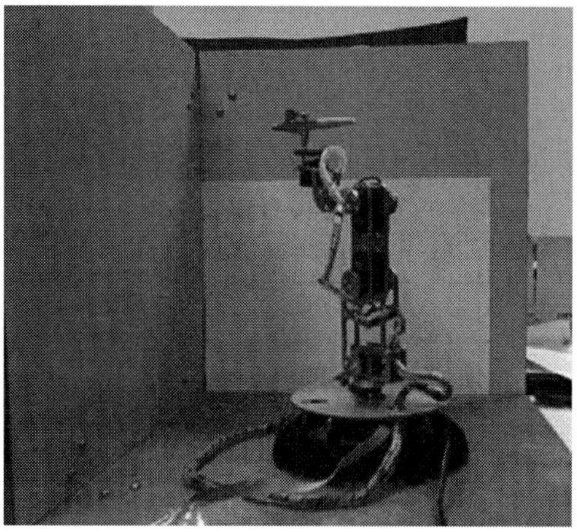

Figure 7 - Model of the painting robot at work

References

[1] Applegate, D. L.; Bixby, R. M.; Chvátal, V.; Cook, W. J. (2006), *The Traveling Salesman Problem*, ISBN 0-691-12993-2.

[2] Paar Ch., Pelzl J., Preneel B.: Understanding Cryptography: *A Textbook for Students and* 2010. *Practitioners. Springer. p. 7.* ISBN 3-642-04100-0.

[3] Thomas H. Cormen et al.,. Introduction to Algorithms. — MIT Press, 2001. — P. 1292. — ISBN 978-0-262-03384-8.

[4] Hazewinkel, M., ed. (2001), "Greedy algorithm", *Encyclopedia of Mathematics*, Springer, ISBN 978-1-55608-010-4

[5] Press, WH; Teukolsky, SA; Vetterling, WT; Flannery, BP (2007). *Numerical Recipes: The Art of Scientific Computing* (3rd ed.). New York: Cambridge University Press. ISBN 978-0-521-88068-8

[6] V. Jarník: O jistém problému minimálním, Práce Moravské Přírodovědecké Společnosti, 6, 1930, pp. 57-63.

[7] Rektorys K. a kol.: Přehled užitné matematiky. Prometheus, Praha 1995, ISBN 80-85849-92-5

[8] Schwefel Hans-Paul. Numerical optimization of computer models (Translation of 1977 Numerische Optimierung von Computor-Modellen mittels der Evolutionsstrategie. — Wiley, 1981. — ISBN 0-471-09988-0.

[9] M. Dorigo & T. Stützle, 2004. Ant Colony Optimization, MIT Press. ISBN 0-262-04219-3

[10] Demel J.: Grafy a jejich aplikace. Academia, Praha 2002, ISBN 80-200-0990-6

[11] Večerka A.: Grafy a grafové algoritmy. Přírodovědecká fakulta, Univerzita Palackého. Olomouc 2007

[12] Stříkací pistole vzduchová a tlaková; čím se liší. *IReceptář* [online]. Praha, 2011 [cit. 2016-01-21]. Dostupné z: http://www.ireceptar.cz/pro-kutily/postupy-a-navody/strikaci-pistole-vzduchova-a-tlakova-cim-se-lisi/

[13] Spray Guns and Applicators Information. *IHS Engineering* [online]. [cit. 2016-01-21]. Dostupné z: http://www.globalspec.com/learnmore/manufacturing_process_equipment/surface_coating_protection/coating_paint_spray_guns

[13] Airless Paint Sprayer Tips for Exterior Paint Jobs. *Building and Construction* [online]. [cit. 2016-01-21]. Dostupné z: http://www.buildconstructpros.com/airless-paint-sprayer-tips-for-exterior-paint-jobs#1

[14] Airless Spray Coating Technology: Including Air-Assisted Airless Spray Technology. *Chreed* [online]. [cit. 2016-01-21]. Dostupné z: http://www.chreed.com/help_pages/airless_guns.pdf

A Point Cloud-Vision Hybrid Approach for 3D Location Tracking of Mobile Construction Assets

Y. Fang[a], J. Chen[b], Y. K. Cho[a], and P. Zhang[c]

[a] School of Civil and Environmental Engineering, Georgia Institute of Technology
[b] School of Electrical and Computer Engineering, Robotics Institute, Georgia Institute of Technology
[c] Department of Construction Management, Tsinghua University
E-mail: yihaifang@gatech.edu, jchen490@gatech.edu, yong.cho@ce.gatech.edu, zhangpy14@mails.tsinghua.edu.cn

Abstract –

Modeling as-is site condition and tracking the three-dimensional (3D) location of mobile assets (e.g., worker, equipment, material) are essential for various construction applications such as progress monitoring, quality control and safety management. Many efforts have been dedicated to vision-based technologies due to their merits in cost-effectiveness and light infrastructure compared to real-time location systems (RTLS). However, a major challenge of vision-based tracking is that it lacks 3D information and thus the results are sensitive to occlusion, illumination conditions and scale variation. To address this problem, this study presents a point cloud-vision hybrid approach to reconstruct and update the area of interest in 3D for scene updating and mobile asset tracking. Baseline 3D geometry information in point cloud is obtained at the start by Structure from Motion (SfM) using Unmanned Aerial Vehicle (UAV), given which mobile and static assets present in the scene are recognized and labeled. Based on 2D aerial isometric images capture by the UAV, labeled assets are automatically recognized and their locations are updated. The proposed approach was implemented in a field test and the results demonstrate that it was able to reconstruct the site and update the location of mobile assets accurately and reliably. Findings in this study indicate the proposed hybrid approach effectively augments the state-of-the-art in site modeling and asset tracking in construction.

Keywords –

Point cloud-vision hybrid approach; Mobile asset tracking; Structure from Motion (SfM); Unmanned Aerial Vehicle (UAV)

1 Introduction

Three-dimensional (3D) location information of construction assets are of great interest to various engineering and management applications including progress monitoring, quality control, operation analysis, safety monitoring and occupational health assessments. Although much research efforts have been dedicated to investigating the merits of Real-time Location System (RTLS) and vision-based methods, limitations in cost-effectiveness, ease of use, and robustness greatly hinder their field deployment. This study focuses on addressing the challenges in obtaining 3D location data of construction assets through mobile camera systems.

Traditional vision-based tracking methods recognize and track the objects of interest from images or video streams captured by cameras at known locations and angles. Such settings can be easily achieved by installing cameras at multiple locations on a construction site or taking advantage of the existing surveillance camera systems. However, cameras at a fixed location inevitably suffer from massive occlusions introduced by ever-changing site conditions such as structure elements, temporary structures, and equipment. In addition, although 3D location can be computed based on the 2D image captured by two cameras setup at known locations [1], this method requires the tracked objects to be present on both images and not fully occluded. Therefore, continuously tracking the 3D location of construction assets is not always practical at many construction sites with cameras at fixed locations. A mobile camera system on an aerial platform such as an Unmanned Aerial Vehicles (UAV) is considered a promising alternative. Recently, UAV technology has drawn much attention from the construction industry for its potential in various applications including maintenance inspection, construction survey, and safety management. Nevertheless, 3D location tracking using a mobile camera faces several major challenges, including estimating the position and orientation of the on-board camera, and transforming the pixel coordinates of recognized objects from the camera frame to the global frame.

To address these challenges, this paper proposes a point cloud-vision hybrid approach for 3D site reconstruction and mobile asset tracking. This paper first reviews current practices in construction asset tracking and the state-of-the-art Structure from Motion (SfM) technology. Then, the point cloud-vision hybrid approach is introduced by a flowchart and the details of the techniques and algorithms used in each step. Results from a case study implementing the proposed method for vehicle location tracking are presented followed by discussion and conclusions.

2 Related Work

2.1 State-of-the-art Construction Asset Tracking Methods

The location information of construction assets such as workers, equipment, and materials is of interest to various construction applications including progress monitoring, quality control, operation analysis, safety monitoring and occupational health assessments. Much efforts have been dedicated to real-time location systems (RTLS) such as Global Positioning System (GPS) [2], Radio Frequency Identification (RFID) [3], and Ultra-wide Band (UWB) [4]. Although varying in tracking accuracy (e.g., meters for GPS and centimeters for UWB), the RTLS technologies provide direct measurement of the 3D location of the tracked objects. However, most RTLS systems require tagging the objects to be tracked, which increases the complexity in applications. In addition to the tags, high-accuracy RTLS such as UWB requires heavy infrastructure deployment, for example the installation of a series of antennas around the tracking site. This results in huge investment in time and cost ($140/m2).

Another research direction for construction asset tracking focuses on computer vision technologies that recognize and track the objects of interest from 2D images or video streams captured by cameras. Compared to RTLS-based tracking methods, vision-based tracking does not require additional sensors and tags and thus has the advantages of simple deployment and low cost [5]. Various vision-based tracking algorithms have been studied and tested in construction scenarios. Generally, tracking algorithms can be categorized into kernel-based [6], contour-based [7], and point-based [8] methods. Different in the means to represent object, contour-based methods use contours or silhouettes that enclose the object region, kernel-based methods use the responses of the object region to selected kernels, and point-based methods use a set of feature points detected in the object region [9]. Compared to the other two methods, the point-based method is more robust to illumination variation and occlusions, which commonly occur in outdoor construction environment. Although the images from a single camera provide only 2D pixel coordinates, images from multiple cameras at multiple known and fixed locations provide 3D metric coordinates through camera calibration, pose estimation, and triangulation [1].

2.2 Obtaining 3D Information by Photogrammetry and Structure from Motion (SfM)

Photogrammetry is an image-based technology that reconstructs 3D objects from 2D photographs. This technology extracts 2D input data from photographs and maps them onto a 3D space. Since constructing a 3D model only requires taking images from different angles, using photogrammetry for 3D data acquisition is flexible, cost-effective, and non-invasive to the survey objects. Structure from Motion (SfM) photogrammetry is an emerging technique that was built upon but fundamentally differs from traditional photogrammetry. In SfM approach, the critical parameters such as camera location/orientation and scene geometry are automatically computed without the need of a series of targets with known locations [10]. Instead, these parameters are computed simultaneously using a highly redundant, iterative bundle adjustment procedure, based on a database of features automatically extracted from a set of multiple overlapping images [11].

The results of SfM or traditional photogrammetry are represented by a dense point cloud comprised of millions of points, each of which contains 3D position (XYZ) and color (RGB) data. The point cloud data is useful in various construction applications such as acquiring as-is geometry data for building component modeling [12], construction progress monitoring and control [13] [14], and construction documentation especially for historical structures [15]. Although the point cloud contains comprehensive 3D geometric data of the objects in the scene, it has been a challenge to track objects using a SfM-generated point cloud. This is mainly because in SfM technique, generating a point cloud from a large amount of images takes time, as it requires massive computation capability. Therefore, noticeable delay between the actual and tracked location makes it impractical to track dynamic objects on construction sites using SfM method alone.

2.3 Alternative Monitoring Method using Unmanned Aerial Vehicles (UAV)

Many vision-based tracking methods are based on the assumption of using the images captured by cameras at fixed and known locations. This is convenient since many construction sites are equipped with surveillance camera systems. However, construction sites are usually very congested, which makes it impossible for static cameras to constantly maintain a clear line-of-sight to the objects to be tracked without the occlusions from structure elements, equipment, and materials present on the site. Recent development of Unmanned Aerial Vehicles (UAV) offers a low-cost alternative for construction monitoring applications such as bridge and road assessment [16] [17], earthwork surveying [18], and safety inspection [19] . Compared to fixed site cameras, the onboard camera on a UAV is more flexible in image capture angles and thus less prone to occlusions. It should be noticed that a major limitation of UAV-based monitoring is the limited time for a single flight (around 30 minutes) due to the battery life.

3 Point cloud-vision Hybrid Approach

To address the challenges in vision-based tracking using mobile cameras, this study proposes a point cloud-vision hybrid approach for tracking of mobile construction assets using SfM and UAV technologies. The flowchart for the proposed hybrid approach is shown in Figure 1.

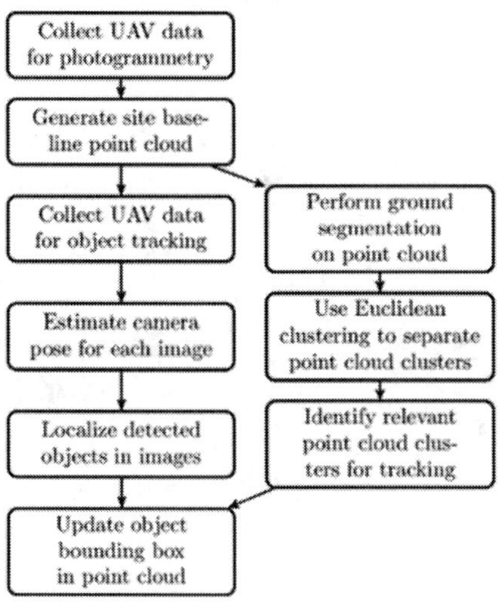

Figure 1: Flowchart for the proposed point cloud-vision hybrid approach

The first step involves collecting aerial images of a construction site from a UAV at multiple viewpoints. The construction site images are highly-overlapping and encircle the site in order to cover the full 3D structure of construction-related entities. Next, the 3D point cloud of the construction site (see Figure 2a) is generated based on the image data using a Structure from Motion (SfM) algorithm adopted from [11]. This algorithm detects common features across each camera frames using Scale Invariant Feature Transform (SIFT). The process works by finding point correspondences between images and solving for point coordinates and camera poses in a bundle adjustment procedure. This establishes a baseline 3D model of the construction site which includes both background elements and the mobile assets. The mobile assets located in the point cloud are separated out using a segmentation and clustering routine. Ground segmentation is first applied to filter out points belonging to the ground which is considered as background. Next, individual point cloud clusters are separated based on neighboring Euclidean distance. Point cloud clusters for objects to be tracked are further identified through a supervised procedure where the user selects a set of clusters corresponding to the interested objects. Bounding boxes are calculated for each identified point cloud cluster which acts as a compact representation of the object.

A series of images is then collected across time from the UAV to specifically track targeted mobile assets. The image data can be either in the form of a video feed or discrete images taken at specific timestamps. Compared to fixed camera setups for object tracking, here it is necessary to solve for the image-specific position and orientation of the camera since the UAV is moving from frame to frame. This is formulated as a perspective-n-point (PnP) problem to recover the complete 6 degree-of-freedom motion (i.e., x, y, z coordinates and yaw, pitch, roll rotations) of the UAV based on the input images. First, an image is synthetically generated from the point cloud projected onto a two-dimensional plane (Figure 2a). Feature points can be calculated from the synthetic image that can be matched to feature points derived from the UAV images. Then a depth buffer is created based on the 3D point cloud data collected from the photogrammetry step. The depth buffer is shown in Figure 2b, where bright points indicate points that are close to the camera while darker points indicate points that are further away from the camera. This enables us to calculate the 3D position of each feature point on the synthetic image. The UAV images are then matched to the synthetic image to obtain a corresponding set of 3D point features. Finally, a camera pose estimate is calculated for each UAV image which minimizes the least-squares re-projection error of 3D point features in the image.

Figure 2: (a) synthetic image and (b) depth buffer generated from point cloud

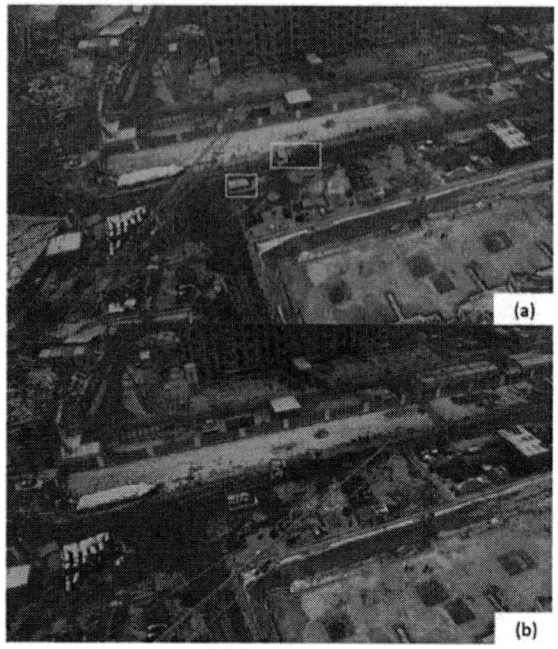

Figure 3: Object detection using feature point matching with a reference image, (a) objects of interest, (b) feature points identified

Figure 4: Projection of image coordinates into 3D space from camera origin

The next step in the processing pipeline is object detection. For each image taken from the UAV in the previous step, the pixel coordinates of objects to be tracked are identified through a point-based method, namely matching of SIFT feature points. This process is semi-automated by the user specifying interested objects in the reference image. As shown in Figure 3, bounding boxes are drawn around the two vehicles to be tracked in the reference image (left image) while the same objects are detected in the tracking image through feature point matching (right image). The process can potentially be fully-automated by having a database of possible objects to be tracked or training an object detection classifier.

In the last step, the location of each detected object in global coordinates is calculated based on the recovered camera pose and its image coordinates. A ray casting method is used where the object location is determined by the intersection of a line formed by an image projection vector originating from the camera with the point cloud surface. For each detected object in the image, a corresponding projection vector is determined based on its pixel coordinates and camera parameters such as focal length and image size. Figure 4 shows the projection of detected objects from image coordinates to 3D space. Successfully matched objects have their bounding boxes updated in the point cloud based on the estimated 3D location.

4 Case Study and Preliminary Results

An ongoing construction project was selected as a case study of the proposed method. Two vehicles (i.e., a concrete mixer truck and a minivan) were chosen as the targeted mobile assets to be tracked. Both vehicles moved in random patterns in an area of 40 m by 20 m. This case study employed an 8-axis UAV (octocopter) equipped with a mirrorless digital camera. In total 169 images at a resolution of 4912 x 3264 were used to generate the 3D point cloud of the site. Table x shows the computation time involved in each step of the proposed pipeline. The step of SfM computation and point cloud generation takes the longest amount of time but only needs to be carried out once at the start of the experiment. On the other hand, the tracking and location updating step involves a trade-off between accuracy and computation time. Using high resolution images for tracking will potentially improve the tracking accuracy since more feature points can be detected but this will also increase the computation time.

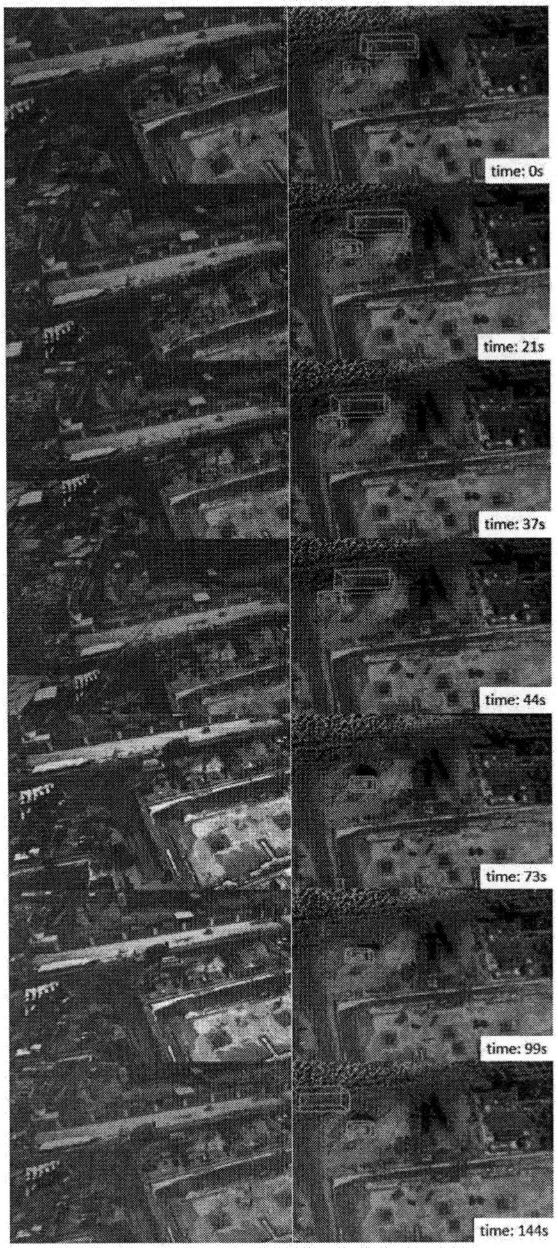

Figure 5: Results of 3D location tracking of two vehicles: image captured from UAV (left) and updated 3D site model (right)

Table 1: Preparation and computation time

Activities	Time
Preparation	~20 min
UAV flight	~3.5 min
SfM computation/Point cloud generation	~150 min
Tracking image by UAV	Up to 30 min
3D location computation	3 min

Figure 5 shows the results obtained from tracking two vehicles over time using images captured from the UAV. The left column shows the time where each image is taken whereas the middle column shows the image captured by the camera. The images are annotated with feature points for each detected object. The right column shows the updated 3D site model (point cloud) corresponding to each captured image. The 3D site model is generated using a top-down view of the site 3D point cloud with bounding boxes formed around each tracked object.

Results from the case study indicate that the proposed method dynamically updates the 3D location of two vehicles in a construction site by using the images captured from a UAV and matching them to a 3D site model in the form of a point cloud. The first four and the last images (time: 0s to 44s and 144s) show the cases of successful tracking where the bounding box for the two vehicles are shown in green. The results at time 73s and 99s show a sequence when the matching algorithm lost track of a vehicle due to insufficient feature points. The corresponding bounding box is drawn based on the previous location estimate but is highlighted in red to indicate uncertainty in location.

5 Discussion

In this study, the tracked objects focuses on large objects exhibiting smooth linear motion such as vehicles. Smaller objects that are prone to occlusion incur difficulty in the feature point matching stage. The method is also limited in terms of automation in the sense that not all dynamic assets are automatically tracked and updated in the 3D point cloud. Instead, the user manually specifies a fixed number of objects to be tracked in the 3D scene update.

In terms of accuracy in location estimation, the method relies on accurately calculating the UAV camera position and orientation for each captured image and correctly identifying the tracked object in each image. The camera pose is derived through a least-squares estimation scheme and is negatively affected by the presence of outliers. In the matching stage with a reference image, outlier points have to be rejected by threshold elimination considering the re-projection error to the image coordinate frame. In the object detection stage, the pixel coordinates of tracked objects can also be incorrectly identified when there exists feature points that are incorrectly matched. Thus, the matching algorithm needs to ensure that a sufficient number of feature points can be identified for each object and that there exists geometric consistency among the matched feature points for objects from each frame to the next frame.

6 CONCLUSION

To address the challenges in 3D location tracking using a mobile camera, this paper proposes a point cloud-vision hybrid approach for 3D site reconstruction and mobile asset tracking. The method involves a processing pipeline with the steps of point cloud generation, camera pose estimation, object detection and object localization. The method updates the baseline 3D scene in the form of point cloud with dynamic bounding boxes for each tracked vehicle, which can be further utilized in site management applications. Preliminary results from a case study show that the proposed method was able to successfully track the 3D location of two vehicles in a construction site by using images captured from a UAV and matching them to a 3D site model in the form of a point cloud. Despite of the limitations in semi-automated processing pipeline and limited UAV flight time, the proposed point cloud-vision hybrid approach enable by SfM and UAV technology shows advantages in flexibility and robustness to occlusions over traditional method using fixed cameras. Findings in this study indicate great potential of the proposed method in 3D location tracking of mobile construction assets in congested construction environment.

The proposed method involves limitations in terms of the number and size of objects that can be tracked. Having a large number of tracked objects or having target objects that are too small will complicate the feature point matching process and reduce the localization accuracy. For future work, the authors would like to experiment with alternative vision tracking methods such as kernel-based and contour-based methods to investigate whether those methods will improve the localization accuracy.

7 ACKNOWLEDGEMENT

This material is based upon work supported by the National Science Foundation (Award #: CMMI-1358176). Any opinions, findings, and conclusions or recommendations expressed on this material are those of the authors and do not necessarily reflect the views of the NSF.

References

[1] M.-W. Park, C. Koch, and I. Brilakis, "Three-Dimensional Tracking of Construction Resources Using an On-Site Camera System," J. Comput. Civ. Eng., vol. 26, no. 4, pp. 541–549, 2012.

[2] N. Pradhananga and J. Teizer, "Automatic spatio-temporal analysis of construction site equipment operations using GPS data," Autom. Constr., vol. 29, pp. 107–122, 2013.

[3] Y. Fang, Y. K. Cho, S. Zhang, and E. Perez, "Case Study of BIM and Cloud–Enabled Real-Time RFID Indoor Localization for Construction Management Applications," J. Constr. Eng. Manag., pp. 1–12, 2016.

[4] Y. K. Cho, J. H. Youn, and D. Martinez, "Error modeling for an untethered ultra-wideband system for construction indoor asset tracking," Autom. Constr., vol. 19, no. 1, pp. 43–54, 2010.

[5] J. Yang, M.-W. Park, P. a. Vela, and M. Golparvar-Fard, "Construction performance monitoring via still images, time-lapse photos, and video streams: Now, tomorrow, and the future," Adv. Eng. Informatics, vol. 29, no. 2, pp. 211–224, 2015.

[6] E. Maggio and A. Cavallaro, "Accurate appearance-based Bayesian tracking for maneuvering targets," Comput. Vis. Image Underst., vol. 113, no. 4, pp. 544–555, 2009.

[7] M. Yokoyama and T. Poggio, "A contour-based moving object detection and tracking," 2005 IEEE Int. Work. Vis. Surveill. Perform. Eval. Track. Surveill., no. 1, pp. 271–276, 2005.

[8] T. Mathes and J. Piater, "Robust non-rigid object tracking using point distribution manifolds," Pattern Recognit., pp. 515–524, 2006.

[9] I. Brilakis, M.-W. Park, and G. Jog, "Automated vision tracking of project related entities," Adv. Eng. Informatics, vol. 25, no. 4, pp. 713–724, Oct. 2011.

[10] M. J. Westoby, J. Brasington, N. F. Glasser, M. J. Hambrey, and J. M. Reynolds, "'Structure-from-Motion' photogrammetry: A low-cost, effective tool for geoscience applications," Geomorphology, vol. 179, pp. 300–314, 2012.

[11] N. Snavely, S. M. Seitz, and R. Szeliski, "Modeling the world from Internet photo collections," Int. J. Comput. Vis., vol. 80, no. 2, pp. 189–210, 2008.

[12] F. Dai and M. Lu, "Assessing the Accuracy of Applying Photogrammetry to Take Geometric Measurements on Building Products," J. Constr. Eng. Manag., vol. 136, no. February, pp. 242–250, 2010.

[13] S. El-Omari and O. Moselhi, "Integrating 3D laser scanning and photogrammetry for progress measurement of construction work," Autom. Constr., vol. 18, no. 1, pp. 1–9, 2008.

[14] M. Golparvar-Fard, J. Bohn, J. Teizer, S. Savarese, and F. Peña-Mora, "Evaluation of image-based

modeling and laser scanning accuracy for emerging automated performance monitoring techniques," Autom. Constr., vol. 20, no. 8, pp. 1143–1155, 2011.
[15] N. Yastikli, "Documentation of cultural heritage using digital photogrammetry and laser scanning," J. Cult. Herit., vol. 8, no. 4, pp. 423–427, 2007.
[16] N. Metni and T. Hamel, "A UAV for bridge inspection: Visual servoing control law with orientation limits," Autom. Constr., vol. 17, no. 1, pp. 3–10, 2007.
[17] S. Rathinam, Z. W. Kim, and R. Sengupta, "Vision-Based Monitoring of Locally Linear Structures Using an Unmanned Aerial Vehicle," J. Infrastruct. Syst., vol. 14, no. 1, pp. 52–63, 2008.
[18] S. Siebert and J. Teizer, "Mobile 3D mapping for surveying earthwork projects using an Unmanned Aerial Vehicle (UAV) system," Autom. Constr., vol. 41, pp. 1–14, 2014.
[19] J. Irizarry, M. Gheisari, and B. N. Walker, "Usability assessment of drone technology as safety inspection tools," Electron. J. Inf. Technol. Constr., vol. 17, no. September, pp. 194–212, 2012.

Investigating Safety Passage Planning for System Shoring Supports with BIM

R.L. Zhuang[a], S. Hung[a,b], Y.C. Shiau[c,*], K.T. Liu[c] and C.C. Liu[c]

[a] Ph.D. Program of Technology Management, Chung Hua University, Taiwan
[b] Department of Digital Media Design, Hwa Hsia Institute of Technology, Taiwan
[c] Department of Construction Management, Chung Hua University, Taiwan
E-mail: luckiness740525@hotmail.com, arion3d@gmail.com, *ycshiau@ms22.hinet.net, dk313419@gmail.com, E10416003@chu.edu.tw

Abstract –

All temporary structures at the construction stage are collectively called falsework. As part of the falsework, the importance of system shoring support (SSS) is often overlooked in construction projects, thus causing increasing accidents from SSS collapse. The 2014 occupational accident statistics of the Council of Labor Affairs show that most occupational accidents occurred in the construction industry, and fall and collapse were the most common accidents in the construction industry in Taiwan. Although the SSS is commonly used in construction, no well-defined instruction for the installation and required safety facilities for SSS have been established to provide a dependable guide. The status of SSS used in falsework of construction projects and the types of potential accidents and relevant examples will be investigated. With building information modeling (BIM), this study defined the relevant SSS facilities with the 3D modeling of BIM. This research drew the actual status of the erection of SSS on the construction site with the Revit from Autodesk to provide a communication platform for engineers, SSS contractors, and labor safety and health personnel. The corresponding preventive measures in terms of safety passages and stairs have been proposed for the reference of engineering personnel in construction site. Through the help of this study, the construction falsework safety can be enhanced and the accidents of SSS can be minimized.

Keywords –

System Shoring Support, Safety Passage, Building Information Modelling.

1 Introduction

1.1 Stages of Construction Projects

In terms of the life of buildings or civil structures, a construction project falls into two main stages: construction and use stages. At the "use stage," main concerns are load factors including seismology, wind resistance, live load, and dead load. At the "construction stage," they are new concrete placing, steel, formwork, workers, and machinery vibration. Nonetheless, the external load pattern of both stages is different [1]. Given that the safety at the "use stage" is the main concern of engineering professionals, the safety requirements of structures at the use stage are the focus of all existing structure design specifications, particularly seismic performance and wind resistance of structures. All temporary structures at the "construction stage" are collectively called "falsework." Its importance is often overlooked as it is demolished after project completion. Kwak and Kim have analyzed shoring systems installed to support applied loads during construction [2]. Currently, there are very few specifications for the reference of falsework design. Most constructors simply calculate the structural strength of falsework based on the design specifications of structures at the "use stage", the accuracy remains inconclusive. Reinfurt et al. have described the design, construction, and monitoring of a temporary shoring support system which permitted the construction of two urban stations for the Metro Link St. Louis Light Rail System [3]. As part of the falsework, the importance of SSS is often overlooked in construction projects, thus causing increasing accidents from SSS collapse. Stauffer et al. have presented the design and construction of an innovative shoring system in Seattle, Washington [4]. As concrete placing for large-area and atrium structures requires many workers working together at the same time, when SSS collapses, this usually causes mass injuries and deaths of workers and massive personal and property damages. Figure 1 shows the formwork SSS collapse occurred at the Beiling Boost Station Service Reservoir Project in Gangshan, Kaohsiung, on December 13, 1997, where seven workers were killed and twelve injured on-site [5]. Figure 2 shows the formwork SSS collapse occurred at the Beishan Interchange Bridge Project of Highway 6 in Nantou on September 30, 2010, where seven were killed and three injured on-site [6].

Figure 1. Formwork SSS collapse at the Beiling Boost Station Service Reservoir Project in Gangshan, Kaohsiung [5]

Figure 2. Formwork SSS collapse at the Beishan Interchange Bridge Project of Highway 6 in Nantou [6]

1.2 Introduction to BIM

Building information modeling (BIM) has become an undeniable trend across the world. As its ability to link 3D models with various analysis tools can improve accuracy, reduce errors, synchronize design, enhance energy efficiency rate, and promote sustainable development, it has been widely applied to architecture, engineering, and construction (AEC) industries and building life-cycle project design and engineering integration [7]. BIM is a 3D building software integration concept and a practical solution. It enables complete linking between fundamental module technology and architectural design databases and allows digital information interchange in the construction process. It also establishes mutually coordinated and consistent computable information for construction projects under design and construction. According to the engineering characteristics and needs to select the appropriate software in BIM products will be save a lot of manpower and resources in construction projects [8]. A BIM system generally refers to a system enabling the establishment, integration, and reuse of construction information and professional knowledge during the life-cycle of buildings [9].

1.3 Research Motivation and Purpose

The visualized interface of BIM can effectively help engineers to convert what were originally 2D plans into 3D elevations. Except for well-trained senior engineers, it is difficult for ordinary people to capture all three-dimensional information from a 2D drawing, and not to mention detecting the conflicts and problems of work items from the drawing. As it is strenuous to erect the SSS, constructors are less likely to erect it again when errors are found after erection. When the problems of SSS were found, best way to solve these issues is to reconsider the whole design of SSS. As they simply strengthen the sections with probably lower strength, industrial safety accidents thus occur [10]. According to CLA's 2011 analysis of major occupational accidents in the construction industry [11], most occupational accidents occurred in the construction industry, and fall and collapse were the most common and second most common occupational accidents among all major occupational accidents in the construction industry, as shown in Figure 3. Effective planning of safety passages and stairs thus become the key to the prevention of occupational accidents. By drawing detailed, actual on-site support conditions on the SSS in falsework of construction projects with BIM, engineers can visually examine the potential problems of SSS and adjust the method or position of erection from the computer to avoid disassembly and re-assembly of SSS when errors are found after erection. The purpose of this research includes:

1. To investigate the status of SSS uses in falsework of construction projects and the types of potential accidents and relevant examples.
2. To draw the actual status of erection of SSS and

relevant safety facilities on a construction site with BIM to provide engineers with a platform to discuss SSS suitability.

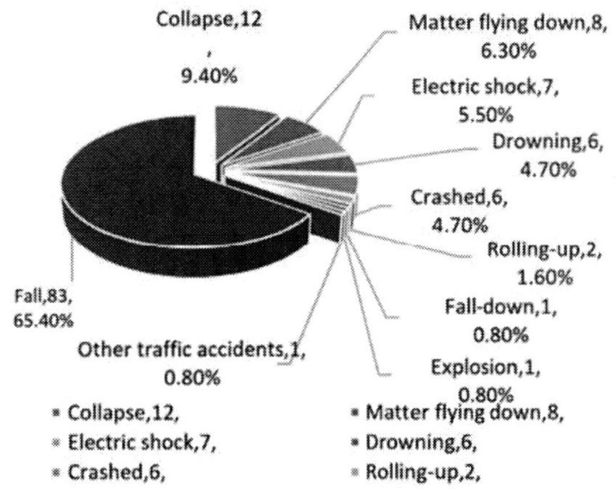

Figure 3. Occupational deaths by accident type in the construction industry [11]

2 Research Results

2.1 Current Status of SSS Safety Facilities

This research drew the status of on-site erection of SSS on the construction site with BIM software to assist in SSS safety passage design and investigated the current status of SSS safety passages by on-site interview and literature collection. Currently in Taiwan, falls likely occur when assembling SSS columns and beams and erecting safety nets due to safety feature inadequacy, personnel passage difficulty, and lack of appropriate stairs. When installing these safety facilities, as their reliability and security are often overlooked, collapse also likely occurs. In the absence of well-defined and dependable guidelines for the installation, choice, and layout of safety facilities for SSS using sites, even if industrial safety personnel are eager to install them, they can only bring a certain amount of safety, and unreliable safety protection cannot bring substantive protection, such as incomplete pavement of footsteps along passages or the absence of guardrails.

2.2 Investigating SSS with BIM

This research drew the `actual status of the erection of SSS on the construction site with the Revit from Autodesk to provide a communication platform for engineers, SSS contractors, and labor safety and health personnel. This research investigated the working space of workers in a construction project with the 3D display of BIM for SSS contractors to prepare adequate quantity of different kinds of elements and safety facilities and reviewed the safety considerations in construction and the conflicts of construction interface among different contractors for a comparatively reasonable and safer work environment.

Based on the current status of a bridge project, this research simulated the status of SSS uses on the construction site and drew up the 3D layout and planning of SSS with BIM software. Figure 4 shows the sectional view of the case. Figure 5 shows the side view of planning. Figure 6 shows the 3D perspective of the entire site.

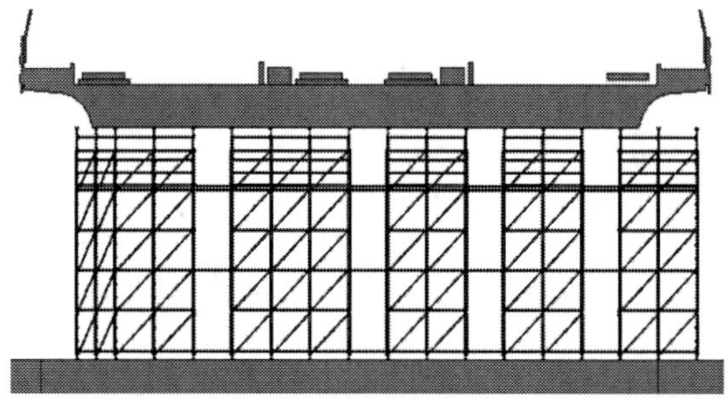

Figure 4. Sectional view of SSS erection

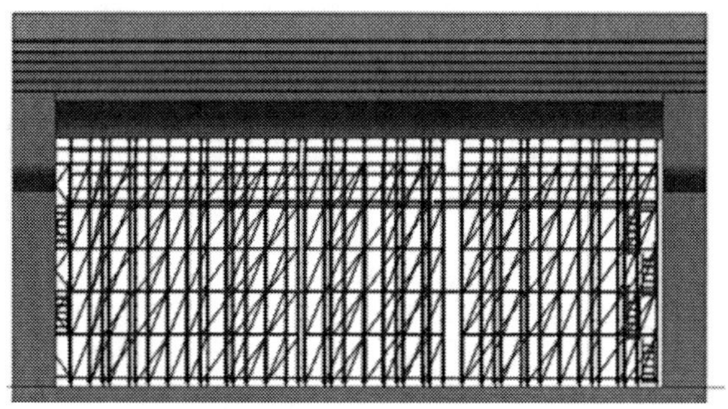

Figure 5. Elevation of SSS erection

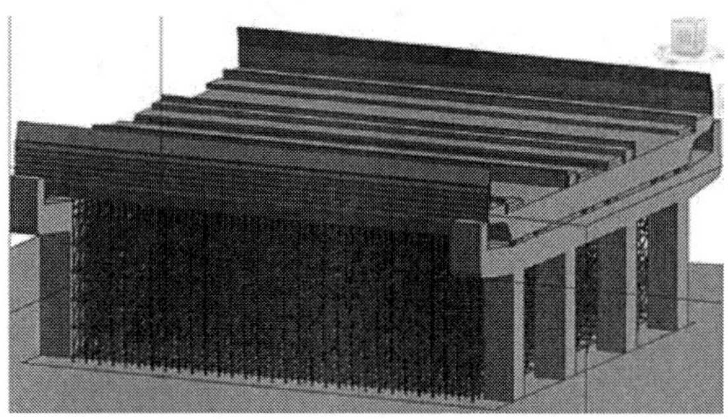

Figure 6. Perspective of SSS erection

2.3 Investigating Footsteps with BIM

This research constructed a 3D model to investigate the viability and safety of SSS safety passage planning. When laying out footsteps based on current practices, passage difficulty or fall often occurs. Although it is difficult to find passage difficulty in SSS from 2D layouts, the 3D drawing in Figure 7 clearly shows the problems in footstep layout on SSS and vertically deployed footsteps must be installed on the left for workers to pass through SSS. Figure 8 shows the need for an additional footstep to prevent any fall of workers, as the interval from the footstep on the right hand side becomes larger after moving the footstep to the left. In addition, as the connection point of the two vertically deployed footsteps is not aligned under this situation, workers may fall when they overlook this, and warning signs should be posted. Figure 9, the aerial view of the entire site, shows that the safety passages are too narrow and the interval between footsteps is too large, thus personnel or objects falling is very likely to occur. From the BIM, this research can adjust the footstep layout (as shown in Figure 10) to improve problems of narrow safety passages and too large intervals between footsteps, minimizing the possibility of personal falls.

Figure 7. Passage difficulty caused by the present footstep layout

Figure 8. Footstep deployment after adjustment to the left

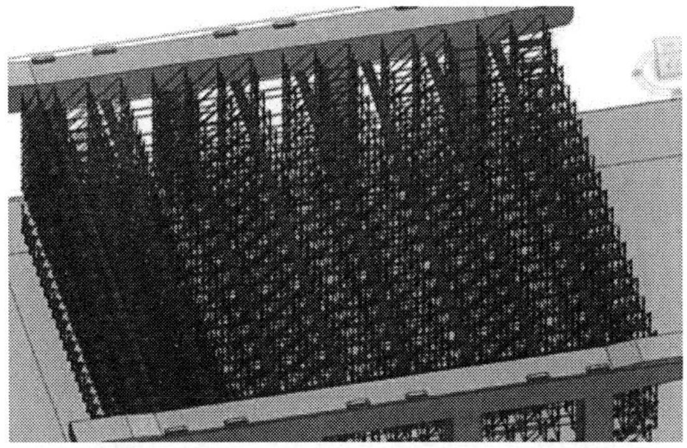

Figure 9. Footstep layout simulation with BIM

Figure 10. Simulation of footstep layout adjustment with BIM

2.4 Investigating Safety Stairs with BIM

The quantity and relevant requirements of SSS safety stairs are well-defined in relevant specifications [12]. In most cases, constructors only need to check the quantity, gradient, and width of safety stairs with reference to relevant standards. The layout of SSS safety stairs is shown in Figure 11 [13], and its BIM simulation is shown in Figure 12.

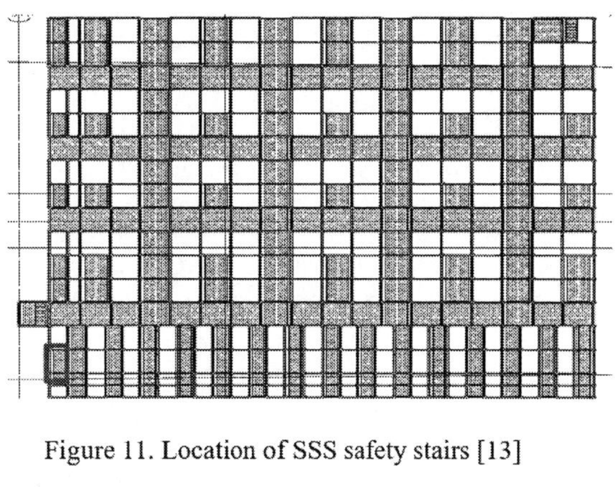

Figure 11. Location of SSS safety stairs [13]

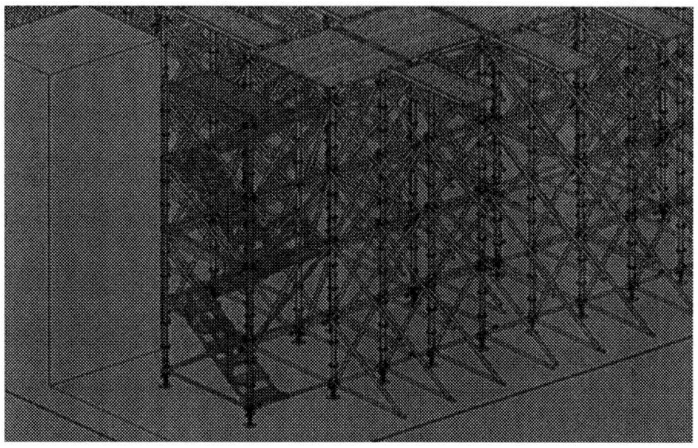

Figure 12. Simulation of SSS safety stairs on the lower left

3 Conclusions

The following conclusions are obtained from simulating SSS in 3D images with BIM.

1. In terms of safety passage, the work platform is the passage of the main work area at top of the SSS. BIM can help engineers to verify passage safety and compliance with the regulatory requirements of work platforms. Support formworks and safety nets should be installed on work platforms where footsteps are inadequate.
2. At the beginning of design, compared to traditional 2D drawings, BIM enables engineers to effectively plan the layout of safety passages, stairs, and safety facilities.
3. If budget allow, a fully paved work platform can always maintain personal safety and prevent personnel falls. This also provides a safer work environment for formwork workers afterwards.

References

[1] Peng R. L., Yen C., Lu L. Z., Zhou D. C., and Chen H. F., Discussion of the Collapse of Engineering Formwork Supports at Beishan Interchange of Freeway 6: A Perspective of Formwork Support Damage I, *Structural Engineering*, 28(3):61-93, 2013.

[2] Kwak H. G. and Kim J. K., Determination of efficient shoring system in RC frame structures, *Building and Environment*, 41(12):1913-1923, 2006.

[3] Reinfurt J. E., Anderson T. C., Reitz P. and Licari T., Temporary shoring support systems in an urban environment, *Geotechnical Special Publication*, 42:24-42, 1994.

[4] Stauffer S. D., Chin K. H. and Byrne J., Design and construction of an innovative shoring system at a challenging urban site in Seattle, Washington, *Geotechnical Special Publication*, 384(208):334-341, 2010.

[5] Pong J. L., Yen C. and Guo C. J., Discussion on the Form Support Safety from Supporting Collapsed of Construction Formwork in Northridge Water Pressurized Station, Labor Safety and Health Studies, 10(3): 252-271, 2022.

[6] http://www.doc88.com/p-0701519196954.html

[7] Ding Y. C., Promotion and Application of BIM Implementation by the Construction and Planning Agency of the Ministry of the Interior, *Journal of the Chinese Institute of Engineers*, 87(5):10-17, 2014.

[8] J. Feng, J. P. Zhang and S. P. Gao, "Choosing suitable BIM software for engineering projects based on the fuzzy neural network evaluation model," *Applied Mechanics and Materials*, 405-408:3348-3351, 2013.

[9] Shiau Y. C., Lu L. T., Lee Y. L., Lu Y. M. and Chan J. C., Establish Traceability System for Construction Projects in BIM Environment, *ICIC Express Letters, Part B: Applications*, 4(4):1159-1166, 2013.

[10] Shiau Y. C., Song S. J., Yang J. C. and Lee G. C., Study on Disaster Prevention of Site-Support Method for Bridge Construction, *Disaster Advances*, 5(4): 1541-1546, 2012.

[11] Council of Labor Affairs, Executive Yuan, Analysis and Inspection Strategy of Major Occupational Accidents in the Construction Industry, 2012.

[12] Council of Labor Affairs, Executive Yuan, Rules for Labor Safety and Health Facilities, amended and announced by the Council of Labor Affairs, Executive Yuan, on October 13, 2009

[13] Institute of Labor, Occupational safety and Health, Council of Labor Affairs, Executive Yuan, Guidelines for Safety Passages and Stairs, and Safety Facilities, for System Shoring Supports, 2013.

Development of House Selling System for Construction Industries

R.L. Zhuang[a], Y.P. Chiu[a,b*], Y.C. Shiau[c,*], and C.J. Guo[c]

[a] Ph.D. Program of Technology Management, Chung Hua University, Taiwan
[b] Department of Digital Media Design, Hwa Hsia Institute of Technology, Taiwan
[c] Department of Construction Management, Chung Hua University, Taiwan
E-mail: luckiness740525@hotmail.com, *frank@cc.hwh.edu.tw, *ycshiau@ms22.hinet.net, E10316015@chu.edu.tw

Abstract –

Mobile information service can provide positioning, query, text, drawings, video and other browser interfaces. It is an indispensable object to most people. Advanced technologies have changed the manner and speed of information delivery. People need to work for a long time for buying a house. Choosing a suitable house is a very important task for homebuyers. This study discusses related important information during housing transactions which are foremost concerns of home buyers. The system analysis is performed to fulfill demands of home-buyers in house transactions. House selling system is developed by using Unity in this study. The basic house data, aerial 3D perspective, site information, plan drawings, hall schematics and landscape are compiled in this system. Real estate agents can communicate with customers by using this system on smart phones, tablet PC and other advanced communication platforms. Home buyers can examine interesting objects at home to handle future affairs such as interior decoration and furnishings. Careful preparations and discussions can prevent human mistakes and render a decent house buying experience through the help of this system. Through the development of this project, real estate agents can prevent lengthy explanation procedure; speed up project sales pace and save considerable manpower expenses.

Keywords –

BIM; Mobile Devices; Smart Phones; House Selling System; 3D Perspective

1 Introduction

1.1 Background

Currently digital information can be fetched on variety platforms. It is an indispensable object to most people. Advanced technologies have changed the manner and speed of information delivery. The screen size becomes bigger and screens are thinner for smart phones and tablet PCs. Some works that can only be executed on the computers can be performed smoothly in smart phones. Suki has revealed that social needs and social influences significantly affect people's dependence on smart phones [1]. Zhou et al. indicated that Smart phones have been developing rapidly over the past decade, more and more people love to use its mobile Application to process the day-to-day affairs [2]. Construction industries have played an important role in the process of economic development in Taiwan. Approximately every seven years, there are considerable gains for real estate prices in the early stage [3]. Since the internationalization of economic market and diversification of investment channels, plenty of money moved outside of Taiwan. The construction industry has very much been challenged since the demand for housing is becoming saturated. In order to survive in this unfavorable environment, using ICT application to promote the service and competitiveness is a key issue for construction industry.

1.2 Unity 3D Software

This study use Unity 3D Software [4] to develop house selling system. Unity 3D is a game development software which provides multiplatform development and exhibits simple operability. It consist of a powerful engine to develop games executable on various products such as Android phones, iOS, tablet PC, PC, Mac, PS3,

Wii, XBox360, and other gaming supports. Unity provides a user-friendly interface, support the PhysX physics engine and particle system, as well as multiplayer features in internet environments. This significantly reduces the game programming threshold, because users do not need to learn complicated languages. This tool decreases the game-designing time and reduces costs of game production. The major features of Unity include:

1. An editor which supports constructing terrain effects efficiently, producing natural-looking displays.
2. Provides NVIDIA PhysX physics engine that offers photorealistic corporeal effects.
3. Offers OpenGL and DirectX illustrations optimization practices.
4. The capability to publish games to various platforms such as Androids, iOS, PC, and Mac.
5. Supporting synchronized multiplayer online games over internet.

1.3 Motivations and Objectives of the Study

In Chinese culture, to buy a house is a major goal for any family. Young people need to work for a long time to pay the down payment for a house. Choosing a suitable house is a very important task for homebuyers. System analysis and software development were used in this study to establish mobile construction house selling system by applying AutoCAD and 3D drawing on Unity platform. The house buyer can rapidly fetch related information such as project profile to make appropriate decisions during the process of buying a house. This can help the buyers to effectively grasp the relevant construction information, shorten time consuming information inquiring process and save the time for house inspection on site. The objectives of this study include:

1. Through literature review and questionnaire to aggregate related information concerning building construction and sales process. The system analysis can then be performed to fulfill demands of home-buyers in house transactions.
2. Collect relative construction drawings and documents and use Unity 3D to develop house selling system.

Input construction project data into system to verify the functionality of the software. Feedbacks from users are consulted to revise the system function.

2 Investigation on Current Studies

The realtors in Taiwan have expanded service that allows users to browse housing item via smart mobile devices. Since the browse interface on smart phone is limited, the provided functionality is inadequate. Wang has discussed related information and introduced human-computer interface design to improve the browsing interface for homebuyer [5]. The provided message include price, square footage, address, house age, and structure types of building. The photos and 360-degree surrounding views are added as reference for realtors to develop the system.

2.1 Interactive Interface

Information asymmetry is a common phenomenon in various trading activities. This trading environment may cause high trading expense. When consumers cannot evaluate the quality of products and services before transactions, they are exposed to certain risks. Hsieh has explored the effects of introducing digital facilities and designing consideration upon the trading environment of information asymmetry in the housing industry [6]. This study has established an interactive interface and designing model as reference for related services.

2.2 Functions of House Sales Management System

House sales management system is an integral part of sales company [7]. Its content is critical for the company's decision-makers. It can provide an appropriate tool for house sales and customer management. It can also improve the quality of house sales service, avoid errors of the leakage of payments received, and improve the company's competitiveness. In this study, the house sales management system for construction industry has been divided into seven modules. It has been decomposition into thirty-three secondary module, and sixteen essential modules were verified through expert questionnaire. This can be used as reference for house sales management system development.

2.3 Data Mining Techniques

Since database and IT have been rapidly developing, many large enterprises have stored lots of data about customer profile, consumer activities and trading records. Using data mining techniques to fetch potential business information is an important issue for these enterprises. Data mining is the key process in which valuable information or knowledge is captured from huge data sets. The business model of real estate is to provide a trading platform that matches sellers and buyers of properties. They make certain profit from the service charges on successful mediation. Therefore, presenting the appropriate properties to the right person in the shortest amount of time is a critical issue on the operation of real estate industries. Lee constructed a recommendation model by employing data mining techniques to predict potential properties trading of a house selling through characteristics of customers [8]. Apriori algorithm in Association Rule is applied on a real estate agency in Taiwan to construct this model. This model can provide managerial implications and improve the decisions making for real estate enterprises. Shi et al. have indicated that these rules represent some universal rules and can provide reference for enterprises [9].

2.4 Space Visualization

The construction companies in Taiwan have provided service of changing the design when consumers buy presale house. Although sample house is provided for customer to catch the possible appearance, it is not an easy task for house buyer to precisely catch the real looks for different building material combinations. There are some relevant design software for interior decoration, such as Space Magician and 3ds MAX. It is inconvenient for consumers and home builders to use these tools. Tsai indicated that a simple and convenient visual system is essential for users to select appropriate materials for presale houses [10]. Photoshop is utilized to cut out visual object such as furniture and bathtub. The system combines database to change visual objects, produce building materials list and building materials confirmation form. The software will show the simulation related to selected decoration material selected by house buyers.

2.5 Implementation of BIM

To use BIM as a tool is an irreversible trend for the construction industry in Taiwan. Many related implementation in the planning, design, construction and maintenance phases utilizing BIM environment for construction projects. Bansal indicated the use of BIM to create the simulation of construction process by linking execution schedule with the 3D model is an important trend [11]. BIM has been used throughout whole life cycle of a building except in the house selling stage. Construction project usually entrust realtor agency to sell the subjects in Taiwan. By applying BIM environment, the information exchange between realtor agency and construction company will be altered from traditional 2D Cad drawing into 3D model. Tseng has established a guide to perform BIM execution for real estate agency [12]. The guide was formulated based on "Facility Owner's Guide for Preparing BIM Guidelines" which is published in 2014 by NTU BIM Center. The simulation result proved that the guide can help real estate agency to establish a BIM execution plan. The team members should be chosen from sales department since these staffs can combine practices and execute information transmissions effectively. Although the investments, such as training and software, for using BIM is relatively low in early phases, the advantages of import BIM can be achieved such as gaining new business and obtaining higher customer satisfaction.

3 Methods

3.1 Functional Analysis

This study has divided the function of the system into following categories (Figure 1):

1. Aerial 3D Perspective

This feature will provide users to view 3D perspective views for all buildings of the project. This study divides this menu into sub-function from Building A to Building F.

2. Site Location

This feature provides related information for users to examine some environment data such as "Soils and Geology", "Neighborhood Traffic" and "Geographic Environment".

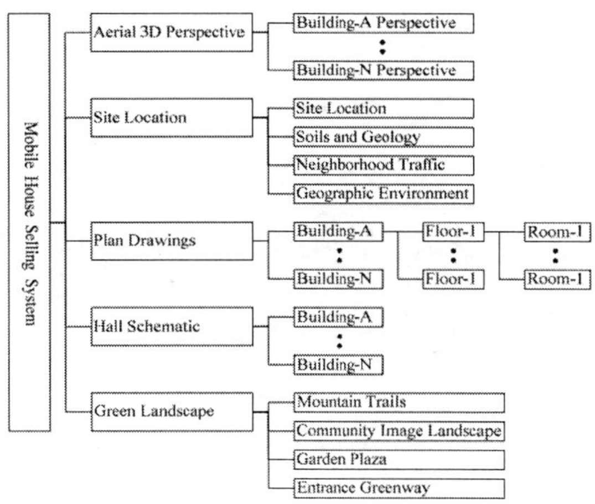

Figure 1. System framework

3. Plan Drawings

This feature allows users to view the plan drawing of interesting objects. The hierarchy of this function includes Building, Floor and Room. When related

"Building" is clicked, all floors belonged to the building will be displayed for further selection. When certain floor is clicked, all rooms belonged to the floor will be displayed for users to select for viewing the plan drawing.

4. Hall Schematic

This function allow user to view the hall perspective for all buildings.

5. Green Landscape

This feature allows homebuyers to watch relative outdoor views of the project such as "Mountain Trails", "Community Image Landscape", "Garden Plaza" and "Entrance Greenway".

3.2 Development of the System

This study uses a construction project as example to develop the mobile house selling system on Unity environment. The AutoCAD drawings are imported to the system to construct the BIM module. The execution display of smart phone is shown in Figure 2. In order to clearly present the program functions, tablet PC version of the system are demonstrated as shown in Figure 3.

Function examination is to confirm that all design functions performed properly. Feedbacks from users should be considered to assure that system can fulfill all needs of the homebuyers. Figure 4 shows the popup menus of the system. A sample construction project was used as an example to implement the system. Figure 5 is the close view of the exterior wall of the building. Figure 6 to Figure 13 is part of the execution display of the system.

Figure 2. Execution display of the system for smart phone version

Figure 3. Execution display of the system for tablet PC version

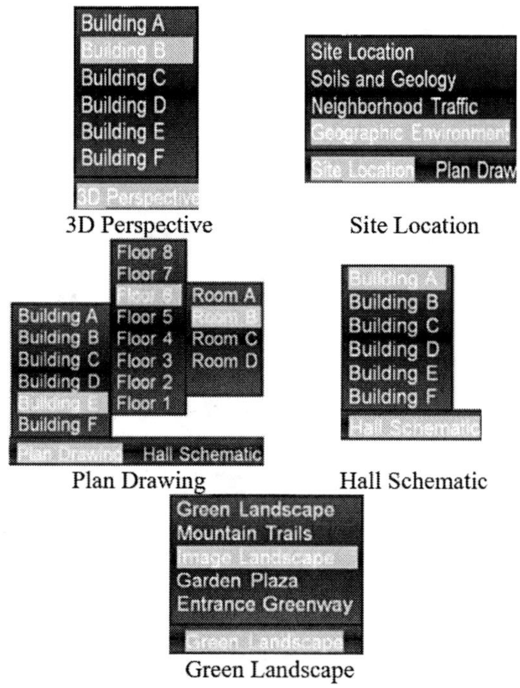

Figure 4. Popup menus of the system

Figure 5. Close view of exterior wall

Figure 6. Soil and geology display

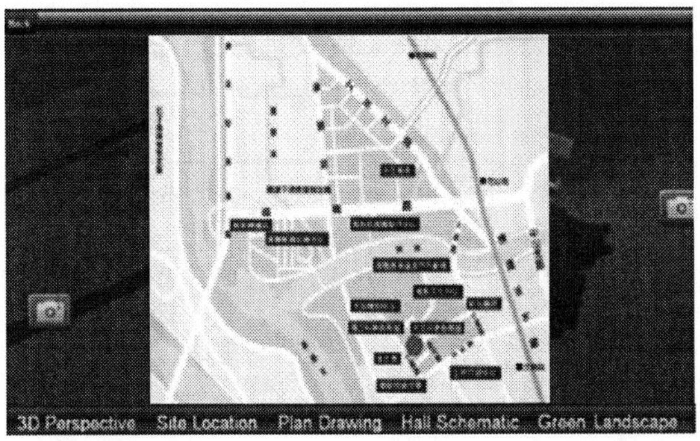

Figure 7. Neighbor traffic display

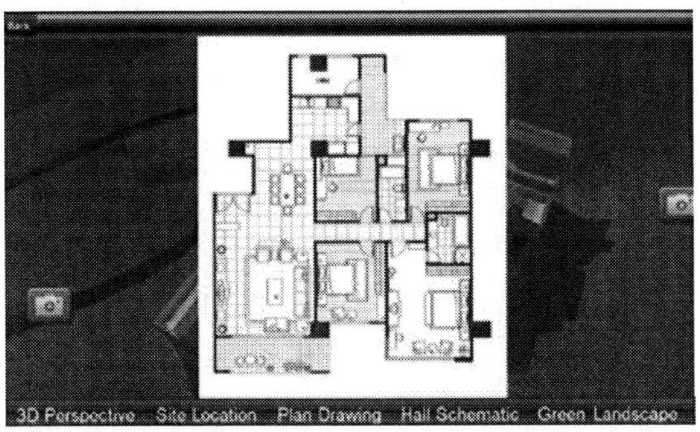

Figure 8. Plan drawing display

Figure 9. Hall schematic display 1

Figure 10. Hall schematic display 2

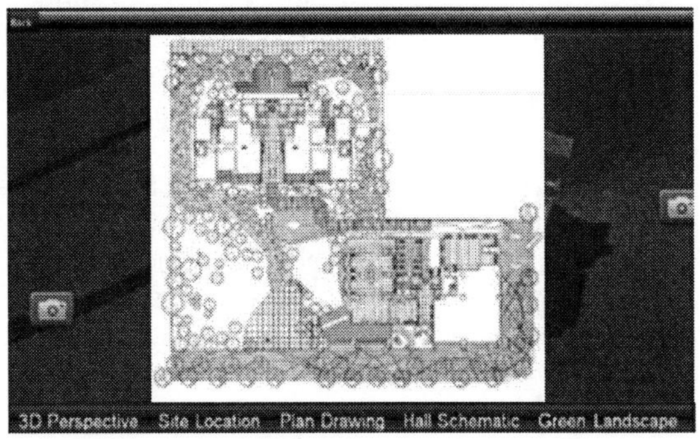

Figure 11. Green landscape display

Figure 12. Image landscape display 1

Figure 13. Image landscape display 2

4 Conclusion

This study uses Unity environment and BIM to develop mobile house selling system. This can bring the following contributions for homebuyers and construction industry:

1. Homebuyers can instantly search information related to the target building and reduce data fetching time through smart mobile applications.
2. Real estate agents can promote business image, endorse transactions and enhance product values by using smart phones, tablet PCs and other advanced communication platform.
3. This system can provide related object information of the house being sold. Staffs of real estate agents can rapidly fetch relative data for customers to reduce unnecessary resources and promote services efficiency.
4. Home buyers can examine interesting objects at home to handle future affairs such as interior decoration and furnishings. In general, employees may have to work several decades to have the opportunity to buy a house. Careful preparations and discussions can prevent human mistakes and render a decent house buying experience through the help of this mobile house selling system.
5. Through the development of this project, real estate agents can prevent lengthy explanation procedures; speed up project sales pace and save considerable manpower expenses.
6. The software can be installed on most smart phone platforms such as IOS, Android, Windows Phone, etc.

References

[1] Suki N. M., Students' dependence on smart phones: The influence of social needs, social influences and convenience, Campus-Wide Information Systems, 30(2):124-34, 2013.

[2] Zhou C. H., Weng Z. Y., Xu C. and Su Z. Z., Integrated Traffic Information Service System for Public Travel Based on Smart Phones Applications: A Case in China, *International Journal of Intelligent Systems and Applications*, 5(12):72-80, 2013.

[3] Shiau Y. C., Study on the Contract Management for Construction Industry, *Construction World*, June 2006.

[4] Unity website: http://www.unity3d.com.

[5] Wang C. H., A Study of the Usability of Mobile App Browsing Interface on Cellular Phone–Use Real Estate Enterprise as an Example, Master Thesis, Department of Information and Communications, Chinese Culture University, 2013.

[6] Hsieh F. Y., Designing Interactions Interface for Information Asymmetry: A Case Study of the Housing Industry, Master Thesis, Department of Industrial Engineering and Management, Ming Chi University of Technology, 2012.

[7] Chen S. F., Study on Functions of House Sales Management System for Construction Industry, Master Thesis, Department of Construction Management, Chung Hua University, 2011.

[8] Lee C. S, A Study of Using Data Mining Technology to Build up House Selling Recommendation Model, Master Thesis, Department of Business Administration, National Taipei University, 2012.

[9] Shi G. Y., Chen P., and Liu S., Application of association rules mining in maritime casualty's analysis, *ICIC Express Letters, Part B: Applications*, 6(1):81-87, 2015.

[10] Tsai Y. T., Space Visualization System for Customized Change in Real Estate Sale, Master Thesis, Department of Civil Engineering, National Chiao-Tung University, 2008.

[11] Bansal V. K., Application of geographic information systems in construction safety planning, *International Journal of Project Management*, 29(1):66-77, 2011.

[12] Tseng Y. J., The Study of Implementing BIM in Real Estate Agency, You-Jung Tseng, Master Thesis, Department of Civil Engineering, National Taiwan University, 2014.

Dynamic Space Conflict Modeling for Construction Operations

Arian Saeedfar[a], Lingguang Song[b] and Satish Gore[c]

[a]Graduate Student, Department of Construction Management, University of Houston, Houston, TX 77204
[b]Associate Professor, Department of Construction Management, University of Houston, 111D T1, Houston, TX 77204 ; Phone (713) 743-4377 ; Fax (713) 743-4032
[c]Architectural Draftsman, SGV International LLC, Houston, TX 77024
Email: asaeedfar@uh.edu, lsong5@uh.edu, sgore@sgvinternational.com

Abstract –

Site planning is critical for efficient and safe construction operations. Current industry practice primarily relies on highly experienced planers and their personal judgment to anticipate and prevent space conflicts during construction operation. Inaccurate and unreliable site space planning not only leads to major loss of time and cost, but also negatively impact safety performance. Recent research has introduced new planning methods during the pre-construction phase to enhance site safety and productivity. This paper proposes a site spatial planning solution called Dynamic Space Conflict Modeling (DSCM), which can model dynamic field operations on a construction job site in a 3D model format based on the space demand of each activity and each resource. DSCM provides a process in which schedulers use time, crew information, and space demand of resources as inputs and capture the live space conflict as the output.

Keywords –

Dynamic Planning; Space Demand; Space Conflict; Safety; Productivity

1 Introduction

Successful delivery of a construction project rely heavily on effective project planning[1]. It is vital to plan the work flow considering various constraints such as time and space to maximize performance and reliability of construction processes[2]. Nonetheless, site space is usually not considered explicitly as a resource like material, equipment, labor, time, and money. Not surprisingly, site space is one of the most overlooked resources in planning[3].

Site congestion can significantly impacts productivity and safety performance, and past research estimated an efficiency loss of up to 65% [3]. This is especially the case for labor productivity since 50% to 60% of the total project cost devotes to labor cost on large construction projects[4]. In addition, numerous past research has confirmed that site congestion and stacking of trades is one of the leading contributors to safety accidents.

Current industry practice primarily relies on a planner's experience and personal judgment to have a feel about site space, visualize it in a 2D drawing format, anticipate conflicts, evaluate the impact on project performance, and resolve these conflicts proactively[1][6]. Due to the highly dynamic and complex nature of construction field operation, this informal and subjective approach in managing space can result in inaccurate and unreliable site planning that further leads to poor productivity and safety performance [7].

With the new advancements in modeling techniques (e.g. 3D, 4D modeling, and building information modeling), sensing (e.g. range camera and laser scanner), and spatial computational methods, site spatial planning is evolving from a 2D format and a manual procedure to a virtual 3D modeling approach with automated data processing that minimizes manual data entry. Virtual construction environment continuously capturing as-built or future site conditions in a digital format enables planners to better visualize and explore dynamic site space for planning purposes[6][6]. It has a great potential to minimize subjectivity in the current planning practice and improve the accuracy and timeliness of recognizing and avoiding conflicts**Error! Reference source not found.**[8].

A key element of this virtual space planning approach is to capture the space demand of construction activities and detect potential conflicts among them, which is the objective of this study. The following section provides a review of relevant past studies. A 3D site planning solution, Dynamic Space Conflict Modeling (DSCM), is then proposed. The proposed method is demonstrated in a prototype sytem.

2 Literature Review

Construction projects consist of a large number of activities and the interactions among the activities are highly dynamic and constantly changing[9]. Traditional planning methods consider space resource implicitly in a 2D drawing format, which is less efficient and accurate to capture the intricacies and dynamics of site conditions[3]. Recent research in the area of construction space planning benefits from the advancement in computer graphics, simulation, and artificial intelligence[10].

2.1 Dynamic Layout Planning

Recognizing and analyzing space demand of construction activities is critical in site space planning. Each activity on the site needs to have a specific amount of space in order to be executed with acceptable safety and productivity performance.

Dynamic site planning uses the project schedule to generate a logical sequence of layouts spanning the entire construction duration. As the condition of the site varies over time and project progresses, schedule can be adjusted to overcome any space availability issue[3]. How to update schedule dynamically to accommodate space demand is the subject of many studies. One of the solutions, MovePlan, assists users to plan the location and handling of resources on site with a 2D graphical user interface [3]. When the user interactively moves resources around until a satisfactory layout is obtained, MovePlan automatically updates the layout plan of the project. [3]This dynamic site planning approach allows users to model approximate site space needs during project scheduling and to create layouts for different stages of construction.

Thabet and Beliveau developed a work-space model that uses a numerical index to measure the degree of congestion at different locations of the site[11][11]. This index is calculated as the ratio of space demand to available space on the construction site. This model helps the planners to identify, prioritize, and response to space conflicts of the project.

Riley and Sanvido presented another work-space model that identifies the space needs of the activities by defining twelve construction-activity space use patterns[12]. This method is applicable to repetitive construction operations, such as multistory building construction. The typical pattern of space demand for each activity and potential interface problems can be predicted using this model, thus supports the logical planning of productive work sequence.

Guo proposed a method which categorizes construction activities based on type and location of the activities along with the project timeline and existing demands to expose overlapping spaces and conflicts on the construction site [13]. This method attempted to automatically resolve conflicts using predefined resolution strategies[13].

2.2 4D Planning and Virtual Construction

4D modeling visualizes the construction process in a time-phased 3D format. 4D CAD enables the planners to analyze construction sequence before the actual start of construction[14]. It helps planners to identify performance issues and avoid conflicts which can save a considerable amount of time and money during the field operation[15]. 4D planning also enables location-based scheduling that integrates the work flow and its performing locations. The planning data can be merged with the spatial data of the project in one single 4D CAD model[2].

Another application in 4D planning is 4D WorkPlanner which aims at overcoming the weakness of the traditional Critical Path Method by representing the spatial requirements of activities and managing the spatial conflicts[16]. This method is able to analyze time-space conflicts and offer customized solutions to planners.

Finally, Virtual Construction Environment (VCE) empowers users to reconstruct a project in a 3D digital model and visualize many different aspects of the project, e.g. the facility, resources, space availability, and work sequence [1]. This environment can be thought of as a test bed for multiple purposes, such as planning and constructability review.

2.3 Spatial Optimization

Construction site layouts can be simulated and optimized with respect to cost and safety of the project. In this method, the user is able to evaluate the site layout efficiently and intelligently, thus save planning time and cost[10]. Dawood and Mallasi developed a method for optimizing temporary facilities of a project. In this method, different types of conflicts are categorized and the productivity rates of activities are assumed to vary from time to time due to uncertainty[17].

2.4 Space Demand

There are many factors affecting the work-space demand on a construction project. It is important to understand the demand and quantify it in the first place[11]. The space demand for each activity consists of the space that the activity occupies on the construction site plus its surrounding area (Equation 1)[11].

$$D_R = S_P + S_S \quad (1)$$

$D_R = Resource\ space\ demand$

$S_P = Space\ needed\ for\ performance$

$S_S = Surrounding\ space\ for\ safety$

The surrounding space is considered as a safety zone and is necessary for the safe movement of resources. This space must be added to the physical space needed for activity performance in order to form a safe and trustworthy space demand[11]. It should be noted that not all types of resources have the same demand:

1. Space demand for equipment and labor resources must consider the mobility and volume of these resources as well as their quantity (Equation 2).
2. Equipment and labor need larger space than material resource does due to performance and safety considerations.
3. The space demand of material is determined based on the physical area that the material occupies on the site (Equation 3)[11].

The physical performance space can be estimated from the facility 3D model as well as scheduling data such as the size of a crew, while the volume of surrounding safety area can be estimated based on past project experience.

$$D_{L\,\&\,E} = \sum Q \times (S_P + S_S) \quad (2)$$

$$D_M = \sum (Q \times S_P) + S_S \quad (3)$$

$D_{L\,\&\,E} = Labor\ \&\ equiment\ space\ Demand$

$D_M = Material\ space\ demand$

$Q = Quantity\ of\ resources$

3 Dynamic Space Conflict Modeling

Most of existing research recognizes and analyzes space demand and space conflict on the construction site and assist the user in taking corrective actions. However, although most of them make the planner aware of the existing conflicts, there is no tool that assists the planner in updating the activity space requirements dynamically while visualizing the conflicts. The proposed method, Dynamic Space Conflict Modeling (DSCM), strives to offer a space planning method which visualizes dynamic space conflicts among activities. Meanwhile, the user has the ability to change the characteristics of each activity such as crew size and space demand per crew member dynamically, compare the results to the original plan, and make optimal decision.

The initial input for the proposed method is the project schedule. The planner needs to plan construction activities and resources (e.g. crew, equipment, and material) as usual before using the DSCM model. The activity data is later used as inputs for visualizing and evaluating space demand and space conflicts. The following section defines the method used in quantifying space conflicts.

3.1 Space Conflict

Space conflict between the activities and resources causes a large amount of problems in different aspects of the project like constructability, safety and productivity. space conflicts with different forms can create various problems in the project life cycle and multiple types of space conflict can exist between two overlapping resources on the jobsite [16]. Space conflict has a close correlation with space demand in the project. As discussed in the literature review, once the space demand for each resource and each activity are determined, different types of space conflicts among resources must be defined so that conflicts and their potential impact can be quantified.

3.1.1 Type A

In this type of conflict, no resource has conflicts with other resources. As shown in Figure 1, solid lines represent the physical performance space and dotted lines represent surrounding area. When this type occurs, the activities can be executed as planned safely.

3.1.2 Type B

In the space conflict Type B, the surrounding area of resource *I* has conflicts with the surrounding area of resource *II*. The risk of implementation of activities with this conflict type is moderate in regards to safety.

3.1.3 Type C

In this type, the surrounding space of resource *II* has conflicts with the physical space of resource *I*. However, the physical spaces of two resources does not have conflicts. The risk of implementing the activities while the resources have conflict Type C is higher than Type B.

3.1.4 Type D

In this type, the physical spaces of both resources have conflicts with each other. The risk of implementing the activities with this type of conflict is high and the planner should adjust the project schedule accordingly.

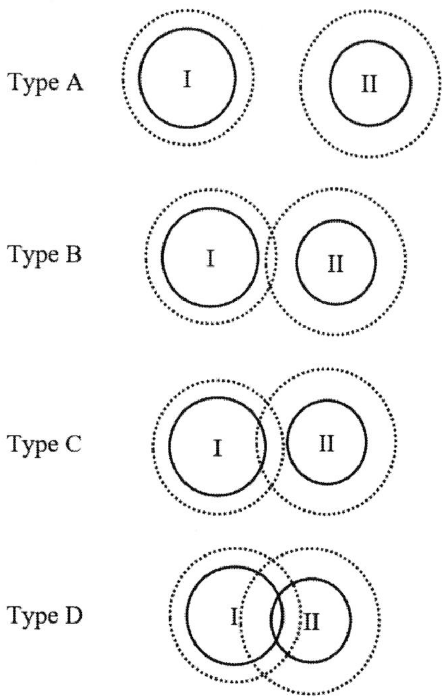

Figure 1. Various types of space conflicts.

3.2 Model Definition

The proposed DSCM model evaluates and visualizes space conflicts among resources. It captures space demand of each resource in each activity and visualizes space conflicts with adjacent resources. In addition to the project schedule, the inputs to this model also include resource type, space demand per unit of resource, and resource quantity. Once all inputs are defined, the model calculates the physical performance and surrounding space demand for each resource based on the resource type. In the next step, the conflicts between the space demand of a resource and its adjacent resources are calculated. The immediate output of the model is the visualization of space conflicts. Additionally, the model defines an index based on the spatial information of the conflicts. Space Conflict Index (SCI) calculates the proportion of space conflicts to the space demand of a resource. SCI assists the planner in making decisions about the feasibility of performing an activity based on the existing conflicts and desired safety and productivity performance.

$$SCI = \frac{Space\ Conflict}{Space\ Demand} \quad (4)$$

However, as different types of conflicts can occur among the resources, different types of SCI can be identified:

1. The space conflict between physical space of resource A and physical space of other adjacent resources divided by physical space of resource A.
2. The space conflict between physical space of resource A and surrounding space of other adjacent resources divided by physical space of resource A.
3. The space conflict between surrounding space of resource A and physical space of other adjacent resources divided by surrounding space of resource A.
4. The space conflict between surrounding space of resource A and surrounding space of other adjacent resources divided by surrounding space of resource A.

SCI represents how much is the proportion of space conflicts to space demand of each resource. Depending on the desired performance, the user must specify the maximum allowed SCI threshold for each resource and each type of SCI in order to detect space conflicts. This threshold value can be determined by the planner and based on historical experience.

As illustrated in Figure 2, when a SCI value is larger than the allowed threshold, the model will flag the conflicts for the planner to take corrective action, such as adjusting activity start time, sequence, and/or crew mix to avoid conflicts.

4 Prototype Implementation

DSCM requires a graphical environment in which planners can update resources, activities, and space demands. Rhinoceros is a 3D CAD application that is used as the platform for the implementation of the DSCM method. BIM facility models can be imported to or exported from Rhinoceros easily using Industry Foundation Classes (IFC) file format. Rhinoceros allows users to update and interact with the 3D model using a visual programming language named Grasshopper, as illustrated in Figure 3[18]. The abilities of visual programming helps the users with limited programming knowledge to develop interactive and dynamic models which could be later updated easily.

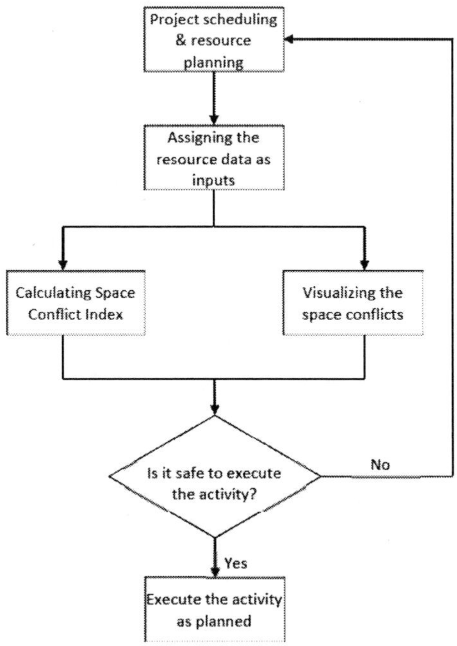

Figure 2. Dynamic Space Conflict Model

For demonstration purpose, the proposed method is implemented on a basic 3D model of a construction site in Rhinoceros 3D and programmed using Grasshopper as described below.

The first step for using the proposed model is to enter the inputs for the resources that will be evaluated, as shown in Figure 4. The initial inputs for each resource are:

1. A base or anchor point on the construction site 3D model that all of the space conflicts can be visualized based on that point.
2. Resource type which identifies the relationship between the physical and surrounding space as discussed before.
3. Space demand of each resource per unit. This input is demonstrated based on the nature of each resource and the planner's knowledge.
4. Quantity of each resource which is already allocated based on the project schedule.
5. The last input is the surrounding factor, which is a floating number between 0 and 1. It identifies the surrounding area devoted to each resource for safety and will be added to the physical space required by each resource and considering the type of each resource.

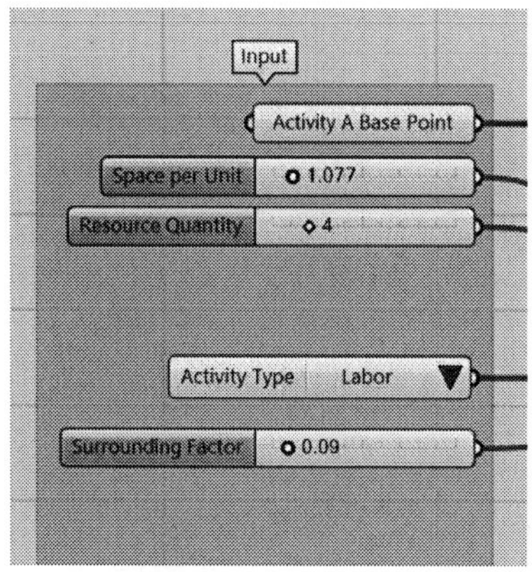

Figure 4. The inputs for one resource in DSCM model.

Once all of the inputs have been recorded, the DSCM model will calculate and visualize the space conflict among resources. At any time the planner updates the input data, the space conflict will be regenerated based on the new input data. Changing these values will trigger the change of space conflict visualization dynamically.

In Figure 5, each red point demonstrates the base point for each resource which is identified by the planner and the area highlighted in green shows the space conflict between the physical space of one resource and surrounding area of the other resource. Similarly, in Figure 6, the area highlighted in green shows the space conflict between physical spaces of two resources.

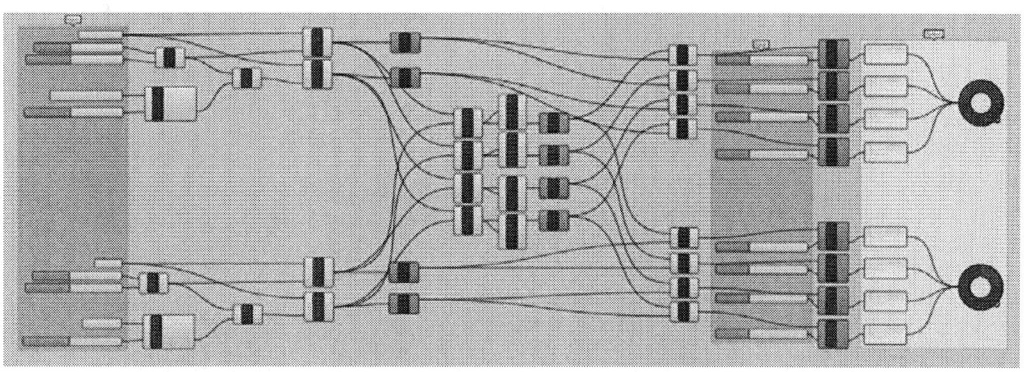

Figure 3. Implementation of DSCM using visual programming language

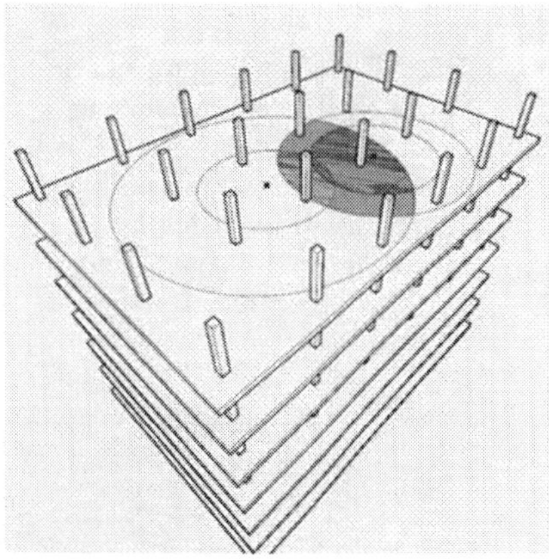

Figure 5. The existing amount of space conflict between the physical space of one resource and surrounding area of the other resource.

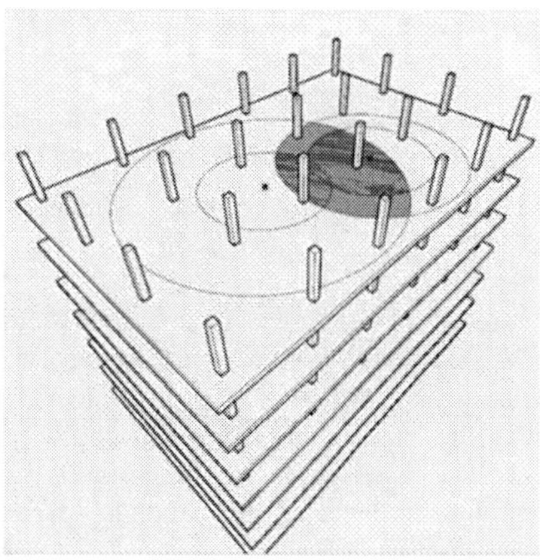

Figure 6. The existing amount of space conflict between physical spaces of two resources.

For automatic space conflict identification, the planner must predefine a maximum allowed value for each 4 types of SCI described in "Model Definition" (see Figure 7). With these threshold values, the DSCM model will first calculate the amount of visualized conflicts and the proportion of existing space conflicts to the space demand (SCI) for all 4 types of SCI, and then compare the existing SCIs to the planner's maximum allowed value. The results will be either "true" or "false" format (see Figure 8). The value "true" indicates that the existing SCI value is less than the maximum allowed value and the resource passes that particular type of SCI test. The value "false" indicate that the existing SCI value is more than the maximum allowed value and the resource fails that particular SCI test. The results of failed and passed SCI tests for each resource then will be presented in a pie chart, as shown in Figure 8. This index makes the planners aware of the existing space conflicts and helps them in determining resolutions. Also, with this index, the planners can make decision whether or not to perform the activity based on the existing condition of the resources and the present space conflicts.

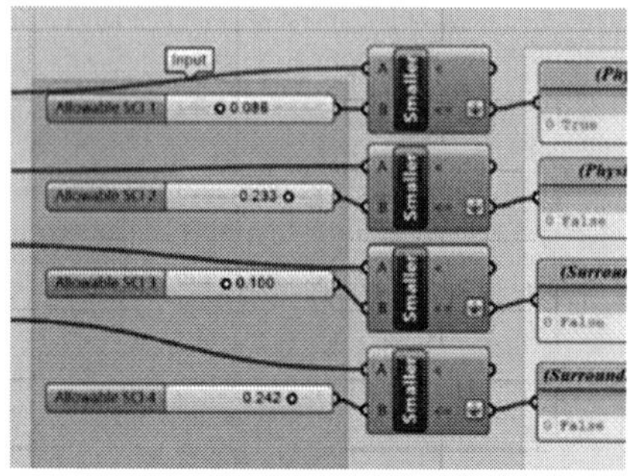

Figure 7. The maximum allowable amount for different types of SCI for a resource.

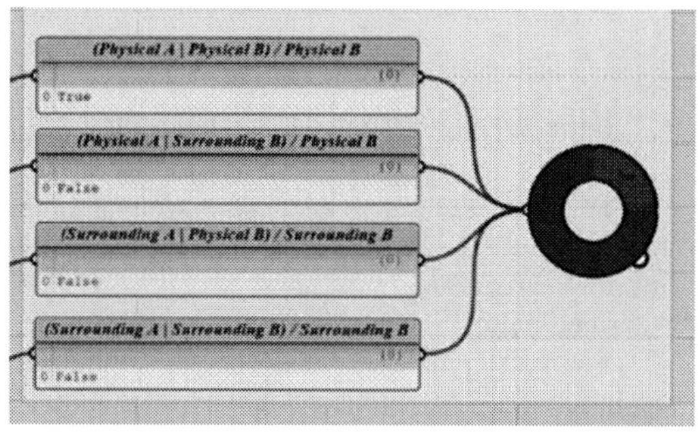

Figure 8. The results of comparing different types of existing and allowable SCIs for each resource.

5 Conclusion

Site space planning is one of the most important areas of construction planning which is inadequately considered by planners. A significant amount of time and cost can be saved with an accurate and reliable site space plan. Existing applications and methods focus on space demand modeling and conflict recognition. However, there is no tool that generates and visualizes the space conflict dynamically. DSCM proposed in this study assists planners in modeling space demand and identifying space conflicts among resources on the construction site. This method visualizes the space conflicts and calculates the amount of conflicts at the same time. With an allowable amount of space conflict defined, the model can compare the conflicts with the allowable threshold and warn the planner any future space conflict that affects productivity and safety performance of the work. The advantage of this model over the existing models is the ability to capture the interactive and live space conflicts in different areas of the project using the resource information as the input. In this model, the planners are able to observe the changes happening to the space conflicts of the project by changing the resource input and also observe the changes in the visualized 3D model at the same time. Traditional methods allow the planner to either observe the changes visually or mathematically.

For future research, the proposed method have several limitations which need to be addressed. The current implementation lacks automation, for example the planner needs to manually transfer scheduling data and evaluate resources individually. Another improvement is to measure space conflict in 3D format so that more type of conflicts (e.g. overhead work) can be analyzed. Finally, our recent work is focused on quantifying the impact of space conflicts on productivity and safety thus minimizing the subjectivity involved in decision making.

References

[1] Waly A.F. and Thabet W.Y. A virtual construction environment for preconstruction planning. *Automation in Construction*, 12(2): 139-154, 2003.

[2] Jongeling R. and Olofsson T. A method for planning of work-flow by combined use of location-based scheduling and 4D CAD. *Automation in Construction,* 16(2): 189-198, 2007.

[3] Tommelein I.D. and Zouein P.P. Interactive dynamic layout planning. *Journal of Construction Engineering and Management*, 119(2): 266-287, 1993.

[4] Elbeltagi E. *Construction site layout planning: Identifying, sizing and locating temporary facilities on construction sites*, volume 1(1). Lap Lambert Academic Publishing, Saarbrücken, Germany, 2014.

[5] Elbeltagi E., Hegazy T. and Eldosouky A. Dynamic layout of construction temporary facilities considering safety. *Journal of Construction Engineering and Management*, 130(4): 534-541, 2004.

[6] Liu C.W. and Kang S.C. A Video-Enabled Dynamic Site Planner. *Computing in Civil and Building Engineering*, 1562-1569, 2014.

[7] Riley D.R. and Sanvido V.E. Space planning method for multistory building construction. *Journal of Construction Engineering and Management*, 123(2): 171-180, 1997.

[8] Lai K.C. and Kang S.C. Collision detection strategies for virtual construction simulation, *Automation in Construction*, 18(6): 724-736, 2008.

[9] Gore S., Song L. and Eldin N. Photo-modeling for construction site space planning. In *Proceedings of Construction Research Congress 2012*, pages 1350-1359, West Lafayette, IN, 2012.

[10] Tawfik, H. and Fernando, T. A simulation environment for construction site planning. In *Proceedings of Fifth International Conference on Information Visualization*, pages 199-204, London, United Kindom, 2001.

[11] Thabet W.Y. and Beliveau Y.J. Modeling work space to schedule repetitive floors in multistory buildings. *Journal of Construction Engineering and Management*, 120(1): 96-116, 1994.

[12] Riley D.R. and Sanvido V.E. Patterns of Construction-space use in multistory buildings. *Journal of Construction Engineering and Management*, 121(4): 464-473, 1995.

[13] Guo S.J. Identification and resolution of work space conflicts in building construction. *Journal of Construction Engineering and Management*, 128(4): 287-295, 2002.

[14] Mallasi Z. and Dawood N. Registering space requirements of construction operations using site-PECASO model. In *Proceedings of Distributing Knowledge In Building, CIB W78 International Conference*, Aarhus School of Architecture, Denmark, 2002.

[15] Heesom D. and Mahdjoubi L. Trends of 4D CAD applications for construction planning. *Construction Management and Economics*, 22(2): 171-182, 2004.

[16] Akinci B. and Fischer M. 4-D workplanner: A prototype system for automated generation of construction spaces and analysis of time-space conflicts. *Computing in Civil and Building Engineering*, 740-747, 2000.

[17] Dawood N. and Mallasi Z. Construction workspace planning: assignment and analysis utilizing 4D visualization technologies. *Computer-Aided Civil and Infrastructure Engineering*, 21(7): 498-513, 2006.

[18] Robert McNeel & Associates. Grasshopper 3D. On-line: http://www.grasshopper3d.com, Accessed: 10/03/2016

Audio Signal Processing for Activity Recognition of Construction Heavy Equipment

Chieh-Feng Cheng[1], Abbas Rashidi[2], Mark A. Davenport[3], and David Anderson[4]

[1] PhD student, School of Electrical and Computer Engineering, Georgia Institute of Technology, Atlanta, GA
Email: ccheng71@gatech.edu
[2] Visiting Assistant Professor, College of Engineering and Information Technology, Georgia Southern University, Statesboro, GA
Email: arashidi@georgiasouthern.edu
[3] Assistant Professor, School of Electrical and Computer Engineering, Georgia Institute of Technology, Atlanta, GA
Email: mdav@gatech.edu
[4] Professor, School of Electrical and Computer Engineering, Georgia Institute of Technology, Atlanta, GA
Email : anderson@gatech.edu

Abstract

Action recognition and tracking of construction heavy equipment is the first step for benchmarking and analyzing the performance of individual machines and evaluating the productivity of a jobsite as a whole. Aside from direct observations, the current approaches for automatically recognizing and tracking various actions of construction heavy equipment includes: 1) using active sensors such as RFID tags, GPS and accelerometers or 2) computer vision-based activity analysis (processing images or videos).

In this paper, we present a novel audio-based approach for activity recognition of construction heavy equipment. Construction machines often produce distinct sound patterns while performing certain activities and it is possible to extract useful information by recording and processing those audio files at construction jobsites. The proposed audio-based framework begins with recording generated sound patterns of construction equipment using commercially available audio recorders. The recorded signal is then fed into a signal enhancement algorithm to reduce background noise commonly found at construction jobsites. The modified audio signal is then converted into a time-frequency representation using the Short-Time Fourier Transform (STFT). A Support Vector Machine (SVM) is then trained to differentiate between the acoustic patterns of the various activities of each machine. The processed audio signal is then finally divided and classified into various activities using a window filtering approach and by setting proper thresholds. We implemented the presented audio-based system at several jobsites as case studies and the results illustrate the efficiency of the system in automatically recognizing various actions of construction heavy equipment.

Keywords–

Audio signals; Construction; Heavy equipment; Activity recognition; Short-Time Fourier Transform

1 Introduction

According to the Lean Construction Institute, productive time in the construction industry is only 43% while it is 88% in manufacturing [1]. For many years, the construction industry has suffered from the lack of real-time performance monitoring, holistic project management, labor efficiency, and waste prevention tools. This has led to cost overruns in almost 90% of construction projects with an average of 28% higher than forecast costs [2]. For these reasons, "effective monitoring and feedback process" has been identified as the third most important attribute (after scheduling deficiencies and construction methods) that affects cost escalations within construction projects [3].

Action recognition and tracking of construction heavy equipment and personnel is the first step for benchmarking, analyzing, and evaluating the productivity of a construction jobsite. A number of monitoring techniques, such as direct observations, activity sampling, computer vision-based activity analysis (images or videos), etc., have been proposed to address this problem, but these proposals are either too inefficient to be practically useful or have other limitations that have limited their adoption by practitioners; hence, the construction industry is still overwhelmed by delays and suffers from frequent cost and time overruns.

To address these existing issues, we propose a novel audio-based activity recognition and performance monitoring system for construction heavy equipment. Construction heavy equipment often generates distinct acoustic patterns while performing various operations and it is feasible to extract useful information by processing the recorded audio data. In addition, a monitoring system based on processing audio data is computationally efficient, non-intrusive, and inexpensive.

The rest of the paper is organized as follows: Section 2 briefly reviews the state of practice and research for activity recognition and performance monitoring of construction equipment. Details of the proposed audio-based performance monitoring system are presented in Section 3. Necessary experimental setup for evaluating the proposed system and the obtained results are summarized and discussed in Section 4. Finally, Section 5 concludes the paper by interpreting the results and providing recommendations for future work.

2 State of Knowledge: Activity Recognition of Construction Equipment

The state-of-practice is based on manual data collection and analysis (watching the operation directly or through real-time video streams). It involves manually filling out timesheets for different equipment, collecting paper documentation like maintenance notes, measuring the amount of accomplished work, etc. The assessment is generally done through the work sampling concept. The concept relies on the principle that the percentage of equipment time spent on value-adding activities is an indirect measure of productivity. The objective is to approximate what is taking place in the field as accurately as possible. An observer records activities at regular intervals and then categorizes them to assess where the time is utilized [4]. This translates to generic and infrequent control activities that are performed off-line and consequently do not enable timely corrective measures to mitigate damage to an ongoing project. The performance is often demonstrated using a combination of graphical tools that represent the operation and the resources involved; for example, equipment balance charts, process charts, and flow diagrams [5].

A substantial need for real-time, automated, and dynamic control systems for on-site equipment has been recognized in both the industry and academia [6-10]. Such systems which provide relevant information to the user by exploiting context are called context aware systems in robotics and telecommunications.

2.1. Activity Recognition of Construction Equipment Using Active Sensors

Location tracking of heavy equipment has been one the first attempts for automating the performance monitoring of construction operations.

The problem of location tracking of construction machines can be currently addressed by several active spatial remote sensing technologies such as GPS, RFID, and Ultra-Wideband (UWB) sensors. They work based on the time-of-arrival principle; specifically, that the propagation time of a signal can be translated directly into distance from the source to the receiver [11]. These sensors are attached to equipment or workers and provide their 3D location in real-time. However, they are not capable of providing any information regarding the type of the ongoing actions. This is addressed in two ways: 1) human observation and work sampling; and 2) using extra sensors such as laser scanners, depth cameras (RGB-D), and a system of RGB cameras that are able to provide a 3D point cloud of the equipment (Fig. 4). This makes the problem more complex. The choice of a particular sensor depends on several factors including the application type, line-of-sight between objects and sensors, required signal strength, calibration requirements, permitted bandwidth, implementation costs, and environmental conditions [11].

Besides location tracking of heavy equipment, certain types of active sensors might be able to provide useful information about the performed operations. For example, in [12] Ahn et al. proposed an activity recognition system based on using an accelerometer. This approach is particularly efficient for earthmoving machines; however it is not general enough to cover a vast range of different types of construction equipment.

2.2. Computer Vision-Based Activity Recognition of Construction Equipment

Video-based techniques present an alternative and appealing solution for the aforementioned active positioning sensors. They are passive (i.e., do not emanate any sorts of energy and only use the natural light in the environment; [13-15]) and capable of providing semi-real-time information at relatively low-cost. A single inexpensive camera could be used to recognize actions of multiple pieces of equipment and minimize the need for sophisticated on-board sensors for each piece of equipment [16]. The recorded videos also offer a reliable documentation for future reviews. The entire process commonly consists of 4 steps, as illustrated in Figure 1: equipment recognition, equipment tracking, action recognition, and estimation of activity performance indicators.

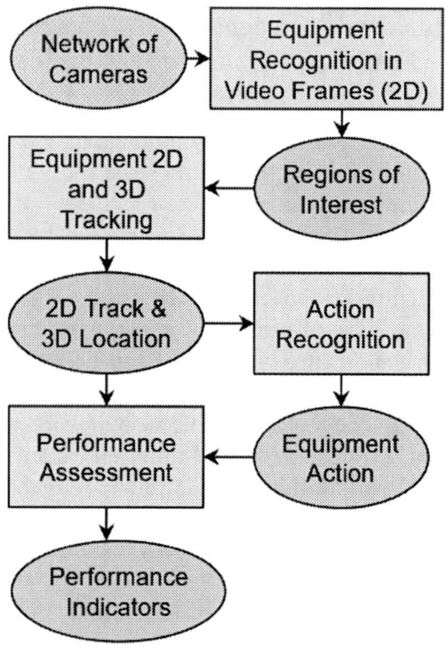

Figure 1. Overall flowchart for computer vision-based performance monitoring of construction equipment.

Equipment recognition, as a subset of object recognition, is a well-studied but challenging task due to several difficulties that arise from variations in illumination, viewpoint, occlusion, scale, articulated shapes, background clutter, and different equipment color [17]. The existing object recognition methods can be classified into three categories: recognition by parts, appearance-based recognition, and feature-based methods. It has been shown that the feature-based approach can achieve the desired performances in complex scenarios by quantizing the descriptors of positive and negative samples in order to be used for training a classifier.

Precise localization and tracking of multiple pieces of equipment in video streams is required for successful action recognition. Tracking the 3D trajectory of the equipment allows one to limit the search space to certain regions in video streams and hence minimizes the effect of noise caused by lateral movements of the camera and dynamic foreground. Contour-based tracking, kernel-based tracking, and feature matching are proposed in the literature for this purpose. Several comparative studies have indicated that the Mean-Shift algorithm outperforms other methods in visually noisy construction environment [18, 19] while the addition of the Kalman filter can further stabilize its performance.

Action recognition consists of identifying the various actions performed by a piece of equipment over time and is the most challenging step for video-based equipment performance monitoring [16]. A more precise terminology has been proposed by Bobick [20] that differentiates between movements, activities, and actions. In this terminology, for example, the action of digging a foundation by an excavator includes different activities like digging, swinging, or dumping. Each activity consists of a sequence of different movements like raising the arm or swinging the bucket. Several video-based equipment and worker action recognition studies have been performed in the construction community. Rezazadeh and McCabe [6] introduced a logical framework that combines object recognition, tracking, and rational events to recognize dirt loading to a dump truck by a hydraulic excavator. Kim and Caldas [21] presented an action recognition method for observing construction workers using interactions between actions and related objects which can be used to measure work rates for labor productivity monitoring. Golparvar-Fard et al. [16] proposed a method that initially represents a video stream as a collection of spatio-temporal visual features and then automatically learns the distributions of these features and action categories using a multi-class support vector machine classifier. More recently, Bugler et al. [22] combined photogrammetry and video analysis for tracking the progress of earthwork processes. They used photogrammetry to determine the volume of the excavated soil in regular intervals while the video analysis was used for generating statistical data regarding the construction activities.

The continuous recording of the data from previous steps enables the documentation of the various equipment actions over a period of time. A statistical analysis of the data would result in activity performance indicators such as active and idle time, cost, quality, productivity, ratio of completed work, etc. These indicators provide proactive resource monitoring capabilities and enable project managers to take corrective actions on performance deviations.

Despite the advantages that video-based performance monitoring provides over the techniques that use active sensing, the vision-based techniques still need to address significant technical and practical challenges. The first is the necessity for an appropriate level of illumination in the scene (neither dark nor direct sunlight). At the very least, this makes it extremely challenging to use video streams for performance monitoring of the tasks that are performed after sunset, which is not negligible especially in urban projects. In the absence of significant artificial lighting, a video-based approach would be nearly impossible. Another challenge is posed by the fact that cameras have a limited field of view and require a clear line-of-sight between the camera and any piece of equipment that is to be characterized. A related challenge is posed by the fact that construction sites are generally crowded with a large number of workers and equipment. Such congestion creates noisy and occluded areas in video streams which eventually disrupts accurate and automatic recognition of equipment actions. An additional key challenge is that the required video processing techniques are generally quite computationally expensive, which hinders their real-time performance and scalability. The last, and perhaps the most important, challenge is privacy, as not everyone is willing to be permanently monitored and recorded by cameras in a jobsite.

2.3. Gaps in Knowledge: Audio vs. Video

Audio processing provides four advantages over the location-based and/or computer vision-based techniques:

1) Some activities or actions are easier to recognize with sound than with other sensors. As a simple example, consider an excavator that is excavating a pit. In general, this action is not associated with a single characteristic gesture. Rather, it involves a series of movements such as digging, rotating, swinging, dumping, etc. that could be performed in a variety of ways and sequences (e.g., swinging in different directions and angles). This is extremely difficult to recognize via computer vision-based systems.
2) The sound produced by each piece of equipment is independent of the operator. An operator can perform an activity in a variety of ways. An action recognition system based on location sensors needs to take this into account, while an audio analysis always yields the same result.
3) The limited field of view of cameras used in vision-based action recognition methods constraints their range of operation from a single spot. A network of cameras is therefore needed to be placed in a circular pattern to cover all the angles in a jobsite (Figure 2). This is not a limitation for audio recording because a recorder can operate in 360 degrees. Moreover, because the sound produced by each piece of equipment can be heard even in the absence of a direct line-of-sight, an audio-based approach is less sensitive to occlusions.

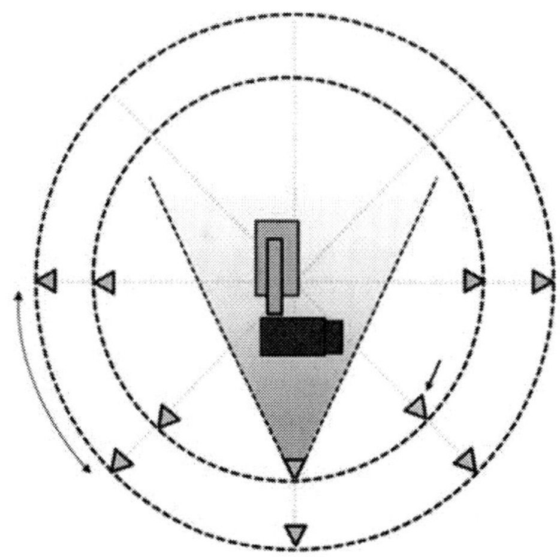

Figure 2. Camera layout in computer vision-based equipment action recognition [16].

4) The data rates that need to be analyzed in each scenario are different for every sensor modality. Table 1 demonstrates the order of magnitude for data rates in different scenarios. Even though the data rates could differ if a specific application is considered, it is obvious that processing audio data is computationally less demanding than video data.

Table 1. Approximate data rates for different sensor modalities

Input Type	Location (NMEA format)	Video	Sound
# Sensors	10	1	1
Sampling Rate	12 Hz	4608 kHz	10 kHz
Resolution	800 bit	8 bit	8 bit
Data Rate	12 kB/s	4608 kB/s	10 kB/s

3 Proposed Methodology: Audio-Based Activity Recognition of Heavy Equipment

To address the challenges associated with active sensors and computer vision-based approaches for activity recognition of construction equipment, we present an innovative activity recognition framework. This framework is based on processing audio signals generated at construction jobsites. The overall procedure is illustrated in Figure 3.

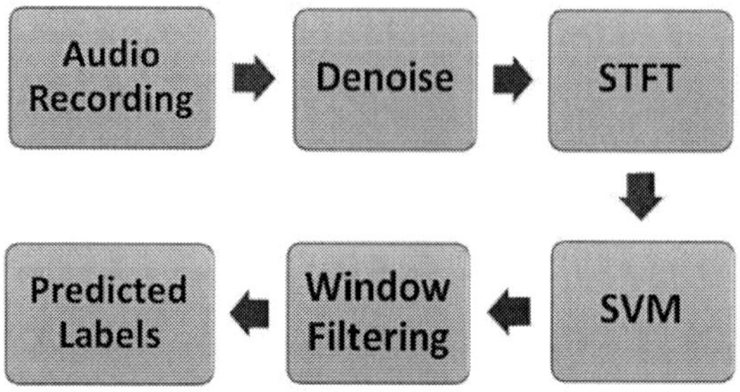

Figure 3. Overall framework for the proposed audio-based activity recognition algorithm.

While performing various tasks, the generated acoustic patterns of construction equipment can be recorded using commercially available microphones (Figure 4). We note that the current study is based on recognizing activities of a single machine and the more general and realistic setting (assuming the entire jobsite as a whole: multiple machines plus multiple microphones) will be considered as part of the future research plans.

Figure 4. An audio recorder (top) and placing the recorder at a jobsite (bottom).

As the next step, each audio recording will go through a number of processing stages as explained below.

3.1. Denoising

The recorded audio is assumed to contain the signal of interest along with environmental and other noise sources. We apply a signal enhancement algorithm developed by Rangachari & Loizou [24] because it has proven to be both efficient and effective. A key aspect of this algorithm is that it can perform noise-estimation in highly non-stationary noise environments such as what might be encountered at a construction job site. An estimate of the noise is continuously updated in every frame using time–frequency smoothing factors computed based on signal-presence probability in each frequency bin of the noisy recording spectrum. More details about this algorithm can be found in [24].

3.2. STFT

The captured audio is then converted to a time-frequency representation using the Short-Time Fourier Transform (STFT). We use Hanning window with size 512, a 1024-point DFT (discrete Fourier transform), and a 50% overlap (256 overlapped samples). The output of the STFT consists of both magnitude and phase matrices, but for single microphone recordings, we only consider the magnitude part.

3.3. SVM

To identify various activities within a captured audio recording, we train a classifier for each piece of machinery. The learning algorithm which has been used in this research is the popular Support Vector Machine (SVM) algorithm. SVM training and operation phases were carried out using the MATLAB SVM package, LIBSVM [23], with a radial basis function kernel. We extract 10 to 20 seconds for each activity as the training data to construct our SVM model. The SVM parameters C and γ were tuned using a grid search over a log-scale ranging from 2^{-6} to 2^{5}, and set differently in the models for each machine. We use 10-fold cross validation to select each combination of C and γ. After building the SVM models for each machine, we randomly extract other periods of the audio recordings to use as the testing data.

3.4. Window Filtering

After building SVM models for each machine, we can identify activities from any part of the audio. However, a label will be assigned for each time bin. In practice what we would prefer is a higher-level labeling of the full period for each of the different activities. Thus, we apply window filtering on the predicted label. Using a small window the algorithm scans through the SVM labels in the time domain and calculates the percentage for each activity. If the percentage for a particular activity is higher than a set threshold, the window is labeled with that certain activity. This procedure is then repeated using a larger window.

4 Experimental Setup and Results

In order to evaluate the performance of the proposed system, 4 different pieces of construction machines operating at various jobsites were selected as case studies: 1) JCB 3CX mini excavator 2) JD 270C Backhoe 3) Volvo L250G Wheel Loader and 4) CAT D5C Dozer.

Each piece of machine was carefully monitored and the generated sounds while performing routine tasks were captured using a commercially available recorder (Tascam DR-05 2 GB). Next, each audio recording was manually labeled based on the various activities that took place during the recording time. Construction heavy equipment usually perform one major task (digging, loading, breaking, etc.) and one or more minor tasks (maneuvering, swinging, moving, etc.) in each cycle so we classified each audio recording based on two activities: Major and minor (or activity 1 and activity 2). For example, the major activity for JD 270C Backhoe is crushing.

Each audio recording was then sent through the audio processing pipeline and divided into activities 1 and 2 considering their manifestations in the STFT domain.

Finally, the performance of the algorithm for each case study was compared to manually generated labels. The comparison results are depicted in Table 3 and Figures 5 and 6.

In the table and figures, the label "Act 1" represents the major activity, while "Act 2" illustrates the minor activities. Table 3 indicates that the performance of the proposed system for automatically recognizing activities of single machines is very promising. For JCB 3CX mini excavator and CAT D5-C Dozer , "Act 1" and "Act 2" have strongly dissimilar patterns in the STFT domain, thus the identification performance can be over 90 percent. Another way to visually evaluate the performance of the system is throughout the comparison charts as shown in Figure 6.

5 Conclusion Remarks and Future Research Plans

In this paper, an innovative system for automatically recognizing construction heavy equipment activities is presented and evaluated. This system utilities audio signals as the major data source. Many tools and equipment used in construction activities produce a distinct sound. Even more, many of the actions are accompanied by a clear and dissimilar sound, be it digging soil or pouring concrete.

Processing audio signals to extract semantic information requires several steps as described in the previous sections. It is necessary to implement a signal enhancement algorithm to reduce the background noise which can potentially produce false activity recognition results. A machine learning procedure (SVM) is also required to train the sounds of each operation and machine.

The presented framework was tested using a number of audio files collected from various jobsites and the results are promising. As the first step of this research project, the proposed system has been able to accurately recognize sound patterns of single machines operating various tasks. As the plan for future research, the authors intend to study a challenging, yet more realistic setting by considering multiple machines operating simultaneously.

Acknowledgments

The presented research has been funded by the U.S. National Science Foundation (NSF) under Grants CMMI-1606034 and CMMI-1537261. The authors gratefully acknowledge NSF's support. Any opinions, findings, conclusions, and recommendations expressed in this paper are those of the authors and do not necessarily reflect the views of the NSF. The authors would also like to thank Chris Andres Sabillon Albarran, graduate student at the College of Engineering and Information Technology at Georgia Southern University, for his assistance in collecting and recording audio files.

References

[1] Aziz, R.F. and Hafez, S.M. Applying lean thinking in construction and performance improvement, *Alexandria Engineering Journal*, 52(4): 679-695, 2013.

[2] Azis, A.A.A., Memon, A.H., Rahman, I.A., and Karim, A.T.A. Controlling cost overrun factors in construction projects in Malaysia. *Research Journal of Applied Sciences, Engineering and Technology*, 5(8):2621-2629, 2012.

[3] Doloi, H. Cost overruns and failure in project management: Understanding the roles of key stakeholders in construction projects. *Journal of Construction Engineering and Management*, 139(3):267-279, 2013.

[4] Joshua, L. and Varghese, K. Accelerometer-based activity recognition in construction. *Journal of Computing in Civil Engineering*, 25(5):370-379, 2011.
[5] Su, Y.Y. Construction crew productivity monitoring supported by location awareness technologies. *PhD dissertation, University of Illinois at Urbana-Champaign*, 2010.
[6] Rezazadeh Azar, E. and McCabe, B. Part based model and spatial–temporal reasoning to recognize hydraulic excavators in construction images and videos. *Automation in Construction*, 24: 194–202, 2012.
[7] Ahn, C., Lee, S., and Peña-Mora, F. Application of low-cost accelerometers for measuring the operational efficiency of a construction equipment fleet. *Journal of Computing in Civil Engineering*, 29(2): 04014042-11, 2015.
[8] Akhavian, R. and Behzadan, A. Knowledge-based simulation modeling of construction fleet operations using multimodal-process data mining. *Journal of Construction Engineering and Management*, 139(11): 04013021-11, 2013.
[9] Akhavian, R. and Behzadan, A. Client-Server Interaction Knowledge Discovery for Operations-level Construction Simulation Using Process Data. *In Proceedings of the Construction Research Congress*, pages 41-50, Atlanta, GA, USA, 2014.
[10] Khosrowpour, A., Nieblesb, J. C., and Golparvar-Fard, M. Vision-based workface assessment using depth images for activity analysis of interior construction operations. *Automation in Construction*, 48:74–87, 2014.
[11] Cheng, T., Venugopal, M., Teizer, J., and Vela, P.A. Performance evaluation of ultra wideband technology for construction resource location tracking in harsh environments. *Automation in Construction*, 20(8):1173-1184, 2011.
[12] Ahn, C.R., Lee, S.H., and Pena-Mora, F. Construction equipment activity recognition from accelerometer data for monitoring operational efficiency and environmental performance. *In Proceedings of the International Conference on Construction Engineering and Project Management*, Anaheim, CA, USA, 2013.
[13] Dai, F., Rashidi, A., Brilakis, I., and Vela, P. Comparison of image-based and time-of-fight-based technologies for three dimensional reconstruction of infrastructure, *Journal of Construction Engineering and Management*, 139(1): 69-79, 2013.
[14] Brilakis, I., Fathi, H., and Rashidi, A. Progressive 3D reconstruction of infrastructure with videogrammetry, *Automation in Construction*, 20(7):_884–895, 2011.
[15] Rashidi, A., Fathi, H., and Brilakis, I. Innovative stereo vision-based approach to generate dense depth map of transportation infrastructure, *Journal of the Transportation Research Board*, 2215:93-99, 2011.
[16] Golparvar-Fard, M., Heydarian, A., and Niebles, J.C. Vision-based action recognition of earthmoving equipment using spatio-temporal features and support vector machine classifiers. *Advanced Engineering Informatics*, 27(4):652–663, 2013.
[17] Rezazadeh Azar, E. 2013. Computer vision-based solution to monitor earth material loading activities. *PhD Dissertation, Department of Civil Engineering, University of Toronto*, 2013.
[18] Gong, J., Caldas, C.H., and Gordon, C. Learning and classifying actions of construction workers and equipment using Bag-of-Video-Feature-Words and Bayesian network models. *Advanced Engineering Informatics*, 25(4):771–782, 2011.
[19] Park, M.W, Makhmalbaf, A., and Brilakis, I. Comparative study of vision tracking methods for tracking of construction site resources. *Automation in Construction*, 20(7): 905-915, 2011.
[20] Bobick, A. and Davis, J. Real-time recognition of activity using temporal templates. *In Proceedings of the Applications of Computer Vision, WACV '96., Proceedings 3rd IEEE Workshop on*, Pages 39-42, Sarasota, FL, USA, 1996.
[21] Kim, J.Y. and Caldas, C.H. Vision-based action recognition in the internal construction site using interactions between worker actions and construction objects. *International Association for Automation and Robotics in Construction. In Proceedings of the 30th ISARC*, Pages 661-668, Montréal, Canada, 2013.
[22] Bügler, M., Ogunmakin, G., Teizer, J., Vela, P.A., and Borrmann, A. A comprehensive methodology for vision-based progress and activity estimation of excavation processes for productivity assessment. *In Proceedings of the EG-ICE Workshop on Intelligent Computing in Engineering*, Pages 1-10, Cardiff, Wales, UK, 2014.
[23] Chang, C.C. and Lin, C.J. LIBSVM: a library for support vector machines. *ACM Transactions on Intelligent Systems and Technology*, 2(3): 1-27, 2011.
[24] Rangachari, S. and Loizou, P.C. A noise-estimation algorithm for highly non-stationary environments. *Speech Communication*, 48:220-231, 2006.

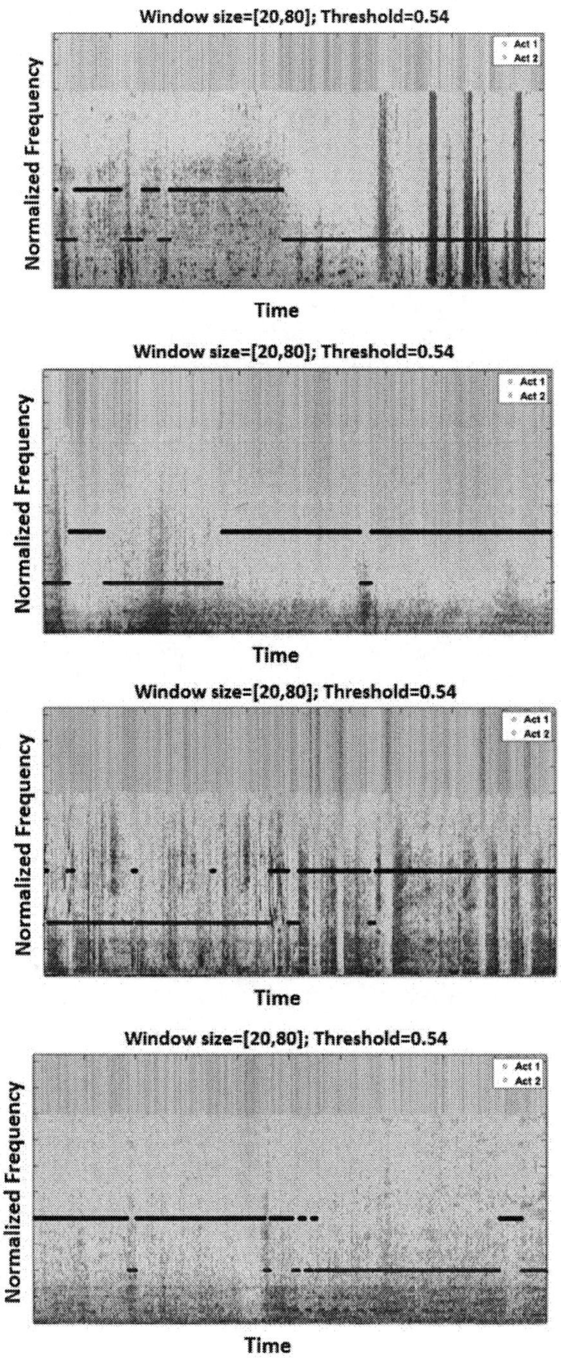

Figure 5. Visual representations of signal spectrums in frequency domain; From top to bottom: JD 270C Backhoe; Volvo L250G Wheel Loader; JCB 3CX mini excavator; CAT D5C Dozer.

Table 3. Comparison results for manually labeled audio files (correct label) versus the results obtained from implementing the audio system (predicted label).

JD 270C Backhoe		Correct Label	
		Act. 1	Act. 2
Predicted Label	Act. 1	0.8375	0.1728
	Act. 2	0.1625	0.8272
Volvo L250G Wheel loader		Correct Label	
		Act. 1	Act. 2
Predicted Label	Act. 1	0.8003	0.1136
	Act. 2	0.1997	0.8864
JCB 3CX mini excavator		Correct Label	
		Act. 1	Act. 2
Predicted Label	Act. 1	0.8326	0.0218
	Act. 2	0.1674	0.9782
CAT D5C Dozer		Correct Label	
		Act. 1	Act. 2
Predicted Label	Act. 1	0.8490	0.0477
	Act. 2	0.1510	0.9523

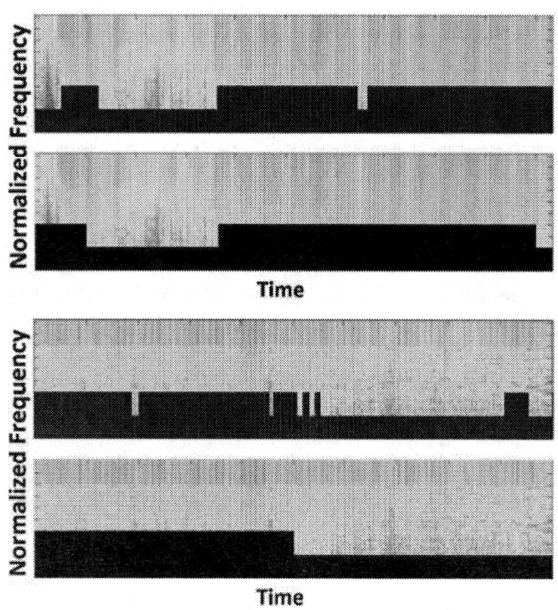

Figure 6. Comparison results between manually labeled audio files (blue) versus the results obtained from implementing the audio system (black).

An Automated GIS based Piezoelectric Potential Assessment: PE Calculator©

Shahaboddin Hashemi Toroghi [a], Mingyang Li [b] and Baabak Ashuri [c]

[a] School of Building Construction & School of City and Regional Planning, Georgia Institute of Technology, USA
[b] School of Civil Engineering & School of City and Regional Planning, Georgia Institute of Technology, USA
[c] School of Building Construction & School of Civil Engineering, Georgia Institute of Technology, USA
E-mail: Toroghi@gatech.edu, mingyli@gatech.edu, baabak.ashuri@coa.gatech.edu

Abstract –

Renewable Sources are at the front line of policy discussions for replacing the fossil fuels around the world. Piezoelectric (PE) material can generate electricity in response to mechanical stress. Knowing that cars put a significant amount of stress onto the roadways every day, the use of piezoelectric transducer for asphalt pavement energy harvesting has gained the attention of researchers in the recent years. Despite the continuing research on improving the efficiency of this new technology, developing a tool to measure the PE potential of the roads, is a necessity for future utilization of piezoelectric materials. Addressing such a need, in this research, we use the python programming language, the scripting language of choice for the ArcGIS platform, to develop a state of the art software – PE Calculator© - which generates the PE potential of roads. The PE Calculator© gives the user the choice of defining the boundary of study, traffic condition and PE coefficient. In addition to visualized PE potential maps using Geographies Information System (GIS), the PE Calculator© generates a detailed table, expressing the PE potential for each road segment. The result of this research can be used by both public and private sectors, including policy makers, transportation engineers, city planners, and etc. to prioritize future PE implementation plans accordingly.

Keywords –

Piezoelectric potential; Python; Geographic Information System (GIS); Energy

1. Introduction

The current need for energy and its increasing consumption trend across the countries in the world encourage researchers to search for a new renewable source of energy. Such an alternative source of energy not only contribute to addressing the issue of scarcity of fossil fuel in the future, but also would have minimal negative impact on the environment. The Piezoelectric (PE) ceramics are an example of such efforts. PE ceramics can transform mechanical impact energy into electrical power [1] [2] [3] [4] [5]. During the past decade, the traffic on the roads in the US have intensely increased, causing several issues, including congestion, which results in waste a massive amount of time, money, and fuel [6]. However, the traffic may not always cause negative consequences. Some previous research focus on the feasibility and efficiency of piezoelectric transducers for the purpose of generating electricity, in form of asphalt pavement energy harvesting [7] [8]. Placing the PE devices five centimetres below the asphalt surface, can generate electricity as the pressure caused by running vehicles would slightly deform the piezoelectric crystals [9]. In 2011, California state assemblyman Mike Gatto proposed Assembly Bill 306 to develop this technology for Californian roads [10]. Although, several research has been conducted on the topic of PE energy generation via pavements on the roads, estimating the energy that can be generated by such a technology, if implemented in a metropolitan area, within any given boundary has not been done to this date. The high sensitivity of PE technology on the road to the traffic level, at each given road, adds additional complexity to the calculation of energy generation. The analysis of energy generation, also highly depends on the type and location of the road/s, where we implement the PE system, as the traffic load varies for each road. The Geographic Information System (GIS) can provide a perfect mean to extract such information in any given scale, from one block to a whole neighbourhood in a city, and even an entire metropolitan area. However, calculating the energy generation using PE, would be a time consuming process and even impossible, as we increase the scope of study. In this research, we develop a state of the art software, PE Calculator©, which is compatible with a GIS program to read the input data from, in addition to the data that user can directly insert through its user friendly interface, and generate the result both in visual format (map) and more detailed version (table). The PE Calculator can be used by policy makers, researchers, city planners, and investors who want to assess the different scenarios for implementing the PE technology on the urban roads. In this research, we limit the scope of PE Calculator the boundaries of the Atlanta metropolitan area, Georgia. However, the PE Calculator can expand its limit to any extend to which the road system network and historical traffic information would be available.

This paper is organized as follows: section 2 describes the method we used for developing the PE Calculator. In section 3, we explain in detail, how the PE Calculator computes the potential energy that the PE technology can generate. Section 4 provides a numerical case example, which we use the PE Calculator to assess the energy potential using PE technology on the road, for two separate neighbourhoods in the Atlanta. Finally, the conclusion is discussed in section 5.

2. Methodology

We consider some assumptions for the calculation of the PE energy potential which are discussed in more detail in this section. Hill and Tong (2013) summarized the recent progress on PE technology, especially focusing on the recent studies on its conversion efficiency from stress to electricity. Assuming that PE materials will be install underneath the roadway to absorb the mechanical stress caused by the passing vehicles, we used the result of previous research on PE technology in the past to calculated the conversion factor which is about 0.0267 kWhr/year/feet/lane for average weighted passenger cars [8] [11]. The total PE energy potential is then a function of the conversion factor, length of the road, and the number of cars running on these roads.

We acquire a detailed road network shapefile in Georgia from TIGER on Census website, which contains not only the shape, length, and number of lanes of each road, but also the classifications of each road segment. After examining the current road network in Georgia, we found that highways, ramps, secondary roads and local roads made up about 97% of all roads in the TIGER shapefile (Table 1). Thus we only use those four road types in the PE Calculator© at this research.

Table 1 – PE Calculator road coverage

Road ID	Description	# of Lane	Proportion of Total
S1100	Highway	8	1.25%
S1200	;econdary Roads	4	6.74%
S1400	Local Roads	2	88.05%
S1630	Ramps to HWY	1.5	0.93%
		Sum	96.97%

To account for the different daily traffic at each road type, we construct different baseline traffic conditions for each of these four toad types. For instance, we define a medium traffic level on highway to be 30,000 vehicle/day, while the medium traffic level for secondary road beings at 3,000. All these numbers are average approximations, based on Georgia Department of Transportation's (GDOT's) annual average daily traffic number [1].

We also incorporate ArcGIS into our PE Calculator. While Python GUI was chosen to build our user interface and the calculation module, we used ArcGIS to deliver the calculation results, due to its capability of generating an easy to read maps, with multiple layers.

3. PE Calculator© – State of the Art software

The goal of our work was to develop a software, compatible with a GIS model, which can assess the potential electricity that could be generated by PE technology on the roads. The PE Calculator© is a standalone software and though its interactive capability with the ArcGIS software makes it possible to read data generated by ArcGIS, such as the shapefile for the study boundaries (defined by the user). The software also generates results in visualized format that can be read by ArcGIS software. The PE calculator analyzes the specification of each road, takes into account the traffic level, and computes the energy potential generated by the embedded PE technology on a road network, based on the PE density coefficient defined by the user. Addition data were predefined and loaded into the software during the development stage.

Figure 1 shows the flowchart of how the PE Calculator computes the energy potential by PE technology on the road. The PE Calculator uses three layers of inputs. We follow two principles for the development of the PE Calculator: simplicity in terms of getting only the necessary information from the user, and accuracy which we only used the data from the official governmental sectors such as Department of Transportation (DOT), or reputable resources such as Google map. We used three layers of data in PE Calculator©:

1- The first layer of input data can be provided by the user through ArcGIS. The user, through the ArcGIS software, defines the boundaries for the energy calculation. This boundary can be as small as one section of a street, or as large as the Atlanta metropolitan area. The PE Calculator then extracts the streets and street types within that boundary, through its pre-saved street network TIGER database.

2- The second layer of input data is gotten directly from the user, via the PE Calculator's User Interface (UI).

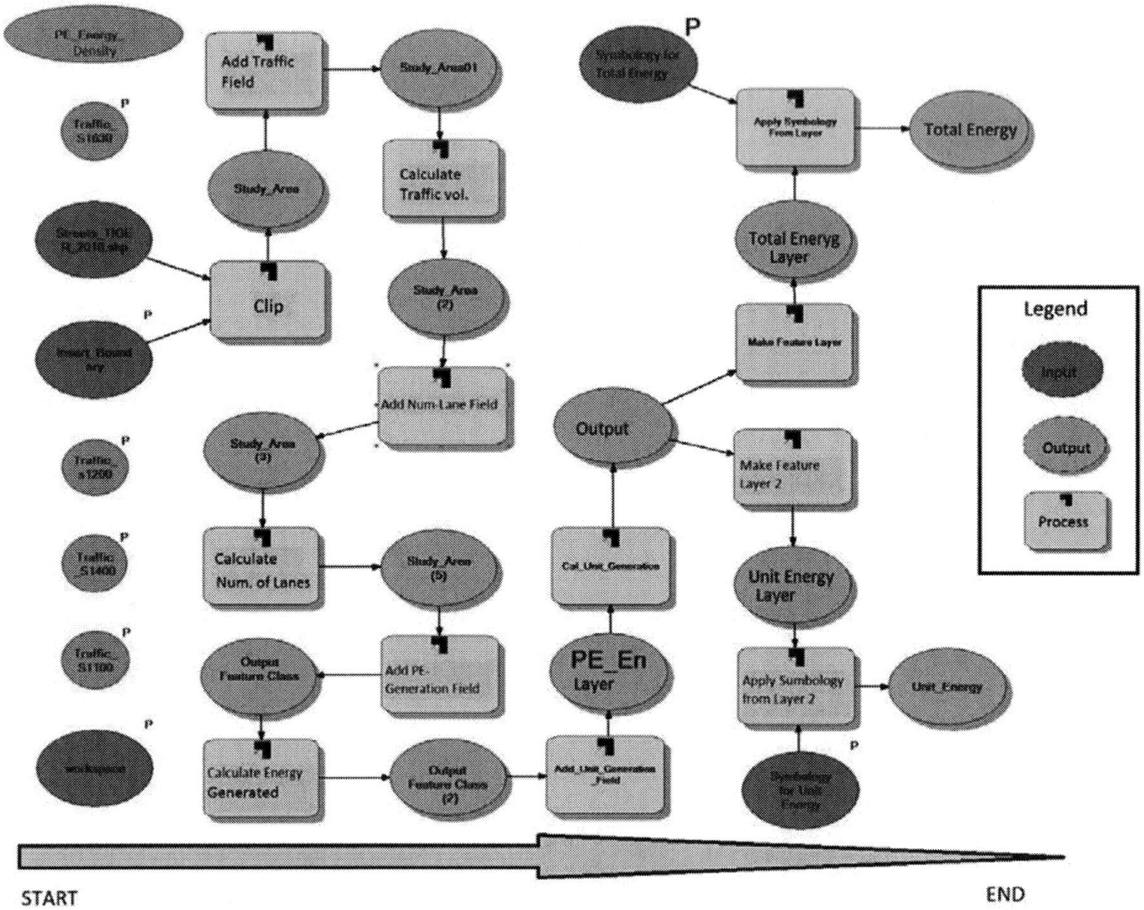

Figure 1. PE Calculator Flowchart.

These data are either necessary to be gained via user, such as the traffic level (high, medium, and low) for the assessment purpose, or can be changed through time such as PE coefficient, which can change as the technology advances. Although we preload the traffic information on each road type from GDOT's Geo-counting website [1]. Here, we provide an extra level of customizability into our software, for the user to investigate the energy that can be generated at different level of traffic.

3- The third layer of input data are those that were preloaded into the software, such as the street network of the Atlanta metropolitan area. These data are either fixed or do not change in a short time span.The PE calculator computes the energy generated by the PE system on the roads, and generates the result, in two formats: map and table. Using the map, the user can visualize the energy generated by the PE within different sections of predefined boundaries. The table provides a more detailed information for the user, such as the exact amount of potential energy, each road segment within the given boundary can generate.

We developed the PE Calculator, using Python programming language. User Interface (UI) of the PE Calculator is simple, while it gets all the required information from a user. Figure 2 shows the UI of the PE Calculator©.

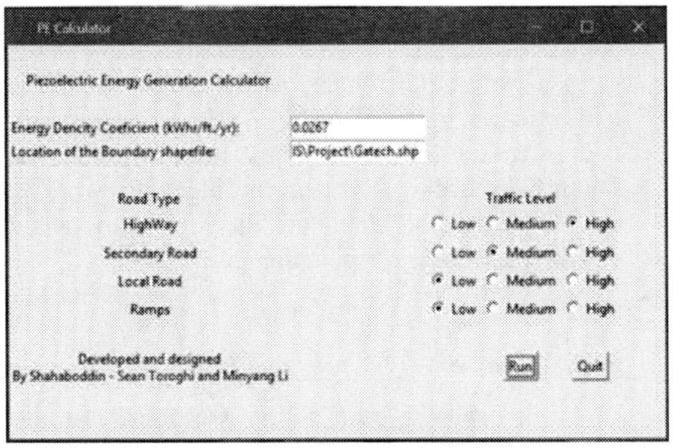

Figure 2. PE Calculator User Interface.

The simple UI provides the user an easy and intuitive environment to enter the inputs. More specifically, user can enter the PE energy conversion factor. As the PE technology advances through time, the efficiency of it can change as well. The user can also enter the location of the saved boundary shapefile on the hard drive of the computer that was created in the past, using any version of ArcGIS software. The traffic level section will take user inputs on the traffic level of different road types. The user will be able to investigate the impact of different traffic levels on energy potential on the roadways. There are three traffic level, Low, Medium, and High, for each of the four road types: Highway, Secondary Roads, Local Roads, and Ramps.

Once the user clicks the Run button, the PE Calculator© computes the PE energy potential on the roadways within the boundaries, and generates and saves the results as the output. The output of the software is in the form of maps and tables, which have the attributes of the energy potential generated by the PE, on different unit scales. The maps are in shapefile format, which can easily read or printed out by the ArcGIS software. Moreover, the shape file provides a unique capability for the user to include or exclude parts if the computed information, based on the desire point of focus. In the next section, we explain in more detail how the PE Calculator© can be beneficial for both public and private sectors, by presenting two case studies, for two neighborhoods in the Atlanta metropolitan area, followed by a comparison and discussion parts.

4. Case Study

In this section we present two case studies, to illustrate the benefits of using our PE Calculator© Software. The first case study illustrates the software's capability to be used in scenario planning: By selecting different traffic level across the four predefined road types in the software, users will be able to get an accurate output that better represent the real world traffic. The second case study focus on the software's ability in performing analysis on multiple boundaries. Such feature enables the users to easily compare the total PE energy potential across different geographic boundaries.

4.1. Case Study 1: Using the Software for Scenario Planning

In the first case study, we firstly create a boundary shapefile for our study area in ArcMap, as shown in Figure 3 below.

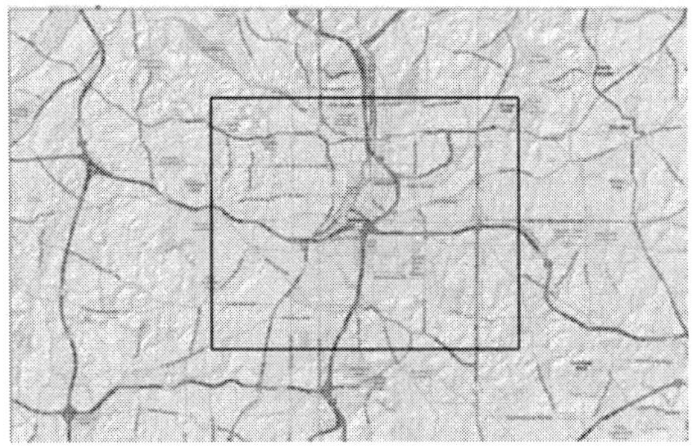

Figure 3. Study Area Boundary.

We first examine the PE energy potential per feet under the scenario that all four predefined road types are experiencing high level of traffic in this study area (Figure 4). The output of unit PE energy potential are exhibit at Figure 5.

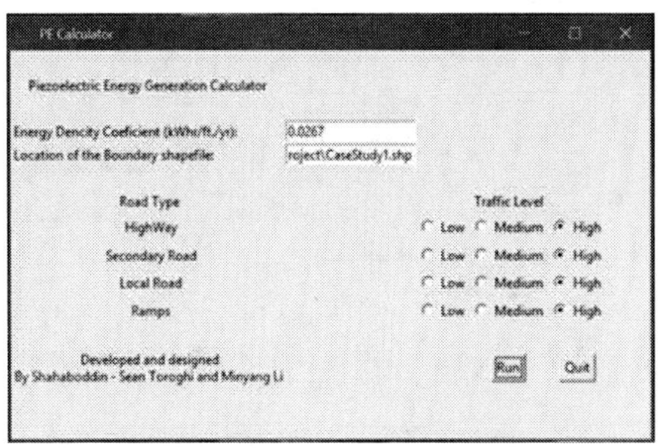

Figure 4. High level of traffic on all four roads in the study area.

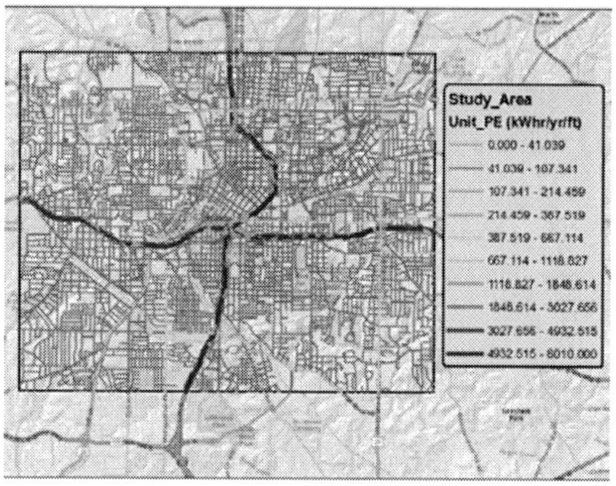

Figure 5. High level of traffic on all four types of roads in the study area.

We then changed the traffic level on highways and ramps from high to low, as shown in Figure 6, while keeping the traffic level on other roads constant. Result is shown in Figure 7.

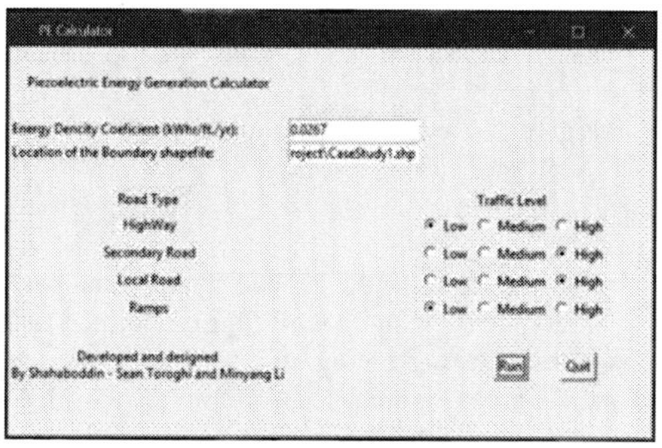

Figure 6. Changing to low traffic level on highways and ramps in the study area.

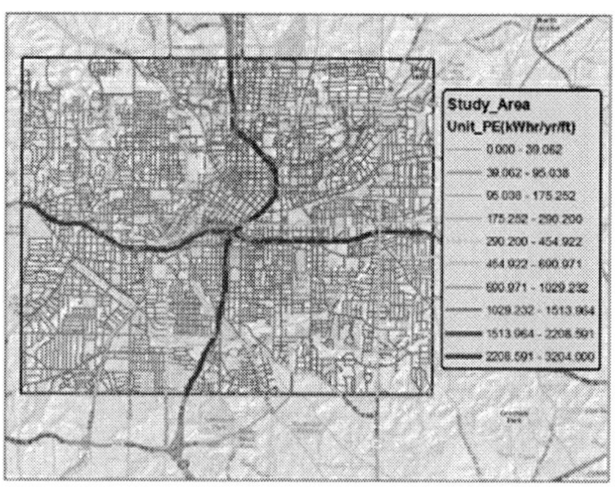

Figure 7. High level of traffic on Secondary and Local Road, Low level of traffic on Highway and Ramps in the study area.

Users can notice the changes in the legends and the consistency in the overall symbology of the output. Our software will automatically adjust the legend according to calculation results. Combining the unified symbology feature with its ability to adjust the traffic level across four pre-defined road types according to real world traffic, the PE Calculator© facilitate the process of analyzing different scenarios for planning purposes.

4.2. Case Study 2

In the second case study, we focus on the PE Calculator's ability to perform analysis on multiple areas simultaneously. The study area consists of two rectangles with the same dimension, as shown in Figure 8 below. The upper rectangle represents Georgia Tech (GT) Campus area, while the lower one represents the Downtown Atlanta.

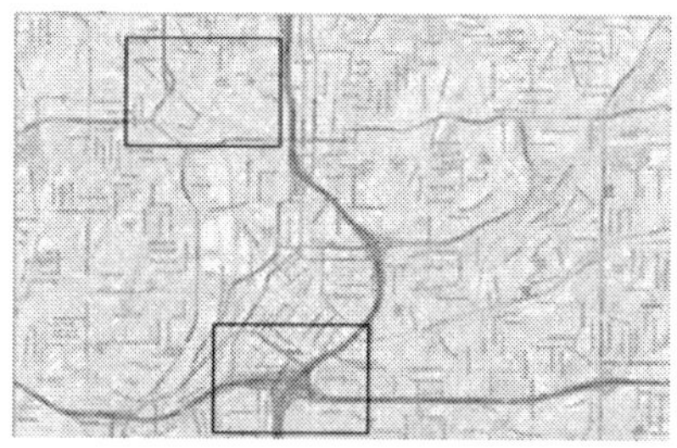

Figure 8.Study area consists of two areas. Upper rectangle being GT campus, and the lower one being Downtown Atlanta.

Using the mentioned two boundaries, we run the PE Calculator© using the boundary shapefile. For demonstration purposes, we selected high level of traffic on all of the predefined road types. The output is shown in Figure 9.

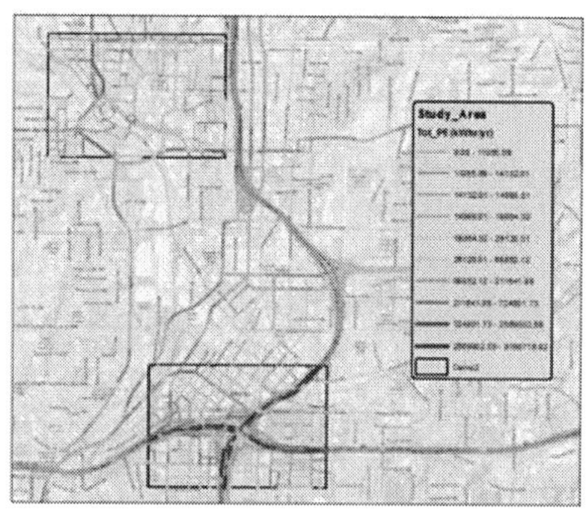

Figure 9.Downtown Atlanta have a higher PE energy potential compared to GT campus

The differences (the red, yellow, and green colors) in potential energy generation can be seen in figure 9, where the Downtown Atlanta has a higher total PE energy potential compared with GT campus area. In fact, PE Calculator also provides the total PE energy potential for each of the areas in the output tables: Downtown Atlanta has about 20 times as much PE energy potential as GT campus. The vast difference is largely due to the different traffic volume on a daily basis, which the PE Calculator© took into account. Downtown Atlanta has a much larger traffic volume, in comparison with the campus area. This unique feature enables the user to easily compare the benefits of implementing the PE technique in different areas side by side, leading to more informed decisions.

The two case studies illustrate how our software, PE Calculator, is capable to perform automatic calculations at different scenarios considering different traffic levels within multiple study areas. Such features of the PE Calculator© software give it an edge compared to other manual or semi-automatic methods in estimating the PE energy potential on roadways.

5. Acknowledgements

This material is based upon work supported by the National Science Foundation under Grants nos. 1300918 and 1441208.

6. Conclusion

Piezoelectric technology opens a new door to scientists and researcher searching for new sources of clean energy. The Piezoelectric (PE) material can transform the mechanical energy to electricity. This technology has already passed the preliminary stage. The use of PE technology inside the road pavement can result in generating electricity, when vehicles move on the road. The viability study shows the vendor-based levelized cost of energy (LCOE) in average is $0.11 /KWh [11]. Although the PE technology shows a promising future in energy harvesting, assess the energy generation in a large scale (neighborhood, city, and etc.) can be too complex and time consuming. Furthermore, for the purpose of comparison of energy harvesting using PE in pavement, researcher and investors would need to take into account other factors that influence the PE energy generation potential, such as traffic level. In this article, we introduce the PE Calculator©, a state-of-art software that is capable of performing automatic calculation of PE energy potential on roadways and generate meaningful and appealing visualizations in ArcGIS. Key features of the software include its ability to incorporate variations in user inputs, assess PE energy potential within the study area, according to the user inputs, and perform the calculation simultaneously on multiple study areas. Such features give it an edge compared to other manual or semi-automatic methods in estimating the PE energy potential on roadways. Potential users from multiple fields, such as scientist working on improving PE technology, city planners, policy makers, traffic engineers, and investment companies, will all benefit from incorporating PE Calculator ©, an easy to use, comprehensive and accurate software for computing the energy potential of PE technology on the roads.

References

[1] Chure, M., Wu, K., Tung, C., Lin, J., and Ma, W. Power Generation Characteristics Of Pzt Piezoelectric Ceramics Using Drop Weight Impact Techniques: Effect Of Dimensional Size. *Ceramics International*, 40(1), 341-345, 2014.

[2] Wang, Xu, Lin, Liwei, John, Sabu, Watkins, Simon, Xiao, Han, Ren, He, & Liang, Xingyu. Dimensionless comparison of electromagnetic and piezoelectric vibration energy harvesters for similarity and duality. *Mechanical Systems and Signal Processing,* Mechanical Systems and Signal Processing. 2013.

[3] Goldfarb, M., and Jones, L. On the Efficiency of Electric Power Generation With Piezoelectric Ceramic. *Journal of Dynamic Systems, Measurement, and Control, 121*(3), 566, 1999.

[4] Starner, T. Human-powered wearable computing. *IBM Systems Journal, 35*(3,4), 618-629, 1996.

[5] Umeda, Mikio, Kentaro Nakamura and Sadayuki Ueha. Analysis of the Transformation of Mechanical Impact Energy to Electric Energy Using Piezoelectric Vibrator. *Japanese Journal of Applied Physics*, Volume 35, Part 1, Number 5B, 1996.

[6] Schrank, David, Bill Eisel, Tim Lomax, and Jim Bak. Urban Mobility Scorecard. The Texas A&M Transportation Institute and INRIX. Aug, 2015.

[7] Yao, Ling, Hongduo Zhao, Zhanyu Dong, Yuanfa Gao. Laboratory Testing of Piezoelectric Bridge Transducers for Asphalt Pavement Energy Harvesting [J]. *Key Engineering materials*, Vol. 492: 172-175, 2011.

[8] Li, C. Road Performance of Common Piezoelectric Transducer for Asphalt Pavement Energy Harvesting. *Applied Mechanics and Materials, 744-746*(Advances in Civil Engineering and Transportation IV), 1491-1494, 2015.

[9] Garland, Rex. Piezoelectric Roads in California. On-line: http://large.stanford.edu/courses/2012/ph240/garland1/ (Accessed Nov. 2015).

[10] Schmidt, Greg, E. Dotson Wilson. Legislative

Index and Table of Sections Affected. California Legislature, 30 Nov 12. On-line: http://large.stanford.edu/courses/2012/ph240/garland1/docs/TOSAIndex1112.pdf (Accessed Nov. 2015)

[11] Hill, D., and N. Tong. Assessment of piezoelectric materials for roadway energy harvesting. California Energy Commission, CEC-500-2013-007, 2013.

[12] Geocounts. Traffic Counts in Georgia. On-line: http://geocounts.com/gdot/ (Access Feb. 2016).

Low Cost Augmented Reality Framework for Construction Applications

Sai Meghasyam Kodeboyina[a] and Koshy Varghese[b]

[a] Graduate Student, Department of Civil Engineering, Indian Institute of Technology Madras, India
[b] Professor, Department of Civil Engineering, Indian Institute of Technology Madras, India

E-mail: ce11b092@smail.iitm.ac.in, koshy@iitm.ac.in

Abstract –

Augmented Reality can be of immense benefit to the construction industry. The oft-cited benefits of AR in construction industry include real time visualization of projects, project monitoring by overlaying virtual models on actual built structures and onsite information retrieval. But this technology is restricted by the high cost and limited portability of the devices. Further, problems with real time and accurate tracking in a construction environment hinder its broader application. To enable utilization of augmented reality on a construction site, a low cost augmented reality framework based on the Google Cardboard visor is proposed. The current applications available for Google cardboard has several limitations in delivering an AR experience relevant to construction requirements. To overcome these limitations Unity game engine, with the help of Vuforia & Cardboard SDK, is used to develop an application environment which can be used for location and orientation specific visualization and planning of work at construction workface. The real world image is captured through the smart-phone camera input and blended with the stereo input of the 3D models to enable a full immersion experience. The application is currently limited to marker based tracking where the 3D models are triggered onto the user's view upon scanning an image which is registered with a corresponding 3D model preloaded into the application. A gaze input user interface is proposed which enables the user to interact with the augmented models. Finally usage of AR app while traversing the construction site is illustrated.

Keywords –

Construction, Augmented Reality, Marker-based Tracking, Information Retrieval, Google Cardboard.

1 Introduction

Augmented Reality (AR) is the enhanced version of reality created by the use of technology to overlay computer generated graphics, sounds and other data on user's view of the real world. It is different from the virtual reality where the user is immersed completely in a virtual environment. Augmented reality has a wide range of applications in various sectors like commerce, military, architecture and construction, medicine, gaming etc. In commerce, it can been used to enable a user to view the contents of a packaged product without actually tearing it open. In military operations AR can aid in projecting critical information on soldiers' goggles. Print and media industries develop apps to display digital content like videos & images that pop up on scanning printed magazines. Travellers can find information about places by pointing their cameras to subjects. Surgeons can image the 3D model of brain on top of the patient's anatomy without actually cutting it open. The simple augmented reality user case is: a person scans an image with camera on his smartphone or tablet and underlying platform identifies a marker in the image, triggering the invocation of a virtual model on that real-world image and is displayed on the camera screen.

The construction industry particularly can realise great benefits with the development of AR technology. AR makes visualization easy. In fact it reduces the effort of architects and designers by a great deal. Real time visualization of projects, upon pointing user's smartphone camera at a specific location or scanning a specific image, will benefit the architect in giving greater insights into details of the project to owners and contractors. Real time monitoring of the project can be done by superimposing the as-built 3D model of the project upon the actual built up structure without having to look at the construction schedules. But bigger challenges are faced in implementing technology at site due to lack of awareness, sophisticated nature of the current AR/VR systems, cost of AR/VR gear and cost of implementation. Also the inherent problems of augmented reality question its implementation to the fullest. The critical problem with any augmented reality systems is the difficulties in real-time and accurate tracking. The major requirement for any augmented information is it has to synchronise with the real world. Slightest of the errors in model synchronization with the real world can be easily caught by the human eye. Any mismatch between augmented objects and real objects can be discomforting and also result in incorrect information being given to the user.

There can be endless discussion on applications of augmented reality but for the purpose of this paper the notable benefits and of this technology are discussed and a specific solution for obtaining a cost effective AR framework is proposed. The solution proposed states the use of Virtual Reality (VR) device Google Cardboard and smartphone for providing AR experience. In Virtual Reality, the user gets immersed into another virtual world and cannot see the real world he actually lives in. So to provide AR experience, it is proposed we capture the real world from the smartphone camera and view the same from Cardboard to provide that immersive experience. Virtual models are now augmented onto the real world input from the camera and also a menu-based user interface has been created to interact with the augmentations

2 Survey on Applications of Augmented Reality in Construction Industry

The technical definition of augmented reality [1] is a system that combines real and virtual world, which is interactive in real time and is registered in three dimensions. Ivan Sutherland is credited largely for the concept of augmented reality as early as 1960s. One of the earliest works of augmented reality in architecture and construction industry is the work done at the Computer Graphics and User Interfaces Lab at Columbia University on so called "Architectural Anatomy" which is overlaying the graphical illustration of the structural systems of a building over the user's view of the room which he currently is in [2]. The other work done at the same labs involves space frame construction. It is a step by step space frame construction process where the AR system directs the user to pick up elements though audio clippings, confirms whether the correct piece has been picked by scanning the barcode and virtual image is displayed next to it showing where and how to install the picked strut.

From this stage the augmented reality application has shifted from indoors to outdoors enabling the user to move around with the heavy equipment. A prototype system developed by the Computer Graphics and User Interfaces Lab, Columbia University acts as an information system in the campus, assisting a person in finding places and allowing to request information about items of interest, like buildings, statues, etc. [3]. The user wears a head-mounted display that gets the inputs over a wireless network from a computer carried in the backpack. GPS is used to track the position of the user while the head mounted display gives the orientation information as the user looks around the campus, the head worn display superimposes textual labels on campus buildings. Also the user can interact with the system and obtain any information about the building on his screen.

The increase in need for information technologies in architecture, engineering, construction and facility management industries is owed to their complex nature that increased demand of information access for evaluation, communication and collaboration [4]. Thus the application of AR technologies are varied and across multiple verticals. AR technologies can benefit the AEC industry through: 1) Visualization of 3D models super imposed on the real world 2) Information Retrieval involving BIM models & drawings for on-site communication and 3) Interaction of virtual models in response to changes in the real world [5]. Some of the specific applications of AR in training, information retrieval and planning are described below.

Virtual Training System for Construction Excavator Operations: A virtual training system (VTS) is developed for construction excavator operators based on a game engine tool [8]. Their study reflects the experience of using a game engine to develop a VTS for an earthmoving excavator in a construction site. They selected Unity3D as a simulator construction tool because of its easy programming and versatility. Unity3D has an advantageous world building tool, user-friendly interface, and lots of target platforms -mobiles, web, PC/MacOS, etc. are supported [6]. Also they employed 3DS Max as modelling software for modelling the excavator and imported into a simulation environment created in Unity 3D. Various scripts are written which enable interaction of various 3D objects say bucket, boom and swing. Similarly soil is also modelled as grey boxes due to limitations in the older version of Unity. Based on this approach the whole training system is developed and used for training where the inputs to the model are given from the joystick connected to the display. This attempt in using game engine to create simulation environment for construction is considered to be an important step in adopting training programs with least cost burdens.

On-site Information Retrieval: A wearable device had been developed that projects the drawings and related information on user need basis [7]. This device is developed to help engineers avoid carrying bulky construction drawings to the site, and to pin point the exact drawing needed instead of going through the pile of drawings. Four modules enable the information from the BIM to the projection.1) Information integration module for transfer of information from BIM to image format to enable on site retrieval 2) Position module to enable users load images based on their location 3) Manipulation module to understand the gestures of the user like touch and turn and then perform changes to images accordingly 4) Display module that links to the projector, scales the images properly and ensure projection to the scale based on the information from other three modules. A test was done at site by enabling engineers to use these helmets and access particular information. Those with the helmets showed increased accuracy in acquiring the information quickly than manual searching through the 2D drawings.

Plan virtual construction site: AR Planner is an AR system that enables the construction worksite planner to position construction materials and machines/equipment, handling devices and the corresponding outing lines in the planned worksite using representative 3D-model model objects in the virtual world [8]. Fiducial markers are used to trigger the 3D objects onto the virtual worksite where they can be moved by the user. There is also a data base that stores all the virtual element set that is modelled and can be used in planning.

Apart from the above mentioned applications there are generalized frameworks which enable integrating AR with Building Information Models for various uses. A conceptual framework that integrates AR with building information modelling (BIM) to detect construction defects has also been presented [9]. In order to visualize the dynamic site operations during construction phase a mobile 3-dimensional AR system has been investigated [10].

Not only in the domain of engineering has AR found its utility, also in teaching especially classroom education. Enhancing a text book with 3D and multimedia virtual information by designing a collaborative Context Aware Mobile Augmented Reality Tool (CAM-ART) in construction and civil engineering curriculum [11]. A group of students were tested on using this CAM-ART by loading an AR application in their smartphones and tablet computers. Performance data was analysed which showed CAM-ART actually had a positive impact on students' learning in both short term and long term.

The benefits of augmented reality to construction industry can be extended from individual construction activities or structural systems invocation to the scale of projects as a whole. Deployment of drones with powerful cameras makes this possible. Once a drone flies over the project area, its cameras scan the whole area and pick up markers that are positioned in the site. When these markers are tracked all the augmentations like as-built 3D site map or project monitoring information can be passed as live feed to the top managers at site or can be recorded into a memory device. This saves a lot of time and gives the whole summary of the project at one go. Also the use of Simultaneous Localization and Mapping (SLAM) helps in generating a site map in real time at different stages of the project which can be compared with the planned site map. When the user walks the site with SLAM app scanning the path along, the determination of the position of the user on the site on a virtual map and 3D mapping of his surrounding environment is done simultaneously. Both the above mentioned benefits of AR are costly investments but when compared to its costs to benefits which are at the scale of project, it is a profitable investment.

3 Cost Effective AR Technology – Hardware and Software

The basic hardware requirements for AR are: input devices, sensors, processor and display. Modern computing devices like smartphones and tablet computers contain these elements which usually include a camera and Micro-Electro-Mechanical-Systems (MEMS) sensors such as accelerometer, GPS, and solid state compass, making them appropriate AR platforms. Head Mounted Displays (HMD) are trending in the current AR and VR display systems. Oculus Rift, Razer's "OSVR", HTC and Valve's "Vive", Samsung Gear VR are considered to be the leading head mounted displays in the virtual reality domain. Google glass, an optical head mounted see through display, uses ubiquitous computing technique that aims at giving people the best AR experience. To assist swift development of Augmented Reality Application, some software development kits (SDK) have emerged. Some of the well-known AR SDKs are offered by Vuforia, ARToolKit, Mobinett AR, Wikitude, Blippar and Layar [12].

Cost of AR application has two main components. The cost of application building and the cost of hardware. The cost of application depends entirely on the creative content added in it and is almost same irrespective of the hardware used. The difference can be made by choosing low cost hardware as the application building cost is almost same across different platforms.

Currently a very ambitious AR device which is dedicated to give high end AR experience to users is the Microsoft HoloLens. It's a mixed technology headset using both augmented and virtual reality where it merges real-world components with virtual images called holograms. The headset is not connected to any PC as it is run on battery with a Windows 10 system built in it. It has a Kinect-style receiving of voice commands and gestures. There is no other device in the making which is as close to HoloLens in giving that high end AR experience. In the VR devices category Oculus Rift is one of the first consumer-targeted device. HTC Vive offers 360 degree tracking and also a "context aware controller" which enables interaction with the virtual world elements. FOVE VR differs from the likes of Rift and PlayStation VR because it enables interactive eye-tracking using an IR sensor inside the headset that monitor's wearer's eyes and controls the rendering on the screen based on where we look. The Gear VR is an Oculus Rift-powered device which has a Samsung Galaxy smartphone as its processor and display. The Galaxy handset simply slots in front of the lenses, into a Micro USB dock. The smartphone's super AMOLED display serves as the screen for the user. It comes in handy for AR experiences if one can enable the phone's camera to capture the real world. Table 1 lists the popular devices various devices currently in the market or in the development stagealong with the market price. It can be seen from the table that the Google Cardboard device is the most economical option. As a smartphone has become ubiquitous, this work focused on exploring the current and extensible capabilities of this platform in providing AR experience for construction applications. Google Cardboard is a virtual reality (VR) and augmented reality (AR) platform developed by Google for use with a head mount for a mobile phone. This is intended to be the low-cost system to boost interest and development in AR and VR applications. Users can build their own cardboard viewer by using low cost materials or purchase from third party with specifications issued by Google.

While choosing the cheapest hardware enables widespread application, the challenges lie in developing a suitable application that can provide a seamless AR experience. Open Source Software Development Kits (SDK) are available to extend and customize the capabilities of the platform. A popular Augmented Reality Software Development Kit (SDK) which enables the creation of AR applications for mobile devices is Vuforia. In Vuforia, Computer Vision technology is used to recognize and help in real time tracking of planar images (Image Targets) and simple3D objects, such as boxes [13]. It is this capability that enables developers to extend capabilities to enable positioning and changing the orientation of virtual objects, such as 3D models and other media, in relation to real world images when these are viewed through the camera of a mobile device.

Table 1. Cost comparison of AR/VR devices

Device	Price
Microsoft HoloLens 	$3,000.00
HTC Vive 	$799.00
Oculus Rift 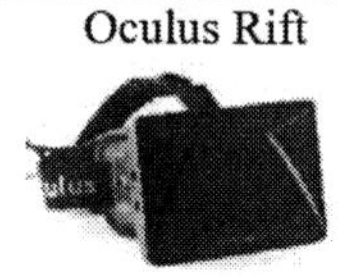	$599.00
FOVE VR 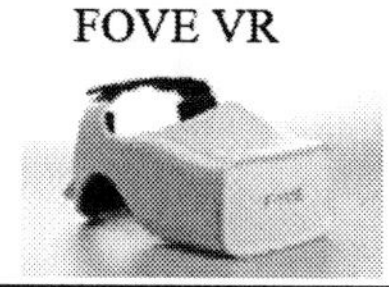	$349.00
Zeiss VR 	$120.00
Samsung Gear VR 	$99.00
Google Cardboard 	$20.00

The Vuforia platform supports both Android and iOS smartphones and tablets as well as digital eyewear. Developers can build Vuforia apps in Eclipse, Xcode and Unity cross-platform game engine. Unity 3D is one of the best professional multiplatform development engine. It is used to develop video games for PC, consoles, mobile devices and websites. There is also a SDK for Cardboard which helps in creation of VR application from scratch or conversion of existing application to VR or enabling switching of application in and out of VR. Therefore, importing Vuforia and Cardboard SDKs (software development kit) on Unity can help in the creation of augmented reality system compatible with Google Cardboard. Taking advantage of this feature a low cost augmented reality solution is presented in this paper.

4 AR Mechanisms

Over the years the AR technologies have evolved rapidly and today we are in the age of smart glasses. The likes of Google Glass and Microsoft HoloLens have emerged which are ubiquitous computing devices that project digital information directly onto the human eye. Figure 1 shows how AR technologies have evolved throughout the years in various levels.

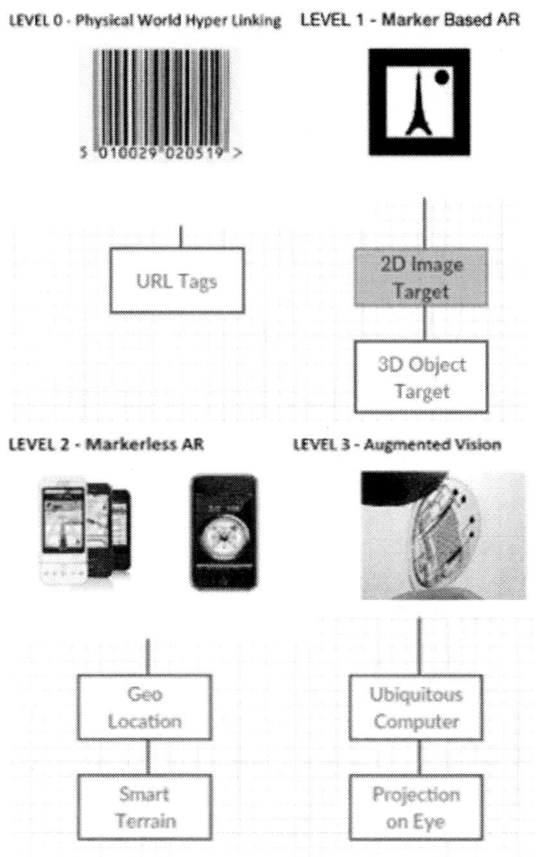

Figure 1. Road Map of AR Technologies

With reference to Figure 1, Level 0 of AR is the simple hyperlinking where we are directed to a particular website upon scanning a barcode tagged with an URL. Level 1 comprises of the major area where our proposed solution lies in. A large number of applications both for Hand-held and Head-mounted devices have been developed for marker based tracking which then evolved into Level 2- as tracking based on the natural features in the environment and geo location. The final level or Level 3 is the currently most concentrated area for development where even contact lenses with AR capability are aimed to be achieved. For the purpose of this discussion we stick to the 2D image target capability on a smartphone with a Cardboard mount.

The simple AR case is the user scans a target in the real world and the virtual model is invoked and aligned itself accordingly to give the augmented experience. There is a triggering event which occurs every time a target is scanned. This target can be an image or an object.

Image targets represent those images which the SDK detects and tracks. The Vuforia SDK usually detects and tracks the features that are found in the image itself by comparing these natural features against a known target resource database. Once the Image Target is recognized from the database which contains correspondence information i.e. the corresponding 3D model to a particular image target, the SDK will track the image as long as the image lies in the camera's field of view.

Object Targets are generated by creating a digital representation of the features and geometry of a real physical object. They are different from image based targets that require the use of a planar source image. Vuforia Object scanner generates object targets by scanning physical 3D objects. Object targets are ideal in cases of building rich interactive experiences with rigid 3D objects. They can be augmentations on toys, manuals overlaid on consumer products or triggering of new events on scanning a consumer product.

For the application in construction site where one would like to augment building models or information onto their displays we have chosen image targets because of its ease of implementation and a variety of images can be created by the user to serve various purposes. One can use QR codes or floor or elevation plans or any other specific purpose images as targets. Figure 2 below shows the buildings being augmented onto the real world once the image (a cover page lying on the desk) is scanned with the AR app.

Figure 2. Difference between normal view and augmented view

4.1 Extended Tracking

Extended tracking is another powerful feature which can be used efficiently to position various models in the real world. It uses the features of the real world environment to improve tracking performance and continue tracking even when the target is no longer in view. When the target goes out of view, other information from the surrounding environment is used by Vuforia to obtain the target position by visually tracking the environment [13]. A map is built around the target for this purpose and therefore assumes that the target and environment are static. Once a model is enabled for extended tracking even if it is not in the view of the camera not only the model still stays but the coordinate system still remains intact with respect to the target.

Figure 3. Models in place even after respective image targets have been removed. This is made possible by enabling extended tracking feature.

If extended tracking is enabled the AR application remembers the reference frame of the first target for an extended time even when the camera encounters a problem. Also the subsequent images that are tracked after the first target maintain their positions with respect to the first target as well as with respect to the surrounding environment. From Figure 3, it can be seen that the models stay in their positions with respect to each other as well as the environment even after there are no image targets. Also for extended tracking to work better the environment should be feature rich and static. This is one of its limitations. The more moving parts or features in an environment the lesser efficient is the positioning of the model with respect to the environment.

4.2 Gaze input user interface

Once the user places a smartphone in the Cardboard slot and mounts it for use, he/she cannot interact with the mobile with touchscreen. So one has to depend on other methods of input. Gaze input user interface is a way of interacting by looking at a certain point. An onscreen cursor appears in the user's view which moves with the head motions of the user.

Wherever the user points his head the onscreen cursor also points in the same direction. So once this onscreen cursor is placed on the User Interface (UI) buttons, a particular function, which is preassigned to the UI buttons is performed.

Figure 4 illustrated how the user points his/her smartphone at the image target (QR code) a menu pops up onto his/her view. In the menu shown there are three UI buttons each associated with a specific function. In order to select the menu options the user has to tilt and orient his/her head such a way that the cursor lies on the UI buttons. Ray casting technique is used to determine when the cursor overlaps with UI objects and the resulting action is triggered. Specific functions for each UI buttons are attached using code scripts.

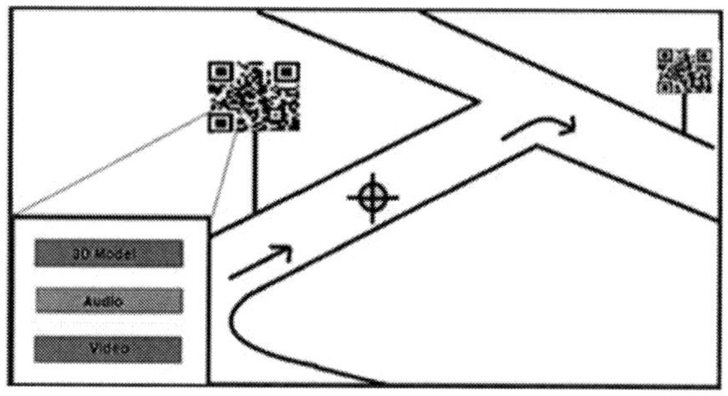

Figure 4. UI buttons and on-screen cursor on user's view

4.3 Integration with Cardboard

Cardboard SDK enables conversion of any application to VR mode. Integrating both Vuforia and Cardboard SDK on Unity game engine can result in the following output shown in the Figure 5 below. It shows a stereo output of the augmentations. The image target in the figure is "TRACKER" which when scanned gives a 3d model augmentation. But since the app had been integrated with Cardboard, for the viewer to get the immersive experience, the stereo output of the augmentation is intended. This integration is done on Unity 3D, where the AR camera from Vuforia SDK is bound to the Left and Right Cardboard Main cameras imported from Cardboard SDK.

Figure 5. Stereo output of AR app

5 Construction Application- Case Illustrations

When the user mounts the cardboard and starts traversing along the site he/she usually picks up the real world camera input until he encounters the image targets. Here the image targets are the QR codes which are specifically positioned at places to ensure the right superimpositions.

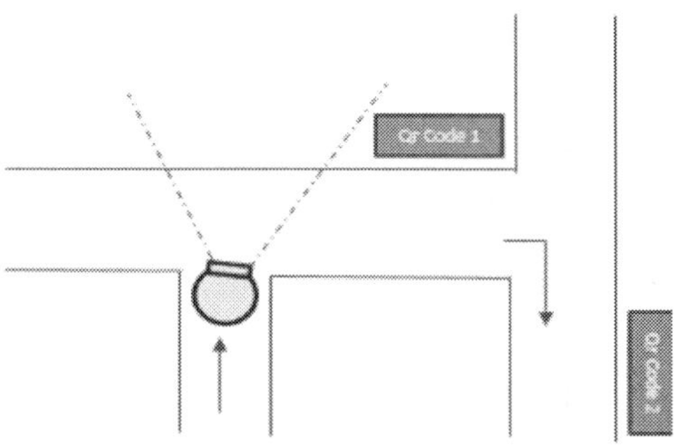

Figure 6. Plan view showing the user's head mounted with Cardboard while traversing the site.

From the Figure 6, 7 and 8 it can be seen that the user traverses the path initially with no augmentations till he encounters QR code 1, then he visualizes a building which is scaled and located at a particular orientation and position with respect to the image target. Also the dotted face in the Figure 7 shows that even when the image target is out of view the model is seen. Hence the first model tracked here is extended tracked as discussed in section 4.1. In an extended tracing system the subsequent models can be as shown in Figure 8, where model 1 stays referenced even though the user is viewing another image target. In this way the entire site can be traversed while positioning the models with respect to the initial image tracked.

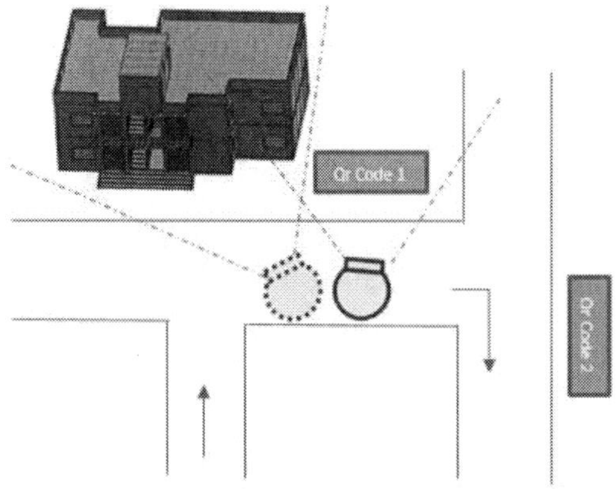

Figure 7. Building model augmented once the user views the QR code.

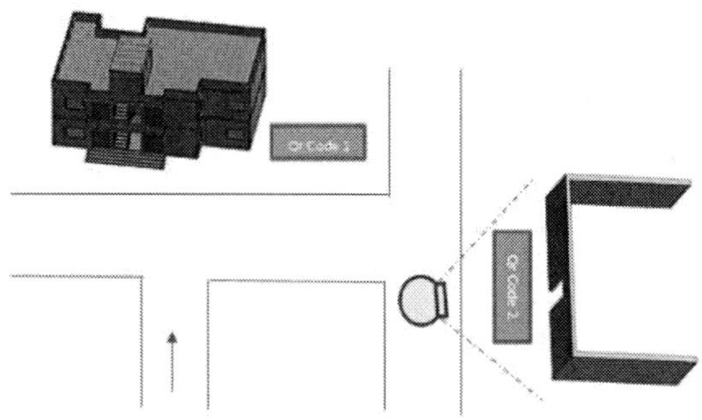

Figure 8. Extended Tracking shows the first image still in position while tracking the next one.

In the modes discussed the models have been invoked as soon as the image target is tracked. But if a menu is invoked instead of the model directly, the user can decide what he wants to view. As shown in Figure 9, a user can use the menu to select whether to view just the 3D model or the reinforcement being superimposed or MEP lines superimposed on the walls.

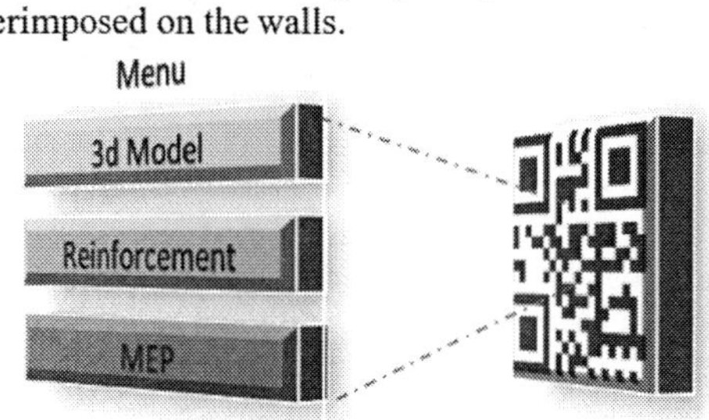

Figure 9. Menu options invoked from QR Code.

Consider the case where the user encounters a QR code as shown in the Figure 10. Once the user mounts the cardboard and views the same image it appears that the menu pops up first onto his view as shown in the Figure 9. And then using the head movements the user has to point the cursor on the desired menu option. In the case below first the user points the cursor onto the "3d Model" option to view the wall with two windows and an entrance. For example, if the QR code to be printed and erected on site is having a width of 30cm, while modelling itself the wall is scaled 10 times to the size of QR code image, so that once 30cm QR code is scanned a 3 meter model wall gets augmented. Also the positioning with respect to the image is done accurately otherwise the slightest of the errors can be picked up by the eye if the model doesn't overlap with the real world as expected.

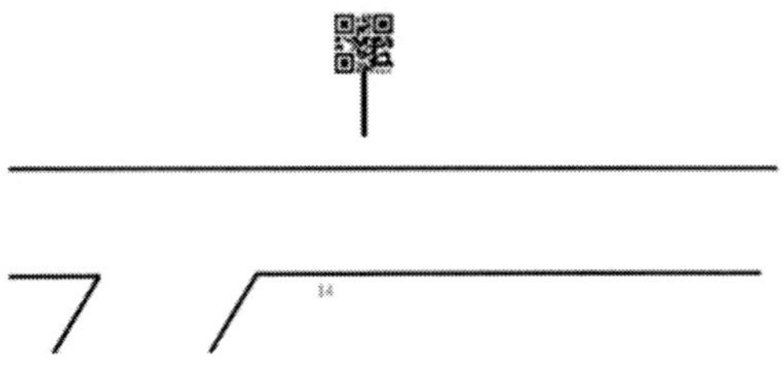

Figure 10. Naked eye view of QR code

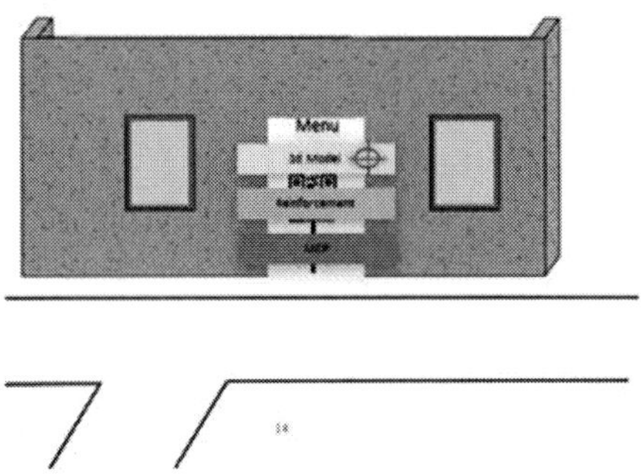

Figure 11. Cursor Invoked 3D model

Similarly the other options are selected by the user by gazing the cursor on the corresponding menu options. Figures 12 and 13 show the super-imposition of the reinforcement and Electrical Lines onto the augmented wall. Also if there is an existing real wall, one can invoke a reinforcement or other attribute display onto it. The visualization of such details of elements helps in understanding where the embedded lines run so that one can exactly pin point a location for maintenance or other activities without actually cutting it open and without going through a large number of drawings.

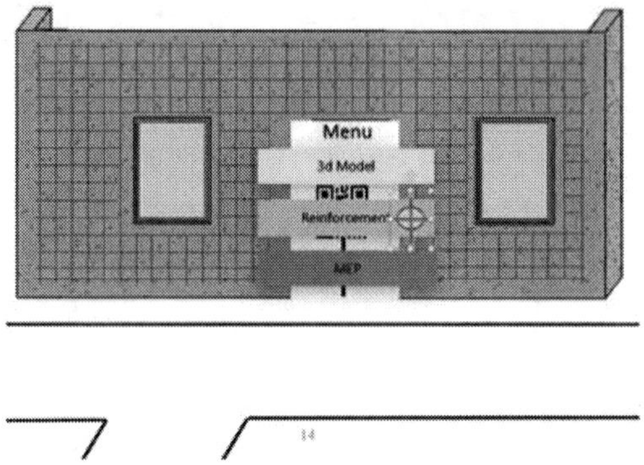

Figure 12. Superimposed Reinforcement

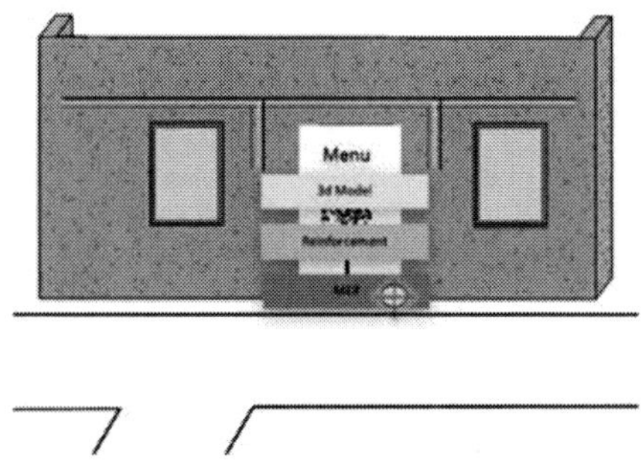

Figure 13. Superimposed Electrical Lines

6 Discussion & Conclusion

Investment in any kind of technology needs a detailed assessment of the value and the potential return on investment. One might deem that implementing advanced technology at construction site would simply increase the cost of construction and early adopters of new technology take a large risk. . Although AR is a technology in its infant stage, the projected utility is enormous that no one can ignore it and preparing for its adoption is a must.

Low investment efforts such as this enables decision makers to experience and evaluate the benefits of AR with minimal financial risk. As full scale deployments of AR technologies in construction becomes feasible the impact on the productivity should be significant.

The cost effective AR platforms is also expected to contribute to the widespread adoption of the technology. While there is need for expensive and sophisticated technologies such as the Microsoft HoloLens, it is the availability for inexpensive and familiar platforms which will propel the widespread usage. Though there are certain limitations to this technology, as a starter these technologies can be used for training workers on site, visualization purposes that gives a clear understanding of the product they are working on. From preliminary initiatives like these the technology can be taken to a full-fledged adoption once it is observed that the results are fruitful.

References

[1] Azuma R. A Survey of Augmented Reality. *Presence: Teleoperators and Virtual Environments*, 6(4), pages 355-385, 1997.

[2] Webster A., Feiner, S., MacIntyre B., Massie W. and Krueger T. Augmented reality in architectural construction, inspection and renovation. In *Proceedings of ASCE Third Congress on Computing in Civil Engineering,* pages 913-919, 1996.

[3] Feiner S., MacIntyre B., Hollerer T. and Webster, A. A touring machine: Prototyping 3d mobile augmented reality systems for exploring the urban environment. ISWC'97: *Proceedings of the 1st IEEE International Symposium on Wearable Computers, strona 74, Washington, DC, USA. IEEE Computer Society,* 1997.

[4] Rankohi S. and Waugh L. Review and analysis of augmented reality literature for construction industry. *Vis Eng*, 1(1), page 9. 2013.

[5] Dong S. and Kamat V. SMART: scalable and modular augmented reality template for rapid development of engineering visualization applications. *Vis Eng*, 1(1), page 1, 2013.

[6] Wang X., Dunston P.S., Proctor R., Hou L. and So J. Reflections on using a game engine to develop a virtual training system for construction excavator operators. In *Proceedings of the 28th International Symposium on Automation and Robotics in Construction (ISARC 2011): Seoul, Korea.* 2012

[7] Yeh K., Tsai M. and Kang S. On-Site Building Information Retrieval by Using Projection-Based Augmented Reality. *Journal of Computing in Civil Engineering*, 26(3), pages 342-355, 2012

[8] Wang X. Using Augmented Reality to Plan Virtual Construction Worksite. *Int J Adv Robotic Sy*, page1. 2007

[9] Park C., Lee D., Kwon O. and Wang X. A framework for proactive construction defect management using BIM, augmented reality and ontology-based data collection template. *Automation in Construction*, 33, pages 61-71.2013.

[10] Behzadan AH., & Kamat VR. Integrated information modelling and visual simulation of engineering operations using dynamic augmented reality scene graphs. *Journal of Information Technology in Construction,* ITcon: 16, pages 259–278, 2011.

[11] Shirazi A. and Behzadan A. Design and Assessment of a Mobile Augmented Reality-Based Information Delivery Tool for Construction and Civil Engineering Curriculum. *J. Prof. Issues Eng. Educ. Pract.*, 141(3), page 04014012. 2015

[12] Augmented Reality. Online. https://en.wikipedia.org/wiki/Augmented_reality, Accessed: 01/03/2016.

[13] Developer.vuforia.com. *Vuforia Developer Portal*. Online. http://developer.vuforia.com, Accessed: 01/03/2016.

Entity Matching across Stereo Cameras for Tracking Construction Workers

Yong-Joo Lee[a], Man-Woo Park[b], Ioannis Brilakis[c]

[a, b] Department of Civil and Environmental Engineering, Myongji University, Republic of Korea
[c] Department of Engineering, University of Cambridge, United Kingdom
E-mail: leetoday@mju.ac.kr, mwpark@mju.ac.kr, ib340@cam.ac.uk

Abstract

While GPS and RFID are dominantly used for tracking construction equipment and materials, 3D vision-based tracking that employs stereo-vision system has been also proposed as an alternative to efficiently track construction resources operating in a congested site. The 3D vision-based tracking requires an entity to be tracked in both cameras of the stereo-vision system so that two of 2D pixel coordinates are generated for each entity. Its 3D coordinate can be calculated by triangulating two projection rays that go through the 2D pixel coordinates. In order to track multiple entities of the same type simultaneously, the triangulation requires an additional process to match entities across the two camera views. This paper proposes an efficient method of entity matching for tracking construction workers. The method is mainly based on epipolar geometry which represents the geometric relation between two camera views. Two additional strategies – distance ratio thresholding and frame grouping – are introduced to reject false matchings and to ultimately increase precision of the 3D tracking. Experiment results demonstrate its suitability for automated entity matching for 3D vision-based tracking of construction workers.

Keywords –

Construction worker; Stereo Vision; Tracking; Detection; Entity Matching

1 Introduction

Any given construction project is composed of a series of processes through the organic and diversified interactions of manpower, equipment and materials. These processes in turn are composed of complicated tasks that are difficult to anticipate. Therefore, it is important to manage the schedules and extent of progress by minimizing uncertainties that can occur in construction sites in order to successfully execute construction projects. Uncertainties ultimately arise from lack of information. With a greater amount of and more diversified information, site managers and decision makers are able to make decisions that are more effective and safer. Data on the location and conditions of the construction equipment, manpower and materials, etc. can be used as important data for the assessment of the status of the construction sites. Research on the utilization of such data have been conducted for the purposes of management of on-site safety and productivity, along with the active execution of research on the position tracking system for each of the types of entities.

Global Positioning System (GPS) is one of the most representative position tracking systems. It is highly reliable since it is possible to secure extremely high accuracy and can be used in all regions provided it is in outdoor settings depending on the composition of the system, and there are many models of the latest heavy machineries with built-in GPS. Recently, Radio Frequency Identification (RFID) [1], Ultra Wide Band (UWB) [2] and Bluetooth [3] formats of position tracking technologies have been introduced to the market. However, the majority of the position tacking methods including GPS need to attach a tag to each entity to be tracked. Accordingly, due to the need for the management of the system such as reading the tag or replacing of the aged tag at prescribed intervals, the cost of position tracking increases rapidly if the number of entities to be tracked increases substantially. However, the image-based position tracking system only uses a camera and processor for the image processing without the need for a separate tag. Due to such advantage, research on the image-based position tracking system has been conducted actively. Research on real-time monitoring of the construction sites by using a 3D imaging sensor and managing the materials by checking the location and conditions of the construction equipment and manpower through the image information and by recognizing the construction materials have been executed [4][5][6]. Image-based tracking system is being recognized as a technology capable of efficiently providing key data for the assessment of the movement paths and patterns of the equipment, manpower and materials in the construction sites.

Majority of the image-based process management methods introduced above were limited to the obtaining of 2D pixel coordinates by assessing the movement and transformation of the entities by using monocular camera. Stereo image is necessary to compute the 3D coordinates in the actual space of particular entity from the 2D images. Park et al. proposed the method of acquiring the actual 3D position coordinate using a stereo vision system [5]. They showed that it is possible to estimate the 3D coordinates of a given entity through the triangulation among the entity photographed and its projected pixel coordinates on two cameras [5]. However, the validation was limited to tracking of a single entity. An additional separate method is necessary in order to track multiple entities. Acquiring 2D coordinates of the entities in each of the images should be followed by the matching of the same entities in the left and right images, which is not needed when tracking a single entity since the matching is straightforward. This paper proposes an efficient method of entity matching for tracking construction workers. The method is mainly based on epipolar geometry which represents the geometric relation between two camera views, and the preliminary experiments showed its suitability for the stereo vision based 3D tracking.

2 Background

GPS and RFID have attracted a lot of attention from construction industry as they can be used for tracking the positions of construction entities such as workers, equipment, materials. As an alternative to GPS and RFID, the vision-based tracking system has also been investigated because of its several potential advantages. For example, it does not require to tag every entity to be tracked, and it can potentially track any entities that appear in the camera view [7]. Continuous research efforts have been made on the image processing and computer vision algorithms to achieve the aforementioned potential advantages. Object detection and object tracking are the main algorithms required to locate the positions of interested objects in the 2D pixel coordinate system.

Chi and Caldas proposed the method of detecting the construction entities by using the background subtraction and the machine learning classifiers [8]. The construction entities were categorized by using the area size, aspect ratio and occupancy percentage, etc. separated by the background subtraction in the artificial neural network. Rezazadeh and McCabe proposed the method that can be used in the progress management through the tracking of the dump truck in the construction sites [9]. The method applied the Haar-like features and the HOG (Histograms of Oriented Gradients) features to detect the dump trucks. Training data are generated by using the images of dump trucks viewed from various directions for detection. They also presented a part-based method to detect articulated shapes of machineries. Focusing on detecting construction workers, Park and Brilakis proposed the method to detect people wearing safety vests of fluorescent colours, which is composed of three steps - background subtraction, HOG + SVM (Support Vector Machine), and colour histogram + k-NN (k-nearest neighbour) [10]. Similarly, Memarzadeh et al. integrated the HOG and hue-saturation-based color histogram features into a single feature [11]. The feature is trained with SVM, and applied to workers, excavators, and dump trucks.

Whereas the methods described above were focused on detecting construction entities, several studies have been conducted on combining detection and tracking methods for continuous position tracking throughout video frames. Rezazadeh Azar et al. [12] employed a HOG based detection method and the KLT (Kanade-Lucas-Tomasi) tracker. The method is capable of tracking excavators and dump trucks even when they are partially occluded. Similarly, Park and Brilakis combined the function of detection and tracking algorithms for locating construction equipment [13] and workers.[14] Their methods stably handled total and partial occlusions by automatically identifying the events of occlusion based on the absence of detection results.

Park et al. [5] proposed the method of acquiring the 3D position coordinates by using a template-based 2D tracker [15] and applying the stereo vision technology. A fixed stereo vision system is used, and the geometric relation between two cameras were estimated by executing the camera calibration process through the use of SIFT (Scale-Invariant Feature Transform) and MAPSAC (Maximum A Posterior SAmple Consensus). The 2D pixel coordinates of the targeted entity is obtained by using a template-based tracking method in the left and right video frames. At each frame, the two 2D pixel coordinates of the entity extracted from the left and right video frames are fed into the triangulation process which calculates the 3D coordinate based on the geometric relation of the two cameras. The method was tested on tracking of a van, a worker, and a steel plate held by a worker. Though the tests showed reasonably high accuracy, each test involved a single target entity. The method is not applicable to tracking more than two entities since it is not able to distinguish the matching pair across the two views. An additional process is required to match the 2D detection/tracking results between the left and right camera views. Therefore, this paper proposes an efficient method for finding the matching pairs across the two views to enable the 3D tracking of multiple construction workers.

3 Eipoplar Geometry based Entity Matching for Multiple Workers

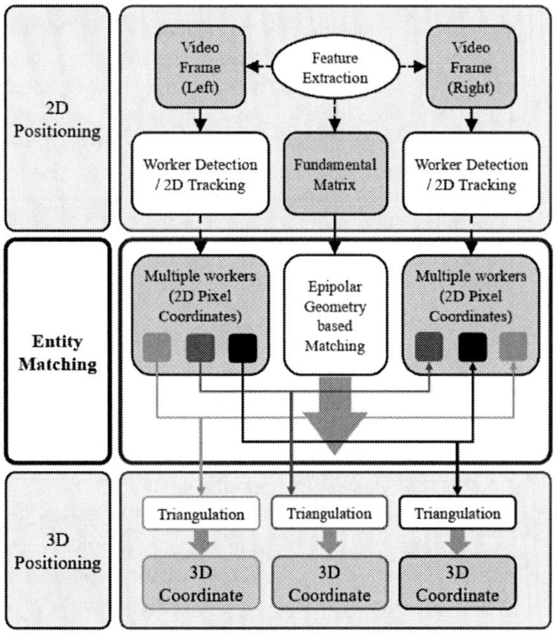

Figure 1. Framework for 3D tracking of multiple construction workers

Figure 1 illustrates the framework for the 3D tracking of multiple workers using a stereo vision system. First, the combination of the detection and 2D tracking methods is executed by receiving input from the two camera views. For this purpose, the method proposed by Park and Brilakis [14] is adopted. This step provides 2D pixel coordinates of the workers in each camera view. If three workers are detected and tracked in both camera views, three of 2D pixel coordinates will be provided continuously throughout the video frames from both views, respectively. Calculating the 3D coordinate of each worker is processed through triangulation which requires a pair of its 2D coordinates – one from the left frame and the other from the right frame. Therefore, the 3D tracking of a worker can be initiated only when the worker entity is found from both views and matched across the two views. As illustrated in Figure 1, the three worker entities found from both views need to be matched each other to determine three pairs of the 2D coordinates, each of which is corresponding to the same worker entity. It should be noted that the matching process is required only once to initiate the 3D tracking and no further matching process afterwards. This paper proposes an efficient method of the entity matching that mainly uses the epipolar geometry information.

3.1 Epipolar geometry based entity matching

Figure 2 shows brief summary of the matching principles based on the epiploar geometry. The epipolar geometry illustrates the geometric correlation between two cameras in terms of the relative position of the camera origins (O_L and O_R) and their view angle difference. For a worker positioned at x in actual 3D space, an epipolar plane is determined as the plane containing three points x, x_L, and x_R, where x_L, and x_R are the 3D-to-2D projections of x onto the left and right image planes, respectively. The left projection line passing through x_L is projected onto the right image plane as l_R which is referred to the epipolar line of x_L. The epipolar line connects x_R and e_R which are the projections of x and O_L onto the right image plane, respectively. This geometric configuration shows that the 2D projection of the same entity onto the right image plane is located on the epipolar line l_R. Therefore, there must be image coordinates, among the pixel coordinates that epipolar line passes through, onto which the targeted entity has been projected. Accordingly, the primary matching principle is established that the matching pair of a 2D position tracked in the left view should be found from its corresponding epipolar line in the right. Complying with this principle, the matching pair of a worker tracked in the left view can be determined as the one appears closest to its epipolar line in the right view. This principle also works in the opposite way. In other words, the matching pair of a worker tracked in the right view can be determined as the one appears closest to its epipolar line in the left view.

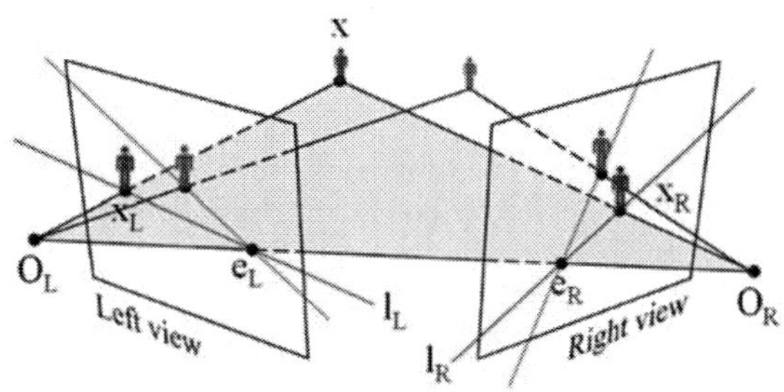

Figure 2. Relation of global 3D points and epipolar lines on each images

Figure 3 shows an example of the matching process. The figure illustrates the process of finding the matching pair of the worker entity which is represented as a bounding box (A) in the right view. In the figure, there are three candidate worker entities detected and tracked in the left view ((a), (b), (c)). The red line drawn in the bottom-left image is the epipolar line that corresponds to the centroid of the bounding box (A). In order to find the matching entity from the left view, the distance from the bounding box centroid to the epipolar line is calculated for each candidate. The candidate (c) is determined as the matching entity since its distance d_c is the smallest.

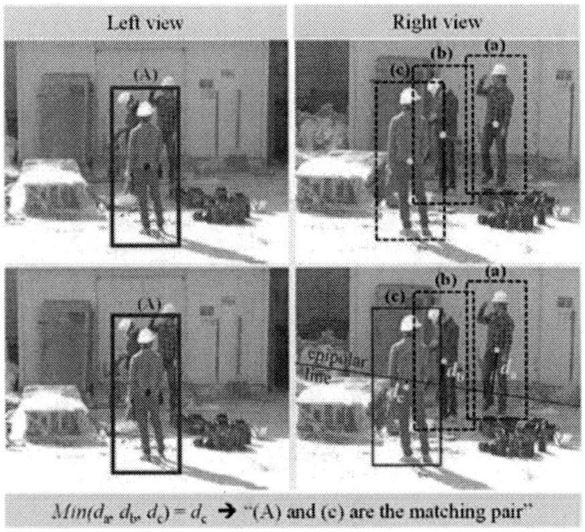

Figure 3. Matching process based on the distance from entity centroids to the epipolar line

The epipolar geometry is generally represented by the 3-by-3 fundamental matrix. The fundamental matrix between the two cameras can be computed by extracting the feature points independently from the images taken by each of the cameras [16]. Since the calculation of the fundamental matrix by using an accurate pair of feature points can accomplish even more accurate entity matching, it is important to use an algorithm that can extract the feature points without being affected by the changes in scale, rotation and distortion of the images. There are diverse methods for the extraction of such feature points. FAST (Feature from Accelerated Segment Test) [17], SURF (Speeded-Up Robust Features) [18] and SIFT [19] are known to be generally outstanding in the area of computer vision are compared. FAST is generally used for mobile applications which require real time processes. In this research, the calculation of the fundamental matrix is processed only once after the camera system has been fixed. Therefore, the real time process is not an important issue. SURF and SIFT would be good options for this research because of their ability to provide a large number of matching points with high accuracy. The number of matching points becomes more important when the image distortion is not calibrated well enough. By using the fundamental matrix computed with SURF or SIFT, the epipolar line corresponding to the centroid of the detected/tracked entities is generated.

3.1.1 Distance ratio thresholding

Finding the tracked entity located closest to the epipolar line does not always give the correct matching. There are several problems to be resolved to implement the epipolar geometry based entity matching. The first problem is the case when the corresponding entity is not detected or tracked in the opposite view. For example, if the entity labelled as (c) is not detected or tracked in the left view, then the matching entity of (A) would be determined as (b) which is the second closest. However, the candidate (b) is not the correct matching and locates much farther than the actual matching entity. In order to remove this type of false matching and minimize the effect of the false results from the detection and 2D tracking methods, this paper introduces the distance threshold. The distance threshold prevents the matching with the entity at a distance too far in this process.

The second problem is the case when two or more entities may be projected onto the single epipolar line. This case can happen if multiple entities are actually on the same epipolar plane. It can also happen even when entities are on the different epipolar plane because of the errors of the detection and 2D tracking methods. Accordingly, it is necessary to introduce an additional procedure to solve the problem. For this purpose, distance ratio threshold was additionally applied. The distance ratio thresholding minimizes the false matchings and maximize precision. The distance ratio (see Equation (1) and Figure 4) illustrates the ratio of the distances of the entities that are the closest and the second closest from an epipolar line. Under the presumption that it is not possible to determine which of the entities is the correct pair if this ratio is too low, the closest entity will be matched to the epipolar line only when the distance ratio is higher than the threshold.

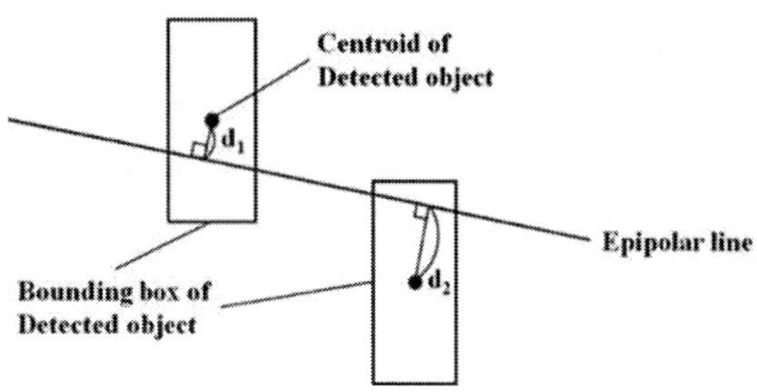

Figure 4. Distance between an epipolar line and the centroids of tracked objects

$$Distance\ ratio = \frac{d_2}{d_1} \qquad (1)$$

The entity matching process is executed by using these two thresholds. Following is the summary of this process for finding the matching pair of each entity tracked in the left camera view. The matching pair is searched in the right camera view.

1. If the pixel distance between the epipolar line and the centroid of the entity in the right camera view is smaller than the distance threshold (T_1), it is deemed to have been matched.
2. If there are more than two entities on a single epipolar line,
   - The closest one is determined to be the matching entity only if the distance ratio is higher than the distance ratio threshold (T_2).
   - If the distance ratio is lower than the distance ratio threshold (T_2), both entities will be excluded from matching.

The above process is repeated in the opposite way – finding the matching pair of the tracked entities in the right camera view. The entities that produced the same results in both ways are matched. While this thresholding process is helpful for reducing false matchings, it also cuts true matching results. In order words, it enhances the matching precision, but degrades the matching recall. It should be noted that the precision is more important than the recall in this research since the matching process is required only once at the beginning of the tracking for each entity. The critical factor is not providing the matching results every frame, but reducing false matchings.

3.1.2 Frame Grouping

Due to the errors that can occur in the worker detection and 2D tracking processes, there are frequent situations in which accurate matching cannot be performed even if various thresholds are mobilized. To enhance the matching precision even further, this paper introduced the frame grouping strategy that determines the matching results for a certain interval, instead of every frame. It integrates the results of several frames in the format of voting in accordance with the set frequency.

Figure 5 illustrates the frame grouping process for searching the matching result of the worker on the farthest left in the image, which is labelled as '1'. From top to bottom, the figure are the results of the individual matchings for the 348th, 349th, and 350th frames, respectively. If the frame grouping is not applied, the method would generate two correctly matched results (#348 and #349) and one wrongfully matched result (#350). However, if frame grouping is applied, the three frames are grouped together and the one false result would be neglected due to the majority votes of the correct results. In other words, instead of providing two correct results and one false results, it generates one true matching result. As such, it is possible to anticipate more accurate results if frame grouping is applied. Although results are obtained more intermittently if the number of the frames grouped together increases, its effect on precision will manifest to greater extent.

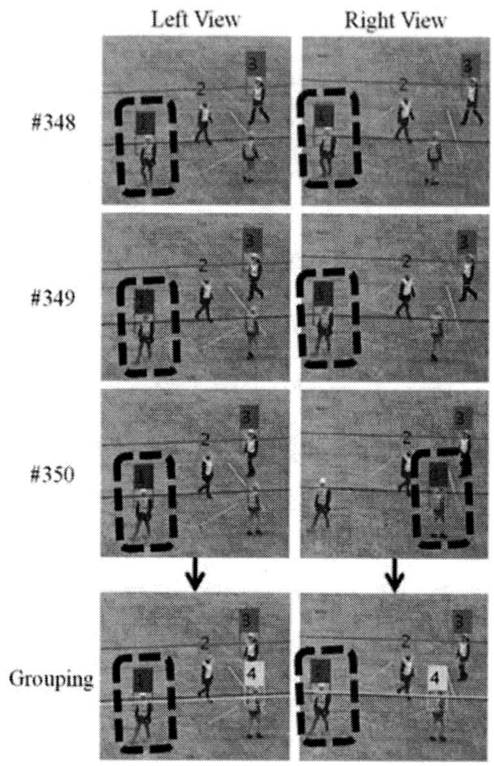

Figure 5. Frame grouping process for the three consecutive frames

4 Experimental Results

For the experimental tests, videos were taken of the 4 persons wearing hardhats and safety vests. These subjects were allowed to walk freely within the prescribed area. The original images in MTS format were edited to the length of approximately 40 seconds and were encoded in MP4 format while maintaining the same number of frames. The propose framework was processed for 10 fps and the total number of the processed frames was 398.

4.1 Fundamental Matrix Calculation

As a first step to implement the proposed matching framework, the fundamental matrix is calculated using the feature point matching results. In this research, no intrinsic calibration is performed, so camera distortion is not corrected. Therefore, a large number of feature matchings are required to obtain more accurate fundamental matrix. A simple experiment is performed to compare the SIFT and the SURF in terms of the number of correct matching points. For both method, RANSAC (RANdom Sample Consensus) [20] is used to remove outliers and to compute the fundamental matrix as accurately as possible. The test showed that the SIFT and the SURF provide 954 and 224 matching pairs, respectively. Accordingly, the SIFT is used for the fundamental matrix calculation. Since the entity matching method proposed in this research uses fixed cameras, it only uses a single fundamental matrix that was computed based on the first frames of the left and the right camera views.

4.2 Epipolar Geometry based Entity Matching with Thresholding

Figure 6. Correct matching results from the 257th frame ($T_1 = 10$, $T_2 = 2.5$)

Firstly, the experiments were conducted without using frame grouping to determine the appropriate value of the distance ratio threshold (T_2). Figure 6 and Figure 7 illustrate some example results. Entities determined to be the same worker are indicated with the same coloured bounding box and ID number. Lines drawn with the same colour used for the bounding box indicate the epipolar line of the corresponding entity. The epipolar line of the entity tracked on the left is drawn on the right video frame and vice versa. Figure 6 illustrates the correctly matched results while the Figure 7 illustrates the results in which the entities labelled as '1' and '4' are falsely matched. If multiple entities are confirmed on the same epipolar line, it may cause a false matching as illustrated.

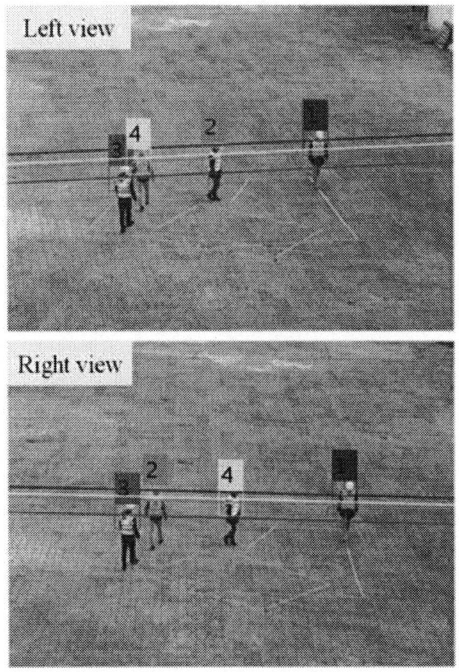

Figure 7. False matching results from the 35th frame (Threshold: distance 10, distance ratio 2.5)

In order to investigate effect of distance ratio thresholding, the recall and the precision are calculated for various threshold values. The precision and the recall are defined as follows.

- Precision = TP / (TP + FP)
- Recall = TP / (TP + FN)
- TP (True Positive) = # of the retrieved matchings which are correct
- FP (False Positive) = # of the retrieved matchings which are incorrect
- FN (False Negative) = # of the matchings which are not retrieved

Table 1. Results of entity matching according to the distance ratio threshold

Distance ratio threshold, (T_2)	Recall (%)	Precision (%)
Not Applied (0.0)	91.21	96.49
Applied (2.5)	80.72	98.10
Applied (4.0)	71.40	98.65

The experiment was conducted by changing the distance ratio threshold from 1.1 to 4.0 at the interval of 0.1 in order to determine the appropriate threshold value T_2. Throughout the experiments, the distance threshold T_1 is maintained equal to 10. Table 1 and Figure 8 illustrate the summary of the results. When the thresholding is not applied (T_2 = 0), the proposed method scored 96.49% precision, which is increased to 98.10% by applying T_2 = 2.5. The precision was found to be higher when the higher threshold was applied although the recall falls to 70%. As mentioned in the previous section, the entity matching method introduced in this research places greater importance on the precision rather than on the recall. Therefore, it was decided to use distance ratio threshold of 4.0 because no further precision enhancement is observed with $T_2 > 4.0$.

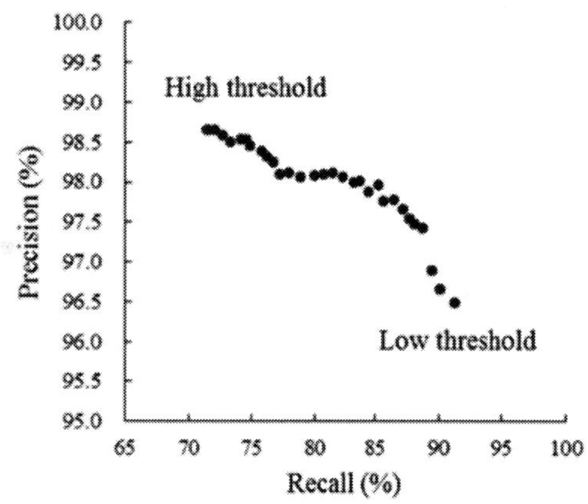

Figure 8. Changes in the recall-precision according to the distance ratio threshold set

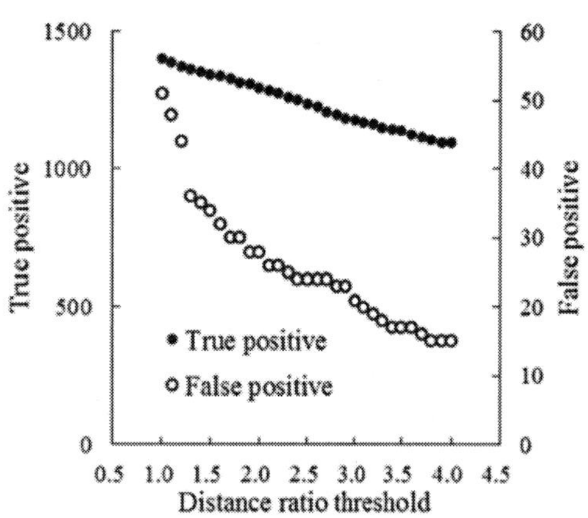

Figure 9. Changes in TP and FP according to the distance ratio threshold set

Figure 9 illustrates the numbers of the true positives and false positives in accordance with T_2. There is rapid improvement of performances in the range of 1.0~1.5 with gradual improvement of performances thereafter. Although the number of entities that are correctly detected decreases with the increase in T_2, the overall precision actually improves as the false matchings are reduced far more than the decrease in the number of the true matchings.

4.2.1 Epipolar Geometry based Entity Matching with Frame Grouping

Table 2 compares the entity matching results with and without using the frame grouping. Both experiments were conducted under the condition of T_2 = 4.0. When applying the frame grouping, the number of the frames to be grouped was increased from 3 to 13 at the interval of 2 frames. Only odd numbers were used since the even numbers can bring the score to a tie. Comparisons were made by using the results at the frequency 13, which has the highest precision among all the groupings. Slightly higher precision is obtained by applying the frame grouping. It exhibits substantially lower recall value which is a predictable result because the frame grouping provides a matching result every 13 frames. According to the results, it is fair to expect the proposed method with frame grouping to provide the matching pair of each entity in a second with 99.17% accuracy.

Table 2. Entity matching results with/without frame grouping (the number of frames in a group = 13)

Frame Grouping	Precision (%)	Correct matchings	Wrong matchings
Not Applied	98.65	1096	15
Applied	99.17	119	1

5 Conclusion

Research efforts have been made on the vision-based tracking for positioning construction entities in construction sites. The 2D localization in a single camera view through detection and 2D tracking methods has been investigated in a series of research works. Furthermore, it was reported that 3D locations of the entities can be obtained by employing stereo vision system. This paper proposes an efficient method of entity matching which is essential for tracking multiple entities using the stereo vision. It was confirmed that the matching between the same entities in the stereo videos is possible by applying the epipolar geometry onto the results of the worker detection and 2D tracking. In order to maximize the precision which is considered as the most critical factor of the matching performance in this research, the distance ratio thresholding and frame grouping are introduced. The preliminary results showed that the distance ratio thresholding could greatly reduce the false matchings which was reflected in the higher precision results. Also, the frame grouping could enhance the precision even further achieving the precision over 99%. The preliminary results showed the potential of the proposed method that can be fitted in the 3D tracking framework to finally enable acquiring 3D positions of multiple workers simultaneously.

Acknowledgement

This material is based upon work supported by the Grant No. 14CTAP-C078828-01 from Infrastructure and Transportation Technology Promotion Research Program funded by the Ministry of Land, Infrastructure and Transport of Korean government. Any opinions, findings, and conclusions or recommendations expressed in this material are those of the authors and do not necessarily reflect the views of the Ministry of Land, Infrastructure and Transport of Korean government.

References

[1] Song J., Caldas C., Ergen E., Haas C. and Akinci B. Field trials of RFID technology for tracking pre-fabricated pipe spools. In Proceedings of the 21th International Symposium on Automation & Robotics in Construction, 2004.

[2] Teizer J., Lao D. and Sofer M. Rapid automated monitoring of construction site activities using ultra-wideband. In Proceedings of the 24th International Symposium on Automation & Robotics in Construction, pages 23-28, 2007.

[3] Hallberg J., Nilsson M. and Synnes K. Positioning with Bluetooth. In Proceedings of the 10th International Conference on Telecommunications 2003, pages 954-958, 2003.

[4] Teizer J., Caldas C. and Haas C. Real-Time Three-Dimensional Occupancy Grid Modeling for the Detection and Tracking of Construction Resources. Journal of Construction Engineering and Management, 133(11):880–888, 2007.

[5] Park M., Koch C. and Brilakis I. Three-Dimensional Tracking of Construction Resources Using an On-Site Camera System. Journal of Computing in Civil Engineering, 26(4):541–549, 2012.

[6] Park M., Makhmalbaf A. and Brilakis I. Comparative study of vision tracking methods for tracking of construction site resources. Automation in Construction, 20(7): 905–915, 2011.

[7] Brilakis I., Park M. and Jog G. Automated vision tracking of project related entities. Advanced Engineering Informatics, 25(4):713–724, 2011.

[8] Chi S. and Caldas C. Automated object identification using optical video cameras on construction sites. Computer-Aided Civil and Infrastructure Engineering, 26(5):398-380, 2010.

[9] Rezazadeh E. and McCabe B. Automated visual recognition of dump trucks in construction videos. Journal of Computing in Civil Engineering, 26(6): 769–781, 2012.

[10] Park M. and Brilakis I. Construction worker detection in video frames for initializing vision trackers. Automation in Construction, 28:15-25, 2012.

[11] Memarzadeh M., Golparvar-Fard M., and Niebles J. C. Automated 2D detection of construction equipment and workers from site video streams using histograms of oriented gradients and colors. Automation in Construction, 32:24-37, 2013.

[12] Rezazadeh Azar E., Dickinson S., and McCabe B. Server-Customer Interaction Tracker: Computer Vision-Based System to Estimate Dirt-Loading Cycles. Journal of Construction Engineering and Management, 139(7):785-794, 2013.

[13] Park M. and Brilakis I. Enhancement of Construction Equipment Detection in Video Frames by Combining with Tracking. In Proceedings of ICCCE 2012, pages 421-428, Clearwater Beach, Florida, United States, 2012.

[14] Park M. and Brilakis I. Improved localization of construction workers in video frames by integrating detection and tracking. The CSCE International Construction Specialty Conference, Vancouver, Canada, 2015.

[15] Ross D., Lim J., Lin R. and Yang M. Incremental learning for robust visual tracking. International Journal of Computer. Vision, 77(1):125–141, 2008.

[16] Luong Q. and Faugeras O. The fundamental matrix: Theory, algorithms, and stability analysis. International journal of computer vision, 17(1):43–75, 1996.

[17] Rosten E. and Drummond T. Machine learning for high-speed corner detection. In Proceedings of the 9th European Conference on Computer Vision, pages 430-443, Graz, Austria, 2006.

[18] Bay H., Tuytelaars T. and Gool L. Speeded-Up Robust Features (SURF). Computer Vision and Image Understanding, 110(3):346–359, 2008.

[19] Lowe D. Distinctive Image Features from Scale-Invariant Keypoints. International journal of computer vision, 60(2):91–110, 2004.

[20] Martin F. and Robert B. Random sample consensus: a paradigm for model fitting with applications to image analysis and automated cartography. Communications of the ACM, 24(6):381-395, 1981.

Analysis of Accuracy Factor for 3D Reconstruction using 2D Image Obtained from Unmanned Aerial Vehicle (UAV)

H.L. Ko[a], S.W. Kwon[b], S.W. Chung[c], D.Y. Moon[d], J.W. Park[e], J.W. Seo[f] and S.M. Kang[g]

[a]Department of Convergence Engineering for Future City, Sungkyunkwan University, Republic of Korea
[b]School of Civil & Architectural Engineering, Sungkyunkwan University, Republic of Korea
[c]Department of Convergence Engineering for Future City, Sungkyunkwan University, Republic of Korea
[d]Department of Convergence Engineering for Future City, Sungkyunkwan University, Republic of Korea
[e]Korea Institute of Civil Engineering and Building Technology, Republic of Korea
[f]Civil Engineering, Hanyang University, Republic of Korea
[g]Korea Plant Maintenance, Republic of Korea
E-mail: kodjmc@nate.com, swkwon@skku.edu, suwanx@nate.com, yoshy17@nate.com, missmr72@gmail.com, jseo@hanyang.ac.kr, sungmin2534@gmail.com

Abstract

Recently UAV has been used in various fields such as construction and civil engineering, disaster, terrain analysis, transportation, etc. In addition, many companies from all over the world are focusing on developing UAV technology. The market of personal UAV has been growing and attracting people's attention year after year due to the low cost of UAV. Especially, in construction industries where people work on large-scale fields, there are many methods to obtain space information and those methods can be done at low cost. However, the existing studies are mostly about utilization of UAV such as flight plan and acquisition 3d image methods. Therefore, it is necessary to study result values according to use of variable value.

This study focuses on the variables which affect the result of 3D reconstruction using 2D image obtained by UAV. First, we fixed values of flight planning of UAV and distortion correction of picture. After that, we took pictures of building and obtained point cloud that allowed us to make 3D model of the building. Second, we analyzed results of the 3D model according to variable of illumination. Then, we found number of point cloud from the previous results. Finally, we identified optimal illumination result to build an effective 3D Modeling. We present about optimal value of illuminance variation to improve current method of developing 3d model. This study can be a more efficient and accurate method to obtain images by UAV.

Keywords –

UAV; Structure from motion; illumination

1 Introduction

Recently, unmanned aerial vehicles (UAVs) are being used in many areas including agriculture, civil engineering, disasters, deliveries, and military operations. The global market for UAVs is steadily growing and is expected to form a large market in the future. In particular, the expansion of the market for spatial data portal services such as Google Earth and MS Virtual Earth is closely associated with the growth of UAV market. The contribution of aerial monitoring to the spatial data areas is also expected to increase gradually. The use of UAVs for the easiness and speed of aerial monitoring is emerging. A mapping system using UAVs allows monitoring with a high frequency because it is efficient in terms of time, cost, and manpower. Furthermore, the applications of UAV-based aerial monitoring systems are gradually increasing as it has now become possible to send data acquired through auto flight and wireless communication technologies to ground stations in real time.

As explained above, the research on UAVs is being conducted actively due to such various advantages of UAVs. Existing studies have focused on data capturing through UAVs and data processing with the acquired data. Various studies have been attempted to convert the acquired data to data that can be measured and managed instead of being explored. Among them, studies on 3D modeling with 2D images are mostly based on the point cloud generation algorithm. In this study, a 2D image environment that is more advantageous for point cloud generation is searched by adjusting the environmental values of 2D image that can directly influence the point cloud generation algorithm. For this purpose, an accuracy analysis is performed by checking the number of point clouds generated by the variation of illumination value.

2 3D Reconstruction using 2D Image

2.1 Literature review

3D modelling with 2D images is generally known as image-based modeling or 3D photography. It is also referred to as structure from motion. Multiview stereo (MVS) and reconstruction are core elements in the process of obtaining 3D modeling through multiple pictures or video images. These technologies are being used in various areas including TV, industries, surveys, cultural assets restoration, and games. Recently, Seitz et al. [1] showed that 1/200-scale objects have higher accuracy in low-resolution images in current MVS algorithms. They can be classified into several grades depending on the object model. First, Voxel-based approaches [2][3][4][5][6]. Faugeras et al present a novel geometric approach for solving the stereo problem for an arbitrary number of images [2]. Pons et al presents a novel formulation for the multiview scene reconstruction problem [3]. Hornung et al present a new volumetric stereo algorithm to reconstruct the 3D shape of an arbitrary object [4]. Vogiatzis et al present a quantitative comparison of several multi-view stereo reconstruction algorithms [5]. Sinha et al formulate multi-view 3D shape reconstruction as the computation of a minimum cut on the dual graph of a semi-regular, multi-resolution, tetrahedral mesh [6]. Second, Algorithms based on deformable polygonal meshes [7][8][9].

Thus, many studies are being conducted on the accuracy and algorithm of structure from motion. This study focused on 2D images that can directly influence point cloud generation.

2.2 2D Image processing for 3D reconstruction

There are various studies on the acquisition of 2D images using UAVs. UAVs are actively used especially for elevation and terrain analysis in construction and civil engineering. Furthermore, UAVs are also used in the management and real-time monitoring of building facilities. Many UAV-related studies can be also found among studies on the aforementioned structure from motion. Darren et al research a technique for geometric correction and mosaicking of UAV photography using feature matching and Structure from Motion (SfM) photogrammetric techniques [10]. Adam et al study explores the use of structure from motion (SfM), a computer vision technique, to model vine canopy structure at a study vineyard [11]. Francesco et al tests the utility of the Structure from Motion (SfM) approach to low-altitude aerial imageries collected by Unmanned Aerial Vehicle (UAV) [12]. Ronald et al demonstrates the use of genetic algorithms in optimized view planning for 3D reconstruction applications using small unmanned aerial vehicles (UAVs) [13].

Many different software applications implementing 3D modeling with 2D images are being developed lately. These applications are classified largely by their use or by modeled object size, such as measurement or land modeling, modeling of building and facilities, and modeling of people or small objects. Among them, Pix4Dmapper and Photoscan are software applications that contain various features and are used widely. These applications extract 3D models by applying option values in line with the object using a variety of image formats. Therefore, they are good for 3D modeling of large and small objects and objects of different environments. However, you must buy the professional version in order to use these various features. Thus, each application has its merits and faults. Pix4Dmapper (Table 1) is one of the most popular software applications. It is specialized for the aerial images of UAVs and is used in many different areas because it offers various features and expressions including project optimization, point cloud editor, area mixing and editing.

Another software application that is often used alongside Pix4Dmapper is Photoscan (Table 2), which has many shared features with Pix4Dmapper. It supports more various extensions for input images and increased the accuracy and shortened the time of camera tracing with the Align photo feature. The overall process is not much different, which consists of picture upload, mask area specification, picture alignment, point cloud building, point cloud editing, transformation into mesh-shaped polygon, mesh editing, high quality texture creation, and exporting as desired output values. However, the cloud editor provides various features such as automatic control by python, 4D processing, DEM data output, and change of the use of meshes. Thus, Photoscan shows good performance in sculptures, buildings, or topography in addition to aerial images. For output formats, OBJ, PLY, XYZ, U3D, ASPRS LAS, and PDF are available.

Autodesk developed a lighter software application called Memento (Table 3) which is more useful for extracting 2D images and point clouds compared to the existing Recap 3D scanner software. Even though it

Table 1. Pix4Dmapper

Software	Feature			
	Input	Process	Cloud editor	Result
Pix4Dmapper	• Edit and import ground control points (.csv, .txt). • Coordinate systems based on local, global, and random points supporting meter and feet units.	• Camera self-correction • Automatic aero triangulation (AAT) and bundle block adjustment (BBA) • Automatic point cloud filtering and modification • Automatic topography/object point cloud classification and DTM extraction (BETA)	• Limitation of size and direction • Project optimization • Point cloud editor • Reflectance map • Multi-area management	• OBJ, PLY, DXF, FBX, Zipped OBJ format • LAS, LAZ, XYZ, PLY • SHP, DXF, PDF • DXF, SHP, DGN, KML

Table 2. Photoscan

Software	Feature			
	Input	Process	Cloud editor	Result
Photoscan	• Automatic image alignment • High accuracy and time shortening of camera tracing with the align photo feature	• Specification of mask area for unnecessary part. • Transformation to mesh-type polygon • Mesh editing • Creation of high-quality texture (original)	• Build dense cloud • Automatic control by python • 4D processing • Creation of texture according to shape	• OBJ, PLY, XYZ text, U3D, ASPRS LAS, PDF

Table 3. Autodesk Memento

Software	Feature			
	Input	Process	Cloud editor	Result
Autodesk Memento	• HD3D mesh modeling based on image such as JPEG, TIFF, PNG, and BMP • Creation of HD3D after scanning the light of the model	• Mesh analysis, rearrangement and modification • Creation of size report • Analysis of differences between the real model and the 3D model	• Removal of unnecessary noises and rearrangement • Measurement of the distance of mesh and checking the size information • Analysis of differences between the real model and the digital model	• OBJ, .STL, .PLY, FBX, .RCM

Table 4. Mobile software

Mobile	SmartMobileVision-SCANN 3D	• Implementation of 3D model using mobile devices • Automatic adjustment of the parameters of different mobile devices • Fast modeling by creating 3D images within 2 minutes
	Fuel 3D	• Automatic conversion of 3D objects to point clouds • Work at a high speed with a low-performance PC by applying video work • Cloud service • 0.1 second processing speed for 3D capture images
	itSeez3D	• Need for purchasing a scan sensor (iSense) device • Scan objects with object sensor or iSense ~ Check the initial 3D model and perform rendering after sending the model to the cloud system ~ Save as a desired file • Specialized for people and small objects

has a disadvantage in the supported formats compared to the existing applications, Memento is more efficient in implementing more vivid 3D objects that are called HD3D. Due to this advantage, Memento is optimized for scanning small detailed objects rather than large objects. The cloud editor shows a high loading speed even for voluminous meshes and offers such features as realistic expression of textures, wire frame mode, isolation mode for mesh arrangement, and analysis of differences between real and digital models, and modifications. Another advantage is the creation of 3D models immediately after saving in STL format for CNC or 3D printers. For output formats, OBJ, STL, PLY, FBX, and RCM are available.

Recently many 3D scanning techniques using mobile technology are being developed (Table 4). Even though their functionality is inferior to the aforementioned applications due to using mobile devices such as smart phone and tablet PC, it is an efficient application for non-professionals performing general tasks because it contains all the essential features. Mobile 3D scan applications offer the advantage of a high processing speed such as quick image scanning and instant modeling. Furthermore, the application is light because they are implemented on mobile devices. They are also good for instant uses because you can scan images anytime, anywhere regardless of time and place. Another mobile scanning

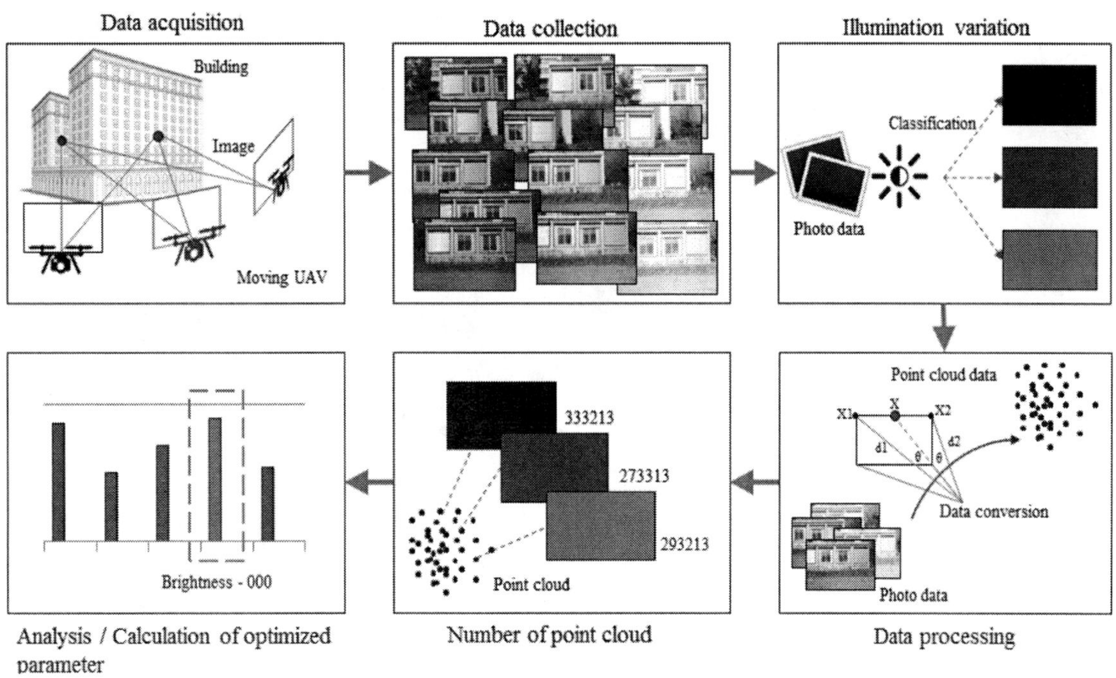

Figure 1. Research process

technique is to buy and mount a scanning device to a tablet PC, which can express 3D objects as effectively as other applications running on PC. Nowadays such various mobile scanning techniques and software applications are being developed.

3 Research methodology and scope

Before conducting this study, existing studies were analyzed through literature review. Most studies on accuracy were about algorithms. In this study, however, we focused on analyzing accuracy based on the changing number of point clouds according to the changing illumination of 2D images. For variations of illumination, camera shutter speed and aperture values were used (Figure 1). First, 2D images are collected using an UAV, and the 2D image data are classified by parameter values. Then the images are converted to point cloud values through data processing. The output values are derived after analysis based on the final values.

4 Implementation, experiments and results

4.1 Test condition

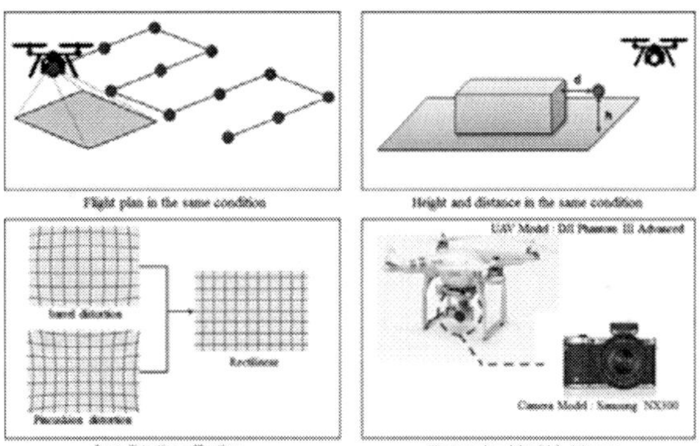

Figure 2. Test condition / Device

Before conducting this study, an identical flight plan for collecting picture data of the same location and angle was established. The distances and heights of buildings were also identical. For this study, two buildings of Sungkyunkwan University in South Korea were selected as subjects of experiment. DJI Phantom III Advanced was used for the UAV model and Samsung NX300 was used for the camera (Figure 2). Picture data are obtained from the identical location using an UAV. The photographed picture data are classified by illumination value. Then the number of point clouds of the data with each different illumination value is obtained using the Photoscan application. The numbers of point clouds by illumination value are analyzed to find the most efficient illumination value. The average illumination value of the pictures in this illumination value is determined. Finally, the illumination value that received the largest number of point clouds is determined.

For photographing, the combinations of aperture and shutter speed of F8 and 1/6000, F13 and 1/2000, F22 and 1/200, and F22 and 1/500 were used. A total of 40 pictures were taken with 10 pictures for each combination in Test1, and a total of 96 pictures with 24 pictures for each combination in Test2. The ground was marked every 3 m and the accurate position of the UAV to the vertical direction was verified. In Test1, the front view of the southward building was photographed to check the variation of illumination. In Test2, another building was photographed in two directions for comparison. The picture data of each environment and different illumination value were classified by aperture and shutter speed values. A total of four illumination values of F8 and 1/6000, F13 and 1/2000, F22 and 1/200, and F22 and 1/500 were used in this study. The average illumination distribution and average illumination value were determined for each group of pictures. The average illumination was converted to brightness value. The point cloud value of the classified picture data was created using the Photoscan application.

4.2 Number of point cloud for performance evaluation

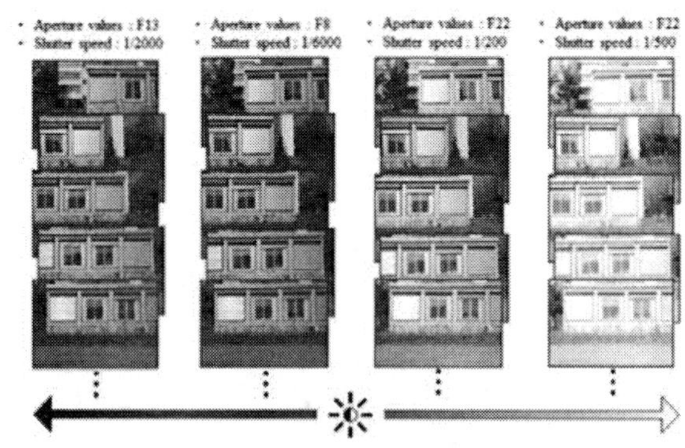

Figure 3. Data classification for data conversion

The brightness was analyzed and presented as lux value to standardize the brightness results according to the aperture and shutter speed values. The brightness values for F13 and 1/2000, F8 and 1/6000, F22 and 1/500, and F22 and 1/200 were 100lx, 120lx, 150lx, and 200lx. The number of cloud point values for each brightness in Test1 was 1269 for F13 and 1/2000, 1059 for F8 and 1/6000, 1180 for F22 and 1/200, and 1349 for F22 and 1/500. In Test2, it was 12339 for F13 and 1/2000, 11320 for F8 and 1/6000, 11639 for F22 and 1/200, and 12050 for F22 and 1/500.

4.3 Results

The final results were analyzed. The number of generated point clouds for the shutter speed 1/2000, aperture value of F13, and the average brightness (lux) of 100lx was 1,269 in Test1 and 12,339 in Test2. The number of generated point clouds for the shutter speed 1/6000, aperture value of F18, and the average brightness (lux) of 100lx was 1,059 in Test1 and 12,339 in Test2. The number of generated point clouds for the shutter speed 1/500, aperture value of F22, and the average brightness (lux) of 150lx was 1,180 in Test1 and 11,639 in Test2. The number of generated point clouds for the shutter speed 1/200, aperture value of F22, and the average brightness (lux) of 200lx was 1,349 in Test1 and 12,505 in Test2 (Table 5).

Consequently, the number of point clouds of 2D image data for 100lx and 200lx were greater by about 11% over the point cloud count by about 11% than the number of point clouds of 2D image data for 120lx and 150 lx.

5 Conclusions

In this study, hypotheses for the 3D modeling of 2D images were defined and verified to analyze accuracy. It was demonstrated that the accuracy of 3D modeling can be increased by analyzing the changing number of point clouds according to the illumination value of 2D images. In conclusion, this study revealed that illumination and point cloud generation are correlated. However, this study was insufficient in the research of buildings and topographies with more diverse conditions, and in the use of more various types of equipment. More in-depth analysis will be possible if more diverse parameters and detailed RGB analysis are added.

6 Acknowledgement

This work is supported by Korea Minister of Ministry of Land, Infrastructure, Transport affairs as Convergence Engineering of Future City Master and Doctoral Grant Program

This research was supported by a grant(14SCIP-B079344-01) from Smart Civil Infrastructure Research Program funded by Ministry of Land, Infrastructure and Transport(MOLIT) of Korea government and Korea Agency for Infrastructure Technology Advancement (KAIA).

Table 5. Results of number of point cloud

Shutter speed (sec)	1/2000	1/6000	1/500	1/200
Aperture values	F13	F8	F22	F22
Average brightness(lux)	100	120	150	200
Pixel	157464	157464	157464	157464
Test 1				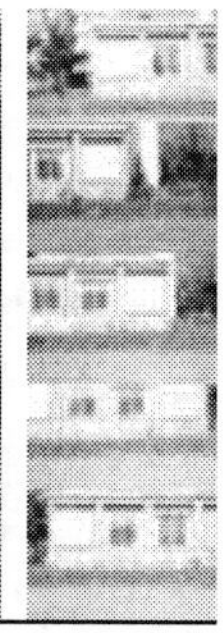
Number of point cloud	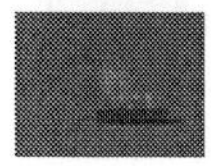 1,269	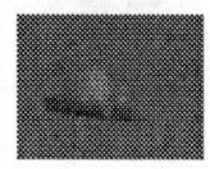 1,059	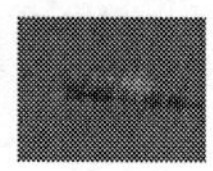 1,180	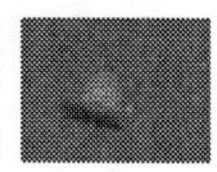 1,349
Test 2	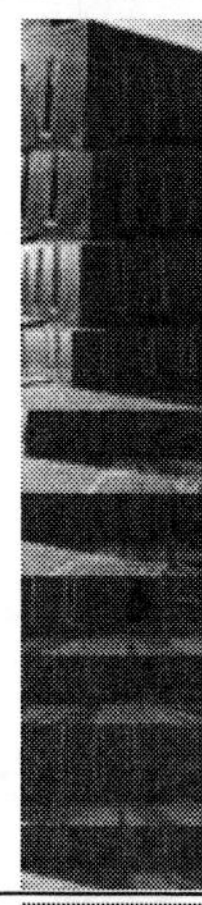			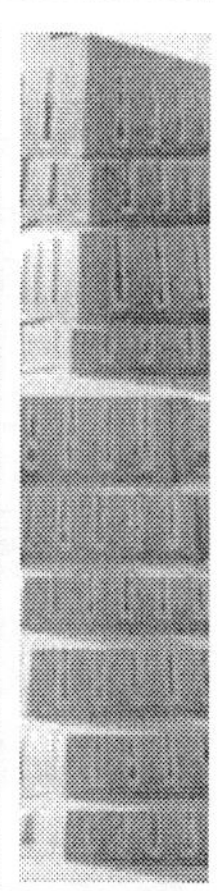
Number of point cloud	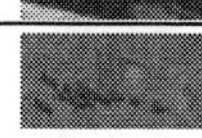 12,339	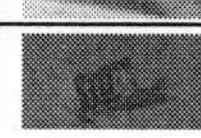 11,320	 11,639	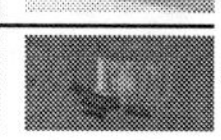 12,050

References

[1] S.M. Seitz, B. Curless, J. Diebel, D. Scharstein, and R. Szeliski, "Multi-View Stereo Evaluation," http://vision.middlebury.edu/mview/, 2010.

[2] O. Faugeras and R. Keriven, "Variational Principles, Surface Evolution, PDE's, Level Set Methods and the Stereo Problem,"IEEE Trans. Image Processing, vol. 7, no. 3, pp. 336-344, Mar. 1998.

[3] J.-P. Pons, R. Keriven, and O.D. Faugeras, "Multi-View Stereo Reconstruction and Scene Flow Estimation with a Global Image-Based Matching Score," Int'l J. Computer Vision, vol. 72, no. 2, pp. 179-193, 2007.

[4] A. Hornung and L. Kobbelt, "Hierarchical Volumetric Multi-View Stereo Reconstruction of Manifold Surfaces Based on Dual Graph Embedding," Proc. IEEE Conf. Computer Vision and Pattern Recognition, 2006.

[5] G. Vogiatzis, P.H. Torr, and R. Cipolla, "Multi-View Stereo Via Volumetric Graph-Cuts," Proc. IEEE Conf. Computer Vision and Pattern Recognition, 2005.

[6] S. Sinha, P. Mordohai, and M. Pollefeys, "Multi-View Stereo via Graph Cuts on the Dual of an Adaptive Tetrahedral Mesh," Proc. Int'l Conf. Computer Vision, 2007.

[7] C. Herna´ndez Esteban and F. Schmitt, "Silhouette and Stereo Fusion for 3D Object Modeling," Computer Vision and Image Understanding, vol. 96, no. 3, pp. 367-392, 2004.

[8] A. Zaharescu, E. Boyer, and R. Horaud, "Transformesh: A Topology-Adaptive Mesh-Based Approach to Surface Evolution," Proc. Asian Conf. Computer Vision, Y. Yagi, S.B. Kang, I.-S. Kweon, and H. Zha, eds., pp. 166-175, 2007.

[9] Y. Furukawa and J. Ponce, "Carved Visual Hulls for Image-Based Modeling," Int'l J. Computer Vision, vol. 81, pp. 53-67, Mar. 2009.

[10] Turner, Darren, Arko Lucieer, and Christopher Watson. "An automated technique for generating georectified mosaics from ultra-high resolution unmanned aerial vehicle (UAV) imagery, based on structure from motion (SfM) point clouds." Remote Sensing 4.5:1392-1410, 2012.

[11] Mathews, Adam J., and Jennifer LR Jensen. "Visualizing and quantifying vineyard canopy LAI using an unmanned aerial vehicle (UAV) collected high density structure from motion point cloud." Remote Sensing 5.5:2164-2183, 2013.

[12] Mancini, Francesco, et al. "Using Unmanned Aerial Vehicles (UAV) for high-resolution reconstruction of topography: The structure from motion approach on coastal environments." Remote Sensing 5.12:6880-6898, 2013.

[13] Martin, Ronald A., et al. "Evolutionary View Planning for Optimized UAV Terrain Modeling in a Simulated Environment." Remote Sensing 8.1: 26, 2015.

Target-Free Automatic Registration of Point Clouds

Pileun Kim[a], Yong Kwon Cho[a], and Jingdao Chen[b]

[a]School of Civil and Environmental Engineering, Georgia Institute of Technology
[b]School of Electrical and Computer Engineering, Robotics Institute, Georgia Institute of Technology
E-mail: pkim45@gatech.edu,yong.cho@ce.gatech.edu, jchen490@gatech.edu

Abstract –

The aim of this paper is to introduce a novel method that automatically registers colored 3D point cloud sets without using targets or any other manual alignment processes. For fully automated point cloud registration without targets or landmarks, our approach utilizes feature detection algorithms used in computer vision. A digital camera and a laser scanner is utilized and the sensor data is merged based on a kinematic solution. The proposed approach is to detect and extract common features not directly from a 3D point cloud but from digital images corresponding to the point clouds. The initial alignment is achieved by matching common SURF features from corresponding digital images. Further alignment is obtained using plane segmentation and matching from the 3D point clouds. The test outcomes show promising results in terms of registration accuracy and processing time.

Keywords –

3D LIDAR; Point cloud; Registration; Feature point; RGB image; Segmentation

1 Introduction

Virtual 3D scenario modelling and mapping is critical for construction applications for understanding the scene of interest, monitoring construction progress and detecting safety hazards. The abundant amount of point cloud data can be used to efficiently model a construction site. In most situations, the construction site has to be scanned from many different viewpoints in order to obtain a complete reconstruction of the site. This is due to the fact that some scans may be affected by occlusion and each scan position has its own local coordinate system. Therefore, all the individual point clouds collected in the local coordinate frame must be transformed to a global coordinate system in a procedure known as point cloud registration. However, most of the current methods used for point cloud registration work properly only if there is a good initial alignment or manually marked correspondences and targets [1]. Therefore, an effective point cloud registration method is an indispensable tool to merge several acquired point cloud data from different scanning points in the as-built construction site modelling pipeline. In this study, a target- free automatic point cloud registration framework is introduced.

2 Literature Review

Target-free automatic point cloud registration methods have been widely studied in the literature to overcome the limitations of target-based registration methods. There are three types of point cloud registration methods in existence: 1) ICP based, 2) feature-based, and 3) geo referencing based.

2.1 Iterative Closest Point (ICP) based

The most popular method for point cloud registration is the iterative closest point (ICP) algorithm developed by Besl and McKay [2], Chen and Medioni [3]. In the ICP algorithm, the closest points in two different scans are used as relative control points. Furthermore, the iterative closest compatible point (ICCP) algorithm has been presented to reduce the search space of the ICP algorithm. In the ICCP algorithm, the minimization of distance is accomplished only between the pairs of points considered compatible on their viewpoint invariant properties[4].In addition, Men and Pochiraju [5] integrated Hue values in the registration process to achieve a 4D ICP algorithm. With the Hue values, the ICP algorithm is able to achieve higher accuracy and faster convergence. However, ICP-based registration methods still incur problems with calculation time due to the heavy computation load involved in the ICP algorithm. Also, the performance may not be reliable depending on the overlapping area and the initial starting points [6].

2.2 Feature based

Feature-based registration can be realized without knowing initial starting points since 2D images are used to assist the recognition of feature points. The feature point extraction based method utilized 2D intensity images with Scale-invariant feature transform (SIFT) [7]. However, this method is very sensitive to the overlapping area size. In addition, a large number of scans are necessary to achieve good performance, and the image feature extraction process can be heavily affected by the environment due to illumination changes. Another significant disadvantage of feature-based registration is that a heavy amount of calculation is involved [8]. Even though thousands of feature points can be extracted from each scan based on geometry or image information, most of them are filtered out because of incorrect matches.

2.3 Geo referencing based

Olsen and Johnstone [9] proposed a registration method where the position of each scan spot is derived from GPS. This method is widely used in outdoor studies, but suffers from a lack of accuracy in the registration process because of the low accuracy of GPS. For indoor registration, an automatic construction of 3D basic-semantic models of inhabited interiors was developed using laser scanners with RFID [10]. This method is applicable only for indoor spaces, and the laser scanner is required to be set up in close proximity to the objects in order to recognize the RFID tags. Thus, geo-referencing based registration using sensors such as GPS and RFID is not suitable for all situations because of the limitations in sensor performance [11].

Therefore, there are significant challenges involved in order to achieve a target-free automation point cloud registration method in complex data collection environments in a rapid and accurate manner.

3 Hybrid 3D LIDAR system

In this study, a robotic hybrid Light Detection And Ranging (LiDAR) system was used, consisting of four SICK LMS511 2D line laser scanners (65 meter working ranges at 25Hz scan speed, 200 sec / 360° scan, 190° for vertical line), and a regular DSLR 2D camera, as shown in Figure 1. The resolution of each line laser is 0.1667 degrees in a vertical direction and 0.072 degrees in a horizontal direction. The customized 3D LiDAR system provides more flexibility in hardware control and software programming compared to a commercial LiDAR scanner.

Figure 1 Hybrid 3D LIDAR system

3.1 Data fusion of point clouds and digital images

A digital camera captures RGB texture data from the surroundings, which can be mapped on the 3D point clouds. In the texture mapping process, a camera calibration step is necessary for the digital camera. There are two kinds of matrices for camera calibration. The camera calibration process is finding internal and external parametric matrix for the camera which affect the image processing process. The internal parametric matrix, also known as intrinsic parameters, consist of focal length, image sensor format, and principal point that could be estimated by the pinhole camera model. The intrinsic parameters can be denoted by Equation (1).

$$K_{int} = \begin{bmatrix} f_x & 0 & c_x \\ 0 & f_y & c_y \\ 0 & 0 & 1 \end{bmatrix} \quad (1)$$

The parametersf_x and $f_y$are associated with the focal length whereas the parameters c_x and c_y representthe principal point [12].

The external parametric matrix, also known as extrinsic parameters, denote the coordinate transformation from 3D world coordinates to 3D camera coordinates. This transformation is necessary since the laser scanned 3D point cloud data is obtained in 3D world coordinates. The extrinsic parameters can be obtained through a kinematic relationship based on the mounting configuration.

$$K_{ext} = [R \quad T] = \begin{bmatrix} R_{11}R_{12}R_{13}T_x \\ R_{21}R_{22}R_{23}T_y \\ R_{31}R_{23}R_{33}T_z \end{bmatrix} \tag{2}$$

Using these intrinsic and extrinsic parameters, the laser scanned 3D point cloud can be transformed to 3D camera coordinates according to Equation (3).

$$\begin{bmatrix} u \\ v \\ w \end{bmatrix} = K_{int}K_{ext} = \begin{bmatrix} f_x & 0 & c_x \\ 0 & f_y & c_y \\ 0 & 0 & 1 \end{bmatrix} \begin{bmatrix} R_{11}R_{12}R_{13}T_x \\ R_{21}R_{22}R_{23}T_y \\ R_{31}R_{23}R_{33}T_z \end{bmatrix} \begin{bmatrix} x_s \\ y_s \\ z_s \\ 1 \end{bmatrix} \tag{3}$$

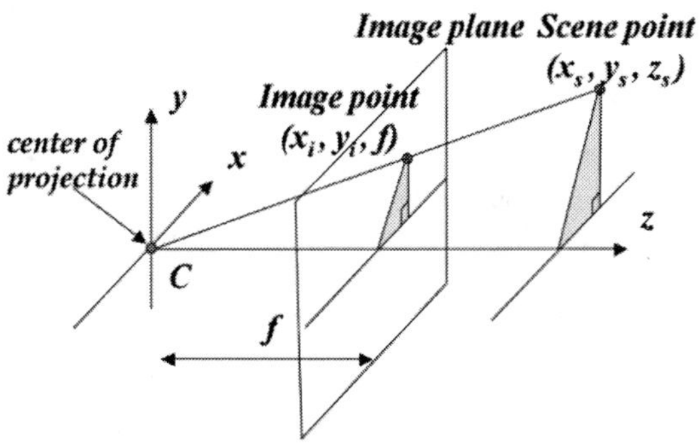

Figure 2 Projection from world coordinate to image plane

Then, the 3D camera coordinates can be transformed to the 2D digital image plane by the Equation (4).

$$\begin{bmatrix} x_i \\ y_i \end{bmatrix} = \begin{bmatrix} \frac{u}{w} \\ \frac{v}{w} \end{bmatrix} \tag{4}$$

Thus, the coordinate systems of 3D point cloud data and RGB image data are aligned using the concept of perspective projection [13]. Figure 2 shows how the point cloud of scene transforms to the image plane. This enables a correct texture mapping between a point cloud and digital camera images.

3.2 RGB feature point based transformation

There are many techniques for detection of points of interests in an image such as Harris corners, SIFT, and SURF. The method of Harris features was proposed by Harris and Stephens in 1988. The notion of corner should only be taken in a general sense as it includes not only corners, but edges and keypoints. The limitation of Harris features is that they are not scale invariant and need to be recomputed for different scales. On the other hand, the Scale Invariant Feature Transform (SIFT) presented by David Lowe detects scale-invariant image feature points, which can easily be matched between images to perform tasks such as object detection and recognition, or to compute geometrical transformations between images. Additionally, the Speeded Up Robust Feature (SURF) is a robust descriptor motivated by the SIFT descriptor, both of them using local gradient histograms. The main difference between the two descriptors is the performance, where SURF decreases the computation time by using integral images for image convolutions and a Hessian matrix-based detector. The standard version of SURF is several times faster than SIFT and is claimed by the authors to be more robust against different image transformations than SIFT [14].

In this study, SURF feature points are used to obtain the initial transformation between point clouds. Once the feature points of each image are extracted, we can track points in the 3D point cloud by matching feature points in the image plane to the RGB-fused point cloud data set.

The Kabsch algorithm (root mean square distance concept) is used to estimate the transformation matrix between point clouds. The algorithm starts with two sets of paired points, P and Q, where each set of points are represented as an N×3 matrix.

$$\mathrm{P} = \begin{bmatrix} x_1 y_1 z_1 \\ x_2 y_2 z_2 \\ \vdots \ \vdots \ \vdots \\ x_n y_n z_n \end{bmatrix} \quad \mathrm{Q} = \begin{bmatrix} x_1' y_1' z_1' \\ x_2' y_2' z_2' \\ \vdots \ \vdots \ \vdots \\ x_n' y_n' z_n' \end{bmatrix}$$

Both sets of points should be translated, so that their centroid corresponds with the origin of the coordinate system. This could be achieved by subtracting from the point coordinates the coordinates of the respective centroid. The next step consists of calculating a covariance matrix A. The optimal rotation matrix U can be computed by singular value decomposition (SVD) of the covariance matrix A.

$$\mathrm{A} = \mathrm{P}^{\mathrm{T}}Q = \mathrm{VSW}^{\mathrm{T}} \tag{5}$$

Next, we decide whether we need to correct our rotation matrix in order to ensure a right-handed coordinate system using the variable *d*.

$$\mathrm{d} = \mathrm{sign}(\det(\mathrm{WV}^{\mathrm{T}})) \tag{6}$$

Finally, we can calculate our optimal rotation matrix, U, as

$$U = W\begin{bmatrix}1 & 0 & 0\\0 & 1 & 0\\0 & 0 & d\end{bmatrix}V^T \quad (7)$$

and the translation matrix D can be obtained by the difference between the centroid of each point cloud [15].

To match each point cloud set, the initial rigid transformation matrix is defined. In this case, the transformation is a perspective projection for 6 degrees of freedom, composed of a rotation matrix and a translation vector in 3 dimensions. This transformation can be written as 3x4 matrix. We can then project a point P where $P = [x\ y\ z\ 1]^T$ simply by applying this transformation matrix to the point:

$$P' = TP = \begin{bmatrix}U_{11}U_{12}U_{13}D_x\\U_{21}U_{22}U_{23}D_y\\U_{31}U_{23}U_{33}D_z\end{bmatrix}\begin{bmatrix}x\\y\\z\\1\end{bmatrix} = \begin{bmatrix}x'\\y'\\z'\end{bmatrix} \quad (8)$$

The summary of finding an initial transformation using RGB feature points is as follows:

1. SURF features are extracted from RGB panorama images using only 2D information
2. Identify coordinates in the image frame for each extracted feature
3. Track the feature points on the corresponding 3D point cloud with RGB-fused point cloud data
4. Compute an initial transformation matrix using Kabsch algorithm
5. Transform a point cloud to the reference point cloud

3.3 Point cloud registration using plane matching and corner points

Two point clouds can be further registered using the method of plane-to-plane matching. This method relies on finding three plane correspondences between the point cloud to be registered and the reference point cloud. The selected planes have to be linearly independent and intersect at a unique point in order to fully recover the transformation parameters. For example, one of the planes can be the ground plane whereas the second plane is a vertical wall in the x-axis whereas the third plane is a vertical wall in the y-axis. First, the Random Sample Consensus (RANSAC) algorithm is used to perform plane segmentation for each point cloud. The RANSAC algorithm works by iteratively sampling points from a given point cloud and estimating a set of plane parameters of the form $ax + by + cz + d = 0$. The best estimate is determined as the set of plane parameters that maximizes the number of points that are considered inliers [16]. The obtained plane parameters are used to segment the original point cloud into points belonging to the plane and the remaining points. This procedure is repeated until three suitable plane candidates are found that satisfy the linear independence criteria. Then, plane correspondences between the input point cloud and the reference point cloud are determined by finding the closest match between normal vectors. Second, the rotation component R of the transformation matrix is calculated using the plane normal vectors found in the previous step. The rotation component is determined such that the normal vectors (n_1, n_2, n_3)in the input point cloud will be transformedin order to match the normal vectors (n_1, n_2, n_3)in the reference point cloud [17]. An intermediate rotation matrix that rotates a vector v_1 to another vector v_2 is derived using Equation (9).

$$R = I + skew(v_1 \times v_2) + skew(v_1 \times v_2)^2 \frac{1 - v_1 \cdot v_2}{\|v_1 \times v_2\|} \quad (9)$$

Three intermediate rotation matrices are calculated for each plane correspondence as shown in Equation 2. The final rotation matrix is then obtained by multiplying the intermediate rotation matrices together.

$$R_1 = getRotation(n_1, n_1')$$
$$R_2 = getRotation(R_1, n_2, n_2')$$
$$R_3 = getRotation(R_2R_1, n_3, n_3')$$
$$R = R_3R_2R_1$$

Third, the translation component T of the transformation matrix is calculated by comparing corner points between the point cloud to be registered and the reference point cloud. A corner point is defined as the unique intersection point between three planes. The corner point can be calculated by solving three plane equations simultaneously for the (x, y, z) values. This in turn can be formulated as a matrix-vector multiplication operation using the corresponding plane parameters. Once the corner point is obtained for each point cloud, the translation vector is determined as the difference between the positions of the two corner points. The calculations involved in this step are shown in detail in Equation (10).

$$\begin{bmatrix}x\\y\\z\end{bmatrix} = \begin{bmatrix}a_1 & b_1 & c_1\\a_2 & b_2 & c_2\\a_3 & b_3 & c_3\end{bmatrix}^{-1}\begin{bmatrix}-d_1\\-d_2\\-d_3\end{bmatrix}$$
$$\begin{bmatrix}x'\\y'\\z'\end{bmatrix} = \begin{bmatrix}a_1' & b_1' & c_1'\\a_2' & b_2' & c_2'\\a_3' & b_3' & c_3'\end{bmatrix}^{-1}\begin{bmatrix}-d_1'\\-d_2'\\-d_3'\end{bmatrix} \quad (10)$$
$$T = \begin{bmatrix}x\\y\\z\end{bmatrix} - \begin{bmatrix}x'\\y'\\z'\end{bmatrix}$$

$$T = \begin{bmatrix} x \\ y \\ z \end{bmatrix} - \begin{bmatrix} x' \\ y' \\ z' \end{bmatrix}$$

Finally, a registered version of the input point cloud is obtained by applying the rotation and translation operations to each point in the point cloud.

4 Test

The data acquisition process for validating the proposed framework is performed at the Bunger-Henry building in Georgia Tech.Figures3 and 4 show the two sets of RGB-fused point clouds scanned at different scan positions with the proposed texture mapping algorithm.

Figure 3 3D RGB-fused point cloud at first scan position

Figure 4 3D RGB-fused point cloud at second scan position

Figure 5 Original point clouds from different scan position

Figure 5 visualizes the two raw point cloud data set from the different scan positions. The two scan data collected as viewed are in the local coordinate system so the origin is the same. In this experiment, the first scan data set will be transformed and the second scan data set will be used as the reference data set. Each scan data collection occurred in an indoor environment where there are fewer visual features. Figure 6 shows the RGB feature extraction results using SURF features for the collected panoramic images. However, not all the feature points are correctly matched between the two panoramic views due to similar feature points in both images. For this reason, the final plane-to-plane matching step is necessary to achieve the desired accuracy for the presented framework.

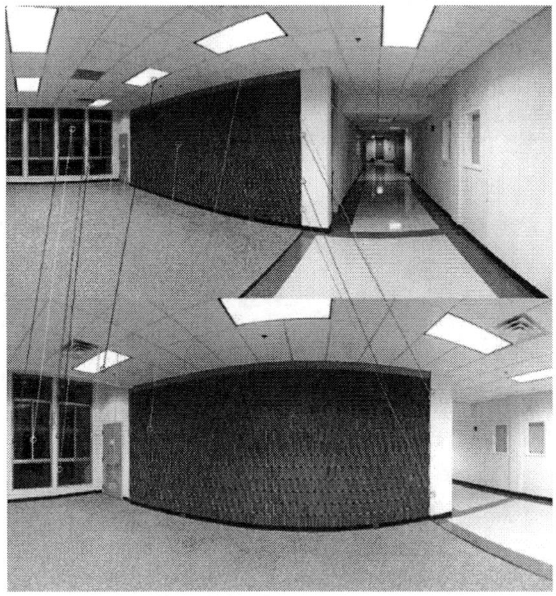

Figure 6 Feature point extraction for panoramic images

Figure 7 Transformed and reference point cloud after feature point matching step

Figure 7 visualizes the point cloud sets after the initial

transformation using RGB feature extraction. As shown in this figure, the point cloud sets are better aligned compared to the raw data set in Figure 5. It can be verified from the reduced deviation angle in Table 1. Then, we can apply the third step which is the plane-to-plane matching algorithm. The initial alignment is sufficient that we can easily find plane correspondences in the final step. Finally, Figure 8 represents the final result for the proposed framework. To verify the result for this test, the second point cloud set is assumed as a true ground and measured the deviation angle from each reference axis at each step of the proposed framework.

Table 1 Deviation angle

Dev. angle measured from	Original point clouds	After initial transform	After final transform
X axis	-17.404	-7.188	0.549
Y axis	-3.155	-1.702	0.342
Z axis	29.899	9.430	-0.811

Figure 8 Transformed and reference point cloud after plane-to-plane matching step

5 Conclusion

In conclusion, a novel method for target-free automatic point cloud registration was demonstrated and validated. A laser scanning system with a digital camera was used to obtain point clouds with mapped RGB texture data. The proposed framework consists of three steps. The Figure 9 indicates the flow diagram for the presented framework. The first step involves constructing RGB-fused point clouds with collected (x, y, z) position data from laser scans and RGB information from digital images. The next step involves obtaining an initial transformation matrix by extracting common SURF feature points in digital images and tracking its corresponding (x, y, z) position. Lastly, plane-to-plane matching algorithm is utilized for accurate registration.

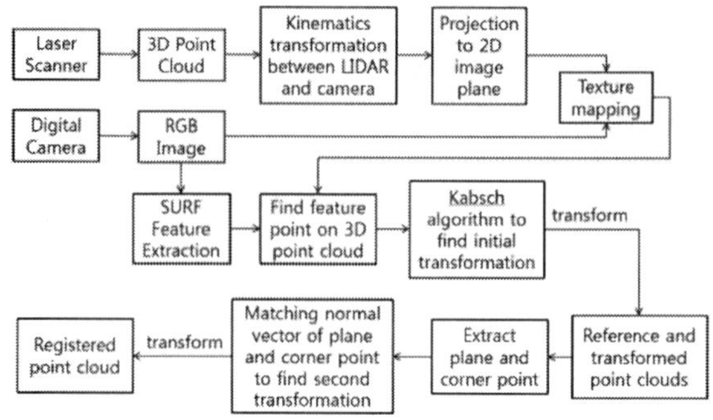

Figure 9 Procedure of proposed framework

Even though the proposed framework should have three plane with one corner point on overlapped area, it achieved automatic point cloud registration without the need for target references and manual adjustments. This framework consist of two sequential methods: RGB feature matching and plane-to-plane matching. The plane-to-plane matching algorithm only works fine, when the RGB feature matching method get a good results. For future work, we focus on how to get more good result for RGB feature point matching and how to apply this framework for more complex and large areas with scans from multiple viewpoints.

6 Acknowledgement

This material is based upon work supported by the National Science Foundation (Award #: CMMI-1358176). Any opinions, findings, and conclusions or recommendations expressed on this material are those of the authors and do not necessarily reflect the views of the NSF. This research was also supported by the Chung-Ang University Research Grants in 2015.

References

[1] H. Xiong, S. Szedmak, and J. Piater. A study of point cloud registration with probability product kernel functions. *In Proc. 2013 Int. Conf. 3D Vision, IEEE*, pp. 207–214, Seattle, USA, 2013.

[2] P.J. Besl, N.D. McKay, A method for registration of 3-d shapes, *IEEE Trans. Pattern Anal. Mach. Intell.* 14 (2):239–256, 1992.

[3] Chen, Y., Medioni, G. Object modelling by registration of multiple range images. *Image Vision Comput.*, 10(3): 145–155, 1992.

[4] Akca, D., Full automatic registration of laser scanner point clouds. *Optical 3-D Measurements VI*, pp. 330–337, 2003.

[5] Men, H.,Gebre, B., Pochiraju, K. Color Point Cloud

Registration with 4D ICP Algorithm. *Int. Conf. Robotics and Automation, IEEE*, pp. 1511-1516, Shanghai, China, 2011.

[6] Wang, C., Cho, Y., and Park, J. Performance Tests for Automatic Geometric Data Registration Technique for Construction Progress Monitoring. *Int. Conf. Computing in Civil and Building Engineering (ICCCBE), ASCE*, pp. 1053-106, Orlando, FL. 2014.

[7] Eo, Y. D., Pyeon, M. W., Kim, S. W., Kim, J. R., Han, D. Y. Coregistration of terrestrial lidar points by adaptive scale-invariant feature transformation with constrained geometry. *Automation in Construction*, 25, 49-58, 2012.

[8] Gai, M., Cho, Y., and Qinghua, X. Target-free automatic point clouds registration using 2D images. *Proceedings of the 2013 ASCE International Workshop on Computing in Civil Engineering*, pp.865-872, Univ. of Southern California, LA, USA. 2013.

[9] Olsen, M. J., Johnstone, E., Kuester, F., Driscoll, N., and Ashford, S. A. New Automated Point-Cloud Alignment for Ground-Based Light Detection and Ranging Data of Long Coastal Sections. *Journal of Surveying Engineering, ASCE*, 137(1), 14-25, 2011.

[10] Valero, E., Adan, A., and Cerrada, C. Automatic Construction of 3D Basic-Semantic Models of Inhabited Interiors Using Laser Scanners and RFID Sensors. *Sensors*, 12(5), 5705-5724, 2012.

[11] Mastin, A., Kepner, J. and Fisher, J. Automatic registration of lidar and optical images of urban scenes. *IEEE Conference on Computer Vision and Pattern Recognition*, pp. 2639–2646, 2009

[12] R. Y. Tsai. A Versatile Camera Calibration Technique for 3D Machine Vision. *IEEE J. Robotics and Automation*.3(4), pp. 323-344, 1987.

[13] O. Faugeras. *Three-Dimensional Computer Vision: A Geometric Approach*. MIT Press, pp. 33-68, 1996

[14] R. Hartley and A. Zisserman. *Multiple View Geometry in Computer Vision*. Cambridge University Press, pp. 138-183, 2000.

[15] Wikipedia. https://en.wikipedia.org/wiki/Kabsch_ algorithm, Accessed: 02/03/2016

[16] R. B. Rusu and S. Cousins, 3D is here: Point Cloud Library (PCL), *IEEE Int. Conf. Robotics and Automation (ICRA)*, 2011.

[17] Mathematics Stack Exchange. http://math. stackexchange.com/questions/180418/calculate-rotation-matrix-to-align-vector-a-to-vector-b-in-3d, Accessed: 05/03/2016

Resource Allocation Methodology of Virtual Desktop Infrastructure in Cloud Computing BIM: Focusing Korean Small and Medium Sized Architectural Design Firm

K.H. Lee [a], S.W. Kwon [b], J.H. Shin [c], D.H. Kim [d], D.Y. Moon [e], S.Y. Lee [f]

[a] Department of Convergence Engineering for Future city of Sungkyunkwan Universicty, Republic of Korea
[b] School of Civil & Architectural Engineering, Sungkyunkwan University, Republic of Korea
[c] Department of Convergence Engineering for Future city of Sungkyunkwan Universicty, Republic of Korea
[d] Department of Convergence Engineering for Future city of Sungkyunkwan Universicty, Republic of Korea
[e] Department of Convergence Engineering for Future city of Sungkyunkwan Universicty, Republic of Korea
[f] Department of Convergence Engineering for Future city of Sungkyunkwan Universicty, Republic of Korea
E-mail: kyup@skku.edu, swkwon@skku.edu, kiwijoa@skku.edu, ih5318@skku.edu, yoshy17@nate.com, 2_songyi@naver.com

Abstract

BIM (Building Information Modelling) technology has been expanded for using in construction project during the past 20 years. However, recently its spread has been delayed than the initial expectations, due to the high-cost of BIM infrastructure development, the lack of regulations and process.

In design phase, especially, collaboration based on BIM system has being a key factor for successful next generation building project. In the many analysis of current research about virtualization and BIM service, convergence of cloud computing and virtual server system can be proposed as a most successful solution for data exchange such as drawing, 3D model, object data, properties,.

Therefore, this study has a purpose DaaS(desktop as a service) for BIM user design phase by phase. By analysis user's desktop resource data from Pre-design phase to Schematic design, Design develop and Construction design phase, we defined criteria to resource distribution on virtual desktop environment.

The goal of the study is to enable the BIM customized cloud computing server. It will bring cost saving, high-performance quality of working environment and cooperation's convenience and high security when doing BIM work in small and medium sized architectural design firm.

Keywords –

VDI (Virtual Desktop Infrastructure); Cloud Computing; BIM; Design Collaboration; Virtualization

1 Introduction

The introduction of building information modelling (BIM) in construction has changed many existing processes in the industry. Communication tools are now available that allow designers and constructors to discuss and interact in the construction environment. This process can reduce the errors and changes associated with construction. Meanwhile, it supports the construction process and reduces costs while minimizing legal conflicts. Ultimately, it facilitates the successful implementation of entire projects.

However, the introduction of BIM at construction sites also causes many problems due to the diversity of environments. Table 1illustrates the types of problems resulting from the BIM introduction process.

Of these problems, this study aims to focus on the systematic shortcomings of BIM-based cooperation. Previous studies related to BIM cooperation include "The Difficulty of Utilizing BIM in a Large-Scale Project (Chi Joo Lee 2009), "The Inadequate Configuration of BIM-Based Cooperation Systems (Jung Wook Park 2009), and "The Cost of Software/Hardware Burdens for the Use of BIM (Smart Market Report 2008), to name a few. Summarizing the aforementioned problems, the current BIM-based cooperation system has important problems resulting from the adequate sharing configuration system. Thus, recent studies have emphasized the importance of research into relevant areas.

Table 1 Diagram of Research Methodology

Problems when Introducing BIM
Lack of standards and guidelines 3.5cm
Compatibility issues with BIM software
BIM process fixation problems 2.5cm
BIM tools utilizing professional shortage
The initial cost of risk when introducing BIM

This paper aims to conduct a study on the design of a virtual desktop infrastructure (VDI) system model for a cloud computing BIM service. This research seeks to make improvements in the aforementioned problems associated with the cooperation system.

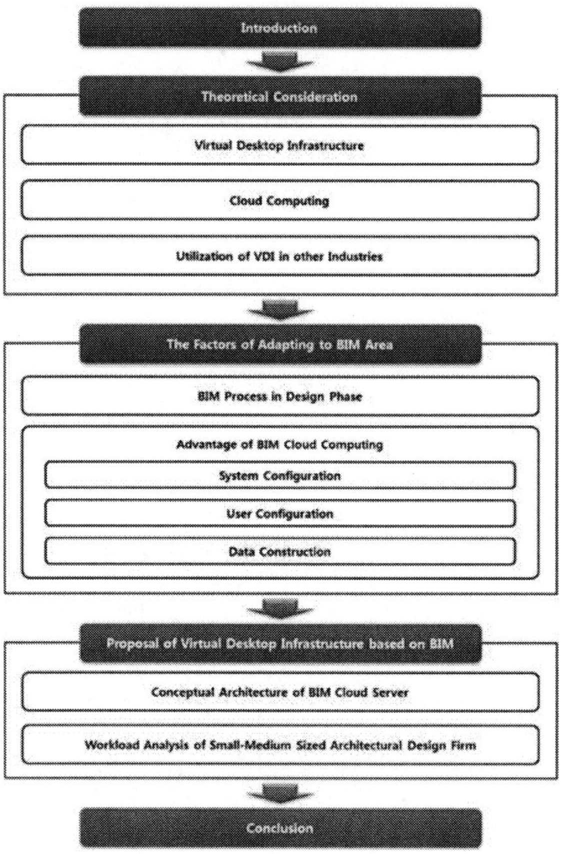

Figure 1 Diagram of Research Methodology

Figure 1 illustrates the procedures performed in this study. First, we investigated small- and medium-sized design firms in the Republic of Korea to perform a theoretical analysis of BIM processes. Then, we prepared a checklist for applying BIM to the context of this study.

Third, this study proposes a virtual desktop infrastructure based on BIM while giving appropriate authoring usage/authorization allowances to users.

2 Theoretical Considerations

2.1 Virtual Desktop Infrastructure (VDI)

VDI services generate virtual machines (virtual PCs) to be operated in a server's cloud data. Moreover, these services allow users to utilize desktop environments anytime and anywhere through a device connected to a network. In the field of VDI, this process is known as "virtualization."

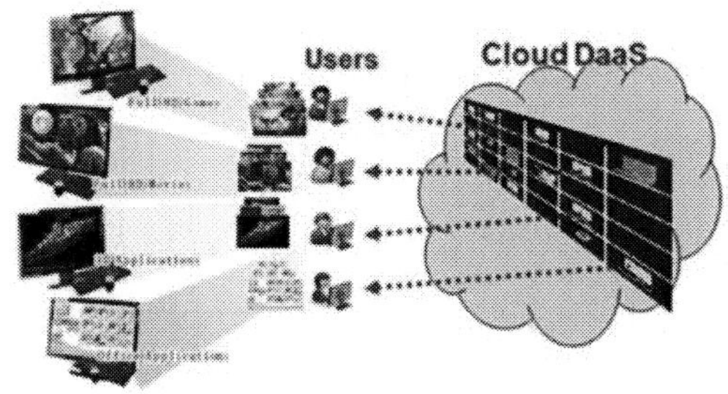

Figure 2 Desktop as a Service Solution

First introduced in 2008, VDI technology has been growing rapidly with virtualization technology, cloud computing, and increased use of mobile devices in the corporate, financial, telecommunication, and public institute computer markets. Currently, the most prominent VDI companies are Citrix and VMware. Approximately 20 VDI companies have released relevant products, including Microsoft, which owns Windows OS, Quest, and Virtual Bridges.

2.2 Cloud Computing

To utilize VDI effectively, one needs a cloud computing service model suited to a given usage. Such services allow users to use a service by accessing the shared pool of computer resources that control an environment. DaaS- (Desktop as a Service) based cloud computing services can be provided as shown in Figure 3 below.

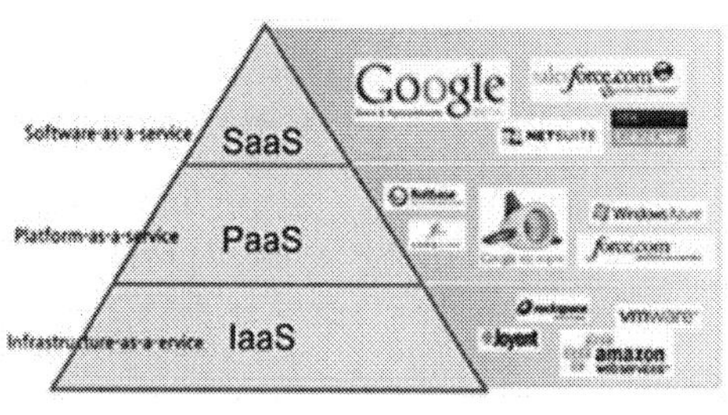

Figure 3 Cloud Service

SaaS (Software as a Service) provides an environment that can develop, distribute, and manage PaaS (Platform as a Service) applications or cloud applications. Prominent companies in this category include Salesforce.com, Google, and Facebook.

IaaS (Infrastructure as a Service) offers an infrastructure environment, such as servers and storage. The most prominent giant in the IaaS field is Amazon, the first company that embarked on the cloud business journey in earnest. Based on this early entry, Amazon has the highest degree of infrastructure reputation and competitiveness in the industry.

2.3 Utilization of VDI in other Industries

Table 2 Application and Utilization in other Industries

Company	Characteristic
1. The Catholic University of Korea, Yeouido ST. Mary' s Hospital	Provides easy solution for building a cloud computing system
	Improves care spaces
2. Seoul National University Hospital	Seeks to minimize power consumption in the 24-hour hospital
	Uses health information systems while moving toward a virtual solution
3. GangNam Severance Hospital	Allows full PC availability, regardless of the device type.
	Provides real-time, delivered health information
4. Kindred Healthcare	Improves stability of the desktop in the data
	Employs a low-cost strategy, implemented using a thin client
5. Bell	Manages virtual desktops easily, more than 4000 VM
	Includes a built-in disaster recovery solution
6. IRMC (Intell Risk Management Corporation)	Saves time and operates hardware
	Provides the additional time-saving of a new PC
7. Samsung Advanced Institute of Technology	Prevents important data leakage (CAD design drawings, etc.)
	Supports tablet devices, such as mobile work
8. Hyundai Heavy Industries	Allows faster speeds (text, Internet, etc.)
	Provides comfortable office space
9. LG CNS	Includes anytime, anywhere high-level security
	Provides integrated services to the storage network
10. SK Telecom	Gives stable resource management
	Provides an all-in-one thin client
11. Korea Workers' Compensation & Welfare Service	Prevents information leaks and protects against cyber terrorism
	Improves speed and usability
12. Court of Korea	Provides the same working environment at all locations
	Offers smart work by VDI
13. Woong-Jin	Enhances eco-friendly corporate image
	Improves the work environment (dust, heat, reduced occurrence of noise)
14. Allianz	Increases convenience for IT management staff
	Switches the old PC into VDI
15. MPC Marketing Technology & Services	Provides WAN, LAN environment support for outsourcing services
	Supports mobile environments
16. University of Incheon	Complements disadvantages of PC (fan noise and heat generated by the PC, complex PC Accessories)
	Provides smart classrooms throughout the VDI
17. National Health Insurance Service	Reduces costs by 83% through zero client installations
18. Korea Institute for Curriculum and Evaluation	Reflects success through the national 2012 proficiency test in English
19. Korean Intellectual Property Office	Secures unpublished patent data
20. Multifunctional Administrative City Construction Agency	Allows data sharing and interactive communication between teachers and students
21. MIRAE Asset	Applies virtualized resources as differential user's tasks and acts

Table 2 illustrates how other industries have used and applied VDI. A total of 21 businesses were surveyed regarding how they introduced and utilized virtual computing. There were 85 application items that we classified into 16 larger units. Classifying VDI into 16 larger units, it became that security was most frequently utilized in other industries, followed by maintenance, efficiency, and mobility. According to previous analyses, other industries actively use security, maintenance, VDI efficiency, and mobile items, verifying its applicability.

The architectural field also takes seriously security, maintenance, efficiency, and mobile items. If such items could properly align with features of the architectural industry, the possibility of success would increase. Therefore, further research into utilization cases in other industries is necessary.

3 The Factors of Adapting to the BIM Area

The introduction of BIM in the field of construction has significantly changed many areas of construction. CAD (computer-aided design) work, which implemented BIM, has resulted in work-time reductions, reutilization of extensive design data, efficient workforce placement, improved design quality, and greater economic profits. It has also brought about significant changes across the construction industry.

Figure 4 Processing Time for Standard PC

As for small-sized buildings, a variety of variables influence them, leading to many difficulties in managing construction sites. However, the introduction of BIM in the remodelling sites of small-sized buildings has resulted in improved quality for buildings. BIM also makes construction more economical by eliminating factors of construction delay, such as design changes and reconstruction.

In particular, designing atypical structures with conventional 2D design methods has caused several floor plan errors because such structures contain many areas that cannot be represented with 2D drawing. However, 3D-based BIM locates inconsistencies between structural members; thus, it allows a smaller number of design errors requiring additional review compared to 2D drawing. As a result, 3D-based BIM is superior in terms of construction costs, responsibility issues, and maintenance and management concerns, all of which are important in the construction process of an atypical structure.

3.1 The BIM Process in the Design Phase

This study limits its scope to the utilization of BIM in the design phase. Therefore, it aims to present clear criterion regarding the design area for construction projects prior to the main text.

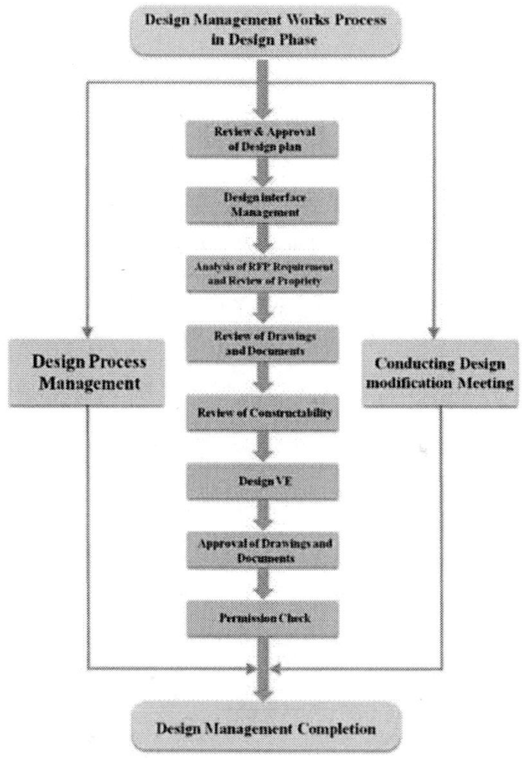

Figure 5 Activity of Design Management

Figure 5 illustrates the specific activity of design management. In general, the design area for construction projects is classified into preliminary or implementation design. The preliminary design phase involves deciding whether to reflect the client's requirements in the project direction. This phase also includes conducting a broad review as to the implementation schedule for design projects and construction productivity (costs, constructability, quality, etc.). The implementation design phase specifies actual work and involves conducting a review on design outcomes. Ultimately, this phase decides whether actual construction can commence. Along with a contractor, this phase provides various support for the design VE task, construction schedule management, and licensing procedures.

3.2 Advantages of BIM Cloud Computing

In general, the reason for introducing VDI is its ability to build an environment allowing one to process tasks anywhere as though using one's own PC.

Users are able to process tasks by accessing personal virtual desktop environments at other departments or sites. They are able to work flexibly by conducting VDI with their PC.

The system considerations for configuring VDI can be classified into the following three factors: system, user, and data configuration.

3.2.1 System Configuration

A construction project typically uses a great deal of software. When utilizing a variety of software, VDI is capable of conducting a batch management function through central management. It can proceed with batch upgrades by using the software distribution function when conducting a security patch in the user's window. This function is based on the software installed in VM (virtual machine), collecting hardware and software inventories. It can be applied equally to the upgrade of office software installed in VM. Even users who do not use high-end programs are able to manage computing resources and conduct efficient application installation and distribution.

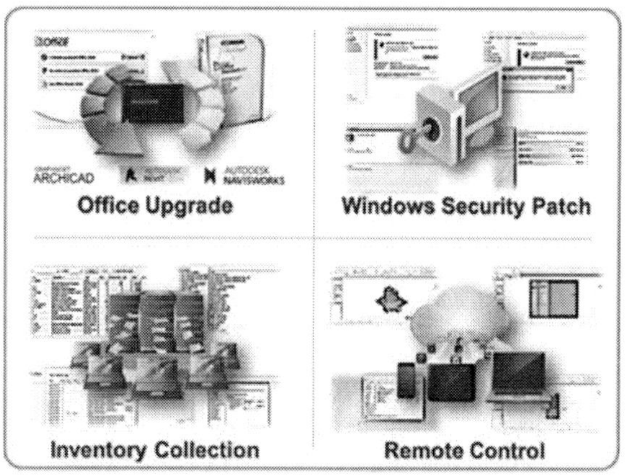

Figure 6 VDI-Based Integrated Management Systems

3.2.2 User Configuration

A variety of users exist in a VDI-based BIM project suitable for construction sites. They can be subdivided into three broad user types. The first type involves a practitioner who performs high-performing modelling tasks. The second type is a cloud access user who performs relatively less-demanding tasks, such as presentation and mark-up. The third type is a user who accesses the private cloud in order to upload simple work journals.

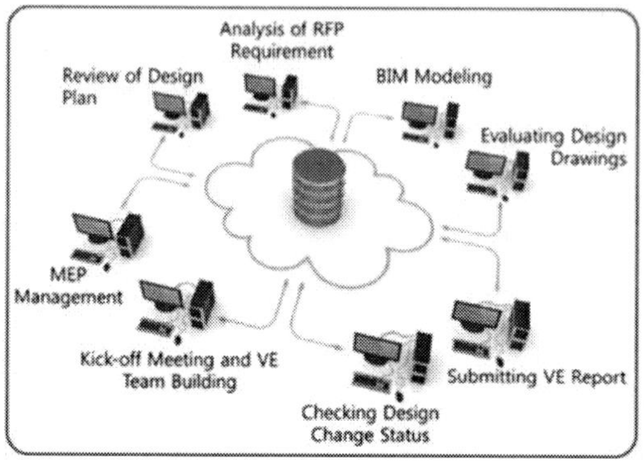

Figure 7 Configuration of VDI User

It is possible to improve work processing capability based upon these three types of practitioners. To do so, one must identify their work environment patterns and distinguish their authorities in each construction environment. In this way, one may also distribute resources more efficiently and design virtual models in advance.

3.2.3 Data Construction

Given the nature of construction projects, there are many occasions to work jointly with partners, involving tow data types: data that needs to be shared together and data that requires security management.

It is possible to configure in such a way that an external user can access a corresponding project server only via virtual desktop after accessing the VPN and desktop when an external user accesses a corresponding project server. In this case, an external user can conduct a task only after accessing a corresponding server via the network, whereas cannot retrieve data from outside via virtual desktop.

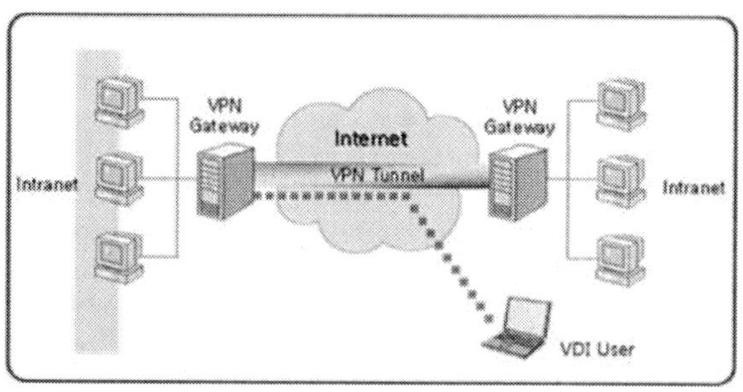

Figure 8 VDI-Based Data Security Design

3.3 Efficient Scale Computation for BIM Cloud Computing

Table 3 Optimal Scale of Server Estimation

Methodology	Characteristic	
Calculating Method	Concept	A method of calculating numerical scale and applying correction values based on elements for calculating scale, including number of users.
	Advantage	Ground of scale estimation that can be clearly presented and it can be estimated more simply than other methods.
	Disadvantage	If the correction value is wrong, the big difference with the desired value has taken place and it is hard to present accurate back-up data for the correction value.
Referencing Method	Concept	See similar scale by comparing rough system scale based on basic data depending on business volume (number of users, DB size).
	Advantage	Relatively safe scale that can be estimated as a comparison with existing established systems.
	Disadvantage	Presented ground (back-up data) is not clear as scale is estimated by comparison, not by calculation.
Simulation Technique	Concept	Scale is estimated by modelling work load for target business and simulating it.
	Advantage	Relatively accurate values can be obtained.
	Disadvantage	Great deals of time money are required.

There are three methods for sizing the cloud server.

First, one can calculate the numerical scale and apply the correction value based on elements for calculating scale including a number of users.

Second, the referencing method involves referring to similar scales by depending on business volume (number of users, DB size). Third, in the simulation technique, scale is estimated by modelling and simulating the work load of the target business. In this study, we focused on the referencing method and calculating method in order to build up the BIM cloud system.

4 Proposed VDI System Based on

The BIM-based information exchange framework proposed in this study is built on the basis of a system model ideal for a VDI supporting both the cloud system and virtual environment simultaneously. Therefore, the proposed BIM service resource allocation methodology for the virtual desktop environment can be described as follows.

The resource allocation methodology for the virtual desktop infrastructure should integrate BIM with the conventional cloud. In so doing, one is able to minimize excessive network construction costs by reflecting the latest cloud technologies and construction tasks while also improving speed and work efficiency through information exchange.

4.1 Conceptual Architecture of BIM Cloud Server

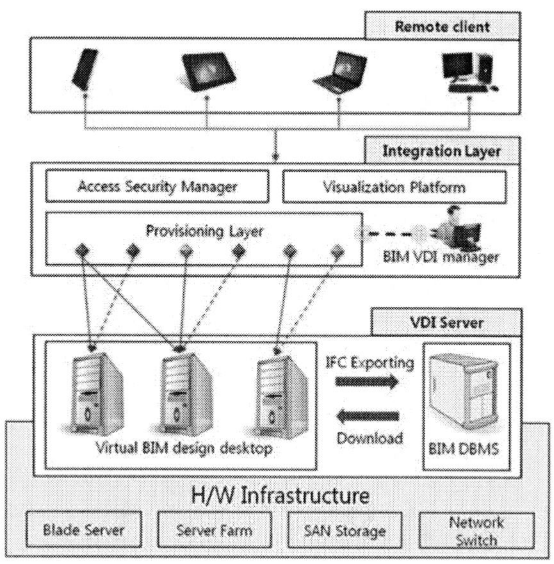

Figure 9 Integrated Management Systems Based on VDI

Figure 9 illustrates the operation method of the VDI-based BIM server. Each user of the design team remotely accesses the virtual BIM work environment through their own IP based on their access permission received through the Xen Hypervisor and Active Directory. The BIM VDI manager is deployed to play the role of resource allocation to each user, network protocol selection, and user grouping through the Xen Studio of Master OS. The BIM model management module or 3D viewer can be connected through the BIM DBMS in the VDI workstation or on an external cloud network. The optimized environment for users can be constructed by installing additional APIs in the Master besides the basic tools. The setup software in the Master instantly provides updated software applications to individual server users through the Virtual Machine distribution settings. Furthermore, the work environment can be integrated through the Snapshot setting (server environment can be restored to a certain point in time).

4.2 Workload Analysis of Small-Medium Sized Architectural Design Firm

A VM performance test must be conducted to prevent any dissatisfaction of VDI users due to delay, and the resource that is inputted to the BIM design organization must be determined through this test.

Table 4 Processing Time by Standard PC

	Revit	IFC viewer	A BIM viewer
Connection Image		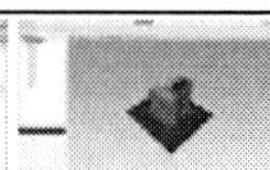	
Access Time	00 :00 :24.72	00 :00 :19.84	00 :00 :22.48
Loading Time	00 :02 :46.30	00 :00 :07.46	00 :00 :38.72
Control Speed	00 :00 :04.37	-	-
Rendering Time	00 :02 :23.30	-	-

The standards of the server environment must be defined to calculate the size of resources. The existing user server environment and VDI server environment were set as in the Value (balanced price and performance) stage to satisfy the general requirements of the BIM tool environment (the recommended system specifications of Autodesk Revit).

Table 5 Processing Time by Virtual Machine

	Revit	IFC viewer	A BIM viewer
Connection Image		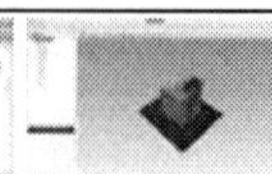	
Access Time	00 :00 :15.72	00 :00 :13.59	00 :00 :18.67
Loading Time	00 :02 :54.48	00 :00 :08.04	00 :00 :42.45
Control Speed	00 :00 :04.48	-	-
Rendering Time	00 :02 :18.30	-	-

The access, loading, and operation speed of Revit and the software applications of two BIM viewer producers were measured based on the test scenario with existing users and VDI users. The measured time values are shown in Table 5 the VDI performance was more stable than the performance of existing users.

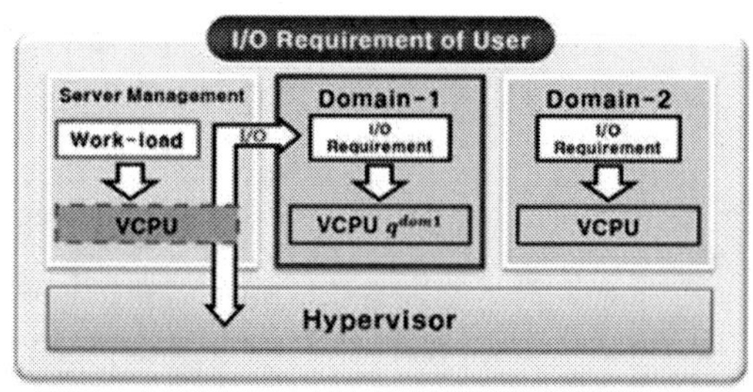

Figure 10 Data Processor of Input-Output Control System

Table 6 Resource Allocation Design Based on VDI

Variable Name	Explain	Variable Name	Explain
i	Request Time	q	Server request Value
j	Request Desktop No.	$\bar{q}$	Average Duration
$\emptyset$	Desktop Performance	C	Total Resource
ρ	Performance Ratio of Desktop	α	Premium Resource

$$q_i = \emptyset 1 \rho i1 + \emptyset 2 \rho i2 + \emptyset 3 \rho i3 + \cdots \quad (1)$$

$$\bar{q} = \frac{1}{n}\sum_{i=0}^{n} q\, i$$

$$C = \bar{q} + \alpha \quad (2)$$

$$\alpha = 20 \sim 30\% \quad (3)$$

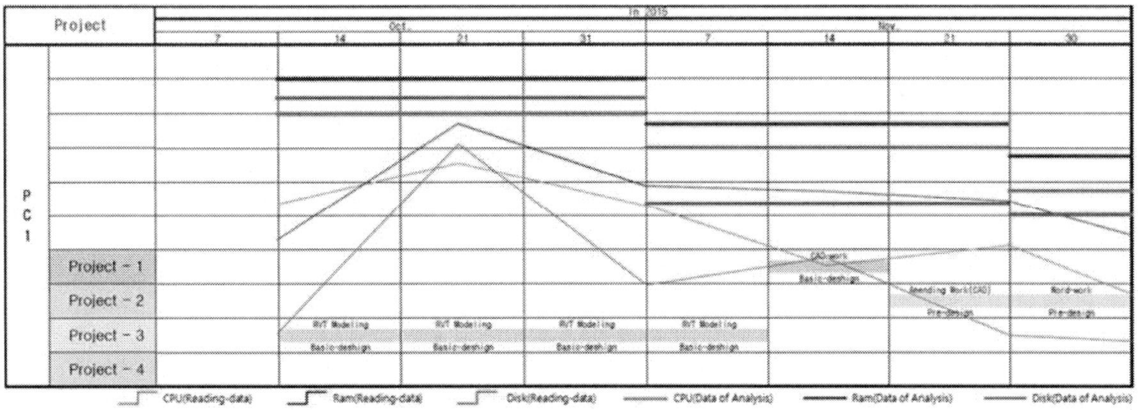

Figure 11 Resource Data in the BIM Environment

Figure 10 shows the work-load request of users in the cloud computing environment according to their resource usage. The average work-load was determined by referring to these work-load data (1), and the total resource was calculated by adding the alpha value to the average value (2). For the alpha value, the allowance rate was determined (3) through an existing server size calculation study (Na 2010). Figure 11 shows the sample data calculated through the calculation method (2). In the next study, we will present the optimum resource data for BIM workers and suggest a method of introducing the data to a VDI system.

5 Conclusion

This paper proposed a BIM cloud system methodology for BIM users. The system proposed in this study would improve many areas of BIM by leveraging VDI in the BIM construction environment process.

However, there may be many adverse effects of this framework if leveraging is performed recklessly in the conventional IT environment and in the general work environment. To address these adverse effects, a thorough investigation must be conducted in advance to ensure the systems are suitable for the construction environment. Moreover, it will be imperative to identify the tasks and environment of the system's users, to consider the characteristics of each construction project, and to conduct a follow-up study in the mid and long term.

As for the results of this study, it is still necessary to regularly conduct development and research regarding BIM data lightening, shape/attribute data parsing technology, and resource allocation for the implementation of the framework, as shown in Figure 11.

Acknowledgement

This research was supported by a grant (13AUDP-C067809-01) from Architecture & Urban Development Research Program funded by Ministry of Land, Infrastructure and Transport of Korean government.

This work is financially supported by Korea Ministry of Land, Infrastructure and Transport(MOLIT) as 「U-City Master and Doctor Course Grant Program」

References

[1] Jung, T.H., Kim, S.W., "The construction management of BIM-based small building remodelling." *Architectural Institute of Korea*, volume 12(1): 191-192, 2012

[2] Lee, S.Y., Ham N.H., "A Study of BIM-based Construction Management Work Process Development in Design Phase." *Korean journal of construction engineering and management*, volume 14(6): 90-101, 2013

[3] David Linthicum "Complex clouds must be managed with automation." *Infoworld Article*, March 2013.

[4] Yoo, S.Y., "Implementation Issues of Virtual Desktop Infrastructure and Its Case Study for a Physician's Round at Seoul National University Bundang Hospital" *The Korean Society of Medical Informatics*, volume 18(4): 259-265, 2012

[5] Tien, D.N., "Energy Management for Mobile Virtual Desktop Infrastructure in Mobile Cloud Environment" *KIISE Transactions on Computing Practices*, volume 20(8): 477-481, 2014

[6] Dillon, T., Wu, C., Chang, E., "Cloud Computing: Issues and Challenges" *International Conference on Advanced Information Networking and Applications*, 27-33, 2010

Using Simulation Applications for Sustainable Design and Construction

Hoda Khoshnevis, Nima Khairdoost

Tabaran Institute of Higher Education, Mashhad, Iran
{khoshnevis, khairdoost}@tabaran.ac.ir

Abstract–

This paper briefly illustrates the application of Building Energy Consumption Simulation (BECS) for sustainable construction education. The curricula of building-related majors have not greatly taken advantage of simulated environments as supplementary tools besides traditional methods. The main purpose of the research project was to determine how a simulation application can be used in a construction curriculum. In order to apply energy consumption optimization in a construction project, there are various suggestions and guidelines that should be carefully incorporated in design and construction phases. BECS is a simulation application that navigates construction students through a set of pre-defined stages and provides necessary information accordingly. BECS puts students in a role of a designer and requires them first to specify the main properties of the construction project such as location, total area, height, orientation, shape, function, and number of levels; and then, interactively specify the characteristics of a building element or construction method in each stage of simulation. The diversity of students' responses were portrayed by 3D avatars accompanied with related calculations and corresponding criteria. Each stage iterates until design and details are satisfactory. The final result, depicting construction details, is shown at the end based on each students design process in accordance with sustainable construction criteria. This study aimed to investigate the effect of using simulation on participants' perceptions of simulation on learning sustainable construction concepts. BECS was tested by a group of 42 undergraduate students in fall 2015. A quantitative method was used in this research. The data retrieved from a retrospective pre and post-test survey were clustered into three main categories. These categories represent sustainable construction contents provided in BECS. A t test was performed to show any statistical difference between pre and post situations. The results support development of more sustainable construction simulation applications and indicate BECS is an effective tool for sustainable construction education.

Keywords –

Simulation; Sustainable Construction; Education; Building Energy Consumption

1 Introduction

Educational systems have been largely changed by the emergence of the Internet, PCs and laptops. Their capabilities are increasingly developed and this has resulted in the development of programming languages and a variety of software, which enhances the design and development of educational materials [1]. Using these tools for education has improved the quality of education and contributed to lifelong and active learning, learning engagement, and virtual education [2]. It is shown that using simulation applications has improved students' engagement [3]. Computer-mediated education methods include computers as tutors, practice and repetition, the inductive method, play method, computer-managed instruction, and finally computer-assisted simulation [4].In educational systems, using simulation applications is an effective method for training in complex situations [5]. Simulation is a tool for creating real phenomena [6]. In simulation, the situation is made as similar as possible to the real position, so that what is learned can be transferred to the real environment [7]. In fact, most simulators are made based on real life descriptions, although, they have been changed for educational purposes. Simulation, as a training method, includes activities that imitate a real environment and has been designed to illustrate processes, decision-making, and critical thinking by using methods, such as playing role and using tools, such as instructional videos and applications [8].Using simulations has been widely increased, as they provide opportunities for students to apply the knowledge gained in the classroom [9]. Learners should develop the concepts and skills necessary to play roles in the target fields to progress through simulated environment. The simulated teaching model is an effective training technique through real activity and subsequent discussions. It also develops various educational targets, such as development of concepts and skills, cooperation and competition, logical thinking, decision-making, conformity, understanding political, social and economic systems, feeling individual usefulness and satisfaction, and awareness of the consequences [10]. Using simulation in training is an application of cybernetic principles that was simulated in the 18thcentury in the form of games [11]. Project-based learning method, which is often a main element of simulation applications, also has a history dating back more than a century [12], however, the history of modern simulation in training goes back to World War II, when it was used to train pilots [13]. Then, modern simulation applications were used to teach sciences, such as nuclear and military fields and in the past few decades it has been increasingly taken into consideration in medical education [14]. Other fields like construction and project management are incorporating subject-oriented simulation in their curricula [15, 16]. Despite the use of simulation application in various fields, it has not widely used in construction. However, literature review shows that due to changes in construction training, using simulation in the training of construction disciplines is gradually expanding [17]. There are still some challenges in using this method practically in construction programs, so that this approach can be facilitated by detecting challenges and effective planning to solve them. According to [18], “there are a number of technical, curricular, political and practical factors that could emerge may inhibit implementation of a project-based construction curriculum”. In this paper, a brief explanation about the types of simulations is given, and, then, the necessity of using simulation in education of construction and challenges associated with the use of this approach are discussed.

1.1 Simulation Steps

In general simulation model has four steps: justification (preparation), participants training, simulation, and asking questions. During the first step, instructor provides some explanations about simulation and general concepts and explains an overview on the play. This step is an important background for the next steps. In the second step, instructor determines a scenario with introducing rules, roles, methods, scoring methods, and simulation purposes to learners, and organizes the learners in various roles. In the third step, learners participate in the simulation and instructor plays his or her role as an educator or arbiter. Finally, in the fourth step, considering purposes, instructor can ask questions from students to contribute in their reflection about the materials [10].

1.2 Challenges of Using Simulation

Despite many benefits of simulation, this learning method also has disadvantages that have limited its use in training. For instance, faculty members are not very familiar with this approach and the way to use it, educators are needed to train, and also there is a resistance against changing the common educational practice [19]. According to [20] many faculty members have not received any training on the use of simulation, and, therefore, are inexperienced in this field.

2 Background

It has been shown that nearly half of the energy consumed in the world is being wasted in buildings. Other consumers in this regard are industry and transport whose activities are closely associated with buildings and their places. Therefore, energy consumption in buildings should be paid attention [21].Sustainable construction includes a wide range of issues, such as optimal energy consumption in various stages (design, construction, operation, and demolition of building) and the use of building materials that cause less damage to the environment, increase the quality of building, and decrease the negative effect of construction on environment around the building to a minimum level [22].It is hard to understand the behavior of buildings in relation to energy consumption and design low-energy building without using tools that can calculate the energy consumption of buildings before constructing. Energy simulation applications are tools used to calculate the energy consumption of buildings and understand their behavior. Finally, building energy consumption before and after changes can be accurately calculated and compared by modeling the dynamics of energy. In this case, except comparing building energy consumption before and after changes, the energy consumption of building before implementing changes and the strengths and weaknesses in terms of energy can be informed [23].

Currently, there are a few simulation applications that are developed with a specific purpose and target. Except the model which is used for calculations in modeling simulation, which determines the accuracy of calculations and static and dynamic modeling, the capabilities of simulation is different. Therefore, construction experts should equip themselves with modeling tools. BECS is an energy modeling simulation application to simulate buildings in terms of energy. The purpose of this application is that the user can apply the necessary changes (in insulation or materials) to optimize consumption energy in a simulated environment, and observe the results calculated by the software through entering building characteristics in the optimization part with suggestions relating to the elements in the "calculations" in optimization part.

Using the energy simulation application, which can be learned and used easily enables practitioners of building-related disciplines such as architecture engineering, civil engineering, mechanical engineering, and construction to model a building from different aspects of design and construction. Such applications, along with instructors' experiences, leads to defining a path by which theoretical and traditional instruction concepts can be understood by everyone.

3 Methodology

Engineering community's efforts to reduce energy consumption in the building sector in Iran have led to the development of national building regulations and also the development of BECS energy simulation application. The main research hypothesis was that BECS application, as a virtual project-based learning tool is an effective learning one for construction students. This study aimed to examine the application of construction students' learning and knowledge, who had no work experience in a real environment, to design and construct sustainable buildings considering saving energy consumption (heating – cooling) in a virtual environment. This was possible by using simulation environment and receiving a quantitative output with suggesting efficient building details to save energy consumption.

Participants of this test included 42 undergraduate students of construction at Tabaran Institute of Higher Education in fall 2015. The evaluation was most about the information sources and learning of construction students in theoretical courses, including elements and construction details, knowledge of building materials, building design courses and building heat transfer calculations. First, a residential building with certain conditions was selected for building energy simulation. The building plans were provided for each participant in the form of files with the format of drawing software such as AutoCAD along with a three-dimensional model. Participants were asked to record the project name and general information, including the number of floors and building's useful surface area. In the next step, the building's location was completed by the target city. Cities were classified by simulation database in terms of different climate zones. Then, the type of applied group of buildings was determined according to local building regulations and standards. In the related table in each group, the type of buildings is classified with more details such as residential, commercial, educational, and so forth. After entering the information related to the building, the thermal energy calculation by the simulator is recommended.

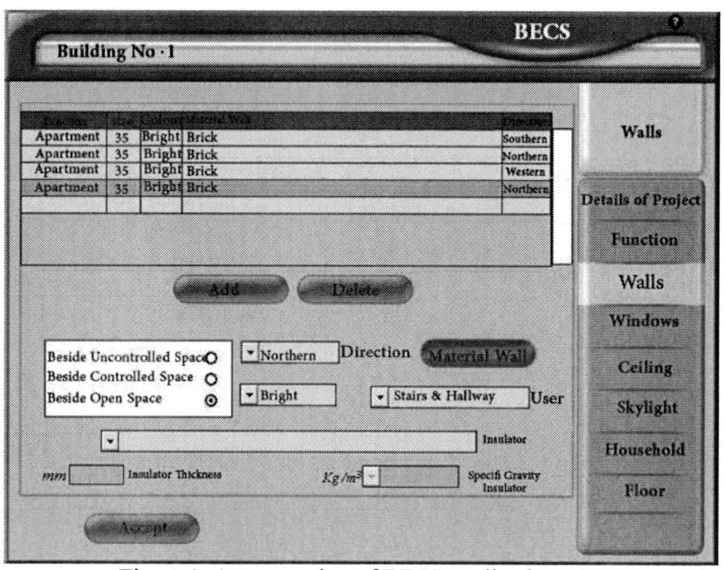

Figure 1: An screenshot of BECS application

In addition, in this simulation application, a part titled DB Viewer, was considered to enter new information (defining new materials, adding a city, defining new username).

When the mentioned format was confirmed, in the next step the type of energy consumption (electric –non-electric) and usage time (continuous –intermittent) and the type of building (villa –non-villa) were determined by the user. Based on the information provided. BECS informs participants the required coefficients to calculate reference heat transfer in different parts of the building in a table. The most important step in working with simulation application was the suggestion part of effective details (Figure 2). These details were portrayed based on the concepts of energy-saving provided in construction curricula. In the suggested resistant materials, participants were asked to design details with the possibility of organizing layers of materials from the outermost to innermost. After selecting materials through which participants entered the suggested layers thickness according to previous knowledge, BECS provided thermal conductivity coefficient and thermal resistance. In the next step, materials were entered. Based on the materials selection, the R value (the overall thermal resistance,m^2.k/w) and U value (the overall heat transfer coefficient, w/m^2. K) were calculated and indicated by the BECS(Figure 3).

At the end, the proposed details were prepared in the form of images. In this case, all building components had to be specified, so that related calculations were performed. The building accountability can be controlled.

In the stage of building condition, and in the case of lacking accountability the details should be replaced so that the building is accountable for thermal needs. To achieve favorite results, participants could repeatedly select different materials to finally suggest the most appropriate details with optimum heat resistance and low thermal coefficient. In every output, the BECS application informs whether or not the proposed details are acceptable for sustainable building thermal load. It was important that, after several decisions about design techniques, best materials were selected to achieve more efficient results. In addition, prediction of building efficiency should be controlled by designing and calculating before the actual construction. Wrong decisions will lead to the building thermal inefficiency.

After several failures in the selection of various materials and their location to meet the thermal requirements of the building, it was possible to evaluate participants' ability to use their knowledge. To evaluate the effectiveness of BECS, a pre and post-test was designed with 21 items in three thematic sections A, B, C. Section A questions consisted of general definition and terms of energy saving. In Section B, questions investigated the effect of using thermal-resistant materials in the design of building details. In this section, thermal insulation and various ways of designing details of shells such as walls, roofs, floors, and facade in different sides of the building were provided. Students could select various materials from the default materials in the application, and, thus, they could compare the

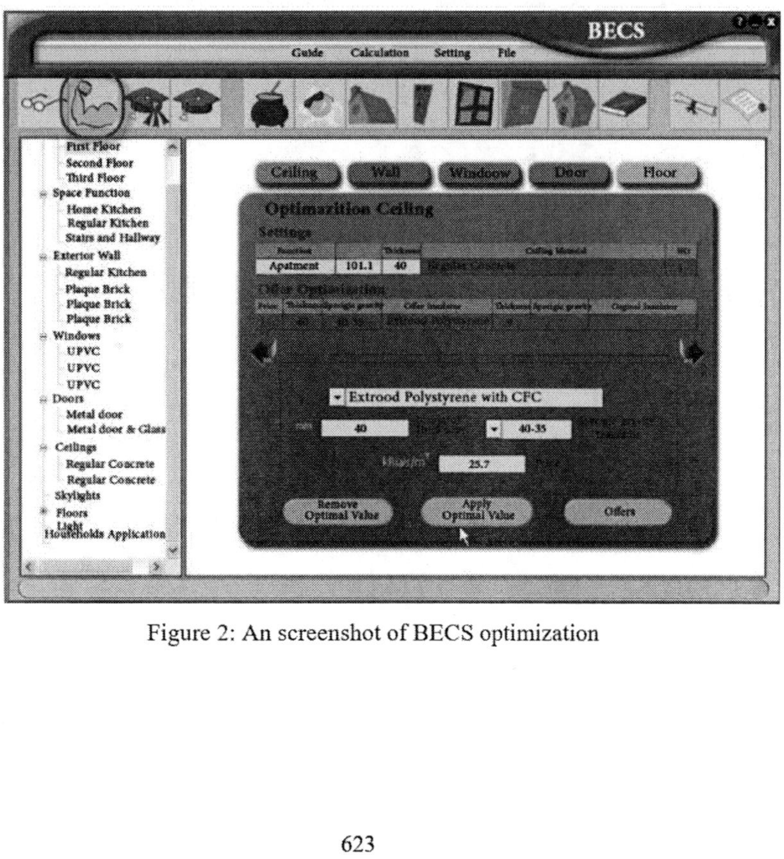

Figure 2: An screenshot of BECS optimization

effect of their selections in the proposed details on thermal resistance. The effect of light diffuser area, the location, calculation of thermal resistance, and profile selection of different materials and thickness of walls were provided in Section C. In order to quantify the participants' perceptions in pre and post-test, a five level Likert scale(very low: 1 to very high: 5) was used. Before playing the BECS, the pre-test questionnaire was distributed to assess students' theoretical energy optimization perceptions. The post-test was conducted at the end of working with the simulation application to measure its effectiveness.

4 Results

Participants included 42 students, 57% male and 43% female students among which 28% of female students and 45% of male students had work experience related to their field of study. Although, none of the students were trained outside the class in the field of energy optimization, six students reported previous experience working with simulation applications. Section A of the survey was designed in accordance with general definitions and terms used in energy saving concepts. Through this section, about 90 percent participants rated their knowledge in calculating thermal coefficient and estimating the efficiency of building energy requirements as low or very low. This indicated the lack of enough understanding about the energy calculations based on building components. In section B, 88% of participants in pre-test rated their knowledge on the effect of using different thermal insulation materials as low or very low. After playing simulation, 85% of participants rated their understanding on subjects, provided in this section, as high or very high. Overall, 80%of students rated the necessity of integrating different types of simulation applications in construction curricula as high or very high. Fifty four percent of participants showed high tendency for replacing the related courses in construction programs with the simulation applications. On average, 69% of participants rated their experience on using BECS application as "satisfied" or "very satisfied".

A paired sample t test was utilized to determine the mean difference between the retrospective pre and post-test scores for each of 21 questions at a significance level of 0.05. As shown in Table 1, there was a statistically significant different between pre and post situations. This indicates that BECS has been an effective simulation application for Sustainable Design and Construction education.

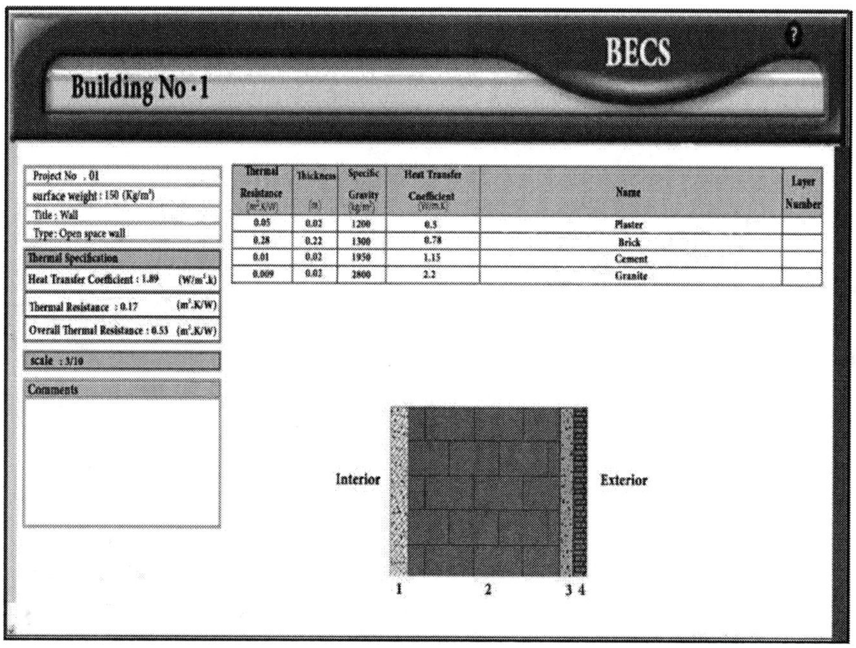

Figure 3: An screenshot of BECS details

Table 1: Paired Sample t-Test

		Paired Differences					t	df	Sig. (2-tailed)
		Mean	Std. Deviation	Std. Error Mean	95% Confidence Interval of the Difference				
					Lower	Upper			
Pair 1	Post 1 - Pre 1	2.024	1.179	.182	1.656	2.391	11.126	41	.000
Pair 2	Post 2 - Pre 2	2.238	1.165	.180	1.875	2.601	12.453	41	.000
Pair 3	Post 3 - Pre 3	2.071	.997	.154	1.761	2.382	13.460	41	.000
Pair 4	Post 4 - Pre 4	2.143	1.181	.182	1.775	2.511	11.763	41	.000
Pair 5	Post 5 - Pre 5	2.310	1.093	.169	1.969	2.650	13.695	41	.000
Pair 6	Post 6 - Pre 6	2.357	1.226	.189	1.975	2.739	12.458	41	.000
Pair 7	Post 7 - Pre 7	2.381	1.125	.174	2.030	2.732	13.714	41	.000
Pair 8	Post 8 - Pre 8	2.000	1.169	.180	1.636	2.364	11.091	41	.000
Pair 9	Post 9 - Pre 9	2.381	1.229	.190	1.998	2.764	12.558	41	.000
Pair 10	Post 10 - Pre 10	2.214	1.353	.209	1.793	2.636	10.605	41	.000
Pair 11	Post 11 - Pre 11	2.095	1.322	.204	1.683	2.507	10.274	41	.000
Pair 12	Post 12 - Pre 12	2.049	1.161	.181	1.682	2.415	11.301	40	.000
Pair 13	Post 13 - Pre 13	2.262	1.149	.177	1.904	2.620	12.759	41	.000
Pair 14	Post 14 - Pre 14	2.238	1.055	.163	1.909	2.567	13.751	41	.000
Pair 15	Post 15 - Pre 15	2.238	1.246	.192	1.850	2.626	11.644	41	.000
Pair 16	Post 16 - Pre 16	2.143	1.299	.200	1.738	2.548	10.694	41	.000
Pair 17	Post 17 - Pre 17	2.167	1.057	.163	1.837	2.496	13.281	41	.000
Pair 18	Post 18 - Pre 18	1.929	1.135	.175	1.575	2.282	11.015	41	.000
Pair 19	Post 19 - Pre 19	2.024	1.239	.191	1.638	2.410	10.583	41	.000
Pair 20	Post 20 - Pre 20	2.214	1.094	.169	1.873	2.555	13.117	41	.000
Pair 21	Post 21 - Pre 21	2.619	1.125	.174	2.268	2.970	15.085	41	.000

Participants were also asked to describe the strengths and weaknesses of the simulation application. The lack of default details in the application was reported as its main disadvantage. The strengths included the ease of calculations, the total estimation of building thermal requirement, control of all changes on the design and details in the calculation of building thermal resistance with only limited knowledge, output of proposed details, and the possibility of designing the building with optimal energy consumption.

5 Conclusion

Several educational learning theories support teaching with the use of simulation in construction education. In recent years the use of simulation in construction education has been growing. The results of studies indicate that this approach has been effective in creating interest and attraction among students and increases their satisfaction and confidence [24]. Compared to traditional models, simulation is less efficient in learning mental principles and skills, but it is very effective in training practical skills, especially when mental concepts and principles were previously taught with other methods. Different types of simulation can be used in construction discipline, which require further studies. There are challenges for using this learning approach that can be solved to some extent with proper planning. Educational institutions can pave the way for the development of simulation applications by using their potentials. Despite challenges, this method has many advantages and is a very effective method in training communication and practical skills, critical thinking and decision-making. Given the special emphasis on cost and operational issues in the construction field, the use of this approach in teaching construction students seems essential and should be pursued more seriously in the construction courses planning. Although the results of this study indicate a statistically meaningful difference between before and after exposure, more research should be conducted to generalize the results. However, construction simulation can be considered as a rich supplementary tool in sustainable construction education.

References

[1] Zandi, S., Abedi, D.,Yousefi, A.,Yamani, N., Kabiri, P., Electronic Learning as a New Educational Technology and its Integration in Medical Education curricula, *Iranian Journal of Medical Education*, 4(1), 61-70, 2004.

[2] Bork, A., Learning with the World Wide Web, *The Internet and Higher Education*, 2(23), 81-85, 2000.

[3] Rokooei, S., Goedert, J., and Fickle, K. Using a Simulation Model for Project Management Education, 7th Conference on Higher Education Pedagogy, Virginia Tech, Blacksburg, 2015.

[4] Ghezelghash, A., Atashzadeh Shurideh, F., AlaviMajid, H., Yaghmaii, F., Comparing methods of lecturing , problem solving and Self-Learning via internet to learn proper interpretation of electrocardiogram among nursing student, *Iranian Journal of Nursing Research*, 3(11), 7-15, 2008.

[5] Sanford, P., Simulation in Nursing Education: A Review of the Research, *The Qualitative Report*, 715(4), 1006-1011, 2010.

[6] Durham, C. and Alden, K., Enhancing patient safety in nursing education through patient simulation, Patient safety and quality: An evidence-based handbook for nurses, 6(3), 221-250, 2008.

[7] Rauen, C. , Simulation as a Teaching Strategy for Nursing Education and Orientation in Cardiac Surgery, *Critical Care Nurse*, 24(3), 46-51, 2004.

[8] Jeffries, P., Getting in S.T.E.P with simulations: Simulations take educator preparation, *Nursing Education Perspectives*, 29(2), 70-73, 2008.

[9] Johnson, M., Graham, C., and Su-Ling H., The effect of instructional use on teaching and learning, Brigham Young University, Department of Instructional , 2006.

[10] Joyce, B., Calhoun, E., and Hopkins, D., *.Models of Learning, Tools for Teaching*, SAMT: Tehran, 2009.

[11] Bremner, M., Aduddell, K., Bennett, D, The use of human patient simulators: Best practices with novice nursing students, *Nurse Education*, 31(4), 170-174, 2006.

[12] Goedert, J., Rokooeisadabad, S., Pawloski, R., A Project-based Simulation Model for Construction Education, 5th Conf. on Higher Education Pedagogy, Virginia Tech, Blacksburg, 2012, 8, 2013.

[13] Sanford, P., Simulation in Nursing Education: A Review of the Research, *The Qualitative Report*, 715(4), 1006-1011, 2010.

[14] Ricketts, B., The role of simulation for learning within pre-registration nursing education — A literature review, *Nurse Education Today*, 31(7), 650 -65, 2011.

[15] Goedert, J., Pawloski, R., Rokooeisadabad, S., Subramaniam, M., Project-Oriented Pedagogical Model for Construction Engineering Education Using Cyber infrastructure Tool, *Journal of Professional Issues in Engineering Education & Practice*, ASCE, 139 (4). 300-309, 2013.

[16] Rokooei, S., Goedert, J., and Weerakoon, A.,.Simulation as an Effective Tool for Gender Education in Construction, 6th Conference on Higher Education Pedagogy, Virginia Tech, Blacksburg, 2014.

[17] Rokooei, S., and Goedert, J. D., Lessons Learned from a Simulation Project in Construction Education, ASEE Annual Conference and Exposition, Seattle, Washington. 10.18260/p.24420, 2015.

[18] Goedert, J., Rokooei, S., and Pawloski, R., Virtual Interactive Construction Education: A Project-Based Pedagogical Model for Construction Engineering and Management, Higher Education Pedagogy Proceedings of the 4th Annual Conf., Virginia Tech University, 73-74, 2012.

[19] Willford, A., and Doyle, T., Integrating simulation training into the nursing curriculum, *British Journal of Nursing*, 15(11), 926-930., 2006.

[20] Childs, J., and Sepples, S., Clinical teaching by simulation: Lessons learned from a complex patient care scenario, *Nursing Education Perspectives*, 27(3), 154-158, 2006.

[21] Ghiasvand, J., Interaction of architecture and new energy forms, *Journal of Road and Building*, 38, 22-27, 2005.

[22] Taghizadeh, K., Sustainable development, sustainable architecture, and building industry, Iranian experts association, (n.d.).

[23] Nasrollahi, Energy Efficient Architecture for Tehran, IFHP World Congress, Urban Technology for Urban Sustainability, Berlin, Germany, 2009.

[24] Goedert, J., Rokooei, S., Project-Based Construction Engineering Education in a Virtual Environment, *International Journal of Construction Education and Research*, 2016.

3D Grid Size Optimization of Automatic Space Analysis for Plant Facility Using Point Cloud Data

Gyu seong Choi [a], S.W. Kwon [b], D. H. Kim [c], D.Y. Moon [d], S.W. Jung [e], S.Y. Lee [f], H.Y. Ko [g]

[a] Department of Convergence Engineering for Future city of Sungkyunkwan Universicty, Republic of Korea
[b] School of Civil & Architectural Engineering, Sungkyunkwan University, Republic of Korea [c] Department of Convergence Engineering for Future city of Sungkyunkwan Universicty, Republic of Korea
[d] Department of Convergence Engineering for Future city of Sungkyunkwan Universicty, Republic of Korea
[e] Department of Convergence Engineering for Future city of Sungkyunkwan Universicty, Republic of Korea
[f] Department of Convergence Engineering for Future city of Sungkyunkwan Universicty, Republic of Korea
[g] Department of Convergence Engineering for Future city of Sungkyunkwan Universicty, Republic of Korea
E-mail: gyupreme@skku.edu , soonwookkwon@skku.edu, ih5318@skku.edu, yoshy17@nate.com, suwanx@nate.com, 2_songyi@naver.com, kodjmc@nate.com

Abstract –

Currently, as plant market grows up, client demands to increase construction quality, technology and to reduce construction time. Due to the vast scale of the construction site and complexity of process, plant project can have some problems of communication among the project participants, the duplication of work, errors and rework. To solve this problem, 3D cloud point data of space and equipment is collected by 3D laser scanning. And the space matching is operated to build as-is environment. In the space matching process, data is simplified by using the 3D grid. It is important to select 3D grid size in the algorithm. Data processing speed and error rates depend on the size of 3D grid. But, still there wasn't the study about optimized 3D grid size. The grid size has been determined only by the user's experience. This study purpose to define optimized grid-size for 3D grid base space analysis. The specific research target of this study is Indoor plant facility. We followed experiment on these Condition. First, classified plant equipment according to the complexity of shape and capacity. Second, we set the classification equipment again by size of grid. Optimization of 3D grid size derive from comparing the volume of 3D grid and real volume of equipment. Third, compare the volume between grid-based model and real object to verify whether it useful or not. Using this method, makes it possible to apply the automatic space analysis algorithm efficiently. And this research applies to automatic space analysis for plant facility. It is expected to be possible to solve the problems and the differences of reactions for the space rearrangement. However, this research only optimizes indoor equipment in plant. So it is necessary to optimization of various equipment.

Keywords –

3D grid size, Space matching, Point cloud, Laser Scanning

1 Introduction

The data acquisition using 3D scanner is widely used in various industries, and its performance and application methods are consistently developed. Accordingly, it is also to be applied and researched for various new areas as well as pre-applied industries such as medical science, construction, and atmosphere measurement. Researches to apply the 3D scanner in the construction areas are also quite active. Various researches using the point cloud are under progress for construction error inspection, BIM modeling, slope analysis, safety management and so on. Recently, the application of 3D scanner to plant project is studied. Because of its own characteristics, the plant project has bigger scale than other construction projects, and its process is complicated as various ranges of equipment are applied. Especially, equipment in various sizes and shapes are installed on site and continuous maintenance of them (extension, repair, reinstall) is an important part of the project. On the maintenance step, addition/exchange of new equipment is included, and they have different requirement of installation space and shapes from existing equipment, so that there happen many difficulties such as shortage in space, space utilization, and so on. To overcome these difficulties, researches about the procurement of installation space for the plant equipment are progressed based on the point cloud.

For the installation space and plant equipment, the point cloud data are acquired using a 3D scanner and acquired data is simplified by 3D space modeling. Comparing the space with the equipment based on the constructed space model, the availability of installation can be decided. In this process, 3D grid method is utilized as a model construction method to secure installation space. In comparison with 3D grids based method, the space matching only with point cloud requires excessive computer memory for data processing and the speed of data processing is also very slow. By comparison, when the data from point cloud are simplified by 3D grid method, the usage of memory decreases and the processing speed is accelerated. In case of using 3D grid method, the number of nodes that is simplified the point cloud data is differentiated by the size of the grid. The smaller grid requires more nodes and detail parts of point cloud can be simplified, but the speed of data processing decreases and memory usage increases. Therefore, this study infers optimal size of the grid through experiments to secure various cases of installation space for equipment.

In this study, existing algorithm to secure installation space of equipment inside the plant based on point cloud data is complemented. To complement the algorithm, a machine room of the building has been laser scanned. In addition, the virtual scenario has been built by scanning equipment to be installed in the machine room and it has been utilized as base data to infer the optimal size of the grid for the equipment by changing its value.

The targeted algorithm of this study, which is to secure installation space of equipment, has been produced for plants or machine rooms of building with complex machine equipment. Therefore, the target space of this study and its data are restricted as the machine room. The target equipment and the data are also restricted to the real one that is used on site to review the availability of installation in the machine room.

2 Theoretical Consideration

2.1 Installation space analysis of plant equipment using laser scanning and acquisition of 3D shape information

The algorithm of previous study, which was to secure installation space of equipment in plant based on the point cloud data, controlled the size of target equipment and indoor space to apply the multi-level cube grid method to the acquired data from 3D laser scanner. It is because each unit size (n) should be same to check interference between indoor space and equipment, while they are subdivided to be small nodes by multi-level cube grid method. After equalizing the unit size of indoor space and equipment, the equipment and space were matched. The machine has been divided into 4 conditions, according to existence of indoor space and equipment data. As the table 1 shows, all data have been regulated to be simplified, except grids with both space data and equipment.

Table 1 Matching condition based space and unit

Matching to availability	Equipment	Space
available	○	○
available	○	●
available	●	○
Impossible	●	●

○ : Data Absence / ●: Data Exists

And then, the data about equipment and space has been arranged with 3D grid composed of nodes of designated size n. The matching result based on data arrangement is as follows.

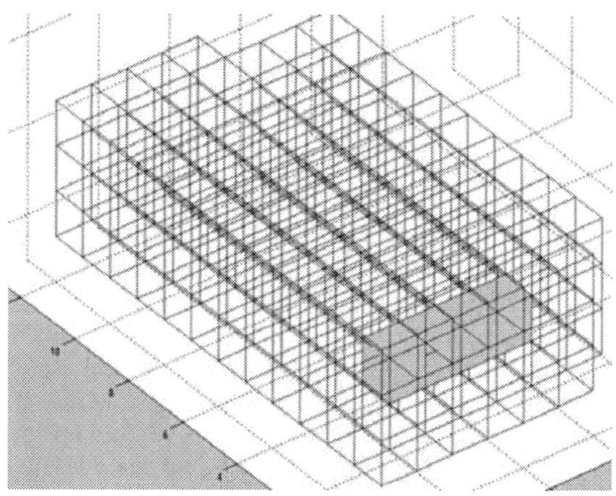

Figure 1 Matching Result

Throughout the process, the algorithm to check the space matching has been produced (Donghyun Kim, 2015)

However, there has been no standard to assign the grid size, so that it depended on decision of users. Thus, detailed parts of equipment and space have not been handled and subsidiary problems appeared. In this study, the decision standard to calculate the grid size and the optimal value are to be defined to solve those problems.

2.2 3D space division method

The 3D space division method can be divided into regular and irregular methods. Regularly divided space can be described by Voxel. The space division method such as Voxel is modeling an object in the 3D space with a set of identical size nodes. If the object to be described has curved boundary, the accuracy can be decided by the size of the cell. So, smaller cell brings higher precision. If the size of a cell is decreased to raise the precision, there will be demerits of increased calculation time and memory usage. (Yonghyun Kim, 2005)

The 3D grid structure is a space division method using cube grid structure. The width, length, and height of nodes have identical size, but the size can be flexibly changed to divide the space. By necessities, multiple grids with diversified levels of size can be generated (Suhui Han, 2012). In addition, the 3D grid structure does not have a hierarchical structure unlike the octree (Seungchan Yang, 2012)

Octree and 3D lattice structure shows a hierarchy difference. Space division method using octree is divided between top and bottom of each node. It has advantage of an index and search operation, etc. (Han, 2013)

However, in this study when the point cloud occupy space, it makes 3d grid structure. Then, it is quickly analysis the occupied space based on 3D grid structure. This study does not require between top and bottom structure of each node. Therefore, 3D grid structure is a more appropriate method for this study.

3 Grid size optimization

3.1 Optimization method by grid size

The goal of optimization is diversification of installation space for equipment using the installation space algorithm. Also, accuracy of space procurement should be increased by using small nodes, when the equipment is small and detailed for large scale equipment, large space considering the size of equipment should be explored and the availability of installation is to be checked rather than increasing the accuracy for detailed parts.

Therefore, the size of grid, which could lead high accuracy, fast processing and various installation cases, has been defined as the optimized value, by analyzing these values for the equipment.

3.2 Equipment classification

The scanning targets for experiments were defined as real machine rooms and equipment to be installed. The equipment has been classified by the shape and divided into the simple type and complex type that consists of two or more shapes. Reasons for the classification of simple and complex types are to decrease the error rate of data simplification according to the shape and fast decision of installation for simple equipment.

The simple type equipment can contribute to accelerate the processing speed and reduce the memory usage using the gird with similar size. Figure 2 shows generated nodes of the simple type equipment. The black node is 0.9m, red one is 1.1m, and blue one is 0.2m as the result of simplification. The smaller node increases the accuracy of shape, but the data processing speed becomes slower and memory usage becomes excessive. On the other hand, when it is simplified with a grid of similar size with the diameter or scale of equipment, the volume of a node is close to real equipment and the data processing time and memory usage decrease. Thus, it is expected to be more effective to generate nodes having similar size with actual equipment for data processing in case of simple type.

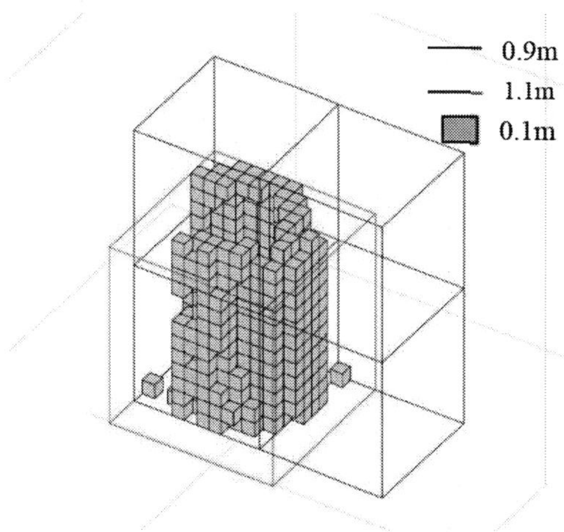

Figure 2 Comparison of node shapes according to node sizes (simple type)

In case of complex type, the application of grid size is different from simple type, since it consists of multiple shapes. The complex type is mixed with two or more different shapes and their size is also different. If the complex equipment is simplified with a single grid, the error rate between the equipment and a grid increases. Therefore, the grid size of complex type has been set to be small to simplify the parts in detail for experiments. (refer to figure 3)

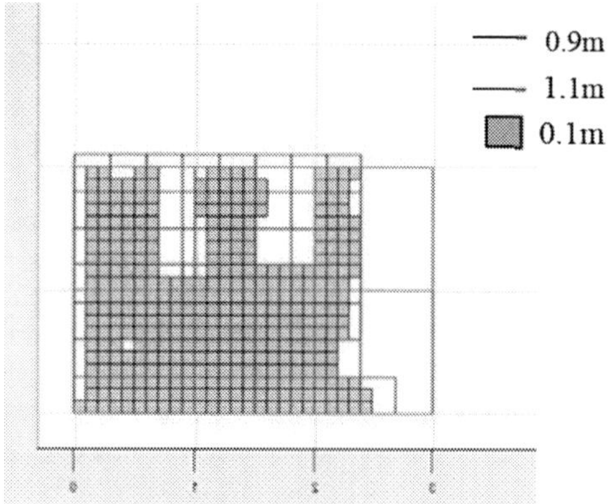

Figure 3. Comparison of node shapes according to node sizes (complex type)

3.3 Setting the range of the grid size

The maximum and minimum size of the grid is to decrease the error rate between the real equipment and simplified nodes in the point cloud. The accuracy of simplified data to real equipment without the maximum value setting, and the distance between points according to the resolution of the scanner becomes bigger than the grid size so that the shape recognition error occurs when the minimum value setting is not defined.

The bigger value between the width or length size of equipment is set to be the maximum value. If the data is simplified with bigger grid size than the maximum value, the error rate between actual device and node data increases. A node with large error rate makes difficult to find accurate space.

The minimum value of grid size is decided by scanner resolution. The minimum value of grid size should be larger than the distance between points (resolution). The resolution differs from the performance of applied scanner, and Scan Station C5 model of Leica that has been applied to the experiment has a resolution of 20 cm by 100 m, when it is scanned with low value. When the scan distance (distance between the scanner and scanning object) is 100 m, there is a gap of 20 cm between points. When the grid size is smaller than 20cm, dedicated space is recognized as a space without objects. The error happens on the matching step, when the space is regarded as one without objects. So, the minimum value of grid size has been set over 20 cm for experiments.

However, the experiments have been progressed indoor, and the scanning distance to object is between 0. 5m ~ 30 m, so that actual resolution was smaller than 20cm. In this study, the scanning distance between installation space and equipment were within 20 m, and the resolution gap was between 0.5cm and 5cm. If the grid size was smaller than 0.5 ~ 5cm, it would be regarded a space without points, so that the node was not to be generated (gray grid on figure 4). Therefore, the minimum size of the grid has been set over 5cm in this study.

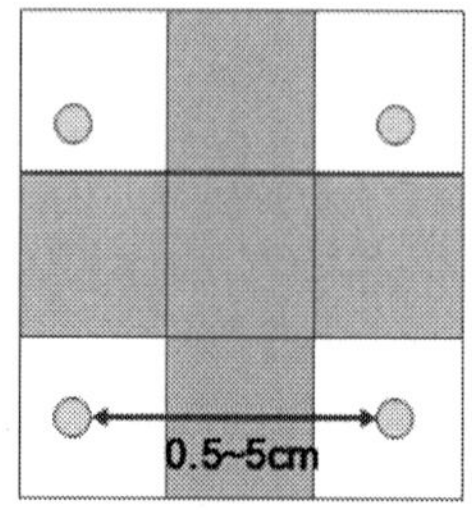

Figure 4 . minimum size of the grid.

4 Experiments and analysis

4.1 Experiments Overview

The scanning targets were the installation space for the space matching and equipment (pump, tank, etc.) to be installed. The point cloud data has been acquired by 3D laser scanner, and the point cloud has been simplified by 3D grid. Three kinds of equipment have been targeted for the experiments and the size of indoor space was 6 m of width, 10.5 m of length, and 2.8 m of height. The size of 3D grid for simplification of point cloud was set between maximum 1 m and minimum 5 cm for various experiments. The result values about changes of grid size were measured by the time lapse of data simplification, number of grids, and number of space matching cases. The system specification for experiment process was the hardware of Intel core i5-4690@ 3.50GHz, RAM 8.0GB. To implement the space matching algorithm, the engineering software 'MATLAB', which has been developed by 'Math Works' and provides numerical analysis and programming conditions, was applied. The MATLAB supports the matrix based calculation, drawing graphs of functions and data, and implementation of algorithms by programming. The specification of scanning equipment is as shown in table 2 and the shape of point cloud data that has been acquired by the scanner was as shown in figure 5 and 6.

Figure 3. Shape of installation space

Table 2. Specification of Data

Terrestrial Laser Scanner	Leica Scan station C5			
	Space	Equipment A	Equipment B	Equipment C
Scanned Objected	Δx = 10.51m Δy = 6.02m Δz = 2.80m	Δx = 4.08m Δy = 0.41m Δz = 2.102m	Δx = 2.38m Δy = 1.01m Δz = 1.934m	Δx = 0.95m Δy = 1.00m Δz = 1.07m
Point cloud	5,339,501	286,901	280,405	63,155
Shape type	-	Complex	Complex	Simple

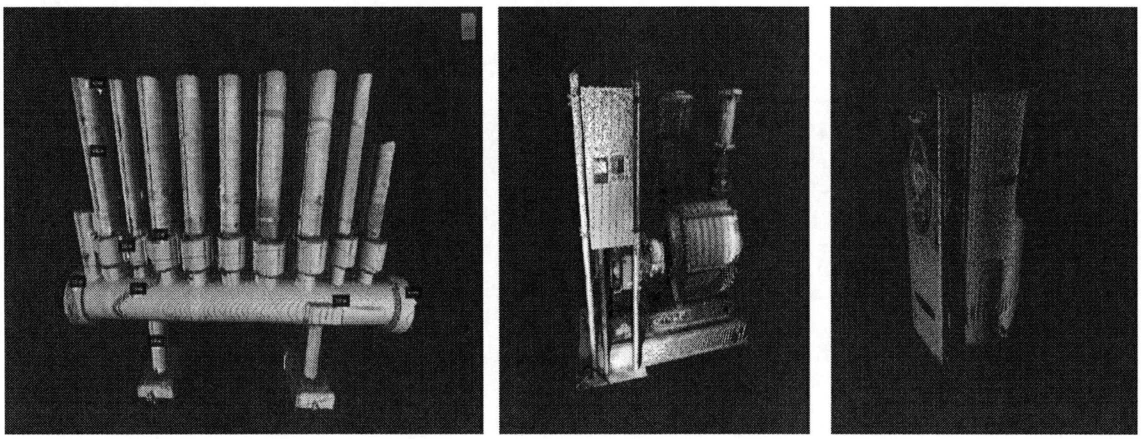

Figure 4. Point cloud data of Equipment (A),(B),(C) (from left)

4.2 Experiment Result

Experiments have been performed as described in the overview, and following results have been acquired. The experiment for equipment A has been progressed with maximum value of 4 m and minimum value of 0.1 m. However, the space matching error happened because the size of the grid was bigger than the height of indoor space, when the maximum value was set to be 4 m. Thus, the maximum value of the experiment has been reset to be 2 m that is the height of the equipment. The speed of simplification process was the fastest with the grid of 2 m, and the grid of 0.1 m required the longest time lapse. The number of grids was 11 with the grid of 2 m, 21 to 1 m, and 2291 to 0.1 m. Grids, of which size were between 1 m to 0.2 m, have shown the deviation of data processing time between 1 and 20 seconds. But it was over 200 seconds for the grid of 0.1 m. Also, the number of nodes drastically increased to be 2291 with the grid of 0.1 m. There were 3 cases of space matching cases for equipment A with the grid of 1 m, 879 to 0.2 m, and 2113 to 0.1m. Experimental results of equipment B and C were also similar to A. When the size of the grid got smaller, the number of gird and the time lapse of data processing increased. For the equipment B, the maximum number of grid was 1,783 with the grid of 0.1 m and the maximum time lapse was 181.9 seconds. There were 4 cases of space matching with the grid of 1 m, 919 cases with 0.2m, and 1911 cases with 0.1m. For the equipment C, the maximum number of grid was 1,783 with the grid of 0.1 m and the maximum time lapse was 6.7 seconds. There were 23 cases of space matching with the grid of 1 m, 1,320 cases with 0.2m, and 3,365 cases with 0.1m. The table 3 shows the time lapse of node generation and the number of grids for each grid size.

Table 3. Result of Time lapse and number of node

Grid Size(m)	Equipment A		Equipment B		Equipment C	
	Time lapse (s)	Number of node (n)	Time lapse (s)	Number of node (n)	Time lapse (s)	Number of node (n)
1	0.713	21	0.504	12	0.077	1
0.9	0.748	23	0.678	18	0.139	6
0.8	0.873	36	1.064	18	0.15	7
0.7	1.258	38	1.391	26	0.158	8
0.6	2.113	53	1.453	40	0.193	12
0.5	2.949	73	1.701	51	0.213	18
0.4	5.783	129	3.944	76	0.336	29
0.3	12.748	223	9.215	173	0.56	53
0.2	35.707	565	25.408	411	1.276	116
0.1	245.314	2291	181.983	1783	6.697	440

4.3 Result Analysis

Figure 7 shows compared results (table 3) of simplification processing speed for each of equipment A, B, and C, according to the grid size. The equipment A and B that is complex type have similar numbers of point cloud, but the difference of node numbers and data processing time have been compared. The nodes of 1 m ~ 0.3 m required 10 seconds for data processing and the deviation among variables were 1~7 seconds However, node generation time for 0.2 m and 0.1 m was 20 times longer. Grid numbers of each size have a deviation of about 1.2 times between 1 m and 0.3 m, but it became 40 times between 0.2 m and 0.1 m. Difference in shapes of equipment A and B brought about 500 gaps of node numbers. From the result, the shape can differentiate the number of grids, though there are similar numbers of point data. The equipment C has about 20% points of other two and its shape is simpler so that it acquires grids 20% less than other two.

Also, the number of space matching cases became similar to previous two results. Because the size of equipment C is 50% smaller than equipment A and B, it has 1.2 ~ 3 times more cases of space matching. As an exception, the simple type does not guarantee to increase cases of space matching, as the size of the grid becomes smaller. When the grid size (figure 2, table 4 – case with a grid of 1 m) is identical to the real equipment, it has been simplified as a grid and there were more cases of space matching than the grid of 0.9 m.

According to the result of experiments, as the size of the grid gets smaller, the error rate between actual equipment and data simplified shape decreases, and the smaller error rate increases cases of space matching.

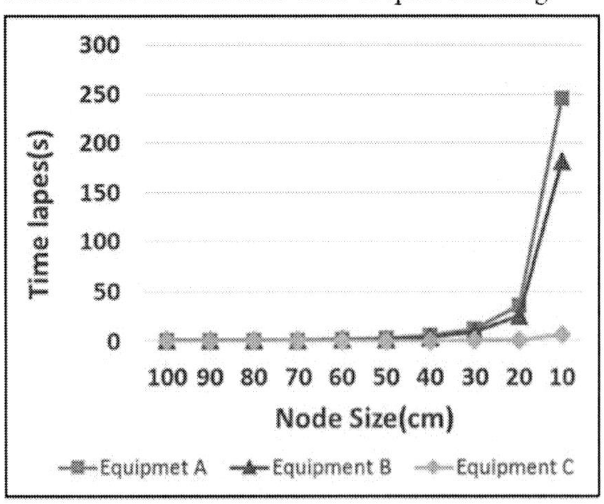

Figure 5. Comparison of Time lapse according to node sizes

However, if the grid size is set to be under 0.1 m to reduce the error rate, the time lapse of data processing will be too excessive for space matching.

The data simplification is not to decrease the error rate to be 0%, but to process data efficiently. So, the grid under 0.1 m has been decided as ineffective size. Faster speed of data processing and more cases of space matching became the standard of grid optimization. According to this study, the optimal size of the grid has been concluded as 0.3 m ~ 0.2 m.

In case of simple type equipment, the larger reference value from the actual equipment contributes to the faster data acquisition, because a node can involve the whole equipment. However, when the space is also set with the node of identical size and it is too large, the large unit node brings about ineffective space application. The time lapse for simple type equipment does not increase significantly, when same grid size with complex type is applied. Therefore, applying the identical method to both simple and complex types has been concluded appropriate.

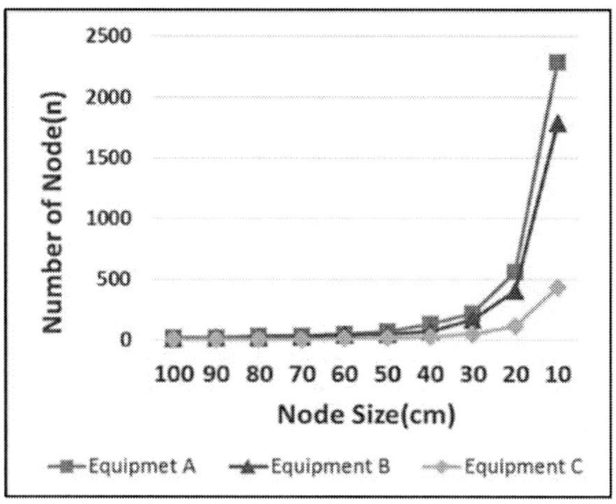

Figure 6. Comparison of number of nodes according to node sizes

Table 4. Result of Space matching

Grid Size(m)	Equipment A	Equipment B	Equipment C
	Number of Space Matching (n)	Number of Space Matching (n)	Number of Space Matching (n)
1	3	4	15
0.9	7	7	9
0.8	67	68	92
0.7	76	81	110
0.6	55	31	164
0.5	123	103	176
0.4	197	235	335
0.3	434	446	590
0.2	879	919	1320
0.1	2113	2211	3165

5 Conclusion

This study was to achieve an optimized value of the grid size, which has not been verified by previous study. Simplifying point cloud acquired from 3D scanner, indoor space and various installation spaces of equipment has been analyzed. For optimization, the equipment has been classified by their shape and experiments have been progressed by applying space matching algorithm of previous study targeting classified equipment. The standard of optimization has been concluded to be the small error rate between actual device and the node, fast data processing, and more cases of space matching, and the optimized node has been inferred with such a grid size. Acquired experiment results were data processing speed, the number of node generation, and the number of available installation cases from the matching between the equipment and space nodes. Among experiment results, the grid of 0.1m has shown the smallest error rate of node and the most cases of space matching, but the data processing speed was ineffective. Comparing to the gird of 0.1 m, the grid of 0.2 m was not quite different in error rate and numbers of space matching cases, and the data processing speed was relatively acceptable. Accordingly, the optimal size of the grid has been decided to be 0.3m ~ 0.2m.

However, the number of target equipment has been restricted to three and the equipment has not varied in size and shape. As a future study, experiments with equipment of more various scales and shapes are necessary and additional verification of defined optimal values of grid size is to be considered. Also, other optimizations will be studied except the methods based on data processing speed or number of space matching cases.

6 Acknowledgement

" This work is financially supported by Korea Ministry of Land, Infrastructure and Transport(MOLIT) as 「U-City Master and Doctor Course Grant Program」 ".

"This research was supported by Basic Science Research Program through the National Research Foundation of Korea(NRF) funded by the Ministry of Science, ICT & Future Planning (NRF-2014R1A2A2A01005893)"

7 Reference

[1] Kim, Y. H., & Ko, S. L. " Development of a Cutting Simulation System using Octree Algorithm ", *Transactions of the Society of CAD/CAM Engineers*, volume 10(2):107-113, 2005

[2] Kim,D.H, Kwon, S.W. "A study on automatic space analysis for facilities based on 3D octree algorithm by using laser scanning information - for remodeling construction project", *Korea Institute of Construction Engineering and Management* (KICEM), 2, 2014

[3] Yang, S. C., Han, S. H., & Heo, J. Massive "3D Point Cloud Visualization by Generating Artificial Center Points from Multi-Resolution Cube Grid Structure" *Journal of the Korean Society of Surveying, Geodesy, Photogrammetry and Cartography*, 30(4), 335-342, 2014

[4] Han, S. H., Lee, S. J., Kim, S. P., Kim, C. J., Heo, J., & Lee, H. B. "A Comparison of 3D R-tree and octree to index large point clouds from a 3D terrestrial laser scanner" *Journal of the Korean Society of Surveying, Geodesy, Photogrammetry and Cartography*, 29(1), 39-46, 2014

[5] Lee, K. H., Woo, H., & Suk, T. "Point data reduction using 3D grids" *The International Journal of Advanced Manufacturing Technology*, 18(3), 201-210, 2014

[6] Han S.H, "Comparison between Main-memory-based Octree and File-based Octree to Index Huge 3D Point Cloud" *The Korean Society for Geo-Spatial Information System, 157-158*

KBimCode-based Applications for the Representation, Definition and Evaluation of Building Permit Rules

Seokyung Park[a] and Jin-Kook Lee[a]

[a]Dept of Interior Architecture Design, Hanyang University, Korea
E-mail: seokyung.park529@gmail.com, designitlab@hanyang.ac.kr

Abstract –

This paper aims to describe applications of KBimCode that enables representation, definition and evaluation of building permit rules for BIM-enabled code compliance checking. KBimCode is a standardized script language. It is developed to represent the Korean Building Act as explicit computable rules. The application of KBimCode can be summarized as: 1) Preparation of natural language rule, 2) representation of rule, 3) definition of rule, 4) execution of rule, and 5) execution result and report. This paper mainly focuses on describing the process and mechanism of each stage of KBimCode application. First, natural language rules are translated into KBimCode through logic rule-based mechanism. KBimCode are generated and managed by logic rule-based system called KBimLogic. Second, we introduce a design strategy and definition of KBimCode syntax. Application of KBimCode database is also handled. Third, execution of KBimCode is demonstrated with BIM assessment software and its result is reported. For the demonstration presented in this paper, Korean Building Act sentences were translated into KBimCode series and evaluated in a BIM assessment tool named KbimAssess-lite. KBimCode is an ongoing project and will contribute to establish automated design quality assessment system for building permit in Korea.

Keywords –

Automated code compliance checking; Automated Building Permit System; Korean Building Act; KBimCode; KBimLogic; BIM (Building Information Modeling)

1 Introduction

Assessment of building design is an important task in every stage of design process to improve the entire quality of design, meeting owner's requirement, or getting a building permit [1]. Various design requirements have been reviewed by building designers in manual. Recently, as building information modelling wields great influence upon architecture, engineering, construction and facility management (AEC-FM) industry, research on BIM-enabled automatic code compliance checking has been explored by domain experts. Some challenging projects and research reported that BIM-enabled code compliance checking benefits conventional assessment process by improving efficiency of the process, reducing errors, etc. [2, 3, 4, 5, 6]. Among various issues related to the BIM-enabled code compliance checking, this paper focuses on interpretation of design rule and its execution. Issues related to building model preparation and rule checking reports are handled with less importance. Design rules are written in natural language, therefore, they need to be translated into a computable format that enables automatic reasoning. This paper points out how design rules can be assessed in a BIM environment centered on KBimCode. KBimCode is a standardized script language that represents natural language rule into a computable format. It was developed as an outcome of ongoing government-funded research and development project for establishing an automated building permit system in Korea [7]. We describe an approach to representing, defining and evaluating building permit requirement in Korea Building Act based on application of KBimCode.

2 Background

2.1 Researches in Rule-based Checking of Building Design

There have been challenging attempts to automate code compliance checking process in the field of the AEC-FM industry. The tracks of research activities in this area date back to the early 1960s when Fenves formalized specifications of the American Institute of Steel Construction (AISC) using decision table for automatic reasoning [8]. Recent studies and projects showed the possibility of various approaches with the issue. The CORENET (COnstruction and Real Estate NETwork) project in Singapore translated the Building Code of Singapore by hardcoding it with a programming language [9]. The USA General Services Administration's courthouse project parameterized circulation rules according to a predefined parametric table [10]. Lee [11] developed a domain-specific language named Building Environment and Rule Analysis (BERA) to analyze and define building circulation and spatial programming rules. The DesignCheck project of Australia formalized the Building Code of Australia and encoded it according to the EDM rule schema based on the EXPRESS language [12]; Uhm et al. analyzed RFPs for large public buildings in Korea and formalized them using four morphemes (object, methods, level of strictness, and others). The formalized rules were translated into the semantic web rule language (SWRL) and translated into the Python script language [13]. Zhang et al. developed algorithms that automatically extract information from the International Building Code (IBC) and transform it into a formalized format using the NLP approach [14, 15].

The existing studies showed various approaches of rule-making in the design assessment. However, there still remain limitations to apply existing approaches to the translation of the Korean Building Act. Korean Building Act inheres complex relationships between sentences, therefore, new approach reflecting such feature is needed. In addition, because it is national binding force and errors in the translation may cause legal conflicts, it should be translated by manual to clarify the responsibility of the translation.

2.2 KBimCode-based Application

Taking advantage of BIM, the Korean government is establishing automated building permit system. As a part of the Korea government-led initiative for enhancing building design productivity as well as overall quality and performance using BIM, automated checking of Korea building permit requirements and rules is the motivation of this paper. The project can be summarized as follows: 1) development of an open BIM based quality verification system, 2) development of an open BIM based construction document optimization standard and applied technology, 3) development of an integrated cooperation work system during the public administrative process [7]. Several software tools that assist in the automated review process are also under development and will soon be publicly available [16]. Among the outcomes, this paper mainly focuses on the following development:

- KBimLogic

A logic rule-based management system that translates Korea Building Act (natural language) into computer-executable rule set files. Following the logic rule-based mechanism, the building code sentences are converted into an intermediate script language called KBimCode. KBimLogic includes 1) logic rule based-meta database and 2) KBimCode authoring tool. The logic rule-based meta database provides regulation-specific information that can be accessed by the KBimCode authoring tool and used for the generation and management of KBimCode.

- KBimCode

KBimCode is developed as a standardized rule set file that aims IFC-based open BIM. Design rules are translated into KBimCode and established as a database that can be reused and executed according to the user's purpose. KBimCode is generated and consistently managed by domain experts. Exported series of KBimCode can be used in various BIM assessment software such as KBimAssess-lite.

KBimCode-based application is a standardized process of rule representation, definition and evaluation. The process can be summarized as five stages.

1. Rules preparation (natural language),
2. Rule representation,
3. Rule definition,
4. Rule execution,
5. Result and report

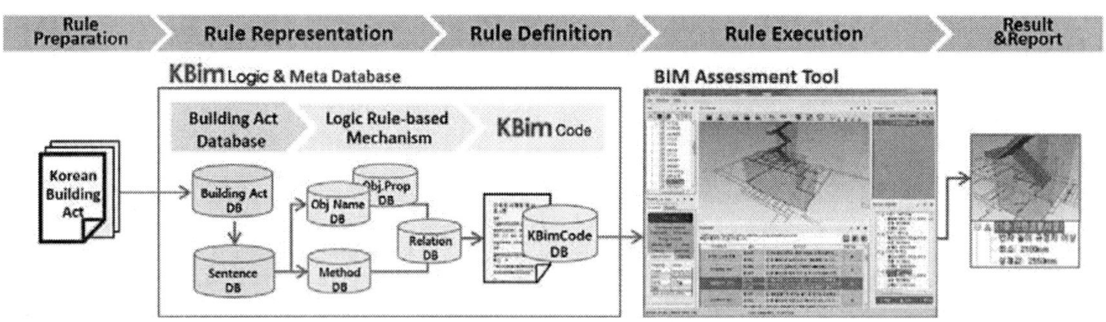

Figure 1. Overview of KBimCode-based applications and use of according software for each stage

Among the five stages of the KBimCode-based application process, this paper mainly deals with mechanism for rule representation, methods for rule definition, implementation issues with rule execution, and type of execution result and reports. Figure 1 illustrates overviews of KBimCode-based applications and use of according software for each stage.

3 Rule Representation

For the rule preparation, target rules are split into sentence unit and stored as a database. The database is used to author KBimCode through the logic rule-based mechanism. This mechanism provides a standardized way of rule representation [17, 18]. It can be explained in three parts according to the structure of a sentence: 1) noun phrases (objects/properties), 2) verb phrases (high-level methods), and 3) relation of sentences (Logic). The following sections describe each part in detail.

3.1 Noun Phrases: objects and properties

Noun phrases represent the key topic of a design rule. By classifying noun phrases, objects and associated properties handled in design assessment can be captured. Each object is developed in an object name database. Based on the name database, object and properties are classified as Figure 2.

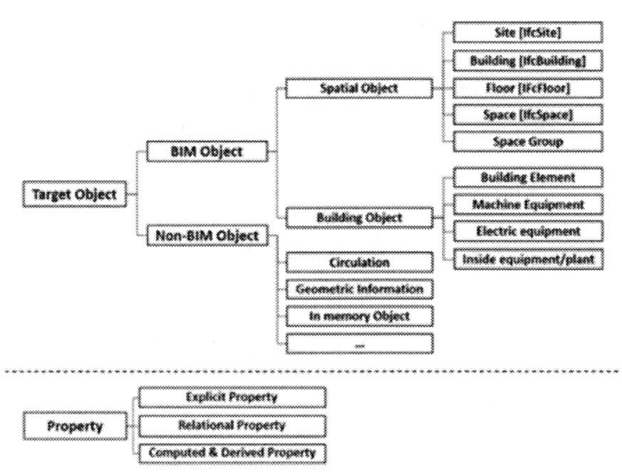

Figure 2. Classification of objects/property

Objects can be divided into two types: target object relevant to assessment of building design and non-target object that is exclusive to the assessment. The target object can be categorized as 1) BIM-enabled object (building objects and spatial object) and 2) non-BIM object derived from BIM data (circulation, geometric information, etc.). Properties are object-specific. They can be categorized into three types: 1) explicit property (e.g. object's name), 2) relational property (specified by two or more objects such as inclusion, distance, etc.), and 3) computed or derived property (e.g. building's gross area, floor number). Classification of objects and properties is mapped with standards such as IFC.

3.2 Verb Phrases: High-level Methods

Verb phrases define the main contents of assessment. Common rules obtained through a classification of verb phrase are defined as implementable methods. The methods are defined in high-level and distinguished from low-level methods for the implementation of BIM assessment tool. The high-level methods aims to represent verb phrases of natural language in intuitive form. The common rule become function and object/property become argument of the high-level methods. According to the object/property, the high-level methods can be categorized into four levels. At the first level, methods are distinguished into two types according to their checking target: 1) Object instance, and 2) property of the object's instance. In the second level, properties are categorized into four types: 1) basic property (object name, window type, etc.), 2) geometric property (door height, space area, etc.), 3) complex property (regulation-specific properties such as fireproof properties), and 4) relational property (inclusion, connectivity, distance, etc.). The third level specifies the content of checking. Contents of review are subdivided in this level. Lastly, methods are determined in the fourth level. There are two types of methods: 1) Representative methods that represent checking the types of the third level, and 2) extended methods subdivided from the representative methods. Extended methods check specific objects while the representative methods can review extensive objects. For example, a representative method 'getObjectArea()' has 'getFloorArea()', 'getWindowArea()', etc. as its extended methods.

Table 1 shows the example of high-level methods and their classification developed from a number of 1,977 Korean Building Act sentences related to building permit. Examples of methods on the fourth level are representative methods.

Table 1 Classification and example of high-level methods

1st Level (Target)	2nd Level (Property Type)	3rd Level (Checking Type)	4th Level (Method)
Object instance(s)		Object Query	`getObject()`
		Object Existence	`isExist()`
		Object Count	`getObjectCount()`
Property of Object	Basic Property	Property Query	`getProperty()`
		Object Type	`getObjectT`

instance(s)			ype()
		Object Material	getObjectMaterial()
		Object Usage	getObjectUsage()
	Geometric Property	Object Height	getObjectHeight()
		Object Width	getObjectWidth()
		Object Area	getObjectArea()
		Object Gradient	getObjectGradient()
	Derived Property	Material Type	getObjectMaterialType()
		Fire Safety	isFireResistantCompartment()
			...
	Relational Property	Inclusion	hasObject()
		Distance	getObjectDistance()
		connectivity	isConnectedTo()
			isAdjacent()
			...
		Path	isAccessible()
		Direction	getObjectDirection()

Specification of high-level methods is provided to software vendors for implementation. The high-level methods are mapped with low-level method library in BIM assessment tool. The detailed method specification for the example methods of Table 1 is available at [19].

3.3 Relation of Sentences: Logic

There are two types of relations pertaining to sentences: 1) relations in a sentence, and 2) relation between sentences.

3.3.1 Relation in a Sentence

Korean Building Act sentence consists of the checking condition and content [23]. Before checking the content, the condition should be satisfied. By clarifying the relation of the condition and content, logical composition of a sentence can be clarified. They are represented with IF-THEN-ELSEIF-ELSE logic. Figure 3 shows the sentence restructuring process of building permit rule to KBimCode. A sentence is split into semantic unit and their relations are figured. Based on the type of relationship, the sentence is restructured and represented as KBimCode.

The detailed description of the restructuring process is as follows:

1. Atomic sentence (AS)

Human can be aware of connected meanings of several sentences. However, computer cannot recognize the connected meaning of sentences in the same way as people do. Therefore, processing of split content of design rule with sentence-basis is needed. In this paper, atomic sentence is a minimum unit of sentence for assessment. It is a declarative sentence that returns either truth or falsity.

2. Translated atomic sentence (TAS)

Translated atomic sentence (TAS) is semantically split unit of atomic sentence. TAS has a subject (S) + verb (V) + object (Object) structure. TAS is obtained through the parsing of AS. With TAS, the checking condition and content of a sentence are clarified.

3. Arithmetic Logic Unit (ALU)

Originally, ALU is a digital electronic circuit that performs arithmetic and bitwise logical operations [20]. In this paper, ALU stands for a declarative clause that acts as the atomic unit for rule checking. The ALU is comprised of left operand, operator, and right operand [21]. The high-level methods and object/properties are placed in the left operand and compared with explicit values in the right operand using comparison operators.

4. Representation of rule (KBimCode)

Reflecting the logical composition of a sentence, ALU are restructured into KBimCode. The details of KBimCode syntax is handled in the next section. Table 2 shows the overall process of restructuring of natural language sentences to KBimCode. Enforcement Decree of Building Act (EDBA), article 64, clause 1 was used as an example

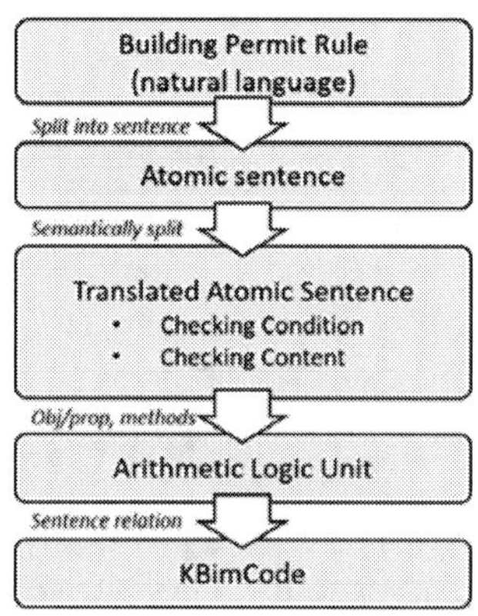

Figure 3. From sentence to KBimCode: Sentence restructuring process based on logic rule-based mechanism.

Table 2. Example of sentence restructuring process based on logic rule-based mechanism

Sentence		[Enforcement Decree of Building Act, article 64] A project owner of a building with six or more floors and a total floor area of 2,000 square meters or more shall have an elevator installed therein. In such cases, the size and structure of elevators shall be prescribed by Ordinance of the Ministry of Land, Infrastructure and Transport. [22]
AS		A building with six or more floors and a total floor area of 2,000 square meters or more shall have an elevator installed.
TAS	Condition	TAS1. A building has six or more floors TAS2. A building has a total floor area of 2,000 square meters or more
	Content	TAS3. A building has an elevator
ALU	Condition	ALU 1. `getBuildingStoriesCount() >= 6` ALU 2. `getGrossFloorArea() >= 2000`
	Content	ALU3. `isExist(Elevator) = TRUE`

KBim Code		`Check(BA_64_1){`
		`IF(getBuildingStoriesCount()>=6`
		`AND getGrossFloorArea()>=2000)`
		`THEN isExist(Elevator)=TRUE`
		`ENDIF`
		`}`

3.3.2 Relation between sentences

Sentences may have relationships with other sentences. For instance, Korea Building Act sentences form complex cross referencing relationships as exemplified in Figure 4.

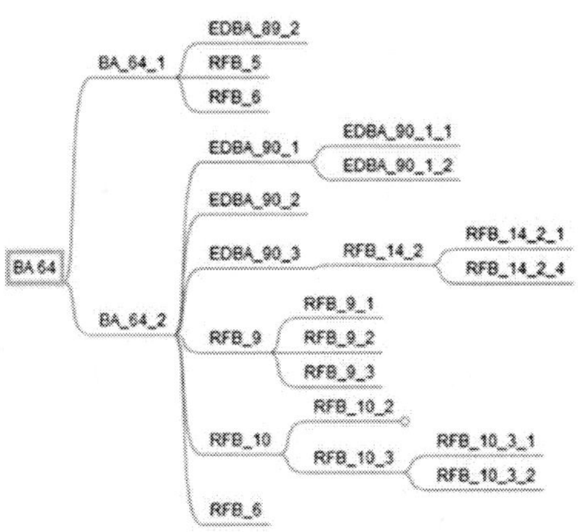

Figure 4. Example of relations between sentences. Each node stands for one sentence. The left-side nodes refer to the right-side nodes. For example, the sentences of the Building Act article 64, clause 1 (BA_64_1) refers to the sentences of the Enforcement Decree of Building Act article 89 clause 2 (EDBA_89_2), and the Regulations for Facility in Building article 5 and 6 (RFB_5, RFB_6).

Relationship between sentences are represented by control methods. For example, the Building Act article 64, clause 1 (BA_64_1) refers to the Enforcement Decree of Building Act 89, clause 2 (EDBA_89_2). Their relationship can be represented in the KBimCode of BA_64_1 as follows:

```
getResult(EDBA_89_2) = TRUE
```

Relations between sentences are developed into a database and directly used in KBimLogic application to logically combine sentences and generate set of KBimCode.

4 Rule Definition

4.1 KBimCode Language Design

This section deals with design strategy for KBimCode language definition. The design of KBimCode Language reflects features of logic rule-based mechanism. The strategy can be described with two points: 1) lexical design, and 2) syntactic design. The lexical design strategy plays as basic tokens and idioms in KBimCode Language. The subject (S) of TAS is specified in a dictionary and used as the fundamental lexicon. They are Korea Building Act-specific building objects and properties (KBimCode Object). The relations between objects and properties are represented as dot-notation. The verb (V) and object (O) structure will be used in a basic form of idioms. They are pre-defined as basic methods such as, getObectCount, getObjectHeight, isFireResistant, hasElement, getResult etc. More details of the predefined methods are depicted in [21].

The basic syntactic form of KBimCode is as follows:

```
check (arg) {
      Statement;
      ...
}
```

The check is a pre-defined method that declares the checking of a certain building code sentence. The non-terminal token arg identifies a target building code sentence. The Statement is a rule defined in building code sentences. There are three types of statements: Arithmetic Logic Units (ALUs), Condition Statement, and KBimCode Object Model (KOM). ALUs is a single ALU or multiple ALUs joined with conjunctions such as

AND, OR. The ALUs is an expressive way of representing the rules. The condition statement is IF-THEN-ELSEIF-ELSE logic including ALUs as its statements. The KBimCode Object Model (KOM) is one of the key features of KBimCode Language. It enables users to define virtual objects with desired rules as constraints.

A set of statements can be grouped as statements group. The syntactic form of the statements group is as follows:

```
var {
   Statement;
   ...
}
```

The statement group is defined outside of the basic syntactic form. The non-terminal token var is a user-defined variable name that represents the statements group. By using the statements group, KBimCode can be written in a clear and simplified way.

4.2 KBimCode Language Definition

This section briefly introduces definition of KBimCode Language syntax and its grammar. The entire syntax of KBimCode is available at [24]. The formal definition described in this section is EBNF-based ANTLR rule [25]. ANTLR is a language parser generator and in its rule, non-terminals starting with lower case are syntactic rules and upper-case non-terminals are lexical definition.

Most of all, a non-terminal syntactic rule kKBimCodeProgram is the starting point of KBimCode Language definition. A lower case 'k' means KBimCode and the rest of the alphabets are simplified token to represent each syntactic component. As mentioned in the previous section, KBimCode Language basically has three key components 1) check declaration (the basic syntactic form), 2) statements group declaration, and 3) KBimCode statement. The check declaration and the statement group declaration are the syntactic form of KBimCode, while KBimCode Statement is used as components of the other two components. These three components are defined as: [2] kCheckDef, [3] kStatGruopDef, and [4] kStatDef. The snippet of the KBimCode definition is as follows:

[1] kKBimCodeProgram

```
kKBimCodeProgram
      :      kCheckDef? kStatObjDef?
      ;
```

[2] kCheckDef

```
kCheckDef
      :      CHECK '(' kCheckParamDef ')'
             '{' kStatDef '}'
      ;
```

[3] kStatObjDef

```
kStatObjDef
      :      kStatObject'{' kStatDef '}'
      ;
```

[4] kStatDef

```
kStatDef
      :      kStatLines
      ;
kStatLines
      :      kStatLine+
      ;
kStatLine
      :      (kALUsStat | kIfThenElseStat |
             kKOMDef)
      ;
```

4.3 KBimCode Database

The KBimCode database provides users with the means to acquire specific-purposed rules, even without knowledge of logic rule-based mechanism or KBimCode syntax. It is accessible through the KBimLogic application. As illustrated in Figure 4, there is an interface for selecting the target design rule (written in natural language). The rules are then logically organized based on pre-defined relationships between sentences. As a result, a KBimCode series can be exported by KBimLogic as a computer-executable rule set file such as JSON [26] and XML [27].

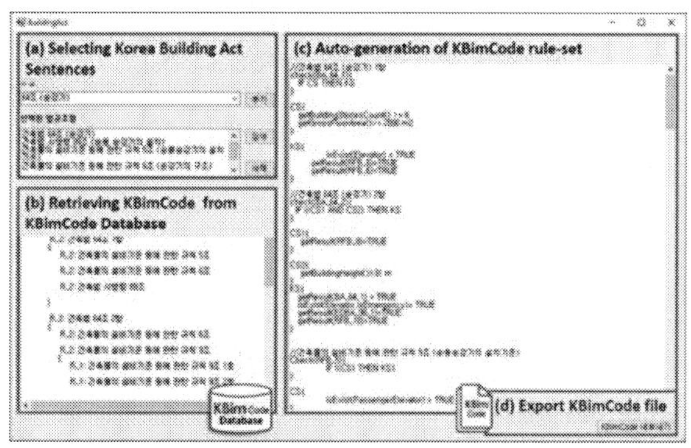

Figure 5. A screenshot of KBimLogic application: by selecting the target design rule (sentences), series of KBimCode can be automatically generated through KBimCode database.

Through the parsing library defined in the BIM assessment software (KBimAssess-lite), the rule set file is executed and compliance checking is performed. Example cases based on the Korean Building Act are tested and demonstrated in the next section.

5 Rule Evaluation

This section describes rule execution and evaluation based on actual demonstration. Specific information about the demonstration is as follows: 1) Design rule: Among Korean Building Act, regulations on the installation of an elevator (exemplified in the table 2) and the installation of escape stair were chosen for the demonstration. 2) BIM model: An IFC model of an actual office building in Seoul, Korea was used as a test model. This building is an office building and has a total floor area of 5,734 m2 with 4 basement floors and 15 ground floors. 3) BIM assessment software: KBimCode was developed to be software-independent and applicable in various BIM assessment tools. Specification of KBimCode is provided to software vendors for the development of syntax analyzer and parser. In this demonstration, one of BIM assessment tools named KBimAssess-lite was used.

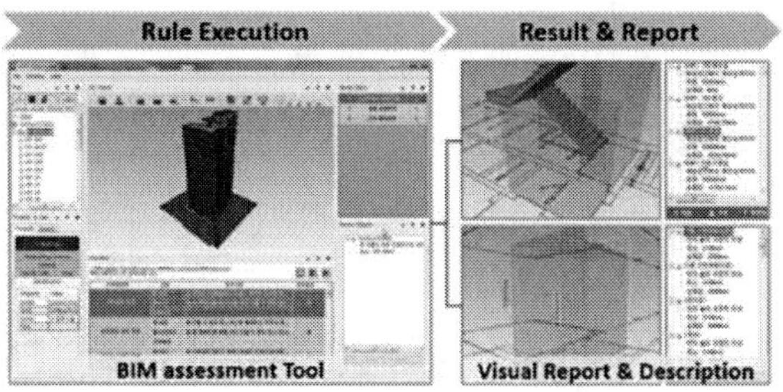

Figure 6. Overview of rule execution, result and report. Visualization as well as a description of the result is reported in KBimAssess-lite.

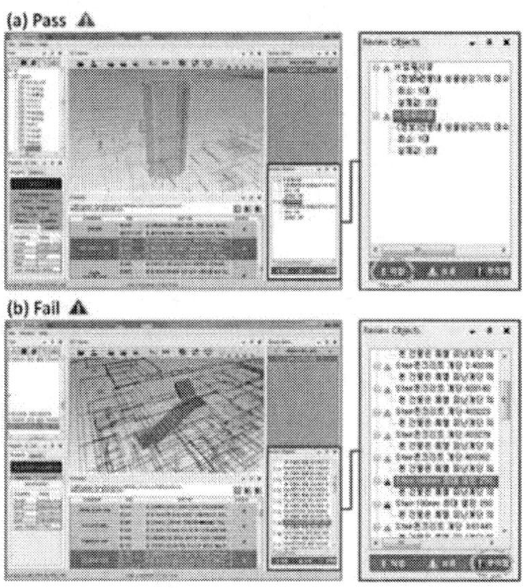

Figure 7. Example of rule execution result: pass case (a) and fail case (b).

KBimCode file is imported in the KBimAssess-lite and the rule is executed. Assessment result is provided with visual report and description. There can be three types of results: 1) pass, 2) fail, and 3) pending. In case of pending, the assessment result should be reviewed by domain expert in manual. Figure 7 shows detailed report of pass case and fail case. Visual report is given in the center of the interface and according description is provided in the lower right corner. The assessment result is reported in object unit. For example, in the (b) fail case of Figure 6, entire stairs in the target building is examined. Stairs with fail result is highlighted and provided with a detailed visual report and description of failing reason.

6 Conclusion

In this paper, we described KBimCode-based applications for rule representation, definition, and evaluation. For rule representation, Korean Building Act sentences decomposed into three parts: 1) noun phrase (object/property), 2) verb phrase (methods), 3) relation (logic). Following the logic rule-based mechanism, sentences are restructured into KBimCode. The definition of KBimCode was explored with two points: 1) lexical/syntactic design strategy, and 2) language definition with EBNF notation. KBimCode is developed into a database and reused for the architect's purpose. Evaluation of KBimcode was demonstrated with a BIM assessment tool named KBimAssess-lite. Result of rule execution can be three types: 1) pass, 2) fail, and 3) pending. Visual report as well as description is provided.

The KBimCode-based application provides a mechanism to generalize translation of natural language rules into a computable format for automated design assessment. In addition, it also suggests specific process for the application of KBimCode. KBimCode is an ongoing project and further verification is in progress. Its application and database will contribute to develop BIM-enabled quality assessment system for building permit in Korea. Although KBimCode was developed based on building permit requirements in Korea Building Act, its application can be extended to various design rules such as design guides, request for proposal etc. for automated code compliance checking.

Acknowledgement

This research was supported by a grant (15AUDP-C067809-03) from Architecture & Urban Development Research Program funded by Ministry of Land, Infrastructure and Transport of Korean government.

References

[1] J. Dimyadi, R. Amor, Automated Building code Compliance Checking Where is it at?, In Proceedings of CIB WBC 2013, Brisbane, Australia, pp. 172-185.

[2] J.Choi, J. Choi, I. Kim, Development of BIM-based evacuation regulation checking system for high-rise and complex buildings, Automation in Construction, 46(2014) 38-49, 2014.

[3] J-K. Lee, J. Lee, Y.-S. Jeong, H. Sheward, P. Sanguinetti, S. Abdelmohsen et al., Development of space database for automated building design review system, Automation in Construction, 24 (2012) 203-212.

[4] C.M. Eastman, P. Teicholz, R. Sacks, K. Liston, BIM Handbook—A guide to Building Information Modeling for Owners, Managers, Designers, Engineers, and Contractors, John Wiley & Sons Inc., Hoboken, NJ, United States of America, 2008.

[5] C. Han, C. J. Kunz, and K. H. Law. Making Automated Building Code Checking a Reality, Facility Management Journal, 1997, pp. 22-28.

[6] D. Greenwood, S. Lockley, S. Malsane, J. Matthews, Automated compliance checking using building information models, In Proceeding of The Construction, Building and Real Estate Research Conference of the Royal Institution of Chartered Surveyors, Paris, France, 2010.

[7] I. Kim, Open BIM based Technological Environment for Building Design Quality Enhancement, Kyung Hee Univ., (Project report), 2015.

[8] W. Solihin, C. Eastman, Classification of rules for automated BIM rule checking development, Automation in Construction, 53(2015): 69–82.

[9] CORENET, CORENET Singapore, Available at: http://www.corenet.gov.sg/

[10] C. Eastman, J.-M. Lee, Y.-S. Jeong, J.-K. Lee, Automatic Rule-based Checking of Building Designs, Automation in Construction, 18(8):1011-1033, 2009.

[11] J-K. Lee, Automated checking of building requirements on circulation over a range of design phase, Ph.D. dissertation, Georgia Institute of Technology, 2012

[12] L. Ding, R. Drogemuller, M. Rosenman, D. Marchant, J. Gero, Automating code checking for building designs, in: K. Brown, K. Hampson, P. Brandon (Eds.), Clients Driving Construction Innovation: Moving Ideas into Practice, CRC for Construction Innovation, Brisbane, Australia, 2006, pp. 113–126.

[13] M. Uhm, G. Lee, Y. Park, S. Kim, J. Jung, J.-K. Lee, Requirements for computational rule checking of requests for proposals(RFPs) for building designs in South Korea, Uhm et al., Advanced Engineering Informatics, 29(3): 602–615, 2015.

[14] J. Zhang, and N.M. El-Gohary, A.M.ASCE, Semantic NLP-based information extraction from construction regulatory documents for automated compliance checking, Journal of Computing in Civil Engineering, 30(2), 2013.

[15] J. Zhang, S.M.ASCE1; and Nora M. El-Gohary, A.M.ASCE2, Automated information transformation for automated regulatory compliance checking in construction, Journal of Computing in Civil Engineering, 29, 2015.

[16] Open BIM based Technological Environment for Building Design Quality Enhancement, On-line: http://dq.kbims.or.kr/

[17] H. Lee, S. Lee, S. Park and J-K. Lee, An Approach to Translate Korea Building Act into Computer-readable Form for Automated Design Assessment, International Association for Automation and Robotics in Construction, 2015.

[18] H. Lee, J-K. Lee, S. Park and I. Kim, An Approach to Translate the Korean Building Act into a Computer-executable Form for Evaluating Building Permit Requirements, Automation in construction, in press.

[19] Method manual, Open BIM based Technological Environment for Building Design Quality Enhancement, On-line: http://designitlab.kr/bim/kbim/method/classification.asp

[20] Wikipedia, Arithmetic Logic Unit, On-line: https://en.wikipedia.org/wiki/Arithmetic_logic_unit

[21] S. Park, H. Lee, S-i. Lee, J. Shin, and J-K. Lee, Rule Checking Method-centered Approach to Represent Building Permit Requirements, International Association for Automation and Robotics in Construction, 2015.

[22] Korea Building Code (English), Statutes of Republic of Korea, On-line: http://elaw.klri.re.kr/kor_service/lawView.do?hseq=32490&lang=ENG

[23] H. Lee, S. Park, I. Kim, J-K. Lee, A Logical Rule-based Approach to the Korea Architecture Code Sentences for BIM-enabled Design Assessment Systems, Korea Design Knowledge Society, 2015.

[24] KBimCode Syntax, On-line: http://designitlab.kr/bim/kbim/kbimcode/KBimCodeLangDef.asp

[25] Parr: 2008, The Definition ANTLR Refernce: "Building Domain-Specific Languages", Pragmatic Bookshelf.

[26] JSON, ECMA-404 The JSON Data Interchange Standard, Online: http://www.ecmainternational.org/publications/files/ECMA-ST/ECMA-404.pdf

[27] XML, W3C, Online: http://www.w3.org/TR/WD-xml-961114.html

OPTIMIZING SURVEILLANCE CAMERA PLACEMENT IN BUILDINGS USING BIM

Ameen Hamza Albahri[a] and Amin Hammad[b]

[a]Ph.D. Candidate, Faculty of Engineering and Computer science, Concordia University, Canada
[b]Professor, Faculty of Engineering and Computer science, Concordia University, Canada
E-mail: a_albah@encs.concordia.ca, hammad@ciise.concodia.ca

Abstract

Surveillance cameras are becoming indispensable tools due to their ability to ensure security in buildings. However, many organizations claim that purchasing, installing and maintaining these cameras are costly. Furthermore, current practices have conducted different 2D and 3D techniques to find the optimum camera placement inside buildings. These techniques have many limitations which can affect the camera coverage. On the other hand, the technology of Building Information Modelling (BIM) is becoming a well-known method that can help in various tasks related to the Architectural, Engineering and Construction (AEC) industry. This study aims to find the optimum placement of a fixed camera inside a BIM model. The proposed method includes: (1) using BIM for specifying constraints; (2) identifying decision variables; and (3) integrating simulation and optimization.

Keywords –

Surveillance cameras; BIM; Optimization; Genetic Algorithm

1 Introduction

The issue of optimizing the camera placement in buildings is becoming significant and it requires more attention by the scientific community. This paper presents a simulation-based method to optimize the placement of a single fixed camera in a building using the technology of Building Information Modeling (BIM). The optimization should ensure the maximum coverage for a specific area. The method is adopting Genetic Algorithm (GA) to address the optimization problem.

The remaining of this work includes the following. Section 2 includes the literature review. Section 3 explains the proposed method. Section 4 implements the proposed method and validates it by conducting a case study. Section 5 summarizes the method and future work.

2 Literature Review

Chelliger [2] outlines that identifying risks and establishing their mitigation standards are the most important features of an efficient security system. Wilkis [3] highlighted that no security system can ensure complete coverage. One of the most significant and serious issues is that surveillance cameras might miss hidden areas behind objects (e.g. columns). These hidden areas can become a vulnerability of an asset. Thus, numerous studies have taken place in order to improve the placement of these surveillance cameras.

The study of Bigdeli et al. [4] was conducted to evaluate the performance of surveillance cameras in railway platforms. The objective of this study is to guarantee clear detection of people during events (e.g. crowded periods). The method showed a real-life trial system that is composed of diverse video analytic systems for detecting important events. The work highlighted that the process of placing various types of surveillance cameras is time consuming.

Furthermore, Kelly [5] mentioned that the placement of surveillance cameras is becoming expensive. Nam and Hong [6] optimized the placement of multiple cameras in indoor spaces taking into consideration the cost constraints. The proposed method includes four steps: space modeling, agent modeling, generating trajectories and selection of optimal camera placement. However, this study did not take into account the importance of a clear segmentation process for the monitoring area, which can facilitate achieving accurate calculation of the camera coverage. Also, their way of finding the important areas by using the path-finding algorithm is inaccurate, since some of these areas might include constraints such as privacy areas.

Yabuta and Kitazawa [7] used a 2D method to find the optimum camera placement in a specific area. Their study specified a number of important parameters related to surveillance cameras as the following: visual distance, Field of View (FOV) and resolution of the camera. Their method showed some weaknesses as the following: (1) the 2D method of calculating the coverage is inefficient in cases where hidden areas are available behind objects because of the variety of elevations; (2) their segmentation process divided the monitoring area into irregular shapes, which will result in inaccurate calculations. This is because small shapes behind obstacles might be missed during the camera coverage process.

Moreover, Chen et al. [8] conducted a study to optimize the real angle of depression of surveillance cameras. The study aimed to validate the applicability of Building Information Modeling (BIM) in improving the performance of surveillance cameras. The study highlighted the role of BIM in detecting obstacles inside the model that are able to affect the camera coverage and placement. The study adopted a 2D method for calculating the camera coverage by using a tool in Revit called "the Fill Region Tool". Although their method proved the ability of BIM in optimizing the real angle of depression for a number of surveillance cameras, it did not consider the optimization of the camera placement.

3 Proposed method

3.1 Using BIM for specifying constraints

Several elements can impact the optimization process as the following:

(1) The placement area of the camera (e.g. ceiling).
(2) The monitored area including the existence of different floor elevations.
(3) The existence of different important areas in the monitored area.
(4) The existence of privacy areas such as bathrooms.
(5) The constraints on the monitored area (e.g. columns) and the placement area (e.g. HVAC system). Other constraints are light sources, shiny surfaces and ceiling signboards which can also affect the visibility of the camera.

BIM is able to identify the locations of different constraints located on the placement or the monitored areas. This ability can help in providing the valid coordinates of the camera placement to the optimization process. This can decrease the time of the optimization process by making the camera placement restricted to valid positions and avoiding the positions of constraints. This study is focusing on four types of constraints:

(1) Geometrical constraints on the ceiling (e.g. ceiling signboards).
(2) Operational constraints (e.g. in places where the camera is exposed to vibration generated by the HVAC system).
(3) Logical constraints (e.g. in places where the camera is facing walls).
(4) Legal constraints (e.g. in case if the monitoring area includes privacy areas).

3.2 Identifying decision variables

The decision variables of the camera placement include the position and the orientation of the camera. The camera position is represented by the values of *X*, *Y* and *Z* coordinates, while the camera orientation is represented by the values of *Pan* and *Tilt* angles. The *Y* variable (the height) will be fixed during the optimization process in a way to ensure the maximum height of the camera. The solution of the GA includes one gene that gives a specific pose of the single camera on the ceiling including five variables *X*, *Y*, *Z*, *Pan* and *Tilt* angles.

Moreover, it is important to note that a bigger population size results in better optimization results and longer computation time, and vice versa. Parameters related to the crossover and mutation processes should be also specified in the GA. For simplicity, this study adopted the rectangular shape for the placement area.

The Preference-Inspired Co-evolutionary Algorithm (PICEA) is used [9]. The general algorithm aims to generate candidate solutions for a specific number of generations. Every generation includes a number of parents which are exposed to crossover and mutation processes in order to produce new offsprings. Parents and offsprings are then combined together and the members of the resulting population are evaluated based on their fitness. The resulting population is then subjected to a truncation selection in order to produce new generation of parents. This process is repeated until the termination criterion is achieved (i.e. maximum number of generations).

3.3 Integrating simulation and optimization

In order to apply GA for the camera placement process, the following steps are applied:

(1) BIM is used to identify the search space of the placement area and exclude the constraints.
(2) These feasible coordinates are then sent to the optimization process in order to generate the decision variables.
(3) As shown in Figure 1, GA will initialize the first generation (*g*) which has specified population size (*P*) of solutions.
(4) The first solution (i.e. gene) of the first generation is sent to a coverage simulation process in order to calculate the fitness. This process will continue until all solutions of the current generation are

evaluated.

(5) GA will then produce the next generation by applying the crossover and mutation. This process will continue until the GA finishes from all generations.

(6) Finally, the optimum solutions are specified and simulated in BIM.

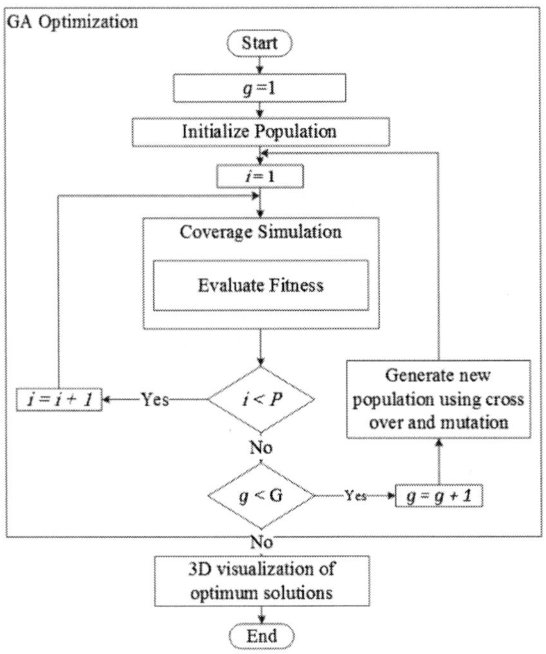

Figure 1 Proposed method

The previous process includes integration between two systems. One represents the optimization process, while the other is responsible of calculating the coverage. In order to calculate the camera coverage, the following steps should be conducted: (1) accessing the data from the BIM, (2) adjusting the camera features (i.e. FOV, the camera height and clipping planes (far and near points)), (3) placing the camera automatically based on the data of each solution in the population, (4) generating cells on the targeted surface (i.e. the floor), and (5) calculating the coverage using Equation 1, [10] and Equation 2 [1]. The output will be sent to the GA to evaluate the fitness as previously discussed. It is also important to mention that the objective function is aiming to maximize the coverage by applying Equations 1 to 7.

$$CC_{a_i} = \sum_{v=1}^{n`} CC_{iv} \times IV_i \qquad \text{Equation 1}$$

$$\text{Coverage of camera } K_x = \sum_{i=1}^{m}\sum_{j=1}^{n}\sum_{v=1}^{n`} \frac{CC_{iv} \times IV_i}{C_{ij} \times IV_i} \qquad \text{Equation 2}$$

Subject to:

$$0 \le X \le L \qquad \text{Equation 3}$$

L is the maximum value of the length

Cameras should be on the ceiling or on the upper part of the wall

$$0 \le Y \le H \qquad \text{Equation 4}$$

H is the maximum value of the height

$$0 \le Z \le W \qquad \text{Equation 5}$$

W is the maximum value of the width

$$0 \le Pan \le 360° \qquad \text{Equation 6}$$

$$0 \le Tilt < 90° \qquad \text{Equation 7}$$

where C_{ij} represents the cell j in important area i, and j = 1:n, IV_i is the importance value assigned all cells in area i, and i = 1:m, CC_{iv} is the covered cell v in area a_i and v = 1:n`, K_x is camera x.

4 Implementation and case study

A case study is applied in order to validate the proposed method. The case study is implemented on the ground floor of the Engineering and Visual Art (EV) building in Concordia University, as shown in Figure 2. This figure shows the current locations of seven surveillance cameras installed inside this model.

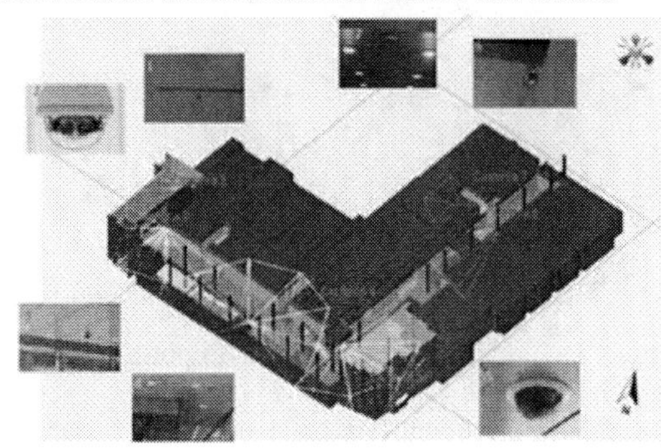

Figure 2 BIM model aligned with FOVs in the game engine [1]

This case study is validating the proposed method using BIM, game engine tool and Matlab for optimizing the placement of single fixed camera in the BIM model. In addition, the floor has some physical constraints such as columns that have an impact on the camera coverage. Autodesk Revit was used to develop the BIM model [11]. The BIM model is sent to Unity3D [12] using FBX format for the purpose of simplifying the programming processes which used C# language.

In Unity3D, the ground floor of the BIM model was divided into important areas as shown in Figure 3. The important areas include cells which have equal cubic shapes. These cells are assigned with different importance values based on their positions inside the model. The importance values can help in defining the optimum position of the camera. These cells can offer accurate results of the camera coverage by adjusting their sizes as discussed in Albahri and Hammad [1]. Also, a grid system is developed in order to facilitate the camera placement process.

As previously discussed, the optimization process includes integration between the GA tool of Matlab [13] and Unity3D. The GA is set to 65 generations. The population size of each generation is 200 candidate solutions. GA will send these generations one by one in the form of a text file to Unity3D.

Inside Unity3D, this text file is processed as the following: (1) Unity3D runs the placement process for the selected 3D camera based on the given solution in the text file; (2) Then, it calculates the coverage of the selected camera; and finally (3) The results of the camera coverage are saved in text file and sent to Matlab in order to continue running the next generations.

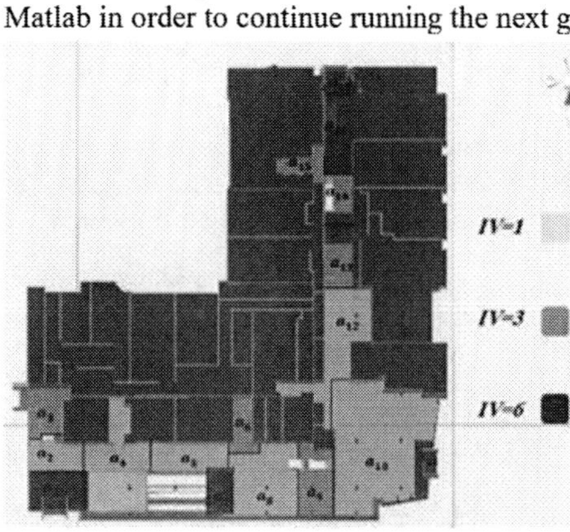

Figure 3 Plan view showing the important areas with their Importance Values (IV) [1]

It is important to mention that the optimization process is conducted on the areas from a_5 - a_8, as shown in Figure 4. These areas were covered by cells that have different importance values (i.e. value of 1 for the lowest importance, value of 3 for the medium importance and value of 6 for the highest importance).

Moreover, the placement surface has the dimensions of 2500 cm on the Z axis and 1400 cm on the X axis. The placement surface is divided by a number of grids that can define the camera placement. In this case study, 56 grids were generated on the X axis, and 100 grids on the Z axis. The selected camera has the FOV = 60°.

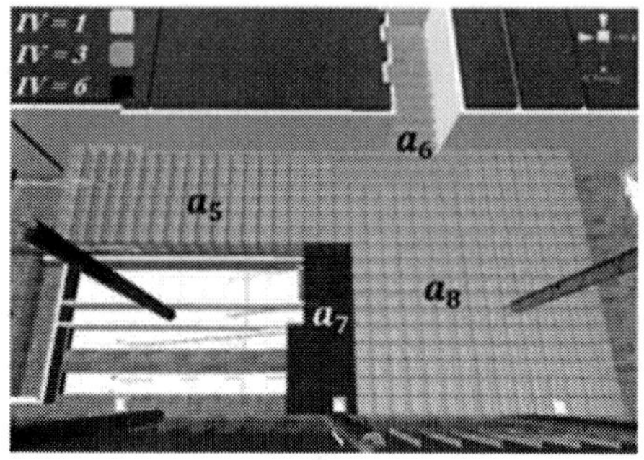

Figure 4 Areas a_5 to a_8 on the monitoring area

Table 1 shows the best 10 solutions with their coverage results. For example, in solution 10, the GA found a camera position at X = 100 cm; Y = 500 cm; Z = 2500 cm; pan angle = 205°; and tilt angle = 61°. This solution offered camera coverage of 65%. Solution 1 has the optimal coverage of 90%, as shown in Figure 5, with coordinates X = 1250 cm, Y=500 cm, Z = 1550 cm, pan angle = 115° and tilt angle = 62°. Figure 6 shows the global fitness, best fitness and the mean fitness curves. The best fitness curve displays a fluctuation in the camera coverage from the 8th generation to the 27th generation. This is because GA is randomly searching for various solutions which lead to various fitness results. The curve faced coverage of 82% in the 28th generation and continued increasing until it reached 90%, which is the global fitness value, in the 58th generation. The GA showed a slight increasing of the coverage from the 54th generation to the 65th generation with coverage values from 88% to 90%, respectively.

Table 1 Best 10 solutions of the optimization process

Solution #	Coverage (%)
1	90
2	89
3	88
4	86
5	84
6	82
7	78
8	74
9	69
10	65

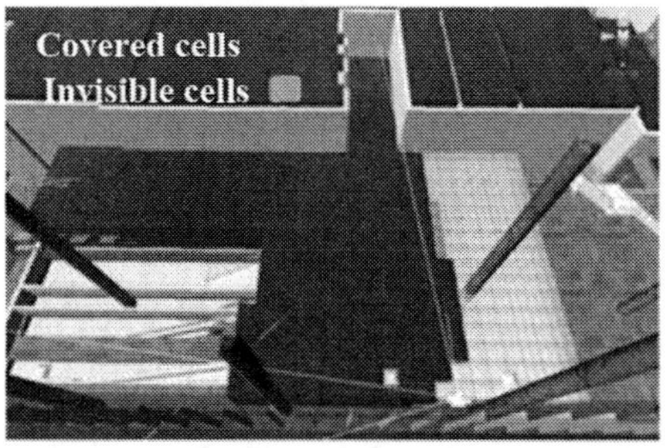

Figure 5 3D Simulation of solution 1

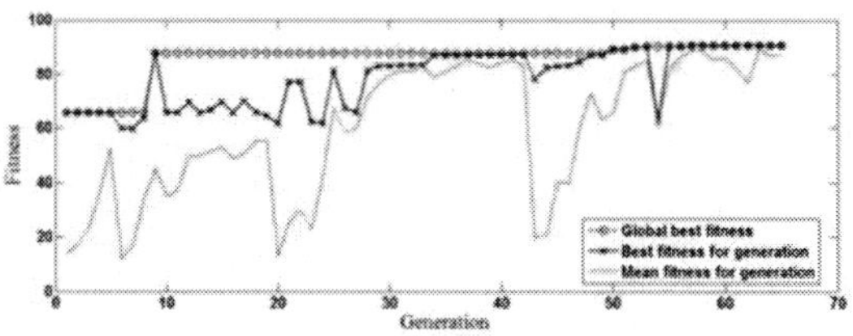

Figure 6 GA fitness progression of the case study

5 Summary and future work

This work presented a simulation-based optimization method for the placement of a single fixed camera inside a BIM model. The optimal placement of the camera aims to achieve the maximum coverage in a specific area inside a building. In addition, BIM is used to define the constraints on the ceiling in order to determine the valid positions of the camera placement. The proposed method adopted integration between BIM and a GA optimization process. The proposed method included the following: (1) using BIM for specifying constraints; (2) identifying decision variables; and (3) integrating simulation and optimization. A case study was developed to validate the method. This method can assist individuals who are responsible of the camera installation to accurately determine the suitable placement in the facility. Future work will include the placement of multiple camera types inside buildings.

Acknowledgement

Special thanks to the Saudi Arabian Cultural Bureau in Canada for their support during the planning and the implementation of this work.

References

[1] Albahri, A., and Hammad, A. (2015). Evaluating the Coverage of Surveillance Cameras Using BIM and Game Engine. In 2nd International Conference on Civil and Building Engineering Informatics (ICCBEI 2015), Tokyo.

[2] Chelliger, D., (2008). From the Ground up: Security for Tall Buildings. CRISB Report. Connecting Research in Security to Practice. In ASIS International. Alexandria, VA, USA.

[3] Wilkis, K. (2015). Who's Watching You? Video Surveillance-User Beware. Extracted on 23 Oct 2015 from: < http://blog.willis.com/2015/05/whos-watching-you-video-surveillance-user-beware/ >.

[4] Bigdeli, A., Lovell, B. C., Sanderson, C., Shan, T., and Chen, S. (2007). Vision Processing in Intelligent CCTV for Mass Transport Security. In IEEE Workshop on Signal Processing Applications for Public Security and Forensics. (SAFE'07), 1-4.

[5] Kelly, H. (2013), After Boston: The Pros and Cons of Surveillance Cameras, CNN Tech. Extracted at 25 Sep 2013 from: <http://www.cnn.com/2013/04/26/tech/innovation/security-cameras-boston-bombings/>.

[6] Nam, Y., and Hong, S. (2014). Optimal Placement of Multiple Visual Sensors Considering Space Coverage and Cost Constraints. Multimedia Tools and Applications, *73*(1): 129-150.

[7] Yabuta, K., and Kitazawa, H. (2008). Optimum Camera Placement Considering Camera Specification for Security Monitoring. IEEE International Symposium on Circuits and Systems (ISCAS), pp. 2114-2117.

[8] Chen, H. T., Wu, S. W., and Hsieh, S. H. (2013). Visualization of CCTV Coverage in Public Building Space Using BIM Technology. Visualization in Engineering, 1(1):1-17.

[9] Purshouse, R.C., Jalba, C. and Fleming, P.J. (2011). Preference-Driven Co-evolutionary Algorithms Show Promise for Many-Objective Optimisation. In Evolutionary Multi-criterion Optimization (EMO), pp. 136-150.

[10] Amriki, K. A., and Atrey, P. K. (2014). Bus Surveillance: How Many and Where Cameras Should be Placed. Multimedia Tools and Applications, 71 (3): 1051-1085.

[11] Autodesk, (2014). Revit. Extracted on 16 Sep 2014 from: <http://www.autodesk.com/products/revit-family/overview>

[12] Unity (2015). Create and Connect With Unity 5. Extracted on: 28 Oct 2015 from: < https://unity3d.com/ >.

[13] De Freitas, A. (2012). Matlab Central: Open Genetic Algorithm Toolbox. Extracted at 15 Oct 2015 from: <http://www.mathworks.com/matlabcentral/fileexchange/37998-open-genetic-algorithm-toolbox>.

Integration of BIM and FEA in Automation of Building and Bridge Engineering Design

F. Fedorik1 and T. Makkonen2 and R. Heikkilä3

Structural Engineering and Construction Technology, University of Oulu, Finland
E-mail: Filip.Fedorik@oulu.fi, Tomi.Makkonen@oulu.fi, Rauno.Heikkila@oulu.fi

Abstract –

Current development in structural engineering allows designers to create efficient and controlled structural systems, which lead into savings in terms of financial cost, time consumption and lack of lucidity in structural designs. Building Information Modeling (BIM) creates a great opportunity for cooperation of individual experts within one project to maximize an efficiency of the design and minimize errors between concrete details of the entire design and management process.

Physical behavior of structures could be currently controlled with high accuracy by applying numerical simulations using Finite Element Analysis (FEA). Full integration of BIM and FEA allows complex monitoring of structural design during the entire development through initial design, manufacturing processes, and construction through thel structures' life-cycle. The presented paper deals with experiences of major bridge and building engineering projects concerning interoperability of BIM and FEA. Future vision of BIM and FEA integration considering current key findings in the visionary model are also presented. Most of BIM software already enable interoperability of BIM and FEA, however certain issues are still not achieved, especially considering extraordinary geometry characteristics of structures. Because of incomplete interoperability of BIM and FEA, engineers apply them separately, which in turn impairs efficiency of the design. Bridge and building engineers play major role in software applicability and development as well as software developers. The errors occurred during a design should be reported to software developers to improve the applicability and minimize errors in the future structural designs.

Keywords –

BIM; FEA; Interoperability; Bridge engineering; Building engineering

1 Introduction

Efforts towards economic and energy efficient designs in construction engineering and other associated professions support the invention and development of new technologies that allow quicker and more efficient design and control of a structure during the entire life-cycle. Building Information Modeling (BIM) represents one of the most important information management systems of construction industry. It allows detailed information about all objects implemented in a project, and also controls their interaction and behavior while they are in use. The aim of the BIM is to connect of teams and experts involved in the project allowing their collaboration and sharing of ideas while minimizing errors caused by e.g. misunderstanding of cooperators and incompatibility of individual objects.

Traditionally, the information is exchanged between project participants through two-dimensional drawings and other paper documents such as cost plans. In this matter, the collaboration is limited and opens space for errors in the design leading to inefficient and unsustainable results and/or unexpected additional costs. The BIM represents digital virtual model where each elements that effects the structural design and/or its life-cycle is described in detail. The BIM is an IT-based information system recently operated by an increasing amount of Architecture, Engineering and Construction (AEC) companies. The process of the design usually consists of architectural design that creates exterior and interior features of the structure, consider renewable energy system (RES) effort and incorporation of energy analysis tools [1]. Then, structural design provides a sustainable design of individual elements and complex structure as a unit with collaboration regarding geological prospects. In the case of residential and non-residential buildings, systematic design is likely included; for instance heating, ventilation and air conditioning (HVAC). Traffic and bridge engineering structures design features of surrounding infrastructural accessibility for vehicles and/or design of roads or rails.

The BIM represents area for specialists, researchers and developers not only in the building engineering and financial resort, but also in an information system (IS) research that can be of significant contribution in current and future BIM development [2]. The collaboration across different applications within BIM requires creation of a multi-disciplinary platform with identified technical features. BIM-server research and development facilitates the collaboration of BIM participants towards intelligent and automated collaboration [3].

One of the most important numerical techniques used by structural designers for physical phenomenon simulation in structures is the Finite Element Analysis (FEA). This method allows designers to simulate the natural behavior of solids, liquids, gases and their interaction. Static and dynamic analysis of common structures, as well as extraordinary structures, can be simulated with high accuracy. The FEA achieves very good results in many professions, such as medicine, mechanical engineering, building and structural engineering, automotive and aeroplane industry but also in gastronomy, etc.

In comparison, the BIM considers every detail of the design including processes in time-schedules. FEA usually simplifies the structural model that leads to saving time and energy. Interoperability of FEA and BIM extends complexity of digitalized designs leading to simulation of structural behavior before, during construction until the end of their life-span.

Integration and interoperability between the BIM and other applications have been studied to promote and develop the BIM, such as BIM and geographic information system GIS [4, 5], architectural design [6, 7], site-linked BIM concept [8], etc.

2 BIM and FEA application

The current market already offers many software applications allowing the cooperation of BIM and FEA. Many companies are experienced in interoperability of BIM and FEA by transferring .DWG and .DXF formats leading to transfer geometrical entities only. The interaction of BIM and FEA is performed via different software that usually causes loss of some information, depending on geometric complexity. This procedure does not usually allow the exchange of material properties and boundary conditions. The importance of full interoperability between BIM and FEA consists in smart data exchange, where certain simplifications are made and the interaction can be performed to both directions. This means that the BIM model transfers digital data into the FEA considering all necessary parts needed for the study and the BIM is updated at the same time as FEA is performed. The process of integration of FEA and BIM can be illustrated as in the Figure 1.

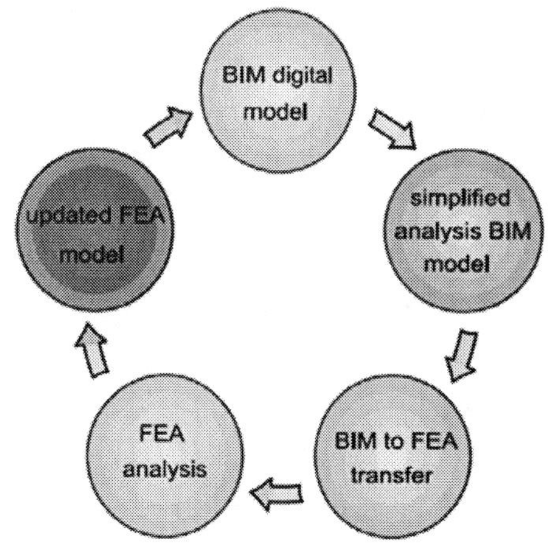

Figure 1. BIM and FEA integration

The current state of the integration of FEA and BIM consists in creating model in BIM, transferring the model to FEA, numerical simulation and evaluation of the obtained results. In the case the results are accepted, the model is not transferred back to the BIM, but the used as such. Otherwise, the model is modified until the goal is achieved and then transferred back to BIM. The difficulties appear especially if detailed and special analysis, such as airflow, thermodynamic, seismic, and other mostly dynamic analyses are needed. The mentioned analyses are usually time consuming and have high requirements for hardware, hence certain simplifications must be used. Structural details, for example bridge equipment, such as handle bars, electrification, lamps, etc. are removed from the FEA model to minimize its size as far as the removed elements don't markedly effect the solution. From this point of view, the FEA should still be considered as individual process requiring detailed professional attention.

In many cases, BIM software allows a certain finite element analyses that can be applied in the BIM directly. On the other hand, FEA specialists prefer specialized multi-physics FEA software where more options of special simulations can be realized. This slows down the BIM process considering difficulties that may occur in transferring the digital model. If the simulation is performed within the BIM software, the integration is usually very good and the probability of occurred errors is minimized. However, extraordinary structures such as bridges, towers, high-rise buildings, etc. quite often represent structures which are significantly affected by dynamic forces, i.e. wind and earthquakes. These kind of analyses are better simulated in multi-physics software in detail, because the correct numerical model definition, solution process and achieving results is likely more accurate and controlled.

The FEA is generally represented by three procedures;

preprocessing, solution and post-processing. The current FEA applications already allow parameterization of the model, which opens space to mathematical techniques that require re-definition in the preprocessor based on results obtained in the post-processing, such as optimization. Hence, the optimization or other loop technique applied in the FEA create their own part of the design. Then, the illustration in the Figure 1 can be updated as follows (Figure 2).

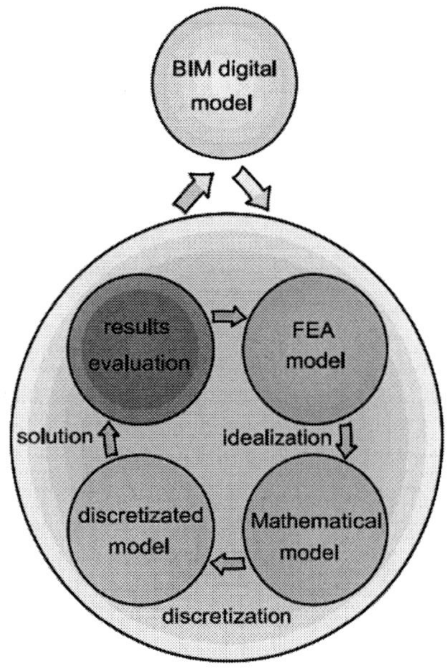

Figure 2. BIM and FEA integration considering loop computing technique

2.1 FEA in bridge and building engineering

The FEA is a numerical technique for finding approximate solutions of partial differential equations as well as of integral equations by variable methods. The method consists in discretization of complex model into small elements. The FEA allows the simulation of a structural behavior close to reality that leads to more accurate and economically efficient designs. The FEA applications have recently applied extensional modulus that allows expanding the structural and mechanical analyses in advanced mathematical techniques supporting efficient designs, for instance optimization. Application of mathematical techniques usually requires a parameterized FEA model. It means the model geometry, loading, material properties and other mechanical and physical features are described in the parameters that represent variables in mathematical expression. This allows the adaptability of the model and its properties to create an optimal design. The FEA represents important part of current designs regardless of the type or characteristics of the structure. The simulation of advanced physical phenomena has been applied in analyses of dynamic behavior of extraordinary structures such as bridges, towers, skyscrapers, wind mills, etc. caused by imposed loading such as wind and seismic activities. The FEA can perform one- to three-dimensional structural problems. Three-dimensional extraordinary structural models represent mathematical challenges that can hardly be performed manually, especially when considering the objects' dynamic behavior. The 3D simulation represents demanding modelling even for the currently available hardware. In many cases, complex models consist of millions of finite elements that represent several equations to be solved, depending on specific features of the analyses (Figure 3).

Figure 3. FEA bridge model

Time-dependent analyses may take several hours or days to solve. This may cause issues in the cooperation between interoperable team members and slows down the entire design process. That is why the FEA models are usually simplified to achieve satisfying accuracy of results but without excess time consumption. Currently the FEA is, among others, also applied in "simple" structures such as residential and non-residential building designs to simulate air and heat flow or mass transfer. This allows control beyond energy efficiency of the structures and health conditions indoor and inside the structural elements to ensure health environment of inhabitants and sustainability of the buildings. Advantages of the FEA consists in accuracy and wide-range of usability in many professional fields. As a disadvantage one could consider the strong dependency of errors and need of experience and judgment of a user – the method is still far from the simplicity required to be used by a regular consumer. In the case of contradictions arise during the preprocessing stage, the results and sequential design may cause fatal errors.

2.2 BIM in bridge and building engineering

Standards lead engineers through parts of BIM that have been developed by the BuildingSmart and are continuously updated. Guidelines determining the requirements for BIM in new building designs, renovation and management facility were published in Common BIM Requirements 2012 (COBIM 2012). COBIM 2012 includes, among others, the requirements for architectural design, structural design, energy analysis, management of a BIM project, use of models in facility management, use of models in construction, etc.

BuildingSMART has estimated a usage of BIM in building engineering sector about 20-30% in building projects. It suggests 20% use in public sector, less than 10% in private sector, 40-50% in large construction companies, less than 10% in small construction companies and around 70% among AEC's. The lack of BIM application is specially caused by private companies that have not found its benefits yet. The discouragement is caused by high initial expenses where the adaptation of structural design requires further education of designers, constructors, financials and customers about BIM.

BIM in bridge designs is also widely discussed and studied topic[9]. The most bridge designs still prefer traditional methods [10] although pilot projects of BIM in bridge designs applications have proven the benefits of the 3D visualization and close collaboration of all involved parties.

Next to the continuously developing guides and standards for BIM application in building engineering, a guide applicable to bridges has been recently provided. BIM Guidelines Applicable to Bridges provides instructions on uniform processes for the BIM-based design, implementation and maintenance of bridges [11]. The guideline includes information of how to proceed during the entire process and life-cycle of BIM-based design of bridges. The investigation and possibilities in BIM bridge designs have been performed in development projects, e.g. Älykäs Silta, 5DSilta, 5DSilta2 and 5DSilta3. This research focused on the development of 3D laser scanning GPR scanning of bridges and transferring the measurement and road geometry to the 3D bridge modelling [12, 13].

3 BIM interoperability

The AEC industry still widely uses geometry-model based applications although a building-specific data model has been on the market for more than 20 years [14]. The issue in commercial vendors' applications is using the internal communication technique that doesn't allow direct collaboration with other programs and applications. The proprietary techniques slow down or disable communication between different object-based building data models. Interoperability between project team members and software applications is a key to success in efficient design in any kind of structural engineering projects.

The neutral platform for the object-based building data model is represented by the Industry Foundation Class (IFC) developed and continuously updated by BuildingSmart Alliance. The BuildingSmart Alliance (International Alliance for Interoperability – IAI) is a global consortium of commercial companies and research organizations founded in 1995 [14]. BuildingSmart drives the transformation of the built environment through entire design, processing and maintaining and defines international standards for OpenBIM. The idea of OpenBIM is the interoperability of various BIM applications regardless on developer or producer of the software. The smooth interaction of team-members participating the BIM project is the key in efficient designs. The OpenBIM welcome software vendors, designers, engineers, constructors, building owners and investors as well as construction managers in collaboration and developing the OpenBIM idea. The interoperability should be achieved by applying IFC. The IFC object model is in some cases considered as too complex and that it may cause speed and accuracy problems. Therefore, effort towards processing a full object-based model requires simplification of the model into subsets or novel use of XML [15, 16].

The IFC represents standardized protocol for interoperability defining rules for data description and representation the buildings. It is a language for the text-file representation of models and adjacent facilities including constructs for a wide range of modelling features, such as geometries, management facilities, electrical, mechanical and other subsystems, processing, maintaining and cost models [17].

4 BIM and FEA integration

The successful integration of BIM and FEA consists in proper communication between the applications to control the efficiency of the design and correct data transfer. Special engineering field discussed in a matter of the BIM and FEA interoperability is a bridge design. Bridges are, in most cases, extraordinary structures of unique shapes and features, where a major role play dynamic loading and infrastructural design.

Interaction between BIM and FEA can be performed via direct- or indirect link allowing access to the information needed to analyze the problem and upgrade all changes in the BIM process. The direct link is applied within a software or from one to another via Application Programming Interface (API). An open API is an interface that enables interaction of two software applications to exchange data and communicate. The interoperability of BIM and FEA in commercially available applications has proven successful in both direct and indirect linking of simple structures, but in the case of complex structures the results are yet merely satisfactory. The direct link allows data transfer of geometry, loads, load groups, material properties. Indirect link (IFC) allows geometry and material properties data transfer only. On the other hand, the exterior data transfer through IFC is satisfactory for wide range of designs, as the direct link needs a further development [9]. Also other techniques, such as hybrid method of BIM and FEA application has already been studied in scattering properties of a dielectric targets above a dielectric rough surface [18].

Difficulties in BIM and FEA interoperability in bridge design may occur in the case of complex geometry, where some geometry entities could be of irregular shapes. The accuracy of the FEA depends on a number of created elements. Hence, if the numerical model isn't expressed by suitable amount of finite elements, the obtained results may affect efficient design of the entire structures. Beside the analyses performed within the numerical model representing the real structure, some analyses require also a model of the surrounding space. For instance, analyzing airflow around the structure or its components requires numerical model of the ambient space and the structure is represented by its boundaries only (Figure 4).

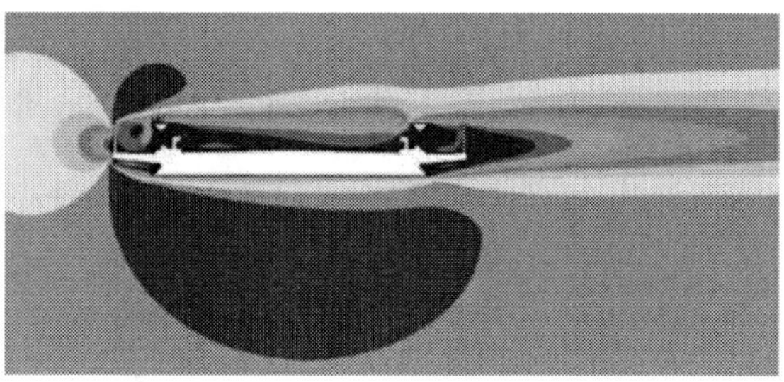

Figure 4. FEA bridge cross-section wind flow

5 Conclusion

Recently, the use of BIM design has increased its popularity. It allows not only 3D visualization of a model but at the same time it provides management information data. Through the BIM, all project participants are connected and share information to achieve more efficient design, manufacturing and construction process and maintenance the structure during entire life-span. The close cooperation minimizes mistakes and errors in the design since any change in the design can be updated by cooperative disciplines.

The idealized full interaction of BIM and FEA saves time and increases efficiency, considering a model is created once in the BIM and linked to the FEA application while any change in the FEA is reflected at the same time back to the BIM model. Although many applications already allow both BIM and FEA, where the data transfer is fully integrated via direct link, the cooperation of project parties is limited by the software they use. The interoperability with no regard to software and applications used can be achieved by IFC that represents open standardized data protocol creating a comprehensive representation of building model [17]. Generally, the interoperability of BIM and FEA cooperates sufficiently up to a certain level. Yet, the further development must be provided in this area.

AEC industry has been developing the BIM continuously to improve the interoperability of all involved disciplines in design, however full integration of FEA has not been developed in detail yet. Construction companies have usually sufficient knowledge about BIM and high proficiency in FEA analyses, but the integrated design knowledge is very limited.

A major progress has been achieved in bridge engineering BIM application as BIM Guidelines Applicable to Bridges has been currently published by Finnish Transport Agency. The guideline defines instructions covering the design phase and content of the modelling [11].

6 Future vision

The aim of the BIM is to create complex control beyond a structural design and management. The complexity consists in digital storage of all data effecting the structural design, structure and management facility throughout the entire life-cycle. Full interoperability of the entire spectrum of involved parties would improve economic conditions, efficiency of the structure and its maintaining to ensure environmental and structural health and extend life-cycle of the construction. Complex BIM design in building and bridge engineering allows controlling behavior of the structure and protect unexpected physical issues before an extended damage. The proper interoperability between BIM and FEA increases the ability to control the structural behavior at its initiation and predict future errors or damage. The importance consists in careful selection of structural elements applied in the FEA.

Essential part of a successful future vision is fully developed IFC that allows collaboration between different software at all four levels; file-level, syntax-level, visualization-level and semantic-level [19]. The benefits from the BIM applications have been already considered, especially in many quite large projects. On the other hand, most of the small companies are not prepared for integrated design yet. The promotion of the BIM application in bridge and building engineering can be performed by arranging workshops and education events with software and engineering vendors. The main issue consists in a fear of investing into the education programs and adapting the design processes into a BIM.

The investors and companies should be aware of the future benefits the systems integration can offer.

References

[1] Gupta A. Developing a BIM-based methodology to support renewable energy assessment of buildings, PhD thesis, Cardiff School of Engineering, Wales, UK, 2013.

[2] Merschbrock C. and Munkvold B. E. A research review on Building Information Modeling in construction - an area ripe for IS research. *Communications of the Association for Information Systems*, 31(10), 2012.

[3] Singh V., Gu N. and Wang X. A theoretical framework of a BIM-based multi-disciplinary collaboration platform, *Automation in Construction*, 20:134-144, 2011.

[4] Karan E., Irizarry J. and Haymaker J. BIM and GIS integration and interoperability based on semantic web technology, *Journal of Computing in Civil Engineering*, 10.1061/(ASCE)CP.1943-5487.0000519, 2015.

[5] Karan E. P. and Irizarry J. Extending BIM interoperability to preconstruction operations using geospatial analyses and semantic web services, *Automation in Construction*, 53: 1-12, 2015.

[6] Arayici Y., Egbu C. and Coates P. Building Information Modelling (BIM) implementation and remote construction projects: Issues, Challenges and Critiques, *Journal of Information Technology in Construction*, 17: 75-92, 2012.

[7] Gustavsen K. Z. Interoperability between architectural BIM models and structural analysis software, MSc thesis, Technical University of Denmark, Department of Civil Engineering, Denmark, 2012.

[8] Wang M. Building Information Modeling (BIM): Site-building interoperability methods, MSc thesis, Worcester Polytechnic Institute, 2011.

[9] Saifi H. Integration of Building Information Modeling (BIM) and Finite Element Analysis (FEA) in Bridge Engineering, MSc thesis, Aalto University, School of Engineering, Finland, 2012.

[10] Johansen J. BIM in bridge design, MSc thesis, Norwegian University of Science and Technology, Department of Structural Engineering, 2013.

[11] Finnish Transport Agency, Guidelines of the Finnish Transport Agency 6eng/2014, ISBN 978-952-317-156-5, Helsinki, 2014.

[12] Heikkilä R. Siltojen rakentamisen, korjaamisen ja kunnossapidon automaation kehittäminen (5D-SILTA2), Liikenneviraston, Tutkimuksia ja selvityksiä, 43/2011, 2011.

[13] Heikkilä R. Siltojen tuotemallintamisen ja rakentamisautomaation kehittäminen (5D-SILTA), Tiehallinnon selvityksiä 22/2008, 2008.

[14] Khemiani L. The IFC Building Model: A Look Under the Hood, Anaylis, Research, and Reviews of AEC Technology, AECbytes Feature, March 30, 2004.

[15] Redmond A. and Smith B. Exchanging partial BIM information through a cloud-based service: testing the efficacy of a major innovation, In *IBEA Conference 2011*, pages 84-95, London, UK, 2011.

[16] Pniewski V. *Building Information Modeling (BIM) interoperablity issues in light of interdisciplinary collaboration*, Collaborative Modeling Ltd, www.collaborativemodeling.com, 2011.

[17] Aranda-Mena G. and Wakefield R. Interoperability of building information: myth or reality?, In *eWork and eBusiness in Architecture, Engineering and Construction. ECPPM 2006: European Conference on Product and Process Modelling 2006 (ECPPM 2006)*, pages 127-135, Valencia, Spain, 2006.

[18] Guo L. X. and Xu R. W. Hybrid FE-BIM for electromagnetic scattering from a dielectric target above a dielectric rough surface, In *Geoscience and Remote Sensing Symposium (IGARSS), 2014 IEEE International*, pages 5215-5218, Quebec City, Canada, 2014.

[19] Steel J., Drogemuller R. and Toth B. Model interoperability in Building Information Modelling, *Software and Systems Modeling*, 11(1):99-109, 2012.

Challenges and Opportunities for Implementation of Laser Scanners in Building Construction

Samad M.E. Sepasgozar[a], Cynthia (Changxin) Wang[a], Sara Shirowzhan[b]
[a] Faculty of the Built Environment, University of New South Wales Australia
[b]School of Civil and Environmental Engineering, University of New South Wales Australia
E-mail: samad.sepasgozar@gmail.com, cynthia.wang@unsw.edu.au, s.shirowzhan@unsw.edu.au

Abstract –

Digital technologies, including hand-held laser scanners and locating systems, are being increasingly introduced to the construction market and have the potential to provide real-time data to create construction information modeling (e.g. shop drawings and as-built). Focusing on the unique advantage of these technologies, this paper identifies the challenges and opportunities for implanting such technologies to create multi-dimensional information modeling based on updated data from construction objects. The results of the paper will be based on implementation of a proposed framework to assist in semi-automated conversion of the raw 3D point clouds from advanced scanners into a compact, semantically rich model aiding in updating construction drawings and developing as-built models. A couple of advanced laser scanners have been deployed within an educational building and have scanned a floor as a sample in order to collect the data. In addition, further data was collected using novel positioning systems, to generate updated data from objectives and labor working in the sample area. The findings show that mobile laser scanners and the novel locating system are efficient tools for updating data to assist construction and maintenance managers to revise the drawings and use updated multi-dimensional models, and are about 50% less expensive than the current technologies. It was found that the practical framework - using mobile laser scanners - enables construction managers to acquire updated information from complex objects, using a real-time communication system. The study provides a step towards understanding the possibility of implementing advanced technologies and internet-of things in construction, and assisting in developing a shop drawing revision modelling for construction purposes.

Keywords –

Laser Scanner, Implementation, Challenges, Ease of Use, Building Information Modelling, Drawing.

1 Introduction

There is a desire to generate digital just-in-time information in construction for management purposes, and to convert all construction documentation to an electronic edition compatible with other software and existing information molding. The emerging technology such as laser scanners coupled with positioning systems provides an opportunity to move forward to a cost-efficient way of generating required information in construction and also in digital as-built creation. Many governments encourage construction engineers to use digital 3-dimensional (3D) documentation in BIM, for example: The United States General Service Administration's Office of the Chief Architect [19], Singapore, the UK, and Australia. However, the current practices' accuracy, possibility of acquiring data from complex objects indoor and outdoor, and skill required to work with the data needs to be improved for as-built documentation purposes [10]. Details and specifications of many construction objects that are left behind walls, covered by another item or buried for another scheduling could not be detected in each stage. Therefore, a tool capable of detecting objects and collecting accurate coordinates of the objects in a timely manner would have a significant value for contractors and contribute to the information flow and modelling in construction.

This paper aims to identify challenges and opportunities for using laser scanners and locating systems in construction. Based on the findings of the paper, a discussion will be provided to the scanner adoption factors, the future of digital technology usage in construction. The main objectives of this paper are: 1) to analyze the result of implementation of mobile scanners to generate accurate data from construction site; 2) to examine the feasibility and challenges of using new scanners and positioning systems to capture data in construction.

The originality of this paper lies in providing an insight to new technology implementation based on experiments applying cutting edge hand-held scanners for real-time indoor data acquisition. The generated data also enables contractors to create and/or update the construction information modeling in a cost-efficient and timely manner.

The paper firstly reviews existing technologies for creating information modeling including as-built models as an example, which are mostly manual processes; and identifies technology gaps and barriers to the automation process. Secondly, the results of applying the advanced technologies to provide independent data sets of a sample educational building at the University of New South Wales, Sydney, Australia are briefly presented. Thereafter, the challenges and opportunities of using such technologies in construction projects will be discussed from a practical sense.

2 Updating the Construction Drawings

In any project, there will be a requirement to update the building objects dimensions where they were changed during construction based on the consultant change orders or as a result of clash management. This mainly is covered by as-built information modelling. The process of as-built information modelling using new technologies can be divided into two main phases: data acquisition and building information modelling. The data acquisition method is the focus of the study.

Several techniques can be used for data acquisition in order to create as-built in construction. However, there are limited techniques for acquiring data with the possibility for creating digital modelling. Corresponding to the challenging problem of reliable data collection of constructed objects in a timely manner, scanners can be proper devices of data collection for as-built creation in construction. The comparison suggests that laser scanners are an accurate and fast solution that will be employed in this paper. In order to solve these problems, we will use a novel mobile equipment to acquire data.

2.1.1 The Current Manual Methods

The manual as-built practices are mainly based on graphical standards for 2D drawings [6]. To develop such illustrations in two or three dimensions, traditional measuring equipment is used. The accuracy of these traditional as-built methods is within the required tolerance [11]. However, this traditional method of data acquisition produces a mass of drawings that makes their management and usage time consuming [15]. For example, Wang and Love [33] explain that traditional site layout method is labor intensive and required many times re-measuring.

Historically, sketches and photos are used to supplement the traditional method to assist CAD operators to enhance the information and accuracy of traditional as-built planning without any ground control [13]. Recently, researchers have tried to use photogrammetry techniques to produce digital and parametric data for as-built information modelling. Photogrammetry refers to geometric information derived from photographs [37]. However, this method has limitations [3,14]. For example, extracting object points from a wide angle shot near an object is difficult [14]. This approach is unable to produce the required information about the topography of irregular shapes in detail, and cannot provide the details of curves and irregular shapes, whereas lidar scanners can capture such details easily. Photogrammetry usually cannot be used independently in creating as-built, and it is not an ideal solution for as-built [17]. Recent studies attempt to integrate digital photogrammetry with lidar scanners [13,16].

2.1.2 Laser Technologies

Light detection and ranging (lidar) is a laser imaging technology that is increasingly employed for capturing scenes with millimeter to centimeter accuracy. Laser scanners collect data in the form of point clouds which are shapes and dimensions of objects in real space converted and represented as a collection of points in a 3D digital space. The fact that laser scanners collect 3D data gives it many advantages over the 2D methods such as using 2D plans.

New research studies intend to use laser scanners for construction purposes [7,35]. For example, several studies attempt to use laser scanners for as-built creation [27,36]. However, there are two main problems in this area. First, geometric information such as lines and surfaces cannot be easily extracted from millions of points data of objects [1], and are recommended as future research [1,4]. Second, a limited number of scanners such as terrestrial scanners are suitable for BIM [36], and the state-of-the-art technologies have not been investigated fully.

2.1.3 Gap in using the generated information

Building Information Modelling (BIM) is a collection of data which digitally represent the relevant characteristics of a building [34]. It is created through a process of modelling which includes practices such as distribution and storage of these data-sets. BIM is a rich data platform, but it would not be very useful if it only represents the design, as it currently occurs. The data platform has the potential to be fully used during construction as well as through the building maintenance life cycle. However, the challenges of generating new information that originated during construction, and updating the data or providing just-in-time information are barriers of the current practice. As-built BIM includes all the changes made on the building during the construction phase; it can be an up-to-date representation model of the building. The widespread use of As-built BIM is prevented by the lack of a time-efficient and accurate method of easily creating BIMs [12]. The capabilities of BIM as a representation model and its requirements for as-built BIMs are well investigated [5,8,9,20]. However, Huber et al. [12] claim that the topic of representing as-built BIM is in its early stages. Furthermore, current procedures rely on manual processes which are labor-intensive, time consuming and susceptible to errors [28]. According to Tang et al [28] existing work focuses on modelling the simplest objects of a building rather than modelling complex objects such as doors and windows. In addition, they show that there is a significant disconnect between surface based modelling and volumetric modelling representations[28].

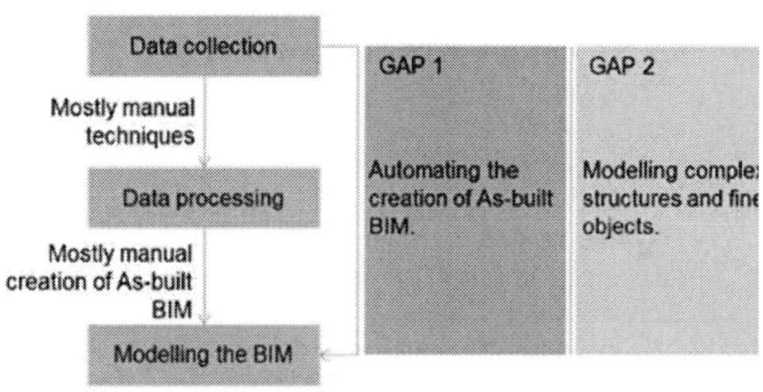

Figure 1 Schematic of the gap in scanner usage

Figure 1 schematically represents the literature divided into three main categories: data collection (input), data processing, and modeling (output). The literature shows that studies examining new technology applications to acquire input data for creating as-built BIM are scarce. However, much study focused on modeling and the output of modeling [2,32,33]. This paper examines a novel mobile laser scanner to create as-built BIM in construction.

In summary, the review reveals four significant challenges for creating as-built BIM. First, previous work mainly used terrestrial scanners and fixed lidar equipment [8,9,12,28]. Using state-of-the-art technologies have been totally ignored. Second, modelling fine and complex objects of buildings have received little attention [12,28]. Third, a procedural approach toward automatic as-built creation of the entire building is still unachievable [8]. Forth, the integration of laser scanner technologies with photography techniques and positioning systems to acquire much rich information model has not been covered in the literature.

3 Research Method

3.1 Case Study for Experiment

The experiment has been carried out in an educational building at the University of New South Wales, Sydney. The sample area chosen is the fourth floor as shown in Figure 2. The area includes different objects such as openings and stairs making it complex enough to explore the accuracy of the work for different building elements. For example, modelling such fine objects and details are still challenging. The general attributes of the case study are represented in Table 1. For consistency in understanding the analysis, pseudonyms are used for the parts such as W4 to refer to the west wing and C4 to refer to the middle corridor of the fourth floor which is being studied.

3.2 Data Acquisition Methods

Using new scanners in corporation, a procedural framework was employed to assist in creating drawings documents. The paper attempts to identify the challenges and opportunities of using the employed scanners and positioning systems by utilizing the scanners to create more accurate as-built with low skill labor in a shorter time. Based on the literature and previous work [22,24,26], the overall process for as-built creation is proposed consisting of scanning, processing, and creation [22].

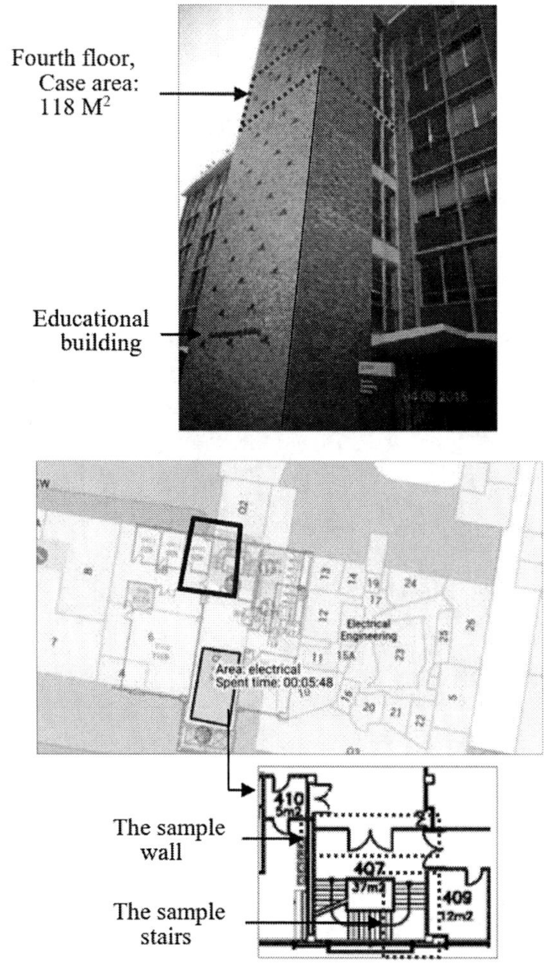

Figure 2 Location of the case study and the sample area (blue area detected by a chosen locating system)

In addition, the algorithm and feature of the new hand-held mobile scanners which are used in this paper dictated the procedure to follow to create the information modelling. The procedure is a part of an on-going study in order to be fully implemented to create as-built models to gain benefit from the advantages of the new scanners. The procedure consists of four stages as follows:

In the first stage, different scanners were used for data capturing, data processing, polygon extraction, making volumetric shapes and creating an initial information model. The implementation of this stage is presented in this paper, and the challenges and suggestions are discussed in the following sections. In the data collection step, one mobile mounted range-sensing system, one mobile hand-held scanner, and two terrestrial scanners were used. Next, the data is registered and obvious noises were removed. Main elements including openings, walls, floors and ceilings were segmented in the next step. Then, the extracted elements were combined as the identified as-built elements. Field work was conducted to assess and verify the level of accuracy obtained using the dense scanner points. In this stage, the results of the mobile scanners were compared and verified by a data set acquired from the terrestrial scanners.

Table 1. The case study details

Items	The building case	The sample area
Location	Kensington	The north-south corridor
Use	Higher education	Crossing corridors
State	Renovated	In use and decorated
Scope	Architectural objects (e.g. window and door)	The partition wall
Materials	Mix	Timber, glass, steel and concrete
Point cloud size	10 to 30 millions	5 to 7 millions

In the second stage, a digital camera was used to collect complementary information about the texture and color. In the data acquisition step, the sample area has been scanned using the scanners.

In the third stage, the information collected from previous phases including point clouds and photos were analyzed and the topology for information modelling was made. In this phase all data should be incorporated to produce a rich model. The model should also update epoch by epoch analysis when the shop drawing needs to be updated.

In the fourth stage, the modelled objects and other information such as colors and positions will be collected in an integrated system to measure the progress. This part of the algorithm is designed but will be examined in the future study.

3.3 Device for Data Acquisition

Four advanced laser scanners including a novel handheld mobile scanner, DPI mobile scanner and Leica scanners were used to implement the framework. The first device is a mobile scanner capturing data by utilizing a handheld mobile mapping device including a lightweight scanner. The utilized device is a 3D sensor system that consists of a rotating and trawling 2D scanner and a MicroStrain miniature internal measurement unit (IMU) mounted on a spring mechanism. This technology was used because, in contrast to stationary terrestrial scanners, it does not need a tripod or a vehicle or skilled operation. Indoor points of the study's extent were scanned using an algorithm of acquiring points that takes advantage of recording points against a trajectory route and other points.

The other data set was collected using a state-of-the-art scan- and multi-station at two locations from less than 3 meters distance from the objects. The maximum distance measurement and the maximum range are 50 and 1000 meters, respectively. Similarly, another terrestrial scanner was also used to collect the third set of the data in order to verify the results. Figure 3 shows two views of the partition wall using Leica C5.

3.4 Data Process

Scanning the indoor environment of the fourth floor of the building using the handheld laser scanner took about 10 minutes. The point cloud segment of the fourth floor is overlaid on the layout plan of the same place. Using segmentation technique, the 3D point cloud of this area was analyzed to get required dimensions for parametric modelling. In the current practice, registration is a semi-automated process. Data processing is also a semi-automated process that includes manual and automated filtering to remove noises and unwanted data, such as points from moving objects and reflections.

The multi-station terrestrial scanners (Figure 3) were required to be adjusted at two locations. The process of data collection took about 30 minutes from two locations including stationing and scanning. This data collected by terrestrial method contains fewer noise points compared to the handheld laser scanner.

The segmentation is used as the first technique for preprocessing. This technique is used in order to get comparable results. The same segment of point clouds including the wall in the corridor, two windows, one door and couple of stairs behind the door were selected and processed. Processing of the data sets consists of conversion of the text files to ply and las files for visualization purpose. Then, detection and reduction of noise (unwanted objects such as people and furniture) was carried out. Thereafter, both processed mobile and terrestrial lidar data sets were imported and processed into Autodesk Revit and converted within the program into a compatible format. Extending previous work [22,24,26], the examples of initial results are shown in Figure 4. Two segmented components: wall (including windows and an opening) and a section of stairs of the sample C_4 were modelled. Levels were allocated in components and a model was created.

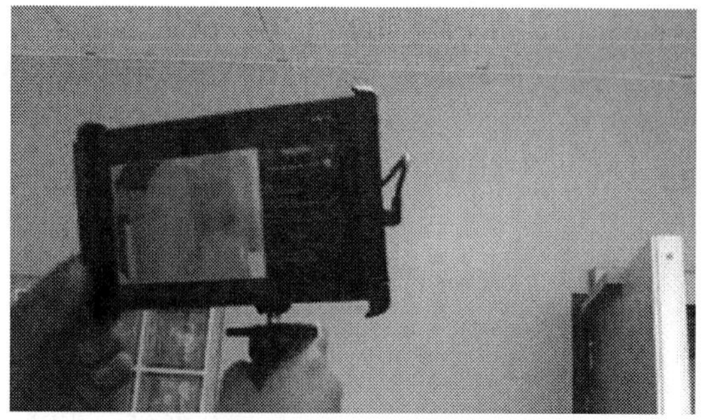

a) Mobile Scanner

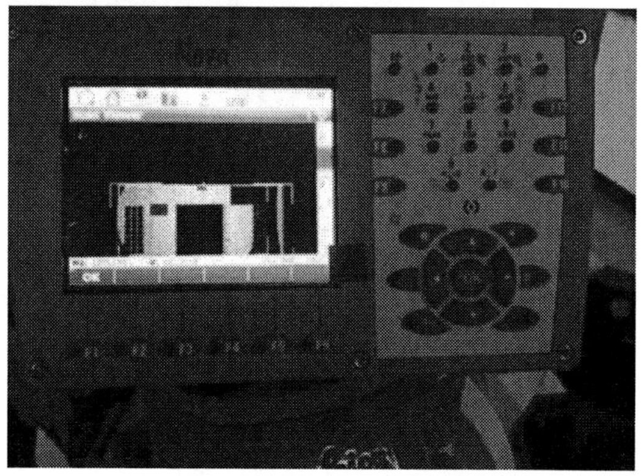

b) Tresstiral Scanner

Figure 3 Selected and Utilized Terrestrial and Mobile Scanners

4 Findings and Discussion

In this section the results of experiments will be used to identify how ease of use is the result of implementation of the advanced technologies in construction. Ease of use is one of the main advantages of a new information technology [29]. The implementation attributes including ease of use of a new technology are critical measurements of technology adoption in construction [21,23,25]. The reason is that the construction industry is a naturally low-skilled labor based industry, and they are less likely to use complicated technologies. However, ease of use has not been yet examined specifically on laser scanners and location systems in construction. Thus, the discussion of this section provides a totally new insight in predicting advanced scanner adoption in construction. The result of the study will appeal to scanner manufacturer and vendors as it shows how ease of use concept can be applied to each type of scanner.

The results of using the advanced laser scanners and a location system in the same area (educational building) were compared and evaluated as follows. Previous studies mostly focused on the feasibility of using a specific terrestrial scanner to acquire data for information modeling. This paper utilized both terrestrial and mobile scanners and compares their performance. The performance evaluation was carried out across five main determinates as shown in Table 2.

Lidar result demonstrability (LRD) – All advanced terrestrial scanners and the DPI mobile scanners used in this study, demonstrate the objects when we were scanning the area. However, the first hand held mobile scanner was not able to demonstrate the three dimensional object when scanner meaning that the operator would not get an understanding of whether the whole area had been scanned or not. But DPI shows what has been detected and the relevant point clouds will be saved and available for further analysis. When we moved the mobile scanner very quickly, the scanner was not able to detect and create point clouds and a message appeared that the object had not been detected.

Lidar output quality (LOQ) – The results show that the terrestrial lidar scanners are more accurate than mobile lidar scanners in most cases, and from visual inspection fewer noise points can been seen in the terrestrial scanner data than in the mobile scanner data. The results show that the accuracy of the result of implementing handheld mobile and terrestrial scanners are 25 mm (i.e. from 5 to 30) and 11 mm (i.e. from 1 to 12) respectively. According to Randall (2011), the accuracy for construction site monitoring and structural analysis and inspection should be less than 10 cm and 1 cm respectively. Thus the output quality of scanners is accurate for construction purposes. Randall (2011) also reports that the distance of scanner to target should be less than 25m and 10m respectively as well. The accuracy of the result of implementations of the frameworks using terrestrial scanner (#2) comparing traditional measures varies from -2% to +2% for openings, while this accuracy for the other experiments vary between -3% to +3% and from 1 mm to 12 mm. All in all, the results show that the terrestrial lidar is more accurate than the mobile lidar in most of the cases.

Lidar result demonstrability (LRD) – All advanced terrestrial scanners and the DPI mobile scanners used in this study, demonstrate the objects when we were scanning the area. However, the first hand held mobile scanner was not able to demonstrate the three dimensional object when scanner meaning that the operator would not get an understanding of whether the whole area had been scanned or not. But DPI shows what has been detected and the relevant point clouds will be saved and available for further analysis. When we moved the mobile scanner very quickly, the scanner was not able to detect and create point clouds and a message appeared that the object had not been detected.

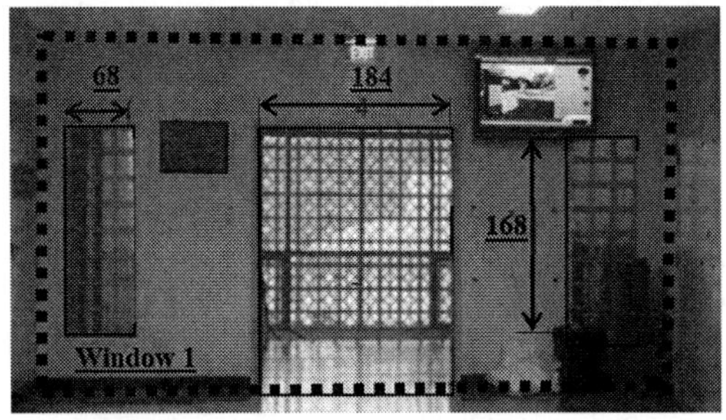

a) Traditional measurements of objects of the sample wall

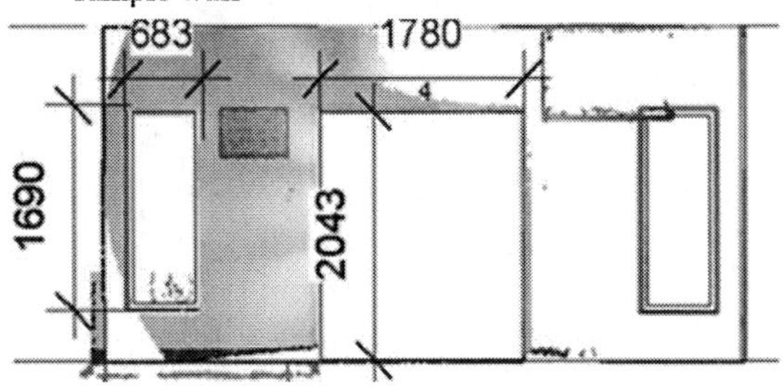

b) Segment A – A in BIM [24]

Figure 4 The segments and modeling of the sample partition

Lidar objective usability (LOU) – the data collection process of the terrestrial scanner is not as fast as the mobile scanners. The results of the mobile scanners also showed that they were not able to scan fine objects in a short time. However, for scanning small objects where high accuracy is required, we recommend the use of terrestrial scanners. For larger objects, where the contractor needs quick results, mobile scanners are more suitable with low cost in operation and ownership. Further analysis applied on the measurements of doors and windows, and errors for the opening dimensions were calculated. Standard Deviation for these three residuals for handheld scanner and terrestrial scanner 1 and 2 are 19.3, 26.3 and 1.9 mm respectively. The findings show that terrestrial scanner 2 has a large difference than others for scanning openings, and gives better results than others in this sample.

One type of scanner cannot be used alone for collecting the data from all construction objects. For example, it is impossible to move a terrestrial scanner inside the building during construction because of many obstacles. The experiments show using both mobile and terrestrial lidar scanners can be more helpful for the shop drawers in order to create a mature model and easily update the documentation during construction.

Table 2. Evaluation of lidars' ease of use

Determinant/ Reference	Definition	Results
Lidar output quality (LOQ) [30]	The degree of the scanner performance for doing the job.	Yes (between -2% to +2% accuracy)
Lidar result demonstrability (LRD) [18]	The degree showing that the results of using a scanner are tangible, observable, and communicable.	Yes/No
Lidar objective usability (LOU) [30]	A "comparison of systems based on the actual level (rather than perceptions) of effort required to completing specific tasks".	Yes/No
Lidar job relevance (LJR) [30]	The degree of applicability to as-built creation.	Yes
Lidar external control (LEC) [31]	The degree of the affordability of availability of resources to support the use of the scanner.	Positive/Negative

Using conventional scanners can barely help us to get the texture and color of each object. In particular, many materials have similar colors. So, using digital photos can assist shop drawers to increase the level of BIM maturity by adding the information collected from photos. However this practice was a manual one and needs further research to develop an automated procedure to decrease the errors of data entry for large projects.

The experiment shows that there are difficulties in scanning indoor objects by terrestrial scanners, because there are many obstacles, particularly during the construction period when many people and machines move around the area.

The experiments show that the process of integration lidar and other technologies (e.g. location systems or photogrammetry) needs much work to be semi-automatic. However, the integration practice is useful for a contractor to collect more information using a unique framework. In addition, the prior robotics applications are typically concerned with navigations and detection algorithm; whereas the feature of mobile lidar equipment enables us to use it in small and limited areas of complex buildings where adjusting terrestrial lidar equipment is not possible.

Lidar job relevance (LJR) – The experiments show that all of these scanner technologies are applicable for updating design drawings and creating as-built drawings. However, as construction sites are a congested area with equipment and a labor force often moving around, using terrestrial scanners provides more noise point clouds compared to mobile scanners, because the user is able to control a mobile scanner to move away from disruptions. In fact, using terrestrial scanner technologies will produce more noise as they cannot be moved when something passes from in front of the scanner while the scanner is automatically detecting the object. Mobile technologies, in contrast, can be moved when there is an object in front and are flexible enough to vary the distance from the object.

Lidar external control (LEC) – Another difference between the mobile and terrestrial lidar is the external factor of cost. The cost of the mobile lidar is three times cheaper than the multi-station scanner. At the same time, using mobile lidar equipment does not need a skilled operator. However, processing the data collected by the mobile lidar needs a skilled expert compared to the terrestrial lidar which is fully commercialized. A summary of the comparison of the two experiments is listed in Table 4.

5 Conclusion

This study aimed to identify the challenges and advantages of terrestrial and mobile scanners for construction purposes. The paper employed a proposed framework consisting of five different determinants of ease of use in order to examine the performance of each scanner in construction. The results of four scanners' experiments were evaluated based on the five determinants.

The results show that lidar output quality (LOQ) for all scanners is acceptable for construction purpose as the accuracy varies and is less than 2.1%. However, Lidar result demonstrability (LRD) criterion has not been met by all scanners. Lidar objective usability (LOU) has proved acceptable for all scanners when the objective is as-built creation, but other construction usability of scanners is arguable. For example, there is much work to be done to implement the scanners in different fields and automatically develop a digital as-built model. Lidar job relevance (LJR) is acceptable for all scanners, while lidar external control (LEC) is arguable and will depends on organization affordability.

The comparison shows that mobile scanners are suitable for indoor buildings and can be used in conjunction with terrestrial scanners which are able to collect both outdoor and indoor data. However, there are many different complex objects in construction and mining, so future studies should examine LOU and LJR more specifically in terms of building complexity, size, volume, location and in extreme weather and conditions in construction, piping, tunneling, and mining.

The limitation of this study is that we did not repeat the experimentations in different conditions such as weather conditions. Future work should consider different factors such as factors that affect the scanners during their placement (e.g. height, distance and extreme weather conditions). As for future work, we suggest to select other types of buildings and construction sites such as tunnels, railways and mining objects. In addition, examining the framework for creating staging as-built during construction is needed as future work. This study shows some significant potential benefits of the scanners from performance perspective (i.e. ease of use), that can be improved by vendors to increase the rate of technology adoption in construction.

Table 3 Cost per hour for each scanner

Scanner	Expenditure cost (A$)	Scanner hire cost per hour (A$)	Operation cost per hour (A$)	Process software cost per hour (A$)	Process cost per hour (A$)	Total cost per hour (A$)
T 1	78,000	260	16.6	5,000	150	83,427
T 2	65,000	216	12.5	5,000	150	70,379
T 3	25,000	83	4.1	5,000	150	30,237
M 1	11,000	36	6.6	5,000	150	16,193
M 2	22,000	73	1.6	-	400	22,475

References

[1] Y. Arayici, An Approach for Real World Data Modelling with the 3D Terrestrial Laser Scanner for Built Environment, Automation in construction 16 (6) (2007) 816-829.

[2] M. Bew, M. Richards, BIM Maturity Model, Construct IT Autumn 2008 Members' Meeting. Brighton, UK, 2008.

[3] A. Bhatla, S.Y. Choe, O. Fierro, F. Leite, Evaluation of Accuracy of As-Built 3D Modeling from Photos Taken by Handheld Digital Cameras, Automation in construction 28 (0) (2012) 116-127.

[4] L. Bornaz, F. Rinaudo, Terrestrial Laser Scanner Data Processing, Proceedings ISPRS Conference Commission V, Istanbul, Citeseer, 2004.

[5] T. Cerovsek, A Review and Outlook for a 'Building Information Model'(BIM): A Multi-Standpoint Framework for Technological Development, Advanced Engineering Informatics 25 (2) (2011) 224-244.

[6] C.A. Cory, Utilization of 2D, 3D, or 4D CAD in construction communication documentation, Information Visualisation, 2001. Proceedings. Fifth International Conference on, IEEE, 2001, pp. 219-224.

[7] M. Dix, A. Abd-Elrahman, B. Dewitt, L. Nash Jr, Accuracy Evaluation of Terrestrial LIDAR and Multibeam Sonar Systems Mounted on a Survey Vessel, Journal of Surveying Engineering 138 (4) (2011) 203-213.

[8] C. Dore, M. Murphy, Semi-Automatic Generation of As-Built BIM Façade Geometry from Laser and Image Data, ITcon, 2014.

[9] B. Giel, R. Issa, Using Laser Scanning to Assess the Accuracy of As-Built BIM, (2011).

[10] B. Giel, R.R.A. Issa, Using Laser Scanning to Access the Accuracy of As-Built BIM, Computing in Civil Engineering (2011), 2011, pp. 665-672.

[11] J. Goedert, P. Meadati, Integrating Construction Process Documentation into Building Information Modeling, Journal of construction engineering and management 134 (7) (2008) 509-516.

[12] D. Huber, B. Akinci, A. Oliver, E. Anil, B. Okorn, X. Xiong, Methods for Automatically Modeling and Representing As-Built Building Information Models, Proceedings of the NSF CMMI Research Innovation Conference, 2011.

[13] K. Jeyapalan, Mobile digital cameras for as-built surveys of roadside features, Photogrammetric engineering and remote sensing 70 (3) (2004) 301-312.

[14] L. Klein, N. Li, B. Becerik-Gerber, Imaged-Based Verification of As-Built Documentation of Operational Buildings, Automation in construction 21 (0) (2012) 161-171.

[15] S. Lagüela, L. Díaz-Vilariño, J. Armesto, P. Arias, Non-Destructive Approach for the Generation and Thermal Characterization of an As-Built BIM, Construction and Building Materials 51 (0) (2014) 55-61.

[16] X. Liu, M. Eybpoosh, B. Akinci, Developing As-Built Building Information Model Using Construction Process History Captured by a Laser Scanner and a Camera, Construction Research Congress 2012, 2012, pp. 1232-1241.

[17] J.D. Markley, J.R. Stutzman, E.N. Harris, Hybridization of photogrammetry and laser scanning technology for as-built 3D CAD models, Aerospace Conference, 2008 IEEE, IEEE, 2008, pp. 1-10.
[18] G.A. Moore, Crossing the chasm: Marketing and selling high-tech products to mainstream consumers, New York: HarperBusiness (1991).
[19] OCA, 3D Laser Scanning in: U.S.G.S. Adminstration (Ed.), GSA's Office of the Chief Architect (OCA), 2013.
[20] T. Randall, Construction Engineering Requirements for Integrating Laser Scanning Technology and Building Information Modeling, Journal of construction engineering and management 137 (10) (2011) 797-805.
[21] S.M. Sepasgozar, S.R. Davis, Modelling the Construction Technology Implementation Framework: an Empirical Study, in: R. Heikkilä (Ed.), The International Symposium on Automation and Robotics in Construction and Mining (ISARC 2015) Oulu, Finland, 2015.
[22] S.M. Sepasgozar, S. Lim, S. Shirowzhan, Implementation of Rapid As-built Building Information Modeling Using Mobile LiDAR, Construction Research Congress 2014, Construction in a Global Network, ASCE, 2014, pp. 209-218.
[23] S.M.E. Sepasgozar, L.E. Bernold, Factors Influencing Construction Technology Adoption, in: A.P.K.M.a.P.K.H. Stephen Kajewski (Ed.), 19th CIB World Building Congress, Brisbane 2013: Construction and Society, Queensland University of Technology, Brisbane, 2013.
[24] S.M.E. Sepasgozar, S. Lim, S. Shirowzhan, Y.M. Kim, Implementation of As-Built Information Modelling Using Mobile and Terrestrial Lidar Systems, The 31st International Symposium on Automation and Robotics in Construction and Mining (ISARC 2014), IAARC, Sydney, 2014.
[25] S.M.E. Sepasgozar, M. Loosemore, S.R. Davis, Conceptualising information and equipment technology adoption in construction: A critical review of existing research, Engineering, Construction and Architectural Management 23 (2) (2016) 158-176.
[26]S. Sepasgozara, S. Limb, S. Shirowzhanc, Y. Kimd, Z.M. Nadoushanie, Utilisation of a New Terrestrial Scanner for Reconstruction of As-built Models: A Comparative Study, ISARC. Proceedings of the International Symposium on Automation and Robotics in Construction, Vol. 32, Vilnius Gediminas Technical University, Department of Construction Economics & Property, 2015, p. 1.
[27] Y. Su, Y. Hashash, L. Liu, Integration of Construction As-Built Data Via Laser Scanning with Geotechnical Monitoring of Urban Excavation, Journal of construction engineering and management 132 (12) (2006) 1234-1241.
[28] P. Tang, D. Huber, B. Akinci, R. Lipman, A. Lytle, Automatic reconstruction of as-built building information models from laser-scanned point clouds: A review of related techniques, Automation in construction 19 (7) (2010) 829-843.
[29] V. Venkatesh, Determinants of Perceived Ease of Use: Integrating Control, Intrinsic Motivation, and Emotion into the Technology Acceptance Model, Information Systems Research 11 (4) (2000) 342-365.
[30] V. Venkatesh, F.D. Davis, A Theoretical Extension of the Technology Acceptance Model: Four Longitudinal Field Studies, Management Science 46 (2) (2000) 186-204.
[31] V. Venkatesh, M.G. Morris, G.B. Davis, F.D. Davis, User acceptance of information technology: Toward a unified view, MIS Quarterly: Management Information Systems 27 (3) (2003) 425-478.
[32] R. Volk, J. Stengel, F. Schultmann, Building Information Modeling (BIM) for existing buildings — Literature review and future needs, Automation in construction 38 (0) (2014) 109-127.
[33] X. Wang, P. Love, BIM+ AR: Onsite information sharing and communication via advanced visualization, Computer Supported Cooperative Work in Design (CSCWD), 2012 IEEE 16th International Conference on, IEEE, 2012, pp. 850-855.
[34] X. Wang, P.E.D. Love, M.J. Kim, C.-S. Park, C.-P. Sing, L. Hou, A Conceptual Framework for Integrating Building Information Modeling with Augmented Reality, Automation in construction 34 (0) (2013) 37-44.
[35] C. Watson, S.-E. Chen, H. Bian, E. Hauser, LiDAR Scan for Blasting Impact Evaluation on a Culvert Structure, Journal of Performance of Constructed Facilities 27 (4) (2011) 460-467.
[36] X. Xiong, A. Adan, B. Akinci, D. Huber, Automatic creation of semantically rich 3D building models from laser scanner data, Automation in construction 31 (0) (2013) 325-337.
[37] Z. Zhu, I. Brilakis, Comparison of Optical Sensor-Based Spatial Data Collection Techniques for Civil Infrastructure Modeling, Journal of Computing in Civil Engineering 23 (3) (2009) 170-177.

Design and simulation of a fuzzy supervisory control system integrated in a small public building

Alessandro Carbonari, Silvia Meschini, Mariangela Benedettelli

Università Politecnica delle Marche, Department of Bulding and Civil Engineering and Architecture, via Brecce Bianche, 60131 Ancona, Italy

alessandro.carbonari@staff.univpm.it ; silviamesc@gmail.com ; m.benedettelli@pm.univpm.it

Abstract –

Public administrations are encouraged to adopt actions for energy requalification of public buildings, which are responsible for considerable amounts of energy consumption. Low-cost technology improvements for energy renovation with high return on investment is a good opportunity when resource shortage hampers the adoption of expensive refurbishment. To this purpose, the integration of advanced control policies may optimize energy use of buildings with minimum hardware enhancement.

In this contribution, we evaluated the performances of a control system based on fuzzy logic, supervising local Proportional-Integrative-Derivative (PID) controllers, through numerical simulation. First, the design process will be described, with particular concern to the proper selection of the rules in the knowledge base and the design of the inference engine. Then, its performances will be compared with the ones provided by the use of just ordinary PID control units. The fuzzy based control system was simulated as an addiction to the heating system of an existing small public building, that is the "San Elpidio a Mare" community clinic located in Italy. The performances were assessed by means of simulations of an energy dynamic model developed in the Dymola™ environment, that had been previously validated with real consumption measurements. In this paper we will show not only what energy saving can be obtained by means of the fuzzy supervisor, but also its satisfactory performances in terms of stability of the control outputs.

Keywords –

Fuzzy logic; advanced control systems; energy saving; heating system; public buildings

1 Introduction

The survey of appropriate renovation actions to improve the energy behaviour of existing buildings can dramatically contribute to the reduction of the overall energy needs. Indeed, the building stock represents one of the biggest energy consumers, and public buildings were shown great potential for improvement, even by means of low cost renovation actions. As shown in the energy analysis of a Malaysian public hospital [1], the typical assessment procedure includes an energy audit process that collects information about the hospital's equipment and its energy consumption. Then, low cost tailored energy saving measures were identified and implemented, such as the integration of variable speed drives. The arrangement of a very detailed energy baseline is particularly critical in the case of complex strategic buildings, such as those ones owned by public administrations, where the development of a detailed consumption breakdown [2] can be quite demanding, due to the shortage of management systems. The only way auditors can accurately survey, diagnose inefficiencies and quantify energy savings that can be achieved by means of enhanced control is by means of tailored simulations. Such evaluations need the adoption of dynamic simulation tools, that are presently under wide consideration by researchers and practitioners [3]. They are more accurate than static calculation procedures traditionally used for plant sizing and they are compliant with suggestions from the European standard series EN 16247 [4].

In this paper, we will evaluate potential improvements determined by the adoption of advanced control systems based on fuzzy logic, focusing on a recurring Italian scenario that is relative to the building stock built around the 1970s and 1980s. In order for the requalification of small hospitals to be cost-effective, they might take advantage of the availability of low cost energy renovation actions.

In this research, one possible application of fuzzy logic will be proposed, whose performances will be compared both with the reference building in its current operation and with those ones provided by ordinary PID control, that was widely used in HVAC systems, as proved in many research works [5],[6]. Sometimes, fuzzy logic was integrated with PID control units in order to adjust their parameters according to control needs [7]. In this paper we will apply fuzzy logic as the supervisor sub-system of several local PID controllers.

2 The case study

San Elpidio a Mare's community clinic is located in Italy and managed by the public health authority's division, whose headquarters are in Fermo (Italy). The building is made of two blocks. The first one was built in the 1970s (A1), while the second block A2 was built in the 1980s (Figure 1). All the wards and offices are located in the older block of the building, whereas staircases and the elevators are accommodated in the other block. The clinic is arranged on seven levels, three of them located below grade, but at the time this study is referred to (i.e. the beginning of 2013) not all the levels were occupied. The main heating plant was built on the third level below grade and consists in a centralized system from which branch out the two main pipes of the secondary circuits that serve the various levels of the two blocks. Energy consumption was available in the form of energy bills and due to gas methane and electrical power. Both of them were invoiced monthly, so it was possible to estimate the actual picture that is provided in Table 1.

However, invoices were relative to the whole amount of electrical power and gas consumption and were not separated into sub-categories. In the case of natural gas consumption, it was split into heating consumption and hot water consumption. To this purpose, it was assumed that from May to September the heating plant was being turned off, so all the consumption in this period was charged to hot water only. The corresponding average value was assumed valid even over winter months. This assumption is confirmed by the Italian Standard D.P.R. 412/93 [8], because it states that all the heating systems must be turned off from 16th April to 31st October, thus supporting the validity of this approach. The difference between the total consumption and the estimated hot water consumption was charged to the heating system, thus estimating the monthly distribution of gas methane requested for heating purposes and shown in the two rightmost columns in Table 1.

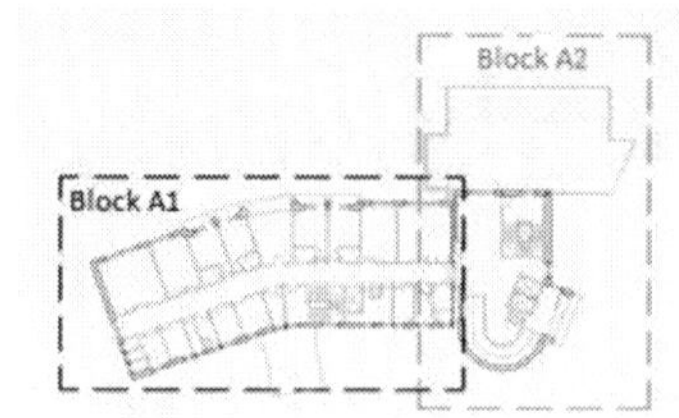

Figure 1. Main plan of the building adopted as a case study.

Table 1 Monthly gas methane consumption

	Total fuel (m^3)		*Primary Energy (kWh)*		*Fuel for heating (m^3)*	
Month	2012	2013	2012	2013	2012	2013
Jan	18818	7538	186125	77811	16879	6026
Feb	19756	11292	195125	113833	17817	9781
Mar	10140	7965	102864	81912	8201	6454
Apr	8731	5313	87893	55015	6792	3802
May	2428	18260	25966	20106	0	0
Jun	1880	1790	20708	19760	0	0
Jul	1826	1474	20192	16728	0	0
Aug	1641	957	18416	11768	0	0
Sep	1919	1050	21083	12660	0	0
Oct	3494	1972	36193	21506	1555	0
Nov	6025	4852	63383	52045	4086	3341
Dec	11027	7945	111370	81720	9088	6434
TOT.	87685	53974	889317	564864	64418	35835

The electrical power consumption was due to "auxiliary consumption for hot water", "auxiliary consumption for heating" and "building", that – in turn - sums up "lighting" and "facilities". A thorough analysis of the building's consumption pattern is presented in the former paper by Pontoni et al. [9]. To sum up, a breakdown of the electrical power consumption was carried out by means of behavioural models and surveys and interviews to technical personnel dedicated to maintenance. Then, all these consumption figures were converted into primary energy, using the current conversion factors for gas methane (i.e. Lower Calorific Value equals to 34.54 MJ/m^3 = 9.59 kWh/m^3) and electrical power (i.e. 1 Kwh_{el} = 2.34 KWh_{prim}), as suggested by the recommendation in AEEG n. 103/03 [10]). As a result, the following figures were derived: total primary energy for heating, total primary energy for hot water, primary energy of the whole building. According to the European Standard UNI EN 15217:2007 [11], the building performance level is assessed through the Energy Performance index (EPi), that in this case is equal to 51.65 kWh/m^3 in 2012 and equals to 38.58 kWh/m^3 in 2013. Hence, both values are much higher than mandatory legislation thresholds (i.e. 14,58 kWh/m^3 in 2012 and 17,71 kWh/m^3 in 2013).

3 The fuzzy controller integrated in the simulation model

3.1 Fuzzy logic

Basically, the approach suggested by fuzzy logic captures the expert's operation design rules in the form of a set of logic rules that would be applied if a human were operating the building control. The main components in a fuzzy control scheme are as follows:

1. the fuzzification interface, that includes the membership functions of each variable considered (both in input and in output) to make possible to convert crisp logic variables into linguistic variables and vice versa;
2. the rule base: simulating the expertise of a knowledgeable operator, it translates the experience of an operator in rules of the kind: "IF premise THEN consequence";
3. the inference engine: it performs decision making in terms of linguistic variables, finding for each crisp variable input those membership functions that are intersected and identifies what rules are active and applicable at each iteration;
4. the defuzzification interface: it turns the linguistic variable values back to crisp logic variable values, which are then applied as a signal to actuators.

A straight way to sum up how fuzzy logic works can be describing it as a procedure. As a first step, actual measurements sent by a monitoring system installed in the building are treated as crisp logic variable values. In the specific case considered in this paper, measurements were relative to ambient temperature and deviation between indoor (i.e. room) air temperature and its set-point (i.e. "error"). These are converted into linguistic variable values by means of previously defined membership functions that were built in the form of trapezoidal or triangular type. The result of this process is that the crisp logic variable values are converted into the corresponding linguistic variable values, that are those ones whose membership functions have the highest value for the measured crisp logic variable value.

As a second step, the actuators must be defined. In the case study, the fuzzy logic supervises the action of the local PID controllers embedded on the actuators, being able to adjust the temperature of the water supplied by the boiler to the primary circuit and by the opening of the three-way valves to the secondary circuit. Thus, three actuators in total, and two actuators per each fuzzy control system, since two separated control systems were designed in each block of the building to overcome the problems of control due to the different behavior of the two blocks of the hospital. All the three actuators were assigned linguistic variables and the corresponding membership functions of the linguistic variables. Then, a set of rules must be defined in order to convert any input linguistic variables combination (whose size depend on the number of controlled variables and actuators) to all possible actuator states. The initial approach for the definition of the rules, and also of the membership functions, was that proposed by Dounis et al [12], then a "trial and error" approach made possible the definition of a set of fuzzy rules relative to every output variable and encoded as a table. Each row in the table represents one rule. To be noticed that each "fuzzy set" represents the value of the respective input/output linguistic variables for the given rule. At this juncture, all the steps needed to implement the inference engine are done. In this case we have a fuzzy system with two inputs and two outputs, that can be written in the compact form as:

$$Y_j = \bigcap_{i=1}^{2} \{X_i \circ R_{ij}\}, j = 1,2 \quad (1)$$

The fuzzy relations R are defined using the rule base according to the formula:

$$R_{ij} = \bigcup_{k=1}^{n} \{X_i(k) \cap Y_j(k)\}, i \text{ and } j = 1,2 \quad (2)$$

Where X and Y are the input and output and the symbol operators mean fuzzy set union and intersection. Finally, the defuzzification method called center of area (COA) was adopted for determining the crisp value that corresponds to linguistic values computed by the inference engine. The COA method defines the crisp value as the center of gravity of the average of the values at which the membership function has maximum values.

3.2 The simulation model

The control approach tested in this paper was targeted to enhance the performance of the heating system of the hospital in San Elpidio a Mare. The energy model was developed in the Dymola™ environment, which is based on the Modelica programing language. This type of modelling is more flexible and accurate than standard simulation tools usually used for whole building simulation, because it allows users to model in detail and control any device affecting the building's consumption. The Modelica "Buildings Library" v1.6 was used, because it is a library made of components specifically tailored to building simulation [13].

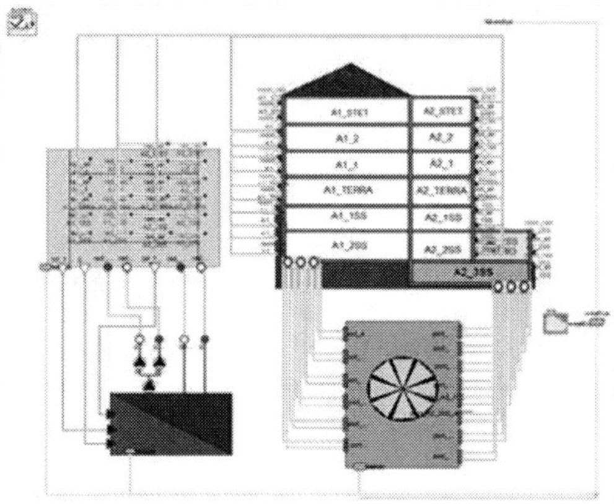

Figure 2. Top layer of the Dymola model.

Modelica is an object-oriented, equation based language to conveniently model complex physical systems containing, e.g. hydraulic, thermal, control and electric power. Figure 2 provides a graphic logic representation of the Dymola model's top layer of the case study. It is made up of four top level components, namely: the main heating plant (bottom left icon on Figure 2), the heating units (top left icon on Figure 2), the indoor space (top right icon on Figure 2) and the ventilation system (bottom right icon on Figure 2). Each of this top level's components was further detailed into lower levels of the simulation model. The indoor space and ventilation system are not illustrated in detail since these units do not include components someway related to the control sub-system. On the contrary, the basic features of the remaining two units will be described here and shown in the following sub-sections.

The heating plant is made of a primary circuit and two secondary circuits. The primary circuits includes the boiler (i.e. RTQ 400 type with 630 W power), which regulates the temperature of supply water by means of a temperature reset control system; one 870 W pump working at constant speed; safety devices imposed by regulations. The two secondary circuits deliver water to the thermal zones of the two building blocks (A1 and A2), whose power is 1350 W and 375 W, serving blocks A1 and A2, respectively. Presently, they work at their maximum rate over the whole heating period. Regulation is provided by two three-way motorized mixing valves, that are installed on the two supply mains and that regulate the temperature of supply water by mixing return water with the hot water supplied by the primary circuit. The valves are activated according to measurements by two temperature sensors installed on the water supply mains.

The heating units were modelled through the elements available in the "Building library" and called "Radiator". Their water content was regulated by three-way motorized valves. Characteristics implemented in the model match with those ones typical of a traditional cast iron radiators. Not all the thermal zones were heated and in operation, so the three-way valves were used either to keep the radiators turned on, or to switch them off, according to real conditions of use.

The indoor space was organized into thermal zones, each modelled through the "Room" component from the library. As a consequence, 14 thermal zones were considered on the overall, and only the heated zones were assigned tailored thermal heat gains due to people, equipment and lighting system, that correspond to approximately 7 W/m^2 on the average.

Each room of the Dymola model was assigned the corresponding thermal characteristics of the envelopes (e.g. thermal transmittance was assumed between 1.232 $W/m^2{\cdot}K$ and 1.443 $W/m^2{\cdot}K$), of the floors (whose thermal transmittance was assumed to range from 1.499 $W/m^2{\cdot}K$ to 2.064 $W/m^2{\cdot}K$) and of the windows (between 2.942 $W/m^2{\cdot}K$ and 6.281 $W/m^2{\cdot}K$).

The ventilation system is the last macro-level unit, which includes both natural air changes (due to air leakages) and the permeability of windows. These values were set as the sum of the contribution from natural air change and window permeability for the heated zones (estimated in 0.366 vol/h), while for the non-heated zones only the windows permeability was considered. Weather conditions were set using real data derived by the "APIwunderground" web site, where the plots from the weather station closest to San Elpidio a Mare (Italy) were retrieved from, relatively to the year 2013.

4 Analysis of performances

4.1 The benchmark

The macro-level component modelling the heating units is shown in Figure 3. Out of all the thermal zones, the three-way valves worked to let hot water flow into 6 thermal zones and preventing that flow from the remaining eight not heated thermal zones. This subdivision and the occupancy patterns were determined according to real conditions of usage. To sum up, most of block A1 was heated, while just a small portion of block A2 was. This situation is expected to change in few years, until the whole hospital will be made fully operational. But the purpose of our study was to accurately model the actual operational conditions. The average internal gains mentioned in paragraph 3.2 were differentiated according to the use of every thermal zone under operation. In every thermal zone, it was given as the sum of the gains generated by people, those ones generated by lighting fixtures and by equipment.

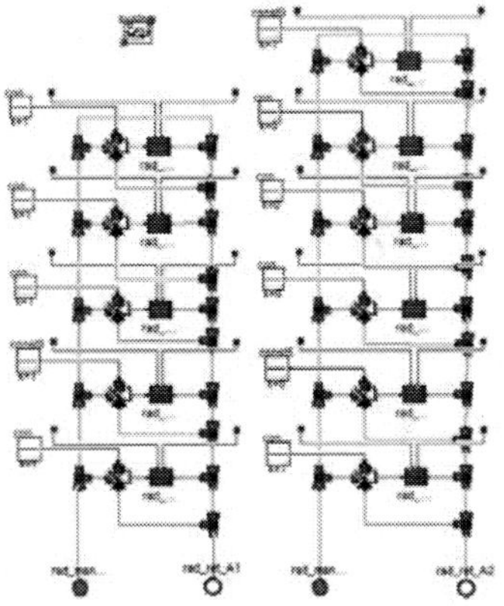

Figure 3. The Heating Units component of the Dymola model shown in Figure 2.

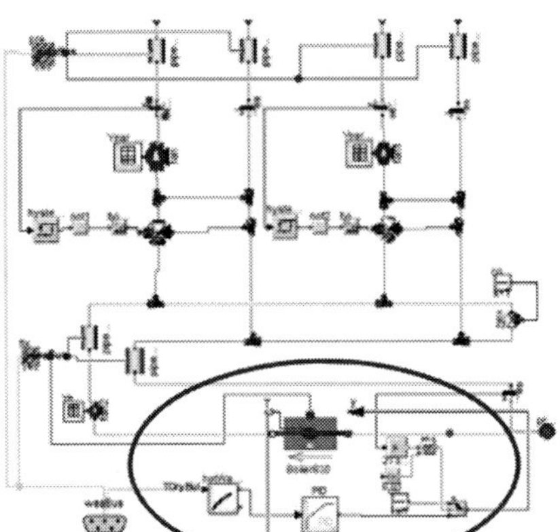

Figure 4. The Heating Plant component of the Dymola model shown in Figure 2.

The Heating Plant depicted in Figure 4 is the last macro-level component that is included in the whole Dymola model. It is equipped with a temperature reset control (circled in red on the figure), that is in charge of adjusting the temperature of the hot water supplied by the boiler and circulated in the primary system towards the distribution system. Such adjustment is varied according to the actual trend of outdoor air temperature (which is sensed in real-time by a sensor). The adjustment parameters were defined as inputs in the "PID_Boi" component displayed at the bottom of Figure 4. The other components between the "PID_Boi" unit and the boiler simulate safety devices that switch off the boiler if either the supply water temperature rises up to 90°C, or the pump in the main circuit is not in operation.

Then, the overall heating energy consumption estimated by this model was compared with the real consumption mentioned in paragraph 3.2 (Figure 5).

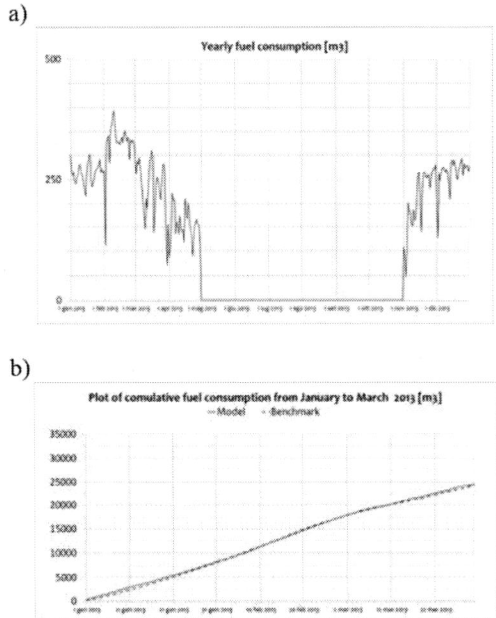

Figure 5: yearly fuel consumption by the boiler (a) and cumulative plots from January to March 2013, used for validation (b).

Table 2: Validation of the benchmark

Date	Measured fuel consum. (m^3)	Estimated fuel consum. (m^3)
2013,Jan 10th	2648.82	2091.46
2013,Jan 25th	3872.47	4226.69
2013, Feb 5th	3020.91	3094.36
2013,Feb 15th	3446.81	3397.00
2013, Feb 25th	3315.41	3172.51
2013, Mar 5th	2405.86	2202.02
2013,Mar 15th	2246.76	2063.37
2013,Mar 25th	2349.09	2318.26
	MBE	-3.28 %
	CV(RMSE)	9.22 %

It can be noticed that no fuel consumption due to heating was experienced during the summer. The validation was performed in the first three months of the year 2013 and was performed by means of the indicators suggested by reference standards. Mean Bias Error (MBE) and Cumulative Variation of Root Mean Squared Error (CVRMSE) were calculated using the formulae [14]:

$$MBE = \frac{\sum_{i=1}^{N_P}(M_i - S_i)}{\sum_{i=1}^{N_P} M_i} \quad (3)$$

$$CVRMSE = \frac{\sqrt{\sum_{i=1}^{N_P}\frac{(M_i-S_i)^2}{N_P}}}{\overline{M_P}} \quad (4)$$

where M_i is the list of measured data, S_i is the set of simulated data, N_P is the size of the database and the average value of the measured data is $\overline{M_P}$. It can be noticed from Table 2 that MBE and CVRMSE are well below the maximum thresholds suggested by ASHRAE, that is 5% and 15%, respectively.

Hence the reference model (i.e. benchmark) was considered as successfully calibrated.

4.2 The PID controlled model

PID controllers are conventional and well-known control systems. As a consequence, a detailed description of their mechanism and tuning methods falls out of the scope of this paper. However, a description of their implementation in the Dymola model for the case study will be provided, in order to clarify the reasons that led to the design and integration of the supervisory fuzzy control.

Besides the PID control that constitutes the temperature reset control at the boiler level and that was embedded even in the benchmark, additional PID controllers were implemented. Two of them were accommodated to adjust the opening rate of the two three-way mixing valves that control water supply temperature values of the two mains in the secondary circuit, and an additional PID controller was added to drive the opening of the three-way valves that regulate the water flow through the radiators in every thermal zone. Figure 6 shows that a set of PID components has been added aside each valves in the heating plant, differently from the benchmark version that is shown in Figure 4.

All the parameters of the PID controllers were estimated by means of the so called "one shot tuning" or "tuning on demand" [15]. This technique can be applied any time the control domain cannot be described by means of a set of mathematical formulas.

As a result, the consumption determined by this new model, whose mixing valves are controlled by PID devices, are lower than the benchmark (Figure 7). Moreover, the comfort conditions are improved even if the local PID controllers are not able to maintain the desired setpoint of 20°C (i.e. required by the mandatory legislations for healthcare buildings) and the internal temperature fluctuates around it as soon as an unexpected external disturb occurs. It was also found that when valves are driven by PID controllers, their activation is smoothed, that is the rate of variation from the position they hold for longer was decreased with respect to the benchmark. In other words, their variations were smoothed. The overall reduction in terms of fuel consumption after the first three months amounts to 5.8 %.

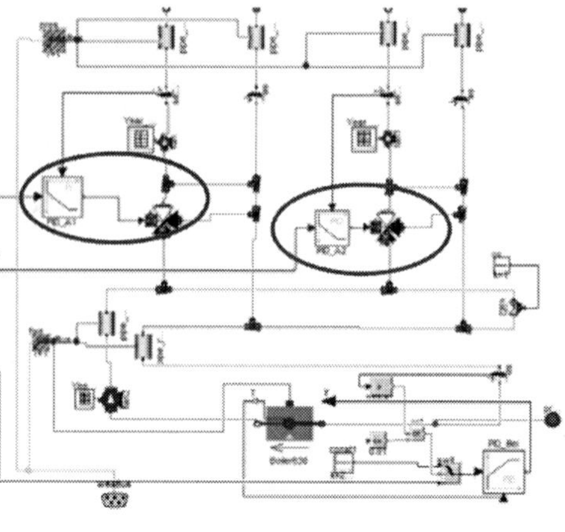

Figure 6: new version of the heating Plant component, where a PID controller was embedded per each mixing valve (circled in red).

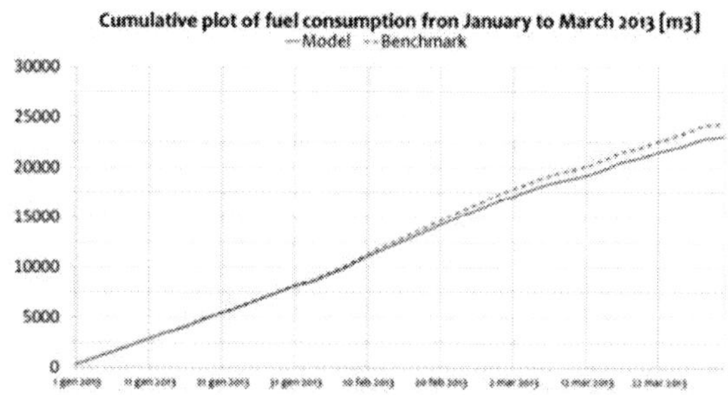

Figure 7: reduction of energy consumption in the first three months of the simulation year.

4.3 The fuzzy controlled model

The fuzzy logic was integrated in the model, in order to assess the potential of a supervisory control, considered as the best solution to overcome the limitations posed by the distribution sub-system of the heating system in block A1. It is served by a two-pipe vertical distribution system, that is unable to differentiate water flow rate and temperature according to the specific needs of each thermal zone. Hence, a conventional local PID control cannot guarantee the best performance, because a trade-off between the several zones can be obtained just in case a supervisor is installed. Moreover, the situation is different in block A2, because its distribution sub-system serves one manifold station on each floor. So, the fuzzy logic was developed as a supervisor, while the local PID controllers in each thermal zone were kept unvaried with respect to what described in sub-section 4.2. Two fuzzy supervisor controls were developed: the first one controlled block A1 and the second one controlled block

A2. Both of them got as input variables the measured crisp variable values: error and ambient temperature. Both fuzzy control units provided the following control signals as outputs: opening rate of the valve located upstream the pumps of the secondary circuits and temperature of the water supplied by the boiler. Considering that two signals were sent by the two fuzzy units at each iteration step, the highest temperature signal was considered as valid to drive the boiler. The linguistic variables associated to the error and ambient temperature are: Negative, Neutral and Positive and Small, Medium, Big respectively. Both the output linguistic variables (PID of the boiler and PID of the mixing valves) were given three states: Small, Medium and Big. A new fuzzy component was integrated in the heating units component of the Dymola model (circled in red in Fig. 8-a), whose architecture is provided in Fig. 8-b.

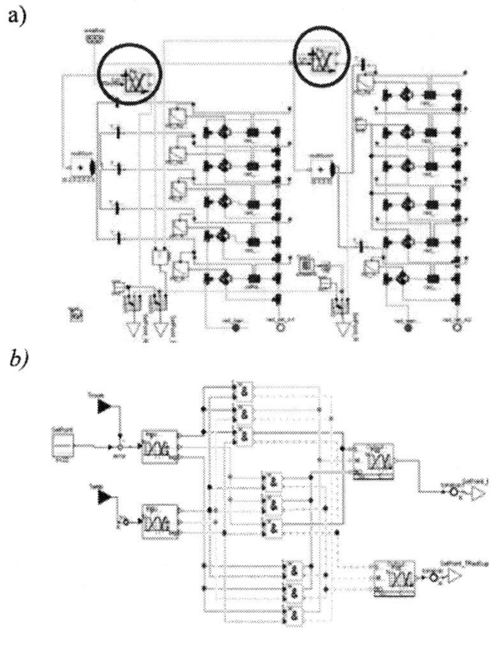

Figure 8: fuzzy component integrated in the Dymola component (a) and its architecture (b).

The component shown in Figure 8-b implemented all the components needed to perform the fuzzy inference (i.e. input and output variables, membership functions and rules) based on the set of rules shown in Figure 9. The defuzzification process was based on the COA approach and the control signals were sent to the heating plant component in Dymola, that was modified accordingly.

As the local PID control in paragraph 4.2 was unable to keep the required setpoint within all the rooms, the comparison was set at a 19°C setpoint, still within the mandatory legislation thresholds. The use of fuzzy logic allows to maintain the desired setpoint without significant variations, achieving a remarkable comfort improvement (Figure 10). Moreover, it determines a reduction in energy consumption with respect to the local PID control estimated as 3.6% (Figure 11).

Output: SP_Boil[illegible]			
T_{amb}	Error		
	Negative	Neutral	Positive
Small	M	B	B
Medium	M	B	B
Big	S	S	M

Output: SP_Rad[illegible]			
T_{amb}	Error		
	Negative	Neutral	Positive
Small	M	B	B
Medium	M	B	B
Big	S	S	M

Output: SP_Boil[illegible]			
T_{amb}	Error		
	Negative	Neutral	Positive
Small	M	B	B
Medium	M	B	B
Big	S	S	M

Output: SP_Rad[illegible]			
T_{amb}	Error		
	Negative	Neutral	Positive
Small	M	B	B
Medium	M	B	B
Big	S	S	M

Figure 9: Fuzzy rules in the inference engine.

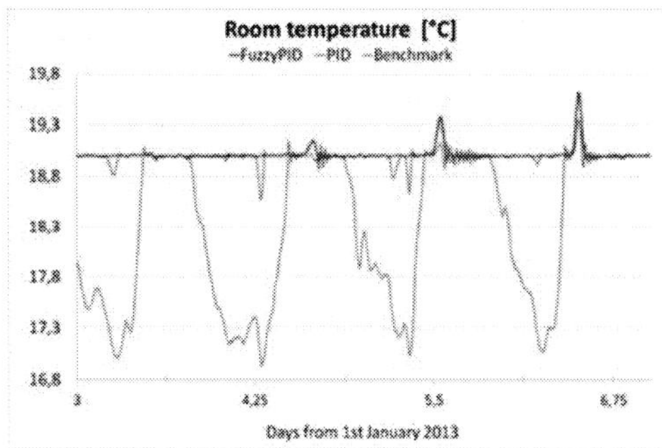

Figure 10: temperature in a representative room determined by fuzzy supervisor (magenta), PID control (blue), the benchmark (green).

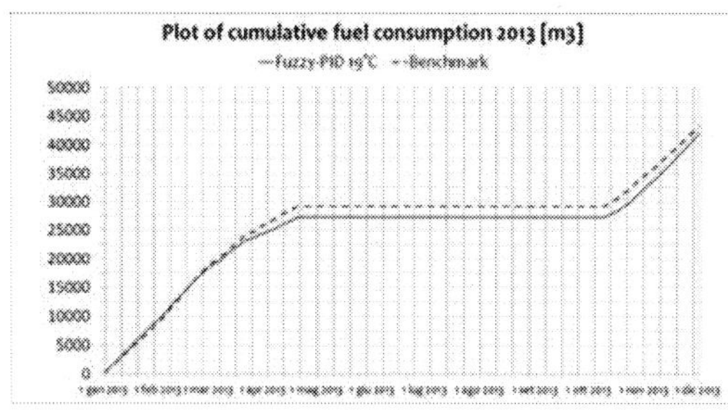

Figure 11: energy consumption reduction due to the fuzzy supervisory control.

5 Conclusions

This paper tackled the problem of HVAC systems improved regulation in order to enhance the energy performances of existing public buildings by means of low-cost hardware improvements and advanced logics.

The benefits of applying a fuzzy supervisory control system were tested on the heating plant of a small hospital, taking advantage of a lumped parameters model simulated in Dymola environment, which has been previously validated through comparison with real consumptions.

The proposed control strategy exploits the fuzzy logic to overcome the limitations pointed out by the implementation of a conventional PID control on a two-pipe vertical distribution system. This kind of plant cannot differentiate water supply temperature in the various thermal zones. The introduction of a fuzzy supervisor actually allows reaching and maintaining the desired thermal comfort, as opposed to the scenario with local PID controllers, which gives back a greater energy saving but also a quite changeable indoor temperature. When fuzzy logic and PID control are compared and applied on the same scenario, that is setting the request internal temperature at 19°C, the fuzzy logic provides a meaningful thermal comfort improvement in addition to energy saving.

The benefits resulting from the implementation of a low-cost control strategy based on logic and simple hardware components as the one shown in this research are easily replicable. Thus, given the large number of old buildings scattered on our territories similar to the one examined, remarkable energy savings could be attained on the overall.

References

[1] Saidur R., Hasanuzzaman M., Yogeswaran S., Mohammed H.A., Hossain M.S. An end-use energy analysis in a Malaysian public hospital. *Energy*, 35: 4780-4785, 2010.

[2] Short C.A., Cook M., Cropper P.C., Al-Maiyah S. Low energy refurbishment strategies for health buildings. *Journal of Building Performance Simulation*, 3(3): 197-216, 2010.

[3] Felgner F., Agustina R., Cladera Bohigas R., Merz R., Litz L. Simulation of Thermal Building Behaviour in Modelica. In *Proceedings of the 2nd International Modelica Conference*, 2002, March 18-19, DLR in Oberpfaffenhofen, Germany.

[4] Technical Standard UNI CEI EN 16247, *Energy audits*.

[5] Malhotra R., Sodhi R. Boiler flow control using PID and fuzzy logic control. *International Journal of Computer Science Engineering and Technology*, 2011, July, 1(6): 315-319.

[6] Shome A., Ashok S.D. Fuzzy Logic Approach for Boiler Temperature & Water Level Control *International Journal of Scientific and Engineering Research*, 2012, June, 3(6).

[7] Xin H., Huang T., Liu X., Tang X, Temperature Control System Based on Fuzzy Self-Adaptive PID Controller, In *Genetic and Evolutionary Computing*, IEEE, 2009, October, 3rd International Conference on: 537-540.

[8] D.P.R. 412/93, *Regolamento recante norme per la progettazione, l'installazione, l'esercizio e la manutenzione degli impianti termici degli edifici ai fini del contenimento dei consumi di energia*, in attuazione dell'art. 4, comma 4, della L. 9 gennaio 1991, n. 10

[9] Pontoni R., Giuliani M., Carbonari A., Vaccarini M. A Lumped Parameter Model for Dynamic Simulation of Energy Control Policies Applied to Small Public Buildings. In *Proceedings of the 32nd ISARC*, 2015, June 15-18, Oulu, Finland.

[10] AEEG n. 103/03, *Linee guida per la preparazione, esecuzione e valutazione dei progetti di cui all'articolo 5, comma 1, dei decreti ministeriali 20 luglio 2004 e per la definizione dei criteri e delle modalità per il rilascio dei titoli di efficienza energetica*.

[11] Standard UNI EN 15217:2007, *Energy performance of buildings – methods for expressing energy performance and for energy certification of buildings*, 2007.

[12] Dounis A.I., Manolakis D.E. Design of a fuzzy system for living space thermal-comfort regulation. *Applied Energy*, 69: 119-144, 2001.

[13] Wetter M. Modelica Library for Building Energy and Control Systems: Buildings 1.0_build2, available at: http://simulationresearch.lbl.gov/publications/recent-developments-modelica, February 2015.

[14] ASHRAE, ASHRAE Guideline 14-2002: Measurement of Energy and Demand Savings, 2002.

[15] Shinskey F.G. Process control systems. Applications, design and tuning. McGraw-Hill, New York, 1988.

Auto-layout of Lighting Objects to Support Lighting Design in the Early Phase of Design

Hyunjung Kim[a], Jinhua Huang[a], Yeon-Sook Hwang[a] and Jin-Kook Lee[a]

[a]Dept of Interior Architecture Design, Hanyang University, Seoul, Republic of Korea
E-mail: bluecleane@gmail.com, sara920319@gmail.com, ysh@hanyang.ac.kr and designit@hanyang.ac.kr

Abstract

This paper aims to propose an approach to automated lighting layout system in indoor space in the perspective of auto-generation of design alternatives. Lighting environment is one of the most important factors in the domain of interior design. Well-designed lighting environment enables occupants to perform their actions safely. Furthermore, it supports occupants' psychological stability. In existing lighting design process, difficultness is that the process is usually depended on experts' experience and it is performed manually based on 2D CAD drawings. Thus, it tends to be time-consuming to create alternatives and error-prone to meet standards in the early phase of design. In this regards, as the way to create alternatives of lighting environment based on quantitative data, this paper suggests an approach to auto-layout system of lighting objects in indoor space in BIM plug-in form. The entire system consists of three major parts: 1) derivation of properties from BIM object and user parameter, 2) calculation of illuminance and the number of lighting fixture according to derived properties for creating alternative which meet standards and 3) suggestion of automatically generated alternative, reports on result value associated with created alternative and visualization of created alternative. In short, this paper proposes a way to create design alternatives of lighting design intuitively using properties included in BIM model by auto-layout of lightings. And it also includes provision of output data associated with result reports from performance analysis.

Keywords –

Automated lighting arrangement system; Building information modeling(BIM); Parametric design; Alternative creation;

1 Introduction

1.1 Research Background and Objectives

The demand for Building information modeling(BIM) is increasing in the domain of Architecture, Engineering, Construction, Facility Management(AEC-FM)[1]. BIM supports efficient utilization of various information within the lifecycle of building based on 3D model[10,18]. And one of applications of BIM, powerful parametric modeling has become possible by taking advantage of object's various properties and its relation with the other object in 3D computational environment[11]. It enables to produce various alternatives intuitively with parameters which are derivable and controllable properties in BIM model.

On the other hand, in the early phase of design, repetition of alternative production, alternative analysis, alternative assessment, alternative selection, decision-making has been existed[12]. Its process tends to be time-consuming and error-prone because it is usually performed based on experts' knowledge by manual. Thus, it is still vulnerable to design changes. And dynamic decision making process which is based on information from performance analysis is still insufficient[3]. Thus, new approach which is flexible to design change and minimizes effort and time consumption that occurs when to find suitable alternative. And it also should meet complex design requirements in the perspective of utilizing information[21].

With this background, this paper aims to suggest an approach that utilizes BIM properties as parameters for creating design alternatives in the domain of lighting design. As one of supportive tools to lighting design, the approach is achieved by auto arrangement of lighting objects in indoor space. The approach is for minimization of trial and error and immediate alternative production, analysis and assessment.

1.2 Research Approach and Scope

This paper suggests automated lighting arrangement system which lays out lighting fixtures in indoor space to meet specific standard by utilizing BIM object's various properties. To design lightings in indoor space to meet standards, various variables associated with lighting fixture and particular space are needed to be considered such as luminous flux, the number of lighting, space area, space height, purpose of space, etc. For the research, we studied preceding researches associated with existing lighting design process. In terms of supplementation and extension of existing mechanism, this paper aims to suggest auto-layout of lighting objects system using BIM object's properties and user input parameter as variables for meeting recommended levels of illuminance according to facility type and space type based on illuminance calculation. And database is also applied in the system for additional parameter according to collected properties. The suggested system is implemented in actual BIM plug-in form. As shown in figure 1, its system is composed of three major parts; 1) Input data part: derives variables related to space and lighting which are selected by user from BIM object's properties and user parameters, 2) Calculation part: using collected information, calculates intensity of illumination to judge whether space's illuminance meets standards or not. In addition, it calculates the number of lighting fixtures according to properties of space object and lighting object and 3) Output data parts: arranges lightings in space according to calculation automatically and provides outputs such as reports associated with designed lighting environment and visualizations of created alternative.

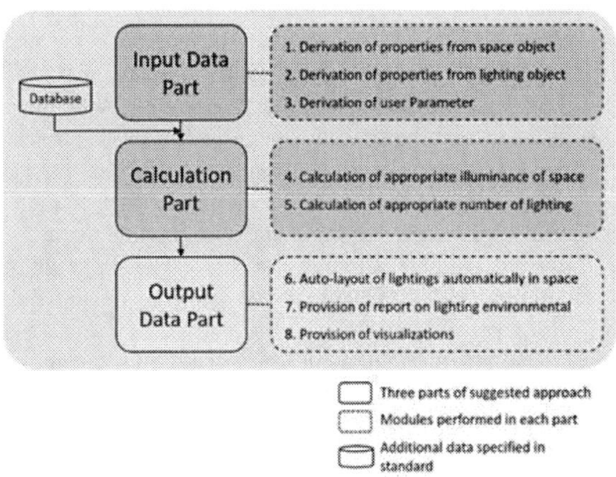

Figure 1. Composition and flow of the approach

After dealing with the three parts of automated lighting layout system, this paper will conduct a demonstration on an actual space. The target space is designed automatically through the implemented BIM plug-in. This paper is based on recommended levels of illumination specified in Korean Industrial Standards(KS) which is similar to USA's American National Standard Institute(ANSI) or Japan's Japanese Industrial Standards. The research progresses according to numeric data and actual standard quantitatively based on lumen method which is general method that calculates illuminance in indoor space and the number of lighting. And esthetic or qualitative factors of lighting environment which affects mentally are excluded in this paper.

2 Lighting Design Process

Lighting design is for providing visually suitable environment which enables occupants to perform their activities safely[22]. Furthermore, it should provide psychological stability[2]. To design suitable lighting environment, accurate illuminance calculation of each space according to its purpose is important. Because it is hard to change lighting design after installation of lighting equipments[4].

The standard associated with recommended levels of illuminance is specified in Korean Industrial Standards(KS)[5]. It specifies recommended standard according to type of facilities such as hospital, public institution as a higher level classification and type of spaces according to actions which performed in space as a lower level classification[6]. To design lighting environment which meets recommended levels of illuminance specified in KS, variables such as lighting system, type of lighting fixtures, luminous flux, space area, purpose of space, reflexibility of space's finishing materials, etc. must be considered[13]. These have been analyzed and derived through reviewing architectural drafting such as 2D-CAD manually[13,14]. Existing lighting design process and various variables for illuminance calculation are shown in fig.2. According to result value of illuminance calculation, appropriate number of lighting fixtures and their arrangement are decided. Thus, consideration of variables for energy-efficient and appropriate lighting environment is essential[21].

On the other hand, 2D-CAD based design process usually has been performed manually and often depended on experts' experience and knowledge in terms of architectural and interior design. Thus, it has been time-consuming and somewhat ad-hoc to make alternatives which meet standards in the design developing process[15]. In 2D-CAD based process, it has general design errors due to error of information, omission of information, etc[16]. In addition, those are connected to static decision-making due to lack of information[17]. Thus, new kinds of design process which enables dynamic decision-making through utilizing rich information in BIM-model is needed[7,9].

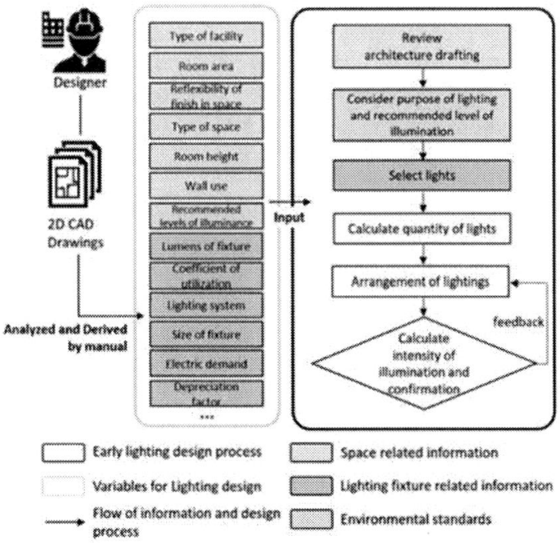

Figure 2. Existing lighting design process and variables for each phase

3 Lighting Design Process

3.1 Parametric Design

Parametric design has become possible though the development of computer operation. It can maximize creativity and economical benefits[8]. It was developed for 'reuse of existing design'. And it sets parameters on variables which define shape and it produces appropriate shapes by controlling relations between parameters[9]. Furthermore, generative parametric design which is mixed concept of mass production and parametric design enables BIM to increase its capability as a creative design tool. Based on the object's properties and relations with other objects, process that producing infinite instances which specific type can support analysis of performance and tracking flow of changes[3]. In the perspective of early phase of design, parametric design is meaningful that idea and intention can be easily realization[23].

3.2 Compatibility of Parametric Design and BIM in the Early Phase of Design

In the early phase of design, various alternatives are created and analyzed. There is repetition of creation, analysis, assessment, selection, decision-making. Thus, design changes frequently exist and this phase has a big impact on decision-making and design quality improvement[12]. In other words, design process without comparison among alternatives based on performance analysis is usually directly connected to static design-making[3,17]. In case of parametric design, it is important that grasp result value of analysis and find superior alternative simultaneously through repetitive process. In this perspective, utilizing properties in BIM model is important for meaningful performance analysis.

4 Lighting Design Process

The automated lighting design system is based on characteristics of BIM and parametric design. In other words, it utilizes information in BIM model as the parameters for producing possible alternatives which meet limited condition. In BIM environment, the system uses properties of space object and lighting object in model for automated lighting design which meets standards. It arranges lighting fixtures automatically in space based on calculation of illuminance and the number of lighting under suitable condition that is specified in Korea Industrial Standard. Figure3 gives an overview of process and information flow in the system.

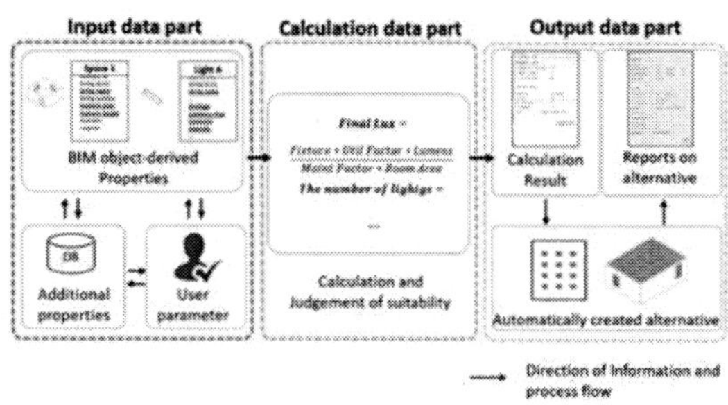

Figure 3. Flow of process and information in auto-layout of lighting objects system

4.1 Input Data Part

4.1.1 Derivation of Space Related Properties

In existing lighting design process, information for designing lighting environment is derived by reviewing architecture drafting, 2D CAD drawings and field survey[19].

In this paper, we suggest to derive this information automatically. In illuminance calculation for auto lighting arrangement, various space related properties are required as shown in Table 1. Type of facility, type of space, area of space, height of space, condition that some object is adjacent to wall or not, reflexibility of finishing in space are fundamental requisites for illuminance calculation[13]. They are properties which are included in BIM model and they can be derived directly from BIM objects. And additional data according to derived BIM properties is able to be collected from database which we designed. For example, recommended range of illumination is collected from database according to facility type and type of space as specified in KS. Table 1 shows this classification of properties.

Table 1 Space related properties for auto lighting arrangement system and classification

	Properties	
BIM properties derived directly from BIM object	Space name	Area of space
	Facility type	Space type
	Height of space	Condition that something is adjacent to wall or not
	Reflexibility of finishing in space	
Additional property	Recommended levels of illumination	

4.1.2 Derivation of Lighting Fixture Related Properties

Lighting fixture related data is also classified into directly BIM-derived properties and additional properties. Information for illuminance calculation, luminous flux of lighting, electric demand can be derived. And additional properties according to derived BIM properties are able to be set default value or collected from database such as depreciation factor, coefficient of utilization. Depreciation factor is property which exists due to expectation illuminance decrease[13]. Thus, it is set to default value because we have no data associated with factors that cause illuminance decrease in the research scope. And coefficient of utilization is the factor that occurs because of expectation of illuminance decrease. Coefficient of utilization is the factor which means incident light of luminous flux on working surface. It is affected by type of lighting system, room ratio and reflexibility of finishing materials in indoor space. Thus, information associated with coefficient of utilization is served as chart so that designer can find appropriate value in actual lighting design process[14]. We mapped this chart into database.

Table 2 Lighting properties for illumination calculation

	Properties	
BIM properties derived directly from BIM object	Lighting name	Electric demand
	Luminous flux	.
Additional properties	Depreciation factor	Coefficient of utilization

4.1.3 Gathering of User Paratemer

After deriving properties from BIM model, additional parameters for illuminance calculation are required. Type of lighting system is needed to be selected for automated lighting design. Because it affects to variables such as distance between lighting fixtures and distance between lighting fixtures and wall[19].

4.2 Calculation Part

4.2.1 Calculation of the Number of Lighting

This module calculates maximum distance between walls and lightings and maximum distance between lightings. When maximum distance between walls and lightings is decided, minimum area in ceiling where lightings can be arranged is decided. It is affected by condition that whether some objects is adjacent to wall or not[13]. On the other hand, maximum distance between lightings is affected by type of lighting system. And both distances are affected by height of space. Consequently, two types of data affect to the number of lighting fixture[19].

4.2.2 Calculation of Illuminance of Space

In this module, illuminance calculation is conducted using parameter which were collected in before parts. For calculate overall illuminance, lumen method is applied in the system. Equation (1) shows formula for calculating illuminance according to lumen method[19]. It is implemented in the system.

$$Final\ illuminance = \frac{Luminous\ flux * The\ number\ of\ lighting * Coefficient\ of\ utilzation}{Space\ area * Depreciation\ factor} \quad (1)$$

4.3 Output Data Part

4.3.1 Suggestion of Result Value

Result value associated with properties of space, lighting fixture and lighting environmental factors of the created alternative are suggested in this module. It can be used as the basis for comparison among alternatives in performance-oriented design process. In addition, it informs whether current state of lighting environment is suitable to standards or not so that feedback can be performed quickly. Table 5 shows providable information in the perspective of performance analysis.

Table 5 Providable information based on performance analysis of created alternative

Properties		
Properties derived directly from BIM model	Space name	Area (ceiling, wall, floor)
	Reflexibility of finishing (ceiling, wall, floor)	Facility type
	Space purpose	Condition that something is adjacent to wall or not
	Recommended illuminance	Recommended range of illuminance
	Lighting fixture name	Electric demand
	Depreciation factor	Luminous flux
Calculated properties of alternative	The number of lighting fixture	Final illuminance
	Maximum distance between lightings	Maximum distance between lighting and wall

4.3.2 Visualization

One of BIM application, visualization based on objects' properties is possible. This module suggests visualization of indoor space which is designed by the system based on actual value of BIM object. Addition to report of result value associated with created alternative, it is meaningful that visualization such as 3D renderings is enable for intuitive comparison among alternatives. In addition, it enables to provide indirect experience that how space going to be changed after lighting design.

5 Demonstration

We implemented the suggested system in BIM plug-in form based on Revit2015 for demonstration. The demonstration was conducted with one of lecture rooms in college of human ecology in Hanyang University as a target space. Figure 4 shows target space which is on fifth floor in college of human ecology. And we choose two different types of lighting objects for auto arrangement in the target space. Two scenarios was conducted for demonstration according to difference between lightings.

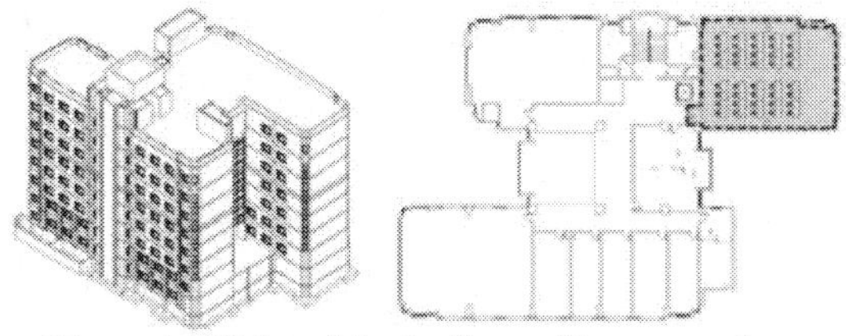

Figure 4. BIM model of college of human ecology building and one target lecture room for the demonstration

5.1.1 Derivation of variables

Figure 5 shows GUI(Graphical user interface) of BIM plug-in when space object of target space and target lighting fixture were selected. As shown in figure 5, properties associated with target space and lighting can be classified in three types of property: 1) properties derived from BIM model automatically, 2) properties derived from predefined database and 3) properties which are additional user parameters. We could derive those properties from two scenarios as shown in Table 6.

Table 6 Derived properties in input part of the system

Property type	Property name	Scenario A	Scenario B
Properties derived directly from BIM model	Space name	Lecture room 507	Lecture room 507
	Area (ceiling)	163.34m^2	163.34m^2
	Area (wall)	194.92m^2	194.92m^2
	Area (floor)	163.34m^2	163.34m^2
	Reflexibility of finishing (ceiling)	75%	75%
	Reflexibility of finishing (wall)	50%	50%
	Reflexibility of finishing (floor)	20%	20%
	Facility type	Universit y	Universit y
	Space type	Lecture room	Lecture room
	Condition of wall	No	No
	Type of lighting	M_Ceilin g light-Linear box (3 lamps)	M_Ceilin g light-Linear box (2 lamps)
	Electric demand	66W	80W
	Luminous flux	9450lm	6300lm

Properties derived from predefined database	Recommended illuminance	400	400
	Recommended range of illuminance	300~600	300~600
	Depreciation factor	140	140
Properties as user parameter	Type of lighting system	Direct (recessed)	Direct (recessed)

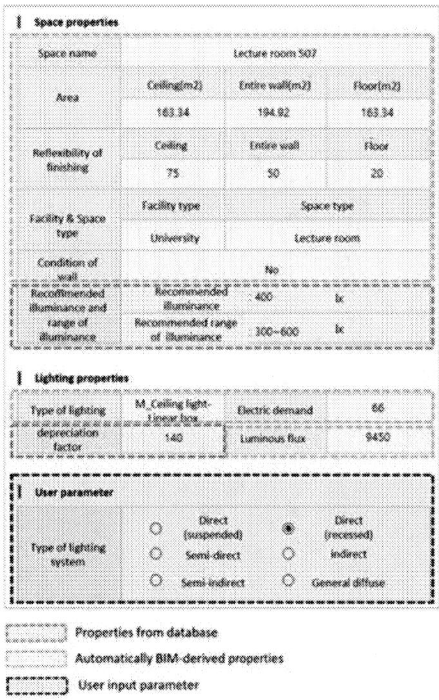

Figure 5 GUI of data input part and its components implemented in BIM plug-in form

According to scenarios which are based on difference of lighting objects, entire properties except lighting related properties are same. And one direct type of lighting system was selected as additional user parameter identically. And then, several calculations implemented in the system could be conducted.

5.1.2 Calculation and provision of data outputs

Figure 6 shows next step of BIM plug-in GUI which informs simple pre-plan of layout and suitability of lighting fixture after calculation as one of data outputs. In other words, it informs properties associated with arrangement of lighting fixture and lighting environmental factors. They could be classified into three types: 1) calculated properties, 2) properties derived from database according to case of formerly derived properties and 3) final suitability decision of alternative whether it meets standards or not. Table 7 shows derived properties from two scenarios based on properties shown in Table 6.

Table 7 Properties in Pre-plan of auto-layout

Property type	Property name	Scenario A	Scenario B
Calculated properties	Minimum number of lighting	30	30
	Maximum distance between lighting and wall	1.33m	1.33m^2
	Maximum distance between lightings	2.38m	2.84m^2
	Final illuminance	520lx	132 lx
	Optimal luminous flux of a lighting	7260lm	7260lm
	Luminous flux of selected lighting	9450lm	2400lm
Properties derived from predefined database	Recommen-ded illuminance	400lx	400lx
	Recommen-ded range of illuminance	300~600lx	300~600lx
Suitability judgement	Suitability (calculation result)	YES	No

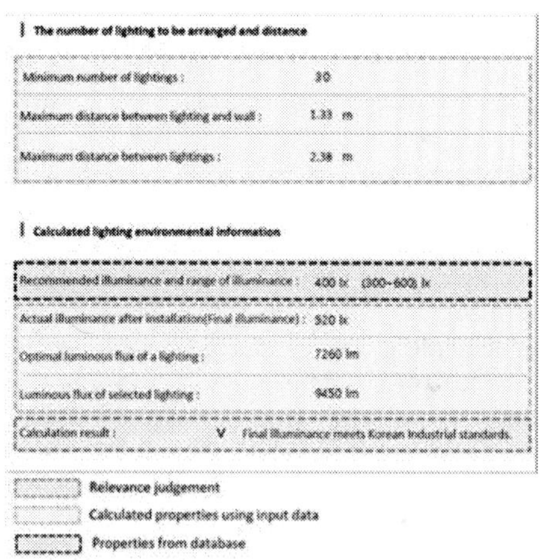

Figure 6. GUI of BIM plug-in as one of data output parts which suggests information related to auto-layout

As shown in Table 7, partly different values were derived from two scenarios. According to the calculation result, it was analyzed that selected lighting fixture's luminous flux is too low to meet standards when arranged in scenario B. When alternative is created, final illuminance of target space is too low to meet standards. On the other hand, in scenario A, it is analyzed that selected lighting's luminous flux is suitable to arrange. When it arranged automatically as many as calculated before, final illuminance of target space is able to be included in recommended range of illuminance. As a result, it was analyzed that only scenario A is allowed to create suitable alternative.

As next phase, the system arranged lighting objects in indoor space and created alternatives were provided as shown in Figure 7. According to properties and calculated values from two scenarios, two different alternatives were created.

And final reports was suggested as shown in Figure 8. It is comprehensive list of properties which was suggest in derivation properties and their calculation phase. Thus, it suggests overall information associated with created alternative. And information could be classified into three parts: 1) properties from calculation, 2) properties from database and 3) properties from created alternative.

Lastly, visualization through 3D rendering images was provided. Figure 9 shows 3D rendering images from two scenarios. It was on fundamental level because we used BIM model with simple material to just observe difference between two lighting environment for demonstration. We found that 3D rendering image of scenario B is visually darker then scenario A as the system analyzed before.

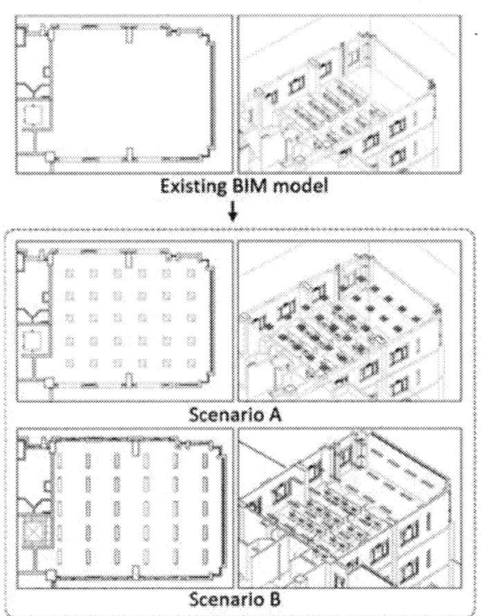

Figure 7. Ceiling plan and 3D view of target space and created alternatives based on scenario through auto layout system

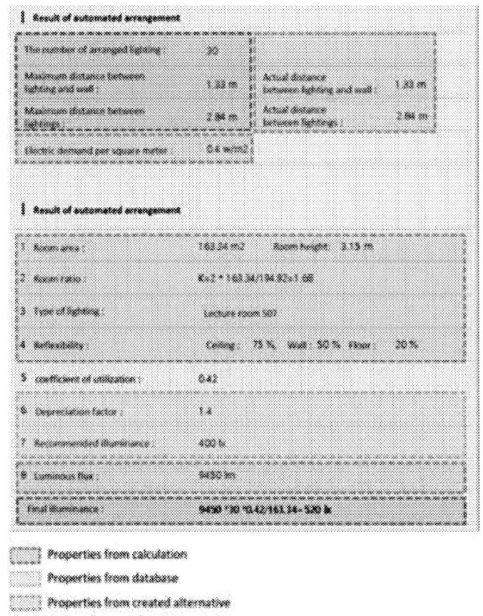

Figure 8. GUI of final report in the system as one of results of data output part.

Figure 9. Suggested 3D rendering images based on two scenarios through the system(left: scenario A, right: scenario B).

6 Conclusion

This paper introduces and demonstrates an approach to auto-arrangement of lighting fixture in indoor space under the condition of recommended levels of illuminance specified in KS. As one of supportive tools to lighting design process, the approach is implemented in BIM plug-in form. The entire system consists of three major parts according to type of data processing. As shown in section 5.Demonstration, arrangement mechanism suggested in this paper enables to create design alternative using various properties in BIM model parametrically to meet recommended standards. Lighting design based on BIM and parametric design has the merit of being updated immediately when design changes occur. And comparison among design alternatives are possible because each alternative is based on performance analysis with quantitative data. To include qualitative and esthetic factors as parameters in auto lighting arrangement system as a supportive tool, we need to develop further study.

Acknowledgement

This work was supported by the National Research Foundation of Korea Grant funded by the Korean Government(NRF-2015R1C1A1A01053497)

References

[1] Jin-Kook Lee, Charles M. Eastman, Yong Cheol Lee. Implementation of a BIM Domain-specific Language for the Building Environment Rule and Analysis. *Journal of Intelligent & Robotic Systems*, DOI 10.1007/s10846-014-0117-7, 2014.

[2] An Seob Choi, Byung Chul Park. Environment-friendly Lightings and Lighting Design. *Journal of the Society of Air0conditioning and Refrigerating Engineers in Korea*, volume 38(6), 2009.

[3] Sung-Ah Kim. Digital Architecture and BIM. *Journal of Architecture Institute of Korea*, volume 56(11), 2012.

[4] Yu-Sin Kim, Byoung-Chul Park, Ken-Young Jeong, An-Seop Choi. A Study on the Coefficients of Utilization of Luminaires for an Efficiency Interior Lighting Design. *Journal of Korean Institute of Architectural Sustainable Environment and Building System,* volume 2(2), 2008

[5] KS A 3011 - Recommended levels of illumination.

[6] Sung-O Cho. A Study on the Development of Database for Lighting Design System. *Journal of Korea Institute of Interior Design*, volume 13(4), 2004.

[7] A. Kiviniemi. *The effects of Integrated BIM in processes and business models*, Distributed Intelligence in Design, Wiley-Blackwell, 2011.

[8] Jung Dae Park. A Study of the Digital-based Form Generation Methodology -Focused on the Parametric Algorithms for the Non-orthogonal buildings-. *Journal of Korea Digital Design Council*, volume 12(1), 2012.

[9] P. Brandon. *Sharing intelligence: The problem of knowledge atrophy*, Distributed Intelligence in Design, Wiley-Blackwell, 2011.

[10] Bilal Succar. Building information modelling framework: A research and delivery foundation for industry stakeholders. *Journal of Automation in Construction*, volume 18(3), 2009.

[11] Rafael Sacks, Charles M. Eastman, Ghang Lee. Parametric 3D modeling in building construction with examples from precast concrete. *Journal of Automation in Construction*, volumn 13(3), 2004.

[12] Do-Young Kim, Sung-Ah Kim. An integrated Design Process of Climate Adaptive Building Shells using Generative Algorithm. *Journal of Architectural Institute of Korea*, volume 30(3), 2014.

[13] Jungsoo Kim, Youngsun Youn, Jong-ho Park, Ki-Young Ko, *Architectural lighting planning,* Kimundang, 2001.

[14] San-Ho Choi, Hong-Bae Kim, Sunduck-Kangm, Si-Bok Nam, *Practice of Architectural Lighitng*, Kimundang, 2013.

[15] Chan Kang. A Study on the Improvement of a Building Remodeling Process Utilizing BIM. Kongju University Graduate School, 2015.

[16] Hyong-Hoon Kim, Eun-Joo Kim, Jae-Jun Kim. A Study to Judge the Importance of Indices to Estimate Architectural Competition on the Apartment for Using BIM. *Journal of Architectural Institute of Korea,* volume 25(2), 2009.

[17] Han-Jong Jun, Digital Design Technology for Design Environment in near Future. *Journal of Architectural Institute of Korea,* volume 56(11), 2012.

[18] Charles M. Eastman, Paul Teicholz, Rafael Sacks, Kathleen Liston, *BIM HANDBOOK-A guide to Building Information Modeling for Owners, Managers, Designers, Engineers and Constructors*, John Wiley & Son. Inc, New Jersy, 2007.

[19] Kyung-Nam Cho, Byoung-Sook Choi, Gang-Wook Shin, Seong-Soo Kim, Hong-Kyoo Choi. A Study on Applying the Utilization Factor for the Room Indix Under 0.7 in Illuminance Calculation by the Lumen Method. *Proceeding of the symposium of The Korean Institute of Illuminating and Electrical Installation Engineers*. 2011.

[20] Ingeun Yi, Sung-Ah Kim. Interaction between BIM model and Physical Model During Conceptual Design Stage. *Journal of The Korean Institute of Communications and Information Science*, volume 38(4), 2013.

[21] Jae-Hong Kim, Sung-Min Yoon, Ki-Taek Jung, Jin-Won Choi, Won-Hwa Hong. A Study on Development of a BIM-based Intelligent Lighting Management System. *Proceeding of the symposium of The Korean Solar Energy Society*, volume 33(1), 2013.

[22] William M.C.Lam, *Perception and Lighting as Formgivers for Architecture*, McGraw-Hill, 1977.

[23] Seung-Yeon Choo, Seung-Hak Woo. Free-form Design and Digital Fabrication for Education of Digital Architecture. *Journal of The Korean Institute of Communications and Information Science,* volume 58(2), 2014.

Automated Layout Planning of Climbing Formwork System Using Genetic Algorithm

Min Ju Kim[a], Taehoon Kim[a], Hyunsu Lim[a], Hunhee Cho[a], and Kyung-In Kang[a]

[a] School of Civil, Environmental, and Architectural Engineering, Korea University, Republic of Korea
E-mail: minju830@korea.ac.kr, kimth0930@korea.ac.kr, iroze00@korea.ac.kr, hhcho@korea.ac.kr, kikang@korea.ac.kr

Abstract –

Layout planning of climbing formwork system is a prerequisite for executing efficient and cost-effective structural framework in a high-rise building construction. However, most of the formwork system layout is operated manually under heuristic approach of the supervisor, and thus is prone to errors that affect constructability of the entire construction. Furthermore, inflexibility of the current method to changes in building design also results in unexpected variability during system utilization. Automated formwork layout method that considers essential factors such as configurations on unit module of the form as an input data, information on building parameters for specifying possible areas of system installation, and configurations on working platform for optimal arrangement of unit systems can be suggested as an alternative to the current manual execution of formwork layout. Therefore, this study proposes an automated layout planning method of climbing formwork system by using genetic algorithm. The method developed in this study is to be utilized for reasonable decision making of the construction worker in control of the automated layout planning.

Keywords –

Automated layout planning; Climbing formwork system; High-rise building; Genetic algorithm

1 Introduction

Climbing formwork system is applied in a high-rise building construction as an appropriate facility for vertical erection of a rigid structure. As structural framework is the single largest cost-related activity in a building construction [1, 2], framing a high-rise structure by attaining economic feasibility becomes a major concern. Equipment and material utilization associated with concrete formwork are prerequisite for a successful construction performance. Thus, layout planning of climbing formwork system must be preceded for executing efficient and cost-effective structural framework in a high-rise building construction.

However, current layout planning method of climbing formwork system encounters several factors in need of improvement. First of all, current system layout is operated under heuristic approach of the supervisor by relying on personal experience in managing climbing formwork system. Not only is manual operation of formwork layout inevitably prone to errors, but also does it consume excessive work duration and workload in actual construction site. Furthermore, current layout method cannot instantly reflect changes in design of the building, and thus results in higher variability in both quality and constructability. Therefore, this study proposes a flexible, automated layout planning method of climbing formwork system that serves as a framework of a decision support model for optimal climbing formwork system layout in high-rise building construction.

2 Theoretical Background on Climbing Formwork System and Related Work

2.1 Climbing Formwork System in High-rise Building Construction

Built environment of a dense area prefers buildings to be erected vertically rather than spread horizontally for efficient construction within limited space. In response to this shift of paradigm on building construction, ensuring rigidity of highly erected structure with minimum cost input on framework construction became a primary concern [3].

Compared with traditionally applied facilities for vertical framework such as gondola and gang form, a climbing formwork system of which scaffold components and wall formwork are all fabricated as one unit can adopt various forms of structure with higher versatility. By executing construction processes with standardized components, system formwork method offers a systemized operation processes with positive effects on improving quality, safety, and constructability [4]. As a result, current trend of high-rise building construction prefers to form a high-rise building structure by operating either crane-lifted-, or self-climbing formwork system during structural framework [5, 6].

Even with the benefits associated with utilizing climbing formwork system in vertical framework construction, the system requires relatively high cost input distributed during three main phases of system utilization - fabrication, erection, and removal [7]. If an error is discovered while comparing the planned layout with applied formwork system to an actual construction site, rework associated with re-adjustment of the system causes variability in both constructability and project cost. Thus, material components of a climbing formwork system should be assembled to minimize workload related to system re-adjustment and also to maximize economic feasibility.

2.2 Major Configurations of Climbing Formwork System

As configurations related to size of climbing formwork system components are adjustable according to site conditions and needs of client, there is no fixed modular form of climbing systems with standardized unit size and specification. For instance, size and elevation level of working platform in a scaffold are adjustable according to the decision-maker in charge of system design. Nevertheless, certain major configurations are chosen to be considered unconditionally in the upmost priority for optimal use of climbing formwork system, and also for its resulting constructability of workers.

The system has maximum allowable length of a working platform that one bracket can endure. Platform is designed to satisfy performance in both work efficiency and safety. Modular aluminium sub-elements of equal size are assembled to form a unitized platform, and combination of the elements for fabrication determine the length of one platform. Combination of aluminium sub-elements are determined under the possibility that elements are cut off in case of need for re-adjustment to a new wall configuration.

One unit of a climbing formwork system is determined by one set of two brackets that supports the working platform and also eventually the entire system. Brackets must be positioned in optimal location with appropriately designed platforms to endure high load-bearing capacity and ensure structural safety of the system. In addition, bracket set is deemed important in formwork layout, since cost on one bracket set accounts for a considerable portion of total cost on climbing formwork system.

Brackets also have a main role of connecting the climbing rail and anchoring shoe for installation of the system to wall surface. Whether the climbing cone can be anchored to a surface insinuates availability of system arrangement within the boundary.

Figure 1 depicts an exemplary climbing formwork system and its major components.

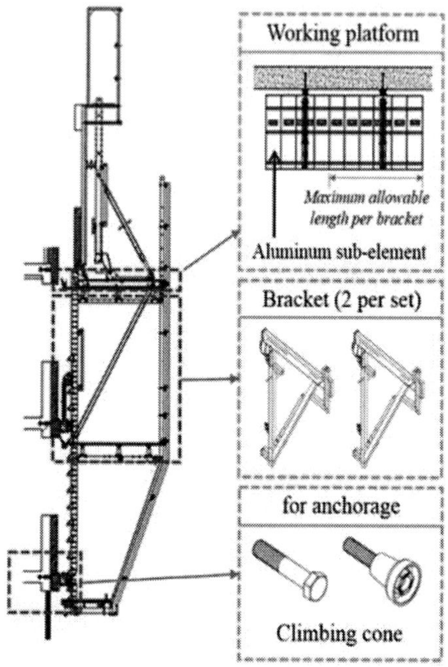

Figure 1. Configuration for climbing formwork system with major components

2.3 Layout Planning of Climbing Formwork System Compared with Other Facilities

Current formwork layout planning method provides alternatives for decision makers, but it does not guarantee whether the chosen alternative is the optimal decision. In a structure with equal parameter, total number of units and associated bracket sets can differ according to the decision made by the supervisor. If the supervisor chooses a layout plan with higher total number of system units to be applied in construction, it means total number of bracket sets becomes proportionally higher. Not only additional installation of bracket results in higher cost, but also re-fabrication of working platforms for adjustment to uneven configurations results in additional workload.

It has been deduced from literature review and interview with experts who majored in building structure that researches on automating layout planning of horizontal formwork and vertical temporary facilities have been previously proposed [4, 8-10]. In case of horizontal formwork, layout planning has been automated based on imported floor plans of building solely by focusing on interior space, not the outer boundary of the building. In case of vertical temporary facilities, few construction corporations have developed their own automated software programs to draw formwork around a building structure and create a bill-of-materials based on the planned layout. However, developed programs for vertical formwork facility are dictated under highly limited conditions, and are applicable to structures with smaller magnitude. Also, the developed programs for automating layout planning of vertical facilities are not intended for 'optimal' layout with targets on specific performances.

In case of climbing formwork system, formwork experts rely on manually planned formwork layout for decision-making. Based on the fixed floor plan, working platform of a system with the longest width are stationed in building parameters and visualized as a layout. In case more than one type of temporary facility must be used in a building construction, decision makers encounter difficulty in finding the optimal set. This insinuates that considerable amount of work hour is consumed for manually planning the layout demonstrated as exemplary on Figure 2.

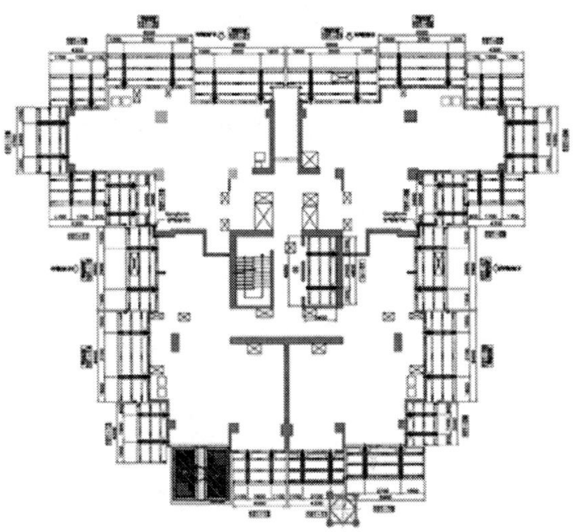

Figure 2. Layout planning of formwork systems

Layout planning dependent on heuristic approach results in higher cost input on labor during planning and design phases. In addition, total cost increases during structural framework of construction phase if any errors occur in any of the formworks that have been fabricated based on the designed layout. Therefore, an automated layout planning model is needed for optimal layout execution of climbing formwork system based on a clear objective in construction performance.

3 Automated Layout Planning of Climbing Formwork System

3.1 Consideration Factors and Assumptions for Optimal Layout Planning

As a decision supporting model for optimizing the use of an appropriate formwork system, automated layout planning method of climbing formwork system has been defined as the objective of this study. The proposed automated method for formwork layout should identify the best arrangement of the systems within the constraints imposed by various conditions. However, simply contributing to cost reduction for formwork construction does not satisfy the conditions of an 'optimal' model.

Since standardization of resources results in operational stabilization in construction, maximizing the use of standardized systems with modular size is efficient in both cost and work constructability. Thus, solution to this optimization problem aims to minimize formwork cost through minimizing material usage for special units or any form of rework associated with system utility. Since cost associated with bracket installation has been determined as the most cost-consuming factor, minimizing quantity of bracket sets by maximizing number to platform installation with minimum variability in modular sub-elements is another definition of the optimization problem. A set of parameters is considered in order to optimize a set of variables set for achieving the specified targets [11].

During this study, the two major components critical for optimal arrangement of the system - length of working platform and distance between two supporting brackets that are connected to the anchoring shoe for system anchorage – are design parameters for configuration of formwork layout and arrangement. In addition to information related to the materials composing a unit module system, information related to construction site acts as the key for space modelling. Imported shop drawings related to the building structure set constraints for system installation. For example, information on location and dimension of window opening and columns serve to determine the resultant possibility of anchorage for formwork system. Also, information on retaining wall and column is also critical for determining availability of platform installation.

Another important controlling factor of climbing formwork system installation is specification of tower crane in construction site. Regardless of whether crane-lifted or hydraulically lifted, the system requires use of tower crane for erection and removal. Also, components of scaffold attached to the system have to be re-adjusted from the previously used modular system if location of tower crane or its turning radius interferes with formwork construction. Although lifting load of tower crane is not additionally reflected in this study, constructability on fabrication, erection, and removal of the system by assistance of tower crane has to be ensured [12, 13]

3.2 Relationship between Structural Safety and Design Parameters of Climbing Formwork System

Optimal solution set should be derived within a reasonable boundary, in both physical workspace perspective and structural safety perspective [14]. Constraint set in this problem requires: (1) a set of brackets to not be installed in window openings, (2) length of working platform to be deduced based on maximum allowable length per bracket, and (3) designed platform to have sufficient capacity to endure the system load.

After the configurations above are fixed, total length of platform is deduced based on the locations of anchorage and their resultant distance between two brackets. However, the two brackets are not always positioned to share the same amount of loading capacity with equal projecting length of platform. Consequently, fitness of anchorage location for two brackets to support a working platform is evaluated in this study by: (1) whether platform length satisfies maximum allowable length of platform per bracket, (2) whether locations of anchorage and capacity of one bracket are compromised to deduce optimal number of aluminium sub-elements of platform for each system arrangement, and (3) whether total length of platform is designed to satisfy the structural limits.

Structural safety verification of the utilized climbing formwork system is applied for satisfying all the above constraints related to the design of platform. Thus, locations of anchorage for brackets should satisfy the maximum deflection of the platform (δ_{max}), and maximum allowable unit stress in bending (F_b).

Considering total vertical load in a platform (W) as summation of dead load and permissive live load under extreme working condition, safe transfer of bearing forces into the building structure is verified under the constraint of maximum allowable length of platform per bracket l_{max} as:

$$\frac{w{l^2}_{max}}{8Z} \leq F_b \tag{1}$$

$$\delta_{max} = \frac{5w{l^2}_{max}}{384EI} \leq {}^{l}/_{240} \tag{2}$$

in which uniform loads(w) are calculated as total vertical load multiplied by maximum length of platform. Maximum deflection of platform should also satisfy the fundamental equation on allowable deflection of an element. The two inequality functions are to be converted and accommodated as a constraint set during search for optimal arrangement of climbing formwork system by using genetic algorithm.

3.3 Genetic Algorithm Formulation for Optimal Solution to Formwork Layout

3.3.1 Framework for Developing Automated Layout Planning

An optimization problem for construction layout has been commonly interpreted as arrangement of a set of predetermined target into appropriate locations. Likewise, developing an automated layout planning of climbing formwork system is not only concerned with providing multiple alternatives of formwork layout. Primary objective of incorporating automation technology for layout planning of climbing formwork system is to provide an optimal solution with definite standards for decision-makings to supervisors.

Development of automated layout planning of climbing formwork system through incorporation of search algorithm contains the following procedures depicted on Figure 3.

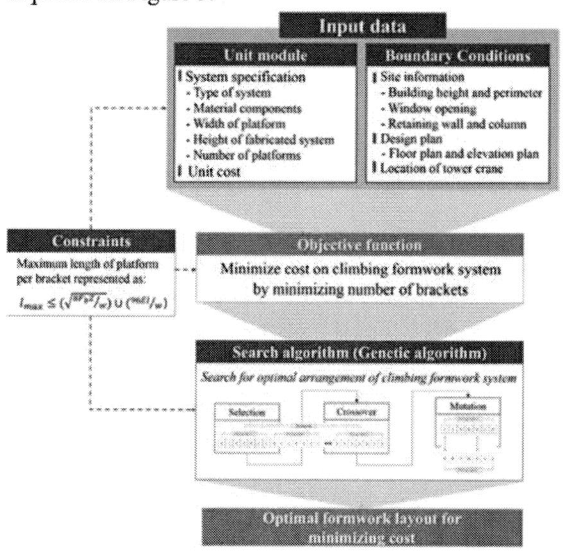

Figure 3. Framework for developing automated layout planning

First, classification of formwork – including size, location of use with different performance requirements, type of climbing formwork system and its material components for high adaptability to all shapes, should be clarified. Unchangeable information on unit module of a climbing formwork system including width of working platform, size of wall formwork, total height of the fabricated system, and number of platforms serve as input data. Unit cost of the climbing system estimated by unit cost of material components is also an input value and a determinant factor for the ultimate output.

Second, shop drawings of construction site with information on building dimensions, including building height and perimeter, are imported for space modelling. Boundary conditions, which in this case interpret possible areas of layout plan, are implemented on a grid with x and y axis. Whether a climbing formwork system can be arranged in an area is determined by availability of anchorage of the brackets to the wall. Location of anchorage considering areas incapable of system installation is first reflected in the space, such as wall surface with openings for window installation. As previously mentioned, imported plans should also include information on number of tower cranes installed along with their locations.

Shop drawing of a building structure is implemented on a grid with information on axis for possible anchorage of climbing formwork system, as demonstrated on Figure 4. Within the boundary space that enables arrangement of the system, a set of optimization algorithm is employed for random search. In this study, locations of anchorage for the two brackets in a unit module of the system are ultimately searched and imported on the grid with (x_i, y_i) and (x_j, y_j) coordinates, respectively. The two coordinates determine distance between two brackets. Prior to derivation of optimal location for anchorage, grid array portraying installation areas of platforms around the building is constructed with length (cl_{ij}) and width (cw_{ij}).

3.3.2 Genetic Algorithm Formulation

Genetic algorithm is applied as a search algorithm to provide optimal solution for arranging climbing formwork system in high-rise building construction. Genetic algorithms employ random search for locating the globally optimal solution through parallel processing [15], and represent schemes that provide alternatives for layout plans in this study. Applying principles of genetic algorithm involves five primary aspects: (1) setting the chromosome structure, (2) deciding the evaluation criteria (objective function), (3) generating an initial population of chromosome for initial solutions, (4) selecting an offspring generation mechanism as a process to generate new potential solutions by mutation, and (5) schematizing the results [16].

Building structure with information on openings, columns, and retaining wall is set as the optimization area for installation of climbing formwork system. Data structure is first established in order to derive the ultimate

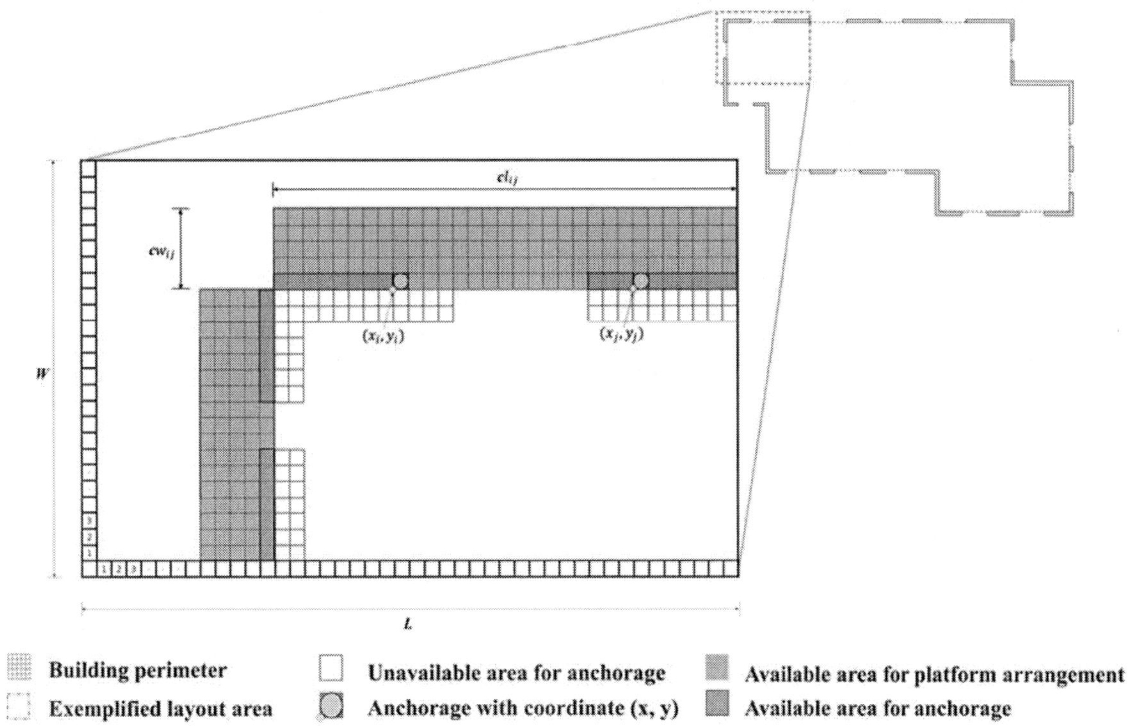

Figure 4. Geometry of allowable area for system installation (reference to [14])

optimal solution [17, 18] based on the available area for installation. Design variables that affect optimization objective are represented as chromosomes with various genes. Prior to the setting of fitness based on objective function, objective function parameters are coded in binary codes.

With reference to Figure 4, possible areas for installation of platforms with predetermined modular width are coordinated around the building surface. Parameter p_{xy} recognizes whether an entire modular platform of unit i with length (cl_{ij}) and width (cw_{ij}) can be placed within the rectangular grid coordinates with length L and width W. Reflecting the binary parameter implemented based on openings or any other areas that disallow anchorage, parameters (a_{xyi}, a_{xyj}) are plotted for possible installation of two anchoring shoes within the platform area coordinated as (x_i, y_i) and (x_j, y_j). Binary parameters $(p_{xy}, a_{xyi}, a_{xyj})$ are given a value of 1 if the system is installed, and 0 if otherwise.

During the search for optimal solution in this study, location of anchorage for system installation to the wall surface is coded, and length of working platform is to be determined based on maximum allowable length of platform per bracket and additional information on anchorage information. Considering that brackets serve to connect anchoring shoe and scaffold in order to support the system, locations of climbing formwork system anchorage convey allowable length of working platform that one bracket can endure.

After continuous search for feasible layout, arrangement of climbing formwork system is executed based on imported boundary conditions of building perimeter and location of the tower crane that confine the boundary set. Parameters on anchoring location converge after searching for optimal locations, and yields to position in which the resultant length of platform satisfies the constraint on maximum allowable length of platform per bracket. Thus, the automated layout planning method of climbing formwork system is operated to search for a solution that minimizes cost input by minimizing use of brackets and associated material components under the following objective function:

$$Minimize \sum_{x=1}^{L}\sum_{y=1}^{W} p_{xy}\left(\sum_{i} a_{xyi} + \sum_{j} a_{xyj}\right) \quad (3)$$

4 Conclusion

As a preliminary study, this paper presented an algorithm that serves as a framework of a decision-support model for optimal layout planning of climbing formwork system. Through an extensive study on formwork layout that takes place during planning and design phases, it has been deduced that optimal layout method for vertical formwork system has not been yet proposed. Multiple factors have been defined to be under consideration, including specification on major components of climbing formwork system such as working platforms and brackets for installation to a wall surface. Based on the derived decision variables that are applied as input data, flow chart defining input data, boundary conditions related to the building structure, and constraints such as maximum allowable length per bracket of a system was produced.

'Optimal' layout of a facility has responded to greater demands for higher-quality construction delivered on lower cost, higher constructability of workers, and enhanced work efficiency. Likewise, optimization problem of this study has been defined as arrangement of climbing formwork system with minimum cost input and maximum material usage during system utilization. Considering the modelled space of the building structure, specification on anchorage with its resultant total length of platform have been determined through random search within constraints. Genetic algorithm has been chosen as the optimal form of algorithm for providing alternatives of formwork layout planning and facilitating construction workers to make a rational decision.

The following conditions should be additionally considered in future study for reliability assessment of the proposed model:

1. The proposed method views the construction space in two dimensions. Since layout planning of climbing formwork system should provide both horizontal and vertical layouts based on shop drawings from various perspectives, providing the solution in three axis (x, y, and z) is expected to provide a clear vision.
2. Some of the conditions that are considered important in actual construction site are not reflected in the current algorithm. Thus, future study is to complement the proposed layout method, and is expected to extend from a framework to an actual automated layout planning that provides optimal solution to the defined problem.

Acknowledgment

This research was supported by a grant (15AUDP-B106327-01) from Architecture & Urban Development Research Program funded by Ministry of Land, Infrastructure and Transport of Korean Government.

References

[1] Peurifoy R.L, and Oberlender G.D. *Formwork for*

Concrete Structures, Fourth Edition. McGraw Hill, New York, 2011

[2] Lab R.H. Think Formwork – Reduce Costs. On-line:, Accessed: 10/02/2016

[3] Lee C., Ham S., Lee G. The Development of Automatic Module for Formwork Layout using the BIM. In *The 3rd International Conference on Construction Engineering and Management/The 6th International Conference on Construction Project Management (ICCEM/ICCPM 2009)*, pp. 1266-1271, Jeju, Korea, 2009

[4] Kim T. Advanced system formwork and construction planning model for tall build construction. Doctorate thesis, Korea University, 2013

[5] PERI GmbH. 2015. Climbing system. On-line: www.peri.de, Accessed: 16/02/2016

[6] Yang S-H, Cho J-Y, Cho J-W, Lee J-H, Kim Y-S. A Study on the Development of Automated Gangform Climbing System for Apartment Housing Construction. Korean Journal of Construction Engineering and Management, 13(1), pp. 53-66, 2012

[7] Mawdesley M., Al-jibouri S., Yang H. Genetic Algorithms for Constriction Site Layout in Project Planning. Journal of Construction Engineering and Management, 128(5), pp. 418-426, 2002

[8] Lee D., Lim H., Kim T., Cho H., Kang K.I. A Formwork Layout Model based on Genetic Algorithm. In *The 31st International Symposium on Automation and Robotics in Construction and Mining (ISARC 2014)*, pages 474–479, Sydney, Australia, 2014

[9] Cha M. A Decision Support Model for Optimal Formwork Layout in High-rise Building Construction. Master's thesis, Korea University, 2012

[10] Kannan M., and Santhi M. Constructability Assessment of Climbing Formwork Systems Using Building Information Modeling. Procedia Engineering. 64, 00. 1129-1138, 2013

[11] Whitley D. A Genetic Algorithm Tutorial. Statistics and Computing, 4, pp. 65-85, 1994

[12] Cai P., Cai Y., Chandrasekaran I., Zheng J. Parallel genetic algorithm based automatic path planning for crane lifting in complex environments. Automation in Construction, 62, pp. 133-147, 2016

[13] Ahn B., Kim J., Kim K., Kim S. A Study on the Optimization of Site Layout in the High-Rise Building Construction. Journal of the Architectural Institute of Korea – Structure & Construction, 15(12), pp. 107-118, 1999

[14] Kim T., Lim H., Lee U.K., Cha M., Cho H., Kang K.I. Advanced formwork method integrated with a layout planning model for tall building construction. Canadian Journal of Civil Engineering. 39:1173-1183, 2012

[15] Conway D., and Venkataramanan M. Genetic Search and the Dynamic Facility Layout Problem. Computers & Operations Research, 21(8), pp. 955-960, 1994

[16] Hegazy T. Optimization of Construction Time-cost Trade-off Analysis using Genetic Algorithm. Canadian Journal of Civil Engineering, 26(6), pp. 685-697, 1999

[17] Koo C., Hong T., Kim S. An Integrated Multi-Objective Optimization Model for Solving the Construction Time-Cost Trade-off Problem. Journal of Civil Engineering and Management, 21(3), pp. 323-333, 2015

[18] Park U-Y, Cho S-S, Kang K-I. A Study on Rebar Detailing Optimization Using Genetic Algorithm. Journal of the Architectural Institute of Korea – Structure & Construction, 19(10), pp. 125-132, 2003

An Extended Ambient Intelligence Implementation for Enhanced Human-Space Interaction

A. Liu Cheng[a,b] and H. H. Bier[a]

[a]Faculty of Architecture and the Built Environment, Delft University of Technology, Delft, The Netherlands
[b]Facultad de Arquitectura y Urbanismo, Universidad Tecnológica Equinoccial, Quito, Ecuador
E-mail: a.liucheng@tudelft.nl, h.h.bier@tudelft.nl

Abstract –

This paper proposes an extended *Ambient Intelligence* (AmI) solution that expresses intelligence with respect to both *Information and Communications Technologies* (ICTs) and spatial reconfiguration in the built-environment. With respect to the former, a solution based on a decentralized yet unified *Wireless Sensor Network* (WSN) is proposed. This is deployed across exterior, interior, and wearable domains, equipped with heterogeneous platforms across embedded and ambulant nodes, and open to a variety of proprietary and non-proprietary communication protocols. With respect to the latter, a corresponding functionally and physically reconfigurable built-environment pertinent to the Adaptive Architecture discourse is revisited. The ICTs component aims to demonstrate the advantages of a cohesive and interoperable heterogeneity distributed along local and web-based proprietary and non-proprietary services over a prevalent locally based homogeneity with respect to both development platforms and communication protocols in a WSN. The architectural component aims to demonstrate that a highly adaptive and transformable built-environment is better suited to complement and to sustain assistive as well as interventive services enabled by said WSN. As a unified solution, the proposal showcases that the merging of technological and architectural considerations in the design of an intelligent environment enables more intuitive solutions that actively adapt to, interact with, intervene on the user to promote comfort and well-being via computational as well as physical feedback-loops.

Keywords –

Ubiquitous Computing, Adaptive Architecture, Wireless Sensor Networks, Ambient Intelligence.

1 Introduction

Present *Ambient Intelligence* (AmI) solutions tend to be situated in discrete structured environments. The salient disadvantage of this tendency is that the functional scope of such solutions does not extend beyond their deployment domain, and can therefore neither extend their services to nor gather valuable physiological data from the user in his/her absence. This is particularly problematic to solutions with components that depend on aggregated human interaction (explicit and/or implicit) and/or physiological data to perform optimally. A brief overview of developing trends and themes in the AmI discourse—e.g., see [1]—is sufficient to suggest that a considerable segment of AmI-related solutions are developing with such considerations.

In order to extend the monitoring capabilities of a system, AmI solutions built on *Wireless Sensor Networks* (WSN) may integrate remote mobile sensor nodes to work in conjunction with local ones to ascertain uninterrupted monitoring [2]. While said remote sensors may be fundamentally developed as project-specific devices and from technologies similar to those employed within the local WSN, a more effective approach would be to render the AmI system interoperable with existing commercial products and solutions. Researchers and developers have typically favored technological homogeneity due to practical considerations, but such an approach limits the reach of the developed system, as no single standard supersedes all others in every aspect. Since some standards are more robust than others depending on particular tasks, heterogeneous solutions that capitalize on this distinction have an advantage over their homogeneous counterparts. However, interoperability among heterogeneous systems is an on-going challenge in the development of AmI solutions (see [3], for example). Researchers have noted that commercial equipment is often closed, making seamless integration with developing AmI systems unfeasible [2]. But in the last five years commercial manufacturers have acted on a vested interest in making their products interoperable with a variety of systems in order to broaden their market. As a result, proprietary *Application*

Program Interfaces (APIs) have enabled the seamless integration of some commercial products with non-proprietary systems. The WSN discussed in this paper is heterogeneous not only with respect to devices but also to communication protocols, and it uses a proprietary API to extend its monitoring abilities to ascertain uninterrupted operation.

In conjunction with technological considerations, another imperative yet generally overlooked consideration pertains to the built-environment within which said WSN is to be deployed. Conventional models tend to labor on a retrofitting strategy, where new services and technologies are made to work within a setting whose built typologies were conceived without ICT-integration in mind. Such models are informed by the belief that the costs of retrofitting the new into the old are less than those associated with late-stage design consolidation of intelligent services and their corresponding production [4, 5]. But this is true insofar as said design consolidation is *late-stage*. Early-stage design consolidation of intelligent services in conjunction with robotically driven production [6], which considers the changes in the structure and infrastructure of the architecture that must be adopted in order to enable robotic environments suitable for ubiquitous systems and service robots [7, 8], do in fact instigate considerable cost reductions [9].

Yet the conventional model prevails due to either industry conservatism or lack of dialogue between researchers in the fields of ICTs and their counterparts in architecture and construction. As a result, contemporary discussions within the AmI discourse focus on ICT-based solutions while continuing to presuppose a static and conventional architecture. Such a tendency inadvertently hinders the development of the AmI discourse by neglecting considerations of consequence to both the pertinence and performance of said ICT-based solutions. For example, present conventional models are not able to accommodate situations where the very architecture of the built-environment altogether prevents or partially hinders the proper deployment and operation of ICT-based services. In order to facilitate a corresponding implementation and proper operation of heterogeneous ICT-based services, the architecture of the built-environment must be considered as part of the overall design strategy in order to express intelligence within it. In this manner, intangible computational services work in conjunction with tangible architectural transformations to mutually inform one another in order to enhance the quality and efficacy of both assistive and interventive services. Recent work in Adaptive Architecture has demonstrated that the immersed experience of built-environment transformations may inform and influence physiological responses—see [10], for example. The present work aims to showcase how ICT-based systems may work in conjunction with transformable architectural systems based on physiological phenomena in order to enhance the quality of living of the user.

2 Concept and Approach

The proposed solution establishes a justified link between sensed physiological user-data with actuated architectural transformations within a built-environment. It does this via a series of interrelated mechatronic devices and architectural mechanisms that conform the WSN (see Figure 1). The architectural transformations explored are intended to promote the well-being of the user as well as enhance his/her overall experience of the space. In order to do this, a set of decision-making criteria to justify the link between sensed input and actuated output is defined (see Section 3). In the present scope, a variety of simple actuation events corresponding to particular scenarios are triggered based on a rudimentary decision-making mechanism.

In the first scenario, the user is assumed to be in a seating position, and his/her heart rate, perspiration levels, temperature, and changes in posture are processed to gauge the period and extent of user-inactivity. Based on medical baselines, the user is tacitly encouraged to change position, stand and/or walk for a variable period of time by the actuating system's gradual shifting of the desk and seat positions. In the second scenario, the user is assumed to be in a standing position. The system gauges the period in which the user has been inactive in such a position and invites him/her to engage in exercises that are only possible with the appearance of specialized transformable components. Such components would tacitly shift to indicate that the activity levels are sufficient while leaving the user the option to continue through expressed persistence. In the third and final scenario, the user is assumed to be absent from the structured environment for the majority of the day. The configuration of the interior environment adjusts itself in preparation for the arrival of its occupant into different combinations depending on the heart rate, and activity averages gathered remotely throughout the day. As a result, the configuration of the space may be more open and ventilated or more enclosed and hermetic.

These basic scenarios are sufficient to demonstrate that transformations in the built-environment are capable of engaging the user in response to his/her physiological state in order to regulate it. Further work to showcase higher degrees of decision-making sophistication is presently under development (see Section 4).

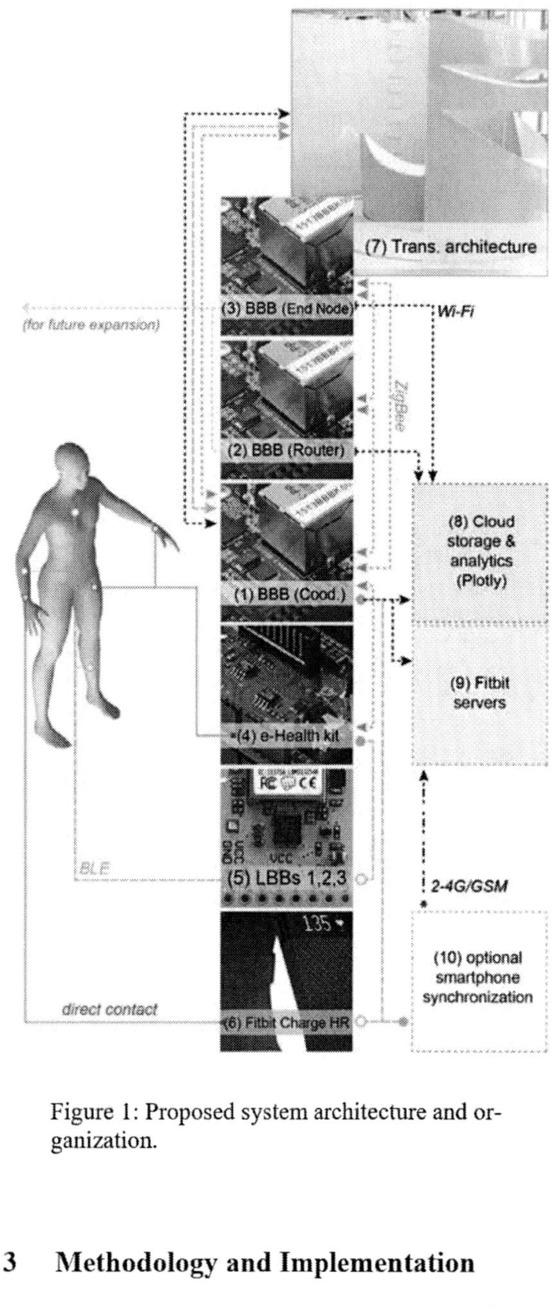

Figure 1: Proposed system architecture and organization.

3 Methodology and Implementation

3.1 Sensor types with respect to scenarios

The WSN developed in this paper builds on the decentralized architecture implemented in [11]. In addition to three *BeagleBone Black*© [12] (BBB) development platforms, i.e., a (1) *Coordinator*, a (2) *Router*, and an (3) *End Node*; an (4) *Arduino UNO*© [13] microcontroller unit (MCU) coupled with *Libelium*®'s *e-Health Sensor Platform V2.0*© [14] (e-Health kit); (5) three *PunchThrough*®'s *LightBlue Bean*©s [15] (LBBs); and (6) *Fitbit*®'s *Charge HR*© [16], have also been integrated into the WSN. Only the *Coordinator* BBB node is used in the present implementation, while the remaining two will be used in subsequent development of the system. Furthermore, in addition to heterogeneity with respect to MCUs, the WSN operates across a variety of communication protocols, i.e., Bluetooth Low Energy (BLE), Wi-Fi, and ZigBee. The present combination of devices serves as a sampler of WSN functionality in exterior, interior, and wearable domains.

The BBB nodes represent the self-healing and meshed backbone of the WSN, while the Arduino UNO and e-Health sensors function as local embedded and principal sensors for the first scenario mentioned in Section 2; the LBB as the local ambulant and principal sensor corresponding to the second scenario; and the *Charge HR* as a remote ambulant and principal sensor corresponding to the third scenario. Although each scenario has a principal sensor system, it should be noted that all sensor systems play a role in each scenario. In the following subsections below a description for each corresponding system is detailed.

3.1.1 Local embedded sensing for scenario 1

The e-Health kit is used in conjunction with an Arduino UNO MCU to gather physiological data—i.e., pulse, peripheral capillary oxygen saturation (SpO2), sweat-levels, temperature—as well as pressure data as measured from a (7) transformable workstation consisting of a desk and a corresponding seat. The galvanic and pulse oximeter sensors are fixed directly unto the fingers of the user while temperature and pressure sensors are placed throughout the seat. These sensors are wired to the MCU. Simultaneously, the LBBs are used to track changes in acceleration, which occur whenever a change in posture takes place. The data gathered is transferred to the MCU via BLE, which is then relayed to the *Coordinator* BBB via ZigBee and made available to all other BBB nodes in case future subsystems require such data. The *Coordinator* node both analyses the data and streams it via Wi-Fi to a cloud-based plotting and analytics service, i.e., (8) *Plotly*® [17]. In this manner, long-term daily datasets are stored in *Plotly* and made accessible to any person and/or AmI system with the proper user-credentials; while short-term datasets are processed locally to trigger immediate corresponding actuation events. Over time, with the long-term statistical analysis, a personalized baseline may be ascertained. Short-term analysis would be set against this baseline to identify deviations from the user's *normal* state. Naturally, personalized baselines must also take into account prescribe medical baselines according to gender, age, weight range, etc.

The overall aim is to ensure well-being and encourage

healthier living. According to the World Health Organization (WHO), physical inactivity is a leading risk factor for global mortality with approximately 3.2 million deaths each year [18]. In order to prevent, for instance, the formation of decubitus ulcers due to persistent applied pressure to a tissue region or to reduce the development of varicose veins [19] sensor-actuators embedded in the built environment will initiate physical movement or activity. Pressure and temperature datasets relative to time and extent may be sufficient to estimate the probabilities of ulcer formations [20]. With respect to varicose veins, when the user has been sitting on one muscle group and/or has remained inactive for extended periods of time, the *Coordinator* BBB will command the seat and the desk component to shift in different directions to tacitly encourage the user to shift his/her weight and/or to exercise his/her lower extremities. The duration and frequency of this interventive service vary and depend on inactivity levels considered against both short-term data as well as long-term personal statistical baselines.

3.1.2 Local ambulant sensing for scenario 2

A series of LBBs are used to provide the user with wearable devices operational within the structured environment. The redundancy in these devices is intended to enhance accuracy and regional coverage—i.e., one LBB represents a wearable attached to the lower extremities, another to the upper extremities, yet another to the core of the upper-body. As in the first scenario, the data is shared across the entire WSN as well as stored and analyzed in *Plotly*. When the extent and frequency of changes in acceleration from these three regions are minimal and scattered far between, the *Coordinator* BBB shifts exercise components towards the region of the user to tacitly invite him/her to increase in activity levels. In the present scope these exercise components are general and hypothetically assumed components, represented as attached to the shifting seat and desk/wall components (see Figure 2). At this stage it is not important to define particular types of exercises as long as the user is moving with frequency and consistency.

The principal objectives in this scenario include the second objective of the previous scenario as well as the intention to proactively engage the muscular and circulatory systems. Aside from these considerations, there is also the intention to engage the user in entertaining tasks, as well-being is not confined strictly to medical concerns. The transformations instantiated in the geometry of the architecture may serve to engage the user in stimulating ways, some of which may have purely physical consequences while others may be both physically and mentally stimulating—for e.g., architecture-embedded Playware [21].

3.1.3 Remote ambulant sensing for scenario 3

A *Charge HR* activity tracker is used as the system's remote sensor node. This wearable device is designed to sense and record data pertaining to steps taken, heart rate, distance covered, calories burned, and floors climbed. The manufacturer's (9) API provides a means for the proposed WSN to connect to their servers in order to download user-specific datasets pertaining to the listed activities. This API is first installed in the *Coordinator* BBB, which is used in a *Python* script expressly written to fetch such datasets. As with the previous two scenarios, the downloaded time-series data is parsed for local analysis as well as *Plotly* storage and analytics. Inside of the structured environment, the Fitbit device synchronizes with the *Coordinator* BBB via BLE, and all collected data is subsequently relayed to Fitbit's servers via Wi-Fi. Outside of the structured environment, the device synchronizes with a smartphone with Fitbit's Application installed, and all collected data may be relayed to Fitbit's servers via (10) cellular communication technologies (e.g., 3G/4G). It should be noted that in the unlikely event that the *Charge HR* should malfunction, Fitbit's iOS / Android application is capable of turning the smartphone into an *ad hoc* tracker, thereby providing a level of justified redundancy.

The principal objective in this scenario is to maintain uninterrupted physiological data tracking with respect to the user even when he/she is outside of the structured environment. While he/she is within it, the Fitbit device may be used in conjunction with the e-Health kit to refine the heart rate analysis. But when he/she is outside the environment, it is in the interest of the WSN to keep gathering physiological data for the generation of a personalized long-term statistical profile. In addition to this objective, this scenario demonstrates how a remote agent may still be capable to influencing his/her home environment while absent. That is to say, depending on the currency of the gathered data *in absentia*, and relying on GPS-tracking services to ascertain user proximity to the structured environment, the WSN may adjust the built-environment's configuration in order to provide a more suitable welcome to the user. If the measured heart rate has been consistently elevated, and the activity levels have been construed as "intense" (under Fitbit's criteria) within a recent period, the built-environment may reconfigure itself to instantiate a more open and highly ventilated space to help balance the user's physiological status.

3.2 Transformable architectural systems

Local and remote sensing inform spatially reconfigurable architectural systems in order to facilitate adaptation and changing use. For instance, *Hyperbody*'s reconfigurable apartment is an experiment into designing a small apartment of 150 m^3 that has all the spatial qualities and functional performances of a standard 300 m^3 apartment. The initial assumption was that when a user is in the kitchen or living room, this user does not use the sleeping room at the same time implying that at one moment of the day large sections of the space could cater to only one or two functions.

Spatial transformation is exploiting material and geometrical properties in order to easily facilitate continuous change of use. Developed as a compliant mechanism the system transfers input force or displacement through elastic deformation. Consisting of a monolithic (single-piece) or jointless components, this adaptive architecture was envisioned to easily populate empty office buildings, which at the time corresponded to six million square meters and the number is—until now—still growing.

Such a reconfigurable apartment could employ WSN/BAN-enabled high-tech intelligence working in conjunction with architectural components in order to coordinate physical / geometrical rearrangements and/or spatial reconfigurations according to a set of conditions established in order to maximize the welfare of the user(s) (see Figure 2). The reconfigurable apartment is intended to cater to quotidian activities such as sleeping, eating, etc.; and the heterogeneous and decentralized WSN is envisioned to continuously gather interior environmental data specific to the respective activities via directly and indirectly attached sensors.

The apartment is envisioned to gradually *learn* that when the user is studying, s/he prefers a more private and intimate space, and accordingly walls, doors, ceiling heights, etc., may be adjusted in conjunction with temperature and illumination settings in order to instantiate just such space. Similarly, if s/he would prefer an open and welcoming space when socializing, the apartment would adapt its physical, geometrical, and spatial properties accordingly.

Furthermore, the apartment would encourage the user to adopt a healthier lifestyle by initiating spatial reconfiguration when physical activity is recommended. It would also adjust climate according to monitored needs. Heating, lighting, and ventilation would be distributed and locally deployed when needed and as needed, increasing levels of comfort while at the same time minimizing energy-waste.

In this context, sensor-actuated environments are not only customisable in order to provide healthier ambiance, but they may also offer solutions for managing the demands associated with rapid population growth, urban densification, and the contemporary inefficient use of built space by enabling multiple, changing uses within reduced timeframes.

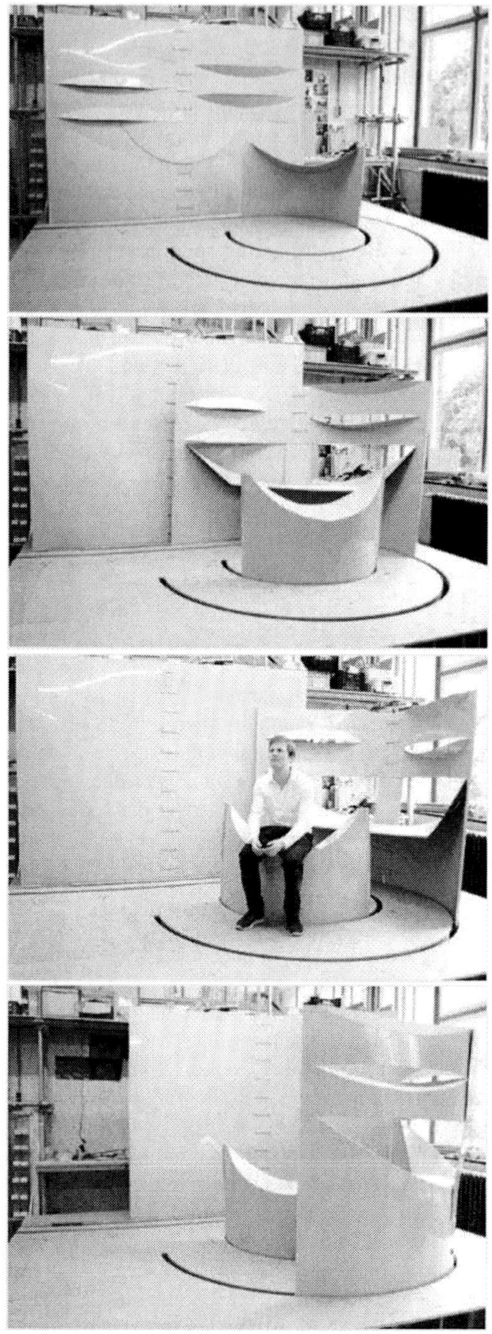

Figure 2: Reconfigurable apartment developed with MSc 2 students and industry partners at *Hyperbody*, TU Delft (Pop-up Apartment, 2013)[6]

4 Discussion and Conclusions

The present proposal attempts to demonstrate the advantages of integrating considerations pertaining to both the ICT as well as the adaptive / transformable architecture discourse into AmI solutions. Such solutions would not be confined to delivering ICT services only, but their very design would enable the means via which to express such services in a tangible manner, thereby increasing effectiveness. On a personal level, users would benefit from the proactive, intuitive, and highly personalized suggestions provided with corresponding means of execution, particularly with respect to quotidian activities. On a societal level, public and private healthcare providers would be alleviated from unnecessary and premature cases of patients whose ailments were exacerbated by sedentary lifestyles. This latter consideration is particularly pressing. For example, according to Espinoza, the *Organization for Economic Cooperation and Development* predicts that the health expenditure in the EU alone is expected to rise by 350% by the year 2050 compared to an economic expansion of only 180% [22]. This reality alone serves to promote extended AmI environments capable of both servicing and providing the means of engaging users to counteract sedentary habits, thereby becoming effective promoters and extenders of health. Although the proposed system is designed to engage and to benefit all age-groups, foreseeable early adopters in the near future are likely to include people with early symptoms related to metabolic syndrome, which—if left untreated—may lead to more serious secondary diseases [23] such as heart attacks, cognitive impairment [25], and even vascular dementia [28]. It should be stressed that although the system does indeed potentially mitigate the extent of health degradation of such target group, it is not intended purely as a mitigating solution. The extended AmI proposal detailed in this paper intends to enhance user-experience, use space intelligently by shifting programmatic functions within the same space throughout the day, and increase meaningful engagement between users and their built-environment in order to promote well-being and health.

In order to more accurately meet the proposal's intended objectives, further work into the extent of spatial transformability and its appropriateness with respect to form, function, and users' needs has to be developed. Furthermore, limitations inherent in the present solution must be resolved. The following three salient limitations may be highlighted: (1) the BLE operational range with respect to the *Coordinator* BBB; (2) the resolution and reliability of low-cost sensors; and (3) the rudimentary character of the decision-making mechanisms.

With respect to the first consideration, a series of routing devices may be scattered throughout the built-environment to ascertain communication between the LBBs and the Arduino UNO MCU and/or any of the BBBs (provided these latter are equipped with pertinent BLE-enabling shields). The feasibility of this solution decreases as the structured environment's size increases. In a relatively extreme scenario, long-range communication protocols may be used instead of BLE. Alternatively, the system could rely on the smartphone's gyroscopic and accelerometer data, broadcasted via *User Datagram Protocol* (UDP) [24]—assuming Wi-Fi coverage throughout the structured environment—whenever the LBBs are out of range.

With respect to the second consideration, the implementation of higher quality and costly sensor systems may be justified if and only if the data resolution, reliability, and performance is considerably greater than that of the low-cost counterparts. If the difference is marginal, then the system should be alternatively configured to operate with lower levels of resolution. With respect to the long term statistical profiling, the loss of some datasets is tolerable. This, of course, depends on what the sensor is measuring and what the actuating mechanism is intended to respond to. Certain actuation events will require higher resolution and system fidelity, while others may tolerate a certain degree of simplification. Therefore, a solution to this consideration must involve a more robust definition and justification of sensed inputs and actuated outputs, which would help to ascertain tolerances as well as to gauge the appropriateness of low-cost sensors. With respect to the third and final consideration, it may be noted that decision-making mechanisms—and corresponding inputs / outputs—are simultaneously the most promising and limiting factor in the development of such cyber-physical AmI solutions. In order for such mechanisms to become the drivers of dynamic and adapting artificial agents, there must be a degree of machine learning involved. This is made manifest in the present literature with respect to the computational side of AmI.

Finally, it needs to be stressed that cyber-physical AmI solutions have to be designed and implemented in such a way as to sustain a new kind of artificial ecosystem, where the environment's components are self-sustaining, and where their development, adaptation, and evolution occur in symbiosis with their corresponding users [26]. In more sophisticated iterations of the present project, architectural components could begin to demonstrate basic yet explicit forms of agency based on and in response to processed user as well as environmental data [27]. The fundamentally and analytically intelligent built-environment effortlessly facilitates an immediate and intimate version of what Oosterhuis has described as a *Society of Building Components*, where the environment's components act and react computationally (i.e., exchange data) and physically (i.e., change forms) towards one another and towards the users [29].

Acknowledgements

This paper has profited from the contribution of *Hyperbody* researchers and students involved in the described project.

References

[1] Hervás, R., Lee, S., Nugent, C., and Bravo, J., Eds. 2014. *Ubiquitous computing and ambient intelligence. Personalisation and user adapted services*. LNCS sublibrary. SL 3, Information systems and applications, incl. Internet/Web, and HCI 8867. Springer, Cham, New York.

[2] Harrison, B. L., Consolvo, S., and Choudhury, T. 2010. Using Multi-modal Sensing for Human Activity Modeling in the Real World. In *Handbook of ambient intelligence and smart environments*, H. Nakashima, H. K. Aghajan and J. C. Augusto, Eds. Springer, New York, 463–478.

[3] Jiménez-Fernández, S., Toledo, P. de, and del Pozo, F. Usability and Interoperability in Wireless Sensor Networks for Patient Telemonitoring in Chronic Disease Management. *IEEE transactions on bio-medical engineering* 60, 12, 3331–3339. 2013.

[4] Blismas, N. and Wakefield, R. Drivers, constraints and the future of offsite manufacture in Australia. *Construction Innovation* 9, 1, 72–83. 2009.

[5] Tam, C. M., Tam, V., Zeng, S. X., and Ng, W. Towards adoption of prefabrication in construction. *Building and Environment* 42, 10, 3642–3654. 2007.

[6] Bier, H. H. Robotically Driven Architectural Production. *Archidoct* 2, 1, 11–16. 2014.

[7] Forlizzi, J. and DiSalvo, C. 2006. Service robots in the domestic environment: a study of the roomba vacuum in the home. In *Proceedings of the 1st ACM SIGCHI/SIGART conference on human-robot interaction*. ACM.

[8] Bier, H. H. 2011. Robotic environments. In *Proceedings of the 28th international symposium on automation and robotics in construction*, 863–868.

[9] Bock, T. and Linner, T. 2015. *Robot-oriented design. Design and management tools for the deployment of automation and robotics in construction*. Cambridge handbooks in construction robotics. Cambridge University Press, New York, NY.

[10] Schnädelbach, H., Slovák, P., Fitzpatrick, G., and Jäger, N. The immersive effect of adaptive architecture. *Pervasive and Mobile Computing* 25, 143–152. 2016.

[11] Liu Cheng, A. 2015. *Design and Implementation of a Novel Cost-effective Fall-Detection and -Intervention System for Independent Living based on Wireless Sensor Network Technologies*. Master of Science thesis, Technische Universität München.

[12] BeagleBoard®. 2015. *BeagleBone Black*. http://beagleboard.org/black. Accessed 10 June 2015.

[13] Arduino®. 2016. *Arduino UNO & Genuino UNO*. https://www.arduino.cc/en/Main/ArduinoBoardUno. Accessed 6 March 2016.

[14] Libelium®. 2016. *e-Health Sensor Platform Complete Kit V2.0 for Arduino, Raspberry Pi and Intel Galileo [Biometric / Medical Applications]*. https://www.cooking-hacks.com/ehealth-sensors-complete-kit-biometric-medical-arduino-raspberry-pi. Accessed 6 March 2016.

[15] PunchThrough®. 2016. *The LightBlue Family. Bean and Bean+: BLE dev boards for the mobile age*. https://punchthrough.com/bean. Accessed 6 March 2016.

[16] Fitbit®. 2016. *Charge HR*. https://www.fitbit.com/chargehr. Accessed 6 March 2016.

[17] Plotly®. 2015. *API Settings*. https://plot.ly/settings/api. Accessed 15 February 2015.

[18] World Health Organization®. 2016. *Health topics: Physical activity*. http://www.who.int/topics/physical_activity/en/. Accessed 11 March 2016.

[19] Milio, G., Siragusa, S., Minà, C., Amato, C., Corrado, E., Grimaudo, S., and Novo, S. Superficial venous thrombosis. Prevalence of common genetic risk factors and their role on spreading to deep veins. *Thrombosis Research* 123, 2, 194–199. 2008.

[20] Angelidis, I. K., Lidman, D., Sjöberg, F., and Thorfinn, J. Decubitus ulcer development. Pressure alone increases tissue temperature. *Eur J Plast Surg* 32, 5, 241–244. 2009.

[21] Lund, H. H. and Marti, P. 2009. Designing modular robotic playware. In *Robot and Human Interactive Communication, 2009. RO-MAN 2009. The 18th IEEE International Symposium on*, 115–121.

[22] Espinoza, J. 2011. *Europe's Failing Health*. http://www.wsj.com/articles/SB10001424052748704893604576200724221948728. Accessed 15 April 2015.

[23] McNeill, A. M., Rosamond, W. D., Girman, C. J., Golden, S. H., Schmidt, M. I., East, H. E., Ballantyne, C. M., and Heiss, G. The Metabolic Syndrome and 11-Year Risk of Incident Cardiovascular Disease in the Atherosclerosis Risk in Communities Study. *Diabetes Care* 28, 2, 385–390. 2005.

[24] Liu Cheng, A. 2014. InTUItive Robot Teleoperation within Ambient Assisted Living Environments via Grasshopper, Firefly, and gHowl. In *Proceedings of the CIB IAARC W119 CIC 2014*

Workshop. Lehrstuhl für Baurealisierung und Baurobotik (BR²), Technische Universität München, Munich, Germany, 33–38.
[25] Ninomiya, J. K., L'Italien, G., Criqui, M. H., Whyte, J. L., Gamst, A., and Chen, R. S. Association of the metabolic syndrome with history of myocardial infarction and stroke in the Third National Health and Nutrition Examination Survey. *Circulation* 109, 1, 42–46. 2003.
[26] Jaskiewicz, T. 2014. Approaching Distributed Architectural Ecosystems. In *ALIVE. Advancements in adaptive architecture*, M. Kretzer and L. Hovestadt, Eds. Applied Virtuality Book Series v.8. Birkhäuser, Basel/Berlin/Boston, 133–139.
[27] Sanchez, J. 2014. Polyomino. The Missing Topology Mechanic. In *ALIVE. Advancements in adaptive architecture*, M. Kretzer and L. Hovestadt, Eds. Applied Virtuality Book Series v.8. Birkhäuser, Basel/Berlin/Boston, 125–128.
[28] Raffaitin, C., Gin, H., Empana, J.-P., Helmer, C., Berr, C., Tzourio, C., Portet, F., Dartigues, J.-F., Alpérovitch, A., and Barberger-Gateau, P. Metabolic syndrome and risk for incident Alzheimer's disease or vascular dementia: the Three-City Study. *Diabetes Care* 32, 1, 169–174. 2009.
[29] Oosterhuis, K. 2014. Caught in the Act. In *ALIVE. Advancements in adaptive architecture*, M. Kretzer and L. Hovestadt, Eds. Applied Virtuality Book Series v.8. Birkhäuser, Basel/Berlin/Boston, 114–119.

Evaluation of Lifting Efficiency of Double-Cage Construction Lift in Supertall Building Construction

Wansoub Kim[a], Dongmin Lee[a], Taehoon Kim[b], Seung Woo Kim[c], Hong-Ku Jung[c], Hunhee Cho[a], Kyung-In Kang[a]

[a]School of Civil, Environmental and Architectural Engineering, Korea University, Republic of Korea
[b]School of Architecture, Chosun University, Republic of Korea
[c]Liftec Corp., Republic of Korea
E-mail: wan0911@korea.ac.kr, ldm1230@korea.ac.kr, thoonkim@chosun.ac.kr, ksw@liftec.co.kr, nayaspy@liftec.co.kr, hhcho@korea.ac.kr, kikang@korea.ac.kr

Abstract

Construction lifts are key equipment used in the vertical transportation of resources in supertall building construction. Also, according to the increase of vertical lifting distance, construction lifts are becoming more important for successful projects completion. However, a limited number of construction lifts were hard to handle the lifting load during peak-time in which the construction resources are concentrated. Also, a large number of stops caused a decrease of lifting efficiency. Consequently, these problems cause an increase of lifting time of the resources and construction delay can occur. Therefore, it is necessary to apply concept of double-deck elevators to construction lift in order to decrease the lifting time of the resources. In this study, two discrete event simulation models using the CYCLONE are developed to evaluate the lifting time of workers, who are required to be located on the upper floor workplace in the morning peak-time. Also, comparative analysis of lifting efficiency between single-cage lift and double-cage lift is conducted. As a result, the proposed simulation model of double-cage lift shows a remarkable decrease of 38.0% compared with single-cage lift. This result could contribute to set the direction of development for double-cage lift. Furthermore, it is expected to support construction manager to find optimal hoisting plan with minimum lifting time in supertall building construction.

Keywords –

Double-cage; Construction Lift; Supertall Building; Lifting Efficiency; Discrete Event Simulation; CYCLONE

1 Introduction

1.1 Background and Objective

As the number of supertall buildings has been increased recently, the heights of buildings have consistently been getting higher[1]. Vertical lifting distance of construction resources, such as material, equipment and labor, has been also getting longer. Thus, the construction lift, which is one of the key lifting facilities, is becoming more important for successful project completion[2].

In supertall building construction, a limited number of construction lifts should move amount of construction resources up to working area on high floors[3], and it cannot afford the vertical loads during peak-time in which the resources are concentrated. This problem can cause an increase of lifting time and construction delay. In the case of the Korea Convention and Exhibition Center, during the peak-time when workers arrive in the morning, it took about 130 minutes to lift 1,200 workers to their work floors[4]. Also, in case of the Burj Kalifa construction project in Dubai, it took approximately 45 minutes as one cycle to transport workers from ground floor to the top floor by using the three-stage lift system [4]. These cases indicate an improvement of lifting efficiency, that is vertical transport capacity of workers at the same time, can shorten the construction time through the reduction of lifting time in whole project's period.

Meanwhile, in the elevator industry, double-deck elevators, each comprising two connected cars that move simultaneously in the same vertical shaft, are developed to reduce the total number of needed elevators and also to minimize the number of stops[5,6]. This elevator has twice the amount of transport capacity compared with single-deck elevator and decreases passenger's waiting time, which insinuates increase in lifting efficiency[7]. Therefore, it is necessary to apply the concept of double-deck elevators to construction lift, in order to save the lifting time of workers.

The aim of this study is to evaluate lifting efficiency between single-cage lift and double-cage lift which is comprised two cages, one located above the other, by calculating lifting time of workers using the discrete event simulation. The result of this research could contribute to setting the direction of double-cage lift's development. Furthermore, it is expected to be utilized for supporting construction manager to make efficient planning of the lifting equipment in supertall buildings construction.

1.2 Research scope and method

Research scope is constrained to supertall buildings, over 100 floors, which is needed to improve lifting efficiency. Also, workers who are rising to work floors in the morning is included only in this study. Because traffic for lifting resources, especially in the number of workers, has been a sharp rise, at 06:30-8:00 am. And, high speed type of construction lift, which is generally applied in supertall building construction, is used in this study.

In this research, we develop a model for calculating worker's lifting time through lift operation process by using the CYCLONE (Cyclic Operation Network) which is given in Halpin(1973). Two simulation models are developed, one for single-cage lift under unlimited operation and the other for double-cage lift under skip operation suitable for peak-time with the maximum lifting load, for comparison of lifting efficiency

2 Literature review

2.1 Research Trend of Construction Lift

Supertall building construction utilized lifts for the vertical transportation of resources, including labor and relatively small materials. This section introduces some related research about construction lift.

Ahn presents a process of a vertical movement plan for labors to improve lifting efficiency of construction using discrete event simulation[3]. Kim et al. suggested the estimation process of proper numbers of construction lifts[14]. Shin et al. proposed a simulation model incorporating genetic algorithms for supporting hoist planners to make an optimal plan with minimal time and effort for high-rise building construction[17]. Moon et al. studied method of optimizing a transfer floor for workers during the morning peak time by using discrete event simulation[18]. Most researches for vertical movement focused on the lifting management rather than development of new equipment for lifting efficiency.

2.2 Types of Double-Cage Lift Operations

Operation types of double-cage lifts are divided into skip operation, restricted operation and unlimited operation. Conceptual diagram is illustrated in Figure 1.

Firstly, skip operation's regulation is the upper cage stops at even number floors and the lower cage stops at odd number floors. Secondly, restricted operation's regulation is registration of target floors by people boarding at the ground floor is restricted to only even number floors for the upper cage and only odd number floors for the lower cage. However, no restrictions apply to workers boarding at other floors. Lastly, unlimited operation regulation is to put in a lift call to any floors.

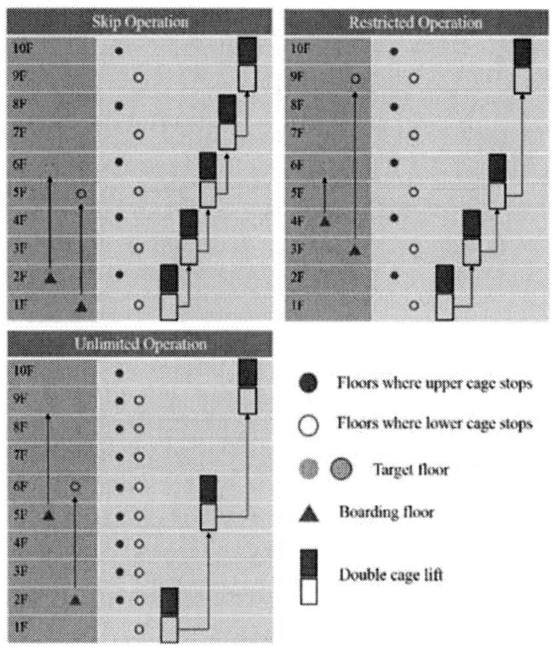

Figure 1. Three Operation Types of Double-Cage Lift

2.3 CYCLONE

The CYCLONE system approach was developed in the early 1970s. This system showed the potential for modeling and simulation of repetitive construction processes. In 1987, Lluch and Halpin further developed the microcomputer version of CYCLONE named MicroCYCLONE[10].

CYCLONE is the most suitable method to measure the actual productivity considering the relationship with resources, duration and work task. The CYCLONE simulation model can provide the actual construction situation simply for the construction engineers to occur problems in construction process. Also, it suggests to derive improvement way which is efficient work process[10]. Table 2 shows CYCLONE modeling elements.

Table 1. CYCLONE Modeling Elements (Halpin and Riggs 1992; Han et al. 2010)

Symbol	Name	Description
	NORMAL	Unconstrained in its starting logic and indicates active processing of (or by) resource entities
	COMBI	Logically constrained in its starting logic, otherwise similar to the NORMAL work task modelling element
	QUEUE	Represents a queuing up, or waiting for the use of passive state resources
	COUNTER	Counts the number of times a key unit passes a particular control point in the network model so that production can be measured
	ARC	Used to model the direction of resource entity flow between the various active-state nodes and the passive-state nodes

3 Simulation Model for Evaluation of Lifting Efficiency

3.1 Operation Process of Construction Lift

Operation process of a construction lift was analyzed in order to establish the simulation model of single-cage lift and double-cage lift. This process begins with the preparation of lifting such as opening door by crew, getting on a lift for workers and closing door by the crew in the ground floor. When its preparation is finished, the construction lift will be starting to move to target floors. And workers are getting off the construction lift which is arrived at target floor. Lastly, lift is descending toward the ground floor. This process is repeated until all workers who were waiting on the ground floor arrive at target floor.

In this research, operation process of construction lift was simply classified into 10 steps, and Table 3 represent it.

Table 2. Operation Process of Construction Lift

Step	Work task	
1	Door opening (ground floor)	Safety door opening
		Lift door opening
2	Getting on the lift (workers)	
3	Door closing (ground floor)	Lift door closing
		Safety door closing
4	Moving to target floor (lift)	
5	Arrival at target floor (lift)	
6	Door opening (Target floor)	Lift door opening
		Footboard lowering
		Safety door opening
7	Getting off the lift (workers)	
8	Door closing (Target floor)	Safety door closing
		Footboard putting
		Lift door closing
9	Moving to ground floor (lift)	
10	Arrival at ground floor (lift)	

3.2 Calculation Method of Hoisting Time

To derive a result from the simulation model, hoisting time of construction lift should be calculated on every travels between ground floor and target floor. For a more accurate estimation of the hoisting time, this study adopted calculation method considering acceleration and deceleration, suggested by Cho et al.[12]. Also, we utilized the acceleration and deceleration time of single-cage lift which was provided by Cho et al.[13]. And we identified the acceleration and deceleration time of double-cage lift by interviewing with the hoist manufacturer who is developed the double-cage lift. Table 4 represents the acceleration and deceleration time of lift

Table 3. Acceleration/Deceleration Time of Lift

	Constant Velocity (m/min, m/sec)	Acceleration time (sec)	Deceleration time (sec)
Single-Cage Lift	100, 1.67	5.5	2.8
Double-Cage Lift	80, 1.33	4.5	4.0

Lifting time(T_m) can be calculated by the sum of lifting time at constant velocity(T_{cv}), acceleration time(S_1), and deceleration time(S_2) (see Equation (1)).

T_{cv} is the sum of the lifting time after the acceleration section, reached to top of the first floor(T_{v1}), the lifting time from bottom of (n-1)floor to starting point of deceleration($T_{v(n-1)}$), and the lifting time of the rest of all at constant velocity(T_v) (see Equation (2),(3)). For the calculation of acceleration distance(h_1) and deceleration distance(h_2), we assumed that acceleration for reaching the constant velocity was uniform. Figure 3, Figure 4 shows velocity-time graph of lift operation.

$$T_m = T_{cv} + S_1 + S_2 \quad (1)$$

$$T_{cv} = T_v + T_{v1} + T_{v(n-1)} \quad (2)$$

$$T_{v1} = \frac{H_1 - h_1}{V}, T_{v(n-1)} = \frac{H_{n-1} - h_2}{V}, T_v = \frac{\sum_{i=2}^{n-2} H_i}{V} \quad (3)$$

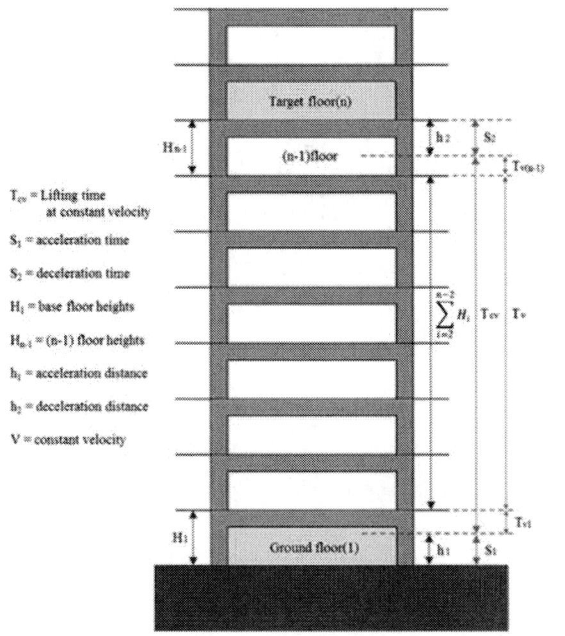

Figure 2. Basic Model for Calculation of Hoisting Time (Cho et al. 2010)

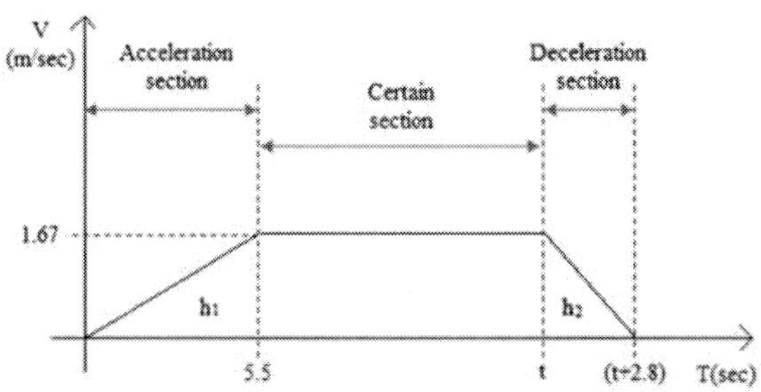

Figure 3. Graph of Single-Cage Lift Operation

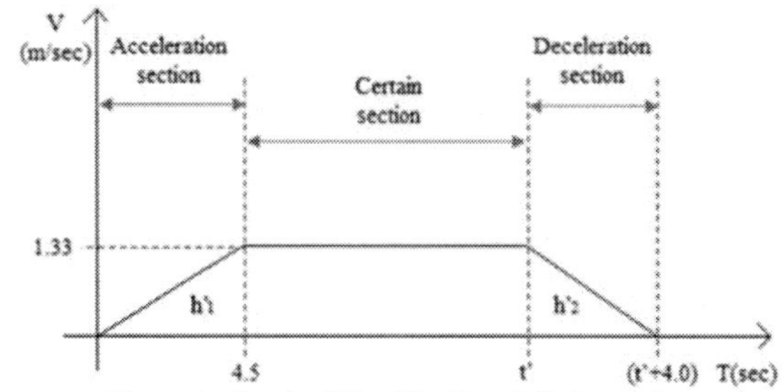

Figure 4. Graph of Double-Cage Lift Operation

3.3 Simulation

3.3.1 Assumptions for Simulation

In this research, some conditions were assumed to estimate the lifting time of workers. The assumptions are described as follows;

1. All workers(1,200) arrive at the construction site simultaneously in the morning, and they are waiting for lifting.
2. Every worker was assumed that weight is 80kg and volume is 0.2m^2. And we determined the maximum number of boarding workers in the cage to select smaller between calculation by weight and calculation by volume.
3. Performance of the cage(size:1.5m * 4.5m * 2.65m, capacity:3,000kg) is same. However the maximum capacity of double-cage lift is set 2,500kg each, considering safety of lift operation.
4. The number of single-cage lift and double-cage lift is one each.
5. Workers are always getting on a lift at ground floor, and they are getting off the lift at 122nd floor, 123rd floor only.
6. The required number of workers on 1st target floor(122nd floor) and 2nd target floor(123rd floor) are assumed 600 workers each.
7. The height of each floor is 3.5m.
8. The sum of workers cannot exceed the maximum number of boarding workers for safety.
9. Lift always operates normally, and any unexpected malfunction was excluded.

Simulation conditions are shown in table 5.

3.3.2 Simulation Model for the Single-Cage Lift

Lifting workers by single-cage lift consisted of 10 work tasks as shown in the table 3. Work tasks were divided into two parts, one is from 'door opening(ground floor)' to 'arrive at target floor', the other is from 'door opening(target floor)' to 'arrive at ground floor'.

Table 4. Simulation Conditions

	Single-Cage Lift	Double-Cage Lift
Total Lifting workers	1,200	
Lift Capacity(kg)	3,000	5,000
Number of Lift	1	
Speed of Lift(m/s)	1.67	1.33
Boarding Floor	1	
1st Target Floor	122	
2nd Target Floor	123	
Floor Height(m)	3.5	

Simulation model for the single-cage lift using the CYCLONE is shown in Figure 5.

Resources which are input the simulation model are a single-cage lift, 3 operating crews who helped with the operating lift each floor, and 33 workers who are required to transport work floor. Table 6 is the details of resource of input data for single-cage lift.

Table 5. Resource of input data for Single-Cage Lift

Node	Work task	Resource	Amount of resource
2	Lift door opening (ground floor)	Operating crew	1 crew
		Single-Cage Lift	1 each
6	Getting on the lift	Worker	33 workers
11	Lift door opening (1st target floor)	Operating crew	1 crew
17	Lift door opening (2nd target floor)	Operating crew	1 crew

Finally, we have to set the duration of work tasks before running the simulation model. In order to get the duration information of 'lift door opening', 'getting on a lift', 'getting off the lift', 'lift door closing', interviews were conducted with engineers who were the experts of construction lift. And then, we input duration information of work tasks by using the uniform distribution. Also, lifting time was calculated by using the Equation (1,2,3), and it is defined as deterministic values. Table 7 is the details of duration of input data for single-cage lift.

Table 6. Duration of input data for Single-Cage Lift

Node	Work task	Value type	Duration (Second)	
			Min	Max
2	Lift door opening (ground floor)	Uniform	3	5
6	Getting on the lift	Uniform	16.5	49.5
8	Lift door closing (ground floor)	Uniform	3	5
9	Moving to 1st target floor	Deter-ministic	259.7	
11	Lift door opening (1st target floor)	Uniform	5	8
13	Getting off the lift	Uniform	8.5	25.5
14	Lift door closing (1st target floor)	Uniform	5	8
15	Moving to 2nd target floor	Deter-ministic	5.14	
17	Lift door opening (2nd target floor)	Uniform	5	8
19	Getting off the lift	Uniform	8	24
20	Lift door closing (2nd target floor)	Uniform	5	8
21	Moving to ground floor	Deter-ministic	261.9	

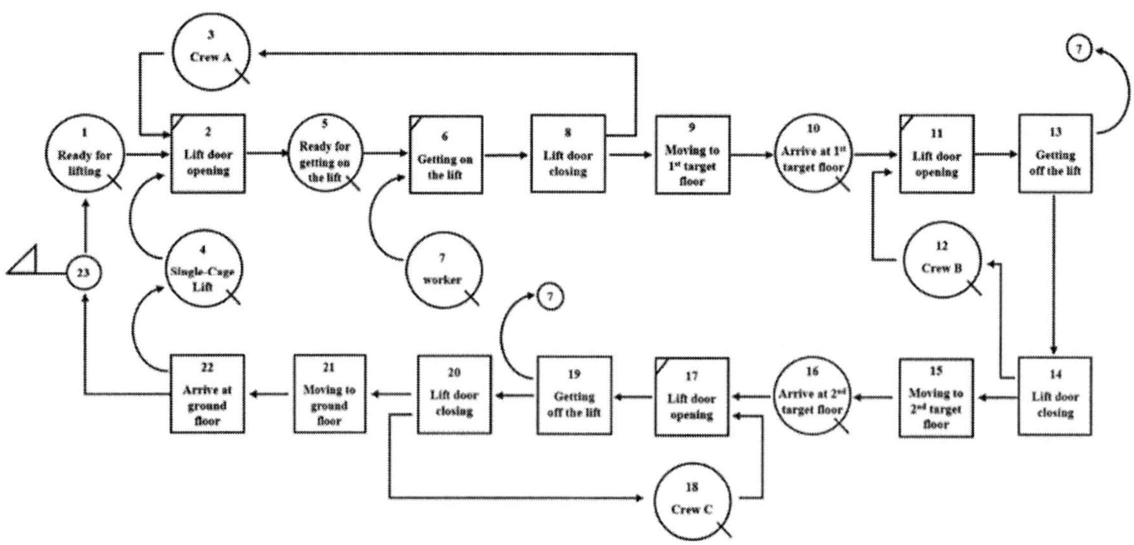

Figure 5. Simulation model for the Single-Cage Lift

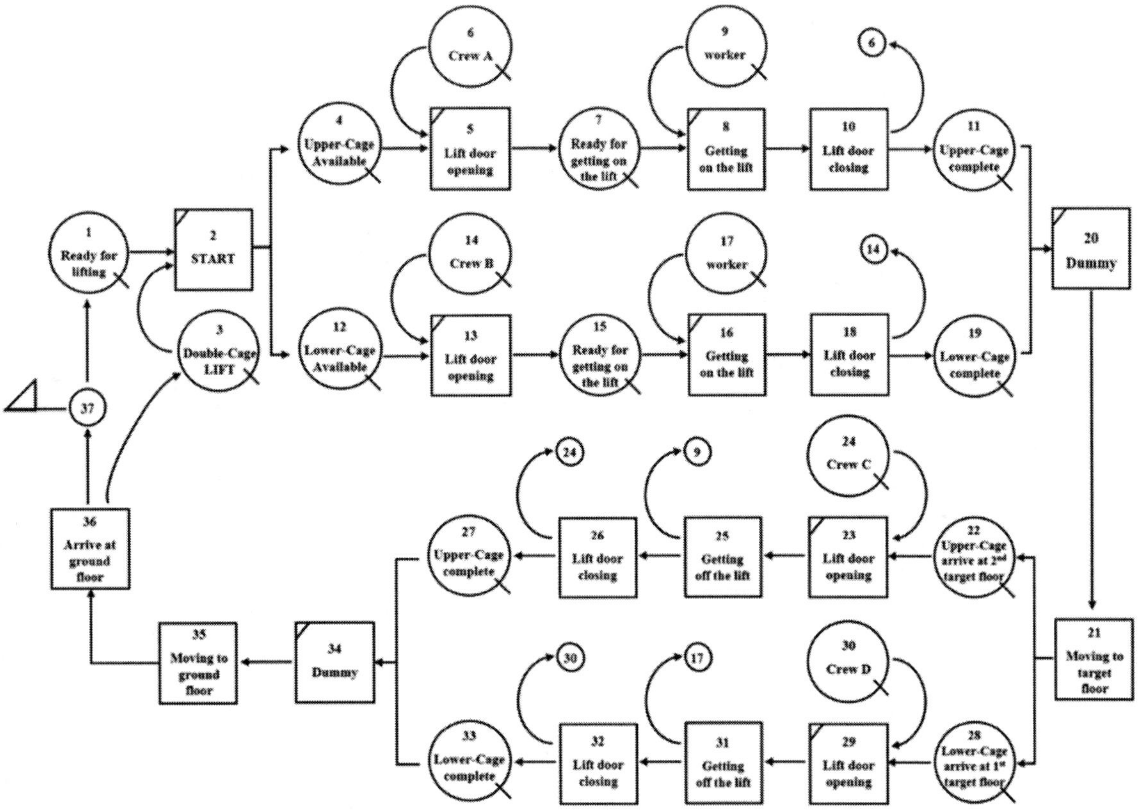

Figure 6. Simulation model for the Double-Cage Lift

3.3.3 Simulation Model for the Double-Cage Lift

Simulation model for the double-cage lift using the CYCLONE is shown in Figure 6. Operation process is very similar between single-cage lift and double-cage lift. However, due to make up with upper-cage and lower-cage in double-cage lift, several work tasks(eg. step 1,2,3,6,7,8 in Table 3) were progressed separately. At this time, when the work tasks of two cages are entirely finished before moving to other floor, it could be possible for work task to progress to the next step.

Resources which are input the simulation model are a double-cage lift, 4 operating crews who helped with the operating lift, and 62 workers who are required to transport work floor. Table 8 is the details of resource of input data for double-cage lift.

Duration of input data for double-cage lift is shown in Table 9. We utilized duration information of single-cage lift, since several work tasks of double-cage lift were similar with single-cage lift. However, lifting time and duration of getting on/off a lift are different by the number of boarding workers and speed of lift. So, we set the duration of them to suit condition of double-cage lift.

Table 7. Resource of input data for Double-Cage Lift

Node	Work task	Resource	Amount of resource
2	Start	Double-Cage Lift	1 each
5	Lift door opening (Upper-Cage/ Ground floor)	Operating crew	1 crew
8	Getting on the lift (Upper-Cage)	Worker	31 workers
13	Lift door opening (Lower-Cage/ Ground floor)	Operating crew	1 crew
16	Getting on the lift (Lower-Cage)	Worker	31 workers
23	Lift door opening (Upper-Cage/ 2nd target floor)	Operating crew	1 crew
29	Lift door opening (Lower-Cage/ 1st target floor)	Operating crew	1 crew

Table 8. Duration of input data for Double-Cage Lift

Node	Work task	Value type	Duration (Second)	
			Min	Max
5	Lift door opening (Upper-Cage/ Ground floor)	Uniform	3	5
8	Getting on a lift (Upper-Cage/ workers)	Uniform	15.5	46.5
10	Lift door closing (Upper-Cage/ Ground floor)	Uniform	3	5
13	Lift door opening (Lower-Cage/ Ground floor)	Uniform	3	5
16	Getting on a lift (Lower-Cage/ workers)	Uniform	15.5	46.5
18	Lift door closing (Lower-Cage/ Ground floor)	Uniform	3	5
21	Moving to target floor	Deter-ministic	319.4	
23	Lift door opening (Upper-Cage/ 2nd target floor)	Uniform	5	8
25	Getting off a lift (Upper-Cage/ workers)	Uniform	15.5	46.5
26	Lift door closing (Upper-Cage/ 2nd target floor)	Uniform	5	8
29	Lift door opening (Lower-Cage/ 1st target floor)	Uniform	5	8
31	Getting off a lift (Lower-Cage/ workers)	Uniform	15.5	46.5
32	Lift door closing (Lower-Cage/ 1st target floor)	Uniform	5	8
35	Moving to ground floor	Deter-ministic	319.4	

4 Results

Table 10 shows the results of single-cage lift simulation. In the simulation test, we conducted a simulation cycle 1,000 times to get more accurate results of cycle time. The total simulation time for 1,000 cycles was 174.19 hours and each cycle time of average was 10.45 minutes. Thus, the total number of lifting cycles required to reach for 1,200 workers to their designated work floors was 37 cycles and it took approximately 6.44 hours. Table 10 shows the results of double-cage lift simulation. Simulation model was carried out 1,000 times, and the total simulation time was 200.67 hours. Also, a cycle time of average was 12.04 minutes. Thus, the total number of repetition cycles to lift 1,200 workers to target floors was 20 cycles and it took about 4.01 hours.

As a result of the comparison, it was estimated that the lifting time of the double-cage lift would decrease by about 38.0% compared with the single-cage lift. The reasons of the reduction lifting time were not only an increase in lift capacity but also minimization of stops. Therefore, this result supports the lifting efficiency for the double-cage lift is better than the single-cage lift.

5 Conclusions and Further Studies

According to the recent increase in the height of tall buildings, more time is needed for the vertical lifting of resources in supertall construction. However, due to the limit of maximum capacity of lift, double-cage lift is demanded for supertall construction project. In this research, we have developed two simulation models for single-cage lift and double-cage lift to evaluate lifting efficiency by using the CYCLONE. The developed models showed comparison value in which lifting time of double-cage lift is more efficient than single-cage lift about 38.0% during the morning peak-time. The result of this study will be utilized for setting the direction of development for double-cage lift. Also, it is expected to support construction manager to make optimal plans with minimum lifting time in supertall building construction. Further studies will need to verify the lifting efficiency of double-cage lift comprehensively by considering production costs, operation types, the number of cages and combination of multiple lifts.

Table 9. Comparison of simulation results of single-cage lift and double-cage lift

	Total Simulation time (hour)	Number of Simulation (cycle)	Productivity (cycle/second)	Cycle Time of Average (min)	Number of Repetition Cycles, Total Lifting Time (cycle, hour)
Single-Cage Lift	174.19	1000	0.001594	10.45	37 , 6.44
Double-Cage Lift	200.67	1000	0.001384	12.04	20 , 4.01

Acknowledgement

This research was supported by a grant(15AUDP-B106327-01) from Architecture & Urban Development Research Program funded by Ministry of Land, Infrastructure and Transport of Korean government.

References

[1] Jason G. Marty C. and Marshall G. 2015 Year in Review report, *CTBUH Journal 1*, 2016 .

[2] Won S. Interactive Decision-making Model of Lift Operation Planning in Tall Building Projects, Doctorate thesis, Kyunghee University, 2011.

[3] Ahn B. A Vertiacl Movement Plan for Labors in High-Rise Building Construction Using Discrete-Event Simulation, *Journals of KICEM*, 5(2):47–54, 2004.

[4] Park M. Ha S. Lee H. Choi Y. Kim H. and Han S. Lifting demand-based zoning for minimizing worker vertical transportation time in high-rise building construction, *Automation in Construction*, 32:88–95, 2013.

[5] Kheir A. Tall Buildings and Elevators: A Review of Recent Technological Advances, *Buildings*, 5(3):1070–1104, 2015.

[6] Joaquin R. Fernandez and Pablo Cortes, A Survey of Elevator Group Control Systems for Vertical Transportation, *IEEE Control Systems*, 35(4):38–55, 2015.

[7] Kim N. A Study of Emergency Operation of Double Deck Elevator, Master's thesis, Yonsei University, 2010.

[8] Simaan M. Daniel W. and James D. State of the art in construction simulation, *Proceedings of the 1992 Winter Simulation Conference*, pages 1271–1277, Arlington, USA, 1992.

[9] Lee I. Yang H. Vertical transportation in Super High Rise Buildings, *The society of air-conditioning and refrigerating engineers of Korea*, 2006.

[10] Han S. and Daniel W. The Use of Simulation for Productivity Estimations based Multiple Regression Analysis, *Proceedings of the 2005 Winter Simulation Conference*, pages 1492–1499, Orlando, USA, 2005.

[11] Han S. Productivity analysis comparison of different types of earthmoving operations by means of various productivity measurements, *Journal of Asian Architecture and Building Engineering*, 9(1):185–192, 2010.

[12] Cho C. Kim J. Cho M. Lee J. Kim Y. and Kwon S. Simulation method of construction hoist operating plan for high rise buildings considering lifting heights and loads, *27th International Symposium on Automation and Robotics in Construction (ISARC 2010)*, pages 22–28, Bratislava, Slovakia, 2010.

[13] Cho C. Shin Y. Won S. Kim J. and Cho M. An Algorithm for Hoisting Time Calculation in Super-tall Building Construction, *Journals of KICEM*, 12(6):120–129, 2011.

[14] Kim S. and Han K. A Study on the Estimation of Proper Numbers of Construction Lifts, *Journal of The Korean Institute of Building Construction*, 8(3):119–125, 2008.

[15] Han C. Lee J. and Won S. A computation model for Resource-based Lifting loads of the lift-cars for super high-rise buildings, *Journals of KICEM*, 13(5):135–143, 2012.

[16] Lee M. Won M. Lee W. Cho H. and Kang K. Integrated Lifting Model of Planning and Operation for Finishing Works in Tall Building, *Journal of The Architectural Institute of Korea Structure & Construction*, 30(5):125–133, 2014.

[17] Shin Y. Cho H. and Kang K. Simulation model incorporating genetic algorithms for optimal temporary hoist planning in high-rise building construction, *Automation in Construction*, 20:550–558, 2011.

[18] Moon J. Park M. Lee H. and Jung M. Study on Optimization for Construction Vertical Lifting with Transfer Operation for Super High-rise Buildings, *Journals of KICEM*, 15(6):53–62, 2014.

Robotic application of foam concrete onto bare wall elements

E. Lublasser[a], J. Brüninghaus[a], A. Vollpracht[b], L. Hildebrand[c] and S. Brell-Cokcan[a,d]

[a]Chair for Individualized Production in Architecture, RWTH Aachen University, Germany
[b]Chair for Environmental Compatibility of Building Materials at the Institute of Building Materials Research, RWTH Aachen University, Germany
[c]Chair for Reuse in Architecture, RWTH Aachen University, Germany
[d]Association for Robots in Architecture, Austria

E-mail: lublasser@ip.rwth-aachen.de, brueninghaus@ip.rwth-aachen.de, vollpracht@ibac.rwth-aachen.de, linda.hildebrand@rwth-aachen.de; brell-cokcan@ip.rwth-aachen.de

Abstract –

In the course of frequently altering energy saving regulations, numerous buildings have to be comprehensively refurbished to meet the rising energy-efficiency standards in order to protect the global environment and to save resources. However, available materials as well as adaptable design concepts for additional energy-saving insulation layers are not yet convincing in terms of their long term recyclability nor variation of shape. Therefore, we investigate the application of foam concrete onto bare walls of existing buildings to gain a façade finish which is highly insulating, easily recyclable and at the same time promises to be individually designable due to the properties of the raw material mixture. To ensure controllable as well as reproducible application and to react to changing working methods in architecture and construction, the research focuses on the automatized application of foam concrete using a robotic setup.

We analyzed manual application strategies of foam concrete, considering parameters of handcraft, used tools as well as the reaction on varying material characteristics during application. Based on the analysis results, we present a concept for the robotic application of foam concrete, including suggestions regarding end effectors, robot programming and surface design planning.

Keywords –

Insulation; Foam Concrete; Robotic application; Automated process; Surface Design

1 Introduction

Rising energy efficiency standards and saving regulations increase the number of building refurbishments [1][2]. The constructional implementation of these energy-oriented refurbishment processes mostly include an optimization of the façade's external insulation often executed as External Thermal Insulation Composite Systems (ETICS) in order to reduce internal thermal loss [3]. At the same time, also the recycling standards for construction materials are constantly intensified attaching importance to separability of construction components into varietal raw material to increase reusability of construction material. However, recycling strategies for ETICS most frequently based on expanded polystyrene (EPS) are not fully developed [4]. And due to the high degree of material compound of ETICS, a material extraction with high purity of variety is expected to be only achievable to a minor degree.

To fulfill both mentioned regulations, we introduce the concept of a mono-material thermal insulation system based on an easily reusable as well as highly insulating mineral material. As basis material for this system we investigate the application of foam concrete using different foam concrete densities according to the layers functions. Figure 1 depicts the structure of the system applied to an existing wall including the following layers: anchor with high density of ≤ 800 kg/m³ (a), insulation with low density of ≤ 150 kg/m³ (b), surface finish with low density between 150 and 250 kg/m³ reinforced with specific cement suspension (c).

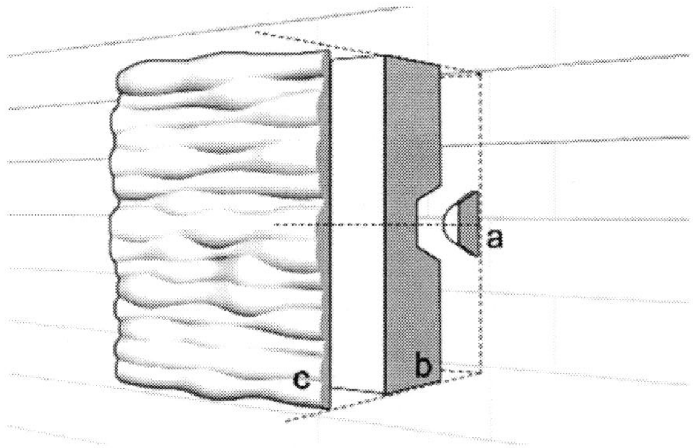

Figure 1. Mono-material insulation system onto bare wall element: anchor (a), insulation (b), designed surface finish (c)

2 Literature Review

2.1 State of the art foam concrete

Foam concrete is proposed as façade insulation layer because it is known for its excellent insulating characteristics due to its low density [5][6]. Furthermore, it is most suitable for façade application because of its higher sound absorbing rate in comparison to dense concrete [7] as well as its acceptable fire resistance [8]. Moreover, foam concrete is ecologically harmless and easily recyclable due to its mineral composition and minimal consumption of aggregate [6][9]. In comparison to similar highly insulating mineral materials such as aerated concrete, foam concrete is strain hardening and therefore also the production process is environmentally friendly due to its low energy consumption [9].

In addition, foam concrete is a competitive material in terms of costs. So far, no reliable study on cost comparison is published. A rough calculation of material costs according to information of local companies indicates a cost saving potential of approximately 15% using foam concrete with a density of 150–200 kg/m³ instead of EPS based systems; even though the layer thickness using foam concrete has to be enlarged to reach the same heat transmission coefficient which EPS provides.

2.2 State of the art foam concrete application

Foam concrete is mainly applied in situ as void fill, bridge abutments and roof insulation or as prefabricated blocks or panels [9][10]. Partly, the blocks are used as insulation material in a constructional setup comparable to external thermal insulation composite systems [11]. While foamed concrete with densities up to 300 kg/m³ is well investigated [6][12] and fabricated as listed above, the manufacturing of building elements made of foam concrete with a low density about 150 kg/m³ is not fully examined nor in common use. However, the lower the density the better foam concrete functions as insulation material due to a decreasing heat transition coefficient [12].

To base the concept of the robotic process on the current foam concrete manufacturing techniques, we analyzed the production process of foam concrete slurry. Mostly, this process consists of the following four sub-steps:

1. Mixing of foam agent and cement slurry separately or mixing of all ingredients at a time
2. Pumping of the foam concrete slurry to its destination of application
3. Pouring of slurry into casting mold or locally predefined in-situ shape
4. Curing

All of these steps but the application via pouring have to be considered for the robotic application concept.

3 Research aim

The overall aim of our research is the development of a concept for the robotic application of foam concrete onto vertical building elements such as bare walls in the context of the explained insulation system. Thereby, the following descriptions are focused only on the outer layer being the designable surface finish of the system.

A particularly robotic instead of manual application is targeted to meet the demands of changing architectural methods and processes not only in the digital design phase but also during the actual building construction and building element fabrication on site. Current trends more frequently establish custom oriented design and adaptable building element generation according to varying structural, physical and aesthetic demands of different parts of one building [14]. In order to be able to feasibly produce and assemble these highly adaptable, parameter based as well as complex formed building elements, the process of building element production has to be generally reconsidered leading to a mass-customized and automated manufacturing process triggering the rapid, affordable and predictable generation of manifold building elements of small batch sizes [14].

However, the commonly used production process of foam concrete via pouring material into molds is not compatible with this development. The great issues thereby are on the one hand the dependency of form generation on the geometric and constructional producibility of the mold and on the other hand the economic competitiveness of small batch sizes using molds which are only reusable to a very limited degree.

Therefore, we investigate suitable application strategies of foam concrete using robotic assistance for direct additive material application without the necessity of expensive or time consuming preparations. Additionally, robots instead of other machines provide the chance of implementing different tools as well as a great variety in path planning throughout different parts of one process [15]. Furthermore, a robotic setup is able to guarantee a reproducible as well as at all times controllable process in comparison to manual application techniques of craftsmen. Using robotic application processes, hence, can lead to a general optimization of the working processes on building sites and an increase in building element accuracy.

4 Research Methodology

The described research started out with an identification of necessary sub-steps for the pursued robotic application concept based on a modification of the described current foam concrete manufacturing techniques (see section 2.2).

1. Material supply including mixing of ingredients as well as pumping of the foam concrete slurry to end effector for application (section 5.1)
2. Development of additive material application technique with focus on generating a flexible method for highly adaptable surface design instead of pouring into casting molds (section 5.2)
3. Curing

In the further course of this interdisciplinary research project our team was focused on the second sub-step of the process. Therefore, the approach as well as the methodology are explained in more detail.

Since this research is related to the outer layer of the foam concrete mono-material system, the distinct and locally adjustable designability of the surface finish within a robotic application process is key factor of the investigation. That is because many factors for this outmost layer greatly vary along the surface area reacting on the different demands of the underlying construction or external circumstances. Some of these factors are exemplary listed below:

- Varying material thickness according to differing thermal insulation demands (especially in the context of refurbishment processes)
- Distinct composition of surface shape for sound insulation of surrounding city noise
- Assurance of rain water discharge
- Aesthetic aspects of architectural design ideas

Harmonizing these aesthetic as well as building physics factors is a highly complex and to a certain extend artistic task of common architectural working methods. Therefore, techniques of the creative industry were used as a research method. Accordingly, manual application experiments were executed enabling an intuitive material handling similar to workflows of sculptors.

In order to find a suitable application technique for the robotic process, the manual application method was used to first, select a tool capable of handling the specific material behaviour and allowing designability during the process. This tool then serves as basis for the development of the end effector. Secondly, the manual application experiments were used to analyse detailed specifications on the path planning parameters of the robotic end effector.

Both experiments were observed not only in terms of the surface designability but as well with the criteria of controllability and reproducibility to gain a fully determinable as well as fully automated process. Furthermore, the analysis of the experiments provides additional information influencing the global robotic setup such as material mixture, processing time, curing progress, order of application, movement speed and direction and amount of applied material.

Based on the experiments and analysis, the robotic application concept was formulated.

5 Experiments

5.1 Material fabrication and supply

The material mixture as well as the material behavior over time greatly influence the application process. In order to provide suitable material conditions, the material fabrication and supply process have to be included in the robotic application concept. In the course of the manual application experiments, we compared two foam concrete manufacturing techniques:

- Preliminary fabrication of foam using a foam branch pipe and mixing of cement slurry plus subsequent folding in of these fractions to gain foam concrete slurry
- Collection of cement ingredients as well as foaming agent in one vessel and afterwards mixing of all fractions at a time with high-performance mixer.

The latter more constantly produced suitable material conditions. Furthermore, this fabrication method had a shorter fabrication time leading to a faster start of application which in turn positively served the materials workability.

5.2 Manual application strategies

The manual experiments were executed using a foam concrete mixture with an approximate density of 200 kg/m³. The underlying vertical wall elements were either made of 4 cm plasterboard or sand-lime bricks. The choice of tools was inspired by the manufacturing techniques of sculptors working with viscous material such as clay. The tools as well as their utilization are listed in Table 1 and further described in the following chapters.

Table 1 Tools for manual application

Utilization	Tools
Spreading	Trowel
Dripping	Spoon
Squirting tubes and drops	Icing bag
Spraying	Spray bottle
Tossing handsize clumps	Hands

5.3 Analysis of application methods

5.3.1 Spraying and tossing method

Experiments with the spray bottle were not carried on because an impeccable material application could not be guaranteed. The concrete ingredients clogged the spraying mechanism. Investigations on tossing were also not carried on because the surface design was not controllable nor reproducible at any time.

5.3.2 Dripping method

Initial tests using the dripping method showed the influence of the relation of applied material amount and speed of tool movement (Figure 2, left).

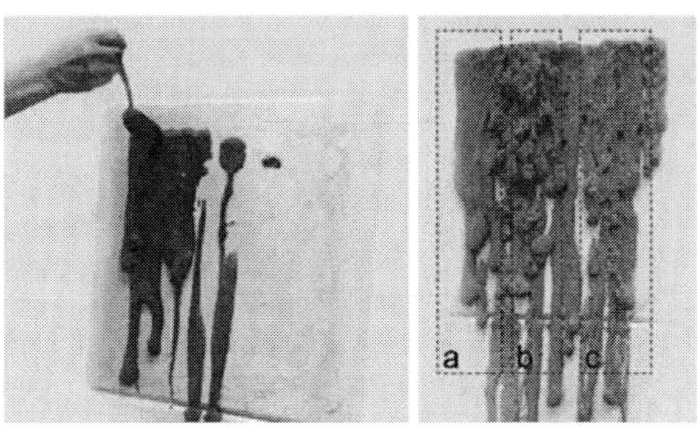

Figure 2. Manual dripping method – application (left) and final result (right)

By quickly dripping the material onto the wall with a volume to contact area ratio of less than 1, the material uncontrollably streams down the wall. This effect increases with a rising fluidity of the foam concrete mixture.

Within the same experiment, the effect of the processing time was observed (Figure 2, right). The depicted foam concrete finish was applied in three time units using the same material mixture. Table 2 displays the observation of the application behavior as well as material quality in relation to the processing time. In summary, the material quality decreases over time due to an increase in foam burst during the storage period. In contrast, the application behavior becomes more controllable over time according to a decrease in fluidity, which is caused by the ongoing curing progress during the storage period.

Table 2 Application behavior and material quality in relation to processing time

Fig.	Time (min)	Application behavior	Foam concrete quality
a	0-4	Strong, smooth streaming	Homogeneous, stable material matrix, good foam distribution
c	4-7	Streaming & fall out of clumps	Inhomogeneous, partly crumbly, foam partly burst
b	7-12	Slight streaming	Inhomogeneous, crumbly, foam mostly burst, material matrix slumped down

Hence, further research has to be focused on the material mixture and additional aggregates to optimize the application process according to curing times and coordinated processing succession.

5.3.3 Spreading method

The application of foam concrete via spreading with a flat, extensive tool represents an advancement of the dripping method. By using the enlarged linear edge of the tool the volume to contact area ratio favorably increases. In this way, an extensive application with a minimum of uncontrollable material streaming becomes increasingly manageable and offers the possibility for the implementation of design strategies such as a wavelike sequence (Figure 3, bottom right) or island like structures with sharp edges and partially smooth surfaces (Figure 3, top right).

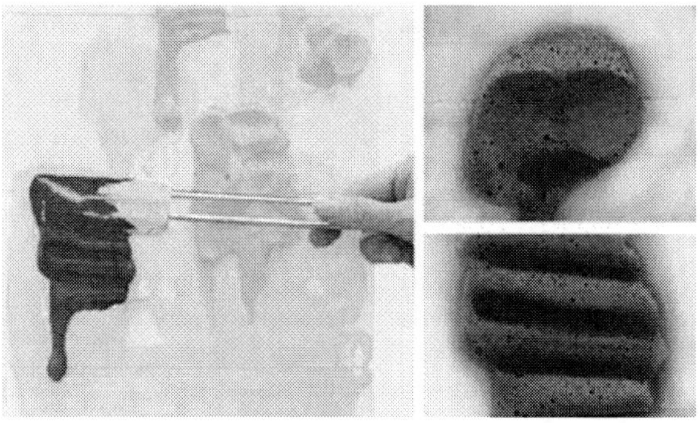

Figure 3. Manual spreading method – application bottom-up (left), final results with smoothed plateau and wave structure (right)

For both methods, dripping as well as spreading, the material supply to the tool ahead of the processing is not yet considered but has to be analyzed and implemented for an automated robotic process.

5.3.4 Squirting method

In contrast to the dripping and spreading method, the results of the squirting method analysis show a higher degree of controllability of the process. The material adhesion to the wall is easily adjustable due to the defined shape and small size of the tool tip as well as the opportunity of adaptable proportioning of material supply. Streaming of the material can be reduced to a minimum by precisely arranging the volume to contact area ratio with this tool (Figure 4, left).

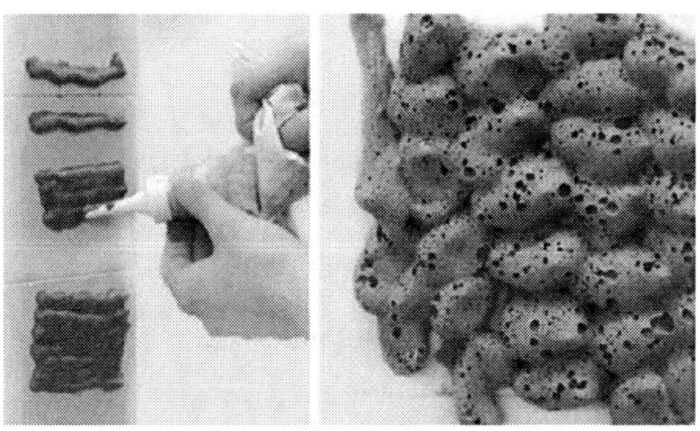

Figure 4. Manual squirting method – application of tubes (left) and final result of drop design (right)

In the course of increasing controllability, the process at the same time guarantees the desired designability of the materials shape. Tube-like and drop-like material application techniques have been successfully tested (Figure 4) which also allow conclusions to be drawn on the path planning for a robot assisted material application (see section 6).

Furthermore, the tools construction of interlinked forming tool and material storage generating a constant material flow provides a conceptual approach for the connection of end-effector to material supply via pipes.

5.4 Conclusion on manual application methods

The analysis shows, that all application methods provide a certain designability with different aesthetic languages of form. In comparison, the squirting method allows for greatest control and reproducibility of form generation during material application of all methods. Moreover, the structure of the squirting tool provides basic connecting factors for the advanced development of a robotic end-effector. Hence, the squirting method is further investigated for robotic foam concrete application and surface design. But since the surface design adjustments are limited to the tool tip size of the squirting mechanism, additionally, the spreading tool is further considered for extensive post-processing of the surface shape without further material supply.

6 Analysis for Path Planning

Suggestions for robot path planning are mostly depending on the tool movement analysis of the different manual application methods. Therefore, additional application experiments were executed for the most suitable methods being spreading and squirting with focus on material behavior according to movement parameters of speed, direction and order.

6.1 Path planning for spreading method

The movement parameters of the trowel for the spreading method had to be investigated more closely in terms of compatibility to the material characteristics. The experiments showed, that material application by spreading performs best when executed bottom-up. By shifting the tool horizontally, unintentional streaming occurs. Whereas top-down movement turned out to be problematic because material adhesion to the tool instead of the wall could not be prevented. As a consequence, the material was dragged along with the tool while only a thin, incalculable layer of foam concrete remained on the wall. In addition, the tilting angle of the tool is important for the spreading method. Any deviation from a parallel position of tool and wall helps to increase material adhesion to the wall instead of tool because of a smaller contact surface area of material and tool.

6.2 Path planning for squirting method

Although the squirting method allows comparatively precise material application anyway, different path orders and directions have been tested. As a result we observed the following path planning factors:

- Linear vertical application works best if performed bottom-up to prevent material drop formation at the end of one application sequence
- For linear horizontal application the speed of the tool movement has to be lower than for the vertical application to prevent minor streaming
- The execution of a combination of horizontal material rows worked best in a bottom-up order. The previously applied row prevents streaming of the following rows and therefore increases shape precision
- For both directions the initial material application has to be executed with a retention time of the tool at the starting position in order to improve material adhesion
- Specific design of the materials shape can be adjusted according to tool speed, distance of tool tip to wall, retention time and force strength of material supply

Moreover, tests on a variation in application order differing from a continuous execution of adjoining rows of material (Figure 5) provide conceptual links for an implementation of curing time in the path planning development, which is expected to increase form precision. The order of application was manipulated to produce every other row of the intended design. With a small time delay, the missing rows were then applied. In the meantime, the curing of the first row set had already proceeded, decreasing the materials high fluidity. This led to a greater form stabilization of these rows and, hence, a decrease in unpredictable material smudging in-between rows was observed.

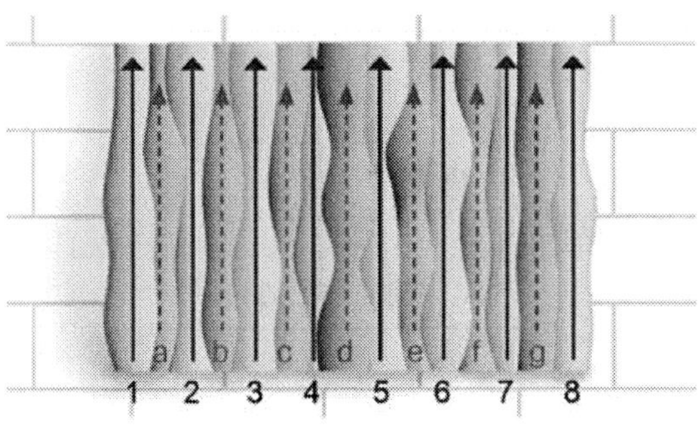

Figure 5. Path order: first material application 1-8, second material application a-g

6.3 Global path planning

The described experiments for path planning analysis were tested in extracts for a material application to cover a surface area of 15 cm x 15 cm. Further research has to investigate the global path planning for entire wall surface areas including a strategy for convenient workflow partitioning and operation order with reference to greatest possible process speed and form precision. Moreover, material characteristics have to be observed to allow smooth transition of material conditions and shape in the course of intended or incidental process interruptions leading to advanced curing progress.

7 Robot assisted application concept

As result of the described experiments we propose the following concept for a robotic application of foam concrete onto bare wall elements, which is also depicted in Figure 6.

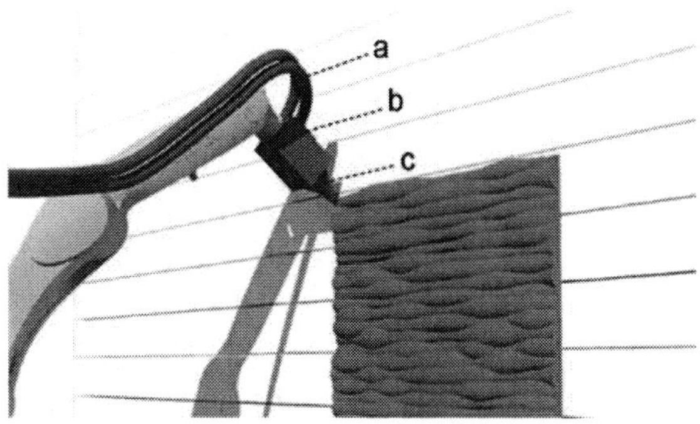

Figure 6. Concept for robot end-effector implementation: (a) separate material supply (b) mixing of material components (c) interchangeable tool tip for varying material shape generation

The concept is based on using a standardized robotic arm with a customized end-effector consisting of the following important elements:

- Separate material supply of foaming agent and cement slurry ingredients via pipes (a)
- High-performance mixing mechanism to produce foam concrete slurry (b)
- Interchangeable tool tip with conical form and varying opening radius for diverse shape generation; directly attached to mixing mechanism (c)
- In addition, a trowel can be attached to the side of the mixing mechanism in order to handle the post-processing of the materials shape with only a slight time delay

The construction of the end-effector elements ensures a fast work progress and a constant material flow from the material supply to the application phase which guarantees best material properties and shape precision of the applied foam concrete.

The suggested process and setup are closely equivalent to 3D printing methods but in contrast to common fields of application of 3D printing, here, a vertical material application is pursued. As part of the ongoing research, the hardware development of this concept can benefit from current 3D concrete printing research. In contrast to the hardware, the robot programming has to be newly developed and implemented because of the modified demands for path-planning of vertical material application as described earlier. Hereby, the main impact for specific application demands is the high fluidity of foam concrete with a low density of about 150 kg/m³ which - if not handled correctly via path planning- leads to unpredictable and uncontrollable material streaming.

8 Conclusion & outlook

The method of analyzing manual application techniques provided a suitable starting point for the development of a robotic concept for the automated application of foam concrete considering a designability of the foam concrete surface shape. The method allowed for intuitive knowledge acquisition concerning the handling of the relatively new material of foam concrete with densities $\leq$ 200kg/m³ as basis for a concept development. The limit of this method was an exact reproducibility of the experiments. Therefore, no quantitative data evaluation was performed.

In the further course of the research project, we are currently developing the hardware and software and verify the conceptual approach.

Figure 7 depicts an initial semi-automated robotic application approach using a modified foam branch pipe. All elements of the conceptual setup considering the end effector are implemented and referenced within the figure (see a-c). This temporary end effector construct is attached to a KUKA iiwa robot and was moved by the manual drive mode instead of a predetermined program. Furthermore, the material supply is achieved via compressed air. At the current state of the ongoing research, material application onto the vertical wall element was in general successful but not yet satisfying in terms of the surface design nor controllability of the process. The influencing factors of the here described research have now to be implemented in more detail to gain the pursued automated robotic application process.

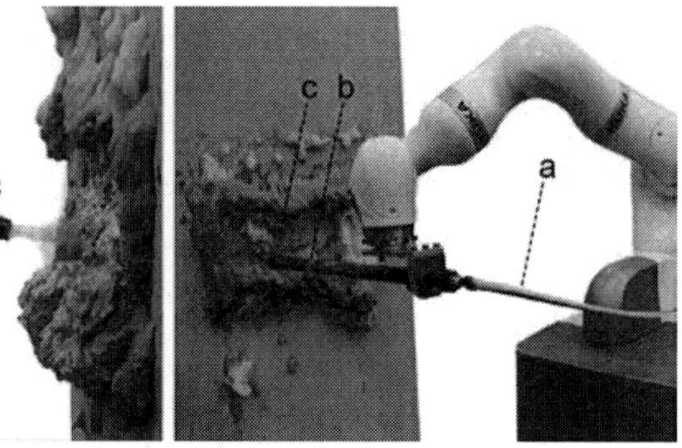

Figure 7. Semi-automated robotic foam concrete application with (a) separate material supply of foam concrete slurry and foaming agent (b) mixing of material components with modified foam branch pipe (c) conical tool tip

Moreover, details of the foam concrete characteristics and manufacturing techniques have to be investigated for improvement of application precision, material adhesion and reproducibility of material properties for constant and consistent material supply.

In order to link the application process to the architectural and constructional working procedures, the robot programming has to be connected to the commonly used software in the field of architecture. Moreover, the concept of using a robotic arm with specific end-effector design has to be extended considering a larger kinematic to allow for material application on the total surface area of a building based on investigations on robotic façade upgrading systems [16]. Therefore, the proposed path planning concept and robotic setup have to be examined for their scalability.

In addition, the application processes for the other layers of the suggested mono-material system have to be developed accordingly.

All in all, we expect a high transferability of the finalized robotic vertical application process for other viscous materials.

References

[1] Bundesministerium für Wirtschaft und Energie (BMWi) *Sanierungsbedarf im Gebäudebestand*. PRpetuum GmbH, Munich, 2014.

[2] Meijer F., Itard L. and Sunikka-Blank M. Comparing European residential building stocks: performance, renovation and policy opportunities. In *Building Research & Information*, 37(5-6): 533-551, 2009

[3] Deutsche Energie.Agentur GmbH The energy refurbishment top five. On-line: http://www.dena.de/en/press-releases/pressemitteilungen/the-energy-

refurbishment-top-five.html, Accessed: 13/03/2016

[4] Albrecht W. and Schwitalla C. *Rückbau, Recycling und Verwertung von WDVS*. Fraunhofer-IRB-Verl., Stuttgart, 2015.

[5] Giannakou A. and Jones M.R. Potentials of foamed concrete to enhance the thermal performance of low rise dwellings. In *Innovations and development in concrete materials and construction*, pages 533–544, London, UK, 2002

[6] Ramamurthy K., Kunhanandan Nambiar E.K. and Indu Siva Ranjani G. A classification of studies on properties of foam concrete. In *Cement & Concrete Composites,* 31: 388–396, 2009

[7] Taylor, W. H. *Concrete technology and practice*. Angus and Robertson, London, 1969.

[8] Sach J. and Seifert H. Foamed Concrete Technology: Possibilities for Thermal Insulation at High Temperatures, *In CFI Ceramic Forum International*, 76(9): 23–30, 1999.

[9] GERTEC GmbH Warum Schaumbeton?. On-line: http://www.gertec-gmbh.de/Aktuelles/Schaumbeton/, Accessed: 21/01/16

[10] Mugahed Amran Y.H., Farzadnia N. and Abang Ali A.A. Properties and applications of foamed concrete; a review. In *Construction and Building Materials*, 101: 990–1005, 2015.

[11] Luca Industries International Herstellung vorgefertigter Isolierplatten: On-line: http://www.dr-Masa-luca.de/pages/de/lithopore_bauweisen/lithopore75_150/index.html, Accessed: 10/12/2015

[12] Hartwich P. et al. Properties of ultra-lightweight concrete based on protein and surfactant foaming agents. In *27th Biennial National Conference of the Concrete Institute of Australia in Conjunction with the 69th RILEM Week*, Melbourne, Australia, 2015.

[13] Shrivastava O.P. Lightweight aerated concrete – a review. In *Indian Concrete Journal*, 51:10–23, 1977.

[14] Reinhardt D. et al. TriVoc - Robotic Manufacturing for Affecting Sound Through Complex Curved Geometries. In *Robotic Fabrication in Architecture, Art and Design*: 163-180, 2014.

[15] Braumann, J. and Brell-Cokcan (eds.), *Robotic Fabrication in Architecture, Art and Design,* Wien, 2012.

[16] Iturralde K., Linner T. and Bock T. Comparison of Automated and Robotic Support Bodies for Building Facade Upgrading. In *Proceedings of the 32st ISARC*, Oulu, Finland, 2015.

Development of a fast and effective solution for on-site building envelope installation

Wen Pan [a], **Thomas Bock** [a], **Thomas Linner** [a] **and Kepa Iturralde** [a]

[a]Chair of Building Realisation and Robotics,TechnischeUniversitätMünchen,Germany
E-mail: {wen.pan, Thomas.Bock, Thomas Linner, Kepa Iturralde}@br2.ar.tum.de

Abstract –

Purpose: building cladding installation process is one of the most important tasks in the construction phase. This paper will focus on the development of a fast and effective method for assembly precast concrete exterior wall panel in high-rise building. The existing method can be described as inefficient, time consuming, expensive and labor intensive. There are various parameters has to take in to consideration when planning and execute the precast concrete wall panel installation task. For instance the size, orientation, weight of the panel, then the ground and weather condition of the site yet the transportation sequence of the panel all need to be carefully planned to ensure smooth operation on-site. Method: the proposed method will investigate the existing procedure from the following aspects; designing, logistical, assembly and management and then provide an alternative lean construction method. The proposed approach will introduce an innovative installation technology that consists of a ground level loading bay, vertical transportation system, positioning system and intergraded management system. The innovative solution requires minimal alteration of an existing constriction tower crane hence with less additional changes of on-site equipments the cost of such system will remain affordable. Results & Discussion: the proposed solution will impose a fast and effective installation method for installing precast concrete panel in high-rise building. It requires further implementation and pilot project in the future to verify the effectiveness and practicality of the system under on-site operation.

Keywords –

Assembly; Construction sequence; Logistics; Continuous flow

1 Introduction

There are many types of cladding systems applied in the construction of high-rise buildings. Usually cladding systems are non-load bearing but structurally sounds itself. Such system can be removed from the structural frame of the building without affecting the structural performance of the building. Precast concrete exterior wall panels are commonly used for cladding multi-storey or high-rise buildings around the world. Yet it has shown many constraints throughout its assembly process. Those constraints often include following aspects; technology availability, tower crane usage, vertical transportation, site labor moment, site management, time, cost, weather, location of the site and the design of the building. Assembly task of the wall panel often marks the completion of the building envelope. With aforementioned constraints this task becomes rather tricky and unpredictable. Therefore, solving problems associated with on-site assembly technique of precast concrete exterior wall panel is essential.

Technology availability varies between different parts of the world. Many developed countries have accomplished standardization, industrialization and some even automation in the construction sector. However, in many developing countries this is still out of reach. For example, in developed countries such as Germany its industrialisation of building production can be traced back into the 1920's and 1930's Bauhaus period. Nowadays, there is profound evidence that European industrialising process in building is widespread and matured. This is not only referring to the technology used in the construction process but also its concept, manufacturing equipment, management and other services are more advanced comparing to many developing countries. Currently, in China the government is urgently promoting the concept of construction industrialization. Stress is given to the improvement and extension of applications of existing industrialized building systems. In the meantime, to find the suitable method to improve work efficiency, engineering quality and save investment become apparent challenge. [1]

Tower crane is one of the most commonly used construction equipment. However there are many drawbacks during the installation of precast concrete exterior wall panel process. When the tower crane is in operation there is only a single lifting task can be executed. Other work station may redundant due to the crane is occupied. Therefore down-time of crane usage can cause unnecessary delay in work schedule. The lifting task requires a high degree of skill and organized planning. Yet weather condition could make this task even more formidable. Heavy wind will cause the hoist cable and the panel to swing. It may cause damage of the panel and the structural frame of the building. In the worst scenario, if the hoist cable snaps it may result injury or casualty of on-site personal.

In many construction sites, the manoeuvre of labor and machinery is often disorganized or even random. This could result unnecessary moment of labor, poor work efficiency on-site and increased health and safety risks.

The building component and the structural frame of the building have to be developed with the consideration of the choice of assembly tools and sequence will be deployed during the on-site assembly process. The size, weight, and logistical strategy have to be compatible with each other to ensure the assembly tasks to be carried out smoothly. [2]

This paper will focus on solving those aforementioned issues and develop a fast, effective method for assembly precast concrete exterior wall panel in high-rise building. The proposed system will be developed with regard of lean construction, design for assembly (DFA) principles. Thoroughly investigates issues that appeared in the conventional method and find the appropriate solution. More detailed description of the proposed system will be mentioned in the paper.

2 Existing method

The conventional installation method of precast concrete exterior wall panel in high-rise building it is not a straightforward process. It is complex, dangerous, time consuming task which requires careful planning and team work among all personals involved.

The overall installation process can be briefly described as follow:

Access route of the building must carefully arranged and the ground condition must capable of taking the live load to ensure safety operation during delivery of the wall panels. Ensure there is adequate height for mobility of machinery and the positioning of the panel in the assembly area.

Centre point of the steel frame structure need to be measured and the position of loading bays are assigned.

Precast wall panel will be lifted from the delivery vehicle or from the stacking area to the assigned loading bay. The lifting clutches will be attached on to the embedded lifting inserts.

Tower crane will lift the wall panel to position gently to avoid swing. Then lift wall panel to assembly location and align the connection joint with the bracket on the steel frame work. Slowly lower and adjusting the wall panel into position.

Tighten the connection bolts and ensure the panel is align both vertically and horizontally. Finally apply joint sealant to the wall panel joints.

There are many challenges exposed during the existing installation process. Initially, the wall panels are off-loaded from the on-site stock yard or directly from the delivery vehicle by tower crane then lifted to the job floor where the panel will be installed into position. This task could be problematic as the tower crane can only lift one piece of panel at a time. While the tower crane is lifting the panel, the delivery vehicle has to remain stable on-site. This can cause two potential issues, first, Down-time of tower crane time and second congestion on-site. [3] The design of the connection joints between the wall panel and the steel building frame can also determine the duration of the assembly time. The movement of the wall panel caused by swinging lifting hoist can make alignment and joining between the wall panel and the steel structural elements extremely difficult. Furthermore, damage on wall panel during assembly process can increase project cost and time delay dramatically. Mechanization during the assembly process is relatively low. Most of the assembly task is conducted by human labor. Dealing with heavy building element while exposed in the environment will inevitably increase safety hazard or cause unpredictable delay. In addition to those challenges, there is an increasing trend in many countries which is aging work force. Demographic change and aging society will have directly impact on the construction sector. Shortage in skilled labor, increased wages and less young population is willing to engage in such physical demanding industry. One of the feasible solution is to increase the degree of mechanization and semi-mechanization with manual operation will flexibly cope with varies conditions and be adopted easily with minimal system development and alteration cost. [2]

2.1 Related work

In the late 1980s, a shortage of skilled labor, as well as an aging society has been a major concern in the Japanese construction industry. Single-task robots have been developed to address this issue while also trying to attract young workers to get involved in the industry.

The example below will demonstrate a multi-purpose material-handling robot developed by Kajima called "Mighty Hand". (Figure 1)

Wall panel installation work is physically demanding as it involves heavy lifting in an enclosed space where it is impossible to use cranes or other heavy lifting devices. Traditionally, workers need to work as a team to lift heavy loads. The task is dirty, dangerous and repetitive. The main goals of this development were to increase productivity and relieve workers from heavy physical exertion. The vertical material lift moves the material to the desired position with little manual assistance. Once in position the material can be attached by hand tools. Preparation material-handling are: transportation of the robot, inspection of the floor surface, accessibility of the floor area, identification of any obstacles, preparation of material for picking up and set-up of the robot. The manipulator commonly has five degrees of freedom: rotating, tilting, shifting, lifting and retraction. The robot is controlled remotely from a small control bow light enough to be carried by a worker. The worker will be responsible for directing the moment of the robot and off-load and position of the building materials.

The robot has increased productivity on-site, reduced manual material handling and the control box can easily carried by humans. Even though interchangeable end effectors allow the robot to perform multi-task jobs, a trained worker must follow the robot at all times, and is responsible for placing materials. Preparation time is long therefore increasing total operation time and total cost. [4]

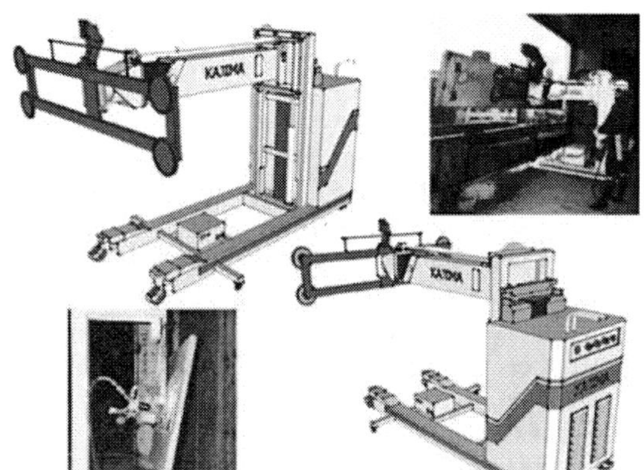

Figure 1: Kajima "Mighty hand" material-handling robot

3 Proposed Method

The proposed system BFIS stands for Building Facade Installation System. The principle design concept is aim to reduce initial R&D cost and manufacturing of the system by utilising existing applied technology in the construction industry or other industrial field. Develop a flexible integrated assembly system that will satisfy vast range of end users demand. BFIS will focus on the design of its on-site logistic, installation, lifting coordination and management technique to solve the aforementioned issues regards to on-site assembly of precast concrete exterior wall panel in high-rise building. The system aim to generate a continuing work flow, reduce down-time in every work station and necessary waste. In terms of design, the degree of automation in the system is carefully applied to be able to achieve reduction in R&D cost. Semi-mechanization combined with manual operation will be able cope varies conditions and be appropriately adopted in many countries. Therefore, the goal of to save investment, reduces expenses, improve the engineering quality and accelerate the overall construction process. [5]

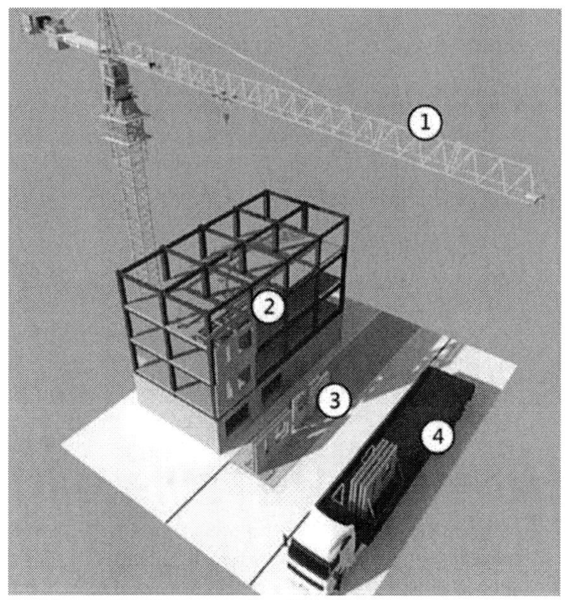

Figure 2: Components of BFIS

3.1 Components of BFIS

The proposed system is designed to be highly modular that the main component of the system consists of three main modules: (Figure 2)

- Number one is the tower crane system. The module is similar to conventional tower crane system, but it equipped with on-board camera, it will connects with the hoist systems and lifts wall panels from the delivery vehicle and unload them to the loading bay. The hoist system will then connects with the BFIS hoist system for final wall panel lifting task while monitor and execute final assembly adjustment.

- Number two is the assembly system module. This is also the key module of the BFIS, it provides support to the wall panel during the assembly process as well as allocation and fine-tuning of the wall panel before it connect with the building structural frame. The assemble module consists of several sub-systems and it will be introduced in detail later in the paper.
- Number three is the loading bay. The wall panel is placed on the loading area by a tower crane from the delivery vehicle. The system will be following a pre-programmed order generated by the construction schedule. A programmable logic controller (PLC) is used for the control of the logistic system. At the prefabrication factory each wall panel is issued a RFID tag. When they arrive on-site, the tag will be scanned and the information read will show the exact assembly sequence and assembly position of each wall component. The control system can be controlled remotely allowing workers to operate the system in case of emergency. [6]
- Number four is the delivery bay. The delivery vehicle will dock into the correct bay following the assembly order generated by the PLC. To adopt just-in-time (JIT) Just-in-sequence (JIS) principle in the construction process will enhance overall productivity on-site.

The assembly system module is the key design feature of the entire system as shown in (Figure 3). The system offers an integrated assembly approach, which include positioning, fine-tuning and fitting of the wall panel. It also provides a self-adjustable approach that allows the system to move along the assembly position. The system can be divided into six modules; main hoist module, tuning hoist module, travel support platform, tuning support module, mobile crane and a trolley system. The system is designed to provide efficient vertical and horizontal transfers of the wall panel. Especially in this high-rise building project, efficient planning for different types of lifting tasks is essential. [7]

The main hoist module connects with the tower crane and the tuning hoist module. It provides vertical lift of the wall panel along with the tuning hoist module. It will lift a set of wall panel with tuning hoist module from the loading bay. Once the tuning hoist module reaches the docking position and securely connected with the turning support arm, the main hoist will be manually detached from the tuning hoist module. The stop damper is designed to minimize the damage may accrue during the lifting of the wall panel. There is an on-board camera will monitor the entire sequence.

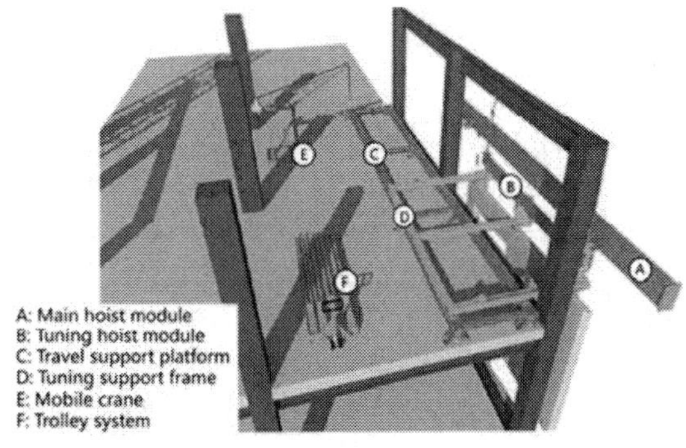

Figure 3: The assembly system module

Once the main hoist module is detached from the tuning hoist module, it will repeat the sequence again.

The tuning hoist module manually operated controlled by assembly operator by a joystick controller. Once it detached from the main hoist module while the wall panel is connected with the tuning hoist supported by the tuning support frame. The hydraulic legs will be inserted into the docking guide on top of the sliding rail. It offers vertical adjustment of the wall panel. The hydraulic legs can be removed from the tuning hoist module and reattached to the correct position based on the size of the wall panel to be assembled. (Figure 4)

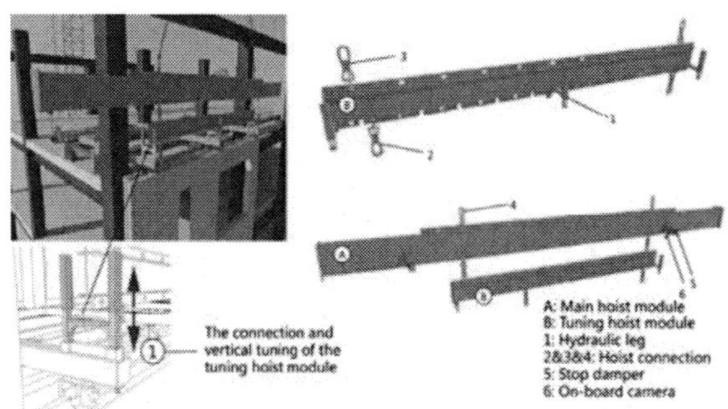

Figure 4: Main and tuning hoist module

There is a motor located in the arm body that will pull and push the docking guide back and forth along the sliding rail. It offers horizontal adjustment of the wall panel. The tuning support adjusting frame will allows the tuning support arm to be fitted in correct position according to the size of the panel. (Figure 7) The tuning support frame is connected with the travel support platform which is linked with the building floor. Both the travel support platform and the tuning support frame are designed as flexible modular components that suitable for on-site assembly and disassembly and easy for transportation across the building site. The mobile crane is for lifting tuning support frame parts and travel platform parts during wall assembly process. The spare parts are temporarily stored on the trolley system.

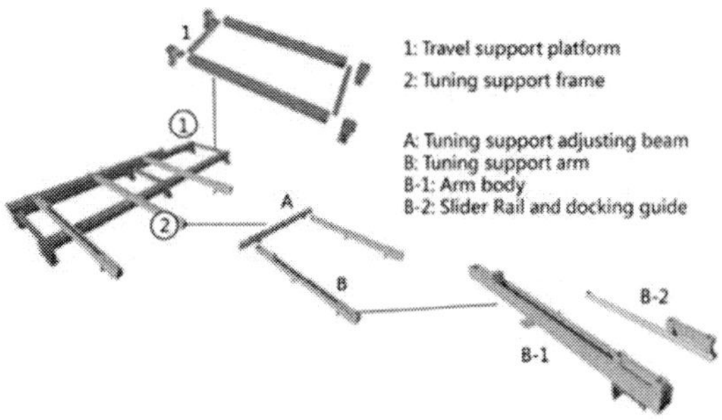

Figure 5: Travel support platform and tuning support frame

3.2 Work flow of BFIS

The wall panels are arrived on-site and to be unloaded from the delivery vehicle on to the loading bay. The tuning hoist module will be connected with the correct wall panel ready to be lifted by the main hoist module that attached on the tower crane. At the same time, on the assembly floor level, the travel support platform and the tuning support module is adjusted and positioned. The panel will be lifted and securely positioned on the tuning support frame along with the tuning hoist module. At this point, the main hoist module will be disconnected from the tuning hoist module and return back to the loading bay for the next load. At the assembly floor level the wall panel is fine-tuned and finely assembled to position by the operators. As shown on (Figure 6) the operators will arrange the assembly tasks of the next section of the travel support platform and the tuning support frame based on the evolution of the assembly process. Those tasks are overlapping according to the installation plan ensures continuous work flow. [8]

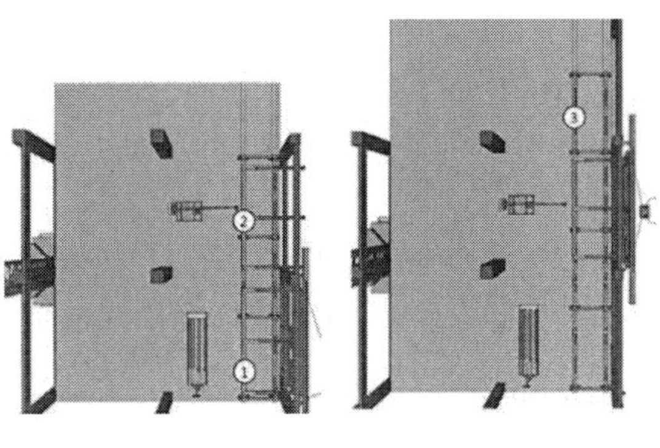

Figure 6: Work set-up of BFIS

Figure 7: Different ranges of hoisting capacity

4 Discussion

In comparison with the conventional installation process, BFIS reflects some advantages as well as limitation.

The advantages are listed as below:

- Time saving –The system is specially designed for the specific task. Due to each task is carefully planned and the overall composition of the system is closely guided by the work flow to ensure work is executed in a parallel fashion. The overall duration of the wall panel assembly process will be reduced.
- Improved on-site logistic – The proposed system enables the detail of the wall panel to be traced from the factory to its final assembly position of the building. On the other hand, the availability of each station and personal is also traceable. This feature will enhance the coordination of work force on-site.
- Stability – The wall panel will be positioned on a static support frame before final assembly. This design will eliminate any swing caused by wind or other external force.
- Cost effective – Semi-mechanization combined with manual operation enable the system to save considerable amount of R&D cost. Most of the system components are available on the existing market and require minimal alteration to achieve the concept specification.

The system limitation and challenges in later implementation stage can be surmised as below:

- Handing variety weight and sizes of the wall panels –The assembly system module is supported by the floor of the building therefore the system weight plus the weight of the wall panel must be considered. Lifting heavy wall panel may cause structural damage of the building. Even through the tuning support adjusting frame and allows the tuning support arm to be fitted in various positions to accommodate different sizes of wall panels. However, because most of dismantle work will be carried out by operators, therefore it will increase overall assembly time.
- Operation Safety – There is not any similar

assembly system available on the market for operators to train or compare. This will increase the potential risk when unfamiliar situation occurs during the assembly process. In addition, the system requires the operators to work closely with the machinery that will increase potential injury caused by malfunction of the machine or misconduct by human error. Therefore, additional training is necessary for all personnel who will be working with the system. An expected learning curve will occur, hence when planning out assembly schedule this period of time should be considered.

- Manufacturing – Even through most of the component can be found on the existing market, however, reconfigure the systems to preform functions as a new system can impose some difficulties. Each party might from a different industry yet has to share each other's technological know-how and expertise. Strong understanding and collaboration between various industries is an essential factor for the success of the proposed concept.

5 Future development

Generally the BFIS still at its concept development stage, the system provides a huge potential for solving issues in assembly of precast concrete exterior wall panel in high-rise building. However, it also faces many changes in terms of system integration, safety and management. Development of such system and improve the assembly processes will face cross disciplinary tasks between the manufacturers, building designers, property investors, and construction contractors. As mentioned earlier, the most optimized working relationships of developing the system would be increased collaboration between all stakeholders. The system has potential to be upgraded with more degree of automation depends on the budgets of the end-user.

6 Conclusion

This paper has given an overview of the concept of the Building Facade Installation System. The concept will impose a fast and cost-effective installation method for installing precast concrete panel in high-rise building. It has demonstrated an innovative approach in reform the assembly process and also has potential to reduce construction time and save overall project investment cost. However, it still requires further research, implementation and pilot project in the future to verify the effectiveness and practicality of the system under on-site operation.

References

[1] Sarja A. a book *OPEN AND INDUSTRIALISED BUILDING*. E&FN Spon, 1998.

[2] Alarcón, L. a book *Lean construction*. CRC Press, 1997.

[3] Cousineau L. and Nobuyasu M. A book, *Construction robots: the search for new building technology in Japan*, ASCE Publications, 1998.

[4] Pan W. M.Sc. Thesis, *Proposed solution for implementing the housing industrialization strategy in China*, Technique University Munich, 2013.

[5] Thomas B. And Thomas L. A journal article. *Robot oriented construction management*, in *Poorang A.E.Piroozfar.et al. eds. Mass customization and personalization in architecture and construction*. Routedge, 2013.

[6] Neelamkavil J. A conference paper. *Automation in the Prefab and modular Construction Industry*. In *Proceeding of the 26th ISARC*, 2009.

[7] Cho C. Kim J. Cho M. Lee J. KimY and Kwon S. A conference paper. *In Simulation method of construction hoist operating plan for high rise buildings considering lifting heights and loads*. In *Proceedings of the 27th ISARC*, p22-28, 2010.

[8] Sacks R and Goldin M. A journal article *Lean management model for construction of high-rise apartment buildings*. In *Journal of construction engineering and Management*, p374-384, 2007.

Accuracy Assessment for 5 Commercial RTK-GNSS Systems using a New Roadlaying Automation Test Center Calibration Track

R. Heikkilä[a], M. Vermeer[b], T. Makkonen[a], P. Tyni[a] and M. Mikkonen[c]

[a]Structures and Construction Technology, University of Oulu, Finland
[b]School of Engineering, Aalto University, Finland
[c]Mitta Oy, Finland

E-mail: rauno.heikkila@oulu.fi, tomi.makkonen@oulu.fi, martin.vermeer@aalto.fi, martti.mikkonen@mitta.fi

Abstract –

In Finland several work sites reported errors of several centimeters between the height results when using different commercial virtual reference station networks and receiver brands. The real time kinematic results are calculated by manufacturer-provided software containing in-house know-how. Further complications for calculations will arise when using physical or especially virtual base stations. In the study five commercial RTK-GNSS systems using a local base station and three virtual reference networks are tested for static accuracy at the OuluZone construction automation center at high precision static GNSS track. Errors on the several-centimeter level were found, and for one manufacturer, even larger gross errors were seen, possibly caused by operator error.

Keywords –

RTK-GNSS, Accuracy of measurement, Virtual reference station

Introduction

Automated 3-D machine control systems have been active research and developments targets during the last decade. The use of the present commercial systems is nowadays very common especially in the Northern countries such as Finland (estimated at over 1500 systems in 2016), Norway and Sweden. These automated machine control systems operate using specific, partly automated, guidance or control methods. When using the guidance method (such as a road grader), the operator drives the machine movements and control blade based on a graphical user interface to the machine control model, where online location and position of the blade are shown. In the control method (such as road grader and bulldozer), the blade is automatically moved according to the machine movements driven by the operator, and the calculated difference of the blade to the machine control model. Continuous accurate 3-D positioning of moving machines and/or blades is the key function for automated 3-D machine control. Two alternative 3-D measurement techniques are typically used on construction sites, i.e. real-time kinematic global navigation satellite system (RTK-GNSS) or kinematic robotic total stations. Robotic total stations need accurate reference points with 200-300 m intervals in order to locate themselves in the site coordinate system used. These instruments typically do not use a model of the local geoid, though error caused by this is likely negligible.

During 2013-2014 in Finland several work sites have reported deviations at the several-centimeter level (even 5-10 cm) between the height results of commercial RTK-GNSS systems with real base stations on site, different commercial RTK correction services and total station systems used. No specific and solvable reasons for the deviations were found by the studies made on the sites. Based on the observations, University of Oulu planned and started a new research project "3D measurement base". The aim was focused to study the improvement of the accuracy and reliability of 3D measurements on road and railway construction sites utilizing RTK-GNSS measurement systems and robotic total stations.

In the literature, the achievable precision of RTK-GNSS using both real base stations and network corrections, has been studied, e.g., Berber & Arslan (2013), Martin & McGovern (2012) and especially Bae et al. (2015). A recurring theme is, that discrepancies with ground truth may be in the 5-10 cm ball park, rather than the 2-3 cm tolerances that are often specified and also assumed by us. So, while not new in principle, our results should sound a warning bell.

Method

The study was carried out in the Ouluzone center located about 30 km North-East of Oulu City. A new reference measurement network with three reference points was built and measured in the Ouluzone center area. The coordinates of the reference points (A, B, C) were measured using the static GNSS measuring method in ETRS-GK26 map co-ordinates [1] and the N2000 height system [2]. The maximum random error after adjustment was found to be 7 mm (horizontal) and 12 mm (vertical). As initial reference points, two points (95M8121, 09M5402) measured earlier by the National Land Survey of Finland were used.. The Mobile 3G network was available for the wireless communication during measurements.

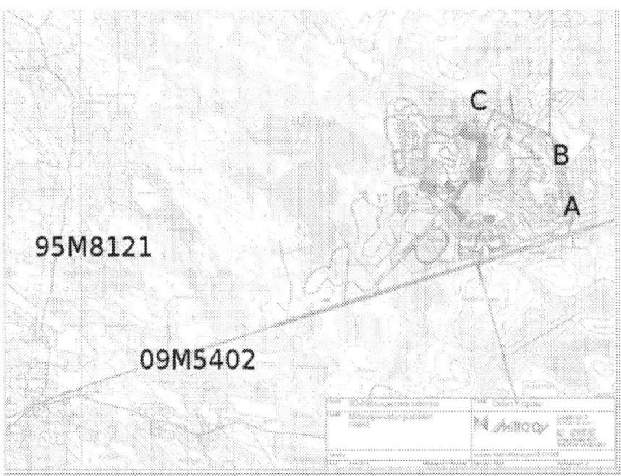

Figure 1. Map of the reference target points (A, B, C) with the global coordinate references (95M6121, 09M5402). Map is approx. 5 km wide.

Figure 2. Setting up a base station.

In total, five different commercial RKT-GNSS systems were found to have been imported and sold in Finland in 2014-2015. All of the companies involved (Leica Geosystems Oy, Geotrim Oy, Sitech Finland Oy, Topgeo Oy, Geolaser Oy and Geostar Oy) were invited to execute measurements of the reference points. University of Oulu was observing and documenting the measurements and observations.

Figure 3. Measurement of a reference point in Ouluzone by a commercial RTK-GNSS system.

In the measurement test, each of the reference points A, B and C were measured using different available measurement modes. The first series was measured using the own base station located on the reference point 09M5402. Position correction information was sent real-time by radio modem or 3G network. Measurements were repeated using sequentially Trimble Trimnet, Leica Smartnet and the correction service of Finnish Geospatial Research Institute FGI. All the GNSS observations above 5 degrees angles above horizon level were accepted into the adjustments. As measurement modes, quick point mode (1 observation), mapping mode (5 observations) and benchmark mode (60 s observations) were used. From the observations, GK26 coordinates with N2000 height coordinate as well as WGS84 coordinates (latitude, longitude) and GRS80 ellipsoidal height were calculated for later analysis.

Results

The measurement results compared to the reference coordinates are presented considering the five different commercial systems available and used in Finland 2014-2015. The tested systems were Trimble SPS985 (Sitech Finland Oy), Trimble R10 (Geotrim Oy), Leica iCON iCG60 (base) and Leica Viva GS14 (rover, Leica Geosystems Oy), Topcon (GR5) and Geomax.

We summarize as examples the measurement results in point A by the Geomax system (table 1) and the Trimble SPS985 system (figures 4-6). The results for all systems are verbally summarized below.

Table 1. Measurement results of the reference point A in Ouluzone by the Geomax system. A means point A, R own base station, S Smartnet, G FGI, V Trimnet, 1 quick point mode, 5 mapping mode, 60 benchmark mode. The *xy* coordinates are, in whole metres: *x* = 73230796 m, *y* = 26503151 m, decimal fractions in table.

Point, mode	x	y	dx	dy	dz	dxy
AR1	.132	.568	-0.004	-0.003	+0.044	0.005
AR5	.137	.572	+0.001	+0.001	+0.048	0.001
AR60	.133	.572	-0.003	+0.001	+0.046	0.003
AS1	.119	.588	-0.017	+0.017	+0.033	0.024
AS5	.112	.588	-0.024	+0.017	+0.013	0.029
AS60	.124	.588	-0.012	+0.017	+0.020	0.021
AG1	.157	.588	+0.021	+0.017	+0.027	0.027
AG5	.167	.588	+0.031	+0.017	+0.036	0.035
AG60	.169	.582	+0.033	+0.011	+0.014	0.035
AV1	.123	.577	-0.013	+0.006	+0.044	0.014
AV5	.115	.575	-0.021	+0.004	+0.030	0.021
AV60	.142	.576	+0.006	+0.005	+0.102	0.008

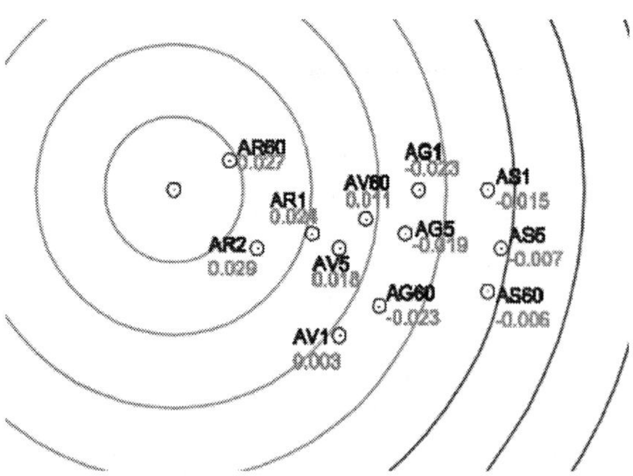

Figure 4. Measurement results at the reference point A (Trimble SPS985). In the dartboard, *xy* deviations were illustrated by circle lines with 5 mm intervals, green lines are inside the tolerance and red lines outside the allowed tolerance. Vertical *z* deviation is presented numerically, where the deviations inside the tolerance (±30 mm) are green and the deviations outside the tolerance red ones.

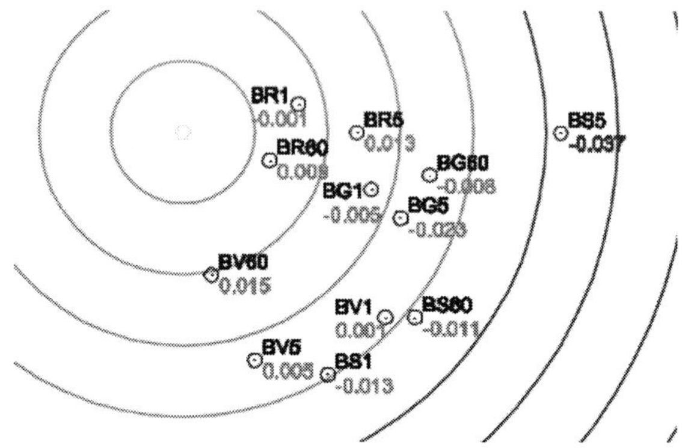

Figure 5. Measurement results at the reference point B (Trimble SPS985).

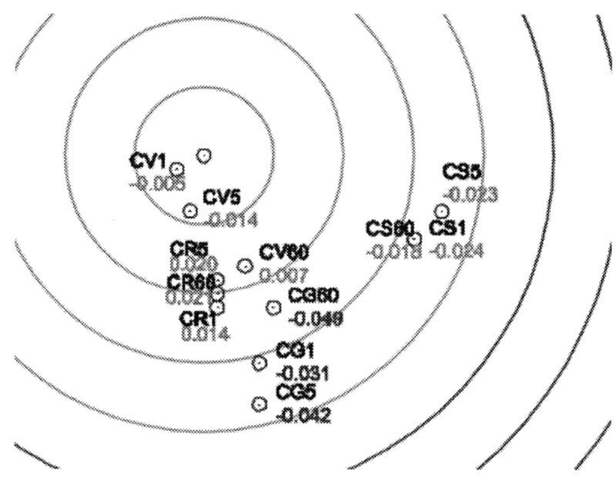

Figure 6. Measurement results at the reference point C (Trimble SPS985).

Considering the **Trimble SPS985** system, all the measurement results with own base station at the points A, B and C were accurate and inside the tolerances. The results using the Trimnet correction service were also accurate at all the points. When using Smartnet corrections, the results were accurate at all points except point A, where a +0.024 m deviation was found. Using FGI corrections all the results were good with the exception of point C, where height deviations were -0,031 m, -0,042 m and -0,049 m.

Considering the **Trimble R10** system, the measurement results using own base station were nearly within tolerances at the A and B points, at C there were *z* deviations of -0.031...-0.033 m. The results using Trimnet were accurate at point B, at point A there were *z* deviations of +0.033...+0.044 m. When using the Smartnet corrections, the point A was measured accurately, at points B and C there were minor deviations outside tolerances like -0.029...+0.022 m. When using FGI corrections, all the points and coordinates were measured accurately.

Considering the **Leica RTK-GNSS system** (Leica iCON iCG60 antenna, iCON CC65 control unit, Viva GS14 performance, Viva CS15 control unit), all the *xy* results with own base station were good and inside tolerances at all the points A, B and C. The *z* deviations at points B and C were -0.032...-0.042 m. The *xy* results when using Trimnet corrections were good at points A and B, but at point C there were *xy* deviations of -0.028...-0.033 m. At points A and B there were *z* deviations of -0.045...+0.058 m. When using Smartnet corrections, point C was measured accurately with all the correction methods, but at point A there were *y* deviations of +0.029...+0.033 m, and at points A and B, +0.033...+0.054 *z* deviations. When using FGI correction, points A and B were measured accurately enough, at point C there were z deviations as -0.042...-0.048 m.

Considering the **Topcon GR5** RTK-GNSS system, the results with own base station were good except at point C, where the *z* deviation was 5-6 cm. The *xy* results when using Trimnet corrections were good, however there were some 4-5 cm deviations especially at points B and C. When using Smartnet or FGI corrections, there were significant deviations as follows: point A – *xy* deviations of 8-9 cm and *z* deviations of 17-23 cm when using Smartnet corrections, *xy* deviations 6-7 cm and *z* deviation 10 cm when using FGI correction. Point B – *xy* deviations of 4-7 cm when using Smartnet corrections. Point C – *xy* deviations 7-12 cm and *z* deviation 12-20 cm when using Smartnet corrections.

Considering the **Geomax RTK-GNSS** system, the results with own base station were inside the tolerances at points B and C. At point A there were +0.044...+0.048 m *z* deviations. The *xy* results with Trimnet correction were good at all the points, but there were +0.044...+0.048 m deviations in *z* direction. When using Smartnet corrections, all the coordinates were quite accurate at points A and B, at point C there were *xy* deviations of +0.024...+0.032 m. When using FGI corrections, the results of point C were within tolerance, at points A and B there were *xy* deviations of +0.021...+0.033 m and *z* deviations of +0.036... -0.047 m.

Considering the **overall results** in terms of worst-case errors per measurement system /correction source combination (figure 7), we found that, with the exception of two Topcon combinations, all combinations perform similarly slightly poorer than the set tolerances, without any clear single cause being implicated.

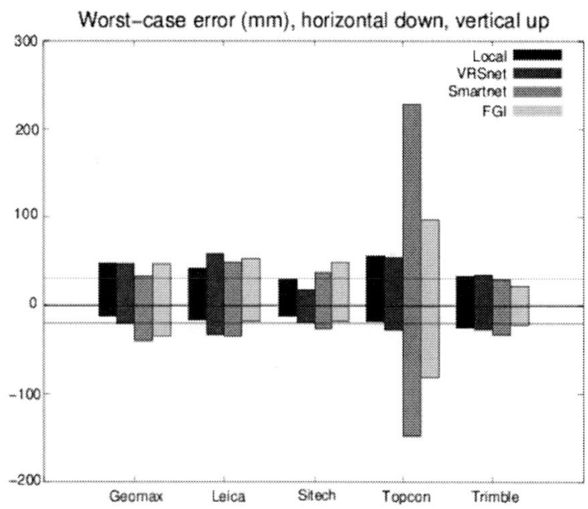

Figure 7. Worst-case error, over nine measurements each for every measurement system / correction source combination, separately for horizontal and vertical errors. Tolerances plotted as red/blue lines.

Conclusions

In the study, five different commercial RTK-GNSS systems used in Finland were used for research experiments at the Ouluzone test center. Also the different RTK correction methods on offer in Finland were used for comparisons in the tests. A significant observation was that the results of the Topcon GR5 system used were quite inaccurate when using the

Smartnet or FGI correction services. No obvious reason was found, but, with complex systems like this, operator error is always possible as an explanation. The best of the tested systems was, according to the test results, the Trimble SPS985 system presented by Sitech Finland, but the difference with the other systems appears not to be statistically significant.

However, even when the most obviously poor results are left outside consideration, there remain largish discrepancies that could not be clearly tied to GNSS system used, RTK correction source used, or observation mode. It would appear that the RTK-GPS technique is not quite as robustly precise as some would like to believe. This is something users would ignore at their peril.

It would appear, and this agrees with literature studies, e.g., Bae et al. (2015), that the RTK-GNSS technique is not quite as robustly precise as some would like to believe. This is something users would ignore at their peril. A solution to this conundrum may be two-fold: often the specified tolerances of 2-3 cm are unnecessarily tight, and may be relaxed. Where this is not possible, the RTK-GNSS technique should not be used but rather, e.g., terrestrial guidance systems.

References

[1] JUHTA. JHS 197, EUREF-FIN - koordinaattijärjestelmät, niihin liittyvät muunnokset ja karttalehtijako. Julkisen Hallinnon Suositukset, URL: http://www.jhs-suositukset.fi/web/guest/jhs/recommendations/197, 2016.

[2] JUHTA. JHS 163, Suomen korkeusjärjestelmä N2000. Julkisen Hallinnon Suositukset. URL: http://www.jhs-suositukset.fi/web/guest/jhs/recommendations/163. 2010.

[3] Heikkilä, R. & Kivimäki, T. (2009) Integrating 5D Product Modelling to On-site 3D Surveying of Bridges. ISARC'2009, The 26th International Symposium on Automation and Robotics in Construction, 24-27 June 2009, Austin, Texas, U.S.A., pp. 445-450.

[4] Makkonen, T. & Heikkilä, R. & Kaaranka, A. (2014) The Applicability of a Geomagnetic Field based 3D Positioning Technique to Underground Tunnels. ISARC'2014, The 31st International Symposium on Automation and Robotics in Construction and Mining, 9-11 July, 2014, Sydney, NSW, Australia, x p.

[5] Tae-Suk Bae, Dorota Grejner-Brzezinska, Gerald Mader, and Michael Dennis: Robust Analysis of Network-Based Real-Time Kinematic for GNSS-Derived Heights. Sensors 2015, 15, 27215-27229; doi:10.3390/s151027215 URL: http://www.ncbi.nlm.nih.gov/pmc/articles/PMC4634509/

[6] Berber, M. & Arslan, N. Network RTK: A case study in Florida. Measurement. Vol. 46:8. 2013. S. 2798—2806. DOI:10.1016/j.measurement.2013.04.078.

[7] Martin, A. & McGovern, E. An Evaluation of the Performance of Network RTK GNSS Services in Ireland. International Federation of Surveyors (FIG) Working week. Rooma, Italia 6.—10.5.2012. FIG. 2012. 19 s. URL: http://arrow.dit.ie/cgi/viewcontent.cgi?article=1001&context=dsiscon

A Case Study: Projecting Images for Designing Interior Panels using Parametric Modeling Tool

HaYan Kim[a], JinHua Huang[a], Jin-Kook Lee[a]

[a]Dept of Interior Architecture Design, Hanyang University, Seoul, Republic of Korea
E-mail: hayaan92@gmail.com, sara920319@gmail.com, designit@hanyang.ac.kr

Abstract

This paper aims to describe a case study of parametric interior design based on BIM (Building Information Modeling). As the practical use of BIM-based design grows, its influence expands into the field of interior architecture design. BIM makes possible to check various design plan, reach decision making in an effective way, and change design plan in an efficient method. Therefore, BIM is also promising field in interior architecture design. However, compared to other fields such as architecture, engineering, and construction (AEC) industry, there have been less research and projects on BIM in the field of interior architecture design. For increasing the feasibility of adapting BIM in interior architecture design, this paper describes a case study-projecting images for designing interior panels using parametric modeling. This process needs elaborate, delicate, and precise steps for harmonious output. For steady using of the building, users look forward to the design plan which can be variable and changeable according to user's preference and the environment. Therefore, demand for parametric design in the interior design part such as panel pattern design for various decoration is growing. The process in this paper deals with an advanced phase which is effective in decreasing time consumption and useless part of the traditional process. Finally, this paper suggests the possibility of using BIM in the interior design process and field where BIM can be applied.

Keywords –

Parametric Modeling; BIM; Interior Design; Automated Design Process

1 Introduction

The word "Interior Design" was meaning of the minimum essential room for human living. [5] These days meaning of "Interior Design" is transformed into the place where residents reveal their personality and characteristic. As a result, according to a different user and changed interest about interior design, remodeling of interior emerged as a main part of the interior design. [6] For remodeling, objects such as furniture or decorations which have variables are in a better position than fixed objects for parametric design. Therefore, a crucial part in interior design of building depends on how much this space is favorable for new use after once occupied. Unlike exterior part, the interior is the abutted part where directly connected with users' life. Interior design must be varied by users' preference and environment. Parametric design is emerging as a practical alternative in changeable and diverse design part in the interior design field.

The parametric design makes designers to change their work and make their own algorithm much easier for effective way of their design work to make final output. As another way of parametric design work, it generates families and collections of families that display different result according to the input value. [1] If there is certain output which user want, objective and mathematic algorithm prevents consequence get too far off from the result that user want. Computer's data calculating process is used for those algorithms. [2]

BIM is used for extensive fields such as architecture, engineering, construction, and facility management (AEC-FM) industry. Naturally, in the field of interior design, BIM has positive effects on toward improving efficiency in design and construction. Nevertheless, BIM in interior design has insufficient researches and practical uses compared with other fields. Because of requisition of an elaborate scheme and relatively small scale project, the interior design process is not familiar with BIM. However, graphical programming environment in BIM authoring tool and relevant algorithm generating function in BIM visual programming environment could be an advantage for growing effectiveness in design and construction of interior design. [7] Accordingly, this paper introduces a way of the practical using method of BIM in the interior design field and suggests the feasibility of BIM application for the supportive design process.

2 Background and Objective

Previous interior design process reflects designer's prompt preference according to circumstances and environments and goes along without standard library of the application object. However, advent of BIM cause emerging of formation about standard definition and library of interior objects such as furniture, window, and door. Interior design of high LoD (Level of Detail) which considered as impossible in BIM modeling tool is now became possible because of High LoD objects in the library. Owners of object -usually design companies- in library gradually upgrading their product's information for usability and coadaptation for various BIM modeling project.

For the useful, smart and variable LoD (Level of Detail) object of interior decoration, parametric property plays a leading part in object property. Parametric property is essential element for represent geometry which generated by parameters and rules. Parameters and rules also can show the relationship between geometry and other object. Therefore, geometry is automatically updated in real time by operation of user or alteration of relationship. LoD about parametric property can be defined by user's needs and status of modeling. For quick modeling, the level of the parametric property will be low, and for the high quality of modeling, the level of the parametric property will be high.

Above all process, this paper is dealing with the improvement of effectiveness in the process of decorating design wall by BIM and parametric modeling. Therefore, simple process for demonstration part of this study is as follows Figure 1. Authoring tool and algorithm generator both are based on BIM. By using two methods, process will start from generating prototype of panel and end by QTO (Quantity Takeoff). Exact example of input in the process will be made by BIM authoring tool as we called BIM modeling environment.

On authoring tool, application object must be selected in the project and parametric property must be defined on the interior panel in the family editor. Algorithm generator accepts input value from authoring tool for node combination which can draw out the final output from the design process. For lead high LoD quality till the end of design construction process, deduction of not just output but also quantity take-off is necessary. In conclusion, this paper aims to describe an effective way of applying BIM on interior design to expanding usability of BIM to small project.

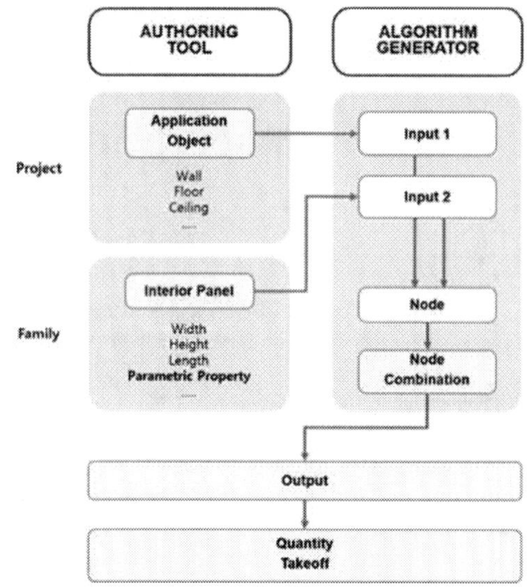

Figure 1. The relationship between BIM authoring tool and algorithm generator in the demonstration.

3 Analysis of preceding research and example

3.1 Tendency of research related to Interior design and BIM

Thanks to BIM extending its influence into interior design field the number of the related research paper is increasing. However, compared with another main part of BIM, such as Architecture, research papers are short of the number. (Figure 2.)

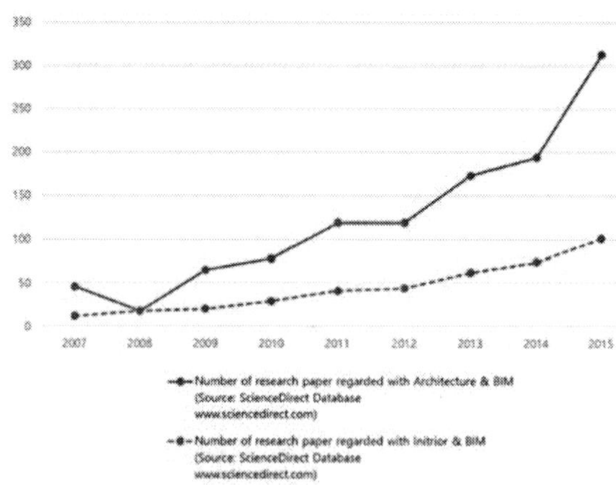

Figure 2. Comparison between the number of research papers of BIM related to architecture and interior (Source: ScienceDirect Database) [3]

There are several main fields which should be deal with large scale in architecture is arousing BIM related studies such as Energy performance assessment, identification of safety hazards and site safety. [8] In the field of interior design, because of BIM' s capabilities of analyzing and simulating system, green energy and design practice. [9] At construction step of interior design, generation of BIM model is expected to play an important role for communication between designer and constructor and improved productivity. [10, 11]

Among interior design process, decorative steps are showing the particularly weak development of BIM research and study. The decorative part in the interior design process are especially easy to interpose creativity and idea where machine and technology yet cannot handle. However, for shortening the construction period and efficient work processes, the introduction of BIM approach seems unavoidable.

3.2 Practical use of parametric modeling in interior design

Decorative interior panel design can be found in lobby or showroom of public facilities such as exhibition area and theatre. In commercial facilities, corporation or retail shop use interior panel design point in their lobby or representative spot to make visitor an impression of their brand image. In residential space, because of waterproof function, this design can be found from mosaic tile decoration of kitchen and bathroom. This interior panel strategy often found at the place where people's attention is converged.

Table 1. Examples of practical use of parametric design in interior design

Project name	Space type	Used software	Application purpose
Gravity Fields [12]	Gallery Lobby	Rhinoceros Grasshopper	Exhibition work
Toolbox Office [13]	Office	Grasshopper	Interior design point
Parametric Furniture [14]	Exhibition Hall	Rhinoceros Grasshopper	Parametric furniture

Interior panel pattern design distorted if even one of the panel is arranged in wrong place. Therefore, the design process must proceed in the exquisite and accurate way. However, in the as-is design processes, which depends on designer's visual perception and sensation from experience, it is easy to arouse error in the final result.

Table 1. describes examples of practical use of parametric design in interior design. Representatively Rhinoceros and Grasshopper used for generating parametric algorithm and geometry. The parametric design seems to be used for exhibition work, design point, and furniture design. Various parts in interior design are using the parametric design for a decorative reason.

4 Projecting images for designing interior panels using parametric modeling

4.1 Process for demonstration of designing interior panel pattern based on 2D image

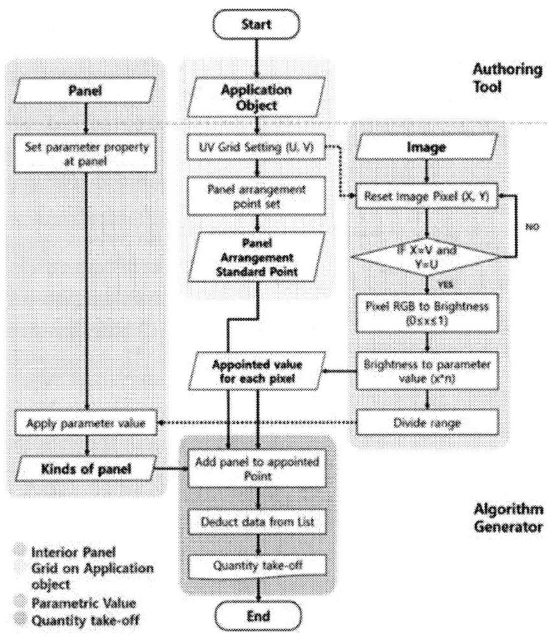

Figure 3. The overall process for a demonstration of pilot tests about the interior panel.

The aim of this process is to make the reduction of time to design and construction and minimization of possibility that error can occur for effective work by projecting a pattern from designated images to design interior panels using a parametric modeling function from BIM authoring tool and algorithm generator.

The as-is design process of interior panel depends on designer's visual perception from image and sensation from experience. This process describes the way of projecting images to design interior panels using parametric modeling. Using BIM application, this process involved design process analysis color and brightness from image automatically. Analyzed result is substituted on the parametric property. According to each value that assigned on the pixel, every interior panel will be placed on the right spot.

Practical use of BIM decreases deficiency during construction and time consumption. Another virtue from BIM is unconstrained quantity take-off after conclusion deducted. Quantity Takeoff is an important part in the design process for connecting design field and construction field. If there is no much communication between both parts, final output will be deteriorated and lost from the output which user expected. To prevent those problems, design process in this paper largely use BIM authoring tool and algorithm generator. Figure 3. is the overall process of demonstration. BIM authoring tool is role as platform for design process and gathering of outputs. BIM algorithm generator starts to emerge for much wider variety of geometric modeling and information management.

4.2 Generating prototype in BIM authoring tool

BIM authoring tool including Autodesk Revit-3D knowledge-rich parametric modeling systems- role as a system for generating a prototype of the interior panel as family and parametric property in the family editor. [4] Before using algorithm generator for panel pattern, three steps need to be done in BIM modeling system. 1) Set of application object 2) Create panel 3) Set parameter property on panel, those steps will be executed by project and family editor in Autodesk Revit.

In this paper, basic model in Revit which used for educational propose is selected for the application object as seen in Figure 4.. Side wall next to hallway in educational facilities is the proper background for panel design because this wall could be seen from the upper floor throughout void between both floors.

A prototype of the panel has a circular hole in the middle of the panel. Figure 5. is different parametric property of panel which the value of the radius of the panel. According to the radius of the hole in the panel, each panel displays correspondent pixel's data. Figure 6. is applied property value in Family types dimension property. Gathering of every panel on assigned spot with different parametric value will show one completed panel pattern design.

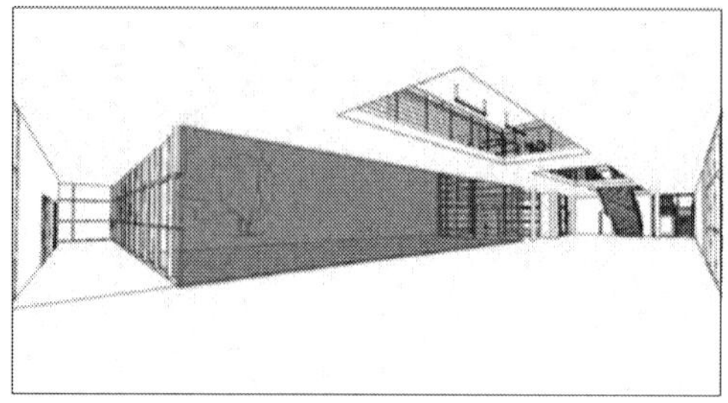

Figure 4. Application objects in facility

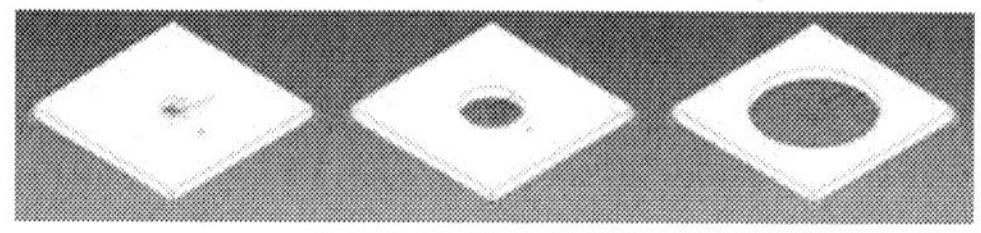

Figure 5. Interior panels with different parameter property (from left, Panel Radius=200, 500, 1000)

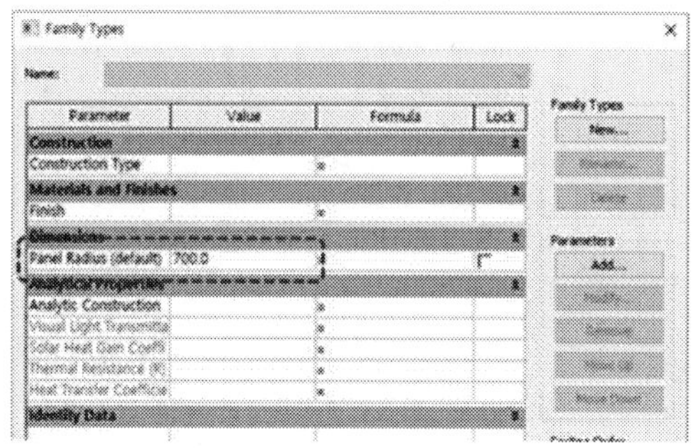

Figure 6. Interior panel's parameter property (Panel Radius) in Family types and value.

4.3 Generating algorithm using graphical programming tool

Each node can expose the result of the process immediately. At the same time, if there are some problems in node combination sequence, problem nodes change their color and show an error's contents. Visual effects during the process make course deduct conclusion much fast and flawless. Accurate result by Dynamo will be drawn by nodes and connections which also operate well only when the proper type of string or formula is entered in. These system makes user know the most direct procedure of Dynamo and finds themselves more effective way to solve the problem.

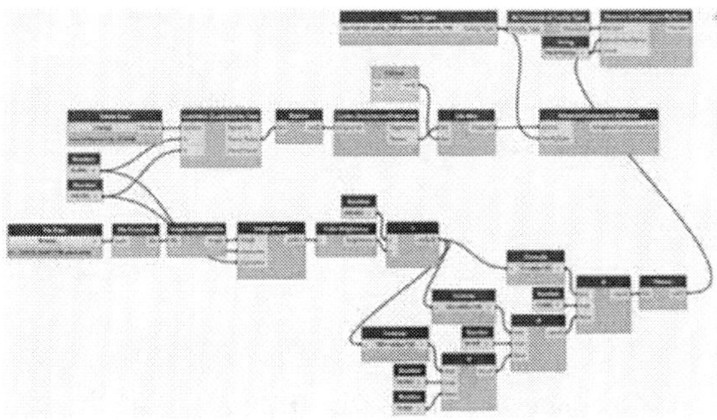

Figure 7. Dynamo, open source graphical programming tool and computational design application for design.

Every node combination and nodes have input and output value as seen in Figure 7., and every value also can be input value of another node. For the arrangement of interior panel node combination largely divided into five parts.

- Set grid on application object: For the arrangement of the panel, the proper number of U and V values are needed to generate a grid for showing the 2D image. If the too small number is entered, it is hard to recognize what user wanted to show because of each panel represent too many pixels.
- Get parametric property in family type: To control the radius of the hole in the panel, its family type must be brought as family type. Through name string, 'Element.SetParameterByName' node find parameter which has same string of name.
- Image's brightness to a parametric degree: This step project 2D image and automatically analyze color value from the pixel on the image and change color value to brightness value between 0 and 1. To deal delicate value of brightness effectively, multiplication node is needed to amplification. Multiplied values are added to parameter property.
- Apply panel to point: Each point of the selected face, parametric panels will be placed with finalized panel radius. Each panel which has their own panel radius locate definite point.

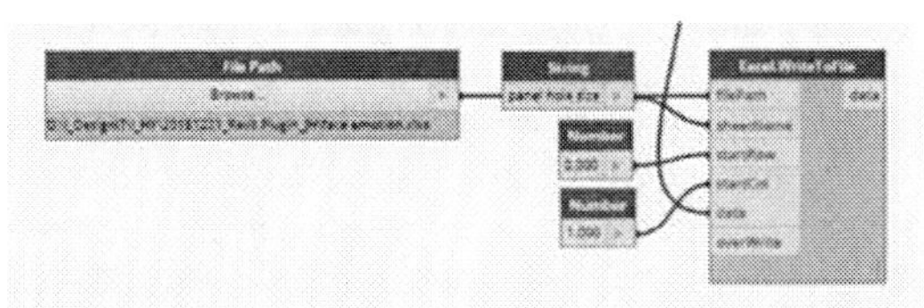

Figure 8. Node combination for quantity take-off in the final process of pilot test.

- Quantity Takeoff: For the effective construction of interior panel design, the data source for the constructor is important for communication with the designer. As seen in Figure 8., node 'Excel.WriteToFile' creates data table depending on selected information. The input of data will be the number of a panel according to kinds of the panel. To arrange panel on the right place, this node creates another sheet for panel plan. The advantage of a data table which set by selected data from user generates plenty kinds of the data sheet for preference of constructor.

5 Parametric interior panel pattern design and arrangement system pilot test

5.1 Different output by input image

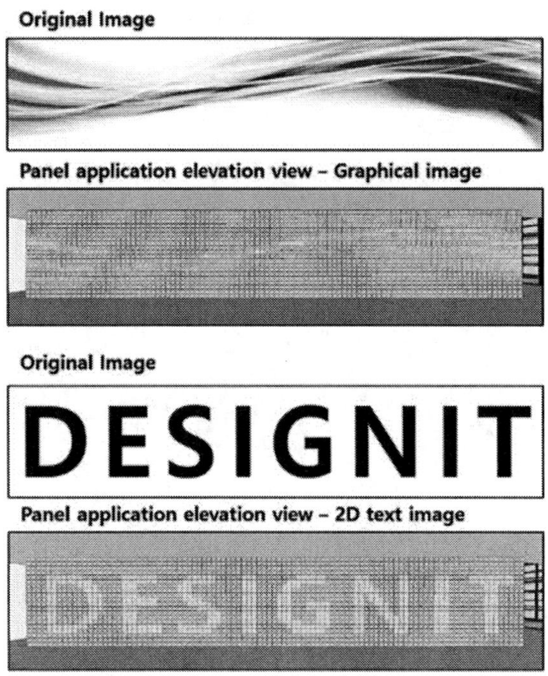

Figure 4. Different output by different entered image by the user.

By different images entered by user draw out the various final result. As Figure 9., the first original image has diverse color in a single image while the second original image has only two colors which derived from 2D text image.

To generate proper output for final, setting kinds of panel step is important. If only two panels arranged for a first original image which has several colors and brightness value, panels could not show softness of wave which the first image tried to reveal. Proper value selection related with image's color and brightness is needed for high quality of the final result.

5.2 Different output by LoD of parameter property

Figure 10. is dealing with parameter property's LoD for a design wall. In the low level of detail, wall used 4 kinds of panel (Panel Radius=10, 30, 50, 70), and high level of detail wall used 7 kinds of panel (Panel Radius=10, 20, 30, 40, 50, 60, 70)

As seen in detail view, detail part of swoosh wave in the original image represent different LoD. For the high quality of the final result, the user needs to enter the large number for kinds of panels.

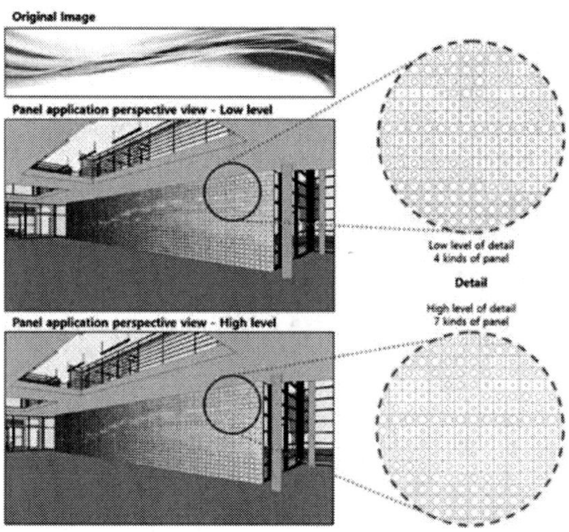

Figure 5. Difference between high level and low level of panels

6 Conclusion

This process has advantages on reducing time consumption and errors which can occur during process. BIM in interior design role as controller and manager of detailed building element. Even if this paper's pilot test is limited on interior panel, there are much feasibility of application of BIM at interior design elements.

Various design tools are coming up for new design paradigms. The trend of design is following parametric design depends on changeable and adaptive geometry. This pilot test is not just for dealing with panels having limited physical characteristics. This paper attempting for further develop for the arrangement of a various object such as space object, a collection of specific objects according to assigned algorithm. For the start of the further study, this case study uses BIM authoring tool and algorithm generator for parametric modeling to propose a new way to design interior panels from an image.

Acknowledgement

This work was supported by the National Research Foundation of Korea Grant funded by the Korean Government (NRF-2013S1A5A8024262)

References

[1] Robert Woodbury, Shane Williamson, Philip Beesley, Parametric Modelling as a Design Representation in Architecture: A Process Account, 159-165, 2006

[2] Ronald Hudson, Strategies for parametric design in architecture. An application of practice led research, PhD Thesis, University of Bath, 21-22, 2010

[3] ScienceDirect Database. On-line: www.sciencedirect.com, Accessed: 06/03/2016

[4] Lee, G., Sacks, R., & Eastman, C. M. (2006). Specifying parametric building object behavior (BOB) for a building information modeling system. Automation in construction, 15(6), 758-776.

[5] Park, S.Y., Kim, K.H., Kim, J.H., Kim, J.J. (2012). Analysis of Business Process for Customizing Interior Design Introducing BIM. Journal of KIBIM, 2(1), 18-26.

[6] Rosenfeld, Y., & Shohet, I. M. (1999). Decision support model for semi-automated selection of renovation alternatives. Automation in Construction, 8(4), 503-510.

[7] Kensek, K. M. (2014). Integration of Environmental Sensors with BIM: case studies using Arduino, Dynamo, and the Revit API.

[8] Yalcinkaya, M., & Singh, V. (2015). Patterns and trends in Building Information Modeling (BIM) research: A Latent Semantic Analysis. Automation in Construction, 59, 68-80.

[9] Lee, Y. S. (2012). Using building information modeling for green interior simulations and analyses. Journal of Interior Design, 37(1), 35-50.

[10] GUDGEL, J. O. H. N. Building information modeling: Transforming design and construction to achieve greater industry productivity. Bedford, MA: McGraw-Hill Construction Research and Analytics, 2008.

[11] SmartMarket Brief: BIM Advancements No. 1. On-line: http://www.smartmarketbrief.com/construction-modeling/, Accessed: 10/02/2016

[12] Noiz acrhitects, Le meridian lobby art "Gravity Fields". On-line: http://noizarchitects.com/projects/le-meridian-lobby-art/, Accessed: 12/02/2016
[13] Caterina Tiazzoldi, Toolbox office lab & coworking. On-line: http://www.tiazzoldi.com/#!m-toolbox/ctyw, Accessed: 10/02/2016
[14] 24Studio Architecture, Parametric furniture. On-line: www.24studioarq, Accessed: 20/02/2016

SeeBridge Information Delivery Manual (IDM) for Next Generation Bridge Inspection

Rafael Sacks[a], Amir Kedar[b], André Borrmann[c], Ling Ma[d], Dominic Singer[e] and Uri Kattel[f]

[a]Faculty of Civil and Environmental Engineering., Technion – Israel Institute of Technology, Israel
[b]Kedmor Engineers, Givatayim, Israel
[c]Computational Modeling and Simulation, Technical University Munich, Germany
[d]National Building Research Institute, Technion – Israel Institute of Technology, Israel
[e]Computational Modeling and Simulation, Technical University Munich, Germany
[f]Faculty of Civil and Environmental Engineering., Technion – Israel Institute of Technology, Israel
E-mail: cvsacks@technion.ac.il, akedar@kedmor.co.il, andre.borrmann@tum.de, lingma@technion.ac.il, dominic.singer@tum.de, uri.kattel@gmail.com

Abstract

Innovative solutions for rapid and intelligent survey and assessment methods are required in maintenance, repair, retrofit and rebuild of enormous numbers of bridges in service throughout the world. Motivated by this need, a next-generation integrated bridge inspection system named SeeBridge is proposed. To frame the system, an Information Delivery Manual (IDM) was compiled to specify the technical components, activities and information exchanges in the SeeBridge process. The IDM supports development of the system by rigorously defining the information and data repositories that structure bridge engineers' knowledge. The SeeBridge process is mapped, parts of the data repositories are presented and the future use of the IDM is discussed.

Keywords –

Information Delivery Manual, Bridge inspection, Building Information Modelling, Semantic enrichment, Remote sensing.

1 Introduction

Highway asset owners face severe problems acquiring status data for their bridges. There are not enough experienced bridge engineers for the extensive work required for inspection of a large number of bridges; bridge inspections mean interruption of transportation and are potentially dangerous activities; and the data available in many Bridge Management Systems (BMS) does not meet the standard of information needed for subsequent bridge repair, retrofit and rebuild work.

Remote sensing technologies are attracting increasing research interest for inspection for health monitoring and valuation for bridges [1-5]. Among the remote sensing technologies, both laser scanning technology and photo- or videogrammetry can produce point clouds from which 3D primitives can be derived. However, the challenge that must be overcome for implementation of remote sensing in bridge inspection is to enable automatic recognition of bridge components from point clouds and make the model semantically rich [6].

To address the challenges, a Semantic Enrichment Engine for Bridges (SeeBridge) is proposed, targeting the development of a comprehensive solution for rapid and intelligent survey and assessment of bridges. The SeeBridge concept is the subject of an EU Infravation research project comprising seven partners in the US, UK, Germany and Israel. In the SeeBridge approach, various advanced remote sensing technologies are used to rapidly and accurately capture the state of a bridge in the format of point cloud data. A bridge model is automatically generated by a point cloud processing system, an expert system that encodes bridge engineers' knowledge for classification of bridge components, and a damage measurement tool that associates the identified defects with the bridge model.

In order to guide and connect the subsystems in the system as a whole, an Information Delivery Manual (IDM) [7] was compiled to formally specify the user requirements and to ensure that the final model would be sufficiently semantically meaningful to provide most of the information needed for decision-making concerning the repair, retrofit or rebuild of a bridge. The IDM approach is outlined in the US National BIM Standard [8] and has been used in numerous BIM interoperability research projects [9-12].

The IDM includes:

- A detailed process map defining the Seebridge process, its component processes and its information exchanges.
- A list of typical bridge elements classified by structure types, their function, shape representation and relative importance in the structure.
- Definition of the possible logical connections between the elements in a bridge structure type.
- A defect table for defects modelling and classification.
- Definition of the required information contents of the exchanges specified in the process map.

The following sections describe the overview and the systematic process of SeeBridge framed by the IDM, explain the information exchange between the component processes, and present parts of the data repositories compiled in the IDM. The conclusion section discusses the need for extensions to the IFC Schema [13] for bridges and the value of the IDM approach to research and development of this kind.

2 SeeBridge Inspection Process

Bridge inspection and management is a part of the bridge life-cycle and is related to the operational and maintenance stage. The data needed for managing the bridge stock within a given defined road network is used for decision making regarding the maintenance, repair, retrofit and rebuild/replacement of the bridges. Bridge inspections are the main source of data regarding the actual condition of a bridge during its life cycle.

Bridge inspection and management methods differ among Departments of Transport (DOT) and authorities in different countries, yet the core innovations of the SeeBridge process are applicable to most if not all. The system integrates four novel technical components to upgrade the traditional bridge inspection process and produce semantically rich BIM models for the inspected bridges. The new components are:

- A bridge data collection system using remote sensing techniques such as terrestrial/mobile laser scanning and photogrammetry/videogrammetry.
- A bridge object detection and classification software for automated compilation of 3D geometry from the remote sensing data using both parametric shape representation and boundary representation.
- A semantic enrichment engine for converting the 3D model to a semantically rich BIM model using forward chaining rules derived from bridge engineers' knowledge.
- A damage detection tool for damage identification, measurement, classification and integration of this information in the BIM model.

Figure 1 shows four bridge types in SeeBridge project.

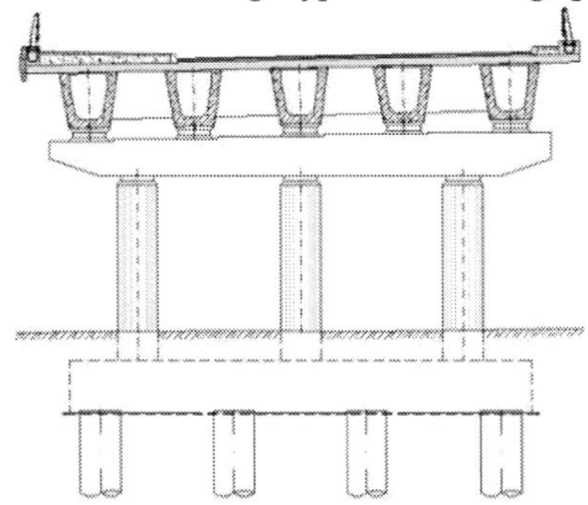

(a) Concrete Beam/Girder Bridge

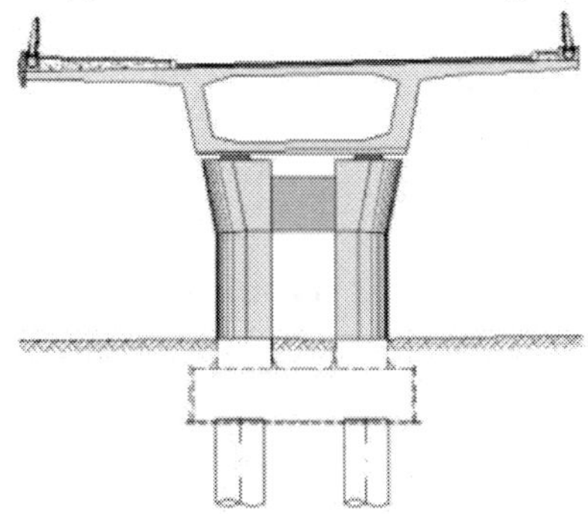

(b) Concrete Box Girder Bridge

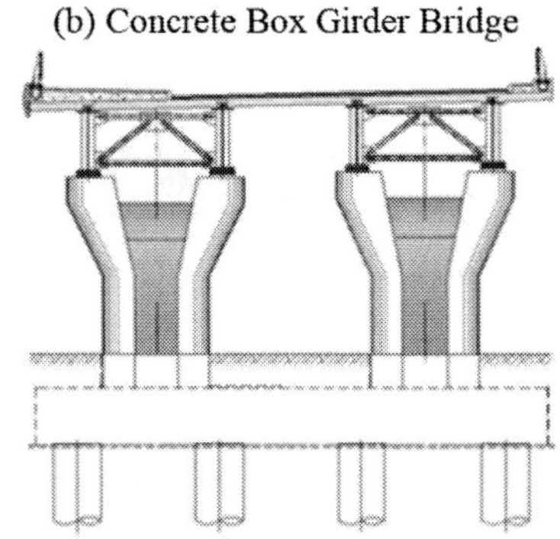

(c) Steel Beam/Girder composite Bridge

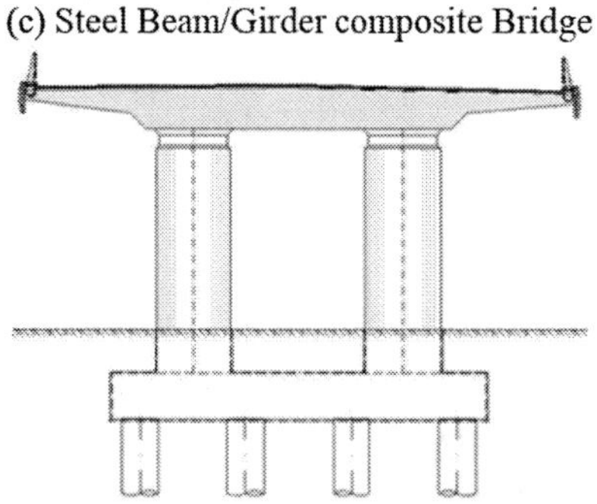

(d) Concrete slab Bridge

Figure 1 SeeBridge Bridge Types

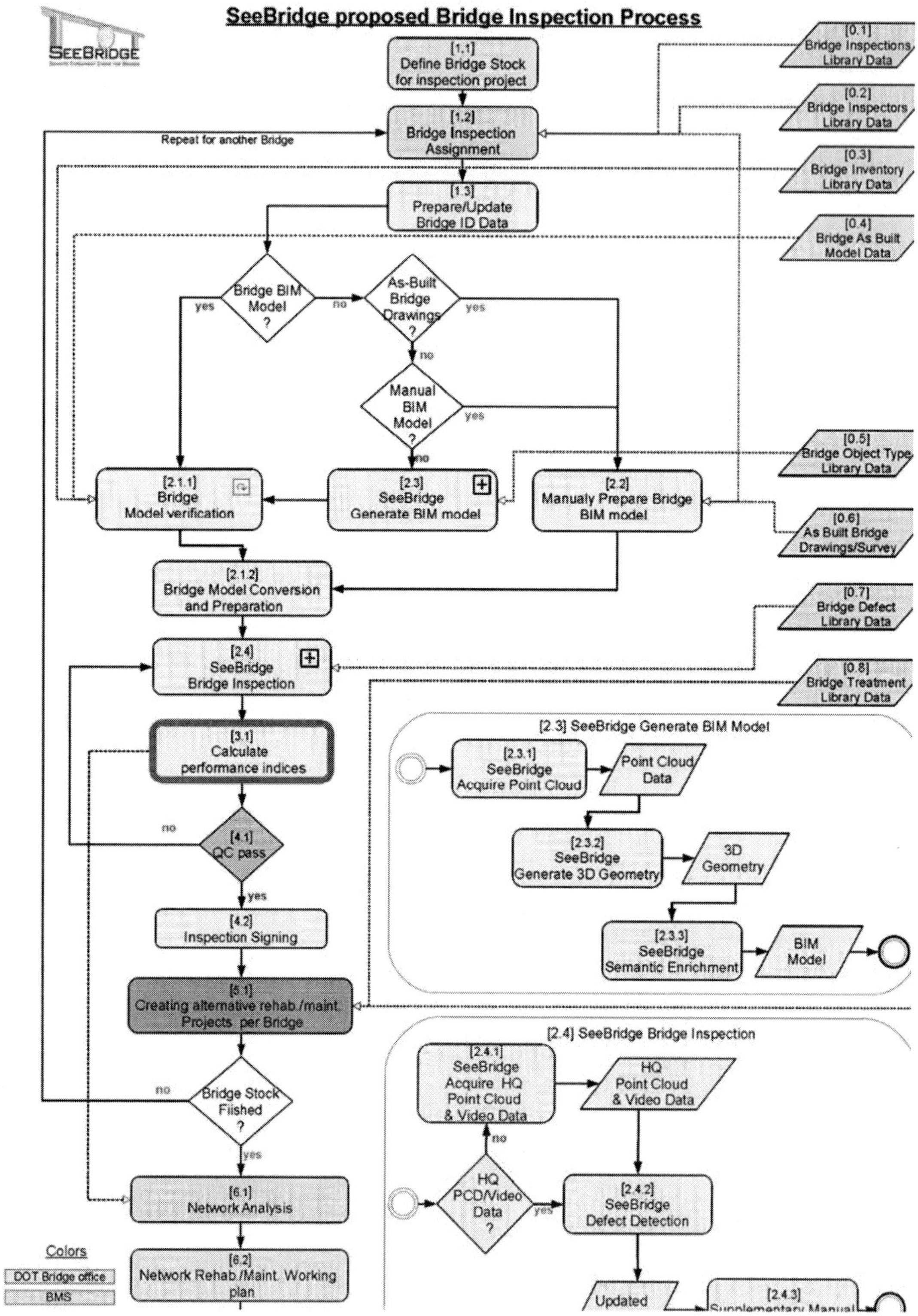

Figure 2 Workflow diagram of proposed SeeBridge Bridge Inspection process.

The workflow of the SeeBridge system is shown in Figure 2 (on the previous page). Incorporating the suggested SeeBridge technical components into an existing bridge inspection and management process should be done with great care as the impact on the existing workflow and on the way the BMS is used to manage the bridge stock may be significant. One of the major changes is the introduction of a BIM model as a database for the bridge inspection and management process. There are three options/situations for incorporating BIM models into the process:

- Using the 'as-built' BIM models of bridges if and where they exist.
- Automatic creation of 'as-is' BIM models of bridges using the SeeBridge technical components numbered 1-3 above (activities 2.3.1, 2.3.2, and 2.3.3 in Figure 2).
- Preparation of 'as-built' BIM models of bridges manually based on drawings.

The second option is the major solution that SeeBridge provides, since most of the existing BMS have not incorporated BIM models. The SeeBridge solution of this aspect should greatly reduce the effort and costs required for BIM model integration into the BMS.

A detailed SeeBridge process map was developed in the IDM using Business Process Modelling Notation (BPMN), which defines the information exchange, including Non Model Exchanges (NME) and BIM Exchange Models (EM), between the activities. Part of the process map is shown in Figure 3.

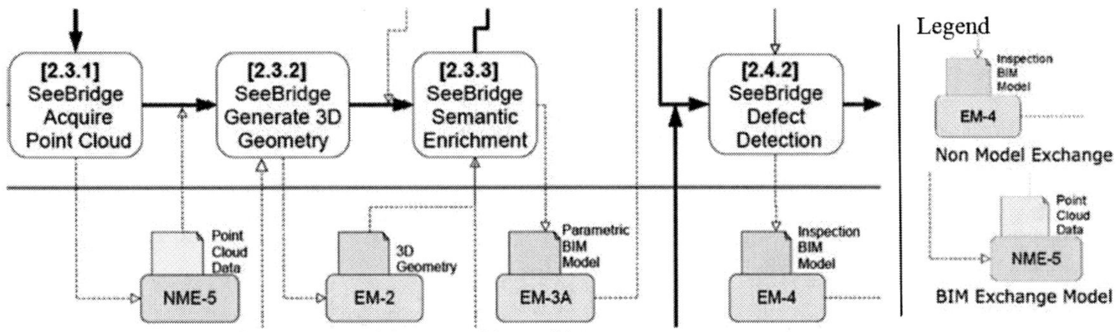

Figure 3 Part of the SeeBridge Bridge Management Process Map

3 Activities and Information Exchange in SeeBridge

The four major activities (technical components) in the SeeBridge system are advanced in the area of survey technology, computer vision, information interoperability and modelling of bridges defects.

3.1 Remote Sensing Technology

The use of these technologies for capture of existing structures is the topic of much research [3, 4]. In activity 2.3.1 shown in Figure 2, the bridge inspector, depending on the bridge type and inspection criteria, selects a proper 3D scanning approach. The options are terrestrial/mobile laser scanning and video/photogrammetry.

In case of laser scanning, the inspector evaluates the site and designs the laser scanning set-points so that they collectively cover the entire bridge structure. The laser scanner is then set at every set-point and a 3D point cloud is captured at each set-point. The individual point clouds are then registered to each other using automated software or manually.

In case of video/photogrammetry, the inspector selects a proper camera resolution based on the project criteria, distance of the camera to the bridge surfaces, and required point cloud resolution. Once the camera is selected, the inspector captures video or takes photographs from the bridge. The important point here is to cover every surface of the bridge from multiple viewpoints. The video or photographs are then input to the processing software. The software automatically estimates camera parameters and trajectory which will lead to the generation of a dense point cloud data (PCD), i.e. the NME-5, as the input of the 2.3.2 activity (as shown in Figure 3).

3.2 Reconstruction of 3D Model from PCD

Current practice for the generation of as-built models from PCD involves manual conversion through user-guided specification of components combined with automated fitting of the components to specified subsets of the point cloud data. In activity 2.3.2 in the SeeBridge process (as shown in Figure 2), the 3D geometry generation engine processes the PCD created in 2.3.1 and generates a geometric model of the infrastructure associated to the PCD. The engine segments the main bridge components by matching the data with a repository of predefined bridge element shapes defined in the IDM. The techniques used employ a surface primitive extraction algorithm and a component detection and classification algorithm. As the detection and classification is based on machine learning, training data is required for learning the proper relationships between surface primitives and integrated components.

Most of the bridge components can be modelled using extruded, prismatic solid shape representations, while others require a BREP approach. To support component detection of extruded area solid elements, a comprehensive set of parametric cross-sections were defined in the IDM, including all of the typical concrete box, double T and girder sections. An example of the SeeBridge Generic Girder Parametric Cross-Section is shown in Figure 4. The parameters are specified in Table 1.

The output of this activity (2.3.2) is a simplified building information model of the sensed bridge with the main bridge components identified and modelled, but with no relationships or other information. Elements that are occluded or that are too small to be discerned due to insufficient scan resolution are not provided. The level of detail satisfies or is superior to LoD 300, but is inferior to LoD 400 [14]. The data format of the output model will be an IFC or equivalent BIM model file with the component objects and their full geometry (defined as EM-2 in Figure 3).

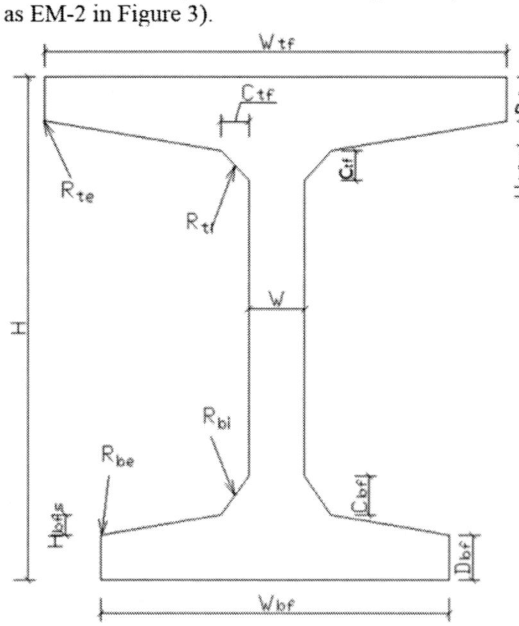

Figure 4 SeeBridge generic girder parametric cross-section

Table 1 Definition of parameters for generic girder parametric cross-section

Parameter	Label	Notes
Height	H	
Top flange depth	D_{tf}	
Top flange slope height	H_{tfs}	
Bottom flange slope height	H_{bfs}	
Top flange chamfer	C_{tf}	Chamfers are all 45°
Bottom flange chamfer	C_{bf}	
Bottom flange depth	D_{bf}	
Top flange width	W_{tf}	
Bottom flange width	W_{bf}	
Web width	W	
Top flange inner filet radius	R_{ti}	These values are only relevant for a small group of bulb tees (e.g. North East and for California bulb tees).
Top flange edge filet radius	R_{te}	
Bottom flange inner filet radius	R_{bi}	
Bottom flange edge filet radius	R_{be}	

3.3 Semantic Enrichment of the 3D Model

In activity 2.3.3, the semantic enrichment engine parses the 3D model and extracts the geometric, topologic and functional characteristics from the model. It then progressively creates, updates or deletes semantically rich model entities (including tangible objects, virtual aggregation containers and objectified relationships of them) following a chain of predefined rules. The rule sets capture the knowledge of bridge engineers concerning the characteristics of the 3D model objects that represent bridge components, including their geometric features (e.g., the parametric cross-sections), their occurrence and the topological and other relationships among them. The general approach to semantic enrichment follows that derived by Belsky et al. [15, 16].

The information derived is structured in the IDM. Some of the examples are shown in Table 2 and Table 3. The output of this activity is a bridge "Pre-Inspection BIM Model" (EM-3A in Figure 3), usually in IFC format, with explicit geometry representation and property sets in a verified LoD similar to LoD 350, but the data must represent 'as-is' conditions (in the same sense as LoD 500 calls for a 'field-verified' model).

Table 2 Part of the IDM Table of Bridge Elements and Occurrence

Element Type		Primary Girders	Slab	Box	Transverse Beam/Diaphragm
Bridge type	**Description**				
Concrete Beam/Girder Bridges	At/Below deck surface	+			+
	Box Girder (exterior & interior)			+	+
Steel Beam/Girder Composite Bridges	At/Below deck surface	+			+
Slab Bridges	Monolithic Slab Bridges	+			

Note:
+ means that this element type always exists in this type of bridge

Table 3 Part of the IDM Table of Spatial Relationships between Elements

Element description		Primary Girders	Box (Box girder)	Slab	Transverse Beam/Diaphragm
Deck/Superstructure	Primary Girders	■			E
	Box (Box girder)		■		E
	Slab			■	
	Transverse Beam/Diaphragm	E	E		■
	Deck Slab - (Concrete Slab)	E	E		P

Note:
E = Exists: normally the elements are in physical contact
P = Possible: the elements may or may not be in physical contact

3.4 Bridge Defects Modeling

A pre-process activity of the damage detection (2.4.2 activity in Figure 3) process is to enable all the elements in the BIM model generated from 2.3.3, i.e., EM-3A, to have boundary shape representation (BREP), because it is much easier to represent defects on the bridge surface when using BREP, which is a composite of faces. Any bridge elements that were only modelled using solid extrusions and CSG in EM-3A maintain both their original representations and BREP in the resulting model - EM-3B. The objects also have high resolution imagery registered with them at this stage (note that EM-3B is not shown in Figure 3 due to space limitations).

The damage detection algorithm (activity 2.4.2 in Figure 3) iterates over every BIM element in EM-3B and analyses the imagery, shape and function in the structure. First, imagery is used solely to localize visually detectable damage groups. Subsequently, these findings are further refined to a specific damage type (structural crack, non-structural crack, spalling, scaling, efflorescence, corrosion, other) using additional extracted properties such as element type, damage position and damage location. The defects' types and possible occurrence in bridge elements are listed in bridge defect occurrence tables that are compiled in the IDM; some examples are shown in Table 4.

Meaningful damage parameters (damage type, absolute and relative size measurements, etc.) are extracted from the findings and embedded into the BIM model. The result is an 'Inspection BIM Model' (EM-4) with defect data attached and located on bridge component surfaces.

The 'Inspection BIM Model' enables automatic calculation of performance indicators of the bridges and automatic classification of the defects based on the defect classification tables, which are compiled in the IDM according to the DOTs/Highway Authorities' regulations. An example of severity levels is shown in Table 5.

4 Conclusion

The proposed IDM establishes the professional knowledge basis of the domain of highway bridges in order to ensure the correct development of the technical components in the proposed SeeBridge system. It specifies the data collection process; details the activities for 3D model reconstruction and the geometric shape representations needed; presents the process of semantic enrichment and the required structured knowledge; and it specifies the defect identification and modeling activities and the defect classifications that facilitate the process.

The proposed IDM was developed and validated with a network of domain experts representing highway departments and DOT's in four countries. It captures general data exchange scenarios relevant to the bridge inspection process in the SeeBridge system, as well as country-specific aspects. It also forms a sound basis for the development of a Model View Definition (MVD), which can be used as an evaluation tool to rigorously validate the comprehensiveness of the bridge information instance models generated when the SeeBridge process is used in the future.

Development of the MVD will also enable review of the currently proposed IfcBridge [17] data model extension. Specification of any new entities, relationships or properties that may be found lacking in the IFC schema will depend heavily on the IDM for their content. For example, there is currently no accepted, consistent or thorough way to represent the defects that may occur in bridges. Definitions for objects that represent defects, defect patches and similar objects will need to be added to the IFC schema.

Use of the IDM approach to modeling the data exchanges has proven to be an effective way of establishing a common basis for the activities of the different research teams engaged in the SeeBridge project. In addition to providing the basis for a Model View Definition (MVD), its development forced the researchers to rigorously confront and solve a range of issues concerning questions of data modeling and the coherence of the process as a whole. As such, the IDM is a central component for R&D of this type.

Table 4 Part of the Bridge Defect Occurrence Table in the IDM

	Defect Group	02 Reinforced & Prestressed Concrete				
				Cracks in reinforced concrete		
	Defect Description	Spalls	Delamination	Cracks likely to affect the stability of the element/ structure	Cracks which do not affect the stability of the element/structure	Cracks in prestressed concrete
Deck/Superstructure	Primary Girders (Concrete Beam/Girders)	+	+	+	+	+
	Primary Girders (Steel Beam/Girders)					
	Box (Box girder)	+	+	+	+	+
	Slab	+	+	+	+	+
	Secondary Deck element - Transverse Beam/Diaphragm	+	+	+	+	+
	Deck Slab (Concrete Beam/Girders, Box Girder, Composite)	+	+	+	+	+

Note: + means normally this type of defect may be identified in this element

Table 5 Part of the Defects Classification Table in the IDM

02 Reinforced & Prestressed Concrete					
	Severity				
Defect	**1**	**2**	**3**	**4**	**5**
Spalls	No spalling	Slight, but clear, local spalling. Partial exposure of the outer reinforcement layer (stirrups in beams, external reinforcement in slabs) usually accompanied by signs of corrosion	Large, discrete spalls, exposing the cross-section of the shear stirrups and/or longitudinal reinforcing bars. Usually accompanied by general corrosion of the exposed bars, with possible local reduction in cross-section of longitudinal bars	Delamination in regions of low bending or shear, with no influence on the stability of the element	The element is no longer structurally functional, as a result of developments described under "Degree of severity 4"

Acknowledgement

This project has received funding from the European Union's Seventh Framework Programme for research, technological development and demonstration under grant agreement no. 31109806.0007. SeeBridge is co-funded by Funding Partners of the ERA-NET Plus Infravation and the European Commission. The Funding Partners of the Infravation 2014 Call are: Ministerie van Infrastructuur en Milieu, Rijkswaterstaat, Bundesministerium für Verkehr, Bau und Stadtentwicklung, Danish Road Directorate, Statens Vegvesen Vegdirektoratet, Trafikverket – Trv, Vegagerðin, Ministere de L'ecologie, du Developpement Durable et de L'energie, Centro para el Desarrollo Tecnologico Industrial, Anas S.P.A., Netivei Israel – National Transport Infrastructure Company Ltd. and Federal Highway Administration USDOT.

References

[1] Jaselskis, E.J., Z. Gao, and R.C. Walters, Improving transportation projects using laser scanning. *Journal of Construction Engineering and Management*, 2005. **131**(3): p. 377-384.

[2] Fuchs, P., G. Washer, S. Chase, et al., Applications of laser-based instrumentation for highway bridges. *Journal of Bridge Engineering*, 2004. **9**(6): p. 541-549.

[3] Tang, P.B. and B. Akinci, Formalization of workflows for extracting bridge surveying goals from laser-scanned data. *Automation in Construction*, 2012. **22**: p. 306-319.

[4] Endsley, A., C. Brooks, D. Harris, et al. Decision support system for integrating remote sensing in bridge condition assessment and preservation. in *SPIE Smart Structures and Materials+ Nondestructive Evaluation and Health Monitoring*. 2012. International Society for Optics and Photonics.

[5] Tang, P., *Extraction of surveying goals from point clouds obtained from laser scanners to support bridge inspection*. 2009, Carnegie Mellon University.

[6] Tang, P., B. Akinci, and J.H. Garrett. Laser scanning for bridge inspection and management. in *IABSE Symposium Report*. 2007. International Association for Bridge and Structural Engineering.

[7] Kedar, A., R. Sacks, L. Ma, et al., *SeeBridge Information Delivery Manual*. 2016.

[8] NIBS, *US National Building Information Model Standard, Version 2.0,* National Institute of Building Sciences. 2012.

[9] Venugopal, M., C.M. Eastman, R. Sacks, et al., Semantics of model views for information exchanges using the industry foundation class schema. *Advanced Engineering Informatics*, 2012. **26**(2): p. 411-428.

[10] Sacks, R., I. Kaner, C.M. Eastman, et al., The Rosewood experiment — Building information modeling and interoperability for architectural precast facades. *Automation in Construction*, 2010. **19**(4): p. 419-432.

[11] Eastman, C.M., R. Sacks, Y.-S. Jeong, et al., *Building Information Modeling (BIM) for Precast Concrete*. 2007, National Institute of Building Sciences: Washington DC, USA. p. 212.

[12] Jeong, Y.S., C.M. Eastman, R. Sacks, et al., Benchmark tests for BIM data exchanges of precast concrete. *Automation in Construction*, 2009. **18**(4): p. 469-484.

[13] BuildingSmart. *Industry Foundation Classes Release 4 (IFC4)*. 2013 2013.

[14] Bedrick, J., A Level of Development Specification for BIM Processes. *AECbytes Viewpoint*, 2013. **68**.

[15] Belsky, M., R. Sacks, and I. Brilakis, Semantic Enrichment for Building Information Modeling. *Computer-Aided Civil and Infrastructure Engineering*, 2015. **31**(4): p. 261-274.

[16] Belsky, M., R. Sacks, and I. Brilakis. A framework for semantic enrichment of ifc building models. in *Proceedings of the 30th CIB W78 International Conference*. 2013. Beijing, China.

[17] Lebegue, E., J. Gual, G. Arthaud, et al., *IFC-Bridge V2 data model*. 2007, Technical Report Edition.

Development of the Scenario-based Evaluation of Indoor Circulation using BIM-enabled Parameters

Jaeyoung Shin [a], Jin-Kook Lee [a] and Jinhua Huang [a]

[a]Dept of Interior Architecture Design, University of Hanyang, Republic of Korea
E-mail: jjyoung311@gmail.com, designit@hanyang.ac.kr, sara920319@gmail.com

Abstract –

This paper aims to develop the scenario-based analysis of indoor circulation using BIM-enabled parameterized quantities of a circulation path computed by BIM-enabled applications such as walking distance, spatial topology, number of turns, etc. A building circulation plays an important factor to plan overall spatial configuration based on specific purposes so that it has been one of priorities even in the beginning of the design. As supported by Building Information Modeling (BIM) and its applications, a design review process has been conducted more efficiently and accurately by using computed quantitative circulation-related data, compared to the conventional process performed in manual. However, qualitative factors of building circulation such as efficient route have been excluded practically as a target of the BIM-based design evaluation.

As motivated by such fundamentals, precedent research discusses design evaluation approach in a BIM environment focusing on qualitative circulation issues such as 'convenient access to a specific space'. Based on the research, this paper develop the scenario based evaluation framework in terms of a generalization of the evaluation process. Furthermore, the framework is demonstrated using a test model with specific scenario and circulation evaluation-purposed application implemented by the authors.

Keywords –

Building Information Modeling (BIM); Indoor circulation; Scenario-based evaluation; Parameterization

1 Background

As supported by Building Information Modeling (BIM), design review process has been conducted more efficiently and accurately by using rich building information compared to the conventional process performed in manual [1, 2]. The BIM-enabled applications have computed series of explicit quantity data, especially for indoor circulation such as metric distance, number of turns, spatial depth, etc [3, 4, 5]. This enables us to measure the explicit egress distance between specific spaces and exits, and establish quantitative indicators based on those data in order to review various requirements and support design decision [3, 6, 7].

In terms of BIM application for indoor circulation analysis, there have been various approaches, including circulation graph representation, rule compliance checking, domain-specific language development for reviewing rules, circulation related data application for design review and so on. Lee, J. K. et al. [6] defined a computational method named as universal circulation network (UCN) to measure explicit walking distance and visualize the circulation path using geometry and spatial topology on a given BIM model. The UCN is implemented as specific software to review circulation-related rules and has been utilized for practical use. Lee, J. M. [8] represented logical, automated checking process of circulation associated rules considering various design development stage from preliminary concept design to the final concept design stage. In addition, the generic method was applied to develop related application system and validated by applying actual courthouse models. As BIM domain-specific programming language, the Building Environment Rule and Analysis (BERA) was implemented for the purpose of review of building model focusing on spatial program and building circulation analysis [3, 9]. With using BERA language, varied building objects including spaces and paths with their properties can be handled by user-defined query. Its applications have presented its utility to enable users to define customized rule and review circulation design based on the user-defined rules. Lee, H. et al. [5, 10] utilized BERA language to derive useful circulation property data and calculated results achieved from BIM models so as to analyze design alternatives of actual remodeling building. Lee, Y. C. et al. [11] extended BERA language for checking accessibility and visibility conditions of spatial requirements. The algorithm for checking accessibility and visibility analyzes the number of turns to reach the end point from the start point. Lee, H. [12] established path object with its related properties and expanded properties which is called as Numeric Data of Building Circulation (NDBC). The path objects with its properties are defined as a BIM-enabled and building-oriented approach in order to resolve building circulation issues.

Current BIM-enabled design analysis and evaluation approaches have handled circulation factors partially that are expressed in explicit way using as-is data. In other words, qualitative factors such as efficient or accessible route have been excluded practically as a target of the BIM-based design evaluation [4, 13]. However, large portion of circulation design requirements described in the RFP or design guidelines (except for building regulatory codes) consist of qualitative factors rather than explicitly-defined ones and those factors influence pedestrian circulation quality from various perspectives [13, 15]. As motivated by such fundamentals, precedent research [14] discussed design evaluation approach in the BIM environment focusing on qualitative circulation issues. To develop a circulation evaluation approach, following strategies was introduced; 1) As-is data derived from BIM model is fundamental sources for definition of evaluation factors. 2) Those data are weighted in accordance with their relative priorities considering design intention.

This paper extends and demonstrates the proposed circulation evaluation framework in terms of a generalization of the evaluation process. In addition, related application is implemented for applying the framework easily by users.

2 Research objective and scope

This research aims to deal with BIM-based approach to circulation evaluation considering qualitative circulation factors. With its objective the scope and process of the research are summarized as follows.

1. Representation of qualitative circulation factors with quantitative data: This part is to summarize developed classification and definition of circulation quality-associated factors that were tackled in precedent research [14]. Those factors are represented with related circulation quantities derived from BIM models.
2. Development of a circulation evaluation framework: The suggested framework is based on the parameterized circulation data that define circulation factors. The framework is composed of three parts and results in final calculated value which is called as PPV (Parameterized Path Value).
3. Implementation of circulation evaluation application: External database and its application software were implemented for demonstration purpose. Using the application circulation evaluation framework can be easily managed.
4. Demonstration of circulation evaluation using the application: In order to verify applicability of the suggested framework case study was conducted by using conceptual office model based on a specific scenario and the implemented application.

3 Representation of Indoor Circulation factors with Quantity Data

In order to define circulation evaluation factors, this paper mainly focuses on BIM-enabled circulation related quantities that have been recognized as quantitative and objective grounds for analyzing circulation design[3, 5, 7, 10, 11, 12, 13, 16]. The scope of circulation factors is defined referring to previous studies in circulation-related design aspects from a pedestrian point of view [16, 17, 18, 19, 20, 21, 22, 23, 24]. They are categorized into 4 types and each factor type is composed of two or three circulation-related quantity data.

3.1 Relative Distance

One of the most effective and fundamental evaluation methods of pedestrian circulation is generally based on metric distance. In general, vertical circulation paths increase burden to move rather than horizontal circulation from the perspective of pedestrians. Especially for people with physical disabilities, circulation with less vertical distance rather than circulation with short distance can be a serious issue. In this regard, relative distance composed of horizontal and vertical distance is a main circulation factor depending on design purposes, etc.

The first type of circulation factors and its related BIM quantities are defined as:

$$D_i = \{VD_i, HD_i\} \quad (1)$$

Where D_i connotes relative distance of P_i, ith path and *VD*, *HD* are vertical distance and horizontal distance each.

3.2 Accessibility of Circulation path

As quantitative measures for assessing the accessibility of circulation, condition of intermediate spaces regarding connectivity is focused in this paper. There is two representation way in this paper; the number of intermediate spaces between start and end spaces and directly accessible spaces to those spaces. In terms of planning circulation with easy access, the number of intermediate spaces has a negative effect on accessibility and the (average) number of directly accessible space of intermediate space, however, would have a positive effect.

The second type of circulation factors and its related BIM quantities are defined as:

$$A_i = \{IS_i, AS_i\} \quad (2)$$

Where A_{ccess} connotes accessibility of P_i, ith path and *IS*, *AS* are the number of intermediate spaces and the average number of directly-accessible spaces in order.

3.3 Simplicity of Circulation path

Influence of circulation pattern has been acknowledged through several researches such as those in the domain of space syntax [21, 23, 24]. Accepting the idea of turning conditions which derive circulation pattern, simplicity of circulation is represented with two elements; the number of turns and the angle of each turn

There are three path objects, path A, B and C instantiated after assigning start and end spaces as described in Figure 1. They each feature different turning conditions;1) path A has no turn, 2) path B has a turn with a small degree of variation and 3) path C has a turn with a large degree of variation. When the three paths have same metric distance, path A is regarded as the most simple and easy path among them and followed by path B.

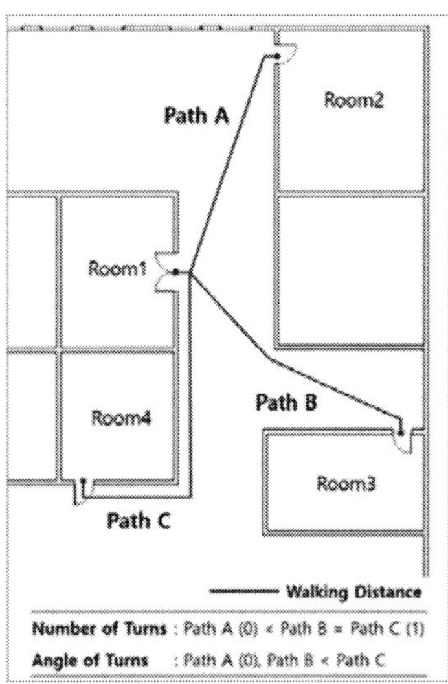

Figure 1. Example of two kinds of BIM data related to clarity of the path; the number of turns and angle of each turn

In this regard, the third type of circulation factors and its related BIM quantities are defined as:

$$C_i = \{NT_i, AT_i\} \quad (3)$$

Where C_i connotes simplicity of circulation of P_i, ith path and *NT*, *AT* are the number of turn and the angle of turn in order.

3.4 Pedestrian friendly path environment

Among the various circulation factors for pedestrians, path environment such as sunlight exposure, spatial scale and surrounding scenery can affect people in building physiologically or physically [12, 17, 18, 19]. In this paper, circulation factor regarding pedestrian friendly path environment has 3 scenarios; 1) window area, 2) passing space volume, 3) ratio of specific space usage such as public spaces and recreational spaces.

The Figure 2 shows two alternatives, path A and path B, which have same start space (Room1) and end space (Room2). Even though path A has short metric distance and no turn, path B is preferable when taking into account comfortable and spacious circulation space with little traffic.

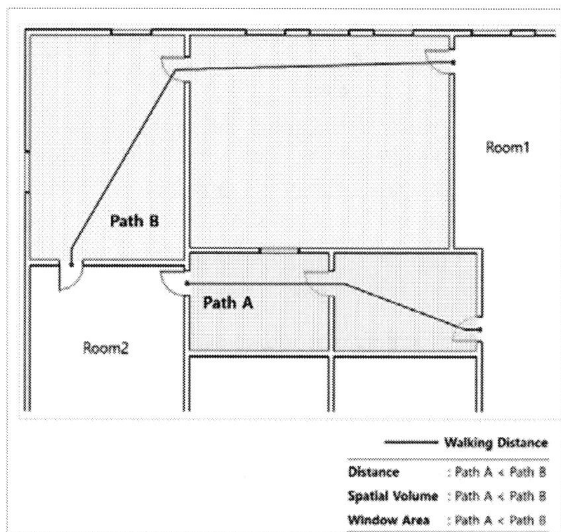

Figure 2. Example of three kinds of BIM data related with pedestrian friendly path; window area, spatial volume, ratio of space usage

The fourth type of circulation factors and its related BIM quantities are defined as:

$$U_i = \{WA_i, VM_i, UG_i\} \quad (4)$$

Where U_i connotes pedestrian friendly path environment of P_i, ith path and *WA*, *VM*, *UG* are window area, spatial volume and ratio of specific space usage in order.

4 BIM-enabled Circulation Evaluation Framework using Parameterized Quantitative Data

4.1 Path generation

The BIM-enabled evaluation process begins with path generation from a BIM model. Unlike static building models such as space object and structural object, path object is dynamically generated by two types of spaces which are the start and end space. Therefore, definition of start and end space (at least one of each) is prerequisite for generating paths in BIM environment. After creating path objects, circulation related quantitative data introduced in the previous section can be obtained from circulation graph-driven path properties such as metric distance and turn-related properties and spatial or architectural object properties which is related the path objects.

4.2 Parameterization of Circulation Data

The driven quantitative data representing four types of pedestrian circulation factors are raw data that play a role of objective source to compare design alternatives and can be used for various applications. In this paper, parameterized circulation data are suggested for design evaluation as application of circulation-related quantitative data. The parameterized circulation data have several concepts as follows.

1. The quantitative circulation data with different unit of measure can be compared after standardization process. When standardizing them, circulation data can be distinguished between positive factors and negative factors of circulation quality.
2. Weighting function has two types including weighting each circulation data within a specific factor and weighting among circulation factors.
3. Weight (*W(X)*, *W(PF)*) implies a relative degree of importance of each factor and total weights is 1 (100%).
4. Weighting of circulation factors depends on users who determine two types of weights individually.
5. The result value (PPV) which is calculated from the circulation data and their assigned weights is the representative value of the individual path object for evaluation.

4.3 Derivation of Parameterized Path Value

On the basis of definition of circulation factors, their component (quantitative data) and two types of weight, parameterized path value derivation process is summarized as follows. The final result value is called as PPV(Parameterized Path Value) and it implies that the path with higher PPV indicates highly evaluated path than others.

- **Definition 1. - 5.**

1. $P = \{P_i | P_i$ is derived path objects from a building model, $i \leq n; n$ is the number of path objects $\}$
2. $PF_i = \{PF_i \, | PF_i$ is a circulation factor of path $Pi\}$ $= \{D_i, A_i, C_i, U_i\}$
3. $X_i = \{X_i \, | X_i$ is quantitative data consisting of $PF_i\}$ $= \{VD_i, HD_i, IS_i, AS_i, NT_i, AT_i, WA_i, VM_i, UG_i\}$
4. $W_X = \{W(X_i) \mid W(X_i)$ is a weight given to X_i, $0 \leq W(X_i) \leq 1\}$
5. $W_{PF} = \{W(PF_i) \mid W(PF_i)$ is a weight of PF_i, $0 \leq W(W_{PF}) \leq 1\}$

- **Process 1.** Standardization of the circulation data [25]

(Positive Factor)

$$(S(X_i)) = \frac{X_i - Xmin}{Xmax - Xmin}$$

(Negative Factor)

$$(S(X_i)) = \frac{Xmax - X_i}{Xmax - Xmin}$$

X_{min}, X_{max} : Minimum/ maximum value among overall path objects

- **Process 2.** Parameterization of path data

  $Prm(PF_i) = \Sigma \, S(X_i) \cdot W(X_i)$

- **Process 3.** Derivation of PPV

  $PPV_i = 1000 \cdot \Sigma \, Prm(PF_i) \cdot W(PF_i)$

5 Implementation of Circulation Evaluation Application

For demonstration of suggested circulation evaluation an external database and application that is connected with database sources are implemented. This application is developed to support scenario-based circulation evaluation with consideration for priority of circulation factor. With the application, user can 1) review the overall path objects and their quantitative data needed for evaluation, 2) set user-defined weights of each circulation factors considering the purpose of design, and 3) evaluate the target paths based on PPV analysis. Figure

3 represents examples of user interface for controlling weight.

In order to support its functions, the application is operated based on an external database of overall circulation quantity data computed by BIM-enabled calculators in advance. Based on the circulation data established in the database, weights assigned to each circulation data and factor are determined by users. Applying user-defined weights with the circulation data that is the process of parameterization of circulation data, calculation for the PPV is done in the back end side of the software. Figure 4 describes the link between the external database for the model information and the application.

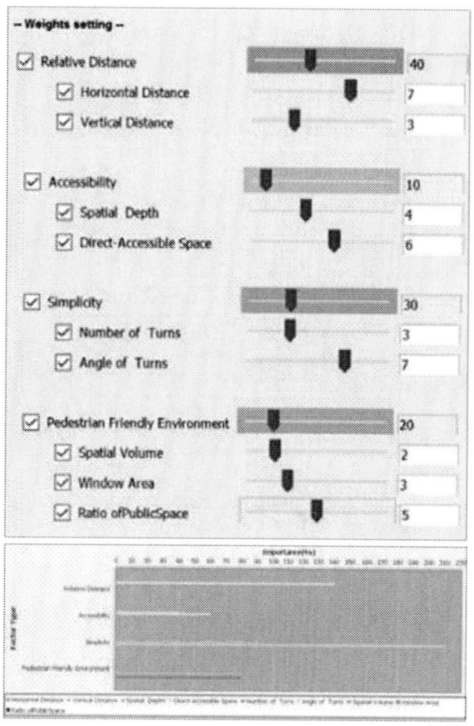

Figure 3. Examples of user interface for controlling weights of the application

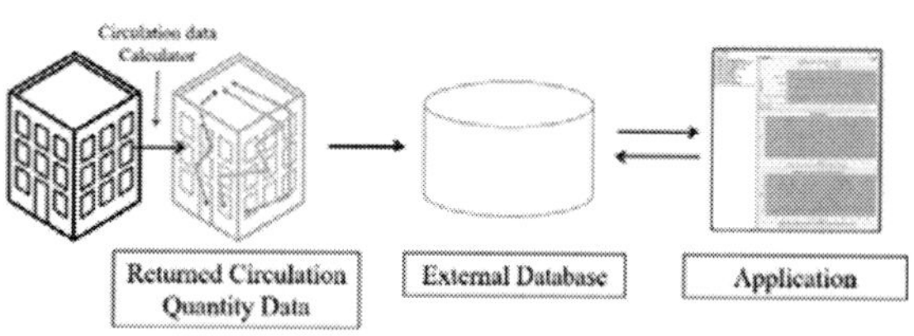

Figure 4. Establishing an external database and its application for circulation evaluation

6 Demonstration

6.1 Scenario of Circulation Evaluation

In this paper, a conceptual office building model which can be modelled in the early phase of design is used to demonstrate design evaluation. The target path objects includes 13 types of paths from different office rooms to a conference room which derived from a given BIM model as represented in Figure 5.

For planning office space, circulation is desirable to be designed in terms of efficient and reasonable spatial allocation and moreover induce workers to communicate formally or informally. In general, informal conversations in work place occur, usually involving people who are physically in close proximity to each other [20]. In terms of circulation, supporting informal communication between workers, this case study focuses two circulation data, the number of directly-accessible spaces of intermediate spaces and ratio of public space to overall passing spaces, as main factors for evaluation. According to its evaluation purpose, weights are assigned to each corresponding factor and data as described in Table 1.

Table 1. Assigning weights to each path objects

Circulation factor	$W(PF_i)$	Circulation data	$W(X_i)$
Relative Distance	0.2	Horizontal Distance	0.3
		Vertical Distance	0.7
Accessibility	0.3	Spatial depth	0.4
		The number of directly-accessible spaces	0.6
Simplicity	0.1	The number of turns	0.3
		Angle of turns	0.7
Pedestrian-friendly environment	0.4	Spatial volume	0.2
		Window area	0.3
		Ratio of public spaces	0.5

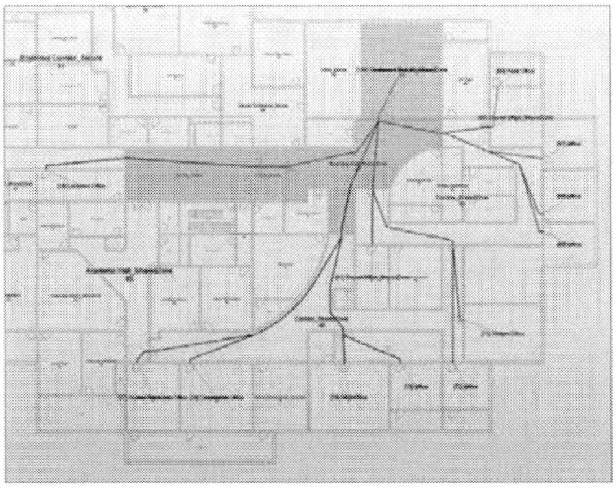

Figure 5. Target 13 path objects for evaluation based on the scenario; yellow (start spaces), red (end space), blue (public spaces)

6.2 Circulation Evaluation Result

Based on the scenario, we conducted circulation evaluation applying the framework through the implemented application. There are two ways of circulation evaluation approaches depending on weighting aspects.

1) Assigning same weight to all circulation factors (that implies circulation factors have the same importance)
2) Assigning different weight to circulation factors (that implies there is a different importance between circulation factors) (Table 1)

Figure 6 describes the result graph generated by the application. The gray-colored graph indicates series of PPV of 13 paths with the same weight and red-colored graph is relevant to PPV of the paths with different weight. WhenPPVs on two graphs are compared, paths with id=2, 10, 13 get highest PPV among alternatives. And most of path rankings stay same, however, 4 paths are changed in priorities. In case of path (id=7) with the distinguishing ranking difference, main reason for change comes from directly accessible spaces of the intermediate spaces and window area. The application reports not only for final evaluation results, but also intermediate result values of each path within the scope of each type of factor before deriving PPV. According to the reporting as shown in Figure 7, paths which are top 3 in PPV (id=2, 10, 13) are above average in almost all factor types.

The scenario-based evaluation approach can be applied diversely to different evaluation cases. For instance, evaluation results of this case study can be expanded to design evaluation case in terms of spatial allocation. By utilizing and comparing PPV and extra analysis information of design alternatives, designers can be supported to find a reasonable location for the conference starting from 13 offices.

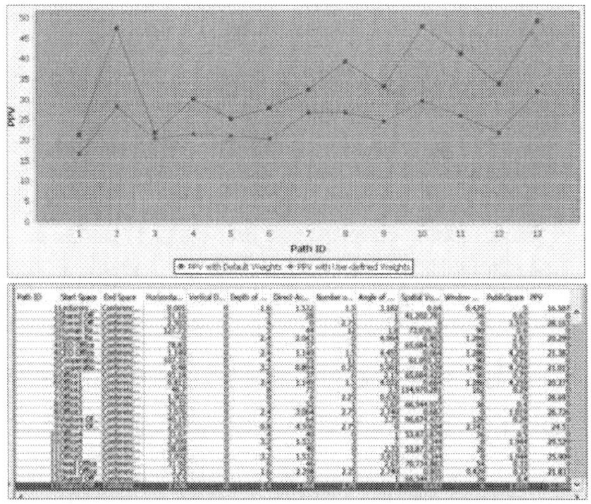

Figure 6. Scenario based circulation evaluation results of target 13 path objects - PPV Graph and Calculation table

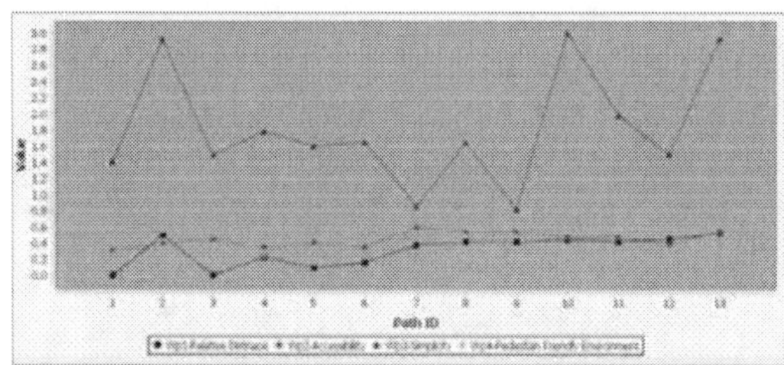

Figure 7. Scenario based circulation evaluation results of target 13 path objects - Intermediate result values graph

7 Conclusion

In terms of design analysis issues, this research aims to study BIM-enabled indoor circulation evaluation approach to dealing with qualitative circulation factors which have been excluded as a target of evaluation in BIM environment. In the early design phase, it is a main task to develop optimal design alternatives considering various design requirements according to building types or usages and building users. Thanks to building design supportive applications facilitated by BIM, explicit information derived from BIM models can play an important role of design analysis in the early design phase. The BIM-enabled design analysis approach has shown its capabilities that evaluate not only explicit factors, but also intrinsic circulation factors by using additional BIM-derived information, including metric distance, number of turns, depth and various quantitative data regarding circulation spaces as well.

This paper focuses on developing scenario-based circulation evaluation framework by applying BIM-enabled circulation data and parameters and demonstrating its applicability based on the idea of parameterizing circulation quantities. Circulation data is mapped with indoor circulation factors for evaluation and parameterized circulation data become adaptable factors depending on evaluation scenarios. The evaluation framework is composed of three parts, 1) path generation 2) parameterization of circulation data and 3) derivation of PPV of each path. The framework is demonstrated using a test model with specific scenario and circulation evaluation-purposed application that enables users to manage the framework easily. The evaluation results can play a role of objective indicators for users to get feedback promptly and effectively in the design development phase.

The suggested framework can be extended as defining additional circulation factors with BIM-enabled parameters. Furthermore, the scenario-based evaluation approach represented in this paper is expected to apply to other design analysis issues such as space program and energy analysis as one of the BIM-enabled evaluation methods.

Acknowledgment

This work was supported by the National Research Foundation of Korea Grant funded by the Korean Government (NRF-2015R1C1A1A01053497)

References

[1] Eastman, C., Eastman, C. M., Teicholz, P., & Sacks, R. *BIM handbook: A guide to building information modeling for owners, managers, designers, engineers and contractors*. John Wiley & Sons, 2011

[2] Eastman, C. Automated assessment of early concept designs. *Architectural Design*, 79(2), 52-57, 2009.

[3] Lee, J. K. *Building Environment Rule and Analysis (BERA) Language: And its Application for Evaluating Building Circulation and Spatial Program*. Ph.D. thesis, College of Architecture, Georgia Institute of Technology, 2011.

[4] Lee, J. K. Need for BIM-enabled Design Quality Assurance from the Early Phase of Building Design. *The BIM*, 8, 25-27, 2012.

[5] Lee, H., Lee, J.K. An approach to the BIM-enabled Building Design Assessment using Quantitative Datasets Derived from Circulation-related Objects & Properties. *Design Convergence Study*, 13(5), 49-66, 2014.

[6] Lee, J. K., Eastman, C. M., Lee, J., Kannala, M., & Jeong, Y. S. Computing walking distances within buildings using the universal circulation network. *Environment and planning. B*, Planning & design, 37(4), 628, 2010.

[7] Solihin, W., & Eastman, C. Classification of rules for automated BIM rule checking development. *Automation in Construction*, 53, 69-82, 2015.

[8] Lee, J. M. *Automated checking of building requirements on circulation over a range of design phases*. Ph.D. thesis, College of Architecture, Georgia Institute of Technology, 2010.

[9] Lee, J. K., Eastman, C. M., and Lee, Y. C. Implementation of a BIM Domain-specific Language for the Building Environment Rule and Analysis. *Journal of Intelligent & Robotic Systems*, 1-16, 2014.

[10] Lee, H., Kim, J., Shin, M., Kim, I. and Lee, J. K. A Demonstration of BIM-enabled Quantitative Circulation Analysis using BERA Language. *Proceedings of the 31st International Symposium on Automation and Robotics in Construction and Minings (IAARC)*, 202-209, 2014.

[11] Lee, Y. C., Eastman, C. M., and Lee, J. K. Automated Rule-Based Checking for the Validation of Accessibility and Visibility of a Building Information Model. *Proceeding, ASCE International workshop on Computing in Civil Engineering*, 572-579, 2015.

[12] Lee, H. *Definition of the Path Object and its Properties for Evaluating Building Circulation Issues using BIM-Enabled Numeric Data*. M.S. thesis, Department of Interior Architecture Design, Graduate School of Hanyang University, 2015.

[13] Uhm, M., Lee, G., Park, Y., Kim, S., Jung, J., and Lee, J.-k., Requirements for computational rule checking of requests for proposals (RFPs) for building designs in South Korea. *Advanced Engineering Informatics*, 29(3), 602-615, 2015.

[14] Shin, J., Lee, H., Lee, S., Park, S., Kim, H. and and Lee, J.K. An Approach to the BIM-enabled Assessment of Building Circulation Using Quantitative Data and Its Weight. *Proceedings of the 32nd International Symposium on Automation and Robotics in Construction and Minings (IAARC)*, 831-837, 2015.

[15] Lee. J. The need for information management system for the Employer Requirements Management. *Architecture & Society*, 18, 150-157, 2009.

[16] Lee, J. K. and Kim, M. J. BIM-enabled Conceptual Modelling and Representation of building circulation. *International Journal of Advanced Robotic Systems*, 2014, 11:X, DOI: 10.5772/58440

[17] Brown, B. B., Werner, C. M., Amburgey, J. W., &

Szalay, C. Walkable route perceptions and physical features converging evidence for en route walking experiences. *Environment and Behavior*, 39(1), 34-61, 2007.

[18] Veitch, J. A., & Galasiu, A. D. The physiological and psychological effects of windows, daylight, and view at home: review and research agenda., 2012.

[19] Mahlum. healthcare design insights- Daylighting. http://www.mahlum.com/pdf/MahlumHDIAutumn2009Issue01.pdf, Accessed: 15/03/2016

[20] Kraut, R. E., Fish, R. S., Root, R. W., & Chalfonte, B. L. Informal communication in organizations: Form, function, and technology. *In Human reactions to technology: Claremont symposium on applied social psychology*, 145-199. 1990.

[21] S. Bafna, Space syntax - a brief introduction to its logic and analytical techniques. Environment and Behavior, 35 (1), 17-29, 2003.

[22] Park, K.H., Lee, S. Y., Shin, J. C. Importance Weight Analysis of Architectural Design Factors Affecting on the Office Valuation. *Seoul Studies*, 12(1), 53-71, 2011.

[23] Alasdair Turner. Angular Analysis. *Proceedings 3rd International Space Syntax Symposium Atlanta* 2001.

[24] Nadia Charalambous, Magda Mavridou. Space Syntax: Spatial Integration Accessibility and Angular Segment Analysis by Metric Distance (ASAMeD). Angela Hull, Cecilia Silva and Luca Bertolini (Eds.) Accessibility Instruments for Planning Practice. COST Office, 57-62.

[25] Ryan Richardson Reeves. *Deriving traverse paths for scientific fieldwork with multicriteria evaluation and path modelling in a geographic information system*. M.S. thesis, the USC graduate school, university of southern California, 2015.

Applying Systems Modeling Approaches to Building Construction

Francisco Valdes

Georgia Tech Research Institute
francisco.valdes@gtri.gatech.edu

Russell Gentry

Georgia Institute of Technology
russell.gentry@coa.gatech.edu

Charles Eastman

Georgia Institute of Technology
charles.eastman@coa.gatech.edu

Stephen Forrest

Maplesoft
sforrest@maplesoft.com

Abstract – Current building modeling environments lack the ability to support coordinated, incremental and systematic description of project requirements, design specifications, and the set of interactions that emerge among them, from different stakeholders, during building design. This paper discusses the prospect of a system-centric modeling framework that integrates a parametric CAD tool with a system modeling application to assess design intent based on construction-specific knowledge. The CAD tool provides strong geometric modeling capabilities, while system modeling allows the description of feature-based design specifications, aligned with construction standards and construction know-how. The ultimate goal of the proposed approach aims for the identification of conflicting interactions between design specifications, from different building material systems and stakeholders, to prevent design errors and on-site rework. The proposed framework will enable collaborative scenarios between Model Based System Engineering and BIM based on parametric, simultaneous, software integration to reduce human-to-data translation errors, improving model consistency among material systems, BIM tools, and project stakeholders.

Keywords –

Building Information Modeling; Model Based Systems Engineering; System Modeling Language; Building Construction; Knowledge-Based Design.

1 Introduction

In building construction, it is common that after the execution of a project, certain stakeholders will not be pleased with the manufacturing accuracy or overall quality of the final product [1]. These situations often occur in buildings due to the construction method was not precise enough, specifications were not properly communicated [3], or, as is often the case, due to the suggested tolerances were unachievable or unrealistic. In addition, design errors arise when design, as documented, reflects the designer's intent, but that intent is flawed due to a lack of information about system implications or due to wrong assumptions about material behavior and assemblies. All these factors produce significant geometric deviations that must be considered and mitigated as part of the construction requirements and specifications development stage.

Likewise, coordination issues among modeling

environments, construction specifications, and knowledge are important sources of geometric variability [4]. These issues are evident when assembling multi-material systems, produced off site, by different vendors with different design workflows. In current practice, the individual material vendors apply their own tolerance specifications during shop-drawing production, based on their tacit knowledge of their own manufacturing and erection processes. Then, these individual models are not re-instantiated into an integrated CAD model to assess the effectiveness of the multi-material tolerance strategy. Therefore, disputes about tolerances incompatibility arise later during erection stages. Also, in any of these cases, as-built geometric deviations obtained in construction are usually much larger than commonly expected [5]. No matter what the reason is, these issues continue to spread in construction projects and contributes to cost and schedule growth due to design changes and errors [6] [7] [4].

Similarly, another common example of geometric variability occurs when designers make late changes to reduce construction costs associated with some building component (e.g. to replace welding in steel connections of a roof structure with bolted connections). While a modification may satisfy the specific construction requirement goal (e.g., reduce installation time), the systems-level implications and long-term side effects are usually not well understood (e.g. bolted connections may allow more movement at the joints, increasing deflection, leading to poor rain drainage, leakage, corrosion, and air infiltrations) and even if the problem is identified qualitatively, there exists no modeling framework in which to assess the implications of the problem quantitatively. Also, for the most common design-bid-build project delivery system, the team includes design professionals, a construction manager or general contractor, and many subcontractors [8]. In the early part of the project, the design team is primary – but in the later stages the general contractor assumes primacy. And so in this case, the responsibility for addressing specifications incompatibility issues is often not clearly defined. In managed contractual systems in which the construction manager does not self-perform the work, field personnel may not be familiar with the manufacturing specifications of the project, and they are also less likely to anticipate tolerance incompatibility problems [9]. In order to reduce these issues, BIM tools are required to represent a building at a whole-system level, capturing the functional and behavioral relationships that span across different domains, material systems, and lifecycle stages. It is in modeling these relationships that the identification of conflicts among design specifications can be facilitated. While early multidisciplinary integration and constant coordination efforts under a BIM-augmented workflow are certainly important means to reduce geometric variability problems [10] [11], they are not sufficient. Current tools and methodologies lack the ability to support coordinated, incremental, and systematic description of design specifications and the set of interactions that emerge across them during building lifecycle.

In contrast, during the past decades, aerospace engineering has largely improved its ability to managing geometric variability and manufacturing-specific knowledge [12]. The aerospace industry has taken advantage of modern computer-aided manufacturing technology to integrate CAD tools with manufacturing processes through a Systems Engineering (SE) approach. It is SE that allows linking together all the disparate elements of a product design into an intelligent product model, which can be continuously validated over its lifecycle. This is the key to enabling true model-based development [13]. Thus, the expansion of SE has facilitated the development of model-centric architectures, and the ability to integrate numerous domain-specific tools. This research advises how the tools promulgated by SE can generate new collaborative environments to facilitate the process of knowledge allocation in construction models. To achieve this goal, this research proposes a system-centric modeling framework that integrates a parametric CAD tool with a MBSE modeling platform. The CAD tool provides robust geometric modeling capabilities, while MBSE allows the modeling of building specifications from a system-level standpoint. Consequently, the identification of system interactions between construction specifications from different domain-specific tools and stakeholders is based on this CAD-MBSE integration.

The present research, which corresponds to number 2 of the following list, belongs to a wider area or investigation to develop an knowledge-integrated modeling framework for construction. Other parallel tracks of this study that will be part of complementary publications include:

1. Review of tolerances and geometric deviations in construction and engineering,
2. Study of the likelihood of using a MBSE approach to model and store reusable manufacturing knowledge and design specifications for construction,
3. Proposal of a model integration and model consistency approach among MBSE models, mathematical engines and BIM (CAD) models, and
4. Development and computational implementation of a system-level tolerances modeling and allocation based on a MBSE approach.

The expected general contribution of this research intends to establish a general MBSE approach for the representations of construction knowledge and tolerances interaction across building sub-systems and stakeholders. A further expected contribution, after the approach is fully developed, is to early identify conflicting design specifications to minimize costly late-discovered construction errors.

2 The Case of Construction Tolerances Issues

The main formal modeling element used to anticipate random (unintended) geometric deviations of manufacturing and construction processes is known as a tolerance. Tolerance has many different meanings based in the field that it applies. For this research, a tolerance is defined as the permissible limit or limits of variation in a physical dimension. Although the concept of tolerance is broadly understood, applicability of construction tolerances have not been adequately established due to the lack of knowledge integration during design stages, and the lack of multidisciplinary coordination among different stakeholders of project. Furthermore, many of the construction specifications cannot be assured from the beginning because they evolve during the course of a project. Early decisions about tolerances are usually made based on improper assumptions, or without an understanding of the "big picture" with respect to system implications. Decisions made late in a design stage are often taken without knowledge or consideration of earlier decisions, or without understanding of the effects that these changes will produce in other building systems. As a result, during project development, the state of knowledge about construction tolerances is diffuse, and no stakeholder has access to the entire knowledge base about what tolerances are realistic to prescribe. As it was discussed previously, there are a number of factors leading to a resulting building that differs geometrically from the nominal, dimensionally perfect, building model. The summary these issues is:

- Multiple material systems with different bodies of manufacturing knowledge [14];
- Geometry does not necessarily comply with manufacturability while being designed and later updates to remedy inconsistencies will increase the likelihood of mismatches with other components;
- Lack of knowledge representation and allocation methods for each material system [15] [14];
- Lack of integrated manufacturing knowledge traceability from specifications to geometry [16];
- Lack of manufacturing and tolerances verification methods [16]; and
- Lack of consistency across different tools and models [17].

The following section will discuss the main limitations of current BIM tools that produce some of the described shortcomings of system coordination and knowledge integration in building design.

3 Limitations of Current BIM

In current practices of architectural design, building engineering, and construction, products and systems are expected to perform at predicted levels. As Friedenthal et al. [18] states: "Competitive pressures demand that these systems leverage technological advances to provide continuously increasing capability at reduced costs and within shorter delivery cycles." In the building industry, this statement usually refers to a highly detailed set of requirements that challenge current methods of design, delivery, and operation of buildings. To successfully produce better buildings, the design and construction industry has to integrate computational tools that shift away from the traditional approach of independently-developed systems and stakeholders requirements. This is Building Information Modeling (BIM). BIM can be defined as a centralized modeling environment that allows connectivity of multiple vectors, including project information, assembly specifications, building operation, and building users [15]. However, the development of BIM, although crucial at the geometry level, has not been equally successful in developing well-defined transactional workflows to eliminate data interoperability issues [19].

Additionally, a building, as any other complex system, is not a static entity. Rather, it changes over time as sub-systems are incorporated or detached during the building lifecycle. These changes result in requirements and behaviors of constituent systems that may not have been anticipated when the system was designed [18]. Furthermore, in building design, multi-functional components are highly common. For example, a building roof covers the space of a building; protects the inner space from weather events; adds thermal protection to the interior; and enables the installation of other systems such as windows or solar panels. Any of these functions requires strict compliance with functional, structural, aesthetic, and economical constraints during building lifecycle. If no proper knowledge and project data integration platform is implemented, presumably, any change of the roof design, which is meant to improve some performance aspect, will result in the detriment or, at least, some change in some other functionality. As a proposed solution to this larger problem, the SE approach, has been extensively recognized in the aerospace and mechanical engineering industry to provide system solutions to technologically challenging and mission-critical problems that require knowledge management, analysis and model coordination [18] [20] [21]. The next section describes the Model Based Systems Engineering methodology and associated modeling language, compares document-centric and model-centric approaches, and gives an overview about how the figure of a systems architect can increase levels of interoperability and consistency among BIM tools and stakeholders.

4 Model Based Systems Engineering and the System Modeling Language

In the SE field, the development of a mature Model Based System Engineering (MBSE) approach allows the management of multiple domains and applications in a progressively complex Information Technology (IT) environment [22] [23] [24]. MBSE is defined as a practice of applying modeling and simulation for implementing the processes and practices of SE [20]. The main characteristic of a MBSE methodology is to link different modeling tools, from different domains, in a central model that allows interoperability and consistency between domains. Use of MBSE has led to the development of a general-purpose system-level architecture that allows multi-disciplinary modeling in different levels of abstraction. The modeling language for MBSE knowledge-modeling environment is the System Modeling Language (SysML), which was established by the Object Management Group (OMG) based on the Unified Modeling Language (UML). SysML is a general-purpose modeling language for systems engineering applications. It supports the specification, analysis, design, verification and validation of a broad range of systems and systems-of-systems [22]. By means of integrating SysML into the construction workflow tools, this research aims for the early identification of conflicting design specifications to minimize costly construction errors. This objective is proposed by programmatically linking a geometric BIM tool with a system engineering tool (based on SysML), and a mathematical simulation engine for analysis calculations. The proposed integration is intended to support the collaborative modeling of a building project as a "system-of-systems," and to provide the computational infrastructure and knowledge necessary to fix conflicts when they are detected.

4.1 Document-Centric Approach for Managing Project Data

One of the important contributions of MBSE has been the development of model-based architectures that have enhanced the ability to share and exchange project data. This contribution, although significant, requires improved knowledge and skills of users to facilitate the adoption of model-based practices. However, even with the development of BIM and system engineering, the current practice of architectural design and construction still relies on the conventional document-centric approach (Figure 1) to deliver and manage building lifecycle data. This method usually emphasizes the generation of individual design documents, in hard copies or electronic file formats with restrictive interoperable capabilities, which are exchanged among the project stakeholders.

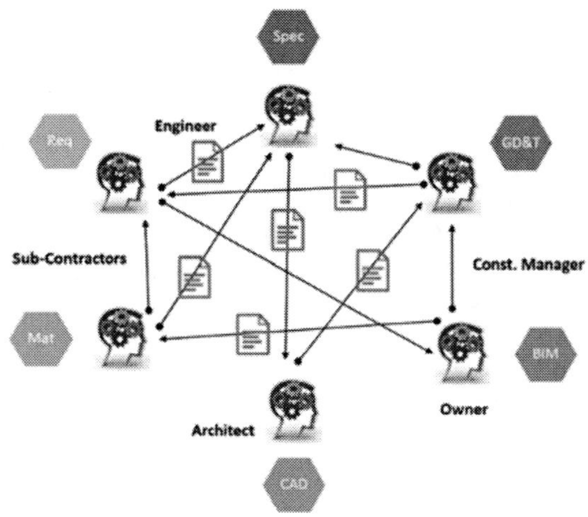

Figure 1, Document-Centric Architecture

Considering this increasingly complex and fragmented IT environment, the document-centric approach requires a significant amount of time to ensure that documentation is valid, complete and consistent. In the classic document-centric approach specifications are depicted in specifications trees. Then, a systems engineering management plan (SEMP) defines how the systems procedure fits in the project, and how all the concurring disciplines come together to develop the documentation necessary to satisfy the requirements in the specification tree [18]. Usually, these kind of relationships will be depicted in design documentation such as flow diagrams. However, flow diagrams of a document-centric approach lack interoperable functionality. Consequently, though a document-centric approach may be quite rigorous, it has a critical limitation when assessing the consistency and completeness of project data. For this approach to be successful, the software architecture of BIM must be transversal to different building stakeholders and tools, and the actor in charge of the document mapping must be consistent and constant in order to maintain a complete model. As [18] points out, The comprehensiveness, consistency, and relationships between requirements, design, analysis, and test data are hard to evaluate due to the fact that information is distributed across several documents. Understanding a particular view of the system and executing the necessary traceability and design-change impact assessments is clearly challenging. Applying this scenario to the AEC domain may lead to a deficient coordination of design requirements, which could subsequently lead to poor knowledge integration regarding material systems and manufacturing processes, and finally to quality issues when the final product is delivered.

4.2 Model-Centric Approach for Managing Project Data

As it has been described above, the document-centric approach for systems engineering –although having many advantages, suffers from an important disadvantage: model inconsistencies. This situation was one of the main motivations to propose this BIM-MBSE approach. With the MBSE approach, many of the intermediate deliverables of the modeling activities seen in the document-centric method can be generated automatically.

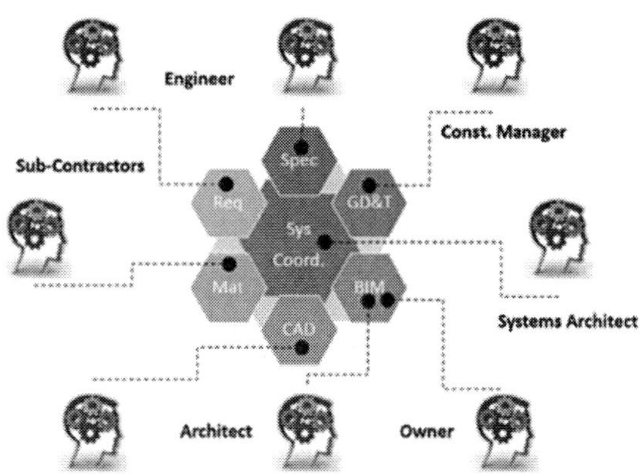

Figure 2, Model-Centric Architecture

As [20] explains, in the model-centric approach, the main product of those activities is an integrated, coherent, and consistent system model, produced through a dedicated systems modeling tool: the System Modeling Language (SysML). All other artifacts are secondary—automatically generated from the system model using the same modeling tool.

One of the important characteristics of a comprehensive BIM model is that it enables stakeholders to take informed decisions. Decisions made within a MBSE framework take place within a central repository, where each design decision is captured by a model element or a relationship among model elements. With the model-centric method, all objects depicted in different BIM or engineering tools are simply views of the underlying system model, they are not the model itself. And that difference is the core of the return on investment (ROI) that MBSE offers over the document-centric approach [20]. In a united MBSE-BIM model, as all modeling elements are programmatically and systemically integrated, any change that is produced will be automatically propagated to the rest of the model. It does not matter if the elements are depicted in a diagram that is user-defined or automatically created, or if the model is too large or complex. After all, the elements of the system models are just views of the real model, which keeps its internal consistency based on the seamlessly integrated nature of its modeling approach.

4.3 The Role of the Systems Architect

Similar to the functions of a BIM manager, an SysML-BIM framework requires an actor that assumes all the responsibilities for the creation of tool-specific integration routines, and proper knowledge allocation during coordination of trades and stakeholders. Also, besides modeling integration, another important function of the systems architect is knowledge integration. This task involves the creation and maintenance of design specifications and standards based on data collected from the different stakeholders. After all domain-specific models have been imported/linked, the systems architect will guarantee that knowledge-based verifications are executed right on time to meet project schedule or to discuss corrective actions with the trade-specific design teams.

In the AEC world, the skill of modelers has been challenged by the implementation of BIM, which is inherently 3D and requires a higher level of modeling skill. Most BIM authoring tools require that modelers assert the relationships between building objects as part of building BIM models, which is an additional challenge, but which makes the building model richer and more useful. The system model includes everything in BIM and adds sub-models for requirements and processes. Thus the modeling complexity is increased even further, leading to the identification of the "systems architect" as a managing entity. The next section presents the proposed general approach for building design, and describes its main functionalities to be developed in a computational implementation.

5 Proposed Design Methodology

In Figure 3, the depicted diagram represents the current approach to inform design in building construction. In this case, domain-specific knowledge is never formally integrated with the assembly geometry. Assumptions about material interactions and components design, rather than formal model-based assessments, create room for inconsistencies between design specifications and design geometry. In this research, a formal link between domain-specific knowledge and geometry is the proposed way to assure the full validation of building requirements and specifications. To implement this approach, the interactions diagram needs to incorporate a new element that will open several other kinds of relations in the process: a geometric constraint.

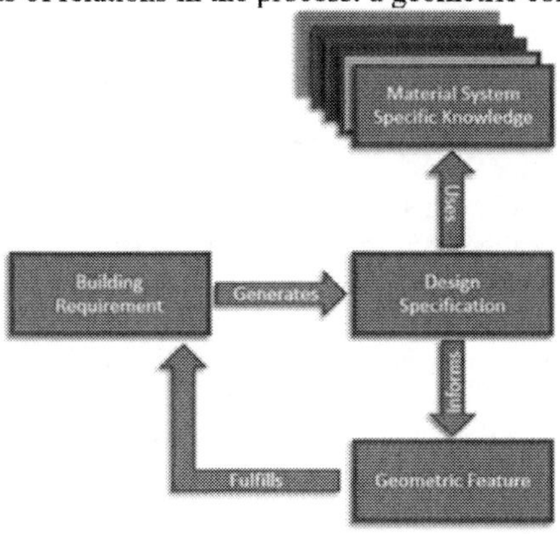

Figure 3, Modeling approach without the proposed framework

A geometric constraint is proposed as the negotiating point between domain-specific construction knowledge and design geometry (Figure 4). Furthermore, a geometric constraint can be the formalization of a piece of manufacturing knowledge. For example, a basic formula to calculate the minimum bending radius of sheet metal is $r = t$, where r is the radius of the bending and t is the thickness of the part. Then, the mathematical expression $r = t$ represents a portion of domain-specific manufacturing knowledge. This basic piece of knowledge can be automatically evaluated in a geometric feature, if CAD parameters and knowledge are linked together. In order to make this geometric constraint operational, most of the exchanges depicted in Figure 4 must be programmatically formalized. For this research, the material system-specific knowledge will be formalized as a specialization of SysML requirements, which will be programmatically linked to their mathematical formalizations as design constraints. Also, another part of the process, which involves data exchange among specifications, geometric constraints, and geometric features, must be automated. As it can be seen in Figure 4, design specifications will inform geometric features, and parameters of geometric features will populate the domain-specific constraint, which will verify that the design specification is in compliance.

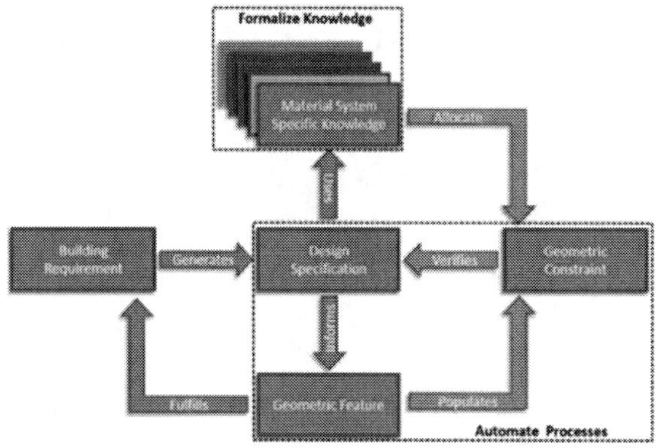

Figure 4, Modeling Approach as proposed

5.1 Functionalities of the proposed modeling framework

In order to implement the proposed modeling framework, the following list of general functionalities is required:

- Model-to-Model Transformation: structural, feature-based decomposition of parametric CAD models into system models.
- Model Integration Approach: parametric, seamless software integration for knowledge allocation, analysis, and verification to reduce human data translation.
- One Truth, multiple model views: centralized project requirements, geometry, and design specifications in an interoperable modeling environment.
- Machine Readable/ Executable: CAD geometry programmatically integrated to manufacturing know-how through knowledge-based mathematical and logical constraints.
- Model Consistency Approach: On-demand model-to-model and tool-to-tool consistency assessment and model data update.

The next sections will convert the previous set of system requirements into specific activities that will be programmatically implemented after full development of this research.

5.2 Required Activities to Implement the Proposed Framework

The methodology to full develop the proposed modeling framework is composed of the following six activities, which will be entirely implemented during future stage of this project:

1. Structural Decomposition: Includes the creation of a feature-based representation of the CAD model in the SysML environment. It follows the project>assembly>part>feature>parameter approach to describe geometry. Also, it creates a data graph based on CAD meta-model.
2. Knowledge Acquisition: Corresponds to the domain-specific knowledge, and its formalization, necessary for specifications compliance analysis or optimization/verification processes of an assembly or section of a building. This process will be carried away by adding specific rules as requirements in the SysML environment. All knowledge created will be stored within a domain-specific knowledge repository in the SysML model. This repository will have requirements that lead to design specifications represented as mathematical expressions.
3. Knowledge Allocation: CAD features decomposed in numeric parameters from CAD data will be linked to mathematical constraints that carry domain-specific knowledge. The allocation process will be executed automatically so that manual data translation is avoided. Then, the application will query the features and material types of the imported CAD, and will offer options to link specific pieces of knowledge that match those details.
4. Parametric Execution: The application will evaluate all the domain-specific knowledge constraints, through a mathematical engine, by using instances data obtained from the CAD models. The results of each parametric evaluation will be stored so that the user can compare them and pick the best analysis scenario for a given analysis context.
5. Specifications Verification: Routines coded for this implementation in SysML will evaluate and verify the consistency between CAD metrics and the formal description of design specifications defined for the specific building project.
6. Knowledge-Compliant Geometry Update: This stage defines a series of functions that will consolidate changes produced in the model on either the CAD or the SysML side. In an integrated framework, changes might be produced in different domain-specific applications. For this framework, if changes that were positively evaluated by the application were produced on the CAD side, there will be an "update SysML model from CAD" command added in the SysML modeling interface. Conversely, if changes were made in the SysML side, there will be an "update CAD model from SysML" command.

At a general level, the main practical functionalities of the proposed application are:

- Adding knowledge-compliant, feature-oriented, case-based design specification to the CAD model;
- Automatically assessing specifications of parts and assemblies to identify possible system conflicts;
- Upgrading "nominal geometry" by adding feature-oriented considerations based on material-system-specific engineering and manufacturing knowledge; and
- Evaluating and validating tolerances and clearances specified for parts and assemblies.

6 Conclusions

This research proposes a system-based, knowledge-aided modeling framework that integrates a parametric CAD tool with a system modeling platform to assess specifications compliance in building construction. Main motivations of this approach are the lack of manufacturing-specific knowledge available for designers in design stages, the lack of manufacturing compliance and verification methods for BIM models, and the lack of multidisciplinary consistency among BIM tools. This research argues that the aforementioned problems can best be addressed in the context of an integrated knowledge-modeling platform. With proper development, the framework proposed by this research could create a new kind of building design paradigm: A modeling environment that virtually and simultaneously brings to the table all domain experts, anytime that building feature is created.

References

[1] P. C. Albert and P. L. Chan an Ada, "Key performance indicators for measuring construction success," Benchmarking: An International Journal, vol. 11, no. 2, pp. 203-221, 2004.

[2] K. Ballast, "Tolerances In Construction: Dimensional Tolerances for Surface Accessibility," Architectural Research Consulting, 2011.

[3] P. Love, P. Davis, J. Ellis and Cheung SO, "Dispute causation: identification of pathogenic influences," Engineering, Construction and Architecture Management, vol. 17, no. 4, pp. 404-423, 2010.

[4] P. Birkeland and L. Westhoff, "Dimensional Tolerances in a Tall Concrete Building," ACI

Journal, vol. 68, pp. 600-607, 1971.
[5] P. Love, "Forensic project management: the underlying causes of rework in construction projects," Civil Environmental Engineering Systems, vol. 12, no. 3, pp. 207-228, 2004.
[6] P. Love, J. Edwards, J. Smith and D. Walker , "Congruence or divergence? A path model of rework in building and civil engineering projects," ASCE Journal of Performance Construction Facilities, vol. 23, no. 6, pp. 480-488, 2009.
[7] C. Juliana and A. Ramirez, "Construction Management/Design-Build," Lorman Seminar, 2005.
[8] A. S. o. C. Contractors, "Reducing the Cost of Tolerances compatibility Problems," Missouri, 2006.
[9] P. Love, D. Edwards, S. Han and Y. Goh, "Design error reduction: toward the effective utilization of building information modeling," Research in Engineering Design, vol. 22, pp. 173-187, 2011.
[10] R. K. Allen, S. N. Pollalis and B. R. Schwegler, "Promise and barriers to technology enabled and open project collaboration," ASCE Journal Prof Issues Engineering Education Practice, vol. 131, no. 4, pp. 301-311, 2005.
[11] National Defense Industrial Association, "21st Century Advanced Manufacturing Modeling & Simulation Roadmap Focus Area Recommendations," Advanced Manufacturing Engineering Capabilities (AMEC) Committee, 2013.
[12] E. Yares, "3D CAD World," Siemems, 7 August 2012. [Online]. Available: http://www.3dcadworld.com/how-was-the-mars-rover-curiosity-designed/. [Accessed 21 8 2015].
[13] American Concrete Institute, Commentary on Standard Specifications for Tolerances for Concrete Construction and Materials (ACI 117R-90),, Farmington Hills, MI: American Concrete Institute, 1990.
[14] C. Eastman, P. Teicholz, R. Sacks and G. Lee, "Development of a Knowledge-Rich CAD System for American Precast Concrete Industry," in Computer Aided Design in Architecture (ACADIA), Muncie, Indiana, 2003.
[15] B. Anselmetti and H. Louati, "Generation of manufacturing Tolerancing based on ISO Standards," International Journal of Machine Tools and Manufacture, vol. 45, no. 10, pp. 1124-1131, 2005.
[16] S. Herzig, A. Qamar, A. Reichwein and C. Paredis, "A conceptual framework for consistency management in model-based systems engineering," in International Design Engineering Technical Conferences & Computer Information in Engineering Conference, 2011.
[17] S. Friedenthal, A. Moore and R. Steiner, A practical Guide to SysML, 3rd Edition, Morgan Kaufman, October 2014.
[18] S. Azhar, "Building Information Modeling (BIM): Trends, Benefits, Risks, and Challenges for the AEC Industry," Leadership Manage. Eng., vol. 11, no. 3, pp. 241-252, 2011.
[19] L. Delligatti, SysML Distilled: A Brief Guide to the System Modeling Language, Addison-Wesley Professional, Nov. 2013.
[20] T. Johnson, C. Paredis and R. Burkhart, "Integrating Models and Simulations of Continuous Dynamics into SysML," in Modelica Conference, Germany, 2008.
[21] Object Direct, sysML Modeling Language explained, Object Direct.
[22] No Magic, SysML Plugin User Guide 18.1, No Magic, 2015.
[23] R. Peak, S. Friedenthal, M. Wilson, M. Bajaj and KIm, "Simulation-based design using SysML," in INCOSE International Symposium , San Diego, 2007.
[24] G. Lee, R. Sacks and C. Eastman, "Specifying parametric building object behavior (BOB) for a building information modeling system," Knowledge Enabled Information System Applications in Construction, vol. 15, no. 6, pp. 758-776, 2006.
[25] C. Adourian and H. Vangheluwe, "Consistency between geometric and dynamic views of a mechanical system," in Summer Computer Simulation Conference, San Diego, 2007.
[26] P. Hehenberger, A. Egyed and K. Zeman, "Consistency Checking of Mechatronic Design Models," in International Design Engineering Technical Conferences & Computers and Information in Engineering Conference , Qebec, 2010.
[27] J. Gausemier, W. Schafer, J. Greenyer, S. Kahl and P. Sascha, "Management of cross domain model consistency during the development of advanced mechatronic systems.," in International Conference of Engineering Design, 2009.
[28] J. Simmonds, R. Van Der Straeten, V. Jonckers and T. Mens, "Maintaining consistency between UML models using description logic," EUZENAT and Bernard, Carre, editors, RSTI serie L'Objet, vol. 10, pp. 231-244, 2004.
[29] T. Mens, R. Van Der Straeten and J. Simmonds, "A framework for managing consistency of evolving UML models," Software Evolution with UML and XML, pp. 1-31, 2005.

Robot-Oriented Design for Production in the context of Building Information Modeling

F.R. Correa[a]

[a]Escola Politecnica, University of Sao Paulo, Brazil
E-mail: fabiano.correa@usp.br

Abstract –

Construction industry must make more extensively use of automation and robots in order to increase its productivity and reduce its impact in the environment. It seems that the first steps in this direction was taken with the Building Information Modeling (BIM) paradigm. Wide adoption of BIM, particularly for the activities of project and construction planning, could be exploited also in trying to introduce more control to the final processes of building construction through a computer interface, from design to production. The main motivation for this research is to study the impact of BIM in bridging the gap between design and construction. Considering the growing interest in applying additive manufacturing technology for future building construction, this paper proposes a computer-aided system that translates a generic architectural project in a set of pieces to be fabricated with 3D printers. The system uses a proposed set of algorithms to process the architectural project: it considers the division of the building in different parts or pieces to be fabricated, based on the work volume of the printer; it also considers the relative position of the part in the bed of the printer, for best results in production; minimum dimension of features to achieve mechanical resistance; and geometric features that would demand support material. IFC standard, in its fourth version, was analyzed for validation in the process to carry all the relevant information produced in design phases to fabrication.

Keywords –

BIM Integration with Robotics; IFC; Additive Manufacture

1 Introduction

Few countries could be regarded as *avant-garde* in the use of automation and robots by the construction industry [1]. In a global perspective, it turns out that the construction industry, albeit its relative importance in the economy of each country is still behind other industries in terms of adopting new technologies to increase its productivity and reduce its impact in the environment. Moreover, this scenario is critical in developing countries like Brazil, which have under qualified and cheap labor, turning more difficult the introduction of costly solutions, even when producing better products.

There are many reasons for this situation, and the cost of implementation is one of them. In that regard, there is a range of technologies that could be exploited by the construction industry, in which the costs varies a lot. One extreme being the use or the investment in research for the use of robots for off-site or on-site pre-fabrication, and on-site assembly.

The other extreme are software developments and IT technologies promoting the automation of many of the common processes, which today are under the Building Information Modeling (BIM) paradigm [2].

Nevertheless, there is a discontinuity between implementations using each of the two extremes, mostly because 3D CAD is still used in some countries in applications that could be solved with BIM technology [3]. Consequently, there is a missed opportunity to capitalize in the integration of both extremes, as occurs in other industries with CAD/CAE/CAM (Computer-Aided Design/Engineering/Manufacturing) approaches.

However, the increasing rate of adoption of BIM, particularly for the activities of project and construction planning, could be exploited also in trying to introduce more control to the final processes of building construction through a computer interface, from design to production. Ultimately, the production could be done by CNC machines and industrial robots [4].

The main motivation for this research is to study the impact of BIM in bridging the gap between design and fabrication, in a future scenario where different robots could be used on-site, for example, for the final assembly of pre-fabricated modules.

To achieve that goal, the focus in this article is in how a computer-assisted design software could make the design process transparent to the architect for production based on a 3D printer. Following the concept of Robot-Oriented Design (ROD) [5], the system will automatically introduces features and changes in the final geometry to be able to produce a functional building that could be 3D printed. The proposed system considers:

- Division of the building in different parts or pieces to be fabricated, based on the work volume of the printer;
- Relative position of the part in the bed of the printer, for best results in production;
- Minimum dimension of features to achieve mechanical resistance;
- Geometric features that would demand support material.

In this case, the integration from design to production or between BIM and Robotics is achieved with the use of the IFC Schema to represent BIM model with building design, and the transformation of that information in an STL file to be used in a 3D printer.

The IFC file format is the de facto standard in the construction industry and the way information is passed between different professionals and software platforms. IFC standard, in its fourth version, was analyzed for validation of the process to carry all the relevant information produced in design phases to fabrication.

It is also discussed the role that additive manufacturing could have in this scenario, when multi-material and control of mechanical properties along different parts of the same component become commonplace.

2 BIM Integration with Robotics

Although BIM has been largely adopted by many countries, and is the responsible for inserting new computational technologies in AEC, it did not bring along direct opportunities for automation and / or the use of robots in construction. In part, this observation is because most BIM implementations are for project or planning, and not for fabrication / construction [6].

BIM could be associated with the CAD / CAM / CAE in mechanical engineering, but the percentage of use of CAM is very limited. There are a few examples in using BIM for fabrication or production, mostly in Japan and China [3].

The interest of manufactures in participate in BIM processes could be viewed in two different scenarios [7]:

1. Producing models/components which represents its products in detail and could be directly used inside BIM models in design or construction;
2. To fully extract rich details of the models and use the information to fabricate certain components, as steel components for HVAC.

For the discussion presented in this article, the focus will be in the second reason.

In working with a 3D printer to fabricate components of an architectural design, the main reason is to collaborate in proposing solutions and identify difficulties in the path for integration of BIM with Robotics.

All the construction information are contained in IFC files, the open BIM standard.

2.1 IFC

In studying the impact of BIM for production, it is important to deal with Industry Foundation Classes (IFC) standard [8]. It is a standard that evolved for almost two decades, and since 2013, is ISO 16739 standard.

IFC was conceived to accommodate construction information regarding all lifecycle of a building.

The main interest in dealing with IFC is to locate all the necessary information to process and analyze each component of the model, and eventually to store information for production inside the model. As an open standard, the solution could be easily ported in a proprietary BIM model.

Although IFC is not the primary file type in authoring BIM tools, it is part of the process when the model needs to exchange information between different applications. So, in the case of fabrication or production, it is necessary to explore if IFC could contain properly all the information necessary to fabrication, information which is used by computer controlled machines and robots. In the case of this article, it would be explored if IFC files could represent STL format which is used in 3D printers.

The main interest in IFC for this particular work regards how the components of an architectural project design could be geometrically represented inside IFC.

2.1.1 Geometric Representation

IFC schema allows the representation of building elements in many different formats. Two are of particular interest:

- Boundary representation: to simplify the approach, it will be used only the *IfcFacetedBrep* (Figure 1) representation instead of the more free-form allowed with *IfcAdvancedBrep*;
- Tessellated: this representation would be valuable to communication with 3D printer and translation to STL format. It uses the *IfcTriangulatedFaceSet* (Figure 2).

In boundary representation, all vertex of the solid are in the end represented as *IfcCartesianPoint*. The triangulated representation could be created directly from the solid resultant from the alterations proposed by the system.

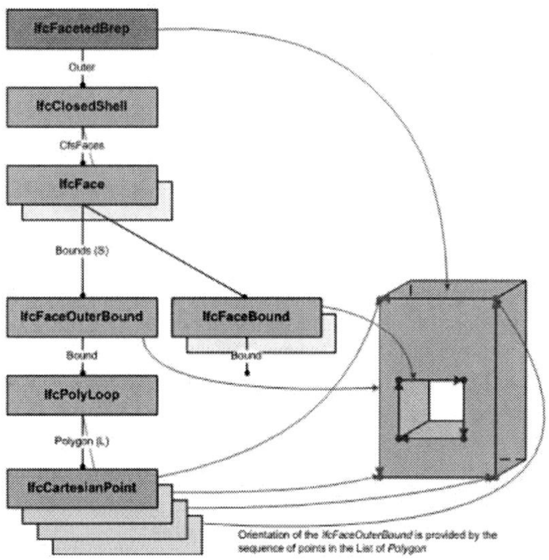

Figure 1. Entities of the IFC schema used to represent a building component with Brep representation [8].

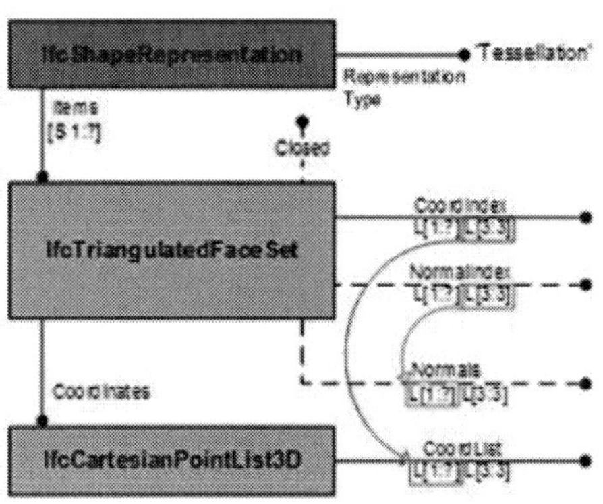

Figure 2. One of three forms of specify a component with *IfcTriangulatedFaceSet* [8].

2.2 Additive Manufacturing in Construction

Additive manufacturing, more popularly known as 3D printing, has been considered a disruptive technology in regarding manufacture processes [9]. It already was successfully applied in the automotive, aviation, and health fields.

The process consists of depositing layer after layer of a material that flows through a channel. In addition, this probe could be placed anywhere in the horizontal plane to draw the intended shape. The material must be hard enough to hold its form before receive another layer of material.

In the last years, it has been used for construction, to build parts of a house or building [10][11].

Independent of the technological barriers that still exists, there is not a single vision in the way this process could be applied to construction. Potential exists to avoid material waste, to use recycled material or in situ material, to control the material properties locally, to produce parts with geometric, and so on.

The benefits of this manufacturing process is the freedom of form, its reduced setup time between different components, its potential capacity of work with different materials. Its negative features is the time to manufacture.

With regard to the constraints applicable to production with additive manufacturing, it could be divided in two different categories: geometry in design, parameters of production.

The parameters of production do not affect the geometry of the component but its mechanical properties. For example, the density of the core of the component could be changed based on the final mechanical resistance necessary for that component.

Using BIM for fabrication and trying to promote the additive manufacturing technology for future building construction and interpreted as a robot, this paper proposes a computer-aided system that translates a generic architectural project in a set of pieces to be fabricated with 3D printers.

3 CAM for Additive Manufacturing

CAM is used in many industries for a long time, existing for the majority of the production system. However, additive manufacturing is a relative novel process that is still under continuously development, primarily because new materials are been conceived to better exploit its potentiality [9].

Some computerized systems exists that deal with Additive Manufacturing [12], but the difference is that it helps in the design process, and the proposed system try to correct the design without the input of the user. Other systems are similar in trying to create a geometry-based grammar to interpret the design [13].

The proposed system consists of a software that processes a BIM model, and extract from it a batch of components to be 3D printed. Normally, the designer of mechanical parts must have a very well understanding of the fabrication process involved in manufacturing the part designed by him. In such scenario inside the construction industry, one architect could design a façade, and send it directly to a CNC controlled machine or a robot for fabrication.

In developing such system, three questions had been

considered:

- To create an abstraction between design and production based on additive manufacturing, so that the architect could concentrate in its intents;
- To promote the entrance of the additive manufacturing in the construction industry;
- To work with the principles of ROD, in which it helps the construction industry to perceive the added value brought by automation and robotics.

For the following discussion, it will be considered that the 3D printer possesses only one head of extrusion, and thus works with a single material, PLA. Unfortunately, it would be difficult to make all the discussions necessary here in using a material appropriate for the construction industry, such as concrete, which is still matter of intensive research. However, the discussion provided and results obtained are still valid for other materials.

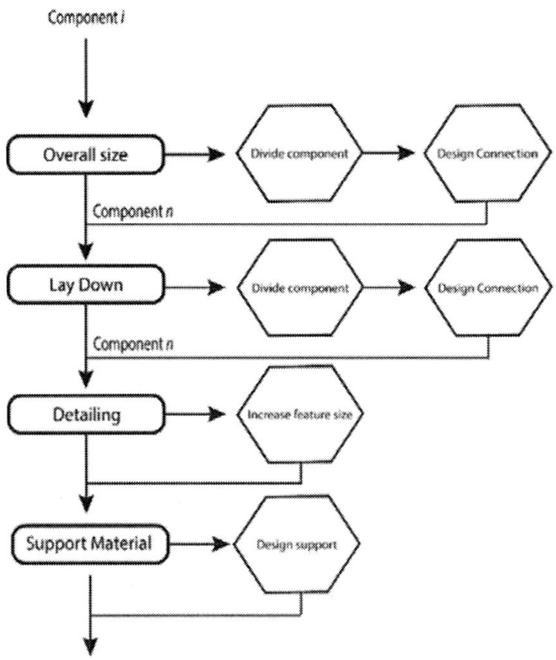

Figure 3. Diagram of the analysis flow for each component of the BIM model.

The methodology of the presented research consists in: 1) Enumerate geometric characteristics or features in building components that would be difficult or impossible to achieve with a standard FDM 3D printer using PLA; 2) To propose a workflow where these features would be automatically and entirely detected in processing a BIM file; 3) Implement algorithms to slightly change those features in each component, so that it could be printed; 4) Print the entire project with the eventually altered parts, and contrast it with initial design intent;

The main objective in developing such system is to study if the designer, i.e. the architect, could have a larger liberty in the design process, relaying some decisions to an expert system, which will most fatefully translate its intent to produce components through 3D printing.

In the following sections, it will be discussed all the characteristics of components that the proposed system verifies. For each component of the model, the analysis workflow is depicted in Figure 3.

Each step verify a specific characteristic (in the diagram, round boxes indicates that a verification occurs). If it passes in the verification, it go for the next step. If not, it is directed to pass through some functions, which could alter the geometry of the component. In some cases, it could create new components from the original one (in the diagram, each hexagon represents an operation on the geometry of the component, altering thus the original geometry of the architect, but not, hopefully its intent).

All the geometric reasoning is based on a voxel representation of the components (Figure 4). It could be implemented using efficient data structures such as octrees, but in this case, as each component have a relative small size, the gains are negligible.

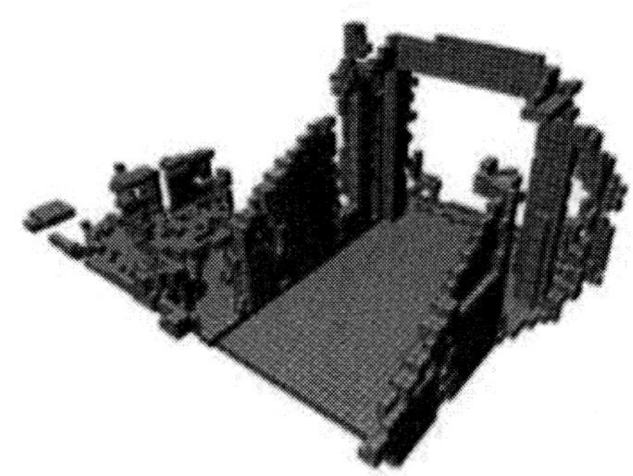

Figure 4. An example of how architectural components could be represented with octrees [14].

The transformation algorithm starts by transforming each building component in a discrete representation composed of voxels. The size of each voxel is chosen to be equal to the diameter of the 3D printer.

Each voxel receives one of three possible values: { 0, 0.5, 1 }. If the voxel is entirely filled by the component, it receives 1; if it is completely empty, 0; and 0.5 when there are voids inside the voxel. Thus, the representation works by indicating when a specific feature of the component is too detailed as to be 3D printed.

Processing the information on the voxel, it is possible

to do all the verification necessary for the proposed system. In addition, it enables the discrete localization in each component for insertion of additional features, such as connectors, and to cut for division of a large component.

All considerations regarding proper design for 3D printing was based on [15] and the own experience of the author. In the following sections, an overview of the workflow is presented, along discussion on how the information generated is stored in IFC files.

3.1 Overall Size

In checking if a component could be printed in one piece, it must not be larger than the workspace of the printer. The bounded box that contains the piece is used in that verification.

In IFC Schema, the entity *IfcBoundingBox* and its attributes *Corner* (*IfcCartesianPoint*), *XDim*, *YDim*, and *ZDim* (*IfcPositiveLengthMeasure*), and *Dim* = 3 (*IfcDimensionCount*) must be filled accordingly.

The volume of the workspace is defined as the size of the bed multiplied by the useful height of the printer. The volume of the bounding box must be smaller than that value:

$$IfcBoundingBox.XDim * IfcBoundingBox.YDim * IfcBoundingBox.ZDim < Workspace\ Volume \quad (1)$$

If the size is larger, it must be divided in more than one part, and certain features will be created which would warrants a consistent assembly. This division must produce as few pieces as necessary to print. If more than two parts are necessary, it must produce parts with similar size. The location of the division is based on the voxel discretization of the component.

It would be interesting to create a library of possible connections; the system implements the connector depicted in Figure 5. Its size must be compatible with the size of the voxel

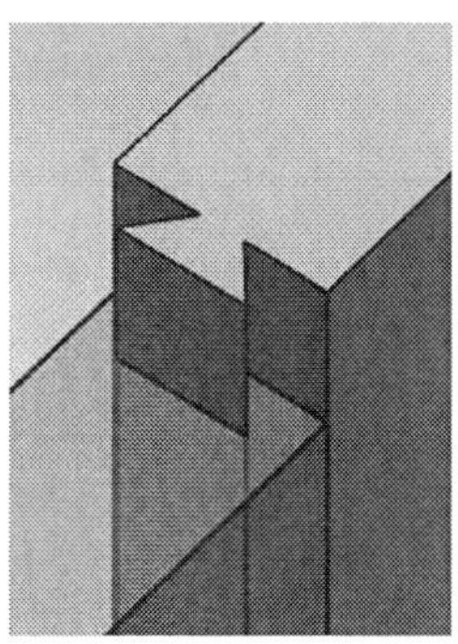

Figure 5. One example of possible parametric connections between components.

3.2 Position in Bed

Checking for best position in laying it out in the printer bed, could avoid the use of support material, shrinkage effects and promote best adhesion . The system searches for the face that have the larger area in the piece. It uses the voxel representation trying to find an external layer that is completely filled.

The final position of each component will be stored on the entity *IfcLocalPlacement*, relative to the coordinates of the printer's bed, of the entity *IfcBuildingElemenProxy*. It is necessary to create a copy of the original component.

If there is not an entirely flat face to position in bed, the system would try to reap off some features or divide the component in different parts to faces entirely in a plane. As occurs with the first analysis, if the component is split in parts, connections would be added to realize the posterior assembly.

The algorithm needed to produce the division in the component to be adequately positioned in bed is different from the algorithm of overall size.

3.3 Detailing

If the component has some fine detailing, it is necessary to known the diameter of the extrusion nozzle, because the printer cannot obtain a detail that is smaller than that parameter.

The algorithm to find and analyse the details of the components must first identify each detail and isolates it. This occurs when the component is transformed in the voxel representation, and could be determined by aggregating neighbouring voxels in the same layer.

3.4 Support Material

Although it must be avoided, to produce certain components directly instead of assembling different components in a final shape, it could be necessary to produce a support material to give shape to the component. Actually, the system do not implement an algorithm to customize the support material, relying instead in the own printer software to do that.

The verification must search for two different type of features: overhang (Figure 6) and bridging (Figure 7), based on the positioning of the component in bed.

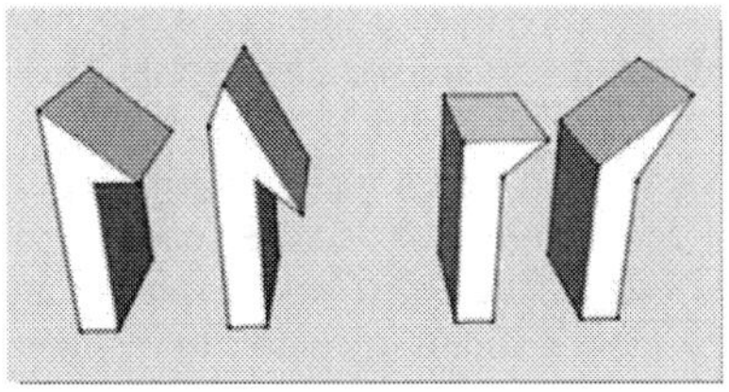

Figure 6. Examples of overhangs: only the two on the right will print correctly without support material [15].

If there are overhangs, and if the inclination of the face is more than 45 degrees from the vertical, it will need support material. If the angle is about 60 degrees, it could be printed but it will be necessary to use the cooler.

Bridging would demand the division of the component in the exact position the that feature.

The support material would be represented inside IFC Schema by creating an *IfcBuildingElementProxy* with attribute *ObjectType* USERDEFINED and attributing to it a geometric representation based on faces.

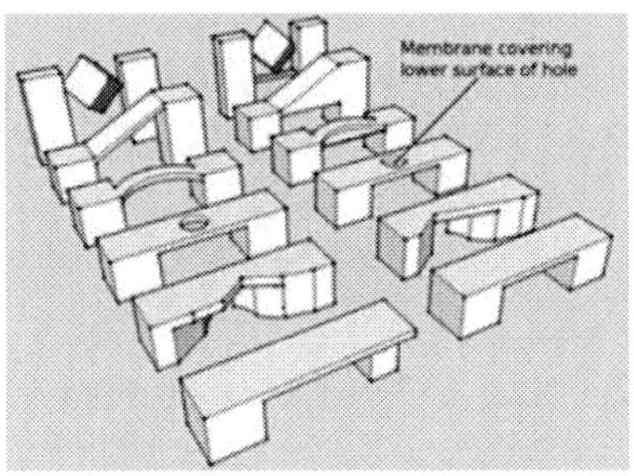

Figure 7. Models in the left will print poorly [15].

3.5 STL Files from IFC

The final step, not depicted in the workflow, would be to generate an STL file for each component. The IFC Schema allows a large number of different geometric representation of the solid. With IFC4, a new form of geometric representation was conceived consisting in transform or in some cases, approximated the geometry of the face of the object using sets of triangles.

For STL format, it is necessary to use the entity *IfcTriangulatedFaceSet* and its attributes *CoordIndex*, *NormalIndex*, and *NumberOfTriangles* = SizeOf(*CoordIndex*).

4 Experiments and Results

The system was implemented using *IfcOpenShell* to parse directly IFC files and written in C++. As *IfcOpenShell* uses *OpenCASCADE* to generate and deal with geometric representation, the voxel representation uses also data structures from *OpenCASCADE*.

Based on a Prusa 3D printer, the configuration parameter for the system are shown in Table 1.

Nozzle diameter affects directly the quality and precision of the final product. No feature with smaller dimension than the nozzle diameter could be produced. Layer's thickness also influences on the final quality of the component and in the time to finish the printing process. The size of the workspace or bed restricts the size of the component to be printed.

Table 1. 3D printer's parameters.

3D Printer's Parameters	Values
Nozzle diameter	0.4 mm
Bed width	20.0 cm
Workspace Height	19.0 cm
Thickness of the layer	0.4 mm
Bed depth	20.0 cm
Diameter of the filament	3.0 mm

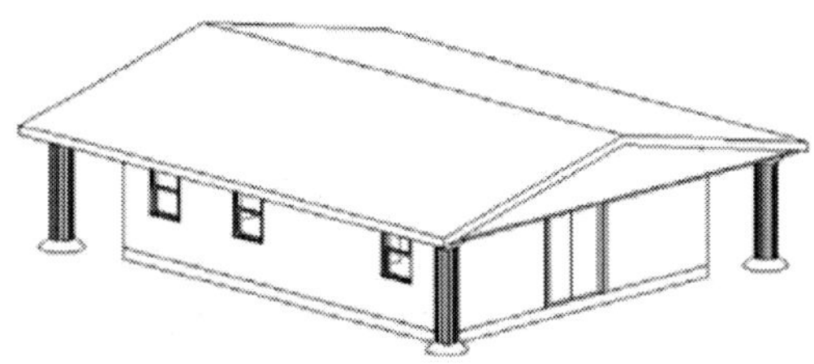

Figure 8. IFC Model.

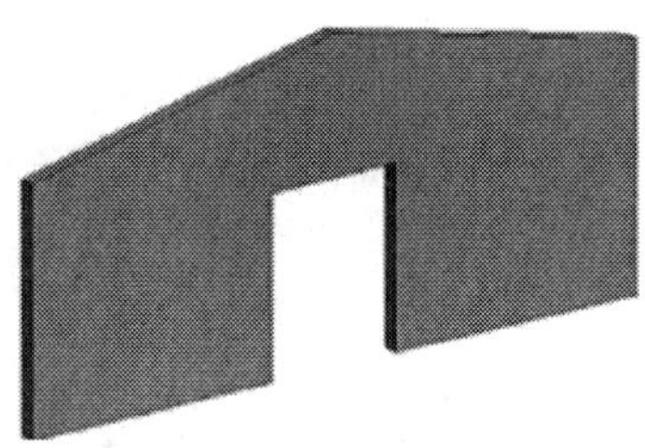

Figure 9. Component to be tested against support material (bridging).

The model used in the experiment is depicted in Figure 8. It will be printed in a 1:200 scale. It generated 11 different components to be analysed, transformed and printed (windows and doors were not considered in this test). The model was generated in Autodesk Revit software and further exported in IFC format, with the built-in exporter, and was given as input to the system.

In the sequence, some components of this building are highlighted to shown the performance of the system in analysing the design for 3D printing.

Figure 9 illustrate one external wall of the house, which because of the main entrance, present the problem of bridging when printed. All walls that have openings (for doors or windows) will present the same problem.

The result of the analysis and transformation of the component generated by the system is illustrated in Figure 10. It shows the division of the wall in three pieces because of its size of 41 cm, and the inclusion of features to connect the different parts.

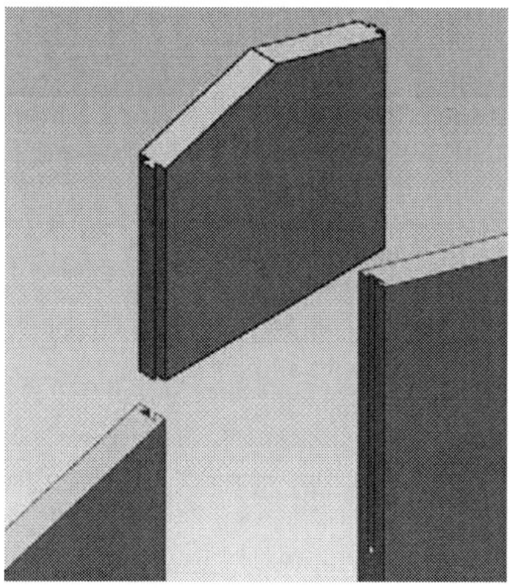

Figure 10. Result from algorithm divide.

In other type of analysis, Figure 11 represents an interior wall whose position in bed for printing is a challenge, without resorting to support material, when considering a larger surface of bed adhesion.

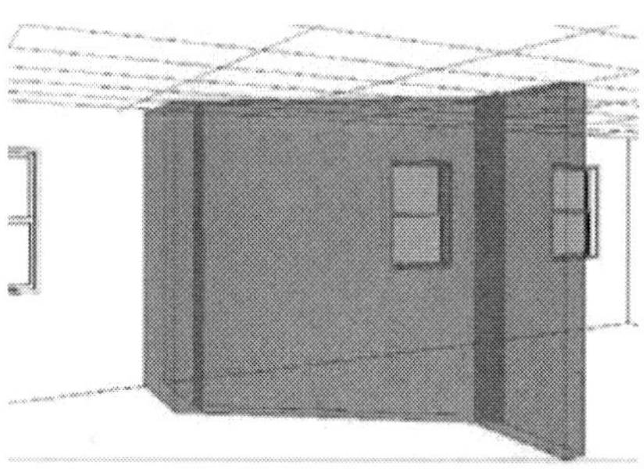

Figure 11. Wall with singular geometry to be tested against lay out position.

As the implementation do not considers a trade-off between size of the area in bed and the necessity of support material or to divide the component (which is not trivial), it instead adopts some priority verification. Thus, the final position to be printed is as depicted in the figure to avoid support material, although this position generate a small area of adhesion.

Based on the voxel representation and the constraint of the size of the nozzle, the depth of all walls were changed from 1 mm to 1.2 mm to accommodate it.

The most difficult operation to realize automatically is to change the feature detailing that could not be printed adequately because of its size. Eliminating such detail is not a solution when the intention of the architect is lost.

Figure 12 illustrate a column with some fine detail that could not be printed as it is. Figure 13 and Figure 14 shows how it was modelled and it is through this analysis that the system could change the number of repetitions along the circle, the size of each detail, and so on.

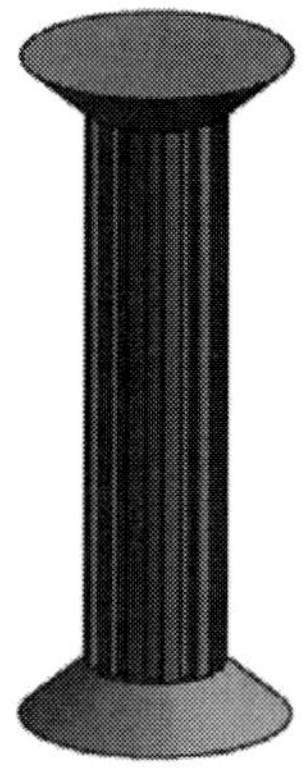

Figure 12. Column with parametric feature.

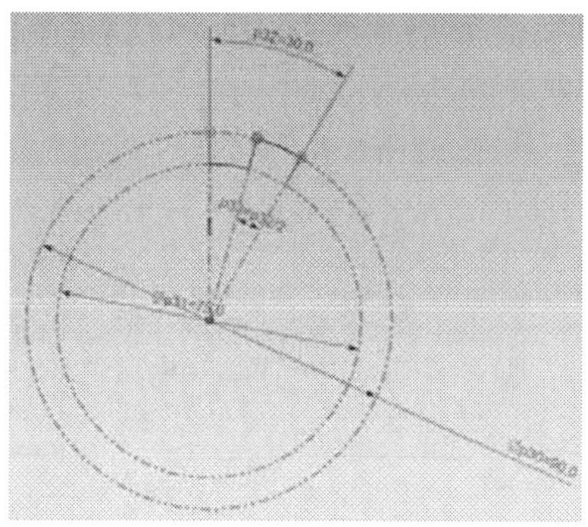

Figure 13. Parametric feature capturing design intent.

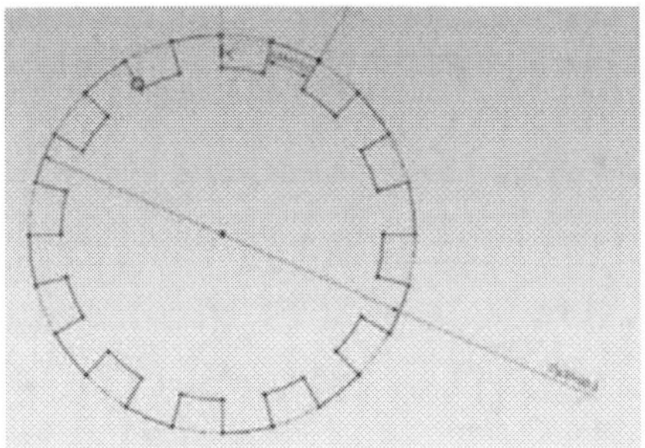

Figure 14. Pattern Curve to create the parametric profile.

To solve this problem some sort of parametric analysis must be provided, but it was not implemented in the current version of the system. However, the IFC Schema in its fourth version could represent such parametric features and models.

5 Conclusion

The initial development of a system that interprets architect's design intent and translate it in a batch of components to be 3D printed was presented. The context in which such system could be useful is in the integration of BIM and robotics.

It was shown the results produced by the system under construction for a BIM model of a house. The IFC file format was used to translate the results of the system's interpretation in the process to transform the initial model attending restrictions and constraints of the proposed fabrication process, i.e. additive manufacture.

It seems that the architect could design its building without worrying too much about the restriction presented by additive manufacturing.

As future work directions, it would be interesting to study how this system could be applied in buildings with geometries very particular, of the kind that could uses all the potential of 3D printers. It certainly would demand the use of a feature-based geometry grammar to deal with the reasoning process by the system [16].

References

[1] Bock T. and Linner T. *Robotic industrialization.* Cambridge University Press, New York, USA, 2015.

[2] Eastman C., Teicholz P., Sacks R., Liston K. *BIM handbook – a guide to Building Information Modeling for owners, managers, designers, engineers, and contractors.* John Wiley & Sons, New Jersey, USA, 2011.

[3] Yamazaki Y., Tabuchi T., Kataoka M., Shimaki D. BIM applications to large-scale complex building projects in Japan. In *International Journal of High-Rise Buildings*, 3(4), 311-323, 2014.

[4] Halabi M. The Sagrada Familia: the starting point of CAD/CAM in architecture. *Scientific Cooperations Journal of Civil Engineerning and Architecture,* 1(2), 12-20, 2016.

[5] Bock T. and Linner T. *Robot-oriented design.* Cambridge University Press, New York, USA, 2015.

[6] McGrawHill Construction. The business value of BIM for construction in major global markets: how contractors around the world are driving innovation with Building Information Modeling. Technical Report, 2014.

[7] Bachmann K. How will BIM benefit product manufactures? On-line: https://www.quora.com/How-will-BIM-benefit-product-manufacturers, Accessed: 14/03/2016.

[8] buildingSMART. IFC4-addendum 1 [final standard]. On-line: http://www.buildingsmart-tech.org/ifc/IFC4/Add1/html/, Acessed: 24/03/2016.

[9] CAMM. Strategic roadmap for the next generation of additive manufacturing materials. Technical Report, PennState University, 2016.

[10] Khoshnevis, B. Automated Construction by Contour Crafting—related robotics and information technologies. *Automation in Construction*, 13 (1): 5–19, 2004.

[11] Kreiger M., MacAllister B., Wilhoit J., Case M. The current state of 3D printing for use in construction. In *Proceedings of the Conference on Autonomous and Robotic Construction of Infrastructure*, pages 149-158, Iowa, United States, 2015.

[12] Rosen D. Multi-scale computer-aided design for additive manufacturing. In *Proceedings of the Intenational Conference on Innovative Design and Manufacturing*, Montreal, Canada, 2014.

[13] Ostrosi E.; Ferney M. A grammar based approach for feature modelling in CAD. *Design computing and cognition,* Springer Netherlands, p. 377-396, 2004

[14] Hornung A., Wurm K., Bennewitz M., Stachniss C., Burgard W. OctoMap: an efficient probabilistic 3D mapping framework based on octrees. *Autonomous Robot*, 34: 189-206, 2013.

[15] Smyth, C. *Functional design for 3D printing – designing 3D printed things for everyday use.* CreateSpace Independent Publishing Platform, 2nd edition, 2015.

[16] Fu Z. and Pennington A. Geometric reasoning based on graph grammar parsing. *Journal of Mechanical Design*, 116(3), 763-769, 1994.

State of Art on Site Space Modeling in Construction Projects

Satish Gore

Architectural Draftsman, SGV International LLC, Houston, TX 77024
Email: sgore@sgvinternational.com

Arian Saeedfar

Graduate Student, Department of Construction Management, University of Houston, Houston, TX 77204
Email: asaeedfar@uh.edu

Lingguang Song

Associate Professor, Department of Construction Management, University of Houston, 111D T1, Houston, TX 77204 ; Phone (713) 743-4377 ; Fax (713) 743-4032
Email: lsong5@uh.edu

Abstract –

Limited space availability is a common issue faced by construction projects and it has great impact on project productivity and safety performance. Effective management of space is challenging due to the highly dynamic and complex nature of construction site environment. Among them is space modeling, i.e. how to capture and visualize the time-phased site space that is constantly evolving. This paper reviews the state-of-art in space modeling studies in thirty publications. These studies are comparatively evaluated based on their methodology, technology basis, applications, maturity, and implementation. Manual processes and techniques have been traditionally used to document as-built site conditions. New space modeling has evolved from this manual, subjective, and 2D format to a more accurate, automated, and integrated approach using technologies, such as photo-modeling, 3D laser scanning, point cloud modeling, and building information modeling. This review provides a better understanding of major shifts of technologies and applications in space modeling research, meanwhile identifies promising areas for future research.

Keywords –

Construction site, space modeling, state of the art, literature review.

1 Introduction

Space is a scarce resource in construction projects. A busy construction site is often characterized by many concurrent activities and continuous movement of workers, equipment, and materials within a limited space [1]. Site congestion can significantly impact productivity, safety and quality performance. Past research estimated an efficiency loss of up to 65% [2]. Therefore, effective management of space as a resource is fundamental to classic construction management functions, such as estimating, scheduling, safety, and quality management. Construction space management involves a continuous effort throughout the project lifecycle in measuring space and its availability, understanding the needs of construction activities and their interaction (e.g. facility, worker, equipment, and material storage), and forecasting and resolving conflicts to ensure effective space usage.

Optimal space management is a challenging task due to the highly dynamic and complex nature of construction. One of such primary challenges is space modeling, i.e. how to capture and visualize the time-phased site space that is constantly evolving. Traditionally, site space is not viewed as a resource, and thus is not managed systematically like other resources, e.g. labor, material, time, and cost. Rather, it is a secondary consideration embedded in other traditional construction management functions. Existing site condition and space availability is captured manually through periodic site walk and personal observation, and later presented in rough 2D sketches for communication and analysis [3] [4] [5]. This manual and sporadic process of capturing space conditions on a constantly changing job site is rather ineffective, costly, and inaccurate [1]. Furthermore, presenting essentially 3D space information on a 2D drawing further reduces the accuracy of information exchange and results in information loss. As a result, the current practice of capturing and visualizing site space is highly dependent on the planner's experience, judgement, intuition, and imagination [6].

The evolution of new reality capturing and visualization technologies has ignited a rethinking of space as a resource and its modeling in construction projects. This research is to review the state-of-the-art on construction site space modeling studies and clarify future research directions. The following section reviews the fundamental technologies employed in space modeling research. This is followed by an overview of the comparison metric we used in evaluating various technologies and modeling techniques. Observations and analysis based on a review of thirty past studies are then presented, such as the major shifts and trends of technologies and applications. The paper also discusses future research needs toward effective space management.

2 Underlying Space Modeling Technologies

There are two aspects of space modeling: capturing space measurements and visualizing them. As discussed previously, the current method of manual and 2D based method is incapable to reflect the dynamic and complex spatial-temporal information, and thus incapable to support the timeliness and accuracy required for project planning needs. Fundamental technologies introduced in recent years to address this challenge can be grouped in two categories: sensing technologies, e.g. photo-modeling and laser scanning, and visualization tools, e.g. Computer Aided Design (CAD) and Building Information Modeling (BIM) which are briefly reviewed in the following sections.

Photo-modeling: Photo-modeling or photogrammetry uses optical cameras to capture 2D photos and later stitches them together to form a 3D point cloud model for measurement and other modeling purposes. This cost-effective and efficient method requires little investment in equipment and time to gather and process data when compared to laser scanning as will be discussed below [7]. It has been used to document as-built site condition in 3D models [8] [7] [9]. However, due to optical distortion, obtaining highly accurate spatial data, e.g. for quality control applications, can be a challenge for photo-modeling [8] [10]. Other limitations, e.g. object occlusion, noise, and capturing large construction site, were also observed [5].

Laser scanning: Laser scanning is primarily used in surveying applications, but it has also been tested for as an alternative method to capture as-built conditions. Based on the reflection of laser pulses and their return time, size and location of objects scanned can be determined by knowing the speed of light [7]. Laser scanning is beneficial in remote monitoring of construction sites [11], recognition of resources on the job site [9], and capturing construction activities and performance data [3] [12]. While producing highly accurate measurements, laser scanning presents some limitations for routine construction management applications [9], such as high initial equipment investment, high operating costs, training requirements, slow scanning speed, and the need of a large number of scans in multiple locations to cover a site [10].

CAD: CAD, a traditional graphic design tool, has also been used to document and communicate site spaces [14]. CAD representations are usually implemented in 3D or even 4-dimensional (i.e. 3D plus time) environments known as 3D CAD and 4D CAD. Comparing with 2D drawing, 3D or 4D CAD model is much more powerful in visualizing site spaces and pinpointing space constraints [14]. It has been used to managing time and space constraints [14] [15], assisting multi-objective optimization [2] [9], automating the process of site monitoring [7], and servicing as a unifying interface for construction management integration [16] [17]. Some limitations were noted in past studies, including limited capability in integrating with other management applications [16], difficulty in generalizing its usage to projects of different types [18] [14], and complexity involved in representing larger numbers of space constraints [19].

BIM: While CAD uses simple geometries to describe an object's shape and location, BIM takes an object-oriented approach in representing individual building elements with both 3D geometric and non-geometric (functional) attributes and relationships. These "intelligent" objects and BIM's open structure make virtual design and construction possible [4]. For space representation, BIM effectively integrates design and construction [20] [7], visualizes workspaces and spatial relationships [19] [18] [21], simplifies design and schedule update [11], and automates data processing and analysis [3].

3 Evaluation Methodology

This section reviews past studies, technologies, and applications which were used for space modeling. Thirty such studies between 1993 and 2015 identified through online search and review of academic journals and conference publications. Nine evaluation criteria were used for the comparison: (1) technology used, (2) model dimensions (2D, 3D, or 4D), (3) level of automation, (4) graphical format (e.g. simple geometry, shape, or surface), (5) software/tools used, (6) model usage (e.g. layout, scheduling, and progress measurement), (7) project stage (e.g. usage during pre-construction or construction), and (8) implementation status (e.g. prototype, pilot study or ready to implement), and (9) system cost (e.g. equipment and operating cost). The objective of this evaluation is to clarify the current status and trend of space modeling as well as dependencies among various system design, technology, and application factors. While some parameters such as hardware used, user-friendliness and level of accuracy were considered as part of the initial analysis, they are not included in the final analysis because either our analysis did not find any unique or significant difference or lack of data for a meaningful analysis.

While most of the criteria are self-explanatory, a few clarifications are necessary. Level of automation refers to the creation process of a space model, including data collection, processing, and visualization. The integration of space models with other tools for a specific construction management function (e.g. safety management) is excluded from the evaluation of the level of automation. Development stage describes the maturity of the developed systems and their readiness for industry implementation.

While most of the evaluation results (e.g. software/tools used) were based on factual data presented in the original publications, certain criteria are evaluated by comparing with the average performance of similar studies as well as authors' judgement. This includes system cost where the following evaluation metric is used: 1) low indicates a low investment on equipment/tools (e.g. in the range of hundreds of dollars), an automated information capture and management procedure requiring little or no user intervention, and minimal training requirement; 2) medium indicates initial investment in the range of thousands of dollars, a semi-automated procedure, and formal user training requirement; and 3) high means initial investment is a magnitude of tenth of thousands of dollars, a manual data processing procedure, and significant user training requirement. The evaluation results are tabulated as shown in Table 1.

Table 1. Evaluation results

Research Paper	Technology Used	Modeling Dimension	Level of Auto.	Graphical Format	Software / Tool Used	Model Usage	Project Stage	Implement. Status	System Cost
Tommelein et al. 1993	MovePlan Model	2D	Manual & Automatic	Simple Line Geometry	Move Plan Model	Layout Planning	During Construction	Pilot Study	High
Thabet et al. 1994	Work-space Constraints Model	2D / 3D	Manual	Simple Surface Geometry	AutoCAD	Scheduling	During Construction	Pilot Study	High
Riley et al. 1997	Detailed Manual Space Planning Method	2D	Manual	Simple Line Geometry	AutoCAD	Space Planning	During Construction	Ready to Implement	High
Akinci et al. 2000	4D WorkPlanner	4D	Manual & Automatic	Simple Surface Geometry	AutoCAD,	Space - Time Conflict	During Construction	Prototype	High
Tawfik et al. 2001	VR, GIS, CAD	2D / 3D	Manual & Automatic	Simple Line Geometry	AutoCAD	Space & Risk Analysis, Site Planning	Pre-construction	Pilot Study	Moderate
Guo 2002	Auto CAD + MS Project	2D	Manual & Automatic	Simple Line Geometry	AutoCAD	Work Space Conflict	Pre-construction	Prototype	High
Waly et al. 2002	VCE, CAVE	3D	Manual & automatic	Simple Surface Geometry	AutoCAD	Virtual Planning, Space Conflict	Pre-construction	Pilot Study	High
Akinci et al. 2002	4D CAD Simulation + Excel	4D	Manual & Automatic	Simple Surface Geometry	AutoCAD	Space - Time Conflict	During Construction	Ready to Implement	High
Sriprasert et al. 2003	Multi-constraint Planning	4D	Manual & Automatic	Advanced / Reality based Geometry	AutoCAD	Site Productivity and Risks	During Construction	Ready to Implement	High
Hesham et al. 2003	2D AutoCAD + Genetic Algorithm	2D	Manual & Automatic	Simple Line Geometry	AutoCAD	Site Layout Planning	During Construction	Ready to Implement	Moderate
Wang et al. 2004	4D-MCPRU, 3D AutoCAD + MS Project	4D	Manual & Automatic	Advanced / Reality based Geometry	AutoCAD	Resource Management and Layout Assessment	Pre-construction & During Construction	Ready to Implement	Moderate
Shih et al. 2006	3D Laser Scanning & Octree Author	4D	Automatic	Point Cloud - Advanced	Laser Scanner	Inspection, Quality Control, Progress Measurements	During Construction	Ready to Implement	High
Dawood et al. 2006	Critical Space-time Analysis (CSA)	4D	Manual & Automatic	Simple Line / Surface Geometry	AutoCAD	Space Planning, Space Conflict	During Construction	Prototype	High
Jongeling et al. 2007	4D CAD Models, Line-of-balance	4D	Manual & Automatic	Advanced / Reality based Geometry	AutoCAD, Archi CAD	Planning of Work-flow	Pre-construction	Prototype	High
Dai et al. 2008	Photogrammetry + Photo modeler software + CAD / Sketch up	3D	Automatic & Semi automatic	Advanced / Reality based Geometry	DSLR, Sketch up	Construction Simulation Visualization	During Construction & Post Construction	Prototype	Low
Choi et al. 2008	Stereo Vision System + AutoCAD / Mat lab	3D	Semi Automatic & Automatic	Simple Line / Surface Geometry	Three Sensor base Line Stereo Camera	Progress Monitoring	During Construction & Post Construction	Prototype	High
El-Omari et al. 2008	LADAR Laser Scanning + Photogrammetry	3D	Automatic	Point Cloud - Advanced	DSLR, Laser Scanner	Progress Monitoring	During Construction	Prototype	High
Golparvar-Fard et al. 2009	4D AR Model, SfM, Algorithms	3D / 4D	Automatic	Point cloud - Moderate	DSLAR	Progress Monitoring, Workspace Logistics, Safety, Construction Productivity	During Construction	Ready to Implement	Moderate

Research Paper	Technology Used	Modeling Dimen.	Level of Auto.	Graphical Format	Software / Tool Used	Model Usage	Project Stage	Implement. Status	System Cost
Benjaoran et al. 2010	4D CAD Model, Rule-based Algorithms	4D	Manual & Automatic	Simple Surface Geometry	AutoCAD	Safety and Construction Management	During Construction	Ready to Implement	High
Tang et al. 2010	BIM + Algorithms	3D	Semi Automated & Automatic	Simple Surface Geometry, Point Cloud - Advanced	BIM, Laser Scanner	Progress Monitoring	Post Construction	Prototype	High
Hyojoo et al. 2010	3D Structural Recognition Model, Stereo Vision	2D / 3D	Semi Automated & Automatic	Advanced / Reality based Geometry	Stereo Vision Camera, Mat lab, AutoCAD	Construction Progress Monitoring	During Construction	Ready to Implement	High
Golparvar-Fard et al. 2011	Image Based Reconstruction and Modeling + 3D Laser Scanning	3D	Automatic & Semi Automatic	Point Cloud - Advanced	DSLR, Laser Scanner	Performance, Progress & Safety Monitoring	During Construction	Ready to Implement	High
Bansal 2011	Matrix Based Tool + Mat lab / 3D GIS	3D / 4D	Semi Automatic & Automatic	Advanced / Reality based Geometry	Mat lab, Arc GIS	Space Planning, Time Space Conflict	Pre-construction & During Construction	Prototype	High
Gore et al. 2012	Photogrammetry	3D	Automatic & Semi Automatic	Point Cloud - Moderate	DSLR Camera	Space Conflicts / Planning	During Construction & Post Construction	Prototype	Low
Moon et al. 2014	4D Simulation, V-CPM, Bounding Box Model, Algorithm, BIM	4D	Semi Automatic & Automatic	Simple Surface Geometry	BIM	Workspace Conflict, Safety	During Construction	Prototype	High
Choi et al. 2014	4D BIM	4D	Manual & Automatic	Advanced / Reality based Geometry	BIM	Quality, Productivity, Safety	Pre-construction	Ready to Implement	High
Su et al. 2014	Arc- GIS 10.0	4D	Manual & Automatic	Simple Surface Geometry	AutoCAD, Arc GIS	Construction Planning	Post Construction	Prototype	High
Kumar et al. 2015	BIM based Dynamic Layout Model	4D	Manual & Automatic	Advanced / Reality based Geometry	Revit	Dynamic Site Layout Modeling	Pre-construction	Prototype	High
Wang et al. 2015	BIM and Firefly Algorithm (FA)	4D	Manual & Automatic	Advanced / Reality based Geometry	BIM	Optimal Tower Crane Layout Planning	Pre-construction & During Construction	Ready to Implement	High
Zhou et al. 2015	4D Activity / Operation Level Models + Algorithms for AR, Simulation	4D	Manual & Automatic	Advanced / Reality based Geometry	BIM	Resource Allocations, Safety & Risk Assessment, Site Layout & Route Planning	Pre-construction & During Construction	Ready to Implement	High

Implement. Status = Implementation Status

- Pilot Study - Used in academic ready for field implementation, not ready for implementation, require case study.
- Prototype - Prototyping demonstration, not ready for implementation.
- Ready to Implement - Multiple case studies performed ready for field implementation.

Modeling Dimen. = Modeling Dimensions

CAD = Computer Aided Design.
VCE = Virtual Construction Environment.
AR = Augmented Reality.
SfM = Structure from Motion.
BIM = Building Information Modeling.
V-CPM = Virtual Construction Project Manager.

Level of Auto. = Level of Automation

- Manual & Automatic – Manual data collection, automation for processing.
- Manual – Manual data collection & processing.
- Automatic – Automatic data collection, high level of automation for processing.
- Automatic & Semi-automated – Automatic data collection & semi-automated processing.
- Semi-automatic & Automatic – Semi-automatic data collection and automatic data processing.

4 Observations and Analysis

Observations and trends identified through the above-mentioned comparative studies are discussed in this section.

4.1 Graphical representation format

Tremendous improvement in the graphical representation of spaces can be observed. The representation format evolved from simple line geometry, surface constructs, point clouds, to advanced reality-based geometrical forms, as shown in Figure 1. Traditionally, simple line geometry was used to represent as-built conditions [22]. Waly and Thabet used surface geometry to create space models [2]. More recent research used point-cloud to represent space model [20] [5]. However, the latest trend shows the use of advanced or reality-based geometry for representing space models. Overall, graphical geometry is transforming towards more complex and reality based representation. This trend not only improves the accuracy in modeling spaces in 3D but also provides a more intuitive way to communicate spatial information to field staff.

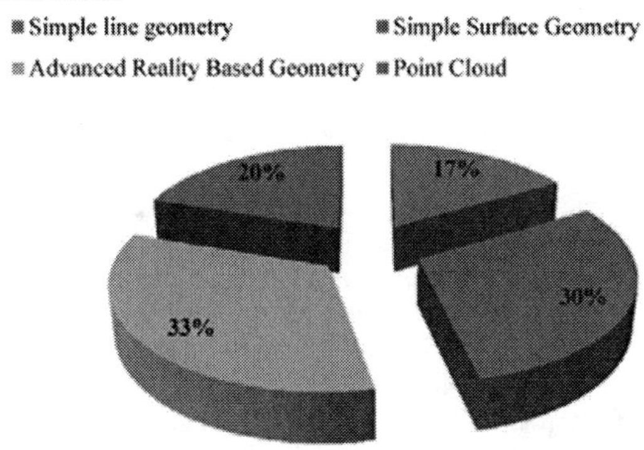

Figure 1. Graphical representation format.

4.2 Model dimension

As mentioned previously, 2D drawings are commonly used for space related construction management tasks [23]. However, space by nature is three dimensional. Not surprisingly, the model dimension in space modeling has evolved from dominantly 2D format to more recent 3D and even 4D format, as shown in Figure 2. Analysis of modeling dimension parameter shows a gradual evolution of space models from two dimensions (2D) to three dimensions (3D) and later to four dimensions (4D). In particular, the 4D format captures not only 3D spaces but also their changes along the project timeline, making it particularly effective for space modeling and analysis in dynamic construction field environment.

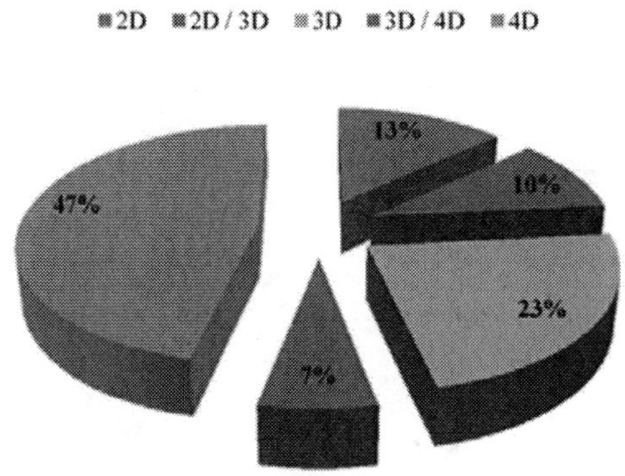

Figure 2. Model dimensions.

4.3 Data capturing, processing, and automation

The analysis of the level of automation suggests two related aspects: 1) capturing raw spatial data, and in many cases, this means capturing as-built site conditions, and 2) processing the raw data for the development of a space model. In early studies, data collection was a manual task through individual observations or measurements, and later these data are manually entered into CAD or BIM tools or in some cases, captured in sketches for further analysis. To improve both accuracy and efficiency, semi-automatic or automatic data collection methods have been introduced. With the two primary technologies, optical camera and 3D laser scanner, field staff can "scan" the job site in multiple locations and later the scans are stitched together to derive a point-cloud model and eventually a 3D site model. The stitching process may be automated such as the case of photo-modeling [5] or may involve users to manually fuse the data sets from multiple scans.

4.4 Model usage and maturity

An analysis was conducted to understand space modeling and its applications in different construction project phases, namely pre-design, design, pre-construction, construction, as-built, and post construction. As shown in Figure 3, majority of space models have been used during construction to capture as-built conditions of a given construction site. Moreover, some are used during pre-construction and some during post-construction for documentation purposes. It is observed that, in early studies, space model tends be a single purpose model, i.e. it is used only for a specific management functions, such as layout planning. The newer reality-based space modeling elevates the accuracy, visualization, and interoperability to a new level. It provides a generic platform for accurate space modeling that supports a multitude of construction management functions. It will not only minimize the time and cost required for collecting, processing, and modeling spatial data, but also streamline the analysis of such data throughout the life cycle of a construction project.

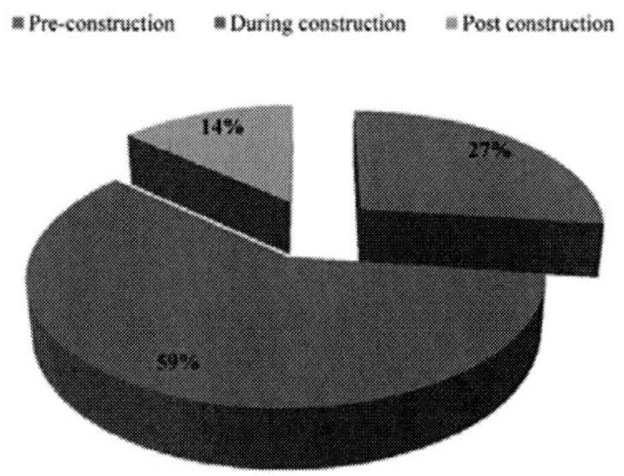

Figure 3. Model usage.

Furthermore, as illustrated in Figure 4, the implementation status of past studies implies that majority of studies are either ready for field implementation after several field tests or prototyped to prove the concept (i.e. not yet ready for field implementation).

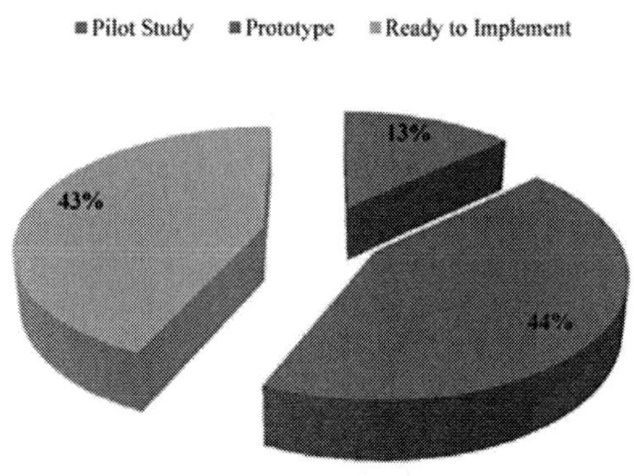

Figure 4. Implementation status.

4.5 User friendliness and system costs

In the case of user-friendliness, majority studies mentioned the requirement of training in implementing space modeling. Some of the training requirements observed were related to information handling, software/tool usage, as well as maintenance of equipment used in capturing site conditions. As shown in Figure 5, system costs space modeling tends to be high. It is worth noting that interdependency exists between level of automation, user-friendliness, and system costs. If training is required while the system has low level of automation, this will drive up the operating costs. Meanwhile, technologies such as photo-modeling, the use of open ended software, and better system integration for automation will be not only help to lower the overall system cost but also make the process more user-friendly.

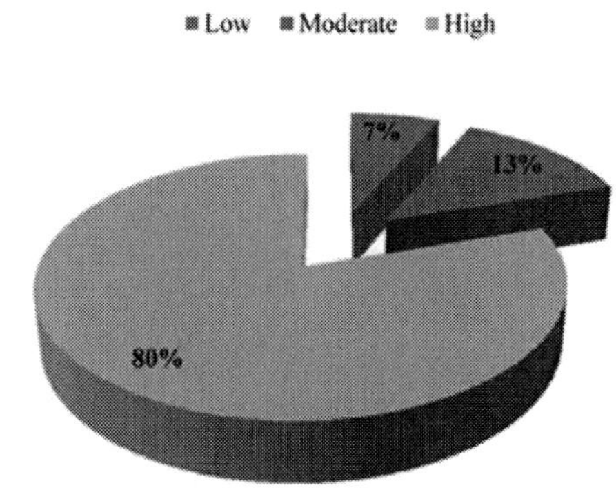

Figure 5. System costs.

4.6 Future research directions

This state-of-art review also offers a peek into future research needs to achieve a comprehensive space management capability. First, regardless of the technologies used, past research efforts suggest further increasing the level of automation in data capturing and handling thus minimizing the amount of manual data input [22] [24]. Second, most of the applications aim at a specific type of construction or a particular level of detail and the space modeling should be more inclusive in regard to the construction type or detail level [3] [19]. Third, to meet multiple project management needs, space modeling must have a standardized and open structure for a better integration with other management functions, e.g. scheduling and project control [7] [5]. As an example, one of our recent research, Dynamic Space Conflict Modeling (DSCM), integrates space modeling with scheduling and productivity and safety analysis. It captures space demand of each construction activity over time and analyses space conflicts and their impact on productivity and safety performance. Finally, expanding the usage of real-time sensing and modeling coupled with immersive visualization technologies such as Augmented Reality (AR) and Virtual Reality (VR) can make space modeling much more user friendly and accessible even for field staff [25] [9]. A notable number of researches have intended to develop the capability to allow users to use real-time monitoring tool to build virtual construction site models. As an example, an active research of the authors involves using Google's Project Tango to sense a space and build a virtual model in real-time. In addition, the new generation of VR equipment (e.g. Oculus Rift, Microsoft HoloLens, and Razer OSVR) along with natural human-computer interface based on gesture tracking can allow users to have an immersive experience of spaces [25][26][9].

5 Conclusion

This paper highlights the importance of space as a resource on a construction site, and more importantly, existing solutions in capturing and visualizing site spaces. We reviewed thirty publications in the emerging field of study in construction site space modeling. These studies are comparatively evaluated based on their methodology, technology basis, applications, maturity, and implementation. Most of the proposed research studies intended to offer innovative solutions for space modeling using latest sensing and visualization technologies. Manual processes and techniques have been used to document as-built site conditions. Our review suggest that sensing, point cloud modeling, and CAD/BIM are widely researched in recent years for a more accurate, automated, and integrated space modeling process. This lays a foundation to achieve a common platform of space modeling to support multiple management functions that must consider space as a resource.

References

[1] Riley D. R. and Sanvido V. E. Space planning method for multistory building construction. *Journal of Construction Engineering and Management,* 123(2):171–180, 1997.

[2] Sriprasert E. and Dawood N. Multi-constraint information management and visualisation for collaborative planning and control in construction. *Electronic Journal of Information Technology in Construction,* (8):341–366, 2003.

[3] Waly A. F. and Thabet W. Y. A virtual construction environment for preconstruction planning. *Automation in construction,* 12(2):139–154, 2003.

[4] Golparvar-Fard M. Bohn J. Teizer J. Savarese S. and Peña-Mora F. Evaluation of image-based modeling and laser scanning accuracy for emerging automated performance monitoring techniques. *Automation in Construction,* 20(8): 1143–1155, 2011.

[5] Gore S. Song L. and Eldin N. Photo-Modeling for Construction Site Space Planning. In *Proceedings of the Construction Research Congress,* pages 1350–1359, West Lafayette, USA, 2012.

[6] Guo S. Identification and Resolution of Work Space Conflicts in Building Construction. *Journal of Construction Engineering and Management,* 128(4):287–295, 2002.

[7] Nuri C. Son H. Changwan K. Changyoon K. and Hyoungkwan K. Rapid 3D object recognition for automatic project progress monitoring using a stereo vision system. In *Proceedings of The 25th International Symposium on Automation and Robotics in Construction,* pages 58–63, Vilnius, Lithuania, 2009.

[8] Dai F. and Lu M. Photo-based 3D modeling of construction resources for visualization of operations simulation: Case of modeling a precast façade. In *Proceedings of the 40th Conference on Winter Simulation*, pages 2439–2446, Piscataway, USA, 2008.

[9] Zhou Y. Ding L. Wang X. Truijens M. and Luo H. Applicability of 4D modeling for resource allocation in mega liquefied natural gas plant construction. *Automation in Construction,* 50:50–63, 2015.

[10] El-Omari S. and Moselhi O. Integrating 3D laser scanning and photogrammetry for progress measurement of construction work. *Automation in construction,* 18(1):1–9, 2008.

[11] Naai-Jung S. and Huang S. 3D scan information management system for construction management. *Journal of construction engineering and management,* 132(2):134–142, 2006.

[12] Tang P. Huber D. Akinci B. Lipman R. and Lytle A. Automatic reconstruction of as-built building information models from laser-scanned point clouds: A review of related techniques. *Automation in construction,* 19(7): 829–843, 2010.

[13] Sarcar M. Rao K. and Narayan K. *Computer aided design and manufacturing*. PHI Learning Pvt. Ltd., New Delhi – 110001, 2008.

[14] Heesom, D. and Mahdjoubi L. Trends of 4D CAD applications for construction planning. *Construction Management and Economics,* 22(2):171–182, 2004.

[15] Akinci B. and Fischer M. 4-D Workplanner: A Prototype System for Automated Generation of Construction Spaces and Analysis of Time-Space Conflicts. In *Proceedings of the 8th International Conference on Computing in Civil and Building Engineering*, pages 740–747, Stanford, USA, 2000.

[16] Wang, H. J. Zhang J. P. Chau K. W. and Anson M.

4D dynamic management for construction planning and resource utilization. *Automation in Construction,* 13(5):575–589, 2004.

[17] Moon H. Dawood N. and Kang L. Development of workspace conflict visualization system using 4D object of work schedule. *Advanced Engineering Informatics,* 28(1):50–65, 2014.

[18] Dawood N. and Zaki M. Construction workspace planning: assignment and analysis utilizing 4D visualization technologies. *Computer-Aided Civil and Infrastructure Engineering,* 21(7):498–513, 2006.

[19] Akinci B. Fischen M. Levitt R. and Carlson R. Formalization and automation of time-space conflict analysis. *Journal of Computing in Civil Engineering,* 16(2):124–134, 2002.

[20] Rogier J. and Olofsson T. A method for planning of work-flow by combined use of location-based scheduling and 4D CAD. *Automation in Construction,* 16(2):189–198, 2007.

[21] Benjaoran V. and Bhokha S. An integrated safety management with construction management using 4D CAD model. *Safety Science,* 48(3):395–403, 2010.

[22] Tommelein I. D. and Zouein P. P. Interactive dynamic layout planning. *Journal of Construction Engineering and Management,* 119(2):266–287, 1993.

[23] Tawfik H. and Fernando T. A simulation environment for construction site planning. In *Proceedings of the Fifth International Conference on Information Visualisation*, pages 199–204, London, United Kingdom, 2001.

[24] Thabet W.Y. and Beliveau Y. J. Modeling work space to schedule repetitive floors in multistory buildings. *Journal of Construction Engineering and Management,* 120(1):96–116, 1994.

[25] Kumar S. and Cheng J. A BIM-based automated site layout planning framework for congested construction sites. *Automation in Construction,* 59: 24–37, 2015.

[26] Wang J. Xuedong Z. Wenchi S. Xiangyu W. Bo X. Kim M. and Wu P. A BIM-based approach for automated tower crane layout planning. *Automation in Construction,* 59:168–178, 2015.

[27] Golparvar-Fard M. Peña-Mora F. and Savarese S. D^4AR–a 4-dimensional augmented reality model for automating construction progress monitoring data collection, processing and communication. *Journal of information technology in construction,* 14(13):129–153, 2009.

[28] Bansal V. K. Use of GIS and Topology in the Identification and Resolution of Space Conflicts. *Journal of Computing in Civil Engineering,* 25(2):159–171, 2011.

[29] Choi B. Lee H. Park M. Cho Y. and Kim H. Framework for work-space planning using four-dimensional BIM in construction projects. *Journal of Construction Engineering and Management,* 140(9):0401–4041, 2014.

[30] Osman H. Georgy M. and Ibrahim M. A hybrid CAD-based construction site layout planning system using genetic algorithms. *Automation in construction,* (12):749–764, 2003.

[31] Son H. and Kim C. 3D structural component recognition and modeling method using color and 3D data for construction progress monitoring. *Automation in construction*, 19(7): 844–854, 2010.

[32] Su X. and Cai H. Enabling Construction 4D Topological Analysis for Effective Construction Planning. *Journal of Computing in Civil Engineering*, 10.1061/(ASCE)CP.1943-5487.0000463, 04014123-01–04014123-10, 2014.

Interoperability with IFC in the automated rebar fabrication

Maciel, Alex R. [a] **and Corrêa, Fabiano R.** [b]

Escola Politécnica, University of Sao Paulo, Brazil
E-mail: [a] alex_roda@usp.br , [b] fabiano.correa@usp.br

Abstract –

Some problems of efficiency in the cut and bent rebar supply chain is related to stakeholder's interaction and the flow of information between designers, constructors and rebar fabricator. In the actual process, distinctive standards are used for exchange of information between design and fabrication. The BVBS format represents a standard widely used for automated cut and bent rebar production while the Industry Foundation Classes (IFC) aims to become the standard for exchange in the AEC industry with the use of Building Information Modeling (BIM). This study aims to explore the use of IFC schema to enable an effective integration of the rebar fabrication process with the BIM workflow, improving data exchange and reducing the need for manual intervention. Initially, the requirements for exchanging reinforcement data laid down in BVBS standard are reviewed. Key requirements to a digital interface are parsed across IFC schema specification in order to ascertain how this information are mapped on IFC. The current way for geometric representation of reinforcing bars, used by some structural BIM tools to export IFC files, is reviewed, in addition with new entities release in IFC4. Some recommendations are provided for improving the interoperability efficiency.

Keywords –

BIM; IFC; Reinforcement bar; Automated bending machine;

1 Introduction

Several authors [1][2][3][4] have studied the interaction between agents involved in the supply chain of reinforcement bars. Although some activities, such as raw material supply, design of required reinforcement and on-site assembly, have a clear actor designated to that role, other activities such as detailing, reckoning and fabrication can be performed by different actors depending of particular supply chain configuration [1].

In the Brazilian cut & bent rebar supply chain, both the design and rebar's detailing activities are performed by the engineering firm and delivered to the constructor or General Contractor (GC) [2]. This configuration features a high dependence among stakeholder's, particularly for the fabricator [1] that carries out the reckoning, produces and deliver the rebar goods to construction site based on the drawings and order form handed in by GC. It is crucial that they received accurate information [2].

The main source of problems in the Brazilian case [4] is related to failures on information flow and communication between the agents involved, especially due to: a) imprecise order and b) poor quality of structural design.

The use of Building Information Modeling (BIM) contributes to improving the quality and integrity of the process. The production of detailing drawings [5] directly from 3D coordinated reinforcement models (with bi-directional associativity) can reduce errors and considerably enhance its consistency. To that end, it is necessary to establish how the information could be exchanged from BIM design environment and the standard accepted by the automated process of cut and bend rebar.

Currently, the main among the available formats used in cut and bent rebar for Cast-in-place (CIP) is the BVBS standard, allowing a digital data exchange with CAM controlled bending machine or Production Planning and Scheduling (PPS) software. However, it is limited to geometry and some references between detailing and placing rebar drawings, and do not allow the addition of information under the contractor's responsibility relating to the order and delivery date.

Another point to be considered is the feasibility of this interface, which demands synergy between the engineering firm and fabricator. Due to the supply chain configuration, such as Brazilian case, these agents does not collaborate directly. This link is effected through the general contractor and he acts as an intermediate.

The importance of improvement of existing interfaces or development of new standards is highlighted by [5] in order to reduce the need for manual intervention by the fabricator. However, it must also consider interfaces that allow the reuse of information created along the design, detailing, planning and procurement phases and enables the collaboration between all stakeholders.

Nowadays, the improvement of existing standards maintaining compatibility with existing cut and bent reinforcement industrial plants can be considered a suitable choice to be held alongside the development of new standards.

For being a consolidated and non-proprietary format for information exchange in the AEC industry, the use of the Industry Foundation Classes (IFC) appears as a good option alongside rebar supply chain.

In this study we analyzed the requirements for exchanging reinforcement data laid down in BVBS standard in order to ascertain how this information are mapped on IFC schema. There is a particular interest in the geometric representation of reinforcing bars, and we provide a comparison of forms present in some structural BIM tools to export IFC files, in addition with the reviewed new entities released in IFC4.

2 Design-production digital interface

Although the CAD-CAM integration have been in use for several years, the majority of standards and file formats have been developed considering a CAD context focused on bending patterns based on 2D drawings **Error! Reference source not found.**. This solution is effective especially when both detailing and production are under fabricator's control as in the United States. Otherwise, it requires a smooth communication among the players.

Table 1. List of some standard files formats used for digital interface

Name	Developer	File format
BVBS	BundesVereinigung der BauSoftwarehäuser E.V.	.abs
ProgressXML	Progress Maschinen & Automation AG.	.pxml
Unitechnik 7.0	Unitechnik Systems GmbH	.uxml
Unitechnik 6.1	Unitechnik Systems GmbH	.cam
Rebar Data Exchange	Applied Systems Associates, Inc. (aSa)	.rdx

There is a variety of standards and files formats on the market, which allows the transcription of the information contained in the reinforcement detailing design in a digital format, as shown in Table 1. Some formats are proprietary while others are created jointly by the cutting and bending rebar supply chain stakeholders.

Amid the available formats the BVBS standard represent the main used on automation of cut and bent rebar manufacture for Cast-in-Place industry. ProgressXML and Unitechnik standards are widely used in precast industry, generally for production of precast wall panels and floor slab.

Some BIM authoring tools have native support for these standards, while others require a third-party plugin to accomplish the design-production digital interface.

3 BVBS interface

The BundesVereinigung der BauSoftwarehäuser standard (BVBS) [6] was developed in consensus by bending machine factories, construction software companies, reinforcement bending works, steel producers and academic institutions to assist the exchange of information between rebar detailing software and CNC controlled bending machines or PPS software without manual intervention.

The BVBS specification features a data structure from the designer's perspective regardless of the fabrication machine which will be used and may also be understandable without authoring CAD/BIM tool. The reinforcement data are exchanged through a ASCII encoded text file and, as other CAD-CAM formats, BVBS is focused on 2D drawings.

These files are composed by a data string divided into blocks, and preceded by a recognition code used to set the shape type group, which can be: two-dimensional rebar (BF2D), three-dimensional rebar (BF3D), spiral links (BFWE), mesh (BFMA) or lattice girders (BFGT). The recognition code also enables the machine to check if it will be able to produce the specified shape.

The blocks in the file are arranged in the following order:

1. Header block (H): provides data about identification and characteristics of the bar;
2. Geometry block (G): describes the rebar's shape bending geometry;
3. Chair mesh block (A): defines the positions of the chair mesh in relation to bars.
4. Bar block (X/Y): used only to mesh for define a diameter, bar origins and length;
5. Private block: used for project or other internal data;
6. Checksum block (C): for a checksum value.

Besides being a standard widely used in rebar's Cast-in-Place industry, the BVBS choice was held because it provides the data required by CNC controlled bending machine and can be used either directly at machine via USB or via barcode as a mass production workflow, via PPS software.

Although this standard encompasses a wide variety of reinforcements types like spiral links, standardized meshes or engineered meshes, this article focused only on 2D and 3D reinforcement bars, and in the following sections the header and geometric blocks will be explored in more details.

3.1 Header block requirements

The requirement of the BVBS header block can be split essentially into three groups of information containing:

- **Identification and document reference:** project number (j), drawing number (r) and revision number index (i);
- **Material and rebar properties:** steel grade (g), bar diameter (d) and bending diameter (s);
- **Quantity Sets:** bar length (l), item quantity (n), weight per bar item (e);

3.2 Geometric block requirements

The coordinate system adopted by BVBS to describe the reinforcement bar geometry is set according to the shape group. While the BF2D (2D rebar) uses polar coordinate, the BF3D (3D rebar) is described in Cartesian coordinates.

The measuring of shape dimensions is based on external length, similarly to the way predicted by some detailing standards as American ACI 315 [7], British BS 8666 [8] or European ISO 3766 [9].

BF2D geometry is described by the leg length (l) and subsequently by angle of the following bend (w). Each bend is determined by the bending diameter indicated in the header block and remains unique for all transitions. In case where distinctive bending diameters are needed, the radius of the bent element (r) must be set on the geometric block.

For defining the BF3D geometry the coordinates of the bar's vertex (X,Y,Z) shall be provided.

Figure 1 illustrates the BVBS specification for a two-dimensional reinforcement bar.

The header block therefore reads:

BF2D@HjTestPDF@r417@ia@p1@l1000@n10@e0.888@d12@g500S@s48@v@

Example 1: (59 characters)

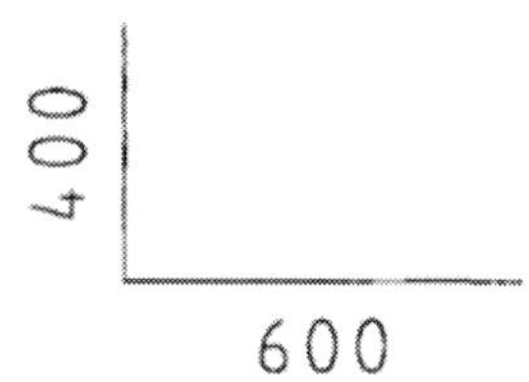

The geometry block reads:

Gl400@w90@l600@w0@

The complete string therefore reads:

BF2D@HjTestPDF@r417@ia@p1@l1000@n10@e0.888@d12@g500S@s48@v@
Gl400@w90@l600@w0@
C88@CRLF (83 characters)

Figure 1. Example of a two-dimensional reinforcement bar (BF2D) according to BVBS specification [6]

4 Interoperability with IFC

Since 2013, the Industry Foundation Classes (IFC) is a ISO standard (ISO 16739:2013) proposed by buildingSMART, and currently is in its IFC4 Add. 1 [10] version. It is a data schema that encodes information related to the entire lifecycle of a building.

In its data architecture, there is a layer of data with many different discipline domains including the Structural Elements Domain, where most of the entities necessary to the workflow described here are modelled.

From the 17 entities included in structural elements domain, 8 are new entities in IFC4, and other 6 were changed following the difference in conceptual modelling proposed by this new version of IFC. It appears to be advocating the use of object types for attribute specifications instead of using the very instance of the object entity, which is a positive change.

The reinforcement of a concrete structural element can be defined in multiple ways when relied upon the production and assembly hierarchy [11] [12]: a) individual bars; b) a set of identical bars distributed on a row pattern (e.g. stirrup) or a group of rebar with same function; or c) a rebar cage.

The way how the reinforcement will be represented in the IFC schema also depend on the way they were defined.

Information related to rebar's geometry and properties are defined at object-type level by means of *IfcReinforcingBarType* entity. Through the mapped representation associated with a single or multiple

IfcMappedItem, individual rebars or sets of same rebars can be represented in the object-level occurrence as shown in Figure 2 .

The *IfcReinforcingBar* (attributes changed in IFC4) and *IfcReinforcingBarType* (new in IFC4) entities are the central objects for the discussion in the following sections that presents how the information regarding the header and geometry blocks of BVBS could be represented inside the IFC specification.

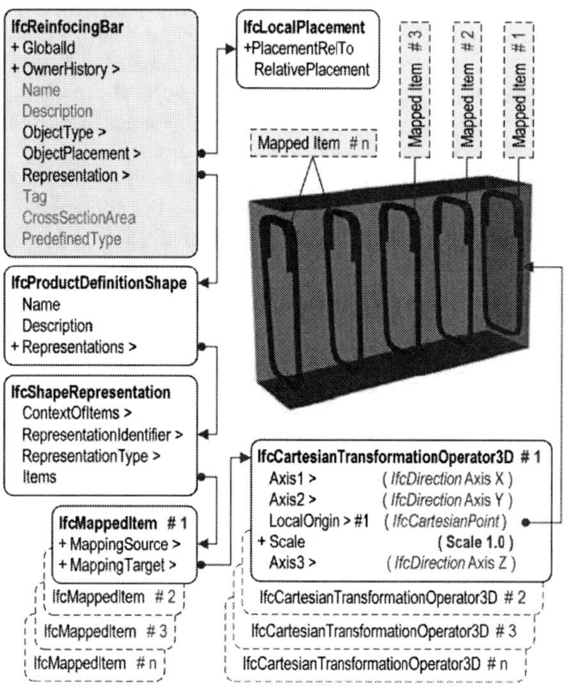

Figure 2. Example of a set of the same rebars (stirrup) represented by multiple mapped entities.

4.1 Identification and document reference

Throughout the structural design, several documents are produced. But despite being BIM-based generated, it has some graphic information that presently remains on the authoring BIM tool.

Currently, the design documentation is focused fundamentally in a paper-based format [13]. However, it is possible to produce detailed and placing drawings directly from the 3D coordinated reinforcement models, often using a 2D orthographic projections. Annotation, schedules, 2D details, notes and symbols used on drawing sheets are not explicitly represented in the IFC model.

In order to fulfill the current reference document requirements of BVBS header block it's recommended that reference information of detailing and placing drawings be referenced in the IFC through the *IfcRelAssociatesDocuments* objectified relationship.

This set of documents, issued in printed or digital formats, are extremely important to allow the tracking and reinforcement assembly on site. Though not stored explicitly in the model, external documents can be described on the IFC by means of *IfcDocumentInformation* or have their location, identification, name and description designated by an *IfcDocumentReference* entity.

An *IfcDocumentInformation* captures external document metadata, providing a set of identification and source information, and control information like status revision and data.

The rebar mark identifier (item number) would be provided by the *IfcElement.Tag* attribute that is inherited by the *IfcReinforcingBar* entity.

4.2 Material and rebar properties

Before the release of IFC4, in the 2x3 version that is still in use throughout the market, the reinforcement bar proprieties such as steel grade and diameter were described in *IfcReinforcingElement* and *IfcReinforcingBar* entities, respectively.

In IFC4, these values were deprecated and currently the diameter is given by *IfcReinforcingBarType*'s nominal diameter attribute and the steel grade turned to be provide by *IfcMaterial* entity.

Besides the steel grade attribute, defined in *IfcMaterial*, an applicable standard can be appointed as an external reference by *IfcExternalReferenceRelationship* entity. If required, the mechanical properties shall be provided by the set of properties specified in the *Pset_MaterialSteel*.

Reinforcing bars have standardized sizes and characteristics determined in accordance with national standard (e.g. NBR 7880 in Brazil, ASTM A615 in the US). Besides defining the bar's properties these standards also define the types steel mill rebars produced by a particular country.

The rebar steel mechanical characteristics are defined based on the steel grade (yield strength level) and all bar properties are set based upon the nominal diameter (or bar designator number) given by the standard adopted.

Currently this information is handled asunder by the IFC schema where inherent to bar properties are defined in the context of instance or element type. The ability to link the bar properties via external reference relationship with an applicable standard (by means of *IfcClassificationReference*) together with an *IfcReinforcementBarProperties* entity are possible but nowadays this entity is used for a specify section rebar proprieties on an early design phase defining a required as-design reinforcement.

This would require disjoin the section's particular properties, as such *TotalCrossSectionArea*, *EffectiveDepth* and *BarCount* attributes, of the standardized rebar's properties. An option could be the creation of a new entity (e.g. *IfcSectionReinforcementBarProperties*) and the subsequent use of *IfcReinforcementBarProperties* entity only to the specific rebar properties.

4.3 Quantity Sets

The quantities associated with the rebar elements are represented by an *IfcElementyQuantity* entity. An objectified relationship between the *IfcReinforcingBar* and the *IfcElementQuantity* is represented by the *IfcRelDefinedByProperties* entity. It is possible to represent one or a row of rebars with the *IfcReinforcingBar*, and the quantities must be calculated accordingly.

Currently, there is a reinforcement quantity definition template in IFC, *Qto_ReinforcingElementBaseQuantities* that predicts the following base quantity measures: count, length and weight. The *IfcElementQuantity* that contains this data is also related to three specific *IfcPhysicalQuantity* (which is an abstract supertype): *IfcQuantityCount*, *IfcQuantityLength*, and *IfcQuantityWeight*.

These quantities are derived from the physical properties of the elements and related with the global identifier of each instance.

4.4 Geometric representation

Although it is recommended [5] [10] the use of a swept disk solid as the rebar section, swept along a three dimensional curve (directrix) to define the rebar geometry, the method of how the segments of this directrix will be made is not standardized by the buildingSMART in IFC.

This swept disk solid can be defined by *IfcSweptDiskSolid* entity or its subtype *IfcSweptDiskSolidPolygonal*. While the directrix attribute is defined in *IfcSweptDiskSolid* by an *IfcCurve* type, *IfcSweptDiskSolidPolygonal* require the use of a *IfcPolyline*.

An *IfcCurve* is an abstract entity used to represent curves in 2D or 3D space. This entity is the supertype of many subtypes like lines, polylines, circles and others primitives used to represent bounded and unbounded curves.

The reinforcement bar shape is a curve with finite length composed by straight and arc segments. Although it is possible to use any subtype of *IfcCurve*, some options demand a combination of other IFC entities to describe this composite curve. This fact can affect the size of the models files, compromising the performance and usability of the model.

The subtype *IfcSweptDiskSolidPolygonal*, included in IFC4, allowed an additional fillet radius attribute applied to all transitions of polyline segments requiring only the definition of straight segments of the directrix for a rebar geometric representation which can be a smart choice, especially in case of rebar shapes with standard bending radius.

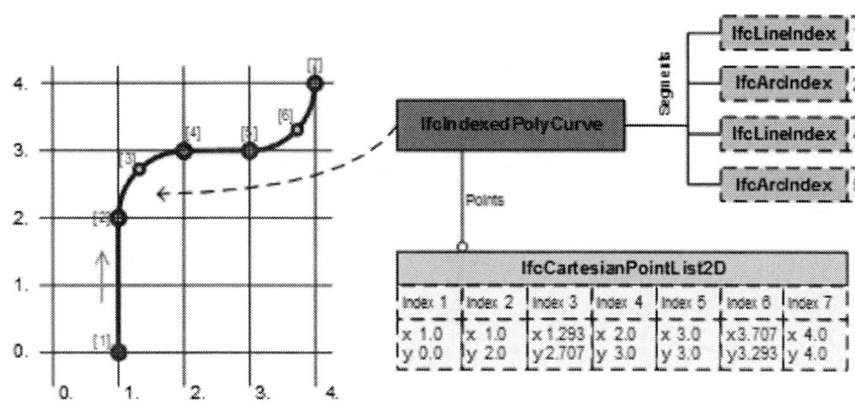

Figure 3. *IfcIndexedPolyCurve* entity included in IFC 4 Addendum 1 [10]

Also, the current IFC release provided a simplified and more compact way for the geometric representation of composite curves using a list of indexed segments with the employment of the *IfcIndexedPolyCurve* entity, a subtype of *IfcBoundedCurve* (Figure 3).

The use of swept disk solid directrix definition to drive a geometric shape on fabrication can be considered as a common point for geometric data exchanging.

In deciding the best approach to model or represent reinforcing bars, one must consider that most of the its shapes are produced with the minimum bent diameter determined in accordance to national structural design standard (e.g. NBR 6118, ACI 318, Eurocode 2) along with reinforcement detailing standards such as ACI 315, BS 8666 or EN ISO 3766. It is generally considered a multiplier of rebar nominal diameter, relative to a structural purpose, to specify the bending machine mandrel diameter.

Due to this, the use of *IfcSweptDiskSolidPolygonal* with fillet radius attribute defined according these standards should be considered, serving almost all cases. This option also favor the bending diameter attribute track down required on BVBS header block.

4.5 Geometric representation in BIM authoring tools

To better illustrate the discussion about geometric representation of structural elements in IFC, a test with three of the most representative authoring BIM tools was conducted: Autodesk Revit 2017, Tekla Structure 2016 and Allplan Engineering 2016.

In order to evaluate the method for geometric representation used by some of the main structural detailing BIM tools, a model was created based on the example of a single reinforcement stirrup present in the IFC4 Add. 1 documentation [14]. In its constant evolution, the new release of the IFC schema included others forms of geometric representation. Nevertheless, these new entities are not supported by the reviewed BIM tools.

The current method used to export reinforcing bars in IFC files, specifically the directrix attribute that establishes the swept disk solid shape, were compared with a new entities released in IFC4 and IFC4 Add.1 with the purpose of verifying their influence in the size of the models files.

Distinctive methods are adopted by the structural detailing BIM tools for define the directrix. In

Table 2 are present the method employed by reviewed BIM tools for a reinforcement bars geometric representation.

Included in the Annex E of IFC4 Add. 1 documentation, the sample E.14.5 uses the new entity *IfcIndexedPolyCurve* to define the swept disk solid directrix. This option allows a simplified and more compact way for describe this stirrup shape, requiring only two lines of an IFC file to describe a composite curve. The first line is used for indexing of the segments (straight and curved) and the second line to define the coordinates of the points, grouped with the use of *IfcCartesianPointList3D*.

The directrix is defined in Autodesk Revit 2017 by *IfcCompositeCurve*, straight segments are defined with *IfcPolyline* and curve segments by *IfcTrimmedCurve* with *IfcCircle* basis curve.

The method adopted by Tekla Structure 2016 also uses an *IfcCompositeCurve* to define the swept disk solid directrix. Both segments which compose the parent curve are defined by *IfcTrimmedCurve* entity by means of *IfcLine* for straight segments and *IfcCircle* for arc segments.

In Allplan Engineering 2016, the directrix is defined only by *IfcPolyline* curve. The segments in curve are discretized in several straight segments with intermediate points of the arc.

Table 2.Geometric representation of reinforcement bars by reviewed BIM tools

Model	IFC4 Add. 1 E.14.5	Autodesk Revit 2017	Tekla Structures 2016	Allplan Engineering 2016
Rebar	*IfcReinforcingBar* *IfcReinforcingBartype*	*IfcReinforcingBar*	*IfcReinforcingBar*	*IfcReinforcingBar*
	IfcSweptDiskSolid	*IfcSweptDiskSolid*	*IfcSweptDiskSolid*	*IfcSweptDiskSolid*
Directrix	*IfcIndexedPolycurve*	*IfcCompositeCurve*	*IfcCompositeCurve*	*IfcPolyline*
Line segment	*IfcLineIndex*	*IfcCompositeCurveSegment* + *IfcPolyline*	*IfcCompositeCurveSegment* + *IfcLine*	*IfcCatersianPoint*
Arc segment	*IfcArcIndex*	*IfcCompositeCurveSegment* + *IfcTrimmedCurve*	*IfcCompositeCurveSegment* + *IfcTrimmedCurve*	
IFC release	IFC 4 Add.1	IFC 2x3 CV	IFC 2x3 CV	IFC 2x3 CV

5 Conclusions

It is necessary to improve the existing interface of information exchange required by the fabricator and to promote the development of a way that allow a collaboration between rebar supply chain players based on IFC model instead of the current flow based 2D drawings.

In order to summarize the results, Figure 4 represents the proposal discussed in this article to map the required information content of the BVBS standard into the IFC data schema, in its 4th version.

The overview of how the information required by BVBS header block are currently mapped on the IFC schema are shown in Table 3.

In this article, a BVBS-IFC digital interface for the workflow of the rebar cut and bend supply chain was proposed. The principal characteristics of the BVBS standard were shown and a possible mapping inside IFC schema was elaborated and discussed. Particular attention was given to the geometric representation of reinforcement bars in IFC and inside some BIM authoring tools.

The review of some structural detailing BIM tools has shown that distinctive methods of geometric representation are currently used for exporting reinforcement bars in IFC.

IFC4 version incorporated many changes in the structural elements domain, when compared to the previous IFC2x3 version, and these changes helped to solve part of the problems existent in the structural design workflow. The use of IFC object-type on reinforcement domain (included in the IFC4) simplify the representation of structural elements and its assemblies, allowing the reuse of common characteristics shared by similar rebar occurrence. But some issues still exist, which must be addressed in the future development of IFC schema.

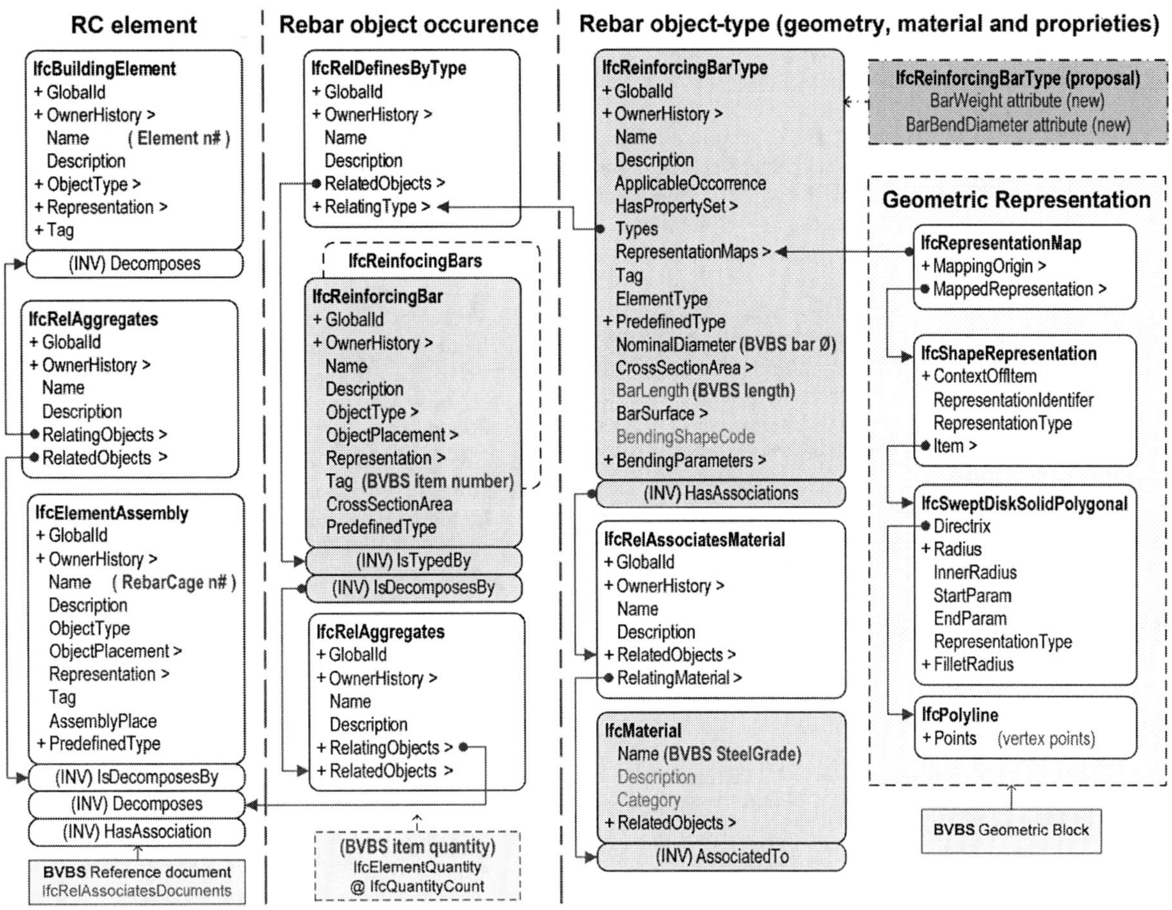

Figure 4. BVBS requirements map in IFC 4 Addendum 1

Table 3. BVBS Header block information in IFC 4 Add.1

Field Identification **H**	Header-Block description	IFC entity		Attribute
j	Project number (optional)	*IfcProject*	# 3	Name
r	Drawing number of the respective drawing	*IfcDocumentInformation*	# 1	Identification
i	Index of the respective drawing	*IfcDocumentInformation*	# 8	Revision
p	re-bar/mesh item number	*IfcReinforcingBar*	# 8	Tag
l	bar/mesh length [mm]	*IfcReinforcingBarType*	# 13	BarLength
n	item quantity	*IfcQuantityCount*	# 4	CountValue
e	weight per mesh/bar item [kg]	—	—	—
d	bar diameter [mm]	*IfcReinforcingBarType*	# 11	NominalDiameter
g	steel grade	*IfcMaterial*	# 1	Name
s	bending diameter [mm]	—	—	—

References

[1] Polat G., Ballard G., Construction supply chains: Turkish supply chain configurations for cut and bent rebar, in: 11th Annual Conference of the International Group for Lean Construction, Virginia, USA, 2003.

[2] Marder T.S., Formoso C.T., Oportunidades de melhorias na cadeia de suprimentos do aço cortado e dobrado para a construção civil, in: I Conferência Latino-Americana Construção Sustentável X Encontro Nacional de Tecnologia do Ambiente Construído, São Paulo, Brasil, 2004: p. 14.

[3] Azambuja M.M.B., Isatto E.L., Marder T.S., Formoso C.T., The importance of commitments management to the integration of make-to-order supply chains in construction industry, in: 14th Annual Conference of the International Group for Lean Construction, Santiago, Chile, 2006: pp. 609–623.

[4] Isatto E.L., Azambuja M., Formoso C.T., The Role of Commitments in the Management of Construction Make-to-Order Supply Chains, J. Manag. Eng. 31 (2015) 04014053. doi:10.1061/(ASCE)ME.1943-5479.0000253.

[5] Aram S., Eastman C., Sacks R., Requirements for BIM platforms in the concrete reinforcement supply chain, Autom. Constr. 35 (2013) 1–17. doi:10.1016/j.autcon.2013.01.013.

[6] BundesVereinigung der BauSoftwarehäuser, BVBS - Guidelines: Exchanging Reinforcement Data - interface description release 2.0, 2.0 ed., BundesVereinigung der BauSoftwarehäuser E.V., Bonn, Germany, 2000.

[7] American Concrete Institute (ACI), ACI 315-99 Details and Detailing of Concrete Reinforcement, ACI Committee 315, Farmington Hills, MI, USA, 1999.

[8] BSi, BS 8666 - Scheduling, dimensioning, bending and cutting of steel reinforcement for concrete - Specification, Britsh, 2005.

[9] ISO, BS EN ISO 3766 - Construction drawings - Simplified representation of concrete reinforcement, EN ISO 3766:2003/AC:2004, UK, 2003.

[10] buildingSMART, Industry Foundation Classes Version 4 - Addendum 1, (2015). On-line: http://www.buildingsmart-tech.org/ifc/IFC4/Add1/html/, Accessed: 15/06/216.

[11] Venugopal M., Eastman C.M., Sacks R., Teizer J., Semantics of model views for information exchanges using the industry foundation class schema, Adv. Eng. Informatics. 26 (2012) 411–428. doi:10.1016/j.aei.2012.01.005.

[12] Venugopal M., Eastman C., Sacks R., Panushev I., Aram V., Engineering semantics of model views for building information model exchanges usinf IFC, in: Proc. CIB W78 2010 27th Int. Conf., Cairo, Egypt, 2010

[13] Eastman C., Teicholz P., Sacks R., Liston K., BIM handbook : a guide to building information modeling for owners, managers, designers, engineers, and contractors, 2nd ed., John Wiley & Sons, Hoboken, New Jersey, 2011.

[14] buildingSMART, IFC4 Add1 Example E.14.5 Reinforcing stirrup, (2015). On-line: http://www.buildingsmart-tech.org/ifc/IFC4/Add1/html/annex/annex-e/reinforcing-stirrup.htm, Accessed: 15/06/216.

Data-Processing Model for Vertical Zoning based on Construction Lift Operation Records for High-rise Building

J.H. Shin[a], S.W. Kwon[b], D.H. Kim[c], D.Y. Moon[d], S.W. Chung[e], G.S. Choi[f] and K.H. Lee[g]

[a]Department of Convergence Engineering for Future City, Sunkyunkwan University, Republic of Korea
[b]School of Civil & Architectural Engineering, Sungkyunkwan University, Republic of Korea
[c]Department of Convergence Engineering for Future City, Sunkyunkwan University, Republic of Korea
[d]Department of Convergence Engineering for Future City, Sunkyunkwan University, Republic of Korea
[e]Department of Convergence Engineering for Future City, Sunkyunkwan University, Republic of Korea
[f]Department of Convergence Engineering for Future City, Sunkyunkwan University, Republic of Korea
[g]Department of Convergence Engineering for Future City, Sunkyunkwan University, Republic of Korea
E-mail: kiwijoa@skku.edu, swkwon@skku.edu, ih5318@nate.com, yoshy17@nate.com, suwanx@nate.com, gyupreme@skku.edu, leekyuhyup@naver.com

Abstract

As recent buildings have been getting higher and larger, vertical construction lift planning and operation are key factor for successful project. Although many studies have been trying to set up a construction lift planning system at early stage before lift installation. But there are not regarding real-time control system for lift during construction stage. In this study, we use sensor module and real-time data storage device named ELIS(Embedded Lift Information System) to collect the lift operation data for improvement of vertical zoning efficiency. Lift operation data linked with finishing work activities is utilized to predict to-be lifting events by data processing. This procedure analyzes the pattern of lift moving. Finally, we propose optimum vertical zoning performance simulation method based on post processed lift operation data from ELIS. This system aims to update and to provide daily optimum vertical zoning. We expect that operation data based decision making for vertical zoning can contribute to finishing work planning in complex and tall building construction over 400m.

Keywords –

Construction lift, Vertical zoning, ELIS(Embedded Lift Information System)

1 Introduction

For the recent past few years, buildings in the world became high-risen. Hereupon, there were various restrictions about the plan and operation of construction. Among them, lift management of the workers and materials for finishing work of skyscrapers have been an important issue as the most important element required when managers considerate productivity and cycle of construction.

The vertical movement of materials in the field has been to be very complicated because the buildings have become high-risen. Especially, as the vertical distance for moving became longer than before, it was very difficult to rationally manage it. It is difficult for individuals who made a decision on the lift management on skyscraper by just using their long accumulated experience and intuition. It causes increase of labor's queue time to use lift in high-rise building.

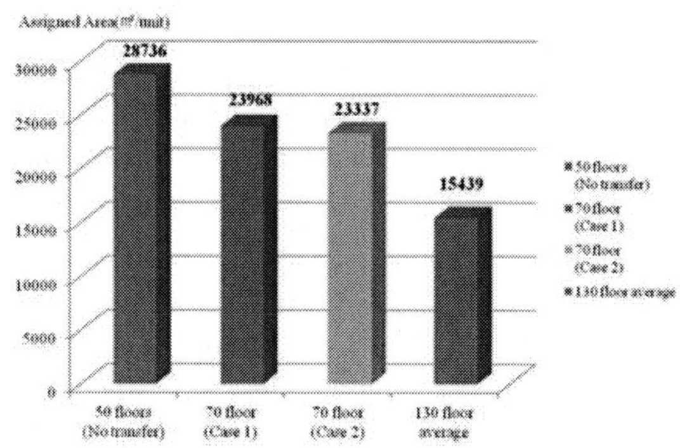

Figure 1. Assigned area per unit cage according to building's number of floor

When establishing the lift management in the stage of construction plan, there are various practical and influential factors that should to be dealt with in the mathematical model consisting of simplified formula or exploratory technique. Especially, number of lift, selection of lift and location are reflected to the period of lift installation, quantity of material and packing method in the early stage. Many kinds of technique have been used to estimate plan of lift equipment. However, currently vertical zoning only has been decided based on labor's commuting time or restrictions during peak-time.

Shin(2011) classified vertical lifting management into two elements; moving plan and movement operation. And he indicated the importance of monitoring or operation management during construction period as important as lift plan. Compare to low-rise building, high-rise building needs additional ration on cost including equipment planning. If different project sites use same lift system, in other word, according to number of floors assigned area per each cage is decreased shown as Fig 1.

2 Research Trend

Sacks et al. developed an automated lifting equipment monitoring system (Sacks et al, 2005). Cho et al. conducted a study on construction lift operation planning in terms of lifting height and loading (Cho, 2011). Further, Shin proposed optimal operation of temporary construction lifts in a super high-rise building based on simulation and genetic algorithm. Before them, most studies focused on the use of tower crane or mobile crane, and other studies on construction lift tended to emphasize lifting planning rather than lifting operation.

As super high-rise construction becomes more popular in Korea, there is a growing need for a systematic construction planning and site management. The government and private corporations are actively undertaking studies on operation planning of construction lifts and tower crane lifting. Kim et al. studied how to calculate a number of necessary construction lifts at super high-rise construction site (2008), and Shin et al. (2010) proposed a construction lift movement planning model for super high-rise construction. Cho et al. (2011) proposed an algorithm that calculates lifting time in consideration of acceleration and deceleration capability of construction lifts (2011). While there are many studies on construction lift planning, few have been conducted on the system, management and algorithm of construction lift operation, with no data based access.

3 Construction Lift Operation Model

3.1 Daily Lifting Pattern on Site

There is a necessity of distinguishing the operation types and patterns to analyze operation states in the field of project. Ahn(2001) and Cho(2011) have divided the daily lifting time into material lifting time, moving time of operators, combined lifting time for employees/materials, lunch time, and peak-time.

Previous studies dealing with characteristics of daily lifting time have distributed the lifting time of materials before and after lifting time of employees. However, according to the result of examining characteristics of actual operation, lifting time of materials was turned out to frequently occur early in the morning or late night due to the expansion of a concept of JIT (Just in Time). Therefore, this study has defined the lifting time of materials into a case other lifting time as shown in the Fig 2.

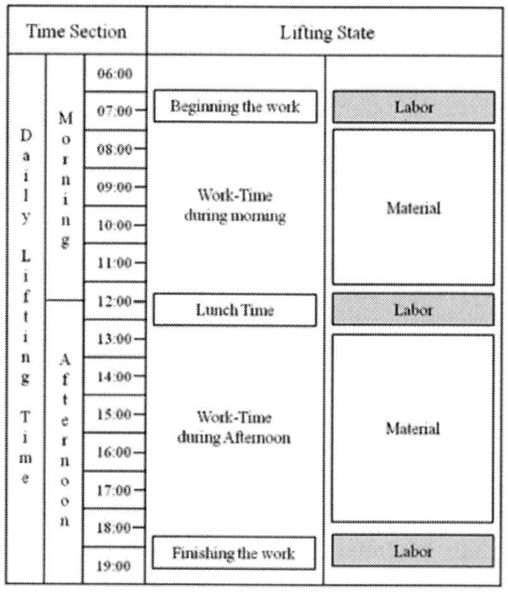

Figure 2. Daily lifting time pattern

3.2 Transfer & Vertical Zoning

A lift applied up to 400m of operation zone has been used for the construction of the highest-risen building in the world at Burj Khalifa. However, a significant construction delay has been caused due to an issue of frequent breakdown of power cable.

Fig. 3 explains the difference between the space zoning type and transfer operation method. Park has suggested optimal space zoning model in reducing the lifting time by setting up the operation zones. Park(2013) has suggested an optimal method based on transfer method.

In addition, as the buildings became high-risen, the daily amount of transported materials has been increased as well. In addition, most of the skyscrapers were built in the form of set-back type due to an issue of slenderness ratio. Therefore, division and setup of lift operation zones have become an inevitable factor.

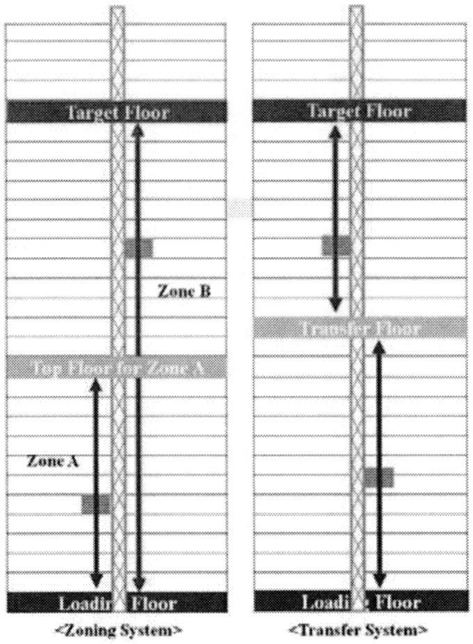

Figure 3. Types of vertical zoning

These studies have setup the commute time of employees as a peak time controlling the lift circulation only during the morning time, increasing available work hours, and preventing the confusion of lift usage. They have utilized result value in the optimized simulation instead of the operation data in the field.

This study is intended to suggest a decision-making process as an alternative of both space zoning method and transfer method depending on the shape of buildings and situations in the field. In addition, this study represents a difference from others aiming to adjust zones based on the database during the operation after installing the lift different from plans for lift operation zones in the milestone for each process, representative finalizing materials, calculation of outcomes per area, and setup for operating group.

3.3 Variables for Operation Management

In order to verify the vertical operation zoning simulation based on the lift operation records suggested in this study, the lift cycle time variables and formulas are used for be reflected on decision-making model by simulation.

Calculation of expected distribution time and entire lift completion for the lift call is performed according to formula (1) to (4). Variables of lift operation are comprised of door-opening time (D_1), door-closing time (D_2), entire moving time (T_{os}), and loading/unloading time (T), and the cycle time consumed for lifting the materials and human resources from the original place to destination is as follows in (1).

$$Cycle = (D_1 + D_2) \times 2 + T_{os} + L \quad (1)$$

$$T_{os} = t_1 + T + t_2 \quad (2)$$

$$T = (H - h_1 - h_2)/V \quad (3)$$

$$t_1 = \sqrt{h_1/a_2} \quad t_2 = \sqrt{h_2/a_2} \quad (4)$$

The entire moving time (T_{os}) is comprised of the total of accelerating time (t_1), moving in a regular speed (T), and decelerating time (t_2). Each of the time factors is stated by standards such as regular speed of distributed lift (V), accelerating/decelerating speed (a_1/a_2), accelerated/decelerated distance (h_1/h_2) and moving distance (H). However, formula (1) indicates 1Cyde comprised of loading or unloading that a total cycle formula for distributing the lift car on the floor given an order is required.

3.4 Embedded Lift Information System

In the current field of construction project, lift management has been taken by field manager preparing for lift plans on a daily basis. However, there is no way to confirm the breakdown of each of the lifts and efficiency of the usage. For this, an effort has actively been exerted to secure a system for collecting information of lift equipment status and operation on a real time basis. Internally installed lift information collecting device in the use of sensor module suggested by Kwon has been proposed for embedded lift information system based on the data on a real time basis. Fig. 4 represents a developed data-box and display to collect operation history data in the ELIS system.

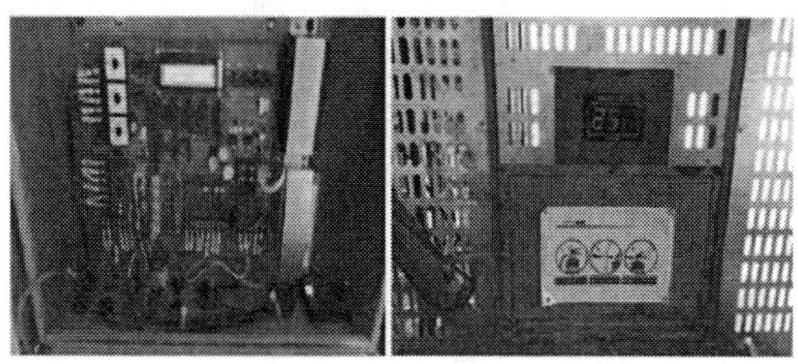

Figure 4. ELIS data-box in construction lift

ELIS system receives location, speed, load, and moving direction on a real time basis from six-tier sensors (encoder, load cell, partitioning converter, current sensor, limit switch, and proximity sensor) installed on each of the lift cars. It saves them on the database for analysis of operation record of each lift during the entire period of project and compares or analyzes data between lifts to establish the operation plans in a near future when reflecting the lift call information received in current time period. Fig. 5 represents data for collecting information of lift on a real time basis.

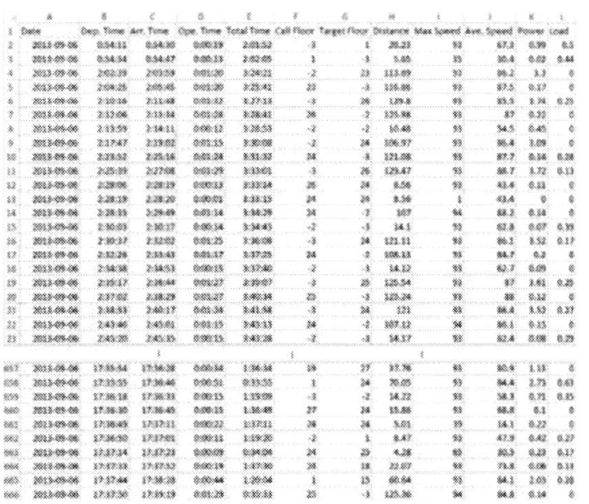

Figure 5. Operation history data-set from ELIS

4 Data-Processing for Performance Analysis of Vertical Zoning

4.1 Integration between Activity & Operation DB

The operation pattern of lift operation has deep relationship with work plan of finishing work. Operation pattern can be derived from just moving history like Fig 5. However, it could not handle dynamic pattern changes caused by the birth and death of finishing work activities during whole project cycle unless lift operation plan doesn't reflect each activity's relevant item, lifting frequency. Table 1 shows a case that 55 floor building's lifting frequency record on materials.

Table 1. Case of lifting frequency record (55 floor building)

Lifting item	Lifting Frequency	Ratio (%)
Fire resistive covering	4,913	13.5
Gypsum board	9,583	26.3
Glass	1,632	4.5
Wall Cover	2,533	6.9
Curtain Box	408	1.1
Light Ceiling-Frame	816	2.2
Ceiling Texture	1,207	3.3
FCU Cover	1,309	3.6
Paint	810	2.2
Door & Frame	2,812	7.7
Stone & Frame	918	2.5
Tile Carpet	1,904	5.2
Finishing Material etc.	7,653	21

In order to this study has displayed the operation patterns of the lift in the current process of project in the diagram as shown in the Fig. 6, and made it convenient for a manager to analyze current situations by systemizing data flow. In addition, vertical zoning simulation inputs were utilized.

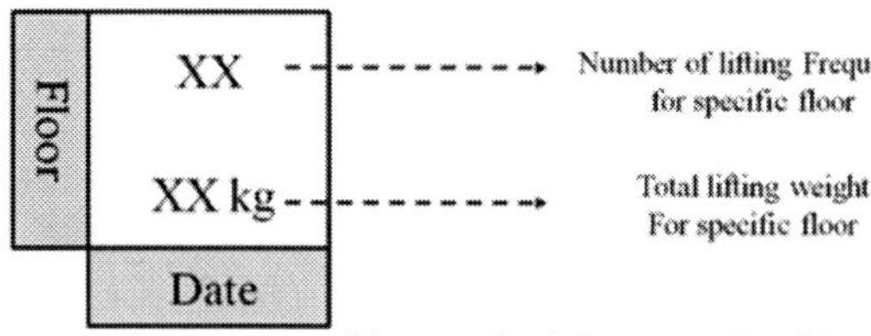

Figure 6. Graphing method for operation history analysis

According to the characteristics of finalizing construction where repetitive processes are vertically connected along with continuity of the work, operation history data module defined as the Fig. 6 can be used as an efficient tool for examination of progress in the present compared to plans by adding objective data-based trend information of a manager on the monitoring of finalizing construction limited to the schedule of finalizing construction.

In finishing work process control, to avoid interference and confliction between finishing work activities some process techniques methods are used such as LOB(Line of Balance) and TACT. Among them, TACT technique has an advantageous to reduce wasting time by confliction. Fig. 7 shows TACT cycle based data composition method.

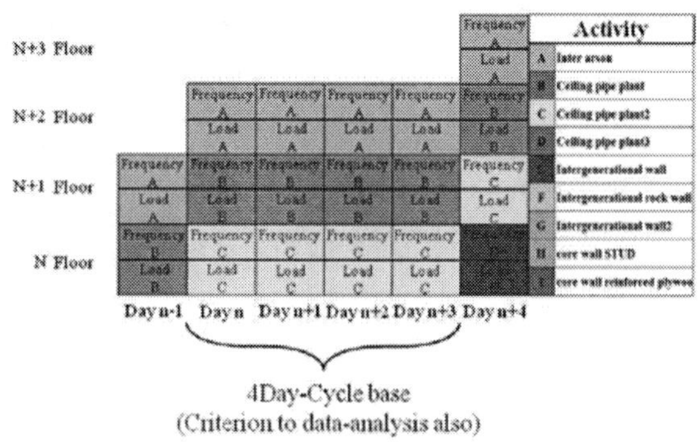

Figure 7. Data composition reflecting TACT-technique

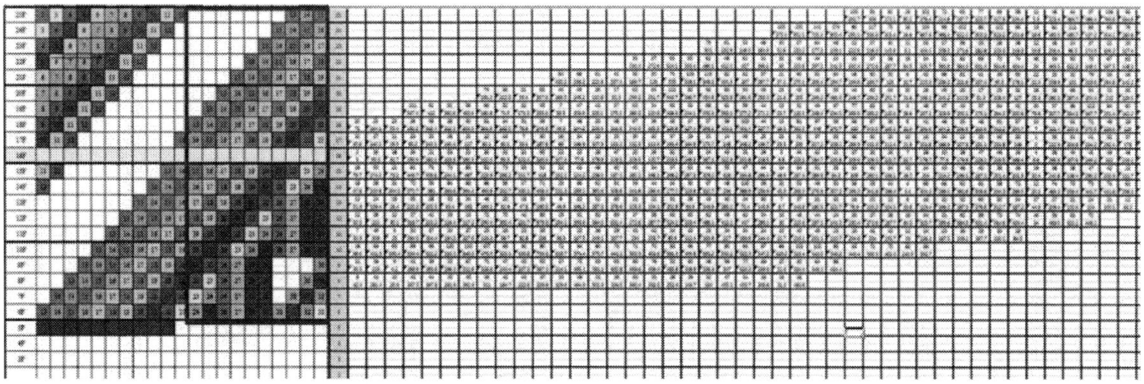

Figure 8. ELIS history data matching to TACT-technique

TACT is performed with the process schedule assuming that all the tact-elements are implemented in the same manner within the consistent cycle. However, a degree of progress in TACT process plans is subject to be changed depending on the work situations in the field. In addition, this is what makes it feasible to find delay factors and modified tact on the plan through operation analytic data and revise them.

Fig. 8 represents the result of matching lift operation records in the field in Korea where ELIS lift operation data system is installed at and tact process schedule in the field. Fields for deriving operation history are the tact process with four-day cycle in principles, and monitoring is performed based on the tact work sheet and tact process minutes. Frequency of operation on each floor and lifting load was setup in accordance with tact format by using the methodology of data in Fig. 6. In addition, a certain format for convenient comparison with a degree of progress for tact process has been derived.

It was analyzed that a progress of finalizing work for core wall side in the field was delayed compared to the plan according to analytic data in Fig. 8. Furthermore, it was confirmed that a schedule of initiating the stud work in the insider wall was made as the tact element number 13. Field manager is able to conveniently identify a progress of tact for finalizing work based on operation record tact data and utilize operation record tact as an index of tact adjustment in the future. Delay factors of tact and issues are reflected on the lift vertical zoning in the field in the future and added on the simulation information based on the date of initiating tact factors.

4.2 Decision Making of Predicting Lifting Event

Generating the predicted lifting event is essential for performance simulation by type of zoning to optimize vertical construction lift zoning. ELIS database is based on data-type of 4×M arrangement. On the row of data arrangement, time, target floor, lifting weight, moving direction linked with storage date are listed. On the column of data arrangement, counted event numbers of specific date are listed through time flow. Fig.9 shows data arrangement of ELIS DB.

	Event 1	Event 2	Event 3	Event 4	• • •	Event M
Signal Time	(1,1)	(1,2)	(1,3)	(1,4)	• • •	(1,M)
Target Floor	(2,1)	(2,2)	(2,3)	(2,4)	• • •	(2,M)
Lifting Weight	(3,1)	(3,2)	(3,3)	(3,4)	• • •	(3,M)
Moving Direction	(4,1)	(3,2)	(4,3)	(4,4)	• • •	(4,M)

Figure 9. Data arrangement on ELIS DB

As shown in Fig. 10, database arrangement described at Fig. 9, algorithm for decision making that able to compute predicted lifting event of next-day. *E* indicates the variable of operation history arrangement stored at ELIS DB. Activity DB stored predicted lifting event information at the time of activity beginning. *A* indicates the variable of predicted lifting data arrangement at the time of activity beginning.

Algorithm for decision making of lifting event prediction needs interaction with manager's recognition about the progress of finishing work because computational algorithm can't check that there is construction delay or not. Hereupon, an input process of variable *n*, *AN(n)* is needed by user.

Compare to last-day, in case that there are additional activities, predicted lifting event has to be added to predicted lifting event arrangement derived from operation history data. And then totally listed event information are re-listed in descending order by time-flow to utilize it as the input data of simulation. *EE* indicates the variable of final predicted lifting event arrangement going to be used on operation simulation. This type of data arrangement is useful to discrete event

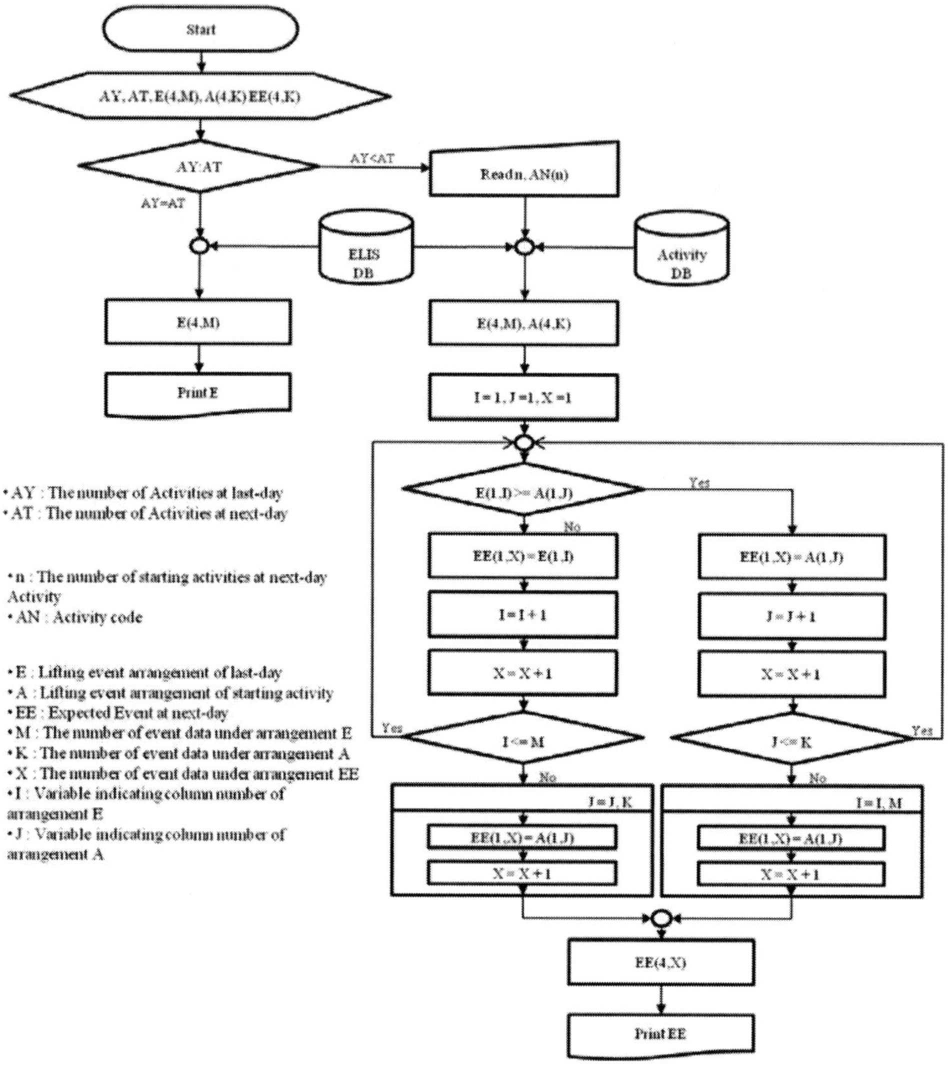

Figure 10. Algorithm flow of decision making for lifting event prediction

modeling based optimization simulation.

Vertical zoning performance simulation for each zoning situation generates many situations by zoning condition of each lift car. As shown in table 2, because each lift car has their own moving section, simulation should compute the result of imaginary operation about lifts in project site.

Table 2. Condition of lift moving section

Situations	Lift A	Lift B	Lift C
Situation 1	1~10	1~20	1 → 10~ Top floor
Situation 2	1~11	1~20	1 → 10~ Top floor
Situation 3	1~12	1~20	1 → 10~ Top floor
⋮	⋮	⋮	⋮
Final Situation	1~ Top floor	1~ Top floor	1~ Top floor

5 Conclusion

As the buildings became high-risen, and the limits on the maximum lifted floors increased, variables applied to the management of finalizing construction became complicated. Hereupon, operation management of the lift is needed for the construction after implementing finalizing process along with a decision-making procedure on the transport-circulation of finishing work materials that are appropriate for flexibly changed situations in the field.

Therefore, in this study, operation history information was saved from sensor modules attached to the lift through development of ELIS(Embedded lift information System), Then it was utilized to draw predicted lifting event using for vertical zoning simulation for future finishing work progress. Based on activity-base process management and TACT technique, we proposed algorithm process for decision making to use an input data on discrete event simulation.

Lastly, this study has served as a starting point of operation system based on the mathematical model and software decision-making process. Therefore, it is anticipated to settle as a system based on unmanned smart lift through verification made in the entire period of research project.

Acknowledgements

This research was supported by a grant(15AUDP-B106327-01) from Architecture & Urban Development Research Program funded by Ministry of Land, Infrastructure and Transport of Korean government.

This work is supported by Korea Minister of Ministry of Land, Infrastructure, Transport affairs as Convergence Engineering of Future City Master and Doctoral Grant Program

References

[1] Lee, H.J., Kim, D.W., Cho, H.H., and Kang, K.I., "Development of the Model Using Queueing Theory for Lifting Planning in Tall Buildings", *Korean journal of Construction Engineering and Management*, volume 11: 635-640, 2008

[2] Ahn, B.J., Kim, J.J., "A Decision-Making Model of Integrated Vertical and Horizontal Move Plan for Finishing Material in High-Rise Building Construction", *Korean journal of Construction Engineering and Management*, volume 2:47-58, 2001

[3] Kim, D.J., Song, Y.S., and Lim, H.C., "Rationalization the Lift car Planning Method in a High-rise Building Project", *Architectural Institute of Korea*, volume 22:483-486, 2002

[4] Kim, S.K., Han, K.K., "A Study on the Estimation of Proper Numbers of Construction Lifts", *The Korea Institute of Building Construction*, volume 8:119-125, 2008

[5] Cho, C.Y., Shin, Y.S, Won, S.K., Kim, J.Y., and Cho, M.Y., "An Algorithm for Hoisting Time Calculation in Super-tall Building Construction" *Korean journal of Construction Engineering and Management*, KICEM, volume 12(6):120-129, 2011

[6] Shin, Y.S., "Construction Lift Planning System in Tall Building Construction", *Architectural Institute of Korea*, volume 27:227-235, 2011

[7] Sacks, R., Navon, R., Brodetskaia, I., and Shapira, A., "Feasibility of Automated Monitoring of Lifting Equipment in Support of Project Control." *J. Constr. Eng. Manage.*, volume 5:604-614, 2005

[8] Lee, J.W., Choi, J.P., "A Study of Algorithm in Developing a Program for Elevator Planning in High-rise Buildings", *Proceeding of annual conference of the architectural institute of Korea*, volume 30:83-84, 2010

[9] Shin, J.H., Kwon, S.W., "Development of Optimum Construction Lift Operation System using Sensing Information for High-rise Building", *Korea Journal of Construction Engineering and Management*, volume 5: 153-163, 2013

[10] P. Cortés, J. Larrañeta, and L. Onieva, "Genetic algorithm for controllers in elevator groups: analysis and simulation during lunchpeak traffic", *Applied Soft Computing*, volume 4:159-174, 2004

Automatic Crane-Related Workflow Control for Nuclear Plant Outages through Computer Vision and Simulation

Cheng Zhang [a], Pingbo Tang [a], Alper Yilmaz [b], Nancy Cooke [c], Verica Buchanan [c], Alan Chasey [a], Ronald Laurids Boring [d], Shawn W. St. Germain [d] , Timothy Vaughn [e], and Samuel Jones [e]

[a] Del E. Webb School of Construction, Arizona State University, USA
[b] Department of Civil, Environment and Geodetic Engineering, The Ohio State University, USA
[c] Human Systems Engineering, The Polytechnic School,, Arizona State University, USA
[d] Idaho National Lab, USA
[e] Arizona Public Service Company, USA
E-mail: Cheng.Zhang.7@asu.edu, tangpingbo@asu.edu, alpery74@gmail.com, Nancy.Cooke@asu.edu, vbuchana@asu.edu, achasey@asu.edu, ronald.boring@inl.gov, shawn.stgermain@inl.gov, Timothy.Vaughn@aps.com, Samuel.Jones@aps.com

Abstract

Nuclear power plant (NPP) outages involve a large number of maintenance activities with a tight schedule and zero-tolerance for accidents. Outage projects thus need real-time control to ensure safety and productivity. During outages, crane lifting is critical for outage control and risk management. An effective outage control method should monitor detailed interactions between human and workspaces, and streamline the workflows of cranes to control both productivity and risks. Unfortunately, current approaches of outage control rely heavily on tedious and error-prone manual inspection that can hardly achieve detailed spatiotemporal monitoring.

This paper presents an automated outage control framework that enables detailed human behavior analysis, automatic comparison of as-planned and actual crane-related operations, and effective decision-making for crane-related workflow control. In this framework, a real-time human tracking algorithm uses 2D/3D imagery to automatically derive the status of workspaces (e.g., waiting, active). Then a change-analysis algorithm detects and diagnoses differences between as-is workflow information against as-planned schedules, and thus enables field managers to implement a close-loop outage control. Preliminary results indicate the potential of this integrated outage control in improving the safety, productivity, and quality of outages, as well as outage project planning.

Keywords

Nuclear power plant; Outages; Crane; Workflow control; Computer vision; Simulation

1 Introduction

In the United States, many nuclear power plants (NPPs) were built for more than 40 years [1] and they require regular maintenance. For example, the management company of Palo Verde Nuclear Power Plant needs to shut down each reactor every 18 months to refuel and repair degraded infrastructures. Such processes are called "outages" at NPPs. NPP refueling outages are challenging because they require tracking and coordinating thousands of activities in a short span of time, usually between twenty to thirty days. Moreover, any delays in the NPP outage processes will cause significant economic losses. If the Palo Verde Nuclear Power Plant is shut down for one more day because of a delay from an outage, the deficiency of power would lead to one to two million dollars' loss for the energy company. Therefore, to prepare for an NPP outage, an outage group in the Work Management department spends months on the outage planning, together with the occasionally long range plans that projects many years into the future. NPP outages require a significant supplemental workforce that consists of thousands of contract personnel, which increases the complexity of communication and information flows [2]. Other challenges, including scheduling, work group coordination, nuclear safety concerns arising from different system configurations, and resource allocation issues, can create delays and schedule overruns, driving up outage costs [3]. All these features of the NPP outage process appeal for an efficient and effective workflow control framework to reduce its cost, duration, labor, and accident rates.

With outage processes, crane-related tasks are critical for maintaining the safety and productivity of related workflows. Crane related tasks often involve lifts of over 30 tons, which are defined as "very heavy lifts" that can significantly increase the damages caused by crane-related accidents [1]. In addition, crane activities often require large moving spaces, increasing the difficulty of coordinating different labors, workspaces and tasks in spatial and temporal domains. Finally, a crane is usually a shared resource for multiple tasks involving different groups of workers, which creates a higher chance of encountering problems of communication, collaboration, and scheduling. Any accidents or delays related to crane are likely to propagate to relevant tasks, affecting the safety and productivity of the entire workflow. Therefore, crane-related tasks are critical for the NPP outage workflow control.

Crane-related safety and productivity are heavily discussed topics in the domain of construction engineering and management. For example, researchers tried to solve this problem by predicting and planning the moving path of the lifted object in order to prevent people from injury under its path [4,5]. Admittedly, previous studies are necessary but cannot handle the problems faced by the nuclear industry. People from both academy and industry still lack of an efficient and effective method to monitor and control human behavior in order to reduce crane-related errors and accidents caused by human factors, which is the major reason of crane-related accidents and near-misses in NPP. Such a deficiency is caused by two reasons. First, people from nuclear industry still lack a systematical understanding about how the features of NPP outage could influence crane-related anomalies and how such anomalies could influence the downstream tasks in NPP outage processes [1]. Any human error in crane-related construction workflows could cause delays that impact productivity and accident rates and such anomaly propagation effect will be amplified by extremely limited time, space, and critical facility resources (e.g. cranes) on the NPP outage jobsite. Second, NPP outages need an efficient and reliable data collection method for human behavior monitoring. In real-world NPP outages, the management team suffers from large amounts of manual data collection and analysis to monitor and maintain the safety and productivity of crane-related tasks and workflows [2]. As lengthy manual data collection and processing could seriously delay the information needed for timely decision-making, manual approaches for crane-related safety and productivity diagnosis cannot meet the requirements of efficient and effective workflow control in outages.

The contribution of this research address this gap through integration of computer-vision-based human behavior monitoring and simulation models that predict how anomalies in workflows escalate into delays and risks. A workflow control methodology that integrates real-time human behavior monitoring and decision making could reduce the labor-intensive work of maintaining the safety of crane-related tasks. The proposed framework consists of two parts. The task-duration anomaly propagation modeling will provide the decision-making methodology according to the as-planned workflow and the monitoring data. The computer-vision based human behavior monitoring algorithm will identify delays of tasks and the anomalies happening on the job site based on the as-planned workflow. This proposed methodology will help prevent the accidents in advance and quickly identify and diagnose the deviations between as-planned and as-is workflows for decision support.

2 Literature Review

Because of the huge size, mass and range of motion of cranes, any anomalies in crane-related activities could cause catastrophic consequences, such as injuries and fatalities. Beavers et al [6] determine the proximal causes and contributing physical factors of accidents by analyzing the Occupational Safety and Health Administration's (OSHA) case files during the years of 1997–2003. However, this approach only analyzed the direct causes of crane-related fatalities (e.g. falls, struck by load, and electrocution) instead of various indirect influential or casual factors (e.g. human misbehaviors, equipment quality issues, etc.). Such direct causal analysis cannot provide suggested adjustments about indirect factors for reducing the high fatality rates related to crane activities. In order to understand the causal factors of the accidents, the authors analyzed the newest accident cases of Fatality and Catastrophe Investigation Summaries from OSHA [7]. In addition, the authors determined the influential tree of an accident to see whether it has multiple indirect causal factors. The result of the influential tree analysis shows that 60% of the accidents have multiple indirect factors in their influential trees. These results indicate that on construction job sites, "domino effects" could enable small anomalies to trigger severe accidents [8]. In addition, human behavior is directly or indirectly related to more than 80% of the reported crane-related accidents in OSHA records.

Currently, few studies focus on the outage control of nuclear plants. Germain et al. [2] stated that the deficiency of communication and collaboration is due to large numbers of personnel and activities on outage job sites. That fact argues for the concept of "advanced outage control center (AOCC)," which is specifically designed to maximize the usefulness of communication and collaboration technologies for outage coordination and problem resolution. However, using the proposed AOCC cannot completely solve the problem of automatically and rapidly detecting the detailed deviation between as-is and as-planned workflows during outage, because detailed spatiotemporal data comparison and analysis is still manual in the current AOCC conceptual framework. Without efficient deviation detection between as-planned and actual workflows, diagnosing indirect impacts of propagative deviations is unrealistic. A report from NEUNG analyzed all accidents related to cranes in all nuclear plant in the U.S. from 1968 to 2002 [1]. According this report, the major reason that caused crane-related accidents in NPP is human behavior, which causes over 80 percent of the accidents. That report, which is specific to NPPs, corroborates the more general crane-related accident database of OSHA.

Researchers have developed many approaches to improve the safety and productivity of crane-related tasks. Tracking the position of crane loads enables nearby workers to keep away from zones in danger of load falls. Fang et al [5] used an inertial measurement unit (IMU) module for measuring load orientation and then predict the load sway trajectories. Yang et al. [9] demonstrated the use of a surveillance camera for measuring the jib angle and the trolley position of a tower crane during a work day. In construction projects, failing to acquire timely, detailed, accurate spatial information for decision-making can cause low project quality, low productivity, and accidents. Computer vision has drawn attention because it is useful for automated and continuous monitoring of construction sites. Seo et al [10] states that technical challenges exist for achieving efficient and effective safety and health monitoring for construction projects using computer vision techniques. These challenges include sufficient high-quality imagery data collection, automatic scene understanding, and activity recognition of equipment/workers. In the domain of human behavior modeling and analysis, Shappell and Wiegmann [11] stated that studies should focus more on human misbehaviors if accidents are to be further reduced, because human misbehavior occupies a significant part in accident causation. Garret et al [12] proposed a general human misbehavior framework originally developed and tested as a tool for investigating and analyzing the human causes of accidents. That research also emphasizes the importance of effective investigation in safety.

However, previous studies have not provided an unified framework for improving the safety and productivity of crane-related tasks considering human behavior and propagations of various errors. In busy and safety-sensitive NPP outages, such a framework is important for timely decision support. In outage processes, perpetuation of errors, delays and other propagative factors will be enlarged by the extremely limited time, space, labor, and availability of cranes and relevant resources. Any delays in an outage process may increase the worker's stress, which increases the misbehavior rate and causes accidents [13]. As a result, the workflow control system should be able to monitor crane-related human behaviors to discover the anomalies as early as possible, while limiting the propagation of anomalies according to the as-designed workflow and real-time monitoring data.

3 Task-duration Anomaly Propagation Model

In order to prevent possible anomalies from influencing the productivity of the entire workflow, researchers and people from industry have applied different methods involving training, planning, and inspection to prevent human misbehaviors at the front end [14] to stabilize the workflow. Such approaches bring a huge amount of manual work in the busy NPP outage projects and rely heavily on the experience of the management team, which is error-prone and inefficient. Therefore, it is very important to have a sensor-based workflow control system. This proposed computer-vision based workflow control system first automatically identifies all anomalies, which are defined as any differences between as-is and as-planned workflow, on the jobsite. Then the system will identify critical anomalies which influence more tasks or causing severer delays to help project managers adjust the schedule in order to limit the influence of anomalies in a controllable range. In nuclear power plant outages, cranes are always a key resource that influences the scheduling of multiple tasks. As a result, crane-related tasks will be the major focus of this workflow control system.

The first step of workflow control is to generate a mathematical model that describes how a task duration anomaly will influence the schedule of other tasks in order to identify these critical anomalies. A task-duration-anomaly propagation model describes how an anomaly in one task influences the starting time of other tasks. This model will show the critical tasks whose duration will possibly influence the productivity of the entire workflow, which provides a mathematical model of real-time decision making for scheduling and resource allocation. In this section, the authors identified three types of basic relationship types between tasks that repeatedly appear in outage projects: Linear, Co-prerequisite, and Resource sharing. A task-duration anomaly propagation model can help the project manager to understand the most influential tasks after an anomaly is discovered, thus optimizing decision-making.

3.1 Relationship Type 1: Chain

Figure 1 shows the structure of tasks that follow the "Chain" relationship. The authors define the duration of Task 1 as D_1, which follows the normal distribution

$N(T_1, \sigma_1^2)$ and so on. Therefore, the finish time of this workflow equals $D_1 + D_2 + D_3$ which still follows the normal distribution:

$$N(T_1 + T_2 + T_3, \sigma_1^2 + \sigma_2^2 + \sigma_3^2) \quad (1)$$

As a result, all tasks that have a chain relationship with each other can be combined into a new task when scheduling.

Figure 1. Structure of tasks that follow "Chain" relationship

3.2 Relationship Type 2: Co-prerequisite

Figure 2 shows a structure of tasks that follow the "Co-prerequisite" relationship. This is a very typical critical path problem. However, outage projects often have a very busy schedule. The float of the non-critical activities is usually very small. The authors define the duration of Task 1 as D_1, which follows the normal distribution $N(T_1, \sigma_1^2)$ and so on: $D_1 : N(T_1, \sigma_1^2)$, $D_2 : N(T_2, \sigma_2^2)$, $D_3 : N(T_3, \sigma_3^2)$. The finish time of this workflow equals:

$$\begin{cases} D_1 + D_3 : \mathrm{N}(T_1 + T_3, \sigma_1^2 + \sigma_3^2), \ \text{if } D_1 \geq D_2 & (2) \\ D_2 + D_3 : \mathrm{N}(T_2 + T_3, \sigma_2^2 + \sigma_3^2), \ \text{if } D_1 < D_2 & (3) \end{cases}$$

In order to identify the probability of (2) happening, we have:

$$P(D_1 \geq D_2) = P(D_1 - D_2 \geq 0) \quad (4)$$

and $D_1 - D_2$ follows:

$$N(T_1 - T_2, \sigma_1^2 + \sigma_2^2) \quad (5)$$

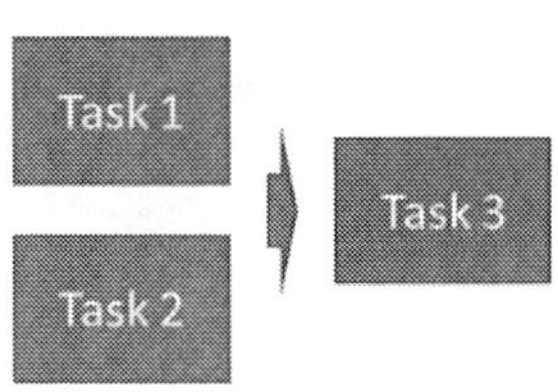

Figure 2. Structure of tasks that follow "Co-prerequisite" relationship

Equation (5) enables us to calculate the probability of changing of critical path. In outage projects, the floats (or slacks) of tasks not on the critical path are usually small. A slight delay of tasks that have a Co-prerequisite relationship with tasks on the critical path (most likely crane-related tasks) may cause the critical path to change and result in hardly predictable durations of the entire workflow.

3.3 Relationship Type 3: Resource Sharing

Figure 3 shows a structure of tasks that follows "Resource sharing" relationship. Tasks with sharing resources cannot be processed at the same time. The tasks with shared resources (Task 2A and Task 2B) follow the "first come, first serve" rule. Using similar signs to represent the duration of the tasks, the authors derived the finishing time of team A (finishing time of Task 3A) as:

$$\begin{cases} D_{1a} + D_{2a} + D_{3a} : \\ \mathrm{N}(T_{1a} + T_{2a} + T_{3a}, \sigma_{1a}^2 + \sigma_{2a}^2 + \sigma_{3a}^2), & (6) \\ \text{if } D_1 \leq D_2 \\ D_{1b} + D_{2b} + D_{2a} + D_{3a} : \\ \mathrm{N}(T_{1b} + T_{2b} + T_{2a} + T_{3a}, \sigma_{1b}^2 + \sigma_{2b}^2 + \sigma_{2a}^2 + \sigma_{3a}^2), & (7) \\ \text{if } D_1 > D_2 \end{cases}$$

Similarly, in order to identify the probability of having (6) happening, we have:

$$P(D_{1a} \leq D_{1b}) = P(D_{1a} - D_{1b} \leq 0) \quad (8)$$

and $D_{1a} - D_{1b}$ follows:

$$N(T_{1a} - T_{1b}, \sigma_{1a}^2 + \sigma_{1b}^2) \quad (9)$$

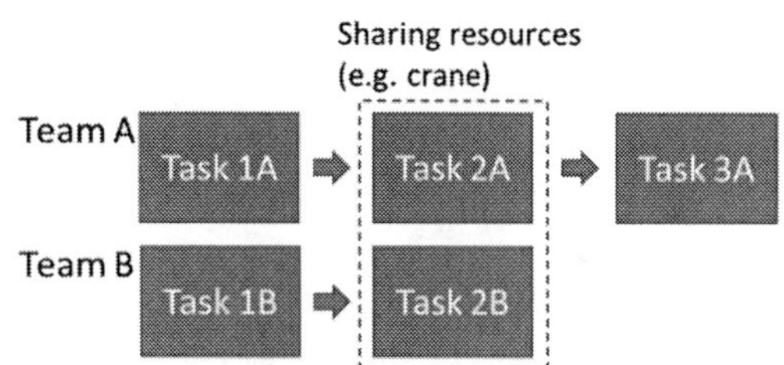

Figure 3. Structure of tasks that follow "Resource sharing" relationship

Resource sharing triggers uncertainties on the sequences of tasks, making the duration of the entire workflow unpredictable and causing the critical path to change. The proposed workflow control method will use simulation to understand the influence of the uncertainty of these "resource sharing" tasks and use computer vision technology to reduce their uncertainty in order to ensure the productivity of the entire workflow.

Table 1 shows the features of the proposed three types of relationships of tasks. In outage projects, tasks can have more complicated relationship types other than the three cases listed above. For example, two tasks can be executed at the same time but with a lower productivity (e.g. two tasks need material from the same tooling trailer). We will leave these more complicated relationship types for future researches.

Table 1. Features of different relationships between tasks

Relationship type	Changing critical path?	Changing task sequence?	Uncertainty bringing to the workflow
Chain	No	No	Small
Co-prerequisite	Yes	No	Medium
Resource sharing	Yes	Yes	Large

3.4 Simulation of the Task-Duration Anomaly Propagation Model

This research uses a simulation to show how different relation types of tasks influence the uncertainty of the duration of a workflow. Figure 4 shows the as-designed workflow of this simulation. First, the workers will be in the waiting area for completing the security check, which takes about 10 minutes. They have a 10% chance of failing the security check, causing a 15-minutes' security rework. After passing the security check, the team will work with a crane operator for about 30 minutes. On the other hand, another crane operation task on a different job site (defined as Site B) is also on the crane operator's schedule, and the crane operator will work on whichever site is ready first. As a result, when the crane operator is called for the box moving task, he or she needs to finish the task on Site B first if he already starts. This "first come, first serve" rule of the crane operator results in uncertainties in the workflow of Site A. The authors run the Monte Carlo simulation of the proposed workflow for 1,000 times, and Figure 5 shows the histogram of the workflow durations for this 1,000 simulations.

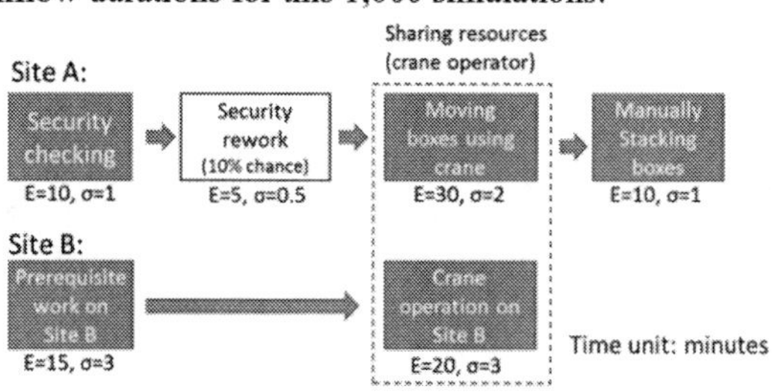

Figure 4. As-designed workflow of this indoor experiment

Ideally this workflow consists of tasks with "Chain" relationships between each other, which will finish in about 50 minutes and its standard deviation is 2.44 minutes. However, the simulation result shows that the average workflow duration is 56.0 minute with a standard deviation of 9.2 minutes and two peaks exist in the histogram. This result means that the "resource sharing" relationship between tasks and the possibility of rework can cause uncertainties to the duration of the entire workflow.

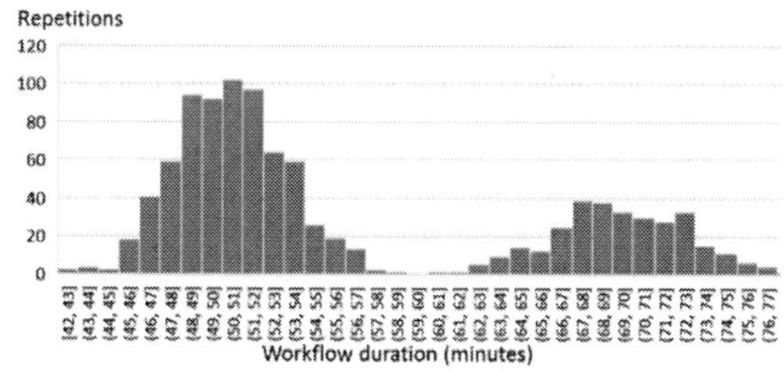

Figure 5. Histogram of the workflow duration

4 Computer-vision Based Real-time Outage Control

The previous section shows that the critical path method does not work well in outage projects because this method only works when the uncertainties of task durations is small. An effective outage control method should inspect the spatiotemporal details of human, workspace (including materials, equipment and environment), and all tasks related to cranes in order to reduce the uncertainties in the workflows. According to the causal factor analysis of OHSA report [7], human behavior is the most frequent cause of crane-related accidents. In outage projects, furthermore, human behavior become more complicated and unpredictable because of stress (nuclear safety), fatigue (working around the clock), and limited time and space (busy site and complicated environment). However, current approaches about outage control rely heavily on manual inspection and analysis, causing less-detailed or delayed job site information and inefficient diagnosis of deviations against the as-planned workflow. Such information deficiency brings difficulties to effective control of crane-related tasks. Furthermore, the systematic approach of human behavior monitoring and understanding in such a busy construction job site is not yet mature.

This method uses an algorithm that automatically detects real-time human activities in certain areas of a NPP outage job site, which can be compared against as-planned workflows in these areas for anomaly identification. This algorithm enables the real-time monitoring of human behavior in critical area of the jobsite, which helps identify the deviations between as-designed and as-is workflow while discovering misbehaviors as early as possible. This algorithm could reduce the uncertainty of the duration of the critical tasks in the workflow and thus expedite the decision-making by limiting the searching space of options.

This research will focus on monitoring human activity patterns in certain areas to reduce the uncertainties in the actual field workflows in a NPP outage project. A real-time and robust human tracking algorithm that uses 2D/3D imagery can automatically observe the critical areas that related to the tasks with co-prerequisite or resource sharing relationships. Then a deviation-detection algorithm will detect differences between as-is workflow information against as-planned workflows and human errors (e.g. queue jumping, chaos when distributing tools, idling around, etc.). Finally, a deviation assessment algorithm will diagnose the potential impacts of detected deviations according to the task-duration anomaly propagation model, thus enables a close-loop control of the workflow schedule. In this section, the authors will introduce two scenes of using computer vision techniques to reduce the uncertainties in various outage workflows.

4.1 Scene 1: Human Counting & Tracking

Outage management teams often struggle with understanding the real-time progress and transportation time of certain tasks. The human counting technique can identify each person, and then label all humans to different groups according to their visual features (e.g. hard hats with different colors) in the neighborhood of a crane, as shown in Figure 6. With this technique, the workflow control system will be able to estimate the transportation time of different groups of workers. Furthermore, human counting & tracking technique provides the possibility of identifying the collaboration status between teams (e.g. well collaborated, quarreling, having difficulties) by analyzing the trajectory of each individual worker, which can reduce the uncertainty caused by human behavior.

4.2 Scene 2: Waiting Line Monitoring

The crane is a critical resource that is always shared by multiple tasks. To predict the finishing time of tasks precisely, the workflow control system needs to monitor workers' activity at a relatively early stage of the workflow. Activities in nuclear power plant often involve strict clearance at its beginning, which leads to waiting lines at the checkpoints. A waiting-line monitoring technique thus enables the workflow control system to reduce the uncertainty of the duration of each task, shown in Figure 7. The waiting line monitoring technique will track each person in the waiting line area, acquiring the line-moving speed and whether people are jumping within a queue, which may indicate the level of experience and training of the team. Moreover, this technique will track the location of certain tools and equipment together with their users. With this information, the workflow control system will automatically identify possible anomalies in the workflow (e.g. shortage of tools, poor-trained team, and accidents), providing efficient indicators to estimate the task duration precisely.

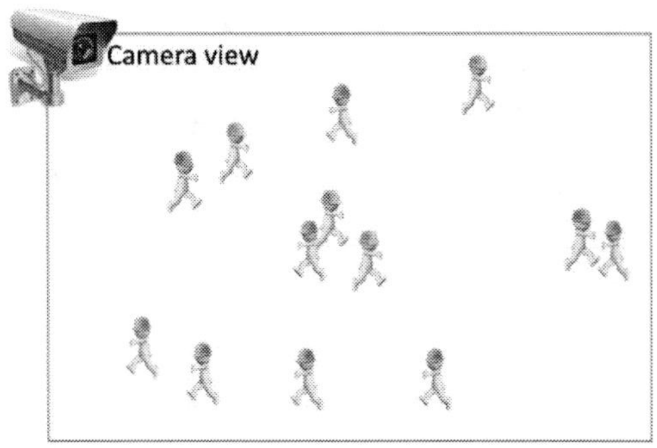

Figure 6. Scene of human counting & tracking

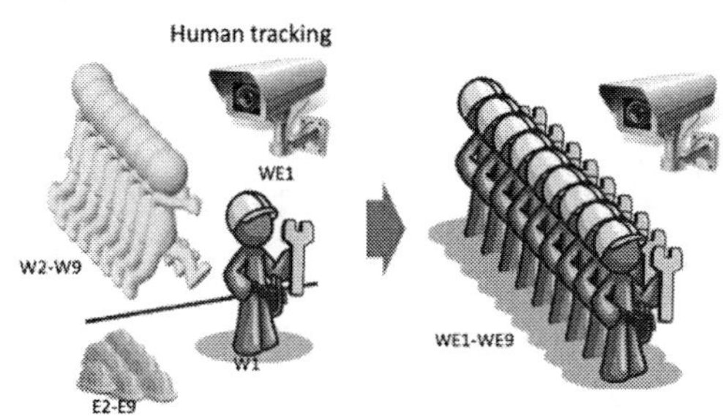

Figure 7. Scene of waiting line monitoring. W=Worker; E=Equipment; WE# = Worker with the No. # Equipment. Barcode or RFID can help identify the equipment.

Within the proposed framework, the 2D/3D imaging data of captured human behaviors of related personnel (e.g. video recording the job site activities related to the crane) will help the management team identify the pattern of anomalies that seems to commonly happen on the job site to trigger other accidents or delays. Also, a real-time and robust object tracking algorithm could use 2D/3D imagery to automatically derive the status of outage workspace, including as-built conditions of structures, materials and equipment, which will be helpful to support the workflow anomaly analyses. In addition, databases that record field activities (especially near-misses, activities that different from the plan, emergencies, etc) can restore the identified anomaly information and its propagation pattern instead of the raw 2D/3D field imageries, which saves data storage space while keeping the detailed jobsite activity information.

5 Experiment and Discussion

5.1 Experiment Setup

The authors designed an indoor experiment to observe how different types of task relationships influence the workflow duration and to test the as-is workflow capturing capability of the proposed workflow control system. This experiment involves three sites: one real site (Site B) in our lab and two virtual sites (Site A and Site C) in computer simulation. Participants will do activities mimicking different tasks in Site B; tasks in Site A and C as well as the transportations between sites will be simulated by people waiting outside the lab for a certain amount of time that is calculated by a simulation algorithm. Figure 8 shows the overall workflow in real and virtual sites. Each site will have the same workflow (shown in Table 2) and share three resources: crane, electricians, and mechanics.

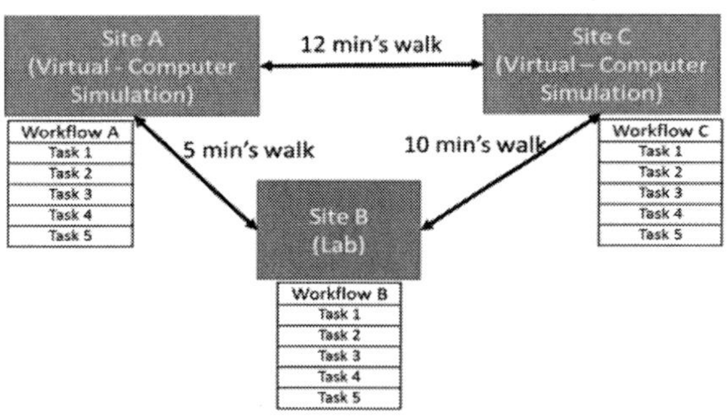

Figure 8. Overall workflow of the experiment

Table 2 Experiment workflow

No.	Task name	Resource
1	Remove the valve	Crane
2	De-term the motor operator	Electricians
3	Perform valve maintenance	Mechanics
4	Re-term the motor operator	Electricians
5	Re-install the valve	Crane

Figure 9 shows the layout of the experimental site (Site B) as well as the position of the sensors wherein light blue areas indicate the view of two Kinect V2 sensors. In site B, volunteers will walk slowly passing the Waiting line area mimicking the security check process and then move and stack some boxes in in the Working area. In this preliminary experiment, only one person will be in the view of the sensors at one time. This experiment uses both GigE Camera and Microsoft Kinect V2 as the sensors of collecting data (shown in Figure 10) for supporting computer vision and spatiotemporal analysis of the scene. Kinect sensors can easily track individual human and capture human action in a relatively short range (maximum 10 meters). On the other hand, GigE camera can monitor a large space with a fast data transfer rates up to 1000Mb/s. The combination of the advantages of these two types of sensors enables the precise tracking of human behavior in an indoor environment.

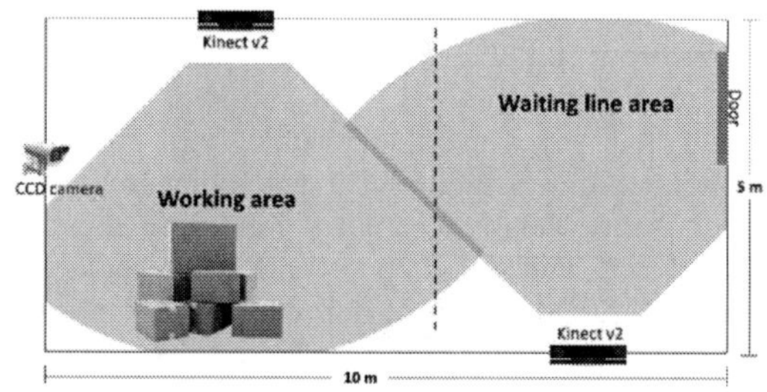

Figure 9. Layout of the experimental site

5.2 Experiment Results and Discussions

Currently, the proposed multi-visual sensor system can capture the human behaviors in the indoor environment. The system can also acquire the start/end time of each task by automatically counting the number of people in both working area and waiting line area in real-time. In the future, the authors plan to focus on estimating potential delays of future tasks according to the human behaviors of current tasks and predictive simulations by enabling multi-human tracking and human-behavior anomaly detection.

Figure 10. Collected Kinect 3D data (left) and video data (right)

In the indoor experiment, the authors observed that one communication mistake (the participant who represented crane did not inform other participants when it finished one of the critical tasks) delayed the overall workflow by 30%. This mistake influenced the starting time of 60% of the following tasks because many tasks are sharing the resource of crane. Due to the page limit, the complete analysis of the experiment will be in another publication. Overall, the results indicate that it is necessary to have a workflow diagnosis and control method to design the topology of the workflow that restrict the propagation of anomalies from one task to the whole workflow, and monitor and control the real-time status of workflows for proactive rescheduling and resource allocation in order to achieve effective and efficient NPP outage control.

6 Conclusion

Crane safety problems in a Nuclear Power Plant (NPP) outage are exascerbated due to the heavy lifts, busy construction site, and dangerous environment with nuclear facilities. Therefore, a reliable and efficient control framework that control human behavior and tasks related to the crane will potentially improve the safe, efficient proceeding of outage project in a nuclear power plant. This need relies on systematic theories and technologies of human error patterns, anomaly propagation patterns in the project, and efficient and effective inspection of the jobsite. This integrated outage control framework will not only improve the safety, productivity, and quality of current outage project, but also provide sufficient data for future outages to reduce excessively long planning stage of an outage project. In addition, the techniques demonstrated for improving crane safety and productivity may be transferable to other outage processes and bottlenecks to further improve overall outage performance.

Acknowledgement

This material is based upon work supported by the Department of Energy, Nuclear Engineering University Program (NEUP) under Award No. DE-NE0008403. DOE's support is gratefully acknowledged. Any opinions and findings presented are those of authors and do not necessarily reflect the views of DOE.

Reference

[1] R. Lloyd, A Survey of Crane Operating Experience at US Nuclear Power Plants from 1968 through 2002 (NUREG-1774), (2003).

[2] S.W. St. Germain, R.K. Farris, A.M. Whaley, H.D. Medema, D.I. Gertman, Guidelines for Implementation of an Advanced Outage Control Center to Improve Outage Coordination, Problem Resolution, and Outage Risk Management, (2014) 77.

[3] K. Le Blanc, J. Oxstrand, Computer-Based Procedures for Field Workers in Nuclear Power Plants: Development of a Model of Procedure Usage and Identification of Requirements, (2012). doi:10.2172/1047193.

[4] X. Luo, F. Leite, M. Asce, W.J.O. Brien, M. Asce, Location-Aware Sensor Data Error Impact on Autonomous Crane Safety Monitoring, 29 (2015) 1–11.doi:10.1061/(ASCE)CP.1943-5487.0000411.

[5] Y. Fang, Y.K. Cho, Crane Load Positioning and Sway Monitoring Using an Inertial Measurement Unit, in: Comput. Civ. Eng. 2015, American Society of Civil Engineers, Reston, VA, 2015: pp. 700–707. doi:10.1061/9780784479247.087.

[6] J.E. Beavers, J.R. Moore, R. Rinehart, W.R. Schriver, Crane-Related Fatalities in the Construction Industry, J. Constr. Eng. Manag. 132 (2006) 901–910. doi:10.1061/(ASCE)0733-9364(2006)132:9(901).

[7] Occupational Safety & Health Administration, (n.d.).

[8] J. Reason, Human error: models and management, BMJ. 320 (2000) 768–770. doi:10.1136/bmj.320.7237.768.

[9] J. Yang, P. Vela, J. Teizer, Z. Shi, Vision-Based Tower Crane Tracking for Understanding Construction Activity, J. Comput. Civ. Eng. 28 (2014) 103–112. doi:10.1061/(ASCE)CP.1943-5487.0000242.

[10] J. Seo, S. Han, S. Lee, H. Kim, Computer vision techniques for construction safety and health monitoring, Adv. Eng. Informatics. 29 (2015) 239–251. doi:10.1016/j.aei.2015.02.001.

[11] S.A. Shappell, D.A. Wiegmann, Applying reason: The human factors analysis and classification system (HFACS), Hum. Factors Aerosp. Saf. (2001).

[12] J.W. Garrett, J. Teizer, Human Factors Analysis Classification System Relating to Human Error Awareness Taxonomy in Construction Safety, J. Constr. Eng. Manag. 135 (2009) 754–763. doi:10.1061/(ASCE)CO.1943-7862.0000034.

[13] V.I. Lohr, C.H. Pearson-mims, G.K. Goodwin, Interior Plants May Improve Worker Productivity and Reduce Stress in a Windowless Environment 1, (1962).

[14] E.J. Jaselskis, S.D. Anderson, J.S. Russell., Strategies for achieving excellence in construction safety performance, J. Constr. Eng. Manag. 122 (1996) 61–70.

Decision Support System for Material Selection based on Supplier Rating

A.A. Akanmu[a] and B. Asfari[b]

[a]Department of Civil and Construction Engineering, Western Michigan University, Michigan, United States
[b]New York City Department of Environmental Protection, NewYork, United States
E-mail: abiola.akanmu@wmich.edu, bushra_ce@hotmail.com

Abstract –

Buildings are continuously recognized as being responsible for significant amount of carbon emission to the environment. A good way to reduce this emission is through the choice of appropriate materials. However, selection of buildings materials is typically based on a number of factors, some of which can either be cost or environmental related. The selection process becomes complicated when designers are faced with several material options of building elements and each option can be supplied by different suppliers whose selection criteria may affect the budgetary and environmental requirements of the project. In order to help decision makers select the most appropriate materials, this paper proposes a decision support system which integrates a modified harmony search algorithm, building information models and supplier performance rating. To illustrate the decision support system, this paper presents a case study of a building in Michigan. The system is capable of producing the cost and environmental implications of different material combinations or building designs.

Keywords –

Harmony Search Optimization; Supplier Rating; Material Selection

1 Introduction

There are a number of benefits to designing buildings that are both cost and environmentally friendly. These includes reduction in carbon-dioxide emission to the environment [1,2], embodied energy in buildings [3,4] and improvement of indoor air quality [5]. When designing buildings, designers are usually faced with the challenge of selecting the most suitable material from different material options or alternatives. This decision becomes more complicated when each material option can be supplied by different suppliers. Furthermore, each supplier may have ratings with different contributions towards the budgetary and environmental requirements of the project in terms of criteria such as price, quality of material and service. Depending on the rating assigned to each criteria, the budgetary and environmental impacts of a project could be affected. For example, if the quality of a material is of more importance to a designer, the cost of the material, hence, the project cost will be higher and if the supplier is selected for his low cost, other criteria such as suppliers material quality, distance and environmental considerations, may be dissatisfying. The later may result in increase in the projects total carbon emission and transportation costs. Although, there has been a number of studies addressing material selection, there is still limited study on the influence of supplier selection criteria on the budget and environmental considerations of a project.

Hence, this paper presents a framework for improving decision making in sustainable construction through appropriate selection of building materials. The framework integrates building information models (BIM), a modified harmony search optimization algorithm and supplier rating. The proposed system adopts two Leadership in Energy and Environmental Design (LEED) requirements pertaining to material selection. To illustrate the functionality of the system, a case study of an office building project in Michigan is discussed.

2 Background

In recent years, material selection has been addressed using different approaches such as ranking methods and optimization techniques. Castro-Lacouture, *et al.* [6] proposed a mixed integer linear programming for material selection using LEED rating. Florez and Castro-Lacouture [7] also utilized mixed integer linear programming but considered objective and subjective factors. Chan and Tong [8] proposed a multi-objective optimization model that uses grey relational analysis to simulate all design criteria and requirements. Zhou, *et al.* [9] developed a multi-objective optimization model for sustainable material selection that includes factors such as process, cost, mechanical properties, performance and environmental impact. Rao [10] proposed an improved compromised ranking method for material selection considering material attributes and their relative importance. Jee and Kang [11] utilized a ranking method to rank materials according to their level of fulfilment of several requirements. Chatterjee, *et al.* [12] explored the capability of the complex proportional assessment and evaluation of mixed data methods for ranking alternative materials according to their capability of meeting predefined design requirements. In spite of these efforts, there has been limited or no study on the effect of the weights of supplier selection criteria on budget and environmental considerations during decision making involving material selections.

3 Proposed System

This section presents an architecture of the decision support system for material selection. The system architecture (shown in Figure 1) integrates BIM, Microsoft database and a modified harmony search algorithm.

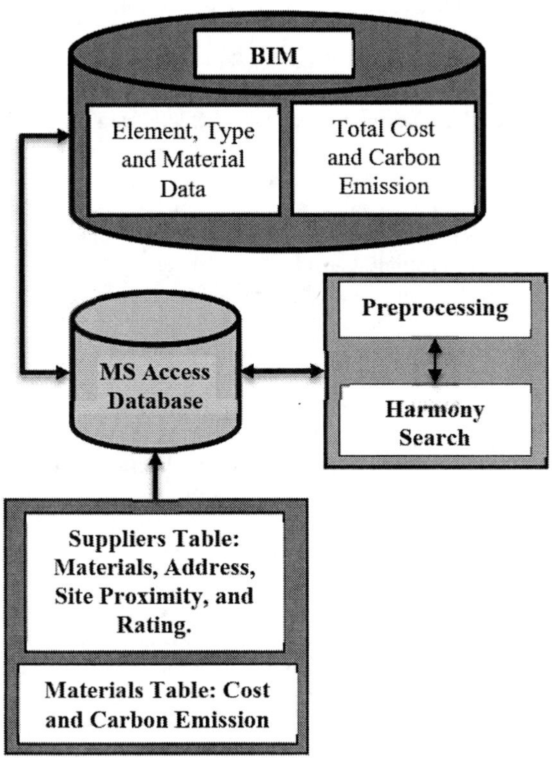

Figure 1. System Architecture for Material Selection

The overview of the information flow among the various applications in the proposed framework is shown in Figure 2. The stages of the model are described below:

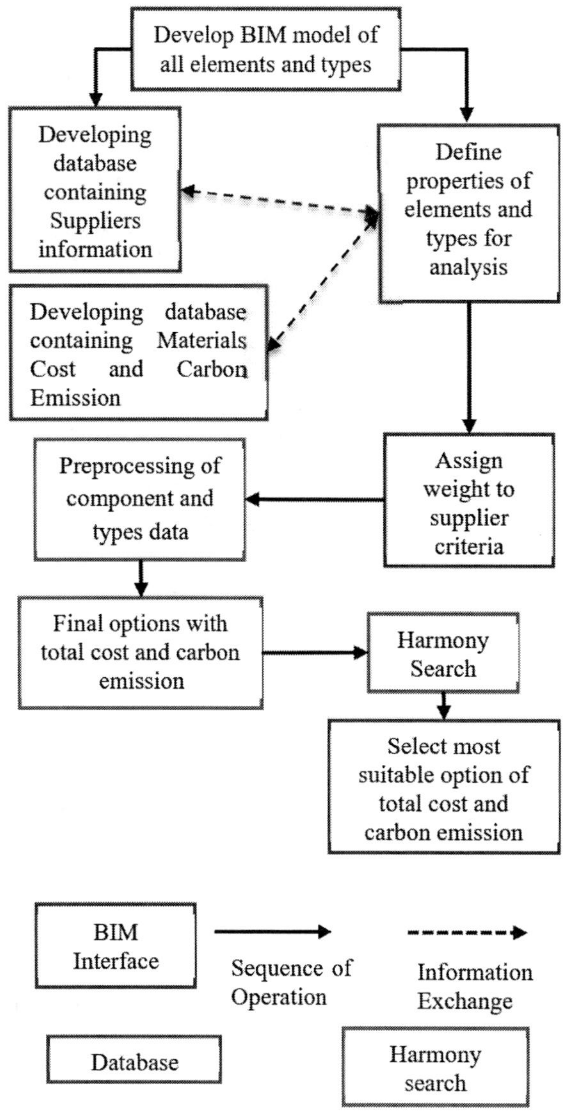

Figure 2. Information flow in Decision Support System.

3.1 Step 1: BIM Module: Defining Building Elements and Properties

In step 1, the building elements and types of each element are defined in the BIM model. The following element properties are also defined: alternatives of each material, the elements to be included in the simulation and elements to be considered for LEED analysis. Autodesk Revit architecture was utilized as the BIM tool. LEED points are also defined in the BIM model. Credit 5 (the regional materials) is implemented for each material option by checking the LEED property (within BIM) for the suppliers in 500 miles range around the project's location.

Table 1: LEED Credits Considered in this Research

Credit	Points	Intent	Requirement
Credit 4	1 point	To increase demand for building products that incorporate recycled content materials, thereby reducing impacts resulting from extraction and processing of virgin materials	Minimum percentage materials recycled for each point threshold = 10% of the total value based on cost
	2 points		Minimum percentage materials recycled for each point threshold = 20% of the total value based on cost
Credit 5	1 point	To increase demand for building materials and products that are extracted and manufactured within the region, thereby supporting the use of indigenous resources and reducing the environmental impacts resulting from transportation	Minimum percentage regional materials for each point threshold = 10% of the total value based on cost
	2 points		Minimum percentage regional materials for each point threshold = 20% of the total value based on cost

3.2 Step 2: BIM-Microsoft Access database

The list of materials, cost and carbon emissions and the supplier information are contained in two separate tables within a Microsoft access database (shown in Tables 1 and 2). The supplier information includes addresses, materials supplied, performance rating and proximity to the construction site. The second table contains a list of building materials, their cost and carbon emission. This can be obtained from published inventories such as the inventory of carbon and energy [18]. The contents of the database and the inputs (from step 1) will be the inputs to the harmony search optimization. A plugin was developed within BIM that enables extraction of supplier data from the supplier database for input to the harmony search optimization.

In order to determine the most suitable supplier of each material alternative, it is important to evaluate and rate the suppliers. To accomplish this, a set of criterions were established, to compare the suppliers. A review of 36 articles on supplier selection criteria (from 1996-2014) was conducted to determine the most influential supplier criterions. From the review, quality was identified as the most popular criterion, followed by cost, while distance and environmental considerations have the lowest percentages. These four criterion were considered in the analysis. Cost refers to the piece-part prices of the type/material of a component as determined by the supplier; distance refers to the distance between the supplier and the jobsite. This affects freight charges, in addition to contributing to the overall carbon emissions of a project. Quality relates to the quality control tool used, defect rates, quality certifications and quality control inspection methods of the supplier. Environmental considerations determines whether the materials offered by the supplier contains recycled or reused content, or whether the supplier has any other environmental certificates for his materials [27]. This is the second LEED credit utilized in the framework (Credit 4: Recycled content).

Likert scale was used as the rating system. Likert scale is based on a 1-5 scale, where 1 means 'not preferred', and 5 means 'mostly preferred' (Table 3). This rating system has previously being adopted by other authors [28,29]. For each supplier, each criterion is assigned a rate in the range of 1 and 5 based on the designer's degree of satisfaction. Furthermore, depending on the designer's needs, the final score of a supplier is determined by assigning each criterion a weight between 0 and 1.

The plugin extracts element properties from BIM, the material and supply information from the database for use in the harmony search optimization. The plugin also extracts the supplier information from the database and displays this on the Revit interface so that the designer can assign suppliers to each material. After assigning suppliers to each material, the plugin extracts these information for the next step.

Table 2: Guideline for Supplier Rates [13]

Point	Grade	Description
5	Exceptional	Demonstrates substantially excellent performance, and has been in the excellence category for last 12 months
4	Excellence	Exceeds company's and customer's expectations, demonstrates extra effort, and is superior to vast majority of suppliers
3	Good	Meets the company's expectations
2	Acceptable	Meets company's minimum requirements
1	Poor	Does not meet the company's and customer's minimum acceptable level

3.3 Step 3: Harmony Search Optimization

Harmony search algorithm is an optimization method for solving both discrete and continuous variable problems. HS algorithm is inspired by the behaviour of musicians whereby each player searches for a better state of harmony. Furthermore, the aesthetic quality is determined by the pitch of each musician's instrument just as the objective function is determined by the set of values assigned to each design variable. Also, the quality of the sound is improved by constant practice just as the value of the objective function is improved by an increase in the number of iterations. HS algorithm requires only few parameters and can be easily implemented. HS has been successfully applied to a variety of practical or real-world optimization on problems, thus, offering more advantages than traditional optimization techniques. These advantages includes utilizing less mathematical requirements, engaging in random or stochastic search and generating a new solution vector, after considering previous vectors. These advantages demonstrate the flexibility of the HS algorithm and potential for producing better solutions. The stages of the harmony search model are described below:

3.3.1 Preprocessing Stage

The pre-processing stage serves two purposes:

1. To arrange the options of each component in an acceptable way to the harmony search stage. This is accomplished by determining the enumerated combinations between each type/material of a component and suppliers. Figure 3 illustrates the pre-processing process of curtain walls. From the figure, the designer has the option of selecting from two types of curtain walls and each curtain wall can be supplied by three suppliers.
2. To compute the cost and carbon emissions for each generated option in a way that incorporates the suppliers' scores using Equations (3) and (4). The results of Equations (3) and (4) are utilized in the HS stage for optimization.

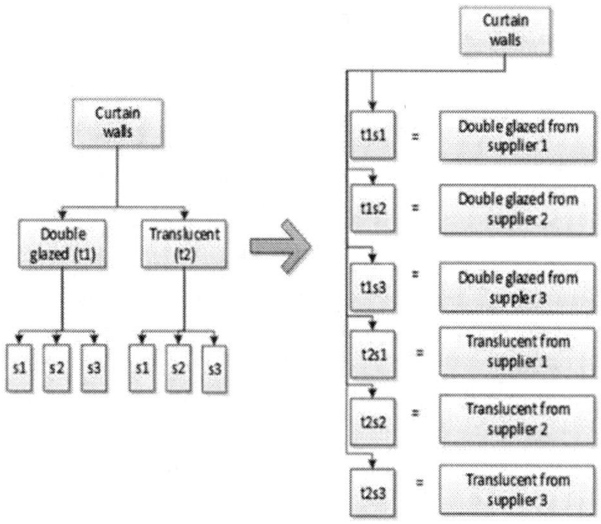

Figure 3. Pre-processing with an example of curtain walls

Computations of Total Cost and Carbon Emission: In computing the cost and carbon emission of buildings, the following phases need to be considered: preconstruction, construction, operations and end of life phases. Based on these phases, the total cost per unit of a material (TC) can be estimated as:

$$TC = C_1 + C_2 + C_3 + C_4 \quad (1)$$

Where, C_1, C_2, C_3, and C_4 are the unit cost of a material in the preconstruction, construction, maintenance and demolition phases respectively.

Likewise the total carbon emissions per unit of a material (TCE) can be estimated as below:

$$TCE = C_1 + C_2 + C_3 + C_4 \quad (2)$$

Where, E_1, E_2, E_3 and E_4 are the unit carbon emissions of a material in the preconstruction, construction, maintenance and demolition phases respectively.

Computations of Modified Total Cost and Carbon Emission: To account for the effect of the supplier rating and selection, Equations (1) and (2) are modified to Equations (3) and (4). These equations can be constructed as a weighted sum of the partial cost/carbon emissions, and are based on the following logics:

- Dividing the initial cost by the cost criterion rate: this means that when a supplier has a good price, the score will be high. This will decrease the initial cost, and in turn reduce the total value. Thus, making this material more probable to be chosen as an optimal solution;
- Dividing the transportation cost by the distance criterion rate: The transportation cost is directly related to the distance of supplier from the jobsite. In the same way, if the supplier is close, he will get a higher rate, which will decrease the transportation cost, and make the material more probable to be chosen as an optimal solution;
- Dividing the total cost by the quality criterion rate: The quality of a material affects the cost. If a material has a good quality, it will probably have a higher cost;
- Dividing the initial carbon emissions by the environmental considerations criterion rate: There is a difference in initial carbon emissions between the virgin and recycled/reused materials. As such, if the supplier provides products with recycled/reused materials, he will get a high score and that will reduce the initial carbon emissions. Hence, this material will probably be selected as an optimal solution.

$$Modified\ TC = \frac{\frac{C_i}{w_c * sc_c} + C_o + \frac{C_t * distance}{w_d * sc_d}}{w_q * sc_q} \quad (3)$$

$$Modified\ TCE = \frac{E_i}{w_e * sc_e} + E_o + \frac{E_t * distance}{w_d * sc_d} \quad (4)$$

Equations (3) and (4) represents components with only one material. Where, C_i is the initial cost of the material, $C_o = C_2 + C_3 + C_4$ and (C_t*distance) is the transportation cost. w_c, w_d, w_q and w_e are the weights assigned to the cost, distance, quality and environmental consideration criteria, respectively. sc_c, sc_d, sc_q and sc_e are the scores given to a supplier for the cost, distance, quality and environmental consideration criteria, respectively. E_i is the initial carbon emissions of a material, $E_o = E_2 + E_3 + E_4$ and (E_t*distance) is the carbon emission from transportation of a material.

In the case of components such as metal studs which consists of multiple materials, the Equations (3) and (4) becomes Equations (5) and (6):

$$Modified\ total\ unit\ cost = \sum_{j=1}^{m} \left(\frac{\frac{initial\ cost}{w_c * sc_c} + C_o + \frac{C_t * distance}{w_d * sc_d}}{w_q * sc_q} \right) j \quad (5)$$

$$Modified\ total\ unit\ carbon\ emissions = \sum_{j=1}^{m} \left(\frac{initial\ carbon\ emissions}{w_e * sc_e} + E_o + \frac{E_t * distance}{w_d * sc_d} \right) j \quad (6)$$

Where, m is the number of materials constituting a type of component

3.3.2 Harmony Search Stage

The objective function of the search process is to minimize the lifecycle cost and carbon emission of a building as shown in equations (7) and (8) respectively. This objective function is evaluated by simulating random values which are initially assigned to the design variables. Depending on the results of the simulation, the design variables are re-assigned new values and another simulation is performed to evaluate the objectives of the new design. The new values of the design variables can be chosen either by random or using the best obtained values which are already stored in harmony memory of the algorithm. In case the new solution is better than the worst solution available in the harmony memory, the worst solution is replaced by the new solution. As optimization process proceeds, the solutions stored in harmony memory becomes better and gradually approaches the optimum solution. The process is continued until a pre-specified maximum number of iterations is reached.

$$\min Cost = \sum_{i=1}^{n} C_i \quad (7)$$

$$\min Carbon\ Emissions = \sum_{i=1}^{n} CE_i \quad (8)$$

Where, n = number of variables, C_i and CE_i are the total cost and carbon emissions of the chosen alternative of a specific variable, respectively.

3.4 Step 4: BIM Module: Selecting the Most Suitable Option

The BIM module aims to present the designer with different designs options and the values of their cost and carbon emissions. Each design will have different combinations of materials. The designer can visualize the different options of total cost and carbon emissions. The selected option is typically the preferred design. However, in order to enable the designer understand the effect of different contributing weights on the supplier criteria, five scenarios were developed. Each scenario represents different weight criterions assigned to each of the supplier selection criteria. The details of the scenarios are shown in Table 3. In this stage, the designer can vary the weights assigned to each criterion depending on the objectives of the design. For example, from scenario 1, the designer can assign equal weights (i.e. 0.25) to each criterion or from scenario 2, he can assign 0.7 to cost and 0.1 to the other criteria. After the harmony search optimization, the designer can select from multiple options of total cost and carbon emissions.

Table 3. Weights assigned to each criterion

Scenarios	Weights
1	Equal weights of 0.25
2	Cost is assigned a higher weight of 0.7, other criterions were given a weight of 0.1
3	Quality was given a higher weight of 0.7, other criterions were given a weight of 0.1
4	Environmental Considerations was given a higher weight of 0.7, other criterions were given a weight of 0.1
5	Distance was given a higher weight of 0.7, other criterions were given a weight of 0.1

4 Case Study

An office building with a total floor area of 34,132 m^2 was used for the case study. The building consists of exterior masonry walls, double glazed curtain walls, precast plank with a concrete topping floor, double glazed curtain walls, exterior doors, hard board wood interior doors, fixed wood framed windows and metal stud-plaster board interior walls. The building was modelled using Autodesk Revit Architecture 2014. A Microsoft access database was developed. The database consist of two tables: One of the tables contains a list of building materials, their cost and carbon emission and the other table contains a list of suppliers, their addresses, materials they supply and their proximity to site. A plugin interface (in C#) was developed to enable interaction between the building components, the database and a harmony search optimization algorithm.

After developing the BIM model, the designer selects elements to be included in the analysis by checking the 'Analysis' property of the elements. The following elements were considered in this analysis: walls, doors, floors, windows, plumbing pipes and HVAC ducts. Table 4 shows the list of elements, types and materials. As part of the properties, alternative materials are assigned to each element (from Table 4) in the BIM model. If the project is to be considered for LEED certification, the 'LEED' property of each element is also checked. The developed plugin is then loaded from the 'add-in' menu of the BIM model. On loading the plugin, the designer selects any model element under the project elements and a list of associated materials (from the properties) are populated in a 'listbox' on the model interface as shown in Figure 5. On selecting any of the materials, the corresponding suppliers (from the database) are also populated on the model interface. The designer can select the potential suppliers of the materials. As each element, material and suppliers are selected from the model, the plugin will upload these information in a new table (in the database) alongside the cost and carbon emission data of the materials.

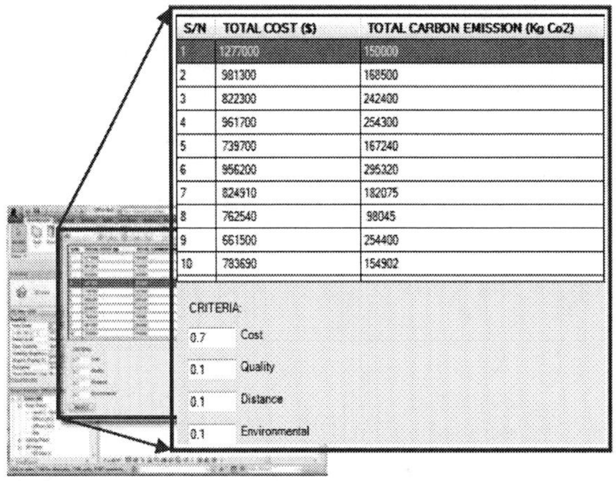

Figure 5. Model Interface showing results.

On selecting the suppliers, it is important to rate them according to their performance in respect to a set of criteria defined in section 2. The scores were assigned randomly, since these information cannot be obtained from the supplier's website. Suppliers whose products consider LEED were assigned the highest score of 5 for environmental considerations while suppliers whose products have any other environmental certificates were assigned a lower score of 4. However, suppliers whose products have no environmental certificate were assigned the lowest score of 1. Suppliers whose location was closer to the project site were assigned a high distance score of 5 and vice versa.

After rating the suppliers, depending on the objective of the design, the designer assigns different weights to each criterion (shown in Table 3). The analysis commences by executing the pre-processing stage. In this stage, the cost and carbon emission of each material are computed using Equations (5) and (6) from section 2.

The HS optimization uses the materials cost and emission data to search for the optimal solutions. The objective functions are shown in equations (1) and (2). The variables are the elements (labelled from X1-X6) in Table 4. The pitch adjustment (PAR) and harmony memory consideration rates (HMCR) helps the algorithm search for a better solution than the worst individual in the harmony memory [14]. The maximum and minimum PAR were set as 0.9 and 0.4 respectively, while the HMCR was 0.8. For each of the scenarios, the plugin can display the total cost and carbon emission of the project on the interface of the model.

Table 4. List of Components, Types and Materials

Components	Types	Materials
X1 = Exterior walls	Wood stud 1 Masonry 2 Metal stud 1	Wood stud + plywood sheathing + steel veneer Brick blocks + plywood sheathing+ steel veneer Metal stud + plywood + brick veneer
X2 = Curtain walls	Single glazed Double glazed Storefront Translucent	
X3 = Exterior doors	Embossed panel steel embossed half glass steel French	
X4 = Plumbing pipes	0.5 in. brass 0.5 in. plastic 0.5 in. stainless steel	
X5 = HVAC ducts	Aluminum alloy under 100 lb Galvanized steel under 200 lb Fibrous-glass 1" thick	

5 Results

This section presents the results of the harmony search optimization. Tables 5 and 6 shows the optimal solutions for the case-study described in Section 3, considering scenarios 1 and 2. Three points have been selected from the optimal solutions of each scenario. Point 1 represents the solution with the highest cost but lowest carbon emission, point 3 represents the solution with lowest cost but highest carbon emission and point 3 is a randomly selected intermediate point. Point 1 can be regarded as the optimal solution for situations where suppliers are selected solely based on cost. The same applies to point 3, which is obtained only when minimizations of carbon emissions is carried out.

Table 5. Optimal Solutions based on Scenario 1.

Variable	Type (Point 1)	Type (Point 2)	Type (Point 3)
X1	Metal Stud 1	Metal Stud 1	Metal Stud 2
X2	Store Front	Store Front	Store Front
X3	Wood Framed 2	Wood Framed 2	Wood Framed 2
X4	Embossed Half Glass Steel	Embossed Panel Steel	Embossed Panel Steel
X5	Wood Bi-Fold	Wood Bi-Fold	Wood Hard Board
Total Cost ($)	301,340.9	278,490.1	226,134.9
Total Carbon Emission (kg)	46,201.45	56,226.6	43,172.2

Table 6. Optimal Solutions based on Scenario 3.

Variable	Type (Point 1)	Type (Point 2)	Type (Point 3)
X1	Metal Stud 2	Metal Stud 2	Metal Stud 2
X2	Store Front	Store Front	Store Front
X3	Wood Framed 2	Wood Framed 4	Wood Framed 2
X4	Embossed Half Glass Steel	Embossed Panel Steel	Embossed Panel Steel
X5	Wood Bi-Fold	Wood Bi-Fold	Wood Hard Board
Total Cost ($)	301,720.3	480,650.0	392,280.5
Total Carbon Emission (kg)	78,585.4	68,520.8	51,593.2

6 Conclusions

This paper presents a framework for selecting building designs that are both cost and environmental effective by integrating BIM, a modified harmony search optimization algorithm and supplier rating. A modified harmony search optimization algorithm that includes supplier selection criteria and rating has been presented.

The search algorithm provides the values of the cost and carbon emissions of material alternatives. The case study illustrates the capabilities of the developed decision support system. The system shows good capability of practical material selection and building design. From a designer's perspective, time and effort could be saved through a BIM-based material selection tool that provides the cost and environmental implications of different design options.

References

[1] Hacker, J.N.; De Saulles, T.P.; Minson, A.J. and Holmes, M.J. Embodied and operational carbon dioxide emissions from housing: A case study on the effects of thermal mass and climate change. *Energy and Buildings, 40*: 375-384, 2008.

[2] Ürge-Vorsatz, D. and Novikova, A. Potentials and costs of carbon dioxide mitigation in the world's buildings. *Energy policy 36*: 642-661, 2008.

[3] Reddy, B.V. and Jagadish, K. Embodied energy of common and alternative building materials and technologies. *Energy and buildings 35*: 129-137, 2003.

[4] Scheuer, C.; Keoleian, G.A. and Reppe, P. Life cycle energy and environmental performance of a new university building: Modeling challenges and design implications. *Energy and buildings 35*: 1049-1064, 2003.

[5] Rackes, A. and Waring, M.S. Using multiobjective optimizations to discover dynamic building ventilation strategies that can improve indoor air quality and reduce energy use. *Energy and Buildings 75*: 272-280, 2014.

[6] Castro-Lacouture, D.; Sefair, J.A.; Flórez, L.; Medaglia, A.L. Optimization model for the selection of materials using a leed-based green building rating system in colombia. *Building and Environment 44:* 1162-1170, 2009

[7] Florez, L. and Castro-Lacouture, D. Optimization model for sustainable materials selection using objective and subjective factors. *Materials & Design 46*: 310-321, 2013.

[8] Chan, J.W. and Tong, T.K. Multi-criteria material selections and end-of-life product strategy: Grey relational analysis approach. *Materials & Design 28*: 1539-1546, 2007.

[9] Zhou, C.-C.; Yin, G.-F. and Hu, X.-B. Multi-objective optimization of material selection for sustainable products: Artificial neural networks and genetic algorithm approach. *Materials & Design 30:* 1209-1215, 2009

[10] Rao, R.V. A decision making methodology for material selection using an improved compromise ranking method. *Materials & Design 29*: 1949-1954, 2008

[11] Jee, D.-H. and Kang, K.-J. A method for optimal material selection aided with decision making theory. *Materials & Design 21*: 199-206, 2000.

[12] Chatterjee, P.; Athawale, V.M. and Chakraborty, S. Materials selection using complex proportional assessment and evaluation of mixed data methods. *Materials & Design 32*: 851-860, 2011.

[13] Egeröd, J.; Nordling, E. Strategic supplier evaluation-considering environmental aspects. **2010**.

[14] Yadav, P.; Kumar, R.; Panda, S.K. and Chang, C. An intelligent tuned harmony search algorithm for optimisation. *Information Sciences 196:* 47-72, 2012.

An Adaptive Automated Monitoring of Construction Activities using Swarm Nodes

A.A. Akanmu[a] and J. Zhang[b]

[a,b]Department of Civil and Construction Engineering, Western Michigan University, Michigan, United States
E-mail: abiola.akanmu@wmich.edu, jiansong.zhang@wmich.edu

Abstract –

Effective progress monitoring of construction activities is always of great value to project stakeholders. It is increasingly being recognized that the success of any project is hinged on the level of awareness of the project status. Although, automated technologies have shown to potentially offer real-time information for quick decision making, there is still resistance to their adoption because of the amount of effort required in obtaining progress data. This calls for investigations into approaches and technologies for creating an adaptive and automated monitoring method. Thus, to address this need, this paper introduces an approach for automatically tracking the status of installed steel components using swarm nodes. The swarm nodes are attached to prefabricated steel components and automatically range to each other in real-time during the steel installation process. In this way, the installation of each steel component and the correctness of the installation are automatically identified and verified in real-time. Particle swarm optimization is used to support an automated arrangement of the swarm nodes for adapting to any steel components configuration. The method of using swarm nodes for data collection, preliminary experiments and results are presented. The potential of the developed method for performance monitoring is also discussed.

Keywords –

Swarm; Monitoring; Adaptive; Optimization

1 Introduction

Recent advances in information technology have triggered investigations into efficient and innovative approaches for automating construction activities. In-spite of these efforts, construction personnel still spend significant amount of time manually tracking and recording project status. Data collected using manual methods are usually incomplete due to the reluctance and inherent human limitations of the workers to record all needed data constantly. Efficient and effective construction progress tracking is crucial for management of construction activities. Although there has been significant improvement to existing progress tracking approaches through the isolated use of radio frequency identification (RFID) tags to the use of integrated approaches such as the integration of RFID tags and virtual models, these approaches still involve manually embedding status information into the tags. As such construction personnel would typically need to spend significant amount of time learning about and using the technology to gather data in addition to accomplishing their required task. For example, in the use of RFID tags for progress tracking by [1] and [2], construction crew need to manually scan installed tagged steel or precast concrete components in order to record their status. For these approaches to be adopted by the industry, construction workers will need to be motivated to collect and record progress or status data, thus reducing the reliability and access to real-time information. Furthermore, the use of information technologies in construction operations is to greatly simplify the complexity and burden associated with existing practices and not to contribute to it [8]. Hence, there is a need for investigations into adaptive and more automated approaches for tracking the status of construction projects.

Thus, this paper introduces an approach that adapts computational resources to the physical construction activities such as to automatically track the status of installed components. We collect component positioning data using swarm nodes. The swarm nodes are arranged on steel components (according to their roles) using particle swarm optimization. The accuracy of the swarm nodes in providing proximity data is presented.

2 Background

Over the years, a number of reasons have necessitated the need to track the progress of construction projects, including tighter schedules, shrinking budgets, environmental influence and increasing competition.

There have been a number of effort developing approaches to tracking construction progress using automated technologies. Among these efforts are the integration of data acquisition technologies (both image and component based technologies) with virtual models. The image based approaches involve the integration of virtual models with laser scanners and digital cameras [3,4]. While these approaches were capable of identifying discrepancies between as-built and as-planned construction models, further processing of the captured data is required before progress data can be attained. Component based approaches involve the integration of RFID tags with the virtual models, [2,5,6] investigated the integration of passive RFID tags and 4D CAD for supply chain and steel management. A higher level of automation in tracking components was illustrated by [5] using active RFID tags. Although, the active RFID tags enables access and update of tag information from distant locations, status information is still updated manually.

3 Research Objective

The research objective is to assist the construction industry in monitoring the progress of construction activities. The purpose of the research is to demonstrate an approach that leverages swarm nodes for automatically tracking the status of installed building components.

4 Swarm Nodes

A swarm node (shown in Figure 1) consists of a swarm bee radio and a microcontroller, both connected using a USB to serial connector.

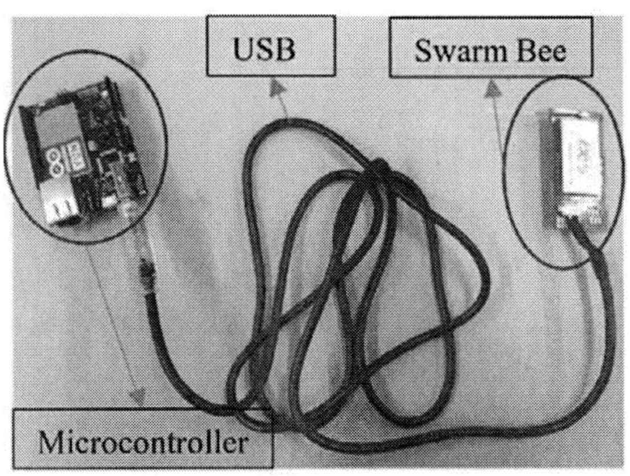

Figure 1. Swarm node

The swarm bee radios developed by Nanotron Technologies, each, has a dimension of 65 x 35 x 12 mm and a weight of 15 grams. The swarm radios have unique ID which distinguishes each node from one another. The swarm bee was powered using the Arduino Yun microcontroller. The microcontroller has a dimension of 73mm and 53mm and a weight of 38 grams. Communication between the swarm bee and the microcontroller is established through a USB to serial connector. The swarm nodes communicate and determine their distance to one another without the need for fixed anchor nodes. The distance between the swarm nodes is determined using the Symmetric Double Sided Two Way Ranging (or SDS-TWR). SDS-TWR estimates the distance using time of flight (ToF) symmetrically from both radios. The ranging methodology and wireless communication are integrated on a single chip in each radio and operates at 2.5 GHz frequency. The range data is extracted from the swarm nodes using a host node. Figure 2 shows interaction between three swarm nodes and a host node.

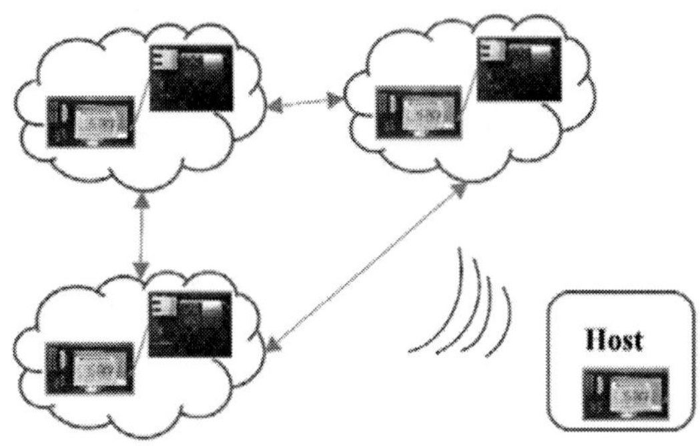

Figure 2. Interaction between swarm nodes

5 Adaptive Automated Steel Installation Monitoring

5.1 Key Considerations

An adaptive automated approach to monitoring the installation of steel members needs to consider the following:

- Steel members/components typically vary in sizes. Installing each member in the right location is important to prevent rework.
- Steel components are erected in specific sequences such as to enable stability of the frames. As such, the proposed approach needs to allow for dependencies between steel members during steel erection.

5.2 Framework

In order to demonstrate the functionality of the swarm nodes, an application was developed for tracking installation of steel components. The framework (Figure 3) is described as follows:

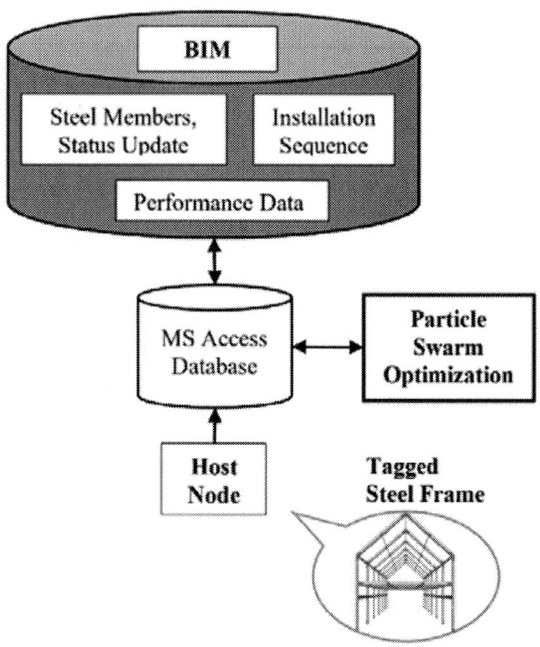

Figure 3. Framework for adaptive automated monitoring system for tracking steel installation

The overview of the information flow among the various applications in the proposed framework is shown in Figure 4. The stages of the model are described below:

5.2.1 Stage 1: BIM Module: Define Steel Members and Properties

At this stage, the steel members are defined and the erection sequence of each member is also defined. This erection sequence is embedded in each member. This erection sequence indicates the dependencies between the steel members (i.e. which members need to be installed before others). This essence of the erection sequence is to avoid the problems that can occur when a member is in line for installation before the supporting members have been erected. The model contains dimensions and proximity data of each steel component, alongside the ID of each component. This is extracted and exported to the database. BIM tools provide a platform for defining and embedding properties of building components. In Autodesk Navisworks, when building elements are defined, erection sequence can also be defined as part of the element properties.

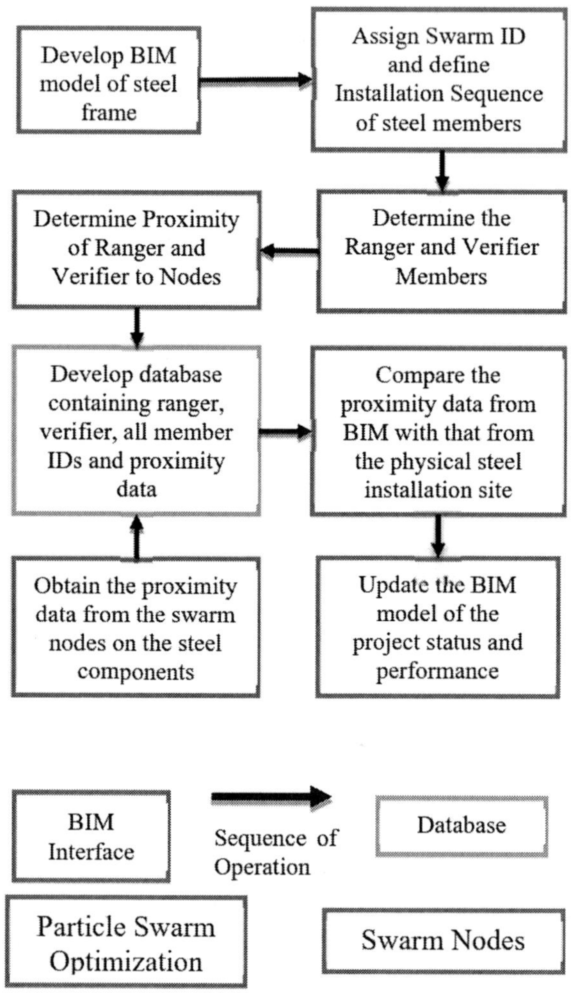

Figure 4. Information Flow for Developed System

5.2.2 Stage 2: Particle Swarm Optimization Module

Swarm nodes range to each other and determine their proximity. To determine the position of a swarm node, it is important to range two fixed swarm nodes to the node with unknown location. These two fixed swarm nodes are assigned the roles of a ranger and a verifier, respectively. The ranger and verifier both range to the unknown swarm node (Figure 5). The minimum number of swarm nodes that can range or have line of sight to the maximum number of nodes are assigned the roles of rangers and verifiers. The rangers and verifiers are determined using the particle swarm optimization (PSO).

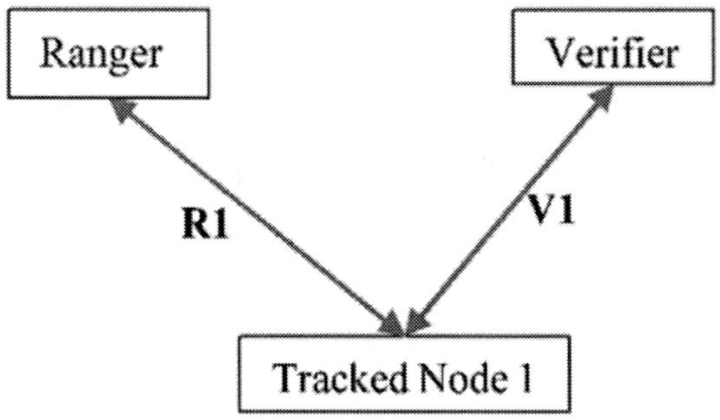

Figure 5. Triangulation to Determine Node Location

PSO is a metaheuristic that makes little or no assumptions regarding the problem being optimized. PSO can also search very large spaces of candidate solutions; this makes it suitable for this research problem. PSO has been widely used for combinatorial optimization and continuous parameter optimization problems. The PSO algorithm can be considered as a swarm intelligence-based multi-agent search method with solutions coded as particles [7]. Each particle keeps track of its location or coordinates in the problem space until the best solution is achieved, at which stage the fitness value is stored. At each step, the PSO involves changing velocity of each particle at its best fitness and global location.

If $X_i = (x_{i1,}x_{i2,} \dots \dots, x_{id})$ is the position of the i th particle in the B-dimension, (5)

$V_i = (v_{i1,}v_{i2,} \dots \dots, v_{id})$ is the velocity representing the search direction, (6)

During the iteration process, each particle keeps the best position that it finds $(pbest)$ and knows the best position searched by the group of particles and changes the velocity based on the two best positions. The standard PSO formula is given below [8]:

$$v_{id}^{k+1} = wv_{id}^{k} + c_1 r_1 \left(p_{id} - x_{id}^{k}\right) + c_2 r_2 \left(p_{gd} - x_{id}^{k}\right) \quad (7)$$

$$x_{id}^{k+1} = x_{id}^{k} + v_{id}^{k} \quad (8)$$

Where

$i = 1,2, \dots \dots, N$

N = Population of the group particles

$d = 1,2, \dots \dots, D$

k = Maximum number of iteration

r_1, r_2 = Random values between [0, 1] used to keep the diversity of the group particles

c_1, c_2 = Acceleration coefficients

v_{id}^{k} = Velocity of particle i in k-th iterating for the d component

x_{id}^{k} = Position of particle i in k-th iterating for the d component

p_{id} = Best position particle i has ever found for the d component

p_{gd} = Best position the group particles have ever found for the d component

The procedure of implementing a standard PSO algorithm is as follows [8]:

1. Initialize the original velocity and position of a particle;
2. Calculate the fitness value of each particle;
3. For each particle, compare the fitness value with the fitness value of *pbest*. If the current value is better, update the position and fitness value with new values;
4. Compute the best particle with the best fitness value. However, if the fitness value is better than the fitness value of *gbest*, update the *gbest* and its fitness value with the position;
5. If the finalizing criterion is satisfied, quit the iteration; otherwise, return to step 2).

5.2.3 Stage 3: BIM-Microsoft Access Database

As shown in Figures 3 and 4, the list of steel members, their IDs and roles, proximity data and installation sequence are exported to a Microsoft access database. These information are obtained from the BIM model and the swarm nodes on the physical steel components. The contents of the database and the inputs defined in stage 1 will be the inputs to the particle swarm optimization and Navisworks plugin, respectively.

5.2.4 Stage 4: Swarm Nodes

The physical steel components are tagged with swarm nodes. The IDs of the swarm nodes are assigned to their corresponding virtual components. The ranger and verifier nodes range to each component and obtain their proximity data for each node. The host node extracts the proximity data from all the verifiers and rangers and sends these to the database.

5.2.5 Stage 5: BIM Module: Visualizing Project Status

The objective of this BIM module is to present to the designer, a visual display of the status of installation of tagged steel components. Navisworks has an open application programming interface (API) which enables developers to easily extend the capabilities of the software. Using the API, a plugin was developed within Navisworks. The plugin extracts and compares proximity data from the database with the data from the swarm nodes. The result of this comparison is used to update the status of steel components. The visual interface of Navisworks serves as a platform for tracking the status of installed/uninstalled steel components (i.e., it can highlight which members have and have not been installed to date).

5.3 Implementation

An experiment was designed to evaluate the performance of the swarm node on a steel construction site in Michigan. The project contains 1200 steel members of which 450 were un-identical. Prior to commencing the experiments, the nodes had to be activated in order to function effectively in the swarm. The nodes were activated using a swarm ranging demo application (from Nanotron technologies). Six nodes was utilized for the experiments. Five steel members were tagged with 5 of the nodes and the remaining node was used as the host. Figure 6 shows a tagged steel component. The swarm node was attached to the steel components using a Velcro. The host node was connected to a computer via a USB cable. The host node interrogates the verifier and ranger nodes and transfers the range data to the computer via the USB.

Figure 6. Steel component tagged with Swarm Node

On executing the swarm demo application, the range data obtained from the host node are populated on the swarm interface. The swarm application was programmed to collect range data every 500ms. Alongside the measurements taken with the swarm node, a Total Station was utilized to record the ground truths of each installed steel component. The Total Station measurements was taken, in order to understand the amount of error associated with the swarm node measurements.

5.4 Experimental Results

The results of the field test is discussed in the following paragraphs:

For each tagged steel, 285 results was collected in order to fully understand the consistency or inconsistency in the generated range data. Table 1 and 2 present a summary of the data collected for each of the tagged components at different placement positions. The average of the range results is shown in Table 1 together with their ground truth collected using Total Station. The error and standard deviation of each sets of range data is represented in Table 2. The error of the swarm range data ($E_{Swarm\ node}$) is determined by computing the absolute difference between the average swarm range data ($R_{Swarm\ node}$) and the range from the (calculated from) Total Station data ($R_{Total\ Station}$).

$$E_{Swarm\ node} = |R_{Swarm\ node} - R_{Total\ Station}| \quad (9)$$

Table 1. Actual and Swarm Range Results

	Ranger (30H6D2)		Verifier (DC107F)	
Comp. ID	Swarm	Actual	Swarm	Actual
7C9524	7.17	6.99	4.69	4.84
2E5DC7	3.25	4.11	6.54	6.96
8F6542	8.35	8.33	4.10	5.36
20F7D7	7.01	5.16	12.63	11.98

Table 2. Error and Standard Deviation of Range Results

	Ranger (30H6D2)		Verifier (DC107F)	
Comp. ID	Error	Std. Dev	Error	Std. Dev
7C9524	0.18	0.36	0.15	0.42
2E5DC7	0.86	0.54	0.42	0.77
8F6542	0.17	0.32	1.26	0.54
20F7D7	1.85	0.14	0.65	0.15

6 Conclusions

This paper has shown that existing approaches to tracking the progress of construction activities has some benefits and limitations. Swarm nodes offer some potential for automated tracking of the progress of construction activities. This paper presents preliminary results on capability of swarm nodes for enhancing tracking automation of steel components installation. The effectiveness of the swarm nodes in providing the range between tagged steel components is discussed. The range data obtained from the swarm nodes is compared with the ground truth obtained from a Total Station. To apply swarm nodes to other components, future research is required to determine the optimal placement of the nodes.

This is because swarm nodes require line of sight in order to range to one another.

Because of the potential of swarm nodes in providing proximity data, further research may leverage swarm nodes to investigate applications for:

- Capturing and documenting near misses and close calls; and
- Delivering context specific information based on social networks.

ACKNOWLEDGEMENT

The authors would like to extend their appreciation to AVB Construction Company for their vital input and cooperation during the case study.

References

[1] Sørensen K.B.; Christiansson P. and Svidt K. Ontologies to support rfid - based link between virtual models and construction components. *Computer - Aided Civil and Infrastructure Engineering, 25:* 285-302, 2010.

[2] Chin S.; Yoon S.; Choi C. and Cho, C. Rfid+ 4d cad for progress management of structural steel works in high-rise buildings. *Journal of Computing in Civil Engineering, 22*:74-89, 2008.

[3] Turkan Y.; Bosche F.; Haas C.T. and Haas, R. Automated progress tracking using 4d schedule and 3d sensing technologies. *Automation in Construction, 22*:414-421, 2012.

[4] Golparvar-Fard M.; Peña-Mora F. and Savarese S. D4ar–a 4-dimensional augmented reality model for automating construction progress monitoring data collection, processing and communication. *Journal of information technology in construction, 14*:129-153, 2009.

[5] Motamedi A. and Hammad, A. Lifecycle management of facilities components using radio frequency identification and building information model. *Journal of Information Technology in Construction, 14*:238-262, 2009.

[6] Akanmu A.A.; Anumba C.J and Messner, J.I. An rtls-based approach to cyber-physical systems integration in design and construction. *International Journal of Distributed Sensor Networks, Article ID 596845*2012.

[7] Wan J.; Yan H.; Suo H. and Li F. Advances in cyber-physical systems research. *TIIS, 5*:1891-1908, 2011.

[8] Mu A.-Q.; Cao D.-X. and Wang X.-H. A modified particle swarm optimization algorithm. *Natural Science, 1:*151-155, 2009.

A Process-Driven Representation Schema for Masonry Wall Assemblies

Andrés Cavieres[a], Russell Gentry[b], and Charles Eastman[b]

[a]University of Oklahoma, [b]Georgia Institute of Technology
E-mail: andres.cavieres@ou.edu, russell.gentry@coa.gatech.edu, charles.eastman@coa.gatech.edu

Abstract – The paper introduces a new approach for the representation of masonry walls in Building Information Modeling applications. The proposed representational scheme addresses the different types and levels of information required to support the design and construction of masonry walls. In particular, the paper proposes the concept of masonry wall "region", as a suitable abstraction to represent the variety of view-dependent features that characterize the life-cycle of masonry walls. At the geometric level, a masonry region works as a surrogate for the description of arbitrary aggregation relations without the cost associated with the explicit propagation of masonry units. In this way, a higher degree of semantic expressiveness can be achieved while keeping the design model flexible and agile. The concept of masonry regions motivates the formulation of a conceptual data model as foundation upon which different masonry-specific applications can be developed in the future, along with the definition of model views necessary to support masonry related data queries and exchanges. The paper outlines the theoretical background behind the concept of masonry regions and its relationship with the Industry Foundation Classes (IFC). Finally the paper introduces a proof-of-concept implementation for a masonry wall schema, and discusses the next steps in the research.

Keywords – BIM; Masonry; Representation; Levels of Development; Design Process; IFC.

1 Introduction

Building design and construction processes are being increasingly facilitated by new Building Information Modeling (BIM) technologies. At the core of these technologies there is a series of data models engineered to enable the exchange of information between different stakeholders. These data models encapsulate and codify industry standard product descriptions, enabling data exchange across software platforms and from heterogeneous stakeholder viewpoints.

In the context of a building material system such as masonry, the conceptualization of the data model should address the representation of more abstract features of assemblies beyond structural aspects such as geometry and material properties. Some of these abstract features may be performance dependent, spatiotemporal, related to cost, ownership, production status, etc. When a given stakeholder viewpoint – which includes the role of the stakeholder and the design activity in question, for example, a structural engineering performing lateral load analysis – is applied to a model, a domain-specific model view needs to be derived from the source model so to support the activity [1].

Currently, the most mature material-specific BIM data models are for structural steel – with the early standardization of steel shapes forming the basis for the steel components used in buildings today [2]. Almost a century after steel shapes were standardized, the first computational data model for structural steel was released as the "Logical Product Model" by CIMSteel [3]. Since then, material-specific models for precast concrete [4] and cast-in-place concrete [5] systems have been developed.

The Building Information Modeling for Masonry Initiative (BIM-M), organized in 2013 in North America, has developed a roadmap for establishing the requirements for masonry-specific data models, to support design, procurement and construction of masonry buildings [6]. The second phase of the roadmap, recently completed, focused on the development of data requirements for masonry units and masonry walls. In addition, the initiative has completed an extensive set of masonry-building case studies, focusing on the information needs of architects, engineers, material suppliers, and mason contractors [7].

As part of the initiative, two preliminary data models have been proposed. The first is a data model for masonry units, which is follows the functionality of databases of hot rolled steel shapes promulgated by AISC and the BSI [8]. Unlike structural steel however, the Masonry Unit Definition model or MUD is extended to include material properties, color, and texture, in addition to geometry. The second data model, called Masonry Wall Definition (MWD) aims towards a specification of masonry wall models that explicitly capture key relationships at the assembly level. As introduced earlier, it is important to point out that the complexity in the representation of masonry assemblies stems not only from the geometry of a wall and the properties of its units, but also from more abstract relationships that need to be formally described in order to provide computationally support to a variety of view-dependent tasks. In particular, one of the main challenges is the development of a flexible representation that could enable incremental levels of design information, geometric or otherwise, without the need for explicit instantiation of individual units. In this way, more informed exploration of alternatives would be facilitated by avoiding oversimplifications embedded in current representational approaches, and without the cost of excessively detailed geometric models.

The present paper focuses on the main guiding principles of the Masonry Wall Definition data model. Brief references to the Masonry Unit Definition (MUD) data model will be provided when needed. More information on the MUD model can be found on Sharif et al. [9].

2 Masonry Wall Definition model

The problem of representing masonry walls in a machine-readable format is considerably different than the representation of masonry units per se. For instance, the MUD model is internally-focused to provide comprehensive information about units, but little information about the context in which the units are applied. Masonry walls on the other hand are defined wholly by their context – the functional, engineering and aesthetic requirements dictate the geometry of the walls and the masonry is configured to fulfil these requirements. The overall geometry of a given masonry wall can be defined in terms of its start and end points, as well as its base and top reference planes. As an assembly however, a masonry wall can be characterized in several different ways simultaneously. Besides the self-evident parts such as the masonry units themselves, and other discrete accessories, a masonry wall can also be characterized by features such as openings, protrusions, niches, cut-outs, indentations, inlays, corners, etc. Furthermore, some of aspects of the assembly may be more abstract and context-dependent. For instance, a particular load-distribution pattern or a special sequence of erection entails different types of relationships. Each of these serves a specific purpose and therefore needs to have its own set of domain-specific properties and attributes.

The representation of these abstract and context-dependent features in turn requires a formal definition of modularity as well as different aggregation patterns that are relevant for masonry construction. In this way the model not only "looks" like a masonry wall, but more importantly, provides a consistent source of information regarding constructability and performance. Currently however, the representation of masonry modularity and aggregation patterns is limited to bonding and coursing. Moreover, these are typically represented in an oversimplified manner, by using a 2D pattern or "hatch" which is applied to the surfaces of wall objects to denote a masonry composition, but without explicit description of abstract and context-dependent features of the entire assembly. This limitation not only reduces the scope of automation that could be implemented otherwise, but affects the ability of teams to detect conflicts in timely manner, and ultimately, to make better design decisions.

By recognizing the importance of this problem, the goal of the Masonry Wall Definition project is to establish the nature of these assembly properties along with the best methods to represent them. For that purpose, information requirements of typical masonry workflows identified by Lee et al. [7] were used as starting point for the definition of those properties. However, the iterative and incremental nature of the design process raises a number of problems.

First, in the early stage of design, the design problem is generally ambiguous and requirements are still ill-defined [10, 11]. During this stage, several candidate building shapes and geometric relationships are explored without committing to any specific semantics of construction or materiality. As the design process moves-on, generic objects are replaced by more specific ones such as walls, doors, and windows. Once masonry is adopted as material of choice, some of these objects have to be moved, sized or deleted from the model. At this point the designer may have little interest in tracking the location of specific masonry units. All that matters is some level of coordination with an underlying modular system that is consistent with the type of masonry chosen.

As the design is refined, the issue of constructability comes into play, imposing the need for more specific information. The concept of Level of Development (LOD) in BIM was introduced to guide the amount and fidelity of information to be added to a building model as design proceeds. According to the LOD Specification [12], model elements are represented with a range of information granularity, from the most schematic representation (LOD 100), to the most detailed (LOD 500). At the early stages of design, walls are represented at LOD 100, without the need to indicate wall thicknesses and material types. At some later stage in the design process, the wall may be identified as a masonry wall, with specific masonry units, bonding patterns and reinforcement information. This may correspond to a LOD 400. At this point, the global geometry of the wall and the local geometry of masonry units have to be resolved in various situations. In particular, the way that masonry units relate to certain boundary conditions of the wall needs to be represented explicitly. This is necessary for example to calculate the number of masonry cuts, custom units or other types of components required to resolve special situations along the wall (e.g. reinforcement, barriers and insulation, etc.).

One possible method to represent such type of conditions is the explicit propagation of masonry components by means of algorithmic procedures and parametric rules [13, 14]. However, this leads to a second problem. BIM parametric applications are known to be computationally-intensive and the performance of any parametric-modeling software degrades as the number of parametric elements in the model increases, which in turn cause negative implications in the design process itself [15]. Since it is likely that a masonry building will have tens or even hundreds of thousands of masonry units, plus all relevant accessories, it is simply not efficient to model each unit, even if the procedure is automated. Furthermore, it is rather questionable that exhaustive modeling of every single unit would be actually an effective approach in supporting conventional design workflows. From a more pragmatic point of view, the aggregation of masonry units according to some functional or aesthetic criteria may provide a more cost-effective approach without losing significant precision or expressiveness.

From these observations it became evident that a different type of representational strategy was needed to model masonry walls more effectively. In particular, it became clear the need for a process-centric representation, so that the iterative and incremental nature of the design process could be not only supported but promoted. This means not only that the representation needs to be flexible enough as to support the use of different levels of information at different stages of the design process, but more importantly, to support the transition between design stages and levels of information. The need for a flexible, process-driven design representation has been discussed extensively in the past from different perspectives [16 - 18].

Within the context of the MWD project, the strategy adopted was the conceptualization of a new type of abstraction. This abstraction is intended to mediate between the monolithic representation of masonry walls and the exhaustive propagation of individual units. Such an intermediate abstraction was called in the research a masonry "region". A masonry region is geometrically represented by a solid object that can be created by decomposition of larger solids – a wall or another region – by means conventional solid modeling operators (e.g. subtraction and intersection). From a semantic perspective however, a region describes a view-dependent aspect of feature of the wall assembly. In particular, a region denotes geometrically an arbitrary aggregation of masonry units. Such an aggregation may include components and accessories connected or functionally associated to the units. In prior work we have discussed how regions may be associated to specific stakeholders as well as with different design or evaluation tasks. Altogether, this information characterizes the context under which region definitions may be created [19].

Part of the motivation behind this formulation is that certain types of information requests between parties could be satisfied through the derivation of regions. In this regard, the conceptualization behind masonry regions is consistent with the principle of targeted interoperability within specific use case scenarios [20, 21]. In the case of masonry walls, and masonry buildings in general, the use cases in which view-dependent information may be requested typically involve analysis of some sort related to structural or energy performance evaluation, detailing, quantity take-off, construction planning, etc.

In the remaining of the paper the main principles behind the concept of masonry regions will be introduced. These principles form the basis for the Masonry Wall Definition data model under development. The requirements for the representation of regions within a wall are enumerated below. We provide specific, and somewhat limiting requirements at this time so to facilitate the implementation of an initial proof-of-concept software application.

2.1 Masonry wall regions

An example of a brick wall with region decomposition is shown in Figure 1. The regions are represented as solid partitions of the wall, each implying a specific aggregation of masonry components. As discussed before, the criterion for the aggregation is view dependent. In this case, boundary conditions for window openings and corners may be necessary for the specification of insulation and water barriers, estimation of unit types (e.g. half-units) and sequence of erection.

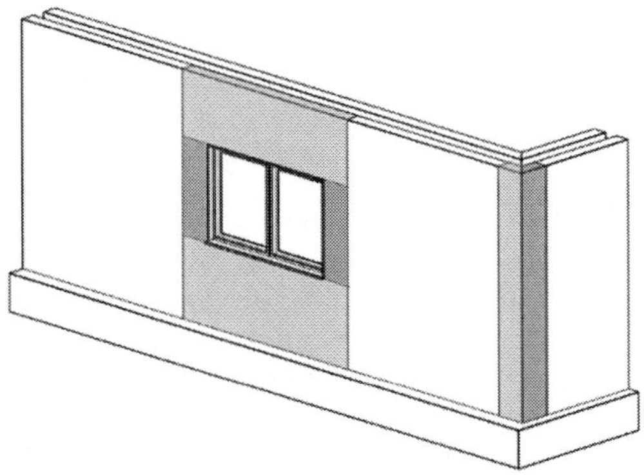

Figure 1 Decomposition of wall into regions associated to opening and corner conditions. The window opening region can be further decomposed into regions for the sill, lintel and jambs.

General definitions and rules for the decomposition of masonry regions are the following:

1. A region is the geometric representation of an arbitrary masonry wall feature. Masonry wall features may be a *bona fide* portion of the wall with a clear, identifiable shape and function (e.g. an arch, a pilaster, etc.), or a *fiat* portion of the wall defined according to some arbitrary criteria (e.g. bricks to be laid by crew per day, damaged bricks to be replaced, etc.). Masonry wall features are discussed later in section 2.3.
2. A masonry region is bound by horizontal lines of masonry courses. The first and last courses are the outermost horizontal boundaries.
3. A masonry region is bound by vertical lines in coordination with the masonry bonding pattern and head joints. Thus, a region may contain only full and half masonry units. When a wall in running bond id represented at LOD 400 or higher, the vertical boundaries are staggered lines that follow the overlapping of the bond. The first and last edges of a wall are the outermost vertical boundaries.
4. The largest possible region within a wall is the wall itself in its whole. In conjunction with the minimal region (see definition 5), the maximal region provides the basis for modular coordination for a given unit type, along with the context for which rule for boundary conditions can be established.
5. The smallest region is the size of a half unit. This is called the minimal region, representing the basic module required for dimensional coordination. In some cases, niches, recesses or other type of feature can be depicted using a minimal region representation.
6. The masonry within a given region must all be laid according to a specific bonding pattern. However, it is possible to decompose a region further into sub-regions, which can have different bonding patterns.
7. Regions may be rectangular, trapezoidal or triangular, so that gables and other forms can be represented.
8. Regions may be defined also through the thickness of the wall, to accommodate walls with multiple wythes of masonry.
9. Within a given region a set of rules can be established that control the placement of masonry-specific components, such as vertical reinforcement, bond beams, grout, wall ties, weeps, etc. A given rule set applies to a region, and if the rules change, a new subdivision is required.

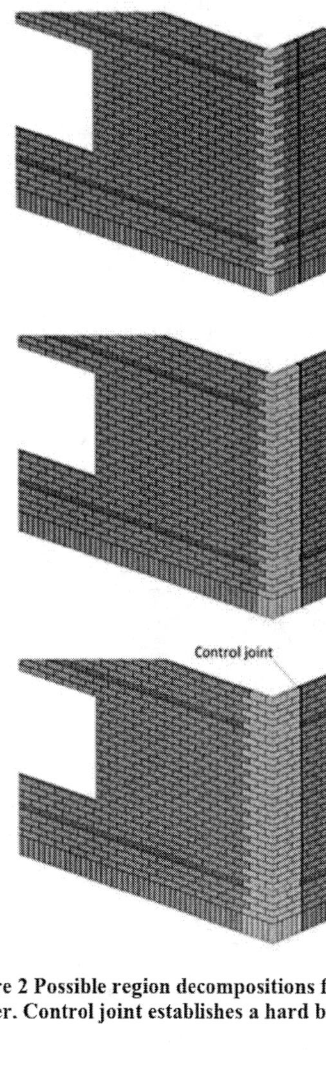

Figure 2 Possible region decompositions for L-type corner. Control joint establishes a hard boundary.

10. Parametric constraints can be associated to regions to control their placement and the geometric behavior of its boundaries, as well as individual parts within the region.
11. Parametric constrains and rules may enforce that a wall be "in-coursing" and "in-bond" in relations to its neighbors, boundary conditions, or other type of requirements. These rules may include geometric tolerance criteria to adjust the thickness of mortar joints to fit masonry units into regions where possible.
12. Masonry wall corners at L- and T- shaped intersections form their own type of region. Therefore a complex sequence of walls can all be controlled geometrically from a single "anchor point", from which the masonry pattern is established. Figure 2 shows an example of an L-type corner region.

2.2 Relationship between regions and LOD

In a design workflow certain activities can only occur if the right type of information is available in an appropriate format. In a BIM enabled workflow, the information required typically involves the exchange of only a subset of the entire model. For that reason, the concept of Levels of Development (LOD) was developed, in order to ensure that the required levels of information are present in the source model to support different design activities across different design stages.

The concept of region representation was developed to support incremental LODs as well as the existence of different LODs simultaneously in the same model. This is particularly relevant for masonry, given that in some circumstances it is necessary to model individual units and accessories explicitly (e.g. virtual mock-ups), while at the same time keeping a more generic description in other areas of the model.

At LOD 100, masonry regions are very generic. By default, only the maximal region may be defined, coinciding with the overall geometry of the wall. However, as soon as LOD 200 is required, the wall can be decomposed into more specific, view-dependent regions. For instance, the insertion of new masonry features into a wall, such as openings, corbels, pilasters, recesses or quoins, to name a few, are all associated to specific forms of region decomposition. This decomposition process is intended to be automatic, similarly to the functionality provided in some pre-cast concrete BIM applications (e.g. IDAT pre-cast module for Revit) [22].

At LOD 200, each region may have associated different masonry unit types, coursing and bonding patterns. Also, as part of this characterization, the behavior of the masonry at different region boundaries must be established in a coordinated manner. Typical examples for vertical boundaries include: "preserve running bond with adjacent regions" and "insert half bricks and establish control joint" (see Fig. 2).

At this point it is possible to generate a custom hatch (2D surface pattern) for each region on the masonry wall. These patterns can be used for manual verification that the masonry wall bonding and coursing are correct. The hatch representation is computationally lightweight – and might well be sufficient for much of the early-stage architectural design process.

The act of placing and correcting the architectural hatch on walls is not trivial. This may mean that overall building dimensions need to be adjusted, or that the size and location of doors and windows need to be changed. Or it may be that alternative masonry units will be specified to meet the overall building geometry. In some situations it may be possible to adjust the width of the head and bed joints – or allocate the dimension mismatch to vertical and horizontal control joints. Once the masonry patterning has been established, and the patterns accepted, the masonry wall can be considered to be at LOD 300. For structural masonry, LOD 350 has a specific definition as outlined in the 2015 BIM Forum Specification [12]. In the structural layer of the walls, the following elements should be included in the model: bond beam and lintels, reinforcing and embedments, and jambs sections. These are key elements included in automated clash detection and trade coordination.

Finally, the propagation of individual masonry units into the BIM model, if required, occurs at LOD 400. The region concept supports the selective placement of masonry units into the model on a region by region basis. Therefore, if certain regions require more complex detailing, then only those regions may be promoted to LOD 400. The masonry units in the MUD can be instantiated in the model and propagated either manually or algorithmically. The specification of placement rules according to local coordinate systems (Fig. 3), allows the masonry units to be merged with the hatch pattern at either the wall face or wall centreline, as appropriate.

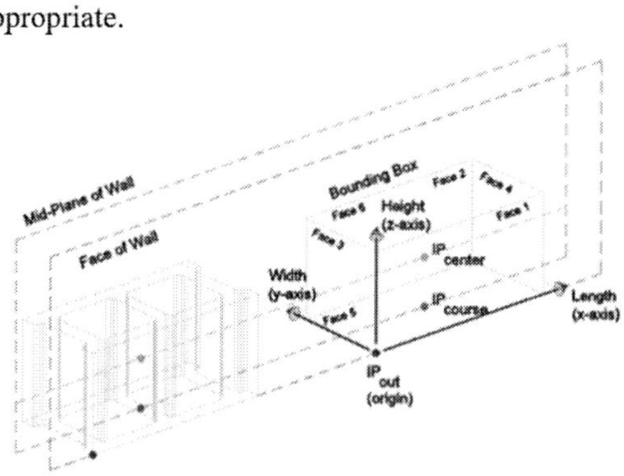

Figure 3 Insertion of masonry units into masonry walls.

The ability to specify different LODs in the same model and to explicitly propagate masonry units into specific sub-regions only when needed is a useful capability (Figure 4). Indeed, we expect that such approach would facilitate a more effective exploration and evaluation of alternatives without significant compromise in computational performance.

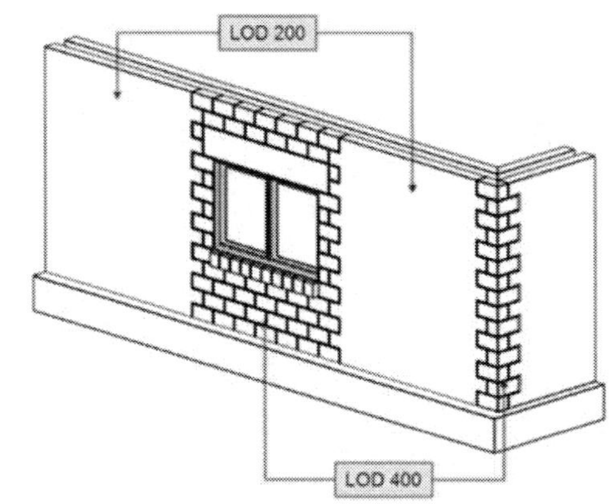

Figure 4 Different LODs assigned to masonry regions.

2.3 Masonry wall features

The implementation of masonry specific BIM applications entails first and foremost the definition of conceptual data models that capture the properties and relationships that are relevant for most use case scenarios and design workflows. As mentioned before, this need has been initially addressed by the BIM-M initiative with the development of the MUD project, which focused primarily on a conceptual model for masonry units. The next step in MUD is to include within the same conceptual framework the definition of masonry components and accessories that are most commonly used in masonry construction.

However, the computational representation of masonry walls requires more than the definition of individual masonry units, components and accessories. In fact, masonry walls need to be characterized as being composed not only of discrete, off-the-shelf collections of parts but also by a series of intermediate aggregations that result from particular arrangements of masonry units. These aggregations conform functional and aesthetic features of masonry walls, such as wythes, veneers, corbels, recesses or inlays to name a few, that that are intrinsic to masonry construction. Figure 5 provides a classification of typical masonry features identified by the research. Since different masonry features imply different combinations of components, sequences of erection, equipment and skill, they also play an important role in construction planning and cost estimation. Therefore, the semantics of masonry features was recognized as key in the specification of the masonry wall data model.

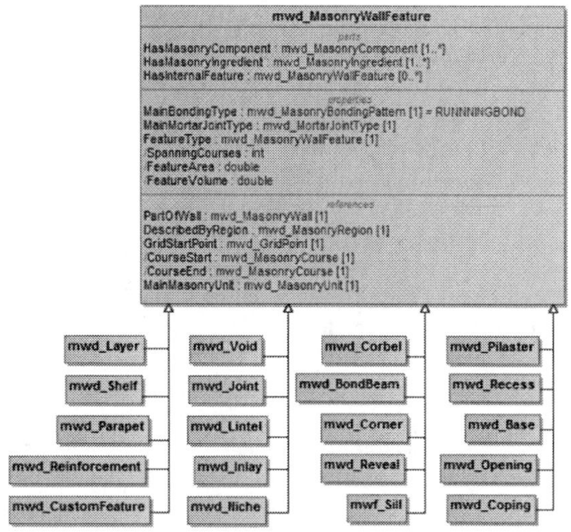

Figure 5 Schema definition for masonry wall feature types.

At the most general level, a masonry feature may be seen as being part of a wall, having a relative position, a shape, a set of internal masonry units as well as internal components and accessories, a bonding pattern and a function. Notice that a masonry feature may contain internal sub-features.

As discussed in section 2.1, regions are geometric abstractions of masonry features, similar to the notion of bounding boxes. By decoupling the representation of masonry features (i.e. the "real-world" entity of interest) from its geometric abstraction, a greater degree of modelling flexibility is provided. In particular, regions play a dual role as both place holders for further addition of geometric detail, as well as implicit form of aggregation of masonry components. This approach is intended to facilitate not only the transition from lower to higher LODS, but also the inverse, from higher to lower LODs. Such a representational capability is considered relevant to support more effective design iterations, by making it easier to designers to go back to a previous stage and explore alternatives.

3 Proof of concept schema

The conceptual data model developed for masonry walls was preliminarily implemented as a XML Schema and imported into SketchUp as a proof-of-concept. The proposed semantics for masonry walls, masonry wall features and masonry regions were formalized using IFC 2x4 definitions as main reference framework [23] This approach aims towards future compatibility with the IFC standard, and the reusability of existing definitions related to walls in general. For example, the semantics of *IfcWallElementedCase* was found to be more appropriate for the description of masonry walls, especially in situation where Levels of Development need to be higher than LOD 100. As a consequence, masonry walls can be treated explicitly as assemblies, while the allowable set of element parts (i.e. *IfcBuildingElementPart*) can be extended and customized further to meet the information requirements of masonry-specific workflows. This provision also facilitated the characterization of masonry wall features as subtype of *IfcBuildingElementPart,* thus keeping the consistency of relations established at different levels of the assembly hierarchy.

Finally, *IfcBuildingElementProxy* provides the basic framework for the semantic characterization of masonry regions. In particular, this is an IFC construct intended to serve as spatial place holders for future allocation of functions and exchange of undefined geometries. As such it can have associations to different placement objects, shape representations and material definitions, as well as spatial containment, element compositions and property sets. This is precisely the goal behind the conceptualization of masonry regions, which may be created during early design stages without a precise meaning or associated function. The process of region generation, by geometric decomposition of larger solid objects is also consistent with the exploratory nature of conceptual design which is arguably top-down.

Within the envisioned software functionality, a region can be assigned a LOD value, the type of masonry wall feature the region denotes (i.e. *isAbstractionOf* relation), along with other inherited and specialized properties (Figure 6).

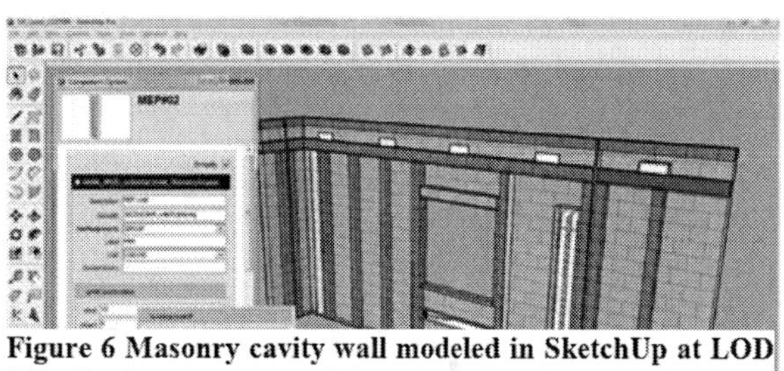

Figure 6 Masonry cavity wall modeled in SketchUp at LOD 350. Regions are created by top-down decomposition and assigned to predefined feature types by means of the "*isAbstractionOf*" relation. In this example the region represents CMU blocks in the back-up wall containing MEP ductwork.

4 Summary and Conclusions

The semantics of masonry construction and particularly of masonry walls are largely missing from current BIM applications. In order to provide better computational support, BIM applications need to go beyond representational approaches that oversimplify the complexity involved in masonry assemblies. Instead, a more sophisticated approach is needed, where information requirements that are unique to masonry workflows can be more effectively satisfied. To do so, the representation of masonry walls needs to be approached from process-centric perspective, with the goal of allowing incremental levels of design information without compromising computational performance nor the ability of designers to explore alternatives. These conditions imply the need for a more expressive representation to cover multiple stakeholders' perspectives. At the same time the representation has to be flexible, so that different Levels of Development (LOD) may co-exist in the same wall model.

This paper outlines the underlying philosophy for a compact but extensible representation of masonry walls intended to address these issues. At the core of this representation lies the concept of masonry region, which is a geometric abstraction for wall features that are meaningful from a domain-specific perspective. Since masonry wall features denote particular aggregations of masonry units, a region serves as geometric proxy for such aggregation. In this way different levels of geometric detail may be added selectively to different features of the wall, independently from their non-geometric aspects.

The paper introduces the early stage of development for a proof-of-concept implementation of a schema for masonry wall using IFC as conceptual framework. The proof-of-concept was encoded as XML Schema and imported into SketchUp for preliminary evaluation. This consisted in a number of masonry cavity wall models that were built at different LODs in order to compare modelling efforts against information content. While the evaluation is still preliminary the exercise provided some valuable insights. For instance, the use of regions facilitated the addition of geometric detail up to LOD 500 (i.e. units and accessories) at selected locations of a masonry wall while keeping overall modularity. This was seen a more efficient way of adding resolution where needed, while keeping the model workable. Similarly, geometric detail could be deleted from specific regions, while keeping semantic consistency with the overall assembly. This is an important functionality because the ability to transition back and forth between LODs may potentially facilitate the exploration of design alternatives, especially in the context of masonry construction.

Future work will focus on further refinement of the proposed masonry wall schema. This will involve development of prototypes, systematic testing and validation. For this purpose collaboration with both the masonry and software industries is critical.

5 References

[1] Lee Y., Eastman C., Solihin W. and See R. Modularized rule-based validation of a BIM model pertaining to model views. *Automation in Construction*, 63: 1-11, 2016.

[2] Standard Specification for Structural Steel 1896. Association of American Steel Manufacturers. Steel Construction Manual Shapes Database, Version 14.1. American Institute of Steel Construction, 2013.

[3] Crowley A. J. and Watson A. S. Representing engineering information for constructional steelwork. *Microcomputers in civil engineering*, 12(1): 69-81, 1997.

[4] Eastman C., Sacks R. and Lee G. Development and Implementation Of Advanced IT in the North American Precast Concrete Industry. *ITcon International Journal of IT in Construction*, 8: 247-262, 2003.

[5] Barak R., Jeong Y. S., Sacks R. and Eastman C. M. Unique requirements of building information modeling for cast-in-place reinforced concrete. *Journal of Computing in Civil Engineering*, 23(2): 64-74, 2009.

[6] Gentry R., Eastman C. and Biggs D. A Roadmap for Developing and Deploying Building Information Modeling (BIM) for the Masonry Industry. Atlanta, Georgia USA, Georgia Institute of Technology, Digital Building Laboratory, 2013.

[7] Lee B., Haymaker J., Gentry R. and Biggs D. Developing a Framework for BIM for Masonry through a Systems Modeling and Case Study Approach. *12th North American Masonry Conference*, Denver, Colorado, USA, 2015. The Masonry Society.

[8] Structural steel sections. Specification for hot-rolled sections. British Standards Institute. BSI 4-1:2005.

[9] Sharif S., Gentry R., Eastman C. and Elder J. Masonry Unit Database Development for BIM-Masonry. *12th North American Masonry Conference*, Denver, Colorado, USA, 2015. The Masonry Society.

[10] Cross N. Designerly ways of knowing. *Design Studies*, 3(4): 221-227, 1982.

[11] Schön D.A. Designing as reflective conversation with the materials of a design situation, *Knowledge-Based Systems*, 5 (1): 3–14, 1992.

[12] Levels of Development Specification for BIM Models, BIM Forum, 2015.

[13] Cavieres A., Gentry T. R. and Al-Haddad T. Knowledge-Based Parametric Tools for Concrete Masonry Walls: Conceptual Design and Preliminary Structural Analysis. *Automation in Construction*, 20(6): 661-740, 2011.

[14] Bonwetsch T., Bärtschi R. and Helmreich M. BrickDesign. Rob | Arch 2012: Robotic Fabrication in Architecture, Art, and Design. S. Brell-Çokcan and J. Braumann. Vienna, Springer Vienna: 102-109.

[15] Eastman C., Teicholz P., Sacks R. and Liston K. BIM Handbook, A Guide to Building Information Modeling for Owners, Managers, Designers, Engineers, and Contractors, John Wiley and Sons Inc, Hoboken, New Jersey, 2011.

[16] Turk T. Phenomenologial foundations of conceptual product modelling in architecture, engineering and construction. *Artificial Intelligence in Engineering*, 15, 83-92, 2001.

[17] Haymaker J., Kunz J., Suter B. and Fischer M. Perspectors: composable, reusable reasoning modules to construct an engineering view from other engineering views. *Advanced Engineering Informatics*, 18, 49-67, 2004.

[18] Mora R., Rivard H. and Bedard C. Computer Representation to Support Conceptual Structural Design within a Building Architectural Context. *Journal of Computing in Civil Engineering*, 20, 76-87, 2006.

[19] Cavieres A. and Gentry R. Masonry Regions: A New Approach for the Representation of Masonry Walls in BIM Applications. In *Proceedings of the 33rd eCAADe Conference*, pages 585-595, Vienna, Austria, 2015.

[20] East E. W., Nisbet N. and Liebich T. Facility Management Handover Model View. *Journal of Computing in Civil Engineering*, 27(1): 61-67, 2013.

[21] Venugopal M., Eastman C., M., Sacks R. and Teizer J. Semantics of Model Views for Information Exchanges using the Industry Foundation Class Schema." *Advanced Engineering Informatics*, 26(2): 411-428, 2012.

[22] IDAT GmbH, RevitPrecast, http://www.idat.de/precast-concrete/products/revit-precast-for-revit-structure/ (last accessed on April 27, 2016).

[23] buildingSmart, IFC 2x4 Schema, http://www.buildingsmart-tech.org/ifc/IFC4/final/html/, (last accessed on April 27, 2016).

LCA of a designed single-family housing in Atlanta under two different energy usage systems

Yifu Shi[a], Tianyu Feng[a]

[a]School of Architecture, Georgia Institute of Technology, USA
E-mail: yshi35@gatech.edu, tyfeng@gatech.edu

Abstract - According to the United Nations Environment Program (UNEP), 87% of the US population will be living in urban areas by 2030 which brings in an increasing demand for residential buildings in metropolitan areas all over the US. High performance houses and net zero house which equipped with photovoltaic (PV) panels are among new generation of buildings emerging and rapidly growing in real estate industry. This study presents the comparison results of an environmental Life Cycle Assessment (LCA) on a pre-designed high performance single-family detached building with a net zero type of it in Atlanta, Georgia. The study covers both embodied and operational energy usage and material of construction, operation, maintenance and demolition phases of the buildings over 70 years of its life span. The relevant Life Cycle Inventory (LCI) was conducted from the project designed by Georgia Tech "High Performance Building" group for the US Department of Energy (DOE) "Zeroes" competition and the LCA software tool, SimaPro, was used to perform the Eco-Indicator 99 impact assessment method. The results indicated that although there are significant energy savings due to the implementation of PV panels, the impact of human health and environment damage, such as carcinogenic effects, radiation damage and mineral resource depletion involved with these technologies which are not usually considered, is highly related to the primary resource of the power generation. Additionally, the environmental and economic damages of a backup battery and the grid stability are other important issues which will eventually arise by the increasing number of high performance and net zero energy buildings.

Keywords: LCA, Residential, High Performance House, Net-zero House, PV, Environmental Impacts

1 Introduction

According to a recent report from Washington Post, the detached, single-family houses are far and away the most common style of housing in major American cities. More specifically, based on new 2014 American Community Survey data on the characteristics of occupied housing, almost 40% of the homes in Atlanta are single family detached houses.

A residential building utilizes a variety of building components in its assemblage designs. The assemblages involve a complex arrangement of material fabricated with various technologies, meeting legislative requirements. The assemblages are produced from a wide range of resources using energy intensive processes, from raw material extraction to final disposal. Energy intensive processes consume a large amount of resources for the generation of power, and produce significant emissions and solid waste. The environmental impacts associated with building assemblages also include operational use of the residence with its heating and cooling, as well as their maintenance and disposal. These systems also have a significant economic cost and social effects. Assessing the environmental impact and cost of a whole building over its lifetime and its' social effects is a complex exercise, as it requires assessment of all its elements and life cycle stages. Life cycle assessment (LCA) approach evaluates the life cycle environmental impacts, life cycle cost and life cycle social effects, respectively [1].

The LCA has been studied in multiple climate zones ranging from hot and humid India to extremely cold Michigan. The energy consumption of the building should include initial embodied energy, operation energy and maintenance energy [2]. A comprehensive LCA analysis on residential houses could be used to make informed budget decision or market purposes [3]. As the PV panel installation get more attention from both public and private sector, it is crucial to analyze economic value, energy conservation and environment impact of the PV panels by integrating it into LCA of residential house [4]. The quality of the input and output data is important to evaluate the result, as such, the SimaPro provides reliable and useful database and result of LCA [5].

This report focuses on all three goals of sustainability which are environmental, economic and social aspects of two residential buildings in Atlanta. The tools, frameworks and processes of LCA of buildings are discussed in the next. The limitations, gaps and assumptions are identified and summarized as well.

Finally, the results, and pros and cons of each type of building is discussed and negotiated.

2 Life Cycle Assessment

Performing an LCA is one way to assess the impacts a product has on the natural environment. The International Organization for Standardization (ISO) has developed international standards that describe how to conduct an LCA (ISO 14040 series). An LCA is a study of the environmental aspects and potential impacts throughout a product's life—from raw material acquisition through production, use and disposal [6]. The ISO standards describe three phases of an LCA. The first phase is an inventory of the inputs and outputs of a product system. The second phase is the assessment of the potential environmental impacts associated with those inputs and outputs. The third phase is the interpretation of the result of the inventory analysis and impact assessment phases in relation to the objectives of the study. These three phases are commonly referred to as (i) life cycle inventory analysis, (ii) life cycle impact assessment, and (iii) life cycle interpretation. In this report, "LCA" will refer to all three phases [7].

2.1 Goal

The purpose of this project was to determine whether a net zero single-family house has an overall environmental, economic and social advantage over a high performance house in Atlanta. It should be an energy conservation guideline for the future residential house. To achieve this goal we used life cycle inventory data to conduct a life cycle assessment of two kinds of houses: net zero house, high performance house equipped with PV panels; the other would be a high performance house using electricity from the power grid.

2.2 Scope

The functional unit in an LCA is defined in ISO 14040 as the quantified performance of a product system [8]. In this case, the functional unit is one house. The system boundary is the interface between a product system and the environment. The system boundary of this study is shown in Figure **1**. It includes the inputs and outputs of energy and material from construction, occupancy, and maintenance. The system boundary excludes human resources, infrastructure, and accidental spills, impacts caused by people, and decomposition of household components after disposal.

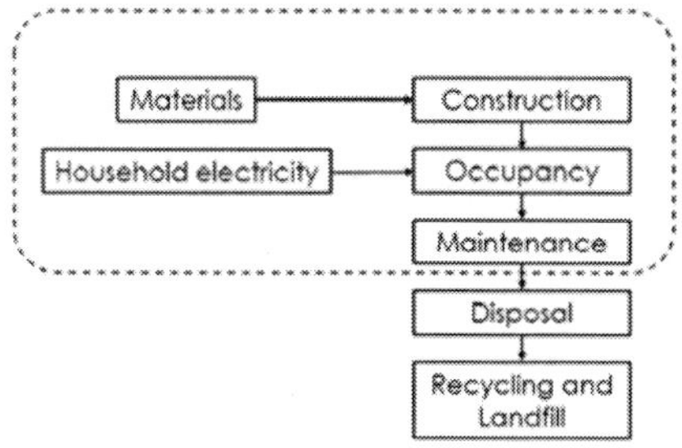

Figure 1 System boundary for house Life Cycle Assessment

The same layout is assumed for both the high performance and the net zero houses. These houses were designed by Georgia Tech "High Performance Building" group for the US Department of Energy (DOE) "Zeroes" competition and all the inventory and materials was extracted from their documents. More detailed description about the buildings and their characteristics are presented in the following section.

2.3 Impact Assessment Methods

The aim of the impact assessment of LCA is to evaluate and assess the importance of the environmental impacts of the two building systems in this study. There are ten main environmental impact categories applied in this research including impacts to human health (carcinogens, respiratory organics and inorganics, climate change, radiation and ozone layer), impacts to ecosystems (eco-toxicity and acidification/ eutrophication, land use) as well as impacts to resources (minerals).

In this project, Sima-Pro was used as the LCA software to analyze our data inventory. Sima-Pro was developed by PRé Environmental Consultants to conduct various steps of the environmental impact assessment simulations with the purpose to perform LCA calculations. In this study, the Eco-indicator 99 system method was used to interpret and weight the LCA results.

3 Methodology – House Description

3.1 Basic Information

The project is a 3 bedroom single family detached residential house assuming that a couple and their future children are using in for its 70 years of life span.

As mentioned earlier, we have developed environmental life cycle inventory data to evaluate the environmental aspects of a high performance house versus a net zero house.

Each house has the same layout but only net zero house is modelled with PV panels. This paper presents a comparison between the designed high performance building and the net zero house of its type.

3.2 Material Flow

The general concept of material usage in this design is to unify the type of material in building, improve the insulation level without compromising the affordability. Both houses are designed in the same method and materials, which could eliminate the differences caused by design. With this design, the base house already reaches a decent quality compared to an average single family housing in current market.

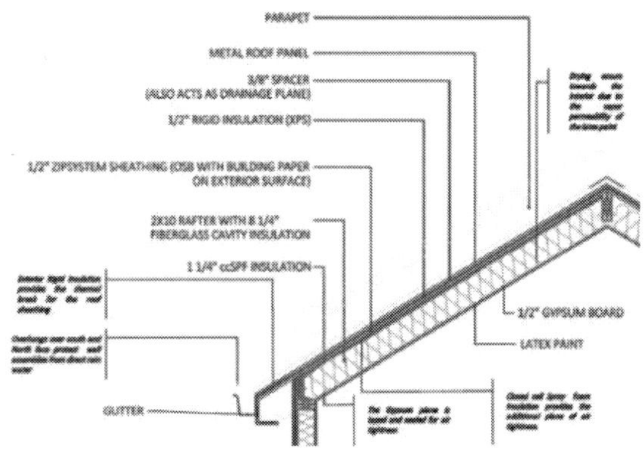

Figure 2 Cross-sectional view of the roof

The design concept is to design a compact roof consisting of a metal panel over an unvented ceiling. The roof ridge is oriented east west and the sloping plane is north-south. Larger portion of roof faces south so that it can receive more solar panels [9]. All the joints in the gypsum board are specified to be taped and sealed. The joint where the roof meets the wall is also taped to maintain the continuity of the interior air barrier. The air sealing is further reinforced with one inch of closed cell spray foam insulation underneath the sheathing. This helps in preventing moist indoor humid air to reach the underside of the roof sheathing where condensation can occur [10].

The roof sheathing is a ZIP System that comes preinstalled with house wrap with optimum permeability, which allows for vapor diffusion to the exterior side. The roof achieves a total thermal resistance of R-41.75 with an R-30 8.25 inch fiber glass-batt cavity insulation and 1.25 inch closed cell spray foam insulation within 2x10 rafters, and 0.5'' XPS board R-3 on exterior sheathing. The half inch XPS rigid insulation controls the thermal bridging that could occur between the sheathing exterior roof panel and the sheathing.

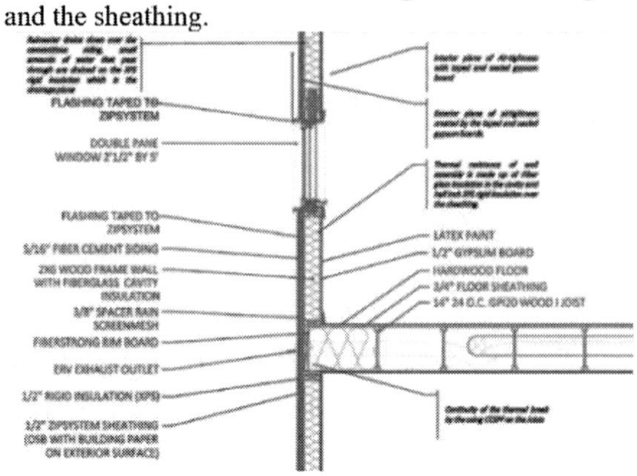

Figure 3 Cross-sectional view of the wall

The design team employed Mixed-humid House Design as introduced in Building Science Corporation [11] as the wall assemblies in this design. The exterior surface of the wall uses the bulk water "source control" strategy. Polyethylene flashing as show in figure 7 installed at horizontal joints to ensure no seepage occurs between those joints. The wall assembly is designed to be air tight from the inside plane of the Gypsum board. Like the roof, the joints of the gypsum board is sealed with taped to make it air tight. All electrical penetrations are sealed to the board. An exterior plane of airtightness is created using taped and sealed rigid insulated sheathing reducing issues with wind washing. Our 2x6 wood advanced frame walls achieve a nominal R value of R-22 using fiberglass cavity batt R-19; 0.5'' XPS board R-3. The XPS rigid insulating sheathing is added to elevate the temperature of the first condensing surface (inner surface of the ZIP system) to reduce the risk of condensation occurring within the assembly. Winter months require lower humidity levels within the spaces to prevent moisture diffusion through the wall assembly. During summer months vapor drive will primarily be from exterior to the interior.

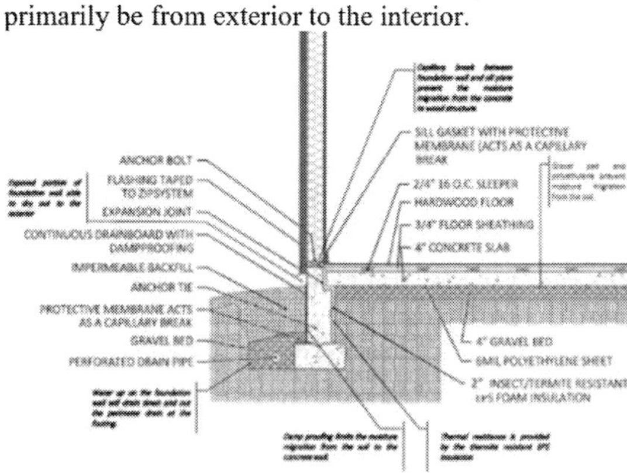

Figure 4 Cross-sectional view of the foundation

The foundation stem walls are cast in place. An impermeable layer of soil is installed next to the foundation to prevent bulk rainwater to flow into the soil. Any amount of water against the stem wall of the foundation is allowed to flow down to drain off into the perimeter drain installed at the footing. This wall is dam proofed to prevent liquid water to flow into it. Moisture migration is controlled between the floor structure and the foundation with the installation of a capillary break on top of the sill plate. A 4-inch deep, ¾ - inch stone bed functions as granular capillary break below the polyethylene. Sill gasket with protective membrane is installed between the floor plate and top of the foundation stem wall.

The stem wall is able to dry out to the exterior through the exposed portion above grade. For exterior soil moisture, the damp proofing on the foundation walls is used to control moisture migrating from the soil into the foundation.

The thermal resistance to the concrete slab is provided by the 2 inches EPS rigid insulation. This comes with a borate coating which prevents insect issues. A thermal break is maintained between the stem wall and the concrete slab using the same.

As it is more clearly states, the whole building is constructed by wood (pinewood, plywood and OSB), and fiber glass for insulation. The usage of concrete and steel are minimized to alleviate the environment impact from material usage. To ease the input process in Sima-pro, we categorize the material inventory as below.

3.3 Energy Usage

The energy usage of these two houses in their 70 years' life span is one important input for life cycle assessment as well as the material flow. For this two houses particularly, all of the energy consumed in the house is coming from electricity.

For the two residential houses, electricity usage can be simply divided into three categories: appliance usage, heating usage and cooling usage. Based on the energy simulation result of the house model, the heating season can be simplified as from October to March and cooling season is from April to September. There are two assumptions in the model: first, only 2.2 people averagely living in the house in the 70 years period and people normally will not be at home during the daytime. During the heating season, the heating load is calculated by subtracting the appliance usage from the weather-related load. On the contrary, the cooling load in the cooling season is estimated by adding the appliance usage from the weather-related load. The appliance electricity usage, heating load and cooling load is obtained from the energy simulation result of the house model.

3.4 Assumptions of Building Load

During a 70 years period, many factors could effect on the house performance. Some basic assumptions are setup to make the model more dynamic. For the house load, the weather-related load is assumed to increase 1% per year since the infiltration efforts and heat loss of the house will increase along with the aging of the house. Secondly, the performance of the house is estimated to improve to the initial stage after the renovation which will occur after 35 years and then the weather-related load will vary based on the assumption 1 in the second 35 years. Thirdly, the appliance load is assumed to increase 2% every 5 years since more equipment will be brought into the home along with the development of technology. The electricity consumed by appliance will also go up correspondingly even the equipment is expecting to operate more efficiently. The detail information is shown in Table 1.

Table 1 Assumptions of building load

Item	Assumption	Reason
Weather-related Load	Plus 1% per year	House aging problem
Appliance Load	Plus 2% per 5 year	Add more equipment
Retrofit	The thermal load drops back to new house level	Replace floor, ceiling, insulation, windows, etc.

3.4.1 Assumptions of HVAC System

The HVAC system is same in the two houses. Based on the design value, the high performance HVAC system can reach a Seasonal Energy Efficiency Ratio (SEER) at 17.8 and a HSPF at 10.7. Additionally, based on the average life span of HVAC system, we setup the life span of HVAC system as 20 years and the HVAC system will be replaced every 20 years. In the entire life span of the house, the HVAC system needs to be replaced 3 times. Besides the equipment initial information, we also make an assumption to estimate the performance of HVAC system in the next 70 years. The detail information is shown in Table 2.

Table 2 Assumptions of HVAC system

Assumptions of HVAC system		
Efficiency	Drops 1% per year	HVAC aging problem

3.4.2 Assumptions of PV panels

According to the house design, the model of designed PV panel is OPT 280-60-4-100, produced by Sunvia®. The product information is shown in

Table 3. Based on weather data of Atlanta area and the amount of building electricity consumption of the first year which we obtained from energy simulation, 18 pieces of the PV modules can provide enough electricity to satisfy the house for the first year, and the annual electricity generation capacity will be 7267.82 kWh [12]. Based on the 80% power output warranty period, we estimate the life span is 25 years. In the entire life span of the house, the PV panels need to be replaced 2 times.

Table 3 PV panel product information

PV panel information	
Power Classification (Pmax)	280W
Module Efficiency (%)	17%
Type of Solar Cell High-efficiency ARTisun Select cells of 156 x 156 mm (6 in.)	Type of Solar Cell High-efficiency ARTisun Select cells of 156 x 156 mm (6 in.)
Cells / Module 60 (6 x 10)	Cells / Module 60 (6 x 10)
Module Dimensions 1652 x 982 mm (65.04 x 38.66 in.)	Module Dimensions 1652 x 982 mm (65.04 x 38.66 in.)
Operating Module Temperature -40°C to +85°C (-40°F to +185°F)	Operating Module Temperature -40°C to +85°C (-40°F to +185°F)
80% Power Output Warranty Period	25yrs
90% Power Output Warranty Period	10yrs

To calculate the power generation in the 70 years period, several assumptions are established for the PV system, which are shown in Table 4.

Table 4 Assumptions of PV system

Item	Assumption	Reason
Efficiency	Power output drops 5% per 5 years	Aging problem
Replacement (every 25 years)	The efficiency increases to the initial efficiency	New product

Based on the assumptions of the building load and energy systems, the total electricity consumption trends are shown in Figure 5. From the figure we can find that PV system cannot provide enough electricity after the first several years which mainly because the increase of the electricity consumption and the decrease of the PV generation efficiency.

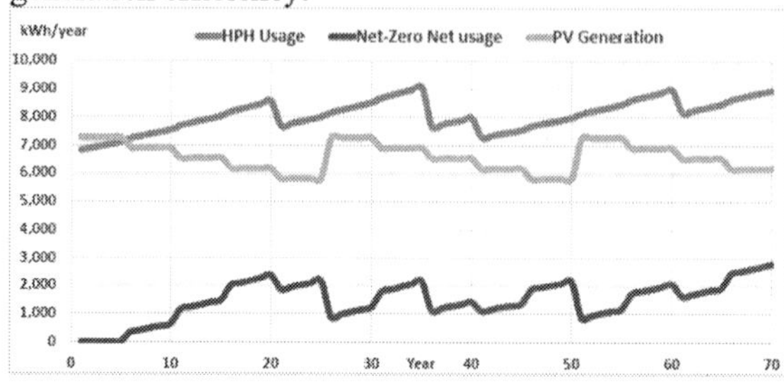

Figure 5 Electricity consumption trend in 70 years

For the high performance house, all of the electricity comes from the electricity grid. For the net zero house, the electricity from the grid is the gap between the house electricity consumption and PV generation. Based on our calculation, the high performance and net zero houses consume the delivered electricity are 564,599kWh and 104,625kWh in the entire 70 years.

4 Result and Discussion

Based on the data we gathered from the house design and the calculated result of the energy consumption, a final inventory in Sima-Pro is shown in Table 5. The inventory includes the total material input during the building construction period and the entire operation and maintenance period. The amount of the material includes the material used for the retrofit and equipment replacement. Georgia electricity generation by resource from EIA is shown in Table 6

Table 5 Final Inventory input in Sima-Pro

Materials/Assemblies/Processes	Quantity	Unit
Material inventory of the basic house model		
Concrete (reinforced) I	5610	kg
Fe360 I	156.51	kg
Pine wood, timber, production mix, at saw mill, 40% water content DE S	8921.63	kg
Plywood, at plywood plant, US PNW/kg/US	15946	kg
Oriented strand board product, US SE/kg/US	3584.4	kg
Glass fiber I	224421.2	kg
Iron and steel, production mix/US	928	kg
Vinyl chloride, at plant/RER S	48184	kg
Vinyl fluoride, at plant/US S	368.18	kg
Flat glass, coated, at plant/RER S	364.2	kg
Concrete roof tile, at plant/CH S	9562.5	kg
Gypsum plaster (CaSO4 alpha hemihydrates) DE S	17.76	kg

Materials/Assemblies/Processes	Quantity	Unit
Galvanized steel sheet, at plant/RNA	288.65	kg
Paint ETH S	40.86	kg
Sealing tape, aluminum/PE, 50 mm wide, at plant/RER U	74000	m
N-olefins, at plant/RER S	850	kg
Copper tube, technology mix, consumption mix, at plant, diameter 15 mm, 1 mm thickness EU-15 S	74	kg
Refrigerant R134a, at plant/RER S	28	kg
Loader Operation (one tractor)	2	month
Net zero house special material inventory		
Photovoltaic panel, single-Si, at plant/RER/I S	90	m2

Table 6 Georgia electricity generation by resource

Resource	Percentage (%)
Gas	43.76
Coal	15.14
Nuclear	33.03
Hydro	3.72
Other	4.35

4.1 Environmental Aspect

To analyze the environmental influence of the two houses, we need to consider the issue from three categories: the impacts on human health, the impacts on ecosystems and the impacts on resources.

Figure 6 shows the comparison result of damage assessment for the two houses. From the comparison, the high performance house has a better performance on Eco-toxicity, Land use, Minerals. Meanwhile, the Net-Zero house has significant less damage on Carcinogens, Respiratory organics, Respiratory Inorganics, Climate Change, Radiation and Acidification/Eutrophication. From the relative percentage result we cannot tell which house has an absolute better performance since neither house performs better in all indicators.

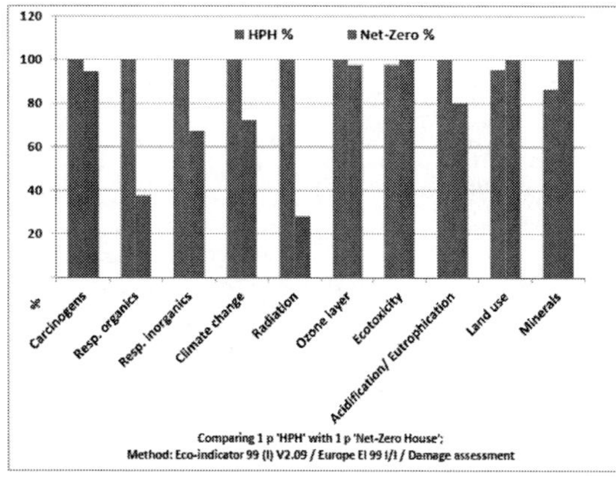

Figure 6 Comparison of damage assessment for the two houses

Figure 7 shows the comparison result of single score. The result shows that comparing with the impact on ecosystems and resources, the impact on human health take the most weighted portion for both houses. And the net zero house has a better performance than the high performance house in this category. The impact on human health should be the priority concern in the comparison since it takes much more points than other two categories.

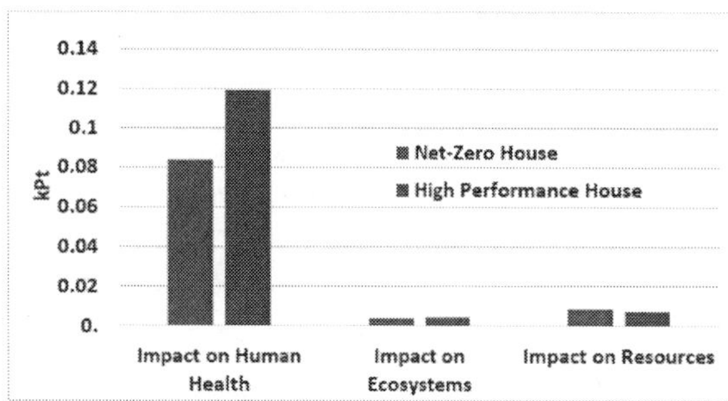

Figure 7 Comparison of single score result

To analyze which indicator has the most impact on human health, the Figure 8 and Figure 9 are developed which shows the single score result of the high performance house and the net zero house. The results show that the climate change and respiratory inorganics are the top two indicators which impact on human health.

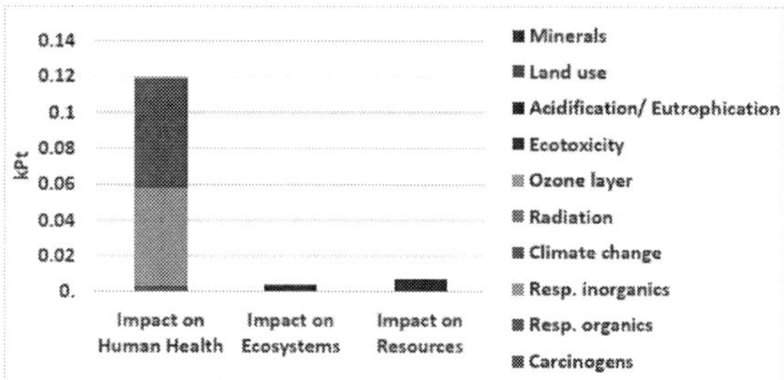

Figure 8 Single Score result of high performance house

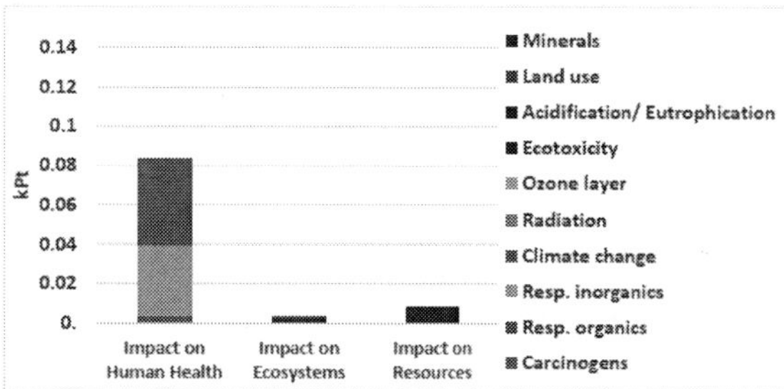

Figure 9 Single Score result of net zero house

Furthermore, the normalization result we got from Sima-Pro tells which material/energy input in the entire life of the house has the most impact on human health. Figure 10 shows that the delivered electricity and glass fiber has the most impact for human health, which mainly effects on respiratory inorganics and climate changes. Figure 11 shows that delivered electricity has much less impact since the total usage in the net zero house is much less than the grid electricity usage in the high performance house.

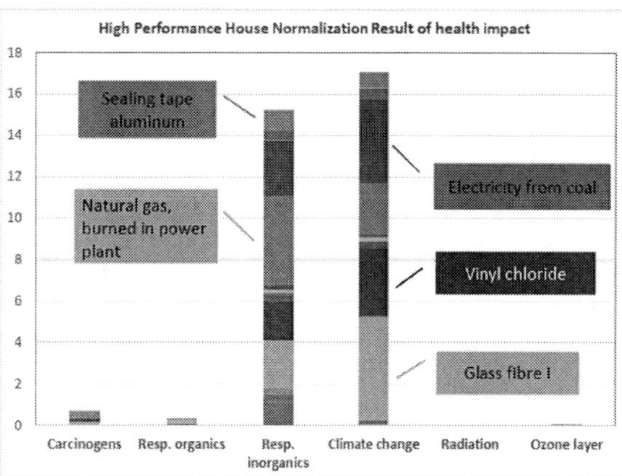

Figure 10 The normalization result of health impact of the high performance house

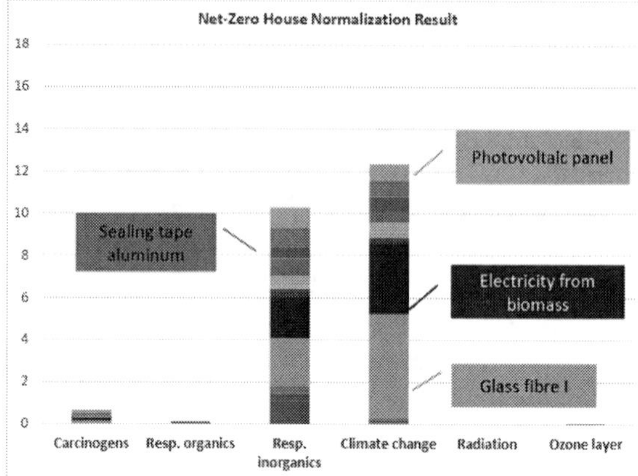

Figure 11 The normalization result of health impact of the net zero house

4.2 Economic Aspect

To simulate the economic perspective of the project, we made several assumptions. The price of electricity on average in Atlanta is 12 cents per kWh and has 3.6% annual increase in next 70 years based on the energy price increase trend in Atlanta. We assume the inflation rate could be stable at 2% during the research period. Based on the warranty information and typical model, the power output of PV panels drop 5% every 5 years, As a result, the PV panels will be replaced every 25 years as well as the HVAC system will be replaced every 20 years based on the information of the life span. As the technology make profess gradually, we assume the efficiency of HVAC system will increase by 5% by next replacement.

The present value of initial cost of net zero energy house is $7,378 higher than the high performance house due to PV panel installation, but the operation cost of the net zero house is significantly lower than a regular high performance house. With a payback calculation, the PV panels on the house could be pay back within 7.82 years without inflation rate and 8.37 years with 2% inflation. One important factor here is that the PV panels have 30% of total system cost incentives from the federal government with the policy of encouraging people using renewable energy, which directly reduce price of the PV panels from $10,540 to $7,378.

From this perspective, the usage of PV panels are beneficial for residences since it save money on energy usage over a long period of time, but it might not be as appealing to builders and contractor since the initial cost are what they are mostly concern.

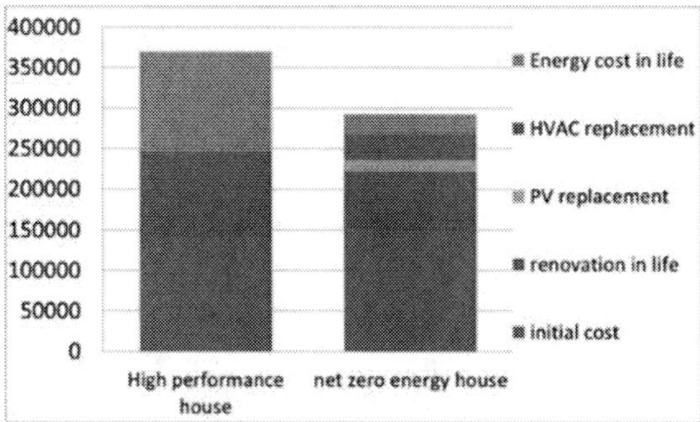

Figure 12 Life cost comparison between the two cases

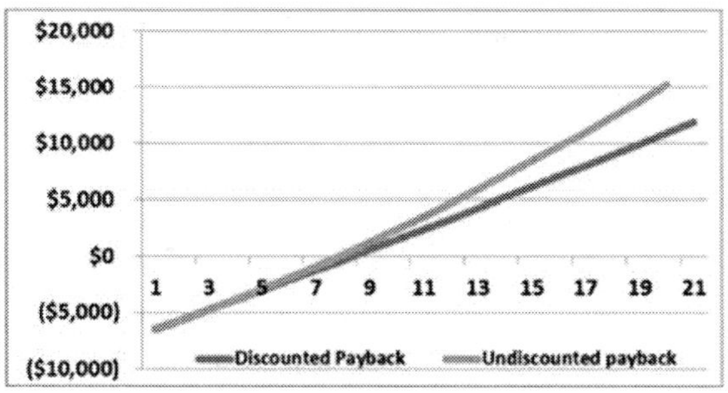

Figure 13 Pay-back year analysis of PV panel

4.3 Social Aspect

We have already covered the environmental and economic aspects of sustainability of these two houses. However, as mentioned before, besides the ecological impacts, there are multiple impacts that are relative to the social environment when new technologies are adopted. In this part, we briefly went over different social aspects of the PV system we have used in the net zero house structure.

4.3.1 PV panels

1. Increase consumer choices: In this case, at least consumers will have choices to choose the electricity resource for their building usage. Hence, the environmental lovers will have the chance to adopt a renewable energy technology for their building's energy systems.
2. Energy independence and security: The more that we reduce the energy consumption of our buildings; the less dependent we are on fossil fuels and foreign sources of energy. Ultimately, this increased independence and security supports peace and prosperity.
3. Significant potential of savings from electricity: Increase potential spending on other aspects of living: leisure, entertainment, etc.
4. Protect communities throughout the supply chain of PV industry and recycling operations
5. Family Wage Jobs: Need to employ highly skilled craftspeople and technicians, which support family wage jobs.

4.4 PV Reuses

As people are focusing more on renewable energy and the technology of PV technology getting more mature, the PV usage has been growing steadily in the past decades. It also brings a new issue of recycling and reusing aging PV panels. As the previous study shows that PV panel generation are energy intensive and create a quite heavy environmental burden. To solve this problem, a recent study has published a promising solution and analysis [13].

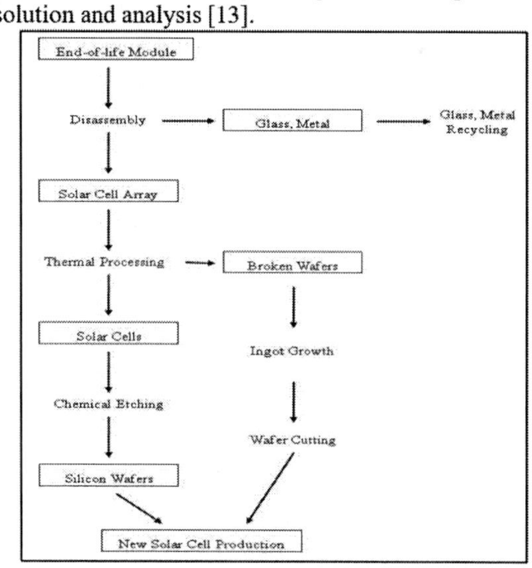

Figure 14 The flow chart for PV panel recycles

The research has stated that the end-of-life module could be disassembled and reused in new solar cell product with a series of process. The important breakthrough of this technique is the energy saving throughout the process. The brand new PV modules need 306 kWh while the recycled modules consume 0 kWh. To recycle the end-of-life module, it only consumes 92 kWh energy which is around ¼ of using new modules. The most prominent advantage of recycled modules is it generate the same amount of energy as the new modules

Table 7 The cost comparison between new wafer and recycled wafer

	modules with new wafter(kWh)	modules with recycled water(kWh)
wafter production	306	0
recycling process	0	92
cell processing	45	45
Module assembly	49	49
Total	400	186
Energy generation	K120(kWh/year)	120(kWh/year)
EPBT	3.3	1.6

5 Conclusion and Future Research

The comparison of the 2 different energy usage systems could be concluded from three aspects. The net zero housing currently is apparently appealing based on current Georgia electricity generation resource distribution, which mainly affected the climate change and Respiratory inorganics. The net zero housing option uses less energy from the primary resource and alleviates the environmental impact from this perspective. From economic perspective, the net zero housing has a more obvious advantage to residences with a roughly 8-year payback period per installation. The 30% incentives from the federal government is a major factor of reaching these benefits. It clearly shows how the power from the policy level could influence the decision making of energy usage. From social level, the net zero housing improves the health condition even from a macro scale, and improve the quality of air and human health. Also, by promoting the renewable energy usage, it could bring up public awareness to the environmental issues and sustainability.

There are several concerns not brought up in this research. The energy storage is not considered, if the house is responsible to store the excessive energy generation, the space usage, and battery production could significantly influence the result. Currently, the energy generated during the day are assumed to be sent back to grid system and taken back when needed. The fluctuation of energy could affect grid stability; especially the user usage pattern and weather condition various constantly. The weight of each environmental impact shifts along with the climate changes. The decision making might be completely differently later.

Acknowledgement

The house design comes from the HouZe GT team for Race to Zero Competition hold by Department of Energy. We thank our classmates from HouZe GT team who provided data and drawings. Also, we thank Professor John Crittenden for the adivse when we form the paper.

References

[1] H. Islam, M. Jollands and S. Setunge, "Life cycle assessment and life cycle cost implication of residential buildings—A review," *Renewable and Sustainable Energy Reviews Volume 42,* p. 129–140, February 2015.

[2] T. Ramesh, "Life Cycle Energy Analysis of a Multifamily Residential House: A Case Study in Indian Context," *Open Journal of Energy Efficiency,* pp. 34-41, 2013.

[3] CCHRC, "Life Cycle Accessment for the Built Environment," Cold Climate Housing Research Center.

[4] W. R. M. Zaki, "Making Economic Sense of Green Energy:Photovoltaic Application in Two Houses, Bangi, Malaysia," *World Applied Sciences Journal*, vol. 6, pp. 474-481, 2009.

[5] Y. Sun, "LIFE CYCLE ASSESSMENT OF A NOVEL BUILDING-INTEGRATED PHOTOVOLTAIC-THERMAL (BIPVT) SYSTEM," columbia university, New York, 2014.

[6] International Organization for Standardization, "Environment Management -- Life Cycly Assessment -- Goal and Scope Definition and Inventory Analysis," 1998.

[7] M. L. Marceau and M. G. VanGeem, "Life cycle assessment of an insulating concrete form house compared to a wood frame house," Portland cement Association, Skokie, Illinois, 2012.

[8] I. O. f. Standardization, "Environmental Management – Life Cycle Assessment – Principles and Framework," 1997.

[9] C. Schumacher, "BSD-149: Unvented Roof Assemblies for All Climates," 24 7 2007. [Online]. Available: http://buildingscience.com/documents/digests/bsd-149-unvented-roof-assemblies-for-all-climates. [Accessed 11 11 2015].

[10] G. Zero, "HouZe GT Volume II," Department of Energy , Denver, CO, 2015.

[11] B. S. Corporation, "Mixed-Humid Climate:

Sandstone 40% Prototype," [Online]. Available: http://buildingscience.com/project/mixed-humid-climate-sandstone-40-prototype. [Accessed 11 11 2015].

[12] NREL, "PVWatts Calculator," 9 July 2015. [Online]. Available: http://pvwatts.nrel.gov/.

[13] N. Weadock, "Recycling Methods for Used Photovoltaic Panels," 9 2011. [Online]. Available: http://2011.solarteam.org/news/recycling-methods-for-used-photovoltaic-panels. [Accessed 11 11 2015].

[14] Bureau of Labor Statistics, U.S Department of Labor, "Average Energy Prices, Atlanta – September 2015," 2015.

[15] J. Crittenden, "Principles of Sustainable or Green," Atlanta, 2015.

[16] V. P. Shah, D. C. Debella and R. J. Ries, "Life cycle assessment of residential heating and cooling systems in four regions in the United States," *Energy and Buildings,* pp. 503-513, 2008.

Challenges of Migrating from Desktop-based BIM in Construction

Samad M.E. Sepasgozar[1], Aaron Costin[2], Cynthia (Changxin) Wang[1]
[1] Faculty of the Built Environment, University of New South Wales Australia
[2] Georgia Institute of Technology, USA
E-mail: samad.sepasgozar@gmail.com, aaron.costin@gatech.edu, cynthia.wang@unsw.edu.au

Abstract –

Recent studies increasingly offer new Building Information Modelling (BIM) solutions and some other studies are modelling BIM adoption and implementation practices. However, the migrating process from the current practice of using desktop-based BIM to a web-based BIM covering broader activities of the entire project lifecycle is complicated. This paper intends to develop the migrating process including main factors that inhibit higher level of web-based BIM usage (e.g. digital autonomous online systems) at construction companies. A case study approach is utilised to collect rich data from three contractors that provided best practices of using BIM in Sydney, Australia. The result of the study shows that there are 'challenges' that hinder pragmatists to migrate from lower level BIM usage to higher level BIM implementation at the construction companies. The key challenges of higher level of BIM usage from contractors' point of view can be divided to technical and managerial challenges. Technical factors include availability of internet infrastructure for web-based BIM and compatibility with other available technologies. Managerial challenges include possibility of changing engineering procedure within contractor's organisation and availability of expert and low skill workers. The study contributes in the body of knowledge by deepening understanding of challenges of BIM implementation for conservatives, which may follow pragmatists in adopting more advanced technologies. The results of the study help the industry policy makers who encourage contractors to use higher level of BIM by resolving the challenges and provide more supportive resources or infrastructure for contractors to cope the challenges. The study will expand to investigate more contractors and other AEC companies to continue to uncover more in-depth challenges. Future recommendations include comparing both the levels and types of challenges faced in different countries. The result can be generalised if many other cases investigated across the country. The study will be extended by focusing on infrastructure projects using more interviews, and developing a roadmap to facilitate the target in each country for web-based BIM implementation.

Keywords –

Desktop-based BIM; web-based BIM; Implementation; Process.

1 Introduction

There is a growing desire to introduce new digital technologies to the construction market. There has been also a vast amount of research on introducing such technologies or dealing with the positive impact of technology in construction [e.g. 1, 2-8]. The concern of innovators and vendors is to increase the rate of up take of the technologies, because the awareness of technologies will not ensure that technology adoption will occur [9]. In the architecture, engineering, and construction (AEC) industry, there has been a paradigm shift to convert all construction documentation into digital representations, including the process of designing, constructing, and operating a facility, known as Building Information Modelling (BIM). Australian government suggests compulsory BIM use for public sector projects. BIM is a digital representation of physical and functional characteristics of a facility, and is a shared knowledge resource for information about that facility forming a reliable basis for decisions from early development to demolition [10]. Being a centralized database of information, BIM has many capabilities and degrees of usage, e.g. 2D drawings, 3D model, material information, analysis model, etc. Since there is confusion of how BIM is used, and to what degree, standardized levels need to be established. Therefore, Bew and Richards [11] developed the BIM Maturity Model consists of three levels (0 to 3). Level 0 is unmanaged CAD probably 2D, with paper (or electronic paper). Level 1 is managed CAD in 2D or 3D format using BS1192:2007 with a collaboration tool providing a common data environment. Level 2 is a "Managed 3D environment held in separate discipline 'BIM' tools with attached data. The approach may utilize 4D program data and 5D cost elements as well as feed operational systems" [11]. Level 3 is defined as "Fully open process and data integration enabled by web services compliant with emerging IFC/IFD standards, managed by a collaborative model server. Could be regarded as iBIM or integrated BIM potentially employing concurrent engineering processes" [11]. The hypothesis is that as the usage of BIM increases, the challenges increase as well (i.e. there is a correlation of increased challenges with increased usage). Therefore, this paper investigates the challenges of implementing level 2 and higher level of BIM usage in construction projects.

It has been identified that BIM has been used for many different purposes such as: designing a 3D model [12], safety [13], co-ordination of models (including clash detection and spatial analysis), real-time construction resource tracking [14], and using BIM for FM/operations [15, 16]. Eadie [17] examines the awareness of the current status of the BIM adoption in the UK. The author concludes that current BIM usage within the UK construction sector revolving around development of 3D models and coordination for clash detection. The study concludes that only a few companies surveyed in their sample has used BIM on all of their projects. However, the question that is raised is 'why other companies do not use full potential of BIM when it is known as a beneficial modelling tool?' The process of the migrating from the current practice to higher level of BIM usage for construction purposes (e.g. including key activities, challenges, and influential factors) have been given less attention in previous research. Eadie [17] suggests that future study should argue the increase in the level of BIM implementation to ensure the UK government will achieve its target.

This paper aims to explore challenges of utilizing higher levels BIM usage in construction projects. The main objectives of this paper are: 1) to understand the current practice of contractors; 2) to understand the main challenges of implementing higher level of BIM usage; and 3) to identify the role/importance of a benchmark using high levels of BIM for mid-sized contractors. This paper will focus on the process of implementation and factors hindering the process.

The originality of this paper lies in investigation of contractors view who are experienced with lower level of BIM implementation compared with government desired level of implementation [18]. This is a step forward to conceptualize a framework for the adoption of high level BIM by contractors, who are identified as the most important players in the adoption process. The presented challenges enable policy makers to facilitate wide spread diffusion of full BIM in the construction industry. In addition, the current practice of the BIM implementation and the designed BIM implementation is compared, which would be useful for the next step of the study for developing high level BIM adoption road map.

The paper firstly reviews existing models of technology adoption and implementation and identifies gaps and barriers to the adoption process. Secondly, the research method that includes an in-depth data collection process from a case study using semi-structure interviews' process is explained. Finally, the result of the implementation challenges is discussed and a novel framework of web-based implementation process is presented.

2 Technology Implementation Process

This section reviews the implementation concepts, general models, and provides a structure for further research. The literature in the area of innovation adoption can be classified into three key perspectives: socio-economic perspective [19]; managerial perspective [20]; and psychological perspective [21, 22]. Studies that take a socio-economic perspective have focused on profiling the users of particular technologies in different disciplines. Rogers [19] envisions technology adoption as occurring within a social system where potential technology adopters communicate with each other and make decisions based on a variety of attitudes towards technology. The formative model includes five categories of adopters, namely: 1) innovators; 2) early adopters; 3) early majority adopters; 4) late majority adopters; and 5) laggards [19]. The key premise of this theory is that it conceptualizes innovation relative to individual action, social relationships, and communication. Studies from the managerial perspective focus on innovation adoption in organizations [23]. Finally, studies using a psychological perspective, focus to a series of mental and behavioral states that a person passes through leading to the adoption or rejection of an innovation [24].

The general framework consists of three overall phases that a potential decision maker may pass through: "pre-adoption", "adoption decision", and "post-adoption"

[7, 20, 23, 25-29]. Although the model is general, and does not refer to any clear activity, it still is helpful as the basis of new relevant research to people involving in any phase of the process.

Technology adoption has been studied in many different areas, and has been defined from various perspectives. Costin, Felkl [30] identify an "innovation gap" between academic research technology and industry adoption, and outlines a roadmap to commercialize university research, e.g. safety technology utilizing BIM. According to Rogers [31], adoption is defined as the steps taken in the process through which the adopter passes to reach a decision to accept or reject a new technology. Implementation has different meanings when viewed from different perspectives. For example, from the diffusion perspective IT implementation defines as "organizational effort directed toward diffusion of appropriate information technology within a user community" [32].

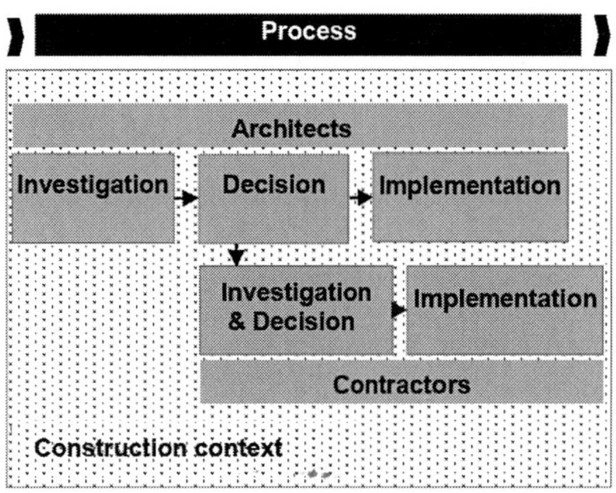

Figure 1 General framework for BIM Implementation

Figure 1 presents a framework for technology implementation process based on previous studies. The framework is an extension of the three overall phases that a potential decision maker may pass through: "pre-adoption", "adoption decision", and "post-adoption" [20, 28, 31]. The general process of pre- and post-adoption is modified by considering consultants (architects) and contractors. It is assumed that consultant implements BIM in three main phases: investigation, adoption decision and implementation. Lately, based on their tendering documentation, contractors are highly encouraged to use BIM in a specific project. Thus, the framework is used to structure the overall research approach.

The literature reveals that the adoption and implementation process is complex and not clearly defined in construction [4, 33]. The complexity originates from the lack of uniformity of decision makers, complexity of new projects, and the variety of differentiated activities involved in any construction project [34, 35]. These coupled with the uniqueness of the construction products [36], the nature of the industry itself [37], project-based organisations and the locations [4, 38] produce a growing number of outsourced project participants [39]. This study considers that the construction industry may affect the adoption process, and so investigate challenges of BIM implementation.

Even with the use of various BIM tools, the exchange between non-interoperable software will require manual data entry to an extent. This lack of interoperability creates impediments for the efficient and seamless transfer of information across a facility's lifecycle. In 2002, Gallaher, O'Connor [40] revealed that the operation and maintenance personnel spent an estimated $4.8 billion (USD) in verifying that documentation accurately represented existing conditions and another $613 million (USD) transferring paper based information into useful electronic format.

3 Research Method

In order to explore the challenges of migrating from desktop-based BIM to a web-based BIM, the case study approach including in-depth interview with the BIM implementer of a construction project has been employed. This method utilized because of its flexibility in obtaining deep understanding of the relative process and influential factors [41]. Qualitative methods were chosen rather than quantitative methods because the literature shows that research is scarce in this area, and so there is a need to investigate and interpret the basic processes occurring, the aim being to produce new insights and understanding of the phenomena concerned [42-46]. Qualitative study is known as a useful research approach to determine richer information about technology adoption [47, 48]. This research strategy enables description of the adoption process, and production of new insight [44].

The semi-structured interview technique was chosen as the best tool to collect data about the vendors and customers' experiences of the technology adoption process, because it enables the researcher to get in-depth data about the process being studied [41, 42, 44, 46]. This type of interview is a flexible tool that allows the researcher to generate rich data to advance understanding and consequently develop empirically and theoretically grounded argument about the process [41, 46]. By interviewing the implementer managers, we can portrait their perspectives into a larger picture and recognize commonalities [45]. The format of semi-structured interview has been applied in many research studies in construction to investigate a process and the associated related factors such as Agapiou [49], Sarshar and Isikdag [50], Bassioni, Price [51] and Samuelson and Björk [52]. For example, Samuelson and Björk [52] investigate factors that affect the decisions to implement different techniques of information technology in construction as well as the actual adoption process. The authors [36] justify their choice of semi-structured interviews because it allows a wider discussion, while the interview is held around defined areas and the selected theoretical framework. A semi-structured interview is a tool to collect rich data, which are open to the participants' decisions about what is important and relevant to discuss. The case company used a middle size construction company that can be fairly a benchmark for other same size construction companies. The case study has been used from one of the University of New South Wales (UNSW) construction contractors of education buildings. The company has been used were carefully choose their contractors from the available construction companies with good track records of quality and safety, because they are working at the university environment surrounded with a large number of students and staff.

At this stage of the study, in order to develop the implementation framework, a BIM implementer of a project was interviewed from a chosen case study. A total number of four case studies and eight interviewees have been planned, which will be conducted in the next phase of the study. We analyzed the data collected from the case study and the interviewee including transcriptions, sketches and the company procedure using thematic analysis technique, as reflected in grounded theory methods [53].

4 Current BIM Practice And Drivers

This section presents the current usage level of BIM in the case study, and compares the current practices of the BIM implementation with the desired level of usage according the literature. Figure 2 shows an example contractor's view from the case study, current practice of BIM usage, and a model of desired BIM usage encouraged by governments and policy makers. Figure 2a shows the interviewees' sketch explaining that 50% of drawings and BIM information will be provided by consultant. The interview shows that the consultant has the main role in decision making to adopt BIM, and contractors had to use BIM as they are committed to in the tendering process. In documented design contracts (DD), consultant provides 70% of information using in BIM. This shows that a contractor has less effort to collect information to use in BIM, even while construction has many different activities occurring on construction sites. The activities, installed materials, and their information are very crucial in the future life cycle of the building, e.g. operations and maintenance [16, 30]. Figure 2b model the current practice of BIM usage in construction based on the case study in Australia. The model is in line with the results of Eadie, Browne [17] study, which show design and build and modeling are the only and most common BIM procurement routes in the UK construction industry. Figure 2c shows the model of desired practice based on the government visions such as the UK [17], Singapore and Australia.

Figure 2 shows there is a clear gap between the current practice of BIM and the desired model for high BIM usage in construction projects. Many associations, champions, and subsequently governments around the world encourage different players in the construction industry to highly use BIM. However, there are still many challenges in using BIM for different purposes other than drawings in construction projects. The challenges are discussed in the next section.

5 Challenges Of BIM Implementation

This section discusses the challenges of migrating from desktop-based to web-based BIM derived from the case study. Then the findings are compared with the existing literature and an overall framework provided for a future study. Figure 3 shows the chasm between early adopters and early majority groups. Chasm refers to differences that exist between innovators and pragmatists in adopting new technologies. The result shows that there should be a large 'challenge' between pragmatists and conservatives in construction. Pragmatists are those companies who are looking for different ways to exploit BIM in order to improve their productivity and situations. Conservatives mostly believe in their traditional way of producing drawings such as deploying AutoCAD for producing 2D documentations. The difference of BIM and any other information systems is that followers (other construction companies such as smaller ones) would not adopt BIM, because early adopters already adopted BIM [54]. The reason is that conservative companies should change some procedures in their company when they are going to use BIM. In addition, the company capability sometimes is not enough to adopt BIM. Therefore, there is a gap between pragmatists who look for new solutions of BIM, and conservative who believe traditional way of drawings are adequate.

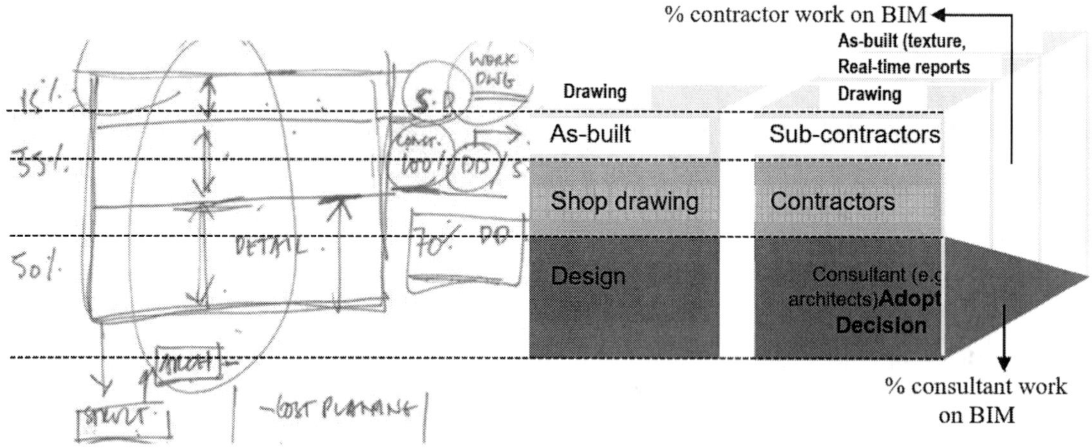

a) Contractor's view (case study example) b) Current practice c) Desired model

Figure 2 Model high level of BIM usage by contractors.

Figure 3 shows two waves of BIM implementation. The interview shows that a mid-sized company in Australia implemented level 1 BIM in the case project. The main driver for adopting BIM for the company was that the architect company uses BIM. The study explained that they are far from Level two. The challenges will be discussed below. The existing of challenges keeps conservatives away from BIM, although they use 2D CAD and other software for drawing (which can be considered as level 0 of BIM). Therefore, innovators and pragmatists use level 1 BIM and others use level 0 BIM. This pattern is shown as Wave 1. The interviewee explains that their company, which is one of early majority companies, tries to improve their situation and productivity. Therefore, they need higher level of BIM, where governments also encouraged them to move up another level of BIM implementation. This can be called Wave 2 in adoption and implementing BIM in construction. So, a new trend should be defined in the literature about new topics and researchers. This study focuses on Wave 2 and identifies challenges. The second wave refers to the high usage of BIM lifecycle.

The study reveals challenges in adopting higher level of BIM implementation. Key challenges are listed as:

- Project scale – projects with smaller scale of budget and size would not allow us to use higher level of BIM;
- Reluctant to record all information – personal are reluctant to measure and record every activity in the site because of privacy, disclosing data in case of fatality;
- Less skilled personal – the level of personal education and skill is not enough for involving BIM, when higher level of implementation is required;
- Full time experts – there are few (if any) full time BIM experts/technicians in many residential projects;
- Resistance to change – conservative contractors would not happy to change the traditional methods and shift to higher level of BIM, which required changes their current procedures and develop new operation procedures in site;
- Massive information – the amount of information is massive in projects, and contractors cannot collect them automatically in digital format, store them and effectively use them;
- Compatibility – the compatibility of new tools and software with each other is a big concern, which are required for collect data and use them;
- Technical support system – lack of availability of supportive system cause delay in construction project, where everything should occur quickly;
- Lack of bench mark projects – contractors need to have benchmarks of BIM implementation to follow them;
- Budget- Some contractors have a small budget to adopt multiple BIM tools, especially since many are not proven to be cost effective.
- Liability issues- Contracts still require 2D print outs for liability. The use of 3D modelling is still not fully supported by owners;
- Interoperability- Software applications that cannot exchange information present an impediment to adopt various software tools.

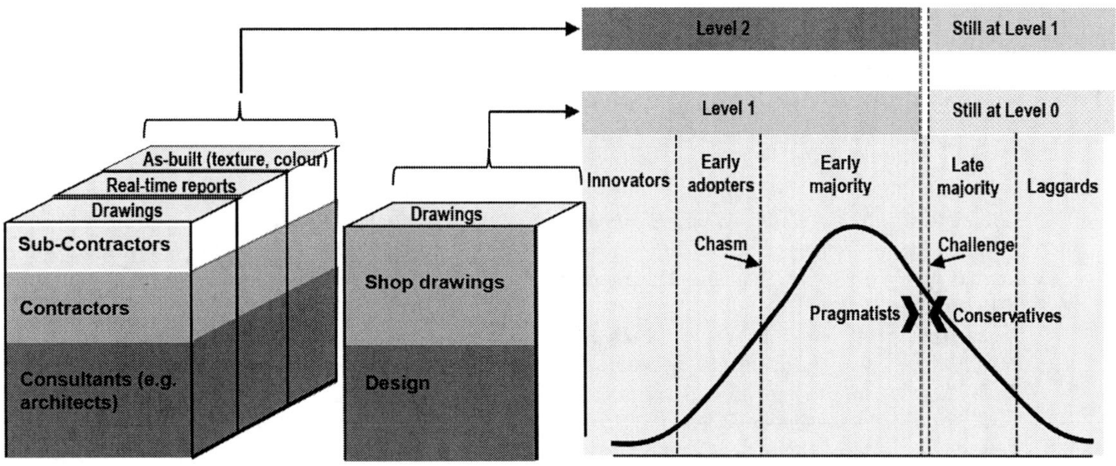

c) Level 2 b) Level 1 a) BIM diffusion life cycle

Figure 3 Two waves of BIM implementation life cycle at two levels.

6 Conclusion

This paper was aimed 1) to understand the current practice of contractors; 2) to understand the main challenges of implementing higher level of BIM usage; and 3) to identify the role/importance of a benchmark using high levels of BIM for mid-sized contractors. The case study approach was used in order to understand the current practice of contractors in Australia, and identify key challenges of BIM implementation in their organization.

The result of the study reveals that consultant provides the main part of BIM information now, and the role of contractor to provide and import information into BIM should be improved. The effort of contractors and sub-contractors in updating BIM information and its richness is critical for maintenance period. In addition, the result shows that there exist many challenges that hinder pragmatists to migrate from level 1 BIM to level 2 BIM implementation. In addition, there are challenges that exist for conservative to follow pragmatists in adopting higher level of BIM in their projects. The results show that as the adoption of BIM goes to a higher level the more complex challenges face, thus proving the hypothesis.

The study contributes in the body of knowledge by deepen understanding of challenges of BIM implementation in construction projects. In addition, the results reveal that there are two waves of BIM adoption. In each wave, conservatives face challenges to follow pragmatists in adopting higher level of adoption. Pragmatists also face nine challenges to shift from level 1 to higher level of BIM usage in construction projects.

The results of the study help the industry policy makers who encourage contractors to use higher level of BIM by resolving the challenges and provide more supportive resources or infrastructure for contractors to cope the challenges. The study will expand to investigate more contractors and other AEC companies to continue to uncover more in-depth challenges. Future recommendations include comparing the levels and types of challenges faced in different countries. Additionally, this research will explore solutions to these problems, such as the use of standards or ontologies to promote interoperability. Furthermore, the study will investigate the recent trends of using the Semantic Web to share ontologically based information over the internet for use in BIM applications.

References

1. O'Connor, J. and L. Yang, *Project Performance versus Use of Technologies at Project and Phase Levels.* Journal of Construction Engineering and Management, 2004. **130**(3): p. 322-329.
2. Kang, Y., et al., *Impact of Information Technologies on Performance: Cross Study Comparison.* Journal of Construction Engineering and Management, 2008. **134**(11): p. 852-863.
3. Goodrum, P.M., D. Zhai, and M.F. Yasin, *Relationship between Changes in Material Technology and Construction Productivity.* Journal of Construction Engineering and Management, 2009. **135**(4): p. 278-287.

4. Sardroud, J., *Perceptions of Automated Data Collection Technology Use in the Construction Industry.* Journal of Civil Engineering and Management, 2014. **21**(1): p. 54-66.
5. Wang, X. and H.-Y. Chong, *Setting New Trends of Integrated Building Information Modelling (BIM) for Construction Industry.* Construction Innovation, 2015. **15**(1): p. 2-6.
6. Skibniewski, M.J., *Information Technology Applications in Construction Safety Assurance.* Journal of Civil Engineering and Management, 2014. **20**(6): p. 778-794.
7. Sepasgozar, S.M. and S.R. Davis. *A Decision Framework for Advanced Construction Technology Adoption.* in *Transportation Research Board 94th Annual Meeting*. 2015.
8. Sepasgozar, S.M.E. and S. Davis, *Diffusion Pattern Recognition of Technology Vendors in Construction*, in *Construction Research Congress 2014*. 2014. p. 2106-2115.
9. Goodrum, P.M., et al., *Model to Predict the Impact of a Technology on Construction Productivity.* Journal of Construction Engineering and Management, 2010. **137**(9): p. 678-688.
10. NIBS, *United States National Building Information Modeling Standard Version 3*, in *National Institute of Building Sciences*. 2015, Northern American Chapter of buildingSMART International (bSI).
11. Bew, M. and M. Richards. *BIM Maturity Model.* in *Construct IT Autumn 2008 Members' Meeting. Brighton, UK.* 2008.
12. Sykes, T.A., V. Venkatesh, and S. Gosain, *Model of acceptance with peer support: A social network perspective to understand employees' system use.* MIS Quarterly: Management Information Systems, 2009. **33**(2): p. 371-393.
13. Zhang, S., et al., *BIM-based fall hazard identification and prevention in construction safety planning.* Safety Science, 2015. **72**: p. 31-45.
14. Costin, A., J. Teizer, and B. Schoner, *RFID and BIM-enabled worker location tracking to support real-time building protocol and data visualization.* 2015, ITcon.
15. Costin, A.M. and J. Teizer, *Fusing passive RFID and BIM for increased accuracy in indoor localization.* Visualization in Engineering, 2015. **3**(1): p. 1-20.
16. Becerik-Gerber, B. and K. Kensek, *Building Information Modeling in Architecture, Engineering, and Construction: Emerging Research Directions and Trends.* Journal of Professional Issues in Engineering Education and Practice. **136**(3): p. 139-147.
17. Eadie, R., et al., *A Survey of Current Status of and Perceived Changes Required for BIM Adoption in the UK.* Built Environment Project and Asset Management, 2015. **5**(1): p. 4-21.
18. Australasia, b., *National Building Information Modelling Initiative.* 2012, buildingSMART Australasia: Australia.
19. Rogers, E.M., *Diffusion of innovations*. 4th ed ed. 1995, New York: Free Press.
20. Damanpour, F. and M. Schneider, *Phases of the Adoption of Innovation in Organizations: Effects of Environment, Organization and Top Managers1.* British Journal of Management, 2006. **17**(3): p. 215-236.
21. Davis, F.D., R.P. Bagozzi, and P.R. Warshaw, *User Acceptance of Computer Technology: A Comparison of Two Theoretical Models.* Management Science, 1989. **35**(8): p. 982-1003.
22. Venkatesh, V. and H. Bala, *Technology Acceptance Model 3 and a Research Agenda on Interventions.* Decision Sciences, 2008. **39**(2): p. 273-315.
23. Sepasgozar, S.M.E., M. Loosemore, and S.R. Davis, *Conceptualising information and equipment technology adoption in construction: A critical review of existing research.* Engineering, Construction and Architectural Management, 2016. **23**(2): p. 158-176.
24. Howard, J.A. and W.L. Moore, *Changes in Consumer Behavior over the Product Life Cycle. Readings in the Management of Innovation* 1988. **2**.
25. Nystrom, P.C., K. Ramamurthy, and A.L. Wilson, *Organizational context, climate and innovativeness: adoption of imaging technology.* Journal of Engineering and Technology Management, 2002. **19**(3-4): p. 221-247.
26. Rogers, E.M., *Diffusion of Innovations*. 2003, New York: Free Press.
27. Hameed, M.A., S. Counsell, and S. Swift, *A Conceptual Model for the Process of IT Innovation Adoption in Organizations.* Journal of Engineering and Technology Management, 2012. **29**(3): p. 358-390.
28. Saeed, K.A. and S. Abdinnour, *Understanding Post-Adoption IS Usage Stages: An Empirical Assessment of Self-Service Information Systems.* Information Systems Journal, 2013. **23**(3): p. 219-244.
29. Sepasgozar, S.M. and S.R. Davis. *Modelling the Construction Technology Implementation Framework: an Empirical Study.* in *The International Symposium on Automation and*

Robotics in Construction and Mining (ISARC 2015) 2015. Oulu, Finland.
30. Costin, A., et al. *Roadmap to Guide Construction Safety Research Commercialization*. in *ISARC. Proceedings of the International Symposium on Automation and Robotics in Construction*. 2014. Vilnius Gediminas Technical University, Department of Construction Economics & Property.
31. Rogers, E.M., *Diffusion of innovations*. 2010, New York: Free press.
32. Cooper, R.B. and R.W. Zmud, *Information Technology Implementation Research: A Technological Diffusion Approach*. Management Science, 1990. **36**(2): p. 123-139.
33. Arayici, Y., et al., *Technology Adoption in the BIM Implementation for Lean Architectural Practice*. Automation in Construction, 2011. **20**(2): p. 189-195.
34. Straub, E.T., *Understanding Technology Adoption: Theory and Future Directions for Informal Learning*. Review of Educational Research, 2009. **79**(2): p. 625-649.
35. Ozorhon, B., C. Abbott, and G. Aouad, *Integration and Leadership as Enablers of Innovation in Construction: A Case Study*. Journal of Management in Engineering, 2013.
36. Munkvold, B.E., *Challenges of IT implementation for supporting collaboration in distributed organizations*. European Journal of Information Systems, 1999. **8**(4): p. 260-272.
37. ABS, *Innovation in Australian Business, 2012-13*, A.B.o. Statistics, Editor. 2014, Australian Bureau of Statistics: Belconnen ACT, Australia.
38. Aouad, G., B. Ozorhon, and C. Abbott, *Facilitating Innovation in Construction: Directions and Implications for Research and Policy*. Construction Innovation: Information, 2010. **10**(4): p. 374 - 394.
39. Mohamed, S. and R.A. Stewart, *An empirical investigation of users' perceptions of web-based communication on a construction project*. Automation in Construction, 2003. **12**(1): p. 43-53.
40. Gallaher, M.P., et al., *Cost Analysis of Inadequate Interoperability in the U.S. Capital Facilities Industry*. 2004, NIST Gaithersburg.
41. Bryman, A., *Social Research Methods*. 2012, NY, US: Oxford University Press.
42. Ritchie, J. and L. Spencer, *Qualitative Data Analysis for Applied Policy Research*. The Qualitative Researcher's Companion, 2002: p. 305-329.
43. Cassell, C. and G. Symon, *Essential Guide to Qualitative Methods in Organizational Research*. 2004, CA, US: Sage Publications
44. Corbin, J. and A. Strauss, *Basics of Qualitative Research: Techniques and Procedures for Developing Grounded Theory*. 2008, Thousand Oaks, CA: Sage Publications.
45. Yin, R.K., *Qualitative Research from Start to Finish*. 2010, New York: Guilford Press.
46. Corbin, J. and A. Strauss, *Basics of Qualitative Research: Techniques and Procedures for Developing Grounded Theory*. 2014, Thousand Oaks, CA: Sage.
47. Lee, Y., K. Kozar, and K. Larsen, *The Technology Acceptance Model: Past, Present, and Future*. Communications of the Association for Information Systems, 2003. **12**(1): p. 50.
48. Hinkka, V. and J. Tätilä, *RFID Tracking Implementation Model for the Technical Trade and Construction Supply Chains*. Automation in Construction, 2013. **35**: p. 405-414.
49. Agapiou, A., *Perceptions of gender roles and attitudes toward work among male and female operatives in the Scottish construction industry*. Construction Management & Economics, 2002. **20**(8): p. 697-705.
50. Sarshar, M. and U. Isikdag, *A survey of ICT use in the Turkish construction industry*. Engineering, Construction and Architectural Management, 2004. **11**(4): p. 238-247.
51. Bassioni, H.A., A.D. Price, and T.M. Hassan, *Building a conceptual framework for measuring business performance in construction: an empirical evaluation*. Construction Management and Economics, 2005. **23**(5): p. 495-507.
52. Samuelson, O. and B.-C. Björk, *Adoption processes for EDM, EDI and BIM technologies in the construction industry*. Journal of Civil Engineering and Management, 2013. **19**(sup1): p. S172-S187.
53. Glaser, B. and A. Strauss, *The Discovery of Grounded Theory*. Weidenfield & Nicolson, London, 1967.
54. Sepasgozar, S.M. and S.R. Davis. *Pioneers, Followers and Interaction Networks in New Technology Adoption*. in *ANZAM* 2014. Sydney.

Data-Driven Scene Parsing Method for Construction Site Monitoring

Hongjo Kim, Kinam Kim, and Hyoungkwan Kim

School of Civil and Environmental Engineering, Yonsei University;
Seoul, Korea
E-mail: hongjo@yonsei.ac.kr, kinam@yonsei.ac.kr, hyoungkwan@yonsei.ac.kr

Abstract –

This paper presents the applicability of a nonparametric scene parsing model for recognizing all objects in an image. The data-driven model labels all the pixels of query images as their object classes using the labels in similar images of the existing dataset. The model has a flexible number of parameters depending on the size of the training data, it is possible to add or remove new data without re-training the whole model. The capability of the nonparametric model would improve the monitoring performance on the construction site with updating the size of the image dataset over time.

Keywords –

Monitoring; Construction Site; Computer Vision; Image Processing; Scene Parsing; Object Recognition

1 Introduction

A lot of efforts have been made for vision-based monitoring in the construction industry [1-5]. Few monitoring methods exist to recognize numerous types of construction objects [6-9]. The previous studies recognizing multiple construction objects used parametric models requiring cumbersome model training processes. Model training processes involve preparing image training data and tuning parameter for which developers spend considerable time and cost. Developer's efforts are proportional to the number of objects of interest; therefore, recognizing holistic construction site scene is elusive.

To address the issues, a data-driven scene parsing method is proposed to recognize various construction object classes. The proposed method is nonparametric in which modeling prior information of target object's features is unnecessary. Furthermore, if a training data size is changed, only an additional data is processed to update the monitoring system, unlike parametric methods involving reconstruction of the whole system.

Benefits of the proposed construction site monitoring system are (1) capability of understanding the whole image area, (2) ease of updating the system with a changing size of training data, and (3) simplicity of tuning model parameters.

2 Data-driven scene parsing

In an image, each pixel has a brightness value within the range of 0–255. A brightness value or cluster has no intuitive meaning of an object to a person. To get object's information, an object boundary in a pixel grid should be properly segmented and then recognized based on features of the pixel cluster. Scene parsing is a term including these processes.

Among several data-driven scene parsing methods, this paper adopts a nonparametric scene parsing method suggested in [10]. The key idea of the method is transferring labels–object identities–of several images to annotate labels of a query image scene. Specifically, every pixel in a query image is annotated by searching the most similar label of retrieved images relevant to a query image from a database. Retrieving similar images to a query image is conducted by two stage filtering. The first filtering stage uses nearest neighbors to scene contents represented by GIST features (refer to [11] for details). The second stage is the dense scene correspondence filtering, which uses SIFT (scale invariant feature transform) flow matching across the filtered images at the first stage. SIFT flow is a concept similar to optical flow, which extracts a SIFT feature at every pixel of an image to calculate a matching score to a query image [12]. The retrieved images are used to transfer their labels to each pixels of a query image using a probabilistic Markov random field model for energy function formulation. In a Markov random field model, an image pixel grid is represented as an undirected graph having edge and node components. An edge represents a relation between neighbor pixels and a node signifies a pixel. The scene parsing procedure is illustrated in Figure 1.

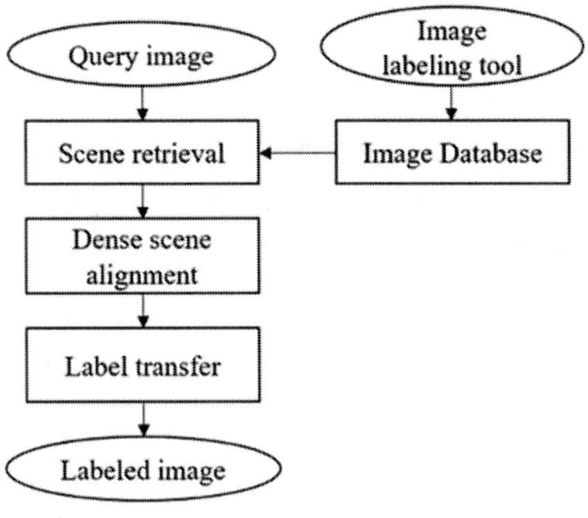

Figure 1. Scene parsing process

3 Experimental settings and results

An experiment was conducted on a desktop computer with Intel i7-4770 CPU, 32 GB RAM in the Windows 10 and 64 bit operating system. This study used the open source code for scene parsing [10] and executed it using a MATLAB programming language. The LabelMe image annotation tool [13] was used to create training image data. The 169 images were used to train the system and 42 images were tested for validation.

The performance of the scene parsing system was measured by the average pixel-wise recognition rate. The evaluation criterion in [10] was selected. The average pixel-wise recognition rate was 81.48% from the 42 test images, with a median value of 84.47% and standard deviation of 13.12%. The average processing time was 7.93 s/query image. Samples of the result are shown in Figure 2.

4 Discussion

The result shows the reasonable performance on recognizing construction site images. However, there has a room for improvement: (1) per-class recognition rates vary in a large degree by object classes. (2) some objects taking a small region in an image were failed to be identified. (3) the long processing time prevents real-time applications. Further studies are required to address these issues.

5 Conclusion

This paper proposed a vision system to identify (scene-parse) the whole image area of construction sites. The system could identify object classes using labels of images that are similar to the query image. The system needed a relatively small number of parameters. The accuracy of the scene parsing was 81.48% based on the average pixel-wise recognition rate. The scene parsing method, in its current version, is capable of recognizing static objects. Further studies, with more training data, would increase the accuracy and the number of construction object classification.

Acknowledgement

This work was supported by the National Research Foundation of Korea (NRF) grant funded by the Korea government (MSIP; Ministry of Science, ICT & Future Planning) (NRF-2014R1A2A1A11052499 and No. 2011-0030040). The authors would like to thank Yonsei University and Doosan Engineering & Construction for granting access to the construction site of the Baekyang-ro renovation project and the tower construction of College of Engineering Hall 1.

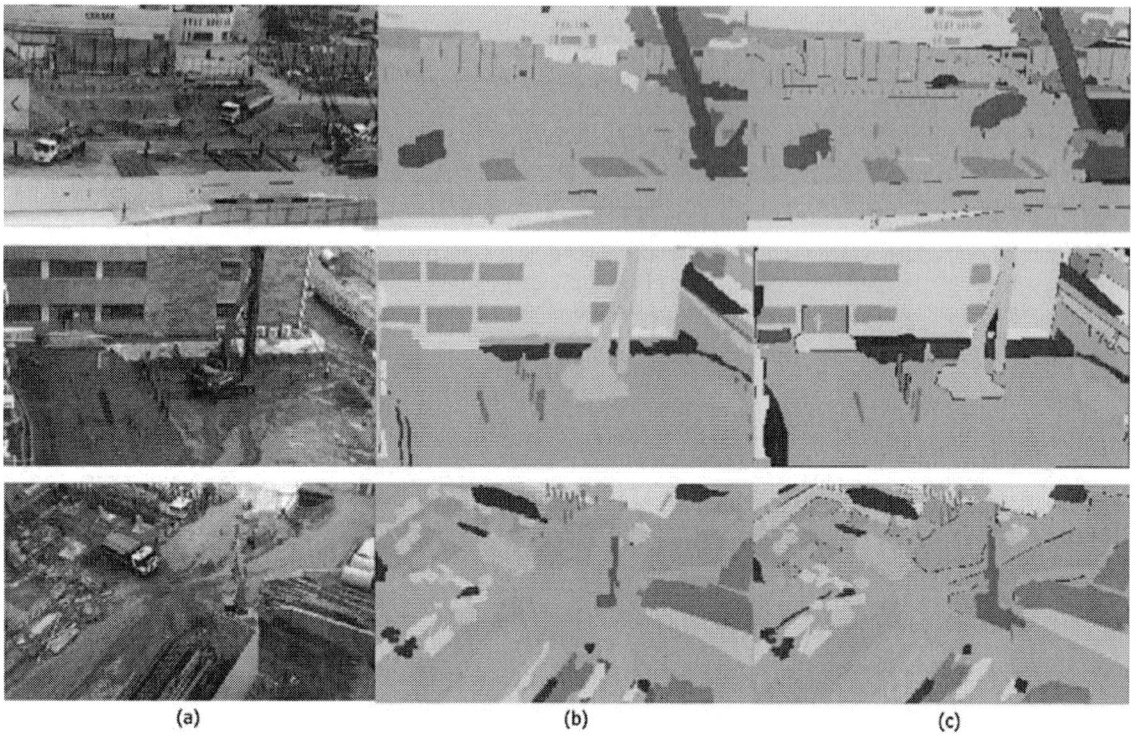

Figure 2. Samples of the scene parsing result. (a) query image, (b) labeled result, and (c) ground-truth label.

References

[1] H. Fathi, F. Dai, M. Lourakis, Automated as-built 3D reconstruction of civil infrastructure using computer vision: Achievements, opportunities, and challenges, Advanced Engineering Informatics 29 (2) (2015) 149-161.

[2] C. Koch, K. Georgieva, V. Kasireddy, B. Akinci, P. Fieguth, A review on computer vision based defect detection and condition assessment of concrete and asphalt civil infrastructure, Advanced Engineering Informatics 29 (2) (2015) 196-210.

[3] H. Son, F. Bosché, C. Kim, As-built data acquisition and its use in production monitoring and automated layout of civil infrastructure: A survey, Advanced Engineering Informatics 29 (2) (2015) 172-183.

[4] J. Teizer, Status quo and open challenges in vision-based sensing and tracking of temporary resources on infrastructure construction sites, Advanced Engineering Informatics 29 (2) (2015) 225-238.

[5] J. Yang, M.-W. Park, P.A. Vela, M. Golparvar-Fard, Construction performance monitoring via still images, time-lapse photos, and video streams: Now, tomorrow, and the future, Advanced Engineering Informatics 29 (2) (2015) 211-224.

[6] S. Chi, C.H. Caldas, Automated Object Identification Using Optical Video Cameras on Construction Sites, Computer-Aided Civil and Infrastructure Engineering 26 (5) (2011) 368-380.

[7] A. Dimitrov, M. Golparvar-Fard, Vision-based material recognition for automated monitoring of construction progress and generating building information modeling from unordered site image collections, Advanced Engineering Informatics 28 (1) (2014) 37-49.

[8] H. Son, C. Kim, N. Hwang, C. Kim, Y. Kang, Classification of major construction materials in construction environments using ensemble classifiers, Advanced Engineering Informatics 28 (1) (2014) 1-10.

[9] H. Tajeen, Z. Zhu, Image dataset development for measuring construction equipment recognition performance, Automation in Construction 48 (2014) 1-10.

[10] C. Liu, J. Yuen, A. Torralba, Nonparametric scene parsing via label transfer, Pattern Analysis and Machine Intelligence, IEEE Transactions on 33 (12) (2011) 2368-2382.

[11] A. Oliva, A. Torralba, Modeling the shape of the scene: A holistic representation of the spatial envelope, International Journal of Computer Vision 42 (3) (2001) 145-175.

[12] C. Liu, J. Yuen, A. Torralba, Sift flow: Dense correspondence across scenes and its applications, Pattern Analysis and Machine Intelligence, IEEE Transactions on 33 (5) (2011) 978-994.

[13] B.C. Russell, A. Torralba, K.P. Murphy, W.T. Freeman, LabelMe: a database and web-based tool for image annotation, International Journal of Computer Vision 77 (1-3) (2008) 157-173.

Large Scale 3D Printing of Complex Geometric Shapes in Construction

Jochen Teizer[1], Alexander Blickle[1], Tobias King[2], Olaf Leitzbach[3] and Daniel Guenther[2]

[1] Ed. Züblin AG, Stuttgart, Germany
[2] voxeljet AG, Friedberg, Germany
[3] MEVA Schalungs-Systeme GmbH, Haiterbach, Germany

Abstract –

3D printing, also known as additive manufacturing, has become an established technology in many industry sectors for the fabrication of three-dimensional (3D) objects. The layered production of scaled prototypes and smaller series typically use automated computer controlled systems that rely on a-priori designed digital 3D models. General principles for 3D modeling, printing, and finishing exist utilizing the cutting, melting, or softening of paper, polymer, or metal materials.

While advantages and limitations of the 3D printing processes require careful review for its final application, the construction industry itself has adopted industrial applications successfully in many examples where, for example, consecutive layers of concrete are combined into a desired structure or form. One of construction's key challenges though is its need for large scale 3D printing of complex geometric shapes on projects where construction time, cost, and quality are the predominant and determining success criteria. While complexity and scale of the planned structure become available at finer detail during the architectural design process, final fabrication of large scale geometric shapes often fails because of constructability issues.

This article introduces conventional construction methods for building large scale and complex geometric structures using on purpose built, automated and robotic 3D printing machines. It therefore contributes the missing link between demanding architectural design that otherwise could only be built at large cost. Commonly known advantages and limitations of existing 3D printing processes, including modeling, printing, and finishing principles are reviewed. Significance of resolution, speed, and quality of materials in 3D printing are explained. Results to an implementation of complex formwork in a major capital construction project are shown. Preliminary benefits and limitations from the perspective of a construction company explain what it takes to advance 3D printing to a field-ready construction method.

Keywords –

3D printing, additive manufacturing, concrete formwork, contour crafting, machine automation and robotics, modularization.

1 Introduction

Three-dimensional printing has been the focus of attention in the media at present and a subject that arouses great expectations. Why then, is this technology, which is often seen superior to conventional methods, so little used in the realm of construction?

One of the roots of additive manufacturing lies in a MIT patent from Sachs et al. from 1989. It describes "a process for making a component comprising the steps of (1) depositing a layer of a powder material in a confined region; (2) applying a further material to one or more selected regions of said layer of powder material which will cause said layer of powder material to become bonded [..]; (3) repeating steps (1) and (2) [..]; (4) removing unbounded powder material [..]".

As a digital means of fabrication, 3D printing offers many advantages. Computer control of the manufacturing sequence affords latitude and shorter production times, often just a matter of days from design to the finished object. Analogue forms of production, in contrast, frequently involve a lot of stages and take much longer. 3D printing is found in many realms of work – in mechanical engineering, in metal pouring, medical technology, research and development, design and education. In architecture, the technology reveals its advantages especially in the creation of 3D building models. Construction companies, however, have yet to develop a strong business case for implementing the technology and its overlaying processes in practice.

2 Background

2.1 Principles of 3D printing

The fundamental principle of 3D printing is that a three-dimensional volume is reduced in the computer to a series of 2D layers. What are initially virtual layers are then built up into a real object, each successive layer being individually worked; for example, by means of laser rays. The process is a generative or additive. The principle is implemented by machines in a number of different forms.

2.1.1 Stereolithography (SLA)

Stereolitography (SLA), for example, is based on a process in which a platform is repeatedly lowered in a bath of synthetic resin (see Figure 1). The very thin layers formed in this way are hardened by exposure to ultraviolet (UV) laser rays. Projecting parts have to be supported, and after hardening, the resin is smoothed. This step is repeated until the construction object is complete. The elements themselves are strong and transparent and have a good resolution. Mainly polymerhybrid resins are used. In part, these contain ceramic additives and have a strength known of 2k cast resins. The largest machines have a capacity of up to two cubic meters. This process is, therefore, particularly suited to the creation of architectural models.

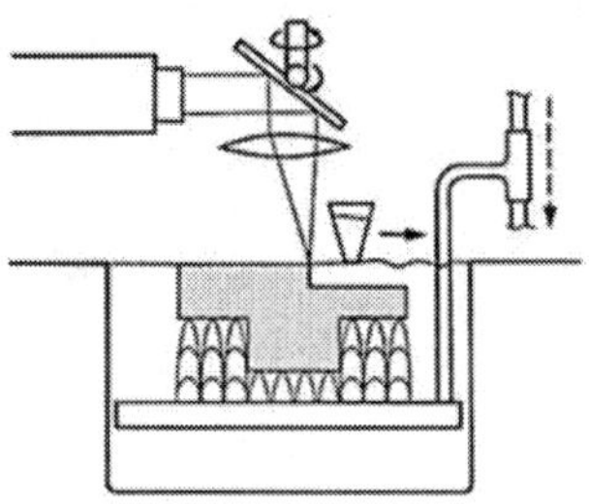

Figure 1. Diagrammatic depiction of the Stereolithography 3D printing system [3]

2.1.2 Selective Laser Sintering (SLS)

Another process is selective laser sintering (SLS). Powder is poured in a thin layer on to a platform, which is moved in a steel container or "box" (see Figure 2). The material is shaped in a series of layers that are fused together by means of a laser ray.

The individual steps are again repeated until the building component is complete as a powder block. After the units have cooled, they are cleaned of loose powder. Overall, the SLS process can lead to a high degree of resolution and excellent strength. The surface quality – often an important criterion for a construction company to get its work approved and paid – will depend on the material used. Normally, particles of medium grain (ca. 50 µm) are used, creating a porous, but not coarse, finish. Metals, such as aluminum, rust-free steel and titanium can be processed, as well as plastics – polyamide mainly. In view of the thermomanagement of the machines, elements not much larger than an eighth of a cubic meter in size are produced.

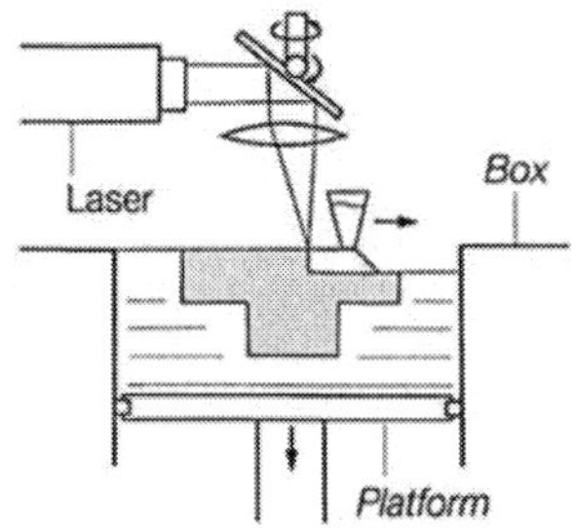

Figure 2. Selective laser sintering (SLS) [3]

2.1.3 Fused Deposition Modeling (FDM)

Fused deposition modeling (FDM) is a process by which plastifiable building material is extruded through a usually heated nozzle (see Figure 3). The latter is moved in relation to the construction platform.

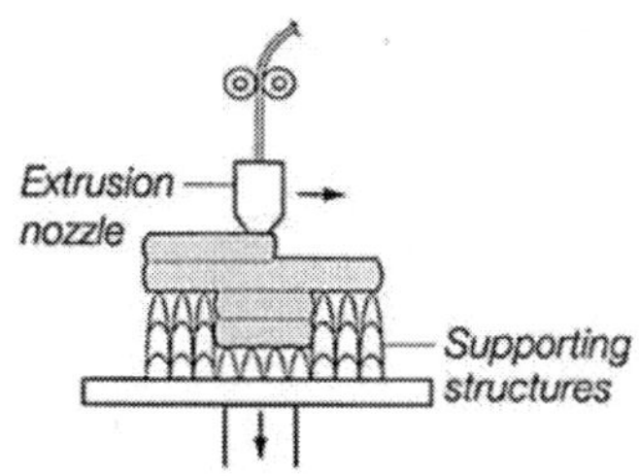

Figure 3. Selective laser sintering (SLS) [3]

The shape of the element is defined in outline and the enclosed area is then filled. Different kinds of material can be used with this method such as plastic, wax, concrete paste or ceramic paste. A variation of this system is known as "contour crafting" [1], whereby an adjustable outer rim prevents the material from flowing out unimpeded. The rim consists of parallel slabs or strips that can be turned about the nozzle. In this way, for example, rapid- hardening vertical concrete wall elements can be formed [2]. This process allows resolution of only low detail. For bigger volumes, very large nozzles are required [3]. Although constructing spaces of more than 10 m^3 becomes available, a major current limitation is the precision in which large and likely free-swinging nozzles form the material. Construction industry experts claim the final surface quality, for example of concrete walls, to be required to be within millimeters to find acceptance by the client.

2.1.4 3D printing with inkjets

3D printing with inkjets is closely related to

superficial printing on paper. In contrast to the techniques described earlier, inkjet print heads, usually with several thousand nozzles, are not drawn along a contour line, but across the construction area (see Figure 4).

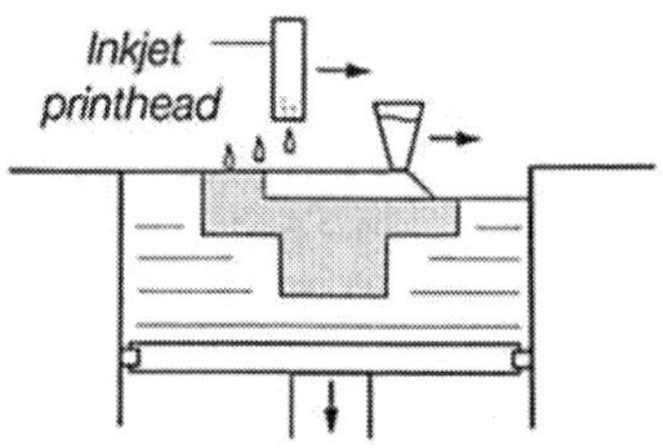

Figure 4. 3D printing with inkjets [3]

The liquid is hardened after leaving the head; for example, through a polymerization process stimulated by a hardener previously mixed into the powder. With non-powder-based applications, the supporting structure and the construction element are printed with different materials – similar to FDM techniques. With powder-based processes, various ground materials can be applied over a large area while the print head selectively prints binder onto the surface where the parts are to be produced and subsequently bonds this area. The process is repeated until the object is complete. Powder-based methods are limited by the grain size of the material, which cannot be reduced at will. The strength of these products is generally low, but it can be increased subsequently by incorporating 2K synthetic resin. The spatial volume of the largest production machine at present is 8 m^3 [3]. An example of its PolyPor system with binder-based on acrylics (a mixture of binder monomers and chemical activator in the powder [4]) is shown in Figure 5.

Figure 5. Binding mixture of particle materials [5]

2.2 Cost of 3D printing

The overall costs of 3D printing include the machines, materials, and labor. The latter occurs mainly at the extraction stage. SLA and SLS are the most expensive 3D printing processes to date, with costs of more than €30,000 per m^3. Powder-based 3D printing with sand costs about €3,000 per m^3, and the FDM process, depending on the type of material, costs less than €2,000 per m^3.

2.3 Direct and indirect use of 3D printing

A distinction is made between the direct use of 3D printed units and the indirect manufacturing of elements by means of 3D printed tools. Many architects, for example, already employ models of buildings created with 3D printing machines for presentation purposes. A special feature of powder-based 3D printing is the use of color (see Figure 6). Obstacles to a wider application in the design process are the high costs and the lack of easy data preparation.

Figure 6. Color-printed building model [Rietveld Architects, New York]

2.4 Techniques in construction applications

Several individual cases exist where engineers applied 3D printing in practice. A few are briefly highlighted. A manufacturing technique known as the "D-shape" process developed by the Italian engineer Enrico Dini and first presented in 2009 used an inorganic liquid solidifying agent that is poured into the powder mix. Part of a large pavilion created by these means is shown in Figure 7.

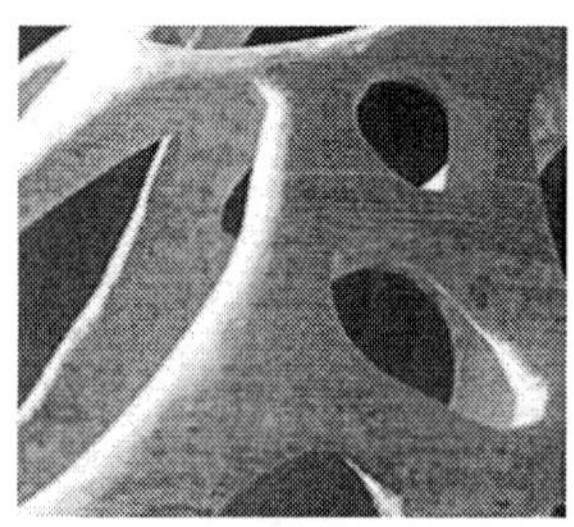

Figure 7. 3D printed architectural object from powder mixture with inorganic solidifying agent

The Chinese company Winsun is developing whole buildings by means of 3D printing. With a technique similar to contour crafting, wall elements are manufactured that are from extruded prismatic bodies; i.e. there is no change to their cross-sectional form over their full height. The rather simple form of the elements created in this way can be assembled using conventional hoisting techniques. They then form constructional components for use in buildings ranging at present from single-family houses (see Figure 8) to a multi-story (5) housing block.

Figure 8. House assembled from printed basic elements, manufactured by Winsun

As an alternative to direct production, indirect processes use forms that have been created as tools. For example, sand forms are already being printed in large numbers for applications with cast metal. They can also be used for mineral castings and concrete. As with metal castings, the mould is usually unfit for use after a single pouring, but the advantage of this system is the possibility of inserting reinforcement in the mold prior to casting. Figure 9 shows a concrete column form for a washbasin cast in two halves. Similar processes have been used to produce sculptures in ultra-high-performance concrete (UHPC) or free-form façade elements.

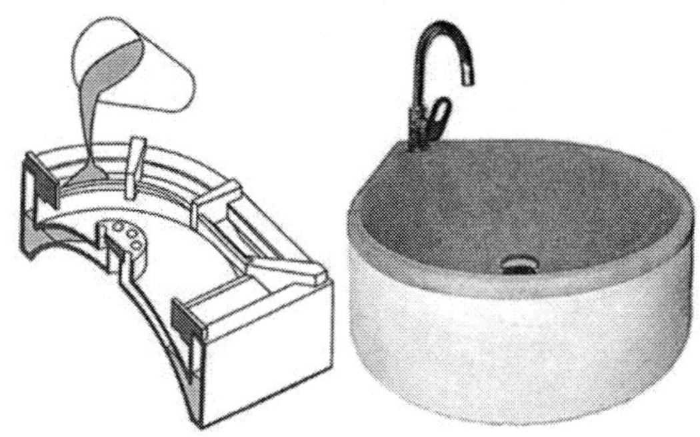

Figure 9. Cast-concrete washbasin: diagrammatic depiction of the casting process in 3D printed form (left); finished washbasin with tap fitting (right) [3]

2.5 Ongoing research activities

A lot of current research is focused on the materials used. Efforts are being made to exploit the properties of UHPC in powder-based 3D printing techniques. Investigations [3] have led, for example, to a cement-based material that, after final treatment, possesses a closed, weatherproof surface with great fire resistance. An example of a screen wall with dimensions of roughly 5 x 2 m is shown in Figure 10). Its strength, however, is not comparable with normal concrete, and it is subsequently not suitable for load-bearing elements.

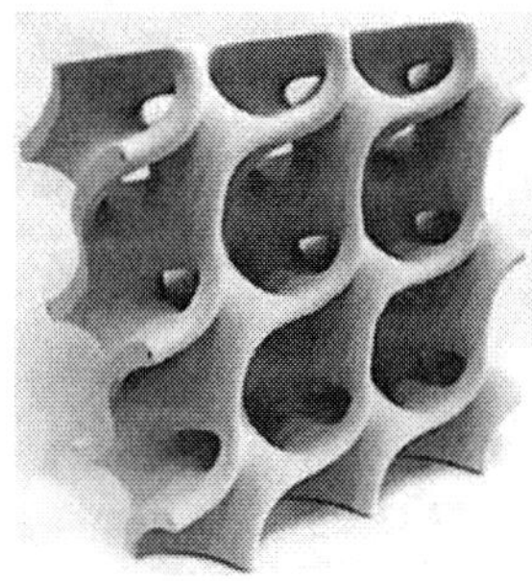

Figure 10. Directly printed building component consisting of cement-bonded material; design: Wieland Schmidt [3]

Many investigations are also concerned with integrating other construction elements in what are essentially monolithic components, things like reinforcement, conduits and cables. The 3D Print Canal House project in the Netherlands is another example. It investigates the use of printing techniques in the realm of architecture. In this project, a large FDM machine has been installed in a container. A synthetic material is used (or externally called "bio-plastic") which is worked in a container close to the working site (see Figure 11). After being printed, the elements can be directly assembled. Alternatively, hollow forms filled with concrete can be used for load-bearing purposes.

Figure 11. Fused deposition modeling process: 3D Print Canal House; DUS architects, Amsterdam [3]

A further example of printing entire houses is the work of the group 3M-FutureLab about Peter Ebner. An extremely compact "micro-home", developed as a student project, was printed out to full scale as a demonstration object. It was produced in two half-shells as a sand model (see Figure 12). This home offers the bare minimum for living with just a few square meters and a height of approximately 3 m.

Figure 12. "Micro-home" printed as sand model; design: Peter Ebner and 3M Future Lab [3]

In mechanical engineering, 3D printing techniques are well established. In construction, however, 3D printing is not commonly encountered despite its many potential benefits it may yield for the society at large. The largest construction components are created at present by extrusion methods, but their monolithic character makes it difficult to incorporate other components such as reinforcement, insulation and many more items. With the use of sintering techniques, it is possible to produce entire building elements, but high costs often prevent a wider application.

Since transformation of well-known industry best practices can and usually take a long time, gradual rather than immediate change is expected to happen in the construction industry. New approaches that are likely to enrich existing construction methods, such as pouring concrete in 3D printed forms, offer an alternative to conventional formwork. Soon they might be incorporated as established technique. The many different techniques and materials used in 3D printing demonstrate that there is a great potential for it use in construction and that the method offers new design scope. For these reasons, research was started to investigate 3D printing in construction in greater detail.

3 Methods

3.1 Team building and setting goals

The companies Ed. Züblin AG (a large construction company belonging to STRABAG SE), voxeljet AG (a manufacturer of 3D printing machines and supplier of large scale 3D print elements), and MEVA Formwork Systems (expertise in formwork systems) teamed up to investigate the understanding and potential of 3D printing in the construction of complex-shaped geometric concrete elements. Ed. Züblin AG's research and development team was in the lead role of the project.

3.2 Work steps

After aligning the project team and setting the goals, the following main work tasks were agreed upon and executed. First, in order to improve any of the existing design, planning and work processes, they had to be analyzed in great detail for potential return on investment (ROI). Building information modeling (BIM) processes were about to be integrated, i.e. the use of digital 3D design techniques that 3D printing machines can read and automatically use in the fabrication process. Second, the team allowed itself the time to understand any market potential and to ensure technical feasibility. Third, digital design began based on a priori known architectural drawings. Tests on smaller scale (reason: small investment and risk) were to demonstrate the feasibility of the selected approach. Meanwhile, choosing a test site for the final field trial fitting the criteria set at the beginning of the project (i.e., answering one of the main research question: where does 3D printing make most sense, economically as well as from construction methods and quality viewpoints?) was selected. Repeating the earlier process of creating a digital 3D design put heavy focus on any potential constructability issue. Second to last, the elements were printed and prepared for the final field trial. The steps can be summarized:

1. Understanding conventional and desired construction processes
2. Survey of market potential and technical feasibility
3. Digital design processes and 3D design generation
4. Test printing of small scale 3D prototypes
5. Selecting a test site and finalizing digital design
6. 3D printing of elements
7. Infiltration of printed elements
8. Final large scale implementation and tests on realistic construction project (in progress)

Some tasks, due to their nature, overlapped with others and were schedule to occur simultaneously. Since 3D printing has yet to become a standardized construction method, all elements included in the preliminary tests and the final field trial, were supposed not to be of the final structure. They were solely built for testing purposes and soon afterwards archived or demolished. Due to the nature of being a research project, some of the work steps are still in progress.

3.3 3D printing technology and process

This project used the voxeljet AG VX4000 3D printing machine (see Figure 13) to produce the sand molds for the elements. The machine's space requirement is 25 x 12 meters and a height of 4.5 m. As a standalone factory, it is one of the largest industrial facilities of its kind. Fabricating one piece at a time, it produces elements of up to 8 m^3 (4,000 x 2,000 x 1,000 mm) using a fully automated layer building process. Depending on size and shape, 3D prints can last several dozen hours (15 mm/h build speed at 600 dpi print resolution), while each piece typically weighs hundreds of kilograms or much more. The process selected for 3D printing is explained in Figure 14.

Figure 13. 3D printing machine VX 4000 [3]

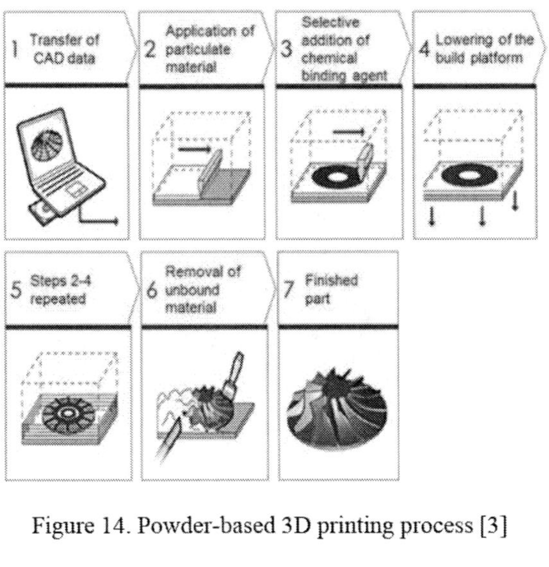

Figure 14. Powder-based 3D printing process [3]

The standard for data exchange is .STL (Standard Tesselation Language (STL), which uses oriented surfaces that are represented by triangles. Their composition is: header (Type (ASCII, binary), name, number of triangles (4 bytes)) and data ((3 vectors of 3 points each (4 Bytes), 1 normal vector). STL allows the identification of sectional planes that are running through all triangles. Since in 3D printing only cut lines are saved (2 points per line) it makes data exchange efficient. The creation of the jetting dataset is based on: real-time conditions while printing (allowing some flexibility), movement data and printing data. Casual events are replenishment of print head and powder and the preparation of data.

4 Experiments and results

Two experiments were performed: printing (1) small scale 3D elements for early feedback on the design, including material tests and (2) large scale 3D elements to determine constructability, including reusability of formwork system elements. First, however, the wider business scope for 3D printing on a large scale capital-intensive construction project is introduced.

4.1 "Stuttgart 21" main central station

The requirements that were set for 3D printing originated from a realistic construction project, called "Stuttgart 21". The heyday of railway stations in the 19th century saw great halls for train terminals in many European cities. Typically, stations were erected outside the historic city center, which soon grew around them, turning the tracks into a hindrance to urban development. This is especially true for the city of Stuttgart in southwest Germany. The existing station is a terminus and difficult to link to the growing European high-speed rail network. New tunnels and an underground station are being built. The most striking impression of the new subterranean station will be its brightness and visual openness, which combines aesthetic with security advantages. It will have an attractive special white and light-weight concrete structure subjected only to compressive loads with minimal construction thickness. The thickness of the optimized shell structure, i.e. of the complex parabolic concrete columns, was reduced to one hundredth of the span, resulting in the use of much less material for construction (see Figure 15).

Figure 15. "Stuttgart 21" main station with parabolic concrete column [ingenhoven architects, Düsseldorf]

These reasons required the reuse of formwork and the prefabrication of modular components which facilitate efficient construction [5,6]. It made it very attractive for exploring the potential of 3D printing.

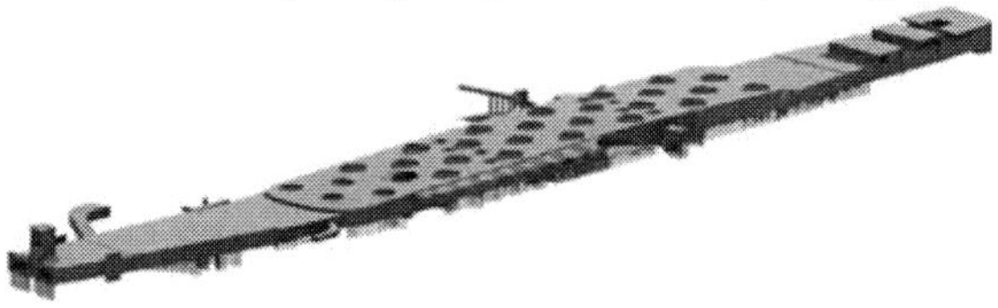

Figure 15. Isometric view on main station model

4.2 Small scale testing

The research method foresaw to experiment initially with small scale test models first. A simplified 3D model of the column was generated and divided into two sand mold segments. These later were joined to function as a formwork system. Aside from developing the general 3D printing process for formwork purposes (see Figure 14), the experiment allowed practical experiences with handling and hardening the sand mold with epoxy resin, pouring the concrete into the molds, and removing the formwork. The success of generating the structure can be seen in the center of Figure 16.

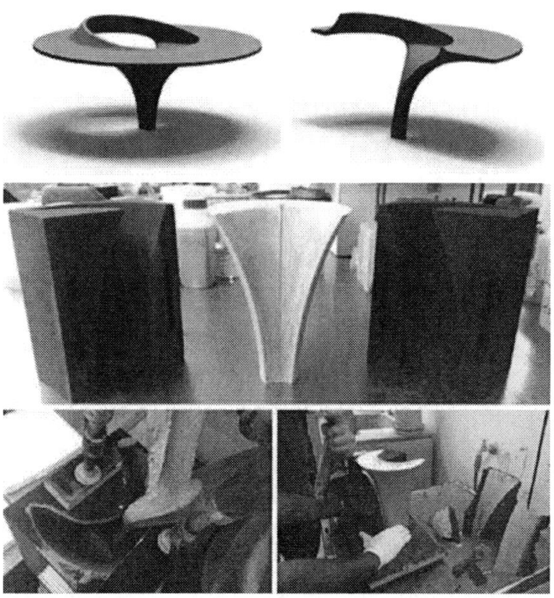

Figure 16. Conceptual 3D model of the parabolic concrete column (top left and cross-sectional cut top right side), printed sand molds (middle image on the left and right side), pouring concrete and removing formwork (bottom image left and right side, respectively) and finished element (center of the middle image)

4.3 Large scale testing

The large scale test involved detailed element and formwork design as well as construability planning (see one segment in Figure 17) of one parabolic concrete column in the main station. Due to the original size of the concrete model, any of the supporting scaffolding was not modeled before the 3D printing process.

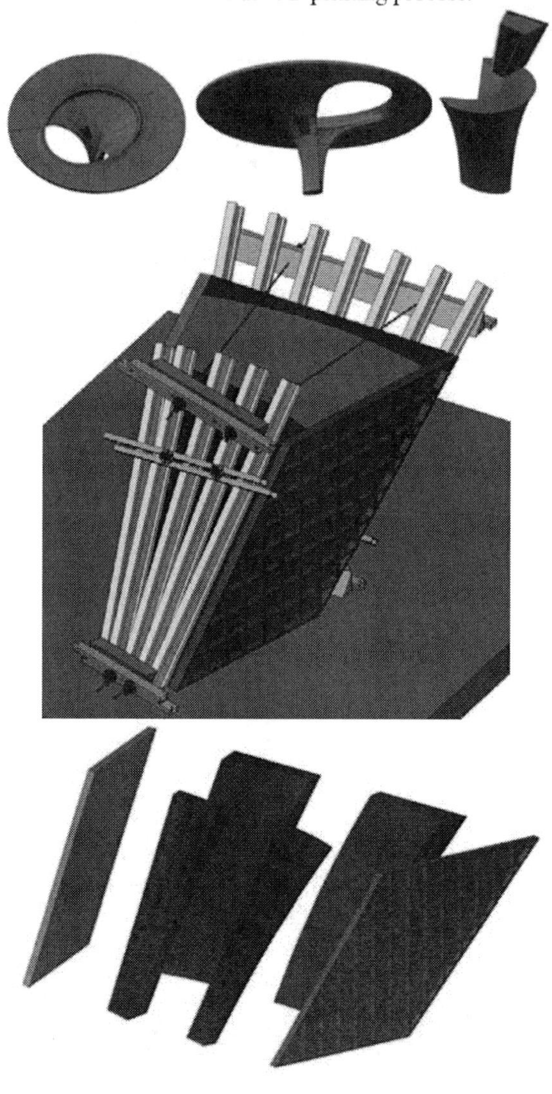

Figure 17. Detailed model (top image), detailed formwork model (middle) and extracted formwork elements due for 3D printing (bottom) [MEVA Schalungs-Systeme GmbH]

The elements were further arranged and optimized for efficient 3D printing in the VX4000 (see Figure 18). The expected total weight of all four elements was about 840 kg. The 3D printing lasted about 1 day. The task thereafter was to harden the material with epoxy resin which took also about 1 work day (see Figure 19). A final task was to erect the 3D printed formwork elements and to pour a layer of concrete. This has yet to be done at the construction site, but has been tested in indoor laboratory-like settings with smaller 3D printed elements already (see Figure 20). As the resulting concrete structure shows, the desired complex geometric shape was successfully built. Structural and further quality performance tests, which have been done for the small scale experiment, have still to be conducted to meet the client's and any regulative demands. Furthermore, efficient processes to repeat the process for all concrete column elements need to be investigated in future work.

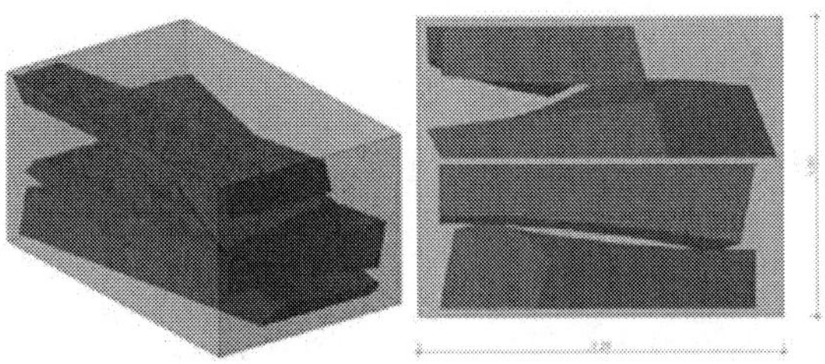

Figure 18. Efficient 3D printing: isometric (left image) and front views (right image) [MEVA Schalungs-Systeme GmbH]

Figure 19. Manual application of epoxy resin on the surfaces of the 3D-printed elements

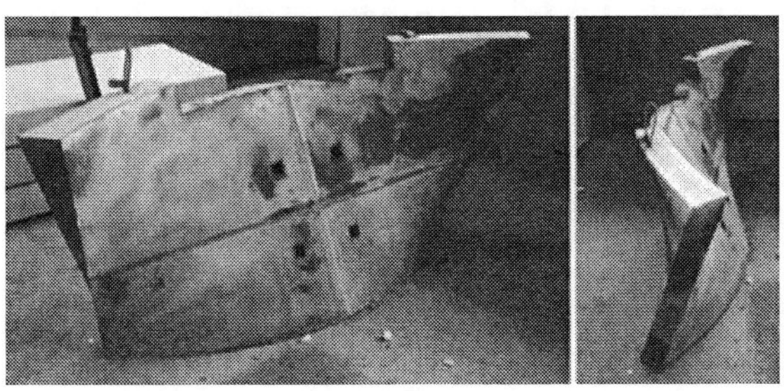

Figure 20. Concrete after removal of 3D printed formwork elements [3]

5 Discussion and Conclusions

Additive manufacturing offers new opportunities in the production and in particular in the fabrication of highly individualized and repetitive modules. It is likely to impact the entire construction supply chain. The current awareness of 3D printing in the media also helps bringing it into a spotlight where construction companies start exploring alternative construction methods. Therefore, the current interest in investing in research and development is growing, albeit many issues remain that demand intelligent solutions. Many of them relate to structural and quality issues (i.e. material properties and finishes) and some, of which many are of high relevance to construction companies, relate to constructability issues.

The presented research gives a brief overview in additive manufacturing and provides a realistic scenario when a construction company investigates alternative construction methods. Examples of small and large scale 3D printing of complex shaped concrete formwork elements demonstrate the successfully developed process along this avenue that likely is transformative.

To achieve the latter, several tasks have to happen before such processes, incl. the application of 3D printing technologies, begin changing the business. They have to be cost efficient and need to follow standardized procedures. For the technology itself it will be important to rely on open data formats, process-oriented design and constructability planning, and flexible material characteristics that yield highest quality work and reduce waste [6] (e.g., rework).

References

[1] Khoshnevis B. Automated construction by contour crafting-related robotics and information technologies, *Aut. in Constr.*, 12:5-19, 2004.

[2] Khoshnevis B., Hwang D., Yao K., Yeh Z. Mega-scale fabrication by contour crafting, *Industrial and Systems Engineering*, 1:301-320, 2006.

[3] Guenther D. 3D printing – the state of the technology and the future of this process, *Detail – Technology*, 596-600, 2015.

[4] Teizer J., Venugopal M., Teizer W. and Felkl J. Nanotechnology and its impact on construction: bridging the gap between researchers and industry professionals, *Construction Engineering and Management*, 138(5):594-604, 2012.

[5] Lim S, Buswell R, Le T, Austin S, Gibb A, Thorpe T. Developments in construction-scale additive manufacturing processes. *Aut. in Constr.*, 21:262-268, 2011.

[6] Oesterle S, Vansteenkiste A and Mirjan A. Zero waste free-form formwork, ICFF2012, 258-267, 2012.

Sensing Workers Gait Abnormality for Safety Hazard Identification

Kanghyeok Yang[a], Changbum R. Ahn[a], Mehmet C. Vuran[b] and Hyunsoo Kim[a]

[a] Charles Durham School of Architectural Engineering and Construction, University of Nebraska-Lincoln, USA
[b] Department of Computer Science and Engineering, University of Nebraska-Lincoln, USA
E-mail: kyang12@huskers.unl.edu, cahn2@unl.edu, mcvuran@cse.unl.edu, hkim13@unl.edu

Abstract –

Ironwork is considered one of the most dangerous construction trades due to its fall-prone working environment. Since safety-hazard identification is fundamental to preventing ironworkers' fall accidents, engineering measures have been applied to eliminate fall hazards or to reduce their associated risks. However, a significant quantity of hazards usually remains unidentified or not well assessed because most current efforts rely on human judgment to identify hazards. To enhance hazard identification efforts, this paper develops a technique for detecting the jobsite safety hazards of ironworkers by analyzing their gait anomalies. Using wearable inertial measurement units (WIMUs) to record kinematic data about ironworkers' gait, this study collected kinematic data while the workers interacted with two types of jobsite hazards. The anomaly level of each gait was modeled using diverse gait-related metrics. Moreover, relationships between safety hazards and worker gait abnormalities were examined through extensive experiment evaluations. The results reveal opportunities for enhancing hazard identification performance by monitoring workers' bodily response.

Keywords – Hazard Identification; Gait Analysis; Inertial Measurement Units; Safety management

1 Introduction

Hazard identification is the critical first step in the construction industry's safety-management programs [1] since unidentified hazards negate the whole safety-management process [2]. However, a significant quantity (on average 10 ~ 57%) of hazards in a construction project still go unidentified or not well-assessed, and these unidentified hazards present the most unmanageable risks [1,3,4]. Poor hazard-identification performance stems from the fact that most current efforts rely on human judgment to identify hazards [1,4,5], a reality that faces twofold challenges: 1) It is unlikely that an individual possesses all the knowledge and experience necessary to identify every potential hazard [1]. 2) Dynamic and unpredictable construction environments compound the complexity for individuals seeking to recognize hazards [1,4]. Furthermore, considering the fact that most hidden hazards are beyond the threshold of human recognition or are not defined with explicit knowledge, there is a critical need to enhance hazard recognition capabilities beyond the threshold of human recognition abilities.

In this context, our study examines whether recording data about workers bodily and behavioral response patterns can enhance hazard-identification capabilities in construction. Thanks to the fact that safety hazards generally cause disruptions in workers' behavior patterns, the analysis of workers behavioral and bodily response patterns provides valuable information that can be used to identify hazards. Toward this end, this paper investigates the feasibility of utilizing workers gait information—specifically, the spatiotemporal features of workers' gaits—to estimate the existence of fall hazards on the surface. In particular, this study develops wearable inertial measurement units (WIMU) -based Gait Abnormality Score (I-GAS) to effectively represent multiple gait features in a single score for detecting the existence of hazard. In turn, this study examines to what extent I-GAS's performance reveals hazards in a laboratory setting. For validation, this study designed and conducted laboratory experiments simulating ironworkers' working environment. Through these experiments, we collected participants' gait data using WIMUs. The results of the experiments demonstrate the potential of using collective gait information to detect fall hazards in construction.

2 Research Background

Most current hazard identification techniques in a construction project rely on individual workers' recognition skills. Previous studies on safety hazard identification have thus focused on enhancing individual workers' hazard-recognition abilities via training programs and virtual environments [1,5]. While these studies have advanced hazard responsiveness, they are still limited by workers' ability to recognize hazards and/or by workers' existing knowledge of known hazards. Furthermore, human recognition abilities are highly subject to diverse environmental factors, such as low light, noise, and jobsite organization. For instance, a slippery/contaminated surface—the destructive causal factor of fall accidents in several trades—is hard to identify in low-light conditions. Moreover, manual material handling (e.g., carrying, pulling and pushing) is generally required during the construction work process, and it often interferes with visibility in the workplace [6].

To mitigate this reality, several studies examined the potential of applying diverse sensing approaches [7–11] to construction safety. Computer vision-based techniques are used to detect workers' unsafe behaviors [7] and hardhat wearing [8] in a construction environment. However, the applications of vision based techniques have several challenges, including the limited sensing range of a camera and visual occlusions.

With the development of wearable sensing devices and technology, workers' physiological responses—such as heart rates, breathing rates, and postures—have been monitored and studied for enhancing safety and productivity in previous studies [9–12]. Our previous studies [13,14] also investigated the feasibility of detecting near-miss falls of ironworkers using sensed bodily response. However, previous studies focused on detecting workers' unsafe behaviors and/or their exposure to recognized jobsite hazards, rather than on identifying and assessing a hidden hazard discovered through sensed data.

It is well known that safety hazards can cause disruptions in workers' behavior patterns, which may then result in undesired events or accidents [15]. Therefore, analyzing the abnormality in workers' behavior patterns provides an opportunity to identify hidden hazards. In particular, by analyzing workers' gaits, researchers can gain valuable insights regarding fall hazards that may cause slips, trips, and falls.

Gait analysis in clinical applications has been widely used to assess the fall risks of health-care patients or elderly individuals who have gait disorders [16,17]. Gait is defined as a particular way of walking [18]. For gait analysis, a maker-based system and a floor sensor-based system have been widely used in previous studies [17,19]. However, these systems are only able to collect gait information in the laboratory setting. Given the need for an ambulatory gait-monitoring system, some studies used WIMU sensors in gait analysis for fall-accident prevention and evaluated these sensors' performance [17, 18]. Using the kinematic data captured via WIMUs, researchers computed spatiotemporal gait features (e.g., stride time and stride distance) to assess abrupt changes in patients' gait patterns and to measure the risk of falls [20,22,23]. While such gait analysis has been used for patients, there are no previous attempts to use it for identifying hazards in construction environments.

Computed gait features may capture the disruption in workers' gait patterns caused by safety hazards, but it is not certain whether each feature is sensitive to all types of hazards. Moreover, each gait feature has a different range of values with different measurement units. Thus, to comprehensively assess the gait abnormalities caused by hazards, it is necessary to represent the deviations of all the gait features from a normal gait in a single score. Expanding upon previous studies, this paper will compute the gait features of workers and use them to analyze the abnormality in workers' gait patterns.

3 Methodology

3.1 Gait Analysis and Gait Feature Computation

During a gait, the sensor coordinate system—the coordinate axes of the WIMU sensors attached to workers' ankles—are not aligned with the human body coordinate system—the coordinate system of the workers' body represented by the anterior-posterior, mediolateral, and vertical axes. This misalignment is due to the difference between the ankle-movement direction (cyclic between the floor, the height of the step, and the foot's forward motion) and the gait direction (i.e., walking forward). Thus, determining the angle between the body coordinate system and the sensor coordinate system (theta in Figure 1) is necessary to calculate the workers' gait features (such as stride length and velocity) using the data garnered from the WIMU attached on the ankle.

As illustrated in Figure 1 (left), the orientation angle between these coordinate systems continues to change by the location of the ankle during a gait. To calculate orientation angle of WIMU, this study utilized the rotation matrix to project the acceleration from the sensor coordinate system to the body coordinate system; this study used the quaternion ($q = q_r + q_i i + q_j j + q_k k$) format, which is available method to represent the rotation of an object using a complex number [24].

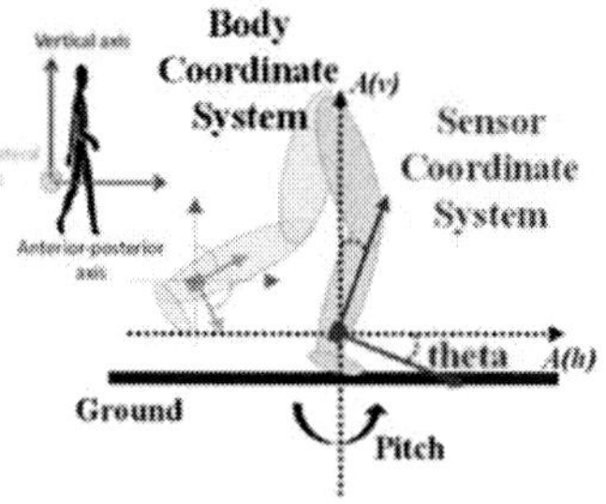

Figure 1. Sensor coordinate system and body coordinate system during gait

Projected acceleration for the gait analysis (i.e., for the body coordinate system) were then computed by multiplying the rotational matrix (Rm)—which we derived from the quaternion (See Equation 1)—and the collected acceleration values from the WIMU sensors.

$$\begin{bmatrix} 1-2{q_j}^2-2{q_k}^2 & 2(q_iq_j-q_kq_r) & 2(q_iq_k+q_jq_r) \\ 2(q_iq_j-q_kq_r) & 1-2{q_i}^2-2{q_k}^2 & 2(q_jq_k-q_iq_r) \\ 2(q_iq_k-q_jq_r) & 2(q_jq_k-q_iq_k) & 1-2{q_i}^2-2{q_j}^2 \end{bmatrix} \quad (1)$$

To compute the horizontal velocity (i.e., velocity of anterior-posterior axis, V_h) and vertical velocity (i.e., velocity of vertical axis, V_v), this study found the integral of the projected horizontal and vertical acceleration values (A_h and A_v in Figure 1). Since acceleration data from WIMU sensors already contain the effect of gravity (9.81 m/s^2), this effect was offset from the rotated acceleration value of the vertical axis.

The next step for computing the gait features is to detect gait events, such as toe off (TO) and heel strike (HS), to define the gait cycle. The gait cycle is defined as the time between two successive occurrences of one of these events during walking [25] (See Figure 2).

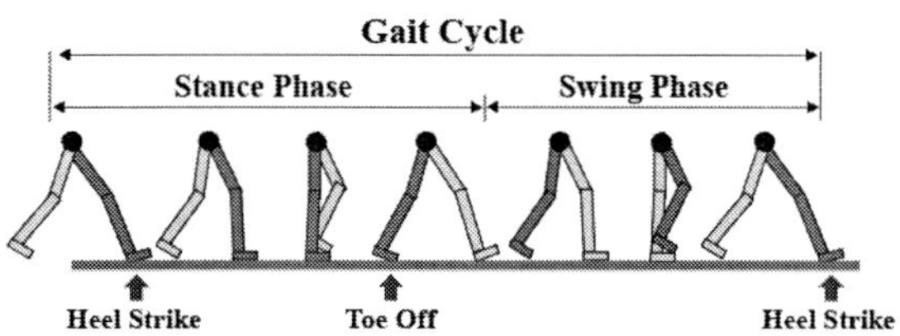

Figure 2. Definition of a gait cycle and gait events

In general, heel strike, the moment when a foot contacts the ground, is widely used to define a gait cycle, although any gait events can be used [26]. Thus, based on other gait analysis studies [15,16,19,20], this study defined a gait cycle as a gait between consecutive heel strikes of one foot. Since this study utilized heel strike as the starting point of the gait cycle, the initial gait cycle before the first heel strike point is neglected in our gait feature computation.

The gaits events (HS and TO) can be directly identified from the pitch angle of a gyroscope [28] inside the WIMU sensor (see Figure 3).

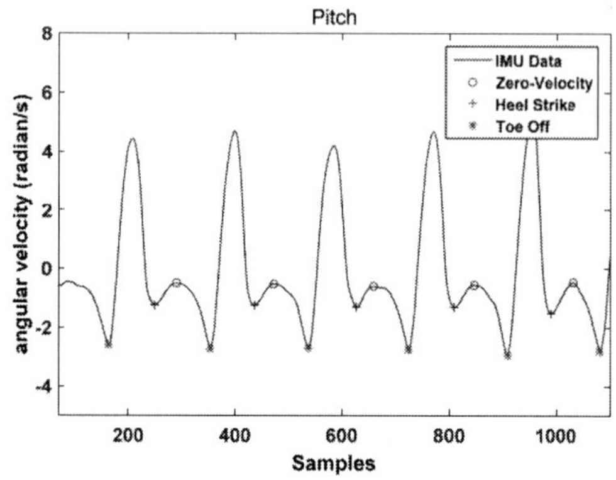

Figure 3. Heel strike, toe off and zero-velocity points with angular velocity in pitch

Based on the defined gait cycle, this study computed the four gait features—stride time, stride distance, average velocity and maximum foot clearance—for each gait cycle as follows:

- Stride Time, ST:
  Time between heel-strike points of one foot.
  $$ST = HS_{i+1} - HS_i \quad (2)$$
- Stride Distance, SD:
  Distance covered during a gait cycle.
  $$\mathrm{SD} = \int_{HS_i}^{HS_{i+1}} V_h(t)\, dt \quad (3)$$
- Average Velocity, $\overline{V_h}$:
  Average of horizontal velocity during a gait cycle.
  $$\overline{V_h} = \frac{1}{HS_{i+1} - HS_i} \int_{HS_i}^{HS_{i+1}} V_h(t)\, dt \quad (4)$$
- Foot Clearance, FC(t)
  Foot height from the ground during gait at time (t), finding an integral of vertical velocity (V_v) at time (t).
  $$\mathrm{FC(t)} = \int_o^t V_v(t)\, dt \quad (5)$$
- Maximum Foot Clearance, FC_{max}:
  The maximum value of foot clearance during a gait cycle.
  $$FC_{max} = \max_{t \in \{HS_i, \cdots, HS_{i+1}\}} FC(t) \quad (6)$$

where HS_i, HS_{i+1} are the time of ith and (i+1)[th] heel strike events.

Sensor drift is another issue in computing gait features from WIMU measurements [29]. This study used the zero-velocity assumption and zero-velocity update (ZUPT) to compensate for drift errors in gait-feature computations. Previous studies [22,28,30] have widely used the zero-velocity assumption to offset sensor measurement errors and drifts; this assumption says that the velocity of the foot is zero when the foot is located horizontally on the ground, which is why this location (the ground) is also known as the zero-velocity point. The zero-velocity update (ZUPT) method compensates for drift errors in velocity computations by updating a zero value for each zero-velocity points during a gait. This study used the pitch angle of the WIMU's gyroscope to identify the zero-velocity point same as gait event identification (see Figure 3).

3.2 Gait Abnormality Score Measurement

The clinical domain develops and uses the Gillette Gait Index (GGI) to measure the gait abnormality of patients in a single score [16]. GGI utilizes 16 gait-related features (e.g., mean pelvic rotation, total range of knee flexion–extension and mean foot progression angle) computed from an optical device with markers (e.g., VICON). Then, it performs principal component analysis, which is a widely used method to remove correlation between features, and measures the distance (i.e., Euclidean) of gait feature values between a specific test subject and a healthy subject group. While GGI has successfully quantified the extent of gait abnormality of Parkinson's [31] or cerebral palsy [28] patients compared to normal subjects, GGI has not been used to quantify a single subject's abnormal gait compared to his/her normal gait to capture the moment when hazards affect his/her gait pattern. Furthermore, GGI requires a total of 16 gait features to compute, which makes GGI hard to apply in a dynamic construction environment.

To respond to the need for such a gait abnormality assessment tool for construction workers, we used WIMU data to develop the IMU based Gait Abnormality Score (I-GAS) to measure the deviation of a single subject's momentary gait stride (i.e., hazard) from his/her normal gait strides. We used the Mahalanobis Distance (MD) to measure the abnormality (i.e., deviations) of gait between normal and hazard. The MD is a distance measurement method that measures the distance between a sample point and group of reference samples while simultaneously considering the distribution of reference samples [33]; this measure is widely used in multivariate outlier detection and clustering algorithms.

Thanks to MD, the correlation between computed gait features can be removed, and measurement unit difference between gait features can also be removed since MD rescales each feature to have unit variance. This paper calculated the I-GAS using Equation 7.

$$I\text{-}GAS = \sqrt{(x-\bar{y})C^{-1}(x-\bar{y})^T} \qquad (7)$$

where, x is an (N x M) vector of the number of observation samples (N) and computed gait features (M), y is a (K x M) vector of number of reference samples (K) and computed gait features (M), $\bar{y}$ is a (1 x M) vector of mean value of reference data (y), and C^{-1} is the inverse covariance matrix of reference data (y).

This paper investigates the usefulness of I-GAS in estimating the existence and the location of a hazard in laboratory experiments.

3.3 Laboratory Experiments

The laboratory experiments involving four humans were designed and conducted to collect subjects' gait data while they were simulating movements of ironworkers on a skeletal steel beam. To accomplish this task, we constructed a 24.4 m (80 feet) long steel beam 10 cm off the ground. Subjects were asked to walkover its surface (15 cm width) at the typical speeds for eight times. To have the same length of walk distance between experiment trials, walking start-point and end-point are assigned by the experiment organizer at the 1.5m (5 feet) and 22.9 m (75 feet). The experiment included either an obstacle (experiment #2 and #3) or a slippery surface (experiment #4 and #5) surface at varying locations on the steel beam to represent hazards that could cause slips and trips (or possibly falls). To have gait data which is not influenced by hazard as a reference for abnormality measure, all subjects also asked to walk on the beam without any installed hazard, called as "normal gait" in this study (See Table 1).

Table 1. Detail of laboratory experiments

Number	Hazard Types	Hazard Location	IMU	Time
1	-	-	Ankle	5 min
2	Obstacle	9.1 m (30 feet)	Ankle	5 min
3	Obstacle	15.2 m (50 feet)	Ankle	5 min
4	Slippery	9.1 m (30 feet)	Ankle	5 min
5	Slippery	15.2 m (50 feet)	Ankle	5 min

For the obstacle, this study installed wood block which has 76 cm (30-inch) length and 10 cm (4-inch) height on the I-beam. The slippery surface was installed to cover 152 cm (60-inch) length of the I-beam and by using liquid (oil) on the plastic. During experiment, all of the subjects were asked to wear a safety harness, a safety hat, and safety boots (See Figure 4).

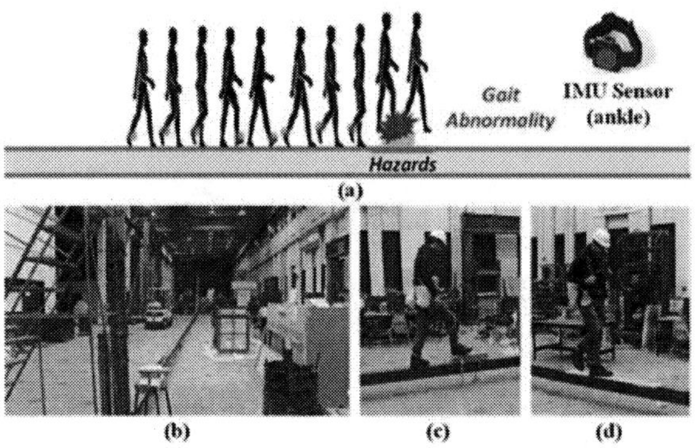

Figure 4. (a) Concept of hazard identification using WIMUs, (b) laboratory experiment environment, (c) obstacle on I-beam, and (d) slippery surface on I-beam

Experiment subjects walked continuously on the steel I-beam (return at the end point). During experiments, a WIMU (Opal, APDM Inc.) was attached to the right ankle of subjects to collect kinematic data. Three axes of acceleration (m/s2), angular velocity (rad/s), and magnetic field (uT) were obtained at a sampling frequency of 128 Hz. The quaternion rotation information of the object was acquired from the WIMU, which used the methodology of Bertrand et al. [24]. In addition to the WIMU, video cameras recorded data about all experiment procedures; these video data were synchronized to have the same timestamp with the WIMU data. These video data were used as a ground truth to identify the gait cycles which are influenced by installed hazards. In general, if there is any overlap between the stride distance of a gait cycle and the boundary of an installed hazard, that gait cycles are considered to include the interaction with the hazard. One to three consecutive gait cycles were thus classified as interacting with each installed hazard.

4 Results and Discussion

4.1 Gait Feature Computations

Gait features were computed using data collected from WIMUs for workers in both normal and hazardous conditions. Figure 5 illustrates the distributions of gait features for two subjects walking without a hazard present (*No Hazard*), walking over an obstacle (*Obstacle*), and walking on a slippery surface (*Slippery Surface*). Hazards caused a certain level of disruptions to gait features (e.g., Figure 5-e and 5-f), a fact that demonstrates the possibility of identifying hazards using worker's gait disruptions. The analysis of the aggregated data from all of the subjects also confirmed that the hazards caused significant differences ($p<0.05$) in several gait feature values.

However, the results also confirmed that such disruptions caused by the hazards varied by subjects and by gait features. For example, the slippery surface caused a change in Subject 1's stride distance (Figure 5-d) but did not create any noticeable change in Subject 2's stride distance (Figure 5-c). The analysis of the aggregated data from all of the subjects confirmed such uncertainty in the effects of hazards on individual gait features; it found that the stride distance at the aggregated data level was susceptible only to the slippery surface, not to the obstacle. These findings demonstrate the need for I-GAS to evaluate anomalies of multiple gait features in a comprehensive way.

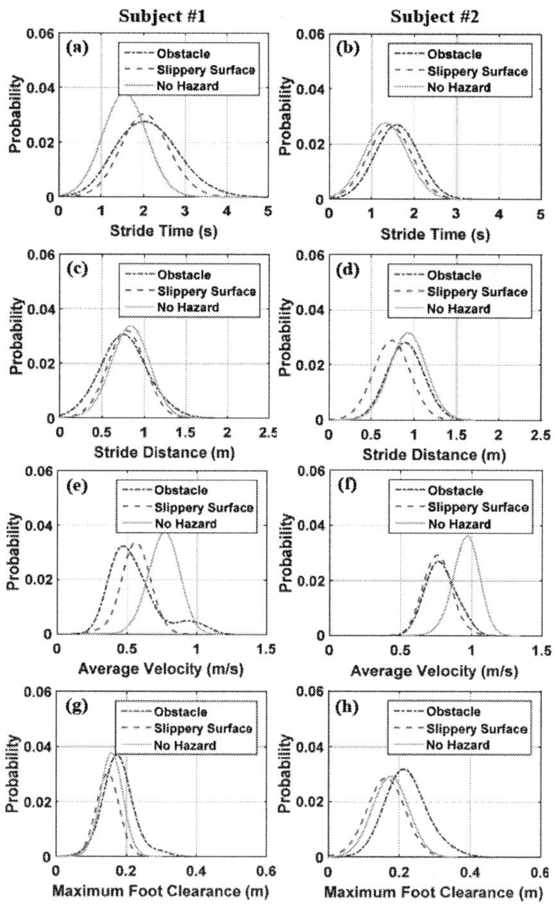

Figure 5. Results of the probability distribution for the computed gait features of Subject #1 and #2: (a, b) stride time; (c, d) stride distance; (e, f) average velocity; and (g, h) maximum foot clearance

4.2 WIMU-based Gait Abnormality Score Computations

Figure 6 shows the distributions of the subjects' combined I-GAS values for the *No Hazard, Obstacle,* and *Slippery Surface* trials (see Figure 6). The results show a clear difference in the I-GAS distributions for hazard and no-hazard conditions, and they demonstrate the usefulness of using IGAS in identifying hazard conditions.

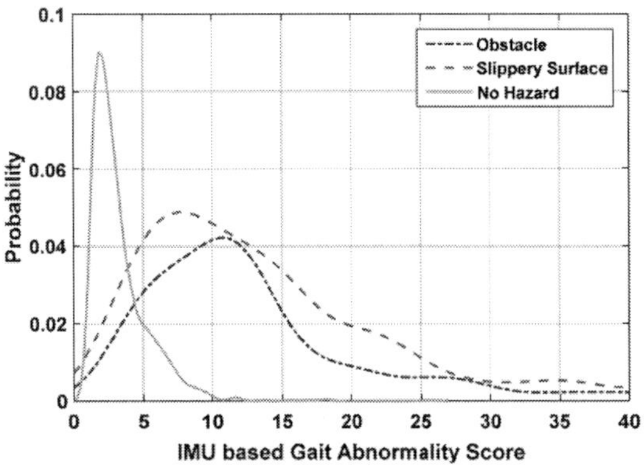

Figure 6. Probability distribution of the subjects' combined I-GAS.

Figure 7 illustrates an experimental application of the I-GAS to identify hazards. The I-GAS value of each gait cycle was mapped on the experimental site according to its estimated location—the location of a gait cycle was determined based on the stride lengths of previous gait cycles. Figure 7-a presents the mapped I-GAS values for individual subjects. The hazard location (at around 15.2 m) has a high I-GAS value, which demonstrates the success of the score at identifying hazards. However, similarly high values are also found in other, no-hazard locations. These noisy values illustrate that subjects continue to change their gait patterns even in no-hazard conditions.

Aggregation of the I-GAS values from multiple subjects (see Figure 7-b) helped filter out such noise. In particular, more samples (e.g., I-GAS values from multiple trials of multiple subjects) help to distinguish hazardous conditions more clearly. To this end, analyzing the collective gait patterns of multiple workers would be necessary to effectively identify hazards in a dynamic construction environment. Future research needs to investigate an efficient way to reveal these hazards with lower data requirements.

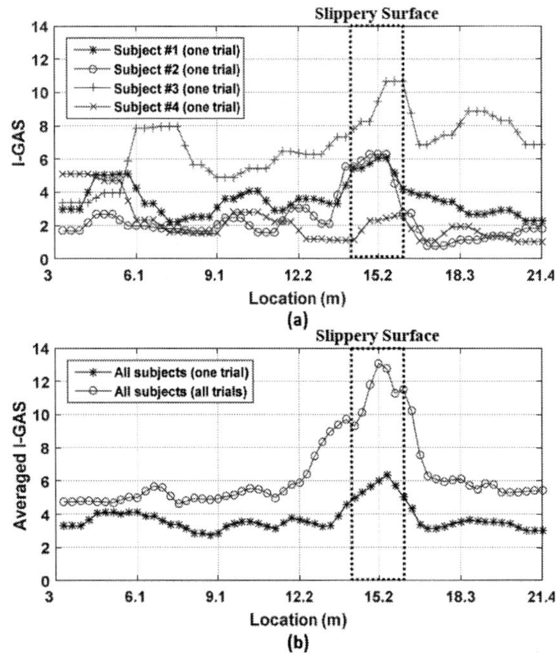

Figure 7. Preliminary results of hazard identification (experiment #5) using I-GAS: (a) I-GAS values from one trial with individual subjects, (b) averaged I-GAS values from one/all trials with all subjects

5 Conclusion

This study investigates the usefulness of gait analysis in identifying safety hazards. Using gait features that were estimated using WIMU data, this study developed the I-GAS to quantify the abnormality of each stride compared to normal strides at an individual level. Laboratory experiment settings that simulated steel connection tasks demonstrated the usefulness of the I-GAS in identifying two types of fall hazards (i.e., obstacles and slippery surfaces). Considering the possibility of ambulatory gait monitoring with WIMUs, the presented approach could help identify various types of fall hazards that cause disruptions in workers' gait and balance.

Our future research will look into the difference in gait features and the I-GAS values between individual workers and will study aggregation methods to evaluate the collective gait abnormality from multiple workers. Also, different types of possible hazards or hidden hazards will be studied to test the developed I-GAS-based technique for automated hazard identification.

Acknowledgement

The authors would like to acknowledge the excellent support and technical assistance from Cory Lyons—Project manager, Davis Erection Co Inc. (Topping Out Inc.) of Lincoln, NE— and Dr. Terry Stentz— Associate Professor, Durham School of Architectural Engineering and Construction, UNL— in designing and conducting the experiment. This study was financially supported by the Nebraska Research Initiative and National Science Foundation grant CIS #1538029 and CNS #1423379. Any opinions, findings, conclusions, or recommendations expressed in this article are those of the authors and do not necessarily reflect the views of the Nebraska Research Initiative and National Science Foundation.

References

[1] G. Carter, S. Smith, Safety Hazard Identification on Construction Projects, Journal of Construction Engineering and Management. 132 (2006) 197–205.

[2] D.L. Goetsch, Occupational Safety and Health in the Age of High Technology : For Technologists, Engineers, and Managers, Prentice Hall, Eaglewood Clifts, New Jersey, 1996.

[3] A. Albert, M.R. Hallowell, B.M. Kleiner, Experimental field testing of a real-time construction hazard identification and transmission technique, Construction Management and Economics. 32 (2014) 1000–1016.

[4] S. Bahn, Workplace hazard identification and management: The case of an underground mining operation, Safety Science. 57 (2013) 129–137.

[5] A. Albert, M. Hallowell, B. Kleiner, Enhancing Construction Hazard Recognition and Communication with Energy-Based Cognitive Mnemonics and Safety Meeting Maturity Model: Multiple Baseline Study, Journal of Construction Engineering and Management. 140 (2014)

[6] H.J. Lipscomb, J.E. Glazner, J. Bondy, K. Guarini, D. Lezotte, Injuries from slips and trips in construction, Applied Ergonomics. 37 (2006) 267–274.

[7] S. Han, S. Lee, A vision-based motion capture and recognition framework for behavior-based safety management, Automation in Construction. 35 (2013) 131–141.

[8] Hardhat-Wearing Detection for Enhancing On-Site Safety of Construction Workers, Journal of Construction Engineering and Management. 141 (2015) 04015024.

[9] T. Cheng, G. Migliaccio, J. Teizer, U. Gatti, Data Fusion of Real-Time Location Sensing and Physiological Status Monitoring for Ergonomics Analysis of Construction Workers, Journal of Computing in Civil Engineering. 27 (2013) 320–335.

[10] U. Gatti, G. Migliaccio, S. Schneider, Wearable Physiological Status Monitors for Measuring and Evaluating Workers' Physical Strain: Preliminary Validation, in: Computing in Civil Engineering (2011), American Society of Civil Engineers, n.d.: pp. 194–201. http://ascelibrary.org/doi/abs/10.1061/41182%28416%2924 (accessed March 11, 2014).

[11] U.C. Gatti, S. Schneider, G.C. Migliaccio, Physiological condition monitoring of construction workers, Automation in Construction. 44 (2014) 227–233.

[12] L. Joshua, K. Varghese, Accelerometer-Based Activity Recognition in Construction, Journal of Computing in Civil Engineering. 25 (2011) 370–379.

[13] K. Yang, S. Aria, T.L. Stentz, C.R. Ahn, Automated Detection of Near-miss Fall Incidents in Iron Workers Using Inertial Measurement Units, in: Construction Research Congress 2014, American Society of Civil Engineers, 2014: pp. 935–944. http://ascelibrary.org/doi/abs/10.1061/9780784413517.096 (accessed August 22, 2014).

[14] K. Yang, H. Jebelli, C.R. Ahn, M.C. Vuran, Threshold-Based Approach to Detect Near-Miss Falls of Iron Workers Using Inertial Measurement Units, in: Computing in Civil Engineering 2015, American Society of Civil Engineers, 2015: pp. 148–155. http://ascelibrary.org/doi/abs/10.1061/9780784479247.019 (accessed April 4, 2016).

[15] V.V. Khanzode, J. Maiti, P.K. Ray, Occupational injury and accident research: A comprehensive review, Safety Science. 50 (2012) 1355–1367.

[16] L.M. Schutte, U. Narayanan, J.L. Stout, P. Selber, J.R. Gage, M.H. Schwartz, An index for quantifying deviations from normal gait, Gait Posture. 11 (2000) 25–31.

[17] T.A. Wren, K.P. Do, R. Hara, F.J. Dorey, R.M. Kay, N.Y. Otsuka, Gillette Gait Index as a gait analysis summary measure: comparison with qualitative visual assessments of overall gait, Journal of Pediatric Orthopaedics. 27 (2007) 765–768.

[18] M.O. Derawi, Accelerometer-based gait analysis, a survey, Nor Informasjonssikkerhetskonferanse NISK. (2010).

[19] A. Muro-de-la-Herran, B. Garcia-Zapirain, A. Mendez-Zorrilla, Gait Analysis Methods: An Overview of Wearable and Non-Wearable Systems, Highlighting Clinical Applications, Sensors. 14 (2014) 3362–3394.

[20] S.R. Hundza, W.R. Hook, C.R. Harris, S.V. Mahajan, P.A. Leslie, C.A. Spani, L.G. Spalteholz,

B.J. Birch, D.T. Commandeur, N.J. Livingston, Accurate and reliable gait cycle detection in Parkinson's disease, Neural Systems and Rehabilitation Engineering, IEEE Transactions on. 22 (2014) 127–137.
[21] N. Kitagawa, N. Ogihara, Estimation of foot trajectory during human walking by a wearable Inertial Measurement Unit mounted to the foot, Gait & Posture. (2016).
[22] S.J.M. Bamberg, A.Y. Benbasat, D.M. Scarborough, D.E. Krebs, J.A. Paradiso, Gait Analysis Using a Shoe-Integrated Wireless Sensor System, IEEE Transactions on Information Technology in Biomedicine. 12 (2008) 413–423.
[23] M. Chen, B. Huang, Y. Xu, Intelligent shoes for abnormal gait detection, in: IEEE International Conference on Robotics and Automation, 2008. ICRA 2008, 2008: pp. 2019–2024.
[24] P.J. Bertrand, A. Anderson, A. Hilbert, D.J. Newman, Feasibility of Spacesuit Kinematics and Human-Suit Interactions, 44th International Conference on Environmental Systems, 2014. https://repositories.tdl.org/ttu-ir/handle/2346/59680 (accessed September 21, 2015).
[25] M.W. Whittle, Gait analysis: an introduction, Butterworth-Heinemann, 2014.
[26] S.-W. Lee, K. Mase, K. Kogure, Detection of spatio-temporal gait parameters by using wearable motion sensors, in: IEEE, 2006: pp. 6836–6839.
[27] K. Aminian, B. Najafi, C. Büla, P.-F. Leyvraz, P. Robert, Spatio-temporal parameters of gait measured by an ambulatory system using miniature gyroscopes, Journal of Biomechanics. 35 (2002) 689–699.
[28] A.M. Sabatini, C. Martelloni, S. Scapellato, F. Cavallo, Assessment of walking features from foot inertial sensing, IEEE Transactions on Biomedical Engineering. 52 (2005) 486–494.
[29] J.C. Alvarez, D. Alvarez, A. López, R.C. González, Pedestrian Navigation Based on a Waist-Worn Inertial Sensor, Sensors. 12 (2012) 10536–10549.
[30] A.R. Jimenez, F. Seco, C. Prieto, J. Guevara, A comparison of Pedestrian Dead-Reckoning algorithms using a low-cost MEMS IMU, in: IEEE International Symposium on Intelligent Signal Processing, 2009. WISP 2009, 2009: pp. 37–42.
[31] M. Galli, V. Cimolin, M.F.D. Pandis, M.H. Schwartz, G. Albertini, Use of the Gait Deviation Index for the Evaluation of Patients With Parkinson's Disease, Journal of Motor Behavior. 44 (2012) 161–167.
[32] M.H. Schwartz, A. Rozumalski, The gait deviation index: A new comprehensive index of gait pathology, Gait & Posture. 28 (2008) 351–357.
[33] R. De Maesschalck, D. Jouan-Rimbaud, D.L. Massart, The Mahalanobis distance, Chemometrics and Intelligent Laboratory Systems. 50 (2000) 1–18.

Integrating Crane Information Models in BIM for Checking the Compliance of Lifting Plan Requirements

Justin K.W. Yeoh[a], J.H. Wong[a] and L. Peng[a]

[a] Department of Civil and Environmental Engineering, National University of Singapore, Singapore
E-mail: justinyeoh@nus.edu.sg, A0088544@u.nus.edu, peng_le@u.nus.edu

Abstract

Cranes typically represent the single biggest equipment investment deployed on a construction site. The success of lifting operations depends on addressing project conditions and adequate site safety management. The Crane Lifting Plan (CLP) is a safety document wherein the information required for planning crane lifting operations are encapsulated and analyzed. A range of regulatory, safety and operational requirements must be complied with. Chief among these considerations is the correct selection and siting of cranes to support the lifting operations. The current practice of developing and checking a CLP is a manual, tedious and potentially error-prone process. Recent advances in Building Information Modeling (BIM) may help to address these difficulties. This paper presents a formalized representation framework for CLP requirements. A parametric Crane Information Model, which enables various regulatory and safety information to be incorporated is also developed. The result of this paper is a step towards automatically checking the compliance of CLP requirements. Finally, a case study of an academic building is used to validate the proposed framework.

Keywords –

Crane Lifting Plans; Crane Information Model; Regulatory Compliance; Construction Requirements

1 Introduction

In Singapore, static cranes are commonly deployed in high-rise projects, of which the two common types are Tower Cranes and Luffing Jib Cranes. The densely built-up urban environment in Singapore constrains the deployment of these construction equipment, creating significant construction challenges to contractors, and compliance issues to the regulatory authorities.

Moreover, cranes typically represent the single biggest equipment investment deployed on a construction site [1], and are also one of the main sources of hazards on site [2]. Statistics from the Workplace Safety and Health Institute (Singapore) [3] show that crane related incidents accounted for almost 20% of all fatalities in the construction industry from 2011 to 2014.

Shapira et al. [4] studied the various factors affecting the safe use of cranes on site, and reported that site safety management, as well as the various project conditions are major determinants. The Crane Lifting Plan (CLP) is a safety document wherein the information required for planning crane lifting operations are encapsulated and analyzed [5]. This involves a range of regulatory, safety and operational requirements which must be complied with to address the project conditions and site safety management issues. Chief among these considerations is the correct selection and layout/siting of cranes to support the lifting operation, subject to the contractor's cost constraints. The current practice of developing a CLP is a manual, tedious and potentially error-prone process [6]. Similarly, checking the CLP for compliance is often difficult, and the consequences of an erroneous CLP could be disastrous, resulting in major accidents and fatalities onsite.

Several research challenges exist which need to be addressed. The first is the need to represent CLP requirements in a form that is usable for analysis. This paper overcomes this challenge by developing a formalized representation of CLP requirements. The second challenge is to capture generic parametric information of a crane. Such a generic information representation schema should allow for different crane types to be modeled. Recent advances in Building Information Modeling (BIM) may help to address this challenge. This paper introduces a Crane Information Model (CIM) which will embed the crane information within BIM for analysis.

This paper starts with a literature review of the current methods for workspace analysis, as well as current crane analysis techniques, and code compliance. The next section details the identification of workspaces from common CLP requirements using a formalized approach. These identified workspaces are then incorporated into a Crane Information Model (CIM), with the purpose of reducing the errors during CLP. The result of this paper is a step towards automatically checking the compliance of CLP requirements. This is validated via a case study.

2 Relevant Literature

2.1 Representing and Analyzing Workspaces for Construction Operations

Various methodologies for modeling construction workspaces exist. These have been proposed for the analysis of spatial conflicts, and the primary method of conflict detection is the analysis of interferences.

Riley and Sanvido [7] identified 13 types of spaces required by construction activities: building components, layout areas, unloading areas, material paths, personnel paths, storage areas, staging areas, prefabrication areas, crew areas, tool and equipment areas, debris paths, protected areas, and hazard areas.

Akinci et al [8] studied the representation of spaces in a 3D CAD environment, and identified the parameters required for representing these spaces. These include: Reference object, volumetric parameters and orientation. A taxonomy of space usage and the resultant conflicts were also developed based on the interactions between spaces [9]. One key contribution is the recognition that not all space interferences constitutes conflict.

Guo [10] analyzed the spatial requirements of movement paths (pathspaces) for workers, equipment and materials on-site. This inclusion of pathspaces, which are abstracted as pathspace requirements (minimum path height and minimum path width) was a separate space type, which could facilitate the verification of the availability of access to work faces.

Chua et al [11] proposed an alternative analysis methodology based on the utilization of space. In their paper, the space taxonomy was differentiated according to the amount of space used. This led to the identification of Product (Dead) spaces, Process/Workspaces and Pathspaces, as well as Interdiction (Clearance) spaces.

Su and Cai [12] have extended the traditional 3D topological analysis, by enhancing the coupling of time analysis with 3D. This results in a 4D topological relationship analysis technique. The technique is then validated using a crane example to test for crane reach and crane hazard spaces. This validation demonstrated the applicability of determining feasible crane operations from a 3D/4D perspective.

The above methodologies aid the visualization of space utilization among the different construction trades. Such visualization helps engineers to identify possible conflicts arising from the detected space collisions. Thus far, analyzing the crane's workspaces and space requirements for lift planning has not been commonly conducted to the authors' knowledge. Instead, most code compliance approaches focus on building design [13], fire engineering [14] or construction quality [15]. One contribution of this paper is the application of workspace analysis for cranes to regulatory conformance. This entails the identification of spaces for cranes, and the recognition of the conditions leading to spatial conflicts during crane operations. These often arise from the interactions of these spaces.

2.2 Analysis of Static Crane Operations

The analysis of static crane operations is also widely studied in the literature. Analyzing static crane operations is one of the fundamental activities in a CLP. In general, there have been several approaches used: Mathematical models, and Simulation.

The mathematical models are often formulated to optimize some quantitative metric like transportation cost or time. Huang et al [16] noted that the crane layout problem involves various nonlinearities, leading to some of these models being computationally expensive. Various metaheuristic models like genetic algorithms and artificial neural networks [17] have been applied. Such approaches have been successfully applied, but requires that a certain amount of crane and project information be 'hard coded' into the optimization problem.

Other approaches attempted to address the crane layout problem by incorporating 3D visualization with simulation. Such approaches typically incorporate 3D crane models into existing 3D building models, and simulate the workflow of the cranes [1,18]. Such simulation models are often difficult to generalize, and may even miss some conflicts if discrete event simulation is employed [19].

Several recent efforts have begun looking at BIM as an information platform. BIM has shown to be extremely useful for providing project information, like sizes and quantities of elements. Irizarry and Karan [20] implemented a GIS-BIM hybrid to reason about suitable crane locations, while Olearczyk et al [21] employed a BIM based methodology for modular construction. Marzouk and Abubakr [22] and Wang et al [23] have also leveraged BIM to mitigate the difficulties of incorporating site information for layout planning.

3 Research Approach: Identifying Information Requirements for Automating CLP

To address the problem of automating a CLP, the research approach adopted in this paper looks to identify the various information requirements needed. Three information requirements are identified (Figure 1): The first information requirement is BIM. BIM should encompass the 3D facility information, the site layout plan, temporary structures, material delivery and construction phasing plan. This information requirement is assumed to be available in this research.

The second information requirement is the identification of the relevant crane codes and regulations. This has to be translated into a formalized representation, so that automation can be conducted. This second requirement forms the basis for the following section.

The third information requirement is the identification of the relevant crane parameters and information. This will include the loading characteristics of the available cranes on the market, which is obtainable from the respective load charts, as well as the relevant crane geometry. This third requirement is addressed in the following sections of this paper.

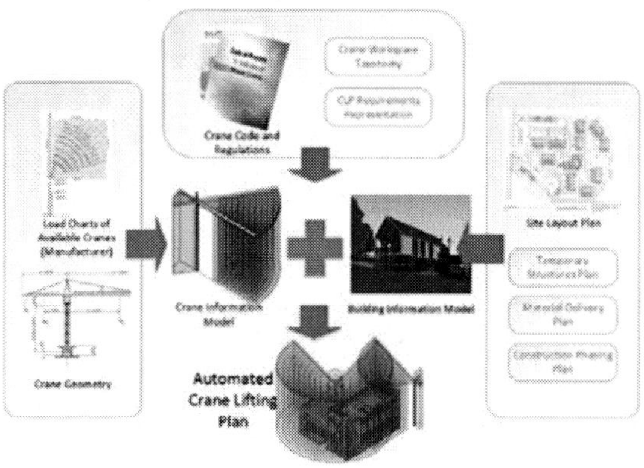

Figure 1. Information Requirements for Automated Crane Lifting Plan

4 Identifying and Modeling Workspaces from Common CLP Requirements

This section discusses the formalized representation of some CLP requirements by starting with a framework for representation, a space taxonomy for crane workspaces, and finally a summary of some common CLP requirements in Singapore.

4.1 A Space Taxonomy for representing CLP Requirements

Song and Chua [24] introduced a formal framework to represent functional construction requirements, and this framework was extended by Yeoh and Chua [25] to incorporate non-functional requirements as well. These construction requirements may be described as capabilities and conditions for which the construction process must conform to. CLP requirements are a subset of these construction requirements.

The representation framework adopted in this paper has been adapted from [25] and includes the following elements: A function (requirement) user, a function (requirement) provider, and the necessary condition (relationship) between the function user and function provider.

Two types of requirements are identified for CLP: Spatial and Parametric. In the spatial requirement (Figure 2), the function user is defined as the requester of the function, while the function provider describes the conditions for fulfilling the requirement. Each function user and function provider is comprised of a set of 3D spaces. In the CLP context, an example of a user and provider could be the Crane and Building respectively. The necessary condition then describes the interaction between both the provider and user via set operators. In particular, this relationship describes the failure conditions of the requirement, i.e. the conditions under which the requirement is not fulfilled.

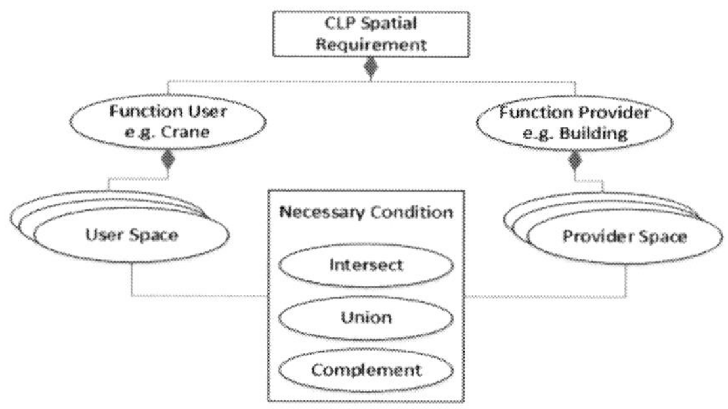

Figure 2. Crane Lifting Plan Spatial Requirements Representation Framework

The second requirement is the parametric requirement (Figure 3). This requirement is similar to that in Figure 2, except that the parameters used to describe the Function User and Function Provider are compared to determine the conditions of failure for the requirement. Mathematical operators are used in place of set operators, as this allows for manipulation and comparison of the parametric values.

Figure 4 shows the space taxonomy related to the requirements. Following the characterization of space utilization in Chua et al [11], three main classes of space are identified: Hazard space, Process space and Product space. The spaces in CLP requirements originate from three sources: Site, Building Facility, and Crane.

The crane and building structure are examples of a product space, and describe the physical elements of the crane and building. The crane working envelope is a type of process space. It is the union of the different possible jib positions of the crane, and its corresponding zone of operation. Building process spaces include the path and activity spaces.

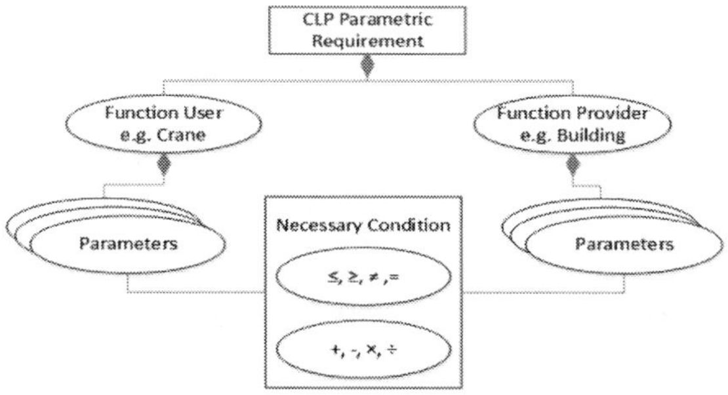

Figure 3. Crane Lifting Plan Parametric Requirements Representation Framework

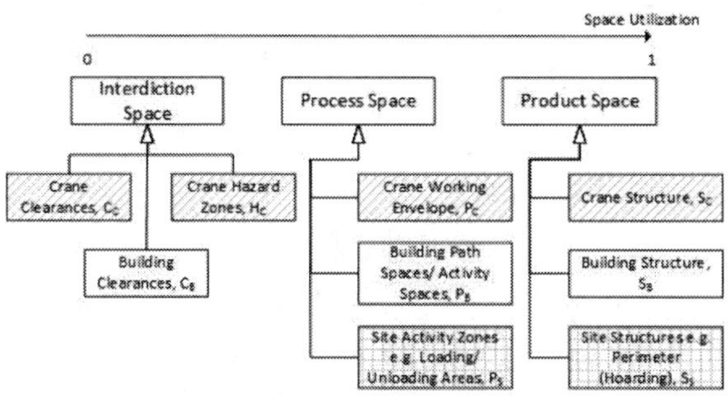

Figure 4. Space Taxonomy for CLP Requirements

The interdiction spaces are spaces where no space utilization is allowed. Two types of interdiction spaces are possible: Clearances and Hazards. Clearance spaces are meant to denote protective zones against physical elements. Clearance spaces include buffer areas around physical elements. Hazard zones (spaces) are used to demarcate areas of hazardous operations. Examples of hazard zone include the zone under the counterweight of the counter-jib.

4.2 Identifying Common CLP Requirements

The above taxonomy aids in allowing common CLP requirements to be represented in a manner useful for analysis. The advantage is that conflict detection in the requirements for crane operations becomes automated.

To illustrate the applicability of the representation framework, the following obvious crane requirement is used: "The crane structure should not be co-located with the building structure". In other words, the crane must not occupy the same space as any of the elements in the building structure.

The Function User is identified as the Crane, while the Function Provider is identified as the Building. Using the space taxonomy provided, the spaces associated with the User and Provider are identified: Crane Structure and Building Structure respectively. To fulfil the intention of the requirement, the Necessary Condition is that both crane structure and building structure must not intersect.

Table 2 presents a list of common CLP requirements found in the relevant Code of Practice in Singapore [5], as well as common operational considerations. This table uses a dot notation to display the spaces or parameters of the Function User and Function Provider, e.g. *User.Space* or *Provider.Parameter*. An ID for each space or parameter is also provided in parenthesis. This notation is then used to describe the Necessary Condition to evaluate for non-compliance.

The result of conducting this formalization of CLP requirements is the ability to analyze crane and building information models, and evaluate non-compliances.

5 Crane Information Model (CIM)

The Crane Information Model (CIM) is developed based on several requirements: Firstly, CIM was intended to be an extension of current BIM systems. This means that elements, and spaces within CIM should be parametrically driven. This will allow CIMs to be easily modified and collaboratively shared. Secondly, CIM is intended to be a repository of crane information. Hence, the relevant crane information should incorporate not only the crane geometry, but the load charts as well.

The following models were created as Autodesk Revit Families, where the parametric crane information were entered as Revit Family parameters to automatically generate the model geometry.

5.1 Mapping Parametric Information to Crane Spaces

Some of the basic parameters of a static crane are presented in the following Table 1. These basic parameters allow static crane information to be easily modified to suit specific project requirements. Additionally, Table 1 provides a schematic mapping showing the relationship between the real crane parameters to the corresponding crane spaces. The source of information for these parameters is also provided in Table 1. Some are user-defined, while others arise from regulatory compliance, crane catalogues, or project specific information, or a combination thereof. These parameters thus drive the geometric and information properties of the crane spaces.

The load chart information of the cranes are also embedded into the CIM via the workspace envelope. The workspace envelope is discretized, with each discrete section having a corresponding load capacity. To illustrate this, an example of a tower crane load chart is shown in Figure 5. The load capacity is typically dependent upon the jib length for the tower cranes. This also typically holds for luffing jib type tower cranes.

Jib Length (m)	m/kg									
	25	30	35	40	45	50	55	60	65	70
70	10220	8230	6820	5760	4940	4280	3750	3300	2920	2600
65	10690	8620	7150	6050	5190	4510	3960	3490	3100	
60	11200	9040	7510	6360	5470	4760	4180	3700		
55	11680	9440	7850	6660	5740	5000	4400			
50	12080	9770	8130	6910	5960	5200				
45	12340	9990	8320	7070	6100					
40	12550	10160	8470	7200						

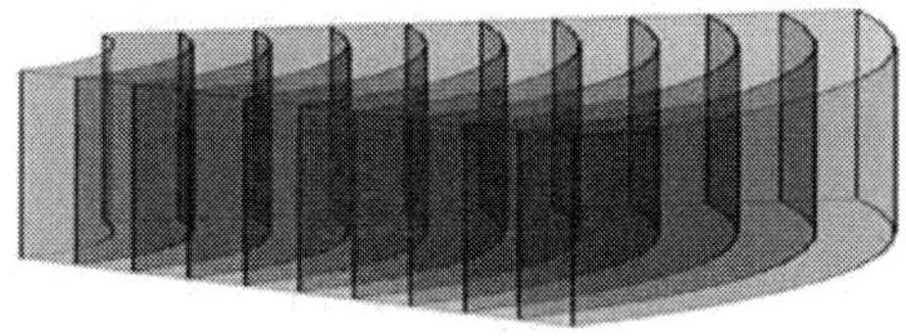

Figure 5. Relating Load Chart information to Workspace Envelope

Table 1. Static Crane Parameters. (*) represents information for Luffing Jib Type Crane

	Parameter	Type	Source*
Crane Structure	Total Height	Length	U/C
	Mast Height	Length	U/C
	Head Height	Length	C
	Hook Height	Length	C
	Jib Length	Length	U
	Counter Jib Length	Length	C
	Mast Size	Length	C
	Jib Size	Length	C
	Jib Angle*	Angle	U/P
Workspace Envelope	Coverage Angle	Angle	U
	Base Angle	Angle	P
	Load Capacity	Weight	C
	Hook Weight	Weight	C
	Load Factor of Safety	Real	U/R
Interdiction (Safety) Space	Jib Clearance Value	Length	R
	Counter Jib Clearance Value	Length	R
	Minimum Building Height Clearance	Length	R
Orientation	True North	Angle	P

U – User Defined, C – Crane Catalogue, P – Project Specific, R - Regulatory

Figure 9 shows the results of the developed CIMs for both Tower Cranes and Luffing Jib Tower Cranes incorporating the various workspace envelopes, crane clearances and hazard areas. Crane parameters are indicated in italics.

6 Validation: Demonstrating the CLP Requirements for an Academic Building

An academic building was used as a case study to validate the formalization of requirements as well as to demonstrate the applicability of CIM. The building comprised two adjoining blocks. The first block is a 10 story structure comprising primarily precast elements, of which there are about 210 pieces per story. The second is a 6 story workshop, with about 70 precast elements per story. The largest element, a post tensioned deep beam situated at the auditorium, weighed about 10 tons. The site was also subject to a height restriction of 80m.

The proposed lifting plan is shown in Figure 6 below. Two tower cranes and a luffing jib crane with specific characteristics as shown in the figure were analyzed.

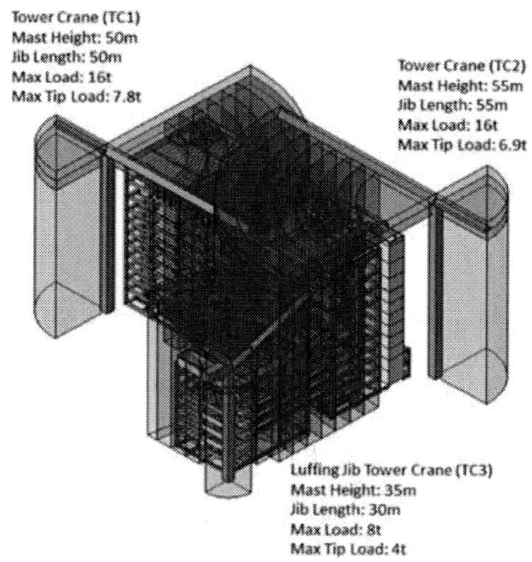

Figure 6. Proposed Layout with Crane Information Model

To validate the proposed approach in this paper, the various corresponding requirements in Table 2 were checked in the case study. These checks may be carried out by visual inspection, or by using interference checking or filtering mechanisms commonly found in BIM applications.

6.1 Requirement 1, 2, 6 and 7: Site Boundary, Counter-Jib Clearance, Crane Coverage and Load Capacity

Figure 7 shows the plan view of the project site. From this plan view, the site boundary has not been exceeded by the proposed workspace envelopes or the various clearances required by the crane.

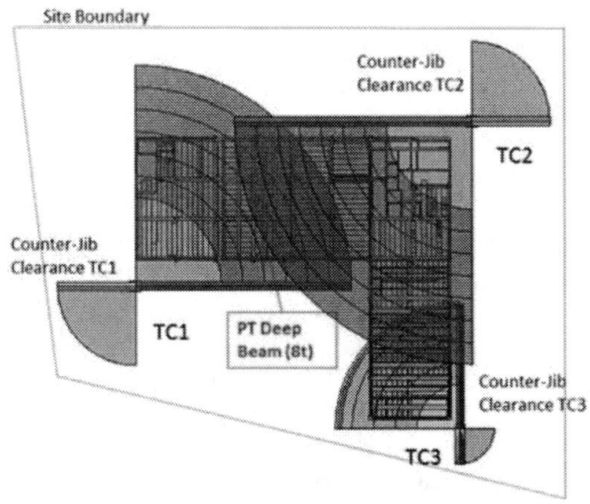

Figure 7. Plan View showing Site Boundary, CIM and Deep Beam position

The crane positions are adequate, with no clashes occurring between the counter-jib clearances. Also, the coverage requirement is met, with all elements within the workspace envelopes.

The 8 ton deep beam is situated at the intersection of the workspace envelopes of TC1 and TC2. An inspection of the load capacities of the discretized workspace envelopes reveals TC2 does not have adequate capacity to hoist. However, TC1 is found to be adequate.

6.2 Requirement 3, 4 and 5: Jib Height Separation, Building Height Separation and Crane Height Limit

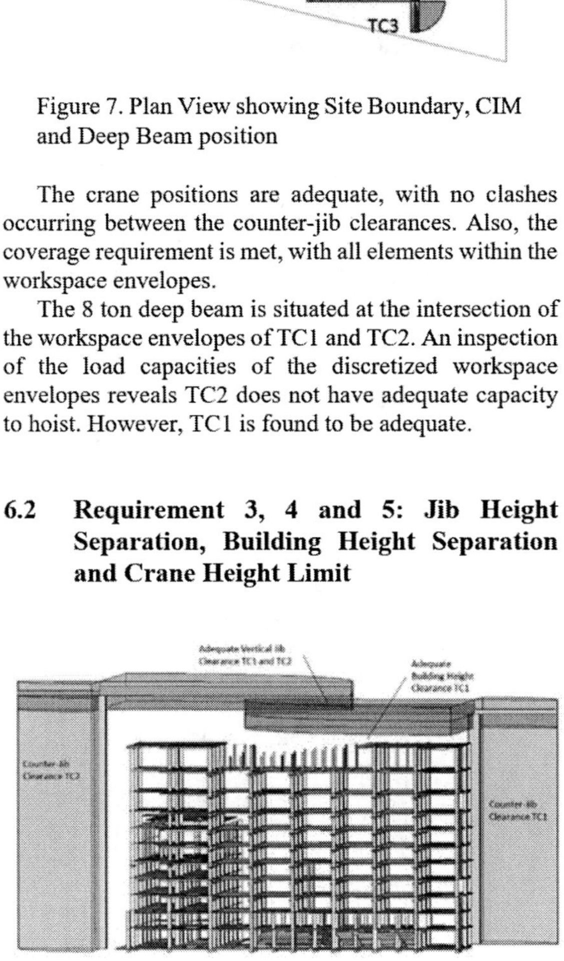

Figure 8. Elevation View

Inspecting the elevation view of the model (Figure 8), TC1 and TC2 fulfills the requirement of minimum jib height separation. Also, TC1 meets the minimum building height separation from the 10 story block. Similarly, TC3 meets the building height separation requirement from the 6 story carpark. The total height of the cranes was also found to be adequately within the permissible site limit.

7 Conclusion

This paper addresses several challenges associated with automating the compliance of Crane Lifting Plan requirements. The research contributions of this paper include the formalization of crane requirements for analysis. Two types of crane requirements are identified: spatial requirements and parametric requirements. Both requirement types encompass the necessary failure conditions for non-compliance of the crane requirement.

The second research contribution is the development of a Crane Information Model. CIM maps the crane geometry to its spaces, and also incorporates information like the load capacity and reach. This model is parametric and easily modifiable, allowing different crane types to be represented. This also allows it to be embedded within existing BIM systems. The advantage is that the crane information need not be 'hard coded', and a certain amount of flexibility is available to quickly evaluate alternative crane configurations.

The value of the developed framework is it provides the foundation for automated checking of crane layout planning requirements. This automation reduces both the time taken, as well as the errors made, providing an advantage for planners as well as regulatory agencies checking the CLP for compliance.

Extensions to the CIM are possible. The requirements in this paper are not exhaustive, and many other requirements exist. One example is the dismantling requirements of the crane, which has not been considered. Also, requirements which cannot be measured, cannot be captured by the framework. These include requirements like maintenance and safety checks. Future work include extending the CIM for mobile cranes, as well as developing an optimization method that integrates the flexibility of the proposed approach to determine optimal crane configurations onsite.

References

[1] M. Al-Hussein, M. Athar Niaz, H. Yu, H. Kim, Integrating 3D visualization and simulation for tower crane operations on construction sites, Autom. Constr. 15 (2006) 554–562. doi:10.1016/j.autcon.2005.07.007.

[2] A. Shapira, M. Simcha, Identification and Analysis of Factors Affecting Safety on Construction Sites with Tower Cranes, J. Constr. Eng. Manag. 135 (2009) 307–318. doi:10.1061/(ASCE)0733-9364(2009)135:4(307).

[3] Workplace Safety and Health Report 2014, A report by the Workplace Safety and Health Institute Singapore, 2015.

[4] A. Shapira, M. Simcha, M. Goldenberg, Integrative model for quantitative evaluation of safety on construction sites with tower cranes, J. Constr. Eng. Manag. 138 (2012) 1281–1293. doi:10.1061/(ASCE)CO.1943-7862.0000537.

[5] Code of Practice on Safe Lifting Operations in the Workplaces, Workplace Safety and Health Council, 2011.

[6] M. Al-Hussein, S. Alkass, O. Moselhi, D-CRANE: a database system for utilization of cranes, Can. J. Civ. Eng. 27 (2000) 1130–1138. doi:10.1139/l00-039.

[7] D.R. Riley, V.E. Sanvido, Patterns of Construction-Space Use in Multistory Buildings, J. Constr. Eng. Manag. 121 (1995) 464–473.

[8] B. Akinci, M. Fischer, J. Kunz, R. Levitt, Representing Work Spaces Generically in Construction Method Models, J. Constr. Eng. Manag. 128 (2002) 296–305. doi:10.1061/(ASCE)0733-9364(2002)128:4(296).

[9] B. Akinci, M. Fischer, R. Levitt, R. Carlson, Formalization and Automation of Time-Space Conflict Analysis, J. Comput. Civ. Eng. 16 (2002) 124–134. doi:10.1061/(ASCE)0887-3801(2002)16:2(124).

[10] S.J. Guo, Identification and resolution of work space conflicts in building construction, J. Constr. Eng. Manag. 128 (2002) 287–295. doi:10.1061/(ASCE)0733-9364(2002)128:4(287).

[11] D.K.H. Chua, K.W. Yeoh, Y. Song, Quantification of Spatial Temporal Congestion in Four-Dimensional Computer-Aided Design., J. Constr. Eng. Manag. 136 (2010) 641–649. doi:10.1061/(ASCE)CO.1943-7862.0000166.

[12] X. Su, H. Cai, Enabling Construction 4D Topological Analysis for Effective Construction Planning, J. Comput. Civ. Eng. (2014) 1–10. doi:10.1061/(ASCE)CP.1943-5487.0000463.

[13] X. Tan, A. Hammad, P. Fazio, Automated Code Compliance Checking for Building Envelope Design, J. Comput. Civ. Eng. 24 (2010) 203–211. doi:10.1061/(ASCE)0887-3801(2010)24:2(203).

[14] S. Malsane, J. Matthews, S. Lockley, P.E.D. Love, D. Greenwood, Development of an object model for automated compliance checking, Autom. Constr. 49 (2015) 51–58. doi:10.1016/j.autcon.2014.10.004.

[15] B.T. Zhong, L.Y. Ding, H.B. Luo, Y. Zhou, Y.Z. Hu, H.M. Hu, Ontology-based semantic modeling of regulation constraint for automated construction quality compliance checking, Autom. Constr. 28 (2012) 58–70. doi:10.1016/j.autcon.2012.06.006.

[16] C. Huang, C.K. Wong, C.M. Tam, Optimization of tower crane and material supply locations in a high-rise building site by mixed-integer linear programming, Autom. Constr. 20 (2011) 571–580. doi:10.1016/j.autcon.2010.11.023.

[17] C.M. Tam, T.K.L. Tong, GA-ANN model for optimizing the locations of tower crane and supply points for high-rise public housing construction, Constr. Manag. Econ. 21 (2003) 257–266. doi:10.1080/0144619032000049665.

[18] W.C. Hornaday, C.T. Haas, J.T. O'Connor, J. Wen, Computer-Aided Planning for Heavy Lifts, J. Constr. Eng. Manag. 119 (1994) 498–515.

[19] K. Tantisevi, B. Akinci, Transformation of a 4D product and process model to generate motion of mobile cranes, Autom. Constr. 18 (2009) 458–468. doi:10.1016/j.autcon.2008.10.008.

[20] J. Irizarry, E.P. Karan, Optimizing location of tower cranes on construction sites through GIS and BIM integration, Electron. J. Inf. Technol. Constr. 17 (2012) 361–366.

[21] J. Olearczyk, M. Al-Hussein, A. Bouferguène, Evolution of the crane selection and on-site utilization process for modular construction multilifts, Autom. Constr. 43 (2014) 59–72. doi:10.1016/j.autcon.2014.03.015.

[22] M. Marzouk, A. Abubakr, Decision support for tower crane selection with building information models and genetic algorithms, Autom. Constr. 61 (2015) 1–15. doi:10.1016/j.autcon.2015.09.008.

[23] J. Wang, X. Zhang, W. Shou, X. Wang, B. Xu, M.J. Kim, et al., A BIM-based approach for automated tower crane layout planning, Autom. Constr. 59 (2015) 168–178. doi:10.1016/j.autcon.2015.05.006.

[24] Y. Song, D.K.H. Chua, Modeling of Functional Construction Requirements for Constructability Analysis, J. Constr. Eng. Manag. 132 (2006) 1314–1326. doi:10.1061/(ASCE)0733-9364(2006)132:12(1314).

[25] K.W. Yeoh, D.K.H. Chua, Representing Requirements of Construction from an IFC Model, in: Comput. Civ. Build. Eng. 2014, American Society of Civil Engineers, 2014: pp. 331–338. doi:doi:10.1061/9780784413616.042.

Table 2. Representation of Common CLP Requirements

	Requirement	Function User.Space/ Function User.Parameter	Function Provider.Space/ Function User. Parameter	Necessary Failure Condition
1	Tower Crane Workspace must be within Site Boundary.	Crane.Workspace Envelope (A)	Site.Boundary (B)	$A \cap B$
2	Hazard Zone of Crane counter-jib must not interfere with Workspace of another Crane.	Crane1.Counter-jib Hazard (C)	Crane2.Workspace Envelope (D)	$C \cap D$
3	When Workspaces of two tower cranes interfere with each other, there must be a minimum 3m height separation between jibs.	Crane1.Jib Clearance (E) Crane1.Jib Clearance Value (G)	Crane2.Jib (F)	$(E \cap F)$ & $(G \geq 3m)$
4	There must be a height separation of at least 3m from the building roof to the lowest point of the hook on the crane.	Crane1.Jib Clearance (H) Crane1.Jib Clearance Value (J)	Building.Structure (I)	$(H \cap I)$ & $(J \geq 3m)$
5	Height Limit of Tower Crane is not to exceed Site Limit.	Crane.Mast Height (K) Crane.Head Height (M)	Site.Limit (L)	$K + M \leq L$
6	Crane Coverage should include all relevant elements	Crane.Workspace Envelope (O)	Building.Structure (N)	$(O \cap N)^C \neq \emptyset$
7	Building Element Weight must not exceed Crane Load Capacity.	Crane.Workspace Envelope.Load Capacity (Q)	Building.Element.Weight (P)	$P \leq Q$

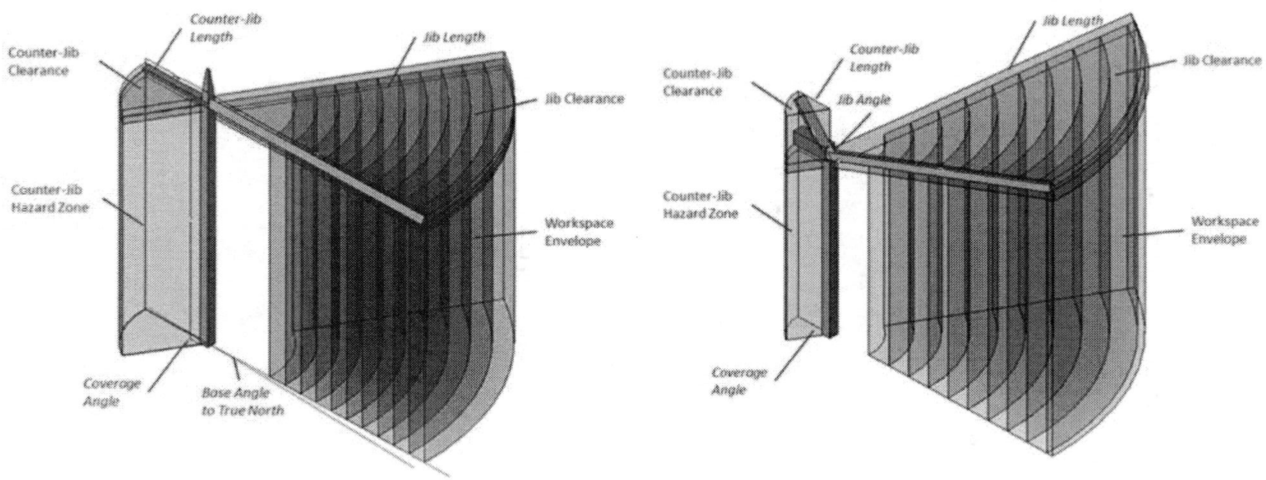

(a) Tower Crane (b) Luffing Jib Tower Crane

Figure 9. Parameters and Spaces in Tower Crane and Luffing Jib Tower Crane CIM

Handheld Simultaneous Localization and Mapping Devices for 3D Scanning

M. Froehlich [a] and S. Azhar [a]

[a] McWhorter School of Building Science, College of Architecture, Design and Construction, Auburn University
E-mail: maf0032@auburn.edu, salman@auburn.edu

Abstract – 3D Scanning has allowed for the scanning of spaces serving as a valuable tool for verifying existing conditions. The cost, setup and processing times associated with procuring this activity limits its use within an active construction site. With setup consisting of multiple scans from triangulated views of one another, achieving an accurate scan can be troublesome. Recent developments of Simultaneous Localization and Mapping (SLAM) technologies have enable indoor tracking of spaces through an array of sensors. The Google® Tango Tablet incorporates SLAM technology within a handheld device. Paired with the tablets depth sensors and software, 3D scanning can be achieved quickly and at substantially lower costs. This paper presents findings of a research study which investigated the 3D scanning applications of Handheld SLAM devices. This research helps us to explore an emerging technology that could allow for significant costs and time saving for the construction industry. This research will be a great aid in determining the value and best applications of handheld SLAM devices.

Keywords –
3D Scanning Technologies; IR Depth Sensors; SLAM;

1 Introduction

There are increasingly a large number of 3D scanners available for many different markets and applications covering, Art, Reverse Engineering, Automotive, Archaeology, Product Design, Architecture, Fashion Design, Body Measurement, BIM, Civil Engineering, Metrology, Forensics, and National Security. There are limitations, advantages and disadvantages to each scanner such as resolution, scan area, mobility, data capturing speed, tolerances [1].

Over the last twenty years, laser-scanning technology has seen a significant shift in applications across multiple disciplines [2]. The use of scanning is growing in this area thanks to steadily declining services costs and better informed clients. Although this area has historically not been quite as lucrative as the civil, infrastructure, industrial plant, and facilities applications, there are numerous successful projects and service providers in this space today [3].

In a case study by C. Clark and J. Liu conducted field testing for drywall installation using laser scanning and BIM. They determined that field scanning requirements for simple rooms such as offices are minimal requiring only one medium resolution scan with image capture for reference; a twenty-minute investment for each office. The accuracy proved that the integration of laser scanning and BIM technology can be used in a specific area without having to make any field measurements. The study additionally concluded that on most construction projects, the window for conducting scans is relatively short. Suggesting a faster process for scanning of an area to further be researched [2].

Becerik-Gerber, et. al. performed research on data acquisition errors caused by target setup, acquisition, and reorientation. They explored how different target types and target layouts affect registration accuracy[4]

Additionally, 3D laser scanning requires stillness in the scene that is captured. Thus, ongoing production activities can prove troublesome. If the geometries in the scene are geometrically complex and numerous, more data captures are required to get comprehensive spatial data of the entire scene [5].

The progression of low cost technology continues with the recent development of the Google® Tango Tablet. Combining Infrared (IR) depth sensors and cameras paired with Simultaneous Localization And Mapping (SLAM) technology allows for scanning on the jobsite quickly with minimal training and prep time by any construction personnel or trade.

The purpose of this study is to test the feasibility of a 3D scanning tablets and their applications for use in construction practice. The research will focus on prep time, scanning time, and accuracy of the scan.

2 Literature Review

Despite recent technological advances, developing a functional mobile AR system to be used in an unprepared environment is a considerable challenge. Solutions require high-accuracy data and precise positioning of the user and their relation to the hardware to ensure correct registration between reality and the data to be augmented [6]. Accurate registration and positioning of virtual objects in the real environment require accuracy in tracking the user's head as well as sensing the locations of real objects in the environment. The biggest single obstacle to building effective AR systems is the requirement of accurate, long-range sensors and trackers [7]. A common approach to indoor localization is to make use of WiFi or Bluetooth sources in the environment [8] [9] [10]. These techniques tend to be sensitive to the WiFi source placement, the accuracy of the signal strength map and the amount of clutter and dynamics in the environment. Additionally, the number of WiFi sources in the environment plays an important role [11]. Zollmann finds tracking currently consist of sensor-based registration techniques that incorporate inertial sensors, camera and Global Positioning System (GPS), and remote localization techniques that only make use of a camera, but depend on a 3-D point cloud [12].

A different approach, widely used in robotics, provides an alternative. So-called simultaneous localization and mapping (SLAM) methods build and update a map of an unknown environment while the same time keeping track of the tracking device's location within the environment (the learned map). SLAM methods can work with various sensors but often optical sensors like cameras, depth cameras or laser range scanners are used. Therefore, SLAM methods combined with mobile hardware, for sensing and mapping the 3D structure of an environment, allow for self-contained tracking of a user's viewpoint [13].

Determining a camera's position and orientation is a key requirement in augmented reality applications, a crucial step in simultaneous localization and mapping (SLAM) algorithms, and overall a major area of focus in computer vision [14][15][16][17]. Absolute pose methods utilize 2D-3D correspondences between image pixels and 3D points in a known scene to determine the full 6-degree-of freedom (d.o.f.) camera pose. These methods are often more efficient than relative pose methods and have been demonstrated to improve accuracy and reduce camera pose jitter in SLAM[18].

The Google® Tango tablet features a color camera and a motion and depth sensing camera. The device's fisheye camera a large field-of-view is observed in each im- age, which significantly simplifies and speeds up the capturing of larger scenes compared to standard lenses [19]. It includes software for SLAM which builds and/or updates a 3D map of the environment. The resulting map is essentially a mesh which can be stored for later use (i.e. later tracking in a known environment) or shared with other users [13].

2.1 Google® Tango Tablet

The recent developments in mobile technology allow advanced interfaces such as AR to be moved from the laboratory into the field. Their ability to present information on site, where it is needed, holds many advantages for professional application [12]. The "Project Tango®" prototype is an Android smartphone-like device which tracks the 3D motion of the device, and creates a 3D model of the environment around it. The Tango® Tablet was developed around, motion tracking, depth perception, and area learning. Project Tango® devices contain customized hardware and software designed to track the full 3D motion of the device, while simultaneously creating a map of the environment. These sensors allow the device to make over a quarter million 3D measurements every second, updating its position and orientation in real-time, combining that data into a single 3D model of the space around you [20].

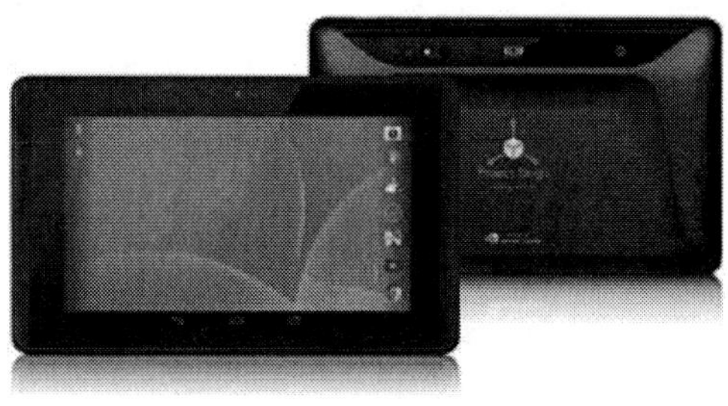

Figure 1. Google® Tango Tablet

Table Google® Tango Tablet

Specification	
Screen	7.02" 1920x1200 HD IPS display (323 ppi) Scratch-resistant Corning glass
Dimensions	119.77 x 196.33 x 15.36mm
Weight	0.82 lbs (370g)
Cameras	4 MP 2mm RGB-IR pixel sensor 1 MP front facing, fixed focus
OS	Android 4.4 KitKat
Wireless	4G LTE

	Dual-band Wi-Fi (2.4GHz/5GHz) WiFi 802.11 a/b/g/n NFC(reader and peer to peer modes)
Audio Output	Dual stereo speakers 3.5mm audio connector (OMTP standard)
Memory	128 GB Storage 4 GB RAM
Ports	Micro HDMI USB 3.0 Micro SD card Nano SIM slot
Processor	NVIDIA® Tegra K1 w/ 192 CUDA cores
Sensors	Motion tracking camera 3D depth sensing Accelerometer Ambient Light Barometer Compass GPS Gyroscope

2.2 Limitations

The tango tablet alike 3D laser scanners are 3D laser scanners are sensitive to the laser light reflected from the scanned object surface. When scanning reflective surfaces with complex geometry, the scanners may record extensive outliers due to undesirable specular reflections. It has been reported that outliers caused by specular reflections can be formed according to the developed mixed reflection and multi-path reflection models [21].

Additionally, the hardware of focus in this research, the Google® Tango Tablet at the current time of this paper is still in the development phase.

3 Research Methodology

The aim of this paper is to determine the feasibility of IR Depth Sensor with SLAM tablets for area scanning. The process of laser scanning requires significant planning and time, of which is not conducive of many construction project schedules. The release of the Google® Tango Tablet has the potential to allow for a faster process for scanning of areas. Preliminary testing of the tablet and its accuracy is carried out though scanning as-built conditions. The as-built area scanned, a small conference room in the Miller Gorrie Center on the campus of Auburn University was chosen. The produced scans were then compared to the as-built drawings and also the as-built Autodesk® Revit model. Time to prep, conduct the scan, and time to process the scan were additionally assessed during this preliminary study.

4 Details of Study Extent

The SLAM technology of the Google® Tango Tablet, and its ability to compute its location, independently, without the the aid of GPS required minimal prep time. Additionally, there is no need for markers to be placed as required by laser scanning. Unlike a fixed scanner, i.e. laser scanners, the Google® Tango Tablet utilizes built-in motion tracking. Of which allows for the user to sweep the area from different angles scanning untethered within a space.

The Google® Tango Constructor application developed for the tablet was used to conduct the scans. Upon starting the tablet calibrates itself requiring the user to extend the tablet out in front of then in a steady position for a few seconds. Once the tablet is calibrated the user has full range to scan spaces within 0.5 to 4 meters (1.5 to 13 feet) from the tablets sensors. Currently the tablet is designed to work best indoors at moderate distances [8]. The application supports functions allowing the user to pause, resume, or reset a scan in progress. This allows scanning to stop or pause and resume scanning of the same area. Additional options allow for the scan to be saved in the native file type for the application, as will as an .OBJ file time.

The results of a scan produce a 3D mesh which can be saved in the .OBJ file format and transferred to a computer for further processing. The file was then imported into Autodesk® Memento Software and exported as an Autodesk® Recap Point Cloud File. The converted scan was then imported into Autodesk® Revit as a point cloud file. Providing the same functionality as laser scanned produced point cloud.

5 Results

During the duration of this research multiple operating (OS) system and application updates were released. With each iteration, both the OS and application significantly improved in functionality as well as overall stability. Scans were repeated to obtain uniformity of the produced 3D models and point clouds. The multiple scans allowed for comparison and analysis in terms of accuracy, preparation time, scan time, processing time, and usability of the produced 3D mesh with regards to its applications.

5.1 Conference Room Scan

The size of the chosen conference room was approximately 13ft x 18ft. With a conference table and chairs centrally located in the room. To test the accuracy of the produced 3D models from the scan, two starting points, point A and point B were determined in the conference room, as marked in figure 2. The scanning path followed the perimeter of the room, clockwise from each starting point.

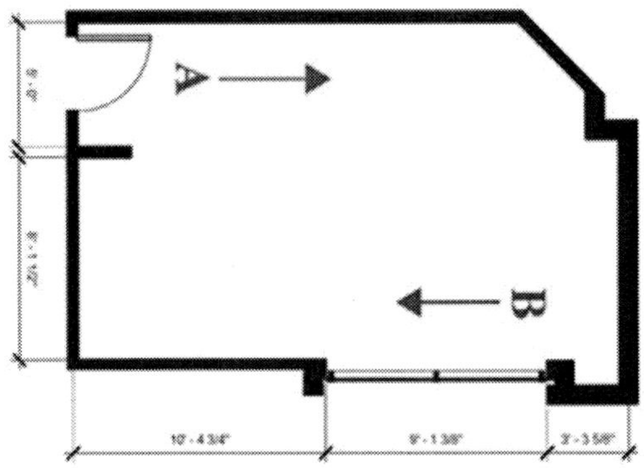

Figure 2: Starting Points of Scan

A total of ten scans were conducted, five scans with a start at point A, and five scans with a start at point B. A live view of the 3D Mesh can be seen on the tablet screen as it scans the area. The view shows the mesh model simultaneously being constructed refining itself as the the tablet sweeps the area. The user has the following options for viewing on the screen: first person, third person or top view. First person viewing was chosen as it displays in scanning in real-time as the user moves the tablet over areas to be scanned.

The multiple scans of the small conference room were able to be completed in a relatively short amount of time. The times of all ten scans were completed within the time range of 5-8 minutes. The scanning times and results are dictated by the user and their movements through the space. Figure 3 illustrates the results of a scan initiated at starting point A, scanning clockwise around the room. Figure 4 illustrates the results of a scan initiated at the starting point B, also scanning clockwise around the room. Figure 5 illustrates the results of a scan also initiated at starting point B.

The scans were conducted in the order as mentioned. There were mixed results produced by this initial test. The 3D models and point clouds varied in quality as well as in accuracy, with similarities shared by scans of the same starting points. Additionally, quality improved as the scans progressed, likely the result of the user's level of comfort in handling the tablet during scanning.

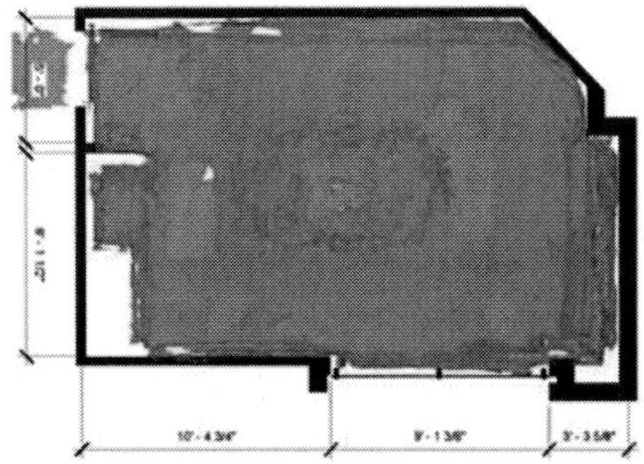

Figure 3: Scan Result Example 1

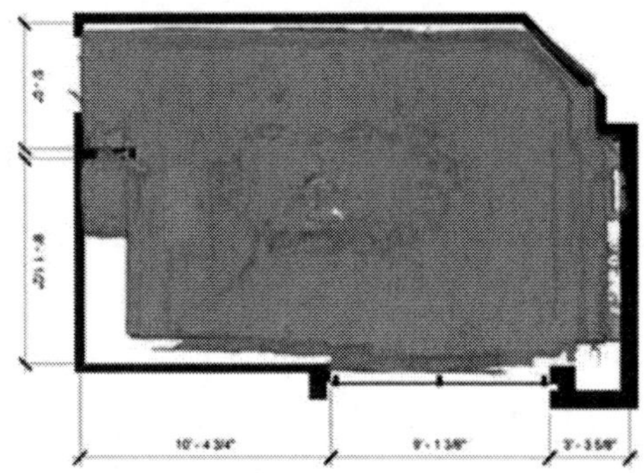

Figure 4: Scan Result Example 2

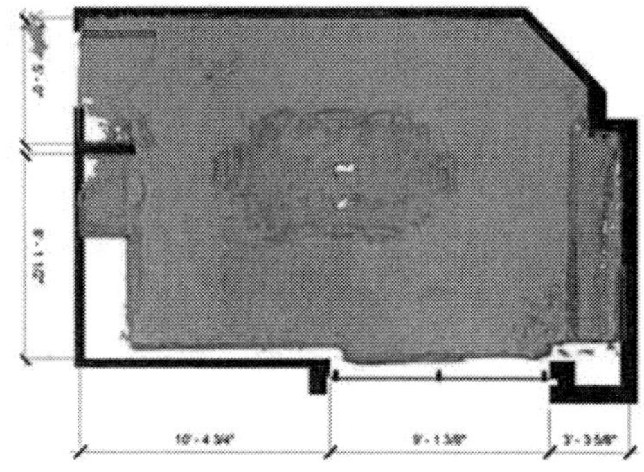

Figure 5: Scan Result Example 3

All the scans initiating at starting point A and starting point B exhibited improved quality and accuracy within the initial trajectory line among commencing from the starting points. Scanning accuracy of the overall model deteriorated with time and its movement along the path. This was due to drifting of the tablet and its ability to maintain its known location in the space. The portions affected by the drift maintain a greater level accuracy based of its new coordinate location within the space.

5.2 Mechanical Room Scan

Next the mechanical room within the Gorrie Center was chosen for scanning. The room is filled with mechanical equipment, piping, and ducts. The small space would be a challenge for laser scanning, and its multiple views required to develop a 3D scan of the space. Figure 6 shows a point cloud of the Mechanical Room produced from the tablet. Figures 6-8 illustrate sections through the room.

The scan of the Mechanical Room gave the tablet more surfaces and objects to reference its to. Overall there was only minor drifting of which occurred along the bottom wall in figure 7. The location of the drifting was poorly lit as well as having only flat surfaces. Both of which are the likely cause of the drifting. The scan the room was completed in 19 minutes. The complexity and tight spaces of the room required additional attention in order to get a usable model.

The existing Revit Model did not include the modelling of the pumps, pipes, ducts and equipment. The point cloud produced from the tablets scan shows the location of those items. The location of these items within the room can easily be modelled from the point cloud if needed.

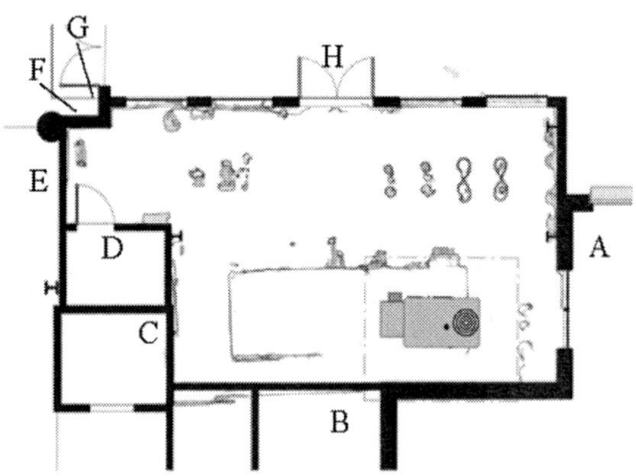

Figure 6: Mechanical Room Plan

Table 1. Mechanical Room scan analyses of As-Built Dimensions and Tablet scanned Dimensions

Wall	As Built Model (Meters)	Tablet Scan Range (Meters)
A	6.28	6.07-6.49
B	9.22	8.79-9.65
C	3.62	3.43-3.81
D	2.5	2.11-2.89
E	2.22	1.29-3.15
F	1.04	1.01-1.07
G	0.46	0.41-0.51
H	10.68	10.46-10.9

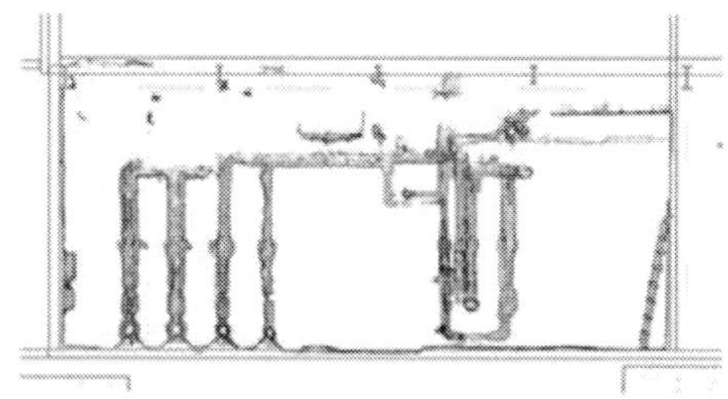

Figure 7: Mechanical Room Section A

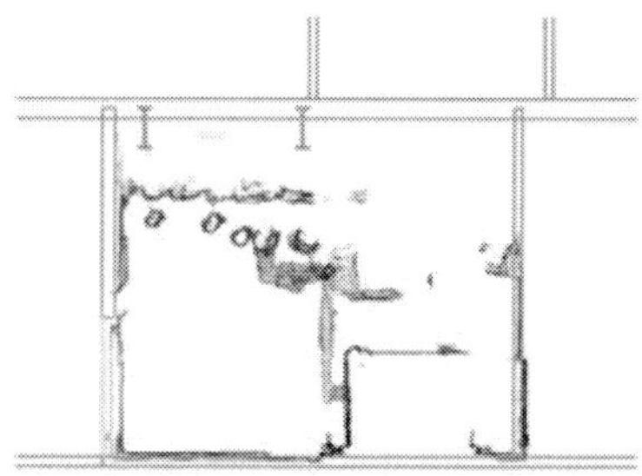

Figure 8: Mechanical Room Section B

Figure 9: Mechanical Room Point Cloud

Figure 10: Mechanical Room Point Cloud

5.3 Large Space Scans

Expanding the scope of areas scanned, an office and connecting hallway were scanned on the second floor of the Gorrie Center. The initial start location is marked by the red dot in Figure 10. The starting office was full of desk, shelves, and equipment, providing ample surfaces for the tablet to reference itself to. Little to no drifting of the scan occurred initially within this first space. Shown in purple of figure 10 is the point cloud aligned to this first space.

As the scan continued into the hallway, the first major drift occurred as seen, in Figure 10. Shown in green is the point cloud aligned to the tablets new coordinate system. As the scan continued down the hallway a third drift of the tablets location occurred, shown in orange, the point cloud is aligned to its location as well. There was a significant amount of drift between the initial point cloud and the drifted aligned point clouds. While there was a significant drift from beginning to end, each segment produced usable scan information. Consistency (the true position is always located inside the estimated uncertainty) cannot be guaranteed in such cases [22]. The longer the trajectory is, the more important the drift becomes [23].

It can also be determined the greater amount of surfaces in a space to be scanned has greater potential in producing a more accurate model or point cloud.

Figure 11: Overlay Illustrating Drifting producing any scans.

6 Discussion and Recommendations

6.1 Area Learning

The Tablet has a feature "Area Learning" in which the tablet familiarizes itself or "localizes" know locations within a space. Every time the tablet scans the area it registers any new points and modifies the "learned area" of that space. At the time of this research, our scans utilizing the constructor app we not able to benefit from this feature. This would significantly benefit and aid to the accuracy of indoor location tracking, as well as producing any scans.

6.2 Integrity of Areas to be scanned

This research has determined there to be a wide range

of factors that influence the integrity of a produced point cloud. For any area in order to be scanned must have surfaces capable of being registered by the tablets sensors. Translucent or Glass material cannot be read by the tablet, as well as shiny or surfaces with a gloss finish. The tablets sensors aren't able to get an accurate depth reading due to the scattering of the transmitted laser and IR signals.

Lighting is also essential to the scan and its ability to scan an area. Too much light, or not enough light caused drifts in the tablets location within the space as well as holes in the produced model or point cloud. Direct sunlight on surfaces, specifically cause holes, as they are not able to be scanned, similar to glass or windows. As this research was conducted indoors in an as-built environment, lighting was detrimental to the overall quality of a scan.

The user's method of scanning a space, in regards to how they handle the tablet, their speed through the space, and ability to thoroughly scan around the space from multiple perspectives. Further research to determine the best practices for scanning using the tablet could be beneficial.

Additionally, through this research it can be determined that the quantity and close proximity of surfaces to one another can produce a more accurate model. The more surfaces the tablet is able to identify and register, the more accurate it can identify its own location within an area to be scanned. While cluttered and/or narrow environments provide many optical features and simplify the use of SLAM methods, they reduce the free walking experience as the resulting resets would disturb the user experience [13].

The accuracy of the scans completed in this research, those of the conference room and of the mechanical room produced different levels of accuracy. With the mechanical room have more surfaces though out the space, the tablet was able to consistently track its own location. Within the mechanical room scan the area of drift occurred in an area consisting of flat surfaces (wall and air handing unit). There were no distinguishable surfaces of which could be registered by the tablet.

6.3 Conclusions

This paper presented results of the use of handheld SLAM for 3D scanning. The results consisted of an array of as-built conditions. The scans (point clouds) were analysed and compared against as-built Building Information Models to test accuracy. Dimensions were analysed in comparison to the tablet scans. Multiple scans were completed giving a range of accuracy as shown in Table 1. While these scans are not of the same quality or high accuracy of terrestrial laser scans, they can serve as a beneficial aid during construction. Time and skill required to complete scans on site are minimal and can serve as a beneficial tool. Additionally, there are significant costs and time savings in comparison to the cost of traditional fixed laser scanners.

Future Research

Incorporating SLAM technologies into user friendly devices has a wide range of applications as well as aid in higher levels of communication amongst those involved in building construction. SLAM allows for a device to know where it is within a space, untethered, deriving location from a array of sensors. The Google® Tango Tablet while still in development shows great promise. This research tested the applications of area scanning for use in the construction industry and has determined the tablet has the great potential for fast area scanning. While scanning of large spaces isn't nearly as accurate as scanning smaller ones, they both offer advantages.

Future research will build upon this study, continuing research with the Google® Tango Tablet, as well as the Microsoft® HoloLens. The HoloLens is a mixed reality headset of which utilizes SLAM technology to overlay information within a space. The headset will release the user of handling a tablet, allowing for a hands-free experience. Like the tablet the headset scans the space around the user, continuously scanning as the user moves around. SLAM technology is an innovative approach that has yet to be fully realized.

References

[1] 3D scanners comparison. Aniwaa 2016. http://www.aniwaa.com/comparison/3d-scanners/ (accessed April 3, 2016).

[2] Caitlin T. Clark, Junshan Liu. A Case Study of Integrating 3D High Definition Laser Scanning Technology and BIM into Drywall Installation. 50th ASC Annual International Conference Proceedings, 2014.

[3] Geoff Jacobs. Uses in Building and Architectural Surveys. Professional Surveyor Magazine 2005.

[4] Becerik-Gerber B, Jazizadeh F, Kavulya G, Calis G. Assessment of target types and layouts in 3D laser scanning for registration accuracy. Automation in Construction 2011;20:649–58. doi:10.1016/j.autcon.2010.12.008.

[5] Humayun Kabir Biswasa D, Boschéa F, Suna M. Planning for Scanning Using Building Information Model: A Novel Approach with Occlusion Handling. Symposium on Automation and Robotics in Construction and Mining (ISARC 2015), vol. 15, 2015, p. 18.

[6] Schall G, Zollmann S, Reitmayr G. Smart Vidente: advances in mobile augmented reality for interactive visualization of underground infrastructure. Personal and Ubiquitous Computing 2013;17:1533–49.

[7] Wang X, Kim MJ, Love PED, Kang S-C. Augmented Reality in built environment: Classification and implications for future research. Automation in Construction 2013;32:1–13. doi:10.1016/j.autcon.2012.11.021.

[8] Biswas J, Veloso MM. Wifi localization and navigation for autonomous indoor mobile robots 2010.

[9] Chintalapudi K, Padmanabha Iyer A, Padmanabhan VN. Indoor localization without the pain. Proceedings of the sixteenth annual international conference on Mobile computing and networking, ACM; 2010, p. 173–84.

[10] Thrun S, Burgard W, Fox D. Probabilistic robotics. Cambridge, Mass: MIT Press; 2005.

[11] Winterhalter W, Fleckenstein F, Steder B, Spinello L, Burgard W. Accurate indoor localization for RGB-D smartphones and tablets given 2D floor plans. Intelligent Robots and Systems (IROS), 2015 IEEE/RSJ International Conference on, IEEE; 2015, p. 3138–43.

[12]S. Zollmann, C. Hoppe, S. Kluckner, C. Poglitsch, H. Bischof, and G. Reitmayr *Augmented Reality for Construction Site Monitoring and Documentation* Proceedings of the IEEE, vol. 102, no. 2, pp. 137–154, Feb. 2014.

[13] Nescher T, Zank M, Kunz A. Simultaneous Mapping and Redirected Walking for ad hoc Free Walking in Virtual Environments, IEEE Virtual Reality Conference 2016, USA; 2016.

[14] Sweeney C, Flynn J, Nuernberger B, Turk M, Hollerer T. Efficient Computation of Absolute Pose for Gravity-Aware Augmented Reality, IEEE; 2015, p. 19–24. doi:10.1109/ISMAR.2015.20.

[15] Landy MS, Movshon JA, editors. Computational models of visual processing. Cambridge, Mass: MIT Press; 1991.

[16] Maybank S. Theory of reconstruction from image motion. Berlin: Springer; 1993.

[17] Kneip L, Scaramuzza D, Siegwart R. A novel parametrization of the perspective-three-point problem for a direct computation of absolute camera position and orientation. Computer Vision and Pattern Recognition (CVPR), 2011 IEEE Conference on, IEEE; 2011, p. 2969–76.

[18] Gauglitz S, Sweeney C, Ventura J, Turk M, Hollerer T. Model Estimation and Selection towardsUnconstrained Real-Time Tracking and Mapping. IEEE Transactions on Visualization and Computer Graphics 2014;20:825–38. doi:10.1109/TVCG.2013.243.

[19] Schops T, Sattler T, Hane C, Pollefeys M. 3D Modeling on the Go: Interactive 3D Reconstruction of Large-Scale Scenes on Mobile Devices, IEEE; 2015, p. 291–9. doi:10.1109/3DV.2015.40.

[20] Google. Project Tango – Google 2015. https://www.google.com/atap/project-tango/hardware/ (accessed September 18, 2015).

[21] Wang Y, Feng H-Y. Modeling outlier formation in scanning reflective surfaces using a laser stripe scanner. Measurement 2014;57:108–21. doi:10.1016/j.measurement.2014.08.010.

[22] Julier SJ, Uhlmann JK. Building a million beacon map. In: McKee GT, Schenker PS, editors., 2001, p. 10–21. doi:10.1117/12.444158.

[23] Bresson G, Aufrère R, Chapuis R. Making visual SLAM consistent with geo-referenced landmarks. Intelligent Vehicles Symposium (IV), 2013 IEEE, IEEE; 2013, p. 553–8.

The Study on the Integrated Control System for Curtain Wall Building Cleaning Robot

M.S. Gil[a], S.H. Kim[b], Y.S. Lee[b], S.H. Lee[c] and C.S. Han[d]

[a]Dept. of Mechatronics, Eng., Hanyang University, Republic of Korea
[b]Dept. of Mechanical, Eng., Hanyang University, Republic of Korea
[c]Dept. of Advanced Mechanical Eng., Eng., Korea Polytechnic University, Republic of Korea
[d]Dept. of Robot, Eng., Hanyang University, Republic of Korea

E-mail: msgill798@gmail.com, shkim8313@gmail.com, pacheon76@gmail.com, leesh@kpu.ac.kr, cshan@hanyang.ac.kr

Abstract –

The industry's interest in building façade maintenance is rising with the growing number of new high-rise buildings in the metropolises. The conventional method of cleaning the façade of high-rise buildings relies on the use of ropes, gondolas, and the winch system involving human labor. Recently the building maintenance unit (BMU) has been developed and has been applied to building maintenance to prevent safety accidents and to ensure high work efficiency. Diverse robot systems are being applied to building maintenance globally, including in Germany, the United States, and France. In South Korea, the built-in guided robot system was developed. This paper deals with an integrated control system of built-in guided robot system to ensure consistent and stable control. This control system is considered its unique characteristics as a building façade cleaning work. The integrated control system proposed in this paper performs cleaning work in three steps: preparation, cleaning, and return to the initial position, with each module consisting of the robot system performing its task sequentially and independently at each step. In addition, the robot system makes up a network composed mainly of VMR to ensure smooth and stable communication among the different modules. The integrated control system proposed in this paper was applied to the built-in guided robot system for performance verification.

Keywords –

Curtain wall building cleaning robot; Integrated control system; Sensor based control; Built-in guide rail

1 Introduction

As the population density has increased and the function of the city has become more complex, the number of high-rise buildings has been increasing. As the façade of such high-rise buildings can be used for advertising and PR purposes, they are usually designed to embody diverse and unique characteristics in terms of their structure. Moreover, the interest in building façade maintenance as opposed to building remodeling is growing as an effective means to extend the life cycle of buildings. The conventional building facade maintenance methods for the cleaning, painting, and crack inspection of high-rise buildings mainly depend on ropes, gondolas, and the winch system. As these methods are highly dependent on manpower, in contrast to the other construction processes that are being automated, their use also incurs high socioeconomic losses due to accidents. The demand for automated maintenance work for high-rise buildings performed in a dangerous environment is gradually increasing [1, 2, 3].

To address the aforementioned demand, many studies on the automation of the building maintenance unit (BMU) and the related robots have been conducted in Europe, Japan, and the U.S [4].

For example, many research institutes and companies, including Fraunhofer IPA and Manntech in Germany and Nihon Bisoh in Japan, introduced robot platforms in the test operation stage. In particular, Nihon Bisoh has already successfully applied robots to the cleaning work of high-rise building landmarks in Tokyo, Osaka, and Yokohama. SIRIUS-c, a BMU developed by Fraunhofer in Germany, is equipped with two sets of linear translation module in addition to the wire-based vertical mobility function along the building façade. The BMU is designed to perform uninterrupted cleaning while maintaining a constant distance from the building façade with the help of this module. Likewise, Skycleaner 3 in Germany is hooked up with the following unit mounted on top of the building via wires while cleaning the building, with the cleaning supplies transported from the ground level by a supporting vehicle [5].

Skycleaner 3 moves along the building façade using pneumatic cylinders as well as multiple suctions pads installed along the sides of the cylinders. Tito 500 developed by RATIOFOREM in the United States moves up and down the building façade along the vertical wires while cleaning the façade, without bumping onto the glass curtain walls, through the use of a wheel mechanism installed inside the robot, which functions as a damper [6]. RobuGlass, developed by Robosoft in France [7] and Gekko Façade, developed by Serbot in Switzerland [8] were designed to perform cleaning independently by sucking onto the building façade only with suction cups, without using wires.

In this paper, a built-in guided robot system mounted with cleaning supplies and that cleans the building façade by gliding along the rails installed around the building will be introduced, along with the integrated control system required to carry out uninterrupted and stable control by considering the robot's cleaning method and cleaning environment.

2 System Configuration and Work Environment

2.1 Work environment

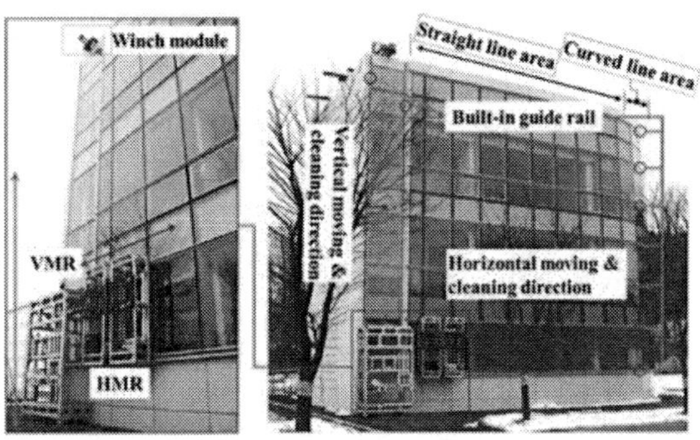

Figure 1. Work environment

The work environment applied to the built-in guided robot system described in this paper is shown in Figure 1. The building on which the robot was tested was a three-story building, with each floor surrounded by curtain walls, each measuring 1,800×1,320 and 900×1,320 mm. The façade of the building to which the built-in guided robot was applied was covered with two sets of rail, one running vertically and the other running horizontally, while the cleaning sections were divided into straight and curved sections depending on the shape of the building.

2.2 System configuration and roles

The built-in guided robot that performs cleaning in the work environment shown in Figure 1 is composed of an HMR (horizontal-moving robot), a VMR (vertical-moving robot), a winch module, and a monitoring module. The VMR performs cleaning while moving along the rail installed horizontally across the façade whereas the HMR plays the role of transporting the VMR to each floor while moving along the rail installed vertically along the façade, as well as of cleaning the façade. The winch module moves the HMR and VMR to their target floors, respectively, while the monitoring module sends the cleaning command, monitors the current position of the cleaning robot, and transmits the robot status information to the control staff and various other systems

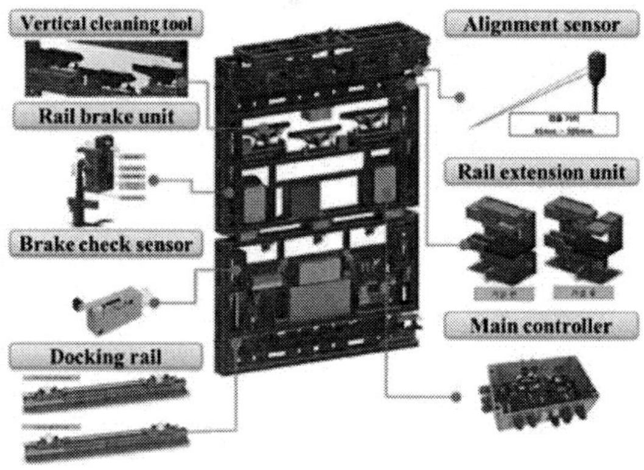

Figure 2. Configuration of VMR

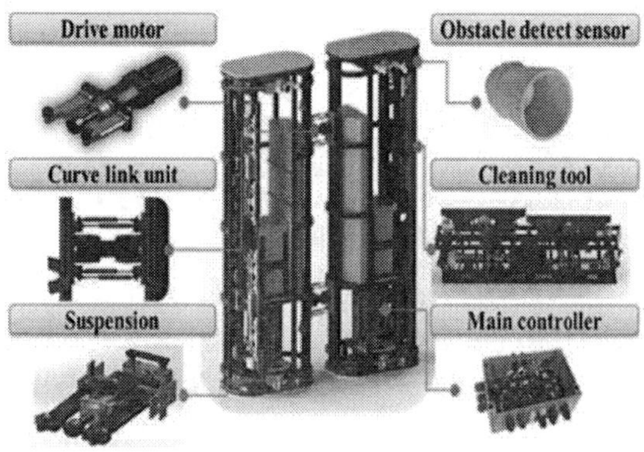

Figure 3. Configuration of HMR

The VMR and HMR, which play the key roles in the

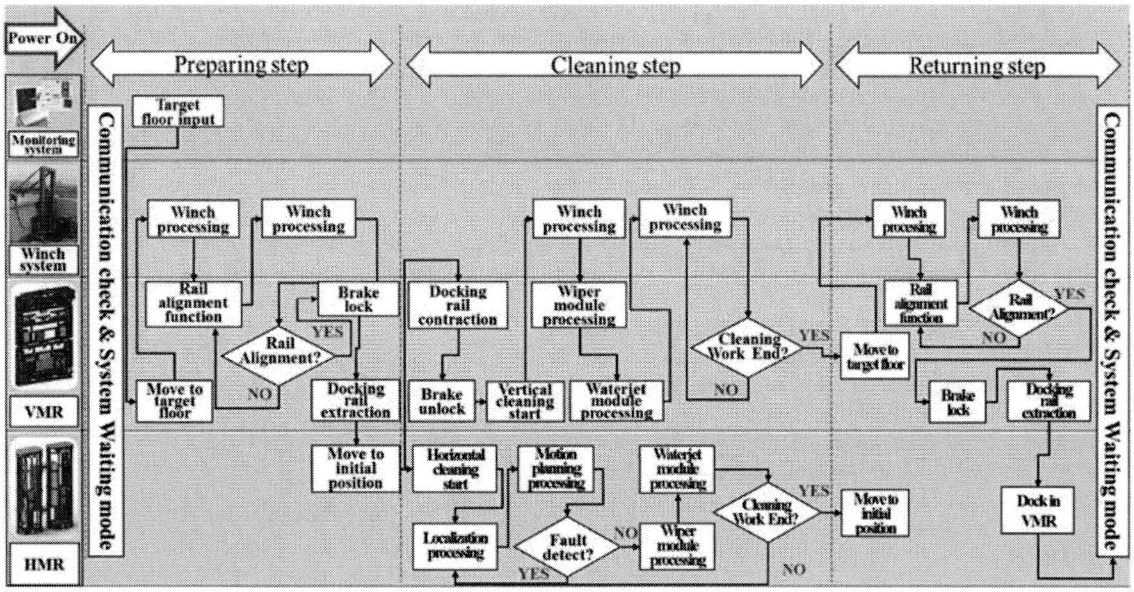

Figure 4. Integrated control system for built-in guided robot

built-in guided robot, are configured as shown in Figure 2 and 3. The VMR contains four brake units and brake check sensors designed to prevent safety accidents due to a shift in its weight, as well as a docking rail, which allows the HMR to dock in the rail running horizontally along the building. The two rail extension units connect the building and the docking rail while the vertical cleaning tool mounted on the VMR is used to clean the curtail walls covered with rails running vertically. The HMR is made up of two units connected with each other via a curb link unit so that it would be able to move along regardless of whether the section to be cleaned is a linear or curved section. Also included in the package is an obstacle-detecting sensor, which detects any obstacle along the path, such as a suspension or a window, to prevent damage due to vibration shock while moving along the horizontal rail along the building. A rail cleaning unit, a small compressor designed to remove the contaminants (fallen leaves, dust, bird nests) inside the horizontal rails, is also included.

3 Integrated control system architecture for the Built-in Guided robot

3.1 Integrated control system architecture

To clean a building, the monitoring system, winch system, VMR, and HMR perform cleaning by either transmitting or receiving the control command sequentially via communication between the modules, or in certain sections, each module performs the given task independently according to the specific control command prescribed for the section. Figure 4 shows the structure of the control system for the built-in guided robot system, including the control flowchart among the modules in accordance with the cleaning work sequence.

All the modules of the built-in guided robot remain in the standby mode after the power is turned on, until the communication check is completed. Upon receiving from the monitoring system (an interface for the initial operator) the command to start the work on the floor, the VMR then decides the target speed and direction of the winch by calculating them with the rail alignment function. Once the VMR and HMR arrive at the target floor, the VMR locks the brake and expands its rails. When the VMR executes the rail expansion command, the HRM moves to the horizontal rail along the building. The HMR's arrival at the initial position located on the horizontal rail of the building marks the completion of the preparation stage by both the VMR and HMR.

Unlike in the preparation stage, where the tasks are performed sequentially via inter-module communication, in the cleaning stage, the HMR and VMR perform their respective tasks independently. In the case of the VMR, the rails contract and the brakes are unlocked once the cleaning work begins. Then the VMR performs cleaning starting from the highest floor and moving all the way down to the lowest floor, by spraying a small amount of water onto the curtain walls and then wiping them dry. In the case of the HMR, it decides its target speed and direction by calculating them with the motion planning function by receiving feedback on the robot's current location based on the localization algorithm. As the HMR performs cleaning along the horizontal rail surrounding the building, it includes a fault detection function, which protects the robot from the windows or other elements.

Both the VMR and HMR embark on the return maneuver to their original locations once the cleaning is over. As in the preparation stage, the VMR locks up the brakes and expands the rails while the HMR moves back to its initial position and remains in the standby mode. Once the signal indicating that the VMR has completed the expansion of the rails is confirmed, the HMR embarks on the dock-in maneuver. Once the HMR completes the dock-in maneuver, the VRM contracts the rails and unlocks the brakes before returning to the standby mode. The built-in guided robot system allows the operator to check the status of each module by sending the status information to the monitoring system once every second throughout the three stages. In the operator mode, the aforementioned sequential commands can be transmitted by the operator in real time to ensure the smooth return of the robot to the initial location in the event of work failure.

3.2 Communication network

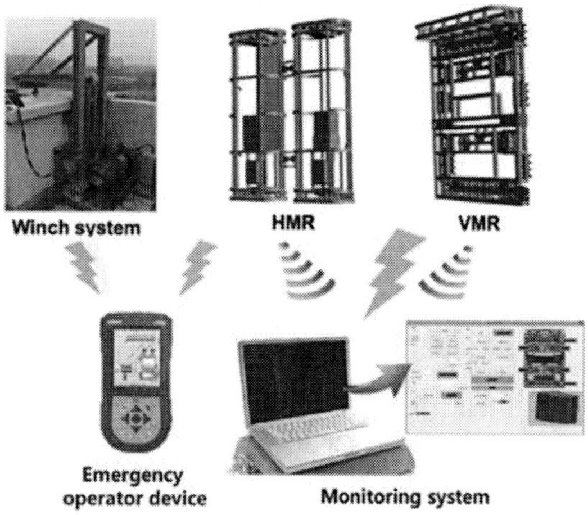

Figure 5. Communication network for robot system

The built-in guided robot system should be able to return to its original location safely in the event of system failure due to an external factor or a system error while performing cleaning work on the building façade. This is where the communication network functioning smoothly between the modules kicks in to improve the cleaning performance and to prevent a safety accident. The built-in guided robot system has a network configured as shown in Figure 5 below. In the built-in guided robot system, which performs the algorithm at regular intervals while sending and receiving various commands, the VMR transmits the control commands that it received from the monitoring system to each module as well as forwards the status information it received from each module and unit back to the monitoring system. Such network configuration can prevent the occurrence of a communication error during the carrying out of the sequential commands. In the event of communication failure between the VMR and the monitoring system during cleaning, the operator shall communicate with each module independently via RF communication, using the handheld operator system.

4 Integrated control system architecture for the Built-in Guided robot

4.1 Motion planning algorithm

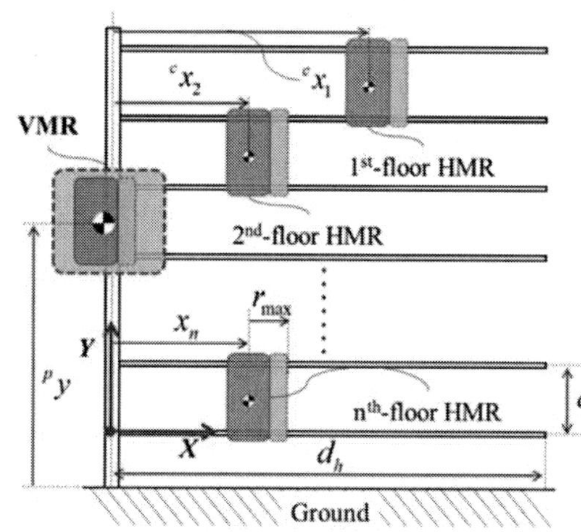

Figure 1. Schematic and coordinate system of the robot system

The coordinate system for indicating the locations of the VMR and HMR is shown in Figure 6. The position of the VMR movement along the vertical rail is Py, and the position from the vertical rail of the HMR along the ith horizontal rail is xi. The length of the horizontal rail is dh, and the vertical spacing between the horizontal rails is dv. The maximum velocity of the VMR is defined as vp_max, and it must always move at this speed during cleaning. In addition, the maximum and minimum velocities of the HMR are defined as vc_max and vc_min, respectively. The maximum value at no contamination is vc_max, and the minimum value at the maximum contamination is vc_min. The velocity of the HMR according to the contamination level between these two speeds has a linear relationship. Here, it is assumed that when the HMR moves at the changing velocity of the contamination level, sufficient cleaning is performed to remove the contaminants at the location. As the contamination level is associated with the distance of each horizontal rail, the value can be expressed as ci(xi). The integral value of the contamination level by the distance from the ith horizontal rail is expressed by si, as shown below.

$$s_i = \int_0^{d_h} c^i(x^i)dx^i \quad (1)$$

As $0 \leq c^i \leq 1$, $0 \leq s^i \leq d_h$ is true. Equation (1) considers the range of the time required for the HMR to perform cleaning according to the contamination level. If equation (1) has the minimum value, $s^i=0$, and the HMR always moves at the maximum speed because the contamination level cannot be measured. The time required for the HMR to return to its original location after cleaning is $2d_h/v_{c\_max}$. On the other hand, if equation (1) has the maximum value, $s^i=d_h$, and the contamination level of all the areas of the horizontal rail is the maximum value (1). Therefore, the range of time that it takes the HMR to perform cleaning on the i_{th} horizontal rail and to return to its original location (t_{i_HMR}) can be expressed as follows:

$$\frac{2d_h}{v_{c\_max}} \leq t_{i_{HMR}} \leq \frac{d_h}{v_{c\_min}} + \frac{d_h}{v_{c\_max}} \quad (2)$$

4.2 Rail alignment algorithm

For the VMR to be able to clean the areas between the floors, it should first move to the inter-floor rails that connect the floors. The rail alignment algorithm controls the precise location and speed of the HMR so that the VMR could dock into the horizontal rail surrounding the building. First, the two laser sensors (Sharp EX-L211) are mounted on the upper/lower half of the HMR's rail extension module, as shown in the following figure, to perform the algorithm and to detect the space and building wall inside the horizontal rail with ON/OFF signals. Assuming the DIO signals from the laser sensor mounted on the upper section as S1 and DIO signals from the sensor mounted on the lower section as S2, the rail alignment algorithm attempts to align the HMR rails with the building's horizontal rails while moving slowly from the lower floors to the upper floors to prevent failure due to the weights of the HMR and VMR. As mentioned earlier, the HMR moves from the lower floors to the upper floors, and if S1 and S2 detect the inner cavities of the horizontal rails of the building one by one while doing so, the HMR defines the rails as aligned. The rail alignment algorithm performs the aforementioned definition in accordance with the following five modes, depending on the locations of the building's horizontal rails and the HMR's rails. Mode 1 signifies an occasion

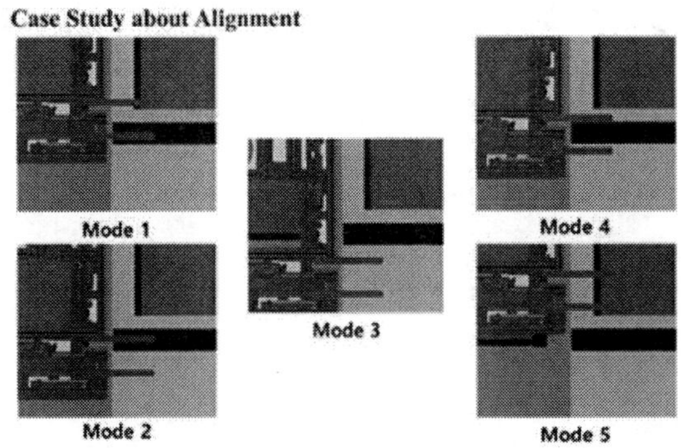

Figure 7. Rail-alignment algorithm

Figure 8 Ultrasonic sensor module for obstacle detecting

where S2 is ON and S1 is OFF, and the HMR glides down slowly until S2 is turned OFF. If S1 is turned ON, the HMR starts to glide up again slowly while the algorithm function stops when S2 is turned OFF. Mode 2 signifies an occasion where S1 is ON and S2 is OFF, and as in Mode 1, the HMR performs its algorithm while slowly gliding up the building. In Mode 3, 4, and 5, both S1 and S2 are OFF, and the process of detecting the rails is included in the algorithm. The survey process in the rail alignment algorithm signifies the process where S2 or S2 detects the inner cavities of the rails while the HMR is gliding up to the upper floors. Mode 3 signifies an algorithm where S1 detects the inner cavities of the building rails, which is followed by the performance of the task in Mode 2. In Mode 4, S2 detects the inner cavities of the building rails, which is followed by the performance of the task in Mode 1. In Mode 5, the robot glides up the building for 15 seconds considering its inter-floor distance and HMR's speed before it glides down again when the S1 and S2 sensors are OFF, and then performs the task in Mode 1.

4.3 Fault detection algorithm

An ultrasonic sensor was used to perceive an obstacle

like the opening of the external wall ventilation window. To prevent the building external wall from being perceived as an obstacle, a vibration- and waterproof ultrasonic wave sensor was used at a place with a narrow ultrasonic-wave directional angle and for industrial use (Figure 8). The control system transmits a trigger signal and sends an ultrasonic wave from the sensor behind 20 ms to transmit an analog signal by ultrasonic-wave strength. The control system was designed to judge a signal larger than a specific size as an obstacle, and to determine the location of an obstacle by the reflection time of the ultrasonic wave.

5 Conclusion

This paper dealt with an integrated controller for the built-in guided robot system, which is used to clean the exterior of curtain walls. The built-in guided robot system is designed to fit into the structure of modern high-rise buildings and consists of four modules: a monitoring system, a winch system, a vertical-moving robot (VMR), and a horizontal-moving robot (HMR). Each module of the built-in guided robot performs cleaning while communicating wirelessly with the other modules on the façade of the building. Therefore, each module of the robot system is designed to return to the ground floor safely in the event of a work or communication error. Therefore, the built-in guided robot system proposed in this paper is designed to perform the cleaning task uninterrupted and stably by enabling each module to be controlled sequentially and independently.

The integrated controller of the built-in guided robot system performs the cleaning task sequentially in three stages: preparation, cleaning, and return to the initial location. In the preparation stage, the integrated controller performs the task sequentially while sending and receiving control commands to/from the different modules. In the cleaning stage, the integrated controller lets the VMR and HMR, the actual devices that clean the building, work independently. In the final stage (return stage), the integrated controller waits for the additional input of the operator while performing sequentially. To communicate as many control commands smoothly, the built-in guided robot's network is configured to forward the control commands inputted by the initial operator to each module via the VMR.

In this paper, the performance of the integrated controller was verified by mounting a controller on a built-in guided robot and then deploying it for an actual building cleaning task. Additional onsite tests of the integrated controller with the built-in guided robot working on the façade of a building would be required for further modification and enhancement of the said controller.

Although the building on which the robot was tested was a three-story building, this robot system and algorithms are designed for high-rise building. But the additional test to safely operate in high-rise building will be progressed in future work.

Acknowledgment

This research was supported by the Industrial Strategic technology development program (No.10052965) funded by the Ministry of Trade, Industry and Energy (MOTIE), KOREA and was supported by a grant(14SCIP-B079344-01) from Smart Civil Infrastructure Research Program funded by Ministry of Land, Infrastructure and Transport(MOLIT) of Korea government and Korea Agency for Infrastructure Technology Advancement(KAIA).

References

[1] Horner R. M. W., El-Haram M. A., and Munns A.K. Building maintenance strategy : a new management approach. *Journal of Quality in Maintenance Engineering*, 3(4):273-280, 1997.

[2] Spedding A. Building maintenance economics and management, Routledge, 5 Howick Place, London. 1987. "

[3] Christer A. H. Modeling inspection policies for building maintenance. *The Journal of the Operational Research Society*, 33(8):723-732, 1982.

[4] Moon S. M., Huh J. M., Lee S. H., Kang S. P., Han C. S. and Hong D. H. A survey on Robot Systems for High-rise Building Wall Maintenance. *Journal of the Korean Society for precision Engineering*, 30(4):359-367, 2013.

[5] Zhang, H., Zhang, J., Liu, R., Wang, W., and Zong, G. Pneumatic Climbing Robots for Glass Wall Cleaning. *In Proceedings of the 7th International Conference on Climbing and Walking Robots*, pages 1061-1069, Madrid, Spain, 2004.

[6] Akinfiev, T., Armada, M., and Nabulsi, S. Climbing Cleaning Robot for Vertical Surfaces. *Industrial Robot: An International Journal*, 36(4):352-357, 2009.

[7] Robosoft robuGLASS. On-line: http://www.robosoft.com/robotic-solutions/cleanliness/glass-roof.html, Accessed: 05/01/2016.

[8] Serbot AG. Gekko Façade. On-line: http://www.serbot.ch/en/products/facade-cleaning/gekko-facade, Accessed: 07/01/2016.

Development of a Sensor System for Detecting Window Contamination for the Building Maintenance Robot System

Y.S. Lee[a], S.H. Kim[a], J.H. Lee[b], M.S. Kang[c], D.I.SUN[c] and C.S. HAN[d]

[a]Dept. of Mechanical, Eng., Hanyang University, Republic of Korea
[b]Dept. of Mechatronics, Eng., Hanyang University, Republic of Korea
[c]Dept. of Interdisciplinary Eng. Systems, Hanyang University, Republic of Korea
[d]Dept. of Robot, Eng., Hanyang University, Republic of Korea

E-mail: pacheon76@gmail.com, shkim83@hanyang.ac.kr, zjagnswkd@hanmail.net, wowmecha@gmail.com, jeniussdi@naver.com, cshan@hanyang.ac.kr

Abstract –

Because of the increasing number of high-rise buildings and large glass facades and the resulting problem of safe and effective maintenance, many studies have been going on progress to develop automated window cleaning systems in the last few years. But the research has not been going on progress to make decision the additional cleaning through the diagnosis of contaminant degree on outer window before or after cleaning. So, the sensor system is proposed to measure the contamination of building's outer-window in this paper. To check the cleaning performance of the window, it is mounted on a Building Maintenance Robot (BMR) which is automated window cleaning system. For detecting the contaminant level of outer window, the sensor module using the IR sensor is proposed. In order to apply the proposed sensor module to BMR system, the mechanism for the horizontal or vertical movement of the sensor module is designed. To improve the diagnostic accuracy of contaminant level, The Algorithm for position control using the proximity sensor is developed to keep a certain distance between sensor and window at the horizontal mechanism. A method detecting the contaminant level is verified by lab test. To validate the mechanism of the proposed sensor module, the field test is executed with the BMR system equipped with the sensor module.

Keywords –

ISARC Window Contamination; Building Maintenance Robot System; IR Sensor

1 Introduction

The number of new high-rises covered with curtain wall glasses is growing along with the improvements in construction technology, with a corresponding increase in the need for their maintenance, repair, and care [1].

The prevalent building maintenance, repair, and care practices largely rely on the conventional means for external wall care, such as the use of a rope, a gondola, and a winch, which are prone to causing safety accidents. As they are more dependent on human labor compared to other construction processes, the safety accident occurrence ratio is also high among them. Several studies are being conducted on ways of automating the process of providing curtain wall care, such as the building maintenance robot (BMR). Several research institutes and companies, including Nihon Bisoh in Japan and Fraunhofer IPA in Germany, are experimenting with the potential application of building maintenance robots for cleaning external wall glasses [2, 3].

Figure 1. State-of-the-art building façade maintenance System

Those robots, however, simply clean the glasses with a constant pace, without paying attention to the varying extent of contamination of the glass surfaces, which again makes it necessary for human staff to check the cleanness of the glass surfaces after the cleaning as the robots are unable to determine their status. Therefore, studies should be conducted on some sensors that could be used to measure the extent of contamination of a glass surface so as to automate the curtain wall glass cleaning system. In this paper, a module-type sensor system capable of measuring the contamination of a glass surface is proposed. The system uses IR sensors mounted on the curtain wall glass cleaning system to survey the extent of contamination on a glass surface.

2 Sensor Module for Detecting Window Contamination

An IR-sensor-based sensor module was developed to detect the contamination of a glass surface. The sensor module was assembled as shown in Figure 2, by combining the light emitter and receiver of the IR sensor. If the light emitter and receiver of the IR sensor are installed at a 45° angle, respectively, part of the light emitted from the light emitter will pass through the glasses, with the rest being reflected back to the light receiver in accordance with the principles of the angle of incidence and angle of reflection. The amount of reflected and inputted light will increase in clean and transparent glasses and will decrease in contaminated glasses, thereby enabling the determination of the extent of glass contamination.

For the IR sensor module developed by these researchers to accurately detect the contamination level, the sensor should be consistently kept at a regular distance from the glasses. If the IR sensor moves back farther from the glasses, both the angle of incidence and angle of reflection will change, thereby making it impossible to detect the contamination of the glasses. If the IR sensor moves too close to the glasses, the angle of incidence and angle of reflection will change again, thereby making it impossible to detect glass contamination or even causing it to bump onto the glasses.

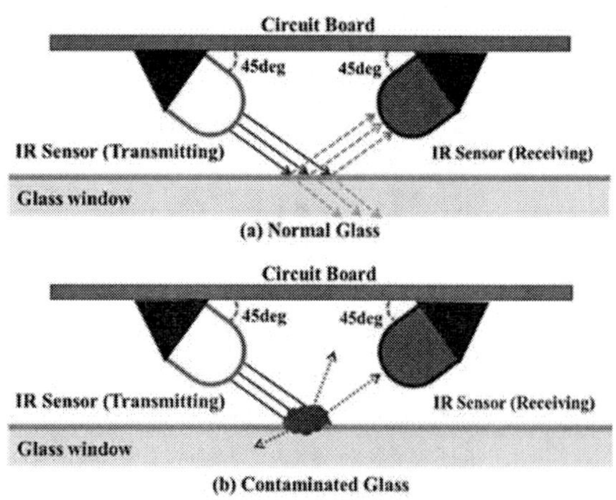

Figure 2. Concept of the IR sensor module

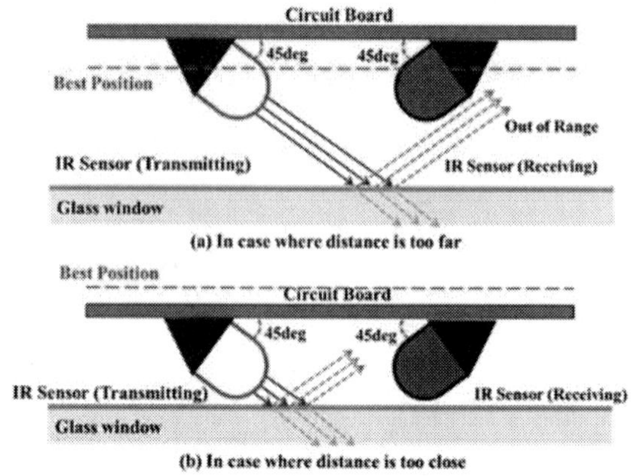

Figure 3. Importance of the distance between the IR sensor module and glass

Therefore, a sensor system with the capacity to maintain a constant distance between the sensor and glasses is required. Chapter 3.1 will introduce a mechanical design for such sensor system while Chapter 3.2 will introduce a control algorithm for the sensor system.

3 Configuration of a Mechanical Design and Control System for Detecting Window Contamination

3.1 Mechanical Design of the Sensor System

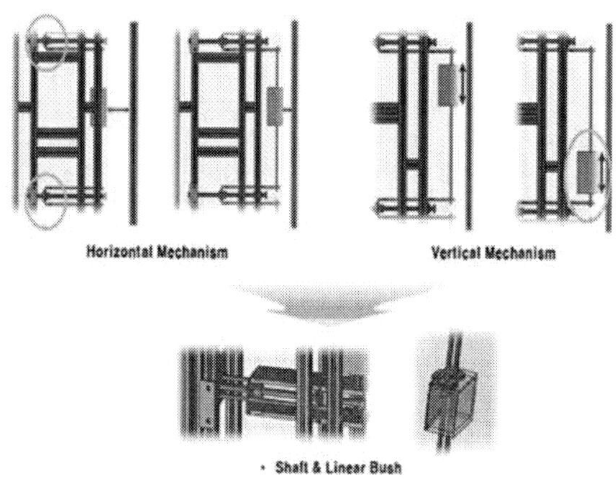

Figure 4. Mechanism design

As described in Chapter 2, a mechanism that maintains a constant distance between a glass surface and a sensor is indispensable for the sensor to possess the specific characteristics required for contamination detection and for improving the reliability and accuracy in the detection of glass contamination.

A guiderail was installed as shown in Figure 4 to enable the forward-backward horizontal movement of the sensor, followed by the installation of a resetting spring equipped with a fail-safe function that enables returning to the default position to prevent a clash between the sensor and the glass surface in case the driving motor is shut off.

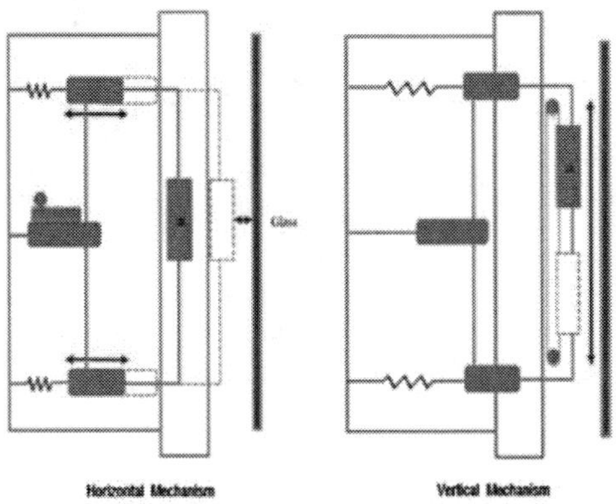

Figure 5. Mechanical design concept

A mechanical design concept equipped with a vertical up-down movement mechanism such as that found in the printer mechanism that was designed to detect the contamination of 1.8m-high large curtain walls using a certain number of IR sensors is proposed herein. Figure 5 shows the mechanical concept that realized vertical up-down movement by employing belts and pulleys.

The belt-and-pulley mechanism was adopted for the following reasons: (1) it is appropriate for a relatively lightweight and long-distance shuttle movement as the movement entails an up-down movement that is longer than the forward-backward movement; and (2) no positioning control is necessary because the sensor is located in the middle of the up-down movement path.

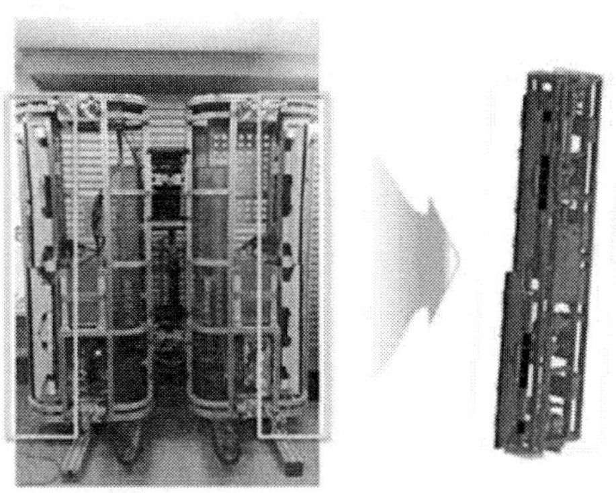

Figure 6. Detachable module-type cleaning tool

As shown in Figure 6, the proposed contamination detector sensor system was designed as a detachable module that can be mounted on the existing curtain wall cleaning robots. The existing curtain wall cleaning robots are designed to remove the contaminants on a glass surface using the cleaning tool mounted on the side opposite the robot's movement (cleaning tools are mounted on the left and right sides of the robot).

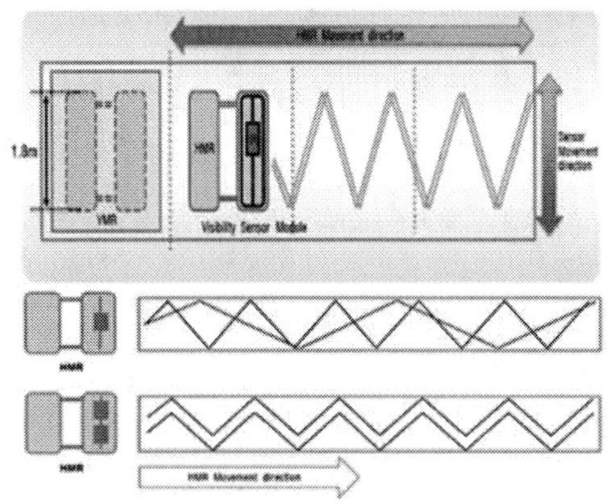

Figure 7. Contamination detection range depending on the number of sensor

The sensor system developed by these researchers was installed on the robot, replacing the robot's right-side cleaning tool. As shown in Figure 7, the sensor system surveys the extent of contamination on a glass surface while the robot installed with the contamination detection sensor moves rightwards, whereas the left-side cleaning tool surveys the cleaned state of the glass surface while the robot is already turning back to its previous position.

The survey range as well as the amount of attainable data from each survey on the contamination of the glass surface can be adjusted as shown in Figure 7, by changing the robot's horizontal movement speed, up-down shuttling speed, and number of sensors.

3.2 Control System and Algorithm for the Sensor System

The proposed system for detecting contamination of a glass surface is composed of eight limit switches, four proximity sensors, four driving motors, and an IR sensor module for controlling the position of the sensor module. Four vertically arranged limit switches and two driving motors control the vertical movement of the sensor module while four horizontally arranged limit switches, along with two driving motors and four proximity sensors, control the horizontal movement.

While the sensor module attached on the belt moves vertically, it measure the contamination on the glass surface at regular intervals. Two proximity sensors capable of detecting the presence of glass are used to enable the sensor module to close in on the glass surface at a regular distance. The proximity sensors with digital output can measure the distance to the glass surface by being mounted on the upper and lower parts of the module at different lengths, respectively. Moreover, the following location-based control algorithm is used to maintain the distance between the sensor module and the glass surface.

The target values in Figure 8 are generated via the following equation:

$$X_{des} = X_{pre} + A * Q_1 * Q_2 \quad (1)$$

where Xdes signifies the target coordinates on which the sensor module will be positioned, Xpre signifies the current coordinates of the sensor module, A signifies the control gain used to change the target coordinates, and Q1 and Q2 signify the output values of the proximity sensor. The initial value of Q1 and Q2 is 1. If proximity sensor 2 detects glass, the value of Q2 changes 1 to 0, and if two proximity sensor detect glass, the value of Q1 changes 1 to -1 and value of Q2 changes 0 to 1. The horizontal movement target value of the sensor module changes in accordance with the values of Q1 and Q2.

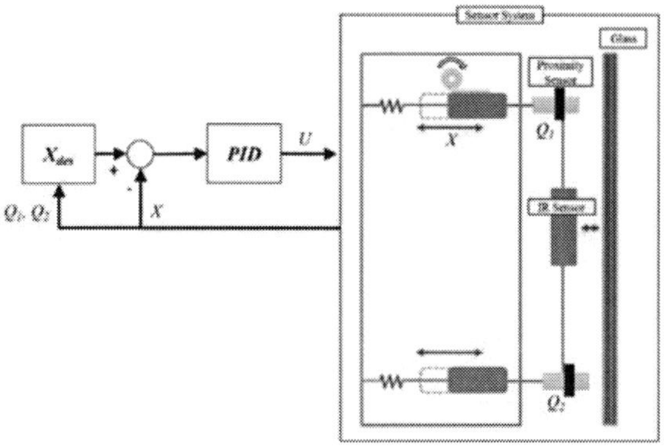

Figure 8. Configuration of the control algorithm

4 Experiment and Result

A field test was performed to drive the sensor system and to detect contamination in an attempt to verify the performance of the proposed system. The proposed system was mounted on a building maintenance robot (BMR), which is an automated window cleaning system.

Figure 9. Sensor system installed on a BMR for detecting glass contamination

The test environment was configured as shown in Figure 10. The experiment was performed in a test building wrapped with 1200mm-wide curtain wall glasses. The BMR robot cleaned the curtain walls twice before the experiment to maintain the contamination of the glass surface at the standard level. An area of the glass curtain wall some 1200 mm behind the initial location of the BMR robot was stained with contaminants before the BMR on which the proposed sensor system was mounted measured the degree of contamination of the glass. The measurement of the contamination was performed at every 3 mm interval since the sensor module embarked on a vertical movement. Over 14,000 contamination data points were measured in the end.

Figure 10. Configuration of the experiment environment

The contamination detection results on the upper and lower curtain wall glasses are shown in Figure 11 and 12, respectively, and a graph indicating the current position of the BMR robot that moved horizontally is shown in Figure 13. The contamination detection results on the upper curtain wall glasses did not show any perceptible change, but values lower than the average output values were observed within the space between 6000 and 9000 msec when the degree of contamination of the lower curtain wall glasses was measured. As such, it was confirmed that those sections whose voltage levels were lower than the average value were contaminated areas. The location of the robot at 6000 msec, where the contamination started, was about 2700 mm, about 1200 mm away from the robot's initial location. The result also shows the location of the contaminated section.

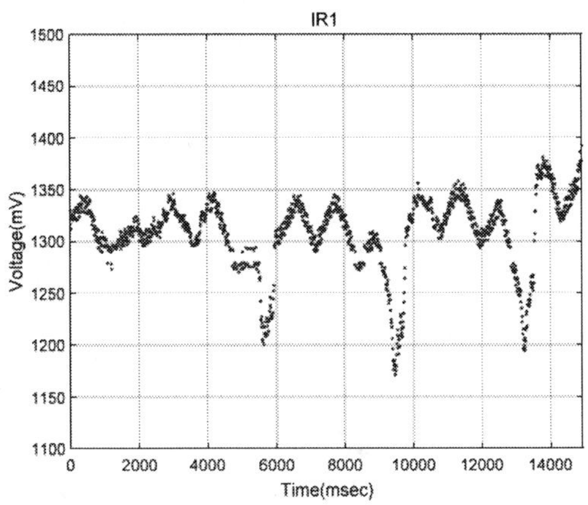

Figure 11. Contamination detection results on the upper curtain wall glasses (IR sensor detection results)

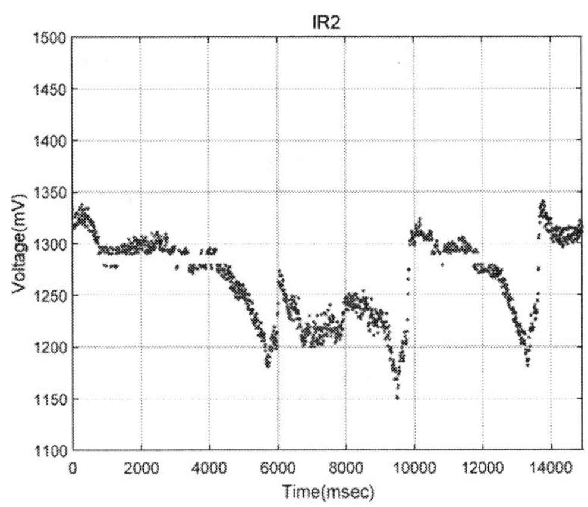

Figure 12. Contamination detection results on the lower curtain wall glass (IR sensor detection results)

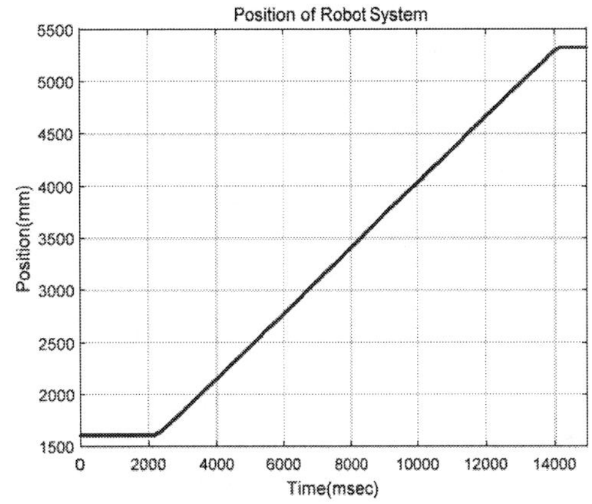

Figure 13. Horizontal movement position of the BMR robot

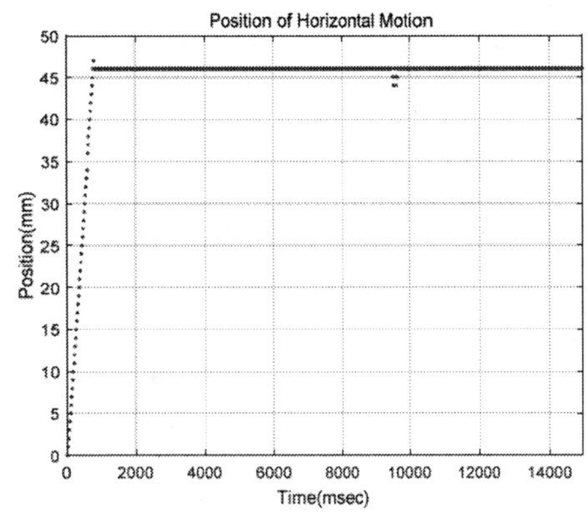

Figure 14. Results of the horizontal movement of the IR sensor module

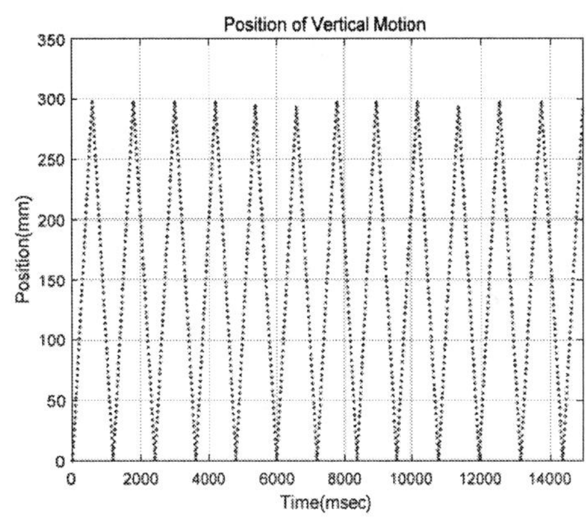

Figure 15. Results of the vertical movement of the IR sensor module

The coordinates of the forward-backward direction of

the IR sensor module required for contamination detection are shown in Figure 14 and 15 while Figure 14 shows the results where the distance between the sensor module and the glass surface was kept constant. It was confirmed from the coordinates of the IR sensor module that the proposed system is capable of maintaining a constant distance from the glass surface as well as of detecting contamination in as many areas as possible.

5 Conclusion

In this paper, a sensor system for detecting contamination on curtain wall glasses for the automated window-cleaning robot is proposed. The sensor system for detecting contamination on windows using an IR sensor requires a subsystem that maintains a constant distance from the glass surfaces given the specific requirements of the sensor system. To address this issue, a mechanism that can move forward-backward and upward-downward is proposed herein. The field test results of the proposed sensor system confirmed that a voltage relatively lower than the average of the sensor values gained from detecting a clean glass surface was observed in the sensor signals gained while detecting contamination on a glass surface.

Acknowledgment

This research was supported by the Industrial Strategic technology development program (No.10052965) funded by the Ministry of Trade, Industry and Energy (MOTIE), KOREA and was supported by a grant(14SCIP-B079344-01) from Smart Civil Infrastructure Research Program funded by Ministry of Land, Infrastructure and Transport(MOLIT) of Korea government and Korea Agency for Infrastructure Technology Advancement(KAIA).

References

[1] Horner R. M. W., El-Haram M. A. and Munns A. K. Building maintenance strategy: a new management approach. *The Journal of Quality in Maintenance Engineering*, 3(4): 273-280, 1997.

[2] NIHON BISOH CO. Automatic Window Washers. On-line: https://www.bisoh.co.jp/en/case/washers/, Accessed: 31/03/2016

[3] Hortig J., Elkmann N., Felsch T., Sack M. and Bohme T. The Sirius robot platform for tool handling on large vertical surface. In *International Symposium on Automation and Robotics in Construction* (ISARC2010), 2010

[4] Albagul A., Asseni A., Jomah O., Omer M. and Farge B. Design and fabrication of an automatic window cleaning robot. In *Proceedings of the 9th WSEAS International Conference on Signal Processing*, 2010

[5] Lee S. H., Kang M. S. and Han C. S., Sensor based motion planning and estimation of high-rise building façade maintenance robot. *International Journal of Precision Engineering and Manufacturing*, 13(12):2127-2134, 2012.

Development of BIM-based Real-time Evacuation and Rescue System for Complex Buildings

Min-Yuan Cheng[a], Kuan-Chang Chiu[b,*], Yo-Ming Hsieh[c], I-Tung Yang[d], and Jui-Sheng Chou[d]

[a] Distinguished Professor, Department of Civil and Construction Engineering, National Taiwan University of Science and Technology, Taiwan
[b] Ph.D. Candidate, Department of Civil and Construction Engineering, National Taiwan University of Science and Technology, Taiwan
[c] Associate Professor, Department of Civil and Construction Engineering, National Taiwan University of Science and Technology, Taiwan
[d] Professor, Department of Civil and Construction Engineering, National Taiwan University of Science and Technology, Taiwan
E-mail: myc@mail.ntust.edu.tw, D10005101@mail.ntust.edu.tw, ymhsieh@mail.ntust.edu.tw, ityang@mail.ntust.edu.tw, jschou@mail.ntust.edu.tw

Abstract –

With the architectural environment is constantly changing such as high-rise buildings, high-ceilinged space, and special space constantly constructed, thus the fire disasters are becoming diverse and difficult to predict. Therefore, the development and application of building's fire disaster prevention integration system become an urgent requirement for prevention the disaster and reach all protection of life and property.

The response system and organization of fire prevention and disaster relief currently in Taiwan, the planning of evacuation and rescue guidance main depends on human judgment to obtain disaster information. This method may easily either cause error conduction associated with information transfer or can't rescue immediately due to the human negligence or human limitations. Furthermore, wireless sensor networks have been widely discussed in many domains of building disaster-prevention management. This study will focus on fire disaster and combine the application of Building Information Modeling (BIM) to construct a BIM-based building fire prevention and disaster relief system, with integration of fire prevention equipment information, bluetooth sensor technology, and route optimization information of evacuation and rescue. In order to create intelligent fire disaster prevention system frame, which applied the BIM model shows the research results of three-dimension (3D) visualization. Under the fire incidents created by information integration platform, it can provide early detection and accurate alarm ability, endangered people can have correct and rapid guide the evacuation, fire-fighters can obtain the correct information and assist the right guidance of the rescue routes, as a result, reached the purposes of safety and disaster reduction.

Keywords –

Wireless Sensor Networks; Building Information Modeling; Bluetooth Sensor; Route Optimization, 3D Visualization

1 Introduction

In recent years, the rapid development in Taiwan's economy, the enhancement of the construction techniques, and the urbanization of population that have led to the intensive and highly land uses of urban. These phenomena will cause the buildings to develop gradually the high-rise and particular types, such as Taipei 101 building, congregate housing, shopping mall, public buildings, Taipei main station, and MRT station in Taiwan. For this reason, the use of inner spaces for high-rise buildings will be more complex than general ones. Besides, due to the highly capacity and centralized goods for these high-rise and complex buildings, there are the difficulties to escape and rescue as a result of the high-rise and the complex routes when the fire disaster is occurred. Hence, in order to ensure the safety of personnel, the users just use the escape facilities or fire extinguishers to save oneself. This condition will cause the higher fire hazards in high-rise and complex buildings then in the general buildings [1]. Thus, building's fire and disaster prevention integration system and application development has not only become an urgent requirement for prevention the disaster, but also to enhance disaster prevention trend.

Furthermore, with the building's environment are constantly changing, the disasters are becoming diverse and difficult to predict. The fire disaster prevention management for buildings has been highly concentrated on. The tasks includes alarm notification [2, 3, 4], the design of fire evacuation guidance in the building [5, 6], and evacuation routes simulation [7, 8, 9], etc. However, the common evacuation routes were usually guided in 2D views, but are not able to advise the exact location within the building in the immediate condition, then to select the proper evacuation exit. In recent years, some applications of Building Information Modeling (BIM) have been addressed to develop on building disaster-prevention management [10, 11, 12, 13]. The use of BIM can help us to display the distinct site-layout in 3D visualization and immediately grasp the status of the fire and disease in the building. BIM is a kind of new concept and has the ability to improve the traditional 2D plane for disaster prevention management. On the other hand, the fire disaster prevention must integrate sensing technology, analysis, judgment, decision, and action ability. Thus, the key aspects for the study will be : (1) "Effective Detection" to provide earlier detection and accurate alarm abilities, and the immediate mechanisms of extinguishing ; (2) with "Correct Refuge" view , endangered people can have correct and rapid guide the refuge ; (3) with "Fire Relief" view provides fire-fighters the correct information and assist the right guidance of the relief route; (4) with "Integration System" to perspective information integration and sharing mechanism , with a BIM information integration platform to integrate the various fire and rescue information. Then, this effectively exert the linkage mechanism to solve the information transmission problems between endangered people and fire-fighters.

This study proposes the application of BIM to construct buildings in 3D massing models, with integration of fire prevention equipment information, bluetooth sensor information, and route optimization information of evacuation and disaster prevention. In order to create intelligent fire and disease prevention system frame, which applied the BIM model shows the research results of visualization. When the bluetooth sensors detect the information of smoke and high temperature earlier, the system can determine the authenticity and sound the alarm to monitor locations and the evacuation/rescue routes which planned by the back-end operation in BIM's 3D model. Furthermore, the system can instantly show which floor occurs the fire disaster and the space plan of the suspected fire area in the 3D model, then connect the users' APP trapped in the fire scene to active a real-time and dynamic evacuation guidance. The proposed integration system will enhance the time effectiveness and safety of evacuation / rescue.

2 Proposed BIM-based fire reduction system

In order to achieve the message passing among the upstream, middlestream and downstream, and integrate all of the on-site fire information so that can share and analyze the real-time data to the users. This study writes an Application Programming Interface (API) through Revit 2015 to integrate the FireSense system which combines the application of smoke detectors with bluetooth wireless sensor networks, evacuation / rescue guiding device for mobiles (see Fig. 1), and the evacuation / rescue routes planning through Dijkstra's algorithm [28] searching the shortest path for BIM-based system. The BIM platform can provide the data storage of fire agency to update, query and share data, furthermore, apply capabilities of 3D visualization to support fire safety management. The BIM-based system consists of four modules: (a) FireSense module, (b) Enviro-Sense/Location module, (c) Evacuation Planning module, (d) Fire Rescue Planning module, and (e) Global Information module. Fig. 2 shows the integrated system framework. The figure presents the inputs, information transmissions, and outputs of each module. Follow subsections will introduce the modules of the BIM-based intelligent fire reduction integrated system.

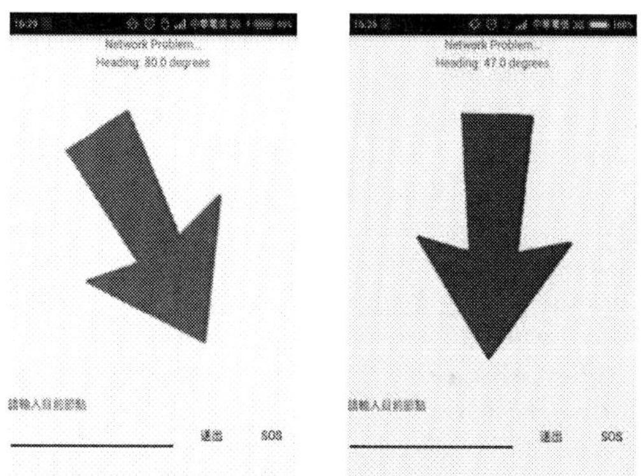

Figure 1. Mobile guiding device

2.1 FireSense module

The FireSense module mainly provides users to monitor the on fire stories of a building. The module combines the multi-functional smoke detector of bluetooth which integrates with temperature sensor, smoke sensor, LED illuminant, power management, System on a Chip (SoC), and power supply (see Fig. 3) to detect the temperature/smoke conditions nearby fire

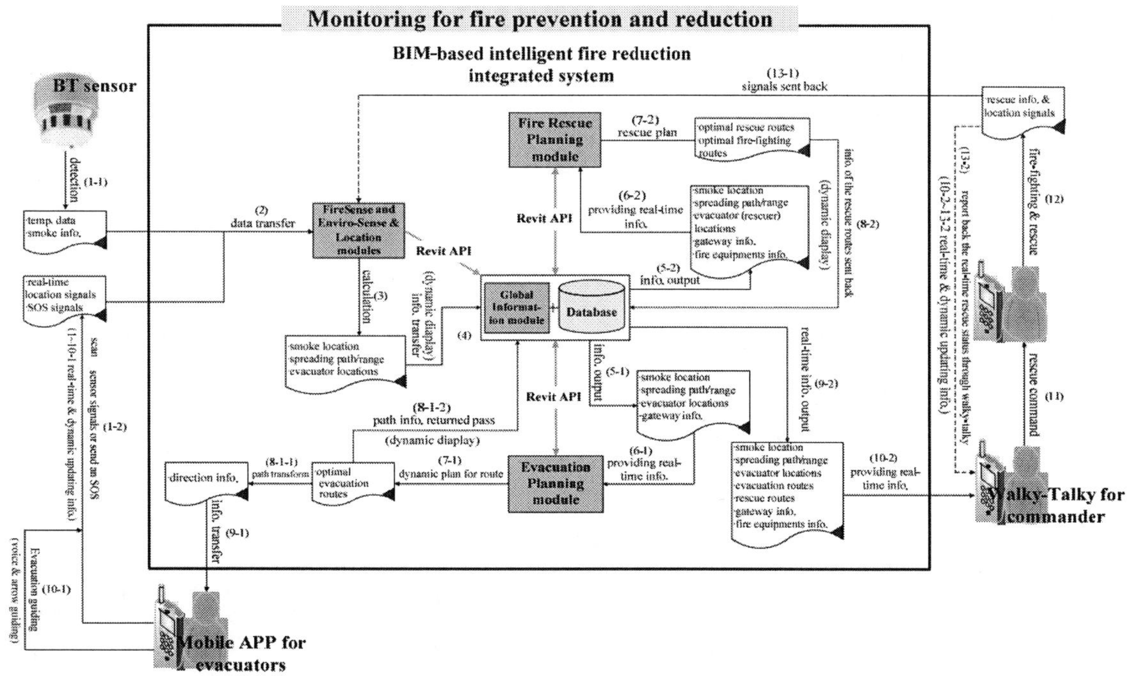

Figure 2. Framework of proposed system

scene and provide the direction guiding of emergency evacuation. Besides, the real-time detected data will be stored in SQL server of the proposed system. When the temperature/smoke are higher than the alert values, the system will highlight the on fire stories of a building and notify the proper fire locations by displaying tips through the FireSense module.

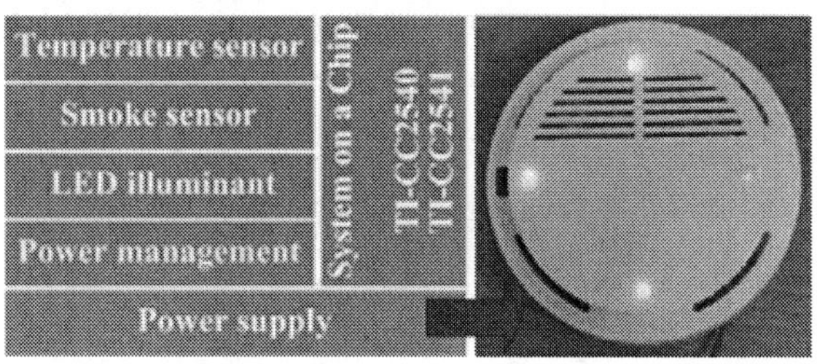

Figure 3. Main components of the FireSense device

2.2 Enviro-Sense/Location module

The FireSense module also combines the multi-functional smoke detector of bluetooth 4.0 to broadcast sensor ID through a radio mechanism of BLE (Bluetooth Low Energy). The broadcast information will be received by smart phones or tablets which support the BLE, and furthers, the smart phone on one's hand can capture the Receiver Signal Strength Indicator (RSSI) through matching up the use of mobile APP to proceed a one's location. Fig. 4 shows the process of personal location. First, the bluetooth MAC of smart phone will be captured by user's APP and provided to the server to track each user's location. Second, the MAC will be stored in a new object of HttpThread to upload the location information. Third, starting the smart phone's bluetooth to scan the multi-functional smoke detectors around the user. The action will last one second to obtain the device ID of the multi-functional smoke detectors around the user and each RSSI. Fourth, the all received device ID and RSSI will be upload to the server and calculated the user's location by the server. Fifth, when completing the uploading process, the all data will be cleared and repeat step3 and step 4 to continue the locating process. Furthermore, the commander can find out the evacuators are in which smoke detector's ID on BIM model through the locating information.

Thus, the module will provide users to query the current real-time information of the fire scene which consists of the temperature/smoke information, quantity of bluetooth sensor charge, and fire/evacuators/rescuers locations. When users click on the sensor components of BIM model from this module, the users can not only query the above mentioned environmental data but also get the detected time and alarm status. On the other hand, if the users click the human shape components on BIM mode, the information of sensor ID which the evacuators/rescuers stay in, number of people, and people's identity can be provided to grasp the latest current situation of fire scene.

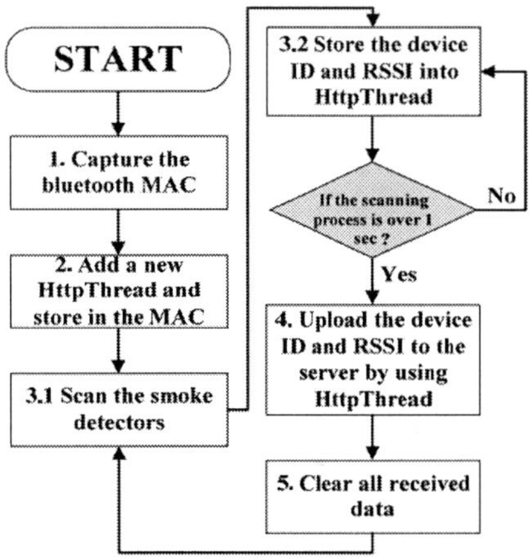

Figure 4. Locating process in back service through integrating smoke detectors and user's APP

2.3 Evacuation Planning module

The evacuation planning module not only provides users to look over the location information of fires, evacuators and rescuers at the same time but also uses the BIM 3D visualization to help preliminarily allocate the evacuation areas of the building to each exit according to the latest locations of fires and evacuators. Besides, the module can real-time calculate the shortest safe path from the start node (evacuator locations) to the end node (exit) on the basis of Dijkstra's algorithm and dynamic display the evacuation routes on the target BIM modeland so that provide the fire commander to grasp the evacuation status of evacuators rapidly. However, the shortest safe path is determined by multiplying penalty factor from the relative paths of on fire nodes to evade the fire areas. Thus, the Dijkstra's algorithm can be defined through the following function [14] :

$$\text{dist}[j] = \min[\text{dist}[j], \text{dist}[k] + c(k, j)] \tag{1}$$

where dist[i] is the shortest path from starting point to i, c(i ,j) presents the distance from i to j. Owing to penalty factor must be a comparatively large value, The study raises a suggestion that takes the penalty factor equation to be set as the longest path from all proper paths, and further times 100 to ensure the relative path of on fire longer than other safe path. This method will be avoided effectively to pass through the fire areas.

On the other hand, an evacuation guiding APP device for mobile (hereafter, mobile APP) is applied by this study owing to its popularity, well performance and easy to carry. After the dynamic evacuation routes are calculated by the proposed module, the results of routes and next target node will be uploaded to database through internet and connect the direction angle on PHP web server at the same time to provide for the use of mobile APP of voice and arrow (shown as Fig. 5). The mobile APP is in combination with the application of gyroscope [15] and orientation sensor [16] which built-in the smart phone to calculate the horizontal azimuth of mobile device and guide direction through an arrow. When the evacuators move to next nodes, the module will recalculate new paths in accordance with the latest location data on BIM model and re-guide through mobile APP to achieve the target of dynamic real-time evacuation guiding and helping evacuators escape to exit rapidly and safely.

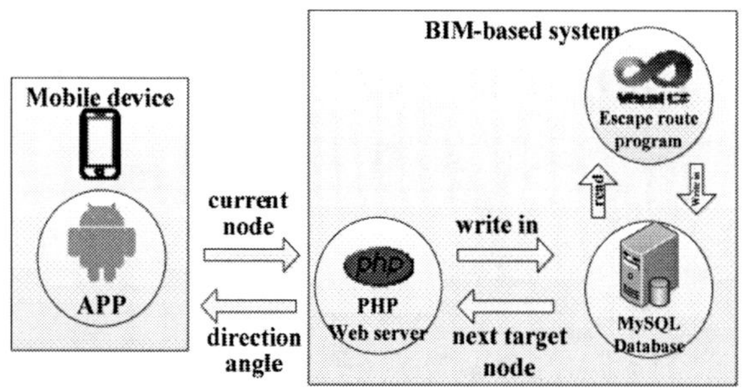

Figure 5. Communications between module and mobile device for evacuation

2.4 Fire Rescue Planning module

The functions of fire rescue planning module are similar to some parts of the evacuation planning module. It also can not only provide commander to look over the location information of fires, evacuators and rescuers, but also display the dynamic real-time fire rescue routes on the BIM 3D visual model. The planned routes are divided into two parts: (a) fire-fighting routes, and (b) rescue routes. In the phase of planning fire-fighting routes, the calculation method is the same as optimal evacuation routes by using of Dijkstra's algorithm. But we won't take the penalty factor into consideration in planning fire-fighting routes and avoid the clash with evacuation routes, owing to fire-fighters must find the shortest path to arrive the fire point and act a fire-fighting. In the phase of planning recue routes, we further import the concept of many (gateways)-to-many (evacuators) into Dijkstra's algorithm and consider the penalty factor equation to calculate the shortest recue routes. The planning results will be uploaded to database through internet to provide for the use of mobile rescue device of voice and arrow (shown as Fig. 5). This model can assist to grasp fire conditions quickly and correctly for on-site commander, help reaching the fire point and the location of victims through the fastest and safest ways for the indoor fire-fighters, and optimize the abilities of fire reduction to enhance the rescue efficiency.

2.5 Global Information module

The global information model not only shows the look of building and fire equipments and locations, etc. on BIM 3D visualization, but also integrates all of fire and disaster reduction information of the first four modules. This function can provide fire commanders to proceed an overall grasp of all real-time information of the fire scene and make quick and correct disaster relief decisions.

3 Case study and system operations

Autodesk Revit Architecture 2015 which opens a full-fledged API called Revit Platform API [17] is utilized herein to develop the proposed BIM model as a result of its popularity and practical utility in Taiwan. Implementing the five modules requires a BIM model with a level of detail (LOD) of 300. An LOD 300 BIM model should contain columns, beams, walls, floorslabs in structure phase, and windows, decoration walls, stairs, doors in architectural phase as objects, along with their size, shape, location, orientation and geometric data (such as length, area, and volume) [18]. The five modules of this BIM-based system are developed on a Windows operation environment and the .NET Framework 4.5, and compiled by using the C# programming, with a Windows 7 computer operating system. Besides, using the SQL SERVER 2010 to manage database and communicate with Revit.

The following subsections describe the results of applying the proposed BIM-based system in a case study.

3.1 Building description

This study concentrates on the high-rise and complex building "International Building (hereafter, IB) of Taiwan Tech" located in northern Taiwan as a case study. The IB building is a building of twelve-storey on the floor with a cockloft designed to act as the uses of office, classroom, performance space, and meeting room, and three-storey below ground as parking lots. The total floor area of this building is about 36535.24 square meters, and the building height is 49.6 meters (see Fig. 6 (a), (b)). Based on the experience of the architect and the fire safety staff, occupants who use gateways of each storey exits will face the most danger because they are in the higher and complex area. On the other hand, since most of the student are mainly gathered up to conduct a class on the third or fourth floors with the more exits, the actions of evacuation and rescue are relative difficult if a fire occurs. Accordingly, in the following sections, where applicable, the area in which occupants use the third floor of IB building as exiting floor will be used to illustrate the simulations and experimental verifications of the proposed modules. The case floor (the third floor) looks like a "L-shape" with six gateways, as displayed in Fig. 7.

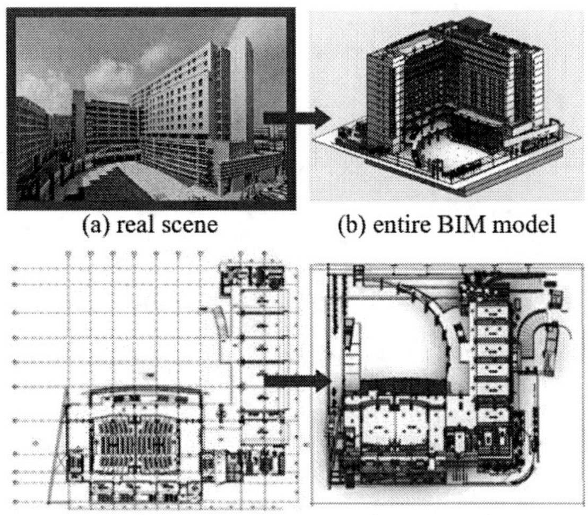

(a) real scene (b) entire BIM model

(c) 2D plane for the 3rd floor (d) 3D model for the 3rd floor

Figure 6. Case building (IB) in NTUST

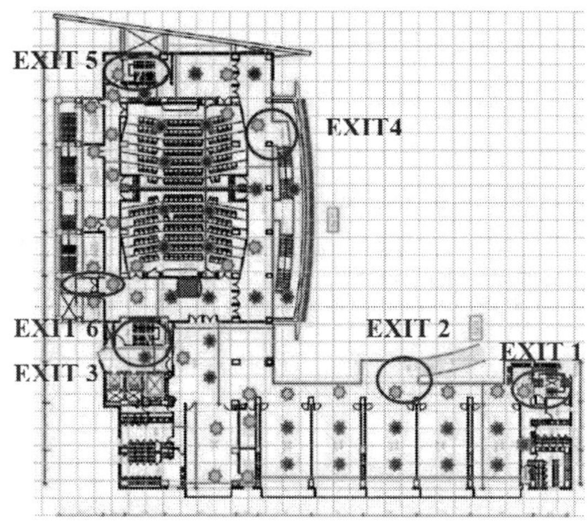

Figure 7. Exits layout of the third floor plane in case building

Besides, this study aims to guide information system into BIM model and be good at using each fire equipment /space data of BIM model to let fire safety staff making a correct decision though the display of

BIM 3D visualization. But before operating the BIM-based system, there must be a constructed BIM model. Therefore, we need to establish the complete BIM model before guiding into integrated system. For this reason, a BIM model of the case building (IB) during operation phase was made available using Autodesk Revit Architecture 2015 (see Fig. 6 (b), (d)).

3.2 Bluetooth sensor layouts and encoding principle

The BIM-based intelligent fire reduction integrated system combines the application of multi-functional smoke detector of bluetooth to detect environmental conditions. In this study, there are 61 bluetooth sensors embedded in the third floor of the IB building as nodes (e.g., located on classrooms, sidewalk, corners, and exits), and the distance of each node equals 5m. We also construct all of the sensor components on the BIM model shown as Fig. 8 (green balls). The follows will describe the principle of bluetooth sensor encoding. The sensors which embedded in interior spaces (including the large meeting rooms and classroom) are numbered by original spatial code (initial number as 301). Thus, the principle of the interior spaces is through the clockwise from the first sensor of upper right to number. For example, Fig. 8 shows that there are four bluetooth sensors embedded in meeting room 301, so the codes are numbered by 3011, 3012, 3013, and 3014 in turns according the principles of top-down and right-left. The similar same as the external sidewalk, the initial value is 300 and the codes are numbered as 3001, 3002, and so on from the first sensor of upper right in Fig. 8 according to right-left and top-down (see Table 1). The codes in this system provide the unique id for bluetooth sensors. The system combines the technique of blutooth wireless sensor networks and writes in/ reads the sensor id and spatial data to real-time locate people/fire in buildings through the display of BIM 3D visual model.

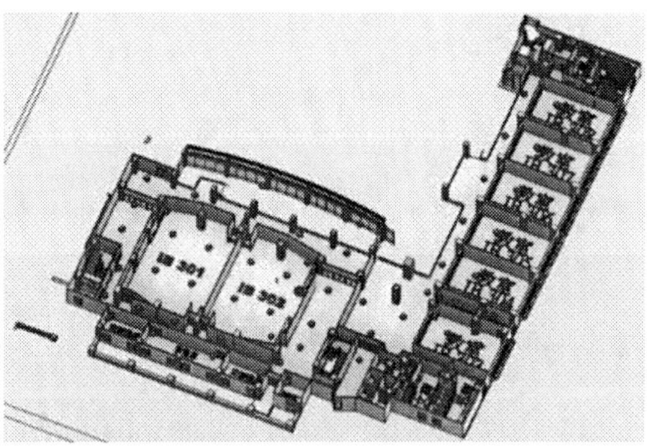

Figure 8. Bluetooth sensor embedded in the 3rd floor of IB building

Table 1. The bluetooth sensor id on the 3rd floor of the IB building

Category	Space No.	Floor	Sensor ID
Meeting room	IB 301	3	**3011**
	IB 301	3	**3012**
	IB 301	3	**3013**
	IB 301	3	**3014**
Meeting room	IB 302	3	**3021**
	IB 302	3	**3022**
	IB 302	3	**3023**
	IB 302	3	**3024**
Classroom	IB 303	3	**3031**
	IB 303	3	**3032**
	⋮		
Classroom	IB 308	3	**3081**
	IB 308	3	**3082**
Sidewalk	-	3	**3001**
	⋮		
Sidewalk	-	3	**30061**

3.3 Operating process of the proposed system

The use of the BIM-based system which established by the communications of BIM model and API, has four operating phases as follows: (a) implementation phase, (b) start phase, (c) update phase, and (d) storage phase. The subsection will introduce the four operating frameworks.

(a) Implementation：When carrying out Autodesk Revit Architecture 2015 to open the target building of BIM model and operate the system, the five modules will be added in the two ribbon panels: "Fire Monitoring" and the "Fire 3D Visualization" of the "Building Fire Prevention and Disaster Relief " tab (see Fig. 9), then pushing each module's button can prepare to start this system.

(b) Start：After implementing, the system will spend a few seconds to capture BIM model and the data from SQL database, and then the dynamic 3D visual data of fire scene will be displayed on BIM model. This current data of visualization model and relative components can be provided to the fire rescuers as references.

(c) Update：When the fire occurs, most of the fire information can be displayed on BIM model of this "BIM-based intelligent fire reduction integrated system" clearly, and "real-time" update the dynamic fire information under a steady interval through compiling the API. The latest dynamic fire data on BIM model can be provided to the relative rescuers.

(d) Storage：In this phase, all of the information which ever displayed on the BIM-based system will be stored and feedback into the SQL database automatically. Furthermore, these past data can be queried or analyzed someday.

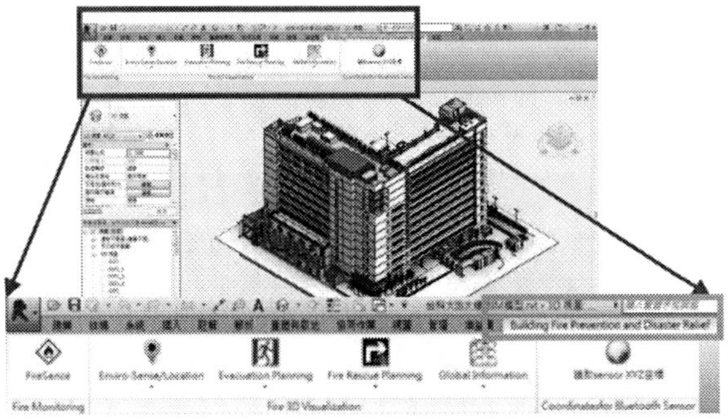

Figure 9. The panels of five modules

3.4 Case simulations

3.4.1 FireSence

First, we start Revit 2015 to open the 3D model of the case building, then click the "FireSense" button within the "Fire Monitoring" tab to prepare monitoring the fire event. In this case study, we assume the fire initial ignites on third floor of IB building. When a fire occurs, the walls of on fire storey will be all highlighted by orange color and moreover shows the dialog box to notice the information of the proper on fire floor and fire locations shown as Fig. 10 (a). When the available fire information obtained in this system, follows users can click the last four module embedded in the tab of "Fire 3D Visualization" to look for another real-time fire disaster information, such as evacuation routes, personnel status,...etc. Fig. 10 (b) shows the user selects the needed 3D sectional drawing of floor to look for relative fire information.

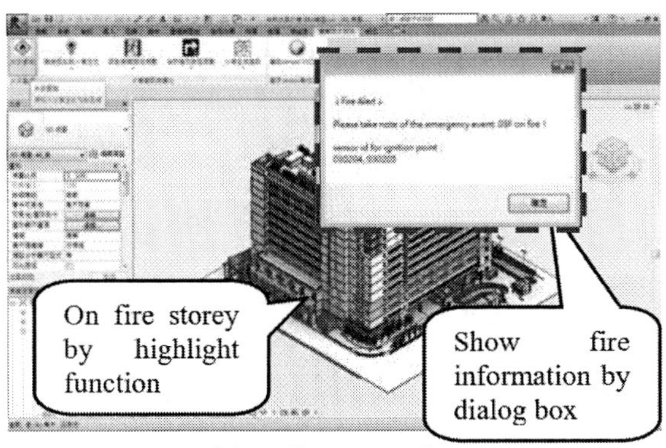

(a) on fire storey alert

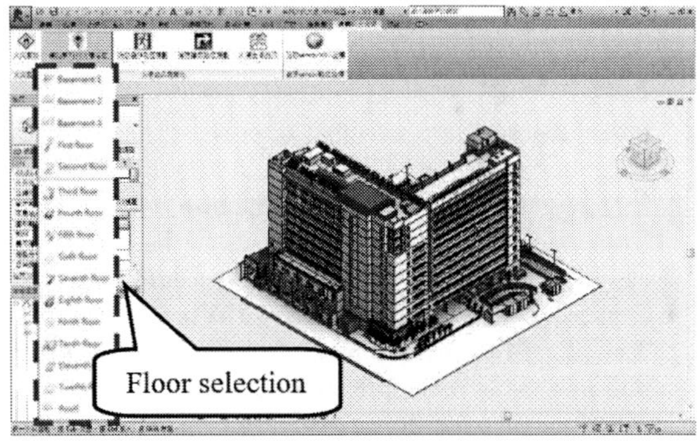

(b) Select floor to open 3D sectional drawing

Figure 10. Functions of FireSense module

3.4.2 Environmental detection and personnel localization

The Enviro-Sense/Location module provides a 3D representation of hazardous/personnel locations, environmental conditions, and the alert for low battery bluetooth sensors. When users carry out this module to look for the fire information, the system will show a alert window to advice users needing to replace battery, and change the sensor component color into red if there is a low battery bluetooth sensor (see Fig. 11 (a)). In addition, users can also click the sensor component button to query its id, detecting time, smoke value, and current temperature.

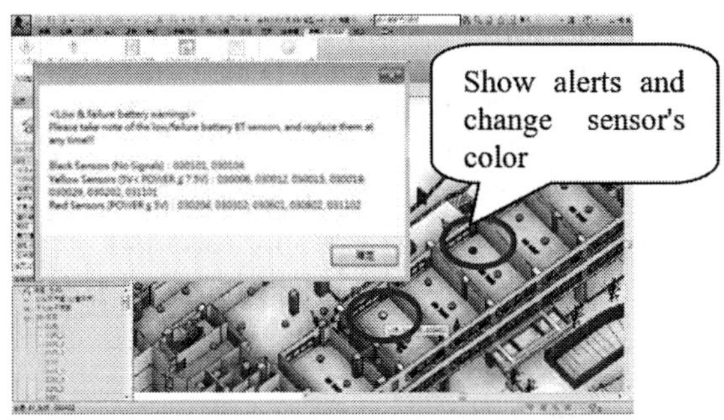

(a) visualization for low battery sensor

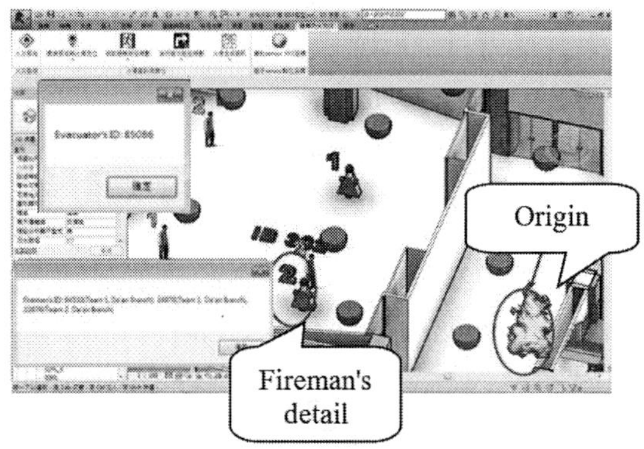

(b) Visualization of personnel locating

Figure 11. Functions of Enviro-Sense/Location module

The second deliverable of this module - fire/personnel locating - is very useful in fire disaster management. Because most high-rise buildings own the more complex indoor spaces and structure designation, so the difficulty of disaster reduction will be increased from fireman's opinions. Based on the results of the assessment using the previously described Enviro-Sense/Location module, Fig. 11 (b) presents a real-time dynamic 3D locating map using the 3rd floor of IB building, as an example. Moreover, this module's query function is used to provide the personnel's identity (device id of smart phone) and organization. And the personnel amounts under each bluetooth will be calculated and displayed on the 3D model of the human shape components. In Fig. 11 (b), the fire symbol presents on fire location, the real human shape component is trapped evacuator, and the component of red head with triangular body presents fire-fighter.

3.4.3 Evacuation/Rescue Planning and Guiding

When a fire first ignites, it may be small enough that you can have an even better chance to escape from the fire scene without risking injury by use of the mobile APP provided evacuation data by our proposed BIM-based system on hand. Initially, the architect of the case building used 2D drawings to make an guiding map. The proposed evacuation planning module uses the 3D BIM model to display the real-time evacuation routes from each personnel location. This study further added an additional rescue guiding map based on BIM model.

In this simulation, we will prior discuss the results of evacuation planning and guiding, owing to the methods of planning rescue routes and guiding are similar to each other. Therefore, in this part, a evacuation planning case study is simulated. We assume the on fire location is No. 030019 and the initial location of two evacuators are No. 030204. The shortest path where these two evacuators can originally escape is from No. 030204, No. 030019, No. 030020 to No. 030029 (see Fig. 12). But in order to avoid the origin fire, the evacuation route is planned from No. 030204, No. 030203, No. 030022 to No. 030013 (Exit) according to the proposed method of planning evacuation presented in Subsection 3.3. Fig. 12 resents the guiding processes between for the interactive relationship between 3D visualization guiding on BIM model and the arrow guiding of the use of mobile APP. of a building to act a disaster relief through fire-fighters.

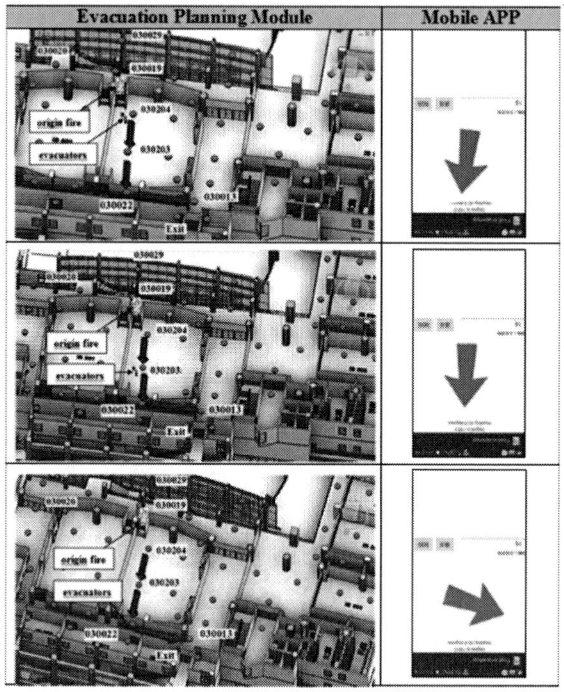

Figure 12 Process of the real-time evacuation guiding

4 Conclusions

Currently, many structures of buildings are most constructed by the fireproofing. When the fire occurs, the people in this kind of buildings still have a heavy casualties and property damages. It can be seen that it is insufficient to be just provided with fire resistance for structure itself. In response stage of the fire, the real-time fire location, the spreading status of smoke, and the occupants' locations must be overall taken into considerations. Furthermore, it is needed for helping evacuators to evacuate from a dangerous area quickly and safely while a emergency event occurs. Besides, saving lives in building fires depends on locating fires quickly and accurately, attacking them aggressively and reaching occupants who may be trapped in the interior For those reasons, This study promotes the usages of BIM spatial and visual information in combinations with "fire alarm sensing system", "location-aided design", "planning of the optimal evacuation and rescue route", and "real-time guiding system of mobile APP" in an BIM-based integrated system for evacuators, fire-fighters, and commanders in the preparedness and the response stages. The proposed integrated system hopes to solve the problems of the error information or without evacuating quickly and rescuing immediately which resulting of the human errors and the human limitations when the fire occurs in buildings. In this study, the numbers of the fire and personnel locations are transmitted by bluetooth sensors which constructed within a building then send back them to the BIM-based integrated system, and combine the database's searching and integrate the relative information of the disaster prevention / reduction (e.g., fire facilities information, etc.) to provide the consultations for planning routes, dynamic guiding, and rescue decision-making.

Applying the "BIM-based intelligent fire reduction integrated system" developed by this study can greatly enhance the accuracy of the fire alarm and real-time monitor the status of disaster areas to plan and show the optimal evacuation / rescue routes by the use of BIM 3D visual model with mobile guiding system. Fortunately, this can not only be used by fire-fighters to quickly locate destinations (e.g., the location of a fire and potential victims) but also reduce response times to accidents and minimize the search times required to find potential trapped evacuators.

Acknowledgments

The authors would like to thank the Taiwan Building Technology Center, NTUST for financially supporting this research. And we thank the Autodesk Company of Taiwan for providing the BIM design software and the Department of Civil and Construction Engineering, NTUST research team for their help in the project.

References

[1] Gang Zhao, "Taiwan Super High-rise Building Fire Disaster Prevention", *China Fire Association Scientific Technology Annual Meeting & Guangdong High-rise Building Fire Safety Management Summit*, Chinese Association of Fire Protection, 2012.

[2] S. Matthews et al., "Field evaluation of two image-based wildland fire detection systems", *Fire Safety Journal*, Vol. 47, pages 54–61, 2012.

[3] F. Yuan, "Video-based smoke detection with histogram sequence of LBP and LBPV pyramids", *Fire Safety Journal*, Vol. 46(3), pages 132–139, 2011.

[4] B.U. T¨oreyin, Y.Dedeoglu, U.G¨ud¨ukbay, A.E. Cetin, "Computer vision based method for real-time fire and flame detection", *Pattern Recognition Lett*, Vol. 27(1), pages 49–58, 2006.

[5] S. Cao et al., "Modeling pedestrian evacuation with guiders based on a multi-grid model", *Physics Letters A*, Vol. 380, pages 540–547, 2016.

[6] H. Ran et al., "Influences of intelligent evacuation guidance system on crowd evacuation in building fire", *Automation in Construction*, Vol. 41, pages 78–82, 2014.

[7] Tanaporn Tanachawengsakul et al., "A simulation study on fire evacuation routes in primary stage for a historic canal residential area" *Procedia -Social and Behavioral Sciences*, Vol. 216, pages 492–502, 2016.

[8] Yu Cai et al., "Study on law of personnel evacuation in deep buried metro station based on the characteristics of fire smoke spreading", *Procedia Engineering*, Vol. 135, pages 543–549, 2016.

[9] Vancho Adjiski et al., "Simulation and optimization of evacuation routes in case of fire in underground mines", *Journal of Sustainable Mining*, Vol. 14(3), pages 133–143, 2015.

[10] Y. Zou et al., "A review of risk management through BIM and BIM-related technologies", *Safety Science, In Press*, 2016.

[11] S.-H. Wang et al., "Applying building information modeling to support fire safety management", *Automation in Construction*, Vol. 59, pages 158–167, 2015.

[12] J. Choi et al., "Development of BIM-based evacuation regulation checking system for high-rise and complex buildings", *Automation in Construction*, Vol. 46, pages 38–49, 2014.

[13] Kun-Chi Wang et al., "Application of Building Information Modeling in Designing Fire Evacuation-a Case Study", *The 31st International Symposium on Automation and Robotics in Construction and Mining (ISARC 2014)*, July 9-July 11, 2014, Sydney, Australia, 2014.

[14] E. W. Dijkstra, "A note on two problems in connexion with graphs", *Numerische Mathematik*, 1959.

[15] Wikipedia, "Gyroscope", 2015.04, <http://tinyurl.com/o7ggnk7>.

[16] Bin-Hua Huang, "Android 4.X App", *GOTOP Information*，pages 10-2 - 10-19, 2013.

[17] Autodesk, Inc., "Revit 2015 API Developer's Guide", Jul.16.2015.

[18] Robert S. Weygent, "BIM Concept Development: Standards, Strategies, and Best Practices", New Jersey: John Wiley & Sons Inc, 2011, pages 79-85.

Modeling the Permanent Deformation Behavior of Asphalt Mixtures Using a Novel Hybrid Computational Intelligence

M.-Y. Cheng[a] and D. Prayogo[a,b]

[a]Dept of Civil and Construction Engineering, National Taiwan University of Science and Technology, Taiwan
[b]Dept. of Civil Engineering, Petra Christian University, Indonesia

E-mail: myc@mail.ntust.edu.tw, doddyprayogo@ymail.com

Abstract –

One of the main causes of pavement rutting is the repetitive action of traffic loads which results in the accumulation of permanent deformations. As a result, it is important to understand the characteristics of the permanent deformation behavior of asphalt mixes under repeated loading and to build the accurate mix model before they are placed in roadways. This study proposed a hybrid computational intelligence system named SOS-LSSVM for modelling the permanent pavement deformation behavior of asphalt mixtures. The SOS-LSSVM fuses Least Squares Support Vector Machine (LSSVM) and Symbiotic Organisms Search (SOS). LSSVM is employed for establishing the relationship model between the flow number, which is obtained from the laboratory test, and the parameters of the asphalt mix design. SOS is used to find the best LSSVM tuning parameters. A total 118 historical cases were used to establish the intelligence prediction model. Obtained results validate the ability of SOS-LSSVM to model the pavement rutting behavior of asphalt mixture with a relatively high accuracy measured by four error indicators. Therefore, the proposed computational intelligence systems can offer a high benefit for road designers and engineers in decision-making processes.

Keywords –

Symbiotic Organisms Search; Least Squares Support Vector Machine; Asphalt concrete mixture; Flow number; Prediction

1 Introduction

Over the last decades, the permanent pavement deformation or rutting of asphalt mixtures has gained great attention in the field of pavement engineering. Rutting progressively develops in asphalt pavements and increases the number of vehicle load cycles. It became a major concern because pavement rutting decreases the useful life service of the pavement and creates serious hazards for highway users. As a result, it is important to understand the characteristics of the permanent deformation behavior of asphalt mixes under repeated loading and consequently construct the accurate asphalt mixture before they are placed in roadways [1].

Recently, dynamic creep test is found to be one of the best methods for assessing the permanent deformation potential of asphalt mixtures [2]. The most important output of the dynamic creep test is the curve of accumulated strain against number of load cycles. The flow number, defined as the loading cycle number where the permanent deformation starts [3], is a good indicator of the rutting resistance of a given asphalt mixture. However, the dynamic creep test is sensitive and costly and cannot always be performed. Therefore, developing a relationship model based on the collectible historical data between the flow number obtained from the dynamic creep test and the parameters from the asphalt mix design leads to considerable savings in construction costs and time.

The prediction of the asphalt mixture performance of asphalt mixtures is a significantly important matter. However, building that accurate relationship model has traditionally been a complicated task. It is because the permanent deformation of asphalt mixtures is influenced by several factors which are also complex and highly nonlinear. The conventional approaches such as linear regression or decision tree are not sufficient to build a satisfactory model in terms of accuracy and computational cost. The artificial intelligence (AI) approaches have been reported to outperform the conventional approaches due to the excellence of their learning features [4-8]. Yet, only a few researches have utilized AI methods to predict the flow number of the asphalt mixtures [1,9,10].

The primary objective of this research work is to build the accurate prediction model using the advanced AI methods for enhancing the permanent deformation analysis. This research proposes a new computational intelligence system called Symbiotic Organisms Search – Least Squares Support Vector Machine (SOS-LSSVM). SOS-LSSVM fuses an accurate prediction technique, Least Squares Support Vector Machine (LSSVM) [11], and a very promising optimization tool, Symbiotic Organisms Search (SOS) [12]. The proposed model will be investigated alongside other prediction methods in building an accurate prediction model of the permanent deformation of asphalt mixtures.

In presenting the SOS-LSSVM as a novel computational intelligence system, this study makes three important contributions. First, SOS-LSSVM is a new variant of the nature-inspired based LSSVM. It integrates a recent promising metaheuristic algorithm (SOS) to the widely used and robust predictive method of LSSVM. Second, this study uses a cross validation technique that was adopted to validate the training and testing process. Third, three performance measurements are also used to evaluate the performance of the methods.

2 Model Construction

2.1 Least Squares Support Vector Machine

LSSVM is a modified version of the Support Vector Machine (SVM) [13]. The LSSVM is a statistical learning theory that adopts a least squares linear system as a loss function instead of the quadratic program in the original support vector machine (SVM) [14]. Notably, in LSSVM training process, a least squares cost function is proposed to obtain a linear set of equations in the dual space. One main difference between LSSVM and the original SVM is that LSSVM requires equality constraints instead of inequality ones while operating with a least squares cost function. The optimization problem and the constraints for LSSVM can be stated by the following formulation.

Minimize: $J_p(w, e) = \frac{1}{2}w^T w + \gamma \frac{1}{2} \sum_{k=1}^{N} e_k^2$ (1)

Subjected to: $y_k = w^T \phi(x_k) + b + e_k, k = 1, \ldots, N$ (2)

where $w \in R$ is an undetermined parameter vector; $\phi(\cdot)$ is the nonlinear function that is usually introduced by a kernel function that maps the input space to a high-dimensional feature space; $e_k \in R$ are error variables; $\gamma > 0$ denotes a regularization constant.

The resulting LS-SVM model for function estimation is expressed as:

$$y(x) = \sum_{k=1}^{N} \alpha_k K(x_k, x_l) + b \quad (3)$$

where α_k and b are the solution to the linear system (4). The kernel function that is often utilized is the Radial Basis Function (RBF) kernel. Description of RBF kernel is given as follows:

$$K(x_k, x_l) = exp\left(\frac{\|x_k - x_l\|^2}{2\sigma^2}\right) \quad (4)$$

where σ is the kernel function parameter.

The accuracy of LSSVM depends on searched problem parameters. The regularization parameter (γ) is a positive cost parameter similar to the C in SVM, while the Kernel parameter (σ) is an additional parameter since LSSVM utilized the RBF kernel function. These two parameters need to be specified in order to find the best prediction model.

2.2 Symbiotic Organisms Search

Symbiotic Organisms Search (SOS) is one of the newly promising metaheuristic algorithms proposed by Cheng and Prayogo [12]. SOS simulates symbiotic interactions to iteratively move a population (ecosystem) of candidate solutions (organisms) to promising areas in the search space during the process of seeking the optimal global solution. Each organism in the ecosystem is associated with a certain fitness value, which reflects a degree of adaptation to the desired objective. The basic steps of the algorithm are summarized below.

1. Generate initial population
2. **Do**
3. Mutualism phase
4. Commensalism phase
5. Parasitism phase
6. Update the best solution
7. **Until** stopping criteria are satisfied

In [12], the performance of SOS has been compared with other metaheuristic techniques such as Genetic Algorithm (GA), Particle Swarm Optimization (PSO), Differential Evolution (DE), and Bees Algorithm (BA) in numerous mathematical test functions and engineering problems. The comparison results indicate that SOS was able to achieve a better performance in terms of effectiveness and efficiency. Furthermore, SOS has been proven to successfully solve various problems in different fields of research [15-24].

2.3 Symbiotic Organisms Search – Least Squares Support Vector Machine

The Symbiotic Organisms Search – Least Squares Support Vector Machine (SOS-LSSVM) model is a hybrid artificial Intelligence (AI) system that combines the two different techniques of SOS and LSSVM. In this system, the LSSVM acts as a supervised-learning-based predictor to build the accurate input-output relationship of the dataset; and the SOS works to optimize the LSSVM parameters, the σ and γ parameters.

The SOS-LSSVM involves eight major steps which are categorized into two phases, beginning with training phase followed by the testing phase. The whole procedure of SOS-LSSVM is shown in Figure 1. An explanation of the major steps involved in SOS-LSSVM is given below:

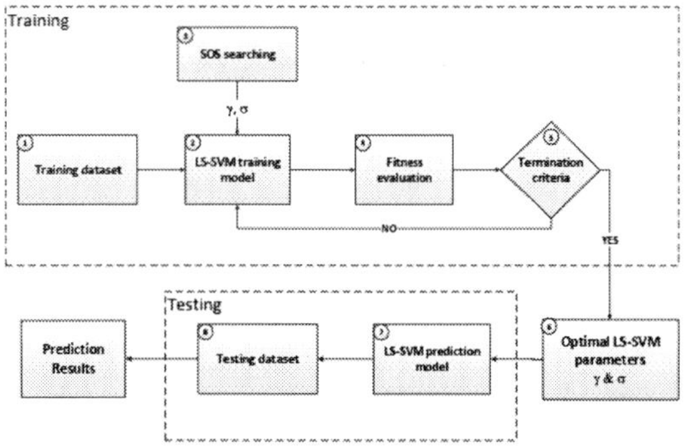

Figure 1. SOS-LSSVM procedure.

1. Training data:
   The data for training are obtained from data collection.
2. LSSVM training model:
   LSSVM addresses the complex relationship between input and output variables. LSSVM requires two parameters, σ and γ, to finish the learning process. LSSVM will gradually obtain the tuning parameters from the optimizer – the SOS algorithm.
3. SOS parameters search:
   In this hybrid AI system, SOS is utilized to explore the various combinations of σ and γ parameters to look for the best set of them. SOS utilizes mutualism, commensalism, and parasitism phase to gradually improve the fitness value of each solution.
4. Fitness evaluation:
   A fitness function is now developed to evaluate accuracy of the learning system. This function has a correlation with the accuracy of the prediction model. The most accurate prediction model is represented by the combination of σ and γ parameters that produces the best fitness value. Instead of randomly splitting the data, it was partitioned into two subsets; learning subset and validation subset. To avoid the sampling bias, the 10-fold cross-validation technique is used in splitting the training data into learning and validation subsets. In this study, the fitness function utilizes mean square error (MSE) of the validation dataset.
5. Termination criteria:
   The process terminates when the termination criterion is satisfied. While still unsatisfied, the model will proceed to the next iteration. As SOS-LSSVM uses SOS, the termination criterion used in this study was the SOS iteration number.
6. Optimal LSSVM parameters:
   The loop stops when the termination criterion is fulfilled. This condition means that the prediction model has identified the input/output mapping relationship with the optimal σ and γ parameters.
7. LSSVM prediction model:
   The optimal LSSVM σ and γ parameters obtained from the training phase are used to establish the prediction model for predicting the testing data.
8. Testing data:
   The data for testing is obtained from data collection. The test set will be used to measure the performance of the LSSVM prediction model.

3 Experimental Results

The experimental dataset was obtained from [1]. A total of 118 samples of uniaxial dynamic creep tests were evaluated from the laboratory experiment at Iran University of Science and Technology, Asphalt Mixtures and Bitumen Research Center. The database includes the percentage of filler to the total aggregate (FP), percentage of binder to the total aggregate (BP), percentage voids in mineral aggregate (VMA), ratio of Marshall stability to Marshall flow (M/F), and the flow number which represents the number of loading cycles where the tertiary deformation starts (Fn). The characteristics of the aggregates, fillers, and the bitumen of the dataset can be seen in Table 1. It also shows the experimental dataset of 5 total input and output variables used in this study.

Table 1 Statistical description of permanent deformation analysis obtained from the dynamic creep test (I = input, O = output)

Variables	Unit	Type	Min	Max	Avg	Std. dev.
FP	%	I	1	10	5.54	3.17
BP	%	I	4	7	5.51	0.81
VMA	%	I	13.2	19.04	16.55	1.41
M/F	-	I	0.61	4.81	2.99	0.74
Fn	cycle	O	22	510	227	143.97

To validate the Symbiotic Organisms Search – Least Squares Support Vector Machine (SOS-LSSVM) performance, another comparison to other predictive models is performed. This includes the Least Squares Support Vector Machine (LSSVM) and the Artificial Neural Network (ANN). In this study, all parameters were set to default for fair comparison.

- For SOS-LSSVM, maximum number of iteration was set to 50; population size was set to 25; and search range for the σ and γ parameters starts from 10^{-5} to 10^{5}.
- For ANN, maximum hidden layers were set to 1; the number of neurons in the hidden layer was set to 4, which is equal to the number of input variables; the number of cycles was set to 1000; and the learning rate was set to 0.1.
- For LSSVM, the parameters of σ and γ were set to 1, as suggested in [13].

Three accuracy measures (coefficient of correlation R, root mean squared error RMSE, and mean absolute percentage error MAPE) were utilized to measure the performance of the predictive techniques. The data was split into 90/10 of training and testing sets. The training and testing processes are performed for each predictive model. The complete experimental results can be seen in Table 2. It can be seen that SOS-LSSVM has performed better in all measurement categories, not only in the training dataset but also in the testing dataset. LSSVM and ANN outputted a similar result in the testing dataset. In terms of R and MAPE, LSSVM earned a slightly better score. Meanwhile, in terms of RMSE, ANN earned a slightly better score.

Table 2 Comparative experimental results among predictive methods.

Performance measure	Predictive methods	Training Result	Testing Result
R	LSSVM	0.9816	0.9234
	ANN	0.9604	0.9161
	SOS-LSSVM	0.9882	0.9564
RMSE	LSSVM	32.37	47.62
(cycle)	ANN	40.91	46.57
	SOS-LSSVM	22.31	35.37
MAPE	LSSVM	23.59	13.88
(%)	ANN	19.09	14.35
	SOS-LSSVM	12.29	8.53

The detailed testing results for the LSSVM, ANN, and SOS-LSSVM are displayed in Table 3. As discussed before, there are 11 cases selected as testing data that represents approximately 10% of the total historical data. After the testing process is finished, the predicted Fn is recorded. Beside the aforementioned performance measures, the differences between the actual and predicted Fn were computed to further illustrate the accuracy of the competing methods. With respect to Table 3, the mean absolute deviation from actual Fn produced by LSSVM, ANN, and SOS-LSSVM are 35.2, 34.7, and 24.9, respectively. It can be concluded that the SOS-LSSVM produced less error than other methods.

Table 3 Detail of testing results.

Case number of testing data	Actual Fn (cycle)	Deviation from Actual Fn (cycle)		
		LSSVM	ANN	SOS-LSSVM
4	260	19.84	61.81	37.31
11	260	-4.79	7.20	-1.97
15	160	75.97	50.57	89.06
19	270	9.94	15.16	14.83
38	190	-12.35	-14.23	-14.99
44	150	37.95	62.40	39.27
57	420	-62.96	-11.77	-8.00
62	420	-25.39	-8.95	1.36
66	380	-107.90	-103.93	-44.88
67	360	-28.37	45.57	19.04
82	60	1.77	-0.04	-3.32

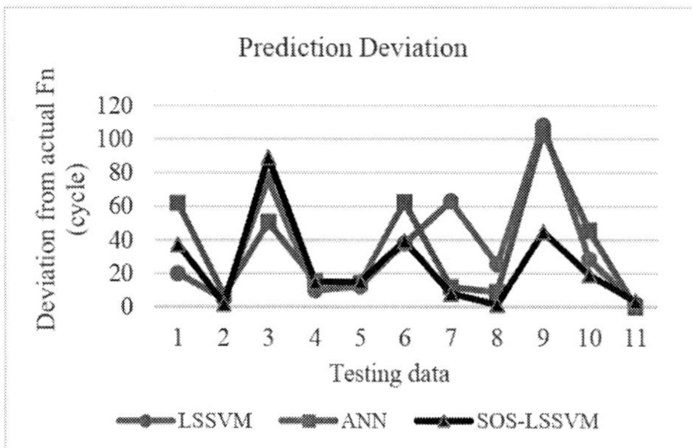

Figure 2. Prediction deviation of LSSVM, ANN, and SOS-LSSVM.

Figure 2 illustrates testing deviations between the actual and predicted values of the three prediction techniques. It can be seen that the SOS-LSSVM produced the best data fit among the other prediction techniques. This further supports that the SOS-LSSVM is the most reliable method for establishing the prediction model.

4 Conclusion

This study developed a new prediction method called Symbiotic Organisms Search – Least Squares Support Vector Machine (SOS-LSSVM) to predict the permanent deformation of asphalt mixtures. To investigate the accuracy of the proposed method, two prediction methods, LSSVM and ANN, were used as benchmarks for the SOS-LSSVM. The experimental data set was acquired from dynamic creep tests of 118 samples. In this investigation, the proposed predictive techniques were applied to the prepared training and testing datasets.

The proposed SOS-LSSVM was further compared for performance outcomes by using three different performance measures (R, RMSE, and MAPE) to obtain a comprehensive comparison of the applied predictive techniques. The findings showed that the proposed SOS-LSSVM achieved the best accuracy for all performance measures. The ANN was in the second place in terms of overall accuracy, while the LSSVM was in the last place.

This study presents a significant contribution to address the importance of the problem of permanent deformation in asphalt mixtures. By accurately predicting the appropriate flow number with the corresponding mixture, the SOS-LSSVM assists the road designers to select the asphalt mixtures accurately within the given specifications. Analytical results indicate that SOS-LSSVM is the most reliable model for building accurate asphalt mixtures based on the specific characteristics of permanent deformation.

References

[1] Alavi A. H., Ameri M., Gandomi A. H., Mirzahosseini M. R. Formulation of flow number of asphalt mixes using a hybrid computational method. *Construction and Building Materials,* 25(3): 1338-1355, 2011.

[2] K.E. Kaloush M. W. W. Tertiary flow characteristics of asphalt mixtures. *Journal of the Association of Asphalt Paving Technologists,* 71: 278–306, 2002.

[3] Witczak M. W., Kaloush K. E., Pellinen T., El-Basyouny M., Von Quintus H. Simple performance test for superpave mix design. *NCHRP Report*: 465, 2002.

[4] Cheng M.-Y., Firdausi P. M., Prayogo D. High-performance concrete compressive strength prediction using Genetic Weighted Pyramid Operation Tree (GWPOT). *Engineering Applications of Artificial Intelligence,* 29: 104-113, 2014.

[5] Cheng M.-Y., Prayogo D., Wu Y.-W. Novel Genetic Algorithm-Based Evolutionary Support Vector Machine for Optimizing High-Performance Concrete Mixture. *Journal of Computing in Civil Engineering,* 28(4): 06014003, 2014.

[6] Cheng M.-Y., Wibowo D. K., Prayogo D., Roy A. F. V. Predicting productivity loss caused by change orders using the evolutionary fuzzy support vector machine inference model. *Journal of Civil Engineering and Management,* 21(7): 881-892, 2015.

[7] Hoang N.-D., Pham A.-D. Hybrid artificial intelligence approach based on metaheuristic and machine learning for slope stability assessment: A multinational data analysis. *Expert Systems with Applications,* 46: 60-68, 2016.

[8] Cheng M.-Y., Hoang N.-D., Wu Y.-W. Hybrid intelligence approach based on LS-SVM and Differential Evolution for construction cost index estimation: A Taiwan case study. *Automation in Construction,* 35: 306-313, 2013.

[9] Gandomi A. H., Alavi A. H., Mirzahosseini M. R., Nejad F. M. Nonlinear Genetic-Based Models for Prediction of Flow Number of Asphalt Mixtures. *Journal of Materials in Civil Engineering,* 23(3): 248-263, 2011.

[10] Mirzahosseini M. R., Aghaeifar A., Alavi A. H., Gandomi A. H., Seyednour R. Permanent deformation analysis of asphalt mixtures using soft computing techniques. *Expert Systems with Applications,* 38(5): 6081-6100, 2011.

[11] Suykens J. A. K., Vandewalle J. Least Squares Support Vector Machine Classifiers. *Neural Process. Lett.,* 9(3): 293-300, 1999.

[12] Cheng M.-Y., Prayogo D. Symbiotic Organisms Search: A new metaheuristic optimization algorithm. *Computers & Structures,* 139: 98-112, 2014.

[13] Suykens J. A. K., Vandewalle J. Least squares support vector machine classifiers. *Neural Processing Letters,* 9(3): 293-300, 1999.

[14] S. K. K., Kim D., S.L.Sekar, Samui P. Model of least square support vector machine (LSSVM) for prediction of fracture parameters of concrete. *International Journal of Concrete Structures and Materials,* 5(1): 29-33, 2011.

[15] Tejani G. G., Savsani V. J., Patel V. K. Adaptive symbiotic organisms search (SOS) algorithm for structural design optimization. *Journal of Computational Design and Engineering*, 2016.

[16] Cheng M.-Y., Prayogo D., Tran D.-H. Optimizing Multiple-Resources Leveling in Multiple Projects Using Discrete Symbiotic Organisms Search. *Journal of Computing in Civil Engineering*: 04015036, 2015.

[17] Tran D.-H., Cheng M.-Y., Prayogo D. A novel Multiple Objective Symbiotic Organisms Search (MOSOS) for time–cost–labor utilization tradeoff problem. *Knowledge-Based Systems*, 94: 132-145, 2016.

[18] Cheng M.-Y., Chiu C.-K., Chiu Y.-F., Wu Y.-W., Syu Z.-L., Prayogo D., Lin C.-H. SOS optimization model for bridge life cycle risk evaluation and maintenance strategies. *Journal of the Chinese Institute of Civil and Hydraulic Engineering*, 26(4): 293-308, 2014.

[19] Duman S. Symbiotic organisms search algorithm for optimal power flow problem based on valve-point effect and prohibited zones. *Neural Computing and Applications*: 1-15, 2016.

[20] Prasad D., Mukherjee V. A novel symbiotic organisms search algorithm for optimal power flow of power system with FACTS devices. *Engineering Science and Technology, an International Journal*, 19(1): 79-89, 2016.

[21] Kamankesh H., Agelidis V. G., Kavousi-Fard A. Optimal scheduling of renewable micro-grids considering plug-in hybrid electric vehicle charging demand. *Energy*, 100: 285-297, 2016.

[22] Verma S., Saha S., Mukherjee V. A novel symbiotic organisms search algorithm for congestion management in deregulated environment. *Journal of Experimental & Theoretical Artificial Intelligence*: 1-21, 2015.

[23] Abdullahi M., Ngadi M. A., Abdulhamid S. i. M. Symbiotic Organism Search optimization based task scheduling in cloud computing environment. *Future Generation Computer Systems*, 56: 640-650, 2016.

[24] Ruskartina E., Yu V. F., Santosa B., Redi A. A. N. P. Symbiotic Organism Search (SOS) for Solving the Capacitated Vehicle Routing Problem. *International Journal of Mechanical, Aerospace, Industrial, Mechatronic and Manufacturing Engineering*, 101: 857-861, 2015.

The Development of a BIM-enabled Inspection Management System for Maintenance Diagnoses of Oil and Gas Plants

H.L. Chi[a] J. Wang[a] J. Chai[a] and X. Wang[a,b]

[a]Australasian Joint Research Centre for Building Information Modelling, Curtin University, Australia
[b]Department of Housing and Interior Design, Kyung Hee University, South Korea
E-mail: hung-lin.chi@curtin.edu.au, jun.wang15@student.curtin.edu.au, jian.chai@postgrad.curtin.edu.au, xiangyu.wang@curtin.edu.au

Abstract –

Current inspection process has productivity issues, such as double handing information, time-consuming data aggregation, lack of information transparency and so on. About inspections for plant maintenance, site managers encounter difficulties in getting global pictures tracing a significant amount of inspection information at management perspective. The decision-making is based on paper-based schedules and updated progress of them, which lacks information reliability and change flexibility. The purpose of the research-industry cooperation is to determine how the integration of inspection management and Building Information Modelling (BIM) influences the performance of the plant inspection and maintenance processes. The research team developed the integration among an online inspection management portal, a mobile client application and an existing BIM platform, Navisworks™, in this study. By using the developed system, the information of inspection tasks, including daily status reports, schedules, and task assignments, can be linked with BIM for better visualization, communication and asset life cycle management. Site managers are able to view the inspection processes through 4D simulation identifying particular inspection tasks and time duration of their interests. The outcomes of a pilot study conducted through industrial partners shows positive comments, such as reducing time spent on aggregating inspection information, going paperless and increasing real-time inspection visibility for managers to make corresponding decisions.

Keywords –

Building Information Modelling; Inspection; Plant life extension; Maintenance

1 Introduction

Inspection is one of the most essential processes [4] in identifying potential issues affecting the execution of process plants in current maintenance practice. There are different purposes for conducting inspections of a plant, such as quality control of under-constructing plant modules, pipe corrosion detection and replacement, functionality analysis of plant instruments, pre-determination of the requirement for shutdown maintenances and so on. Each of these requires accurate and efficient evaluations, a perpetual cycle of inspections and capital renewal actions [3] in order to make the maintenance diagnoses with sufficient supportive evidences and to be judged to maintain the serviceability of the plant facility. It usually requires a significant amount of field crews' efforts, management and effective inspection tools to accomplish such missions, especially in a large-scaled and complex facility.

As can be seen in state-of-the-art inspection processes, many issues exist and obstacle the productivity gained at every stage of processes. The transformation of the inspection information is a manual process between the locations of site offices and fields, which results in data aggregation problems, such as double handling information, misinterpretation, human errors and risks of document loss. These problems sometimes let the inspection be seriously dragged on. In addition, error and insufficient information increases the difficulty for managers to make flexible and appropriate decisions, resulting in cost blowout and significant delay in the completion of the inspection tasks. Furthermore, current inspection practice lacks the information transparency among site managers and field crews, increasing the difficulty of communication and the opportunities to get global pictures of the entire inspection statuses simultaneously. The above-mentioned issues need to be solved urgently given the circumstance that the complexity of current project is getting higher and higher these days.

Previous research work focuses on improving current inspection processes through the innovations at management and technical aspects. Take technologies adoption for an instance, Choi and Park [8] designed and implemented a field inspection management system based on wireless sensor networks (WSN). Yang *et al.* [9] conducted an application research on the intellectual inspection system of transmission line based on the mobile information platform. Regarding management innovations, Akanni and Alonge [5] proposed an effective implementation of risk based inspection (RBI) approach in asset integrity management of oil and gas facilities.

An observation from the previous research work shows that researchers are working on integrating emerging technologies to create a digitalized environment for reducing the use of paper-based documentary and manual effort of processing. Besides potential hardware that can be used to facilitate current inspections, new considerations of management also help improve their efficiency and effectiveness. However, the way to view and organize the information for a large-scaled and complex project still requires managers' effort in comparing and discovering deviations among planned and actual information. There should be an intuitive approach to show clear status of the entire process and facilitate appropriate decisions to be made for further maintenances actions and predictions.

To further eliminate the above-mentioned issues between current inspection systems and their users, the researchers introduce Building Information Modelling (BIM) technologies to be integrated as a user interaction layer for providing sufficient visualization capabilities to site managers. BIM is utilized in the entire life cycle of a facility which contains various information potentially benefits to discover the progress issues of the inspection. The extra information, such as cost, existing schedules and component properties, can be further referenced by managers to easily perform the inspection data comparisons and organizations. The objective of this research-industrial cooperation is to develop a BIM-enabled inspection management system so that site managers can have better communication and understanding of the status of the inspections, through the visualization of the 3D models. It is expected to help create a paper-less environment minimizing data aggregation error and waste, and to provide necessary information for further decision making.

The rest of sections are written as the follows: Section 2 describes a typical inspection process and its critical decision points needed to be aware of; Section 3 is written to introduce the architecture of the proposed BIM-enabled inspection system. The BIM-enabled functionalities are presented as well; Section 4 shows the pilot study of the research and the summary of the discussion with industrial partners regarding the use of the proposed system; Finally, section 5 presents the conclusion and potential future work of this study.

2 Inspection Process

Considering a typical inspection process, it always comes out with a decision about whether maintenance actions are needed depending on the result of the inspection. After a consultation with industrial experts, a general and complete life cycle of a typical inspection process is summarized and can be seen as Figure 1. It goes through six stages: Planning, Execution, Review and Engineering, Implementation, Examination, and Closeout. At the beginning stage, it always needs to plan inspection tasks based on the facility's maintenance strategy. The tasks will be scheduled and issued for execution. In cases of oil and gas plants' inspections, these stages always represent a significant amount of planning effort, large scaled survey activities and frequent coordination in order not to influence with the oil and gas production processes. It can be even a decision for the preparation of a shutdown maintenance which closes the entire plant for essential process components' repair and replacement.

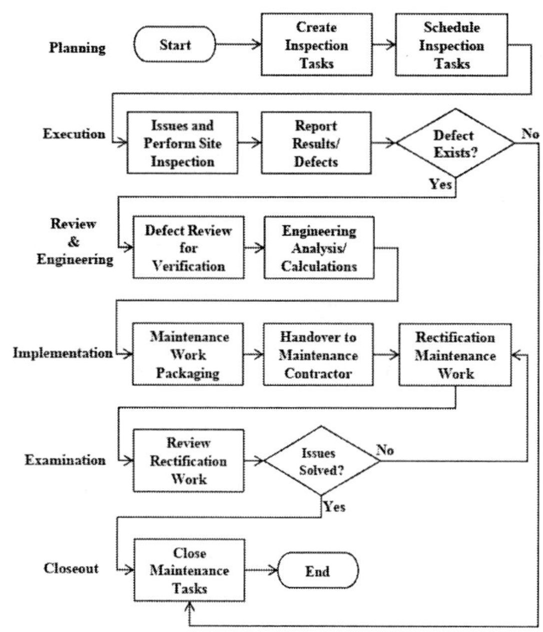

Figure 1. A general life cycle of the inspection process for infrastructures, buildings and industrial facilities

After the execution of the planned inspection task, here comes to the first decision point: is any defect or urgent fatal discovered and needed to be responded immediately? It the answer is *No*, the inspection task is considered to be completed and closed. Otherwise, a review process should be adopt with more verification, engineering analysis and calculation in order to identify the properties of the defect in more details and for scoping necessary maintenance tasks.

After the maintenance work packages are defined and scheduled, the developed maintenance tasks will be contracted and the rectification will be executed. The following review process is heading to the second decision point: have the issues discovered during the inspection task been solved through the maintenance actions? If it has been judged as a *Yes*, consider the maintenance task is done and the case will be closed. If it is not the case, further rectification is necessary in order to fix remaining problems.

As can be seen in the above-mentioned process, the entire inspection process is a complex mission with a considerable amount of effort expected on planning, coordination, judgement, communication and verification [3]. It is especially difficult to be well-managed in a complicated and large-scaled facility like oil and gas plant [6, 7]. Thus, BIM, as a management data base with visualization capabilities, is integrated to serve for the better communication among decision makers. It is done by providing visualizations of the entire inspection situations, coordinative functions of the inspection management and historical inspection simulations. These features can enhance the efficiency and effective of the inspection process during Planning, Review and Engineering, and Examination stages. A BIM-enabled inspection system is proposed and developed in this study. The following section describes its overall architecture and elaborates its' functions for better management.

3 BIM-enabled Inspection System

In this study, a BIM-enabled inspection system which integrates a BIM visualization platform, Navisworks™ [1], a web-based inspection management portal, Inspectivity Portal™, and a mobile inspection application, Inspectivity Go™, [2] has been developed. Multiple inspection managers can access the inspection information through the web interface and see the visualization results on the BIM platforms of their own computers simultaneously, as long as the system keeps the data synchronized. The details of the developed system architecture and functionalities can be referred to the following two sub-sections:

3.1 System Architecture

The overall system architecture is illustrated as Figure 2. Basically, the system can be divided into three main modules: Web-based management, Mobile computing and Visualization module. The web-based management module is the foundation of the inspection system, which contains two sub-modules: Inspection and Issue management sub-module.

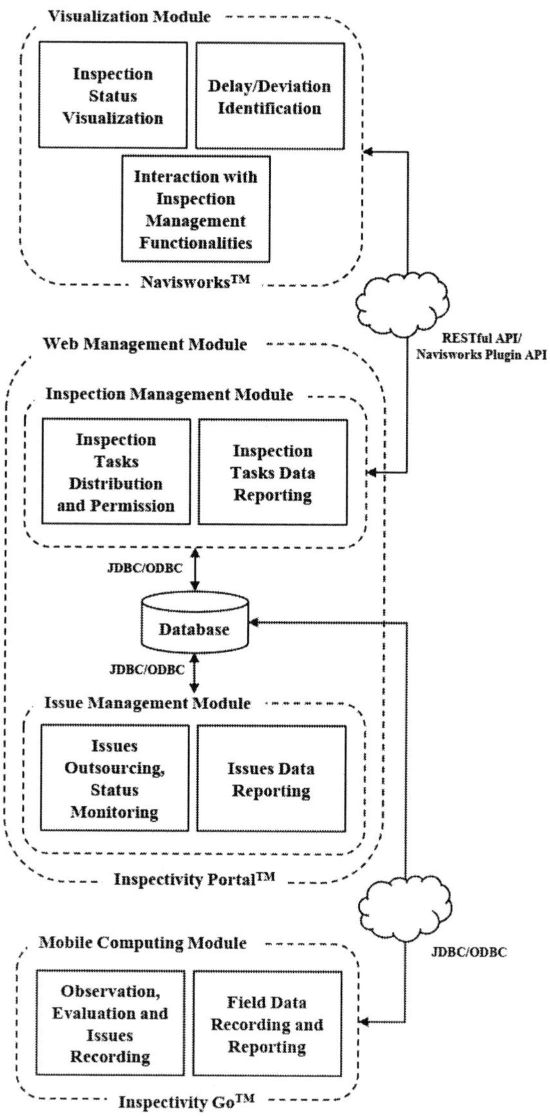

Figure 2. System architecture of the developed BIM-enabled inspection management system

The inspection management sub-module is a web portal allowing user to coordinate every inspection tasks. The coordination options include creating, assigning, approving, modifying and monitoring inspection tasks through web-based interface; As for the issue management sub-module, it basically shares the same information of a database with the inspection management sub-module. Once an issue has been raised and reported through mobile computing module used at field, the work flow of handling the issue can be created through this sub-module and the created process will be tracked and updated once any further examination been conducted. Both of the sub-modules can be accessed through Internet by office computers for site managers or planners to know the latest information of the inspection.

Mobile computing module is an application installed in mobile devices, such as tablets, which can be utilized by site inspectors to perform any inspection and examination tasks. Firstly, the field inspectors need to download the information of the inspection tasks assigned to them, from the web-based management module to their mobile application. The list of inspection tasks is varies depending on the assignments and different login authorizations for the individual field inspector. Once any latest assignment is announced, the notification messages will be sent out to the account which is already logged in. The application further provides inspection related functions, such as inspection form editing, check option designing, field photo capturing, annotating and commenting and so on, for field inspectors to record the inspection results. Those results can be uploaded to the database and update the information to the web-based management module, once the field inspectors have headed back to the office or any places with Internet capacities.

By utilizing the above-mentioned two main modules, the data of inspection tasks can be transferred digitally which reduces potential manual error of data collection and organization. However, it still lacks interpretabilities for better communication among managers to make effective decisions. Due to all the format of the information is in text- or list-based, it is difficult to get a global picture without significant effort on data referencing and comparing, in order to determine what the appropriate maintenance action the managers should go for at the next step. Regarding the effective visualization which helps managers to create maintenance plans with more confidents, the study is particularly focused on the development of the connection between the web-based management module and visualization module.

The visualization module is developed on an existing BIM platform to show the 3D facility model related to inspections in the virtual environment. The facility information is also associated with the corresponding components on the model. Furthermore, schedule information linked between different work activities and corresponded components are also provided. Thus, the inspection status of different components can be easily retrieved by calling RESTful web application programming interface (API) [10] to synchronize the latest information from the web-based management module. It is also easy to identify any delay or deviation of inspection tasks by comparing the planned schedule with the actual reported one.

For the implementation of the system in this study, the researchers utilized commercial software, Inspectivity Portal™ and Go™, as the web-based management module and the mobile computing module individually. They are all synchronized with the same database. As for the visualization module, the researchers utilized Autodesk Navisworks™ as the BIM platform and developed a plugin by using Navisworks plugin API and RESTful web API to communicate with the Inspectivity Portal™. The details of the BIM-enabled functionalities can be seen in the following sub-section.

3.2 BIM-enabled Functionalities

The purpose of integrating BIM capabilities in the visualization module of the system is to increase the data interpretabilities and communications during decision-making processes. The overall BIM-enabled functionalities contain three main sets of functions: (1) data synchronization, (2) status identification and animation, and (3) task manipulation. They are elaborated accordingly in the following sub-sections:

3.2.1 Data Synchronization

Data synchronization functions help user connect the relationships between the component of 3D geometrical facility and its related inspection task in the inspection management sub-module. It can be demonstrated in Figure 3 and 4. Figure 3 shows the user interface of the inspection management sub-module. As can be seen in the figure, a tree hierarchy of the facility's components (left-side of the figure) is generated. It is referenced to the uploaded information of 3D components. It executes in real-time by pressing a menu item of the plugin (as can be seen in Figure 4) and a batch uploading process can be executed. By generating the tree hierarchy automatically, user can plan and create related inspection tasks on the individual item on the tree, including the detail task descriptions, schedule, assignee, and supervisors.

The defined inspection tasks can be further performed by field crews through the use of mobile computing module to report the latest outcomes back to the inspection management sub-module. Once the inspection results are captured at sites, the data will be uploaded as long as crews go back to the office and submit it with Internet capability. It is a necessary procedure similar with that of manual reporting approach instead of the higher accuracy and efficiency. It helps rapid reactions against urgent task changes. Figure 4 shows the user interface of the visualization module. As can be seen in the figure, detail inspection information (properties and outcomes) from the inspection management sub-module can be further downloaded to the visualization module, and updated and displayed as properties of the 3D components (right-side of the figure). In addition, related inspection schedule of the corresponding component can also be generated automatically (bottom of the figure).

It should be noted that in this study the synchronization is dominated by the information from the inspection management sub-module. BIM models and inspection tasks are bounded by identifiers and cannot be modified by normal users. The proper approach in handling synchronization issues is expected and will be investigated in the future studies.

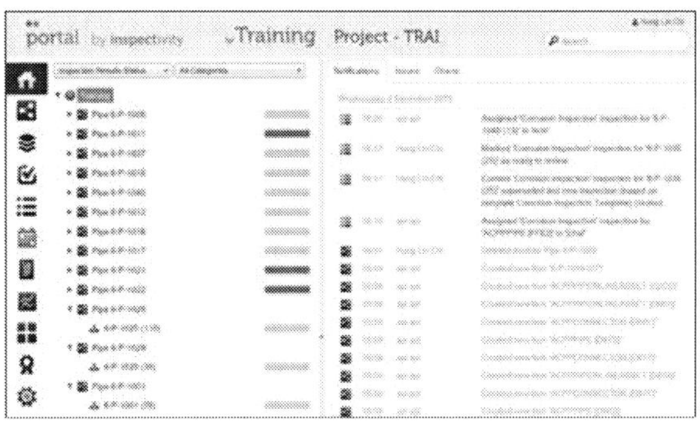

Figure 3. The user interface of the inspection management sub-module (Inspectivity Portal™)

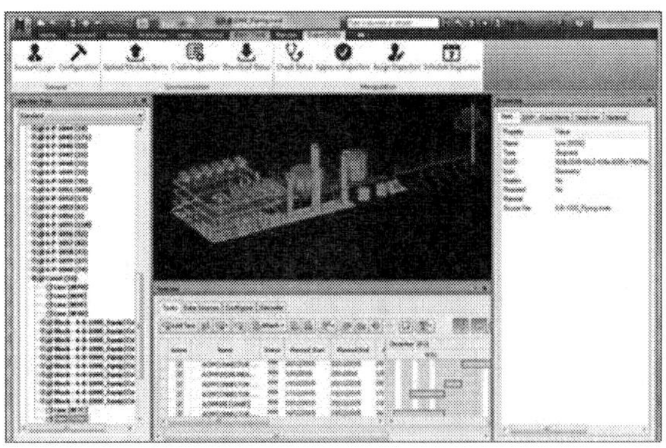

Figure 4. The user interface of the BIM visualization module (Navisworks™)

3.2.2 Status Identification and Animation

Once all the latest inspection information is synchronized with the corresponding ones in the visualization module, typical BIM simulation functions become useful. As can be seen in Figure 5, color-coded inspection statuses of different pipes in an oil and gas plant are presented based on the synchronized data. It is easy for user to click on any color-coded component and get its related inspection details right away on the properties list (right side of the user interface shown in Figure 5).

For example, managers may want to check the delay reasons if there is a task wasn't finished on time. It can be done by simply comparing the actual progress with the planned inspection schedule. It should be noted that the researchers defined a set of general inspection statuses in the plugin: Draft, Review, Assignable, Assigned, In Progress, In Progress at Field, Performed, Completed and Archived. As shown in Figure 6, different inspection tasks show their status in different colors at the specific tracking period. The representative colors of different inspection status can be edited through configuration function as shown in the right side of Figure 6.

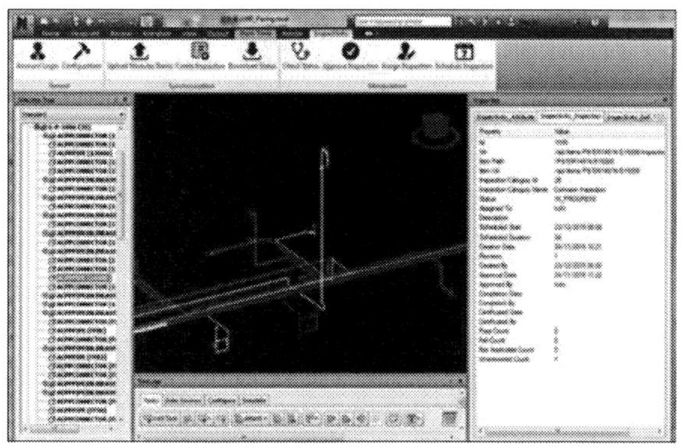

Figure 5. The color-coded inspection statuses of different pipes in an oil and gas plant

Besides showing the overall inspection status to users, a 4D (3D plus schedule) animation function is implemented to show the different inspection status with time goes by. It also helps managers evaluate how reliable the current inspection plans against actual implementations at field, which would facilitates re-planning process to be performed on the web-based management module or modifications in the visualization module right away. The re-planning process of inspection tasks allows better control of the inspection efficiency and reflects actual situations of field activities. The re-planning functions are described in the next sub-section.

It should be noted that only CAD representations of the existing oil and gas facility could be enough for the status identification because of the sufficient information of geometrical visualization. However, further related information for supporting decision-making, such as cost analysis of maintenance, quantity information and animation, which BIM can provide would better achieve the goals of inspection monitoring. This is the reason why the researchers adopt BIM platform in the proposed system development.

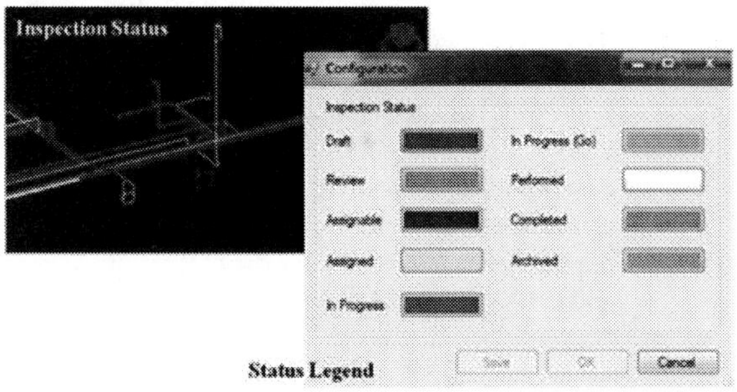

Figure 6. The configuration of the color-coded inspection status

3.2.3 Task Manipulation

In the visualization module, partial inspection management functions have been realized for modifying the details of inspection tasks, as the same ones in the inspection management sub-module. It is implemented because managers may need to rapidly re-plan the current inspection tasks based on the actual inspection progress reported from fields. These inspection management functions include approving certain inspection tasks, assigning them to related field crews, and re-scheduling specific tasks. Approving inspection tasks is a function that helps perform quick reactions while managers observed any permission haven't been dealt with. It can be observed by viewing the visualization of inspection statuses; Assigning function is designed with similar intentions, which speeds up the inspection progress when the similar delays in the work arrangement are discovered; As for the re-scheduling of inspection tasks (as shown in Figure 7), manager can easily reference to the current progress of the specific task and evaluate the actual inspection effort, to adjust its schedule plan. It would help better control the field resources and shorten the actual inspection duration of finishing all tasks, which will benefit to complex and large-scaled inspection events such as the preparation of the shutdown maintenance.

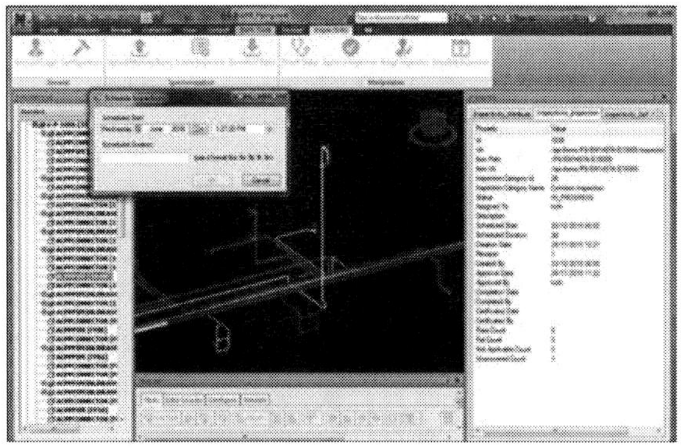

Figure 7. The re-scheduling function for the specific inspection task

All BIM-enabled functionalities have been implemented and packaged as a plugin on Navisworks™ platform. Different roles of project crews in the inspection management process require different account with different authorities to access this plugin. The plugin can be logged in by users and provide different functions depending on what role the user is, such as inspection manager, inspector, site crews, supervisors and so on.

After the implementation of the proposed system, the next section describes the details of a pilot study through a discussion with industrial partners. Some comments from the industrial partners regarding the use of the proposed system have been presented as well.

4 Pilot Study and Discussion

For evaluating the usefulness of the proposed BIM-enabled inspection system, the researchers have conducted a pilot study by (1) cooperating with a software company professionalized in the inspection management of the oil and gas plant; (2) introducing the developed system to an oil and gas industrial company and (3) seeking for their professional comments regarding the adoption feasibility of the system in practice.

The oil and gas company is currently executing a life extension project for a 30+ years old Liquefied Natural Gas (LNG) plant. In this project, a significant amount of inspection tasks involving establish of scaffoldings, utilization of corrosion detection instruments and massive evaluations are planned and scheduled. And the inspection activities are performed simultaneously with the regular LNG production processes. So the inspection task can only be executed in one limited region at one time, trying to minimize effects to the production. Therefore, accurate and flexible inspection plans and effective task monitoring are very important for the project so that the time and cost spent on inspections can be optimized for further maintenance processes.

By presenting the developed system to industrial inspection managers, the following comments and benefits has been summarized and identified:

- *Data aggregation approach of the proposed system is efficient and effective*: Inspection managers generally agree that the proposed system is useful to create a paper-less environment which eliminated the potential information loss and error during the data aggregation processes. In the conventional way, field crews need to collect and remember all inspection information and record them through papers, photos and computers manually. The system helps reduce the chances of encountering uncertain situations such as double handling the data, manual input error and documents missing.
- *Status identification of the proposed system help support decision making*: Inspection managers used to identify all the inspection statuses and display them with different colors on the 3D model, just as what the developed system did instead of the entire process is manual. It usually takes time in weeks to come out with a decent drawing for decision makings. With the use of the system, the time spent on the view generation can be significantly reduced and it will indeed buy more reaction time while the actual situation going far from the expectation.
- *More configurations of the inspection status are needed*: Inspection managers suggest that the function of the status identification and animation should be extended, so that users can define the names of different inspection status by themselves to include different outcomes of the inspection. In a pipe inspection task, for example, managers would like to know how serious the corrosion condition of a specific segment of the pipe is and how the corrosions will grow with time through the collected historical data. It can facilitate making decisions for prioritizing the maintenance work sequence due to the limited budget. The growing pattern of the corrosion can be also a good reference to be considered in further predictive maintenance considerations.
- *Issue management also requires visualization capabilities*: Regarding the issues discovered during inspection processes, it is also necessary to keep tracking the progress of the corresponding maintenance. Inspection managers suggest a new synchronization function to be realized for monitoring when the issues can be solved and how to do so, by viewing the global view of the issues' status in the system. Technically, such function can be implemented by cooperating with the software company to provide another RESTful web API for the issue management sub-module. It can be then integrated into the current developed plugin. However, how to make the data structure of inspection tasks co-existing with that of issues and to be visualized individually still needs further design effort and examinations.
- *Further validation is required as well as the experimental adoption*: In despite of positive feedbacks received in the discussion, inspection managers raised several concerns to be noted, such as culture changes of implementing new technologies among field crews against the conventional approaches. The actual benefit of the developed system must be identified through more evaluations in practice, in order to validate the expectations summarized from the discussion. It should be done at the next stage of the study.

5 Summary and Conclusion

A BIM-enabled inspection management system is developed in this study in order to increase the productivity of inspection tasks and get better control of uncertainties. The developed system contains three main modules: web-based management, mobile computing and visualization module. They are sharing the same database and can be synchronized or linked through web API. A plugin in the BIM platform, Navisworks™, is developed to support the visualization module. It also provides three sets of functions: data synchronization, status identification and animation, and task manipulation for users to easily get ideas about inspection details. Pilot study including discussions with inspection experts, from an oil and gas industrial partner, has been conducted and the feedbacks are generally good for the feasibility assessment of the system adoption in practice.

The contributions of the research-industry coordination include that: (1) the proposed approach extended BIM visualization capacities, such as color-coded status identification and 4D animations, from oil and gas plant planning and construction stages to maintenance and operation stages; (2) the system provides efficiency and effectiveness in inspection data aggregations as well as inspection status tracking. In addition, it helps managers know the deviations of inspection tasks rapidly in order for better communication and decision making; (3) through this study, BIM model will have inspection information as another dimension of the management, which facilitates the asset management at the entire life cycle of the oil and gas facility in the future.

For the future plan of the research-industry cooperation, evaluations will be conducted in order to further validate the true values of the developed system. Quantitative results collected in a simulated inspection environment will be used to support the claims in this study. Furthermore, the developed system will be integrated with more features based on the comments provided from the industrial partners, such as adding issues' status monitoring functions, historical data animation functions for inspection outcomes, the freedom of naming different inspection status and so on.

Acknowledgement

This research was undertaken with the benefit of a grant from Australian Research Council Linkage Program (Grant No. LP130100451). The researchers would also like to acknowledge company Inspectivity Ltd. for their contribution of the software, web API, inspection expertise and mobile devices.

References

[1] Autodesk Navisworks | Project Planning. On-line: http://www.autodesk.com.au/products/navisworks/overview, Accessed: 15/03/2016.

[2] Inspectivity Commissioning Completion Software. On-line: http://www.inspectivity.com, Accessed: 15/03/2016.

[3] Hegazy T. and Gad W. Dynamic System for Prioritizing and Accelerating Inspections to Support Capital Renewal of Buildings. *Journal of Computing in Civil Engineering*, 29(5): 04014058, 2015.

[4] Laofor C. and Peansupap V. Defect Detection and Quantification System to Support Subjective Visual Quality Inspection via a Digital Image Processing: A Tiling Work Case Study. *Automation in Construction*, 24: 160-174, 2012.

[5] Akanni J. and Alonge O. Effective Implementation of Risk Based Inspection (RBI) Approach in Asset Integrity Management of Oil and Gas Facilities. In *Proceeding of ASME 2015 Pressure Vessels and Piping Conference*, Boston, MA, 2015.

[6] Chandima Ratnayake R. M. Challenges in Inspection Planning for Maintenance of Static Mechanical Equipment on Ageing Oil and Gas Production Plants: The State of the Art. In *Proceeding of ASME 2012 31st International Conference on Ocean, Offshore and Arctic Engineering*, Rio de Janeiro, Brazil, 2012.

[7] PharmOut White Paper: 10 Common GMP Challenges Facing Maintenance Departments in Pharmaceutical Plants. On-line: https://pharmout.net/downloads/white-paper-10-problems-with-pharma-plant-maintenance.pdf, Accessed: 24/05/2016.

[8] Choi S. H. and Park H. K. Design and Implementation of a Field Inspection Management System Based on Wireless Sensor Networks. *International Journal of Distributed Sensor Networks*, 2014:1-4, 2014.

[9] Yang Z. W., Yi W. and Jin Y. Application Research on the Intellectual Inspection System of Transmission Line Based on the "Mobile Information Platform." *Applied Mechanics and Materials*, 596: 719-722, 2014.

[10] Pautasso C., Wilde E. and Alarcon R. *REST: Advanced Research Topics and Practical Applications*. Springer-Verlag, New York, 2014.

A Framework for Establishing Early Warning System for Working in Hot Environments

W. Yi[a], J. Zhu[b], X. Liu[b], X. Wang[b] and A.P.C. Chan[a]

[a]Department of Building and Real Estate, The Hong Kong Polytechnic University, Hong Kong, China
[b]School of Built Environment, University of Curtin, Australia
E-mail: yiwen96@163.com, junxiang.zhu@postgrad.curtin.edu.au, xin.liu@curtin.edu.au, xiangyu.wang@curtin.edu.au, albert.chan@polyu.edu.hk

Abstract

Heat stress is a well-known occupational hazard, and climate change together with the increased frequency and intensity of extreme heat events has made risks more severe and widespread. The liquefied natural gas (LNG) construction industry is found to be more susceptible to heat stress than other industries because LNG workers are subjected to heat stress not only from outdoor strenuous physical work but also in confined working spaces which could be even worse. Traditional management of heat stress are by and large some "dos and don'ts" and are not based on scientific measurements. The continuing high frequency of heat related incidents in the construction LNG industry calls for better approaches to detect impending attacks of heat stress. The purpose of study is to develop a smart work site against hot climates using modern technologies. The proposed system could be automated by information communication technology, sensor technology, Radio Frequency Identification (RFID) technology, to protect the wellbeing for those who have to work in hot environments. A Bayesian networks model could be established to evaluate the heat strain of workers. 3D environment map of a LNG site could be simulated by the readings of environmental sensors and trilinear interpolation. The work duration, location, and workplace heat exposure of workers could be monitored during the work and warning alerts would be sent to the workers when certain threshold exceeds. The proposed smart work site for early warning of heat stress would be of tremendous value in better safeguarding workers' health and safety by reducing the occurrences of heat stress on site.

Keywords –

Heat stress; Health and safety; Liquefied natural gas (LNG) construction industry; Early-warning system

1 Introduction

Liquefied natural gas (LNG) industry is one of the most rapid growing energy market trades. As a large, dynamic, and complex sector, it creates employment for millions of people worldwide. It is estimated that the LNG workforce in Queensland could increase to 17,000 workers in 2024 based on 59,000 wells and six LNG trains [1]. Construction projects of LNG plant involving assembling and installing of pipes, pumps, vessels, storage tanks, cooler and electronic system is complicated, which would cause a higher risk of work-related illnesses and accidents. Considering the complicated nature and the large scale, health and safety of LNG workers needs to be paid greater attention.

Heat stress is recognized as major occupational hazard in many workers [2]. In a broader sense of comparison between the East and the West, an average of 13.8 workers per year died from heat related disorders in Japan between 1991 and 2001 [3]. A cohort study among 58,495 Thai workers indicated that approximately 20% of them experienced occupational heat stress that significantly related to injuries and illnesses [4]. In the United States, 423 workers in agricultural and non-agricultural industries were reported to have died from exposure to environmental heat between 1992 and 2006 [5]. Many occupational settings are commonly encountered in extreme hot environments, including outdoor cleaning and horticulture, iron manufacturing, mining, firefighting, military, and catering, and glass factories. Liquefied natural gas (LNG) workers are at high risk of heat stress not only from outdoor strenuous task, but also in confined working spaces, which could be even worse. More importantly, the four existing and six proposed LNG facilities in Australia are built in Zone 1 characterised as high humid summer proposed by the Building Code of Australia. Zone 1 represents a climate with hot humid summer and warm winter. As the number of oil and gas developments increase in a hot and humid environmental condition in Australia, heat-related disorders and heat stress prevention are of critical importance for the Australian LNG industry.

To protect workers from the risk of heat stress, a number of guidelines and initiatives have been put in place to promote health and safety in the workplace including proper work-rest schedule, ventilation facilities at work or rest places, provision of portable water, personal cooling system [6]. However, these recommendations are largely based on regulations, which is considered as a reactive approach to manage heat stress. No systematic studies has been initiated to early warning and safety evaluation of heat stress recognition, assessment, control and management of individual behaviour. Advanced technologies and devices have been regarded as an effective and proactive method to enhance on-site safety performance. For example, Wu et al. argued that by using RFID sensor networks for tracking near-misses, onsite accidents can be avoided, hence improving safety performance [7]. Cheng et al. adopted physiological status monitoring (PSM) technology to monitor safe and unsafe behaviour of workers [8]. Therefore, the research aims to develop an early-warning system for heat stress using modern technologies and Bayesian network as the analytical algorithm.

2 Framework for Early-warning System of Heat Stress

The aim of the early-warning system is to reduce the occurrence of heat-related incidents on site. The research will combine the assessment of workers' heat strain, simulation of 3D site environment, and establishment of early-warning system. The framework of the early-warning system of heat stress in shown in Figure 1.

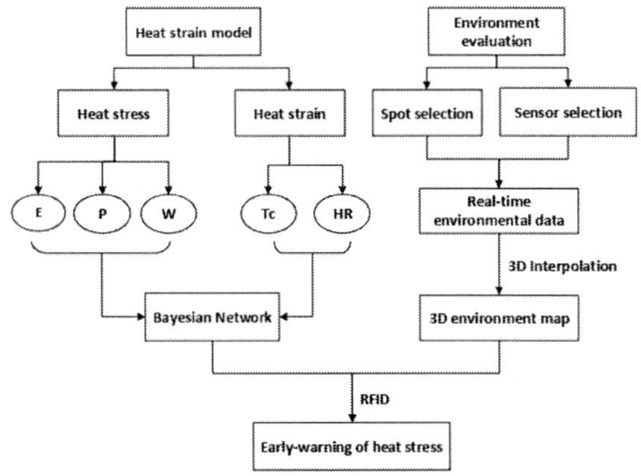

Figure 1 Framework of the early-warning for heat stress

Note: E is environmental data (i.e., air temperature, relative humidity, wind speed, solar radiation); P is personal information (i.e., age, percentage of body fat, smoking habit, alcohol drinking habit); W is work-related data (i.e., work duration, work intensity); Tc is core temperature (°C); HR is heart rate (bpm).

2.1 Stage 1: Heat Strain Model

Heat stress is defined as "the net heat load to which a worker may be exposed from the combined contributions of metabolic cost of work, environmental factors and clothing requirements" [9]. The effects of working in hot weather can potentially have a severe influence on the welfare of workers if the risks have not been properly managed or considered. It may result in an increase in heart rate, body core temperature, and sweating. When subjected to a hot environment, heart rate will increase to move blood and dissipate heat from the heart, lungs, and other vital organs to the skin [10] and [11]. If the excessive heat is not fully dissipated from the body, the body will store heat and as a result core temperature will rise. The common heat-related disorders are heat fainting, heat exhaustion, and heat stroke. The heat strain of workers is assessed by their body core temperature and heart rate. Heat exhaustion is a moderate heat illness in which core temperature is between 38°C and 40°C [12]. Heat stroke is the most severe heat illness when core temperatures exceed 40°C [13]. Earlier studies reveal that human heart rate is in the range of 130 and 160 bpm when he/she performs exercise in hot environments. The American Conference of Governmental Industrial Hygienists (ACGIH) has recommended the threshold limit values (TLVs) for heat stress and suggested the limit for heart rate is 180 bpm minus the age of an individual lasting no more than three minutes [14]. Furthermore, physiological researches have demonstrated heat strain are associated with personal, work-related, and environmental factors [15].

Field studies will be conducted during summer time in different work sites. Outdoor workers without flu in the week prior to participation, without a history of diagnosed major health problem such as diabetes, hypertension, cardiovascular disease and regular medication intake will be invited to participate in the field study. Subjects will be asked to provide personal information on age, alcohol drinking habit, sleeping habit. Their height, weight, percentage of body fat, resting heart rate, blood pressure, and thermal insulation of their clothing will be measured. During the test, the workers will perform tasks as their usual daily routine. Wet bulb globe temperature, integrating all four elements of the thermal environment (i.e., air temperature, relative humidity, air velocity, and solar radiation), will be measured by sensors mounted on safety helmet at regular time intervals. Heart rate and energy consumption will be recorded every 10s by a heart rate belt (Polar Vantage XL, Finland). Core temperature will be measured by an ingestible telemetric temperature system, which consists of a commercial temperature pill and a compact data receiver (CorTrack™, HQInc., Palmetto, Florida, USA). A sensor inside the temperature pill could transmit a continuous radio wave that varies with temperature. The data logger will receive and store this signal accordingly.

RFID integrated with mobile communication system will be adopted to track subjects' positions in outdoor environment. Subjects' positions will be sent back to a tracking server at defined time interval. Subjects' real-time positions will be displayed in a Geographic Information System (GIS) system. The position, physiological and environmental information collected by RFID, CorTrack™ and environmental sensors, respectively, will be stored in a geospatial database for further analysis together with site structure and development drawings, plans. The relationship between heat strain (core temperature and heart rate) and factors affecting heat strain (e.g., personal characteristics, environmental variable, work-related data, and work location) will be established by the Bayesian network model.

2.2 Stage 2: Environment Evaluation

Environmental conditions in various work locations are at different levels of heat stress within a work site, the aim of stage 2 is to simulate the environment condition of a LNG site. WBGT is calculated by the display of three thermometers on globe temperature, wet bulb temperature, and dry bulb temperature. Sampling locations for WBGT will be optimized by Kriging to better capture the spatial variation of the environmental condition [16]. To monitor the WBGT at different locations of a LNG site, several WBGT sensors packed by senor module (that involved globe temperature, wet bulb temperature, and dry bulb temperature sensors), Global System for Mobile Communications (GSM) module, and microcontroller will be deployed [17]. The microcontroller converts the signal from analog to digital form, issues the data to the GSM module, and orchestrates the routing of data through the network. Thus, real-time environmental data sampled from WBGT sensor can be transmitted to a server via GSM. Based on the timely updated WBGT data at different locations, the 3D environmental map of a LNG site will be generated through the trilinear interpolation.

2.3 Stage 3: Early-warning System

The purpose of the early-warning system is to safeguard the wellbeing of LNG workers while working in hot environment. The proposed early warning system is capable of collecting data on (a) instant WBGT at different zones of LNG sites; (b) location of workers; (c) work duration; and (d) personal information of workers. It could be automated by integrating smart sensor technology, RFID technology, and information communication technology. It is used to monitor the environmental condition of the whole LNG site, track the location of LNG workers, and alert workers the risks of heat stress by alarming sounds. Attendance of LNG workers are recorded automatically when workers walk through the entrance and enter the site with their helmets. Basic information (e.g., age, percentage of body fat, smoking habit, alcohol drinking habit, job nature, etc.) could be transmitted to the site office. The real-time location of workers in different zones of the LNG site could be monitored by the RFID. According to location of workers, the level of heat stress the workers exposed to can be obtained by the 3D environmental map. Based on these personal, work-related, and environmental data, workers' heat strain can be evaluated according to the Bayesian network model.

3 A Bayesian Network Approach to Assess Heat Stress

3.1 Concepts of a Bayesian Network

A Bayesian Network is a network used to model the uncertainty relationships of the actual world with probabilities [18]. It has been applied to numerous studies involving human physiology, such as predicting individuals' physiological responses in an interactive virtual environment [19], estimating complex multitask operators' workload from their physiological and psychological features [20], and predicting personalised optimal comfort temperature for individuals [21]. Two parts are contained in a Bayesian Network: a body and the knowledge. The body of a Bayesian network is the structure made up of two components. One is nodes which represent variables of interest and the other is arcs pointing from one node to another that indicate causal relationships among nodes. Nodes with arcs orienting from them to other nodes are called parents nodes and those with arcs pointing to them are called children nodes. Also, a node is a root node when it has no parent nodes, and one node are called leaf node when it doesn't have any children nodes. A Bayesian Network's body formed by nodes and arcs is also referred to as a Directed Acyclic Graph (DAG), in which you can never get back to a previous node following the direction of arcs [22, 23].

The knowledge of a Bayesian Network is the posterior probability distribution of each node in the network. When the data supporting the Bayesian Network are discrete, i.e. the variables have several definite statuses, the posterior probability distribution comes in the form of a Conditional Probability Table (CPT), which records the probabilities of every possible combination of its parent nodes' statuses. In practice, if the data are continuous, they tend to undergo discretization before they can be used to build a Bayesian Network. This is because the continuous-data situation is far more complicated than a discrete one to compute and to understand. The equation for the joint probability distribution of all variables in a Bayesian Network is shown as follows (Neapolitan, 2004):

$$P(X_1, X_2, \dots, X_n) = \prod_{i}^{n} P(X_i|Parents(X_i)) \quad (1)$$

Where $P(X_1, X_2, \dots, X_n)$ refers to the joint probability distribution of node $X_1, X_2, \dots, X_n$, $Parents(X_i)$ means all the parents nodes of node X_i. $P(X_i|Parents(X_i))$ stands for the conditional probability distribution of X_i, provided its parents [18, 24-26].

3.2 Bayesian Network for Heat Strain

A number of International Organization for Standardization (ISO) recommended indicators of heat strain as body core temperature and heart rate [27]. Evaluation of heat stress is generally determined through meteorological parameters that enable the estimation of the influence of several environmental factors on thermal comfort and physiological ability [28]. WBGT, integrating the effects of temperature, humidity, air velocity, and solar radiation, is by far the most widely used index of heat stress. Apart from the meteorological parameter, which can be measured by a comprehensive environmental index, heat strain is also highly correlated with work-related factors. It includes the type of exercise habit and other personal factors such as age, physique, percentage of body fat, and hydration level (Chan et al. 2012). Previous studies also demonstrated that the effects of aging on body fat [29]. Therefore, a Bayesian Networks model including these information was built to assess heat strain of workers (Figure 2).

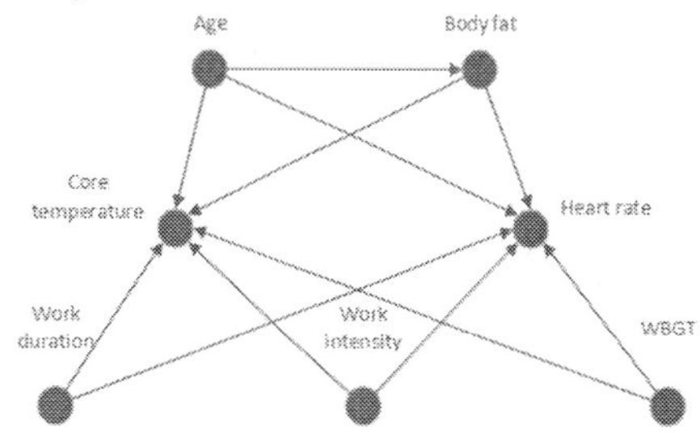

Figure 2 Bayesian Network in heat strain of workers

4 3D Mapping of Work Environments

In reality, environmental information can only be collected at particular spots. Those data are scattered point data. However, people work and live in a 3D space. As a useful approach to derive values for unknown points from a discrete set of known values, 3D-interpolation is employed to obtain spatially continuous and gridded data.

Many interpolation methods have been developed, including nearest neighbour, averaging, linear, and so on. However, for spatial data like temperature, Kriging, which is applied in Geosciences like Geology, Soil Science, Geography etc., has been mathematically proven to be the most stable linear unbiased estimation technique [30, 31], as traditional interpolation methods are mathematical only, while the Kriging is built on a model of stochastic spatial variation that fits well with reality [32] . The basic Kriging equation is shown as

$$\hat{Z}(\mathrm{p}) = \sum_{i=1}^{n} w_i Z_i(p_i) \quad (2)$$

Where $\hat{Z}(\mathrm{p})$ is the unknown value that needs to be estimated at the position p with known points p_i and corresponding known values $Z_i(p_i)$ as its neighborhood. n stands for the number of neighbouring known points used to krige the estimate, and w_i is the weight for point p_i.

Equations (3) and (4) make $\hat{Z}$ the best linear unbiased estimator [32, 33].

$$E(\hat{Z} - Z) = 0 \quad (3)$$

$$E(\hat{Z} - Z)^2 \text{ is minimum} \quad (4)$$

$\hat{Z} - Z$, the estimation error, indicates the similarity between $\hat{Z}$ and Z, $E(\hat{Z} - Z)^2$ is the mean square error and E is the expectation.

Figure 3 shows an example of scattered temperature measurement on site. The points in the 3D space represent the positions of sampling collected by the smart sensors. Different colours stand for various degrees of temperature. Environment conditions at different locations are varying. Points in different colours denote different WBGT values. Through 3D interpolation, as shown in Figure 4, we transform those irregular points into spatially continuous WBGT data, which is much easier to perceive and understand.

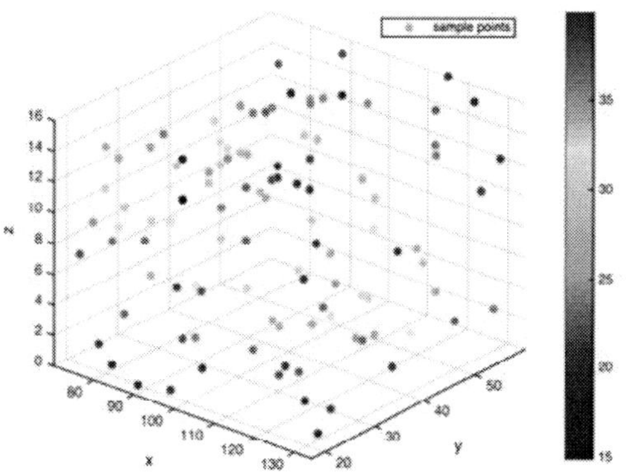

Figure 3 Sample of temperature measures in a 3D space

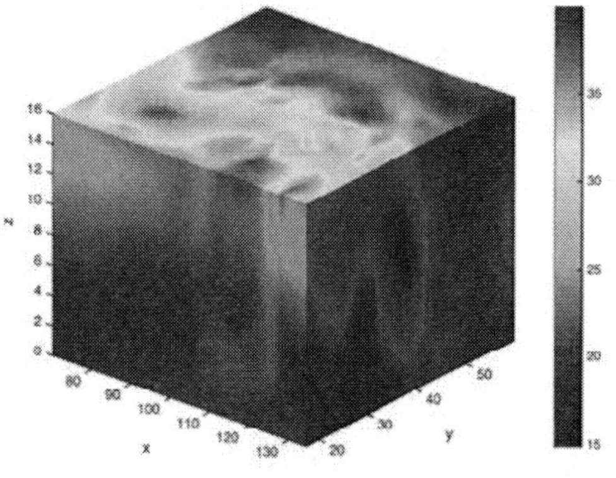

Figure 4 3D temperature field interpolated from scattered temperature samples

5 Conclusion

LNG workers are often exposed to severe environmental conditions, especially in North Australia, where most nature resource companies are located and the occupational settings are commonly in extreme hot environments. This increases the risk of heat stroke, injury and death. To protect workers from the risk of heat stress, a number of guidelines and initiatives have been put in place to promote health and safety in the workplace. However, most of these guidelines are largely depend on the specific regulations of workers' own behaviours. This study proposed an automatic way to monitor and early-warn any potential heat stress risk. The significant contributions of this study include: (1) a framework of interactive heat strain and environment evaluation model based on sensors technologies and mobile communication system; (2) a Bayesian Network in heat strain of workers; and (3) 3D interpolation model of environment condition on the work site.

References

[1] Queensland, E.S., *Queensland CSG to LNG industry workforce plan: operations and maintenance 2014-2034.* 2013.

[2] Chan, A.P. and Yi, W., *Heat stress and its impacts on occupational health and performance.* Indoor and Built Environment, **25**(1): p. 3-5, 2016.

[3] Inaba, R. and Mirbod, S.M., *Comparison of subjective symptoms and hot prevention measures in summer between traffic control workers and construction workers in Japan.* Industrial health, **45**(1): p. 91-99, 2007.

[4] Tawatsupa, B., et al., *Association between heat stress and occupational injury among Thai workers: Findings of the Thai Cohort Study.* Industrial health, **51**(1): p. 34-46, 2013.

[5] Control, C.f.D. and Prevention, *Heat-related deaths among crop workers--United States, 1992--2006.* MMWR: Morbidity and mortality weekly report, **57**(24): p. 649-653, 2008.

[6] Council, C.I., *Guidelines on site safety measures for working in hot weather.* Version 2, 2013.

[7] Wu, W., et al., *Towards an autonomous real-time tracking system of near-miss accidents on construction sites.* Automation in Construction, **19**(2): p. 134-141, 2010.

[8] Cheng, T., et al., *Data fusion of real-time location sensing and physiological status monitoring for ergonomics analysis of construction workers.*

Journal of Computing in Civil Engineering, **27**(3): p. 320-335, 2012.

[9] Di Corleto, R., Firth, I., and Coles, G., *Heat stress standard & documentation developed for use in the Australian environment.* Australian Institute of Occupational Hygienists, 2003.

[10] Brouha, L. and Maxfield, M.E., *Practical evaluation of strain in muscular work and heat exposure by heart rate recovery curves*.* Ergonomics, **5**(1): p. 87-92, 1962.

[11] Wang, F., et al., *Can the PHS model (ISO7933) predict reasonable thermophysiological responses while wearing protective clothing in hot environments?* Physiological measurement, **32**(2): p. 239, 2010.

[12] Howe, A.S. and Boden, B.P., *Heat-related illness in athletes.* The American Journal of Sports Medicine, **35**(8): p. 1384-1395, 2007.

[13] McDermott, B.P., et al., *Exertional heat stroke basics: What strength and conditioning coaches need to know.* Strength & Conditioning Journal, **30**(3): p. 29-32, 2008.

[14] Worldwide, A., *TLVs and BEIs—Threshold limit values for chemical substances and physical agents biological exposure indices.* ACGIH Worldwide ISBN: p. 1-882417, 2002.

[15] Chan, A.P., et al., *Determining an optimal recovery time after exercising to exhaustion in a controlled climatic environment: Application to construction works.* Building and Environment, **56**: p. 28-37, 2012.

[16] Liu, X. and Wang, X. *Development of a smart work site for early warning of heat stress.* in *20th International Symposium on Advancement of Construction Management and Real Estate.* 2015. Hangzhou, China.

[17] Yi, W., et al., *Development of an early-warning system for site work in hot and humid environments: A case study.* Automation in Construction, **62**: p. 101-113, 2016.

[18] Wagner, M.M., Moore, A.W., and Aryel, R.M., *Handbook of biosurveillance.* Academic Press, 2011.

[19] McQuiggan, S.W., Lee, S., and Lester, J.C., *Predicting User Physiological Response for Interactive Environments: An Inductive Approach.* AIIDE, **2006**: p. 60-65, 2006.

[20] Besson, P., et al., *Effectiveness of Physiological and Psychological Features to Estimate Helicopter Pilots' Workload: A Bayesian Network Approach.* Intelligent Transportation Systems, IEEE Transactions on, **14**(4): p. 1872-1881, 2013.

[21] Auffenberg, F., Stein, S., and Rogers, A., *A personalised thermal comfort model using a Bayesian network.* 2015.

[22] Bang-Jensen, J. and Gutin, G.Z., *Digraphs: theory, algorithms and applications.* Springer Science & Business Media, 2008.

[23] Jensen, F.V., *An introduction to Bayesian networks.* Vol. 210. UCL press London, 1996.

[24] Jouffe, S.C.L., *Bayesian Networks and BayesiaLab—A Practical Introduction for Researchers.* USA. 383, 2015.

[25] Korb, K.B. and Nicholson, A.E., *Bayesian artificial intelligence.* CRC press, 2010.

[26] Liao, Y., et al., *Risk assessment of human neural tube defects using a Bayesian belief network.* Stochastic Environmental Research and Risk Assessment, **24**(1): p. 93-100, 2010.

[27] ISO, I., *Evaluation of Thermal Strain by Physiological Measurements.* International Organisation for Standardisation, 1992.

[28] Moran, D., et al., *Evaluation of the environmental stress index for physiological variables.* Journal of thermal biology, **28**(1): p. 43-49, 2003.

[29] Silver, A.J., et al., *Effect of aging on body fat.* Journal of the American Geriatrics Society, **41**(3): p. 211-213, 1993.

[30] Matheron, G., *Principles of geostatistics.* Economic geology, **58**(8): p. 1246-1266, 1963.

[31] Stytz, M.R. and Parrott, R.W., *Using kriging for 3D medical imaging.* Computerized Medical Imaging and Graphics, **17**(6): p. 421-442, 1993.

[32] Oliver, M.A. and Webster, R., *Kriging: a method of interpolation for geographical information systems.* International Journal of Geographical Information System, **4**(3): p. 313-332, 1990.

[33] Cressie, N., *The origins of kriging.* Mathematical geology, **22**(3): p. 239-252, 1990.

An Integrated INS-GPS-Raspberry Pi System Using the Time-Sphere Model for Real-Time Identification of Struck-by-Equipment Hazard

J. Wang[a], S. Du[a] and S. Razavi[a]

[a] Department of Civil Engineering, McMaster University, Canada
E-mail: wang786@mcmaster.ca, dus5@mcmaster.ca, razavi@mcmaster.ca

Abstract

Struck-by-equipment hazard, i.e, contact collision between worker-on-foot and construction equipment or between equipment and equipment, is a major cause of injuries and fatalities in construction. A variety of smart technologies and systems have been developed and utilized to alleviate the risks of this type of hazard. However, most of the existing technologies and systems perform with a high rate of false alarms which have critically impeded their implementations on real construction sites. Therefore, a system performing more accurately and in real-time needs to be developed to enhance construction safety. In this context, a real-time system integrating with the 4D Time-Sphere model is developed to timely identify and alarm struck-by-equipment hazards with reduced false alarms. This system is comprised of Global Positioning System (GPS), Inertial Navigation System (INS), Raspberry Pi micro-processor and the developed Time-Sphere model. Corresponding actuation will be triggered if an unsafe situation is identified, i.e., struck-by-equipment hazard(s) are or to be presented. A controlled field experiment was conducted to evaluate the feasibility of the integrated system and the effectiveness of the Time-Sphere model in reducing false alarms. The obtained results show that average 53% of the alarms generated by the prevalent method can be reduced by the Time-Sphere model. All of the reduced alarms are false positives. The average false positive rate of the prevalent method is 29%. Moreover, it is demonstrated that the developed system is feasible and effective in identifying struck-by-equipment hazards in real time and potentially enhancing construction safety.

Keywords

Real-Time Hazard Identification; Construction Safety; Technological Innovation; Raspberry Pi; Struck by; False Alarm; 4D Model; Inertial Navigation System

1 Introduction

Construction industry has the largest number of fatal occupational injuries against to all of other sectors in many countries such as the U.S., Canada and others [1,2]. The Construction Focus Four Training program initiated by the OSHA provides safety training particularly on the top four types of hazards, i.e., falls, struck by, caught in or between, and electrocution. The largest proportion (38%) of fatal injuries in the category of struck by occurred as workers struck by equipment or falling objects [3]. Therefore, exploring an effective approach to avoid struck-by-equipment hazards has attracted attention of both industry and academia.

In recent years, a variety of smart technologies and proximity detection systems have been developed for construction to prevent struck-by-equipment hazards [4-6]. However, only a limited of them have been indeed adopted and implemented on real-world jobsites. One major reason restricting their wide adoptions in construction is the high number of false alarms (false positive and false negative) generated [7].

As thus, a more accurate and effective hazard identification method, i.e., with reduced false alarms, needs to be proposed and developed. At the same time, a capable platform which supports real-time data collection, transmission and analysis/calculation also needs to be established to enable the implementation of the developed hazard identification method in construction jobsites. Therefore, a formalized system having all of the above mentioned functions needs to be developed to enhance construction safety.

This paper introduces the feasibility of an integrated micro-processor system for real-time struck-by-equipment hazard identification. The developed integrated system uses (i) a GPS aided Inertial Navigation System (INS-GPS) for construction entities' state information collection (3D position, velocity and orientation), (ii) a Raspberry Pi coded with the developed Time-Sphere model for real-time and more accurate data analysis/calculation, i.e., reduced false alarms, (iii) ZigBee modules for real-time data transmission, and (iv) the function of notifying the involved entities of the potential safety risks. The feasibility of the developed integrated system and the effectiveness of the Time-Sphere model in reducing false alarms have been evaluated through a controlled field experiment.

2 Literature Review

Technologies such as radio-frequency identification (RFID), ultrasonic, radar, infrared and others have been utilized and developed as proximity warning systems for construction to prevent collisions. For instance, ultrasonic and pulsed radar were adopted specifically for back-over safety practices in construction work zones. The performances of the developed sensing systems (ultrasonic and pulsed radar respectively) were evaluated under different conditions including sensor installation, static test, dynamic test, and dirty sensor test [4]. In another work, Bluetooth based wireless sensing technology (iBeacon) was utilized for the detection of workers who breached into hazardous areas around equipment, along with auxiliary components to support sound alert, vibration and visualization. A wheel loader and a dump truck were used to assess the system detection distance which was the horizontal distance between beacons and receiver [5]. Magnetic field sensing and actuation technology was also applied to alert workers from being too close to heavy equipment in real time [6].

In the application of unsafe proximity identification, the measurement and determination of the hazardous areas around equipment is essential [8]. A framework integrated with six major steps for creating the hazard zones around construction equipment was presented in the work [9]. A user interface was also developed to generate the hazard zone automatically for a specific piece of equipment. However, the implementation of the developed framework is limited to the jobsites and equipment that are at a horizontal grade. In a study of Teizer [10], by analyzing the point cloud data collected from a laser scan, the equipment blind spots can be detected and determined automatically. In addition, an autonomous pro-active real-time proximity safety alert system was developed by Teizer et al. for construction equipment and workers [11].

Wireless systems have attracted extensive interests and been broadly developed and used in multiple fields in civil engineering. Crane safety and efficiency can be controlled and improved by using a wireless communication prototype that encompassed RFID, GPS, the Electronic Construction Kiosk (eCKiosk), wireless IP camera and Bluetooth intercom [12]. A wireless sensor network was established to acquire both environmental and energy data in real time for monitoring the state of a metro station [13]. In conclusion, a robust, reliable and capable wireless system plays a critical role to promote the effective management of civil projects.

It is noteworthy that most of the current solutions to unsafe proximity detection problem have a major and common shortcoming, i.e., focusing on mere distance between entities. Moreover, generally the hazardous distance or zone (i.e., the threshold) around equipment is defined in 2D space [4,9]. Also in some studies, the thresholds were defined as constant values without updating over time [4]. A dynamic workspace generation method was developed and employed to construction equipment by Vahdatikhaki and Hammad [14]. Nonetheless, not all of the detected collisions were real safety threats as the method did not consider equipment's motion and geometry along with the vertical axis. The identified drawbacks cause such prevalent sensing systems to generate a high rate of false alarms which greatly impedes their real implementations in the real world [7]. In chorus, a capable and robust wireless platform that supports the execution of a more accurate hazard identification method in a real-time manner is needed. Therefore, this paper focuses on proposing and developing effective and timely solutions to the identified gaps and needs elaborated above.

3 Research Objectives

Developing a real-time and reliable struck-by-equipment hazard identification system is the main goal of this study. Therefore, the following three objectives are included: (1) developing a more effective proximity detection method which can detect hazards and reduce false alarms; (2) establishing an integrated platform for real-time data collection, communication and analysis; and (3) conducting a field experiment to check the feasibility and effectiveness of the integrated system developed herein.

4 Methodology

4.1 Integrated INS-GPS-Raspberry Pi System

An integrated INS-GPS-Raspberry Pi system using the Time-Sphere model for real-time identification of struck-by-equipment hazard is proposed and developed in this paper. It is worth mentioning that selection and utilization of Raspberry Pi in the development of this application is attributed to its computation capabilities, small size, high integration and low cost. Raspberry Pi which acts as a micro-processor can achieve real-time sensor data acquisition, entity state analysis and system actuation.

The framework of the integrated system is illustrated in Figure 1 and the major components ((a)-(f)) are explained as below:

(a) Each entity (construction equipment and workers-on-foot) is equipped with an integrated INS-GPS module, a ZigBee module and a Raspberry Pi;

(b) The INS-GPS module [36mm(L), 49mm(W), and 25mm(H)] is used to collect entities' 3D position, orientation and velocity;

(c) The collected state information is input into the Time-Sphere model for hazard identification with a low false alarm rate;

(d) The Time-Sphere model is embedded in the Raspberry Pi in the data processing unit to analyze the gathered state information from multiple entities by taking advantage of the processing capabilities of Raspberry Pi;

(e) Corresponding actuations will be triggered if an unsafe situation is identified, i.e., struck-by-equipment hazard(s) are or to be presented, e.g., the LED lights on respective Raspberry Pi will turn on (more reliable and advanced actuation mechanisms will be explored in the next step of this study);

(f) ZigBee modules are responsible for wireless data transmission between multiple Raspberry Pis.

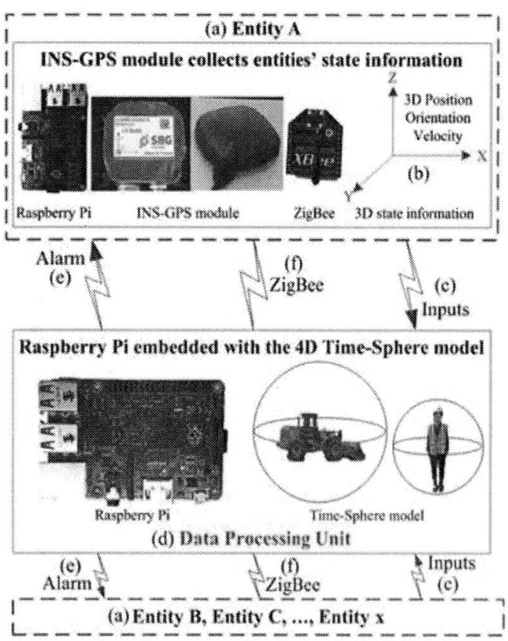

Figure 1. The framework of the integrated INS-GPS-Raspberry Pi system (images by authors)

For workers-on-foot, the developed integrated system needs to be mounted at the elevation of workers knee (e.g. on knee pads) or slightly higher. In the next step of this study, the authors will endeavor to make the device further compacted in order to be more comfortably used by workers-on-foot.

4.2 4D Model: Time-Sphere Model

As identified in the Literature Review section, generation of high false alarms is a major limitation of most of current proximity sensing systems. Therefore, the Time-Sphere model is developed and specialized to reduce false alarms through providing comprehensive solutions [15]. The hazard identification process of the Time-Sphere model is illustrated in Figure 2. The Time-Sphere model can be applied to multiple construction entities to prevent collisions. In this paper, only two entities are used to explain the model's development and evaluation.

The distinctions of the Time-Sphere model are summarized as follow:

- Along with 3D distance between entities, entities' velocity and orientation are also taken into consideration;
- Two thresholds, i.e., alert distance and warning distance, are used for hazard justifications. Alert zone around equipment and workers-on-foot is defined as a sphere in 3D space which will be used to quantify alert distance; Warning distance is dynamically adjusted over time fully in accordance with entities' moving characteristics and personnel reaction and execution time;
- The unsafe proximity query rules developed in the Time-Sphere model not only can detect the actual intersections of 3D alert zones, but also can identify the impending collisions in 3D space in a proactive manner;
- Entities' movements are not limited on a horizontal plane. Vertical motion, different altitudes, site terrain and others are also taken into account in the Time-Sphere model.

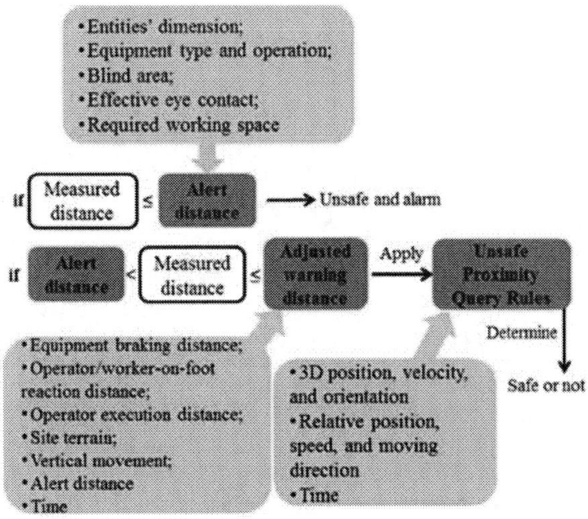

Figure 2. Illustration of the Time-Sphere model

The distance between entities is measured by mounting a sensing module on individual entities. The sensing module will collect entities' state information. Generally the distance between two sensing modules can be approximated as the distance between entities in real-world applications. To be more accurate, the distance between two sphere centers is the distance between entities. The state information of the sphere center can be calculated by using the sensor's state information with the geometry of the sensor's installation position in the sphere. Linear and angular velocity along with entities' orientation need to be considered in the calculation of the sphere center's velocity.

As the Time-Sphere model is established by considering construction entities' 3D kinematics with time, thus it can also be extended and utilized to prevent several types of collisions including materials transported in air, temporal and permanent site facilities and other dynamic elements on sites.

The three major parts involved in the development of the Time-Sphere model (see Figure 2) are explained in detail in the sections 4.2.1-4.2.3.

4.2.1 Identification of Alert Zone

Alert zone is the hazardous zone around a construction entity and is represented using a sphere in this paper. The definition is fairly self-explanatory, i.e., no entities can enter hazardous zones without authorizations. Diverse 2D or 3D shapes such as circle, cylinder and others have been used to denote the hazardous area around a construction element depending on their applications [16].

It is worth pointing out that in the study of struck by hazard prevention, the definition of a situation that is truly unsafe (true positive) or safe (true negative) is crucial. Even though the overwhelming majority of existing studies defined an unsafe situation as the collision of the 2D or 3D zones around the objects. However, no studies conclude a certain 2D or 3D shape that can represent the hazardous zone for all or one specific object. For different objects and applications, the shape that is the closest to the abstraction of the object's hazardous zone is different.

In this study sphere is adopted (see Figure 1) for three reasons: (i) the scope of this paper is to identify and predict collisions in 3D space; (ii) the unsafe proximity query rules which can reduce the generation of false alarms will use the 3D alert zone's 3D position, velocity and orientation (described in section 4.2.3); and (iii) the computational efficiency of sphere contributes to implementing the proposed system in real time. More 3D shapes with corresponding unsafe proximity query rules for reducing false alarms will be studied in the future.

The radius of the sphere for one specific piece of equipment is determined by Equation (1). The quantification of the buffer distance Δ needs to consider equipment type and operation, equipment blind spots, and effective eye contacts between operator and workers-on-foot [17].

$$\text{sphere radius(equipment)} = 0.5 \times \text{equipment length} + \Delta \quad (1)$$

For workers-on-foot, the diameter of the sphere is determined as 2 meters which is slightly higher than the average height of an adult male for safety reasons [18].

Therefore, alert distance is the sum of the radius of the involved spheres. If the distance between entities is smaller than the corresponding alert distance, the situation will be identified as unsafe.

4.2.2 Quantification of Warning Distance

Warning distance is another threshold used in the Time-Sphere model and is dynamically adjusted over time based on entities' kinematics and personnel execution and reaction time. Due to the multiple uncertainties and dynamics on jobsites, it is difficult to calculate the exact warning distance needed by the entities to avoid collisions when they come to a complete stop. The existing methods were endeavoring to quantify it as close to the needed distance as possible by taking more factors into account [9,14].

In this paper, the major factors considered in the quantification include equipment braking distance, operator reaction distance, workers-on-foot reaction distance, alert distance, site terrain (e.g., slopes and different altitudes), entities' orientation and vertical motion (see Equations (2)-(4)). Accurate quantification of warning distance helps to reduce false alarms. The Equations (2) and (3) depict situations without vertical motions. The Equation (4) represents the situation in which the worker has vertical movement.

$$\text{Warning Distance} = \text{sqrt}\{[[(\text{sphere raduis} + \text{reaction distance} + \text{braking distance})\text{of E1} \times \cos(\text{Pitch1})] + [(\text{sphere raduis} + \text{reaction distance} + \text{braking distance})\text{of E2} \times \cos(\text{Pitch2})]]\text{^}2 + [[(\text{sphere raduis} + \text{reaction distance} + \text{braking distance})\text{of E1} \times \sin(\text{Pitch1})] + [(\text{sphere raduis} + \text{reaction distance} + \text{braking distance})\text{of E2} \times \sin(\text{Pitch2})]]\text{^}2\} \quad (2)$$

$$\text{Warning Distance} = \text{sqrt}\{[[(\text{sphere raduis} + \text{reaction distance})\text{of W} \times \cos(\text{Pitch1})] + [(\text{sphere raduis} + \text{reaction distance} + \text{braking distance})\text{of E} \times \cos(\text{Pitch2})]]\text{^}2 + [[(\text{sphere raduis} + \text{reaction distance})\text{of W} \times \sin(\text{Pitch1})] + [(\text{sphere raduis} + \text{reaction distance} + \text{braking distance})\text{of E} \times \sin(\text{Pitch2})]]\text{^}2\} \quad (3)$$

$$\text{Warning Distance} = \text{sqrt}\{[[\text{sphere raduis of W}] + [(\text{sphere raduis} + \text{reaction distance} + \text{braking distance})\text{of E } \times \cos(\text{Pitch2})]]^2 + [(\text{sphere raduis} + \text{reaction distance})\text{of W} + (\text{sphere raduis} + \text{reaction distance} + \text{braking distance})\text{of E} \times \sin(\text{Pitch2})]\text{^}2\} \quad (4)$$

Where: E means moving equipment and W is worker-on-foot; Pitch is one dimension of entities' orientation; the detailed calculations of equipment reaction distance and braking distance, and workers-on-foot reaction distance are explained in the work [15]. Equation (2) expresses the distance for two pieces of equipment and Equation (3) and (4) are for one piece of equipment and one worker-on-foot.

Based on the collected 3D velocity, the system can determine whether the studied entity has vertical movement or not. Different from the prevalent methods in which an alarm will be triggered once the warning distance is violated, the Time-Sphere model will apply the unsafe proximity query rules to further confirm whether a real safety threat exists.

4.2.3 Unsafe Proximity Query Rules

The unsafe proximity query rules are developed especially for the use in struck by hazards identification and are another effective solution to reduce false alarms. When the distance between entities is smaller than the updated warning distance, the rules will be applied to further confirm whether an alarm is needed. The unsafe proximity query rules use relative position, speed and moving direction to identify (i) the occurred collisions of spheres and (ii) the imminent collisions. The identification of the occurred and upcoming intersections between spheres is a three-dimensional problem which is converted to three two-dimensional problems. The 3D coordinate system adopted is the geographic east-north-up (E-N-U) system. The overall process of the unsafe proximity query rules is presented in Figure 3. As shown in Figure 3, a situation can be concluded as hazardous only when all three 2D planes are identified as unsafe. A 2D plane is identified as unsafe if the circles projected from spheres are overlapped or to be overlapped in the future. The approach developed to judge whether two circles will overlap in the future is illustrated in Figure 4 and is explained in detail at the end of this section. Execution of spheres collision identification by implementing the developed unsafe proximity query rules on 2D planes has two major advantages:

(1) Speed up the spheres collision detection as the computations on a 2D plane is effortless and the judgment process will stop as long as one 2D plane is confirmed as safe. In addition, consideration of the real operations on jobsites and the scope of this study, if the minimum altitude of sphere A is lower than the maximum altitude of sphere B and the maximum altitude of sphere A is higher than the minimum altitude of sphere B, only the calculation on the geographic east-north plane needs to be conducted. Therefore, whether the situation presents struck by risks or not depends on whether the geographic east-north plane is hazardous.

(2) Detect the upcoming collisions by analyzing entities' relative position, speed and moving direction, without assuming a forecast time interval and generating the corresponding 3D shapes within the time interval. Predicting the 3D shapes within an assumed time interval to check clashes is commonly used in current studies [19,20]. Taking the east-north plane as an example, the identification of two circles that are to overlap is displayed in Figure 4.

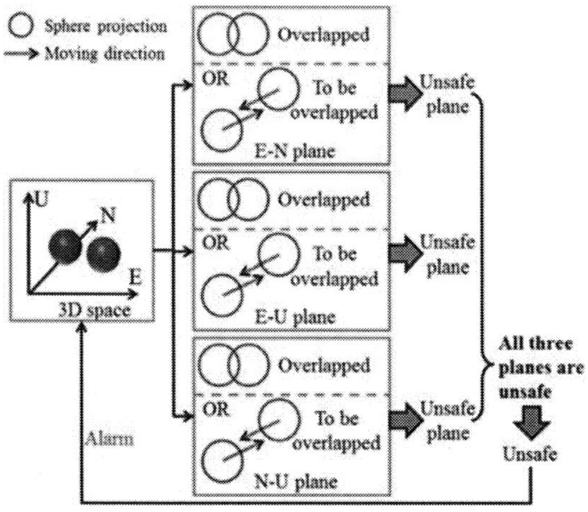

Figure 3. Unsafe proximity query rules

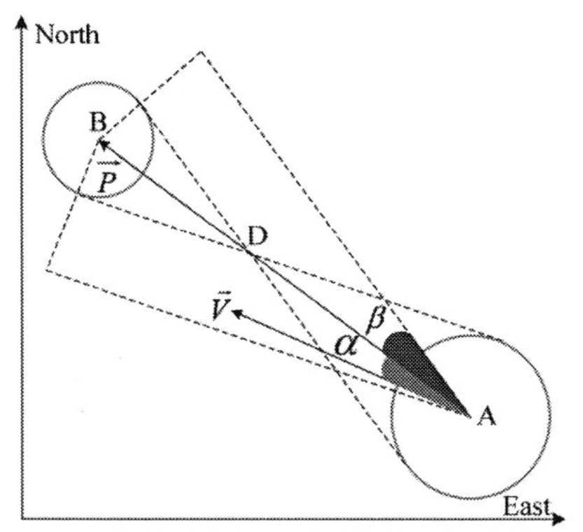

Figure 4. Identification of two circles (sphere projections) which are to overlap

In Figure 4, D is the distance between two circle centers; $\vec{P}$ is the relative position of B with respect to A; $\vec{V}$ is the relative velocity of A with respect to B; α is the angle between the vectors $\vec{P}$ and $\vec{V}$; 2β is the movement range within which two circles will overlap. Therefore, the two circles will intersect with each other if α is smaller than β.

5 Experiment and Results Analysis

A controlled field experiment was conducted to assess the feasibility of the developed integrated INS-GPS-Raspberry Pi System and the effectiveness of the Time-Sphere model in reducing false alarms.

5.1 Experiment Setting

The controlled field experiment was completed on a parking lot by using a vehicle [4.3m(L), 1.8m(W), 1.5(H)] as equipment and a person as worker-on-foot (see Figure 5(a)). Five scenarios were designed and conducted to test the integrated INS-GPS-Raspberry Pi System. It should be noted that at the current stage the developed system was only applied to two entities in each scenario to check the system's feasibility and evaluate its effectiveness. Implementation of the integrated system to multiple entities on real construction jobsites is the next step of this study.

Three out of the five scenarios were designed as "equipment struck by equipment" and the remaining two scenarios on "worker-on-foot struck by equipment". The trajectory of each entity in each scenario was roughly planned by researchers in advance. The trajectories in one scenario ("equipment struck by equipment") are shown in Figure 5(b) by uploading the collected 3D positions to the *Google Earth*.

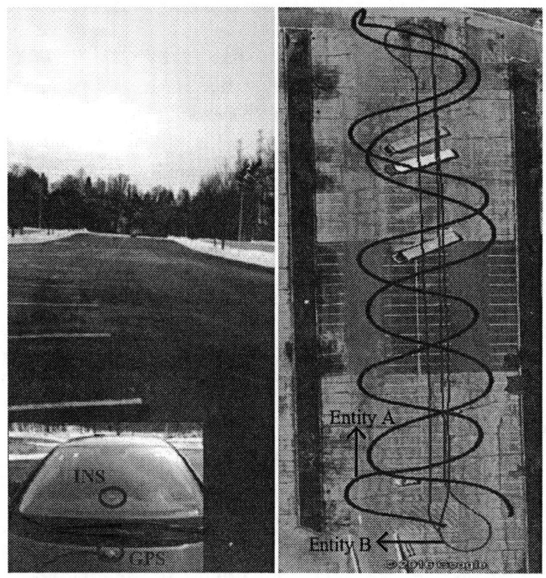

Figure 5. (a) Experiment site and the vehicle installed with the INS-GPS module; and (b) obtained trajectories in one scenario

5.2 Results Analysis and Discussion

In each scenario, entities' state information was collected by the INS-GPS module and analyzed using the Time-Sphere model which was embedded in the Raspberry Pi, i.e., the data processing unit in Figure 1. The accuracy of the INS-GPS module in localization was also assessed in advance on the parking lot. The average localization error obtained is less than 0.5m.

If the situation at any given time was identified as unsafe by the Time-Sphere model, the LED lights on the corresponding entities would turn on. As discussed, in the future work of this study, more reliable and advanced actuations will be developed and implemented to alarm entities of the potential safety risks, such as, flashing of large LEDs mounted on equipment, audible alarms and others.

During the experiment, all of the entities' state information with corresponding hazard identification result (i.e., generation of alarm or not at each moment) was recorded for further evaluations. In addition, a numerical model of the current prevalent unsafe proximity warning method (i.e., merely relying on the distance between entities to identify hazards) was also applied to the state information collected at each moment. The hazard identification results obtained by the prevalent method were also recorded. All of the saved information was used for the assessment of the Time-Sphere model's performance through two major types of analysis:

(1) Verification of the Time-Sphere model by simulation:

- For the collected state information that the prevalent method identified the situation as unsafe while the Time-Sphere model identified it as safe, the results show that the involved spheres neither intersected at the moment nor being intersected in the future (checked through simulation), even though the distance between them appeared to be smaller than the warning distance;
- For the collected state information that the Time-Sphere model identified the situation as unsafe, the results show that the involved spheres either intersected at the moment or being intersected in the future (checked through simulation), indicating that the generated alarms are rightly needed.

(2) Evaluation of the effectiveness of the Time-Sphere model in reducing false alarms

As explained earlier in the section 4.2.1, the definitions of the true positive and true negative vary depending on the application. As the Time-Sphere model was verified, it can be concluded that the alarms reduced by the Time-Sphere model are all false positives generated by the prevalent method. The obtained false positive rates (FPR) of the prevalent method are shown in Table 1.

Another indicator, i.e., reduced alarm percentage (RAP), also is used to specifically evaluate the effectiveness of the developed Time-Sphere unsafe proximity query rules in reducing alarms (Equation (5)). RAP denotes the percentage of alarms generated by the prevalent method can be avoided by the Time-Sphere model. All of the reduced alarms are false positives.

$$\text{RAP} = \frac{a_P - a_{TS}}{a_P} \quad (5)$$

Where: a_P and a_{TS} is the total number of alarms triggered by the prevalent method and Time-Sphere model respectively.

Table 1. Results of the five scenarios in experiment

Scenario	Total scans	False positives	FPR	RAP
Equip.-Equip. 1	1252	196	0.23	0.33
Equip.-Equip. 2	1138	154	0.13	0.58
Equip.-Equip. 3	1245	418	0.41	0.64
Equip.-Worker 1	1272	156	0.14	0.51
Equip.-Worker 2	1076	421	0.53	0.60
Average			0.29	0.53

According to Table 1, (1) the average FPR of the prevalent method is 29%, and (2) average 53% of the alarms generated by the prevalent method can be reduced by the Time-Sphere model.

6 Limitations and Future Work

In the conducted experiment, only two entities without vertical movements were included in each scenario. Applying the developed integrated INS-GPS-Raspberry Pi system to multiple entities for more types of struck by hazards identification on real construction jobsites is left for the future work. Performance evaluation on real-world construction jobsites is another area of future research. As such, the performance of the developed system (e.g., robustness, timeliness and accuracy) in dealing with different types of movement and altitudes will be checked. In addition, more reliable and advanced actuation mechanisms will be explored and tested on jobsites.

An accurate abstraction of the hazardous zone around entities can improve the efficiency of using the limited jobsite space and further reduce the generation of false alarms. In this way, more comprehensive analysis on false positive rate, false negative rate, sensitivity and specificity can be performed. Therefore, using other 3D shapes to develop more accurate hazard identification models also is a perspective research direction.

As the integrated system can collect entities' state information and perform hazard identification in real time, it serves as a potentially valuable platform to provide data and information for the risk analysis of struck by hazards.

7 Concluding Remarks

To detect struck-by-equipment hazards in a timely and more accurate (i.e., a low false alarm rate) manner, an integrated INS-GPS-Raspberry Pi system with Time-Sphere model was proposed and developed in this paper. The feasibility of the system and the effectiveness of the Time-Sphere model in reducing false alarms have been evaluated through a controlled field experiment. The obtained results positively indicate that the developed integrated system has the potential to be employed to multiple types of struck by hazards on jobsites. The study presented in this paper lays a foundation for the safety and risk analysis of struck by hazard and further enhancing construction safety.

References

[1] Bureau of Labor Statistics. Injuries, Illnesses, and Fatalities. Online: http://www.bls.gov/iif/ Accessed: 01/03/2016.

[2] Association of Workers' Compensation Boards of Canada. Fatalities, by Industry and Jurisdiction. Online: http://awcbc.org/?page_id=14 Accessed: 01/03/2016.

[3] Bureau of Labor Statistics. National census of fatal occupational injuries in 2014. Online: http://www.bls.gov/news.release/pdf/cfoi.pdf. Accessed: 01/03/2016.

[4] Choe S., Leite F., Seedah D., and Caldas C. Evaluation of sensing technology for the prevention of backover accidents in construction work zones. *ITcon*, 19:1–19, 2014.

[5] Park J., Cho Y., and Suryanto W. Bluetooth Low Energy Sensing Technology for Proximity Construction Applications. In *Proceedings of the 2015 Conference on Autonomous and Robotic Construction of Infrastructure*, pages 171–179, Ames, U.S., 2015.

[6] Teizer J. Magnetic Field Proximity Detection and Alert Technology for Safe Heavy Construction Equipment Operation. In *Proceedings of the ISARC*, pages 460–468, Oulu, Finland, 2015.

[7] Ruff T.M. Overview of proximity warning and technologies and approaches. In *NIOSH Workshop on Proximity Warning Systems for Mining Equipment*, Charleston, U.S., 2010.

[8] Wang J. and Razavi S. Low false alarm rate model for unsafe-proximity detection in construction. *J. Comput. Civ. Eng.*, 30(2): 04015005,1–13, 2016.

[9] Shen X., Marks E., Pradhananga N., and Cheng T. Hazardous Proximity Zone Design for Heavy Construction Excavation Equipment. *J. Constr. Eng. Manage.*, 10.1061/(ASCE)CO.1943-7862.0001108, 2016.

[10] Teizer J., Allread B., and Mantripragada U. Automating the blind spot measurement of construction equipment. *Autom. Constr.*, 19(4):491–501, 2010.

[11] Teizer J., Allread B. S., Fullerton C. E., and Hinze J. Autonomous proactive real-time construction worker and equipment operator proximity safety alert system. *Autom. Constr.*, 19(5), 630–640, 2010.

[12] Zekavat P. and Bernold L. Embedded Wireless Communication Platform Addresses Crane Safety and Efficiency. CRC, pages 309-318, Atlanta, Georgia, 2014.

[13] Nurchis M., Valta M., Vaccarini M., and Carbonari A. A Wireless System for Real-time Environmental and Energy Monitoring of a Metro Station: Lessons Learnt from a Three-year Research Project. In Proceedings of the ISARC, pages 416–423, Oulu, Finland, 2015.

[14] Vahdatikhaki F. and Hammad A. Dynamic equipment workspace generation for improving earthwork safety using real-time location system. *Adv. Eng. Inf.*, 29(3):459–471, 2015.

[15] Wang J. and Razavi S. Two 4D Models Effective in Reducing False Alarms for Struck-by-Equipment Hazard Prevention. *J. Comput. Civ. Eng.*,2016. (Accpted)

[16] Roofigari Esfahan N., Wang J., and Razavi S. An integrated framework to prevent unsafe proximity hazards in construction by optimizing spatio-temporal constraints. *Proc., 5th Inter./11th Constr. Specialty Conf.*, 126,1-10. Vancouver, Canada, 2015.

[17] Ray S. J., and Teizer J. Computing 3D blind spots of construction equipment: implementation and evaluation of an automated measurement and visualization method utilizing range point cloud data. *Autom. Constr.*, 36, 95–107, 2013.

[18] Dagan D., and Isaac S. Planning safe distances between workers on construction sites. *Autom. Constr.*, 50, 64-71, 2015.

[19] Moon H., Nawood N. and Kang L. Development of workspace conflict visualization system using 4D object of work schedule. *Adv. Eng. Inf.*, 28(31):50–65, 2014.

[20] Moon H., Kim H., Kim C. and Kang L. Development of a schedule-workspace interference management system simultaneously considering the overlap level of parallel schedules and workspaces. *Autom. Constr.*, 39, 93-105, 2014.

Applications of Supernumerary Robotic Limbs to Construction Works: Case Studies

Wooseok Seo[1], Chang-Yeob Shin[2], Juheon Choi[3], and Daehie Hong[3*], Chang Soo Han[4]

[1]Graduate School of Division of Mechanical Engineering, Korea University, Seoul, Korea
[2]Department of Mechanical Engineering, University of California, Los Angeles, USA
[3]Department of Mechanical Engineering, Korea University, Seoul, Korea
[4]Department of Robot Engineering, Hanyang University, An-san, Korea
E-mail: prime@korea.ac.kr, shinhujune@ucla.edu, wngjs3@korea.ac.kr, dhhong@korea.ac.kr*, cshan@hanyang.ac.kr

Abstract

In construction sites, most of manual works require at least two workers cooperating each other to complete a task properly. If it is possible to make a single worker carry out these kinds of tasks with the assistance of a robot, the overall construction efficiency can be increased. This study deals with the application of a supernumerary robotic limb (SRL) to the construction tasks.

In this paper, two typical construction tasks that can be improved by supports of SRL are considered. One includes the tasks that require a worker to hold an object like a panel in a desired position with an appropriate force while the other worker finishes main processes. The other contains moving or holding construction materials which are difficult to be held by one hand not because of its weight but because of its shape or size. For the first case, the SRL that sits on shoulder or waist with position based force controller can be utilized. For the second case, the robotic limb should be able to assist the worker in holding and balancing the object while the one carries it with one hand. As for the case studies, two types of extra robotic limbs were used. One was mounted on the shoulder and the other was attached on the wrist. Both types were applied in the situation of ceiling works in different ways. From many experiments, it was shown that the SRL can be effectively utilized in construction sites for diverse works.

Keywords –

Supernumerary robotic limbs; Wearable robot; Construction work

1 Introduction

There have been much research on the applications of automation and robotics in building construction. Robots can be utilized to substitute human works in a dangerous environment or to increase the construction performance. One way to improve the efficiency of overall construction works is to enable a single worker to complete a task that originally needs two or more workers to collaborate. Most of such manual works in the construction site require at least two people to cooperate each other to complete the job properly. In this case, the main process is usually done by one person and the others perform supporting tasks. Thus, if robots are able to assist the main process autonomously, the manual works can be carried out by one worker and the construction efficiency will be improved.

Various kinds of robots for different purposes have been researched and among them, wearable ones have emerged as one of the promising robots with diverse capabilities. A typical wearable robot is the exoskeleton which is normally used to support human in the aspect of power [1]. Separately with this, a new type of robot named SRL (Supernumerary Robotic Limbs) has been developed [2]. The main difference between the exoskeleton and the SRL is that the SRL have ability to be operated independently of human arms. Therefore, the SRL can be applied to a work that requires multitasking rather than power supporting. With this point, the SRLs are able to help a single worker to perform the main process and supporting tasks simultaneously which will result to improvement in efficiencies of the construction works.

In this study, in order to investigate applicability of the SRL to construction works in terms of multitasking, two different types of SRL were designed and manufactured. As case studies, by conducting many experiments in the assumption that the SRL are applied to assist ceiling work, the performance and feasibility of the SRL will be examined.

2 Case Analysis

Among many construction tasks which are done by two or more people, two typical cases are introduced in this study. The first case includes paneling, ceiling, and pilling works. The common ground of these works is that the main job like drilling or hammering, etc. (jobs for fixing objects in place) is performed by one worker, and the other worker have to hold construction materials such as panels or stake in the desired position and orientation until the main process is completed. By substituting the entire manual holding task with the operation of SRL, a single worker can play a role of two people at the same time. For this purpose, SRL first have to apply forces to the desired point of objects with appropriate magnitudes. In addition, since the base of SRL is attached to the human body, SRL should be able to maintain its contact point and contact force on materials while the worker moves and finish the main process.

The second case is about moving or holding construction materials which are difficult to be held by one hand not because of its weight but because of its shape or size, and thus need to be supported by two hands. If both 2 hands are used only to carry objects, the person is not able to do any other works. However, by making one hand free from holding the material, some simple tasks which are easily dealt with the other hand can be carried out simultaneously, and therefore, multitasking is available. To achieve this objective, the SRL should be able to assist the worker in holding and balancing the building materials such as panels or pipes while the worker moves around or perform other tasks. For each case and corresponding requirements mentioned above, two different types of SRL were applied to actual experiments.

3 Hardware and Control

Specifications of the SRL (such as dimensions, attachment point on human body, or kinds of end-effectors, etc.) are various depending on the practical task in construction site. For instance, the SRL can be installed on waist [2-4] and the end-effectors in arc-shape might be suitable for grabbing stakes in case of the pilling work. If SRL assist panelling or ceiling works, the attachment point of SRL can be on back or shoulders [5], and the end-effectors should be flat, or round shaped to press and hold panels effectively. In this study, two types of SRL for different cases were utilized with intention of examining both types in the situation of ceiling works in separate ways.

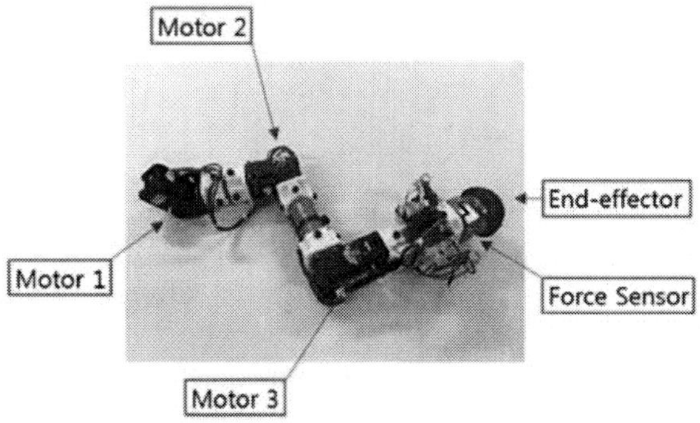

Figure 1. Manipulator structure of SRL for ceiling work

3.1 SRL for Ceiling Work

3.1.1 Hardware Specification

For the first case of applications, SRL with two identical arms are used for ceiling work. The entire SRL is divided into manipulators and shoulder mount (base of limbs) [6, 7]. Figure 1 illustrates the structure of manipulator. The end-effector of the SRL is made of hemisphere-shaped rubber material and it is fixed to the manipulator. The hemisphere structure makes a point-contact possible between the end-effector and panels regardless of posture of the manipulator. Rubber material prevents the contact point from slipping on the panel. Force sensor is connected with end-effector to calculate forces acting on the end-effector in 3-axis. Each manipulator has three motors (three-DOFs) to press panels on the ceiling and to compensate for horizontal movement of the worker while the drilling task is in progress. Shoulder mount which is described in Figure 2 consists of manipulator holder, back holder and vest. An IMU (inertial measurement unit) is attached to one of the manipulator holder to measure rotational movement of human body. Each manipulator is fixed to manipulator holders and two manipulator holders are connected to the 1 back holder by hinge structure to make rotating of the worker in yaw direction comfortable. The back holder is installed on back shoulder of vest and the worker can collaborates with the SRL by wearing the vest [6, 7].

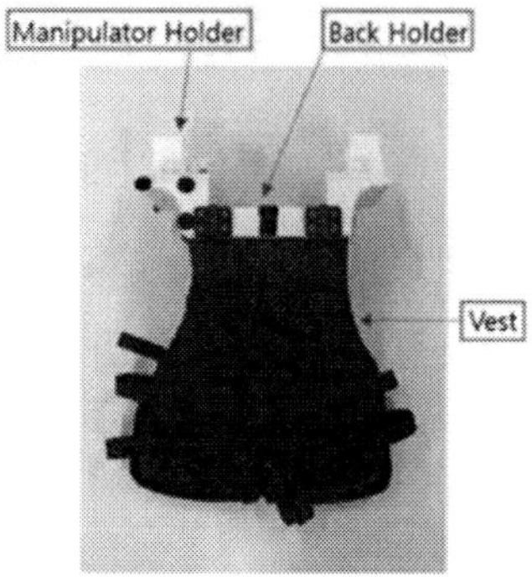

Figure 2. Shoulder mount of SRL for ceiling work

3.1.2 Control Algorithm

To maintain contact point and contact force, position based force controller is applied [8, 9]. Once the end-effector of SRL are set on the desired point of panel, force in Z-direction is controlled first according to preset magnitude to hold the panel stably. After that, force sensors and IMU continuously measure forces in X, Y, Z-directions, roll and pitch. If worker moves during the work, relative position of end-effector with respect to the human body should be adjusted to keep supporting panels. Since values from force sensors and IMU varies depending on translational and rotational movements of the worker, the controller can estimate motion of the worker. Based on these values, the controller calculate required position of the end-effector and required rotation angles of motors to compensate for movement of the worker. According to this process, three motors are regulated by the controller to keep contact point and contact force [6, 7]. Overall algorithm is represented in Figure 3.

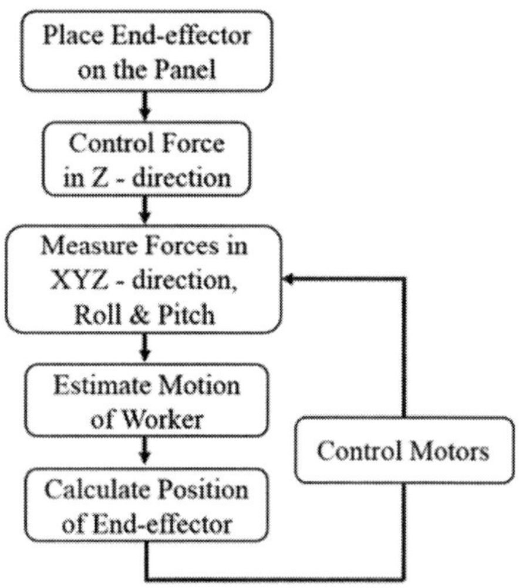

Figure 3. Control algorithm for ceiling work

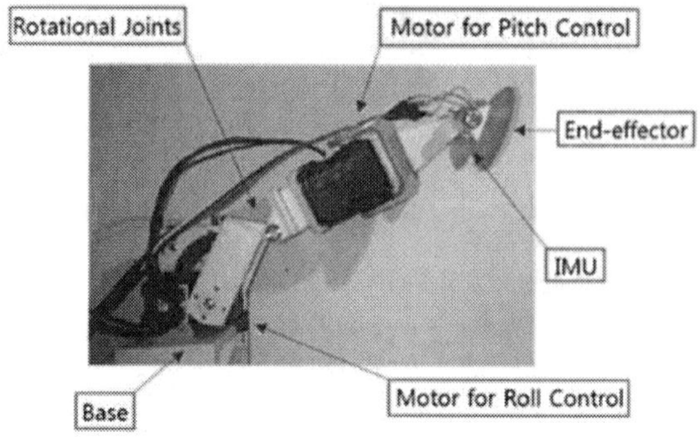

Figure 4. Structure of SRL for holding object

3.2 SRL for Holding Object

3.2.1 Hardware Specification

For the second case of applications which is for holding or carrying construction materials with one hand, SRL with one arm is applied. Overall structure of the SRL represented in Figure 4 is composed of manipulator and base. The end-effector of SRL which is flat-shaped and covered with rubber is connected to the motor by free rotational joint. Shape of plane makes supporting panels stable and rubber material is used to slips between the end-effector and object. An IMU is installed on the end-effector to measure slopes of the construction material. The number of motors for manipulator was determined according to required number of degrees of freedom. To prevent an object from falling from the hand, roll and pitch of the object should be adjusted, and this account for two motors. A free rotational joint in pitch direction is applied to control elevation of the object effectively and this joint is supported by torsion spring to maintain initial posture of the SRL. When values of roll and pitch change at the same time, it causes change in value of yaw. The value of yaw seems uninfluential to balancing, however, if yaw of the object is varied with respect to human body, it causes twisting and pains at wrist. Therefore, another free rotational joint in yaw direction is adopted between two motors not to make changes in yaw apply uncomfortable forces to wrist of the worker. Manipulator is assembled with the base and the base can be fixed on wrist or forearm by band. Consequently, SRL is attached on human body when it operates and it is illustrated in Figure 5.

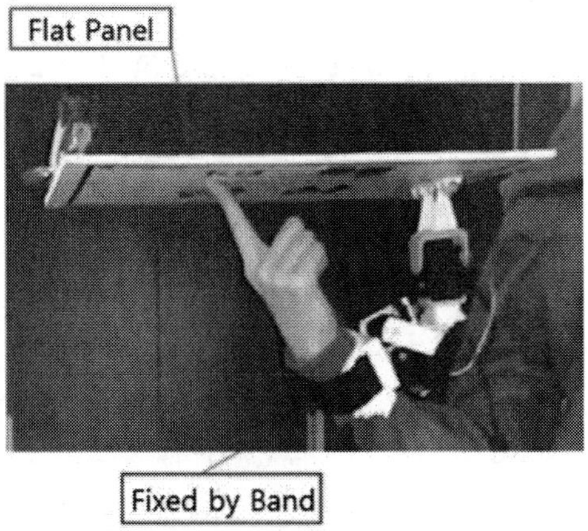

Figure 5. SRL for holding object in operation

3.2.2 Control Algorithm

During the work, construction material is held by two contact areas which are divided into approximately. One is between the SRL and the object. The other is made between hand of the worker and the object. Position and orientation of the former is regulated autonomously by the SRL, and therefore, overall balance of the material is maintained. Figure 6 shows control algorithm of the SRL. Once it starts operating, values of roll and pitch of the object are measured. If the object is inclined a lot to certain direction, it is easy to be dropped since it is not fixed to the hand or SRL. Hence, angular velocities of motors have to be large enough to avoid fall of the object. On the other hand, if slope of the object is small, the angular velocities of motors should be rather small to prevent the object from moving by its inertia near almost stable position. Based on this, the controller determines the velocities of motors which is proportional to the values of IMU. According to this process, motors are regulated to balance the object and to make both roll and pitch values become zero.

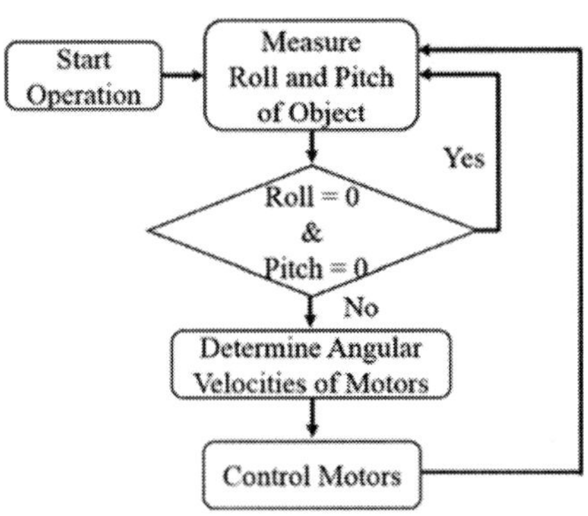

Figure 6. Control algorithm for holding object

4 Experiment

Among various kinds of task to which the SRL can be applied, ceiling work was selected as a task to be carried out in experiment to verify performance of the SRL. The SRL for holding object as well as SRL for ceiling work was examined in situation of the ceiling work in a way of holding panels with one hand.

4.1 SRL for Ceiling Work

The most important performance for this type of SRL is ability to support panels until drilling work is finished. Therefore, maintaining contact force and contact point between the end-effector and object should be verified while a worker who wears the SRL moves. For this purpose, position and forces of the end-effector were measured in the experiment. The end-effector was placed on certain point of structure for experiment, and contacting force was regulated. The person then made translational movements in directions of three-axis and rotational movements as indicated in Figure 7. Measured displacement of the end-effector and forces according to the position of the end-effector are described in Figure 8, and Figure 9 respectively. From the results, it was observed that compared to position change of human, which was approximately 170mm in maximum, position change of the end-effector was less than 20mm which is small enough to maintain contact point, and contacting forces in each axis were also maintained. After that, separately from the measurement experiment, the ceiling work process which imitated the actual tasks in construction site was performed successively.

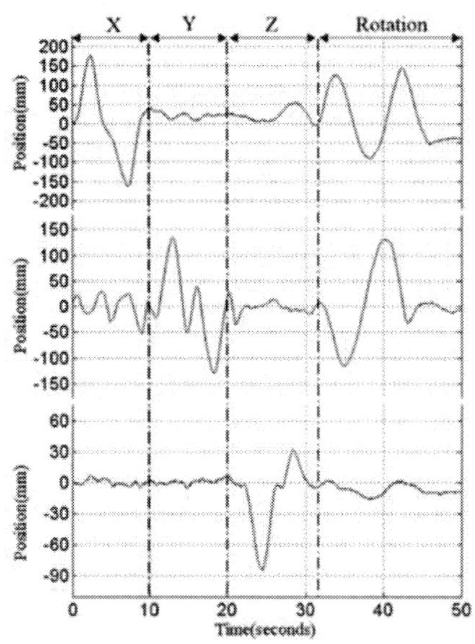

Figure 7. Input movement of person

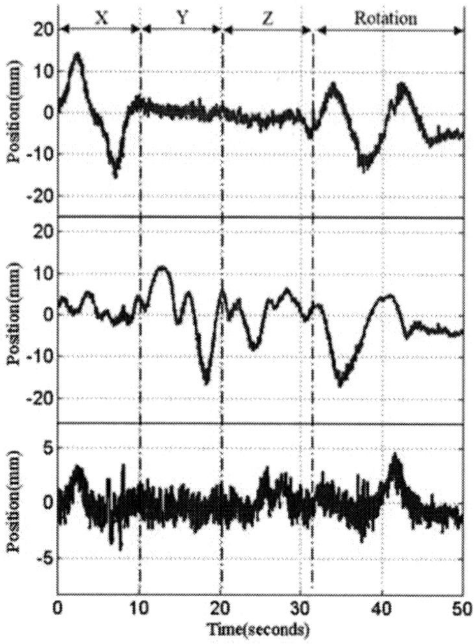

Figure 8. Output movement of end-effector

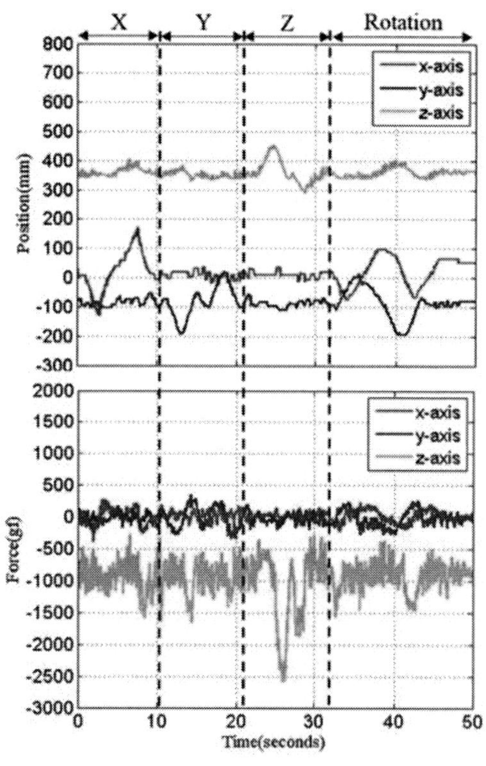

Figure 9. Force data according to position of end-effector

4.2 SRL for Holding Object

The main performance of SRL for holding object is making a worker be able to hold or carry construction materials with one hand. Accordingly, the ability to balance an object should be demonstrated. In the experiment, a flat panel was placed on the SRL. After that, input slope angle which would have been angle of the panel if SRL had not work was given in terms of roll and pitch angle separately. The result output roll and pitch angles which were compensated by the SRL are represented in Figure 10, and Figure 11 respectively. From the experiment result, it was examined that input angle up to 30 degrees in which the object is easy to fall can be reduced to output angle of 5 degrees stably. To verify performance of supporting object in situation similar to actual work environment (worker proceeds certain task with the other hand), ceiling work was carried out with his SRL. In this case, the task was proceeded in a way of completing the ceiling work by holding panel by one hand with assist of SRL, and conducting drilling with the other hand.

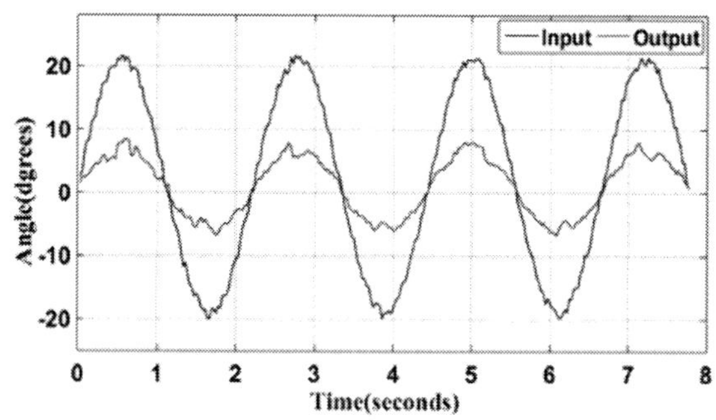

Figure 10. Input and output roll angle

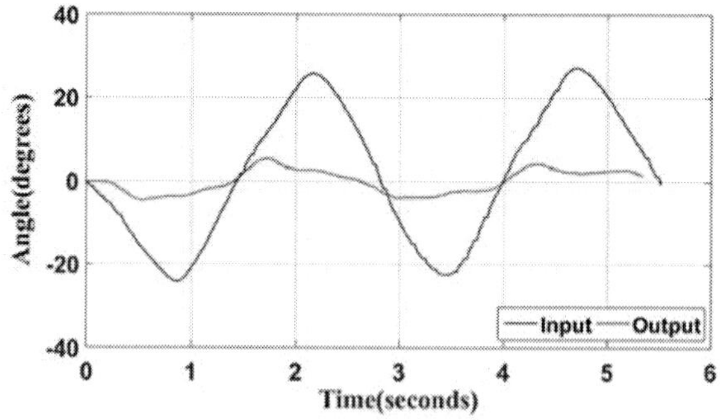

Figure 11. Input and output pitch angle

5 Conclusion

This study suggests ways to apply new type of wearable robot which is called Supernumerary Robotic Limbs (SRL) to construction works. From the fact that most of manual construction works require collaboration of more than two workers, the study focus on one of the advantages of the SRL that it can operate independently of human limbs. By making use of this point, substituting supports of additional workers with operation of SRL can be achieved, and therefore, some kind of works are able to be dealt with by one person or one person becomes able to handle several tasks simultaneously, which result to improvement in efficiency. As specific kind of works, ceiling work, and holding or carrying construction material with one hand were selected. Two different types of SRL were utilized for each case. Experiments were conducted to verify performances of the SRL, and ceiling works were carried out according to the procedure of practical ceiling work. From many experiments, it was demonstrated that Supernumerary Robotic Limbs can be effectively utilized in construction site for diverse works.

References

[1] Zoss AB., Kazerooni H. and Chu A. Biomechanical design of the Berkeley lower extremity exoskeleton (BLEEX). *IEEE/ASME Transactions on Mechatronics*, Vol. 11(2): 128–138, 2006.

[2] Llorens-Bonillla B., Parietti F. and Asada H. H. Demonstration-based control of supernumerary robotic limbs. In *IEEE/RSJ International Conference on Intelligent Robots and Systems (IROS)*, pages 3936–3942, Vilamoura, 2012.

[3] Parietti H., Chan K. and Asada H. H. Bracing the human body with supernumerary robotic limbs for physical assistance and load reduction. In *IEEE International Conference on Robotics and Automation (ICRA)*, pages 141–148, Hong Kong, 2014.

[4] Parietti, F. and Asada H. H. Supernumerary robotic limbs for aircraft fuselage assembly: body stabilization and guidance by bracing. In *IEEE International Conference on Robotics and Automation (ICRA)*, pages 1176–1183, Hong Kong, 2014.

[5] Llorens-Bonilla H. and Asada H. H. A robot on shoulder: Coordinated human-wearable robot control using coloured petri nets and partial least squares predictions. In *IEEE International Conference on Robotics and Automation (ICRA)*, pages 119–125, Hong Kong, 2014.

[6] Shin C-Y., Bae J. and Hong D. Ceiling work scenario based hardware design and control algorithm of supernumerary robotic limbs. In *International Conference on Control, Automation and Systems (ICCAS)*, pages 1228–1230, Busan, 2015.

[7] Shin C-Y. 'Position holding and force regulation control of supernumerary robotic limbs for ceiling work.' MS Thesis, Korea University, 2015.

[8] De Schutter J. and Van Brussel H. Compliant robot motion II. A control approach based on external control loops. *The International Journal of Robotics Research*, Vol. 7(4): 18-33, 1988.

[9] Roy J. and Whitcomb L. L. Adaptive force control of position/velocity controlled robots: theory and experiment. *IEEE Transactions on Robotics and Automation*, Vol. 18(2): 121-137, 2002.

Acknowledgement

This work was jointly supported by the Industrial Strategic Technology Development Program (No.10052965) funded by the Ministry of Trade, Industry and Energy (MOTIE), KOREA, the Human Resources Development of the Korea Institute of Energy Technology Evaluation and Planning (KETEP) grant funded by the Korea government Ministry of Trade, industry & Energy (No. 20144010200770), and the Technological Innovation R&D Program (S2357158) funded by the Small and Medium Business Administration (SMBA, Korea).

Technical term similarity model for natural language based data retrieval in civil infrastructure projects

T. Le and H. D. Jeong

Department of Civil, Construction and Environmental Engineering, Iowa State University, United States
E-mail: ttle@iastate.edu, djeong@iastate.edu

Abstract –

Recent advances in data and information technologies have enabled extensive digital datasets to be available to decision makers during the life cycle of a civil infrastructure project. However, much of the data is not yet fully reused due to the challenging and time consuming process of extracting the desired data for a specific purpose. Digital datasets are presented only in computer-readable formats and they are mostly complicated. In order to accurately extract a required subset of data, end users need to have deep understanding of the structure of the data schema, the meaning of each data entity and a query language. Thus, to truly facilitate the reuse of digital project data, a computational platform is needed to allow users to present their data needs in natural language. One of the critical requirements for a computer to perform this task is the ability to understand and interpret users' natural language inputs where keywords are a basic linguistic component. This research aims to collect technical terms commonly used in the civil infrastructure domain and develop a semantic similarity model that can measure the meaning relatedness/similarity between terms. Natural Language Processing (NLP) techniques and C-value method are used to automatically extract terms from text documents. A machine learning model called Skip-gram model is then employed to learn the semantic relatedness between technical terms using an unlabeled highway corpora as the input data. The input corpus includes 10 million words mainly collected from roadway design guidelines across the U.S. The model is evaluated by comparing the mapping results performed by a computer and a human.

Keywords –

Civil infrastructure project, Landxml, Data retrieval, Natural language interface, NLP, Vector space model

1 Introduction

The advanced computerized technologies such as 3D modeling and Geographic Information System (GIS) throughout the life cycle of a civil infrastructure project has allowed a large portion of project data to become available in digital format. In order to enable digital data exchange between proprietary software applications, several neutral data standards, for instance, LandXML [1] and TransXML [2], have been developed. However, these schemas are presented in machine-readable format and so complicated that it is difficult for end users to extract the desired properties [3]. The end user is required to have considerable programming skills and properly understand the structure and the meaning of each entity or attribute included in the source data schema. Thus, there have been apparent demands for an automatic data extraction means that would eliminate manual processing.

To address the above demand, a considerable amount of research efforts has been undertaken in both the building and transportation sectors. The most commonly adopted method is Model View Definition (MVD) which defines a subset of data for a specific business process. Some examples from this line of efforts include Construction to Operation Building Information Exchange (Cobie) [4] for IFC schema and InfraModel subsets for LandXML schema. Although a large number of MVDs have been proposed, they have not yet kept up with the dynamic demand from the industry since the current method for MVD development is based on a manual process which is highly time consuming [5,6,7]. Moreover, the business processes are dynamic and tend to change over time, hence MVDs must be periodically maintained and tailored to reflect the changes from industry practices. Therefore, there is a need to change the current practice of model view definition from the ad-hoc approach to a more rigorous methodology [5].

A natural language interface that can allow for human-computer interaction in natural language would enable digital data retrieval to overcome the bottle-neck of MVDs and remove the current burden on the end user.

One fundamental requirement for such a system is the ability to understand technical terms/keywords since they are a basic unit of natural language and users prefer to use them for obtaining data [8]. One of the major obstacles to fulfill the above requirement is the ambiguity issue of technical terms. A technical term in a domain specific document implicitly refers to something that only experts in that field can correctly understand. For example, the term 'roadway type', in general context, can mean the classification of roadways in terms of either material, function or location; but in the highway context, it refers to roadway functional classification. Another issue related to term ambiguity is that two different terms may be used to represent the same concept. For instance, the concept of longitudinal centerline of a roadway has a variety of terms including 'profile', 'crest', 'grade-line' and 'vertical alignment'. Addressing those issues will provide a foundation for natural language interfaces to fast and exactly extract data from the complicated sets of data with minimized human intervention and costs.

To fulfill that need, this research aims to propose a novel model that can be used to measure the semantic similarity between technical terms in the civil infrastructure domain. In order to achieve that goal, Natural Language Processing (NLP) techniques and C-Value method [9] are employed to process domain-specific guidelines and extract technical terms commonly used in the civil sector. A matching algorithm implementing the result from the previous step is developed to automatically look for the nearest entities and attributes in the Landxml schema for a certain keyword. The proposed semantic similarity model and the data mapping algorithm are evaluated by comparing the automatically retrieved data with the results performed by a human for performance assessment.

2 Related research

2.1 Partial digital model extraction

Methods for extracting partial models for specific use cases can be classified into the following three groups ordered by the degree of ease of use for end users: (1) developing a query language specifically for Building Information Modeling (BIM) models, (2) ontology-based query approaches, and (3) user-oriented query methods. The first group aims to tailor the conventional query languages (e.g., SQL, Object Orientation) for extracting information from BIM models. The major focus is on developing spatial filter strategies. Examples of these efforts include the Spatial query language [3], QL4BIM Spatio-semantic query language [10]; graph-based BIM retrieval [11], and topological querying [12]. The second group is to enhance the human-readability of data schema by utilizing an ontology approach to transform relations among data entities from implicit to explicit. With these semantic representations, it is easier for end users to read and comprehend a complicated data schema. An extensive number of studies based on this approach have been carried out for various use cases including ontology-driven construction information retrieval for tunnel projects [13], ontology partial BIM model extraction for building projects [14], ontology-based extraction of construction information [15] and ontology based querying over linked life cycle data spaces [16]. The last class of partial model query approaches moves a step further in terms of enhancing the ease of data extraction by providing query tools that require less effort from users. For example, Won et al. (2013) [17] proposed a non-schema algorithm that allows for the extraction of IFC instances without using IFC schema or MVD. In addition, a visual BIM query [18] was also established to visualize query codes. Although significant research efforts have been conducted, there is still a lack of natural language interface platforms that can enable computers to understand and interpret the end user's data interests in the civil infrastructure domain.

2.1.1 Semantic data label matching

In the construction industry, research efforts are currently focusing on standardizing the data structure format. There are very few studies that dealt with the issue of sense ambiguity. Zhang and El-Gohary (2015) [19] proposed an algorithm called ZESeM aiming to match a certain keyword to the most semantically relevant IFC entity. The algorithm includes two sequential steps including term-based matching and semantic relation based matching. Since the algorithm accepts matches from the label-based matching step, disambiguation still remains where the same word form is used for different senses. In addition, since ZESeM relies on Wordnet which is a generic lexicon, the applicability would be limited. Lin et al. (2015) [20] developed an IFD based framework for BIM information retrieval. IFD Library (International Framework for Dictionaries library), which is developed and maintained by the international buildingSMART, is a dictionary of BIM data terminology that assigns the same ID to synonyms. The integration or exchange of data using IDs rather than data names would eliminate semantic mismatching. However, since IFD is a hand-made electronic vocabulary, constructing this e-dictionary is time consuming and therefore it is still very limited considering the large collection of terms used in the construction industry.

2.2 Natural Language Processing

NLP is a collection of techniques that can analyze and extract information from natural language like text and speech. The major applications of NLP include translation, information extraction, and opinion mining [21]. These applications are supported by a combination of several techniques such as Named Entity Recognition (NER), Part-of-Speech (POS) tagging [22,23], tokenization (or word segmentation) [24,25], relation extraction, sentence parsing, word sense disambiguation [26,27,28], etc. NLP methods can be classified into two main groups: (1) rule-based and (2) machine-learning (ML) based methods. Since the former group is based solely on hand-coded rules, these systems are not able to cover all the complicated set of human grammars [29]; and therefore they do not perform well. NLP research has shifted from rule-based analysis to ML-based methods [21]. ML models are able to learn patterns from training examples and predict the output, hence they are independent to languages, linguistic grammars and consequently human resource cost would be reduced [30].

2.3 Word vector representation

Vector space model is a popular method used for estimating word similarity when digital dictionaries are not available. This method is based on the distributional model which represents meanings of words through their contexts (surrounding words) in the corpus [31]. The distributional model stands on the distributional hypothesis that states that two similar terms would occur in the same context [32]. The outcome of this approach is a Vector Space Model (VSM) in which words are converted into vectors and the similarity between two vectors represents the context similarity between the corresponding words [31]. VSM outperforms the dictionary-based method in terms of time saving as the semantic model can be automatically obtained from text corpus and collecting of these corpus is much easier than manually constructing a digital dictionary [33]. Among the methods to develop VSM, Skip-Gram model [34], which is an un-supervised machine-learning model, outperforms other statistical computational methods in various performance aspects such as accuracy and degree of computational complexity [34]. This machine-learning model learns the semantic similarity between two technical terms through their context similarity. The outcome of the training process is a set of representation vectors for technical terms.

3 Highway term space model development

The ultimate goal of this research is to build a highway vector space model (H-VSM) that can support the disambiguation task for advancing the computer-human interaction in partial model extraction. For addressing ambiguity, there are several methods including thesaurus based, ontology based and distributional method. The first two methods require a full lexicon or ontology consisting of semantic descriptions for all relevant concepts. These methods would be ideal for the disambiguation task if domain related thesauruses are available. However, since building those dictionaries requires a huge amount of empirical work, they can cover only limited vocabulary. Wordnet is one of the largest lexicons available containing 117,000 synsets [35], but it is generic and not suitable for the highway domain. For this reason, this research employs an unsupervised machine learning method called Skip-gram to train unlabeled data and learn the meanings of words by analysing their context words. The process of developing the H-VSM includes the following steps: (1) text document collection, (2) technical term extraction, and (3) semantic similarity training. The sub-sections below discuss the detailed procedures for each step.

3.1.1 Data collection

In this study, a highway corpus has been built using technical documents including textbooks, and highway engineering manuals collected from the Federal Department of Transportation (DOT) and from 18 State DOTs. The focus of the highway corpora in this research is on three project phases including design, construction and asset management. Technical terms in a guidance document in the engineering field are organized in various formats such as plain text, tables, and equations. Since tables and equations are not yet supported by the state-of-the-art NLP techniques, they are removed from the text corpora. The result of data collection is a plain text corpora consisting of approximately 10 million words. This dataset is utilized to extract highway related technical terms which are then trained and converted into vectors.

3.1.2 Technical term detection

The first step of data processing to construct the H-VSM is to detect highway related technical terms. In order to achieve this goal, a set of NLP techniques including tokenization, part of speech tagging are utilized to identify the POS tag for each word in the highway corpora. The OpenNLP library is used to perform this task. The linguistic process, as illustrated in Figure 1, includes the following steps:

- ***Word Tokenization:*** In this step, text is broken down into individual units (called tokens).
- ***POS tagging:*** The purpose of this step is to determine the part of speech tag (e.g., noun, adjective, verb, etc.) for each token.
- ***Noun phrase detection:*** Linguists argue that a technical term is either a noun (e.g., road) or a noun phrase (NP) (e.g., right of way) that frequently occurs in the domain text documents [36]. This research utilizes the following patterns $(A|N)_1{}^*N_1$ Prep(of) $(A|N)_2{}^*N_2$ or $(A|N)_1{}^*N_1(A|N)_2N_2$ to detect good candidates for technical terms. In the patterns above, A is adjective and N is noun.
- ***Termhood measurement and concept extraction:*** The C-Value algorithm, proposed by Frantzi et al. (2000) [9], is employed to rank the candidates extracted from the previous step. C-value represents the degree of termhood and is computed based on the frequency of occurrence and the length of terms. Equation 1 below presents the C-value measurement of termhood.

$$C\_value(a) = \begin{cases} log_2|a|f(a), & a \text{ is not nested} \\ log_2|a|f(a) - \frac{1}{P(Ta)}\sum_{b \in Ta} f(b), & otherwise \end{cases} \quad (1)$$

Where: a is a candidate noun phrase, f is the frequency of a in the corpus, Ta is the set of extracted noun phrases that contains a, and P(Ta) is the number of these candidate terms.

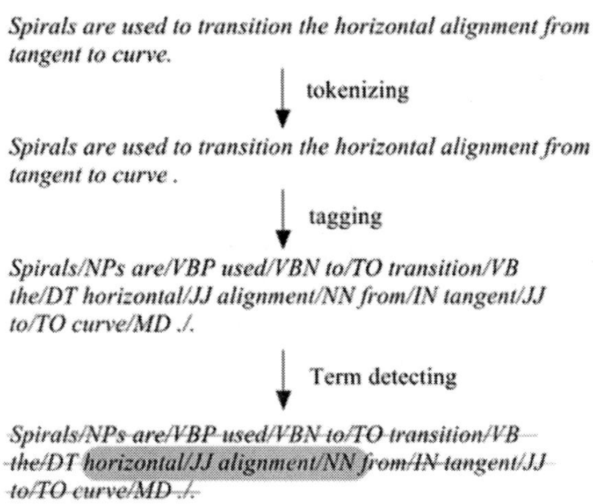

Figure 1. Term detection procedure

3.1.3 Data training and results

The tagged text corpora from the previous phases will serve as a data source for developing the semantic similarity model. Before collecting the training datasets, multi-word terms in the corpus are replaced with connected blocks of their word members so that they can be treated as single tokens. For instance, 'vertical alignment' will become 'vertical_alignment'. To train the highway corpus, this research employs the skip-gram neural network training model which was developed by [37]. The Skip-Gram model, as illustrated in Figure 2, requires a set of training data in which the input data is a linguistic unit (noun or noun phrase) and the output data is a set of context words. In order to collect this training dataset, the tagged text corpus is scanned to collect instances of terms and their corresponding context words. Each occurrence of a technical term will correspondingly generate a data point in the training dataset.

The semantic similarity is trained using the word2vec package developed based on the Skip-gram neural network model. Parameters required for word2vec include frequency threshold values, hidden layers, and context window. The parameters used for the training model are presented in Table 1.

Table 1. Skip-gram model parameters

Parameter	Value
Frequency threshold	50
Hidden layer size	300
Context window size	10

Figure 3 presents the term space model developed from the training process. In this model, each technical term collected from technical documents is represented as a vector in a high dimensional space; and the distance between them represents the semantic similarity. The preliminary term space presented in this paper consists of more than six thousand technical keywords.

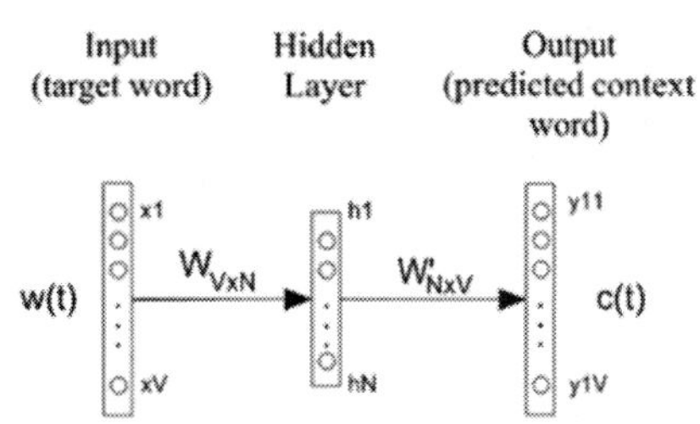

Figure 2. Skip-gram model

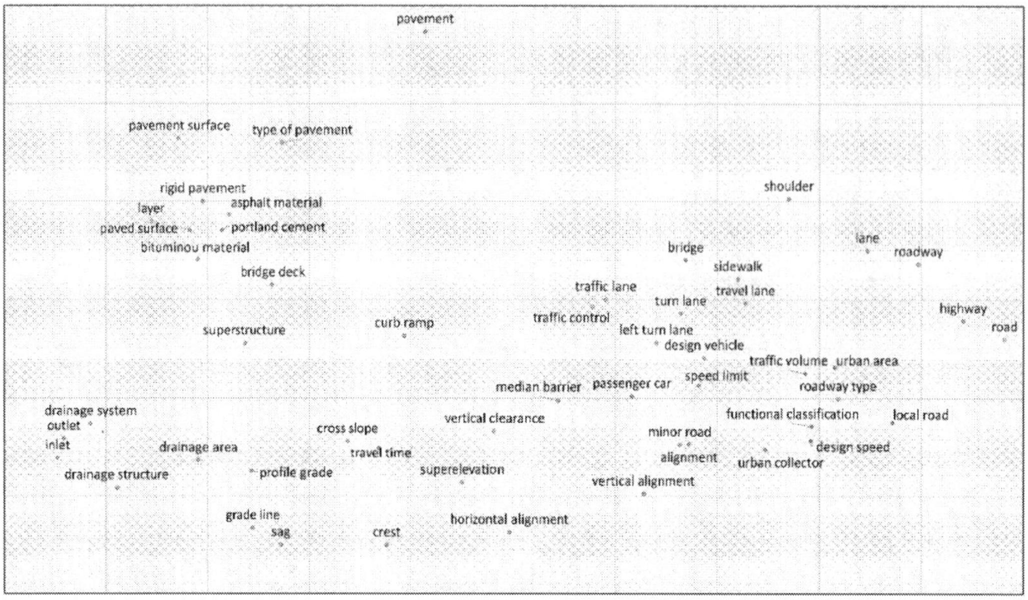

Figure 3. Highway term space model (H-VSM)

4 Test and evaluation

4.1.1 LandXML entity search

The proposed semantic model has been tested on the ability to search for data entities in the Landxml schema. As part of the test, an algorithm (see Figure 4) was developed for semantically searching for equivalent entities/attributes in Landxml schema. The algorithm is a two-stage procedure. In the first stage, a list of the nearest terms of the keyword input is generated by utilizing the vector space model developed in this research. A string-based similarity algorithm is then applied to find the entity in the Landxml schema that has the most similar name for each synonym in the list. The string similarity between a word in the nearest list and a Landxml entity or an attribute is computed using the Levenshtein algorithm [38] which is an edit-distance matching based method. The final matches are ranked by the similarity score.

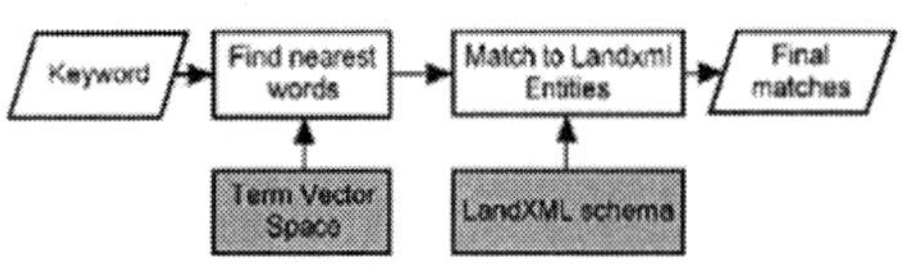

Figure 4. Landxml entities search algorithm

4.1.2 Evaluation

An evaluation experiment was conducted to evaluate the performance of H-VSM and the searching algorithm. In this experiment, a graduate student was asked to look for the data entities and attributes in the Landxml schema that are equivalent or the most related to each of 37 randomly selected keywords. Meanwhile, a prototype built upon the developed algorithm was applied to automatically generate the most matched Landxml entities. The results from the two methods were used to calculate the accuracy of the searching algorithm.

$$Recall = \frac{correctly\ matched\ keywords}{total\ keywords} \quad (2)$$

$$Precision = \frac{correctly\ matched\ keywords}{total\ matched\ keywords} \quad (3)$$

$$F_{measure} = \frac{2. Precision. Recall}{Precision + Recall} \quad (4)$$

Table 2 presents a portion of the experiment result and Table 3 shows the evaluation result using the criteria presented in Equations 2, 3 and 4. As presented in Table 3, the system shows a 37 percent precision which is relatively low. This is possibly due to the training data size. The searching algorithm accuracy highly relies on the capacity of finding synonyms which is based on the vector space model. This model is currently based on a data training set consisting of only approximately 10 million words. In order to enhance the accuracy, the data training set needs to be extended. Future research will be conducted to extend the training data set.

Table 2. A portion of the experiment result

Keyword	Top matched Landxml entities	Y/N
Drainage system	Outlet OutletStruct	Yes
Vertical alignment	roadTerrain pointGeometry	No
Pavement type	pavementSurfaceType stateType Project type	Yes
Roadway type	Classification Roadsign type	Yes

Table 3. Evaluation result

Recall (%)	Precision (%)	F-measure (%)
27	37	31

5 Conclusions

Digital project data is now widely available throughout the project life cycle in the civil infrastructure sector. However, the data collected and generated in the early project development stages are not typically reusable in the downstream phases. This is due to the interoperability issue when digital data from the original data creator is not readable or correctly understandable by the data receiver. This research developed a framework that semantically searches for the desired data from a transferred data file. The framework is composed of two components including (1) a term space model which represents highway related concepts extracted from the highway corpora in vectors and (2) a searching algorithm that can search for entities in the Landxml schema based on their semantic similarity instead of string based similarity.

The framework was evaluated by testing on a randomly selected set of input keywords. The result shows the accuracy of over 30 percent. The accuracy is low due to the size of the training data. Future research will be conducted to increase the data size.

The developed semantic similarity model is expected to serve as a fundamental resource for data integration and query systems to eliminate the issue of data terminology inconsistency. The model also has a broad impact in text mining for the infrastructure sector. Although digital data is increasingly available, text documents are still a major data and information communication channel among project stakeholders. The ability to understand the meanings of technical terms will enable computer systems to fast and exactly extract value information from project documents such as contracts or reports. Thanks to that, laborious work of reading and manipulating data and information in paper format would be eliminated.

References

[1] landxml.org, About landxml.org, On-line: http://www.landxml.org/, Accessed: 10/11/2005.

[2] Scarponcini P., Methodology for selection and development of transxml schemas, *Transportation Research Record: Journal of the Transportation Research Board* (1) 107-115, 2008.

[3] Borrmann A., Rank E., *Query support for bims using semantic and spatial conditions*, Handbook of Research on Building Information Modeling and Construction Informatics: Concepts and Technologies, 2009.

[4] East E.W., Construction operations building information exchange (cobie), *Tech. rep., DTIC Document*, 2007.

[5] Venugopal M., Eastman C. M., R. Sacks, J. Teizer, Semantics of model views for information exchanges using the industry foundation class schema, *Advanced Engineering Informatics* 26 (2) 411-428, 2012.

[6] Eastman C., The future of ifc: Rationale and design of a sem ifc layer, *presentation at the IDDS workshop*, 2012.

[7] Hu H., Development of interoperable data protocol for integrated bridge project delivery, Ph.d., *UMI Dissertations Publishing 2014*, 2014.

[8] Shekarpour S., Auer S., Ngomo A.-C. N., Gerber D., Hellmann S., Stadler C., Keyword-driven sparql query generation leveraging background knowledge, in: *Web Intelligence and Intelligent Agent Technology (WIIAT)*, Vol. 1, IEEE, pp. 203-210, 2011.

[9] Frantzi K., Ananiadou S., Mima H., Automatic recognition of multi-word terms: the c-value/nc-value method, *International Journal on Digital Libraries* 3 (2) 115, 2000.

[10] Daum S., Borrmann A., Langenhan C., Petzold F., Automated generation of building fingerprints using a spatio-semantic query language for building information models, eWork and eBusiness in *Architecture, Engineering and Construction*: ECPPM 2014 87, 2014.

[11] Langenhan C., Weber M., Liwicki M., Petzold F., Dengel A., Graph-based retrieval of building information models for supporting the early design stages, *Advanced Engineering Informatics* 27 (4) 413-42, 2013.

[12] Khalili A., Chua D., Ifc-based graph data model for topological queries onbuilding elements, *Journal of Computing in Civil Engineering* 0 (0) 04014046,

2013.
[13] Min H., Zhewen H., Ontology-driven tunnel construction information retrieval and extraction, in: *Control and Decision Conference* (2014 CCDC), The 26th Chinese, IEEE, 14, pp. 4741-4746, 2014.
[14] Zhang L., Issa R. R., Ontology-based partial building information model extraction, *Journal of Computing in Civil Engineering* 27 (6) 576-584, 2012.
[15] Nepal M. P., Staub-French S., Pottinger R., Zhang J., Ontology-based feature modeling for construction information extraction from a building information model, *Journal of Computing in Civil Engineering* 27 (5) 555-569, 2012
[16] Le T., Jeong H. D., Interlinking life-cycle data spaces to support decision making in highway asset management, *Automation in Construction* 64, pp. 54-64, 2016.
[17] Won J., Lee G., Cho C., No-schema algorithm for extracting a partial model from an ifc instance model, *Journal of Computing in Civil Engineering* 27 (6) 585-592, 2013.
[18] Wülfing A., Windisch R., Scherer R., A visual bim query language, eWork and eBusiness in *Architecture, Engineering and Construction*: ECPPM 2014 157, 2014.
[19] Zhang J., El-Gohary N., A semantic similarity-based method for semi-automated ifc extension, 2015.
[20] Lin J., Hu Z., Zhang J., Yu F., A natural-language-based approach to intelligent data retrieval and representation for cloud bim, *Computer-Aided Civil and Infrastructure Engineering*, 2015.
[21] Cambria E., White B., Jumping NLP curves: a review of natural language processing research [review article], *Computational Intelligence Magazine*, IEEE 9 (2) 48-57, 2014.
[22] Toutanova K., Klein D., Manning C. D., Singer Y., Feature-rich part-of-speech tagging with a cyclic dependency network, in: *Proceedings of the 2003 Conference of the North American Chapter of the Association for Computational Linguistics on Human Language Technology*-Volume 1, Association for Computational Linguistics, pp. 173-180, 2003
[23] Cunningham H., Maynard D., Bontcheva K., Tablan V., Gate: an architecture for development of robust hlt applications, in: *Proceedings of the 40th annual meeting on association for computational linguistics*, Association for Computational Linguistics, pp. 168-175, 2002.
[24] Webster J. J., Kit C., Tokenization as the initial phase in nlp, in: *Proceedings of the 14th conference on Computational linguistics*-Volume 4, Association for Computational Linguistics, pp. 1106-1110, 1992.
[25] Zhao H., Kit C., Integrating unsupervised and supervised word segmentation: *The role of goodness measures, Information Sciences* 181 (1) 163-183, 2011.
[26] Lesk M., Automatic sense disambiguation using machine readable dictionaries: how to tell a pine cone from an ice cream cone, in: *Proceedings of the 5th annual international conference on Systems documentation*, ACM, pp. 24-26, 1986.
[27] Yarowsky D., Unsupervised word sense disambiguation rivaling supervised methods, in: *Proceedings of the 33rd annual meeting on Association for Computational Linguistics*, Association for Computational Linguistics, pp. 189-196, 1995.
[28] Navigli R., Word sense disambiguation: A survey, *ACM Computing Surveys (CSUR)* 41 (2) 10, 2009.
[29] Marcus M., New trends in natural language processing: statistical natural language processing, *Proceedings of the National Academy of Sciences* 92 (22) 10052-10059, 1995.
[30] Costa-Jussa M. R., Farrús M., Mariño J. B., Fonollosa J. A., Study and comparison of rule-based and statistical catalan-spanish machine translation systems, *Computing and Informatics* 31 (2) 245-270, 2012.
[31] Erk K., Vector space models of word meaning and phrase meaning: A survey, *Language and Linguistics Compass* 6 (10) 635-653, 2012.
[32] Harris Z. S., Distributional structure, Word, 1954.
[33] Turney P. D., Pantel P., From frequency to meaning: Vector space models of semantics, *Journal of artificial intelligence research* 37 (1) 141-188, 2010.
[34] Mikolov T., Sutskever I., Chen K., Corrado G. S., Dean J., Distributed representations of words and phrases and their compositionality, in: *Advances in neural information processing systems*, pp. 3111-3119, 2013.
[35] Miller G. A., Wordnet: a lexical database for english, *Communications of the ACM* 38 (11) 39-41, 1995.
[36] Justeson J. S., Katz S. M., Technical terminology: some linguistic properties and an algorithm for identification in text, *Natural Language Engineering* 1 (01) 9-27, 1995.
[37] Mikolov T., Chen K., Corrado G., Dean J., Eficient estimation of word representations in vector space, *arXiv preprint arXiv:1301*.3781, 2013.
[38] Gale W. A., Church K. W., A program for aligning sentences in bilingual corpora, *Computational linguistics* 19 (1) 75-102, 1993.

Consolidated Exchange Models for Implementing Precast Concrete Model View Definition

Kereshmeh Afsari[a] and Charles M. Eastman[a]

[a]School of Architecture, Georgia Institute of Technology, USA
E-mail: keresh@gatech.edu, charles.eastman@coa.gatech.edu

Abstract –

The US National BIM Standard proposes facilitating information exchanges through Model View Definitions (MVDs). An MVD, defines a subset of the IFC schema that is needed to satisfy one or many exchange requirements of the AEC industry. In the MVD definition for precast concrete domain specified in precast concrete Information Delivery Manual (IDM), four sets of Exchange Models (EMs) have been specified for four different processes i.e. architectural precast, precast lead project, precast detailer as subcontractor, and precast fabrication and erection processes. This approach identifies 47 exchange models for the lifecycle of a precast concrete project. However, there are many similar exchanges among these EMs that can be combined. Two approaches of Exchange Model consolidation has been proposed and this paper identifies the implementation issues of these consolidation strategies. The main objective of this paper is to define the MVD with limited number of EMs that addresses the problems of existing consolidation strategies to facilitate the MVD implementation and execution. This study suggests combining 47 original Exchange Models to twelve EMs by combining them based on two major aspects. First aspect is with regard to the MVD concepts to ensure that the exchange models are being combined based on similar data they include. Second aspect is the phases of the project that the data exchange is happening considering their exchange disciplines to ensure that the consolidated exchange models are related to similar phase/phases of the project with the same exchange purpose.

Keywords –

Exchange Model, MVD, Precast Concrete, IFC, BIM

1 Introduction

The US National BIM Standard [1] proposes facilitating information exchanges through Model View Definitions (MVDs). A Model View Definition that is specific to an Industry Foundation Classes (IFC) release, defines a subset of the IFC schema that is needed to satisfy one or many exchange requirements of the AEC industry [2]. In fact, MVD consists of one or multiple information exchange data and each information exchange is the data that must be provided by one application to support work in one or more other applications. For the IFC work, two types of exchanges can be distinguished: model-based exchanges and all others, non-model exchanges. The unique strength of IFC is its support for model-based exchanges. This paper investigates the model-based exchanges known as Exchange Models (EMs).

IFC MVD provides implementation guidance and agreements for all IFC concepts such as classes, attributes, relationships, property sets, and quantity definitions used within this MVD subset. Therefore, it represents the software requirement specification for the implementation of an IFC interface to satisfy the exchange requirements [3]. In the Information Delivery Manual (IDM) for precast concrete domain [4], four sets of Exchange Models (EMs) have been specified for four different precast business context. These are architectural precast, precast lead project, precast detailer as subcontractor, and precast fabrication and erection processes.

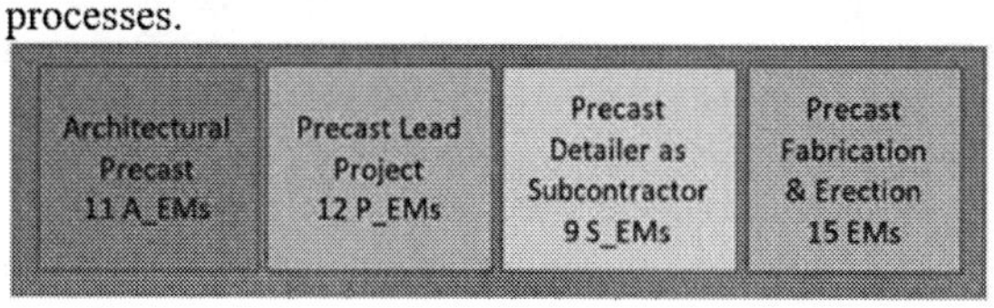

Figure 1 Four sets of Exchange Models identified in precast concrete IDM

As shown in Figure 1, the IDM identified 47 distinct exchange models for the lifecycle of a precast concrete project including 11 exchange models for the architectural precast process (i.e. A_EM), 12 exchange models for precast lead project process (i.e. P_EM), 9 exchange models for precast detailer as subcontractor process (i.e. S_EM), and 15 exchange models for precast fabrication and erection process (i.e. EM) [4, 5].

However, among these EMs there are many similar exchanges [4, 5, 6] such as A_EM.1, P_EM.1, and S_EM.1 that are all identical [4] referring to architectural and engineering concept model. Most importantly, the implementation of precast concrete MVD- either for automated validation of BIM Exchange Models or for implementing in software applications- faces a major challenge when dealing with 47 exchange models. In fact, for implementing precast concrete MVD, limited number of EMs are required to make the software implementation manageable and the functional distinctions between the EMs discernible. Also, consolidation of precast concrete Exchange Models are previously recommended [4, 6]. Therefore, the main objective of this paper is to define the MVD with limited number of EMs in order to facilitate its implementation and execution. This paper, first reviews and compares the existing recommendations to point out the inconsistencies and challenges that exist in current consolidated Exchange Models for precast concrete domain. Then, to address the existing problems, this paper indicates how 47 originally defined exchange models should be combined to address the existing challenges and to facilitate the implementation of precast concrete EMs within a wide range of contexts of an MVD domain.

2 Existing Consolidation Strategies

The percentage degrees of differences between 47 precast exchange models are identified in the IDM for precast concrete [4] and the effort in combining similar Exchange Models have been reviewed in [4] and [6]. Combining Exchange Models in these two efforts is done by reviewing the overlaps, inconsistencies, and redundancies in the exchange models through a detailed analysis. In this analysis all of the exchange models were compared in order to remove inconsistencies and also to identify opportunities for consolidation of exchange models to reduce their number [4, 6]. For this analysis, a Visual Basic Macro was used to scan each field of each EM and compare it to the parallel field of every other EM [4]. Wherever the degree of difference was less than 10%, the EMs were compared critically with a view to unifying them [4, 6]. Table 1 compares the consolidation results in these two efforts as approach "A" [4] and approach "B" [6].

In the first approach [4] (i.e. approach A), some of the similar EMs have not been identified. For instance, A_EM.2, S_EM.2 both define the same exchange in the preliminary project description stage. The exchange description for A_EM.2 and S_EM specifies that each one is related to the engineering concept model which provides information about the structural grid, structural system, major precast connections and issues, interfaces between precast and other structural and curtain wall systems [4]. This similarity is identified in the second approach [6] (i.e. approach B).

Table 1 Comparison of consolidation approach "A" [4] and approach "B" [6]

Consolidated Exchange Model A [4]	Consolidated Exchange Model B [6]	Original Exchange Models A [4]	Original Exchange Models B [6]
BC_EM (Building Concept)	BC_EM (Building Concept)	A_EM.1 P_EM.1 S_EM.1 P_EM.2	A_EM.1, P_EM.1, S_EM.1, P_EM.2, P_EM.3
-	Engineering Concept	-	A_EM.2, S_EM.2, S_EM.3
PC_EM (Precast Concept)	PC_EM (Precast Concrete Concept)	A_EM.5 A_EM.6	A_EM.5, A_EM.6, S_EM.3
-	Engineering design development	-	P_EM.4, P_EM.5, P_EM.6, P_EM.7, P_EM.8
ASC_EM (Architectural/ Structural Contract)	AC_EM (Architectural)	S_EM.4 P_EM.9	A_EM.4, P_EM.9, P_EM.11, S_EM.4, S_EM.6, A_EM.11, EM.51
-	ECO_EM (Engineering Contract)	-	P_EM.10, S_EM.5
PDC_EM (Precast Detailed Coordination)	PDC_EM (Precast Concrete Detailed Coordination)	A_EM.7 A_EM.10	A_EM.7, A_EM.8, A_EM.9, A_EM.10, P_EM.12, S_EM.9, EM.53, EM.54, EM.55, EM.60
-	SRC_EM (Structural review and coordination)	-	S_EM.7, EM.61, EM.62
-	EAR_EM (Engineering analysis results)	-	EM.52, EM.55, EM.56
-	PF_EM (Precast concrete)	-	S_EM.8, EM.58, EM.59
-	PR_EM (Production data)	-	EM.57, EM.64, EM.65

Another similarity that is not identified in approach A is with regard to P_EM.10 and S_EM.5 that both define the same exchange models in the construction documentation phase. Each of these EMs specifies that the exchange model is the engineering contract model focused on the structural design and integrates the structural layout with other building systems. It includes structural elements, connections and details. In this exchange, both the precast and other structural systems are fully designed. The exchange is prepared as a construction drawing set or construction-level model [4]. This similarity has not been identified in approach A but indicated in approach B.

In the second approach [6] (i.e. approach B), while all original EMs are compared then merged into a limited number of EMs i.e. eleven Exchange Models, some issues exist in the consolidation strategy. For instance, as can be seen in Table 2, S_EM3 is duplicated in two exchange models i.e. Engineering Concept and Precast Concrete Concept. Moreover, in approach B although eleven exchange models were defined in the resulting IDM, only six EMs out of eleven have been utilized and further explained.

Another example of the issues of inconsistencies in approach B can be seen in AC_EM which is specified as Architectural Exchange Model. In this exchange, A_EM4 and EM 51 are happening in two different stages of the project with two different purpose. A_EM4 happens in Design Development Phase and EM.51 occurs in Product Development Phase. A_EM4 is specified as the architectural contract model [4] that integrates the building layout of all precast pieces with all other building systems. It includes precast layout of surface finishes, molding, reveals and other decorative features. Other systems interacting with precast are also passed. The pass-off exchange is prepared to support production of a construction level model. EM 51 [4] on the other hand, passes back to the precast fabricator a report of the design intent issues identified by the architect for precast assembly-level piece layout, based on information supplied by the precast fabricator. In this exchange "design constraints of buildings and spaces are indicated, where relevant. Product information that raises issues about the design intent are reported, including layout, shape, material types, geometry and material of finishes of products, both in the piece and assembly level, and assembly relation of the pieces and connections. Openings and opening frames may be identified. Detailed information of different types of products may be included. Facade layout and grid geometry may be designated; slab topping thickness, material and surface treatment may be returned. For load-bearing and non-load bearing pieces, assembly and joint relations may be identified as problems. Characteristics of thermal and acoustic insulation may be referenced Nested relations and details of joints specifications may be referenced. Finally, other building parts affecting precast pieces specifications and systems may be indicated" [4]. A detail analysis of the information groups that A_EM4 and EM 51 include in the IDM for precast concrete [4] shows that A_EM4 includes sets of data that are not included in EM 51 and vice versa. For instance, A_EM4 requires all information groups related to the project and spatial hierarchy data to be provided in the model. However, in EM 51 these information groups are not included except the project name and the unit definition. Also, in A_EM4 all data related to grids such as grid axis assignment and placement relative to grids are required. But in EM 51 these information groups are excluded. In EM 51, the data related to reinforcing bar assignment and aggregation, embeds and joints are required while such data is not included in A_EM4. If these two Exchange Models are combined, the resulting EM will include a lot of information groups that are required in one original EM and not in another.

Given this review, two significant points can be derived. In addition to similarities in exchange description, consolidated set of exchange models should be defined with regard to two major aspects: a) project stage that each exchange is happening considering the exchange disciplines b) detailed specification and information groups each exchange model includes. This ensures that the exchange models are similarly aligned for consolidation.

3 Exchange Requirements and IFC Concepts

In defining the MVD for precast concrete, exchange requirements are combined into a set of information modules [6]. In fact, contents in different exchanges but within similar domains are often replicated. This creates the notion of data exchange modules that could be reused. These reusable modules represent semantic units that map the exchange pieces to an information model schema that is most often an IFC sub-schema [5].

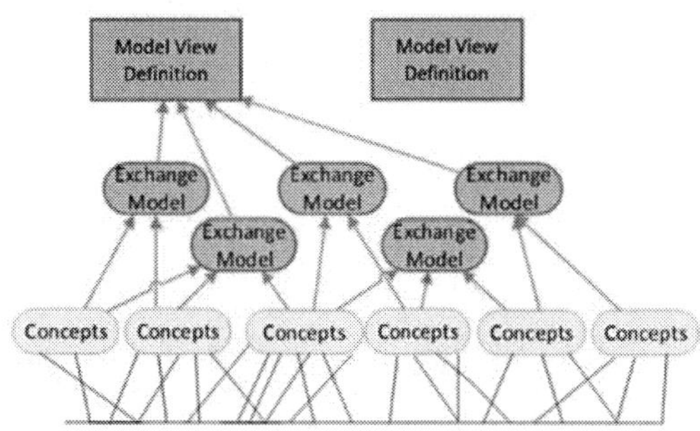

Figure 2 MVD concepts structure

The reusable modules are called MVD concepts or IFC concepts. They are aggregated into Exchange Models that are subsets of MVD concepts specifying the information for a specific workflow exchange. Figure 2 diagrams the concept structure. These concepts are the proposed basis for the Exchange Models representing the semantic and functional knowledge of an industry domain, such as precast concrete [5, 7]. The relationship of IDM requirements and reusable MVD concepts is represented in the work of [8].

The MVD implementation of precast concrete identified 93 reusable concepts while each of the 47 EMs consists of a set of concepts. An example of precast concepts is shown in Figure 3. In this concept, IFC entities are specified to define the data for placement of pieces relative to building elements. Another example of IFC concepts can be the IFC entities that define the geometry definition such as extrusion or B-rep.

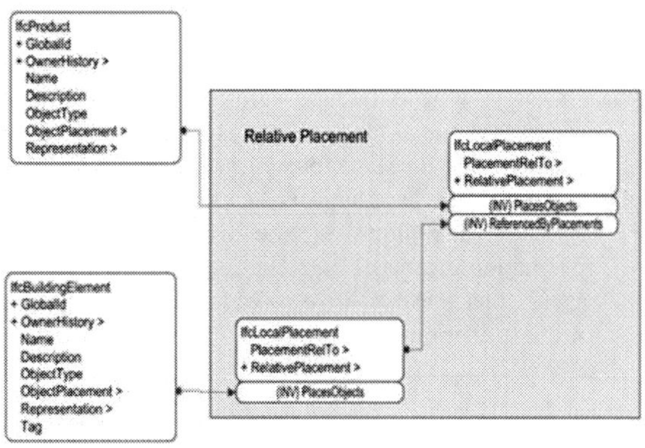

Figure 3 PCI-063 concept for placement of pieces to building element

Comparing original 47 precast concrete EMs with regard to the MVD concepts that each include reveals which Exchange Models have the same set of exchange requirements. Therefore, in this paper the consolidation is mainly based on a detailed review of the MVD concepts in each original Exchange Models. This ensures that the exchange models are being combined with regard to similar exchange data they include. Also, the investigation of the project stages that each exchange is happening (discussed in the next section) is another critical aspect of consolidation to ensure that the consolidated EMs are related to similar phase/phases of the project.

As an example, a detail study on A_EM.1, P_EM.1, S_EM.1, S_EM.2, and A_EM.2 shows that these exchanges are all happening in the preliminary stages of the project [4] with the purpose of architectural concept model or engineering concept model being passed to detailer for further preliminary precast structural and fabrication detailing. In addition, these exchange models all include the same set of IFC concepts such as project and spatial hierarchy, geometry representation as extrusion, metadata including status and approvals concepts. Hence, combining these models ensures that the consolidated model introduces an EM with the same purpose and the same set of concept definition as five original EMs.

Table 2 Precast concrete containment concepts (columns) and the consolidated Exchange Models (rows)

	PCI-040	PCI-042	PCI-043
	Building Element Aggregation	Site Contained in Project	Building Contained in Site
EMPC1	R		R
EMPC2	R		
EMPC3	R	R	R
EMPC4	R	R	R
EMPC5	R	R	R
EMPC6	R	R	R
EMPC7	R	R	R
EMPC8	R	R	R
EMPC9	O		
EMPC10	R	R	R
EMPC11a	R		
EMPC11b	R		

Table 2 shows 12 consolidated Exchange Models developed in this paper associated with some examples of IFC concepts (i.e. PCI-040, PCI-042, and PCI-043). These IFC concepts are defined for addressing containment data in precast concrete National BIM Standard [4]. In this table, if the concept is not included in the exchange it is shown as blank, if the concept is required it is marked with "R", and if the concept is optional in the exchange it is marked as "O". Complete table as EMs/Concepts matrix can be accessed on http://dcom.arch.gatech.edu/pci/MVD/Concept-Mapping which associates each of the resulting twelve consolidated Exchange Models with 93 precast concrete IFC concepts.

4 Project Stages and Merging EMs

Seven project stages have been identified in IDM for precast concrete [4] as: Preliminary Project Description, Design Development, Construction Documentation, Procurement, Product Development, Fabrication, and

Erection Phase. Also, the data exchange disciplines are identified as: Architecture, Structural Engineering, Building Product Manufacturing which is the precast detailer, and General Contracting.

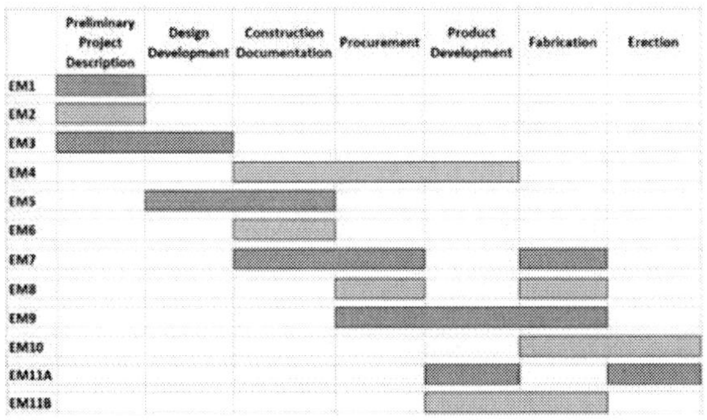

Figure 4 Consolidated Exchange Models (rows) within project stages (columns)

The consolidated Exchange Models developed in this study are illustrated in Figure 4. This figure demonstrates new EMs with regard to the project stages that the exchange is happening. In addition, exchange disciplines and the list of the original EMs are shown in Table 3.

Table 3 Consolidated EMs with associated exchange disciplines and mapping to old EMs

Consolidated Exchange Models	Exchange Disciplines	Original EMs
EMPC1	Architecture Structural Engineering Precast Engineer	A_EM.1 P_EM.1 S_EM.1 S_EM.2 A_EM.2
EMPC2	Architecture Precast Engineer	A_EM.3 P_EM.2
EMPC3	Architecture Structural Engineering Precast Engineer General Contractor	A_EM.5 A_EM.6 S_EM.3 P_EM.3 P_EM.4 P_EM.5 P_EM.6 P_EM.7 P_EM.8
EMPC4	Architecture Structural Engineering Precast Engineer General Contractor	P_EM.12 A_EM.7 A_EM.8 EM.53
EMPC5	Architecture Structural Engineering Precast Engineer General Contractor	A_EM.4 A_EM.11 P_EM.9 P_EM.11 S_EM.4 S_EM.6
EMPC6	Architecture Structural Engineering Precast Engineer General Contractor	P_EM.10 S_EM.5
EMPC7	Architecture Structural Engineering Precast Engineer General Contractor	A_EM.9 A_EM.10 S_EM.9 EM.60
EMPC8	Structural Engineering Precast Engineer Plant Management	EM.59 S_EM.8 EM.57
EMPC9	Structural Engineering Precast Engineer General Contractor	S_EM.7 EM.61 EM.62 EM.52
EMPC10	Precast Engineer General Contractor Plant Management	EM.64 EM.58
EMPC11A	Precast Engineer General Contractor Plant Management	EM.54 EM.63 EM.66
EMPC11B	Architecture Precast Engineer	EM.51 EM.55 EM.56

EM11A and EM11B both include the similar sets of concepts but the purpose of the exchanges and the exchange disciplines are different. Thus, they are kept separate from each other. In EM11A the fabricator passes the model of precast pieces and assemblies to the general contractor for coordination and then during the erection phase, the general contractor sends orders for piece delivery to plant manager. EM11B transfers coordination action items to the fabricator from the architect for piece detailing. This exchange passes back to the precast fabricator a report of the design intent issues identified by the architect for precast assembly-level piece layout, based on information supplied by the precast fabricator for the architects' review/approval. Therefore, these EMs are considered as separate exchanges.

5 Precast Concrete Consolidated Exchange Models

Based on the analysis of the exchange description as well as two major aspects described earlier which are: project stages and concept definition for each of the Exchange Models, 47 original EMs defined in precast concrete National BIM Standard are combined to twelve Exchange Models (i.e. EMPCs) listed below. Mapping to original EMs are listed in Table 3.

1. *EMPC1* (Building Concept- BC) consists of concept design layout of precast pieces optionally composed into assemblies. Geometry is nominal, without camber or twisting. It does not include surface or structural detailing. It includes structural- and other grid-controls, if used. It optionally includes major architectural finishes, and site information. It identifies interfaces with other structural elements and curtain wall systems. Extrusion is used as the geometry representation.
2. *EMPC2* (Precast Concept- PC) is the precast detailer's review on Building Concept model from architects and structural engineers. It specifies major architectural/structural precast components. This may deal with precast structural system, panelization, architectural finishes and site logistics. Extrusion is used as the geometry representation.
3. *EMPC3* (Precast Contract Development- PCD) provides precast design intent dealing with both structural and architectural intent. It defines the structural requirements of the building. It may include loads reactions, precast connection designs, precast-to-structural steel connection design, foundation design, and connection element capacities. Precast finishes may be defined and optionally doors, windows, interior wall partitions, and curtain wall systems embedded in or related to the precast. It is passed between different parties for review to ensure the building design intent and the structural adequacy is preserved. It is further refinement of the concept model, providing basis for the precast cost estimate based on early schematic design models. General contractor adds budget, schedule and specifications for the entire building received from several precast detailers/subcontractors to be passed to the owner/architect group to make a go/no-go decision about the project. Main geometry representation in this exchange is extrusion.
4. *EMPC4* (Engineering Design Development- EDD) is the detailed precast design model. It includes high-level description of precast piece detailing and all connection details. It provides assembly and piece layout and panelization for review to the architect and engineer. Architect's response then identifies those aspects and parts of the design where design intent has not been met to ensure consistency between the architectural design and precast detailing models. The general contractor can use this model for bid preparation or for coordination merged with other trade models. Main geometry representation is B-rep.
5. *EMPC5* (Architectural Contract- AC) integrates the building layout of all precast pieces with all other building systems. It identifies the shape and logical connectivity of all precast pieces. It includes the layout of surface finishes, molding, reveals and other decorative features. Other systems interacting with precast are also represented. Based on the architectural and engineering designs, this exchange model is used for coordination of all precast components includes precast slabs, beams, columns, and connections. It conveys detailed model descriptions of all precast structural elements, using B-rep geometry. The model together with the drawings and specifications are also submitted to the general contractor in order to be assembled with other models and used for the bid preparation. Main geometry representation is B-rep.
6. *EMPC6* (Engineering Contract- EC) is prepared as a construction drawing set or construction-level model. It is focused on the structural design and integrates the structural layout with other building systems. The model includes structural elements, connections and details. Both the precast and other structural systems are fully designed. Main geometry representation in this exchange is extrusion.
7. *EMPC7* (Precast Detailed Coordination- PDC) is general purpose multi-workflow exchange model defined by diverse sources for different recipients for detailed coordination. It may be used for the total building cost estimate based on the early schematic design models. It includes descriptions of all connection details, finishes, joints, embeds, reinforcing, tensioning cable layout, pre-tensioned pieces, and lifting hooks for lifting and transporting. Structural design of logical connections are specified. This model also conveys the results of structural design and reinforcement review of the engineer of record to the precast fabricator during the fabrication phase with information about design constraints, design loads and structural design. Main geometry representation is B-rep.
8. *EMPC8* (Structural Review & Coordination- SRC) includes geometry and assembly relations of buildings. Common categories of information for various types of products are included like layout, related shape and material information; both in the piece and assembly level. Connection relations of the pieces except for non-load bearing pieces are specified. Assembly and nested relations except for connections, and non-load bearing pieces are included. Related identification information and concrete mixes are included. Layout and grid geometry of facades, slab toppings, and reinforcement specifications are designated. More low level, detailed information about products is included. Characteristics of thermal and acoustic insulation are defined. Nested relations of both field

applied and plant applied connections are specified. Finally, related specifications of other building parts and systems are included. It includes detailed description of precast piece detailing, descriptions of all connection details, finishes, joints, embeds, reinforcing, tensioning cable layout, pre-tensioned pieces, and lifting hooks for lifting and transporting. Connections, design constraints, design loads and structural design are defined, using B-rep geometry.

9. *EMPC9* (Engineering Analysis Results- EAR) includes all structural precast elements. Slab layout and topping are defined. Assembly, nested and connection relations of load bearing and voided pieces are specified. Assembly and nested relations of logical connections and both field and plant applied connections are defined. Related identification information and concrete mixes are included. Reinforcement specifications and layout are designated. Structural design for load-bearing pieces and design loads for slabs are specified. Important common categories of information are included such as layout, shape, and material types and surface treatment, both in the piece and assembly level. Openings and opening frames are defined. Detailed information for some types of products is included. Layout and grid geometry of facades are designated. For load-bearing, non-load bearing and voided pieces, joint and connection relations are specified too. Logical and physical connections are defined. Lifting devices are indicated. Thermal and acoustic insulation characteristics are defined. Extrusion is used as the geometry representation.
10. *EMPC10* (Final Precast Detailing & Coordination- FPCD) includes fully detailed information about products and their assembled composition in project- layout, shape, geometry and finishes of all precast products. Assembly relations of the pieces and connections are specified. Connections with other systems, including embeds, are included. Openings and opening frames are defined (not opening fillers). Identification and related production information for different pieces are included. Reinforcement specifications are defined. Relevant information for different types of products is provided. Facade layout and grid geometry are defined. Voided pieces, nested, connection and joint relations are specified. Nested relations of both field applied and plant applied connections are specified. Specifications of other related building parts and systems are included. Concrete mixes and finish material types are defined. Lifting devices are included. Surface treatment areas are included. Main geometry representation is B-rep.
11. *EMPC11A* (Production and Erection Data- PED) provides important common categories of information including layout, shape, material types, and information about product finishes both at the piece and assembly level. Also, assembly relations of products except for foundation parts are specified. The piece marks for identification are included. Detailed information for some types of products is included. Layout and grid geometry of facades are designated and slab topping thickness, material and surface treatment are defined. For load-bearing and non-load bearing pieces, assembly, nested, joint and connection relations are specified. Relevant information about reinforcement is included. Nested and assembly relations of both field applied and plant applied connections are specified. Affecting specifications of other building parts and systems like lifting devices are indicated. Main geometry representation is B-rep.
12. *EMPC11B* (Architectural Review and Coordination- ARC) in which design constraints of buildings and spaces are indicated, where relevant. Product information that raises issues about the design intent are reported, including layout, shape, material types, geometry and material of finishes of products, both in the piece and assembly level. Also, assembly and connection relations of pieces are specified. For load-bearing and non-load bearing pieces, assembly and joint relations may be identified. The specifications of joints are defined. Nested and assembly relations of both field applied and plant applied connections are specified. The piece marks for identification are included. Facade layout and grid geometry may be designated. Slab topping thickness, material and surface treatment may be returned. Related specifications of other building parts and systems are indicated. Main geometry representation is B-rep.

These twelve Exchange Models cover the whole lifecycle of the precast concrete project from preliminary design stage to erection phase. They address all the exchange information requirements listed in National BIM Standard for precast concrete projects.

6 Discussion and Conclusion

This paper highlights that while there are many similar exchanges among the EMs defined in the Information Delivery Manual for precast concrete domain [4], the existing recommendations for consolidating 47 originally defined exchange models described in the work of [4] and [6] have not been effective. In the first approach [4], some of the similar EMs have not been identified ending up with too many Exchange Models. In the second approach [6], even though all original EMs are combined to limited number of EMs (i.e. eleven), some issues with its consolidation strategy are analyzed in this study which points out the challenges. This analysis indicates the existence of duplicated EMs and inconsistencies in exchange requirements of the combined EMs. Such problems make the implementation of precast concrete MVD challenging. The MVD requires its Exchange Models to be limited in numbers so that it can be managed within software implementation. Most importantly the MVD should be exhaustive and definite. Exhaustive means every model-based exchange in the lifecycle of the project must belong to an EM, and definitely means each exchange may only belong to one EM. In addition, if two original EMs are supposed to be combined in a resulting EM, the exchange requirements (i.e. the sets of data required in each EM) should be similar, otherwise the consolidation is not practical.

The consolidation methodology proposed in this paper addresses the existing issues. This is with regard to model description as well as IFC concepts that each EM includes, and project stages that each exchange is happening with related exchange disciplines that are involved. A detailed review of the MVD concepts identified the IFC concepts each EM includes and specifies the IFC concepts that similar EMs have in common. This ensures the exchange models are being combined based on similar exchange data they include. In addition, common project phases that similar EMs occur along with relevant exchange disciplines ensures that the consolidated exchange models are related to similar stages of the project. As a result, twelve exchange models are identified as Building Concept, Precast Concept, Precast Contract Development, Engineering Design Development, Architectural Contract, Engineering Contract, Precast Detailed Coordination, Structural Review and Coordination, Engineering Analysis result, Final Precast Detailing and Coordination, Production and Erection Data, and Architectural Review and Coordination.

This methodology describes how to effectively combine what is found to be too many originally defined precast concrete exchange models. Thus, the main contribution of this paper is to make the implementation of precast concrete MVD manageable by introducing a limited number of exchange models in the life cycle of the precast projects. The methodology has been applied in the implementation of the MVD for automated validation of precast concrete IFC instance files [9].

The identification of these new Exchange Models suggests future revision to National BIM Standard for precast concrete Information Delivery Manual. Additionally, the methodology of combining exchange models described in this paper can be utilized in other domains to facilitate the implementation of their MVDs.

7 Acknowledgement

The work presented here was funded by the Prestress/Precast Concrete Institute (PCI). All information presented is that of the authors alone.

References

[1] NBS, National BIM Standard- United States, Version 3, National Institute of Building Sciences buildingSMART alliance, 2015.

[2] buildingSMART, Model View Definition Summary, Retrieved April 2016 from http://www.buildingsmart-tech.org/specifications/ifc-view-definition, 2016

[3] Venugopal M., Eastman C., Sacks R., Teizer J., Semantics of model views for information exchanges using the industry foundation class schema, Advanced Engineering Informatics, Volume 26, Issue 2, April 2012, Pages 411–428, 2012

[4] Eastman C., Sacks R., Panushev I., Aram S., Yagmur E., The National Building Information Modeling Standards: Information Delivery Manual for Precast Concrete, The Charles Pankaw Foundation and PCI, 2009.

[5] Panushev, I., Eastman, C.M., Sacks, R., Venugopal, M., and Aram, V., 'Development of the National BIM Standard (NBIMS) For Precast/Prestressed Concrete', Applications of IT in the AEC Industry, CIB W78 Workshop Proceedings, Cairo, Egypt, November 2010, Virginia Tech, 2010.

[6] Belsky M., Eastman C., Sacks R., Venugopal M., Aram S., Yang D., Interoperability for precast concrete building models, PCI Journal;Spring2014, Vol. 59 Issue 2, pp. 144, 2014

[7] Eastman, C., Jeong, Y., Sacks, R., and Kaner, I., Exchange Model and Exchange Object Concepts for Implementation of National BIM Standards. J. Comput. Civ. Eng., 10.1061/(ASCE)0887-3801(2010)24:1(25), 25-34, 2010.

[8] Aram S., Eastman C., Sacks R., Panushev I., Venugopal M., Introducing a new methodology to develop the information delivery manual for AEC projects, Proceedings of the CIB W78 2010: 27th International Conference –Cairo, Egypt, 16-18 November, 2010

[9] Lee Y. and Eastman C., The Validation Logic and Structures of a Building Information Model Pertaining to the Model View Definition, Proceedings of CIB W78 Conference, Netherlands, University of Technology Eindhoven, 23-25 October, 2015

A Model for Real-Time Control of Industrial Robots

Fereshteh Shahmiri [a] and [b] Jeremy Ficca

[a] School of Architecture, Georgia Institute of Technology, USA
[b] School of Architecture, Carnegie Mellon University, USA
E-mail: [a] Fereshteh_shahmiri@gatech.edu, [b] jficca@cmu.edu

Abstract –

In this study a user-friendly model is developed for architects and designers without advanced programming skills who need to operate their projects in a real-time workflow from parametric design to robotic fabrication. Based on the implemented network capabilities in proposed model, not only one but also multiple robotic arms can interact and collaborate with design model data. In the network features of this model, a mediated interface is inserted between design software and robot interface that lets the user choose different parametric modeling software as the origin of the workflow. A practical experiment is developed in this research to validate the accuracy of the proposed model.

Keywords –

Real-time, Networking, Socket Communication, TCP/IP, UDP, Robotic Fabrication

1 Introduction

Many reasons necessitate running the live operation of industrial robots in design to fabrication processes. First, regarding the education of architecture, the live control of industrial robots can lead to creative products with exploration of unpredictable features [1][2]. Second, it is beneficial in the detection of potential errors existing in workflows and their correction inside the process with a fast and strong feedback loop originating from design action and fabrication reaction [3]. Moreover, some existing design variables in parametric modeling are not fixed values. So the implementation of the fabrication workflow has to be flexible enough to be responsive to the instantaneous changes of those variables while robot is operated [4]. Finally, there are situations in which multiple robots have to interact with one unique design model database, in these situations, robots' synchronization will be problematic if they are not operated real-time.

The present paper responds to key questions by addressing some challenges in the live control of robotic arms such as connection and compatibility of different interfaces, programming languages, connectivity between software and hardware layers and the complexity of networking. Those key questions are: first, what are the issues and interfaces involved in a real-time robotic workflow? Second, how is the implementation of a user-friendly application for such a workflow? Third, how is a real-time data streaming to both a virtual and an actual robotic arm? At the end, this research claims that the developed model is applicable in many live operations of industrial robots relative to the architecture applications.

2 Research Background

Many different researches and projects are conducted around the concept of live controlling of industrial robots in the area of architecture. A common software relative to parametric modeling in architecture is grasshopper [5]. Different adds-on for robotic fabrication is developed for this environment such as HAL [6] and KUKA|prc [7]. They presented different components to easily program industrial robots where the user generates a movement path with common CAD software and the programming tool generates the robotic execution program [2]. They have specifically developed components to generate and control a real-time workflow. However, in some cases, users need something different in compare with what already exists in those mentioned grasshopper add-ons. First, users need a clear understanding about the process that could simply demonstrate how is the implementation of a real-time workflow. Grasshopper components seem as black boxes to the users. Error checking and debugging is complicated when they are involved in design processes. This study tries to demonstrate how is the communication and networking in a live operation of the industrial robot. Second, mentioned plug-ins are developed just for grasshopper. However, all users don't necessarily work with this software. This study has pipelined the workflow among three interfaces: first, grasshopper as a design software, second, a middle layer application written in python as a mediator between design and fabrication software, third, RobotStudio [8] and IRC5 robot controller [9] as fabrication platform. It needs to be noted that although in this study grasshopper is considered as the default design modeling environment, the proposed model is flexible enough to detach the first interface and insert other parametric modeling software into the pipelined workflow as long as it is able to support network constraints.

3 Research Scope

Controlling the industrial robots and robotic fabrication are entwined in some topics such as robot mechanism, kinematic solvers, robot end-effector set up, robot programming languages, workspace analysis, positioning and motion programming. Dealing with these topics is required in all types of robotic workflows including live operations. However, the concentration area of this research is mainly on networking issues among existing interfaces.

As figure 1 illustrates, in the scope of this research, the real-time workflow is originated from grasshopper to RobotStudio for a live control of virtual robot and to IRC5 robot controller for real-time running of actual robotic arm. In order to streamline the workflow, the technical knowledge is restricted to coding in RAPID (Robot Application Programming Interface Dialog) [10] and Python script, networking in LAN (Local Area Network) and WAN (Wide-Area Network), socket programming for using UDP (User Datagram Protocol) and TCP (Transport Control Protocol)/ IP (Internet Protocols).

This study is conducted in Carnegie Mellon University; digital Fabrication lab (dFab) [11] and Computational Design Lab (CoDe Lab) [12] in 2014. Involved industrial robots are ABB IRB-4400 & IRB-6640. It is an ongoing research project and will continue at Georgia Tech; Digital Fabrication lab, with Kuka industrial robots.

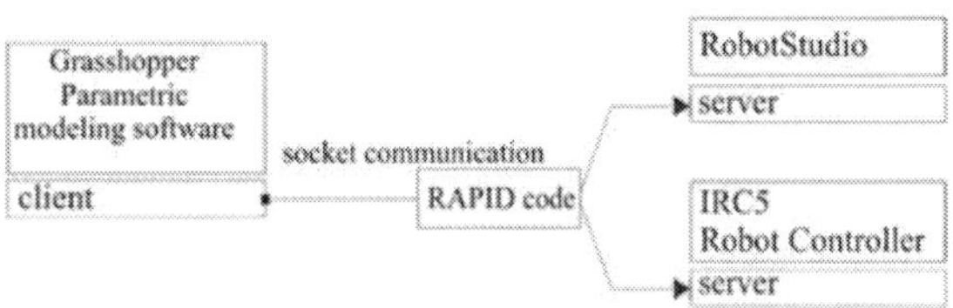

Figure 1. Focus areas in the research scope

4 Methodology

This research project is organized in two levels. In top level, it establishes a robust and reliable method of live communication among design and fabrication interfaces in order to run an industrial robot. To achieve such a reliable real-time workflow, client-server applications are developed. In these applications, UDP and TCP protocols are used for data streaming in LAN and WAN. Implemented method has the potential of operating not only one but also multiple robotic arms simultaneously. In second level, developed model of live controlling the robots, is applied to a practical experiment to validate the accuracy of the research result. This experiment aims to deal with a common challenge in the area of architecture for designing the surface patterns.

4.1 Create a Client-Server Application

In order to operate a live communication between design software and robot software and controller, data is streamed from grasshopper to RobotStudio and IRC5 robot controller. For data streaming, a Client-Server application is required in which grasshopper acts as a client and RobotStudio as a server. Such a Client-Server application is necessary to establish a network between two existing platforms. The connection between client and server in this network is either over LAN or WAN such as Internet. This network service and data streaming is provided by transport layer which comprises two types of protocols; UDP and TCP. Socket programming is used as an interface for these protocols. Figure 2 demonstrates how sockets are created in mentioned client-server application. As a result, it is essential to determine the type of network and protocol to stream data from design to fabrication platform.

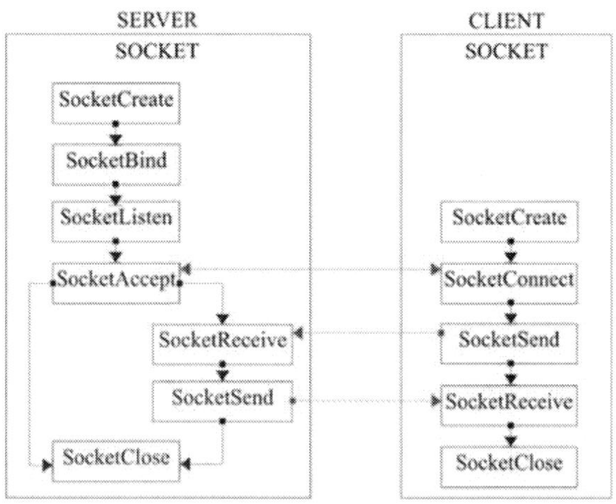

Figure 2. Socket communication between server and client [13]

4.2 UDP and TCP Protocols

Choosing either TCP or UDP as a transport layer is

depend on each existing platform. It is possible that some design model data sources and environments like grasshopper don't support socket programming over TCP protocol. Grasshopper can stream data over UDP [14]. UDP is a protocol that sends independent packets of data, called datagrams, from one computer to another with no guarantees about arrival and sequencing [15]. Missing and ordering problem of datagrams is mostly because of the network delays. As long as, data transmission is on a LAN, network delays are insignificant and missing and ordering the data won't be an issue. From the other side, IRC5 robot controller is only programmed to work with TCP. TCP is a connection-oriented protocol that provides a reliable flow of data. Since two existing platforms are not able to communicate by a mutual protocol, third interface is inserted in the workflow as a mediator.

4.3 Python Script: A Mediator Interface

A mediator interface between design software and robot software and controller is developed in python. According to figure 3, this middle layer interface aims to create two separate networks and acts both as a server and a client. In first network, it is a server in LAN for receiving data over UDP from grasshopper and in the second one, it is a client to stream data over TCP/IP to RobotStudio and IRC5 Robot Controller.

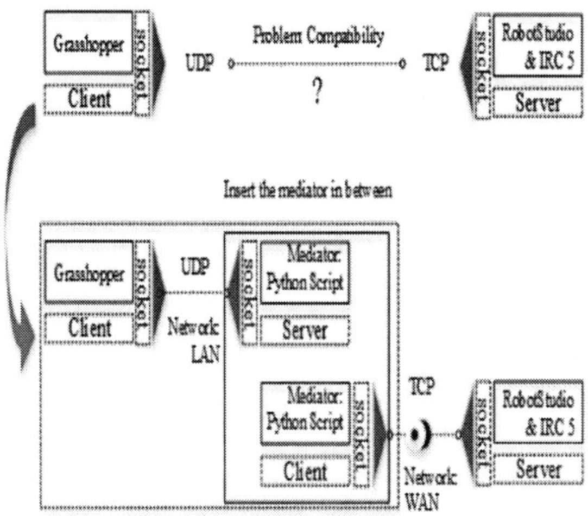

Figure 3. socket programming in mediator

There are key benefits of inserting such a mediator in the workflow. It aims users to customize the workflow depend on their own demands and makes data streaming to fabrication interface flexible enough to be independent of the design software. That means users have freedom in choosing different design software as long as the selected software is able to support either UDP or TCP protocols. Even if the software would support data streaming over TCP/IP, keeping the middle layer interface assists those users who are not interested in coding in RAPID. Figure 4 and 5 illustrate how python script and RAPID code interact with each other via programmed sockets as client and server.

```
import socket
Import time
HOST1='localhost'
#after receiving data by localhost on UDP, data
should be sent by host IP over TCP connection.
PORT1 =8000
S =socket.socket(socket.AF_INET,
socket.SOCK_DGRAM)
S.bind(HOST1, PORT1)
S1 = socket.socket(socket.AF_INET,
socket.SOCK_STREAM)
// if stream to multiple robot, second HOST and
PORT is assigned
S1.connect(HOST1,PORT1)
while 1:
        data, addr = s.recvfrom(1024)
        s1.sendall(data)
        time.sleep(2)
```

Figure 4. Insert middle layer interface to stream data between design and fabrication platforms

```
MODULE RobTalk
VAR socketdev ServSock; !Socket for server
VAR socketdev InSock; !Socket for client
VAR string received_string;
VAR string stRecieved; ! String from socket
VAR string clientIP;
VAR bool listen:= FALSE;
PROC RobComm()
    SocketCreate ServSock;
    S = socket.socket(socket.AF_INET,
    socket.SOCK_STREAM)
    SocketBind ServSock, "128.2.109.20",4997 ;
    SocketListen ServSock;
    SocketAccept ServSock, InSock;
    WHILE TRUE DO
        !variable data inserted here
        SocketRecieve InSock
        !str:=stRecieved;
        TPWRITE stReceived;

        ENDWHILE
        ENDPROC
        ENDMODULE
```

Figure 5. socket programming in RAPID code

4.4 Communication with Multiple Robots

There are different architecture and design applications in which multiple robotic arms are required in doing a specified task. According to the present research methodology, the middle layer interface is able to interact with multiple servers. In this case, multiple WAN networks relative to the number of robots are established. According to figure 6, by assigning unique IP addresses and port numbers to each server, manipulated data in design model can interact with each robot independently. Figure 8 depicts how two ABB robots interact with a common data source simultaneously.

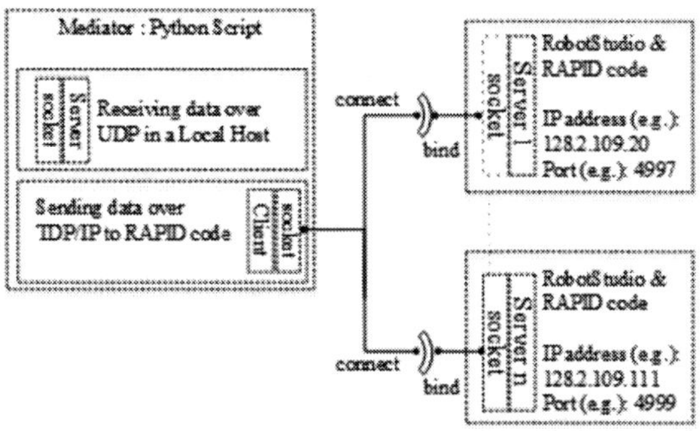

Figure 6. Data streaming to multiple servers

5 Result and Discussion

Figure 7 is the proposed model for a live control of robots and illustrates how network issues and real-time data streaming is solved between grasshopper, RobotStudio and ABB industrial robots. Based on the developed model, design and fabrication software have the potential to be customized according to the users' decision. There are key points in this developed model that users have to consider. First, for those design software without the capability of communication over TCP/IP like grasshopper, UDP is an option as long as first and second interfaces remain in LAN. If another design software support TCP protocol, the mediator interface could be removed and this model remains valid in connection between first and last interfaces. In this case, client socket that is implemented in the first interface has to be modified relevant to TCP protocol requirements. In the future, this research will develop the model to support both TCP and UDP in the first interface. Second, sockets are programed in different ways. If transmitting additional acknowledgement is required between interfaces, it is possible to program the sockets as both client and server in each interface. This optional concept will be added to the developed model in the future. Third, there are two advantages of inserting the middle layer interface as a mediator in the workflow. It creates separate sockets and networks to make connections over both UDP and TCP protocols. It also, can be a platform for the implementation of an external memory when users need to deal with large data set.

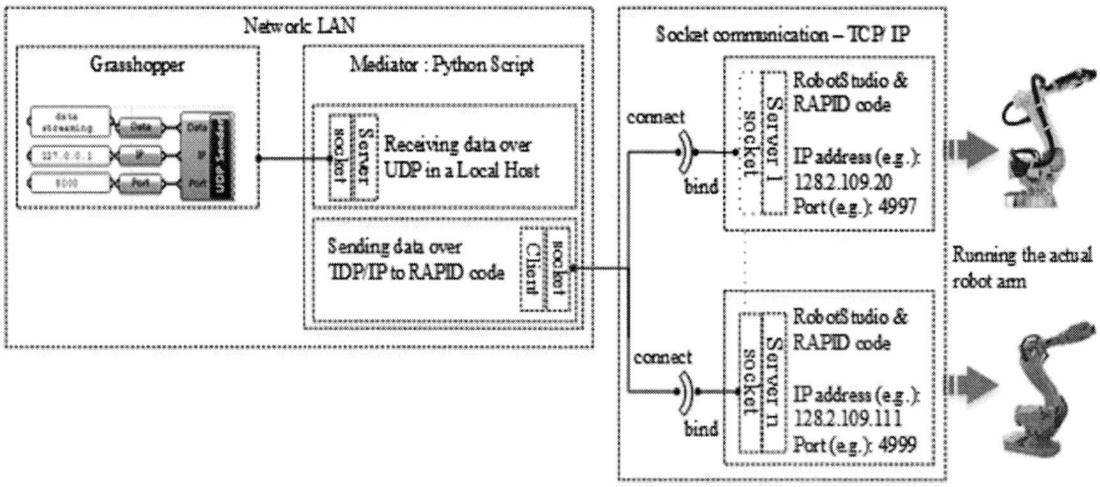

Figure 7. a diagrammatic view of final application

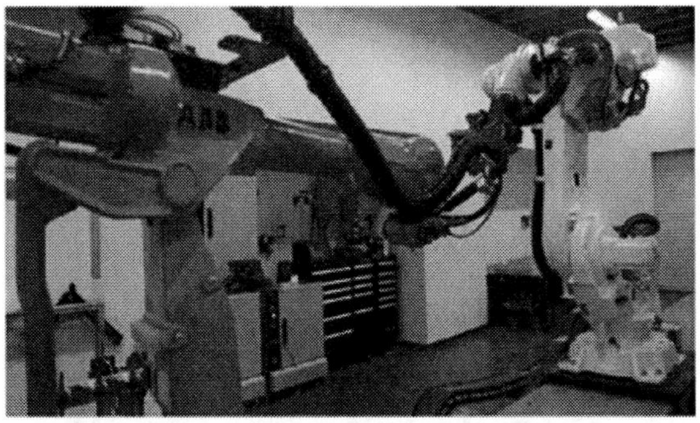

Figure 8. based on the proposed model, two ABB robots interact real-time with one unique design model database. (CMU dFab)

6 Validation

To validate the developed model and methodology, a practical experiment is conducted in order to simulate a variety of unpredictable patterns on a surface. For a long time designing the surface patterns has been a challenging issue for architects and designers. Specifically, when they are seeking nature patterns with many variable inputs that influence on the patterns formation. In this experiment, patterns are generated from water flow particles and tracing on a freeform surface. This simulation is done in Kangaroo [16]. The real-time operation of the robotic arm applies instantaneous changes in virtual model to the physical prototype synchronously. As figure 9 illustrates, variable inputs are particles' position and their non-linear velocity in each defined time frame. Data streaming in this step is done by using "UDP sender" component in Firefly [17]. To deal with such a large data set, an external memory out of grasshopper is required. This external memory is created in the mediator interface in the proposed model in figure 7. In this memory, data is saved and sifted according to the relation between patterns resolution and memory size. This manipulated data is streamed to RAPID code and led to the live operation of the robot. Figure 10, 11 and 12, demonstrates how this experiment is developed.

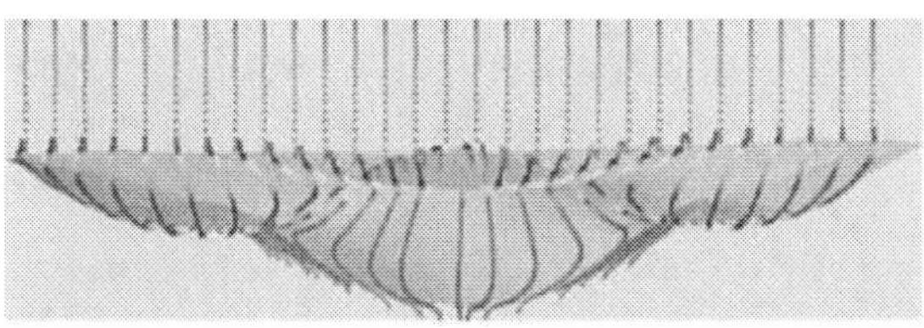

Figure 9. Particles have non-linear velocities in each position.

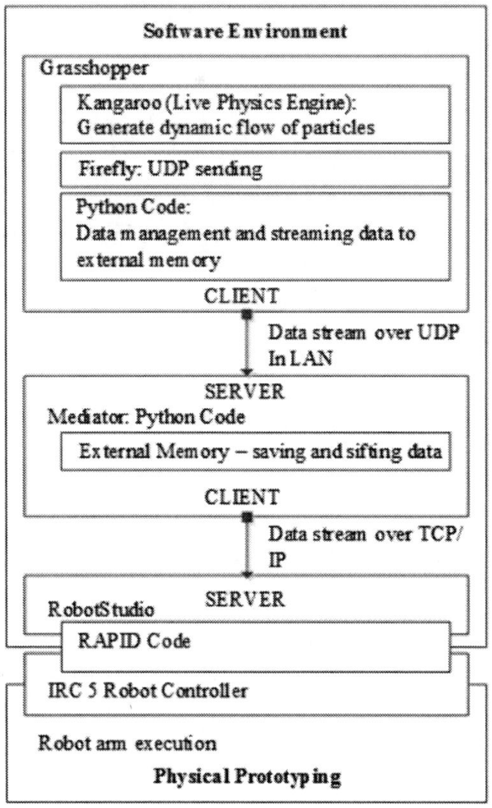

Figure 10. workflow for a live simulation of water flow patterns on freeform surfaces.

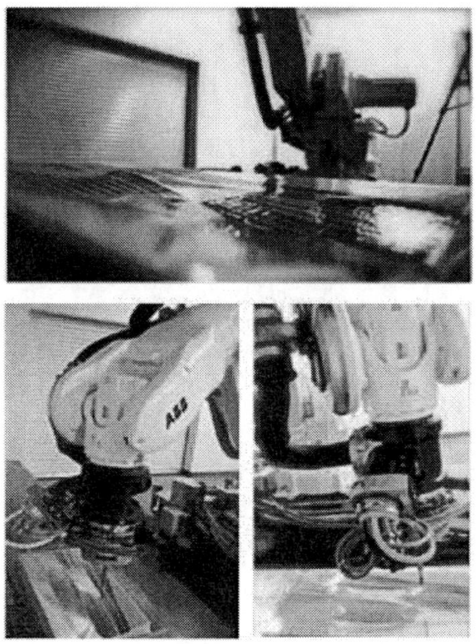

Figure 11. Physical prototyping – real-time workflow synchronized with virtual model.

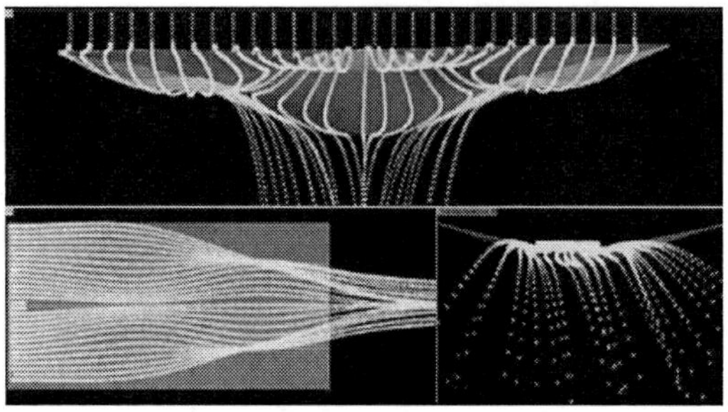

Figure 12. Water flow particles' simulation in virtual model

7 Summary and Conclusion

This research has introduced the existing challenges in a real-time control of ABB industrial robots via an architecture software like grasshopper. It has established a robust and reliable method of live communication among design and fabrication interfaces. To achieve such a live communication, technical issues related to the networking and socket programming are solved. In the proposed methodology, client-server applications are developed. UDP and TCP protocols are used for data streaming in LAN and WAN and the possibility of live control of multiple robot arms is tested and validated. This study has verified the accuracy and efficiency of the methodology by presenting a practical experiment.

There are advantages in the proposed methodology. First, connection and compatibility of the interfaces are open which provides a clear understanding about the workflow for the users. Second, the users also can choose grasshopper as with any other design software as long as it could support network constraints. The methodology has the potential to apply different types of industrial robots into the workflow.

The main conclusions drawn in this research are to investigate several issues. first; developed model has to be customized for other types of industrial robots such as Kuka. Second, the integration of the proposed methodology in to the other devices such as vision system and sensory devices is part of the future development of the research. Third, the more exploration of real-time applications is required. Regarding the developed project, one helpful application is to explore how different patterns generated from a particle system in virtual space can lead to a variety of unpredicted but adjustable patterns in physical model. Variables like different shapes of freeform surfaces and the density of the particles are some of those explorations. Such differentiation between virtual and physical models would be created and controlled by different sensors applied to the design to fabrication process.

References

[1] Yoshida H, "Robotic aggregation of materials with unpredictable geometry". On-line: http://www.dfab.ch/portfolio/robotic-aggregations-of-materials-with-unpredictable-geometry/, Accessed Jun 2016

[2] Elashry K, and Glynn R. "An Approach to Automated Construction Using Adaptive Programing." *Robotic Fabrication in Architecture, Art and Design 2014*. Springer International Publishing: 51-66, 2014.

[3] Braumann J, and Brell-Cokcan S. "Real-time robot simulation and control for architectural design." *30th eCAADe Conference* pages 469-476, Prague. 2012.

[4] Vasey, L. Maxwell, I. Pigram, D. "Adaptive Part Variation; A Near Real-Time Approach to Construction Tolerances." *Robotic Fabrication in Architecture, Art and Design 2014*. Springer International Publishing: 291-304, 2014.

[5] Rutten, D. "A graphical algorithm editor integrated with Rhino's 3-D modeling tools". On-line: http://www.grasshopper3d.com/, Accessed Jun 2016

[6] Schwartz, T. "Application of Robot Control and programming in architecture and construction industry". On-line: http://www.hal-robotics.com/

[7] Accessed Jun 2016

[8] Braumann J, and Brell-Cokcan S. "Parametric Robot Control and Simulation". On-line: http://www.robotsinarchitecture.org/kuka-prc, Accessed Jun 2016

[9] "ABB's fifth generation robot controller". On-line: http://new.abb.com/products/robotics/controllers/irc5 , Accessed Jun 2016

[10] ABB Robotics, Application Manual, PC SDK. on-line:https://library.e.abb.com/public/124d6b59313ed85fc125793400410c5b/3HAC036957-en.pdf , Accessed Jun 2016

[11] "Off-line programming software for ABB Robots". On-line: http://new.abb.com/products/robotics/robotstudio , Accessed Jun 2016

[12] Carnegie Mellon University, Digital Fabrication

Lab. On-line: http://soa.cmu.edu/dfab/ , Accessed Jun 2016

[13] Carnegie Mellon University, Code Lab. On-line: http://code.arc.cmu.edu/ , Accessed Jun 2016

[14] Szabolcs, C. " Comparison between two methods of robot control in the production of prismatic parts". *nonconventional technologies review16.2* (2012).

[15] Afsari K., Swarts M.E., Gentry T.R. Integrated generative technique for interactive design of brickworks, *Journal of Information Technology in Construction (ITcon)*, Vol.19, pg. 225-247, 2014.

[16] Xue, M. and Zhu, C. "The socket programming and software design for communication based on client/server." *Circuits, Communications and Systems*, 2009. PACCS'09. Pacific-Asia Conference on. IEEE, 2009.

[17] Piker, D. "Kangaroo; live physics for Rhino/grasshopper" On-line: http://api.ning.com/files/yPjjd3IjSzDhL3dzJkMTSiPHFBFe8m2SxAX-pvUfidJ*2W82Ij0BG-2z*M*LELCkzVSTxX2UToz-sJgBVHa6MnPmf-Gvdppk/KangarooManualGrasshopperversion_.htm , Accessed Jun 2016

[18] Payne, A. and Johnson, J." Firefly; connection between grasshopper and Arduino." on-line: http://www.fireflyexperiments.com/ , Accessed Jun 2016

Cloud-based BIM Data Transmission: Current Status and Challenges

Kereshmeh Afsari[a], Charles M. Eastman[a] and Dennis R. Shelden[a]

[a]School of Architecture, Georgia Institute of Technology, USA
E-mail: keresh@gatech.edu, charles.eastman@coa.gatech.edu, dennis.shelden@coa.gatech.edu

Abstract –

In the architecture, engineering and construction (AEC) industry model-based data exchange methods are mainly based on manual file transfer, data conversion, and data-merge. File-based Building Information Model (BIM) data exchange is either in vendor specific file formats or neutral format using Industry Foundation Classes (IFC). IFC Model View Definitions (MVDs) proposed by the US National BIM Standard can assist cross-platform BIM data exchange. Since BIM applications are steadily moving to the Cloud, the study of Cloud-based BIM data transmission techniques become significant. The main objective of this paper is to investigate how building data transmission can be managed in current Cloud-BIM applications and what challenges exist in the current systems. Therefore, in this study, methodologies for Cloud-BIM data integration are investigated. Features of each technique is specified. The strengths and weaknesses of current systems are indicated with regard to data transmission requirements for Cloud-based BIM applications. In addition, the study investigates the challenges to cross-platform BIM data transfer in making multiple Cloud-BIM applications interoperate. The challenges in current data transmission approaches highlight the need for an effective network-based BIM data exchange to address a collaborative BIM work flow in the Cloud.

Keywords –

BIM; Cloud Computing; Data Exchange; IFC

1 Introduction

As a rapidly emerging technology, Cloud-BIM has become a new research area in architecture, engineering and construction (AEC) industry since 2010 [1, 2]. It is believed that Cloud-BIM could provide project partners and design disciplines with the capabilities to share and exchange design data requirements and solutions [3]. Cloud computing can perform as an effective means to overcome current BIM challenges by providing real-time access to a pool of data, on-demand access to computing resources (e.g. storage, servers) and applications, and potentially better interoperability [4]. The question is whether existing Cloud-BIM interoperability efforts have been successful in achieving this vision or not. The objective of this research is to investigate current Cloud-based BIM data transfer methodologies and to study the advantages and disadvantages of each approach to outline existing challenges. Therefore, in this paper, the significance of Cloud-based BIM development is discussed and existing Cloud-BIM data integration solutions are investigated. Moreover, the paper provides a comparison of current Cloud-based BIM interoperability architectures.

2 BIM and Cloud Computing

Cloud-based BIM technology is known as a cost effective alternative to current state of data exchange and storage [4]. Since potential values of Cloud-based applications such as efficiency and low-cost has been recognized, the combination of BIM and Cloud computing is believed to be a promising trend [2, 4]. As a result, BIM applications are gradually moving to the Cloud and BIM web services and Cloud-based apps are becoming more and more popular [5, 6, 7]. Some examples of the Cloud-BIM services are GRAPHISOFT's BIM Explorer (BIMx) and BIMcloud, AutoCAD WS, Autodesk's A360 and BIM360, BIMServer, ONUMA System [1, 5], Trimble Quadri^{DCM} [18] and Trimble Connect [25], as well as third party packages such as Assemble Insight by Assemble Systems [23], and BIM Assure by invicara [24]. With the growing development of Cloud-based BIM technologies, there is a need to reconsider the approach to interoperability of new Cloud-BIM services [8].

In addition, Cloud-BIM can facilitate a distributed system environment for multi-user collaborative interaction [9]. Cloud-BIM is anticipated to change the AEC industry although the technology is still relatively new [1]. The development of Cloud-based BIM services has already created a new direction in BIM implementation to support collaborative BIM data generation and consumption among project partners [5]. Therefore, Cloud-BIM technology is believed to provide higher levels of information interaction further cross-disciplinary collaboration [1, 9].

Most importantly, interoperability is the key to the success of Cloud services implementation [1]. However, the challenge is that making multiple Cloud-BIM applications interoperate would be very difficult when they are developed by different vendors running on different Cloud platforms [10]. Therefore, the issue of data exchange and interoperability for Cloud-based BIM applications needs further studies [5]. Thus this paper investigates interoperability architectures in current Cloud-BIM data integration solutions.

3 BIM Interoperability Standards

Information sharing which is the basis for collaboration requires applications to be able to exchange data regardless of vendors and data formats. To achieve this, in the AEC industry, building data is described in Industry Foundation Classes (IFC) specification to support a neutral data format and to facilitate cross-platform BIM interoperability [11, 12]. IFC is in fact the international openBIM standard [13].

Besides, Model View Definitions (MVDs) proposed by the US National BIM Standard (NBIMS) is supposed to assist BIM data exchange [14] in cross-platform collaborations. The NBIMS process [14] shown in Figure 1, first organizes the team that will participate in defining data exchanges for a specific domain or stage of the project and determines the functional requirements for the identified exchanges. It is the fundamental step that defines the functionality of exchanges to be supported later by MVDs. The result of the works such as forming a workgroup, developing a process map for the identified work flow, defining the set of use case exchanges being addressed by the workgroup, describing the activities involved especially the exchange requirements creates the basis of the Information Delivery Manual (IDM) for a given NBIMS project. IDM [13] captures the user needs and specification of the exchanges in a form that can be translated into technical exchange specifications. The design phase of NBIMS creates the MVD binding and specification required to implement the exchanges defined in the IDM [14].

An MVD or what is known as an IFC view definition, specifies a subset of the IFC schema that is needed to satisfy one or many exchange requirements of the AEC industry [13]. MVD consists of one or multiple exchange requirements which must be provided by the sender of data to support work in the receiving application. Sender uses an MVD to generate the exported IFC model in a BIM application and then passes the file to the receiver of data. Upon receiving the exported IFC model, the receiver of data imports the model in the receiving application which uses an IFC importer module to translate the IFC file to native binding. This process is illustrated in Figure 2.

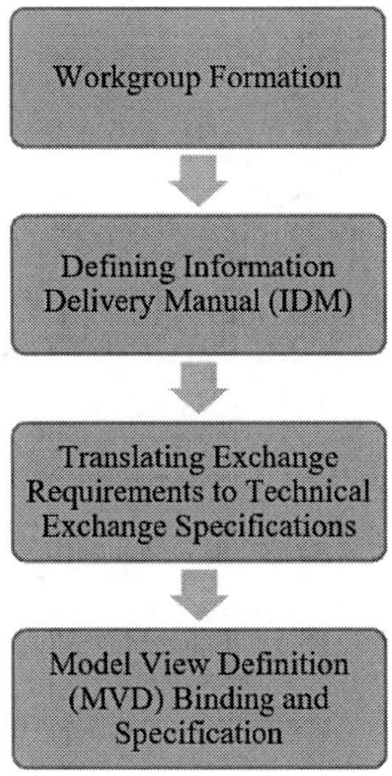

Figure 1 NBIMS development process for cross-platform BIM data exchange

In this process, the request for data, exporting the BIM model from the sending application and importing it in the receiver application should be done manually, usually through emails or other methods of correspondence.

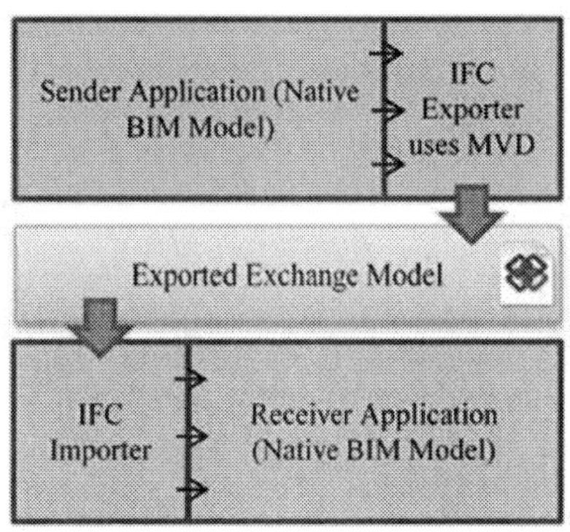

Figure 2 IFC-based BIM data exchange process

On the other hand, as stated in NIST Cloud Computing Technology Roadmap, interoperability relates to communication and data transfer between different systems. Interoperability decreases the need for manual intervention or providing the same information to multiple systems. Therefore, Cloud providers need to deliver standards-based systems to allow the provisioning of data to the users [15]. BIMSie project [6] has worked towards introducing a standardized Service Interface for Cloud-BIM solutions. This standard Service Interface is supposed to automate interaction between Cloud-BIM applications by introducing an Application programming Interface (API) for Cloud-BIM applications with standardized methods [6]. In the BIMSie initiative when services are connected, the data exchange relies on file-based data using industry standards such as IFC files.

4 Cloud-BIM Data Integration

Available Cloud-BIM data integration solutions on the market mainly address the idea of centralization of BIM data. The emergence of server-based BIM solutions have provided a central BIM service to all project members [2, 7]. These model server technologies establish a single-sourced data server [2] that uses data directly from the models (i.e. sub-models) and with a consolidated model it can improve multidisciplinary collaboration [7]. BIM server implementations are still limited [7] but BIM server technology has changed BIM work sharing with a database driven approach [5, 16, 7].

Current BIM server technologies on the market provide functionalities like querying BIM models as well as graphical interfaces for sharing and viewing BIM models in a team on a centralized platform. Examples of these BIM server technologies are GRAPHISOFT® BIMcloud® for design process [17], Trimble® Connect for project collaboration [25], Autodesk® A360 for design delivery [19] and BIM360 for construction project delivery [20]. Also Autodesk® Revit® Server is the server application for Revit® Architecture, Revit® Structure, and Revit® MEP performs as a server-based worksharing platform for Revit® projects. In addition to proprietary BIM server technologies, BIMServer.org as an open source technology has been developed by TNO and the University of Eindhoven [16]. BIMServer.org architecture is shown in Figure 3. BIMserver centralizes the information of a project with a core that is based on IFC and shares its information to client applications through some interfaces. This solution is not a file server but its architecture interprets IFC data from a file and stores it in a database. Therefore, it can merge and query the model and it eventually generates IFC files [21]. Other examples of IFC-based BIM server implementation are IFC model server developed by VTT Building and Transport and SECOM Co., Ltd. And also EDM Model Server which is an IFC model server based on EDM developed by Jotne EPM Technology [2].

These Cloud-based BIM servers could be theoretically a good solution however in practice they have faced major challenges [2] such as scalability and robustness [5]. BIM server technology overcomes the issue with full-size model synchronization [5] while the integration of BIM models in the Cloud provides a network-based data exchange [1]. However, the technology is not robust [5]. In this type of Cloud-based collaboration some data management issues exist especially in combining models [1] and in sub-model extraction [2]. Also, these solutions require more powerful Web-based operating systems, file-sharing platforms and hardware controllers [1].

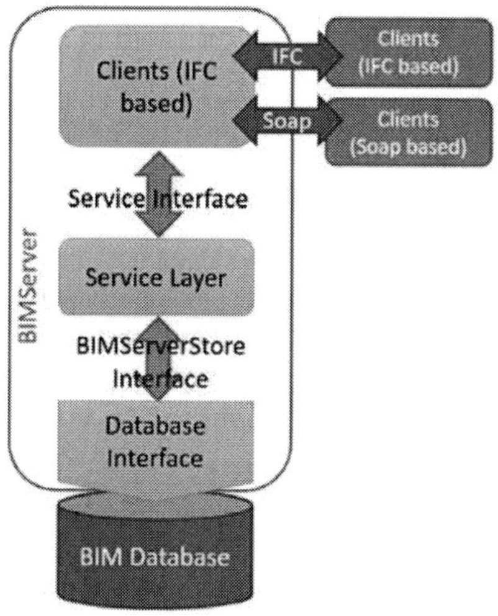

Figure 3 BIMServer.org architecture

Studies identified that existing BIM servers require functional and performance improvements [2, 7] as well as coordination with vendor specific format data [2].

Figure 4 Architecture of a BIM server solution

Most importantly, the input and output of these systems, as shown in Figure 4, is heavily based on files. For instance, input and output of BIMServer.org is based on IFC files. Also, the existing BIM servers cannot work with the decentralized, heterogeneous and dynamic design data and as a result cannot support the whole lifecycle of the project [2].

In addition to BIM server solutions, Flux project [22] which was started in late-2010 at Google[x], Google's research lab, is a cloud-based collaboration tools for design process to assist architects, engineers, and contractors with exchanging data. Unlike conventional file-based data transfer, Flux acts as an interchange point for sharing project data such as design, analysis, and schedules. Flux plugins should be installed on design software applications to automate data transfer to and from Flux. Currently, Flux works with limited number of applications such as Rhino/Grasshopper, Excel, Revit/Dynamo, and SketchUp. Figure 5 illustrates the data flow between applications and Flux.

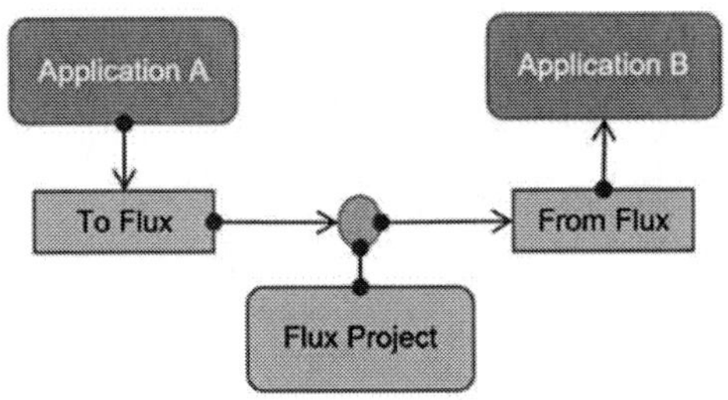

Figure 5 Flux data flow for design data exchange

5 Comparison of Data Integration Approaches

The study of the existing Cloud-BIM data integration methodologies suggests that there are three main categories of BIM data integration in the Cloud that allows cross-platform data exchange.

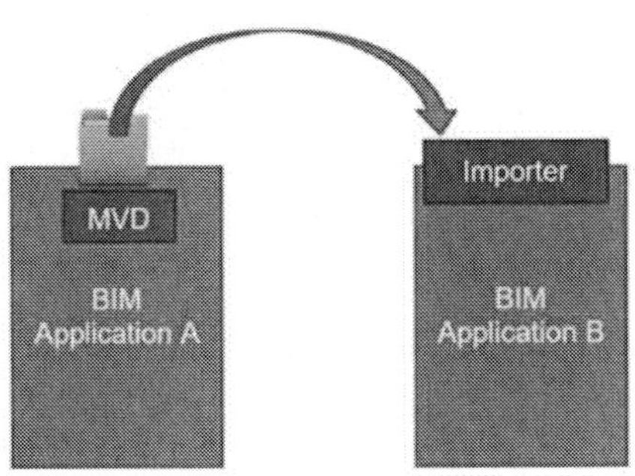

Figure 6 Data flow for manual file transfer

1. **Manual File Transfer**: Manual file transfer can be based on BIM open standard as well as vendor specific formats. Using neutral file format based on IFC data model and established MVDs can be potentially advantageous to help AEC industry stick to a common language to improve collaboration. However, file-based data transfer is a one-way communication that should be repeated for each design iteration to include constant design changes. In addition, in this process, there are methods to validate the exported model against the initial exchange requirements but how to validate the imported model is still vague. This data flow is shown in Figure 6. Existing file-based data transfer technologies are not suitable for BIM applications because of incapability of managing data redundancy and inconsistencies [2].
2. **BIM Server Technology**: BIM server centralizes BIM data in a database and can improve collaboration in an integrated model. But the technology is limited and has performance and scalability issues in dealing with complex and big projects.

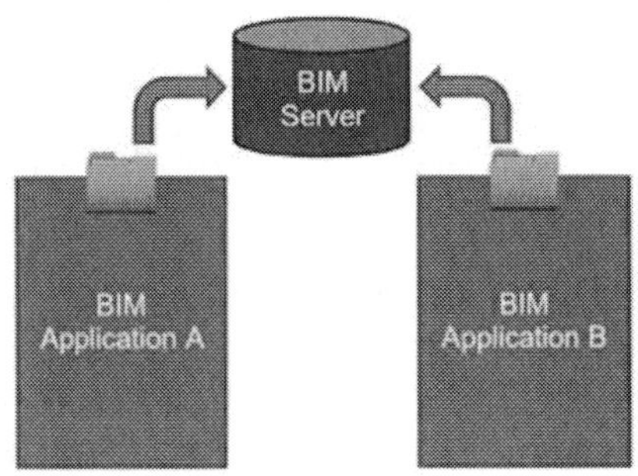

Figure 7 Integration data flow for BIM server technology in data transmission between applications developed by different vendors.

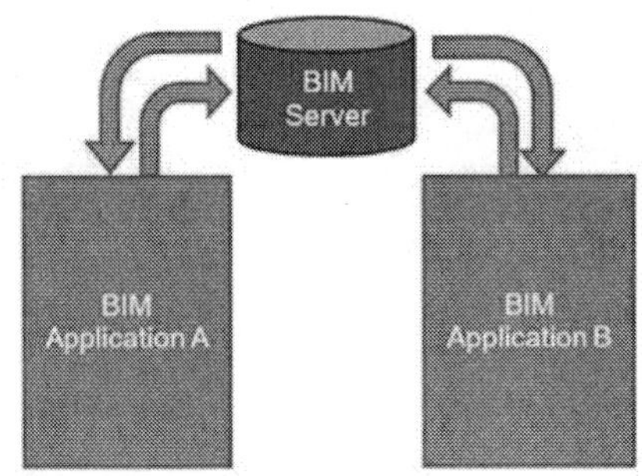

Figure 8 Integration data flow for BIM server technology in data transmission between applications developed by one vendor.

The integration data flow for this methodology is shown in Figure 7 and 8. BIM server technology provides a centralized and accessible repository for the project and can provide access to project data almost anytime and anywhere throughout the building lifecycle. However, since these solutions only support a limited number of data formats as inputs and outputs (i.e. mainly data formats supported by one vendor), the centralized model can be disconnected from the origin of data if the data is provided by a different vendor whose format is not supported. In this case data transfer should be tackled with exporting and importing files (Figure 7). This makes the integration to rely on manual file transfer. In BIMServer.org for instance, a user (e.g. an architect) should check in a file-based IFC instance model and the model will communicate with server in the Graphical User Interface (GUI). Similarly, the outputs of the service are based on files as well if needed to be transferred after analysis, simulation, etc.

3. **Data Interchange Hub**: DIH technology such as Flux project can automate data flow between certain applications. This solution can reflect the changes in real-time while each user and application controls when to synchronize data with the project. Data Interchange solution allows users to work in isolation and share their changes when they are ready [22]. But this solution currently supports very few design applications to exchange data. In addition, the applications are dependent on Flux although the model can be updated on each platform on its own. The integration data flow for Data Interchange Hub solution is shown in Figure 9.

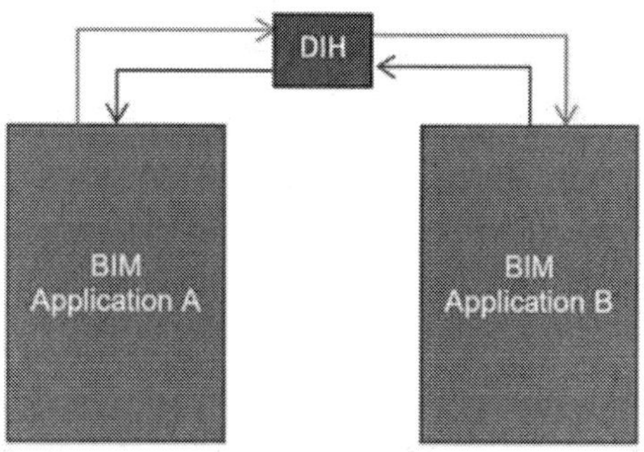

Figure 9 Integration data flow for Data Interchange Hub (i.e. Flux) solution

Comparison of these three methodologies are summarized in Table 1. Advantages and disadvantages of each method are specified.

In addition to these BIM data integration solutions that are developed to enable cross-platform data exchange and to provide cross-disciplinary collaboration solutions, there are other third party solutions developed for a specific purpose. These cloud-based third party solutions extract data from the BIM models in BIM authoring tools through some plug-ins to publish the model in a new environment so that they can provide additional specialized services such as analysis.

Table 1 Comparison of Cloud-BIM data integration methodologies

Cloud-BIM Data Integration Technique	Pros	Cons
Manual File Transfer	Could be based on established standards and MVDs	One-way data transfer and data import issues
	Can use neutral data format	Repeats for each design iteration
BIM Server Technologies	Centralizes BIM data	Scalability and performance issues
	Improves collaboration in an integrated Model	Depends on the server platform and not completely connected to the origin of data
Data Interchange Hub	Automates data flow	Supports very few applications
	Reflects design changes in real-time within each BIM application	Depends on the interchange platform

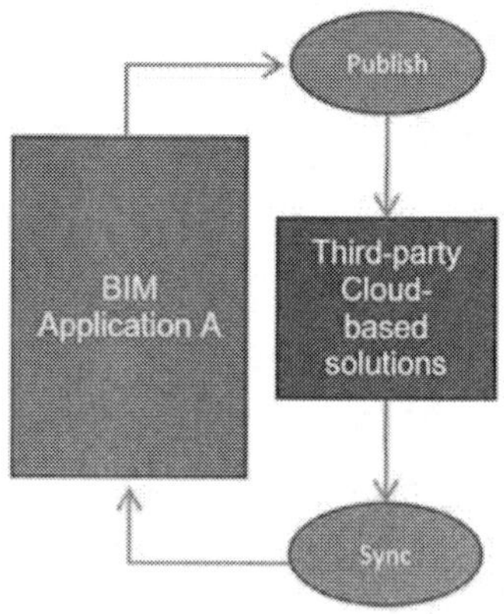

Figure 10 Data flow for third-party Cloud-based BIM solutions

As distributed services, these Cloud-based solutions can eventually synchronize the modifications applied in the model back to the original BIM authoring tool. For instance, BIM Assure [24] provides BIM model checking service for validating the model accuracy and completeness. Also, Assemble Insight [23] provides cost estimation solutions and analyses such as value engineering. These solutions are currently limited in exchanging and mapping of data structures in different formats and can only deal with BIM data from very few BIM authoring tools and data formats, mainly supporting one vendor specific data structure. Since these type of Cloud-based BIM solutions are not considered as consolidation products and are not supposed to provide integration platforms, cross-platform data transfer solutions or solutions to connect applications, they are excluded from the comparison in this research. The overall data flow for these Cloud-based solutions is illustrated in Figure 10.

6 Key Challenges in Cloud-BIM Data Exchange

As can be seen in Table 1, Cloud-BIM data transmission for cross-platform collaboration faces several challenges.

- Standardization- There is a lack of Cloud specific standards for BIM interoperability. With the growing number of Cloud-BIM services developed by several providers, standardization among these service providers become important. Open standards like IFC data schema should be expanded to address the requirements of Cloud-based applications.
- Data interdependency- Current BIM integration solutions that deal with Cloud interoperability are either addressing model federation in a centralized platform or interconnecting a limited number of design applications on premise through a new Cloud-based platform with the help of plug-ins. These solutions integrate two or more systems to a third new system, instead of creating a loosely coupled aggregation, shown in Figure 11, where each system remains self-contained.
- Data access and security- Collaborative nature of Cloud-BIM data integration causes security challenges such as liability and BIM model ownership [1, 4]. User authentication and authorization is key to successful deployment of Cloud-BIM and therefore Cloud identity management and role-based user access becomes significant which require advancement in trust and privacy preserving techniques [4]. Providing a safe service for Cloud-BIM applications and integration solutions is critical. The issues of dealing with the data security, ownership and stability for Cloud-BIM technologies are still open research areas.

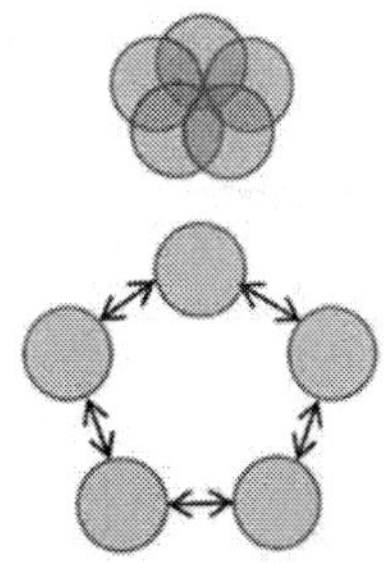

Figure 11 Tight coupling (top) vs. loose coupling (bottom) between Cloud applications

7 Discussion and Conclusion

Existing architectures for Cloud-BIM data integration face several challenges in model-based data exchange and have not fully exploit the potential of the Cloud towards a loosely coupled integration. Therefore, there is a need to develop a new architecture that redefines the data flow utilizing web-based technologies as major enablers of the Cloud. For instance, Cloud computing is based on the Transmission Control Protocol/Internet Protocol (TCP/IP). Such data transfer protocols should be considered for cross-platform BIM data exchange architecture. Therefore, future study will likely propose an architecture considering web technologies and elements of the Cloud interoperability such as Cloud APIs.

In addition, while vendor specific data formats are quite diverse, these are based on multiple and different data schemas. On the other hand though, data standards for Cloud-based cross-platform data exchange purposes are limited. IFC data model which describes building data provides a means to define building components and processes in a publicly available data schema. So far, the specification of IFC standard is provided in EXPRESS (using the STEP physical file structure) and XSD schema (using the XML document structure). As an industry-wide open standard, IFC schema definition should become the basis for Cloud-BIM integration solutions to ensure a common understanding of building data across the applications and disciplines. Future work should investigate the possibility of representation of IFC schema in data encoding suitable for web-based data transfer. In addition, the NBIMS process should be revisited considering the requirements of Cloud-based data transmission and data integration work flow with Cloud-based use cases.

References

[1] Wong J., Wang X., Li H., Chan G., Li H. A Review of Cloud-based BIM Technology in the Construction Sector, *Journal of Information Technology in Construction (ITcon)*, Special Issue BIM Cloud-Based Technology in the AEC Sector: Present Status and Future Trends, Vol. 19, pg. 281-291, 2014.

[2] Zhang J., Liu Q., Yu F., Hu Z. and Zhao W. A Framework of Cloud-Computing-Based BIM Service for Building Lifecycle, *Computing in Civil and Building Engineering*, 1514-1521, 2014.

[3] Redmond, A., Hore, A., Alshawi, M. and West, R. Exploring How Information Exchange Can Be Enhanced through Cloud BIM. *Automation in Construction*, 24, 175-183, 2012.

[4] Mahamadu A-M., Mahdjoubi L. and Booth C. Challenges to BIM-cloud integration: Implication of security issues on secure collaboration. *Proceedings of the 5th IEEE International Conference on Cloud Computing Technology and Science*, 209-214, 2013.

[5] Wu, W., and Issa, R. Leveraging Cloud-BIM for LEED Automation. *Journal of Information Technology in Construction*. Vol. 17, 367, 2012.

[6] Berlo L. BIM Service interface exchange (BIMSie). On-line: https://www.nibs.org/?page=bsa_bimsie, Accessed 31-May-2016.

[7] Shafiq MT, Matthews J, Lockley SR. A study of BIM collaboration requirements and available features in existing model collaboration systems. *Journal of Information Technology in Construction*. Vol. 18, 148-16, 2013.

[8] Curry, E., O'Donnell, J., Corry, E., Hasan, S., Keane, M., & O'Riain, S. Linking building data in the cloud: Integrating cross-domain building data using linked data. *Advanced Engineering Informatics*. 206–219, 2013.

[9] Juan, D. and Zheng, Q. Cloud and Open BIM-Based Building Information Interoperability Research. *Journal of Service Science and Management*, 7, 47-56, 2014.

[10] Yang, J., Anand, R., Hobson, S., Lee, J., Wang, Y., & Xu, J. Data Service Portal for Application Integration in Cloud Computing In *8th International Conference & Expo on Emerging Technologies for a Smarter World (CEWIT)*. New York, NY: IEEE. 1-3, 2011.

[11] Eastman, C., Jeong, Y., Sacks, R., and Kaner, I. Exchange Model and Exchange Object Concepts for Implementation of National BIM Standards. *Journal of Computing in Civil Engineering*, 10.1061/(ASCE)0887-3801(2010)24:1(25), 25-34, 2010

[12] buildingSMART. IFC Overview summary. On-line: buildingSMART: http://www.buildingsmart-tech.org/specifications/ifc-overview, Accessed 17-June-2016

[13] buildingSMART. An Integrated Process for Delivering IFC Based Data Exchange, buildingSMART International User Group, 2012.

[14] NBS. National BIM Standard- United States. Version 3. National Institute of Building Sciences, buildingSMART alliance, 2015.

[15] NIST. *US Government Cloud Computing Technology Roadmap*, Volum I and II, Special Publication 500-293. National Institute of Standards and Technology, 2014.

[16] Beetz, J., Berlo, L., Laat, R., & Bonsma, P. Advances in the development and application of an open source model server for building information. *CIB W078 – Information Technology for Construction*. Sophia Antipolis, France, 2011.

[17] GRAPHISOFT. BIMcloud, on-line: http://www.graphisoft.com/bimcloud/, Accessed 17-June-2016.

[18] Trimble. Design and Construct with Confidence, Civil Engineering and Construction, on-line: http://construction.trimble.com/sites/construction.trimble.com/files/marketing_material/Trimble_CEC_solutions_brochure_MEDIUM.pdf, Accessed 17-June-2016.

[19] Autodesk, A360, on-line: https://a360.autodesk.com/, Accessed June 2016.

[20] Autodesk, BIM360: Collaborative construction management software, on-line: http://www.autodesk.com/products/bim-360/overview, Accessed 17-June-2016.

[21] Beetz, J., Berlo, L., Laat, R., Halem P. BIMSERVER.ORG - An Open Source IFC Model Serer. In *Proceedings of the CIB W78: 27th International Conference* –Cairo. Egypt, 2010.

[22] Flux. Flux Overview, on-line: https://community.flux.io/content/kbentry/1258/flux-overview.html, Accessed June 2016.

[23] Assemble Systems. White paper: Project Collaboration- How to Use Assemble to Track Level of Development to Deliver on Your BIM Execution Plan, on-line: http://assemblesystems.com/project-collaboration-level-of-development-tracking/, Accessed 17-June-2016.

[24] invicara. BIM Assure product overview, on-line: http://bimassure.com/wp-content/uploads/2016/06/BIM-Assure-Product-Overview-2016-03.pdf, Accessed 17-June-2016.

[25] Trimble. Trimble Connect Overview, on-line: http://bpmidamerica.com/wp-content/uploads/2016/03/Trimble-Connect-Overview.pdf, Accessed 17-June-2016.

Development of a modular and integrated product-manufacturing-installation system kit for the automation of the refurbishment process in the research project BERTIM

K. Iturralde 1[a], T. Linner 2[a] T. Bock 3[a]

[a] Chair of Building Realization and Robotics, Technische Universität München, Germany
E-mail: kepa.iturralde@br2.ar.tum.de, thomas.linner@ br2.ar.tum.de, thomas.bock@br2.ar.tum.de

Abstract

The manufacture-installation of prefabricated modules for the refurbishment market needs to be lifted to a more integrated, systemized and automated level, in order to transform the existing buildings stock towards near zero energy consumption at reasonable cost. In building renovation, prefabricated modules have to be industrially customized to each building, and therefore, standard prefabrication and refurbishing approaches have to be re-conceived. In that sense, the automotive industry is a good example of combining modular products and manufacturing processes. Within the research EU-funded research project BERTIM, research institutes and companies are developing an interdisciplinary, cross-country approach for this challenge. In this research project, the goal of a dedicated work package is to ameliorate the existing holistic renovation process, by a combination an integration of robotic, automation and augmented reality technologies. Basically, the final objective is to reduce time and cost of the current process of the manufacturing and installation of 2D and 3D modules significantly (by 30% or more). For the development of this research, the Axiomatic Design method has been used, which is a matrix based method. Here, the main requirements and design parameters are expressed mathematically. The objective of applying this method is to define a scientific approach to the re-conception of the prefab modules and the manufacturing and installation processes, and therefore find an optimal solution for every decomposed requirement.

Keywords: refurbishing, customization, automated, installation, prefabrication, axiomatic design

1 INTRODUCTION and PROBLEM STATEMENT

Achieving a zero energy consumer building stock is a goal of the European Union [1]. For that purpose, there have been several publicly financed projects [2,3] that have been working with several solutions in order to ameliorate traditional way of gathering a holistic energetic building renovation. This paper presents some of the achievements made within the BERTIM project [4] that was launched in June 2016. The BERTIM project scopes the use of timber-frame based 2D and 3D prefabricated modules for the building refurbishment. This research will be carried out in 48 months. The research project´s main goals have been pointed out as:

1. *``define a general methodology for the efficient mass* (customized*) manufacturing process of prefabricated modules in the timber manufacturing industry''*
2. *``The methodology will allow installation time reductions of at least 30% compared to typical renovation''*

Building renovation needs bespoke solutions all the time, and therefore it is necessary to conceive a highly customizable module that can be adapted to the majority of the targeted building typologies. Besides, the Off-site manufacturing process needs to be (re)adapted to this circumstances as well as the On-site installation process. Therefore, there can be defined three main Sub-systems: the (A) 2D and 3D modules configuration, (B) the manufacturing process and the (C) installation process. For achieving these goals, it is necessary an overall perspective, it cannot only be based on product improvement, or only in the manufacturing process or only in the installation process. It must be a general solution, but also flexible for being implemented in different situations [5]. A key question is also how to accomplish this objective in several prefabrication companies, which act in different environments and markets and currently employ different degrees of prefabrication and automation. Therefore, the final solution needs to be adaptable to various construction scenarios, existing manufacturing/installation and automation levels and investment capabilities. In that context, the manufacturing-installing process has to be co-adapted with the design and modularity of the prefabricated high-level components. Besides, it will be considered the insertion of new technology such as manufacturing workstations with variable automation degrees, module installation with cooperative robots, object recognition, assistive devices (e.g., smart glasses) and other data acquisition and tracking solutions. This adaptation process is carried out in collaboration with the industry partners (companies), as well as with the more research-oriented partners. In a first phase of the project, the proposed solution for the manufacturing-installation process will be tested by simulation, small-scale usability tests and expert inclusion. In a later stage of the project the key aspects of the proposed will be tested and evaluated in operational environment (3 real world use cases for buildings to be renovated).

On the first 10 months of the research project three different manufacturers have participated. The manufacturers are from three European countries located in different climatic regions (Company 1, Company 2 and Company 3). The first two companies have participated actively on the research project. The third company participated only on the very first phase. And a fourth will start working in month 11 (Company 4). Different information has been received from the companies, either because they don´t have enough time for answering or either because the required data hasn't been monitored. This situation can be considered as uncertain, since a very marked company profile isn´t defined. Therefore, it must be taken a decision under an unclear definition of the manufacturer companies. Dealing with this uncertainty in principle is good, as the BERTIM system should be adaptable to a high variety of manufacturers. In other words, the three Sub-systems must be designed with incomplete information. For all the reasons explained, it is necessary to define a clear modular system that integrates aspects of the prefab module, manufacturing and installation processes. The reminder of this paper is to explain the development of this complex and multi-aspect system by applying a matrix based methodology such as Axiomatic Design [6].

2 DEVELOPMENT OF THE MODULAR PRODUCT AND PRODUCTION KIT SYSTEM

In this chapter there are 5 main parts. First, methodology is explained. Second, there is an Analysis of the participant companies. Following, a common background of the companies has been subtracted where the main similarities regarding to product, manufacturing and installation have been pointed out. After that, Axiomatic design is implemented into the main three Sub-systems. And finally, one aspect within one of the Sub-systems is furtherly detailed. This aspect is related to the placement of the connectors within the buildings.

2.1 Methodology

In this complex situation, it is a must to avoid subjectivity when adopting a design decision and criteria. Therefore, a step-by-step methodology that guides the research has been defined:

1. Analysis of the manufacturing and installing system of the participating companies. For this purpose, a questionnaire has been provided to the companies. They could define and quantify the manufacturing and installation Providing factory layouts, product details, and the direction where the research should be directed to. The analysis has been completed with a visit to the factories.
2. Analysis of solutions in some other industries, such as car manufacturing, for gathering and identifying suitable technology that could be applied to the production and installation process.
3. Questionnaires have been handed out to the companies asking about the suitability of the identified technologies
4. Preliminary set of solutions for the Proposal of the manufacturing process. This proposal must be adapted to each case, it has to be flexible and modular.
5. First prototypes and tests.
6. Final definition.

In phase 4, and 5, it´ll be applied the so called Axiomatic design method for reorganizing the whole proposal. As a short explanation of the axiomatic design, it can be said that there are four main domains: Customer domain, Functional domain, Physical domain and Process domain. These domains are strongly interconnected. In the Customer domain, the Customer Attributes or needs are defined (CA). The Functional Requirements (FR) are part of the functional domain. In these domain, constraints such as economic feasibility are also underlined. Once it is known what to achieve, it must be asked how to accomplish it. The Physical domain is for conceiving the Design Parameters (DP) or physical artefacts. But is it feasible the adopted solution regarding to achievability with existing technologies? On the Process Domain the Process Variables (PV) are defined in order to assure that the Design Parameters are realizable. The Axiomatic Design offers a Design Matrix for interrelating the CAs, FRs, DPs and PVs. Moreover, for such complex research developments as in our case, the CAs, FRs, DPs and PVs haven been decomposed into smaller units and hierarchized in order to make the problem solving issue affordable and achievable. Also, each of the decomposed CAs, FRs, DPs and PVs must remain independent (Independence axiom).

For the interconnection of the higher degree of DPs and lower degrees, a zig-zagging procedure must be carried out. In this very moment, the research will enter the phase

E. For now, the companies have agreed on modifying and adapting the modules, the manufacturing process and the installation process within some limits. Here it'll start the approach of the solutions for the decomposed FR. The inventiveness of each decomposed problem would also need to be guided by specific methods that facilitate the problem solving during design and development phases such as TRIZ. This method has already been used in previous and parallel research phases [7]. The research process hasn´t advance till this point yet, but it must be considered that the best solution will be always the one with higher probability of success (Information axiom).

2.2 Analysis of the Existing Systems

Prior to the System Development, the information about the companies' characteristics has been gathered. This was apprehended by structured questionnaires and overall information request, interviews with experts, and factory visits. As a résumé, the main qualitative aspects are pointed out in the next table.

Table 1. Main qualitative characteristics of companies

Company 1	
Product standardization	High degree of standardization. Catalogue based products
Production system	Manufacturing line
Main product	Timber-frame
Installation process	No
Customization	Limited
Adaptable manufacturing process	Limited
Manufacturing automation degree	Medium
Building refurbishment	No
Prefab degree	70%
Manufacturing range	Sawed timber-> prefab wall
Main handling device during manufacturing process	Rolling table +Hanging conveyor
Company 2	
Product standardization	Low degree of standardization
Production system	Non connected work stations
Main product	CLT and Timber frame
Installation process	Yes
Customization	High
Adaptable manufacturing process	Highly adaptable
Manufacturing automation degree	Low
Building refurbishment	Yes
Prefab degree	30-90
Manufacturing range	Lumber->3D module installation
Main handling device during manufacturing process	Bridge crane + Forklift
Company 3	
Product standardization	
Production system	Non connected work stations
Main product	CLT and Timber-frame
Installation process	Yes
Customization	High
Adaptable manufacturing process	Highly adaptable
Manufacturing automation degree	Medium
Building refurbishment	Yes
Prefab degree	30-90%
Manufacturing range	Forest -> Building finalization
Main handling device during manufacturing process	Bridge crane. Conveyor in the future
Company 4	
NO INFO	

If we look for quantitative aspects, the main data obtained for the manufacturing process and installation process of 2D modules are shown in table2. The data in this case refer to the manufacturing and installation of new buildings. In the case of Company 1´s installation process, the data are given by the subcontracted company. Besides, some companies don´t work with 3D modules; in this phase of the research we will focus on 2D modules.

The productivity of the processes shows that there is a big difference between companies 1-2 and 3 referring to the installation process.

Table 2. Main quantitative characteristics of companies

Manufacturing	Production staff	2D module	2D m. production
Company 1	26	17,04 m2/hour	0,65 m2 /worker-hour
Company 2	18	10 m2/hour	0,55 m2 /worker-hour
Company 3	9	11,25 m2/hour	**1,25 m2 /worker-hour**
Company 4	No info	No info	No info
Installation	**Staff 2D module**	**2D module**	**Time for Rework**
Company 1	9 operators	1,6 m2 /worker-hour	?
Company 2	4 workers	2,2 m2 /worker-hour	0,08 m2 /worker-hour
Company 3	5 workers	**7,6 m2 /worker-hour**	?
Company 4	No info	No info	No info

The big gap between companies 1-2 and 3 referring to the installation process might be due to the special installation system used by Company 3, which consists of a similar on-site factory as used by NCC Komplett and Skanska´s Flying factories [8]. Anyhow, this system wouldn´t be usable for Building Renovation and therefore we can´t take this data as a benchmark. In order to broaden the sample about this issue, a request has been made [9] to a none-participant company. This company states that they install 60m2 per day with 3 workers. Therefore, they have a ratio of 2,8 m^2 installed per worker-hour. In this example the test has been monitored with a low building and using one aerial work platform, one mobile crane and one forklift crane. Therefore, we consider that reasonable benchmarks for 2D modules therefore are:

1. 0,55 m^2/ worker-hour for manufacturing
2. 2,5 m^2/ worker-hour for installation.

The different degrees of Quality and complexity of the 2D modules that the companies offer is also a parameter must be taken into account. Company 2 and 3´s product quality and complexity is higher than Company 1´s.

2.3 Similarities and common background

As said before, this research will define a timber based 2D and 3D module type that will be adopted by the companies. We have considered to bring out the main similarities of the modules that are produced in the companies, in order to define the common platform [5] that the future BERTIM module must have. This very simple scheme will be the base of the BERTIM module:

1. Base structure: timber-frame with infill insulation (Alternatively CLT).
2. Rigidizing board (alternatively CLT).
3. External insulation placed onto the board.
4. Services will run through the timber frame.
5. Flexible finishing system.

It must be said that within this similarity there are still some peculiarities in each company, for instance, the dimension of primary elements (stud and raster dimension, profiles, insulation types, windows…) that uses each manufacturer.

If we focus on the Similarity of the manufacturing process, we can see that for obtaining the final highly prefabricated 2D module, between 25 and 35 different tasks (e.g. stud cutting, timber-framing, insulation placement, window fixation etc.) must be carried out for the production of elements and the assembly of the modules at the factories. This 25-35 task do not include the simultaneous works needed in parallel production lines for the manufacturing of supplies such as windows. The proposed product needs to be adapted to this line and manufacturing timing. For instance, one of the Functional Requirements of the Company 1 is that any of the proposed Design Parameters must be limited to the Time-Tracking of the production line which means to produce a 2D module every 9-22 minutes.

Regarding to the similarities of the installation process, there is a higher variety of solutions. Only Companies 2 and 3 perform directly installation activities. Company 1 doesn´t install their modules, the client directly contracts an installer company. Company 2 is the one that works more for building refurbishment. Company 1 hasn´t ever worked in building renovation and Company 3 has worked only in the installation of 3D modules on the top of buildings. About the equipment and support devices, mobile cranes, aerial work platforms and even scaffolding are mainly used.

2.4 System Development

As said in the beginning, we´ve determined three main Sub-Systems: the 2D and 3D modules configuration (A), the manufacturing process (B) and the installation process (C). These three different sub-systems must be interrelated and therefore integrated in a unique system. But there must be the choice to implement independently. The issue here is that what traditionally has been considered as PVs (Manufacturing + Installation process) are part of the FR. But the primary goal of an efficient manufacturing and reduction of installation time directs as to consider as FR. Due to the heterogeneous type of manufacturers, the adopted solution must be adapted to any type of company. Even more, each company can be flexible and decide either they use the whole set or only some of the decomposed FRs and DPs. Therefore, the independence axiom is really a must in this case.

In other words, there is uncertainty if the DPs certainly will be implemented totally or independently. We need to assure that they remain independent. So, applying the axiomatic design to our system and A, B and C sub-systems we can have the next matrix:

$$FR = A * DP$$

$$\begin{Bmatrix} FR_1 \\ FR_2 \\ FR_3 \end{Bmatrix} = \begin{Bmatrix} x & A_{12} & A_{13} \\ A_{21} & x & A_{23} \\ A_{31} & A_{32} & x \end{Bmatrix} * \begin{Bmatrix} DP_1 \\ DP_2 \\ DP_3 \end{Bmatrix}$$

(A) $FR_1 = \textit{Customize the existing 2D modules for building refurbishment}$
(B) $FR_2 = \textit{Maximize the offsite manufacturing process of existing facilities}$
(C) $FR_3 = \textit{Minimize onsite Installation time and cost of the modules}$

(A) $DP_1 = \textit{Adaptable 2D modules}$
(B) $DP_2 = \textit{Modular assembly workstation kit}$
(C) $DP_3 = \textit{Rapid installation system}$

By decomposing and Hierarchizing the FRs and DPs, we can get the matrixes explained on the next sub-chapters.

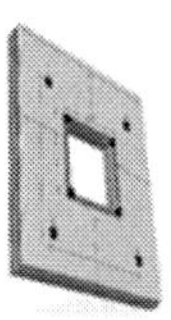

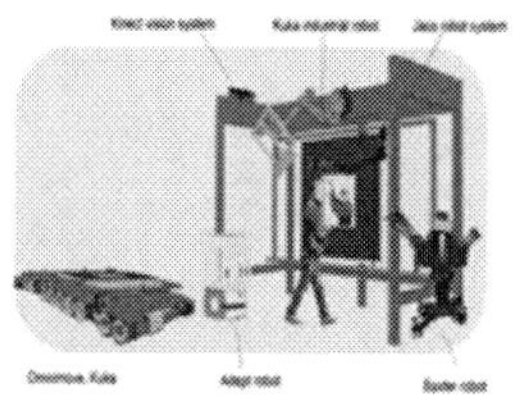

Figure1(left). Definition of the adaptable module(A).
Figure 2(middle). Preliminary idea of the WorkStation (B).
Figure 3 (right). Rapid installation system, a first proposal (C).

In this first step of the definition of the FRs and DPs, there has been an iterative and collaborative process with the companies to check that the offered preliminary DPs are correct and suitable for a further definition and decomposition process. Furthermore, on the matrix besides the diagonal elements the rest should be zero in order to gain a robust solution, or at least the elements on the upper part should be zero to keep no interference among the FRs and DPs.

2.4.1 Decomposed FR1 and DP1: Customization of the 2D modules by Conceiving an adaptable module.

As said before, the modules need to be highly adaptable. The modules that produce the companies should be reconfigured in order to obtain easily customizable modules to a high variety of Building typologies that have been already defined within the BERTIM project: concrete or steel structure buildings where the modules could be supported hanging from the building. In principle, self-supporting modules will be avoided as there would be a need for inserting dedicated foundations.

The next matrix shows the definition on the first decomposed Functional Requirements and Design Parameters of the modules:

$$FR_1 = A_1 * DP_1$$

$$\begin{Bmatrix} FR_{11} \\ FR_{12} \\ FR_{13} \\ FR_{14} \\ FR_{15} \\ FR_{16} \\ FR_{17} \end{Bmatrix} = \begin{Bmatrix} x & A_{12} & A_{13} & A_{14} & A_{15} & A_{16} & A_{17} \\ A_{21} & x & A_{23} & A_{24} & A_{25} & A_{26} & A_{27} \\ A_{31} & A_{32} & x & A_{34} & A_{35} & A_{36} & A_{37} \\ A_{41} & A_{42} & A_{43} & x & A_{45} & A_{46} & A_{47} \\ A_{51} & A_{52} & A_{53} & A_{54} & x & A_{56} & A_{57} \\ A_{61} & A_{62} & A_{63} & A_{64} & A_{65} & x & A_{67} \\ A_{71} & A_{72} & A_{73} & A_{74} & A_{75} & A_{76} & A_{77} \end{Bmatrix} * \begin{Bmatrix} DP_{11} \\ DP_{12} \\ DP_{13} \\ DP_{14} \\ DP_{15} \\ DP_{16} \\ DP_{17} \end{Bmatrix}$$

FR_{11} = *Adapt to existing building configuration*
FR_{12} = *Adapt to existing building minor elements(cornices, friezes ...)*
FR_{13} = *Adapt to different manufacturers common elements*
FR_{14} = *Adapt to different climatic regions*
FR_{15} = *Adapt to different aesthetic options*
FR_{16} = *Adapt to variable services*
FR_{17} = *Adapt to structural needs (hanging modules, self supported etc.)*

DP_{11} = *Convex and concave and vertical and horizontal corner solutions*
DP_{12} = *Separate module from the facade in order absorb minor elements*
DP_{13} = *Module System based on common interfaces (connectors, plugs...) and "geometries" instead of particular elements*
DP_{14} = *Variable thicknes of Insulation*
DP_{15} = *A system that allows variable finishings*
DP_{16} = *Flexible Criteria for placing the services within the 2D modules*
DP_{17} = *Adaptble timberframe and connector configuration*

In principle, and within this decomposition level, we can state the independence axiom of the matrix. Besides the proposed FRs, we might need to include more such as Airtightness, Moisture Barrier and Economic Feasibility of the proposed solution. But these would maybe be included as generic Constrains. Finally, it´s being discussed if the module could be dismountable. Taking into account all the requirements, there has been a preliminary approximation on the Design Parameters. For further development, some simulation and prototypes will be accomplished.

2.4.2 Decomposed FR2 and DP2: Maximize off-site manufacturing process of the modules within the existing facilities by a Modular assembly workstation kit

The goal here is to improve the Off-site Manufacturing process. The proposed solution should all the time implementable to any kind of manufacturing facilities, providing agility and to fulfil the needs of the costumer of BERTIM modules. The modular workstation kit must accomplish some decomposed requirements. The design matrix is solved like this:

$$FR_2 = A_2 * DP_2$$

$$\begin{Bmatrix} FR_{21} \\ FR_{22} \\ FR_{23} \\ FR_{24} \\ FR_{25} \\ FR_{26} \\ FR_{27} \\ FR_{28} \end{Bmatrix} = \begin{Bmatrix} x & A_{12} & A_{13} & A_{14} & A_{15} & A_{16} & A_{17} & A_{18} \\ A_{21} & x & A_{23} & A_{24} & A_{25} & A_{26} & A_{27} & A_{28} \\ A_{31} & A_{32} & x & A_{34} & A_{35} & A_{36} & A_{37} & A_{38} \\ A_{41} & A_{42} & A_{43} & x & A_{45} & A_{46} & A_{47} & A_{48} \\ A_{51} & A_{52} & A_{53} & A_{54} & x & A_{56} & A_{57} & A_{58} \\ A_{61} & A_{62} & A_{63} & A_{64} & A_{65} & x & A_{67} & A_{68} \\ A_{71} & A_{72} & A_{73} & A_{74} & A_{75} & A_{76} & x & A_{78} \\ A_{81} & A_{82} & A_{83} & A_{84} & A_{85} & A_{86} & A_{87} & x \end{Bmatrix} * \begin{Bmatrix} DP_{21} \\ DP_{22} \\ DP_{23} \\ DP_{24} \\ DP_{25} \\ DP_{26} \\ DP_{27} \\ DP_{28} \end{Bmatrix}$$

FR_{21} = *Adapt to different degrees of automation*
FR_{22} = *Perform different assembly tasks*
FR_{23} = *Support the 2D module Vertically and Horizontally during assemly*
FR_{24} = *Adapt to various Handling systems (hanging conveyor, carrier,*
FR_{25} = *Adapt to various assembly processes configurations*
FR_{26} = *Adaptble factory reconfigurations*
FR_{27} = *Adapt to various supplying systems*
FR_{28} = *Operators must reach easily the parts of the 2D module*

DP_{21} = *Adaptable ergonomic framework where either manual processes or robotic activities can be carried out*
DP_{22} = *Possibility to Host different tool types for assembly*
DP_{23} = *Possibility to Host 2D module supporting elements*
DP_{24} = *Integration of the exisitng handling system within the workstation*
DP_{25} = *Rejoinable workstation kit*
DP_{26} = *Movable workstation kit*
DP_{27} = *Dedicated area(s)or passes for supplies*
DP_{28} = *Posibility to include various (mobile)platforms*

The preliminary conception of the workstation has already been presented to the companies and the feedback has been positive (Figure 2). But some issues need to be solved, the Information axiom is not fulfilled. For instance, it affects directly to the FR1 since the work-station is not appropriate for a variable size of 2D modules (it can´t host longer modules than 6 meters) and it is neither usable for the assembly of 3D modules. This has been considered and on the second phase and the problem has been solved. The solution was a more open work-station, with higher degree of flexibility and possibility for being joined in order to create production lines (Figure 4).

2.4.3 Decomposed FR3 and DP3: Minimize on-site Installation time and cost of the modules by a Rapid installation system

How to accomplish a rapid installation process? Nowadays, too many support devices are needed during the installation of these panels. On the next matrix, we have pointed out the Functional Requirements and Design Parameters that reach the installation process to a rapider task.

$$FR_3 = A_3 * DP_3$$

$$\begin{Bmatrix} FR_{31} \\ FR_{32} \\ FR_{33} \\ FR_{34} \end{Bmatrix} = \begin{Bmatrix} x & A_{12} & A_{13} & A_{14} \\ A_{21} & x & A_{23} & A_{24} \\ A_{31} & A_{32} & x & A_{34} \\ A_{41} & A_{42} & A_{43} & x \end{Bmatrix} * \begin{Bmatrix} DP_{31} \\ DP_{32} \\ DP_{33} \\ DP_{34} \end{Bmatrix}$$

FR_{31} = *Avoid time comsuming settingout of the connectors*
FR_{32} = *Minimize time during the placement and fixation of modules*
FR_{33} = *Minimize duration of the accesory support devices on_site*
FR_{34} = *Minimize rework*
DP_{31} = *Connector with interface that offers higher tolerances*
DP_{32} = *Fast placement and fixation system*
DP_{33} = *Easy and stable handling system*
DP_{34} = *Fully or Highly prefabricated modules*

The BERTIM project considers that previous to the Installation process there would be a 3D data acquisition of the existing building that would provide enough data to define the location of the connectors and the accurate size of the panels. This rapid installation process will gather the use of fewer handling and support devices.

On the very first attempts, we´re considering the use of only one single device and no other support system, similar to the handling of cargo containers. For that purpose, the joinery system must be accurately positioned. Once we have defined tall the decomposed FRs and DPs, we can check the independence axiom and if there is any interference between them.

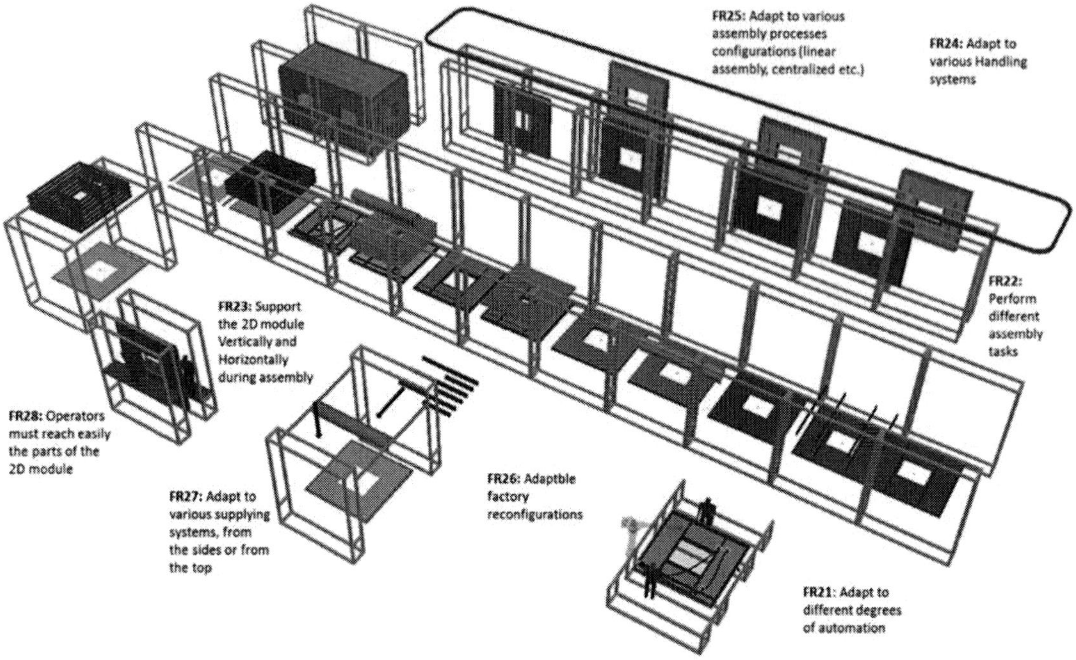

Figure 4. Second development of the work-station modular kit, implementation of the FRs.

2.4.4 Master matrix and information axiom

A master matrix has been configured to check the independence axiom of each of the decomposed FRs and DPs of every Sub-System. As said before, it is non-accurate to state that the decomposed FRs and DPs don´t interfere with each other, more detailed and decomposed solution are needed. In next sub-chapter 2.5 one decomposed FR and DP will be furtherly decomposed in order to gain more definition of the solution.

2.5 Detailed FR31 and DP31 development

The FR_{31} refers to ``Avoid time consuming setting out of the connectors onto existing building´´. How traditionally have been set out the connectors? For none prefabricated solutions such as the installation of rain-screen (or ventilated facade), the procedure is normally as follows:

1. Fixation of the connector plate into the existing wall. The plates are placed with a laser alignment system, normally separated every 600 mm vertically.
2. Vertical guides are placed onto the connectors. Onto this, we might need an extra horizontal guide, depending on the product type. This way, we get a planar situation and the guides are located in a known range of distance.
3. The external envelope modules are cut to the right measure and placed onto the guides.

For the installation of prefabricated elements onto existing buildings, first, a data acquisition of the geometry is necessary. This Data acquisition of the existing building is carried out using 3D laser scanning, photogrammetry or/and theodolites. This data acquisition is normally considered as sufficient for manufacturing the prefab modules. But, in this research, it will be considered that after this overall measurement, we must accomplish strategies for assuring the exact location of the connector. During the installation of any prefabricated element, the accuracy of the joinery system is primordial [2,3]. Some prefabricated timber based modules are being installed following the traditional procedure. Besides, in some other proposals [10] the connector is ``punctual´´, there is no guide or rail for obtaining a planar situation.

In principle, this process using punctual connectors is faster than the previous process. But the connector must be fixed very accurately. For this cases, the connector is composed by at least two elements: the part onto the existing building, Part1, and the part that goes within the module, Part 2. The position on the module is dictated by the position on the wall and vice versa, their coordinates need to be coordinated. The irregularities of a wall might be 50 mm in a concrete and rigid structure building, which means that the joinery system must absorb these irregularities.

Normally, the process of manufacturing the modules and the installation process is carried out in parallel or at least in a subsequent process. This means that when part 1 are being fixed on the wall of the existing building, the 2D modules are being manufactured off-site. Besides, we are considering to use a fast clipping system with latches or similar techniques; that means we have an extra constrain, the maximum tolerance between the connector part in the wall and the module must less than 5 mm. This tolerance must be defined as the allowable variation of $DP_{31}\Delta$.

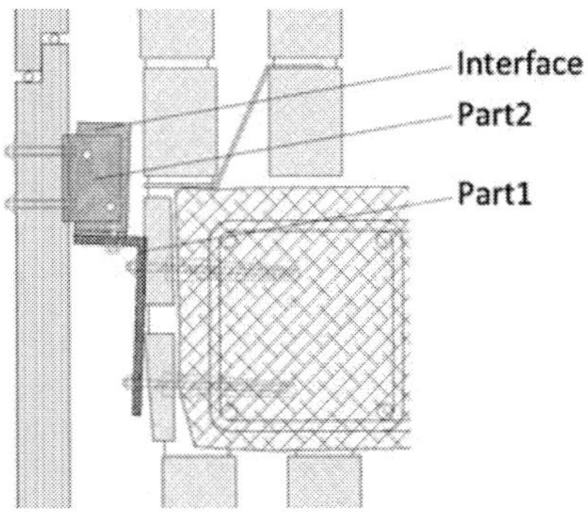

Figure 5. First approaches regarding the connector type. Cross-section of the module and the existing building. We can see that Part 1 is not parallel to the vertical plan. Here, an interface was chosen in order to absorb the irregularities.

We will consider that the Companies manufacture the 2D modules also with that tolerance. On the first approach, we found out two main strategies.

Strategy 1: Place Part 1 of the connector with low tolerance

This means to obtain accuracy on-site, placing the connector part1 on its exact position, based on a given position of the connector by the designer. For achieving this highly accurate placing, we have two main options:

DP3.1a-Use of a multi-hole pattern that would allow us to place the connectors accurately with a degree of flexibility in case of impossibility of making a hole due to a rod bar or some other inconvenience. This way, we could achieve the necessary accuracy for placing the module on its place. The patter can be physical or laser-made:

DP3.1.aa-Using a physical pattern. This technique has been traditionally used in the Japanese construction procedure due to the need of using prefabricated and rigid solutions. Disadvantage: This solution would mean to work with large physical elements on a vertical plan and therefore either a scaffolding or an Aerial Work Platform would be needed.

DP3.1.ab- Using Laser Spatial Positioning system. Disadvantage: it can be used only at night or during cloudy days. Besides, the system might not be appropriate for tall buildings.

Both solutions wouldn't take into account the none-planar irregularities of the existing wall and the part1 might be placed on a non-parallel to vertical plan.

Strategy 2: Place Part 1 of the connector with high tolerance

With this approach, we can place the connectors with a traditional laser alignment system, taking into account that the whole set of connectors might not be in the same plan, meaning parallel with the same distance to the existing wall. After that, we must measure their accurate location. For facilitating this purpose, target-reflectors should be embedded on the connector. Once we know the exact position of the part 1, we can accurately correct the, or better said, modify some parts within the module. Within this strategy 2, we have two main options. One is to use an interface, and the other is to mill accurately some parts on a CNC mill.

DP3.1b- Adapt the connector´s interface by using a modifiable element. For this case we have two options.

DP3.1.ba- Using a 3D printed peg or dowel. The dowel needs to be hard enough to absorb the forces on the connector. Disadvantage: This solution might be very expensive and even time consuming.

DP3.1.bb- Making holes in different positions in a standardized interface. Disadvantage: The interface must be hold in an accurate location while doing the holes.

DP3.1c.- Modify the 2D modules connection point or surface by milling accurately in a CNC the placement of the part 2 within the module.

DP3.1.ca- Mill the stud (a single element) where the connector will be placed. Disadvantage: if the stud is milled before the assembly process of the module, any variance on the position of the stud within the module during the manufacturing process might create inaccuracies.

DP3.1.cb- Once the module is finished, insert the whole module into the CNC and make the necessary milling. Disadvantage: The company must have a large CNC milling station.

As a result of this first research, we have found six different DP_{31} variants for satisfying FR_{31}. We can see that all solutions present an impediment that should be solved on the next decomposition and zig-zagging step. So far, we have gathered several Design Variables. How to validate Design Variables? In the following phase of the research, we must keep the independence axiom, each of the FRs needs to be satisfied without affecting any other. Only this way we will gather a robust solution, an ideal uncoupled design. It has to be pointed out that within this detail level, it cannot be stated if the future solution will fulfil the independence axiom since the definition is still vague. About the Second axiom, we know that the design with highest probability of success is the best design. For choosing the best design variables, we might need to take into consideration the different requirements of the companies. We have entered a phase where a prototyping of the solutions will be needed. The companies will provide the choice of proving some of the DP31 variants. We can still foresee some risks, for instance, due to timber´s unstable physical properties in differential humidity and temperature situation, the module´s geometrical properties might vary and therefore the location of the Part 2 might move. This might jeopardize the installation process; Part1 and Part2 might not fit.

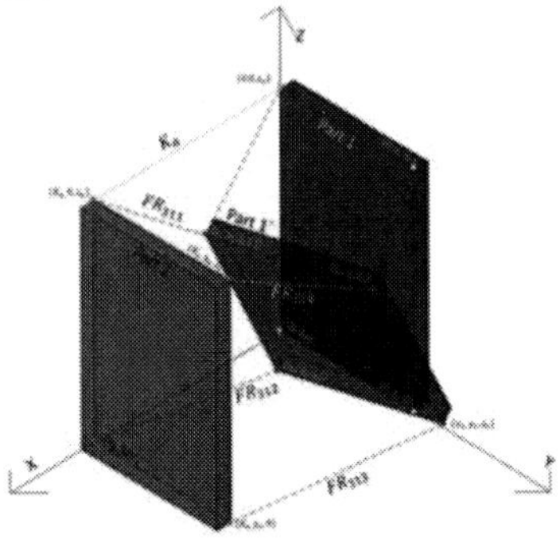

Figure 6. Abstract geometrical definition of FR_{31n}.

Currently, the research is focused in Strategy 2, placing the connectors with high tolerance and specifically using an interface to absorb the variation. This decision has been taken co-ordinately with the expertise in the companies. In that sense, as guidance to future work, we can decompose the FR31 in equations that relate Part1 and Part 2. For that purpose, we will only need to measure the coordinates in Part1 and generate FR_{311}, FR_{312} and FR_{313} by gathering the line´s equation (FRL_{31n}) and the distance (FRD_{31n}). After that, the DPs will be adapted accordingly.

$$FRL_{31n} = \frac{(x - K_a)}{(x_n - x_a)} = \frac{(y - y_a)}{(y_n - y_a)} = \frac{(z - z_a)}{(z_n - z_a)}$$

$$FRD_{31n} = \sqrt{(x_n - K_a)^2 + (y_n - y_a)^2 + (z_n - z_a)^2}$$

We will suppose Ka as constant, assuming that this constant will depend on the thickness of interface type. Either we use a custom made dowel or a customized hole locating system, with three decomposed FR311, FR312 and FR313 we could allocate Part2. In principle, with these equations the independence axiom is fulfilled, but as said before, it will depend on the final taken solutions, the DPs.

3 CONCLUSION

In this phase of the research, we have gathered the main method or design frame for achieving the goal of the project. The Axiomatic Design approach has been applied and the complexity of the overall process has been handled. Gathering a solution for a *modular and integrated product-manufacturing-installation system kit* is a complex process and therefore it needs to be decomposed in affordable and multiple sub-systems. On the next phases and within the Axiomatic Design method, issues such as Weighting factors, Uncertainty, Reduction of variance and System analysis will be analysed. Besides, we must remember that the approximation made on this step will be valid for a robotic installation system of modules onto existing buildings. It must be taken into consideration that the biggest technology gap in terms of automation is within the installation process of the modules and therefore, the main time reduction could be achieved on this phase. For that purpose, prototyping and testing will be necessary. Following the FR31 and DP31 development, the different connector solutions are being tested currently. The authors believe that this is a key issue for achieving a more automated robotic installation process of prefabricated modules for building refurbishment.

Acknowledgments

This paper has been drafted thanks to the funding from the European Union's Horizon 2020 research and innovation programme under grant agreement No 636984. The content of this paper does not reflect the opinion of the European Union.

References

[1] European Commission, "A Roadmap for moving to a competitive low carbon economy in 2050", Brussels, 2011

[2] Larsen, K. E., et al., "Surveying and digital workflow in energy performance retrofit projects using prefabricated elements", Automation in construction, 2011.

[3] Various Authors, "Großelement-Dämmtechnik mit Vakuumdämmung", Institut Wohnen und Umwelt GmbH, Endbericht, Germany, 2007

[4] Webpage visited the 15/03/2016: http://bertim.eu

[5] Waltl H, Wildemann H., Modularisierung der Produktion in der Automobilindustrie, TCW, Transfer-Zentrum, 2014.

[6] Suh, Nam P. "Axiomatic Design: Advances and Applications (The Oxford Series on Advanced Manufacturing)." (2001).

[7] Iturralde, K., Linner, T., Bock, T., Development and preliminary Evaluation of a concept for a Modular End-Effector for automated/robotic Facade Panel Installation in Building Renovation, 10th Conference on Advanced Building Skins. 4-5 October 2015, Bern, Switzerland

[8] Bock, T. ,Linner., T., Volume 3: Site Automation: Automated/Robotic On-Site Factories, Cambridge Handbooks on Construction Robotics, Cambridge University Press, 2016

[9] Source cannot be specified due to privacy policy

[10] Webpage visited 15/03/2016: https://www.lindner-group.com

Lean-based diagnosis and improvement for offsite construction factory manufacturing facilities

YouYi Zhang[a], Gordon Fan[a], Zhen Lei[a], SangHyeok Han[b], Claudio Raimondi[c], Mohamed Al-Hussein[a], Ahmed Bouferguene[d]
[a]Hole School of Construction Engineering, University of Alberta, Edmonton, Canada
[b]-Concordia University, Montreal, Canada
[c]Fortis LGS Structures Inc., Edmonton, Canada
[d]Campus Saint-Jean, University of Alberta, Edmonton, Canada

Abstract –

In current practice, manufacturing companies have daily outputs that are expected to meet the company's targeted production level. In construction, there is an increasing trend toward buildings being manufactured off-site rather than through traditional stick-built methods on site. This trend has created a demand for construction companies to subcontract work and to seek new efficiencies in their operations. In reference to the production line, regardless of how hard an employee works, there are certain limitations that impede a worker's productivity. To reduce or remove these limitations, many companies have begun to implement Lean methodologies in order to reduce waste in the production process. However, before improving the productivity of the production line, an analysis of the entire manufacturing process must be conducted. This paper introduces the methodologies that can be used to evaluate the production process on both weekly and monthly bases to ensure a consistent productivity rate; this research can be implemented in a variety of manufacturing settings. As a case study, it is implemented at Fortis LGS Structures Inc., an Edmonton-based offsite construction company that produces light gauge steel wall panels; Fortis has been working closely with researchers at the University of Alberta to identify areas of waste and improvements which can be implemented to help increase the productivity of future projects.

Keywords –

Offsite construction; Light gauge steel; Modular construction; Lean manufacturing; Waste elimination.

1 Introduction

Modular construction derives its root from the manufacturing industry: entire stick-built construction projects are broken down into components that are easy to manufacture on factory production lines. Various concepts and knowledge have been borrowed from the manufacturing industry and implemented into modular construction over the past decades. Different levels of automation can be observed in the construction field to achieve high efficiency. Bock (2015) provides a review of the past and current tendencies in construction automation. Despite some technology limits, automation has been seen and proven in various construction activities: robotic infrastructure production; robotic transportation systems; general manufacturing industry, and so on. A concept of building component manufacturing (BCM) is introduced in Bock's work, which refers to a robotic industrialization trend based on concrete, brickwork, wood, and steel as building materials and large-scale prefabrication to deliver complex building components and products. Similar review and survey is also provided by Everett and Slocum (1994) and Vaha et al. (2013). For onsite construction, robotic-oriented thinking is also being targeted to reduce human components and errors in the process, such as a steel beam automatic assembly system (Chu et al. 2013; Jung et al. 2013).

Apart from the academic value of industrialization in construction, there is also an economical potential behind this trend, particularly in areas where construction labor is expensive and the outdoor construction environment is not comfortable due to the harsh weather. As an example, modular construction is being implemented for industrial construction projects in remote areas in Canada (Lei et al. 2013); and in the residential and commercial industry, the industrialization concept is being used to manufacture wall panels or modules for multi-story buildings/single-family houses (Yu et al. 2013; Sang et al. 2011). Wood and light gauge steel are two common materials used in this area. However, compared to European practice in modularization, the North American market is smaller due to the low level of acceptance of this method by the market and the technology readiness in North America. Some leading technologies in modular construction can be seen in Europe, such as WEINMANN and Randek machinery, which are attempting to maximize the usage of machinery in factory construction. To fully maximize the potential of machinery, other technologies/knowledge are required (particularly as the scale of production increases, the integration between different systems/technologies becomes necessary). "Industry 4.0" originates from the strategic project by the German government and was first used in 2011, which emphasizes a number of integrated automation, data exchange, and manufacturing technologies. A Cyber-Physical System (CPS) concept was developed to describe transformative theologies for managing interconnected systems between physical assets and capabilities (Brettel et al. 2014; Lee et al. 2015). On a more detailed level, the direct implementation of information-driven systems in modular construction is observed. For example, Altaf et al. (2015) introduce an RFID-based online simulation system for simulating a wood wall-panel production line. CAD-based production has also become a trend to achieve the communication between the design and the machines. In addition, industrial engineering concepts are a valuable contribution to modular construction. Various Lean-based improvements are being carried out in the construction industry. Fortis LGS Structures Inc. is a light gauge steel modular construction company located in Edmonton, Alberta, Canada; over the past decade, Fortis has developed its own method of wall panel production and unique structure detail designs. Their projects mainly focus on residential and commercial buildings. In recent years, Fortis has been seeking higher efficiency and a more cost-effective method in delivering quality products to the construction market, and bottlenecks have been noticed in the production causing ineffectiveness and waste. This paper introduces an initiative which began in 2015, marking the beginning of the company's industrial engineering journey, which involves the development of a method to capture and analyze the company's production line using some Lean-thinking ideas. This also leads to an implementation of a bar-code based tracking system into the factory for better material inventory and management. Case studies will be discussed in this paper to illustrate this transition.

2 Light Gauge Steel Panelization Process

Fortis LGS Structures Inc. has adopted a modular construction method for constructing their buildings using light gauge steel material. Before producing the panels in the factory, the design drawings (e.g., architecture, structure drawings, etc.) are converted into shop drawings, which can be easily understood by the workers in the factory; and the production line is divided into several stations (see Figure 1). Currently the production line consists of the four stations as follows. (1) Assembly Table: where all the raw light gauge steel materials (steel studs, headers, and footers) are sorted and assembled together in bundles and passed to the next station. (2) Framing Table: where the light gauge steel materials are put together as wall components. On this table, semi-automated machines are used to assist the workers to achieve high efficiency and reduce the human components. (3) Sheathing Table: in this case the walls are exterior walls, thus drywall is sheathed to the wall frames with automated nailing guns. (4) Panel Racks: the sheathed walls are hung on the racks for exterior finishing such as stucco or bricks. After this station is completed, the wall panels are shipped to site for onsite assembly. Washroom units are also pre-assembled with plumbing finishing in the factory before shipment (Figure 2 pictures the products).

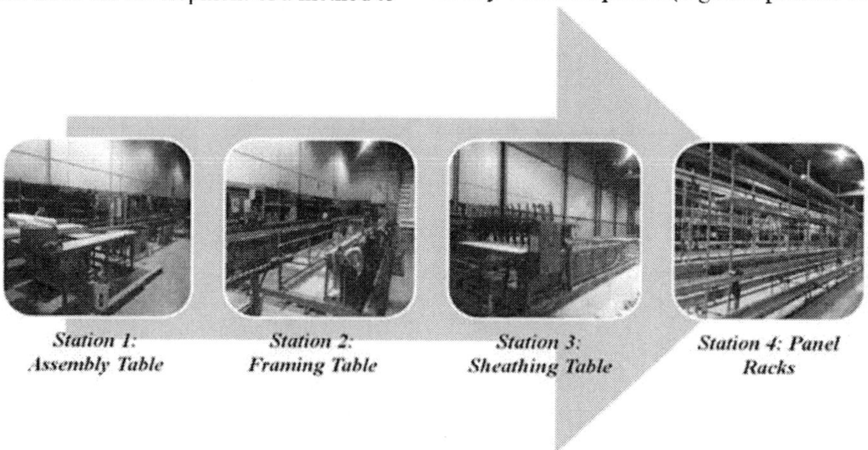

Figure 1. Light gauge steel production line

(*i*) LGS panel wall	(*ii*) Washroom unit

Figure 2. (*i*) Light gauge steel wall, and (*ii*) unit product

3 Production Line Diagnosis Approach

Production analysis for the current state based on the time study is the first step toward identifying/quantifying the company's existing problems. For offsite construction manufacturing—panelized or modular—there exist various approaches to achieve this. Yu et al. (2009) apply Value Stream Mapping (VSM) as the tool for current stage study based on the collected data base for house construction. Yet, the construction utilizes the stick-built methodology where the lead time between activities plays a large role on overall operation time. Shafai (2012) implements a time study for panelized manufacturing by collecting targeted work stations cycle time. The process efficiency is then studied by simulation using a regression model which is generated by statistical analysis. However, in order to achieve this, the quality of the collected data becomes critical. For factories that have limited automated data-collection systems, the procurement of accurate data for each individual operation process still poses a challenge. Considering the aforementioned issues, a less sophisticated but effective way for data-collection is used in this study. Although there is no large data set involved, the level of data collection correlates to the purpose of identifying the key issues on the current production line.

The first objective of the study is to diagnose Fortis's production line and expose the bottlenecks and potential areas for improvement. The study is carried out during a peak production period when the factory is active with a high production volume. The diagnosis procedure consists of the following steps:

- Process mapping and time study: to gather information to obtain the entire picture of the production; based on the process map, processing time is collected for each particular working station; the value stream map is created based on collected data.
- Waste identification and improvement action plan: to identify the waste in production from a high-level and detailed level; potential improvements are then proposed based on the defined wastes/bottlenecks.

3.1 Time study and process mapping

In this study, a floor process diagram is first created, as shown in Figure 3, to describe the production flow. The diagram shows the floor layout of the production facility with description. A time study is then conducted for each individual station to collect processing time. Figure 4 shows a typical time study sheet used for data collection. The key is to relate the time data with characteristics of the product being produced (e.g., dimensions of walls, number of door/window openings) and find the time distribution for each station. After the time study, the VSM is used to document, analyze, and improve the flow of information or product in a production process. However, the VSM in this study focuses on the manufacturing production flow, which is the cycle time for each work station, other than the process between different activities.

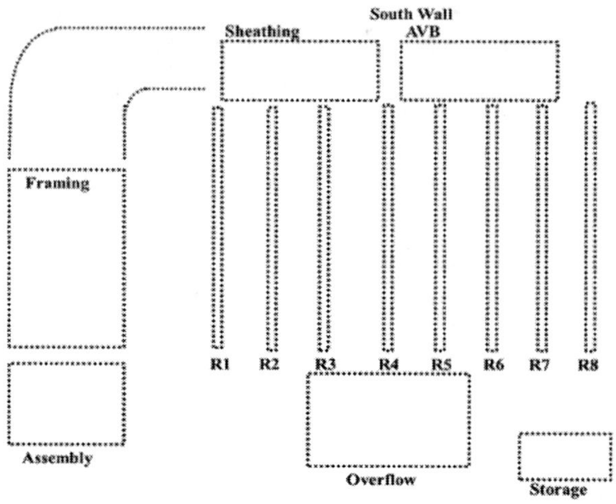

Production process order	Key material	Key tools	General crew required
Assembly	steel, clips, screws, insulation	...	3
Framing	screws	...	3
Sheathing	drywall, screws	...	3
AVB	cement, meshes	...	2
Foam (on racks)	foam	...	2
Rasping (on racks)	-	...	2
Basecoat (on racks)	cement	...	2
Primer (on racks)	-	...	2
Finish coat (on racks)	finishing material (stucco)	...	2
Loading	-	...	2

Figure 3. Process diagram and description

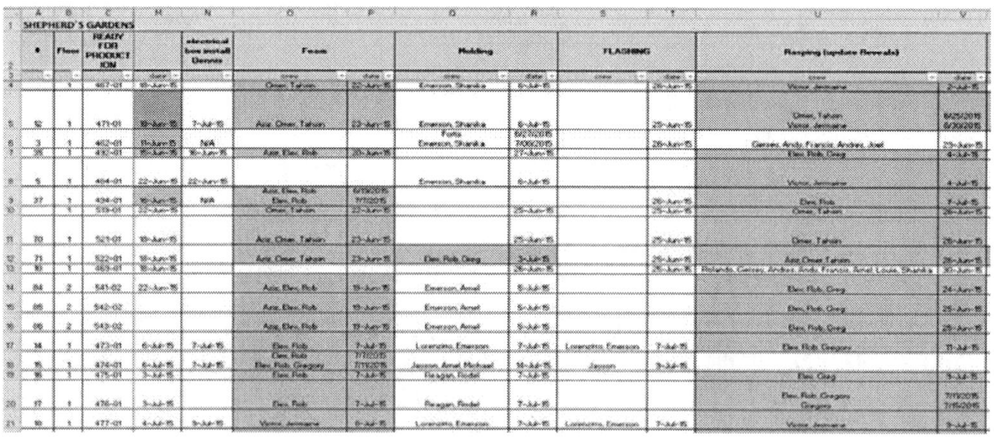

Figure 4. A sample time study sheet

3.2 Waste identification

To ensure the success of Lean implementation, it is necessary to quantify the waste/bottlenecks in the production by setting up some indicators. In this study, indicators are used to primarily capture the waste time in terms of lost products and wasted working hours. The "lost products" are defined as the products lost due to the production inefficiency compared to the calculated takt time. Material waste is also measured in terms of waste percentage. In addition, another typical waste identified in this study is over-production due to non-balanced production lines.

4 Implementation and Results

4.1 Existing problems identification

The first step toward implementing Lean manufacturing is to obtain a commitment from the company's management team. This is achieved by holding a meeting with the management team and providing fundamental Lean training. The goal is to engage the management and assist them in realizing the need for change and improvement. Following this, a two-week plan is provided for the company in order to conduct all the necessary analysis. Figure 5(i) shows a time study sheet filled out by the workers at the assembly station to record the processing time for that station. The collected data is correlated with various product parameters; for example, Figure 5(ii) shows the relationship between the produced wall surface area and its processing time on the assembly table. This can help the management team to identify products that take relatively more time to build, and obtain a rough idea about their production rate. The key findings from these scatter figures include:

(1) There is a wide range of variation in the working durations, which indicates a non-standardized working procedure.
(2) The size of the openings on the panel influences the time duration for assembly, framing, and sheathing stations, yet has very little impact at the AVB and foaming stations.
(3) Approximately 50% of wall panels are designed with specific length and working area range with a similar duration; however, the remaining 50% perform with a large range of variation.

Based on the collected data, a VSM is created to represent the current state of the production line, as shown in Figure 6; and the following drawbacks of the production line are summarized:

(1) Large standard deviation contributes to an unstable production line flow. Variations in work duration on each station indicate the necessity of standardized work procedure.
(2) Work stations in a manufacturing plant should all target a constant cycle time to meet the customer requirement. However, the VSM shows a non-balanced production line where the AVB and potential foaming station interfere the flow of the line.
(3) Lean principles focus on value adding as well as eliminating waste. Regardless of the work efficiency on the station itself, the assembly station can be identified as a non-value added activity which consume close to 30 minutes for every panel that must be built.

The findings from VSM affirm the observations in the factory; areas are identified in the production line that potentially contribute to over-production. For example, the conveyor between the framing station and sheathing station is selected as a target area. The idle time of the products on this conveyor is collected, and an average of 20 hr of idle time is found for the products that are considered as waste from a Lean perspective. Figure 7 depicts that the wall panels are stacked on top of each other due to the non-balance of the production line, which causes the workers to spend time and effort to locate the correct panel to process. This is a typical situation known as a "push" system rather than a "pull" system. The workers' productivity is also measured by calculating the wasted working time, which is conducted by means of monitoring from the company's camera. At 60 minute intervals, an average of wasted time due to non-value added activities (e.g., walking around searching for material) is recorded, and according to the collected data, an average of 33% wasted working time is usual in a daily production.

4.2 Potential solutions and future improvement

Previous findings in this study lead to various Lean strategies, including 5S and Kaizen, in order to achieve a clean and organized work environment. Regarding the implementation of 5S, this study reveals that specific target areas should be selected for improvement, and that it is easier to gain employee buy-in by beginning in an easy-to-improve area; the employees are more likely to be motivated if results are visual (Figure 8 shows an "after" state of the assembly station where 5S was performed). After obtaining employee buy-in, Kaizen sessions are conducted to identify the waste, and solutions are generated through brainstorm sessions. Some of the identified waste includes: (1) unnecessary inventories between framing and sheathing stations; (2) defects occur often, which result from improper material dimension and low production accuracy; (3) shop drawings are not supplied to the factory in a timely manner; and (4) workers are called off their work station by the factory manager from time to time. In general, the management has realized that the root of these causes arises from miscommunications and the solution is to assist employees in realizing their value and making them accountable for their work. From another perspective, Standard Operation Procedures (SOPs) are necessary for producing quality panels. Figure 9 shows a sample SOP created for the assembly station. From the production line flow perspective, the goal is to create work stations with congruent cycle times. In consideration of these observations, two solutions are proposed: (1) eliminate the work at the assembly station by proper machinery support or beforehand planning. Therefore, one extra station is being created. It is also important not to increase the cycle time at the second work station in order to maintain the production line balance; and (2) as the last two activities each use double the cycle time of the previous work station, work should be broken down further to allow the cycle time to be maintained at an approximately constant level. Cycle time, based on the working hours divided by the number of work stations, can be calculated to shorten the current duration, either to decrease the working hours with manpower or machine support, or increase the number of work stations by creating space from non-value added activities.

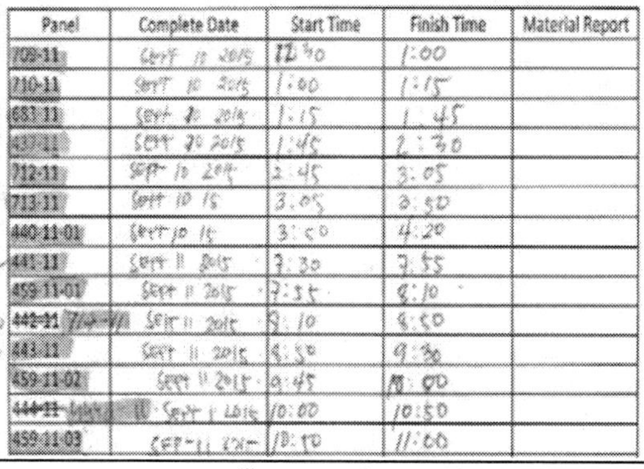

Panel	Complete Date	Start Time	Finish Time	Material Report
709-11	Sept 10 2015	12:40	1:00	
710-11	Sept 10 2015	1:00	1:15	
683-11	Sept 10 2015	1:15	1:45	
437-11	Sept 10 2015	1:45	2:30	
712-11	Sept 10 2015	2:45	3:05	
713-11	Sept 10 15	3:05	3:50	
440-11-01	Sept 10 15	3:50	4:20	
441-11	Sept 11 2015	7:30	7:55	
459-11-01	Sept 11 2015	7:55	8:10	
442-11	Sept 11 2015	8:10	8:50	
443-11	Sept 11 2015	8:50	9:30	
459-11-02	Sept 11 2015	9:45	10:00	
444-11	Sept 11 2015	10:00	10:50	
459-11-03	Sept 11 2015	10:50	11:00	

(i) Time study

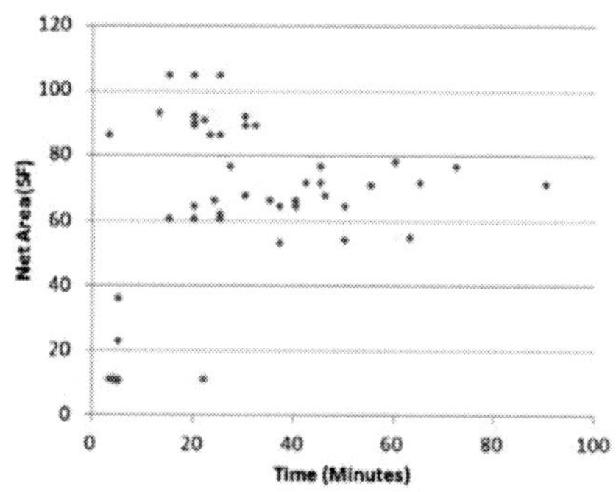

(ii) Relation of the wall SF and processing time

Figure 5. Time study and data analysis

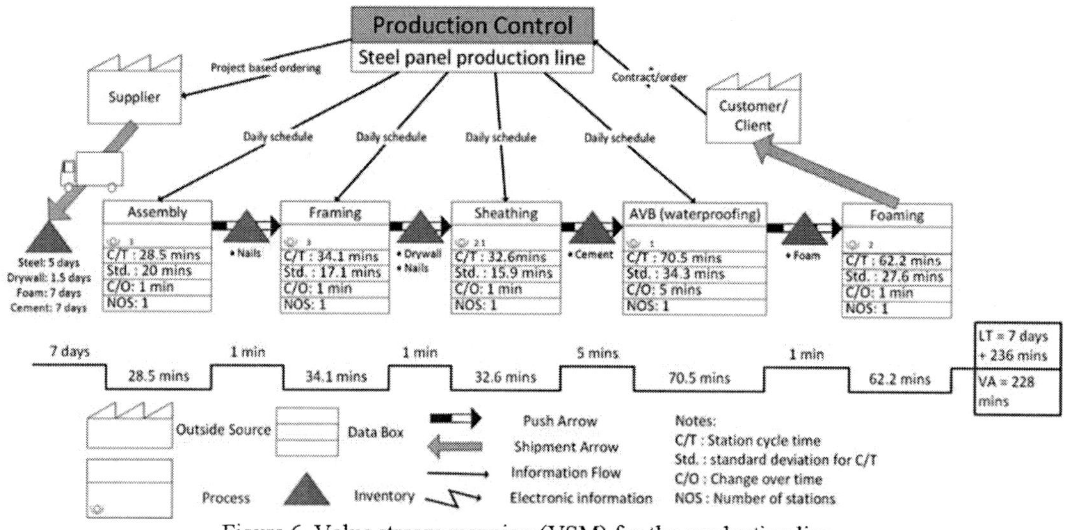

Figure 6. Value stream mapping (VSM) for the production line

Figure 7. Workers searching for the correct panel due to non-balanced line

Figure 8. 5S implementation results

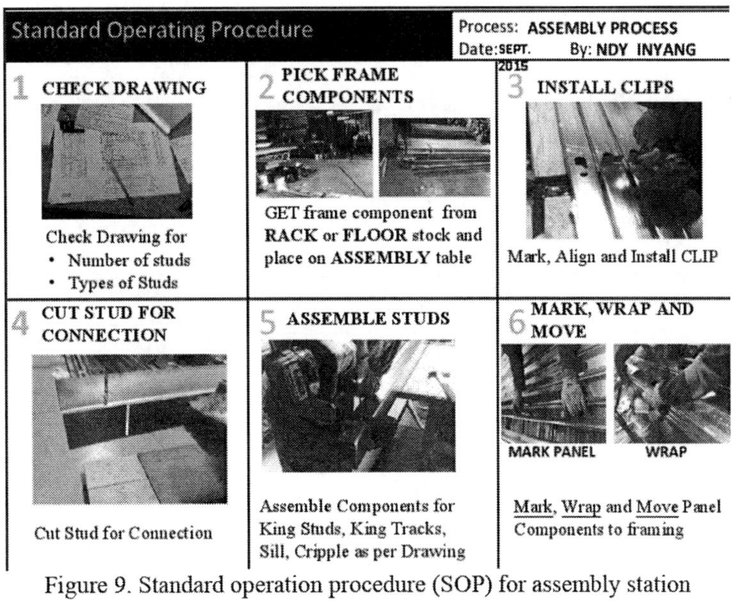

Figure 9. Standard operation procedure (SOP) for assembly station

5 Conclusions and Future Work

This study identifies the values of using Lean manufacturing concepts in a factory-based production line. The objective is to assist the company to realize the areas of waste in their production, and identify key areas for improvement. Time study is proven as the key for the analysis, based on which VSM is created to visualize the bottlenecks of production. The Lean tools are applied to the factory production line to assist in adding value to the production process. Typical tools include 5S, Kaizen, and JIT, etc. This research reveals that it is important to gain the employee/management buy-in in order to achieve Lean improvement. A culture change of the company is required for a continuous improvement process. For future work, the authors see the potential of implementing a real-time tracking system to the production line in order to better control the inventory and production. To achieve a highly balanced flow, immediate information feedback from the floor to the office is mandatory. This can be achieved through many available technologies, such as the barcode system or Radio-frequency identification (RFID) system (Altaf et al. 2015). Digital Kanban systems can also be helpful to the productivity improvement and defects control.

6 Acknowledgements

The authors would like to appreciate the financial support from the Natural Sciences and Engineering Research Council (NSERC); also, the feedback from our collaborator, Fortis LGS Structures Inc., is well acknowledged.

References

[1] Altaf, M. S., Liu, H., Yu, H., and Al-Hussein, M. Online simulation modeling of prefabricated wall panel production using RFID system. Proceedings of 2015 Winter Simulation Conference, pages 3379 – 3390, Huntington Beach, California, USA, 2015.

[2] Bock, T. The future of construction automation: technological disruption and the upcoming ubiquity of robotics. Automation in Construction, 59, 113 – 121, 2015.

[3] Brettel, M., Friederishsen, N., Keller, M., and Rosenberg, M. How virtualization, decentralization and network building change the manufacturing landscape: an industry 4.0 perspective. International Journal of Mechanical, 8(1), 37 – 44, 2014.

[4] Chu, B., Jung, K., Lim, M., and Hong, D. Robot-based construction automation: an application to steel beam assembly (Part I). Automation in Construction, 32, 46 – 61, 2013.

[5] Everett, J. G., and Slocum, A. H. Automation and robotics opportunities: construction versus manufacturing. Journal of Construction Engineering and Management, 120(2), 443 – 452, 1994.

[6] Hyeok, S., Al-hussein, M., Al-jibouri, S., and Yu, H. (2012). "Automation in Construction Automated post-simulation visualization of modular building production assembly line." *Automation in Construction*, Elsevier B.V., 21, 229–236.

[7] Jung, K., Chu, B., and Hong, D. Robot-based construction automation: an application to steel beam assembly (Part II). Automation in Construction, 32, 62 – 79, 2013.

[8] Lee, J., Bagheri, B., and Kao, H. A cyber-physical systems architecture for Industry 4.0-based manufacturing systems. Manufacturing Letters, 3, 18 – 23, 2015.

[9] Lei, Z., Taghaddos, H., Ph, D., Olearczyk, J., Ph, D., Al-hussein, M., Ph, D., Asce, M., and Hermann, U. (2013). "Automated Method for Checking Crane Paths for Heavy Lifts in Industrial Projects." *Journal of Construction Engineering and Management*, 5(2003), 1–9.

[10] Randek, http://www.randek.com/, accessed on April 4th, 2016.

[11] Shafai, L. (2012). "Simulation Based Process Flow Improvement for Wood Framing Home Building Production Lines By."

[12] Vaha, P., Heikkila, T., Kilpelainen, P., Jarviluoma, M., and Gambao, E. Extending automation of building construction – Survey on potential sensor technologies and robotic applications. Automation in Construction, 36, 168 -178, 2013.

[13] WEINMANN machinery, http://www.weinmann-partner.com/, accessed on April 4th 2016.

[14] Yu, H., Al-hussein, M., Asce, M., Al-jibouri, S., and Telyas, A. (2013). "Lean Transformation in a Modular Building Company : A Case for Implementation." *Journal of Management in Engineering,* 29(January), 103–111.

[15] Yu, H., Tweed, T., Al-Hussein, M., and Nasseri, R. (2009). "Development of Lean Model for House Construction Using Value Stream Mapping." *Journal of Construction Engineering and Management*, 135(8), 782–790.

AUTHOR INDEX

AUTHOR INDEX

AUTHOR INDEX

AUTHOR INDEX